INDEX KEWENSIS

SUPPLEMENT TWENTY ONE

NAMES OF SEED-BEARING PLANTS AT THE RANK OF FAMILY

AND BELOW

PUBLISHED BETWEEN

JANUARY 1996 AND THE END OF 2000

WITH SOME OMISSIONS FROM EARLIER YEARS

COMPILED BY

K. M. CHALLIS

R. A. DAVIES

EDITED BY

R. A. DAVIES

AT THE HERBARIUM OF THE ROYAL BOTANIC GARDENS

KEW

ROYAL BOTANIC GARDENS, KEW

2002

First published 2002

Page make-up by Media Resources
Information Services Department, Royal Botanic Gardens, Kew

ISBN 1-84246-041-2

Printed by Bookcraft, Midsomer Norton, Nr Bath

FOREWORD

Supplement XXI of Index Kewensis follows a similar format to Supplement XX. We have scanned books and journals at Kew and the Natural History Museum (BM). Names of all taxonomic ranks from family downwards are included. Names at ranks above genus are arranged alphabetically under families in a separate sequence at the beginning. Names at the rank of genus downwards are arranged alphabetically under genus headings. These are followed by an appendix containing a list of new genera in alphabetic order under family. At each rank there are three categories of names: new taxon, new name (nom. nov.) and new combination. An entry can be divided up as follows: Name, Author, Bibliographic details, and either geography if a new taxon of specific or infraspecific rank, or replaced synonym if a new name, or basionym if a new combination.

Name

The spelling of names is changed, if they are incorrect, in accordance with the appropriate Article of the International Code of Botanical Nomenclature (ICBN). Each genus is assigned to a family according to *R.K. Brummitt, Vascular Plant Families and Genera* (1992), ISBN 0-947643-43-5.

Authorship

The author of the basionym appears in brackets before the publishing author in entries of new combinations. Since its publication in 1992 *R.K. Brummitt & C.E. Powell, Authors of Plant Names*, ISBN 0-947643-44-3, has been used as a standard for the abbreviation of names of basionym authors. However, if a surname of a co-author is given as, for example, 'Smith', it is not feasible to establish which 'Smith' is in question and these names are left as given. The name of the publishing author is always given in full.

Bibliographic details

The same journal abbreviations are used as are found in the *Kew Record of Taxonomic Literature relating to Vascular Plants*. New journal title abbreviations are published regularly in *Kew Record* www.rbgkew.org.uk/bibliographies/KR/KRHomeExt.html.

Nomenclatural status

A note is sometimes added after the bibliographic details, indicating the following:

(1) the original spelling of the epithet if it has been changed;
(2) the status of a name, if incorrectly stated by the author, such as when an author writes 'sp. nov.' where 'nom. nov.' or 'comb. nov.' is appropriate;
(3) the term 'nom. nov.', used to indicate an avowed substitute for a previously validly published name according to Arts. 7.3 and 33.3 of ICBN;
(4) details which might affect nomenclatural validity, as defined by the ICBN;
(5) the term 'isonym', used to indicate a new combination which has already been made by another author; we do not go out of our way to find this information, but it is included if known.

Geography

The countries listed are those from which specimens are cited in the protologue, and may not always coincide with the full extent of the distribution of the taxon as stated by the author. The country of the type collection is given first. Geography is not stated for ranks above species level. *S. Hollis & R.K. Brummitt, World Geographical Scheme for Recording Plant Distribution* (1992), ISBN 0-913196-56-8 has been used as a standard since its publication, using country names at level 4 and where necessary levels 3 and 4.

Basionym/replaced synonym

Sometimes the basionyms are names published some years ago when symbols were used instead of the name of the rank, e.g. **** or letters of the Greek alphabet. Where possible, we reproduce these as they appear in the original publication.

Corrected entries

A dagger (†) at the end of an entry indicates the corrected re-entry of a name cited in an earlier volume.

Acknowledgments

We are grateful to Dr. R.K. Brummitt for nomenclatural advice, the Kew Computer Section for processing the data and Margaret Newman and Dom Costello for page make-up.

Future of Index Kewensis in printed form

Readers may be aware that the whole of the Index Kewensis dataset is now freely available in electronic form via the Internet as part of the International Plant Names Index (IPNI, see www.ipni.org), a joint venture involving the Harvard University Herbaria and the Centre for Plant Biodiversity Research, Canberra as well as the Royal Botanic Gardens, Kew. As levels of usage of IPNI increase, we anticipate that there may be a corresponding decline in the demand for hard copy supplements to Index Kewensis. We would like to learn of readers' views on this issue. Please write to us at Index Kewensis, Herbarium, Royal Botanic Gardens, Kew, Richmond, Surrey, TW9 3AE, United Kingdom or email ik@rbgkew.org.uk

Royal Botanic Gardens
Kew (www.rbgkew.org.uk)
January 2002

Katherine Challis,
Compiler, Index Kewensis
Rosemary Davies,
Compiler and Editor, Index Kewensis
Dr Eimear Nic Lughadha,
Project Manager,
International Plant Names Index
Professor Peter R. Crane FRS,
Director

INDEX KEWENSIS

SUPPLEMENT TWENTY ONE
NAMES OF ALL RANKS ABOVE GENUS
UP TO AND INCLUDING FAMILY

ACMOPYLEACEAE
fam. **Acmopyleaceae** (Pilg.) A.P. Melikyan & A.V. Bobrov in Proc. Internat. Conf. Pl. Anat. & Morphology, St. Petersburg, 1997: 93 (1997): subfam. Acmopyleoideae.

ALLIACEAE
subfam. **Tulbaghioideae** (Endl. ex Meisn.) M.F. Fay & M.W. Chase in Taxon, 45(3): 448 (1996): tribus Tulbaghieae.

AMARYLLIDACEAE
subtribus **Boophoninae** D. Müller-Doblies & U. Müller-Doblies in Feddes Repert., 107(5–6): S.c.3 (1996).
tribus **Calostemmateae** D. Müller-Doblies & U. Müller- in Feddes Repert., 107(5–6): S.c.7 (1996).
tribus **Clinantheae** A.W. Meerow in Syst. Bot., 25(4): 722 (2000).
subtribus **Cliviinae** D. Müller-Doblies & U. Müller-Doblies in Feddes Repert., 107(5–6): S.c.3 (1996).
subtribus **Cyrtanthinae** D. Müller-Doblies & U. Müller-Doblies in Feddes Repert., 107(5–6): S.c.3 (1996).
subtribus **Griffiniinae** (Ravenna) D. Müller-Doblies & U. Müller-Doblies in Feddes Repert., 107(5–6): S.c.6 (1996): tribus Griffinieae.
tribus **Hymenocallideae** (D. Müll.-Doblies & U. Müll.-Doblies) A.W. Meerow in K. Kubitzki (ed.), Fam. & Gen. Vasc. Pl., 3: 105 (1998), without exact basionym page: subtribus Hymenocallidinae.
subtribus **Hymenocallidinae** D. Müller-Doblies & U. Müller-Doblies in Feddes Repert., 107(5–6): S.c.7 (1996).
subtribus **Stenomessinae** D. Müller-Doblies & U. Müller-Doblies in Feddes Repert., 107(5–6): S.c.7 (1996).
tribus **Tocantinieae** P. Ravenna in Onira, Bot. Leafl., 5(3): 12 (2000).
subtribus **Traubiinae** D. Müller-Doblies & U. Müller-Doblies in Feddes Repert., 107(5–6): S.c.6 (1996).

ANEMARRHENACEAE
fam. **Anemarrhenaceae** J.G. Conran, M.W. Chase & P.J. Rudall in Kew Bull., 52(4): 995 (1997).

APOSTASIACEAE
subtribus **Neuwiediinae** (Burns-Bal. & V.A. Funk) D.L. in Fragm. Flor. Geobot., Supp. 3: 9 (1995): Orchidaceae subfam. Neuwiedioideae.

ARACEAE
subfam. **Orontioideae** S.J. Mayo, J. Bogner & P.C. Boyce, The Genera of Araceae: 346 (1997).

ASCLEPIADACEAE
tribus **Cryptolepideae** H.J.T. Venter in Taxon, 46(4): 713 (1997).

ASCLEPIADACEAE
subtribus **Gonolobinae** (G. Don) S. Liede in Taxon, 46(2): 238 (1997): tribus Gonolobeae.
tribus **Gymnanthereae** H.J.T. Venter in Taxon, 46(4): 712 (1997).

AVETRACEAE
fam. **Avetraceae** A. Takhtajan, Diversity & Classific. Flow. Pl.: 524 (1997).

BAXTERIACEAE
fam. **Baxteriaceae** A.L. Takhtajan in Bot. Zhurn., 81(2): 85 (1996).

BEHNIACEAE
fam. **Behniaceae** J.G. Conran, M.W. Chase & P.J. Rudall in Kew Bull., 52(4): 996 (1997).

BEHNIACEAE
fam. **Behniaceae** R. Dahlgren ex J.L. Reveal in Phytologia, 82(4): 273 (1997 publ. 1998), homonym; see Conran et al. in Kew Bull., 52(4): 996 (1997).

BEMBICIACEAE
fam. **Bembiciaceae** R. Keating & A.L. Takhtajan in Bot. Zhurn., 81(2): 85 (1996).

BORYACEAE
fam. **Boryaceae** M.W. Chase, P.J. Rudall & J.G. Conran in Kew Bull., 51(4): 676 (1996), without latin descr. or ref. to latin descr.

BORYACEAE
fam. **Boryaceae** (Baker) M.W. Chase, P.J. Rudall & J.G. Conran in Kew Bull., 52(2): 416 (1997): tribus Boryeae.

BRACTEOCARPACEAE
fam. **Bracteocarpaceae** A.V.F.Ch. Bobrov & A.P. Melikian in Bot. Zhurn., 85(7): 60 (2000).

BROMELIACEAE
tribus **Dyckieae** H. Robinson & D.C. Taylor in Harvard Pap. Bot., 4(1): 198 (1999).

BURCHARDIACEAE
fam. **Burchardiaceae** A.L. Takhtajan in Bot. Zhurn., 81(2): 85 (1996).

BURSERACEAE
subtribus **Boswelliinae** D.C. Daly in World Pollen & Spore Fl. (ed. S. Nilsson), 20: 4 (1995).
subtribus **Pseudodacryodiinae** R. Pierlot in Bull. Jard. Bot. Nation. Belg., 66(3–4): 185 (1997).

CACTACEAE
subtribus **Astrophytinae** (Horan.) A.B. Doweld in Novosti Sist. Vȳssh. Rast., 32: 115 (2000): tribus Astrophyteae.
subtribus **Bravocactinae** A.B. Doweld in Sukkulentȳ (Moscow), 1998(1): 19 (1998).
subtribus **Cochemieinae** A.B. Doweld in Sukkulentȳ (Moscow), 3(1–2): 13 (2000).
subtribus **Copiapoinae** A.B. Doweld in Sukkulentȳ (Moscow), 3(1–2): 63 (2000).
tribus **Cylindropuntieae** A.B. Doweld in Sukkulentȳ (Moscow), 1999(1): 25 (1999).
subtribus **Epithelanthinae** A.B. Doweld in Novosti Sist. Vȳssh. Rast., 32: 116 (2000).
subtribus **Escobariinae** A.B. Doweld in Sukkulentȳ (Moscow), 3(1–2): 22 (2000).
tribus **Maihuenieae** A.B. Doweld in Sukkulentȳ (Moscow), 1999(1): 25 (1999).
subfam. **Maihuenioideae** P.J. Fearn, Rev. Origins Cactus & Search for System of Classif.: [67] (1996).
subtribus **Notocactinae** Buxb. ex A.B. Doweld in Sukkulentȳ (Moscow), 3(1–2): 54 (2000).
subtribus **Pediocactinae** A.B. Doweld in Sukkulentȳ (Moscow), 1999(1): 24 (1999).
subtribus **Pelecyphorinae** A.B. Doweld in Sukkulentȳ (Moscow), 3(1–2): 18 (2000).
tribus **Pereskiopsideae** A.B. Doweld in Sukkulentȳ (Moscow), 1999(1): 26 (1999).
tribus **Pterocacteae** A.B. Doweld in Sukkulentȳ (Moscow), 1999(1): 26 (1999).
subtribus **Sclerocactinae** A.B. Doweld in Sukkulentȳ (Moscow), 1998(1): 16 (1998).
tribus **Tephrocacteae** A.B. Doweld in Sukkulentȳ (Moscow), 1999(1): 26 (1999).
subtribus **Turbinicarpinae** A.B. Doweld in Sukkulentȳ (Moscow), 1998(1): 23 (1998).

CAMPANULACEAE
subfam. **Cyphocarpoideae** (Miers) M.H.G. Gustafsson, Phylogen. Stud. Asterales sensu lato: 34 (1996), basionym not validly publ.: fam. Cyphocarpaceae.
subfam. **Nemacladoideae** (Nutt.) M.H.G. Gustafsson, Phylogen. Stud. Asterales sensu lato: 34 (1996): fam. Nemacladaceae.
subfam. **Nemacladoideae** T.G. Lammers in Novon, 8(1): 37 (1998), isonym; see M.H.G. Gustafsson, Phylogen. Stud. Asterales sensu lato: 34 (1996).
subfam. **Ostrowskioideae** (Fed.) A. Takhtajan, Diversity & Classific. Flow. Pl.: 408 (1997): tribus Ostrowskieae.

CHIONOGRAPHIDACEAE
fam. **Chionographidaceae** A.L. Takhtajan in Bot. Zhurn., 81(2): 85 (1996).

COMPOSITAE
subtribus **Abrotanellinae** H. Robinson, G.D. Carr, R.M. King & A.M. Powell in Ann. Missouri Bot. Gard., 84(4): 896 (1997 publ. 1998
subtribus **Astranthiinae** G.L. Nesom in Sida, 19(2): 265 (2000).
subtribus **Boltoniinae** G.L. Nesom in Sida, 19(2): 266 (2000).
subtribus **Centrapalinae** H. Robinson in Proc. Biol. Soc. Wash., 112(1): 223 (1999).
subtribus **Chaetopappinae** G.L. Nesom in Sida, 19(2): 264 (2000).
subtribus **Chrestinae** H. Robinson in Smithsonian Contrib. Bot., 89: 11 (1999).
subtribus **Crassocephalinae** Y.R. Ling in Bull. Bot. Res. North-East. Forest. Univ., 17(1): 14 (1997), without latin descr.
subtribus **Doronicinae** Y.R. Ling in Bull. Bot. Res. North-East. Forest. Univ., 17(1): 13 (1997), without latin descr.
subtribus **Erlangeinae** H. Robinson in Proc. Biol. Soc. Wash., 112(1): 222 (1999).
subtribus **Gymnantheminae** H. Robinson in Proc. Biol. Soc. Wash., 112(1): 224 (1999).
subtribus **Lagenophorinae** Y.R. Ling in Bull. Bot. Res. North-East. Forest. Univ., 17(1): 12 (1997), without latin descr.
subtribus **Leiboldiinae** H. Robinson in Smithsonian Contrib. Bot., 89: 8 (1999).
subtribus **Liguláriinae** Y.R. Ling in Bull. Bot. Res. North-East. Forest. Univ., 17(1): 14 (1997), without latin descr.
subtribus **Nannoglottidinae** Y.R. Ling in Bull. Bot. Res. North-East. Forest. Univ., 17(1): 14 (1997), as 'Nannoglottisinae', without latin descr.
subtribus **Pacourininae** H. Robinson in Proc. Biol. Soc. Wash., 112(1): 218 (1999).
subtribus **Pentachaetinae** G.L. Nesom in Sida, 19(2): 264 (2000).
subtribus **Plateileminae** B.G. Baldwin in Syst. Bot., 25(3): 536 (2000).
subtribus **Psathyrotinae** B.G. Baldwin in Syst. Bot., 25(3): 535 (2000).
subtribus **Pseudojacobinae** R. Mathur in Fl. India, 13: 303 (1995), as 'Pseudojacobineae'.
subtribus **Sipolisiinae** H. Robinson in Smithsonian Contrib. Bot., 89: 13 (1999).
subtribus **Stokesiinae** H. Robinson in Proc. Biol. Soc. Wash., 112(1): 216 (1999).

COROKIACEAE
fam. **Corokiaceae** Kapil ex A. Takhtajan, Diversity & Classific. Flow. Pl.: 374 (1997).

CRASSULACEAE
tribus **Kalanchoeae** (A. Berger) H. 't Hart in H. 't Hart & U. Eggli (eds.), Evol. & Syst. Crassulac.: 167 (1995): subfam. Kalanchoideae.
tribus **Sedeae** H. 't Hart in H. 't Hart & U. Eggli (eds.), Evol. & Syst. Crassulac.: 167 (1995).
subtribus **Sedinae** H. 't Hart in H. 't Hart & U. Eggli (eds.), Evol. & Syst. Crassulac.: 169 (1995).
subtribus **Telephinae** H. 't Hart in H. 't Hart & U. Eggli (eds.), Evol. & Syst. Crassulac.: 167 (1995).

CUNONIACEAE
subfam. **Aphanopetaloideae** A. Takhtajan, Diversity & Classific. Flow. Pl.: 254 (1997), contrary to Art. 41.1 ICBN (1994).

CUPRESSACEAE
tribus **Widdringtoneae** Z.P. Jiang in Acta Phytotax. Sin., 35(3): 244 (1997).

CYCADACEAE
subfam. **Microcycadoideae** V.M. Tarbaeva in Byull. Glavn. Bot. Sada (Moscow), 173: 111 (1996), as 'Microcycaideae'.

CYPHOCARPACEAE
fam. **Cyphocarpaceae** (Miers) J.L. Reveal & R.D. Hoogland in Phytologia, 79(2): 72 (1995 publ. 1996): subfam. Cyphocarpoideae.

DACRYCARPACEAE
fam. **Dacrycarpaceae** A.V.F.Ch. Bobrov & A.P. Melikian in Bot. Zhurn., 85(7): 59 (2000).

DACRYDIACEAE
fam. **Dacrydiaceae** A.V.F.Ch. Bobrov & A.P. Melikian in Bot. Zhurn., 85(7): 60 (2000).

DECAISNEACEAE
fam. **Decaisneaceae** (Takht. ex H.N. Qin & Y.C. Tang) H. Loconte in U. Jensen & J.W. Kadereit (eds.), Syst. Evol. Ranunculiflorae (Pl. Syst. Evol., Supp. 9): 105 (1995): tribus Decaisneeae.

DIERVILLACEAE
fam. **Diervillaceae** (Raf.) N. Pyck in Taxon, 47(3): 658 (1998): Diervilloideae.

DUABANGACEAE
fam. **Duabangaceae** A.L. Takhtajan, Florist. Reg. World: 332 (1986).

ENGELHARDTIACEAE
fam. **Engelhardtiaceae** J.L. Reveal & A.B. Doweld in Novon, 9(4): 552 (1999).

EPACRIDACEAE
tribus **Archerieae** D.M. Crayn & C.J. Quinn in Austral. Syst. Bot., 11(1): 23 (1998).

EXBUCKLANDIACEAE
fam. **Exbucklandiaceae** J.L. Reveal & A.B. Doweld in Novon, 9(4): 552 (1999).

FALCATIFOLIACEAE
fam. **Falcatifoliaceae** A.V.F.Ch. Bobrov & A.P. Melikian in Bot. Zhurn., 85(7): 61 (2000).

GRAMINEAE
subfam. **Leptaspidoideae** (Tzvelev) C.O. Morales in Sendtnera, 5: 244 (1998): tribus Leptaspideae.
subtribus **Parianinae** V.C. Hollowell in G.P. Chapman (ed.), Bamboos (Linn. Soc. Symp. Ser. 19): 54 (1997).
subfam. **Pharoideae** (Stapf) L.G. Clark & E.J. Judziewicz in Taxon, 45(4): 643 (1996): tribus Phareae.
subfam. **Puelioideae** L.G. Clark, M. Kobayashi, S. Mathews, R.E. Spangler & E.A. Kellogg in Syst. Bot., 25(2): 184 (2000).
subtribus **Sinobambusinae** Z.P. Wang in J. Bamboo Res., 16(4): 5 (1997).

HALOCARPACEAE
fam. **Halocarpaceae** A.V.F.Ch. Bobrov & A.P. Melikian in Bot. Zhurn., 85(7): 60 (2000).

HOPKINSIACEAE
fam. **Hopkinsiaceae** B.G. Briggs & L.A.S. Johnson in Telopea, 8(4): 484 (2000).

HYACINTHACEAE
subfam. **Chlorogaloideae** F. Speta in Phyton (Austria), 38(1): 51 (1998).

HYACINTHACEAE
subtribus **Lachenaliinae** U. Müller-Doblies & D. Müller-Doblies in Feddes Repert., 108(1–2): 65 (1997).

HYACINTHACEAE
subtribus **Ledebouriinae** U. Müller-Doblies & D. Müller-Doblies in Feddes Repert., 108(1–2): 56 (1997).
subfam. **Ornithogaloideae** F. Speta in Phyton (Austria), 38(1): 53 (1998).
subfam. **Oziroeoideae** F. Speta in Phyton (Austria), 38(1): 52 (1998).
subfam. **Urgineoideae** F. Speta in Phyton (Austria), 38(1): 52 (1998).

HYDRASTIDACEAE
subfam. **Glaucidioideae** (Tamura) H. Loconte in U. Jensen & J.W. Kadereit (eds.), Syst. Evol. Ranunculiflorae (Pl. Syst. Evol., Supp. 9): 105 (1995): fam. Glaucidiaceae.

JAPONOLIRIACEAE
fam. **Japonoliriaceae** A.L. Takhtajan in Bot. Zhurn., 81(2): 85 (1996).

KALIPHORACEAE
fam. **Kaliphoraceae** A.L. Takhtajan in Bot. Zhurn., 81(2): 86 (1996).

LARDIZABALACEAE
subtribus **Boquilinae** H.N. Qin in Cathaya, 8–9: 170 (1997).
subfam. **Sinofranchetioideae** (H.N. Qin & Y.C. Tang) H. Loconte in U. Jensen & J.W. Kadereit (eds.), Syst. Evol. Ranunculiflorae (Pl. Syst. Evol., Supp. 9): 105 (1995): tribus Sinofranchetieae.

LEGUMINOSAE
tribus **Diplotropideae** G.P. Yakovlev, Bobovȳe zemnogo shara: 67 (1991).
subtribus **Erinaceinae** S. Talavera in An. Jard. Bot. Madrid, 57(1): 217 (1999).
subtribus **Erophacinae** S. Talavera in An. Jard. Bot. Madrid, 57(1): 220 (1999).
tribus **Hypocalypteae** (Yakovlev) A.L. Schutte in Novon, 8(2): 180 (1998): subtribus Hypocalyptinae.
subtribus **Hypocalyptiinae** G.P. Yakovlev, Bobovȳe zemnogo shara: 87 (1991).
tribus **Myrospermeae** G.P. Yakovlev, Bobovȳe zemnogo shara: 65 (1991).
subtribus **Ononidinae** (Hutch.) G.P. Yakovlev, Bobovȳe zemnogo shara: 117 (1991): tribus Ononideae.
tribus **Ormosieae** G.P. Yakovlev, Bobovȳe zemnogo shara: 72 (1991).
subtribus **Ornithopinae** S. Talavera in Lagascalia, 21(1): 210 (1999).
subtribus **Parochetiinae** G.P. Yakovlev, Bobovȳe zemnogo shara: 117 (1991).
subtribus **Parochetinae** L.B. Chaudhary & M. Sanjappa in Taxon, 47(4): 829 (1998).
tribus **Poiretieae** (Burkart) Hiroyoshi Ohashi in Sci. Rep. Tohoku Univ., Ser. 4, 40(3): 194 (1999): subtribus Poiretiinae.
subtribus **Xiphothecinae** A.L. Schutte in Pl. Syst. Evol., 209(1–2): 26 (1998).

LEPIDOTHAMNACEAE
fam. **Lepidothamnaceae** A.V.F.Ch. Bobrov & A.P. Melikian in Bot. Zhurn., 85(7): 63 (2000).

LILIACEAE
subtribus **Arnocrinineae** G.J. Keighery in Flora (Germany), 175(2): 107 (1984).
subfam. **Medeoloideae** (Benth.) M.N. Tamura in K. Kubitzki (ed.), Fam. & Gen. Vasc. Pl., 3: 350 (1998): tribus Medeoleae.

LINNAEACEAE
fam. **Linnaeaceae** (Raf.) A. Backlund in Taxon, 47(3): 658 (1998): Linnaeoideae.

LOASACEAE
subfam. **Petalonychoideae** M. Weigend, Nasa & Conquest S. America - Syst. Rearrang. in Loasac.: 217 (1997).

LYGINIACEAE
fam. **Lyginiaceae** B.G. Briggs & L.A.S. Johnson in Telopea, 8(4): 488 (2000).

LYTHRACEAE
subfam. **Duabangoideae** (Takht.) S.A. Graham, R.F. Thorne & J.L. Reveal in Taxon, 47(2): 436 (1998): fam. Duabangaceae.
subfam. **Punicoideae** (Horan.) S.A. Graham, R.F. Thorne & J.L. Reveal in Taxon, 47(2): 436 (1998): fam. Punicaceae.
subfam. **Sonneratioideae** (Engl. & Gilg) S.A. Graham, R.F. Thorne & J.L. Reveal in Taxon, 47(2): 436 (1998): fam. Sonneratiaceae.

MAESACEAE
fam. **Maesaceae** (A.DC.) A.A. Anderberg, B. Ståhl & M. Kaüllersjoü in Taxon, 49(2): 185 (2000): tribus Maeseae.

MAGNOLIACEAE
subtribus **Alcimandriinae** Y.W. Law in Acta Phytotax. Sin., 34(1): 87 (Feb. 1996); et in Fl. Reipubl. Popul. Sin., 30(1): 270 (May 1996).

MICROCACHRYACEAE
fam. **Microcachryaceae** A.V.F.Ch. Bobrov & D.S. Kostrikin in 14. Symp. Biodivers. & Evolutionsbiol. Jena 1999 (ed. H. Manitz & F.H. Hellwig): 19 (1999).

MICROCACHRYDACEAE
fam. **Microcachrydaceae** A.B. Doweld & J.L. Reveal in Phytologia, 84(5): 365 (1998 publ. 1999).

MORACEAE
subfam. **Cecropioideae** (C.C. Berg) S. Romaniuc-Neto in Albertoa, n.s., 4: 13 (1999): fam. Cecropiaceae.

MUNTINGIACEAE
fam. **Muntingiaceae** C. Bayer, M.W. Chase & M.F. Fay in Taxon, 47(1): 38 (1998).

NARTHECIACEAE
tribus **Japonolirieae** (Takht.) M.N. Tamura in K. Kubitzki (ed.), Fam. & Gen. Vasc. Pl., 3: 390 (1998): fam. Japonoliriaceae.
subfam. **Narthecioideae** (Benth.) M.N. Tamura in K. Kubitzki (ed.), Fam. & Gen. Vasc. Pl., 3: 390 (1998): tribus Narthecieae.
subfam. **Tofieldioideae** (Takht.) M.N. Tamura in K. Kubitzki (ed.), Fam. & Gen. Vasc. Pl., 3: 388 (1998): fam. Tofieldiaceae.

OLEACEAE
subtribus **Fraxininae** (Vest) E. Wallander & V.A. Albert in Amer. J. Bot., 87(12): 1841 (2000): fam. Fraxinaceae.
subtribus **Schreberinae** (Wight) E. Wallander & V.A. Albert in Amer. J. Bot., 87(12): 1841 (2000): subfam. Schreberoideae.

ORCHIDACEAE
subtribus **Acanthephippiinae** S.C. Chen, Z.H. Tsi & G.H. Zhu in Acta Phytotax. Sin., 37(2): 115 (1999).
tribus **Adrorhizeae** (Schltr.) D.L Szlachetko in Fragm. Flor. Geobot., Supp. 3: 57 (1995): subtribus Adrorhizinae.

ORCHIDACEAE
subtribus **Agrostophyllinae** D.L. Szlachetko in Fragm. Flor. Geobot., Supp. 3: 65 (1995).
subtribus **Anthogoniinae** S.C. Chen, Z.H. Tsi & G.H. Zhu in Acta Phytotax. Sin., 37(2): 116 (1999).
subtribus **Bolusiellinae** D.L. Szlachetko in Fragm. Flor. Geobot., Supp. 3: 91 (1995).
subtribus **Caleaninae** D.L. Szlachetko in Fragm. Flor. Geobot., Supp. 3: 24 (1995).
subtribus **Calyptrochilinae** D.L. Szlachetko in Fragm. Flor. Geobot., Supp. 3: 92 (1995).
subtribus **Chiloglottidinae** D.L. Szlachetko in Fragm. Flor. Geobot., Supp. 3: 23 (1995).
subtribus **Chilopogoninae** D.L. Szlachetko in Fragm. Flor. Geobot., Supp. 3: 65 (1995).
subtribus **Claderiinae** D.L. Szlachetko in Fragm. Flor. Geobot., Supp. 3: 80 (1995).
tribus **Codonorchideae** P.J. Cribb in Lindleyana, 15(3): 169 (2000).
subtribus **Coeliopsidinae** D.L. Szlachetko in Fragm. Flor. Geobot., Supp. 3: 97 (1995).
tribus **Cryptostylideae** (Schltr.) D.L. Szlachetko in Fragm. Flor. Geobot., Supp. 3: 18 (1995): subtribus Cryptostylidinae.
subtribus **Cyclopogoninae** D.L. Szlachetko in Fragm. Flor. Geobot., Supp. 3: 43 (1995).
subtribus **Deceptorinae** D.L. Szlachetko in Fragm. Flor. Geobot., Supp. 3: 88 (1995).
subtribus **Determanniinae** D.L. Szlachetko in Fragm. Flor. Geobot., Supp. 3: 104 (1995).
subtribus **Diceratostelinae** (Dressler) D.L. Szlachetko in Fragm. Flor. Geobot., Supp. 3: 47 (1995): tribus Diceratosteleae.
subtribus **Diplocentrinae** D.L. Szlachetko in Fragm. Flor. Geobot., Supp. 3: 90 (1995).
subtribus **Duckeellinae** D.L. Szlachetko in Fragm. Flor. Geobot., Supp. 3: 52 (1995).
tribus **Elleantheae** D.L. Szlachetko in Fragm. Flor. Geobot., Supp. 3: 71 (1995).
subtribus **Eriopsidinae** D.L. Szlachetko in Fragm. Flor. Geobot., Supp. 3: 94 (1995).
subtribus **Gastrochilinae** D.L. Szlachetko in Fragm. Flor. Geobot., Supp. 3: 89 (1995).
subtribus **Hintonellinae** D.L. Szlachetko in Fragm. Flor. Geobot., Supp. 3: 100 (1995).
subtribus **Leochilinae** D.L. Szlachetko in Fragm. Flor. Geobot., Supp. 3: 104 (1995).

ORCHIDACEAE :-
subtribus **Listrostachyinae** D.L. Szlachetko in Fragm. Flor. Geobot., Supp. 3: 92 (1995).
subtribus **Ludisiinae** D.L. Szlachetko in Fragm. Flor. Geobot., Supp. 3: 41 (1995).
subtribus **Microtidinae** (Szlach.) D.L. Szlachetko in Fragm. Flor. Geobot., Supp. 3: 20 (1995): tribus Microtideae.
subtribus **Monophyllorchidinae** D.L. Szlachetko in Fragm. Flor. Geobot., Supp. 3: 54 (1995).
tribus **Ornithocephaleae** D.L. Szlachetko in Fragm. Flor. Geobot., Supp. 3: 99 (1995).
subtribus **Pelatantheriinae** D.L. Szlachetko in Fragm. Flor. Geobot., Supp. 3: 90 (1995).
subtribus **Phalaenopsidinae** D.L. Szlachetko in Fragm. Flor. Geobot., Supp. 3: 89 (1995).
subtribus **Ponerorchidinae** D.L. Szlachetko & P. Rutkowski in Acta Bot. Fenn., 169: 125 (2000).
subtribus **Raycadencoinae** K. Senghas in Schlechter Orchideen, I/B(31): 1916 (1995).
subtribus **Rhaesteriinae** D.L. Szlachetko in Fragm. Flor. Geobot., Supp. 3: 92 (1995).
tribus **Rhizanthelleae** (R.S. Rogers) D.L. Szlachetko in Fragm. Flor. Geobot., Supp. 3: 20 (1995): subtribus Rhizanthellinae.
subtribus **Risleyinae** S.C. Chen, Z.H. Tsi & G.H. Zhu in Acta Phytotax. Sin., 37(2): 115 (1999).
subtribus **Roeperocharidinae** D.L. Szlachetko & P. Rutkowski in Acta Bot. Fenn., 169: 149 (2000).
subtribus **Stenorrhynchidinae** D.L. Szlachetko in Fragm. Flor. Geobot., Supp. 3: 43 (1995).
subtribus **Stereosandrinae** S.C. Chen, Z.H. Tsi & G.H. Zhu in Acta Phytotax. Sin., 37(2): 114 (1999), as 'Sterosanadrinae', isonym; see Dressler in Telopea, 2(4): 422 (1983).
subtribus **Taeniophyllinae** D.L. Szlachetko in Fragm. Flor. Geobot., Supp. 3: 91 (1995).
subfam. **Tropidioideae** (Pfitzer) D.L. Szlachetko in Fragm. Flor. Geobot., Supp. 3: 35 (1995): subtribus Tropidiinae.
subfam. **Vanilloideae** (Lindl.) D.L. Szlachetko in Fragm. Flor. Geobot., Supp. 3: 48 (1995): fam. Vanillaceae.
subtribus **Warreinae** D.L. Szlachetko in Fragm. Flor. Geobot., Supp. 3: 96 (1995).
subtribus **Xerorchidinae** D.L. Szlachetko in Fragm. Flor. Geobot., Supp. 3: 72 (1995).
subtribus **Yoaniinae** S.C. Chen, Z.H. Tsi & G.H. Zhu in Acta Phytotax. Sin., 37(2): 114 (1999), isonym; see Szlach. in Fragm. Flor. Geobot., Supp. 3: 69 (1995).
subtribus **Yoaniinae** D.L. Szlachetko in Fragm. Flor. Geobot., Supp. 3: 69 (1995).

OROBANCHACEAE
subtribus **Aeginetiinae** E.S. Teryokhin in Opred. Zarazikhovўkh Fl. SSSR (ed. L.Yu. Budantsev): 34 (1993).
subtribus **Boschniakiinae** E.S. Teryokhin in Opred. Zarazikhovўkh Fl. SSSR (ed. L.Yu. Budantsev): 34 (1993).
subtribus **Conopholidinae** E.S. Teryokhin in Opred. Zarazikhovўkh Fl. SSSR (ed. L.Yu. Budantsev): 35 (1993).
subtribus **Epifaginae** E.S. Teryokhin in Opred. Zarazikhovўkh Fl. SSSR (ed. L.Yu. Budantsev): 35 (1993).
subtribus **Mannagettaeinae** E.S. Teryokhin in Opred. Zarazikhovўkh Fl. SSSR (ed. L.Yu. Budantsev): 36 (1993).
tribus **Orobancheae** E.S. Teryokhin in Opred. Zarazikhovўkh Fl. SSSR (ed. L.Yu. Budantsev): 34 (1993).
subtribus **Orobanchinae** E.S. Teryokhin in Opred. Zarazikhovўkh Fl. SSSR (ed. L.Yu. Budantsev): 34 (1993).
subfam. **Orobanchoideae** (Vent.) E.S. Teryokhin in Opred. Zarazikhovўkh Fl. SSSR (ed. L.Yu. Budantsev): 34 (1993): fam. Orobanchaceae.
tribus **Phacellantheae** E.S. Teryokhin in Opred. Zarazikhovўkh Fl. SSSR (ed. L.Yu. Budantsev): 35 (1993).
subtribus **Phacellanthinae** E.S. Teryokhin in Opred. Zarazikhovўkh Fl. SSSR (ed. L.Yu. Budantsev): 36 (1993).
subtribus **Phelipanchinae** E.S. Teryokhin in Opred. Zarazikhovўkh Fl. SSSR (ed. L.Yu. Budantsev): 35 (1993).

PALMAE
subtribus **Eleiodoxinae** W.C. McClatchey, Revis. Gen. Metroxylon sect. Coelococcus (Ph.D. Thesis): 80 (1996).
tribus **Eugeissoneae** W.J. Baker & J. Dransfield in Syst. Bot., 25(2): 318 (2000).
tribus **Raphieae** (Hook. f.) W.C. McClatchey, Revis. Gen. Metroxylon sect. Coelococcus (Ph.D. Thesis): 80 (1996), as 'Raphiinae': subtribus Raphiinae.

PAPAVERACEAE
subfam. **Glaucioideae** (Mill.) H. Loconte in U. Jensen & J.W. Kadereit (eds.), Syst. Evol. Ranunculiflorae (Pl. Syst. Evol., Supp. 9): 105 (1995), with incorrect basionym ref.: gen. Glaucium.

PARASITAXACEAE
fam. **Parasitaxaceae** A.V.F.Ch. Bobrov & A.P. Melikian in Bot. Zhurn., 85(7): 61 (2000).

PINACEAE
tribus **Pseudolariceae** L.Chu Li in Acta Phytotax. Sin., 33(5): 430 (1995).
subfam. **Pseudolaricoideae** L.Chu Li in Acta Phytotax. Sin., 33(5): 429 (1995).

POLEMONIACEAE
tribus **Acanthogilieae** V. Grant in Amer. J. Bot., 85(6): 744 (1998).
subfam. **Acanthogilioideae** J.M. Porter & L.A. Johnson in Aliso, 19(1): 60 (2000).
tribus **Gilieae** J.M. Porter & L.A. Johnson in Aliso, 19(1): 63 (2000).
tribus **Leptodactyloneae** V. Grant in Amer. J. Bot., 85(6): 746 (1998).
tribus **Loeselieae** J.M. Porter & L.A. Johnson in Aliso, 17(1): 84 (8 June 1998).
tribus **Loeselieae** V. Grant in Amer. J. Bot., 85(6): 744 (17 June 1998), isonym; see J.M. Porter & L.A. Johnst. in Aliso, 17(1): 84 (8 June 1998).
tribus **Phlogieae** (Rchb.) J.M. Porter & L.A. Johnson in Aliso, 17(1): 84 (1998): Gruppe Phloginae.

PRIONIACEAE
fam. **Prioniaceae** S.L. Munro & H.P. Linder in Syst. Bot., 23(1): 51 (1998).

PROTEACEAE
subfam. **Bellendenoideae** P.H. Weston in Fl. Australia, 16: 472 (1995).
subfam. **Eidotheoideae** A.W. Douglas & B.P.M. Hyland in Fl. Australia, 16: 472 (1995).

PRUMNOPITYACEAE
fam. **Prumnopityaceae** A.V.F.Ch. Bobrov & A.P. Melikian in Bot. Zhurn., 85(7): 58 (2000).

RANUNCULACEAE
subfam. **Delphinioideae** (Rchb.) A.N. Luferov in Sosud. Rast. Sovet. Dal'nego Vostoka, 7: 37 (1995): Ranunculaceae β Delphinieae.
subtribus **Eranthinae** U. Jensen, S.B. Hoot, J.T. Johansson & K. Kosuge in U. Jensen & J.W. Kadereit (eds.), Syst. Evol. Ranunculiflorae (Pl. Syst. Evol., Supp. 9): 278 (1995).

RHAMNACEAE
tribus **Ampeloziziphеae** J.E. Richardson in Kew Bull., 55(2): 335 (2000).
tribus **Bathiorhamneae** J.E. Richardson in Kew Bull., 55(2): 335 (2000).
tribus **Doerpfeldieae** J.E. Richardson in Kew Bull., 55(2): 335 (2000).

ROSACEAE
subfam. **Coleogynoideae** (J. Agardh) E.I. Golubkova in Biol. Nauki (Moscow), 3(327): 107 (1991): fam. Coleogyneae.

RUTACEAE
tribus **Galipeeae** J.A. Kallunki in Kew Bull., 53(2): 257 (1998), without latin descr.
subtribus **Galipeinae** J.A. Kallunki in Kew Bull., 53(2): 257 (1998), without latin descr.

SARCOBATACEAE
fam. **Sarcobataceae** H.-D. Behnke in Taxon, 46(3): 503 (1997).

SAXEGOTHAEACEAE
fam. **Saxegothaeaceae** Gaussen ex A.B. Doweld & J.L. Reveal in Phytologia, 84(5): 365 (1998 publ. 1999).

SCHLEGELIACEAE
fam. **Schlegeliaceae** (A.H. Gentry) J.L. Reveal in Phytologia, 79(2): 74 (1995 publ. 1996): tribus Schlegelieae.

SCOLIOPACEAE
fam. **Scoliopaceae** A.L. Takhtajan in Bot. Zhurn., 81(2): 86 (1996).

SCYTOPETALACEAE
subfam. **Rhaptopetaloideae** O. Appel in Bot. J. Linn. Soc., 121(3): 225 (1996).
subfam. **Scytopetaloideae** O. Appel in Bot. J. Linn. Soc., 121(3): 225 (1996).

SETCHELLANTHACEAE
fam. **Setchellanthaceae** H.H. Iltis in Taxon, 48(2): 260 (1999).

SOLANACEAE
tribus **Benthamielleae** (Hunz.) A.T. Hunziker in Kurtziana, 28(1): 61 (2000): subtribus Benthamiellinae.
tribus **Capsiceae** W.G. D'Arcy in Phytologia, 80(4): 273 (1996).
subtribus **Iochrominae** (Miers) A.T. Hunziker in Kurtziana, 28(1): 66 (2000), without basionym ref.: tribus Iochromeae.
subfam. **Juanulloideae** (Hunz.) A.T. Hunziker in Kurtziana, 28(1): 59 (2000): tribus Juanulloeae.
subtribus **Nicotianinae** A.T. Hunziker in Fl. Fan. Argentina, 56(256:8): 3 (1998).
tribus **Physaleae** W.G. D'Arcy in Phytologia, 80(4): 274 (1996).
subtribus **Physalinae** (Miers) A.T. Hunziker in Kurtziana, 28(1): 66 (2000), without basionym ref.: tribus Physaleae.
subfam. **Salpiglossidoideae** (Benth.) A.T. Hunziker in Kurtziana, 28(1): 56 (2000), as 'Salpiglossoideae': tribus Salpiglossideae.
subfam. **Schizanthoideae** (Miers) A.T. Hunziker in Kurtziana, 28(1): 56 (2000): tribus Schizantheae.
subtribus **Witheringinae** (Miers) A.T. Hunziker in Kurtziana, 28(1): 65 (2000), without basionym ref.: tribus Witheringeae.

SPERGULACEAE
fam. **Spergulaceae** (Dumort.) N.N. Tzvelev in Novosti Sist. Vȳssh. Rast., 32: 26 (2000): Caryophyllaceae tribus Sperguleae.

TAXACEAE
tribus **Amentotaxeae** W.C. Cheng & C.D. Chu in Acta Phytotax. Sin., 13(4): 87 (1975), in I.K. Supp. 16 under 'Taxodiaceae' in error.†

TAXODIACEAE
subfam. **Athrotaxidoideae** C.J. Quinn in Amer. J. Bot., 87(7): 1056 (2000).
subfam. **Cunninghamioideae** (Hayata) C.J. Quinn in Amer. J. Bot., 87(7): 1056 (2000), isonym; see Silba in J. Int. Conifer Preserv. Soc., 6(1): 10, 9 (1999): fam. Cunninghamiaceae.
subfam. **Cunninghamioideae** J. Silba in J. Int. Conifer Preserv. Soc., 6(1): 10, 9 (1999), as 'Cunninghamiaceae'.
subfam. **Nageioideae** (D.Z. Fu) J. Silba in J. Int. Conifer Preserv. Soc., 6(1): 11 (1999), as 'Nageiaceae': fam. Nageiaceae.
subfam. **Sciadopityoideae** (Luerss.) J. Silba in J. Int. Conifer Preserv. Soc., 6(1): 10 (1999), as 'Sciadopityaceae': fam. Sciadopityaceae.
subfam. **Sequoioideae** (Luerss.) C.J. Quinn in Amer. J. Bot., 87(7): 1056 (2000): fam. Sequoiaceae.

TAXODIACEAE: -
subfam. **Taiwanioideae** (Hayata) C.J. Quinn in Amer. J. Bot., 87(7): 1056 (2000), isonym; see L.Chu Li in Acta Bot. Yunnanica, 11(2): 123 (1989): fam. Taiwaniaceae.
subfam. **Wollemioideae** J. Silba in J. Int. Conifer Preserv. Soc., 6(1): 11 (1999), as 'Wollemiaceae'.

TETRADICLIDACEAE
fam. **Tetradiclidaceae** (Engl.) A.L. Takhtajan, Florist. Reg. World: 333 (1986): subfam. Tetradiclidoideae.

THYMELAEACEAE
tribus **Passerineae** (Domke) C.L. Bredenkamp & A.E. van Wyk in Grana, 35(6): 345 (1996 publ. 1997): subtribus Passerininae.

TILIACEAE
subfam. **Craigioideae** S.J. Xu & H.H. Hsue in J. Trop. Subtrop. Bot., 8(1): 15 (2000).

TRICYRTIDACEAE
fam. **Tricyrtidaceae** A. Takhtajan, Diversity & Classific. Flow. Pl.: 482 (1997).

UMBELLIFERAE
tribus **Aciphylleae** M.F. Watson & S.R. Downie in Amer. J. Bot., 87(2): 288 (2000).
tribus **Heteromorpheae** M.F. Watson & S.R. Downie in Amer. J. Bot., 87(2): 289 (2000).
tribus **Pleurospermeae** M.F. Watson & S.R. Downie in Amer. J. Bot., 87(2): 289 (2000).

XERONEMATACEAE
fam. **Xeronemataceae** M.W. Chase, P.J. Rudall & M.F. Fay in Kew Bull., 55(4): 869 (2000).

XEROPHYLLACEAE
fam. **Xerophyllaceae** A.L. Takhtajan in Bot. Zhurn., 81(2): 86 (1996).

ZYGOPHYLLACEAE
subfam. **Larreoideae** M.C. Sheahan & M.W. Chase in Bot. J. Linn. Soc., 122(4): 299 (1996).
subfam. **Seetzenioideae** M.C. Sheahan & M.W. Chase in Bot. J. Linn. Soc., 122(4): 299 (1996).

INDEX KEWENSIS

SUPPLEMENT TWENTY-ONE
NAMES AT GENERIC RANK AND BELOW

A

ABAREMA (Legumin.).
abbottii (Rose & Leonard) R.C. Barneby & J.W. Grimes, Silk Tree, Guanacaste, Monkey's Earring (Mem. New York Bot. Gard., 74(1)): 105 (1996): Pithecellobium abbottii.
acreana (J.F. Macbr.) L. Rico in Novon, 9(4): 555 (1999): Pithecellobium acreanum.
adenophora (Ducke) R.C. Barneby & J.W. Grimes, Silk Tree, Guanacaste, Monkey's Earring (Mem. New York Bot. Gard., 74(1)): 74 (1996): Pithecellobium adenophorum.
alexandri (Urb.) R.C. Barneby & J.W. Grimes, Silk Tree, Guanacaste, Monkey's Earring (Mem. New York Bot. Gard., 74(1)): 101 (1996): Pithecellobium jupunba var. alexandri.
var. troyana (Urb.) R.C. Barneby & J.W. Grimes, Silk Tree, Guanacaste, Monkey's Earring (Mem. New York Bot. Gard., 74(1)): 103 (1996): Pithecellobium alexandri var. troyanum.
asplenifolia (Griseb.) R.C. Barneby & J.W. Grimes, Silk Tree, Guanacaste, Monkey's Earring (Mem. New York Bot. Gard., 74(1)): 97 (1996): Pithecellobium asplenifolium.
auriculata (Benth.) R.C. Barneby & J.W. Grimes, Silk Tree, Guanacaste, Monkey's Earring (Mem. New York Bot. Gard., 74(1)): 62 (1996): Pithecellobium auriculatum.
barbouriana (Standl.) R.C. Barneby & J.W. Grimes, Silk Tree, Guanacaste, Monkey's Earring (Mem. New York Bot. Gard., 74(1)): 70 (1996): Pithecellobium barbourianum.
var. arenaria (Ducke) R.C. Barneby & J.W. Grimes, Silk Tree, Guanacaste, Monkey's Earring (Mem. New York Bot Gard., 74(1)): 72 (1996): Pithecellobium arenarium.
brachystachya (DC.) R.C. Barneby & J.W. Grimes, Silk Tree, Guanacaste, Monkey's Earring (Mem. New York Bot. Gard., 74(1)): 91 (1996): Inga brachystachya.
callejasii R.C. Barneby & J.W. Grimes, Silk Tree, Guanacaste, Monkey's Earring (Mem. New York Bot. Gard., 74(1)): 85 (1996)—Colombia.
campestris (Benth.) R.C. Barneby & J.W. Grimes, Silk Tree, Guanacaste, Monkey's Earring (Mem. New York Bot. Gard., 74(1)): 82 (1996): Pithecellobium campestre.
centiflora R.C. Barneby & J.W. Grimes, Silk Tree, Guanacaste, Monkey's Earring (Mem. New York Bot. Gard., 74(1)): 109 (1996)—Bolivia.
cochleata (Willd.) R.C. Barneby & J.W. Grimes, Silk Tree, Guanacaste, Monkey's Earring (Mem. New York Bot. Gard., 74(1)): 83 (1996): Inga cochleata.
var. moniliformis (Ducke) R.C. Barneby & J.W. Grimes, Silk Tree, Guanacaste, Monkey's Earring (Mem. New York Bot. Gard., 74(1)): 85 (1996): Pithecellobium moniliforme.

ABAREMA:-
cochliacarpos (Gomes) R.C. Barneby & J.W. Grimes, Silk Tree, Guanacaste, Monkey's Earring (Mem. New York Bot. Gard., 74(1)): 94 (1996): Mimosa cochliacarpos.
commutata R.C. Barneby & J.W. Grimes, Silk Tree, Guanacaste, Monkey's Earring (Mem. New York Bot. Gard., 74(1)): 65 (1996)—Guyana, Venezuela, Surinam.
curvicarpa (H.S. Irwin) Barneby & J.W. Grimes
var. rodriguesii R.C. Barneby & J.W. Grimes, Silk Tree, Guanacaste, Monkey's Earring (Mem. New York Bot. Gard., 74(1)): 57 (1996)—Brazil North (Amazonas).
floribunda (Benth. R.C. Barneby & J.W. Grimes, Silk Tree, Guanacaste, Monkey's Earring (Mem. New York Bot. Gard., 74(1)): 48 (1996): Pithecellobium floribundum.
gallorum R.C. Barneby & J.W. Grimes, Silk Tree, Guanacaste, Monkey's Earring (Mem. New York Bot. Gard., 74(1)): 51 (1996)—French Guiana.
ganymedea R.C. Barneby & J.W. Grimes, Silk Tree, Guanacaste, Monkey's Earring (Mem. New York Bot. Gard., 74(1)): 59 (1996)—Colombia.
glauca (Urb.) R.C. Barneby & J.W. Grimes, Silk Tree, Guanacaste, Monkey's Earring (Mem. New York Bot. Gard., 74(1)): 98 (1996): Pithecellobium glaucum.
idiopoda (S.F. Blake) R.C. Barneby & J.W. Grimes, Silk Tree, Guanacaste, Monkey's Earring (Mem. New York Bot. Gard., 74(1)): 53 (1996): Pithecellobium idiopodum.
josephi R.C. Barneby & J.W. Grimes, Silk Tree, Guanacaste, Monkey's Earring (Mem. New York Bot. Gard., 74(1)): 109 (1996)—Colombia.
jupunba (Willd.) Britton & Killip
var. trapezifolia (Vahl) R.C. Barneby & J.W. Grimes, Silk Tree, Guanacaste, Monkey's Earring (Mem. New York Bot. Gard., 74(1)): 69 (1996): Mimosa trapezifolia.
killipii (Britton & Rose) R.C. Barneby & J.W. Grimes, Silk Tree, Guanacaste, Monkey's Earring (Mem. New York Bot. Gard., 74(1)): 89 (1996): Punjuba killipii.
laeta (Benth.) R.C. Barneby & J.W. Grimes, Silk Tree, Guanacaste, Monkey's Earring (Mem. New York Bot. Gard., 74(1)): 79 (1996): Pithecellobium laetum.
langsdorfii (Benth.) R.C. Barneby & J.W. Grimes, Silk Tree, Guanacaste, Monkey's Earring (Mem. New York Bot. Gard., 74(1)): 95 (1996): Pithecellobium langsdorfii.
lehmannii (Britton & Rose) R.C. Barneby & J.W. Grimes, Silk Tree, Guanacaste, Monkey's Earring (Mem. New York Bot. Gard., 74(1)): 88 (1996): Punjuba lehmannii.
leucophylla (Benth.) R.C. Barneby & J.W. Grimes, Silk Tree, Guanacaste, Monkey's Earring (Mem. New York Bot. Gard., 74(1)): 76 (1996): Pithecellobium leucophyllum.

ABAREMA:-
leucophylla (Benth.) R.C. Barneby & J.W. Grimes
var. vaupesensis R.C. Barneby & J.W. Grimes, Silk Tree, Guanacaste, Monkey's Earring (Mem. New York Bot. Gard., 74(1)): 78 (1996)—Colombia.
levelii (R.S. Cowan) R.C. Barneby & J.W. Grimes, Silk Tree, Guanacaste, Monkey's Earring (Mem. New York Bot. Gard., 74(1)): 106 (1996): Pithecellobium levelii.
longipedunculata (H.S. Irwin) R.C. Barneby & J.W. Grimes, Silk Tree, Guanacaste, Monkey's Earring (Mem. New York Bot. Gard., 74(1)): 52 (1996): Pithecellobium longipedunculatum.
macradenia (Pittier) R.C. Barneby & J.W. Grimes, Silk Tree, Guanacaste, Monkey's Earring (Mem. New York Bot. Gard., 74(1)): 57 (1996): Pithecellobium macradenium.
maestrensis (Urb.) M. Bässler in Fl. Repúbl. Cuba, Ser. A, 2: 121 (1998): Pithecellobium maestrense.
mataybifolia (Sandwith) R.C. Barneby & J.W. Grimes, Silk Tree, Guanacaste, Monkey's Earring (Mem. New York Bot. Gard., 74(1)): 78 (1996): Pithecellobium mataybifolium.
microcalyx (Benth.) R.C. Barneby & J.W. Grimes, Silk Tree, Guanacaste, Monkey's Earring (Mem. New York Bot. Gard., 74(1)): 63 (1996): Pithecellobium microcalyx.
var. enterolobioides R.C. Barneby & J.W. Grimes, Silk Tree, Guanacaste, Monkey's Earring (Mem. New York Bot. Gard., 74(1)): 64 (1996)—Brazil North (Pará).
var. parauaquarae (Ducke) R.C. Barneby & J.W. Grimes, Silk Tree, Guanacaste, Monkey's Earring (Mem. New York Bot. Gard., 74(1)): 65 (1996): Pithecellobium parauaquarae.
nipensis (Britton) R.C. Barneby & J.W. Grimes, Silk Tree, Guanacaste, Monkey's Earring (Mem. New York Bot. Gard., 74(1)): 104 (1996): Pithecellobium nipense.
obovalis (A. Rich.) R.C. Barneby & J.W. Grimes, Silk Tree, Guanacaste, Monkey's Earring (Mem. New York Bot. Gard., 74(1)): 99 (1996): Inga obovalis.
obovata (Benth.) R.C. Barneby & J.W. Grimes, Silk Tree, Guanacaste, Monkey's Earring (Mem. New York Bot. Gard., 74(1)): 91 (1996): Calliandra obovata.
oppositifolia (Urb.) R.C. Barneby & J.W. Grimes, Silk Tree, Guanacaste, Monkey's Earring (Mem. New York Bot. Gard., 74(1)): 103 (1996): Pithecellobium oppositifolium.
oxyphyllidia R.C. Barneby & J.W. Grimes, Silk Tree, Guanacaste, Monkey's Earring (Mem. New York Bot. Gard., 74(1)): 107 (1996)—Honduras.
racemiflora (Donn. Sm.) R.C. Barneby & J.W. Grimes, Silk Tree, Guanacaste, Monkey's Earring (Mem. New York Bot. Gard., 74(1)): 90 (1996): Pithecellobium racemiflorum.
turbinata (Benth.) R.C. Barneby & J.W. Grimes, Silk Tree, Guanacaste, Monkey's Earring (Mem. New York Bot. Gard., 74(1)): 94 (1996): Pithecellobium turbinatum.

ABAREMA:-
villifera *(Ducke) R.C. Barneby & J.W. Grimes, Silk (Mem. NewTree, Guanacaste, Monkey's Earring (Mem. New York Bot. Gard., 74(1)):* 108 (1996): Pithecellobium villiferum.
zolleriana *(Standl. & Steyerm.) R.C. Barneby & J.W. Grimes, Silk Tree, Guanacaste, Monkey's Earring (Mem. New York Bot. Gard., 74(1)):* 75 (1996), as 'zollerana': Pithecellobium zollerianum.

ABELIA (Caprifoliac.).
biflora Turcz.
f. minor *(Nakai) C.F. Fang in Clavis Pl. Chinae Bor.-Or. ed. 2 (ed. P.Y. Fu):* 617 (1995): A. biflora var. minor.
grandifolia *J.A. Villarreal Q. in Brittonia,* 52(2): 174 (2000)—Mexico Northeast (QuereéAtaro).
mexicana *J.A. Villarreal Q. in Brittonia,* 52(2): 172 (2000)—Mexico Southwest (Oaxaca).
occidentalis *J.A. Villarreal Q. in Brittonia,* 49(1): 84 (1997)—Mexico Northeast (Durango), Mexico Southwest (Jalisco).

ABELMOSCHUS (Malvac.).
cancellatus *(L.f.) J.O. Voigt, Hort. Suburb. Calcutt.:* 119 (1845): Hibiscus cancellatus.
chinensis *(Roxb.) J.O. Voigt, Hort. Suburb. Calcutt.:* 119 (1845): Hibiscus chinensis.
pungens *(Roxb.) J.O. Voigt, Hort. Suburb. Calcutt.:* 119 (1845): Hibiscus pungens.

ABIES (Pinac.).
beshanzuensis M.H. Wu
var. ziyuanensis *(L.K. Fu & S.L. Mo) L.K. Fu & Nan Li in Novon,* 7(3): 261 (1997): A. ziyuanensis.
cilicica (Antoine & Kotschy) Carrière
var. isaurica *(Coode & Cullen) J. Silba in J. Int. Conifer Preserv. Soc.,* 7(1): 18 (2000), as 'issaurica': A. cilicica subsp. isaurica.
delavayi Franch.
subsp. fansipanensis *(Q.P. Xiang, L.K. Fu & Nan Li) K. Rushforth in Int. Dendrol. Soc. Yearbook,* 1998: 62 (1999): A. fansipanensis.
fansipanensis *Q.P. Xiang, L.K. Fu & Nan Li in Acta Phytotax. Sin.,* 35(4): 356 (1997)—Vietnam.
ferreana Bordères & Gaussen
var. longibracteata *L.K. Fu & Nan Li in Novon,* 7(3): 261 (1997)—China South-central (Yunnan).
guatemalensis Rehder
var. ixtepejiensis *J. Silba in J. Int. Conifer Preserv. Soc.,* 7(1): 19 (2000)—Mexico Southwest (Oaxaca).
var. rushforthii *J. Silba in J. Int. Conifer Preserv. Soc.,* 7(1): 19 (2000)—Honduras.
var. tamaulipasensis *J. Silba in J. Int. Conifer Preserv. Soc.,* 5: 45 (1998 publ. 1997)—Mexico Northeast (Tamaulipas).
koreana E.H. Wilson
var. yuanbaoshanensis *(Y.J. Lu & L.K. Fu) J. Silba in J. Int. Conifer Preserv. Soc.,* 5: 53 (1998 publ. 1997), not effectively publ., see Art. 30 ICBN (1994): A. yuanbaoshanensis.
nephrolepis (Trautv.) Maxim.
var. kamchatkensis *J. Silba in J. Int. Conifer Preserv. Soc.,* 5: 61 (1998 publ. 1997), not effectively publ., see Art. 30 ICBN (1994); *et in J. Int. Conifer Preserv. Soc.,* 7(1): 20 (2000)—Kamchatka.
f. nigricarpa *Y.N. Lee, Fl. Korea:* 1156 (1996), without type—Locality not stated.
pindrow (Royle ex D. Don) Royle
subsp. gamblei *(Hickel) K. Rushforth in Int. Dendrol. Soc. Yearbook,* 1998: 63 (1999): A. gamblei.
religiosa (Humb., Bonpl. & Kunth) Schltdl. & Cham.
var. colimensis *(Rushforth & Narave) J. Silba in J. Int. Conifer Preserv. Soc.,* 7(1): 20 (2000): A. colimensis.
var. perotensis *J. Silba in J. Int. Conifer Preserv. Soc.,* 7(1): 21 (2000)—Mexico Gulf (Veracruz).
sibirica Ledeb.
var. alpina *(Poljakov ex Kom.) J. Silba in J. Int. Conifer Preserv. Soc.,* 7(1): 21 (2000): A. sibirica f. alpina.

ABILDGAARDIA (Cyperac.).
hispidula (Vahl) Lye
subsp. longespicata *(Lye) J.-P. Lebrun & A.L. Stork, Ënum. Pl. Fleurs Afr. Trop.,* 4: 597 (1997): Bulbostylis hispidula subsp. longespicata.
subsp. macroglumis *(Lye) J.-P. Lebrun & A.L. Stork, Ënum. Pl. Fleurs Afr. Trop.,* 4: 597 (1997): Bulbostylis hispidula subsp. macroglumis.
papillosa *R. Kral & M.T. Strong in Sida,* 18(3): 842 (1999)—Brazil Northeast (Bahia).

ABROTANELLA (Compos.).
rostrata *U. Swenson in New Zealand J. Bot.,* 34(1): 48 (1996)—New Zealand South I.

ABRUS (Legumin.).
melanospermus Hassk.
subsp. suffruticosus *(Boutique) D.K. Harder in Novon,* 10(2): 124 (2000): A. suffruticosus.
subsp. tenuiflorus *(Benth.) D.K. Harder in Novon,* 10(2): 124 (2000): A. tenuiflorus.

ABUTILON (Malvac.).
anglosomaliae *Cufod. ex M. Thulin in Kew Bull.,* 53(4): 1013 (1998), nom. nov.: A. molle *Baker.*
fraseri (Hook.) Walp.
subsp. diplotrichum *(F. Muell.) R.M. Barker in Muelleria,* 9: 113 (1996): A. diplotrichum.
subumbellatum *D. Philcox in M.D. Dassanayake & W.D. Clayton, eds., Revised Handb. Fl. Ceylon,* 11: 341 (1997)—Sri Lanka.

ACACIA (Legumin.).
abbatiana *L. Pedley in Austrobaileya,* 5(2): 313 (1999)—Queensland.
acanthaster *B.R. Maslin in Nuytsia,* 12(3): 312 (1999)—Western Australia.
acanthoclada F. Muell.
subsp. glaucescens *B.R. Maslin in Nuytsia,* 12(3): 314 (1999)—Western Australia.
acoma *B.R. Maslin in Nuytsia,* 12(3): 315 (1999)—Western Australia.
aculeiformis *B.R. Maslin in Nuytsia,* 12(3): 317 (1999)—Western Australia.
acuminata Benth.
subsp. burkitii *(F. Muell. ex Benth.) P.G. Kodela & M.D. Tindale in Telopea,* 7(4): 415 (1998): A. burkittii.
adinophylla *B.R. Maslin in Nuytsia,* 12(3): 318 (1999)—Western Australia.
aemula *B.R. Maslin in Nuytsia,* 10(2): 169 (1995)—Western Australia.
subsp. muricata *B.R. Maslin in Nuytsia,* 10(2): 171 (1995)—Western Australia.
alata R. Br.
var. tetrantha *B.R. Maslin in Nuytsia,* 10(2): 157 (1995)—Western Australia.
amputata *B.R. Maslin in Nuytsia,* 12(3): 493 (1999), nom. nov.: A. pulchella var. subsessilis *Maslin.*
amyctica *R.S. Cowan & B.R. Maslin in Nuytsia,* 10(2): 222 (1995)—Western Australia.
anasilla *A.S. George in J. Roy. Soc. West. Austral.,* 82(2): 67 (1999)—Western Australia.
ancistrophylla C.R.P. Andrews
var. perarcuata *R.S. Cowan & B.R. Maslin in Nuytsia,* 10(2): 227 (1995)—Western Australia.
angustissima (Mill.) Kuntze
var. leucothrix *(Standl.) B.L. Turner in Phytologia,* 81(1): 14 (1996 publ. 1997): A. leucothrix.
var. oaxacana *B.L. Turner in Phytologia,* 81(1): 14 (1996 publ. 1997)—Southwest (Oaxaca), Mexico Central (Puebla).
anthochaera *B.R. Maslin in Nuytsia,* 10(2): 183 (1995)—Western Australia.
applanata *B.R. Maslin in Nuytsia,* 10(2): 158 (1995)—Western Australia.
aprica *B.R. Maslin & A.R. Chapman in Nuytsia,* 12(3): 471 (1999)—Western Australia.
arbiana *L. Pedley in Austrobaileya,* 5(2): 307 (1999)—Queensland.
arcuatilis *R.S. Cowan & B.R. Maslin in Nuytsia,* 12(3): 472 (1999)—Western Australia.
argyrotricha *L. Pedley in Austrobaileya,* 5(2): 310 (1999)—Queensland.

ACACIA:-
aristulata *B.R. Maslin in Nuytsia,* 12(3): 320 (1999)—Western Australia.
aroma Gillies ex Hook. & Arn.
var. huarango *(Ruiz ex J.F. Macbr.) J.E. Ebinger, D.S. Seigler & H.D. Clarke in Syst. Bot.,* 25(4): 602 (2000): Acacia huarango.
asepala *B.R. Maslin in Nuytsia,* 12(3): 321 (1999)—Western Australia.
assimilis S. Moore
subsp. atroviridis *R.S. Cowan & B.R. Maslin in Nuytsia,* 10(2): 240 (1995)—Western Australia.
ataxiphylla Benth.
subsp. magna *B.R. Maslin in Nuytsia,* 12(3): 324 (1999)—Western Australia.
aulacophylla *R.S. Cowan & B.R. Maslin in Nuytsia,* 10(2): 241 (1995)—Western Australia.
auratiflora *R.S. Cowan & B.R. Maslin in Nuytsia,* 12(3): 414 (1999)—Western Australia.
aureocrinita *B.J. Conn & T.M. Tame in Austral. Syst. Bot.,* 9(6): 851 (1996)—New South Wales.
auripila *R.S. Cowan & B.R. Maslin in Nuytsia,* 12(3): 415 (1999)—Western Australia.
balsamea *R.S. Cowan & B.R. Maslin in Nuytsia,* 12(3): 417 (1999)—Western Australia.
barakulensis *L. Pedley in Austrobaileya,* 5(2): 308 (1999)—Queensland.
barbinervis Benth.
subsp. borealis *B.R. Maslin in Nuytsia,* 12(3): 327 (1999)—Western Australia.
bifaria *B.R. Maslin in Nuytsia,* 10(2): 160 (1995)—Western Australia.
blaxellii *B.R. Maslin in Nuytsia,* 12(3): 328 (1999)—Western Australia.
bracteolata *B.R. Maslin in Nuytsia,* 12(3): 329 (1999)—Western Australia.
brumalis *B.R. Maslin in Nuytsia,* 10(2): 185 (1995)—Western Australia.
burdekensis *L. Pedley in Austrobaileya,* 5(2): 313 (1999)—Queensland.
capillaris *A.S. George in J. Roy. Soc. West. Austral.,* 82(2): 69 (1999)—Western Australia.
carens *B.R. Maslin in Nuytsia,* 10(2): 172 (1995)—Western Australia.
carnosula *B.R. Maslin in Nuytsia,* 12(3): 331 (1999)—Western Australia.
castanostegia *B.R. Maslin in Nuytsia,* 12(3): 332 (1999)—Western Australia.
cavealis *R.S. Cowan & B.R. Maslin in Nuytsia,* 12(3): 454 (1999)—Western Australia.
celsa *M.D. Tindale in Austral. Syst. Bot.,* 13(1): 34 (2000)—Queensland.
cerastes *B.R. Maslin in Nuytsia,* 10(2): 173 (1995)—Western Australia.
cernua *M. Thulin & A.S. Hassan in Nordic J. Bot.,* 16(3): 303 (1996 publ. 1997)—Somalia.
chamaeleon *B.R. Maslin in Nuytsia,* 10(2): 189 (1995)—Western Australia.
chapmanii *R.S. Cowan & B.R. Maslin in Nuytsia,* 12(3): 455 (1999)—Western Australia.
subsp. australis *R.S. Cowan & B.R. Maslin in Nuytsia,* 12(3): 457 (1999)—Australia.
clandullensis *B.J. Conn & T.M. Tame in Austral. Syst. Bot.,* 9(6): 849 (1996)—New South Wales.
cochlocarpa Meisn.
subsp. velutinosa *B.R. Maslin & A.R. Chapman in Nuytsia,* 12(3): 475 (1999)—Western Australia.
colei Maslin & L. Thomson
var. ileocarpa *M.W. McDonald & B.R. Maslin in Nuytsia,* 11(2): 219 (1997)—Western Australia, Northern Territory.
concolorans *B.R. Maslin in Nuytsia,* 12(3): 334 (1999)—Western Australia.
congesta Benth.
subsp. cliftoniana *(W. Fitzg.) B.R. Maslin in Nuytsia,* 12(3): 338 (1999): A. cliftoniana.
subsp. wonganensis *B.R. Maslin in Nuytsia,* 12(3): 338 (1999)—Western Australia.
consanguinea *R.S. Cowan & B.R. Maslin in Nuytsia,* 10(2): 243 (1995)—Western Australia.
convallium *L. Pedley in Austrobaileya,* 5(2): 312 (1999)—Northern Territory.
cordifolia *R.Br. ex N. Lémon in Ann. Fl. Pomone (or J. Jardins),* 1836-37: 378 (1837)—Cultivation.

ACACIA:-
cracentis *R.S. Cowan & B.R. Maslin in Nuytsia,* 12(3): 476 (1999)—Western Australia.
cremiflora *B.J. Conn & T.M. Tame in Austral. Syst. Bot.,* 9(6): 853 (1996)—New South Wales.
crenulata *R.S. Cowan & B.R. Maslin in Nuytsia,* 12(3): 418 (1999)—Western Australia.
cummingiana *B.R. Maslin in Nuytsia,* 10(2): 175 (1995)—Western Australia.
cuneifolia *B.R. Maslin in Nuytsia,* 12(3): 339 (1999)—Western Australia.
deficiens *B.R. Maslin in Nuytsia,* 12(3): 340 (1999)—Western Australia.
diaphana *R.S. Cowan & B.R. Maslin in Nuytsia,* 12(3): 420 (1999)—Western Australia.
diminuta *B.R. Maslin in Nuytsia,* 12(3): 342 (1999)—Western Australia.
disparrima *M.W. McDonald & B.R. Maslin in Austral. Syst. Bot.,* 13(1): 46 (2000)—Queensland, New South Wales.
subsp. calidestris *M.W. McDonald & B.R. Maslin in Austral. Syst. Bot.,* 13(1): 52 (2000)—Queensland.
dissona *R.S. Cowan & B.R. Maslin in Nuytsia,* 10(2): 209 (1995)—Western Australia.
var. indoloria *R.S. Cowan & B.R. Maslin in Nuytsia,* 10(2): 211 (1995)—Western Australia.
donaldsonii *R.S. Cowan & B.R. Maslin in Nuytsia,* 12(3): 458 (1999)—Western Australia.
dorsenna *B.R. Maslin in Nuytsia,* 10(2): 192 (1995)—Western Australia.
elachantha *M.W. McDonald & B.R. Maslin in Austral. Syst. Bot.,* 10(3): 311 (1997)—Queensland, Western Australia, Northern Territory, South Australia.
emilioana *R.H. Fortunato & A.M. Cialdella in Candollea,* 51(1): 217 (1996)—Paraguay, Bolivia.
enervia Maiden & Blakely
subsp. explicata *R.S. Cowan & B.R. Maslin in Nuytsia,* 10(2): 232 (1995)—Western Australia.
ericksoniae *B.R. Maslin in Nuytsia,* 12(3): 343 (1999)—Western Australia.
errabunda *B.R. Maslin in Nuytsia,* 12(3): 345 (1999)—Western Australia.
euthyphylla *B.R. Maslin in Nuytsia,* 12(3): 347 (1999)—Western Australia.
evenulosa *B.R. Maslin in Nuytsia,* 12(3): 348 (1999)—Western Australia.
faucium *L. Pedley in Austrobaileya,* 5(2): 314 (1999)—Queensland.
filipes *L. Pedley in Austrobaileya,* 5(2): 315 (1999)—Northern Territory.
fodinalis *L. Pedley in Austrobaileya,* 5(2): 316 (1999)—Queensland.
formidabilis *C.A. Gardner ex R.S. Cowan & B.R. Maslin in Nuytsia,* 12(3): 460 (1999)—Western Australia.
gelasina *B.R. Maslin in Nuytsia,* 10(2): 194 (1995)—Western Australia.
gemina *R.S. Cowan & B.R. Maslin in Nuytsia,* 12(3): 421 (1999)—Western Australia.
glaucissima *B.R. Maslin Nuytsia,* 12(3): 350 (1999)—Western Australia.
gloeotricha *A.R. Chapman & B.R. Maslin in Nuytsia,* 12(3): 487 (1999)—Western Australia.
grahamii *E. Vajravelu in N.C. Nair & A.N. Henry, Fl. Tamil Nadu, India, Ser. 1* 1: Addenda ii (1983), nom. nov.: A. pennata var. canescens *Baker.*
graniticola *B.R. Maslin in Nuytsia,* 12(3): 351 (1999)—Western Australia.
guanacastensis *(H.D. Clarke, Seigler & Ebinger) J.E. Ebinger, D.S. Seigler & H.L. Clarke in Syst. Bot.,* 25(4): 610 (2000): Acacia farnesiana var. guanacastensis.
hadrophylla *R.S. Cowan & B.R. Maslin in Nuytsia,* 10(2): 214 (1995)—Western Australia.
hendersonii *L. Pedley in Austrobaileya,* 5(2): 309 (1999)—Queensland.
heteroclita Meisn.
subsp. valida *R.S. Cowan & B.R. Maslin in Nuytsia,* 12(3): 424 (1999)—Western Australia.
hopperiana *B.R. Maslin in Nuytsia,* 12(3): 495 (1999)—Western Australia.

ACACIA:-
hypermeces *A.S. George in J. Roy. Soc. West. Austral.,* 82(2): 70 (1999)—Western Australia.
hystrix *B.R. Maslin in Nuytsia,* 12(3): 353 (1999)—Western Australia.
subsp. continua *B.R. Maslin in Nuytsia,* 12(3): 356 (1999)—Western Australia.
imitans *B.R. Maslin in Nuytsia,* 12(3): 356 (1999)—Western Australia.
imparilis *B.R. Maslin in Nuytsia,* 12(3): 358 (1999)—Western Australia.
improcera *B.R. Maslin in Nuytsia,* 12(3): 359 (1999)—Western Australia.
incanicarpa *A.R. Chapman & B.R. Maslin in Nuytsia,* 12(3): 489 (1999)—Western Australia.
inceana Domin
subsp. conformis *R.S. Cowan & B.R. Maslin in Nuytsia,* 10(2): 234 (1995)—Western Australia.
insolita E. Pritz.
subsp. efoliolata *B.R. Maslin in Nuytsia,* 12(3): 362 (1999)—Western Australia.
subsp. recurva *B.R. Maslin in Nuytsia,* 12(3): 363 (1999)—Western Australia.
isoneura *B.R. Maslin & A.R. Chapman in Nuytsia,* 12(3): 478 (1999)—Western Australia.
subsp. nimia *B.R. Maslin & A.R. Chapman in Nuytsia,* 12(3): 479 (1999)—Western Australia.
johannis *L. Pedley in Austrobaileya,* 5(2): 311 (1999)—Queensland.
kalgoorliensis *R.S. Cowan & B.R. Maslin in Nuytsia,* 10(2): 215 (1995)—Western Australia.
lacertensis *L. Pedley in Austrobaileya,* 5(2): 316 (1999)—Northern Territory.
lanceolata *B.R. Maslin in Nuytsia,* 12(3): 363 (1999)—Western Australia.
laricina Meisn.
var. crassifolia *B.R. Maslin in Nuytsia,* 12(3): 367 (1999)—Western Australia.
latipes Benth.
subsp. licina *R.S. Cowan & B.R. Maslin in Nuytsia,* 12(3): 464 (1999)—Western Australia.
leptalea *B.R. Maslin in Nuytsia,* 12(3): 367 (1999)—Western Australia.
lineolata Benth.
subsp. multilineata *(W. Fitzg.) R.S. Cowan & B.R. Maslin in Nuytsia,* 10(2): 236 (1995): A. multilineata.
lirellata *B.R. Maslin & A.R. Chapman in Nuytsia,* 12(3): 480 (1999)—Western Australia.
subsp. compressa *B.R. Maslin & A.R. Chapman in Nuytsia,* 12(3): 482 (1999)—Western Australia.
lullfitziorum *B.R. Maslin in Nuytsia,* 12(3): 369 (1999)—Western Australia.
midgleyi *M.W. McDonald & B.R. Maslin in Austral. Syst. Bot.,* 13(1): 61 (2000)—Queensland.
mitodes *A.S. George in J. Roy. Soc. West. Austral.,* 82(2): 71 (1999)—Western Australia.
multistipulosa *M.D. Tindale & M. Bedward in Austral. Syst. Bot.,* 9(6): 859 (1996)—Northern Territory.
mutabilis *B.R. Maslin in Nuytsia,* 12(3): 371 (1999)—Western Australia.
subsp. angustifolia *B.R. Maslin in Nuytsia,* 12(3): 373 (1999)—Western Australia, South Australia.
subsp. incurva *B.R. Maslin in Nuytsia,* 12(3): 374 (1999)—Western Australia.
subsp. rhynchophylla *B.R. Maslin in Nuytsia,* 12(3): 375 (1999)—Western Australia.
subsp. stipulifera *B.R. Maslin in Nuytsia,* 12(3): 376 (1999)—Western Australia.
nigripilosa Maiden
subsp. latifolia *B.R. Maslin in Nuytsia,* 12(3): 378 (1999)—Western Australia.
nivea *R.S. Cowan & B.R. Maslin in Nuytsia,* 10(2): 250 (1995)—Western Australia.
obesa *R.S. Cowan & B.R. Maslin in Nuytsia,* 10(2): 252 (1995)—Western Australia.
octandra *Jacques in Ann. Fl. Pomone,* 1837–38: 164 (1838)—India.
olivensana *G.P. Lewis in Kew Bull.,* 51(2): 372 (1996)—Brazil Northeast (Bahia).

ACACIA:-
ophiolithica *R.S. Cowan & B.R. Maslin in Nuytsia,* 10(2): 246 (1995)—Western Australia.
oviedoensis *R.G. García & M.M. Mejía in Moscosoa,* 11: 7 (2000)—Dominican Republic.
pachyphloia W. Fitzg.
subsp. brevipinnula *M.D. Tindale & P.G. Kodela in Austral. Syst. Bot.,* 9(3): 314 (1996)—Western Australia.
pachyphylla *B.R. Maslin in Nuytsia,* 12(3): 379 (1999), nom. nov.: A. gonophylla var. crassifolia *Benth.*
papulosa *R.S. Cowan & B.R. Maslin in Nuytsia,* 10(2): 219 (1995)—Western Australia.
pedina *P.G. Kodela & T.M. Tame in Telopea,* 8(3): 305 (1999)—New South Wales.
pelophila *R.S. Cowan & B.R. Maslin in Nuytsia,* 12(3): 428 (1999)—Western Australia.
peregrina *M.W. McDonald & B.R. Maslin in Austral. Syst. Bot.,* 13(1): 67 (2000), homonym; *see Willd., Sp. Pl.,* 4: 1073 (1806)—New Guinea (Papua New Guinea, Irian Jaya).
pinguiculosa *R.S. Cowan & B.R. Maslin in Nuytsia,* 12(3): 429 (1999)—Western Australia.
subsp. teretifolia *R.S. Cowan & B.R. Maslin in Nuytsia,* 12(3): 431 (1999)—Western Australia.
plautella *B.R. Maslin in Nuytsia,* 12(3): 381 (1999)—Western Australia.
polyphylla DC.
var. giganticarpa *G.P. Lewis in Kew Bull.,* 51(3): 591 (1996)—Brazil Northeast (Bahia), Brazil Southeast (Minas Gerais).
profusa *B.R. Maslin in Nuytsia,* 12(3): 383 (1999)—Western Australia.
proiantha *L. Pedley in Austrobaileya,* 5(2): 318 (1999)—Northern Territory.
pterocaulon *B.R. Maslin in Nuytsia,* 10(2): 165 (1995)—Western Australia.
puncticulata *B.R. Maslin in Nuytsia,* 12(3): 384 (1999)—Western Australia.
pusilla *B.R. Maslin in Nuytsia,* 12(3): 386 (1999)—Western Australia.
qandalensis *M. Thulin in Nordic J. Bot.,* 18(5): 513 (1998)—Somalia.
quinquenervia *B.R. Maslin in Nuytsia,* 12(3): 387 (1999)—Western Australia.
recurvata *R.S. Cowan & B.R. Maslin in Nuytsia,* 12(3): 432 (1999)—Western Australia.
repens *A.S. George in J. Roy. Soc. West. Austral.,* 82(2): 71 (1999)—Western Australia, Northern Territory.
resinosa *R.S. Cowan & B.R. Maslin in Nuytsia,* 12(3): 433 (1999)—Western Australia.
rhamphophylla *B.R. Maslin in Nuytsia,* 12(3): 389 (1999)—Western Australia.
rigescens *M.D. Tindale & M. Bedward in Austral. Syst. Bot.,* 9(6): 864 (1996)—Northern Territory.
rigida *B.R. Maslin in Nuytsia,* 12(3): 390 (1999)—Western Australia.
rostellata *B.R. Maslin in Nuytsia,* 12(3): 392 (1999)—Western Australia.
rubricola *L. Pedley in Austrobaileya,* 5(2): 309 (1999)—Queensland.
sabulosa *B.R. Maslin in Nuytsia,* 12(3): 393 (1999)—Western Australia, Northern Territory.
santosii *G.P. Lewis in Kew Bull.,* 51(2): 371 (1996)—Brazil Northeast (Bahia).
scalena *B.R. Maslin in Nuytsia,* 12(3): 396 (1999)—Western Australia.
scleroclada *B.R. Maslin in Nuytsia,* 10(2): 196 (1995)—Western Australia.
sclerophylla Lindl.
var. pilosa *R.S. Cowan & B.R. Maslin in Nuytsia,* 12(3): 440 (1999)—Western Australia.
scopulorum *L. Pedley in Austrobaileya,* 5(2): 319 (1999), as 'scopularum'—Northern Territory.
sedifolia Maiden & Blakely
subsp. pulvinata *B.R. Maslin in Nuytsia,* 12(3): 398 (1999)—Western Australia.
sirissa *(Roxb.) Buch.-Ham. ex J.O. Voigt, Hort. Suburb. Calcutt.:* 261 (1845): Mimosa sirissa.
smeringa *A.S. George in J. Roy. Soc. West. Austral.,* 82(2): 73 (1999)—Western Australia.
solenota *L. Pedley in Austrobaileya,* 5(2): 319 (1999)—Queensland.
speckii *R.S. Cowan & B.R. Maslin in Nuytsia,* 12(3): 465 (1999)—Western Australia.

ACACIA:-
sphacelata Benth.
subsp. recurva *B.R. Maslin in Nuytsia,* 12(3): 401 (1999)—Western Australia.
subsp. verticillata *B.R. Maslin in Nuytsia,* 12(3): 402 (1999)—Western Australia.
sphenophylla *B.R. Maslin in Nuytsia,* 12(3): 403 (1999)—Western Australia.
suaveolens (Sm.) Willd.
subsp. grampianensis *D.A. Morrison & A.J. Rupp in Austral. Syst. Bot.,* 8(6): 1025 (1995)—Victoria.
subsp. montana *D.A. Morrison & A.J. Rupp in Austral. Syst. Bot.,* 8(6): 1025 (1995)—New South Wales.
subsp. myallensis *D.A. Morrison & A.J. Rupp in Austral. Syst. Bot.,* 8(6): 1024 (1995)—New South Wales.
subsp. prostrata *D.A. Morrison & A.J. Rupp in Austral. Syst. Bot.,* 8(6): 1024 (1995)—New South Wales, Tasmania.
subflexuosa Maiden
subsp. capillata *R.S. Cowan & B.R. Maslin in Nuytsia,* 12(3): 443 (1999)—Western Australia.
subrigida *B.R. Maslin in Nuytsia,* 10(2): 198 (1995)—Western Australia.
subsessilis *A.R. Chapman & B.R. Maslin in Nuytsia,* 12(3): 490 (1999)—Western Australia.
suma *(Roxb.) Buch.-Ham. ex J.O. Voigt, Hort. Suburb. Calcutt.:* 260 (1845): Mimosa suma.
tephrophylla *M. Thulin in Nordic J. Bot.,* 18(5): 515 (1998)—Somalia.
tetraneura *B.R. Maslin & A.R. Chapman in Nuytsia,* 12(3): 483 (1999)—Western Australia.
tetraptera *B.R. Maslin in Nuytsia,* 12(3): 405 (1999)—Western Australia.
thomsonii *B.R. Maslin & M.W. McDonald in Nuytsia,* 10(3): 444 (1996)—Western Australia, Northern Territory, Queensland.
tingoorensis *L. Pedley in Austrobaileya,* 5(2): 320 (1999), nom. nov.: A. longispicata subsp. velutina *Pedley.*
trinalis *R.S. Cowan & B.R. Maslin in Nuytsia,* 12(3): 444 (1999)—Western Australia.
truculenta *B.R. Maslin in Nuytsia,* 12(3): 407 (1999)—Western Australia.
trulliformis *R.S. Cowan & B.R. Maslin in Nuytsia,* 12(3): 446 (1999)—Australia.
tuberculata *B.R. Maslin in Nuytsia,* 12(3): 408 (1999)—Western Australia.
undosa *R.S. Cowan & B.R. Maslin in Nuytsia,* 10(2): 220 (1995)—Western Australia.
valida *M.D. Tindale & P.G. Kodela in Austral. Syst. Bot.,* 9(3): 307 (1996)—Northern Territory, Western Australia.
vittata *R.S. Cowan & B.R. Maslin in Nuytsia,* 12(3): 448 (1999)—Western Australia.
whibleyana *R.S. Cowan & B.R. Maslin in Nuytsia,* 10(2): 228 (1995)—South Australia.
wilcoxii *B.R. Maslin in Nuytsia,* 10(2): 200 (1995)—Western Australia.
wilsonii *R.S. Cowan & B.R. Maslin in Nuytsia,* 12(3): 449 (1999)—Western Australia.
xerophila W. Fitzg.
var. brevior *(E. Pritz.) B.R. Maslin in Nuytsia,* 12(3): 410 (1999): A. fitzgeraldii var. brevior.
zatrichota *A.S. George in J. Roy. Soc. West. Austral.,* 82(2): 73 (1999)—Western Australia.

ACAENA (Rosac.).
sect. Patagonicae *A. Marticorena in Novon,* 9(2): 227 (1999).
patagonica *A. Marticorena in Novon,* 9(2): 227 (1999)—Chile South (Magallanes), Argentina South (Santa Cruz, Tierra del Fuego).

ACALYPHA (Euphorbiac.).
burquezii *V.W. Steinmann & R.S. Felger in Aliso,* 16(1): 11 (1997)—Mexico Northwest (Sonora, Sinaloa).
filiformis Poir.
var. goudotiana *(Baill.) R. Govaerts in R. Govaerts, D.G. Frodin & A. Radcliffe-Smith, World Checklist & Bibliogr.* Euphorbiaceae, 1: 62 (2000): A. goudotiana.
var. ovalifolia *(Baill.) R. Govaerts in R. Govaerts, D.G. Frodin & A. Radcliffe-Smith, World Checklist & Bibliogr.* Euphorbiaceae, 1: 62 (2000): A. ovalifolia.

ACALYPHA:-
filiformis Poir.
var. pervilleana *(Baill.) R. Govaerts in R. Govaerts, D.G. Frodin & A. Radcliffe-Smith, World Checklist & Bibliogr.* Euphorbiaceae, 1: 62 (2000): A. pervilleana.
subsp. rubra *(Müll. Arg.) R. Govaerts in R. Govaerts, D.G. Frodin & A. Radcliffe-Smith, World Checklist & Bibliogr.* Euphorbiaceae, 1: 62 (2000): A. reticulata var. rubra.
var. urophylla *(Baill.) R. Govaerts in R. Govaerts, D.G. Frodin & A. Radcliffe-Smith, World Checklist & Bibliogr.* Euphorbiaceae, 1: 63 (2000): A. urophylla.
var. urophylloides *(Pax & K. Hoffm.) R. Govaerts in R. Govaerts, D.G. Frodin & A. Radcliffe-Smith, World Checklist & Bibliogr.* Euphorbiaceae, 1: 63 (2000): A. reticulata var. urophylloides.
insulana Müll. Arg.
var. anisodonta *(Müll. Arg.) R. Govaerts in R. Govaerts, D.G. Frodin & A. Radcliffe-Smith, World Checklist & Bibliogr.* Euphorbiaceae, 1: 68 (2000): A. anisodonta.
leicesterfieldiensis *A. Radcliffe-Smith & R. Govaerts in Kew Bull.,* 52(2): 477 (1997), nom. nov.: A. jamaicensis *Britton.*
longipetiolata *J.M. Cardiel in An. Jard. Bot. Madrid,* 57(1): 57 (1999)—Venezuela.
medibracteata *A. Radcliffe-Smith & R. Govaerts in Kew Bull.,* 52(2): 477 (1997), nom. nov.: A. gagnepainii *Leandri.*
var. calcicola *(Leandri) A. Radcliffe-Smith & R. Govaerts in Kew Bull.,* 52(2): 477 (1997): A. gagnepainii var. calcicola.
muelleriana *I. Urban, Symb. Antill.,* 1: 338 (1899)—Venezuela.
paxii *J.B. Palacký, Cat. Pl. Madagasc.,* 2: 25 (1907), nom. nov: A. urophylla *Pax.*
polynesiaca *A. Radcliffe-Smith & R. Govaerts in Kew Bull.,* 52(2): 477 (1997), nom. nov.: A. stokesii *F. Br.*
rubrinervis *Q.C.B. Cronk in Bull. Nat. Hist. Mus. (London), Bot.,* 25(1): 98 (1995), nom. nov.: A. rubra *Roxb.*
stellata *J.M. Cardiel in Novon,* 10(4): 362 (2000)—Ecuador.
taiwanensis *S.S. Ying in Mem. Coll. Agric. Nation. Taiwan Univ.,* 31(1): 32 (1991), contrary to Art. 37.4 ICBN (1994)—Taiwan.
ulei *A. Radcliffe-Smith & R. Govaerts in Kew Bull.,* 52(2): 478 (1997), nom. nov.: A. macrophylla *Ule.*
venezuelica *J.M. Cardiel in An. Jard. Bot. Madrid,* 57(1): 59 (1999)—Venezuela.
websteri *J.M. Cardiel in Novon,* 10(4): 360 (2000)—Ecuador.

ACAMPE (Orchidac.).
thailandica *G. Seidenfaden, Contrib. Orchid Fl. Thailand XIII:* 49 (1997)—Thailand.

ACANTHOCALYCIUM (Cactac.).
klimpelianum (Weidlich & Werderm.) Backeb.
var. macranthum *(Rausch) J.G. Lambert, Cactus d'Argentine, ed. 2:* 9 (1998): Lobivia spiniflora var. macrantha.
thionanthum (Speg.) Backeb.
var. munitum *(Rausch) J.G. Lambert, Cactus d'Argentine, ed. 2:* 16 (1998): Lobivia thionantha var. munita.

ACANTHOLIMON (Plumbaginac.).
jarmilae *J.J. Halda in Acta Mus. Richnov. Sect. Nat.,* 7(2): 71 (2000)—Kirgizistan.

ACANTHOPHYLLUM (Caryophyllac.).
coloratum Schischk.
var. eglandulosum *G.A. Lazkov in Bot. Zhurn.,* 85(8): 117 (2000)—Kirgizistan.

ACANTHOSYRIS (Santalac.).
annonagustata *C.U. Ulloa & P.M. Jørgensen in Novon,* 8(1): 84 (1998)—Ecuador.
asipapote *M. Nee in Brittonia,* 48(4): 574 (1996 publ. 1997)—Bolivia.

ACANTHUS (Acanthac.).
platyphyllus *(Murb.) D. Prain in Index Kewensis,* Suppl. 4: 2 (1913): A. mollis subsp. platyphyllus.

ACANTHUS:-
xiamenensis *R.T. Zhang in Wuyi Sci. J.,* 5: 239 (1985)—China Southeast (Fujian).

ACER (Acerac.).
sect. Apetala *N.N. Tzvelev in Fl. Vostochnoĭ Evropy̆,* 9: 343 (1996).
amamiense *T. Yamazaki in J. Jap. Bot.,* 75(5): 282 (2000)—Nansei-shoto.
bodinieri H. Lév.
var. nayongense *(W.P. Fang) K. Rushforth Int. Dendrol. Soc. Yearbook,* 1998: 72 (1999): A. nayongense.
macropterum *T.Z. Hsu & H. Sun in Acta Bot. Yunnanica,* 19(1): 29 (1997)—Xizang.
medogense *T.Z. Hsu & Z.K. Zhou in Acta Bot. Yunnanica,* 19(1): 29 (1997)—Xizang.
miyabei Maxim.
f. shibatae *(Nakai) K. Ogata in Fl. Japan,* 2c: 70 (1999): A. shibatae.
nipponicum Hara
var. australe *T. Yamazaki in J. Jap. Bot.,* 75(2): 113 (2000)—Japan.
var. koshinense *T. Yamazaki in J. Jap. Bot.,* 75(2): 114 (2000)—Japan.
subsp. orientale *T. Yamazaki in J. Jap. Bot.,* 75(2): 113 (2000)—Japan.
var. orientale *T. Yamazaki in J. Jap. Bot.,* 75(2): 113 (2000)—Japan.
platanoides L.
f. drummondii *(Drummond ex Schwer.) D. Geerinck in Belgian J. Bot.,* 130(2): 120 (1998): A. platanoides [var.] drummondii.
f. meyeringii *D. Geerinck in Belgian J. Bot.,* 130(2): 120 (1998)—Cultivation.
pluridens *T.Z. Hsu & H. Sun in Acta Bot. Yunnanica,* 19(1): 31 (1997)—Xizang.
sieboldianum Miq.
f. dissectum *Baba ex T. Shimizu, Fl. Nagano Pref.:* 1509 (1997)—Japan.
verhaegheanum *D. Geerinck in Nat. Belg.,* 78(1): 9 (1997)—Cultivation. (A. cappadocicum × platanoides).

ACETOSA (Polygonac.).
pratensis Mill.
subsp. fontanopaludosa *(Kalela) N.N. Tzvelev in Novosti Sist. Vy̆ssh. Rast.,* 32: 183 (2000): Rumex fontanopaludosus.

ACHARAGMA *(N.P. Taylor) A. D. Zimmerman ex C.E. Glass, Guia Identific. Cact. Amenazadas México,* 1: Ac/ag (1998 publ. 1997): Escobaria sect. Acharagma. (Cactac.).
aguirreana *(Glass & R.A. Foster) C.E. Glass, Guía Identific. Cact. Amenazadas México,* 1: Ac/ag (1998 publ. 1997): Gymnocactus aguirreanus.
roseana *(Boed.) E.F. Anderson in Cact. Succ. J. (U.S.A.),* 71(6): 323 (1999): Echinocactus roseanus.

ACHILLEA (Compos.).carpatica *Błocki ex O.N. Dubovik in Novosti Sist. Vy̆ssh. Nizsh. Rast. (Kiev),* 1974: 93 (1974)—Ukraina.
clavennae L.
f. velebitica *Ž. Bjelčić in Acta Bot. Croat.,* 43: 76 (1984)—Yugoslavia.
millefolium L.
subsp. ceretanica *(Sennen) O. de Bolòs & J. Vigo, Fl. Països Catalans,* 3: 796 (1995): A. millefolium proles ceretanica.
oxyloba (DC.) Sch. Bip.
subsp. barrelieri *(Ten.) F. Conti in Bocconea,* 10: 14, 172 (1998): Anthemis barrelieri.
setacea Waldst. & Kit.
subsp. asiatica *(Serg.) V.N. Voroschilov in A.K. Skvortsov (ed.), Florist. issl. v razn. raĭonakh SSSR:* 195 (1985): A. asiatica.
wilhelmsii K. Koch
subsp. santolinoides *(Lag.) K.M. Valant-Vetschera & A. Kästner in Feddes Repert.,* 109(7–8): 504 (1998): A. santolinoides.

ACHNATHERUM (Gramin.).
sect. Achnatheropsis *(Tzvelev) J. Holub in Preslia,* 70(2): 99 (1998), isonym; *see Prob. in Sosud. Rast. Sovet. Dal'nego Vostoka,* 1: 343 (1985): Stipa sect. Achnatheropsis.

ACHNATHERUM:-
brandisii *(Mez) Z.L. Wu in Acta Phytotax. Sin.*, 34(2): 154 (1996): Stipa brandisii.
caudatum *(Trin.) S.W.L. Jacobs & J. Everett in Telopea*, 6(4): 582 (1996): Stipa caudata.
lobatum *(Swallen) Z.L. Wu in Acta Phytotax. Sin.*, 34(2): 154 (1996), isonym; *see Barkworth in Phytologia*, 74(1): 9 (1993): Stipa lobata.
mesatlasica *(Quézel) M. Ibn Tattou in Bocconea*, 8: 212 (1998): Calamagrostis argentea subsp. mesatlasica.
petriei *(Buchanan) S.W.L. Jacobs & J. Everett in Telopea*, 6(4): 582 (1996): Stipa petriei.
stillmanii *(Bol.) Z.L. Wu in Acta Phytotax. Sin.*, 34(2): 154 (1996), isonym; *see Barkworth in Phytologia*, 74(1): 14 (1993): Stipa stillmanii.
wallowaensis *J. Maze & K.A. Robson in Madroño*, 43(3): 401 (1996)—Oregon.

ACHYRANTHES (Amaranthac.).
bidentata Blume
var. katsuudakemontana *(Tawada) S. Hatusima in S. Hatusima & T. Amano, Fl. Ryukyus, S. of Amami Is., ed. 2:* 46 (1994), without basionym ref.: Basionym not stated.
var. villosa *J.L. Liu in Acta Phytotax. Sin.*, 34(3): 275 (1996)—China South-central (Sichuan).
lanuginosa *T. Nuttall in Trans. Amer. Philos. Soc., n.s.*, 5: 166 (1835)—Arkansas.

ACHYROCLINE (Compos.).
gaudens *V.M. Badillo & M. González Sánchez in Ernstia*, 8(2–3): 3 (1998)—Venezuela.

ACHYROSPERMUM (Labiat.).
purpureum *P.B. Phillipson in Fl. Madagascar*, fam. 175: 48 (1998)—Madagascar.

ACIANTHUS (Orchidac.).
sect. Caudatus *P.J. Kores in Allertonia*, 7(3): 194 (1995).
sect. Cyrtostylis *(R. Br.) P.J. Kores in Allertonia*, 7(3): 181 (1995): gen. Cyrtostylis.
sect. Macropetalus *P.J. Kores in Allertonia*, 7(3): 149 (1995).
subgen. Univiscidiatus *P.J. Kores in Allertonia*, 7(3): 148 (1995).
sect. Univiscidiatus *P.J. Kores in Allertonia*, 7(3): 163 (1995).
halleanus *P.J. Kores in Allertonia*, 7(3): 149 (1995)—New Caledonia.

ACIDOSASA *C.D. Chu & C.S. Chao ex P.C. Keng in J. Bamboo Res.*, 1(2): 165 (1982); *see Taxon*, 46(1): 106 (1997). *GRAMINEAE*
bilamina *W.T. Lin & Z.M. Wu in J. S. China Agric. Univ.*, 14(3): 113 (1993)—China Southeast (Guangdong).
chinensis *C.D. Chu & C.S. Chao ex P.C. Keng in J. Bamboo Res.*, 1(2): 165 (1982); *see Taxon*, 46(1): 106 (1997)—China Southeast (Guangdong).
denigrata *W.T. Lin in J. S. China Agric. Univ.*, 14(1): 48 (1993)—China Southeast (Guangdong).
hirtiflora *Z.P. Wang & G.H. Ye ex C.S. Chao & C.D. Chu in Acta Phytotax. Sin.*, 29(6): 524 (1991); *see Taxon*, 46(1): 106 (1997)—China Southeast (Guangxi).
longiligula
var. amara *(T.H. Wen) D. Ohrnberger, Bamboos of the World Intro. 2:* 10 (1996): Indosasa longiligula var. amara.
venusta *(McClure) Z.P. Wang & G.H. Ye ex C.S. Chao & C.D. Chu in Acta Phytotax. Sin.*, 29(6): 524 (1991); *see Taxon*, 46(1): 106 (1997): Semiarundinaria venusta.

ACILEPIS (Compos.).
aspera *(Buch.-Ham.) H. Robinson in Proc. Biol. Soc. Wash.*, 112(1): 226 (1999): Vernonia aspera.
clivorum *(Hance) H. Robinson in Proc. Biol. Soc. Wash.*, 112(1): 226 (1999): Vernonia clivorum.
dalzelliana *(J.R. Drumm. & Hutch.) H. Robinson in Proc. Biol. Soc. Wash.*, 112(1): 226 (1999): Vernonia dalzelliana.
dendigulensis *(DC.) H. Robinson in Proc. Biol. Soc. Wash.*, 112(1): 226 (1999): Decaneurum dendigulense.

ACILEPIS:-
nantcianensis *(Pamp.) H. Robinson in Proc. Biol. Soc. Wash.*, 112(1): 226 (1999): Vernonia bracteata var. nantcianensis.
saligna *(DC.) H. Robinson in Proc. Biol. Soc. Wash.*, 112(1): 226 (1999): Vernonia saligna.
scariosa *(DC.) H. Robinson in Proc. Biol. Soc. Wash.*, 112(1): 227 (1999): Decaneurum scariosum.
silhetensis *(DC.) H. Robinson in Proc. Biol. Soc. Wash.*, 112(1): 227 (1999): Decaneurum silhetense.
spirei *(Gand.) H. Robinson in Proc. Biol. Soc. Wash.*, 112(1): 227 (1999): Vernonia spirei.

ACINOS (Labiat.).
alpinus (L.) Moench.
f. albiflorus *E. Brandis ex Č. Šilić, Monografija Satureja, Calamintha, Micromeria, Acinos i Clinopodium fl.* Jugoslavije: 281 (1979)—Yugoslavia.
var. dinaricus *Č. Šilić, Monografija Satureja, Calamintha, Micromeria, Acinos i Clinopodium fl.* Jugoslavije: 274 (1979)—Yugoslavia.
subsp. dinaricus *Č. Šilić, Monografija Satureja, Calamintha, Micromeria, Acinos i Clinopodium fl.* Jugoslavije: 274 (1979)—Yugoslavia.
f. hirsutus *(Pant.) Č. Šilić, Monografija Satureja, Calamintha, Micromeria, Acinos i Clinopodium fl.* Jugoslavije: 281 (1979): A. alpinus var. hirsutus.
arvensis (Lam.) Dandy
var. argutus *(Rchb.) Č. Šilić, Monografija Satureja, Calamintha, Micromeria, Acinos i Clinopodium fl.* Jugoslavije: 331 (1979): A. villosus var. argutus.
var. lancifolius *(Murb.) Č. Šilić, Monografija Satureja, Calamintha, Micromeria, Acinos i Clinopodium fl.* Jugoslavije: 327 (1979): Calamintha acinos var. lancifolia.
hungaricus *(Simonk.) Č. Šilić, Monografija Satureja, Calamintha, Micromeria, Acinos i Clinopodium fl.* Jugoslavije: 296 (1979): Melissa hungarica.
f. albiflorus *(K. Malý) Č. Šilić, Monografija Satureja, Calamintha, Micromeria, Acinos i Clinopodium fl.* Jugoslavije: 312 (1979): Satureja majoranifolia f. albiflora.
f. coloratus *(Bornm.) Č. Šilić, Monografija Satureja, Calamintha, Micromeria, Acinos i Clinopodium fl.* Jugoslavije: 313 (1979): Calamintha patavina f. colorata.
f. cuneatus *(Simonk.) Č. Šilić, Monografija Satureja, Calamintha, Micromeria, Acinos i Clinopodium fl.* Jugoslavije: 313 (1979): Calamintha hungarica var. cuneata.
var. doiranensis *(Bornm.) Č. Šilić, Monografija Satureja, Calamintha, Micromeria, Acinos i Clinopodium fl.* Jugoslavije: 313 (1979): Calamintha patavina var. doiranensis.
f. kindingeri *(Adamović) Č. Šilić, Monografija Satureja, Calamintha, Micromeria, Acinos i Clinopodium fl.* Jugoslavije: 313 (1979): Calamintha patavina var. kindigeri.
var. marginatus *(Borbás) Č. Šilić, Monografija Satureja, Calamintha, Micromeria, Acinos i Clinopodium fl.* Jugoslavije: 314 (1979): Calamintha alpina var. marginata.
f. ozrensis *(Ritter-Studn.) Č. Šilić, Monografija Satureja, Calamintha, Micromeria, Acinos i Clinopodium fl.* Jugoslavije: 312 (1979): Satureja majoranifolia f. ozrensis.
f. rosiflorus *(K. Malý) Č. Šilić, Monografija Satureja, Calamintha, Micromeria, Acinos i Clinopodium fl.* Jugoslavije: 312 (1979): Satureja majoranifolia f. rosiflora.
suaveolens *(Sibth. & Sm.) G. Don in J.C. Loudon, Hort. Brit.*: 239 (1830): Thymus suaveolens.†
f. acuminatus *(Velen.) Č. Šilić, Monografija Satureja, Calamintha, Micromeria, Acinos i Clinopodium fl.* Jugoslavije: 340 (1979): Calamintha suaveolens var. acuminata.
subsp. bulgaricus *J. Chrtek in Čas. Nár. Muz. Praze Přír.*, 157(1–4): 143 (1988 publ. 1991)—Bulgaria.
subsp. kalampacanus *J. Chrtek in Čas. Nár. Muz. Praze Přír.*, 157(1–4): 144 (1988 publ. 1991)—Greece.
subsp. thessalicus *J. Chrtek in Čas. Nár. Muz. Praze Přír.*, 157(1–4): 144 (1988 publ. 1991)—Greece.

ACINOS:-
suaveolens *(Sibth. & Sm.) G. Don*
f. tinctus *(Bornm.) Č. Šilić, Monografija Satureja, Calamintha, Micromeria, Acinos i Clinopodium fl.* Jugoslavije: 340 (1979): Calamintha suaveolens f. tincta.
var. viridis *(Heldr. & Hausskn.) Č. Šilić, Monografija Satureja, Calamintha, Micromeria, Acinos i Clinopodium fl.* Jugoslavije: 340 (1979): Calamintha suaveolens var. viridis.

ACION *B.G. Briggs & L.A.S. Johnson in Telopea*, 7(4): 353 (1998). *RESTIONACEAE*
hookeri *(D.I. Morris) B.G. Briggs & L.A.S. Johnson in Telopea*, 8(1): 22 (1998): Restio hookeri.
monocephalum *(R. Br.) B.G. Briggs & L.A.S. Johnson in Telopea*, 7(4): 355 (1998): Restio monocephalus.

ACIOTIS (Melastomatac.).
alata *(Beurl.) F. Almeda in Novon*, 7(4): 335 (1997 publ. 1998): Spennera alata.

ACISANTHERA (Melastomatac.).
leptalea *F. Almeda in Brittonia*, 51(2): 159 (1999)—Bolivia.

ACMISPON (Legumin.).
sect. Anisolotus *(Bernh.) D.D. Sokoloff in Ann. Bot. Fenn.*, 37(2): 129 (2000): gen. Anisolotus.
brachycarpus *(Benth.) D.D. Sokoloff in Ann. Bot. Fenn.*, 37(2): 130 (2000): Hosackia brachycarpa.
denticulatus *(Drew) D.D. Sokoloff in Ann. Bot. Fenn.*, 37(2): 130 (2000): Hosackia denticulata.
maritimus *(Nutt.) D.D. Sokoloff in Ann. Bot. Fenn.*, 37(2): 129 (2000): Hosackia maritima.
parviflorus *(Benth.) D.D. Sokoloff in Ann. Bot. Fenn.*, 37(2): 129 (2000): Hosackia parviflora.
rubriflorus *(H. Sharsm.) D.D. Sokoloff in Ann. Bot. Fenn.*, 37(2): 130 (2000): Lotus rubriflorus.
subpinnatus *(Lag.) D.D. Sokoloff in Ann. Bot. Fenn.*, 37(2): 130 (2000): Lotus subpinnatus.
wrangelianus *(Fisch. & C.A. Mey.) D.D. Sokoloff in Taxon*, 48(1): 58 (1999): Lotus wrangelianus.

ACOMIS (Compos.).
bella *A.E. Holland in Austrobaileya*, 4(4): 609 (1997)—Queensland.

ACONITUM (Ranunculac.).
subsect. Catenata *(Steinb. ex Riedl) A.N. Luferov in Bot. Zhurn.*, 85(10): 91 (2000): Aconitum sect. Catenata.
sect. Rotundifolia *Steinb. ex Y. Kadota in J. Jap. Bot.*, 74(5): 287 (1999).
altissimum Mill.
subsp. penninum *(Ser.) J. Holub in Preslia*, 70(2): 99 (1998): A. lycoctonum var. penninum.
barbatum Pers.
subsp. pekinense *(Vorosch.) I.A. Gubanov in Fl. Vostoch. Khangaya (MNR):* 127 (1983): A. pekinense.
confertiflorum (DC.) Gáyer
var. armenum *Vorosch. ex A.N. Luferov in Bot. Zhurn.*, 85(10): 91 (2000)—Transcaucasus (Armenia).
var. versicolor *(Steven ex Ser.) A.N. Luferov in Bot. Zhurn.*, 85(10): 91 (2000): Aconitum anthora var. versicolor.
degenii
nothosubsp. lippertianum *W. Starmühler in Fritschiana*, 7: 11 (1996)—Switzerland. (A. degenii subsp. paniculatum × A. degenii subsp. rhaeticum).
subsp. rhaeticum *W. Starmühler in Fritschiana*, 7: 2 (1996)—Switzerland.
delphinifolium DC.
subsp. kuzenevae *(Vorosch.) V.N. Voroschilov in A.K. Skvortsov (ed.), Florist. issl. v razn. raĭonakh SSSR:* 168 (1985): A. kuzenevae.
subsp. pseudokusnezowii *(Vorosch.) V.N. Voroschilov in A.K. Skvortsov (ed.), Florist. issl. v razn. raĭonakh SSSR:* 168 (1985): A. pseudokusnezowii.

ACONITUM :-
fengii W.T. Wang
var. crispulum *Q.E. Yang in Acta Phytotax. Sin.*, 37(6): 572 (1999)—China South-central (Yunnan).
fukutomei Hayata
var. formosanum *(Tamura) T.Y.A. Yang & T.C. Huang in Taiwania*, 41(2): 119 (1996): A. formosanum.
heterophyllum Wall. ex Royle
var. roylei *L.B. Chaudhary & R.R. Rao in Feddes Repert.*, 109(7–8): 532 (1998)—Jammu-Kashmir.
iidemontanum *Y. Kadota, Kita & Ueda in J. Jap. Bot.*, 74(5): 290 (1999)—Japan.
jaluense Kom.
var. paniculigerum *(Nakai) S.X. Li in Clavis Pl. Chinae Bor.-Or. ed. 2 (ed. P.Y. Fu)*: 182 (1995): A. paniculigerum.
kamelinii *A.A. Solovjev in Turczaninowia*, 1(2): 5 (1998), as 'kamelini'—Mongolia.
karafutense Miyabe & Nakai
subsp. baburinii *(Vorosch.) V.N. Voroschilov in A.K. Skvortsov (ed.), Florist. issl. v razn. raĭonakh SSSR*: 168 (1985): A. karafutense var. baburinii.
kirghistanicum *Y. Kadota in J. Jap. Bot.*, 74(5): 287 (1999)—Kirgizstan.
korshinskyi *N.N. Tzvelev in Bot. Zhurn.*, 81(12): 113 (1996 publ. 1997)—Russia East.
kusnezoffii Rchb.
var. birobidshanicum *(Vorosch.) S.X. Li in Clavis Pl. Chinae Bor.-Or. ed. 2 (ed. P.Y. Fu)*: 183 (1995): A. birobidshanicum.
lasiocarpum (Rchb.) Gáyer
subsp. kotulae *(Pawł.) W. Starmühler & J. Mitka in Acta Soc. Bot. Pol.*, 69(2): 150 (2000): A. variegatum subsp. kotulae.
nagarum Stapf
var. acaule *(Finet & Gagnep.) Q.E. Yang in Acta Phytotax. Sin.*, 37(6): 557 (1999): A. napellus var. acaule.
napellus L.
nothosubsp. hinterhuberi *W. Starmühler in Fritschiana*, 18: 12 (1999)—Austria. (A. napellus subsp. formosum × napellus subsp. napellus).
nothosubsp. polatschekii *Mucher ex W. Starmühler in Fritschiana*, 18: 13 (1999)—Austria. (A. napellus subsp. formosum × napellus subsp. lobelii).
nothosubsp. seitzii *Mucher ex W. Starmühler in Fritschiana*, 18: 14 (1999)—Austria. (A. napellus subsp. lobelii × napellus subsp. napellus).
nasutum Fisch. ex Rchb.
subsp. tuscheticum *(N. Busch) A.N. Luferov in Bot. Zhurn.*, 85(10): 93 (2000): Aconitum caucasicum var. tuscheticum.
neosachalinense H. Lév.
subsp. miyabei *(Nakai) V.N. Voroschilov in A.K. Skvortsov (ed.), Florist. issl. v razn. raĭonakh SSSR*: 168 (1985): A. miyabei.
okuyamae Nakai
var. wagaense *Y. Kadota in J. Jap. Bot.*, 74(5): 292 (1999)—Japan.
× pawlowskii *J. Mitka & W. Starmühler in Acta Soc. Bot. Pol.*, 69(2): 151 (2000)—Poland, Czechoslovakia. (A. lasiocarpum × variegatum).
prutense *(Zapał.) N.N. Tzvelev in Bot. Zhurn.*, 81(12): 115 (1996 publ. 1997): A. paniculatum var. prutense.
pulcherrimum Nakai
subsp. birobidshanicum *(Vorosch.) V.N. Voroschilov in A.K. Skvortsov (ed.), Florist. issl. v razn. raĭonakh SSSR*: 168 (1985): A. birobidshanicum.
racemulosum Franch.
var. austrokoreense *(Koidz.) Y.N. Lee, Fl. Korea*: 1158 (1996): A. austrokoreense.
ranunculoides Turcz. ex Ledeb.
subsp. ajanense *(Steinb.) V.N. Voroschilov in A.K. Skvortsov (ed.), Florist. issl. v razn. raĭonakh SSSR*: 169 (1985): A. ajanense.
subsp. gigas *(H. Lév. & Vaniot) V.N. Voroschilov in A.K. Skvortsov (ed.), Florist. issl. v razn. raĭonakh SSSR*: 169 (1985): A. gigas.
subsp. puchonroenicum *(Uyeki & Sakata) V.N. Voroschilov in A.K. Skvortsov (ed.), Florist. issl. v razn. raĭonakh SSSR*: 169 (1985): A. puchonroenicum.

ACONITUM :-
ranunculoides Turcz. ex Ledeb.
subsp. umbrosum *(Korsh.) V.N. Voroschilov in A.K. Skvortsov (ed.), Florist. issl. v razn. raĭonakh SSSR*: 169 (1985): A. lycoctonum f. umbrosum.
rilongense *Y. Kadota in J. Jap. Bot.*, 74(5): 283 (1999)—China South-central (Sichuan).
tauricum Wulfen
var. eustachyum *(Rchb.) W. Starmühler in Hladnikia*, 6: 40 (1996): A. eustachyum.
subsp. latemarense *(Degen & Gáyer) W. Starmühler in W. Maurer, Fl. Steiermark*, 1: 84 (1996): A. latemarense.
tuguancunense *Q.E. Yang in Acta Phytotax. Sin.*, 33(6): 572 (1995)—China South-central (Yunnan).
variegatum L.
var. carniolicum *W. Starmühler in Fritschiana*, 10: 1 (1997)—Yugoslavia, Austria.
villosum Rchb.
var. daxinganlinense *(Y.Z. Zhao) S.X. Li in Clavis Pl. Chinae Bor.-Or. ed. 2 (ed. P.Y. Fu)*: 182 (1995): A. daxinganlinense.
subsp. tschangbaischanense *(S.X. Li & Y.H. Huang) S.X. Li in Clavis Pl. Chinae Bor.-Or. ed. 2 (ed. P.Y. Fu)*: 182 (1995): A. tschangbaischanense.

ACONOGONON (Polygonac.).
× fennicum *Reiersen in Nordic J. Bot.*, 19(4): 386 (1999)—Norway. (A. alpinum × weyrichii).
molle (D. Don) H. Hara
var. paniculatum *(Blume) K. Yonekura & Hiroyoshi Ohashi in J. Jap. Bot.*, 72(3): 157 (1997): Polygonum paniculatum.
weyrichii (F. Schmidt) H. Hara
subsp. savatieri *(Nakai) N.N. Tzvelev in Novosti Sist. Vÿssh. Rast.*, 32: 183 (2000): Polygonum savatieri.

ACOURTIA (Compos.).
ciprianoi *J.L. Panero & J.L. Villaseñor in Brittonia*, 51(1): 87 (1999)—Mexico Southwest (Oaxaca).

ACRACHNE (Gramin.).
sundararajii *P. Umamaheswari, S.A. Muthukumar & P. Daniel in Kew Bull.*, 52(4): 1007 (1997)—India (Tamil Nadu).

ACRITOPAPPUS (Compos.).
catolesensis *D.J.N. Hind & H.P. Bautista in Kew Bull.*, 55(4): 950 (2000)—Brazil Northeast (Bahia).
diamantinicus *H.P. Bautista, S. Ortiz & J. Rodríguez-Oubiña in Nordic J. Bot.*, 20(2): 173 (2000)—Brazil Northeast (Bahia).
pintoi *H.P. Bautista & D.J.N. Hind in Kew Bull.*, 55(4): 951 (2000)—Brazil Northeast (Bahia).

ACRODON (Aizoac.).
purpureostylus *(L. Bolus) P.M. Burgoyne in Aloe*, 35(2): 61 (1998): Mesembryanthemum purpureostylum.
quarcicola *H.E.K. Hartmann in Bradleya*, 14: 33 (1996)—Cape Province.

ACRONEMA (Umbellif.).
edosmioides *(H. Boissieu) M.G. Pimenov & E.V. Kljuykov in Feddes Repert.*, 110(7–8): 481 (1999): Pimpinella edosmioides.

ACROPTILON (Compos.).
repens (L.) DC.
subsp. australe *(Iljin) I.A. Gubanov, Konsp. Fl. Vneshnei Mongolii (Sosud. Rast.)*: 95 (1996): A. australe.

ACTAEA (Ranunculac.).
sect. Dichanthera *(P.K. Hsiao) J.A. Compton in Taxon*, 47(3): 623 (1998): Cimicifuga sect. Dichanthera.
sect. Oligocarpae *(Tamura) J.A. Compton in Taxon*, 47(3): 618 (1998): Cimicifuga sect. Oligocarpae.
sect. Pityrosperma *(Siebold & Zucc.) J.A. Compton in Taxon*, 47(3): 619 (1998): gen. Pityrosperma.
sect. Podocarpae *J.A. Compton in Taxon*, 47(3): 620 (1998).

ACTAEA :-
sect. Souliea *(Franch.) J.A. Compton in Taxon*, 47(3): 613 (1998): gen. Souliea.
acuminata Wall. ex Royle
subsp. asiatica *(H. Hara) A.N. Luferov in Komarovia*, 1: 61 (1999): A. asiatica.
arizonica *(S. Watson) J.A. Compton in Taxon*, 47(3): 618 (1998): Cimicifuga arizonica.
bifida *(Nakai) J.A. Compton in Taxon*, 47(3): 624 (1998): Cimicifuga heracleifolia var. bifida.
brachycarpa *(P.K. Hsiao) J.A. Compton in Taxon*, 47(3): 622 (1998): Cimicifuga brachycarpa.
cimicifuga L.
var. racemosa *(Regel) J.A. Compton in Taxon*, 47(3): 623 (1998): Cimicifuga foetida var. racemosa.
europaea *(Schipcz.) J.A. Compton in Taxon*, 47(3): 623 (1998): Cimicifuga europaea.
heracleifolia *(Kom.) J.A. Compton in Taxon*, 47(3): 624 (1998): Cimicifuga heracleifolia.
japonica Thunb.
var. acutiloba *(H. Hara) J.A. Compton in Taxon*, 47(3): 620 (1998): Cimicifuga japonica var. acutiloba.
kashmiriana *(J. Compton & Hedd.) J.A. Compton in Taxon*, 47(3): 622 (1998): Cimicifuga kashmiriana.
laciniata *(S. Watson) J.A. Compton in Taxon*, 47(3): 621 (1998).Cimicifuga laciniata.
mairei *(H. Lév.) J.A. Compton in Taxon*, 47(3): 622 (1998): Cimicifuga mairei.
var. foliolosa *(P.K. Hsiao) J.A. Compton in Taxon*, 47(3): 622 (1998): Cimicifuga foetida var. foliolosa.
matsumurae *(Nakai) J.A. Compton & T.A.J. Heddersen in Taxon*, 47(3): 625 (1998): Cimicifuga foetida var. matsumurae.
purpurea *(P.K. Hsiao) J.A. Compton in Taxon*, 47(3): 619 (1998): Cimicifuga acerina f. purpurea.
racemosa L.
var. dissecta *(A. Gray) J.A. Compton in Taxon*, 47(3): 614 (1998): Cimicifuga racemosa var. dissecta.
taiwanensis *J.A. Compton, T.A.J. Heddersen & T.Y. Yang in Taxon*, 47(3): 624 (1998)—Taiwan.
vaginata *(Maxim.) J.A. Compton in Taxon*, 47(3): 613 (1998): Isopyrum vaginatum.
yesoensis *(Nakai) J.A. Compton & T.A.J. Heddersen in Taxon*, 47(3): 624 (1998): Cimicifuga simplex var. yesoensis.
yunnanensis *(P.K. Hsiao) J.A. Compton in Taxon*, 47(3): 621 (1998).Cimicifuga yunnanensis.

ACTINIDIA (Actinidiac.).
arguta (Siebold & Zucc.) Planch. ex Miq.
var. purpurea *(Rehder) C.F. Liang ex Q.Q. Chang in Wuyi Sci. J.*, 2: 27 (1982): A. purpurea.

ACTINOSERIS (Compos.).
arenaria *(Baker) N. Roque in Kew Bull.*, 52(1): 199 (1997): Trichocline arenaria.

ACTINOTUS (Umbellif.).
laxus *G.J. Keighery in Nuytsia*, 13(1): 23 (1999)—Western Australia.
periculosus *M.J. Henwood in Telopea*, 8(4): 456 (2000)—Queensland.
whicheranus *G.J. Keighery in Nuytsia*, 13(1): 24 (1999)—Western Australia.

× **ADAPASIA** *hort. in Orchid Rev.*, 105(1217): 318 (1997). *ORCHIDACEAE* (Ada × Aspasia).

ADELIA (Euphorbiac.).
retusa *J. Graham, Cat. Pl. Bombay*: 185 (1839)—India (Karnataka).

ADENANTHE (Ochnac.).
ciliata *C. Sastre in Novon*, 8(3): 301 (1998)—Guyana.

ADENANTHERA (Legumin.).
aculeata *(Roxb.) W. Hunter in Asiat. Res.*, 6: 66 (1799): Prosopis aculeata.

ADENARAKE (Ochnac.).
macrocarpa *C. Sastre in Novon*, 8(3): 301 (1998)—Venezuela.

ADENIA (Passiflorac.).
lapiazicola *M. Bardot-Vaucoulon in Adansonia,* 19(1): 156 (1997)—Madagascar.
lobata (Jacq.) Engl.
subsp. rumicifolia *(Engl.) K.A. Lye in Lidia,* 4(3): 92 (1998): A. rumicifolia.
subsp. schweinfurthii *(Engl.) K.A. Lye in Lidia,* 4(3): 92 (1998): A. schweinfurthii.

ADENIUM (Apocynac.).
obesum
f. albiflorum *Lodé, Fichier Encycl. Cact. Autres Succ.,* Ser. 15: no. 1394 (1995), without latin descr. or type; *cf. Repert. Pl. Succ. (I.O.S.),* 46: 5 (1995 publ. 1996)—Locality not stated.

ADENOCALYMMA (Bignoniac.).
magnoalatum *V.V. Scudeller in Novon,* 10(3): 234 (2000)—Brazil Southeast (Minas Gerais).
ubatubensis *M.A. Assis & J. Semir in Novon,* 9(2): 136 (1999)—Brazil Southeast (São Paulo).

ADENOCARPUS (Legumin.).
battandieri *(Maire) S. Talavera in An. Jard. Bot. Madrid,* 57(1): 216 (1999): Cytisus battandieri.
complicatus (L.) J. Gay
subsp. bracteatus *(Font Quer & Pau) S. Talavera & P.E. Gibbs in Lagascalia,* 18(2): 267 (1996): A. bracteatus.
subsp. parvifolius *(DC.) R. García Adá, G. López & P. Vargas in Candollea,* 51(2): 376 (1996): A. parvifolius.
desertorum *S. Castroviejo in An. Jard. Bot. Madrid,* 57(1): 43 (1999)—Spain.
gibbsianus *S. Castroviejo & S. Talavera in An. Jard. Bot. Madrid,* 56(1): 177 (1998)—Spain.
lainzii *(Castrov.) S. Castroviejo in An. Jard. Bot. Madrid,* 57(1): 43 (1999): A. complicatus subsp. lainzii.

ADENOPHORA (Campanulac.).
sect. Basiphyllae *P.F. Tu & G.J. Xu in Acta Bot. Bor.-Occid. Sin.,* 18(4): 615 (1998).
subsect. Liliifoliae *P.F. Tu & G.J. Xu in Acta Bot. Bor.-Occid. Sin.,* 18(4): 619 (1998).
subsect. Ovatifoliae *P.F. Tu & G.J. Xu in Acta Bot. Bor.-Occid. Sin.,* 18(4): 616 (1998).
sect. Remotiflorae *P.F. Tu & G.J. Xu in Acta Bot. Bor.-Occid. Sin.,* 18(4): 618 (1998).
subsect. Tricuspidatae *P.F. Tu & G.J. Xu in Acta Bot. Bor.-Occid. Sin.,* 18(4): 617 (1998).
curvidens Nakai
f. alba *Y.N. Lee, Fl. Korea:* 1161 (1996), without type—Locality not stated.
erecta *S. Lee, Joongku Lee & S. Kim in J. Pl. Res.,* 110(1097): 77 (1997)—South Korea.
grandiflora Nakai
f. albiflora *Y.N. Lee, Fl. Korea:* 1161 (1996), without type—Locality not stated.
izuensis *H. Ohba & S. Watanabe in J. Jap. Bot.,* 73(2): 81 (1998)—Japan.
microsperma *Y.Y. Qian in Guihaia,* 18(1): 9 (1998)—China South-central (Yunnan).
ningxianica *S. Ge & D.Y. Hong in Novon,* 9(1): 46 (1999)—Inner Mongolia (Ningxia), China North-central (Gansu).
pereskiifolia (Fisch. ex Roem. & Schult.) G. Don
subsp. sublata *(Kom.) V.N. Voroschilov in A.K. Skvortsov (ed.), Florist. issl. v razn. raïonakh SSSR:* 193 (1985): A. sublata.
polydentata *P.F. Tu & G.J. Xu in J. China Pharm. Univ.,* 21(1): 16 (1990)—China North-central (Shanxi).
potaninii Korsh.
var. bockiana *(Diels) S.W. Liu in Fl. Qinghaiica,* 3: 317 (1996): A. bockiana.
racemosa *J. Lee & S. Lee in Korean J. Pl. Taxon.,* 20(2): 122 (1990)—Korea.
remotiflora (Siebold & Zucc.) Miq.
var. hirticalyx *S. Lee, Y. Chung & Joongku Lee in Korean J. Pl. Taxon.,* 20(3): 192 (1990)—South Korea.
stenanthina
subsp. xifengensis *P.F. Tu & Y.S Zhou in J. China Pharm. Univ.,* 21(1): 16 (1990)—China North-central (Gansu).

ADENOPHORA:-
stricta
subsp. henanica *P.F. Tu & G.J. Xu in J. China Pharm. Univ.,* 21(1): 17 (1990)—China Southeast (Henan).
var. nanjingensis *P.F. Tu & G.J. Xu in J. China Pharm. Univ.,* 21(1): 17 (1990)—China Southeast (Jiangsu).
var. qinglongshanica *P.F. Tu & G.J. Xu in J. China Pharm. Univ.,* 21(1): 17 (1990)—China Southeast (Jiangsu).
suolunensis *P.F. Tu & X.F. Zhao in Acta Bot. Bor.-Occid. Sin.,* 18(4): 616 (1998)—Inner Mongolia (Nei Mongol).
tetraphylla (Thunb.) Fisch.
f. angustifolia *(Regel) C.Y. Li in Clavis Pl. Chinae Bor.-Or. ed. 2 (ed. P.Y. Fu):* 631 (1995): A. verticillata f. angustifolia.
triphylla (Thunb.) A.DC.
f. albiflora *Y.N. Lee, Fl. Korea:* 1161 (1996), without type—Locality not stated.
verticillata Fisch.
var. primorskensis *A.E. Kozhevnikov in Sosud. Rast. Sovet. Dal'nego Vostoka,* 8: 280 (1996)—Primorye.
xifengensis *(P.F. Tu & Y.S. Zhou) P.F. Tu & Y.S. Zhou in Acta Bot. Bor.-Occid. Sin.,* 18(4): 617 (1998): A. stenanthina subsp. xifengensis.

ADENOSTEMMA (Compos.).
lavenia (L.) Kuntze
var. angustifolium *(Arn.) B.P. Uniyal in Fl. India,* 12: 347 (1995): A. angustifolium.

ADESMIA (Legumin.).
candida Hook. f.
var. cabrerae *(Burkart) E.A. Ulibarri & A. Burkart in Darwiniana,* 38(1–2): 96 (2000): A. cabrerae.
comechingona *C.A. Bianco & F. Weberling in Feddes Repert.,* 110(7–8): 515 (1999)—Argentina Northeast (Córdoba).
cordobensis Burkart
var. appendiculata *E.A. Ulibarri & A. Burkart in Darwiniana,* 38(1–2): 84 (2000)—Argentina Northwest (San Luis).
corymbosa Clos
var. sinepenna *E.A. Ulibarri & A. Burkart in Darwiniana,* 38(1–2): 81 (2000)—Argentina South (Santa Cruz).
globosa *M. Davyt & P. Izaguirre in Parodiana,* 9(1–2): 99 (1996)—Uruguay, Argentina Northeast (Entre Ríos).
incana Vogel
var. oblata *M. Davyt & P. Izaguirre in Parodiana,* 9(1–2): 104 (1996)—Uruguay.
minor (Hook. & Arn.) Burkart
var. caespitosa *(Phil.) E.A. Ulibarri & A. Burkart in Darwiniana,* 38(1–2): 117 (2000): A. caespitosa.
var. riojana *(Burkart) E.A. Ulibarri & A. Burkart in Darwiniana,* 38(1–2): 117 (2000): A. riojana.
punctata (Poir.) DC.
var. sessiliflora *M. Davyt & P. Izaguirre in Parodiana,* 9(1–2): 109 (1996)—Uruguay.

ADINANDRA (Theac.).
auriformis *L.K. Ling & S.X. Liang in Fl. Reipubl. Popul. Sin.,* 50(1): 190 (1998)—China Southeast (Guangxi).
collettiana *T.K. Paul in Bull. Bot. Surv. India,* 35(1–4): 126 (1993 publ. 1997)—Burma.
grandis *L.K. Ling in Fl. Reipubl. Popul. Sin.,* 50(1): 190 (1998)—China South-central (Yunnan).
hirta Gagnep.
var. macrobracteata *L.K. Ling in Fl. Reipubl. Popul. Sin.,* 50(1): 191 (1998)—China South-central (Guizhou), China Southeast (Guangxi).
incornuta *(Y.C. Wu) T.L. Ming in Fl. Yunnanica,* 8: 345 (1997): Cleyera incornuta.
japonica *(Thunb.) T.L. Ming in Fl. Yunnanica,* 8: 344 (1997): Cleyera japonica.
var. wallichiana *(DC.) T.L. Ming in Fl. Yunnanica,* 8: 344 (1997): Cleyera ochnacea var. wallichiana.

ADINANDRA:-
lancipetala *L.K. Ling in Fl. Reipubl. Popul. Sin.,* 50(1): 192 (1998)—China Southeast (Guangxi), China South-central (Yunnan).
latifolia *L.K. Ling in Fl. Reipubl. Popul. Sin.,* 50(1): 189 (1998)—China South-central (Yunnan).
nigroglandulosa *L.K. Ling in Fl. Reipubl. Popul. Sin.,* 50(1): 193 (1998)—China South-central (Yunnan).
pingbianensis *L.K. Ling in Fl. Reipubl. Popul. Sin.,* 50(1): 188 (1998)—China South-central (Yunnan).

ADONANTHE (Ranunculac.).
apennina *(L.) A.N. Sennikov in Novosti Sist. Vyssh. Rast.,* 31: 83 (1998): Adonis apennina.
brevistyla *(Franch.) A.N. Sennikov in Novosti Sist. Vyssh. Rast.,* 31: 83 (1998): Adonis brevistyla.
chrysocyatha *(Hook. f. & Thomson) A.N. Sennikov in Novosti Sist. Vyssh. Rast.,* 31: 84 (1998): Adonis chrysocyatha.
coerulea *(Maxim.) A.N. Sennikov in Novosti Sist. Vyssh. Rast.,* 31: 83 (1998): Adonis coerulea.
cyllenea *(Boiss., Heldr. & Orph.) A.N. Sennikov in Novosti Sist. Vyssh. Rast.,* 31: 84 (1998): Adonis cyllenea.
davidii *(Franch.) A.N. Sennikov in Novosti Sist. Vyssh. Rast.,* 31: 83 (1998): Adonis davidii.
distorta *(Ten.) A.N. Sennikov in Novosti Sist. Vyssh. Rast.,* 31: 84 (1998): Adonis distorta.
sutchuenensis *(Franch.) A.N. Sennikov in Novosti Sist. Vyssh. Rast.,* 31: 83 (1998): Adonis sutchuenensis.

ADONIS (Ranunculac.).
subsect. Aestivales *(W.T. Wang) A.N. Luferov in Bot. Zhurn.,* 84(8): 110 (1999): A. ser. Aestivales.
subsect. Amurenses *(Poschk.) M.H. Hoffmann in Pl. Syst. Evol.,* 211(1–2): 45 (1998): A. ser. Amurenses.
subsect. Dentatae *(W.T. Wang) A.N. Luferov in Bot. Zhurn.,* 84(8): 110 (1999): A. ser. Dentatae.
amurensis Regel & Radde
f. argentatus *Y.N. Lee, Fl. Korea:* 1157, 182 (1996), as 'algentatus', contrary to Arts. 37.4, 37.5 ICBN (1994)—South Korea.
subsp. nanus *Y.N. Lee, Fl. Korea:* 1157 (1996), contrary to Arts. 37.4, 37.5 ICBN (1994)—South Korea.
f. viridescensicalyx *Y.N. Lee, Fl. Korea:* 1157 (1996), contrary to Arts. 37.4, 37.5 ICBN (1994)—South Korea.

ADRIANA (Euphorbiac.).
tomentosa Gaudich.
var. hookeri *(F. Muell.) C.L. Gross & M.A. Whalen in Austral. Syst. Bot.,* 9(5): 767 (1996): Trachycaryon hookeri.

AECHMEA (Bromeliac.).
aculeatosepala *(Rauh & Barthlott) E.M.C. Leme, Canistrum - Bromeliads Atlantic Forest:* 95 (1997): Neoregelia aculeatosepala.
amorimii *E.M.C. Leme in J. Bromeliad Soc.,* 46(1): 19 (1996)—Brazil Northeast (Bahia).
andersonii *H.E. Luther & E.M.C. Leme in J. Bromeliad Soc.,* 48(3): 127 (1998)—Brazil Northeast (Bahia).
aquilega (Salisb.) Griseb.
f. alba *F. Oliva-Esteve in J. Bromeliad Soc.,* 49(2): 77 (1999)—Venezuela.
bruggeri *E.M.C. Leme in Bromélia,* 3(3): 24 (1996)—Brazil Southeast (Minas Gerais).
canaliculata *E.M.C. Leme & H.E. Luther in Selbyana,* 19(2): 184 (1998 publ. 1999)—Brazil Northeast (Bahia).
confusa *H.E. Luther in Selbyana,* 19(2): 224 (1998 publ. 1999), nom. nov.: Greigia amazonica *L.B. Sm.*
echinata *(Leme) E.M.C. Leme, Canistrum - Bromeliads Atlantic Forest:* 79 (1997): Wittrockia echinata.
glandulosa *E.M.C. Leme in Harvard Pap. Bot.,* 4(1): 144 (1999)—Brazil Northeast (Bahia).
incompta *E.M.C. Leme & H.E. Luther in Selbyana,* 19(2): 186 (1998 publ. 1999)—Brazil Northeast (Bahia).
leonard-kentiana *H.E. Luther & E.M.C. Leme in Bromélia,* 3(3): 10 (1996)—Brazil Northeast (Bahia).

AECHMEA:-
lingulatoides *E.M.C. Leme & H.E. Luther in Selbyana,* 19(2): 187 (1998 publ. 1999)—Brazil Northeast (Bahia).
linharesiorum *E.M.C. Leme in Harvard Pap. Bot.,* 4(1): 145 (1999), as 'linharesii'—Brazil Northeast (Bahia).
maasii *E. Gouda & W. Till in Bromélia,* 4(1): 4 (1997)—Brazil Southeast (Espírito Santo).
manzanaresiana *H.E. Luther in Selbyana,* 19(2): 222 (1998 publ. 1999)—Ecuador.
paradoxa *(Leme) E.M.C. Leme, Canistrum - Bromeliads Atlantic Forest:* 81 (1997): Wittrockia paradoxa.
patriciae *H.E. Luther in J. Bromeliad Soc.,* 49(4): 160 (1999)—Ecuador.
reclinata *C. Sastre & R. Brithmer in Biogeographica,* 75(1): 43 (1999)—Windward Is. (Martinique).
roeseliae *H.E. Luther Selbyana,* 19(1): 86 (1998)—Ecuador.
vasquezii *H.E. Luther in Selbyana,* 21(1–2): 129 (2000)—Peru.
weberi *(E. Pereira & Leme) E.M.C. Leme, Canistrum - Bromeliads Atlantic Forest:* 80 (1997): Nidularium weberi.
wuelfinghoffii *E. Gross in Bot. Jahrb.,* 1998(1): 13 (1998)—Ecuador.

× **AECHOPSIS** *D. Butcher in J. Bromeliad Soc.,* 49(1): 14 (1999). *BROMELIACEAE* (Aechmea × Canistropsis).

AEGILOPS (Gramin.).
biuncialis Vis.
subsp. archipelagica *(Eig) T. Raus in Bot. Chronika,* 12: 45 (1996): A. biuncialis var. archipelagica.
× insulae-cypri *H. Scholz in Willdenowia,* 30(1): 62 (2000)—Cyprus. (A. geniculata × peregrina).

AEGIPHILA (Verbenac.).
elata Sw.
var. macrophylla *(Humb., Bonpl. & Kunth) S. López-Palacios in Pittieria,* 5: 15 (1973): A. macrophylla.
meridensis *S. López-Palacios in Pittieria,* 5: 35 (1973)—Venezuela.
moldenkeana *S. López-Palacios in Pittieria,* 5: 27 (1973)—Venezuela.
ternifolia Humb., Bonpl. & Kunth
f. oppositifolia *S. López-Palacios in Pittieria,* 5: 47 (1973)—Venezuela.
ulei *(Hayek) B. Wallnöfer in Linz. Biol. Beitr.,* 28(2): 1053 (1996): Clerodendrum ulei.

AEGOPODIUM (Umbellif.).
podagraria L.
subsp. nadeshdae *N.V. Stepanov, Florogenet. Analiz,* 1: 99 (1994)—Krasnoyarsk.

AENHENRYA (Orchidac.).
rotundifolia *(Blatt.) C. Sathish Kumar & F.N. Rasmussen in Novon,* 7(1): 81 (1997): Odontochilus rotundifolius.

AEONIUM (Crassulac.).
sect. Greenovia *(Webb & Berthel.) T.H.M. Mes in H. 't Hart & U. Eggli (eds.), Evol. & Syst. Crassulac.:* 41 (1995): gen. Greenovia.
× aguajilvense *Á. Bañares in Vieraea,* 25: 182 (1996)—Canary Is. (A. castello-paivae × gomerense).
aizoon *(Bolle) T.H.M. Mes in H. 't Hart & U. Eggli (eds.), Evol. & Syst. Crassulac.:* 41 (1995): Greenovia aizoon.
appendiculatum *Á. Bañares Baudet in Willdenowia,* 29(1–2): 98 (1999)—Canary Is.
aureum *(C. Sm. ex Hornem.) T.H.M. Mes in H. 't Hart & U. Eggli (eds.), Evol. & Syst. Crassulac.:* 41 (1995): Sempervivum aureum.
× cabrerae *(A. Santos) Á. Bañares in Vieraea,* 25: 184 (1996), basionym not validly publ.: × Greenonium cabrerae.
diplocyclum *(Webb ex Bolle) T.H.M. Mes in H. 't Hart & U. Eggli (eds.), Evol. & Syst. Crassulac.:* 41 (1995): Greenovia diplocycla.
dodrentale *(Willd.) T.H.M. Mes in H. 't Hart & U. Eggli (eds.), Evol. & Syst. Crassulac.:* 41 (1995): Sempervivum dodrentale.
× holospathulatum Bañares
nothovar. sanchezii *(Bañares) Á. Bañares in Vieraea,* 25: 184 (1996): A. × sanchezii.

AEONIUM:-
× isorense *Á. Bañares in Vieraea,* 25: 184 (1996)—Canary Is. (A. arboreum var. holochrysum × hierrense).
× nogalesii Bañares
nothovar. dasyphyllum *Á. Bañares & J.M. Macarrón in Vieraea,* 25: 188 (1996)—Canary Is. (A. canariense var. canariense × sedifolium).
× orbelindense *Á. Bañares in Vieraea,* 25: 188 (1996)—Canary Is. (A. canariense var. canariense × pseudourbicum).
× pseudohawbicum *Á. Bañares in Vieraea,* 25: 190 (1996)—Canary Is. (A. haworthii × pseudourbicum).
× tijarafense *(A. Santos) Á. Bañares in Vieraea,* 25: 190 (1996), basionym not validly publ.: × Greenonium tijarafense.
urbicum (C. Sm. ex Hornem.) Webb & Berthel.
var. meridionale *Á. Bañares Baudet in Willdenowia,* 29(1–2): 98 (1999)—Canary Is.
× voggenreiteri *Á. Bañares in Vieraea,* 25: 192 (1996), nom.nov., as 'voggenreiterii', without replaced synonym date: × Aeoniogreenovia lambii *Voggenr.*
volkeri *E. Hernández & Á. Bañares in Vieraea,* 25: 160 (1996), as 'volkerii'—Canary Is.

AEQUATORIUM (Compos.).
subgen. Praegynoxys *(Cuatrec.) B. Nordenstam in Compositae Newsl.,* 31: 2 (1997): Senecio sect. Praegynoxys.
lepidotum *B. Nordenstam in Compositae Newsl.,* 31: 6 (1997)—Ecuador.
limonense *B. Nordenstam in Compositae Newsl.,* 31: 14 (1997)—Ecuador.
repandiforme *B. Nordenstam in Compositae Newsl.,* 31: 9 (1997)—Ecuador.
tatamanum *S. Díaz-Piedrahita & A. Correa in Rev. Acad. Colomb. Cienc. Exact. Fis. Nat.,* 88: 332 (1999)—Colombia.
venezuelanum *V.M. Badillo in Ernstia,* 10(1): 16 (2000)—Venezuela.

AERANGIS (Orchidac.).
bouarensis *G. Chiron in Orchid Rev.,* 106(1222): 229 (1998)—Central African Republic.
maireae *I.F. la Croix & J. Stewart Orchid Rev.,* 106(1222): 227 (1998)—Tanzania.
rusituensis *W. Fibeck & M. Dare in S. Afr. Orchid J.,* 28(3): 73 (1997)—Zimbabwe.
verdickii (De Wild.) Schltr.
var. rusituensis *(Fibeck & Dare) I. la Croix & P.J. Cribb in Fl. Zambes.,* 11(2): 391 (1998): A. rusituensis.

AERIDES (Orchidac.).
multiflora *W. Roxburgh, Pl. Coromandel,* 3: 68 (1820)—Bangladesh.†

AESCHYNANTHUS (Gesneriac.).
amboinensis *(Merr.) M. Mendum in Edinburgh J. Bot.,* 55(3): 363 (1998): Trichosporum amboinense.
argentii *M. Mendum Edinburgh J. Bot.,* 56(2): 265 (1999)—Borneo (Sabah).
batakiorum *M. Mendum & Madulid in Edinburgh J. Bot.,* 56(2): 269 (1999)—Philippines.
flavidus *M. Mendum & P.J.B. Woods in Edinburgh J. Bot.,* 54(2): 254 (1997)—Borneo (Sarawak).
intraflavus *M. Mendum in Edinburgh J. Bot.,* 55(3): 359 (1998)—Moluccas.
pseudohybridus *M. Mendum in Edinburgh J. Bot.,* 56(2): 268 (1999)—Borneo (Sabah, Sarawak).
roseoflorus *M. Mendum in Edinburgh J. Bot.,* 55(3): 361 (1998)—Moluccas.
tirapensis *U.C. Bhattacharyya in Bull. Bot. Surv. India,* 35(1–4): 131 (1993 publ. 1997)—Assam (Arunachal Pradesh).

AESCHYNOMENE (Legumin.).
ser. Sclerosae *Afr. Fernandes, O Táxon Aeschynomene no Brasil:* 116 (1996), without type.
benthamii *(Rudd) Afr. Fernandes, O Táxon Aeschynomene no Brasil:* 113 (1996), without basionym ref.: A. mollicula var. benthamii.
emerus *Aubl., Hist. Pl. Guiane:* 2: 775, table des noms p.1 (1775)—Locality not stated.

AESCHYNOMENE:-
fructipendula *M.de L. A. A. de Oliveira in Iheringia, Bot.,* 46: 22 (1995)—Brazil South (Rio Grande do Sul, Santa Catarina).
gilbertoi *M. Brandão in Daphne,* 1(2): 5 (1991)—Brazil Southeast (Minas Gerais).
laca-buendiana *M. Brandão in Daphne,* 1(3): 5 (1991)—Brazil Southeast (Minas Gerais).
lewisiana *Afr. Fernandes, O Táxon Aeschynomene no Brasil:* 78 (1996)—Brazil Northeast (Bahia).
matosii *Afr. Fernandes, O Táxon Aeschynomene no Brasil:* 97 (1996)—Brazil Northeast (Piauí).
pawekiae *B. Verdcourt in Fl. Zambes.,* 3(6): 88 (2000)—Malawi, Mozambique.
priscoana *Afr. Fernandes, O Táxon Aeschynomene no Brasil:* 86 (1996)—Brazil Northeast (Piauí).
venulosa *Afr. Fernandes, O Táxon Aeschynomene no Brasil:* 111 (1996), homonym; *see Verdc. in Kew Bull.,* 24: 8 (1970)—Brazil Northeast (Alagoas, Bahia).

AESCULUS (Hippocastanac.).
turbinata Blume
f. pubescens *(Rehder) Ohwi ex Y. Endo in Fl. Japan,* 2c: 76 (1999): A. turbinata var. pubescens.

AETHEOLAENA (Compos.).
decipiens *(Benoist) B. Nordenstam in Compositae Newsl.,* 30: 47 (1997): Senecio decipiens.
ledifolia *(Kunth) B. Nordenstam in Compositae Newsl.,* 30: 47 (1997): Culcitium ledifolium.
otophora (Wedd.) B. Nord.
var. christophori *(Cuatrec.) B. Nordenstam in Compositae Newsl.,* 30: 47 (1997): Senecio otophorus var. christophori.
var. microcephala *(Hieron.) B. Nordenstam in Compositae Newsl.,* 30: 48 (1997): Senecio otophorus var. microcephalus.
pichinchensis *(Cuatrec.) B. Nordenstam in Compositae Newsl.,* 30: 48 (1997): Culcitium pichinchense.

AETHIONEMA (Crucif.).
dumanii *M. Vural & N. Adıgüzel in Turkish J. Bot.,* 19(4): 481 (1995)—Turkey.
karamanicum *K. Ertuğrul & O. Beyazoğlu in Turkish J. Bot.,* 21(2): 99 (1997), as 'karamanicus'—Turkey.
semnanensis *V. Mozaffarian in Iranian J. Bot.,* 7(1): 128 (1996)—Iran.

AFRAMOMUM (Zingiberac.).
spiroligulatum *A.D. Poulsen & J.M. Lock in Kew Bull.,* 52(3): 608 (1997)—Uganda, Rwanda.
uniflorum *J.M. Lock & A.D. Poulsen in Kew Bull.,* 52(3): 606 (1997)—Uganda, Zaire.

AGALMYLA (Gesneriac.).
bicolor *O.M. Hilliard & B.L. Burtt in Blumea,* 44(2): 382 (1999)—Sulawesi.
chalmersii *(F. Muell.) B.L. Burtt in Blumea,* 44(2): 383 (1999): Dichrotrichum chalmersii.
elongata *(Blume) B.L. Burtt in Blumea,* 44(2): 383 (1999): Tromsdorffia elongata.
exannulata *O.M. Hilliard & B.L. Burtt in Blumea,* 44(2): 384 (1999)—Sulawesi.
macrocalyx *O.M. Hilliard & B.L. Burtt in Blumea,* 44(2): 385 (1999)—Borneo (Sarawak).
manuselae *O.M. Hilliard & B.L. Burtt in Blumea,* 44(2): 386 (1999)—Moluccas.
remotidentata *O.M. Hilliard & B.L. Burtt in Blumea,* 44(2): 387 (1999)—Sulawesi.
samarica *O.M. Hilliard & B.L. Burtt in Blumea,* 44(2): 388 (1999)—Philippines.
triflora *(Valeton) B.L. Burtt in Blumea,* 44(2): 389 (1999): Dichrotrichum triflorum.

× **AGANANTHES** *hort. Orchid Rev.,* 106(1224): 380 (1998). *ORCHIDACEAE* (Aganisia × Cochleanthes).

× **AGANAX** *hort. in Orchid Rev.,* 104(1207): 37 (1996). *ORCHIDACEAE* (Aganisia × Colax).

AGANISIA (Orchidac.).
graminea *(Lindl.) N.E. Brown, Suppl. Johnson's Gard. Dict.:* 866 (1882): Maxillaria graminea.

AGAPETES (Ericac.).
gongshanensis *H. Li in Fl. Dulongjian Reg.*: 230 (1993), without latin descr. or type—China South-central (Yunnan).
guangxiensis *D. Fang in Acta Phytotax. Sin.*, 36(3): 276 (1998)—China Southeast (Guangxi).
mondangensis *H. Li in Fl. Dulongjian Reg.*: 230 (1993)—China South-central (Yunnan).
subsessilifolia *S.H. Huang, H. Sun & Z.K. Zhou in Novon*, 8(4): 378 (1998)—Xizang.

AGATHIS (Araucariac.).
macrophylla (Lindl.) Mast.
var. vitiensis *(Seem.) J. Silba in J. Int. Conifer Preserv. Soc.*, 3: 19 (1996): Dammara vitiensis.
robusta (C. Moore ex F. Muell.) F.M. Bailey
var. palmerstonii *(F. Muell.) J. Silba in J. Int. Conifer Preserv. Soc.*, 3: 29 (1996): Dammara palmerstonii.

AGAVE (Agavac.).
sect. Yuccaefoliae *(A. Terracc.) B. Ullrich in Piante Grasse*, 16(1): 29 (1996): A. subser. Yuccaefoliae.
apedicellata *J. Thiede & U. Eggli in Kakt. And. Sukk.*, 50(5): 111 (1999), nom. nov.: Bravoa sessiliflora *Hemsl.*
× blissii *(Worsley) J. Thiede & U. Eggli in Kakt. And. Sukk.*, 50(5): 111 (1999): Polianthes × blissii.
brittoniana Trel.
subsp. brachypus *(Trel.) A. Álvarez de Zayas in Fontqueria*, 44: 121 (1996): A. brittoniana var. brachypus.
subsp. sancti-spirituensis *A. Álvarez de Zayas in Fontqueria*, 44: 125 (1996)—Cuba.
bulliana *(Baker) J. Thiede & U. Eggli in Kakt. And. Sukk.*, 50(5): 112 (1999): Bravoa bulliana.
× bundrantii *(T.M. Howard) J. Thiede & U. Eggli in Kakt. And. Sukk.*, 50(5): 111 (1999): Polianthes × bundrantii.
chamelensis *(E.J. Lott & Verh.-Will.) J. Thiede & U. Eggli in Kakt. And. Sukk.*, 50(5): 110 (1999): Manfreda chamelensis.
confertiflora *J. Thiede & U. Eggli in Kakt. And. Sukk.*, 50(5): 111 (1999), nom. nov.: Bravoa densiflora *B.L. Rob. & Fernald.*
delamateri *W.C. Hodgson & L. Slauson in Haseltonia*, 3: 133 (1995)—Arizona.
dolichantha *J. Thiede & U. Eggli in Kakt. And. Sukk.*, 50(5): 111 (1999), nom. nov.: Polianthes longiflora *Rose.*
duplicata *J. Thiede & U. Eggli in Kakt. And. Sukk.*, 50(5): 111 (1999), nom. nov.: Bravoa geminiflora *Lex.*
subsp. clivicola *(McVaugh) J. Thiede & U. Eggli in Kakt. And. Sukk.*, 50(5): 111 (1999): Polianthes geminiflora var. clivicola.
subsp. graminifolia *(Rose) J. Thiede & U. Eggli in Kakt. And. Sukk.*, 50(5): 112 (1999): Polianthes graminifolia.
gentryi *B. Ullrich in Succulenta*, 69(10): 211 (1990)—Mexico Northeast (Nuevo León).
gracilis *A. García-Mendoza & E. Martínez Salas in Sida*, 18(1): 227 (1998), homonym; *see Jacobi in Abh. Schles. Ges. Vaterl. Cult., Abth. Naturwiss.*, 1869/72: 150 (1872)—Mexico Southwest (Guerrero, Oaxaca).
howardii *(Verh.-Will.) J. Thiede & U. Eggli in Kakt. And. Sukk.*, 50(5): 112 (1999): Polianthes howardii.
involuta *(McVaugh) J. Thiede & U. Eggli in Kakt. And. Sukk.*, 50(5): 110 (1999): Manfreda involuta.
longibracteata *(Verh.-Will.) J. Thiede & U. Eggli in Kakt. And. Sukk.*, 50(5): 110 (1999): Manfreda longibracteata.
michoacana *(M. Cedano, R. Delgad. & Enciso) J. Thiede & U. Eggli Kakt. And. Sukk.*, 50(5): 112 (1999): Polianthes michoacana.
montana *J.A. Villarreal Q. in Sida*, 17(1): 191 (1996)—Mexico Northeast (Nuevo León).
nanchititlensis *(Matuda) J. Thiede & U. Eggli in Kakt. And. Sukk.*, 50(5): 110 (1999), without exact basionym page: Manfreda nanchititlensis.
neonelsonii *J. Thiede & U. Eggli in Kakt. And. Sukk.*, 50(5): 112 (1999), nom. nov.: Polianthes nelsonii *Rose.*
neopringlei *J. Thiede & U. Eggli in Kakt. And. Sukk.*, 50(5): 112 (1999), nom. nov.: Polianthes pringlei *Rose.*

AGAVE:-
palustris *(Rose) J. Thiede & U. Eggli in Kakt. And. Sukk.*, 50(5): 112 (1999): Polianthes palustris.
petrophila *A. García-Mendoza & E. Martínez Salas in Sida*, 18(2): 627 (1998), nom. nov. A. gracilis *García-Mend. & E. Martínez.*
platyphylla *(Rose) J. Thiede & U. Eggli in Kakt. And. Sukk.*, 50(5): 112 (1999): Polianthes platyphylla.
producta *J. Thiede & U. Eggli in Kakt. And. Sukk.*, 50(5): 112 (1999), nom. nov.: Polianthes elongata *Rose.*
rosei *J. Thiede & U. Eggli in Kakt. And. Sukk.*, 50(5): 112 (1999), nom. nov.: Polianthes montana *Rose.*
sileri *(Verh.-Will.) J. Thiede & U. Eggli in Kakt. And. Sukk.*, 50(5): 111 (1999): Manfreda sileri.
stictata *J. Thiede & U. Eggli in Kakt. And. Sukk.*, 50(5): 111 (1999), nom. nov.: Polianthes maculata *Mart.*
tuberosa *(L.) J. Thiede & U. Eggli in Kakt. And. Sukk.*, 50(5): 112 (1999): Polianthes tuberosa.
victoriae-reginae
subsp. swobodae *J.J. Halda in Acta Mus. Richnov. Sect. Nat.*, 7(2): 71 (2000)—Mexico Northeast (Coahuila).

AGELAEA (Connarac.).
palmata *C.C.H. Jongkind in Bull. Jard. Bot. Nation. Belg.*, 66(1–2): 149 (1997)—Gabon.

AGELANTHUS (Loranthac.).
sect. Acranthemum *(Tiegh.) R.M. Polhill & D. Wiens, Mistletoes of Africa*: 139 (1998): gen. Acranthemum.
sect. Erectilobi *(Sprague) R.M. Polhill & D. Wiens, Mistletoes of Africa*: 143 (1998): Loranthus sect. Erectilobi.
sect. Longiflori *(Engl.) R.M. Polhill & D. Wiens, Mistletoes of Africa*: 138 (1998): Loranthus group Longiflori.
sect. Obtectiflori *(Engl.) R.M. Polhill & D. Wiens, Mistletoes of Africa*: 140 (1998): Loranthus group Obtectiflori.
sect. Purpureiflori *(Engl.) R.M. Polhill & D. Wiens, Mistletoes of Africa*: 142 (1998): Loranthus group Purpureiflori.
atrocoronatus *R.M. Polhill & D. Wiens, Mistletoes of Africa*: 179 (1998)—Tanzania.
bipartitus *Balle ex R.M. Polhill & D. Wiens in Novon*, 7(3): 274 (1997)—Tanzania, Malawi.
crassifolius *(Wiens) R.M. Polhill & D. Wiens, Mistletoes of Africa*: 172 (1998): Tapinanthus crassifolius.
flammeus *R.M. Polhill & D. Wiens, Mistletoes of Africa*: 169 (1998)—Zimbabwe.
gracilis *(Toelken & Wiens) R.M. Polhill & D. Wiens, Mistletoes of Africa*: 149 (1998): Tapinanthus gracilis.
guineensis *R.M. Polhill & D. Wiens, Mistletoes of Africa*: 159 (1998)—Guinea.
kraussianus *(Meisn.) R.M. Polhill & D. Wiens, Mistletoes of Africa*: 169 (1998): Loranthus kraussianus.
lancifolius *R.M. Polhill & D. Wiens, Mistletoes of Africa*: 170 (1998)—Zimbabwe.
microphyllus *R.M. Polhill & D. Wiens, Mistletoes of Africa*: 150 (1998)—Tanzania.
pilosus *R.M. Polhill & D. Wiens in Novon*, 7(3): 276 (1997)—Malawi.
prunifolius *(E. Mey. ex Harv.) R.M. Polhill & D. Wiens, Mistletoes of Africa*: 168 (1998): Loranthus prunifolius.
sansibarensis (Engl.) Polhill & Wiens
subsp. montanus *R.M. Polhill & D. Wiens, Mistletoes of Africa*: 166 (1998)—Kenya.
songeensis *Balle ex R.M. Polhill & D. Wiens, Mistletoes of Africa*: 175 (1998)—Tanzania.
toroensis *(Sprague) R.M. Polhill & D. Wiens, Mistletoes of Africa*: 158 (1998): Loranthus toroensis.
validus *R.M. Polhill & D. Wiens, Mistletoes of Africa*: 163 (1998)—Tanzania.

AGERATINA (Compos.).
appendiculata *V.M. Badillo in Ernstia*, 10(1): 7 (2000)—Venezuela.
aracaiensis *V.M. Badillo in Ernstia*, 10(1): 8 (2000), contrary to Art. 37.4 ICBN (1994)—Venezuela.

AGERATINA:-
areolaris *(DC.) D. Gage ex B.L. Turner in Phytologia Mem.*, 11: 19 (1997): Eupatorium areolare.
ayerscottiana *B.L. Turner in Phytologia*, 79(4): 309 (1995 publ. 1996)—Mexico Northeast (Chihuahua).
bobjansenii *B.L. Turner in Phytologia*, 81(2): 118 (1996 publ. 1997)—Mexico Northeast (Durango).
capazica *V.M. Badillo in Ernstia*, 10(1): 9 (2000)—Venezuela.
dorrii *V.M. Badillo in Ernstia*, 10(1): 10 (2000), as 'dorri'—Venezuela.
haageana *(Regel & Körn.) H. Robinson in Phytologia*, 84(5): 347 (1998 publ. 1999): Eupatorium haageanum.
jaliscensis *(B.L. Rob.) D. Gage ex B.L. Turner in Phytologia Mem.*, 11: 31 (1997): Piptothrix jaliscensis.
jalpana *B.L. Turner in Phytologia*, 81(5): 343 (1996 publ. 1997)—Mexico Northeast (Zacatecas).
jocotepecana *B.L. Turner in Phytologia*, 80(2): 128 (1996)—Mexico Southwest (Jalisco).
josepaneroi *B.L. Turner in Phytologia*, 81(5): 344 (1996 publ. 1997)—Mexico Southwest (Guerrero).
juxtlahuacensis *J.L. Panero & J.L. Villaseñor in Lundellia*, 1: 72 (1998)—Mexico Southwest (Oaxaca).
leiocarpa *(B.L. Rob.) D. Gage ex B.L. Turner in Phytologia Mem.*, 11: 32 (1997): Eupatorium areolare var. leiocarpum.
miahuatlana *B.L. Turner in Phytologia*, 79(4): 310 (1995 publ. 1996)—Mexico Southwest (Oaxaca).
palmeri *(A. Gray) B.L. Turner in Phytologia Mem.*, 11: 38 (1997): Piptothrix palmeri.
pendula *J.L. Panero & J.L. Villaseñor in Brittonia*, 48(4): 498 (1996 publ. 1997)—Mexico Southwest (Oaxaca).

AGLAIA (Meliac.).
canarana *(Turcz.) C.J. Saldanha, Fl. Karnataka*, 2: 229 (1996): Oraoma canarana.
edulis *(Roxb.) N. Wallich, Report on Calcutta Bot. Gard. to G.A. Bushby*: 26 (1840): Milnea edulis.
mackiana *C.M. Pannell in Kew Bull.*, 52(3): 715 (1997)—New Guinea (Papua New Guinea).
saxonii *W. Takeuchi in Sida*, 19(1): 11 (2000)—New Guinea (Papua New Guinea).
tamilnadensis *N.C. Nair & R. Rajan in N.C. Nair & A.N. Henry, Fl. Tamil Nadu, India, Ser. 1*, 1: 66 (1983), nom. nov.: Oraoma canarana *Turcz.*

AGLAONEMA (Arac.).
chermsiriwattanae *D. Sookchaloem in Thai Forest Bull., Bot.*, 25: 54 (1997)—Thailand.
flemingianum *A. Hay in Gard. Bull. Singapore*, 50(1): 1 (1998)—Malaya (Peninsular Malaysia).

AGONANDRA (Opiliac.).
brasiliensis
subsp. racemigera *P. Hiepko in Bot. Jahrb.*, 117(4): 498 (1995)—Colombia.
goldbergiana *P. Hiepko in Willdenowia*, 27(1–2): 225 (1997), nom. nov.: Izabalaea excelsa *Lundell.*
obtusifolia Standl.
subsp. conzattii *(Standl.) P. Hiepko in Willdenowia*, 27(1–2): 226 (1997): A. conzattii.
peruviana *P. Hiepko in Bot. Jahrb.*, 117(4): 495 (1995)—Peru.

AGRIMONIA (Rosac.).
pilosa Ledeb.
var. succapitata *N. Naruhashi in J. Phytogeogr. Taxon.*, 44(1–2): 82 (1996)—Japan.

AGROPYRON (Gramin.).
sinkiangense *D.F. Cui in Fl. Xinjiangensis*, 6: 602 (1996); *et in Acta Bot. Bor.-Occid. Sin.*, 18(2): 284 (1998)—Xinjiang.

AGROSTEMMA (Caryophyllac.).
× hybrida *Pépin in Ann. Fl. Pomone*, 1833–34: 81 (1833)—Cultivation. (A. coronaria × flos-jovis).

AGROSTIS (Gramin.).
barceloi *L. Sáez & J.A. Rosselló in Bot. J. Linn. Soc.*, 133(3): 361 (2000)—Baleares.

AGROSTIS:-
buchtarmensis *Yu.A. Kotukhov in Turczaninowia,* 1(1): 12 (1998)—Altay.
capillaris L.
subsp. oreophila *(O. Schwarz) K. Senghas in K. Senghas & S. Seybold, Schmeil & Fitschen Fl. Deutschland u. angrenz.* Länd., 91. Aufl.: 798 (2000), isonym; *see Soják in Čas. Nár. Muz. (Prague),* 150(3–4): 137 (1982 publ. 1982): A. tenuis subsp. oreophila.
collicola *(D.I. Morris) A.J. Brown & N.G. Walsh in Muelleria,* 14: 80 (2000): Agrostis billardierei var. collicola.
infirma Buse
var. arisan-montana *(Ohwi) J.F. Veldkamp in Blumea,* 41(2): 408 (1996): A. arisan-montana.
var. borneensis *(Stapf) J.F. Veldkamp in Blumea,* 41(2): 408 (1996): A. canina var. borneensis.
var. diffusissima *(Ohwi) J.F. Veldkamp in Blumea,* 41(2): 408 (1996): A. reinwardtii var. diffusissima.
var. formosana *(Hack.) J.F. Veldkamp in Blumea,* 41(2): 408 (1996): A. canina var. formosana.
var. fukuyamae *(Ohwi) J.F. Veldkamp in Blumea,* 41(2): 408 (1996): A. fukuyamae.
var. kinabaluensis *(Ohwi) J.F. Veldkamp in Blumea,* 41(2): 409 (1996): A. kinabaluensis.
var. remota *(Buse) J.F. Veldkamp in Blumea,* 41(2): 409 (1996): A. stricta var. remota.
megathyrsa Keng f.
var. angustispicata *D.Z. Ma & J.N. Li in Fl. Ningxiaensis,* 2: 522 (1988 publ. 1990)—Inner Mongolia (Ningxia).
musjidii *S. Rajeswari, R.R. Rao & Arti Garg in J. Ind. Bot. Soc.,* 74(1–4): 333 (1995)—Jammu-Kashmir.
petelotii *(Hitchc.) H.J. Noltie in Edinburgh J. Bot.,* 56(3): 386 (1999): Aulacolepis petelotii.
punicea *A.J. Brown & N.G. Walsh in Muelleria,* 14: 83 (2000), nom. nov.: Agrostis billardierei var. setifolia *Hook. f.*
var. filifolia *(Vickery) A.J. Brown & N.G. Walsh Muelleria,* 14: 85 (2000): Agrostis billardierei var. filifolia.
reuteri Boiss.
subsp. botelhoi *M.L. Rocha Afonso & J. do Amaral Franco in Silva Lusitana,* 5(1): 141 (1997)—Azores.
robusta *(Vickery) A.J. Brown & N.G. Walsh in Muelleria,* 14: 82 (2000): Agrostis billardierei var. robusta.
stolonifera L.
var. calcicola *(Philipson) P.D. Sell in P.D. Sell & G. Murrell, Fl. Great Britain & Ireland,* 5: 363 (1996): A. stolonifera ecas calcicola.
var. marina *(Gray) P.D. Sell in P.D. Sell & G. Murrell, Fl. Great Britain & Ireland,* 5: 363 (1996), isonym; *see Kerguélen in Lejeunia,* 75: 68 (1975): Vilfa stolonifera var. marina.
ushae *H.J. Noltie in Edinburgh J. Bot.,* 56(3): 388 (1999)—Sikkim.

AGROSTOPHYLLUM (Orchidac.).
flavidum *S. Phukan in J. Ind. Bot. Soc.,* 69(1–2): 209 (1990)—Assam (Meghalaya).
stipulatum (Griff.) Schltr.
subsp. bicuspidatum *(J.J. Sm.) A. Schuiteman in Orchid Monogr.,* 8: 14 (1997): A. bicuspidatum.

× **AGROTRIGIA** (Gramin.).
berelica *Yu.A. Kotukhov in Turczaninowia,* 1(1): 15 (1998)—Altay. (Agropyron krylovianum × Elytrigia geniculata).
urunchaica *Yu.A. Kotukhov in Turczaninowia,* 1(1): 14 (1998)—Altay. (Agropyron tarbagataicum × Elytrigia repens var. cristatum).

AIDIA (Rubiac.).
sect. Randiella *(Baill.) C.E. Ridsdale in Blumea,* 41(1): 154 (1996): Randia sect. Randiella.
sect. Xylanthorandia *(Baill.) C.E. Ridsdale in Blumea,* 41(1): 159 (1996): Genipa sect Xylanthorandia.
acutipetala *C.E. Ridsdale in Blumea,* 41(1): 160 (1996)—Borneo (Sarawak).
auriculata *(Wall.) C.E. Ridsdale in Blumea,* 41(1): 161 (1996): Webera auriculata.
var. indigiriensis *C.E. Ridsdale in Blumea,* 41(1): 163 (1996)—Sumatera.

AIDIA:
bakeri *(Merr.) C.E. Ridsdale in Blumea,* 41(1): 165 (1996): Randia bakeri.
beccariana *(Baill.) C.E. Ridsdale in Blumea,* 41(1): 159 (1996): Randia beccariana.
borneensis *C.E. Ridsdale in Blumea,* 41(1): 147 (1996)—Borneo (Sarawak, Sabah, Kalimantan, Brunei).
bracteata *C.E. Ridsdale in Blumea,* 41(1): 171 (1996)—Sulawesi.
brisipensis *C.E. Ridsdale in Blumea,* 41(1): 155 (1996)—Solomon Is. (South Solomons).
congestum *(Schltr. & E.H.L. Krause) C.E. Ridsdale in Blumea,* 41(1): 154 (1996): Rhopalobrachium congestum.
dilleniacea *(Baill.) C.E. Ridsdale in Blumea,* 41(1): 165 (1996): Randia dilleniacea.
endertii *C.E. Ridsdale in Blumea,* 41(1): 175 (1996)—Borneo (Kalimantan).
foveata *C.E. Ridsdale in Blumea,* 41(1): 166 (1996)—Borneo (Sabah, Sarawak, Kalimantan), Philippines.
glabra *(Valeton) C.E. Ridsdale in Blumea,* 41(1): 156 (1996): Randia gynopachis var. glabra.
var. rubiginosa *(Valeton) C.E. Ridsdale in Blumea,* 41(1): 157 (1996): Randia gynopachis var. rubiginosa.
halleri *C.E. Ridsdale in Blumea,* 41(1): 169 (1996)—Borneo (Sarawak).
heterophylla *C.E. Ridsdale in Blumea,* 41(1): 171 (1996)—Sulawesi.
impressinervis *(King & Gamble) C.E. Ridsdale in Blumea,* 41(1): 168 (1996): Randia impressinervis.
jambosoides *(Valeton) C.E. Ridsdale in Blumea,* 41(1): 159 (1996): Randia jambosoides.
kinabaluensis *C.E. Ridsdale in Blumea,* 41(1): 175 (1996)—Borneo (Sabah).
longiflora *C.E. Ridsdale in Blumea,* 41(1): 160 (1996)—Borneo (Sarawak).
magnifolia *C.E. Ridsdale in Blumea,* 41(1): 174 (1996)—Borneo (Kalimantan).
moluccana *C.E. Ridsdale Blumea,* 41(1): 173 (1996)—Moluccas.
paiei *C.E. Ridsdale in Blumea,* 41(1): 174 (1996)—Borneo (Sarawak, Sabah, Kalimantan).
polystachya *(Valeton) C.E. Ridsdale in Blumea,* 41(1): 157 (1996): Randia polystachya.
pseudospicata *C.E. Ridsdale in Blumea,* 41(1): 149 (1996)—Assam.
pulcherrima *(Merr.) C.E. Ridsdale in Blumea,* 41(1): 173 (1996): Randia pulcherrima.
shweliensis *(J. Anthony) W.C. Chen in Fl. Reipubl. Popul. Sin.,* 71(1): 350 (1999): Randia shweliensis.
solomonensis *C.E. Ridsdale in Blumea,* 41(1): 158 (1996)—Solomon Is. (South Solomons).
tomentosa *(Blume) C.E. Ridsdale in Blumea,* 41(1): 169 (1996): Gynopachis tomentosa.
var. sundaica *C.E. Ridsdale in Blumea,* 41(1): 170 (1996)—Lesser Sunda Is.
vieillardii *(Baill.)C.E. Ridsdale in Blumea,* 41(1): 154 (1996): Randia vieillardii.
waugia *C.E. Ridsdale in Blumea,* 41(1): 156 (1996)—New Guinea (Papua New Guinea).
zippeliana *(Scheff.) C.E. Ridsdale in Blumea,* 41(1): 158 (1996): Gynopachis zippeliana.

AIDIOPSIS (Rubiac.).
orophila *(Miq.) C.E. Ridsdale in Blumea,* 41(1): 176 (1996): Stylocoryna orophila.
lancangensis *Y.Y. Qian in J. Trop. Subtrop. Bot.,* 8(2): 161 (2000)—China South-central (Yunnan).

AIOUEA (Laurac.).
opaca *H. van der Werff in Brittonia,* 47(4): 372 (1995)—French Guiana.

AIPHANES (Palmae).
hirsuta Burret
subsp. fosteriorum *(H.E. Moore) F. Borchsenius & R. Bernal in Fl. Neotrop.,* 70: 65 (1996): A. fosteriorum.
subsp. intermedia *F. Borchsenius & R. Bernal in Fl. Neotrop.,* 70: 64 (1996)—Colombia.
subsp. kalbreyeri *(Burret) F. Borchsenius & R. Bernal in Fl. Neotrop.,* 70: 63 (1996): A. kalbreyeri.
spicata *F. Borchsenius & R. Bernal in Fl. Neotrop.,* 70: 78 (1996)—Peru.

AIRA (Gramin.).
uniaristata Lag. & Rodr.
var. biaristata *(Emb. & Maire) J. Lambinon in Soc. Éch. Pl. Vasc. Eur. Bassin Médit. Bull.,* 27: 88 (1998): A. pulchella var. biaristata.

AIZOPSIS (Crassulac.).
sect. Villosae *(Boriss.) S.B. Gontcharova in Bull. Nation. Sci. Mus., B (Tokyo),* 25(2): 55 (1999): Sedum ser. Villosae.
kurilensis *(Vorosch.) S.B. Gontcharova in Bull. Nation. Sci. Mus., B (Tokyo),* 25(2): 54 (1999): Sedum kurilense.
maximowiczii *(Regel) S.B. Gontcharova in Bull. Nation. Sci. Mus., B (Tokyo),* 25(2): 54 (1999): Sedum maximowiczii.
middendorfiana (Maxim.) Grulich
subsp. sichotensis *(Vorosch.) S.B. Gontcharova in Bull. Nation. Sci. Mus., B (Tokyo),* 25(2): 54 (1999): Sedum sichotense.

AJANIA (Compos.).
qiraica *C.H. An & Dilixiat in Fl. Xinjiangensis,* 5: 476 (1999)—Xinjiang.
shiwogiku
var. kinokumiensis *(Shimot. & Kitam.) R. Govaerts, World Checklist Seed Plants,* 3(1): 9 (1999), as 'kinokumiense': Chrysanthemum shiwogiku var. kinokumiense.

AJUGA (Labiat.).
decaryana *Danguy ex R.A. Clement in Fl. Madagascar,* fam. 175: 14 (1998)—Madagascar.

AKEBIA (Lardizabalac.).
trifoliata (Thunb.) Koidz.
subsp. longisepala *H.N. Qin in Cathaya,* 8–9: 71 (1997)—China North-central (Gansu).

ALA *D.L. Szlachetko in Fragm. Flor. Geobot.,* Supp. 3: 113 (1995), contrary to Art. 20.2 ICBN (1994). *ORCHIDACEAE*
aethiopica *(S. Thomas & P.J. Cribb) D.L. Szlachetko & T.S. Olszewski in Fl. Cameroun,* 34: 104 (1998), genus name not validly publ.: Habenaria aethiopica.
decorata *(Hochst. ex A. Rich.) D.L. Szlachetko in Fragm. Flor. Geobot.,* Supp. 3: 113 (1995), genus name not validly publ.: Habenaria decorata.
tricruris *(A. Rich.) D.L. Szlachetko in Fragm. Flor. Geobot.,* Supp. 3: 113 (1995), genus name not validly publ.: Platanthera tricruris.

ALAFIA (Apocynac.).
erythrophthalma *(K. Schum.) A.J.M. Leeuwenberg in Novon,* 6(3): 271 (1996): Tabernaemontana erythrophthalma.
falcata *A.J.M. Leeuwenberg in Novon,* 6(3): 271 (1996)—Gabon.
velutina *A.J.M. Leeuwenberg in Novon,* 6(3): 271 (1996)—Gabon.

ALANGIUM (Alangiac.).
qingchuanense *M.Y. He in J. Sichuan Univ., Nat. Sci. Ed.,* 33(6): 739 (1996)—China South-central (Sichuan).

ALATAVIA *G.I. Rodionenko in Signa,* 58: 3026 (1997), without type; *et in Bot. Zhurn.,* 84(7): 103 (1999), with type. *IRIDACEAE*
kolpakowskiana *(Regel) G.I. Rodionenko in Signa,* 58: 3026 (1997), generic name not validly publ.; *et in Bot. Zhurn.,* 84(7): 104 (1999): Iris kolpakowskiana.
winkleri *(Regel) G.I. Rodionenko in Signa,* 58: 3026 (1997), generic name not validly publ.; *et in Bot. Zhurn.,* 84(7): 104 (1999): Iris winkleri.

ALBERTISIA (Menispermac.).
triplinervis *L.L. Forman in Checklist Flow. Pl. & Gymn. Brunei Darussalam (ed. M.J.E. Coode et al.):* 439 (1996)—Borneo (Brunei).

ALBIZIA (Legumin.).
sect. Arthrosamanea *(Britton & Rose) R.C. Barneby & J.W. Grimes, Silk Tree, Guanacaste, Monkey's Earring (Mem. New York Bot. Gard., 74(1)):* 206 (1996): gen. Arthrosamanea.

ALBIZIA:-
ser. Arthrosamanea *(Britton & Rose) R.C. Barneby & J.W. Grimes, Silk Tree, Guanacaste, Monkey's Earring (Mem. New York Bot. Gard., 74(1)):* 229 (1996): gen. Arthrosamanea.
ser. Inundatae *R.C. Barneby & J.W. Grimes, Silk Tree, Guanacaste, Monkey's Earring (Mem. New York Bot. Gard., 74(1)):* 238 (1996).
ser. Multiflorae *R.C. Barneby & J.W. Grimes, Silk Tree, Guanacaste, Monkey's Earring (Mem. New York Bot. Gard., 74(1)):* 234 (1996).
ser. Paniculatae *(Benth.) R.C. Barneby & J.W. Grimes, Silk Tree, Guanacaste, Monkey's Earring (Mem. New York Bot. Gard., 74(1)):* 208 (1996): Pithecellobium ser. Paniiculatae.
buntingii *R.C. Barneby & J.W. Grimes, Silk Tree, Guanacaste, Monkey's Earring (Mem. New York Bot. Gard., 74(1)):* 223 (1996)—Venezuela.
burkartiana *R.C. Barneby & J.W. Grimes, Silk Tree, Guanacaste, Monkey's Earring (Mem. New York Bot. Gard., 74(1)):* 211 (1996)—Brazil South (Santa Catarina, Paraná).
decandra *(Ducke) R.C. Barneby & J.W. Grimes, Silk Tree, Guanacaste, Monkey's Earring (Mem. New York Bot. Gard., 74(1)).* 237 (1996): Pithecellobium decandrum.
duckeana *L. Rico in Kew Bull.,* 55(2): 404 (2000), nom. nov.: Pithecellobium elegans *Ducke.*
edwallii *(Hoehne) R.C. Barneby & J.W. Grimes, Silk Tree, Guanacaste, Monkey's Earring (Mem. New York Bot. Gard., 74(1)).* 209 (1996): Pithecellobium edwallii.
elegans *(Ducke) L. Rico in Novon,* 9(4): 556 (1999), homonym; *see Kurz, Prelim. Rep. Forest Pegu:* App. B, 47 (1875): Pithecellobium elegans.
inundata *(Mart.) R.C. Barneby & J.W. Grimes, Tree, Guanacaste, Monkey's Earring (Mem. New York Bot. Gard., 74(1)):* 238 (1996): Acacia inundata.
leonardii *Britton & Rose ex R.C. Barneby & J.W. Grimes, Silk Tree, Guanacaste, Monkey's Earring (Mem. New York Bot. Gard., 74(1)):* 216 (1996)—Haiti.
mainaea *J.-F. Villiers in Novon,* 7(4): 420 (1997 publ. 1998)—Madagascar.
multiflora *(Kunth) R.C. Barneby & J.W. Grimes, Silk Tree, Guanacaste, Monkey's Earring (Mem. New York Bot. Gard., 74(1)):* 234 (1996): Acacia multiflora.
var. sagasteguii *R.C. Barneby & J.W. Grimes, Silk Tree, Guanacaste, Monkey's Earring (Mem. New York Bot. Gard., 74(1)):* 237 (1996)—Peru.
niopoides (Benth.) Burkart
var. colombiana *(Britton) R.C. Barneby & J.W. Grimes, Silk Tree, Guanacaste, Monkey's Earring (Mem. New York Bot. Gard., 74(1)):* 222 (1996): Albizia colombiana.
pedicellaris *(DC.) L. Rico in Novon,* 9(4): 555 (1999): Inga pedicellaris.
pistaciifolia *(Willd.) R.C. Barneby & J.W. Grimes, Silk Tree, Guanacaste, Monkey's Earring (Mem. New York Bot. Gard., 74(1)):* 230 (1996): Mimosa pistaciifolia.
subdimidiata *(Splitg.) R.C. Barneby & J.W. Grimes, Silk Tree, Guanacaste, Monkey's Earring (Mem. New York Bot. Gard., 74(1)):* 232 (1996): Acacia subdimidiata.
var. minor *R.C. Barneby & J.W. Grimes, Silk Tree, Guanacaste, Monkey's Earring (Mem. New York Bot. Gard., 74(1))* 234 (1996)—Guyana.

ALBUCA (Hyacinthac.).
sect. Trianthera *U. Müller-Doblies in Feddes Repert.,* 106(5–8): 355 (1995 publ. 1996).
aspera *U. Müller-Doblies in Feddes Repert.,* 106(5–8): 355 (1995 publ. 1996)—Cape Province.
bontebokensis *U. Müller-Doblies in Feddes Repert.,* 106(5–8): 358 (1995 publ. 1996)—Cape Province.
clanwilliamae-gloria *U. Müller-Doblies in Feddes Repert.,* 106(5–8): 358 (1995 publ. 1996)—Cape Province.
cremnophila *E. van Jaarsveld & A.E. van Wyk in Aloe,* 36(4): 72 (1999)—Cape Province.
decipiens *U. Müller-Doblies in Feddes Repert.,* 106(5–8): 358 (1995 publ. 1996)—Cape Province.
dinteri *U. Müller-Doblies in Feddes Repert.,* 106(5–8): 363 (1995 publ. 1996)—Namibia.

ALBUCA:-
etesiogaripensis *U. Müller-Doblies in Feddes Repert.,* 106(5–8): 363 (1995 publ. 1996)—Namibia.
foetida *U. Müller-Doblies in Feddes Repert.,* 106(5–8): 363 (1995 publ. 1996)—Cape Province.
goswinii *U. Müller-Doblies in Feddes Repert.,* 106(5–8): 363 (1995 publ. 1996)—Cape Province.
hesquaspoortensis *U. Müller-Doblies in Feddes Repert.,* 106(5–8): 366 (1995 publ. 1996)—Cape Province.
jacquinii *U. Müller-Doblies in Feddes Repert.,* 106(5–8): 366 (1995 publ. 1996)—Cape Province.
papyracea *J.C. Manning & P. Goldblatt in Strelitzia,* 9: 709 (2000)—Cape Province.
robertsoniana *U. Müller-Doblies in Feddes Repert.,* 106(5–8): 367 (1995 publ. 1996)—Cape Province.
viscosella *U. Müller-Doblies in Feddes Repert.,* 106(5–8): 369 (1995 publ. 1996)—Cape Province.
weberlingiorum *U. Müller-Doblies in Feddes Repert.,* 106(5–8): 369 (1995 publ. 1996)—Cape Province.

ALCANTAREA (Bromeliac.).
benzingii *E.M.C. Leme in Bromélia,* 2(3): 19 (1995)—Brazil Southeast (Espírito Santo).
brasiliana *(L.B. Sm.) J.R. Grant in Trop. Subtrop. Pflanzenwelt,* 91: 12 (1995): Vriesea brasiliana.
burle-marxii *(Leme) J.R. Grant in Bromélia,* 2(3): 26 (1995): Vriesea burle-marxii.
duarteana *(L.B. Sm.) J.R. Grant in Trop. Subtrop. Pflanzenwelt,* 91: 13 (1995): Vriesea duarteana.
edmundoi *(Leme) J.R. Grant in Bromélia,* 2(3): 26 (1995): Vriesea edmundoi.
extensa *(L.B. Sm.) J.R. Grant in Trop. Subtrop. Pflanzenwelt,* 91: 13 (1995): Vriesea extensa.
farneyi *(Martinelli & And. Costa) J.R. Grant in Trop. Subtrop. Pflanzenwelt,* 91: 13 (1995): Vriesea farneyi.
geniculata *(Wawra) J.R. Grant in Trop. Subtrop. Pflanzenwelt,* 91: 13 (1995): Platystachys geniculata.
glaziouana *(Lem.) E.M.C. Leme in Bromélia,* 4(2): 35 (1997): Vriesea glaziouana.
hatschbachii *(L.B. Sm. & Read) E.M.C. Leme in Bromélia,* 2(3): 22 (1995): Vriesea hatschbachii.
nahoumii *(Leme) J.R. Grant in Trop. Subtrop. Pflanzenwelt,* 91: 13 (1995): Vriesea nahoumii.
nevaresii *E.M.C. Leme in Bromélia,* 2(3): 15 (1995)—Brazil Southeast (Rio de Janeiro).
odorata *(Leme) J.R. Grant in Trop. Subtrop. Pflanzenwelt,* 91: 13 (1995): Vriesea odorata.
roberto-kautskyi *E.M.C. Leme in Harvard Pap. Bot.,* 4(1): 148 (1999)—Brazil Southeast (Espírito Santo).
vinicolor *(E. Pereira & Reitz) J.R. Grant in Trop. Subtrop. Pflanzenwelt,* 91: 14 (1995): Vriesea vinicolor.

ALCHEMILLA (Rosac.).
subser. Appressipilae *V.N. Tikhomirov in Byull. Mosk. Obshch. Ispȳt. Prir., Biol.,* 103(2): 62 (1998).
subser. Commixtae *V.N. Tikhomirov in Byull. Mosk. Obshch. Ispȳt. Prir., Biol.,* 103(2): 61 (1998).
sect. Coriaceae *S. Fröhner in Gleditschia,* 14(1): 41 (1986).†
sect. Decumbentes *S. Fröhner in Gleditschia,* 14(1): 36 (1986).†
subser. Denudatae *V.N. Tikhomirov in Byull. Mosk. Obshch. Ispȳt. Prir., Biol.,* 103(2): 62 (1998).
sect. Flabellatae *S. Fröhner in Gleditschia,* 14(1): 44 (1986).†
ser. Flabellatae *(S.E. Fröhner) V.N. Tikhomirov in Byull. Mosk. Obshch. Ispȳt. Prir., Biol.,* 103(2): 60 (1998): A. sect. Flabellatae.
subser. Glabricaules *V.N. Tikhomirov in Byull. Mosk. Obshch. Ispȳt. Prir., Biol.,* 103(2): 62 (1998).
sect. Glaciales *S. Fröhner in Gleditschia,* 14(1): 39 (1986).†
ser. Glaucescentes *V.N. Tikhomirov in Byull. Mosk. Obshch. Ispȳt. Prir., Biol.,* 103(2): 60 (1998).
subser. Heptagonae *V.N. Tikhomirov in Byull. Mosk. Obshch. Ispȳt. Prir., Biol.,* 103(2): 61 (1998).
ser. Hirsuticaules *V.N. Tikhomirov in Byull. Mosk. Obshch. Ispȳt. Prir., Biol.,* 103(2): 60 (1998).
subser. Pastorales *V.N. Tikhomirov in Byull. Mosk. Obshch. Ispȳt. Prir., Biol.,* 103(2): 61 (1998).
ser. Plicatae *(S.E. Fröhner) V.N. Tikhomirov in Byull. Mosk. Obshch. Ispȳt. Prir., Biol.,* 103(2): 60 (1998): A. ser. Plicatae.

ALCHEMILLA:-
sect. Plicatae *S. Fröhner in Gleditschia,* 14(1): 45 (1986).†
angustata *S.E. Fröhner in An. Jard. Bot. Madrid,* 56(2): 406 (1998)—Spain.
angustiserrata *S. Fröhner in An. Jard. Bot. Madrid,* 55(2): 242 (1997)—Spain.
atriuscula *S. Fröhner in An. Jard. Bot. Madrid,* 53(2): 173 (1995 publ. 1996)—Spain.
austroitalica *S. Brullo, F. Scelsi & G. Spampinato in Willdenowia,* 27(1–2): 69 (1997)—Italy.
czamsinensis *V.N. Tikhomirov in Byull. Mosk. Obshch. Ispȳt. Prir., Biol.,* 101(5): 81 (1996)—Russia Central.
fageti *S.E. Fröhner in An. Jard. Bot. Madrid,* 56(2): 405 (1998)—Spain.
hypercycla *S. Fröhner in An. Jard. Bot. Madrid,* 53(2): 175 (1995 publ. 1996)—Spain.
legionensis *S. Fröhner in An. Jard. Bot. Madrid,* 55(2): 235 (1997)—Spain.
lunaria *S. Fröhner An. Jard. Bot. Madrid,* 55(2): 238 (1997)—Spain.
macrochira *S. Fröhner in An. Jard. Bot. Madrid,* 53(2): 174 (1995 publ. 1996)—Spain.
melanoscytos *S. Fröhner in An. Jard. Bot. Madrid,* 55(2): 236 (1997)—Spain.
microcephala *S. Fröhner in An. Jard. Bot. Madrid,* 55(2): 241 (1997)—Spain.
montenegrina *A. Plocek in Candollea,* 53(2): 317 (1998), homonym; *see Gand., Contrib. Fl. Terr. Slav. Merid.,* 1: 11 (1883)—Yugoslavia.
nieto-felineri *S. Fröhner in An. Jard. Bot. Madrid,* 55(2): 243 (1997)—Spain.
rubidula *A. Plocek in Candollea,* 53(2): 311 (1998)—Yugoslavia.
rugulosa *S. Fröhner in An. Jard. Bot. Madrid,* 53(2): 171 (1995 publ. 1996)—Spain.
spectabilior *S. Fröhner in An. Jard. Bot. Madrid,* 53(2): 175 (1995 publ. 1996)—Spain.
tenerifolia *S. Fröhner in An. Jard. Bot. Madrid,* 55(2): 240 (1997)—France, Spain.
tenerrima *S. Fröhner in An. Jard. Bot. Madrid,* 55(2): 239 (1997)—France, Italy.
trabzonica *S. Hayırlıoğlu-Ayaz & O. Beyazoğlu in Ann. Bot. Fenn.,* 34(2): 111 (1997)—Turkey.
vetteri *R. Buser in C. Bicknell, Fl. Bordighera:* 99 (1896)—Italy.
vincekii *A. Plocek in Candollea,* 53(2): 310 (1998)—Yugoslavia.
vizcayensis *S. Fröhner in An. Jard. Bot. Madrid,* 53(2): 172 (1995 publ. 1996)—Spain.

ALCHORNEA (Euphorbiac.).
anamariae *R. Secco in Rev. Brasil. Bot.,* 22(2): 141 (1999)—Bolivia.
fluviatilis *R. de S. Secco in Bol. Mus. Paraense Emílio Goeldi, Bot.,* 9(1): 60 (1993)—Brazil Northeast (Maranhão), Brazil North (Amapá, Amazonas, Pará), Venezuela, French Guiana.
pubescens *(Britton) R. Secco in Rev. Brasil. Bot.,* 22(2): 142 (1999), homonym; *see Merr. in Philipp. J. Sci.,* 20(4): 399 (1922): Conceveiba pubescens.
rugosa (Lour.) Müll. Arg.
var. pubescens *(Pax & K. Hoffm.) H.S. Kiu in Fl. Reipubl. Popul. Sin.,* 44(2): 69 (1996): A. hainanensis var. pubescens.
tachirensis *R. de S. Secco in Bol. Mus. Paraense Emílio Goeldi, Bot.,* 14(1): 81 (1998 publ. 1999)—Venezuela, Colombia.

ALDINA (Legumin.).
macrophylla Spruce ex Benth.
var. yapacanensis *(R.S. Cowan) B. Stergios in Fl. Venezuelan Guayana (gen. eds. J.A. Steyermark et al.),* 5: 248 (1999): A. yapacanensis.
paulberryi *G. Aymard C. in Novon,* 8(4): 330 (1998)—Venezuela.

ALECTRA (Scrophulariac.).
alba *(Hepper) B.L. Burtt Edinburgh J. Bot.,* 56(3): 460 (1999): Harveya alba.
fruticosa *Eb. Fischer in Bull. Mus. Nation. Hist. Nat., B, Adansonia,* Ser. 4, 18(1–2): 58 (1996)—Madagascar.
hildebrandtii *Eb. Fischer in Bull. Mus. Nation. Hist. Nat., B, Adansonia,* Ser. 4, 18(1–2): 55 (1996)—Madagascar.
humbertii *Eb. Fischer in Bull. Mus. Nation. Hist. Nat., B, Adansonia,* Ser. 4, 18(1–2): 56 (1996)—Madagascar.

ALECTRA:-
ibityensis *Eb. Fischer in Bull. Mus. Nation. Hist. Nat., B, Adansonia,* Ser. 4, 18(1–2): 56 (1996)—Madagascar.
excelsus Gaertn.
subsp. grandis *(Cheeseman) P.J. de Lange & E.K. Cameron in New Zealand J. Bot.,* 37(1): 9 (1999): A. excelsus var. grandis.

ALEISANTHIOPSIS *C. Tange in Nordic J. Bot.,* 16(6): 571 (1996 publ. 1997). *RUBIACEAE*
distantiflora *(Merr.) C. Tange in Nordic J. Bot.,* 16(6): 573 (1996 publ. 1997): Greenea distantiflora.
multiflora *C. Tange in Nordic J. Bot.,* 16(6): 574 (1996 publ. 1997)—Borneo (Kalimantan).

ALEURITES (Euphorbiac.).
rockinghamensis *(Baill.) P.I. Forster in Muelleria,* 9: 8 (1996): A. moluccana var. rockinghamensis.

× **ALGASTOLOBA** *D.M. Cumming in Haworthiad,* 13(1): 20 (1999); *cf. Repert. Pl. Succ. (I.O.S.),* 50: 7 (1999 publ. 2000). *ALOACEAE* (Aloe × Astroloba × Gasteria).

ALIBERTIA (Rubiac.).
premontana *C.M. Taylor in Novon,* 6(2): 215 (1996)—Costa Rica, Panamá.utleyorum *(Dwyer) C.M. Taylor in Novon,* 6(2): 215 (1996): Duroia utleyorum.

ALICIELLA (Polemoniac.).
sect. Giliandra *(A. Gray) J.M. Porter in Aliso,* 17(1): 27 (1998): Gilia sect. Giliandra.
subgen. Gilmania *(H. Mason & A.D. Grant) J.M. Porter in Aliso,* 17(1): 43 (1998): Gilia subgen. Gilmania.
subsect. Subnuda *J.M. Porter in Aliso,* 17(1): 31 (1998).
caespitosa *(A. Gray) J.M. Porter in Aliso,* 17(1): 34 (1998): Gilia caespitosa.
formosa *(Greene ex Brand) J.M. Porter in Aliso,* 17(1): 33 (1998): Gilia formosa.
haydenii *(A. Gray) J.M. Porter in Aliso,* 17(1): 31 (1998): Gilia haydenii.
subsp. crandallii *(Rydb.) J.M. Porter in Aliso,* 17(1): 32 (1998): Gilia crandallii.
heterostyla *(S. Cochrane & A.G. Day) J.M. Porter in Aliso,* 17(1): 37 (1998): Gilia heterostyla.
humillima *(Brand) J.M. Porter in Aliso,* 17(1): 41 (1998): A. triodon var. humillima.
hutchinsifolia *(Rydb.) J.M. Porter in Aliso,* 17(1): 35 (1998): Gilia hutchinsifolia.
latifolia *(S. Watson) J.M. Porter in Aliso,* 17(1): 43 (1998): Gilia latifolia.
subsp. imperialis *(S.L. Welsh) J.M. Porter in Aliso,* 17(1): 44 (1998): Gilia latifolia var. imperialis.
leptomeria *(A. Gray) J.M. Porter in Aliso,* 17(1): 38 (1998): Gilia leptomeria.
lottiae *(A. Day) J.M. Porter in Aliso,* 17(1): 40 (1998): Gilia lottiae.
mcvickerae *(M.E. Jones) J.M. Porter in Aliso,* 17(1): 28 (1998): Gilia mcvickerae.
micromeria *(A. Gray) J.M. Porter in Aliso,* 17(1): 40 (1998): Gilia micromeria.
nyensis *(Reveal) J.M. Porter in Aliso,* 17(1): 36 (1998): Gilia nyensis.
pentstemonoides *(M.E. Jones) J.M. Porter in Aliso,* 17(1): 30 (1998): Gilia pentstemonoides.
pinnatifida *(Nutt. ex A. Gray) J.M. Porter in Aliso,* 17(1): 27 (1998): Gilia pinnatifida.
ripleyi *(Barneby) J.M. Porter in Aliso,* 17(1): 45 (1998): Gilia ripleyi.
sedifolia *(Brandegee) J.M. Porter in Aliso,* 17(1): 29 (1998): Gilia sedifolia.
stenothyrsa *(A. Gray) J.M. Porter in Aliso,* 17(1): 30 (1998): Gilia stenothyrsa.
subacaulis *(Rydb.) J.M. Porter & L.A. Johnson in Aliso,* 19(1): 70 (2000): Gilia subacaulis.
subnuda *(Torr. ex A. Gray) J.M. Porter in Aliso,* 17(1): 33 (1998): Gilia subnuda.
tenuis *(F.J. Sm. & Neese) J.M. Porter in Aliso,* 17(1): 35 (1998): Gilia tenuis.

ALJELLA (Compos.).
iminouakensis *(Emb.) A. Dobignard & D. Jeanmonod in Candollea,* 52(1): 143 (1997): Phagnalon iminouakense.

ALKANNA (Boraginac.).
mughlae *H. Duman, A. Güner & H. Şağban in Bot. J. Linn. Soc.,* 129(4): 340 (1999)—Turkey.

ALLEXIS (Violac.).
zygomorpha *G. Achoundong & J.-M. Onana in Kew Bull.,* 53(4): 1009 (1998)—Cameroon.

ALLIUM (Alliac.).
sect. Avulsea *F.O. Khassanov in M.A. Öztürk, Ö. Seçmen & G. Görk (eds.), Pl. Life S.W. &* Central Asia, 1: 149 (1996).
sect. Brevidentia *F.O. Khassanov & S.C. Iengalycheva in M.A. Öztürk, Ö. Seçmen & G. Görk (eds.), Pl. Life S.W. &* Central Asia, 1: 147 (1996).
sect. Coerulea *(Omelczuk) F.O. Khassanov in M.A. Öztürk, Ö. Seçmen & G. Görk (eds.), Pl. Life S.W. &* Central Asia, 1: 149 (1996): A. ser. Coerulea.
sect. Costulatae *F.O. Khassanov & S.C. Iengalycheva in M.A. Öztürk, Ö. Seçmen & G. Görk (eds.), Pl. Life S.W. &* Central Asia, 1: 147 (1996).
sect. Crystallina *F.O. Khassanov & S.C. Iengalycheva in M.A. Öztürk, Ö. Seçmen & G. Görk (eds.), Pl. Life S.W. &* Central Asia, 1: 147 (1996).
sect. Eremoprasa *(Kamelin) F.O. Khassanov in O.A. Ashurmetov et al. (eds.), Pl. Life S.-W. & C. Asia Vth Internat.* Symp. Tashkent 1998: 110 (2000), without basionym ref.: Basionym not stated.
subsect. Filidentia *(Tsag.) F.O. Khassanov & S.C. Iengalycheva in M.A. Öztürk, Ö. Seçmen & G. Görk (eds.), Pl. Life S.W. &* Central Asia, 1: 147 (1996): A. ser. Filidentia.
subsect. Filiformia *F.O. Khassanov & S.C. Iengalycheva in M.A. Öztürk, Ö. Seçmen & G. Görk (eds.), Pl. Life S.W. &* Central Asia, 1: 147 (1996).
ser. Globoprason *J. Radić in Acta Biokovica,* 5: 57 (1989), without type.
sect. Kopetdagia *F.O. Khassanov in M.A. Öztürk, Ö. Seçmen & G. Görk (eds.), Pl. Life S.W. &* Central Asia, 1: 150 (1996).
sect. Longivaginata *(Kamelin) F.O. Khassanov in O.A. Ashurmetov et al. (eds.), Pl. Life S.-W. & C. Asia Vth Internat.* Symp. Tashkent 1998: 109 (2000), without basionym ref.: Basionym not stated.
sect. Minuta *F.O. Khassanov in M.A. Öztürk, Ö. Seçmen & G. Görk (eds.), Pl. Life S.W. &* Central Asia, 1: 150 (1996).
sect. Multicaulea *F.O. Khassanov & S.C. Iengalycheva in M.A. Öztürk, Ö. Seçmen & G. Görk (eds.), Pl. Life S.W. &* Central Asia, 1: 148 (1996).
ser. Neuroprason *J. Radić in Acta Biokovica,* 5: 46 (1989), without type.
sect. Pallasia *(Tsag.) F.O. Khassanov in O.A. Ashurmetov et al. (eds.), Pl. Life S.-W. & C. Asia Vth Internat.* Symp. Tashkent 1998: 109 (2000), without basionym ref.: Basionym not stated.
ser. Platoprason *J. Radić in Acta Biokovica,* 5: 61 (1989), without type.
ser. Proionoprason *J. Radić in Acta Biokovica,* 5: 38 (1989), without type.
sect. Spathulata *F.O. Khassanov & R.M. Fritsch in O.A. Ashurmetov et al. (eds.), Pl. Life S.-W. & C. Asia Vth Internat.* Symp. Tashkent 1998: 109 (2000).
aegilicum *D. Tzanoudakis in Bot. Chronika,* 13: 81 (2000)—Greece.
amethystinum Tausch *var.* antiquorum *J. Radić in Acta Biokovica,* 5: 65 (1989), contrary to Art. 37.3 ICBN (1994)—Yugoslavia.
ampeloprasum L.
f. scaberrimum *J. Radić in Acta Biokovica,* 5: 57 (1989), contrary to Art. 37.3 ICBN (1994)—Yugoslavia.
f. urceolatum *J. Radić in Acta Biokovica,* 5: 59 (1989), contrary to Art. 37.3 ICBN (1994)—Yugoslavia.
anzalonei *S. Brullo, P. Pavone & C. Salmeri in Sendtnera,* 4: 34 (1997)—Italy.
archeotrichon *S. Brullo, P. Pavone & C. Salmeri in Nordic J. Bot.,* 19(1): 42 (1999)—East Aegean Is.
assadii *A. Seisums in Iranian J. Bot.,* 8(2): 230 (2000)—Iran.

ALLIUM:-
atrosanguineum Schrenk
var. fedschenkoanum *(Regel) G.H. Zhu & N.J. Turland in Novon,* 10(2): 181 (2000): A. fedschenkoanum.
var. tibeticum *(Regel) G.H. Zhu & N.J. Turland in Novon,* 10(2): 182 (2000): A. monadelphum var. tibeticum.
austrokyushuense *M. Hotta in Acta Phytotax. Geobot.,* 49(1): 64 (1998)—Japan.
brulloi *C. Salmeri in Willdenowia,* 28(1–2): 71 (1998)—East Aegean Is.
calabrum *(N. Terracc.) S. Brullo, P. Pavone & C. Salmeri in Giorn. Bot. Ital.,* 128(1): 246 (1994): A. pulchellum var. calabrum.
carinatum L.
subsp. biokovense *J. Radić in Acta Biokovica,* 5: 55 (1989), contrary to Art. 37.3 ICBN (1994)—Yugoslavia.
cupuliferum Regel
subsp. nuratavicum *R.M. Fritsch & Beshko in O.A. Ashurmetov et al. (eds.), Pl. Life S.-W. & C. Asia Vth Internat.* Symp. Tashkent 1998: 64 (2000)—Uzbekistan.
daninianum *S. Brullo, P. Pavone & C. Salmeri Willdenowia,* 26(1–2): 239 (1996)—Palestine (Israel, Jordan), Lebanon, Syria.
egorovae *M.V. Agababian & M.E. Oganesian in Willdenowia,* 30(1): 95 (2000)—Transcaucasus (Nakhichevan)
flavum L.
subsp. lignosum *J. Radić in Acta Biokovica,* 5: 53 (1989), contrary to Art. 37.3 ICBN (1994)—Yugoslavia.
fritschii *F.O. Khassanov & S.S. Yengalycheva in Bot. Jahrb.,* 119(2): 270 (1997)—Uzbekistan.
goekyigitii *T. Ekim, H. Duman & A. Güner in Karaca Arbor. Mag.,* 5(1): 27 (1999)—Turkey.
guttatum Steven
subsp. dilatatum *(Zahar.) B. Mathew, Review Allium sect. Allium:* 136 (1996): A. dilatatum.
haneltii *F.O. Khassanov & R.M. Fritsch in Linz. Biol. Beitr.,* 30(1): 282 (1998)—Uzbekistan.
hindukuschense *R.V. Kamelin & A.G. Seisums in Novosti Sist. Vyssh. Rast.,* 30: 31 (1996)—Afghanistan.
incensiodorum *J. Radić in Acta Biokovica,* 5: 29, 37 (1989), contrary to Art. 37.3 ICBN (1994)—Yugoslavia.
isakulii *R.M. Fritsch & F.O. Khassanov in O.A. Ashurmetov et al. (eds.), Pl. Life S.-W. & C. Asia Vth Internat.* Symp. Tashkent 1998: 65 (2000)—Tadzhikistan.
subsp. balkhanicum *R.M. Fritsch & F.O. Khassanov in O.A. Ashurmetov et al. (eds.), Pl. Life S.-W. & C. Asia Vth Internat.* Symp. Tashkent 1998: 65 (2000)—Turkmenistan.
subsp. nuratense *(Kamelin) R.M. Fritsch & F.O. Khassanov in O.A. Ashurmetov et al. (eds.), Pl. Life S.-W. & C. Asia Vth Internat.* Symp. Tashkent 1998: 65 (2000): A. iliense subsp. nuratense.
subsp. subkopetdagense *R.M. Fritsch & F.O. Khassanov in O.A. Ashurmetov et al. (eds.), Pl. Life S.-W. & C. Asia Vth Internat.* Symp. Tashkent 1998: 65 (2000)—Turkmenistan.
kuramense *F.O. Khassanov & N.V. Friesen in Linz. Biol. Beitr.,* 30(1): 287 (1998)—Uzbekistan.
kurdaicum *M.S. Bajtenov in Izv. Natsion. Akad. Nauk Resp. Kazakhstan, Ser. Biol.,* 1995(6): 81 (1995 publ. 1996)—Kazakhstan.
ledebourianum Roem. & Schult.
var. maximowiczii *(Regel) Q.S. Sun in Fl. Plant. Herb. Chinae Bor.-Or.,* 12: 130 (1998): A. maximowiczii.
lehmannii Lojac.
subsp. castellanense *F. Garbari, P. Miceli & F.M. Raimondo in Atti Soc. Tosc. Sci. Nat., Mem., B,* 101: 141 (1994 publ. 1995)—Sicilia.
leucanthum K. Koch
var. tridentatum *M.E. Oganesian in Fl. Rastitel'nost' Rast. Res. Arm.,* 12: 12 (1999)—Transcaucasus (Armeniya).
monanthum Maxim.
var. floribundum *Z.J. Zhong & X.T. Huang in Bull. Bot. Res. North-East. Forest. Univ.,* 17(1): 53 (1997)—Manchuria (Jilin).
nabelekii *R.V. Kamelin & A.G. Seisums in Novosti Sist. Vyssh. Rast.,* 30: 29 (1996)—Turkey.

ALLIUM:-
neapolitanum Cirillo
subsp. philippi *J. Radić in Acta Biokovica,* 5: 42 (1989), contrary to Art. 37.3 ICBN (1994)—Yugoslavia.
neriniflorum (Herb.) Baker
f. albiflorum *(Kitag.) Q.S. Sun in Fl. Plant. Herb. Chinae Bor.-Or.,* 12: 139 (1998): Nothoscordum neriniflorum f. albiflorum.
oporinanthum *S. Brullo, P. Pavone & C. Salmeri in An. Jard. Bot. Madrid,* 55(2): 297 (1997)—Spain, France.
pallens L.
var. admirabile *J. Radić in Acta Biokovica,* 5: 52 (1989), contrary to Art. 37.3 ICBN (1994)—Yugoslavia.
subvar. arenae-maris *J. Radić in Acta Biokovica,* 5: 52 (1989), contrary to Art. 37.3 ICBN (1994)—Yugoslavia.
subsp. christophori *J. Radić in Acta Biokovica,* 5: 51 (1989), contrary to Art. 37.3 ICBN (1994)—Yugoslavia.
var. grandiflorum *(Maire & Weiller) J. Pastor & B. Valdés in Lagascalia,* 18(2): 330 (1996): A. paniculatum var. grandiflorum.
subvar. stoloniferum *J. Radić in Acta Biokovica,* 5: 52 (1989), contrary to Art. 37.3 ICBN (1994)—Yugoslavia.
paniculatum L.
subsp. jacobi *J. Radić in Acta Biokovica,* 5: 48 (1989), contrary to Art. 37.3 ICBN (1994)—Yugoslavia.
pseudoalbidum *N. Friesen & N. Özhatay in Feddes Repert.,* 109(1–2): 25 (1998)—Turkey.
pseudowinklerianum *R.M. Fritsch & F.O. Khassanov in O.A. Ashurmetov et al. (eds.), Pl. Life S.-W. & C. Asia Vth Internat.* Symp. Tashkent 1998: 66 (2000)—Kirgizstan.
rhodopeum Velen.
subsp. turcicum *S. Brullo, A. Guglielmo & M.C. Terrasi in Pl. Biosystems,* 132(1): 64 (1998)—Turkey-in-Europe, Turkey, Greece.
ritsi *G. Iatroú & D. Tzanoudakis in Phyton (Austria),* 35(2): 248 (1995), as 'ritsii'—Greece.
roseum L.
var. coacervatum *J. Radić in Acta Biokovica,* 5: 38 (1989), without type—Yugoslavia.
rotundum L.
subsp. jajlae *(Vved.) B. Mathew, Review Allium sect. Allium:* 101 (1996): A. jajlae.
rubellum M. Bieb.
subsp. syntamanthum *(K. Koch) M.E. Oganesian in Fl. Rastitel'nost' Rast. Res. Arm.,* 12: 12 (1999): A. syntamanthum.
runemarkii *P. Trigas & D. Tzanoudakis in Nordic J. Bot.,* 20(1): 89 (2000)—Greece.
salsuginis *J. Radić in Acta Biokovica,* 5: 11, 18 (1989), contrary to Art. 37.3 ICBN (1994)—Yugoslavia.
f. piscatorum *J. Radić in Acta Biokovica,* 5: 23 (1989), contrary to Art. 37.3 ICBN (1994)—Yugoslavia.
f. spiritus-procellarum *J. Radić in Acta Biokovica,* 5: 21 (1989), contrary to Art. 37.3 ICBN (1994)—Yugoslavia.
schchianae *M.E. Oganesian in Fl. Rastitel'nost' Rast. Res. Arm.,* 12: 10 (1999)—Transcaucasus (Armeniya).
spathulatum *F.O. Khassanov & R.M. Fritsch in Linz. Biol. Beitr.,* 30(1): 286 (1998)—Kirgizstan.
sphaerocephalon L.
var. stipitatum *J. Radić in Acta Biokovica,* 5: 62 (1989), contrary to Art. 37.3 ICBN (1994)—Yugoslavia.
struzlianum *M.E. Oganesian in Fl. Rastitel'nost' Rast. Res. Arm.,* 12: 11 (1999)—Transcaucasus (Armeniya).
subhirsutum L.
f. bistonense *J. Radić in Acta Biokovica,* 5: 45 (1989), contrary to Art. 37.3 ICBN (1994)—Yugoslavia.
subf. convexum *J. Radić in Acta Biokovica,* 5: 45 (1989), contrary to Art. 37.3 ICBN (1994)—Yugoslavia.
subf. minimum *J. Radić in Acta Biokovica,* 5: 45 (1989), contrary to Art. 37.3 ICBN (1994)—Yugoslavia.
subf. rubellum *J. Radić in Acta Biokovica,* 5: 45 (1989), contrary to Art. 37.3 ICBN (1994)—Yugoslavia.

ALLIUM:-
taeniopetalum Popov & Vved.
subsp. mogoltavicum *(Vved.) R.M. Fritsch & F.O. Khassanov in Linz. Biol. Beitr.,* 30(1): 290 (1998): A. mogoltavicum.
subsp. turakulovii *R.M. Fritsch & F.O. Khassanov in Linz. Biol. Beitr.,* 30(1): 290 (July 1998), contrary to Art. 37.5 ICBN (1994); *et in Linz. Biol. Beitr.,* 30(2): 843 (Dec. 1998)—Kirgizstan.
taquetii H. Lév. & Vaniot
f. albiflorum *Y.N. Lee, Fl. Korea:* 1163 (1996), without type—Locality not stated.
tschulaktavicum *M.S. Bajtenov & N.V. Nelina in Izv. Natsion. Akad. Nauk Resp. Kazakhstan, Ser. Biol.,* 1993(4): 79 (1993 publ. 1994)—Xinjiang.
urmiense *R.V. Kamelin & A.G. Seisums in Novosti Sist. Vyssh. Rast.,* 30: 30 (1996)—Iraq.
valdesianum *S. Brullo, P. Pavone & C. Salmeri in Sendtnera,* 3: 96 (1996)—Spain.
vasilevskajae *M.E. Oganesian in Willdenowia,* 30(1): 96 (2000)—Transcaucasus (Armeniya).
virgunculae F. Maek. & Kitam.
var. yakushimense *M. Hotta in Acta Phytotax. Geobot.,* 49(1): 66 (1998)—Japan.
yamarakkyo *M. Honda in Bot. & Zool.,* 8: 97 (1940)—Japan.

ALLOCHEILOS (Gesneriac.).
guangxiensis *H.Q. Wen, Y.G. Wei & S.H. Zhong in Acta Phytotax. Sin.,* 38(3): 297 (2000)—China Southeast (Guangxi).

ALLOMAIETA (Melastomatac.).
caucana *G. Lozano C. in Rev. Acad. Colomb. Cienc. Exact. Fis. Nat.,* 86: 11 (1999), without latin descr.—Colombia.
ebejicosana *G. Lozano C. in Rev. Acad. Colomb. Cienc. Exact. Fis. Nat.,* 86: 17 (1999), without latin descr.—Colombia.
hirsuta *(Gleason) G. Lozano C. in Rev. Acad. Colomb. Cienc. Exact. Fis. Nat.,* 86: 12 (1999): Cyphostyla hirsuta.
pancurana *G. Lozano C. in Rev. Acad. Colomb. Cienc. Exact. Fis. Nat.,* 86: 12 (1999), without latin descr.—Colombia.
strigosa *(Gleason) G. Lozano C. in Rev. Acad. Colomb. Cienc. Exact. Fis. Nat.,* 86: 11 (1999): Cyphostyla strigosa.
villosa *(Gleason) G. Lozano C. in Rev. Acad. Colomb. Cienc. Exact. Fis. Nat.,* 86: 11 (1999): Cyphostyla villosa.
zenufanasana *G. Lozano C. in Rev. Acad. Colomb. Cienc. Exact. Fis. Nat.,* 86: 12 (1999), without latin descr.—Colombia.

ALLOMARKGRAFIA (Apocynac.).
campanulata *(Markgr.) J.F. Morales in Brittonia,* 49(3): 339 (1997): Mandevilla campanulata.
ecuatoriana *J.F. Morales in Brittonia,* 49(3): 340 (1997)—Ecuador.
insignis *J.F. Morales in Brittonia,* 49(3): 341 (1997)—Costa Rica.

ALLONEURON (Melastomatac.).
sect. Meiandra *(Markgr.) B. Wallnöfer in Ann. Naturhist. Mus. Wien,* 98B(Suppl.): 454 (1996): gen. Meiandra.
liron *B. Wallnöfer in Ann. Naturhist. Mus. Wien,* 98B(Suppl.): 457 (1996)—Peru.
ronliesneri *B. Wallnöfer in Ann. Naturhist. Mus. Wien,* 101B: 595 (1999)—Peru.

ALLOPHYLLUM (Polemoniac.).
sect. Kelloggia *(H. Mason & A.D. Grant) A.G. Day & V. Grant in Phytologia,* 84(5): 375 (1998 publ. 1999): G. subgen. Kelloggia.
capillare *(Kellogg) A.G. Day & V. Grant in Phytologia,* 84(5): 375 (1998 publ. 1999): Gilia capillaris.
leptaleum *(A. Gray) A.G. Day & V. Grant in Phytologia,* 84(5): 375 (1998 publ. 1999): Collomia leptalea.
subsp. bicolor *(H. Mason & A.D. Grant) A.G. Day & V. Grant in Phytologia,* 84(5): 376 (1998 publ. 1999): Gilia leptalea subsp. bicolor.

ALLOPHYLLUM:-
nemophilophyllum *J.M. Porter & L.A. Johnson in Aliso,* 19(1): 64 (2000)—Mexico Northwest (Baja California).
sinistrum *(M.E. Jones) A.G. Day & V. Grant in Phytologia,* 84(5): 376 (1998 publ. 1999): Gilia sinistra.
subsp. pinnatisectum *(H. Mason & A.D. Grant) A.G. Day & V. Grant in Phytologia,* 84(5): 376 (1998 publ. 1999): Gilia leptalea subsp. pinnatisecta.
tenerrimum *(A. Gray) V. Grant in Phytologia,* 84(2): 79 (1998 publ. 1999): Gilia tenerrima.

ALLOPHYLUS (Sapindac.).
africanus P. Beauv.
var. griseotomentosus *(Gilg) B. Verdcourt in Fl. Trop. E. Afr., Sapindac.:* 82 (1998): A. griseotomentosus.
borbonicus *(J.F. Gmel.) F. Friedmann in Fl. Mascar.,* 76: 6 (1997): Ornitrophe borbonica.
delicatulus *B. Verdcourt in Fl. Trop. E. Afr., Sapindac.:* 84 (1998)—Tanzania.
ferrugineus Taub.
var. stipitatus *B. Verdcourt in Fl. Trop. E. Afr., Sapindac.:* 87 (1998)—Tanzania.
grotei *F.G. Davies & B. Verdcourt in Fl. Trop. E. Afr., Sapindac.:* 93 (1998)—Tanzania.
paniger *F.R. Fosberg in Allertonia,* 7(2): 61 (1993), nom. nov.: Aporetica ternata *J.R. Forst.*
rubifolius (A. Rich.) Engl.
var. alnifolius *(Baker) I. Friis & K. Vollesen in Kew Bull.,* 51(4): 802 (1996): Schmidelia alnifolia.
var. dasystachys *(Gilg) B. Verdcourt in Fl. Trop. E. Afr., Sapindac.:* 90 (1998): A. dasystachys.
var. rhusiphyllus *(Balf. f.) I. Friis & K. Vollesen in Kew Bull.,* 51(4): 802 (1996): Allophylus rhusiphyllus.
tanzaniensis *F.G. Davies in Fl. Trop. E. Afr., Sapindac.:* 87 (1998)—Tanzania.
vestitus *F.G. Davies in Fl. Trop. E. Afr., Sapindac.:* 93 (1998)—Tanzania.

ALLOPLECTUS (Gesneriac.).
martinianus *J.F. Smith in Novon,* 9(3): 419 (1999)—Ecuador.
penduliflorus *M. Freiberg in Phyton (Austria),* 37(1): 134 (1997)—Ecuador.

ALNUS (Betulac.).
brembana *L. Rota, Prosp. fl. Bergamo:* 102 (1853)—Italy.†
viridis (Chaix) DC.
subsp. brembana *(Rota) P.A. Schmidt in Mitt. Deutsch. Dendrol. Ges.,* 82: 33 (1996): A. brembana.

ALOCASIA (Arac.).
balgooyi *A. Hay in Gard. Bull. Singapore,* 50(2): 291 (1998)—Sulawesi.
boyceana *A. Hay in Gard. Bull. Singapore,* 51(1): 9 (1999)—Philippines.
clypeolata *A. Hay in Gard. Bull. Singapore,* 51(1): 12 (1999)—Philippines.
flemingiana *Yuzammi & A. Hay in Gard. Bull. Singapore,* 50(2): 292 (1998)—Jawa.
kerinciensis *A. Hay in Gard. Bull. Singapore,* 50(2): 322 (1998)—Sumatera.
melo *A. Hay, P.C. Boyce & K.M. Wong in Curtis's Bot. Mag.,* 14(2): 84 (1997)—Borneo (Sabah).
minuscula *A. Hay in Gard. Bull. Singapore,* 50(2): 320 (1998)—Borneo (Sarawak).
nebula *A. Hay in Curtis's Bot. Mag.,* 17(1): 16 (2000)—Borneo (Sarawak)?
pangeran *A. Hay in Gard. Bull. Singapore,* 50(2): 273 (1998)—Borneo (Sabah).
principiculus *A. Hay in Gard. Bull. Singapore,* 50(2): 277 (1998)—Borneo (Sabah, Kalimantan).
puteri *A. Hay in Gard. Bull. Singapore,* 50(2): 275 (1998)—Borneo (Sabah).
ramosii *A. Hay in Gard. Bull. Singapore,* 51(1): 27 (1999)—Philippines.
reginula *A. Hay in Gard. Bull. Singapore,* 50(2): 258 (1998)—Cultivation.
ridleyi *A. Hay in Gard. Bull. Singapore,* 50(2): 263 (1998)—Borneo (Sarawak).
scalprum *A. Hay in Gard. Bull. Singapore,* 51(1): 30 (1999)—Philippines.

ALOCASIA:-
suhirmaniana *Yuzammi & A. Hay in Telopea,* 7(4): 303 (1998)—Sulawesi.
venusta *A. Hay in Gard. Bull. Singapore,* 50(2): 281 (1998)—Borneo (Sarawak).
wongii *A. Hay in Gard. Bull. Singapore,* 50(2): 270 (1998)—Borneo (Sabah).

ALOE (Aloac.).
sect. Aristatae *(A. Berger) H.F. Glen & D.S. Hardy in Fl. S. Afr.,* 5(1:1): 31 (2000): A. ser. Aristatae.
sect. Asperifoliae *(A. Berger) H.F. Glen & D.S. Hardy in Fl. S. Afr.,* 5(1:1): 76 (2000): A. ser. Asperifoliae.
sect. Chabaudia *H.F. Glen & D.S. Hardy in Fl. S. Afr.,* 5(1:1): 88 (2000).
sect. Haemanthifoliae *(A. Berger) H.F. Glen & D.S. Hardy in Fl. S. Afr.,* 5(1:1): 23 (2000): A. ser. Haemanthifoliae.
sect. Latebracteatae *(A. Berger) H.F. Glen & D.S. Hardy Fl. S. Afr.,* 5(1:1): 85 (2000): A. ser. Latebracteatae.
sect. Lomatophyllum *G.D. Rowley in Excelsa,* 17: 59 (1996).
sect. Longistylae *(A. Berger) H.F. Glen & D.S. Hardy in Fl. S. Afr.,* 5(1:1): 25 (2000): A. ser. Longistylae.
sect. Macrifoliae *(Haw.) H.F. Glen & D.S. Hardy in Fl. S. Afr.,* 5(1:1): 92 (2000): A. ser. Macrifoliae.
sect. Ortholophae *(Christian) H.F. Glen & D.S. Hardy in Fl. S. Afr.,* 5(1:1): 121 (2000): A. subsect. Ortholophae.
sect. Pachythamnos *H.F. Glen & D.S. Hardy in Fl. S. Afr.,* 5(1:1): 126 (2000).
sect. Principales *(A. Berger) H.F. Glen & D.S. Hardy in Fl. S. Afr.,* 5(1:1): 114 (2000): A. ser. Principales.
sect. Superpositae *(Pole-Evans) H.F. Glen & D.S. Hardy in Fl. S. Afr.,* 5(1:1): 73 (2000): A. ser. Superpositae.
× abhaica *J.J. Lavranos & I.S. Collenette in Cact. Succ. J. (U.S.A.),* 72(2): 87 (2000), without latin descr.—Saudi Arabia. (A. edentata × pseudorubroviolacea).
ahmarensis *P. Favell, M.B. Miller & A.N. Al-Gifri in Cact. Succ. J. (U.S.A.),* 71(5): 257 (1999)—South Yemen.
aldabrensis *(Marais) L.E. Newton & G.D. Rowley in Excelsa,* 17: 59 (1996): Lomatophyllum aldabrense.
anivoranoensis *(Rauh & Hebding) L.E. Newton & G.D. Rowley in Bradleya,* 16: 114 (1998): Lomatophyllum anivoranoense.
ankaranensis *W. Rauh & R.D. Mangelsdorff in Kakt. And. Sukk.,* 51(10): 274, 273 (2000)—Madagascar.
ankoberensis *M.G. Gilbert & Sebsebe Demissew in Kew Bull.,* 52(1): 146 (1997)—Ethiopia.
antsingyensis *(Leandri) L.E. Newton & G.D. Rowley in Excelsa,* 17: 59 (1996): Lomatophyllum antsingyense.
armatissima *J.J. Lavranos & I.S. Collenette in Cact. Succ. J. (U.S.A.),* 72(1): 22 (2000)—Saudi Arabia.
ballii Reynolds
var. makurupiniensis *A.F.N. Ellert in Cact. Succ. J. (U.S.A.),* 70(3): 130 (1998)—Zimbabwe/ Mozambique border.
belavenokensis *(Rauh & Gerold) L.E. Newton & G.D. Rowley in Excelsa,* 17: 59 (1996): Lomatophyllum belavenokense.
berevoana *J.J. Lavranos in Kakt. And. Sukk.,* 49(7): 161 (1998)—Madagascar.
bernadettae *J.-B. Castillon in Adansonia,* 22(1): 136 (2000)—Madagascar.
bertemariae *Sebsebe Demissew & M. Dioli in Kew Bull.,* 55(3): 679 (2000)—Ethiopia.
bosseri *J.-B. Castillon in Adansonia,* 22(1): 135 (2000)—Madagascar.
brunneodentata *J.J. Lavranos & I.S. Collenette in Cact. Succ. J. (U.S.A.),* 72(2): 86 (2000)—Saudi Arabia.
capmanambatoensis *W. Rauh & R. Gerold in Kakt. And. Sukk.,* 51(11): 294, 293 (2000)—Madagascar.
carnea *S. Carter in Kew Bull.,* 51(4): 784 (1996)—Zimbabwe, Mozambique.
cephalophora *J.J. Lavranos & I.S. Collenette in Cact. Succ. J. (U.S.A.),* 72(1): 20 (2000)—Saudi Arabia.

ALOE:-
citrea *(Guillaumin) L.E. Newton & G.D. Rowley in Excelsa,* 17: 61 (1996): Lomatophyllum citreum.
collina *S. Carter in Kew Bull.,* 51(4): 781 (1996)—Zimbabwe.
compressa H. Perrier
var. paucituberculata *J.J. Lavranos in Kakt. And. Sukk.,* 49(7): 158 (1998)—Madagascar.
cyrtophylla *J.J. Lavranos in Kakt. And. Sukk.,* 49(7): 160 (1998)—Madagascar.
dichotoma Masson
var. ramosissima *(Pillans) H.F. Glen & D.S. Hardy in Fl. S. Afr.,* 5(1:1): 142 (2000): A. ramosissima.
edentata *J.J. Lavranos & I.S. Collenette in Cact. Succ. J. (U.S.A.),* 72(2): 86 (2000)—Saudi Arabia.
eumassawana *S. Carter, M.G. Gilbert & Sebsebe Demissew in Kew Bull.,* 51(4): 776 (1996)—Ethiopia.
fimbrialis *S. Carter in Kew Bull.,* 51(4): 779 (1996)—Zambia.
fleuretteana *W. Rauh & R. Gerold in Kakt. And. Sukk.,* 51(5): 122 (2000)—Madagascar.
friisii *Sebsebe Demissew & M.G. Gilbert in Kew Bull.,* 55(3): 683 (2000)—Ethiopia.
gilbertii Sebsebe & Brandham
subsp. megalacanthoides *M.G. Gilbert & Sebsebe Demissew in Kew Bull.,* 52(1): 151 (1997)—Ethiopia.
heybensis *J.J. Lavranos in Cact. Succ. J. (U.S.A.),* 71(3): 159 (1999)—Somalia.
hijazensis *J.J. Lavranos & I.S. Collenette in Cact. Succ. J. (U.S.A.),* 72(1): 23 (2000)—Saudi Arabia.
inyangensis Christian
var. kimberleyana *S. Carter in Kew Bull.,* 51(4): 777 (1996)—Zimbabwe, Mozambique.
kefaensis *M.G. Gilbert & Sebsebe Demissew in Kew Bull.,* 52(1): 140 (1997)—Ethiopia.
lindenii *J.J. Lavranos in Cact. Succ. J. (U.S.A.),* 69(3): 149 (1997)—Somalia.
lucile-allorgeae *W. Rauh in Bradleya,* 16: 97 (1998)—Madagascar.
megalacantha Baker
subsp. alticola *M.G. Gilbert & Sebsebe Demissew in Kew Bull.,* 52(1): 150 (1997)—Ethiopia.
megalocarpa *J.J. Lavranos in Kakt. And. Sukk.,* 49(7): 162 (1998)—Madagascar.
menyharthii Baker
subsp. ensifolia *S. Carter in Kew Bull.,* 51(4): 783 (1996)—Mozambique.
microstigma Salm-Dyck
subsp. framesii *(L. Bolus) H.F. Glen & D.S. Hardy in Fl. S. Afr.,* 5(1:1): 107 (2000): A. framesii.
namorokaensis *(Rauh) L.E. Newton & G.D. Rowley in Bradleya,* 16: 114 (1998): Lomatophyllum namorokaense.
occidentalis *(H. Perrier) L.E. Newton & G.D. Rowley in Excelsa,* 17: 61 (1996): Lomatophyllum occidentale.
orientalis *(H. Perrier) L.E. Newton & G.D. Rowley in Excelsa,* 17: 61 (1996): Lomatophyllum orientale.
parvicapsula *J.J. Lavranos & I.S. Collenette in Cact. Succ. J. (U.S.A.),* 72(2): 84 (2000)—Saudi Arabia.
parvicoma *J.J. Lavranos & I.S. Collenette in Cact. Succ. J. (U.S.A.),* 72(1): 21 (2000)—Saudi Arabia.
pembana *L.E. Newton in Cact. Succ. J. (U.S.A.),* 70(1): 27 (1998)—Tanzania.
porphyrostachys *J.J. Lavranos & I.S. Collenette in Cact. Succ. J. (U.S.A.),* 72(1): 18 (2000)—Saudi Arabia.
propagulifera *(Rauh & Razaf.) L.E. Newton & G.D. Rowley in Bradleya,* 16: 114 (1998): Lomatophyllum propaguliferum.
prostrata *(H. Perrier) L.E. Newton & G.D. Rowley in Excelsa,* 17: 61 (1996): Lomatophyllum prostratum.
subsp. pallida *W. Rauh & R.D. Mangelsdorff in Kakt. And. Sukk.,* 51(6): 159 (2000)—Madagascar.
pseudorubroviolacea *J.J. Lavranos & I.S. Collenette in Cact. Succ. J. (U.S.A.),* 72(1): 17 (2000)—Saudi Arabia.

ALOE:-
pulcherrima *M.G. Gilbert & Sebsebe Demissew in Kew Bull.,* 52(1): 147 (1997)—Ethiopia.
× qaharensis *J.J. Lavranos & I.S. Collenette in Cact. Succ. J. (U.S.A.),* 72(2): 87 (2000), without latin descr.—Saudi Arabia. (A. fleurentiniorum × woodii).
rosea *(H. Perrier) L.E. Newton & G.D. Rowley in Excelsa,* 17: 61 (1996): Lomatophyllum roseum.
ruffingiana *W. Rauh & H. Petignat in Kakt. And. Sukk.,* 50(11): 271 (1999)—Madagascar.
schilliana *L.E. Newton & G.D. Rowley in Excelsa,* 17: 61 (1996), nom. nov.: Lomatophyllum viviparum *H. Perrier.*
shadensis *J.J. Lavranos & I.S. Collenette in Cact. Succ. J. (U.S.A.),* 72(2): 82 (2000)—Saudi Arabia.
socialis *(H. Perrier) L.E. Newton & G.D. Rowley in Excelsa,* 17: 61 (1996): Lomatophyllum sociale.
steffaniana *W. Rauh in Kakt. And. Sukk.,* 51(3): 73, 72 (2000)—Madagascar.
tewoldei *M.G. Gilbert & Sebsebe Demissew in Kew Bull.,* 52(1): 143 (1997)—Ethiopia.
tormentorii *(Marais) L.E. Newton & G.D. Rowley in Excelsa,* 17: 61 (1996): Lomatophyllum tormentorii.
trichosantha A. Berger
subsp. longiflora *M.G. Gilbert & Sebsebe Demissew in Kew Bull.,* 52(1): 142 (1997)—Ethiopia, Somalia.
vulcanica *J.J. Lavranos & I.S. Collenette in Cact. Succ. J. (U.S.A.),* 72(2): 88 (2000), nom. provis., without latin descr. or type—Saudi Arabia.
woodii *J.J. Lavranos & I.S. Collenette in Cact. Succ. J. (U.S.A.),* 72(2): 82 (2000)—Saudi Arabia.
zombitsiensis *W. Rauh & M. Teissier in Kakt. And. Sukk.,* 51(8): 203, 201 (2000)—Madagascar.

ALOITIS (Gentianac.).
azurea *(Bunge) S. Omer in Fl. Pakistan,* 197: 109 (1995): Gentiana azurea.
smithii *S. Omer in Fl. Pakistan,* 197: 121 (1995)—Pakistan.

ALONSOA (Scrophulariac.).
honoraria *J. Grau in Sendtnera,* 6: 110 (1999)—Chile North (Coquimbo).

ALOPECURUS (Gramin.).
alpinus *D. Villars, Hist. Pl. Dauphiné,* 1: 306 (1786)—France.
borealis Trin.
subsp. glaucus *(Less.) M. Doğan in Turkish J. Bot.,* 23(4): 259 (1999): A. glaucus.
gerardii Vill.
subsp. cassius *(Boiss.) M. Doğan in Turkish J. Bot.,* 23(4): 257 (1999): A. cassius.
myosuroides Huds.
subsp. tonsus *Blanche ex Boiss.) M. Doğan in Turkish J. Bot.,* 23(4): 250 (1999): A. agrestis var. tonsus.
× winklerianus *Asch. & Graebn., Syn. Mitteleur. Fl.,* 2(1): 139 (1899)—Poland. (A. fulvus × pratensis).

ALOYSIA *Palau, Parte Práct. Bot.,* 1: 767 (1784). *VERBENACEAE*†
citriodora *Palau, Parte Práct. Bot.,* 1: 768 (1784)—Southern South America.†

ALPHONSEA (Annonac.).
havilandii *P.J.A. Kessler in Bot. Jahrb.,* 118(1): 91 (1996)—Borneo (Sarawak).
malayana *P.J.A. Kessler in Bot. Jahrb.,* 118(1): 97 (1996)—Malaya (Peninsular Malaysia).
orthopetala *H. Okada in Acta Phytotax. Geobot.,* 47(1): 1 (1996)—New Guinea (Papua New Guinea).
siamensis *P.J.A. Kessler in Bot. Jahrb.,* 118(1): 100 (1996)—Thailand.
tonquinensis Aug. DC.
var. coriacea *P.J.A. Kessler in Bot. Jahrb.,* 118(1): 101 (1996)—Vietnam, Laos.

ALPINIA (Zingiberac.).
breviligulata *(Gagnep.) Gagnep. in Lecomte, Fl. Indo-Chine,* 6: 88 (1908): A. calcarata var. breviligulata.

ALPINIA:-
calcicola *Q.B. Nguyen & M.F. Newman in Gard. Bull. Singapore*, 52(1): 211 (2000)—Vietnam.
celebica *H.N. Ridley in J. Straits Branch Roy. Asiat. Soc.*, 34: 98 (1900)—Sulawesi.
conghuaensis *J.P. Liao & T.L. Wu in Acta Phytotax. Sin.*, 38(6): 563 (2000)—China Southeast (Guangdong).
henryi K. Schum.
var. densihispida *H. Dong & G.J. Xu in Acta Phytotax. Sin.*, 35(6): 568 (1997)—Hainan.
jianganfeng *T.L. Wu in Novon*, 7(4): 441 (1997 publ. 1998)—China Southeast (Guangdong, Guangxi, Hong Kong, Hunan, Jiangxi).
malaccensis (Burm.) Roscoe
var. nobilis *(Ridl.) I.M. Turner in Novon*, 6(2): 223 (1996): A. nobilis.
menghaiensis *S.Q. Tong & Y.M. Xia in Bull. Bot. Res. North-East. Forest. Univ.*, 18(2): 142 (1998)—China South-central (Yunnan).
pectinata *H.N. Ridley in J. Straits Branch Roy. Asiat. Soc.*, 34: 97 (1900)—Sulawesi.
stachyoides Hance
var. yangchunensis *Z.L. Zhao & L.S. Xu in Acta Phytotax. Sin.*, 36(6): 551 (1998)—China Southeast (Guangdong).
vitellina (Lindl.) Ridley
var. cannaefolia *(Ridl.) I.M. Turner in Novon*, 6(2): 223 (1996): A. cannaefolia.

ALSINE (Caryophyllac.).
bocconi *Scheele in Flora*, 26: 431 (1843)—Locality not stated.
postii *(Holmboe) J. Holub in Preslia*, 70(2): 100 (1998): Stellaria media subsp. postii.

ALSTONIA (Apocynac.).
annamensis *(Monach.) K. Sidiyasa in Blumea*, Suppl. 11: 106 (1998): A. angustifolia var. annamensis.
beatricis *K. Sidiyasa in Blumea*, 41(1): 29 (1996)—New Guinea (Irian Jaya).
breviloba *K. Sidiyasa in Blumea*, Suppl. 11: 118 (1998)—New Guinea (Papua New Guinea).
penangiana *K. Sidiyasa in Blumea*, Suppl. 11: 162 (1998)—Malaya (Peninsular Malaysia).
rubiginosa *K. Sidiyasa in Blumea*, Suppl. 11: 171 (1998)—New Guinea (Papua New Guinea).

ALSTROEMERIA (Alstroemeriac.).
sect. Andesine *P. Ravenna in Onira, Bot. Leafl.*, 4(10): 45 (2000).
ser. Magnificae *P. Ravenna in Onira, Bot. Leafl.*, 4(10): 45 (2000).
ser. Mutabiles *P. Ravenna in Onira, Bot. Leafl.*, 5(8): 41 (2000).
achirae *M. Muñoz Schick & Brinck in Gayana, Bot.*, 57(1): 56 (2000)—Chile Central (Maule).
altoparadisea *P. Ravenna in Onira, Bot. Leafl.*, 5(5): 17 (2000)—Brazil West-Central (Goiás).
anajeana *P. Ravenna in Onira, Bot. Leafl.*, 4(10): 37 (2000)—Brazil Northeast (Bahia).
annapolina *P. Ravenna in Onira, Bot. Leafl.*, 5(5): 18 (2000)—Brazil West-Central (Goiás).
aquidauanica *P. Ravenna in Onira, Bot. Leafl.*, 5(8): 35 (2000)—Brazil West-Central (Mato Grosso).
arnicana *P. Ravenna in Onira, Bot. Leafl.*, 5(8): 36 (2000)—Brazil West-Central (Goiás).
atrorubra *P. Ravenna in Onira, Bot. Leafl.*, 4(10): 38 (2000)—Brazil West-central (Mato Grosso).
aulica *P. Ravenna in Onira, Bot. Leafl.*, 4(10): 41 (2000)—Chile North (Valparaíso).
bahiensis *P. Ravenna in Onira, Bot. Leafl.*, 4(10): 37 (2000)—Brazil Northeast (Bahia).
caiaponica *P. Ravenna in Onira, Bot. Leafl.*, 5(8): 36 (2000)—Brazil West-Central (Goiás).
caudiculata *P. Ravenna in Onira, Bot. Leafl.*, 4(10): 36 (2000)—Brazil Southeast (Espírito Santo).
cuiabana *P. Ravenna in Onira, Bot. Leafl.*, 5(5): 18 (2000)—Brazil West-Central (Mato Grosso).
cultrifolia *P. Ravenna in Onira, Bot. Leafl.*, 5(8): 37 (2000)—Brazil West-Central (Brazilia Distrito Federal).
curralensis *P. Ravenna in Onira, Bot. Leafl.*, 4(10): 35 (2000)—Brazil Southeast (Minas Gerais).
decora *P. Ravenna in Onira, Bot. Leafl.*, 4(10): 42 (2000)—Chile North (Valparaíso)

ALSTROEMERIA:-
discolor *P. Ravenna in Onira, Bot. Leafl.*, 4(10): 44 (2000)—Chile North (Valparaíso).
douradensis *P. Ravenna in Onira, Bot. Leafl.*, 5(8): 37 (2000)—Brazil West-Central (Goiás).
espigonensis *P. Ravenna in Onira, Bot. Leafl.*, 5(5): 19 (2000)—Brazil Northeast (Bahia).
firmulifolia *P. Ravenna in Onira, Bot. Leafl.*, 5(5): 20 (2000)—Brazil Southeast (Minas Gerais).
fuscovinosa *P. Ravenna Onira, Bot. Leafl.*, 4(10): 34 (2000)—Brazil Southeast (Minas Gerais).
glaucandra *P. Ravenna in Onira, Bot. Leafl.*, 4(10): 41 (2000)—Chile North (Valparaíso).
hygrophila *A.W. Meerow, A.F.C. Tombolato & F.K. Meyer in Brittonia*, 51(4): 440 (1999)—Brazil West-central (Brasília Distrito Federal).
ibitipocae *P. Ravenna in Onira, Bot. Leafl.*, 4(10): 35 (2000)—Brazil Southeast (Minas Gerais).
igarapavica *P. Ravenna in Onira, Bot. Leafl.*, 4(10): 36 (2000)—Brazil Southeast (São Paulo).
itabiritensis *P. Ravenna in Onira, Bot. Leafl.*, 5(5): 20 (2000)—Brazil Southeast (Minas Gerais).
itatiaica *P. Ravenna in Onira, Bot. Leafl.*, 5(8): 38 (2000)—Brazil Southeast (Rio de Janeiro).
jequitiana *P. Ravenna in Onira, Bot. Leafl.*, 5(5): 21 (2000)—Brazil Southeast (Minas Gerais).
lactilutea *P. Ravenna & Brinck in Onira, Bot. Leafl.*, 5(5): 23 (2000)—Chile North (Tarapacá).
ligtu L.
f. carnea *P. Ravenna in Onira, Bot. Leafl.*, 5(8): 42 (2000)—Chile Central (Bíobío).
f. haemantha *(Ruiz & Pav.) P. Ravenna in Onira, Bot. Leafl.*, 5(8): 42 (2000): Alstroemeria haemantha.
f. nivea *P. Ravenna in Onira, Bot. Leafl.*, 5(8): 42 (2000)—Chile Central (Bíobío).
litterata *P. Ravenna in Onira, Bot. Leafl.*, 5(8): 39 (2000)—Brazil West-Central (Mato Grosso).
lutea *M. Muñoz Schick in Gayana, Bot.*, 57(1): 55 (2000)—Chile North (Tarapacá).
nervosa *P. Ravenna in Onira, Bot. Leafl.*, 5(8): 39 (2000), without type—Brazil Southeast (Minas Gerais).
subsp. multiradiata *P. Ravenna in Onira, Bot. Leafl.*, 5(8): 40 (2000), species name not validly publ.—Brazil Southeast (Minas Gerais).
orchidioides *A.W. Meerow, A.F.C. Tombolato & F.K. Meyer in Brittonia*, 51(4): 442 (1999)—Brazil West-central (Goiás).
pubiflora *P. Ravenna in Onira, Bot. Leafl.*, 5(5): 21 (2000)—Brazil West-Central (Goiás).
pudica *P. Ravenna in Onira, Bot. Leafl.*, 4(10): 43 (2000)—Chile North (Santiago).
punctata *P. Ravenna in Onira, Bot. Leafl.*, 4(10): 33 (2000)—Brazil West-central (Brasília Distrito Federal).
roseoviridis *P. Ravenna in Onira, Bot. Leafl.*, 5(5): 22 (2000)—Brazil West-Central (Mato Grosso do Sul).
talcaensis *P. Ravenna in Onira, Bot. Leafl.*, 4(10): 40 (2000)—Chile North (Maule).
umbrosa *P. Ravenna in Onira, Bot. Leafl.*, 4(10): 39 (2000)—Brazil West-central (Goiás).
virginalis *P. Ravenna & Brinck in Onira, Bot. Leafl.*, 5(8): 40 (2000)—Chile Central (Coquimbo).
viridiflora *P. Ravenna in Onira, Bot. Leafl.*, 4(10): 34 (2000), homonym; *see Warm. in Vidensk. Meddel. Dansk Naturhist. Foren. Kjøbenhavn*, 1872: 118 (1872)—Brazil Southeast (Minas Gerais).
werdermannii Ehr. Bayer
var. flavicans *M. Muñoz Schick in Gayana, Bot.*, 57(1): 57 (2000)—Chile North (Atacama).
xavantinensis *P. Ravenna in Onira, Bot. Leafl.*, 4(10): 39 (2000)—Brazil West-central (Mato Grosso).

ALTERNANTHERA (Amaranthac.).
collina *T.M. Pedersen in Adansonia*, 19(2): 218 (1997)—Argentina Northwest (Santiago del Estero).
hirtula (Mart.) R.E. Fr.
var. hirsuta *T.M. Pedersen in Bonplandia*, 10(1–4): 84 (2000)—Paraguay.
var. nitens *T.M. Pedersen in Bonplandia*, 10(1–4): 84 (2000)—Paraguay.
inaccessa *T.M. Pedersen in Adansonia*, 19(2): 219 (1997)—Paraguay.

ALTERNANTHERA:-
kurtzii Pedersen
subsp. sclerosperma *T.M. Pedersen in Bonplandia*, 10(1–4): 86 (2000)—Paraguay.
malmeana R.E. Fr.
var. straminea *(Chodat) T.M. Pedersen in Adansonia*, 19(2): 220 (1997): Telanthera rosea e straminea.
paronichyoides A. St.-Hil.
f. aberrans *(Pedersen) T.M. Pedersen in Zuloaga & Morrone (eds.), Cat. Pl. Vasc. Rep. Arg. 2 (Monogr. Syst. Bot.* Missouri Bot. Gard. 74): 1245 (1999): Alternanthera paronichyoides f. aberrans.
pennelliana *Mears ex T.M. Pedersen in Adansonia*, 19(2): 221 (1997), nom. nov.: Telanthera geniculata *S. Moore.*
philoxeroides (Mart.) Griseb.
f. acutifolia *(Moq.) T.M. Pedersen in Adansonia*, 19(2): 221 (1997): Telanthera philoxeroides β acutifolia.
subf. philoxerina *(Suess.) T.M. Pedersen in Adansonia*, 19(2): 222 (1997): A. philoxerina.
piptantha *T.M. Pedersen in Bonplandia*, 10(1–4): 88 (2000)—Paraguay.
praelonga A.St.-Hil.
var. australis *T.M. Pedersen in Adansonia*, 19(2): 222 (1997)—Brazil South (Rio Grande do Sul, Santa Catarina).
ramosissima (Mart.) Chodat
var. missionum *T.M. Pedersen in Adansonia*, 19(2): 223 (1997)—Argentina Northeast (Misiones).
var. reptans *T.M. Pedersen in Adansonia*, 19(2): 224 (1997)—Brazil South (Santa Catarina).
serpens *T.M. Pedersen in Adansonia*, 19(2): 224 (1997)—Paraguay.

ALTHAEA (Malvac.).
sect. Hirsutae *Iljin ex L.G. Olyanitskaya & N.N. Tzvelev in Fl. Vostochnoĭ Evropy̆*, 9: 250 (1996).

ALUTA *B.L. Rye & M.E. Trudgen in Nuytsia*, 13(2): 347 (2000). *MYRTACEAE*
appressa *(C.R.P. Andrews) B.L. Rye & M.E. Trudgen in Nuytsia*, 13(2): 351 (2000): Thryptomene appressa.
aspera *(E. Pritz.) B.L. Rye & M.E. Trudgen in Nuytsia*, 13(2): 353 (2000): Thryptomene aspera.
subsp. hesperia *B.L. Rye & M.E. Trudgen in Nuytsia*, 13(2): 355 (2000)—Western Australia.
subsp. localis *B.L. Rye & M.E. Trudgen in Nuytsia*, 13(2): 357 (2000)—Western Australia.
maisonneuvei *(F. Muell.) B.L. Rye & M.E. Trudgen in Nuytsia*, 13(2): 358 (2000): Thryptomene maisonneuvei.
subsp. auriculata *(F. Muell.) B.L. Rye & M.E. Trudgen in Nuytsia*, 13(2): 360 (2000): Thryptomene auriculata.
quadrata *B.L. Rye & M.E. Trudgen in Nuytsia*, 13(2): 361 (2000)—Western Australia.
teres *B.L. Rye & M.E. Trudgen in Nuytsia*, 13(2): 364 (2000)—Western Australia.

ALYSICARPUS (Legumin.).
longifolius (Rottler ex Spreng.) Wight & Arn.
var. major *D.S. Pokle in Reinwardtia*, 11(4): 292 (1999)—India (Maharashtra).
var. pygmaeus *D.S. Pokle in Reinwardtia*, 11(4): 290 (1999)—India (Maharashtra).
mahbubnagarensis *S.R. Raghava Rao, T. Rajagopal & S.T. Ramachandra Chary in Asian J. Pl. Sci.*, 1: 97 (1989)—India (Andhra Pradesh).
naikianus *D.S. Pokle in Reinwardtia*, 11(4): 285 (1999)—India (Maharashtra).
saplianus *D.S. Pokle in Reinwardtia*, 11(4): 288 (1999)—India (Maharashtra).

ALYSISTYLES *N.E. Br. ex R.A. Dyer, Gen. S. Afr. Fl. Pl.*, 1: 275 (1975), orth. variant of 'Alistilus'. *LEGUMINOSAE*

ALYSSOIDES (Crucif.).
bulgarica (Sagorski) Assenov
var. macedonica *K. Micevski, Fl. Republ. Macedonia,* 1(3): 772 (1995 publ. 1996)—Yugoslavia.

ALYSSUM (Crucif.).
campestre (L.) L.
subsp. strigosum *(Banks & Sol.) J. Jalas in Ann. Bot. Fenn.,* 33(4): 283 (1996): A. strigosum.
debarense *K. Micevski in Prilozi Oddel. Biol. Med. Nauki Makedonska Akad. Nauk. Umet.,* 15(1–2): 50 (1994), as 'debarensis'—Yugoslavia.
gevgelicense *K. Micevski in Prilozi Oddel. Biol. Med. Nauki Makedonska Akad. Nauk. Umet.,* 15(1–2): 55 (1994), as 'gevgelicensis'—Yugoslavia.
f. multiradiatum *K. Micevski in Prilozi Oddel. Biol. Med. Nauki Makedonska Akad. Nauk. Umet.,* 15(1–2): 56 (1994), as 'multiradiata'—Yugoslavia.
kavadarcense *K. Micevski in Prilozi Oddel. Biol. Med. Nauki Makedonska Akad. Nauk. Umet.,* 15(1–2): 52 (1994), as 'kavadarcensis'—Yugoslavia.
obovatum (C.A. Mey.) Turcz.
subsp. baicalicum *(Nyár.) G.A. Peschkova, Stepnaya Fl. Baĭkal'skoĭ Sibiri:* 61 (1972): A. baicalicum.
subsp. orbiculare *(Regel) G.A. Peschkova, Stepnaya Fl. Baĭkal'skoĭ Sibiri:* 61 (1972): A. alpestre var. orbiculare.
pulvinare Velen.
var. mariovoense *K. Micevski & V. Matevski in K. Micevski, Fl. Republ. Macedonia,* 1(3): 772 (1995 publ. 1996), as 'mariovoensis'—Yugoslavia.
serpentinum *K. Micevski in Prilozi Oddel. Biol. Med. Nauki Makedonska Akad. Nauk. Umet.,* 15(1–2): 46 (1994)—Yugoslavia.
skopjense *K. Micevski in Prilozi Oddel. Biol. Med. Nauki Makedonska Akad. Nauk. Umet.,* 15(1–2): 43 (1994), as 'skopjensis'—Yugoslavia.
trichostachyum Rupr.
var. stenophyllum *(Halácsy) K. Micevski, Fl. Republ. Macedonia,* 1(3): 666 (1995 publ. 1996): A. trichostachyum f. stenophyllum.

ALYXIA (Apocynac.).
celebica *D.J. Middleton in Blumea,* 45(1): 38 (2000)—Sulawesi.
fosbergii *J. Florence in Novon,* 7(1): 27 (1997)—Pitcairn Is.
globosa *D.J. Middleton in Blumea,* 45(1): 51 (2000)—Sulawesi.
graciliflora *D.J. Middleton in Blumea,* 45(1): 53 (2000)—New Guinea (Papua New Guinea).
kwalotabaa *D.J. Middleton in Blumea,* 45(1): 61 (2000)—Solomon Is (North Solomons, South Solomons), Bismarck Archipelago.
lackii *D.J. Middleton in Blumea,* 45(1): 64 (2000)—Sulawesi.
longiloba *D.J. Middleton in Blumea,* 45(1): 67 (2000)—New Guinea (Papua New Guinea).
manusiana *D.J. Middleton in Blumea,* 45(1): 71 (2000)—Bismarck Archipelago.
minutiflora *D.J. Middleton in Blumea,* 45(1): 76 (2000)—Sulawesi.
muguma *D.J. Middleton in Blumea,* 45(1): 78 (2000)—Philippines.
papuana *D.J. Middleton in Blumea,* 45(1): 87 (2000)—New Guinea (Papua New Guinea).
uniflora *D.J. Middleton in Blumea,* 45(1): 130 (2000)—Sulawesi.
vera *D.J. Middleton in Blumea,* 45(1): 132 (2000)—Moluccas.

AMARANTHUS (Amaranthac.).
sect. Acanthochiton *(Torr.) S.L. Mosyakin & K.R. Robertson in Ann. Bot. Fenn.,* 33(4): 277 (1996): gen. Acanthochiton.
sect. Acnida *(L.) S.L. Mosyakin & K.R. Robertson in Ann. Bot. Fenn.,* 33(4): 277 (1996), with incorrect basionym ref.: gen. Acnida.
nothosect. Dubia *S.L. Mosyakin & K.R. Robertson in Ann. Bot. Fenn.,* 33(4): 278 (1996). (A. sect. Amaranthus × sect. Centrusa).
sect. Goerziella *(Urb.) S.L. Mosyakin & K.R. Robertson in Ann. Bot. Fenn.,* 33(4): 280 (1996): gen. Goerziella.

AMARANTHUS:-
subsect. Hybrida *S.L. Mosyakin & K.R. Robertson in Ann. Bot. Fenn.,* 33(4): 278 (1996).
sect. Pentamorion *(Beck) S.L. Mosyakin & K.R. Robertson in Ann. Bot. Fenn.,* 33(4): 280 (1996): Euxolus sect. Pentamorion.
sect. Saueranthus *S.L. Mosyakin & K.R. Robertson in Ann. Bot. Fenn.,* 33(4): 277 (1996).
crassipes Schltdl.
var. warnockii *(I.M. Johnst.) J. Henrickson in Sida,* 18(3): 788 (1999): A. warnockii.
hybridus L.
var. bouchonii *(Thell.) J. Lambinon in Candollea,* 52(2): 273 (1997): A. bouchonii.
hypochondriacus L.
var. powellii *(S. Watson) T.M. Pedersen in Zuloaga & Morrone (eds.), Cat. Pl. Vasc. Rep. Arg. 2 (Monogr. Syst. Bot.* Missouri Bot. Gard. 74): 1245 (1999): Amaranthus powellii.
ividus L.
f. rubens *(Moq.) S. Hatusima in S. Hatusima & T. Amano, Fl. Ryukyus, S. of Amami Is., ed. 2:* 47 (1994), without basionym ref.: Basionym not stated.
retroflexus L.
subsp. delilei *(Richt. & Loret) N.N. Tzvelev in Novosti Sist. Vÿssh. Rast.,* 32: 183 (2000): A. delilei.
silvestris *Vill., Cat. Jard. Pl. Strasbourg:* 111 (1807)—France.
tamariscinus *T. Nuttall in Trans. Amer. Philos. Soc., n.s.,* 5: 165 (1835)—Arkansas.
tamaulipensis *J. Henrickson in Sida,* 18(3): 800 (1999)—Mexico Northeast (Tamaulipas, San Luis Potosí), Mexico Southeast (Tabasco), Texas.
× texensis *J. Henrickson in Sida,* 18(3): 791 (1999)—Texas, Mexico Northeast (Tamaulipas). (A. crassipes × scleropoides).
× tucsonensis *J. Henrickson in Sida,* 18(3): 804 (1999)—Arizona, Mexico Northwest (Sonora).

AMARYLLIS (Amaryllidac.)
paradisicola *D.A. Snijman in Bothalia,* 28(2): 193 (1998)—Cape Province.
starkiorum *I.S. Nelson & H.P. Traub in Plant Life,* 24: 49 (1968), as 'starkii'—Bolivia.†
aunieri *M. Gandoger, Dec. Pl. Nov.,* 1: 31 (1875)—France.

AMELANCHIER (Rosac.).
fallens *M. Gandoger, Dec. Pl. Nov.,* 1: 31 (1875)—France.
mucronata *M. Gandoger, Dec. Pl. Nov.,* 1: 30 (1875)—France.
ovalis Medik.
subsp. comafredensis *(O. Bolòs & Vigo) A.M. Romo, Fl. Silvestres Baleares:* 130 (1994): A. ovalis var. comafredensis.
parviflora Boiss.
subsp. chelmea *(Halácsy) J. Zieliński in Willdenowia,* 30(2): 239 (2000): Amelanchier cretica var. chelmea.
utahensis Koehne
var. covillei *(Standl.) N.H. Holmgren in A. Cronquist et al., Intermountain Fl., Vasc. Pl. Intermountain West,* U.S.A., 3A: 153 (1997): A. covillei.
vallis-clausae *M. Gandoger, Dec. Pl. Nov.,* 1: 31 (1875)—France.
vapincis *M. Gandoger, Dec. Pl. Nov.,* 1: 31 (1875)—France.

AMENTOTAXUS (Taxac.).
hatuyenensis *N.T. Hiêp in Fl. Cambodge, Laos & Vietnam,* 28: 126 (1996)—Vietnam.

AMEROGLOSSUM *Eb. Fischer, S. Vogel & A.V. Lopes Feddes Repert.,* 110(7–8): 529 (1999). *SCROPHULARIACEAE*
pernambucense *Eb. Fischer, S. Vogel & A.V. Lopes in Feddes Repert.,* 110(7–8): 529 (1999)—Brazil Northeast (Pernambuco).

AMESIELLA (Orchidac.).
minor *K. Senghas in J. Orchideenfreund,* 6(2): 121 (1999)—Philippines.
monticola *J.E. Cootes & D.P. Banks in Orchids Austral.,* 10(5): 26 (1998)—Philippines.

AMISCHOTOLYPE (Commelinac.).
griffithii *(C.B. Clarke) I.M. Turner in Novon,* 6(2): 221 (1996): Forrestia griffithii.
irritans *(Ridl.) I.M. Turner in Novon,* 6(2): 221 (1996): Forrestia irritans.
laxiflora *(Merr.) R.B. Faden in Checklist Flow. Pl. & Gymn. Brunei Darussalam (ed. M.J.E. Coode et al.):* 434 (1996): Forrestia laxiflora.
mollissima (Blume) Hassk.
var. glabrata *(Hassk.) R.S. Rao in Maharashtra Vidnyan Mandir Patrika,* 6(2): 53 (1971): Campelia glabrata.
var. marginata *(Blume) R.S. Rao in Maharashtra Vidnyan Mandir Patrika,* 6(2): 53 (1971): Campelia mollissima.
monosperma *(C.B. Clarke) I.M. Turner in Novon,* 6(2): 221 (1996): Forrestia monosperma.
sphagnorrhiza *E.J. Cowley in Checklist Flow. Pl. & Gymn. Brunei Darussalam (ed. M.J.E. Coode et al.):* 434 (1996)—Borneo (Brunei).
tenuis *(C.B. Clarke) R.S. Rao in Maharashtra Vidnyan Mandir Patrika,* 6(2): 53 (1971): Buforrestia tenuis.

AMITOSTIGMA (Orchidac.).
thailandicum *G. Seidenfaden & Thaithong in G. Seidenfaden, Contrib. Orchid Fl. Thailand XIII:* 8 (1997)—Thailand.

AMOMUM (Zingiberac.).
angustipetalum *S. Sakai & H. Nagamasu in Edinburgh J. Bot.,* 55(1): 49 (1998)—Borneo (Sarawak).
bilabiatum *S. Sakai & H. Nagamasu in Edinburgh J. Bot.,* 55(1): 57 (1998)—Borneo (Sarawak).
borealiborneense *I.M. Turner in Sandakania,* 12: 25 (1998), nom. nov.: A. sylvestre *Ridl.*
botryoideum *J. Cowley in Kew Bull.,* 55(3): 674 (2000)—Borneo (Brunei).
calyptratum *S. Sakai & H. Nagamasu in Edinburgh J. Bot.,* 55(1): 52 (1998)—Borneo (Sarawak).
deorianum *D.P. Dam & N. Dam in Bull. Bot. Surv. India,* 34(1–4): 212 (1992 publ. 1997)—Assam (Meghalaya).
dulongense *S.Q. Tong & H. Li in Fl. Dulongjian Reg.:* 314 (1993), without latin descr. or type—China South-central (Yunnan).
durum *S. Sakai & H. Nagamasu in Edinburgh J. Bot.,* 55(1): 55 (1998)—Borneo (Sarawak).
gagnepainii *T.L. Wu, K. Larsen & N.J. Turland in Novon,* 10(1): 90 (2000), nom. nov.: A. thyrsoideum *Gagnep.*
garoense *Sunil Tripathi & Ved Prakash in Rheedea,* 9(2): 177 (1999), as 'garoensis'—Assam (Meghalaya).
hedyosmum *I.M. Turner in Asian J. Trop. Biol.,* 4(1): 18 (2000), nom. nov.: Amomum trilobum *Ridl.*
jainii *S. Tripathi & V. Prakash in Nordic J. Bot.,* 19(5): 609 (1999)—Assam (Meghalaya).
neoaurantiacum *T.L. Wu, K. Larsen & N.J. Turland in Novon,* 10(1): 90 (2000), nom. nov.: A. aurantiacum *H.T. Tsai & S.W. Zhao.*
petaloideum *(S.Q. Tong) T.L. Wu in Novon,* 7(4): 441 (1997 publ. 1998): Paramomum petaloideum.
roseisquamosum *H. Nagamasu & S. Sakai in Edinburgh J. Bot.,* 53(1): 39 (1996)—Borneo (Sarawak).
somniculosum *S. Sakai & H. Nagamasu in Edinburgh J. Bot.,* 55(1): 53 (1998)—Borneo (Sarawak).
subcapitatum *Y.M. Xia in Acta Phytotax. Sin.,* 35(3): 259 (1997)—China South-central (Yunnan).
terminale *H.N. Ridley in J. Straits Branch Roy. Asiat. Soc.,* 34: 98 (1900)—Bismarck Archipelago.
vermanum *S. Tripathi & V. Prakash in Edinburgh J. Bot.,* 57(2): 257 (2000)—Assam (Meghalaya).

AMORPHOPHALLUS (Arac.).
abyssinicus (A. Rich.) N.E. Br.
subsp. akeassii *S. Ittenbach in Willdenowia,* 27(1–2): 159 (1997)—Ivory Coast, Ghana, Nigeria.
ankarana *W.L.A. Hetterscheid, S. Ittenbach & J. Bogner in Bot. Jahrb.,* 121(1): 2 (1999)—Madagascar.

AMORPHOPHALLUS:-
antsingyensis *J. Bogner, W.L.A. Hetterscheid & S. Ittenbach in Bot. Jahrb.*, 121(1): 8 (1999)—Madagascar.
barthlottii *S. Ittenbach & W. Lobin in Willdenowia*, 27(1–2): 147 (1997)—Ivory Coast, Liberia.
calabaricus N.E. Br.
subsp. mayoi *S. Ittenbach in Willdenowia*, 27(1–2): 159 (1997)—Uganda, Kenya, Zaire.
canaliculatus *S. Ittenbach, W.L.A. Hetterscheid & W. Lobin in Willdenowia*, 27(1–2): 150 (1997)—Gabon.
hetterscheidii *S. Ittenbach & W. Lobin in Willdenowia*, 27(1–2): 152 (1997)—Gabon, Central African Republic, Zaire.
impressus *S. Ittenbach in Willdenowia*, 27(1–2): 154 (1997)—Tanzania, Malawi.
margretiae *S. Ittenbach in Willdenowia*, 27(1–2): 156 (1997), as 'margretae'—Zaire.
richardsiae *S. Ittenbach in Willdenowia*, 27(1–2): 157 (1997)—Zambia, Tanzania.
taurostigma *S. Ittenbach, W.L.A. Hetterscheid & J. Bogner in Bot. Jahrb.*, 121(1): 12 (1999)—Madagascar.
zengianus *C.L. Long & H. Li in Novon*, 10(2): 125 (2000)—China South-central (Yunnan).

AMPELOCALAMUS (Gramin.).
mianningensis *(Q. Li & Xin Jiang) D.Z. Li & C.M.A. Stapleton in Kew Bull.*, 51(4): 811 (1996): Dendrocalamus mianningensis.

AMPELOCISSUS (Vitac.).
butoensis *C.L. Li in Chinese J. Appl. Environ. Biol.*, 2(1): 52 (1996)—China South-central (Sichuan).
hoabinhensis *C.L. Li in Acta Phytotax. Sin.*, 35(6): 554 (1997)—Vietnam, China South-central (Yunnan).
javalensis *(Seem.) W.D. Stevens & A. Pool in Novon*, 9(3): 424 (1999): Vitis javalensis.
xizangensis *C.L. Li in Acta Phytotax. Sin.*, 35(6): 552 (1997)—Xizang.

AMPELOPSIS (Vitac.).
chondisensis *(Vass. & Vassil.) M. Tulyaganova in Opred. Rast. Sred. Azii*, 7: 93 (1983): A. aegirophylla var. chondisensis.
delavayana Planch.
var. glabra *(Diels & Gilg) C.L. Li in Chinese J. Appl. Environ. Biol.*, 2(1): 48 (1996): A. aconitifolia var. glabra.
var. setulosa *(Diels & Gilg) C.L. Li in Chinese J. Appl. Environ. Biol.*, 2(1): 48 (1996): A. aconitifolia var. setulosa.
var. tomentella *(Diels & Gilg) C.L. Li in Chinese J. Appl. Environ. Biol.*, 2(1): 48 (1996): A. aconitifolia var. tomentella.
gongshanensis *C.L. Li in Chinese J. Appl. Environ. Biol.*, 2(1): 48 (1996)—China South-central (Yunnan).
heterophylla Siebold & Zucc.
var. brevipedunculata *(Regel) C.L. Li in Chinese J. Appl. Environ. Biol.*, 2(1): 47 (1996): Cissus humulifolia var. brevipedunculata.
var. kulingensis *(Rehder) C.L. Li in Chinese J. Appl. Environ. Biol.*, 2(1): 48 (1996): A. brevipedunculata var. kulingensis.
hypoglauca *(Hance) C.L. Li in Chinese J. Appl. Environ. Biol.*, 2(1): 48 (1996): Hedera hypoglauca.
megalophylla Diels & Gilg
var. jiangxiensis *(W.T. Wang) C.L. Li in Chinese J. Appl. Environ. Biol.*, 2(1): 48 (1996): A. jiangxiensis.
tomentosa Planch.
var. glabrescens *C.L. Li in Chinese J. Appl. Environ. Biol.*, 2(1): 48 (1996)—China South-central (Yunnan).

AMPHIBOLIA (Aizoac.).
hutchinsonii *(L. Bolus) H.E.K. Hartmann in Bradleya*, 16: 46 (1998): Ruschia hutchinsonii.
obscura *H.E.K. Hartmann in Bradleya*, 16: 47 (1998)—Cape Province, Namibia.
rupis-arcuatae *(Dinter) H.E.K. Hartmann in Bradleya*, 14: 34 (1996): Mesembryanthemum rupis-arcuatae.
saginata *(L. Bolus) H.E.K. Hartmann in Bradleya*, 16: 48 (1998): Ruschia saginata.
succulenta *(L. Bolus) H.E.K. Hartmann in Bradleya*, 16: 48 (1998): Ruschia succulenta.

AMPHIGLOSSA (Compos.).
celans *M. Koekemoer in Bothalia*, 29(1): 68 (1999)—Cape Province.
grisea *M. Koekemoer in Bothalia*, 29(1): 70 (1999)—Cape Province.
rudolphii *M. Koekemoer in Bothalia*, 29(1): 71 (1999)—Cape Province.
susannae *M. Koekemoer in Bothalia*, 29(1): 71 (1999)—Cape Province.
tecta *(Brusse) M. Koekemoer in Bothalia*, 29(1): 72 (1999): Pterothrix tecta.
thuja *(Merxm.) M. Koekemoer in Bothalia*, 29(1): 73 (1999): Pterothrix thuja.

AMPHITHALEA (Legumin.).
bowiei *(Benth.) A.L. Schutte in Taxon*, 47(1): 62 (1998): Coelidium bowiei.
bullata *(Benth.) A.L. Schutte in Taxon*, 47(1): 62 (1998): Coelidium bullatum.
cedarbergensis *(Granby) .L. Schutte in Taxon*, 47(1): 62 (1998): Coelidium cedarbergense.
cymbifolia *(C.A. Sm.) A.L. Schutte in Taxon*, 47(1): 62 (1998): Coelidium cymbifolium.
dahlgrenii *(Granby) A.L. Schutte in Taxon*, 47(1): 62 (1998): Coelidium dahlgrenii.
esterhuyseniae *(Granby) A.L. Schutte in Taxon*, 47(1): 62 (1998): Coelidium esterhuyseniae.
flava *(Granby) A.L. Schutte in Taxon*, 47(1): 62 (1998): Coelidium flavum.
minima *(Granby) A.L. Schutte in Taxon*, 47(1): 63 (1998): Coelidium minimum.
monticola *A.L. Schutte in Taxon*, 47(1): 63 (1998), nom. nov.: Coelidium humile *Schltr.*
muirii *(Granby) A.L. Schutte Taxon*, 47(1): 63 (1998): Coelidium muirii.
muraltioides *(Benth.) A.L. Schutte in Taxon*, 47(1): 63 (1998): Coelidium muraltioides.
obtusiloba *(Granby) A.L. Schutte in Taxon*, 47(1): 63 (1998): Coelidium obtusilobum.
pageae *(L. Bolus) A.L. Schutte in Taxon*, 47(1): 63 (1998): Coelidium pageae.
parvifolia *(Thunb.) A.L. Schutte in Taxon*, 47(1): 63 (1998), homonym; *see E. Phillips, Gen. S. Afr. Fl. Pl., ed. 2*: 402 (1951): Crotalaria parvifolia.
purpurea *(Granby) A.L. Schutte in Taxon*, 47(1): 64 (1998): Coelidium purpureum.
spinosa *(Harv.) A.L. Schutte in Taxon*, 47(1): 64 (1998), homonym; *see Steud., Nomencl. Bot., ed. 2*, 1: 80 (1840): Coelidium spinosum.
vlokii *(A.L. Schutte & B.-E. van Wyk) A.L. Schutte in Taxon*, 47(1): 64 (1998): Coelidium vlokii.

AMYREA (Euphorbiac.).
celastroides *A. Radcliffe-Smith in Kew Bull.*, 53(2): 440 (1998)—Madagascar.
eucleoides *A. Radcliffe-Smith in Kew Bull.*, 53(2): 440 (1998)—Madagascar.
gracillima *A. Radcliffe-Smith in Kew Bull.*, 53(2): 442 (1998)—Madagascar.
grandifolia *A. Radcliffe-Smith in Kew Bull.*, 53(2): 442 (1998)—Madagascar.
lancifolia *A. Radcliffe-Smith in Kew Bull.*, 53(2): 445 (1998)—Madagascar.
maprouneifolia *A. Radcliffe-Smith in Kew Bull.*, 53(2): 447 (1998)—Madagascar.
myrtifolia *A. Radcliffe-Smith in Kew Bull.*, 53(2): 447 (1998)—Madagascar.
remotiflora *A. Radcliffe-Smith in Kew Bull.*, 53(2): 449 (1998)—Madagascar.
stenocarpa *A. Radcliffe-Smith in Kew Bull.*, 53(2): 451 (1998)—Madagascar.

AMYRIS (Rutac.).
oblanceolata *A. Pool in Novon*, 8(1): 61 (1998)—Nicaragua.

ANACAMPSEROS (Portulacac.).
mallei *(G. Will.) G. Williamson in Aloe*, 34(1–2): 45 (1997): Avonia mallei.
pisina *G. Williamson in Aloe*, 35(1): 18 (1998)—Cape Province.
prominens *G. Williamson in Aloe*, 34(1–2): 45 (1997)—Cape Province.
recurvata *S. Schönland in Rec. Albany Mus.*, 1(1): 52 (1903)—Locality not stated.

ANACAMPTIS (Orchidac.).
boryi *(Rchb. f.) R.M. Bateman, A.M. Pridgeon & M.W. Chase in Lindleyana*, 12(3): 120 (1997): Orchis boryi.
champagneuxii *(Barnéoud) R.M. Bateman, A.M. Pridgeon & M.W. Chase in Lindleyana*, 12(3): 120 (1997): Orchis champagneuxii.
collina *(Banks & Sol. ex Russell) R.M. Bateman, A.M. Pridgeon & M.W. Chase in Lindleyana*, 12(3): 120 (1997): Orchis collina.
coriophora *(L.) R.M. Bateman, A.M. Pridgeon & M.W. Chase in Lindleyana*, 12(3): 120 (1997): Orchis coriophora.
subsp. fragrans *(Pollini) R.M. Bateman, A.M. Pridgeon & M.W. Chase in Lindleyana*, 12(3): 120 (1997): Orchis fragrans.
israelitica *(H. Baumann & Dafni) R.M. Bateman, A.M. Pridgeon & M.W. Chase in Lindleyana*, 12(3): 120 (1997): Orchis israelitica.
laxiflora *(Lam.) .M. Bateman, A.M. Pridgeon & M.W. Chase in Lindleyana*, 12(3): 120 (1997): Orchis laxiflora.
longicornu *(Poir.) R.M. Bateman, A.M. Pridgeon & M.W. Chase in Lindleyana*, 12(3): 120 (1997): Orchis longicornu.
morio *(L.) R.M. Bateman, A.M. Pridgeon & M.W. Chase in Lindleyana*, 12(3): 120 (1997): Orchis morio.
palustris *(Jacq.) R.M. Bateman, A.M. Pridgeon & M.W. Chase in Lindleyana*, 12(3): 120 (1997): Orchis palustris.
subsp. elegans *(Heuff.) R.M. Bateman, A.M. Pridgeon & M.W. Chase in Lindleyana*, 12(3): 120 (1997): Orchis elegans.
subsp. robusta *(T. Stephenson) R.M. Bateman, A.M. Pridgeon & M.W. Chase in Lindleyana*, 12(3): 120 (1997): Orchis palustris var. robusta.
papilionacea *(L.) R.M. Bateman, A.M. Pridgeon & M.W. Chase in Lindleyana*, 12(3): 120 (1997): Orchis papilionacea.
pyramidalis (L.) Rich.
var. nivea *P. Delforge in Nat. Belg.*, 81(4): 396 (2000)—Greece.
f. sanguinea *(Druce) P.D. Sell in P.D. Sell & G. Murrell, Fl. Great Britain & Ireland*, 5: 364 (1996): Orchis pyramidalis var. or subvar. sanguinea.
sancta *(L.) R.M. Bateman, A.M. Pridgeon & M.W. Chase in Lindleyana*, 12(3): 120 (1997): Orchis sancta.
syriaca *(Boiss. ex H. Baumann & Künkele) R.M. Bateman, A.M. Pridgeon & M.W. Chase in Lindleyana*, 12(3): 120 (1997): Orchis syriaca.

× **ANACAMPTORCHIS** (Orchidac.).
lesbiensis *B. Biel in J. Eur. Orchideen*, 30(2): 276 (1998)—East Aegean Is. (Anacamptis pyramidalis × Orchis sancta).

ANADENDRUM (Arac.).
ellipticum *D.S. Widyartini & E.A. Widjaja in Reinwardtia*, 11(4): 282 (1999)—Jawa, Malaya (Peninsular Malaysia, Singapore), Borneo (Kalimantan, Sabah, Brunei), Sulawesi.

ANAGALLIS (Primulac.).
deccanensis *J. Graham, Cat. Pl. Bombay:* 166 (1839)—India (Maharashtra).

× **ANAGELIA** *E.L. Smith in J. Bromeliad Soc.*, 33(2): 72 (1983). *BROMELIACEAE* (Ananas × Neoregelia).

× **ANANANANAS** *D.A. Beadle, Prelim. Listing Cultivar Grex Names Bromel.*: 20 (1991). *BROMELIACEAE* (Ananas × Pseudananas).

ANAPHALIOIDES (Compos.).
alpina *(Cockayne) D. Glenny in New Zealand J. Bot.*, 35(4): 467 (1997): Helichrysum alpinum.
bellidioides *(G. Forst.) D. Glenny in New Zealand J. Bot.*, 35(4): 467 (1997): Xerothemum bellidioides.
mariae *(F. Muell.) D. Glenny in New Zealand J. Bot.*, 35(4): 472 (1997): Anaphalis mariae.
papuana *(Lauterb.) D. Glenny in New Zealand J. Bot.*, 35(4): 473 (1997): Anaphalis nubigena var. papuana.

ANAPHALIS (Compos.).
cooperi *A.J.C. Grierson & L.S. Springate in Edinburgh J. Bot.*, 57(3): 397 (2000)—Bhutan.
kokonorica *(Y. Ling) V.I. Grubov in Novosti Sist. Vyssh. Rast.*, 31: 249 (1998): A. bicolor var. kokonorica.

ANATHEROSTIPA *(Hack. ex Kuntze) P. Peñailillo B. in Gayana, Bot.*, 53(2): 278 (1996): Stipa sect. Anatherostipa. (Gramin.).
bomanii *(Hauman) P. Peñailillo B. in Gayana, Bot.*, 53(2): 279 (1996): Stipa bomanii.
hans-meyeri *(Pilg.) P. Peñailillo B. in Gayana, Bot.*, 53(2): 279 (1996): Stipa hans-meyeri.
henrardiana *(Parodi) P. Peñailillo B. in Gayana, Bot.*, 53(2): 279 (1996): Stipa henrardiana.
obtusa *(Nees & Meyen) P. Peñailillo B. in Gayana, Bot.*, 53(2): 279 (1996): Piptatherum obtusum.
rigidiseta *(Pilg.) P. Peñailillo B. in Gayana, Bot.*, 53(2): 279 (1996): Oryzopsis rigidiseta.
saltensis *(Kuntze) P. Peñailillo B. in Gayana, Bot.*, 53(2): 279 (1996): Stipa saltensis.
venusta *(Phil.) P. Peñailillo B. in Gayana, Bot.*, 53(2): 279 (1996): Stipa venusta.

ANAXAGOREA (Annonac.).
inundata *P.E. Berry & R.B. Miller in Syst. Bot.*, 24(4): 506 (1999 publ. 2000)—Venezuela.

ANCHUSA (Boraginac.).
crispa Viv.
subsp. maritima *(Vals.) F. Selvi & M. Bigazzi in Pl. Biosystems*, 132(2): 136 (1998): A. maritima.
formosa *F. Selvi, M. Bigazzi & G. Bacchetta in Pl. Biosystems*, 131(2): 104 (1997)—Sardegna.
samothracica *M. Bigazzi & F. Selvi in Nordic J. Bot.*, 20(2): 141 (2000)—Greece.
sardoa *(Illario) F. Selvi & M. Bigazzi in Pl. Biosystems*, 132(2): 136 (1998): A. crispa var. sardoa.

ANCHUSELLA *M. Bigazzi, E. Nardi & F. Selvi in Pl. Syst. Evol.*, 205(3–4): 253 (1997). *BORAGINACEAE*
cretica *(Mill.) M. Bigazzi, E. Nardi & F. Selvi in Pl. Syst. Evol.*, 205(3–4): 257 (1997): Anchusa cretica.
variegata *(L.) M. Bigazzi, E. Nardi & F. Selvi in Pl. Syst. Evol.*, 205(3–4): 254 (1997): Lycopsis variegata.

ANCISTROCACTUS (Cactac.).
mathssonii *(Berge ex K. Schum.) A.B. Doweld in Sukkulenty (Moscow)*, 1999(1): 29 (1999): Echinocactus mathssonii.
scheeri
subsp. tobuschii *(W.T. Marshall) A.B. Doweld in Sukkulenty (Moscow)*, 1999(1): 29 (1999): Mammillaria tobuschii.
grandiflorus *M. Cheek in Kew Bull.*, 55(4): 878 (2000)—Cameroon.

ANCISTROCLADUS (Ancistrocladac.).
tanzaniensis *M.R. Cheek & C. Frimodt-Møller in Kew Bull.*, 55(1): 207 (2000)—Tanzania.

ANDERBERGIA *B. Nordenstam in Ann. Naturhist. Mus. Wien*, 98B (Suppl.): 407 (1996). *COMPOSITAE*
elsiae *B. Nordenstam in Ann. Naturhist. Mus. Wien*, 98B(Suppl.): 411 (1996)—Cape Province.
epaleata *(Hilliard & B.L. Burtt) B. Nordenstam in Ann. Naturhist. Mus. Wien*, 98B(Suppl.): 409 (1996): Petalacte epaleata.
fallax *B. Nordenstam in Ann. Naturhist. Mus. Wien*, 98B(Suppl.): 416 (1996)—Cape Province.
rooibergensis *B. Nordenstam in Ann. Naturhist. Mus. Wien*, 98B: 410 (1996)—Cape Province.
ustulata *B. Nordenstam in Ann. Naturhist. Mus. Wien*, 98B(Suppl.): 417 (1996)—Cape Province.
vlokii *(Hilliard) B. Nordenstam in Ann. Naturhist. Mus. Wien*, 98B(Suppl.): 413 (1996): Petalacte vlokii.

ANDINIA *(Luer) C.A. Luer, Ic. Pleurothallid. XX (Monogr. Syst. Bot. Missouri Bot. Gard., 79)*: 5 (2000): Salpistele subgen. Andinia. (Orchidac.).

ANDINIA:-
dielsii *(Mansf.) C.A. Luer, Ic. Pleurothallid. XX (Monogr. Syst. Bot. Missouri Bot. Gard., 79)*: 6 (2000): Lepanthes dielsii.
pensilis *(Schltr.) C.A. Luer, Ic. Pleurothallid. XX (Monogr. Syst. Bot. Missouri Bot. Gard., 79)*: 6 (2000): Lepanthes pensilis.
tervequinata *R.T. Pennington, G. Aymard & N. Cuello in Novon*, 7(1): 72 (1997)—Venezuela.

ANDRACHNE (Euphorbiac.).
cerebroides *Petra Hoffmann in Adansonia*, 22(1): 125 (2000)—Madagascar.
gracilipes *Petra Hoffmann in Adansonia*, 22(1): 126 (2000)—Madagascar.
nana *(P.T. Li) R. Govaerts in R. Govaerts, D.G. Frodin & A. Radcliffe-Smith, World Checklist & Bibliogr.* Euphorbiaceae, 1: 167 (2000): Leptopus nanus.
pachyphylla *(X.X. Chen) R. Govaerts in R. Govaerts, D.G. Frodin & A. Radcliffe-Smith, World Checklist & Bibliogr.* Euphorbiaceae, 1: 167 (2000): Leptopus pachyphyllus.
robinsonii *(Airy Shaw) R. Govaerts in R. Govaerts, D.G. Frodin & A. Radcliffe-Smith, World Checklist & Bibliogr.* Euphorbiaceae, 1: 168 (2000): Leptopus robinsonii.
yunnanensis *(P.T. Li) R. Govaerts in R. Govaerts, D.G. Frodin & A. Radcliffe-Smith, World Checklist & Bibliogr.* Euphorbiaceae, 1: 170 (2000): Leptopus yunnanensis.

ANDREADOXA *J.A. Kallunki in Brittonia*, 50(1): 59 (1998). *RUTACEAE*
flava *J.A. Kallunki in Brittonia*, 50(1): 59 (1998)—Brazil Northeast (Bahia).

ANDRIANA *B.-E. van Wyk in Taxon*, 48(4): 739 (1999). *UMBELLIFERAE*
coursii *(Humbert) B.-E. van Wyk in Taxon*, 48(4): 740 (1999): Heteromorpha coursii.
marojejyensis *(Humbert) B.-E. van Wyk in Taxon*, 48(4): 740 (1999): Heteromorpha marojejyensis.
tsaratananensis *(Humbert) B.-E. van Wyk in Taxon*, 48(4): 740 (1999): Heteromorpha tsaratananensis.

ANDROCORYS (Orchidac.).
oxysepalus *K.Y. Lang in Guihaia*, 16(2): 106 (1996)—China South-central (Yunnan).
pugioniformis *(Lindl. ex Hook. f.) K.Y. Lang in Guihaia*, 16(2): 105 (1996): Herminium pugioniforme.

ANDROCYMBIUM (Colchicac.).
ser. Eghimocymbia *U. Müller-Doblies & D. Müller-Doblies in Feddes Repert.*, 109(7–8): 560 (1998).
subser. Gradatocymbia *U. Müller-Doblies & D. Müller-Doblies in Feddes Repert.*, 109(7–8): 554 (1998).
ser. Gradatocymbia *U. Müller-Doblies & D. Müller-Doblies in Feddes Repert.*, 109(7–8): 553 (1998).
subsect. Gradatocymbium *U. Müller-Doblies & D. Müller-Doblies in Feddes Repert.*, 109(7–8): 553 (1998).
subser. Leistnerocymbia *U. Müller-Doblies & D. Müller-Doblies in Feddes Repert.*, 109(7–8): 557 (1998).
ser. Melittocymbia *U. Müller-Doblies & D. Müller-Doblies in Feddes Repert.*, 109(7–8): 567 (1998).
ser. Myiocymbia *U. Müller-Doblies & D. Müller-Doblies in Feddes Repert.*, 109(7–8): 564 (1998).
subser. Pachystyla *U. Müller-Doblies & D. Müller-Doblies in Feddes Repert.*, 109(7–8): 559 (1998).
ser. Schlechterocymbia *U. Müller-Doblies & D. Müller-Doblies in Feddes Repert.*, 109(7–8): 571 (1998).
subser. Swazicymbia *U. Müller-Doblies & D. Müller-Doblies in Feddes Repert.*, 109(7–8): 556 (1998).
ser. Therocymbia *U. Müller-Doblies & D. Müller-Doblies in Feddes Repert.*, 109(7–8): 565 (1998).
ser. Trifoliata *U. Müller-Doblies & D. Müller-Doblies in Feddes Repert.*, 109(7–8): 559 (1998).
albanense Schönland
subsp. clanwilliamense *J. Pedrola Monfort, N. Membrives Fernández & J.M. Montserrat Martí in Fontqueria*, 54(2): 8 (1999)—Cape Province.

ANDROCYMBIUM:-
austrocapense *U. Müller-Doblies & D. Müller-Doblies in Feddes Repert.*, 109(7–8): 568 (1998)—Cape Province.
eghimocymbion *U. Müller-Doblies & D. Müller-Doblies in Feddes Repert.*, 109(7–8): 563 (1998)—Cape Province.
hughocymbion *U. Müller-Doblies & D. Müller-Doblies in Feddes Repert.*, 109(7–8): 560 (1998)—Cape Province.
huntleyi *J. Pedrola Monfort, N. Membrives Fernández, J.M. Montserrat Martí & J.* Caujapé Castells *in Fontqueria*, 53: 1 (1999)—Cape Province.
kunkelianum *U. Müller-Doblies, P. Hirsch, Stearn & D. Müller-Doblies in Feddes Repert.*, 109(7–8): 559 (1998)—Cape Province.
leistneri *U. Müller-Doblies & D. Müller-Doblies in Feddes Repert.*, 109(7–8): 557 (1998)—Cape Province.
melanthioides Willd.
subsp. australe *U. Müller-Doblies & D. Müller-Doblies in Feddes Repert.*, 109(7–8): 570 (1998)—Cape Province.
subsp. transvaalense *U. Müller-Doblies & D. Müller-Doblies in Feddes Repert.*, 109(7–8): 570 (1998)—Transvaal.
orienticapense *U. Müller-Doblies & D. Müller-Doblies in Feddes Repert.*, 109(7–8): 571 (1998)—Cape Province.
swazicum *U. Müller-Doblies & D. Müller-Doblies in Feddes Repert.*, 109(7–8): 556 (1998)—Swaziland.
undulatum *U. Müller-Doblies & D. Müller-Doblies in Feddes Repert.*, 109(7–8): 562 (1998)—Cape Province.
vanjaarsveldii *U. Müller-Doblies, L.V. Hähnlin, U.U. Müler-Doblies & D. Müller*-Doblies *in Feddes Repert.*, 109(7–8): 563 (1998)—Cape Province.
walteri *J. Pedrola Monfort, N. Membrives Fernández & J.M. Montserrat Martí in Fontqueria*, 54(2): 7 (1999)—Cape Province.
worsonense *U. Müller-Doblies & D. Müller-Doblies in Feddes Repert.*, 109(7–8): 559 (1998)—Cape Province.

ANDROGRAPHIS (Acanthac.).
subgen. Indoneesiella *(Sreem.) L.H. Cramer in Kew Bull.*, 51(3): 555 (1996): gen. Indoneesiella.
longipedunculata *(Sreem.) L.H. Cramer in Kew Bull.*, 51(3): 555 (1996): Indoneesiella longipedunculata.
macrobotrys Nees
var. ceylanica *(Nees) L.H. Cramer in M.D. Dassanayake & W.D. Clayton, eds., Revised Handb. Fl. Ceylon*, 12: 100 (1998): A. ceylanica.

× **ANDROLAECHMEA** *C. Chevalier in Bull. Bromeliad Soc.*, 1(3): 24 (1951). *BROMELIACEAE* (Aechmea × Androlepis).

ANDROPOGON (Gramin.).
crucianus *S.A. Renvoize, Gramineas de Bolivia*: 596 (1998)—Bolivia.
gerardii Vitman
nothosubsp. chrysocomus *(Nash) J.K. Wipff in Phytologia*, 80(5): 344 (1996 publ. 1997): A. chrysocomus.
subsp. hallii *(Hack.) J.K. Wipff in Phytologia*, 80(5): 343 (1996 publ. 1997): A. hallii.
glomeratus (Walter) Britton, Sterns & Poggenb.
subsp. reinoldii *(Léon) L. Catasús Guerra in Fontqueria*, 44: 144 (1996): A. reinoldii.
hondurensis *(R.W. Pohl) J.K. Wipff in Phytologia*, 80(5): 346 (1996 publ. 1997): A. gerardii var. hondurensis.
multiflorus *S.A. Renvoize, Gramineas de Bolivia*: 596 (1998)—Bolivia.

ANDROSACE (Primulac.).
centriberica *(Kress) A. Kress, Primulac.-Stud.*, 15: 2 (1999): A. vitaliana var. centriberica.
subsp. assoana *(M. Laínz) A. Kress, Primulac.-Stud.*, 15: 2 (1999): Vitaliana primuliflora subsp. assoana.
subsp. maragatorum *A. Kress, Primulac.-Stud.*, 15: 2 (1999)—Spain.

ANDROSACE:-
congesta *(O. Schwarz) A. Kress, Primulac.-Stud.*, 15: 2 (1999): Vitaliana congesta.
hohxilensis *R.F. Huang & S.K. Wu in S.G. Wu (S.K. Wu) & Z.J. Feng (eds.), Biol. & Human Physiol. Hoh Xil* Region: 69 (1996)—Qinghai.
kuczerovii *M.S. Knjasev in Bot. Zhurn.*, 83(3): 137 (1998)—Russia East.
× schachtii *A. Kress, Primulac.-Stud.*, 14: 1 (1999)—Cultivation. (A. puberula × pyrenaica).
tauscheri *M. Gandoger, Dec. Pl. Nov.*, 2: 32 (1876)—Hungary.
torrepandoi *M. Gandoger, Dec. Pl. Nov.*, 2: 31 (1876)—Spain.
tribracteata *R.F. Huang in S.G. Wu (S.K. Wu) & Z.J. Feng (eds.), Biol. & Human Physiol. Hoh Xil* Region: 71 (1996)—Qinghai.
vitaliana (L.) Lapeyr.
var. assoana *(M. Laínz) A. Kress, Primulac.-Stud.*, 13: 10 (1997): Vitaliana primuliflora subsp. assoana.
subsp. aurelii *M. Luceño in An. Jard Bot. Madrid*, 56(1): 165 (1998), nom. nov.: A. vitaliana var. centriberica *Kress*.
var. centriberica *A. Kress, Primulac.-Stud.*, 13: 10 (1997)—Spain.
subsp. flosjugorum *A. Kress, Primulac.-Stud.*, 13: 9 (1997)—Spain.
var. nevadensis *(Chiarugi) A. Kress, Primulac.-Stud.*, 13: 10 (1997): Vitaliana primuliflora var. nevadensis.
subsp. nevadensis *(Chiarugi) M. Luceño in An. Jard. Bot. Madrid*, 56(1): 165 (1998): Vitaliana primuliflora var. nevadensis.

ANDROSAEMUM (Guttif.).
calycinum *(L.) T.G. Lestiboudois, Botanogr. Belg.*, 2: 346 (1827): Hypericum calycinum.

ANEILEMA (Commelinac.).
trispermum *R.B. Faden in Brittonia*, 48(4): 544 (1996 publ. 1997)—Somalia.

ANEMANTHELE *J.F. Veldkamp in Acta Bot. Neerl.*, 34(1): 107 (1985). *GRAMINEAE*
lessoniana *(Steud.) J.F. Veldkamp Acta Bot. Neerl.*, 34(1): 108 (1985): Agrostis lessoniana.

ANEMONE (Ranunculac.).
biflora DC.
var. eranthioides *(Regel) S. Ziman in Thaiszia*, 8(1): 68 (1998): A. eranthioides.
var. petiolulosa *(Juz.) S. Ziman in Thaiszia*, 8(1): 67 (1998): A. petiolulosa.
flaccida
var. tagawae *(Ohwi) M. Tamura in Acta Phytotax. Geobot.*, 46(2): 210 (1995 publ. 1996): A. tagawae.
× gokayamensis *M. Sugimoto, Tak. Sato & N. Naruhashi in J. Phytogeogr. Taxon.*, 46(1): 103 (1998)—Japan. (A. debilis × pseudoaltaica).
narcissiflora L.
var. crimea *S. Ziman & M. Fedoronchuk S. Ziman et al., Taxon. & Evol. Anemone narcissiflora complex:* 34 (1997), contrary to Art. 37.3 ICBN (2000)—Krym.
var. fasciculata *(L.) S. Ziman & M. Fedoronchuk in S. Ziman et al., Taxon. & Evol. Anemone narcissiflora complex:* 34 (1997): Anemone fasciculata.
subsp. fasciculata *(L.) S. Ziman & M. Fedoronchuk in S. Ziman et al., Taxon. & Evol. Anemone narcissiflora complex:* 34 (1997): Anemone fasciculata.
subsp. protracta *(Ulbr.) S. Ziman & M. Fedoronchuk in S. Ziman et al., Taxon. & Evol. Anemone narcissiflora complex:* 34 (1997): Anemone narcissiflora var. protracta.
subsp. sachalinensis *(Miyabe & T. Miyake) S. Ziman & M. Fedoronchuk in S. Ziman et al., Taxon. & Evol. Anemone narcissiflora complex:* 34 (1997): Anemone narcissiflora var. sachalinensis.
var. speciosa *(Adams) S. Ziman & M. Fedoronchuk in S. Ziman et al., Taxon. & Evol. Anemone narcissiflora complex:* 34 (1997), with incorrect basionym ref.: Anemone speciosa.

ANEMONE:-
obtusiloba D. Don
var. leiovaria *J.Q. Fu in Fl. Loess-Plat. Sin.*, 1: 611 (2000), contrary to Art. 37.5 ICBN (1994)—Qinghai.
pseudoaltaica H. Hara
f. prolifera *Y. Kadota in J. Jap. Bot.*, 73(5): 289 (1998)—Japan.
pulsatilla L.
subsp. gotlandica *(Johanss.) Lindell in Nordic J. Bot.*, 16(1): 7 (1996): Pulsatilla vulgaris var. gotlandica.
tuberosa Rydb.
var. texana *M. Enquist & B. Crozier in Phytologia*, 78(6): 441 (1995)—Texas, New Mexico, Mexico Northeast (Coahuila, Nuevo León).
vitifolia Buch.-Ham.
var. takasagomontana *(Masam.) S.S. Ying in Mem. Coll. Agric. Nation. Taiwan Univ.*, 29(2): 54 (1989): A. takasagomontana.

ANETHOLEA *Peter G. Wilson in Austral. Syst. Bot.*, 13(3): 434 (2000)—*MYRTACEAE*
anisata *(Vickery) Peter G. Wilson in Austral. Syst. Bot.*, 13(3): 434 (2000): Backhousia anisata.

ANGELICA (Umbellif.).
angelicifolia *(Franch.) E.V. Kljuykov in Feddes Repert.*, 110(7–8): 482 (1999): Ligusticum angelicifolium.
anomala Avé-Lall.
var. glabra *(Koidz.) H. Ohba in Fl. Japan*, 2c: 300 (1999): A. matsumurae var. glabra.
f. pubescens *(T. Yamaz.) H. Ohba in Fl. Japan*, 2c: 300 (1999): A. sachalinensis var. pubescens.
subsp. sachalinensis *(Maxim.) H. Ohba in Fl. Japan*, 2c: 299 (1999): A. sachalinensis.
var. sachalinensis *(Maxim.) H. Ohba in Fl. Japan*, 2c: 299 (1999): A. sachalinensis.
balangshanensis *R.H. Shan & F.T. Pu in Acta Phytotax. Sin.*, 33(5): 476 (1995)—China South-central (Sichuan).
cryptotaeniifolia Kitag.
var. kyushiana *T. Yamazaki in J. Jap. Bot.*, 71(2): 120 (1996)—Japan.
gmelinii (DC.) Pimenov
subsp. saxatilis *(Turcz. ex Ledeb.) V.N. Voroschilov in A.K. Skvortsov (ed.), Florist. issl. v razn. rinakh SSSR:* 184 (1985): A. saxatilis.
kangdingensis *R.H. Shan & F.T. Pu in Acta Phytotax. Sin.*, 33(5): 478 (1995)—China South-central (Sichuan).
lignescens *J.-P. Reduron & P. Danton in Acta Bot. Gallica*, 144(1): 186 (1997 publ. 1998)—Azores.
nakaiana *(Kitag.) M.G. Pimenov in Feddes Repert.*, 110(7–8): 482 (1999): Homopterix nakaiana.
pinnatiloba *R.H. Shan & F.T. Pu in Acta Phytotax. Sin.*, 33(5): 481 (1995)—China South-central (Sichuan).
songpanensis *R.H. Shan & F.T. Pu in Acta Phytotax. Sin.*, 33(5): 480 (1995)—China South-central (Sichuan).
tenuisecta (Makino) Makino
var. furcijuga *(Kitag.) H. Ohba in Fl. Japan*, 2c: 295 (1999)—Angelica furcijuga.
var. mayebarana *(Koidz.) H. Ohba in Fl. Japan*, 2c: 295 (1999)—Peucedanum mayebaranum.

ANGELONIA (Scrophulariac.).
alternifolia *V.C. Souza in Bradea*, 8(7): 37 (1997)—Brazil North (Tocantins).

ANGELPHYTUM (Compos.).
apense *(Chodat) A.A. Sáenz in Zuloaga & Morrone (eds.), Cat. Pl. Vasc. Rep. Arg. 2 (Monogr. Syst. Bot.* Missouri Bot. Gard. 74): 1245 (11 June 1999), as 'apensis': Aspilia apensis.
apense *(Chodat) J.F. Pruski in Compositae Newsl.*, 34: 2 (21 Dec. 1999): Aspilia apensis.
herzogii *(Hassl.) J.F. Pruski in Compositae Newsl.*, 34: 2 (1999): Zexmenia herzogii.

ANGINON (Umbellif.).
fruticosum *I. Allison & B.-E. van Wyk in Nordic J. Bot.*, 17(6): 567 (1997 publ. 1998)—Cape Province.
intermedium *I. Allison & B.-E. van Wyk in Nordic J. Bot.*, 17(6): 576 (1997 publ. 1998)—Cape Province.

ANGINON:-
pumilum *I. Allison & B.-E. van Wyk in Nordic J. Bot.*, 17(6): 569 (1997 publ. 1998)—Cape Province.
streyi *(Merxm.) I. Allison & B.-E. van Wyk in Nordic J. Bot.*, 17(6): 574 (1997 publ. 1998): Sonderina streyi.
tenuius *I. Allison & B.-E. van Wyk in Nordic J. Bot.*, 17(6): 569 (1997 publ. 1998), as 'tenuior'—Cape Province.
ternatum *I. Allison & B.-E. van Wyk in Nordic J. Bot.*, 17(6): 566 (1997 publ. 1998)—Cape Province.

ANGOLLUMA (Asclepiadac.).
lenewtonii *J.J. Lavranos in Asklepios*, 73: 15 (1998)—Kenya.

ANGOPHORA (Myrtac.).
exul *K.D. Hill in Telopea*, 7(2): 100 (1997)—New South Wales.
inopina *K.D. Hill in Telopea*, 7(2): 97 (1997)—New South Wales.

ANGOSTURA (Rutac.).
alipes *J.A. Kallunki in Kew Bull.*, 53(2): 261 (1998)—Ecuador.
bracteata *(Nees & Mart.) J.A. Kallunki in Kew Bull.*, 53(2): 263 (1998): Sciuris bracteata.
granulosa *(Kallunki) J.A. Kallunki in Kew Bull.*, 53(2): 263 (1998): Galipea granulosa.
longiflora *(K. Krause) J.A. Kallunki in Kew Bull.*, 53(2): 263 (1998): Galipea longiflora.
quinquefolia *J.A. Kallunki in Kew Bull.*, 53(2): 263 (1998)—Brazil North (Pará), Brazil Northeast (Maranhão).
simplex *J.A. Kallunki in Kew Bull.*, 53(2): 264 (1998)—Brazil North (Amazonas), Peru.

ANGRAECOPSIS (Orchidac.).
cryptantha *P.J. Cribb in Kew Bull.*, 51(2): 361 (1996)—Cameroon.

ANGRAECUM (Orchidac.).
aloifolium *J. Hermans & P.J. Cribb in Orchid Rev.*, 105(1214): 108 (1997)—Madagascar.
cephalotes *(Rchb. f.) N.E. Brown, Suppl. Johnson's Gard. Dict.*: 873 (1882): Listrostachys cephalotes.
chimanimaniense *G. Williamson in Kew Bull.*, 51(3): 557 (1996)—Zimbabwe.
dollii *K. Senghas in J. Orchideenfreund*, 4(1): 16 (1997)—Madagascar.
ringens *(Rchb. f.) N.E. Brown, Suppl. Johnson's Gard. Dict.*: 873 (1882): Listrostachys ringens.
sedenii *(Rchb. f.) N.E. Brown, Suppl. Johnson's Gard. Dict.*: 873 (1882): Listrostachys sedenii.

ANGULOA (Orchidac.).
× acostae *H.F. Oakeley in Orchids (Tokyo)*, 38: 30 (10 Feb. 1999), holotype not indicated; *et in Orquideologia*, 21(2): 185 (June 1999), holotype not indicated—Colombia. (A. eburnea × hohenlohii).
cliftonii Rolfe
var. alba *H.F. Oakeley in Orchids (Tokyo)*, 38: 32 (10 Feb. 1999); *et in Orquideologia*, 21(2): 189 (June 1999)—Colombia.
var. concolor *H.F. Oakeley in Orchids (Tokyo)*, 38: 32 (10 Feb. 1999); *et in Orquideologia*, 21(2): 189 (June 1999)—Cultivation.
hohenlohii C. Morren
var. macroglossa *(Schltr.) H.F. Oakeley in Orchids (Tokyo)*, 38: 28, 34 (10 Feb. 1999), correct basionym pagination on p. 34; *et in Orquideologia*, 21(2): 183 (June 1999): A. macroglossa.
tognettiae *H.F. Oakeley in Orchids (Tokyo)*, 38: 26 (10 Feb. 1999); *et in Orchidee*, 50(1): 28 (24 March 1999); *et in Orquideologia*, 21(2): 173 (June 1999)—Venezuela.
virginalis *Linden ex R. Schlechter, Orchideen:* 404 (1914)—Colombia.
var. turneri *(B.S. Williams) H.F. Oakeley in Orchids (Tokyo)*, 38: 24 (10 Feb. 1999); *et in Orchidee*, 50(1): 27 (24 March 1999); *et in Orquideologia*, 21(2): 173 (June 1999): A. turneri.

ANISACANTHUS (Acanthac.).
nicaraguensis *L.H. Durkee in Novon*, 9(4): 503 (1999)—Nicaragua.

ANISACHNE (Gramin.).
gracilis Keng
var. multinodis *Y.Y. Qian in Bull. Bot. Res. North-East. Forest. Univ.*, 18(4): 398 (1998)—China South-central (Yunnan).

ANISOCARPUS (Compos.).
scabridus *(Eastw.) B.G. Baldwin in Novon*, 9(4): 463 (1999): Raillardella scabrida.

ANISOCHILUS (Labiat.).
henryi *K. Ravikumar & V. Lakshmanan in Rheedea*, 9(1): 72 (1999)—India (Tamil Nadu).

ANISOLOTUS (Legumin.).
subpinnatus *(Lag.) P. Lassen in Svensk Bot. Tidskr.*, 91(5): 249 (1997 publ. 1998): Lotus subpinnatus.

ANISOPAPPUS (Compos.).
abercornensis G. Taylor
subsp. anemonifolius *(DC.) S. Ortiz, J.A.R. Paiva & J. Rodríguez-Oubiña in An. Jard. Bot. Madrid*, 54(1): 388 (1996): Epallage anemonifolia.
chinensis Hook. & Arn.
subsp. buchwaldii *(O. Hoffm.) S. Ortiz, J.A.R. Paiva & J. Rodríguez-Oubiña in An. Jard. Bot. Madrid*, 54(1): 383 (1996): Sphacophyllum buchwaldii.
var. dentatus *(DC.) S. Ortiz, J.A.R. Paiva & J. Rodríguez-Oubiña in An. Jard. Bot. Madrid*, 54(1): 385 (1996): Epallage dentata.
var. macrocephala *(Humbert) S. Ortiz, J.A.R. Paiva & J. Rodríguez-Oubiña in An. Jard. Bot. Madrid*, 54(1): 383 (1996): Epallage dentata var. macrocephala.
subsp. oliveranus *(Wild) S. Ortiz, J.A.R. Paiva & J. Rodríguez-Oubiña in An. Jard. Bot. Madrid*, 54(1): 385 (1996): A. oliveranus.
subsp. paucidentatus *(Wild) S. Ortiz, J.A.R. Paiva & J. Rodríguez-Oubiña in An. Jard. Bot. Madrid*, 54(1): 386 (1996): A. paucidentatus.
var. rupestris *(DC.) S. Ortiz, J.A.R. Paiva & J. Rodríguez-Oubiña in An. Jard. Bot. Madrid*, 54(1): 383 (1996): Epallage rupestris.
subsp. scrophulariifolius *(Baker) S. Ortiz, J.A.R. Paiva & J. Rodríguez-Oubiña in An. Jard. Bot. Madrid*, 54(1): 386 (1996): Temnolepis scrophulariifolia.
davyi S. Moore
var. bampsianus *(Lisowski) S. Ortiz, J.A.R. Paiva & J. Rodríguez-Oubiña in An. Jard. Bot. Madrid*, 54(1): 387 (1996): A. bampsianus.
subsp. pumilus *(Hiern) S. Ortiz, J.A.R. Paiva & J. Rodríguez-Oubiña in An. Jard. Bot. Madrid*, 54(1): 386 (1996): Sphacophyllum pumilum.
var. pumilus *(Hiern) S. Ortiz, J.A.R. Paiva & J. Rodríguez-Oubiña in An. Jard. Bot. Madrid*, 54(1): 387 (1996): Sphacophyllum pumilum.

ANNELIESIA (Orchidac.).
cuneata *(Lindl.) K. Senghas & Lückel in Schlechter Orchideen, ed. 3*, I/C(33–36): 2284 (1997): Miltonia cuneata.
kayasimae *(Pabst) K. Senghas & Lückel in Schlechter Orchideen, ed. 3*, I/C(33–36): 2284 (1997): Miltonia kayasimae.
russeliana *(Lindl.) K. Senghas & Lückel in Schlechter Orchideen, ed. 3*, I/C(33–36): 2284 (1997): Miltonia russeliana.

ANNESLEA (Theac.).
fragrans Wall.
var. rubriflora *(Hu & H.T. Chang) L.K. Ling in Fl. Reipubl. Popul. Sin.*, 50(1): 180 (1998): A. rubriflora.

ANNONA (Annonac.).
× atemoya *D.J. Mabberley, Paradisus:* 143 (1998)—Cultivation. (A. cherimola × squamosa).
warmingiana *R. de Mello-Silva & J.R. Pirani in Kew Bull.*, 54(1): 237 (1999), nom. nov.: A. coriacea var. pygmaea *Warm.*

ANODENDRON (Apocynac.).
borneense *(King & Gamble) D.J. Middleton in Blumea*, 41(1): 46 (1996): Cleghornia borneensis.
gracile *(King & Gamble) D.J. Middleton in Blumea*, 41(1): 50 (1996): Cleghornia gracilis.

ANODENDRON:-
seramense *D.J. Middleton in Blumea*, 41(1): 59 (1996)—Moluccas.
whitmorei *D.J. Middleton in Blumea*, 41(1): 63 (1996)—Solomon Is. (South Solomons), Moluccas, New Guinea (Papua New Guinea).

ANOECTOCHILUS (Orchidac.).
acalcaratus *L.V. Averyanov in Bot. Zhurn.*, 81(10): 74, 73 (1996)—Vietnam.
calcareus *L.V. Averyanov in Bot. Zhurn.*, 81(10): 75 (1996)—Vietnam.
candidus *(T.P. Lin & C.C. Hsu) K.Y. Lang in Fl. Reipubl. Popul. Sin.*, 17: 217 (1999): Odontochilus candidus.
gengmanensis *K.Y. Lang in Acta Phytotax. Sin.*, 34(5): 554 (1996)—China South-central (Yunnan).
pingbianensis *K.Y. Lang in Acta Phytotax. Sin.*, 34(5): 556 (1996)—China South-central (Yunnan).
roxburghii (Wall.) Lindl.
var. baotingensis *K.Y. Lang in Acta Phytotax. Sin.*, 34(5): 557 (1996)—Hainan.
subregularis *(Rchb. f.) P. Ormerod in Austral. Orchid Rev.*, 61(4): 36 (August 1996): Georchis subregularis.
subregularis *(Rchb. f.) A. Schuiteman in Blumea*, 41(2): 400 (27 Nov. 1996), isonym; *see P. Ormerod in Austral. Orchid Rev.*, 61(4): 36 (August 1996): Georchis subregularis.

ANOMOSTACHYS (Euphorbiac.).
lastellei *(Baill.) R.C. Kruijt in Biblioth. Bot.*, 146: 11 (1996): Stillingia lastellei.

ANOPLOCARYUM (Boraginac.).
helenae *K.A. Volotovskii in Novosti Sist. Vyssh. Rast.*, 30: 147 (1996)—Yakutiya.

ANREDERA (Basellac.).
aspera *C.R. Sperling in Phytologia*, 79(1): 3 (1995 publ. 1996)—Bolivia.
boliviensis *(Hauman) R. Govaerts, World Checklist Seed Plants*, 2(1–2): 9 (1996): Boussingaultia ramosa var. boliviensis.
brachystachys *(Moq.) C.R. Sperling in Phytologia*, 79(1): 3 (1995 publ. 31 Jan. 1996): Tandonia brachystachys.
brachystachys *(Moq.) R. Govaerts, World Checklist Seed Plants*, 2(1–2): 9 (8 Nov. 1996), isonym; *see Sperling in Phytologia*, 79(1): 3 (31 Jan. 1996): Tandonia brachystachys.
cordifolia (Ten.) Steenis
subsp. gracilis *(Miers) C. Xifreda & Argimón in Zuloaga & Morrone (eds.), Cat. Pl. Vasc. Rep. Arg. 2 (Monogr. Syst. Bot.* Missouri Bot. Gard. 74): 1245 (1999): Boussingaultia gracilis.
densiflora *C.R. Sperling in Phytologia*, 79(1): 2 (1995 publ. 1996)—Peru, Ecuador.
floribunda *(Moq.) C.R. Sperling in Phytologia*, 79(1): 2 (1995 publ. 31 Jan. 1996): Boussingaultia floribunda.
floribunda *(Moq.) R. Govaerts, World Checklist Seed Plants*, 2(1–2): 9 (8 Nov. 1996), isonym; *see Sperling in Phytologia*, 79(1): 2 (31 Jan. 1996): Boussingaultia floribunda.
krapovickasii *(Villa) C.R. Sperling in Phytologia*, 79(1): 3 (1995 publ. 31 Jan. 1996): Boussingaultia krapovickasii.
krapovickasii *(Villa) R. Govaerts, World Checklist Seed Plants*, 2(1–2): 9 (8 Nov. 1996), isonym; *see Sperling in Phytologia*, 79(1): 3 (31 Jan. 1996): Boussingaultia krapovickasii.
obovata *(Kunth) R. Govaerts, World Checklist Seed Plants*, 2(1–2): 9 (1996): Basella obovata.
ramosa (Moq.) Eliasson
subsp. aurea *R. Eriksson in Fl. Ecuador*, 55(40): 83 (1996)—Ecuador.
tucumanensis *(Lillo & Hauman) C.R. Sperling in Phytologia*, 79(1): 3 (1995 publ. 1996): Boussingaultia tucumanensis.
volkesii *(Ulbr.) R. Govaerts, World Checklist Seed Plants*, 2(1–2): 9 (1996): Boussingaultia volkesii.

ANTENNARIA (Compos.).
alpina (L.) Gaertn.
subsp. canescens *(Lange) J.G. Chmielewski in Rhodora*, 100(901): 61 (1998): A. alpina var. canescens.

ANTENNARIA:-
alpina (L.) Gaertn.
subsp. porsildii *(E. Ekman) J.G. Chmielewski in Rhodora*, 100(901): 63 (1998): A. porsildii.
media Greene
subsp. compacta *(Malte) J.G. Chmielewski in Brittonia*, 49(3): 324 (1997): A. compacta.
subsp. fusca *(E.W. Nelson) J.G. Chmielewski in Brittonia*, 49(3): 325 (1997): A. fusca.
subsp. pulchella *(Greene) J.G. Chmielewski in Brittonia*, 49(3): 321 (1997): A. pulchella.

ANTENORON (Polygonac.).
filiforme (Thunb.) Roberty & Vautier
var. neofiliforme *(Nakai) A.J. Li in Fl. Reipubl. Popul. Sin.*, 25(1): 108 (1998): Polygonum neofiliforme.

ANTERIORCHIS (Orchidac.).
coriophora (L.) E. Klein & Strack
subsp. fragrans *(Pollini) P. Jacquet, Rectificatif à la Répartition des Orchidées Sauvages de France* (Orchidophile, 125): III (1997), without basionym ref.: Orchis coriophora subsp. fragans.
subsp. fragrans *(Pollini) P. Jacquet in Orchidophile*, 127: 134 (1997): Orchis fragrans.
subsp. martrinii *(Timb.-Lagr.) P. Jacquet, Rectificatif à la Répartition des Orchidées Sauvages de France* (Orchidophile, 125): III (1997), without basionym ref.: Orchis coriophora subsp. martrinii.
subsp. martrinii *(Timb.-Lagr.) P. Jacquet in Orchidophile*, 127: 134 (1997): Orchis martrinii

ANTHAENANTIA (Gramin.).
villaregalis *(McVaugh & R. Guzmán) A. Espejo Serna & A.R. López-Ferrari in Acta Bot. Mex.*, 51: 62 (2000): Leptocoryphium villaregalis.

ANTHEMIS (Compos.).
abylaea *(Font Quer & Maire) C. Oberprieler in Bocconea*, 9: 227 (1998): A. punctata var. abylaea.
arvensis L.
subsp. glabra *(Rouy) D. Jeanmonod in J. Gamisans & D. Jeanmonod, Asteraceae I: Compl. Prodr. Fl. Corse (ser.* eds. Jeanm. & Burdet): 227 (1998): A. arvensis subvar. glabra.
cotula L.
subsp. lithuanica *(DC.) N.N. Tzvelev in Novosti Sist. Vyssh. Rast.*, 32: 185 (2000): Maruta cotula var. lithuanica.
cretica L.
var. macrocephala *(Jeanb. & Timb.-Lagr.) O. de Bolòs & J. Vigo, Fl. Països Catalans*, 3: 786 (1995): A. gerardiana var. macrocephala.
dicksoniae *A. Ghafoor in Feddes Repert.*, 108(5–6): 319 (1997)—Saudi Arabia.
maritima L.
subsp. pseudopunctata *C. Oberprieler in Bocconea*, 9: 248 (1998)—Algeria.
maroccana Batt. & Pit.
subsp. aguilarii *(Maire & Sennen) C. Oberprieler in Bocconea*, 9: 174 (1998): A. boveana var. aguilarii.
mauritiana Maire & Sennen
*subsp.*faurei *(Maire) C. Oberprieler in Bocconea*, 9: 129 (1998): A. pedunculata var. faurei.
parviceps *Dobrocz. & Fed. ex M.V. Klokov in Novosti Sist. Vyssh. Nizsh. Rast. (Kiev)*, 1974: 115 (1974)—Krym.
pedunculata Desf.
subsp. atlantica *(Pomel) C. Oberprieler in Bocconea*, 9: 282 (1998): A. atlantica.
subsp. clausonis *(Pomel) C. Oberprieler in Bocconea*, 9: 278 (1998): A. clausonis.*var.* discoidea *(Boiss.) C. Oberprieler in Bocconea*, 9: 273 (1998): A. tuberculata var. discoidea.
subsp. turolensis *(Pau ex Caball.) C. Oberprieler in Bocconea*, 9: 255 (1998): A. turolensis.
peregrina L.
subsp. heracleotica *(Boiss. & Heldr.) O. Georgiou in Flora Medit.*, 7: 103 (1997): A. peregrina var. heracleotica.plutonia Meikle*var.* artemisioides *(Holmboe) C. Oberprieler & R. Vogt in Bocconea*, 11: 99 (1999): A. tricolor var. artemisioides.

ANTHEMIS:-
punctata Vahl.
subsp. kabylica *(Batt.) C. Oberprieler in Bocconea,* 9: 293 (1998): A. montana var. kabylica.sheilae *A. Ghafoor & T.A. Al-Turki in Edinburgh J. Bot.,* 56(1): 55 (1999)—Saudi Arabia.
stiparum Pomel
subsp. intermedia *C. Oberprieler in Bocconea,* 9: 142 (1998)—Algeria, Morocco.
subsp. sabulicola *(Pomel) C. Oberprieler in Bocconea,* 9: 145 (1998): A. sabulicola.
tenuisecta Ball
subsp. jahandiezii *(Maire) C. Oberprieler in Bocconea,* 9: 183 (1998): A. boveana var. jahandiezii.
triumfetti (L.) DC.
subsp. aligulata *(Losa) J. Sánchez ex O. de Bolòs & J. Vigo, Fl. Països Catalans,* 3: 783 (1995): Cota aligulata.
zaianica *C. Oberprieler in Bocconea,* 9: 149 (1998)—Morocco.

ANTHERICUM (Anthericac.).
argentinense *(Hauman) E.R. Guaglianone in Hickenia,* 2(31): 137 (1995): A. ciliatum var. argentinense.

ANTHOCERCIS (Solanac.).
sylvicola *T.D. Macfarlane & G. Wardell-Johnson in Nuytsia,* 11(1): 71 (1996)—Western Australia.

ANTHONOTHA (Legumin.).
hallei *(Aubrév.) J. Léonard in Biodiversity African Pl. (Proc. XIV AETFAT Congr.), eds. L.J.G. van der* Maesen et al.: 446 (1996): Isomacrolobium hallei.

ANTHOPTERUS (Ericac.).
subgen. Gonandra *J.L. Luteyn in Brittonia,* 48(4): 607 (1996 publ. 1997).
costaricensis *J.L. Luteyn in Brittonia,* 48(4): 609 (1996 publ. 1997)—Costa Rica.
gentryi *J.L. Luteyn in Fl. Ecuador,* 54(147): 390 (1996)—Ecuador.
molaui *(Luteyn) J.L. Luteyn in Fl. Ecuador,* 54(147): 392 (1996): Themistoclesia molaui.
pterotus *(A.C. Sm.) J.L. Luteyn in Brittonia,* 48(4): 609 (1996 publ. 1997): Themistoclesia pterota.
revolutus *(Wilbur & Luteyn) J.L. Luteyn in Fl. Ecuador,* 54(147): 387 (1996): Themistoclesia revoluta.
schultzeae *(Sleumer) J.L. Luteyn in Fl. Ecuador,* 54(147): 388 (1996): Themistoclesia schultzeae.

ANTHOSPERMOPSIS *(K. Schum.) J.H. Kirkbride in Brittonia,* 49(3): 373 (1997): Staelia sect. Anthospermopsis. (Rubiac.)
catechosperma *(K. Schum.) J.H. Kirkbride in Brittonia,* 49(3): 373 (1997): Staelia catechosperma.

ANTHOTIUM (Goodeniac.).
odontophyllum *L.W. Sage in Nuytsia,* 12(2): 229 (1998)—Western Australia.

ANTHOXANTHUM (Gramin.).
davidsei *(R.W. Pohl) G.C. Tucker in Harvard Pap. Bot.,* 9: 66 (1996), isonym; *see Veldkamp in Blumea,* 30(2): 347 (1985): Hierochloe davidsei.
hirtum (Schrank) Y. Schouten & Veldkamp
subsp. arcticum *(J. Presl) G.C. Tucker in Harvard Pap. Bot.,* 9: 66 (1996): Hierochloe arctica.
maderense *H. Teppner in Phyton (Austria),* 38(2): 309 (1998)—Madeira.
monticola (Bigelow) Veldkamp
subsp. orthanthum *(T.J. Sørensen) G.C. Tucker in Harvard Pap. Bot.,* 9: 66 (1995): Hierochloe orthantha.
nitens (Weber) Y. Schouten & Veldkamp
subsp. balticum *(G. Weim.) G.C. Tucker in Harvard Pap. Bot.,* 9: 66 (1995): Hierochloe odorata subsp. baltica.
occidentale *(Buckley) G.C. Tucker in Harvard Pap. Bot.,* 9: 66 (1996), isonym; *see Veldkamp in Blumea,* 30(2): 348 (1985): Hierochloe occidentalis.

ANTHRISCUS (Umbellif.).
sylvestris (L.) Hoffm.
subsp. fumarioides *(Waldst. & Kit.) K. Spalik in Pol. Bot. Stud.,* 13: 58 (1997): Scandix fumarioides.

ANTHRISCUS:-
sylvestris (L.) Hoffm.
f. hirtifructus *(Ohwi) H. Ohba in Fl. Japan,* 2c: 278 (1999): A. nemorosus var. hirtifructus.

ANTHURIUM (Arac.).
balslevii *T.B. Croat & J. Rodríguez de Salvador in Aroideana,* 18: 54 (1995 publ. 1996)—Ecuador.
bromelicola *S.J. Mayo & L.P. Félix in Aroideana,* 23: 89 (2000)—Brazil Northeast (Pernambuco).
subsp. bahiense *S.J. Mayo, J.G. Jardim & A.M. Carvalho in Aroideana,* 23: 95 (2000)—Brazil Northeast (Bahia).
cabuyalense *T.B. Croat & J. Rodríguez de Salvador in Aroideana,* 18: 57 (1995 publ. 1996)—Ecuador.
clidemioides Standl.
subsp. pacificum *T.B. Croat & M.H. Grayum in Phytologia,* 82(1): 31 (1997)—Costa Rica.
cupulispathum *T.B. Croat & J. Rodríguez de Salvador in Aroideana,* 18: 63 (1995 publ. 1996)—Ecuador.
hebetatilaminum *T.B. Croat & J. Rodríguez de Salvador in Aroideana,* 18: 73 (1995 publ. 1996)—Ecuador.
jaramilloi *T.B. Croat & J. Rodríguez de Salvador in Aroideana,* 18: 80 (1995 publ. 1996)—Ecuador.
jimenae *T.B. Croat in Aroideana,* 18: 83 (1995 publ. 1996)—Ecuador.
limonense *M.H. Grayum in Phytologia,* 82(1): 32 (1997)—Costa Rica.
magnifolium *T.B. Croat & J. Rodríguez de Salvador in Aroideana,* 18: 93 (1995 publ. 1996)—Ecuador.
maricense *M.A. Nadruz Coelho & S.J. Mayo in Aroideana,* 23: 82 (2000)—Brazil Southeast (Rio de Janeiro).
minarum *C.M. Sakuragui & S.J. Mayo in Feddes Repert.,* 110(7–8): 535 (1999)—Brazil Southeast (Minas Gerais, Rio de Janeiro, São Paulo).
montezumense *C.M. Sakuragui in Aroideana,* 23: 73 (2000)—Brazil Southeast (Minas Gerais).
nigropunctatum *T.B. Croat & J. Rodríguez de Salvador in Aroideana,* 18: 101 (1995 publ. 1996)—Ecuador.
obtusum *(Engl.) M.H. Grayum in Phytologia,* 82(1): 35 (1997): A. trinerve var. obtusum.
subsp. puntarenense *M.H. Grayum in Phytologia,* 82(1): 35 (1997)—Costa Rica.
propinquum Sodiro
var. albispadix *T.B. Croat & J. Rodríguez de Salvador in Aroideana,* 18: 118 (1995 publ. 1996)—Ecuador.
pulverulentum Sodiro
var. adsimile *(Sodiro) T.B. Croat & J. Rodríguez de Salvador in Aroideana,* 18: 122 (1995 publ. 1996): A. adsimile.
rodrigueziae *T.B. Croat in Aroideana,* 18: 126 (1995 publ. 1996)—Ecuador.
silanchense *T.B. Croat & J. Rodríguez de Salvador in Aroideana,* 18: 130 (1995 publ. 1996)—Ecuador.
thompsoniae *I. Arias Granda in Rev. Jard. Bot. Nacion. Univ. Habana,* 17–18: 5 (1996–97 publ. 1998), as 'thompsonii'—Cuba.

ANTHYLLIS (Legumin.).
subgen. Barba-jovis *V.N. Tikhomirov & D.D. Sokoloff in Byull. Mosk. Obshch. Ispȳt. Prir., Biol.,* 101(1): 66 (1996).
subgen. Cornicina *(DC.) Akulova ex V.N. Tikhomirov & D.D. Sokoloff in Byull. Mosk. Obshch. Ispȳt. Prir., Biol.,* 101(1): 70 (1996): A. sect. Cornicina.
sect. Sericeae *V.N. Tikhomirov & D.D. Sokoloff in Byull. Mosk. Obshch. Ispȳt. Prir., Biol.,* 101(1): 68 (1996).
subgen. Terniflora *V.N. Tikhomirov & D.D. Sokoloff in Byull. Mosk. Obshch. Ispȳt. Prir., Biol.,* 101(1): 69 (1996).
sect. Terniflora *(V.N. Tikhom. & D.D. Sokoloff) C. Benedí González in An. Jard. Bot. Madrid,* 56(2): 280 (1998): A. subgen. Terniflora.
lagascana *C. Benedí González in An. Jard. Bot. Madrid,* 53(2): 283 (1995 publ. 1996), nom. nov.: A. sericea *Lag.*
langei *(Jalas) G.H. Loos in Flor. Rundbr.,* 30(2): 152 (1996): A. vulneraria var. langei.
pseudovulneraria *(Sagorski) G.H. Loos in Flor. Rundbr.,* 30(2): 152 (1996): A. vulneraria Unterrasse pseudovulneraria.

ANTHYLLIS
subsimplex Pomel
subsp. valentina *(Esteve) G. Mateo Sanz & M.B. Crespo Villalba, Fl. Abrev. Comun. Valenciana:* 430 (1995): A. sericea subsp. valentina.
tejedensis Boiss.
subsp. plumosa *(Cullen ex E. Domínguez) C. Benedí González in An. Jard. Bot. Madrid,* 56(2): 288 (1998): A. plumosa.
versicolor *(Sagorski) G.H. Loos in Flor. Rundbr.,* 30(2): 152 (1996): A. dillenii var. versicolor.
vulneraria L.
subsp. arenaria *(Rupr.) N.N. Tzvelev in Novosti Sist. Vȳssh. Rast.,* 32: 184 (2000): Vulneraria rustica f. arenaria.
subsp. colorata *(Juz.) N.N. Tzvelev in Novosti Sist. Vȳssh. Rast.,* 32: 184 (2000): A. colorata.
subsp. insularum *(Rothm.) A.M. Romo, Fl. Silvestres Baleares:* 153 (1994): A. insularum.
subsp. microcephala *(Willk.) C. Benedí González in An. Jard. Bot. Madrid,* 56(2): 291 (1998): A. webbiana var. microcephala.
subsp. schiwereckii *(DC.) N.N. Tzvelev in Novosti Sist. Vȳssh. Rast.,* 32: 184 (2000): A. vulneraria var. schiwereckii.

ANTIDAPHNE (Eremolepidac.).
hondurensis *J. Kuijt in Novon,* 8(4): 402 (1998), as 'hondurense'—Honduras.

ANTIDESMA (Euphorbiac.).
brevipes *P. Hoffmann in Kew Bull.,* 54(2): 348 (1999)—Sulawesi.
bunius (L.) Spreng.
var. pubescens *P. Hoffmann in Kew Bull.,* 54(2): 350 (1999)—Thailand.
edule Merr.
var. apoense *P. Hoffmann in Kew Bull.,* 54(2): 350 (1999)—Philippines.
elbertii *P. Hoffmann in Kew Bull.,* 54(2): 352 (1999)—Sulawesi, Lesser Sunda Is, Moluccas.
excavatum Miq.
var. indutum *(Airy Shaw) P. Hoffmann in Kew Bull.,* 54(2): 355 (1999): A. moluccanum var. indutum.
madagascariense Lam.
var. hildebrandtii *(Pax & K. Hoffm.) Petra Hoffmann in Kew Bull.,* 54(4): 883 (1999 publ. 2000): A. hildebrandtii.
montanum Blume
var. microphyllum *(Hemsl.) P. Hoffmann in Kew Bull.,* 54(2): 357 (1999): A. microphyllum.
var. salicinum *(Ridl.) P. Hoffmann in Kew Bull.,* 54(2): 359 (1999): A. salicinum.
var. wallichii *(Tul.) P. Hoffmann in Kew Bull.,* 54(2): 357 (1999): A. oblongifolium var. wallichii.
neurocarpum Miq.
var. hosei *(Pax & K. Hoffm.) P. Hoffmann in Kew Bull.,* 54(2): 360 (1999): A. hosei.
var. linearifolium *(Pax & K. Hoffm.) P. Hoffmann in Kew Bull.,* 54(2): 360 (1999): A. linearifolium.
pradoshii *T. Chakrabarty & M. Gangopadhyay in J. Econ. Taxon. Bot.,* 21(2): 479 (1997)—Sumatera.
riparium Airy Shaw
subsp. ramosum *P. Hoffmann in Kew Bull.,* 54(2): 355 (1999)—New Guinea (Irian Jaya).
tomentosum Blume
var. stenocarpum *(Airy Shaw) P. Hoffmann in Kew Bull.,* 54(2): 361 (1999): A. stenocarpum.
venosum Tul.
subsp. membranaceum *(Müll. Arg.) K.A. Lye in Lidia,* 4(3): 91 (1998): A. membranaceum.

ANTIMIMA (Aizoac.).
addita *(L. Bolus) H.E.K. Hartmann in Bothalia,* 28(1): 70 (1998): Ruschia addita.
amoena *(Schwantes) H.E.K. Hartmann in Bothalia,* 28(1): 70 (1998): Ruschia amoena.
androsacea *(Marloth & Schwantes) H.E.K. Hartmann in Bothalia,* 28(1): 70 (1998): Ruschia androsacea.
argentea *(L. Bolus) H.E.K. Hartmann in Bothalia,* 28(1): 70 (1998): Ruschia argentea.
aurasensis *H.E.K. Hartmann in Kakt. And. Sukk.,* 47(11): 231 (1996)—Namibia.

ANTIMIMA:-
biformis *(N.E. Br.) H.E.K. Hartmann in Bothalia,* 28(1): 70 (1998): Mesembryanthemum biforme.
bina *(L. Bolus) H.E.K. Hartmann in Bothalia,* 28(1): 70 (1998): Ruschia bina.
bracteata *(L. Bolus) H.E.K. Hartmann in Bothalia,* 28(1): 70 (1998): Ruschia bracteata.
brevicarpa *(L. Bolus) H.E.K. Hartmann in Bothalia,* 28(1): 70 (1998): Ruschia brevicarpa.
brevicollis *(N.E. Br.) H.E.K. Hartmann in Bothalia,* 28(1): 71 (1998): Mesembryanthemum brevicolle.
buchubergensis *(Dinter) H.E.K. Hartmann in Bothalia,* 28(1): 71 (1998): Ruschia buchubergensis.
compacta *(L. Bolus) H.E.K. Hartmann in Bothalia,* 28(1): 71 (1998): Ruschia compacta.
compressa *(L. Bolus) H.E.K. Hartmann in Bothalia,* 28(1): 71 (1998): Ruschia compressa.
concinna *(L. Bolus) H.E.K. Hartmann in Bothalia,* 28(1): 71 (1998): Ruschia concinna.
condensa *(N.E. Br.) H.E.K. Hartmann in Bothalia,* 28(1): 71 (1998): Mesembryanthemum condensum.
crassifolia *(L. Bolus) H.E.K. Hartmann in Bothalia,* 28(1): 71 (1998): Ruschia crassifolia.
dasyphylla *(Schltr.) H.E.K. Hartmann in Bothalia,* 28(1): 71 (1998): Mesembryanthemum dasyphyllum.
defecta *(L. Bolus) H.E.K. Hartmann in Bothalia,* 28(1): 71 (1998): Ruschia defecta.
dekenahi *(N.E. Br.) H.E.K. Hartmann in Bothalia,* 28(1): 72 (1998): Mesembryanthemum dekenahi.
distans *(L. Bolus) H.E.K. Hartmann in Bothalia,* 28(1): 72 (1998): Mesembryanthemum distans.
dolomitica *(Dinter) H.E.K. Hartmann in Bothalia,* 28(1): 72 (1998): Mesembryanthemum dolomiticum.
eendornensis *(Dinter) H.E.K. Hartmann in Bothalia,* 28(1): 72 (1998): Mesembryanthemum eendornense.
elevata *(L. Bolus) H.E.K. Hartmann in Bothalia,* 28(1): 72 (1998): Ruschia elevata.
emarcescens *(L. Bolus) H.E.K. Hartmann in Bothalia,* 28(1): 72 (1998): Ruschia emarcescens.
erosa *(L. Bolus) H.E.K. Hartmann in Bothalia,* 28(1): 72 (1998): Ruschia erosa.
evoluta *(N.E. Br.) H.E.K. Hartmann in Bothalia,* 28(1): 72 (1998): Mesembryanthemum evolutum.
excedens *(L. Bolus) C. Klak in Bothalia,* 30(1): 35 (2000): Ruschia excedens.
exsurgens *(L. Bolus) H.E.K. Hartmann in Bothalia,* 28(1): 73 (1998): Ruschia exsurgens.
fenestrata *(L. Bolus) H.E.K. Hartmann in Bothalia,* 28(1): 73 (1998): Ruschia fenestrata.
fergusoniae *(L. Bolus) H.E.K. Hartmann in Bothalia,* 28(1): 73 (1998): Ruschia fergusoniae.
gracillima *(L. Bolus) H.E.K. Hartmann in Bothalia,* 28(1): 73 (1998): Ruschia gracillima.
granitica *(L. Bolus) H.E.K. Hartmann in Bothalia,* 28(1): 73 (1998): Mesembryanthemum graniticum.
hallii *(L. Bolus) H.E.K. Hartmann in Bothalia,* 28(1): 73 (1998): Ruschia hallii.
hamatilis *(L. Bolus) H.E.K. Hartmann in Bothalia,* 28(1): 73 (1998): Ruschia hamatilis.
herrei *(Schwantes) H.E.K. Hartmann in Bothalia,* 28(1): 73 (1998): Ruschia herrei.
intervallaris *(L. Bolus) H.E.K. Hartmann in Bothalia,* 28(1): 74 (1998): Ruschia intervallaris.
ivori *(N.E. Br.) H.E.K. Hartmann in Bothalia,* 28(1): 74 (1998): Mesembryanthemum ivori.
karroidea *(L. Bolus) H.E.K. Hartmann in Bothalia,* 28(1): 74 (1998): Ruschia karroidea.
klaverensis *(L. Bolus) H.E.K. Hartmann in Bothalia,* 28(1): 74 (1998): Mesembryanthemum klaverense.
koekenaapensis *(L. Bolus) H.E.K. Hartmann in Bothalia,* 28(1): 74 (1998): Ruschia koekenaapensis.
komkansica *(L. Bolus) H.E.K. Hartmann in Bothalia,* 28(1): 74 (1998): Ruschia komkansica.
lawsonii *(L. Bolus) H.E.K. Hartmann in Bothalia,* 28(1): 74 (1998): Mesembryanthemum lawsonii.
leipoldtii *(L. Bolus) H.E.K. Hartmann in Bothalia,* 28(1): 74 (1998): Ruschia leipoldtii.
leucanthera *(L. Bolus) H.E.K. Hartmann in Bothalia,* 28(1): 75 (1998): Mesembryanthemum leucantherum.

ANTIMIMA:-
limbata *(N.E. Br.) H.E.K. Hartmann in Bothalia,* 28(1): 75 (1998): Mesembryanthemum limbatum.
lodewykii *(L. Bolus) H.E.K. Hartmann in Bothalia,* 28(1): 75 (1998): Ruschia lodewykii.
loganii *(L. Bolus) H.E.K. Hartmann in Bothalia,* 28(1): 75 (1998): Ruschia loganii.
lokenbergensis *(L. Bolus) H.E.K. Hartmann in Bothalia,* 28(1): 75 (1998): Ruschia lokenbergensis.
luckhoffii *(L. Bolus) H.E.K. Hartmann in Bothalia,* 28(1): 75 (1998): Ruschia luckhoffii.
maleolens *(L. Bolus) H.E.K. Hartmann in Bothalia,* 28(1): 75 (1998): Ruschia maleolens.
maxwellii *(L. Bolus) H.E.K. Hartmann in Bothalia,* 28(1): 75 (1998): Ruschia maxwellii.
menniei *(L. Bolus) H.E.K. Hartmann in Bothalia,* 28(1): 75 (1998): Ruschia menniei.
mesklipensis *(L. Bolus) H.E.K. Hartmann in Bothalia,* 28(1): 76 (1998): Ruschia mesklipensis.
meyerae *(Schwantes) H.E.K. Hartmann in Bothalia,* 28(1): 76 (1998): Ruschia meyerae.
minima *(Tischer) H.E.K. Hartmann in Bothalia,* 28(1): 76 (1998): Cheiridopsis minima.
minutifolia *(L. Bolus) H.E.K. Hartmann in Bothalia,* 28(1): 76 (1998): Ruschia minutifolia.
modesta *(L. Bolus) H.E.K. Hartmann in Bothalia,* 28(1): 76 (1998): Ruschia modesta.
mucronata *(Haw.) H.E.K. Hartmann in Bothalia,* 28(1): 76 (1998): Mesembryanthemum mucronatum.
mutica *(L. Bolus) H.E.K. Hartmann in Bothalia,* 28(1): 76 (1998): Ruschia mutica.
nobilis *(Schwantes) H.E.K. Hartmann in Bothalia,* 28(1): 77 (1998): Ruschia nobilis.
nordenstamii *(L. Bolus) H.E.K. Hartmann in Bothalia,* 28(1): 77 (1998): Ruschia nordenstamii.
oviformis *(L. Bolus) H.E.K. Hartmann in Bothalia,* 28(1): 77 (1998): Ruschia oviformis.
papillata *(L. Bolus) H.E.K. Hartmann in Bothalia,* 28(1): 77 (1998): Ruschia papillata.
paucifolia *(L. Bolus) H.E.K. Hartmann in Bothalia,* 28(1): 77 (1998): Ruschia paucifolia.
pauper *(L. Bolus) H.E.K. Hartmann in Bothalia,* 28(1): 77 (1998): Ruschia pauper.
peersii *(L. Bolus) H.E.K. Hartmann in Bothalia,* 28(1): 77 (1998): Ruschia peersii.
perforata *(L. Bolus) H.E.K. Hartmann in Bothalia,* 28(1): 77 (1998): Ruschia perforata.
persistens *H.E.K. Hartmann in Bothalia,* 28(1): 78 (1998), nom. nov., as comb. nov.: Ruschia persistens *L. Bolus (1963)*.
pilosula *(L. Bolus) H.E.K. Hartmann in Bothalia,* 28(1): 78 (1998): Ruschia pilosula.
piscodora *(L. Bolus) H.E.K. Hartmann in Bothalia,* 28(1): 78 (1998): Ruschia piscodora.
prolongata *(L. Bolus) H.E.K. Hartmann in Bothalia,* 28(1): 78 (1998): Ruschia prolongata.
propinqua *(N.E. Br.) H.E.K. Hartmann in Bothalia,* 28(1): 78 (1998): Mesembryanthemum propinquum.
prostrata *(L. Bolus) H.E.K. Hartmann in Bothalia,* 28(1): 78 (1998): Ruschia prostrata.
pumila *(Fedde & C. Schust.) H.E.K. Hartmann in Bothalia,* 28(1): 78 (1998): Mesembryanthemum pumilum.
pusilla *(Schwantes) H.E.K. Hartmann in Bothalia,* 28(1): 79 (1998): Ruschia pusilla.
pygmaea *(Haw.) H.E.K. Hartmann in Bothalia,* 28(1): 79 (1998): Mesembryanthemum pygmaeum.
quarzitica *(Dinter) H.E.K. Hartmann in Bothalia,* 28(1): 79 (1998): Mesembryanthemum quarziticum.
roseola *(N.E. Br.) H.E.K. Hartmann in Bothalia,* 28(1): 79 (1998): Mesembryanthemum roseolum.
saturata *(L. Bolus) H.E.K. Hartmann in Bothalia,* 28(1): 79 (1998): Ruschia saturata.
saxicola *(L. Bolus) H.E.K. Hartmann in Bothalia,* 28(1): 79 (1998): Ruschia saxicola.
schlechteri *(Schwantes) H.E.K. Hartmann in Bothalia,* 28(1): 79 (1998): Ruschia schlechteri.
simulans *(L. Bolus) H.E.K. Hartmann in Bothalia,* 28(1): 79 (1998): Ruschia simulans.
sobrina *(N.E. Br.) H.E.K. Hartmann in Bothalia,* 28(1): 80 (1998): Mesembryanthemum sobrinum.
solida *(L. Bolus) H.E.K. Hartmann in Bothalia,* 28(1): 80 (1998): Mesembryanthemum solidum.
stayneri *(L. Bolus) H.E.K. Hartmann in Bothalia,* 28(1): 80 (1998): Ruschia stayneri.

ANTIMIMA:-
stokoei *(L. Bolus) H.E.K. Hartmann in Bothalia,* 28(1): 80 (1998): Ruschia stokoei.
subtruncata *(L. Bolus) H.E.K. Hartmann in Bothalia,* 28(1): 80 (1998): Ruschia subtruncata.
triquetra *(L. Bolus) H.E.K. Hartmann in Bothalia,* 28(1): 80 (1998): Ruschia triquetra.
tuberculosa *(L. Bolus) H.E.K. Hartmann in Bothalia,* 28(1): 80 (1998): Ruschia tuberculosa.
turneriana *(L. Bolus) H.E.K. Hartmann in Bothalia,* 28(1): 80 (1998): Ruschia turneriana.
vanzylii *(L. Bolus) H.E.K. Hartmann in Bothalia,* 28(1): 81 (1998): Ruschia vanzylii.
varians *(L. Bolus) H.E.K. Hartmann in Bothalia,* 28(1): 81 (1998): Ruschia varians
ventricosa *(L. Bolus) H.E.K. Hartmann in Bothalia,* 28(1): 81 (1998): Mesembryanthemum ventricosum.
verruculosa *(L. Bolus) H.E.K. Hartmann in Bothalia,* 28(1): 81 (1998): Ruschia verruculosa.
watermeyeri *(L. Bolus) H.E.K. Hartmann in Bothalia,* 28(1): 81 (1998): Ruschia watermeyeri.
wittebergensis *(L. Bolus) H.E.K. Hartmann in Bothalia,* 28(1): 81 (1998): Mesembryanthemum wittebergense.

ANTIRHEA (Rubiac.).
triflora *J.H. Kirkbride in Biollania,* ed. esp. 6 (J.J. Wurdack Festschr.): 396 (1997)—French Guiana.

ANTIRRHINUM (Scrophulariac.).
subsect. Hispanica *(Rothm.) F.J. Fernández Casas in Fontqueria,* 48: 197 (1997 publ. 1998 A. ser. Hispanica.
sect. Kickxiella *(Rothm.) F.J. Fernández Casas in Fontqueria,* 48: 197 (1997 publ. 1998): A. subsect. Kickxiella.
subsect. Sempervirentia *(Rothm.) F.J. Fernández Casas in Fontqueria,* 48: 197 (1997 publ. 1998): A. ser. Sempervirentia.
sect. Sempervirentia *(Rothm.) F.J. Fernández Casas in Fontqueria,* 48: 197 (1997 publ. 1998): A. ser. Sempervirentia.
subsect. Sicula *(Rothm.) F.J. Fernández Casas in Fontqueria,* 48: 197 (1997 publ. 1998): A. ser. Sicula.
sect. Streptosepalum *(Rothm.) F.J. Fernández Casas in Fontqueria,* 48: 199 (1997 publ. 1998): A. subsect. Streptosepalum.
subsect. Valentina *(Rothm.) F.J. Fernández Casas in Fontqueria,* 48: 197 (1997 publ. 1998): A. ser. Valentina.

ANTISTROPHE (Myrsinac.).
glabra *A.G. Pandurangan & V.J. Nair in Nordic J. Bot.,* 17(4): 367 (1997)—India (Kerala).

ANULOCAULIS (Nyctaginac.).
leiosolenus (Torr.) Standl.
var. gypsogenus *(Waterf.) R. Spellenberg & T. Wootten in Sida,* 18(4): 992 (1999): A. gypsogenus.
var. howardii *R. Spellenberg & T. Wootten in Sida,* 18(4): 989 (1999)—New Mexico.

AORCHIS (Orchidac.).
cyclochila *(Franch. & Sav.) T. Hashimoto in Proc. 12th World Orchid Conf., 1987 (eds. K. Saito & R. Tanaka)*: 119 (1987): Habenaria cyclochila.

AOSA *M. Weigend, Nasa & Conquest S. America - Syst. Rearrang. in Loasac.*: 214 (1997). *LOASACEAE*
gilgiana *(Urb.) M. Weigend, Nasa & Conquest S. America - Syst. Rearrang. in Loasac.*: 218 (1997): Loasa gilgiana.
parviflora *(DC.) M. Weigend, Nasa & Conquest S. America - Syst. Rearrang. in Loasac.*: 214, 218 (1997): Loasa parviflora.
plumieri *(Urb.) M. Weigend, Nasa & Conquest S. America - Syst. Rearrang. in Loasac.*: 218 (1997): Loasa plumieri.
rostrata *(Urb.) M. Weigend, Nasa & Conquest S. America - Syst. Rearrang. in Loasac.*: 218 (1997): Basionym not stated.
rupestris *(Gardner) M. Weigend, Nasa & Conquest S. America - Syst. Rearrang. in Loasac.*: 218 (1997): Loasa rupestris.

AOSA:-
sigmoidea *M. Weigend in Nordic J. Bot.*, 19(2): 213 (1999)—Brazil Southeast (Minas Gerais).
uleana *(Urb. & Gilg) M. Weigend, Nasa & Conquest S. America - Syst. Rearrang. in Loasac.*: 218 (1997), without basionym date: Loasa uleana.

APATOPHYLLUM (Celastrac.).
flavovirens *A.R. Bean & L.W. Jessup in Austrobaileya*, 5(4): 693 (2000)—Queensland.
teretifolium *A.R. Bean & L.W. Jessup in Austrobaileya*, 5(4): 693 (2000)—Queensland.

APETAHIA (Campanulac.).
seigelii *J. Florence in Allertonia*, 7(4): 248 (1997)—Marquesas.

APHANACTIS (Compos.).
antisanensis *H. Robinson in Brittonia*, 49(1): 73 (1997)—Ecuador.
barclayae *H. Robinson in Brittonia*, 49(1): 74 (1997)—Ecuador.
boliviana *H. Robinson in Brittonia*, 49(1): 75 (1997)—Bolivia.
hutchisonii *H. Robinson in Phytologia*, 82(1): 58 (1997)—Peru.
ollgaardii *H. Robinson in Brittonia*, 49(1): 76 (1997)—Ecuador.

APHANES (Rosac.).
cotopaxiensis *K. Romoleroux & P. Frost-Olsen in Nordic J. Bot.*, 16(5): 473 (1996 publ. 1997)—Ecuador.
lusitanica *P. Frost-Olsen in An. Jard. Bot. Madrid*, 55(1): 195 (1997)—Portugal.

APHANOCALYX (Legumin.).
subgen. Antherodontus *J.J. Wieringa in Wageningen Agric. Univ. Pap.*, 99(4): 174 (1999).
hedinii *(A. Chev.) J.J. Wieringa in Wageningen Agric. Univ. Pap.*, 99(4): 175 (1999): Cynometra hedinii.
heitzii *(Pellegr.) J.J. Wieringa in Wageningen Agric. Univ. Pap.*, 99(4): 179 (1999): Monopetalanthus heitzii.
jenseniae *(Gram) J.J. Wieringa in Wageningen Agric. Univ. Pap.*, 99(4): 128 (1999): Monopetalanthus jenseniae.
ledermannii *(Harms) J.J. Wieringa in Wageningen Agric. Univ. Pap.*, 99(4): 131 (1999): Monopetalanthus ledermannii.
libellula *J.J. Wieringa in Wageningen Agric. Univ. Pap.*, 99(4): 184 (1999)—Angola.
microphyllus *(Harms) J.J. Wieringa in Wageningen Agric. Univ. Pap.*, 99(4): 140 (1999): Monopetalanthus microphyllus.
subsp. compactus *(Hutch. ex Lane-Poole) J.J. Wieringa in Wageningen Agric. Univ. Pap.*, 99(4): 148 (1999): Monopetalanthus compactus.
obscurus *J.J. Wieringa in Wageningen Agric. Univ. Pap.*, 99(4): 152 (1999)—Gabon, Zaire.
pectinatus *(A. Chev.) J.J. Wieringa in Wageningen Agric. Univ. Pap.*, 99(4): 157 (1999): Monopetalanthus pectinatus.
pteridophyllus *(Harms) J.J. Wieringa in Wageningen Agric. Univ. Pap.*, 99(4): 161 (1999): Monopetalanthus pteridophyllus.
richardsiae *(J. Léonard) J.J. Wieringa in Wageningen Agric. Univ. Pap.*, 99(4): 165 (1999): Monopetalanthus richardsiae.
trapnellii *(J. Léonard) J.J. Wieringa in Wageningen Agric. Univ. Pap.*, 99(4): 171 (1999): Monopetalanthus trapnellii.

APHELANDRA (Acanthac.).
albinotata *D.C. Wasshausen in Nordic J. Bot.*, 16(4): 405 (1996 publ. 1997)—Ecuador.
anderssonii *D.C. Wasshausen in Nordic J. Bot.*, 16(4): 394 (1996 publ. 1997)—Ecuador.
azuayensis *D.C. Wasshausen in Nordic J. Bot.*, 16(4): 394 (1996 publ. 1997)—Ecuador.
longibracteolata *G. Lindau in Bull. Herb. Boiss.*, 3(8): 367 (1895)—Bolivia.
loxensis *D.C. Wasshausen in Nordic J. Bot.*, 16(4): 398 (1996 publ. 1997)—Ecuador.
modesta *(Mildbr.) D.C. Wasshausen in Nordic J. Bot.*, 16(4): 401 (1996 publ. 1997): Aphelandrella modesta.

APHELANDRA:-
mollissima *D.C. Wasshausen in Nordic J. Bot.*, 16(4): 405 (1996 publ. 1997)—Peru, Ecuador.
neillii *D.C. Wasshausen in Nordic J. Bot.*, 16(4): 400 (1996 publ. 1997)—Ecuador, Peru.
rosulata *(Lindau) D.C. Wasshausen in Nordic J. Bot.*, 16(4): 403 (1996 publ. 1997): Rhombochlamys rosulata.
var. albinotata *D.C. Wasshausen in Nordic J. Bot.*, 16(4): 403 (1996 publ. 1997)—Ecuador.
scabra (Vahl) Sm.
var. angustifolia *L.H. Durkee in Novon*, 9(4): 505 (1999)—Panama.
simplex *G. Lindau in Bull. Herb. Boiss.*, 3(8): 366 (1895), in Supp. 1 with incorrect ref.—Bolivia.†
tomentosa *G. Lindau in Bull. Herb. Boiss.*, 3(8): 369 (1895), in Supp. 1 with incorrect ref.—Venezuela.†
tumbecensis *D.C. Wasshausen in Nordic J. Bot.*, 16(4): 396 (1996 publ. 1997)—Peru, Ecuador.
villonacensis *D.C. Wasshausen in Nordic J. Bot.*, 16(4): 392 (1996 publ. 1997)—Ecuador.
zamorensis *D.C. Wasshausen in Nordic J. Bot.*, 16(4): 398 (1996 publ. 1997)—Ecuador.

APHRAGMUS (Crucif.).
nepalensis *(H. Hara) I.A. Al-Shehbaz in Harvard Pap. Bot.*, 5(1): 112 (2000): Staintoniella nepalensis.
przewalskii *(Maxim.) A.L. Ebel in Turczaninowia*, 1(4): 25 (1998): Eutrema przewalskii.

APHYLLODIUM (Legumin.).
australiense *(Schindl.) Horoyoshi Ohashi in Taiwania*, 42(2): 143 (1997): Dicerma biarticulatum var. australiense.
glossocarpum *L. Pedley in Austrobaileya*, 5(2): 214 (1999)—Western Australia.
hispidum *(Schindl.) Hiroyoshi Ohashi in Taiwania*, 42(2): 144 (1997): Dicerma hispidum.
var. collettii *(Schindl.) Hiroyoshi Ohashi in Taiwania*, 42(2): 145 (1997): Dicerma biarticulatum var. collettii.
latifolium *L. Pedley in Austrobaileya*, 5(2): 212 (1999)—Queensland.
novoguineense *(Schindl.) Hiroyoshi Ohashi in Taiwania*, 42(2): 146 (1997): Dicerma novoguineense.
parvifolium *L. Pedley in Austrobaileya*, 5(2): 212 (1999)—Western Australia.
schindleri *L. Pedley in Austrobaileya*, 5(2): 213 (1999)—Queensland, Western Australia, Northern Territory, New Guinea (Papua New Guinea).
stylosanthoides *L. Pedley in Austrobaileya*, 5(2): 215 (1999)—Western Australia, Northern Territory.

APHYLLON (Scrophulariac.).
purpureum *(A. Heller) J. Holub in Preslia*, 70(2): 100 (1998): Thalesia purpurea.

APHYLLORCHIS (Orchidac.).
annamensis *L.V. Averyanov in Bot. Zhurn.*, 81(10): 82 (1996)—Vietnam.
sanguinea *N. Pearce & P.J. Cribb in Edinburgh J. Bot.*, 56(2): 273 (1999)—Bhutan.

APIUM (Umbellif.).
nodiflorum
subsp. mairei *J.A. Molina Abril & S. Sardinero in Lagascalia*, 18(2): 278 (1996)—Morocco, Algeria.

APODANTHERA (Cucurbitac.).
anatuyana *(Mart. Crov.) R. Pozner in Hickenia*, 2(40): 186 (1996): Melothria anatuyana.

APODASMIA *B.G. Briggs & L.A.S. Johnson in Telopea*, 7(4): 371 (1998). *RESTIONACEAE*
brownii *(Hook. f.) B.G. Briggs & L.A.S. Johnson in Telopea*, 7(4): 371 (1998): Leptocarpus brownii.
chilensis *(Gay) B.G. Briggs & L.A.S. Johnson in Telopea*, 8(1): 22 (1998): Schoenodum chilense.
similis *(Edgar) B.G. Briggs & L.A.S. Johnson in Telopea*, 8(1): 22 (1998): Leptocarpus similis.

APONOGETON (Aponogetonac.).
elongatus F. Muell. ex Benth.
subsp. fluitans *C.B. Hellquist & S.W.L. Jacobs in Telopea*, 8(1): 18 (1998)—Queensland.

APONOGETON:-
euryspermus *C.B. Hellquist & S.W.L. Jacobs in Telopea*, 8(1): 16 (1998)—Western Australia, Northern Territory.
kimberleyensis *C.B. Hellquist & S.W.L. Jacobs in Telopea*, 8(1): 15 (1998)—Western Australia.
lancesmithii *C.B. Hellquist & S.W.L. Jacobs in Telopea*, 8(1): 12 (1998)—Queensland.
madagascariensis (Mirb.) H. Bruggen
var. henkelianus *(Baum) H.W.E. van Bruggen in Aqua Planta*, 1998(1): 9 (1998): A. henkelianus.
var. major *(Baum) H.W.E. van Bruggen in Aqua Planta*, 1998(1): 11 (1998): A. fenestralis var. major.
proliferus *C.B. Hellquist & S.W.L. Jacobs in Telopea*, 8(1): 11 (1998)—Queensland.
vanbruggenii *C.B. Hellquist & S.W.L. Jacobs in Telopea*, 8(1): 17 (1998)—Queensland, Northern Territory.

APOROCACTUS (Cactac.).
× moennighoffii *(D.F. Fisch.) A.J.S. McMillan in Brit. Cact. Succ. J.*, 16(4): 205 (1998): Cereus moennighoffii.(A. flagelliformis × martianus).

APOROHELIOCEREUS *hort. ex H.K. Airy Shaw in Willis, Dict. Flow. Pl. Ferns, ed. 7*: 80 (1966). *CACTACEAE* Aporocactus × Heliocereus

APORUSA (Euphorbiac.).
merrilliana *R. Govaerts & A. Radcliffe-Smith in Kew Bull.*, 51(1): 175 (1996), nom. nov.: A. acuminatissima *Merr.*

APPENDICULA (Orchidac.).
bilobulata *J.J. Wood in Contrib. Univ. Mich. Herb.*, 21: 315 (1997)—Borneo (Sarawak).
gracilis *L.V. Averyanov in Bot. Zhurn.*, 82(3): 137 (1997)—Vietnam.

APPENDICULOPSIS *(Schltr.) D.L. Szlachetko in Fragm. Flor. Geobot.*, Supp. 3: 119 (1995): Agrostophyllum sect. Appendiculopsis. (Orchidac.).
bicuspidata *(J.J. Sm.) D.L. Szlachetko in Fragm. Flor. Geobot.*, Supp. 3: 120 (1995): Agrostophyllum bicuspidatum.
celebica *(Schltr.) D.L. Szlachetko in Fragm. Flor. Geobot.*, Supp. 3: 120 (1995): Agrostophyllum celebicum.
confusa *(J.J. Sm.) D.L. Szlachetko in Fragm. Flor. Geobot.*, Supp. 3: 120 (1995): Agrostophyllum confusum.
costata *(J.J. Sm.) D.L. Szlachetko in Fragm. Flor. Geobot.*, Supp. 3: 120 (1995): Agrostophyllum costatum.
var. concava *(Schltr.) D.L. Szlachetko in Fragm. Flor. Geobot.*, Supp. 3: 120 (1995): Agrostophyllum costatum var. concavum.
elongata *(Ridl.) D.L. Szlachetko in Fragm. Flor. Geobot.*, Supp. 3: 120 (1995): Appendicula elongata.
stipulata *(Griff.) D.L. Szlachetko in Fragm. Flor. Geobot.*, Supp. 3: 120 (1995): Appendicula stipulata.
sumatrana *(Schltr. & J.J. Sm.) D.L. Szlachetko in Fragm. Flor. Geobot.*, Supp. 3: 120 (1995): Agrostophyllum sumatranum.
trifida *(Schltr.) D.L. Szlachetko in Fragm. Flor. Geobot.*, Supp. 3: 120 (1995): Agrostophyllum trifidum.

APTERANTHES (Asclepiadac.).
burchardii *(N.E. Br.) D.C.H. Plowes in Haseltonia*, 3: 61 (1995): Caralluma burchardii.
europaea *(Guss.) D.C.H. Plowes in Haseltonia*, 3: 59 (1995), isonym; *see Murb. in Acta Univ. Lund.*, Afd. 2, 34(7): 2 (1898): Stapelia europaea.
var. affinis *(De Wild.) D.C.H. Plowes in Haseltonia*, 13: 59 (1995): Caralluma affinis.
var. albotigrina *(Maire) D.C.H. Plowes in Haseltonia*, 3: 61 (1995): Caralluma europaea var. albotigrina.
var. barrueliana *(Maire) D.C.H. Plowes in Haseltonia*, 3: 60 (1995): Caralluma europaea var. barrueliana.

APTERANTHES:-
europaea *(Guss.) D.C.H. Plowes*
var. confusa *(Font Quer) D.C.H. Plowes in Haseltonia,* 3: 61 (1995): Caralluma confusa.
var. decipiens *(Maire) D.C.H. Plowes in Haseltonia,* 3: 61 (1995): Caralluma europaea var. decipiens.
var. gattefossei *(Maire) D.C.H. Plowes in Haseltonia,* 3: 61 (1995): Caralluma maroccana var. gattefossei.
subsp. gusssoneana *(Mikan) D.C.H. Plowes in Haseltonia,* 3: 59 (1995), with incorrect basionym ref.: Basionym not stated.
var. judaica *(Zohary) D.C.H. Plowes in Haseltonia,* 3: 61 (1995): Caralluma europaea var. judaica.
var. marmaricensis *(A. Berger) D.C.H. Plowes in Haseltonia,* 3: 61 (1995): Caralluma europaea var. marmaricensis.
subsp. maroccana *(Hook. f.) D.C.H. Plowes in Haseltonia,* 3: 60 (1995): Boucerosia maroccana.
var. micrantha *(Maire) D.C.H. Plowes in Haseltonia,* 3: 61 (1995): Caralluma europaea var. micrantha.
var. schmuckiana *(Gattef. & Maire) D.C.H. Plowes in Haseltonia,* 3: 59 (1995): Caralluma europaea var. schmuckiana.
var. simonis *(A. Berger) D.C.H. Plowes in Haseltonia,* 3: 61 (1995): Caralluma europaea var. simonis.
var. tristis *(Maire) D.C.H. Plowes in Haseltonia,* 3: 59 (1995): Caralluma europaea var. tristis.
joannis *(Maire) D.C.H. Plowes in Haseltonia,* 3: 61 (1995): Caralluma joannis.
negevensis *(D. Zohary) D.C.H. Plowes in Haseltonia,* 3: 61 (1995): Caralluma negevensis.

AQUILEGIA (Ranunculac.).
subsect. Colchicae *I.M. Vassiljeva in Novosti Sist. Vÿssh. Rast.,* 30: 15 (1996).
sect. Darwasianae *I.M. Vassiljeva in Novosti Sist. Vÿssh. Rast.,* 30: 27 (1996).
subsect. Flabellatae *I.M. Vassiljeva in Novosti Sist. Vÿssh. Rast.,* 30: 18 (1996).
sect. Glandulosae *I M. Vassiljeva in Novosti Sist. Vÿssh. Rast.,* 30: 19 (1996).
sect. Lactiflorae *I.M. Vassiljeva in Novosti Sist. Vÿssh. Rast.,* 30: 24 (1996).
subsect. Leptocerates *I.M. Vassiljeva in Novosti Sist. Vÿssh. Rast.,* 30: 17 (1996).
sect. Parviflorae *I.M. Vassiljeva in Novosti Sist. Vÿssh. Rast.,* 30: 21 (1996).

ARABIDOPSIS (Crucif.).
arenosa (L.) Lawalrée
subsp. borbasii *(Zapał.) S.L. O'Kane & I.A. Al-Shehbaz in Novon,* 7(3): 325 (1997): Arabis arenosa subsp. borbasii.
cebennensis *(DC.) S.L. O'Kane & I.A. Al-Shehbaz in Novon,* 7(3): 325 (1997): Arabis cebennensis.
croatica *(Schott) S.L. O'Kane & I.A. Al-Shehbaz in Novon,* 7(3): 325 (1997): Arabis croatica.
halleri *(L.) S.L. O'Kane & I.A. Al-Shehbaz in Novon,* 7(3): 325 (1997): Arabis halleri.
subsp. gemmifera *(Matsum.) S.L. O'Kane & I.A. Al-Shehbaz in Novon,* 7(3): 325 (1997): Cardamine gemmifera.
subsp. ovirensis *(Wulfen) S.L. O'Kane & I.A. Al-Shehbaz in Novon,* 7(3): 325 (1997): Arabis ovirensis.
lyrata *(L.) S.L. O'Kane & I.A. Al-Shehbaz in Novon,* 7(3): 325 (1997): Arabis lyrata.
subsp. kamchatica *(Fisch. ex DC.) S.L. O'Kane & I.A. Al-Shehbaz in Novon,* 7(3): 326 (1997): Arabis lyrata var. kamchatica.
subsp. petraea *(L.) S.L. O'Kane & I.A. Al-Shehbaz in Novon,* 7(3): 326 (1997): Cardamine petraea.
mongolica *(Botsch.) J. Měsíček & J. Soják in Folia Geobot. Phytotax.,* 30(4): 448 (1995): Arabis mongolica.
neglecta *(Schult.) S.L. O'Kane & I.A. Al-Shehbaz in Novon,* 7(3): 326 (1997): Arabis neglecta.
pedemontana *(Boiss.) S.L. O'Kane & I.A. Al-Shehbaz in Novon,* 7(3): 326 (1997): Arabis pedemontana.
qiranica *Z.X. An in Fl. Xinjiangensis,* 2(2): 376 (1995)—Xinjiang.
trichocarpa *R.F. Huang in Fl. Qinghaiica,* 1: 509 (1997)—Qinghai.

ARABIS (Crucif.).
alpina L.
var. formosana *Masam. ex S.F. Huang in Taiwania,* 40(4): 386 (1995)—Taiwan.
boivinii *G.A. Mulligan in Rhodora,* 97(890): 157 (1995 publ. 1996)—Saskatchewan, Québec, Montana.
calderi *G.A. Mulligan in Rhodora,* 97(890): 144 (1995 publ. 1996)—British Columbia, Alberta, Northwest Territories (Mackenzie), Yukon.
codyi *G.A. Mulligan in Rhodora,* 97(890): 151 (1995 publ. 1996)—Yukon, British Columbia.
flagellosa Miq.
var. kawachiensis *S. Fujii in Bull. Osaka Mus. Nat. Hist.,* 53: 50 (1999)—Japan.
formosana *(Masam. ex S.F. Huang) Liu & S.S. Ying in Fl. Taiwan, ed. 2,* 2: 746 (1996): A. alpina var. formosana.
glauca H. Boissieu
subsp. pseudoauriculata *(H. Boissieu) V.N. Voroschilov in A.K. Skvortsov (ed.), Florist. issl. v razn. raĭonakh SSSR:* 173 (1985): A. pseudoauriculata.
hirshbergiae *S. Boyd in Aliso,* 17(2): 203 (1998)—California.
holboellii Hornem.
var. consanguinea *(Greene) G.A. Mulligan in Rhodora,* 97(890): 155 (1995 publ. 1996): A. consanguinea.
lycia *G. Parolly & P. Hein in Willdenowia,* 30(2): 294 (2000)—Turkey.
murrayi *G.A. Mulligan in Rhodora,* 97(890): 151 (1995 publ. 1996)—Yukon, Alberta, British Columbia, Washington.
pamirica *Y.C. Lan & Z.X. An in Fl. Xinjiangensis,* 2(2): 375 (1995)—Xinjiang.
stelleri DC.
subsp. japonica *(A. Gray) V.N. Voroschilov in A.K. Skvortsov (ed.), Florist. issl. v razn. raĭonakh SSSR:* 173 (1985): A. alpina var. japonica.

ARACHNOTHRYX (Rubiac.).
affinis *(Hemsl.) A. Borhidi in Acta Bot. Hung.,* 38(1–4): 139 (1993–94 publ. 1995): Rondeletia affinis.
atravesadensis *(Lorence) A. Borhidi in Acta Bot. Hung.,* 38(1–4): 139 (1993–94 publ. 1995): Rondeletia atravesadensis.
chaconis *(Lorence) A. Borhidi in Acta Bot. Hung.,* 38(1–4): 140 (1993–94 publ. 1995): Rondeletia chaconis.
chiriquiana *(Lorence) A. Borhidi in Acta Bot. Hung.,* 40(1–4): 15 (1996–97 publ. 1997): Rondeletia chiriquiana.
dwyeri *(Lorence) A. Borhidi in Acta Bot. Hung.,* 38(1–4): 140 (1993–94 publ. 1995): Rondeletia dwyeri.
evansii *(Lorence) A. Borhidi in Acta Bot. Hung.,* 40(1–4): 15 (1996–97 publ. 1997): Rondeletia evansii.
ginetteae *(Lorence) A. Borhidi in Acta Bot. Hung.,* 38(1–4): 140 (1993–94 publ. 1995): Rondeletia ginetteae.
guerrerensis *(Lorence) A. Borhidi in Acta Bot. Hung.,* 38(1–4): 140 (1993–94 publ. 1995): Rondeletia guerrerensis.
manantiensis *(Lorence) A. Borhidi in Acta Bot. Hung.,* 38(1–4): 140 (1993–94 publ. 1995): Rondeletia manantiensis.
monteverdensis *(Lorence) A. Borhidi in Acta Bot. Hung.,* 38(1–4): 140 (1993–94 publ. 1995): Rondeletia monteverdensis.
povedae *(Lorence) A. Borhidi in Acta Bot. Hung.,* 38(1–4): 140 (1993–94 publ. 1995): Rondeletia povedae.
pringlei *(Lorence) A. Borhidi in Acta Bot. Hung.,* 40(1–4): 16 (1996–97 publ. 1997): Rondeletia pringlei.
purpurea *(Lorence) A. Borhidi in Acta Bot. Hung.,* 38(1–4): 140 (1993–94 publ. 1995): Rondeletia purpurea.
ricoi *(Lorence) A. Borhidi in Acta Bot. Hung.,* 38(1–4): 140 (1993–94 publ. 1995): Rondeletia ricoi.
rzedowskii *(Lorence) A. Borhidi in Acta Bot. Hung.,* 38(1–4): 140 (1993–94 publ. 1995): Rondeletia rzedowskii.

ARACHNOTHRYX:-
scoti *(Lorence) A. Borhidi in Acta Bot. Hung.,* 38(1–4): 140 (1993–94 publ. 1995): Rondeletia scoti.
tayloriae *(Lorence) A. Borhidi in Acta Bot. Hung.,* 38(1–4): 140 (1993–94 publ. 1995): Rondeletia tayloriae.
tenorioi *(Lorence) A. Borhidi in Acta Bot. Hung.,* 38(1–4): 140 (1993–94 publ. 1995): Rondeletia tenorioi.

ARAEOCOCCUS (Bromeliac.).
montanus *E.M.C. Leme in Harvard Pap. Bot.,* 4(1): 142 (1999)—Brazil Northeast (Bahia).

ARALIA (Araliac.).
houheensis *W.X. Wang, W.Y. Guo & Y.S. Fu in Abstr. XV Internat. Bot. Congr., Yokohama, Japan 1993:* 250 (1993)—China South-central (Hubei).
houheensis *W.X. Wang & Y.S. Fu in J. Wuhan Bot. Res.,* 13(3): 207 (1995)—China South-central (Hubei).
nantouensis *S.S. Ying in Mem. Coll. Agric. Nation. Taiwan Univ.,* 28(2): 44 (1988)—Taiwan.
ryukyuensis *(J. Wen) T. Yamazaki in J. Jap. Bot.,* 71(1): 55 (1996): A. elata var. ryukyuensis.
var. inermis *(Yanagita) T. Yamazaki in J. Jap. Bot.,* 71(1): 55 (1996): A. hypoleuca var. inermis.

ARAUCARIA (Araucariac.).
angustifolia (Bert.) Kuntze
var. vinacea *J.R. de Mattos in Loefgrenia,* 111: 1 (1997)—Brazil South (Santa Catarina).
bernieri Buchholz
var. pumilio *J. Silba in J. Int. Conifer Preserv. Soc.,* 7(1): 21 (2000)—New Caledonia.

ARCEUTHOBIUM (Viscac.).
sect. Americana *D.L. Nickrent in F.G. Hawksworth & D. Wiens, Dwarf Mistletoes (USDA Forest Serv. Agric.* Handb. 709): 170 (1996).
sect. Azorica *D.L. Nickrent in F.G. Hawksworth & D. Wiens, Dwarf Mistletoes (USDA Forest Serv. Agric.* Handb. 709): 170 (1996).
sect. Chinense *D.L. Nickrent in F.G. Hawksworth & D. Wiens, Dwarf Mistletoes (USDA Forest Serv. Agric.* Handb. 709): 170 (1996).
ser. Globosa *D.L. Nickrent in F.G. Hawksworth & D. Wiens, Dwarf Mistletoes (USDA Forest Serv. Agric.* Handb. 709): 170 (1996).
sect. Penda *D.L. Nickrent in F.G. Hawksworth & D. Wiens, Dwarf Mistletoes (USDA Forest Serv. Agric.* Handb. 709): 170 (1996).
sect. Pusilla *D.L. Nickrent in F.G. Hawksworth & D. Wiens, Dwarf Mistletoes (USDA Forest Serv. Agric.* Handb. 709): 171 (1996).
vaginatum (Willd.) J. Presl
var. cryptopodum *(Engelm.) A. Cronquist in A. Cronquist et al., Intermountain Fl., Vasc. Pl. Intermountain West,* U.S.A., 3A: 253 (1997): A. cryptopodum.

ARCHIBACCHARIS (Compos.).
almedana *G.L. Nesom in Phytologia,* 84(1): 51 (1998)—Mexico Southwest (Guerrero).
breedlovei *G.L. Nesom & B.L. Turner in Phytologia,* 84(1): 50 (1998)—Mexico Southwest (Oaxaca).
lewisii *H. Robinson in Biollania,* ed. esp. 6 (J.J. Wurdack Festschr.): 502 (1997)—Bolivia.
trichotoma *(Klatt) G.L. Nesom in Sida,* 19(1): 85 (2000): Baccharis trichotoma.

ARCHIDENDRON (Legumin.).
kanisii *R.S. Cowan in Nuytsia,* 11(1): 18 (1996)—Queensland.

ARCHIHYOSCYAMUS *A.M. Lu in Adansonia,* 19(1): 136 (1997). *SOLANACEAE*
leptocalyx *(Stapf) A.M. Lu in Adansonia,* 19(1): 138 (1997): Hyoscyamus leptocalyx.

ARCHILEPTOPUS *P.T. Li in J. S. China Agric. Univ.,* 12(3): 38 (1991). *EUPHORBIACEAE*
fangdingianus *P.T. Li in J. S. China Agric. Univ.,* 12(3): 39 (1991)—China Southeast (Guangxi).

ARCHIVEA *E.A. Christenson & R. Jenny in Orchids,* 65(5): 497 (1996). *ORCHIDACEAE*
kewensis *E.A. Christenson & R. Jenny in Orchids,* 65(5): 497 (1996)—Brazil.

ARCTIUM (Compos.).
sect. Lappacea *(Bunge) H. Duistermaat, Monogr. Arctium L. (Asterac.):* 105 (1996), as 'Lappaceum': Cousinia sect. Lappacea.
sect. Nanarctium *(Tscherneva) H. Duistermaat, Monogr. Arctium L. (Asterac.):* 108 (1996): Cousinia sect. Nanarctium.
sect. Pseudarctium *(Juz.) H. Duistermaat, Monogr. Arctium L. (Asterac.):* 112 (1996): Cousinia sect. Pseudarctium.
× dualis *(Juz.) H. Duistermaat, Monogr. Arctium L. (Asterac.):* 126 (1996): Cousinia × cualis.
pseudarctium *(Bornm.) H. Duistermaat, Monogr. Arctium L. (Asterac.):* 114 (1996): Cousinia pseudarctium.
× semiconstrictum *H. Duistermaat, Monogr. Arctium L. (Asterac.):* 124 (1996), contrary to Art. 37.4 ICBN (1994)—Iran.(A. minus × pa ladini).

ARCTOSTAPHYLOS (Ericac.).
columbiana Piper
subsp. nortensis *P.V. Wells in Four Seasons,* 8(1): 50 (1988)—California.
incognita *J.E. Keeley, A. Massihi, J. Delgadillo Rodríguez & S.A. Hirales in Madroño,* 44(2): 138 (1997 publ. 1998)—Mexico Northwest (Baja California).
manzanita Parry
subsp. glaucescens *P.V. Wells in Four Seasons,* 8(1): 51 (1988)—California.
parryana Lemmon
subsp. deserticum *J.E. Keeley, L. Boykin & A. Massihi in Madroño,* 44(3): 265 (1997 publ. 1998)—California.
subsp. tumescens *J.E. Keeley, L. Boykin & A. Massihi in Madroño,* 44(3): 264 (1997 publ. 1998)—California.
uva-ursi (L.) Spreng.
*subsp.*crassifolia *(Braun-Blanq.) Rivas Mart. ex A. De la Torre, F. Alcaraz & M.B. Crespo in Lazaroa,* 16: 154 (1995 publ. 1996): A. uva-ursi var. crassifolia.

ARCUATOPTERUS (Umbellif.).
harae *(Pimenov) M.G. Pimenov & T.A. Ostroumova in Feddes Repert.,* 111(7–8): 557 (2000): Angelica harae.
ramosissimus *(DC.) M.G. Pimenov & T.A. Ostroumova in Feddes Repert.,* 111(7–8): 557 (2000): Palimbia ramosissima.
sikkimensis *(C.B. Clarke) M.G. Pimenov & T.A. Ostroumova in Feddes Repert.,* 111(7–8): 557 (2000): Peucedanum sikkimense.

ARCYTOPHYLLUM (Rubiac.).
serpyllaceum *(Schltdl.) E.E. Terrell in Novon,* 9(2): 263 (1999): Hedyotis serpyllacea.

ARDISIA (Myrsinac.).
aciphylla Pit.
var. touranensis *C.M. Hu & J.E. Vidal in J. Trop. Subtrop. Bot.,* 4(4): 2 (1996)—Vietnam.
alajuelae *(Lundell) J.J. Pipoly & J.M. Ricketson in Sida,* 18(2): 511 (1998): Icacorea alajuelae.
albipedicellata *J.J. Pipoly & J.M. Ricketson in Sida,* 18(2): 511 (1998), nom. nov.: Icacorea parvifolia *Lundell.*
albipetala *(Lundell) J.J. Pipoly & J.M. Ricketson in Sida,* 18(2): 515 (1998): Icacorea albipetala.
albisepala *(Lundell) J.J. Pipoly & J.M. Ricketson in Sida,* 18(2): 511 (1998): Auriculardisia albisepala.
amherstiana A.DC.
var. pubescens *(H.R. Fletcher) K. Larsen & C.M. Hu in Fl. Thailand,* 6(2): 127 (1996): A. attenuata var. pubescens.
apodophylla *J.F. Morales in Phytologia,* 83(2): 111 (1997 publ. 1998), nom. nov.: Auriculardisia sessilifolia *Lundell.*
applanata *C.M. Hu & J.E. Vidal in J. Trop. Subtrop. Bot.,* 4(4): 3 (1996)—Vietnam.
asymmetrica *C.M. Hu & J.E. Vidal in J. Trop. Subtrop. Bot.,* 4(4): 4 (1996)—Vietnam.
azaharensis *J.J. Pipoly & J.M. Ricketson in Sida,* 18(2): 511 (1998), nom. nov.: Auriculardisia microcalyx *Lundell.*

ARDISIA:-
bartletii Lundell
subsp. lilacina *(Lundell) J.J. Pipoly & J.M. Ricketson in Sida,* 18(2): 466 (1998): A. lilacina.
baviensis *C.M. Hu & J.E. Vidal in J. Trop. Subtrop. Bot.,* 4(4): 4 (1996)—Vietnam.
bekomiensis *(Lundell) J.F. Morales in Phytologia,* 83(2): 111 (1997 publ. 1998): Icacorea bekomiensis.
bristanii *J.J. Pipoly & J.M. Ricketson in Sida,* 18(2): 511 (1998), nom. nov.: Auriculardisia parviflora *Lundell.*
caloneura *C.M. Hu & J.E. Vidal in J. Trop. Subtrop. Bot.,* 4(4): 5 (1996)—Laos.
chinensis Benth.
var. stenophylla *C.M. Hu & J.E. Vidal in J. Trop. Subtrop. Bot.,* 4(4): 6 (1996)—Vietnam.
chiriquiana *(Lundell) J.J. Pipoly & J.M. Ricketson in Sida,* 18(2): 511 (1998): Auriculardisia chiriquiana.
clemensii *C.M. Hu & J.E. Vidal in J. Trop. Subtrop. Bot.,* 4(4): 6 (1996)—Vietnam.
cornudentata Mez
subsp. morrisonensis *(Hayata) Yuen P. Yang in Fl. Taiwan, ed. 2,* 4: 42 (1998): A. morrisonensis.
var. morrisonensis *(Hayata) Yuen P. Yang in Fl. Taiwan, ed. 2,* 4: 42 (1998): A. morrisonensis.
var. stenosepala *(Hayata) Yuen P. Yang in Fl. Taiwan, ed. 2,* 4: 42 (1998): A. stenosepala.
crenata Sims
subsp. crassinervosa *(E. Walker) C.M. Hu & J.E. Vidal in J. Trop. Subtrop. Bot.,* 4(4): 7 (1996): A. crassinervosa.
subsp. mouretii *(Pit.) C.M. Hu & J.E. Vidal in J. Trop. Subtrop. Bot.,* 4(4): 8 (1996): A. mouretii.
crotonifolia *C.M. Hu & J.E. Vidal in J. Trop. Subtrop. Bot.,* 4(4): 8 (1996)—Vietnam.
darlingii Merr.
subsp. podadenia *(B.C. Stone) C.M. Hu in Blumea,* 44(2): 398 (1999): A. podadenia.
demissa Miq.
var. bambusetorum *(King & Gamble) C.M. Hu in Blumea,* 44(2): 394 (1999): A. bambusetorum.
dressleri *J.J. Pipoly & J.M. Ricketson in Sida,* 18(2): 511 (1998), nom. nov.: Auriculardisia roseiflora *Lundell.*
dryeri *J.J. Pipoly & J.M. Ricketson in Sida,* 18(2): 512 (1998), nom. nov.: Auriculardisia micrantha *Lundell.*
dumicola *C.M. Hu & J.E. Vidal in J. Trop. Subtrop. Bot.,* 4(4): 8 (1996)—Vietnam.
duripetala *(Lundell) J.J. Pipoly & J.M. Ricketson in Sida,* 18(2): 515 (1998): Icacorea duripetala.
eciliata *(Lundell) J.J. Pipoly & J.M. Ricketson in Sida,* 18(2): 512 (1998): Zunilia eciliata.
ellipticifolia *J.J. Pipoly & J.M. Ricketson in Sida,* 18(2): 512 (1998): Amatlania elliptica *Lundell.*
eucuneata *(Lundell) J.J. Pipoly & J.M. Ricketson in Sida,* 18(2): 512 (1998): Auriculardisia eucuneata.
eurubiginosa *(Lundell) J.F. Morales in Phytologia,* 83(2): 111 (1997 publ. 1998): Auriculardisia eurubiginosa.
fasciculiflora *C.M. Hu & J.E. Vidal in J. Trop. Subtrop. Bot.,* 4(4): 9 (1996)—Vietnam.
feniana *J.J. Pipoly & J.M. Ricketson in Sida,* 18(2): 512 (1998), nom. nov.: Zunilia purpusii *Lundell.*
filipendula *C.M. Hu & J.E. Vidal in J. Trop. Subtrop. Bot.,* 4(4): 10 (1996)—Vietnam.
flavida *J.J. Pipoly in Sida,* 17(2): 451 (1996)—Ecuador.
forbesii S. Moore
var. cataractorum *(B.C. Stone) C.M. Hu in Blumea,* 44(2): 399 (1999): A. synneura var. cataractorum.
fulva King & Gamble
var. multiflora *(B.C. Stone) C.M. Hu in Blumea,* 44(2): 403 (1999): A. junghuhniana var. multiflora.
glauciflora *Urb., Symb. Antill.,* 1: 382 (1899), in I.K. Supp. 2 as 'glaucifolia' in error—Puerto Rico.†
glomeriflora *J.F. Morales in Phytologia,* 83(2): 109 (1997 publ. 1998)—Costa Rica.
gracilenta *C.M. Hu & J.E. Vidal in J. Trop. Subtrop. Bot.,* 4(4): 11 (1996)—Vietnam.
guanacastensis *(Lundell) J.J. Pipoly & J.M. Ricketson in Sida,* 18(2): 512 (1998): Icacorea guanacastensis.

ARDISIA:-
guinealensis *(Lundell) J.J. Pipoly & J.M. Ricketson in Sida,* 18(2): 512 (1998): Icacorea guinealensis.
hatoana *(Lundell) J.J. Pipoly & J.M. Ricketson in Sida,* 18(2): 515 (1998): Icacorea hatoana.
heterotricha *(Lundell) J.J. Pipoly & J.M. Ricketson in Sida,* 18(2): 512 (1998): Auriculardisia heterotricha.
hornitoana *J.J. Pipoly & J.M. Ricketson in Sida,* 18(2): 513 (1998), nom. nov.: Icacorea reflexa *Lundell.*
hugonensis *(Lundell) J.J. Pipoly & J.M. Ricketson in Sida,* 18(2): 513 (1998): Auriculardisia hugonensis.
illicioides *C.M. Hu & J.E. Vidal in J. Trop. Subtrop. Bot.,* 4(4): 11 (1996)—Vietnam.
imperialis K. Schum.
var. novoguineensis *(Mez) C.M. Hu in Blumea,* 44(2): 403 (1999): A. novoguineensis.
interjacens *C.M. Hu & J.E. Vidal in J. Trop. Subtrop. Bot.,* 4(4): 12 (1996)—Vietnam.
ixcanensis *(Lundell) J.J. Pipoly & J.M. Ricketson in Sida,* 18(2): 513 (1998): Icacorea ixcanensis.
jaliscensis *(Lundell) J.J. Pipoly & J.M. Ricketson in Sida,* 18(2): 513 (1998): Icacorea jaliscensis.
knappii *(Lundell) J.J. Pipoly & J.M. Ricketson in Sida,* 18(2): 513 (1998): Auriculardisia knappii.
lajana *(Lundell) J.J. Pipoly & J.M. Ricketson in Sida,* 18(2): 515 (1998): Icacorea lajana.
lamdongensis *C.M. Hu & J.E. Vidal in J. Trop. Subtrop. Bot.,* 4(4): 13 (1996)—Vietnam.
latisepala *(Lundell) J.J. Pipoly & J.M. Ricketson in Sida,* 18(2): 513 (1998): Auriculardisia latisepala.
leptopoda *(Lundell) J.J. Pipoly & J.M. Ricketson in Sida,* 18(2): 513 (1998): Auriculardisia leptopoda.
lucidula *C.M. Hu & J.E. Vidal in J. Trop. Subtrop. Bot.,* 4(4): 14 (1996)—Vietnam.
macrosepala
var. schmidii *C.M. Hu & J.E. Vidal in J. Trop. Subtrop. Bot.,* 5(1): 1 (1997)—Vietnam.
merrillii
var. rosea *C.M. Hu & J.E. Vidal in J. Trop. Subtrop. Bot.,* 5(1): 1 (1997)—Laos.
mesoamericana *J.J. Pipoly & J.M. Ricketson in Sida,* 18(2): 515 (1998), nom. nov.: Icacorea latisepala *Lundell.*
monteverdeana *(Lundell) J.J. Pipoly & J.M. Ricketson in Sida,* 18(2): 513 (1998): Icacorea monteverdeana.
moonii C.B. Clarke
var. serratifolia *B.M. Wadhwa in M.D. Dassanayake & W.D. Clayton, eds., Revised Handb. Fl. Ceylon,* 13: 145 (1999)—Sri Lanka.
nebulosa *(Lundell) J.J. Pipoly & J.M. Ricketson in Sida,* 18(2): 513 (1998): Auriculardisia nebulosa.
neomirandae *J.J. Pipoly & J.M. Ricketson in Sida,* 18(2): 514 (1998), nom. nov.: Zunilia mirandae *Lundell.*
nhatrangensis *C.M. Hu & J.E. Vidal in J. Trop. Subtrop. Bot.,* 5(1): 2 (1997)—Vietnam.
var. glaucescens *C.M. Hu & J.E. Vidal in J. Trop. Subtrop. Bot.,* 5(1): 2 (1997)—Vietnam.
var. neurophylla *C.M. Hu & J.E. Vidal in J. Trop. Subtrop. Bot.,* 5(1): 3 (1997)—Vietnam.
niambiensis *J.J. Pipoly & A. Cogollo P.in Sida,* 18(2): 390 (1998)—Colombia.
oaxacana *(Lundell) J.J. Pipoly & J.M. Ricketson in Sida,* 18(2): 514 (1998): Icacorea oaxacana.
obovata Desv.
var. blancheana *(Stehlé) J. Fournet, Fl. Illustr. Phan. Guadeloupe & Martinique:* 1050 (1978), without basionym ref.: A. guadalupensis var. blancheana.
obtusa Mez
var. montana *(Pit.) C.M. Hu & J.E. Vidal in J. Trop. Subtrop. Bot.,* 5(1): 3 (1997): A. pseudopedunculosa var. montana.
subsp. pachyphylla *(Dunn) J.J. Pipoly & C. Chen in Novon,* 5(4): 359 (1995): A. pachyphylla.
opegrapha Oerst.
subsp. paquitensis *(Lundell) J.J. Pipoly & J.M. Ricketson in Sida,* 18(2): 457 (1998): A. paquitensis.
subsp. wagneri *(Mez) J.J. Pipoly & J.M. Ricketson in Sida,* 18(2): 452 (1998): A. wagneri.
paradoxa *C.M. Hu & J.E. Vidal in J. Trop. Subtrop. Bot.,* 5(1): 3 (1997)—Vietnam.

ARDISIA:-
parvidenticulata *J.J. Pipoly & J.M. Ricketson in Sida,* 18(2): 514 (1998), nom. nov.: Icacorea denticulata *Lundell.*
parvipunctata *(Lundell) J.J. Pipoly & J.M. Ricketson in Sida,* 18(2): 514 (1998): Icacorea parvipunctata.
perpendicularis E. Walker
var. balansana *(Yuen P. Yang) C.M. Hu in J. Trop. Subtrop. Bot.,* 5(1): 4 (1997): A. balansana.
pitardii *C.M. Hu & J.E. Vidal in J. Trop. Subtrop. Bot.,* 5(1): 4 (1997)—Vietnam.
polysticta Miq.
subsp. punctipetala *(Merr.) C.M. Hu in Blumea,* 44(2): 395 (1999): A. punctipetala.
premontana *J.J. Pipoly in Sida,* 17(2): 446 (1996)—Peru, Ecuador.
prionata E. Walker
var. linearifolia *(Pit.) C.M. Hu & J.E. Vidal in J. Trop. Subtrop. Bot.,* 5(1): 5 (1997): A. quinquegona var. linearifolia.
prolifera *C.M. Hu & J.E. Vidal in J. Trop. Subtrop. Bot.,* 5(1): 5 (1997): Laos.
prunifolia *C.M. Hu & J.E. Vidal in J. Trop. Subtrop. Bot.,* 5(1): 6 (1997)—Vietnam.
pubicalyx Miq.
var. collinsiae *(H.R. Fletcher) C.M. Hu in Blumea,* 44(2): 404 (1999): A. collinsiae.
quadrata *(Lundell) J.F. Morales in Phytologia,* 83(2): 111 (1997 publ. 1998): Auriculardisia quadrata.
quangnamensis *C.M. Hu & J.E. Vidal in J. Trop. Subtrop. Bot.,* 5(1): 7 (1997)—Vietnam.
quinquegona Blume
var. salicifolia *(E. Walker) C.M. Hu & J.E. Vidal in J. Trop. Subtrop. Bot.,* 5(1): 8 (1997): A. salicifolia.
rapaneifolia *C.M. Hu & J.E. Vidal in J. Trop. Subtrop. Bot.,* 5(1): 8 (1997)—Vietnam.
ravida *C.M. Hu & J.E. Vidal in J. Trop. Subtrop. Bot.,* 5(1): 9 (1997)—Vietnam.
rhodochroa *C.M. Hu & J.E. Vidal in J. Trop. Subtrop. Bot.,* 5(1): 10 (1997)—Vietnam.
rubescens Pit.
var. oblongifolia *C.M. Hu & J.E. Vidal in J. Trop. Subtrop. Bot.,* 5(1): 10 (1997)—Vietnam.
var. puberula *C.M. Hu & J.E. Vidal in J. Trop. Subtrop. Bot.,* 5(1): 11 (1997)—Vietnam.
samalana *(Lundell) J.J. Pipoly & J.M. Ricketson in Sida,* 18(2): 514 (1998): Icacorea samalana.
silvestris
var. appressa *C.M. Hu & J.E. Vidal in J. Trop. Subtrop. Bot.,* 5(1): 11 (1997)—Vietnam, Hainan.
sordida *(Lundell) J.F. Morales in Phytologia,* 83(2): 111 (1997 publ. 1998): Auriculardisia sordida.
squamata *(Lundell) J.F. Morales in Phytologia,* 83(2): 111 (1997 publ. 1998): Auriculardisia squamata.
steinii *J.J. Pipoly & J.M. Ricketson in Sida,* 18(2): 514 (1998), nom. nov.: Auriculardisia baruana *Lundell.*
stevensii *(Lundell) J.J. Pipoly & J.M. Ricketson in Sida,* 18(2): 514 (1998): Gentlea stevensii.
suboppositifolia *C.M. Hu & J.E. Vidal in J. Trop. Subtrop. Bot.,* 5(1): 11 (1997)—Vietnam.
symplocifolia *(C. Chen) K. Larsen & C.M. Hu in Fl. Thailand,* 6(2): 141 (1996): A. maculosa var. symplocifolia.
tahanica King & Gamble
subsp. eucalyptifolia *(B.C. Stone) C.M. Hu in Blumea,* 44(2): 404 (1999): A. eucalyptifolia.
ternatensis Scheff.
var. forstensii *(Scheff.) C.M. Hu in Blumea,* 44(2): 397 (1999): A. forstensii.
toroana *(Lundell) J.J. Pipoly & J.M. Ricketson in Sida,* 18(2): 514 (1998): Auriculardisia toroana.
triangula *(Lundell) J.J. Pipoly & J.M. Ricketson in Sida,* 18(2): 514 (1998): Icacorea triangula.
utleyi *(Lundell) J.J. Pipoly & J.M. Ricketson in Sida,* 18(2): 515 (1998): Icacorea utleyi.
verapazensis Donn. Sm.
subsp. cucullata *(Lundell) J.J. Pipoly & J.M. Ricketson in Sida,* 18(4): 1156 (1999): A. cucullata.
vidalii *C.M. Hu in J. Trop. Subtrop. Bot.,* 5(1): 12 (1997)—Vietnam.
var. orbicularis *C.M. Hu in J. Trop. Subtrop. Bot.,* 5(1): 13 (1997)—Vietnam.

ARDISIA:-
vietnamensis *C.M. Hu & J.E. Vidal in J. Trop. Subtrop. Bot.,* 5(1): 13 (1997)—Vietnam.
warburgiana *Mez in Engl., Pflanzenr.,* IV, 236: 143 (1902)—Philippines.
warneri *J.J. Pipoly & J.M. Ricketson in Sida,* 18(2): 515 (1998), nom. nov.: Icacorea brevipes *Lundell.*
websteri *J.J. Pipoly in Sida,* 17(2): 454 (1996), as 'websterii'—Ecuador.
wendtii *(Lundell) J.J. Pipoly & J.M. Ricketson in Sida,* 18(2): 515 (1998): Ibarraea wendtii.
zakii *J.J. Pipoly in Sida,* 17(2): 449 (1996)—Ecuador.

ARECA (Palmae).
bifaria *D.R. Hodel in Palm J.,* 136: 7 (1997)—Thailand.
recurvata *D.R. Hodel in Palm J.,* 134: 28 (1997)—Thailand.

ARENARIA (Caryophyllac.).
bryophylla Fernald
var. brevipedicella *R.F. Huang & S.K. Wu in S.G. Wu (S.K. Wu) & Z.J. Feng (eds.), Biol. & Human Physiol. Hoh Xil* Region: 30 (1996)—Qinghai.
chamdoensis *C.Y. Wu ex L.H. Zhou in Fl. Qinghaiica,* 1: 506 (1997)—Xizang, Qinghai.
conferta Boiss.
var. macedonica *K. Micevski in Prilozi Oddel. Biol. Med. Nauki Makedonska Akad. Nauk. Umet.,* 9(1–2): 47 (1988 publ. 1991)—Yugoslavia.
forrestii Diels
f. cernua *(F.N. Williams) C.Y. Wu in Fl. Reipubl. Popul. Sin.,* 26: 204 (1996): A. glanduligera var. cernua.
f. micrantha *(F.N. Williams) C.Y. Wu in Fl. Reipubl. Popul. Sin.,* 26: 204 (1996): A. glanduligera var. micrantha.
geladaindongensis *R.F. Huang & S.K. Wu in S.G. Wu (S.K. Wu) & Z.J. Feng (eds.), Biol. & Human Physiol. Hoh Xil* Region: 33 (1996)—Qinghai.
haitzeshanensis *Tsui ex L.H. Zhou in Acta Biol. Plateau Sin.,* 13: 1 (1995–97 publ. 1997)—China South-central (Sichuan), Xizang.
iochanensis *C.Y. Wu in Fl. Reipubl. Popul. Sin.,* 26: 241 (1996), nom. nov.: Cerastium mairei *H. Lév.*
latisepala *R.F. Huang & S.K. Wu in S.G. Wu (S.K. Wu) & Z.J. Feng (eds.), Biol. & Human Physiol. Hoh Xil* Region: 31 (1996)—Qinghai.
martrinii *N.N. Tzvelev in Novosti Sist. Vyssh. Rast.,* 32: 34 (2000), nom. nov.: A. patula *Martrin-Donos.*
modesta Dufour
subsp. africana *(Pau) A. Dobignard in Candollea,* 52(1): 127 (1997): A. modesta var. africana.
nigricans Hand.-Mazz.
var. zhengkangensis *(C.Y. Wu ex L.H. Zhou) C.Y. Wu in Fl. Reipubl. Popul. Sin.,* 26: 209 (1996): A. zhengkangensis.
runemarkii *D. Phitos in A. Strid & Kit Tan (eds.), Fl. Hellenica,* 1: 166 (1997)—East Aegean Is.
serpyllifolia L.
subsp. lloydii *(Jord.) H.E. Weber in Abh. Naturwiss. Ver. Bremen,* 43(2): 246 (1996): Arenaria lloydii.
taiwanensis *S.S. Ying in Mem. Coll. Agric. Nation. Taiwan Univ.,* 30(2): 56 (1990), as 'taiwanense'—Taiwan.
takasagomontana *(Masam.) S.S. Ying in Fl. Taiwan, ed. 2,* 2: 346 (1996): Cerastium takasagomontana.
xerophila W.W. Sm.
var. xiangchengensis *(L.H. Zhou) C.Y. Wu in Fl. Reipubl. Popul. Sin.,* 26: 210 (1996): A. xiangchengensis.

ARENIFERA (Aizoac.).
pungens *H.E.K. Hartmann in Bradleya,* 14: 37 (1996)—Cape Province.
spinescens *(L. Bolus) H.E.K. Hartmann in Bradleya,* 14: 38 (1996): Ruschia spinescens.
stylosa *(L. Bolus) H.E.K. Hartmann in Bradleya,* 14: 38 (1996): Ruschia stylosa.

ARGOMUELLERA (Euphorbiac.).
pierlotiana *J. Léonard in Bull. Jard. Bot. Nation. Belg.,* 65(1–2): 31 (1996)—Zaire.

ARGOSTEMMA (Rubiac.).
laeve Benn.
subsp. setosum *(E.T. Geddes) K. Sridith in Nordic J. Bot.,* 19(2): 178 (1999): A. setosum.
lobulatum Craib
var. variabile *K. Sridith in Nordic J. Bot.,* 19(2): 177 (1999), sp. author given as 'Geddes' in error—Thailand.
monophyllum *K. Sridith in Nordic J. Bot.,* 19(2): 171 (1999), homonym; *see Ridl. in J. Bot.,* 65: 27 (1927)—Thailand.
puffii *K. Sridith in Nordic J. Bot.,* 19(2): 175 (1999)—Thailand.
rotundicalyx *K. Sridith in Nordic J. Bot.,* 19(2): 173 (1999)—Thailand.
taiwanense *S.S. Ying in Mem. Coll. Agric. Nation. Taiwan Univ.,* 29(2): 56 (1989)—Taiwan.
thaithongiae *K. Sridith in Nordic J. Bot.,* 19(2): 174 (1999), as 'thaithongae'—Thailand.

ARGYRANTHEMUM (Compos.).
pinnatifidum *(L.f.) R.T. Lowe, Man. Fl. Madeira:* 460 (1868): Chrysanthemum pinnatifidum.†
subsp. montanum *Ø.H. Rustan in Bocagiana,* 55: 15 (1981)—Madeira.

ARGYREIA (Convolvulac.).
elongata *L.L. Forman in Checklist Flow. Pl. & Gymn. Brunei Darussalam (ed. M.J.E. Coode et al.):* 435 (1996)—Borneo (Brunei).

ARGYRODERMA (Aizoac.).
theartii *E. Van Jaarsveld in Aloe,* 34(1–2): 12 (1997)—Cape Province.

ARGYROLOBIUM (Legumin.).
angustissimum *(E. Mey.) T.J. Edwards in Bothalia,* 26(1): 45 (1996): Chasmone angustissima.
barikotense *K.H. Rechinger in Fl. Iran.,* 157: 32 (1984)—Pakistan.
itremoense *D.J. Du Puy & J.-N. Labat in Bull. Mus. Nation. Hist. Nat., B, Adansonia,* Ser. 4, 17(3–4): 179 (1995)—Madagascar.
pulvinatum *K.H. Rechinger in Fl. Iran.,* 157: 33 (1984)—Pakistan.
robustum *T.J. Edwards in S. Afr. J. Bot.,* 63(5): 271 (1997)—Transvaal, Natal, Swaziland.
roseum (Cambess.) Jaub. & Spach
var. album *(U.C. Bhattach.) L.B. Chaudhary in J. Econ. Taxon. Bot.,* 21(1): 220 (1997): A. album.
var. elongatum *(Ali) L.B. Chaudhary in J. Econ. Taxon. Bot.,* 21(1): 220 (1997): A. roseum subsp. elongatum.

ARIDARIA (Aizoac.).
noctiflora (L.) Schwantes
subsp. defoliata *(Haw.) M. Gerbaulet in Bot. Jahrb.,* 118(1): 50 (1996): Mesembryanthemum defoliatum.
subsp. straminea *(Haw.) M. Gerbaulet in Bot. Jahrb.,* 118(1): 52 (1996): Mesembryanthemum stramineum.

ARIDARUM (Arac.).
borneense *(M. Hotta) J. Bogner & A. Hay in Telopea,* 9(1): 185 (2000): Heteroaridarum borneense.
incavatum *H. Okada & Y. Mori in Acta Phytotax. Geobot.,* 51(1): 1 (2000)—Borneo (Kalimantan).
rostratum *J. Bogner & A. Hay in Telopea,* 9(1): 191 (2000)—Borneo (Kalimantan).

ARIOCARPUS (Cactac.).
sect. Encephalocarpus *(A. Berger) J.J. Halda in Acta Mus. Richnov. Sect. Nat.,* 5(1): 38 (1998): gen. Encephalocarpus.
subgen. Obregonia *(Frič) J.J. Halda in Acta Mus. Richnov. Sect. Nat.,* 5(1): 37 (1998): gen. Obregonia.
subgen. Pelecyphora *(Ehrenb.) J.J. Halda in Acta Mus. Richnov. Sect. Nat.,* 5(1): 37 (1998): gen. Pelecyphora.
sect. Pelecyphora *(Ehrenb.) J.J. Halda in Acta Mus. Richnov. Sect. Nat.,* 5(1): 38 (1998): gen. Pelecyphora.
subgen. Strombocactus *(Britton & Rose) J.J. Halda in Acta Mus. Richnov. Sect. Nat.,* 5(1): 37 (1998): gen. Strombocactus.

ARIOCARPUS:-
bravoanus H.M. Hern. & E.F. Anderson
subsp. hintonii *(Stuppy & N.P. Taylor) E.F. Anderson & W.A. Fitz Maurice in Haseltonia,* 5: 11 (1997 publ. 1998): A. fissuratus var. hintonii.
confusus *J.J. Halda & Horácek in Cactaceae etc.,* 7(1): 1, 3–4 (1997); *cf. Repert. Pl. Succ. (I.O.S.),* 48: 7 (1997 publ. 1998)—Locality not stated.
disciformis (DC.) W.T. Marshall
subsp. jarmilae *(Halda) J.J. Halda in Acta Mus. Richnov. Sect. Nat.,* 5(1): 37 (1998): Strombocactus jarmilae.
× drabi *J.J. Halda & J. Sladkovský in Acta Mus. Richnov. Sect. Nat.,* 7(1): 33 (2000)—Mexico Northeast (Nuevo León).(A. kotschoubeyanus × retusus).
elongatus *(Salm-Dyck) R. von Wettstein in Kakteenkunde,* 1933(1): 13 (1933): Anhalonium elongatum.†
subsp. horacekii *(Halda) J.J. Halda in Acta Mus. Richnov. Sect. Nat.,* 5(1): 35 (1998): A. trigonus var. horacekii.
fissuratus
subsp. bravoanus *(H.M. Hern. & E.F. Anderson) J.M. Lüthy in Kakt. And. Sukk.,* 50(11): 278 (1999): A. bravoanus.
subsp. hintonii *(Stuppy & N.P. Taylor) J.J. Halda in Acta Mus. Richnov. Sect. Nat.,* 5(1): 36 (1998): A. fissuratus var. hintonii.
subsp. pailanus *J.J. Halda in Acta Mus. Richnov. Sect. Nat.,* 5(5): 184 (1998)—Mexico Northeast (Coahuila).
kotschoubeyanus (Lem.) K. Schum.
subsp. agavoides *(Castañeda) J.J. Halda in Acta Mus. Richnov. Sect. Nat.,* 5(1): 35 (1998): Neogomesia agavoides.
subsp. albiflorus *(Backeb.) C.E. Glass, Guía Identific. Cact. Amenazadas México,* 1: Ar/ko ssp. albiflorus (1998 publ. 1997): A. kotschoubeyanus var. albiflorus.
subsp. bravoanus *(H.M. Hern. & E.F. Anderson) J.J. Halda in Acta Mus. Richnov. Sect. Nat.,* 5(1): 35 (1998): A. bravoanus.
subsp. elephantidens *J.J. Halda in Acta Mus. Richnov. Sect. Nat.,* 5(5): 185 (1998), as comb. nov., with latin descr. & type—Mexico Northeast (Querétaro).
subsp. macdowellii *(Backeb.) J.J. Halda in Acta Mus. Richnov. Sect. Nat.,* 5(5): 184 (1998), without basionym ref., basionym author as "Marsh. ex Krainz": Basionym not stated.
subsp. sladkovskyi *J.J. Halda & L. Horacek in Acta Mus. Richnov. Sect. Nat.,* 5(5): 185 (1998)—Mexico Northeast (San Luis Potosí).
subsp. tulensis *J.J. Halda in Acta Mus. Richnov. Sect. Nat.,* 5(1): 35 (1998); *et in Acta Mus. Richnov. Sect. Nat.,* 5(5): 185 (1998), nom. nov.: Roseocactus kotschoubeyanus var. albiflorus *Backeb.*
pulcherrimus *(Halda) J.J. Halda in Acta Mus. Richnov. Sect. Nat.,* 5(1): 37 (1998): Strombocactus pulcherrimus.
retusus
subsp. confusus *(Halda & Horacek) J.M. Lüthy in Kakt. And. Sukk.,* 50(11): 278 (1999): A. confusus.
subsp. horacekii *J.J. Halda & P. Panarotto in Acta Mus. Richnov. Sect. Nat.,* 5(4): 163 (1998), as 'horaceki'—Mexico Northeast (San Luis Potosí).
subsp. jarmilae *J.J. Halda, L. Horacek & P. Panarotto in Acta Mus. Richnov. Sect. Nat.,* 5(4): 162 (1998)—Mexico Northeast (Nuevo León).
subsp. panarottoi *J.J. Halda & L. Horacek in Acta Mus. Richnov. Sect. Nat.,* 5(4): 163 (1998)—Mexico Northeast (Tamaulipas).
subsp. scapharostroides *J.J. Halda & Horácek in Cactaceae etc.,* 7(3): 97–98 (1997); *cf. Repert. Pl. Succ. (I.O.S.),* 48: 7 (1997 publ. 1998)—Locality not stated.
subsp. sladkovskyi *J.J. Halda & P. Kupčák in Acta Mus. Richnov. Sect. Nat.,* 7(1): 33 (2000)—Mexico Northeast (Coahuila).
subsp. trigonus *(F.A.C. Weber) E.F. Anderson & W.A. Fitz Maurice in Haseltonia,* 5: 18 (1997 publ. 1998): Anhalonium trigonum.

ARIOCARPUS:-
scapharostrus
var. swobodae *J.J. Halda, L. Horacek & P. Panarotto in Acta Mus. Richnov. Sect. Nat.,* 5(4): 163 (1998)—Mexico Northeast (Nuevo León).
trigonus
var. horacekii *J.J. Halda in Cactaceae etc.,* 7(3): 97–99 (1997); *cf. Repert. Pl. Succ. (I.O.S.),* 48: 7 (1997 publ. 1998)—Locality not stated.

ARIPUANA *L. Struwe, P.J.M. Maas & V.A. Albert in Harvard Pap. Bot.,* 2(2): 236 (1997). *GENTIANACEAE*
cullmaniorum *L. Struwe, P.J.M. Maas & V.A. Albert in Harvard Pap. Bot.,* 2(2): 236 (1997)—Brazil North (Amazonas).

ARISAEMA (Arac.).
bogneri *P. Boyce & H. Li in Acta Bot. Yunnanica,* Supp. 11: 55 (1999)—China South-central (Yunnan), Xizang.
burmaense *P. Boyce & H. Li in Acta Bot. Yunnanica,* Supp. 11: 59 (1999)—Burma.
ciliatum H. Li
var. liubaense *G. Gusman & E.J. Gouda in Blumea,* 44(1): 39 (1999)—China South-central (Sichuan).
cochleatum *Stapf ex H. Li in Kew Bull.,* 55(2): 417 (2000)—China North-central (Shanxi).
condaoense *V.D. Nguyen in Aroideana,* 23: 36 (2000), as 'condaoensis'—Vietnam.
consanguineum Schott
subsp. kelung-insularis *(Hayata) G. Gusman in Bull. Jard. Bot. Nation. Belg.,* 67(1–4): 223 (1999): A. kelung-insularis.
echinoides *H. Li in Kew Bull.,* 55(2): 425 (2000)—China South-central (Yunnan).
guixiense *S.Y. Liu in Guihaia,* 15(4): 302 (1995)—China Southeast (Guangxi).
ilanense *J.C. Wang in Bot. Bull. Acad. Sin. (Taipei),* 37(1): 71 (1996)—Taiwan.
jethompsonii *J.G. Thiyagaraj & P. Daniel in Kew Bull.,* 54(2): 483 (1999)—India (Tamil Nadu).
sivadasanii *S.R. Yadav, K.S. Patil & M.K. Janarthanam in Aroideana,* 20: 53 (1997 publ. 1998)—India (Maharashtra).
tengtsungense H. Li
var. pentaphyllum *H. Li in Acta Bot. Yunnanica,* Supp. 11: 58 (1999)—China South-central (Yunnan).
thunbergii
subsp. autumnale *J.C. Wang, J. Murata & Hiroyoshi Ohashi in Bot. Bull. Acad. Sin. (Taipei),* 37(1): 75 (1996)—Taiwan.
victoriae *V.D. Nguyen in Aroideana,* 23: 38 (2000)—Vietnam.
wumengense *H. Li, Q.T. Zhang & L.S. Xie in Kew Bull.,* 55(2): 421 (2000)—China South-central (Yunnan, Sichuan).
xuanweiense *H. Li in Kew Bull.,* 55(2): 419 (2000)—China South-central (Yunnan).
zhui *H. Li in Kew Bull.,* 55(2): 423 (2000)—China South-central (Yunnan).

ARISTEA (Iridac.).
cantharophila *P. Goldblatt & J.C. Manning in Novon,* 7(2): 138 (1997)—Cape Province.
fimbriata *P. Goldblatt & J.C. Manning in Novon,* 7(4): 361 (1997 publ. 1998)—Cape Province.
inaequalis *P. Goldblatt & J.C. Manning in Novon,* 7(4): 357 (1997 publ. 1998)—Cape Province.
ranomafana *P. Goldblatt in Bull. Mus. Nation. Hist. Nat., B, Adansonia,* Ser. 4, 17(3–4): 160 (1995)—Madagascar.
rupicola *P. Goldblatt & J.C. Manning in Novon,* 7(4): 360 (1997 publ. 1998)—Cape Province.
teretifolia *P. Goldblatt & J.C. Manning in Novon,* 7(2): 141 (1997)—Cape Province.

ARISTIDA (Gramin.).
fredscholzii *H. Scholz & Kürschner in Willdenowia,* 30(2): 273 (2000)—Oman.
guayllabambensis *S. Laegaard in Fl. Ecuador,* 57(214:1): 49 (1997)—Ecuador.
gypsophila Beetle
f. diffusa *K.W. Allred & J. Valdés-Reyna in Brittonia,* 49(1): 65 (1997)—Mexico Northeast (Chihuaha, San Luis Potosí).

ARISTIDA:-
gypsophila Beetle
f. gypsophiloides *K.W. Allred & J. Valdés-Reyna in Brittonia,* 49(1): 65 (1997)—Mexico Northeast (Chihuahua, Coahuila, San Luis Potosí), Texas.
humidicola *S.M. Phillips in Kew Bull.,* 55(1): 203 (2000)—Tanzania.
pansa Wooton & Standl.
f. contracta *K.W. Allred & J. Valdés-Reyna in Brittonia,* 49(1): 63 (1997)—Mexico Northeast (Coahuila, Nuevo León, Tamaulipas, Zacatecas).
f. dissita *(I.M. Johnst.) K.W. Allred & J. Valdés-Reyna in Brittonia,* 49(1): 62 (1997): A. dissita.
venustula Arechav.
var. venustuloides *(Caro) H.M. Longhi-Wagner in Bol. Inst. Bot. (São Paulo),* 12: 152 (1999): A. venustuloides.

ARISTOGEITONIA (Euphorbiac.).
magnistipula *A. Radcliffe-Smith in Kew Bull.,* 51(4): 801 (1996)—Tanzania.
uapacifolia *A. Radcliffe-Smith in Kew Bull.,* 53(4): 978 (1998)—Madagascar.

ARISTOLOCHIA (Aristolochiac.).
ser. Hexandrae *(Duch.) F.A. González Garavito in Fl. Colombia,* 12: 78 (1990): A. § Hexandrae.†
subsect. Hexandrae *(Duch.) F.A. González Garavito in Fl. Colombia,* 12: 44 (1990): A. § § Hexandrae.
subser. Hexandrae *(Duch.) F.A. González Garavito in Fl. Colombia,* 12: 82 (1990): A. § Hexandrae.†
andina *F. González & I. Vargas in Kew Bull.,* 55(4): 911 (2000)—Bolivia.
baenzigeri *B. Hansen & L. Phuphathanaphong in Nordic J. Bot.,* 19(5): 575 (1999)—Thailand.
bahiensis *F. González in Brittonia,* 50(1): 8 (1998)—Brazil Northeast (Bahia).
baseri *H. Malyer & S. Erken in Turkish J. Bot.,* 21(6): 381 (1997)—Turkey.
bilabiata L.
subsp. maestrensis *R. Rankin Rodríguez in Rev. Jard. Bot. Nacion. Univ. Habana,* 13: 49 (1992 publ. 1995)—Cuba.
chiapensis *J.F. Ortega Ortiz & R.V. Ortega Ortiz in Acta Bot. Mex.,* 44: 2 (1998)—Mexico Southeast (Chiapas).
colombiana *F. González in Rev. Acad. Colomb. Cienc. Exact. Fis. Nat.,* 90: 52 (2000)—Colombia.
dalyi *F. González in Brittonia,* 50(1): 6 (1998)—Brazil North (Acre), Peru.
davilae *I. Calzada J., G. Flores F. & O. Téllez V.in Contrib. Univ. Mich. Herb.,* 21: 143 (1997)—Mexico Southwest (Nayarit).
dinghoui *Favio González & O. Poncy in Brittonia,* 51(4): 452 (1999)—Thailand.
goliathiana *Michael J. Parsons in Bot. J. Linn. Soc.,* 120(3): 212 (1996)—New Guinea (Papua New Guinea), Solomon Is. (South Solomons).
holostylis *F. González in Rev. Acad. Colomb. Cienc. Exact. Fis. Nat.,* 88: 338 (1999), nom. nov.: Holostylis reniformis *Duch.*
longeracemosa *B. Hansen & L. Phuphathanaphong in Nordic J. Bot.,* 19(5): 577 (1999)—Thailand.
longispathulata *F. González in Novon,* 10(4): 371 (2000)—Brazil Northeast (Bahia).
luzmariana *F.J. Santana-Michel in Bol. Inst. Bot. (Guadalajara),* 3(1–3): 87 (1995 publ. 1997)—Mexico Southwest (Jalisco).
nauseifolia *Michael J. Parsons in Bot. J. Linn. Soc.,* 120(3): 208 (1996)—Queensland, Northern Territory.
occidentalis *F.J. Santana-Michel & S. Lemus Juarez in Acta Bot. Mex.,* 36: 54 (1996)—Mexico Southwest (Colima, Jalisco).
ophioides *L. Marión Heredia in Moscosoa,* 11: 3 (2000)—Dominican Republic.
reniformis *(Duch.) F. González in Caldasia,* 19(1–2): 124 (1997), homonym; *see Vell., Fl. Flumin.,* 9: t. 100 (1827): Holostylis reniformis.
rojasiana *(Chodat & Hassl.) F. González in Caldasia,* 19(1–2): 123 (1997): Euglypha rojasiana.
schottii *L. Marión Heredia in Moscosoa,* 11: 1 (2000)—Dominican Republic.

ARMATOCEREUS (Cactac.).
mataranus
subsp. ancashensis *(F. Ritter) C. Ostolaza Nano in Cactaceae Consensus Initiatives,* 6: 8 (1998): A. mataranus var. ancashensis.
rauhii
subsp. balsasensis *(F. Ritter) C. Ostolaza Nano in Cactaceae Consensus Initiatives,* 6: 8 (1998): A. balsasensis.

ARMENIACA (Rosac.).
hypotrichodes *(Cardot) C.L. Li & S.Y. Jiang in Acta Phytotax. Sin.,* 36(4): 367 (1998): Prunus hypotrichodes.
limeixing *J.Y. Zhang & Z.Ming Wang in Acta Phytotax. Sin.,* 37(1): 107 (1999)—Manchuria (Liaoning).
vulgaris Lam.
var. zhidanensis *(C.Z. Qiao & Y.P. Zhu) L.T. Lu in Acta Phytotax. Sin.,* 38(3): 281 (2000): A. zhidanensis.
zhengheensis *J.Y. Zhang & M.N. Lu in Acta Phytotax. Sin.,* 37(1): 105 (1999)—China Southeast (Fujian).

ARMERIA (Plumbaginac.).
alliacea (Cav.) Hoffmanns. & Link
var. bilbilitana *(Bernis) O. de Bolòs & J. Vigo, Fl. Països Catalans,* 3: 97 (1995): A. maritima var. bilbilitana.
subsp. loscosii *A.M. Romo in Estud. Mus. Cienc. Nat. Alava,* 10–11: 198 (1995–96 publ. 1996)—Spain.
aspromontana *S. Brullo, F. Scelsi & G. Spampinato in Edinburgh J. Bot.,* 54(1): 91 (1997)—Italy.
filicaulis (Boiss.) Boiss.
subsp. nevadensis *G. Nieto Feliner, J.A. Rosselló & J. Fuertes in An. Jard. Bot. Madrid,* 56(1): 163 (1998)—Spain.
masguindalii *(Pau) G. Nieto Feliner in Lagascalia,* 18(2): 236 (1996): A. plantaginea var. masguindalii.
rivasmartinezii *S. Sardinero & G. Nieto Feliner in Collect. Bot.,* 23: 98 (1997 publ. 1998)—Spain.
rumelica Boiss.
f. temskyana *(Degen & Dörfl.) K. Micevski, Fl. Republ. Macedonia,* 1(3): 478 (1995 publ. 1996), without basionym ref.: Basionym not stated.
scabra Pall. ex Roem. & Schult.
var. glabricalyx *N.N. Tzvelev in Fl. Vostochnoĭ Evropy̆,* 9: 163 (1996)—Russia North.
var. labradorica *(Wallr.) N.N. Tzvelev in Fl. Vostochnoĭ Evropy̆,* 9: 163 (1996): A. labradorica.

ARNICA (Compos.).
dealbata *(A. Gray) B.G. Baldwin in Novon,* 9(4): 460 (1999): Whitneya dealbata.

ARNOGLOSSUM (Compos.).
album *L.C. Anderson in Sida,* 18(2): 378 (1998)—Florida.

ARPOPHYLLUM (Orchidac.).
giganteum Hartw. ex Lindl.
subsp. alpinum *(Lindl.) R.L. Dressler in Novon,* 10(3): 193 (2000): A. alpinum.
subsp. medium *(Rchb. f.) R.L. Dressler in Novon,* 10(3): 193 (2000): A. medium.

ARRHENATHERUM (Gramin.).
baeticum *(Romero Zarco) S. Brullo, P. Minissale & G. Spampinato in Lagascalia,* 19(1–2): 908 (1997): A. elatius subsp. baeticum.
nebrodense *S. Brullo, P. Minissale & G. Spampinato in Lagascalia,* 19(1–2): 903 (1997)—Sicilia.
sardoum *(Em. Schmid) S. Brullo, P. Minissale & G. Spampinato in Lagascalia,* 19(1–2): 908 (1997): A. elatius var. sardoum.

ARROJADOA (Cactac.).
dinae
subsp. eriocaulis *(Buining & Brederoo) N.P. Taylor & D.C. Zappi in Cactaceae Consensus Initiatives,* 3: 7 (1997): A. eriocaulis.
heimenii *W. van Heek & W. Strecker in Kakt. And. Sukk.,* 50(6): 137 (1999)—Brazil Southeast (Minas Gerais).

ARTEMISIA (Compos.).
subsect. Annua *(Rydb.) I.M. Krasnoborov in Fl. Sibir. (Asterac.),* 13: 116 (1997 publ. 1998), without basionym ref.: Basionym not stated.
subsect. Auratae *(Poljakov) I.M. Krasnoborov in Fl. Sibir. (Asterac.),* 13: 116 (1997 publ. 1998), without basionym ref.: Basionym not stated.
abbreviata *(Krasch. ex Korobkov) I.M. Krasnoborov in Fl. Sibir. (Asterac.),* 13: 114 (1997 publ. 1998): A. lagopus subsp. abbreviata.
alba Turra
subsp. glabrescens *(Willk.) E. Valdés-Bermejo in Soc. Ech. Pl. Vasc. Eur. Bassin Médit. Bull.,* 20: 55 (1985): A. incanescens var. glabrescens.
bhutanica *A.J.C. Grierson & L.S. Springate in Edinburgh J. Bot.,* 57(3): 400 (2000)—Bhutan.
borealis Pall.
var. bargusinensis *(Spreng.) V.N. Voroschilov in A.K. Skvortsov (ed.), Florist. issl. v razn. raĭonakh SSSR:* 195 (1985): A. bargusinensis.
desertorum Spreng.
var. lineatoangustissima *G.Yi Zhang in G.Y. Zhang et al., Monogr. Artemisia (Compos.) N.E. China:* 154 (1996), as 'lineato-angustissima', without type—Manchuria (Heilongjiang).
var. longiflora *G.Yi Zhang in G.Y. Zhang et al., Monogr. Artemisia (Compos.) N.E. China:* 154 (1996), as 'longiffora', without type—Inner Mongolia (Nei Mongol).
feddei H. Lév. & Vaniot
subsp. arschantinica *(Darijma) I.A. Gubanov & R.V. Kamelin in I.A. Gubanov, Konsp. Fl. Vneshnei Mongolii (Sosud. Rast.):* 97 (1996): A. arschantinica.
filatovae *A.N. Kuprianov in Bot. Zhurn.,* 80(7): 83 (1995)—Kazakhstan.
finita *M. Kitagawa in Rep. Inst. Sci. Res. Manchoukuo,* 6(4): 124 (1942)—Manchuria.
frigida Willd.
f. atropurpurea *(Pamp.) W. Wang & C.Y. Li in Clavis Pl. Chinae Bor.-Or. ed. 2 (ed. P.Y. Fu):* 666 (1995): A. frigida var. atropurpurea.
furcata M. Bieb.
subsp. insulana *(Krasch.) V.N. Voroschilov in A.K. Skvortsov (ed.), Florist. issl. v razn. raĭonakh SSSR:* 196 (1985): A. insulana.
gallica Willd.
subsp. densiflora *(Viv.) J. Gamisans in J. Gamisans & D. Jeanmonod, Asteraceae I: Compl. Prodr. Fl. Corse (ser.* eds. Jeanm. & Burdet): 246 (1998): A. densiflora.
var. gargantae *(Vallès-Xirau & Seoane-Camba) O. de Bolòs & J. Vigo, Fl. Països Catalans,* 3: 826 (1995), without exact basionym page: A. caerulescens subsp. gargantae.
gmelinii Web. ex Stechm.
var. minor *(Ledeb.) B.D. Naithani in Fl. India,* 12: 24 (1995): A. sacrorum var. minor.
var. vestita *(Wall. ex DC.) B.D. Naithani in Fl. India,* 12: 24 (1995): A. vestita.
herba-alba Asso
var. communis *A. Ferchichi in Acta Bot. Gallica,* 144(1): 153 (1997 publ. 1998)—Tunisia.
var. desertii *A. Ferchichi in Acta Bot. Gallica,* 144(1): 154 (1997 publ. 1998)—Tunisia.
integrifolia L.
var. latifolia *G.Yi Zhang in G.Y. Zhang et al., Monogr. Artemisia (Compos.) N.E. China:* 153 (1996)—Inner Mongolia (Nei Mongol).
f. sublata *G.Yi Zhang in G.Y. Zhang et al., Monogr. Artemisia (Compos.) N.E. China:* 154 (1996)—Manchuria (Heilongjiang).
iwayomogi Kitam.
var. discolor *(Torr. & A. Gray) Y.N. Lee, Fl. Korea:* 1162 (1996), without basionym date: A. discolor.
jordanica *A. Danin in Willdenowia,* 29(1–2): 147 (1999)—Palestine (Jordan), Saudi Arabia, Iraq.
kotuchovii *A.N. Kuprianov in Bot. Zhurn.,* 84(4): 114 (1999)—Kazakhstan.
latifolia Ledeb.
subsp. maximowiczii *(F. Schmidt) V.N. Voroschilov in A.K. Skvortsov (ed.), Florist. issl. v razn. raĭonakh SSSR:* 196 (1985): A. latifolia var. maximowiczii.
marschalliana Spreng.
f. gracilis *H.V. Savczenko in Ukr. Bot. Zhurn.,* 56(3): 284 (1999)—Ukraina.

ARTEMISIA:-
obtusiloba Ledeb.
subsp. altaiensis *(Krasch.) I.M. Krasnoborov in Fl. Sibir. (Asterac.),* 13: 124 (1997 publ. 1998): A. altaiensis.
subsp. martjanovii *(Krasch. ex Poljakov) I.M. Krasnoborov in Fl. Sibir. (Asterac.),* 13: 124 (1997 publ. 1998): A. martjanovii.
subsp. subviscosa *(Turcz. ex Besser) I.M. Krasnoborov in Fl. Sibir. (Asterac.),* 13: 126 (1997 publ. 1998): A. subviscosa.
orthobotrys *M. Kitagawa in Rep. Inst. Sci. Res. Manchoukuo,* 6(4): 125 (1942)—North Korea, Manchuria.
pineticola *A.N. Kuprianov in Bot. Zhurn.,* 84(4): 115 (1999)—Altay.
pronutans *M. Kitagawa in Rep. Inst. Sci. Res. Manchoukuo,* 6(4): 126 (1942)—Manchuria, North Korea.
radicans *A.N. Kuprianov in Bot. Zhurn.,* 80(7): 82 (1995)—Kazakhstan.
sacrorum Ledeb.
subsp. iwayomogi *(Kitam.) V.N. Voroschilov in Byull. Mosk. Obshch. Ispy̆t. Prir., Biol.,* 102(1): 68 (1997): A. iwayomogi.
santolinifolia Turcz. ex Besser
var. scheludjakoviae *(Korobkov) I.M. Krasnoborov in Fl. Sibir. (Asterac.),* 13: 107 (1997 publ. 1998): A. gmelinii subsp. scheludjakoviae.
sieversiana Ehrh. ex Willd.
var. koreana *(Nakai) W. Wang & C.Y. Li in Clavis Pl. Chinae Bor.-Or. ed. 2 (ed. P.Y. Fu):* 663 (1995): A. koreana.
stolonifera *(Maxim.) V.L. Komarov, Fl. Manshur.,* 3: 676 (1907): A. vulgaris var. stolonifera.
var. laciniata *G.Yi Zhang in G.Y. Zhang et al., Monogr. Artemisia (Compos.) N.E. China:* 153 (1996)—Manchuria (Heilongjiang).
vulgaris L.
var. articulatopilosa *G.A. Peschkova in Fl. Tsentral'noi Sibiri,* 2: 859 (1979)—Irkutsk.
wallichiana Besser
f. nitida *(Pamp.) B.D. Naithani in Fl. India,* 12: 46 (1995): A. moorcroftiana f. nitida.
× wurzellii *C.M. James & C.A. Stace in Watsonia,* 23(1): 145 (2000)—Great Britain. (A. verlotiorum × vulgaris).

ARTHROCEREUS (Cactac.).
melanurus
subsp. magnus *N.P. Taylor & D.C. Zappi in Cactaceae Consensus Initiatives,* 3: 7 (1997)—Brazil Southeast (Minas Gerais).
subsp. odorus *(F. Ritter) N.P. Taylor & D.C. Zappi in Cactaceae Consensus Initiatives,* 3: 7 (1997): A. odorus.

ARTHROCHILUS (Orchidac.).
huntianus (F. Muell.) Blaxell
subsp. nothofagicola *D.L. Jones in Austral. Orchid Res.,* 3: 4 (1998)—Tasmania.

ARTHROPOGON (Gramin.).
rupestris *T.S. Filgueiras in Nordic J. Bot.,* 16(1): 69 (1996)—Brazil West-central (Goiás).

ARTHROSTYLIDIUM (Gramin.).
auriculatum *X. Londoño & L.G. Clark in Novon,* 8(4): 410 (1998)—Colombia.
canaliculatum *S.A. Renvoize, Gramineas de Bolivia:* 33 (1998)—Bolivia.
chiribiquetense *X. Londoño & L.G. Clark in Novon,* 8(4): 412 (1998), as 'chiribiquetensis'—Colombia.
punctulatum *X. Londoño & L.G. Clark in Novon,* 8(4): 415 (1998)—Colombia.
virolinense *X. Londoño & L.G. Clark in Novon,* 8(4): 417 (1998), as 'virolinensis'—Colombia.

ARTOCARPUS (Morac.).
anisophyllus
var. sessilifolius *K.M. Kochummen in Gard. Bull. Singapore,* 50(2): 200 (1998)—Borneo (Sabah, Sarawak).
corneri *K.M. Kochummen in Gard. Bull. Singapore,* 50(2): 197 (1998)—Borneo (Sarawak).
jarrettiae *K.M. Kochummen in Gard. Bull. Singapore,* 50(2): 198 (1998)—Borneo (Sabah, Sarawak).

ARTOCARPUS:-
primackii *K.M. Kochummen in Gard. Bull. Singapore,* 50(2): 199 (1998)—Borneo (Sabah, Sarawak).

ARUM (Arac.).
campanulatum *W. Roxburgh, Pl. Coromandel,* 3: 68 (1820)—India (West Bengal).†
majoricum Chodat
subsp. lucanum *(Bonafé) A.M. Romo, Fl. Silvestres Baleares:* 391 (1994), basionym not validly publ: A. lucanum.
pictum L.f.
subsp. sagittifolium *J.A. Rosselló & L. Sáez in Acta Bot. Barcin.,* 44: 170 (1997)—Baleares.

ARUNCUS (Rosac.).
dioicus (Walter) Fernald
subsp. vulgaris *(Raf.) N.N. Tzvelev in Novosti Sist. Vyssh. Rast.,* 32: 184 (2000): A. vulgaris.

ARUNDINARIA (Gramin.).
bicorniculata *W.T. Lin & Z.J. Feng in J. S. China Agric. Univ.,* 14(1): 50 (1993)—China Southeast (Guangdong).
conghuaensis *W.T. Lin in J. Bamboo Res.,* 19(4): 2 (2000)—China Southeast (Guangdong).
hamadae *(Hatus.) J.-P. Demoly in Bambou,* 21: 14 (1995): Pseudosasa hamadae.
herklotsii *(McClure) J.-P. Demoly, Bambous en France:* 76 (1996): Indocalamus herklotsii.
heterolodicula *(W.T. Lin & Z.J. Feng) W.T. Lin in J. Bamboo Res.,* 19(1): 6 (2000): Oligostachyum heterolodiculum.
kodzumae *(Makino) J.-P. Demoly in Bambou,* 21: 14 (1995): Pleioblastus kodzumae.
multinervis *W.T. Lin & Z.M. Wu in J. S. China Agric. Univ.,* 11(3): 48 (1990)—China Southeast (Guangdong).
munsuensis *Y.N. Lee in Korean J. Pl. Taxon.,* 28(1): 33 (1998)—South Korea.
orthotropoides *(W.T. Lin) W.T. Lin in J. Bamboo Res.,* 19(1): 8 (2000): Oligostachyum orthotropoides.
tenuivagina *W.T. Lin & Z.M. Wu in J. S. China Agric. Univ.,* 11(3): 48 (1990), as 'tenui-vagina'—China Southeast (Guangdong).
yangshanensis *W.T. Lin in J. Bamboo Res.,* 16(3): 26 (1997)—China Southeast (Guangdong).

ARUNDINELLA (Gramin.)
dagana *H.J. Noltie in Edinburgh J. Bot.,* 56(3): 396 (1999)—Bhutan.

ASARUM (Aristolochiac.).
caudigerum Hance
var. leptophyllum *(Hayata) S.S. Ying in Mem. Coll. Agric. Nation. Taiwan Univ.,* 30(2): 64 (1990): A. leptophyllum.
var. triangulare *(Hayata) S.S. Ying in Mem. Coll. Agric. Nation. Taiwan Univ.,* 31(1): 33 (1991): A. leptophyllum var. triangulare.
ikegamii *(F. Maek. ex Y. Maek.) T. Sugawara in Acta Phytotax. Geobot.,* 49(1): 15 (1998): Heterotropa ikegamii.
var. fujimakii *T. Sugawara in Acta Phytotax. Geobot.,* 49(1): 15 (1998)—Japan.
misandrum *B.U. Oh & J.G. Kim in Korean J. Pl. Taxon.,* 27(4): 493 (1997)—South Korea.
mitoanum *T. Sugawara in Acta Phytotax. Geobot.,* 47(2): 136 (1996 publ. 1997)—Japan.
taiwanense *S.S. Ying in Mem. Coll. Agric. Nation. Taiwan Univ.,* 30(2): 54 (1990)—Taiwan.
unzen (F. Maek.) Kitam. & Murata
f. luteoviride *K. Yonekura & H. Ohashi in J. Jap. Bot.,* 72(5): 309 (1997)—Japan.
yoshikawae *T. Sugawara in Acta Phytotax. Geobot.,* 49(1): 14 (1998)—Japan.

ASCLEPIAS (Asclepiadac.).
jorgeana *M. Fishbein & S.P. Lynch in Novon,* 9(2): 179 (1999)—Mexico Northeast (Chihuahua, Durango, Guanajuato), Mexico Northwest (Sonora).
lynchiana *M. Fishbein in Novon,* 7(3): 234 (1997), nom. nov.: A. grandiflora *E. Fourn.*
tingens *W. Roxburgh, Pl. Coromandel,* 3: 34 (1815)—Burma.†
viridiflora *(E. Mey.) D.J. Goyder in Kew Bull.,* 52(1): 247 (1997), homonym; *see Raf. in Med. Repos.,* Hex. 2, 5: 360 (1808): Pachycarpus viridiflorus.

ASCOCENTROPSIS *K. Senghas & H. Schildhauer in J. Orchideenfreund,* 7(4): 289 (2000). *ORCHIDACEAE*
pusilla *(Aver.) K. Senghas & H. Schildhauer in J. Orchideenfreund,* 7(4): 290 (2000), as 'pusillum': Ascocentrum pusillum.

ASIASARUM (Aristolochiac.).
patens *K. Yamaki in J. Jap. Bot.,* 71(1): 3 (1996)—South Korea.
sieboldii (Miq.) F. Maek.
var. versicolor *K. Yamaki in J. Jap. Bot.,* 71(1): 1 (1996)—South Korea.

ASIPHONIA *W. Griffith in Proc. Linn. Soc.,* 1(22): 218 (1844). *ARISTOLOCHIACEAE*†
piperiformis *W. Griffith in Proc. Linn. Soc.,* 1(22): 218 (1844)—Malaya (Peninsular Malaysia).†

ASPARAGUS (Asparagac.).
acutifolius
var. achhalii *B. Valdés in Lagascalia,* 18(2): 327 (1996)—Morocco.
aggregatus *(Oberm.) A.C. Fellingham & N.L. Meyer in Bothalia,* 25(2): 206 (1995): Protasparagus aggregatus.
angusticladus *(Jessop) A.C. Fellingham & N.L. Meyer in Bothalia,* 25(2): 206 (Oct. 1995), isonym; *see J.-P. Lebrun & Stork, Enum. Pl. Fleurs Afr. Trop.,* 3: 60 (Feb. 1995): A. aethiopicus var. angusticladus.
bayeri *(Oberm.) A.C. Fellingham & N.L. Meyer in Bothalia,* 25(2): 206 (1995): Protasparagus bayeri.
biflorus *(Oberm.) A.C. Fellingham & N.L. Meyer in Bothalia,* 25(2): 206 (1995): Protasparagus biflorus.
clareae *(Oberm.) A.C. Fellingham & N.L. Meyer in Bothalia,* 25(2): 206 (1995): Protasparagus clareae.
coddii *(Oberm.) A.C. Fellingham & N.L. Meyer in Bothalia,* 25(2): 206 (1995): Protasparagus coddii.
devenishii *(Oberm.) A.C. Fellingham & N.L. Meyer in Bothalia,* 25(2): 206 (1995): Protasparagus devenishii.
divaricatus *(Oberm.) A.C. Fellingham & N.L. Meyer in Bothalia,* 25(2): 206 (1995): Protasparagus divaricatus.
edulis *(Oberm.) A.C. Fellingham & N.L. Meyer in Bothalia,* 25(2): 206 (Oct. 1995), isonym; *see J.-P. Lebrun & Stork, Enum. Pl. Fleurs Afr. Trop.,* 3: 61 (Feb. 1995): Protasparagus edulis.
exsertus *(Oberm.) A.C. Fellingham & N.L. Meyer in Bothalia,* 25(2): 206 (1995): Protasparagus exsertus.
f. ecklonii *(Baker) A.C. Fellingham & N.L. Meyer in Bothalia,* 25(2): 207 (1995): A. ecklonii.
filicladus *(Oberm.) A.C. Fellingham & N.L. Meyer in Bothalia,* 25(2): 207 (1995): Protasparagus filicladus.
flavicaulis *(Oberm.) A.C. Fellingham & N.L. Meyer in Bothalia,* 25(2): 207 (1995): Protasparagus flavicaulis.
subsp. setulosus *(Oberm.) A.C. Fellingham & N.L. Meyer in Bothalia,* 25(2): 207 (1995): Protasparagus flavicaulis subsp. setulosus.
fouriei *(Oberm.) A.C. Fellingham & N.L. Meyer in Bothalia,* 25(2): 207 (1995): Protasparagus fouriei.
fractiflexus *(Oberm.) A.C. Fellingham & N.L. Meyer in Bothalia,* 25(2): 207 (1995): Protasparagus fractiflexus.
graniticus *(Oberm.) A.C. Fellingham & N.L. Meyer in Bothalia,* 25(2): 207 (1995): Protasparagus graniticus.
hajrae *S.Y. Kamble in J. Econ. Taxon. Bot.,* 20(3): 688 (1996)—Assam (Meghalaya).
intricatus *(Oberm.) A.C. Fellingham & N.L. Meyer in Bothalia,* 25(2): 207 (1995): Protasparagus intricatus.
lynetteae *(Oberm.) A.C. Fellingham & N.L. Meyer in Bothalia,* 25(2): 207 (1995): Protasparagus lynetteae.
mariae *(Oberm.) A.C. Fellingham & N.L. Meyer in Bothalia,* 25(2): 207 (1995): Protasparagus mariae.

ASPARAGUS:-
mollis *(Oberm.) A.C. Fellingham & N.L. Meyer in Bothalia,* 25(2): 207 (1995): Protasparagus mollis.
natalensis *(Baker) A.C. Fellingham & N.L. Meyer in Bothalia,* 25(2): 208 (Oct. 1995), isonym; *see J.-P. Lebrun & Stork, Enum. Pl. Fleurs Afr. Trop.,* 3: 61 (Feb. 1995): A. aethiopicus var. natalensis.
nodulosus *(Oberm.) A.C. Fellingham & N.L. Meyer in Bothalia,* 25(2): 208 (Oct. 1995), isonym; *see J.-P. Lebrun & Stork, Enum. Pl. Fleurs Afr. Trop.,* 3: 208 (Feb. 1995): Protasparagus nodulosus.
officinalis L.
subsp. polyphyllus *(Steven) N.N. Tzvelev in Novosti Sist. Vyssh. Rast.,* 32: 183 (2000): A. polyphyllus.
oliveri *(Oberm.) A.C. Fellingham & N.L. Meyer in Bothalia,* 25(2): 208 (1995): Protasparagus oliveri.
pendulus *(Oberm.) A.C. Fellingham & N.L. Meyer in Bothalia,* 25(2): 208 (Oct. 1995), isonym; *see J.-P. Lebrun & Stork, Enum. Pl. Fluers Afr. Trop.,* 3: 61 (Feb. 1995): Protasparagus pendulus.
polyphyllus *C. von Steven in Bull. Soc. Imp. Naturalistes Mosc.,* 30(3): 91 (1857)—Krym.†
recurvispinus *(Oberm.) A.C. Fellingham & N.L. Meyer in Bothalia,* 25(2): 208 (1995): Protasparagus recurvispinus.
sekukuniensis *(Oberm.) A.C. Fellingham & N.L. Meyer in Bothalia,* 25(2): 208 (1995): Protasparagus sekukuniensis.
transvaalensis *(Oberm.) A.C. Fellingham & N.L. Meyer in Bothalia,* 25(2): 208 (1995): Protasparagus transvaalensis.

ASPERULA (Rubiac.).
subsect. Graveolentes *(Klokov) S.K. Pyatunina in Byull. Mosk. Obshch. Ispyt. Prir., Biol.,* 100(4): 75 (1995): A. ser. Graveolentes.
cynanchica L.
subsp. rumelica *(Boiss.) S.K. Pyatunina in Byull. Mosk. Obshch. Ispyt. Prir., Biol.,* 100(4): 71 (1995): A. rumelica.
exasperata V.I. Krecz. ex Klokov
subsp. petraea *(V.I. Krecz. ex Klokov) S.K. Pyatunina in Byull. Mosk. Obshch. Ispyt. Prir., Biol.,* 100(4): 73 (1995): A. petraea.
subsp. tephrocarpa *(Czern. ex Popov & Chrshan.) S.K. Pyatunina in Byull. Mosk. Obshch. Ispyt. Prir., Biol.,* 100(4): 72 (1995): A. tephrocarpa.
graveolens M. Bieb. ex Schult. & Schult. f.
subsp. danilewskiana *(Basiner) S.K. Pyatunina in Byull. Mosk. Obshch. Ispyt. Prir., Biol.,* 100(4): 77 (1995): A. danilewskiana.
subsp. leiograveolens *(Popov & Chrshan.) S.K. Pyatunina in Byull. Mosk. Obshch. Ispyt. Prir., Biol.,* 100(4): 76 (1995): A. leiograveolens.
naufraga *F. Ehrendorfer & W. Gutermann in Bot. Chronika,* 13: 62 (2000)—Greece.
paui Font Quer
subsp. dianensis *(Font Quer) A. De la Torre, F. Alcaraz & M.B. Crespo in Lazaroa,* 16: 154 (1995 publ. 1996): A. paui var. dianensis.
supina M. Bieb.
subsp. caespitans *(Juz.) S.K. Pyatunina in Byull. Mosk. Obshch. Ispyt. Prir., Biol.,* 100(4): 75 (1995): A. caespitans.
taurina L.
subsp. caucasica *(Woronow ex Pobed.) A.G. Jelenevsky & S.K. Pyatunina in Byull. Mosk. Obshch. Ispyt. Prir., Biol.,* 100(4): 70 (1995), isonym; *see Ehrend. in Notes Roy. Bot. Gard. Edinburgh,* 37(2): 248 (1979): A. caucasica.
tephrocarpa *Czern. ex M.G. Popov & V.G. Khrzhanovskiĭ in Byull. Moskovsk. Obshch. Isp. Prir., Otd. Biol.,* 50(5-6): 96 (1945)—Ukraina.

ASPHODELINE (Asphodelac.).
sect. Appendicigera *E. Tuzlacı in Candollea,* 53(2): 424 (1998).
sertachiae *E. Tuzlacı in Candollea,* 53(2): 424 (1998), as 'sertachae'—Turkey.
turcica *E. Tuzlacı in Candollea,* 53(2): 426 (1998)—Turkey.

ASPHODELUS (Asphodelac.).
albus
subsp. carpetanus *Z. Díaz Lifante & B. Valdés in Lagascalia,* 17(2): 274 (1994)—Spain.

ASPHODELUS:-
albus Mill.
subsp. delphinensis *(Gren. & Godr.) Z. Díaz Lifante & B. Valdés in Boissiera,* 52: 95 (1996): A. delphinensis.
subsp. occidentalis *(Jord.) Z. Díaz Lifante & B. Valdés in Boissiera,* 52: 99 (1996): A. occidentalis.
bento-rainhae P. Silva
subsp. salmanticus *Z. Díaz Lifante & B. Valdés in Boissiera,* 52: 105 (1996)—Spain.
fistulosus L.
subsp. cirerae *(Sennen) A.M. Romo, Fl. Silvestres Baleares:* 323 (1994): A. cirerae.
lusitanicus Cout.
var. ovoideus *(Merino) Z. Díaz Lifante & B. Valdés in Boissiera,* 52: 71 (1996): A. chambeironi var. ovoideus.
macrocarpus Parl.
var. arrondeaui *(J. Lloyd) Z. Díaz Lifante & B. Valdés in Boissiera,* 52: 126 (1996): A. arrondeaui.
subsp. rubescens *Z. Díaz Lifante & B. Valdés in Boissiera,* 52: 129 (1996)—Spain, Morocco.
ramosus L.
var. africanus *(Jord.) Z. Díaz Lifante & B. Valdés in Boissiera,* 52: 56 (1996): A. africanus.
subsp. distalis *Z. Díaz Lifante & B. Valdés in Boissiera,* 52: 61 (1996)—Spain, Portugal, Canary Is., Morocco.
var. nervosus *(Pomel) Z. Díaz Lifante & B. Valdés in Boissiera,* 52: 59 (1996): A. nervosus.

ASPIDISTRA (Convallariac.).
ser. Acetabuliformes *G.Z. Li in Guihaia,* 20(3): 214 (2000).
ser. Austrosinenses *G.Z. Li in Guihaia,* 20(3): 214 (2000).
ser. Cavicolae *G.Z. Li in Guihaia,* 20(3): 212 (2000).
ser. Cruciformes *G.Z. Li in Guihaia,* 20(3): 215 (2000).
ser. Fimbriatae *G.Z. Li in Guihaia,* 20(3): 215 (2000).
ser. Flaviflorae *G.Z. Li in Guihaia,* 20(3): 213 (2000).
ser. Fungiliformes *G.Z. Li in Guihaia,* 20(3): 212 (2000).
ser. Leyeenses *G.Z. Li in Guihaia,* 20(3): 216 (2000).
ser. Longilobae *G.Z. Li in Guihaia,* 20(3): 217 (2000).
ser. Luodianenses *G.Z. Li in Guihaia,* 20(3): 216 (2000), as 'Luodianensis'.
ser. Luridae *G.Z. Li in Guihaia,* 20(3): 215 (2000).
sect. Luridae *G.Z. Li in Guihaia,* 20(3): 214 (2000).
ser. Minutiflorae *G.Z. Li in Guihaia,* 20(3): 212 (2000).
ser. Oblanceifoliae *G.Z. Li in Guihaia,* 20(3): 214 (2000).
ser. Oblongifoliae *G.Z. Li in Guihaia,* 20(3): 212 (2000), as 'Oblogifoliae'.
ser. Patentilobae *G.Z. Li in Guihaia,* 20(3): 216 (2000).
sect. Primihenae *G.Z. Li in Guihaia,* 20(3): 211 (2000).
ser. Retusae *G.Z. Li in Guihaia,* 20(3): 213 (2000).
ser. Sexexcavatae *G.Z. Li in Guihaia,* 20(3): 213 (2000).
sect. Trilineares *G.Z. Li in Guihaia,* 20(3): 211 (2000).
ser. Yingjiangenses *G.Z. Li in Guihaia,* 20(3): 213 (2000).
ebianensis *K.Y. Lang & Z.Y. Zhu in Acta Phytotax. Sin.,* 37(5): 492 (1999)—China South-central (Sichuan).
fasciaria *G.Z. Li in Acta Phytotax. Sin.,* 37(5): 484 (1999)—China Southeast (Guangxi).
fenghuangensis *K.Y. Lang in Acta Phytotax. Sin.,* 37(5): 494 (1999)—China Southeast (Hunan).
hekouensis *H. Li, C.L. Long & J. Bogner in Sendtnera,* 5: 15 (1998)—China South-central (Yunnan).
oblongifolia *F.T. Wang & K.Y. Lang in Acta Phytotax. Sin.,* 37(5): 476 (1999)—China Southeast (Guangxi).

ASPIDOGYNE (Orchidac.).
decora *(Rchb. f.) L.A. Garay & G.A. Romero-González in Harvard Pap. Bot.,* 3(1): 53 (1998): Physurus decorus.

ASPIDOGYNE:-
utriculata *(Dressler) D.L. Szlachetko in Fragm. Flor. Geobot.,* Supp. 3: 115 (1995): Erythrodes utriculata.

ASPIDOPTERYS (Malpighiac.).
glabriuscula
var. lohitensis *R.C. Srivastava in Nation. Acad. Sci. Lett.,* 6(7): 223 (1983)—Assam (Arunachal Pradesh).
microcarpa *H.W. Li ex S.K. Chen in Acta Bot. Yunnanica,* 18(4): 405 (1996)—China Southeast (Guangxi), Vietnam.

ASPIDOSPERMA (Apocynac.).
subgen. Coutinia *(Vell.) W. Marcondes-Ferreira in Rev. Brasil. Bot.,* 19(2): 213 (1996 publ. 1997): gen. Coutinia.
sect. Excelsa *W. Marcondes-Ferreira in Rev. Brasil. Bot.,* 19(2): 212 (1996 publ. 1997).
sect. Inundata *W. Marcondes-Ferreira in Rev. Brasil. Bot.,* 19(2): 211 (1996 publ. 1997).
sect. Nobilia *(Woodson) W. Marcondes-Ferreira in Rev. Brasil. Bot.,* 19(2): 210 (1996 publ. 1997): A. ser. Nobiles.
sect. Polyneura *(Woodson) W. Marcondes-Ferreira in Rev. Brasil. Bot.,* 19(2): 211 (1996 publ. 1997): A. ser. Polyneura.
sect. Pungentia *(Pichon) W. Marcondes-Ferreira in Rev. Brasil. Bot.,* 19(2): 207 (1996 publ. 1997): A. ser. Pungentia.
sect. Ramiflora *(Woodson) W. Marcondes-Ferreira in Rev. Brasil. Bot.,* 19(2): 210 (1996 publ. 1997): A. ser. Ramiflora.
sect. Rigida *(Woodson) W. Marcondes-Ferreira in Rev. Brasil. Bot.,* 19(2): 212 (1996 publ. 1997): A. ser. Rigida.
sect. Schultesia *W. Marcondes-Ferreira in Rev. Brasil. Bot.,* 19(2): 212 (1996 publ. 1997).
thomasii *W. Marcondes-Ferreira in Brittonia,* 51(1): 74 (1999)—Brazil Northeast (Bahia).

ASPILIA (Compos.).
albuquerquei *J.U.M. Santos in Bol. Mus. Paraense Emílio Goeldi, Bot.,* 9(2): 209 (1993 publ. 1994)—Brazil North (Tocantins).
cearensis *J.U.M. Santos in Bol. Mus. Paraense Emílio Goeldi, Bot.,* 9(2): 210 (1993 publ. 1994)—Brazil Northeast (Ceará).
discolor *J.U.M dos Santos in Bol. Mus. Paraense Emílio Goeldi, Bot.,* 10(1): 113 (1994 publ. 1995)—Brazil West-central (Mato Grosso).
goiazensis *J.U.M. dos Santos in Bol. Mus. Paraense Emílio Goeldi, Bot.,* 10(1): 109 (1994 publ. 1995)—Brazil West-central (Goiás).
hatschbachii *J.U.M. dos Santos in Bol. Mus. Paraense Emílio Goeldi, Bot.,* 10(1): 106 (1994 publ. 1995)—Brazil West-central (Mato Grosso do Sul).
helianthoides *(Schumach. & Thonn.) D. Oliver & W.P. Hiern in Oliv., Fl. Trop. Afr.,* 3: 381 (1877): Coronocarpus helianthoides.†
ioletae *J.U.M. Santos in Bol. Mus. Paraense Emílio Goeldi, Bot.,* 9(2): 208 (1993 publ. 1994)—Brazil North (Tocantins).
itabaianensis *J.U.M. Santos in Bol. Mus. Paraense Emílio Goeldi, Bot.,* 9(2): 212 (1993 publ. 1994)—Brazil Northeast (Sergipe).
matogrossensis *J.U.M. dos Santos in Bol. Mus. Paraense Emílio Goeldi, Bot.,* 10(1): 111 (1995)—Brazil West-central (Mato Grosso).
subscandens *J.U.M. Santos in Bol. Mus. Paraense Emílio Goeldi, Bot.,* 9(2): 213 (1993 publ. 1994)—Brazil Northeast (Bahia).

ASPLUNDIA (Cyclanthac.).
gigantea *D. Tuberquia in Caldasia,* 19(1–2): 182 (1997)—Colombia.
trilobulata *D. Tuberquia in Caldasia,* 19(1–2): 179 (1997)—Colombia.

ASPRELLA (Gramin.).
hexandra *(Sw.) P. Beauv., Ess. Agrostogr.:* 2, 153 (1812): Leersia hexandra.
lenticularis *(Michx.) P. Beauv., Ess. Agrostogr.:* 2, 153 (1812): Leersia lenticularis.
mexicana *(Kunth) J.J. Roemer & J.A. Schultes, Syst. Veg.,* 2: 267 (1817): Leersia mexicana.
monandra *(Sw.) J.J. Roemer & J.A. Schultes, Syst. Veg.,* 2: 267 (1817): Leersia monandra.

ASPRELLA:-
virginica *(Willd.) P. Beauv., Ess. Agrostogr.:* 2, 153 (1812): Leersia virginica.

ASTER (Compos.).
subgen. Canadanthus *(G.L. Nesom) J.C. Semple in Univ. Waterloo Biol. Ser.,* 38: 36 (1996): gen. Canadanthus.
sect. Polyliguli *(Semple & Brouillet) J.C. Semple in Univ. Waterloo Biol. Ser.,* 38: 84 (1996): Lasallea sect. Polyliguli.
bachtiaricus *V. Mozaffarian in Iranian J. Bot.,* 7(1): 130 (1996)—Iran.
ciliatus
subsp. angustus *(Lindl.) R. Govaerts, World Checklist Seed Plants,* 2(1–2): 9 (1996): Tripolium angustum.
subsp. laurentianus *(Fernald) R. Govaerts, World Checklist Seed Plants,* 2(1–2): 9 (1996): A. laurentianus.
coronopifolia *(Nutt.) N.E. Brown, Suppl. Johnson's Gard. Dict.:* 879 (1882): Dieteria coronopifolia.
fallax Tamamsch.
var. brachyglossus *(Onno) G.A. Peschkova in Fl. Tsentral'noi Sibiri,* 2: 819 (1979): A. alpinus var. brachyglossus.
formosissimus *Lémon in Ann. Fl. Pomone,* 1833–34: 79 (1833)—Southeastern U.S.A.
hispidus Thunb.
var. tubulosus *K. Asano ex T. Shimizu, Fl. Nagano Pref.:* 1510 (1997)—Japan.
hualiensis *S.S. Ying in Mem. Coll. Agric. Nation. Taiwan Univ.,* 28(2): 39 (1988)—Taiwan.
magnus *Y.N. Lee & C. Kim in Korean J. Pl. Taxon.,* 28(1): 31 (1998)—South Korea.
f. albiflorus *Y.N. Lee & C. Kim in Korean J. Pl. Taxon.,* 28(1): 33 (1998)—South Korea.
menthodorus *(Benth.) R. Govaerts, World Checklist Seed Plants,* 2(1–2): 9 (1996), as 'methodorus': Brachyactis menthodora.
microcephalus (Miq.) Franch. & Sav.
var. ovatus *(Franch. & Sav.) A. Soejima & Mot. Ito in Acta Phytotax. Geobot.,* 49(2): 151 (1998 publ. 1999): A. trinervius var. ovatus.
var. yezoensis *(Kitam. & H. Hara) A. Soejima & Mot. Ito in Acta Phytotax. Geobot.,* 49(2): 152 (1998 publ. 1999): A. ageratoides var. yezoensis.
muliensis *Hand.-Mazz.in Oesterr. Bot. Z.,* 88(4): 309 (1939)—China South-Central (Sichuan).
novi-belgii L.
var. crenifolius *(Fernald) J. Labrecque & L. Brouillet in Canad. J. Bot.,* 74(2): 185 (1996): A. foliaceus var. crenifolius.
ontarionis *K.M. Wiegand in Rhodora,* 30(357): 179 (1928), in Supp. 8 as 'ontariensis'—New York, Ontario.†
var. glabratus *J.C. Semple in Univ. Waterloo Biol. Ser.,* 38: 60 (1996)—Ontario, Québec.
ovatus (Franch. & Sav.) Mot. Ito & Soejima
var. angustifolius *(Franch. & Sav.) K. Asano ex T. Shimizu, Fl. Nagano Pref.:* 1510 (1997): A. trinervius var. angustifolius.
ruoqiangensis *Y. Wei & C.H. An in Fl. Xinjiangensis,* 5: 475 (1999)—Xinjiang.
spathulifolius Maxim.
var. oharae *(Nakai) Y.N. Lee, Fl. Korea:* 1162 (1996), as 'oharai': A. oharae.
tripolium L.
subsp. minoricensis *A.M. Romo, Fl. Silvestres Baleares:* 294 (1994), as 'minoricense'—Baleares.
umbrosus *(Benth.) R. Govaerts, World Checklist Seed Plants,* 2(1–2): 9 (1996): Brachyactis umbrosa.

ASTERIDEA (Compos.).
archeri *P.S. Short in Austral. Syst. Bot.,* 13(5): 739 (2000)—Western Australia.
morawana *P.S. Short in Austral. Syst. Bot.,* 13(5): 741 (2000)—Western Australia.

× **ASTERIGERON** *N.N. Tzvelev in Fl. Evropeĭskoĭ Chasti SSSR,* 7: 185 (1994), homonym; *see Rydb., Fl. Rocky Mts.:* 891 (1917). *COMPOSITAE* (Aster × Erigeron).

ASTERISCUS (Compos.).
aquaticus (L.) Less.
subsp. nanus *(Boiss.) O. de Bolòs & J. Vigo, Fl. Països Catalans,* 3: 761 (1995): A. aquaticus var. nanus.

ASTERISCUS:-
cyrenaicus *(Alavi) A. Dobignard in Candollea,* 52(1): 147 (1997): Pallenis cyrenaica.
daltonii (Webb) Walp.
subsp. vogelii *(Webb) W. Greuter in Flora Medit.,* 7: 46 (1997): Odontospermum vogelii.
graveolens (Forssk.) Less.
subsp. odorus *(Schousb.) W. Greuter in Flora Medit.,* 7: 46 (1997): Buphthalmum odorum.
subsp. stenophyllus *(Link) W. Greuter in Flora Medit.,* 7: 46 (1997): Buphthalmum stenophyllum.
spinosus (L.) Sch. Bip.
var. gymnesicus *(O. Bolòs & P. Monts.) O. de Bolòs & J. Vigo, Fl. Països Catalans,* 3: 760 (1995): Pallenis spinosa var. gymnesica.
subsp. microcephalus *(Halácsy) O. de Bolòs & J. Vigo, Fl. Països Catalans,* 3: 760 (1995): Pallenis spinosa f. microcephala.
teknensis *A. Dobignard & F. Jacquemoud in Candollea,* 52(1): 145 (1997)—Morocco.

ASTEROLASIA (Rutac.).
pallida Benth.
subsp. hyalina *Paul G. Wilson in Nuytsia,* 12(1): 84 (1998)—Western Australia.
rivularis *Paul G. Wilson in Nuytsia,* 12(1): 85 (1998)—New South Wales.

ASTEROPEIA (Asteropeiac.).
labatii *G.E. Schatz, P.P. Lowry & A.-E. Wolf in Adansonia,* 21(2): 260 (1999)—Madagascar.
matrambody *(Capuron) G.E. Schatz, P.P. Lowry & A.-E. Wolf in Adansonia,* 21(2): 262 (1999): A. amblyocarpa var. matrambody.
mcphersonii *G.E. Schatz, P.P. Lowry & A.-E. Wolf in Adansonia,* 21(2): 263 (1999)—Madagascar.

ASTEROSTIGMA (Arac.).
cryptostylum *J. Bogner in Sendtnera,* 4: 7 (1997)—Brazil West-central (Goiás).
lombardii *E.G. Gonçalves in Aroideana,* 22: 30 (1999)—Brazil Southeast (Minas Gerais).

ASTOMAEA (Umbellif.).
galiocarpa *(Korovin) M.G. Pimenov & E.V. Kljuykov in Opred. Rast. Sred. Azii,* 7: 194 (1983): Astomatopsis galiocarpa.

ASTONIA *S.W.L. Jacobs in Telopea,* 7(2): 140 (1997). *ALISMATACEAE*
australiensis *(Aston) S.W.L. Jacobs in Telopea,* 7(2): 141 (1997): Limnophyton australiense.

ASTRACANTHA (Legumin.).
sicula *(Biv.) Greuter ex U. Reer & D. Podlech in Mitt. Bot. Staatssamml. München,* 22: 544 (31 Dec. 1986): Astragalus siculus.

ASTRAGALUS (Legumin.).
sect. Aberrantes *D. Podlech in Sendtnera,* 6: 136 (1999).
subsect. Bakerophaca *D. Podlech in Fl Iran.,* 174: 78 (1999).
subsect. Chronopus *(Bunge) D. Podlech in Fl. Iran.,* 174: 123 (1999): A. sect. Chronopus.
subsect. Concordi *S.L. Welsh in Great Basin Nat.,* 58(4): 386 (1998).
subsect. Hymenocoleus *(Bunge) D. Podlech & S. Zarre M. in Sendtnera,* 3: 261 (1996): A. sect. Hymenocoleus.
subsect. Microcymbi *S.L. Welsh in Great Basin Nat.,* 58(1): 50 (1998).
sect. Mikrophyton *D. Podlech in Sendtnera,* 6: 161 (1999).
sect. Monotheca *N. Ulziykhutag in Novosti Sist. Vyssh. Rast.,* 30: 109 (1996).
sect. Panicula *L.R. Xu & Z.H. Wu in Acta Bot. Bor.-Occid. Sin.,* 17(2): 257 (1997).
subsect. Phyllolobium *(Fisch.) D. Podlech in Fl. Iran.,* 174: 79 (1999): gen. Phyllolobium.
sect. Rechingeria *D. Podlech in Fl. Iran.,* 174: 80 (1999).
adsurgens Pall.
subsp. oreogenus *(Jurtzev) V.N. Voroschilov in A.K. Skvortsov (ed.), Florist. issl. v razn. raĭonakh SSSR:* 178 (1985): A. inopinatus subsp. oreogenus.

ASTRAGALUS:-
ahmad-parsae *A.A. Maassoumi in Iranian J. Bot.,* 7(2): 222 (1998), as 'ahmad-parsai'—Iran.
akhanii *D. Podlech in Sendtnera,* 6: 142 (1999)—Iran.
akmanii *Z. Aytaç & H. Duman in Turkish J. Bot.,* 19(4): 477 (1995)—Turkey.
aktauensis *N.F. Gontscharow in Bot. Mater. Gerb. Bot. Inst. Komarova Akad. Nauk SSSR,* 9: 96 (1941)—Uzbekistan.
aktiubensis *A.K. Sytin in Novosti Sist. Vyssh. Rast.,* 32: 92 (2000)—Kazakhstan.
albispinus Širj. & Bornm.
subsp. esfandiarii *(Širj. & Rech. f.) S. Zarre-Mobarakeh, Syst. Revis. Astragalus sect. Adiaspastus, sect. Macrophyllium & sect.* Pterophorus (Fabac.): 45 (1998): A. esfandiarii.
alexeenkoi *N.F. Gontscharow in Fl. Tadzhikistana,* 5: 669 (1937)—Tadzhikistan.
altiusculus *A.A. Maassoumi & F. Ghahremani-nejad in Iranian J. Bot.,* 8(1): 36 (1999)—Iran.
ampullarioides *(S.L. Welsh) S.L. Welsh in Great Basin Nat.,* 58(1): 51 (1998): A. eremiticus var. ampullarioides.
andreji-sytinii *D. Podlech in Sendtnera,* 6: 172 (1999), nom. nov.: A. eugenii *Grossh.*
antalyensis *A. Duran & D. Podlech in Sendtnera,* 6: 143 (1999)—Turkey.
antoninae *J.S. Grigorjev in Fl. Tadzhikistana,* 5: 673 (1937)—Tadzhikistan.
apiculatus *N.F. Gontscharow in Fl. Tadzhikistana,* 5: 668 (1937)—Tadzhikistan.
arakliensis *D. Podlech in Sendtnera,* 6: 162 (1999)—Turkey.
arasbaranensis *A.A. Maassoumi & M.M. Ranjbar in Iranian J. Bot.,* 7(1): 40 (1996)—Iran.
arnoldii Hemsl. & H. Pearson
f. albiflorus *Y.H. Wu in Acta Bot. Yunnanica,* 20(1): 39 (1998), as 'albiflous'—Qinghai.
auratus *N.F. Gontscharow in Bot. Mater. Gerb. Bot. Inst. Komarova Akad. Nauk SSSR,* 9: 102 (1941)—Turkmenistan.
australis (L.) Lam.
var. aboriginorum *(Richardson) S.L. Welsh in Great Basin Nat.,* 58(1): 48 (1998): A. aboriginorum.
var. cottonii *(M.E. Jones) S.L. Welsh in Great Basin Nat.,* 58(1): 48 (1998): A. cottonii.
var. glabriusculus *(Hook.) D. Isely in Iowa State J. Res.,* 59(2): 130 (1984): Phaca glabriuscula.
var. lepagei *(Hultén) S.L. Welsh in Great Basin Nat.,* 58(1): 47 (1998): A. lepagei.
var. muriei *(Hultén) S.L. Welsh in Great Basin Nat.,* 58(1): 47 (1998): A. aboriginorum var. muriei.
austroiranicus *D. Podlech in Sendtnera,* 6: 172 (1999), nom. nov.: A. minutissimus *Freyn & Bornm.*
austrouralensis *P.V. Kulikov in Novosti Sist. Vyssh. Rast.,* 32: 90 (2000)—West Siberia.
avajensis *D. Podlech in Sendtnera,* 6: 145 (1999)—Iran.
badachschanicus *A.G. Borissova in Fl. Tadzhikistana,* 5: 654 (1937)—Tadzhikistan.
bakirdaghensis *D. Podlech in Sendtnera,* 6: 157 (1999)—Turkey.
balkaricus *A.K. Sytin in Novosti Sist. Vyssh. Rast.,* 30: 103 (1996)—North Caucasus (Kabardino-Balkariya).
barboides *S. Zarre & H. Duman in Edinburgh J. Bot.,* 55(3): 352 (1998)—Turkey.
belangeri *(Kuntze) D. Podlech in Sendtnera,* 6: 172 (1999): Tragacantha belangeri.
beypazaricus *D. Podlech & Z. Aytaç in Sendtnera,* 5: 266 (1998)—Turkey.
binaludensis *A.A. Maassoumi & M. Pakravan in Iranian J. Bot.,* 7(1): 47 (1996)—Iran.
bischkendicus *N.F. Gontscharow in Fl. Tadzhikistana,* 5: 660 (1937)—Tadzhikistan.
bowes-lyonii *D. Podlech in Sendtnera,* 6: 140 (1999)—Pakistan.
brachycalyx Fisch. ex Kotschy
subsp. eriostylus *(Boiss. & Hausskn.) S. Zarre-Mobarakeh, Syst. Revis. Astragalus sect. Adiaspastus, sect. Macrophyllium & sect.* Pterophorus (Fabac.): 57 (1998): A. eriostylus.
brevidentatus *D. Podlech in Sendtnera,* 6: 151 (1999), homonym; *see C.H. Wright in Kew Bull.,* 1906: 200 (1906)—Turkey.

ASTRAGALUS:-
breviflorus DC.
subsp. elongatus *(D.F. Chamb. & V.A. Matthews) S. Zarre-Mobarakeh, Syst. Revis. Astragalus sect. Adiaspastus, sect. Macrophyllium & sect.* Pterophorus (Fabac.): 59 (1998): A. eriocephalus subsp. elongatus.
burchan-buddaicus *N. Ulziykhutag in Novosti Sist. Vyssh. Rast.,* 30: 104 (1996)—Qinghai.
cartilagineus *N.F. Gontscharow in Bot. Mater. Gerb. Bot. Inst. Komarova Akad. Nauk SSSR,* 9: 99 (1941)—Turkmenistan.
caspicus M. Bieb.
subsp. pseudocaspius *(Fisch.) S. Zarre-Mobarakeh, Syst. Revis. Astragalus sect. Adiaspastus, sect. Macrophyllium & sect.* Pterophorus (Fabac.): 65 (1998): A. pseudocaspius.
cedreticola *A. Duran & D. Podlech in Sendtnera,* 6: 149 (1999)—Turkey.
chehreganii *S. Zarre M. & D. Podlech in Sendtnera,* 3: 264 (1996)—Iran.
chorinensis Bunge
var. popovii *G.A. Peschkova in Fl. Tsentral'noi Sibiri,* 2: 599 (1979)—Buryatiya, Irkutsk.
concordius *S.L. Welsh in Great Basin Nat.,* 58(4): 387 (1998)—Utah.
confertus Benth. ex Bunge
f. albiflorus *R.F. Huang & Y.H. Wu in Bull. Bot. Res. North-East. Forest. Univ.,* 17(1): 36 (1997)—Qinghai.
conjunctus S. Watson
var. rickardii *S.L. Welsh, F. Caplow & K. Beck in Great Basin Nat.,* 57(4): 354 (1997)—Washington.
cutleri *(Barneby) S.L. Welsh in Great Basin Nat.,* 58(1): 51 (1998): A. preussii var. cutleri.
dabanshanicus *Y.H. Wu in Acta Bot. Yunnanica,* 20(1): 35 (1998)—Qinghai.
danicus Retz.
subsp. dasyglottis *(Fisch.) V.N. Voroschilov in A.K. Skvortsov (ed.), Florist. issl. v razn. raĭonakh SSSR:* 178 (1985): A. dasyglottis.
delutulus *A.A. Maassoumi in Iranian J. Bot.,* 7(2): 225 (1998), nom. nov.: A.dilutus *Bornm.*
demirizii *R. Kramer & D. Podlech in Sendtnera,* 6: 163 (1999)—Turkey.
densifolius Lam.
subsp. amasiensis *(Freyn & Bornm.) Z. Aytaç & Ekim in Turkish J. Bot.,* 21(1): 40 (1997), without basionym ref.: Basionym not stated.
devesae *S. Talavera, A. González & G. López in Lagascalia,* 21(1): 194 (1999)—Spain.
discessiflorus *N.F. Gontscharow in Fl. Tadzhikistana,* 5: 676 (1937)—Tadzhikistan.
diversipilosus *D. Podlech in Sendtnera,* 6: 145 (1999)—Iran.
diyarbakirensis *D. Podlech in Sendtnera,* 6: 138 (1999)—Turkey.
drasianus *H.J. Chowdhery, B.P. Uniyal & B. Balodi in Bull. Bot. Surv. India,* 34(1–4): 209 (1992 publ. 1997)—Jammu-Kashmir.
dsharfi *B.A. Fedtschenko in Fl. Tadzhikistana,* 5: 675 (1937)—Tadzhikistan.
dulanensis *Y.H. Wu in Bull. Bot. Res. North-East. Forest. Univ.,* 17(1): 34 (1997)—Qinghai.
eastwoodiae M.E. Jones
var. debequaeus *(S.L. Welsh) D. Isely, Native & Natural. Legumin. (Fabac.) U.S. (excl. Alaska & Hawaii):* 278 (1998): A. debequaeus.
ekimii *S. Zarre & H. Duman in Edinburgh J. Bot.,* 55(3): 355 (1998)—Turkey.
ephemeretorum *N.F. Gontscharow in Fl. Tadzhikistana,* 5: 658 (1937)—Tadzhikistan.
eskishehiricus *D. Podlech in Sendtnera,* 6: 164 (1999)—Turkey.
fallacinus *D. Podlech in Sendtnera,* 6: 148 (1999)—Turkey.
ferganensis *(Popov) B.A. Fedtschenko in Fl. Tadzhikistana,* 5: 414 (1937): A. erioceras var. ferganensis.
floridus Benth.
var. multipilus *Y.H. Wu in Bull. Bot. Res. North-East. Forest. Univ.,* 17(1): 36 (1997)—Qinghai.
frigidus (L.) A. Gray
subsp. secundus *(DC.) V.N. Voroschilov in A.K. Skvortsov (ed.), Florist. issl. v razn. raĭonakh SSSR:* 178 (1985): A. secundus.

ASTRAGALUS:-
gigantirostratus *A.A. Maassoumi, A. Ghahreman, F. Ghahremani-Nejad & F. Matin in Willdenowia,* 29(1–2): 222 (1999)—Iran.
gigantostegius *D. Podlech in Sendtnera,* 6: 154 (1999)—Turkey.
gines-lopezii *S. Talavera, D. Podlech, J.A. Devesa & F.M. Vázquez Pardo in An. Jard. Bot. Madrid,* 57(1): 219 (1999)—Spain.
glaberrimus *D. Podlech in Sendtnera,* 6: 146 (1999), homonym; *see Širj. & Rech. f. in Anz. Österr. Akad. Wiss., Math.-Naturwiss.* Kl., 90: 180 (1953)—Iran.
glaucopsiformis *A.A. Maassoumi in Iranian J. Bot.,* 8(2): 311 (2000)—Iran.
golmuensis Y.C. Ho
var. paucipilus *Y.H. Wu in Bull. Bot. Res. North-East. Forest. Univ.,* 17(1): 36 (1997)—Qinghai.
guinanicus *Y.H. Wu in Bull. Bot. Res. North-East. Forest. Univ.,* 17(1): 33 (1997)—Qinghai.
hakkariensis *D. Podlech in Sendtnera,* 6: 168 (1999)—Turkey.
harpocarpus *V.V. Meffert in Fl. Tadzhikistana,* 5: 657 (1937)—Tadzhikistan.
huiningensis Y.C. Ho
var. psilocarpus *K.T. Fu in Fl. Loess-Plat. Sin.,* 2: 511 (1992), with two types cited—Inner Mongolia (Ningxia), China North-central (Gansu, Shaanxi).
hyalolepidoides *A.P. Khokhrjakov in Byull. Glavn. Bot. Sada (Moscow),* 172: 24 (1995)—Transcaucasus (Adzhariya).
hymenocystis Fisch. & C.A. Mey.
subsp. confiniorum *S. Zarre M. & D. Podlech in Sendtnera,* 3: 273 (1996)—Iran, Turkey.
insignis *N.F. Gontscharow in Fl. Tadzhikistana,* 5: 663 (1937)—Tadzhikistan.
jacobsii *D. Podlech in Sendtnera,* 6: 158 (1999)—Iran.
jamzadiae *A.A. Maassoumi in Iranian J. Bot.,* 8(2): 312 (2000), as 'jamzadae'—Iran.
juratzkanus Freyn & Sint.
subsp. brotherusii *(Freyn & Sint.) D. Podlech in Sendtnera,* 6: 173 (1999): A. xanthoxiphidium subsp. brotherusii.
kaghysmani *N.F. Gontscharow in Bot. Mater. Gerb. Bot. Inst. Komarova Akad. Nauk SSSR,* 9: 100 (1941)—Transcaucasus (Armeniya).
karasarensis *D. Podlech in Sendtnera,* 6: 155 (1999)—Turkey.
karategini *N.F. Gontscharow in Fl. Tadzhikistana,* 5: 660 (1937)—Tadzhikistan.
kashanensis *D. Podlech in Sendtnera,* 6: 159 (1999)—Iran.
kentrophyllus *D. Podlech in Sendtnera,* 6: 156 (1999)—Iran.
khatamsaziae *A.A. Maassoumi in Iranian J. Bot.,* 8(2): 313 (2000), as 'khatamsazae'—Iran.
kirilovii *D. Podlech in Sendtnera,* 6: 172 (1999), nom. nov.: A. intermedius *Kar. & Kir.*
konyaensis *D. Podlech in Sendtnera,* 6: 165 (1999)—Turkey.
kordloricus *S. Zarre-Mobarakeh, Syst. Revis. Astragalus sect. Adiaspastus, sect. Macrophyllium & sect.* Pterophorus (Fabac.): 86 (1998)—Iran.
kudrjaschovii *A.S. Koroleva in Fl. Tadzhikistana,* 5: 677 (1937)—Tadzhikistan.
kukunoricus *N. Ulziykhutag in Novosti Sist. Vyssh. Rast.,* 30: 107 (1996)—Qinghai.
laccoliticus *(M.E. Jones) S.L. Welsh in Great Basin Nat.,* 58(1): 53 (1998): A. cicadae var. laccoliticus.
lackschewitzii *M. Lavin & H. Marriott in Syst. Bot.,* 22(2): 214 (1997)—Montana.
laconicus *G. Iatroú & Kit Tan in Willdenowia,* 29(1–2): 56 (1999)—Greece.
langtangensis *D. Podlech in Sendtnera,* 6: 136 (1999)—Nepal, India (Uttar Pradesh).
latianicus *A.A. Maassoumi & M.M. Ranjbar in Iranian J. Bot.,* 7(1): 41 (1996)—Iran.
latus *S. Zarre-Mobarakeh, Syst. Revis. Astragalus sect. Adiaspastus, sect. Macrophyllium & sect.* Pterophorus (Fabac.): 87 (1998), homonym; *see M.E. Jones in Zoe,* 4: 272 (1893)—Iraq.
laxmannii Jacq.
subsp. robustior *(Hook.) D. Podlech in Sendtnera,* 6: 173 (1999): A. adsurgens var. robustior.
var. robustior *(Hook.) R.C. Barneby & S.L. Welsh in Great Basin Nat.,* 56(1): 85 (1996): A. adsurgens var. robustior.

ASTRAGALUS:-
laxmannii Jacq.
var. tananaicus *(Hultén) R.C. Barneby & S.L. Welsh in Great Basin Nat.,* 56(1): 85 (1996): A. tananaicus.
subsp. viciifolius *(S.L. Welsh) D. Podlech in Sendtnera,* 6: 173 (1999): A. adsurgens subsp. viciifolius.
lepsensis Bunge
var. leduensis *Y.H. Wu in Acta Bot. Yunnanica,* 20(1): 37 (1998)—Qinghai.
linczevskii *N.F. Gontscharow in Fl. Tadzhikistana,* 5: 667 (1937)—Tadzhikistan.
longiracemosus *N. Ulziykhutag in Novosti Sist. Vyssh. Rast.,* 30: 114 (1996)—Qinghai.
longisubulatus *D. Podlech in Sendtnera,* 6: 151 (1999)—Turkey.
longivexillatus *D. Podlech & Ekici in Sendtnera,* 6: 170 (1999)—Turkey.
luteiflorus *N. Ulziykhutag in Novosti Sist. Vyssh. Rast.,* 30: 112 (1996)—Qinghai.
maabudii *M.M. Ranjbar in Iranian J. Bot.,* 7(1): 43 (1996)—Iran.
mahoschanicus Hand.-Mazz.
var. multipilosus *Y.H. Wu in Acta Bot. Yunnanica,* 20(1): 39 (1998)—Qinghai.
var. mengdaensis *Y.H. Wu in Acta Bot. Yunnanica,* 20(1): 40 (1998)—Qinghai.
malatyaensis *D. Podlech in Sendtnera,* 6: 147 (1999)—Turkey.
maniaticus *Kit Tan & A. Strid in Pl. Syst. Evol.,* 206(1–4): 47 (1997)—Greece.
mattam Tsai & Yu
var. macroflorus *Y.H. Wu in Acta Bot. Yunnanica,* 20(1): 39 (1998)—Qinghai.
meiliensis *C. Chen & Zi G. Qian in Acta Bot. Yunnanica,* 19(3): 236 (1997)—China South-central (Yunnan).
membranaceus (Fisch.) Bunge
f. albiflorus *Y.N. Lee, Fl. Korea:* 1159 (1996), without type—Locality not stated.
f. purpurinus *(Y.C. Ho) K.T. Fu in Fl. Loess-Plat. Sin.,* 2: 392 (1992): A. membranaceus var. purpurinus.
membranostipulus *A.A. Maassoumi in Iranian J. Bot.,* 8(2): 313 (2000)—Iran.
mercklinii Boiss. & Buhse
subsp. farsicus *D. Podlech in Sendtnera,* 6: 152 (1999)—Iran.
mikrophytoides *D. Podlech in Sendtnera,* 6: 161 (1999)—Afghanistan.
molybdenus Barneby
var. shultziorum *(Barneby) S.L. Welsh in Great Basin Nat.,* 58(1): 46 (1998): A. shultziorum.
mongholicus Bunge
var. dahuricus *(DC.) D. Podlech in Sendtnera,* 6: 172 (1999): Phaca alpina var. dahurica.
montis-varvashti *D. Podlech in Sendtnera,* 6: 169 (1999)—Iran.
montosus *A.A. Maassoumi in Iranian J. Bot.,* 8(2): 314 (2000)—Iran.
moussavii *A.A. Maassoumi, F. Ghahremani-nejad & A. Ghahreman in Nordic J. Bot.,* 20(3): 353 (2000)—Iran.
naignichus *Y.H. Xu & Q.R. Liu in Acta Bot. Yunnanica,* 19(2): 190 (1997), nom. nov.: A. chingianus *P.C. Li.*
nathaliae *V.V. Meffert in Fl. Tadzhikistana,* 5: 678 (1937)—Tadzhikistan.
neocarpus *E. Gómez-Sosa in Brittonia,* 49(4): 537 (1997)—Argentina South (Chubut).
neomozaffarianii *A.A. Maassoumi in Iranian J. Bot.,* 8(2): 315 (2000), as 'neo-mozaffarianii'—Iran.
nepalensis *D. Podlech in Sendtnera,* 6: 137 (1999)—Nepal.
nevadensis Boiss.
subsp. muticus *(Pau) S. Zarre-Mobarakeh & D. Podlech in S. Zarre-Mobarakeh, Syst. Revis. Astragalus sect. Adiaspastus, sect.* Macrophyllium & sect. Pterophorus (Fabac.): 96 (1998): A. muticus.
nevshehiricus *D. Podlech in Sendtnera,* 6: 153 (1999)—Turkey.
neyshaburensis *D. Podlech in Sendtnera,* 6: 147 (1999)—Iran.
nigdeanus *D. Podlech & Ekici in Sendtnera,* 6: 167 (1999)—Turkey.
nigrifructus *D. Podlech & Z. Aytaç in Sendtnera,* 5: 265 (1998)—Turkey.

ASTRAGALUS:-
nigrocalycinus *D. Podlech in Sendtnera,* 6: 139 (1999)—Turkey.
nigrocalyx *Slobodov ex J.S. Grigorjev in Fl. Tadzhikistana,* 5: 674 (1937)—Tadzhikistan.
nydeggeri *S. Zarre & H. Duman in Edinburgh J. Bot.,* 55(3): 356 (1998)—Turkey.
ochranthus *N.F. Gontscharow in Fl. Tadzhikistana,* 5: 659 (1937)—Tadzhikistan.
ovalifoliolatus *A.A. Maassoumi & M.M. Ranjbar in Iranian J. Bot.,* 7(1): 42 (1996)—Iran.
ovczinnikovii *A.G. Borissova in Fl. Tadzhikistana,* 5: 655 (1937)—Tadzhikistan.
oxypterus *A.G. Borissova in Fl. Tadzhikistana,* 5: 655 (1937)—Tadzhikistan.
parnassi Boiss.
subsp. calabricus *(Fisch.) S. Zarre-Mobarakeh, Syst. Revis. Astragalus sect. Adiaspastus, sect. Macrophyllium & sect.* Pterophorus (Fabac.): 164 (1998): A. calabricus.
passargadensis *A.A. Maassoumi in Iranian J. Bot.,* 8(2): 316 (2000)—Iran.
pauxillis *A.A. Maassoumi & F. Ghahremani-nejad in Iranian J. Bot.,* 8(1): 37 (1999)—Iran.
peltopsis *D. Podlech & K.H. Rechinger in Fl. Iran.,* 174: 98 (1999), nom. nov.: A. distinctissimus *Rech. f. & Edelb.*
pereshkhoranicus *A.A. Maassoumi & F. Ghahremani-nejad in Iranian J. Bot.,* 8(1): 38 (1999)—Iran.
petrovii *N. Ulziykhutag in Novosti Sist. Vyssh. Rast.,* 30: 109 (1996)—China North-central (Gansu).
peymanii *A.A. Maassoumi in Iranian J. Bot.,* 8(2): 317 (2000)—Iran.
pineticola *D. Podlech in Sendtnera,* 6: 166 (1999)—Turkey.
polhillii *D. Podlech in Sendtnera,* 6: 153 (1999)—Turkey.
polyanthus Bunge
subsp. vedicus *(Takht.) S. Zarre-Mobarakeh, Syst. Revis. Astragalus sect. Adiaspastus, sect. Macrophyllium & sect.* Pterophorus (Fabac.): 105 (1998): A. vedicus.
polycladus Bureau & Franch.
var. glabricarpus *Y.H. Wu in Acta Bot. Yunnanica,* 20(1): 39 (1998)—Qinghai.
var. magniflorus *Y.H. Wu in Bull. Bot. Res. North-East. Forest. Univ.,* 17(1): 35 (1997), as 'magniflors'—Qinghai.
polystachys *A.A. Maassoumi in Iranian J. Bot.,* 8(2): 318 (2000)—Iran.
pseudanthylloides *N.F. Gontscharow in Bot. Mater. Gerb. Bot. Inst. Komarova Akad. Nauk SSSR,* 9: 101 (1941)—Uzbekistan.
pseudoamygdalinus *M.G. Popov in Fl. Tadzhikistana,* 5: 661 (1937)—Tadzhikistan.
pseudoeremophysa *M.G. Popov in Fl. Tadzhikistana,* 5: 671 (1937)—Tadzhikistan.
pseudolanuginosus *N.F. Gontscharow in Fl. Tadzhikistana,* 5: 657 (1937)—Tadzhikistan.
pseudorhacodes *N.F. Gontscharow in Bot. Mater. Gerb. Bot. Inst. Komarova Akad. Nauk SSSR,* 9: 103 (1941)—Tadzhikistan.
roschanicus *B.A. Fedtschenko in Fl. Tadzhikistana,* 5: 673 (1937)—Tadzhikistan.
rubellus *N.F. Gontscharow in Bot. Mater. Gerb. Bot. Inst. Komarova Akad. Nauk SSSR,* 9: 94 (1941)—Uzbekistan.
rumpens *V.V. Meffert in Fl. Tadzhikistana,* 5: 678 (1937)—Tadzhikistan.
sabulosus M.E. Jones
var. vehiculus *S.L. Welsh in Great Basin Nat.,* 58(1): 53 (1998)—Utah.
saralensis *N.F. Gontscharow in Bot. Mater. Gerb. Bot. Inst. Komarova Akad. Nauk SSSR,* 9: 92 (1941)—Krasnoyarsk.
sarzehensis *M.M. Ranjbar in Iranian J. Bot.,* 7(1): 40 (1996)—Iran.
scapiger *M. Ranjbar & A.A. Maassoumi in Iranian J. Bot.,* 7(2): 236 (1998)—Iran.
serpentinicola *H. Duman & T. Ekim in Willdenowia,* 27(1–2): 181 (1997)—Turkey.
shahbazanicus *D. Podlech in Sendtnera,* 6: 172 (1999), nom. nov.: A. drepanophorus *Bornm.*
slobodovii *N.F. Gontscharow & A.S. Koroleva in Fl. Tadzhikistana,* 5: 676 (1937)—Tadzhikistan.
spongocarpus *V.V. Meffert in Fl. Tadzhikistana,* 5: 656 (1937)—Tadzhikistan.

ASTRAGALUS:-
subcaracugensis *G.T. Sitpaeva in Izv. Min. Nauki Akad. Nauk Resp. Kazakhstan, Ser. Biol. Med.*, 1997(5–6): 32 (1997 publ. 1998)—Kazakhstan.
subcinereus A. Gray
var. sileranus *(M.E. Jones) S.L. Welsh in Great Basin Nat.*, 58(1): 49 (1998): A. sileranus.
subpenicillatus *D. Podlech in Fl. Iran.*, 174: 259 (1999)—Afghanistan.
sultan-bulaghensis *D. Podlech in Sendtnera*, 6: 160 (1999)—Iran.
sumarensis *A.A. Maassoumi in Iranian J. Bot.*, 8(2): 319 (2000)—Iran.
sungpanensis E. Peter
f. albiflorus *Y.H. Wu in Bull. Bot. Res. North-East. Forest. Univ.*, 17(1): 35 (1997)—Qinghai.
tadshikorum *A.S. Koroleva in Fl. Tadzhikistana*, 5: 679 (1937)—Tadzhikistan.
tegetarioides M.E. Jones
var. anxius *(Meinke & Kaye) S.L. Welsh in Great Basin Nat.*, 58(1): 51 (1998): A. anxius.
tetuanensis *D. Podlech in Sendtnera*, 6: 170 (1999)—Morocco.
thomsonii *D. Podlech in Sendtnera*, 6: 140 (1999)—Xizang, India (Himachal Pradesh), Jammu-Kashmir.
tongolensis Ulbr.
var. longibracteatus *Y.H. Wu in Bull. Bot. Res. North-East. Forest. Univ.*, 17(1): 35 (1997)—Qinghai.
trachycarpus *N.F. Gontscharow in Fl. Tadzhikistana*, 5: 672 (1937)—Tadzhikistan.
tricholobus DC.
subsp. hohenackeri *(Boiss.) S. Tietz in Sendtnera*, 3: 181 (1996): A. hohenackeri.
tuna-ekimii *N. Adıgüzel in Ann. Bot. Fenn.*, 36(4): 231 (1999)—Turkey.
ui-eilakensis *F. Ghahremani-Nejad in Feddes Repert.*, 111(3–4): 181 (2000), nom. nov.: A. caudiculosus *Kom.*
ulziykhutagii *A.K. Sytin in Kew Bull.*, 51(2): 376 (1996), nom. nov.: A. alexandri *N. Ulziykh.*
vachanicus *A.G. Borissova & A.S. Koroleva in Fl. Tadzhikistana*, 5: 655 (1937)—Tadzhikistan.
valerii *N. Ulziykhutag in Novosti Sist. Vyssh. Rast.*, 30: 111 (1996)—Qinghai.
varius S.G. Gmel.
subsp. eupatoricus *A.K. Sytin in Bot. Zhurn.*, 84(12): 121 (1999)—Krym.
varzobicus *N.F. Gontscharow in Fl. Tadzhikistana*, 5: 666 (1937)—Tadzhikistan.
venulosus Boiss.
subsp. iraqensis *D. Podlech in Sendtnera*, 6: 167 (1999)—Iraq.
villosidentatus *D. Podlech in Soc. Éch. Pl. Vasc. Eur. Bassin Médit. Bull.*, 27: 94 (1998)—Turkey.
villosulus *D. Podlech in Sendtnera*, 6: 141 (1999)—Afghanistan.
voburus *Y.H. Xu & Q.R. Liu in Acta Bot. Yunnanica*, 19(2): 190 (1997), nom. nov.: A. grubovii *Cheng f. ex P.C. Li.*
wiedemannianus *F.B. Fischer in Bull. Soc. Imp. Naturalistes Moscou*, 26(4): 395 (1853 publ. 1854)—Turkey.†
xiqingshanicus *Y.H. Wu in Acta Bot. Yunnanica*, 20(1): 36 (1998)—Qinghai.
yangii *C. Chen & Zi G. Qian in Acta Bot. Yunnanica*, 19(3): 237 (1997)—China South-central (Yunnan).
zaprjagaevii *N.F. Gontscharow in Fl. Tadzhikistana*, 5: 670 (1937)—Tadzhikistan.
zarjabadensis *M.M. Ranjbar in Iranian J. Bot.*, 7(1): 41 (1996)—Iran.

ASTROLOBA (Aloac.).
corrugata *N.L. Meyer & G.F. Smith in Bothalia*, 28(1): 61 (1998)—Cape Province.
rubriflora *(L. Bolus) G.F. Smith & J.C. Manning in Bothalia*, 30(1): 53 (2000): Apicra rubriflora.

ASTROPHYTUM (Cactac.).
ser. Anastrophytum *A.B. Doweld in Sukkulenty (Moscow)*, 3(1–2): 36 (2000).
capricorne
subsp. senile *(Frič) A.B. Doweld in Sukkulenty (Moscow)*, 3(1–2): 36 (2000): Astrophytum senile.

ASTROPHYTUM:-
mirum *J.J. Halda & P. Panarotto in Acta Mus. Richnov. Sect. Nat.*, 7(1): 33 (2000)—Mexico Northeast (Tamaulipas).

ASTROTRICHA (Araliac.).
brachyandra *A.R. Bean in Austrobaileya*, 5(1): 65 (1997)—Queensland.
umbrosa *A.R. Bean in Austrobaileya*, 5(1): 66 (1997)—Queensland.

ASTROTRICHILIA (Meliac.).
diegoensis *J.-F. Leroy & M. Lescot in Bull. Mus. Nation. Hist. Nat., B, Adansonia*, Ser. 4, 18(1–2): 30 (1996)—Madagascar.
elegans *J.-F. Leroy & M. Lescot in Bull. Mus. Nation. Hist. Nat., B, Adansonia*, Ser. 4, 18(1–2): 33 (1996)—Madagascar.
masoalensis *J.-F. Leroy in Bull. Mus. Nation. Hist. Nat., B, Adansonia*, Ser. 4, 18(1–2): 25 (1996)—Madagascar.
parvifolia *J.-F. Leroy & M. Lescot in Bull. Mus. Nation. Hist. Nat., B, Adansonia*, Ser. 4, 18(1–2): 28 (1996)—Madagascar.
procera *J.-F. Leroy in Bull. Mus. Nation. Hist. Nat., B, Adansonia*, Ser. 4, 18(1–2): 32 (1996)—Madagascar.
rakodomena *J.-F. Leroy in Bull. Mus. Nation. Hist. Nat., B, Adansonia*, Ser. 4, 18(1–2): 22 (1996)—Madagascar.
thouvenotii *J.-F. Leroy in Bull. Mus. Nation. Hist. Nat., B, Adansonia*, Ser. 4, 18(1–2): 24 (1996)—Madagascar.
valiandra *J.-F. Leroy in Bull. Mus. Nation. Hist. Nat., B, Adansonia*, Ser. 4, 18(1–2): 29 (1996)—Madagascar.
voamatata *J.-F. Leroy in Bull. Mus. Nation. Hist. Nat., B, Adansonia*, Ser. 4, 18(1–2): 21 (1996)—Madagascar.
zombitsyensis *J.-F. Leroy & M. Lescot in Bull. Mus. Nation. Hist. Nat., B, Adansonia*, Ser. 4, 18(1–2): 26 (1996)—Madagascar.

ASYNEUMA (Campanulac.).
compactum (Boiss. & Heldr.) Damboldt
var. eriocarpum *G. Parolly in Willdenowia*, 30(1): 72 (2000)—Turkey.
dulongense *H. Li in Fl. Dulongjian Reg.*: 282 (1993), without latin descr. or type—China South-central (Yunnan).
junceum *G. Parolly in Willdenowia*, 30(1): 71 (2000), nom. nov.: Campanula juncea *Wettst.*

ASYSTASIA (Acanthac.).
albiflora *Ensermu Kelbessa in Kew Bull.*, 53(4): 930 (1998)—Tanzania, Zaire, Zambia.
gangetica (L.) T. Anderson
var. mendeliana *S.R. Kundu & Mahua Pal in J. Econ. Taxon. Bot.*, 21(2): 486 (1997)—India (West Bengal, Kerala, Tamil Nadu).
richardsiae *Ensermu Kelbessa in Kew Bull.*, 53(4): 929 (1998)—Tanzania.

ATELEIA (Legumin.).
acunae *A. Barreto Valdés in Fontqueria*, 44: 247 (1996)—Cuba.
apetala Griseb.
var. pubescens *A. Barreto Valdés in Fontqueria*, 44: 247 (1996)—Cuba.
var. trinitensis *A. Barreto Valdés in Fontqueria*, 44: 250 (1996)—Cuba.
baracoensis *A. Barreto Valdés in Fontqueria*, 44: 250 (1996)—Cuba.
bissei *A. Barreto Valdés in Fontqueria*, 44: 250 (1996)—Cuba.
cubensis Griseb.
var. marianensis *A. Barreto Valdés in Fontqueria*, 44: 251 (1996)—Cuba.
var. orientensis *A. Barreto Valdés in Fontqueria*, 44: 251 (1996)—Cuba.
roigii *A. Barreto Valdés in Fontqueria*, 44: 251 (1996)—Cuba.
salicifolia Mohlenbr.
var. pubescens *A. Barreto Valdés in Fontqueria*, 44: 252 (1996)—Cuba.

ATRACTOCARPUS (Rubiac.).
benthamianus *(F. Muell.) C.F. Puttock in Austral. Syst. Bot.*, 12(2): 277 (1999): Randia benthamiana.

ATRACTOCARPUS:-
benthamianus *(F. Muell.) C.F. Puttock*
subsp. glaber *C.F. Puttock in Austral. Syst. Bot.*, 12(2): 281 (1999)—New South Wales, Queensland.
carolinensis *(Valeton) C.F. Puttock in Austral. Syst. Bot.*, 12(2): 300 (1999): Randia carolinensis.
chartaceus *(F. Muell.) C.F. Puttock in Austral. Syst. Bot.*, 12(2): 282 (1999): Gardenia chartacea.
crosbyi *(Burkill) C.F. Puttock in Austral. Syst. Bot.*, 12(2): 301 (1999): Randia crosbyi.
decorus *(Valeton) C.F. Puttock in Austral. Syst. Bot.*, 12(2): 301 (1999): Randia decora.
fitzalanii *(F. Muell.) C.F. Puttock in Austral. Syst. Bot.*, 12(2): 285 (1999): Gardenia fitzalanii.
subsp. tenuipes *C.F. Puttock in Austral. Syst. Bot.*, 12(2): 288 (1999)—Queensland.
hirtus *(F. Muell.) C.F. Puttock in Austral. Syst. Bot.*, 12(2): 289 (1999): Gardenia hirta.
longipes *(A.C. Sm.) C.F. Puttock in Austral. Syst. Bot.*, 12(2): 302 (1999): Sukunia longipes.
macarthurii *(F. Muell.) C.F. Puttock in Austral. Syst. Bot.*, 12(2): 303 (1999): Randia macarthurii.
merikin *(F.M. Bailey) C.F. Puttock in Austral. Syst. Bot.*, 12(2): 292 (1999): Gardenia merikin.
obscurinervius *(Merr.) C.F. Puttock in Austral. Syst. Bot.*, 12(2): 304 (1999): Gardenia obscurinervia.
pentagonioides *(Seem.) C.F. Puttock in Austral. Syst. Bot.*, 12(2): 305 (1999): Gardenia pentagonioides.
pterocarpon *(Guillaumin) C.F. Puttock in Austral. Syst. Bot.*, 12(2): 305 (1999): Franciella pterocarpon.
sessilis *(F. Muell.) C.F. Puttock in Austral. Syst. Bot.*, 12(2): 294 (1999): Randia sessilis.
tahitiensis *(Nadeaud) C.F. Puttock in Austral. Syst. Bot.*, 12(2): 306 (1999): Randia tahitiensis.
tenuiflorus *(A.C. Sm.) C.F. Puttock in Austral. Syst. Bot.*, 12(2): 306 (1999): Randia tenuiflora.
versteegii *(Valeton) C.F. Puttock in Austral. Syst. Bot.*, 12(2): 307 (1999): Randia versteegii.

ATRACTYLIS (Compos.).
arbuscula *E.R. Sventenius & P. Michaelis in Ind. Sem. Hort. Acclim. Pl. Arautapae*, 1968: 50 (1969)—Canary Is.
delvarii *V. Mozaffarian in Iranian J. Bot.*, 7(1): 130 (1996)—Iran.

ATRIPLEX (Chenopodiac.).
subsect. Heterospermae *(Aellen) N.A. Medvedeva in Fl. Vostochnoĭ Evropy*, 9: 48 (1996): A. ser. Heterospermae.
subsect. Latifoliae *N.A. Medvedeva in Fl. Vostochnoĭ Evropy*, 9: 50 (1996).
subsect. Patulae *(Aellen) N.A. Medvedeva in Fl. Vostochnoĭ Evropy*, 9: 48 (1996): A. sect. Patulae.
subsect. Suffruticosae *N.A. Medvedeva in Fl. Vostochnoĭ Evropy*, 9: 52 (1996).
alaschanica *Y.Z. Zhao in Acta Phytotax. Sin.*, 35(3): 257 (1997)—Inner Mongolia (Nei Mongol).
altaica *A.P. Suchorukow in Feddes Repert.*, 111(3–4): 176 (2000)—Altay.
dimorphostegia Kar. & Kir.
f. collenettea *T.A. Al-Turki, S. Omer & A. Ghafoor in Feddes Repert.*, 111(5–6): 278 (2000)—Saudi Arabia.
erecticaulis *H.C. Stutz, G.L. Chu & S.C. Sanderson in Madroño*, 44(1): 89 (1997)—California.
hollowayi *P.J. de Lange & D.A. Norton in New Zealand J. Bot.*, 38(4): 553 (2000)—New Zealand North.
leucoclada Boiss.
var. mandavillea *T.A. Al-Turki, S. Omer & A. Ghafoor in Feddes Repert.*, 111(5–6): 281 (2000)—Saudi Arabia.
longitrichoma *H.C. Stutz, G.L. Chu & S.C. Sanderson in Madroño*, 45(2): 128 (1998 publ. 1999)—Nevada, California.
pachypoda *H.C. Stutz & G.L. Chu in Madroño*, 44(3): 277 (1997 publ. 1998)—Colorado, New Mexico.
sagittiformis *(Aellen) G.L. Chu in Higher Plants China*, 4: 327 (2000): Atriplex dimorphostegia var. sagittiformis.
subtilis *H.C. Stutz & G.L. Chu in Madroño*, 44(2): 184 (1997 publ. 1998)—California.

ATROPA (Solanac.).
lutea *(Döll) A.S. Al-Nowaihi & M.M. Mourad in Taeckholmia,* 19(2): 175 (1999), without basionym ref.: A. belladonna var. lutea.

ATTALEA (Palmae).
barreirensis *S.F. Glassman, Taxon. Treatment Palm Subtribe Attaleinae (Tribe Cocoeae) (Illinois Biol.* Monogr., 59): 25 (1999)—Brazil Northeast (Bahia).
brasiliensis *S.F. Glassman, Taxon. Treatment Palm Subtribe Attaleinae (Tribe Cocoeae) (Illinois Biol.* Monogr., 59): 65 (1999)—Brazil West-central (Goiás).
salvadorensis *S.F. Glassman, Taxon. Treatment Palm Subtribe Attaleinae (Tribe Cocoeae) (Illinois Biol.* Monogr., 59): 63 (1999)—Brazil Northeast (Bahia).
seabrensis *S.F. Glassman, Taxon. Treatment Palm Subtribe Attaleinae (Tribe Cocoeae) (Illinois Biol.* Monogr., 59): 50 (1999)—Brazil Northeast (Bahia).
× voeksii *Noblick ex S.F. Glassman, Taxon. Treatment Palm Subtribe Attaleinae (Tribe Cocoeae) (Illinois Biol.* Monogr., 59): 191 (1999)—Brazil Northeast (Bahia).(A. funifera × humilis).

AUBLETIANA *J. Murillo-A. in Rev. Acad. Colomb. Cienc. Exact. Fis. Nat.,* 24(92): 360 (2000). *EUPHORBIACEAE*
leptostachys *(Breteler) J. Murillo-A. in Rev. Acad. Colomb. Cienc. Exact. Fis. Nat.,* 24(92): 361 (2000): Conceveiba leptostachys.
macrostachys *(Breteler) J. Murillo-A. in Rev. Acad. Colomb. Cienc. Exact. Fis. Nat.,* 24(92): 361 (2000): Conceveiba macrostachys.

AUGUSTA (Rubiac.).
subgen. Lindenia *(Benth.) J.H. Kirkbride in Brittonia,* 49(3): 358 (1997): gen. Lindenia.
sect. Lindenia *(Benth.) J.H. Kirkbride in Brittonia,* 49(3): 358 (1997): gen. Lindenia.
sect. Pacifica *(S.P. Darwin) J.H. Kirkbride in Brittonia,* 49(3): 358 (1997): Lindenia sect. Pacifica.
austrocaledonica *(Brongn.) J.H. Kirkbride in Brittonia,* 49(3): 359 (1997): Lindenia austrocaledonica.
longifolia (Spreng.) Rehder
var. parvifolia *(Pohl) P.G. Delprete in Brittonia,* 49(4): 495 (1997), as 'parviflora': A. parvifolia.
rivalis *(Benth.) J.H. Kirkbride in Brittonia,* 49(3): 358 (1997): Lindenia rivalis.
f. glabra *(S.P. Darwin) J.H. Kirkbride in Brittonia,* 49(3): 358 (1997): Lindenia rivalis f. glabra.
vitiensis *(Seem.) J.H. Kirkbride in Brittonia,* 49(3): 358 (1997): Lindenia vitiensis.

AULACOCALYX (Rubiac.).
jasminiflora Hook. f.
subsp. kivuensis *E. Figueiredo in Kew Bull.,* 52(3): 649 (1997)—Zaire.
pallens *(Hiern) D.M. Bridson & E. Figueiredo in Kew Bull.,* 52(3): 653 (1997): Randia pallens.
subsp. letestui *(Pellegr.) E. Figueiredo in Kew Bull.,* 52(3): 655 (1997): Dorothea letestui.
subulata *(N. Hallé) E. Figueiredo in Kew Bull.,* 52(3): 643 (1997): A. lujae var. subulata.
subsp. glabra *E. Figueiredo in Kew Bull.,* 52(3): 645 (1997)—Gabon.

AULACOSPERMUM (Umbellif.).
ikonnikovii *R.V. Kamelin in Novosti Sist. Vyssh. Rast.,* 32: 131 (2000)—Tadzhikistan.

AULONEMIA (Gramin.).
bogotensis *L.G. Clark, X. Londoño & M. Kobayashi in Brittonia,* 49(4): 504 (1997)—Colombia.
boliviana *S.A. Renvoize, Gramineas de Bolivia:* 39 (1998)—Bolivia.
longipedicellata *S.A. Renvoize, Gramineas de Bolivia:* 41 (1998)—Bolivia.
patriae *R.W. Pohl in Ann. Missouri Bot. Gard.,* 68(1): 225 (1981), with type—Costa Rica.
tremula *S.A. Renvoize, Gramineas de Bolivia:* 39 (1998)—Bolivia.

AURANTICARPA *L.W. Cayzer, M.D. Crisp & I.R.H. Telford in Austral. Syst. Bot.,* 13(6): 904 (2000). *PITTOSPORACEAE*
edentata *L.W. Cayzer, M.D. Crisp & I.R.H. Telford in Austral. Syst. Bot.,* 13(6): 908 (2000)—Queensland.
ilicifolia *L.W. Cayzer, M.D. Crisp & I.R.H. Telford in Austral. Syst. Bot.,* 13(6): 912 (2000)—Queensland.
melanosperma *(F. Muell.) L.W. Cayzer, M.D. Crisp & I.R.H. Telford in Austral. Syst. Bot.,* 13(6): 914 (2000): Pittosporum melanospermum. (Pittosporac.).
papyracea *L.W. Cayzer, M.D. Crisp & I.R.H. Telford in Austral. Syst. Bot.,* 13(6): 909 (2000)—Queensland.
resinosa *(Domin) L.W. Cayzer, M.D. Crisp & I.R.H. Telford in Austral. Syst. Bot.,* 13(6): 915 (2000): Pittosporum resinosum.
rhombifolia *(A. Cunn. ex Hook.) L.W. Cayzer, M.D. Crisp & I.R.H. Telford in Austral. Syst. Bot.,* 13(6): 906 (2000): Pittosporum rhombifolium.

AURELIANA (Solanac.).
angustifolia *R.C. Almeida-Lafetá in Novon,* 10(3): 187 (2000)—Brazil Southeast (Minas Gerais).

AUSTRALLUMA *D.C.H. Plowes in Haseltonia,* 3: 54 (1995). *ASCLEPIADACEAE*
peschii *(Nel) D.C.H. Plowes in Haseltonia,* 3: 54 (1995): Caralluma peschii.

AUSTROBUXUS (Euphorbiac.).
megacarpus *P.I. Forster in Austrobaileya,* 4(4): 622 (1997)—Queensland.

AUSTROCYLINDROPUNTIA (Cactac.).
hirschii *(Backeb.) E.F. Anderson in Cact. Succ. J. (U.S.A.),* 71(6): 324 (1999): Tephrocactus hirschii.
punta-caillan *(Rauh & Backeb.) E.F. Anderson in Cact. Succ. J. (U.S.A.),* 71(6): 324 (1999): Tephrocactus punta-caillan.
yanganucensis *(Rauh & Backeb.) E.F. Anderson in Cact. Succ. J. (U.S.A.),* 71(6): 324 (1999): Tephrocactus yanganucensis.

AUSTRODANTHONIA *H.P. Linder in Telopea,* 7(3): 269 (1997). *GRAMINEAE*
acerosa *(Vickery) H.P. Linder in Telopea,* 7(3): 270 (1997): Danthonia acerosa.
alpicola *(Vickery) H.P. Linder in Telopea,* 7(3): 270 (1997): Danthonia alpicola.
auriculata *(J.M. Black) H.P. Linder in Telopea,* 7(3): 270 (1997): Danthonia auriculata.
biannularis *(Zotov) H.P. Linder in Telopea,* 7(3): 270 (1997): Notodanthonia biannularis.
bipartita *(Link) H.P. Linder in Telopea,* 7(3): 270 (1997): Avena bipartita.
bonthainica *(Jansen) H.P. Linder in Telopea,* 7(3): 270 (1997): Danthonia pilosa var. bonthainica.
caespitosa *(Gaudich.) H.P. Linder in Telopea,* 7(3): 271 (1997): Danthonia caespitosa.
carphoides *(Benth.) H.P. Linder in Telopea,* 7(3): 271 (1997): Danthonia carphoides.
clavata *(Zotov) H.P. Linder in Telopea,* 7(3): 271 (1997): Notodanthonia clavata.
diemenica *(D.I. Morris) H.P. Linder in Telopea,* 7(3): 271 (1997): Danthonia diemenica.
duttoniana *(Cashmore) H.P. Linder in Telopea,* 7(3): 271 (1997): Danthonia duttoniana.
eriantha *(Lindl.) H.P. Linder in Telopea,* 7(3): 271 (1997): Danthonia eriantha.
fulva *(Vickery) H.P. Linder in Telopea,* 7(3): 271 (1997): Danthonia linkii var. fulva.
geniculata *(J.M. Black) H.P. Linder in Telopea,* 7(3): 271 (1997): Danthonia geniculata.
induta *(Vickery) H.P. Linder in Telopea,* 7(3): 272 (1997): Danthonia induta.
laevis *(Vickery) H.P. Linder in Telopea,* 7(3): 272 (1997): Danthonia laevis.
mera *(Connor & Edgar) H.P. Linder in Telopea,* 7(3): 272 (1997): Rytidosperma merum.
monticola *(Vickery) H.P. Linder in Telopea,* 7(3): 272 (1997): Danthonia monticola.
occidentalis *(Vickery) H.P. Linder in Telopea,* 7(3): 272 (1997): Danthonia occidentalis.

AUSTRODANTHONIA:-
oreophila *(H.P. Linder & N.G. Walsh) H.P. Linder in Telopea,* 7(3): 272 (1997): Rytidosperma oreophilum.
penicillata *(Labill.) H.P. Linder in Telopea,* 7(3): 272 (1997): Arundo penicillata.
pilosa *(R. Br.) H.P. Linder in Telopea,* 7(3): 272 (1997): Danthonia pilosa.
popinensis *(D.I. Morris) H.P. Linder in Telopea,* 7(3): 273 (1997): Danthonia popinensis.
racemosa *(R. Br.) H.P. Linder in Telopea,* 7(3): 273 (1997): Danthonia racemosa.
var. obtusata *(Benth.) H.P. Linder in Telopea,* 7(3): 273 (1997): Danthonia racemosa var. obtusata.
remota *(D.I. Morris) H.P. Linder in Telopea,* 7(3): 273 (1997): Danthonia remota.
richardsonii *(Cashmore) H.P. Linder in Telopea,* 7(3): 273 (1997): Danthonia richardsonii.
setacea *(R. Br.) H.P. Linder in Telopea,* 7(3): 273 (1997): Danthonia setacea.
tenuior *(Steud.) H.P. Linder in Telopea,* 7(3): 273 (1997): Plinthanthesis tenuior.

AUSTRODRIMYS *A.B. Doweld in Novosti Sist. Vyssh. Rast.,* 32: 38 (2000). *WINTERACEAE*
lanceolata *(Poir.) A.B. Doweld in Novosti Sist. Vyssh. Rast.,* 32: 38 (2000): Winterania lanceolata.

AUSTROMUELLERA (Proteac.).
valida *B.P.M. Hyland in Fl. Australia,* 17B: 396 (1999)—Queensland.

AUSTROSTEENISIA (Legumin.).
blackii (F. Muell.) R. Geesink
var. astipella *D.J. Dixon in Austrobaileya,* 5(1): 85 (1997)—Queensland.
mollitricha *D.J. Dixon in Austrobaileya,* 5(1): 87 (1997)—Queensland.

AUSTROSTIPA *S.W.L. Jacobs & J. Everett in Telopea,* 6(4): 582 (1996). *GRAMINEAE*
subgen. Arbuscula *S.W.L. Jacobs & J. Everett in Telopea,* 6(4): 591 (1996).
subgen. Aulax *S.W.L. Jacobs & J. Everett in Telopea,* 6(4): 591 (1996).
subgen. Bambusina *S.W.L. Jacobs & J. Everett in Telopea,* 6(4): 593 (1996).
subgen. Ceres *S.W.L. Jacobs & J. Everett in Telopea,* 6(4): 592 (1996).
subgen. Eremophilae *S.W.L. Jacobs & J. Everett in Telopea,* 6(4): 592 (1996).
subgen. Falcatae *S.W.L. Jacobs & J. Everett in Telopea,* 6(4): 592 (1996).
subgen. Lancea *S.W.L. Jacobs & J. Everett in Telopea,* 6(4): 591 (1996).
subgen. Lanterna *S.W.L. Jacobs & J. Everett in Telopea,* 6(4): 591 (1996).
subgen. Lobatae *S.W.L. Jacobs & J. Everett in Telopea,* 6(4): 592 (1996).
subgen. Longiaristatae *S.W.L. Jacobs & J. Everett in Telopea,* 6(4): 591 (1996).
subgen. Petaurista *S.W.L. Jacobs & J. Everett in Telopea,* 6(4): 592 (1996).
subgen. Tuberculatae *S.W.L. Jacobs & J. Everett in Telopea,* 6(4): 592 (1996).
acrociliata *(Reader) S.W.L. Jacobs & J. Everett in Telopea,* 6(4): 584 (1996): Stipa acrociliata.
aphylla *(Rodway) S.W.L. Jacobs & J. Everett in Telopea,* 6(4): 584 (1996): Stipa pubescens var. aphylla.
aquarii *(Vickery, S.W.L. Jacobs & J. Everett) S.W.L. Jacobs & J. Everett in Telopea,* 6(4): 584 (1996): Stipa aquarii.
aristiglumis *(F. Muell.) S.W.L. Jacobs & J. Everett in Telopea,* 6(4): 584 (1996): Stipa aristiglumis.
bigeniculata *(Hughes) S.W.L. Jacobs & J. Everett in Telopea,* 6(4): 584 (1996): Stipa bigeniculata.
blackii *(C.E. Hubb.) S.W.L. Jacobs & J. Everett in Telopea,* 6(4): 584 (1996): Stipa blackii.
blakei *(Vickery, S.W.L. Jacobs & J. Everett) S.W.L. Jacobs & J. Everett in Telopea,* 6(4): 584 (1996): Stipa blakei.
breviglumis *(J.M. Black) S.W.L. Jacobs & J. Everett in Telopea,* 6(4): 584 (1996): Stipa breviglumis.
campylachne *(Nees) S.W.L. Jacobs & J. Everett in Telopea,* 6(4): 584 (1996): Stipa campylachne.

AUSTROSTIPA
centralis *(Vickery, S.W.L. Jacobs & J. Everett) S.W.L. Jacobs & J. Everett in Telopea*, 6(4): 584 (1996): Stipa centralis.
compressa *(R. Br.) S.W.L. Jacobs & J. Everett in Telopea*, 6(4): 584 (1996): Stipa compressa.
crinita *(Gaudich.) S.W.L. Jacobs & J. Everett in Telopea*, 6(4): 585 (1996): Stipa crinita.
curticoma *(Vickery) S.W.L. Jacobs & J. Everett in Telopea*, 6(4): 585 (1996): Stipa curticoma.
densiflora *(Hughes) S.W.L. Jacobs & J. Everett in Telopea*, 6(4): 585 (1996): Stipa densiflora.
dongicola *(Vickery, S.W.L. Jacobs & J. Everett) S.W.L. Jacobs & J. Everett in Telopea*, 6(4): 585 (1996): Stipa dongicola.
drummondii *(Steud.) S.W.L. Jacobs & J. Everett in Telopea*, 6(4): 585 (1996): Stipa drummondii.
echinata *(Vickery, S.W.L. Jacobs & J. Everett) S.W.L. Jacobs & J. Everett in Telopea*, 6(4): 585 (1996): Stipa echinata.
elegantissima *(Labill.) S.W.L. Jacobs & J. Everett in Telopea*, 6(4): 585 (1996): Stipa elegantissima.
eremophila *(Reader) S.W.L. Jacobs & J. Everett in Telopea*, 6(4): 585 (1996): Stipa eremophila.
exilis *(Vickery) S.W.L. Jacobs & J. Everett in Telopea*, 6(4): 585 (1996): Stipa exilis.
feresetacea *(Vickery, S.W.L. Jacobs & J. Everett) S.W.L. Jacobs & J. Everett in Telopea*, 6(4): 585 (1996): Stipa feresetacea.
flavescens *(Labill.) S.W.L. Jacobs & J. Everett in Telopea*, 6(4): 585 (1996): Stipa flavescens.
geoffreyi *S.W.L. Jacobs & J. Everett in Telopea*, 6(4): 585 (1996)—Western Australia.
gibbosa *(Vickery) S.W.L. Jacobs & J. Everett in Telopea*, 6(4): 586 (1996): Stipa gibbosa.
hemipogon *(Benth.) S.W.L. Jacobs & J. Everett in Telopea*, 6(4): 586 (1996): Stipa hemipogon.
juncifolia *(Hughes) S.W.L. Jacobs & J. Everett in Telopea*, 6(4): 586 (1996): Stipa juncifolia.
lanata *(Vickery, S.W.L. Jacobs & J. Everett) S.W.L. Jacobs & J. Everett in Telopea*, 6(4): 586 (1996): Stipa lanata.
macalpinei *(Reader) S.W.L. Jacobs & J. Everett in Telopea*, 6(4): 587 (1996): Stipa macalpinei.
metatoris *(J. Everett & S.W.L. Jacobs) S.W.L. Jacobs & J. Everett in Telopea*, 6(4): 587 (1996): Stipa metatoris.
mollis *(R. Br.) S.W.L. Jacobs & J. Everett in Telopea*, 6(4): 587 (1996): Stipa mollis.
muelleri *(Tate) S.W.L. Jacobs & J. Everett in Telopea*, 6(4): 587 (1996): Stipa muelleri.
multispiculis *(J.M. Black) S.W.L. Jacobs & J. Everett in Telopea*, 6(4): 587 (1996): Stipa multispiculis.
mundula *(J.M. Black) S.W.L. Jacobs & J. Everett in Telopea*, 6(4): 587 (1996): Stipa mundula.
nitida *(Summerh. & C.E. Hubb.) S.W.L. Jacobs & J. Everett in Telopea*, 6(4): 587 (1996): Stipa nitida.
nivicola *(J.H. Willis) S.W.L. Jacobs & J. Everett in Telopea*, 6(4): 587 (1996): Stipa nivicola.
nodosa *(S.T. Blake) S.W.L. Jacobs & J. Everett in Telopea*, 6(4): 587 (1996): Stipa nodosa.
nullanulla *(J. Everett & S.W.L. Jacobs) S.W.L. Jacobs & J. Everett in Telopea*, 6(4): 587 (1996): Stipa nullanulla.
nullarborensis *(Vickery, S.W.L. Jacobs & J. Everett) S.W.L. Jacobs & J. Everett in Telopea*, 6(4): 587 (1996): Stipa nullarborensis.
oligostachya *(Hughes) S.W.L. Jacobs & J. Everett in Telopea*, 6(4): 587 (1996): Stipa oligostachya.
petraea *(Vickery) S.W.L. Jacobs & J. Everett in Telopea*, 6(4): 587 (1996): Stipa petraea.
pilata *(S.W.L. Jacobs & J. Everett) S.W.L. Jacobs & J. Everett in Telopea*, 6(4): 587 (1996): Stipa pilata.
platychaeta *(Hughes) S.W.L. Jacobs & J. Everett in Telopea*, 6(4): 587 (1996): Stipa platychaeta.
plumigera *(Hughes) S.W.L. Jacobs & J. Everett in Telopea*, 6(4): 588 (1996): Stipa plumigera.
puberula *(Steud.) S.W.L. Jacobs & J. Everett in Telopea*, 6(4): 588 (1996): Stipa puberula.
pubescens *(R. Br.) S.W.L. Jacobs & J. Everett in Telopea*, 6(4): 588 (1996): Stipa pubescens.
pubinodis *(Trin. & Rupr.) S.W.L. Jacobs & J. Everett in Telopea*, 6(4): 588 (1996): Stipa pubinodis.
pycnostachya *(Benth.) S.W.L. Jacobs & J. Everett in Telopea*, 6(4): 588 (1996): Stipa pycnostachya.

AUSTROSTIPA:-
ramosissima *(Trin.) S.W.L. Jacobs & J. Everett in Telopea*, 6(4): 588 (1996): Urachne ramosissima.
rudis *(Spreng.) S.W.L. Jacobs & J. Everett in Telopea*, 6(4): 588 (1996): Stipa rudis.
subsp. australis *(J. Everett & S.W.L. Jacobs) S.W.L. Jacobs & J. Everett in Telopea*, 6(4): 588 (1996): Stipa rudis
subsp. nervosa *(Vickery) S.W.L. Jacobs & J. Everett in Telopea*, 6(4): 588 (1996): Stipa nervosa var. nervosa.
scabra *(Lindl.) S.W.L. Jacobs & J. Everett in Telopea*, 6(4): 588 (1996): Stipa scabra.
subsp. falcata *(Hughes) S.W.L. Jacobs & J. Everett in Telopea*, 6(4): 588 (1996): Stipa falcata.
semibarbata *(R. Br.) S.W.L. Jacobs & J. Everett in Telopea*, 6(4): 588 (1996): Stipa semibarbata.
setacea *(R. Br.) S.W.L. Jacobs & J. Everett in Telopea*, 6(4): 589 (1996): Stipa setacea.
stipoides *(Hook. f.) S.W.L. Jacobs & J. Everett in Telopea*, 6(4): 589 (1996): Dichelachne stipoides.
stuposa *(Hughes) S.W.L. Jacobs & J. Everett in Telopea*, 6(4): 589 (1996): Stipa stuposa.
tenuifolia *(Steud.) S.W.L. Jacobs & J. Everett in Telopea*, 6(4): 589 (1996): Stipa tenuifolia.
trichophylla *(Benth.) S.W.L. Jacobs & J. Everett in Telopea*, 6(4): 589 (1996): Stipa trichophylla.
tuckeri *(F. Muell.) S.W.L. Jacobs & J. Everett in Telopea*, 6(4): 589 (1996): Stipa tuckeri.
variabilis *(Hughes) S.W.L. Jacobs & J. Everett in Telopea*, 6(4): 589 (1996): Stipa variabilis.
velutina *(Vickery, S.W.L. Jacobs & J. Everett) S.W.L. Jacobs & J. Everett in Telopea*, 6(4): 589 (1996): Stipa velutina.
verticillata *(Nees ex Spreng.) S.W.L. Jacobs & J. Everett in Telopea*, 6(4): 589 (1996): Stipa verticillata.
vickeryana *(J. Everett & S.W.L. Jacobs) S.W.L. Jacobs & J. Everett in Telopea*, 6(4): 589 (1996): Stipa vickeryana.
wakoolica *(Vickery, S.W.L. Jacobs & J. Everett) S.W.L. Jacobs & J. Everett in Telopea*, 6(4): 589 (1996): Stipa wakoolica.

AUTONOE *(Webb & Berthel.) F. Speta in Phyton (Austria)*, 38(1): 93 (1998): Scilla sect. Autonoe. (Hyacinthac.).
berthelotii *(Webb & Berthel.) F. Speta in Phyton (Austria)*, 38(1): 93 (1998): Scilla berthelotii.
dasyantha *(Webb & Berthel.) F. Speta in Phyton (Austria)*, 38(1): 93 (1998): Scilla dasyantha.
haemorrhoidalis *(Webb & Berthel.) F. Speta in Phyton (Austria)*, 38(1): 93 (1998): Scilla haemorrhoidalis.
iridifolia *(Webb & Berthel.) F. Speta in Phyton (Austria)*, 38(1): 95 (1998): Scilla iridifolia.
latifolia *(Willd.) F. Speta in Phyton (Austria)*, 38(1): 95 (1998): Scilla latifolia.
madeirensis *(Menezes) F. Speta in Phyton (Austria)*, 38(1): 95 (1998): Scilla madeirensis.

AVENA (Gramin.).
sect. Triaristata *C. Romero Zarco in Lagascalia*, 18(2): 185 (1996).
sect. Ventricosa *B.R. Baum ex C. Romero Zarco in Lagascalia*, 18(2): 194 (1996).
parlatorei *Joseph Woods, Tourist's fl.*: 405 (1850), as 'parlatorii'—Italy.†
sativa L.
subsp. byzantina *(K. Koch) C. Romero Zarco in Lagascalia*, 18(2): 180 (1996): A. byzantina.
sterilis L.
nothosubsp. malzevii *H. Scholz in Ber. Bayer. Bot. Ges.*, 66–67: 229 (1996)—Italy.(A. fatua subsp. fatua × sterilis subsp. ludoviciana).

AVENOCHLOA (Gramin.).
albinervis (Boiss.) Holub
subsp. occidentalis *C. Gervais in Denkschr. Schweiz. Naturforsch. Ges.*, 88: 122 (1973)—Portugal.
× praeusta *(Rchb.) C. Gervais in Denkschr. Schweiz. Naturforsch. Ges.*, 88: 103 (1973): Avena praeusta.
vasconica *(Sennen) C. Gervais in Denkschr. Schweiz. Naturforsch. Ges.*, 88: 95 (1973): Avena vasconica.

AVENULA (Gramin.).
aenigmatica *(D. Lange) J. Holub in Preslia*, 70(2): 100 (1998): Helictotrichon blavii subsp. aenigmatica.
arundana *(Romero Zarco) J. Holub in Preslia*, 70(2): 100 (1998): A. gervaisii subsp. arundana.
cintrana *(Röser) J. Holub in Preslia*, 70(2): 100 (1998): Helictotrichon cintranum.
hackelii (Henriq.) Holub
subsp. stenophylla *(Franco) J. do Amaral Franco in Silva Lusitana*, 5(1): 141 (1997): A. occidentalis subsp. stenophylla.
lodunensis (Delastre) Kerguélen
subsp. cintrana *(Röser) C. Cebolla Lozano, J.Á. López Rodríguez & M.A. Rivas Ponce in Fontqueria*, 49: 12 (1997): Helictotrichon cintranum.
subsp. marginata *(Lowe) J. Holub in Preslia*, 70(2): 100 (1998): Avena marginata.
lusitanica *(Romero Zarco) J. Holub in Preslia*, 70(2): 100 (1998): A. pratensis subsp. lusitanica.
pseudoviolacea *(Dalla Torre) J. Holub in Preslia*, 70(2): 100 (1998): Avena pseudoviolacea.
sulcata (Boiss.) Dumort.
subsp. reuteri *(Romero Zarco) J. do Amaral Franco in Silva Lusitana*, 5(1): 141 (1997): A. sulcata var. reuteri.

AVERRHOIDIUM (Sapindac.).
spondioides *(Standl.) P. Acevedo-Rodríguez & M.S. Ferrucci in Novon*, 8(2): 105 (1998): Matayba spondioides.

AVONIA (Portulacac.).
prominens *(G. Will.) G. Williamson in Brit. Cact. Succ. J.*, 15(4): 208 (1997): Anacampseros prominens.
recurvata (Schönland) G.D. Rowley
subsp. buderiana *(Poelln.) G. Williamson in Aloe*, 35(1): 22 (1998): Anacampseros buderiana.
variabilis *(Poelln.) G. Williamson in Aloe*, 35(1): 19 (1998): Anacampseros variabilis.

AVONSERA *F. Speta in Phyton (Austria)*, 38(1): 95 (1998). *HYACINTHACEAE*
convallarioides *(H. Perrier) F. Speta in Phyton (Austria)*, 38(1): 95 (1998): Ornithogalum convallarioides.
lachenalioides *(Baker) F. Speta in Phyton (Austria)*, 38(1): 96 (1998): Scilla lachenalioides.

AXONOPUS (Gramin.).
boliviensis *S.A. Renvoize, Gramineas de Bolivia*: 496 (1998)—Bolivia.
morronei *D. Giraldo-Cañas in Caldasia*, 20(2): 88 (1998)—Colombia.
piccae *D. Giraldo-Cañas in Caldasia*, 21(2): 133 (1999)—Colombia, Venezuela.
zuloagae *D. Giraldo-Cañas in Caldasia*, 20(2): 89 (1998)—Colombia.

AYAPANOPSIS (Compos.).
beckii *H. Robinson in Biollania*, ed. esp. 6 (J.J. Wurdack Festschr.): 509 (1997)—Bolivia.
luteynii *H. Robinson & Pruski in Biollania*, ed. esp. 6 (J.J. Wurdack Festschr.): 510 (1997)—Ecuador.
wurdackiana *H. Robinson in Biollania*, ed. esp. 6 (J.J. Wurdack Festschr.): 511 (1997)—Peru.

AYENIA (Sterculiac.).
latifolia *C.L. Cristóbal in Bonplandia*, 9(3–4): 214 (1997)—Brazil Northeast (Bahia), Brazil West-central (Goiás).
mansfeldiana *(Herter) C.L. Cristóbal in Bonplandia*, 9(1–2): 149 (1996): Tragia mansfeldiana.
saligna *L.J. Dorr in Brittonia*, 48(2): 213 (1996)—Colombia.

AZALEASTRUM (Ericac.).
albiflorum Rydb.
subsp. warrenii *(A. Nelson) W.A. Weber in Phytologia*, 79(2): 65 (1995 publ. 1996): A. warrenii.

AZORELLA (Umbellif.).
diversifolia Clos
var. antillanca *J.C. Zech in Madroño*, 44(2): 193 (1997 publ. 1998)—Chile South (Los Lagos).

B

BABINGTONIA (Myrtac.).
angusta *A.R. Bean in Austrobaileya,* 5(2): 163 (1999)—New South Wales, Queensland.
behrii *(Schltdl.) A.R. Bean in Austrobaileya,* 4(4): 637 (1997): Camphoromyrtus behrii.
bidwillii *A.R. Bean in Austrobaileya,* 5(2): 161 (1999)—Queensland.
brachypoda *A.R. Bean in Austrobaileya,* 5(2): 168 (1999)—Queensland.
collina *A.R. Bean in Austrobaileya,* 5(2): 167 (1999)—Queensland, New South Wales.
crassa *A.R. Bean in Austrobaileya,* 5(2): 166 (1999)—New South Wales.
crenulata *(F. Muell.) A.R. Bean in Austrobaileya,* 4(4): 635 (1997): Camphoromyrtus crenulata.
cunninghamii *(Schauer) A.R. Bean in Austrobaileya,* 4(4): 638 (1997): Harmogia cunninghamii.
granitica *A.R. Bean in Austrobaileya,* 4(4): 642 (1997)—Queensland.
jucunda *(S.T. Blake) A.R. Bean in Austrobaileya,* 4(4): 639 (1997): Baeckea jucunda.
leratii *(Schltr.) A.R. Bean in Austrobaileya,* 4(4): 633 (1997): Baeckea leratii.
odontocalyx *A.R. Bean in Austrobaileya,* 4(4): 644 (1997)—New South Wales.
papillosa *A.R. Bean in Austrobaileya,* 5(2): 169 (1999)—Queensland.
pinifolia *(Labill.) A.R. Bean in Austrobaileya,* 4(4): 633 (1997): Leptospermum pinifolium.
pluriflora *(F. Muell.) A.R. Bean in Austrobaileya,* 5(2): 165 (1999), without basionym page: Camphoromyrtus pluriflora.
procera *(J.W. Dawson) A.R. Bean in Austrobaileya,* 4(4): 633 (1997): Baeckea procera.
prominens *A.R. Bean in Austrobaileya,* 4(4): 639 (1997)—New South Wales.
silvestris *A.R. Bean in Austrobaileya,* 4(4): 641 (1997)—Queensland, New South Wales.
similis *A.R. Bean in Austrobaileya,* 5(2): 162 (1999)—Queensland, New South Wales.
squarrulosa *(Domin) A.R. Bean in Austrobaileya,* 4(4): 642 (1997): Baeckea squarrulosa.
taxifolia *(Merr.) A.R. Bean in Austrobaileya,* 4(4): 632 (1997): Baeckea taxifolia.
tozerensis *A.R. Bean in Austrobaileya,* 4(4): 634 (1997)—Queensland.

BACCAUREA (Euphorbiac.).
carinata *R.M.A.P. Haegens in Blumea,* Suppl. 12: 103 (2000)—New Guinea (Papua New Guinea).
dulcis *(Jack) Wall. ex J.O. Voigt, Hort. Suburb. Calcutt.:* 95 (1845): Pierardia dulcis.
microcarpa *(Airy Shaw) R.M.A.P. Haegens in Blumea,* Suppl. 12: 127 (2000): B. nesophila var. microcarpa.
mollis *R.M.A.P. Haegens in Blumea,* Suppl. 12: 131 (2000)—Borneo (Sarawak).
parviflora (Müll. Arg.) Müll. Arg.
var. ptychopyxis *(Airy Shaw) T. Chakrabarty & M. Gangopadhyay in J. Econ. Taxon. Bot.,* 21(3): 531 (1997 publ. 1998): B. ptychopyxis.
pierardi *N. Wallich, Report on Calcutta Bot. Gard. to G.A. Bushby:* 31 (1840), nom. nov.: Pierardia sapida *Roxb.*
purpurea *R.M.A.P. Haegens in Blumea,* Suppl. 12: 164 (2000)—New Guinea (Papua New Guinea).
reniformis *T. Chakrabarty & M. Gangopadhyay in J. Econ. Taxon. Bot.,* 19(2): 449 (1995)—Malaya (Peninsular Malaysia).
simaloerensis *R.M.A.P. Haegens in Blumea,* Suppl. 12: 179 (2000)—Sumatera.

BACCHARIS (Compos.).
brachylaenoides
var. polycephala *(I.L. Teodoro) R. Govaerts, World Checklist Seed Plants,* 2(1–2): 9 (1996), basionym not validly publ.: Pseudobaccharis polycephala.
var. polycephala *R. Govaerts, World Checklist Seed Plants,* 3(1): 9 (1999), nom. nov., replaced syn. not validly publ.: B. polycephala *Sch. Bip.*

BACCHARIS (Compos.).
brevipappa *(McVaugh) G.L. Nesom in Phytologia,* 84(1): 46 (1998): multiflora var. brevipappa.
crassicuneata *G.L. Nesom in Phytologia,* 84(1): 43 (1998)—Mexico Northeast (Coahuila, Nuevo León, Tamaulipas).
johnwurdackiana *H. Robinson in Biollania,* ed. esp. 6 (J.J. Wurdack Festschr.): 506 (1997)—Peru.
longii *R. Govaerts, World Checklist Seed Plants,* 2(1–2): 9 (1996), nom. nov.: B. rufescens var. alpestris *Baker.*
malibuensis *R.M. Beauchamp & J. Henrickson in Aliso,* 14(3): 202 (1995 publ. 1996)—California.
markoi *H. Robinson in Biollania,* ed. esp. 6 (J.J. Wurdack Festschr.): 503 (1997)—Bolivia.
potosiensis *H. Robinson in Biollania,* ed. esp. 6 (J.J. Wurdack Festschr.): 504 (1997)—Bolivia.
pseudotenuifolia I.L. Teodoro
var. leptophylla *(DC.) D. Giuliano in Zuloaga & Morrone (eds.), Cat. Pl. Vasc. Rep. Arg. 2 (Monogr. Syst. Bot.* Missouri Bot. Gard. 74): 1245 (1999): Baccharis leptophylla.
semiserrata
var. elaeagnoides *(Steud. ex Baker) R. Govaerts, World Checklist Seed Plants,* 2(1–2): 9 (1996), as 'elaegnoides': B. elaeagnoides.
serranoi *H. Robinson in Biollania,* ed. esp. 6 (J.J. Wurdack Festschr.): 505 (1997)—Bolivia.
squamulosa *(A. Gray) M.E. Jones in Contr. W. Bot.,* 18: 71 (1933): Brickellia squamulosa.

BACCHAROIDES (Compos.).
adoensis (Sch. Bip. ex Walp.) H. Rob.
var. kotschyana *(Sch. Bip. ex Walp.) M.A. Isawumi, G. El-Ghazaly & B. Nordenstam in Grana,* 35(4): 219 (1996): Vernonia kotschyana.
var. mossambiquensis *(Steetz) M.A. Isawumi, G. El-Ghazaly & B. Nordenstam in Grana,* 35(4): 219 (1996): Ascaricida mossambiquensis.
ballyi *(C. Jeffrey) M.A. Isawumi, G. El-Ghazaly & B. Nordenstam in Grana,* 35(4): 219 (1996): Vernonia ballyi.
bracteosa *(O. Hoffm.) M.A. Isawumi, G. El-Ghazaly & B. Nordenstam in Grana,* 35(4): 219 (1996): Vernonia bracteosa.
calvoana *(Hook. f.) M.A. Isawumi in Feddes Repert.,* 104(5–6): 312 (1993), without basionym ref.: Stengelia calvoana.
calvoana *(Hook. f.) M.A. Isawumi, G. El-Ghazaly & B. Nordenstam in Grana,* 35(4): 226 (1996): Stengelia calvoana.
var. acuta *(C.D. Adams) M.A. Isawumi in Feddes Repert.,* 104(5–6): 317 (1993), species name not validly publ.; *et in Grana,* 35(4): 226 (1996): Vernonia leucacalyx var. acuta.
subsp. adolfi-friderici *(Muschl.) M.A. Isawumi, G. El-Ghazaly & B. Nordenstam in Grana,* 35(4): 227 (1996): Vernonia adolfi-friderici.
var. hymenolepis *(A. Rich.) M.A. Isawumi in Feddes Repert.,* 104(5–6): 317 (1993), species name not validly publ.: Vernonia hymenolepis.
var. insignis *(Hook. f.) M.A. Isawumi in Feddes Repert.,* 104(5–6): 317 (1993), without basionym ref., species name not validly publ.: Stengelia insignis.
subsp. leucocalyx *(O. Hoffm.) M.A. Isawumi, G. El-Ghazaly & B. Nordenstam in Grana,* 35(4): 227 (1996): Vernonia leucocalyx.
subsp. meridionalis *(Wild) M.A. Isawumi, G. El-Ghazaly & B. Nordenstam in Grana,* 35(4): 226 (1996): Vernonia hymenolepis subsp. meridionalis.
var. microcephala *(C.D. Adams) M.A. Isawumi in Feddes Repert.,* 104(5–6): 317 (1993), species name not validly publ.: Vernonia calvoana var. microcephala.
subsp. mokaensis *(Mildbr. & Mattf.) M.A. Isawumi, G. El-Ghazaly & B. Nordenstam in Grana,* 35(4): 227 (1996): Vernonia mokaensis.
subsp. oehleri *(Muschl.) M.A. Isawumi, G. El-Ghazaly & B. Nordenstam in Grana,* 35(4): 227 (1996): Vernonia oehleri.
subsp. ruwenzoriensis *(C. Jeffrey) M.A. Isawumi, G. El-Ghazaly & B. Nordenstam in Grana,* 35(4): 227 (1996): Vernonia calvoana subsp. ruwenzoriensis.

BACCHAROIDES (Compos.).
calvoana *(Hook. f.)*
subsp. ulugurensis *(O. Hoffm.) M.A. Isawumi, G. El-Ghazaly & B. Nordenstam in Grana,* 35(4): 227 (1996): Vernonia ulugurensis.
subsp. usambarensis *(C. Jeffrey) M.A. Isawumi, G. El-Ghazaly & B. Nordenstam in Grana,* 35(4): 227 (1996): Vernonia calvoana subsp. usambarensis.
cardiolepis *(O. Hoffm.) M.A. Isawumi, G. El-Ghazaly & B. Nordenstam in Grana,* 35(4): 219 (1996): Vernonia cardiolepis.
dumicola *(S. Moore) M.A. Isawumi, G. El-Ghazaly & B. Nordenstam in Grana,* 35(4): 228 (1996): Vernonia dumicola.
filigera *(Oliv. & Hiern.) M.A. Isawumi, G. El-Ghazaly & B. Nordenstam in Grana,* 35(4): 228 (1996): Vernonia filigera.
filipendula *(Hiern) M.A. Isawumi, G. El-Ghazaly & B. Nordenstam in Grana,* 35(4): 220 (1996): Vernonia filipendula.
guineensis (Benth.) H. Rob.
var. cameroonica *(C.D. Adams) M.A. Isawumi in Feddes Repert.,* 104(5–6): 321 (1993): Vernonia guineensis var. cameroonica.
var. procera *(O. Hoffm.) M.A. Isawumi in Feddes Repert.,* 104(5–6): 321 (1993): Vernonia procera.
hymenolepis *(A. Rich.) M.A. Isawumi, G. El-Ghazaly & B. Nordenstam in Grana,* 35(4): 228 (1996): Vernonia hymenolepis.
incompta *(S. Moore) M.A. Isawumi, G. El-Ghazaly & B. Nordenstam in Grana,* 35(4): 226 (1996): Vernonia incompta.
kirungae *(R.E. Fr.) M.A. Isawumi, G. El-Ghazaly & B. Nordenstam in Grana,* 35(4): 222 (1996): Vernonia kirungae.
lasiopus (O. Hoffm.) H. Rob.
var. acuta *(C. Jeffrey) M.A. Isawumi, G. El-Ghazaly & B. Nordenstam in Grana,* 35(4): 224 (1996): Vernonia lasiopus var. acuta.
var. caudata *(C. Jeffrey) M.A. Isawumi, G. El-Ghazaly & B. Nordenstam in Grana,* 35(4): 224 (1996): Vernonia lasiopus var. caudata.
var. grandiceps *(C. Jeffrey) M.A. Isawumi, G. El-Ghazaly & B. Nordenstam in Grana,* 35(4): 224 (1996): Vernonia lasiopus var. grandiceps.
var. iodocalyx *(O. Hoffm.) M.A. Isawumi, G. El-Ghazaly & B. Nordenstam in Grana,* 35(4): 224 (1996): Vernonia iodocalyx.
longipedunculata *(De Wild.) M.A. Isawumi, G. El-Ghazaly & B. Nordenstam in Grana,* 35(4): 220 (1996): Vernonia longipedunculata.
var. manikensis *(De Wild.) M.A. Isawumi, G. El-Ghazaly & B. Nordenstam in Grana,* 35(4): 222 (1996): Vernonia manikensis.
var. retusa *(R.E. Fr.) M.A. Isawumi, G. El-Ghazaly & B. Nordenstam in Grana,* 35(4): 222 (1996): Vernonia retusa.
nimbaensis *(C.D. Adams) M.A. Isawumi, G. El-Ghazaly & B. Nordenstam in Grana,* 35(4): 222 (1996): Vernonia nimbaensis.
prolixa *(S. Moore) M.A. Isawumi, G. El-Ghazaly & B. Nordenstam in Grana,* 35(4): 220 (1996): Vernonia prolixa.
pumila *(Kotschy & Peyr.) M.A. Isawumi in Feddes Repert.,* 104(5–6): 321 (1993): Vernonia pumila.
ringoetii *(De Wild.) M.A. Isawumi, G. El-Ghazaly & B. Nordenstam in Grana,* 35(4): 226 (1996): Vernonia ringoetii.
schimperi *(DC.) M.A. Isawumi, G. El-Ghazaly & B. Nordenstam in Grana,* 35(4): 222 (1996): Vernonia schimperi.
stenostegia *(Stapf) M.A. Isawumi in Feddes Repert.,* 104(5–6): 311 (1993): Candidea stenostegia.
sunzuensis *(Wild) M.A. Isawumi, G. El-Ghazaly & B. Nordenstam in Grana,* 35(4): 228 (1996): Vernonia sunzuensis.
tenoreana *(Oliv.) M.A. Isawumi in Feddes Repert.,* 104(5–6): 311 (1993): Vernonia tenoreana.
tolypophora *(Mattf.) M.A. Isawumi, G. El-Ghazaly & B. Nordenstam in Grana,* 35(4): 228 (1996): Vernonia tolypophora.

BACOPA (Scrophulariac.).
sect. Alternifoliae *R. Mielcarek in Fragm. Flor. Geobot.,* 41(1): 34 (1996).
subsect. Bracteolatae *R. Mielcarek in Fragm. Flor. Geobot.,* 41(1): 34 (1996).

BACOPA:-
subsect. Ebracteolatae *R. Mielcarek in Fragm. Flor. Geobot.*, 41(1): 40 (1996).

BACTRIS (Palmae).
acanthocarpa Mart.
var. trailiana *(Barb. Rodr.) A.J. Henderson in Fl. Neotrop.*, 79: 28 (2000): B. trailiana.
ana-juliae *A. Cascante in Palms*, 44(3): 146 (2000)—Costa Rica.
bahiensis *Noblick ex A.J. Henderson in Fl. Neotrop.*, 79: 32 (2000)—Brazil Northeast (Bahia), Brazil Southeast (Espírito Santo, Rio de Janeiro).
charnleyae *G. de Nevers, Andrew Henderson & M.H. Grayum in Proc. Calif. Acad. Sci.*, 49(7): 176 (1996)—Panama.
chaveziae *A.J. Henderson in Fl. Neotrop.*, 79: 49 (2000), nom. nov.: B. concinna var. sigmoidea *A.J. Hend.*
dahlgreniana *R. Govaerts, Checklist Seed Plants*, 2(1–2): 9 (1996), nom. nov.: Guilielma microcarpa *Huber.*
gasipaes Kunth
var. chichagui *(H. Karst.) A.J. Henderson in Fl. Neotrop.*, 79: 73 (2000): B. speciosa var. chichagui.
glandulosa Oerst.
var. baileyana *(H.E. Moore) G. de Nevers in Proc. Calif. Acad. Sci.*, 49(7): 183 (1996): B. baileyana.
glassmanii *Med.-Costa & Noblick ex A.J. Henderson in Fl. Neotrop.*, 79: 78 (2000)—Brazil Northeast (Bahia, Alagoas, Pernambuco).
grayumii *G. de Nevers & Andrew Henderson in Proc. Calif. Acad. Sci.*, 49(7): 188 (1996)—Costa Rica, Nicaragua.
halmoorei *A.J. Henderson in Fl. Neotrop.*, 79: 86 (2000)—Brazil North (Amazonas, Acre), Peru.
hatschbachii *Noblick ex A.J. Henderson in Fl. Neotrop.*, 79: 87 (2000)—Brazil South (Paraná), Brazil Southeast (São Paulo).
herrerana *A. Cascante in Palms*, 44(3): 148 (2000)—Costa Rica.
hirta Mart.
var. jenmanii *A.J. Henderson in Fl. Neotrop.*, 79: 92 (2000)—Guyana.
var. lakoi *(Burret) A. Henderson in Fl. Neotrop.*, 79: 92 (2000): B. lakoi.
var. spruceana *(Trail) A.J. Henderson in Fl. Neotrop.*, 79: 92 (2000): B. pectinata var. spruceana.
horridispatha *Noblick ex A.J. Henderson in Fl. Neotrop.*, 79: 98 (2000)—Brazil Northeast (Bahia).
kunorum *G. de Nevers & M.H. Grayum in Proc. Calif. Acad. Sci.*, 49(7): 195 (1996)—Panama, Colombia.
maraja Mart.
var. trichospatha *(Trail) A.J. Henderson in Fl. Neotrop.*, 79: 119 (2000): B. trichospatha.
martiana *A.J. Henderson in Fl. Neotrop.*, 79: 120 (2000)—Peru, Colombia, Ecuador, Brazil North (Acre, Amazonas).
mexicana Mart.
var. trichophylla *(Burret) Andrew Henderson in Proc. Calif. Acad. Sci.*, 49(7): 204 (1996): B. trichophylla.
neomilitaris *G. de Nevers & A.J. Henderson in Brittonia*, 51(1): 77 (1999)—Panama, Costa Rica.
panamensis *G. de Nevers & M.H. Grayum in Proc. Calif. Acad. Sci.*, 49(7): 205 (1996)—Panama.
polystachya *H. Wendl. ex M.H. Grayum in Phytologia*, 84(4): 308 (1998 publ. 1999)—Costa Rica.
soeiroana *Noblick ex A.J. Henderson in Fl. Neotrop.*, 79: 154 (2000)—Brazil Northeast (Bahia).
timbuiensis *H.Q.B. Fernandes in Bol. Mus. Biol. Mello Leitão, Nova Sér.*, 5: 4 (1996)—Brazil Southeast (Espírito Santo).

BAECKEA (Myrtac.).
kandos *A.R. Bean in Telopea*, 7(3): 260 (1997)—New South Wales.
latifolia *(Benth.) A.R. Bean in Telopea*, 7(3): 263 (1997): B. gunniana var. latifolia.
omissa *A.R. Bean in Telopea*, 7(3): 256 (1997)—New South Wales, Queensland.
trapeza *A.R. Bean in Telopea*, 7(3): 255 (1997)—Queensland.

BAEOMETRA (Colchicac.).
breyniana *(L.) J.O. Voigt, Hort. Suburb. Calcutt.*: 647 (1845): Tulipa breyniana.

BAFUTIA (Compos.).
tenuicaulis C.D. Adams
var. zapfackiana *H. Beentje & B.J. Pollard in M. Cheek et al., Pl. Mount Oku & Ijim Ridge, Cameroon, Conserv. Checklist*: 92 (2000)—Cameroon.

BAIMASHANIA *I.A. Al-Shehbaz in Novon*, 10(4): 321 (2000). *CRUCIFERAE*
pulvinata *I.A. Al-Shehbaz in Novon*, 10(4): 321 (2000)—China South-Central (Yunnan).
wangii *I.A. Al-Shehbaz in Novon*, 10(4): 322 (2000)—Qinghai.

× **BALAGUERARA** *hort. in Orchid Rev.*, 105(1213): 62 (1997). *ORCHIDACEAE* (Broughtonia × Epidendrum × Laeliopsis × Tetramicra).

BALAKA (Palmae).
streptostachys *D. Fuller & J.L. Dowe in Palms*, 43(1): 10 (1999)—Fiji.

BALAKATA *H.-J. Esser in Blumea*, 44(1): 154 (1999). *EUPHORBIACEAE*
baccata *(Roxb.) H.-J. Esser in Blumea*, 44(1): 155 (1999): Sapium baccatum.
luzonica *(S. Vidal) H.-J. Esser in Blumea*, 44(1): 157 (1999): Myrica luzonica.

BALANOPHORA (Balanophorac.).
affinis *W. Griffith in Proc. Linn. Soc.*, 1(22): 220 (1844); *et in Trans. Linn. Soc.*, 20(1): 94 (1846)—India.†
alveolata *W. Griffith in Proc. Linn. Soc.*, 1(22): 220 (1844); *et in Trans. Linn. Soc.*, 20(1): 94 (1846)—India.†
burmannica *W. Griffith in Proc. Linn. Soc.*, 1(22): 219 (1844); *et in Trans. Linn. Soc.*, 20(1): 93 (1846), in IK Vol. 1 as 'burmanniana'—Burma.†
indica (Arn.) Griff.
var. globosa *(Jungh.) N.T. Ban in Konsp. Sosud. Rast. Fl. V'etnama*, 2: 45 (1996): B. globosa.
lancangensis *Y.Y. Qian in J. Trop. Subtrop. Bot.*, 4(2): 12 (1996)—China South-central (Yunnan).
picta *W. Griffith in Proc. Linn. Soc.*, 1(22): 220 (1844); *et in Trans. Linn. Soc.*, 20(1): 94 (1846)—India?†
polyandra *W. Griffith in Proc. Linn. Soc.*, 1(22): 220 (1844); *et in Trans. Linn. Soc.*, 20(1): 94 (1846)—India.†
saxicola *F.W. Xing & Z.X. Li in Acta Bot. Austro Sinica*, 10: 17 (1995)—Hainan.

BALFOURIA *(H. Ohba) H. Ohba in H. 't Hart & U. Eggli (eds.), Evol. & Syst. Crassulac.*: 156 (1995), homonym; *see R. Br., Asclepiadeae*, 59 (1810): Sedum subgen. Balfouria. (Crassulac.)
balfourii *(Raym.-Hamet) H. Ohba in H. 't Hart & U. Eggli (eds.), Evol. & Syst. Crassulac.*: 156 (1995): Sedum balfourii.

BALFOURODENDRON (Rutac.).
molle *(Miq.) J.R. Pirani in Brittonia*, 50(3): 359 (1998): Esenbeckia mollis.

BALINOTELLA *J. Soják in Zprávy Krajsk. Vlastiv. Muz. Olomouci*, 215: 3 (1982). *UMBELLIFERAE*
buchtormensis *(Spreng.) J. Soják in Zprávy Krajsk. Vlastiv. Muz. Olomouci*, 215: 4 (1982): Bubon buchtormensis.
calycina *(Korovin) J. Soják in Zprávy Krajsk. Vlastiv. Muz. Olomouci*, 215: 4 (1982): Libanotis calycina.
krylovii *(V.N. Tikhom.) J. Soják in Zprávy Krajsk. Vlastiv. Muz. Olomouci*, 215: 3 (1982): Libanotis krylovii.
libanotis *(L.) J. Soják in Zprávy Krajsk. Vlastiv. Muz. Olomouci*, 215: 3 (1982): Athamanta libanotis.
subsp. leiocarpa *(Heuff.) J. Soják in Zprávy Krajsk. Vlastiv. Muz. Olomouci*, 215: 3 (1982): Libanotis montana var. leiocarpa.

BALINOTELLA:-
libanotis
subsp. pyrenaica *(L.) J. Soják in Zprávy Krajsk. Vlastiv. Muz. Olomouci*, 215: 3 (1982): Crithmum pyrenaicum.
subsp. sibirica *(L.) J. Soják in Zprávy Krajsk. Vlastiv. Muz. Olomouci*, 215: 3 (1982): Athamanta sibirica.
schrenkiana *(C.A. Mey. ex Schischk.) J. Soják in Zprávy Krajsk. Vlastiv. Muz. Olomouci*, 215: 4 (1982): Libanotis schrenkiana.
seseloides *(Turcz.) J. Soják in Zprávy Krajsk. Vlastiv. Muz. Olomouci*, 215: 4 (1982): Libanotis seseloides.
transcaucasica *(Schischk.) J. Soják in Zprávy Krajsk. Vlastiv. Muz. Olomouci*, 215: 4 (1982): Libanotis transcaucasica.

BALIOSPERMUM (Euphorbiac.).
yui *Y.T. Chang ex R. Govaerts, World Checklist Seed Plants*, 2(1–2): 10 (1996)—China South-central (Yunnan).

BALIZIA *R.C. Barneby & J.W. Grimes, Silk Tree, Guanacaste, Monkey's Earring (Mem. New York Bot. Gard., 74(1))*: 34 (1996). *LEGUMINOSAE*
sect. Leucosamanea *R.C. Barneby & J.W. Grimes, Silk Tree, Guanacaste, Monkey's Earring (Mem. New York Bot. Gard., 74(1))*: 36 (1996).
elegans *(Ducke) R.C. Barneby & J.W. Grimes, Silk Tree, Guanacaste, Monkey's Earring (Mem. New York Bot. Gard., 74(1))*: 40 (1996): Pithecellobium elegans.
leucocalyx *(Britton & Rose) R.C. Barneby & J.W. Grimes, Silk Tree, Guanacaste, Monkey's Earring (Mem. New York Bot. Gard., 74(1))*: 36 (1996): Samanea leucocalyx.
pedicellaris *(DC.) R.C. Barneby & J.W. Grimes, Silk Tree, Guanacaste, Monkey's Earring (Mem. New York Bot. Gard., 74(1))*: 37 (1996): Inga pedicellaris.

BALLOTA (Labiat.).
deserti *(Noë) S.L. Jury, M. Rejdali & A.J.K. Griffiths in Willdenowia*, 28(1–2): 168 (1998): Sideritis deserti.

BALLYANTHUS *P.V. Bruyns in Aloe*, 37(4): 76 (2000). *ASCLEPIADACEAE*
prognathus *(P.R.O. Bally) P.V. Bruyns in Aloe*, 37(4): 76 (2000): Stapelia prognatha.

BALOGHIA (Euphorbiac.).
pininsularis *Guillaumin ex G. McPherson & C. Tirel in Fl. N. Caléd. & Dépend.*, 14: 58 (1987)—New Caledonia.

BALOSKION (Restionac.).
australe *(R. Br.) B.G. Briggs & L.A.S. Johnson Telopea*, 8(1): 23 (1998): Restio australis.
fimbriatum *(L.A.S. Johnson & O.D. Evans) B.G. Briggs & L.A.S. Johnson in Telopea*, 8(1): 23 (1998): Restio fimbriatus.
gracile *(R. Br.) B.G. Briggs & L.A.S. Johnson in Telopea*, 8(1): 23 (1998): Restio gracilis.
longipes *(L.A.S. Johnson & O.D. Evans) B.G. Briggs & L.A.S. Johnson in Telopea*, 8(1): 23 (1998): Restio longipes.
pallens *(R. Br.) B.G. Briggs & L.A.S. Johnson in Telopea*, 8(1): 23 (1998): Restio pallens.
stenocoleum *(L.A.S. Johnson & O.D. Evans) B.G. Briggs & L.A.S. Johnson in Telopea*, 8(1): 23 (1998): Restio stenocoleus.
tenuiculme *(S.T. Blake) B.G. Briggs & L.A.S. Johnson in Telopea*, 8(1): 23 (1998): Restio tenuiculmis.
tetraphyllum *(Labill.) B.G. Briggs & L.A.S. Johnson in Telopea*, 8(1): 23 (1998): Restio tetraphyllus.
subsp. meiostachyum *(L.A.S. Johnson & O.D. Evans) B.G. Briggs & L.A.S. Johnson in Telopea*, 8(1): 24 (1998): Restio tetraphyllus subsp. meiostachyus.

BALSAMIFLUA (Salicac.).
euphratica (Olivier) Kimura
f. pruinosa *(Schrenk) N. Chao & J. Liu in J. Sichuan Forest. Sci. & Techn.*, 18(2): 3 (1997): Populus pruinosa.

BALSAMIFLUA:-
mexicana *(Wesm.) N. Chao & G.T. Gong in J. Sichuan Forest. Sci. & Techn.*, 18(2): 3 (1997): Populus mexicana.
subsp. dimorpha *(Brandegee) N. Chao & G.T. Gong in J. Sichuan Forest. Sci. & Techn.*, 18(2): 4 (1997): Populus dimorpha.

BALSAMODENDRUM (Burserac.).
roxburghianum *(Wight & Arn.) Wall. ex J.O. Voigt, Hort. Suburb. Calcutt.*: 150 (1845): Protium roxburghianum.

BAMBUSA (Gramin.).
albostriata *(McClure) D. Ohrnberger, Bamboos of the World Intro. 4:* 18 (1997): B. textilis var. albostriata.
basisolida *W.T. Lin in J. Bamboo Res.*, 16(3): 23 (1997)—China Southeast (Guangdong).
chungii McClure
var. barbellata *(Q.H. Dai) D. Ohrnberger, Bamboos of the World Intro. 4:* 18 (1997): Lingnania chungii var. barbellata.
var. petilla *(T.H. Wen) D. Ohrnberger, Bamboos of the World Intro. 4:* 18 (1997): Lingnania chungii var. petilla.
comillensis *M.K. Alam in Bangladesh J. Pl. Taxon.*, 3(2): 41 (1996)—Bangladesh.
concava *W.T. Lin in J. Bamboo Res.*, 16(3): 24 (1997)—Hainan.
emeiensis L.C. Chia & H.L. Fung
f. chrysotricha *(J.R. Xue & T.P. Yi) D. Ohrnberger, Bamboos of the World Intro. 4:* 18 (1997): Sinocalamus affinis f. chrysotricha.
f. flavidovirens *(T.P. Yi) D. Ohrnberger, Bamboos of the World Intro. 4:* 18 (1997), as 'flavidorivens': Sinocalamus affinis var. flavidovirens.
f. viridiflava *(T.P. Yi) D. Ohrnberger, Bamboos of the World Intro. 4:* 18 (1997): Sinocalamus affinis var. viridiflavus.
glaucescens (Lam.) Siebold ex Munro
var. floribunda *(Büse ex Miq.) S. Hatusima in S. Hatusima & T. Amano, Fl. Ryukyus, S. of Amami Is., ed. 2:* 242 (1994), without basionym ref.: Basionym not stated.
glaucophylla *E.A. Widjaja in Reinwardtia*, 11(2): 59 (1997)—Jawa.
grandis *(Q.H. Dai & X.L. Tao ex Keng f.) D. Ohrnberger, Bamboos of the World Intro. 4:* 18 (1997): Dendrocalamopsis grandis.
jacobsii *E.A. Widjaja in Reinwardtia*, 11(2): 60 (1997)—Jawa.
lako *E.A. Widjaja in Reinwardtia*, 11(2): 61 (1997)—Lesser Sunda Is. (East Timor).
maculata *E.A. Widjaja in Reinwardtia*, 11(2): 63 (1997)—Moluccas.
membranacea *(Munro) C.M.A. Stapleton & N.H. Xia in Kew Bull.*, 52(1): 238 (1997): Dendrocalamus membranaceus.
multiplex (Lour.) Raeusch. ex Schult. & Schult. f.
var. incana *B.M. Yang in J. Hunan Sci. Technol. Univ.*, 1(1): 115 (1985)—China Southeast (Hunan).
f. silverstripe *(R.A. Young) T.P. Yi, Bamboos Fl. Sichuan:* 54 (1997); *et in Fl. Sichuanica*, 12: 29 (1997 publ. 1998), basionym not validly publ.: B. multiplex var. silverstripe.
f. yellowstripe *(L.C. Chia & C.Y. Sia) T.P. Yi, Bamboos Fl. Sichuan:* 54 (1997); *et in Fl. Sichuanica*, 12: 29 (1997 publ. 1998), basionym not validly publ.: B. glaucescens cv. Yellowstripe.
nguyenii *D. Ohrnberger, Bamboos of the World Intro. 4:* 19 (1997), nom. nov.: B. vulgaris var. latiflora *Balansa*.
papillata *(Q.H. Dai) Q.H. Dai in Fl. Reipubl. Popul. Sin.*, 9(1): 129 (1996), isonym; *see K.M. Lan in Fl. Guizhouensis*, 5: 281 (1988): Lingnania papillata.
papillata *(Q.H. Dai) D. Ohrnberger, Bamboos of the World Intro. 4:* 19 (1997), isonym; *see K.M. Lan in Fl. Guizhouensis*, 5: 281 (1988): Lingnania papillata.
papillatoides *Q.H. Dai & D.Y. Huang in Acta Phytotax. Sin.*, 36(3): 279 (1998)—China Southeast (Guangxi).

BAMBUSA:-
pervariabilis McClure
var. multistriata *W.T. Lin in J. Bamboo Res.*, 16(3): 25 (1997)—China Southeast (Guangdong).
riauensis *E.A. Widjaja in Reinwardtia*, 11(2): 65 (1997)—Sumatera.
ritchiei *W. Munro in Trans. Linn. Soc. London*, 26(1): 113, 157 (1868), as 'ritcheyi'—India (Maharashtra).†
rugata *(W.T. Lin) D. Ohrnberger, Bamboos of the World Intro. 4:* 19 (1997): Lingnania rugata.
salarkhanii *M.K. Alam in Bangladesh J. Pl. Taxon.*, 3(2): 44 (1996)—Bangladesh, Nepal.
scabriculma *W.T. Lin in J. S. China Agric. Univ.*, 14(1): 46 (1993)—China Southeast (Guangdong).
semitecta *W.T. Lin & Z.M. Wu in J. S. China Agric. Univ.*, 14(3): 111 (1993)—China Southeast (Guangdong).
surrecta *(Q.H. Dai) Q.H. Dai in Fl. Reipubl. Popul. Sin.*, 9(1): 119 (1996): Lingnania surrecta.
surrecta *(Q.H. Dai) D. Ohrnberger, Bamboos of the World Intro. 4:* 19 (1997), isonym; *see Q.H. Dai in Fl. Reipubl. Popul. Sin.*, 9(1): 119 (1996): Lingnania surrecta.
tengchongensis *D.Z. Li & N.H. Xia in Acta Bot. Yunnanica*, 22(1): 44 (2000), nom. nov.: Neosinocalamus yunnanensis *J.R. Xue & Jia R. Xue*
textilis McClure
var. pubescens *B.M. Yang in J. Hunan Sci. Technol. Univ.*, 1(1): 116 (1985)—China Southeast (Hunan).
f. purpurascens *(N.H. Xia) T.P. Yi, Bamboos Fl. Sichuanica:* 63 (1997); *et in Fl. Sichuanica*, 12: 39 (1997 publ. 1998): B. textilis var. purpurascens.
valida *(Q.H. Dai) D. Ohrnberger, Bamboos of the World Intro. 4:* 19 (1997), isonym; *see W.T. Lin in Guihaia*, 10(1): 15 (1990): Dendrocalamopsis validus.
varioaurita *W.T. Lin & Z.J. Feng in J. S. China Agric. Univ.*, 14(1): 47 (1993)—China Southeast (Guangdong).
viridis *E.A. Widjaja in Reinwardtia*, 11(2): 67 (1997)—New Guinea (Irian Jaya).
wenchouensis *(T.H. Wen) Q.H. Dai in Fl. Reipubl. Popul. Sin.*, 9(1): 117 (1996): Lingnania wenchouensis.
wenchouensis *(T.H. Wen) D. Ohrnberger, Bamboos of the World Intro. 4:* 19 (1997), isonym; *see Q.H. Dai in Fl. Reipubl. Popul. Sin.*, 9(1): 117 (1996): Lingnania wenchouensis.
xueana *D. Ohrnberger, Bamboos of the World Intro. 4:* 19 (1997), nom. nov. Neosinocalamus yunnanensis *J.R. Xue & Jia R. Xue.*

BANARA (Flacourtiac.).
caymanensis *G.R. Proctor in Kew Bull.*, 51(3): 491 (1996)—Cayman Is.

BANISTERIOPSIS (Malpighiac.).
carolina *W.R. Anderson in Contrib. Univ. Mich. Herb.*, 22: 1 (1999)—French Guiana.
mariae *W.R. Anderson in Contrib. Univ. Mich. Herb.*, 22: 2 (1999)—Brazil Northeast (Bahia).

BANKSIA (Proteac.).
subser. Acclives *K. Thiele in Austral. Syst. Bot.*, 9(5): 721 (1996).
sect. Coccinea *(A.S. George) T.L. Maguire, M. Sedgley & J.G. Conran in Austral. Syst. Bot.*, 9(6): 890 (1996): B. ser. Coccineae.
subser. Cratistyles *K. Thiele in Austral. Syst. Bot.*, 9(5): 720 (1996).
subser. Ericifoliae *K. Thiele in Austral. Syst. Bot.*, 9(5): 721 (1996).
subser. Integrifoliae *K. Thiele in Austral. Syst. Bot.*, 9(5): 721 (1996).
subser. Leptophyllae *K. Thiele in Austral. Syst. Bot.*, 9(5): 720 (1996).
ser. Lindleyanae *K. Thiele in Austral. Syst. Bot.*, 9(5): 719 (1996).
subser. Longistyles *K. Thiele in Austral. Syst. Bot.*, 9(5): 720 (1996).
subser. Nutantes *K. Thiele in Austral. Syst. Bot.*, 9(5): 720 (1996).
subser. Occidentales *K. Thiele in Austral. Syst. Bot.*, 9(5): 721 (1996).

BANKSIA:-
ser. Ochraceae *K. Thiele in Austral. Syst. Bot.*, 9(5): 719 (1996).
subser. Sphaerocarpae *K. Thiele in Austral. Syst. Bot.*, 9(5): 720 (1996).
subser. Spinulosae *K. Thiele in Austral. Syst. Bot.*, 9(5): 721 (1996).
ser. Tricuspidae *A.S. George in Fl. Australia*, 17B: 397 (1999).
aquilonia *(A.S. George) A.S. George in Nuytsia*, 11(1): 22 (1996): B. integrifolia var. aquilonia.
brevidentata *(A.S. George) K. Thiele in Austral. Syst. Bot.*, 9(5): 719 (1996): B. gardneri var. brevidentata.
conferta A.S. George
subsp. penicillata *(A.S. George) A.S. George in Nuytsia*, 11(1): 22 (1996): B. conferta var. penicillata.
dolichostyla *(A.S. George) K. Thiele in Austral. Syst. Bot.*, 9(5): 719 (1996): B. sphaerocarpa var. dolichostyla.
ericifolia L.f.
subsp. macrantha *(A.S. George) A.S. George in Nuytsia*, 11(1): 22 (1996): B. ericifolia var. macrantha.
hiemalis *(A.S. George) K. Thiele in Austral. Syst. Bot.*, 9(5): 719 (1996): B. gardneri var. hiemalis.
meisneri Lehm.
subsp. ascendens *(A.S. George) A.S. George in Nuytsia*, 11(1): 22 (1996): B. meisneri var. ascendens.
paludosa R.Br.
subsp. astrolux *A.S. George in Nuytsia*, 11(1): 23 (1996)—New South Wales.
penicillata *(A.S. George) K. Thiele in Austral. Syst. Bot.*, 9(5): 719 (1996): B. conferta var. penicillata.

× **BAPTICIDIUM** *hort. in Orchid Rev.*, 107(1228): 255 (1999). *ORCHIDACEAE* (Baptistonia × Oncidium).

BAPTISIA (Legumin.).
× variicolor *M.A. Kosnik, G.M. Diggs, P.A. Redshaw & B.L. Lipscomb in Sida*, 17(2): 498 (1996)—Texas. (B. australis × sphaerocarpa).

BARBACENIA (Velloziac.).
lymansmithii *R. de Mello-Silva & N.L. de Menezes in Harvard Pap. Bot.*, 4(1): 271 (1999)—Brazil Southeast (Minas Gerais).

BARBAMINE *A.P. Khokhrjakov in Byull. Glavn. Bot. Sada (Moscow)*, 175: 50 (1997). *CRUCIFERAE*
ketzkhovelii *(Mardal.) A.P. Khokhrjakov in Byull. Glavn. Bot. Sada (Moscow)*, 175: 50 (1997): Barbarea ketzkhovelii.
procumbens *A.P. Khokhrjakov in Byull. Glavn. Bot. Sada (Moscow)*, 175: 51 (1997)—Turkey.
sachokiana *(N. Busch) A.P. Khokhrjakov in Byull. Glavn. Bot. Sada (Moscow)*, 175: 50 (1997): Cardamine sachokiana.

BARBAREA (Crucif.).
sect. Strictae *V.I. Dorofeev in Bot. Zhurn.*, 81(3): 132 (1996).
hongii *I.A. Al-Shehbaz & G. Yang in Acta Phytotax. Sin.*, 38(1): 71 (2000)—Manchuria (Jilin).
macrocarpa *(Boiss.) I.A. Al-Shehbaz & F. Jacquemoud in Candollea*, 55(1): 202 (2000): Nasturtium macrocarpum.

BARBOSELLA (Orchidac.).
subgen. Cardinella *C.A. Luer, Ic. Pleurothallid. XX (Monogr. Syst. Bot. Missouri Bot. Gard., 79)*: 10 (2000).
macaheensis *(Cogn.) C.A. Luer, Ic. Pleurothallid. XX (Monogr. Syst. Bot. Missouri Bot. Gard., 79)*: 23 (2000): Pleurothallis macaheensis.
ricii *C.A. Luer & R. Vásquez in C.A. Luer, Ic. Pleurothallid. XX (Monogr. Syst. Bot. Missouri Bot. Gard.*, 79): 26 (2000)—Bolivia.
vasquezii *C.A. Luer, Ic. Pleurothallid. XX (Monogr. Syst. Bot. Missouri Bot. Gard., 79)*: 29 (2000)—Bolivia.

BARLERIA (Acanthac.).
sect. Cavirostrata *M.-J. Balkwill in J. Biogeogr.*, 25(1): 110 (1998).

BARLERIA:-
sect. Chrysothrix *M.-J. Balkwill in J. Biogeogr.*, 25(1): 110 (1998).
sect. Fissimura *M.-J. Balkwill in J. Biogeogr.*, 25(1): 110 (1998).
sect. Stellatohirta *M.-J. Balkwill in J. Biogeogr.*, 25(1): 110 (1998).
lancifolia T. Anderson
subsp. charlesii *(Benoist) J.-P. Lebrun & Monod in J.-P. Lebrun & A.L. Stork, Enum. Pl. Fleurs Afr. Trop.*, 4: 470 (1997): B. hereroensis var. charlesii.

BARNADESIA (Compos.).
subgen. Bacasia *(Ruiz & Pav.) E. Urtubey in Ann. Missouri Bot. Gard.*, 86(1): 106 (1999): gen. Bacasia.
lehmannii Hieron.
var. villosa *(I.C. Chung) E. Urtubey in Ann. Missouri Bot. Gard.*, 86(1): 94 (1999): B. polyacantha var. villosa.

BARNARDIA (Hyacinthac.).
alboviridis *(Hand.-Mazz.) F. Speta in Phyton (Austria)*, 38(1): 96 (1998): Scilla alboviridis.
bispatha *(Hand.-Mazz.) F. Speta in Phyton (Austria)*, 38(1): 96 (1998): Scilla bispatha.
borealijaponica *(M. Kikuchi) F. Speta in Phyton (Austria)*, 38(1): 97 (1998): Scilla boreali-japonica.
pulchella *(Kitag.) F. Speta in Phyton (Austria)*, 38(1): 97 (1998): Scilla thunbergii var. pulchella.
sinensis *(Lour.) F. Speta in Phyton (Austria)*, 38(1): 97 (1998): Ornithogalum sinense.

BARNEBYDENDRON *J.H. Kirkbride in Sida*, 18(3): 817 (1999), nom. nov.: gen. Phyllocarpus *Riedel ex Tul.* (Legumin.).
riedelii *(Tul.) J.H. Kirkbride in Sida*, 18(3): 817 (1999): Phyllocarpus riedelii.

BARONIELLA (Asclepiadac.).
capillacea *J. Klackenberg in Candollea*, 52(2): 406 (1997)—Madagascar.
ensifolia *J. Klackenberg in Candollea*, 52(2): 395 (1997)—Madagascar.
longicornis *J. Klackenberg in Candollea*, 52(2): 397 (1997)—Madagascar.

BARRINGTONIA (Lecythidac.).
josephstaalensis *W. Takeuchi in Sida*, 19(1): 9 (2000)—New Guinea (Papua New Guinea).
terengganuensis *P. Chantaranothai in Gard. Bull. Singapore*, 48(1–2): 201 (1996 publ. 1998)—Malaya (Peninsular Malaysia).

BARROSOA (Compos.).
grossedentata *V.M. Badillo in Ernstia*, 5(3): 135 (1995)—Venezuela.

BARTERIA (Passiflorac.).
solida *F.J. Breteler in Adansonia*, 21(2): 316 (1999)—Cameroon, Congo, Zaire, Gabon, Nigeria.

BARTHLOTTIA *Eb. Fischer in Bull Mus. Nation. Hist. Nat., B, Adansonia*, Ser. 4, 18(3–4): 352 (1996). *SCROPHULARIACEAE*
madagascariensis *Eb. Fischer in Bull. Mus. Nation. Hist. Nat., B, Adansonia*, Ser. 4, 18(3–4): 352 (1996)—Madagascar.

BARTLETTINA (Compos.).
calderonii *(B.L. Turner) B.L. Turner in Phytologia*, 82(6): 385 (1997 publ. 1998): Eupatorium calderonii.
lanicaulis *(B.L. Rob.) B.L. Turner in Phytologia*, 82(6): 386 (1997 publ. 1998): Eupatorium lanicaule.
xalapana *(B.L. Turner) B.L. Turner in Phytologia*, 82(6): 386 (1997 publ. 1998): Eupatorium xalapanum.

BARTSCHELLA (Cactac.).
ser. Glassella *A.B. Doweld in Sukkulentÿ (Moscow)*, 3(1–2): 38 (2000).
ser. Napinae *(Keizer) A.B. Doweld in Sukkulentÿ (Moscow)*, 3(1–2): 38 (2000): Mammillaria ser. Napinae.

BARTSCHELLA:-
ser. Phellosperma *(Britton & Rose) A.B. Doweld in Sukkulentÿ (Moscow)*, 3(1–2): 38 (2000): gen. Phellosperma.
ser. Probartschella *A.B. Doweld in Sukkulentÿ (Moscow)*, 3(1–2): 38 (2000).
ser. Zephyranthoides *(E. Kuhn & Hofman) A.B. Doweld in Sukkulentÿ (Moscow)*, 3(1–2): 38 (2000): Mammillaria ser. Zephyranthoides.
albicans *(Britton & Rose) A.B. Doweld in Sukkulentÿ (Moscow)*, 3(1–2): 37 (2000): Neomammillaria albicans.
subsp. fraileana *(Britton & Rose) A.B. Doweld in Sukkulentÿ (Moscow)*, 3(1–2): 37 (2000): Neomammillaria fraileana.
armillata *(K. Brandegee) A.B. Doweld in Sukkulentÿ (Moscow)*, 3(1–2): 37 (2000): Mammillaria armillata.
boolii *(G.E. Linds.) A.B. Doweld in Sukkulentÿ (Moscow)*, 3(1–2): 37 (2000): Mammillaria boolii.
cerralboa *(Britton & Rose) A.B. Doweld in Sukkulentÿ (Moscow)*, 3(1–2): 37 (2000): Neomammillaria cerralboa.
deherdtiana *(Farwig) A.B. Doweld in Sukkulentÿ (Moscow)*, 3(1–2): 37 (2000): Mammillaria deherdtiana.
subsp. dodsonii *(Bravo) A.B. Doweld in Sukkulentÿ (Moscow)*, 3(1–2): 37 (2000): Mammillaria dodsonii.
glassii *(R.A. Foster) A.B. Doweld in Sukkulentÿ (Moscow)*, 3(1–2): 37 (2000): Mammillaria glassii.
subsp. ascensionis *(Repp.) A.B. Doweld in Sukkulentÿ (Moscow)*, 3(1–2): 37 (2000): Mammillaria ascensionis.
subsp. nominis-dulcis *(A.B. Lau) A.B. Doweld Sukkulentÿ (Moscow)*, 3(1–2): 38 (2000): Mammillaria glassii var. nominis-dulcis.
subsp. siberiensis *(A.B. Lau) A.B. Doweld in Sukkulentÿ (Moscow)*, 3(1–2): 38 (2000): Mammillaria glassii var. siberiensis.
guelzowiana *(Werderm.) A.B. Doweld in Sukkulentÿ (Moscow)*, 3(1–2): 38 (2000): Mammillaria guelzowiana.
heidiae *(Krainz) A.B. Doweld in Sukkulentÿ (Moscow)*, 3(1–2): 38 (2000): Mammillaria heidiae.
hernandezii *(Glass & R.A. Foster) A.B. Doweld in Sukkulentÿ (Moscow)*, 3(1–2): 38 (2000): Mammillaria hernandezii.
hutchisoniana *(H.E. Gates) A.B. Doweld in Sukkulentÿ (Moscow)*, 3(1–2): 38 (2000): Neomammillaria hutchisoniana.
subsp. louisae *(G.E. Linds.) A.B. Doweld in Sukkulentÿ (Moscow)*, 3(1–2): 38 (2000), as 'louisiae': Mammillaria louisae
insularis *(H.E. Gates) A.B. Doweld in Sukkulentÿ (Moscow)*, 3(1–2): 38 (2000): Mammillaria insularis.
multidigitata *(G.E. Linds.) A.B. Doweld in Sukkulentÿ (Moscow)*, 3(1–2): 38 (2000): Mammillaria multidigitata.
napina *(J.A. Purpus) A.B. Doweld in Sukkulentÿ (Moscow)*, 3(1–2): 38 (2000): Mammillaria napina.
palmeri *(J.M. Coult.) A.B. Doweld in Sukkulentÿ (Moscow)*, 3(1–2): 38 (2000): Cactus palmeri
tetrancistra *(Engelm.) A.B. Doweld in Sukkulentÿ (Moscow)*, 3(1–2): 38 (2000): Mammillaria tetrancistra.
yaquensis *(R.T. Craig) A.B. Doweld in Sukkulentÿ (Moscow)*, 3(1–2): 38 (2000): Mammillaria yaquensis.
zephyranthoides *(Scheidw.) A.B. Doweld in Sukkulentÿ (Moscow)*, 3(1–2): 38 (2000): Mammillaria zephyranthoides.

BARTSIELLA *M. Bolliger in Willdenowia*, 26(1–2): 76 (1996). *SCROPHULARIACEAE*
rameauana *(Emb.) M. Bolliger in Willdenowia*, 26(1–2): 76 (1996): Odontites rameauanus.

BASANANTHE (Passiflorac.).
subsessilicarpa *J.B. Gillett ex B. Verdcourt in Kew Bull.*, 53(4): 989 (1998)—Kenya.

BASHANIA (Gramin.).
abietina *T.P. Yi & Lin Yang in J. Bamboo Res.*, 17(4): 1 (1998)—China South-central (Sichuan).
baoxingensis *T.P. Yi in J. Bamboo Res.*, 19(1): 9 (2000)—China South-central (Sichuan).

BASIPHYLLAEA (Orchidac.).
wrightii *(Acuña) M.A. Nir, Orchidaceae Antillanae*: 35 (2000): Bletia wrightii.

BASKERVILLA (Orchidac.).
stenopetala *R.L. Dressler in Bol. Inst. Bot. (Guadalajara)*, 5(1–3): 70 (1998)—Panama.

BASSECOIA *B.L. Burtt in R.K. Tandon & Prithipalsingh (eds.), Biodivers., Taxon., & Ecol.*: 137 (1999). *DIPSACACEAE*
bretschneideri *(Batalin) B.L. Burtt in R.K. Tandon & Prithipalsingh (eds.), Biodivers., Taxon., & Ecol.*: 137 (1999): Scabiosa bretschneideri.
siamensis *(Craib) B.L. Burtt in R.K. Tandon & Prithipalsingh (eds.), Biodivers., Taxon., & Ecol.*: 138 (1999): Scabiosa siamensis.

BASSIA (Chenopodiac.).
scoparia
var. subvillosa *(Moq.) J. Lambinon in Soc. Éch. Pl. Vasc. Eur. Bassin Médit. Bull.*, 28: 33 (2000): Kochia scoparia var. subvillosa.

BASTARDIOPSIS (Malvac.).
eggersii *(Baker f.) Fuertes & P.A. Fryxell in Fl. St. John (Mem. New York Bot. Gard., 78)*: 310 (1996): Sida eggersii.

BATHYSA (Rubiac.).
bathysoides *(Steyerm.) P.G. Delprete in Brittonia*, 48(1): 42 (1996)—Chimarrhis bathysoides.
bracteosa *(Wedd.) P.G. Delprete in Brittonia*, 49(4): 481 (1997): Schizocalyx bracteosus.
perijaensis *(Steyerm.) P.G. Delprete in Brittonia*, 48(1): 43 (1996): Chimarrhis perijaensis.

BATOPILASIA *G.L. Nesom & R.D. Noyes in Sida*, 19(1): 81 (2000). *COMPOSITAE*
byei *(S.D. Sundb. & G.L. Nesom) G.L. Nesom & R.D. Noyes in Sida*, 19(1): 81 (2000): Erigeron byei.

BATRACHIUM (Ranunculac.).
subsect. Circinata *V.I. Krecz. ex N.N. Tzvelev in Novosti Sist. Vÿssh. Rast.*, 31: 80 (1998).
subsect. Litoralia *V.I. Krecz. ex N.N. Tzvelev in Novosti Sist. Vÿssh. Rast.*, 31: 73 (1998).
subsect. Peltata *V.I. Krecz. ex N.N. Tzvelev in Novosti Sist. Vÿssh. Rast.*, 31: 73 (1998).
subsect. Trichophylla *(Dumort.) N.N. Tzvelev in Novosti Sist. Vÿssh. Rast.*, 31: 77 (1998): B. sect. Trichophylla.
baudotii Godr.
subsp. marinum *(Fr.) N.N. Tzvelev in Novosti Sist. Vÿssh. Rast.*, 32: 183 (2000): B. marinum.
fucoides *(Freyn) J. Holub in Preslia*, 70(2): 100 (1998): Ranunculus fucoides.
mongolicum (Krylov) V.I. Krecz.
var. suberadicatum *N.N. Tzvelev Novosti Sist. Vÿ ssh. Rast.*, 31: 76 (1998)—Russia North.
nevensis *N.N. Tzvelev in Novosti Sist. Vÿssh. Rast.*, 31: 75 (1998)—Russia North.
trichophyllum (Chaix ex Vill.) Bosch
var. jingpoense *(G.Y. Chang et al.) W.T. Wang in Phytologia*, 79(5): 387 (1995 publ. 1996): B. jingpoense.

BAUHINIA (Legumin.).
aherniana Perkins
var. subglabra *(Merr.) K. Larsen & S.S. Larsen in Nordic J. Bot.*, 13(6): 661 (1993): B. subglabra.
ampla Span.
var. schlechteri *(Harms) K. Larsen & B. Sunarno in Nordic J. Bot.*, 13(6): 658 (1993): B. schlechteri.
anguina *W. Roxburgh, Pl. Coromandel*, 3: 82 (1820)—Bangladesh.†
angulosa Vogel
var. bahiana *A.S. da Fonseca Vaz Brittonia*, 47(4): 378 (1995)—Brazil Northeast (Bahia).
var. meridionalis *(Hoehne) A.S.F. Vaz in Albertoa*, 4(5): 55 (1995): B. splendens f. meridionalis.
bakeriana *S.S. Larsen in Grana*, 14(2–3): 126 (1974), nom. nov., without replaced synonym ref.: B. rufa *Graham ex Baker.*

BAUHINIA:-
bassacensis Pierre ex Gagnep.
var. backeri *(de Wit) S.S. Larsen in Grana,* 14(2–3): 122 (1974), as 'bakeri': Phanera bassacensis var. backeri.
bidentata Jack
var. breviflora *(Ridl.) K. Larsen & S.S. Larsen in Nordic J. Bot.,* 13(6): 663 (1993): B. breviflora.
var. fraseri *(de Wit) K. Larsen & S.S. Larsen in Nordic J. Bot.,* 13(6): 663 (1993): Phanera bidentata var. fraseri.
var. gracilipes *(Merr.) K. Larsen & S.S. Larsen in Nordic J. Bot.,* 13(6): 663 (1993): B. gracilipes.
var. monticola *(Ridl.) K. Larsen & S.S. Larsen in Nordic J. Bot.,* 13(6): 663 (1993): B. monticola.
brevipedicellata *J.K. Jarvie in Kew Bull.,* 53(2): 495 (1998)—Borneo (Kalimantan).
calciphila *D.X. Zhang & T.C. Chen in Nordic J. Bot.,* 18(2): 141 (1998)—China Southeast (Guangxi).
carvalhoi *A.S. da Fonseca Vaz in Brittonia,* 47(4): 376 (1995)—Brazil Northeast (Bahia).
erythropoda Hayata
var. guangxiensis *D.X. Zhang & T.C. Chen in Nordic J. Bot.,* 18(2): 145 (1998)—China Southeast (Guangxi).
finlaysoniana (Graham ex Benth.) Baker
var. amoena *(de Wit) K. Larsen & S.S. Larsen in Nordic J. Bot.,* 13(6): 663 (1993): Phanera finlaysoniana var. amoena.
var. leptopus *(Perkins) K. Larsen & S.S. Larsen in Nordic J. Bot.,* 13(6): 663 (1993): B. leptopus.
var. montana *(de Wit) K. Larsen & S.S. Larsen in Nordic J. Bot.,* 13(6): 663 (1993): Phanera finlaysoniana var. montana.
foraminifera Gagnep.
var. falcata *K. Larsen & S.S. Larsen in Nordic J. Bot.,* 13(6): 661 (1993)—Borneo (Brunei).
gilva *(F.M. Bailey) R. Govaerts, World Checklist Seed Plants,* 2(1–2): 10 (1996): B. cunninghamii f. gilva.
gilva *(F.M. Bailey) A.S. George in Fl. Australia,* 12: 197 (1998), isonym; *see R. Govaerts, World Checklist Seed Plants,* 2(1–2): 10 (1996): B. cunninghamii f. gilva.
glabrifolia (Benth.) Baker
var. maritima *K. Larsen & S.S. Larsen in Thai Forest Bull., Bot.,* 25: 14 (1997), without type—Locality not stated.
glauca (Wall. ex Benth.) Benth.
var. caterviflora *(H.Y. Chen) K. Larsen & S.S. Larsen ex D.X. Zhang in Acta Bot. Austro Sinica,* 10: 64 (1995), without basionym ref.: B. caterviflora.
var. hupehana *(Craib) K. Larsen & S.S. Larsen ex D.X. Zhang in Acta Bot. Austro Sinica,* 10: 64 (1995), without basionym ref.: B. hupehana.
var. pernervosa *(H.Y. Chen) K. Larsen & S.S. Larsen ex D.X. Zhang in Acta Bot. Austro Sinica,* 10: 64 (1995), without basionym ref.: B. pernervosa.
integrifolia Roxb.
var. nymphaeifolia *(Perkins) K. Larsen & S.S. Larsen in Nordic J. Bot.,* 13(6): 658 (1993): B. nymphaeifolia.
kockiana Korth.
var. angustifolia *K. Larsen & S.S. Larsen in Nordic J. Bot.,* 13(6): 664 (1993)—Borneo (Sabah).
var. bakoensis *K. Larsen & S.S. Larsen in Nordic J. Bot.,* 13(6): 664 (1993)—Borneo (Sarawak).
var. beccarii *K. Larsen & S.S. Larsen in Nordic J. Bot.,* 13(6): 664 (1993)—Borneo (Sabah).
var. brevipedicellata *K. Larsen & S.S. Larsen in Nordic J. Bot.,* 13(6): 664 (1993)—Borneo (Kalimantan).
var. calcicola *K. Larsen & S.S. Larsen in Nordic J. Bot.,* 13(6): 664 (1993)—Borneo (Sarawak).
var. scarlatina *(Backer ex Cammerl.) K. Larsen & S.S. Larsen in Nordic J. Bot.,* 13(6): 665 (1993): B. scarlatina.
var. sericeinervia *(de Wit) K. Larsen & S.S. Larsen in Nordic J. Bot.,* 13(6): 665 (1993): Phanera kockiana var. sericeinervia.
var. velutina *(de Wit) K. Larsen & S.S. Larsen in Nordic J. Bot.,* 13(6): 665 (1993): Phanera kockiana var. velutina.

BAUHINIA:-
lingua DC.
var. antipolana *(Perkins) K. Larsen & S.S. Larsen in Nordic J. Bot.,* 13(6): 660 (1993): B. antipolana.
var. riedelii *(Baker) K. Larsen & S.S. Larsen in Nordic J. Bot.,* 13(6): 660 (1993): B. riedelii.
longicuspis
var. bicuspidata *(Benth.) R. Govaerts, World Checklist Seed Plants,* 2(1–2): 10 (1996): B. bicuspidata.
merrilliana Perkins
var. borneensis *K. Larsen & S.S. Larsen in Nordic J. Bot.,* 13(6): 658 (1993)—Borneo (Sarawak, Sabah).
pachyphylla Merr.
var. wenzelii *K. Larsen & S.S. Larsen in Nordic J. Bot.,* 13(6): 659 (1993)—Philippines.
pottsii
var. mollissima *(Wall. ex Prain) R. Govaerts, World Checklist Seed Plants,* 2(1–2): 10 (1996): B. mollissima.
semibifida Roxb.
var. acuminata *K. Larsen & S.S. Larsen in Nordic J. Bot.,* 13(6): 660 (1993)—Borneo (Sabah, Kalimantan).
var. bruneiana *K. Larsen & S.S. Larsen in Nordic J. Bot.,* 13(6): 660 (1993)—Borneo (Brunei, Sarawak).
var. longebracteata *K. Larsen & S.S. Larsen in Nordic J. Bot.,* 13(6): 660 (1993)—Borneo (Kalimantan).
var. perkinsiae *(Merr.) K. Larsen & S.S. Larsen in Nordic J. Bot.,* 13(6): 661 (1993): B. perkinsiae.
var. stenostachya *(Baker) K. Larsen & S.S. Larsen in Nordic J. Bot.,* 13(6): 661 (1993): B. stenostachya.
sirindhorniae *K. Larsen & S.S. Larsen in Nordic J. Bot.,* 17(2): 113 (1997)—Thailand.
stipularis Korth.
var. brachystylus *K. Larsen & S.S. Larsen in Nordic J. Bot.,* 13(6): 659 (1993)—Sumatera.
tianlinensis *T.C. Chen & D.X. Zhang in Nordic J. Bot.,* 18(2): 141 (1998)—China Southeast (Guangxi).
wrayi Prain
var. blumeana *K. Larsen & S.S. Larsen in Nordic J. Bot.,* 13(6): 662 (1993)—Sumatera.
var. borneensis *K. Larsen & S.S. Larsen in Nordic J. Bot.,* 13(6): 662 (1993)—Borneo (Sabah).
var. cancellata *(Ridl.) K. Larsen & S.S. Larsen in Nordic J. Bot.,* 13(6): 662 (1993): B. cancellata.
var. cardiophylla *(Merr.) K. Larsen & S.S. Larsen in Nordic J. Bot.,* 13(6): 662 (1993): B. cardiophylla.
var. moultonii *(Merr.) K. Larsen & S.S. Larsen in Nordic J. Bot.,* 13(6): 662 (1993): B. moultonii.
var. rubella *(de Wit) K. Larsen & S.S. Larsen in Nordic J. Bot.,* 13(6): 662 (1993): Phanera wrayi var. rubella.
wuzhengyii *S.S. Larsen in Novon,* 9(4): 526 (1999)—China South-central (Yunnan).

BAYABUSUA *W.J.J.O. de Wilde in Sandakania,* 13: 1 (1999). *CUCURBITACEAE*
clarkei *(King) W.J.J.O. de Wilde in Sandakania,* 13: 2 (1999): Zanonia clarkei.

× **BAYERARA** *D.M. Cumming in Haworthiad,* 13(1): 20 (1999); *cf. Repert. Pl. Succ. (I.O.S.),* 50: 7 (1999 publ. 2000). *ALOACEAE* (Aloe × Gasteria × Haworthia).

BAYNESIA *P.V. Bruyns in Novon,* 10(4): 354 (2000). *ASCLEPIADACEAE*
lophophora *P.V. Bruyns in Novon,* 10(4): 354 (2000)—Namibia.

BEAUFORTIA (Myrtac.).
aestiva *K.J. Brooks in Nuytsia,* 12(2): 164 (1998)—Western Australia.
sprengelioides *(DC.) L.A. Craven in Taxon,* 48(1): 54 (1999): Melaleuca sprengelioides.

BECCARINDA (Gesneriac.).
sumatrana *B.L. Burtt in Beitr. Biol. Pfl.,* 70(2–3): 378 (1997–98 publ. 1998)—Sumatera.

BECHIUM (Compos.).
nudicaule *(Less.) H. Robinson in Proc. Biol. Soc. Wash.,* 112(1): 227 (1999): Vernonia nudicaulis.
rhodolepis *(Baker) H. Robinson in Proc. Biol. Soc. Wash.,* 112(1): 228 (1999): Vernonia rhodolepis.

BECIUM (Labiat.).
grandiflorum (Lam.) Pic. Serm.
var. cordatum *(A.J. Paton) J.-P. Lebrun & A.L. Stork, Énum. Pl. Fleurs Afr. Trop.,* 4: 534 (1997): B. obovatum subsp. cordatum.
var. crystallinum *(A.J. Paton) J.-P. Lebrun & A.L. Stork, Énum. Pl. Fleurs Afr. Trop.,* 4: 534 (1997): B. obovatum subsp. crystallinum.
var. densiflorum *(A.J. Paton) J.-P. Lebrun & A.L. Stork, Énum. Pl. Fleurs Afr. Trop.,* 4: 534 (1997): B. grandiflorum subsp. densiflorum.
var. punctatum *(Baker) J.-P. Lebrun & A.L. Stork, Énum. Pl. Fleurs Afr. Trop.,* 4: 534 (1997): Ocimum punctatum.

BECLARDIA (Orchidac.).
grandiflora *J. Bosser in Adansonia,* 19(2): 185 (1997)—Madagascar.

BEGONIA (Begoniac.).
ser. Cathayanae *T.C. Ku in Fl. Reipubl. Popul. Sin.,* 52(1): 402 (1999).
subsect. Cathayanae *T.C. Ku in Fl. Reipubl. Popul. Sin.,* 52(1): 402 (1999).
subsect. Crassirostres *T.C. Ku in Fl. Reipubl. Popul. Sin.,* 52(1): 401 (1999).
ser. Crassirostres *T.C. Ku in Fl. Reipubl. Popul. Sin.,* 52(1): 401 (1999).
ser. Cylindricae *T.C. Ku in Fl. Reipubl. Popul. Sin.,* 52(1): 401 (1999).
ser. Floribundae *T.C. Ku in Fl. Reipubl. Popul. Sin.,* 52(1): 402 (1999).
subsect. Hemsleyanae *T.C. Ku in Fl. Reipubl. Popul. Sin.,* 52(1): 402 (1999).
ser. Leprosae *T.C. Ku in Fl. Reipubl. Popul. Sin.,* 52(1): 401 (1999).
ser. Ornithophyllae *T.C. Ku in Fl. Reipubl. Popul. Sin.,* 52(1): 402 (1999).
sect. Pleiothece *T.C. Ku in Fl. Reipubl. Popul. Sin.,* 52(1): 401 (1999).
subsect. Summoglabrae *T.C. Ku in Fl. Reipubl. Popul. Sin.,* 52(1): 401 (1999).
ser. Umbraculifoliae *T.C. Ku in Fl. Reipubl. Popul. Sin.,* 52(1): 402 (1999).
ser. Wilsonianae *T.C. Ku in Fl. Reipubl. Popul. Sin.,* 52(1): 401 (1999).
subsect. Wilsonianae *T.C. Ku in Fl. Reipubl. Popul. Sin.,* 52(1): 401 (1999).
acutitepala *K.Y. Guan & D.K. Tian in Acta Bot. Yunnanica,* 22(2): 129 (2000)—China South-central (Yunnan).
almedana *K. Burt-Utley & J.F. Utley in Novon,* 9(4): 485 (1999)—Mexico Southeast (Chiapas).
arborescens Raddi
var. oxyphylla *(A.DC.) S.F. Smith in Selbyana,* 20(1): 28 (1999): B. oxyphylla.
awongii *M.J.S. Sands in Checklist Flow. Pl. & Gymn. Brunei Darussalam (ed. M.J.E. Coode et al.):* 432 (1996)—Borneo (Brunei).
bahakensis *M.J.S. Sands in Checklist Flow. Pl. & Gymn. Brunei Darussalam (ed. M.J.E. Coode et al.):* 432 (1996)—Borneo (Brunei).
balansana Gagnep.
var. rubropilosa *S.H. Huang & Y.M. Shui in Acta Bot. Yunnanica,* 21(1): 12 (1999)—China South-central (Yunnan).
biflora *T.C. Ku in Acta Phytotax. Sin.,* 35(1): 43 (1997)—China South-central (Yunnan).
brevicaulis *T.C. Ku in Acta Phytotax. Sin.,* 35(1): 53 (1997), homonym; *see A.DC. in Ann. Sci. Nat. (Paris),* Ser. 4, 11: 135 (1859)—China South-central (Yunnan).
bruneiana *M.J.S. Sands in Checklist Flow. Pl. & Gymn. Brunei Darussalam (ed. M.J.E. Coode et al.):* 433 (1996)—Borneo (Brunei).
subsp. angustifolia *M.J.S. Sands in Checklist Flow. Pl. & Gymn. Brunei Darussalam (ed. M.J.E. Coode et al.):* 433 (1996)—Borneo (Brunei).
subsp. labiensis *M.J.S. Sands in Checklist Flow. Pl. & Gymn. Brunei Darussalam (ed. M.J.E. Coode et al.):* 433 (1996)—Borneo (Brunei).

BEGONIA:-
bruneiana
subsp. retakensis M.J.S. Sands in Checklist Flow. Pl. & Gymn. Brunei Darussalam (ed. M.J.E. Coode et al.): 433 (1996)—Borneo (Brunei).
cacauicola L.B. Sm. ex S.F. Smith & D.C. Wasshausen in Selbyana, 20(1): 16 (1999)—Brazil Northeast (Bahia).
ceratocarpa S.H. Huang & Y.M. Shui in Acta Bot. Yunnanica, 21(1): 13 (1999)—China South-central (Yunnan).
chlorandra M.J.S. Sands in Checklist Flow. Pl. & Gymn. Brunei Darussalam (ed. M.J.E. Coode et al.): 432 (1996)—Borneo (Brunei).
chloroneura P. Wilkie & M.J.S. Sands in New Plantsman, 6(3): 132 (1999)—Philippines.
cyanescens M.J.S. Sands in Checklist Flow. Pl. & Gymn. Brunei Darussalam (ed. M.J.E. Coode et al.): 433 (1996)—Borneo (Brunei).
daxinensis T.C. Ku in Acta Phytotax. Sin., 35(1): 45 (1997)—China Southeast (Guangxi).
euryphylla L.B. Sm. ex S.F. Smith & D.C. Wasshausen in Selbyana, 20(1): 16 (1999)—Brazil Northeast (Bahia).
eutricha M.J.S. Sands in Checklist Flow. Pl. & Gymn. Brunei Darussalam (ed. M.J.E. Coode et al.): 434 (1996)—Borneo (Brunei)
faustinoi K. Burt-Utley & J.F. Utley in Novon, 9(4): 486 (1999)—Mexico Southeast (Chiapas).
floribunda T.C. Ku in Acta Phytotax. Sin., 35(1): 48 (1997), homonym; see Carrière in Rev. Hortic.: 420 (1875)—China Southeast (Guangxi).
foliosa Kunth
var. major (L.B. Sm.) L.J. Dorr in Harvard Pap. Bot., 4(1): 260 (1999): B. microphylla var. major.
var. microphylla (Klotzsch) L.J. Dorr in Harvard Pap. Bot., 4(1): 261 (1999): Lepsia microphylla.
fuscisetosa M.J.S. Sands in Checklist Flow. Pl. & Gymn. Brunei Darussalam (ed.M.J.E. Coode et al.): 433 (1996)—Borneo (Brunei).
gomantongensis R. Kiew in Gard. Bull. Singapore, 50(2): 164 (1998)—Borneo (Sabah).
gongshenensis C.Y. Wu in Fl. Dulongjian Reg.: 106 (1993), without latin descr. or type—China South-central (Yunnan).
guangxiensis C.Y. Wu in Acta Phytotax. Sin., 35(1): 45 (1997)—China Southeast (Guangxi).
hekouensis S.H. Huang in Acta Bot. Yunnanica, 21(1): 21 (Feb. 1999); et in Fl. Reipubl. Popul. Sin., 52(1): 226 (Oct. 1999), nom. nov.: B. gesnerioides S.H. Huang & Y.M. Shui.
hexaptera M.J.S. Sands in Checklist Flow. Pl. & Gymn. Brunei Darussalam (ed. M.J.E. Coode et al.): 434 (1996)—Borneo (Brunei).
hondurensis K. Burt-Utley & J.F. Utley in Novon, 9(4): 483 (1999)—Honduras.
jarmilae J.J. Halda in Acta Mus. Richnov. Sect. Nat., 7(2): 71 (2000)—Mexico Northeast (Nuevo León).
jureiensis S.J. Gomes da Silva & M.C.H. Mamede in Novon, 10(1): 24 (2000)—Brazil Southeast (São Paulo).
keithii R. Kiew in Gard. Bull. Singapore, 50(2): 189 (1998)—Borneo (Sabah).
laccophora M.J.S. Sands in Checklist Flow. Pl. & Gymn. Brunei Darussalam (ed. M.J.E. Coode et al.): 434 (1996)—Borneo (Brunei).
lancangensis S.H. Huang in Acta Bot. Yunnanica, 21(1): 13 (1999)—China South-central (Yunnan).
lazat R. Kiew & Reza Azmi in Gard. Bull. Singapore, 50(1): 45 (1998)—Borneo (Sabah).
leucochlora M.J.S. Sands in Checklist Flow. Pl. & Gymn. Brunei Darussalam (ed. M.J.E. Coode et al.): 432 (1996)—Borneo (Brunei).
leucotricha M.J.S. Sands in Checklist Flow. Pl. & Gymn. Brunei Darussalam (ed. M.J.E. Coode et al.): 434 (1996)—Borneo (Brunei).
linearifolia J. Sierra Calzado in Rev. Jard. Bot. Nacion. Univ. Habana, 12: 43 (1991 publ. 1993); et in Rev. Jard. Bot. Nacion. Univ. Habana, 19: 10 (1998), nom. nov.: B. stenophylla J. Sierra.
lipolepis L.B. Sm.
var. luteyniorum (L.B. Sm. & Wassh.) L.J. Dorr in Harvard Pap. Bot., 4(1): 254 (1999): B. brevipetala var. luteyniorum.
longanensis C.Y. Wu in Acta Phytotax. Sin., 35(1): 54 (1997)—China Southeast (Guangxi).

BEGONIA:-
longialata K.Y. Guan & D.K. Tian in Acta Bot. Yunnanica, 22(2): 132 (2000)—China South-central (Yunnan).
longicarpa K.Y. Guan & D.K. Tian in Acta Bot. Yunnanica, 22(2): 131 (2000)—China South-central (Yunnan).
luzhaiensis T.C. Ku in Acta Phytotax. Sin., 37(3): 287 (1999)—China Southeast (Guangxi).
manhaoensis S.H. Huang & Y.M. Shui in Acta Bot. Yunnanica, 21(1): 21 (1999)—China South-central (Yunnan).
matogrossensis L.B. Sm. ex S.F. Smith & D.C. Wasshausen in Selbyana, 20(1): 17 (1999)—Brazil West-central (Mato Grosso).
mattos-silvae L.B. Sm. ex S.F. Smith & D.C. Wasshausen in Selbyana, 20(1): 19 (1999)—Brazil Northeast (Bahia).
nantoensis M.J. Lai & N.J. Chung in Quart. J. Exp. Forest. Nation. Taiwan Univ., 6(1): 60 (1992)—Taiwan.
nevadensis L.J. Dorr in Harvard Pap. Bot., 4(1): 264 (1999), nom. nov.: B. trapa var. pilosa L.B. Sm. & Wassh.
obliquifolia S.H. Huang & Y.M. Shui in Acta Bot. Yunnanica, 21(1): 21 (1999), as 'obliquefolia'—China South-central (Yunnan).
papyraptera M.J.S. Sands in Checklist Flow. Pl. & Gymn. Brunei Darussalam (ed. M.J.E. Coode et al.): 432 (1996)—Borneo (Brunei).
paucilobata C.Y. Wu
var. maguanensis (S.H. Huang & Y.M. Shui) T.C. Ku in Fl. Reipubl. Popul. Sin., 52(1): 261 (1999): B. maguanensis.
pinheironis L.B. Sm. ex S.F. Smith & D.C. Wasshausen in Selbyana, 20(1): 21 (1999)—Brazil Northeast (Bahia).
pluvialis L.B. Sm. ex S.F. Smith & D.C. Wasshausen in Selbyana, 20(1): 22 (1999)—Brazil South (Paraná).
polygonifolia A.DC.
var. differens L.B. Sm. ex S.F. Smith & D.C. Wasshausen in Selbyana, 20(1): 24 (1999)—Brazil Northeast (Bahia).
postarii R. Kiew in Gard. Bull. Singapore, 50(2): 165 (1998)—Borneo (Sabah).
russelliana L.B. Sm. ex S.F. Smith & D.C. Wasshausen in Selbyana, 20(1): 25 (1999)—Brazil Northeast (Bahia).
salesopolensis S.J. Gomes da Silva & M.C.H. Mamede in Novon, 10(1): 22 (2000)—Brazil Southeast (São Paulo).
serranegrae L.B. Sm. ex S.F. Smith & D.C. Wasshausen in Selbyana, 20(1): 25 (1999)—Brazil South (Paraná).
setulosopeltata C.Y. Wu in Acta Phytotax. Sin., 35(1): 48 (1997), as 'setuloso-peltata'—China Southeast (Guangxi).
sibutensis M.J.S. Sands in Checklist Flow. Pl. & Gymn. Brunei Darussalam (ed. M.J.E. Coode et al.): 433 (1996)—Borneo (Brunei).
siccacaudata J. Doorenbos in Blumea, 45(2): 400 (2000)—Sulawesi.
sino-vietnamica C.Y. Wu in Acta Phytotax. Sin., 35(1): 50 (1997)—China Southeast (Guangxi).
sinobrevicaulis T.C. Ku in Acta Phytotax. Sin., 37(2): 193 (1999), nom. nov.: B. brevicaulis T.C. Ku.
sinofloribunda L.J. Dorr in Harvard Pap. Bot., 4(1): 265 (1999), nom. nov.: B. floribunda T.C. Ku.
stenogyna M.J.S. Sands in Checklist Flow. Pl. & Gymn. Brunei Darussalam (ed. M.J.E. Coode et al.): 432 (1996)—Borneo (Brunei).
subhowii S.H. Huang in Acta Bot. Yunnanica, 21(1): 20 (1999)—China South-central (Yunnan).
× taipeiensis C.-I Peng in Bot. Bull. Acad. Sin. (Taipei), 41(2): 151 (2000)—Taiwan.(B. aptera × formosana).
temburongensis M.J.S. Sands in Checklist Flow. Pl. & Gymn. Brunei Darussalam (ed. M.J.E. Coode et al.): 432 (1996)—Borneo (Brunei).
teysmanniana (Miq.) M.C. Tebbitt in Brittonia, 52(1): 116 (2000): Platycentrum teysmannianum.
thelmae L.B. Smith & D.C. Wasshausen in Begonian, 48(5): 114 (1981), vol. no. 49 in error—Cultivation ex Brazil.
turrialbae K. Burt-Utley & J.F. Utley in Novon, 9(4): 488 (1999)—Costa Rica.

BEGONIA:-
windischii L.B. Sm. ex S.F. Smith & D.C. Wasshausen in Selbyana, 20(1): 27 (1999)—Brazil Southeast (São Paulo).
yingjiangensis S.H. Huang in Acta Bot. Yunnanica, 21(1): 18 (1999)—China South-central (Yunnan).
yishanensis T.C. Ku in Acta Phytotax. Sin., 37(3): 285 (1999)—China Southeast (Guangxi).

BEILSCHMIEDIA (Laurac.).
durifolia (Mez) R. Govaerts, World Checklist Seed Plants, 2(1–2): 10 (1996): Persea durifolia.
glaucoides (H.W. Li) H.W. Li in Acta Bot. Yunnanica, 18(1): 54 (1996): B. glauca var. glaucoides.
hexanthera H. van der Werff in Brittonia, 47(4): 374 (1995)—French Guiana.
immersinervis Sachiko Nishida in Ann. Missouri Bot. Gard., 86(3): 678 (1999)—Costa Rica.
latifolia (Nees) Sachiko Nishida in Ann. Missouri Bot. Gard., 86(3): 680 (1999): Hufelandia latifolia.
linharensis Sachiko Nishida & H. van der Werff in Ann. Missouri Bot. Gard., 86(3): 681 (1999)—Brazil Northeast (Bahia), Brazil Southeast (Espírito Santo).
manantlanensis R. Cuevas G. & T.S. Cochrane in Novon, 9(1): 18 (1999)—Mexico Southwest (Jalisco).
moratii H. van der Werff in Novon, 6(4): 463 (1996)—Madagascar.
ovalioides Sachiko Nishida in Ann. Missouri Bot. Gard., 86(3): 686 (1999)—Mexico Southwest (Oaxaca), Mexico Southeast (Chiapas).
tilaranensis Sachiko Nishida in Ann. Missouri Bot. Gard., 86(3): 694 (1999)—Costa Rica, Panama.
tovarensis (Meisn.) Sahiko Nishida in Ann. Missouri Bot. Gard., 86(3): 696 (1999)—Hufelandia tovarensis.
tsangii Merr.
var. delicata (S.K. Lee & Y.T. Wei) J. Li & H.W. Li in Acta Bot. Yunnanica, 18(1): 54 (1996): B. delicata.

BELLEVALIA (Hyacinthac.).
sitiaca Z. Kypriotakis & D. Tzanoudakis in Bot. Helvetica, 109(1): 85 (1999)—Kriti.

BELLIS (Compos.).
annua
subsp. repens (Lam.) R. Govaerts, World Checklist Seed Plants, 2(1–2): 10 (1996): B. repens.
croatica M. Gandoger, Dec. Pl. Nov., 2: 3 (1876)—Yugoslavia.
decipiens M. Gandoger, Dec. Pl. Nov., 2: 4 (1876)—France.
latifolia M. Gandoger, Dec. Pl. Nov., 2: 4 (1876)—Algeria.
minor M. Gandoger, Dec. Pl. Nov., 2: 4 (1876)—France.
subserrata M. Gandoger, Dec. Pl. Nov., 2: 3 (1876)—France.
validula M. Gandoger, Dec. Pl. Nov., 2: 4 (1876)—Yugoslavia.

BELLOA (Compos.).
wurdackiana V.M. Badillo in Biollania, ed. esp. 6 (J.J. Wurdack Festschr.): 235 (1997)—Venezuela.

BELLUCIA (Melastomatac.).
villosa G. Lozano-Contreras & L.M. Quiñones in Rev. Acad. Colomb. Cienc. Exact. Fis. Nat., 77: 244 (1996)—Colombia.

BELOGLOTTIS (Orchidac.).
sect. Helonoma (Garay) D.L. Szlachetko in Fragm. Flor. Geobot., 41(2): 702 (1996): gen. Helonoma.
americana (C. Schweinf. & Garay) D.L. Szlachetko in Fragm. Flor. Geobot., 41(2): 702 (1996): Manniella americana.
bifida (Ridl.) D.L. Szlachetko in Fragm. Flor. Geobot., 41(2): 702 (1996): Spiranthes bifida.
chiropterae D.L. Szlachetko in Fragm. Flor. Geobot., 41(2): 703 (1996)—Venezuela.

BELONOPHORA (Rubiac.).
coffeoides Hook. f.
subsp. hypoglauca *(Welw. ex Hiern) S.E. Dawson & M.R. Cheek in Kew Bull.*, 55(1): 77 (2000): Coffea hypoglauca.
ongensis *S.E. Dawson & M.R. Cheek in Kew Bull.*, 55(1): 75 (2000)—Cameroon.

BENTHAMIA (Orchidac.).
chlorantha *(Spreng.) L.A. Garay & G.A. Romero-González in Harvard Pap. Bot.*, 3(1): 53 (1998): Habenaria chlorantha.
drummondii *(H.P. Linder & G. Will.) D.L. Szlachetko & P. Rutkowski in Acta Bot. Fenn.*, 169: 380 (2000): Oligophyton drummondii.

BENTHAMIDIA (Cornac.).
sinensis *(Nakai) T. Yamazaki in J. Jap. Bot.*, 75(6): 374 (2000): Cynoxylon sinense.

BERBERIS (Berberidac.).
annamica *(Gagnep.) J.E. Laferrière in Bot. Zhurn.*, 82(9): 96 (19 Nov. 1997); *et in Fragm. Flor. Geobot.*, 42(2): 350 (19 Dec. 1997); *et in Acta Bot. Indica*, 25(2): 243 (Dec. 1997): Mahonia annamica.
aquifolium
var. lyallii *(Ahrendt) J.S. Marroquín & J.E. Laferrière in J. Ariz.-Nev. Acad. Sci.*, 30(1): 53 (1997): Mahonia aquifolium var. lyallii.
var. nutkana *(DC.) J.S. Marroquín & J.E. Laferrière in J. Ariz.-Nev. Acad. Sci.*, 30(1): 53 (1997): Mahonia aquifolium var. nutkana.
aristulata *(Ahrendt) J.E. Laferrière in Bot. Zhurn.*, 82(9): 96 (19 Nov. 1997); *et in Fragm. Flor. Geobot.*, 42(2): 350 (19 Dec. 1997); *et in Acta Bot. Indica*, 25(2): 243 (Dec. 1997): Mahonia aristulata.
assamana *J.E. Laferrière in Bot. Zhurn.*, 82(9): 96 (19 Nov. 1997); *et in Fragm. Flor. Geobot.*, 42(2): 350 (19 Dec. 1997); *et in Acta Bot. Indica*, 25(2): 243 (Dec. 1997), nom. nov.: Mahonia simonsii *Takeda*.
atroviridiana *T.S. Ying in Acta Phytotax. Sin.*, 37(4): 336 (1999)—Xizang.
batangensis *T.S. Ying in Acta Phytotax. Sin.*, 37(4): 344 (1999)—China South-central (Sichuan).
bealei
var. planifolia *(Hook. f. & Thomson) J.E. Laferrière in Bot. Zhurn.*, 82(9): 96 (19 Nov. 1997); *et in Fragm. Flor. Geobot.*, 42(2): 350 (19 Dec. 1997); *et in Acta Bot. Indica*, 25(2): 243 (Dec. 1997): Mahonia bealei var. planifolia.
beijingensis *T.S. Ying in Acta Phytotax. Sin.*, 37(4): 324 (1999)—China North-central (Hebei, Shandong).
berberidifolia *(P.K. Hsiao & Y.S. Wang) J.E. Laferrière in Bot. Zhurn.*, 82(9): 96 (19 Nov. 1997); *et in Fragm. Flor. Geobot.*, 42(2): 350 (19 Dec. 1997); *et in Acta Bot. Indica*, 25(2): 243 (Dec. 1997): Mahonia berberidifolia.
bodinieri *(Gagnep.) J.E. Laferrière in Bot. Zhurn.*, 82(9): 96 (19 Nov. 1997); *et in Fragm. Flor. Geobot.*, 42(2): 350 (19 Dec. 1997); *et in Acta Bot. Indica*, 25(2): 243 (Dec. 1997), homonym; *see H. Lév. in Repert. Spec. Nov. Regni Veg.*, 9: 454 (1911): Mahonia bodinieri.
borealis *(Takeda) J.E. Laferrière in Bot. Zhurn.*, 82(9): 96 (19 Nov. 1997); *et in Fragm. Flor. Geobot.*, 42(2): 350 (19 Dec. 1997); *et in Acta Bot. Indica*, 25(2): 243 (Dec. 1997): Mahonia borealis.
var. parryi *(Ahrendt) J.E. Laferrière in Bot. Zhurn.*, 82(9): 96 (19 Nov. 1997); *et in Fragm. Flor. Geobot.*, 42(2): 350 (19 Dec. 1997); *et in Acta Bot. Indica*, 25(2): 244 (Dec. 1997): Mahonia borealis var. parryi.
bracteolata *(Takeda) J.E. Laferrière in Bot. Zhurn.*, 82(9): 96 (19 Nov. 1997); *et in Fragm. Flor. Geobot.*, 42(2): 350 (19 Dec. 1997); *et in Acta Bot. Indica*, 25(2): 244 (Dec. 1997): Mahonia bracteolata.
var. zhongdianensis *(S.Y. Bao) J.E. Laferrière in Bot. Zhurn.*, 82(9): 96 (19 Nov. 1997); *et in Fragm. Flor. Geobot.*, 42(2): 350 (19 Dec. 1997); *et in Acta Bot. Indica*, 25(2): 244 (Dec. 1997): Mahonia bracteolata var. zhongdianensis.
breviracema *(Y.S. Wang & P.K. Hsiao) J.E. Laferrière in Bot. Zhurn.*, 82(9): 96 (19 Nov. 1997); *et in Fragm. Flor. Geobot.*, 42(2): 350 (19 Dec. 1997); *et in Acta Bot. Indica*, 25(2): 244 (Dec. 1997): Mahonia breviracema.

BERBERIS:-
caelicolor *(S.Y. Bao) J.E. Laferrière in Bot. Zhurn.*, 82(9): 96 (19 Nov. 1997); *et in Fragm. Flor. Geobot.*, 42(2): 350 (19 Dec. 1997); *et in Acta Bot. Indica*, 25(2): 244 (Dec. 1997): Mahonia caelicolor.
caesia *(C.K. Schneid.) J.E. Laferrière in Bot. Zhurn.*, 82(9): 96 (19 Nov. 1997); *et in Fragm. Flor. Geobot.*, 42(2): 350 (19 Dec. 1997); *et in Acta Bot. Indica*, 25(2): 244 (Dec. 1997): Mahonia caesia.
calamicaulis *(Spare & C.E.C.Fisch.) J.E. Laferrière in Bot. Zhurn.*, 82(9): 96 (19 Nov. 1997); *et in Fragm. Flor. Geobot.*, 42(2): 350 (19 Dec. 1997); *et in Acta Bot. Indica*, 25(2): 244 (Dec. 1997): Mahonia calamicaulis.
chilensis Gillies ex Hook.
var. brachybotria *(Gay) L.R. Landrum in Ann. Missouri Bot. Gard.*, 86(4): 805 (1999): B. brachybotria.
chunanensis *T.S. Ying in Acta Phytotax. Sin.*, 37(4): 315 (1999), contrary to Art. 37.4 ICBN (1994)—China Southeast (Zhejiang).
confusa *(Sprague) J.E. Laferrière in Bot. Zhurn.*, 82(9): 97 (19 Nov. 1997); *et in Fragm. Flor. Geobot.*, 42(2): 350 (19 Dec. 1997); *et in Acta Bot. Indica*, 25(2): 244 (Dec. 1997): Mahonia confusa.
var. bournei *(Ahrendt) J.E. Laferrière in Bot. Zhurn.*, 82(9): 97 (19 Nov. 1997); *et in Fragm. Flor. Geobot.*, 42(2): 351 (19 Dec. 1997); *et in Acta Bot. Indica*, 25(2): 244 (Dec. 1997): Mahonia confusa var. bournei.
contracta *T.S. Ying in Acta Phytotax. Sin.*, 37(4): 322 (1999)—China South-central (Yunnan).
× convoluta *(Ahrendt) J.S. Marroquín & J.E. Laferrière in J. Ariz.-Nev. Acad. Sci.*, 30(1): 53 (1997), basionym not validly publ.: Mahonia × convoluta.
daiana *T.S. Ying in Acta Phytotax. Sin.*, 37(4): 345 (1999)—China South-central (Sichuan).
daochengensis *T.S. Ying in Acta Phytotax. Sin.*, 37(4): 336 (1999)—China South-central (Sichuan).
decipiens *(C.K. Schneid.) J.E. Laferrière in Bot. Zhurn.*, 82(9): 97 (19 Nov. 1997); *et in Fragm. Flor. Geobot.*, 42(2): 351 (19 Dec. 1997); *et in Acta Bot. Indica*, 25(2): 244 (Dec. 1997): Mahonia decipiens.
derongensis *T.S. Ying in Acta Phytotax. Sin.*, 37(4): 333 (1999)—China South-central (Sichuan).
discolorifolia *(Ahrendt) J.E. Laferrière in Bot. Zhurn.*, 82(9): 97 (19 Nov. 1997); *et in Fragm. Flor. Geobot.*, 42(2): 351 (19 Dec. 1997); *et in Acta Bot. Indica*, 25(2): 244 (Dec. 1997): Mahonia discolorifolia.
dolichostylis *(Takeda) J.E. Laferrière in Bot. Zhurn.*, 82(9): 97 (19 Nov. 1997); *et in Fragm. Flor. Geobot.*, 42(2): 351 (19 Dec. 1997); *et in Acta Bot. Indica*, 25(2): 244 (Dec. 1997): Mahonia dolichostylis.
dongchuanensis *T.S. Ying in Acta Phytotax. Sin.*, 37(4): 312 (1999)—China South-central (Yunnan).
duclouxiana *(Gagnep.) J.E. Laferrière in Bot. Zhurn.*, 82(9): 97 (19 Nov. 1997); *et in Fragm. Flor. Geobot.*, 42(2): 351 (19 Dec. 1997); *et in Acta Bot. Indica*, 25(2): 244 (Dec. 1997): Mahonia duclouxiana.
var. hilaica *(Ahrendt) J.E. Laferrière in Bot. Zhurn.*, 82(9): 97 (19 Nov. 1997); *et in Fragm. Flor. Geobot.*, 42(2): 351 (19 Dec. 1997); *et in Acta Bot. Indica*, 25(2): 244 (Dec. 1997): Mahonia duclouxiana var. hilaica.
eurybracteata *(Fedde) J.E. Laferrière in Bot. Zhurn.*, 82(9): 97 (19 Nov. 1997); *et in Fragm. Flor. Geobot.*, 42(2): 351 (19 Dec. 1997); *et in Acta Bot. Indica*, 25(2): 244 (Dec. 1997): Mahonia eurybracteata.
fargesii *(Takeda) J.E. Laferrière in Bot. Zhurn.*, 82(9): 97 (19 Nov. 1997); *et in Fragm. Flor. Geobot.*, 42(2): 351 (19 Dec. 1997); *et in Acta Bot. Indica*, 25(2): 244 (Dec. 1997): Mahonia fargesii.
feddei *(Ahrendt) J.E. Laferrière in Bot. Zhurn.*, 82(9): 97 (19 Nov. 1997); *et in Fragm. Flor. Geobot.*, 42(2): 351 (19 Dec. 1997); *et in Acta Bot. Indica*, 25(2): 244 (Dec. 1997): Mahonia feddei.
flavida *(C.K. Schneid.) J.E. Laferrière in Bot. Zhurn.*, 82(9): 97 (19 Nov. 1997); *et in Fragm. Flor. Geobot.*, 42(2): 350 (19 Dec. 1997); *et in Acta Bot. Indica*, 25(2): 244 (Dec. 1997): Mahonia flavida.

BERBERIS:-
flavida *(C.K. Schneid.)*
var. integrifoliola *(Hand.-Mazz.) J.E. Laferrière in Bot. Zhurn.*, 82(9): 97 (19 Nov. 1997); *et in Fragm. Flor. Geobot.*, 42(2): 351 (19 Dec. 1997); *et in Acta Bot. Indica*, 25(2): 244 (Dec. 1997): Mahonia flavida var. integrifoliola.
fordii *(C.K. Schneid.) J.E. Laferrière in Bot. Zhurn.*, 82(9): 97 (19 Nov. 1997); *et in Fragm. Flor. Geobot.*, 42(2): 351 (19 Dec. 1997); *et in Acta Bot. Indica*, 25(2): 244 (Dec. 1997): Mahonia fordii.
fortunei
var. szechuanica *(Ahrendt) J.E. Laferrière in Bot. Zhurn.*, 82(9): 97 (19 Nov. 1997); *et in Fragm. Flor. Geobot.*, 42(2): 351 (19 Dec. 1997); *et in Acta Bot. Indica*, 25(2): 244 (Dec. 1997): Mahonia fortunei var. szechuanica.
gautamae *J.E. Laferrière in Bot. Zhurn.*, 82(9): 97 (19 Nov. 1997); *et in Fragm. Flor. Geobot.*, 42(2): 351 (19 Dec. 1997); *et in Acta Bot. Indica*, 25(2): 244 (Dec. 1997), nom. nov.: Mahonia sikkimensis *Takeda*.
griffithii *(Takeda) J.E. Laferrière in Bot. Zhurn.*, 82(9): 97 (19 Nov. 1997); *et in Fragm. Flor. Geobot.*, 42(2): 351 (19 Dec. 1997); *et in Acta Bot. Indica*, 25(2): 244 (Dec. 1997): Mahonia griffithii.
guizhouensis *T.S. Ying in Acta Phytotax. Sin.*, 37(4): 320 (1999)—China South-central (Guizhou).
hancockiana *(Takeda) J.E. Laferrière in Bot. Zhurn.*, 82(9): 97 (19 Nov. 1997); *et in Fragm. Flor. Geobot.*, 42(2): 351 (19 Dec. 1997); *et in Acta Bot. Indica*, 25(2): 244 (Dec. 1997): Mahonia hancockiana.
haoi *T.S. Ying in Acta Phytotax. Sin.*, 37(4): 339 (1999)—China North-central (Gansu).
henryi *J.E. Laferrière in Bot. Zhurn.*, 82(9): 97 (19 Nov. 1997); *et in Fragm. Flor. Geobot.*, 42(2): 351 (19 Dec. 1997); *et in Acta Bot. Indica*, 25(2): 244 (Dec. 1997), nom. nov.: Mahonia conferta *Takeda*.
× herveyi *(Ahrendt) J.S. Marroquín & J.E. Laferrière in J. Ariz.-Nev. Acad. Sci.*, 30(1): 54 (1997), basionym not validly publ.: Mahonia × herveyi.
hicksii *(Ahrendt) J.E. Laferrière in Bot. Zhurn.*, 82(9): 97 (19 Nov. 1997); *et in Fragm. Flor. Geobot.*, 42(2): 351 (19 Dec. 1997); *et in Acta Bot. Indica*, 25(2): 244 (Dec. 1997): Mahonia hicksii.
huiliensis *(Hand.-Mazz.) J.E. Laferrière in Bot. Zhurn.*, 82(9): 97 (19 Nov. 1997); *et in Fragm. Flor. Geobot.*, 42(2): 352 (19 Dec. 1997); *et in Acta Bot. Indica*, 25(2): 244 (Dec. 1997): Mahonia huiliensis.
integripetala *T.S. Ying in Acta Phytotax. Sin.*, 37(4): 334 (1999)—China North-central (Gansu).
intermedia *(Dupuis) J.E. Laferrière in Bot. Zhurn.*, 82(9): 97 (19 Nov. 1997); *et in Fragm. Flor. Geobot.*, 42(2): 352 (19 Dec. 1997); *et in Acta Bot. Indica*, 25(2): 244 (Dec. 1997): Mahonia intermedia.
japonica
var. trifurca *(Loudon) J.E. Laferrière in Bot. Zhurn.*, 82(9): 97 (19 Nov. 1997), with incorrect basionym ref.; *et in Fragm. Flor. Geobot.*, 42(2): 352 (19 Dec. 1997), with incorrect basionym ref.; *et in Acta Bot. Indica*, 25(2): 244 (Dec. 1997), with incorrect basionym ref.: B. trifurca.
jaunsarensis *(Ahrendt) J.E. Laferrière in Bot. Zhurn.*, 82(9): 97 (19 Nov. 1997); *et in Fragm. Flor. Geobot.*, 42(2): 352 (19 Dec. 1997); *et in Acta Bot. Indica*, 25(2): 244 (Dec. 1997): Mahonia jaunsarensis.
jinfoshanensis *T.S. Ying in Acta Phytotax. Sin.*, 37(4): 316 (1999)—China South-central (Sichuan).
jingguensis *G.S. Fan & X.W. Li in J. Trop. Subtrop. Bot.*, 5(3): 1 (1997)—China South-central (Yunnan).
jiulongensis *T.S. Ying in Acta Phytotax. Sin.*, 37(4): 320 (1999), holotype not indicated—China South-central (Sichuan).
kangdingensis *T.S. Ying in Acta Phytotax. Sin.*, 37(4): 349 (1999)—China South-central (Sichuan).
keikoe *J. Laferrière in Fragm. Flor. Geobot.*, 42(2): 352 (19 Dec. 1997), nom. nov. superfl.; *see Laferr. in Bot. Zhurn.*, 82(9): 99 (19 Nov. 1997): Mahonia mairei *Takeda*.

BERBERIS:-
klossii *(Baker f.) J.E. Laferrière in Bot. Zhurn.*, 82(9): 97 (19 Nov. 1997); *et in Fragm. Flor. Geobot.*, 42(2): 352 (19 Dec. 1997); *et in Acta Bot. Indica*, 25(2): 244 (Dec. 1997): Mahonia klossii.
laojunshanensis *T.S. Ying in Acta Phytotax. Sin.*, 37(4): 318 (1999)—China South-central (Hubei).
leboensis *T.S. Ying in Acta Phytotax. Sin.*, 37(4): 328 (1999)—China South-central (Sichuan).
leptodonta *(Gagnep.) J.E. Laferrière in Bot. Zhurn.*, 82(9): 97 (19 Nov. 1997); *et in Fragm. Flor. Geobot.*, 42(2): 352 (19 Dec. 1997); *et in Acta Bot. Indica*, 25(2): 244 (Dec. 1997): Mahonia leptodonta.
leschenaultii *Wall. ex Wight & Arn., Prodr. Fl. Ind. Orient.*: 16 (1834)—India (Tamil Nadu).
leveilleana *(C.K. Schneid.) J.E. Laferrière in Bot. Zhurn.*, 82(9): 97 (19 Nov. 1997); *et in Fragm. Flor. Geobot.*, 42(2): 352 (19 Dec. 1997); *et in Acta Bot. Indica*, 25(2): 244 (Dec. 1997): Mahonia leveilleana.
lomariifolia *(Takeda) J.E. Laferrière in Bot. Zhurn.*, 82(9): 98 (19 Nov. 1997); *et in Fragm. Flor. Geobot.*, 42(2): 352 (19 Dec. 1997); *et in Acta Bot. Indica*, 25(2): 245 (Dec. 1997): Mahonia lomariifolia.
var. estylis *(C.Y. Wu) J.E. Laferrière in Bot. Zhurn.*, 82(9): 98 (19 Nov. 1997); *et in Fragm. Flor. Geobot.*, 42(2): 352 (19 Dec. 1997); *et in Acta Bot. Indica*, 25(2): 245 (Dec. 1997): Mahonia lomariifolia var. estylis.
longibracteata *(Takeda) J.E. Laferrière in Bot. Zhurn.*, 82(9): 98 (19 Nov. 1997); *et in Fragm. Flor. Geobot.*, 42(2): 352 (19 Dec. 1997); *et in Acta Bot. Indica*, 25(2): 245 (Dec. 1997): Mahonia longibracteata.
longipes *(Standl.) J.S. Marroquín & J.E. Laferrière in J. Ariz.-Nev. Acad. Sci.*, 30(1): 54 (1997): Odostemon longipes.
longlinensis *(Y.S. Wang & P.K. Hsiao) J.E. Laferrière in Bot. Zhurn.*, 82(9): 98 (19 Nov. 1997); *et in Fragm. Flor. Geobot.*, 42(2): 352 (19 Dec. 1997); *et in Acta Bot. Indica*, 25(2): 245 (Dec. 1997): Mahonia longlinensis.
luhuoensis *T.S. Ying in Acta Phytotax. Sin.*, 37(4): 323 (1999)—China South-central (Sichuan).
magnifica *(Ahrendt) J.E. Laferrière in Bot. Zhurn.*, 82(9): 98 (19 Nov. 1997); *et in Fragm. Flor. Geobot.*, 42(2): 352 (19 Dec. 1997); *et in Acta Bot. Indica*, 25(2): 245 (Dec. 1997): Mahonia magnifica.
manipurensis *(Takeda) J.E. Laferrière in Bot. Zhurn.*, 82(9): 98 (19 Nov. 1997); *et in Fragm. Flor. Geobot.*, 42(2): 352 (19 Dec. 1997); *et in Acta Bot. Indica*, 25(2): 245 (Dec. 1997): Mahonia manipurensis.
medogensis *T.S. Ying in Acta Phytotax. Sin.*, 37(4): 350 (1999)—Xizang.
mianningensis *T.S. Ying in Acta Phytotax. Sin.*, 37(4): 347 (1999)—China South-central (Sichuan).
micropetala *T.S. Ying in Acta Phytotax. Sin.*, 37(4): 313 (1999), homonym; *see C.K. Schneid. in Repert. Spec. Nov. Regni Veg.*, 46: 266 (1939)—China South-central (Yunnan).
miethkeana *(L.W. Melander & Eade) J.E. Laferrière in Bot. Zhurn.*, 82(9): 98 (19 Nov. 1997); *et in Fragm. Flor. Geobot.*, 42(2): 352 (19 Dec. 1997); *et in Acta Bot. Indica*, 25(2): 245 (Dec. 1997): × Mahoberberis miethkeana.
monyulensis *(Ahrendt) J.E. Laferrière in Bot. Zhurn.*, 82(9): 98 (19 Nov. 1997); *et in Fragm. Flor. Geobot.*, 42(2): 352 (19 Dec. 1997); *et in Acta Bot. Indica*, 25(2): 245 (Dec. 1997): Mahonia monyulensis.
× moseri *(Ahrendt) J.S. Marroquín & J.E. Laferrière in J. Ariz.-Nev.Acad. Sci.*, 30(1): 54 (1997), basionym not validly publ.: Mahonia × moseri.
multiovula *T.S. Ying in Acta Phytotax. Sin.*, 37(4): 309 (1999)—China South-central (Sichuan).
napaulensis *(DC.) J.E. Laferrière in Bot. Zhurn.*, 82(9): 98 (19 Nov. 1997); *et in Fragm. Flor. Geobot.*, 42(2): 352 (19 Dec. 1997); *et in Acta Bot. Indica*, 25(2): 245 (Dec. 1997), isonym; *see Spreng., Syst. Veg.*, 2: 120 (1825): Mahonia napaulensis.
var. pycnophylla *(Fedde) J.E. Laferrière in Bot. Zhurn.*, 82(9): 98 (19 Nov. 1997); *et in Fragm. Flor. Geobot.*, 42(2): 353 (19 Dec. 1997); *et in Acta Bot. Indica*, 25(2): 245 (Dec. 1997): Mahonia napaulensis var. pycnophylla.

BERBERIS:-
nitens *(C.K. Schneid.) J.E. Laferrière in Bot. Zhurn.*, 82(9): 98 (19 Nov. 1997); *et in Fragm. Flor. Geobot.*, 42(2): 353 (19 Dec. 1997); *et in Acta Bot. Indica*, 25(2): 245 (Dec. 1997), homonym; *see Schrad. in Linnaea*, 12: 382 (1838): Mahonia nitens.
nivea *(C.K. Schneid.) J.E. Laferrière in Bot. Zhurn.*, 82(9): 98 (19 Nov. 1997); *et in Fragm. Flor. Geobot.*, 42(2): 353 (19 Dec. 1997); *et in Acta Bot. Indica*, 25(2): 245 (Dec. 1997): Mahonia nivea.
oiwakensis *(Hayata) J.E. Laferrière in Bot. Zhurn.*, 82(9): 98 (19 Nov. 1997); *et in Fragm. Flor. Geobot.*, 42(2): 353 (19 Dec. 1997); *et in Acta Bot. Indica*, 25(2): 245 (Dec. 1997): Mahonia oiwakensis.
pachakshirensis *(Ahrendt) J.E. Laferrière in Bot. Zhurn.*, 82(9): 98 (19 Nov. 1997); *et in Fragm. Flor. Geobot.*, 42(2): 353 (19 Dec. 1997); *et in Acta Bot. Indica*, 25(2): 245 (Dec. 1997): Mahonia pachakshirensis.
paucijuga *(C.Y. Wu) J.E. Laferrière in Bot. Zhurn.*, 82(9): 98 (19 Nov. 1997); *et in Fragm. Flor. Geobot.*, 42(2): 353 (19 Dec. 1997); *et in Acta Bot. Indica*, 25(2): 245 (Dec. 1997): Mahonia paucijuga.
paxii *(Fedde) J.S. Marroquín & J.E. Laferrière in J. Ariz.-Nev. Acad. Sci.*, 30(1): 54 (1997): Mahonia paxii.
philippinensis *(Takeda) J.E. Laferrière in Bot. Zhurn.*, 82(9): 98 (19 Nov. 1997); *et in Fragm. Flor. Geobot.*, 42(2): 353 (19 Dec. 1997); *et in Acta Bot. Indica*, 25(2): 245 (Dec. 1997): Mahonia philippinensis.
pingwuensis *T.S. Ying in Acta Phytotax. Sin.*, 37(4): 339 (1999)—China South-central (Sichuan).
pinnata
var. hortensis *(Fedde) J.S. Marroquín & J.E. Laferrière in J. Ariz.-Nev. Acad. Sci.*, 30(1): 54 (1997): Mahonia pinnata var. hortensis.
polyodonta *(Fedde) J.E. Laferrière in Bot. Zhurn.*, 82(9): 98 (19 Nov. 1997); *et in Fragm. Flor. Geobot.*, 42(2): 353 (19 Dec. 1997); *et in Acta Bot. Indica*, 25(2): 245 (Dec. 1997): Mahonia polyodonta.
pomensis *(Ahrendt) J.E. Laferrière in Bot. Zhurn.*, 82(9): 98 (19 Nov. 1997); *et in Fragm. Flor. Geobot.*, 42(2): 353 (19 Dec. 1997); *et in Acta Bot. Indica*, 25(2): 245 (Dec. 1997): Mahonia pomensis.
pseudoamoena *T.S. Ying in Acta Phytotax. Sin.*, 37(4): 331 (1999), contrary to Art. 37.4 ICBN (1994)—China South-central (Sichuan).
puca *J.E. Laferrière in Bot. Zhurn.*, 82(9): 98 (19 Nov. 1997); *et in Fragm. Flor. Geobot.*, 42(2): 353 (19 Dec. 1997); *et in Acta Bot. Indica*, 25(2): 245 (Dec. 1997), nom. nov.: Mahonia morrisonensis *Takeda*.
quinquefolia *(Standl.) J.S. Marroquín & J.E. Laferrière in J. Ariz.-Nev. Acad. Sci.*, 30(1): 54 (1997): Odostemon quinquefolius.
racemosa *(A.R. Molina) J.S. Marroquín & J.E. Laferrière in J. Ariz.-Nev. Acad. Sci.*, 30(1): 54 (1997): Mahonia racemosa.
repens
var. macrocarpa *(Jouin) J.S. Marroquín & J.E. Laferrière in J. Ariz.-Nev. Acad. Sci.*, 30(1): 54 (1997): Mahonia repens var. macrocarpa.
var. rotundifolia *(Fedde) J.S. Marroquín & J.E. Laferrière in J. Ariz.-Nev. Acad. Sci.*, 30(1): 54 (1997): Mahonia repens var. rotundifolia.
reticulinerva *(C.Y. Wu) J.E. Laferrière in Bot. Zhurn.*, 82(9): 98 (19 Nov. 1997); *et in Fragm. Flor. Geobot.*, 42(2): 352 (19 Dec. 1997); *et in Acta Bot. Indica*, 25(2): 245 (Dec. 1997): Mahonia reticulinerva.
reticulinervis *T.S. Ying in Acta Phytotax. Sin.*, 37(4): 305 (1999)—Xizang, China South-central (Sichuan).
var. brevipedicellata *T.S. Ying in Acta Phytotax. Sin.*, 37(4): 307 (1999)—China North-central (Gansu).
retinervia *J.J. Triana & J.E. Planchon in Ann. Sci. Nat. Bot.*, Ser. 4, 17: 53 (1862)—Colombia, Ecuador.†

BERBERIS:-
retinervis *(Y.S. Wang & P.K. Hsiao) J.E. Laferrière in Bot. Zhurn.*, 82(9): 98 (19 Nov. 1997); *et in Fragm. Flor. Geobot.*, 42(2): 353 (19 Dec. 1997); *et in Acta Bot. Indica*, 25(2): 245 (Dec. 1997), homonym; *see Triana & Planch. in Ann. Sci. Nat. Bot.*, Ser. 4, 17: 53 (1862): Mahonia retinervis.
retusa *T.S. Ying in Acta Phytotax. Sin.*, 37(4): 338 (1999)—China South-central (Sichuan).
roxburghii *(DC.) J.E. Laferrière in Bot. Zhurn.*, 82(9): 98 (19 Nov. 1997); *et in Fragm. Flor. Geobot.*, 42(2): 353 (19 Dec. 1997); *et in Acta Bot. Indica*, 25(2): 245 (Dec. 1997): Mahonia napaulensis var. roxburghii.
russellii *(N.P. Taylor) J.S. Marroquín & J.E. Laferrière in J. Ariz.-Nev. Acad. Sci.*, 30(1): 54 (1997): Mahonia russellii.
salweenensis *(Ahrendt) J.E. Laferrière in Bot. Zhurn.*, 82(9): 98 (19 Nov. 1997); *et in Fragm. Flor. Geobot.*, 42(2): 353 (19 Dec. 1997); *et in Acta Bot. Indica*, 25(2): 245 (Dec. 1997): Mahonia salweenensis.
schochii *(Hand.-Mazz.) J.E. Laferrière in Bot. Zhurn.*, 82(9): 98 (19 Nov. 1997); *et in Fragm. Flor. Geobot.*, 42(2): 353 (19 Dec. 1997); *et in Acta Bot. Indica*, 25(2): 245 (Dec. 1997): Mahonia schochii.
setosa *(Gagnep.) J.E. Laferrière in Bot. Zhurn.*, 82(9): 98 (19 Nov. 1997); *et in Fragm. Flor. Geobot.*, 42(2): 353 (19 Dec. 1997); *et in Acta Bot. Indica*, 25(2): 245 (Dec. 1997): Mahonia setosa.
shenii *(Chun) J.E. Laferrière in Bot. Zhurn.*, 82(9): 99 (19 Nov. 1997); *et in Fragm. Flor. Geobot.*, 42(2): 354 (19 Dec. 1997); *et in Acta Bot. Indica*, 25(2): 245 (Dec. 1997): Mahonia shenii.
sheridaniana *(C.K. Schneid.) J.E. Laferrière in Bot. Zhurn.*, 82(9): 99 (19 Nov. 1997); *et in Fragm. Flor. Geobot.*, 42(2): 354 (19 Dec. 1997); *et in Acta Bot. Indica*, 25(2): 245 (Dec. 1997): Mahonia sheridaniana.
sibirica Pall.
var. glauconigrescens *G.A. Peschkova in Fl. Sibir. (Berberidac.-Grossulariac.)*, 7: 10 (1994), contrary to Art. 37.5 ICBN (1994)—Altay.
sichuanica *T.S. Ying in Acta Phytotax. Sin.*, 37(4): 329 (1999)—China South-central (Sichuan, Yunnan).
standleyi *J.S. Marroquín & J.E. Laferrière in J. Ariz.-Nev. Acad. Sci.*, 30(1): 54 (1997), nom. nov.: Mahonia glauca *Standl. & L.O. Williams*.
subimbricata *(Chun & F. Chun) J.E. Laferrière in Bot. Zhurn.*, 82(9): 99 (19 Nov. 1997); *et in Fragm. Flor. Geobot.*, 42(2): 354 (19 Dec. 1997); *et in Acta Bot. Indica*, 25(2): 245 (Dec. 1997): Mahonia subimbricata.
sumatrensis *(Merr.) J.E. Laferrière in Bot. Zhurn.*, 82(9): 99 (19 Nov. 1997); *et in Fragm. Flor. Geobot.*, 42(2): 354 (19 Dec. 1997); *et in Acta Bot. Indica*, 25(2): 245 (Dec. 1997): Mahonia sumatrensis.
tarokoensis *S.Y. Lu & Yuen P. Yang in Fl. Taiwan, ed. 2*, 2: 581 (1996)—Taiwan.
tenuipedicellata *T.S. Ying in Acta Phytotax. Sin.*, 37(4): 343 (1999)—China South-central (Sichuan).
tianshuiensis *T.S. Ying in Acta Phytotax. Sin.*, 37(4): 341 (1999)—China North-central (Gansu).
tibetensis *J.E. Laferrière in Bot. Zhurn.*, 82(9): 99 (19 Nov. 1997); *et in Fragm. Flor. Geobot.*, 42(2): 354 (19 Dec. 1997); *et in Acta Bot. Indica*, 25(2): 245 (Dec. 1997), nom. nov.: Mahonia taronensis *Hand.-Mazz.*
tikushiensis *(Hayata) J.E. Laferrière in Bot. Zhurn.*, 82(9): 99 (19 Nov. 1997); *et in Fragm. Flor. Geobot.*, 42(2): 354 (19 Dec. 1997); *et in Acta Bot. Indica*, 25(2): 245 (Dec. 1997): Mahonia tikushiensis.
tsailunii *J.E. Laferrière in Bot. Zhurn.*, 82(9): 99 (19 Nov. 1997); *et in Acta Bot. Indica*, 25(2): 246 (Dec. 1997), nom. nov.: Mahonia mairei *Takeda*.
tsienii *T.S. Ying in Acta Phytotax. Sin.*, 37(4): 307 (1999)—China South-central (Guizhou).
undulata *(Ahrendt) J.S. Marroquín & J.E. Laferrière in J. Ariz.-Nev. Acad. Sci.*, 30(1): 54 (1997), basionym not validly publ.: Mahonia undulata.

BERBERIS:-
veitchiorum
var. kingdon-wardiana *(Ahrendt) J.E. Laferrière in Bot. Zhurn.*, 82(9): 99 (19 Nov. 1997); *et in Fragm. Flor. Geobot.*, 42 (2): 354 (19 Dec. 1997); *et in Acta Bot. Indica*, 25(2): 246 (Dec. 1997): Mahonia veitchiorum var. kingdon-wardiana.
volcania *(Standl. & Steyerm.) J.S. Marroquín & J.E. Laferrière in J. Ariz.-Nev. Acad. Sci.*, 30(1): 54 (1997): Mahonia volcania.
weiningensis *T.S. Ying in Acta Phytotax. Sin.*, 37(4): 326 (1999)—China South-central (Guizhou).
xingwenensis *T.S. Ying in Acta Phytotax. Sin.*, 37(4): 311 (1999)—China South-central (Sichuan).
yui *T.S. Ying in Acta Phytotax. Sin.*, 37(4): 309 (1999), as 'yuii'—China South-central (Sichuan).
zimapana *(Fedde) J.S. Marroquín & J.E. Laferrière in J. Ariz.-Nev. Acad. Sci.*, 30(1): 54 (1997): Mahonia zimapana.
ziyunensis *P.K. Hsiao & Z.Yu. Li in Acta Bot. Yunnanica*, 21(1): 30 (1999)—China South-central (Guizhou).

BERCHEMIA (Rhamnac.).
flavescens *(Wall. ex Roxb.) A.T. Brongniart, Mém. Fam. Rhamnées:* 50 (1826): Ziziphus flavescens.†
floribunda *(Wall. ex Roxb.) A.T. Brongniart, Mém. Fam. Rhamnées:* 50 (1826): Ziziphus floribunda.†

BERGIA (Elatinac.).
koganii *V.V. Nikitin ex M. Tulyaganova in Opred. Rast. Sred. Azii*, 7: 369 (1983)—Turkmenistan.

BERLANDIERA (Compos.).
pumila (Michx.) Nutt.
var. scabrella *G.L. Nesom & B.L. Turner in Sida*, 18(2): 499 (1998)—Texas.

BERNARDIA (Euphorbiac.).
brasiliensis *Müll. Arg. in Linnaea*, 34: 174 (1865)—Brazil.
dodecandra *(Sessé ex Cav.) R. Govaerts in R. Govaerts, D.G. Frodin & A. Radcliffe-Smith, World Checklist & Bibliogr.* Euphorbiaceae, 1: 250 (2000): Adelia dodecandra.

BERRYA (Tiliac.).
ammonilla *W. Roxburgh, Pl. Coromandel*, 3: 60 (1820)—Sri Lanka.†

BERTEROA (Crucif.).
potaninii Maxim.
var. latifolia *Z.X. An in Fl. Xinjiangensis*, 2(2): 375 (1995)—Xinjiang.

BERTIERA (Rubiac.).
bracteosa *(Donn. Sm.) B. Ståhl & L. Andersson in Fl. Ecuador*, 62(162:3): 127 (1999): Gonzalea bracteosa.
pubiflora *(Steyerm.) L. Andersson & B. Ståhl in Fl. Ecuador*, 62(162:3): 125 (1999): B. guianensis subsp. pubiflora.

BESCHORNERIA (Agavac.).
yuccoides
subsp. dekosteriana *(K. Koch) R. Govaerts, World Checklist Seed Plants*, 2(1–2): 10 (1996), isonym; *see García-Mendoza in Espejo Serna & López-Ferrari, Monocot. Mexicanas* Sinopsis Florística, 1(1): 30 (1993): B. dekosteriana.

BETA (Chenopodiac.).
maritima
subsp. marcosii *(O. Bolòs & Vigo) A. Juan & M.B. Crespo in Acta Bot. Malacitana*, 24: 226 (1999): B. vulgaris var. marcosii.

BETULA (Betulac.).
× alpestris
nothovar. sukaczevii *(Soczava) R. Govaerts, World Checklist Seed Plants*, 2(1–2): 10 (1996): B. × sukaczevii.
brunnea *(J.X. Huang) J.X. Huang in J. Hebei Agric. Univ.*, 13(2): 43 (1990), without basionym ref.: B. platyphylla var. brunnea.

BETULA:-
ceratoptera *G.H. Liu & Y.C. Ma in J. Inner Mongolia Forest. Coll.*, 1: ? (1988), without type; *cf. Collected Works of Ma Yu-Chuan:* 237 (1995)—Inner Mongolia (Nei Mongol).
fruticosa Pall.
subsp. extremiorientalis *(Kuzen. & V.N. Vassil.) V.N. Voroschilov in A.K. Skvortsov (ed.), Florist. issl. v razn. raĭonakh SSSR:* 162 (1985): B. extremiorientalis.
subsp. fusca *(Pall. ex Georgi) M.A. Shemberg in Fl. Sibir. (Salicac.-Amaranthac.)*, 5: 67 (1992): B. fusca.
subsp. montana *M.A. Shemberg in Fl. Sibir. (Salicac.-Amaranthac.)*, 5: 67 (1992)—Altay.
gmelinii Bunge
var. zyzyphifolia *(Z. Wang & S.L. Tung) G.H. Liu & E.W. Ma in J. Inner Mongolia Forest. Coll.*, 1: ? (1988); *cf. Collected Works of Ma Yu-Chuan:* 235 (1995): B. zyzyphifolia.
mandshurica (Regel) Nakai
subsp. tauschii *(Regel) Kamelin ex I.A. Gubanov, Konsp. Fl. Vneshnei Mongolii (Sosud. Rast.):* 36 (1996): B. alba var. tauschii.
nana
subsp. rotundifolia *(Spach) L.I. Malyschev in L.I. Malyschev (ed.), Vŷsokogornaya Flora Stanovogo Nagor'ya:* 82 (1972): B. rotundifolia.
platyphylla Sukaczev
subsp. kamtschatica *(Regel) V.N. Voroschilov in A.K. Skvortsov (ed.), Florist. issl. v razn. raĭonakh SSSR:* 162 (1985): B. alba var. kamtschatica.
pubescens Ehrh.
f. columnaris *T. Ulvinen in Ann. Bot. Fenn.*, 36(1): 34 (1999)—Finland.
var. pumila *(L.) R. Govaerts, World Checklist Seed Plants*, 2(1–2): 10 (1996): B. alba var. pumila.

BIARUM (Arac.).
mendax *P.C. Boyce in Aroideana*, 22: 90 (1999)—Spain.

BIDARIA (Asclepiadac.).
foetida *(Tsiang) P.T. Li in J. S. China Agric. Univ.*, 12(3): 39 (1991): Gymnema foetidum.
var. mairei *(Tsiang) P.T. Li in J. S. China Agric. Univ.*, 12(3): 40 (1991): Gymnema foetidum var. mairei.
hainanensis *(Tsiang) P.T. Li in J. S. China Agric. Univ.*, 12(3): 39 (1991): Gymnema hainanense.
latifolia *(Wall. ex Wight) P.T. Li in J. S. China Agric. Univ.*, 12(3): 39 (1991): Gymnema latifolium.
longiretinaculata *(Tsiang) P.T. Li in J. S. China Agric. Univ.*, 12(3): 40 (1991): Gymnema longiretinaculatum.
yunnanensis *(Tsiang) P.T. Li in J. S. China Agric. Univ.*, 12(3): 39 (1991): Gymnema yunnanense.

BIDENS (Compos.).
connata
f. ambiversa *(Fassett) E.G. Voss in Mich. Bot.*, 34(4): 139 (1995 publ. 1996): B. connata var. ambiversa.
f. anomala *(Farw.) E.G. Voss in Mich. Bot.*, 34(4): 139 (1995 publ. 1996): B. connata var. anomala.
hassleriana *(Chodat) A.A. Sáenz in Zuloaga & Morrone (eds.), Cat. Pl. Vasc. Rep. Arg. 2 (Monogr. Syst. Bot.* Missouri Bot. Gard. 74): 1245 (1999): Bidens riedelii var. hassleriana.
leptophylla *C.H. An in Fl. Xinjiangensis*, 5: 476 (1999)—Xinjiang.
radiata Thuill.
var. microcephala *C.H. An in Fl. Xinjiangensis*, 5: 476 (1999)—Xinjiang.
triparta L.
var. quinqueloba *C.H. An in Fl. Xinjiangensis*, 5: 476 (1999)—Xinjiang.

BIDWILLIA (Anthericac.).
glauca *(Ruiz & Pav.) G. Don in Loudon, Encycl. Pl.:* 1344 (1855): Anthericum glaucum.

BIEBERSTEINIA (Geraniac.).
sect. Achillaeophyllum *Vved. ex R.V. Kamelin, Mater. Istorii Fl. Azii (Altaisk. Gorn. Strana):* 15 (1998).
sect. Heterostemonia *R.V. Kamelin, Mater. Istorii Fl. Azii (Altaisk. Gorn. Strana):* 15 (1998).

× **BIFRANISIA** *hort. in Orchid Rev.*, 105(1216): 254 (1997). *ORCHIDACEAE* (Aganisia × Bifrenaria).

BIFRENARIA (Orchidac.).
barbosae *V.P. Castro Neto in Proc. 15th World Orchid Conf., Rio de Janeiro 1996 (ed. C.E.B. Pereira):* 377, 380 (1998), nom. nov. superfl.: B. calcarata *Barb. Rodr.*
calcarata *(Vell.) V.P. Castro Neto in Proc. 15th World Orchid Conf., Rio de Janeiro 1996 (ed. C.E.B. Pereira):* 377 (1998), homonym; *see Barb. Rodr., Gen. Spec. Orchid.*, 2: 213 (1881): Epidendrum calcaratum.
silvana *V.P. Castro Neto in Bol. CAOB*, 3(4): 41 (1991)—Brazil Northeast (Bahia).
stefanae *V.P. Castro Neto in Bol. CAOB*, 3(4): 42 (1991)—Brazil Southeast (São Paulo)
verboonenii *G.A. Romero & V.P. Castro in Harvard Pap. Bot.*, 5(1): 187 (2000)—Brazil Southeast (Minas Gerais).

BIGNONIA (Bignoniac.).
suberosa *W. Roxburgh, Pl. Coromandel*, 3: 11 (1811)—Locality not stated.†

BIKINIA *J.J. Wieringa in Wageningen Agric. Univ. Pap.*, 99(4): 187 (1999). *LEGUMINOSAE*
aciculifera *J.J. Wieringa in Wageningen Agric. Univ. Pap.*, 99(4): 191 (1999)—Gabon.
breynei *(Bamps) J.J. Wieringa in Wageningen Agric. Univ. Pap.*, 99(4): 195 (1999): Monopetalanthus breynei.
congensis *J.J. Wieringa in Wageningen Agric. Univ. Pap.*, 99(4): 200 (1999)—Zaire, Congo.
coriacea *(J. Morel ex Aubrév.) J.J. Wieringa in Wageningen Agric. Univ. Pap.*, 99(4): 204 (1999): Monopetalanthus coriaceus.
durandii *(F. Hallé & Normand) J.J. Wieringa in Wageningen Agric. Univ. Pap.*, 99(4): 207 (1999): Monopetalanthus durandii.
evrardii *(Bamps) J.J. Wieringa in Wageningen Agric. Univ. Pap.*, 99(4): 213 (1999): Monopetalanthus evrardii.
grisea *J.J. Wieringa in Wageningen Agric. Univ. Pap.*, 99(4): 218 (1999)—Gabon, Equatorial Guinea.
le-testui *(Pellegr.) J.J. Wieringa in Wageningen Agric. Univ. Pap.*, 99(4): 222 (1999): Monopetalanthus le-testui.
subsp. mayumbensis *J.J. Wieringa in Wageningen Agric. Univ. Pap.*, 99(4): 229 (1999)—Congo.
media *J.J. Wieringa in Wageningen Agric. Univ. Pap.*, 99(4): 230 (1999)—Gabon.
pellegrinii *(A. Chev.) J.J. Wieringa in Wageningen Agric. Univ. Pap.*, 99(4): 234 (1999): Monopetalanthus pellegrinii.

BILLBERGIA (Bromeliac.).
brachysiphon
var. breviflora *H.E. Luther in Selbyana*, 19(2): 224 (1998 publ. 1999)—Ecuador.
dasilvae *E.M.C. Leme in Harvard Pap. Bot.*, 4(1): 147 (1999)—Brazil North (Rondônia).
mohammadii *R. Vásquez & P.L. Ibisch in Bromelie*, 2000(1): 11 (2000)—Bolivia.
rufa *(Gaudich.) N.E. Brown, Suppl. Johnson's Gard. Dict.:* 883 (1882): Quesnelia rufa.

× **BILLMEA** *K. Williams in J. Bromeliad Soc.*, 24(1): 26 (1974). *BROMELIACEAE* (Aechmea × Billbergia).

BIONDIA (Asclepiadac.).
microcentra *(Tsiang) P.T. Li in J. S. China Agric. Univ.*, 12(3): 39 (1991): Adelostemma microcentrum.

BIOPHYTUM (Oxalidac.).
congestiflorum *E. Govindarajalu in J. Econ. Taxon. Bot.*, 20(2): 311 (1996)—India (Kerala).
longipedunculatum *E. Govindarajalu in J. Econ. Taxon. Bot.*, 20(2): 312 (1996)—India (Kerala).
uzungwaensis *C. Frimodt-Møller in Kew Bull.*, 54(1): 185 (1999)—Tanzania.
veldkampii *A.E. Shanavas Khan, E.S. Santhosh Kumar, S. Binu & P. Pushpangadan in Rheedea*, 8(1): 79 (1998)—India (Kerala).

BISCUTELLA (Crucif.).
austriaca *A. Jordan, Diagn. esp. nouv.*: 293 (1864)—Austria.†
var. rudavensis *M. Peniašteková in Acta Bot. Slov. Acad. Sci. Slov., A, Taxon. Geobot.*, 10: 76 (1987)—Czechoslovakia.
var. trebejensis *M. Peniašteková in Acta Bot. Slov. Acad. Sci. Slov., A, Taxon. Geobot.*, 10: 76 (1987)—Czechoslovakia.
bilbilitana *G. Mateo & M.B. Crespo in Bot. J. Linn. Soc.*, 132(1): 2 (2000)—Spain.
conquensis *G. Mateo & M.B. Crespo in Bot. J. Linn. Soc.*, 132(1): 8 (2000)—Spain.
ebusitana *J.A. Rosselló, N. Torres & L. Sáez in Bot. J. Linn. Soc.*, 129(2): 157 (1999)—Baleares.
laevigata L.
subsp. australis *M. Raffaelli & L. Baldoin in Webbia*, 52(1): 115 (1997)—Italy.
subsp. hispidissima *(Posp.) M. Raffaelli & L. Baldoin in Webbia*, 52(1): 109 (1997): B. laevigata var. hispidissima.
subsp. ossolana *M. Raffaelli & L. Baldoin in Webbia*, 52(1): 107 (1997)—Italy.
subsp. prinzerae *M. Raffaelli & L. Baldoin in Webbia*, 52(1): 112 (1997)—Italy.
prealpina *M. Raffaelli & L. Baldoin in Webbia*, 52(1): 118 (1997)—Italy.
segurae *G. Mateo & M.B. Crespo in Bot. J. Linn. Soc.*, 132(1): 5 (2000)—Spain.
valentina
var. controversa *(Boreau) R. Govaerts, World Checklist Seed Plants*, 2(1–2): 10 (1996): B. controversa.
subsp. coronopifolia *(L.) R. Govaerts, World Checklist Seed Plants*, 2(1–2): 10 (1996): B. coronopifolia.
subsp. cuneata *(Font Quer) R. Govaerts, World Checklist Seed Plants*, 2(1–2): 10 (1996): B. laevigata f. cuneata.
var. divionensis *(Jord.) R. Govaerts, World Checklist Seed Plants*, 2(1–2): 10 (1996): B. divionensis.
var. gracilis *(Mach.-Laur.) R. Govaerts, World Checklist Seed Plants*, 2(1–2): 10 (1996): B. laevigata subsp. gracilis.
var. guestphalica *(Mach.-Laur.) R. Govaerts, World Checklist Seed Plants*, 2(1–2): 11 (1996): B. laevigata subsp. guestphalica.
var. guillonii *(Jord.) R. Govaerts, World Checklist Seed Plants*, 2(1–2): 11 (1996): B. guillonii.
var. hungarica *(Soó) R. Govaerts, World Checklist Seed Plants*, 2(1–2): 11 (1996): B. laevigata subsp. hungarica.
var. intricata *(Jord.) R. Govaerts, World Checklist Seed Plants*, 2(1–2): 11 (1996): B. intricata.
var. lamottei *(Jord.) R. Govaerts, World Checklist Seed Plants*, 2(1–2): 11 (1996): B. lamottei.
subsp. leptophylla *(Pau) G. Mateo Sanz & M.B. Crespo Villalba, Fl. Abrev. Comun. Valenciana*: 430 (1995): B. leptophylla.
var. lucida *(Balb. ex DC.) R. Govaerts, World Checklist Seed Plants*, 2(1–2): 11 (1996): B. lucida.
var. montenegrina *(Rohlena) R. Govaerts, World Checklist Seed Plants*, 2(1–2): 11 (1996): B. laevigata subsp. montenegrina.
var. neustriaca *(Bonnet) R. Govaerts, World Checklist Seed Plants*, 2(1–2): 11 (1996): B. neustriaca.
subsp. rotgesii *(Foucaud) R. Govaerts, World Checklist Seed Plants*, 2(1–2): 11 (1996): B. rotgesii.
var. sclerocarpa *(Revel) R. Govaerts, World Checklist Seed Plants*, 2(1–2): 11 (1996): B. sclerocarpa.
var. tenuifolia *(Bluff & Fingerh.) R. Govaerts, World Checklist Seed Plants*, 2(1–2): 11 (1996): B. ambigua var. tenuifolia.
var. varia *(Dumort.) R. Govaerts, World Checklist Seed Plants*, 2(1–2): 11 (1996): B. varia.
subsp. variegata *(Boiss. & Reut.) R. Govaerts, World Checklist Seed Plants*, 2(1–2): 11 (1996): B. variegata.

BISNAGA (Cactac.).
haematacantha *(Monv. ex Salm-Dyck) A.B. Doweld in Sukkulentÿ (Moscow)*, 1999(1): 28 (1999): Echinocactus electracanthus var. haematacanthus.

BISNAGA:-
haematacantha
subsp. sinuata *(A. Dietr.) A.B. Doweld in Sukkulentÿ (Moscow)*, 1999(1): 28 (1999): Echinocactus sinuatus.
histrix *(DC.) A.B. Doweld in Sukkulentÿ (Moscow)*, 1999(1): 28 (1999): Echinocactus histrix.
latispina *(Haw.) A.B. Doweld in Sukkulentÿ (Moscow)*, 3(1–2): 36 (2000): Cactus latispinus.
subsp. spiralis *(Karw. ex Pfeiff.) A.B. Doweld in Sukkulentÿ (Moscow)*, 3(1–2): 36 (2000): Echinocactus spiralis.
macrodisca *(Mart.) A.B. Doweld in Sukkulentÿ (Moscow)*, 1999(1): 29 (1999): Echinocactus macrodiscus.
subsp. septentrionalis *(J. Meyrán) A.B. Doweld in Sukkulentÿ (Moscow)*, 1999(1): 29 (1999), without basionym date: Ferocactus macrodiscus var. septentrionalis.
pueblensis *A.B. Doweld in Sukkulentÿ (Moscow)*, 1999(2): 43 (1999), nom. nov.: Echinocactus electracanthus var. haematacanthus *Monv. ex Salm-Dyck*.
recurva
subsp. latispina *(Haw.) A.B. Doweld in Sukkulentÿ (Moscow)*, 1999(1): 29 (1999): Cactus latispinus.

BISTORTA (Polygonac.).
alopecuroides (Turcz. ex Meisn.) Kom.
var. attenuata *(Kom.) I.A. Gubanov, Konsp. Fl. Vneshnei Mongolii (Sosud. Rast.)*: 38 (1996): B. attenuata.
coriacea *(Sam.) K. Yonekura & Hiroyoshi Ohashi in J. Jap. Bot.*, 72(3): 157 (1997): Polygonum coriaceum.
honanensis *(H.W. Kung) K. Yonekura & Hiroyoshi Ohashi in J. Jap. Bot.*, 72(3): 157 (1997): Polygonum honanense
macrophylla (D. Don) Soják
var. stenophylla *(Meisn.) F. Miyamoto in Bull. Nation. Sci. Mus., B (Tokyo)*, 26(3): 97 (2000): Polygonum stenophyllum.
paleacea *(Wall. ex Hook. f.) K. Yonekura & Hiroyoshi Ohashi in J. Jap. Bot.*, 72(3): 157 (1997): Polygonum paleaceum.
purpureonervosa *(A.J. Li) K. Yonekura & Hiroyoshi Ohashi in J. Jap. Bot.*, 72(3): 158 (1997): Polygonum purpureonervosum.
rubra *K. Yonekura & H. Ohashi in J. Jap. Bot.*, 74(6): 338 (1999)—Nepal, India (Uttar Pradesh).
sinomontana *(Sam.) F. Miyamoto in Bull. Nation. Sci. Mus., B (Tokyo)*, 25(4): 153 (1999): Polygonum sinomontanum.
taipaishanensis *(H.W. Kung) K. Yonekura & Hiroyoshi Ohashi in J. Jap. Bot.*, 72(3): 158 (1997): Polygonum taipaishanense.
tenuicaulis (Bisset & S. Moore) Nakai
var. chionophila *K. Yonekura & H. Ohashi in J. Jap. Bot.*, 73(1): 7 (1998)—Japan.

BITUMINARIA (Legumin.).
bituminosa
var. brachycarpa *(Feldman) A. Danin in Willdenowia*, 28(1–2): 141 (1998): Psoralea bituminosa var. brachycarpa.

BLACKWELLIA (Flacourtiac.).
cochinchinensis *(Lour.) R. Br. ex J.O. Voigt, Hort. Suburb. Calcutt.*: 64 (1845): Astranthus cochinchinensis.

BLAKEA (Melastomatac.).
coloradensis *F. Almeda in Novon*, 10(4): 304 (2000)—Panama.
darcyana *F. Almeda in Novon*, 10(4): 307 (2000)—Panama.
perforata *F. Almeda in Novon*, 10(4): 311 (2000)—Panama.
storkii *(Standl.) F. Almeda in Novon*, 10(4): 314 (2000): Topobea storkii.
tapantiana *G. Umaña Dodero & F. Almeda in Novon*, 5(4): 305 (1995)—Costa Rica.
wilsoniorum *F. Almeda in Novon*, 10(4): 316 (2000)—Costa Rica, Panama.

BLANCHETIODENDRON *R.C. Barneby & J.W. Grimes, Silk Tree, Guanacaste, Monkey's Earring (Mem. New York Bot. Gard., 74(1))*: 127 (1996). *LEGUMINOSAE*
blanchetii *(Benth.) R.C. Barneby & J.W. Grimes, Silk Tree, Guanacaste, Monkey's Earring (Mem. New York Bot. Gard., 74(1))*: 129 (1996): Enterolobium blanchetii.

BLEPHARIS (Acanthac.).
sect. Biflora *K. Vollesen, Blepharis (Acanthac.) Taxon. Revis.*: 155 (2000).
subgen. Ebracteata *K. Vollesen, Blepharis (Acanthac.) Taxon. Revis.*: 156 (2000).
sect. Inopinata *K. Vollesen, Blepharis (Acanthac.) Taxon. Revis.*: 231 (2000).
sect. Scorpioidea *Oberm. ex K. Vollesen, Blepharis (Acanthac.) Taxon. Revis.*: 236 (2000).
aequisepala *K. Vollesen, Blepharis (Acanthac.) Taxon. Revis.*: 126 (2000)—Cape Province.
boranensis *K. Vollesen, Blepharis (Acanthac.) Taxon. Revis.*: 110 (2000)—Ethiopia, Djibouti, Somalia, Kenya.
burundiensis *K. Vollesen, Blepharis (Acanthac.) Taxon. Revis.*: 180 (2000)—Burundi, Zaire.
cuanzensis S. Moore
var. parvispina *K. Vollesen, Blepharis (Acanthac.) Taxon. Revis.*: 216 (2000)—Zaire.
diplodonta *K. Vollesen, Blepharis (Acanthac.) Taxon. Revis.*: 153 (2000)—Somalia.
drummondii *K. Vollesen, Blepharis (Acanthac.) Taxon. Revis.*: 77 (2000)—Zimbabwe.
dunensis *K. Vollesen, Blepharis (Acanthac.) Taxon. Revis.*: 272 (2000)—Mozambique.
duvigneaudii *K. Vollesen, Blepharis (Acanthac.) Taxon. Revis.*: 283 (2000)—Zaire.
fenestralis *K. Vollesen, Blepharis (Acanthac.) Taxon. Revis.*: 54 (2000)—Transvaal.
flava *K. Vollesen, Blepharis (Acanthac.) Taxon. Revis.*: 212 (2000)—Angola.
gazensis *K. Vollesen, Blepharis (Acanthac.) Taxon. Revis.*: 56 (2000)—Mozambique.
hildebrandtii Lindau
subsp. phillipsiae *(Rendle) K. Vollesen, Blepharis (Acanthac.) Taxon. Revis.*: 167 (2000): B. phillipsiae.
huillensis *K. Vollesen, Blepharis (Acanthac.) Taxon. Revis.*: 140 (2000)—Angola.
inflata *K. Vollesen, Blepharis (Acanthac.) Taxon. Revis.*: 252 (2000)—Zambia, Zaire.
inopinata *K. Vollesen, Blepharis (Acanthac.) Taxon. Revis.*: 233 (2000)—Tanzania, Zambia.
itigiensis *K. Vollesen, Blepharis (Acanthac.) Taxon. Revis.*: 251 (2000)—Tanzania.
kenyensis *K. Vollesen, Blepharis (Acanthac.) Taxon. Revis.*: 263 (2000)—Kenya.
laevifolia *K. Vollesen, Blepharis (Acanthac.) Taxon. Revis.*: 81 (2000)—Transvaal.
macra *(Nees) K. Vollesen, Blepharis (Acanthac.) Taxon. Revis.*: 122 (2000): Acanthodium macrum.
meyeri *K. Vollesen, Blepharis (Acanthac.) Taxon. Revis.*: 133 (2000)—Namibia.
montana *K. Vollesen, Blepharis (Acanthac.) Taxon. Revis.*: 261 (2000)—Zaire.
obermeyerae *K. Vollesen, Blepharis (Acanthac.) Taxon. Revis.*: 84 (2000)—Transvaal, Mozambique.
ogadenensis *K. Vollesen, Blepharis (Acanthac.) Taxon. Revis.*: 55(4): 167 (2000)—Somalia, Ethiopia, Kenya.
petalidioides *K. Vollesen, Blepharis (Acanthac.) Taxon. Revis.*: 143 (2000)—Botswana.
pusilla *K. Vollesen, Blepharis (Acanthac.) Taxon. Revis.*: 243 (2000)—Tanzania.
reekmansii *K. Vollesen, Blepharis (Acanthac.) Taxon. Revis.*: 294 (2000)—Burundi.
richardsiae *K. Vollesen, Blepharis (Acanthac.) Taxon. Revis.*: 262 (2000)—Zambia.
scandens *K. Vollesen, Blepharis (Acanthac.) Taxon. Revis.*: 189 (2000)—Somalia.
sericea *K. Vollesen, Blepharis (Acanthac.) Taxon. Revis.*: 256 (2000)—Natal.
somaliensis *K. Vollesen, Blepharis (Acanthac.) Taxon. Revis.*: 188 (2000)—Somalia.
spinipes *K. Vollesen, Blepharis (Acanthac.) Taxon. Revis.*: 130 (2000)—Transvaal.
subglabra *K. Vollesen, Blepharis (Acanthac.) Taxon. Revis.*: 53 (2000)—Natal.

BLEPHARIS:-
swaziensis *K. Vollesen, Blepharis (Acanthac.) Taxon. Revis.*: 184 (2000)—Natal, Swaziland, Mozambique.
tanganyikensis *(Napper) K. Vollesen, Blepharis (Acanthac.) Taxon. Revis.*: 217 (2000): B. cuanzensis subsp. tanganyikensis.
tanzaniensis *K. Vollesen, Blepharis (Acanthac.) Taxon. Revis.*: 273 (2000)—Tanzania.
torrei *K. Vollesen, Blepharis (Acanthac.) Taxon. Revis.*: 269 (2000)—Tanzania, Mozambique.
trifida *K. Vollesen, Blepharis (Acanthac.) Taxon. Revis.*: 155 (2000)—Somalia.
turkanae *K. Vollesen, Blepharis (Acanthac.) Taxon. Revis.*: 195 (2000)—Kenya.
uzondoensis *K. Vollesen, Blepharis (Acanthac.) Taxon. Revis.*: 299 (2000)—Tanzania.

BLEPHARODON (Asclepiadac.).
manicatum *(Decne.) J. Fontella Pereira in Bradea*, 5(23): 262 (1989): Ditassa manicata.
pictum *(Vahl) W.D. Stevens in Novon*, 10(3): 242 (2000): Cynanchum pictum.

BLETIA (Orchidac.).
antillana *M.A. Díaz Dumas & V. Sosa in Brittonia*, 49(1): 81 (1997)—Cuba.
× tamayoana *S. Rosillo ex R. Soltero Q.in Bol. Inst. Bot. (Guadalajara)*, 1(5): 341 (1993)—Mexico Southwest (Jalisco). (B. amabilis × coccinea).

× **BLETIAGLOTTIS** *hort. in Orchid Rev.*, 107(1228): 255 (1999). *ORCHIDACEAE* (Bletia × Spathoglottis).

BLETILLA (Orchidac.).
elegantula *(Kraenzl.) L.A. Garay & G.A. Romero-González in Harvard Pap. Bot.*, 3(1): 53 (1998): Coelogyne elegantula.
formosana (Hayata) Schltr.
f. rubrolabella *S.S. Ying in Mem. Coll. Agric. Nation. Taiwan Univ.*, 30(2): 68 (1990)—Taiwan.

BLOTIA (Euphorbiac.).
leandriana *Petra Hoffmann & G. McPherson in Novon*, 7(3): 249 (1997)—Madagascar.
mimosoides *(Baill.) Petra Hoffmann & G. McPherson in Adansonia*, 20(2): 253 (1998): Savia mimosoides.

BLUMEA (Compos.).
adamsii *J.-P. Lebrun & A.L. Stork, Énum. Pl. Fleurs Afr. Trop.*, 4: 261 (1997), nom. nov.: B. alata var. gracilis *O. Hoffm. & Muschl.*
braunii *(Vatke) J.-P. Lebrun & A.L. Stork, Énum. Pl. Fleurs Afr. Trop.*, 4: 261 (1997): Laggera braunii.
crispata (Vahl) Merxm.
var. montana *(C.D. Adams) J.-P. Lebrun & A.L. Stork, Énum. Pl. Fleurs Afr. Trop.*, 4: 262 (1997): Laggera alata var. montana.
emeiensis *Z.Y. Zhu in Guihaia*, 17(1): 16 (1997)—China South-central (Sichuan).
gomphrena *(Walp.) Sch. Bip. ex Hand.-Mazz. in Oesterr. Bot. Z.*, 88(4): 310 (1939): Vernonia gomphrena.
lecomteana *(O. Hoffm. & Muschl.) J.-P. Lebrun & A.L. Stork, Énum. Pl. Fleurs Afr. Trop.*, 4: 262 (1997): Laggera lecomteana.
linearis *C.I Peng & W.P. Leu in Bot. Bull. Acad. Sin. (Taipei)*, 40(1): 53 (1999)—Taiwan.
volkensii *(O. Hoffm.) J.-P. Lebrun & A.L. Stork, Énum. Pl. Fleurs Afr. Trop.*, 4: 262 (1997): Laggera volkensii.

BLUMENBACHIA (Loasac.).
exalata *M. Weigend, Nasa & Conquest S. America - Syst. Rearrang. in Loasac.*: 218 (1997)—Brazil South (Rio Grande do Sul).

BLYSMUS (Cyperac.).
subgen. Blysmopsis *(Oteng-Yeb.) N.N. Tzvelev in Bot. Zhurn.*, 84(7): 113 (1999): gen. Blysmopsis.
compressus (L.) Panz. ex Link
subsp. subulifolius *A.P. Khokhrjakov in Byull. Mosk. Obshch. Ispŷt. Prir., Biol.*, 102(4): 46 (1997)—Turkey.

BOBGUNNIA *J.H. Kirkbride & J.H. Wiersema in Brittonia*, 49(1): 1 (1997), nom. nov.: Swartzia sect. Fistuloides *Benth.* (Legumin.).
fistuloides *(Harms) J.H. Kirkbride & J.H. Wiersema in Brittonia*, 49(1): 3 (1997): Swartzia fistuloides.
madagascariensis *(Desv.) J.H. Kirkbride & J.H. Wiersema in Brittonia*, 49(1): 7 (1997): Swartzia madagascariensis.

BOBROVIA *A.P. Khokhrjakov in Novosti Sist. Vŷssh. Rast.*, 31: 137 (1998). *LEGUMINOSAE*
alpina *(L.) A.P. Khokhrjakov in Novosti Sist. Vŷssh. Rast.*, 31: 137 (1998): Trifolium alpinum.
pilzii *(Adamović) A.P. Khokhrjakov in Novosti Sist. Vŷssh. Rast.*, 31: 137 (1998): Trifolium pilzii.
polyphylla *(C.A. Mey.) A.P. Khokhrjakov in Novosti Sist. Vŷssh. Rast.*, 31: 137 (1998): Trifolium polyphyllum.

BOCCONIA (Papaverac.).
hintoniorum *B.L. Turner in Phytologia*, 79(4): 313 (1995 publ. 1996)—Mexico Southwest (Oaxaca).

BOECHERA (Crucif.).
pallidifolia *(Rollins) W.A. Weber in Phytologia*, 79(2): 65 (1995 publ. 1996): Arabis pallidifolia.

BOEHMERIA (Urticac.).
balslevii *I. Friis & C.M. Wilmot-Dear in Cat. Vasc. Pl. Ecuador (Monogr. Syst. Bot. Missouri Bot. Gard., 75)*: 956 (1999), nom. nov.: B. dolichostachys *Friis & Wilmot-Dear.*
bullata Kunth
subsp. coriacea *(Killip) I. Friis & C.M. Wilmot-Dear in Op. Bot.*, 129: 32 (1996): B. coriacea.
subsp. notobullata *I. Friis & C.M. Wilmot-Dear in Op. Bot.*, 129: 34 (1996)—Peru.
dolichostachys *I. Friis & C.M. Wilmot-Dear in Op. Bot.*, 129: 35 (1996)—Ecuador.
macrophylla Hornem.
subsp. canescens *(Wedd.) G. Panigrahi & S.K. Murti, Fl. Bilaspur Distr.*, 2: 558 (1999): B. platyphylla var. canescens.
var. rotundifolia *(D. Don) W.T. Wang in Fl. Reipubl. Popul. Sin.*, 23(2): 337 (1995): B. rotundifolia.
var. zeylanica *(Wedd.) B.M. Wadhwa in M.D. Dassanayake & W.D. Clayton, eds., Revised Handb. Fl. Ceylon*, 13: 265 (1999): B. platyphylla var. zeylanica.
malabarica Wedd.
var. leioclada *(W.T. Wang) W.T. Wang in Fl. Reipubl. Popul. Sin.*, 23(2): 326 (1995): B. glomerulifera var. leioclada.
nivea (L.) Gaudich.
f. concolor *(Makino) S. Hatusima in S. Hatusima & T. Amano, Fl. Ryukyus, S. of Amami Is., ed. 2*: 37 (1994), without basionym ref.: Basionym not stated.
penduliflora Wedd. ex D.G. Long
var. loochooensis *(Wedd.) W.T. Wang in Fl. Reipubl. Popul. Sin.*, 23(2): 355 (1995): B. platyphylla var. loochooensis.

BOENNINGHAUSENIA (Rutac.).
albiflora (Hook.) Meisn.
var. japonica *(Nakai ex Makino & Nemoto) H. Ohba in Fl. Japan*, 2c: 33 (1999), basionym not validly publ.: B. japonica.

BOERHAVIA (Nyctaginac.).
ambigua *(Meikle) R. Govaerts, World Checklist Seed Plants*, 2(1–2): 11 (1996): Commicarpus ambiguus.
arabica *(Meikle) R. Govaerts, World Checklist Seed Plants*, 2(1–2): 11 (1996): Commicarpus arabicus.
australis *(Meikle) R. Govaerts, World Checklist Seed Plants*, 2(1–2): 11 (1996): Commicarpus australis.
brandegeei *(Standl.) R. Govaerts, World Checklist Seed Plants*, 2(1–2): 12 (1996): Commicarpus brandegeei.
chinensis
subsp. natalensis *(Meikle) R. Govaerts, World Checklist Seed Plants*, 2(1–2): 12 (1996): Commicarpus chinensis subsp. natalensis.
crassifolia *(Heimerl) R. Govaerts, World Checklist Seed Plants*, 2(1–2): 12 (1996): Commicarpus crassifolius.

BOERHAVIA:-
decipiens *(Meikle) R. Govaerts, World Checklist Seed Plants*, 2(1–2): 12 (1996): Commicarpus decipiens.
greenwayi *(Meikle) R. Govaerts, World Checklist Seed Plants*, 2(1–2): 12 (1996): Commicarpus greenwayi.
heronensis *R. Govaerts, World Checklist Seed Plants*, 2(1–2): 12 (1996), nom. nov.: Commicarpus insularum *Meikle.*
hiranensis *(Thulin) R. Govaerts, World Checklist Seed Plants*, 2(1–2): 12 (1996): Commicarpus hiranensis.
lantsangensis *(D.Q. Lu) R. Govaerts, World Checklist Seed Plants*, 2(1–2): 12 (1996): Commicarpus lantsangensis.
mista *(Thulin) R. Govaerts, World Checklist Seed Plants*, 2(1–2): 12 (1996): Commicarpus mistus.
montana *(Miré, H. Gillet & Quézel) R. Govaerts, World Checklist Seed Plants*, 2(1–2): 12 (1996): Commicarpus montanus.
parviflora *(Thulin) R. Govaerts, World Checklist Seed Plants*, 2(1–2): 12 (1996): Commicarpus parviflorus.
pilosa *(Heimerl) R. Govaerts, World Checklist Seed Plants*, 2(1–2): 12 (1996): Commicarpus fallacissimus f. pilosus.
ramosissima *(Thulin) R. Govaerts, World Checklist Seed Plants*, 2(1–2): 12 (1996): Commicarpus ramosissimus.
raynalii *(J.-P. Lebrun & Meikle) R. Govaerts, World Checklist Seed Plants*, 2(1–2): 12 (1996): Commicarpus raynalii.
somalensis *(Chiov.) R. Govaerts, World Checklist Seed Plants*, 2(1–2): 12 (1996): Selinocarpus somalensis.
traubae *R. Spellenberg in Madroño*, 46(4): 208 (1999 publ. 2000)—Mexico Northwest (Sonora).
tsarisbergensis *R. Govaerts, World Checklist Seed Plants*, 2(1–2): 12 (1996), nom. nov.: Commicarpus fruticosus *Pohnert.*

BOESENBERGIA (Zingiberac.).
armeniaca *J. Cowley in Kew Bull.*, 55(3): 669 (2000)—Borneo (Brunei, Sabah).
bruneiana *J. Cowley in Kew Bull.*, 53(3): 626 (1998)—Borneo (Brunei).
gelatinosa *K. Larsen in Nordic J. Bot.*, 17(4): 361 (1997)—Thailand.
hosensis *J. Cowley in Kew Bull.*, 55(3): 672 (2000)—Borneo (Sarawak).
jangarunii *J. Cowley in Kew Bull.*, 53(3): 623 (1998)—Borneo (Brunei).
loerzingii *(Valeton) K. Larsen in Nordic J. Bot.*, 17(4): 365 (1997), without basionym ref.: Gastrochilus loerzingii.
malaccana *(Baker) R. Govaerts, World Checklist Seed Plants*, 3(1): 9 (1999): Scaphochlamys malaccana.
phyllostachya *(Gagnep.) K. Larsen in Nordic J. Bot.*, 17(4): 363 (1997), without basionym ref.: Gastrochilus phyllostachyus.
roseopunctata *(Ridl.) I.M. Turner in Asian J. Trop. Biol.*, 4(1): 27 (2000): Gastrochilus roseopunctatus.
trangensis *K. Larsen in Nordic J. Bot.*, 17(4): 363 (1997)—Thailand.

× **BOGARDARA** *hort. in Orchid Rev.*, 104(1211): 276 (1996). *ORCHIDACEAE* (Ascocentrum × Phalaenopsis × Vanda × Vandopsis).

BOKKEVELDIA (Amaryllidac.).
aestivalis *(Snijman) D. Müller-Doblies & U. Müller-Doblies in Feddes Repert.*, 107(5–6): S.c.5 (1996): Strumaria aestivalis.
perryae *(Snijman) D. Müller-Doblies & U. Müller-Doblies in Feddes Repert.*, 107(5–6): S.c.4 (1996). Strumaria perryae.

BOLANTHUS (Caryophyllac.).
chelmicus Phitos
subsp. meteoricus *D. Phitos in A. Strid & Kit Tan (eds.), Fl. Hellenica*, 1: 328 (1997)—Greece.

BOLBOSCHOENUS (Cyperac.).
fernaldii *(C. Bicknell) Soó ex R. Govaerts, World Checklist Seed Plants*, 2(1–2): 14 (1996): Scirpus fernaldii.
novae-angliae *(Britton) S.G. Smith in Brittonia*, 47(4): 434 (1995): Scirpus novae-angliae.

BOLIVICACTUS *A.B. Doweld in Sukkulentý (Moscow)*, 3(1–2): 62 (2000). *CACTACEAE*
aureicentrus *(Backeb.) A.B. Doweld in Sukkulentý (Moscow)*, 3(1–2): 62 (2000): Parodia aureicentra.
columnaris *(Cárdenas) A.B. Doweld in Sukkulentý (Moscow)*, 3(1–2): 62 (2000): Parodia columnaris.
comarapanus *(Cárdenas) A.B. Doweld in Sukkulentý (Moscow)*, 3(1–2): 62 (2000): Parodia comarapana.
commutans *(F. Ritter) A.B. Doweld in Sukkulentý (Moscow)*, 3(1–2): 62 (2000): Parodia commutans.
hausteinianus *(Rausch) A.B. Doweld in Sukkulentý (Moscow)*, 3(1–2): 62 (2000): Parodia hausteiniana.
maassii *(Heese) A.B. Doweld in Sukkulentý (Moscow)*, 3(1–2): 62 (2000): Parodia maassii.
mairananus *(Cárdenas) A.B. Doweld in Sukkulentý (Moscow)*, 3(1–2): 62 (2000): Parodia mairanana.
ocampoi *(Cárdenas) A.B. Doweld in Sukkulentý (Moscow)*, 3(1–2): 62 (2000): Parodia ocampoi.
procerus *(F. Ritter) A.B. Doweld in Sukkulentý (Moscow)*, 3(1–2): 62 (2000): Parodia procera.
ritteri *(Buining) A.B. Doweld in Sukkulentý (Moscow)*, 3(1–2): 62 (2000): Parodia ritteri.
saint-pieanus *(Backeb.) A.B. Doweld in Sukkulentý (Moscow)*, 3(1–2): 62 (2000): Parodia saint-pieana.
schwebsianus *(Werderm.) A.B. Doweld in Sukkulentý (Moscow)*, 3(1–2): 62 (2000): Echinocactus schwebsianus.
stuemeri *(Werderm.) A.B. Doweld in Sukkulentý (Moscow)*, 3(1–2): 63 (2000): Echinocactus stuemeri.
subterraneus *(F. Ritter) A.B. Doweld in Sukkulentý (Moscow)*, 3(1–2): 63 (2000): Parodia subterranea.
taratensis *(Cárdenas) A.B. Doweld in Sukkulentý (Moscow)*, 3(1–2): 63 (2000): Parodia taratensis.
tilcarensis *(Werderm. & Backeb.) A.B. Doweld in Sukkulentý (Moscow)*, 3(1–2): 63 (2000): Echinocactus stuemeri var. tilcarensis.
tuberculatus *(Cárdenas) A.B. Doweld in Sukkulentý (Moscow)*, 3(1–2): 63 (2000): Parodia tuberculata.

BOLLEA (Orchidac.).
hirtzii *E. Waldvogel in Orchidee*, 33(4): 143 (1982)—Ecuador.

BOLUSIA (Legumin.).
polhilliana *S. Lisowski in Fragm. Flor. Geobot.*, 40(2): 908 (1995)—Zaire.

BOLUSIELLA (Orchidac.).
lebeliana *G. Delepierre & D. Geerinck in Belgian J. Bot.*, 130(2): 135 (1998)—Rwanda.

BOMAREA (Alstroemeriac.).
subgen. Leontochir *(Phil.) P. Ravenna in Onira, Bot. Leafl.*, 5(8): 45 (2000): gen. Leontochir.
subgen. Surgentia *P. Ravenna in Onira, Bot. Leafl.*, 5(8): 45 (2000).
cachimbensis *P. Ravenna in Onira, Bot. Leafl.*, 5(8): 44 (2000)—Brazil North (Pará).
ovallei *(Phil.) P. Ravenna in Onira, Bot. Leafl.*, 5(8): 45 (2000), as 'ovalleii': Leontochir ovallei.
paradoxa *P. Ravenna in Onira, Bot. Leafl.*, 5(8): 44 (2000)—Brazil West-Central (Mato Grosso do Sul).

BOMBYCILAENA (Compos.).
californica *(Fisch. & C.A. Mey.) J. Holub in Preslia*, 70(2): 100 (1998): Micropus californicus.

BONA (Legumin.).
johannis *(Tamamsch.) A.K. Stankevich & Yu.R. Roskov in Kew Bull.*, 53(3): 732 (1998): Vicia johannis.
turkestanica *(Vassilkovsk.) A.K. Stankevich & Yu.R. Roskov in Kew Bull.*, 53(3): 732 (1998): Vicia turkestanica.

BONAMIA (Convolvulac.).
apurensis *D.F. Austin in Fl. Venezuela*, 8(3): 40 (1982)—Venezuela.
tsivory *T. Deroin in Candollea*, 53(1): 93 (1998)—Madagascar.

BONIA (Gramin.).
amplexicaulis *(L.C. Chia, H.L. Fung & Y.L. Yang) N.H. Xia in Kew Bull.*, 51(3): 568 (1996): Monocladus amplexicaulis.
levigata *(L.C. Chia, H.L. Fung & Y.L. Yang) N.H. Xia in Kew Bull.*, 51(3): 568 (1996): Monocladus levigatus.
saxatilis *(L.C. Chia, H.L. Fung & Y.L. Yang) N.H. Xia in Kew Bull.*, 51(3): 567 (1996): Monocladus saxatilis.
var. solida *(C.D. Chu & C.S. Chao) D.Z. Li in Acta Bot. Yunnanica*, 22(1): 44 (2000): Indocalamus solidus.
solida *(C.D. Chu & C.S. Chao) N.H. Xia in Kew Bull.*, 51(3): 568 (1996): Indocalamus solidus.

BONINIA (Rutac.).
grisea Planch.
var. crassifolia *(Nakai) T. Yamaz. ex H. Ohba in Fl. Japan*, 2c: 34 (1999): B. glabra var. crassifolia.

BONNETIA (Guttif.).
ahogadoi *(Steyerm.) A.L. Weitzman & P.F. Stevens in Biollania*, ed. esp. 6 (J.J. Wurdack Festschr.): 559 (1997): Acopanea ahogadoi.
fasciculata *P.F. Stevens & A.L. Weitzman in Biollania*, ed. esp. 6 (J.J. Wurdack Festschr.): 559 (1997), nom. nov.: Neogleasonia wurdackii *Maguire*.
rubicunda *(Sastre) A.L. Weitzman & P.F. Stevens in Biollania*, ed. esp. 6 (J.J. Wurdack Festschr.): 558 (1997): Neblinaria rubiciunda.

BOREALLUMA *D.C.H. Plowes in Haseltonia*, 3: 63 (1995). *ASCLEPIADACEAE*
munbyana *(Decne.) D.C.H. Plowes in Haseltonia*, 3: 63 (1995): Boucerosia munbyana.
plicatiloba *(Lavranos) D.C.H. Plowes in Haseltonia*, 3: 63 (1995): Caralluma plicatiloba.
staintonii *(H. Hara) D.C.H. Plowes in Haseltonia*, 3: 63 (1995): Caralluma staintonii.
tuberculata *(N.E. Br.) D.C.H. Plowes in Haseltonia*, 3: 63 (1995): Caralluma tuberculata

BORINDA (Gramin.).
albocerea *(Jia R. Xue & T.P. Yi) C.M.A. Stapleton in Kew Bull.*, 53(2): 455 (1998): Fargesia albocerea.
edulis *(Jia R. Xue & T.P. Yi) C.M.A. Stapleton in Kew Bull.*, 53(2): 455 (1998): Fargesia edulis.
fansipanensis *(T.Q. Nguyen) C.M.A. Stapleton in Kew Bull.*, 53(2): 455 (1998): Fargesia fansipanensis.
frigida *(T.P. Yi) C.M.A. Stapleton in Kew Bull.*, 53(2): 456 (1998), as 'frigidorum' originally 'frigidis': Fargesia frigida.
fungosa *(T.P. Yi) C.M.A. Stapleton in Kew Bull.*, 53(2): 456 (1998): Fargesia fungosa.
hsuehiana *(T.P. Yi) C.M.A. Stapleton in Kew Bull.*, 53(2): 456 (1998), as 'hsuehana': Fargesia hsuehiana.
lushuiensis *(Jia R. Xue & T.P. Yi) C.M.A. Stapleton in Kew Bull.*, 53(2): 457 (1998): Fargesia lushuiensis.
papyrifera *(T.P. Yi) C.M.A. Stapleton in Kew Bull.*, 53(2): 457 (1998): Fargesia papyrifera.
perlonga *(Jia R. Xue & T.P. Yi) C.M.A. Stapleton in Kew Bull.*, 53(2): 457 (1998): Fargesia perlonga.
schmidiana *(A. Camus) C.M.A. Stapleton in Kew Bull.*, 53(2): 457 (1998): Arundinaria schmidiana.

BORNEOSICYOS *W.J.J.O. de Wilde in Reinwardtia*, 11(3): 224 (1998). *CUCURBITACEAE*
simplex *W.J.J.O. de Wilde in Reinwardtia*, 11(3): 224 (1998)—Borneo (Sabah).

BORNMUELLERA (Crucif.).
kiyakii *Z. Aytać & A. Aksoy in Bot. J. Linn. Soc.*, 134(3): 487 (2000)—Turkey.

BORONIA (Rutac.).
sect. Alatae *M.F. Duretto in Muelleria*, 12(1): 13 (1999).
sect. Algidae *M.F. Duretto in Muelleria*, 12(1): 16 (1999).
subsect. Bowmaniae *M.F. Duretto in Muelleria*, 12(1): 37 (1999).

BORONIA:-
ser. Erianthae *M.F. Duretto in Muelleria*, 12(1): 42 (1999).
ser. Ericifoliae *M.F. Duretto in Muelleria*, 12(1): 34 (1999).
subser. Filicifoliae *M.F. Duretto in Muelleria*, 12(1): 110 (1999).
ser. Fraseriae *M.F. Duretto in Muelleria*, 12(1): 51 (1999).
subsect. Grandisepalae *M.F. Duretto in Muelleria*, 12(1): 88 (1999).
ser. Grandisepalae *M.F. Duretto in Muelleria*, 12(1): 92, 88 (1999).
subser. Grandisepalae *M.F. Duretto in Muelleria*, 12(1): 95, 88 (1999).
subser. Jucundae *M.F. Duretto in Muelleria*, 12(1): 107 (1999).
ser. Lanuginosae *M.F. Duretto in Muelleria*, 12(1): 102 (1999).
subser. Lanuginosae *M.F. Duretto in Muelleria*, 12(1): 103, 102 (1999).
ser. Quadrilatae *M.F. Duretto in Muelleria*, 12(1): 90 (1999).
ser. Rupicolae *M.F. Duretto in Muelleria*, 12(1): 54 (1999).
ser. Ternatae *M.F. Duretto in Muelleria*, 12(1): 24 (1999).
subsect. Ternatae *M.F. Duretto in Muelleria*, 12(1): 24 (1999).
subser. Verecundae *M.F. Duretto in Muelleria*, 12(1): 93 (1999).
acanthoclada *Paul G. Wilson in Nuytsia*, 12(1): 146 (1998)—Western Australia.
amplectens *M.F. Duretto in Austral. Syst. Bot.*, 10(2): 287 (1997)—Northern Territory.
anceps *Paul G. Wilson in Nuytsia*, 12(1): 132 (1998)—Western Australia.
anemonifolia A. Cunn.
subsp. aurifodina *P.G. Neish in Muelleria*, 14: 9 (2000)—Victoria.
subsp. variabilis *(Hook.) P.G. Neish in Muelleria*, 14: 11 (2000)—Boronia variabilis.
subsp. wadbilligensis *P.G. Neish in Muelleria*, 14: 10 (2000)—New South Wales.
angustisepala *M.F. Duretto in Muelleria*, 12(1): 63 (1999)—New South Wales.
anomala *M.F. Duretto in Muelleria*, 12(1): 22 (1999)—Western Australia.
baeckeacea F. Muell.
subsp. patula *Paul G. Wilson in Nuytsia*, 12(1): 148 (1998)—Western Australia.
bella *M.F. Duretto in Austrobaileya*, 5(2): 287 (1999)—Queensland.
corynophylla *Paul G. Wilson in Nuytsia*, 12(1): 143 (1998)—Western Australia.
crenulata Sm.
var. angustifolia *Paul G. Wilson in Nuytsia*, 12(1): 124 (1998)—Western Australia.
subsp. obtusa *Paul G. Wilson in Nuytsia*, 12(1): 125 (1998)—Western Australia.
subsp. pubescens *(Benth.) Paul G. Wilson in Nuytsia*, 12(1): 125 (1998): B. crenulata var. pubescens.
subsp. viminea *(Lindl.) Paul G. Wilson in Nuytsia*, 12(1): 125 (1998): B. viminea.
decumbens *M.F. Duretto in Nuytsia*, 11(3): 323 (1997)—Northern Territory.
duiganiae *M.F. Duretto in Austrobaileya*, 5(2): 292 (1999)—Queensland.
excelsa *M.F. Duretto in Austrobaileya*, 5(2): 284 (1999)—Queensland.
exilis *Paul G. Wilson in Nuytsia*, 12(1): 134 (1998)—Western Australia.
fabianoides (Diels) Paul G. Wilson
subsp. rosea *Paul G. Wilson in Nuytsia*, 12(1): 150 (1998)—Western Australia.
fastigiata Bartl.
subsp. tenuior *(Benth.) Paul G. Wilson in Nuytsia*, 12(1): 137 (1998): B. fastigiata var. tenuior.
foetida *M.F. Duretto in Austrobaileya*, 5(2): 285 (1999)—Queensland.
forsteri *M.F. Duretto in Austrobaileya*, 5(2): 280 (1999)—Queensland.
grandisepala F. Muell.
subsp. acanthophida *M.F. Duretto in Austral. Syst. Bot.*, 10(2): 278 (1997)—Northern Territory.
hoipolloi *M.F. Duretto in Austrobaileya*, 5(2): 288 (1999)—Queensland.

BORONIA:-
humifusa *Paul G. Wilson in Nuytsia,* 12(1): 126 (1998)—Western Australia.
jensziae *M.F. Duretto in Austrobaileya,* 5(2): 282 (1999)—Queensland.
jucunda *M.F. Duretto in Nuytsia,* 11(3): 328 (1997)—Western Australia.
juncea Bartl.
subsp. laniflora *(Bartl.) Paul G. Wilson in Nuytsia,* 12(1): 138 (1998): B. laniflora.
subsp. micrantha *(Bartl.) Paul G. Wilson in Nuytsia,* 12(1): 138 (1998): B. laniflora var. micrantha.
subsp. minima *Paul G. Wilson in Nuytsia,* 12(1): 138 (1998)—Western Australia.
kalumburuensis *M.F. Duretto in Nuytsia,* 11(3): 334 (1997)—Western Australia.
laxa *M.F. Duretto in Austral. Syst. Bot.,* 10(2): 279 (1997)—Northern Territory.
minutipinna *M.F. Duretto in Nuytsia,* 11(3): 335 (1997)—Western Australia.
odorata *M.F. Duretto in Austrobaileya,* 5(2): 294 (1999)—Queensland.
palasepala *M.F. Duretto in Austrobaileya,* 5(2): 280 (1999)—Queensland.
prolixa *M.F. Duretto in Austral. Syst. Bot.,* 10(2): 283 (1997)—Northern Territory.
purdieana Diels
subsp. calcicola *Paul G. Wilson in Nuytsia,* 12(1): 128 (1998)—Western Australia.
quadrilata *M.F. Duretto in Austral. Syst. Bot.,* 10(2): 297 (1997)—Northern Territory.
quinkanensis *M.F. Duretto in Austrobaileya,* 5(2): 291 (1999)—Queensland.
ramosa (Lindl.) Benth.
subsp. lesueurana *Paul G. Wilson in Nuytsia,* 12(1): 151 (1998)—Western Australia.
rupicola *M.F. Duretto in Nuytsia,* 11(3): 336 (1997)—Northern Territory.
scabra Lindl.
subsp. attenuata *Paul G. Wilson in Nuytsia,* 12(1): 129 (1998)—Western Australia.
subsp. condensata *Paul G. Wilson in Nuytsia,* 12(1): 130 (1998)—Western Australia.
splendida *M.F. Duretto in Austrobaileya,* 5(2): 278 (1999)—Queensland.
squamipetala *M.F. Duretto in Austrobaileya,* 5(2): 295 (1999)—Queensland.
suberosa *M.F. Duretto in Austral. Syst. Bot.,* 10(2): 288 (1997)—Northern Territory.
ternata Endl.
var. austrofoliosa *M.F. Duretto in Muelleria,* 12(1): 32 (1999)—Western Australia.
var. promiscua *M.F. Duretto in Muelleria,* 12(1): 28 (1999)—Western Australia.
tetragona *Paul G. Wilson in Nuytsia,* 12(1): 140 (1998)—Western Australia.
tolerans *M.F. Duretto in Nuytsia,* 11(3): 326 (1997)—Northern Territory.
verecunda *M.F. Duretto in Austral. Syst. Bot.,* 10(2): 291 (1997)—Northern Territory.
viridiflora *M.F. Duretto in Austral. Syst. Bot.,* 10(2): 295 (1997)—Northern Territory.
westringioides *Paul G. Wilson in Nuytsia,* 12(1): 152 (1998)—Western Australia.
wilsonii *(F. Muell. ex Benth.) M.F. Duretto in Nuytsia,* 11(3): 320 (1997): B. artemisiifolia var. wilsonii.
xanthastrum *M.F. Duretto in Austral. Syst. Bot.,* 10(2): 292 (1997)—Northern Territory.

BORRERIA (Rubiac.).
ser. Brachystemonoides *N.M. Bacigalupo & E.L. Cabral in E. Robbrecht et al. (eds.), 2nd Internat. Rubiac. Conf. Proc. (Op. Bot.* Belg., 7): 306 (1996).
subgen. Dasycephala *(DC.) N.M. Bacigalupo & E.L. Cabral in E. Robbrecht et al. (eds.), 2nd Internat. Rubiac. Conf. Proc. (Op. Bot.* Belg., 7): 306 (1996): Diodia sect. Dasycephala.
ser. Laeves *N.M. Bacigalupo & E.L. Cabral in E. Robbrecht et al. (eds.), 2nd Internat. Rubiac. Conf. Proc. (Op. Bot.* Belg., 7): 306 (1996).
subsect. Latifoliae *(K. Schum.) N.M. Bacigalupo & E.L. Cabral in E. Robbrecht et al. (eds.), 2nd Internat. Rubiac. Conf. Proc. (Op. Bot.* Belg., 7): 306 (1996): B. ser. Latifoliae.

BORRERIA:-
burchellii *E.L. Cabral & N.M. Bacigalupo in Bonplandia,* 10(1–4): 126 (2000), nom. nov.: Borreria tenella var. pumila *K. Schum.*
crispata *(K. Schum.) E.L. Cabral & N.M. Bacigalupo in Bonplandia,* 10(1–4): 126 (2000): Borreria tenella var. crispata.
dasycephala *(Cham. & Schltdl.) N.M. Bacigalupo & E.L. Cabral in E. Robbrecht et al. (eds.), 2nd Internat. Rubiac. Conf. Proc. (Op. Bot.* Belg., 7): 306 (1996): Diodia dasycephala.
delicatula *E.L. Cabral in Hickenia,* 3(7): 21 (1999), nom. nov.: B. filiformis *E.L. Cabral.*
densiflora DC.
var. pilosa *N.M. Bacigalupo in A.L. Cabrera, Fl. Prov. Jujuy (Inst. Nac. Tecn. Agropec., 13),* 9: 413 (1993)—Argentina Northwest (Jujuy, Salta).
dimorpha *J.H. Kirkbride in Brittonia,* 49(3): 373 (1997)—Brazil West-central (Goiás).
eryngioides Cham. & Schltdl.
var. ostenii *(Standl.) E.L. Cabral & N.M. Bacigalupo in E. Robbrecht et al. (eds.), 2nd Internat. Rubiac. Conf. Proc. (Op. Bot.* Belg., 7): 317 (1996): B. ostenii.
filiformis *E.L. Cabral in Bonplandia,* 9(1–2): 35 (1996), as 'filiforme', homonym; *see Hutch. & Dalziel, Fl. W. Trop. Afr.,* 2: 135 (1931)—Brazil North (Rondônia).
flavovirens *N.M. Bacigalupo & E.L. Cabral in Hickenia,* 2(56): 261 (1998)—Argentina Northeast (Entre Ríos, Chaco, Corrientes, Misiones), Brazil Southeast (Minas Gerais, São Paulo), Brazil South (Paraná, Rio Grande do Sul), Brazil North (Rondônia), Paraguay.
flexuosa *E.L. Cabral in Bonplandia,* 9(1–2): 36 (1996)—Brazil West-central (Mato Grosso).
gardneri *(K. Schum.) N.M. Bacigalupo & E.L. Cabral in E. Robbrecht et al. (eds.), 2nd Internat. Rubiac. Conf. Proc. (Op. Bot.* Belg., 7): 307 (1996): Diodia gardneri.
hyssopifolia *(Roem. & Schult.) N.M. Bacigalupo & E.L. Cabral in E. Robbrecht et al. (eds.), 2nd Internat. Rubiac. Conf. Proc. (Op. Bot.* Belg., 7): 307 (1996): Spermacoce hyssopifolia.
incognita *E.L. Cabral in E. Robbrecht et al. (eds.), 2nd Internat. Rubiac. Conf. Proc. (Op. Bot.* Belg., 7): 322 (1996)—Brazil West-central (Goiás), Brazil North (Tocantins).
irwiniana *E.L. Cabral in Bonplandia,* 9(1–2): 36 (1996)—Brazil West-central (Goiás).
latifolia (Aubl.) K. Schum.
var. bartlingiana *(DC.) J. Fournet, Fl. Illustr. Phan. Guadeloupe & Martinique:* 1182 (1978), without basionym ref.: B. bartlingiana.
multiflora *(DC.) N.M. Bacigalupo & E.L. Cabral in E. Robbrecht et al. (eds.), 2nd Internat. Rubiac. Conf. Proc. (Op. Bot.* Belg., 7): 307 (1996): Diodia multiflora.
ocymifolia *(Roem. & Schult.) N.M. Bacigalupo & E.L. Cabral in E. Robbrecht et al. (eds.), 2nd Internat. Rubiac. Conf. Proc. (Op. Bot.* Belg., 7): 307 (1996): Spermacoce ocymifolia.
ocymoides (Burm. f.) DC.
var. dussii *(Standl.) J. Fournet, Fl. Illustr. Phan. Guadeloupe & Martinique:* 1181 (1978), without basionym ref.: B. dussii.
palustris *(Cham. & Schltdl.) N.M. Bacigalupo & E.L. Cabral in Hickenia,* 2(56): 264 (1998): Diodia palustris.
paraensis *E.L. Cabral & N.M. Bacigalupo in Darwiniana,* 37(3–4): 268 (1999)—Brazil North (Pará), Brazil Northeast (Bahia).
paranaensis *E.L. Cabral & N.M. Bacigalupo in Bol. Soc. Argent. Bot.,* 34(3–4): 149 (2000)—Brazil South (Paraná), Brazil Southeast (São Paulo).
paulista *E.L. Cabral & N.M. Bacigalupo in Acta Bot. Brasil.,* 11(1): 46 (1997)—Brazil Southeast (São Paulo).
pulchristipula *(Bremek.) N.M. Bacigalupo & E.L. Cabral in Bol. Soc. Argent. Bot.,* 34(3–4): 151 (2000): Diodia pulchristipula.
remota *(Lam.) N.M. Bacigalupo & E.L. Cabral in Darwiniana,* 37(3–4): 334 (1999): Spermacoce remota.
rosmarinifolia *E.L. Cabral & N.M. Bacigalupo in Acta Bot. Brasil.,* 11(1): 48 (1997), as 'rosmariinifolia'—Brazil Southeast (Minas Gerais).

BORRERIA:-
semiamplexicaulis *E.L. Cabral in Bonplandia,* 9(1–2): 37 (1996), as 'semiamplexicaule'—Brazil North (Pará).
spicata *(Miq.) N.M. Bacigalupo & E.L. Cabral in E. Robbrecht et al. (eds.), 2nd Internat. Rubiac. Conf. Proc. (Op. Bot.* Belg., 7): 307 (1996): Diodia spicata.

BORYA (Anthericac.).
inopinata *P.I. Forster & E.J. Thompson in Austrobaileya,* 4(4): 597 (1997)—Queensland.

BORZICACTUS (Cactac.).
aureispinus
f. cristatus *P.V. Heath in Calyx,* 5(4): 137 (1997)—Cultivation.

BOSSIAEA (Legumin.).
cucullata *J.H. Ross in Muelleria,* 11: 8 (1998)—Western Australia.
halophila *J.H. Ross in Muelleria,* 11: 5 (1998)—Western Australia.

BOSWELLIA (Burserac.).
glabra *W. Roxburgh, Pl. Coromandel,* 3: 4 (1811)—India (Tamil Nadu).†

BOTHRIOCHLOA (Gramin.).
panormitana *(Parl.) S. Brullo in Giorn. Bot. Ital.,* 129(2): 173 (1995), isonym; *see Pilg. in Engl. & Prantl, Nat. Pflanzenfam. ed. 2,* 14e: 161 (1940): Andropogon panormitanus.

BOTHRIOCLINE (Compos.).
bampsii *S. Lisowski in Fragm. Flor. Geobot.,* 41(1): 487 (1996)—Zaire.
leonardiana *S. Lisowski in Fragm. Flor. Geobot.,* 41(1): 488 (1996)—Zaire.
pauwelsii *S. Lisowski in Fragm. Flor. Geobot.,* 41(1): 488 (1996)—Zaire.

BOUCEROSIA (Asclepiadac.).
dentata *(Forssk.) A. Deflers in Mém. Inst. Egypt.,* 3: 273 (1896); see also M. G. Gilbert *in Bradleya,* 7: 80 (1989) and F.N. Hepper & I. Friis, *Pl. Pehr Forsskal's Fl. Aegypt.-Arab.:* 79 (1994): Stapelia dentata.
indica *(Wight & Arn.) D.C.H. Plowes in Haseltonia,* 3: 59 (1995): Hutchinia indica.
nilagiriana *(Kumari & Subba Rao) D.C.H. Plowes in Haseltonia,* 3: 59 (1995): Caralluma nilagiriana.
procumbens *(Gravely & Mayur.) D.C.H. Plowes in Haseltonia,* 3: 59 (1995): Caralluma procumbens.

BOURRERIA (Boraginac.).
grandicalyx *J.S. Miller & B. Sirot in Novon,* 7(4): 395 (1997 publ. 1998)—Costa Rica.
rinconensis *J.S. Miller in Novon,* 9(2): 230 (1999)—Costa Rica.

BOUTELOUA (Gramin.).
barbata
var. major *(Vasey) R. Govaerts, World Checklist Seed Plants,* 2(1–2): 13 (1996): B. polystachya var. major.
bracteata *(McVaugh) J.T. Columbus in Aliso,* 18(1): 63 (1999): Opizia bracteata.
chihuahuana *(M.C. Johnst.) J.T. Columbus in Aliso,* 14(3): 227 (1995 publ. 1996): B. rigidiseta var. chihuahuana.
dactyloides *(Nutt.) J.T. Columbus in Aliso,* 18(1): 63 (1999): Sesleria dactyloides.
dimorpha *J.T. Columbus in Aliso,* 18(1): 63 (1999), nom. nov.: Opizia stolonifera *J. Presl.*
diversispicula *J.T. Columbus in Aliso,* 18(1): 63 (1999), nom. nov.: Cathestecum brevifolium *Swallen.*
erecta *(Vasey & Hack.) J.T. Columbus in Aliso,* 18(1): 63 (1999): Cathestecum erectum.
griffithsii *J.T. Columbus in Aliso,* 18(1): 63 (1999), nom. nov.: Cathestecum prostratum *J. Presl.*
hirsuta Lag.
subsp. pectinata *(Feath.) J.K. Wipff & S.D. Jones in Sida,* 17(1): 109 (1996): B. pectinata.
mexicana *(Scribn.) J.T. Columbus in Aliso,* 18(1): 63 (1999): Fourniera mexicana.
multifida *(Griffiths) J.T. Columbus in Aliso,* 18(1): 63 (1999): Cathestecum multifidum.

BOUTELOUA:-
polymorpha *(E. Fourn.) J.T. Columbus in Aliso,* 18(1): 63 (1999): Atheropogon polymorphus.
reederorum *J.T. Columbus in Aliso,* 18(1): 64 (1999), nom. nov.: Atheropogon stolonifer *E. Fourn.*
scabra *(Kunth) J.T. Columbus in Aliso,* 18(1): 64 (1999): Pentarrhaphis scabra.
swallenii *J.T. Columbus in Aliso,* 18(1): 64 (1999), nom. nov.: Pentarrhaphis annua *Swallen.*
varia *(Swallen) J.T. Columbus in Aliso,* 18(2): 64 (1999): Cathestecum varium.

BRACHIONIDIUM (Orchidac.).
alpestre *C.A. Luer & R. Vásquez in C.A. Luer, Ic. Pleurothallid. XII (Monogr. Syst. Bot. Missouri Bot. Gard.,* 57): 12 (1995)—Bolivia.
andreettae *C.A. Luer & Hirtz in C.A. Luer, Ic. Pleurothallid. XII (Monogr. Syst. Bot. Missouri Bot. Gard.,* 57): 14 (1995)—Ecuador.
arethusa *C.A. Luer, Ic. Pleurothallid. XII (Monogr. Syst. Bot. Missouri Bot. Gard., 57):* 16 (1995)—Peru.
ballatrix *C.A. Luer & Hirtz in C.A. Luer, Ic. Pleurothallid. XII (Monogr. Syst. Bot. Missouri Bot. Gard.,* 57): 18 (1995)—Ecuador.
calypso *C.A. Luer, Ic. Pleurothallid. XII (Monogr. Syst. Bot. Missouri Bot. Gard., 57):* 24 (1995)—Panama.
capillare *C.A. Luer & Hirtz in C.A. Luer, Ic. Pleurothallid. XII (Monogr. Syst. Bot. Missouri Bot. Gard.,* 57): 26 (1995)—Ecuador.
dentatum *C.A. Luer & Dressler in C.A. Luer, Ic. Pleurothallid. XII (Monogr. Syst. Bot. Missouri Bot. Gard.,* 57): 34 (1995)—Panama.
dodsonii *C.A. Luer, Ic. Pleurothallid. XII (Monogr. Syst. Bot. Missouri Bot. Gard., 57):* 38 (1995)—Ecuador.
dressleri *C.A. Luer, Ic. Pleurothallid. XII (Monogr. Syst. Bot. Missouri Bot. Gard., 57):* 40 (1995)—Panama, Costa Rica.
ephemerum *C.A. Luer & Hirtz in C.A. Luer, Ic. Pleurothallid. XII (Monogr. Syst. Bot. Missouri Bot. Gard.,* 57): 46 (1995)—Ecuador.
filamentosum *C.A. Luer & Hirtz in C.A. Luer, Ic. Pleurothallid. XII (Monogr. Syst. Bot. Missouri Bot. Gard.,* 57): 50 (1995)—Ecuador, Colombia, Panama.
fornicatum *C.A. Luer & Hirtz in C.A. Luer, Ic. Pleurothallid. XII (Monogr. Syst. Bot. Missouri Bot. Gard.,* 57): 54 (1995)—Ecuador, Colombia.
furfuraceum *C.A. Luer, Ic. Pleurothallid. XII (Monogr. Syst. Bot. Missouri Bot. Gard., 57):* 56 (1995)—Peru.
galeatum *C.A. Luer & Hirtz in C.A. Luer, Ic. Pleurothallid. XII (Monogr. Syst. Bot. Missouri Bot. Gard.,* 57): 58 (1995)—Ecuador.
haberi *C.A. Luer, Ic. Pleurothallid. XII (Monogr. Syst. Bot. Missouri Bot. Gard., 57):* 60 (1995)—Costa Rica.
imperiale *C.A. Luer & R. Escobar in C.A. Luer, Ic. Pleurothallid. XII (Monogr. Syst. Bot. Missouri Bot. Gard.,* 57): 64 (1995), with two types cited—Colombia, Ecuador.
ingramii *C.A. Luer & Dalström in C.A. Luer, Ic. Pleurothallid. XIV (Monogr. Syst. Bot. Missouri Bot. Gard.,* 61(4)): 1 (1996); *see also C.A. Luer, Ic. Pleurothallid. XV (Monogr. Syst. Bot. Missouri Bot.* Gard., 64): 122 (1997)—Ecuador.
jesupiae *C.A. Luer, Ic. Pleurothallid. XII (Monogr. Syst. Bot. Missouri Bot. Gard., 57):* 66 (1995)—Ecuador.
lehmannii *C.A. Luer, Ic. Pleurothallid. XII (Monogr. Syst. Bot. Missouri Bot. Gard., 57):* 72 (1995)—Ecuador.
loxense *C.A. Luer, Ic. Pleurothallid. XII (Monogr. Syst. Bot. Missouri Bot. Gard., 57):* 76 (1995)—Ecuador.
lucanoideum *C.A. Luer, Ic. Pleurothallid. XII (Monogr. Syst. Bot. Missouri Bot. Gard., 57):* 78 (1995)—Panama.
minusculum *C.A. Luer & Dressler in C.A. Luer, Ic. Pleurothallid. XII (Monogr. Syst. Bot. Missouri Bot. Gard.,* 57): 82 (1995)—Costa Rica.
operosum *C.A. Luer & Hirtz in C.A. Luer, Ic. Pleurothallid. XII (Monogr. Syst. Bot. Missouri Bot. Gard.,* 57): 86 (1995)—Ecuador.
peltarion *C.A. Luer, Ic. Pleurothallid. XII (Monogr. Syst. Bot. Missouri Bot. Gard., 57):* 92 (1995)—Panama.

BRACHIONIDIUM:-
piuntzae *C.A. Luer, Ic. Pleurothallid. XII (Monogr. Syst. Bot. Missouri Bot. Gard., 57):* 96 (1995)—Ecuador.
polypodium *C.A. Luer, Ic. Pleurothallid. XII (Monogr. Syst. Bot. Missouri Bot. Gard., 57):* 98 (1995)—Panama, Costa Rica.
puraceense *C.A. Luer, Ic. Pleurothallid. XII (Monogr. Syst. Bot. Missouri Bot. Gard., 57):* 104 (1995)—Colombia.
renzii *C.A. Luer, Ic. Pleurothallid. XII (Monogr. Syst. Bot. Missouri Bot. Gard., 57):* 108 (1995), without type; *et in Lindleyana,* 11(2): 118 (1996), with type—Guyana.
satyreum *C.A. Luer, Ic. Pleurothallid. XII (Monogr. Syst. Bot. Missouri Bot. Gard., 57):* 112 (1995)—Panama.
stellare *C.A. Luer & Hirtz in C.A. Luer, Ic. Pleurothallid. XII (Monogr. Syst. Bot. Missouri Bot. Gard.,* 57): 120 (1995)—Ecuador.
syme-morrisii *C.A. Luer, Ic. Pleurothallid. XII (Monogr. Syst. Bot. Missouri Bot. Gard., 57):* 122 (1995)—Jamaica.
uxorium *C.A. Luer & R. Vásquez in C.A. Luer, Ic. Pleurothallid. XII (Monogr. Syst. Bot. Missouri Bot. Gard.,* 57): 130 (1995)—Bolivia.
zunagense *C.A. Luer & Hirtz in C.A. Luer, Ic. Pleurothallid. XII (Monogr. Syst. Bot. Missouri Bot. Gard.,* 57): 136 (1995)—Ecuador.

BRACHYACTIS (Compos.).
alpina *Y. Wei & C.H. An in Fl. Xinjiangensis,* 5: 475 (1999), as 'alpinus'—Xinjiang.

BRACHYCORYTHIS (Orchidac.).
lisowskiana *D.L. Szlachetko & T.S. Olszewski in Fl. Cameroun,* 34: 76 (1998)—Congo.
ovata Lindl.
var. schweinfurthii *(Rchb. f.) D.L. Szlachetko & T.S. Olszewski in Fl. Cameroun,* 34: 72 (1998): B. schweinfurthii.
var. welwitschii *(Rchb. f.) D.L. Szlachetko & T.S. Olszewski in Fl. Cameroun,* 34: 73 (1998): B. welwitschii.

BRACHYLAENA (Compos.).
discolor DC.
var. rotundata *(S. Moore) H.J. Beentje in Kew Bull.,* 55(1): 27 (2000): B. rotundata.
var. transvaalensis *(E. Phillips & Schweick.) H.J. Beentje in Kew Bull.,* 55(1): 25 (2000): B. transvaalensis.

BRACHYLOMA (Epacridac.).
nguba *R.J. Cranfield in Nuytsia,* 12(2): 180 (1998)—Western Australia.

BRACHYOTUM (Melastomatac.).
incrassatum *E. Cotton in Nordic J. Bot.,* 20(2): 185 (2000)—Ecuador.
johannes-julii *E. Cotton in Nordic J. Bot.,* 20(2): 187 (2000)—Ecuador.
russatum *E. Cotton in Nordic J. Bot.,* 20(2): 187 (2000)—Ecuador.

BRACHYPODIUM (Gramin.).
sylvaticum (Huds.) P. Beauv.
var. gaditanum *(Talavera) A. Galán de Mera in Lagascalia,* 18(1): 56 (1995): B. gaditanum.
wilsonis *Sennen in Ann. Junta Cièn. Nat.,* 2: 658 (1917)—Spain.

BRACHYSCIAS *J.M. Hart & M.J. Henwood in Austral. Syst. Bot.,* 12(2): 176 (1999). *UMBELLIFERAE*
verecundus *J.M. Hart & M.J. Henwood in Austral. Syst. Bot.,* 12(2): 176 (1999)—Western Australia.

BRACHYSOLA *B.L. Rye in Nuytsia,* 13(2): 332 (2000), nom. nov.: Chloanthes sect. Brachysolenia *F. Muell.* (Labiat.).
coerulea *(F. Muell. & Tate) B.L. Rye in Nuytsia,* 13(2): 335 (2000): Chloanthes coerulea.
halganiacea *(F. Muell.) B.L. Rye in Nuytsia,* 13(2): 337 (2000): Chloanthes halganiacea.

BRACHYSTEGIA (Legumin.).
astlei *A.C. Hoyle in Kew Bull.,* 54(1): 155 (1999)—Zambia.

BRACHYSTEGIA:-
michelmorei *A.C. Hoyle in Kew Bull.,* 54(1): 156 (1999)—Zambia.
torrei *A.C. Hoyle in Kew Bull.,* 54(1): 159 (1999)—Mozambique, Zimbabwe.

BRACHYSTELE (Orchidac.).
bicrinita *D.L. Szlachetko in Fragm. Flor. Geobot.,* 41(2): 849 (1996)—Brazil South (Santa Catarina).
hintoniorum *(Todzia) A. Espejo Serna & A.R. López-Ferrari in Phytologia,* 82(2): 79 (1997): Galeotiella hintoniorum.
luzmariana *D.L. Szlachetko & R. González Tamayo in Bol. Inst. Bot. (Guadalajara),* 5(1–3): 390 (1998)—Mexico Southwest (Jalisco).
maasii *D.L. Szlachetko in Fragm. Flor. Geobot.,* 41(2): 847 (1996)—Colombia.
markowskiana *(Szlach.) A. Espejo Serna & A.R. López-Ferrari in Phytologia,* 82(2): 79 (1997): Galeotiella markowskiana.
oxyanthos *D.L. Szlachetko in Fragm. Flor. Geobot.,* 41(2): 846 (1996)—Surinam.
pappulosa *D.L. Szlachetko in Fragm. Flor. Geobot.,* 41(2): 848 (1996)—Uruguay.
waldemarii *D.L. Szlachetko in Fragm. Flor. Geobot.,* 41(2): 850 (1996)—Uruguay.

BRACHYSTELMA (Asclepiadac.).
albipilosum *A. Lancaster ex R. Peckover in Cact. Succ. J. (U.S.A.),* 69(3): 155 (1997)—Zimbabwe.
bikitaense *R. Peckover in Aloe,* 32(3–4): 78 (1995), as 'bikitaensis'—Zimbabwe.
bracteolatum *U. Meve in Kew Bull.,* 52(3): 711 (1997)—Nigeria.
burchellii *(Decne.) R. Peckover in Aloe,* 33(2–3): 43 (1996), without basionym date: Macropetalum burchellii.
chloranthum *(Schltr.) R. Peckover in Aloe,* 33(2–3): 43 (1996): Tenaris chlorantha.
codonanthum *P.V. Bruyns in Bothalia,* 25(2): 157 (1995)—Namibia.
filifolium *(Schltr.) R. Peckover in Aloe,* 33(2–3): 43 (1996), without basionym date: Macropetalum filifolium.
gymnopodum *(Schltr.) P.V. Bruyns in Bothalia,* 25(2): 161 (1995): Ceropegia gymnopoda.
maritae *R. Peckover in Cact. Succ. J. (U.S.A.),* 68(1): 3 (1996)—Tanzania.
megasepalum *R. Peckover in Kakt. And. Sukk.,* 47(12): 250 (1996)—Tanzania, Zambia.
molaventi *R.G. Peckover & A.E. van Wyk in Aloe,* 36(2–3): 46 (1999)—Natal.
naorojii *P. Tetali, D.K. Kulkarni, S. Tetali & M.S. Kumbhojkar in Rheedea,* 8(1): 75 (1998)—India (Maharashtra).
nepalense *(Radcl.-Sm.) U. Meve in Kew Bull.,* 52(4): 1012 (1997): Riocreuxia nepalensis.
pellacibellum *L.E. Newton in Bradleya,* 14: 97 (1996)—Kenya.
recurvatum *P.V. Bruyns in Bothalia,* 25(2): 156 (1995)—Namibia.
richardsii *R. Peckover in Excelsa,* 17: 33 (1996)—Zimbabwe.
rubellum *(E. Mey.) R. Peckover in Aloe,* 33(2–3): 43 (1996): Tenaris rubella.
schultzei *(Schltr.) P.V. Bruyns in Bothalia,* 25(2): 162 (1995): Kinepetalum schultzei.

BRACHYSTEPHANUS (Acanthac.).
africanus S. Moore
var. madagascariensis *E. Figueiredo in Kew Bull.,* 51(4): 757 (1996)—Madagascar.
var. velutinus *(De Wild.) E. Figueiredo in Kew Bull.,* 51(4): 756 (1996): B. velutinus.
densiflorus *E. Figueiredo in Kew Bull.,* 51(4): 758 (1996)—Madagascar.
glaberrimus *D. Champluvier in Bull. Jard. Bot. Nation. Belg.,* 66(3–4): 194 (1997)—Uganda, Zaire, Rwanda.

BRACHYTOME (Rubiac.).
kachinensis *R. Govaerts, World Checklist Seed Plants,* 2(1–2): 13 (1996), nom. nov.: B. wardii *Deb & M.G. Gangop.*

BRACTEANTHA (Compos.).
palustris *C. Flann in Muelleria,* 11: 97 (1998)—Victoria, Tasmania.

BRACTEOCARPUS *A.V. Bobrov & A.P. Melikyan in Byull. Mosk. Obshch. Ispŷt. Prir., Biol.,* 103(1): 58 (1998). *PODOCARPACEAE*
cinctus *(Pilg.) A.V. Bobrov & A.P. Melikyan in Byull. Mosk. Obshch. Ispŷt. Prir., Biol.,* 103(1): 58 (1998): Podocarpus cinctus.
compactus *(Wasscher) A.V. Bobrov & A.P. Melikyan in Byull. Mosk. Obshch. Ispŷt. Prir., Biol.,* 103(1): 58 (1998): Podocarpus compactus.
cumingii *(Parl.) A.V. Bobrov & A.P. Melikyan in Byull. Mosk. Obshch. Ispŷt. Prir., Biol.,* 103(1): 58 (1998): Podocarpus cumingii.
dacrydiifolius *(Wasscher) A.V. Bobrov & A.P. Melikyan in Byull. Mosk. Obshch. Ispŷt. Prir., Biol.,* 103(1): 59 (1998): Podocarpus dacrydiifolius.
expansus *(de Laub.) A.V. Bobrov & A.P. Melikyan in Byull. Mosk. Obshch. Ispŷt. Prir., Biol.,* 103(1): 59 (1998): Dacrycarpus expansus.
imbricatus *(Blume) A.V. Bobrov & A.P. Melikyan in Byull. Mosk. Obshch. Ispŷt. Prir., Biol.,* 103(1): 58 (1998): Podocarpus imbricatus.
kawaii *(Hayata) A.V. Bobrov & A.P. Melikyan in Byull. Mosk. Obshch. Ispŷt. Prir., Biol.,* 103(1): 59 (1998): Podocarpus kawaii.
kinabaluensis *(Wasscher) A.V. Bobrov & A.P. Melikyan in Byull. Mosk. Obshch. Ispŷt. Prir., Biol.,* 103(1): 59 (1998): Podocarpus imbricatus var. kinabaluensis.
leptophyllus *(Wasscher) A.V. Bobrov & A.P. Melikyan in Byull. Mosk. Obshch. Ispŷt. Prir., Biol.,* 103(1): 59 (1998): Podocarpus leptophyllus.
papuanus *(Ridl.) A.V. Bobrov & A.P. Melikyan in Byull. Mosk. Obshch. Ispŷt. Prir., Biol.,* 103(1): 59 (1998): Podocarpus papuanus.
steupii *(Wasscher) A.V. Bobrov & A.P. Melikyan in Byull. Mosk. Obshch. Ispŷt. Prir., Biol.,* 103(1): 59 (1998): Podocarpus steupii.

BRADBURIA (Compos.).
pilosa *(Nutt.) J.C. Semple in Univ. Waterloo Biol. Ser.,* 37: 7 (1996): Chrysopsis pilosa.

BRAHEA (Palmae).
sarukhanii *H.J. Quero in Palms,* 44(3): 110 (2000)—Mexico Southwest (Nayarit, Jalisco).

BRASILICACTUS (Cactac.).
ser. Ritterocactoides *A.B. Doweld in Sukkulentŷ (Moscow),* 3(1–2): 59 (2000).
alacriportanus *(Backeb. & Voll) A.B. Doweld in Sukkulentŷ (Moscow),* 1999(2): 24 (1999): Parodia alacriportana.
brevihamatus *(W. Haage ex Backeb.) A.B. Doweld in Sukkulentŷ (Moscow),* 1999(2): 24 (1999): Parodia brevihamata.
buenekeri *(Buining) A.B. Doweld in Sukkulentŷ (Moscow),* 1999(2): 24 (1999): Parodia buenekeri.
catarinensis *A.B. Doweld in Sukkulentŷ (Moscow),* 1999(2): 24 (1999)—Locality not stated.
rechensis *(Buining) A.B. Doweld in Sukkulentŷ (Moscow),* 1999(2): 24 (1999): Notocactus rechensis.

BRASILOCYCNIS *G. Gerlach & M.W. Whitten in J. Orchideenfreund,* 6(3): 188 (1999). *ORCHIDACEAE*
breviloba *(Summerh.) G. Gerlach & M.W. Whitten in J. Orchideenfreund,* 6(3): 188 (1999): Polycycnis breviloba.

BRASSAIOPSIS (Araliac.).
phanrangensis *C.B. Shang in Adansonia,* 19(1): 80 (1997)—Vietnam, Laos.
rufosetosa *(Ridl.) M.H.P. Jebb in Glasra (n.s.),* 3(2): 108 (1998): Trevesia rufosetosa.

BRASSIA (Orchidac.).
subgen. Glumacea *(Lindl.) K. Senghas in Schlechter Orchideen, ed. 3,* I/C(33–36): 2097 (1997): B. sect. Glumacea.
andreettae *(Dodson) K. Senghas in Schlechter Orchideen, ed. 3,* I/C(33–36): 2097 (1997): Ada andreettae.
bennettiorum *(Dodson) K. Senghas in Schlechter Orchideen, ed. 3,* I/C(33–36): 2097 (1997): Ada bennettiorum.

BRASSIA:-
mendozae *(Dodson) K. Senghas in Schlechter Orchideen, ed. 3,* I/C(33–36): 2097 (1997): Ada mendozae.
pozoi *(Dodson & N.H. Williams) K. Senghas in Schlechter Orchideen, ed. 3,* I/C(33–36): 2097 (1997): Ada pozoi.
rolandoi *(D.E. Benn. & Christenson) K. Senghas in Schlechter Orchideen, ed. 3,* I/C(33–36): 2097 (1997), with incorrect basionym ref.: Ada rolandoi.

BRASSICA (Crucif.).
subgen. Brassicaria *(Godr.) C. Gómez Campo in An. Jard. Bot. Madrid,* 56(2): 379 (1998): Diplotaxis sect. Brassicaria.
sect. Erucastroides *N.N. Tzvelev, Opred. Sosud. Rast. Severo-Zap. Rossii:* 679 (2000).
subsect. Junceae *V.I. Dorofeev in Bot. Zhurn.,* 81(2): 84 (1996).
sect. Nasturtiops *(Pomel) Salmeen ex C. Gómez Campo & J. Martínez Laborde in An. Jard. Bot. Madrid,* 56(2): 379 (1998): gen. Nasturtiops.
sect. Rapa *(Mill.) Salmeen ex C. Gómez Campo & J. Martínez Laborde in An. Jard. Bot. Madrid,* 56(2): 379 (1998): gen. Rapa.
brevirostrata *Z.X. An in Fl. Xinjiangensis,* 2(2): 374 (1995)—Xinjiang.
juncea (L.) Czern.
subsp. napiformis *(Paillieux & Bois) T. Gladis in Feddes Repert.,* 103(7–8): 474 (1992): Sinapis juncea var. napiformis.
subsp. tsatsai *(Mao) T. Gladis in Feddes Repert.,* 103(7–8): 474 (1992): B. juncea var. tsatsai.
nantouensis *S.S. Ying in Mem. Coll. Agric. Nation. Taiwan Univ.,* 29(2): 47 (1989), as 'nantouense'—Taiwan.
nigra (L.) W.D.J. Koch
subsp. hispida *(O.E. Schulz) T. Gladis in Feddes Repert.,* 103(7–8): 490 (1992): B. nigra f. hispida.
var. dissecta *(O.E. Schulz) T. Gladis in Feddes Repert.,* 103(7–8): 500 (1992): B. campestris var. dissecta.
var. pandurata *(V.G. Sun) T. Gladis in Feddes Repert.,* 103(7–8): 496 (1992), basionym without latin descr.: B. chinensis var. pandurata.
repanda (Willd.) DC.
subsp. glabrescens *(Poldini) C. Gómez Campo in An. Jard. Bot. Madrid,* 56(2): 379 (1998): B. glabrescens.
rupestris Raf.
subsp. brevisiliqua *F.M. Raimondo & P. Mazzola in Lagascalia,* 19(1–2): 834 (1997)—Sicilia.
subsp. hispida *F.M. Raimondo & P. Mazzola in Lagascalia,* 19(1–2): 832 (1997)—Sicilia.
villosa Biv.
subsp. bivoniana *(Mazzola & Raimondo) F.M. Raimondo & P. Mazzola in Lagascalia,* 19(1–2): 832 (1997): B. bivoniana.
subsp. drepanensis *(Caruel) F.M. Raimondo & P. Mazzola in Lagascalia,* 19(1–2): 832 (1997): Eruca drepanensis.
subsp. tinei *(Lojac.) F.M. Raimondo & P. Mazzola in Lagascalia,* 19(1–2): 832 (1997): B. tinei.

BRAVOCACTUS *A.B. Doweld in Sukkulentŷ (Moscow),* 1998(1): 22 (1998). *CACTACEAE*
horripilus *(Lem.) A.B. Doweld in Sukkulentŷ (Moscow),* 1998(1): 22 (1998): Mammillaria horripila.

BRAZORIA (Labiat.).
truncata (Benth.) Engelm. & A. Gray
var. pulcherrima *(Lundell) M.W. Turner in Pl. Syst. Evol.,* 203(1–2): 77 (1996): B. pulcherrima.

BREDIA (Melastomatac.).
longiradiosa *C. Chen ex R. Govaerts, World Checklist Seed Plants,* 2(1–2): 13 (1996), nom. nov.: Barthea cavaleriei *H. Lév.*

BRENANDENDRON *H. Robinson in Proc. Biol. Soc. Wash.,* 112(1): 244 (1999). *COMPOSITAE*
donianum *(DC.) H. Robinson in Proc. Biol. Soc. Wash.,* 112(1): 244 (1999): Vernonia doniana.

BRENANDENDRON:-
frondosum *(Oliv. & Hiern) H. Robinson in Proc. Biol. Soc. Wash.,* 112(1): 244 (1999): Vernonia frondosa.
titanophyllum *(Brenan) H. Robinson in Proc. Biol. Soc. Wash.,* 112(1): 244 (1999): Vernonia titanophylla.

BRENANIODENDRON *J. Léonard in Bull. Jard. Bot. Nation. Belg.,* 67(1–4): 16 (1999). *LEGUMINOSAE*
carvalhoi *(Harms) J. Léonard in Bull. Jard. Bot. Nation. Belg.,* 67(1–4): 18 (1999): Cynometra carvalhoi.

BREWCARIA (Bromeliac.).
brocchinioides *(L.B. Sm.) B.K. Holst in Fl. Venezuelan Guayana (gen. eds. J.A. Steyermark et al.),* 3: 569 (1997): Navia brocchinioides.
hechtioides *(L.B. Sm.) B.K. Holst in Fl. Venezuelan Guayana (gen. eds. J.A. Steyermark et al.),* 3: 571 (1997): Navia hechtioides.
hohenbergioides *(L.B. Sm.) B.K. Holst in Fl. Venezuelan Guayana (gen. eds. J.A. Steyermark et al.),* 3: 571 (1997): Navia hohenbergioides.
reflexa *(L.B. Sm.) B.K. Holst in Fl. Venezuelan Guayana (gen. eds. J.A. Steyermark et al.),* 3: 571 (1997): Navia reflexa.

BREYNIA (Euphorbiac.).
subangustifolia *Nguyen Nghia Thin, Euphorbiac. Vietnam:* 50 (1995 publ. 1996)—Vietnam.
subindochinensis *Nguyen Nghia Thin, Euphorbiac. Vietnam:* 50 (1995 publ. 1996)—Vietnam.

BRICKELLIA (Compos.).
hinckleyi Standl.
var. terlinguensis *(Flyr) B.L. Turner in Phytologia Mem.,* 11: 69 (1997): B. brachyphylla var. terlinguensis.
villarrealii *(R.M. King & H. Rob.) R. Govaerts, World Checklist Seed Plants,* 2(1–2): 13 (1996): Brickelliastrum villarrealii.

BRIDELIA (Euphorbiac.).
cathartica G. Bertol.
f. fischeri *(Pax) A. Radcliffe-Smith in Kew Bull.,* 51(2): 302 (1996): B. fischeri.
var. lingelsheimii *(Gehrm.) A. Radcliffe-Smith in Kew Bull.,* 51(2): 302 (1996): B. lingelsheimii.
f. melanthesoides *(Baill.) A. Radcliffe-Smith in Fl. Zambes.,* 9(4): 17 (1996): Pentameria melanthesoides.
var. melanthesoides *(Baill.) A. Radcliffe-Smith in Fl. Zambes.,* 9(4): 16 (1996): Pentameria melanthesoides.
f. niedenzui *(Gehrm.) A. Radcliffe-Smith in Kew Bull.,* 51(2): 302 (1996): B. niedenzui.
f. pubescens *A. Radcliffe-Smith in Kew Bull.,* 51(2): 302 (1996)—Mozambique.
erapensis *S. Dressler in Kew Bull.,* 51(3): 601 (1996)—New Guinea (Papua New Guinea).
finalis *P.I. Forster in Austrobaileya,* 5(3): 408 (1999)—Queensland.
glauca Blume
var. acuminatissima *(Merr.) S. Dressler in Blumea,* 41(2): 314 (1996): B. acuminatissima.
var. sosopodonica *(Airy Shaw) S. Dressler in Blumea,* 41(2): 315 (1996): B. sosopodonica.
insulana Hance
var. subnuda *(K. Schum. & Lauterb.) S. Dressler in Blumea,* 41(2): 320 (1996): B. subnuda.
scleroneura Müll. Arg.
subsp. angolensis *(Welw. ex Müll. Arg.) A. Radcliffe-Smith in Kew Bull.,* 51(2): 303 (1996): B. angolensis.

BRIGGSIA (Gesneriac.).
dulongensis *H. Li in Fl. Dulongjian Reg.:* 300 (1993), without latin descr. or type—China South-central (Yunnan).

BRILLANTAISIA (Acanthac.).
riparia *(Vollesen & Brummitt) K. Sidwell in Bull. Nat. Hist. Mus. (London), Bot.,* 28(2): 86 (1998): B. pubescens var. riparia.
stenopteris *K. Sidwell in Bull. Nat. Hist. Mus. (London), Bot.,* 28(2): 89 (1998)—Tanzania.

BROCCHINIA (Bromeliac.).
cataractarum *(Sandwith) B.K. Holst in Biollania,* ed. esp. 6 (J.J. Wurdack Festschr.) 377 (1997): Navia cataractarum.
rupestris *(Gleason) B.K. Holst in Biollania,* ed. esp. 6 (J.J. Wurdack Festschr.): 377 (1997): Tofieldia rupestris.
wurdackiana *B.K. Holst in Biollania,* ed. esp. 6 (J.J. Wurdack Festschr.): 375 (1997)—Venezuela.

BROMELIA (Bromeliac.).
flemingii *I.M. Ramirez & G. Carnevali in J. Bromeliad Soc.,* 46(5): 206 (1996)—Venezuela.
granvillei *L.B. Smith & E.J. Gouda in Bromélia,* 3(4): 9 (1996)—French Guiana.

BROMHEADIA (Orchidac.).
borneensis J.J. Sm.
var. longiflora *H.J. van Scheindelen & E.F. de Vogel in Orchid Monogr.,* 8: 88 (1997)—Singapore, Borneo (Kalimantan).
cecieliae *J. Kruizinga in Orchid Monogr.,* 8: 101 (1997)—Borneo (Sarawak, Sabah).
coomansii *J.J. Sm. ex J. Kruizinga & E.F. de Vogel in Orchid Monogr.,* 8: 102 (1997)—Borneo (Sarawak, Sabah, Kalimantan).
devogelii *J. Kruizinga in Orchid Monogr.,* 8: 102 (1997)—Borneo (Sarawak), Philippines.
gracilis *J. Kruizinga & E.F. de Vogel in Orchid Monogr.,* 8: 104 (1997)—Borneo (Sabah).
graminea *J. Kruizinga & E.F. de Vogel in Orchid Monogr.,* 8: 105 (1997)—Borneo (Sarawak, Kalimantan).
grandiflora *J. Kruizinga & E.F. de Vogel in Orchid Monogr.,* 8: 106 (1997)—Borneo (Sarawak, Brunei, Kalimantan), Malaya (Peninsular Malaysia).
humilis *J. Kruizinga & E.F. de Vogel in Orchid Monogr.,* 8: 107 (1997)—Borneo (Sabah).
latifolia *J. Kruizinga & E.F. de Vogel in Orchid Monogr.,* 8: 107 (1997)—Borneo (Kalimantan).
lohaniensis *J. Kruizinga & E.F. de Vogel in Orchid Monogr.,* 8: 108 (1997)—Borneo (Sabah).
longifolia *J. Kruizinga & E.F. de Vogel in Orchid Monogr.,* 8: 109 (1997)—Borneo (Sabah, Brunei).
pendek *E.F. de Vogel in Orchid Monogr.,* 8: 93 (1997)—Borneo (Brunei).
robusta *J. Kruizinga & E.F. de Vogel in Orchid Monogr.,* 8: 110 (1997)—Borneo (Kalimantan, Sarawak).
srilankensis *J. Kruizinga & E.F. de Vogel in Orchid Monogr.,* 8: 112 (1997)—Sri Lanka.

BROMOPSIS (Gramin.).
arctica (Shear ex Scribn. & Merr.) Holub
var. alpina *(Malyschev) G.A. Peschkova in Fl. Tsentral'noi Sibiri,* 1: 125 (1979): Bromus sibiricus f. alpinus.
pumpelliana (Scribn.) Holub
subsp. ircutensis *(Kom.) N.N. Tzvelev in Novosti Sist. Vyssh. Rast.,* 32: 182 (2000): Bromus irkutensis.
subsp. sibirica *(Drobow) N.N. Tzvelev in Novosti Sist. Vyssh. Rast.,* 32: 182 (2000): Bromus sibiricus.
subsp. taimyrensis *(Roshev.) N.N. Tzvelev in Novosti Sist. Vyssh. Rast.,* 32: 182 (2000): Bromus sibiricus var. taimyrensis.

BROMUS (Gramin.).
bidentatus *G. Holmström & H. Scholz in Bot. Jahrb.,* 122(2): 196 (2000), contrary to Art. 8.2 ICBN (2000)—Cyprus.
bolivianus *Hack. ex S.A. Renvoize, Gramineas de Bolivia:* 247 (1998)—Bolivia.
cabrerensis *C. Acedo & F. Llamas in Willdenowia,* 27(1–2): 48 (1997)—Spain.
catharticus Vahl
var. rupestris *(Speg.) A.M. Planchuelo & P.M. Peterson in Novon,* 8(1): 54 (1998): B. unioloides var. rupestris.
danthoniae Trin.
subsp. pseudodanthoniae *(Drobow) H. Scholz in Willdenowia,* 28(1–2): 146 (1998): B. pseudodanthoniae.
subsp. rogersii *C.E. Hubb. ex H. Scholz in Willdenowia,* 28(1–2): 147 (1998)—Afghanistan, Iran.
elidis *H. Scholz in Feddes Repert.,* 108(1–2): 99 (1997)—Greece.

BROMUS:-
kunkelii *(H. Scholz) H. Scholz in Mitteilungsb. Botaniker-Tagung Wien (Sept. 1984):* 137 (1984): B. madritensis subsp. kunkelii.
nervosus *C. Acedo & F. Llamas in Willdenowia,* 27(1–2): 50 (1997)—Portugal.
optimae *H. Scholz in Bocconea,* 11: 86 (1999)—Cyprus.
rubens L.
subsp. kunkelii *(H. Scholz) H. Scholz in Willdenowia,* 25(2): 578 (1996): B. madritensis subsp. kunkelii.
secalinus L.
subsp. decipiens *W. Bomble & H. Scholz in Feddes Repert.,* 110(5–6): 427 (1999)—Germany, Sweden, Denmark, Great Britain, Belgium, France, Poland, Czechoslovakia, Hungary, Austria, Yugoslavia, Italy, Greece, Palestine (Israel), Transcaucasus (Abkhasiya).
setifolius
var. minor *(Desv.) R. Govaerts, World Checklist Seed Plants,* 2(1–2): 13 (1996): B. macranthus var. minor.
tomentellus Boiss.
subsp. nivalis *(Bornm.) H. Scholz & A.J. Byfield in Turkish J. Bot.,* 24(4): 263 (2000): B. tomentellus var. nivalis.
turcomanicus *H. Scholz in Willdenowia,* 28(1–2): 148 (1998)—Turkmenistan.

BRONGNIARTIA (Legumin.).
guiengolensis *O. Dorado & L. Torres-Colín in Novon,* 6(1): 9 (1996)—Mexico Southwest (Oaxaca).
riesebergii *O. Dorado in Brittonia,* 51(4): 398 (1999)—Mexico Northeast (Durango).

BROSIMUM (Morac.).
multinervium *C.C. Berg in Novon,* 6(3): 230 (1996)—Ecuador.

BROWALLIA (Solanac.).
mirabilis *S. Leiva González in Arnaldoa,* 3(2): 14 (1995 publ. 1996)—Peru.

BROWNANTHUS (Aizoac.).
fraternus *C. Klak in Bothalia,* 30(1): 37 (2000)—Cape Province.
glareicola *C. Klak in Bothalia,* 30(1): 37 (2000)—Cape Province.
neglectus *S.M. Pierce & M. Gerbaulet in Aloe,* 34(1–2): 43 (1997)—Namibia, Cape Province.
nucifer *(Ihlenf. & Bittrich) S.M. Pierce & M. Gerbaulet in Aloe,* 34(1–2): 44 (1997): Pseudobrownanthus nucifer.
pseudoschlichtianus *S.M. Pierce & M. Gerbaulet in Aloe,* 34(1–2): 43 (1997)—Cape Province, Namibia.

BROWNEA (Legumin.).
chocoana *L.M. Quiñones in Rev. Acad. Colomb. Cienc. Exact. Fis. Nat.,* 80: 225 (1997)—Colombia.
enricii *L.M. Quiñones in Caldasia,* 18(86): 17 (1995)—Colombia.
santanderensis *L.M. Quiñones in Caldasia,* 18(86): 19 (1995)—Colombia.

BRUGMANSIA (Solanac.).
cubensis *(V.R. Fuentes) J.M.H. Shaw in New Plantsman,* 6(3): 150 (1999), isonym; *see V.R. Fuentes in Rev. Jard. Bot. Nacion. Univ. Habana,* 11(2–3): 113 (1990 publ. 1993): Datura cubensis.
sanguinea
subsp. vulcanicola *(A.S. Barclay) R. Govaerts, World Checklist Seed Plants,* 2(1–2): 13 (1996): Datura vulcanicola.
subsp. vulcanicola *(A.S. Barclay) Lockwood ex J.M.H. Shaw in New Plantsman,* 5(1): 56 (1998), isonym; *see R. Govaerts, World Checklist Seed Plants,* 2(1–2): 13 (1996): Datura vulcanicola.

BRUGUIERA (Rhizophorac.).
gymnorrhiza (L.) Lam.
var. nana *S. Hatusima in S. Hatusima & T. Amano, Fl. Ryukyus, S. of Amami Is., ed. 2:* 148 (1994), without latin descr. or type—Locality not stated.

BRUNFELSIA (Solanac.).
brasiliensis (Spreng.) L.B. Sm. & Downs
subsp. macrocalyx *(Dusén) T.C. Plowman in Fieldiana, Bot.,* 39: 66 (1998): B. hopeana var. macrocalyx.
hydrangeiformis (Pohl) Benth.
subsp. capitata *(Benth.) T.C. Plowman in Fieldiana, Bot.,* 39: 86 (1998): B. capitata.
lindeniana *(Planch.) N.E. Brown, Suppl. Johnson's Gard. Dict.:* 886 (1882): Franciscea lindeniana.

BRUNIA (Bruniac.).
noduliflora *P. Goldblatt & J.C. Manning in Strelitzia,* 9: 706 (2000)—Cape Province.

BRUNONIELLA (Acanthac.).
neocaledonica *(Heine) E.C. Moylan in Kew Bull.,* 55(2): 480 (2000): Hemigraphis neocaledonica.

BRYANTIELLA *J.M. Porter in Aliso,* 19(1): 70 (2000). *POLEMONIACEAE*
glutinosa *(Phil.) J.M. Porter in Aliso,* 19(1): 71 (2000): Gilia glutinosa.
palmeri *(S. Watson) J.M. Porter in Aliso,* 19(1): 71 (2000): Gilia palmeri.

BRYASPIS (Legumin.).
psittacorhyncha *(Webb) R. Govaerts, World Checklist Seed Plants,* 2(1–2): 13 (1996): Soemmeringia psittacorhyncha.

BRYONIA (Cucurbitac.).
cretica
subsp. marmorata *(E. Petit) R. Govaerts, World Checklist Seed Plants,* 2(1–2): 13 (1996): B. marmorata.

BRYOPHYLLUM (Crassulac.).
bergeri *(Raym.-Hamet & H. Perrier) R. Govaerts, World Checklist Seed Plants,* 2(1–2): 13 (1996): Kalanchoe bergeri.
calcicola *(H. Perrier) V.V. Byalt in Novosti Sist. Vyssh. Rast.,* 32: 51 (2000): Kalanchoe pinnata var. calcicola.
jongmansii *(Raym.-Hamet & H. Perrier) R. Govaerts, World Checklist Seed Plants,* 2(1–2): 13 (1996): Kalanchoe jongmansii.
laxiflorum *(Baker) R. Govaerts, World Checklist Seed Plants,* 2(1–2): 13 (1996): Kalanchoe laxiflora.
marnierianum *(H. Jacobsen) R. Govaerts, World Checklist Seed Plants,* 2(1–2): 13(1996): Kalanchoe marnieriana.
marnierianum *(H. Jacobsen) V.V. Byalt in Novosti Sist. Vyssh. Rast.,* 32: 51 (2000), isonym; *see Govaerts, World Checklist Seed Plants,* 2(1–2): 13 (1996): Kalanchoe marnieriana.
poincarei *(Raym.-Hamet & H. Perrier) R. Govaerts, World Checklist Seed Plants,* 2(1–2): 13 (1996): Kalanchoe poincarei.
poincarei *(Raym.-Hamet & H. Perrier) V.V. Byalt in Novosti Sist. Vyssh. Rast.,* 32: 51 (2000), isonym; *see Govaerts, World Checklist Seed Plants,* 2(1–2): 13 (1996): Kalanchoe poincarei.
pseudocampanulatum *(Mannoni & Boiteau) R. Govaerts, World Checklist Seed Plants,* 2(1–2): 13 (1996): Kalanchoe pseudocampanulatum.
pubescens *(Baker) R. Govaerts, World Checklist Seed Plants,* 2(1–2): 13 (1996): Kalanchoe pubescens.
rolandi-bonapartei *(Raym.-Hamet & H. Perrier) R. Govaerts, World Checklist Seed Plants,* 2(1–2): 13 (1996): Kalanchoe rolandi-bonapartei.

BUCEPHALANDRA (Arac.).
magnifolia *H. Okada & Y. Mori in Acta Phytotax. Geobot.,* 51(1): 4 (2000)—Borneo (Kalimantan).

BUCHANANIA (Anacardiac.).
angustifolia *W. Roxburgh, Pl. Coromandel,* 3: 58 (1820)—India (Tamil Nadu).†

BUCHNERA (Scrophulariac.).
subsect. Compactiflorae *R. Mielcarek in Fragm. Flor. Geobot.,* 41(1): 120 (1996).
subsect. Distantiflorae *R. Mielcarek in Fragm. Flor. Geobot.,* 41(1): 99 (1996).

BUCHNERA:-
sect. Nuttii *R. Mielcarek in Fragm. Flor. Geobot.*, 41(1): 161 (1996).
sect. Prorepentes *R. Mielcarek in Fragm. Flor. Geobot.*, 41(1): 97 (1996).
bampsiana *R. Mielcarek in Fragm. Flor. Geobot.*, 41(1): 159 (1996)—Zaire.
decandollei *R. Govaerts, World Checklist Seed Plants*, 2(1–2): 14 (1996), nom. nov.: B. capitata *Benth.*
dilungensis *R. Mielcarek in Fragm. Flor. Geobot.*, 41(1): 125 (1996)—Zaire.
libenii *R. Mielcarek in Fragm. Flor. Geobot.*, 41(1): 133 (1996)—Burundi.
philcoxii *R. Mielcarek in Fragm. Flor. Geobot.*, 41(1): 103 (1996)—Zaire.
robynsii *R. Mielcarek in Fragm. Flor. Geobot.*, 41(1): 157 (1996)—Zaire.
symoensiana *R. Mielcarek in Fragm. Flor. Geobot.*, 41(1): 130 (1996)—Zaire.
vandenberghenii *R. Mielcarek in Fragm. Flor. Geobot.*, 41(1): 140 (1996)—Zaire.

BUDDLEJA (Buddlejac.).
ser. Anchoenses *E.M. Norman in Fl. Neotrop.*, 81: 104 (2000).
ser. Brachiatae *E.M. Norman in Fl. Neotrop.*, 81: 118 (2000).
ser. Cordatae *E.M. Norman in Fl. Neotrop.*, 81: 148 (2000).
ser. Globosae *(Benth.) E.M. Norman in Fl. Neotrop.*, 81: 96 (2000): B. subsect. Globosae.
ser. Glomeratae *(Benth.) E.M. Norman in Fl. Neotrop.*, 81: 104 (2000): B. subsect. Glomeratae.
ser. Lanatae *E.M. Norman in Fl. Neotrop.*, 81: 128 (2000).
ser. Oblongae *E.M. Norman in Fl. Neotrop.*, 81: 83 (2000).
ser. Scordioides *E.M. Norman in Fl. Neotrop.*, 81: 132 (2000).
ser. Stachyoides *(Benth.) E.M. Norman in Fl. Neotrop.*, 81: 83 (2000): B. subsect. Stachyoideae.
ser. Thyrsoides *(Benth.) E.M. Norman in Fl. Neotrop.*, 81: 75 (2000): B. subsect. Thyrsoideae.
ser. Verticillatae *(Benth.) E.M. Norman in Fl. Neotrop.*, 81: 144 (2000): B. subsect. Verticillatae.
cardenasii *Standl. ex E.M. Norman in Fl. Neotrop.*, 81: 163 (2000)—Bolivia.
cordata Kunth
subsp. ovandensis *(Lundell ex E.M. Norman) E.M. Norman in Fl. Neotrop.*, 81: 174 (2000): B. ovandensis.
skutchii C.V. Morton
subsp. costaricensis *E.M. Norman in Fl. Neotrop.*, 81: 176 (2000)—Panama, Costa Rica.
striata *Zhi Y. Zhang ex R. Govaerts, World Checklist Seed Plants*, 2(1–2): 14 (1996)—Locality not stated.
vexans *Kraenzl. & Loes. ex E.M. Norman in Fl. Neotrop.*, 81: 157 (2000)—Peru.

BUESIELLA (Orchidac.).
ornata *W. Königer in Arcula*, 9: 250 (1999)—Peru.

BUFONIA (Caryophyllac.).
sect. Calyculatae *J. Chrtek & B. Křísa in Acta Univ. Carol., Biol.*, 43(2): 81 (1999).
sect. Enervis *J. Chrtek & B. Křísa in Acta Univ. Carol., Biol.*, 43(2): 82 (1999).
sect. Leptoclada *J. Chrtek & B. Křísa in Acta Univ. Carol., Biol.*, 43(2): 83 (1999).
sect. Loliopsis *J. Chrtek & B. Křísa in Acta Univ. Carol., Biol.*, 43(2): 82 (1999).
sect. Micrantha *J. Chrtek & B. Křísa in Acta Univ. Carol., Biol.*, 43(2): 83 (1999).
subsect. Oliveriana *J. Chrtek & B. Křísa in Acta Univ. Carol., Biol.*, 43(2): 81 (1999).
anatolica *J. Chrtek & B. Křísa in Acta Univ. Carol., Biol.*, 43(2): 94 (1999)—Turkey.
var. elatior *J. Chrtek & B. Křísa in Acta Univ. Carol., Biol.*, 43(2): 95 (1999)—Turkey.
ephedrina Rech. f.
var. minor *J. Chrtek & B. Křísa in Acta Univ. Carol., Biol.*, 43(2): 98 (1999)—Syria, Lebanon.
subsp. musilii *J. Chrtek & B. Křísa in Acta Univ. Carol., Biol.*, 43(2): 98 (1999)—Saudi Arabia.
kotschyana Boiss.
subsp. calycina *(Boiss. & Hausskn.) J. Chrtek & B. Křísa in Acta Univ. Carol., Biol.*, 43(2): 106 (1999): B. calycina.

BUFONIA:-
kotschyana Boiss.
subsp. densa *J. Chrtek & B. Křísa in Acta Univ. Carol., Biol.*, 43(2): 105 (1999)—Iran.
leptoclada *Rech. f. ex R. Govaerts, World Checklist Seed Plants*, 2(1–2): 14 (1996)—Locality not stated.
pabotii *J. Chrtek & B. Křísa in Acta Univ. Carol., Biol.*, 43(2): 99 (1999)—Lebanon-Syria (Syria).
paniculata Dubois
subsp. orientalis *J. Chrtek & B. Křísa in Acta Univ. Carol., Biol.*, 43(2): 87 (1999)—Turkey, Lebanon, Syria.

BUGLOSSOIDES (Boraginac.).
goulandriorum *(Rech. f.) R. Govaerts, World Checklist Seed Plants*, 2(1–2): 14 (1996): Lithospermum goulandriorum.
subsp. thessalica *(Aldén) R. Govaerts, World Checklist Seed Plants*, 2(1–2): 14 (1996), as 'thessalicum': Lithospermum goulandriorum subsp. thessalicum.

BULBINE (Asphodelac.).
bruynsii *S.A. Hammer in Cactus & Co.*, 2(4): 6–7 (1998); *cf. Repert. Pl. Succ. (I.O.S.)*, 49: 13 (1998 publ. 1999)—Locality not stated.
canaliculata *G. Williamson in Bradleya*, 18: 37 (2000)—Cape Provinces (Eastern Cape Province).
cremnophila *E. van Jaarsveld in Aloe*, 36(4): 72 (1999)—Cape Province.
disimilis *G. Williamson in Aloe*, 34(3–4): 70 (1997)—Cape Province.
fragilis *G. Williamson in Haseltonia*, 4: 13 (1996)—Cape Province.
francescae *G. Williamson & H. Baijnath in S. Afr. J. Bot.*, 61(6): 312 (1995)—Namibia.
lamprophylla *G. Williamson in Bradleya*, 14: 84 (1996)—Cape Province.
lavrani *G. Williamson & H. Baijnath in Aloe*, 36(2–3): 28 (1999)—Cape Province.
muscicola *G. Williamson in Bradleya*, 18: 36 (2000)—Cape Provinces (Western Cape Province).
pendens *G. Williamson & H. Baijnath in S. Afr. J. Bot.*, 61(6): 316 (1995)—Cape Province.
quartzicola *G. Williamson in Haseltonia*, 4: 19 (1996)—Cape Province.
rupicola *G. Williamson in Bradleya*, 18: 36 (2000)—Cape Provinces (Eastern Cape Province).
torsiva *G. Williamson in Haseltonia*, 4: 21 (1996)—Cape Province.
truncata *G. Williamson in Haseltonia*, 4: 19 (1996)—Cape Province.
undulata *G. Williamson in Bradleya*, 18: 33 (2000)—Cape Provinces (Western Cape Province).
vitrea *G. Williamson & H. Baijnath in S. Afr. J. Bot.*, 61(6): 314 (1995)—Cape Province.
vittatifolia *G. Williamson in Haseltonia*, 4: 21 (1996)—Cape Province.

BULBINELLA (Asphodelac.).
latifolia Kunth
subsp. denticulata *P.L. Perry in Strelitzia*, 8: 28 (1999)—Cape Provinces (Western Cape Province).
subsp. doleritica *(P.L. Perry) P.L. Perry in Strelitzia*, 8: 28 (1999): Bulbinella latifolia var. doleritica.
subsp. toximontana *P.L. Perry in Strelitzia*, 8: 31 (1999)—Cape Provinces (Western Cape Province).
nutans (Thunb.) T. Durand & Schinz
subsp. turfosicola *(P.L. Perry) P.L. Perry in Strelitzia*, 8: 23 (1999): Bulbinella nutans var. turfosicola.
potbergensis *P.L. Perry in Strelitzia*, 8: 36 (1999)—Cape Provinces (Western Cape Province).

BULBOPHYLLUM (Orchidac.).
agastor *L.A. Garay, F. Hamer & E. Siegerist in Lindleyana*, 11(4): 224 (1996)—New Guinea (Papua New Guinea).
aithorhachis *J.J. Vermeulen in Blumea*, 41(2): 349 (1996)—Borneo (Brunei).
anjozorobeense *J. Bosser in Adansonia*, 22(2): 170 (2000)—Madagascar.

BULBOPHYLLUM:-
arcuatilabium *L.V. Averyanov in Lindleyana*, 14(4): 214 (1999)—Vietnam.
bidi *R. Govaerts, World Checklist Seed Plants*, 2(1–2): 14 (1996), nom. nov.: B. pumilio *Ridl.*
blepharochilum *L.A. Garay in Harvard Pap. Bot.*, 4(1): 301 (1999)—Cameroon.
bosseri *K. Lemcke in Orchidee*, 50(6): 663 (1999), nom. nov.: B. inauditum *Schltr. (1925)*.
burfordiense *L.A. Garay, F. Hamer & E. Siegerist in Lindleyana*, 11(4): 224 (1996)—New Guinea (Papua New Guinea).
cameronense *L.A. Garay, F. Hamer & E. Siegerist in Lindleyana*, 11(4): 226 (1996)—Malaya (Peninsular Malaysia).
camptochilum *J.J. Vermeulen in Blumea*, 41(2): 355 (1996)—Borneo (Brunei).
carrianum *J.J. Vermeulen in Gard. Bull. Singapore*, 52(2): 279 (2000)—Borneo (Sabah).
cavipes *J.J. Vermeulen in Blumea*, 41(2): 356 (1996)—Sumatera.
cephalophorum *L.A. Garay, F. Hamer & E. Siegerist in Lindleyana*, 11(4): 227 (1996)—Philippines.
cerebellum *J.J. Vermeulen in Blumea*, 41(2): 350 (1996)—Borneo (Sabah).
× cipoense *E.L. Borba & J. Semir in Lindleyana*, 13(2): 114 (1998)—Brazil Southeast (Minas Gerais). (B. involutum × weddellii).
congolanum Schltr.
subsp. leucorrhachis *F. Perez-Vera in Orchidées Cult. Protect.*, 24bis: 8 (1996), without latin descr. or type—Ivory Coast.
cootesii *M.A. Clements in Austral. Orchid Rev.*, 64(6): 6 (1999–2000 publ. 1999)—Philippines.
crassiusculifolium *L.V. Averyanov in Bot. Zhurn.*, 82(3): 138 (1997)—Vietnam.
cryptophoranthoides *L.A. Garay in Harvard Pap. Bot.*, 4(1): 302 (1999), homonym; *see Kraenzl. in Bot. Jahrb. Syst.*, 34: 251 (1904)—Philippines.
cryptophoranthus *L.A. Garay in Harvard Pap. Bot.*, 4(2): 475 (1999), nom. nov.: B. cryptophoranthoides *Garay*.
cyanotriche *J.J. Vermeulen in Blumea*, 41(2): 348 (1996)—Borneo (Sabah).
cylindrocarpum Frapp. ex Cordem.
var. andringitrense *J. Bosser in Adansonia*, 22(2): 169 (2000)—Madagascar.
dewildei *J.J. Vermeulen in Blumea*, 41(2): 364 (1996)—Sumatera.
dracunculus *J.J. Vermeulen in Gard. Bull. Singapore*, 52(2): 276 (2000)—Borneo (Sabah).
dulongense *H. Li in Fl. Dulongjian Reg.*: 338 (1993), without latin descr. or type—China South-central (Yunnan).
erythroglossum *J. Bosser in Adansonia*, 22(2): 179 (2000)—Madagascar.
facetum *L.A. Garay, F. Hamer & E. Siegerist in Lindleyana*, 11(4): 227 (1996)—Philippines.
falcatum
var. bufo *(Lindl.) R. Govaerts, World Checklist Seed Plants*, 2(1–2): 14 (1996): Megaclinium bufo.
foraminiferum *J.J. Vermeulen in Blumea*, 41(2): 357 (1996)—Borneo (Sarawak).
fraudulentum *L.A. Garay, F. Hamer & E. Siegerist in Lindleyana*, 11(4): 229 (1996)—New Guinea (Irian Jaya).
fulgens *J.J. Vermeulen in Blumea*, 41(2): 361 (1996)—Sumatera.
gemma-reginae *J.J. Vermeulen in Blumea*, 41(1): 21 (1996)—Borneo (Sabah).
gomesii *C.N. de Fraga in Bradea*, 8(24): 136 (1999)—Brazil Southeast (Espírito Santo).
guttifilum *G. Seidenfaden in Nordic J. Bot.*, 16(2): 141 (1996)—Thailand.
hapalanthos *L.A. Garay in Harvard Pap. Bot.*, 4(1): 302 (1999)—Madagascar.
hashimotoi *T. Yukawa & K. Karasawa in Ann. Tsukuba Bot. Gard.*, 16: 17 (1997)—New Guinea (Papua New Guinea).
hornense *R. Govaerts, World Checklist Seed Plants*, 2(1–2): 14 (1996), nom. nov., as 'hornensis': B. stenophyllum *Schltr.*
involutum *E.L. Borba, J. Semir & F. de Barros in Novon*, 8(3): 225 (1998)—Brazil Southeast (Minas Gerais), Brazil Northeast (Bahia).
jiewhoei *J.J. Vermeulen & P. O'Byrne in Gard. Bull. Singapore*, 52(2): 293 (2000)—Locality not stated.

BULBOPHYLLUM:-
kautskyi *A.L.V. Toscano de Brito in Lindleyana,* 15(3): 184 (2000)—Brazil Southeast (Espírito Santo), Brazil Northeast (Bahia).
kegelii *F. Hamer & L.A. Garay in Bol. Inst. Bot. (Guadalajara),* 3(1–3): 9 (1995 publ. 1997)—Brazil North (Amazonas), Surinam, Trinidad-Tobago.
kettridgei *(Garay, Hamer & Siegerist) J.J. Vermeulen in Nordic J. Bot.,* 16(4): 381 (1996 publ. 1997): Synarmosepalum kittredgei.
laetum *J.J. Vermeulen in Blumea,* 41(2): 365 (1996)—Borneo (Sarawak).
lemurense *J. Bosser & P.J. Cribb in D. Du Puy, P. Cribb, J. Bosser, J. Hermans & C. Hermans, Orchids of* Madagascar: 4, 77 (1999), nom. nov.: B. clavigerum *H. Perrier.*
leptobulbon *J.J. Vermeulen in Blumea,* 41(2): 358 (1996)—New Guinea (Papua New Guinea).
lilacinum Ridl.
var. sorocianum *M. Ahmed, M.K. Pasha & M.A. Aziz Khan in J. Bombay Nat. Hist. Soc.,* 91(1): 79 (1994)—Bangladesh.
luciphilum *T. Stévart in Syst. Geogr. Fl.,* 70(1): 143 (2000)—Gulf of Guinea Is. (São Tomé).
mayombeense *L.A. Garay in Harvard Pap. Bot.,* 4(1): 303 (1999)—Congo.
nasseri *L.A. Garay in Harvard Pap. Bot.,* 4(1): 303 (1999)—Borneo.
ngoclinhensis *L.V. Averyanov in Bot. Zhurn.,* 82(3): 140 (1997)—Vietnam.
obtusiangulum *Z.H. Tsi in Acta Phytotax. Sin.,* 33(6): 578 (1995)—Hainan.
papulosum *L.A. Garay in Harvard Pap. Bot.,* 4(1): 304 (1999)—Philippines.
pectinatum Finet
var. transarisanense *(Hayata) S.S. Ying in Mem. Coll. Agric. Nation. Taiwan Univ.,* 29(2): 51 (1989): B. transarisanense.
pleurothallidanthum *L.A. Garay in Harvard Pap. Bot.,* 4(1): 304 (1999)—Sumatera.
polycyclum *J.J. Vermeulen in Blumea,* 41(2): 351 (1996)—Borneo (Kalimantan, Sabah).
proculcastris *J.J. Vermeulen in Gard. Bull. Singapore,* 52(2): 281 (2000)—Borneo (Sabah).
pseudotrias *J.J. Vermeulen in Blumea,* 41(2): 359 (1996)—New Guinea (Papua New Guinea).
purpureifolium *L.V. Averyanov in Bot. Zhurn.,* 82(3): 141 (1997)—Vietnam.
recurvilabre *L.A. Garay in Harvard Pap. Bot.,* 4(1): 304 (1999)—Philippines.
reflexiflorum H. Perrier
subsp. pogonochilum *(Summerh.) J. Bosser in Adansonia,* 22(2): 174 (2000): Bulbophyllum pogonochilum.
saronae *L.A. Garay in Harvard Pap. Bot.,* 4(1): 305 (1999)—New Guinea (Papua New Guinea).
schmidii *L.A. Garay in Harvard Pap. Bot.,* 4(1): 305 (1999)—New Guinea (Irian Jaya).
setuliferum *J.J. Vermeulen & L.G. Saw in Gard. Bull. Singapore,* 52(2): 289 (2000)—Malaya (Peninsular Malaysia).
siederi *L.A. Garay in Harvard Pap. Bot.,* 4(1): 306 (1999)—Sumatera.
signatum *J.J. Vermeulen in Blumea,* 41(2): 362 (1996)—Borneo (Sarawak).
smitinandii *G. Seidenfaden & C. Thorut in Nordic J. Bot.,* 16(3): 287 (1996 publ. 1997)—Thailand.
spissum *J.J. Vermeulen in Blumea,* 41(2): 360 (1996)—Sumatera.
spongiola *J.J. Vermeulen in Blumea,* 41(2): 361 (1996)—New Guinea (Papua New Guinea).
subbullatum *J.J. Vermeulen in Blumea,* 41(2): 347 (1996)—Borneo (Sarawak).
sulawesii *L.A. Garay, F. Hamer & E. Siegerist in Lindleyana,* 11(4): 230 (1996)—Sulawesi.
sumatranum *L.A. Garay, F. Hamer & E. Siegerist in Lindleyana,* 11(4): 230 (1996), nom. nov.: B. lobbii var. breviflorum *J.J. Sm.*
taeter *J.J. Vermeulen in Blumea,* 41(2) 363 (1996)—Borneo (Brunei).
tseanum *(S.Y. Hu & Barretto) Z.H. Tsi in Fl. Reipubl. Popul. Sin.,* 19: 239 (1999): Cirrhopetalum tseanum.
uviflorum *P. O'Byrne in Malayan Orchid Rev.,* 33: 46 (1999), as 'uvaeflorum'—Sulawesi.
wakoi *N.H.S. Howcroft in Sida,* 18(4): 1019 (1999)—New Guinea (Papua New Guinea).

BULBOPHYLLUM:-
woelfliae *L.A. Garay, K. Senghas & K. Lemcke in J. Orchideenfreund,* 3(1): 37 (1996)—Philippines.
xenosum *J.J. Vermeulen in Blumea,* 41(2): 353 (1996)—Borneo (Brunei).
xiphion *J.J. Vermeulen in Blumea,* 41(2): 354 (1996)—Borneo (Kalimantan).

BULBOSTYLIS (Cyperac.).
carajana *R. Kral & M.T. Strong in Sida,* 18(3): 844 (1999)—Brazil North (Pará).
eleocharoides *R. Kral & M.T. Strong in Sida,* 18(3): 846 (1999)—Bolivia, Paraguay.
hispidula (Vahl) R.W. Haines
subsp. longespicata *K.A. Lye in Lidia,* 3(6): 207 (1996)—Ethiopia.
subsp. macroglumis *K.A. Lye in Willdenowia,* 25(2): 596 (1996)—Somalia.
latifolia *R. Kral & M.T. Strong in Sida,* 18(3): 848 (1999)—Brazil West-central (Goiás).
lombardii *R. Kral & M.T. Strong in Sida,* 18(3): 850 (1999)—Brazil Southeast (Minas Gerais).
pachypoda *R. Kral & M.T. Strong in Sida,* 18(3): 852 (1999)—Brazil Southeast (Minas Gerais), Brazil West-central (Mato Grosso).
rugosa *M.G. López in Bonplandia,* 9(1–2): 29 (1996)—Argentina Northeast (Corrientes), Paraguay.
scirpoides *R. Kral & M.T. Strong in Sida,* 18(3): 855 (1999)—Brazil West-central (Goiás, Brasília Distrito Federal).
somaliensis *K.A. Lye in Edinburgh J. Bot.,* 53(2): 223 (1996)—Somalia.
subsp. confusa *K.A. Lye in Edinburgh J. Bot.,* 53(2): 225 (1996)—Somalia.
subsp. microcarpa *(Chiov.) K.A. Lye in Edinburgh J. Bot.,* 53(2): 227 (1996): Fimbristylis cioniana var. microcarpa.
spectabilis *R. Kral & M.T. Strong in Sida,* 18(3): 857 (1999)—Brazil South (Paraná).
splendens *M.T. Strong in Brittonia,* 48(1): 96 (1996)—Brazil North (Rondônia).

BUNCHOSIA (Malpighiac.).
caroli *W.R. Anderson in Bol. Inst. Bot. (Guadalajara),* 5(1–3): 110 (1998)—Mexico Southwest (Oaxaca, Guerrero).
cauliflora *W.R. Anderson in Contrib. Univ. Mich. Herb.,* 21: 37 (1997)—Ecuador.
luzmariae *W.R. Anderson in Bol. Inst. Bot. (Guadalajara),* 5(1–3): 108 (1998)—Mexico Southwest (Jalisco, Nayarit).
pernambucana *W.R. Anderson in Contrib. Univ. Mich. Herb.,* 21: 39 (1997)—Brazil Northeast (Pernambuco).
petraea *W.R. Anderson in Contrib. Univ. Mich. Herb.,* 21: 40 (1997)—Venezuela.
thaumatothrix *W.R. Anderson in Contrib. Univ. Mich. Herb.,* 21: 43 (1997)—Venezuela.

BUNIUM (Umbellif.).
balearicum *(Sennen) G. Mateo Sanz & S. López Udias in An. Jard. Bot. Madrid,* 57(1): 229 (1999): Bulbocastanum balearicum.
bulbocastanum L.
var. peucedanoides *(Desf.) J.M. Montserrat in Lagascalia,* 21(2): 367 (2000): Laserpitium peucedanoides.
elegans (Fenzl) Freyn
var. involucratum *Ö. Saya in Turkish J. Bot.,* 16(3): 425 (1992)—Turkey.
var. latilobum *Ö. Saya in Turkish J. Bot.,* 16(3): 426 (1992)—Turkey.
var. pachypus *Ö. Saya in Turkish J. Bot.,* 16(3): 424 (1992)—Turkey.
macuca Boiss.
subsp. nivale *(Boiss.) G. Mateo Sanz & S. López Udias in An. Jard. Bot. Madrid,* 57(1): 231 (1999): B. nivale.
pinnatifolium *E.V. Kljuykov in Bot. Zhurn.,* 83(1): 139 (1998)—Turkey.

BUPLEURUM (Umbellif.).
ajanense *(Regel) I.M. Krasnoborov in Abstr. XV Internat. Bot. Congr., Yokohama, Japan 1993:* 255 (1993): B. ranunculoides var. ajanense.
var. alpinum *(Rupr.) I.M. Krasnoborov in Bot. Zhurn.,* 83(2): 124 (1998): B. triradiatum var. alpinum.

BUPLEURUM:-
ajanense
subsp. purpureum *(Blank.) I.M. Krasnoborov in Bot. Zhurn.,* 83(2): 124 (1998): B. purpureum.
arcticum *(Regel) I.M. Krasnoborov in Abstr. XV Internat. Bot. Congr., Yokohama, Japan 1993:* 255 (1993); *et in Bot. Zhurn.,* 83(2): 124 (1998): B. ranunculoides var. arcticum.
atlanticum
subsp. mairei *(Panel. & Vindt) R. Govaerts, World Checklist Seed Plants,* 2(1–2): 14 (1996): B. mairei.
bicaule Helm
f. latifolium *(Y.C. Chu) Y.C. Chu in Clavis Pl. Chinae Bor.-Or. ed. 2 (ed. P.Y. Fu):* 465 (1995): B. bicaule var. latifolium.
falcatum L.
subsp. komarovianum *(Lincz.) V.N. Voroschilov in A.K. Skvortsov (ed.), Florist. issl. v razn. raĭonakh SSSR:* 183 (1985): B. komarovianum.
gansuense *S.L. Pan & P.S. Hsu in Rheedea,* 5(2): 145 (1995)—China North-central (Gansu).
linczevskii *M.G. Pimenov & Sdobnina in Opred. Rast. Sred. Azii,* 7: 371 (1983)—Tadzhikistan.
longiradiatum Turcz.
subsp. shikotanense *(M. Hiroe) V.N. Voroschilov in A.K. Skvortsov (ed.), Florist. issl. v razn. raĭonakh SSSR:* 183 (1985): B. shikotanense.
rosulare *Korovin ex M.G. Pimenov & Sdobnina in Opred. Rast. Sred. Azii,* 7: 372 (1983)—Kazakhstan.
sibiricum Vest
subsp. americanum *(J.M. Coult. & Rose) I.M. Krasnoborov in Bot. Zhurn.,* 83(2): 125 (1998): B. americanum.
wightii P.K. Mukh.
var. ramosissimum *(Wight & Arn.) M. Chandrabose in N.C. Nair & A.N. Henry, Fl. Tamil Nadu, India, Ser. 1,* 1: 178 (1983): B. ramosissimum.

BURMANNIA (Burmanniac.).
filamentosa *D.X. Zhang & R.M.K. Saunders in Nordic J. Bot.,* 20(4): 392 (2000)—China Southeast (Guangdong).
larseniana *D.X. Zhang & R.M.K. Saunders in Nordic J. Bot.,* 19(2): 243 (1999)—Thailand.

BURMEISTERA (Campanulac.).
racemiflora *T.G. Lammers in Brittonia,* 50(2): 258 (1998)—Ecuador.
venezuelensis *T.G. Lammers in Brittonia,* 50(2): 260 (1998)—Venezuela.

BURRETIOKENTIA (Palmae).
dumasii *J.-C. Pintaud & D.R. Hodel in Principes,* 42(3): 160 (1998)—New Caledonia.
grandiflora *J.-C. Pintaud & D.R. Hodel in Principes,* 42(3): 162 (1998)—New Caledonia.
koghiensis *J.-C. Pintaud & D.R. Hodel in Principes,* 42(3): 164 (1998)—New Caledonia.

BURSARIA (Pittosporac.).
calcicola *L.W. Cayzer, M.D. Crisp & I.R.H. Telford in Austral. Syst. Bot.,* 12(1): 137 (1999)—New South Wales.
reevesii *L.W. Cayzer, M.D. Crisp & I.R.H. Telford in Austral. Syst. Bot.,* 12(1): 139 (1999)—Queensland.
spinosa Cav.
subsp. lasiophylla *(E.M. Benn.) L.W. Cayzer, M.D. Crisp & I.R.H. Telford in Austral. Syst. Bot.,* 12(1): 129 (1999): B. lasiophylla.

BURSERA (Burserac.).
laurihuertae *J. Rzedowski & G. Calderón de Rzedowski in Acta Bot. Mex.,* 52: 75 (2000)—Mexico Southwest (Oaxaca).
macvaughiana *R. Cuevas G. & J. Rzedowski in Acta Bot. Mex.,* 46: 78 (1999)—Mexico Southwest (Jalisco, Colima).
palaciosii *J. Rzedowski & G. Calderón de Rzedowski in Acta Bot. Mex.,* 50: 47 (2000)—Mexico Southwest (Jalisco).
ribana *J. Rzedowski & G. Calderón de Rzedowski in Acta Bot. Mex.,* 50: 51 (2000)—Mexico Southwest (Jalisco, Michoacan).

BURSERA:-
vazquezyanesii *J. Rzedowski & G. Calderón de Rzedowski in Acta Bot. Mex.*, 50: 55 (2000)—Mexico Southwest (Jalisco).

BUTIA (Palmae).
yatay (Mart.) Becc.
subsp. paraguayensis *(Barb. Rodr.) C.C. Xifreda & A.M. Sanso in Hickenia*, 2(43): 207 (1996): Cocos paraguayensis.

BUXUS (Buxac.).
bissei *Egon Köhler in Feddes Repert.*, 109(5–6): 351 (1998)—Cuba.
kitashimae *Y. Yanagita in J. Soc. Forest.*, 15: 931 (1933), as 'kitashimai'—Japan.
leivae *Egon Köhler in Feddes Repert.*, 109(5–6): 353 (1998)—Cuba.
microphylla Siebold & Zucc.
var. kitashimae *(Yanagita) H. Ohba in Fl. Japan*, 2c: 107 (1999): B. kitashimae.
moctezumae *Egon Köhler, R. Fernández & S. Zamudio in Feddes Repert.*, 104(5–6): 295 (1993)—Mexico Northeast (Querétaro).
pseudaneura *Egon Köhler in Feddes Repert.*, 109(5–6): 355 (1998)—Cuba.

BYBLIS (Byblidac.).
aquatica *A. Lowrie & J.G. Conran in Nuytsia*, 12(1): 62 (1998)—Northern Territory.
rorida *A. Lowrie & J.G. Conran in Nuytsia*, 12(1): 70 (1998)—Western Australia.

BYRSONIMA (Malpighiac.).
basiliana *W.R. Anderson in Contrib. Univ. Mich. Herb.*, 21: 45 (1997)—Venezuela.
cordifolia *W.R. Anderson in Contrib. Univ. Mich. Herb.*, 22: 4 (1999)—Brazil West-central (Goiás).
formosa *W.R. Anderson in Contrib. Univ. Mich. Herb.*, 21: 46 (1997)—Guyana, Brazil North (Amazonas).
hirsuta *W.R. Anderson in Contrib. Univ. Mich. Herb.*, 22: 5 (1999)—Brazil North (Pará).
huberi *W.R. Anderson in Contrib. Univ. Mich. Herb.*, 21: 48 (1997)—Venezuela.
lanulosa *W.R. Anderson in Contrib. Univ. Mich. Herb.*, 22: 6 (1999)—Bolivia, Brazil West-central (Mato Grosso), Brazil North (Rondônia).
pedunculata *W.R. Anderson in Contrib. Univ. Mich. Herb.*, 21: 51 (1997)—Brazil Northeast (Pernambuco).
piresii *W.R. Anderson in Contrib. Univ. Mich. Herb.*, 21: 52 (1997)—Brazil North (Rondônia, Amazonas), Brazil West-central (Mato Grosso).
riparia *W.R. Anderson in Contrib. Univ. Mich. Herb.*, 22: 8 (1999)—Bolivia, Brazil West-central (Mato Grosso), Brazil North (Rondônia).

BYTTNERIA (Sterculiac.).
cristobaliana *L.J. Dorr in Kew Bull.*, 54(4): 991 (1999 publ. 2000)—Brazil Northeast (Bahia).
fontis *C.L. Cristóbal in Bonplandia*, 9(3–4): 213 (1997)—Bolivia.
jaramilloana *L.J. Dorr in Biollania*, ed. esp. 6 (J.J. Wurdack Festschr.): 305 (1997)—Ecuador.

C

CAATINGANTHUS *H. Robinson in Smithsonian Contrib. Bot.*, 89: 15 (1999). *COMPOSITAE*
harleyi *H. Robinson in Smithsonian Contrib. Bot.*, 89: 16 (1999)—Brazil Northeast (Bahia).
rubropappus *(Soares Nunes) H. Robinson in Smithsonian Contrib. Bot.*, 89: 16 (1999): Stilpnopappus rubropappus.

CABOBANTHUS *H. Robinson in Proc. Biol. Soc. Wash.*, 112(1): 228 (1999). *COMPOSITAE*
bullulatus *(S. Moore) H. Robinson in Proc. Biol. Soc. Wash.*, 112(1): 228 (1999): Vernonia bullulata.

CABOBANTHUS:-
polysphaerus *(Baker) H. Robinson in Proc. Biol. Soc. Wash.*, 112(1): 228 (1999): Vernonia polysphaera.

CACALIA (Compos.).
auriculata DC.
subsp. praetermissa *(Pojark.) V.N. Voroschilov in A.K. Skvortsov (ed.), Florist. issl. v razn. raĭonakh SSSR:* 196 (1985): Hasteola praetermissa.
var. praetermissa *(Poljakov) W. Wang & C.Y. Li in Clavis Pl. Chinae Bor. Or. ed. 2 (ed. P.Y. Fu)*: 685 (1995): Hasteola praetermissa.
eupatorifolia *(DC.) Kuntze, Revis. Gen. Pl.*, 2: 970 (1891): Vernonia eupatorifolia.
paucicapitata *B.L. Robinson & J.M. Greenman in Amer. J. Sci.*, Ser. 3, 50(296): 158 (1895), in Supp. 1 as 'paucispicata'—Mexico Southwest (Oaxaca).

CACCINIA (Boraginac.).
macranthera
var. glauca *(Savi) R. Govaerts, World Checklist Seed Plants*, 3(1): 9 (1999): C. glauca.

CADETIA (Orchidac.).
maliliensis *(J.J. Sm.) A. Schuiteman in Blumea*, 43(1): 116 (1998): Dendrobium maliliense.
siewhongii *P. O'Byrne in Malayan Orchid Rev.*, 30: 73 (1996)—Sulawesi.

CAESALPINIA (Legumin.).
balansae *E.A. Ulibarri in Darwiniana*, 34(1–4): 307 (1996), nom. nov.: Hoffmannseggia parviflora *Micheli*.
bonduc (L.) Roxb.
var. caymanensis *(Millsp.) G.R. Proctor in Kew Bull.*, 51(3): 494 (1996): C. caymanensis.
erythrocarpa *L. Pedley in Austrobaileya*, 5(1): 98 (1997)—Queensland, New Guinea (Papua New Guinea).
exostemma DC.
subsp. tampicoana *(Britton & Rose) G.P. Lewis, Caesalpinia: Revis. Poincianella-Erythrostemon group:* 72 (1998): Poincianella tampicoana.
hughesii *G.P. Lewis, Caesalpinia: Revis. Poincianella-Erythrostemon group:* 73 (1998)—Mexico Southwest (Oaxaca, Colima, Guerrero, Michoacan).
nelsonii *(Britton & Rose) J.L.R. Contreras Jiménez, Contrib. Conocim. Gen. Caesalpinia Estado de Guerrero, Mexico:* 91 (1991): Poincianella nelsonii.
nicaraguensis *G.P. Lewis, Caesalpinia: Revis. Poincianella-Erythrostemon group:* 86 (1998)—Nicaragua.
nitens *(F. Muell. ex Benth.) L. Pedley in Austrobaileya*, 5(1): 99 (1997): Pterolobium nitens.
pauciflora *(Griseb.) C. Wright in An. Acad. Ci. Méd. Habana*, 5: 404 (1869): Lebidibia pauciflora.
pinnata (Griseb.) C. Wright
subsp. oblongifolia *(Urb.) A. Barreto Valdés & Á. Beyra Matos in Fontqueria*, 48: 182 (1997 publ. 1998): C. oblongifolia.
pluviosa DC.
var. cabraliana *G.P. Lewis, Caesalpinia: Revis. Poincianella-Erythrostemon group:* 148 (1998)—Brazil Northeast (Bahia).
var. intermedia *G.P. Lewis, Caesalpinia: Revis. Poincianella-Erythrostemon group:* 146 (1998)—Brazil Northeast (Bahia), Brazil Southeast (Minas Gerais).
var. paraensis *(Ducke) G.P. Lewis, Caesalpinia: Revis. Poincianella-Erythrostemon group:* 150 (1998): C. paraensis.
var. peltophoroides *(Benth.) G.P. Lewis, Caesalpinia: Revis. Poincianella-Erythrostemon group:* 146 (1998): C. peltophoroides.
var. sanfranciscana *G.P. Lewis, Caesalpinia: Revis. Poincianella-Erythrostemon group:* 151 (1998)—Brazil Northeast (Bahia), Brazil Southeast (Minas Gerais).
robinsoniana *(Britton & Rose) G.P. Lewis, Caesalpinia: Revis. Poincianella-Erythrostemon group:* 42 (1998): Poincianella robinsoniana.
traceyi *L. Pedley in Austrobaileya*, 5(1): 99 (1997)—Queensland.

CAESALPINIA:-
yucatanensis Greenm.
subsp. chiapensis *G.P. Lewis, Caesalpinia: Revis. Poincianella-Erythrostemon group:* 85 (1998)—Mexico Southeast (Chiapas), Guatemala.
subsp. hondurensis *G.P. Lewis, Caesalpinia: Revis. Poincianella-Erythrostemon group:* 86 (1998)—Honduras.

CAJANUS (Legumin.).
volubilis *(Blanco) R.D. Gaur, Fl. Distr. Garhwal N.W. Himalaya:* 259 (1999), isonym; *see Blanco, Fl. Filip., ed. 2:* 417 (1845): Cytisus volubilis.

CAKILE (Crucif.).
maritima L.
subsp. islandica *(Gand.) R. Elven in Nordic J. Bot.*, 16(1): 8 (1996): C. maritima f. islandica.

CALADENIA (Orchidac.).
subgen. Calonema *(Benth.) S.D. Hopper & A.P. Brown in Lindleyana*, 15(2): 124 (2000): C. sect. Calonema.
subgen. Drakonorchis *S.D. Hopper & A.P. Brown in Lindleyana*, 15(2): 124 (2000).
subgen. Elevata *S.D. Hopper & A.P. Brown in Lindleyana*, 15(2): 124 (2000).
subgen. Phlebochilus *(Benth.) S.D. Hopper & A.P. Brown in Lindleyana*, 15(2): 126 (2000): C. sect. Phlebochilus.
actensis *D.L. Jones & M.A. Clements in Orchadian*, 12(11): 522 (1999)—New South Wales (Australian Capital Territory).
amnicola *D.L. Jones in Orchadian*, 12(4): 166 (1997)—New South Wales.
anthracina *D.L. Jones in Austral. Orchid Res.*, 3: 21 (1998)—Tasmania.
atradenia *D.L. Jones, B.P.J. Molloy & M.A. Clements in Orchadian*, 12(5): 221 (1997)—New Zealand North I., New Zealand South I.
atrochila *D.L. Jones in Austral. Orchid Res.*, 3: 23 (1998)—Tasmania.
attenuata *(Brinsley) D.L. Jones in Orchadian*, 13(6): 256 (2000): Caladenia carnea var. attenuata.
bartlettii *(Hatch) D.L. Jones, B.P.J. Molloy & M.A. Clements in Orchadian*, 12(5): 227 (1997): C. carnea var. bartlettii.
callitrophila *D.L. Jones in Orchadian*, 13(1): 5 (1999)—New South Wales.
campbellii *D.L. Jones in Austral. Orchid Res.*, 3: 25 (1998)—Tasmania.
capillata *D.L. Jones in Orchadian*, 13(6): 255 (2000), nom. nov.: Caladenia filamentosa var. tentaculata *R.S. Rogers*.
chamaephylla *D.L. Jones in Orchadian*, 13(1): 6 (1999)—Queensland.
chlorostyla *D.L. Jones, B.P.J. Molloy & M.A. Clements in Orchadian*, 12(5): 223 (1997)—New Zealand South I., New Zealand North I.
cracens *D.L. Jones in Muelleria*, 9: 46 (1996)—Tasmania.
cruciformis *D.L. Jones in Orchadian*, 13(1): 10 (1999)—Victoria.
dienema *D.L. Jones in Austral. Orchid Res.*, 3: 28 (1998)—Tasmania.
gracillima *(Rupp) D.L. Jones in Orchadian*, 13(6): 256 (2000): Caladenia carnea var. gracillima.
maritima *D.L. Jones in Orchadian*, 13(1): 11 (1999)—Victoria.
mentiens *D.L. Jones in Austral. Orchid Res.*, 3: 35 (1998)—Tasmania.
nothofageti *D.L. Jones, B.P.J. Molloy & M.A. Clements in Orchadian*, 12(5): 226 (1997)—New Zealand North I., New Zealand South I.
ornata *(Nicholls) D.L. Jones in Orchadian*, 13(6): 256 (2000): Caladenia carnea var. ornata.
pilotensis *D.L. Jones in Orchadian*, 13(1): 15 (1999)—Victoria.
porphyrea *D.L. Jones in Orchadian*, 13(1): 16 (1999)—New South Wales.
rileyi *D.L. Jones in Orchadian*, 12(4): 164 (1997)—New South Wales.
saggicola *D.L. Jones in Austral. Orchid Res.*, 3: 39 (1998)—Tasmania.
sanguinea *D.L. Jones in Orchadian*, 13(1): 17 (1999)—South Australia.
subtilis *D.L. Jones in Orchadian*, 13(1): 21 (1999)—New South Wales.

CALADENIA:-
sylvicola *D.L. Jones in Austral. Orchid Res.*, 3: 40 (1998)—Tasmania.
tonellii *D.L. Jones in Austral. Orchid Res.*, 3: 41 (1998)—Tasmania.
transitoria *D.L. Jones in Austral. Orchid Res.*, 3: 42 (1998)—Tasmania, Victoria, New South Wales.

CALAMAGROSTIS (Gramin.).
abnormis *(Hook.f.) U. Shukla, Grasses N.-East. India:* 45 (1996): Deyeuxia abnormis.
accedens *(Vickery) R. Govaerts, World Checklist Seed Plants,* 3(1): 9 (1999): Deyeuxia accedens.
acuminata *(Vickery) R. Govaerts, World Checklist Seed Plants,* 3(1): 9 (1999): Deyeuxia acuminata.
affinis *(M. Gray) R. Govaerts, World Checklist Seed Plants,* 3(1): 9 (1999): Deyeuxia affinis.
alpicola *R. Govaerts, World Checklist Seed Plants,* 3(1): 9 (1999), nom. nov., homonym; *see Ohwi in Bot. Mag. (Tokyo),* 44: 569 (1930): Agrostis monticola *Roem. & Schult.*
aniselytron *R. Govaerts, World Checklist Seed Plants,* 3(1): 9 (1999), nom. nov.: Aniselytron agrostoides *Merr.*
appressa *(Vickery) R. Govaerts, World Checklist Seed Plants,* 3(1): 9 (1999): Deyeuxia appressa.
austrodensa *R. Govaerts, World Checklist Seed Plants,* 3(1): 9 (1999), nom. nov.: Deyeuxia densa *Benth.*
austroscaberula *R. Govaerts, World Checklist Seed Plants,* 3(1): 9 (1999), nom. nov.: Deyeuxia scaberula *Vickery.*
benthamiana *(Vickery) R. Govaerts, World Checklist Seed Plants,* 3(1): 9 (1999) Deyeuxia benthamiana.
brachyanthera *(Stapf) R. Govaerts, World Checklist Seed Plants,* 3(1): 9 (1999): Dichelachne brachyanthera.
canadensis
var. dorei *B. Boivin in Provancheria* 12: 40 (1981)—Eastern Canada.
carchiensis *S. Laegaard in Novon,* 8(1): 23 (1998)—Ecuador.
carinata *(Vickery) R. Govaerts, World Checklist Seed Plants,* 3(1): 10 (1999): Deyeuxia carinata.
chrysophylla *(Phil.) R. Govaerts, World Checklist Seed Plants,* 3(1): 10 (1999): Deyeuxia chrysophylla.
clemensiae *(Hitchc.) R. Govaerts, World Checklist Seed Plants,* 3(1): 10 (1999), as 'clemensae': Aulacolepis clemensiae.
cordechii *R. Govaerts, World Checklist Seed Plants,* 3(1): 10 (1999), nom. nov.: Deyeuxia ciliata *Rúgolo & Villav.*
crassiuscula *(Vickery) R. Govaerts, World Checklist Seed Plants,* 3(1): 10 (1999): Deyeuxia crassiuscula.
crispa *(Rúgolo & Villav.) R. Govaerts, World Checklist Seed Plants,* 3(1): 10 (1999): Deyeuxia crispa.
curtoides *(Rúgolo & Villav.) R. Govaerts, World Checklist Seed Plants,* 3(1): 10 (1999): Deyeuxia curtoides.
cuzcoensis *O. Tovar in Publ. Mus. Hist. Nat. Javier Prado, B,* 33: 11 (1985)—Peru.
decipiens *(R. Br.) R. Govaerts, World Checklist Seed Plants,* 3(1): 10 (1999): Agrostis decipiens.
drummondii *(Steud.) R. Govaerts, World Checklist Seed Plants,* 3(1): 10 (1999): Pentapogon drummondii.
ecuadoriensis *S. Laegaard in Novon,* 8(1): 25 (1998)—Ecuador.
epileuca *(Stapf) R. Govaerts, World Checklist Seed Plants,* 3(1): 10 (1999): Deyeuxia epileuca.
fibrovaginata *S. Laegaard in Novon,* 8(1): 29 (1998), nom. nov.: C. coarctata *(Kunth) Steud.*
fulgida *S. Laegaard in Novon,* 8(1): 26 (1998)—Ecuador.
gigantea *T. Nuttall in Trans. Amer. Philos. Soc., n.s.,* 5: 143 (1835)—Arkansas.
hirta *(Sodiro ex Mille) S. Laegaard in Novon,* 8(1): 29 (1998): Deyeuxia hirta.
imbricata *(Vickery) R. Govaerts, World Checklist Seed Plants,* 3(1): 10 (1999): Deyeuxia imbricata.
inaequalis *(Vickery) R. Govaerts, World Checklist Seed Plants,* 3(1): 10 (1999): Deyeuxia inaequalis.
inexpansa *A. Gray, N. Amer. Gram. & Cyper.,* 1: 20 (1834)—New York, New Hampshire, Vermont, Maine, Newfoundland.†

CALAMAGROSTIS:-
japonica *(Hack.) R. Govaerts, World Checklist Seed Plants,* 3(1): 10 (1999): Aulacolepis japonica.
kashmiriana *(Bor) R. Govaerts, World Checklist Seed Plants,* 3(1): 10 (1999): Deyeuxia kashmiriana.
korotkyi Litv.
subsp. monticola *(Petrov ex Kom.) V.N. Voroschilov in A.K. Skvortsov (ed.), Florist. issl. v razn. Raĭonakh SSSR:* 148 (1985): C. monticola.
lawrencei *(Vickery) R. Govaerts, World Checklist Seed Plants,* 3(1): 10 (1999): Deyeuxia lawrencei.
linifolia *R. Govaerts, World Checklist Seed Plants,* 3(1): 10 (1999), nom. nov.: Deyeuxia angustifolia *Vickery.*
llanganatensis *S. Laegaard in Novon,* 8(1): 26 (1998)—Ecuador.
mckiei *(Vickery) R. Govaerts, World Checklist Seed Plants,* 3(1): 11 (1999): Deyeuxia mckiei.
menhoferi *R. Govaerts, World Checklist Seed Plants,* 3(1): 11 (1999), nom. nov.: Deyeuxia hirsuta *Rúgolo & Villav.*
mesanthera *(Vickery) R. Govaerts, World Checklist Seed Plants,* 3(1): 11 (1999): Deyeuxia mesanthera.
microseta *(Vickery) R. Govaerts, World Checklist Seed Plants,* 3(1): 11 (1999): Deyeuxia microseta.
neglecta (Ehrh.) P. Gaertn., B. Mey. & Scherb.
subsp. holmii *(Lange) V.N. Voroschilov in A.K. Skvortsov (ed.), Florist. issl. v razn. Raĭonakh SSSR:* 149 (1985): C. holmii.
neocontracta *R. Govaerts, World Checklist Seed Plants,* 3(1): 11 (1999), nom. nov.: Agrostis contracta *F. Muell.*
ningxiaensis *D.Z. Ma & J.N. Li in Fl. Ningxiaensis,* 2: 521 (1988 publ. 1990)—Inner Mongolia (Ningxia).
nudiflora *(Vickery) R. Govaerts, World Checklist Seed Plants,* 3(1): 11 (1999): Deyeuxia nudiflora.
petelotii *(Hitchc.) R. Govaerts, World Checklist Seed Plants,* 3(1): 11 (1999): Aulacolepis petelotii.
porteri A. Gray
subsp. insperata *(Swallen) C.W. Greene in Amer. J. Bot.,* 71(3): 285 (1984): C. insperata.
pseudopoa *(Jansen) R. Govaerts, World Checklist Seed Plants,* 3(1): 11 (1999): Deyeuxia pseudopoa.
purpurascens R. Br.
subsp. sesquiflora *(Trin.) V.N. Voroschilov in A.K. Skvortsov (ed.), Florist. issl. v razn. Raĭonakh SSSR:* 149 (1985): Trisetum sesquiflorum.
purpurea (Trin.) Trin.
subsp. tenuis *(V.N. Vassil.) N.N. Tzvelev in Novosti Sist. Vyssh. Rast.,* 32: 182 (2000): C. tenuis.
reflexa *(Vickery) R. Govaerts, World Checklist Seed Plants,* 3(1): 11 (1999): Deyeuxia reflexa.
rodwayi *(Vickery) R. Govaerts, World Checklist Seed Plants,* 3(1): 11 (1999): Deyeuxia rodwayi.
stricta (Timm) Koeler
subsp. inexpansa *(A. Gray) C.W. Greene in Amer. J. Bot.,* 71(3): 286 (1984): C. inexpansa.
tatarica *(Hook. f.) D.F. Cui in Fl. Xinjiangensis,* 6: 277 (1996): C. littorea var. tatarica.
teretifolia *S. Laegaard in Novon,* 8(1): 27 (1998)—Ecuador.
treutleri *(Kuntze) U. Shukla, Grasses N.-East. India:* 51 (1996): Milium treutleri.
treutleri *(Kuntze) R. Govaerts, World Checklist Seed Plants,* 3(1): 11 (1999), isonym; *see U. Shukla, Grasses N.-East. India:* 51 (1996): Milium treutleri.

CALAMINTHA (Labiat.).
sect. Macromelissa *(Benth.) R. Morales in An. Jard. Bot. Madrid,* 55(2): 272 (1997): Melissa sect. Macromelissa.
glandulosa (Req.) Benth.
f. albiflora *Č. Šilić, Monografija Satureja, Calamintha, Micromeria, Acinos i Clinopodium fl.* Jugoslavije: 142 (1979)—Yugoslavia.
var. canescens *(J. Presl) Č. Šilić, Monografija Satureja, Calamintha, Micromeria, Acinos i Clinopodium fl.* Jugoslavije: 142 (1979): C. canescens.
nepeta (L.) Savi
subsp. sylvatica *(Bromf.) R. Morales in An. Jard. Bot. Madrid,* 55(2): 271 (1997): C. sylvatica.

CALAMINTHA:-
sylvatica Bromf.
f. boveana *(K. Malý) Č. Šilić, Monografija Satureja, Calamintha, Micromeria, Acinos i Clinopodium fl.* Jugoslavije: 131 (1979): Satureja sylvatica var. boveana.
f. villosissima *(Benth.) Č. Šilić, Monografija Satureja, Calamintha, Micromeria, Acinos i Clinopodium fl.* Jugoslavije: 131 (1979): C. officinalis var. villosissima.
vardarensis *Č. Šilić, Monografija Satureja, Calamintha, Micromeria, Acinos i Clinopodium fl.* Jugoslavije: 154 (1979)—Yugoslavia.

× **CALAMMOPHILA** (Gramin.).
don-hensonii *A.A. Reznicek & E.J. Judziewicz in Mich. Bot.,* 35(2): 36 (1996 publ. 1997); *see also Mich. Bot.,* 35(3–4): 119 (1996 publ. 1998)—Michigan. (Ammophila breviligulata × Calamagrostis canadensis).

CALAMUS (Palmae).
bimaniferus *T. Evans, K. Sengdala, O.V. Viengkham, B. Thammavong & J. Dransfield in Kew Bull.,* 55(4): 936 (2000)—Laos.
bousigonii Becc.
subsp. smitinandii *J. Dransfield in Kew Bull.,* 55(3): 713 (2000)—Thailand.
laoensis *T. Evans, K. Sengdala, O.V. Viengkham, B. Thammavong & J. Dransfield in Kew Bull.,* 55(4): 929 (2000)—Laos.
loeiensis *D.R. Hodel in Palm J.,* 139: 54 (1998)—Thailand.
maiadum *J. Dransfield, Rattans Brunei Darussalam:* 193 (1997 publ. 1998)—Borneo (Brunei).
mogeae *J. Dransfield in Kew Bull.,* 55(3): 717 (2000)—Sumatera.
neelagiricus *C. Renuka in Rheedea,* 7(2): 69 (1997)—India (Kerala).
occidentalis *J.R. Witono & J. Dransfield in Kew Bull.,* 53(3): 747 (1998)—Jawa.
pauciflorus *T. Evans, K. Sengdala, O.V. Viengkham, B. Thammavong & J. Dransfield in Kew Bull.,* 55(4): 935 (2000), homonym; *see Ridl., Fl. Malay Penins.,* 5: 57 (1925)—Laos, Thailand.
solitarius *T. Evans, K. Sengdala, O.V. Viengkham, B. Thammavong & J. Dransfield in Kew Bull.,* 55(4): 932 (2000)—Laos, Thailand.
temburongii *J. Dransfield, Rattans Brunei Darussalam:* 194 (1997 publ. 1998)—Borneo (Brunei).
yunnanensis *R. Govaerts, World Checklist Seed Plants,* 3(1): 11 (1999)—China South-central (Yunnan).

CALANDRINIA (Portulacac.).
arenicola *S.T. Syeda in Proc. Linn. Soc. New S. Wales,* 116: 153 (1996)—Queensland.
discolor *H.A. Schrader in Index Seminum Horti Academici Gottingensis:* 2 (1831)—Locality not stated.†
tumida *S.T. Syeda in Proc. Linn. Soc. New S. Wales,* 116: 156 (1996)—Queensland.

CALANTHE (Orchidac.).
alpina Hook. f.
var. keshabii *(Lucksom) R.C. Srivastava in Novon,* 8(2): 203 (1998): C. keshabii.
carrii *R. Govaerts, World Checklist Seed Plants,* 3(1): 11 (1999), nom. nov.: C. pusilla *Carr.*
davidii Franch.
var. bungoana *(Ohwi) T. Hashimoto in Proc. 12th World Orchid Conf., 1987 (eds. K. Saito & R. Tanaka):* 124 (1987): C. bungoana.
dulongensis *H. Li in Fl. Dulongjian Reg.:* 340 (1993), without latin descr. or type—China South-central (Yunnan).
fragilis *R. Govaerts, World Checklist Seed Plants,* 3(1): 11 (1999), nom. nov.: C. tenuis *Gilli.*
lyroglossa Rchb.f.
var. longibracteata *P. O'Byrne in Malayan Orchid Rev.,* 31: 29 (1997)—Jawa.
nicolae *P. O'Byrne in Malayan Orchid Rev.,* 30: 69 (1996)—Sumatera.
otuhanica *C.L. Chan & T.J. Barkman in Sandakania,* 9: 29 (1997)—Borneo (Sabah).
rhodochila Schltr.
var. reconditiflora *(J.J. Sm.) S. Thomas in Blumea,* 44(1): 29 (1999): C. reconditiflora.

CALANTHE:-
sylvatica (Thouars) Lindl.
var. geerinckiana *T. Stévart in Syst. Geogr. Pl.*, 70(1): 147 (2000)—Gulf of Guinea Is. (São Tomé).
triplicata (Willemet) Ames
var. gracillima *(Lindl.) N. Hallé in Fragm. Flor. Geobot.*, 43(1): 41 (1998): C. gracillima.
yuksomnensis *S.Z. Lucksom in J. Bombay Nat. Hist. Soc.*, 95(2): 319 (1998)—Sikkim.

CALATHEA (Marantac.).
annae *H. Kennedy & J. Marcelo in Phytologia*, 82(2): 96 (1997)—Brazil.
caquetensis *S. Suárez & G. Galeano in Caldasia*, 22(1): 9 (2000)—Colombia.
compacta *S. Suárez & G. Galeano in Caldasia*, 22(1): 12 (2000)—Colombia.
fatimae *H. Kennedy & J. Marcelo in Phytologia*, 82(2): 94 (1997)—Brazil Southeast (Rio de Janeiro).
gloriana *H. Kennedy in Selbyana*, 18(1): 35 (1997)—Costa Rica.
grazielae *H. Kennedy & J. Marcelo in Phytologia*, 82(2): 101 (1997)—Brazil.
hammelii *H. Kennedy in Selbyana*, 18(1): 33 (1997)—Costa Rica.
hylaeanthoides *H. Kennedy in Canad. J. Bot.*, 75(8): 1356 (1997)—Costa Rica.
incompta *H. Kennedy in Canad. J. Bot.*, 75(8): 1361 (1997)—Costa Rica.
misantlensis *M. Lascurain in Novon*, 6(4): 385 (1996)—Mexico Gulf (Veracruz).
regalis *(Rollison ex Lem.) H. Kennedy in Acta Hort. (Netherlands)*, 413: 172 (1995): Maranta regalis.
retroflexa *H. Kennedy in Canad. J. Bot.*, 75(8): 1357 (1997)—Costa Rica.
singularis *Helen Kennedy in Novon*, 9(1): 61 (1999)—Brazil Southeast (Espírito Santo).

CALATHIANA (Gentianac.).
bavarica
subsp. subacaulis *(Gaudin) J. Holub in Preslia*, 70(2): 100 (1998): Gentiana bavarica β subacaulis.
fenetii *(Litard. & Maire) J. Holub in Preslia*, 70(2): 101 (1998): Gentiana verna subsp. fenetii.
magellensis *(Ronniger) J. Holub in Preslia*, 70(2): 101 (1998): Gentiana verna f. magellensis.
sierrae *(Briq.) J. Holub in Preslia*, 70(2): 101 (1998): Gentiana sierrae.
verna
subsp. delphinensis *(Beauverd) J. Holub in Preslia*, 70(2): 101 (1998): Gentiana delphinensis.

CALCARATOLOBELIA *R.L. Wilbur in Sida*, 17(3): 561 (1997). *CAMPANULACEAE*
aurita *(Brandegee) R.L. Wilbur in Sida*, 17(3): 563 (1997): Heterotoma aurita.
cordifolia *(Hook. & Arn.) R.L. Wilbur in Sida*, 17(3): 562 (1997): Lobelia cordifolia.
flexuosa (C. Presl) Wilbur
var. intermedia *(Hemsl.) R.L. Wilbur in Sida*, 17(3): 563 (1997): Heterotoma intermedia.
gibbosa *(S. Watson) R.L. Wilbur in Sida*, 17(3): 563 (1997): Heterotoma gibbosa.
goldmanii *(Fernald) R.L. Wilbur in Sida*, 17(3): 563 (1997): Heterotoma goldmanii.
knoblochii *(T.J. Ayers) R.L. Wilbur in Sida*, 17(3): 562 (1997): Lobelia knoblochii.
macrocentron *(Benth.) R.L. Wilbur in Sida*, 17(3): 562 (1997): Heterotoma macrocentron.
margarita *(E. Wimm.) R.L. Wilbur in Sida*, 17(3): 564 (1997): Lobelia margarita.
mcvaughii *(T.J. Ayers) R.L. Wilbur in Sida*, 17(3): 562 (1997): Lobelia mcvaughii.
pringlei *(B.L. Rob.) R.L. Wilbur in Sida*, 17(3): 564 (1997): Heterotoma pringlei.
tenella *(Turcz.) R.L. Wilbur in Sida*, 17(3): 563 (1997): Heterotoma tenella.
villaregalis *(T.J. Ayers) R.L. Wilbur in Sida*, 17(3): 562 (1997): Lobelia villaregalis.

CALCEOLARIA (Scrophulariac.).
sect. Tenella *C. Ehrhart in Biblioth. Bot.*, 153: 70 (2000).
aiseniana *C. Ehrhart in Sendtnera*, 4: 63 (1997)—Chile South (Aisén). (Scrophulariac.).

CALCEOLARIA:-
alicahuensis *M. Muñoz Schick in Gayana, Bot.*, 52(2): 61 (1995)—Chile North (Valparaíso).
arachnoidea Graham
subsp. nubigena *(Poepp.) C. Ehrhart in Biblioth. Bot.*, 153: 152 (2000): Calceolaria nubigena.
ascendens Lindl.
subsp. glandulifera *(Witasek) C. Ehrhart in Biblioth. Bot.*, 153: 118 (2000): Calceolaria glandulifera.
subsp. pristiphylla *(Phil.) C. Ehrhart in Biblioth. Bot.*, 153: 119 (2000): Calceolaria pristiphylla.
borsinii *R.A. Rossow in Fl. Patagonica (Collecc. Cient. 8)*, 6: 313 (1999)—Argentina South (Neuquén).
caerulescens *Jacques in Ann. Fl. Pomone*, 1835–36: 155 (1836)—Locality not stated.
collina Phil.
subsp. conferta *(Witasek) C. Ehrhart in Biblioth. Bot.*, 153: 113 (2000): Calceolaria conferta.
subsp. subincisa *(Benth.) C. Ehrhart in Biblioth. Bot.*, 153: 113 (2000): Calceolaria dentata var. subincisa.
corymbosa Ruiz & Pav.
subsp. floccosa *(Witasek) C. Ehrhart in Biblioth. Bot.*, 153: 130 (2000): Calceolaria floccosa.
subsp. mimuloides *(Clos) C. Ehrhart in Biblioth. Bot.*, 153: 130 (2000): Calceolaria mimuloides.
subsp. montana *(Cav.) C. Ehrhart in Biblioth. Bot.*, 153: 130 (2000): Calceolaria montana.
subsp. santiagina *C. Ehrhart in Biblioth. Bot.*, 153: 129 (2000)—Chile Central (Santiago, Valparaíso).
subsp. tetraphylla *(Phil.) C. Ehrhart in Biblioth. Bot.*, 153: 130 (2000): Calceolaria tetraphylla.
densifolia Phil.
subsp. laxa *C. Ehrhart in Biblioth. Bot.*, 153: 222 (2000)—Chile Central (Santiago, Valparaíso).
dentata Ruiz & Pav.
subsp. araucana *(Phil.) C. Ehrhart in Biblioth. Bot.*, 153: 210 (2000): Calceolaria araucana.
subsp. cummingiana *(Witasek) C. Ehrhart in Biblioth. Bot.*, 153: 209 (2000): Calceolaria cummingiana.
filicaulis Clos
subsp. luxurians *(Witasek) C. Ehrhart in Biblioth. Bot.*, 153: 167 (2000): Calceolaria luxurians.
flavovirens *C. Ehrhart in Biblioth. Bot.*, 153: 225 (2000)—Chile Central (Coquimbo), Chile North (Atacama).
glandulosa Poepp. ex Benth.
subsp. alicahuensis *(Muñoz-Schick) C. Ehrhart in Biblioth. Bot.*, 153: 249 (2000): Calceolaria alicahuensis.
meyeniana Phil.
subsp. cheiranthoides *(Witasek) C. Ehrhart in Biblioth. Bot.*, 153: 267 (2000): Calceolaria cheiranthoides.
subsp. glabrata *(Phil.) C. Ehrhart in Biblioth. Bot.*, 153: 267 (2000): Calceolaria glabrata.
subsp. nahuelbutae *(Phil.) C. Ehrhart in Biblioth. Bot.*, 153: 267 (2000): Calceolaria nahuelbutae.
pinnata *L. in Kongl. Vetensk. Acad. Handl.*, 31: 289 (1770)—Peru.†
rinconada *C. Ehrhart in Sendtnera*, 4: 62 (1997)—Chile North (Atacama, Antofagasta).
tripartita Ruiz & Pav.
subsp. yuparum *(Steyerm.) J.A. Steyermark in Pittieria*, 7: 29 (1978): C. yuparum.
valdiviana Phil.
subsp. crassifolia *(Phil.) C. Ehrhart in Biblioth. Bot.*, 153: 202 (2000): Calceolaria crassifolia.

CALCHAS *P.V. Heath in Calyx*, 5(4): 160 (1997), nom. nov.: Coleus sect. Solenostemoides *Briq.* (Labiat.).
acuminatus *(Benth.) P.V. Heath in Calyx*, 5(4): 160 (1997): Coleus acuminatus.
atropurpureus *(Benth.) P.V. Heath in Calyx*, 5(4): 160 (1997): Coleus atropurpureus.
autranii *(Briq.) P.V. Heath in Calyx*, 5(4): 160 (1997): Coleus autranii.
bernieri *(Briq.) P.V. Heath in Calyx*, 5(4): 160 (1997): Coleus bernieri.
betonicoides *(Hiern) P.V. Heath in Calyx*, 5(4): 160 (1997): Coleus betonicoides.
bojeri *(Benth.) P.V. Heath in Calyx*, 5(4): 160 (1997): Coleus bojeri.

CALCHAS:-
briquetii *(Baker) P.V. Heath in Calyx*, 5(4): 160 (1997): Coleus briquetii.
caeruleus *(Gürke) P.V. Heath in Calyx*, 5(4): 160 (1997): Coleus caeruleus.
chevalieri *(Briq.) P.V. Heath in Calyx*, 5(4): 160 (1997): Coleus chevalieri.
claessensii *(De Wild.) P.V. Heath in Calyx*, 5(4): 160 (1997): Coleus claessensii.
collinus *(Lebrun & Touss.) P.V. Heath in Calyx*, 5(4): 160 (1997): Coleus collinus.
copiosiflorus *(Briq.) P.V. Heath in Calyx*, 5(4): 160 (1997): Coleus copiosiflorus.
crispipilus *(Merr.) P.V. Heath in Calyx*, 5(4): 160 (1997): Coleus crispipilus.
decumbens *(Hook.) P.V. Heath in Calyx*, 5(4): 160 (1997): Plectranthus decumbens.
dewevrei *(Briq.) P.V. Heath in Calyx*, 6(2): 51 (1999): Coleus dewevrei.
latidens *(S. Moore) P.V. Heath in Calyx*, 6(2): 51 (1999): Coleus latidens
parvifolius *(Benth.) P.V. Heath in Calyx*, 6(2): 51 (1999): Coleus parvifolius.
platystomoides *(Robyns & Lebrun) P.V. Heath in Calyx*, 6(2): 51 (1999): Coleus platystomoides.
schizophyllus *(Baker) P.V. Heath in Calyx*, 6(2): 51 (1999): Plectranthus schizophyllus.
scutellarioides *(L.) P.V. Heath in Calyx*, 6(2): 51 (1999): Ocimum scutellarioides.
var. angustifolia *(Benth.) P.V. Heath in Calyx*, 6(2): 51 (1999): Coleus scutellarioides var. angustifolia.
var. crispipilus *(Merr.) P.V. Heath in Calyx*, 6(2): 51 (1999): Coleus macranthus var. crispipilus.
var. limnophila *(Benth.) P.V. Heath in Calyx*, 6(2): 51 (1999): Coleus scutellarioides var. limnophila.
seretii *(De Wild.) P.V. Heath in Calyx*, 6(2): 51 (1999): Coleus seretii.
shirensis *(Gürke) P.V. Heath in Calyx*, 6(2): 51 (1999): Coleus shirensis.

CALEA (Compos.).
sect. Haplocalea *(Less.) J.F. Pruski in Kew Bull.*, 53(3): 683 (1998): C. subgen. Haplocalea.
sect. Monanthocalea *(Less.) J.F. Pruski in Kew Bull.*, 53(3): 684 (1998): C. subgen. Monanthocalea.
harlingii *H. Robinson in Phytologia*, 82(1): 59 (1997)—Ecuador.
hatschbachii *J.F. Pruski & D.J.N. Hind in Kew Bull.*, 53(3): 695 (1998)—Brazil Southeast (Minas Gerais).
lucidivenia Gleason & S.F. Blake
var. cardonae *(Maguire & Wurdack) J.F. Pruski in Fl. Venezuelan Guayana (gen. eds. J.A. Steyermark et al.)*, 3: 227 (1997): C. cardonae.
var. orientalis *(Maguire & Wurdack) J.F. Pruski in Fl. Venezuelan Guayana (gen. eds. J.A. Steyermark et al.)*, 3: 227 (1997): C. cardonae var. orientalis.
neblinensis *(Maguire & Wurdack) J.F. Pruski in Fl. Venezuelan Guayana (gen. eds. J.A. Steyermark et al.)*, 3: 228 (1997): C. cardonae subsp. neblinensis.
nematophylla *J.F. Pruski in Kew Bull.*, 53(3): 684 (1998)—Bolivia.
robinsoniana *J.F. Pruski in Kew Bull.*, 53(3): 688 (1998)—Brazil West-central (Goiás).
semirii *J.F. Pruski & D.J.N. Hind in Kew Bull.*, 53(3): 698 (1998)—Brazil Southeast (Minas Gerais).
solidaginea Humb., Bonpl. & Kunth
subsp. deltophylla *(R.S. Cowan) J.F. Pruski in Fl. Venezuelan Guayana (gen. eds. J.A. Steyermark et al.)*, 3: 229 (1997): C. deltophylla.
verticillata *(Klatt) J.F. Pruski in Kew Bull.*, 53(3): 689 (1998): Tridax verticillata.

CALENDULA (Compos.).
suffruticosa
subsp. maderensis *(DC.) R. Govaerts, World Checklist Seed Plants*, 3(1): 11 (1999): C. maderensis.

CALIBRACHOA (Solanac.).
dusenii *(R.E. Fr.) J.R. Stehmann & J. Semir in Novon*, 7(4): 419 (1997 publ. 1998): Petunia dusenii.

CALIBRACHOA:-

eglandulata *J.R. Stehmann & J. Semir in Novon,* 7(4): 417 (1997 publ. 1998)—Brazil South (Santa Catarina).

elegans *(Miers) J.R. Stehmann & J. Semir in Novon,* 7(4): 419 (1997 publ. 1998): Petunia elegans.

humilis *(R.E. Fr.) J.R. Stehmann & J. Semir in Novon,* 7(4): 419 (1997 publ. 1998): Petunia humilis.

micrantha *(R.E. Fr.) J.R. Stehmann & J. Semir in Novon,* 7(4): 419 (1997 publ. 1998): Petunia micrantha.

ovalifolia *(Miers) J.R. Stehmann & J. Semir in Novon,* 7(4): 419 (1997 publ. 1998): Petunia ovalifolia.

sendtneriana *(R.E. Fr.) J.R. Stehmann & J. Semir in Novon,* 7(4): 419 (1997 publ. 1998): Petunia sendtneriana.

serrulata *(L.B. Sm. & Downs) J.R. Stehmann & J. Semir in Novon,* 7(4): 419 (1997 publ. 1998): Petunia serrulata.

spathulata *(L.B. Sm. & Downs) J.R. Stehmann & J. Semir in Novon,* 7(4): 419 (1997 publ. 1998): Petunia spathulata.

thymifolia *(A. St.-Hil.) J.R. Stehmann & J. Semir in Novon,* 7(4): 419 (1997 publ. 1998): Fabiana thymifolia.

CALLERYA (Legumin.).

atropurpurea (Wall.) Schot

var. pubescens *(Craib) P.K. Lôc in Bot. Zhurn.,* 81(10): 98 (1996): Padbruggea pubescens.

bonatiana *(Pamp.) P.K. Lôc in Bot. Zhurn.,* 81(10): 99 (1996): Millettia bonatiana.

championii *(Benth.) P.K. Lôc in Bot. Zhurn.,* 81(10): 97 (1996), without basionym ref.: Millettia championii.

dielsiana *(Harms) P.K. Lôc in Bot. Zhurn.,* 81(10): 99 (1996), without basionym page: Millettia dielsiana.

CALLIANDRA (Legumin.).

sect. Acistegia *R.C. Barneby, Silk Tree, Guanacaste, Monkey's Earring (Mem. New York Bot. Gard., 74(3)):* 139 (1998).

sect. Acroscias *R.C. Barneby, Silk Tree, Guanacaste, Monkey's Earring (Mem. New York Bot. Gard., 74(3)):* 146 (1998).

ser. Ambivalentes *R.C. Barneby, Silk Tree, Guanacaste, Monkey's Earring (Mem. New York Bot. Gard., 74(3)):* 103 (1998).

ser. Androcallis *R.C. Barneby, Silk Tree, Guanacaste, Monkey's Earring (Mem. New York Bot. Gard., 74(3)):* 21 (1998).

sect. Androcallis *R.C. Barneby, Silk Tree, Guanacaste, Monkey's Earring (Mem. New York Bot. Gard., 74(3)):* 21 (1998).

ser. Biflorae *R.C. Barneby, Silk Tree, Guanacaste, Monkey's Earring (Mem. New York Bot. Gard., 74(3)):* 98 (1998).

ser. Chilenses *R.C. Barneby, Silk Tree, Guanacaste, Monkey's Earring (Mem. New York Bot. Gard., 74(3)):* 100 (1998).

ser. Comosae *R.C. Barneby, Silk Tree, Guanacaste, Monkey's Earring (Mem. New York Bot. Gard., 74(3)):* 196 (1998).

ser. Hymenaeodeae *R.C. Barneby, Silk Tree, Guanacaste, Monkey's Earring (Mem. New York Bot. Gard., 74(3)):* 134 (1998).

ser. Leptopodae *R.C. Barneby, Silk Tree, Guanacaste, Monkey's Earring (Mem. New York Bot. Gard., 74(3)):* 199 (1998).

ser. Longipedes *R.C. Barneby, Silk Tree, Guanacaste, Monkey's Earring (Mem. New York Bot. Gard., 74(3)):* 137 (1998).

ser. Microcallis *R.C. Barneby, Silk Tree, Guanacaste, Monkey's Earring (Mem. New York Bot. Gard., 74(3)):* 198 (1998).

sect. Microcallis *R.C. Barneby, Silk Tree, Guanacaste, Monkey's Earring (Mem. New York Bot. Gard., 74(3)):* 197 (1998).

ser. Pauciflorae *R.C. Barneby, Silk Tree, Guanacaste, Monkey's Earring (Mem. New York Bot. Gard., 74(3)):* 101 (1998).

ser. Tsugoideae *R.C. Barneby, Silk Tree, Guanacaste, Monkey's Earring (Mem. New York Bot. Gard., 74(3)):* 190 (1998).

CALLIANDRA:-

ser. Virgatae *R.C. Barneby, Silk Tree, Guanacaste, Monkey's Earring (Mem. New York Bot. Gard., 74(3)):* 189 (1998).

antioquiae *R.C. Barneby, Silk Tree, Guanacaste, Monkey's Earring (Mem. New York Bot. Gard., 74(3)):* 123 (1998)—Colombia.

bahiana Renvoize

var. erythematosa *R.C. Barneby, Silk Tree, Guanacaste, Monkey's Earring (Mem. New York Bot. Gard., 74(3)):* 151 (1998)—Brazil Northeast (Bahia).

carrascana *R.C. Barneby, Silk Tree, Guanacaste, Monkey's Earring (Mem. New York Bot. Gard., 74(3)):* 55 (1998)—Brazil Southeast (Minas Gerais).

chulumania *R.C. Barneby, Silk Tree, Guanacaste, Monkey's Earring (Mem. New York Bot. Gard., 74(3)):* 85 (1998)—Bolivia.

coccinea Renvoize

var. trimera *R.C. Barneby, Silk Tree, Guanacaste, Monkey's Earring (Mem. New York Bot. Gard., 74(3)):* 168 (1998)—Brazil Northeast (Bahia).

colimae *R.C. Barneby, Silk Tree, Guanacaste, Monkey's Earring (Mem. New York Bot. Gard., 74(3)):* 88 (1998)—Mexico Southwest (Colima).

concinna *R.C. Barneby, Silk Tree, Guanacaste, Monkey's Earring (Mem. New York Bot. Gard., 74(3)):* 71 (1998)—Brazil Southeast (Minas Gerais).

duckei *R.C. Barneby, Silk Tree, Guanacaste, Monkey's Earring (Mem. New York Bot. Gard., 74(3)):* 95 (1998)—Brazil Northeast (Pernambuco).

dysantha Benth.

var. macrocephala *(Benth.) R.C. Barneby, Silk Tree, Guanacaste, Monkey's Earring (Mem. New York Bot. Gard., 74(3)):* 63 (1998): C. macrocephala.

var. opulenta *R.C. Barneby, Silk Tree, Guanacaste, Monkey's Earring (Mem. New York Bot. Gard., 74(3)):* 63 (1998)—Paraguay.

var. turbinata *(Benth.) R.C. Barneby, Silk Tree, Guanacaste, Monkey's Earring (Mem. New York Bot. Gard., 74(3)):* 64 (1998): C. turbinata.

fasciculata Benth.

var. bracteosa *(Benth.) R.C. Barneby, Silk Tree, Guanacaste, Monkey's Earring (Mem. New York Bot. Gard., 74(3)):* 158 (1998): C. bracteosa.

fernandesii *R.C. Barneby, Silk Tree, Guanacaste, Monkey's Earring (Mem. New York Bot. Gard., 74(3)):* 67 (1998)—Brazil Northeast (Piauí).

ganevii *R.C. Barneby, Silk Tree, Guanacaste, Monkey's Earring (Mem. New York Bot. Gard., 74(3)):* 173 (1998)—Brazil Northeast (Bahia).

germana *R.C. Barneby, Silk Tree, Guanacaste, Monkey's Earring (Mem. New York Bot. Gard., 74(3)):* 175 (1998)—Brazil Northeast (Bahia).

glomerulata H. Karst.

var. parvifolia *(Benth.) R.C. Barneby, Silk Tree, Guanacaste, Monkey's Earring (Mem. New York Bot. Gard., 74(3)):* 26 (1998): C. laxa var. parvifolia.

goldmanii *Rose ex R.C. Barneby, Silk Tree, Guanacaste, Monkey's Earring (Mem. New York Bot. Gard., 74(3)):* 35 (1998)—Mexico Southeast (Chiapas).

haematocephala Hassk.

var. boliviana *(Britton) R.C. Barneby, Silk Tree, Guanacaste, Monkey's Earring (Mem. New York Bot. Gard., 74(3)):* 110 (1998): C. boliviana.

haematomma (DC.) Benth.

var. colletioides *(Griseb.) R.C. Barneby, Silk Tree, Guanacaste, Monkey's Earring (Mem. New York Bot. Gard., 74(3)):* 141 (1998): C. colletioides.

var. correllii *R.C. Barneby, Silk Tree, Guanacaste, Monkey's Earring (Mem. New York Bot. Gard., 74(3)):* 142 (1998)—Bahamas.

var. locoensis *(R.G. García & Kolt.) R.C. Barneby, Silk Tree, Guanacaste, Monkey's Earring (Mem. New York Bot. Gard., 74(3)):* 144 (1998): C. locoensis.

CALLIANDRA:-

haematomma (DC.) Benth.

var. rivularis *(Urb. & Ekman) R.C. Barneby, Silk Tree, Guanacaste, Monkey's Earring (Mem. New York Bot. Gard., 74(3)):* 142 (1998): C. rivularis.

var. tortuensis *(Alain) R.C. Barneby, Silk Tree, Guanacaste, Monkey's Earring (Mem. New York Bot. Gard., 74(3)):* 143 (1998): C. tortuensis.

hintonii *R.C. Barneby, Silk Tree, Guanacaste, Monkey's Earring (Mem. New York Bot. Gard., 74(3)):* 88 (1998)—Mexico Central (México State).

hirtiflora Benth.

var. ripicola *R.C. Barneby, Silk Tree, Guanacaste, Monkey's Earring (Mem. New York Bot. Gard., 74(3)):* 165 (1998)—Brazil Northeast (Bahia).

houstoniana (Mill.) Standl.

var. acapulcensis *(Britton & Rose) R.C. Barneby, Silk Tree, Guanacaste, Monkey's Earring (Mem. New York Bot. Gard., 74(3)):* 180 (1998): Anneslia acapulcensis.

subsp. alamosensis *(Britton & Rose) D.J. Macqueen & H.M. Hernández in Kew Bull.,* 52(1): 34 (1997): Anneslia alamosensis.

var. anomala *(Kunth) R.C. Barneby, Silk Tree, Guanacaste, Monkey's Earring (Mem. New York Bot. Gard., 74(3)):* 179 (1998): Inga anomala.

var. calothyrsus *(Meisn.) R.C. Barneby, Silk Tree, Guanacaste, Monkey's Earring (Mem. New York Bot. Gard., 74(3)):* 180 (1998): C. calothyrsus

var. colomasensis *(Britton & Rose) R.C. Barneby, Silk Tree, Guanacaste, Monkey's Earring (Mem. New York Bot. Gard., 74(3)):* 180 (1998): Anneslia colomasensis.

subsp. stylesii *D.J. Macqueen & H.M. Hernández in Kew Bull.,* 52(1): 36 (1997)—Mexico Southeast (Tabasco, Campeche, Chiapas), Mexico Northeast (Hidalgo, Querétaro, San Luis Potosí, Tamaulipas), Mexico Southwest (Oaxaca), Mexico Central (Puebla), Guatemala.

humilis Benth.

var. gentryana *R.C. Barneby, Silk Tree, Guanacaste, Monkey's Earring (Mem. New York Bot. Gard., 74(3)):* 47 (1998)—Mexico Northwest (Sinaloa).

iligna *R.C. Barneby, Silk Tree, Guanacaste, Monkey's Earring (Mem. New York Bot. Gard., 74(3)):* 171 (1998)—Brazil Southeast (Minas Gerais).

imperialis *R.C. Barneby, Silk Tree, Guanacaste, Monkey's Earring (Mem. New York Bot. Gard., 74(3)):* 70 (1998)—Brazil Northeast (Piauí).

inermis

subsp. alamosensis *(Britton & Rose) R. Govaerts, World Checklist Seed Plants,* 3(1): 11 (1999): Anneslia alamosensis.

subsp. stylesii *(Macqueen & H.M. Hern.) R. Govaerts, World Checklist Seed Plants,* 3(1): 12 (1999): C. houstoniana subsp. stylesii.

iselyi *B.L. Turner in Lundellia,* 3: 17 (2000)—Texas.

jariensis *R.C. Barneby, Silk Tree, Guanacaste, Monkey's Earring (Mem. New York Bot. Gard., 74(3)):* 120 (1998)—Brazil North (Pará).

laxa (Willd.) Benth.

var. stipulacea *(Benth.) R.C. Barneby, Silk Tree, Guanacaste, Monkey's Earring (Mem. New York Bot. Gard., 74(3)):* 32 (1998): C. stipulacea.

var. urimana *R.C. Barneby, Silk Tree, Guanacaste, Monkey's Earring (Mem. New York Bot. Gard., 74(3)):* 33 (1998)—Venezuela.

lintea *R.C. Barneby, Silk Tree, Guanacaste, Monkey's Earring (Mem. New York Bot. Gard., 74(3)):* 149 (1998)—Brazil Northeast (Bahia).

longipedicellata *(McVaugh) D.J. Macqueen & H.M. Hernández in Kew Bull.,* 52(1): 40 (1997): C. grandiflora var. longipedicellata.

macqueenii *R.C. Barneby, Silk Tree, Guanacaste, Monkey's Earring (Mem. New York Bot. Gard., 74(3)):* 131 (1998)—Mexico Southwest (Guerrero).

macrocalyx Harms

var. aucta *R.C. Barneby, Silk Tree, Guanacaste, Monkey's Earring (Mem. New York Bot. Gard., 74(3)):* 66 (1998)—Brazil Northeast (Bahia).

CALLIANDRA:-
magdalenae (DC.) Benth.
var. colombiana *(Britton & Killip) R.C. Barneby, Silk Tree, Guanacaste, Monkey's Earring (Mem. New York Bot. Gard., 74(3)):* 83 (1998): C. colombiana.
nebulosa *R.C. Barneby, Silk Tree, Guanacaste, Monkey's Earring (Mem. New York Bot. Gard., 74(3)):* 150 (1998)—Brazil Northeast (Bahia).
paterna *R.C. Barneby, Silk Tree, Guanacaste, Monkey's Earring (Mem. New York Bot. Gard., 74(3)):* 172 (1998)—Brazil Northeast (Bahia).
pauciflora *(A. Rich.) Griseb., Pl. Wright.:* 180 (1860): Acacia pauciflora.†
pittieri Standl.
var. polyphylla *(Harms) R.C. Barneby, Silk Tree, Guanacaste, Monkey's Earring (Mem. New York Bot. Gard., 74(3)):* 23 (1998): C. polyphylla.
pityophila *R.C. Barneby, Silk Tree, Guanacaste, Monkey's Earring (Mem. New York Bot. Gard., 74(3)):* 87 (1998)—Mexico Southwest (Guerrero).
renvoizeana *R.C. Barneby, Silk Tree, Guanacaste, Monkey's Earring (Mem. New York Bot. Gard., 74(3)):* 169 (1998), nom. nov.: C. gracilis *Renvoize.*
samik *R.C. Barneby, Silk Tree, Guanacaste, Monkey's Earring (Mem. New York Bot. Gard., 74(3)):* 77 (1998)—Peru.
santosiana *Glaz. ex R.C. Barneby, Silk Tree, Guanacaste, Monkey's Earring (Mem. New York Bot. Gard., 74(3)):* 169 (1998)—Brazil Southeast (Minas Gerais).
semisepulta *R.C. Barneby, Silk Tree, Guanacaste, Monkey's Earring (Mem. New York Bot. Gard., 74(3)):* 174 (1998)—Brazil Northeast (Bahia).
staminea *(Thunb.) R.C. Barneby, Silk Tree, Guanacaste, Monkey's Earring (Mem. New York Bot. Gard., 74(3)):* 92 (1998): Mimosa staminea.
stelligera *R.C. Barneby, Silk Tree, Guanacaste, Monkey's Earring (Mem. New York Bot. Gard., 74(3)):* 166 (1998)—Brazil Northeast (Bahia).
tergemina (L.) Benth.
var. emarginata *(Willd.) R.C. Barneby, Silk Tree, Guanacaste, Monkey's Earring (Mem. New York Bot. Gard., 74(3)):* 129 (1998): Inga emarginata.
var. arborea *(Standl.) R.C. Barneby, Silk Tree, Guanacaste, Monkey's Earring (Mem. New York Bot. Gard., 74(3)):* 117 (1998): C. arborea.
var. carbonaria *(Benth.) R.C. Barneby, Silk Tree, Guanacaste, Monkey's Earring (Mem. New York Bot. Gard., 74(3)):* 115 (1998): C. carbonaria.
var. paniculans *R.C. Barneby, Silk Tree, Guanacaste, Monkey's Earring (Mem. New York Bot. Gard., 74(3)):* 117 (1998)—Peru.
var. peruicola *R.C. Barneby, Silk Tree, Guanacaste, Monkey's Earring (Mem. New York Bot. Gard., 74(3)):* 115 (1998)—Peru.
var. pilosifolia *(R.S. Cowan) R.C. Barneby, Silk Tree, Guanacaste, Monkey's Earring (Mem. New York Bot. Gard., 74(3)):* 116 (1998): C. pilosifolia.
var. stenocylix *R.C. Barneby, Silk Tree, Guanacaste, Monkey's Earring (Mem. New York Bot. Gard., 74(3)):* 117 (1998)—Brazil North (Amazonas).
vaupesiana R.S. Cowan
var. oligandra *R.C. Barneby, Silk Tree, Guanacaste, Monkey's Earring (Mem. New York Bot. Gard., 74(3)):* 193 (1998)—Venezuela.

CALLICARPA (Verbenac.).
badipilosa *S. Atkins in Kew Bull.,* 52(1): 228 (1997)—Borneo (Brunei).
cinnamomea *(Hallier f.) R. Govaerts, World Checklist Seed Plants,* 3(1): 12 (1999): Geunsia cinnamomea.
glabrifolia *S. Atkins in Kew Bull.,* 52(1): 227 (1997)—Borneo (Brunei, Sarawak).
grandiflora *(Hallier f.) R. Govaerts, World Checklist Seed Plants,* 3(1): 12 (1999): Geunsia grandiflora.
homoeophylla *(Hallier f.) R. Govaerts, World Checklist Seed Plants,* 3(1): 12 (1999): Geunsia homoeophylla.

CALLICARPA:-
pullei *(H.J. Lam) R. Govaerts, World Checklist Seed Plants,* 3(1): 12 (1999): Geunsia pullei.
quaternifolia *(Hallier f.) R. Govaerts, World Checklist Seed Plants,* 3(1): 12 (1999): Geunsia quaternifolia.
ramosii *(Moldenke) R. Govaerts, World Checklist Seed Plants,* 3(1): 12 (1999): Geunsia ramosii.
scandens *(Moldenke) R. Govaerts, World Checklist Seed Plants,* 3(1): 12 (1999): Geunsia scandens.
serrulata *(Hallier f.) R. Govaerts, World Checklist Seed Plants,* 3(1): 12 (1999): Geunsia serrulata.
shikokiana *T. Makino in Bot. Mag. Tokyo,* 18(206): 46 (1904)—Japan.†
sinuata *A.L. Budantzev & V.X. Phuong in Konsp. Sosud. Rast. Fl. V'etnama,* 2: 228 (1996)—Vietnam.
subternata *(Hallier f.) R. Govaerts, World Checklist Seed Plants,* 3(1): 12 (1999): Geunsia subternata.

CALLIGONUM (Polygonac.).
aphyllum (Pall.) Gürke
var. dentatum *N.N. Tzvelev in Fl. Vostochnoĭ Evropy̆,* 9: 157 (1996), contrary to Art. 37.5 ICBN (1994)—Locality not stated.

CALLIPHYSA (Polygonac.).
zakirovii *P. Khalkuziev, Rodstv. Svyaz. Nekotor. Semeĭstv Rast. Pust. Oblast.:* 85 (1990)—Uzbekistan.

CALLISTEMON (Myrtac.).
genofluvialis *W. Molyneux in Muelleria,* 10: 57 (1997)—Victoria.

CALLITRICHE (Callitrichac.).
exalata *(Tzvelev) N.N. Tzvelev, Opred. Sosud. Rast. Severo-Zap. Rossii:* 680 (2000): Callitriche hermaphroditica subsp. exalata.
hermaphroditica L.
subsp. exalata *N.N. Tzvelev in Bot. Zhurn.,* 82(12): 96 (1997)—Russia Northwest.
palustris L.
var. elegans *(Petrov) N.N. Tzvelev in Sosud. Rast. Sovet. Dal'nego Vostoka,* 8: 246 (1996), isonym; *see Y.L. Chang in Fl. Plant. Herb. Chinae bor.-or.,* 6: 53 (1977): C. elegans.

CALLITRIS (Cupressac.).
columellaris
var. microcarpa *(Benth.) R. Govaerts, World Checklist Seed Plants,* 3(1): 12 (1999): Frenela robusta var. microcarpa.
cupressoides *(L.) Jacques in Ann. Fl. Pomone,* 1836–37: 267 (1837): Thuja cupressoides.
gracilis R.T. Baker
subsp. murrayensis *(J. Garden) K.D. Hill in Fl. Australia,* 48: 716 (1998): C. preissii subsp. murrayensis.
oblonga Rich.
subsp. corangensis *K.D. Hill in Fl. Australia,* 48: 716 (1998)—New South Wales.
subsp. parva *K.D. Hill in Fl. Australia,* 48: 717 (1998)—New South Wales.

× **CALOMITRA** *hort. in Orchid Rev.,* 106(1224): 380 (1998). *ORCHIDACEAE* (Calochilus × Thelymitra).

CALOPHYLLUM (Guttif.).
multitudinis *P.F. Stevens in Checklist Flow. Pl. & Gymn. Brunei Darussalam (ed. M.J.E. Coode et al.):* 438 (1996)—Borneo (Brunei).

CALOTROPIS (Asclepiadac.).
gigantea (L.) Dryand. ex W.T. Aiton
var. procera *(Aiton) P.T. Li in J. S. China Agric. Univ.,* 12(3): 39 (1991): Asclepias procera.

CALPURNIA (Legumin.).
reflexa *A.J. Beaumont in Bothalia,* 29(1): 13 (1999), as 'reflexus'—Lesotho.

CALTHA (Ranunculac.).
palustris L.
subsp. minor *(Mill.) V.N. Voroschilov in A.K. Skvortsov (ed.), Florist. issl. v razn. raĭonakh SSSR:* 167 (1985): C. minor.

CALTHA:-
palustris L.
subsp. polypetala *(Turcz.) V.N. Voroschilov in A.K. Skvortsov (ed.), Florist. issl. v razn. raĭonakh SSSR:* 167 (1985): C. palustris var. polypetala.

CALYCOLPUS (Myrtac.).
australis *L.R. Landrum in Novon,* 8(3): 244 (1998)—Brazil Southeast (Minas Gerais).

CALYCOPHYLLUM (Rubiac.).
papillosum *J.H. Kirkbride in Brittonia,* 49(3): 354 (1997)—Brazil Southeast (Espírito Santo).
tefense *J.H. Kirkbride in Brittonia,* 49(3): 355 (1997)—Brazil North (Amazonas).

CALYCORECTES (Myrtac.).
ambivalens *M. Sobral in Daphne,* 3(4): 5 (1993)—Brazil Southeast (Minas Gerais).
belemii *J.R. Mattos in Loefgrenia,* 106: 1 (1995)—Brazil Northeast (Bahia).
dominicanus *J.R. de Mattos in Loefgrenia,* 112: 1 (1998)—Dominican Republic.
ferrugineus *J.R. Mattos in Loefgrenia,* 106: 2 (1995)—Brazil West-central (Goiás).
guyanensis *J.R. de Mattos in Loefgrenia,* 108: 1 (1996)—French Guiana.
heringerianus *J.R. de Mattos in Loefgrenia,* 101: 1 (1992)—Brazil Southeast (Minas Gerais).
minutifolius *J.R. de Mattos in Loefgrenia,* 113: 1 (1999)—Dominican Republic.
orlandoi *J.R. de Mattos in Loefgrenia,* 102: 1 (1993)—Brazil Northeast (Bahia).
paranaensis *J.R. de Mattos in Loefgrenia,* 102: 2 (1993)—Brazil South (Paraná).
pirataquinensis *J.R. de Mattos in Loefgrenia,* 107: 1 (1995)—Brazil Northeast (Bahia).
psidiiflorus
var. legrandii *(Mattos) J.R. de Mattos in Loefgrenia,* 102: 3 (1993): C. legrandii.
var. schultzianus *(Mattos) J.R. de Mattos in Loefgrenia,* 108: 2 (1996): C. schultzianus.
var. triflora *J.R. de Mattos in Loefgrenia,* 102: 3 (1993)—Brazil South (Paraná).
rogersianus *J.R. de Mattos in Loefgrenia,* 112: 2 (1998)—Colombia.
schultzianus *J.R. de Mattos in Loefgrenia,* 101: 2 (1992)—Brazil South (Rio Grande do Sul, Santa Catarina).
teixeireanus *J.R. de Mattos in Loefgrenia,* 109: 1 (1996)—Brazil Southeast (Rio de Janeiro).

CALYCOSIA (Rubiac.).
mamosei *W. Takeuchi in Sida,* 19(1): 14 (2000)—New Guinea (Papua New Guinea).

CALYPTRANTHERA *J. Klackenberg in Novon,* 6(1): 27 (1996). *ASCLEPIADACEAE*
baronii *J. Klackenberg in Adansonia,* 19(1): 32 (1997)—Madagascar.
brevicaudata *J. Klackenberg in Adansonia,* 19(1): 33 (1997)—Madagascar.
caudiclava *(Choux) J. Klackenberg in Novon,* 6(1): 27 (1996): Toxocarpus caudiclavus.
gautieri *J. Klackenberg in Candollea,* 53(2): 396 (1998)—Madagascar.
grandiflora *J. Klackenberg in Adansonia,* 19(1): 30 (1997)—Madagascar.
pubipetala *J. Klackenberg in Adansonia,* 19(1): 36 (1997)—Madagascar.
schatziana *J. Klackenberg in Adansonia,* 22(1): 34 (2000)—Madagascar.

CALYPTRANTHES (Myrtac.).
amarulenta *B.K. Holst in Novon,* 9(4): 517 (1999)—Nicaragua.
dryadica *M.L. Kawasaki in Novon,* 8(4): 386 (1998)—Brazil Southeast (São Paulo).
fusiformis *M.L. Kawasaki in Brittonia,* 48(4): 508 (1996 publ. 1997)—Brazil Southeast (São Paulo).
garciae *Alain & M. Mejía in Moscosoa,* 9: 8 (1997)—Dominican Republic.
proctorii *P. Acevedo-Rodríguez in Brittonia,* 51(2): 168 (1999), nom. nov.: C. uniflora *Proctor.*
ursina *G.M. Barroso & A.L. Peixoto in Acta Bot. Brasil.,* 10(1): 80 (1996)—Brazil Southeast (Rio de Janeiro).

CALYPTROGYNE (Palmae).
herrerae *M.H. Grayum in Phytologia*, 84(4): 309 (1998 publ. 1999)—Costa Rica.

CALYSTEGIA (Convolvulac.).
× scanica *R.K. Brummitt in Kew Bull.*, 35(2): 333 (1980), as 'scania' in error in I.K. Supp. 17—Sweden.(C. pulchra × sepium supsp. sepium).†
sepium (L.) R.Br.
f. alba *Y.N. Lee, Fl. Korea:* 1160 (1996), without type—Locality not stated.
silvatica (Kit.) Griseb.
subsp. disjuncta *R.K. Brummitt in Lagascalia*, 18(2): 339 (1996)—Italy, Morocco, Spain, Baleares, France.
spectabilis *(Brummitt) N.N. Tzvelev, Opred. Sosud. Rast. Severo-Zap. Rossii:* 680 (2000): Calystegia sepium subsp. spectabilis.

CAMBESSEDEA (Anacardiac.).
oppositifolia *(Roxb.) Wight & Arn. ex J.O. Voigt, Hort. Suburb. Calcutt.:* 272 (1845): Mangifera oppositifolia.†

CAMBESSEDESIA (Melastomatac.).
sect. Pyramia *(Cham.) A.B. Martins in Acta Bot. Brasil.*, 9(1): 148 (1995): gen. Pyramia.
cambessedesioides *(Wurdack) A.B Martins in Acta Bot. Brasil.*, 9(1): 148 (1995): Pyramia cambessedesioides.
espora
subsp. ilicifolia *(DC.) A.B. Martins in Acta Bot. Brasil.*, 9(1): 148 (1995): C. espora var. ilicifolia.
pityrophylla *(DC.) A.B. Martins in Acta Bot. Brasil.*, 9(1): 148 (1995): Osbeckia pityrophylla.
salviaefolia *(Cham.) A.B. Martins in Acta Bot. Brasil.*, 9(1): 148 (1995): Pyramia salviaefolia.
striatella *(Naudin) A.B. Martins in Acta Bot. Brasil.*, 9(1): 148 (1995): Pyramia striatella.

CAMELINA (Crucif.).
sect. Flexuosae *V.I. Dorofeev in Bot. Zhurn.*, 81(8): 98 (1996).
ser. Microcarpae *V.I. Dorofeev in Bot. Zhurn.*, 81(8): 98 (1996).
hispida Boiss.
var. stiefelhagenii *(Bornm.) Ş. Yıldırımlı in Ot Sist. Bot. Dergisi*, 5(2): 4 (1998 publ. 2000): C. stiefelhagenii.

CAMELLIA (Theac.).
sect. Pleurocarpus *Hung T. Chang in Acta Sci. Nat. Univ. Sunyatseni*, 35(1): 79 (1996).
sect. Protocamellia *Hung T. Chang in Acta Sci. Nat. Univ. Sunyatseni*, 35(1): 79 (1996).
achrysantha *Hung T. Chang & S Ye Liang in Guangxi Forest. Sci.*, 23(1): 52 (1994)—China Southeast (Guangxi).
assamica (J.W. Mast.) Hung T. Chang
var. kucha *(Hung T. Chang & Wang) Hung T. Chang in Fl. Reipubl. Popul. Sin.*, 49(3): 136 (1998): C. sinensis var. kucha.
var. polyneura *(Chang, Y.J. Tan & P.S. Wang) Hung T. Chang in Fl. Reipubl. Popul. Sin.*, 49(3): 135 (1998): C. polyneura.
brevistyla (Hayata) Cohen-Stuart
var. microphylla *(Merr.) T.L. Ming in Acta Bot. Yunnanica*, 21(2): 158 (1999): Thea microphylla.
chunii Hung T. Chang
var. pentaphylax *(Hung T. Chang) Hung T. Chang in Fl. Reipubl. Popul. Sin.*, 49(3): 65 (1998): C. pentaphylax.
cordifolia (F.P. Metcalf) Nakai
var. glabrisepala *T.L. Ming in Acta Bot. Yunnanica*, 19(3): 222 (1997)—China South-central (Yunnan).
crassicolumna H.T. Chang
var. shangbaensis *F.C. Zhang in Acta Bot. Yunnanica*, 19(4): 437 (1997)—China South-central (Yunnan).
crassiphylla *Tran Ninh & N. Hakoda in Int. Camellia J.*, 30: 76 (1998)—Vietnam.
cucphuongensis *Tran Ninh & J.-C. Rosmann in Int. Camellia J.*, 30: 71 (1998)—Vietnam.
cupiformis *T.L. Ming in Acta Bot. Yunnanica*, 19(3): 222 (1997)—China South-central (Yunnan).

CAMELLIA:-
cuspidata (Kochs) hort.
var. trichandra *(Hung T. Chang) T.L. Ming in Acta Bot. Yunnanica*, 21(2): 153 (1999): C. trichandra.
euryoides Lindl.
var. nokoensis *(Hayata) T.L. Ming in Acta Bot. Yunnanica*, 21(2): 154 (1999): C. nokoensis.
flavida Hung T. Chang
var. patens *(S.L. Mo & Y.C. Zhong) T.L. Ming in Acta Bot. Yunnanica*, 21(2): 152 (1999): C. longgangensis var. patens.
fluviatilis Hand.-Mazz.
var. megalantha *(Hung T. Chang) T.L. Ming in Acta Bot. Yunnanica*, 21(2): 157 (1999): C. kissi var. megalantha.
forrestii (Diels) Cohen-Stuart
var. pentamera *(Hung T. Chang) T.L. Ming in Acta Bot. Yunnanica*, 21(2): 153 (1999): C. pentamera.
furfuracea (Merr.) Cohen-Stuart
var. latipetiolata *(C.W. Chi) T.L. Ming in Acta Bot. Yunnanica*, 21(2): 155 (1999): C. latipetiolata.
subsp. megacarpa *(Elmer) T.L. Ming in Acta Bot. Yunnanica*, 21(2): 155 (1999): Thea megacarpa.
var. shanglinensis *T.L. Ming in Acta Bot. Yunnanica*, 21(2): 159 (1999)—China Southeast (Guangxi).
glabricostata *T.L. Ming in Acta Bot. Yunnanica*, 21(2): 159 (1999)—China Southeast (Guangxi).
henryana Cohen-Stuart
var. pilocarpa *T.L. Ming in Acta Bot. Yunnanica*, 19(3): 222 (1997)—China South-central (Yunnan).
var. trichocarpa *(Hung T. Chang) T.L. Ming in Acta Bot. Yunnanica*, 21(2): 155 (1999): C. trichocarpa.
japonica L.
var. rusticana *(Honda) T.L. Ming in Acta Bot. Yunnanica*, 20(2): 132 (1998): C. rusticana.
kissi Wall.
var. confusa *(Craib) T.L. Ming in Fl. Yunnanica*, 8: 300 (1997): T. confusa.
lanceolata (Blume) Seem.
var. mollis *(Merr.) T.L. Ming in Acta Bot. Yunnanica*, 21(2): 158 (1999): Thea lanceolata var. mollis.
laotica *(Gagnep.) T.L. Ming in Acta Bot. Yunnanica*, 21(2): 153 (1999): Thea laotica.
liberofilamenta *Hung T. Chang & C.H. Yang in Guihaia*, 17(4): 290 (1997); *et in Int. Camellia J.*, 30: 80 (1998)—China South-central (Guizhou).
lutchuensis T. Itô
var. minutiflora *(Hung T. Chang) T.L. Ming in Acta Bot. Yunnanica*, 21(2): 154 (1999): C. minutiflora.
mairei (H. Lév.) Melch.
f. alba *(Hung T. Chang) T.L. Ming in Acta Bot. Yunnanica*, 20(2): 142 (1998): C. mairei var. alba.
megasepala *Hung T. Chang & Trin Ninh in Acta Sci. Nat. Univ. Sunyatseni*, 35(1): 79 (1996); *et in Int. Camellia J.*, 30: 80 (1998)—Vietnam.
mileensis *T.L. Ming in Acta Bot. Yunnanica*, 19(3): 221 (1997)—China South-central (Yunnan).
var. microphylla *T.L. Ming in Acta Bot. Yunnanica*, 19(3): 222 (1997)—China South-central (Yunnan).
murauchii *Tran Ninh & N. Hakoda in Int. Camellia J.*, 30: 77 (1998)—Vietnam.
nokoensis Hayata
var. transnokoensis *(Hayata) S.S. Ying in Mem. Coll. Agric. Nation. Taiwan Univ.*, 35(2): 140 (1995): C. transnokoensis.
pitardii Cohen-Stuart
var. cryptoneura *(Hung T. Chang) T.L. Ming in Acta Bot. Yunnanica*, 21(2): 157 (1999): C. cryptoneura.
var. variabilis *(Hung T. Chang & X.K. Wen) T.L. Ming in Acta Bot. Yunnanica*, 20(2): 137 (1998): C. compressa var. variabilis.
polyodonta F.C. How ex Hu
var. longicaudata *(Hung T. Chang & S.Ye Liang) T.L. Ming in Acta Bot. Yunnanica*, 20(2): 144 (1998): C. longicaudata.
reticulata Lindl.
f. albescens *(Hung T. Chang) T.L. Ming in Fl. Yunnanica*, 8: 308 (1997): C. albescens.

CAMELLIA:-
rosmannii *Tran Ninh in Int. Camellia J.*, 30: 74 (1998)—Vietnam.
rubriflora *Tran Ninh & N. Hakoda in Int. Camellia J.*, 30: 77 (1998)—Vietnam.
sealyana *T.L. Ming in Acta Bot. Yunnanica*, 21(2): 158 (1999)—Laos.
semiserrata C.W. Chi
f. albiflora *(Hu & T.C. Huang) T.L. Ming in Acta Bot. Yunnanica*, 20(2): 133 (1998): C. semiserrata var. albiflora.
sinensis (L.) Kuntze
var. kucha *Hung T. Chang & Wang in Acta Sci. Nat. Univ. Sunyatseni*, 1984(1): 10 (1984)—China South-central (Yunnan).
skogiana *C.X. Ye in Acta Sci. Nat. Univ. Sunyatseni*, 35(4): 127 (1996)—China South-central (Yunnan).
synaptica Sealy
var. parviovata *(Hung T. Chang) T.L. Ming in Acta Bot. Yunnanica*, 21(2): 154 (1999): C. parviovata.
tachangensis F.S. Zhang
var. remotiserrata *(Hung T. Chang, H.S. Wang & P.S. Wang) T.L. Ming in Fl. Yunnanica*, 8: 276 (1997), without basionym ref.; *et in Acta Bot. Yunnanica*, 21(2): 152 (1999), with basionym ref.: C. remotiserrata.
tsingpienensis Hu
var. macrophylla *T.L. Ming in Acta Bot. Yunnanica*, 19(3): 222 (1997)—China South-central (Yunnan).
vidalii *J.C. Rosmann in Adansonia*, 21(2): 319 (1999)—Vietnam.
viridicalyx Hung T. Chang
var. linearifolia *T.L. Ming in Acta Bot. Yunnanica*, 21(2): 158 (1999)—China South-central (Guizhou).
wardii Kobuski
var. muricatula *(Hung T. Chang) T.L. Ming in Fl. Yunnanica*, 8: 292 (1997): C. muricatula.
yunnanensis (Pit. ex Diels) Cohen-Stuart
var. camellioides *(Hu) T.L. Ming in Acta Bot. Yunnanica*, 21(2): 155 (1999): Pyrenaria camellioides.
var. trichocarpa *(Hung T. Chang) T.L. Ming in Fl. Yunnanica*, 8: 289 (1997): C. trichocarpa.

CAMISSONIA (Onagrac.).
boothii (Douglas ex Lehm.) P.H. Raven
var. alyssoides *(Hook. & Arn.) N.H. Holmgren & P.K. Holmgren in A. Cronquist et al., Intermountain Fl., Vasc. Pl. Intermountain West*, U.S.A., 3A: 187 (1997): Oenothera alyssoides.

CAMPANULA (Campanulac.).
sect. Heterophylla *(Witasek) N.N. Tzvelev, Opred. Sosud. Rast. Severo-Zap. Rossii:* 680 (2000): Campanula grex Heterophylla.
calycialata *V. Randjelović & B. Zlatković in Flora Medit.*, 8: 87 (1998)—Yugoslavia.
chinalug *R. Govaerts, World Checklist Seed Plants*, 3(1): 12 (1999), nom. nov.: C. ciliata *Steven*.
csepeliensis *M. Gandoger, Dec. Pl. Nov.*, 2: 22 (1876)—Hungary.
frutetorum *M. Gandoger, Dec. Pl. Nov.*, 2: 23 (1876)—France.
gansuensis *L.C. Wang & D.Y. Hong in Bot. Bull. Acad. Sin. (Taipei)*, 41(2): 159 (2000)—China North-central (Gansu).
glomerata L.
subsp. krylovii *M.V. Olonova in Turczaninowia*, 2(3): 8 (1999)—West Siberia.
grandis Fisch. & C.A. Mey.
subsp. rizeensis *(Güner) T.G. Lammers in Novon*, 9(3): 385 (1999): C. latiloba subsp. rizeensis.
immodesta *T.G. Lammers in Novon*, 8(1): 34 (1998), nom. nov.: C. modesta *Hook. f. & Thomson*.
× iserana *M. Kovanda in Thaiszia*, 9(1): 16 (1999)—Czechoslovakia. (C. rhomboidalis × rotundifolia).
koyuncui *H. Duman in Edinburgh J. Bot.*, 56(3): 357 (1999)—Turkey.
langsdorffiana *(DC.) E.R. von Trautvetter & C.A. Meyer in Middend., Reise Sibir. (Florula Ochotensis Phaenogama):* 60 (1856): C. linifolia δ langsdorffiana.

CAMPANULA:-
pangea *P. Hartvig in Willdenowia,* 28(1–2): 65 (1998)—Greece.
punctata Lam.
var. takesimana *(Nakai) Y.N. Lee, Fl. Korea:* 1161 (1996): C. takesimana.
rotundifolia L.
subsp. langsdorffiana *(Fisch. ex Trautv. & C.A. Mey.) N.S. Vodopianova in L.I. Malyschev (ed.), Vÿsokogornaya Flora Stanovogo Nagor'ya:* 139 (1972): C. langsdorffiana.
sibirica L.
subsp. caucasica *(M. Bieb.) V.P. Victorov in Novosti Sist. Vÿssh. Rast.,* 32: 168 (2000): C. caucasica.
subsp. daghestanica *(Fomin) V.P. Victorov in Novosti Sist. Vÿssh. Rast.,* 32: 168 (2000): C. daghestanica.
subsp. komarovii *(Maleev) V.P. Victorov in Novosti Sist. Vÿssh. Rast.,* 32: 166 (2000): C. komarovii.
subsp. longistyla *(Fomin) V.P. Victorov in Novosti Sist. Vÿssh. Rast.,* 32: 165 (2000): C. longistyla.
yaltirikii *H. Duman in Edinburgh J. Bot.,* 56(3): 355 (1999)—Turkey.

CAMPNOSPERMA (Anacardiac.).
lepidotum *Capuron ex A. Randrianasolo & James S. Miller in Adansonia,* 20(2): 286 (1998)—Madagascar.
parvifolium *Capuron ex J.S. Miller & A. Randrianasolo in Novon,* 8(2): 170 (1998)—Madagascar.
schatzii *A. Randrianasolo & James S. Miller in Adansonia,* 20(2): 291 (1998)—Madagascar.

CAMPOMANESIA (Myrtac.).
discolor O. Berg
var. paranensis *(D. Legrand) J.R. de Mattos in Loefgrenia,* 110: 1 (1997): C. paranensis.
erosa *(Miq.) R. Govaerts, World Checklist Seed Plants,* 3(1): 12 (1999): Psidium erosum.
phaea (O. Berg) Landr.
var. lauroana *(Mattos) J.R. de Mattos in Loefgrenia,* 110: 1 (1997): Paivaea langsdorffii var. lauroana.
simulans *M.L. Kawasaki in Brittonia,* 52(2): 188 (2000)—Brazil Southeast (São Paulo, Minas Gerais).

CAMPSIANDRA (Legumin.).
angustifolia Spruce ex Benth.
var. rosea *(Poepp. & Endl.) B. Stergios in Novon,* 6(4): 437 (1996): C. rosea.
aymardii *B. Stergios in Novon,* 6(4): 438 (1996)—Venezuela.
casiquiarensis *B. Stergios in Novon,* 6(4): 438 (1996)—Venezuela.
chigo-montero *B. Stergios in Novon,* 6(4): 439 (1996)—Venezuela, Bolivia, Brazil North (Roraima, Pará).
comosa Benth.
var. surinamensis *(Kleinh.) B. Stergios in Novon,* 6(4): 442 (1996): C. surinamensis.
curaara *B. Stergios in Novon,* 6(4): 442 (1996)—Venezuela.
emonensis *B. Stergios in Novon,* 6(4): 443 (1996)—Venezuela.
ferruginea *B. Stergios in Novon,* 6(4): 444 (1996)—Venezuela.
gomez-alvareziana *B. Stergios in Novon,* 6(4): 445 (1996)—Venezuela, Brazil North (Amazonas).
var. pusilliflora *B. Stergios in Novon,* 6(4): 447 (1996)—Venezuela, Colombia.
guayanensis *B. Stergios in Novon,* 6(4): 447 (1996)—Venezuela.
implexicaulis *B. Stergios in Novon,* 6(4): 448 (1996)—Venezuela, Brazil North (Pará).
macrocarpa R.S. Cowan
var. alveolata *B. Stergios in Novon,* 6(4): 452 (1996)—Venezuela.
var. grandifolia *B. Stergios in Novon,* 6(4): 452 (1996)—Venezuela.
nutans *B. Stergios in Novon,* 6(4): 454 (1996)—Venezuela, Brazil North (Pará).
pasibensis *B. Stergios in Novon,* 6(4): 455 (1996)—Venezuela.
steyermarkiana *B. Stergios in Novon,* 6(4): 456 (1996)—Venezuela.

CAMPSIANDRA:-
taphornii *B. Stergios in Novon,* 6(4): 457 (1996)—Venezuela.
velutina *B. Stergios in Novon,* 6(4): 458 (1996)—Venezuela.
wurdackiana *B. Stergios in Biollania,* ed. esp. 6 (J.J. Wurdack Festschr.): 528 (1997)—Venezuela.

CAMPTOCARPUS (Asclepiadac.).
sect. Harpanema *(Decne.) J. Klackenberg in Bot. Jahrb.,* 120(1): 72 (1998): gen. Harpanema.
acuminatus *(Choux) H.J.T. Venter in Taxon,* 46(4): 712 (1997): Tanulepis acuminata.
cornutus *J. Klackenberg in Bot. Jahrb.,* 120(1): 80 (1998)—Madagascar.
decaryi *(Choux) H.J.T. Venter in Taxon,* 46(4): 712 (1997): Tanulepis decaryi.
lanceolatus *J. Klackenberg in Bot. Jahrb.,* 120(1): 75 (1998)—Madagascar.
madagascariensis *(Schltr.) H.J.T. Venter in Taxon,* 46(4): 712 (1997): Symphytonema madagascariense.
semihastatus *J. Klackenberg in Bot. Jahrb.,* 120(1): 72 (1998), nom. nov.: Harpanema acuminatum *Decne.*
sphenophyllus *(Balf. f.) H.J.T. Venter in Taxon,* 46(4): 712 (1997): Tanulepis sphenophylla.

CAMPTOSEMA (Legumin.).
ellipticum *(Desv.) R. Govaerts, World Checklist Seed Plants,* 3(1): 12 (1999): Rudolphia elliptica.

CAMPTOTHECA (Cornac.).
lowreyana *S.Y. Li in Bull. Bot. Res. North-East. Forest. Univ.,* 17(3): 348 (1997)—China Southeast (Guangdong, Fujian, Guangxi, Hunan), China South-central (Sichuan).

CAMPYLANDRA (Convallariac.).
annulata *(H. Li & J.L. Huang) M.N. Tamura, S. Yun Liang & N.J. Turland in Novon,* 10(2): 159 (2000): Tupistra annulata.
chinensis *(Baker) M.N. Tamura, S. Yun Liang & N.J. Turland in Novon,* 10(2): 159 (2000): Tupistra chinensis.
delavayi *(Franch.) M.N. Tamura, S. Yun Liang & N.J. Turland in Novon,* 10(2): 159 (2000): Tupistra delavayi.
emeiensis *(Z.Y. Zhu) M.N. Tamura, S. Yun Liang & N.J. Turland in Novon,* 10(2): 159 (2000): Tupistra emeiensis.
ensifolia *(F.T. Wang & T. Tang) M.N. Tamura, S. Yun Liang & N.J. Turland in Novon,* 10(2): 159 (2000): Tupistra ensifolia.
fimbriata *(Hand.-Mazz.) M.N. Tamura, S. Yun Liang & N.J. Turland in Novon,* 10(2): 159 (2000): Tupistra fimbriata.
jinshanensis *(Z.L. Yang & X.G. Luo) M.N. Tamura, S. Yun Liang & N.J. Turland in Novon,* 10(2): 159 (2000): Tupistra jinshanensis.
liangshanensis *(Z.Y. Zhu) M.N. Tamura, S. Yun Liang & N.J. Turland in Novon,* 10(2): 159 (2000): Tupistra liangshanensis.
lichuanensis *(Y.K. Yang, J.K. Wu & D.T. Peng) M.N. Tamura, S. Yun Liang & N.J. Turland in Novon,* 10(2): 160 (2000): Tupistra lichuanensis.
longipedunculata *(F.T. Wang & S. Yun Liang) M.N. Tamura, S. Yun Liang & N.J. Turland in Novon,* 10(2): 160 (2000): Tupistra longipedunculata.
tui *(F.T. Wang & T. Tang) M.N. Tamura, S. Yun Liang & N.J. Turland in Novon,* 10(2): 160 (2000): Rohdea tui.
urotepala *(Hand.-Mazz.) M.N. Tamura, S. Yun Liang & N.J. Turland in Novon,* 10(2): 160 (2000): Rohdea urotepala.
verruculosa *(Q.H. Chen) M.N. Tamura, S. Yun Liang & N.J. Turland in Novon,* 10(2): 160 (2000): Tupistra verruculosa.
yunnanensis *(F.T. Wang & S. Yun Liang) M.N. Tamura, S. Yun Liang & N.J. Turland in Novon,* 10(2): 160 (2000): Tupistra yunnanensis.

CAMPYLANTHUS (Scrophulariac.).
anisotrichus *(A.G. Mill.) M. Hjertson & A.G. Miller in Edinburgh J. Bot.,* 57(2): 224 (2000): C. incanus var. anisotrichus.

CAMPYLANTHUS:-
glaber Benth.
subsp. spathulatus *(A. Chev.) C. Brochmann, N. Kilian, W. Lobin & Ø.H. Rustan in Sommerfeltia,* 24: 277 (1997): C. spathulatus.
parviflorus *M. Hjertson & A.G. Miller in Edinburgh J. Bot.,* 57(2): 221 (2000)—Somalia.

CAMPYLOCENTRUM (Orchidac.).
helorrhizum *(Dod) M.A. Nir, Orchidaceae Antillanae:* 59 (2000): Dendrophylax helorrhiza.
natalieae *G. Carnevali & I.M. Ramirez in Biollania,* ed. esp. 6 (J.J. Wurdack Festschr.): 282 (1997)—Venezuela.

CAMPYLOSPERMUM (Ochnac.).
dybovskii *Tiegh. in Ann. Sci. Nat. Bot.,* ser. 8, 16: 298 (1902)—Gabon.
monticola *(Gilg) M. Cheek in Pl. Mount Cameroon: Conserv. Checklist (eds. S. Cable & M. Cheek):* 92 (1998): Ouratea monticola.

CAMPYLOTROPIS (Legumin.).
alba *Schindl. ex Y. Iokawa & Hiroyoshi Ohashi in J. Jap. Bot.,* 72(3): 139 (1997)—China.
pinetorum (Kurz) Schindl.
var. albopubescens *Y. Iokawa & Hiroyoshi Ohashi in J. Jap. Bot.,* 72(3): 141 (1997)—China South-central (Yunnan).

CANARIUM (Burserac.).
acutifolium (DC.) Merr.
var. pioriverensis *W. Takeuchi in Sida,* 19(2): 237 (2000)—New Guinea (Papua New Guinea).
pimeloides *R. Govaerts, World Checklist Seed Plants,* 3(1): 12 (1999), nom. nov.: Pimela nigra *Lour.*

CANISTROPSIS *(Mez) E.M.C. Leme, Canistropsis - Bromeliads of Atlantic Forest:* 20 (1998): Nidularium subgen. Canistropsis. (Bromeliac.).
albiflora *(L.B. Sm.) H.E. Luther & E.M.C. Leme in E.M.C. Leme, Canistropsis - Bromeliads of Atlantic Forest:* 31 (1998): Neoregelia albiflora.
billbergioides *(Schult. f.) E.M.C. Leme, Canistropsis - Bromeliads of Atlantic Forest:* 45 (1998): Hohenbergia billbergioides.
f. azurea *(E. Pereira & Leme) E.M.C. Leme, Canistropsis - Bromeliads of Atlantic Forest:* 50 (1998): Nidularium billbergioides f. azureum.
burchellii *(Baker) E.M.C. Leme, Canistropsis - Bromeliads of Atlantic Forest:* 26 (1998): Aechmea burchellii.
correia-araujoi *(E. Pereira & Leme) E.M.C. Leme, Canistropsis - Bromeliads of Atlantic Forest:* 32 (1998): Nidularium correia-araujoi.
elata *(E. Pereira & Leme) E.M.C. Leme, Canistropsis - Bromeliads of Atlantic Forest:* 51 (1998): Nidularium microps f. elatum.
exigua *(E. Pereira & Leme) E.M.C. Leme, Canistropsis - Bromeliads of Atlantic Forest:* 53 (1998): Canistrum exiguum.
marceloi *(E. Pereira & Moutinho) E.M.C. Leme, Canistropsis - Bromeliads of Atlantic Forest:* 24 (1998): Neoregelia marceloi.
microps *(E. Morren ex Mez) E.M.C. Leme, Canistropsis - Bromeliads of Atlantic Forest:* 35 (1998): Nidularium microps.
f. bicensis *(Ule) E.M.C. Leme, Canistropsis - Bromeliads of Atlantic Forest:* 40 (1998): Nidularium microcephalum var. bicense.
f. pallida *(L.B. Sm.) E.M.C. Leme, Canistropsis - Bromeliads of Atlantic Forest:* 39 (1998): Nidularium microps var. pallidum.
pulcherrima *(E. Pereira & Leme) E.M.C. Leme, Canistropsis - Bromeliads of Atlantic Forest:* 40 (1998): Nidularium pulcherrimum.
seidelii *(L.B. Sm. & Reitz) E.M.C. Leme, Canistropsis - Bromeliads of Atlantic Forest:* 42 (1998): Nidularium seidelii.
selloana *(Baker) E.M.C. Leme, Canistropsis - Bromeliads of Atlantic Forest:* 55 (1998): Quesnelia selloana.
simulans *(E. Pereira & Leme) E.M.C. Leme, Canistropsis - Bromeliads of Atlantic Forest:* 29 (1998): Nidularium simulans.

CANISTRUM (Bromeliac.).
subgen. Cucullatanthus *E.M.C. Leme, Canistrum - Bromeliads Atlantic Forest:* 25 (1997).

CANISTRUM:-
guzmanioides *E.M.C. Leme in Bradea,* 8(21): 116 (1999)—Brazil Northeast (Bahia).
montanum *E.M.C. Leme, Canistrum - Bromeliads Atlantic Forest:* 33 (1997)—Brazil Northeast (Bahia).
sandrae *E.M.C. Leme in Bradea,* 8(21): 117 (1999)—Brazil Northeast (Bahia).

× **CANMEA** *Racine C. Foster & M.B. Foster in J. Bromeliad Soc.,* 23(5): 175 (1973). *BROMELIACEAE* (Aechmea × Canistrum).

CANNA (Cannac.).
amabilis *T. Koyama & Nobuyuki Tanaka in Bull. Nation. Sci. Mus., B (Tokyo),* 26(1): 11 (2000)—Argentina Northeast (Chaco).
jacobiniflora *T. Koyama & Nobuyuki Tanaka in Bull. Nation. Sci. Mus., B (Tokyo),* 26(1): 9 (2000)—Argentina Northeast (Corrientes).
plurituberosa *T. Koyama & Nobuyuki Tanaka in J. Jap. Bot.,* 75(2): 89 (2000)—Argentina Northwest (Jujuy).
stenantha *Nobuyuki Tanaka in Bull. Nation. Sci. Mus., B (Tokyo),* 26(1): 7 (2000)—Argentina Northeast (Misiones), Paraguay.
variegatifolia *M.M. Ciciarelli in Rev. Mus. La Plata, Bot.,* 14(103): 334 (1995)—Argentina Northeast (Santa Fé).

CANNABOIDES *B.-E. van Wyk in Taxon,* 48(4): 740 (1999). *UMBELLIFERAE*
andohahelensis *(Humbert) B.-E. van Wyk in Taxon,* 48(4): 742 (1999): Heteromorpha andohahelensis.
var. denudata *(Humbert) B.-E. van Wyk in Taxon,* 48(4): 742 (1999): Heteromorpha andohahelensis var. denudata.
betsileensis *(Humbert) B.-E. van Wyk in Taxon,* 48(4): 742 (1999): Heteromorpha betsileensis.

CANNAEORCHIS *M.A. Clements & D.L. Jones in Lasianthera,* 1(3): 132 (1997 publ. 1998). *ORCHIDACEAE*
atractoglossa *(N. Hallé) M.A. Clements & D.L. Jones in Lasianthera,* 1(3): 136 (1997 publ. 1998), as 'atractoglossum': Dendrobium polycladium var. atractoglossum.
cymatoleguum *(Schltr.) M.A. Clements & D.L. Jones in Lasianthera,* 1(3): 136 (1997 publ. 1998): Dendrobium cymatoleguum.
delumbis *(Kraenzl.) M.A. Clements & D.L. Jones in Lasianthera,* 1(3): 139 (1997 publ. 1998), as 'delumbe': Dendrobium delumbe.
deplanchei *(Rchb. f.) M.A. Clements & D.L. Jones in Lasianthera,* 1(3): 139 (1997 publ. 1998): Dendrobium deplanchei.
fractiflexa *(Finet) M.A. Clements & D.L. Jones in Lasianthera,* 1(3): 140 (1997 publ. 1998), as 'fractiflexum': Dendrobium fractiflexum.
megalorhiza *(Kraenzl.) M.A. Clements & D.L. Jones in Lasianthera,* 1(3): 140 (1997 publ. 1998), as 'megalorhizum': Dendrobium megalorhizum.
polycladium *(Rchb. f.) M.A. Clements & D.L. Jones in Lasianthera,* 1(3): 141 (1997 publ. 1998): Dendrobium polycladium.
sarcochilus *(Finet) M.A. Clements & D.L. Jones in Lasianthera,* 1(3): 141 (1997 publ. 1998): Dendrobium sarcochilus.
steatoglossa *(Rchb. f.) M.A. Clements & D.L. Jones in Lasianthera,* 1(3): 142 (1997 publ. 1998), as 'steatoglossum': Dendrobium steatoglossum.
vandifolia *(Finet) M.A. Clements & D.L. Jones in Lasianthera,* 1(3): 142 (1997 publ. 1998), as 'vandifolium': Dendrobium vandifolium.
verrucifera *(Rchb. f.) M.A. Clements & D.L. Jones in Lasianthera,* 1(3): 146 (1997 publ. 1998), as 'verruciferum': Dendrobium verruciferum.

CANTUA (Polemoniac.).
coerulea *(Brand) J.M. Porter & L.A. Johnson in Aliso,* 19(1): 62 (2000): Huthia coerulea.
longiflora *(Brand) J.M. Porter & L.A. Johnson in Aliso,* 19(1): 62 (2000), homonym; *see Torr. in Ann. Lyceum Nat. Hist. New York,* 2: 221 (1828): Huthia longiflora.

× **CANULARIUM** *D.A. Beadle, Prelim. Listing Cultivar Grex Names Bromel.:* 36 (1991). *BROMELIACEAE* (Canistrum × Nidularium).

CAPEOBOLUS *J. Browning in S. Afr. J. Bot.,* 65(3): 218 (1999). *CYPERACEAE*
brevicaulis *(C.B. Clarke) J. Browning in S. Afr. J. Bot.,* 65(3): 218 (1999): Costularia brevicaulis.

CAPILLIPEDIUM (Gramin.).
leucotrichum *(A. Camus) M. Schmid ex J.F. Veldkamp in Austrobaileya,* 5(3): 527 (1999): Chrysopogon leucotrichus.

CAPITANOPSIS (Labiat.).
albida *(Baker) I.C. Hedge in Fl. Madagascar,* fam. 175: 258 (1998): Plectranthus albidus.

CAPNOPHYLLUM (Umbellif.).
leiocarpon *(Sond.) J.C. Manning & P. Goldblatt in Strelitzia,* 9: 705 (2000): C. africanum var. leiocarpon.

CAPPARIS (Capparac.).
flexuosa (L.) L.
subsp. polyantha *(Triana & Planch.) H.H. Iltis in Fl. Venezuelan Guayana (gen. eds. J.A. Steyermark et al.),* 4: 139 (1998): C. polyantha.
hastata Jacq.
f. coccolobifolia *(Mart. ex Eichler) H.H. Iltis & A. Dugand in Fl. Venezuelan Guayana (gen. eds. J.A. Steyermark et al.),* 4: 141 (1998): C. coccolobifolia.
himalayensis *S.M.H. Jafri in Pakistan J. Forest.,* 6(3): 197 (1956)—India (Himachal Pradesh).
kollimalayana *M.B. Viswanathan in Kew Bull.,* 55(1): 245 (2000)—India (Tamil Nadu).
muco *H.H. Iltis, L.J. Cumana C., R.E. Delgado M. & G.C. Aymard in Novon,* 6(4): 379 (1996)—Venezuela.
sepiaria L.
var. boscioides *(Pax) L.E. Kers in Kew Bull.,* 54(1): 220 (1999): C. boscioides.
velutina *P.I. Forster in Austrobaileya,* 5(1): 107 (1997)—Queensland, New South Wales.

CAPSELLA (Crucif.).
bursa-pastoris (L.) Medik.
subsp. orientalis *(Klokov) N.N. Tzvelev in Novosti Sist. Vyssh. Rast.,* 32: 184 (2000): C. orientalis.

CAPSICUM (Solanac.).
baccatum L.
var. umbilicatum *(Vell.) A.T. Hunziker & G.E. Barboza in Kurtziana,* 26: 27 (1998): C. umbilicatum.
brachypodum *(Dunal) O. Kuntze, Rev. Gen.,* 2: 450 (1891): Bassovia brachypoda.
ciliatum *(Humb., Bonpl. & Kunth) O. Kuntze, Rev. Gen.,* 2: 450 (1891): Witheringia ciliata.
crassiflorum *(Dunal) O. Kuntze, Rev. Gen.,* 2: 450 (1891): Bassovia crassiflora.
dimorphum *(Miers) O. Kuntze, Rev. Gen.,* 2: 449 (1891): Brachistus dimorphus.
diversifolium *(Klotzsch) O. Kuntze, Rev. Gen.,* 2: 450 (1891): Witheringia diversifolia.
dumetorum *(Dunal) O. Kuntze, Rev. Gen.,* 2: 450 (1891): Witheringia dumetora.
dunalii *O. Kuntze, Rev. Gen.,* 2: 450 (1891), nom nov.: Fregirardia angustifolia *Dunal.*
eriolarynx *(Dunal) O. Kuntze, Rev. Gen.,* 2: 450 (1891): Fregirardia eriolarynx.
fasciculatum *(Vell.) O. Kuntze, Rev. Gen.,* 2: 450 (1891), later homonym see Sturtevant *in Bull Torrey Bot. Club,* 15: 133 (1888): Solanum fasciculatum.
glomuliflorum *(Sendtn.) O. Kuntze, Rev. Gen.,* 2: 450 (1891): Aureliana glomuliflora.
hebephyllum *(Miers) O. Kuntze, Rev. Gen.,* 2: 449 (1891): Brachistus hebephyllus.
hebepodum *(Dunal) O. Kuntze, Rev. Gen.,* 2: 450 (1891): Bassovia hebepoda.
hirsutum *(Gardner) O. Kuntze, Rev. Gen.,* 2: 450 (1891): Witheringia hirsuta.
hookerianum *(Miers) O. Kuntze, Rev. Gen.,* 2: 449 (1891): Brachistus hookerianus.
laeve *(Dunal) O. Kuntze, Rev. Gen.,* 2: 450 (1891): Bassovia laevis.
lanceaefolium *(Miers) O. Kuntze, Rev. Gen.,* 2: 449 (1891): Brachistus lanceaefolius.
leptocladum *(Dunal) O. Kuntze, Rev. Gen.,* 450 (1891): Fregirardia leptoclada.

CAPSICUM:-
leptopodum *(Dunal) O. Kuntze, Rev. Gen.,* 2: 449 (1891): Bassovia leptopoda.
ligustrinum *(Dunal) O. Kuntze, Rev. Gen.,* 2: 450 (1891): Fregirardia leptoclada.
lindenii *(Dunal) O. Kuntze, Rev. Gen.,* 2: 450 (1891): Fregirardia lindenii.
lucidum *(Moric.) O. Kuntze, Rev. Gen.,* 449 (1891): Solanum lucidum.
molle *(Humb., Bonpl. & Kunth) O. Kuntze, Rev. Gen.,* 2: 450 (1891): Witheringia mollis.
oblongifolium *(Miers) O. Kuntze, Rev. Gen.,* 2: 449 (1891): Brachistus oblongifolius.
pyraster *(Dunal) O. Kuntze, Rev. Gen.,* 2: 450 (1891): Bassovia pyraster.
rhomboideum *(Humb., Bonpl. & Kunth) O. Kuntze, Rev. Gen.,* 448 (1891), as 'rhombodeum': Wetheringia rhomboidea.
richardii *(Dunal) O. Kuntze, Rev. Gen.,* 2: 450 (1891): Bassovia richardii.
riparium *(Humb., Bonpl. & Kunth) O. Kuntze, Rev. Gen.,* 2: 450 (1891): Witheringia riparia.
solanaceum *(L'Hér.) O. Kuntze, Rev. Gen.,* 2: 448 (1891): Witheringia solanacea.
spina-alba *(Dunal) O. Kuntze, Rev. Gen.,* 2: 448 (1891): Fregirardia spina-alba.
stramonifolium *(Humb., Bonpl. & Kunth) O. Kuntze, Rev. Gen.,* 2: 450 (1891): Witheringia stramonifolia.
sylvaticum *(Aubl.) O. Kuntze, Rev. Gen.,* 2: 450 (1891): Bassovia sylvatica.
tomentosum *(Sendtn.) O. Kuntze, Rev. Gen.,* 2: 450 (1891): Aureliana tomentosa.
vargasii *(Dunal) O. Kuntze, Rev. Gen.,* 2: 450 (1891): Fregirardia vargasii.
velutinum *(Sendtn.) O. Kuntze, Rev. Gen.,* 2: 448 (1891): Aureliana velutina.

CARAGANA (Legumin.).
[unranked] Brachypodae *(Sanchir) Ch. Sanchir in Acta Sci. Nat. Univ. Neimongol,* 30(4): 510 (1999), contrary to Art. 35.1 ICBN (1994): C. subser. Brachypodae.
ser. Brachypodae *(Sanchir) Ch. Sanchir in Novosti Sist. Vyssh. Rast.,* 32: 87 (2000): C. subser. Brachypodae.
[unranked] Bracteolatae *(Kom.) Ch. Sanchir in Acta Sci. Nat. Univ. Neimongol,* 30(4): 504 (1999), contrary to Art. 35.1 ICBN (1994): C. ser Bracteolatae.
subser. Bracteolatae *(Kom.) Ch. Sanchir in Acta Sci. Nat. Univ. Neimongol,* 30(4): 504 (1999); *et in Novosti Sist. Vyssh. Rast.,* 32: 79 (2000): C. ser. Bracteolatae.
sect. Bracteolatae *(Kom.) M.L. Zhang in Acta Bot. Yunnanica,* 19(4): 339 (1997): C. ser. Bracteolatae.
[unranked] Dasyphyllae *(Pojark.) Ch. Sanchir in Acta Sci. Nat. Univ. Neimongol,* 30(4): 508 (1999), contrary to Art. 35.1 ICBN (1994): Basionym not stated.
sect. Dasyphyllae *(Pojark.) Ch. Sanchir in Novosti Sist. Vyssh. Rast.,* 32: 84 (2000): C. ser. Dasyphyllae.
[unranked] Densae *(Sanchir) Ch. Sanchir in Acta Sci. Nat. Univ. Neimongol,* 30(4): 511 (1999), contrary to Art. 35.1 ICBN (1994): C. subser. Densae.
sect. Densae *(Sanchir) Ch. Sanchir in Novosti Sist. Vyssh. Rast.,* 32: 88 (2000): C. subser. Densae.
[unranked] Erinacanthae *(Kom.) Ch. Sanchir in Acta Sci. Nat. Univ. Neimongol,* 30(4): 506 (1999), contrary to Art. 35.1 ICBN (1994): C. ser. Erinacanthae.
sect. Erinacanthae *(Kom.) Ch. Sanchir in Novosti Sist. Vyssh. Rast.,* 32: 82 (2000): C. ser. Erinacanthae.
subser. Frutescentes *Ch. Sanchir in Acta Sci. Nat. Univ. Neimongol,* 30(4): 509 (1999).
sect. Gerardianae *Ch. Sanchir in Acta Sci. Nat. Univ. Neimongol,* 30(4): 505 (1999); *et in Novosti Sist. Vyssh. Rast.,* 32: 80 (2000).
ser. Gerardianae *Ch. Sanchir in Acta Sci. Nat. Univ. Neimongol,* 30(4): 505 (1999); *et in Novosti Sist. Vyssh. Rast.,* 32: 81 (2000).
[unranked] Leucophloeae *(Gorbunova) Ch. Sanchir in Acta Sci. Nat. Univ. Neimongol,* 30(4): 511 (1999), contrary to Art. 35.1 ICBN (1994): C. ser. Leucophloeae.

CARAGANA:-
sect. Leucophloeae *(Gorbunova) Ch. Sanchir in Novosti Sist. Vÿssh. Rast.,* 32: 88 (2000): C. subsect. Leucophloeae.
[unranked] Occidentales *(Kom.) Ch. Sanchir in Acta Sci. Nat. Univ. Neimongol,* 30(4): 506 (1999), contrary to Art. 35.1 ICBN (1994): C. ser. Occidentales.
sect. Occidentales *(Kom.) Ch. Sanchir in Novosti Sist. Vÿssh. Rast.,* 32: 82 (2000): C. ser. Occidentales.
subser. Opulentes *Ch. Sanchir in Acta Sci. Nat. Univ. Neimongol,* 30(4): 509 (1999); *et in Novosti Sist. Vÿssh. Rast.,* 32: 86 (2000).
[unranked] Pruinosae *(Gorbunova) Ch. Sanchir in Acta Sci. Nat. Univ. Neimongol,* 30(4): 508 (1999), contrary to Art. 35.1 ICBN (1994): C. subsect. Pruinosae.
sect. Pruinosae *(Gorbunova) Ch. Sanchir in Novosti Sist. Vÿssh. Rast.,* 32: 84 (2000): C. subsect. Pruinosae.
[unranked] Pygmaeae *(Kom.) Ch. Sanchir in Acta Sci. Nat. Univ. Neimongol,* 30(4): 510 (1999), contrary to Art. 35.1 ICBN (1994): C. ser. Pygmaeae.
sect. Pygmaeae *(Kom.) Ch. Sanchir in Novosti Sist. Vÿssh. Rast.,* 32: 87 (2000): C. ser. Pygmaeae.
[unranked] Roseae *(Sanchir) Ch. Sanchir in Acta Sci. Nat. Univ. Neimongol,* 30(4): 512 (1999), contrary to Art. 35.1 ICBN (1994): C. subser. Roseae.
ser. Roseae *(Sanchir) Ch. Sanchir in Novosti Sist. Vÿssh. Rast.,* 32: 90 (2000): C. subser. Roseae.
subser. Sukienses *Ch. Sanchir in Acta Sci. Nat. Univ. Neimongol,* 30(4): 505 (1999); *et in Novosti Sist. Vÿssh. Rast.,* 32: 80 (2000).
sect. Tanguticae *Ch. Sanchir in Acta Sci. Nat. Univ. Neimongol,* 30(4): 507 (1999).
ser. Tanguticae *Ch. Sanchir in Acta Sci. Nat. Univ. Neimongol,* 30(4): 507 (1999); *et in Novosti Sist. Vÿssh. Rast.,* 32: 83 (2000).
ser. Tibeticae *Ch. Sanchir in Acta Sci. Nat. Univ. Neimongol,* 30(4): 506 (1999); *et in Novosti Sist. Vÿssh. Rast.,* 32: 81 (2000).
ser. Versicolores *Ch. Sanchir in Acta Sci. Nat. Univ. Neimongol,* 30(4): 511 (1999); *et in Novosti Sist. Vÿssh. Rast.,* 32: 89 (2000).
maimanensis *K.H. Rechinger in Fl. Iran.,* 157: 88 (1984)—Afghanistan.
media *Ch. Sanchir in Novosti Sist. Vÿssh. Rast.,* 32: 72 (2000)—Kazakhstan.
var. fuscata *Ch. Sanchir in Novosti Sist. Vÿssh. Rast.,* 32: 74 (2000)—Kazakhstan.
var. macrocalyx *Ch. Sanchir in Novosti Sist. Vÿssh. Rast.,* 32: 75 (2000)—Kazakhstan.
microphylla Lam.
var. potaninii *(Kom.) Y.X. Liou ex L.Z. Shue in Fl. Loess-Plat. Sin.,* 2: 368 (1992): C. potaninii.
occidentalis *(Kamelin & Yakovlev) Ch. Sanchir in Acta Sci. Nat. Univ. Neimongol,* 30(4): 510 (1999), without basionym ref.: Basionym not stated.
tangutica Maxim. ex Kom.
var. yushuensis *Y.H. Wu in Fl. Qinghaiica,* 2: 433 (1999)—Qinghai.
zaissanica *Ch. Sanchir in Novosti Sist. Vÿssh. Rast.,* 32: 75 (2000)—Altay.

CARALLIA (Rhizophorac.).
integrifolia *J. Graham, Cat. Pl. Bombay:* 68 (1839)—India (Maharashtra).

CARALLUMA (Asclepiadac.).
aucheriana *(Decne.) A.P. Jagtap & N.P. Singh in Fasc. Fl. India,* 24: 208 (1999), isonym; *see N.E. Br. in Gard. Chron.,* Ser. 3, 12: 369 (1892): Boucerosia aucheriana.
burchardii N.E. Br.
subsp. maura *(Maire) U. Meve & F. Albers in Nordic J. Bot.,* 15(5): 465 (1995 publ. 1996): C. burchardii var. maura.
flavovirens *L.E. Newton in Asklepios,* 74: 25 (1998)—Kenya.
munbyana (Decne.) N.E. Br.
subsp. hispanica *(Coincy) M.B. Crespo & Mateo in Acta Bot. Malacitana,* 20: 285 (1995): Boucerosia munbyana var. hispanica.

CARALLUMA:-
turneri E.A. Bruce
subsp. ukambensis *(P.R.O. Bally) L.E. Newton in Asklepios,* 72: 9 (1997 publ. 1998): C. dicapuae subsp. ukambensis.

CARDAMINE (Crucif.).
abchasica *R. Govaerts, World Checklist Seed Plants,* 3(1): 12 (1999), nom. nov.: Dentaria microphylla *Willd.*
amara L.
subsp. austriaca *K. Marhold in Bot. Helvetica,* 109(1): 77 (1999)—Austria, Czechoslovakia, Germany, Italy.
subsp. balcanica *K. Marhold, M.E. Ančev & Kit Tan in Ann. Bot. Fenn.,* 33(3): 201 (1996)—Bulgaria, Greece.
astoniae *I.R. Thompson in Muelleria,* 9: 156 (1996)—Victoria, New South Wales, Tasmania.
changbaiana *I.A. Al-Shehbaz in Novon,* 10(4): 323 (2000)—Manchuria (Jilin).
cheotaiyienii *I.A. Al-Shehbaz & G. Yang in Harvard Pap. Bot.,* 3(1): 73 (1998), nom. nov.: Hilliella alatipes var. macrantha *Y.H. Zhang.*
dichondroides *(Speg.) R. Govaerts, World Checklist Seed Plants,* 3(1): 12 (1999): C. rostrata var. dichondroides.
dubia *H. Li in Fl. Dulongjian Reg.:* 75 (1993), without latin descr. or type—China South-central (Yunnan).
dulongensis *H. Li in Fl. Dulongjian Reg.:* 75 (1993), without latin descr. or type—China South-central (Yunnan).
fargesiana *I.A. Al-Shehbaz in Novon,* 10(4): 324 (2000)—China South-Central (Sichuan).
forsteri *R. Govaerts, World Checklist Seed Plants,* 3(1): 13 (1999), nom. nov.: Sisymbrium heterophyllum *G. Forst.*
franklinensis *I.R. Thompson in Muelleria,* 9: 152 (1996)—Australian Capital Territory, New South Wales, Victoria, Tasmania.
georgiana *I.A. Al-Shehbaz & S.I. Warwick in Rhodora,* 101(907): 275 (1999), nom. nov.: Dentaria microphylla *Willd.*
gongshanensis *H. Li in Fl. Dulongjian Reg.:* 75 (1993), without latin descr. or type—China South-central (Yunnan).
heterandra *J.Z. Sun & K.L. Chang in Taiwania,* 41(2): 113 (1996)—China North-central (Gansu).
lihengiana *I.A. Al-Shehbaz in Novon,* 10(4): 326 (2000)—China South-Central (Yunnan).
lineariloba *I.R. Thompson in Muelleria,* 9: 163 (1996)—Victoria, South Australia.
lojanensis *I.A. Al-Shehbaz in Novon,* 7(1): 6 (1997)—Ecuador.
longipedicellata *Z.M. Tan & G.H. Chen in Acta Phytotax. Sin.,* 34(6): 650 (1996), homonym; *see Rollins in J. Arnold Arbor.,* 21: 395 (1940)—China South-central (Sichuan).
macrocephala *Z.M. Tan & S.C. Zhou in J. Sichuan Univ., Nat. Sci. Ed.,* 33(5): 601 (1996)—China South-central (Sichuan).
macrophylla Willd.
subsp. acutidentata *V.B. Kuvaev & A.E. Sonnikova in Byull. Mosk. Obshch. Ispÿt. Prir., Biol.,* 103(4): 68 (1998)—Siberia.
microphylla Adams
subsp. blaisdellii *(Eastw.) D.F. Murray & S. Kelso in Rhodora,* 99(897): 40 (1997): C. blaisdellii.
microthrix *I.R. Thompson in Muelleria,* 9: 165 (1996)—Victoria, New South Wales, South Australia.
moirensis *I.R. Thompson in Muelleria,* 9: 167 (1996)—Victoria, New South Wales.
papillata *I.R. Thompson in Muelleria,* 9: 161 (1996)—Victoria, South Australia, New South Wales, Tasmania.
penzesii *M.E. Ančev & K. Marhold in Ann. Bot. Fenn.,* 36(3): 172 (1999), nom. nov.: C. tuberosa *Pénzes & Vida.*
pratensis
subsp. submatthiolii *N.N. Tzvelev, Opred. Sosud. Rast. Severo-Zap. Rossii:* 679 (2000)—Northwest European Russia.
prorepens Fisch.
subsp. roseiflora *(Vorosch.) V.N. Voroschilov in A.K. Skvortsov (ed.), Florist. issl. v razn. raionakh SSSR:* 172 (1985): C. yesoensis subsp. roseiflora.

CARDAMINE:-
pulchella *(Hook. f. & Thomson) I.A. Al-Shehbaz & G. Yang in Harvard Pap. Bot.,* 3(1): 77 (1998): Loxostemon pulchellus.
purpurascens *(O.E. Schulz) I.A. Al-Shehbaz, T.Y. Cheo, L.L. Lou & G. Yang in Novon,* 10(4): 324 (2000): Cardamine microzyga var. purpurascens.
robusta *I.R. Thompson in Muelleria,* 9: 151 (1996)—New South Wales.
scutata Thunb.
var. manshurica *(Kom.) P.Y. Fu in Clavis Pl. Chinae Bor.-Or. ed. 2 (ed. P.Y. Fu):* 244 (1995): C. flexuosa var. manshurica.
sibirica *(O.E. Schulz) R. Govaerts, World Checklist Seed Plants,* 3(1): 13 (1999): C. glandulosa subsp. sibirica.
violifolia O.E. Schulz
var. pilosa *K.L. Chang & H.L. Huang in Taiwania,* 41(2): 115 (1996)—China North-central (Gansu).

CARDARIA (Crucif.).
sect. Cyclocarpae *V.I. Dorofeev in Bot. Zhurn.,* 81(4): 93 (1996).

CARDIANDRA (Hydrangeac.).
densifolia *C.F. Wei in Acta Bot. Austro Sinica,* 10: 9 (1995)—Taiwan.
moellendorffii (Hance) Migo
var. laxiflora *(H.L. Li) C.F. Wei in Acta Bot. Austro Sinica,* 10: 8 (1995): C. laxiflora.

CARDIOSPERMUM (Sapindac.).
cuchujaquense *M.S. Ferrucci & P. Acevedo-Rodríguez in Novon,* 8(3): 235 (1998)—Mexico Northwest (Sonora).

CARDUUS (Compos.).
subsect. Onopordioides *N.N. Portenier in Bot. Zhurn.,* 82(9): 101 (1997).
bourgeanus Boiss. & Reut.
subsp. ibicensis *(Devesa & Talavera) A.M. Romo, Fl. Silvestres Baleares:* 313 (1994), as 'ibizenses', name not new. At bas. ref. name stated is: C. bourgeanus subsp. ibicensis.
defloratus L.
subsp. paui *(Devesa & Talavera) O. de Bolòs & J. Vigo, Fl. Països Catalans,* 3: 881 (1995): C. paui.
kirghisicus *B.A. Sultanova in Bot. Zhurn.,* 85(11): 103 (2000)—Kirgizistan.
laciniatus Ledeb.
subsp. novorossicus *(Porten.) A.S. Zernov, Rast. Severo-Zapad. Zakavkaz'ya (Pl. Transcaucas. Bor.-Occ.):* 111 (2000): C. novorossicus.
× leridanus Devesa & Talavera
nothosubsp. mercadaliae *G. Mateo Sanz, C. Fabregat Llueca & S. López Udias in An. Biol. Fac. Biol. Univ. Murcia,* 20: 103 (1995), as 'mercadalii'—Spain.(C. carlinifolius subsp. paui × nutans subsp. nutans).
novorossicus *N.N. Portenier in Bot. Zhurn.,* 82(1): 102 (1997)—North Caucasus (Krasnodar).
nutans L.
subsp. attenuatus *(Klokov) N.N. Tzvelev in Novosti Sist. Vÿssh. Rast.,* 32: 185 (2000): C. attenuatus.
nyassanus (S. Moore) R.E. Fr.
subsp. kikuyorum *(R.E. Fr.) C. Jeffrey in Fl. Trop. E. Afr., Compos. (Part 1):* 49 (2000): Carduus kikuyorum.
schimperi Sch. Bip.
subsp. nanus *(R.E. Fr.) C. Jeffrey in Fl. Trop. E. Afr., Compos. (Part 1):* 47 (2000): Carduus nanus.
subsp. platyphyllus *(R.E. Fr.) C. Jeffrey in Fl. Trop. E. Afr., Compos. (Part 1):* 47 (2000): Carduus platyphyllus.

CAREX (Cyperac.).
subsect. Bostrychostigmaticae *T.V. Egorova, Sedges (Carex L.) of Russia & Adj. States:* 206 (1999).
subsect. Chordorrhizae *(Meinsh.) T.V. Egorova, Sedges (Carex L.) of Russia & Adj. States:* 519 (1999): C. sect. Chordorrhizae.
subsect. Curaicae *(T.V. Egorova) T.V. Egorova, Sedges (Carex L.) of Russia & Adj. States:* 502 (1999): C. ser. Curaicae.

CAREX:-

subsect. Distichae *(T.V. Egorova) T.V. Egorova, Sedges (Carex L.) of Russia & Adj. States:* 501 (1999): C. ser. Distichae.
subsect. Divisae *(H. Christ ex Kük.) T.V. Egorova, Sedges (Carex L.) of Russia & Adj. States:* 518 (1999): C. sect. Divisae.
subsect. Pseudocuraicae *(T.V. Egorova) T.V. Egorova, Sedges (Carex L.) of Russia & Adj. States:* 504 (1999): C. ser. Pseudocuraicae.
sect. Setigerae *I. Kukkonen in Ann. Naturhist. Mus. Wien,* 98B(Suppl.): 92 (1996).
subsect. Silvaticae *T.V. Egorova, Sedges (Carex L.) of Russia & Adj. States:* 206 (1999).
*subsect.*Strigosae *(Andersson) T.V. Egorova, Sedges (Carex L.) of Russia & Adj. States:* 210 (1999): C. γ Strigosae.
subsect. Williamsianae *T.V. Egorova, Sedges (Carex L.) of Russia & Adj. States:* 345 (1999).
acocksii *C. Archer in S. Afr. J. Bot.,* 63(6): 342 (1997)—Cape Province.
albata *Boott ex Franch. & Sav., Enum. Pl. Jap.,* 2: 553 (1878)—Japan.†
alpina
var. lehmanii *(Drejer) R.C. Srivastava in Novon,* 8(2): 203 (1998): C. lehmanii.
andicola *G.A. Wheeler in Hickenia,* 2(41): 196 (1996)—Argentina South (Neuquén, Chubut, Rio Negro, Santa Cruz), Argentina Northeast (Córdoba), Chile North (Coquimbo, O'Higgins).
angustiutricula *Wang & Tang ex L.K. Dai in Acta Phytotax. Sin.,* 37(2): 182 (1999)—China South-central (Sichuan).
anningensis *Wang & Tang ex P.C. Li in Acta Phytotax. Sin.,* 37(2): 171 (1999)—China South-central (Yunnan).
aterrima *Hoppe, Caricolog. Germ.,* 51 (1826)—Austria.†
atlasica *(H. Lindb.) M. Ibn Tattou in Bocconea,* 8: 198 (1998): C. leporina subsp. atlasica.
austrosinensis *Tang & Wang ex S.Yun Liang in Acta Phytotax. Sin.,* 33(5): 488 (1995)—China Southeast (Guangdong).
bigelowii Torr. ex Schwein.
subsp. paishanensis *(Nakai) V.N. Voroschilov in A.K. Skvortsov (ed.), Florist issl. v razn. raĭonakh SSSR:* 153 (1985): C. paishanensis.
brunnea Thunb.
subsp. occidentalis *K.A. Lye in Bull. Mus. Nation. Hist. Nat., B, Adansonia,* Ser. 4, 18(3–4): 235 (1996)—Somalia, Ethiopia.
caespitosa L.
subsp. minuta *(Franch.) V.N. Voroschilov in A.K. Skvortsov (ed.), Florist. issl. v razn. raĭonakh SSSR:* 153 (1985): C. minuta.
capillaris L.
subsp. ledebouriana *(C.A. Mey. ex Trevir.) V.N. Voroschilov in A.K. Skvortsov (ed.), Florist. issl. v razn. raĭonakh SSSR:* 154 (1985): C. ledebouriana.
caudispicata *Wang & Tang ex P.C. Li in Acta Phytotax. Sin.,* 37(2): 170 (1999)—China South-central (Yunnan).
cheniana *T. Tang & F.T. Wang ex S.Yun Liang in Acta Phytotax. Sin.,* 36(6): 532 (1998)—China Southeast (Zhejiang).
chuiana *Wang & Tang ex P.C. Li in Acta Phytotax. Sin.,* 37(2): 167 (1999)—China South-central (Sichuan).
collumanthus *(Steyerm.) L.E. Mora-Osejo in Acta Biol. Colomb.,* 1(1): 40, 41, 31 (1982): Vesicarex collumanthus.
communis L.H. Bailey
var. amplisquama *(F.J. Herm.) J.H. Rettig in Novon,* 6(1): 117 (1996): C. amplisquama.
concolor R. Br.
var. minuscula *V.B. Kuvaev in V.B. Kuvaev et al., Rast. pokrov ostrova Sibiryakova:* 52 (1994)—Krasnoyarsk.
congestiflora *A.A. Reznicek & M.S. González-Elizondo in Contrib. Univ. Mich. Herb.,* 22: 125 (1999)—Mexico Southeast (Chiapas).
deciduisquama *Wang & Tang ex P.C. Li in Acta Phytotax. Sin.,* 37(2): 160 (1999)—China South-central (Yunnan).
densifimbriata T. Tang & F.T. Wang
var. hirsuta *P.C. Li in Acta Phytotax. Sin.,* 37(2): 162 (1999)—China Southeast (Hunan).

CAREX:-

digitata L.
subsp. pallens *(Fristedt) N.N. Tzvelev in Novosti Sist. Vyssh. Rast.,* 32: 183 (2000): C. digitata var. pallens.
dolomitica *P.B. Heenan & P.J. de Lange in New Zealand J. Bot.,* 35(4): 423 (1997)—New Zealand South I.
elata All.
subsp. mauritanica *(Boiss. & Reut.) A. Galán de Mera in Lagascalia,* 18(1): 57 (1995): C. mauritanica.
ensifolia Turcz.
subsp. lugens *(Holm) V.N. Voroschilov in A.K. Skvortsov (ed.), Florist. issl. v razn. raĭonakh SSSR:* 154 (1985): C. lugens.
subsp. soczavaeana *(Gorodkov) V.N. Voroschilov in A.K. Skvortsov (ed.), Florist. issl. v razn. raĭonakh SSSR:* 154 (1985): C. soczavaeana.
evadens *S. González Elizondo & A.A. Reznicek in Acta Bot. Mex.,* 40: 19 (1997)—Mexico Northeast (Querétaro).
flava L.
subsp. bergrothii *(Palmgr.) P.D. Sell in P.D. Sell & G. Murrell, Fl. Great Britain & Ireland,* 5: 363 (1996): C. bergrothii.
subsp. jemtlandica *(Palmgr.) P.D. Sell in P.D. Sell & G. Murrell, Fl. Great Britain & Ireland,* 5: 363 (1996): C. lepidocarpa subsp. jemtlandica.
subsp. oedocarpa *(Andersson) P.D. Sell in P.D. Sell & G. Murrell, Fl. Great Britain & Ireland,* 5: 363 (1996): C. oederi * oedocarpa.
subsp. pulchella *(Lönnr.) P.D. Sell in P.D. Sell & G. Murrell, Fl. Great Britain & Ireland,* 5: 363 (1996): C. oederi * pulchella.
subsp. scotica *(E.W. Davies) P.D. Sell in P.D. Sell & G. Murrell, Fl. Great Britain & Ireland,* 5: 363 (1996): C. lepidocarpa subsp. scotica.
subsp. serotina *(Mérat) P.D. Sell in P.D. Sell & G. Murrell, Fl. Great Britain & Ireland,* 5: 363 (1996): C. serotina.
fujimakii *M. Kikuchi in Ann. Rep. Gakugei Fac. Iwate Univ.,* 13(2): 47 (1958)—Japan.
funhuangshanica *Wang & Tang ex P.C. Li in Acta Phytotax. Sin.,* 37(2): 168 (1999)—China Southeast (Guangxi).
funingensis *T. Tang & F.T. Wang ex S.Yun Liang in Acta Phytotax. Sin.,* 36(6): 534 (1998)—China South-central (Yunnan).
fuscula D'Urv.
subsp. catharinensis *(Boeck.) M. Luceño & M. Alves in An. Jard. Bot. Madrid,* 57(1): 173 (1999): C. catharinensis.
gentilis Franch.
var. intermedia *Tang & Wang ex L.K. Dai in Acta Phytotax. Sin.,* 37(2): 187 (1999), homonym; *see Y.C. Yang in Fl. Xizangica,* 5: 408 (1987)—China South-central (Sichuan).
var. macrocarpa *Tang & Wang ex L.K. Dai in Acta Phytotax. Sin.,* 37(2): 188 (1999)—China South-central (Sichuan).
gibertii *G.A. Wheeler in Sida,* 17(2): 475 (1996), nom. nov.: C. phalaroides var. crassiflora *Kük.*
herteri *G.A. Wheeler in Sida,* 17(2): 471 (1996)—Uruguay.
houghtoniana *Torr. ex Dewey in Amer. J. Sci. Arts,* 30: 63 (1836), in IK Vol. 1 as 'C. houghtonii Torr.'—Minnesota.†
impexa *K.A. Ford in New Zealand J. Bot.,* 36(4): 587 (1998)—New Zealand South I.
indiciformis *Wang & Tang ex P.C. Li in Acta Phytotax. Sin.,* 37(2): 156 (1999), as 'indicaeformis'—Hainan.
jinfoshanensis *Tang & Wang ex S.Yun Liang in Acta Phytotax. Sin.,* 37(2): 191 (1999)—China South-central (Sichuan).
kaoi *T. Tang & F.T. Wang ex S.Yun Liang in Acta Phytotax. Sin.,* 36(6): 530 (1998)—China Southeast (Guangdong).
klaphakei *K.L. Wilson in Telopea,* 6(4): 569 (1996)—New South Wales.
kuchunensis *T. Tang & F.T. Wang ex S.Yun Liang in Acta Phytotax. Sin.,* 36(6): 536 (1998)—China Southeast (Guangxi).
kunlunsanensis *N.R. Cui in Fl. Xinjiangensis,* 6: 604 (1996), as 'kunlunsannsis', holotype not indicated—Xinjiang.

CAREX:-

kwangsiensis *Wang & Tang ex P.C. Li in Acta Phytotax. Sin.,* 37(2): 164 (1999)—China Southeast (Guangxi).
lancangensis *S.Yun Liang in Acta Phytotax. Sin.,* 37(2): 189 (1999)—China South-central (Yunnan).
lanceolata Boott
var. pseudolanceolata *(V.I. Krecz.) V.N. Voroschilov in A.K. Skvortsov (ed.), Florist. issl. v razn. raĭonakh SSSR:* 155 (1985): C. pseudolanceolata.
lancisquamata *L.K. Dai in Acta Phytotax. Sin.,* 37(2): 181 (1999)—China South-central (Yunnan).
lepidocarpa Tausch
subsp. nevadensis *(Boiss. & Reut.) M. Luceño in An. Jard. Bot. Madrid,* 57(1): 176 (1999): C. nevadensis.
leucochlora Bunge
var. horikawae *(K. Okamoto) T. Katsuyama in Bull. Kanagawa Prefect. Mus., Nat. Sci.,* 22: 59 (1993): C. horikawae.†
liqingii *Tang & Wang ex S.Yun Liang in Acta Phytotax. Sin.,* 33(5): 489 (1995)—China Southeast (Guangxi).
lonchocarpa *Willd. ex Spreng., Syst. Veg.,* 3: 817 (1826)—North Carolina.
longerostrata C.A. Mey.
var. hoi *Chü ex S.Yun Liang in Acta Phytotax. Sin.,* 36(6): 537 (1998)—China Southeast (Zhejiang).
longii Mack.
subsp. meridionalis *(Kük.) M. Luceño & M. Alves in An. Jard. Bot. Madrid,* 57(1): 173 (1999): C. albolutescens var. meridionalis.
longpanlaensis *S.Yun Liang in Acta Phytotax. Sin.,* 37(2): 191 (1999)—China South-central (Yunnan).
malanshanensis *R.F. Huang in S.G. Wu (S.K. Wu) & Z.J. Feng (eds.), Biol. & Human Physiol. Hoh Xil* Region: 105 (1996), without latin descr.—Qinghai.
manca Boott
subsp. jiuhuaensis *(S.W. Su) S.Yun Liang in Acta Phytotax. Sin.,* 36(6): 537 (1998): C. jiuhuaensis.
subsp. wichurae *(Boeck.) S.Yun Liang in Acta Phytotax. Sin.,* 36(6): 537 (1998), as 'wichurai': C. wichurae.
marahuacana *A.A. Reznicek in Novon,* 6(4): 419 (1996)—Venezuela.
maximowiczii Miq.
subsp. suifunensis *(Kom.) V.N. Voroschilov in A.K. Skvortsov (ed.), Florist. issl. v razn. raĭonakh SSSR:* 155 (1985): C. suifunensis.
mckittrickensis *P.W. Ball in Novon,* 8(3): 221 (1998)—Texas.
melinacra Franch.
var. changningensis *S.Yun Liang in Acta Phytotax. Sin.,* 37(2): 193 (1999)—China South-central (Yunnan).
minuscula *(Kuvaev) O.V. Rebristaya in Bot. Zhurn.,* 82(7): 33 (1997): C. concolor var. minuscula.
molesta *Mack. ex J. Bright in Trillia,* 9: 4, 20 (1930), as 'modesta'—Pennsylvania.
molestiformis *A.A. Reznicek & P.E. Rothrock in Contrib. Univ. Mich. Herb.,* 21: 300 (1997)—Oklahoma, Arkansas, Missouri.
nanchuanensis *Chü ex S.Yun Liang in Acta Phytotax. Sin.,* 36(6): 532 (1998)—China South-central (Sichuan).
nanella Ohwi
var. callitrichos *(V.I. Krecz.) V.N. Voroschilov in A.K. Skvortsov (ed.), Florist. issl. v razn. raĭonakh SSSR:* 155 (1985): C. callitrichos.
neblinensis *A.A. Reznicek in Novon,* 6(4): 421 (1996)—Brazil North (Amazonas), Venezuela.
nevadensis Boiss. & Reut.
subsp. alpina *(Kneuck.) D. Podlech in Soc. Éch. Pl. Vasc. Eur. Bassin Médit. Bull.,* 28: 113 (2000): Carex flava var. alpina.
nitidiutriculata *L.K. Dai in Acta Phytotax. Sin.,* 37(2): 186 (1999)—China South-central (Yunnan).
obovatosquamata *Wang & Y.L. Chang ex P.C. Li in Acta Phytotax. Sin.,* 37(2): 175 (1999)—China South-central (Yunnan), Xizang.
okamotoi Ohwi
f. variegata *Y.N. Lee, Fl. Korea:* 1164 (1996), without type—Locality not stated.

CAREX:-

ophiolithica *P.J. de Lange & P.B. Heenan in New Zealand J. Bot.*, 35(4): 429 (1997)—New Zealand North I.

orbicularinucis *L.K. Dai in Acta Phytotax. Sin.*, 37(2): 183 (1999)—China South-central (Yunnan).

oregonensis *Olney ex L.H. Bailey in Proc. Amer. Acad. Arts*, 22(1): 73 (1886)—Oregon.†

ozarkana *P.E. Rothrock & A.A. Reznicek in Brittonia*, 48(1): 104 (1996)—Arkansas, Oklahoma, Texas.

pamirica (O. Fedtsch.) O. Fedtsch. & B. Fedtsch.

subsp. dichroa *(Freyn) T.V. Egorova, Sedges (Carex L.) of Russia & Adj. States:* 183 (1999): C. pulla subsp. dichroa.

pediformis C.A. Mey.

var. macroura *(Meinsh.) V.N. Voroschilov in A.K. Skvortsov (ed.), Florist. issl. v razn. raĭonakh SSSR:* 155 (1985): C. macroura.

var. reventa *(V.I. Krecz.) V.N. Voroschilov in A.K. Skvortsov (ed.), Florist. issl. v razn. raĭonakh SSSR:* 155 (1985): C. reventa.

peliosanthifolia *Wang & Tang ex P.C. Li in Acta Phytotax. Sin.*, 37(2): 162 (1999)—China Southeast (Guangxi).

pellita *Willd., Sp. Pl.*, 4: 302 (1805)—Pennsylvania.†

phalaroides Kunth

subsp. crassiflora *(Kük.) M. Luceño & M. Alves in An. Jard. Bot. Madrid*, 57(1): 174 (1999): C. phalaroides var. crassiflora.

subsp. moesta *(Kunth) M. Luceño & M. Alves in An. Jard. Bot. Madrid*, 57(1): 174 (1999): C. moesta.

subsp. paraguayensis *(Maury) M. Luceño & M. Alves in An. Jard. Bot. Madrid*, 57(1): 174 (1999): C. paraguayensis.

pigra *R.F.C. Naczi in Novon*, 7(1): 67 (1997)—Mississippi, Alabama, Florida, Georgia, North Carolina, South Carolina, Tennessee, Virginia.

pinophila *A.A. Reznicek & M.S. González Elizondo in Contrib. Univ. Mich. Herb.*, 21: 294 (1997)—Mexico Northeast (Querétaro).

planispicata *R.F.C. Naczi in J. Kentucky Acad. Sci.*, 60(1): 37 (1999)—Alabama.

podocarpa R. Br.

var. koraginensis *(Meinsh.) V.N. Voroschilov in A.K. Skvortsov (ed.), Florist. issl. v razn. raĭonakh SSSR:* 155 (1985): C. reventa.

porrecta *A.A. Reznicek & K. Camelbeke in Novon*, 6(4): 423 (1996)—Ecuador, Costa Rica, Venezuela, Colombia.

pseudobiwensis *M. Kitagawa in Rep. Inst. Sci. Res. Manchoukuo*, 6(4): 119 (1942), as 'pseudo-biwensis'—Manchuria.

pseudohumilis *Wang & Y.L. Chang ex P.C. Li in Acta Phytotax. Sin.*, 37(2): 172 (1999)—China South-central (Yunnan).

pseudolaticeps *T. Tang & F.T. Wang ex S.Yun Liang in Acta Phytotax. Sin.*, 36(6): 536 (1998), as 'pseudo-laticeps'—China Southeast (Hong Kong).

pseudomacloviana *G.A. Wheeler in Hickenia*, 2(41): 193 (1996)—Argentina Northwest (Salta, Tucumán).

pseudosupina *Y.C. Tang ex L.K. Dai in Acta Phytotax. Sin.*, 37(2): 177 (1999)—China South-central (Sichuan).

purpureosquamata *L.K. Dai in Acta Phytotax. Sin.*, 37(2): 183 (1999)—China South-central (Yunnan).

putuoensis *S.Yun Liang in Acta Phytotax. Sin.*, 33(5): 486 (1995)—China Southeast (Zhejiang).

queretarensis *A.A. Reznicek & M.S. González Elizondo in Contrib. Univ. Mich. Herb.*, 21: 291 (1997)—Mexico Northeast (Querétaro).

rariflora (Wahlenb.) Smith

var. pluriflora *(Hultén) V.N. Voroschilov in A.K. Skvortsov (ed.), Florist. issl. v razn. raĭonakh SSSR:* 156 (1985): C. pluriflora.

rhizina *Blytt ex A.E. Lindblom in Bot. Not.*, 1839(8): 98 (1839)—Locality not stated.

rupestris All.

subsp. alticola *(Popl. ex Sukaczev) V.N. Voroschilov in A.K. Skvortsov (ed.), Florist. issl. v razn. raĭonakh SSSR:* 156 (1985): C. alticola.

CAREX:-

rupestris All.

subsp. argunensis *(Turcz. ex Trevir.) V.N. Voroschilov in A.K. Skvortsov (ed.), Florist. issl. v razn. raĭonakh SSSR:* 156 (1985): C. argunensis.

rupicola *(Pedersen) G.A. Wheeler in Hickenia*, 2(39): 180 (1996): C. bonariensis var. rupicola.

sartwellii *Dewey in Amer. J. Sci. Arts*, 43: 90 (1842), in IK Vol. 1 as 'sartwelliana'—New York.†

scaposa C.B. Clarke

var. hirsuta *P.C. Li in Acta Phytotax. Sin.*, 37(2): 162 (1999)—China Southeast (Guangdong, Fujian).

scirpoidea Michx.

subsp. convoluta *(Kük.) D.A. Dunlop in Novon*, 7(4): 355 (1997 publ. 1998): C. scirpoidea var. convoluta.

subsp. pseudoscirpoidea *(Rydb.) D.A. Dunlop in Novon*, 7(4): 355 (1997 publ. 1998): C. pseudoscirpoidea.

serotina Mérat

var. fennica *(Palmgr.) T.V. Egorova, Sedges (Carex L.) of Russia & Adj. States:* 278 (1999): C. oederi subsp. fennica.

serpenticola *P.F. Zika in Madroño*, 45(3): 261 (1998 publ. 1999)—California, Oregon.

shaanxiensis *Wang & Tang ex P.C. Li in Acta Phytotax. Sin.*, 37(2): 174 (1999)—China North-central (Shaanxi, Gansu).

shanghaiensis *S.X. Qian & Y.Q. Liu in Acta Phytotax. Sin.*, 37(1): 103 (1999)—China Southeast (Jiangsu).

shanghangensis *S.Yun Liang in Acta Phytotax. Sin.*, 34(1): 94 (1996)—China Southeast (Fujian).

shuangbaiensis *L.K. Dai in Acta Phytotax. Sin.*, 37(2): 187 (1999)—China South-central (Yunnan).

sichouensis *P.C. Li in Acta Phytotax. Sin.*, 37(2): 173 (1999)—China South-central (Yunnan).

simulans C.B. Clarke

var. densiflora *T. Tang & F.T. Wang ex S.Yun Liang in Acta Phytotax. Sin.*, 36(6): 537 (1998)—China South-central (Sichuan).

sororia Kunth

subsp. uruguensis *(Boeck.) M. Luceño & M. Alves in An. Jard. Bot. Madrid*, 57(1): 173 (1999): C. uruguensis.

subpumila *Tang & Wang ex L.K. Dai in Acta Phytotax. Sin.*, 37(2): 182 (1999)—China North-central (Hebei).

superata *R.F.C. Naczi, A.A. Reznicek & B.A. Ford in Amer. J. Bot.*, 85(3): 443 (1998)—Alabama, Florida, Georgia, Kentucky, Mississippi, South Carolina, Tennessee, Virginia.

tenejapensis *A.A. Reznicek & M.S. González-Elizondo in Contrib. Univ. Mich. Herb.*, 22: 121 (1999)—Mexico Southeast (Chiapas).

tenuiculmis *(Petrie) P.B. Heenan & P.J. de Lange in New Zealand J. Bot.*, 35(2): 159 (1997): C. secta var. tenuiculmis.

tenuipaniculata *P.C. Li in Acta Phytotax. Sin.*, 37(2): 158 (1999)—China South-central (Yunnan).

tenuispicula *T. Tang ex S.Yun Liang in Acta Phytotax. Sin.*, 34(1): 92 (1996)—China Southeast (Fujian).

thanikaimoniana *E. Govindarajalu in J. Econ. Taxon. Bot.*, 20(2): 305 (1996)—India (Tamil Nadu).

tiogana *D.W. Taylor & J. Mastrogiuseppe in Novon*, 9(1): 120 (1999)—California.

tonsa (Fernald) Bicknell

var. rugosperma *(Mack.) W.J. Crins in Novon*, 6(1): 118 (1996): C. rugosperma.

tristis M. Bieb.

subsp. stenocarpa *(Turcz. ex V.I. Krecz.) T.V. Egorova, Sedges (Carex L.) of Russia & Adj. States:* 147 (1999): C. stenocarpa.

tsaiana *Wang & Tang ex P.C. Li in Acta Phytotax. Sin.*, 37(2): 169 (1999)—China South-central (Yunnan).

tuckermanii *C. Dewey in Amer. J. Sci. Arts*, 49: 48 (1845)—Locality not stated.

tucumanensis *G.A. Wheeler in Hickenia*, 2(41): 190 (1996)—Argentina Northwest (Tucumán).

umbrosa

var. rostrata *(Maxim.) R. Govaerts, World Checklist Seed Plants*, 3(1): 13 (1999): C. pediformis var. rostrata.

CAREX:-

vesiculosa

var. pallida *(Kük.) R.C. Srivastava in Fl. Sikkim*, 1: 213 (1996); *et in Novon*, 8(2): 203 (1998): C. vesiculosa f. pallida.

viridula Michx.

subsp. bergrothii *(Palmgr.) N.N. Tzvelev in Novosti Sist. Vyssh. Rast.*, 32: 183 (2000): C. bergrothii.

wawuensis *Chü ex S.Yun Liang in Acta Phytotax. Sin.*, 37(2): 192 (1999)—China South-central (Sichuan).

wenshanensis *L.K. Dai in Acta Phytotax. Sin.*, 37(2): 179 (1999)—China South-central (Yunnan).

wui *Chü ex L.K. Dai in Acta Phytotax. Sin.*, 37(2): 179 (1999)—China South-central (Sichuan).

wushanensis *S.Yun Liang in Acta Phytotax. Sin.*, 33(5): 484 (1995)—China North-central (Gansu).

wuyishanensis *S.Yun Liang in Acta Phytotax. Sin.*, 34(1): 94 (1996)—China Southeast (Fujian).

yangshuoensis *Tang & Wang ex S.Yun Liang in Acta Phytotax. Sin.*, 33(5): 488 (1995)—China Southeast (Guangxi).

CARICA (Caricac.).

palandensis *V.M. Badillo, V. Van den Eynden & P. Van Damme in Novon*, 10(1): 4 (2000)—Ecuador.

CARINIANA (Lecythidac.).

parvifolia *S.A. Mori, G.T. Prance & Menandro in Bol. Bot. (São Paulo)*, 14: 8 (1995)—Brazil Southeast (Espírito Santo).

CARISSA (Apocynac.).

pichoniana *A.J.M. Leeuwenberg in Wageningen Agric. Univ. Pap.*, 97(2): 101 (1997), nom. nov.: C. verticillata *Pichon*.

stolonifera *(F.M. Bailey) E. Perrot & E. Vogt in Trav. Lab. Mat. Médic.*, 9: 112 (1913), name not accepted as a species by authors: C. ovata var. stolonifera.†

CARLINA (Compos.).

fennica *(Meusel & Kästner) N.N. Tzvelev in Novosti Sist. Vyssh. Rast.*, 31: 264 (1998): C. biebersteinii var. fennica.

× vayredae

nothosubsp. szaferi *(Jasiewicz & Pawł.) R. Govaerts, World Checklist Seed Plants*, 3(1): 13 (1999): C. × szaferi.

CARLOWRIGHTIA (Acanthac.).

ecuadoriana *T.F. Daniel & D.C. Wasshausen in Nordic J. Bot.*, 13(6): 653 (1993)—Ecuador.

sulcata *(Nees) C. Ezcurra in A.L. Cabrera, Fl. Prov. Jujuy (Inst. Nac. Tecn. Agropec., 13)*, 9: 319 (1993): Jacobinia sulcata.

CARLQUISTIA *B.G. Baldwin in Novon*, 9(4): 463 (1999). *COMPOSITAE*

muirii *(A. Gray) B.G. Baldwin in Novon*, 9(4): 463 (1999): Raillardella muirii.

× **CARMICHAELARA** *hort. in Orchid Rev.*, 107(1227): 191 (1999). *ORCHIDACEAE* (Brassavola × Broughtonia × Laelia).

CARMICHAELIA (Legumin.).

carmichaeliae *(Hook. f.) P.B. Heenan in New Zealand J. Bot.*, 36(1): 56 (1998): Notospartium carmichaeliae.

glabrescens *(Petrie) P.B. Heenan in New Zealand J. Bot.*, 36(1): 56 (1998): Notospartium glabrescens.

× hutchinsii *(M.D. Griffiths) P.B. Heenan in New Zealand J. Bot.*, 36(1): 56 (1998): × Carmispartium hutchinsii.

muritai *(A.W. Purdie) P.B. Heenan in New Zealand J. Bot.*, 36(1): 56 (1998): Chordospartium muritai.

stevensonii *(Cheeseman) P.B. Heenan in New Zealand J. Bot.*, 36(1): 56 (1998): Chordospartium stevensonii.

torulosa *(Kirk) P.B. Heenan in New Zealand J. Bot.*, 36(1): 56 (1998): Notospartium torulosum.

vexillata *P.B. Heenan in New Zealand J. Bot.*, 33(4): 471 (1995)—New Zealand South I.

CARNARVONIA (Proteac.).
araliifolia F. Muell.
var. montana *B.P.M. Hyland in Fl Australia,* 16: 496 (1995)—Queensland.

CARPHEPHORUS (Compos.).
subtropicanus *K.R. DeLaney, N. Bissett & J.D. Weidenhamer in Bot. Explor.,* 1: 2 (1999)—Florida.

CARPINUS (Corylac.).
dulongensis *H. Li in Fl. Dulongjian Reg.:* 156 (1993), without latin descr. or type—China South-central (Yunnan).

CARPODIPTERA (Tiliac.).
cubensis Griseb.
subsp. ophiticola *(Bisse) A. Rodríguez Fuentes in Fl. Repúbl. Cuba,* 3(5): 24 (2000): Carpodiptera ophiticola.

CARUM (Umbellif.).
carvi L.
subsp. rosellum *(Woronow) V.N. Voroschilov in A.K. Skvortsov (ed.), Florist. issl. v razn. raĭonakh SSSR:* 184 (1985): C. rosellum.
villosum *Haines, Bot. Bihar Orissa,* 3: 408 (1922)—India.

CARYOCHLOA (Gramin.).
brasiliensis *Trinius, Gram. Panic.:* 248 (1826)—Brazil.†

CARYODAPHNOPSIS (Laurac.).
bilocellata *H. van der Werff & N.K. Dao in Novon,* 9(4): 584 (1999)—Vietnam.

CARYOTA (Palmae).
gigas *W.J. Hahn ex D.R. Hodel in Palm J.,* 139: 51 (1998), without diagnostic latin descr.—Thailand.
kiriwongensis *D.R. Hodel in Palm J.,* 139: 53 (1998), without diagnostic latin descr.—Thailand.
ophiopellis *J.L. Dowe in Austral. Syst. Bot.,* 9(1): 20 (1996)—Vanuatu.
zebrina *G.G. Hambali, R.A. Maturbongs, C.D. Heatubun & J. Dransfield in Palms,* 44(4): 171 (2000)—New Guinea (Irian Jaya, Papua New Guinea).

CASEARIA (Flacourtiac.).
canziala *(Buch.-Ham.) N. Wallich, Report on Calcutta Bot. Gard. to G.A. Bushby:* 31 (1840): Samyda canziala.
glomerata Roxb. & DC.
var. sikkimensis *(N. Mukh.) R.C. Srivastava, Fl. Sikkim (Ranunculac.-Moringac.):* 128 (1998): C. sikkimensis.
grewiifolia
var. hexagona *(Decne.) R. Govaerts, World Checklist Seed Plants,* 3(1): 13 (1999): C. hexagona.
sylvestris Sw.
subsp. myricoides *(Griseb.) J.E. Gutiérrez Amaro in Fl. Repúbl. Cuba,* ser. A, 5(1): 52 (2000): Casearia sylvestris var. myricoides.
tomentosa Roxb.
subsp. reducta *B. Verdcourt in M.D. Dassanayake & W.D. Clayton, eds., Revised Handb. Fl. Ceylon,* 10: 234 (1996)—Sri Lanka.

CASSINE (Celastrac.).
brachycremastra *(Guillaumin) I.H. Müller in Fl. N. Caléd.,* 20: 17 (1996): Elaeodendron brachycremastron.
bupleuroides *(Guillaumin) I.H. Müller in Fl. N. Caléd.,* 20: 28 (1996): Gymnosporia bupleuroides.
kamerunensis *(Loes.) R. Govaerts, World Checklist Seed Plants,* 3(1): 13 (1999): Elaeodendron glaucum var. kamerunense.
parvifolia *I.H. Müller in Fl. N. Caléd.,* 20: 26 (1996), homonym; *see Sond. in Harv. & Sond., Fl. Capensis,* 1: 466 (1860)—New Caledonia.
peragua L.
subsp. affinis *(Sond.) R.H. Archer in S. Afr. J. Bot.,* 63(3): 153 (1997): C. affinis.
subsp. barbara *(L.) R.H. Archer in S. Afr. J. Bot.,* 63(3): 152 (1997): C. barbara.

CASSINE:-
pininsularis *(Hürl.) I.H. Müller in Fl. N. Caléd.,* 20: 22 (1996): Elaeodendron pininsulare.
subsp. poyaensis *I.H. Müller in Fl. N. Caléd.,* 20: 24 (1996)—New Caledonia.
schinoides *(Spreng.) R.H. Archer in S. Afr. J. Bot.,* 63(3): 155 (1997): Elaeodendron schinoides.

CASSIOPE (Ericac.).
membranifolia *R.C. Fang in Novon,* 9(2): 162 (1999)—China South-central (Yunnan).

CASSIPOUREA (Rhizophorac.).
gummiflua Tul.
subsp. ugandensis *(Stapf) K.A. Lye in Lidia,* 4(3): 92 (1998): Dactylopetalum ugandense.

CASSYTHA (Laurac.).
pondoensis Engl.
var. schliebenii *(Robyns & R. Wilczek) M.A. Diniz in Kirkia,* 16(1): 59 (1996 publ. Aug. 1997); *et in Fl. Zambes.,* 9(2): 49 (Sept. 1997): C. schliebenii.

CASTANOPSIS (Fagac.).
damingshanensis *S.L. Mo ex C.C. Huang & Y.T. Chang in Guihaia,* 16(4): 300 (1996)—China Southeast (Guangxi).
globigemmata *Chun & C.C. Huang in Guihaia,* 16(4): 300 (1996)—China South-central (Yunnan).
× kuchugouzhui *C.C. Huang & Y.T. Chang in Guihaia,* 16(4): 301 (1996)—China Southeast (Hunan).(C. sclerophylla × tibetana).
ledongensis *C.C. Huang & Y.T. Chang in Guihaia,* 16(4): 301 (1996)—Hainan.
nigrescens *Chun & C.C. Huang in Guihaia,* 16(4): 301 (1996)—China Southeast (Jiangxi).
purpurella (Miq.) N.P. Balakr.*subsp.*laotica *(Hickel & A. Camus) R. Govaerts in R. Govaerts & D.G. Frodin, World Checklist & Bibliogr. Fagales:* 132 (1998): C. laotica.
subuliformis *Chun & C.C. Huang in Guihaia,* 16(4): 301 (1996)—China Southeast (Guangdong).

CASTILLEJA (Scrophulariac.).
chambersii *M. Egger & R.J. Meinke in Brittonia,* 51(4): 445 (1999)—Oregon.

CASTRILANTHEMUM *R. Vogt & C. Oberprieler in An. Jard. Bot. Madrid,* 54(1): 342 (1996). *COMPOSITAE*
debeauxii *(Degen, Hervier & E. Rev.) R. Vogt & C. Oberprieler in An. Jard. Bot. Madrid,* 54(1): 342 (1996): Pyrethrum debeauxii.

CATACOLEA *B.G. Briggs & L.A.S. Johnson in Telopea,* 7(4): 346 (1998). *RESTIONACEAE*
enodis *B.G. Briggs & L.A.S. Johnson in Telopea,* 7(4): 347 (1998)—Western Australia.

CATALEPIDIA *P.H. Weston in Fl. Australia,* 16: 499 (1995). *PROTEACEAE*
heyana *(F.M. Bailey) P.H. Weston in Fl. Australia,* 16: 499 (1995): Helicia heyana.

CATANTHERA (Melastomatac.).
nummularia *P.C. Bygrave in Kew Bull.,* 53(2): 475 (1998)—Borneo (Brunei).

CATAPODIUM (Gramin.).
balearicum *(Willk.) H. Scholz in Bot. Chronika,* 13: 96 (2000): Desmazeria balearica.
borgesii *H. Scholz in Bot. Chronika,* 13: 102 (2000)—Kriti.

CATASETUM (Orchidac.).
charlesworthii *(Mansf.) R. Jenny in Caesiana,* 11: 37 (1998): C. naso var. charlesworthii.
cucullatum *M.F.F. da Silva & A.T. de Oliveira in Bol. Mus. Paraense Emílio Goeldi, Bot.,* 14(1): 64 (1998 publ. 1999)—Brazil North (Amazonas).
dodsonianum *(E. Aguirre) P.F. Hunt in Orchid Rev.,* 106(1220): 126 (1998): Clowesia dodsoniana.
franchinianum *K.G. de Lacerda in Bradea,* 8(16): 89 (1998)—Brazil West-central (Goiás).
galeatum *K.G. de Lacerda in Bradea,* 8(16): 85 (1998)—Brazil North (Rondônia).

CATASETUM:-
gladiatorium *K.G. de Lacerda in Bradea,* 8(16): 90 (1998)—Brazil West-central (Mato Grosso, Goiás).
maranhense *K.G. de Lacerda & J.B.F. da Silva in Bradea,* 8(13): 69 (1998)—Brazil Northeast (Maranhão, Piauí).
maroaense *G.A. Romero & C. Gómez in Harvard Pap. Bot.,* 5(1): 179 (2000)—Venezuela.
mojuense *A.T. de Oliveira & J.B.F. da Silva in Bol. Mus. Paraense Emílio Goeldi, Bot.,* 14(2): 110 (1998 publ. 1999)—Brazil North (Pará).
osculatum *K.G. de Lacerda & V.P. Castro Neto in Bradea,* 6(44): 381 (1995)—Brazil West-central (Mato Grosso), Brazil North (Rondôa).
× pohlianum *P. Castro & Campacci in Bol. CAOB,* 5(1): 26 (1993)—Brazil Southeast (Rio de Janeiro).(C. hookeri × trulla).
sanguineum Lindl.
var. viride *(T. Moore) R. Jenny in Caesiana,* 11: 30 (1998): C. naso var. viride.
spitzii
var. album *L.C. Menezes in Bol. CAOB,* 28: 60 (1997)—Brazil West-central (Goiás).
var. sanguineum *L.C. Menezes in Bol. CAOB,* 28: 60 (1997)—Brazil West-central (Goiás).
tenebrosum Kraenzl.
f. smaragdinum *D.E. Bennett, E.A. Christenson & B. Collantes M.in Orchids,* 68(5): 478 (1999)—Peru.
tricorne *P. Ortiz V. in Orquideologia,* 20(3): 314 (1997)—Colombia.
tucuruiense *A.T. de Oliveira & J.B.F. da Silva in Bol. Mus. Paraense Emílio Goeldi, Bot.,* 14(2): 112 (1998 publ. 1999)—Brazil North (Pará).
vinaceum
var. album *(L.C. Menezes) L.C. Menezes in Bol. CAOB,* 28: 60 (1997): C. taguariense var. album.
var. splendidum *L.C. Menezes in Bol. CAOB,* 28: 61 (1997)—Brazil West-central (Mato Grosso do Sul).
yavitaense *G.A. Romero-González & C. Gómez in Harvard Pap. Bot.,* 3(1): 54 (1998)—Venezuela.

CATHARIOSTACHYS *S. Dransfield in Kew Bull.,* 53(2): 388 (1998). *GRAMINEAE*
capitata *(Kunth) S. Dransfield in Kew Bull.,* 53(2): 391 (1998): Nastus capitatus.
madagascariensis *(A. Camus) S. Dransfield in Kew Bull.,* 53(2): 394 (1998): Cephalostachyum madagascariense.

CATOLESIA *D.J.N. Hind in Kew Bull.,* 55(4): 942 (2000). *COMPOSITAE*
mentiens *D.J.N. Hind in Kew Bull.,* 55(4): 943 (2000)—Brazil Northeast (Bahia).

CATTLEYA (Orchidac.).
subgen. Aclandia *C.L. Withner, Cattleyas & Relatives,* 1: 21 (1988).
subgen. Aclandia *(Pfitzer) C.L. Withner, Cattleyas & Relatives,* 6: 168 (2000), isonym; *see Withner, Cattleyas & Relatives,* 1: 21 (1988): Cattleya sect. Aclandia.
sect. Aurantiacae *C.L. Withner, Cattleyas & Relatives,* 1: 21 (1988).
subgen. Bicolorea *(Pfitzer) C.L. Withner, Cattleyas & Relatives,* 6: 168 (2000): Cattleya sect. Bicolorea.
subgen. Circumvolva *C.L. Withner, Cattleyas & Relatives,* 1: 21 (1988).
subgen. Falcatae *C.L. Withner, Cattleyas & Relatives,* 1: 21 (1988).
sect. Granulosae *C.L. Withner, Cattleyas & Relatives,* 1: 21 (1988).
sect. Guttatae *C.L. Withner, Cattleyas & Relatives,* 1: 21 (1988).
subgen. Intermediae *C.L. Withner, Cattleyas & Relatives,* 1: 21 (1988).
subgen. Intermediae *(Cogn.) C.L. Withner, Cattleyas & Relatives,* 6: 168 (2000), without full basionym ref.: Cattleya sect. Intermediae.
subgen. Laelioidea *C.L. Withner, Cattleyas & Relatives,* 1: 21 (1988).
subgen. Leopoldia *(Pfitzer) C.L. Withner, Cattleyas & Relatives,* 6: 168 (2000): Cattleya sect. Leopoldia.

CATTLEYA:-
sect. Maximae *C.L. Withner, Cattleyas & Relatives,* 1: 21 (1988).
sect. Moradae *C.L. Withner, Cattleyas & Relatives,* 1: 21 (1988), as 'subgen.'.
subgen. Rhizantha *C.L. Withner, Cattleyas & Relatives,* 1: 21 (1988).
subgen. Rhizanthemum *(Cogn.) C.L. Withner, Cattleyas & Relatives,* 6: 168 (2000), without full basionym ref.: Cattleya subsect. Rhizanthemum.
subgen. Schomburgkoidea *C.L. Withner, Cattleyas & Relatives,* 1: 21 (1988).
subgen. Stellatae *C.L. Withner, Cattleyas & Relatives,* 1: 21 (1988).
sect. Xantheae *C.L. Withner, Cattleyas & Relatives,* 1: 21 (1988).
araguaiensis Pabst
f. alba *(L.C. Menezes) E.A. Christenson in Lindleyana,* 11(1): 14 (1996): C. araguaiensis var. alba.
bicolor Lindl.
subsp. canastrensis *L.C. Menezes & G.J. Braem in Schlechteriana,* 3(2): 54 (1992)—Brazil Southeast (Minas Gerais).
× colnagiana *L.C. Menezes in Orquidário,* 10(1): 9 (1996)—Brazil Northeast (Bahia).
dormaniana Rchb. f.
f. alba *(L.C. Menezes) E.A. Christenson in Lindleyana,* 11(1): 14 (1996): C. dormaniana var. alba.
× extricata *J. Gonzalez Raposo in Schlechteriana,* 3(2): 48 (1992)—Brazil South (Santa Catarina).(C. intermedia × leopoldii).
× gransabanensis *K. Senghas in J. Orchideenfreund,* 6(1): 58 (1999)—Venezuela.(C. jenmanii × lawrenceana).
guttata
var. caerulea *L.C. Menezes in Bol. CAOB,* 5(1): 22, 20 (1993)—Brazil West-central (Brasília Distrito Federal).
× joaquiniana *F.E. Miranda in Bradea,* 8(22): 128 (1999)—Brazil Southeast (Minas Gerais).(C. bicolor × walkeriana).
× mesquitae *L.C. Menezes in Bol. CAOB,* 26: 24 (1996); *et in Orchid Dig.,* 62(1): 22 (1998)—Brazil West-central (Goiás).(C. nobilior × walkeriana var. princeps).
× mixta *L.C. Menezes in Bol. CAOB,* 4(2): 34 (1992)—Brazil Southeast (Espírito Santo).(C. guttata × schofieldiana).
rex
var. flammea *S.A.A. Oliveira in Bol. CAOB,* 4(1): 26 (1992), without latin descr. or type—Peru.
violacea (Kunth) Rolfe
f. alba *(Rolfe) E.A. Christenson in Lindleyana,* 11(1): 14 (1996): C. superba var. alba.
walkeriana
var. princeps *L.C. Menezes in Bol. CAOB,* 5(3): 14 (1993)—Brazil West-central (Goiás).

CATUNAREGAM (Rubiac.).
longispina *(Link) D.D. Tirvengadum in Edinburgh J. Bot.,* 53(1): 95 (1996): Posoqueria longispina.

CAUCASALIA *B. Nordenstam in Pl. Syst. Evol.,* 206(1–4): 22 (1997). *COMPOSITAE*
macrophylla *(M. Bieb.) B. Nordenstam in Pl. Syst. Evol.,* 206(1–4): 23 (1997): Cacalia macrophylla.
parviflora *(M. Bieb.) B. Nordenstam in Pl. Syst. Evol.,* 206(1–4): 29 (1997): Cineraria parviflora.
platyphylloides *(Sommier & H. Lév.) B. Nordenstam in Pl. Syst. Evol.,* 206(1–4): 25 (1997): Senecio platyphylloides.
similiflora *(Kolak.) B. Nordenstam in Pl. Syst. Evol.,* 206(1–4): 27 (1997): Senecio similiflorus.

CAUDANTHERA *D.C.H. Plowes in Haseltonia,* 3: 58 (1995). *ASCLEPIADACEAE*
mireillae *(Lavranos) D.C.H. Plowes in Haseltonia,* 3: 58 (1995): Caralluma mireillae.
sinaica *(Decne.) D.C.H. Plowes in Haseltonia,* 3: 58 (1995): Boucerosia sinaica.

CAULIPSOLON *C. Klak in Bot. Jahrb.,* 120(3): 364 (1998). *AIZOACEAE*
rapaceum *(Jacq.) C. Klak in Bot. Jahrb.,* 120(3): 364 (1998): Mesembryanthemum rapaceum.

CAUSTIS (Cyperac.).
blakei Kük.
subsp. macrantha *M.E. Johnston & J.H. Webber in Austrobaileya,* 4(4): 615 (1997)—Queensland.

CAVENDISHIA (Ericac.).
antioquiensis *J.L. Luteyn & D.S. Sylva S.in Brittonia,* 51(3): 282 (1999)—Colombia.
awa *J.L. Luteyn in Fl. Ecuador,* 54(147): 104 (1996)—Ecuador.
davidsei *J.L. Luteyn in Brittonia,* 48(2): 241 (1996)—Costa Rica, Panama.
gomezii *J.L. Luteyn in Brittonia,* 48(2): 244 (1996)—Panama.
herrerae *J.L. Luteyn & J.F. Morales in Brittonia,* 48(4): 517 (1996 publ. 1997)—Costa Rica.
limonensis *J.L. Luteyn in Brittonia,* 48(2): 244 (1996)—Costa Rica.
linearifolia *J.L. Luteyn & J.F. Morales in Brittonia,* 48(4): 516 (1996 publ. 1997)—Costa Rica.
luteynii *J.F. Morales in Brittonia,* 48(4): 518 (1996 publ. 1997)—Costa Rica.
osaensis *J.L. Luteyn & J.F. Morales in Brittonia,* 48(4): 514 (1996 publ. 1997)—Costa Rica.
parviflora *J.L. Luteyn in Fl. Ecuador,* 54(147): 103 (1996)—Ecuador.
viridiflora *J.L. Luteyn & D.S. Sylva S. in Brittonia,* 51(3): 288 (1999)—Colombia.

CAXAMARCA *M.O. Dillon & A. Sagástegui Alva in Novon,* 9(2): 156 (1999).*COMPOSITAE*
sanchezii *M.O. Dillon & A. Sagástegui Alva in Novon,* 9(2): 156 (1999)—Peru.

CAYAPONIA (Cucurbitac.).
longiloba *A.K. Monro in Novon,* 6(1): 82 (1996)—Mexico Southwest (Oaxaca), Mexico Southeast (Chiapas), Guatemala, Belize, Honduras, Nicaragua, Costa Rica.

CAYRATIA (Vitac.).
albifolia *C.L. Li in Chinese J. Appl. Environ. Biol.,* 2(1): 51 (1996)—China South-central (Sichuan, Guizhou, Hubei, Yunnan), China Southeast (Fujian, Guangxi, Hunan, Jiangxi, Zhejiang).
var. glabra *(Gagnep.) C.L. Li in Chinese J. Appl. Environ. Biol.,* 2(1): 52 (1996): C. oligocarpa f. glabra.
cordifolia *C.Y. Wu ex C.L. Li in Chinese J. Appl. Environ. Biol.,* 2(1): 49 (1996)—China South-central (Yunnan).
daliensis *C.L. Li in Guihaia,* 17(2): 98 (1997)—China South-central (Yunnan).
fugongensis *C.L. Li in Guihaia,* 17(2): 97 (1997)—China South-central (Yunnan).
japonica (Thunb. ex Murray) Gagnep.
var. pseudotrifolia *(W.T. Wang) C.L. Li in Chinese J. Appl. Environ. Biol.,* 2(1): 51 (1996): C. pseudotrifolia.
medoensis *C.L. Li in Chinese J. Appl. Environ. Biol.,* 2(1): 50 (1996)—Xizang.
menglaensis *C.L. Li in Acta Bot. Yunnanica,* 19(3): 219 (1997)—China South-central (Yunnan).
mollissima (Wall.) Gagnep.
var. lanceolata *C.L. Li in Chinese J. Appl. Environ. Biol.,* 2(1): 50 (1996)—Hainan.
timoriensis *(DC.) C.L. Li in Chinese J. Appl. Environ. Biol.,* 2(1): 50 (1996): Cissus timoriensis.
var. mekongensis *(C.Y. Wu ex W.T. Wang) C.L. Li in Chinese J. Appl. Environ. Biol.,* 2(1): 50 (1996): C. mekongensis.

CEARANTHES *P. Ravenna in Onira, Bot. Leafl.,* 5(3): 11 (2000). *AMARYLLIDACEAE*
fuscoviolacea *P. Ravenna in Onira, Bot. Leafl.,* 5(3): 11 (2000)—Brazil Northeast (Ceará).

CEBALLOSIA *G. Kunkel ex H. Förther in Sendtnera,* 5: 129 (1998). *BORAGINACEAE*
fruticosa *(L.f.) G. Kunkel ex H. Förther in Sendtnera,* 5: 130 (1998): Messerschmidia fruticosa.
var. angustifolia *(Lam.) G. Kunkel ex H. Förther in Sendtnera,* 5: 130 (1998): Messerschmidia angustifolia.

CECROPIA (Cecropiac.).
annulata *C.C. Berg & P. Franco Rosselli in Novon,* 6(3): 245 (1996)—Bolivia.

CECROPIA:-
heterochroma *C.C. Berg & P. Franco Rosselli in Novon,* 6(3): 246 (1996)—Panama, Colombia.
puberula *C.C. Berg & P. Franco Rosselli in Novon,* 6(3): 248 (1996)—Peru.
tacuna *C.C. Berg & P. Franco Rosselli in Novon,* 6(3): 248 (1996)—Peru.

CEIBA (Bombacac.).
chodatii *(Hassl.) P. Ravenna in Onira, Bot. Leafl.,* 3(15): 44 (1998): Chorisia chodatii.
crispiflora *(Humb., Bonpl. & Kunth) P. Ravenna in Onira, Bot. Leafl.,* 3(15): 45 (1998): Chorisia crispiflora.
incana *(A. Robyns) P. Ravenna in Onira, Bot. Leafl.,* 3(15): 47 (1998): Chorisia incana.
integrifolia *(Ulbr.) P. Ravenna in Onira, Bot. Leafl.,* 3(15): 46 (1998): Chorisia integrifolia.
jaibana *P. Ravenna in Onira, Bot. Leafl.,* 3(15): 48 (1998)—Brazil Southeast (Minas Gerais).
mythica *P. Ravenna in Onira, Bot. Leafl.,* 3(15): 47 (1998)—Peru.
soluta *(Donn. Sm.) P. Ravenna in Onira, Bot. Leafl.,* 3(15): 47 (1998): Chorisia soluta.
speciosa *(A. St.-Hil., A. Juss. & Cambess.) P. Ravenna in Onira, Bot. Leafl.,* 3(15): 46 (1998): Chorisia speciosa.
ventricosa *(Nees & Mart.) P. Ravenna in Onira, Bot. Leafl.,* 3(15): 47 (1998): Chorisia ventricosa.

CELASTRUS (Celastrac.).
tristyla *J.O. Voigt, Hort. Suburb. Calcutt.:* 167 (1845), nom. nov.: C. trigyna *Roxb.*

CELIANTHA *B. Maguire in Mem. New York Bot. Gard.,* 32: 382 (1981). *GENTIANACEAE*
bella *B. Maguire & J.A. Steyermark in Mem. New York Bot. Gard.,* 32: 383 (1981)—Venezuela.
chimantensis *(Steyerm. & Maguire) B. Maguire in Mem. New York Bot. Gard.,* 32: 387 (1981): Lisianthus chimantensis.
imthurniana *(Oliv.) B. Maguire in Mem. New York Bot. Gard.,* 32: 387 (1981): Lisianthus imthurnianus.

CELMISIA (Compos.).
costiniana *M. Gray & D.R. Given in Austral. Syst. Bot.,* 12(2): 204 (1999)—New South Wales, Victoria.
latifolia *(F. Muell. ex Benth.) M. Gray & D.R. Given in Austral. Syst. Bot.,* 12(2): 201 (1999): C. longifolia var. latifolia.
pugioniformis *M. Gray & D.R. Given in Austral. Syst. Bot.,* 12(2): 203 (1999)—New South Wales (New South Wales, Australian Capital Territory), Victoria.
tomentella *M. Gray & D.R. Given in Austral. Syst. Bot.,* 12(2): 202 (1999)—Victoria, New South Wales (Australian Capital Territory, New South Wales).

CELTIS (Ulmac.).
bungeana Blume
var. deqinensis *Xiang W. Li & G.S. Fan in Acta Bot. Yunnanica,* 20(2): 160 (1998), as 'deqiensis'—China South-central (Yunnan).
gongshanensis *Xiang W. Li & G.S. Fan in Acta Bot. Yunnanica,* 20(2): 160 (1998)—China South-central (Yunnan).
planchoniana *K.I. Christensen in A. Strid & Kit Tan (eds.), Fl. Hellenica,* 1: 54 (1997), nom. nov.: C. glabrata *Steven ex Planch.*
roxburghii *J.O. Voigt, Hort. Suburb. Calcutt.:* 294 (1845), nom. nov.: C. trinervia *Roxb.*

CENCHRUS (Gramin.).
sect. Africani *J. Chrtek & J. Osbornová in Acta Univ. Carol., Biol.,* 39(2): 90 (1995 publ. 1996).
sect. Biflori *J. Chrtek & J. Osbornová in Acta Univ. Carol., Biol.,* 39(2): 90 (1995 publ. 1996).
sect. Caliculati *J. Chrtek & J. Osbornová in Acta Univ. Carol., Biol.,* 39(2): 90 (1995 publ. 1996).
subsect. Caribaei *J. Chrtek & J. Osbornová in Acta Univ. Carol., Biol.,* 39(2): 89 (1995 publ. 1996).
sect. Cenchropsis *(Nash) J. Chrtek & J. Osbornová in Acta Univ. Carol., Biol.,* 39(2): 90 (1995 publ. 1996): gen. Cenchropsis.

CENCHRUS:-
sect. Pacifici *J. Chrtek & J. Osbornová in Acta Univ. Carol., Biol.*, 39(2): 91 (1995 publ 1996).
subsect. Paucisetosi *J. Chrtek & J. Osbornová in Acta Univ. Carol., Biol.*, 39(2): 90 (1995 publ. 1996).
sect. Pennisetopsis *J. Chrtek & J. Osbornová in Acta Univ. Carol., Biol.*, 39(2): 90 (1995 publ. 1996).
subsect. Pennisetopsis *J. Chrtek & J. Osbornová in Acta Univ. Carol., Biol.*, 39(2): 90 (1995 publ. 1996).
subgen. Pennisetopsis *J. Chrtek & J. Osbornová in Acta Univ. Carol., Biol.*, 39(2): 90 (1995 publ. 1996).
subsect. Pilosi *J. Chrtek & J. Osbornová in Acta Univ. Carol., Biol.*, 39(2): 89 (1995 publ. 1996).
subgen. Setigerus *J. Chrtek & J. Osbornová in Acta Univ. Carol., Biol.*, 39(2): 91 (1995 publ. 1996).
sect. Spinosi *J. Chrtek & J. Osbornová in Acta Univ. Carol., Biol.*, 39(2): 89 (1995 publ. 1996).
subsect. Spinosi *J. Chrtek & J. Osbornová in Acta Univ. Carol., Biol.*, 39(2): 89 (1995 publ. 1996).
ciliaris L.
subsp. ibrahimii *J. Chrtek & J. Osbornová in Acta Univ. Carol., Biol.*, 39(2): 99 (1995 publ. 1996)—North Yemen, Egypt, Sinai.
pennisetiformis Hochst. & Steud.
subsp. glabrata *J. Chrtek & J. Osbornová in Acta Univ. Carol., Biol.*, 39(2): 104 (1995 publ. 1996)—Egypt, Sudan.

CENTAUREA (Compos.).
sect. Africanae *M.V. Agababian in Lagascalia*, 19(1–2): 892 (1997).
subsect. Africanae *M.V. Agababian in Lagascalia*, 19(1–2): 892 (1997).
subsect. Aralocaspicae *M.V. Agababian in Lagascalia*, 19(1–2): 894 (1997).
sect. Arpenses *A.D. Mikheev in Bot. Zhurn.*, 84(9): 108 (1999).
sect. Fischerianae *(Czerep.) A.D. Mikheev in Bot. Zhurn.*, 84(9): 104 (1999): C. subsect. Fischerianae.
subsect. Heterocarpa *F.J. Fernández Casas in Fontqueria*, 48: 205 (1997 publ. 1998).
subsect. Iranicae *M.V. Agababian in Lagascalia*, 19(1–2): 897 (1997).
subgen. Pseudoseridia *(Wagenitz) E.Tz. Gabrieljan in Fl. Armenii*, 9: 421 (1995): C. sect. Pseudoseridia.
subsect. Ruthenicae *M.V. Agababian in Lagascalia*, 19(1–2): 894 (1997).
subsect. Turkestanicae *M.V. Agababian in Lagascalia*, 19(1–2): 894 (1997).
subsect. Vicentinae *M.V. Agababian in Lagascalia*, 19(1–2): 891 (1997).
sect. Vicentinae *M.V. Agababian in Lagascalia*, 19(1–2): 891 (1997).
adusta *M. Gandoger, Obs. Centaurea decipiens etc.*: 14 (1873)—France.
aeolica Guss. ex Lojac.
subsp. pandataria *(Fiori & Bég.) B. Anzalone in Boll. Soc. Sarda Sci. Nat.*, 30: 512 (1994–95 publ. 1995): C. cineraria var. pandataria.
aksamica *A. Soldano in Compositae Newsl.*, 32: 17 (1998), nom. nov.: C. veneris *B.L. Burtt & P.H. Davis.*
alba L.
subsp. sterilis *(Steven) A.D. Mikheev in Bot. Zhurn.*, 85(3): 120 (2000): C. sterilis.
argentea L.
var. chionantha *N.J. Turland & L. Chilton in Bot. Chronika*, 13: 77 (2000)—Kriti.
subsp. macrothysana *(Rech. f.) N.J. Turland & L. Chilton in Bot. Chronika*, 13: 76 (2000): C. argentea var. macrothysana.
boissieri DC.
var. saguntina *(Mateo & M.B. Crespo) O. de Bolòs & J. Vigo, Fl. Països Catalans*, 3: 933 (1995): C. saguntina.
var. segariensis *(Figuerola, Peris & Stübing) O. de Bolòs & J. Vigo, Fl. Països Catalans*, 3: 934 (1995): C. segariensis.
cheiranthifolia Willd.
subsp. willdenowii *(Czerep.) A.D. Mikheev in Bot. Zhurn.*, 84(9): 104 (1999): C. willdenowii.
choulettei *M. Gandoger, Dec. Pl. Nov.*, 1: 28 (1875)—Algeria.
corsica *M. Gandoger, Dec. Pl. Nov.*, 1: 27 (1875)—Corse.

CENTAUREA:-
crocata *M. Gandoger, Obs. Centaurea decipiens etc.*: 13 (1873)—France.
dalmatica A. Kern.
subsp. lubenicensis *I. Trinajstić & Z. Pavletić in Nat. Croat.*, 8(1): 58 (1999)—Yugoslavia.
dealbescens *M. Gandoger, Obs. Centaurea decipiens etc.*: 10 (1873)—France.
demirkapiensis *K. Micevski in Prilozi Oddel. Biol. Med. Nauki Makedonska Akad. Nauk. Umet.*, 8(1–2): 49 (1987 publ. 1990)—Yugoslavia.
eriocaulon *M. Gandoger, Obs. Centaurea decipiens etc.*: 12 (1873)—France.
fallacina *M. Gandoger, Obs. Centaurea decipiens etc.*: 16 (1873)—France.
flexilis *M. Gandoger, Obs. Centaurea decipiens etc.*: 11 (1873)—France.
foliosa *M. Gandoger, Obs. Centaurea decipiens etc.*: 11 (1873)—France.
galicicae *K. Micevski in Acta Bot. Croat.*, 44: 83 (1985)—Yugoslavia.
gerhardii *M.V. Agababian in Lagascalia*, 19(1–2): 897 (1997)—Afghanistan.
golestanica *H. Akhani & G. Wagenitz in Edinburgh J. Bot.*, 56(1): 6 (1999)—Iran.
gracilescens *M. Gandoger, Obs. Centaurea decipiens etc.*: 13 (1873)—France.
grandifolia *M. Gandoger, Obs. Centaurea decipiens etc.*: 7 (1873)—France.
hadimensis *G. Wagenitz, K. Ertugrul & H. Dural in Willdenowia*, 28(1–2): 157 (1998)—Turkey.
incurvata *M. Gandoger, Obs. Centaurea decipiens etc.*: 8 (1873)—France.
intermixta *M. Gandoger, Obs. Centaurea decipiens etc.*: 7 (1873)—France.
jacea L.
subsp. microptilon *(Godr.) E. Sauer, Gefässpfl. des Saarlandes (Natur u. Landschaft Saarland, Sonderb. 5)*: 149 (1993), without basionym ref.: Basionym not stated.
subsp. substituta *(Czerep.) A.D. Mikheev in Bot. Zhurn.*, 84(9): 105 (1999): C. substituta.
kavadarensis *K. Micevski in Prilozi Oddel. Biol. Med. Nauki Makedonska Akad. Nauk. Umet.*, 8(1–2): 49 (1987 publ. 1990)—Yugoslavia.
kimeridzei *M.K. Kimeridze in Bot. Zhurn.*, 80(2): 76 (1995)—Transcaucasus (Gruziya).
kunkelii *N. Garcia-Jacas in Ann. Bot. Fenn.*, 35(3): 165 (1998)—Spain.
littorea *M. Gandoger, Dec. Pl. Nov.*, 1: 27 (1875)—France.
lucescens *M. Gandoger, Obs. Centaurea decipiens etc.*: 14 (1873)—France.
lyratifolia *M. Gandoger, Obs. Centaurea decipiens etc.*: 11 (1873)—France.
messenicolasiana *T. Georgiadis, G. Dimitrellos & E. Routsi in Willdenowia*, 25(2): 563 (1996)—Greece.
micrantha S.G. Gmel.
subsp. kemulariae *(Dumbadze) A.D. Mikheev in Bot. Zhurn.*, 85(3): 118 (2000): C. kemulariae.
monticola Boiss. ex DC.
subsp. orthoacantha *(Pau & Font Quer) M. Ibn Tattou in Bocconea*, 8: 29 (1998): C. boissieri var. orthocantha.
subsp. perplexans *(Emb. & Maire) M. Ibn Tattou in Bocconea*, 8: 29 (1998): C. dissecta var. perplexans.
myrioclada *M. Gandoger, Obs. Centaurea decipiens etc.*: 9 (1873)—France.
occasus *F.J. Fernández Casas in Fontqueria*, 48: 221 (1997 publ. 1998)—Portugal.
oriolii-bolosii *F.J. Fernández Casas in Fontqueria*, 48: 220 (1997 publ. 1998)—Morocco.
ornata Willd.
var. galianoi *(Fern. Casas & Susanna) F.J. Fernández Casas in Fontqueria*, 48: 221 (1997 publ. 1998): C. galianoi.
ovina Pall. ex Willd.
subsp. caprina *(Steven) A.D. Mikheev in Bot. Zhurn.*, 85(3): 119 (2000): C. caprina.
subsp. gulissaschwilii *(Dumbadze) A.D. Mikheev in Bot. Zhurn.*, 85(3): 119 (2000): C. gulissashwilii.
subsp. hohenackeri *(Steven) A.D. Mikheev in Bot. Zhurn.*, 85(3): 119 (2000): C. hohenackeri.
× paredensis *(G. López) G. Mateo Sanz & V.J. Arán Redó in Flora Montiberica*, 16: 11 (2000): C. × bilbilitana nm. paredensis.

CENTAUREA:-
phrygia L.
subsp. salicifolia *(M. Bieb. ex Willd.) A.D. Mikheev in Bot. Zhurn.*, 84(9): 105 (1999): C. salicifolia.
poeltiana *D. Puntillo in Flora Medit.*, 6: 219 (1996)—Italy.
pseudomicroptilon *M. Gandoger, Obs. Centaurea decipiens etc.*: 8 (1873)—France.
pseudophrygia C.A. Mey.
subsp. abnormis *(Czerep.) A.D. Mikheev in Bot. Zhurn.*, 84(9): 106 (1999): C. abnormis.
subsp. alutacea *(Dobrocz.) A.D. Mikheev in Bot. Zhurn.*, 84(9): 106 (1999): C. alutacea.
pseudoscabiosa Boiss. & Buhse
subsp. ossethica *(Sosn.) E.Tz. Gabrieljan in Fl. Armenii*, 9: 416 (1995), as 'ossetica': C. ossethica.
pseudosquarrosa *A.D. Mikheev in Bot. Zhurn.*, 85(3): 119 (2000), nom. nov., without replaced synonym ref.: C. squarrosa *Willd.*
pulmonariifolia *M. Gandoger, Obs. Centaurea decipiens etc.*: 8 (1873)—France.
radichii *M. Plazibat in Nat. Croat.*, 9(Suppl. 1): 79 (2000), nom. nov.: C. cuspidata subsp. elegantissima *Radić.*
reflexa Lam.
subsp. salonitana *(Vis.) A.D. Mikheev in Bot. Zhurn.*, 85(3): 116 (2000): C. salonitana.
subsp. sosnovskyi *(Grossh.) A.D. Mikheev in Bot. Zhurn.*, 85(3): 116 (2000): C. sosnovskyi.
sabauda *M. Gandoger, Obs. Centaurea decipiens etc.*: 10 (1873)—France, Italy.
scabiosa L.
subsp. apiculata *(Ledeb.) A.D. Mikheev in Bot. Zhurn.*, 85(3): 117 (2000): C. apiculata.
var. araratica *(Azn.) A.D. Mikheev in Bot. Zhurn.*, 85(3): 117 (2000), without basionym ref.: Basionym not stated.
var. glehnii *(Trautv.) A.D. Mikheev in Bot. Zhurn.*, 85(3): 117 (2000): C. glehnii.
var. ossethica *(Sosn.) A.D. Mikheev in Bot. Zhurn.*, 85(3): 117 (2000), as 'ossetica': C. ossethica.
subsp. pseudoscabiosa *(Boiss. & Buhse) A.D. Mikheev in Bot. Zhurn.*, 85(3): 117 (2000): C. pseudoscabiosa.
scripczinskyi *A.D. Mikheev in Bot. Zhurn.*, 82(11): 86 (1997)—North Caucasus (Stavropol).
× segimonensis *S. Pajarón Sotomayor & F.J. Fernández Casas in Fontqueria*, 48: 223 (1997 publ. 1998)—Spain. (C. gabrielis-blancae × granatensis).
seminutans *M. Gandoger, Obs. Centaurea decipiens etc.*: 9 (1873)—France.
skopjensis *K. Micevski in Prilozi Oddel. Biol. Med. Nauki Makedonska Akad. Nauk. Umet.*, 8(1–2): 52 (1987 publ. 1990)—Yugoslavia.
stevenii M. Bieb.
subsp. armeniaca *E.Tz. Gabrieljan in Fl. Armenii*, 9: 421 (1995), contrary to Art. 37.4 ICBN (1994)—Transcaucasus (Armeniya).
subcana *M. Gandoger, Obs. Centaurea decipiens etc.*: 15 (1873)—France.
subconica *M. Gandoger, Obs. Centaurea decipiens etc.*: 12 (1873)—France.
tamanianiae *M.V. Agababjan in Biol. Zhurn. Arm.*, 42(3): 187 (1989)—Transcaucasus (Armeniya).
tomorosii *K. Micevski in Acta Bot. Croat.*, 44: 84 (1985)—Yugoslavia.
torrida *M. Gandoger, Obs. Centaurea decipiens etc.*: 15 (1873)—France.
transcaucasica Sosn. ex Grossh.
subsp. alexandri *(Bordz.) A.D. Mikheev in Bot. Zhurn.*, 85(3): 121 (2000): C. alexandri.
subsp. georgica *(Klokov) A.D. Mikheev in Bot. Zhurn.*, 85(3): 121 (2000): C. georgica.
treskana *K. Micevski in Prilozi Oddel. Biol. Med. Nauki Makedonska Akad. Nauk. Umet.*, 8(1–2): 47 (1987 publ. 1990)—Yugoslavia.
trinervia Willd.
subsp. kobstanica *(Tzvelev) A.D. Mikheev in Bot. Zhurn.*, 84(9): 108 (1999): C. kobstanica.
valida *M. Gandoger, Obs. Centaurea decipiens etc.*: 15 (1873)—France.
visianiana *M. Plazibat in Nat. Croat.*, 9(Suppl. 1): 81 (2000), nom. nov.: C. cuspidata subsp. visianii *Radić.*

CENTAUREA:-
wagenitzii *K. Micevski in Prilozi Oddel. Biol. Med. Nauki Makedonska Akad. Nauk. Umet.*, 8(1–2): 49 (1987 publ. 1990), homonym; *see Hub.-Mor. in Bauhinia*, 3: 315 (1967)—Yugoslavia.
yozgatensis *G. Wagenitz in Ann. Naturhist. Mus. Wien*, 98B(Suppl.): 176 (1996)—Turkey.

CENTAURIUM (Gentianac.).
clementii *(Domin) L.G. Adams in Fl. Australia*, 28: 315 (1996): Erythraea clementii.
littorale (Turner) Gilmour
subsp. compressum *(Hayne) J. Kirschner in Preslia*, 72(1): 87 (2000): Erythraea compressa.
pulchellum (Sw.) Druce
var. meyeri *(Bunge) S. Omer in Fl. Pakistan*, 197: 10 (1995): Erythraea meyeri.
quadrifolium (L.) G. López & C.E. Jarvis
var. barrelieri *(Dufour) O. de Bolòs & J. Vigo, Fl. Països Catalans*, 3: 116 (1995): Erythraea barrelieri.

CENTELLA (Umbellif.).
annua *M.T.R. Schubert & B.-E. van Wyk in Nordic J. Bot.*, 17(3): 309 (1997)—Cape Province.
brachycarpa *M.T.R. Schubert & B.-E. van Wyk in Nordic J. Bot.*, 18(4): 467 (1998)—Cape Province.
calcaria *M.T.R. Schubert & B.-E. van Wyk in Nordic J. Bot.*, 17(3): 310 (1997)—Cape Province.
cryptocarpa *M.T.R. Schubert & B.-E. van Wyk in Nordic J. Bot.*, 19(4): 433 (1999)—Cape Province.
dolichocarpa *M.T.R. Schubert & B.-E. van Wyk in Nordic J. Bot.*, 18(4): 468 (1998)—Cape Province.
glauca *M.T.R. Schubert & B.-E. van Wyk in Bothalia*, 26(1): 50 (1996)—Cape Province.
gymnocarpa *M.T.R. Schubert & B.-E. van Wyk in Nordic J. Bot.*, 19(4): 433 (1999)—Cape Province.
longifolia *(Adamson) M.T.R. Schubert & B.-E. van Wyk in Nordic J. Bot.*, 19(4): 432 (1999): C. montana var. longifolia.
pilosa *M.T.R. Schubert & B.-E. van Wyk in Nordic J. Bot.*, 18(4): 469 (1998)—Cape Province.
tridentata (L.f.) Drude ex Domin
var. dregeana *(Sond.) M.T.R. Schubert & B.-E. van Wyk in Nordic J. Bot.*, 17(3): 314 (1997): Hydrocotyle dregeana.
var. hermanniifolia *(Eckl. & Zeyh.) M.T.R. Schubert & B.-E. van Wyk in Nordic J. Bot.*, 17(3): 313 (1997): Hydrocotyle hermanniifolia.
var. litoralis *(Eckl. & Zeyh.) M.T.R. Schubert & B.-E. van Wyk in Nordic J. Bot.*, 17(3): 313 (1997): Hydrocotyle litoralis.

CENTRANTHUS (Valerianac.).
amazonum *A. Fridlender & A. Raynal-Roques in Adansonia*, 20(2): 328 (1998)—Sardegna.
angustifolius (Mill.) DC.
subvar. dertosensis *O. de Bolòs & J. Vigo, Fl. Països Catalans*, 3: 624 (1995)—Spain.

CENTRAPALUS (Compos.).
acrocephalus *(Klatt) H. Robinson in Proc. Biol. Soc. Wash.*, 112(1): 236 (1999): Vernonia acrocephala.
africanus *(Sond.) H. Robinson in Proc. Biol. Soc. Wash.*, 112(1): 236 (1999): Vernonella africana.
chthonocephalus *(O. Hoffm.) H. Robinson in Proc. Biol. Soc. Wash.*, 112(1): 236 (1999): Vernonia chthonocephala.
denudatus *(Hutch. & Burtt) H. Robinson in Proc. Biol. Soc. Wash.*, 112(1): 236 (1999): Vernonia denudata.
kirkii *(Oliv. & Hiern) H. Robinson in Proc. Biol. Soc. Wash.*, 112(1): 236 (1999): Vernonia kirkii.
pauciflorus *(Willd.) H. Robinson in Proc. Biol. Soc. Wash.*, 112(1): 236 (1999): Conyza pauciflora.
praemorsus *(Muschl.) H. Robinson in Proc. Biol. Soc. Wash.*, 112(1): 236 (1999): Vernonia praemorsa.
purpureus *(Sch. Bip. ex Walp.) H. Robinson in Proc. Biol. Soc. Wash.*, 112(1): 236 (1999): Vernonia purpurea.
subaphyllus *(Baker) H. Robinson in Proc. Biol. Soc. Wash.*, 112(1): 236 (1999): Vernonia subaphylla.

CENTRATHERUM (Compos.).
fruticosum *S. Vidal, Revis. Pl. Vasc. Filip.*: 159 (1886)—Philippines.

CENTROMADIA (Compos.).
parryi (Greene) Greene
subsp. australis *(D.D. Keck) B.G. Baldwin in Novon*, 9(4): 466 (1999): Hemizonia parryi subsp. australis.
subsp. congdonii *(B.L. Rob. & Greenm.) B.G. Baldwin in Novon*, 9(4): 466 (1999): Hemizonia congdonii.
subsp. rudis *(Greene) B.G. Baldwin in Novon*, 9(4): 467 (1999): C. rudis.
pungens (Hook. & Arn.) Greene
subsp. laevis *(D.D. Keck) B.G. Baldwin in Novon*, 9(4): 467 (1999): Hemizonia pungens subsp. laevis.
subsp. maritima *(Greene) B.G. Baldwin in Novon*, 9(4): 467 (1999): C. maritima.
subsp. septentrionalis *(D.D. Keck) B.G. Baldwin in Novon*, 9(4): 467 (1999): Hemizonia pungens subsp. septentrionalis.

CENTRONIA (Melastomatac.).
ruizii *G. Lozano-Contreras in Rev. Acad. Colomb. Cienc. Exact. Fis. Nat.*, 23(Suplemento Especial): 80 (1999 publ. 2000)—Colombia.

CENTROPOGON (Campanulac.)
sect. Niveopsis *T.G. Lammers in Brittonia*, 50(2): 257 (1998).
arachnocalyx *T.G. Lammers in Brittonia*, 50(2): 246 (1998)—Colombia.
ayavacensis *(Willd. ex Schult.) T.G. Lammers in Novon*, 9(3): 386 (1999): Lobelia ayavacensis.
subsp. cylindricus *(Gleason) T.G. Lammers in Novon*, 9(3): 386 (1999): Siphocampylus cylindricus.
connatilobatus *T.G. Lammers in Brittonia*, 50(2): 248 (1998)—Venezuela.
david-smithii *T.G. Lammers in Brittonia*, 50(2): 255 (1998)—Peru.
dianae *T.G. Lammers in Brittonia*, 50(2): 257 (1998)—Peru.
dillonii *T.G. Lammers in Brittonia*, 50(2): 243 (1998)—Peru.
exsertus *T.G. Lammers in Brittonia*, 50(2): 248 (1998)—Colombia.
jeppesenii *T.G. Lammers in Brittonia*, 50(2): 252 (1998)—Ecuador.
knoxii *T.G. Lammers in Brittonia*, 50(2): 249 (1998)—Peru.
simulans *T.G. Lammers in Brittonia*, 50(2): 251 (1998)—Peru.
steinii *T.G. Lammers in Brittonia*, 50(2): 245 (1998)—Ecuador.
wilburii *T.G. Lammers in Brittonia*, 50(2): 253 (1998)—Mexico Southwest (Oaxaca).

CENTROSEMA (Legumin.).
lucia-helenae *M. Brandão & M.L. Gavilanes in Daphne*, 6(3): 7 (1996)—Brazil Southeast (Minas Gerais).
macrocarpum Benth.
var. andinum *R. Schultze-Kraft & J. Belalcázar in Caldasia*, 18(86): 45 (1995)—Colombia.
platycarpum
var. magnificum *(Afr. Fern. & P. Bezerra) Afr. Fernandes in Albertoa, n.s.*, 7: 7 (2000): Clitoria magnifica.
teresae *M. Brandão & N.M. Sousa Costa in Daphne*, 3(1): 5 (1993)—Brazil Southeast (Minas Gerais).

CEODES (Nyctaginac.).
grandis *(R. Br.) D.Q. Lu in Fl. Reipubl. Popul. Sin.*, 26: 3 (1996): Pisonia grandis.

CEPHALANTHERA (Orchidac.).
alpicola Fukuy.
var. shizuoi *(F. Maek.) T. Hashimoto in Proc. 12th World Orchid Conf., 1987 (eds. K. Saito & R. Tanaka)*: 120 (1987): C. shizuoi.
rubra (L.) Rich.
f. comosa *(Tineo) K. Robatsch in Wulfenia*, 7: 104 (2000): Cephalanthera comosa.
taiwaniana *S.S. Ying in Mem. Coll. Agric. Nation. Taiwan Univ.*, 28(2): 37 (1988)—Taiwan.

CEPHALANTHEROPSIS (Orchidac.).
gracilis (Lindl.) S.Y. Hu
var. halconensis *(Ames) Kohji Karasawa, Orchid Atlas*, 5: 199 (1997), without basionym ref.: Phaius halconensis.

CEPHALANTHEROPSIS:-
gracilis (Lindl.) S.Y. Hu
var. venusta *(Schltr.) Kohji Karasawa, Orchid Atlas*, 5: 199 (1997), without basionym ref.: Calanthe venusta.
laciniata *P. Ormerod in Orchid Dig.*, 62(4): 157 (1998)—Malaya (Peninsular Malaysia).
longipes *(Hook. f.) P. Ormerod in Orchid Dig.*, 62(4): 156 (1998): Calanthe longipes.
obcordata *(Lindl.) P. Ormerod in Orchid Dig.*, 62(4): 157 (1998): Bletia obcordata.

CEPHALARIA (Dipsacac.).
decipiens *M. Gandoger, Dec. Pl. Nov.*, 1: 43 (1875)—Caucasus.
ekimiana *R.S. Göktürk & H. Sümbül in Ann. Bot. Fenn.*, 34(3): 153 (1997)—Turkey.
fanourii *D.K. Perdetzoglou & Kit Tan in Ann. Bot. Fenn.*, 32(4): 228 (1995)—Greece.
squamiflora (Sieber) Greuter
subsp. ebusitana *(O. Bolòs & Vigo) A.M. Romo, Fl. Silvestres Baleares*: 289 (1994), as 'squamifolia'; isonym see O. de Bolòs *in But. Inst. Catalana Hist. Nat. 59 Bot.*, 8: 145 (1991): C. squamiflora var. ebusitana.

CEPHALOCEREUS (Cactac.).
royenii (L.) Britton & Rose
f. cristatus *P.V. Heath in Calyx*, 5(3): 104 (1996)—Locality not stated.

CEPHALOPHILON (Polygonac.).
chinense *(L.) A.E. Borodina-Grabovskaya in Konsp. Sosud. Rast. Fl. V'etnama*, 2: 145 (1996): Polygonum chinense.
malaicum *(Danser) A.E. Borodina-Grabovskaya in Konsp. Sosud. Rast. Fl. V'etnama*, 2: 145 (1996): Polygonum malaicum.
palmatum *(Dunn) A.E. Borodina-Grabovskaya in Konsp. Sosud. Rast. Fl. V'etnama*, 2: 146 (1996): Polygonum palmatum.

CEPHALOPODUM (Umbellif.).
hissaricum *M.G. Pimenov in Opred. Rast. Sred. Azii*, 7: 374 (1983)—Tadzhikistan.

CEPHALORRHYNCHUS (Compos.).
sect. Pterorhynchus *A.N. Sennikov in Bot. Zhurn.*, 82(5): 113 (1997).
sect. Quasibrassica *A.N. Sennikov in Bot. Zhurn.*, 82(2): 110 (1997).

CEPHALOSTACHYUM (Gramin.).
mannii *(Gamble) C.M.A. Stapleton & D.Z. Li in Kew Bull.*, 52(3): 700 (1997): Arundinaria mannii.
scandens *J.R. Hsueh & C.M. Hui in Acta Phytotax. Sin.*, 35(6): 562 (1997), homonym; *see Bor in Kew Bull.*, 12(3): 419 (1957 publ. 1958)—China South-central (Yunnan).

CEPHALOSTEMON (Rapateac.).
flavus *(Link) R. Govaerts, World Checklist Seed Plants*, 3(1): 13 (1999): Schoenus flavus.

CEPHALOTAXUS (Cephalotaxac.).
harringtonia (Knight ex J. Forbes) K. Koch
var. latifolia *(W.C. Cheng & L.K. Fu) J. Silba in J. Int. Conifer Preserv. Soc.*, 7(1): 22 (2000): C. sinensis var. latifolia.
var. thailandensis *J. Silba in J. Int. Conifer Preserv. Soc.*, 7(1): 22 (2000)—Thailand.
latifolia *W.C. Cheng & L.K. Fu ex L.K. Fu & R.R. Mill in Novon*, 9(2): 185 (1999)—China South-central (Sichuan).
sinensis (Rehder & E.H. Wilson) H.L. Li
var. wilsoniana *(Hayata) L.K. Fu & Nan Li in Novon*, 7(3): 263 (1997): C. wilsoniana.

CERAPADUS (Rosac.).
glandulifolia *(Rupr. & Maxim.) V.A. Nedoluzhko in Sosud. Rast. Sovet. Dal'nego Vostoka*, 8: 239 (1996): Prunus glandulifolia.

CERARIA (Portulacac.).
pygmaea *(Pillans) G.D. Rowley in Bradleya*, 14: 82 (1996): Portulacaria pygmaea.

CERASTIUM (Caryophyllac.).
sect. Caespitosa *(Pax & K. Hoffm.) I.V. Sokolova in Novosti Sist. Vyssh. Rast.*, 30: 37 (1996): C. subsect. Caespitosa.
sect. Fugacia *(Fenzl ex Pax & K. Hoffm.) I.V. Sokolova in Novosti Sist. Vyssh. Rast.*, 30: 40 (1996): C. subsect. Fugacia.
ser. Pumila *M.V. Klokov in Novosti Sist. Vyssh. Nizsh. Rast. (Kiev)*, 1974: 25 (1974).
ser. Semidecandra *M.V. Klokov in Novosti Sist. Vyssh. Nizsh. Rast. (Kiev)*, 1974: 34 (1974).
arvense L.
f. calcicola *(Schur) M. Niketić in Glasn. Prir. Muz. Beogradu, B*, 49–50: 49 (1995–98 publ. 1999), as 'calcicolum': C. arvense var. calcicola.
f. capitatum *(T. Georgiev) M. Niketić in Glasn. Prir. Muz. Beogradu, B*, 49–50 50 (1995–98 publ. 1999): C. arvense var. capitatum.
f. petrovicii *M. Niketić in Glasn. Prir. Muz. Beogradu, B*, 49–50: 50 (1995–98 publ. 1999)—Yugoslavia.
banaticum (Rochel) Heuff.
f. adenophorum *Halácsy ex M. Niketić in Glasn. Prir. Muz. Beogradu, B*, 49–50: 53 (1995–98 publ. 1999)—Greece.
var. alpinum *(Boiss.) M. Niketić in Glasn. Prir. Muz. Beogradu, B*, 49–50: 53 (1995–98 publ. 1999): C. grandiflorum var. alpinum.
var. balcanicum *(Vandas) M. Niketić in Glasn. Prir. Muz. Beogradu, B*, 49–50: 53 (1995–98 publ. 1999): C. balcanicum.
f. diklicii *M. Niketić in Glasn. Prir. Muz. Beogradu, B*, 49–50: 51 (1995–98 publ. 1999)—Yugoslavia.
var. grebenscikovii *M. Niketić in Glasn. Prir. Muz. Beogradu, B*, 49–50: 52 (1995–98 publ. 1999)—Yugoslavia.
subsp. hirsutum *(Fenzl ex Griseb.) M. Niketić in Glasn. Prir. Muz. Beogradu, B*, 49–50: 52 (1995–98 publ. 1999): C. grandiflorum var. hirsutum.
var. ilicii *M. Niketić in Glasn. Prir. Muz. Beogradu, B*, 49–50: 52 (1995–98 publ. 1999)—Yugoslavia.
f. kellereri *(T. Georgiev) M. Niketić in Glasn. Prir. Muz. Beogradu, B*, 49–50: 52 (1995–98 publ. 1999): C. banaticum var. kellereri.
var. kosaninii *(T. Georgiev) K. Micevski in Prilozi Oddel. Biol. Med. Nauki Makedonska Akad. Nauk. Umet.*, 9(1–2): 48 (1988 publ. 1991), as 'kosanini': C. banaticum subsp. kosaninii.
f. nikolicii *M. Niketić in Glasn. Prir. Muz. Beogradu, B*, 49–50: 52 (1995–98 publ. 1999)—Yugoslavia.
f. pancicii *M. Niketić in Glasn. Prir. Muz. Beogradu, B*, 49–50: 52 (1995–98 publ. 1999)—Yugoslavia.
beeringianum Cham. & Schltdl.
subsp. jenisejense *(Hultén) V.N. Voroschilov in A.K. Skvortsov (ed.), Florist. issl. v razn. raĭonakh SSSR:* 166 (1985): C. jenisejense.
brachypetalum Desp. ex Pers.
subsp. balearicum *A.M. Romo, Fl. Silvestres Baleares:* 214 (1994)—Baleares.
cerastoides (L.) Britton
var. foliosum *Yu.P. Kozhevnikov in Rast. Tsentral. Azii*, 11: 43 (1994 publ. 1995)—Qinghai.
f. glandulosa *K. Micevski in Prilozi Oddel. Biol. Med. Nauki Makedonska Akad. Nauk. Umet.*, 9(1–2): 47 (1988 publ. 1991)—Yugoslavia.
crassiusculum *M.V. Klokov in Novosti Sist. Vyssh. Nizsh. Rast. (Kiev)*, 1974: 26 (1974)—Krym.
decalvans Schloss. & Vuk.
f. adamovicii *(Velen.) M. Niketic in Glasn. Prir. Muz. Beogradu, B*, 49–50: 46 (1995–98 publ. 1999): C. moesiacum var. adamovicii.
f. albanicum *(Bald.) M. Niketić in Glasn. Prir. Muz. Beogradu, B*, 49–50: 42 (1995–98 publ. 1999): C. grandiflorum var. albanicum.
var. albanicum *(Bald.) M. Niketić in Glasn. Prir. Muz. Beogradu, B*, 49–50: 42 (1995–98 publ. 1999): C. grandiflorum var. albanicum.
f. alpicola *(Hausskn.) M. Niketić in Glasn. Prir. Muz. Beogradu, B*, 49–50: 43 (1995–98 publ. 1999), as 'alpicolum': C. lanigerum var. alpicola.

CERASTIUM:-
decalvans Schloss. & Vuk.
var. anisophyllum *(Bornm.) M. Niketić in Glasn. Prir. Muz. Beogradu, B*, 49–50: 42 (1995–98 publ. 1999): C. balcanicum var. anisophyllum.
f. belassicum *(T. Georgiev) M. Niketić in Glasn. Prir. Muz. Beogradu, B*, 49–50: 46 (1995–98 publ. 1999): C. adamovicii var. belassicum.
var. bosniacum *(Beck) M. Niketić in Glasn. Prir. Muz. Beogradu, B*, 49–50: 42 (1995–98 publ. 1999): C. tomentosum f. bosniacum.
f. doerflerianum *(T. Georgiev) M. Niketić in Glasn. Prir. Muz. Beogradu, B*, 49–50: 45 (1995–98 publ. 1999): C. lanigerum f. doerflerianum.
var. durmitoreum *(Rohlena) M. Niketić in Glasn. Prir. Muz. Beogradu, B*, 49–50: 44 (1995–98 publ. 1999): C. lanigerum var. durmitoreum.
f. glabrum *(T. Georgiev) M. Niketić in Glasn. Prir. Muz. Beogradu, B*, 49–50: 44 (1995–98 publ. 1999): C. lanigerum f. glabrum.
f. glandulosum *(K. Malý) M. Niketić in Glasn. Prir. Muz. Beogradu, B*, 49–50: 47 (1995–98 publ. 1999): C. lanigerum f. glandulosum.
var. halacsyi *(Borza) M. Niketić in Glasn. Prir. Muz. Beogradu, B*, 49–50: 44 (1995–98 publ. 1999): C. moesiacum f. halacsyi.
var. histrio *(Correns ex Prain) M. Niketić in Glasn. Prir. Muz. Beogradu, B*, 49–50: 44 (1995–98 publ. 1999): C. histrio.
f. isophyllum *(Bornm.) M. Niketić in Glasn. Prir. Muz. Beogradu, B*, 49–50: 42 (1995–98 publ. 1999): C. lanigerum var. isophyllum.
f. leontopodium *(Stoj. & Stef.) M. Niketić in Glasn. Prir. Muz. Beogradu, B*, 49–50: 45 (1995–98 publ. 1999): C. banaticum var. leontopodium.
f. lindtneri *M. Niketić in Glasn. Prir. Muz. Beogradu, B*, 49–50: 47 (1995–98 publ. 1999)—Yugoslavia.
f. macedonicum *(T. Georgiev) M. Niketić in Glasn. Prir. Muz. Beogradu, B*, 49–50: 45 (1995–98 publ. 1999): C. macedonicum.
f. minutum *(T. Georgiev) M. Niketić in Glasn. Prir. Muz. Beogradu, B*, 49–50: 43 (1995–98 publ. 1999): C. lanigerum subsp. minutum.
f. nikolovii *(T. Georgiev) M. Niketić in Glasn. Prir. Muz. Beogradu, B*, 49–50: 45 (1995–98 publ. 1999): C. lanigerum subsp. nikolovii.
var. oehmii *M. Niketić in Glasn. Prir. Muz. Beogradu, B*, 49–50: 45 (1995–98 publ. 1999)—Yugoslavia.
var. orbelicum *(Velen.) M. Niketić in Glasn. Prir. Muz. Beogradu, B*, 49–50: 45 (1995–98 publ. 1999): C. orbelicum.
f. orbelicum *(Velen.) M. Niketić in Glasn. Prir. Muz. Beogradu, B*, 49–50: 46 (1995–98 publ. 1999): C. orbelicum.
var. pauciflorum *(Beck) M. Niketić in Glasn. Prir. Muz. Beogradu, B*, 49–50: 46 (1995–98 publ. 1999): C. lanigerum f. pauciflorum.
f. pauciflorum *(Beck) M. Niketić in Glasn. Prir. Muz. Beogradu, B*, 49–50: 46 (1995–98 publ. 1999): C. lanigerum f. pauciflorum.
f. prilepanum *(T. Georgiev) M. Niketić in Glasn. Prir. Muz. Beogradu, B*, 49–50: 45 (1995–98 publ. 1999): C. lanigerum f. prilepanum.
var. pseudolanatum *(K. Malý) M. Niketić in Glasn. Prir. Muz. Beogradu, B*, 49–50: 46 (1995–98 publ. 1999): C. lanigerum var. pseudolanatum.
f. robustum *(Beck) M. Niketić in Glasn. Prir. Muz. Beogradu, B*, 49–50: 43 (1995–98 publ. 1999), as 'robustrum': C. lanigerum f. robustum.
f. scardicum *T. Georgiev ex M. Niketić in Glasn. Prir. Muz. Beogradu, B*, 49–50: 42 (1995–98 publ. 1999)—Yugoslavia.
f. semiglabrum *(Beck) M. Niketić in Glasn. Prir. Muz. Beogradu, B*, 49–50: 44 (1995–98 publ. 1999): C. lanigerum f. semiglabrum.
f. stojanovii *(T. Georgiev) M. Niketić in Glasn. Prir. Muz. Beogradu, B*, 49–50: 46 (1995–98 publ. 1999): C. lanigerum subsp. stojanovii.
f. stribrnyi *M. Niketić in Glasn. Prir. Muz. Beogradu, B*, 49–50: 46 (1995–98 publ. 1999)—Bulgaria.
subsp. sylvaticum *(K. Malý) M. Niketić in Glasn. Prir. Muz. Beogradu, B*, 49–50: 47 (1995–98 publ. 1999): C. lanigerum var. sylvaticum.

CERASTIUM:-
dinaricum Beck & Szyszył.
var. hekuravense *(Jáv.) M. Niketić in Glasn. Prir. Muz. Beogradu, B*, 49–50: 55 (1995–98 publ. 1999): C. hekuravense.
eriophorum Kit.
f. glutinosum *(W.D.J. Koch) M. Niketić in Glasn. Prir. Muz. Beogradu, B*, 49–50: 40 (1995–98 publ. 1999): C. alpinum var. glutinosum.
f. macranthum *(Stoj. & T. Georgiev) M. Niketić in Glasn. Prir. Muz. Beogradu, B*, 49–50: 41 (1995–98 publ. 1999): C. lanatum f. macranthum.
f. rotundifolium *(Schur) M. Niketić in Glasn. Prir. Muz. Beogradu, B*, 49–50: 41 (1995–98 publ. 1999): C. lanatum f. rotundifolium.
fischerianum Ser.
subsp. albimarginatum *(Vorosch.) V.N. Voroschilov in A.K. Skvortsov (ed.), Florist. issl. v razn. raĭonakh SSSR:* 166 (1985): C. fischerianum var. albimarginatum.
fontanum Baumg.
subsp. membranaceum *(Edgew. & Hook. f.) M. Sharma in Geobios New Rep.*, 5(2): 159 (1986): C. vulgatum var. membranaceum.
glabratum
subsp. glaberrimum *(Lapeyr.) R. Govaerts, World Checklist Seed Plants*, 3(1): 13 (1999): C. glaberrimum.
hintoniorum *B.L. Turner in Phytologia*, 79(5): 340 (1995 publ. 1996)—Mexico Southwest (Oaxaca).
latifolium
subsp. pyrenaicum *(J. Gay) R. Govaerts, World Checklist Seed Plants*, 3(1): 13 (1999): C. pyrenaicum.
malyi *(T. Georgiev) M. Niketić in Glasn. Prir. Muz. Beogradu, B*, 49–50: 47 (1995–98 publ. 1999): C. orbelicum subsp. malyi.
f. midzorense *M. Niketić in Glasn. Prir. Muz. Beogradu, B*, 49–50: 48 (1995–98 publ. 1999)—Yugoslavia.
f. neolindtneri *M. Niketić in Glasn. Prir. Muz. Beogradu, B*, 49–50: 48 (1995–98 publ. 1999)—Yugoslavia.
f. orientale *M. Niketić in Glasn. Prir. Muz. Beogradu, B*, 49–50: 48 (1995–98 publ. 1999)—Yugoslavia.
var. orientale *M. Niketić in Glasn. Prir. Muz. Beogradu, B*, 49–50: 48 (1995–98 publ. 1999)—Yugoslavia.
subsp. serpentini *(Novák) M. Niketić in Glasn. Prir. Muz. Beogradu, B*, 49–50: 48 (1995–98 publ. 1999): C. moesiacum f. serpentini.
moesiacum Friv.
subsp. glutinosum *(Halácsy) A. Strid in A. Strid & Kit Tan (eds.), Fl. Hellenica*, 1: 202 (1997): C. alpinum var. glutinosum.
neoscardicum *M. Niketić in Glasn. Inst. Bot. Bot. Bašte Univ. Beogradu*, 26–27: 64 (1994)—Yugoslavia.
f. adenotrichum *M. Niketić in Glasn. Inst. Bot. Bot. Bašte Univ. Beogradu*, 26–27: 68 (1994)—Yugoslavia.
f. glandulosum *M. Niketić in Glasn. Inst. Bot. Bot. Bašte Univ. Beogradu*, 26–27: 68 (1994)—Yugoslavia.
nigrescens (H.C. Watson) Edmondston ex H.C. Watson
var. laxum *(Lindblom) A.K. Brysting & R. Elven in Taxon*, 49(2): 212 (2000): C. latifolium var. laxum.
odessanum *M.V. Klokov in Novosti Sist. Vyssh. Nizsh. Rast. (Kiev)*, 1974: 29 (1974)—Ukraina.
stenopetalum Fenzl ex Gren.
var. nanum *G. Barberis, N. Bechi & P. Miceli in Flora Medit.*, 7: 128 (1997)—Corse.
vagans
subsp. azoricum *(Hochst. ex Seub.) R. Govaerts, World Checklist Seed Plants*, 3(1): 13 (1999): C. azoricum.

CERASUS (Rosac.).
apetala (Siebold & Zucc.) H. Ohba
var. monticola *(Kawas. & H. Koyama) H. Ohba in J. Jap. Bot.*, 73(2): 117 (1998): Prunus apetala var. monticola.
avium (L.) Moench
var. duracina *(L.) R. Cinovskis in Fl. Baltiĭskikh Respublik*, 2: 122 (1996): Prunus cerasus λ duracina.

CERASUS:-
avium (L.) Moench
var. juliana *(L.) R. Cinovskis in Fl. Baltiĭskikh Respublik,* 2: 122 (1996): Prunus cerasus ε juliana.
cerasoides
var. campanulata *(Maxim.) X.R. Wang & C.B. Shang in J. Nanjing Forest. Univ.,* 22(4): 61 (1998): Prunus campanulata.
dolichadenia *(Cardot) X.R. Wang & C.B. Shang in J. Nanjing Forest. Univ.,* 22(4): 60 (1998): Prunus dolichadenia.
× furuseana (Ohwi) H. Ohba
nothovar. pseudoaffinis *(Kawas.) H. Ohba in J. Jap. Bot.,* 73(2): 117 (1998): Prunus × pseudoaffinis.
glandulosa (Thunb.) Loisel.
var. trichostyla *(Koehne) J.X. Yang in Fl. Loess-Plat. Sin.,* 2: 305 (1992): Prunus glandulosa var. trichostyla.
jamasakura *(Siebold ex Koidz.) X.R. Wang & C.B. Shang in J. Nanjing Forest. Univ.,* 22(4): 61 (1998), isonym; *see H. Ohba in J. Jap. Bot.,* 67(5): 278 (1992): Prunus jamasakura.
× katonis *H. Ohba in J. Jap. Bot.,* 73(2): 116 (1998)—Japan.(C. maximowiczii × sargentii).
laxiflora *(Koehne) C.L. Li & S.Y. Jiang in Acta Phytotax. Sin.,* 36(4): 368 (1998): Prunus laxiflora.
nipponica *(Matsum.) V.A. Nedoluzhko in Sosud. Rast. Sovet. Dal'nego Vostoka,* 8: 240 (1996), isonym; *see Ohle in Mansfeld, Verz. Landwirtsch. u. Gartn. Kulturpfl., Auf. 2, ed.* Schultze-Motel, 1: 417 (1986): Prunus nipponica.
× oneyamensis *(Hayashi) H. Ohba in J. Jap. Bot.,* 73(2): 117 (1998): Prunus × oneyamensis.
nothovar. takasawana *(H. Kubota & Funatsu) H. Ohba in J. Jap. Bot.,* 73(2): 118 (1998): Prunus × takasawana.
rehderiana *(Koehne) X.R. Wang & C.B. Shang in J. Nanjing Forest. Univ.,* 23(6): 62 (1999): Prunus rehderiana.
sargentii (Rehder) H. Ohba
var. akimotoi *H. Ohba & Mas. Saito in J. Jap. Bot.,* 75(6): 370 (2000)—Japan (Kyushu).
serrulata (Lindl.) G. Don ex Loudon
var. taishanensis *Y. Zhang & C.D. Shi in Acta Phytotax. Sin.,* 37(1): 87 (1999)—China North-central (Shandong).
subhirtella (Miq.) S.Ya. Sokolov
var. ascendens *(Makino) X.R. Wang & C.B. Shang in J. Nanjing Forest. Univ.,* 24(6): 45 (2000): Prunus pendula var. ascendens.
tomentosa (Thunb.) Wall.
var. pendula *B.Y. Feng & S.M. Xie in Acta Phytotax. Sin.,* 35(3): 268 (1997)—China North-central (Shanxi).
variabilis *(Koehne) X.R. Wang & C.B. Shang in J. Nanjing Forest. Univ.,* 23(6): 62 (1999): Prunus variabilis.
vulgaris Mill.
var. frutescens *(Neilr.) R. Cinovskis in Fl. Baltiĭskikh Respublik,* 2: 123 (1996): Prunus cerasus var. frutescens.
yaoiana *W.L. Cheng in Acta Phytotax. Sin.,* 38(2): 195 (2000)—Xizang.

CERATOCHLOA (Gramin.).
sitchensis *(Trin.) T.A. Cope & T.B. Ryves in Watsonia,* 21(1): 135 (1996): Bromus sitchensis.

CERATOLIMON *M.B. Crespo & M.D. Lledó in Bot. J. Linn. Soc.,* 132(2): 169 (2000). *PLUMBAGINACEAE*
feei *(Girard) M.B. Crespo & M.D. Lledó in Bot. J. Linn. Soc.,* 132(2): 170 (2000): Bubania feei.
var. grandiflorum *(Maire & Wilczek) M.B. Crespo & M.D. Lledó in Bot. J. Linn. Soc.,* 132(2): 171 (2000): Limoniastrum feei var. grandiflorum.
migiurtinum *(Chiov.) M.B. Crespo & M.D. Lledó in Bot. J. Linn. Soc.,* 132(2): 171 (2000): Bubania migiurtina.
rechingeri *(J.R. Edm.) M.B. Crespo & M.D. Lledó in Bot. J. Linn. Soc.,* 132(2): 171 (2000): Limoniastrum rechingeri.
weygandiorum *(Maire & Wilczek) M.B. Crespo & M.D. Lledó in Bot. J. Linn. Soc.,* 132(2): 172 (2000): Limoniastrum weygandiorum.

CERATOSTEMA (Ericac.).
auriculatum *J.L. Luteyn in Fl. Ecuador,* 54(147): 241 (1996)—Ecuador.
bracteolatum *J.L. Luteyn in Fl. Ecuador,* 54(147): 215 (1996)—Ecuador.
cutucuense *J.L. Luteyn in Fl. Ecuador,* 54(147): 249 (1996)—Ecuador.
glans *J.L. Luteyn in Fl. Ecuador,* 54(147): 228 (1996)—Ecuador.
macbrydeorum *J.L. Luteyn in Fl. Ecuador,* 54(147): 242 (1996), as 'macbrydiorum'—Ecuador.
madisonii *J.L. Luteyn in Fl. Ecuador,* 54(147): 227 (1996)—Ecuador.
megalobum *J.L. Luteyn in Fl. Ecuador,* 54(147): 221 (1996)—Ecuador.
oellgaardii *J.L. Luteyn in Fl. Ecuador,* 54(147): 224 (1996)—Ecuador.
pedunculatum *J.L. Luteyn in Fl. Ecuador,* 54(147): 230 (1996)—Ecuador.

CERATOSTYLIS (Orchidac.).
eria *R. Govaerts, World Checklist Seed Plants,* 3(1): 13 (1999), nom. nov.: Eria pygmaea *Hook. f.*
hainanensis *Z.H. Tsi in Acta Phytotax. Sin.,* 33(6): 582 (1995), nom. nov.: C. caespitosa *(Rolfe) T. Tang & F.T. Wang.*
incognita *J.T. Atwood & J. Beckner in Selbyana,* 19(2): 265 (1998 publ. 1999)—Philippines.

CERATOZAMIA (Zamiac.).
alvarezii *M.A. Perez-Farrera, A.P. Vovides & C. Iglesias in Novon,* 9(3): 410 (1999)—Mexico Southeast (Chiapas).
mixeorum *J. Chemnick, T.J. Gregory & S. Salas-Morales in Phytologia,* 83(1): 47 (1997 publ. 1998)—Mexico Southwest (Oaxaca).
morettii *M. Vázquez-Torres & A.P. Vovides in Novon,* 8(1): 87 (1998)—Mexico Gulf (Veracruz).
whitelockiana *J. Chemnick & T.J. Gregory in Phytologia,* 79(1): 51 (1995 publ. 1996)—Mexico Southwest (Oaxaca).

CERBERA (Apocynac.).
laeta *A.J.M. Leeuwenberg in Wageningen Agric. Univ. Pap.,* 98(3): 18 (1999)—New Guinea (Papua New Guinea).

CERCIS (Legumin.).
siliquastrum
nothosubsp. yaltirikii *(Ponert) R. Govaerts, World Checklist Seed Plants,* 3(1): 13 (1999): C. × yaltirikii.

CEREUS (Cactac.).
fernambucensis
subsp. sericifer *(F. Ritter) N.P. Taylor & D.C. Zappi in Cactaceae Consensus Initiatives,* 3: 7 (1997): Piptanthocereus sericifer.
hildmannianus
subsp. uruguayanus *(R. Kiesling) N.P. Taylor in Cactaceae Consensus Initiatives,* 6: 15 (1998): C. uruguayanus.
jamacaru
subsp. calcirupicola *(F. Ritter) N.P. Taylor & D.C. Zappi in Cactaceae Consensus Initiatives,* 3: 7 (1997): Piptanthocereus calcirupicola.
phatnospermus K. Schum.
subsp. adelmarii *(Rizzini & A. Mattos) P.J. Braun & E. Esteves Pereira in Kakt. And. Sukk.,* 48(6): 142 (1997): Monvillea adelmarii.
subsp. kroenleinii *(N.P. Taylor) P.J. Braun & E. Esteves Pereira in Kakt. And. Sukk.,* 48(6): 142 (1997): C. kroenleinii.

CERISCOIDES (Rubiac.).
howii *H.S. Lo in Bull. Bot. Res. North-East. Forest. Univ.,* 18(3): 281 (1998)—Hainan.

CEROCHLAMYS (Aizoac.).
gemina *(L. Bolus) H.E.K. Hartmann in Bradleya,* 16: 53 (1998): Ruschia gemina.
purpureostyla *(L. Bolus) H.E.K. Hartmann in Bradleya,* 16: 56 (1998): Mesembryanthemum purpureostylum.

CEROPEGIA (Asclepiadac.).
adrienneae *W. Rauh & R. Gerold in Succulentes,* 1998(3): 7 (1998), without latin descr.; *et in Succulentes,* 1998(4): 9 (1998), contrary to Art. 45.1 ICBN (1994)—Madagascar.

CEROPEGIA:-
ahmarensis *P.S. Masinde in Kew Bull.,* 55(1): 225 (2000)—Somalia.
ambovombensis *W. Rauh & Gerold in Succulentes,* 1997(2): 6, 3 (1997)—Madagascar.
andamanica *P.V. Sreekumar, K. Veenakumari & M. Prashanth in Blumea,* 43(1): 215 (1998)—Andaman Is.
dichotoma Haw.
subsp. fusca *(Bolle) G.D. Rowley in Asklepios,* 75: 36 (1998): C. fusca.
gikyi *W. Rauh & R. Gerold in Succulentes,* 1998(3): 6 (1998)—Madagascar.
hermannii *W. Rauh & M. Teissier in Succulentes,* 1996(4): 22 (1996)—Madagascar.
inornata *P.R.O. Bally ex P.S. Masinde in Kew Bull.,* 53(4): 949 (1998)—Kenya, Ethiopia.
intracolor *L.E. Newton in Bradleya,* 13: 35 (1995)—Kenya.
keniensis *P.S. Masinde in Kew Bull.,* 54(2): 477 (1999)—Kenya.
kituloensis *P.S. Masinde & F. Albers in Bot. Jahrb.,* 122(2): 162 (2000)—Tanzania.
konasita *P.S. Masinde in Cact. Succ. J. (U.S.A.),* 71(3): 146 (1999)—Kenya.
robivelonae *W. Rauh & Gérold in Succulentes,* 1998(4): 3 (1998)—Madagascar.
striata *U. Meve & P.S. Masinde in Novon,* 8(1): 38 (1998)—Madagascar.
verticillata *P.S. Masinde in Cact. Succ. J. (U.S.A.),* 72(3): 155 (2000)—Kenya.

CESTRUM (Solanac.).
sect. Sessea *(Ruiz & Pav.) L.Á.F. de Carvalho & A. Schnoor in Rodriguésia,* 45–49(71–75): 16 (1993–97 publ. 1998): gen. Sessea.
capsulare *L.Á.F. de Carvalho & A. Schnoor in Rodriguésia,* 45–49(71–75): 17 (1993–97 publ. 1998), nom. nov.: Sessea regnellii *Taub.*
darcyanum *C. Benítez de Rojas & N.W. Sawyer in Brittonia,* 51(2): 163 (1999)—Venezuela.
honduronicaraguense *C.H. Nelson Sutherland in An. Jard. Bot. Madrid,* 57(2): 423 (1999 publ. 2000), nom. nov., as 'honduro-nicaraguense': C. laxiflorum *Francey.*
jaramillanum *C. Benítez de Rojas & W.G. D'Arcy in Novon,* 5(4): 311 (1995)—Ecuador, Venezuela.
milciomejiae *T.A. Zanoni in Ann. Carnegie Mus. Pittsburgh,* 64(4): 263 (1995)—Dominican Republic.
ruizteranianum *C. Benítez de Rojas & W.G. D'Arcy in Novon,* 5(4): 313 (1995)—Venezuela.
tillettii *C. Benítez de Rojas & W.G. D'Arcy in Novon,* 5(4): 315 (1995)—Venezuela.
toledii *L.Á.F. de Carvalho & A. Schnoor in Rodriguésia,* 45–49(71–75): 17 (1993–97 publ. 1998), nom. nov.: Sessea brasiliensis *J.T. Toledo.*

CHAENORHINUM (Scrophulariac.).
crassifolium (Cav.) Lange
subsp. pityusicum *(Font Quer) A.M. Romo, Fl. Silvestres Baleares:* 276 (1994): C. crassifolium var. pityusicum.
granatense *(Willk.) J. Holub in Preslia,* 70(2): 101 (1998): Linaria granatensis.
montserratiorum *J. Holub in Preslia,* 70(2): 101 (1998), nom. nov., as 'montserratorum': C. origanifolium subsp. cotiellae *P. Monts. & J.M. Monts.*
semiglabrum *(Loidi & A. Galán) J.A. Alejandre Sáenz, J.A. Arizaleta Urarte & J. Benito Ayuso in Flora Montiberica,* 12: 45 (1999): C. segoviense subsp. semiglabrum.

CHAEROPHYLLUM (Umbellif.).
sect. Bulbosa *(Koso-Pol.) N.N. Tzvelev, Opred. Sosud. Rast. Severo-Zap. Rossii:* 680 (2000): Chaerophyllum ser. Bulbosa.
aksekiense *A. Duran & H. Duman in Edinburgh J. Bot.,* 56(1): 47 (1999)—Turkey.
posofianum *S. Erik & N. Demirkuş in Willdenowia,* 28(1–2): 153 (1998)—Turkey.

CHAETANTHERA (Compos.).
glabrata *(DC.) Meigen in Bot. Jahrb. Syst.,* 23: 456 (1894), as 'glabraia' in error in I.K. Supp.1—Tylloma glabrata†

CHAETANTHUS (Restionac.).
aristatus *(R. Br.) B.G. Briggs & L.A.S. Johnson in Telopea,* 8(1): 24 (1998): Leptocarpus aristatus.
tenellus *(Nees) B.G. Briggs & L.A.S. Johnson in Telopea,* 8(1): 24 (1998): Restio tenellus.

CHAETOBROMUS (Gramin.).
involucratus (Schrad.) Nees
subsp. dregeanus *(Nees) G.A. Verboom in Nordic J. Bot.,* 18(1): 74 (1998): C. dregeanus.
subsp. sericeus *(Nees) G.A. Verboom in Nordic J. Bot.,* 18(1): 72 (1998): C. involucratus var. sericeus.

CHAETOCALYX (Legumin.).
longiloba *V.E. Rudd in Novon,* 6(1): 119 (1996)—Peru.

CHAETOSERIS (Compos.).
hirsuta *(Franch.) C. Shih in Fl. Reipubl. Popul. Sin.,* 80(1): 282 (1997): Lactuca hirsuta.
qiliangshanensis *S.W. Liu & T.N. Ho in Fl. Qinghaiica,* 3: 512 (1996)—Qinghai, China North-central (Gansu).

CHAETOSTOMA (Melastomatac.).
albiflorum *(Naudin) C. Koschnitzke & A.B. Martins in Novon,* 9(2): 202 (1999): C. pungens var. albiflorum.
canastrense *Rosana Romero & A.B. Martins in Candollea,* 54(2): 449 (1999)—Brazil Southeast (Minas Gerais).
cupressinum *(D. Don) C. Koschnitzke & A.B. Martins in Novon,* 9(2): 202 (1999): Microlicia cupressina.
flavum *C. Koschnitzke & A.B. Martins in Novon,* 9(2): 204 (1999)—Brazil West-central (Goiás).
selagineum *(Naudin) C. Koschnitzke & A.B. Martins in Novon,* 9(2): 202 (1999): Microlicia selaginea.
stenocladon *(Naudin) C. Koschnitzke & A.B. Martins in Novon,* 9(2): 202 (1999): Microlicia stenocladon.

CHAMAECRISTA (Legumin.).
amorimii *R.C. Barneby in Brittonia,* 51(3): 331 (1999)—Brazil Northeast (Bahia).
arachiphylla *R.C. Barneby in Brittonia,* 48(2): 181 (1996)—Paraguay.
aspleniifolia *(H.S. Irwin & Barneby) R. Govaerts, World Checklist Seed Plants,* 3(1): 13 (1999), isonym; *see H.S. Irwin & Barneby in Brittonia,* 31(1): 155 (1979): Cassia aspleniifolia.
biddulphiana *L. Pedley in Fl. Australia,* 12: 196 (1998)—Queensland.
caspariifolia *R.C. Barneby in Brittonia,* 51(3): 334 (1999)—Brazil Southeast (Minas Gerais).
chiquitana *R.C. Barneby in Brittonia,* 51(3): 337 (1999)—Bolivia.
concinna *(Benth.) L. Pedley in Fl. Australia,* 12: 196 (1998): Cassia concinna.
conferta (Benth.) H.S. Irwin & Barneby
var. gurgueiana *Afr. Fernandes & E.P. Nunes in Albertoa, n.s.,* 7: 4 (2000)—Brazil Northeast (Piauí).
coradinii *R.C. Barneby in Brittonia,* 48(2): 180 (1996), isonym; *see Irwin & Barneby in Brittonia,* 44(2): 227 (1992)—Brazil Northeast (Bahia).
cytisoides
var. blanchetii *(Benth.) R. Govaerts, World Checklist Seed Plants,* 3(1): 13 (1999): Cassia blanchetii.
var. brachystachya *(Benth.) R. Govaerts, World Checklist Seed Plants,* 3(1): 13 (1999): Cassia brachystachya.
deserti *L. Pedley in Fl. Australia,* 12: 196 (1998)—Northern Territory, Western Australia, Queensland.
desvauxii (Collad.) Killip
var. latistipula *(Benth.) M. Luckow in Baileya,* 23(4): 208 (1996), isonym; *see G.P. Lewis, Legumes of Bahia:* 77 (1987): Cassia latistipula.
exigua *L. Pedley in Fl. Australia,* 12 196 (1998)—Queensland.
var. minor *L. Pedley in Fl. Australia,* 12: 196 (1998)—Queensland.
fulgida *R.C. Barneby in Brittonia,* 51(3): 336 (1999)—Brazil West-central (Goiás).

CHAMAECRISTA:-
grisea *L. Pedley in Fl. Australia,* 12: 197 (1998)—Northern Territory, Queensland.
harneyi *(Specht) R. Govaerts, World Checklist Seed Plants,* 3(1): 13 (1999): Cassia harneyi.
linearifolia (G. Don) H.S. Irwin & Barneby
var. latifolia *R.C. Barneby in Brittonia,* 51(3): 336 (1999)—Brazil Southeast (Minas Gerais).
lineata
var. arduinervis *(Urb.) R. Govaerts, World Checklist Seed Plants,* 3(1): 13 (1999): Cassia arduinervis.
longipes *(Domin) L. Pedley in Fl. Australia,* 12: 197 (1998): Cassia longipes.
maritima *L. Pedley in Fl. Australia,* 12: 196 (1998)—Queensland, New South Wales.
molinae *G. Flores Franco & M. Sousa S.in An. Inst. Biol. Univ. Nac. Autón. Méx., Bot.,* 71(1): 20 (2000)—Costa Rica, Honduras.
moorei *L. Pedley in Fl. Australia,* 12: 197 (1998)—Western Australia.
neesiana
var. goyazensis *(Taub. ex Ule) R. Govaerts, World Checklist Seed Plants,* 3(1): 14 (1999): Cassia goyazensis.
var. laxiracemosa *(Harms) R. Govaerts, World Checklist Seed Plants,* 3(1): 14 (1999): Cassia laxiracemosa.
var. subnitida *(Taub. ex Harms) R. Govaerts, World Checklist Seed Plants,* 3(1): 14 (1999): Cassia subnitida.
nomame (Siebold) H. Ohashi
var. grandiflora *L. Pedley in Fl. Australia,* 12: 195 (1998)—Queensland, Western Australia, Northern Territory.
olesiphylla *(Vogel) R. Govaerts, World Checklist Seed Plants,* 3(1): 14 (1999): Cassia olesiphylla.
pteropoda *R.C. Barneby in Brittonia,* 51(3): 333 (1999)—Brazil Northeast (Bahia).
rotundata *(Vogel) R. Govaerts, World Checklist Seed Plants,* 3(1): 14 (1999): Cassia rotundata.
var. grandistipula *(Vogel) R. Govaerts, World Checklist Seed Plants,* 3(1): 14 (1999): Cassia grandistipula.
var. interstes *R. Govaerts, World Checklist Seed Plants,* 3(1): 14 (1999)—Brazil Southeast (Minas Gerais).
symonii *L. Pedley in Fl. Australia,* 12: 197 (1998)—Northern Territory, Western Australia, Queensland.
verruculosa *Afr. Fernandes & E.P. Nunes in Albertoa, n.s.,* 7: 2 (2000)—Brazil Northeast (Piauí).
vestita *(Vogel) R. Govaerts, World Checklist Seed Plants,* 3(1): 14 (1999): Cassia vestita.
xanthadena *(Benth.) R. Govaerts, World Checklist Seed Plants,* 3(1): 14 (1999): Cassia exsudans var. xanthadena.

CHAMAECYTISUS (Legumin.).
gueneri *H. Duman, K.H.C. Başer & H. Malyer in Karaca Arbor. Mag.,* 4(1): 13 (1997)—Turkey.
triflorus (Lam.) Skalická
subsp. ciliatus *(Wahlenb.) J. Holub in Fl. Slovenska,* 4(4): 38 (1988): Cytisus ciliatus.
× versicolor *(Dippel) T. Karlsson in Svensk Bot. Tidskr.,* 91(5): 249 (1997 publ. 1998): Cytisus × versicolor.

CHAMAEDOREA (Palmae).
binderi *D.R. Hodel in Principes,* 40(4): 215 (1996)—Costa Rica.
christinae *D.R. Hodel in Novon,* 7(1): 36 (1997)—Colombia.
coralliformis *D.R. Hodel in Principes,* 40(4): 212 (1996)—Costa Rica.
hodelii *M.H. Grayum in Phytologia,* 84(4): 312 (1998 publ. 1999)—Costa Rica.
ponderosa *D.R. Hodel in Novon,* 7(1): 35 (1997)—Panama.

CHAMAEGASTRODIA (Orchidac.).
exigua *(Rolfe) F. Maek. ex P. Ormerod in Austral. Orchid Rev.,* 63(4): 10 (1998): Spiranthes exigua.

CHAMAELEON (Compos.).
cuneatus *(Boiss.) M. Dittrich in Boissiera,* 51: 74 (1996): Atractylis cuneata.
speciosus *(DC.) M. Dittrich in Boissiera,* 51: 74 (1996): Atractylis speciosa.

CHAMAELEORCHIS *K. Senghas & Lückel in Schlechter Orchideen, ed. 3,* I/C(33–36): 2305 (1997). *ORCHIDACEAE*
warscewiczii *(Rchb. f.) K. Senghas & Lückel in Schlechter Orchideen, ed. 3,* I/C(33–36): 2305 (1997): Miltonia warscewiczii.

CHAMAEMELUM (Compos.).
nobile (L.) All.
var. aureum *(L.) M. Ladero Älvarez, C.J. Valle Gutiérrez, M.T. Santos Bobillo, T. Ruiz* Téllez & M.I. Fernández-Arias *in Stud. Bot. (Salamanca),* 4: 185 (1985), without basionym ref.: Anacyclus aureus.

CHAMAERHODOS (Rosac.).
grandiflora (Pall. ex Schult.) Bunge
var. nilssonii *B.A. Jurtzev in Arktich. Fl. SSSR,* 9(1): 238 (1984)—Yakutiya.

CHAMAESPARTIUM (Legumin.).
undulatum *(Ern) S. Talavera & L. Sáez in An. Jard. Bot. Madrid,* 57(1): 208 (1999): Genistella sagittalis subsp. undulata.

CHAMAESYCE (Euphorbiac.).
bombaiensis *(Santapau) R.R. Rao in J. Mysore Univ., B,* 24: 71 (1971–72 publ. 1972): Euphorbia bombaiensis.
carissoides *(F.M. Bailey) D.C. Hassall ex P.I. Forster & R.J.F. Henderson in Novon,* 5(4): 323 (1995): Euphorbia carissoides.
centralis *(B.G. Thomson) P.I. Forster & R.J.F. Henderson in Novon,* 5(4): 323 (1995): Euphorbia centralis.
coghlanii *(F.M. Bailey) D.C. Hassall ex P.I. Forster & R.J.F. Henderson in Novon,* 5(4): 323 (1995): Euphorbia coghlanii.
eleanoriae *D.H. Lorence & W.L. Wagner in Novon,* 6(1): 68 (1996)—Hawaiian Is.
fimbriata *(B. Heyne ex Roth) R.R. Rao & B.A. Razi in J. Mysore Univ., B,* 23(1–2): 53 (1969–70 publ. 1970): Euphorbia fimbriata.
forsskaolii *(J. Gay) E. Figueiredo in Fl. Cabo Verde,* 54: 48 (1996): Euphorbia forsskaolii.
fosbergii *J. Florence in Bull. Mus. Nation. Hist. Nat., B, Adansonia,* Ser. 4, 18(3–4): 243 (1996)—Society Is., Tubuai Is., Tuamotu Is.
inaequilatera *(Sond.) M. Fennane in Bocconea,* 8: 95 (1998), isonym; *see Soják in Čas. Nár. Muz. (Prague),* 140(3–4): 169 (1972): Euphorbia inaequilatera.
jamesonii *(Boiss.) G.L. Webster in Cat. Vasc. Pl. Ecuador (Monogr. Syst. Bot. Missouri Bot. Gard., 75):* 954 (1999): Euphorbia jamesonii.
melanocarpa *(Boiss.) G.L. Webster in Cat. Vasc. Pl. Ecuador (Monogr. Syst. Bot. Missouri Bot. Gard., 75):* 954 (1999): Euphorbia melanocarpa.
mitchelliana *(Boiss.) D.C. Hassall in Austral. J. Bot.,* 24(5): 640 (1976): Euphorbia mitchelliana.
nana *(Klotzsch & Garcke) M.P. Simmons & W.J. Hayden in Brittonia,* 49(2): 166 (1997): Anisophyllum nanum.
obliqua *(Endl.) J. Florence in Bull. Mus. Nation. Hist. Nat., B, Adansonia,* Ser. 4, 18(3–4): 241 (1996): Euphorbia obliqua.
ocellata (Durand & Hilg.) Millsp.
var. arenicola *(Parish) N.H. Holmgren & P.K. Holmgren in A. Cronquist et al., Intermountain Fl., Vasc. Pl. Intermountain West,* U.S.A., 3A: 278 (1997): Euphorbia arenicola.
ophiolitica *P.I. Forster in Austrobaileya,* 5(4): 711 (2000)—Queensland.
petala *(Ewart & L.R. Kerr) P.I. Forster & R.J.F. Henderson in Novon,* 5(4): 323 (1995): Euphorbia petala.
psammogeton *(P.S. Green) P.I. Forster & R.J.F. Henderson in Novon,* 5(4): 323 (1995): Euphorbia psammogeton.
quitensis *(Boiss.) G.L. Webster in Cat. Vasc. Pl. Ecuador (Monogr. Syst. Bot. Missouri Bot. Gard., 75):* 954 (1999): Euphorbia quitensis.
ruiziana *(Klotzsch & Garcke) G.L. Webster in Cat. Vasc. Pl. Ecuador (Monogr. Syst. Bot. Missouri Bot. Gard., 75):* 954 (1999): Anisophyllum ruizianum.
sachetiana *J. Florence in Bull. Mus. Nation. Hist. Nat., B, Adansonia,* Ser. 4, 18(3–4): 247 (1996)—Marquesas.

CHAMAESYCE:-
scordifolia *(Jacq.) M. Fennane in Bocconea*, 8: 95 (1998), isonym; *see Croizat in Degener, Fl. Hawaiiensis*, fam. 190, Chamaesyce, leafl. 3 (1937): Euphorbia scordifolia.
serpens (Kunth) Small
subsp. fissistipula *(Thell.) G. Hügin in Feddes Repert.*, 109(3–4): 223 (1998): Euphorbia serpens var. fissistipula.
setosa *(Boiss.) M.P. Simmons & W.J. Hayden in Brittonia*, 49(2): 176 (1997): Euphorbia selloi var. setosa.
tamanduana *(Boiss.) M.P. Simmons & W.J. Hayden in Brittonia*, 49(2): 177 (1997): Euphorbia tamanduana.
viscoides *(Boiss.) M.P. Simmons & W.J. Hayden in Brittonia*, 49(2): 178 (1997): Euphorbia viscoides.
vulgaris *J.I. Prokhanov in Trudy̆ Kuĭby̆shevsk. Bot. Sada*, 1: 8 (1941), nom. nov.: Euphorbia chamaesyce *L.*

CHAMERION (Onagrac.).
angustifolium (L.) Holub
var. canescens *(A.W. Wood) N.H. Holmgren & P.K. Holmgren in A. Cronquist et al., Intermountain Fl., Vasc. Pl. Intermountain West*, U.S.A., 3A: 228 (1997): Epilobium angustifolium β canescens.
subsp. circumvagum *(Mosquin) P.C. Hoch in Fl. Japan*, 2c: 241 (1999): Epilobium angustifolium subsp. circumvagum.
× prantlii *(Dalla Torre & Sarnth.) J. Holub in Preslia*, 70(2): 101 (1998): Epilobium × prantlii.

CHAMPEREIA (Opiliac.).
manillana (Blume) Merr.
var. longistaminea *(W.Z. Li) H.S. Kiu in J. Trop. Subtrop. Bot.*, 5(3): 3 (1997): Melientha longistaminea.

CHANGIOSTYRAX *Tao Chen in Guihaia*, 15(4): 290 (1995). *STYRACACEAE*
dolichocarpus *(C.J. Qi) Tao Chen in Guihaia*, 15(4): 291 (1995): Sinojackia dolichocarpa.

CHAPMANNIA (Legumin.).
gracilis *(Balf. f.) M. Thulin in Nordic J. Bot.*, 19(5): 604 (1999): Arthrocarpum gracile.
prismatica *(Sessé & Moç.) M. Thulin in Nordic J. Bot.*, 19(5): 600 (1999): Hedysarum prismaticum.
reghidensis *M. Thulin & D. McKean in Nordic J. Bot.*, 19(5): 605 (1999)—Socotra.
sericea *M. Thulin & D. McKean in Nordic J. Bot.*, 19(5): 602 (1999)—Socotra.
somalensis *(Hillc. & J.B. Gillett) M. Thulin in Nordic J. Bot.*, 19(5): 601 (1999): Arthrocarpum somalense.
tinireana *M. Thulin in Nordic J. Bot.*, 19(5): 603 (1999)—Socotra.

CHAPTALIA (Compos.).
chapadensis *D.J.N. Hind in Kew Bull.*, 54(4): 933 (1999 publ. 2000)—Brazil Northeast (Bahia).
hermogenis *M.D. de Moraes in Novon*, 8(2): 173 (1998)—Brazil Southeast (São Paulo).

CHARADRANAETES *J.P. Janovec & H. Robinson in Novon*, 7(2): 165 (1997). *COMPOSITAE*
durandii *(Klatt) J.P. Janovec & H. Robinson in Novon*, 7(2): 165 (1997): Senecio durandii.

CHARYBDIS *F. Speta in Phyton (Austria)*, 38(1): 58 (1998), nom. nov.: gen. Squilla *Steinh.* (Hyacinthac.)
aphylla *(Forssk.) F. Speta in Phyton (Austria)*, 38(1): 60 (1998): Anthericum aphyllum.
elata *(Andrews) F. Speta in Phyton (Austria)*, 38(1): 60 (1998): Ornithogalum elatum.
hesperia *(Webb & Berthel.) F. Speta in Phyton (Austria)*, 38(1): 60 (1998): Urginea hesperia.
maritima *(L.) F. Speta in Phyton (Austria)*, 38(1): 60 (1998): Scilla maritima.
maura *(Maire) F. Speta in Phyton (Austria)*, 38(1): 60 (1998): Urginea maura.
numidica *(Jord. & Fourr.) F. Speta in Phyton (Austria)*, 38(1): 60 (1998): Squilla numidica.
pancration *(Steinh.) F. Speta in Phyton (Austria)*, 38(1): 60 (1998): Squilla pancration.
simensis *(Hochst. ex A. Rich.) F. Speta in Phyton (Austria)*, 38(1): 60 (1998): Scilla simensis.

CHASSALIA (Rubiac.).
laikomensis *M. Cheek in Kew Bull.*, 55(4): 884 (2000)—Cameroon.
petitiana *F. Piesschaert in Syst. Bot.*, 24(3): 315 (1999), nom. nov.: Psychotria epiphytica *Mildbr.*
pteropetala *(K. Schum.) M. Cheek in Kew Bull.*, 55(4): 887 (2000): Psychotria pteropetala.
subcordatifolia *(De Wild.) F. Piesschaert in Syst. Geogr. Pl.*, 69(2): 190 (1999): Psychotria subcordatifolia.

× **CHAUBARDIANTHES** *hort.in Orchid Rev.*, 108(1235): new orchid hybrids 14 (2000). *ORCHIDACEAE* (Chaubardiella × Cochleanthes).

CHEESEMANIA (Crucif.).
stellata *(Allan) P.B. Heenan & P.J. Garnock-Jones in New Zealand J. Bot.*, 37(2): 235 (1999): C. fastigiata var. stellata.

CHEIRIDOPSIS (Aizoac.).
amabilis *S.A. Hammer in Piante Grasse*, 15(4 Suppl.): 78 (1995 publ. 1996)—Cape Province.
brevissima *R. Govaerts, World Checklist Seed Plants*, 3(1): 14 (1999), nom. nov.: C. brevis *L. Bolus.*
campanulata *G. Williamson in Aloe*, 37(1): 4 (2000)—Cape Province.
gamoepensis *S.A. Hammer in Piante Grasse*, 15(4 Suppl.): 78 (1995 publ. 1996)—Cape Province.
glomerata *S.A. Hammer in Piante Grasse*, 15(4 Suppl.): 79 (1995 publ. 1996)—Cape Province.
ponderosa *S.A. Hammer in Piante Grasse*, 15(4 Suppl.): 82 (1995 publ. 1996)—Cape Province.

CHEIROLOPHUS (Compos.).
burchardii *A. Susanna in Fontqueria*, 44: 32 (1996), nom. nov.: Centaurea canariensis var. subexpinnata *Burchard.*
crassifolius *(Bertol.) A. Susanna in Pl. Syst. Evol.*, 214(1–4): 157 (1999): Centaurea crassifolia.
grandifolius *(Font Quer) G. Stübing, J.B. Peris, A. Olivares & J. Martín in An. Jard. Bot. Madrid*, 55(1): 171 (1997): Centaurea intybacea var. grandifolia.
lagunae *A. Olivares, J.B. Peris, G. Stübing & J. Martín in An. Jard. Bot. Madrid*, 53(2): 262 (1995 publ. 1996)—Spain.
mansanetianus *G. Stübing, J.B. Peris, A. Olivares & J. Martín in An. Jard. Bot. Madrid*, 55(1): 172 (1997)—Spain.
sempervirens (L.) Pomel
subsp. mauritanicus *(Font Quer) M. Ibn Tattou in Bocconea*, 8: 31 (1998): Centaurea sempervirens var. mauritanica.
tananicus *(Maire) M. Ibn Tattou in Bocconea*, 8: 31 (1998), isonym; *see Holub in Preslia*, 45(2): 143 (1973): Centaurea tananica.

CHEIROSTYLIS (Orchidac.).
bipunctata *L.V. Averyanov in Bot. Zhurn.*, 81(10): 78 (1996)—Vietnam.
derchiensis *S.S. Ying in Mem. Coll. Agric. Nation. Taiwan Univ.*, 31(1): 20 (1991)—Taiwan.
divina (Guinea) Summerh.
var. ochyrae *D.L. Szlachetko & T.S. Olszewski in Fl. Cameroun*, 34: 252 (1998)—Cameroon.
eglandulosa *L.V. Averyanov in Bot. Zhurn.*, 81(10): 82, 80 (1996)—Vietnam.
gunnarii *A. Nageswara Rao in Nordic J. Bot.*, 18(1): 23 (1998)—Assam (Arunachal Pradesh).
marmorata *L.V. Averyanov in Lindleyana*, 14(4): 216 (1999), homonym; *see Lem. in Fl. Serres Jard. Eur.*, Ser. 1, 4: 370 (1848)—Vietnam.
merrillii *(Ames & Quisumb.) P. Ormerod in Austral. Orchid Rev.*, 63(4): 10 (1998): Adenostylis merrillii.
notialis *D.L. Jones in Muelleria*, 10: 80 (1997)—New South Wales, Queensland.
orobanchoides *(F. Muell.) D.L. Jones & M.A. Clements in Orchadian*, 12(8): 350 (1998): Gastrodia orobanchoides.
pingbianensis *K.Y. Lang in Acta Phytotax. Sin.*, 34(6): 635 (1996)—China South-central (Yunnan).
sherriffii *N. Pearce & P.J. Cribb in Edinburgh J. Bot.*, 56(2): 275 (1999)—Bhutan.
tabiyahanensis *(Hayata) N. Pearce & P.J. Cribb in Edinburgh J. Bot.*, 56(2): 278 (1999): Zeuxine tabiyahanensis.

CHELONANTHUS (Gentianac.).
longistylus *(J.G.M. Pers. & Maas) L. Struwe & V.A. Albert in Harvard Pap. Bot.*, 3(1): 70 (1998): Irlbachia alata subsp. longistyla.
matogrossensis *(J.G.M. Pers. & Maas) L. Struwe & V.A. Albert in Harvard Pap. Bot.*, 3(1): 70 (1998): Irlbachia alata subsp. matogrossensis.
purpurascens *(Aubl.) L. Struwe, S. Nilsson & V.A. Albert in Harvard Pap. Bot.*, 3(1): 70 (1998): Lisianthus purpurascens.

CHELONISTELE (Orchidac.).
laetitia-reginae *E.F. de Vogel in Blumea*, 41(1): 23 (1996)—Borneo (Sarawak).

CHELYORCHIS *R.L. Dressler & N.H. Williams in G.A. Romero & G. Carnevali, Orchids of Venezuela, ed. 2*: 1130 (2000). *ORCHIDACEAE*
ampliata *(Lindl.) R.L. Dressler & N.H. Williams in G.A. Romero & G. Carnevali, Orchids of Venezuela, ed. 2*: 1130 (2000): Oncidium ampliatum.

CHENOPODIUM (Chenopodiac.).
subsect. Adenois *(Moq.) L.E. Simón in An. Jard. Bot. Madrid*, 54(1): 138 (1996): Ambrina sect. Adenois.
sect. Adenois *(Moq.) L.E. Simón in An. Jard. Bot. Madrid*, 54(1): 138 (1996): Ambrina sect. Adenois.
subsect. Capitata *Kowal ex S.L. Mosyakin & S.E. Clemants in Novon*, 6(4): 399 (1996).
subsect. Cicatricosa *(Aellen) S.L. Mosyakin & S.E. Clemants in Novon*, 6(4): 402 (1996): C. ser. Cicatricosa.
subsect. Favosa *(Aellen) S.L. Mosyakin & S.E. Clemants in Novon*, 6(4): 401 (1996): C. ser. Favosa.
subsect. Foliosa *Kowal ex S.L. Mosyakin & S.E. Clemants in Novon*, 6(4): 398 (1996).
subsect. Fremontiana *(Standl.) S.E. Clemants & S.L. Mosyakin in Novon*, 6(4): 401 (1996): C. VII. Fremontiana.
subsect. Leptophylla *(Standl.) S.E. Clemants & S.L. Mosyakin in Novon*, 6(4): 400 (1996): C. IV. Leptophylla.
subsect. Polysperma *Kowal ex S.L. Mosyakin & S.E. Clemants in Novon*, 6(4): 400 (1996).
subsect. Roubieva *(Moq.) L.E. Simón in An. Jard. Bot. Madrid*, 54(1): 138 (1996): gen. Roubieva.
subsect. Standleyana *S.L. Mosyakin & S.E. Clemants in Novon*, 6(4): 402 (1996).
subsect. Undata *Aellen & Iljin ex S.L. Mosyakin & S.E. Clemants in Novon*, 6(4): 400 (1996).
subsect. Urbica *(Standl.) S.L. Mosyakin & S.E. Clemants in Novon*, 6(4): 400 (1996): C. I. Urbica.
acuminatum Willd.
var. minimum *(W. Wang & P.Y. Fu) Z.S. Qin in Clavis Pl. Chinae Bor.-Or. ed. 2 (ed. P.Y. Fu)*: 164 (1995): C. minimum.
aristatum L.
f. muticum *J.Q. Fu in Fl. Loess-Plat. Sin.*, 1: 610 (2000), with two types cited—China North-central (Shaanxi, Shanxi, Gansu), Inner Mongolia (Ningxia).
cyanifolium *S.C. Pandeya, G. Singhal & A.K. Bhatnagar in J. Bombay Nat. Hist. Soc.*, 95(3): 480 (1998)—India (Uttar Pradesh).
× dvorakianum *S.L. Mosyakin in Ukr. Bot. Zhurn.*, 52(4): 459 (1995), nom. nov.: C. laciniatum *Murr.*
fasciculosum Aellen
var. schimperi *(Asch.) M.G. Gilbert in Nordic J. Bot.*, 19(4): 411 (1999): C. murale var. schimperi.
gubanovii *A.P. Suchorukow in Feddes Repert.*, 110(7–8): 493 (1999)—Mongolia.
lobodontum *H. Scholz in Verh. Bot. Ver. Berlin Brandenburg*, 132: 8 (1999)—Germany.
minimum *W. Wang & P.Y. Fu in Fl. Plant. Herb. Chinae bor.-or.*, 2: 111 (1959)—Inner Mongolia (Nei Mongol).
× morarii *(Prodán) S.L. Mosyakin in Ukr. Bot. Zhurn.*, 52(4): 459 (1995): C. album f. morarii.
prostratum Bunge
subsp. karoi *(Murr) M.N. Lomonosova in Fl. Sibir. (Salicac.-Amaranthac.)*, 5: 147 (1992), without basionym ref.: C. album subsp. karoi.
pseudopulifolium *(Scholz) J. Murr in I.M. Hayward & G.C. Druce, Advent. Fl. Tweedside*: 190 (1919): C. album var. pseudopulifolium.

CHENOPODIUM:-
santoshei *S.C. Pandeya, G. Singhal & A.K. Bhatnagar in J. Bombay Nat. Hist. Soc.*, 95(3): 484 (1998)—India (Uttar Pradesh).
strictum Roth
var. angustifolium *F. Dvořák in Feddes Repert.*, 104(7–8): 447 (1993)—Czechoslovakia.

CHERSODOMA (Compos.).
deltoidea *A. Sagástegui-Alva & M.O. Dillon in Brittonia*, 48(4): 598 (1996 publ. 1997)—Peru.
glabriuscula *(Cabrera) M.O. Dillon & A. Sagástegui-Alva in Brittonia*, 48(4): 600 (1996 publ. 1997): Senecio diclinus var. glabriusculus.

CHESNEYA (Legumin.).
hissarica *A.G. Borissova in Fl. Tadzhikistana*, 5: 684 (1937)—Tadzhikistan.
intermedia *(Yakovlev) Z.G. Qian in Acta Bot. Yunnanica*, 20(4): 401 (1998): Spongiocarpella intermedia.
nubigena (D. Don) Ali
subsp. purpurea *(P.C. Li) X.Y. Zhu in Acta Phytotax. Sin.*, 34(5): 561 (1996): C. purpurea.
paucifoliolata *(Yakovlev) Z.G. Qian in Acta Bot. Yunnanica*, 20(4): 400 (1998): Spongiocarpella paucifoliolata.
potaninii *(Yakovlev) R. Govaerts, World Checklist Seed Plants*, 3(1): 14 (1999): Spongiocarpella potaninii.
yunnanensis *(Yakovlev) Z.G. Qian in Acta Bot. Yunnanica*, 20(4): 400 (1998): Spongiocarpella yunnanensis.

CHICHICASTE *M. Weigend, Nasa & Conquest S. America - Syst. Rearrang. in Loasac.*: 215 (1997). *LOASACEAE*
grandis *(Standl.) M. Weigend, Nasa & Conquest S. America - Syst. Rearrang. in Loasac.*: 215 (1997): Loasa grandis.

CHILIADENUS (Compos.).
antiatlanticus *(Emb. & Maire) F. Gómiz in An. Jard. Bot. Madrid*, 58(1): 199 (2000): Jasonia glutinosa var. antiatlantica.

CHILOGLOTTIS (Orchidac.).
jeanesii *D.L. Jones in Muelleria*, 10: 64 (1997)—Victoria.
triceratops *D.L. Jones in Austral. Orchid Res.*, 3: 66 (1998)—Tasmania.

CHIMANTAEA (Compos.).
cinerea (Gleason & S.F. Blake) Maguire, Steyerm. & Wurdack
subsp. similis *(Maguire, Steyerm. & Wurdack) J.F. Pruski & O. Huber in Fl. Venezuelan Guayana (gen. eds. J.A. Steyermark et al.)*, 3: 242 (1997): C. similis.

CHIMAPHILA (Ericac.).
monticola Andres
var. taiwaniana *(Masam.) S.S. Ying in Mem. Coll. Agric. Nation. Taiwan Univ.*, 31(1): 29 (1991): C. taiwaniana.

CHIMARRHIS (Rubiac.).
subgen. Pseudochimarrhis *(Ducke) P.G. Delprete in Fl. Neotrop.*, 77: 169 (1999): gen. Pseudochimarrhis.
clausicorollata *J.H. Kirkbride in Biollania*, ed. esp. 6 (J.J. Wurdack Festschr.): 394 (1997)—Guyana.
duckeana *P.G. Delprete in Fl. Neotrop.*, 77: 179 (1999)—Brazil North (Amazonas)
gentryana *P.G. Delprete in Brittonia*, 48(1): 37 (1996)—Peru, Ecuador, Brazil North (Amazonas).
speciosa *(Steyerm.) P.G. Delprete in Fl. Neotrop.*, 77: 167 (1999): C. microcarpa var. speciosa.

CHIMONOBAMBUSA (Gramin.).
tianquanensis *T.P. Yi in J. Bamboo Res.*, 19(1): 11 (2000)—China South-central (Sichuan).

CHIMONOCALAMUS (Gramin.).
bicorniculatus *Sheng Feng Li & Z.F. Wang in Acta Phytotax. Sin.*, 33(6): 614 (1995)—China South-central (Yunnan).

CHIMONOCALAMUS:-
burmaensis *(C.S. Chao & Renvoize) D. Ohrnberger, Bamboos of the World Intro. 2:* 10 (1996), isonym; *see D.Z. Li in Acta Bot. Yunnanica*, 16(1): 40 (1994): Sinarundinaria burmaensis.
longispiculatus *(C.S. Chao & Renvoize) D. Ohrnberger, Bamboos of the World Intro. 2:* 10 (1996), homonym; *see R.B. Majumdar in Karth. et al., Fl. Ind. Enumerat. - Monocot.*: 276 (1989): Sinarundinaria longispiculata.
lushaiensis *D. Ohrnberger, Bamboos of the World Intro. 3:* 14 (1996), nom. nov.: Sinarundinaria longispiculata *C.S. Chao & Renvoize.*

CHIONANTHUS (Oleac.).
amblirrhinus *P.S. Green in Kew Bull.*, 51(4): 765 (1996)—Thailand.
beccarii *(Stapf) R. Kiew in Blumea*, 43(2): 471 (1998): Linociera beccarii.
brachystachys *(Schltr.) P.S. Green in Fl. N. Caléd.*, 22: 49 (1998): Notelaea brachystachys.
brassii *(Kobuski) R. Kiew in Blumea*, 43(2): 471 (1998): Linociera brassii.
caymanensis Stearn
var. longipetala *G.R. Proctor in Kew Bull.*, 51(3): 501 (1996)—Cayman Is.
clementis *(Quisumb. & Merr.) R. Kiew in Blumea*, 43(2): 472 (1998): Linociera clementis.
cordifolius *J.-N. Labat, Marc Pignal & O. Pascal in Novon*, 9(1): 68 (1999)—Comoros (Mayotte).
coriaceus *(S. Vidal) Yuen P. Yang & S.Y. Lu in Fl. Taiwan, ed. 2*, 4: 129 (1998): Linociera coriacea.
decipiens *P.S. Green in Kew Bull.*, 51(4): 766 (1996)—Thailand, Burma.
eriorachis *(Kerr) P.S. Green in Kew Bull.*, 51(4): 766 (1996): Linociera eriorachis.
gigas *(Lingelsh.) R. Kiew in Blumea*, 43(2): 472 (1998): Linociera gigas.
hahlii *(Rech.) R. Kiew in Blumea*, 43(2): 472 (1998): Linociera hahlii.
insularis *J.-N. Labat, Marc Pignal & O. Pascal in Novon*, 9(1): 69 (1999)—Comoros (Mayotte, Comoros).
kajewskii *(Sleumer) R. Kiew in Blumea*, 43(2): 473 (1998): Linociera kajewskii.
mala-elengi (Dennst.) P.S. Green
subsp. linocieroides *(Wight) P.S. Green in Kew Bull.*, 51(4): 767 (1996): Olea linocieroides.
subsp. terniflorus *(Wall. ex G. Don) P.S. Green in Kew Bull.*, 51(4): 767 (1996): Linociera terniflora.
maxwellii *P.S. Green in Kew Bull.*, 55(1): 194 (2000)—Thailand.
microbotrys *(Kerr) P.S. Green in Kew Bull.*, 51(4): 768 (1996): Linociera microbotrys.
microstigma *(Gagnep.) P.S. Green in Kew Bull.*, 51(4): 768 (1996): Linociera microstigma.
nitidus *(Merr.) R. Kiew in Blumea*, 43(2): 473 (1998): Linociera nitida.
pedunculatus *P.S. Green in Fl. N. Caléd.*, 22: 47 (1998), nom. nov.: Notelaea paniculata *Guillaumin.*
remotinervius *(Merr.) R. Kiew in Blumea*, 43(2): 474 (1998): Linociera remotinervia.
riparius *(Lingelsh.) R. Kiew in Blumea*, 43(2): 474 (1998): Linociera riparia.
rupicola *(Lingelsh.) R. Kiew in Blumea*, 43(2): 475 (1998): Linociera rupicola.
salicifolius *(Lingelsh.) R. Kiew in Blumea*, 43(2): 476 (1998): Linociera salicifolia.
sessiliflorus *(Hemsl.) F.R. Fosberg in Micronesica*, 25(2): 197 (1992): Linociera sessiliflora.
sessiliflorus *(Hemsl.) R. Kiew in Blumea*, 43(2): 476 (1998), isonym; *see Fosberg in Micronesica*, 25(2): 197 (1992): Linociera sessiliflora.
stenurus *(Merr.) R. Kiew in Blumea*, 43(2): 477 (1998): Linociera stenura.
sutepensis *(Kerr) P.S. Green in Kew Bull.*, 51(4): 769 (1996): Linociera sutepensis.
thorelii *(Gagnep.) P.S. Green in Kew Bull.*, 51(4): 769 (1996): Linociera thorelii.
velutinus *(Kerr) P.S. Green in Kew Bull.*, 51(4): 770 (1996): Linociera velutina.

CHIRITA (Gesneriac.).
atroglandulosa *W.T. Wang in Novon*, 7(4): 423 (1997 publ. 1998)—China Southeast (Guangxi).
baishouensis *Y.G. Wei, H.Q. Wen & S.H. Zhong in Acta Phytotax. Sin.*, 38(3): 299 (2000)—China Southeast (Guangxi).

CHIRITA:-
demissa *(Hance) W.T. Wang in Novon*, 7(4): 424 (1997 publ. 1998): Didymocarpus demissus.
halongensis *R. Kiew & T.H. Nguyen in Gard. Bull. Singapore*, 52(1): 189 (2000)—Vietnam.
hiepii *R. Kiew in Gard. Bull. Singapore*, 52(1): 192 (2000)—Vietnam.
modesta *R. Kiew & T.H. Nguyen in Gard. Bull. Singapore*, 52(1): 194 (2000)—Vietnam.
mollifolia *D. Fang, Y.G. Wei & J. Murata in Acta Phytotax. Sin.*, 37(6): 592 (1999)—China Southeast (Guangxi).
napoensis *Z.Yu Li in Novon*, 7(4): 424 (1997 publ. 1998)—China Southeast (Guangxi).
puerensis *Y.Y. Qian in Acta Phytotax. Sin.*, 34(2): 216 (1996)—China South-central (Yunnan).
pungentisepala *W.T. Wang in Novon*, 7(4): 425 (1997 publ. 1998)—China Southeast (Guangxi).
pycnantha *W.T. Wang in Novon*, 7(4): 425 (1997 publ. 1998)—China South-central (Yunnan).
shouchengensis *Z.Yu Li in Novon*, 7(4): 426 (1997 publ. 1998)—China Southeast (Guangxi).
shuii *Z.Yu Li in Novon*, 7(4): 427 (1997 publ. 1998)—China South-central (Yunnan).
skogiana *Z.Yu Li in Novon*, 7(4): 427 (1997 publ. 1998)—China North-central (Gansu).
tamiana *B.L. Burtt in Gloxinian*, 49(4): 20 (1999)—Vietnam.
varicolor *D. Fang & D.H. Qin in Acta Phytotax. Sin.*, 37(6): 591 (1999)—China Southeast (Guangxi).
wangiana *Z.Yu Li in Novon*, 7(4): 428 (1997 publ. 1998)—China Southeast (Guangxi).

CHIROPETALUM (Euphorbiac.).
astroplethos *(J.W. Ingram) A. Radcliffe-Smith & R. Govaerts in Kew Bull.*, 52(2): 478 (1997): Argythamnia astroplethos.
pilosistylum *(Allem & Irgang) A. Radcliffe-Smith & R. Govaerts in Kew Bull.*, 52(2): 478 (1997): Argyrothamnia pilosistyla.
ramboi *(Allem & Irgang) A. Radcliffe-Smith & R. Govaerts in Kew Bull.*, 52(2): 478 (1997): Argyrothamnia ramboi.

CHLAMYDACANTHUS (Acanthac.).
rupestris *(Nees) R. Govaerts, World Checklist Seed Plants*, 3(1): 14 (1999): Aetheilema rupestre.

CHLORIS (Gramin.).
boliviensis *S.A. Renvoize, Gramineas de Bolivia*: 347 (1998)—Bolivia.
sagraeana A. Rich.
subsp. cubensis *(Hitchc. & Ekman) L. Catasús Guerra in Fontqueria*, 44: 143 (1996): C. cubensis.
verticillata *T. Nuttall in Trans. Amer. Philos. Soc., n.s.*, 5: 150 (1835)—Arkansas.

CHLOROLEUCON (Legumin.).
acacioides *(Ducke) R.C. Barneby & J.W. Grimes, Silk Tree, Guanacaste, Monkey's Earring (Mem. New York Bot. Gard., 74(1))*: 141 (1996): Pithecellobium acacioides.
chacoense *(Burkart) R.C. Barneby & J.W. Grimes, Silk Tree, Guanacaste, Monkey's Earring (Mem. New York Bot. Gard., 74(1))*: 140 (1996): Pithecellobium chacoense.
eurycyclum *R.C. Barneby & J.W. Grimes, Silk Tree, Guanacaste, Monkey's Earring (Mem. New York Bot. Gard., 74(1))*: 140 (1996)—Venezuela.
extortum *R.C. Barneby & J.W. Grimes, Silk Tree, Guanacaste, Monkey's Earring (Mem. New York Bot. Gard., 74(1))*: 142 (1996)—Brazil Northeast (Bahia).
mangense (Jacq.) Britton & Rose
var. lentiscifolium *(A. Rich.) R.C. Barneby & J.W. Grimes, Silk Tree, Guanacaste, Monkey's Earring (Mem. New York Bot. Gard., 74(1))*: 156 (1996): Acacia lentiscifolia.
var. leucospermum *(Brandegee) R.C. Barneby & J.W. Grimes, Silk Tree, Guanacaste, Monkey's Earring (Mem. New York Bot. Gard., 74(1))*: 154 (1996): Pithecellobium leucospermum.
var. mathewsii *(Benth.) R.C. Barneby & J.W. Grimes, Silk Tree, Guanacaste, Monkey's Earring (Mem. New York Bot. Gard., 74(1))*: 152 (1996): Pithecellobium mathewsii.

CHLOROLEUCON:-
mangense (Jacq.) Britton & Rose
var. tetrazyx *R.C. Barneby & J.W. Grimes, Silk Tree, Guanacaste, Monkey's Earring (Mem. New York Bot. Gard., 74(1)):* 157 (1996)—Venezuela.
var. vincentis *(Benth.) R.C. Barneby & J.W. Grimes, Silk Tree, Guanacaste, Monkey's Earring (Mem. New York Bot. Gard., 74(1)):* 154 (1996): Pithecellobium vincentis.
tenuiflorum *(Benth.) R.C. Barneby & J.W. Grimes, Silk Tree, Guanacaste, Monkey's Earring (Mem. New York Bot. Gard., 74(1)):* 145 (1996): Pithecellobium tenuiflorum.
tortum *(Mart.) R.C. Barneby & J.W. Grimes, Silk Tree, Guanacaste, Monkey's Earring (Mem. New York Bot. Gard., 74(1)):* 146 (1996), isonym; *see Pittier in Bol. Cient. Tecn. Mus. Comerc. Venez.*, 1: 16 (1927): Pithecellobium tortum.

CHLOROPHYTUM (Anthericac.).
angustissimum *(Poelln.) I. Nordal in Fl. Trop. E. Afr., Anthericac.:* 42 (1997): Anthericum angustissimum.
cameronii (Baker) Kativu
var. grantii *(Baker) I. Nordal in Fl. Trop. E. Afr., Anthericac.:* 32 (1997): Anthericum grantii.
var. pterocaulon *(Baker) I. Nordal in Fl. Trop. E. Afr., Anthericac.:* 32 (1997): Anthericum pterocaulon.
var. purpuratum *(Rendle) R. Govaerts, World Checklist Seed Plants,* 3(1): 14 (1999): Anthericum purpuratum.
filipendulum Baker
subsp. amaniense *(Engl.) I. Nordal & A.D. Poulsen in Fl. Trop. E. Afr., Anthericac.:* 59 (1997): C. amaniense.
hirsutum *A.D. Poulsen & I. Nordal in Kew Bull.*, 54(4): 941 (1999 publ. 2000)—Uganda, Zaire, Burundi.
jacquinii *R. Govaerts, World Checklist Seed Plants,* 3(1): 14 (1999), nom. nov.: Anthericum undulatum *Jacq.*
longiphyllum *R. Govaerts, World Checklist Seed Plants,* 3(1): 14 (1999), nom. nov.: Anthericum longistylum *Baker.*
occultum *A.D. Poulsen & I. Nordal in Kew Bull.*, 54(4): 944 (1999 publ. 2000)—Uganda, Zaire.
sphacelatum (Baker) Kativu
var. hockii *(De Wild.) I. Nordal in Fl. Trop. E. Afr., Anthericac.:* 35 (1997): Anthericum hockii.
var. milanjianum *(Rendle) I. Nordal in Fl. Trop. E. Afr., Anthericac.:* 35 (1997): Anthericum milanjianum.
sylvestre *M. Bardot-Vaucoulon in Adansonia,* 19(1): 152 (1997)—Madagascar.
zingiberastrum *I. Nordal & A.D. Poulsen in Fl. Trop. E. Afr., Anthericac.:* 48 (1997); *et in Kew Bull.*, 53(4): 937 (1998)—Malawi, Tanzania, Zambia, Mozambique.

CHOMELIA (Rubiac.).
bahiae *J.H. Kirkbride in Brittonia,* 49(3): 364 (1997)—Brazil Northeast (Bahia).
laxiflora *(Standl.) R. Govaerts, World Checklist Seed Plants,* 3(1): 14 (1999): Anisomeles laxiflora.
parvifolia *(Standl.) R. Govaerts, World Checklist Seed Plants,* 3(1): 15 (1999): Anisomeles parvifolia.
rudis *(Standl.) D.H. Lorence in Novon,* 7(1): 48 (1997): Anisomeris rudis.

× **CHONDRANTHES** *hort.in Orchid Rev.*, 106(1223): 318 (1998). *ORCHIDACEAE* (Chondrorhyncha × Cochleanthes).

CHONDROPHYLLA (Gentianac.).
prostrata *(Haenke) J.P. Anderson in Proc. Iowa Acad. Sci.*, 25: 445 (1918): Gentiana prostrata.

CHONDRORHYNCHA (Orchidac.).
lankesteriana *F. Pupulin in Lindleyana,* 15(1): 21 (2000)—Costa Rica.
luerorum *R. Vásquez Ch. & C.H. Dodson in Rev. Soc. Boliv. Bot.*, 2(1): 1 (1998)—Bolivia.
manzurii *P. Ortiz V.in Rev. Acad. Colomb. Cienc. Exact. Fis. Nat.*, 90: 57 (2000)—Colombia.

CHONDROSCAPHE (Orchidac.).
embreei *(Dodson & Neudecker) C. Rungius in Orchidee,* Suppl. 3: 16 (1996): Chondrorhyncha embreei.
escobariana *(Dodson & Neudecker) C. Rungius in Orchidee,* Suppl. 3: 16 (1996): Chondrorhyncha escobariana.
gentryi *(Dodson & Neudecker) C. Rungius in Orchidee,* Suppl. 3: 16 (1996): Chondrorhyncha gentryi.

CHORDIFEX *B.G. Briggs & L.A.S. Johnson in Telopea,* 7(4): 356 (1998). *RESTIONACEAE*
abortivus *(Nees) B.G. Briggs & L.A.S. Johnson in Telopea,* 8(1): 24 (1998): Restio abortivus.
amblycoleus *(F. Muell.) B.G. Briggs & L.A.S. Johnson in Telopea,* 8(1): 24 (1998): Restio amblycoleus.
chaunocoleus *(F. Muell.) B.G. Briggs & L.A.S. Johnson in Telopea,* 8(1): 24 (1998): Restio chaunocoleus.
crispatus *(R. Br.) B.G. Briggs & L.A.S. Johnson in Telopea,* 8(1): 24 (1998): Restio crispatus.
gracilior *(F. Muell. ex Benth.) B.G. Briggs & L.A.S. Johnson in Telopea,* 8(1): 25 (1998): Restio gracilior.
isomorphus *(K.W. Dixon & Meney) B.G. Briggs & L.A.S. Johnson in Telopea,* 8(1): 25 (1998): Restio isomorphus.
laxus *(R. Br.) B.G. Briggs & L.A.S. Johnson in Telopea,* 8(1): 25 (1998): Restio laxus.
leucoblepharus *(Gilg) B.G. Briggs & L.A.S. Johnson in Telopea,* 8(1): 25 (1998): Restio leucoblepharus.
ornatus *(Steud.) B.G. Briggs & L.A.S. Johnson in Telopea,* 8(1): 25 (1998): Restio ornatus.
sphacelatus *(R. Br.) B.G. Briggs & L.A.S. Johnson in Telopea,* 8(1): 25 (1998): Restio sphacelatus.
stenandrus *B.G. Briggs & L.A.S. Johnson in Telopea,* 7(4): 356 (1998)—Western Australia.

CHORISEPALUM (Gentianac.).
ovatum Gleason
var. sipapoanum *B. Maguire in Mem. New York Bot. Gard.*, 32: 344 (1981)—Venezuela.
psychotrioides Ewan
var. acuminatum *(Steyerm.) B. Maguire in Mem. New York Bot. Gard.*, 32: 349 (1981): Chorisepalum acuminatum.
sipapoanum *(Maguire) L. Struwe & V.A. Albert in Fl. Venezuelan Guayana (gen. eds. J.A. Steyermark et al.)*, 5: 491 (1999): C. ovatum var. sipapoanum.

CHORISPORA (Crucif.).
tashkorganica *I.A. Al-Shehbaz, T.Y. Cheo, L.L. Lu & G. Yang in Novon,* 10(2): 106 (2000)—Xinjiang.
tianschanica *Z.X. An in Fl. Xinjiangensis,* 2(2): 378 (1995)—Xinjiang.

CHOSENIA (Salicac.).
cardiophylla *(Trautv. & C.A. Mey.) N. Chao in J. Sichuan Forest. Sci. Techn.*, 17(1): 14 (1996): Salix cardiophylla.
maximowiczii *(Kom.) N. Chao in J. Sichuan Forest. Sci. Techn.*, 17(1): 14 (1996): Salix maximowiczii.
urbaniana *(Seemen) N. Chao in J. Sichuan Forest. Sci. Techn.*, 17(1): 14 (1996): Salix urbaniana.

CHOUARDIA *F. Speta in Phyton (Austria)*, 38(1): 97 (1998). *HYACINTHACEAE*
lakusicii *(Šilić) F. Speta in Phyton (Austria)*, 38(1): 98 (1998), basionym may not be effectively publ.: Scilla lakusicii.
litardierei *(Breistr.) F. Speta in Phyton (Austria)*, 38(1): 98 (1998): Scilla litardierei.

CHOUXIA (Sapindac.).
bijugata *G.E. Schatz, R.E. Gereau & P.P. Lowry in Adansonia,* 21(1): 53 (1999)—Madagascar.
borealis *G.E. Schatz, R.E. Gereau & P.P. Lowry in Adansonia,* 21(1): 54 (1999)—Madagascar.
macrophylla *G.E. Schatz, R.E. Gereau & P.P. Lowry in Adansonia,* 21(1): 56 (1999)—Madagascar.
mollis *G.E. Schatz, R.E. Gereau & P.P. Lowry in Adansonia,* 21(1): 58 (1999)—Madagascar.

CHOUXIA:-
saboureaui *Capuron ex G.E. Schatz, R.E. Gereau & P.P. Lowry in Adansonia,* 21(1): 59 (1999)—Madagascar.

CHRISTOLEA (Crucif.).
niyaensis *Z.X. An in Fl. Xinjiangensis,* 2(2): 376 (1995)—Xinjiang.
pinnatifida *R.F. Huang in Acta Phytotax. Sin.*, 35(6): 556 (Dec. 1997); *et in Fl. Qinghaiica,* 1: 508 (Dec. 1997)—Qinghai.

CHROMOLAENA (Compos.).
guiengolensis *(L. Torres & Villaseñor) B.L. Turner in Phytologia,* 82(6): 386 (1997 publ. 1998): Eupatorium guiengolense.
tamaulipasensis *(R.M. King & H. Rob.) B.L. Turner in Phytologia,* 82(6): 386 (1997 publ. 1998): Viereckia tamaulipasensis.

CHRYSACTINIA (Compos.).
luzmariae *J. Rzedowski & G. Calderón de Rzedowski in Bol. Inst. Bot. (Guadalajara)*, 5(1–3): 61 (1998)—Mexico Northeast (Guanajuato).

CHRYSACTINIUM (Compos.).
wurdackii *M.F. Zermoglio & V.A. Funk in Biollania,* ed. esp. 6 (J.J. Wurdack Festschr.): 568 (1997)—Peru.

CHRYSANTHEMUM (Compos.).
coreanum (H. Lév. & Vaniot) Nakai
subsp. maximowiczii *(Kom.) V.N. Voroschilov in A.K. Skvortsov (ed.), Florist. issl. v razn. raĭonakh SSSR:* 195 (1985): C. maximowiczii.
× intermedium *Y.N. Lee, Fl. Korea:* 1162 (1996), without diagnostic latin descr., contrary to Arts. 37.4, 37.5 ICBN (1994)—South Korea.(C. zawadskii subsp. acutilobum × subsp. latilobum).
makinoi
var. wakasaense *(Shimot. ex Kitam.) R. Govaerts, World Checklist Seed Plants,* 3(1): 15 (1999): C. wakasaense.
vulgare (Lam.) Gaterau
subsp. boreale *(Fisch. ex DC.) V.N. Voroschilov in A.K. Skvortsov (ed.), Florist. issl. v razn. raĭonakh SSSR:* 195 (1985): Tanacetum boreale.
zawadskii Herbich
var. alpinum *(Nakai) Y.N. Lee, Fl. Korea:* 1162 (1996): C. sibiricum var. alpinum.
subsp. coreanum *(Nakai) Y.N. Lee, Fl. Korea:* 1162 (1996): C. coreanum.
var. leiophyllum *(Nakai) Y.N. Lee, Fl. Korea:* 1162 (1996): C. leiophyllum.
subsp. lucidum *(Nakai) Y.N. Lee, Fl. Korea:* 1162 (1996): C. lucidum.
subsp. naktongense *(Nakai) Y.N. Lee, Fl. Korea:* 1162 (1996): C. naktongense.
var. tenuisectum *(Nakai) Y.N. Lee, Fl. Korea:* 1162 (1996), with incorrect basionym ref.: Basionym not stated.
subsp. yezoense *(Maek.) Y.N. Lee, Fl. Korea:* 1162 (1996): C. arcticum var. yezoense.

CHRYSOCEPHALUM (Compos.).
serpens *J. Everett in Telopea,* 8(3): 311 (1999)—New South Wales.

CHRYSOCHLAMYS (Guttif.).
allenii *(Maguire) B.E. Hammel in Novon,* 9(3): 363 (1999): Tovomitopsis allenii.
angustifolia *(Maguire) B.E. Hammel in Novon,* 9(3): 363 (1999): Tovomitopsis angustifolia.
balboa *B.E. Hammel in Cat. Vasc. Pl. Ecuador (Monogr. Syst. Bot. Missouri Bot. Gard., 75):* 953 (30 July 1999); *et in Novon,* 9(3): 373 (15 Sept. 1999), nom. nov.: Balboa membranacea *Planch. & Triana.*
grandifolia *(L.O. Williams) B.E. Hammel in Novon,* 9(3): 365 (1999): Tovomita grandifolia.
membrillensis *(D'Arcy) B.E. Hammel in Novon,* 9(3): 366 (1999): Tovomitopsis membrillensis.
silvicola *(Hammel) B.E. Hammel in Novon,* 9(3): 368 (1999): Tovomitopsis silvicola.
skutchii *B.E. Hammel in Novon,* 9(3): 369 (1999)—Costa Rica.
tenuis *B.E. Hammel in Novon,* 9(3): 371 (1999)—Costa Rica, Panama.

CHRYSOCYATHUS (Ranunculac.).
sect. Consiligo *(DC.) J. Holub in Preslia,* 70(2): 101 (1998): Adonis sect. Consiligo.
subsect. Vernales *(Poschk.) J. Holub in Preslia,* 70(2): 102 (1998): Adonis subsect. Vernales.
amurensis *(Regel & Radde) J. Holub in Preslia,* 70(2): 102 (1998): Adonis amurensis.
bobrovianus *(Simonov.) J. Holub in Preslia,* 70(2): 102 (1998): Adonis bobroviana.
davidii *(Franch.) J. Holub in Preslia,* 70(2): 102 (1998): Adonis davidii.
× hybridus *(Simonk.) J. Holub in Preslia,* 70(2): 102 (1998): Adonis hybrida.
leiosepalus *(Butkov) J. Holub in Preslia,* 70(2): 102 (1998): Adonis leiosepala.
mongolicus *(Simonov.) J. Holub in Preslia,* 70(2): 102 (1998): Adonis mongolica.
multiflorus *(Nishikawa & Koji Ito) J. Holub in Preslia,* 70(2): 102 (1998): Adonis multiflora.
pseudoamurensis *(W.T. Wang) J. Holub in Preslia,* 70(2): 102 (1998): Adonis pseudoamurensis.
ramosus *(Franch.) J. Holub in Preslia,* 70(2): 102 (1998): Adonis ramosa.
sibiricus *(Ledeb.) J. Holub in Preslia,* 70(2): 102 (1998): Adonis sibirica.
sutchuensis *(Franch.) J. Holub in Preslia,* 70(2): 103 (1998): Adonis sutchuensis.
tianshanicus *(Adolf) J. Holub in Preslia,* 70(2): 103 (1998): Adonis turkestanica var. tianshanica.
turkestanicus *(Korsh.) J. Holub in Preslia,* 70(2): 103 (1998): Adonis appenina var. turkestanica.
vernalis *(L.) J. Holub in Preslia,* 70(2): 103 (1998): Adonis vernalis.
villosus *(Ledeb.) J. Holub in Preslia,* 70(2): 103 (1998): Adonis villosa.
volgensis *(Steven) J. Holub in Preslia,* 70(2): 103 (1998): Adonis volgensis.

CHRYSOGLOSSUM (Orchidac.).
ensigerum *W. van der Burgh & E.F. de Vogel in Orchid Monogr.,* 8: 142 (1997)—Sumatera.

CHRYSOPHYLLUM (Sapotac.).
ambrense *(Aubrév.) G.E. Schatz & L. Gautier in Novon,* 6(4): 426 (1996): Donella ambrensis.
analalavense *(Aubrév.) G.E. Schatz & L. Gautier in Novon,* 6(4): 426 (1996): Donella analalavensis.
boukokoense *(Aubrév. & Pellegr.) L. Gautier in J.-P. Lebrun & A.L. Stork, Enum. Pl. Fleurs Afr. Trop.,* 4: 27 (1997): Gambeya boukokoensis.
boukokoense *(Aubrév. & Pellegr.) R. Govaerts, World Checklist Seed Plants,* 3(1): 15 (1999), isonym; *see L. Gaut. in J.-P. Lebrun & Stork, Enum. Pl. Fleurs Afr. Trop.,* 4: 27 (1997): Gambeya boukokoensis.
capuronii *G.E. Schatz & L. Gautier in Novon,* 6(4): 426 (1996)—Madagascar.
delphinense *(Aubrév.) G.E. Schatz & L. Gautier in Novon,* 6(4): 428 (1996): Donella delphinensis.
edule *A. Sagástegui Alva & M.O. Dillon in Arnaldoa,* 5(1): 68 (1998), homonym; *see Hoyle in Kew Bull.,* 1932: 269 (1932)—Peru.
fenerivense *(Aubrév.) G.E. Schatz & L. Gautier in Novon,* 6(4): 428 (1996): Donella fenerivensis.
guerelianum *(Aubrév.) G.E. Schatz & L. Gautier in Novon,* 6(4): 428 (1996): Austrogambeya guereliana.
masoalense *(Aubrév.) G.E. Schatz & L. Gautier in Novon,* 6(4): 428 (1996): Donella masoalensis.
nyangense *(Pellegr.) R. Govaerts, World Checklist Seed Plants,* 3(1): 15 (1999): Gambeya nyangensis.
perrieri *(Lecomte) G.E. Schatz & L. Gautier in Novon,* 6(4): 428 (1996): Donella perrieri.
roxburghianum *Wall. ex G.T.F.S. Barlow Speede, New Indian Gardener:* 282 (1848), nom. nov.: C. acuminatum *Roxb.*
ubangiense *(De Wild.) R. Govaerts, World Checklist Seed Plants,* 3(1): 15 (1999), as 'ubanguiensis': Mimusops ubangiensis.
ubangiense *(De Wild.) D.J. Harris in Kew Bull.,* 55(1): 229 (2000), isonym: Mimusops ubangiensis.

CHRYSOPOGON (Gramin.).
castaneus *J.F. Veldkamp & C.B. Salunkhe in Rheedea,* 10(1): 59 (2000)—India (Maharashtra).
celebicus *J.F. Veldkamp in Austrobaileya,* 5(3): 511 (1999)—Sulawesi.

CHRYSOPOGON:-
festucoides *(C. Presl) J.F. Veldkamp in Austrobaileya,* 5(3): 512 (1999): Andropogon festucoides.
fulvibarbis *(Trin.) J.F. Veldkamp in Austrobaileya,* 5(3): 525 (1999): Andropogon fulvibarbis.
intercedens *J.F. Veldkamp in Austrobaileya,* 5(3): 515 (1999)—New Guinea (Papua New Guinea).
lawsonii *(Hook. f.) J.F. Veldkamp in Austrobaileya,* 5(3): 515 (1999): Andropogon lawsonii.
micrantherus *J.F. Veldkamp in Austrobaileya,* 5(3): 516 (1999)—New Guinea (Papua New Guinea).
nigritanus *(Benth.) J.F. Veldkamp in Austrobaileya,* 5(3): 526 (1999): Andropogon nigritanus.
oliganthus *J.F. Veldkamp in Austrobaileya,* 5(3): 526 (1999), nom. nov.: Vetiveria pauciflora *S.T. Blake.*
rigidus *(B.K. Simon) J.F. Veldkamp in Austrobaileya,* 5(3): 527 (1999): Vetiveria rigida.

CHRYSOSPLENIUM (Saxifragac.).
japonicum (Maxim.) Makino
var. cuneifolium *X.H. Guo & X.P. Zhang in Acta Phytotax. Sin.,* 36(4): 365 (1998)—China Southeast (Anhui).
pseudofauriei H. Lév.
var. nipponense *M. Wakabayashi in Acta Phytotax. Geobot.,* 48(2): 142 (1997 publ. 1998)—Japan.
pseudopilosum *M. Wakabayashi & H. Takahashi in Acta Phytotax. Geobot.,* 50(1): 2 (1999)—Japan.
var. divaricatistylosum *M. Wakabayashi & H. Takahashi in Acta Phytotax. Geobot.,* 50(1): 4 (1999)—Japan.

× **CHUATIANARA** *hort. in Orchid Rev.,* 108(1232): new orchid hybrids 14 (2000). *ORCHIDACEAE* (Neofinetia × Renanthera × Rhynchostylis × Vanda).

CHUQUIRAGA (Compos.).
raimondiana *A. Granda Paucar in Kurtziana,* 25: 151 (1997)—Peru.

CHUSQUEA (Gramin.).
subgen. Rettbergia *(Raddi) L.G. Clark in G.P. Chapman (ed.), Bamboos (Linn. Soc. Symp. Ser. 19):* 41 (1997): gen. Rettbergia.
subgen. Swallenochloa *(McClure) L.G. Clark in G.P. Chapman (ed.), Bamboos (Linn. Soc. Symp. Ser. 19):* 42 (1997): gen. Swallenochloa.
antioquensis *L.G. Clark & X. Londoño in Novon,* 8(4): 423 (1998)—Colombia.
arachniformis *L.G. Clark & X. Londoño in Novon,* 8(4): 425 (1998), as 'arachniforme'—Colombia.
bahiana *L.G. Clark in Brittonia,* 48(2): 250 (1996)—Brazil Northeast (Bahia).
bradei *L.G. Clark in Brittonia,* 48(2): 254 (1996)—Brazil Southeast (Espírito Santo), Brazil Northeast (Bahia).
galeottiana *Rupr. ex W. Munro in Trans. Linn. Soc.,* 26(1): 59 (1868)—Mexico Southwest (Oaxaca).†
gigantea *J.-P. Demoly in Bambou,* 30: 8 (1999)—Chile.
leonardiorum *L.G. Clark in Brittonia,* 48(2): 256 (1996)—Ecuador.
loxensis *L.G. Clark in Brittonia,* 48(2): 260 (1996)—Ecuador.
perotensis *L.G. Clark, G. Cortés R. & M. Cházaro B. in Syst. Bot.,* 22(2): 225 (1997)—Mexico Gulf (Veracruz), Mexico Southwest (Oaxaca).

CHYSIS (Orchidac.).
addita *R.L. Dressler in Novon,* 10(3): 194 (2000)—Mexico Southeast (Chiapas).
orichalcea *R.L. Dressler in Novon,* 10(3): 194 (2000)—El Salvador.

CHYTRANTHUS (Sapindac.).
longibracteatus *F.G. Davies in Fl. Trop. E. Afr., Sapindac.:* 67 (1998)—Tanzania.

CHYTROGLOSSA (Orchidac.).
seehaweri *I. Bock in Orchidee,* 48(5): 180 (1997); et *in Orquidário,* 12(1): 10 (1998)—Brazil Southeast (Rio de Janeiro).

CICER (Legumin.).
sect. Annua *(Maesen) I.V. Seferova in Bot. Zhurn.,* 80(8): 101 (1995): C. ser. Annua.
ser. Arietina *Lincz. ex I.V. Seferova in Bot. Zhurn.,* 80(8): 101 (1995).

CICER:-
sect. Cuneicicer *I.V. Seferova in Bot. Zhurn.,* 80(8): 101 (1995).
ser. Flexuosa *Lincz. ex I.V. Seferova in Bot. Zhurn.,* 80(8): 98 (1995).
ser. Graeca *I.V. Seferova in Bot. Zhurn.,* 80(8): 97 (1995).
ser. Macracantha *Lincz. ex I.V. Seferova in Bot. Zhurn.,* 80(8): 99 (1995).
sect. Nanopolycicer *(Popov) I.V. Seferova in Bot. Zhurn.,* 80(8): 98 (1995), as 'Nano-polycicer': C. subsect. Nanopolycicer.
ser. Pinnatifida *I.V. Seferova in Bot. Zhurn.,* 80(8): 102 (1995).
ser. Pungentia *Lincz. ex I.V. Seferova in Bot. Zhurn.,* 80(8): 99 (1995).
ser. Subaphylla *I.V. Seferova in Bot. Zhurn.,* 80(8): 100 (1995).
ser. Tragacanthoidea *Lincz. ex I.V. Seferova in Bot. Zhurn.,* 80(8): 100 (1995).
macracanthum Popov
subsp. acanthophyllum *(Boriss.) I.V. Seferova in Bot. Zhurn.,* 80(8): 103 (1995): C. acanthophyllum.
mogoltavicum *(Popov) A.S. Koroleva in Fl. Tadzhikistana,* 5: 600 (1937): C. flexuosum var. mogoltavicum.

CICERBITA (Compos.).
sect. Azurella *A.N. Sennikov in Bot. Zhurn.,* 82(5): 112 (1997).
sect. Poicilachena *(Kirp.) A.N. Sennikov in Bot. Zhurn.,* 82(5): 112 (1997): C. subgen. Poicilachena.
sect. Tianschania *A.N. Sennikov in Bot. Zhurn.,* 82(5): 112 (1997).
azurea (Ledeb.) Beauverd
var. glabra *A.N. Sennikov in Bot. Zhurn.,* 82(5): 112 (1997)—Xinjiang.

CICUTA (Umbellif.).
virosa L.
f. latisecta *(Čelak.) Y.C. Chu in Clavis Pl. Chinae Bor.-Or. ed. 2 (ed. P.Y. Fu):* 467 (1995): C. virosa var. latisecta.

CIENFUEGOSIA (Malvac.).
glauca *M. Brandão & J.P. Laca-Buendia in Daphne,* 3(4): 35 (1993), without latin descr.—Brazil Southeast (Minas Gerais).
ituiutabensis *M. Brandão & J.P. Laca-Buendia in Daphne,* 3(4): 35 (1993), without latin descr.—Brazil Southeast (Minas Gerais).
longifolia *M. Brandão & J.P. Laca-Buendia in Daphne,* 3(4): 38 (1993), without latin descr.—Brazil Southeast (Minas Gerais).
uberabensis *M. Brandão & J.P. Laca-Buendia in Daphne,* 3(4): 40 (1993), without latin descr.—Brazil Southeast (Minas Gerais).

CIMICIFUGA (Ranunculac.).
foetida L.
var. mairei *(H. Lév.) W.T. Wang & Zh. Wang in Acta Phytotax. Sin.,* 37(3): 212 (1999): C. mairei.
kashmiriana *J.A. Compton & T.A.J. Hedderson in Bot. J. Linn. Soc.,* 123(1): 16 (1997)—Jammu-Kashmir, Pakistan, Xizang.
mairei H. Lév.
var. foliolosa *(P.K. Hsiao) J.A. Compton & T.A.J. Hedderson in Bot. J. Linn. Soc.,* 123(1): 20 (1997): C. foetida var. foliolosa.
purpurea *(P.K. Hsiao) C.W. Park & H.W. Lee in Novon,* 6(1): 93 (1996): C. acerina f. purpurea.

CIMINALIS (Gentianac.).
clarkei *(Kusn.) S.S. Ikonnikov in Bot. Zhurn.,* 83(5): 41 (1998): Gentiana clarkei.

CINCHONA (Rubiac.).
antioquiae *L. Andersson in Mem. New York Bot. Gard.,* 80: 36 (1998)—Colombia.
fruticosa *L. Andersson in Mem. New York Bot. Gard.,* 80: 57 (1998)—Peru.
krauseana *L. Andersson in Mem. New York Bot. Gard.,* 80: 46 (1998), nom. nov.: Ladenbergia coriacea *K. Krause.*
pyrifolia *L. Andersson in Mem. New York Bot. Gard.,* 80: 50 (1998)—Peru.

CINERARIA (Compos.).
argillacea *G.V. Cron in S. Afr. J. Bot.*, 63(6): 402 (1997)—Cape Province.
cyanomontana *G.V. Cron in S. Afr. J. Bot.*, 63(6): 400 (1997)—Transvaal.
lyratiformis *G.V. Cron in S. Afr. J. Bot.*, 65(4): 287 (1999), nom. nov.: C. lyrata *DC.*
tomentolanata *R. Govaerts, World Checklist Seed Plants*, 3(1): 15 (1999), nom. nov., as 'tomentolanatus': Senecio lanatus *L.f.*

CINNA (Gramin.).
arkansana *(Nutt.) G.C. Tucker in Harvard Pap. Bot.*, 9: 62 (1996): Greenia arkansana.

CINNAMOMUM (Laurac.).
birmanicum *A.J.G.H. Kostermans in Reinwardtia*, 11(3): 213 (1998)—Burma.
brachythyrsum *J. Li in Acta Bot. Yunnanica*, 18(1): 53 (1996)—China South-central (Yunnan).
bractefoliaceum *F.G. Lorea-Hernández in Acta Bot. Mex.*, 40: 4 (1997)—Mexico Northeast (Querétaro, San Luis Potosí, Tamaulipas).
concinnum *F.G. Lorea-Hernández in Acta Bot. Mex.*, 40: 2 (1997), nom. nov.: Phoebe elegans *van der Werff.*
cupulatum *A.J.G.H. Kostermans in Reinwardtia*, 11(3): 202 (1998)—Burma.
ellipticifolium *A.J.G.H. Kostermans in Reinwardtia*, 11(3): 211 (1998)—Burma.
glauciphyllum *A.J.G.H. Kostermans in Reinwardtia*, 11(3): 210 (1998)—Burma.
glossophyllum *F.G. Lorea-Hernández in Acta Bot. Mex.*, 40: 6 (1997)—Mexico Southwest (Nayarit).
grisebachii *F.G. Lorea-Hernández in Acta Bot. Mex.*, 40: 2 (1997), nom. nov.: Phoebe triplinervis *Griseb.*
hkinlumense *A.J.G.H. Kostermans in Reinwardtia*, 11(3): 209 (1998)—Burma.
kingdon-wardii *A.J.G.H. Kostermans in Reinwardtia*, 11(3): 197 (1998)—Burma.
leptophyllum *F.G. Lorea-Hernández in Acta Bot. Mex.*, 40: 8 (1997)—Mexico Gulf (Veracruz).
micranthum (Hayata) Hayata
f. kanehirae *(Hayata) S.S. Ying, Coloured Illustr. Fl. Taiwan*, 1: 309 (March 1985); *et in Mem. Coll. Agric. Nation. Taiwan Univ.*, 25(1): 108 (April 1985): Machilus kanehirae.†
nalingway *A.J.G.H. Kostermans in Reinwardtia*, 11(3): 204 (1998)—Burma.
pachyphyllum *A.J.G.H. Kostermans in Reinwardtia*, 11(3): 206 (1998)—Thailand, Burma.
velveti *F.G. Lorea-Hernández in Acta Bot. Mex.*, 40: 10 (1997)—Mexico Southwest (Guerrero).
zapatae *F.G. Lorea-Hernández in Acta Bot. Mex.*, 40: 13 (1997)—Mexico Southwest (Guerrero, Oaxaca), Guatemala.

CINTIA *K. Kníže & J. Říha in Kaktusy (Brno)*, 31(2): 37 (1995). *CACTACEAE*
knizei *J. Říha in Kaktusy (Brno)*, 31(2): 38 (1995)—Bolivia.

CIONOMENE *B.A. Krukoff in Phytologia*, 41(4): 241 (1979). *MENISPERMACEAE*
javariensis *B.A. Krukoff in Phytologia*, 41(4): 241 (1979)—Brazil North (Amazonas).

CIPOCEREUS (Cactac.).
fissuratus *Lodé, Fichier Encycl. Cact. Autres Succ.*, Sér. 17: no. 1587 (1996), nom. provis.; *cf. Repert. Pl. Succ. (I.O.S.)*, 47: 8 (1996 publ. 1997)—Locality not stated.
laniflorus *N.P. Taylor & D.C. Zappi in Cactaceae Consensus Initiatives*, 3: 7 (1997)—Brazil Southeast (Minas Gerais).

CIRRHOPETALUM (Orchidac.).
annamense *L.A. Garay in Harvard Pap. Bot.*, 4(1): 306 (1999)—Vietnam.
cercanthum *L.A. Garay, F. Hamer & E. Siegerist in Lindleyana*, 11(4): 231 (1996)—Borneo.
curtisii Hook.f.
var. lutescens *L.A. Garay in Harvard Pap. Bot.*, 4(1): 307 (1999)—Thailand.
var. purpureum *(Ridl.) L.A. Garay in Harvard Pap. Bot.*, 4(1): 307 (1999): C. concinnum var. purpureum.

CIRRHOPETALUM:-
flabellum-veneris *(J. König) G. Seidenfaden & P. Ormerod in G. Seidenfaden, Descr. Epidendrorum J.G. König 1791*: 39 (1995), as 'flabelloveneris': Epidendrum flabellum-veneris.
minutiflorum *L.A. Garay, F. Hamer & E. Siegerist in Lindleyana*, 11(4): 231 (1996)—Borneo.
panigrahianum *(S. Misra) S. Misra in J. Orchid Soc. India*, 11(1–2): 54 (1997): Bulbophyllum panigrahianum.
pseudopicturatum *L.A. Garay in Harvard Pap. Bot.*, 4(1): 308 (1999)—Thailand.

CIRSIUM (Compos.).
abukumense *Y. Kadota in Mem. Nation. Sci. Mus. (Tokyo)*, 29: 96 (1996)—Japan.
amplexifolium Kitam.
f. albiflorum *Y. Kadota in Bull. Nation. Sci. Mus., B (Tokyo)*, 24(4): 152 (1998)—Japan.
ashiuense *S. Yokoyama & T. Shimizu in J. Phytogeogr. Taxon.*, 44(1–2): 19 (1996)—Japan.
bactiacense *M. Gandoger, Dec. Pl. Nov.*, 1: 41 (1875)—Switzerland.
dirmilense *R.M. Burton in Karaca Arbor. Mag.*, 3(3): 123 (1996)—Turkey.
× fontqueri *A.M. Romo in Estud. Mus. Cienc. Nat. Alava*, 10–11: 200 (1995–96 publ. 1996)—Spain.(C. pyrenaicum var. paniculatum × acaule subsp. acaule).
haberianum *M. Gandoger, Dec. Pl. Nov.*, 1: 41 (1875)—France.
hidakamontanum *Y. Kadota in Bull. Nation. Sci. Mus., B (Tokyo)*, 25(3): 95 (1999)—Japan.
hillii
f. albiflorum *E.G. Voss in Mich. Bot.*, 18(4): 139 (1995 publ. 1996), nom. nov., as comb. nov.: C. pumilum f. candidum *B. Boivin.*
ishizuchiense *(Kitam.) Y. Kadota in Mem. Nation. Sci. Mus. (Tokyo)*, 32: 132 (2000): C. tenue var. ishizuchiense.
kamtschaticum Ledeb.
subsp. weyrichii *(Maxim.) V.N. Voroschilov in A.K. Skvortsov (ed.), Florist. issl. v razn. raionakh SSSR*: 198 (1985): C. weyrichii.
komarovii Schischk.
subsp. schischkinii *(Serg.) O.S. Zhirova in Fl. Sibir. (Asterac.)*, 13: 219 (1997 publ. 1998): C. schischkinii.
longirameum *M. Gandoger, Dec. Pl. Nov.*, 1: 41 (1875)—France.
macranthum *(Schrenk) G. Panigrahi in Fl. U.S.S.R. (English translation)*, 28: 172 (1998): Chamaepeuce macrantha.
occidentalinipponense *Y. Kadota in Bull. Nation. Sci. Mus., B (Tokyo)*, 23(3): 115 (1997)—Japan.
ochrocentrum A. Gray
subsp. martinii *P. Barlow-Irick in Novon*, 9(3): 320 (1999)—New Mexico, Arizona.
rhizocephalum
subsp. frickii *(Fisch. & C.A. Mey.) R. Govaerts, World Checklist Seed Plants*, 3(1): 15 (1999): C. frickii.
schantarense Trautv. & C.A. Mey.
f. albiflorum *Y.N. Lee, Fl. Korea*: 1162 (1996), without type—Locality not stated.
setidens Nakai
f. albiflorum *Y.N. Lee, Fl. Korea*: 1163 (1996), without type—Locality not stated.
umezawanum *Y. Kadota in Bull. Nation. Sci. Mus., B (Tokyo)*, 24(4): 147 (1998)—Japan.
zawoense *Y. Kadota in Bull. Nation. Sci. Mus., B (Tokyo)*, 25(3): 100 (1999)—Japan.

CISCHWEINFIA (Orchidac.).
popowiana *W. Königer in Arcula*, 7: 191 (1997)—Ecuador.

CISSUS (Vitac.).
acreensis *J.A. Lombardi in Novon*, 6(2): 195 (1996)—Brazil North (Acre).
amapaensis *J.A. Lombardi in Novon*, 6(2): 195 (1996)—Brazil North (Amapá).
anisophylla *J.A. Lombardi in Novon*, 7(2): 182 (1997)—Ecuador, Panama.
apendiculata *J.A. Lombardi in Novon*, 6(2): 197 (1996)—Brazil North (Pará, Tocantins), Brazil Northeast (Maranhão).

CISSUS:-
araguainensis *J.A. Lombardi in Novon*, 6(2): 197 (1996)—Brazil North (Tocantins, Pará), Brazil Northeast (Maranhão).
bahiensis *J.A. Lombardi in Brittonia*, 48(2): 195 (1996)—Brazil Northeast (Bahia, Pernambuco).
boliviana *J.A. Lombardi in Brittonia*, 50(1): 19 (1998)—Bolivia.
bracteosa *J.A. Lombardi in Brittonia*, 50(1): 22 (1998)—Peru, Colombia, Ecuador.
cervii *A. Dunaiski in Acta Biol. Parana.*, 21(1–4): 135 (1992)—Brazil South (Paraná).
colombiensis *J.A. Lombardi in Novon*, 6(2): 191 (1996)—Colombia, Brazil North (Amazonas), Bolivia.
compressiflora *J.A. Lombardi in Brittonia*, 48(2): 200 (1996)—Peru, Ecuador.
decidua *J.A. Lombardi in Brittonia*, 48(2): 197 (1996)—Brazil Northeast (Paraíba, Alagoas, Bahia, Pernambuco, Piauí).
descoingsii *J.A. Lombardi in Candollea*, 51(2): 370 (1996)—French Guiana.
flavifolia *J.A. Lombardi in Candollea*, 51(2): 365 (1996)—Ecuador, Peru.
fusifolia *J.A. Lombardi in Candollea*, 51(2): 368 (1996)—Ecuador, Colombia, Peru.
glaucotricha *J.A. Lombardi in Brittonia*, 48(2): 200 (1996)—Peru, Ecuador, Brazil North (Acre).
javana DC.
var. pubescens *C.L. Li in Chinese J. Appl. Environ. Biol.*, 2(1): 49 (1996)—China South-central (Yunnan).
luzoniensis *(Merr.) C.L. Li in Chinese J. Appl. Environ. Biol.*, 2(1): 49 (1996): C. repens var. luzoniensis.
macrocarpa *A. Dunaiski in Acta Biol. Parana.*, 21(1–4): 136 (1992)—Brazil South (Paraná).
mirabilis *(Urb. & Ekman) J.A. Lombardi in Taxon*, 46(3): 427 (1997): Pterocissus mirabilis.
osaensis *J.A. Lombardi in Candollea*, 52(1): 105 (1997)—Costa Rica, Guatemala.
paraensis *J.A. Lombardi in Novon*, 6(2): 199 (1996)—Brazil North (Pará, Acre, Amazonas).
paucinervia *J.A. Lombardi in Brittonia*, 48(2): 199 (1996)—Brazil Northeast (Bahia), Brazil Southeast (Espírito Santo).
peruviana *J.A. Lombardi in Brittonia*, 48(2): 202 (1996)—Peru.
pingtungensis *S.S. Ying in Mem. Coll. Agric. Nation. Taiwan Univ.*, 36(1): 16 (1996)—Taiwan.
pseudofuliginea *J.A. Lombardi in Brittonia*, 48(2): 204 (1996)—Peru, Colombia, Venezuela, Ecuador.
pseudoverticillata *J.A. Lombardi in Novon*, 7(2): 182 (1997)—Peru, Bolivia.
repanda Vahl
var. subferruginea *(Merr. & Chun) C.L. Li in Chinese J. Appl. Environ. Biol.*, 2(1): 49 (1996): Parthenocissus subferruginea.
rubropilosa *J.A. Lombardi in Candollea*, 51(2): 367 (1996)—Bolivia, Peru.
ursina *J.A. Lombardi in Brittonia*, 48(2): 206 (1996)—Peru, Ecuador, Brazil North (Acre).
verticillata
subsp. colombiana *J.A. Lombardi in Novon*, 6(2): 193 (1996)—Colombia, Venezuela.
subsp. micrantha *(Poir.) J.A. Lombardi in Taxon*, 46(3): 430 (1997): C. micrantha.
subsp. oblongolanceolata *(Krug & Urb.) J.A. Lombardi in Taxon*, 46(3): 430 (1997): C. sicyoides f. oblongolanceolata.
wenshanensis *C.L. Li in Acta Bot. Yunnanica*, 19(3): 218 (1997)—China South-central (Yunnan).

CISTANCHE (Scrophulariac.).
sect. Salsae *T.I. Kravtsova in Opred. Zarazikhovýkh Fl. SSSR (ed. L.Yu. Budantsev)*: 36 (1993).

CISTANTHE (Portulacac.).
tovarii *A. Galán de Mera in An. Jard. Bot. Madrid*, 53(2): 253 (1995 publ. 1996)—Peru.

CISTUS (Cistac.).
× bornetianus *J.-P. Demoly in An. Jard. Bot. Madrid*, 54(1): 253 (1996)—Cultivation.(C. albidus × lauriflorus).
× canescens Sweet
nothof. albens *J.-P. Demoly in Biocosme Mésogéen*, 14(2–3): 115 (1998)—Cultivation.(C. albidus (f. albus?) × canescens f. albus).

CISTUS:-
clusii Dunal
var. multiflorus *(Demoly) A.M. Romo, Fl. Silvestres Baleares:* 172 (1994): C. clusii subsp. multiflorus.
× conradiae *J.-P. Demoly in Biocosme Mésogéen,* 13(1): 19 (1996)—Corse.(C. creticus × monspeliensis).
× escartianus *J.-P. Demoly in Biocosme Mésogéen,* 14(2–3): 116 (1998)—Cultivation.(C. creticus × heterophyllus).
× fernandesiae P. Silva
nothovar. latifolius *J.-P. Demoly in An. Jard. Bot. Madrid,* 54(1): 253 (1996), contrary to Art. 37.3 ICBN (1994)—Cultivation.(C. crispus × ladanifer var. sulcatus).
inflatus *Pourr. ex J.-P. Demoly in Acta Bot. Gallica,* 144(1): 42 (1997 publ. 1998), nom. nov.: C. hirsutus *Lam. (1786).*
× lucasii *J.-P. Demoly in Biocosme Mésogéen,* 14(2–3): 116 (1998)—Cultivation.(C. chinamadensis × creticus).
× pauranthus *J.-P. Demoly in An. Jard. Bot. Madrid,* 54(1): 252 (1996)—Cyprus, East Aegean Is.(C. parviflorus × salviifolius).
× purpureus Lam.
nothof. holorhodos *J.-P. Demoly in Biocosme Mésogéen,* 14(2–3): 117 (1998), nom. nov.: C. × purpureus var. argenteus *Demoly.*
× sammonsii *J.-P. Demoly in An. Jard. Bot. Madrid,* 54(1): 253 (1996)—Cultivation.(C. creticus × laurifolius).
× tephreus *J.-P. Demoly in Biocosme Mésogéen,* 14(2–3): 118 (1998)—Cultivation.(C. canescens × heterophyllus).
× timbalii *J.-P. Demoly in An. Jard. Bot. Madrid,* 54(1): 252 (1996)—Cultivation.(C. albidus × salviifolius).

CITHAREXYLUM (Verbenac.).
grandiflorum *G.A. Aymard & R.M. Rueda in Novon,* 7(2): 95 (1997)—Ecuador.
rigidum *(Briq.) H.N. Moldenke in Phytologia,* 1: 17 (1933): C. myrianthum var. rigidum.†

× **CITROFORTUNELLA** (Rutac.).
madurensis *(Lour.) D. Rivera Núñez et al., Varied. Trad. Frut. Cuenca Río Segura Cat. Etnobot.:* 191 (1998): Citrus madurensis.

CITRUS (Rutac.).
aurantifolia (Christm.) Swingle
subsp. latifolia *(Tanaka) S. Ríos Ruiz, D. Rivera Núñez & C. Obón de Castro in D. Rivera et al., Varied. Trad. Frut. Cuenca Río Segura Cat. Etnobot.:* 101 (1998): C. latifolia.
subsp. murgetana *García Lidón, D. Rivera Núñez, S. Ríos Ruiz & C. Obón de Castro in D. Rivera et al., Varied. Trad. Frut. Cuenca Río Segura Cat. Etnobot.:* 98 (1998)—Spain.
bergamia Risso & Poit.
subsp. mellarosa *(Risso) D. Rivera Núñez et al., Varied. Trad. Frut. Cuenca Río Segura Cat. Etnobot.:* 108 (1998): C. mellarosa.
× floridana *(J.W. Ingram & H.E. Moore) D.J. Mabberley in Telopea,* 7(4): 337 (1998): × Citrofortunella floridana.
glauca *(Lindl.) I.H. Burkill in Gard. Bull. Straits Settlem.,* 5(Index): 3 (1932): Triphasia glauca.
gracilis *D.J. Mabberley in Telopea,* 7(4): 340 (1998)—Northern Territory.
hindsii *(Champ. ex Benth.) R. Govaerts, World Checklist Seed Plants,* 3(1): 15 (1999): Sclerostylis hindsii.
limetta Risso
subsp. murcica *S. Ríos Ruiz, D. Rivera Núñez, García-Lidón & C. Obón de Castro in D. Rivera et al., Varied. Trad. Frut. Cuenca Río Segura Cat. Etnobot.:* 133 (1998)—Spain.
× limodulcis *D. Rivera Núñez, C. Obón de Castro, F. Méndez Colmenero & S. Ríos* Ruiz *in D. Rivera et al., Varied. Trad. Frut. Cuenca Río Segura Cat. Etnobot.:* 125 (1998)—Spain.(C. limetta subsp. limetta × C. limon?).
× limorosea *D. Rivera Núñez, F. Méndez Colmenero, C. Obón de Castro & S. Ríos* Ruiz *in D. Rivera et al., Varied. Trad. Frut. Cuenca Río Segura Cat. Etnobot.:* 129 (1998), as 'limoroseus'—Spain.(C. aurantifolia × bergamia?).

CITRUS:-
macrosperma *T.C. Guo & Y.M. Ye in Acta Phytotax. Sin.,* 35(4): 353 (1997)—China Southeast (Guangxi).
pampelmos *A. Risso, Hist. nat. prod. Eur. mérid.,* 1: 412 (1826)—Locality not stated.†
peretta *A. Risso, Hist. nat. prod. Eur. mérid.,* 1: 418 (1826)—Locality not stated.†
polytrifolia *R. Govaerts, World Checklist Seed Plants,* 3(1): 15 (1999), nom. nov.: Poncirus polyandra *S.Q. Ding et al.*
reticulata Blanco
subsp. deliciosa *(Ten.) D. Rivera Núñez et al., Varied. Trad. Frut. Cuenca Río Segura Cat. Etnobot.:* 187 (1998): C. deliciosa.
subsp. tachibana *(Tanaka) D. Rivera Núñez et al., Varied. Trad. Frut. Cuenca Río Segura Cat. Etnobot.:* 189 (1998): C. tachibana.
subsp. unshiu *(Marcow.) D. Rivera Núñez et al., Varied. Trad. Frut. Cuenca Río Segura Cat. Etnobot.:* 181 (1998): C. unshiu.
sinensis (L.) Osbeck
subsp. crassa *(Risso) D. Rivera Núñez et al., Varied. Trad. Frut. Cuenca Río Segura Cat. Etnobot.:* 179 (1998), as 'crassum': C. aurantium var. crassa.
subsp. fetifera *(Risso) D. Rivera Núñez et al., Varied. Trad. Frut. Cuenca Río Segura Cat. Etnobot.:* 176 (1998), as 'fetiferum': C. aurantium var. fetifera.
subsp. hierochuntica *(Risso) D. Rivera Núñez et al., Varied. Trad. Frut. Cuenca Río Segura Cat. Etnobot.:* 168 (1998), as 'hierochunticum': C. aurantium var. hierochuntica.
subsp. lusitanica *(Risso) D. Rivera Núñez et al., Varied. Trad. Frut. Cuenca Río Segura Cat. Etnobot.:* 164 (1998), as 'lusitanicum': C. aurantium var. lusitanica.
× virgata *D.J. Mabberley in Telopea,* 7(4): 339 (1998)—Cultivation.(C. australasica × australis).
wanshanensis *S.S. Ying in Mem. Coll. Agric. Nation. Taiwan Univ.,* 30(2): 61 (1990), contrary to Arts. 37.4, 37.5 ICBN (1994)—Taiwan.
wintersii *D.J. Mabberley in Telopea,* 7(4): 342 (1998), nom. nov.: Microcitrus papuana *Winters*

CLADIUM (Cyperac.).
mariscus
subsp. californicum *(S. Watson) R. Govaerts, World Checklist Seed Plants,* 3(1): 15 (1999): C. mariscus var. californicum.
subsp. chinense *(Nees) R. Govaerts, World Checklist Seed Plants,* 3(1): 15 (1999): C. chinense.

CLADOPUS (Podostemac.).
austro-osumiensis *Y. Kadono & N. Usui in Acta Phytotax. Geobot.,* 46(2): 133 (1995 publ. 1996)—Japan.

CLADOSTIGMA (Convolvulac.).
nigistiae *Sebsebe Demissew in Kew Bull.,* 51(2): 410 (1996)—Ethiopia.

CLADRASTIS (Legumin.).
kentukea (Dum. Cours.) Rudd
f. tomentosa *(Steyerm.) S.A. Spongberg in Int. Dendrol. Soc. Yearbook,* 1996: 29 (1997): C. lutea f. tomentosa.

CLAOXYLON (Euphorbiac.).
sect. Rostratae *S.R.M. Susila Rani & N.P. Balakrishnan in Rheedea,* 5(2): 138 (1995).

× **CLARKEARA** *hort.in Orchid Rev.,* 107(1227): 191 (1999). *ORCHIDACEAE* (Brassavola × Cattleya × Diacrium × Laelia × Sophronitis).

CLARKIA (Onagrac.).
subgen. Godetia *(Spach) N.N. Tzvelev, Opred. Sosud. Rast. Severo-Zap. Rossii:* 680 (2000): gen. Godetia.
mildrediae (A. Heller) H.F. Lewis & M.R. Lewis
subsp. lutescens *L.D. Gottlieb & L.P. Janeway in Madroño,* 44(3): 250 (1997 publ. 1998)—California.

CLAUSENA (Rutac.).
smyrelliana *P.I. Forster in Austrobaileya,* 5(4): 716 (2000)—Queensland.

CLAUSIA (Crucif.).
agideliensis *M.S. Knjasev in Bot. Zhurn.,* 82(7): 125 (1997)—Russia East.

CLAVIJA (Theophrastac.).
laplanadae *B. Ståhl in Novon,* 5(4): 370 (1995)—Colombia.
obtusifolia *B. Ståhl in Novon,* 5(4): 373 (1995)—Peru.
subandina *B. Ståhl in Novon,* 5(4): 370 (1995)—Ecuador.

CLEIDION (Euphorbiac.).
spiciflorum (Burm. f.) Merr.
var. moniliflorum *(Airy Shaw) A. Radcliffe-Smith & R. Govaerts in Kew Bull.,* 52(2): 478 (1997): C. moniliflorum.
taynguyenense *Nguyen Nghia Thin, Euphorbiac. Vietnam:* 47 (1995 publ. 1996)—Vietnam.

CLEIOSTOMA *Raf., Fl. Tellur.,* 4: 80 (1838), as 'Cleisostoma' in I.K. vol. 1 in error. *CONVOLVULACEAE*†

CLEISOSTOMA (Orchidac.).
platystele *P. Ormerod in Austral. Orchid Rev.,* 61(4): 37 (1996)—New Guinea (Papua New Guinea).
stimulatrix *P. Ormerod in Austral. Orchid Rev.,* 61(4): 37 (1996)—New Guinea (Papua New Guinea).
striatum *(Rchb. f.) N.E. Brown, Suppl. Johnson's Gard. Dict.:* 901 (1882): Echinoglossum striatum.

CLEISTANTHUS (Euphorbiac.).
polystachyus Planch.
subsp. milleri *(Dunkley) A. Radcliffe-Smith in Kew Bull.,* 51(2): 303 (1996): C. milleri.

CLEISTES (Orchidac.).
carautae *A.L.V. Toscano de Brito & L.S. Leoni in Pabstia,* 8(5): 1 (1997)—Brazil Southeast (Minas Gerais).
tamboana *C.H. Dodson & Carnevali in Orquideologia,* 20(3): 271 (1997)—Ecuador.
triflora *(C. Schweinf.) G. Carnevali & I. Ramírez in G.A. Romero & G. Carnevali, Orchids of Venezuela, ed. 2:* 1130 (2000): Pogonia triflora.

CLEISTOCACTUS (Cactac.).
acanthurus (Vaupel) D.R. Hunt
var. faustianus *(Backeb.) C. Ostolaza Nano in Brit. Cact. Succ. J.,* 14(4): 171 (1996): Borzicactus faustianus.
subsp. faustianus *(Backeb.) C. Ostolaza Nano in Cactaceae Consensus Initiatives,* 6: 8 (1998): Borzicactus faustianus.
subsp. pullatus *(Rauh & Backeb.) C. Ostolaza Nano in Cactaceae Consensus Initiatives,* 6: 8 (1998): Loxanthocereus pullatus.
var. pullatus *(Rauh & Backeb.) C. Ostolaza Nano in Brit. Cact. Succ. J.,* 14(4): 171 (1996): Loxanthocereus pullatus.
baumannii
subsp. horstii *(P.J. Braun) N.P. Taylor in Cactaceae Consensus Initiatives,* 6: 15 (1998): C. horstii.
clavispinus *(Rauh & Backeb.) C. Ostolaza Nano in Brit. Cact. Succ. J.,* 16(3): 133 (1998): Loxanthocereus clavispinus.
fieldianus
subsp. samnensis *(F. Ritter) C. Ostolaza Nano in Cactaceae Consensus Initiatives,* 6: 8 (1998): Borzicactus samnensis.
subsp. tessellatus *(Akers & Buining) C. Ostolaza Nano in Cactaceae Consensus Initiatives,* 6: 8 (1998): Borzicactus tessellatus.
hystrix *(Rauh & Backeb.) C. Ostolaza Nano in Brit. Cact. Succ. J.,* 16(3): 133 (1998): Loxanthocereus hystrix.
pachycladus *(Rauh & Backeb.) C. Ostolaza Nano in Cactaceae Consensus Initiatives,* 6: 8 (1998): Loxanthocereus pachycladus.
peculiaris *(Rauh & Backeb.) C. Ostolaza Nano in Brit. Cact. Succ. J.,* 16(3): 129 (1998): Loxanthocereus peculiaris.
plagiostoma *(Vaupel) D.R. Hunt in Cactaceae Consensus Initiatives,* 2: 3 (1997): Cereus plagiostoma.

CLEISTOCACTUS:-
roseiflorus *(Buining) G.D. Rowley in Brit. Cact. Succ. J.*, 18(4): 208 (2000): Akersia roseiflora.
xylorhizus *(F. Ritter) C. Ostolaza Nano in Brit. Cact. Succ. J.*, 14(4): 170 (1996): Loxanthocereus xylorhizus.

CLEISTOCALYX (Myrtac.).
eugenioides *E.D. Merrill & L.M. Perry in J. Arnold Arbor.*, 18: 330 (1937), nom. nov.Calyptranthes eugenioides *Seem.*†
nervosum (DC.) Kosterm.
var. paniala *(Roxb.) J. Parnell & P. Chantaranothai in Novon*, 6(2): 201 (1996): Eugenia paniala.

× **CLEISTOCANA** (Cactac.).
mirabilis *(Buining) D.R. Hunt in Cactaceae Consensus Initiatives*, 4: 10 (1997): Matucana mirabilis.

CLEISTOPETALUM *H. Okada in Acta Phytotax. Geobot.*, 47(1): 4 (1996). *ANNONACEAE*
borneense *H. Okada in Acta Phytotax. Geobot.*, 47(1): 4 (1996)—Borneo (Kalimantan).
sumatranum *H. Okada in Acta Phytotax. Geobot.*, 47(1): 5 (1996)—Sumatera.

CLEMATIS (Ranunculac.).
subsect. Acerifoliae *W.T. Wang in Acta Phytotax. Sin.*, 36(2): 161 (1998).
subsect. Aethusifoliae *(Serov) M. Johnson, Släktet Klematis:* 288 (1997): C. sect. Aethusifoliae.
ser. Afoliatae *W.T. Wang in Acta Phytotax. Sin.*, 38(6): 512 (2000).
subsect. Africanae *M. Johnson, Släktet Klematis:* 416 (1997), without type.
subsect. Angustifoliae *(Tamura) M. Johnson, Släktet Klematis:* 603 (1997): C. ser. Angustifoliae.
sect. Angustifoliae *(Prantl) W.A. Brandenburg, Meclatis in Clematis: Yellow flow. Clematis sp.:* 38 (2000), without basionym ref.: Basionym not stated.
sect. Aristatae *(Prantl) W.A. Brandenburg, Meclatis in Clematis: Yellow flow. Clematis sp.:* 38 (2000), without basionym ref.: Basionym not stated.
ser. Aristatae *W.T. Wang in Acta Phytotax. Sin.*, 38(6): 510 (2000).
subsect. Aristatae *(Prantl) W.T. Wang in Acta Phytotax. Sin.*, 38(6): 509 (2000): C. [unranked] Aristatae.
ser. Armandianae *W.T. Wang in Acta Phytotax. Sin.*, 36(2): 158 (1998).
sect. Baldwinianae *(R.O. Erickson) T.Y.A. Yang in Syst. Geogr. Pl.*, 68(1–2): 302 (1999), basionym not validly publ.: C. subsect. Baldwinianae.
ser. Baominianae *W.T. Wang in Acta Phytotax. Sin.*, 36(2): 157 (1998).
subsect. Brachyblasti *M. Johnson, Släktet Klematis:* 172 (1997), without type.
subsect. Chinenses *(Tamura) M. Johnson, Släktet Klematis:* 603 (1997): C. ser. Chinenses.
sect. Connatae *(Koehne) M. Johnson, Släktet Klematis:* 287 (1997): C. subsect. Connatae.
subsect. Crassifoliae *(Tamura) M. Johnson, Släktet Klematis:* 604 (1997): C. ser. Crassifoliae.
sect. Crassifoliae *(Tamura) W.A. Brandenburg, Meclatis in Clematis: Yellow flow. Clematis sp.:* 38 (2000), without basionym ref.: Basionym not stated.
sect. Dioicae *(Prantl) W.A. Brandenburg, Meclatis in Clematis: Yellow flow. Clematis sp.:* 38 (2000), without basionym ref.: Basionym not stated.
ser. Dioicae *W.T. Wang in Acta Phytotax. Sin.*, 38(6): 507 (2000).
subsect. Dioicae *(Prantl) W.T. Wang in Acta Phytotax. Sin.*, 38(6): 506 (2000): Clematis c. Dioicae.
ser. Dissectae *W.T. Wang in Acta Phytotax. Sin.*, 38(5): 423 (2000).
subsect. Fasciculiflorae *(Tamura) W.T. Wang in Acta Phytotax. Sin.*, 38(6): 503 (2000): Clematis ser. Fasciculiflorae.
subsect. Fasciculiflorae *(Tamura) M. Johnson, Släktet Klematis:* 604 (1997): C. ser. Fasciculiflorae.
ser. Fasciculiflorae *W.T. Wang in Acta Phytotax. Sin.*, 36(2): 162 (1998).

CLEMATIS:-
sect. Fruticella *(Tamura) W.A. Brandenburg, Meclatis in Clematis: Yellow flow. Clematis sp.:* 38 (2000), without basionym ref.: Basionym not stated.
subsect. Fuscae *(Tamura) M. Johnson, Släktet Klematis:* 538 (1997): C. ser. Fuscae.
ser. Grandiflorae *W.T. Wang in Acta Phytotax. Sin.*, 38(4): 334 (2000).
subsect. Hastatae *(W.T. Wang) W.T. Wang in Acta Phytotax. Sin.*, 38(6): 503 (2000): Clematis ser. Hastatae.
ser. Hastatae *W.T. Wang in Acta Phytotax. Sin.*, 36(2): 159 (1998).
subsect. Henryanae *(Tamura) M. Johnson, Släktet Klematis:* 288 (1997): Basionym not stated.
subsect. Hexapetalae *(Prantl) W.T. Wang in Acta Phytotax. Sin.*, 38(6): 510 (2000): Clematis d. Hexapetalae.
ser. Hexapetalae *W.T. Wang in Acta Phytotax. Sin.*, 38(6): 512 (2000).
subsect. Heynianae *W.T. Wang in Acta Phytotax. Sin.*, 38(5): 402 (2000).
sect. Hirsutissimae *(R.O. Erickson) T.Y.A. Yang in Syst. Geogr. Pl.*, 68(1–2): 302 (1999), basionym not validly publ.: C. subsect. Hirsutissimae.
subsect. Huchouenses *W.T. Wang in Acta Phytotax. Sin.*, 36(2): 163 (1998).
subsect. Insidiosae *W.T. Wang in Acta Phytotax. Sin.*, 38(4): 315 (2000).
ser. Insidiosae *W.T. Wang in Acta Phytotax. Sin.*, 38(6): 513 (2000).
sect. Integrifoliae *(R.O. Erickson) T.Y.A. Yang in Syst. Geogr. Pl.*, 68(1–2): 302 (1999), basionym not validly publ.: C. subsect. Integrifoliae.
subgen. Integrifoliae *(R.O. Erickson) T.Y.A. Yang in Syst. Geogr. Pl.*, 68(1–2): 302 (1999), basionym not validly publ.: C. subsect. Integrifoliae.
subsect. Ispahanicae *(Serov) M. Johnson, Släktet Klematis:* 652 (1997): Basionym not stated.
subsect. Lanuginosae *M. Johnson, Släktet Klematis:* 668 (1997), contrary to Art. 37.4 ICBN (1994).
sect. Lasiantha *(Tamura) W.A. Brandenburg, Meclatis in Clematis: Yellow flow. Clematis sp.:* 38 (2000), without basionym ref.: Basionym not stated.
subsect. Lasianthae *(Tamura) W.T. Wang in Acta Phytotax. Sin.*, 38(6): 507 (2000): C. sect. Lasianthae.
ser. Liboenses *W.T. Wang in Acta Phytotax. Sin.*, 38(5): 420 (2000).
subsect. Liboenses *W.T. Wang in Acta Phytotax. Sin.*, 36(2): 164 (1998).
ser. Loureirianae *W.T. Wang in Acta Phytotax. Sin.*, 38(5): 420 (2000).
subsect. Macgregorianae *W.T. Wang in Acta Phytotax. Sin.*, 38(5): 421 (2000).
ser. Maratae *W.T. Wang in Acta Phytotax. Sin.*, 38(6): 512 (2000).
ser. Marmorariae *W.T. Wang in Acta Phytotax. Sin.*, 38(6): 512 (2000).
subsect. Meyenianae *(Tamura) M. Johnson, Släktet Klematis:* 605 (1997): C. ser. Meyenianae.
subsect. Microphyllae *W.T. Wang in Acta Phytotax. Sin.*, 38(6): 507 (2000).
ser. Millefoliolatae *W.T. Wang in Acta Phytotax. Sin.*, 38(6): 507 (2000).
sect. Naraveliopsis *(Hand.-Mazz.) W.A. Brandenburg, Meclatis in Clematis: Yellow flow. Clematis sp.:* 38 (2000), without basionym ref.: Basionym not stated.
sect. Novae-zeelandiae *M. Johnson, Släktet Klematis:* 159 (1997), nom. nov.: C. d. Hexapetalae *Prantl.*
sect. Papuasicae *(H. Eichler) W.A. Brandenburg, Meclatis in Clematis: Yellow flow. Clematis sp.:* 38 (2000), without basionym ref.: Basionym not stated.
subsect. Phlebanthae *M. Johnson, Släktet Klematis:* 653 (1997), contrary to Art. 37.4 ICBN (1994).
ser. Pickeringianae *W.T. Wang in Acta Phytotax. Sin.*, 38(6): 510 (2000).
sect. Pierotianae *(Tamura) W.A. Brandenburg, Meclatis in Clematis: Yellow flow. Clematis sp.:* 38 (2000), without basionym ref.: Basionym not stated.
subsect. Pinnatae *W.T. Wang in Acta Phytotax. Sin.*, 36(2): 159 (1998).

CLEMATIS:-
ser. Pogonandrae *W.T. Wang in Acta Phytotax. Sin.*, 36(2): 170 (1998).
subsect. Potaninianae *W.T. Wang in Acta Phytotax. Sin.*, 36(2): 162 (1998), as 'Potanianae'; *see also Acta Phytotax. Sin.*, 38(5): 401 (2000).
subsect. Potaninianae *M. Johnson, Släktet Klematis:* 412 (1997), without type.
ser. Pseudopogonandrae *W.T. Wang in Acta Phytotax. Sin.*, 36(2): 171 (1998).
sect. Pterocarpa *(Tamura) W.A. Brandenburg, Meclatis in Clematis: Yellow flow. Clematis sp.:* 38 (2000), without basionym ref.: Basionym not stated.
sect. Rectae *(Prantl) W.A. Brandenburg, Meclatis in Clematis: Yellow flow. Clematis sp.:* 38 (2000), without basionym ref.: Basionym not stated.
ser. Rutoides *W.T. Wang in Acta Phytotax. Sin.*, 38(6): 513 (2000).
ser. Sichotealinenses *W.T. Wang in Acta Phytotax. Sin.*, 38(6): 504 (2000).
subsect. Smilacifoliae *W.T. Wang in Acta Phytotax. Sin.*, 36(2): 163 (1998); et *in Acta Phytotax. Sin.*, 38(5): 414 (2000).
subsect. Songaricae *(Serov) M. Johnson, Släktet Klematis:* 653 (1997): C. ser. Songaricae.
ser. Tongluenses *W.T. Wang in Acta Phytotax. Sin.*, 38(6): 501 (2000).
subgen. Tubulosae *T.Y.A. Yang in Syst. Geogr. Pl.*, 68(1–2): 301 (1999), nom. nov.: C. § Tubulosae *Decne.*
subsect. Uncinatae *(Tamura) M. Johnson, Släktet Klematis:* 606 (1997): C. ser. Uncinatae.
subsect. Wightianae *(Prantl) W.T. Wang in Acta Phytotax. Sin.*, 38(4): 321 (2000): C. [unranked] Wightianae.
ser. Wightianae *W.T. Wang in Acta Phytotax. Sin.*, 38(5): 421 (2000).
subsect. Williamsianae *M. Johnson, Släktet Klematis:* 366 (1997), contrary to Art. 37.4 ICBN (1994).
ser. Zemuenses *W.T. Wang in Acta Phytotax. Sin.*, 38(6): 505 (2000).
actinostemmatifolia *W.T. Wang in Acta Phytotax. Sin.*, 38(5): 421 (2000)—Comoros.
angolana *M. Johnson, Släktet Klematis:* 145 (1997), nom. nov.: Clematopsis speciosa *Hutch.*
angustifolia Jacq.
var. elliptica *(S.Y. Hu) M. Johnson, Släktet Klematis:* 626 (1997), without basionym page: C. hexapetala var. elliptica.
var. insularis *(S.Y. Hu) M. Johnson, Släktet Klematis:* 626 (1997), without basionym page: C. hexapetala var. insularis.
var. smithiana *(S.Y. Hu) M. Johnson, Släktet Klematis:* 626 (1997), without basionym page: C. hexapetala var. smithiana.
angustifoliola *W.T. Wang in Guihaia*, 17(1): 14 (1997)—China South-central (Yunnan).
armandii Franch.
var. farquhariana *(Rehder & E.H. Wilson) W.T. Wang in Acta Phytotax. Sin.*, 36(2): 158 (1998): C. armandii f. farquhariana.
baominiana *W.T. Wang in Acta Phytotax. Sin.*, 36(2): 157 (1998), nom. nov.: C. villosa *B.M. Yang.*
bojeri Hook.
var. anethifolia *(Hook.) M. Johnson, Släktet Klematis:* 147 (1997): C. anethifolia.
var. macrophylla *(J. Raynal) M. Johnson, Släktet Klematis:* 147 (1997): Clematopsis bojeri var. macrophylla.
var. oligophylla *(Hook.) M. Johnson, Släktet Klematis:* 147 (1997): C. oligophylla.
var. pseudoscabiosifolia *(H. Perrier) M. Johnson, Släktet Klematis:* 147 (1997): C. pseudoscabiosifolia.
brachystemon *Gunn ex W.T. Wang in Acta Phytotax. Sin.*, 38(6): 507 (2000)—Tasmania.
canescens (Turcz.) W.T. Wang & M.C. Chang
var. viridis *(W.T. Wang & M.C. Chang) W.T. Wang in Acta Phytotax. Sin.*, 36(2): 161 (1998): C. canescens subsp. viridis.
caudigera *W.T. Wang in Acta Phytotax. Sin.*, 36(2): 165 (1998)—Xinjiang.
chinensis Osb.
var. vestita *(Rehder & E.H. Wilson) W.T. Wang in Acta Phytotax. Sin.*, 36(2): 158 (1998): C. chinensis f. vestita.

CLEMATIS:-
chrysocarpa Welw. ex Oliv.
subsp. bijuga *R.K. Brummitt in Kew Bull.*, 55(1): 102 (2000)—Malawi, Mozambique, Tanzania, Zaire, Zambia.
cinnamomoides *W.T. Wang in Acta Phytotax. Sin.*, 38(5): 409 (2000)—Vietnam.
commutata Kuntze
var. glabrisepala *W.T. Wang in Acta Phytotax. Sin.*, 38(5): 423 (2000)—Cape Province.
comoresensis *W.T. Wang in Acta Phytotax. Sin.*, 38(4): 326 (2000)—Comoros.
connata DC.
var. trullifera *(Franch.) W.T. Wang in Acta Phytotax. Sin.*, 36(2): 170 (1998): C. buchananiana var. trullifera.
cruttwellii *H. Eichler ex W.T. Wang in Acta Phytotax. Sin.*, 38(4): 320 (2000)—New Guinea (Papua New Guinea).
delicata *H. Eichler ex W.T. Wang in Acta Phytotax. Sin.*, 38(6): 509 (2000)—Western Australia.
eichleri *(Tamura) M. Tamura in Thai Forest Bull., Bot.*, 25: 70 (1997): Naravelia eichleri.
fauriei *(H. Boissieu) M. Johnson, Släktet Klematis:* 192 (1997): C. alpina var. fauriei.
fengii *W.T. Wang in Acta Phytotax. Sin.*, 38(5): 418 (2000)—China South-central (Yunnan).
fusca Turcz.
var. amurensis *(Kuntze) M. Johnson, Släktet Klematis:* 590 (1997): Basionym not stated.
var. obtusifoliola *(Kuntze) M. Johnson, Släktet Klematis:* 594 (1997): Basionym not stated.
fusijamana *(Kuntze) M. Johnson, Släktet Klematis:* 202 (1997): C. alpina var. fusijamana.
gracilifolia Rehder & E.H. Wilson
var. lasiocarpa *W.T. Wang in Acta Phytotax. Sin.*, 38(6): 501 (2000)—Tibet.
var. trifoliolata *M. Johnson, Släktet Klematis:* 393 (1997)—China South-central (Sichuan).
grandifolia *(Staner & J. Léonard) M. Johnson, Släktet Klematis:* 150 (1997): Clematopsis grandifolia.
grossa Benth.
var. striatula *(DC.) W.T. Wang in Acta Phytotax. Sin.*, 38(4): 314 (2000): C. sericea var. striatula.
henryi Oliv.
var. trifolia *R.J. Wang in J. Trop. Subtrop. Bot.*, 7(1): 28 (1999)—China Southeast (Fujian).
heracleifolia DC.
f. albiflora *Y.N. Lee, Fl. Korea:* 1157 (1996), contrary to Arts. 37.4, 37.5 ICBN (1994)—South Korea.
heynei *M. Johnson, Släktet Klematis:* 432 (1997), nom. nov., isonym; *see M.A. Rau in B.D. Sharma et al., Fl. India,* 1: 67 (1993): C. triloba *B. Heyne ex Roth.*
hirsuta Perr. & Guill.
var. incisodentata *(A. Rich.) W.T. Wang in Acta Phytotax. Sin.*, 38(4): 325 (2000), as 'inciso-dentata': C. incisodentata.
var. junodii *(Burtt Davy) W.T. Wang in Acta Phytotax. Sin.*, 38(4): 325 (2000) C. oweniae var. junodii.
ianthina Koehne
var. kuripoensis *M. Johnson, Släktet Klematis:* 597 (1997), without type—South Korea.
intricata Bunge
var. phaeantha *(Rehder) M. Johnson, Släktet Klematis:* 340 (1997): C. glauca f. phaeantha.
katangensis *(Hutch.) M. Johnson, Släktet Klematis:* 151 (1997): Clematopsis katangensis.
khasiana *(Brühl) W.T. Wang in Acta Phytotax. Sin.*, 38(6): 503 (2000): Clematis montana var. khasiana.
kirilowii Maxim.
var. latisepala *(M.C. Chang) W.T. Wang in Acta Phytotax. Sin.*, 36(2): 158 (1998): C. terniflora var. latisepala.
koreana Kom.
var. fragrans *M. Johnson, Släktet Klematis:* 210 (1997), without type—South Korea.
var. lutea *(Rehder) M. Johnson, Släktet Klematis:* 210 (1997): C. koreana f. lutea.
leariloba *(Hutch. & Summerh.) W.T. Wang in Acta Phytotax. Sin.*, 38(5): 428 (2000), homonym; *see DC., Syst. Nat.*, 1: 155 (1817): Clematopsis leariloba.
macrophylla *(J. Raynal) W.T. Wang in Acta Phytotax. Sin.*, 38(5): 425 (2000): Clematopsis bojeri var. macrophylla.

CLEMATIS:-
manipurensis *(Brühl) W.T. Wang in Acta Phytotax. Sin.*, 38(6): 501 (2000): Clematis montana var. manipurensis.
var. lasioclada *W.T. Wang in Acta Phytotax. Sin.*, 38(6): 503 (2000)—East Himalaya (Bhutan).
mankiuensis *Y.N. Lee, Fl. Korea:* 1157 (1996), contrary to Arts. 37.4, 37.5 ICBN (1994)—South Korea.
meyeniana Walp.
var. uniflora *W.T. Wang in Acta Phytotax. Sin.*, 36(2): 158 (1998)—China Southeast (Fujian).
minggangiana *W.T. Wang in Phytologia,* 79(5): 387 (1995 publ. 1996)—China South-central (Yunnan).
montana Buch.-Ham. ex DC.
var. longipes *W.T. Wang in Acta Phytotax. Sin.*, 38(4): 306 (2000)—China South-central (Sichuan, Guizhou, Hubei, Yunnan), China North-central (Gansu), Xizang.
var. urosepala *W.T. Wang in Acta Phytotax. Sin.*, 38(6): 500 (2000)—Myanmar.
nagaensis *W.T. Wang in Acta Phytotax. Sin.*, 38(6): 513 (2000)—Assam (Nagaland).
nainitalensis *W.T. Wang in Acta Phytotax. Sin.*, 38(4): 330 (2000)—India (Uttar Pradesh).
napoensis *W.T. Wang in Acta Phytotax. Sin.*, 37(3): 217 (1999)—China Southeast (Guangxi).
novocaledoniaensis *W.T. Wang in Acta Phytotax. Sin.*, 38(6): 510 (2000)—New Caledonia.
oliveri *(Hutch.) M. Johnson, Släktet Klematis:* 152 (1997), homonym; *see Kuntze in Verh. Bot. Vereins Prov. Brandenburg,* 26: 165 (1885): Clematopsis oliveri.
orientalis L.
var. sinorobusta *W.T. Wang in Acta Phytotax. Sin.*, 36(2): 167 (1998), nom. nov.: C. orientalis var. robusta *W.T. Wang.*
parviloba Gardner & Champ.
var. bartlettii *(Yamam.) W.T. Wang in Acta Phytotax. Sin.*, 38(5): 405 (2000): C. bartlettii.
subsp. bartlettii *(Yamam.) T.Y.A. Yang & T.C. Huang in Taiwania,* 40(3): 235 (1995): C. bartlettii.
var. longianthera *W.T. Wang in Acta Phytotax. Sin.*, 36(2): 156 (1998)—China South-central (Sichuan).
var. rhombicoelliptica *W.T. Wang in Acta Phytotax. Sin.*, 38(5): 405 (2000), as 'rhombico-elliptica'—China South-central (Yunnan).
var. suboblonga *W.T. Wang in Acta Phytotax. Sin.*, 38(5): 405 (2000)—China South-central (Sichuan).
patens C. Morren & Decne.
var. tientaiensis *(M.Y. Fang) W.T. Wang in Acta Phytotax. Sin.*, 36(2): 163 (1998): C. patens subsp. tientaiensis.
peterae Hand.-Mazz.
var. lishanensis *(T.Y.A. Yang & T.C. Huang) W.T. Wang in Acta Phytotax. Sin.*, 36(2): 155 (1998): C. gouriana subsp. lishanensis.
pimpinellifolia Hook.
var. pseudoscabiosifolia *(H. Perrier) W.T. Wang in Acta Phytotax. Sin.*, 38(5): 427 (2000): C. pseudoscabiosifolia.
puberula Hook. f. & Thomson
var. ganpiniana *(H. Lév. & Vaniot) W.T. Wang in Acta Phytotax. Sin.*, 38(5): 407 (2000): C. vitalba var. ganpiniana.
var. subsericea *(Rehder & E.H. Wilson) W.T. Wang in Acta Phytotax. Sin.*, 38(5): 407 (2000): C. brevicaudata var. subsericea.
var. tenuisepala *(Maxim.) W.T. Wang in Acta Phytotax. Sin.*, 38(5): 406 (2000): C. brevicaudata var. tenuisepala.
pubipes *(W.T. Wang) W.T. Wang in Acta Phytotax. Sin.*, 38(5): 416 (2000): C. crassipes var. pubipes.
rigoi *W.T. Wang in Acta Phytotax. Sin.*, 38(4): 327 (2000)—Italy.
rutoides *W.T. Wang in Acta Phytotax. Sin.*, 38(4): 316 (2000)—Madagascar.
salsuginea *Bunge ex W.T. Wang in Acta Phytotax. Sin.*, 38(5): 411 (2000)—Mongolia.
sclerophylla *W.T. Wang in Acta Phytotax. Sin.*, 38(4): 320 (2000)—New Guinea (Papua New Guinea, Irian Jaya).
sibirica (L.) Mill.
var. iliensis *(Y.S. Hou & W.H. Hou) J.G. Liu in Fl. Xinjiangensis,* 2(1): 288 (1994): C. iliensis.

CLEMATIS:-
sibirica (L.) Mill.
var. tianschanica *(Pavlov) M. Johnson, Släktet Klematis:* 236 (1997): Atragene tianschanica.
sikkimensis *(Hook. f. & Thomson) J.R. Drumm. ex I.H. Burkill in Rec. Bot. Surv. India,* 10(2): 229 (1925): C. acuminata var. sikkimensis.
var. clarkei *(Kuntze) W.T. Wang in Acta Phytotax. Sin.*, 36(2): 168 (1998): C. acuminata β clarkei.
var. monantha *W.T. Wang & L.Q. Li in Acta Phytotax. Sin.*, 37(3): 218 (1999)—China South-central (Yunnan).
smilacifolia Wall.
var. peltata *(W.T. Wang) W.T. Wang in Acta Phytotax. Sin.*, 38(5): 415 (2000): C. loureiriana var. peltata.
staintonii *W.T. Wang in Acta Phytotax. Sin.*, 38(4): 333 (2000)—Nepal.
subfalcata Pei ex M.Y. Fang
var. pubipes *W.T. Wang in Acta Phytotax. Sin.*, 36(2): 169 (1998)—China South-central (Yunnan).
var. stenophylla *(Hand.-Mazz.) W.T. Wang in Acta Phytotax. Sin.*, 36(2): 169 (1998): C. clarkeana var. stenophylla.
taiwaniana
var. ryukiuensis *(Tamura) M. Tamura in Acta Phytotax. Geobot.*, 46(2): 210 (1995 publ. 1996): Clematis grata var. ryukiuensis.
tamurae *T.Y.A. Yang & T.C. Huang in Taiwania,* 40(3): 239 (1995)—Taiwan.
tibetana Kuntze
var. leariloba *W.T. Wang in Acta Phytotax. Sin.*, 36(2): 164 (1998)—Xizang, China South-central (Sichuan).
subsp. tangutica *(Maxim.) W.A. Brandenburg, Meclatis in Clematis: Yellow flow. Clematis sp.:* 169 (2000), without basionym ref.: C. orientalis var. tangutica.
var. vernayi *(C.E.C. Fisch.) W.T. Wang in Acta Phytotax. Sin.*, 36(2): 164 (1998), isonym; *see Grey-Wilson in Kew Bull.*, 44(1): 48 (1989): C. vernayi.
tongluensis (Brühl) Tamura
var. mollisepala *W.T. Wang in Acta Phytotax. Sin.*, 38(6): 503 (2000)—Tibet.
trifida Hook.
var. lanceolata *W.T. Wang in Acta Phytotax. Sin.*, 38(5): 424 (2000)—Madagascar.
triloba Thunb.
var. congensis *(A. Chev.) M. Johnson, Släktet Klematis:* 469 (1997): C. thunbergii var. congensis.
tripartita *W.T. Wang in Acta Phytotax. Sin.*, 38(6): 500 (2000)—Tibet, Nepal.
tuaensis *H. Eichler ex W.T. Wang in Acta Phytotax. Sin.*, 38(4): 317 (2000)—New Guinea (Papua New Guinea).
turkestanica *M. Johnson, Släktet Klematis:* 239 (1997)—Pakistan.
var. kirgizica *M. Johnson, Släktet Klematis:* 242 (1997), without type—Kirgizstan.
villosa DC.
subsp. kirkii *(Oliv.) R.K. Brummitt in Kew Bull.*, 55(1): 104 (2000): C. kirkii.
subsp. oliveri *(Hutch.) R.K. Brummitt in Kew Bull.*, 55(1): 104 (2000): Clematopsis oliveri.
xinhuiensis *R.J. Wang in J. Trop. Subtrop. Bot.*, 7(1): 26 (1999)—China Southeast (Guangdong).
zandaensis *W.T. Wang in Acta Phytotax. Sin.*, 36(2): 165 (1998)—Xizang.
zhejiangensis *R.J. Wang in J. Trop. Subtrop. Bot.*, 7(1): 28 (1999)—China Southeast (Zhejiang).

CLEOME (Capparac.).
chapalaensis *H.H. Iltis in Bol. Inst. Bot. (Guadalajara)*, 5(1–3): 428 (1998)—Mexico Southwest (Michoacan, Jalisco).
subsp. albipetala *H.H. Iltis in Bol. Inst. Bot. (Guadalajara)*, 5(1–3): 434 (1998)—Mexico Southwest (Michoacan).
fimbriata
subsp. brachystyla *(Deflers) R. Govaerts, World Checklist Seed Plants,* 3(1): 15 (1999): C. brachystyla.
rutidosperma DC.
var. hainanensis *J.L. Shan in Bull. Bot. Res. North-East. Forest. Univ.*, 19(2): 126 (1999)—Hainan.

CLEOME
torticarpa *H.H. Iltis & T. Ruiz Zapata in Novon,* 7(4): 367 (1997 publ. 1998)—Venezuela.
viscosa L.
var. deglabrata *(Backer) B.S. Sun in Fl. Reipubl. Popul. Sin.,* 32: 539 (1999): Polanisia viscosa var. deglabrata.

CLEOMELLA (Capparac.).
jaliscensis *E. Villegas Flores & R. Ramírez Delgadillo in Bol. Inst. Bot. (Guadalajara),* 6(2–3): 179 (1998 publ. 2000)—Mexico Southwest (Jalisco).

CLERODENDRUM (Verbenac.).
abilioi *R.B. Fernandes in Mem. Soc. Brot.,* 30: 7 (1998)—Mozambique.
aurantiacum Baker
f. faulkneri *(Moldenke) R.B. Fernandes in Mem. Soc. Brot.,* 30: 34 (1998): C. faulkneri.
calundense *R.B. Fernandes in Mem. Soc. Brot.,* 30: 34 (1998)—Angola.
cephalanthum Oliv.
var. torrei *R.B. Fernandes in Mem. Soc. Brot.,* 30: 10 (1998)—Mozambique.
coulteri *(A. Gray) R. Govaerts, World Checklist Seed Plants,* 3(1): 15 (1999): Tetraclea coulteri.
cyaneum *R.B. Fernandes in Mem. Soc. Brot.,* 30: 36 (1998)—Zambia, Zimbabwe, Angola.
glabrum E. Mey.
f. pubescens *R.B. Fernandes in Mem. Soc. Brot.,* 30: 11 (1998)—Angola, Mozambique, Transvaal.
izuinsulare *K. Inoue, Masami Hasegawa & Shiro Kobayashi in J. Jap. Bot.,* 72(2): 119 (1997)—Japan.
luembense De Wild.
var. malawiense *R.B. Fernandes in Mem. Soc. Brot.,* 30: 45 (1998)—Malawi.
subsp. niassense *R.B. Fernandes in Mem. Soc. Brot.,* 30: 45 (1998)—Mozambique.
mendesii *R.B. Fernandes in Mem. Soc. Brot.,* 30: 46 (1998)—Angola.
myricoides (Hochst.) Vatke
f. alatipetiolatum *R.B. Fernandes in Mem. Soc. Brot.,* 30: 72 (1998)—Mozambique.
f. angustilobatum *R.B. Fernandes in Mem. Soc. Brot.,* 30: 73 (1998)—Angola, Zimbabwe.
f. brevilobatum *R.B. Fernandes in Mem. Soc. Brot.,* 30: 74 (1998)—Mozambique.
var. capiriense *R.B. Fernandes in Mem. Soc. Brot.,* 30: 55 (1998)—Zambia.
f. cubangense *R.B. Fernandes in Mem. Soc. Brot.,* 30: 83 (1998)—Angola.
var. dumale *(Hiern) R.B. Fernandes in Mem. Soc. Brot.,* 30: 77 (1998): Siphonanthus dumalis.
var. eleanoriae *R.B. Fernandes in Mem. Soc. Brot.,* 30: 88 (1998), as 'eleanorae'—Malawi.
f. lanceolatilobatum *R.B. Fernandes in Mem. Soc. Brot.,* 30: 55 (1998)—Zimbabwe.
f. lobulatum *R.B. Fernandes in Mem. Soc. Brot.,* 30: 74 (1998)—Mozambique.
var. moldenkei *R.B. Fernandes in Mem. Soc. Brot.,* 30: 89 (1998)—Zimbabwe, Zambia, Botswana.
subsp. namibiense *R.B. Fernandes in Mem. Soc. Brot.,* 30: 93 (1998)—Namibia.
subsp. napperae *B. Verdcourt in Fl. Trop. E. Afr., Verbenac.:* 134 (1992), in Supp. 20 as 'napperi'—Tanzania.†
f. reflexilobatum *R.B. Fernandes in Mem. Soc. Brot.,* 30: 75 (1998)—Zambia.
panigrahianum *A. Rajendran & P. Daniel in J. Bombay Nat. Hist. Soc.,* 95(1): 99 (1998)—Assam (Arunachal Pradesh).
robustum Klotzsch
var. macrocalyx *R.B. Fernandes in Mem. Soc. Brot.,* 30: 17 (1998)—Mozambique.
sansibarense Gürke
var. eratense *R.B. Fernandes in Mem. Soc. Brot.,* 30: 103 (1998)—Mozambique.
f. tomentellum *R.B. Fernandes in Mem. Soc. Brot.,* 30: 102 (1998)—Tanzania.
silvanum Henriq.
f. botryodes *(Hiern) R.B. Fernandes in Mem. Soc. Brot.,* 30: 21 (1998): Siphonanthus botryodes.
var. brevitubum *R.B. Fernandes in Mem. Soc. Brot.,* 30: 22 (1998)—Angola.

CLERODENDRUM:-
silvanum Henriq.
f. caulanthum *(Exell) R.B. Fernandes in Mem. Soc. Brot.,* 30: 22 (1998): Clerodendrum caulanthum.
ternatum Schinz
f. villosum *R.B. Fernandes in Mem. Soc. Brot.,* 30: 26 (1998), nom. nov.: C. lanceolatum *Gürke.*
triphyllum (Harv.) H. Pearson
f. hirsutum *(Hochst.) R.B. Fernandes in Mem. Soc. Brot.,* 30: 104 (1998): Cyclonema hirsutum.
urticifolium *(Roxb.) Wall. ex J.O. Voigt, Hort. Suburb. Calcutt.:* 466 (1845): Volkameria urticifolia.
verdcourtii *R.B. Fernandes in Mem. Soc. Brot.,* 30: 105 (1998)—Zimbabwe, Mozambique.
wildii Moldenke
f. glabrum *R.B. Fernandes in Mem. Soc. Brot.,* 30: 107 (1998)—Mozambique, Botswana, Zambia, Zimbabwe, Malawi.

CLETHRA (Clethrac.).
chiapensis *L.M. González Villarreal in Bol. Inst. Bot. (Guadalajara),* 5(1–3): 138 (1998)—Mexico Southeast (Chiapas).
conzattiana *L.M. González Villarreal in Bol. Inst. Bot. (Guadalajara),* 5(1–3): 141 (1998)—Mexico Southwest (Oaxaca).
luzmariae *L.M. González Villarreal in Bol. Inst. Bot. (Guadalajara),* 5(1–3): 145 (1998)—Mexico Southwest (Oaxaca).

CLEYERA (Theac.).
japonica Thunb.
var. longicarpa *(Yamam.) L.K. Ling & C.F. Hsieh in Fl. Taiwan, ed. 2,* 2: 676 (1996): Sakakia longicarpa.
yangchunensis *L.K. Ling in Fl. Reipubl. Popul. Sin.,* 50(1): 194 (1998)—China Southeast (Guangdong).

CLIBADIUM (Compos.).
websteri *H. Robinson in Phytologia,* 82(1): 60 (1997)—Ecuador.

CLIDEMIA (Melastomatac.).
longipedunculata *E. Cotton in Nordic J. Bot.,* 20(2): 183 (2000)—Ecuador.
tetragona *(DC.) Fournet ex C. Sastre in Biogeographica,* 72(4): 183 (1996): Sagraea tetragona.

CLINANTHUS (Amaryllidac.).
callacallensis *(Ravenna) A.W. Meerow in Syst. Bot.,* 25(4): 723 (2000): Stenomesson callacallense.
campodensis *(Ravenna) A.W. Meerow in Syst. Bot.,* 25(4): 723 (2000): Stenomesson campodense.
caracensis *(Ravenna) A.W. Meerow in Syst. Bot.,* 25(4): 723 (2000): Stenomesson caracense.
chihuanhuayu *(Cárdenas) A.W. Meerow in Syst. Bot.,* 25(4): 723 (2000): Haylockia chihuanhuayu.
coccineus *(Ruiz & Pav.) A.W. Meerow in Syst. Bot.,* 25(4): 723 (2000): Pancratium coccineum.
croceus *(Savigny) A.W. Meerow in Syst. Bot.,* 25(4): 723 (2000): Pancratium croceum.
elwesii *(Baker) A.W. Meerow in Syst. Bot.,* 25(4): 723 (2000): Callithauma viridiflorum var. elwesii.
flammidus *(Ravenna) A.W. Meerow in Syst. Bot.,* 25(4): 723 (2000), as 'flammidis': Stenomesson flammidum.
fulvus *(Herb.) A.W. Meerow in Syst. Bot.,* 25(4): 723 (2000): Coburghia fulva.
glareosus *(Ravenna) A.W. Meerow in Syst. Bot.,* 25(4): 723 (2000): Stenomesson glareosum.
humilis *(Herb.) A.W. Meerow in Syst. Bot.,* 25(4): 723 (2000): Clitanthes humilis.
imasumacc *(Vargas) A.W. Meerow in Syst. Bot.,* 25(4): 723 (2000), as 'imasumac': Stenomesson imasumacc.
incarnatus *(Kunth) A.W. Meerow in Syst. Bot.,* 25(4): 723 (2000): Pancratium incarnatum.
incarum *(Kraenzl.) A.W. Meerow in Syst. Bot.,* 25(4): 723 (2000), as 'incarus': Stenomesson incarum.
macleanicus *(Herb.) A.W. Meerow in Syst. Bot.,* 25(4): 723 (2000): Clitanthes macleanica.
microstephium *(Ravenna) A.W. Meerow in Syst. Bot.,* 25(4): 723 (2000), as 'microstephus': Stenomesson microstephium.

CLINANTHUS:-
mirabilis *(Ravenna) A.W. Meerow in Syst. Bot.,* 25(4): 723 (2000): Stenomesson mirabile.
recurvatus *(Ruiz & Pav.) A.W. Meerow in Syst. Bot.,* 25(4): 723 (2000): Pancratium recurvatum.
sunchubambae *(Ravenna) A.W. Meerow in Syst. Bot.,* 25(4): 723 (2000): Stenomesson sunchubambae.
variegatus *(Ruiz & Pav.) A.W. Meerow in Syst. Bot.,* 25(4): 723 (2000): Pancratium variegatum.
viridiflorus *(Ruiz & Pav.) A.W. Meerow in Syst. Bot.,* 25(4): 724 (2000): Pancratium viridiflorum.

× **CLINOMICROMERIA** *R. Govaerts, World Checklist Seed Plants,* 3(1): 15 (1999). *LABIATAE* (Clinopodium × Micromeria).
hostii *(Caruel) R. Govaerts, World Checklist Seed Plants,* 3(1): 15 (1999): Satureja × hostii.
nothosubsp. narentana *(K. Malý ex Dörfl.) R. Govaerts, World Checklist Seed Plants,* 3(1): 16 (1999): Satureja × narentana.

CLINOPODIUM (Labiat.).
acutifolium *(Benth.) R. Govaerts, World Checklist Seed Plants,* 3(1): 16 (27 July 1999): Gardoquia acutifolia.
acutifolium *(Benth.) R.M. Harley in Cat. Vasc. Pl. Ecuador (Monogr. Syst. Bot. Missouri Bot. Gard., 75):* 954 (30 July 1999), isonym; *see Govaerts, World Checklist Seed Plants,* 3(1): 16 (27 July 1999): Gardoquia acutifolia.
alpestre *(Urb.) R.M. Harley in Kew Bull.,* 55(4): 919 (2000): Micromeria alpestris.
alpinum
subsp. albanicum *(Kümmerle & Jáv.) R. Govaerts, World Checklist Seed Plants,* 3(1): 16 (1999): Calamintha alpina f. albanica.
subsp. hungaricum *(Simonk.) R. Govaerts, World Checklist Seed Plants,* 3(1): 16 (1999): Melissa hungarica.
subsp. majoranifolium *(Mill.) R. Govaerts, World Checklist Seed Plants,* 3(1): 16 (1999): Melissa majoranifolia.
subsp. meridionale *(Nyman) R. Govaerts, World Checklist Seed Plants,* 3(1): 16 (1999), as 'meridionalis': Calamintha alpina subsp. meridionalis.
subsp. orontium *(K. Malý) R. Govaerts, World Checklist Seed Plants,* 3(1): 16 (1999): Satureja orontia.
subsp. pyrenaeum *(Braun-Blanq.) R. Govaerts, World Checklist Seed Plants,* 3(1): 16 (1999): Satureja alpina subsp. pyrenaea.
subsp. sardoum *(Asch. & Levier) R. Govaerts, World Checklist Seed Plants,* 3(1): 16 (1999): Calamintha alpina var. sardoa.
amissum *(Epling & Játiva) R.M. Harley in Kew Bull.,* 55(4): 919 (2000), as 'amissa': Satureja amissa.
argenteum *(Kunth) R. Govaerts, World Checklist Seed Plants,* 3(1): 16 (1999): Gardoquia argentea.
axillare *(Rusby) R.M. Harley in Kew Bull.,* 55(4): 919 (2000): Bystropogon axillaris.
brevicalyx *(Epling) R.M. Harley & A. Granda Paucar in Kew Bull.,* 55(4): 919 (2000): Satureja brevicalyx.
breviflorum *(Benth.) R. Govaerts, World Checklist Seed Plants,* 3(1): 16 (1999): Gardoquia breviflora.
bucheri *(P. Wilson) R.M. Harley in Kew Bull.,* 55(4): 920 (2000): Micromeria bucheri.
caroli-henricanum *(Kit Tan & Sorger) R. Govaerts, World Checklist Seed Plants,* 3(1): 16 (1999): Calamintha caroli-henricana.
cercocarpoides *(Epling) R. Govaerts, World Checklist Seed Plants,* 3(1): 16 (1999): Satureja cercocarpoides.
chandleri *(Brandegee) P.D. Cantino & S.J. Wagstaff in Brittonia,* 50(1): 69 (1998): Calamintha chandleri.
chilense *(Benth.) R. Govaerts, World Checklist Seed Plants,* 3(1): 16 (1999): Gardoquia chilensis.
clivorum *(Epling) R. Govaerts, World Checklist Seed Plants,* 3(1): 16 (1999): Satureja clivorum.
corsicum *(Pers.) R. Govaerts, World Checklist Seed Plants,* 3(1): 16 (1999): Thymus corsicus.
cylindristachys *(Epling & Játiva) R. Govaerts, World Checklist Seed Plants,* 3(1): 17 (1999): Satureja cylindristachys.

CLINOPODIUM:-
domingense *(Urb. & Ekman) R. Govaerts, World Checklist Seed Plants,* 3(1): 17 (1999): Satureja domingensis.
ekmanianum *(Epling & Alain) R.M. Harley in Kew Bull.,* 55(4): 921 (2000): Satureja ekmaniana.
fasciculatum *(Benth.) R. Govaerts, World Checklist Seed Plants,* 3(1): 17 (27 July 1999): Gardoquia fasciculata.
fasciculatum *(Benth.) R.M. Harley in Cat. Vasc. Pl. Ecuador (Monogr. Syst. Bot. Missouri Bot. Gard., 75):* 954 (30 July 1999), isonym; *see Govaerts, World Checklist Seed Plants,* 3(1): 17 (27 July 1999): Gardoquia fasciculata.
flabellifolium *(Epling & Játiva) R. Govaerts, World Checklist Seed Plants,* 3(1): 17 (1999): Satureja flabellifolia.
foliolosum *(Benth.) R. Govaerts, World Checklist Seed Plants,* 3(1): 17 (1999): Gardoquia foliolosa.
ganderi *(Epling) R. Govaerts, World Checklist Seed Plants,* 3(1): 17 (1999): Satureja ganderi.
grandiflorum
subsp. baborense *(Batt.) R. Govaerts, World Checklist Seed Plants,* 3(1) 17 (1999): Calamintha baborensis.
graveolens
subsp. rotundifolium *(Pers.) R. Govaerts, World Checklist Seed Plants,* 3(1): 17 (1999): Acinos rotundifolius.
griseum *(Epling) R.M. Harley in Kew Bull.,* 55(4): 921 (2000): Satureja grisea.
heterotrichum *(Boiss. & Reut.) R. Govaerts, World Checklist Seed Plants,* 3(1): 17 (1999): Calamintha heterotricha.
hintoniorum *(B.L. Turner) R. Govaerts, World Checklist Seed Plants,* 3(1): 17 (1999): Satureja hintoniorum.
hispidulum *(Boiss. & Reut.) R. Govaerts, World Checklist Seed Plants,* 3(1): 17 (1999): Calamintha hispidula.
insulare *(P. Candargy) R. Govaerts, World Checklist Seed Plants,* 3(1): 17 (1999), as 'insularis': Micromeria insularis.
jaliscanum *(McVaugh & R. Schmid) R. Govaerts, World Checklist Seed Plants,* 3(1): 17 (1999): Satureja jaliscana.
jamesonii *(Benth.) R. Govaerts, World Checklist Seed Plants,* 3(1): 17 (27 July 1999): Gardoquia jamesonii.
jamesonii *(Benth.) R.M. Harley in Cat. Vasc. Pl. Ecuador (Monogr. Syst. Bot. Missouri Bot. Gard., 75):* 954 (30 July 1999), isonym *see Govaerts, World Checklist Seed Plants,* 3(1): 17 (27 July 1999): Gardoquia jamesonii.
junctionis *(Epling & Játiva) R. Govaerts, World Checklist Seed Plants,* 3(1): 17 (1999): Satureja junctionis.
kunashirense *N.S. Probatova in Sosud. Rast. Sovet. Dal'nego Vostoka,* 7: 369 (1995)—Kuril Is.
laxiflorum *(Hayata) K. Mori in G. Masamune (ed.), Short Fl. Formosa:* 183 (1936): Calamintha laxiflora.†
loeseneriаnum *(Mansf.) A. Granda Paucar in Kew Bull.,* 55(4): 922 (2000): Satureja loeseneriana.
maderense *(Henr.) R. Govaerts, World Checklist Seed Plants,* 3(1): 17 (1999): Satureja maderensis.
matthewsii *(Briq.) R. Govaerts, World Checklist Seed Plants,* 3(1): 17 (1999): Satureja matthewsii.
menthifolium
subsp. ascendens *(Jord.) R. Govaerts, World Checklist Seed Plants,* 3(1): 17 (1999): Calamintha ascendens.
subsp. hirtum *(Briq.) R. Govaerts, World Checklist Seed Plants,* 3(1): 18 (1999): Satureja calamintha var. hirsuta.
mexicanum *(Benth.) R. Govaerts, World Checklist Seed Plants,* 3(1): 18 (1999): Gardoquia mexicana.
micromerioides *(Hemsl.) R. Govaerts, World Checklist Seed Plants,* 3(1): 18 (1999): Gardoquia micromerioides.
mutabile *(Epling) R.M. Harley in Cat. Vasc. Pl. Ecuador (Monogr. Syst. Bot. Missouri Bot. Gard., 75):* 954 (1999): Satureja mutabilis.
nanum *(P.H. Davis & Doroszenko) R. Govaerts, World Checklist Seed Plants,* 3(1): 18 (1999): Acinos nanus.

CLINOPODIUM:-
nepeta
subsp. glandulosum *(Req.) R. Govaerts, World Checklist Seed Plants,* 3(1): 18 (1999): Thymus glandulosus.
nothosubsp. subisidoratum *(Borbás) R. Govaerts, World Checklist Seed Plants,* 3(1): 18 (1999): Calamintha × subisidorata.
obovatum *(Ruiz & Pav.) R. Govaerts, World Checklist Seed Plants,* 3(1): 18 (1999): Gardoquia obovata.
odorum *(Griseb.) R.M. Harley in Kew Bull.,* 55(4): 923 (2000): Xenopoma odorum.
pallidum *(Epling) R. Govaerts, World Checklist Seed Plants,* 3(1): 18 (1999): Satureja pallida.
pamphylicum *(Boiss. & Heldr.) R. Govaerts, World Checklist Seed Plants,* 3(1): 18 (1999): Calamintha pamphylica.
subsp. davisii *(Contandr. & Quézel) R. Govaerts, World Checklist Seed Plants,* 3(1): 18 (1999): Calamintha davisii.
subsp. piperelloides *(Stapf) R. Govaerts, World Checklist Seed Plants,* 3(1): 18 (1999): Calamintha piperelloides.
× pillichianum *(J. Wagner) R. Govaerts, World Checklist Seed Plants,* 3(1): 18 (1999): Calamintha × pillichiana.
plicatulum *(Epling) R. Govaerts, World Checklist Seed Plants,* 3(1): 18 (1999): Satureja plicatula.
procumbens *(Greenm.) R.M. Harley in Kew Bull.,* 55(4): 924 (2000): Satureja procumbens.
pulchellum *(Kunth) R. Govaerts, World Checklist Seed Plants,* 3(1): 18 (1999): Gardoquia pulchella.
revolutum *(Ruiz & Pav.) R. Govaerts, World Checklist Seed Plants,* 3(1): 18 (1999): Gardoquia revoluta.
rouyanum *(Briq.) R. Govaerts, World Checklist Seed Plants,* 3(1): 18 (1999): Satureja rouyana.
schusteri *(Urb.) R. Govaerts, World Checklist Seed Plants,* 3(1): 19 (1999): Hyptis schusteri.
selerianum *(Loes.) R. Govaerts, World Checklist Seed Plants,* 3(1): 19 (1999): Satureja seleriana.
sericeum *(C. Presl ex Benth.) R. Govaerts, World Checklist Seed Plants,* 3(1): 19 (27 July 1999): Gardoquia sericea.
sericeum *(C. Presl ex Benth.) R.M. Harley in Cat. Vasc. Pl. Ecuador (Monogr. Syst. Bot. Missouri Bot. Gard., 75):* 954 (30 July 1999), isonym; *see Govaerts, World Checklist Seed Plants,* 3(1): 19 (27 July 1999): Gardoquia sericea.
sericifolium *(Epling & Játiva) R. Govaerts, World Checklist Seed Plants,* 3(1): 19 (1999): Satureja sericifolia.
speciosum *(Hook.) R. Govaerts, World Checklist Seed Plants,* 3(1): 19 (1999): Stachys speciosa.
sphenophyllum *(Epling) R. Govaerts, World Checklist Seed Plants,* 3(1): 19 (1999): Satureja sphenophylla.
striatum *(Ruiz & Pav.) R. Govaerts, World Checklist Seed Plants,* 3(1): 19 (1999): Gardoquia striata.
tauricola *(P.H. Davis) R. Govaerts, World Checklist Seed Plants,* 3(1): 19 (1999), as 'tauricolum': Calamintha tauricola.
taxifolium *(Kunth) R. Govaerts, World Checklist Seed Plants,* 3(1): 19 (27 July 1999): Gardoquia taxifolia.
taxifolium *(Kunth) R.M. Harley in Cat. Vasc. Pl. Ecuador (Monogr. Syst. Bot. Missouri Bot. Gard., 75):* 954 (30 July 1999), isonym; *see Govaerts, World Checklist Seed Plants,* 3(1): 19 (27 July 1999): Gardoquia taxifolia.
tenellum *(Epling) R.M. Harley in Cat. Vasc. Pl. Ecuador (Monogr. Syst. Bot. Missouri Bot. Gard., 75):* 954 (1999): Satureja tenella.
tomentosum *(Kunth) R. Govaerts, World Checklist Seed Plants,* 3(1): 19 (27 July 1999): Gardoquia tomentosa.
tomentosum *(Kunth) R.M. Harley in Cat. Vasc. Pl. Ecuador (Monogr. Syst. Bot. Missouri Bot. Gard., 75):* 954 (30 July 1999), isonym; *see Govaerts, World Checklist Seed Plants,* 3(1): 19 (27 July 1999): Gardoquia tomentosa.
troodi *(Post) R. Govaerts, World Checklist Seed Plants,* 3(1): 19 (1999): Calamintha troodi.
subsp. grandiflorum *(Hartvig & Strid) R. Govaerts, World Checklist Seed Plants,* 3(1): 19 (1999): Acinos troodi subsp. grandiflorus.

CLINOPODIUM:-
troodi
subsp. vardaranum *(Leblebici) R. Govaerts, World Checklist Seed Plants,* 3(1): 19 (1999): Acinos troodi subsp. vardaranus.
vanum *(Epling) R.M. Harley & A. Granda Paucar in Kew Bull.,* 55(4): 926 (2000): Satureja vana.
vardarense *(Šilić) R. Govaerts, World Checklist Seed Plants,* 3(1): 19 (1999), as 'varadense': Calamintha vardarensis.
vargasii *(Epling & Mathias) R. Govaerts, World Checklist Seed Plants,* 3(1): 19 (1999): Satureja vargasii.
vulgare L.
f. glabratum *(Vandas) Č. Šilić, Monografija Satureja, Calamintha, Micromeria, Acinos i Clinopodium fl.* Jugoslavije: 362 (1979): C. vulgare f. glabratum.
f. parviflorum *(Rohlena) Č. Šilić, Monografija Satureja, Calamintha, Micromeria, Acinos i Clinopodium fl.* Jugoslavije: 362 (1979): C. vulgare var. parviflorum.
f. rubrotinctum *Č. Šilić, Monografija Satureja, Calamintha, Micromeria, Acinos i Clinopodium fl.* Jugoslavije: 364 (1979)—Yugoslavia.
f. unicephalum *Č. Šilić, Monografija Satureja, Calamintha, Micromeria, Acinos i Clinopodium fl.* Jugoslavije: 364 (1979)—Yugoslavia.
weberbaueri *(Mansf.) R. Govaerts, World Checklist Seed Plants,* 3(1): 19 (1999): Satureja weberbaueri.

CLINTONIA (Convallariac.).
dulongensis *H. Li in Fl. Dulongjian Reg.:* 316 (1993), without latin descr. or type—China South-central (Yunnan).

CLITORIA (Legumin.).
arborea Hoffm. ex Benth.
var. pseudoamazonica *P.R. Fantz in Sida,* 16(4): 721 (1995)—Brazil North (Amazonas, Acre).
epetiolata
var. angustissima *P.R. Fantz in Sida,* 16(4): 722 (1995)—Brazil South (Paraná).
falcata Lam.
var. aurantiaca *(Benth.) P.R. Fantz in Sida,* 16(4): 723 (1995): C. glycinoides var. aurantiaca.
var. latifolia *(Rizzini) P.R. Fantz in Sida,* 16(4): 724 (1995): C. rubiginosa f. latifolia.
guianensis (Aubl.) Benth.
var. chapadensis *(Malme) P.R. Fantz in Sida,* 16(4): 725 (1995): C. chapadensis.
var. macrocleistogama *P.R. Fantz in Sida,* 16(4): 726 (1995)—Brazil Southeast (Minas Gerais).
heterophylla
var. pedunculata *(Bojer ex Benth.) P.R. Fantz in Sida,* 16(4): 726 (1995): C. pedunculata.
leptostachya
var. fruticosa *P.R. Fantz in Sida,* 16(4): 727 (1995)—Brazil North (Pará).
mariana
var. pubescentia *P.R. Fantz in Sida,* 16(4): 727 (1995)—Florida.
polystachya Benth.
var. congesta *P.R. Fantz in Sida,* 16(4): 728 (1995)—Mexico Central (México State).
var. pringlei *P.R. Fantz in Sida,* 16(4): 729 (1995)—Mexico Central (Morelos).
pozuzoensis J.F. Macbr.
var. schunkei *P.R. Fantz in Sida,* 16(4): 729 (1995)—Peru.
stipularis Benth.
var. latifolia *(Rizzini) P.R. Fantz in Sida,* 16(4): 729 (1995): C. stipularis f. latifolia.

CLIVIA (Amaryllidac.).
miniata *(Lindl.) J.F.W. Bosse, Vollst. Handb. Bl.-Gärtn., ed. 3,* 1: 768 (1859): Vallota miniata.†

CLOWESIA (Orchidac.).
amazonica *K.G. de Lacerda & V.P. Castro Neto in Bradea,* 6(44): 386 (1995)—Brazil North (Amazonas).

CLUSIA (Guttif.).
sect. Havetia *(Kunth) J.J. Pipoly in Sida,* 17(4): 766 (1997): gen. Havetia.

CLUSIA:-
sect. Havetiopsis *(Planch. & Triana) J.J. Pipoly in Fl. Venezuelan Guayana (gen. eds. J.A. Steyermark et al.)*, 4: 269 (1998): gen. Havetiopsis.
sect. Oedematopus *(Planch. & Triana) J.J. Pipoly in Fl. Venezuelan Guayana (gen. eds. J.A. Steyermark et al.)*, 4: 269 (1998): gen. Oedematopus.
sect. Pilosperma *(Planch. & Triana) J.J. Pipoly in Sida*, 17(4): 766 (1997): gen. Pilosperma.
sect. Quapoya *(Aubl.) J.J. Pipoly in Fl. Venezuelan Guayana (gen. eds. J.A. Steyermark et al.)*, 4: 269 (1998): gen. Quapoya.
aristeguietae *(Maguire) J.J. Pipoly in Sida*, 18(2): 408 (1998): Oedematopus aristeguietae.
asymmetrica *J.J. Pipoly in Sida*, 16(4): 657 (1995)—Venezuela.
bernardoi *J.J. Pipoly & A. Cogollo P. in Sida*, 18(2): 402 (1998)—Colombia.
burle-marxii *V. Bittrich in Bol. Bot. (São Paulo)*, 15: 76 (1996)—Brazil Northeast (Bahia).
caudata *(Planch. & Triana) J.J. Pipoly in Sida*, 17(4): 766 (1997): Pilosperma caudatum.
colombiana *J.J. Pipoly in Sida*, 17(4): 767 (1997), nom. nov.: Havetia laurifolia *Kunth*.
comans *(Mart.) J.J. Pipoly in Fl. Venezuelan Guayana (gen. eds. J.A. Steyermark et al.)*, 4: 272 (1998), with incorrect basionym and basionym ref.; *et in Sida*, 18(2): 408 (1998), with correct basionym ref.: Schweiggeria comans.
deminuta *J.J. Pipoly in Sida*, 18(4): 1175 (1999)—Colombia.
diamantina *V. Bittrich in Bol. Bot. (São Paulo)*, 15: 73 (1996)—Brazil Southeast (Minas Gerais).
divaricata *(Cuatrec.) J.J. Pipoly in Sida*, 18(2): 408 (1998): Oedematopus divaricatus.
engleriana *J.J. Pipoly in Sida*, 17(4): 766 (1997), nom. nov.: Oedematopus congestiflorus *Engl.*
epiphytica *(Cuatrec.) J.J. Pipoly in Sida*, 18(2): 408 (1998): Oedematopus epiphyticus.
fabiolae *J.J. Pipoly in Sida*, 16(4): 740 (1995)—Venezuela.
flavida *(Benth.) J.J. Pipoly in Fl. Venezuelan Guayana (gen. eds. J.A. Steyermark et al.)*, 4: 272 (1998): Havetia flavida.
grammadenioides *J.J. Pipoly in Sida*, 16(4): 653 (1995)—Venezuela, Colombia.
guayanae *J.J. Pipoly in Sida*, 16(4): 674 (1995)—Guyana, Venezuela.
hammeliana *J.J. Pipoly in Fl. Venezuelan Guayana (gen. eds. J.A. Steyermark et al.)*, 4: 273 (1998), nom. nov.: Rengifa peruviana *Poepp.*
huberi *J.J. Pipoly in Fl. Venezuelan Guayana (gen. eds. J.A. Steyermark et al.)*, 4: 274 (1998), nom. nov.: Oedematopus duidae *Gleason*.
hylaeae *J.J. Pipoly in Sida*, 17(4): 766 (1997), nom. nov.: Oedematopus weberbaueri *Engl.*
immersa *C.M. Vieira in Bradea*, 8(4): 21 (1996)—Brazil Southeast (Rio de Janeiro).
maguireana *J.J. Pipoly in Sida*, 16(4): 669 (1995)—Venezuela, Guyana.
mirandensis *(Cuatrec.) J.J. Pipoly in Sida*, 18(2): 408 (1998): Oedematopus mirandensis.
multilineata *J.J. Pipoly in Sida*, 16(4): 655 (1995)—Venezuela.
niambiensis *J.J. Pipoly, A. Cogollo P. & M.S. González in Sida*, 18(2): 396 (1998)—Colombia, Ecuador.
obdeltifolia *V. Bittrich in Bol. Bot. (São Paulo)*, 15: 78 (1996)—Brazil Northeast (Bahia), Brazil Southeast (Minas Gerais).
obovata *(Spruce ex Planch. & Triana) J.J. Pipoly in Fl. Venezuelan Guayana (gen. eds. J.A. Steyermark et al.)*, 4: 275 (1998): Oedematopus obovatus.
octandra *(Poepp.) J.J. Pipoly in Fl. Venezuelan Guayana (gen. eds. J.A. Steyermark et al.)*, 4: 276 (1998): Havetia octandra.
paisarum *J.J. Pipoly in Sida*, 18(4): 1179 (1999)—Colombia.
parvula *(Maguire) J.J. Pipoly in Fl. Venezuelan Guayana (gen. eds. J.A. Steyermark et al.)*, 4: 277 (1998): Oedematopus parvulus.
polyandra *(Vesque) J.J. Pipoly in Sida*, 17(4): 766 (1997): Oedematopus polyandrus.
ptaritepuiensis *(Steyerm.) J.J. Pipoly in Fl. Venezuelan Guayana (gen. eds. J.A. Steyermark et al.)*, 4: 277 (1998): Oedematopus ptaritepuiensis.

CLUSIA:-
sipapoana *(Maguire) J.J. Pipoly in Fl. Venezuelan Guayana (gen. eds. J.A. Steyermark et al.)*, 4: 279 (1998): Quapoya sipapoana.
tetragona *J.J. Pipoly & A. Cogollo P. in Sida*, 18(2): 404 (1998)—Colombia.
wurdackiana *J.J. Pipoly in Sida*, 16(4): 747 (1995), nom. nov.: C. sessilis *Klotzsch ex Engl.*

CLUSIELLA (Guttif.).
impressinervis *B.E. Hammel in Novon*, 9(3): 354 (1999)—Brazil North (Amazonas), Venezuela.
isthmensis *B.E. Hammel in Novon*, 9(3): 354 (1999)—Costa Rica, Panama.

CLUTIA (Euphorbiac.).
govaertsii *A. Radcliffe-Smith in Kew Bull.*, 52(2): 478 (1997), nom. nov.: C. thunbergii var. vaccinioides *Pax & K. Hoffm.*

CLYPEA (Menispermac.).
glabra *(Roxb.) Wight & Arn. ex J.O. Voigt, Hort. Suburb. Calcutt.*: 330 (1845): Cissampelos glabra.

CNEORIDIUM (Rutac.).
dumosum *(Nutt.) H.E. Baillon, Hist. Pl.*, 4: 498 (1873): Pitavia dumosa.†

CNICUS (Compos.).
bulgaricus *P.P. Panov in Fitologiya*, 48: 9 (1996)—Bulgaria.

CNIDIUM (Umbellif.).
atropurpureum *M.F. Watson in Edinburgh J. Bot.*, 53(1): 128 (1996), homonym; *see Spreng. in Roem. & Schult., Syst. Veg.*, 6: 418 (1820)—Bhutan.
bhutanicum *M.F. Watson in Edinburgh J. Bot.*, 55(3): 373 (1998), nom. nov.: C. atropurpureum *M.F. Watson*.

CNIDOSCOLUS (Euphorbiac.).
regina *(Léon) A. Radcliffe-Smith & R. Govaerts in Kew Bull.*, 52(1): 183 (1997): Jatropha regina.
urens (L.) Arthur
var. stimulosus *(Michx.) R. Govaerts in R. Govaerts, D.G. Frodin & A. Radcliffe-Smith, World Checklist & Bibliogr.* Euphorbiaceae, 1: 398 (2000): Jatropha stimulosa.

COBAEA (Cobaeac.).
sect. Pachysepalae *L.A. Prather, Syst. Cobaea (Polemoniac.) (Syst. Bot. Monog., 57)*: 47 (1999).
sect. Triovulatae *L.A. Prather, Syst. Cobaea (Polemoniac.) (Syst. Bot. Monog., 57)*: 31 (1999).
flava *L.A. Prather in Brittonia*, 48(1): 114 (1996)—Peru.
paneroi *L.A. Prather in Brittonia*, 48(1): 116 (1996)—Mexico Northeast (Hidalgo), Mexico Central (Puebla).
rotundiflora *L.A. Prather in Brittonia*, 48(1): 117 (1996)—Mexico Southeast (Chiapas), Guatemala.

COCCOCHRONDRA (Rubiac.).
laevis (Steyerm.) Rauschert
subsp. maigualidae *J.H. Kirkbride in Biollania*, ed. esp. 6 (J.J. Wurdack Festschr.): 398 (1997)—Venezuela.

COCCOTHRINAX (Palmae).
boschiana *M. Mejía & Ricardo García in Moscosoa*, 9: 1 (1997)—Dominican Republic.
pumila *A. Borhidi & J.A. Hernandez Valdés in Acta Bot. Hung.*, 38(1–4): 195 (1993–94 publ. 1995)—Cuba.

COCHEMIEA (Cactac.).
ser. Barbatae *A.B. Doweld in Sukkulentÿ (Moscow)*, 3(1–2): 39 (2000).
ser. Cochemieoidei *A.B. Doweld in Sukkulentÿ (Moscow)*, 3(1–2): 39 (2000).
sect. Mamillopsis *(E. Morren) A.B. Doweld in Sukkulentÿ (Moscow)*, 3(1–2): 39 (2000): Mammillaria sect. Mamillopsis.
sect. Protocochemiea *A.B. Doweld in Sukkulentÿ (Moscow)*, 3(1–2): 39 (2000).

COCHEMIEA:-
ser. Saboae *A.B. Doweld in Sukkulentÿ (Moscow)*, 3(1–2): 39 (2000).
ser. Seniles *A.B. Doweld in Sukkulentÿ (Moscow)*, 3(1–2): 39 (2000).
barbata *(Engelm.) A.B. Doweld in Sukkulentÿ (Moscow)*, 3(1–2): 38 (2000): Mammillaria barbata.
capensis *(H.E. Gates) A.B. Doweld in Sukkulentÿ (Moscow)*, 3(1–2): 39 (2000): Neomammillaria capensis.
dioica *(K. Brandegee) A.B. Doweld in Sukkulentÿ (Moscow)*, 3(1–2): 39 (2000): Mammillaria dioica.
subsp. angelensis *(R.T. Craig) A.B. Doweld in Sukkulentÿ (Moscow)*, 3(1–2): 39 (2000): Mammillaria angelensis.
subsp. estebanensis *(G.E. Linds.) A.B. Doweld in Sukkulentÿ (Moscow)*, 3(1–2): 39 (2000): Mammillaria estebanensis.
grahamii *(Engelm.) A.B. Doweld in Sukkulentÿ (Moscow)*, 3(1–2): 39 (2000): Mammillaria grahamii.
phitauiana *(E.M. Baxter) A.B. Doweld in Sukkulentÿ (Moscow)*, 3(1–2): 39 (2000): Neomammillaria phitauiana.
saboae *(Glass) A.B. Doweld in Sukkulentÿ (Moscow)*, 3(1–2): 39 (2000): Mammillaria saboae.
subsp. goldii *(Glass & R.A. Foster) A.B. Doweld in Sukkulentÿ (Moscow)*, 3(1–2): 39 (2000): Mammillaria goldii.
subsp. haudeana *(A.B. Lau & K. Wagner) A.B. Doweld in Sukkulentÿ (Moscow)*, 3(1–2): 39 (2000): Mammillaria haudeana.
swinglei *(Britton & Rose) A.B. Doweld in Sukkulentÿ (Moscow)*, 3(1–2): 39 (2000): Neomammillaria swinglei.
theresae *(Cutak) A.B. Doweld in Sukkulentÿ (Moscow)*, 3(1–2): 39 (2000): Mammillaria theresae.
wrightii *(Engelm. & Bigelow) A.B. Doweld in Sukkulentÿ (Moscow)*, 3(1–2): 39 (2000): Mammillaria wrightii.
subsp. wilcoxii *(Toumey ex K. Schum.) A.B. Doweld in Sukkulentÿ (Moscow)*, 3(1–2): 39 (2000): Mammillaria wilcoxii.

COCHLEARIA (Crucif.).
sect. Archaeocochlearia *M. Koch in Pl. Syst. Evol.*, 216(3–4): 227 (1999), without latin descr.
officinalis L.
subsp. aestuaria *(Lloyd) I. Nordal & M.M. Laane in Symb. Bot. Upsal.*, 31(3): 56 (1996): C. officinalis var. aestuaria.

COCHLIDIOSPERMA (Scrophulariac.).
trichadena *(Jord. & Fourr.) R. Govaerts, World Checklist Seed Plants*, 3(1): 19 (1999): Veronica trichadena.

CODIAEUM (Euphorbiac.).
variegatum (L.) Blume
var. cavernicola *R. Kiew & P. van Welzen in Gard. Bull. Singapore*, 50(1): 32 (1998)—Borneo (Sabah).

CODONANTHE (Gesneriac.).
cordifolia *A. Chautems in Candollea*, 52(1): 159 (1997)—Brazil Southeast (São Paulo, Rio de Janeiro, Minas Gerais), Brazil South (Santa Catarina).
venosa *A. Chautems in Candollea*, 52(1): 162 (1997)—Brazil Southeast (São Paulo, Rio de Janeiro).

CODONOPSIS (Campanulac.).
grey-wilsonii *J.M.H. Shaw in New Plantsman*, 3(2): 93 (1996), nom. nov.: C. nepalensis *Grey-Wilson*.
hirsuta *(D.Y. Hong) K.E. Morris & T.G. Lammers in Novon*, 9(3): 387 (1999), homonym; *see D.Y. Hong & L.M. Ma in Fl. Sichuanica*, 10: 546 (1992): Leptocodon hirsutus.

COELACHNE (Gramin.).
simpliciuscula (Wight & Arn. ex Steud) Munro ex Benth.
var. spicata *(Hook. f.) G. Panigrahi in S.K. Murti & G. Panigrahi, Fl. Bilaspur Distr.*, 2: 741 (1999): C. pulchella var. spicata.

COELOGYNE (Orchidac.).
arunachalensis *H.J. Chowdhery & G.D. Pal in Nordic J. Bot.*, 17(4): 369 (1997)—Assam (Arunachal Pradesh).
gongshanensis *H. Li ex S.C. Chen in Fl. Reipubl. Popul. Sin.*, 18: 412 (1999)—China South-Central (Yunnan).
gongshanensis *H. Li in Fl. Dulongjian Reg.*: 341 (1993), without latin descr. or type—China South-central (Yunnan).
hajrae *S. Phukan in Orchid Rev.*, 105(1214): 94 (1997)—Assam (Arunachal Pradesh).
lockii *L.V. Averyanov in Lindleyana*, 15(2): 73 (2000)—Vietnam.
longiana *L.V. Averyanov in Bot. Zhurn.*, 82(3): 136, 135 (1997)—Vietnam.
malipoensis *Z.H. Tsi in Bull. Bot. Res. North-East. Forest. Univ.*, 15(1): 43 (1995)—China South-central (Yunnan).
marthae *S.E.C. Sierra in Blumea*, 45(2): 299 (2000)—Borneo (Sarawak).
motleyi *Rolfe ex J.J. Wood, D.A. Clayton & C.L. Chan in Sandakania*, 11: 38 (1998)—Borneo (Kalimantan, Sabah, Sarawak).
speciosa (Blume) Lindl.
subsp. fimbriata *(J.J. Sm.) B. Gravendeel in Blumea*, 44(2): 301 (1999): C. speciosa var. fimbriata.
subsp. incarnata *B. Gravendeel in Blumea*, 44(2): 303 (1999)—Sumatera.
tommii *B. Gravendeel & P. O'Byrne in Blumea*, 44(2): 310 (1999)—Locality not stated.
verrucosa *S.E.C. Sierra in Blumea*, 45(2): 309 (2000)—Borneo (Sabah).

COELOPLEURUM (Umbellif.).
multisectum (Maxim.) Kitag.
var. trichocarpum *(H. Hara) H. Ohba in Fl. Japan*, 2c: 288 (1999): Angelica trichocarpa.

COELORACHIS (Gramin.).
laevispica *(Keng) P.C. Keng in Pl. Shanghai*, 1: 753 (1999): Rottboellia laevispica.

COFFEA (Rubiac.).
dactylifera *E. Robbrecht & P. Stoffelen in Syst. Geogr. Pl.*, 69(1): 121 (1999)—Zaire.
heterocalyx *P. Stoffelen in Belgian J. Bot.*, 129(1): 72 (1996 publ. 1997)—Cameroon.
leonimontana *P. Stoffelen in Belgian J. Bot.*, 129(1): 72 (1996 publ. 1997)—Cameroon.
leroyi *A.P. Davis in Kew Bull.*, 55(2): 411 (2000)—Madagascar.
liaudii *J.-F. Leroy ex A.P. Davis in Kew Bull.*, 55(2): 409 (2000)—Madagascar.
magnistipula *P. Stoffelen & E. Robbrecht in Taxon*, 46(1): 39 (1997)—Gabon, Cameroon.
manombensis *A.P. Davis in Kew Bull.*, 55(2): 406 (2000)—Madagascar.
montekupensis *P. Stoffelen in Kew Bull.*, 52(4): 990 (1997)—Cameroon.

COHNIELLA (Orchidac.).
aguirrei *(Königer) W. Königer in Arcula*, 10: 280 (2000): Stilifolium aguirrei.
ascendens *(Lindl.) E.A. Christenson in Lindleyana*, 14(4): 177 (1999): Oncidium ascendens.
cebolleta *(Jacq.) E.A. Christenson in Lindleyana*, 14(4): 177 (1999): Dendrobium cebolleta.
jonesiana *(Rchb. f.) E.A. Christenson in Lindleyana*, 14(4): 177 (1999): Oncidium jonesianum.
marvraganii *(Lückel) E.A. Christenson in Lindleyana*, 14(4): 177 (1999): Stilifolium marvraganii.
nuda *(Bateman ex Lindl.) E.A. Christenson in Lindleyana*, 14(4): 177 (1999): Oncidium nudum.
osteniana *(Schltr.) E.A. Christenson in Lindleyana*, 14(4): 177 (1999): Oncidium ostenianum.
pongratziana *(Königer) W. Königer in Arcula*, 10: 280 (2000): Stilifolium pongratzianum.
sprucei *(Lindl.) W. Königer & Pongr. in Arcula*, 10: 280 (2000): Oncidium sprucei.
stacyi *(Garay) E.A. Christenson in Lindleyana*, 14(4): 177 (1999): Oncidium stacyi.
stipitata *(Lindl.) E.A. Christenson in Lindleyana*, 14(4): 177 (1999): Oncidium stipitatum.
teres *(Ames & C. Schweinf.) E.A. Christenson in Lindleyana*, 14(4): 177 (1999): Oncidium teres.

COINCYA (Crucif.).
tournefortii *(Gouan) F. Alcaraz, T.E. Díaz, S. Rivas-Martínez & P. Sánchez-Gómez in Itinera Geobot.*, 2: 108 (1989): Brassica tournefortii.

COIX (Gramin.).
chinensis Tod.
var. formosana *(Ohwi) L. Liou in Fl. Reipubl. Popul. Sin.*, 10(2): 291 (1997): C. lacryma-jobi var. formosana.
gigantea Koenig ex Roxb.
subsp. aquatica *(Roxb.) Bhattacharya in S. Moulik, Grasses & Bamboos India*, 1: 175 (1996): C. aquatica.

COJOBA (Legumin.).
arborea (L.) Britton & Rose
var. angustifolia *(Rusby) R.C. Barneby & J.W. Grimes, Silk Tree, Guanacaste, Monkey's Earring (Mem. New York Bot. Gard., 74(2))*: 44 (1997): Pithecellobium sophorocarpum var. angustifolium.
var. cubensis *(Bisse) R.C. Barneby & J.W. Grimes, Silk Tree, Guanacaste, Monkey's Earring (Mem. New York Bot. Gard., 74(2))*: 44 (1997): Pithecellobium cubense.
bahorucensis *J.W. Grimes & R. García in Barneby & Grimes, Silk Tree, Guanacaste, Monkey's Earring (Mem. New York* Bot. Gard., 74(2)): 59 (1997)—Dominican Republic.
beckii *R.C. Barneby & J.W. Grimes, Silk Tree, Guanacaste, Monkey's Earring (Mem. New York Bot. Gard., 74(2))*: 46 (1997)—Bolivia.
filipes *(Vent.) R.C. Barneby & J.W. Grimes, Silk Tree, Guanacaste, Monkey's Earring (Mem. New York Bot. Gard., 74(2))*: 55 (1997): Inga filipes.
zanonii *(Barneby) R.C. Barneby & J.W. Grimes, Silk Tree, Guanacaste, Monkey's Earring (Mem. New York Bot. Gard., 74(2))*: 57 (1997): Obolinga zanonii.

COLCHICUM (Colchicac.).
actupii *A. Fridlender in Bull. Mens. Soc. Linn. Lyon*, 68(6): 195 (1999)—Sardegna.
arenasii *A. Fridlender in Acta Bot. Gallica*, 146(2): 158 (1999)—Corse.
bakhtiaricum *F. Matin & M. Iranshahr in Iranian J. Bot.*, 8(2): 252 (2000)—Iran.
chalcedonicum Azn.
subsp. punctatum *K. Persson in Candollea*, 53(2): 405 (1998)—Turkey.
chimonanthum *K. Persson in Pl. Syst. Evol.*, 217(1–2): 56 (1999)—Greece.
confusum *K. Persson in Pl. Syst. Evol.*, 217(1–2): 60 (1999)—Greece.
davisii *C.D. Brickell in New Plantsman*, 5(1): 15 (1998)—Turkey.
dolichantherum *K. Persson in Edinburgh J. Bot.*, 56(1): 126 (1999)—Turkey.
euboeum *(Boiss.) K. Persson in Candollea*, 53(2): 400 (1998): C. latifolium var. euboeum.
fharii *A. Fridlender in Bull. Mens. Soc. Linn. Lyon*, 68(9): 268 (1999)—Morocco.
goharíae *E.Tz. Gabrieljan in Fl. Rastitel'nost' Rast. Res. Arm.*, 12: 16 (1999), as 'goharae'—Transcaucasus (Armeniya).
heldreichii *K. Persson in Edinburgh J. Bot.*, 56(1): 98 (1999)—Turkey.
imperatoris-friderici *Siehe ex K. Persson in Edinburgh J. Bot.*, 56(1): 129 (1999)—Turkey.
inundatum *K. Persson in Edinburgh J. Bot.*, 56(1): 99 (1999)—Turkey.
lingulatum Boiss. & Spruner
subsp. rigescens *K. Persson in Candollea*, 53(2): 411 (1998)—Turkey.
micaceum *K. Persson in Edinburgh J. Bot.*, 56(1): 95 (1999)—Turkey.
minutum *K. Persson in Edinburgh J. Bot.*, 56(1): 90 (1999)—Turkey.
munzurense *K. Persson in Edinburgh J. Bot.*, 56(1): 86 (1999)—Turkey.
paschei *K. Persson in Edinburgh J. Bot.*, 56(1): 110 (1999)—Turkey.
rausii *K. Persson in Pl. Syst. Evol.*, 217(1–2): 58 (1999)—Greece.
sanguicolle *K. Persson in Edinburgh J. Bot.*, 56(1): 92 (1999)—Turkey.

COLDENIA (Boraginac.).
darwinii *(Hook. f.) A. Gray in Proc. Amer. Acad. Arts*, 5: 341 (1862): Galapagoa darwinii.
fusca *(Hook. f.) A. Gray in Proc. Amer. Acad. Arts*, 5: 341 (1862): Galapagoa fusca.

COLEBROOKEA (Labiat.).
ternifolia *W. Roxburgh, Pl. Coromandel*, 3: 40 (1815)—India (Karnataka).†

COLEOCEPHALOCEREUS (Cactac.).
buxbaumianus
subsp. flavisetus *(F. Ritter) N.P. Taylor & D.C. Zappi in Cactaceae Consensus Initiatives*, 3: 7 (1997): C. flavisetus.
fluminensis
subsp. decumbens *(F. Ritter) N.P. Taylor & D.C. Zappi in Cactaceae Consensus Initiatives*, 3: 7 (1997): C. decumbens.

COLEOGETON *(Rchb.) D.H. Les & R.R. Haynes in Novon*, 6(4): 389 (1996): Potamogeton [subgen.] Coleogeton. (Potamogetonac.).
filiformis *(Pers.) D.H. Les & R.R. Haynes in Novon*, 6(4): 390 (1996): Potamogeton filiformis.
subsp. alpinus *(Blytt) D.H. Les & R.R. Haynes in Novon*, 6(4): 390 (1996): Potamogeton marinus f. alpinus.
subsp. occidentalis *(J.W. Robbins) D.H. Les & R.R. Haynes in Novon*, 6(4): 390 (1996): Potamogeton marinus var. occidentalis.
pectinatus *(L.) D.H. Les & R.R. Haynes in Novon*, 6(4): 390 (1996): Potamogeton pectinatus.
striatus *(Ruiz & Pav.) D.H. Les & R.R. Haynes in Novon*, 6(4): 390 (1996): Potamogeton striatus.
vaginatus *(Turcz.) D.H. Les & R.R. Haynes in Novon*, 6(4): 390 (1996): Potamogeton vaginatus.

COLEUS (Labiat.).
vaalae *(Forssk.) Deflers in Revue d'Égypte*: 423 (1894); *and as reprint, Esquisses géogr. bot.*: 47 (1894): Ocimum vaalae.

COLLABIUM (Orchidac.).
acuticalcar *W. van der Burgh & E.F. de Vogel in Orchid Monogr.*, 8: 150 (1997)—Sumatera.
carinatum *E.F. de Vogel in Orchid Monogr.*, 8: 152 (1997)—New Guinea (Papua New Guinea).
evrardii *(Gagnep.) L.V. Averyanov in Bot. Zhurn.*, 84(10): 127 (1999): Tainia evrardii.

COLLARE-STUARTENSE *K. Senghas & L. Bockemühl in J. Orchideenfreund*, 4(2): 73 (1997). *ORCHIDACEAE*
astranthum *(Linden & Rchb. f.) K. Senghas & L. Bockemühl in J. Orchideenfreund*, 4(2): 74 (1997): Odontoglossum astranthum.
cariniferum *(Rchb. f.) K. Senghas & L. Bockemühl in J. Orchideenfreund*, 4(2): 75 (1997): Odontoglossum cariniferum.
digitatum *(C. Schweinf.) K. Senghas & L. Bockemühl in J. Orchideenfreund*, 4(2): 74 (1997): Odontoglossum digitatum.
dracoceps *(Dalström) K. Senghas in J. Orchideenfreund*, 7(2): 135 (2000): Odontoglossum dracoceps.
micklowii *(Dalström) K. Senghas & L. Bockemühl in J. Orchideenfreund*, 4(2): 74 (1997): Odontoglossum micklowii.
multistellare *(Rchb. f.) K. Senghas & L. Bockemühl in J. Orchideenfreund*, 4(2): 74 (1997): Odontoglossum multistellare.

COLLETIA (Rhamnac.).
infesta *(Kunth) A.T. Brongniart, Mém. Fam. Rhamnées*: 59 (1826): Ceanothus infesta.†
pubescens *A.T. Brongniart, Mém. Fam. Rhamnées*: 59 (1826)—Australia.†
tetragona *A.T. Brongniart, Mém. Fam. Rhamnées*: 59 (1826)—Peru.†

COLOBANTHUS (Caryophyllac.).
squarrosus Cheeseman
subsp. drucei *B.V. Sneddon in New Zealand J. Bot.*, 37(2): 200 (1999)—New Zealand South I.

COLOCASIA (Arac.).
gaoligongensis *H. Li & C.L. Long in Feddes Repert.*, 110(5–6): 423 (1999)—China South-central (Yunnan).

COLOCASIA:-
gongii *C.L. Long & H. Li in Feddes Repert.*, 111(7–8): 559 (2000)—China South-Central (Yunnan).
oresbia *A. Hay in Sandakania*, 7: 39 (1996)—Borneo (Sabah).

COLUBRINA *Rich. ex A.T. Brongniart, Mém. Fam. Rhamnées:* 61 (1826). *RHAMNACEAE*†
asiatica *(L.) A.T. Brongniart, Mém. Fam. Rhamnées:* 62 (1826): Ceanothus asiatica.†
cubensis *(Jacq.) A.T. Brongniart, Mém. Fam. Rhamnées:* 62 (1826): Rhamnus cubensis.†
ferruginosa *A.T. Brongniart, Mém. Fam. Rhamnées:* 62 (1826), nom. nov.: Rhamnus colubrina *Jacq.*†
granulosa *(Ruiz & Pav.) A.T. Brongniart, Mém. Fam. Rhamnées:* 62 (1826): Ceanothus granulosa.†
reclinata *(L'Hér.) A.T. Brongniart, Mém. Fam. Rhamnées:* 62 (1826): Ceanothus reclinata.†

COLUMNEA (Gesneriac.).
filifera *(Wiehler) L.E. Skog & L.P. Kvist in Novon*, 7(4): 413 (1997 publ. 1998): Dalbergaria filifera.
katzensteinii *(Wiehler) L.E. Skog & L.P. Kvist in Novon*, 7(4): 413 (1997 publ. 1998): Pentadenia katzensteinii.
mastersonii *(Wiehler) L.E. Skog & L.P. Kvist in Novon*, 7(4): 414 (1997 publ. 1998): Dalbergaria mastersonii.
medicinalis *(Wiehler) L.E. Skog & L.P. Kvist in Novon*, 7(4): 414 (1997 publ. 1998): Dalbergaria medicinalis.
ornata *(Wiehler) L.E. Skog & L.P. Kvist in Novon*, 7(4): 414 (1997 publ. 1998): Dalbergaria ornata.
tandapiana *(Wiehler) L.E. Skog & L.P. Kvist in Novon*, 7(4): 414 (1997 publ. 1998): Pentadenia tandapiana.

COLUTEA (Legumin.).
hispanica *S. Talavera & M. Arista in An. Jard. Bot. Madrid*, 56(2): 412 (1998)—Spain.

COMASTOMA (Gentianac.).
jigzhiense *T.N. Ho & J.Q. Liu in Novon*, 6(2): 187 (1996)—Qinghai.
urnigera *(E. Aitken & D.G. Long) J. Holub in Preslia*, 70(2): 103 (1998): Gentianella urnigera.

COMBESIA (Crassulac.).
muscosa
var. variegata *P.V. Heath in Calyx*, 6(2): 61 (1999)—Locality not stated.

COMBRETUM (Combretac.).
sect. Calopyxis *(Tul.) C.C.H. Jongkind in Bull. Mus. Nation. Hist. Nat., B, Adansonia*, Ser. 4, 17(3–4): 193 (1995): gen. Calopyxis.
albidum G. Don
var. cooperi *(Haines) H.O. Saxena & M. Brahmam, Fl. Orissa*, 2: 666 (1995): C. ovalifolium var. cooperi.
albiflorum *(Tul.) C.C.H. Jongkind in Bull. Mus. Nation. Hist. Nat., B, Adansonia*, Ser. 4, 17(3–4): 192 (1995): Poivrea albiflora.
boinensis *C.C.H. Jongkind in Bull. Mus. Nation. Hist. Nat., B, Adansonia*, Ser. 4, 17(3–4): 197 (1995)—Madagascar.
capuronii *C.C.H. Jongkind in Bull. Mus. Nation. Hist. Nat., B, Adansonia*, Ser. 4, 17(3–4): 198 (1995)—Madagascar.
clarense *C.C.H. Jongkind in Fl. Gabon*, 35: 24 (1999), as 'clarensis'—Gabon.
coursianum *(H. Perrier) C.C.H. Jongkind in Bull. Mus. Nation. Hist. Nat., B, Adansonia*, Ser. 4, 17(3–4): 194 (1995): Calopyxis coursiana.
decaryi *C.C.H. Jongkind in Bull. Mus. Nation. Hist. Nat., B, Adansonia*, Ser. 4, 17(3–4): 198 (1995), nom. nov.: Poivrea obscura var. occidentalis *H. Perrier.*
eriogynum *(H. Perrier) C.C.H. Jongkind in Bull. Mus. Nation. Hist. Nat., B, Adansonia*, Ser. 4, 17(3–4): 194 (1995): Calopyxis eriogyna.
erosum *C.C.H. Jongkind in Fl. Gabon*, 35: 37 (1999)—Gabon.
evisceratum *(H. Perrier) C.C.H. Jongkind in Bull. Mus. Nation. Hist. Nat., B, Adansonia*, Ser. 4, 17(3–4): 194 (1995): Calopyxis eviscerata.

COMBRETUM:-
indicum *(L.) C.C.H. Jongkind in Fl. Gabon*, 35: 48 (1999): Quisqualis indica.
longicollum *C.C.H. Jongkind in Bull. Mus. Nation. Hist. Nat., B, Adansonia*, Ser. 4, 17(3–4): 195 (1995), nom. nov.: Calopyxis trichophylla *Baker.*
macrocalyx *(Tul.) C.C.H. Jongkind in Bull. Mus. Nation. Hist. Nat., B, Adansonia*, Ser. 4, 17(3–4): 195 (1995): Poivrea macrocalyx.
meridionalis *(H. Perrier) C.C.H. Jongkind in Bull. Mus. Nation. Hist. Nat., B, Adansonia*, Ser. 4, 17(3–4): 199 (1995): Poivrea obscura var. meridionalis.
octagonum *(H. Perrier) C.C.H. Jongkind in Bull. Mus. Nation. Hist. Nat., B, Adansonia*, Ser. 4, 17(3–4): 199 (1995): Poivrea obscura var. octagona.
oxygonium *(Tul.) C.C.H. Jongkind in Bull. Mus. Nation. Hist. Nat., B, Adansonia*, Ser. 4, 17(3–4): 196 (1995): Calopyxis oxygonia.
polyanthum *C.C.H. Jongkind in Fl. Gabon*, 35: 77 (1999)—Gabon.
sphaeroides *(Tul.) C.C.H. Jongkind in Bull. Mus. Nation. Hist. Nat., B, Adansonia*, Ser. 4, 17(3–4): 196 (1995): Calopyxis sphaeroides.
subumbellatum *(Baker) C.C.H. Jongkind in Bull. Mus. Nation. Hist. Nat., B, Adansonia*, Ser. 4, 17(3–4): 196 (1995): Calopyxis subumbellata.
villosum *(Tul.) C.C.H. Jongkind in Bull. Mus. Nation. Hist. Nat., B, Adansonia*, Ser. 4, 17(3–4): 193 (1995): Poivrea villosa.
violaceum *(Tul.) C.C.H. Jongkind in Bull. Mus. Nation. Hist. Nat., B, Adansonia*, Ser. 4, 17(3–4): 193 (1995): Poivrea violacea.
wallichii
var. ternatum *(C.B. Clarke) R. Govaerts, World Checklist Seed Plants*, 3(1): 20 (1999): C. chinense var. ternatum.

COMMELINA (Commelinac.).
africana
var. karooica *(C.B. Clarke) R. Govaerts, World Checklist Seed Plants*, 3(1): 20 (1999): C. karooica.
arenicola *R.B. Faden in Brittonia*, 48(4): 547 (1996 publ. 1997)—Somalia.
erecta L.
f. dielsii *(Herter) N.M. Bacigalupo in Hickenia*, 2(31): 136 (1995): C. dielsii.
f. roseopurpurea *(Herter) N.M. Bacigalupo in Hickenia*, 2(31): 136 (1995): C. roseopurpurea.
frutescens *R.B. Faden in Brittonia*, 48(4): 548 (1996 publ. 1997)—Somalia.
nivea *A.R. López-Ferrari, A. Espejo Serna & J. Ceja Romero in Acta Bot. Mex.*, 41: 10 (1997)—Mexico Northeast (Durango).
rzedowskii *A.R. López-Ferrari, A. Espejo Serna & J. Ceja Romero in Acta Bot. Mex.*, 41: 12 (1997)—Mexico Central (México State).

COMMICARPUS (Nyctaginac.).
coctoris *N.A. Harriman in Sida*, 18(3): 679 (1999)—Mexico Southwest (Oaxaca).
praetermissus *N.A. Harriman in Sida*, 18(3): 681 (1999)—Mexico Southwest (Michoacan).

COMMIDENDRUM (Compos.).
robustum (Roxb.) DC.
subsp. gummiferum *(Roxb.) Q.C.B. Cronk in Bull. Nat. Hist. Mus. (London), Bot.*, 25(1): 98 (1995): Conyza gummifera.

COMMIPHORA (Burserac.).
africana
var. venosa *(Mattick) R. Govaerts, World Checklist Seed Plants*, 3(1): 20 (1999): C. pilosa var. venosa.
arenaria *M. Thulin in Nordic J. Bot.*, 20(4): 395 (2000)—Somalia.
chiovendana *J.B. Gillett ex M. Thulin in Nordic J. Bot.*, 20(4): 405 (2000)—Somalia.
gardoensis *J.B. Gillett ex M. Thulin in Nordic J. Bot.*, 20(4): 399 (2000)—Somalia.
kucharii *M. Thulin in Nordic J. Bot.*, 20(4): 409 (2000)—Somalia.
lobatospathulata *J.B. Gillett ex M. Thulin in Nordic J. Bot.*, 20(4): 403 (2000)—Somalia.
multifoliolata *J.B. Gillett ex M. Thulin in Nordic J. Bot.*, 20(4): 407 (2000)—Somalia.

COMMIPHORA:-
murraywatsonii *J.B. Gillett ex M. Thulin in Nordic J. Bot.*, 20(4): 408 (2000)—Somalia.
quercifoliola *J.B. Gillett ex M. Thulin in Nordic J. Bot.*, 20(4): 402 (2000)—Somalia.
spinulosa *J.B. Gillett ex M. Thulin in Nordic J. Bot.*, 20(4): 399 (2000)—Somalia.
stellatopubescens *J.B. Gillett ex M. Thulin in Nordic J. Bot.*, 20(4): 397 (2000)—Somalia.

COMOCLADIA (Anacardiac.).
pinnatifolia *Linnaeus, Syst. Nat. ed. 10*, 2: 861 (1759), in IK Vol.1 as 'pinnatifida'; *see also Linnaeus, Syst. Nat. ed. 12*, 2: 74 (1767)—Jamaica.†

CONANTHERA (Tecophilaeac.).
parvula *(Phil.) M. Muñoz-Schick in Not. Mens. Mus. Nacion. Hist. Nat. (Chile)*, 343: 26 (2000): Cummingia parvula.

CONCEVEIBA (Euphorbiac.).
maynasensis *R. de S. Secco in Bol. Mus. Paraense Emílio Goeldi, Bot.*, 14(1): 84 (1998 publ. 1999)—Peru.
prealta *(Croizat) Punt ex J. Murillo-A.in Rev. Acad. Colomb. Cienc. Exact. Fis. Nat.*, 24(92): 365 (2000): Cleidion prealtum.
santanderensis *J. Murillo-A.in Rev. Acad. Colomb. Cienc. Exact. Fis. Nat.*, 24(92): 366 (2000)—Colombia.
tristigmata *J. Murillo-A.in Rev. Acad. Colomb. Cienc. Exact. Fis. Nat.*, 24(92): 368 (2000)—Colombia.

CONCHOCARPUS (Rutac.).
acuminatus *(Pilg.) J.A. Kallunki & J.R. Pirani in Kew Bull.*, 53(2): 275 (1998): Cusparia acuminata.
adenantherus *(Rizzini) J.A. Kallunki & J.R. Pirani in Kew Bull.*, 53(2): 275 (1998): Angostura adenanthera.
bellus *J.A. Kallunki in Kew Bull.*, 53(2): 275 (1998)—Brazil Southeast (Espírito Santo).
cauliflorus *J.R. Pirani in Kew Bull.*, 53(2): 276 (1998)—Brazil Southeast (Espírito Santo).
concinnus *J.A. Kallunki in Kew Bull.*, 53(2): 278 (1998)—Brazil Northeast (Bahia).
cuneifolius Nees & Mart.
var. confertus *J.A. Kallunki in Kew Bull.*, 53(2): 282 (1998)—Brazil Northeast (Bahia).
cyrtanthus *J.A. Kallunki in Kew Bull.*, 53(2): 285 (1998)—Brazil Northeast (Bahia), Brazil Southeast (Espírito Santo).
dasyanthus *J.A. Kallunki in Kew Bull.*, 53(2): 286 (1998)—Brazil Northeast (Bahia).
diadematus *J.R. Pirani in Kew Bull.*, 53(2): 290 (1998)—Brazil Northeast (Bahia), Brazil Southeast (Rio de Janeiro).
elegans *(A. St.-Hil.) J.A. Kallunki & J.R. Pirani in Kew Bull.*, 53(2): 291 (1998): Galipea elegans.
fanshawei *(Sandwith) J.A. Kallunki & J.R. Pirani in Kew Bull.*, 53(2): 292 (1998): Cusparia fanshawei.
fissicalyx *J.R. Pirani in Kew Bull.*, 53(2): 292 (1998)—Brazil Northeast (Bahia).
fontanesianus *(A. St.-Hil.) J.A. Kallunki & J.R. Pirani in Kew Bull.*, 53(2): 294 (1998): Galipea fontanesiana.
furcatus *J.A. Kallunki in Kew Bull.*, 53(2): 294 (1998)—Brazil Southeast (Espírito Santo).
gaudichaudianus *(A. St.-Hil.) J.A. Kallunki & J.R. Pirani in Kew Bull.*, 53(2): 295 (1998): Galipea gaudichaudiana.
subsp. bahiensis *J.A. Kallunki in Kew Bull.*, 53(2): 295 (1998)—Brazil Northeast (Bahia).
grandiflorus *(Engl.) J.A. Kallunki & J.R. Pirani in Kew Bull.*, 53(2): 299 (1998): Cusparia grandiflora.
grandis *J.A. Kallunki in Kew Bull.*, 53(2): 299 (1998)—Brazil North (Pará, Amazonas).
guyanensis *(Pulle) J.A. Kallunki & J.R. Pirani in Kew Bull.*, 53(2): 300 (1998): Almeidea guyanensis.
heterophyllus *(A. St.-Hil.) J.A. Kallunki & J.R. Pirani in Kew Bull.*, 53(2): 300 (1998): Galipea heterophylla.
hirsutus *J.R. Pirani in Kew Bull.*, 53(2): 301 (1998)—Brazil Northeast (Bahia).
inopinatus *J.R. Pirani in Kew Bull.*, 53(2): 303 (1998)—Brazil Northeast (Bahia).

CONCHOCARPUS:-
insignis *J.R. Pirani in Kew Bull.*, 53(2): 305 (1998)—Brazil Southeast (Espírito Santo), Brazil Northeast (Bahia).
larensis *(Tamayo & Croizat) J.A. Kallunki & J.R. Pirani in Kew Bull.*, 53(2): 308 (1998): Raputia larensis.
longifolius *(A. St.-Hil.) J.A. Kallunki & J.R. Pirani in Kew Bull.*, 53(2): 308 (1998): Galipea longifolia.
longipes *J.A. Kallunki in Kew Bull.*, 53(2): 308 (1998)—Brazil Northeast (Bahia).
macrocarpus *(Engl.) J.A. Kallunki & J.R. Pirani in Kew Bull.*, 53(2): 309 (1998): Cusparia macrocarpa.
marginatus *(Rizzini) J.A. Kallunki & J.R. Pirani in Kew Bull.*, 53(2): 310 (1998): Angostura marginata.
mastigophorus *J.A. Kallunki in Kew Bull.*, 53(2): 310 (1998)—Brazil Northeast (Bahia).
modestus *J.A. Kallunki in Kew Bull.*, 53(2): 312 (1998)—Brazil Northeast (Bahia).
nicaraguensis *(Standl. & L.O. Williams) J.A. Kallunki & J.R. Pirani in Kew Bull.*, 53(2): 314 (1998): Galipea nicaraguensis.
obovatus *(Nees & Mart.) J.A. Kallunki & J.R. Pirani in Kew Bull.*, 53(2): 314 (1998): Diglottis obovata.
odoratissimus *(Lindl.) J.A. Kallunki & J.R. Pirani in Kew Bull.*, 53(2): 314 (1998): Galipea odoratissima.
oppositifolius *J.A. Kallunki in Kew Bull.*, 53(2): 315 (1998)—Brazil Northeast (Bahia).
ovatus *(A. St.-Hil. & Tul.) J.A. Kallunki & J.R. Pirani in Kew Bull.*, 53(2): 317 (1998): Galipea ovata.
paniculatus *(Engl.) J.A. Kallunki & J.R. Pirani in Kew Bull.*, 53(2): 317 (1998): Cusparia paniculata.
pentandrus *(A. St.-Hil.) J.A. Kallunki & J.R. Pirani in Kew Bull.*, 53(2): 317 (1998): Galipea pentandra.
punctatus *J.A. Kallunki in Kew Bull.*, 53(2): 318 (1998)—Brazil Northeast (Bahia).
ramiflorus *(Spruce ex Engl.) J.A. Kallunki & J.R. Pirani in Kew Bull.*, 53(2): 319 (1998): Cusparia ramiflora.
santosii *J.R. Pirani & J.A. Kallunki in Kew Bull.*, 53(2): 319 (1998)—Brazil Northeast (Bahia).
silvestris *(Nees & Mart.) J.A. Kallunki & J.R. Pirani in Kew Bull.*, 53(2): 320 (1998): Lasiostemum silvestre.
sordidus *J.A. Kallunki in Kew Bull.*, 53(2): 320 (1998)—Brazil Southeast (Rio de Janeiro).
toxicarius *(Spruce ex Engl.) J.A. Kallunki & J.R. Pirani in Kew Bull.*, 53(2): 322 (1998): Cusparia toxicaria.
transitionalis *(Rizzini) J.A. Kallunki & J.R. Pirani in Kew Bull.*, 53(2): 323 (1998): Angostura transitionalis.
ucayalinus *(Huber) J.A. Kallunki & J.R. Pirani in Kew Bull.*, 53(2): 323 (1998): Cusparia ucayalina.

CONDALIA (Rhamnac.).
thomasiana *J.L. Fernández-Alonso in Caldasia*, 19(1–2): 102 (1997)—Colombia.

CONDAMINEA (Rubiac.).
elegans *P.G. Delprete in Fl. Neotrop.*, 77: 114 (1999)—Peru.

CONDYLOPODIUM (Compos.).
hyalinifolium *S. Díaz-Piedrahita & G.P. Méndez-Ramírez in Rev. Acad. Colomb. Cienc. Exact. Fis. Nat.*, 90: 46 (2000)—Colombia.

CONGEA (Verbenac.).
villosa *(Roxb.) J.O. Voigt, Hort. Suburb. Calcutt.*: 473 (1845): Roscoea villosa.

CONIOSELINUM (Umbellif.).
papyraceum *(C.B. Clarke) M.G. Pimenov & E.V. Kljuykov in Bot. Zhurn.*, 84(3): 91 (1999): Selinum papyraceum.

CONNARUS (Connarac.).
peltatus *L.L. Forman in Checklist Flow. Pl. & Gymn. Brunei Darussalam (ed. M.J.E. Coode et al.)*: 435 (1996)—Borneo (Brunei).
striatus *J. Nimmo in J. Graham, Cat. Pl. Bombay*: 247 (1839)—India (Maharashtra).

CONNELLIA (Bromeliac.).
nahoumii *E.M.C. Leme in Bromélia*, 3(2): 13 (1996)—Venezuela.

CONOCLINIUM (Compos.).
mayfieldii *T.F. Patterson in Phytologia*, 80(2): 104 (1996)—Mexico Northeast (Tamaulipas, Chihuahua, Durango).

CONOPHYTUM (Aizoac.).
sect. Batrachia *M.R. Opel & S.A. Hammer in Cact. Succ. J. (U.S.A.)*, 71(3): 153 (1999).
bachelorum S.A. Hammer
subsp. sponsaliorum *S.A. Hammer in Piante Grasse*, 15(4 Suppl.): 38 (1995 publ. 1996)—Cape Province.
bruynsii *S.A. Hammer in Cactus & Co.*, 2(4): 5–6 (1998); *cf. Repert. Pl. Succ. (I.O.S.)*, 49: 7 (1998 publ. 1999)—Locality not stated.
chrisocruxum *S.A. Hammer in Cact. Succ. J. (U.S.A.)*, 69(3): 127 (1997)—Cape Province.
chrisolum *S.A. Hammer in Cact. Succ. J. (U.S.A.)*, 69(3): 127 (1997)—Cape Province.
cubicum *P. Pavelka in Kaktusy (Brno)*, 35(Speciál 1): 27 (1999)—Cape Province.
danielii *P. Pavelka in Kaktusy (Brno)*, 35(Speciál 1): 30 (1999)—Cape Province.
devium G.D. Rowley
subsp. stiriiferum *S.A. Hammer & C. Barnhill in Bradleya*, 15: 44 (1997)—Cape Province.
hammeri *G. Williamson & H.C. Kennedy in Cact. Succ. J. (U.S.A.)*, 69(4): 205 (1997)—Cape Province.
hanae *P. Pavelka in Kaktusy (Brno)*, 35(Speciál 1): 29 (1999)—Cape Province.
irmae *S.A. Hammer & C. Barnhill in Bradleya*, 15: 43 (1997)—Cape Province.
jarmilae *J.J. Halda in Acta Mus. Richnov. Sect. Nat.*, 5(4): 164 (1998)—Cape Province.
obscurum N.E. Br.
subsp. barbatum *(L. Bolus) S.A. Hammer in Piante Grasse*, 15(4 Suppl.): 34 (1995 publ. 1996): C. barbatum.
roodiae N.E. Br.
subsp. corrugatum *T.C. Smale in Brit. Cact. Succ. J.*, 18(2): 106 (2000)—Cape Province.
subsp. cylindratum *(Schwantes) T.C. Smale in Brit. Cact. Succ. J.*, 18(2): 106 (2000): C. cylindratum.
subsp. sanguineum *(S.A. Hammer) T.C. Smale in Brit. Cact. Succ. J.*, 18(2): 106 (2000): C. rugosum subsp. sanguineum.
tantillum N.E. Br.
subsp. amicorum *S.A. Hammer & C. Barnhill in Bradleya*, 15: 42 (1997)—Cape Province.
tomasii *J.J. Halda in Acta Mus. Richnov. Sect. Nat.*, 5(5): 184 (1998), as 'tomasi', contrary to Art. 37.5 ICBN (1994)—Cape Province.

CONOPODIUM (Umbellif.).
bunioides (Boiss.) Calest.
var. aranii *S. López Udias & G. Mateo Sanz in An. Jard. Bot. Madrid*, 57(2): 470 (1999 publ. 2000)—Spain.
subsp. atlanticum *(Humbert & Maire) Molero ex J.M. Montserrat in Lagascalia*, 21(1): 249 (1999): Conopodium bunioides var. atlanticum.
var. gredensis *(Pau) S. López Udias & G. Mateo Sanz in An. Jard. Bot. Madrid*, 57(2): 469 (1999 publ. 2000): Butinia bunioides var. gredensis.
majus (Gouan) Loret
subsp. marizianum *(Samp.) S. López Udias & G. Mateo Sanz in An. Jard. Bot. Madrid*, 57(2): 473 (1999 publ. 2000): C. marizianum.

CONOSPERMUM (Proteac.).
sect. Axillaria *(Meisn.) E.M. Bennett in Fl. Australia*, 16: 481 (1995): C. subsect. Axillaria.
subgen. Chilurus *(R. Br.) E.M. Bennett in Fl. Australia*, 16: 481 (1995): C. sect. Chilurus.
sect. Corymbocephala *(Meisn.) E.M. Bennett in Fl. Australia*, 16: 481 (1995), contrary to Art. 22.2 ICBN (1994): C. subsect. Corymbocephala.
sect. Eriostachya *(Meisn.) E.M. Bennett in Fl. Australia*, 16: 482 (1995): C. § Eriostachya.
subsect. Foliosa *E.M. Bennett in Fl. Australia*, 16: 482 (1995).

CONOSPERMUM:-
subgen. Isomerium *(R. Br.) E.M. Bennett in Fl. Australia*, 16: 481 (1995): C. sect. Isomerium.
subsect. Multibracteata *E.M. Bennett in Fl. Australia*, 16: 482 (1995).
sect. Paniculata *(Meisn.) E.M. Bennett in Fl. Australia*, 16: 482 (1995): C. subsect. Paniculata.
subsect. Pedunculata *E.M. Bennett in Fl. Australia*, 16: 482 (1995).
subsect. Ramosa *E.M. Bennett in Fl. Australia*, 16: 482 (1995).
sect. Scaposa *(Meisn.) E.M. Bennett in Fl. Australia*, 16: 482 (1995): C. subsect. Scaposa.
subsect. Sericea *E.M. Bennett in Fl. Australia*, 16: 482 (1995).
sect. Teretifolia *E.M. Bennett in Fl. Australia*, 16: 482 (1995).
acerosum Lindl.
subsp. hirsutum *E.M. Bennett in Fl. Australia*, 16: 482 (1995)—Western Australia.
amoenum Meisn.
subsp. cuneatum *E.M. Bennett in Fl. Australia*, 16: 483 (1995)—Western Australia.
boreale *E.M. Bennett in Fl. Australia*, 16: 483 (1995)—Western Australia.
subsp. ascendens *E.M. Bennett in Fl. Australia*, 16: 483 (1995)—Western Australia.
caeruleum R. Br.
subsp. contortum *E.M. Bennett in Fl. Australia*, 16: 483 (1995)—Western Australia.
subsp. debile *(Kippist ex Meisn.) E.M. Bennett in Fl. Australia*, 16: 483 (1995): C. debile.
subsp. marginatum *(Meisn.) E.M. Bennett in Fl. Australia*, 16: 483 (1995): C. marginatum.
subsp. oblanceolatum *E.M. Bennett in Fl. Australia*, 16: 483 (1995)—Western Australia.
subsp. spathulatum *(Benth.) E.M. Bennett in Fl. Australia*, 16: 483 (1995): C. caeruleum var. spathulatum.
canaliculatum Meisn.
subsp. apiculatum *E.M. Bennett in Fl. Australia*, 16: 483 (1995)—Western Australia.
capitatum R. Br.
subsp. glabratum *E.M. Bennett in Fl. Australia*, 16: 483 (1995)—Western Australia.
subsp. velutinum *E.M. Bennett in Fl. Australia*, 16: 484 (1995)—Western Australia.
cinereum *E.M. Bennett in Fl. Australia*, 16: 484 (1995)—Western Australia.
coerulescens F. Muell.
subsp. adpressum *E.M. Bennett in Fl. Australia*, 16: 484 (1995)—Western Australia.
subsp. dorrienii *(Domin) E.M. Bennett in Fl. Australia*, 16: 484 (1995): C. dorrienii.
densiflorum Lindl.
subsp. unicephalatum *E.M. Bennett in Fl. Australia*, 16: 484 (1995)—Western Australia.
elongatum *E.M. Bennett in Fl. Australia*, 16: 484 (1995)—Victoria.
filifolium Meisn.
subsp. australe *E.M. Bennett in Fl. Australia*, 16: 484 (1995)—Western Australia.
flexuosum R. Br.
subsp. laevigatum *(Meisn.) E.M. Bennett in Fl. Australia*, 16: 484 (1995): C. flexuosum var. laevigatum.
galeatum *E.M. Bennett in Fl. Australia*, 16: 485 (1995)—Western Australia.
hookeri *(Meisn.) E.M. Bennett in Fl. Australia*, 16: 485 (1995): C. taxifolium var. hookeri.
leianthum (Benth.) Diels
subsp. orientale *E.M. Bennett in Fl. Australia*, 16: 485 (1995)—Western Australia.
microflorum *E.M. Bennett in Fl. Australia*, 16: 485 (1995)—Western Australia.
multispicatum *E.M. Bennett in Fl. Australia*, 16: 485 (1995)—Western Australia.
paniculatum *E.M. Bennett in Fl. Australia*, 16: 485 (1995)—Western Australia.
quadripetalum *E.M. Bennett in Fl. Australia*, 16: 486 (1995)—Western Australia.
sigmoideum *E.M. Bennett in Fl. Australia*, 16: 486 (1995)—Western Australia.
spectabile *E.M. Bennett in Fl. Australia*, 16: 486 (1995)—Western Australia.
stoechadis Endl.
subsp. sclerophyllum *(Lindl.) E.M. Bennett in Fl. Australia*, 16: 486 (1995)—C. sclerophyllum.

CONOSPERMUM:-
unilaterale *E.M. Bennett in Fl. Australia,* 16: 486 (1995)—Western Australia.
wycherleyi *E.M. Bennett in Fl. Australia,* 16: 486 (1995)—Western Australia.
subsp. glabrum *E.M. Bennett in Fl. Australia,* 16: 486 (1995)—Western Australia.

CONSOLEA (Cactac.).
microcarpa *(K. Schum.) E.F. Anderson in Cact. Succ. J. (U.S.A.),* 71(6): 324 (1999), nom. illegit., see Art. 52.1 ICBN (1994): Opuntia microcarpa.
moniliformis (L.) A. Berger
subsp. guantanamana *A.E. Areces-Mallea in Brittonia,* 48(2): 225 (1996)—Cuba.
nashii (Britton) A. Berger
subsp. gibarensis *A.E. Areces-Mallea in Brittonia,* 48(2): 231 (1996)—Cuba.
picardae *(Urb.) A.E. Areces-Mallea in Cact. Succ. J. (U.S.A.),* 72(1): 42 (2000): Opuntia picardae.

CONSOLIDA (Ranunculac.).
ajacis *(L.) R. Schrödinger in Abh. K.K. Zool.-Bot. Ges. Wien,* 4(5): 7, 62 (1909): Delphinium ajacis.
anthoroidea *(Boiss.) R. Schrödinger in Abh. K.K. Zool.-Bot. Ges. Wien,* 4(5): 10, 62 (1909): Delphinium anthoroideum.
armeniaca *(Stapf ex Huth) R. Schrödinger in Abh. K.K. Zool.-Bot. Ges. Wien,* 4(5): 9, 62 (1909): Delphinium armeniacum.
axilliflora *(DC.) R. Schrödinger in Abh. K.K. Zool.-Bot. Ges. Wien,* 4(5): 8, 62 (1909): Delphinium axilliflorum.
barbata *(Bunge) R. Schrödinger in Abh. K.K. Zool.-Bot. Ges. Wien,* 4(5): 16, 62 (1909): Delphinium barbatum.
divaricata *(Ledeb.) R. Schrödinger in Abh. K.K. Zool.-Bot. Ges. Wien,* 4(5): 62 (1909): Delphinium divaricatum.
fasciola *S. Mobayen, Fl. Iran Vasc. Pl.,* 3: 43 (1985), without latin descr. or type—Iran.
hispanica *(Willk. ex Costa) R. Schrödinger in Abh. K. K. Zool.-Bot. Ges. Wien,* 4(5): 62 (1909): Delphinium hispanicum.
oliveriana *(DC.) R. Schrödinger in Abh. K.K. Zool.-Bot. Ges. Wien,* 4(5): 27, 62 (1909): Delphinium oliverianum.
orientalis *(J. Gay) R. Schrödinger in Abh. K.K. Zool.-Bot. Ges. Wien,* 4(5): 27, 62 (1909): Delphinium orientale.
persica *(Boiss.) R. Schrödinger in Abh. K.K. Zool.-Bot. Ges. Wien,* 4(5): 17, 62 (1909): Delphinium persicum.
pubescens (DC.) Soó
var. dissitiflora *(Coss.) A. Dubuis in Soc. Ech. Pl. Vasc. Eur. Bassin Médit. Bull.,* 20: 45 (1985): Delphinium pubescens var. dissitiflorum.
pusilla *(Labill.) R. Schrödinger in Abh. K.K. Zool.-Bot. Ges. Wien,* 4(5): 9, 62 (1909): Delphinium pusillum.
raveyi *(Boiss.) R. Schrödinger in Abh. K.K. Zool.-Bot. Ges. Wien,* 4(5): 7, 62 (1909): Delphinium raveyi.
regalis Gray
var. glanduligera *(Peterm.) W. Starmühler in Hladnikia,* 6: 43 (1996): Delphinium consolida var. glanduligerum.
var. sparsiflora *(Vis.) W. Starmühler in Hladnikia,* 6: 43 (1996): Delphinium consolida var. sparsiflorum.
thirkeana *(Boiss.) R. Schrödinger in Abh. K.K. Zool.-Bot. Ges. Wien,* 4(5): 10, 62 (1909): Delphinium thirkeanum.
tomentosa *(Boiss.) R. Schrödinger in Abh. K.K. Zool.-Bot. Ges. Wien,* 4(5): 9, 62 (1909): Delphinium tomentosum.
trigonelloides *(Boiss.) R. Schrödinger in Abh. K.K. Zool.-Bot. Ges. Wien,* 4(5): 9, 62 (1909): Delphinium trigonelloides.

CONSTANCEA *B.G. Baldwin in Madroño,* 46(3): 159 (1999 publ. 2000). *COMPOSITAE*
nevinii *(A. Gray) B.G. Baldwin in Madroño,* 46(3): 160 (1999 publ. 2000): Eriophyllum nevinii.

CONVOLVULUS (Convolvulac.).
afanassievii *A.N. Luferov in Komarovia,* 1: 59 (1999)—Uzbekistan.

CONVOLVULUS:-
althaeoides L.
f. austrovalentinus *O. de Bolòs & J. Vigo, Fl. Països Catalans,* 3: 148 (1995)—Spain.
bidrensis *Sebsebe Demissew in Kew Bull.,* 54(1): 67 (1999)—Ethiopia.
gilbertii *Sebsebe Demissew in Kew Bull.,* 54(1): 74 (1999)—Ethiopia.
grigorjevii *R.V. Kamelin in Bot. Zhurn.,* 82(7): 124 (1997)—Kirgizstan.
leptosepalus *M. Gandoger, Dec. Pl. Nov.,* 2: 15 (1876)—France.
longipilis *M. Gandoger, Dec. Pl. Nov.,* 2: 14 (1876)—France.
ruspolii Dammer ex Hallier f.
var. pilosa *Sebsebe Demissew in Kew Bull.,* 54(1): 71 (1999)—Ethiopia, Somalia.
villiflorus *M. Gandoger, Dec. Pl. Nov.,* 2: 15 (1876)—Hungary.
vollesenii *Sebsebe Demissew in Kew Bull.,* 54(1): 69 (1999)—Ethiopia.

CONYZA (Compos.).
groegeri *V.M. Badillo in Ernstia,* 10(1): 5 (2000)—Venezuela.
laevigata *(Rich.) J.F. Pruski in Brittonia,* 50(4): 475 (1998): Erigeron laevigatus.
newii *Oliv. & Hiern in Fl. Trop. Afr.,* 3: 317 (1877), in orig. I.K. as 'neuri' in error—Tanzania.†

COOKTOWNIA *D.L. Jones in Austrobaileya,* 5(1): 74 (1997). *ORCHIDACEAE*
robertsii *D.L. Jones in Austrobaileya,* 5(1): 74 (1997)—Queensland.

COPIAPOA (Cactac.).
calderana
subsp. longistaminea *(F. Ritter) N.P. Taylor in Cactaceae Consensus Initiatives,* 6: 15 (1998): C. longistaminea.
cinerea
subsp. dealbata *(F. Ritter) R. Slaba in Kaktusy (Brno),* 33(speciál): 3 (1997): C. dealbata.
subsp. gigantea *(Backeb.) R. Slaba in Kaktusy (Brno),* 33(speciál): 3 (1997): Copiapoa gigantea.
subsp. haseltoniana *(Backeb.) N.P. Taylor in Cactaceae Consensus Initiatives,* 3: 8 (1997): C. haseltoniana.
var. krainziana *(F. Ritter) G.J. Charles in Brit. Cact. Succ. J.,* 16(1): 15 (1998): C. krainziana.
subsp. krainziana *(F. Ritter) R. Slaba in Kaktusy (Brno),* 33(speciál): 3 (1997): Copiapoa krainziana.
var. longistaminea *(F. Ritter) R. Slaba in Kaktusy (Brno),* 33(speciál): 3 (1997): C. longistaminea.
var. longistaminea *(F. Ritter) G.J. Charles in Brit. Cact. Succ. J.,* 16(1): 15 (1998), isonym; *see Slaba in Kaktusy (Brno),* 33(speciál): 3 (1997): C. longistaminea.
var. tenebrosa *(F. Ritter) R. Slaba in Kaktusy (Brno),* 33(speciál): 3 (1997), isonym; *see A.E. Hoffm., Cactaceas en la flora silvestre de Chile:* 106 (1989): Copiapoa tenebrosa.
humilis
var. tenuissima *(F. Ritter) G.J. Charles in Brit. Cact. Succ. J.,* 16(1): 15 (1998): C. tenuissima.
var. tocopillana *(F. Ritter) G.J. Charles in Brit. Cact. Succ. J.,* 16(1): 15 (1998): C. tocopillana.
var. variispinata *(F. Ritter) G. Charles, Copiapoa (Cactus File Handbook, 4):* 48 (1998): C. variispinata.
hypogaea
var. montana *(F. Ritter) G.J. Charles in Brit. Cact. Succ. J.,* 16(1): 15 (1998): C. montana.
knizei *(Kníže & Říha) J.J. Halda in Cactaceae etc.,* 8(4): 134 (1998); *cf. Repert. Pl. Succ. (I.O.S.),* 49: 15 (1998 publ. 1999): Cintia knizei.
montana
subsp. grandiflora *(F. Ritter) N.P. Taylor in Cactaceae Consensus Initiatives,* 3: 8 (1997): C. grandiflora.

COPROSMA (Rubiac.).
sect. Leptostigma *(Arn.) M.J. Heads in Candollea,* 51(2): 389 (1996): gen. Leptostigma.
subgen. Lucidae *C.J. Webb in New Zealand J. Bot.,* 34(2): 282 (1996).

COPROSMA:-
sect. Lucidae *C.J. Webb in New Zealand J. Bot.,* 34(2): 282 (1996).
sect. Moorei *M.J. Heads in Candollea,* 51(2): 390 (1996).
sect. Nertera *(Banks & Sol. ex Gaertn.) M.J. Heads in Candollea,* 51(2): 386 (1996): gen. Nertera.
arnottiana *(Walp.) M.J. Heads in Candollea,* 51(2): 390 (1996): Leptostigma arnottianum.
balfouriana *(Cockayne) M.J. Heads in Candollea,* 51(2): 388 (1996): Nertera balfouriana.
boliviana *M.J. Heads in Candollea,* 51(2): 390 (1996), nom. nov.: Mitchella pilosa *Benth.*
decurva *M.J. Heads in New Zealand J. Bot.,* 36(1): 65 (1998)—New Zealand South I., New Zealand North I.
dentata *(Elmer) M.J. Heads in Candollea,* 51(2): 389 (1996): Nertera dentata.
dichondrifolia *(A. Cunn.) M.J. Heads in Candollea,* 51(2): 389 (1996): Geophila dichondrifolia.
granadensis *(L.f.) M.J. Heads in Candollea,* 51(2): 388 (1996): Gomozia granadensis.
holmboei *(Christoph.) M.J. Heads in Candollea,* 51(2): 389 (1996): Nertera holmboei.
longiflora *(Standl.) M.J. Heads in Candollea,* 51(2): 390 (1996): Corynula longiflora.
nertera F. Muell.
var. papuana *(Valeton) M.J. Heads in Candollea,* 51(2): 388 (1996): Nertera depressa var. papuana.
nigricarpa *(Hayata) M.J. Heads in Candollea,* 51(2): 389 (1996): Nertera nigricarpa.
patrickii *M.J. Heads in Candollea,* 51(2): 389 (1996), nom. nov.: Nertera ciliata *Kirk.*
pedicellata *B.P.J. Molloy, P.J. de Lange & B.D. Clarkson in New Zealand J. Bot.,* 37(3): 384 (1999)—New Zealand South I, New Zealand North I.
philipsonii *M.J. Heads in Candollea,* 51(2): 388 (1996), nom. nov.: Nertera cunninghamii *Hook. f.*
pseudocuneata *P.J. Garnock-Jones & R. Elder in New Zealand J. Bot.,* 34(1): 140 (1996)—New Zealand North I.
scapanioides *(Lange) M.J. Heads in Candollea,* 51(2): 389 (1996): Nertera scapanioides.
sinensis *(Hemsl.) M.J. Heads in Candollea,* 51(2): 389 (1996): Nertera sinensis.
taiwaniana *(Masam.) M.J. Heads in Candollea,* 51(2): 389 (1996): Nertera taiwaniana.
villosa *(B.H. Macmill. & R. Mason) M.J. Heads in Candollea,* 51(2): 389 (1996): Nertera villosa.
weberbaueri *(Fosberg) M.J. Heads in Candollea,* 51(2): 390 (1996): Leptostigma weberbaueri.

COPTIS (Ranunculac.).
japonica (Thunb.) Makino
var. anemonifolia *(Siebold & Zucc.) H. Ohba in J. Jap. Bot.,* 75(6): 369 (2000): Coptis anemonifolia.

COPTOPHYLLUM (Rubiac.).
sylvestre *(Ridl.) I.M. Turner in Novon,* 6(2): 222 (1996): Pomazota sylvestris.

CORALLORHIZA (Orchidac.).
bentleyi *J.V. Freudenstein in Novon,* 9(4): 511 (1999)—West Virginia.
maculata (Raf.) Raf.
var. mexicana *(Lindl.) J.V. Freudenstein in Harvard Pap. Bot.,* 10: 45 (1997): C. mexicana.
odontorhiza (Willd.) Poir.
var. pringlei *(Greenm.) J.V. Freudenstein in Harvard Pap. Bot.,* 10: 24 (1997): C. pringlei.
f. pringlei *(Greenm.) J.V. Freudenstein in Harvard Pap. Bot.,* 10: 24 (1997): C. pringlei.
f. radia *J.V. Freudenstein in Harvard Pap. Bot.,* 10: 26 (1997)—Mexico Southwest (Guerrero, Michoacan, Oaxaca), Mexico Southeast (Chiapas), Mexico Central (Morelos), Guatemala.
striata Lindl.
var. involuta *(Greenm.) J.V. Freudenstein in Harvard Pap. Bot.,* 10: 15 (1997): C. involuta.
trifida Châtel.
subsp. virescens *(Drejer) B. Løjtnant in Flora Fauna (Århus),* 101(2): 71 (1996): C. virescens.

CORCHORUS (Tiliac.).
neocaledonicus Schltr.
var. estellatus *C. Tirel in Bull. Mus. Nation. Hist. Nat., B, Adansonia,* Ser. 4, 18(1–2): 42 (1996)—New Caledonia.

CORDIA (Boraginac.).
andreana *J. Estrada Sánchez in Fl. Colombia,* 14: 76 (1995)—Colombia.
barbata *J. Estrada Sánchez in Fl. Colombia,* 14: 138 (1995)—Colombia.
bullata
var. globosa *(Jacq.) R. Govaerts, World Checklist Seed Plants,* 3(1): 20 (1999): Varronia globosa.
bullulata *Killip ex J. Estrada Sánchez & García-Barr. in Fl. Colombia,* 14: 70 (1995)—Colombia.
fuertesii *J. Estrada Sánchez in Fl. Colombia,* 14: 125 (1995)—Colombia.
linearicalycina *Killip ex J. Estrada Sánchez in Fl. Colombia,* 14: 119 (1995)—Colombia.
llanense *Killip ex J. Estrada Sánchez in Fl. Colombia,* 14: 122 (1995)—Colombia.
platystachya *Killip ex J. Estrada Sánchez in Fl. Colombia,* 14: 111 (1995)—Colombia.
ramirezii *J. Estrada Sánchez in Fl. Colombia,* 14: 147 (1995)—Colombia.
resinosa *J. Estrada Sánchez in Fl. Colombia,* 14: 133 (1995)—Colombia.
rhombifolia *J. Estrada Sánchez in Fl. Colombia,* 14: 114 (1995)—Colombia.
rubescens *J. Estrada Sánchez in Fl. Colombia,* 14: 143 (1995)—Colombia.
wurdackiana *C. Feuillet in Biollania,* ed. esp. 6 (J.J. Wurdack Festschr.): 331 (1997)—Guyana.

CORDYLINE (Agavac.).
fruticosa (L.) A. Chev.
var. ferrea *(Baker) R.R. Fernandez, Trees of Mumbai (Bombay):* 259 (1999): C. terminalis var. ferrea.

COREOCARPUS (Compos.).
arizonicus (A. Gray) S.F. Blake
var. sanpedroensis *(E.B. Sm.) B.L. Turner in Phytologia,* 80(2): 136 (1996): C. sanpedroensis.
sonoranus Sherff
var. libranus *B.L. Turner in Phytologia,* 80(2): 137 (1996)—Mexico Northwest (Sonora).

COREOPSIS (Compos.).
bolanosana *J.L. Panero & J.L. Villaseñor in Brittonia,* 51(1): 89 (1999)—Mexico Southwest (Jalisco).
crawfordii *Mesfin Tadesse in Nordic J. Bot.,* 19(2): 235 (1999)—Mexico Northeast (Querétaro).
mutica A.DC.
var. miahuatlana *B.L. Turner in Phytologia,* 80(4): 293 (1996)—Mexico Southwest (Oaxaca).

CORIFLORA *W.A. Weber in Phytologia,* 79(2): 66 (1995 publ. 1996), nom. nov.: gen. Viorna *Spach.* (Ranunculac.).
addisonii *(Britton ex Vail) W.A. Weber in Phytologia,* 79(2): 66 (1995 publ. 1996): Clematis addisonii.
albicoma *(Wherry) W.A. Weber in Phytologia,* 79(2): 66 (1995 publ. 1996): Clematis albicoma.
baldwinii *(Torr. & A. Gray) W.A. Weber in Phytologia,* 79(2): 66 (1995 publ. 1996): Clematis baldwinii.
beadlei *(Small) W.A. Weber in Phytologia,* 79(2): 66 (1995 publ. 1996): Viorna beadlei.
bigelovii *(Torr.) W.A. Weber in Phytologia,* 79(2): 66 (1995 publ. 1996): Clematis bigelovii.
crispa *(L.) W.A. Weber in Phytologia,* 79(2): 66 (1995 publ. 1996): Clematis crispa.
fremontii *(J. James) W.A. Weber in Phytologia,* 79(2): 66 (1995 publ. 1996): Clematis ochroleuca var. fremontii.
gattingeri *(Small) W.A. Weber in Phytologia,* 79(2): 66 (1995 publ. 1996): Clematis gattingeri.
glaucophylla *(Small) W.A. Weber in Phytologia,* 79(2): 66 (1995 publ. 1996) Clematis glaucophylla.
hirsutissima *(Pursh) W.A. Weber in Phytologia,* 79(2): 66 (1995 publ. 1996) Clematis hirsutissima.

CORIFLORA:-
integrifolia *(L.) W.A. Weber in Phytologia,* 79(2): 66 (1995 publ. 1996): Clematis integrifolia.
morefieldii *(Kral) W.A. Weber in Phytologia,* 79(2): 66 (1995 publ. 1996): Clematis morefieldii.
ochroleuca *(Aiton) W.A. Weber in Phytologia,* 79(2): 66 (1995 publ. 1996): Clematis ochroleuca.
palmeri *(Rose) W.A. Weber in Phytologia,* 79(2): 66 (1995 publ. 1996): Clematis palmeri.
pitcheri *(Torr. & A. Gray) W.A. Weber in Phytologia,* 79(2): 67 (1995 publ. 1996): Clematis pitcheri.
reticulata *(Walter) W.A. Weber in Phytologia,* 79(2): 67 (1995 publ. 1996): Clematis reticulata.
scottii *(Porter) W.A. Weber in Phytologia,* 79(2): 67 (1995 publ. 1996): Clematis scottii.
texensis *(Buckley) W.A. Weber in Phytologia,* 79(2): 67 (1995 publ. 1996): Clematis texensis.
versicolor *(Small ex Britton) W.A. Weber in Phytologia,* 79(2): 67 (1995 publ. 1996): Clematis versicolor.
viorna *(L.) W.A. Weber in Phytologia,* 79(2): 67 (1995 publ. 1996): Clematis viorna.
viticaulis *(E.S. Steele) W.A. Weber in Phytologia,* 79(2): 67 (1995 publ. 1996): Clematis viticaulis.

CORIS (Primulac.).
monspeliaca
subsp. rivasiana *F. Alcaraz, T.E. Díaz, S. Rivas-Martínez & P. Sánchez-Gómez in Itinera Geobot.,* 2: 108 (1989)—Spain.
subsp. somaliensis *R. Govaerts, World Checklist Seed Plants,* 3(1): 20 (1999), nom. nov.: C. monspeliensis var. longinqua *Airy Shaw.*

CORISPERMUM (Chenopodiac.).
subsect. Aralocaspica *S.L. Mosyakin in Thaiszia,* 7(1): 12 (1997).
subsect. Canescentia *(Klokov) S.L. Mosyakin in Thaiszia,* 7(1): 10 (1997): C. ser. Canescentia.
subsect. Crassifolia *S.L. Mosyakin in Thaiszia,* 7(1): 11 (1997).
subsect. Lehmanniana *S.L. Mosyakin in Thaiszia,* 7(1): 12 (1997).
subsect. Nitida *(Klokov) S.L. Mosyakin in Thaiszia,* 7(1): 10 (1997): C. ser. Nitida.
subsect. Pallasiana *S.L. Mosyakin in Thaiszia,* 7(1): 13 (1997).
subsect. Platyptera *S.L. Mosyakin in Thaiszia,* 7(1): 12 (1997).
americanum *(Nutt.) T. Nuttall in Trans. Amer. Philos. Soc., n.s.,* 5: 165 (1835): C. hyssopifolium β americanum.
var. rydbergii *S.L. Mosyakin in Novon,* 5(4): 351 (1995)—Mexico Northeast (Chihuahua), Colorado.
bardunovii *Popov ex M.N. Lomonosova in Fl. Sibir. (Salicac.-Amaranthac.),* 5: 168 (1992), without type—Tuva.
hookeri *S.L. Mosyakin in Novon,* 5(4): 349 (1995)—Saskatchewan, Alberta, British Columbia, Manitoba, Northwest Territories (Mackenzie), Ontario.
var. pseudodeclinatum *S.L. Mosyakin in Novon,* 5(4): 350 (1995)—British Columbia.
× klokovii *S.L. Mosyakin in Ukr. Bot. Zhurn.,* 52(4): 460 (1995), as 'nom. nov.'—Ukraina.(C. marschallii × nitidum).
× modestii *S.L. Mosyakin in Ukr. Bot. Zhurn.,* 52(4): 460 (1995), as 'nom. nov.', without type—Germany, Ukraina.(C. marschallii × pallasii).
navicula *S.L. Mosyakin in Novon,* 5(4): 349 (1995)—Colorado.
nitidum Kit. ex Schult.*subsp.*microspermum *(Host) S.L. Mosyakin in Ukr. Bot. Zhurn.,* 52(4): 459 (1995): C. microspermum.
ochotense Ignatov
var. alaskanum *S.L. Mosyakin in Novon,* 5(4): 351 (1995)—Alaska.
pacificum *S.L. Mosyakin in Novon,* 5(4): 345 (1995)—Washington, Idaho, Oregon.
pallasii Steven
subsp. membranaceum *(Iljin) N.N. Tzvelev in Novosti Sist. Vyssh. Rast.,* 32: 183 (2000): C. membranaceum.
pallidum *S.L. Mosyakin in Novon,* 5(4): 347 (1995)—Washington.

CORISPERMUM:-
stauntonii Moq.
subsp. elongatum *(Bunge) V.N. Voroschilov in A.K. Skvortsov (ed.), Florist. issl. v razn. raionakh SSSR:* 165 (1985): C. elongatum.
subsp. sibiricum *(Iljin) V.N. Voroschilov in A.K. Skvortsov (ed.), Florist. issl. v razn. raionakh SSSR:* 165 (1985): C. sibiricum.
welshii *S.L. Mosyakin in Novon,* 5(4): 348 (1995)—Utah, Wyoming.

CORNERIA *A.V.F.Ch. Bobrov & A.P. Melikian in Bot. Zhurn.,* 85(7): 62 (2000). *PODOCARP-ACEAE*
comosa *(Corner) A.V.F.Ch. Bobrov & A.P. Melikian in Bot. Zhurn.,* 85(7): 62 (2000): Dacrydium comosum.
cornwalliana *(de Laub.) A.V.F.Ch. Bobrov & A.P. Melikian in Bot. Zhurn.,* 85(7): 62 (2000): Dacrydium cornwallianum.
elata *(Roxb.) A.V.F.Ch. Bobrov & A.P. Melikian in Bot. Zhurn.,* 85(7): 62 (2000): Juniperus elata.
ericoides *(de Laub.) A.V.F.Ch. Bobrov & A.P. Melikian in Bot. Zhurn.,* 85(7): 62 (2000), as 'ericoidea': Dacrydium ericoides.
gibbsiae *(Stapf) A.V.F.Ch. Bobrov & A.P. Melikian in Bot. Zhurn.,* 85(7): 62 (2000), as 'gibbsii': Dacrydium gibbsiae.
gracilis *(de Laub.) A.V.F.Ch. Bobrov & A.P. Melikian in Bot. Zhurn.,* 85(7): 62 (2000), as 'gracile': Dacrydium gracile.
leptophylla *(Wasscher) A.V.F.Ch. Bobrov & A.P. Melikian in Bot. Zhurn.,* 85(7): 62 (2000): Podocarpus leptophyllus.
magna *(de Laub.) A.V.F.Ch. Bobrov & A.P. Melikian in Bot. Zhurn.,* 85(7): 62 (2000): Dacrydium magnum.
media *(de Laub.) A.V.F.Ch. Bobrov & A.P. Melikian in Bot. Zhurn.,* 85(7): 62 (2000): Dacrydium medium.
nausoriensis *(de Laub.) A.V.F.Ch. Bobrov & A.P. Melikian in Bot. Zhurn.,* 85(7): 62 (2000): Dacrydium nausoriense.
nidula *(de Laub.) A.V.F.Ch. Bobrov & A.P. Melikian in Bot. Zhurn.,* 85(7): 62 (2000): Dacrydium nidulum.
novoguineensis *(Gibbs) A.V.F.Ch. Bobrov & A.P. Melikian in Bot. Zhurn.,* 85(7): 62 (2000): Dacrydium novoguineense.
pectinata *(de Laub.) A.V.F.Ch. Bobrov & A.P. Melikian in Bot. Zhurn.,* 85(7): 62 (2000): Dacrydium pectinatum.
pierrei *(Hickel) A.V.F.Ch. Bobrov & A.P. Melikian in Bot. Zhurn.,* 85(7): 62 (2000): Dacrydium pierrei.
spathoides *(de Laub.) A.V.F.Ch. Bobrov & A.P. Melikian in Bot. Zhurn.,* 85(7): 62 (2000), as 'spathoidea': Dacrydium spathoides.
xanthandra *(Pilg.) A.V.F.Ch. Bobrov & A.P. Melikian in Bot. Zhurn.,* 85(7): 62 (2000): Dacrydium xanthandrum.

CORNUKAEMPFERIA *J. Mood & K. Larsen in Nat. Hist. Bull. Siam Soc.,* 45(2): 217 (1997). *ZINGIBERACEAE*
aurantiflora *J. Mood & K. Larsen in Nat. Hist. Bull. Siam Soc.,* 45(2): 218 (1997).Thailand.
longipetiolata *J. Mood & K. Larsen in New Plantsman,* 6(4): 198 (1999)—Thailand.

CORNUS (Cornac.).
sect. Umbellicrania *Z.E. Murrell in Syst. Bot.,* 21(3): 286 (1996 publ. 1997).
kousa F. Buerger & Hance
f. minor *(Nakai) Y.N. Lee, Fl. Korea:* 1159 (1996), with incorrect basionym ref.: Basionym not stated.
f. viridis *(Nakai) Y.N. Lee, Fl. Korea:* 1159 (1996), with incorrect basionym ref.: Basionym not stated.
yaeyamensis *S. Hatusima in S. Hatusima & T. Amano, Fl. Ryukyus, S. of Amami Is., ed. 2:* 160 (1994), without latin descr. or type—Locality not stated.

CORONILLA (Legumin.).
repanda (Poir.) Guss.
subsp. montserratii *(O. Bolòs & Vigo) A.M. Romo, Fl. Silvestres Baleares:* 154 (1994): C. repanda var. montserratii.

CORONILLA:-
varia L.
subsp. hirta *(Bunge ex Boiss.) K.H. Rechinger in Fl. Iran.*, 157: 361 (1984): C. varia var. hirta.

CORREA (Rutac.).
backhouseana Hook.
var. coriacea *(Paul G. Wilson) Paul G. Wilson in Nuytsia*, 12(1): 90 (1998): C. reflexa var. coriacea.
var. orbicularis *Paul G. Wilson in Nuytsia*, 12(1): 91 (1998)—South Australia.
calycina J.M. Black
var. halmaturorum *Paul G. Wilson in Nuytsia*, 12(1): 92 (1998)—South Australia.
eburnea *Paul G. Wilson in Nuytsia*, 12(1): 93 (1998)—South Australia.
glabra Lindl.
var. leucoclada *(Lindl.) Paul G. Wilson in Nuytsia*, 12(1): 95 (1998): C. leucoclada.
var. turnbullii *(Ashby) Paul G. Wilson in Nuytsia*, 12(1): 95 (1998): C. turnbullii.
lawrenceana Hook.
var. grampiana *Paul G. Wilson in Nuytsia*, 12(1): 98 (1998)—Victoria.
var. latrobeana *(F. Muell. ex Hannaford) Paul G. Wilson in Nuytsia*, 12(1): 97 (1998): C. latrobeana.
reflexa (Labill.) Vent.
var. angustifolia *Paul G. Wilson in Nuytsia*, 12(1): 102 (1998)—Victoria.
var. insularis *Paul G. Wilson in Nuytsia*, 12(1): 103 (1998)—South Australia.
var. lobata *Paul G. Wilson in Nuytsia*, 12(1): 100 (1998)—Victoria.
var. scabridula *Paul G. Wilson in Nuytsia*, 12(1): 101 (1998)—South Australia, Victoria.
var. speciosa *(Donn ex Andrews) Paul G. Wilson in Nuytsia*, 12(1): 101 (1998): C. speciosa.
speciosa *Donn ex H.C. Andrews, Bot. Repos.*, 10: t. 653 (1812)—Australia.

CORRIGIOLA (Molluginac.).
sect. Paniculatae *S.S. Ikonnikov in Bot. Zhurn.*, 80(12): 117 (1995 publ. 1996).
africana *(Turrill) S.S. Ikonnikov in Bot. Zhurn.*, 80(12): 116 (1995 publ. 1996): C. litoralis subsp. africana.
perez-larae *(Chaudhri, Muñoz Garm. & Pedrol) S.S. Ikonnikov in Bot. Zhurn.*, 80(12): 116 (1995 publ. 1996): C. litoralis subsp. perez-larae.
poeppigii *(Chaudhri) S.S. Ikonnikov in Bot. Zhurn.*, 80(12): 117 (1995 publ. 1996): C. squamosa var. poeppigii.
vulcanica *S.S. Ikonnikov in Bot. Zhurn.*, 80(12): 115 (1995 publ. 1996)—Italy.

CORRYOCACTUS (Cactac.).
brevistylus
subsp. puquiensis *(Rauh & Backeb.) C. Ostolaza Nano in Cactaceae Consensus Initiatives*, 6: 8 (1998): C. puquiensis.
chachapoyensis *Ochoa & Backeb. ex D.R. Hunt in Cactaceae Consensus Initiatives*, 7: 32 (1999)—Peru.

CORSIA (Corsiac.).
purpurata L.O. Williams
var. wiakabui *W. Takeuchi & J.J. Pipoly in Sida*, 18(1): 164 (1998)—Bismarck Archipelago.

CORSIOPSIS *D.X. Zhang, R.M.K. Saunders & C.M. Hu in Syst. Bot.*, 24(3): 313 (1999). *CORSIACEAE*
chinensis *D.X. Zhang, R.M.K. Saunders & C.M. Hu in Syst. Bot.*, 24(3): 313 (1999)—China Southeast (Guangdong).

CORTADERIA (Gramin.).
boliviensis *M. Lyle in Novon*, 6(1): 72 (1996), nom. nov.: C. bifida var. grandiflora *Henrard*.

CORTIELLA (Umbellif.).
cortioides *(C. Norman) M.F. Watson in Edinburgh J. Bot.*, 53(1): 130 (1996): Selinum cortioides.
lamondiana *F. Fullarton & M.F. Watson in Edinburgh J. Bot.*, 53(1): 130 (1996)—Nepal.

CORYANTHES (Orchidac.).
cavalcantei *M.F.F. da Silva & A.T. de Oliveira in Bol. CAOB*, 25: 21 (1996)—Brazil North (Pará).
elianae *M.F.F. da Silva & A.T. de Oliveira in Bol. Mus. Paraense Emílio Goeldi, Bot.*, 14(1): 44 (1998 publ. 1999)—Brazil North (Amazonas).
gomezii *G.A. Romero & G. Gerlach in Harvard Pap. Bot.*, 5(1): 180 (2000)—Venezuela.
miuaensis *M.F.F. da Silva & A.T. de Oliveira in Bol. Mus. Paraense Emílio Goeldi, Bot.*, 14(1): 47 (1998 publ. 1999)—Brazil North (Amazonas).
thivii *M. Kropf & H.-G. Seeger in Orchidee*, 50(6): 653 (1999)—Bolivia.

CORYBAS (Orchidac.).
iridescens *J.B. Irwin & B.P.J. Molloy in New Zealand J. Bot.*, 34(1): 1 (1996)—New Zealand North I.
karoensis *J.B. Comber & J. Dransfield in Malayan Orchid Rev.*, 29: 71 (1995)—Sumatera.
papa *B.P.J. Molloy & J.B. Irwin in New Zealand J. Bot.*, 34(1): 5 (1996)—New Zealand North I.
scutelliferus *J.B. Comber & J. Dransfield in Malayan Orchid Rev.*, 29: 74 (1995)—Sumatera.
stenotribonos *J.B. Comber & J. Dransfield in Malayan Orchid Rev.*, 29: 73 (1995)—Sumatera.

CORYDALIS (Papaverac.).
ser. Adianta *M. Lidén in Willdenowia*, 26(1–2): 25 (1996).
sect. Asterostigmata *Fedde ex C.Y. Wu & H. Chuang in Fl. Reipubl. Popul. Sin.*, 32: 541, 103 (1999), isonym; *see Fedde in Engl. & Prantl, Nat. Pflanzenfam. ed. 2*, 17b: 135 (1936).
subsect. Benecinctae *(Fedde) M. Lidén in Willdenowia*, 26(1–2): 33 (1996): C. [unranked] Benecinctae.
subsect. Brevinectaria *M. Lidén in Willdenowia*, 26(1–2): 25 (1996).
subgen. Calocapnos *(Spach) N.N. Tzvelev, Opred. Sosud. Rast. Severo-Zap. Rossii:* 679 (2000): gen. Calocapnos.
subsect. Cheilanthifoliae *(Lidén) C.Y. Wu & Z.Y. Su in Fl. Reipubl. Popul. Sin.*, 32: 428 (1999): C. sect. Cheilanthifoliae.
sect. Chinenses *(Gorovoj & Bassargin) C.Y. Wu & Z.Y. Su in Fl. Reipubl. Popul. Sin.*, 32: 395 (1999): C. subsect. Chinenses.
subgen. Chremnocapnos *(Wendelbo) T. Fukuhara & M. Lidén in Pl. Syst. Evol.*, 20(1–4): 419 (1997): C. sect. Chremnocapnos.
ser. Coronatae *M. Lidén in Willdenowia*, 26(1–2): 30 (1996).
ser. Davidianae *C.Y. Wu, H. Chuang & Z.Y. Su in Fl. Reipubl. Popul. Sin.*, 32: 172 (1999).
sect. Elatae *Fedde ex C.Y. Wu & H. Chuang in Fl. Reipubl. Popul. Sin.*, 32: 541, 134 (1999).
ser. Elegantes *M. Lidén in Willdenowia*, 26(1–2): 30 (1996).
sect. Flaccidae *C.Y. Wu in Acta Bot. Yunnanica*, 18(4): 399 (1996).
sect. Geraniifoliae *C.Y. Wu in Acta Bot. Yunnanica*, 18(4): 399 (1996).
ser. Helicosyne *M. Lidén in Willdenowia*, 26(1–2): 23 (1996).
sect. Kingiae *C.Y. Wu & Z.Y. Su in Acta Bot. Yunnanica*, 18(4): 399 (1996).
ser. Laeliae *C.Y. Wu in Fl. Reipubl. Popul. Sin.*, 32: 184 (1999).
sect. Latiflorae *C.Y. Wu & Z.Y. Su in Acta Bot. Yunnanica*, 18(4): 398 (1996).
sect. Linstowianae *C.Y. Wu & Z.Y. Su in Acta Bot. Yunnanica*, 18(4): 400 (1996); *et in Acta Bot. Yunnanica*, 19(3): 227 (1997).
ser. Macrocentrae *Popov ex M. Lidén in Willdenowia*, 26(1–2): 30 (1996).
subsect. Monstruosa *M. Lidén in Willdenowia*, 26(1–2): 26 (1996).
subsect. Numullaria *M. Lidén in Willdenowia*, 26(1–2): 25 (1996).
subsect. Officinales *M. Lidén in Willdenowia*, 26(1–2): 25 (1996).
sect. Oreocapnos *Popov ex C.Y. Wu & H. Chuang in Fl. Reipubl. Popul. Sin.*, 32: 542, 259 (1999).
sect. Oreocapnos *Popov ex M.A. Mikhailova in Bot. Zhurn.*, 85(7): 180 (2000), isonym; *see C.Y. Wu & H. Chuang in Fl. Reipubl. Popul. Sin.*, 32: 542 (1999).

CORYDALIS:-
ser. Petiolatae *M. Lidén in Willdenowia*, 26(1–2): 30 (1996).
sect. Polyphyllae *Z.Y. Su in Acta Bot. Yunnanica*, 18(4): 398 (1996).
ser. Raphanituber *(A.P. Khokhr.) M. Lidén in Willdenowia*, 26(1–2): 25 (1996): C. sect. Raphanituber.
sect. Rapiferae *Fedde ex C.Y. Wu & H. Chuang in Fl. Reipubl. Popul. Sin.*, 32: 542, 213 (1999).
subgen. Sophorocapnos *(Turcz.) T. Fukuhara & M. Lidén in Pl. Syst. Evol.*, 206(1–4): 419 (1997): gen. Sophorocapnos.
ser. Tenerrimae *C.Y. Wu in Fl. Reipubl. Popul. Sin.*, 32: 183 (1999).
sect. Vermiculares *Z.Y. Su & M. Lidén in Acta Bot. Yunnanica*, 19(3): 227 (1997).
ser. Yunnanenses *C.Y. Wu in Fl. Reipubl. Popul. Sin.*, 32: 179 (1999).
acuminata Franch.
subsp. hupehensis *C.Y. Wu in Acta Bot. Yunnanica*, 18(4): 401 (1996)—China South-central (Hubei, Guizhou), China North-central (Shaanxi).
adunca Maxim.
subsp. microsperma *M. Lidén & Z.Y. Su in Fl. Reipubl. Popul. Sin.*, 32: 545, 406 (1999)—China South-central (Yunnan).
subsp. scaphopetala *(Fedde) C.Y. Wu & Z.Y. Su in Fl. Reipubl. Popul. Sin.*, 32: 406 (1999): C. scaphopetala.
afghanica Gilli
subsp. elegans *M. Lidén in Willdenowia*, 26(1–2): 31 (1996)—Afghanistan.
subsp. tenuis *M. Lidén in Willdenowia*, 26(1–2): 31 (1996)—Afghanistan.
aitchisonii Popov
subsp. kamelinii *(Kurbanov) M. Lidén in Willdenowia*, 26(1–2): 31 (1996): C. kamelinii.
albipetala *B.U. Oh in Korean J. Pl. Taxon.*, 26(3): 213 (1996)—South Korea.
ambigua Cham. & Schltdl.
subsp. amurensis *(Maxim.) V.N. Voroschilov in A.K. Skvortsov (ed.), Florist. issl. v razn. raĭonakh SSSR:* 171 (1985): C. ambigua var. amurensis.
f. lacticolora *Y.N. Lee, Fl. Korea:* 1158 (1996), without type—Locality not stated.
anaginova *M. Lidén & Z.Y. Su in Edinburgh J. Bot.*, 54(1): 61 (1997)—Xizang.
aspleniifolia *M. Lidén & Z.Y. Su in Edinburgh J. Bot.*, 54(1): 69 (1997)—China South-central (Sichuan), China North-central (Gansu).
auriculata *M. Lidén & Z.Y. Su in Edinburgh J. Bot.*, 54(1): 71 (1997)—Xizang.
benecincta W.W. Sm.
subsp. trilobipetala *(Hand.-Mazz) M. Lidén in Willdenowia*, 26(1–2): 34 (1996): C. trilobipetala.
bibracteolata *Z.Y. Su in Fl. Reipubl. Popul. Sin.*, 32: 545, 411 (1999)—Xinjiang.
birmanica *M. Lidén in Edinburgh J. Bot.*, 55(3): 345 (1998)—Burma.
blanda Schott
subsp. olympica *M. Lidén in Willdenowia*, 26(1–2): 29 (1996)—Greece.
subsp. oxelmanii *M. Lidén in Willdenowia*, 26(1–2): 29 (1996), as 'oxelmannii'—Greece.
subsp. parnassica *(Orph. & Heldr.) M. Lidén in Willdenowia*, 26(1–2): 29 (1996): C. parnassica.
bokuensis *L.H. Zhou in Fl. Qinghaiica*, 1: 507 (1997)—Qinghai.
brevipedunculata *(Z.Y. Su) Z.Y. Su & M. Lidén in Fl. Reipubl. Popul. Sin.*, 32: 442 (1999): C. foetida var. brevipedunculata.
bulleyana Diels
subsp. muliensis *M. Lidén & Z.Y. Su in Fl. Reipubl. Popul. Sin.*, 32: 543, 124 (1999)—China South-central (Sichuan).
caput-medusae *Z.Y. Su & M. Lidén in Edinburgh J. Bot.*, 54(1): 66 (1997)—China South-central (Yunnan).
caseana
subsp. aquae-gelidae *(M. Peck & W.C. Wilson) H. Zetterlund & M. Lidén in Willdenowia*, 26(1–2): 34 (1996): C. aquae-gelidae.
cashmeriana Royle
var. longicalcarata *(D.G. Long) R.C. Srivastava, Fl. Sikkim (Ranunculac.-Moringac.):* 72 (1998): C. ecristata var. longicalcarata.

CORYDALIS:-
caucasica DC.
subsp. abantensis *M. Lidén in Willdenowia,* 26(1–2): 26 (1996)—Turkey.
chingii Fedde
var. shansiensis *W.T. Wang ex C.Y. Wu & Z.Y. Su in Fl. Reipubl. Popul. Sin.,* 32: 544, 300 (1999)—China North-central (Shanxi)
chionophila Czerniak.
subsp. firouzii *(Wendelbo) M. Lidén in Willdenowia,* 26(1–2): 31 (1996): C. firouzii.
subsp. parviflora *M. Lidén in Willdenowia,* 26(1–2): 31 (1996)—Afghanistan.
crassirhizomata *(C.Y. Wu) C.Y. Wu in Fl. Reipubl. Popul. Sin.,* 32: 149 (1999): C. schusteriana var. crassirhizomata.
curviflora Maxim.
subsp. altecristata *(C.Y. Wu & H. Chuang) C.Y. Wu in Fl. Reipubl. Popul. Sin.,* 32: 223 (1999): C. curviflora var. altecristata.
subsp. minuticristata *(Fedde) C.Y. Wu in Fl. Reipubl. Popul. Sin.,* 32: 223 (1999): C. curviflora var. minuticristata.
subsp. pseudosmithii *(Fedde) C.Y. Wu in Fl. Reipubl. Popul. Sin.,* 32: 223 (1999): Corydalis curviflora var. pseudosmithii.
subsp. rosthornii *(Fedde) C.Y. Wu in Fl. Reipubl. Popul. Sin.,* 32: 221 (1999): C. curviflora var. rosthornii.
cytisiflora *(Fedde) Lidén ex C.Y. Wu, H. Chuang & Z.Y. Su in Fl. Reipubl. Popul. Sin.,* 32: 219 (1999): C. curviflora var. cytisiflora.
decumbens Pers.
f. albiflora *Y.N. Lee, Fl. Korea:* 1158 (1996), without type—Locality not stated.
diphylla Wall.
subsp. murreeana *(Jafri) M. Lidén in Willdenowia,* 26(1–2): 30 (1996), as 'murreana': C murreeana.
subsp. occidentalis *M. Lidén in Willdenowia,* 26(1–2): 30 (1996)—Pakistan.
ecristata (Prain) D.G. Long
subsp. longicalcarata *(D.G. Long) C.Y. Wu in Fl. Reipubl. Popul. Sin.,* 32: 272 (1999): C. ecristata var. longicalcarata.
elata Bureau & Franch.
subsp. ecristata *C.Y. Wu in Fl. Reipubl. Popul. Sin.,* 32: 544, 147 (1999)—China South-central (Sichuan).
ellipticarpa C.Y. Wu & Z.Y. Su
var. taipaica *C.Y. Wu in Fl. Reipubl. Popul. Sin.,* 32: 543, 144 (1999)—China North-central (Shaanxi).
enantiophylla *M. Lidén in Edinburgh J. Bot.,* 55(3): 343 (1998)—Burma.
eugeniae Fedde
subsp. fissibracteata *(Fedde) M. Lidén in Fl. Reipubl. Popul. Sin.,* 32: 251 (1999): C. linarioides var. fissibracteata.
farreri *M. Lidén in Edinburgh J. Bot.,* 55(3): 346 (1998)—Burma.
flexuosa Franch.
subsp. balsamiflora *(Prain) C.Y. Wu in Fl. Reipubl. Popul. Sin.,* 32: 118 (1999): C. balsamiflora.
f. bulbillifera *C.Y. Wu in Acta Bot. Yunnanica,* 18(4): 400 (1996)—China South-central (Sichuan).
subsp. gemmipara *(H. Chuang) C.Y Wu in Fl. Reipubl. Popul. Sin.,* 32: 117 (1999): C. gemmipara.
subsp. kuanhsiensis *C.Y. Wu in Acta Bot. Yunnanica,* 18(4): 400 (1996)—China South-central (Sichuan).
subsp. microflora *(C.Y. Wu & H. Chuang) C.Y. Wu in Fl. Reipubl. Popul. Sin.,* 32: 118 (1999): C. flexuosa var. microflora.
subsp. mucronipetala *(C.Y. Wu & H. Chuang) C.Y. Wu in Fl. Reipubl. Popul. Sin.,* 32: 118 (1999): C. flexuosa var. mucronipetala.
subsp. omeiana *(C.Y. Wu & H. Chuang) C.Y. Wu in Fl. Reipubl. Popul. Sin.,* 32: 118 (1999): C. flexuosa var. omeiana.
subsp. pinnatibracteata *(C.Y. Wu & H. Chuang) C.Y. Wu in Fl. Reipubl. Popul. Sin.,* 32: 119 (1999): C. flexuosa var. pinnatibracteata.
subsp. pseudoheterocentra *(Fedde) Lidén ex C.Y. Wu, H. Chuang & Z.Y. Su in Fl. Reipubl. Popul. Sin.,* 32: 119 (1999): C. pseudoheterocentra.

CORYDALIS:-
foliaceobracteata *C.Y. Wu & Z.Y. Su in Acta Bot. Yunnanica,* 19(3): 232 (1997)—China South-central (Sichuan), China North-central (Gansu).
fukuharae *M. Lidén in Willdenowia,* 26(1–2): 27 (1996)—Japan.
fumariifolia Maxim.
subsp. azurea *M. Lidén & H. Zetterlund in Willdenowia,* 26(1–2): 27 (1996)—Japan.
geocarpa *Harry Sm. ex M. Lidén in Willdenowia,* 26(1–2): 34 (1996)—China South-central (Sichuan).
glaucissima *M. Lidén & Z.Y. Su in Edinburgh J. Bot.,* 54(1): 78 (1997)—China South-central (Yunnan)/Xizang border.
gorodkovii Karav.
subsp. magadanica *(A.P. Khokhr.) V.N. Voroschilov in A.K. Skvortsov (ed.), Florist. issl. v razn. raĭonakh SSSR:* 171 (1985): C. magadanica.
gracillima *C.Y. Wu ex R. Govaerts, World Checklist Seed Plants,* 3(1): 20 (1999), nom. nov.: C. gracilis *Franch.*
gyrophylla *M. Lidén in Willdenowia,* 26(1–2): 33 (1996)—China South-central (Sichuan).
heteropetala Otsch.
subsp. angusticalcarea *M.A. Michajlova in Bot. Zhurn.,* 84(9): 113 (1999)—Transcaucasus (Gruziya).
subsp. bayerniana *M.A. Michajlova in Bot. Zhurn.,* 84(9): 112 (1999)—Transcaucasus (Gruziya).
subsp. grandiflora *M.A. Michajlova in Bot. Zhurn.,* 84(9): 112 (1999)—North Caucasus (Kabardino-Balkariya).
imbricata *Z.Y. Su & M. Lidén in Edinburgh J. Bot.,* 54(1): 59 (1997)—Xizang.
incisa (Thunb.) Pers.
f. albiflora *Y.N. Lee, Fl. Korea:* 1158 (1996), without type—Locality not stated.
kingii Prain
var. megalantha *C.Y. Wu & Z.Y. Su in Acta Bot. Yunnanica,* 18(4): 404 (1996)—Xizang.
kiukiangensis *C.Y. Wu, Z.Y. Su & M. Lidén in Edinburgh J. Bot.,* 54(1): 67 (1997)—China South-central (Yunnan).
lasiocarpa *M. Lidén & Z.Y. Su in Edinburgh J. Bot.,* 54(1): 82 (1997)—Xizang.
latiflora Hook. f. & Thomson
subsp. gerdae *(Fedde) Lidén ex C.Y. Wu, H. Chuang & Z.Y. Su in Fl. Reipubl. Popul. Sin.,* 32: 277 (1999): C. gerdae.
latiloba
var. tibetica *Z.Y. Su & M. Lidén in Acta Bot. Yunnanica,* 19(3): 234 (1997)—Xizang.
var. wumungensis *C.Y. Wu & Z.Y. Su in Acta Bot. Yunnanica,* 19(3): 234 (1997)—China South-central (Yunnan).
leptantha *M. Lidén in Edinburgh J. Bot.,* 55(3): 346 (1998)—Bhutan.
leptophylla *M. Lidén in Edinburgh J. Bot.,* 55(3): 347 (1998)—Burma.
lhasaensis *C.Y. Wu & Z.Y. Su in Fl. Reipubl. Popul. Sin.,* 32: 280 (1999), nom. nov.: C. meifolia var. ecristata *C.Y. Wu & Z.Y. Su.*
liana *M. Lidén & Z.Y. Su in Edinburgh J. Bot.,* 54(1): 64 (1997)—China South-central (Yunnan).
linjiangensis *Z.Y. Su ex M. Lidén in Willdenowia,* 26(2): 28 (1996)—Manchuria (Jilin).
livida Maxim.
var. denticulatocristata *Z.Y. Su in Fl. Reipubl. Popul. Sin.,* 32: 544, 296 (1999), as 'denticulato-cristata'—China South-central (Sichuan).
longipes DC.
var. pubescens *(C.Y. Wu & H. Chuang) C.Y. Wu in Fl. Reipubl. Popul. Sin.,* 32: 341 (1999): C. pubescens.
lydica *M. Lidén in Willdenowia,* 26(1–2): 32 (1996)—Turkey.
mayae Hand.-Mazz.
var. stenophylla *(Fedde) C.Y. Wu in Fl. Yunnanica,* 8: 114 (1997), without basionym ref.; *et in Fl. Reipubl. Popul. Sin.,* 32: 241 (1999), with basionym ref.: C. delavayi var. stenophylla.
megalosperma *Z.Y. Su in Acta Bot. Yunnanica,* 19(3): 230 (1997)—China South-central (Yunnan).

CORYDALIS:-
ohii *M. Lidén in Willdenowia,* 26(1–2): 28 (1996)—South Korea.
oreocoma *M. Lidén & Z.Y. Su in Edinburgh J. Bot.,* 54(1): 77 (1997)—Xizang.
ornata *M. Lidén & H. Zetterlund in Willdenowia,* 26(1–2): 29 (1996)—Primorye.
pallida
var. chanetii *(H. Lév.) R. Govaerts, World Checklist Seed Plants,* 3(1): 20 (1999): C. chanetii.
parviflora *Z.Y. Su & M. Lidén in Edinburgh J. Bot.,* 54(1): 55 (1997)—China Southeast (Guangxi), China South-central (Guizhou, Yunnan).
pauciflora (Stephan. ex Willd.) Pers.
subsp. arctica *(Popov) V.N. Voroschilov in A.K. Skvortsov (ed.), Florist. issl. v razn. raĭonakh SSSR:* 171 (1985): C. arctica.
var. foliosa *L.H. Zhou in Fl. Qinghaiica,* 1: 508 (1997)—Qinghai.
peltata *M. Lidén & Z.Y. Su in Edinburgh J. Bot.,* 54(1): 71 (1997)—China South-central (Yunnan).
petrodoxa *M. Lidén & Z.Y. Su in Edinburgh J. Bot.,* 54(1): 81 (1997)—Xizang.
philippi *M.A. Mikhailova in Bot. Zhurn.,* 85(7): 182 (2000)—Mongolia.
podlechii *M. Lidén in Willdenowia,* 26(1–2): 30 (1996)—Afghanistan.
praecipitorum *C.Y. Wu, Z.Y. Su & M. Lidén in Edinburgh J. Bot.,* 54(1): 83 (1997)—China North-central (Gansu).
pseudobalfouriana *M. Lidén & Z.Y. Su in Edinburgh J. Bot.,* 54(1): 77 (1997)—China South-central (Yunnan).
pseudodensispica *Z.Y. Su & M. Lidén in Edinburgh J. Bot.,* 54(1): 74 (1997)—China South-central (Sichuan).
pseudofilisecta *M. Lidén & Z.Y. Su in Edinburgh J. Bot.,* 54(1): 79 (1997)—Xizang.
pseudoincisa *C.Y. Wu, Z.Y. Su & M. Lidén in Edinburgh J. Bot.,* 54(1): 73 (1997)—China North-central (Shaanxi, Gansu).
pseudomicrophylla *Z.Y. Su in Fl. Reipubl. Popul. Sin.,* 32: 545, 411 (1999)—Xinjiang.
pseudomucronata *C.Y. Wu, Z.Y. Su & M. Lidén in Acta Bot. Yunnanica,* 18(4): 400 (1996)—China South-central (Sichuan).
var. cristata *C.Y. Wu in Acta Bot. Yunnanica,* 18(4): 401 (1996)—China South-central (Sichuan).
pseudomucronata *C.Y. Wu, Z.Y. Su & M. Lidén in Edinburgh J. Bot.,* 54(1): 68 (1997)—China South-central (Sichuan).
pseudorupestris *M. Lidén & Z.Y. Su in Acta Bot. Yunnanica,* 19(3): 234 (1997)—China South-central (Sichuan).
pseudoscaberula *M. Lidén & Z.Y. Su in Edinburgh J. Bot.,* 54(1): 82 (1997)—Xizang.
pseudoweigoldii *Z.Y. Su in Fl. Reipubl. Popul. Sin.,* 32: 544, 387 (1999)—China South-central (Sichuan).
punicea *C.Y. Wu ex R. Govaerts, World Checklist Seed Plants,* 3(1): 20 (1999), nom. nov.: C. rosea *Maxim.*
qinghaiensis *Z.Y. Su & M. Lidén in Edinburgh J. Bot.,* 54(1): 62 (1997)—Qinghai.
rockiana *C.Y. Wu, M. Lidén & Z.Y. Su in Edinburgh J. Bot.,* 54(1): 80 (1997)—Xizang.
rorida *H. Chuang in Acta Bot. Yunnanica,* 18(4): 402 (1996)—China South-central (Sichuan).
saccata *Z.Y. Su in Acta Bot. Yunnanica,* 19(3): 231 (1997)—Manchuria (Liaoning).
seisumsiana *M. Lidén in Willdenowia,* 26(1–2): 32 (1996)—Transcaucasus (Nakhichevan).
shensiana *Lidén ex C.Y. Wu, H. Chuang & Z.Y. Su in Fl. Reipubl. Popul. Sin.,* 32: 219 (1999), nom. nov.: C. curviflora var. giraldii *Fedde.*
shimienensis *C.Y. Wu & Z.Y. Su in Fl. Reipubl. Popul. Sin.,* 32: 543, 110 (1999)—China South-central (Sichuan).
sibirica (L.f.) Pers.
subsp. impatiens *(Pall.) I.A. Gubanov, Konsp. Fl. Vneshnei Mongolii (Sosud. Rast.):* 52 (1996): Fumaria impatiens.
sigmantha *Z.Y. Su & C.Y. Wu in Acta Bot. Yunnanica,* 19(3): 228 (1997)—China North-central (Gansu).
solida (L.) Clairv.
subsp. longicarpa *M. Lidén in Willdenowia,* 26(1–2): 27 (1996)—Greece.

CORYDALIS:-
solida (L.) Clairv.
subsp. subremota *Popov ex M. Lidén & H. Zetterlund in Willdenowia*, 26(1–2): 26 (1996)—Krasnoyarsk.
subsp. subremota *(Popov) G.A. Peschkova in Fl. Sibir. (Berberidac.-Grossulariac.)*, 7: 41 (1994), basionym not validly publ.: C. halleri subsp. subremota.
squamigera *Z.Y. Su in Fl. Reipubl. Popul. Sin.*, 32: 543, 145 (1999)—China South-central (Sichuan).
straminoides *C.Y. Wu & Z.Y. Su in Acta Bot. Yunnanica*, 19(3): 229 (1997)—Xizang.
striatocarpa *H. Chuang in Acta Bot. Yunnanica*, 18(4): 402 (1996)—China South-central (Sichuan).
tangutica Peschkova
subsp. bullata *(Lidén) Z.Y. Su in Fl. Reipubl. Popul. Sin.*, 32: 450 (1999): C. tianzhuensis subsp. bullata.
temulifolia Franch.
subsp. aegopodioides *(H. Lév. & Vaniot) C.Y. Wu in Fl. Yunnanica*, 8: 76 (1997), without basionym ref.; *et in Fl. Reipubl. Popul. Sin.*, 32: 107 (1999), with basionym ref.: C. aegopodioides.
tenerrima *C.Y. Wu in Acta Bot. Yunnanica*, 18(4): 401 (1996)—China South-central (Yunnan).
tenerrima *C.Y. Wu, Z.Y. Su & M. Lidén in Edinburgh J. Bot.*, 54(1): 64 (1997)—China South-central (Yunnan).
ternatifolia *C.Y. Wu, Z.Y. Su & M. Lidén in Edinburgh J. Bot.*, 54(1): 70 (1997)—China South-central (Hubei, Sichuan).
tianzhuensis M.S. Yan & C.J. Wang
subsp. bullata *M. Lidén in Willdenowia*, 26(1–2): 33 (1996)—Xizang.
trachycarpa Maxim.
var. leucostachya *(C.Y. Wu & H. Chuang) C.Y. Wu in Fl. Reipubl. Popul. Sin.*, 32: 197 (1999): C. leucostachya.
var. octocornuta *(C.Y. Wu) C.Y. Wu in Fl. Reipubl. Popul. Sin.*, 32: 197 (1999): C. octocornuta.
tsangensis *M. Lidén & Z.Y. Su in Edinburgh J. Bot.*, 54(1): 63 (1997)—Xizang.
tuberipisiformis *Z.Y. Su in Acta Bot. Yunnanica*, 19(3): 232 (1997)—China North-central (Sichuan).
turtschaninovii Besser
f. albiflora *Y.N. Lee, Fl. Korea:* 1158 (1996), without type—Locality not stated.
f. rotundiloba *(Maxim.) Y.C. Chu in Clavis Pl. Chinae Bor.-Or. ed. 2 (ed. P.Y. Fu)*: 230 (1995): C. remota var. rotundiloba.
subsp. vernyi *(Franch. & Sav.) M. Lidén in Willdenowia*, 26(1–2): 28 (1996): C. vernyi.
uranoscopa *M. Lidén in Edinburgh J. Bot.*, 55(3): 347 (1998)—India (Uttar Pradesh).
uvaria *M. Lidén & Z.Y. Su in Edinburgh J. Bot.*, 54(1): 78 (1997)—China South-central (Sichuan).
vermicularis *M. Lidén & Z.Y. Su in Edinburgh J. Bot.*, 54(1): 56 (1997)—China South-central (Sichuan).
verna *Z.Y. Su & M. Lidén in Edinburgh J. Bot.*, 54(1): 80 (1997)—Xizang.
verticillaris DC.
subsp. parviflora *M. Lidén in Willdenowia*, 26(1–2): 32 (1996)—Iran.
wandoensis *Y.N. Lee in Korean J. Pl. Taxon.*, 28(1): 26 (1998)—South Korea.
watanabei *M. Kitagawa in Rep. Inst. Sci. Res. Manchoukuo*, 6(4): 122 (1942)—Manchuria (Jilin).
wendelboi Lidén
subsp. congesta *M. Lidén & H. Zetterlund in Willdenowia*, 26(1–2): 26 (1996)—Turkey.
wuzhengyiana *Z.Y. Su & M. Lidén in Edinburgh J. Bot.*, 54(1): 59 (1997)—Xizang, China South-central (Sichuan).
zhongdianensis *C.Y. Wu in Acta Bot. Yunnanica*, 18(4): 403 (1996), homonym; *see Z.Y. Su & Lidén in Rheedea*, 5(1): 16 (1995)—China South-central (Yunnan, Sichuan).

CORYMBIA *K.D. Hill & L.A.S. Johnson in Telopea*, 6(2–3): 214 (1995). *MYRTACEAE*
abbreviata *(Blakely & Jacobs) K.D. Hill & L.A.S. Johnson in Telopea*, 6(2–3): 344 (1995): Eucalyptus abbreviata.

CORYMBIA:-
abergiana *(F. Muell.) K.D. Hill & L.A.S. Johnson in Telopea*, 6(2–3): 244 (1995): Eucalyptus abergiana.
aparrerinja *K.D. Hill & L.A.S. Johnson in Telopea*, 6(2–3): 453 (1995)—Northern Territory, Western Australia, Queensland.
arafurica *K.D. Hill & L.A.S. Johnson in Telopea*, 6(2–3): 409 (1995)—Northern Territory.
arenaria *(Blakely) K.D. Hill & L.A.S. Johnson in Telopea*, 6(2–3): 274 (1995): Eucalyptus arenaria.
arnhemensis *(D.J. Carr & S.G.M. Carr) K.D. Hill & L.A.S. Johnson in Telopea*, 6(2–3): 271 (1995): Eucalyptus arnhemensis.
subsp. monticola *K.D. Hill & L.A.S. Johnson in Telopea*, 6(2–3): 273 (1995)—Queensland.
aspera *(F. Muell.) K.D. Hill & L.A.S. Johnson in Telopea*, 6(2–3): 448 (1995): Eucalyptus aspera.
aureola *(Brooker & A.R. Bean) K.D. Hill & L.A.S. Johnson in Telopea*, 6(2–3): 374 (1995): Eucalyptus aureola.
bella *K.D. Hill & L.A.S. Johnson in Telopea*, 6(2–3): 411 (1995)—Northern Territory, Western Australia, Queensland.
blakei *K.D. Hill & L.A.S. Johnson in Telopea*, 6(2–3): 437 (1995)—Queensland.
subsp. rasilis *K.D. Hill & L.A.S. Johnson in Telopea*, 6(2–3): 438 (1995)—Queensland.
bleeseri *(Blakely) K.D. Hill & L.A.S. Johnson in Telopea*, 6(2–3): 288 (1995): Eucalyptus bleeseri.
bloxsomei *(Maiden) K.D. Hill & L.A.S. Johnson in Telopea*, 6(2–3): 372 (1995): Eucalyptus bloxsomei.
brachycarpa *(D.J. Carr & S.G.M. Carr) K.D. Hill & L.A.S. Johnson in Telopea*, 6(2–3): 281 (1995): Eucalyptus brachycarpa.
bunites *(Brooker & A.R. Bean) K.D. Hill & L.A.S. Johnson in Telopea*, 6(2–3): 373 (1995): Eucalyptus bunites.
byrnesii *(D.J. Carr & S.G.M. Carr) K.D. Hill & L.A.S. Johnson in Telopea*, 6(2–3): 331 (1995): Eucalyptus byrnesii.
cadophora *K.D. Hill & L.A.S. Johnson in Telopea*, 6(2–3): 345 (1995), nom. nov.: Eucalyptus perfoliata *R.Br. ex Benth.*
subsp. pliantha *K.D. Hill & L.A.S. Johnson in Telopea*, 6(2–3): 348 (1995)—Western Australia.
calophylla *(Lindl.) K.D. Hill & L.A.S. Johnson in Telopea*, 6(2–3): 240 (1995): Eucalyptus calophylla.
candida *K.D. Hill & L.A.S. Johnson in Telopea*, 6(2–3): 440 (1995)—Western Australia.
subsp. dipsodes *K.D. Hill & L.A.S. Johnson in Telopea*, 6(2–3): 444 (1995)—Western Australia, Northern Territory.
subsp. lautifolia *K.D. Hill & L.A.S. Johnson in Telopea*, 6(2–3): 442 (1995)—Western Australia.
capricornia *(D.J. Carr & S.G.M. Carr) K.D. Hill & L.A.S. Johnson in Telopea*, 6(2–3): 299 (1995): Eucalyptus capricornia.
catenaria *K.D. Hill & L.A.S. Johnson in Telopea*, 6(2–3): 376 (1995)—Queensland.
chartacea *K.D. Hill & L.A.S. Johnson in Telopea*, 6(2–3): 353 (1995)—Northern Territory.
chillagoensis *K.D. Hill & L.A.S. Johnson in Telopea*, 6(2–3): 433 (1995)—Queensland.
chippendalei *(D.J. Carr & S.G.M. Carr) K.D. Hill & L.A.S. Johnson in Telopea*, 6(2–3): 313 (1995): Eucalyptus chippendalei.
chlorolampra *K.D. Hill & L.A.S. Johnson in Telopea*, 6(2–3): 238 (1995)—Western Australia.
citriodora *(Hook.) K.D. Hill & L.A.S. Johnson in Telopea*, 6(2–3): 388 (1995): Eucalyptus citriodora.
subsp. variegata *(F. Muell.) A.R. Bean & M.W. McDonald in Austrobaileya*, 5(4): 735 (2000): Eucalyptus variegata.
clandestina *(A.R. Bean) K.D. Hill & L.A.S. Johnson in Telopea*, 6(2–3): 282 (1995): Eucalyptus clandestina.
clarksoniana *(D.J. Carr & S.G.M. Carr) K.D. Hill & L.A.S. Johnson in Telopea*, 6(2–3): 259 (1995): Eucalyptus clarksoniana.
clavigera *(A. Cunn. ex Schauer) K.D. Hill & L.A.S. Johnson in Telopea*, 6(2–3): 413 (1995): Eucalyptus clavigera.

CORYMBIA:-
cliftoniana *(W. Fitzg.) K.D. Hill & L.A.S. Johnson in Telopea*, 6(2–3): 337 (1995): Eucalyptus cliftoniana.
collina *(W. Fitzg.) K.D. Hill & L.A.S. Johnson in Telopea*, 6(2–3): 290 (1995): Eucalyptus collina.
confertiflora *(F. Muell.) K.D. Hill & L.A.S. Johnson in Telopea*, 6(2–3): 428 (1995): Eucalyptus confertiflora.
curtipes *(D.J. Carr & S.G.M. Carr) K.D. Hill & L.A.S. Johnson in Telopea*, 6(2–3): 332 (1995): Eucalyptus curtipes.
dallachiana *(Benth.) K.D. Hill & L.A.S. Johnson in Telopea*, 6(2–3): 451 (1995): Eucalyptus tessellaris var. dallachiana.
dampieri *(D.J. Carr & S.G.M. Carr) K.D. Hill & L.A.S. Johnson in Telopea*, 6(2–3): 329 (1995): Eucalyptus dampieri.
dendromerinx *K.D. Hill & L.A.S. Johnson in Telopea*, 6(2–3): 431 (1995)—Western Australia.
deserticola *(D.J. Carr & S.G.M. Carr) K.D. Hill & L.A.S. Johnson in Telopea*, 6(2–3): 365 (1995): Eucalyptus deserticola.
subsp. mesogeotica *K.D. Hill & L.A.S. Johnson in Telopea*, 6(2–3): 368 (1995)—Northern Territory.
dichromophloia *(F. Muell.) K.D. Hill & L.A.S. Johnson in Telopea*, 6(2–3): 295 (1995): Eucalyptus dichromophloia.
dimorpha *(Brooker & A.R. Bean) K.D. Hill & L.A.S. Johnson in Telopea*, 6(2–3): 380 (1995): Eucalyptus peltata subsp. dimorpha.
disjuncta *K.D. Hill & L.A.S. Johnson in Telopea*, 6(2–3): 423 (1995)—Northern Territory, Western Australia, Queensland, New Guinea (Papua New Guinea).
dolichocarpa *(D.J. Carr & S.G.M. Carr) K.D. Hill & L.A.S. Johnson in Telopea*, 6(2–3): 267 (1995): Eucalyptus dolichocarpa.
drysdalensis *(D.J. Carr & S.G.M. Carr) K.D. Hill & L.A.S. Johnson in Telopea*, 6(2–3): 297 (1995): Eucalyptus drysdalensis.
dunlopiana *K.D. Hill & L.A.S. Johnson in Telopea*, 6(2–3): 354 (1995)—Northern Territory.
ellipsoidea *(D.J. Carr & S.G.M. Carr) K.D. Hill & L.A.S. Johnson in Telopea*, 6(2–3): 306 (1995): Eucalyptus ellipsoidea.
eremaea *(D.J. Carr & S.G.M. Carr) K.D. Hill & L.A.S. Johnson in Telopea*, 6(2–3): 309 (1995): Eucalyptus eremaea.
subsp. oligocarpa *(Blakely & Jacobs) K.D. Hill & L.A.S. Johnson in Telopea*, 6(2–3): 311 (1995): Eucalyptus polycarpa var. oligocarpa.
erythrophloia *(Blakely) K.D. Hill & L.A.S. Johnson in Telopea*, 6(2–3): 304 (1995): Eucalyptus erythrophloia.
eximia *(Schauer) K.D. Hill & L.A.S. Johnson in Telopea*, 6(2–3): 383 (1995): Eucalyptus eximia.
ferriticola *(Brooker & Edgecombe) K.D. Hill & L.A.S. Johnson in Telopea*, 6(2–3): 446 (1995): Eucalyptus ferriticola.
subsp. sitiens *L.A.S. Johnson & K.D. Hill in Telopea*, 6(2–3): 447 (1995)—Western Australia.
ferruginea *(Schauer) K.D. Hill & L.A.S. Johnson in Telopea*, 6(2–3): 339 (1995): Eucalyptus ferruginea.
subsp. stypophylla *K.D. Hill & L.A.S. Johnson in Telopea*, 6(2–3): 343 (1995)—Western Australia, Northern Territory.
ficifolia *(F. Muell.) K.D. Hill & L.A.S. Johnson in Telopea*, 6(2–3): 245 (1995): Eucalyptus ficifolia.
flavescens *K.D. Hill & L.A.S. Johnson in Telopea*, 6(2–3): 455 (1995)—Northern Territory, Western Australia, Queensland.
foelscheana *(F. Muell.) K.D. Hill & L.A.S. Johnson in Telopea*, 6(2–3): 333 (1995): Eucalyptus foelscheana.
gilbertensis *(Maiden & Blakely) K.D. Hill & L.A.S. Johnson in Telopea*, 6(2–3): 435 (1995): Eucalyptus clavigera var. gilbertensis.
grandifolia *(R. Br. ex Benth.) K.D. Hill & L.A.S. Johnson in Telopea*, 6(2–3): 457 (1995): Eucalyptus grandifolia.
subsp. lamprocardia *L.A.S. Johnson in Telopea*, 6(2–3): 461 (1995)—Western Australia, Northern Territory.
subsp. longa *L.A.S. Johnson in Telopea*, 6(2–3): 460 (1995)—Northern Territory, Western Australia.

CORYMBIA:-
greeniana *(D.J. Carr & S.G.M. Carr) K.D. Hill & L.A.S. Johnson in Telopea,* 6(2–3): 330 (1995): Eucalyptus greeniana.
gummifera *(Gaertn.) K.D. Hill & L.A.S. Johnson in Telopea,* 6(2–3): 233 (1995): Metrosideros gummifera.
haematoxylon *(Maiden) K.D. Hill & L.A.S. Johnson in Telopea,* 6(2–3): 237 (1995): Eucalyptus haematoxylon.
hamersleyana *(D.J. Carr & S.G.M. Carr) K.D. Hill & L.A.S. Johnson in Telopea,* 6(2–3): 314 (1995): Eucalyptus hamersleyana.
hendersonii *K.D. Hill & L.A.S. Johnson in Telopea,* 6(2–3): 279 (1995)—Queensland.
henryi *(S.T. Blake) K.D. Hill & L.A.S. Johnson in Telopea,* 6(2–3): 396 (1995): Eucalyptus henryi.
hylandii *(D.J. Carr & S.G.M. Carr) K.D. Hill & L.A.S. Johnson in Telopea,* 6(2–3): 283 (1995): Eucalyptus hylandii.
subsp. peninsularis *K.D. Hill & L.A.S. Johnson in Telopea,* 6(2–3): 284 (1995)—Queensland.
inobvia *K.D. Hill & L.A.S. Johnson in Telopea,* 6(2–3): 436 (1995)—Queensland.
intermedia *(R.T. Baker) K.D. Hill & L.A.S. Johnson in Telopea,* 6(2–3): 247 (1995): Eucalyptus intermedia.
jacobsiana *(Blakely) K.D. Hill & L.A.S. Johnson in Telopea,* 6(2–3): 225 (1995): Eucalyptus jacobsiana.
karelgica *K.D. Hill & L.A.S. Johnson in Telopea,* 6(2–3): 430 (1995)—Western Australia.
kombolgiensis *(Brooker & Dunlop) K.D. Hill & L.A.S. Johnson in Telopea,* 6(2–3): 414 (1995): Eucalyptus kombolgiensis.
lamprophylla *(Brooker & A.R. Bean) K.D. Hill & L.A.S. Johnson in Telopea,* 6(2–3): 287 (1995): Eucalyptus lamprophylla.
latifolia *(F. Muell.) K.D. Hill & L.A.S. Johnson in Telopea,* 6(2–3): 327 (1995): Eucalyptus latifolia.
leichhardtii *(F.M. Bailey) K.D. Hill & L.A.S. Johnson in Telopea,* 6(2–3): 370 (1995): Eucalyptus leichhardtii.
lenziana *(D.J. Carr & S.G.M. Carr) K.D. Hill & L.A.S. Johnson in Telopea,* 6(2–3): 312 (1995): Eucalyptus lenziana.
leptoloma *(Brooker & A.R. Bean) K.D. Hill & L.A.S. Johnson in Telopea,* 6(2–3): 369 (1995): Eucalyptus leptoloma.
ligans *K.D. Hill & L.A.S. Johnson in Telopea,* 6(2–3): 263 (1995)—Queensland.
subsp. burdekinensis *K.D. Hill & L.A.S. Johnson in Telopea,* 6(2–3): 265 (1995)—Queensland.
subsp. novocastrensis *K.D. Hill & L.A.S. Johnson in Telopea,* 6(2–3): 265 (1995)—Queensland.
maculata *(Hook.) K.D. Hill & L.A.S. Johnson in Telopea,* 6(2–3): 393 (1995): Eucalyptus maculata.
maritima *K.D. Hill & L.A.S. Johnson in Telopea,* 6(2–3): 261 (1995)—Queensland.
nesophila *(Blakely) K.D. Hill & L.A.S. Johnson in Telopea,* 6(2–3): 242 (1995): Eucalyptus nesophila.
novoguinensis *(D.J. Carr & S.G.M. Carr) K.D. Hill & L.A.S. Johnson in Telopea,* 6(2–3): 257 (1995): Eucalyptus novoguinensis.
oocarpa *(D.J. Carr & S.G.M. Carr) K.D. Hill & L.A.S. Johnson in Telopea,* 6(2–3): 293 (1995): Eucalyptus oocarpa.
opaca *(D.J. Carr & S.G.M. Carr) K.D. Hill & L.A.S. Johnson in Telopea,* 6(2–3): 318 (1995): Eucalyptus opaca.
opacula *L.A.S. Johnson in Telopea,* 6(2–3): 325 (1995)—Western Australia.
pachycarpa *K.D. Hill & L.A.S. Johnson in Telopea,* 6(2–3): 359 (1995)—Western Australia, Northern Territory.
subsp. glabrescens *K.D. Hill & L.A.S. Johnson in Telopea,* 6(2–3): 360 (1995)—Northern Territory.
papillosa *K.D. Hill & L.A.S. Johnson in Telopea,* 6(2–3): 362 (1995)—Northern Territory.
subsp. globifera *K.D. Hill & L.A.S. Johnson in Telopea,* 6(2–3): 363 (1995)—Western Australia.
papuana *(F. Muell.) K.D. Hill & L.A.S. Johnson in Telopea,* 6(2–3): 405 (1995): Eucalyptus papuana.
paracolpica *K.D. Hill & L.A.S. Johnson in Telopea,* 6(2–3): 407 (1995)—Queensland.

CORYMBIA:-
paractia *K.D. Hill & L.A.S. Johnson in Telopea,* 6(2–3): 462 (1995)—Western Australia.
pauciseta *K.D. Hill & L.A.S. Johnson in Telopea,* 6(2–3): 426 (1995)—Northern Territory.
pedimontana *L.A.S. Johnson in Telopea,* 6(2–3): 335 (1995)—Western Australia.
peltata *(Benth.) K.D. Hill & L.A.S. Johnson in Telopea,* 6(2–3): 381 (1995): Eucalyptus peltata.
petalophylla *(Brooker & A.R. Bean) K.D. Hill & L.A.S. Johnson in Telopea,* 6(2–3): 375 (1995): Eucalyptus petalophylla.
plena *K.D. Hill & L.A.S. Johnson in Telopea,* 6(2–3): 268 (1995)—Queensland.
pocillum *(D.J. Carr & S.G.M. Carr) K.D. Hill & L.A.S. Johnson in Telopea,* 6(2–3): 303 (1995): Eucalyptus pocillum.
polycarpa *(F. Muell.) K.D. Hill & L.A.S. Johnson in Telopea,* 6(2–3): 254 (1995): Eucalyptus polycarpa.
polysciada *(F. Muell.) K.D. Hill & L.A.S. Johnson in Telopea,* 6(2–3): 417 (1995): Eucalyptus polysciada.
porphyritica *K.D. Hill & L.A.S. Johnson in Telopea,* 6(2–3): 307 (1995)—Queensland.
porrecta *(S.T. Blake) K.D. Hill & L.A.S. Johnson in Telopea,* 6(2–3): 270 (1995): Eucalyptus porrecta.
ptychocarpa *(F. Muell.) K.D. Hill & L.A.S. Johnson in Telopea,* 6(2–3): 250 (1995): Eucalyptus ptychocarpa.
subsp. aptycha *K.D. Hill & L.A.S. Johnson in Telopea,* 6(2–3): 251 (1995)—Northern Territory.
punkapitiensis *K.D. Hill & L.A.S. Johnson in Telopea,* 6(2–3): 449 (1995)—Western Australia.
rhodops *(D.J. Carr & S.G.M. Carr) K.D. Hill & L.A.S. Johnson in Telopea,* 6(2–3): 276 (1995): Eucalyptus rhodops.
rubens *K.D. Hill & L.A.S. Johnson in Telopea,* 6(2–3): 301 (1995)—Western Australia.
scabrida *(Brooker & A.R. Bean) K.D. Hill & L.A.S. Johnson in Telopea,* 6(2–3): 382 (1995): Eucalyptus scabrida.
semiclara *K.D. Hill & L.A.S. Johnson in Telopea,* 6(2–3): 316 (1995)—Western Australia.
setosa *(Schauer) K.D. Hill & L.A.S. Johnson in Telopea,* 6(2–3): 356 (1995): Eucalyptus setosa.
subsp. pedicellaris *K.D. Hill & L.A.S. Johnson in Telopea,* 6(2–3): 358 (1995)—Queensland.
sphaerica *K.D. Hill & L.A.S. Johnson in Telopea,* 6(2–3): 351 (1995)—Northern Territory.
stockeri *(D.J. Carr & S.G.M. Carr) K.D. Hill & L.A.S. Johnson in Telopea,* 6(2–3): 286 (1995): Eucalyptus stockeri.
terminalis *(F. Muell.) K.D. Hill & L.A.S. Johnson in Telopea,* 6(2–3): 323 (1995): Eucalyptus terminalis.
tessellaris *(F. Muell.) K.D. Hill & L.A.S. Johnson in Telopea,* 6(2–3): 402 (1995): Eucalyptus tessellaris.
torelliana *(F. Muell.) K.D. Hill & L.A.S. Johnson in Telopea,* 6(2–3): 385 (1995): Eucalyptus torelliana.
torta *K.D. Hill & L.A.S. Johnson in Telopea,* 6(2–3): 418 (1995)—Western Australia.
subsp. allanii *K.D. Hill & L.A.S. Johnson in Telopea,* 6(2–3): 420 (1995)—Western Australia.
subsp. mixtifolia *K.D. Hill & L.A.S. Johnson in Telopea,* 6(2–3): 421 (1995)—Western Australia.
trachyphloia *(F. Muell.) K.D. Hill & L.A.S. Johnson in Telopea,* 6(2–3): 227 (1995): Eucalyptus trachyphloia.
subsp. amphistomatica *K.D. Hill & L.A.S. Johnson in Telopea,* 6(2–3): 230 (1995)—New South Wales, Queensland.
subsp. carnarvonica *K.D. Hill & L.A.S. Johnson in Telopea,* 6(2–3): 231 (1995)—Queensland.
tumescens *K.D. Hill & L.A.S. Johnson in Telopea,* 6(2–3): 321 (1995)—New South Wales, Queensland, South Australia.
umbonata *(D.J. Carr & S.G.M. Carr) K.D. Hill & L.A.S. Johnson in Telopea,* 6(2–3): 300 (1995): Eucalyptus umbonata.
variegata *(F. Muell.) K.D. Hill & L.A.S. Johnson in Telopea,* 6(2–3): 389 (1995): Eucalyptus variegata.

CORYMBIA:-
watsoniana *(F. Muell.) K.D. Hill & L.A.S. Johnson in Telopea,* 6(2–3): 378 (1995): Eucalyptus watsoniana.
subsp. capillata *(Brooker & A.R. Bean) K.D. Hill & L.A.S. Johnson in Telopea,* 6(2–3): 379 (1995): Eucalyptus watsoniana subsp. capillata.
xanthope *(A.R. Bean & Brooker) K.D. Hill & L.A.S. Johnson in Telopea,* 6(2–3): 277 (1995): Eucalyptus xanthope.
zygophylla *(Blakely) K.D. Hill & L.A.S. Johnson in Telopea,* 6(2–3): 349 (1995): Eucalyptus zygophylla.

CORYMBIUM (Compos.).
africanum L.
var. fourcadei *(Hutch.) F.M. Weitz in S. Afr. J. Bot.,* 55(6): 606 (1989): C. fourcadei.
var. gramineum *(Burm. f.) F.M. Weitz in S. Afr. J. Bot.,* 55(6): 606 (1989): C. gramineum.
subsp. scabridum *(P.J. Bergius) F.M. Weitz in S. Afr. J. Bot.,* 55(6): 605 (1989): C. scabridum.
var. scabridum *(P.J. Bergius) F.M. Weitz in S. Afr. J. Bot.,* 55(6): 605 (1989): C. scabridum.
elsiae *F.M. Weitz in S. Afr. J. Bot.,* 55(6): 627 (1989)—Cape Province.
glabrum L.
var. rogersii *(Markötter) F.M. Weitz in S. Afr. J. Bot.,* 55(6): 618 (1989): C. rogersii.
var. rogersii *(Markötter) R. Govaerts, World Checklist Seed Plants,* 3(1): 20 (1999), isonym; *see Weitz in S. Afr. J. Bot.,* 55(6): 618 (1989): C. rogersii.
laxum Compton
subsp. bolusii *F.M. Weitz in S. Afr. J. Bot.,* 55(6): 627 (1989)—Cape Province.

CORYMBORKIS (Orchidac.).
minima *P.J. Cribb in Kew Bull.,* 51(2): 355 (1996)—Cameroon.

CORYPHA (Palmae).
taliera *W. Roxburgh, Pl. Coromandel,* 3: 51 (1820)—India (West Bengal).†

CORYPHANTHA (Cactac.).
glassii *R.F. Dicht & A.D. Lüthy in Kakt. And. Sukk.,* 51(1): 2 (2000)—Mexico Northeast (San Luis Potosí).
hintoniorum *R.F. Dicht & A.D. Lüthy in Kakt. And. Sukk.,* 50(1): 15, 13 (1999)—Mexico Northeast (Nuevo León).
macromeris
subsp. runyonii *(Britton & Rose) N.P. Taylor in Cactaceae Consensus Initiatives,* 6: 15 (1998): C. runyonii.
macrothele *(Mart. ex Pfeiff.) H. Kümmler in Mitteilungsbl. AfM,* 22(2): 109 (1998); *cf. Repert. Pl. Succ. (I.O.S.),* 49: 15 (1998 publ. 1999): Mammillaria macrothele.
muehlenpfordtii Britton & Rose
subsp. robustispina *(Schott ex Engelm.) R.F. Dicht in Kakt. And. Sukk.,* 47(5): 98 (1996), without basionym page; *et in Kakt. And. Sukk.,* 47(12): 252 (1996), with basionym page: Mammillaria robustispina.
subsp. uncinata *(L.D. Benson) R.F. Dicht in Kakt. And. Sukk.,* 47(5): 98 (1996): C. scheeri var. uncinata.
pallida Britton & Rose
subsp. calipensis *(Bravo ex S. Arias, Gama & U.Guzman) R.F. Dicht & A.D. Lüthy in J. Mammillaria Soc.,* 40(4): 50 (2000): Coryphantha calipensis.
panarottoi *J.J. Halda & L. Horáček in Acta Mus. Richnov. Sect. Nat.,* 6(3): 235 (1999)—Mexico Northeast (Nuevo León).
pycnacantha (Mart.) Lem.
var. calipensis *L.U. Guzmán Cruz in Fl. Valle Tehuacán-Cuicatlán,* 14(Cactac.): 12 (1997), as comb. nov. with incorrect basionym ref.—Mexico Central (Puebla).
ramillosa Cutak
subsp. santarosa *R.F. Dicht & A.D. Lüthy in Kakt. And. Sukk.,* 51(6): 141 (2000)—Mexico Northeast (Coahuila).
robustispina (Engelm.) Britton & Rose
subsp. scheeri *(Lem.) N.P. Taylor in Cactaceae Consensus Initiatives,* 6: 18 (1998): C. scheeri.

CORYPHANTHA:-
robustispina (Engelm.) Britton & Rose
subsp. uncinata *(L.D. Benson) N.P. Taylor in Cactaceae Consensus Initiatives,* 6: 18 (1998): C. scheeri var. uncinata.
salinensis *(Poselg.) R.F. Dicht & A.D. Lüthy in Kakt. And. Sukk.,* 49(11): 257 (1998): Echinocactus salinensis.
werdermannii
subsp. unguispina *J.J. Halda, P. Kupčák & J. Sladkovský in Acta Mus. Richnov. Sect. Nat.,* 7(1): 34 (2000)—Mexico Northeast (Coahuila).

CORYTOPLECTUS (Gesneriac.).
cutucuensis *H. Wiehler in Gesneriana,* 1(1): 35 (1995)—Ecuador.

COSMIANTHEMUM (Acanthac.).
guangxiense *H.S. Lo & D. Fang in Guihaia,* 17(1): 42 (1997)—China Southeast (Guangxi).
longiflorum *D. Fang & H.S. Lo in Guihaia,* 17(1): 43 (1997)—China Southeast (Guangxi), Hainan, Vietnam.
viriduliflorum *(C.Y. Wu & H.S. Lo) H.S. Lo in Guihaia,* 17(1): 42 (1997): Graptophyllum viriduliflorum.

COSMOS (Compos.).
juxtlahuacensis *J.L. Panero & J.L. Villaseñor in Brittonia,* 48(1): 79 (1996)—Mexico Southwest (Oaxaca).

COSTUS (Costac.).
foliaceus *J.M. Lock & A.D. Poulsen in Kew Bull.,* 52(3): 614 (1997)—Uganda.
osae *P.J.M. Maas & H. Maas-van de Kamer in Brittonia,* 49(2): 277 (1997)—Costa Rica.
ricus *P.J.M. Maas & H. Maas-van de Kamer in Brittonia,* 49(2): 274 (1997)—Costa Rica.
speciosus (Retz.) Sm.
var. dilnavaziae *M.R. Almeida & S.M. Almeida in J. Bombay Nat. Hist. Soc.,* 96(2): 356 (1999), as 'dilnavazii'—India (Maharashtra).

COTINUS (Anacardiac.).
carranzae *J. Rzedowski & G. Calderón de Rzedowski in Acta Bot. Mex.,* 47: 24 (1999)—Mexico Northeast (Querétaro).
chiangii *(D.A. Young) J. Rzedowski & G. Calderón de Rzedowski in Acta Bot. Mex.,* 47: 28 (1999): Rhus chiangii.

COTONEASTER (Rosac.).
ser. Arbusculi *G. Klotz in Mitt. Deutsch. Dendrol. Ges.,* 82: 78 (1996).
ser. Coriaceii *G. Klotz in Mitt. Deutsch. Dendrol. Ges.,* 82: 76 (1996).
acutifolius Turcz.
var. lucidus *(Schltdl.) L.T. Lu in Acta Phytotax. Sin.,* 38(3): 277 (2000): C. lucidus.
albokermesinus *J. Fryer & B. Hylmö in New Plantsman,* 5(3): 132 (1998)—China South-central (Sichuan).
angustus *(T.T. Yu) G. Klotz in Mitt. Deutsch. Dendrol. Ges.,* 82: 71 (1996): C. salicifolius var. angustus.
arvernensis *M. Gandoger, Dec. Pl. Nov.,* 1: 7 (1875)—France.
bilokonii *A.T. Grevtsova, Atlas Cotoneasters - Cotoneaster (Medic.) Bauhin:* 304 (1999), without type—Tadzhikistan.
cuspidatus *Marquand ex J. Fryer & B. Hylmö in New Plantsman,* 5(3): 138 (1998)—Burma/Xizang border.
dammeri C.K. Schneid.
subsp. songmingensis *C.Y. Wu & L.H. Zhou in Acta Bot. Yunnanica,* 22(4): 380 (2000)—China South-Central (Yunnan, Sichuan, Guizhou).
daralagesicus *A.T. Grevtsova, Atlas Cotoneasters - Cotoneaster (Medic.) Bauhin:* 304 (1999), without type—Transcaucasus (Armenia).
flinckii *J. Fryer & B. Hylmö in New Plantsman,* 5(3): 140 (1998)—China North-central (Gansu).
ganghobaensis *J. Fryer & B. Hylmö in New Plantsman,* 5(3): 136 (1998)—China South-central (Yunnan).
hsingshangensis *J. Fryer & B. Hylmö in Watsonia,* 21(4): 335 (1997)—China South-central (Hubei).

COTONEASTER:-
hummelii *J. Fryer & B. Hylmö in Watsonia,* 21(4): 336 (1997)—China North-central (Gansu).
ichangensis *G. Klotz in Mitt. Deutsch. Dendrol. Ges.,* 82: 72 (1996)—China South-central (Hubei).
induratus *J. Fryer & B. Hylmö in Watsonia,* 21(4): 339 (1997)—China South-central (Yunnan).
kirgizicus *A.T. Grevtsova, Atlas Cotoneasters - Cotoneaster (Medic.) Bauhin:* 305 (1999), without type—Kirgizistan.
logginovae *A.T. Grevtsova, Atlas Cotoneasters - Cotoneaster (Medic.) Bauhin:* 305 (1999), without type—Chita.
magnificus *J. Fryer & B. Hylmö in New Plantsman,* 5(3): 138 (1998)—China South-central (Sichuan).
mathonnetii *M. Gandoger, Dec. Pl. Nov.,* 1: 8 (1875)—France.
melanotrichus *(Franch.) G. Klotz in Mitt. Deutsch. Dendrol. Ges.,* 82: 65 (1996): C. buxifolius f. melanotrichus.
pritzelii *(C.K. Schneid.) G. Klotz in Mitt. Deutsch. Dendrol. Ges.,* 82: 75 (1996): C. rugosus var. pritzelii.
pseudoambiguus *J. Fryer & B. Hylmö in Watsonia,* 21(4): 337 (1997)—China South-central (Sichuan).
pyrenaicus *M. Gandoger, Dec. Pl. Nov.,* 1: 8 (1875)—France.
raboutensis *K.E. Flinck, J. Fryer, L. Garraud, B. Hylmö & J. Zeller in Bull. Mens. Soc. Linn. Lyon,* 67(10): 276 (1998)—France.
rannensis *B. Hylmö & J. Fryer in Acta Bot. Fenn.,* 162: 183 (1999)—Baltic States (Estoniya), Russia Northwest.
russanovii *A.T. Grevtsova, Atlas Cotoneasters - Cotoneaster (Medic.) Bauhin:* 306 (1999), without type—Tadzhikistan.
sargentii *G. Klotz in Mitt. Deutsch. Dendrol. Ges.,* 82: 72 (1996)—China South-central (Sichuan).
shansiensis *J. Fryer & B. Hylmö in New Plantsman,* 5(3): 135 (1998)—China North-central (Shanxi).
smithii *G. Klotz in Mitt. Deutsch. Dendrol. Ges.,* 82: 77 (1996)—China South-central (Yunnan).
sordidus *G. Klotz in Mitt. Deutsch. Dendrol. Ges.,* 82: 72 (1996)—China South-central (Sichuan).
suboblongus *M. Gandoger, Dec. Pl. Nov.,* 1: 8 (1875)—France.
tardiflorus *J. Fryer & B. Hylmö in New Plantsman,* 5(3): 139 (1998)—China South-central (Sichuan).
tengyuehensis *J. Fryer & B. Hylmö in Watsonia,* 21(4): 338 (1997)—China South-central (Yunnan).
tkatschenkoi *A.T. Grevtsova, Atlas Cotoneasters - Cotoneaster (Medic.) Bauhin:* 306 (1999), without type—Kirgizistan.
uralensis *B. Hylmö & J. Fryer in Acta Bot. Fenn.,* 162: 181 (1999)—Russia North.
uzbetzicus *A.T. Grevtsova, Atlas Cotoneasters - Cotoneaster (Medic.) Bauhin:* 307 (1999), without type—Tadzhikistan.
zaprjagaevae *A.T. Grevtsova, Atlas Cotoneasters - Cotoneaster (Medic.) Bauhin:* 307 (1999), without type—Tadzhikistan.

COTTENDORFIA (Bromeliac.).
tillandsioides *(L.B. Sm.) L.B. Smith in Phytologia,* 7(4): 171 (1960), in Supp. 13 as 'Tillandsia' in error: Lindmania tillandsioides.†

COTULA (Compos.).
nigellifolia (DC.) K. Bremer & Humphries
var. tenuior *(DC.) P.P.J. Herman in Bothalia,* 25(2): 244 (1995): Matricaria nigellifolia var. tenuior.
rotundata *(Cheeseman) D.G. Lloyd in New Zealand J. Bot.,* 10(2): 314 (1972), in Supp. 16 as 'rotunda': C. dioica var. rotundata.†
traillii Kirk
subsp. pulchella *(Kirk) D.G. Lloyd in New Zealand J. Bot.,* 10(2): 324 (1972), in Supp. 16 as 'trailii': C. pulchella.†

COTYLEDON (Crassulac.).
eliseae *E. van Jaarsveld in Bradleya,* 15: 65 (1997)—Cape Province.

COTYLEDON:-
platyphylla *(Schrenk) N.E. Brown, Suppl. Johnson's Gard. Dict.:* 907 (1882): Umbilicus platyphyllus.
turkestanica *(Regel & C. Winkl.) N.E. Brown, Suppl. Johnson's Gard. Dict.:* 907 (1882): Umbilicus turkestanicus.

COUEPIA (Chrysobalanac.).
hondurasensis *G.T. Prance in Kew Bull.,* 54(1): 110 (1999)—Honduras.
leitaofilhoi *G.T. Prance in Kew Bull.,* 54(1): 108 (1999)—Brazil Southeast (São Paulo).

COUSINIA (Compos.).
balchanica *O.V. Tscherneva in Novosti Sist. Vyssh. Rast.,* 30: 153 (1996)—Turkmenistan.
dalahuensis *F. Attar & A. Ghahreman in Iranian J. Bot.,* 8(2): 264 (2000)—Iran.
mobayenii *A. Ghahreman & F. Attar in Iranian J. Bot.,* 8(2): 260 (2000)—Iran.
omissa *O.V. Tscherneva in Novosti Sist. Vyssh. Rast.,* 30: 155 (1996)—Kirgizstan.
parsana *A. Ghahreman, M. Iranshahr & F. Attar in Iranian J. Bot.,* 8(1): 16 (1999)—Iran.
shebliensis *A. Ghahreman, M. Iranshahr & F. Attar in Iranian J. Bot.,* 8(1): 19 (1999)—Iran.

COUSSAPOA (Cecropiac.).
david-smithii *C.C. Berg in Novon,* 6(3): 250 (1996)—Bolivia.
villosa Poepp. & Endl.
subsp. polycephala *C.C. Berg in Novon,* 6(3): 251 (1996)—Colombia.

COUSSAREA (Rubiac.).
acuminata *(Ruiz & Pav.) D.C. Zappi in Kew Bull.,* 55(1): 166 (2000): Coffea acuminata.
amplifolia *C.M. Taylor in Fl. Ecuador,* 62(162:3): 248 (1999)—Ecuador, Colombia.
cephaeloides *C.M. Taylor in Fl. Ecuador,* 62(162:3): 251 (1999)—Ecuador.
dulcifolia *D. Neill, C.E. Cerón & C.M. Taylor in Fl. Ecuador,* 62(162:3): 253 (1999)—Ecuador.
ecuadorensis *C.M. Taylor in Fl. Ecuador,* 62(162:3): 255 (1999)—Ecuador.
granvillei *P.G. Delprete & B.M. Boom in Brittonia,* 51(4): 403 (1999)—French Guiana.
linearis *C.M. Taylor in Novon,* 5(4): 379 (1995)—Peru, Colombia.
loftonii (Dwyer & M.V. Hayden) Dwyer
subsp. calimana *C.M. Taylor in Novon,* 9(3): 431 (1999)—Colombia.
pilosula *C.M. Taylor in Fl. Ecuador,* 62(162:3): 266 (1999)—Ecuador, Brazil North (Acre), Peru.
resinosa *C.M. Taylor in Fl. Ecuador,* 62(162:3): 269 (1999)—Peru, Ecuador, Colombia.
sancti-ciprianii *C.M. Taylor in Novon,* 9(3): 433 (1999)—Colombia.
speciosa *K. Schum. ex M. Gomes in Reserva Ecol. Macaé de Cima: Aspectos Flor. Espéc. Vasc.,* 2: 412 (1996)—Brazil Southeast (Rio de Janeiro).
spiciformis *C.M. Taylor in Fl. Ecuador,* 62(162:3): 270 (1999)—Ecuador.

CRAMBE (Crucif.).
hispanica L.
subsp. abyssinica *(Hochst. ex R.E. Fr.) A. Prina in Bot. J. Linn. Soc.,* 133(4): 517 (2000): C. abyssinica.
var. meyeri *(O.E. Schulz) A. Prina in Bot. J. Linn. Soc.,* 133(4): 519 (2000): C. abyssinica var. meyeri.
santosii *D. Bramwell in Bot. Macaronésica,* 22: 111 (1995), nom. nov., without exact replaced synonym page: C. gigantea *(Ceballos & Ortuño) Bramwell.*
wildpretii *A.O. Prina & D. Bramwell in Ann. Bot. Fenn.,* 37(4): 301 (2000)—Canary Is.

CRANICHIS (Orchidac.).
cochleata *R.L. Dressler in Bol. Inst. Bot. (Guadalajara),* 5(1–3): 70 (1998)—Mexico Southeast (Chiapas), Mexico Gulf (Veracruz), Guatemala.
notata *R.L. Dressler in Bol. Inst. Bot. (Guadalajara),* 5(1–3): 73 (1998)—Mexico Southeast (Chiapas), Honduras.
talamancana *R.L. Dressler in Bol. Inst. Bot. (Guadalajara),* 5(1–3): 75 (1998)—Costa Rica.

CRANIOSPERMUM (Boraginac.).
ser. Diploloma *(Schrenk) S.V. Ovczinnikova in Bot. Zhurn.*, 85(12): 83 (2000): gen. Diploloma.
subsect. Diploloma *(Schrenk) S.V. Ovczinnikova in Bot. Zhurn.*, 85(12): 83 (2000): gen. Diploloma.
subsect. Floccosa *S.V. Ovczinnikova in Bot. Zhurn.*, 85(12): 83 (2000).
sect. Leucolachnos *S.V. Ovczinnikova in Bot. Zhurn.*, 85(12): 79 (2000).
ser. Tuvinica *S.V. Ovczinnikova in Bot. Zhurn.*, 85(12): 84 (2000).
tuvinicum *S.V. Ovczinnikova in Bot. Zhurn.*, 85(12): 84 (2000)—Tuva.

CRASPEDIA (Compos.).
robusta *(Hook.f.) Cockayne in Trans & Proc. New Zealand Inst.*, 45: 262 (1913): C. fimbriata var. robusta

CRASSULA (Crassulac.).
ausensis
subsp. titanopsis *P. Pavelka in Kaktusy (Brno)*, 34(2): 39 (1998)—Namibia.
cremnophila *E. van Jaarsveld & A.E. van Wyk in Aloe*, 36(4): 71 (1999)—Cape Province.
setulosa
nothovar. schlechteri *(Schönland) R. Govaerts, World Checklist Seed Plants*, 3(1): 20 (1999): C. × schlechteri.

CRATAEGUS (Rosac.).
ser. Baroussanae *J.B. Phipps in Sida, Bot. Misc.*, 15: 14 (1997).
ser. Cerrones *J.B. Phipps in Canad. J. Bot.*, 76(11): 1872 (1998 publ. 1999).
ser. Grandifoliae *J.B. Phipps in Sida, Bot. Misc.*, 15: 14 (1997).
ser. Greggianae *J.B. Phipps in Sida, Bot. Misc.*, 15: 14 (1997).
ser. Madrenses *J.B. Phipps in Sida, Bot. Misc.*, 15: 14 (1997).
ser. Purpureofructi *J.B. Phipps & R.J. O'Kennon in Sida*, 18(1): 184 (1998).
ser. Sulfureae *J.B. Phipps in Sida, Bot. Misc.*, 15: 14 (1997).
alemanniensis Cinovskis
var. heterodonta *(Pojark.) R. Cinovskis in Fl. Baltiĭskikh Respublik*, 2: 119 (1996): C. heterodonta.
var. orientobaltica *(Cinovskis) R. Cinovskis in Fl. Baltiĭskikh Respublik*, 2: 119 (1996): C. orientobaltica.
var. subborealis *(Cinovskis) R. Cinovskis in Fl. Baltiĭskikh Respublik*, 2: 119 (1996): C. subborealis.
araneosa *R. Doll in Nat. Naturschutz Mecklenburg*, 13: 11 (1975)—Germany.
aurescens *J.B. Phipps in Sida, Bot. Misc.*, 15: 54 (1997)—Mexico Northeast (Nuevo León, Coahuila).
baroussana Eggl.
var. jamensis *J.B. Phipps in Sida, Bot. Misc.*, 15: 76 (1997)—Mexico Northeast (Coahuila).
× bergiana *R. Doll in Nat. Naturschutz Mecklenburg*, 13: 13 (1975)—Germany.(C. alemanniensis × lindmanii).
× borealoides *R. Doll in Nat. Naturschutz Mecklenburg*, 13: 13 (1975)—Germany.(C. orientobaltica × subborealis).
× brandenii *R. Doll in Nat. Naturschutz Mecklenburg*, 13: 13 (1975)—Germany.(C. lindmanii × oxyacantha).
cuprina *J.B. Phipps in Sida, Bot. Misc.*, 15: 77 (1997)—Mexico Northeast (Nuevo León, Coahuila).
× curvisepaloides *R. Doll in Nat. Naturschutz Mecklenburg*, 13: 15 (1975)—Germany.(C. curvisepala × orientobaltica).
dabelowa *R. Doll in Nat. Naturschutz Mecklenburg*, 13: 15 (1975)—Germany.
× daghestanica *V.N. Gladkova in Novosti Sist. Vyssh. Rast.*, 30: 97 (1996)—North Caucasus (Dagestan). (C. microphylla × pseudoheterophylla).
× eriocalyx *(Sanio) R. Cinovskis in Fl. Baltiĭskikh Respublik*, 2: 110 (1996): C. oxyacantha f. eriocalyx.
nothovar. villosa *(Sanio) R. Cinovskis in Fl. Baltiĭskikh Respublik*, 2: 110 (1996): C. oxyacantha f. villosa.

CRATAEGUS:-
× feldbergii *R. Doll in Nat. Naturschutz Mecklenburg*, 13: 16 (1975)—Germany.(C. monoli × orientobaltica).
× glauca *R. Doll in Nat. Naturschutz Mecklenburg*, 13: 18 (1975), homonym; *see G. Don, Gen. Hist.*, 2: 598 (1832)—Germany.(C. curcina × monogyna).
gracilior *J.B. Phipps in Sida, Bot. Misc.*, 15: 26 (1997), nom. nov.: Mespilus pubescens *Humb., Bonpl. & Kunth.*
grandifolia *J.B. Phipps in Sida, Bot. Misc.*, 15: 88 (1997)—Mexico Northeast (Coahuila).
var. potosina *J.B. Phipps in Sida, Bot. Misc.*, 15: 91 (1997)—Mexico Northeast (Nuevo León).
greggiana Eggl.
var. pepo *J.B. Phipps in Sida, Bot. Misc.*, 15: 63 (1997)—Mexico Northeast (Coahuila, Nuevo León).
× integerrima *R. Doll in Nat. Naturschutz Mecklenburg*, 13: 18 (1975)—Germany.(C. dabelowa × oxyacantha).
johnstonii *J.B. Phipps in Sida, Bot. Misc.*, 15: 79 (1997)—Mexico Northeast (Coahuila).
laevigata (Poir.) DC.
f. ellipticifolia *(Cinovskis) R. Cinovskis in Fl. Baltiĭskikh Respublik*, 2: 110 (1996): C. oxyacantha f. ellipticifolia.
× luzinii *R. Doll in Nat. Naturschutz Mecklenburg*, 13: 19 (1975)—Germany. (C. lindmanii × orientobaltica).
× macrocarpa
nothovar. calycina *(Peterm.) R. Govaerts, World Checklist Seed Plants*, 3(1): 20 (1999): C. × calycina.
× mechowa *R. Doll in Nat. Naturschutz Mecklenburg*, 13: 21 (1975)—Germany.(C. curvisepala × oxyacantha).
× media Bechst.
nothovar. intermixta *(Wenz.) R. Cinovskis in Fl. Baltiĭskikh Respublik*, 2: 110 (1996): Mespilus oxyacantha ε intermixta.
monogyna Jacq.
var. poloniensis *(Cinovskis) R. Cinovskis in Fl. Baltiĭskikh Respublik*, 2: 117 (1996): C. poloniensis.
nananixonii *J.B. Phipps & R.J. O'Kennon in Sida*, 17(3): 569 (1997)—Texas.
okanaganensis *J.B. Phipps & R.J. O'Kennon in Sida*, 18(1): 178 (1998)—British Columbia, Idaho, Washington.
okennonii *J.B. Phipps in Sida*, 18(1): 170 (1998)—British Columbia, Idaho, Montana, Washington.
× oxyacanthoides *R. Doll in Nat. Naturschutz Mecklenburg*, 13: 21 (1975), homonym; *see Thuill., Fl. Env. Paris, ed. 2:* 245 (1799)—Germany.(C. orientobaltica × oxyacantha).
× petiolata *R. Doll in Nat. Naturschutz Mecklenburg*, 13: 23 (1975)—Germany.(C. calycina × monogyna).
phippsii *R.J. O'Kennon in Sida*, 18(1): 185 (1998)—British Columbia, Montana, Washington.
× pseudoalemanniensis *R. Doll in Nat. Naturschutz Mecklenburg*, 13: 23 (1975), as 'pseudo-alemanniensis'—Germany.(C. alemanniensis × lindmanii).
× rosacea *R. Doll in Nat. Naturschutz Mecklenburg*, 13: 23 (1975)—Germany.(C. oxyacantha × subborealis).
rosei Eggl.
var. amoena *J.B. Phipps in Sida, Bot. Misc.*, 15: 39 (1997)—Mexico Northeast (Nuevo León, Coahuila).
var. mahindae *J.B. Phipps in Sida, Bot. Misc.*, 15: 38 (1997)—Mexico Northeast (Tamaulipas, San Luis Potosí).
subsp. parryana *(Eggl.) J.B. Phipps in Sida, Bot. Misc.*, 15: 42 (1997): C. parryana.
serratissima *J.B. Phipps in Sida, Bot. Misc.*, 15: 70 (1997)—Mexico Northeast (Hidalgo, Querétaro, San Luis Potosí).
× subsphaericea Gand.
nothovar. domicensis *(Hrabětova) K.I. Christensen in Feddes Repert.*, 108(1–2): 104 (1997): C. domicensis.
sulfurea *J.B. Phipps in Sida, Bot. Misc.*, 15: 81 (1997)—Mexico Northeast (Coahuila, Nuevo Leon).
tkatschenkoi *K.I. Christensen in Willdenowia*, 25(2): 633 (1996), nom. nov.: C. trilobata *Tkatsch.*

CRATAEGUS:-
tracyi Ashe ex Eggl.
var. coahuilensis *J.B. Phipps in Sida, Bot. Misc.*, 15: 49 (1997)—Mexico Northeast (Coahuila).
var. madrensis *J.B. Phipps in Sida, Bot. Misc.*, 15: 50 (1997)—Mexico Northeast (Coahuila, Nuevo León).
× uhrovae Soó
nothovar. subcurvisepala *(Cinovskis) R. Cinovskis in Fl. Baltiĭskikh Respublik*, 2: 112 (1996): C. pseudooxyacantha var. subcurvisepala.
ulotricha *Pojark. ex V.N. Gladkova in Novosti Sist. Vyssh. Rast.*, 30: 96 (1996)—Transcaucasus (Armeniya).
× werdana *R. Doll in Nat. Naturschutz Mecklenburg*, 13: 26 (1975)—Germany.(C. lindmanii × poloniensis).
zagrica *M. Khatamsaz in Iranian J. Bot.*, 8(2): 182 (2000)—Iran.

CRATEVA (Capparac.).
formosensis *(Jacobs) B.S. Sun in Fl. Reipubl. Popul. Sin.*, 32: 489 (1999): C. adansonii subsp. formosensis.
simplicifolia *J.S. Miller in Novon*, 8(2): 168 (1998)—Madagascar.
trifoliata *(Roxb.) B.S. Sun in Fl. Reipubl. Popul. Sin.*, 32: 489 (1999): Capparis trifoliata.

CRATOXYLUM (Guttif.).
arborescens (Vahl) Blume
var. borneense *A.C. Church & P.F. Stevens in Blumea*, 42(2): 403 (1997)—Borneo (Kalimantan, Sarawak).

CRATYLIA (Legumin.).
intermedia *(Hassl.) L.P. de Queiroz & R. Monteiro in Bol. Bot. (São Paulo)*, 16: 108 (1997): C. floribunda var. intermedia.

CRAWFURDIA (Gentianac.).
lobatilimba *W.L. Cheng in Acta Phytotax. Sin.*, 36(5): 461 (1998)—Xizang.
nyingchiensis *K. Yao & W.L. Cheng in Acta Phytotax. Sin.*, 36(5): 459 (1998)—Xizang.

CREMANTHODIUM (Compos.).
microglossum *S.W. Liu in Novon*, 6(2): 185 (1996)—Qinghai, China South-central (Yunnan), China North-central (Gansu).

CREMASTRA (Orchidac.).
aphylla *T. Yukawa in Ann. Tsukuba Bot. Gard.*, 18: 59 (1999)—Japan.

CRENULLUMA *D.C.H. Plowes in Haseltonia*, 3: 66 (1995). *ASCLEPIADACEAE*
adenensis *(Deflers) D.C.H. Plowes in Haseltonia*, 3: 66 (1995): Boucerosia adenensis.
arabica *(N.E. Br.) D.C.H. Plowes in Haseltonia*, 3: 66 (1995): Caralluma arabica.
aucheriana *(Decne.) D.C.H. Plowes in Haseltonia*, 3: 66 (1995): Boucerosia aucheriana.
awdeliana *(Deflers) D.C.H. Plowes in Haseltonia*, 3: 66 (1995): Boucerosia awdeliana.
dolichocarpa *(O. Schwartz) D.C.H. Plowes in Haseltonia*, 3: 67 (1995): Caralluma dolichocarpa.
flava *(N.E. Br.) D.C.H. Plowes in Haseltonia*, 3: 67 (1995): Caralluma flava.
kalmbacheriana *(Lavranos) D.C.H. Plowes in Haseltonia*, 3: 67 (1995): Caralluma kalmbacheriana.
lavranii *(Rauh & Wertel) D.C.H. Plowes in Haseltonia*, 3: 67 (1995): Caralluma lavranii.
petraea *(Lavranos) D.C.H. Plowes in Haseltonia*, 3: 67 (1995): Caralluma petraea.
rauhii *(Lavranos) D.C.H. Plowes in Haseltonia*, 3: 67 (1995): Caralluma rauhii.

CREPIDIASTRUM (Compos.).
sect. Crepidopsis *(Babc. & Stebbins) A.N. Sennikov in Bot. Zhurn.*, 82(5): 115 (1997): Youngiasect. Crepidopsis.
denticulatum (Houtt.) Pak & Kawano
var. cornutum *A.N. Sennikov in Bot. Zhurn.*, 82(5): 114 (1997)—South Korea.
f. pinnatipartitum *(Makino) A.N. Sennikov in Bot. Zhurn.*, 82(5): 114 (1997): Lactuca denticulata f. pinnatipartita.

CREPIDIASTRUM:-

humifusum *(Dunn) A.N. Sennikov in Bot. Zhurn.*, 82(5): 115 (1997): Lactuca humifusa.

sonchifolium (Bunge) Pak & Kawano

var. elegans *(Franch.) A.N. Sennikov in Bot. Zhurn.*, 82(5): 114 (1997): Lactuca elegans.

tenuicaule *(Babc. & Stebbins) A.N. Sennikov in Bot. Zhurn.*, 82(5): 115 (1997): Youngia tenuifolia subsp. tenuicaulis.

tenuifolium *(Willd.) A.N. Sennikov in Bot. Zhurn.*, 82(5): 115 (1997): Crepis tenuifolia.

CREPIDIUM (Orchidac.).

sect. Commelinoides *(Schltr.) D.L. Szlachetko in Fragm. Flor. Geobot.*, Supp. 3: 123 (1995): Microstylis sect. Commelinoides.

subgen. Pseudoliparis *(Finet) D.L. Szlachetko in Fragm. Flor. Geobot.*, Supp. 3: 123 (1995): gen. Pseudoliparis.

acuminatum *(D. Don) D.L. Szlachetko in Fragm. Flor. Geobot.*, Supp. 3: 123 (1995): Malaxis acuminata.

acutangulum *(Hook. f.) D.L. Szlachetko in Fragm. Flor. Geobot.*, Supp. 3: 123 (1995): Microstylis acutangula.

alagense *(Ames) M.A. Clements & D.L. Jones in Lasianthera*, 1(1): 34 (1996): Microstylis alagensis.

amplectens *(J.J. Sm.) D.L. Szlachetko in Fragm. Flor. Geobot.*, Supp. 3: 124 (1995): Microstylis amplectens.

apollinis *(P.F. Hunt) M.A. Clements & D.L. Jones in Lasianthera*, 1(1): 34 (1996): Malaxis apollinis.

arachnoideum *(Schltr.) D.L. Szlachetko in Fragm. Flor. Geobot.*, Supp. 3: 124 (1995): Microstylis arachnoidea.

arietinum *(Ames) D.L. Szlachetko in Fragm. Flor. Geobot.*, Supp. 3: 124 (1995): Malaxis arietina.

aschistum *(Seidenf.) D.L. Szlachetko in Fragm. Flor. Geobot.*, Supp. 3: 124 (1995): Malaxis aschista.

atratum *(Schltr.) D.L. Szlachetko in Fragm. Flor. Geobot.*, Supp. 3: 124 (1995): Microstylis atrata.

atrobrachiatum *(Ridl.) D.L. Szlachetko in Fragm. Flor. Geobot.*, Supp. 3: 124 (1995): Microstylis atrobrachiatum.

balabacense *(Ames) D.L. Szlachetko in Fragm. Flor. Geobot.*, Supp. 3: 124 (1995): Malaxis balabacensis.

bancanoides *(Ames) D.L. Szlachetko in Fragm. Flor. Geobot.*, Supp. 3: 124 (1995): Malaxis bancanoides.

bancanum *(Ridl.) D.L. Szlachetko in Fragm. Flor. Geobot.*, Supp. 3: 124 (1995): Microstylis bancana.

bataanense *(Ames) M.A. Clements & D.L. Jones in Lasianthera*, 1(1): 35 (1996): Malaxis bataanensis.

bellum *(Rchb. f.) D.L. Szlachetko in Fragm. Flor. Geobot.*, Supp. 3: 124 (1995): Microstylis bella.

bengkulense *(J.J. Sm.) D.L. Szlachetko in Fragm. Flor. Geobot.*, Supp. 3: 124 (1995): Microstylis bengkulensis.

biauritum *(Lindl.) D.L. Szlachetko in Fragm. Flor. Geobot.*, Supp. 3: 124 (1995): Microstylis biaurita.

bilobum *(Lindl.) D.L. Szlachetko in Fragm. Flor. Geobot.*, Supp. 3: 124 (1995), with incorrect basionym ref.: Microstylis biloba.

binabayense *(Ames) D.L. Szlachetko in Fragm. Flor. Geobot.*, Supp. 3: 124 (1995): Microstylis binabayensis.

bispiriferum *(J.J. Sm.) D.L. Szlachetko in Fragm. Flor. Geobot.*, Supp. 3: 124 (1995): Microstylis bispirifera.

blumei *(Boerl. & J.J. Sm.) D.L. Szlachetko in Fragm. Flor. Geobot.*, Supp. 3: 124 (1995): Microstylis blumei.

brachycaulos *(Schltr.) D.L. Szlachetko in Fragm. Flor. Geobot.*, Supp. 3: 124 (1995): Microstylis brachycaulos.

brachyodontum *(Schltr.) D.L. Szlachetko in Fragm. Flor. Geobot.*, Supp. 3: 124 (1995): Microstylis brachyodonta.

bracteosum *(Ames) D.L. Szlachetko in Fragm. Flor. Geobot.*, Supp. 3: 125 (1995): Malaxis bracteosa.

brevidentatum *(C. Schweinf.) M.A. Clements & D.L. Jones in Lasianthera*, 1(1): 35 (1996): Malaxis brevidentata.

CREPIDIUM:-

breviscapum *(Schltr.) D.L. Szlachetko in Fragm. Flor. Geobot.*, Supp. 3: 125 (1995): Microstylis breviscapa.

burbidgei *(Rchb. f. ex Ridl.) D.L. Szlachetko in Fragm. Flor. Geobot.*, Supp. 3: 125 (1995): Microstylis burbidgei.

calcareum *(Schltr.) D.L. Szlachetko in Fragm. Flor. Geobot.*, Supp. 3: 125 (1995): Microstylis calcarea.

calophyllum *(Rchb. f.) D.L. Szlachetko in Fragm. Flor. Geobot.*, Supp. 3: 125 (1995): Microstylis calophylla.

caricoides *(Schltr.) D.L. Szlachetko in Fragm. Flor. Geobot.*, Supp. 3: 125 (1995): Microstylis caricoides.

carinatifolium *(J.J. Sm.) D.L. Szlachetko in Fragm. Flor. Geobot.*, Supp. 3: 125 (1995): Microstylis carinatifolia.

carrii *(Seidenf. & J.J. Wood) M.A. Clements & D.L. Jones in Lasianthera*, 1(1): 35 (1996): Microstylis carrii.

celebicum *(Schltr.) D.L. Szlachetko in Fragm. Flor. Geobot.*, Supp. 3: 125 (1995): Microstylis celebica.

chlorophrys *(Rchb. f.) D.L. Szlachetko in Fragm. Flor. Geobot.*, Supp. 3: 125 (1995): Microstylis chlorophrys.

circaeum *(Ridl.) D.L. Szlachetko in Fragm. Flor. Geobot.*, Supp. 3: 125 (1995): Microstylis circaea.

comans *(C. Schweinf.) M.A. Clements & D.L. Jones in Lasianthera*, 1(1): 35 (1996): Malaxis comans.

comberi *D.L. Szlachetko & H.B. Margońska in Fragm. Flor. Geobot.*, 43(1): 9 (1998)—Sumatera.

commelinifolium *(Zoll. & C. Morren) D.L. Szlachetko in Fragm. Flor. Geobot.*, Supp. 3: 125 (1995): Microstylis commelinifolia.

concavum *(Seidenf.) D.L. Szlachetko in Fragm. Flor. Geobot.*, Supp. 3: 125 (1995): Malaxis concava.

copelandii *(Ames) D.L. Szlachetko in Fragm. Flor. Geobot.*, Supp. 3: 125 (1995): Malaxis copelandii.

cordiglottis *(J.J. Sm.) M.A. Clements & D.L. Jones in Lasianthera*, 1(1): 35 (1996): Microstylis cordiglottis.

cribbianum *D.L. Szlachetko & H.B. Margońska in Adansonia*, 20(2): 346 (1998)—Borneo (Sarawak).

cucullatum *(Schltr.) D.L. Szlachetko in Fragm. Flor. Geobot.*, Supp. 3: 125 (1995): Microstylis cucullata.

cuneipetalum *(Ames) D.L. Szlachetko in Fragm. Flor. Geobot.*, Supp. 3: 125 (1995): Malaxis cuneipetala.

curvatulum *(Schltr.) D.L. Szlachetko in Fragm. Flor. Geobot.*, Supp. 3: 125 (1995): Microstylis curvatula.

cyanobrachium *(Schltr.) D.L. Szlachetko in Fragm. Flor. Geobot.*, Supp. 3: 125 (1995): Microstylis cyanobrachia.

davanense *(Ames) D.L. Szlachetko in Fragm. Flor. Geobot.*, Supp. 3: 126 (1995): Malaxis davanensis.

decumbens *(Schltr.) D.L. Szlachetko in Fragm. Flor. Geobot.*, Supp. 3: 126 (1995): Microstylis decumbens.

dentatum *(Ames) D.L. Szlachetko in Fragm. Flor. Geobot.*, Supp. 3: 126 (1995): Malaxis dentata.

dewildeanum *D.L. Szlachetko & H.B. Margońska in Fragm. Flor. Geobot.*, 43(2): 184 (1998)—Sumatera.

diploceras *(Schltr.) D.L. Szlachetko in Fragm. Flor. Geobot.*, Supp. 3: 126 (1995): Microstylis diploceras.

distans *(Schltr.) D.L. Szlachetko in Fragm. Flor. Geobot.*, Supp. 3: 126 (1995): Microstylis distans.

dolichostachyum *(Schltr.) D.L. Szlachetko in Fragm. Flor. Geobot.*, Supp. 3: 126 (1995): Microstylis dolichostachya.

dresslerianum *H.B. Margońska & D.L. Szlachetko in Adansonia*, 22(2): 265 (2000)—New Guinea (Irian Jaya).

dryadum *(Schltr.) D.L. Szlachetko in Fragm. Flor. Geobot.*, Supp. 3: 126 (1995): Microstylis dryadum.

elmeri *(Ames) D.L. Szlachetko in Fragm. Flor. Geobot.*, Supp. 3: 126 (1995): Malaxis elmeri.

epidendrum *M.A. Clements & D.L. Jones in Lasianthera*, 1(1): 35 (1996), nom. nov.: Malaxis epiphytica *Ames*.

CREPIDIUM:-

epiphyticum *(Schltr.) D.L. Szlachetko in Fragm. Flor. Geobot.*, Supp. 3: 126 (1995): Microstylis epiphytica.

euanthum *(Schltr.) M.A. Clements & D.L. Jones in Lasianthera*, 1(1): 35 (1996): Microstylis euantha.

fasciatum *(Schltr.) D.L. Szlachetko in Fragm. Flor. Geobot.*, Supp. 3: 126 (1995): Microstylis fasciata.

fimbriatum *(Lavarack) D.L. Szlachetko in Fragm. Flor. Geobot.*, Supp. 3: 126 (1995): Malaxis fimbriata.

fissum *(Schltr.) D.L. Szlachetko in Fragm. Flor. Geobot.*, Supp. 3: 126 (1995): Microstylis fissa.

flammeum *D.L. Szlachetko & H.B. Margońska in Fragm. Flor. Geobot.*, 43(2): 186 (1998)—Sumatera.

flavovirens *D.L. Jones & M.A. Clements in Novon*, 7(4): 376 (1997 publ. 1998)—Queensland.

floscularium *(J.J. Sm.) M.A. Clements & D.L. Jones in Lasianthera*, 1(1): 36 (1996): Microstylis floscularia.

fulvum *(Schltr.) M.A. Clements & D.L. Jones in Lasianthera*, 1(1): 36 (1996): Microstylis fulva.

gibbosum *(J.J. Sm.) D.L. Szlachetko in Fragm. Flor. Geobot.*, Supp. 3: 126 (1995): Microstylis gibbosa.

glabrum *(Kraenzl.) M.A. Clements & D.L. Jones in Lasianthera*, 1(1): 36 (1996): Goodyera glabra.

godefroyi *(Rchb. f.) D.L. Szlachetko in Fragm. Flor. Geobot.*, Supp. 3: 126 (1995): Microstylis godefroyi.

graciliscapum *(Ames & C. Schweinf.) D.L. Szlachetko in Fragm. Flor. Geobot.*, Supp. 3: 126 (1995): Malaxis graciliscapa.

gracillimum *M.A. Clements & D.L. Jones in Lasianthera*, 1(1): 36 (1996), nom. nov.: Microstylis graciliscapa *Schltr.*

graminifolium *(Schltr.) D.L. Szlachetko in Fragm. Flor. Geobot.*, Supp. 3: 126 (1995): Microstylis graminifolia.

grandiflorum *(J.J. Sm.) D.L. Szlachetko in Fragm. Flor. Geobot.*, Supp. 3: 126 (1995): Microstylis grandiflora.

grandifolium *(Schltr.) D.L. Szlachetko in Fragm. Flor. Geobot.*, Supp. 3: 127 (1995): Microstylis grandifolia.

heliophilum *(Schltr.) D.L. Szlachetko in Fragm. Flor. Geobot.*, Supp. 3: 127 (1995): Microstylis heliophila.

heliophobum *(J.J. Sm.) D.L. Szlachetko in Fragm. Flor. Geobot.*, Supp. 3: 127 (1995): Microstylis heliophoba.

holttumianum *D.L. Szlachetko in Fragm. Flor. Geobot.*, Supp. 3: 127 (1995), nom. nov.: Malaxis stenophylla *Holttum*.

hutchinsonianum *(Ames) D.L. Szlachetko in Fragm. Flor. Geobot.*, Supp. 3: 127 (1995): Microstylis hutchinsoniana.

hydrophilum *(J.J. Sm.) D.L. Szlachetko in Fragm. Flor. Geobot.*, Supp. 3: 127 (1995): Microstylis hydrophila.

imthurnii *(Rolfe) M.A. Clements & D.L. Jones in Lasianthera*, 1(1): 36 (1996): Microstylis imthurnii.

incurvum *(J.J. Sm.) D.L. Szlachetko in Fragm. Flor. Geobot.*, Supp. 3: 127 (1995): Microstylis incurva.

inexpectatum *(J.J. Sm.) D.L. Szlachetko in Fragm. Flor. Geobot.*, Supp. 3: 127 (1995): Microstylis inexpectata.

integrilabium *(Schltr.) D.L. Szlachetko in Fragm. Flor. Geobot.*, Supp. 3: 127 (1995): Microstylis integrilabium.

irregularis *(J.J. Sm.) D.L. Szlachetko in Fragm. Flor. Geobot.*, Supp. 3: 127 (1995): Microstylis irregularis.

junghuhnii *(J.J. Sm.) D.L. Szlachetko in Fragm. Flor. Geobot.*, Supp. 3: 127 (1995): Microstylis junghuhnii.

kabense *(J.J. Sm.) D.L. Szlachetko in Fragm. Flor. Geobot.*, Supp. 3: 127 (1995): Microstylis kabensis.

kempfii *(Schltr.) M.A. Clements & D.L. Jones in Lasianthera*, 1(1): 36 (1996): Microstylis kempfii.

kerintjiense *(J.J. Sm.) D.L. Szlachetko in Fragm. Flor. Geobot.*, Supp. 3: 127 (1995): Microstylis kerintjiensis.

kerstigianum *(Schltr.) D.L. Szlachetko in Fragm. Flor. Geobot.*, Supp. 3: 127 (1995): Microstylis kerstigiana.

CREPIDIUM:-

khasianum *(Hook. f.) D.L. Szlachetko in Fragm. Flor. Geobot.*, Supp. 3: 127 (1995): Microstylis khasiana.

kinabaluense *(Rolfe) D.L. Szlachetko in Fragm. Flor. Geobot.*, Supp. 3: 127 (1995): Microstylis kinabaluensis.

klabatense *(Schltr.) D.L. Szlachetko in Fragm. Flor. Geobot.*, Supp. 3: 127 (1995): Microstylis klabatensis.

kobi *(J.J. Sm.) M.A. Clements & D.L. Jones in Lasianthera*, 1(1): 36 (1996): Microstylis kobi.

kordersii *(J.J. Sm.) D.L. Szlachetko in Fragm. Flor. Geobot.*, Supp. 3: 127 (1995): Microstylis kordersii.

laeve *(Schltr.) D.L. Szlachetko in Fragm. Flor. Geobot.*, Supp. 3: 128 (1995), as 'laevis': Microstylis laevis.

lamii *(J.J. Sm.) M.A. Clements & D.L. Jones in Lasianthera*, 1(1): 37 (1996): Microstylis lamii.

langkawiense *(Seidenf. & J.J. Wood) D.L. Szlachetko in Fragm. Flor. Geobot.*, Supp. 3: 128 (1995): Malaxis langkawiensis.

latilabre *(Schltr.) M.A. Clements & D.L. Jones in Lasianthera*, 1(1): 37 (1996), as 'latilabrum': Microstylis latilabris.

latipetalum *(J.J. Sm.) D.L. Szlachetko in Fragm. Flor. Geobot.*, Supp. 3: 128 (1995): Microstylis latipetala.

latisegmentum *(C. Schweinf.) M.A. Clements & D.L. Jones in Lasianthera*, 1(1): 37 (1996): Malaxis latisegmenta.

latisepalum *(Rolfe) M.A. Clements & D.L. Jones in Lasianthera*, 1(1): 37 (1996): Microstylis latisepala.

latum *(J.J. Sm.) D.L. Szlachetko in Fragm. Flor. Geobot.*, Supp. 3: 128 (1995): Microstylis lata.

lawleri *(Lavarack & B. Gray) D.L. Szlachetko in Fragm. Flor. Geobot.*, Supp. 3: 128 (1995): Malaxis lawleri.

laxum *(Ridl.) D.L. Szlachetko in Fragm. Flor. Geobot.*, Supp. 3: 128 (1995): Microstylis laxa.

ledermannii *(Schltr.) D.L. Szlachetko in Fragm. Flor. Geobot.*, Supp. 3: 128 (1995): Microstylis ledermannii.

leucodon *(Schltr.) D.L. Szlachetko in Fragm. Flor. Geobot.*, Supp. 3: 128 (1995): Microstylis leucodon.

lilacinum *(Ames) M.A. Clements & D.L. Jones in Lasianthera*, 1(1): 37 (1996): Malaxis lilacina.

lobatocallosum *(J.J. Sm.) D.L. Szlachetko in Fragm. Flor. Geobot.*, Supp. 3: 128 (1995): Microstylis lobatocallosa.

lokonense *(Schltr.) M.A. Clements & D.L. Jones in Lasianthera*, 1(1): 37 (1996): Microstylis lokonensis.

longifolium *(Rolfe) M.A. Clements & D.L. Jones in Lasianthera*, 1(1): 37 (1996): Microstylis longifolia.

longispicum *(Schltr.) D.L. Szlachetko in Fragm. Flor. Geobot.*, Supp. 3: 128 (1995): Microstylis longispica.

lowii *(E. Morren) D.L. Szlachetko in Fragm. Flor. Geobot.*, Supp. 3: 128 (1995): Microstylis lowii.

lunatum *(Schltr.) M.A. Clements & D.L. Jones in Lasianthera*, 1(1): 38 (1996): Microstylis lunata.

luniferum *(J.J. Sm.) D.L. Szlachetko in Fragm. Flor. Geobot.*, Supp. 3: 128 (1995): Microstylis lunifera.

lyroglossum *D.L. Szlachetko & H.B. Margońska in Adansonia*, 20(2): 342 (1998)—Lesser Sunda Is.

maaikeae *H.B. Margonska, D.L. Szlachetko & T. Kubala in Orchidee*, 50(6): 651 (1999)—New Guinea (Papua New Guinea).

maboroensis *(Schltr.) D.L. Szlachetko in Fragm. Flor. Geobot.*, Supp. 3: 128 (1995): Microstylis maboroensis.

macgregorii *(Ames) M.A. Clements & D.L. Jones in Lasianthera*, 1(1): 38 (1996) Malaxis macgregorii.

mackinnonii *(Duthie) D.L. Szlachetko in Fragm. Flor. Geobot.*, Supp. 3: 128 (1995): Microstylis mackinnonii.

macrochilum *(Rolfe) D.L. Szlachetko in Fragm. Flor. Geobot.*, Supp. 3: 128 (1995): Microstylis macrochila.

macrophyllum *(Schltr.) D.L. Szlachetko in Fragm. Flor. Geobot.*, Supp. 3: 128 (1995): Microstylis macrophylla.

CREPIDIUM:-

macrotis *(Kraenzl.) D.L. Szlachetko in Fragm. Flor. Geobot.*, Supp. 3: 128 (1995): Liparis macrotis.

magnifoveum *(J.J. Sm.) D.L. Szlachetko in Fragm. Flor. Geobot.*, Supp. 3: 129 (1995): Microstylis magnifovea.

marginatum *(J.J. Sm.) D.L. Szlachetko in Fragm. Flor. Geobot.*, Supp. 3: 129 (1995): Microstylis marginata.

mariae *H.B. Margońska & D.L. Szlachetko in Ann. Bot. Fenn.*, 37(4): 303 (2000)—Solomon Is. (South Solomons).

marsupichilum *(Upton) D.L. Szlachetko in Fragm. Flor. Geobot.*, Supp. 3: 129 (1995): Malaxis marsupichila.

matsudae *(Yamam.) D.L. Szlachetko in Fragm. Flor. Geobot.*, Supp. 3: 129 (1995): Microstylis matsudae.

maximowiczianum *(King & Pantl.) D.L. Szlachetko in Fragm. Flor. Geobot.*, Supp. 3: 129 (1995): Microstylis maximowicziana.

megalanthum *(Schltr.) D.L. Szlachetko in Fragm. Flor. Geobot.*, Supp. 3: 129 (1995): Microstylis megalantha.

melanophyllum *(Schltr.) D.L. Szlachetko in Fragm. Flor. Geobot.*, Supp. 3: 129 (1995): Microstylis melanophylla.

merapiense *(Schltr.) D.L. Szlachetko in Fragm. Flor. Geobot.*, Supp. 3: 129 (1995), as 'merapiensis': Microstylis merapiensis.

merrillii *(Ames) D.L. Szlachetko in Fragm. Flor. Geobot.*, Supp. 3: 129 (1995): Microstylis merrillii.

metallicum *(Rchb. f.) D.L. Szlachetko in Fragm. Flor. Geobot.*, Supp. 3: 129 (1995): Microstylis metallica.

micranthum *(Hook. f.) D.L. Szlachetko in Fragm. Flor. Geobot.*, Supp. 3: 129 (1995): Microstylis micrantha.

microhybos *(Schltr.) D.L. Szlachetko in Fragm. Flor. Geobot.*, Supp. 3: 129 (1995): Microstylis microhybos.

mindaenaense *(Ames) M.A. Clements & D.L. Jones in Lasianthera*, 1(1): 38 (1996): Malaxis mindaenaensis.

mindorense *(Ames) M.A. Clements & D.L. Jones in Lasianthera*, 1(1): 38 (1996): Microstylis mindorensis.

multiflorum *(Ames & C. Schweinf.) D.L. Szlachetko in Fragm. Flor. Geobot.*, Supp. 3: 129 (1995): Malaxis multiflora.

myosotis *M.A. Clements & D.L. Jones in Lasianthera*, 1(1): 32 (1996)—New Guinea (Papua New Guinea).

negrosianum *(Ames) M.A. Clements & D.L. Jones in Lasianthera*, 1(1): 38 (1996): Malaxis negrosiana.

nemorale *(Ridl.) D.L. Szlachetko in Fragm. Flor. Geobot.*, Supp. 3: 129 (1995): Microstylis nemoralis.

neoebudicum *(Ames) M.A. Clements & D.L. Jones in Lasianthera*, 1(1): 38 (1996): Malaxis neoebudica.

nephroglossum *(Schltr.) M.A. Clements & D.L. Jones in Lasianthera*, 1(1): 38 (1996): Microstylis nephroglossa.

nigrescens *(J.J. Sm.) D.L. Szlachetko in Fragm. Flor. Geobot.*, Supp. 3: 129 (1995): Microstylis nigrescens.

nitidum *(Schltr.) D.L. Szlachetko in Fragm. Flor. Geobot.*, Supp. 3: 129 (1995): Microstylis nitida.

novoguineense *H.B. Margonska, D.L. Szlachetko & T. Kubala in Orchidee*, 50(5): 510 (1999)—New Guinea (Papua New Guinea).

obovatum *(J.J. Sm.) D.L. Szlachetko in Fragm. Flor. Geobot.*, Supp. 3: 129 (1995): Microstylis obovata.

ochyranum *D.L. Szlachetko & H.B. Margońska in Adansonia*, 20(2): 346 (1998), as 'ochyreanum'—New Guinea (Papua New Guinea).

octodentatum *(Seidenf.) D.L. Szlachetko in Fragm. Flor. Geobot.*, Supp. 3: 130 (1995): Malaxis octodentata.

oculatum *(Rchb. f.) D.L. Szlachetko in Fragm. Flor. Geobot.*, Supp. 3: 130 (1995): Microstylis oculata.

oliganthum *(Schltr.) D.L. Szlachetko in Fragm. Flor. Geobot.*, Supp. 3: 130 (1995): Microstylis oligantha.

CREPIDIUM:-

olivaceum *(Schltr.) D.L. Szlachetko in Fragm. Flor. Geobot.*, Supp. 3: 130 (1995): Microstylis olivacea.

ophrydis *(J. König) M.A. Clements & D.L. Jones in Lasianthera*, 1(1): 38 (1996): Epidendrum ophrydis.

orbiculare *(W.W. Sm. & Jeffrey) G. Seidenfaden, Contrib. Orchid Fl. Thailand XIII*: 18 (1997), as 'orbicularum': Microstylis orbicularis.

oreocharis *(Schltr.) D.L. Szlachetko in Fragm. Flor. Geobot.*, Supp. 3: 130 (1995): Microstylis oreocharis.

ovalisepalum *(J.J. Sm.) D.L. Szlachetko in Fragm. Flor. Geobot.*, Supp. 3: 130 (1995): Microstylis ovalisepala.

paguroides *(Schltr.) D.L. Szlachetko in Fragm. Flor. Geobot.*, Supp. 3: 130 (1995): Microstylis paguroides.

palawense *(Schltr.) D.L. Szlachetko in Fragm. Flor. Geobot.*, Supp. 3: 130 (1995): Microstylis palawensis.

partitilobum *(J.J. Sm.) D.L. Szlachetko in Fragm. Flor. Geobot.*, Supp. 3: 130 (1995): Microstylis partitiloba.

pectinatum *(J.J. Sm.) D.L. Szlachetko in Fragm. Flor. Geobot.*, Supp. 3: 130 (1995): Microstylis pectinata.

pedicellare *(Rchb. f.) D.L. Szlachetko & H.B. Margońska in Adansonia*, 20(2): 342 (1998), as 'pedicellaris': Microstylis pedicellaris.

perakense *(Ridl.) D.L. Szlachetko in Fragm. Flor. Geobot.*, Supp. 3: 130 (1995): Microstylis perakensis.

petiolare *(Schltr.) M.A. Clements & D.L. Jones in Lasianthera*, 1(1): 39 (1996): Microstylis petiolaris.

plantagineum *(Hook. & Arn.) M.A. Clements & D.L. Jones in Lasianthera*, 1(1): 39 (1996): Pterochilus plantaginea.

platychilum *(Rchb.f.) M.A. Clements & D.L. Jones in Lasianthera*, 1(1): 39 (1996): Microstylis platychila.

pleistanthum *(Schltr.) D.L. Szlachetko in Fragm. Flor. Geobot.*, Supp. 3: 130 (1995): Microstylis pleistantha.

polyodon *(Hook. f.) D.L. Szlachetko in Fragm. Flor. Geobot.*, Supp. 3: 130 (1995): Microstylis polyodon.

polyphyllum *(Ridl.) M.A. Clements & D.L. Jones in Lasianthera*, 1(1): 39 (1996): Microstylis polyphylla.

prasinum *(Ridl.) D.L. Szlachetko in Fragm. Flor. Geobot.*, Supp. 3: 130 (1995): Microstylis prasina.

productum *(J.J. Sm.) D.L. Szlachetko in Fragm. Flor. Geobot.*, Supp. 3: 130 (1995): Microstylis producta.

propinquum *(Ames) D.L. Szlachetko in Fragm. Flor. Geobot.*, Supp. 3: 130 (1995): Malaxis propinqua.

protractum *(J.J. Sm.) D.L. Szlachetko in Fragm. Flor. Geobot.*, Supp. 3: 130 (1995): Microstylis protracta.

puberulum *(J.J. Sm.) D.L. Szlachetko in Fragm. Flor. Geobot.*, Supp. 3: 130 (1995): Microstylis puberula.

pubicallosum *(Schltr.) D.L. Szlachetko in Fragm. Flor. Geobot.*, Supp. 3: 131 (1995): Microstylis pubicallosa.

punctatum *(J.J. Wood) D.L. Szlachetko in Fragm. Flor. Geobot.*, Supp. 3: 131 (1995): Malaxis punctata.

purpureiflorum *(Ames & Quisumb.) M.A. Clements & D.L. Jones in Lasianthera*, 1(1): 39 (1996): Malaxis purpureiflora.

purpureonervosum *(J.J. Sm.) D.L. Szlachetko in Fragm. Flor. Geobot.*, Supp. 3: 131 (1995): Microstylis purpureonervosa.

purpureum *(Lindl.) D.L. Szlachetko in Fragm. Flor. Geobot.*, Supp. 3: 131 (1995): Microstylis purpurea.

quadridens *(Schltr.) D.L. Szlachetko in Fragm. Flor. Geobot.*, Supp. 3: 131 (1995): Microstylis quadridens.

quadridentatum *(Ames) D.L. Szlachetko in Fragm. Flor. Geobot.*, Supp. 3: 131 (1995): Microstylis quadridentata.

quadrilobum *(Ames) D.L. Szlachetko in Fragm. Flor. Geobot.*, Supp. 3: 131 (1995): Malaxis quadriloba.

CREPIDIUM:-
radicicola *(Rolfe) M.A. Clements & D.L. Jones in Lasianthera,* 1(1): 39 (1996): Microstylis radicicola.
ramosii *(Ames) D.L. Szlachetko in Fragm. Flor. Geobot.*, Supp. 3: 131 (1995): Malaxis ramosii.
ranauense *(J.J. Sm.) D.L. Szlachetko in Fragm. Flor. Geobot.*, Supp. 3: 131 (1995): Microstylis ranauensis.
reineckeanum *(Kraenzl.) M.A. Clements & D.L. Jones in Lasianthera,* 1(1): 39 (1996): Microstylis reineckeana.
repens *(Rolfe) D.L. Szlachetko in Fragm. Flor. Geobot.*, Supp. 3: 131 (1995): Microstylis repens.
resupinatum *(G. Forst.) D.L. Szlachetko in Fragm. Flor. Geobot.*, Supp. 3: 131 (1995): Epidendrum resupinatum.
retusum *(J.J. Sm.) D.L. Szlachetko & H.B. Margońska in Adansonia,* 20(2): 342 (1998): Microstylis retusa.
rhabdophyllum *(Ridl.) D.L. Szlachetko in Fragm. Flor. Geobot.*, Supp. 3: 131 (1995): Microstylis rhabdophylla.
rhinoceros *(J.J. Sm.) D.L. Szlachetko in Fragm. Flor. Geobot.*, Supp. 3: 131 (1995): Microstylis rhinoceros.
ridleyanum *(P.F. Hunt) M.A. Clements & D.L. Jones in Lasianthera,* 1(1): 40 (1996): Malaxis ridleyana.
ridleyi *(J.J. Sm.) D.L. Szlachetko in Fragm. Flor. Geobot.*, Supp. 3: 131 (1995): Microstylis ridleyi.
riparium *(J.J. Sm.) D.L. Szlachetko in Fragm. Flor. Geobot.*, Supp. 3: 131 (1995): Microstylis riparia.
rizalense *(Ames) M.A. Clements & D.L. Jones in Lasianthera,* 1(1): 40 (1996): Malaxis rizalensis.
saccatum *(Ridl.) D.L. Szlachetko in Fragm. Flor. Geobot.*, Supp. 3: 131 (1995): Microstylis saccata.
sagittatum *(J.J. Sm.) D.L. Szlachetko in Fragm. Flor. Geobot.*, Supp. 3: 131 (1995): Microstylis sagittata.
sarawakense *D.L. Szlachetko & H.B. Margońska in Adansonia,* 20(2): 343 (1998)—Borneo (Sarawak).
schlechteri *(Rolfe) M.A. Clements & D.L. Jones in Lasianthera,* 1(1): 40 (1996): Microstylis schlechteri.
schumannianum *(Schltr.) D.L. Szlachetko in Fragm. Flor. Geobot.*, Supp. 3: 131 (1995): Microstylis schumanniana.
sciaphilum *(Schltr.) D.L. Szlachetko in Fragm. Flor. Geobot.*, Supp. 3: 132 (1995): Microstylis sciaphila.
segaarense *(Kraenzl.) M.A. Clements & D.L. Jones in Lasianthera,* 1(1): 40 (1996): Microstylis segaarensis.
seleniglossum *(Schltr.) D.L. Szlachetko in Fragm. Flor. Geobot.*, Supp. 3: 132 (1995): Microstylis seleniglossa.
setipes *(Schltr.) D.L. Szlachetko in Fragm. Flor. Geobot.*, Supp. 3: 132 (1995): Microstylis setipes.
smithianum *D.L. Szlachetko in Fragm. Flor. Geobot.*, Supp. 3: 132 (1995), nom. nov.: Microstylis laxa *J.J. Sm.*
soleiforme *(J.J. Sm.) D.L. Szlachetko in Fragm. Flor. Geobot.*, Supp. 3: 132 (1995): Microstylis soleiformis.
sordidum *(J.J. Sm.) D.L. Szlachetko in Fragm. Flor. Geobot.*, Supp. 3: 132 (1995): Microstylis sordida.
sororium *(J.J. Sm.) D.L. Szlachetko in Fragm. Flor. Geobot.*, Supp. 3: 132 (1995): Microstylis sororia.
stenophyllum *(Schltr.) D.L. Szlachetko in Fragm. Flor. Geobot.*, Supp. 3: 132 (1995): Microstylis stenophylla.
stenostachys *(Schltr.) D.L. Szlachetko in Fragm. Flor. Geobot.*, Supp. 3: 132 (1995): Microstylis stenostachys.
stolleanum *(Schltr.) D.L. Szlachetko in Fragm. Flor. Geobot.*, Supp. 3: 132 (1995): Microstylis stolleana.
sumatrense *(Ridl.) D.L. Szlachetko in Fragm. Flor. Geobot.*, Supp. 3: 132 (1995): Microstylis sumatrensis.
sundaicum *D.L. Szlachetko & H.B. Margońska in Fragm. Flor. Geobot.*, 43(1): 7 (1998)—Lesser Sunda Is.
taurinum *(Rchb. f.) D.L. Szlachetko in Fragm. Flor. Geobot.*, Supp. 3: 132 (1995): Microstylis taurina.
taylorii *(Ames) D.L. Szlachetko in Fragm. Flor. Geobot.*, Supp. 3: 132 (1995): Malaxis taylorii.

CREPIDIUM:-
tenggerense *(J.J. Sm.) M.A. Clements & D.L. Jones in Lasianthera,* 1(1): 40 (1996): Microstylis tenggerensis.
tetralobum *(Schltr.) M.A. Clements & D.L. Jones in Lasianthera,* 1(1): 40 (1996): Microstylis tetraloba.
tixieri *(Seidenf.) D.L. Szlachetko in Fragm. Flor. Geobot.*, Supp. 3: 132 (1995): Malaxis tixieri.
torricellense *(Schltr.) D.L. Szlachetko in Fragm. Flor. Geobot.*, Supp. 3: 132 (1995): Microstylis torricellensis.
trichopodum *(Schltr.) D.L. Szlachetko in Fragm. Flor. Geobot.*, Supp. 3: 132 (1995): Microstylis trichopoda.
trinerve *(Ridl.) D.L. Szlachetko in Fragm. Flor. Geobot.*, Supp. 3: 133 (1995): Microstylis trinervis.
truncicola *(Schltr.) D.L. Szlachetko in Fragm. Flor. Geobot.*, Supp. 3: 133 (1995): Microstylis truncicola.
tubulosum *(J.J. Sm.) D.L. Szlachetko in Fragm. Flor. Geobot.*, Supp. 3: 133 (1995): Microstylis tubulosa.
umbonatum *(Schltr.) D.L. Szlachetko in Fragm. Flor. Geobot.*, Supp. 3: 133 (1995): Microstylis umbonata.
umbraticola *(Schltr.) D.L. Szlachetko in Fragm. Flor. Geobot.*, Supp. 3: 133 (1995): Microstylis umbraticola.
uncatum *(Ames) M.A. Clements & D.L. Jones in Lasianthera,* 1(1): 41 (1996): Malaxis uncata.
undulatum *(Schltr.) D.L. Szlachetko in Fragm. Flor. Geobot.*, Supp. 3: 133 (1995): Microstylis undulata.
variabile *(Ames & C. Schweinf.) D.L. Szlachetko in Fragm. Flor. Geobot.*, Supp. 3: 133 (1995): Malaxis variabilis.
vermeulenianum *D.L. Szlachetko & H.B. Margońska in Fragm. Flor. Geobot.*, 43(2): 186 (1998)—Sumatera.
verruculosum *(J.J. Sm.) D.L. Szlachetko in Fragm. Flor. Geobot.*, Supp. 3: 133 (1995): Microstylis verruculosa.
vinicolor *(Schltr.) D.L. Szlachetko in Fragm. Flor. Geobot.*, Supp. 3: 133 (1995): Microstylis vinicolor.
vinosum *(Schltr.) D.L. Szlachetko in Fragm. Flor. Geobot.*, Supp. 3: 133 (1995): Microstylis vinosa.
vitiense *(Rolfe) M.A. Clements & D.L. Jones in Lasianthera,* 1(1): 41 (1996): Microstylis vitiensis.
wappeanum *(J.J. Sm.) D.L. Szlachetko in Fragm. Flor. Geobot.*, Supp. 3: 133 (1995): Microstylis wappeana.
warapussae *(Schltr.) D.L. Szlachetko in Fragm. Flor. Geobot.*, Supp. 3: 133 (1995): Microstylis warapussae.
warianum *(Schltr.) D.L. Szlachetko in Fragm. Flor. Geobot.*, Supp. 3: 133 (1995): Microstylis wariana.
wenzelii *(Ames) D.L. Szlachetko in Fragm. Flor. Geobot.*, Supp. 3: 133 (1995): Malaxis wenzelii.
werneri *(Schltr.) D.L. Szlachetko in Fragm. Flor. Geobot.*, Supp. 3: 133 (1995): Microstylis werneri.
williamsii *(Ames) D.L. Szlachetko in Fragm. Flor. Geobot.*, Supp. 3: 133 (1995): Malaxis williamsii.
witkowskianum *D.L. Szlachetko & H.B. Margońska in Ann. Bot. Fenn.*, 37(4): 305 (2000)—Bismarck Archipelago.
woodianum *D.L. Szlachetko & H.B. Margońska in Fragm. Flor. Geobot.*, 43(1): 9 (1998)—Borneo (Kalimantan).
xanthochilum *(Schltr.) D.L. Szlachetko in Fragm. Flor. Geobot.*, Supp. 3: 133 (1995): Microstylis xanthochila.
zippelii *(J.J. Sm.) D.L. Szlachetko in Fragm. Flor. Geobot.*, Supp. 3: 133 (1995): Microstylis zippelii.

CREPIDORHOPALON (Scrophulariac.).
affinis *(De Wild.) Eb. Fisch. ex R. Govaerts, World Checklist Seed Plants,* 3(1): 20 (1999): Torenia affinis.
microcarpaeoides *(Bonati) Eb. Fischer in Bull. Mus. Nation. Hist. Nat., B, Adansonia,* Ser. 4, 17(3–4): 252 (1995): Lindernia microcarpaeoides.
parviflorus *(Philcox) Eb. Fischer in Bull. Jard. Bot. Nation. Belg.*, 67(1–4): 378 (1999): Torenia latibracteata subsp. parviflora.

CREPIDORHOPALON:-
robynsii *Eb. Fischer in Bull. Jard. Bot. Nation. Belg.*, 67(1–4): 371 (1999)—Zaire.
rupestris *(Engl.) Eb. Fischer in Bull. Jard. Bot. Nation. Belg.*, 67(1–4): 376 (1999): Lindernia rupestris.
scaettae *(Staner) Eb. Fischer in Bull. Jard. Bot. Nation. Belg.*, 66(1–2): 78 (1997): Stemodiopsis scaettae.
tanzanicus *Eb. Fischer in Bull. Jard. Bot. Nation. Belg.*, 67(1–4): 374 (1999)—Tanzania.

CREPIS (Compos.).
sect. Ruderalia *A.N. Sennikov in Komarovia,* 1: 79 (1999).
amphisboena *M. Gandoger, Dec. Pl. Nov.*, 1: 37 (1875)—France.
aspromontana *S. Brullo, F. Scelsi & G. Spampinato in Flora Medit.*, 5: 59 (1995 publ. 1996)—Italy.
autaretica *M. Gandoger, Dec. Pl. Nov.*, 1: 25 (1875)—France.
bifurcata *(Babc. & Stebbins) Hand.-Mazz.in Oesterr. Bot. Z.*, 88(4): 312 (1939): Youngia bifurcata.
burejensis F. Schmidt
subsp. hokkaidoensis *(Babc.) V.N. Voroschilov in A.K. Skvortsov (ed.), Florist. issl. v razn. raĭonakh SSSR:* 199 (1985): C. hokkaidoensis.
chrysantha (Ledeb.) Turcz.
subsp. minor *(Ledeb.) M.N. Lomonosova in Fl. Sibir. (Asterac.)*, 13: 302 (1997 publ. 1998): C. chrysantha var. minor.
cichoroides *M. Gandoger, Dec. Pl. Nov.*, 1: 13 (1875)—France.
crocea (Lam.) Babc.
subsp. dahurica *M.N. Lomonosova in Fl. Sibir. (Asterac.)*, 13: 303 (1997 publ. 1998)—Chita.
czerepanovii *N.N. Tzvelev in Bot. Zhurn.*, 81(4): 102 (1996)—Russia Northwest.
deloynei *M. Gandoger, Dec. Pl. Nov.*, 1: 14 (1875)—France.
flexuosa (Ledeb.) C.B. Clarke
var. tenuifolia *C.H. An in Fl. Xinjiangensis,* 5: 481 (1999)—Xinjiang.
jacutica *M.N. Lomonosova in Fl. Sibir. (Asterac.)*, 13: 304 (1997 publ. 1998)—Yakutiya.
kurdistanica *R.K. Askerova in Dokl. Akad. Nauk. Azerb. SSR,* 14(11): 901 (1958)—Transcaucasus (Azerbaydzhan).
melanthera *C.H. An in Fl. Xinjiangensis,* 5: 481 (1999), as 'melantherus'—Xinjiang.
newii Oliv. & Hiern
subsp. oliveriana *(Kuntze) C. Jeffrey & H.J. Beentje in Fl. Trop. E. Afr., Compos. (Part 1):* 70 (2000): Hieraciodes oliverianum.
novoana *S. Ortiz, X. Soñora & J. Rodríguez Oubiña in Bot. J. Linn. Soc.*, 123(2): 148 (1997)—Spain.
oxyphylla *M. Gandoger, Dec. Pl. Nov.*, 1: 12 (1875)—France.
perrieri *M. Gandoger, Dec. Pl. Nov.*, 1: 37 (1875)—France.
royi *M. Gandoger, Dec. Pl. Nov.*, 1: 25 (1875)—France.
subglabrescens *M. Gandoger, Dec. Pl. Nov.*, 1: 26 (1875)—France.
trichosticta *M. Gandoger, Dec. Pl. Nov.*, 1: 13 (1875)—Italy.
vendeana *M. Gandoger, Dec. Pl. Nov.*, 1: 13 (1875)—France.
zacintha *(L.) J.L.A. Loiseleur-Deslongchamps, Fl. Gall.*, 2: 527 (1807): Lapsana zacintha.

CRIBBIA (Orchidac.).
confusa *P.J. Cribb in Kew Bull.*, 51(2): 359 (1996)—Cameroon, Liberia, Ivory Coast, Gulf of Guinea Is. (São Tomé).
pendula *I. la Croix & P.J. Cribb in Kew Bull.*, 52(3): 747 (1997)—Gulf of Guinea Is. (São Tomé).
thomensis *I. la Croix & P.J. Cribb in Kew Bull.*, 52(3): 745 (1997)—Gulf of Guinea Is. (São Tomé).

× **CRIMOCHARIS** *D.J. Lehmiller in Herbertia,* 54: 125 (1999 publ. 2000). *AMARYLLIDACEAE* (Ammocharis × Crinum).
hardyi *D.J. Lehmiller in Herbertia,* 54: 126 (1999 publ. 2000), as 'hardyii'—Cultivation. (Ammocharis nerinoides × Crinum baumii).

CRINITINA *J. Soják in Zprávy Krajsk. Vlastiv. Muz. Olomouci,* 215: 1 (1982). *COMPOSITAE*
fominii *(Kem.-Nath.) J. Soják in Zprávy Krajsk. Vlastiv. Muz. Olomouci,* 215: 2 (1982): Linosyris fominii.
linosyris *(L.) J. Soják in Zprávy Krajsk. Vlastiv. Muz. Olomouci,* 215: 1 (1982): Chrysocoma linosyris.
subsp. armoricana *(Rouy) J. Holub in Preslia,* 70(2): 103 (1998): Aster linosyris proles armoricanus.
pontica *(Lipsky) J. Soják in Zprávy Krajsk. Vlastiv. Muz. Olomouci,* 215: 2 (1982): Linosyris vulgaris var. pontica.
tatarica *(Less.) J. Soják in Zprávy Krajsk. Vlastiv. Muz. Olomouci,* 215: 2 (1982): Chrysocoma tatarica.
villosa *(L.) J. Soják in Zprávy Krajsk. Vlastiv. Muz. Olomouci,* 215: 2 (1982): Chrysocoma villosa.

CRINUM (Amaryllidac.).
giessii *D.J. Lehmiller in Herbertia,* 52: 50 (1997)—Namibia.
minus *Jacques in Ann. Fl. Pomone,* 1832–33: 59 (1832)—Cultivation.
verdoorniae *D.J. Lehmiller in Herbertia,* 52: 57 (1997)—Zambia, Caprivi Strip.

CRISTARIA (Malvac.).
aspera Gay
var. formosula *(I.M. Johnst.) M. Muñoz-Schick in Bol. Mus. Nacion. Hist. Nat. (Chile),* 45: 56 (1995 publ. 1996): C. formosula.
var. sadae *(Phil.) M. Muñoz-Schick in Bol. Mus. Nacion. Hist. Nat. (Chile),* 45: 56 (1995 publ. 1996): C. sadae.
calderana *M. Muñoz-Schick in Bol. Mus. Nacion. Hist. Nat. (Chile),* 45: 56 (1995 publ. 1996)—Chile North (Atacama).
dissecta Hook. & Arn.
var. glandulosa *(Phil.) M. Muñoz-Schick in Bol. Mus. Nacion. Hist. Nat. (Chile),* 45: 59 (1995 publ. 1996): C. glandulosa.
var. inconspicua *(F. Phil.) M. Muñoz-Schick in Bol. Mus. Nacion. Hist. Nat. (Chile),* 45: 59 (1995 publ. 1996): C. inconspicua.
glaucophylla Cav.
var. eriantha *(Hook. & Arn.) M. Muñoz-Schick in Bol. Mus. Nacion. Hist. Nat. (Chile),* 45: 61 (1995 publ. 1996): C. eriantha.
insularis F. Phil.
var. johowii *(Skottsb.) M. Muñoz-Schick in Bol. Mus. Nacion. Hist. Nat. (Chile),* 45: 63 (1995 publ. 1996): C. johowii.
integerrima Phil.
var. lobulata *(Phil.) M. Muñoz-Schick in Bol. Mus. Nacion. Hist. Nat. (Chile),* 45: 64 (1995 publ. 1996): C. lobulata.
ovata *M. Muñoz-Schick in Bol. Mus. Nacion. Hist. Nat. (Chile),* 45: 67 (1995 publ. 1996)—Chile North (Atacama, Antofagasta).
tenuissima *M. Muñoz-Schick in Bol. Mus. Nacion. Hist. Nat. (Chile),* 45: 67 (1995 publ. 1996)—Chile North (Antofagasta).
viridi-luteola Gay
var. pinnata *(Phil.) M. Muñoz-Schick in Bol. Mus. Nacion. Hist. Nat. (Chile),* 45: 69 (1995 publ. 1996): C. pinnata.

CRITESION (Gramin.).
brachyantherum *(Nevski) M.E. Barkworth & D.R. Dewey in Amer. J. Bot.,* 72(5): 772 (1985): Hordeum brachyantherum.
brachyantherum *(Nevski) N.N. Tzvelev in Bot. Zhurn.,* 84(7): 113 (1999): as 'brachyanthemum', isonym: Hordeum brachyantherum.
marinum
subsp. gussoneanum *(Parl.) M.E. Barkworth & D.R. Dewey in Amer. J. Bot.,* 72(5): 772 (1985): Hordeum gussoneanum.
nevskianum *(Bowden) N.N. Tzvelev in Bot. Zhurn.,* 84(7): 113 (1999): Hordeum nevskianum.
roshevitzii *(Bowden) N.N. Tzvelev in Bot. Zhurn.,* 84(7): 113 (1999): Hordeum roshevitzii.
turkestanicum *(Nevski) N.N. Tzvelev in Bot. Zhurn.,* 84(7): 113 (1999): Hordeum turkestanicum.

CRITONIA (Compos.).
microdon *(B.L. Rob.) B.L. Turner in Phytologia,* 82(6): 386 (1997 publ. 1998): Eupatorium microdon.
paneroi *B.L. Turner in Phytologia,* 80(3): 145 (1996)—Mexico Southwest (Guerrero).

CRITONIOPSIS (Compos.).
nonoensis *(Benoist) H. Robinson in Smithsonian Contrib. Bot.,* 89: 64 (1999): Vernonia nonoensis.
oblongifolia *A. Sagástegui Alva & M.O. Dillon in Arnaldoa,* 5(1): 20 (1998)—Peru.
solorzanoana *(Rzed. & Calderón) H. Robinson in Smithsonian Contrib. Bot.,* 89: 64 (1999): Vernonia solorzanoana.
villaregalis *(Carvajal) H. Robinson in Smithsonian Contrib. Bot.,* 89: 65 (1999): Vernonia villaregalis.

CROCUS (Iridac.).
biflorus
subsp. fibroannulatus *H. Kerndorff & E. Pasche in Linz. Biol. Beitr.,* 29(1): 594 (1997)—Turkey.
goulimyi Turrill
subsp. leucanthus *(B. Mathew) B. Mathew in Bot. Chronika,* 13: 257 (2000): C. goulimyi var. leucanthus.
hadriaticus Herb.
subsp. parnassicus *(B. Mathew) B. Mathew in Bot. Chronika,* 13: 260 (2000), without exact basionym page, with incorrect basionym vol. & year; *et in Bulb Newsl.,* 32: 7 (2000), with correct basionym ref.: C. hadriaticus f. parnassicus.
subsp. parnonicus *B. Mathew in Bot. Chronika,* 13: 259 (2000)—Greece.
× paulineae *E. Pasche & H. Kerndorff in New Plantsman,* 6(1): 44 (1999)—Turkey.(C. abantensis × ancyrensis).
serotinus
f. albus *B. Mathew in New Plantsman,* 5(1): 12 (1998)—Spain.
vernus
nothosubsp. fritschii *(Derganc) R. Govaerts, World Checklist Seed Plants,* 3(1): 20 (1999): C. × fritschii.

CROIZATIA (Euphorbiac.).
brevipetiolata *(Secco) L.J. Dorr in Sida,* 18(3): 832 (1999): Pseudosagotia brevipetiolata.
brevipetiolata *(Secco) R. Govaerts in R. Govaerts, D.G. Frodin & A. Radcliffe-Smith, World Checklist & Bibliogr.* Euphorbiaceae, 1: 415 (2000), isonym; *see Dorr in Sida,* 18(3): 832 (1999): Pseudosagotia brevipetiolata.

CROSSANDRA (Acanthac.).
acutiloba *K. Vollesen in Kew Bull.,* 52(2): 397 (1997)—Madagascar.
benoistii *K. Vollesen in Kew Bull.,* 52(2): 397 (1997)—Madagascar.
cinnabarina *K. Vollesen in Kew Bull.,* 52(2): 409 (1997)—Madagascar.
flavicaulis *K. Vollesen in Kew Bull.,* 52(2): 388 (1997)—Madagascar.
infundibuliformis (L.) Nees
subsp. arida *L.H. Cramer in Kew Bull.,* 51(3): 554 (1996), nom. nov.: C. indundibuliformis var. crocea *Trimen*.
subsp. axillaris *(Nees) L.H. Cramer in Kew Bull.,* 51(3): 554 (1996): C. axillaris.
isaloensis *K. Vollesen in Kew Bull.,* 52(2): 403 (1997)—Madagascar.
leucodonta *K. Vollesen in Kew Bull.,* 55(4): 965 (2000)—Tanzania.
longehirsuta *K. Vollesen in Kew Bull.,* 52(2): 395 (1997)—Madagascar.
multidentata *K. Vollesen in Kew Bull.,* 52(2): 393 (1997)—Madagascar.
pilosa *(Benoist) K. Vollesen in Kew Bull.,* 52(2): 395 (1997): C. citrina var. pilosa.
tsingyensis *K. Vollesen in Kew Bull.,* 52(2): 390 (1997)—Madagascar.

CROSSOGLOSSA (Orchidac.).
aurantilineata *F. Pupulin in Lindleyana,* 15(1): 23 (2000)—Costa Rica.
bifida *R.L. Dressler in Orquideologia,* 20(3): 255 (1997)—Panama.
elliptica *R.L. Dressler in Orquideologia,* 20(3): 256 (1997)—Panama.

CROSSOPETALUM (Celastrac.).
enervium *B.E. Hammel in Novon,* 7(2): 153 (1997)—Costa Rica.
uragoga *(Jacq.) O. Kuntze, Revis. gen. pl.:* 1: 116 (1891) Myginda uragoga

CROSSOSPERMA *T.G. Hartley in Adansonia,* 19(2): 204 (1997). *RUTACEAE*
cauliflora *T.G. Hartley in Adansonia,* 19(2): 206 (1997)—New Caledonia.
velutina *(Guillaumin) T.G. Hartley in Adansonia,* 19(2): 208 (1997): Melicope velutina.

CROTALARIA (Legumin.).
martiana Benth.
subsp. mohlenbrockii *(Windler & S.G. Skinner) A.M. Planchuelo in Candollea,* 53(2): 465 (1998): C. mohlenbrockii.
prostrata Rottler ex Willd.
var. jinpingensis *(C.Y. Yang) C.Y. Yang in Fl. Reipubl. Popul. Sin.,* 42(2): 373 (1998): C. jinpingensis.
socotrana *(Balf. f.) M. Thulin in Kew Bull.,* 53(2): 460 (1998): Priotropis socotrana.
tanety *D.J. Du Puy, J.-N. Labat & H. Ireland in Kew Bull.,* 52(2): 445 (1997), nom. nov.: C. cytisoides *Hils. & Bojer ex Benth.*

CROTON (Euphorbiac.).
*subgen.*Crotonopsis *(Michx.) A. Radcliffe-Smith & R. Govaerts in Kew Bull.,* 52(1): 183 (1997): gen. Crotonopsis.
subgen. Eremocarpus *(Benth.) A. Radcliffe-Smith & R. Govaerts in Kew Bull.,* 52(1): 184 (1997): gen. Eremocarpus.
subgen. Julocroton *(Mart.) A. Radcliffe-Smith & R. Govaerts in Kew Bull.,* 52(1): 184 (1997): gen. Julocroton.
agrestis *(Pax & K. Hoffm.) A. Radcliffe-Smith & R. Govaerts in Kew Bull.,* 52(1): 184 (1997): Julocroton agrestis.
ambovombensis *A. Radcliffe-Smith & R. Govaerts in Kew Bull.,* 52(1): 186 (1997), nom. nov.: C. divaricatus *Leandri*.
apostolon *A. Radcliffe-Smith & R. Govaerts in Kew Bull.,* 52(1): 184 (1997), nom. nov.: Julocroton vulpinus *Croizat*.
araracuarae *J. Murillo, P.E. Berry & M.V. Arbeláez in Novon,* 10(1): 64 (2000)—Colombia.
argentealbidus *A. Radcliffe-Smith & R. Govaerts in Kew Bull.,* 52(1): 184 (1997), nom. nov.: Julocroton gardneri *Müll. Arg.*
bilocularis *J. Murillo-A.in Caldasia,* 21(2): 156 (1999)—Colombia, Peru.
catamarcensis *L.Z. Ahumada in Darwiniana,* 37(1–2): 167 (1999)—Argentina Northwest (Catamarca).
cinerascens *A. Radcliffe-Smith & R. Govaerts in Kew Bull.,* 52(1): 186 (1997), nom. nov.: C. villosus *Müll. Arg.*
cnidophyllus *A. Radcliffe-Smith & R. Govaerts in Kew Bull.,* 52(1): 186 (1997), nom. nov.: C. urticifolius *Y.T. Chang & Q.H. Chen.*
var. dui *(Y.T. Chang) A. Radcliffe-Smith & R. Govaerts in Kew Bull.,* 52(1): 186 (1997): C. urticifolius var. dui.
coccymelophyllus *A. Radcliffe-Smith & R. Govaerts in Kew Bull.,* 52(1): 186 (1997), nom. nov.: C. prunifolius *Airy Shaw.*
colombianus *J. Murillo-A.in Caldasia,* 21(2): 151 (1999)—Colombia.
cooperianus *(Croizat) A. Radcliffe-Smith & R. Govaerts in Kew Bull.,* 52(1): 184 (1997): Julocroton cooperianus.
cordobensis *L.Z. Ahumada in Darwiniana,* 37(1–2): 170 (1999)—Argentina Northeast (Córdoba), Argentina Northwest (Santiago del Estero, San Luis).
corymbulosoides *A. Radcliffe-Smith & R. Govaerts in Kew Bull.,* 52(1): 186 (1997), nom. nov.: C. eremophilus *Wooton & Standl.*
curuguatyensis *L.Z. Ahumada in Candollea,* 53(1): 122 (1998), nom. nov.: C. pohlianus var. macrophyllus *Chodat & Hassl.*
dinghuensis *H.S. Kiu in J. Trop. Subtrop. Bot.,* 6(2): 101 (1998), contrary to Art. 37.5 ICBN (1994)—China Southeast (Guangdong).
eskuchei *L.Z. Ahumada in Darwiniana,* 37(1–2): 172 (1999)—Argentina Northeast (Corrientes), Paraguay.

CROTON:-
grandifructus *A. Radcliffe-Smith & R. Govaerts in Kew Bull.*, 52(1): 187 (1997), nom. nov.: C. macrocarpus *Ridl.*
guildingioides *A. Radcliffe-Smith & R. Govaerts in Kew Bull.*, 52(1): 187 (1997), nom. nov.: C. subglaber *Urb.*
guizhouensis *H.S. Kiu in J. Trop. Subtrop. Bot.*, 6(2): 103 (1998), nom. nov.: C. urticifolius *Y.T. Chang & Q.H. Chen.*
herzogianus *(Pax & K. Hoffm.) A. Radcliffe-Smith & R. Govaerts in Kew Bull.*, 52(1): 184 (1997): Julocroton herzogianus.
holodiscus *(Croizat) A. Radcliffe-Smith & R. Govaerts in Kew Bull.*, 52(1): 185 (1997): Julocroton holodiscus.
huajuapanensis *M. Martínez Gordillo & R. Cruz Durán in An. Inst. Biol. Univ. Nac. Autón. Méx., Bot.*, 71(1): 26 (2000)—Mexico Southwest (Oaxaca).
integer *(Chodat) A. Radcliffe-Smith & R. Govaerts in Kew Bull.*, 52(1): 185 (1997): Julocroton integer.
iteophyllus *A. Radcliffe-Smith & R. Govaerts in Kew Bull.*, 52(1): 187 (1997), nom. nov.: C. salicifolius *Gagnep.*
janeirensis *A. Radcliffe-Smith & R. Govaerts in Kew Bull.*, 52(1): 187 (1997), nom. nov.: C. macrocalyx *Mart. ex Baill.*
jorgei *J. Murillo-A.in Caldasia*, 21(2): 147 (1999)—Colombia.
lamdongensis *Nguyen Nghia Thin, Euphorbiac. Vietnam:* 48 (1995 publ. 1996)—Vietnam.
langsonensis *Nguyen Nghia Thin, Euphorbiac. Vietnam:* 47 (1995 publ. 1996)—Vietnam.
lauioides *A. Radcliffe-Smith & R. Govaerts in Kew Bull.*, 52(1): 187 (1997), nom. nov.: C. olivaceus *Y.T. Chang & P.T. Li.*
laureltyanus *L.Z. Ahumada in Darwiniana*, 37(1–2): 174 (1999)—Argentina Northeast (Corrientes).
lepidus *(S. Moore) A. Radcliffe-Smith & R. Govaerts in Kew Bull.*, 52(1): 185 (1997): Julocroton lepidus.
lissophyllus *A. Radcliffe-Smith & R. Govaerts in Kew Bull.*, 52(1): 187 (1997), nom. nov.: C. laevifolius *Hook. f.*
lithrifolius *(Croizat) A. Radcliffe-Smith & R. Govaerts in Kew Bull.*, 52(1): 185 (1997): Julocroton lithrifolius.
longipedicellatus J. Léonard
var. glabrescens *A. Radcliffe-Smith in Kew Bull.*, 51(2): 304 (1996)—Tanzania.
macrolepis *A. Radcliffe-Smith & R. Govaerts in Kew Bull.*, 52(1): 187 (1997), nom. nov.: C. squamulosus *Müll. Arg.*
malvoides *(Croizat) A. Radcliffe-Smith & R. Govaerts in Kew Bull.*, 52(1): 185 (1997): Julocroton malvoides.
martinianus *V.W. Steinmann in Novon*, 8(1): 81 (1998)—Mexico Northwest (Sonora, Sinaloa), Mexico Southwest (Jalisco).
microphyllinus *A. Radcliffe-Smith & R. Govaerts in Kew Bull.*, 52(1): 188 (1997), nom. nov.: Timandra erythroxyloides *Klotzsch.*
mocinoi *A. Radcliffe-Smith & R. Govaerts in Kew Bull.*, 52(1): 188 (1997), nom. nov.: C. trilobatus *Sessé & Moç.*
multirameus *L.Z. Ahumada in Darwiniana*, 37(1–2): 176 (1999)—Argentina Northeast (Corrientes).
nicaraguensis *A. Radcliffe-Smith & R. Govaerts in Kew Bull.*, 52(1): 188 (1997), nom. nov.: C. ceanothifolius *Standl. & L.O. Williams.*
nigricans *(Mart. ex Schltdl.) A. Radcliffe-Smith & R. Govaerts in Kew Bull.*, 52(1): 185 (1997): Julocroton nigricans.
pavonianus *A. Radcliffe-Smith & R. Govaerts in Kew Bull.*, 52(1): 188 (1997), nom. nov.: C. tomentosus *Pav. ex Baill.*
pedersenii *L.Z. Ahumada in Candollea*, 53(1): 120 (1998)—Paraguay, Brazil West-central (Mato Grosso do Sul).
pyrosoma *(Croizat) A. Radcliffe-Smith & R. Govaerts in Kew Bull.*, 52(1): 185 (1997): Julocroton pyrosoma.
robustior *(L.B. Sm. & Downs) A. Radcliffe-Smith & R. Govaerts in Kew Bull.*, 52(1): 185 (1997): Julocroton humilis var. robustior.

CROTON:-
roxburghii *N. Wallich, Report on Calcutta Bot. Gard. to G.A. Bushby:* 20 (1840), nom. nov.: C. polyandrum *Roxb.*
ruderalis *L.Z. Ahumada in Darwiniana*, 37(1–2): 178 (1999)—Argentina Northeast (Entre Ríos).
scutatus *P.E. Berry in Syst. Bot.*, 23(2): 171 (1998)—Venezuela, Colombia.
serratoideus *A. Radcliffe-Smith & R. Govaerts in Kew Bull.*, 52(1): 188 (1997), nom. nov.: Timandra dichotoma *Klotzsch.*
spissirameus *A. Radcliffe-Smith & R. Govaerts in Kew Bull.*, 52(1): 185 (1997), nom. nov.: Julocroton crassirameus *Croizat.*
stipuliformis *J. Murillo-A.in Caldasia*, 21(2): 159 (1999)—Colombia, Costa Rica.
tatacuensis *L.Z. Ahumada in Darwiniana*, 37(1–2): 180 (1999)—Argentina Northeast (Corrientes).
thellungianus *(Herter ex Arechav.) A. Radcliffe-Smith & R. Govaerts in Kew Bull.*, 52(1): 185 (1997): Julocroton thellungianus.
thoii *Nguyen Nghia Thin, Euphorbiac. Vietnam:* 47 (1995 publ. 1996)—Vietnam.
tiglium L.
var. xiaopadou *Y.T. Chang & S.Z. Huang in Wuyi Sci. J.*, 2: 23 (1982), with two types cited—China Southeast (Guangdong, Guangxi), China South-central (Guizhou).
tocantinsensis *A. Radcliffe-Smith & R. Govaerts in Kew Bull.*, 52(1): 185 (1997), nom. nov.: Julocroton glaziouі *Croizat.*
trichophilus *(Pax & K. Hoffm.) A. Radcliffe-Smith & R. Govaerts in Kew Bull.*, 52(1): 186 (1997): Julocroton trichophilus.
troncosoi *L.Z. Ahumada in Darwiniana*, 37(1–2): 182 (1999)—Argentina Northeast (Entre Ríos).
venezuelensis *A. Radcliffe-Smith & R. Govaerts in Kew Bull.*, 52(1): 188 (1997), nom. nov.: C. confusus *Pittier.*
verrucosus *A. Radcliffe-Smith & R. Govaerts in Kew Bull.*, 52(1): 188 (1997), nom. nov.: C. echinocarpus *Müll. Arg.*
vietnamensis *A. Radcliffe-Smith & R. Govaerts in Kew Bull.*, 52(1): 188 (1997), nom. nov.: C. longipes *Gagnep.*
villosissimus *(Chodat & Hassl.) A. Radcliffe-Smith & R. Govaerts in Kew Bull.*, 52(1): 186 (1997): Julocroton villosissimus.
viridulus *(Croizat) A. Radcliffe-Smith & R. Govaerts in Kew Bull.*, 52(1): 186 (1997): Julocroton viridulus.
xiaopadou *Y.T. Chang & S.Z. Huang ex H.S. Kiu in J. Trop. Subtrop. Bot.*, 6(2): 103 (1998), as comb. nov., with latin descr. & type—China Southeast (Guangdong, Guangxi, Hunan), China South-central (Guizhou).
yecorensis *V.W. Steinmann & R.S. Felger in Novon*, 8(2): 207 (1998)—Mexico Northwest (Sonora).

CROTONOGYNOPSIS (Euphorbiac.).
akeassii *J. Léonard in Bull. Jard. Bot. Nation. Belg.*, 65(3–4): 342 (1996)—Ivory Coast, Ghana.

CROWEA (Rutac.).
exalata F. Muell.
subsp. revoluta *Paul G. Wilson in Nuytsia*, 11(3): 430 (1997)—Victoria.

CRUCIATA (Rubiac.).
sect. Aculeolatae *A.G. Jelenevsky & N.G. Kuranova in Novosti Sist. Vyssh. Rast.*, 32: 159 (2000).
sect. Pseudovalantia *A.G. Jelenevsky & N.G. Kuranova in Novosti Sist. Vyssh. Rast.*, 32: 159 (2000).
sect. Vernae *(Pobed.) A.G. Jelenevsky & N.G. Kuranova in Novosti Sist. Vyssh. Rast.*, 32: 158 (2000): Galium ser. Vernae.
glabra (L.) Ehrend.
subsp. hirticaulis *(Beck) A. Natali & D. Jeanmonod, Rubiaceae: Compl. Prodr. Fl. Corse (ser. ed. D. Jeanmonod):* 29 (2000): Galium vernum var. hirticaule.
subsp. krylovii *(Iljin) E.G. Naumova in Fl. Sibir. (Solanac.-Lobeliac.)*, 12: 124 (1996 publ. 1997): Galium krylovii.
subsp. krylovii *(Iljin) A.G. Jelenevsky & N.G. Kuranova in Novosti Sist. Vyssh. Rast.*, 32: 159 (2000), isonym; *see E.G. Naumova in Fl. Sibir (Solanac.-Lobeliac.)*, 12: 124 (1996 publ. 1997): Galium krylovii.

CRUCIHIMALAYA *I.A. Al-Shehbaz, S.L. O'Kane & R.A. Price in Novon*, 9(3): 298 (1999). *CRUCIFERAE*
axillaris *(Hook. f. & Thomson) I.A. Al-Shehbaz, S.L. O'Kane & R.A. Price in Novon*, 9(3): 301 (1999): Sisymbrium axillare.
himalaica *(Edgew.) I.A. Al-Shehbaz, S.L. O'Kane & R.A. Price in Novon*, 9(3): 301 (1999): Arabis himalaica.
kneuckeri *(Bornm.) I.A. Al-Shehbaz, S.L. O'Kane & R.A. Price in Novon*, 9(3): 302 (1999): Sisymbrium kneuckeri.
lasiocarpa *(Hook. f. & Thomson) I.A. Al-Shehbaz, S.L. O'Kane & R.A. Price in Novon*, 9(3): 300 (1999): Sisymbrium lasiocarpum.
mollissima *(C.A. Mey.) I.A. Al-Shehbaz, S.L. O'Kane & R.A. Price in Novon*, 9(3): 299 (1999): Sisymbrium mollissimum.
mongolica *(Botsch.) I.A. Al-Shehbaz, S.L. O'Kane & R.A. Price in Novon*, 9(3): 300 (1999): Arabis mongolica.
ovczinnikovii *(Botsch.) I.A. Al-Shehbaz, S.L. O'Kane & R.A. Price in Novon*, 9(3): 300 (1999): Arabidopsis ovczinnikovii.
stricta *(Cambess.) I.A. Al-Shehbaz, S.L. O'Kane & R.A. Price in Novon*, 9(3): 300 (1999): Malcolmia stricta.
wallichii *(Hook. f. & Thomson) I.A. Al-Shehbaz, S.L. O'Kane & R.A. Price in Novon*, 9(3): 301 (1999): Sisymbrium wallichii.

CRUPINA (Compos.).
etrusca *M. Gandoger, Dec. Pl. Nov.*, 1: 40 (1875)—Italy.
insularis *M. Gandoger, Dec. Pl. Nov.*, 1: 40 (1875)—Corse.
villosa *M. Gandoger, Dec. Pl. Nov.*, 1: 40 (1875)—Corse.

CRUSEA (Rubiac.).
andersoniorum *D.H. Lorence in Novon*, 7(1): 57 (1997)—Mexico Southwest (Oaxaca).

CRYOSOPHILA (Palmae).
mocini *(Kunth) R.R. Fernandez, Trees of Mumbai (Bombay):* 269 (1999), as 'mocinni': Chamaerops mocini.

× **CRYPTANANAS** *D.A. Beadle, Prelim. Listing Cultivar Grex Names Bromel.:* 36 (1991). *BROMELIACEAE* (Ananas × Cryptanthus).

CRYPTANDRA (Rhamnac.).
apetala Ewart & Jean White
var. anomala *B.L. Rye in Nuytsia*, 10(2): 257 (1995)—Western Australia.
arbutiflora Fenzl
var. borealis *B.L. Rye in Nuytsia*, 10(2): 260 (1995)—Western Australia.
var. intermedia *B.L. Rye in Nuytsia*, 10(2): 260 (1995)—Western Australia.
var. pygmaea *B.L. Rye in Nuytsia*, 10(2): 262 (1995)—Western Australia.
aridicola *B.L. Rye in Nuytsia*, 10(2): 264 (1995)—Western Australia.
congesta *B.L. Rye in Nuytsia*, 10(2): 265 (1995)—Western Australia.
distigma *B.L. Rye in Nuytsia*, 10(2): 267 (1995)—Western Australia.
graniticola *B.L. Rye in Nuytsia*, 10(2): 270 (1995)—Western Australia.
intonsa *B.L. Rye in Nuytsia*, 10(2): 272 (1995)—Western Australia.
minutifolia *B.L. Rye in Nuytsia*, 10(2): 273 (1995), nom. nov.: C. parvifolia *Turcz.*
subsp. brevistyla *B.L. Rye in Nuytsia*, 10(2): 274 (1995)—Western Australia.
monticola *B.L. Rye & M.E. Trudgen in Nuytsia*, 10(2): 307 (1995)—Western Australia.
nola *B.L. Rye in Nuytsia*, 10(2): 275 (1995)—Western Australia.
polyclada Diels
subsp. aequabilis *B.L. Rye in Nuytsia*, 10(2): 276 (1995)—Western Australia.
pyramidalis *R. Br. ex A.T. Brongniart, Mém. Fam. Rhamnées:* 66 (1826)—Australia.†
recurva *B.L. Rye in Nuytsia*, 10(2): 278 (1995)—Western Australia.
wilsonii *B.L. Rye in Nuytsia*, 10(2): 280 (1995)—Western Australia.

CRYPTANTHA (Boraginac.).
hispidula *Greene ex A. Brand in Engl., Pflanzenr.*, IV, 252(97): 60 (1931)—Northwestern U.S.A., Southwestern U.S.A.

CRYPTANTHUS (Bromeliac.).
caulescens *I.M. Ramírez M.in Harvard Pap. Bot.*, 3(2): 216 (1998)—Brazil Southeast (Espírito Santo), Brazil Northeast (Bahia).
correia-araujoi *E.M.C. Leme in Bromélia*, 2(3): 7 (1995)—Brazil Southeast (Espírito Santo).
diamantinensis *E.M.C. Leme in Harvard Pap. Bot.*, 4(1): 137 (1999)—Brazil Northeast (Bahia).
dorothyae *E.M.C. Leme in Cryptanthus Soc. J.*, 11(1): 6 (1996)—Brazil Southeast (Espírito Santo).
fernseeoides *E.M.C. Leme in Cryptanthus Soc. J.*, 11(3): 9 (1996)—Brazil Southeast (Espírito Santo).
latifolius *E.M.C. Leme in Cryptanthus Soc. J.*, 6(2): 14 (1991)—Brazil Southeast (Espírito Santo).
leuzingerae *E.M.C. Leme in Bradea*, 8(21): 115 (1999)—Brazil Southeast (Espírito Santo).
lutherianus *I.M. Ramírez M.in Harvard Pap. Bot.*, 3(2): 218 (1998)—Brazil Southeast (Espírito Santo).
lyman-smithii *E.M.C. Leme in Harvard Pap. Bot.*, 4(1): 135 (1999)—Brazil Northeast (Bahia).
microglazioui *I.M. Ramírez M.in Harvard Pap. Bot.*, 3(2): 219 (1998)—Brazil Southeast (Espírito Santo).
sergipensis *I.M. Ramírez M.in Harvard Pap. Bot.*, 3(2): 219 (1998)—Brazil Northeast (Sergipe).
ubairensis *I.M. Ramírez M.in Harvard Pap. Bot.*, 3(2): 221 (1998)—Brazil Northeast (Bahia).

× **CRYPTMEA** *E.L. Smith in J. Bromeliad Soc.*, 33(2): 72 (1983). *BROMELIACEAE* (Aechmea × Cryptanthus).

CRYPTOCARYA (Laurac.).
rolletii *H. Wang & H. Zhu in Guihaia*, 19(3): 197 (1999)—China South-central (Yunnan).

CRYPTOCORYNE (Arac.).
annamica *M. Serebryanyi in Aqua Planta*, 1991(3): 98 (1991)—Vietnam.

CRYPTOLEPIS (Asclepiadac.).
arbuscula *(Radcl.-Sm.) H.J.T. Venter in Taxon*, 46(4): 713 (1997): Mitolepis arbuscula.
intricata *(Balf. f.) H.J.T. Venter in Taxon*, 46(4): 713 (1997): Mitolepis intricata.
macrophylla *(Radcl.-Sm.) H.J.T. Venter in Taxon*, 46(4): 713 (1997): Curroria macrophylla.
socotrana *(Balf. f.) H.J.T. Venter in Taxon*, 46(4): 713 (1997): Cochlanthus socotranus.
yemenensis *H.J.T. Venter & R.L. Verhoeven in Bot. J. Linn. Soc.*, 131(4): 420 (1999)—South Yemen.

CRYPTOLLUMA *D.C.H. Plowes in Haseltonia*, 13: 57 (1995). *ASCLEPIADACEAE*
edulis *(Edgew.) D.C.H. Plowes in Haseltonia*, 3: 57 (1995): Boucerosia edulis.

CTENOPSIS (Gramin.).
cynosuroides *(Desf.) Paunero ex A.T. Romero García in Lagascalia*, 18(2): 321 (1996): Festuca cynosuroides.

CUATRESIA (Solanac.).
hunzikeriana *(Benítez & M. Martínez) N.W. Sawyer in Kurtziana*, 27(1): 250 (1999): Larnax hunzikeriana.

CUCURBITA (Cucurbitac.).
pepo L.
var. flogra *H. Teppner in Phyton (Austria)*, 40(1): 35 (2000)—Cultivation.
var. georgica *H. Teppner in Phyton (Austria)*, 40(1): 35 (2000)—Cultivation.
subsp. gumala *H. Teppner in Phyton (Austria)*, 40(1): 34 (2000)—Cultivation.

CULLEN (Legumin.).
australasicum *(Schltdl.) J.W. Grimes in Muelleria*, 9: 195 (1996): Psoralea australasica.
balsamicum *(F. Muell.) J.W. Grimes in Austral. Syst. Bot.*, 10(4): 636 (1997): Psoralea balsamica.

CULLEN:-
candidum *J.W. Grimes in Austral. Syst. Bot.*, 10(4): 624 (1997)—Western Australia.
cinereum *(Lindl.) J.W. Grimes in Muelleria*, 9: 195 (1996): Psoralea cinerea.
corallum *J.W. Grimes in Austral. Syst. Bot.*, 10(4): 626 (1997)—Western Australia, Northern Territory.
cuneatum *(W. Fitzg.) J.W. Grimes in Austral. Syst. Bot.*, 10(4): 633 (1997): Psoralea cuneata.
discolor *(Domin) J.W. Grimes in Muelleria*, 9: 195 (1996): Psoralea discolor.
gaudichaudianum *(Decne.) J.W. Grimes in Austral. Syst. Bot.*, 10(4): 623 (1997): Psoralea gaudichaudiana.
graveolens *(Domin) J.W. Grimes in Austral. Syst. Bot.*, 10(4): 611 (1997): Psoralea graveolens.
lachnostachys *(F. Muell.) J.W. Grimes in Austral. Syst. Bot.*, 10(4): 615 (1997): Psoralea lachnostachys.
leucanthum *(F. Muell.) J.W. Grimes in Austral. Syst. Bot.*, 10(4): 628 (1997): Psoralea leucantha.
leucochaites *J.W. Grimes in Austral. Syst. Bot.*, 10(4): 616 (1997)—Western Australia.
martinii *(F. Muell.) J.W. Grimes in Austral. Syst. Bot.*, 10(4): 634 (1997): Psoralea martinii.
microcephalum *(Rchb. ex Kunze) J.W. Grimes in Muelleria*, 9: 195 (1996): Psoralea microcephala.
pallidum *(N.T. Burb.) J.W. Grimes in Muelleria*, 9: 195 (1996): Psoralea pallida.
parvum *(F. Muell.) J.W. Grimes in Muelleria*, 9: 195 (1996): Psoralea parva.
patens *(Lindl.) J.W. Grimes in Muelleria*, 9: 196 (1996): Psoralea patens.
plumosum *J.W. Grimes in Austral. Syst. Bot.*, 10(4): 612 (1997), nom. nov.: Psoralea plumosa *F. Muell.*
pogonocarpum *J.W. Grimes in Austral. Syst. Bot.*, 10(4): 596 (1997)—Western Australia.
praeruptorum *J.W. Grimes in Austral. Syst. Bot.*, 10(4): 619 (1997)—Western Australia.
pustulatum *(F. Muell.) J.W. Grimes in Austral. Syst. Bot.*, 10(4): 613 (1997): Psoralea pustulata.
stipulaceum *(Decne.) J.W. Grimes in Austral. Syst. Bot.*, 10(4): 620 (1997): Psoralea stipulacea.
tenax *(Lindl.) J.W. Grimes in Muelleria*, 9: 196 (1996): Psoralea tenax.
tomentosum *(Thunb.) J.W. Grimes in Austral. Syst. Bot.*, 10(4): 589 (1997): Trigonella tomentosa.
virens *(W. Fitzg.) J.W. Grimes in Austral. Syst. Bot.*, 10(4): 629 (1997): Psoralea virens.
walkingtonii *(F. Muell.) J.W. Grimes in Austral. Syst. Bot.*, 10(4): 622 (1997): Psoralea walkingtonii.

× **CUMMINGARA** *G.D. Rowley in Haworthiad*, 13(3): 115 (1999), nom. nov.; *cf. Repert. Pl. Succ. (I.O.S.)*, 50: 7 (1999 publ. 2000): × Smithara *D.M. Cumming.* (Gasteria × Haworthia × Poellnitzia). (Aloac.).

CUMULOPUNTIA (Cactac.).
chichensis *(Cárdenas) E.F. Anderson in Cact. Succ. J. (U.S.A.)*, 71(6): 324 (1999): Tephrocactus chichensis.
corotilla *(K. Schum. ex Vaupel) E.F. Anderson in Cact. Succ. J. (U.S.A.)*, 71(6): 324 (1999): Opuntia corotilla.
dactylifera *(Vaupel) E.F. Anderson in Cact. Succ. J. (U.S.A.)*, 71(6): 324 (1999): Opuntia dactylifera.
fulvicoma *(Rauh & Backeb.) E.F. Anderson in Cact. Succ. J. (U.S.A.)*, 71(6): 324 (1999): Tephrocactus fulvicomus.
mistiensis *(Backeb.) E.F. Anderson in Cact. Succ. J. (U.S.A.)*, 71(6): 324 (1999): Tephrocactus mistiensis.
sphaerica *(C.F. Först.) E.F. Anderson in Cact. Succ. J. (U.S.A.)*, 71(6): 324 (1999): Opuntia sphaerica.

CUNILA (Labiat.).
crenata *M.R. García-Peña & P. Tenorio Lezama in An. Inst. Biol. Univ. Nac. Autón. Méx., Bot.*, 68(1): 2 (1997)—Mexico Northeast (Durango).

CUNNINGHAMIA (Taxodiac.).
lanceolata (Lamb.) Hook.
var. unicanaliculata *(D.Y. Wang & H.L. Liu) J. Silba in J. Int. Conifer Preserv. Soc.*, 7(1): 23 (2000): C. unicanaliculata.

CUNONIA (Cunoniac.).
aoupiniensis *R.D. Hoogland in Adansonia*, 19(1): 16 (1997)—New Caledonia.
cerifera *R.D. Hoogland in Adansonia*, 19(1): 13 (1997)—New Caledonia.
nervosa *R.D. Hoogland in Adansonia*, 19(1): 13 (1997)—New Caledonia.
rupicola *R.D. Hoogland in Adansonia*, 19(1): 17 (1997)—New Caledonia.
varijuga *R.D. Hoogland in Adansonia*, 19(1): 17 (1997)—New Caledonia.

CUPANIA (Sapindac.).
fluminensis *P. Acevedo-Rodríguez in Biollania*, ed. esp. 6 (J.J. Wurdack Festschr.): 146 (1997), nom. nov.: C. laxiflora *Radlk.*
impressinervia *P. Acevedo-Rodríguez in Biollania*, ed. esp. 6 (J.J. Wurdack Festschr.): 147 (1997), nom. nov.: C. revoluta *Radlk.*
radlkoferi *P. Acevedo-Rodríguez in Biollania*, ed. esp. 6 (J.J. Wurdack Festschr.): 147 (1997), nom. nov.: C. verrucosa *Radlk.*
seemannii Triana & Planch.
var. fosteri *(Croat) P. Acevedo-Rodríguez in Biollania*, ed. esp. 6 (J.J. Wurdack Festschr.): 148 (1997): C. sylvatica var. fosteri.

CUPHEA (Lythrac.).
apurensis *A. Lourteig in Bradea*, 8(5): 31 (1996)—Venezuela.
avigera B.L. Rob. & Seaton
var. pulcherrima *(R.C. Foster) S.A. Graham, Revis. Cuphea sect. Diploptychia (Lythrac.) (Syst. Bot. Monog., 53)*: 41 (1998): C. pulcherrima.
beckiana *A. Lourteig in Bradea*, 8(5): 30 (1996)—Bolivia.
chiribiquetea *A. Lourteig in Bradea*, 8(5): 27 (1996)—Colombia.
congesta *S.A. Graham & T.B. Cavalcanti in Brittonia*, 51(1): 26 (1999)—Venezuela.
curiosa Lourteig
var. oresbia *A. Lourteig in Bradea*, 8(5): 28 (1996), contrary to Art. 37.5 ICBN (1994); *et in Bradea*, 8(5): 28 (1996 publ. 1997), substitute issue—Venezuela.
cyrilli-nelsonii *R. Zúniga in Ceiba*, 35(1): 63 (1994)—Honduras.
galeatocalcarata *A. Lourteig in Bradea*, 8(12): 65 (1997), as 'galeato-calcarata'—Venezuela.
hookeriana Walp.
var. ixodes *(Hemsl.) S.A. Graham, Revis. Cuphea sect. Diploptychia (Lythrac.) (Syst. Bot. Monog., 53)*: 63 (1998): C. ixodes.
iguazuensis *A. Lourteig in Bradea*, 8(5): 30 (1996)—Brazil South (Paraná).
insolita *A. Lourteig in Bradea*, 8(5): 29 (1996)—Guyana.
luteola *S.A. Graham & T.B. Cavalcanti in Brittonia*, 51(3): 303 (1999)—Bolivia.
maigualidensis *A. Lourteig in Bradea*, 8(5): 28 (1996)—Venezuela.
splendida Lourteig
var. viridiflava *S.A. Graham & T.B. Cavalcanti in Brittonia*, 51(1): 29 (1999)—Bolivia.
xanthopetala *S.A. Graham & T.B. Cavalcanti in Brittonia*, 51(1): 24 (1999)—Brazil West-central (Mato Grosso).

CUPRESSUS (Cupressac.).
dupreziana A. Camus
var. atlantica *(Gaussen) J. Silba in J. Int. Conifer Preserv. Soc.*, 5(2): 29 (1998): C. atlantica.

CUPULANTHUS (Legumin.).
bracteolosus *(F. Muell.) R. Govaerts, World Checklist Seed Plants*, 3(1): 20 (1999): Brachysaema bracteolosum.

CURAREA (Menispermac.).
crassa *R.C. Barneby in Brittonia*, 48(1): 22 (1996)—Brazil Northeast (Bahia).

CURCULIGO (Hypoxidac.).
pilosa (Schumach. & Thonn.) Engl.
subsp. major *(Baker) J. Wiland in Fragm. Flor. Geobot.*, 42(1): 19 (1997): C. gallabatensis var. major.

CURCULIGO:-
pilosa (Schumach. & Thonn.) Engl.
subsp. minor *(Guinea) J. Wiland in Fragm. Flor. Geobot.*, 42(1): 16 (1997): C. minor.
strobiliformis *D. Fang & D.H. Qin in Guihaia*, 16(1): 3 (1996)—China Southeast (Guangxi).

CURCUMA (Zingiberac.).
sect. Nontuberosa *K.C. Velayudhan, V.A. Amalraj & V.K. Muralidharan in J. Econ. Taxon. Bot.*, 20(2): 381 (1996), without type.
sect. Stolonifera *K.C. Velayudhan, V.A. Amalraj & V.K. Muralidharan in J. Econ. Taxon. Bot.*, 20(2): 382 (1996).
sect. Tuberosa *K.C. Velayudhan, V.A. Amalraj & V.K. Muralidharan in J. Econ. Taxon. Bot.*, 20(2): 380 (1996), without type.
kwangsiensis S.K. Lee & C.F. Liang
var. affinis *Y.H. Chen in Acta Pharm. Sin.*, 16(5): 387 (1981)—China Southeast (Guangxi).
var. puberula *Y.H. Chen in Acta Pharm. Sin.*, 16(5): 387 (1981)—China Southeast (Guangxi).
rhabdota *P. Sirirugsa & M.F. Newman in New Plantsman*, 7(4): 196 (2000)—Laos.
wenyujin *Y.H. Chen & C. Ling in Acta Pharm. Sin.*, 16(5): 387, 388 (1981)—China.

CURIO *P.V. Heath in Calyx*, 5(4): 136 (1997), nom. nov.: Senecio sect. Rowleyani *C. Jeffrey*. (Compos.).
acaulis *(L.) P.V. Heath in Calyx*, 5(4): 136 (1997): Cacalia acaulis.
var. burchellii *(DC.) P.V. Heath in Calyx*, 5(4): 136 (1997): Kleinia acaulis α burchellii.
var. ecklonis *(DC.) P.V. Heath in Calyx*, 5(4): 136 (1997): Kleinia acaulis β ecklonis.
archeri *(Compton) P.V. Heath in Calyx*, 5(4): 136 (1997): Kleinia archeri.
articulatus *(L.) P.V. Heath in Calyx*, 5(4): 136 (1997): Cacalia articulatus.
var. globosus *(G.D. Rowley) P.V. Heath in Calyx*, 5(4): 136 (1997): Senecio articulatus f. globosus.
f. variegatus *P.V. Heath in Calyx*, 6(2): 54 (1999)—Locality not stated.
avasimontanus *(Dinter) P.V. Heath in Calyx*, 5(4): 136 (1997): Senecio avasimontanus.
cicatricosus *(Sch. Bip.) P.V. Heath in Calyx*, 5(4): 136 (1997): Senecio cicatricosus.
citriformis *(G.D. Rowley) P.V. Heath in Calyx*, 5(4): 136 (1997): Senecio citriformis.
crassulifolius *(DC.) P.V. Heath in Calyx*, 6(2): 54 (1999): Kleinia crassulifolia.
cuneifolius *(L.) P.V. Heath in Calyx*, 6(2): 54 (1999): Cacalia cuneifolia.
ficoides *(L.) P.V. Heath in Calyx*, 6(2): 54 (1999): Cacalia ficoides.
hallianus *(G.D. Rowley) P.V. Heath in Calyx*, 6(2): 54 (1999): Senecio hallianus.
hanburyanus *(Dinter) P.V. Heath in Calyx*, 6(2): 54 (1999): Senecio hanburyanus.
herreanus *(Dinter) P.V. Heath in Calyx*, 6(2): 54 (1999): Senecio herreanus.
humbertii *(Guillaumin) P.V. Heath in Calyx*, 6(2): 54 (1999): Kleinia humbertii.
ovoideus *(Compton) P.V. Heath in Calyx*, 5(4): 136 (1997): Kleinia ovoidea.
× peregrinus *(G. Kunkel) P.V. Heath in Calyx*, 6(2): 54 (1999), basionym not validly publ.: Kleinia × peregrina.
pinguifolius *(DC.) P.V. Heath in Calyx*, 6(2): 55 (1999): Kleinia pinguifolia.
radicans *(L.) P.V. Heath in Calyx*, 6(2): 55 (1999): Cacalia radicans.
repens *(L.) P.V. Heath in Calyx*, 5(4): 136 (1997): Cacalia repens.
rowleyanus *(H. Jacobsen) P.V. Heath in Calyx*, 6(2): 55 (1999), as 'roeleanus': Senecio rowleyanus.
f. marmoratus *P.V. Heath in Calyx*, 6(2): 55 (1999)—Locality not stated.
sulcicalyx *(N.E. Br.) P.V. Heath in Calyx*, 6(2): 55 (1999): Senecio sulcicalyx.
talinoides *(DC.) P.V. Heath in Calyx*, 6(2): 55 (1999): Kleinia talinoides.
var. aizoides *(DC.) P.V. Heath in Calyx*, 6(2): 55 (1999): Kleinia aizoides.
var. chordifolius *(Hook.) P.V. Heath in Calyx*, 6(2): 55 (1999): Senecio chordifolius.
var. mandraliscae *(Tineo) P.V. Heath in Calyx*, 6(2): 55 (1999): Kleinia mandraliscae.

CURTIA (Gentianac.).
quadrifolia *B. Maguire in Mem. New York Bot. Gard.*, 32: 352 (1981)—Venezuela.

CUSCUTA (Convolvulac.).
castroviejoi *M.A. García in Ann. Bot. Fenn.*, 36(3): 167 (1999)—Ethiopia.
rausii *M.A. García in Ann. Bot. Fenn.*, 35(3): 172 (1998)—Greece.

CUVIERA (Rubiac.).
wernhamii *M. Cheek in Pl. Mount Cameroon: Conserv. Checklist (eds. S. Cable & M. Cheek)*: 106 (1998), nom. nov.: Globulostylis minor *Wernham*.

CYANANTHUS (Campanulac.).
sect. Annui *(Y.S. Lian) K.K. Shrestha in Acta Phytotax. Sin.*, 35(5): 425 (1997): C. subsect. Annui.
subsect. Lichiangenses *K.K. Shrestha in Acta Phytotax. Sin.*, 35(5): 426 (1997).
subgen. Micranthus *K.K. Shrestha in Acta Phytotax. Sin.*, 35(5): 430 (1997).
sect. Suffruticulosi *K.K. Shrestha in Acta Phytotax. Sin.*, 35(5): 415 (1997).
incanus Hook. f. & Thomson
subsp. orientalis *K.K. Shrestha in Acta Phytotax. Sin.*, 35(5): 407 (1997)—Nepal.
leiocalyx (Franch.) Cowan
subsp. lucidus *K.K. Shrestha in Acta Phytotax. Sin.*, 35(5): 409 (1997)—Xizang.
macrocalyx Franch.
subsp. spathulifolius *(Nannf.) K.K. Shrestha in Acta Phytotax. Sin.*, 35(5): 412 (1997): C. spathulifolius.
microphyllus Edgew.
subsp. williamsonii *K.K. Shrestha in Acta Phytotax. Sin.*, 35(5): 419 (1997)—Nepal.
pilosus *(C. Marquand) K.K. Shrestha in Acta Phytotax. Sin.*, 35(5): 405 (1997): C. macrocalyx var. pilosus.

CYANEA (Campanulac.).
calycina *(Cham.) T.G. Lammers in Novon*, 8(1): 32 (1998): Lobelia calycina.
cylindrocalyx *(Rock) T.G. Lammers in Novon*, 8(1): 31 (1998): C. grimesiana var. cylindrocalyx.
habenata *(H. St. John) T.G. Lammers in Novon*, 8(1): 33 (1998): Delissea habenata.
kahiliensis *(H. St. John) T.G. Lammers in Novon*, 8(1): 33 (1998): Delissea kahiliensis.
kuhihewa *T.G. Lammers in Brittonia*, 48(2): 238 (1996)—Hawaiian Is.
lobata H. Mann
subsp. baldwinii *(C.N. Forbes & G.C. Munro) T.G. Lammers in Novon*, 9(3): 387 (1999): C. baldwinii.
mauiensis *(Rock) T.G. Lammers in Novon*, 8(1): 31 (1998): C. grimesiana var. mauiensis.
munroi *(Hosaka) T.G. Lammers in Novon*, 8(1): 31 (1998): C. grimesiana var. munroi.
sessilifolia *(O. Deg.) T.G. Lammers in Novon*, 8(1): 32 (1998): Rollandia sessilifolia.

CYANICULA *S.D. Hopper & A.P. Brown in Lindleyana*, 15(2): 120 (2000). *ORCHIDACEAE*
amplexans *(A.S. George) S.D. Hopper & A.P. Brown in Lindleyana*, 15(2): 120 (2000): Caladenia amplexans.
ashbyae *S.D. Hopper & A.P. Brown in Lindleyana*, 15(2): 120 (2000)—Western Australia.
caerulea *(R. Br.) S.D. Hopper & A.P. Brown in Lindleyana*, 15(2): 121 (2000): Caladenia caerulea.
deformis *(R.Br.) S.D. Hopper & A.P. Brown in Lindleyana*, 15(2): 121 (2000): Caladenia deformis.
fragrans *S.D. Hopper & A.P. Brown in Lindleyana*, 15(2): 121 (2000)—Western Australia.
gemmata *(Lindl.) S.D. Hopper & A.P. Brown in Lindleyana*, 15(2): 123 (2000): Caladenia gemmata.
gertrudae *(Ostenf.) S.D. Hopper & A.P. Brown in Lindleyana*, 15(2): 123 (2000): Caladenia gertrudae.
ixioides *(Lindl.) S.D. Hopper & A.P. Brown in Lindleyana*, 15(2): 123 (2000): Caladenia ixioides.
nikulinskyae *S.D. Hopper & A.P. Brown in Lindleyana*, 15(2): 123 (2000)—Western Australia.
sericea *(Lindl.) S.D. Hopper & A.P. Brown in Lindleyana*, 15(2): 124 (2000): Caladenia sericea.

CYANONEURON *C. Tange in Nordic J. Bot.*, 18(2): 147 (1998). *RUBIACEAE*
cyaneum *(Hallier f.) C. Tange in Nordic J. Bot.*, 18(2): 148 (1998): Myrioneuron cyaneum.
depauperatum *(Merr.) C. Tange in Nordic J. Bot.*, 18(2): 152 (1998): Myrioneuron depauperatum.
grandiflorum *C. Tange in Nordic J. Bot.*, 18(2): 153 (1998)—Borneo (Sarawak).
pedunculatum *C. Tange in Nordic J. Bot.*, 18(2): 153 (1998)—Borneo (Sarawak, Kalimantan).
pubescens *(Valeton) C. Tange in Nordic J. Bot.*, 18(2): 151 (1998): Myrioneuron pubescens.

CYANTHILLIUM (Compos.).
albicans *(DC.) H. Robinson in Proc. Biol. Soc. Wash.*, 112(1): 229 (1999): Vernonia albicans.
conyzoides *(DC.) H. Robinson in Proc. Biol. Soc. Wash.*, 112(1): 229 (1999): Vernonia conyzoides.
hookerianum *(Arn.) H. Robinson in Proc. Biol. Soc. Wash.*, 112(1): 229 (1999): Vernonia hookeriana.
vernonioides *(Muschl.) H. Robinson in Proc. Biol. Soc. Wash.*, 112(1): 229 (1999): Erlangea vernonioides.

CYATHOCHAETA (Cyperac.).
equitans *K.L. Wilson in Nuytsia*, 11(2): 269 (1997)—Western Australia.
stipoides *K.L. Wilson in Nuytsia*, 11(2): 271 (1997)—Western Australia.

CYATHOCLINE (Compos.).
manilaliana *C. Prabhakar Raju & R.R. Venkata Raju in Rheedea*, 9(2): 151 (1999)—India (Andhra Pradesh).

CYATHOCOMA (Cyperac.).
bachmannii *(Kük.) C. Archer in S. Afr. J. Bot.*, 63(4): 170 (1997): Tetraria bachmannii.
hexandra *(Nees) J. Browning in S. Afr. J. Bot.*, 62(5): 250 (1996): Carpha hexandra.

CYATHODES (Epacridac.).
platystoma *C.M. Weiller in Austral. Syst. Bot.*, 9(4): 502 (1996)—Tasmania.

CYATHOSTEMMA (Annonac.).
glabrum *(Span.) Jessup ex T.M.A. Utteridge in Blumea*, 45(2): 385 (2000): Uvaria glabra.
siamense *T.M.A. Utteridge in Blumea*, 45(2): 389 (2000), as 'siamensis'—Thailand.

CYBIANTHUS (Myrsinac.).
anthuriophyllus *J.J. Pipoly in Sida*, 18(1): 89 (1998)—Ecuador, Peru.
cenepensis *J.J. Pipoly in Sida*, 18(1): 106 (1998)—Peru, Ecuador.
comperuvianus *J.J. Pipoly in Sida*, 18(1): 47 (1998)—Brazil West-central (Mato Grosso), Brazil North (Rondônia), Ecuador, Peru, Bolivia.
croatii *J.J. Pipoly in Sida*, 18(1): 71 (1998)—Ecuador, Colombia.
cuatrecasasii *J.J. Pipoly in Caldasia*, 18(3): 285 (1996)—Colombia.
flavovirens *J.J. Pipoly in Sida*, 18(1): 134 (1998)—Peru.
fosteri *J.J. Pipoly in Sida*, 18(1): 144 (1998)—Peru.
grandezii *J.J. Pipoly in Sida*, 18(1): 139 (1998)—Peru.
granulosus *J.J. Pipoly in Sida*, 18(1): 133 (1998)—Peru.
guyanensis (A.DC.) Miq.
subsp. pseudoicacoreus *(Miq.) J.J. Pipoly in Sida*, 18(1): 49 (1998): Ardisia pseudoicacorea.
huampamiensis *J.J. Pipoly in Sida*, 18(1): 130 (1998)—Peru.
idroboi *J.J. Pipoly in Caldasia*, 18(3): 288 (1996)—Colombia.
incognitus *J.J. Pipoly in Sida*, 18(1): 125 (1998)—Peru.
jensonii *J.J. Pipoly in Sida*, 18(1): 142 (1998)—Peru.
kayapii *(Lundell) J.J. Pipoly in Sida*, 18(1): 84 (1998): Weigeltia kayapii.
liesneri *J.J. Pipoly & J.M. Ricketson in Sida*, 18(4): 1168 (1999)—Venezuela.
nestorii *J.J. Pipoly in Sida*, 18(1): 59 (1998)—Peru.
prevostiae *J.J. Pipoly in Brittonia*, 51(2): 128 (1999)—French Guiana.
pseudolongifolius *J.J. Pipoly in Sida*, 18(1): 101 (1998)—Peru.

CYBIANTHUS:-
timanae *J.J. Pipoly in Sida,* 18(1): 53 (1998)—Peru, Ecuador.
vasquezii *J.J. Pipoly in Sida,* 18(1): 103 (1998)—Peru.

CYCAS (Cycadac.).
ser. Angulatosae *K.D. Hill in Telopea,* 7(1): 29 (1996).
ser. Asiorientales *(J. Schust.) D.Yue Wang in Cycads in China (eds. F.X. Wang & H.B. Liang):* 27 (1996): C. sect. Asiorientales.
ser. Cairnsianosae *K.D. Hill in Telopea,* 7(1): 26 (1996).
ser. Calcicolosae *K.D. Hill in Telopea,* 7(1): 42 (1996).
sect. Endemicae *(J. Schust.) D.Yue Wang in Cycads in China (eds. F.X. Wang & H.B. Liang):* 28 (1996): C. subsect. Endemicae.
ser. Furfuraceosae *K.D. Hill in Telopea,* 7(1): 55 (1996).
ser. Multipinnata *D.Yue Wang in Cycads in China (eds. F.X. Wang & H.B. Liang):* 26 (1996).
sect. Panzhihuaenses *D.Yue Wang in Cycads in China (eds. F.X. Wang & H.B. Liang):* 26 (1996).
subgen. Panzhihuaenses *D.Yue Wang in Cycads in China (eds. F.X. Wang & H.B. Liang):* 26 (1996).
ser. Panzhihuaenses *D.Yue Wang in Cycads in China (eds. F.X. Wang & H.B. Liang):* 27 (1996).
subsect. Panzhihuaenses *D.Yue Wang in Cycads in China (eds. F.X. Wang & H.B. Liang):* 26 (1996).
subgen. Pectinata *D.J. de Laubenfels in Blumea,* 43(2): 375 (1998).
subgen. Revoluta *D.J. de Laubenfels in Blumea,* 43(2): 382 (1998).
sect. Rumphiae *(K.D. Hill) D.Yue Wang in Cycads in China (eds. F.X. Wang & H.B. Liang):* 28 (1996): C. subsect. Rumphiae.
*subsect.*Stangerioides *(Smitinand) D.Yue Wang in Cycads in China (eds. F.X. Wang & H.B. Liang):* 26 (1996): C. sect. Stangerioides.
ser. Stangerioides *(Smitinand) D.Yue Wang in Cycads in China (eds. F.X. Wang & H.B. Liang):* 26 (1996).C. sect. Stangerioides.
subgen. Truncata *D.J. de Laubenfels in Blumea,* 43(2): 367 (1998).
subsect. Wadeanae *D.Yue Wang in Cycads in China (eds. F.X. Wang & H.B. Liang):* 27 (1996), without latin descr.
ser. Yorkianosae *K.D. Hill in Telopea,* 7(1): 16 (1996).
acuminatissima *Hung T. Chang, Y.C. Zhong & Z.F. Lu in Acta Sci. Nat. Univ. Sunyatseni,* 37(4): 6 (1998)—China Southeast (Guangxi).
arnhemica K.D. Hill
subsp. muninga *S.K. Chirgwin & K.D. Hill in Telopea,* 7(1): 44 (1996)—Northern Territory.
subsp. natja *K.D. Hill in Telopea,* 7(1): 46 (1996)—Northern Territory.
badensis *K.D. Hill in Telopea,* 7(1) 20 (1996)—Queensland.
brevipinnata *Hung T. Chang, Y.Y. Huang & Y.C. Zhong in Acta Sci. Nat. Univ. Sunyatseni,* 37(4): 8 (1998)—China Southeast (Guangxi).
chamaoensis *K.D. Hill in Brittonia,* 51(1): 58 (1999)—Thailand.
changjiangensis *N. Liu in Acta Phytotax. Sin.,* 36(6): 552 (1998)—Hainan.
clivicola *K.D. Hill in Brittonia,* 51(1): 62 (1999)—Thailand.
subsp. lutea *K.D. Hill in Brittonia,* 51(1): 64 (1999)—Thailand, Cambodia, Vietnam.
crassipes *Hung T. Chang, Y.C. Zhong & Z.F. Lu in Acta Sci. Nat. Univ. Sunyatseni,* 38(3): 121 (1999)—China Southeast (Guangxi).
curranii *(J. Schust.) K.D. Hill in Proc. Third Internat. Conf. Cycad Biol. (ed. P. Vorster):* 150 (1995): C. circinalis var. curranii.
debaoensis *Y.C. Zhong & C.J. Chen in Acta Phytotax. Sin.,* 35(6): 571 (1997)—China Southeast (Guangxi).
diannanensis *Z.T. Guan & G.D. Tao in Sichuan Forest. Surv. Design,* 1995(4): 1 (1995)—China South-central (Yunnan).
edentata *D.J. de Laubenfels in Blumea,* 43(2): 373 (1998)—Philippines.

CYCAS:-
elonga *(Leandri) D.Yue Wang in Cycads in China (eds. F.X. Wang & H.B. Liang):* 51 (1996): C. pectinata var. elonga.
fairylakea *D.Yue Wang in Cycads in China (eds. F.X. Wang & H.B. Liang):* 54 (1996)—China Southeast (Guangdong).
falcata *K.D. Hill in Kew Bull.,* 54(1): 209 (1999)—Sulawesi.
guizhouensis K.M. Lan & R.F. Zou
f. tenuifolia *C.Y. Deng & D.Yue Wang ex Z.T. Guan, L. Zhou & K.S. Hsu in Cycads of China (eds. Z.T. Guan & L. Zhou):* 16 (1996), without type—China South-central (Yunnan, Guizhou), China Southeast (Guangxi).
hongheensis *S.Y. Yang & S.L. Yang ex D.Yue Wang in Cycads in China (eds. F.X. Wang & H.B. Liang):* 62 (1996)—China South-central (Yunnan, Sichuan), China Southeast (Guangdong).
javana *(Miq.) de Laub. ex D.Yue Wang in Cycads in China (eds. F.X. Wang & H.B. Liang):* 65 (1996): C. circinalis var. javana.
javana *(Miq.) D.J. de Laubenfels in Blumea,* 43(2): 381 (1998), isonym; *see D.Yue Wang in Cycads in China:* 65 (1996): C. circinalis var. javana.
lindstromii *S.L. Yang, K.D. Hill & N.T. Hiep in Novon,* 7(2): 213 (1997)—Vietnam.
litoralis *K.D. Hill in Brittonia,* 51(1): 70 (1999)—Thailand, Sumatera, Malaya (Peninsular Malaysia).
longiconifera *Hung T. Chang, Y.C. Zhong & Y.Y. Huang in Acta Sci. Nat. Univ. Sunyatseni,* 37(4): 6 (1998)—China Southeast (Guangxi).
longipetiolula *D.Yue Wang in Cycads in China (eds. F.X. Wang & H.B. Liang):* 68 (1996)—China South-central (Yunnan), China Southeast (Fujian, Guangdong).
longisporophylla *F.N. Wei in Guihaia,* 17(3): 209 (1997)—China Southeast (Guangxi).
longlinensis *H.T. Chang & Y.C. Zhong in Acta Sci. Nat. Univ. Sunyatseni,* 36(3): 68 (1997)—China Southeast (Guangxi).
maconochiei *S.K. Chirgwin & K.D. Hill in Telopea,* 7(1): 48 (1996)—Northern Territory.
subsp. lanata *K.D. Hill in Telopea,* 7(1): 49 (1996)—Northern Territory.
subsp. viridis *K.D. Hill in Telopea,* 7(1): 51 (1996)—Northern Territory.
media R. Br.
subsp. banksii *K.D. Hill in Telopea,* 7(1): 15 (1996): Queensland.
subsp. ensata *K.D. Hill in Telopea,* 7(1): 14 (1996)—Queensland.
micholitzii Dyer
f. distichus *Z.T. Guan in Cycads of China (eds. Z.T. Guan & L. Zhou):* 20 (1996), without type—Locality not stated.
subsp. multipinnata *(C.J. Chen & S.Y. Yang) A. Lindstrom in Biol. & Conserv. Cycads (ed. C.J. Chen):* 266 (1999): C. multipinnata.
f. stonensis *(S.L. Yang) Z.T. Guan in Cycads of China (eds. Z.T. Guan & L. Zhou):* 20 (1996), basionym not validly publ.: C. micholitzii var. stonensis.
multifida *H.T. Chang & Y.C. Zhong in Acta Sci. Nat. Univ. Sunyatseni,* 36(3): 70 (1997)—China Southeast (Guangxi).
multifrondis *D.Yue Wang in Cycads in China (eds. F.X. Wang & H.B. Liang):* 80 (1996)—China Southeast (Fujian, Guangdong).
multiovula *D.Yue Wang in Cycads in China (eds. F.X. Wang & H.B. Liang):* 83 (1996)—China South-central (Yunnan).
multipinnata C.J. Chen & S.Y. Yang
f. latilobus *Z.T. Guan in Cycads of China (eds. Z.T. Guang & L. Zhou):* 20 (1996), without latin descr. or type—Locality not stated.
nongnoochiae *K.D. Hill in Brittonia,* 51(1): 60 (1999)—Thailand.
palmatifida *Hung T. Chang, Y.Y. Huang & Y.C. Zhong in Acta Sci. Nat. Univ. Sunyatseni,* 37(4): 7 (1998)—China Southeast (Guangxi).
parvula *S.L. Yang ex D.Yue Wang in Cycads in China (eds. F.X. Wang & H.B. Liang):* 93 (1996)—China South-central (Yunnan, Sichuan), China Southeast (Fujian, Guangdong).

CYCAS:-
pectinata Buch.-Ham.
f. hongheensis *(S.Y. Yang & S.L. Yang) Z.T. Guan in Cycads of China (eds. Z.T. Guan & L. Zhou):* 18 (1996), basionym not validly publ.: C. hongheensis.
subsp. manhaoensis *C. Chen & P. Yun in Acta Bot. Yunnanica,* 17(4): 400 (1995)—China South-central (Yunnan).
pranburiensis *S.L. Yang, W. Tang, K.D. Hill & P. Vatcharakorn in Brittonia,* 51(1): 44 (1999)—Thailand.
semota *K.D. Hill in Telopea,* 7(1): 23 (1996)—Queensland.
septemsperma *Hung T. Chang, Y.Y. Huang & H.X. Zheng in Acta Sci. Nat. Univ. Sunyatseni,* 37(4): 8 (1998)—China Southeast (Guangxi).
sexseminifera *F.N. Wei in Guihaia,* 16(1): 1 (1996)—China Southeast (Guangxi).
shiwandashanica *H.T. Chang & Y.C. Zhong in Acta Sci. Nat. Univ. Sunyatseni,* 36(3): 67 (1997)—China Southeast (Guangxi).
simplicipinna *(Smitinand) K.D. Hill in Proc. Third Internat. Conf. Cycad Biol. (ed. P. Vorster):* 150 (1995): C. micholitzii var. simplicipinna.
spiniformis *J.Y. Liang in Guihaia,* 17(3): 211 (1997)—China Southeast (Guangxi).
tanqingii *D.Yue Wang in Cycads in China (eds. F.X. Wang & H.B. Liang):* 134 (1996)—China South-central (Yunnan).
tansachana *K.D. Hill & S.L. Yang in Brittonia,* 51(1): 65 (1999)—Thailand.
tuckeri *K.D. Hill in Telopea,* 7(1): 20 (1996)—Queensland.
xilingensis *H.T. Chang & Y.C. Zhong in Acta Sci. Nat. Univ. Sunyatseni,* 36(3): 69 (1997)—China Southeast (Guangxi).
xipholepis *K.D. Hill in Telopea,* 7(1): 32 (1996)—Queensland.
yorkiana *K.D. Hill in Telopea,* 7(1): 18 (1996)—Queensland.

CYCLACHAENA (Compos.).
ambrosiifolia *(A. Gray) P.A. Rydberg in N.L. Britton et al., N. Amer. Fl.,* 33: 10 (1922): Euphrosyne ambrosiifolia.†

CYCLAMEN (Primulac.).
cilicium Boiss. & Heldr.
f. album *R. Frank & M. Koenen in Quart. Bull. Alp. Gard. Soc.,* 51(2): 150 (1983)—Turkey.
coum Mill.
f. albissimum *R.H. Bailey in Cyclamen Soc. J.,* 13(1): 27 (1989)—Turkey.
subsp. elegans *(Boiss. & Buhse) C. Grey-Wilson, Cyclamen:* 174 (1997): C. elegans.
f. pallidum *C. Grey-Wilson, Cyclamen:* 174 (1997)—Cultivation.
creticum (Dörfl.) Hildebr.
var. pallide-roseum *C. Grey-Wilson, Cyclamen:* 174 (1997)—Kriti.
× drydeniae *C. Grey-Wilson, Cyclamen:* 174 (1997), as 'drydenii'—Cultivation.(C. coum × trochopteranthum).
graecum Link
subsp. mindleri *(Heldr.) A.P. Davis & R. Govaerts in A.P. Davis, CITES bulb checklist:* 131 (1999): Cyclaminos mindleri.
hederifolium Aiton
f. albiflorum *(Jord.) C. Grey-Wilson, Cyclamen:* 174 (1997): C. albiflorum.
var. confusum *C. Grey-Wilson, Cyclamen:* 174 (1997)—Kriti.
× meiklei *C. Grey-Wilson, Cyclamen:* 174 (1997)—Cultivation.(C. creticum × repandum).
mirabile Hildebr.
f. niveum *Jill White & C. Grey-Wilson in C. Grey-Wilson, Cyclamen:* 174 (1997)—Turkey.
parviflorum Pobed.
var. subalpinum *C. Grey-Wilson, Cyclamen:* 174 (1997)—Cultivation.
persicum Mill.
f. albidum *(Jord.) C. Grey-Wilson, Cyclamen:* 174 (1997): C. albidum.
var. autumnale *C. Grey-Wilson, Cyclamen:* 174 (1997)—Cultivation.
f. puniceum *(Glasau) C. Grey-Wilson, Cyclamen:* 174 (1997): C. aleppicum subsp. puniceum.

CYCLAMEN:-
pseudibericum Hildebr.
f. roseum *C. Grey-Wilson, Cyclamen:* 174 (1997)—Cultivation.
purpurascens Mill.
f. album *C. Grey-Wilson, Cyclamen:* 174 (1997)—Cultivation.
repandum Sm.
f. album *C. Grey-Wilson, Cyclamen:* 175 (1997), contrary to Art. 37.3 ICBN (1994)—Cultivation.
var. peloponnesiacum *(Grey-Wilson) C. Grey-Wilson, Cyclamen:* 174 (1997): C. repandum f. peloponnesiacum.
var. vividum *(Grey-Wilson) C. Grey-Wilson, Cyclamen:* 174 (1997): C. repandum f. vividum.
× saundersiae *C. Grey-Wilson, Cyclamen:* 175 (1997), as 'saundersii'—Cultivation.(C. balearicum × repandum).
× schwarzii *C. Grey-Wilson, Cyclamen:* 175 (1997)—Cultivation.(C. libanoticum × pseudibericum).
trochopteranthum O. Schwarz
f. leucanthum *C. Grey-Wilson, Cyclamen:* 175 (1997)—Cultivation.
vernum Sweet
var. abchasicum *(Medw. ex Kusn.) Ju.L. Menitsky in Bot. Zhurn.,* 85(6): 166 (2000): C. coum var. abchasicum.
× whiteae *C. Grey-Wilson, Cyclamen:* 175 (1997), as 'whitei'—Cultivation.(C. graecum × hederifolium).

CYCLANTHERA (Cucurbitac.).
jeffreyi *R. Lira & I. Rodríguez-Arévalo in Acta Bot. Mex.,* 48: 12 (1999)—Mexico Northeast (Querétaro).

CYCLEA (Menispermac.).
ochiaiana *(Yamam.) S.F. Huang & T.C. Huang in Fl. Taiwan, ed. 2,* 2: 594 (1996): Cissampelos ochiaiana.

CYCLOBALANOPSIS (Fagac.).
albicaulis *(Chun & W.C. Ko) Y.C. Hsu & H.W. Jen in J. Beijing Forest. Univ.,* 15(4): 45 (1993): Quercus albicaulis.
argyrotricha *(A. Camus) Chun & Chang ex Y.C. Hsu & H.W. Jen in J. Beijing Forest. Univ.,* 15(4): 45 (1993): Quercus argyrotricha.
bambusifolia *(Hance) Chun ex Y.C. Hsu & H.W. Jen in J. Beijing Forest. Univ.,* 15(4): 44 (1993), with incorrect basionym ref.: Quercus bambusifolia.
bella *(Chun & Tsiang) Chun ex Y.C. Hsu & H.W. Jen in J. Beijing Forest. Univ.,* 15(4): 45 (1993): Quercus bella.
camusiae *(Trel. ex Hickel & A. Camus) Y.C. Hsu & H.W. Jen in J. Beijing Forest. Univ.,* 15(4): 44 (1993): Quercus camusiae.
chevalieri *(Hickel & A. Camus) Y.C. Hsu & H.W. Jen in J. Beijing Forest. Univ.,* 15(4): 45 (1993), with incorrect basionym ref.: Quercus chevalieri.
chingsiensis *(Y.T. Chang) Y.T. Chang in Acta Phytotax. Sin.,* 34(3): 339 (1996): Quercus chingsiensis.
dinghuensis *(C.C. Huang) Y.C. Hsu & H.W. Jen in J. Beijing Forest. Univ.,* 15(4): 44 (1993): Quercus dinghuensis.
dongfangensis *(C.C. Huang, F.W. Xing & Z.X. Li) Y.T. Chang in Acta Phytotax. Sin.,* 34(3): 340 (1996): Quercus dongfangensis.
fulvisericea *Y.C. Hsu & D.M. Wang in Acta Bot. Yunnanica,* 5(4): 341 (1983), as 'fulvieriacus' in Supp. 18—China South-central (Yunnan).†
hui *(Chun) Chun ex Y.C. Hsu & H.W. Jen in J. Beijing Forest. Univ.,* 15(4): 45 (1993): Quercus hui.
hypargyrea *(Seemen) Y.C. Hsu & H.W. Jen in J. Beijing Forest. Univ.,* 15(4): 46 (1993): Quercus glauca var. hypargyrea.
kontumensis *(A. Camus) Y.C. Hsu & H.W. Jen ex Y.T. Chang in Acta Phytotax. Sin.,* 34(3): 339 (1996): Quercus kontumensis.
lamelloides *(C.C. Huang) Y.T. Chang in Acta Phytotax. Sin.,* 34(3): 340 (1996): Quercus lamelloides.
motuoensis *(C.C. Huang) Y.C. Hsu & H.W. Jen in J. Beijing Forest. Univ.,* 15(4): 46 (1993): Quercus motuoensis.

CYCLOBALANOPSIS:-
nanchuanica *(C.C. Huang) Y.T. Chang in Acta Phytotax. Sin.,* 34(3): 340 (1996): Quercus nanchuanica.
patelliformis *(Chun) Y.C. Hsu & H.W. Jen in J. Beijing Forest. Univ.,* 15(4): 45 (1993): Quercus patelliformis.
shennongii *(C.C. Huang & S.H. Fu) Y.C. Hsu & H.W. Jen in J. Beijing Forest. Univ.,* 15(4): 45 (1993): Quercus shennongii.
subhinoidea *(Chun & W.C. Ko) Y.C. Hsu & H.W. Jen ex Y.T. Chang in Acta Phytotax. Sin.,* 34(3): 339 (1996): Quercus subhinoidea.
xanthotricha *(A. Camus) Y.C. Hsu & H.W. Jen in J. Beijing Forest. Univ.,* 15(4): 45 (1993): Quercus xanthotricha.
yonganensis *(L.K. Ling & C.C. Huang) Y.C. Hsu & H.W. Jen in J. Beijing Forest. Univ.,* 15(4): 45 (1993), without exact basionym page: Quercus yonganensis.

CYCLOCODON (Campanulac.).
celebicus *(Blume) D.Y. Hong in Acta Phytotax. Sin.,* 36(2): 109 (1998): Campanumoea celebica.
lancifolius
subsp. celebicus *(Blume) K.E. Morris & T.G. Lammers in Novon,* 9(3): 387 (1999): Campanumoea celebica.

CYCLOPHYLLUM (Rubiac.).
balansae *(Baill.) A. Guillaumin in Arch. Bot., Mém., Caen,* 3(5): 22 (1929 publ. 1930): Canthium balansae.
pancheri *(Baill.) A. Guillaumin in Arch. Bot., Mém, Caen,* 3(5): 22 (1929 publ. 1930): Canthium pancheri.
subulatum *(Baill.) A. Guillaumin in Arch. Bot., Mém, Caen,* 3(5): 22 (1929 publ. 1930): Canthium subulatum.
vieillardii *(Baill.) A. Guillaumin in Arch. Bot., Mém., Caen,* 3(5): 22 (1929 publ. 1930): Canthium vieillardii.

CYCLOPIA (Legumin.).
sect. Aequalis *A.L. Schutte in Edinburgh J. Bot.,* 54(2): 139 (1997).
sect. Marsupium *A.L. Schutte in Edinburgh J. Bot.,* 54(2): 149 (1997).
sect. Praegnans *A.L. Schutte in Edinburgh J. Bot.,* 54(2): 161 (1997).
sect. Truncatae *A.L. Schutte in Edinburgh J. Bot.,* 54(2): 144 (1997).
alopecuroides *A.L. Schutte in Edinburgh J. Bot.,* 54(2): 164 (1997)—Cape Province.
alpina *A.L. Schutte in Edinburgh J. Bot.,* 54(2): 156 (1997), nom. nov.: C. genistoides var. ovalifolia *Kies.*
glabra *(Hofmeyr & E. Phillips) A.L. Schutte in Edinburgh J. Bot.,* 54(2): 166 (1997): C. montana var. glabra.

CYCLOPOGON (Orchidac.).
cranichoides
f. albolabia *(P.M. Br. & C.L. McCartney) P.M. Brown in North Amer. Native Orchid J.,* 4(1): 51 (1998), without basionym date: Beadlea cranichoides f. albolabia.

CYCLORHIZA (Umbellif.).
edosmoides *(H. Boissieu) M.L. Sheh in Fl. Yunnanica,* 7: 451 (1997): Pimpinella edosmoides.
peucedanifolia *(Franch.) L. Constance in Edinburgh J. Bot.,* 54(1): 101 (1997): Arracacia peucedanifolia.

× **CYCNANDRA** *hort. in Orchid Rev.,* 105(1213): 62 (1997). *ORCHIDACEAE* (Cycnoches × Galeandra).

CYCNIUM (Scrophulariac.).
sect. Integrifolia *R. Mielcarek in Fragm. Flor. Geobot.,* 41(1): 165 (1996).
sect. Tenuisecta *R. Mielcarek in Fragm. Flor. Geobot.,* 41(1): 163 (1996).

CYCNOCHES (Orchidac.).
barthiorum *G.F. Carr & E.A. Christenson in Orchids,* 68(6): 600 (1999)—Colombia.
carrii *E.A. Christenson in Orchid Dig.,* 63(4): 173 (1999)—Peru.

CYCNOCHES:-
lusiae *G.A. Romero-Gonzalez & L.A. Garay in Harvard Pap. Bot.,* 4(2): 476 (1999)—Venezuela.
manoelae *P. Castro & Campacci in Bol. CAOB,* 5(1): 25 (1993)—Brazil North (Pará).

× **CYCSELLIA** *hort.in Orchid Rev.,* 108(1234): new orchid hybrids 15 (2000). *ORCHIDACEAE* (Ansellia × Cycnoches).

CYDISTA (Bignoniac.).
aequinoctialis
var. sarmentosa *(Bertol.) R. Govaerts, World Checklist Seed Plants,* 3(1): 20 (1999): Bignonia sarmentosa.

CYDONIA (Rosac.).
microcarpa *A. Risso, Hist. nat. prod. Eur. mérid.,* 2: 298 (1826)—Locality not stated.

CYLINDRILLUMA *D.C.H. Plowes in Haseltonia,* 3: 62 (1995). *ASCLEPIADACEAE*
solenophora *(Lavranos) D.C.H. Plowes in Haseltonia,* 3: 63 (1995): Caralluma solenophora.

CYLINDROPUNTIA (Cactac.).
acanthocarpa (Engelm. & J.M. Bigelow) F.M. Knuth
var. coloradensis *(L.D. Benson) D.J. Pinkava in J. Ariz.-Nev. Acad. Sci.,* 32(1): 42 (1999): Opuntia acanthocarpa var. coloradensis.
var. major *(Engelm.) D.J. Pinkava in J. Ariz.-Nev. Acad. Sci.,* 32(1): 42 (1999): Opuntia echinocarpa var. major.
anteojoensis *(Pinkava) E.F. Anderson in Cact. Succ. J. (U.S.A.),* 71(6): 324 (1999), without basionym ref.; *et in Cact. Succ. J. (U.S.A.),* 72(1): 46 (2000), with basionym ref.: Opuntia anteojoensis.
× congesta *(Griffiths) D.J. Pinkava in J. Ariz.-Nev. Acad. Sci.,* 32(1): 42 (1999), isonym; *see F.M. Knuth in Backeb. & F.M. Knuth, Kaktus-ABC:* 123 (1935): Opuntia congesta.
× deserta *(Griffiths) D.J. Pinkava in J. Ariz.-Nev. Acad. Sci.,* 32(1): 42 (1999): Opuntia deserta.

CYLISTA (Legumin.).
tomentosa *W. Roxburgh, Pl. Coromandel,* 3: 16 (1811)—India (Karnataka).†

CYMBALARIA (Scrophulariac.).
glutinosa *M. Bigazzi & M. Raffaelli in Webbia,* 54(2): 201 (2000)—Italy.
subsp. brevicalcarata *M. Bigazzi & M. Raffaelli in Webbia,* 54(2): 202 (2000)—Italy.

CYMBIDIUM (Orchidac.).
acuminatum *M.A. Clements & D.L. Jones in Lasianthera,* 1(1): 28 (1996)—New Guinea (Papua New Guinea).
ensifolium
subsp. haematodes *(Lindl.) Du Puy & P.J. Cribb ex R. Govaerts, World Checklist Seed Plants,* 3(1): 20 (1999): C. haematodes.
var. xiphiifolium *(Lindl.) S.S. Ying in Mem. Coll. Agric. Nation. Taiwan Univ.,* 30(1): 38 (1990): C. xiphiifolium.
faberi Rolfe
f. viridiflorum *(S.S. Ying) S.S. Ying in Mem. Coll. Agric. Nation. Taiwan Univ.,* 30(1): 41 (1990): C. tortisepalum var. viridiflorum.
goeringii Rchb. f.
f. albiflorum *(S.S. Ying) S.S. Ying in Mem. Coll. Agric. Nation. Taiwan Univ.,* 30(1): 33, 35 (1990): C. formosanum f. albiflorum.
var. formosanum *(Hayata) S.S. Ying in Mem. Coll. Agric. Nation. Taiwan Univ.,* 30(1): 31 (1990): C. formosanum.
var. gracillimum *(Fukuy.) R. Govaerts, World Checklist Seed Plants,* 3(1): 20 (1999): C. gracillimum.
lancifolium Hook. f.
var. papuanum *(Schltr.) S.S. Ying in Mem. Coll. Agric. Nation. Taiwan Univ.,* 30(1): 22 (1990): C. papuanum.
lowianum
var. kalawensis *(Colyear) R. Govaerts, World Checklist Seed Plants,* 3(1): 21 (1999): C. grandiflorum var. kalawensis.

CYMBIDIUM:-
maguanense *F.Y. Liu in Acta Bot. Yunnanica,* 18(4): 412 (1996)—China South-central (Yunnan).
rigidum *Z.J. Liu & S.C. Chen in Acta Phytotax. Sin.,* 38(6): 570 (2000), homonym; *see Willd., Sp. Pl.,* 4(1): 106 (1805)—China South-Central (Yunnan).
sinense (Jacks.) Willd.
f. pallidiflorum *(S.S. Ying) S.S. Ying in Mem. Coll. Agric. Nation. Taiwan Univ.,* 30(1): 45 (1990): C. sinense var. pallidiflorum.
f. taiwanianum *(S.S. Ying) S.S. Ying in Mem. Coll. Agric. Nation. Taiwan Univ.,* 30(1): 45 (1990): C. sinense var. taiwanianum.

CYMBOPOGON (Gramin.).
sect. Citrati *(Stapf) L. Liou in Fl. Reipubl. Popul. Sin.,* 10(2): 196 (1997): C. ser. Citrati.
sect. Rusae *(Stapf) L. Liou in Fl. Reipubl. Popul. Sin.,* 10(2): 192 (1997): C. ser. Rusae.
bhutanicus *H.J. Noltie in Edinburgh J. Bot.,* 56(3): 398 (1999)—Bhutan.
munroi *(C.B. Clarke) H.J. Noltie in Edinburgh J. Bot.,* 56(3): 400 (1999): Andropogon munroi.

CYMOPTERUS (Umbellif.).
constancei *R.L. Hartman in Brittonia,* 52(2): 136 (2000)—Wyoming, Arizona, Colorado, New Mexico, Utah.
trotteri *(S.L. Welsh & Goodrich) A. Cronquist in A. Cronquist et al., Intermountain Fl., Vasc. Pl. Intermountain West,* U.S.A., 3A: 380 (1997): Oreoxis trotteri.

CYNANCHUM (Asclepiadac.).
sect. Formosum *S. Liede in Novon,* 7(2): 174 (1997).
sect. Macbridea *(Raf.) S. Liede in Novon,* 7(2): 176 (1997): gen. Macbridea.
sect. Microphyllum *S. Liede in Novon,* 7(2): 176 (1997).
sect. Telminostelma *(E. Fourn.) S. Liede in Novon,* 7(2): 177 (1997): gen. Telminostelma.
aculeatum *(Desc.) S. Liede & U. Meve in Novon,* 6(1): 59 (1996): Prosopostelma aculeata.
amplexicaule (Siebold & Zucc.) Hemsl.
f. castaneum *(Makino) C.Y. Li in Clavis Pl. Chinae Bor.-Or. ed. 2 (ed. P.Y. Fu):* 529 (1995): C. amplexicaule var. castaneum.
anderssonii *G. Morillo in Pittieria,* 23: 36 (1995)—Ecuador.
andinum *S. Liede in R. Govaerts, World Checklist Seed Plants,* 3(1): 21 (1999), nom. nov.: Cynoctonum undulatum *Decne.*
ansamalense *S. Liede in Bull. Mus. Nation. Hist. Nat., B, Adansonia,* Ser. 4, 18(1–2): 112 (1996)—Madagascar.
appendiculatopsis *S. Liede in Bull. Mus. Nation. Hist. Nat., B, Adansonia,* Ser. 4, 18(1–2): 114 (1996)—Madagascar.
argentinense *S. Liede in R. Govaerts, World Checklist Seed Plants,* 3(1): 21 (1999), nom. nov.: Metastelma trifurcatum *Griseb.*
atacamense *S. Liede in R. Govaerts, World Checklist Seed Plants,* 3(1): 21 (1999), nom. nov.: Cynoctonum viride *Phil.*
balense *S. Liede in Ann. Missouri Bot. Gard.,* 83(3): 299 (1996)—Ethiopia.
belloi *P.T. Li in J. S. China Agric. Univ.,* 14(1): 57 (1993), nom. nov.: Metastelma lineare *Bello.*
bernardii *G. Morillo in Pittieria,* 23: 37 (1995)—Venezuela.
bifidum *S. Liede & U. Meve in Novon,* 7(1): 38 (1997)—Ecuador.
blyttioides *S. Liede in Ann. Missouri Bot. Gard.,* 83(3): 301 (1996)—Somalia.
brasiliense *(Morillo) S. Liede in Novon,* 7(2): 178 (1997): Metalepis brasiliense.
bricenoi *G. Morillo in Pittieria,* 23: 39 (1995)—Venezuela.
clavidens N.E. Br.
subsp. hastifolium *(N.E. Br.) S. Liede in Ann. Missouri Bot. Gard.,* 83(3): 305 (1996): C. hastifolium.
crassiantherae *S. Liede in Ann. Missouri Bot. Gard.,* 83(3): 307 (1996)—Somalia.
cyathiforme *(Sundell) W.D. Stevens in Novon,* 10(3): 242 (2000): C. racemosum var. cyathiforme.

CYNANCHUM:-
fimbricorona *S. Liede in R. Govaerts, World Checklist Seed Plants,* 3(1): 21 (1999), nom. nov.: C. fimbriatum *Choux.*
fimbricoronum *P.T. Li in J. S. China Agric. Univ.,* 14(1): 58 (1993), nom. nov.: C. fimbriatum *Choux.*
folotsioides *S. Liede & U. Meve in Novon,* 6(1): 60 (1996)—Madagascar.
galgalense *S. Liede in Ann. Missouri Bot. Gard.,* 83(3): 312 (1996)—Somalia.
gentryi *(Morillo) S. Liede in Novon,* 7(2): 178 (1997): Metalepis gentryi.
gilbertii *S. Liede in Novon,* 7(2): 178 (1997), nom. nov.: Metalepis prevostiae *Morillo.*
gobicum *V.I. Grubov in Novosti Sist. Vyssh. Rast.,* 32: 135 (2000), nom. nov.: Antitoxicum lanceolatum *Grubov.*
graminiforme *S. Liede in Bull. Mus. Nation. Hist. Nat., B, Adansonia,* Ser. 4, 18(1–2): 103 (1996)—Madagascar.
hardyi *S. Liede & U. Meve in Novon,* 6(1): 63 (1996)—Madagascar.
hatusimae *P.T. Li in J. S. China Agric. Univ.,* 12(3): 41 (1991), nom. nov., as 'hatusimai': C. crassifolium *Hatus.*
hoedimeerium Bakh. f.
subsp. timorense *S. Liede in Blumea,* 44(2): 481 (1999)—Lesser Sunda Is (East Timor).
jaramilloi *G. Morillo in Pittieria,* 23: 41 (1995)—Ecuador.
lechleri *G. Morillo in Pittieria,* 23: 43 (1995)—Peru.
lineare N.E. Br.
subsp. keraudreniae *S. Liede in Bull. Mus. Nation. Hist. Nat., B, Adansonia,* Ser. 4, 18(1–2): 108 (1996)—Madagascar, Comoros.
luteifluens (Jum. & H. Perrier) Desc.
var. longicoronae *S. Liede in Bull. Mus. Nation. Hist. Nat., B, Adansonia,* Ser. 4, 18(1–2): 119 (1996)—Madagascar.
maasii *G. Morillo in Pittieria,* 23: 45 (1995)—Peru.
mazzuchii *(Speg.) S. Liede in Novon,* 7(2): 175 (1997): Dicarpophora mazzuchii.
mevei *S. Liede in Bull. Mus. Nation. Hist. Nat., B, Adansonia,* Ser. 4, 18(1–2): 120 (1996)—Madagascar.
mexicoense *P.T. Li in J. S. China Agric. Univ.,* 14(1): 58 (1993), nom. nov.: C. pallidum *Sessé & Moç.*
morilloi *P.T. Li in J. S. China Agric. Univ.,* 14(1): 57 (1993), nom. nov.: Ditassa angustifolia *Decne.*
nanellum *P.T. Li in J. S. China Agric. Univ.,* 14(1): 58 (1993), nom. nov.: C. nanum *Skottsb.*
nanissimum *S. Liede in R. Govaerts, World Checklist Seed Plants,* 3(1): 21 (1999), nom. nov.: C. nanum *Skottsb.*
nematostemma *S. Liede in Bull. Mus. Nation. Hist. Nat., B, Adansonia,* Ser. 4, 18(1–2): 123 (1996), nom. nov.: Nematostemma perrieri *Choux.*
nielsenii *G. Morillo in Pittieria,* 23: 46 (1995)—Ecuador.
paralias *(Decne.) P.T. Li in J. S. China Agric. Univ.,* 14(1): 57 (1993): Metastelma paralias.
paramorum *G. Morillo in Pittieria,* 23: 47 (1995)—Venezuela.
petignatii *S. Liede & Rauh in Bull. Mus. Nation. Hist. Nat., B, Adansonia,* Ser. 4, 18(1–2): 126 (1996)—Madagascar.
phillipsonianum *S. Liede & Meve in Bull. Mus. Nation. Hist. Nat., B, Adansonia,* Ser. 4, 18(1–2): 109 (1996)—Madagascar.
pietrangelii *G. Morillo in Pittieria,* 23: 48 (1995)—Venezuela.
pusilum *V.I. Grubov in Novosti Sist. Vyssh. Rast.,* 32: 137 (2000)—Xinjiang.
rotiforme *P.T. Li in J. S. China Agric. Univ.,* 14(1): 58 (1993), nom. nov.: C. rotatum *Sessé & Moç.*
rubricoronae *S. Liede in Ann. Missouri Bot. Gard.,* 83(3): 333 (1996)—Somalia.
sessei *P.T. Li in J. S. China Agric. Univ.,* 14(1): 58 (1993), nom. nov.: C. hirsutum *Sessé & Moç.*
sigridiae *U. Meve & M. Teissier in Bradleya,* 14: 10 (1996)—Madagascar.
stelligerum *P.T. Li in J. S. China Agric. Univ.,* 14(1): 58 (1993), nom. nov.: C. stellatum *Pobed.*
trollii *S. Liede & U. Meve in Novon,* 7(1): 39 (1997)—Bolivia.
umtalense *S. Liede in Ann. Missouri Bot. Gard.,* 83(3): 341 (1996)—Zimbabwe, Malawi.
ventensis *G. Morillo in Pittieria,* 23: 49 (1995)—Venezuela.

CYNAROSPERMUM *K. Vollesen in Kew Bull.,* 54(1): 171 (1999). *ACANTHACEAE*
asperrimum *(Nees) K. Vollesen in Kew Bull.,* 54(1): 172 (1999): Blepharis asperrima.

CYNOGLOSSUM (Boraginac.).
amabile Stapf & J.R. Drumm.
f. leucanthum *X.D. Dong in Bull. Bot. Res. North-East. Forest. Univ.,* 17(2): 141 (1997)—China South-central (Yunnan).
f. ruberum *X.D. Dong in Bull. Bot. Res. North-East. Forest. Univ.,* 17(2): 141 (1997)—China South-central (Yunnan).
hintoniorum *B.L. Turner in Phytologia,* 79(4): 306 (1995 publ. 1996)—Mexico Southwest (Oaxaca).
kandavanensis *(Bornm. & Gauba) H. Akhani in Stapfia,* 53: 80 (1998): Lindelofia kandavanensis.
nebrodense Guss.
subsp. pustulatum *(Boiss.) O. de Bolòs & J. Vigo, Fl. Països Catalans,* 3: 207 (1995): C. pustulatum.
obtusicalyx *E. Retief & A.E. van Wyk in S. Afr. J. Bot.,* 62(3): 169 (1996)—Cape Province.
sabirense *(R.R. Mill & A.G. Mill.) J.R.I. Wood, Handbook Yemen Fl.:* 241 (1997): Paracynoglossum sabirense.
semnanicum *M. Khatamsaz in Iranian J. Bot.,* 8(1): 2 (1999)—Iran.
yemenense *(R.R. Mill & A.G. Mill.) J.R.I. Wood, Handbook Yemen Fl.:* 241 (1997): Paracynoglossum yemenense.

CYNOMETRA (Legumin.).
floretii *J.-N. Labat & O. Pascal in Kew Bull.,* 54(1): 165 (1999)—Comoros (Mayotte).
mayottensis *J.-N. Labat & O. Pascal in Kew Bull.,* 54(1): 164 (1999)—Comoros (Mayotte).
polyandra *W. Roxburgh, Pl. Coromandel,* 3: 83 (1820)—Bangladesh.†

CYNORKIS (Orchidac.).
subgen. Bicornella *(Lindl.) D.L. Szlachetko & P. Rutkowski in Acta Bot. Fenn.,* 169: 380 (2000): gen. Bicornella.
bardotiana *J. Bosser in Adansonia,* 20(2): 281 (1998)—Madagascar.
hetaerioides *(Summerh.) D.L. Szlachetko in Fragm. Flor. Geobot.,* Supp. 3: 114 (1995): Habenaria hetaerioides.
ochyrae *D.L. Szlachetko & T.S. Olszewski in Fl. Cameroun,* 34: 92 (1998)—Zaire.
rosellata *(Thouars) J. Bosser in Adansonia,* 19(2): 188 (1997): Satyrium rosellatum.

CYNOXYLON (Cornac.).
subgen. Discocrania *(Harms) J. Holub in Preslia,* 70(2): 103 (1998): Cornus sect. Discocrania.

CYPERUS (Cyperac.).
aggregatus (Willd.) Endl.
var. gigas *(Lindm.) E.R. Guaglianone in Hickenia,* 2(31): 137 (1995): Mariscus flavus var. gigas.
alatus (Nees) F. Muell.
subsp. albus *(Nees) K.A. Lye in Lidia,* 3(5): 172 (1995): Kyllinga alba.
subsp. ascolepidioides *(Cherm.) K.A. Lye in Lidia,* 3(5): 172 (1995): Kyllingia ascolepidioides.
subsp. controversus *(Steud.) K.A. Lye in Lidia,* 3(5): 172 (1995): Kyllinga controversa.
altomicroglumis *K.A. Lye in Bull. Jard. Bot. Nation. Belg.,* 65(1–2): 249 (1996)—Somalia.
aureoalatus *K.A. Lye in Lidia,* 3(5): 171 (1995)—Uganda.
baobab *K.A. Lye in Nordic J. Bot.,* 16(4): 371 (1996 publ. 1997)—Somalia.
bifolius *K.A. Lye in Lidia,* 3(6): 206 (1996)—Ethiopia.
brunneofibrosus *K.A. Lye in Candollea,* 51(2): 423 (1996)—Somalia.
densibulbosus *K.A. Lye in Nordic J. Bot.,* 16(4): 369 (1996 publ. 1997)—Somalia.
erectus
subsp. jubensis *(Chiov.) K.A. Lye in Candollea,* 51(2): 424 (1996): Kyllinga jubensis.
floribundus *(Kük.) R. Carter & S.D. Jones in Rhodora,* 99(900): 330 (1997 publ. 1998): C. uniflorus var. floribundus.

CYPERUS:-
gypsophilus *K.A. Lye in Nordic J. Bot.*, 16(4): 374 (1996 publ. 1997)—Somalia.
ligusticus *M. Gandoger, Dec. Pl. Nov.*, 2: 28 (1876)—Italy.
macrocephalus Liebm.
var. eggersii *(Boeck.) S.D. Jones, J.K. Wipff & R. Carter in Phytologia*, 80(4): 289 (1996): C. eggersii.
maculatus Boeck.
subsp. ogadensis *K.A. Lye in Lidia*, 3(6): 206 (1996)—Ethiopia.
micromedusaeus *K.A. Lye in Nordic J. Bot.*, 16(4): 373 (1996 publ. 1997)—Somalia.
micropelophilus *K.A. Lye in Willdenowia*, 26(1–2): 233 (1996)—Somalia.
obtusiusculus *M. Gandoger, Dec. Pl. Nov.*, 2: 29 (1876)—France.
odoratus L.
var. engelmannii *(Steud.) R. Carter, S.D. Jones & J.K. Wipff in Phytologia*, 80(4): 289 (1996): C. engelmannii.
var. squarrosus *(Britton) S.D. Jones, J.K. Wipff & R. Carter in Phytologia*, 80(4): 289 (1996): C. speciosus var. squarrosus.
oliganthus *M. Gandoger, Dec. Pl. Nov.*, 2: 37 (1876)—France.
ossicaulis *K.A. Lye in Kew Bull.*, 51(1): 205 (1996)—Somalia.
pluricephalus *K.A. Lye in Nordic J. Bot.*, 16(4): 371 (1996 publ. 1997)—Somalia.
pseudothyrsiflorus *(Kük.) R. Carter & S.D. Jones in Rhodora*, 99(900): 330 (1997 publ. 1998): C. uniflorus var. pseudothyrsiflorus.
recurvispicatus *K.A. Lye in Nordic J. Bot.*, 16(4): 376 (1996 publ. 1997)—Somalia.
retroflexus Buckley
var. pumilus *(Britton) R. Carter & S.D. Jones in Rhodora*, 99(900): 328 (1997 publ. 1998): C. uniflorus var. pumilus.
robustior *M. Gandoger, Dec. Pl. Nov.*, 2: 37 (1876)—France.
rotundus L.
subsp. divaricatus *K.A. Lye in Nordic J. Bot.*, 16(4): 367 (1996 publ. 1997)—Somalia.
scabricaulis *K.A. Lye in Nordic J. Bot.*, 16(4): 369 (1996 publ. 1997)—Somalia.
somalidunensis *K.A. Lye in Nordic J. Bot.*, 16(4): 374 (1996 publ. 1997)—Somalia.
tibialis *(Ledeb.) R. Govaerts, World Checklist Seed Plants*, 3(1): 21 (1999): Kyllinga tibialis.

CYPHOMANDRA (Solanac.).
meridensis *Pittier ex J.A. Steyermark & Rojas in Pittieria*, 7: 19 (1978)—Venezuela.

CYPHOSTEMMA (Vitac.).
cornus-africani *M. Thulin in Nordic J. Bot.*, 19(3): 341 (1999)—Somalia.
rubroglandulosum *E. Retief & A.E. van Wyk in S. Afr. J. Bot.*, 62(4): 183 (1996)—Natal, Cape Province.

CYPHOSTIGMA (Zingiberac.).
kandariense *(K. Schum.) K. Schum.in Engl., Pflanzenr.*, IV.46: 273 (1904): Amomum kandariense.
stoloniflorum *(K. Schum.) K. Schum.in Engl., Pflanzenr.*, IV.46: 273 (1904): Amomum stoloniflorum.
surculosum *(K. Schum.) K. Schum.in Engl., Pflanzenr.*, IV.46: 273 (1904): Amomum surculosum.

CYPRIPEDIUM (Orchidac.).
sect. Bifolia *(Lindl.) S.C. Chen in Proc. 12th World Orchid Conf., 1987 (eds. K. Saito & R. Tanaka)*: 145 (1987): C. group 4. Bifolia.
subsect. Enantiopedilum *(Pfitzer) S.C. Chen in Proc. 12th World Orchid Conf., 1987 (eds. K. Saito & R. Tanaka)*: 145 (1987): C. sect. Enantiopedilum.
sect. Flabellinervia *(Pfitzer) Hennessy ex P.J. Cribb, Genus Cypripedium*: 225 (1997): C. ser. Flabellinervia.
subsect. Flabellinervia *(Pfitzer) S.C. Chen in Proc. 12th World Orchid Conf., 1987 (eds. K. Saito & R. Tanaka)*: 145 (1987): C. ser. Flabellinervia.

CYPRIPEDIUM:-
sect. Irapeana *P.J. Cribb, Genus Cypripedium*: 120 (1997).
subsect. Macrantha *(Kraenzl.) P.J. Cribb, Genus Cypripedium*: 190 (1997): C. group 2. Macrantha.
sect. Obtusipetala *(Pfitzer) P.J. Cribb, Genus Cypripedium*: 135 (1997): C. subsect. Obtusipetala.
sect. Retinervia *(Pfitzer) S.C. Chen in Proc. 12th World Orchid Conf., 1987 (eds. K. Saito & R. Tanaka)*: 145 (1987): C. ser. Retinervia.
× catherinae *L.V. Averyanov in Turczaninowia*, 2(2): 34 (1999)—Primorye.(C. macranthon × shanxiense).
cheniae *G. Torelli in Caesiana*, 12: 45 (1999), as 'chenii'—China South-central (Guizhou).
froschii *H. Perner in Orchidee*, Suppl. 6: 10 (1999)—China South-central (Yunnan).
guttatum Sw.
f. albiflorum *L.V. Averyanov in Turczaninowia*, 2(2): 30 (1999)—West Siberia.
f. albiflorum *Y.N. Lee, Fl. Korea*: 1164 (1996), contrary to Arts. 37.4, 37.5 ICBN (1994)—South Korea.
kentuckiense C.F. Reed
f. pricei *P.M. Brown in North Amer. Native Orchid J.*, 4(1): 45 (1998)—Arkansas.
lentiginosum *P.J. Cribb & S.C. Chen in Quart. Bull. Alp. Gard. Soc.*, 67(2): 155 (1999)—China South-central (Yunnan).
ludlowii *P.J. Cribb, Genus Cypripedium*: 204 (1997)—Xizang.
macranthon Sw.
nothof. alboroseum *L.V. Averyanov in Turczaninowia*, 2(2): 23 (1999), as 'alboroseum'—Primorye. (C. macranthon var. album × var. macranthon).
nothof. albostriatum *L.V. Averyanov in Turczaninowia*, 2(2): 23 (1999), as 'albostriatum'—Primorye. (C. macranthon var. album × var. atropurpureum).
var. atropurpureum *L.V. Averyanov in Turczaninowia*, 2(2): 18 (1999)—Amur.
nothof. flavoroseum *L.V. Averyanov in Turczaninowia*, 2(2): 23 (1999), as 'flavoroseum'—Primorye.(C. macranthon var. flavum × var. macranthon).
rubronerve *W. Cavestro in Rhôone-Alpes Orchidées*, 25: 14 (10 May 2000), preprint—Manchuria (Heilongjiang).
taibaiense *G.H. Zhu & S.C. Chen in Novon*, 9(3): 454 (1999)—China North-central (Shaanxi).
× ventricosum Sw.
f. bicolor *L.V. Averyanov in Turczaninowia*, 2(2): 26 (1999)—Primorye.
× wenqingiae *H. Perner in Orchidee*, 49(5): 201 (1998)—China South-central (Sichuan).(C. farreri × tibeticum).

CYRTANDRA (Gesneriac.).
erythrotricha *B.L. Burtt in Checklist Flow. Pl. & Gymn. Brunei Darussalam (ed. M.J.E. Coode et al.)*: 436 (1996)—Borneo (Sarawak).
glomeruliflora *B.L. Burtt in Checklist Flow. Pl. & Gymn. Brunei Darussalam (ed. M.J.E. Coode et al.)*: 436 (1996)—Borneo (Brunei).
integerrima *B.L. Burtt in Checklist Flow. Pl. & Gymn. Brunei Darussalam (ed. M.J.E. Coode et al.)*: 436 (1996)—Borneo (Brunei).
kalimantana *B.L. Burtt in Edinburgh J. Bot.*, 56(3): 374 (1999)—Borneo (Kalimantan).
neiothiantha *B.L. Burtt in Checklist Flow. Pl. & Gymn. Brunei Darussalam (ed. M.J.E. Coode et al.)*: 436 (1996)—Borneo (Brunei).
paragibbsiae *B.L. Burtt in Checklist Flow. Pl. & Gymn. Brunei Darussalam (ed. M.J.E. Coode et al.)*: 436 (1996)—Borneo (Brunei).
poulsenii *B.L. Burtt in Checklist Flow. Pl. & Gymn. Brunei Darussalam (ed. M.J.E. Coode et al.)*: 436 (1996)—Borneo (Brunei).
prolata *B.L. Burtt in Checklist Flow. Pl. & Gymn. Brunei Darussalam (ed. M.J.E. Coode et al.)*: 437 (1996)—Borneo (Sarawak).

CYRTANDROMOEA (Scrophulariac.).
pterocaulis *D.D. Tao, X.D. Li & X. Yang in Acta Bot. Yunnanica*, 17(2): 156 (1995)—China South-central (Yunnan).

CYRTANTHUS (Amaryllidac.).
leptosiphon *D.A. Snijman in Bothalia*, 29(2): 259 (1999)—Cape Province.
wellandii *D.A. Snijman in Bothalia*, 29(2): 261 (1999)—Cape Province.

CYRTOCHILUM (Orchidac.).
aureum *(Lindl.) K. Senghas in Schlechter Orchideen, ed. 3*, I/C(33–36): 2205 (1997): Oncidium aureum.
pastorellii *(Dodson & D.E. Benn.) K. Senghas in Schlechter Orchideen, ed. 3*, I/C(33–36): 2205 (1997): Oncidium pastorellii.

CYRTOCHLOA *S. Dransfield in Kew Bull.*, 53(4): 861 (1998). *GRAMINEAE*
fenixii *(Gamble) S. Dransfield in Kew Bull.*, 53(4): 864 (1998): Schizostachyum fenixii.
hirsuta *S. Dransfield in Kew Bull.*, 53(4): 869 (1998)—Philippines.
luzonica *(Gamble) S. Dransfield in Kew Bull.*, 53(4): 871 (1998): Schizostachyum luzonicum.
puser *S. Dransfield in Kew Bull.*, 53(4): 867 (1998)—Philippines.
toppingii *(Gamble) S. Dransfield in Kew Bull.*, 53(4): 862 (1998): Schizostachyum toppingii.

CYRTOPODIUM (Orchidac.).
braemii *L.C. Menezes in Bol. CAOB*, 5(3): 26 (1993)—Brazil West-central (Goiás).
cachimboense *L.C. Menezes in Bol. CAOB*, 26: 22 (1996)—Brazil North (Pará).
gonzalezii *L.C. Menezes in Bol. CAOB*, 6(1): 9 (1995)—Brazil West-central (Brasília Distrito Federal).
graniticum *G.A. Romero-Gonzalez & Carnevali in Harvard Pap. Bot.*, 4(2): 512 (1999)—Venezuela.
inaldianum *L.C. Menezes in Bol. CAOB*, 23: 12 (1996)—Brazil Northeast (Paraíba).
latifolium *L. de Bem Bianchetti & J.A.N. Batista in Lindleyana*, 15(4): 222 (2000)—Brazil West-Central (Brazilia Distrito Federal, Mato Grosso).
macrobulbon *(La Llave & Lex.) G.A. Romero-González & G. Carnevali Fernández-Concha in Harvard Pap. Bot.*, 4(1): 331 (1999), as 'macrobulbum': Epidendrum macrobulbon.
f. salvadorense *(Hamer & Garay) G.A. Romero-González & G. Carnevali Fernández-Concha in Harvard Pap. Bot.*, 4(1): 336 (1999): C. punctatum var. salvadorense.
paludicola *F.C. Hoehne, Com. Lin. Telegr., Bot.*, 4: 24 (1912), as 'paludicolum'—Brazil West-central (Mato Grosso).
var. goiandirense *L.C. Menezes in Bol. CAOB*, 27: 18 (1997)—Brazil West-central (Goiás).
polyphyllum *(Vell.) Pabst ex F. Barros in Acta Bot. Brasil.*, 8(1): 12 (1994): Epidendrum polyphyllum.
polyphyllum *(Vell.) L.C. Menezes in Orquidário*, 12(4): 103 (1999), isonym; *see F. Barros in Acta Bot. Brasil.*, 8(1): 12 (1994): Epidendrum polyphyllum.
var. pickelii *(Hoehne) L.C. Menezes in Orquidário*, 12(4): 104 (1999), basionym not validly publ.: C. paranaense var. pickelii.
roraimense *L.C. Menezes in Orchidee*, 50(1): 19 (1999); *et in Orquidário*, 12(4): 101 (1999)—Brazil North (Roraima).
sanguineum *(Lindl.) N.E. Brown, Suppl. Johnson's Gard. Dict.*: 912 (1882): Cyrtopera sanguinea.
sarneyanum *L.C. Menezes in Orquidário*, 13(3–4): 85 (1999)—Brazil North (Rondônia).
withneri *L.C. Menezes in Orchid Dig.*, 60(1): 13 (1996), as 'withnerii'—Brazil West-central (Brasília Distrito Federal).

CYSTIDOSPERMUM (Euphorbiac.).
petiolatum *(Banks & Sol.) T. Nasimova in Opred. Rast. Sred. Azii*, 7: 77 (1983): Euphorbia petiolata.

CYSTORCHIS (Orchidac.).
ogurae *(Tuyama) P. Ormerod & P.J. Cribb in Kew Bull.*, 55(1): 234 (2000): Hetaeria ogurae.

CYTISUS (Legumin.).
sect. Verzinum *(Raf.) S. Talavera in An. Jard. Bot. Madrid*, 57(1): 214 (1999): gen. Verzinum.

CYTISUS:-
aeolicus *G. Gussone, Fl. Sicul. Prodr. Suppl.*: 221 (1834)—Sicilia.
balansae (Boiss.) Ball
var. galianoi *(Talavera & P.E. Gibbs) G. López González in An. Jard. Bot. Madrid*, 57(2): 450 (1999 publ. 2000): C. galianoi.
galianoi *S. Talavera & P. Gibbs in Bot. J. Linn. Soc.*, 125(4): 338 (1997)—Spain.
grandiflorus (Brot.) DC.
subsp. cabezudoi *S. Talavera in An. Jard. Bot. Madrid*, 57(1): 213 (1999)—Spain.
malacitanus Boiss.
subsp. moleroi *(Fern. Casas) Á. Lora González, J.E. Hernández Bermejo & J. Prados Ligero in Acta Bot. Malacitana*, 23: 235 (1998): C. moleroi.
scoparius (L.) Link
subsp. maurus *(Humbert & Maire) S. Talavera in An. Jard. Bot. Madrid*, 57(1): 212 (1999): C. maurus.
transiens *(Maire) S. Talavera in An. Jard. Bot. Madrid*, 57(1): 215 (1999): C. arboreus var. transiens.
valdesii *S. Talavera & P. Gibbs in Bot. J. Linn. Soc.*, 125(4): 340 (1997), nom. nov.: C. balansae var. atlanticus *Ball*.

CYTOGONIDIUM *B.G. Briggs & L.A.S. Johnson in Telopea*, 7(4): 362 (1998). *RESTIONACEAE*
leptocarpoides *(Benth.) B.G. Briggs & L.A.S. Johnson in Telopea*, 7(4): 363 (1998): Restio leptocarpoides.

D

DABOECIA (Ericac.).
cantabrica (Huds.) K. Koch
f. albiflora *(D.C. McClint.) E.C. Nelson in New Plantsman*, 3(2): 85 (1996): D. azorica f. albiflora.
f. eburnea *E.C. Nelson in New Plantsman*, 3(2): 85 (1996), nom. nov.: D. × scotica f. albiflora *D.C. McClint.*
nothosubsp. scotica *(D.C. McClint.) E.C. Nelson in New Plantsman*, 3(2): 84 (1996): D. × scotica.

DACRYDIUM (Podocarpac.).
nidulum de Laub.
var. vitiensis *J. Silba in J. Int. Conifer Preserv. Soc.*, 7(1): 23 (2000)—Fiji.
papuanum *(de Laub.) T.C. Whitmore in T.C. Whitmore, I.G.M. Tantra & U. Sutisna (eds.), Tree Flora Indonesia*, Check List Irian Jaya: 233 (1997): Falcatifolium papuanum.

DACRYODES (Burserac.).
bampsiana *R. Pierlot in Bull. Jard. Bot. Nation. Belg.*, 65(3–4): 366 (1996)—Zaire.
leonardiana *R. Pierlot in Bull. Jard. Bot. Nation. Belg.*, 65(3–4): 363 (1996)—Zaire.

DACTYLADENIA (Chrysobalanac.).
floretii *F.J. Breteler in Adansonia*, 22(1): 117 (2000)—Gabon.
jongkindii *F.J. Breteler in Adansonia*, 22(1): 118 (2000)—Gabon.
ndjoleensis *F.J. Breteler in Syst. Geogr. Pl.*, 69(1): 111 (1999)—Gabon.

DACTYLIS (Gramin.).
glomerata L.
subsp. nestorii *J.A. Rosselló & L. Sáez in An. Jard. Bot. Madrid*, 56(2): 396 (1998), nom. nov.: D. glomerata subsp. ibizensis *Stebbins & D. Zohary*.
metlesicsii *P. Schönfelder & D. Ludwig in Willdenowia*, 26(1–2): 218 (1996)—Canary Is.

× **DACTYLITELLA** (Orchidac.).
berninaensis *W. Schmid in O. Gerbaud & W. Schmid, Hybrides Genres Nigritella et/ou Pseudorchis* (Cahiers S.F.O., 5): 26 (1999)—Switzerland.(Dactylorhiza majalis × Nigritella rhellicani).

DACTYLOCTENIUM (Gramin.).
aegyptium (L.) Willd.
subsp. aristatum *(Link) M. Sharma in Geobios New Rep.*, 16(1): 34 (1997); *et in J. Econ. Taxon. Bot.*, 21(3): 555 (1997 publ. 1998): D. aristatum.

× **DACTYLODENIA** (Orchidac.).
aravardii *M. Gerbaud & O. Gerbaud in Orchidophile*, 126: 64, 61 (1997)—France. (Dactylorhiza traunsteineri × Gymnadenia odoratissima).

× **DACTYLOGLOSSUM** (Orchidac.).
ponticum *R. Kohlmüller in Ber. Arbeitskr. Heim. Orch.*, 16(1): 74 (1999)—Turkey. (Coeloglossum viride × Dactylorhiza urvilleana).

× **DACTYLOGYMNADENIA** (Orchidac.).
gracilis (A. Camus) Soó
nothovar. major *H. Kümpel in Mitt. Arbeitskr. Heim. Orch. Zentr. Fachausschusses Bot. Kulturb. DDR*, 8: 41 (1978), as nothomorph—Germany. (Dactylorhiza fuchsii subsp. fuchsii × Gymnadenia conopsea subsp. densiflora).

DACTYLORHIZA (Orchidac.).
aristata (Fisch. ex Lindl.) Soó
f. perbracteata *(Lepage) P.M. Brown in North Amer. Native Orchid J.*, 2(1): 68 (1996), isonym; *see Catling in Nat. Canad.*, 109(2): 278 (1982): Orchis aristata f. perbracteata.
× braunii (Halácsy) Borsos & Soó
nothovar. lilacina *F. Procházka in Preslia*, 54(4): 291 (1982), as'nothosubsp.' in Supp. 18—Czechoslovakia.(D. fuchsii subsp. fuchsii × majalis subsp. turfosa).†
nothosubsp. lilacina *(F. Proch.) J. Holub in Preslia*, 70(2): 103 (1998): D. × braunii nm. lilacina.
nothosubsp. smitakii *P. Batousek in J. Eur. Orchideen*, 29(4): 643 (1997)—Czechoslovakia. (D. fuchsii subsp. sooana × majalis subsp. majalis).
carpatica *(Batousek & Kreutz) P. Delforge in Nat. Belg.*, 81(4): 396 (2000): Dactylorhiza traunsteineri subsp. carpatica.
comosa *(Scop.) P.D. Sell in P.D. Sell & G. Murrell, Fl. Great Britain & Ireland*, 5: 364 (1996): Orchis comosa.
subsp. cambrensis *(R.H. Roberts) P.D. Sell in P.D. Sell & G. Murrell, Fl. Great Britain & Ireland*, 5: 365 (1996): Dactylorchis majalis subsp. cambrensis.
f. kerryensis *(Wilmott) P.D. Sell in P.D. Sell & G. Murrell, Fl. Great Britain & Ireland*, 5: 365 (1996): Orchis kerryensis.
subsp. majalis *(Rchb.) P.D. Sell in P.D. Sell & G. Murrell, Fl. Great Britain & Ireland*, 5: 365 (1996): Orchis majalis.
f. occidentalis *(Pugsley) P.D. Sell in P.D. Sell & G. Murrell, Fl. Great Britain & Ireland*, 5: 365 (1996): Orchis majalis var. occidentalis.
subsp. occidentalis *(Pugsley) P.D. Sell in P.D. Sell & G. Murrell, Fl. Great Britain & Ireland*, 5: 364 (1996): Orchis majalis var. occidentalis.
subsp. scotica *(E. Nelson) P.D. Sell in P.D. Sell & G. Murrell, Fl. Great Britain & Ireland*, 5: 364 (1996): D. majalis subsp. scotica.
subsp. turfosa *(F. Proch.) J. Holub in Preslia*, 70(2): 103 (1998): D. majalis subsp. turfosa.
× daunia *W. Rossi, P. Arduino, R. Cianchi, N. Piemontese & L. Bullini in Webbia*, 50(1): 11 (1995)—Italy. (D. romana × saccifera).
ebudensis *(Wief. ex R.M. Bateman & Denholm) P. Delforge in Nat. Belg.*, 81(4): 397 (2000): Dactylorhiza majalis var. ebudensis.
elata (Poir.) Soó
var. brennensis *(E. Nelson) P. Delforge in Nat. Belg.*, 81(4): 397 (2000): Dactylorhiza elata subsp. brennensis.
fuchsii (Druce) Soó
var. trilobata *(Bréb.) P.D. Sell in P.D. Sell & G. Murrell, Fl. Great Britain & Ireland*, 5: 364 (1996): Orchis maculata var. trilobata.

DACTYLORHIZA:-
ilgazica *C.A.J. Kreutz in Eurorchis*, 9: 49 (1997)—Turkey.
× juennensis *M. Perko in Carinthia II*, 187/107(1): 91 (1997)—Austria.(D. lapponica × maculata subsp. fuchsii).
× kernerorum (Soó) Soó
nothosubsp. variablis *(Hesl.-Harr.) P.D. Sell in P.D. Sell & G. Murrell, Fl. Great Britain & Ireland*, 5: 364 (1996): Orchis variabilis.
× kuehnensis *H. Presser & A. Riechelmann in J. Eur. Orchideen*, 32(1): 109 (2000)—Germany. (D. majalis × ruthei).
× uuskiae *E. Breiner & R. Breiner in Ber. Arbeitskr. Heim. Orch.*, 15(2): 34 (1998), as 'kuuskae'—Baltic States (Estoniya). (D. baltica × ruthei).
maculata (L.) Soó
var. elodes *(Griseb.) L.V. Averyanov in Turczaninowia*, 3(1): 49 (2000): Orchis elodes.
subsp. podesta *(Landwehr) C.A.J. Kreutz in C.A.J. Kreutz & H. Dekker, Orchid. van Nederland*: 179 (2000): D. maculata var. podesta.
var. transsilvanica *(Schur) P. Delforge in Nat. Belg.*, 81(4): 397 (2000): Orchis transsilvanica.
parvimajalis *D. Tyteca & J.-L. Gathoye in J. Eur. Orchideen*, 32(3–4): 504 (2000), contrary to Art. 37.6 ICBN (2000)—France.
pontica *(H. Fleischm. & Hand.-Mazz.) P. Delforge in Nat. Belg.*, 81(4): 397 (2000): Orchis pontica.
praetermissa (Druce) Soó
f. junialis *(Verm.) P.D. Sell in P.D. Sell & G. Murrell, Fl. Great Britain & Ireland*, 5: 365 (1996): Orchis latifolia var. junialis.
× reitaluae *M. Hennecke, E. Breiner & R. Breiner in Ber. Arbeitskr. Heim. Orch.*, 15(2): 36 (1998)—Baltic States (Estoniya). (D. incarnata × ruthei).
× silvae-gabretae *F. Procházka & V. Čurn in Preslia*, 71(3): 236 (1999)—Czechoslovakia. (D. fuchsii subsp. fuchsii × traunsteineri).
× transiens (Druce) Soó
nothosubsp. corylensis *(Hesl.-Harr.) P.D. Sell in P.D. Sell & G. Murrell, Fl. Great Britain & Ireland*, 5: 364 (1996): Orchis corylensis.
traunsteineri (Saut.) Soó
subsp. carpatica *P. Batoušek & C.A.J. Kreutz in J. Eur. Orchideen*, 31(3): 684, 683 (1999)—Czechoslovakia.
var. curvifolia *(F. Nyl.) L.V. Averyanov in Turczaninowia*, 3(1): 50 (2000): Orchis curvifolia.
var. russowii *(Klinge) L.V. Averyanov in Turczaninowia*, 3(1): 50 (2000), without basionym ref.: Orchis angustifolia var. russowii.
× venusta (T. Stephenson & T.A. Stephenson) Soó
nothosubsp. hebridella *(Wilmott) P.D. Sell in P.D. Sell & G. Murrell, Fl. Great Britain & Ireland*, 5: 364 (1996): Orchis hebridella.
viridis *(L.) R.M. Bateman, A.M. Pridgeon & M.W. Chase in Lindleyana*, 12(3): 129 (1997): Satyrium viride.
× wisniewskii *E. Hemke in Mitt. Arbeitskr. Heim. Orch. Zentr. Fachausschusses Bot. Kulturb. DDR*, 8: 51 (1978), contrary to Art. 37.3 ICBN (1994)—Germany.(D. incarnata var. haematodes × majalis).

DAHLIA (Compos.).
parvibracteata *D.E. Saar & P.D. Sørensen in Novon*, 10(4): 407 (2000)—Mexico Southwest (Guerrero).
sorensenii *H.V. Hansen & J.P. Hjerting in Nordic J. Bot.*, 16(4): 454 (1996 publ. 1997)—Mexico Central (Mexico Distrito Federal).

DALBERGARIA (Gesneriac.).
albiflora *(L.P. Kvist & L.E. Skog) H. Wiehler in Gesneriana*, 1(1): 121 (1995): Columnea albiflora.
albovinosa *M. Freiberg in Phyton (Austria)*, 37(1): 134 (1997)—Ecuador.
cappilosa *(L.P. Kvist & L.E. Skog) H. Wiehler in Gesneriana*, 1(1): 121 (1995): Columnea cappilosa.
fuscihirta *(L.P. Kvist & L.E. Skog) H. Wiehler in Gesneriana*, 1(1): 121 (1995): Columnea fuscihirta.
gigantifolia *(L.P. Kvist & L.E. Skog) H. Wiehler in Gesneriana*, 1(1): 121 (1995): Columnea gigantifolia.

DALBERGARIA
incredibilis *(L.P. Kvist & L.E. Skog) H. Wiehler in Gesneriana,* 1(1): 43 (1995): Columnea incredibilis.
longinervosa *(L.P. Kvist & L.E. Skog) H. Wiehler in Gesneriana,* 1(1): 121 (1995): Columnea longinervosa.
nematoloba *(L.P. Kvist & L.E. Skog) H. Wiehler in Gesneriana,* 1(1): 121 (1995): Columnea nematoloba.
purpurimarginata *(L.P. Kvist & L.E. Skog) H. Wiehler in Gesneriana,* 1(1): 121 (1995): Columnea purpurimarginata.
robusta *H. Wiehler in Gesneriana,* 1(1): 48 (1995)—Colombia.
rubribracteata *(L.P. Kvist & L.E. Skog) H. Wiehler in Gesneriana,* 1(1): 122 (1995): Columnea rubribracteata.
variabilis *H. Wiehler in Gesneriana,* 1(1): 51 (1995)—Ecuador, Colombia.

DALBERGIA (Legumin.).
sect. Pseudecastaphyllum *A.M. de Carvalho in Brittonia,* 49(1): 88 (1997).
abrahamii *J. Bosser & R. Rabevohitra in Bull. Mus. Nation. Hist. Nat., B, Adansonia,* Ser. 4, 18(3–4): 190 (1996)—Madagascar.
andapensis *J. Bosser & R. Rabevohitra in Bull. Mus. Nation. Hist. Nat., B, Adansonia,* Ser. 4, 18(3–4): 201 (1996)—Madagascar.
assamica Benth.
var. laccifera *(Eberh. & Dubard) C. Niyomdham in Bull. Mus. Nation. Hist. Nat., B, Adansonia,* Ser. 4, 18(1–2): 138 (1996): D. hupeana var. laccifera.
aurea *J. Bosser & R. Rabevohitra in Bull. Mus. Nation. Hist. Nat., B, Adansonia,* Ser. 4, 18(3–4): 172 (1996)—Madagascar.
bintuluensis *B. Sunarno & Hiroyoshi Ohashi in J. Jap. Bot.,* 72(4): 202 (1997)—Borneo (Sarawak).
brachystachya *J. Bosser & R. Rabevohitra in Bull. Mus. Nation. Hist. Nat., B, Adansonia,* Ser. 4, 18(3–4): 182 (1996)—Madagascar.
calderonii Standl.
var. molinae *V.E. Rudd in Novon,* 5(4): 368 (1995)—Honduras, Nicaragua.
cana Graham ex Kurz
var. kurzii *(Prain) C. Niyomdham in Bull. Mus. Nation. Hist. Nat., B, Adansonia,* Ser. 4, 18(1–2): 139 (1996): D. kurzii.
capuronii *J. Bosser & R. Rabevohitra in Bull. Mus. Nation. Hist. Nat., B, Adansonia,* Ser. 4, 18(3–4): 176 (1996)—Madagascar.
comorensis *J. Bosser & R. Rabevohitra in Bull. Mus. Nation. Hist. Nat., B, Adansonia,* Ser. 4, 18(3–4): 204 (1996)—Comoros.
cultrata *T.S. Ralph, Icon. Carpolog.:* 22 (1849)—Locality not stated.
darlacensis *P.H. Hô & C. Niyomdham in Bull. Mus. Nation. Hist. Nat., B, Adansonia,* Ser. 4, 18(1–2): 139 (1996)—Vietnam, Cambodia, Thailand, China Southeast (Hong Kong), Hainan.
davidii *J. Bosser & R. Rabevohitra in Bull. Mus. Nation. Hist. Nat., B, Adansonia,* Ser. 4, 18(3–4): 202 (1996)—Madagascar.
delphinensis *J. Bosser & R. Rabevohitra in Bull. Mus. Nation. Hist. Nat., B, Adansonia,* Ser. 4, 18(3–4): 188 (1996)—Madagascar.
dialoides *(Pierre) C. Niyomdham in Bull. Mus. Nation. Hist. Nat., B, Adansonia,* Ser. 4, 18(1–2): 141 (1996): Coroya dialoides.
elegans *A.M. de Carvalho in Brittonia,* 49(1): 95 (1997)—Brazil Southeast (Espírito Santo).
emirnensis Benth.
var. decaryi *J. Bosser & R. Rabevohitra in Bull. Mus. Nation. Hist. Nat., B, Adansonia,* Ser. 4, 18(3–4): 209 (1996)—Madagascar.
erubescens *J. Bosser & R. Rabevohitra in Bull. Mus. Nation. Hist. Nat., B, Adansonia,* Ser. 4, 18(3–4): 180 (1996)—Madagascar.
foliosa *T.S. Ralph, Icon. Carpolog.:* 22 (1849)—Locality not stated.
foliosa *(Benth.) A.M. de Carvalho in Brittonia,* 49(1): 100 (1997), homonym; *see Ralph, Ic. Carpolog.:* 22 (1849): Ecastaphyllum foliosum.
glaberrima *J. Bosser & R. Rabevohitra in Bull. Mus. Nation. Hist. Nat., B, Adansonia,* Ser. 4, 18(3–4): 175 (1996)—Madagascar.

DALBERGIA:-
glaberrima *J. Bosser & R. Rabevohitra*
subsp. ankaranensis *J. Bosser & R. Rabevohitra in Bull. Mus. Nation. Hist. Nat., B, Adansonia,* Ser. 4, 18(3–4): 176 (1996)—Madagascar.
glabra (Mill.) Standl.
var. chontalensis *(Standl. & L.O. Williams) V.E. Rudd in Novon,* 5(4): 368 (1995): D. chontalensis.
var. paucifoliolata *(Lundell) V.E. Rudd in Novon,* 5(4): 368 (1995): D. paucifoliolata.
glaucocarpa *J. Bosser & R. Rabevohitra in Bull. Mus. Nation. Hist. Nat., B, Adansonia,* Ser. 4, 18(3–4): 182 (1996)—Madagascar.
grandistipula *A.M. de Carvalho in Brittonia,* 49(1): 102 (1997)—Brazil North (Rondônia).
guttembergii *A.M. de Carvalho in Brittonia,* 49(1): 104 (1997)—Brazil North (Roraima).
hildebrandtii Vatke
var. scorpioides *(Baker) J. Bosser & R. Rabevohitra in Bull. Mus. Nation. Hist. Nat., B, Adansonia,* Ser. 4, 18(3–4): 211 (1996): D. scorpioides.
hirticalyx *J. Bosser & R. Rabevohitra in Bull. Mus. Nation. Hist. Nat., B, Adansonia,* Ser. 4, 18(3–4): 204 (1996)—Madagascar.
intermedia *A.M. de Carvalho in Brittonia,* 49(1): 101 (1997), nom. nov.: Ecastaphyllum tomentosum *Spruce ex Benth.*
kostermansii *B. Sunarno & Hiroyoshi Ohashi in J. Jap. Bot.,* 72(4): 209 (1997)—Borneo (Kalimantan, Sarawak).
lanceolaria L.f.
var. errans *(Craib) C. Niyomdham in Bull. Mus. Nation. Hist. Nat., B, Adansonia,* Ser. 4, 18(1–2): 141 (1996): D. errans.
var. lakhonensis *(Gagnep.) C. Niyomdham & P.H. Hô in Bull. Mus. Nation. Hist. Nat., B, Adansonia,* Ser. 4, 18(1–2): 142 (1996): D. lakhonensis.
lemurica *J. Bosser & R. Rabevohitra in Bull. Mus. Nation. Hist. Nat., B, Adansonia,* Ser. 4, 18(3–4): 195 (1996)—Madagascar.
madagascariensis Vatke
subsp. antongilensis *J. Bosser & R. Rabevohitra in Bull. Mus. Nation. Hist. Nat., B, Adansonia,* Ser. 4, 18(3–4): 206 (1996)—Madagascar.
var. poolii *(Baker) J. Bosser & R. Rabevohitra in Bull. Mus. Nation. Hist. Nat., B, Adansonia,* Ser. 4, 18(3–4): 208 (1996): D. poolii.
maritima R. Vig.
var. pubescens *J. Bosser & R. Rabevohitra in Bull. Mus. Nation. Hist. Nat., B, Adansonia,* Ser. 4, 18(3–4): 208 (1996)—Madagascar.
mollis *J. Bosser & R. Rabevohitra in Bull. Mus. Nation. Hist. Nat., B, Adansonia,* Ser. 4, 18(3–4): 211 (1996), nom. nov.: D. stenocarpa *R. Vig.*
var. menabeensis *(R. Vig.) J. Bosser & R. Rabevohitra in Bull. Mus. Nation. Hist. Nat., B, Adansonia,* Ser. 4, 18(3–4): 211 (1996): D. stenocarpa var. menabeensis.
monticola *J. Bosser & R. Rabevohitra in Bull. Mus. Nation. Hist. Nat., B, Adansonia,* Ser. 4, 18(3–4): 198 (1996)—Madagascar.
neoperrieri *J. Bosser & R. Rabevohitra in Bull. Mus. Nation. Hist. Nat., B, Adansonia,* Ser. 4, 18(3–4): 194 (1996)—Madagascar.
nigrescens Kurz
var. anomala *(Pierre) C. Niyomdham in Bull. Mus. Nation. Hist. Nat., B, Adansonia,* Ser. 4, 18(1–2): 143 (1996): Dalbergia anomala.
normandii *J. Bosser & R. Rabevohitra in Bull. Mus. Nation. Hist. Nat., B, Adansonia,* Ser. 4, 18(3–4): 174 (1996)—Madagascar.
oojeinensis *W. Roxburgh, Fl. Indica,* 3: 220 (1832), in original I.K. as 'ougeinensis'—India.
orientalis *J. Bosser & R. Rabevohitra in Bull. Mus. Nation. Hist. Nat., B, Adansonia,* Ser. 4, 18(3–4): 184 (1996)—Madagascar.
peltieri *J. Bosser & R. Rabevohitra in Bull. Mus. Nation. Hist. Nat., B, Adansonia,* Ser. 4, 18(3–4): 212 (1996), nom. nov.: D. microcarpa *R. Vig.*
retusa Hemsl.
var. cuscatlanica *(Standl.) V.E. Rudd in Novon,* 5(4): 368 (1995): Amerimnon cuscatlanicum.
var. hypoleuca *(Pittier) V.E. Rudd in Novon,* 5(4): 368 (1995): D. hypoleuca.

DALBERGIA:-
retusa Hemsl.
var. lineata *(Pittier) V.E. Rudd in Novon,* 5(4): 368 (1995): D. lineata.
var. pacifica *(Standl. & Steyerm.) V.E. Rudd in Novon,* 5(4): 368 (1995): D. pacifica.
richardsii *B. Sunarno & Hiroyoshi Ohashi in J. Jap. Bot.,* 72(4): 212 (1997)—Borneo (Sarawak).
sandakanensis *B. Sunarno & Hiroyoshi Ohashi in J. Jap. Bot.,* 72(4): 217 (1997)—Borneo (Sabah).
suthepensis *C. Niyomdham in Bull. Mus. Nation. Hist. Nat., B, Adansonia,* Ser. 4, 18(1–2): 143 (1996)—Thailand.
teijsmannii *B. Sunarno & H. Ohashi in J. Jap. Bot.,* 71(5): 246 (1996)—Sulawesi.
tilarana *N. Zamora in Novon,* 10(2): 175 (2000)—Costa Rica, Nicaragua, Panama.
tricolor Drake
var. breviracemosa *J. Bosser & R. Rabevohitra in Bull. Mus. Nation. Hist. Nat., B, Adansonia,* Ser. 4, 18(3–4): 210 (1996)—Madagascar.
tsaratananensis *J. Bosser & R. Rabevohitra in Bull. Mus. Nation. Hist. Nat., B, Adansonia,* Ser. 4, 18(3–4): 186 (1996)—Madagascar.
urschii *J. Bosser & R. Rabevohitra in Bull. Mus. Nation. Hist. Nat., B, Adansonia,* Ser. 4, 18(3–4): 192 (1996)—Madagascar.
velutina Benth.
var. annamensis *C. Niyomdham in Bull. Mus. Nation. Hist. Nat., B, Adansonia,* Ser. 4, 18(1–2): 147 (1996)—Vietnam.
var. succirubra *(Gagnep. & Craib) C. Niyomdham in Bull. Mus. Nation. Hist. Nat., B, Adansonia,* Ser. 4, 18(1–2): 146 (1996): Dalbergia succirubra.
var. verrucosa *(Craib) C. Niyomdham in Bull. Mus. Nation. Hist. Nat., B, Adansonia,* Ser. 4, 18(1–2): 147 (1996): D. verrucosa.
vietnamensis *P.H. Hô & C. Niyomdham in Bull. Mus. Nation. Hist. Nat., B, Adansonia,* Ser. 4, 18(1–2): 147 (1996)—Vietnam, Cambodia.
viguieri *J. Bosser & R. Rabevohitra in Bull. Mus. Nation. Hist. Nat., B, Adansonia,* Ser. 4, 18(3–4): 189 (1996)—Madagascar.
villosa (Benth.) Benth.
var. barretoana *(Hoehne) A.M. de Carvalho in Brittonia,* 49(1): 94 (1997): D. barretoana.
xerophila *J. Bosser & R. Rabevohitra in Bull. Mus. Nation. Hist. Nat., B, Adansonia,* Ser. 4, 18(3–4): 183 (1996)—Madagascar.
ximengensis *Y.Y. Qian in J. Trop. Subtrop. Bot.,* 7(2): 151 (1999)—China South-central (Yunnan).

DALEA (Legumin.).
occidentalis *(A. Heller) D. Isely, Native & Natural. Legumin. (Fabac.) U.S. (excl. Alaska & Hawaii):* 520 (1998), homonym; *see L. Riley in Kew Bull.,* 1923: 337 (1923): Kuhnistera occidentalis.

DALECHAMPIA (Euphorbiac.).
sect. Brevicolumnae *W.S. Armbruster in Syst. Bot.,* 21(2): 232 (1996).
brevicolumna *W.S. Armbruster in Syst. Bot.,* 21(2): 232 (1996)—French Guiana, Surinam.
burgeriana *J. Gómez-Laurito in Novon,* 10(4): 368 (2000)—Costa Rica.
gentryi *W.S. Armbruster in Syst. Bot.,* 21(2): 223 (1996)—Peru, Brazil North (Amazonas).
heterobractea *W.S. Armbruster in Syst. Bot.,* 21(2): 226 (1996)—Guyana, Surinam, Venezuela, French Guiana, Brazil North (Amapá).
purpurata *I. Cordeiro in Kew Bull.,* 53(2): 467 (1998)—Brazil Northeast (Bahia).

DAMNACANTHUS (Rubiac.).
guangxiensis *Y.Z. Ruan in Fl. Reipubl. Popul. Sin.,* 71(2): 320 (1999)—China Southeast (Guangxi).
hainanensis *(H.S. Lo) Y.Z. Ruan in Fl. Reipubl. Popul. Sin.,* 71(2): 176 (1999): D. henryi subsp. hainanensis.

DANHATCHIA *L.A. Garay & E.A. Christenson in Orchadian,* 11(10): 469 (1995). *ORCHIDACEAE*
australis *(Hatch) L.A. Garay & E.A. Christenson in Orchadian,* 11(10): 470 (1995): Yoania australis.

DANTHONIA (Gramin.).
boliviensis S.A. Renvoize, Gramineas de Bolivia: 260 (1998)—Bolivia.
chilensis E. Desv.
var. aureofulva (E. Desv.) C.M. Baeza P. in Sendtnera, 3: 32 (1996): D. aureofulva.
holm-nielsenii S. Laegaard in Fl. Ecuador, 57(214:1): 17 (1997)—Ecuador.
rugoloana A.A. Sulekic in Darwiniana, 37(3–4): 341 (1999)—Argentina Northwest (Salta).
schneiderii Pilger
var. minor (Hooker) U. Shukla, Grasses N.-East. India: 246 (1996), without basionym ref.: Basionym not stated.
secundiflora J. Presl
subsp. mattheii C.M. Baeza P. in Sendtnera, 3: 55 (1996)—Brazil South (Paraná), Brazil Southeast (Rio de Janeiro, São Paulo), Bolivia.

DAPHNE (Thymelaeac.).
sect. Bella (Halda) J.J. Halda in Acta Mus. Richnov. Sect. Nat., 5(4): 141 (1998): D. subgen. Bella.
sect. Brachystachyae (S.C. Huang) J.J. Halda in Acta Mus. Richnov. Sect. Nat., 6(3): 198 (1999): Wikstroemia ser. Brachystachyae.
sect. Championi J.J. Halda in Acta Mus. Richnov. Sect. Nat., 7(2): 42 (2000).
subgen. Diplomorpha (Meisn.) J.J. Halda in Acta Mus. Richnov. Sect. Nat., 6(3): 198 (1999): gen. Diplomorpha.
subgen. Eriosolena (Blume) J.J. Halda in Acta Mus. Richnov. Sect. Nat., 5(4): 143 (1998): gen. Eriosolena.
subsect. Esquirollianae J.J. Halda in Acta Mus. Richnov. Sect. Nat., 7(1): 1 (2000).
sect. Graciles J.J. Halda in Acta Mus. Richnov. Sect. Nat., 7(1): 2 (2000).
subgen. Grueningianae J.J. Halda in Acta Mus. Richnov. Sect. Nat., 7(2): 41 (2000).
sect. Haoianae (S.C. Huang) J.J. Halda in Acta Mus. Richnov. Sect. Nat., 6(3): 198 (1999): Wikstroemia ser. Haoianae.
subgen. Herbaceae J.J. Halda in Acta Mus. Richnov. Sect. Nat., 7(2): 41 (2000).
subgen. Holosericeae J.J. Halda in Acta Mus. Richnov. Sect. Nat., 7(2): 41 (2000).
sect. Keisslera J.J. Halda in Acta Mus. Richnov. Sect. Nat., 5(4): 139 (1998).
sect. Lichiangense (S.C. Huang) J.J. Halda in Acta Mus. Richnov. Sect. Nat., 6(3): 198 (1999): Wikstroemia ser. Lichiangense.
sect. Micranthae (S.C. Huang) J.J. Halda in Acta Mus. Richnov. Sect. Nat., 6(3): 198 (1999): Wikstroemia ser. Micranthae.
sect. Parviflorae (S.C. Huang) J.J. Halda in Acta Mus. Richnov. Sect. Nat., 6(3): 198 (1999): Wikstroemia ser. Parviflorae.
subgen. Penicillatae J.J. Halda in Acta Mus. Richnov. Sect. Nat., 7(2): 41 (2000).
sect. Pseudolaurus J.J. Halda in Acta Mus. Richnov. Sect. Nat., 5(4): 140 (1998).
sect. Pseudomezereum J.J. Halda in Acta Mus. Richnov. Sect. Nat., 5(4): 137 (1998).
subgen. Pseudostellera J.J. Halda in Acta Mus. Richnov. Sect. Nat., 7(1): 1 (2000).
sect. Pseudowikstroemia J.J. Halda in Acta Mus. Richnov. Sect. Nat., 7(1): 1 (2000).
subsect. Pseudowikstroemia J.J. Halda in Acta Mus. Richnov. Sect. Nat., 7(1): 1 (2000).
sect. Rehdera J.J. Halda in Acta Mus. Richnov. Sect. Nat., 5(4): 142 (1998).
subgen. Rhamnoneuron (Gilg) J.J. Halda in Acta Mus. Richnov. Sect. Nat., 6(3): 199 (1999): gen. Rhamnoneuron.
sect. Scytophyllae (S.C. Huang) J.J. Halda in Acta Mus. Richnov. Sect. Nat., 6(3): 199 (1999): Wikstroemia ser. Scytophyllae.
subgen. Wangeanae J.J. Halda in Acta Mus. Richnov. Sect. Nat., 7(2): 42 (2000).
subgen. Wikstroemia (Endl.) J.J. Halda in Acta Mus. Richnov. Sect. Nat., 6(3): 196 (1999): gen. Wikstroemia.
sect. Wikstroemia (Endl.) J.J. Halda in Acta Mus. Richnov. Sect. Nat., 6(3): 198 (1999): gen. Wikstroemia.
abbreviata J.J. Halda in Acta Mus. Richnov. Sect. Nat., 6(3): 202 (1999), nom. nov.: Wikstroemia acuminata Merr.

DAPHNE:-
× adamantina J.J. Halda in Acta Mus. Richnov. Sect. Nat., 7(1): 2 (2000)—Cultivation. (D. arbuscula × petraea).
alba (Hand.-Mazz.) J.J. Halda in Acta Mus. Richnov. Sect. Nat., 6(3): 202 (1999): Wikstroemia alba.
subsp. anhuiensis (D.C. Zhang & X.P. Zhang) J.J. Halda in Acta Mus. Richnov. Sect. Nat., 6(3): 202 (1999): Wikstroemia anhuiensis.
alpestrivaga M. Gandoger, Dec. Pl. Nov., 1: 12 (1875)—France.
altaica
subsp. fasciculiflora (T.Z. Hsu) J.J. Halda in Acta Mus. Richnov. Sect. Nat., 4(2): 67 (1997): D. fasciculiflora.
alternifolia (Batalin) J.J. Halda in Acta Mus. Richnov. Sect. Nat., 6(3): 202 (1999): Wikstroemia alternifolia.
var. multiflora (Lecomte) J.J. Halda in Acta Mus. Richnov. Sect. Nat., 7(1): 2 (2000): Wikstroemia alternifolia var. multiflora.
× ambrosiaca J.J. Halda in Acta Mus. Richnov. Sect. Nat., 7(1): 2 (2000)—Cultivation. (D. malyana × petraea).
× ampla J.J. Halda in Acta Mus. Richnov. Sect. Nat., 7(1): 3 (2000)—Cultivation. (D. sericea × striata).
androsaemifolia (Decne.) J.J. Halda in Acta Mus. Richnov. Sect. Nat., 6(3): 202 (1999): Wikstroemia androsaemifolia.
angustissima (Merr.) J.J. Halda in Acta Mus. Richnov. Sect. Nat., 6(3): 202 (1999): Wikstroemia angustissima.
austrocochinchinensis J.J. Halda in Acta Mus. Richnov. Sect. Nat., 7(1): 3 (2000)—Vietnam.
autaretica M. Gandoger, Dec. Pl. Nov., 1: 20 (1875)—France.
axillaris (Merr. & Chun) Chun & C.F. Wei
subsp. rhynchocarpa (C.Yung Chang) J.J. Halda in Acta Mus. Richnov. Sect. Nat., 4(2): 67 (1997): D. rhynchocarpa.
baimashanensis (S.C. Huang) J.J. Halda in Acta Mus. Richnov. Sect. Nat., 6(3): 202 (1999): Wikstroemia baimashanensis.
balansae (Drake) J.J. Halda in Acta Mus. Richnov. Sect. Nat., 6(3): 202 (1999): Wikstroemia balansae.
× beata J.J. Halda in Acta Mus. Richnov. Sect. Nat., 7(1): 3 (2000)—Cultivation. (D. oleoides × petraea).
bholua Buch.-Ham. ex D. Don
subsp. emeiensis (C.Yung Chang) J.J. Halda in Acta Mus. Richnov. Sect. Nat., 4(2): 67 (1997): D. emeiensis.
bicornuta (Hillebr.) J.J. Halda in Acta Mus. Richnov. Sect. Nat., 6(3): 203 (1999): Wikstroemia bicornuta.
× blanda J.J. Halda in Acta Mus. Richnov. Sect. Nat., 7(1): 4 (2000)—Cultivation. (D. cneorum × rodriguezii).
brachyantha (Merr.) J.J. Halda in Acta Mus. Richnov. Sect. Nat., 6(3): 203 (1999): Wikstroemia brachyantha.
brevipaniculata (Rehder) J.J. Halda in Acta Mus. Richnov. Sect. Nat., 6(3): 203 (1999): Wikstroemia brevipaniculata.
brigantiaca M. Gandoger, Dec. Pl. Nov., 1: 11 (1875)—France.
canescens Wall.
subsp. capitatoracemosa (S.C. Huang) J.J. Halda in Acta Mus. Richnov. Sect. Nat., 6(3): 203 (1999): Wikstroemia capitatoracemosa.
capitata (Rehder) J.J. Halda in Acta Mus. Richnov. Sect. Nat., 6(3): 203 (1999): Wikstroemia capitata.
capitellata (H. Hara) J.J. Halda in Acta Mus. Richnov. Sect. Nat., 6(3): 203 (1999): Wikstroemia capitellata.
chamaebuxus M. Gandoger, Dec. Pl. Nov., 1: 11 (1875)—France.
chamaedaphne (Meisn.) J.J. Halda in Acta Mus. Richnov. Sect. Nat., 6(3): 203 (1999): Wikstroemia chamaedaphne.
chui (Merr.) J.J. Halda in Acta Mus. Richnov. Sect. Nat., 6(3): 203 (1999), as 'chuii': Wikstroemia chui.
cneorum L.
subsp. julia (Koso-Pol.) J.J. Halda in Acta Mus. Richnov. Sect. Nat., 5(5): 165 (1998): D. julia.

DAPHNE:-
× colorata J.J. Halda in Acta Mus. Richnov. Sect. Nat., 7(1): 4 (2000)—Cultivation. (D. glomerata × tangutica).
confusa J.J. Halda in Acta Mus. Richnov. Sect. Nat., 7(1): 4 (2000), nom. nov.: Wikstroemia coriacea Seem.
cunninghamii (Meisn.) J.J. Halda in Acta Mus. Richnov. Sect. Nat., 6(3): 203 (1999): Wikstroemia cunninghamii.
× cupida J.J. Halda in Acta Mus. Richnov. Sect. Nat., 7(1): 4 (2000)—Cultivation. (D. glomerata × sericea).
delavayi (Lecomte) J.J. Halda in Acta Mus. Richnov. Sect. Nat., 6(3): 203 (1999): Wikstroemia delavayi.
dolichantha (Diels) J.J. Halda in Acta Mus. Richnov. Sect. Nat., 6(3): 203 (1999): Wikstroemia dolichantha.
var. effusa (Rehder) J.J. Halda in Acta Mus. Richnov. Sect. Nat., 6(3): 204 (1999): Wikstroemia effusa.
endlicheriana J.J. Halda in Acta Mus. Richnov. Sect. Nat., 6(3): 204 (1999), nom. nov.: Wikstroemia australis Endl.
× eschmannii C.D. Brickell & A.R. White in New Plantsman, 7(4): 240 (2000)—Cultivation. (D. blagayana × cneorum).
esquirolii H. Lév.
subsp. pedunculata (H.F. Zhou ex C. Yung Chang) J.J. Halda in Acta Mus. Richnov. Sect. Nat., 4(2): 68 (1997): D. pedunculata.
× fabulosa J.J. Halda in Acta Mus. Richnov. Sect. Nat., 7(1): 5 (2000)—Cultivation. (D. cneorum × malyana).
fargesii (Lecomte) J.J. Halda in Acta Mus. Richnov. Sect. Nat., 7(1): 5 (2000): Stellera fargesii.
floribunda (Merr.) J.J. Halda in Acta Mus. Richnov. Sect. Nat., 7(1): 5 (2000): Wikstroemia floribunda.
forbesii (Skottsb.) J.J. Halda in Acta Mus. Richnov. Sect. Nat., 6(3): 204 (1999): Wikstroemia forbesii.
× fortunata J.J. Halda in Acta Mus. Richnov. Sect. Nat., 7(1): 5 (2000)—Cultivation. (D. petraea × stapfii).
furcata (Hillebr.) J.J. Halda in Acta Mus. Richnov. Sect. Nat., 6(3): 204 (1999): Wikstroemia sandwicensis var. furcata.
ganpi (Siebold & Zucc.) J.J. Halda in Acta Mus. Richnov. Sect. Nat., 6(3): 204 (1999): Passerina ganpi.
genkwa Siebold & Zucc.
subsp. jinzhaiensis (D.C. Zhang & J.Z. Shao) J.J. Halda in Acta Mus. Richnov. Sect. Nat., 4(2): 67 (1997): D. jinzhaiensis.
subsp. leishanensis (H.F. Zhou ex C. Yung Chang) J.J. Halda in Acta Mus. Richnov. Sect. Nat., 4(2): 67 (1997): D. leishanensis.
glabra (W.C. Cheng) J.J. Halda in Acta Mus. Richnov. Sect. Nat., 6(3): 204 (1999): Wikstroemia glabra.
f. purpurea (W.C. Cheng) J.J. Halda in Acta Mus. Richnov. Sect. Nat., 6(3): 204 (1999): Wikstroemia glabra f. purpurea.
gnidium L.
subsp. mauritanica (Nieto Fel.) J.J. Halda in Acta Mus. Richnov. Sect. Nat., 7(1): 6 (2000): D. mauritanica.
grayana J.J. Halda in Acta Mus. Richnov. Sect. Nat., 6(3): 204 (1999), nom. nov.: Wikstroemia retusa A. Gray.
gynopoda (Maxim. ex Makino) J.J. Halda in Acta Mus. Richnov. Sect. Nat., 7(1): 6 (2000): Wikstroemia gynopoda.
hainanensis (Merr.) J.J. Halda in Acta Mus. Richnov. Sect. Nat., 6(3): 204 (1999): Wikstroemia hainanensis.
hanalei (Wawra) J.J. Halda in Acta Mus. Richnov. Sect. Nat., 6(3): 204 (1999): Wikstroemia hanalei.
haoi (Domke) J.J. Halda in Acta Mus. Richnov. Sect. Nat., 6(3): 204 (1999), as 'haoii': Wikstroemia haoi.
helvetica M. Gandoger, Dec. Pl. Nov., 1: 20 (1875)—Switzerland.
hemsleyana (H. Lév.) J.J. Halda in Acta Mus. Richnov. Sect. Nat., 6(3): 205 (1999): Wikstroemia hemsleyana.

DAPHNE:-

hemsleyi *J.J. Halda in Acta Mus. Richnov. Sect. Nat.*, 6(3): 204 (1999), nom. nov.: Wikstroemia angustifolia *Hemsl.*

hillebrandii *J.J. Halda in Acta Mus. Richnov. Sect. Nat.*, 6(3): 205 (1999), nom. nov., as 'hillebranti': Wikstroemia villosa *Hillebr.*

huidongensis *(C.Yung Chang) J.J. Halda in Acta Mus. Richnov. Sect. Nat.*, 6(3): 205 (1999): Wikstroemia huidongensis.

indica L.

subsp. linearifolia *(Elmer) J.J. Halda in Acta Mus. Richnov. Sect. Nat.*, 6(3): 205 (1999): Wikstroemia linearifolia.

subsp. novae-caledoniae *(Gand.) J.J. Halda in Acta Mus. Richnov. Sect. Nat.*, 6(3): 205 (1999): Wikstroemia novae-caledoniae.

subsp. sericea *(Christoph.) J.J. Halda in Acta Mus. Richnov. Sect. Nat.*, 7(1): 6 (2000), basionym as 'Daphne sericea': Wikstroemia sericea.

jarmilae *J.J. Halda in Acta Mus. Richnov. Sect. Nat.*, 7(1): 6 (2000)—Burma.

jasminea Sibth. & Sm.

subsp. jarmilae *J.J. Halda in Acta Mus. Richnov. Sect. Nat.*, 5(4): 143 (1998)—Libya.

johnplewsii *(W.L. Wagner & Lorence) J.J. Halda in Acta Mus. Richnov. Sect. Nat.*, 7(1): 6 (2000): Wikstroemia johnplewsii.

× juraseki *J.J. Halda in Acta Mus. Richnov. Sect. Nat.*, 4(2): 69 (1997)—Caucasus. (D. circassica × glomerata).

kamtschatica Maxim.

subsp. jezoensis *(Maxim.) V.N. Voroschilov in A.K. Skvortsov (ed.), Florist. issl. v razn. raĭonakh SSSR:* 182 (1985): D. jezoensis.

× kazbali *J.J. Halda in Acta Mus. Richnov. Sect. Nat.*, 4(2): 69 (1997)—Cultivation. (D. petraea × sericea).

kingdon-wardii *J.J. Halda in Acta Mus. Richnov. Sect. Nat.*, 7(1): 7 (2000)—Assam.

kudoi *(Makino) J.J. Halda in Acta Mus. Richnov. Sect. Nat.*, 6(3): 205 (1999): Wikstroemia kudoi.

lamatsoensis *(Hamaya) J.J. Halda in Acta Mus. Richnov. Sect. Nat.*, 6(3): 205 (1999): Wikstroemia lamatsoensis.

lanceolata *(Merr.) J.J. Halda in Acta Mus. Richnov. Sect. Nat.*, 6(3): 205 (1999): Wikstroemia lanceolata.

× lepida *J.J. Halda in Acta Mus. Richnov. Sect. Nat.*, 7(1): 7 (2000)—Cultivation. (D. rodriguezii × sericea).

leptophylla *(W.W. Sm.) J.J. Halda in Acta Mus. Richnov. Sect. Nat.*, 6(3): 205 (1999): Wikstroemia leptophylla.

var. atroviolacea *(Hand.-Mazz.) J.J. Halda in Acta Mus. Richnov. Sect. Nat.*, 6(3): 205 (1999): Wikstroemia leptophylla var. atroviolacea.

liangii *(Merr. & Chun) J.J. Halda in Acta Mus. Richnov. Sect. Nat.*, 6(3): 205 (1999): Wikstroemia liangii.

lichiangensis *(W.W. Sm.) J.J. Halda in Acta Mus. Richnov. Sect. Nat.*, 6(3): 205 (1999): Wikstroemia lichiangensis.

ligustrina *(Rehder) J.J. Halda in Acta Mus. Richnov. Sect. Nat.*, 6(3): 206 (1999): Wikstroemia ligustrina.

linoides *(Hemsl.) J.J. Halda in Acta Mus. Richnov. Sect. Nat.*, 6(3): 206 (1999): Wikstroemia linoides.

longilobata Turrill

subsp. purpurascens *(S.C. Huang) J.J. Halda in Acta Mus. Richnov. Sect. Nat.*, 6(3): 206 (1999): D. purpurascens.

× lutetiana *Jacques in Ann. Fl. Pomone*, 1835–36: 253 (1836)—Cultivation. (D. altaica × collina).

× luxuriosa *J.J. Halda in Acta Mus. Richnov. Sect. Nat.*, 7(1): 8 (2000)—Cultivation. (D. blagayana × glomerata).

× mauerbachii *C.D. Brickell & A.R. White in New Plantsman*, 7(1): 9 (2000)—Cultivation. (D. caucasica × petraea).

mauritanica *G. Nieto Feliner in An. Jard. Bot. Madrid*, 53(2): 192 (1995 publ. 1996)—Algeria, Morocco, Tunisia.

× medfordensis *J.J. Halda in Acta Mus. Richnov. Sect. Nat.*, 4(2): 68 (1997)—Cultivation. (D. arbuscula × sericea).

DAPHNE:-

meisneriana *J.J. Halda in Acta Mus. Richnov. Sect. Nat.*, 6(3): 206 (1999), nom. nov.: Wikstroemia chinensis *Meisn.*

merrilliana *J.J. Halda in Acta Mus. Richnov. Sect. Nat.*, 6(3): 206 (1999), nom. nov.: Wikstroemia crassifolia *Merr. ex Domke.*

meyeniana *(Warb.) J.J. Halda in Acta Mus. Richnov. Sect. Nat.*, 6(3): 206 (1999): Wikstroemia meyeniana.

micrantha *(Hemsl.) J.J. Halda in Acta Mus. Richnov. Sect. Nat.*, 6(3): 206 (1999): Wikstroemia micrantha.

subsp. longipaniculata *(S.C. Huang) J.J. Halda in Acta Mus. Richnov. Sect. Nat.*, 6(3): 206 (1999): Wikstroemia longipaniculata.

subsp. lungtzeensis *(S.C. Huang) J.J. Halda in Acta Mus. Richnov. Sect. Nat.*, 6(3): 206 (1999): Wikstroemia lungtzeensis.

subsp. paniculata *(H.L. Li) J.J. Halda in Acta Mus. Richnov. Sect. Nat.*, 6(3): 207 (1999): Wikstroemia paniculata.

monnula *(Hance) J.J. Halda in Acta Mus. Richnov. Sect. Nat.*, 6(3): 207 (1999): Wikstroemia monnula.

var. vaccinium *(H. Lév.) J.J. Halda in Acta Mus. Richnov. Sect. Nat.*, 6(3): 207 (1999): Lonicera vaccinium.

var. xiuningensis *(D.C. Zhang & J.Z. Shao) J.J. Halda in Acta Mus. Richnov. Sect. Nat.*, 6(3): 207 (1999): Wikstroemia monnula var. xiuningensis.

mononectaria *(Hayata) J.J. Halda in Acta Mus. Richnov. Sect. Nat.*, 6(3): 207 (1999): Wikstroemia mononectaria.

mucronata Royle

subsp. linearifolia *(Hart) J.J. Halda in Acta Mus. Richnov. Sect. Nat.*, 5(5): 165 (1998): D. linearifolia.

subsp. turcica *J.J. Halda in Acta Mus. Richnov. Sect. Nat.*, 7(1): 8 (2000)—Turkey.

× nobilis *J.J. Halda in Acta Mus. Richnov. Sect. Nat.*, 7(1): 8 (2000)—Cultivation. (D. cneorum × glomerata).

nutans *(Champ. ex Benth.) J.J. Halda in Acta Mus. Richnov. Sect. Nat.*, 6(3): 207 (1999): Wikstroemia nutans.

var. brevior *(Hand.-Mazz.) J.J. Halda in Acta Mus. Richnov. Sect. Nat.*, 6(3): 207 (1999): Wikstroemia nutans var. brevior.

oahuensis *(A. Gray) J.J. Halda in Acta Mus. Richnov. Sect. Nat.*, 6(3): 207 (1999): Wikstroemia foetida var. oahuensis.

var. austrocochinchinensis *(Lecomte) J.J. Halda in Acta Mus. Richnov. Sect. Nat.*, 6(3): 207 (1999), basionym not validly publ.: Wikstroemia elongata var. austrocochinchinensis.

odora Thunb.

subsp. acutiloba *(Rehder) J.J. Halda in Acta Mus. Richnov. Sect. Nat.*, 5(5): 165 (1998): D. acutiloba.

oleoides Schreb.

subsp. baksanica *(Pobed.) J.J. Halda in Acta Mus. Richnov. Sect. Nat.*, 5(5): 165 (1998): D. baksanica.

subsp. transcaucasica *(Pobed.) J.J. Halda in Acta Mus. Richnov. Sect. Nat.*, 5(5): 165 (1998): D. transcaucasica.

orogenes *M. Gandoger, Dec. Pl. Nov.*, 1: 12 (1875)—France.

ovata *(C.A. Mey.) J.J. Halda in Acta Mus. Richnov. Sect. Nat.*, 6(3): 207 (1999): Wikstroemia ovata.

pampaninii *(Rehder) J.J. Halda in Acta Mus. Richnov. Sect. Nat.*, 6(3): 208 (1999): Wikstroemia pampaninii.

subsp. cochlearifolia *(S.C. Huang) J.J. Halda in Acta Mus. Richnov. Sect. Nat.*, 6(3): 208 (1999): Wikstroemia cochlearifolia.

subsp. subcyclolepidota *(L.P. Liu & Y.S. Lian) J.J. Halda in Acta Mus. Richnov. Sect. Nat.*, 6(3): 208 (1999): Wikstroemia subcyclolepidota.

papyracea Wall. ex Steud.

var. grandiflora *(Meisn. ex Diels) C.Yung Chang in Fl. Reipubl. Popul. Sin.*, 52(1): 376 (1999): D. papyracea f. grandiflora.

subsp. jinyunensis *(C.Yung Chang) J.J. Halda in Acta Mus. Richnov. Sect. Nat.*, 4(2): 68 (1997): D. jinyunensis.

DAPHNE:-

papyracea Wall. ex Steud.

var. longituba *(C.Yung Chang) J.J. Halda in Acta Mus. Richnov. Sect. Nat.*, 4(2): 68 (1997): D. longituba.

var. xichouensis *(H.F. Zhou ex C.Yung Chang) J.J. Halda in Acta Mus. Richnov. Sect. Nat.*, 7(2): 42 (2000): Daphne xichouensis.

subsp. yunnanensis *(H.F. Zhou ex C. Yung Chang) J.J. Halda in Acta Mus. Richnov. Sect. Nat.*, 4(2): 68 (1997): D. yunnanensis.

parviflora *(S.C. Huang) J.J. Halda in Acta Mus. Richnov. Sect. Nat.*, 6(3): 208 (1999): Wikstroemia parviflora.

paxiana *(H.J.P. Winkl.) J.J. Halda in Acta Mus. Richnov. Sect. Nat.*, 6(3): 208 (1999): Wikstroemia paxiana.

× perfecta *J.J. Halda in Acta Mus. Richnov. Sect. Nat.*, 7(1): 8 (2000)—Cultivation. (D. sericea × tangutica).

phillyreifolia *(A. Gray) J.J. Halda in Acta Mus. Richnov. Sect. Nat.*, 6(3): 208 (1999), as 'phillyreaefolia': Wikstroemia phillyreifolia.

× poetica *J.J. Halda in Acta Mus. Richnov. Sect. Nat.*, 7(1): 9 (2000)—Cultivation. (D. cneorum × oleoides).

poilanei *(Leandri) J.J. Halda in Acta Mus. Richnov. Sect. Nat.*, 6(3): 208 (1999): Wikstroemia poilanei.

polyantha *(Merr.) J.J. Halda in Acta Mus. Richnov. Sect. Nat.*, 6(3): 208 (1999): Wikstroemia polyantha.

pseudoretusa *(Koidz.) J.J. Halda in Acta Mus. Richnov. Sect. Nat.*, 6(3): 208 (1999): Wikstroemia pseudoretusa.

× pudica *J.J. Halda in Acta Mus. Richnov. Sect. Nat.*, 7(1): 9 (2000)—Cultivation. (D. petraea × rodriguezii).

pulcherrima *(Skottsb.) J.J. Halda in Acta Mus. Richnov. Sect. Nat.*, 6(3): 208 (1999): Wikstroemia pulcherrima.

raiateensis *(J.W. Moore) J.J. Halda in Acta Mus. Richnov. Sect. Nat.*, 7(1): 9 (2000): Wikstroemia raiateensis.

ridleyi *(Gamble) J.J. Halda in Acta Mus. Richnov. Sect. Nat.*, 6(3): 208 (1999): Wikstroemia ridleyi.

× rollsdorfii *C.D. Brickell & A.R. White in New Plantsman*, 7(1): 8 (2000)—Cultivation. (D. collina × petraea).

rubriflora *(C.Y. Wu ex S.C. Huang) J.J. Halda in Acta Mus. Richnov. Sect. Nat.*, 6(3): 208 (1999): Rhamnoneuron rubriflorum.

sandwicensis *(Meisn.) J.J. Halda in Acta Mus. Richnov. Sect. Nat.*, 6(3): 209 (1999): Wikstroemia sandwicensis.

× schlyteri *C.D. Brickell & A.R. White in New Plantsman*, 7(4): 237 (2000)—Cultivation. (D. arbuscula × cneorum).

scytophylla *(Diels) J.J. Halda in Acta Mus. Richnov. Sect. Nat.*, 6(3): 209 (1999): Wikstroemia scytophylla.

subsp. mekongensis *(W.W. Sm.) J.J. Halda in Acta Mus. Richnov. Sect. Nat.*, 6(3): 209 (1999): Wikstroemia mekongensis.

sericea Vahl

subsp. circassica *(Woronow ex Pobed.) J.J. Halda in Acta Mus. Richnov. Sect. Nat.*, 5(5): 165 (1998): D. circassica.

subsp. pseudosericea *(Pobed.) J.J. Halda in Acta Mus. Richnov. Sect. Nat.*, 5(5): 165 (1998): D. pseudosericea.

sikokiana *(Franch. & Sav.) J.J. Halda in Acta Mus. Richnov. Sect. Nat.*, 6(3): 209 (1999): Wikstroemia sikokiana.

skottsbergiana *(Sparre) J.J. Halda in Acta Mus. Richnov. Sect. Nat.*, 6(3): 209 (1999): Wikstroemia skottsbergiana.

sophia Kalen.

subsp. taurica *(Kotov) J.J. Halda in Acta Mus. Richnov. Sect. Nat.*, 5(5): 165 (1998): D. taurica.

× splendida *J.J. Halda in Acta Mus. Richnov. Sect. Nat.*, 7(1): 9 (2000)—Cultivation. (D. arbuscula × tangutica).

stenantha *(Hemsl.) J.J. Halda in Acta Mus. Richnov. Sect. Nat.*, 6(3): 209 (1999): Wikstroemia stenantha.

DAPHNE:-
stenophylla *(E. Pritz.) J.J. Halda in Acta Mus. Richnov. Sect. Nat.*, 6(3): 209 (1999): Wikstroemia stenophylla.
var. ziyangensis *(C.Y. Yu) J.J. Halda in Acta Mus. Richnov. Sect. Nat.*, 6(3): 209 (1999): Wikstroemia stenophylla var. ziyangensis.
× susannae *C.D. Brickell in New Plantsman*, 7(1): 17 (2000)—Cultivation. (D. arbuscula × collina).
taiwanensis *(Chang) J.J. Halda in Acta Mus. Richnov. Sect. Nat.*, 6(3): 209 (1999): Wikstroemia taiwanensis.
taylorii *J.J. Halda in Acta Mus. Richnov. Sect. Nat.*, 7(1): 10 (2000)—Xizang.
tenuiramis *(Miq.) J.J. Halda in Acta Mus. Richnov. Sect. Nat.*, 6(3): 209 (1999): Wikstroemia tenuiramis.
× transatlantica *C.D. Brickell & A.R. White in New Plantsman*, 7(4): 245 (2000)—Cultivation. (D. caucasica × collina).
trichotoma *(Thunb.) J.J. Halda in Acta Mus. Richnov. Sect. Nat.*, 6(3): 210 (1999): Queria trichotoma.
subsp. techinensis *(S.C. Huang) J.J. Halda in Acta Mus. Richnov. Sect. Nat.*, 6(3): 210 (1999): Wikstroemia techinensis.
uva-ursi *(A. Gray) J.J. Halda in Acta Mus. Richnov. Sect. Nat.*, 6(3): 210 (1999): Wikstroemia uva-ursi.
var. monticola *(Skottsb.) J.J. Halda in Acta Mus. Richnov. Sect. Nat.*, 6(3): 210 (1999): Wikstroemia monticola.
vaccinium *(H. Lév.) J.J. Halda in Acta Mus. Richnov. Sect. Nat.*, 7(1): 10 (2000): Lonicera vaccinium.
venosa *(Merr. & L.M. Perry) J.J. Halda in Acta Mus. Richnov. Sect. Nat.*, 6(3): 210 (1999): Wikstroemia venosa.
wangeana *(Hamaya) J.J. Halda in Acta Mus. Richnov. Sect. Nat.*, 7(1): 10 (2000): D. holosericea var. wangeana.
× whiteorum *C.D. Brickell in New Plantsman*, 7(1): 13 (2000)—Cultivation. (D. jasminea × petraea).
yatabeana *J.J. Halda in Acta Mus. Richnov. Sect. Nat.*, 6(3): 210 (1999), nom. nov.: Wikstroemia albiflora *Yatabe*.
zephyrina *J.J. Halda in Acta Mus. Richnov. Sect. Nat.*, 6(3): 210 (1999), nom. nov.: Wikstroemia pauciflora *Franch. & Sav.*
var. yakusimensis *(Makino) J.J. Halda in Acta Mus. Richnov. Sect. Nat.*, 6(3): 210 (1999): Wikstroemia pauciflora var. yakusimensis.
zhouana *J.J. Halda in Acta Mus. Richnov. Sect. Nat.*, 6(3): 210 (1999), nom. nov.: Wikstroemia linearifolia *H.F. Zhou ex C.Yung Chang.*

DAPHNIPHYLLUM (Daphniphyllac.).
ceramense *(T.C. Huang) T.C. Huang in Blumea*, 41(1): 237 (1996): D. glaucescens subsp. ceramense.
dichotomum *(T.C. Huang) T.C. Huang in Blumea*, 41(1): 238 (1996): D. glaucescens subsp. dichotomum.
gracile Gage
var. newirelandum *T.C. Huang in Blumea*, 41(1): 239 (1996)—Bismarck Archipelago.
papuanum Hallier f.
var. tuberculatum *(T.C. Huang) T.C. Huang in Blumea*, 41(1): 240 (1996): D. gracile var. tuberculatum.
sumatraense *(T.C. Huang) T.C. Huang in Blumea*, 41(1): 241 (1996): D. glaucescens subsp. sumatraense.
timorianum *(T.C. Huang) T.C. Huang in Blumea*, 41(1): 242 (1996): D. glaucescens subsp. timorianum.

DAPHNOPSIS (Thymelaeac.).
aemygdioi *P.P.H. Laclette in Bol. Mus Nac. Rio de Janeiro, Bot.*, 46: 3 (1977), with two types cited—Brazil Southeast (Rio de Janeiro).
bissei *A. Noa Monzón in Rev. Jard. Bot. Nacion. Univ. Habana*, 13: 35 (1992 publ. 1995)—Cuba.
cuneata Radlk.
subsp. maestrensis *A. Noa Monzón in Rev. Jard. Bot. Nacion. Univ. Habana*, 13: 36 (1992 publ. 1995)—Cuba.

DAPSILANTHUS *B.G. Briggs & L.A.S. Johnson in Telopea*, 7(4): 369 (1998). *RESTIONACEAE*
disjunctus *(Mast.) B.G. Briggs & L.A.S. Johnson in Telopea*, 8(1): 25 (1998): Leptocarpus disjunctus.
elatior *(R. Br.) B.G. Briggs & L.A.S. Johnson in Telopea*, 7(4): 371 (1998): Leptocarpus elatior.
ramosus *(R. Br.) B.G. Briggs & L.A.S. Johnson in Telopea*, 8(1): 25 (1998): Leptocarpus ramosus.
spathaceus *(R. Br.) B.G. Briggs & L.A.S. Johnson in Telopea*, 8(1): 26 (1998): Leptocarpus spathaceus.

DARCYA *A.T. Hunziker in Bol. Soc. Argent. Bot.*, 35(1–2): 167 (2000), homonym; *see B.L. Turner & C.C. Cowan in Phytologia*, 74(3): 268 (1993). *SOLANACEAE*
spruceana *(Hunz.) A.T. Hunziker in Bol. Soc. Argent. Bot.*, 35(1–2): 167 (2000): Physalis spruceana.

DARCYANTHUS *A.T. Hunziker in Bol. Soc. Argent. Bot.*, 35(3–4): 345 (2000), nom. nov., without exact replaced syn. page: Darcya *Hunz.* (Solanac.).
spruceanus *(Hunz.) A.T. Hunziker in Bol. Soc. Argent. Bot.*, 35(3–4): 345 (2000), without exact basionym page: Basionym not stated.

DARMERA *A. Voss in Gärt. Zentral-Bl.*, 1: 645 (1899); *see Madroño*, 24(2): 68-74 (1977). *SAXIFRAGACEAE*
peltata *(Torr. ex Benth.) A. Voss in Gärt. Zentral-Bl.*, 1: 646 (1899); *see Madroño*, 24(2): 68-74 (1977): Saxifraga peltata.

DASYGLOSSUM (Orchidac.).
edwardii *(Rchb. f.) W. Königer & H. Schildhauer in Arcula*, 5: 152 (1996): Odontoglossum edwardii.
graminifolium *(C. Schweinf.) W. Königer & H. Schildhauer in Arcula*, 5: 152 (1996): Odontoglossum graminifolium.

DASYLIRION (Dracaenac.).
gentryi *D.J. Bogler in Brittonia*, 50(1): 71 (1998)—Mexico Northwest (Sonora), Mexico Northeast (Chihuahua).
longissimum Lem.
var. treleasei *D.J. Bogler in Brittonia*, 50(1): 85 (1998)—Mexico Northeast (San Luis Potosí).
miquihuanense *D.J. Bogler in Brittonia*, 50(1): 76 (1998)—Mexico Northeast (Tamaulipas, Nuevo León).
sereke *D.J. Bogler in Brittonia*, 50(1): 74 (1998)—Mexico Northeast (Chihuahua).

DASYPHYLLUM (Compos.).
varians *(Gardner) A.L. Cabrera in Candollea*, 52(1): 216 (1997): Flotovia varians.

DASYSTEPHANA (Gentianac.).
vodopjanovae *V.V. Zuev in Fl. Sibir. (Pyrolac.-Lamiac.)*, 11: 66 (1997)—Chita.

DATURA (Solanac.).
stramonium L.
f. godronii *(Danert) D. Geerinck & É. Walravens in Nat. Belg.*, 79(4): 271 (1998); *et in Nat. Belg.*, 80(4): 434 (1999), basionym not validly publ.: D. stramonium var. godronii.
f. tatula *(L.) D. Geerinck & É. Walravens in Nat. Belg.*, 79(4): 270 (1998): D. tatula.

DAUBENYA (Hyacinthac.).
angustifolia *(L.f.) A.M. van der Merwe & J.C. Manning in Strelitzia*, 9: 713 (2000): Massonia angustifolia.
capensis *(Schltr.) A.M. van der Merwe & J.C. Manning in Strelitzia*, 9: 713 (2000): Androsiphon capensis.
namaquensis *(Schltr.) J.C. Manning & P. Goldblatt in Strelitzia*, 9: 713 (2000), as 'namaquana': Neobakeria namaquensis.
stylosa *(W.F. Barker) A.M. van der Merwe & J.C. Manning in Strelitzia*, 9: 713 (2000): Amphisiphon stylosus.

DAUCUS (Umbellif.).
arcanus *F. García Martín & S. Silvestre in Lagascalia*, 15(2): 263 (1990)—Spain.

DAVIDSONIA (Davidsoniac.).
jerseyana *(F. Muell. ex F.M. Bailey) G.J. Harden & J.B. Williams in Telopea*, 8(4): 420 (2000): D. pruriens var. jerseyana.
johnsonii *J.B. Williams & G.J. Harden in Telopea*, 8(4): 423 (2000)—South Wales, Queensland.

DAVIESIA (Legumin.).
apiculata *M.D. Crisp in Austral. Syst. Bot.*, 8(6): 1170 (1995)—Western Australia.
argillacea *M.D. Crisp in Austral. Syst. Bot.*, 8(6): 1171 (1995)—Western Australia.
articulata *M.D. Crisp in Austral. Syst. Bot.*, 8(6): 1173 (1995)—Western Australia.
audax *M.D. Crisp in Austral. Syst. Bot.*, 8(6): 1175 (1995)—Western Australia.
benthamii Meisn.
subsp. acanthoclona *(F. Muell.) M.D. Crisp in Austral. Syst. Bot.*, 8(6): 1175 (1995): D. acanthoclona.
bursarioides *M.D. Crisp in Austral. Syst. Bot.*, 8(6): 1178 (1995)—Western Australia.
campephylla *M.D. Crisp in Austral. Syst. Bot.*, 8(6): 1180 (1995)—Western Australia.
chapmanii *M.D. Crisp in Austral. Syst. Bot.*, 8(6): 1181 (1995)—Western Australia.
crassa *M.D. Crisp in Austral. Syst. Bot.*, 8(6): 1183 (1995)—Western Australia.
cunderdin *M.D. Crisp & G.T. Chandler in Austral. Syst. Bot.*, 10(3): 322 (1997)—Western Australia.
decipiens *(E. Pritz.) M.D. Crisp in Austral. Syst. Bot.*, 8(6): 1185 (1995): D. pectinata var. decipiens.
dilatata *M.D. Crisp in Austral. Syst. Bot.*, 8(6): 1187 (1995)—Western Australia.
elongata Benth.
subsp. implexa *M.D. Crisp in Austral. Syst. Bot.*, 8(6): 1191 (1995)—Western Australia.
emarginata *(Miq.) M.D. Crisp in Austral. Syst. Bot.*, 8(6): 1191 (1995): Fusanus emarginatus.
euryloba *M.D. Crisp & G.T. Chandler in Austral. Syst. Bot.*, 10(3): 327 (1997)—Western Australia.
glossosema *M.D. Crisp in Austral. Syst. Bot.*, 8(6): 1194 (1995)—Western Australia.
grossa *M.D. Crisp in Austral. Syst. Bot.*, 8(6): 1196 (1995)—Western Australia.
hamata *M.D. Crisp in Austral. Syst. Bot.*, 8(6): 1198 (1995)—Western Australia.
incrassata Sm.
subsp. reversifolia *(F. Muell.) M.D. Crisp in Austral. Syst. Bot.*, 8(6): 1201 (1995): D. reversifolia.
subsp. teres *M.D. Crisp in Austral. Syst. Bot.*, 8(6): 1203 (1995)—Western Australia.
intricata *M.D. Crisp in Austral. Syst. Bot.*, 8(6): 1204 (1995)—Western Australia.
subsp. xiphophylla *M.D. Crisp in Austral. Syst. Bot.*, 8(6): 1205 (1995)—Western Australia.
lineata *M.D. Crisp in Austral. Syst. Bot.*, 8(6): 1207 (1995)—Western Australia.
major *(Benth.) M.D. Crisp in Austral. Syst. Bot.*, 8(6): 1208 (1995): D. hakeoides var. major.
megacalyx *M.D. Crisp in Austral. Syst. Bot.*, 8(6): 1209 (1995)—Western Australia.
microcarpa *M.D. Crisp in Austral. Syst. Bot.*, 8(6): 1211 (1995)—Western Australia.
nudiflora Meisn.
subsp. amplectens *M.D. Crisp in Austral. Syst. Bot.*, 8(6): 1214 (1995)—Western Australia.
subsp. drummondii *(Meisn.) M.D. Crisp in Austral. Syst. Bot.*, 8(6): 1214 (1995): D. drummondii.
subsp. hirtella *M.D. Crisp in Austral. Syst. Bot.*, 8(6): 1216 (1995)—Western Australia.
oxyclada *M.D. Crisp in Austral. Syst. Bot.*, 8(6): 1217 (1995)—Western Australia.
oxylobium *M.D. Crisp in Austral. Syst. Bot.*, 8(6): 1218 (1995)—Western Australia.
pleurophylla *M.D. Crisp in Austral. Syst. Bot.*, 8(6): 1221 (1995)—Western Australia.
pseudaphylla *M.D. Crisp in Austral. Syst. Bot.*, 8(6): 1222 (1995)—Western Australia.
pteroclada *M.D. Crisp in Austral. Syst. Bot.*, 8(6): 1223 (1995)—Western Australia.
ramosissima *M.D. Crisp in Austral. Syst. Bot.*, 8(6): 1226 (1995)—Western Australia.
retrorsa *M.D. Crisp in Austral. Syst. Bot.*, 8(6): 1228 (1995)—Western Australia.

DAVIESIA:-
rhizomata *M.D. Crisp in Austral. Syst. Bot.*, 8(6): 1230 (1995)—Western Australia.
rubiginosa *M.D. Crisp in Austral. Syst. Bot.*, 8(6): 1230 (1995)—Western Australia.
sarissa *M.D. Crisp in Austral. Syst. Bot.*, 8(6): 1233 (1995)—Western Australia.
subsp. redacta *M.D. Crisp in Austral. Syst. Bot.*, 8(6): 1234 (1995)—Western Australia.
scoparia *M.D. Crisp in Austral. Syst. Bot.*, 8(6): 1235 (1995)—Western Australia.
sejugata *G.T. Chandler & M.D. Crisp in Austral. Syst. Bot.*, 10(1): 33 (1997)—Tasmania, South Australia.
smithiorum *M.D. Crisp in Austral. Syst. Bot.*, 8(6): 1235 (1995)—Western Australia.
speciosa *M.D. Crisp in Austral. Syst. Bot.*, 8(6): 1237 (1995)—Western Australia.
tortuosa *M.D. Crisp in Austral. Syst. Bot.*, 8(6): 1239 (1995)—Western Australia.
ulicifolia Andrews
subsp. aridicola *G.T. Chandler & M.D. Crisp in Austral. Syst. Bot.*, 10(1): 39 (1997)—South Australia, Western Australia, New South Wales.
subsp. incarnata *G.T. Chandler & M.D. Crisp in Austral. Syst. Bot.*, 10(1): 40 (1997)—South Australia.
subsp. pilligensis *G.T. Chandler & M.D. Crisp in Austral. Syst. Bot.*, 10(1): 45 (1997)—New South Wales.
subsp. ruscifolia *(A. Cunn. ex Benth.) G.T. Chandler & M.D. Crisp in Austral. Syst. Bot.*, 10(1): 44 (1997): D. ruscifolia.
subsp. stenophylla *G.T. Chandler & M.D. Crisp in Austral. Syst. Bot.*, 10(1): 42 (1997)—New South Wales, Queensland.
umbonata *M.D. Crisp & G.T. Chandler in Austral. Syst. Bot.*, 10(3): 324 (1997)—Western Australia.
uncinata *M.D. Crisp in Austral. Syst. Bot.*, 8(6): 1243 (1995)—Western Australia.

DAVILLA (Dilleniac.).
kubitzkii *G.A. Aymard C. in Brittonia*, 50(1): 51 (1998)—Brazil North (Pará).
morii *G.A. Aymard C. in Brittonia*, 50(1): 53 (1998)—Brazil Northeast (Bahia).

DAYIA *J.M. Porter in Aliso*, 19(1): 71 (2000). *POLEMONIACEAE*
grantii *J.M. Porter in Aliso*, 19(1): 72 (2000)—Mexico Northwest (Baja California Sur).
scabra *(Brandegee) J.M. Porter in Aliso*, 19(1): 71 (2000): Gilia scabra.

DECALEPIS (Asclepiadac.).
arayalpathra *(J. Joseph & V. Chandras.) H.J.T. Venter in Taxon*, 46(4): 712 (1997): Janakia arayalpathra.
nervosa *(Wight & Arn.) H.J.T. Venter in Taxon*, 46(4): 712 (1997): Brachylepis nervosa.

DECASPERMUM (Myrtac.).
belense *E.D. Merrill & L.M. Perry in J. Arnold Arbor.*, 23: 234 (1942)—New Guinea (Irian Jaya).†

DECLIEUXIA (Rubiac.).
decumbens *J.H. Kirkbride in Brittonia*, 49(3): 368 (1997)—Brazil Southeast (Minas Gerais).

DEGUELIA (Legumin.).
utilis *(A.C. Sm.) A.M.G. de Azevedo Tozzi in Rev. Brasil. Biol.*, 58(3): 512 (1998): Lonchocarpus utilis.

DEHERAINIA (Theophrastac.).
lageniformis *J. Gómez-Laurito & N. Zamora in Novon*, 8(2): 141 (1998)—Costa Rica.

DEINANDRA (Compos.).
arida *(D.D. Keck) B.G. Baldwin in Novon*, 9(4): 467 (1999): Hemizonia arida.
bacigalupii *B.G. Baldwin in Madroño*, 46(1): 55 (1999)—California.
clementina *(Brandegee) B.G. Baldwin in Novon*, 9(4): 468 (1999): Hemizonia clementina.
conjugens *(D.D. Keck) B.G. Baldwin in Novon*, 9(4): 468 (1999): Hemizonia conjugens.

DEINANDRA:-
corymbosa *(DC.) B.G. Baldwin in Novon*, 9(4): 468 (1999): Hartmannia corymbosa.
subsp. macrocephala *(Nutt.) B.G. Baldwin in Novon*, 9(4): 468 (1999): Hemizonia macrocephala.
frutescens *(A. Gray) B.G. Baldwin in Novon*, 9(4): 468 (1999): Hemizonia frutescens.
greeneana *(Rose) B.G. Baldwin in Novon*, 9(4): 468 (1999): Hemizonia greeneana.
subsp. peninsularis *(Moran) B.G. Baldwin in Novon*, 9(4): 468 (1999): Hemizonia greeneana subsp. peninsularis.
halliana *(D.D. Keck) B.G. Baldwin in Novon*, 9(4): 468 (1999): Hemizonia halliana.
increscens *(H.M. Hall ex D.D. Keck) B.G. Baldwin in Novon*, 9(4): 468 (1999): Hemizonia paniculata subsp. increscens.
subsp. foliosa *(Hoover) B.G. Baldwin in Novon*, 9(4): 468 (1999): Hemizonia paniculata subsp. foliosa.
subsp. villosa *(Tanowitz) B.G. Baldwin in Novon*, 9(4): 469 (1999): Hemizonia increscens subsp. villosa.
martirensis *(D.D. Keck) B.G. Baldwin in Novon*, 9(4): 469 (1999): Hemizonia martirensis.
minthornii *(Jeps.) B.G. Baldwin in Novon*, 9(4): 469 (1999): Hemizonia minthornii.
mohavensis *(D.D. Keck) B.G. Baldwin in Novon*, 9(4): 469 (1999): Hemizonia mohavensis.
pallida *(D.D. Keck) B.G. Baldwin in Novon*, 9(4): 469 (1999): Hemizonia pallida.
palmeri *(Rose) B.G. Baldwin in Novon*, 9(4): 469 (1999): Hemizonia palmeri.
pentactis *(D.D. Keck) B.G. Baldwin in Novon*, 9(4): 469 (1999): Hemizonia lobbii subsp. pentactis.
streetsii *(A. Gray) B.G. Baldwin in Novon*, 9(4): 469 (1999): Hemizonia streetsii.

DEINBOLLIA (Sapindac.).
borbonica Scheff.
f. subcordata *B. Verdcourt in Fl. Trop. E. Afr., Sapindac.*: 72 (1998)—Tanzania.
kilimandscharica Taub.
var. adusta *(Radlk.) B. Verdcourt in Fl. Trop. E. Afr., Sapindac.*: 71 (1998): D. adusta.
mezilii *D.W. Thomas & D.J. Harris in Kew Bull.*, 54(4): 954 (1999 publ. 2000)—Cameroon.

DEIREGYNE (Orchidac.).
pallens *(Szlach.) A. Espejo Serna & A.R. López-Ferrari in Phytologia*, 82(2): 80 (1997): Oestlundorchis pallens.
pterygodium *D.L. Szlachetko in Orchidee*, Suppl. 3: 18 (1996)—Mexico Central (México State).
sheviakiana *(Szlach.) A. Espejo Serna & A.R. López-Ferrari in Phytologia*, 82(2): 80 (1997): Oestlundorchis sheviakiana.
tamayoi *D.L. Szlachetko in Fragm. Flor. Geobot.*, 41(2): 851 (1996)—Mexico Southwest (Jalisco).

DELISSEA (Campanulac.).
kuhihewa *(Lammers) R. Govaerts, World Checklist Seed Plants*, 3(1): 21 (1999): Cyanea kuhihewa.
lanaiensis *(Rock) T.G. Lammers in Novon*, 8(1): 34 (1998): D. sinuata var. lanaiensis.
niihauensis H. St. John
subsp. kauaiensis *(Lammers) T.G. Lammers in Novon*, 9(3): 388 (1999): D. undulata subsp. kauaiensis.

DELOSPERMA (Aizoac.).
deilanthoides *S.A. Hammer in Bradleya*, 15: 46 (1997)—Transvaal.
dolomitica *E. van Jaarsveld & A.E. van Wyk in Aloe*, 36(4): 73 (1999)—Transvaal.

DELPHINIUM (Ranunculac.).
ser. Cossoniana *C. Blanché, J. Molero & J. Simon P.in Lagascalia*, 19(1–2): 76 (1997).
aktoense *W.T. Wang in Acta Phytotax. Sin.*, 37(3): 216 (1999)—Xinjiang.
alpinum Waldst. & Kit.
var. hebecarpum *P.V. Kulikov in Bot. Zhurn.*, 85(4): 133 (2000)—Russia East.
austriacum *(Pawł.) W. Starmühler in W. Maurer, Fl. Steiermark*, 1: 84 (1996): D. elatum subsp. austriacum.

DELPHINIUM:-
austriacum
nothosubsp. drescheri *W. Starmühler in Mitt. Naturwiss. Ver. Steiermark*, 126: 102 (1996)—Austria. (D. austriacum subsp. austriacum × subsp. stiriacum).
subsp. stiriacum *(Pawł.) W. Starmühler in W. Maurer, Fl. Steiermark*, 1: 85 (1996): D. elatum var. stiriacum.
caeruleum Jacq. ex Cambess.
var. crassicalcaratum *W.T. Wang & M.J. Warnock in Guihaia*, 17(1): 9 (1997)—China South-central (Yunnan).
delavayi Franch.
var. baoshanense *(W.T. Wang) W.T. Wang in Phytologia*, 79(5): 384 (1995 publ. 1996): D. baoshanense.
elatum L.
subsp. alpinum *(Waldst. & Kit.) N.N. Tzvelev in Novosti Sist. Vyssh. Rast.*, 32: 183 (2000): D. alpinum.
subsp. polatschekii *W. Starmühler in A. Polatschek, Fl. Nordtirol, Osttirol & Vorarlberg*, 3: 701 (2000)—Austria.
subsp. villosum *(Steven ex DC.) N.N. Tzvelev in Novosti Sist. Vyssh. Rast.*, 32: 183 (2000): D. villosum.
favargeri *C. Blanché, J. Molero & J. Simon P.in Lagascalia*, 19(1–2): 75 (1997)—Morocco.
grandiflorum L.
f. roseum *G.A. Peschkova, Stepnaya Fl. Baĭkal'skoĭ Sibiri*: 57 (1972), without type—Buryatiya.
grandilimbum *W.T. Wang & M.J. Warnock in Guihaia*, 17(1): 10 (1997)—China South-central (Yunnan).
hispanicum *Willk. ex Costa in Anales Soc. Esp. Hist. Nat.*, 2: 27 (1873)—Spain.
huangzhongense *W.T. Wang in Acta Phytotax. Sin.*, 37(3): 216 (1999)—Qinghai.
kansuense W.T. Wang
var. villosiusculum *W.T. Wang & M.J. Warnock in Phytologia*, 79(5): 383 (1995 publ. 1996)—Qinghai.
motingshanicum *W.T. Wang & M.J. Warnock in Guihaia*, 17(1): 3 (1997)—China South-central (Yunnan).
nurguschense *P.V. Kulikov in Bot. Zhurn.*, 85(4): 134 (2000)—West Siberia.
palasianum *R.A. Rafiq in Novon*, 6(3): 295 (1996)—Pakistan.
pentagynum
subsp. formenteranum *N. Torres, L. Sáez, J.A. Rosselló & C. Blanché in Bot. J. Linn. Soc.*, 133(3): 372 (2000)—Baleares.
pseudothibeticum *W.T. Wang & M.J. Warnock in Guihaia*, 17(1): 8 (1997)—China South-central (Yunnan).
pseudoyunnanense *W.T. Wang & M.J. Warnock in Phytologia*, 79(5): 382 (1995 publ. 1996)—China South-central (Yunnan).
shawurense W.T. Wang
var. pseudoaemulans *(C.Y. Yang & B. Wang) W.T. Wang in Phytologia*, 79(5): 384 (1995 publ. 1996): D. pseudoaemulans.
staphisagria L.
subsp. pictum *(Willd.) A.M. Romo, Fl. Silvestres Baleares*: 104 (1994): D. pictum.
subcuneatum *N.N. Tzvelev in Bot. Zhurn.*, 81(12): 118 (1996 publ. 1997)—Russia East.
umbrosum Hand.-Mazz.
var. drepanocentrum *(Bruehl ex Huth) W.T. Wang & M.J. Warnock in Phytologia*, 79(5): 384 (1995 publ. 1996): D. altissimum var. drepanocentrum.
yongningense *W.T. Wang & M.J. Warnock in Guihaia*, 17(1): 7 (1997)—China South-central (Yunnan).
zhangii *W.T. Wang in Acta Phytotax. Sin.*, 37(3): 214 (1999)—Xinjiang.

DENDRANTHEMA (Compos.).
sect. Sinuata *N.M. Boldyreva in Fl. Sibir. (Asterac.)*, 13: 86 (1997 publ. 1998).
foliaceum *G.F. Peng, C. Shih & S.Q. Zhang in Acta Phytotax. Sin.*, 37(6): 600 (1999)—China North-central (Shandong).
longibracteatum *C. Shih, G.F. Peng & S.Y. Jin in Acta Phytotax. Sin.*, 37(6): 598 (1999)—China North-central (Shandong).

DENDRANTHEMA:-
nankingense *(Hand.-Mazz.) P.S. Hsu in Pl. Shanghai,* 1: 625 (1999), isonym; *see X.D. Cui in Fl. Tsinlingensis,* 1(5): 252 (1985): Chrysanthemum nankingense.
ornatum (Hemsl.) Kitam.
var. tokarense *M. Hotta & Y. Hirai in Acta Phytotax. Geobot.,* 47(1): 103 (1996), as 'tokarensis'—Nansei-Shoto.
zawadskii (Herbich) Tzvelev
subsp. peleiolepis *(Trautv.) N.M. Boldyreva in Fl. Sibir. (Asterac.),* 13: 88 (1997 publ. 1998): Leucanthemum sibiricum var. peleiolepis.

DENDROBIUM (Orchidac.).
sect. Tetrodon *(Kraenzl.) P. Ormerod in Austral. Orchid Rev.,* 61(4): 39 (1996): Eria sect. Tetrodon.
aurantiacum Rchb. f.
var. denneanum *(Kerr) Z.H. Tsi in Fl. Reipubl. Popul. Sin.,* 19: 89 (1999): Dendrobium denneanum.
var. zhaojuense *(S.C. Sun & L.G. Xu) Z.H. Tsi in Fl. Reipubl. Popul. Sin.,* 19: 89 (1999): Dendrobium zhaojuense.
aurantiflammeum *J.J. Wood in Orchid Rev.,* 106(1224): 340 (1998)—Borneo (Sabah, Brunei, Sarawak).
auyongii *T. Yukawa in Lindleyana,* 13(1): 28 (1998)—Borneo (Sarawak).
ayubii *J.B. Comber & J.J. Wood in Orchid Rev.,* 107(1226): 90 (1999)—Sumatera.
balzerianum *H. Fessel & E. Lückel in Orchidee,* 48(4): 173 (1997)—Philippines.
chittimae *G. Seidenfaden, Contrib Orchid Fl. Thailand XIII:* 29 (1997)—Locality not stated.
ejirii *T. Yukawa in Ann. Tsukuba Bot. Gard.,* 16: 25 (1997)—Vietnam.
indivisum (Blume) Miq.
var. fuscum *P. O'Byrne in Malayan Orchid Rev.,* 31: 21 (1997)—Malaya (Peninsular Malaysia).
jennae *P. O'Byrne in Malayan Orchid Rev.,* 30: 24 (1996)—Sulawesi.
judithiae *P. O'Byrne in Malayan Orchid Rev.,* 33: 43 (1999)—Sulawesi.
khanhoaense *L.V. Averyanov in Lindleyana,* 14(4): 216 (1999)—Vietnam.
kurashigei *T. Yukawa in Lindleyana,* 13(1): 28 (1998)—Borneo (Sabah, Kalimantan).
lanyaiae *G. Seidenfaden in Nordic J. Bot.,* 16(2): 139 (1996)—Laos.
milaniae *H. Fessel & E. Lückel in Orchidee,* 47(3): A131 (1996)—Philippines.
militare *P.J. Cribb in Kew Bull.,* 51(2): 380 (1996), nom. nov.: D. brevimentum *P.J. Cribb.*
mitriferum *L.V. Averyanov in Lindleyana,* 15(2): 76 (2000), homonym; *see J.J. Sm. in Bull. Dép. Agric. Indes Néerl.,* 39: 10 (1910)—Vietnam.
mussauense *P. Ormerod in Austral. Orchid Rev.,* 62(3): 13 (1997)—Bismarck Archipelago.
orbilobulatum *H. Fessel & E. Lückel in Orchidee,* 47(5): A232 (1996)—Philippines.
praetermissum *G. Seidenfaden, Contrib. Orchid Fl. Thailand XIII:* 34 (1997)—Burma.
pseudoequitans *H. Fessel & E. Lückel in Orchidee,* 51(1): 83 (2000)—Philippines.
setigerum *Ames ex M.A. Clements in Austral. Orchid Rev.,* 64(6): 7 (1999–2000 publ. 1999)—Philippines.
shompenii *B.K. Sinha & P.S.N. Rao in Nordic J. Bot.,* 18(1): 27 (1998)—Nicobar Is.
soriense *N.H.S. Howcroft in Orchadian,* 11(12): 580 (1996)—New Guinea (Papua New Guinea).
stockeri *T. Yukawa in Lindleyana,* 15(3): 189 (2000)—New Guinea (Papua New Guinea).
sulawesiense *J. Erfkamp & O. Gruss in Orchidee,* 51(1): 57 (2000)—Sulawesi.
torajaense *P. O'Byrne in Malayan Orchid Rev.,* 33: 45 (1999)—Sulawesi.
× usitae *T. Yukawa in Ann. Tsukuba Bot. Gard.,* 14: 43 (1995)—Philippines. (D. bullenianum × goldschmidtianum).
vesiculosum *M.A. Clements & D.L. Jones in Lasianthera,* 1(1): 9 (1996)—New Guinea (Papua New Guinea).

DENDROCALAMOPSIS (Gramin.).
basihirsuta *(McClure) P.C. Keng & W.T. Lin in Fl. Reipubl. Popul. Sin.,* 9(1): 143 (1996): Bambusa basihirsuta.

DENDROCALAMOPSIS:-
daii *P.C. Keng in Fl. Reipubl. Popul. Sin.,* 9(1): 149 (1996), nom. nov.: D. grandis *Q.H. Dai & X.L. Tao.*

DENDROCALAMUS (Gramin.).
bengkalisensis *E.A. Widjaja in Reinwardtia,* 11(2): 69 (1997)—Sumatera.
buar *E.A. Widjaja in Reinwardtia,* 11(2): 70 (1997)—Sumatera.
copelandii *(Gamble ex Brandis) N.H. Xia & C.M.A. Stapleton in Kew Bull.,* 52(2): 484 (1997): Bambusa copelandii.
exilis *N.H. Xia & L.C. Chia in J. Trop. Subtrop. Bot.,* 4(1): 23 (1996)—China South-central (Guizhou).
guiyangensis *N.H. Xia & L.C. Chia in J. Trop. Subtrop. Bot.,* 4(1): 25 (1996)—China South-central (Guizhou).
hait *E.A. Widjaja in Reinwardtia,* 11(2): 72 (1997)—Sumatera.
minor
f. amoenus *(Q.H. Dai & C.F. Huang) D. Ohrnberger, Bamboos of the World Intro. 3:* 14 (1996): Sinocalamus minor var. amoenus.

DENDROCHILUM (Orchidac.).
sect. Convoluta *H.A. Pedersen in Op. Bot.,* 130: 22 (1997).
sect. Cruciformia *J.J. Wood & H.A. Pedersen in Op. Bot.,* 130: 30 (1997).
sect. Eurybrachium *Carr ex J.J. Wood, H.A. Pedersen & J.B. Comber in Op. Bot.,* 130: 31 (1997).
*sect.*Falsiloba *J.J. Wood & H.A. Pedersen in Op. Bot.,* 130: 25 (1997).
sect. Heterantha *H.A. Pedersen in Op. Bot.,* 130: 25 (1997).
sect. Longicolumna *J.J. Wood in Op. Bot.,* 130: 40 (1997).
sect. Luzonorchis *H.A. Pedersen in Op. Bot.,* 130: 59 (1997).
sect. Mammosa *J.J. Wood & H.A. Pedersen in Op. Bot.,* 130: 41 (1997).
sect. Mindanaorchis *H.A. Pedersen in Op. Bot.,* 130: 41 (1997).
subgen. Pseudacoridium *(Ames) Butzin ex H.A. Pedersen in Op. Bot.,* 130: 59 (1997): D. sect. Pseudacoridium.
sect. Replicata *H.A. Pedersen in Op. Bot.,* 130: 59 (1997).
ambangense *H.A. Pedersen in Nordic J. Bot.,* 15(4): 391 (1995 publ. 1996)—Sulawesi.
amesianum *H.A. Pedersen in Nordic J. Bot.,* 15(4): 381 (1995 publ. 1996)—Philippines.
cinnabarinum Pfitzer
var. sanguineum *(Ames) H.A. Pedersen in Op. Bot.,* 130: 22 (1997): Acoridium sanguineum.
citrinum *H.A. Pedersen in Nordic J. Bot.,* 15(4): 392 (1995 publ. 1996)—Sulawesi.
convallariaeforme Schauer
var. minor *(Ames) H.A. Pedersen in Op. Bot.,* 130: 34 (1997): D. bicallosum var. minor.
cootesii *H.A. Pedersen in Lindleyana,* 12(4): 205 (1997)—Philippines.
erectilabium *H.A. Pedersen in Nordic J. Bot.,* 15(4): 394 (1995 publ. 1996)—Sulawesi.
exiguum *(Ames) H.A. Pedersen in Op. Bot.,* 130: 60 (1997): Pseudacoridium exiguum.
eymae *H.A. Pedersen in Nordic J. Bot.,* 15(4): 395 (1995 publ. 1996)—Sulawesi.
flexuosum *H.A. Pedersen in Nordic J. Bot.,* 15(4): 383 (1995 publ. 1996)—Philippines.
flos-susannae *J.J. Wood in Orchid Rev.,* 105(1215): 142 (1997)—Borneo (Sarawak).
galeatum *H.A. Pedersen in Nordic J. Bot.,* 15(4): 396 (1995 publ. 1996)—Lesser Sunda Is.
latifolium Lindl.
var. macranthum *(Schltr.) H.A. Pedersen in Op. Bot.,* 130: 51 (1997): D. macranthum.
longipedicellatum *H.A. Pedersen in Nordic J. Bot.,* 15(4): 399 (1995 publ. 1996)—Sulawesi.
louisianum *H.A. Pedersen in Nordic J. Bot.,* 15(4): 384 (1995 publ. 1996)—Philippines.
parvipapillatum *H.A. Pedersen in Nordic J. Bot.,* 15(4): 386 (1995 publ. 1996)—Philippines.
parvulum (Ames) Pfitzer
var. strictiforme *(Ames) H.A. Pedersen in Op. Bot.,* 130: 20 (1997): Acoridium strictiforme.

DENDROCHILUM:-
perplexum (Ames) L.O. Williams
var. montanum *(Ames) H.A. Pedersen in Op. Bot.,* 130: 20 (1997): Acoridium montanum.
philippinense (Ames) Pfitzer
var. purpureum *(Ames) H.A. Pedersen in Op. Bot.,* 130: 20 (1997): D. purpureum.
plocoglottoides *H.A. Pedersen in Nordic J. Bot.,* 15(4): 387 (1995 publ. 1996)—Philippines.
pseudoscriptum *T.J. Barkman & J.J. Wood in Orchid Rev.,* 104(1209): 179 (1996)—Borneo (Sabah).
pumilum Rchb. f.
var. recurvum *(Ames) H.A. Pedersen in Op. Bot.,* 130: 24 (1997): Acoridium recurvum.
quisumbingianum *H.A. Pedersen in Nordic J. Bot.,* 15(4): 389 (1995 publ. 1996)—Philippines.
saccatum *J.J. Wood in Op. Bot.,* 130: 41 (1997), nom. nov.: Pholidota gracilis *L.O. Williams.*
tetradactyliferum *H.A. Pedersen in Nordic J. Bot.,* 15(4): 390 (1995 publ. 1996)—Philippines.
uncatum Rchb. f.
var. longispicatum *(Ames) H.A. Pedersen in Op. Bot.,* 130: 58 (1997): D. longispicatum.

DENDROCNIDE (Urticac.).
cordifolia *(L.S. Sm.) B.R. Jackes & M. Hurley in Austrobaileya,* 5(1): 122 (1997): Laportea cordifolia.

DENDROLOBIUM (Legumin.).
arbuscula *(Domin) Hiroyoshi Ohashi in Taiwania,* 42(2): 137 (1997): D. arbuscula.
cheelii *(C.A. Gardner) L. Pedley in Austrobaileya,* 5(2): 221 (1999): Alysicarpus cheelii.
geesinkii *Hiroyoshi Ohashi in J. Jap. Bot.,* 73(5): 251 (1998)—Philippines.
multiflorum *L. Pedley in Austrobaileya,* 5(2): 220 (1999)—Northern Territory.
papuacola *Hiroyoshi Ohashi & Tomoyuki Nemoto in J. Jap. Bot.,* 74(1): 16 (1999)—New Guinea (Papua New Guinea).
polyneurum *(S.T. Blake) L. Pedley in Austrobaileya,* 5(2): 221 (1999): Desmodium polyneurum.
umbellatum (L.) Benth.
var. hirsutum *(DC.) L. Pedley in Austrobaileya,* 5(2): 223 (1999): Desmodium umbellatum var. hirsutum.

DENDROPANAX (Araliac.).
affinis *(Marchal) J.C. Gamerro & F.O. Zuloaga in Darwiniana,* 35(1–4): 163 (1998): Gilibertia affinis.
hoi *C.B. Shang in Acta Phytotax. Sin.,* 37(6): 607 (1999), nom. nov.: D. macrocarpus *C.N. Ho.*
langbianensis *C.B. Shang in Adansonia,* 19(1): 82 (1997)—Vietnam, Cambodia.

DENDROPHORBIUM (Compos.).
amplexicaule *(Kunth) B. Nordenstam in Compositae Newsl.,* 29: 47 (1996): Senecio amplexicaulis.
angelense *(Domke) B. Nordenstam in Compositae Newsl.,* 29: 47 (1996): Senecio angelensis.
argutidentatum *(Cuatrec.) H. Beltran in Compositae Newsl.,* 34: 51 (1999): Senecio argutidentatus.
balsapampae *(Cuatrec.) B. Nordenstam in Compositae Newsl.,* 29: 48 (1996): Senecio balsapampae
chingualense *(Cuatrec.) S. Díaz Piedrahita & J. Cuatrecasas, Asterac. Fl. Colombia, Senecioneae-1:* 19 (1999): Senecio chingualensis.
cosnipatense *(Cabrera) H. Beltran in Compositae Newsl.,* 34: 51 (1999): Senecio cosnipatense.
dodsonii *(H. Rob. & Cuatrec.) B. Nordenstam in Compositae Newsl.,* 29: 48 (1996): Pentacalia dodsonii.
gesnerifolium *(Cuatrec.) B. Nordenstam in Compositae Newsl.,* 29: 48 (1996): Senecio gesnerifolius.
goodspeedii *(Cuatrec.) H. Beltran in Compositae Newsl.,* 34: 51 (1999): Senecio goodspeedii.
ingens *(Benoist) B. Nordenstam in Compositae Newsl.,* 29: 48 (1996): Senecio ingens.
kleinioides *(Kunth) B. Nordenstam in Compositae Newsl.,* 30: 48 (1997): Cacalia kleinioides.
llewelynii *(Cuatrec.) H. Beltran in Compositae Newsl.,* 34: 51 (1999): Senecio llewelynii.

DENDROPHORBIUM:-
onae (Cuatrec.) C. Jeffrey
var. leonis *(Cuatrec.) B. Nordenstam in Compositae Newsl.*, 29: 48 (1996): Senecio onae var. leonis.
pericaule *(Greenm.) B. Nordenstam in Compositae Newsl.*, 31: 22 (1997): Senecio pericaulis.
solisii *(Cuatrec.) B. Nordenstam in Compositae Newsl.*, 29: 48 (1996): Senecio solisii.
tipocochense *(Domke) B. Nordenstam in Compositae Newsl.*, 29: 48 (1996): Senecio tipocochensis.
toreadoris *(Cuatrec.) B. Nordenstam in Compositae Newsl.*, 29: 49 (1996): Senecio toreadoris.
trigynum *(Cuatrec.) H. Beltran in Compositae Newsl.*, 34: 51 (1999): Senecio trigynus.
vanillodorum *(Cabrera) H. Beltran in Compositae Newsl.*, 34: 51 (1999): Senecio vanillodorus.
vargasii *(Cabrera) H. Beltran in Compositae Newsl.*, 34: 51 (1999): Senecio vargasii.

DENDROPHTHORA (Viscac.).
amalfiensis *J. Kuijt in Bot. Jahrb.*, 122(2): 180 (2000)—Colombia.
argentea *J. Kuijt in Bot. Jahrb.*, 122(2): 182 (2000)—Peru.
avenia *(Trel.) J. Kuijt in Bot. Jahrb.*, 122(2): 174 (2000): Phoradendron avenium.
corynarthron *(Eichler) J. Kuijt in Bot. Jahrb.*, 122(2): 176 (2000): Phoradendron corynarthron.
crispula *(Rizzini) J. Kuijt in Bot. Jahrb.*, 122(2): 176 (2000): Phoradendron crispulum.
densifrons *(Ule) J. Kuijt in Novon*, 6(1): 33 (1996): Phoradendron densifrons.
elegantissima *J. Kuijt in Bot. Jahrb.*, 122(2): 183 (2000)—Peru.
enckeifolia *(Rizzini) J. Kuijt in Bot. Jahrb.*, 122(2): 177 (2000): Phoradendron enckeifolium.
fanshawei *(Maguire) J. Kuijt in Novon*, 6(1): 33 (1996): Phoradendron fanshawei.
filiformis *C.T. Rizzini in Revista Fac. Agron. (Maracay)*, 8(3): 83 (1975)—Venezuela.
gracilipes *J. Kuijt in Bot. Jahrb.*, 122(2): 184 (2000)—Peru.
haberi *J. Kuijt in Novon*, 6(1): 33 (1996)—Costa Rica.
heterophylla *(Rizzini) J. Kuijt in Bot. Jahrb.*, 122(2): 177 (2000): Phoradendron heterophyllum.
intermedia *(Rizzini) J. Kuijt in Novon*, 6(1): 35 (1996): Phoradendron intermedium.
lacryma-jobi *E.A. Kellogg in Novon*, 6(1): 35 (1996)—Venezuela.
longepedunculata *C.T. Rizzini in Revista Fac. Agron. (Maracay)*, 8(3): 84 (1975)—Venezuela.
macbridei *(Standl.) J. Kuijt in Bot. Jahrb.*, 122(2): 177 (2000): Phoradendron macbridei.
oligantha *J. Kuijt in Novon*, 6(1): 38 (1996)—Venezuela.
palaeformis *J. Kuijt in Bot. Jahrb.*, 122(2): 186 (2000)—Peru.
perfurcata *(Rizzini) J. Kuijt in Bot. Jahrb.*, 122(2): 178 (2000): Phoradendron perfurcatum.
purpurea *J. Kuijt in Bot. Jahrb.*, 122(2): 186 (2000)—Bolivia.
roseantha *J. Kuijt in Bot. Jahrb.*, 122(2): 189 (2000)—Peru.
scopulata *J. Kuijt in Novon*, 6(1): 39 (1996)—Honduras.
steyermarkii *C.T. Rizzini in Revista Fac. Agron. (Maracay)*, 8(3): 84 (1975)—Venezuela.
sulcata *J. Kuijt in Bot. Jahrb.*, 122(2): 190 (2000)—Colombia.

DENDROPHYLAX (Orchidac.).
constanzensis *(Garay) M.A. Nir, Orchidaceae Antillanae:* 83 (2000), as 'constanzense': Campylocentrum constanzense.
serpentilingua *(Dod) M.A. Nir, Orchidaceae Antillanae:* 86 (2000): Campylocentrum serpentilingua.

DENDROPORTULACA *U. Eggli in Adansonia*, 19(1): 49 (1997). *PORTULACACEAE*
mirabilis *U. Eggli in Adansonia*, 19(1): 50 (1997)—Madagascar.

DENMOZA (Cactac.).
rhodacantha
var. diamantina *R. Slaba in Kaktusy (Brno)*, 32(4): 98 (1996)—Argentina Northwest (Mendoza).

DEPPEA (Rubiac.).
miahuatlanica *D.H. Lorence in Novon*, 7(1): 48 (1997)—Mexico Southwest (Oaxaca).

DEPREA (Solanac.).
bitteriana *(Werderm.) N.W. Sawyer & C.E. Benítez de Rojas in Brittonia*, 50(4): 527 (1998): Athenaea bitteriana.
ecuatoriana *A.T. Hunziker & G. Barboza in Darwiniana*, 34(1–4): 109 (1996)—Ecuador.

× **DEROSAARA** *hort. in Orchid Rev.*, 104(1209): 166 (1996). *ORCHIDACEAE* (Aspasia × Brassia × Miltonia × Odontoglossum).

DERRIS (Legumin.).
araripensis *(Benth.) N.F. Mattos in Loefgrenia*, 98: 3 (1989): Lonchocarpus araripensis.
campestris *(Mart. ex Benth.) N.F. Mattos in Loefgrenia*, 98: 2 (1989): Lonchocarpus campestris.
costata *(Benth.) N.F. Mattos in Loefgrenia*, 98: 3 (1989): Lonchocarpus costatus.
dasycalyx *(Harms) N.F. Mattos in Loefgrenia*, 98: 4 (1989): Lonchocarpus dasycalyx.
filipes *(Benth.) N.F. Mattos in Loefgrenia*, 98: 2 (1989): Lonchocarpus filipes.
glaziovii *(Taub.) N.F. Mattos in Loefgrenia*, 98: 4 (1989): Lonchocarpus glaziovii.
guilleminiana (Tul.) J.F. Macbr.
var. pubigera *(Tul.) N.F. Mattos in Loefgrenia*, 98: 2 (1989): Lonchocarpus guilleminianus var. pubigerus.
mollis *(Benth.) N.F. Mattos in Loefgrenia*, 98: 3 (1989): Lonchocarpus mollis.
nitidula *(Benth.) N.F. Mattos in Loefgrenia*, 98: 3 (1989): Lonchocarpus nitidulus.
obtusus *(Benth.) N.F. Mattos in Loefgrenia*, 98: 3 (1989): Lonchocarpus obtusus.
paniculata *(Ducke) N.F. Mattos in Loefgrenia*, 98: 5 (1989), homonym; *see Benth. in J. Proc. Linn. Soc., Bot.*, 4(Supp.): 105 (1860): Lonchocarpus paniculatus.
peckoltii *(Wawra) N.F. Mattos in Loefgrenia*, 98: 4 (1989): Lonchocarpus peckoltii.
praecox *(Mart. ex Benth.) N.F. Mattos in Loefgrenia*, 98: 3 (1989): Lonchocarpus praecox.
virgilioides *(Benth.) N.F. Mattos in Loefgrenia*, 98: 2 (1989): Lonchocarpus virgilioides.

DESCHAMPSIA (Gramin.).
cespitosa (L.) P. Beauv.
var. hispanica *(Vivant) C. Cebolla Lozano, J.Á. López Rodríguez & M.A. Rivas Ponce in Fontqueria*, 49: 23 (1997): D. cespitosa subsp. hispanica.
subsp. sikkimensis *H.J. Noltie in Edinburgh J. Bot.*, 56(3): 382 (1999)—Sikkim, Xizang.
sukatschewii (Popl.) Roshev.
subsp. borealis *(Trautv.) N.N. Tzvelev in Novosti Sist. Vyssh. Rast.*, 32: 182 (2000): Aira caespitosa var. borealis.

DESMANTHODIUM (Compos.).
hintoniorum *B.L. Turner in Phytologia*, 79(5): 317 (1995 publ. 1996)—Mexico Southwest (Oaxaca).

DESMIDORCHIS (Asclepiadac.).
edithiae *(N.E. Br.) D.C.H. Plowes in Haseltonia*, 3: 58 (1995): Caralluma edithiae.
foetida *(E.A. Bruce) D.C.H. Plowes in Haseltonia*, 3: 58 (1995): Caralluma foetida.
penicillata *(Deflers) D.C.H. Plowes in Haseltonia*, 3: 58 (1995): Boucerosia penicillata.
somalica *(N.E. Br.) D.C.H. Plowes in Haseltonia*, 3: 59 (1995): Caralluma somalica.
speciosa *(N.E. Br.) D.C.H. Plowes in Haseltonia*, 3: 59 (1995): Sarcodon speciosa.

DESMOCLADUS (Restionac.).
asper *(Nees) B.G. Briggs & L.A.S. Johnson in Telopea*, 8(1): 26 (1998): Calorophus asper.
fasciculatus *(R. Br.) B.G. Briggs & L.A.S. Johnson in Telopea*, 8(1): 26 (1998): Restio fasciculatus.
flexuosus *(R. Br.) B.G. Briggs & L.A.S. Johnson in Telopea*, 8(1): 26 (1998): Restio flexuosus.
glomeratus *K.W. Dixon & K.A. Meney in Telopea*, 6(4): 649 (1996)—Western Australia.

DESMOCLADUS:-
myriocladus *(Gilg) B.G. Briggs & L.A.S. Johnson in Telopea*, 8(1): 27 (1998): Loxocarya myrioclada.
virgatus *(Benth.) B.G. Briggs & L.A.S. Johnson in Telopea*, 8(1): 27 (1998): Loxocarya virgata.

DESMODIUM (Legumin.).
subgen. Acanthocladum *L. Pedley in Austrobaileya*, 5(2): 225 (1999).
ser. Arillata *L. Pedley in Austrobaileya*, 5(2): 226 (1999).
sect. Desmodiopsis *(Schindl.) L. Pedley in Austrobaileya*, 5(2): 226 (1999): Alysicarpus sect. Desmodiopsis.
ser. Sagotia *(Duchass. & Walp.) L. Pedley in Austrobaileya*, 5(2): 226 (1999): gen. Sagotia.
ser. Stenostachys *(Schindl.) L. Pedley in Austrobaileya*, 5(2): 226 (1999): D. sect. Stenostachys.
elongatum *Wall. ex Wight & Arn., Prodr. Fl. Ind. Orient.*: 226 (1834), nom. nov.: Hedysarum diffusum *Roxb.*
flexuosum Wall. ex Benth.
var. cordifoliolatum *(P.C. Li) P.H. Huang in Bull. Bot. Res. North-East. Forest. Univ.*, 18(1): 5 (1998): D. cordifoliolatum.
glareosum *L. Pedley in Austrobaileya*, 5(2): 242 (1999)—Northern Territory, Western Australia.
griffithianum Benth.
var. leiocarpum *X.F. Gao & C. Chen in Acta Bot. Yunnanica*, 20(3): 270 (1998)—China South-central (Sichuan, Yunnan).
lempirae *C. Nelson Sutherland in Fontqueria*, 44: 106 (1996), nom. nov.: D. arbuscula *Standl. & L.O. Williams.*
luteolum *H. Ohashi & T. Nemoto in J. Jap. Bot.*, 73(2): 86 (1998), homonym; *see Standl. in Publ. Field Mus. Nat. Hist., Bot. Ser.*, 17(4): 368 (1938)—China South-central (Yunnan).
pendenticarpum *C.Z. Gao & Q.R. Lai ex Hiroyoshi Ohashi in J. Jap. Bot.*, 70(2): 115 (1995), in syn.—China Southeast (Guangxi).
pullenii *L. Pedley in Austrobaileya*, 5(2): 243 (1999)—Northern Territory, Western Australia, Queensland, Lesser Sunda Is.
pycnotrichum *L. Pedley in Austrobaileya*, 5(2): 236 (1999): Northern Territory, Western Australia.
sericophyllum Schltdl.
var. strobilaceum *(Schltdl.) N. Zamora in Novon*, 10(2): 179 (2000): D. strobilaceum.
strigillosum Schindl.
var. pendenticarpum *P.H. Huang in Bull. Bot. Res. North-East. Forest. Univ.*, 18(1): 5 (1998), as 'comb. nov.' but bas. not valid; 2 refs. give validation as new taxon; Ohashi ref. to type—China Southeast (Guangxi).
tiwiense *L. Pedley in Austrobaileya*, 5(2): 248 (1999)—Northern Territory.
whitfordii *(Schindl.) L. Pedley in Austrobaileya*, 5(2): 247 (1999): D. nemorosum subvar. whitfordii.

DESMOS (Annonac.).
chinensis Lour.
var. laevigatus *(Hook. f. & Thomson) Debika Mitra in Bull. Bot. Surv. India*, 35(1–4): 117 (1993 publ. 3 July 1997); *et in Fl. West Bengal*, 1: 147 (15 Aug. 1997), as 'laevigata': Unona discolor var. laevigata.
evrardii *(Ast) P.T. Li in J. S. China Agric. Univ.*, 14(3): 155 (1993): Dasymaschalon evrardii.
lomentaceus *(Finet & Gagnep.) P.T. Li in J. S. China Agric. Univ.*, 14(3): 155 (1993): Dasymaschalon lomentaceum.
macrocalyx *(Finet & Gagnep.) P.T. Li in J. S. China Agric. Univ.*, 14(3): 155 (1993): Dasymaschalon macrocalyx.
robinsonii *(Ast) P.T. Li in J. S. China Agric. Univ.*, 14(3): 155 (1993): Dasymaschalon robinsonii.

DEUTEROCOHNIA (Bromeliac.).
bracteosa *W. Till & L. Hromadnik in Bromélia*, 4(1): 19 (1997)—Bolivia.

DEUTZIA (Hydrangeac.).
hatusimae *H. Ohba, L.M. Niu & T. Minamitani in J. Jap. Bot.*, 75(5): 292 (2000)—Japan (Kyushu).

DEYEUXIA (Gramin.).
sect. Chamaecalamus *(Pilg.) Z. Rúgolo de Agrasar & X. Villavicencio in X. Villavicencio, Revis. Gattung Deyeuxia Bolivien:* 98 (1995), without exact basionym page: Calamagrostis sect. Chamaecalamus.
sect. Stylagrostis *(Mez) Z. Rúgolo de Agrasar & X. Villavicencio in X. Villavicencio, Revis. Gattung Deyeuxia Bolivien:* 56 (1995): gen. Stylagrostis.
sect. Viridiflavescentia *Z. Rúgolo de Agrasar & X. Villavicencio in X. Villavicencio, Revis. Gattung Deyeuxia Bolivien:* 90 (1995).
alaica *(Litv.) Y.H. Wu in Y.H. Wu & Q.J. Wang, Grasses of Karakorum & Kunlun Mts.:* 112 (1999): Calamagrostis alaica.
boliviensis *(Hack.) X. Villavicencio, Revis. Gattung Deyeuxia Bolivien:* 131 (1995): Calamagrostis boliviensis.
brevifolia J. Presl
var. expansa *Z.E. Rúgolo de Agrasar & X. Villavicencio in Bol. Soc. Argent. Bot.*, 31(1–2): 125 (1995)—Bolivia, Peru
cabrerae (Parodi) Parodi
var. aristulata *Z.E. Rúgolo de Agrasar & X. Villavicencio in Bol. Soc. Argent. Bot.*, 31(1–2): 126 (1995)—Argentina Northwest (Jujuy), Bolivia.
chrysantha J. Presl
var. phalaroides *(Wedd.) X. Villavicencio, Revis. Gattung Deyeuxia Bolivien:* 63 (1995): D. phalaroides.
ciliata *Z.E. Rúgolo de Agrasar & X. Villavicencio in Bol. Soc. Argent. Bot.*, 31(1–2): 126 (1995)—Bolivia.
var. glabrescens *Z.E. Rúgolo de Agrasar & X. Villavicencio in Bol. Soc. Argent. Bot.*, 31(1–2): 128 (1995)—Bolivia.
crispa *Z.E. Rúgolo de Agrasar & X. Villavicencio in Bol. Soc. Argent. Bot.*, 31(1–2): 128 (1995)—Argentina Northwest (Jujuy, Catamarca, La Rioja, Salta), Bolivia, Chile North (Antofagasta, Tarapacá), Peru.
curtoides *Z.E. Rúgolo de Agrasar & X. Villavicencio in Bol. Soc. Argent. Bot.*, 31(1–2): 132 (1995)—Bolivia.
deserticola Phil.
var. breviaristata *Z.E. Rúgolo de Agrasar & X. Villavicencio in Bol. Soc. Argent. Bot.*, 31(1–2): 134 (1995)—Bolivia, Chile North (Tarapacá), Argentina Northwest (Catamarca, Jujuy, La Rioja).
eminens J. Presl
var. discreta *Z.E. Rúgolo de Agrasar & X. Villavicencio in Bol. Soc. Argent. Bot.*, 31(1–2): 135 (1995)—Peru, Bolivia, Chile North (Tarapacá).
filifolia Wedd.
var. festucoides *(Wedd.) Z. Rúgolo de Agrasar & X. Villavicencio in X. Villavicencio, Revis. Gattung Deyeuxia Bolivien:* 201 (1995): D. festucoides.
hirsuta *Z.E. Rúgolo de Agrasar & X. Villavicencio in Bol. Soc. Argent. Bot.*, 31(1–2): 136 (1995)—Bolivia.
kokonorica *(Tzvelev) S.L. Lu in Fl. Qinghaiica*, 4: 137 (1999): Calamagrostis kokonorica.
lacustris *E. Edgar & H.E. Connor in New Zealand J. Bot.*, 37(1): 68 (1999)—New Zealand South I.
minima *(Pilg.) Z. Rúgolo de Agrasar in X. Villavicencio, Revis. Gattung Deyeuxia Bolivien:* 116 (1995): Calamagrostis vicunarum var. minima.
nitidula *(Pilg.) Z. Rúgolo de Agrasar in X. Villavicencio, Revis. Gattung Deyeuxia Bolivien:* 239 (1995): Calamagrostis nitidula.
ovata J. Presl
var. nivalis *(Wedd.) X. Villavicencio, Revis. Gattung Deyeuxia Bolivien:* 75 (1995): D. nivalis.
polygama (Griseb.) Parodi
subsp. filifolia *Z.E. Rúgolo de Agrasar & X. Villavicencio in Bol. Soc. Argent. Bot.*, 31(1–2): 139 (1995)—Bolivia.
rupestris *(Trin.) Z. Rúgolo de Agrasar in X. Villavicencio, Revis. Gattung Deyeuxia Bolivien:* 95 (1995): Calamagrostis rupestris.
scleranitha *(Hack.) Z. Rúgolo de Agrasar in X. Villavicencio, Revis. Gattung Deyeuxia Bolivien:* 134 (1995): Calamagrostis sclerantha.

DEYEUXIA:-
spicigera J. Presl
var. cephalotes *(Wedd.) Z. Rúgolo de Agrasar & X. Villavicencio in X. Villavicencio, Revis. Gattung Deyeuxia Bolivien:* 172 (1995): D. cephalotes.
srilankensis *(Davidse) J.F. Veldkamp in Blumea*, 41(2): 410 (1996): Calamagrostis srilankensis.
swallenii *(Tovar) Z. Rúgolo de Agrasar in X. Villavicencio, Revis. Gattung Deyeuxia Bolivien:* 128 (1995): Calamagrostis swallenii.
tarmensis (Pilg.) Sodiro
var. tarijensis *(Pilg.) X. Villavicencio, Revis. Gattung Deyeuxia Bolivien:* 224 (1995): Calamagrostis tarijensis.
violacea Wedd.
var. puberula *Z.E. Rúgolo de Agrasar & X. Villavicencio in Bol. Soc. Argent. Bot.*, 31(1–2): 139 (1995)—Bolivia, Argentina Northwest (Jujuy).

DIACHROA *T. Nuttall in Trans. Amer. Philos. Soc., n.s.*, 5: 147 (1835). *GRAMINEAE*
procumbens *(Muhl.) T. Nuttall in Trans. Amer. Philos. Soc., n.s.*, 5: 147 (1835): Festuca procumbens.

DIALIUM (Legumin.).
indum L.
var. bursa *(de Wit) J.P. Rojo in Fl. Males.*, Ser. 1, 12(2): 612 (1996), isonym; *see P.J.A. Kessler & Sidiyasa, Trees of Balikpapan-Samarinda Area (Tropenbos*, 7): 71 (1994): D. laurinum var. bursa.
kunstleri Prain
var. trifoliolatum *(de Wit) J.P. Rojo in Fl. Males.*, Ser. 1, 12(2): 614 (1996): D. trifoliolatum.

DIANDRANTHUS *L. Liou in Fl. Reipubl. Popul. Sin.*, 10(2): 10 (1997). *GRAMINEAE*
brevipilus *(Hand.-Mazz.) L. Liou in Fl. Reipubl. Popul. Sin.*, 10(2): 16 (1997): Miscanthus brevipilus.
eulalioides *(Keng) L. Liou in Fl. Reipubl. Popul. Sin.*, 10(2): 15 (1997): Miscanthus eulalioides.
nepalensis *(Trin.) L. Liou in Fl. Reipubl. Popul. Sin.*, 10(2): 11 (1997): Eulalia nepalensis.
nudipes *(Griseb.) L. Liou in Fl. Reipubl. Popul. Sin.*, 10(2): 15 (1997): Erianthus nudipes.
taylorii *(Bor) L. Liou in Fl. Reipubl. Popul. Sin.*, 10(2): 14 (1997): Miscanthus taylorii.
yunnanensis *(A. Camus) L. Liou in Fl. Reipubl. Popul. Sin.*, 10(2): 14 (1997): Miscanthus nudipes subsp. yunnanensis.

DIANELLA (Phormiac.).
ensifolia (L.) DC.
f. aureovariegata *(K. Nakaj.) S. Hatusima in S. Hatusima & T. Amano, Fl. Ryukyus, S. of Amami Is., ed. 2:* 302 (1994), without basionym ref.: Basionym not stated.

DIANTHERA (Acanthac.).
pohliana *(Nees) N.E. Brown, Suppl. Johnson's Gard. Dict.:* 917 (1882): Amphiscopia pohliana.

DIANTHUS (Caryophyllac.).
subsect. Arenarii *M.M. Fedoronchuk in Ukr. Bot. Zhurn.*, 55(1): 21 (1998).
subsect. Capitati *M.M. Fedoronchuk in Ukr. Bot. Zhurn.*, 55(1): 21 (1998).
sect. Chinenses *N.N. Tzvelev, Opred. Sosud. Rast. Severo-Zap. Rossii:* 679 (2000).
subsect. Floribundi *M.M. Fedoronchuk in Ukr. Bot. Zhurn.*, 55(1): 21 (1998).
subsect. Leptopetali *M.M. Fedoronchuk in Ukr. Bot. Zhurn.*, 55(1): 21 (1998).
subsect. Superbi *M.M. Fedoronchuk in Ukr. Bot. Zhurn.*, 55(1): 22 (1998).
aphanoneurus *(Rech. f.) M.L. Kuzmina in Bot. Zhurn.*, 81(8): 82 (1996): D. orientalis subsp. aphanoneurus.
aydogdui *Y. Menemen & E. Hamzaoglu in Ann. Bot. Fenn.*, 37(4): 285 (2000), as 'aydogduii'—Turkey
brachyzonus Borbás & Formánek
subsp. villiger *(Bornm.) K. Micevski in Prilozi Oddel. Biol. Med. Nauki Makedonska Akad. Nauk. Umet.*, 8(1–2): 45 (1987 publ. 1990): D. calocephalus var. villiger.

DIANTHUS:-
chinensis L.
var. jingpoensis *G.Y. Zhang & X.Y. Yuan in Bull. Bot. Res. North-East. Forest. Univ.*, 18(1): 10 (1998), holotype not indicated—Manchuria (Heilongjiang).
subsp. repens *(Willd.) V.N. Voroschilov in A.K. Skvortsov (ed.), Florist. issl. v razn. raĭonakh SSSR:* 167 (1985): D. repens.
subsp. versicolor *(Fisch. ex Link) V.N. Voroschilov in A.K. Skvortsov (ed.), Florist. issl. v razn. raĭonakh SSSR:* 167 (1985): D. versicolor.
cinnamomeus Sm.
subsp. naxensis *Runemark ex A. Strid in A. Strid & Kit Tan (eds.), Fl. Hellenica*, 1: 352 (1997)—Greece.
desideratus *A. Strid in A. Strid & Kit Tan (eds.), Fl. Hellenica*, 1: 371 (1997)—Greece.
dmanissianus *M.L. Kuzmina in Bot. Zhurn.*, 81(8): 80 (1996)—Transcaucasus (Gruziya).
elegans d'Urv.
var. gramineus *R.M. Burton in Karaca Arbor. Mag.*, 3(4): 184 (1997)—Turkey.
galicicae *K. Micevski in Prilozi Oddel. Biol. Med. Nauki Makedonska Akad. Nauk. Umet.*, 8(1–2): 40 (1987 publ. 1990)—Yugoslavia.
haematocalyx Boiss. & Heldr.
subsp. phitosianus *T. Constantinidis in Phyton (Austria)*, 39(2): 279 (1999)—Greece.
hispanicus Asso
subsp. fontqueri *(O. Bolòs & Vigo) A. Barber i Vallès, M.B. Crespo & Mateo in A. Barber i Vallès, Contrib. Coneix. Fl. Fitogeogr. Lit. Comarca Marina* Alta: 77 (1999): D. pungens subsp. fontqueri.
jablanicensis *K. Micevski in Prilozi Oddel. Biol. Med. Nauki Makedonska Akad. Nauk. Umet.*, 8(1–2): 31 (1987 publ. 1990)—Yugoslavia.
jacupicensis *(Košanin) K. Micevski in Prilozi Oddel. Biol. Med. Nauki Makedonska Akad. Nauk. Umet.*, 8(1–2): 41 (1987 publ. 1990): D. pindicola var. jacupicensis.
jugoslavicus *K. Micevski in Prilozi Oddel. Biol. Med. Nauki Makedonska Akad. Nauk. Umet.*, 8(1–2): 33 (1987 publ. 1990)—Yugoslavia.
juzeptchukii *M.L. Kuzmina in Bot. Zhurn.*, 81(8): 80 (1996)—Transcaucasus (Gruziya).
klokovii *M.S. Knjasev in Bot. Zhurn.*, 82(11): 80 (1997)—Russia East, Kazakhstan.
macedonicus *K. Micevski in Prilozi Oddel. Biol. Med. Nauki Makedonska Akad. Nauk. Umet.*, 8(1–2): 35 (1987 publ. 1990)—Yugoslavia.
martuniensis *M.L. Kuzmina in Bot. Zhurn.*, 81(8): 81 (1996)—Transcaucasus (Armeniya).
microlepis Boiss.
subsp. degenii *(Stoj. & Acht.) D. Peev & J. Zlatkova in Phytologia Balcanica*, 1: 38 (1995): D. microlepis var. degenii.
ohridanus *K. Micevski in Prilozi Oddel. Biol. Med. Nauki Makedonska Akad. Nauk. Umet.*, 8(1–2): 34 (1987 publ. 1990)—Yugoslavia.
pinifolius Sm.
subsp. tenuicaulis *(Turrill) A. Strid in A. Strid & Kit Tan (eds.), Fl. Hellenica*, 1: 368 (1997): D. tenuicaulis.
prilepensis *K. Micevski in Prilozi Oddel. Biol. Med. Nauki Makedonska Akad. Nauk. Umet.*, 8(1–2): 39 (1987 publ. 1990)—Yugoslavia.
pygmaeus Hayata
f. albiflorus *(S.S. Ying) S.S. Ying in Fl. Taiwan, ed. 2*, 2: 356 (1996): D. pygmaeus var. albiflorus.
rupicola Biv.
subsp. bocchoriana *(L. Llorens & Gradaille) A.M. Romo, Fl. Silvestres Baleares:* 218 (1994), name not new. At bas. ref. name stated is: D. rupicola subsp. bocchoriana.
scardicus Wettst.
var. incisus *K. Micevski in Prilozi Oddel. Biol. Med. Nauki Makedonska Akad. Nauk. Umet.*, 8(1–2): 44 (1987 publ. 1990)—Yugoslavia.
shinanensis (Yatabe) Makino
f. alpinus *Hid. Takah. ex T. Shimizu, Fl. Nagano Pref.:* 1505 (1997)—Japan.
skopjensis *K. Micevski in Prilozi Oddel. Biol. Med. Nauki Makedonska Akad. Nauk. Umet.*, 8(1–2): 36 (1987 publ. 1990)—Yugoslavia.
superbus L.
f. albiflorus *Y.N. Lee, Fl. Korea:* 1156 (1996), without type—Locality not stated.

DIANTHUS:-
superbus-chinensis *Y.N. Lee, Fl. Korea:* 1156 (1996), contrary to Arts. 37.4, 37.5, H.10.3 ICBN (1994)—South Korea.
sylvestris Wulfen
f. albiflorus *K. Micevski in Prilozi Oddel. Biol. Med. Nauki Makedonska Akad. Nauk. Umet.,* 8(1–2): 44 (1987 publ. 1990)—Yugoslavia.
var. alpestris *K. Micevski in Prilozi Oddel. Biol. Med. Nauki Makedonska Akad. Nauk. Umet.,* 8(1–2): 42 (1987 publ. 1990)—Yugoslavia.
subsp. kozjakensis *K. Micevski in Prilozi Oddel. Biol. Med. Nauki Makedonska Akad. Nauk. Umet.,* 8(1–2): 43 (1987 publ. 1990)—Yugoslavia.
versicolor Fisch. ex Link
var. ninelli *G.A. Peschkova in Fl. Tsentral'noi Sibiri,* 1: 334 (1979)—Buryatiya.
viscidus Chaub. & Bory
var. glandulosus *K. Micevski in Prilozi Oddel. Biol. Med. Nauki Makedonska Akad. Nauk. Umet.,* 8(1–2): 44 (1987 publ. 1990)—Yugoslavia.
vodnensis *K. Micevski in Prilozi Oddel. Biol. Med. Nauki Makedonska Akad. Nauk. Umet.,* 8(1–2): 38 (1987 publ. 1990)—Yugoslavia.

DIAPHANANTHE (Orchidac.).
delepierreana *J.-P. Lebel & D. Geerinck in Belgian J. Bot.,* 130(2): 136 (1998)—Rwanda.

DIASCIA (Scrophulariac.).
esterhuyseniae *K.E. Steiner in S. Afr. J. Bot.,* 65(3): 223 (1999)—Cape Province.

DICELLA (Malpighiac.).
aciculifera *W.R. Anderson in Contrib. Univ. Mich. Herb.,* 21: 54 (1997)—Costa Rica.

DICELLOSTYLES (Malvac.).
zizyphifolia *(Griff.) L. Phuphathanaphong in Thai Forest Bull., Bot.,* 21: 124 (1994): Kydia zizyphifolia.

DICHAEA (Orchidac.).
alcantarae *D.E. Bennett & E.A. Christenson in Lindleyana,* 13(1): 31 (1998)—Peru.
delcastilloi *D.E. Bennett & E.A. Christenson in Lindleyana,* 13(1): 33 (1998)—Peru.
galeata *C.H. Dodson in Orquideologia,* 20(3): 272 (1997)—Ecuador.
latifolia
var. longa *(Schltr.) J.P. Folsom in Orchid Dig.,* 60(4): 154 (1996), without basionym date: D. longa.
tachirensis *G.A. Romero & G. Carnevali, Orchids of Venezuela, ed. 2:* 1130 (2000)—Venezuela.

DICHAPETALUM (Dichapetalac.).
korupinum *F.J. Breteler in Kew Bull.,* 51(1): 201 (1996)—Cameroon.
moralesii *G.T. Prance in Kew Bull.,* 52(1): 216 (1997)—Costa Rica.
morenoi *G.T. Prance in Kew Bull.,* 52(1): 216 (1997)—Nicaragua, Costa Rica.

DICHELACHNE (Gramin.).
lautumia *E. Edgar & H.E. Connor in New Zealand J. Bot.,* 37(1): 67 (1999)—New Zealand South I.

DICHOCARPUM (Ranunculac.).
nipponicum (Franch.) W.T. Wang & P.K. Hsiao
f. didymocalyx *H. Ohba in J. Jap. Bot.,* 75(6): 370 (2000)—Japan (Honshu).

DICHORISANDRA (Commelinac.).
amabilis *J.R. Grant in Novon,* 10(2): 117 (2000)—Costa Rica, Mexico Southeast (Chiapas), Mexico Southwest (Oaxaca), Mexico Gulf (Veracruz), Guatemala, El Salvador, Honduras, Nicaragua, Panama.

DICHOSTYLIS (Cyperac.).
castanea *(Willd.) Palla in Bot. Jahrb. Syst.,* 10(4): 296 (4 Dec. 1888): Cyperus castaneus.
cuspidata *(Kunth) Palla in Bot. Jahrb. Syst.,* 10(4): 296 (4 Dec. 1888): Cyperus cuspidatus.
nitens *(Retz.) Palla in Bot. Jahrb. Syst.,* 10(4): 296 (4 Dec. 1888): Cyperus nitens.

DICHOSTYLIS:-
patens *(Vahl) Palla in Bot. Jahrb. Syst.,* 10(4): 296 (4 Dec. 1888): Cyperus patens.
squarrosa *(L.) Palla in Bot. Jahrb. Syst.,* 10(4): 296 (4 Dec. 1888): Cyperus squarrosus.

DICHROA (Hydrangeac.).
yunnanensis S.M. Huang
var. microcarpa *H. Li in H. Li, H.J. Guo & Z.L. Dao (eds.), Fl. Gaoligong Mount.:* 632 (2000), without latin descr. or type—China South-central (Yunnan).

DICLIDANTHERA (Polygalac.).
wurdackiana *G. Aymard & P.E. Berry in Biollania,* ed. esp. 6 (J.J. Wurdack Festschr.): 191 (1997)—Venezuela.

DICLIPTERA (Acanthac.).
hirta *K. Balkwill in Kew Bull.,* 51(1): 50 (1996)—Zimbabwe, Zambia, Mozambique, Botswana.
leistneri *K. Balkwill in Kew Bull.,* 51(1): 42 (1996)—Orange Free State, Transvaal, Cape Province.
magaliesbergensis *K. Balkwill in Kew Bull.,* 51(1): 39 (1996)—Transvaal.
minor C.B. Clarke
subsp. pratis-manna *K. Balkwill in Kew Bull.,* 51(1): 38 (1996)—Transvaal.
neesii *(Trimen) L.H. Cramer in M.D. Dassanayake & W.D. Clayton, eds., Revised Handb. Fl. Ceylon,* 12: 124 (1998): D. zeylanica var. neesii.
spinulosa *Hochst. ex K. Balkwill in Kew Bull.,* 51(1): 53 (1996)—Sudan, Burkina, Ghana, Nigeria, Central African Republic, Ethiopia, Kenya, Tanzania, Angola, Malawi, Mozambique, Zimbabwe, Namibia, Transvaal, Madagascar.

DICOMA (Compos.).
aethiopica *S. Ortiz & J. Rodríguez Oubiña in Nordic J. Bot.,* 16(3): 279 (1996 publ. 1997)—Ethiopia.
anomala Sond.
subsp. attenuata *(S. Moore) S. Ortiz & J. Rodríguez-Oubiña in Nordic J. Bot.,* 16(6): 584 (1996 publ. 1997): D. anomala var. attenuata.
subsp. gerrardii *(Harv. ex F.C. Wilson) S. Ortiz & J. Rodríguez-Oubiña in Nordic J. Bot.,* 16(6): 584 (1996 publ. 1997): D. gerrardii.
popeana *S. Ortiz & J. Rodríguez Oubiña in Nordic J. Bot.,* 16(3): 277 (1996 publ. 1997)—Somalia.
schimperi (DC.) Baill. ex O. Hoffm.
subsp. cinerea *(S. Ortiz & Rodr. Oubiña) S. Ortiz & J. Rodríguez-Oubiña in Ann. Missouri Bot. Gard.,* 85(3): 451 (1998): D. cinerea.
sessiliflora Harv.
var. membranacea *(S. Moore) S. Ortiz & J. Rodriguez-Oubiña in Bothalia,* 27(1): 49 (1997): D. membranacea.
swazilandica *S. Ortiz, Rodr. Oubiña & I. Pulgar in Nordic J. Bot.,* 20(3): 347 (2000)—Swaziland.
thuliniana *S. Ortiz, J. Rodríguez-Oubiña & Mesfin Tadesse in Ann. Missouri Bot. Gard.,* 85(3): 455 (1998)—Somalia.

DICRANOPYGIUM (Cyclanthac.).
arusisense *D. Tuberquia in Caldasia,* 19(1–2): 184 (1997)—Colombia.
odoratum *D. Tuberquia in Caldasia,* 19(1–2): 187 (1997)—Colombia.

DICRANOSTIGMA (Papaverac.).
erectum *K.F. Günther in Flora (Germany),* 164(2–3): 233 (1975)—China South-central (Yunnan).
fimbrilligerum *(Boiss.) C.H. An in Fl. Xinjiangensis,* 5: 481 (1999): Glaucium fimbrilligerum.
henanense *S.Y. Wang & L.H. Wu in Bull. Bot. Res. North-East. Forest. Univ.,* 17(1): 43 (1997), as 'henanensis'—China Southeast (Henan).

DICRASTYLIS (Verbenac.).
maritima *B.L. Rye & M.E. Trudgen in Nuytsia,* 12(2): 219 (1998)—Western Australia.
soliparma *B.L. Rye & M.E. Trudgen in Nuytsia,* 12(2): 225 (1998)—Western Australia.

DICTYANTHUS (Asclepiadac.).
altatensis *(Brandegee) W.D. Stevens in Novon,* 10(3): 243 (2000): Gonolobus altatensis.
asper *(Mill.) W.D. Stevens in Novon,* 9(4): 564 (1999): Cynanchum asperum.
eximius *(W.D. Stevens) W.D. Stevens in Novon,* 10(3): 243 (2000): Matelea eximia.
hamatus *(W.D. Stevens) W.D. Stevens in Novon,* 10(3): 243 (2000): Matelea hamata.
lautus *(W.D. Stevens) W.D. Stevens in Novon,* 10(3): 243 (2000): Matelea lauta.
macvaughianus *(W.D. Stevens) W.D. Stevens in Novon,* 10(3): 243 (2000): Matelea macvaughiana.
sepicola *(W.D. Stevens) W.D. Stevens in Novon,* 10(3): 243 (2000): Matelea sepicola.
suffruticosus *(W.D. Stevens) W.D. Stevens in Novon,* 10(3): 243 (2000): Matelea suffruticosa.

DIDISSANDRA (Gesneriac.).
brachycarpa *A. Weber & B.L. Burtt in Beitr. Biol. Pfl.,* 70(2–3): 217 (1997–98 publ. 1998)—Sumatera.
elongata (Jack) C.B. Clarke
subsp. burleyi *A. Weber & B.L. Burtt in Beitr. Biol. Pfl.,* 70(2–3): 211 (1997–98 publ. 1998)—Sumatera.
subsp. minor *(Ridl.) A. Weber & B.L. Burtt in Beitr. Biol. Pfl.,* 70(2–3): 211 (1997–98 publ. 1998): D. minor.
subsp. montana *(Blume) A. Weber & B.L. Burtt in Beitr. Biol. Pfl.,* 70(2–3): 213 (1997–98 publ. 1998): Ellobum montanum.
ternata *(Miq.) A. Weber & B.L. Burtt in Beitr. Biol. Pfl.,* 70(2–3): 216 (1997–98 publ. 1998): Didymocarpus ternatus.
wildeana *A. Weber & B.L. Burtt in Beitr. Biol. Pfl.,* 70(2–3): 218 (1997–98 publ. 1998)—Sumatera.

DIDYMOCARPUS (Gesneriac.).
pteronema *B.L. Burtt in Edinburgh J. Bot.,* 56(3): 376 (1999)—Burma.
subpalmatinervis *W.T. Wang in Novon,* 7(4): 428 (1997 publ. 1998)—China South-central (Yunnan).

DIDYMOPLEXIOPSIS *G. Seidenfaden, Contrib. Orchid Fl. Thailand XIII:* 13 (1997). *ORCHIDACEAE*
khiriwongensis *G. Seidenfaden, Contrib. Orchid Fl. Thailand XIII:* 13 (1997)—Thailand.

DIDYMOPLEXIS (Orchidac.).
vietnamica *P. Ormerod in Oasis,* 1(4): 15 (2000)—Vietnam.

DIEFFENBACHIA (Arac.).
beachiana *T.B. Croat & Grayum in Novon,* 9(4): 491 (1999)—Panama, Costa Rica.
concinna *T.B. Croat & Grayum in Novon,* 9(4): 492 (1999)—Costa Rica, Nicaragua.
grayumiana *T.B. Croat in Novon,* 9(4): 494 (1999)—Costa Rica, Colombia, Panama.
hammelii *T.B. Croat & Grayum in Novon,* 9(4): 496 (1999)—Costa Rica, Nicaragua.
tonduzii *T.B. Croat & Grayum in Novon,* 9(4): 497 (1999)—Panama, Costa Rica, Nicaragua, Colombia, Ecuador.

DIELSIA (Restionac.).
stenostachya *(W. Fitzg.) B.G. Briggs & L.A.S. Johnson in Telopea,* 8(1): 27 (1998): Restio stenosytachyus.

DIENIA (Orchidac.).
benguetensis *(Ames) M.A. Clements & D.L. Jones in Lasianthera,* 1(1): 41 (1996): Malaxis benguetensis.
curranii *(Ames) M.A. Clements & D.L. Jones in Lasianthera,* 1(1): 41 (1996): Malaxis curranii.
latifolia *(Sm.) M.A. Clements & D.L. Jones in Lasianthera,* 1(1): 41 (1996): Malaxis latifolia.
montana *(Blume) M.A. Clements & D.L. Jones in Lasianthera,* 1(1): 41 (1996): Gastroglottis montana.
ophrydis *(J. König) G. Seidenfaden, Contrib. Orchid Fl. Thailand XIII:* 18 (1997): Epidendrum ophrydis.
seidenfadeniana *D.L. Szlachetko, H.B. Margońska & P. Rutkowski in Adansonia,* 21(2): 221 (1999)—Sumatera, New Guinea (Irian Jaya).

DIENIA:-
truncicola *(Schltr.) M.A. Clements & D.L. Jones in Lasianthera,* 1(1): 41 (1996): Microstylis truncicola.
volkensii *(Schltr.) M.A. Clements & D.L. Jones in Lasianthera,* 1(1): 42 (1996): Microstylis volkensii.

DIGITALIS (Scrophulariac.).
minor L.
subsp. palaui *(Garcias Font & Marcos) A.M. Romo, Fl. Silvestres Baleares:* 278 (1994): D. dubia subsp. palaui.

DIGITARIA (Gramin.).
effusa *J.F. Veldkamp in Blumea,* 44(2) 447 (1999)—Vietnam.
filiformis (L.) Koeler
var. dolichophylla *(Henrard) J.K. Wipff in Phytologia,* 80(5): 348 (1996 publ. 1997): D. dolichophylla.
var. laeviglumis *(Fernald) J.K. Wipff in Phytologia,* 80(5): 348 (1996 publ. 1997): D. laeviglumis.

DILLENIA (Dilleniac.).
crenatifolia *Hoogland ex D.J. Mabberley in Gard. Bull. Singapore,* 51(2): 311 (1999), nom. nov.: Wormia crenata *A.C. Sm.*

DILLWYNIA (Legumin.).
crispii *P.C. Jobson & P.H. Weston in Telopea,* 8(3): 363 (1999)—New South Wales.
glaucula *P.C. Jobson & P.H. Weston in Telopea,* 8(1): 1 (1998)—New South Wales.
palustris *P.C. Jobson & P.H. Weston in Telopea,* 8(3): 366 (1999)—New South Wales.

DIMERIA (Gramin.).
copei *N. Ravi in Blumea,* 41(1): 251 (1996)—India (Kerala).
falcata Hack.
var. taiwaniana *(Ohwi) S.L. Chen & G.Y. Sheng in Fl. Reipubl. Popul. Sin.,* 10(2): 179 (1997): D. taiwaniana.
kalavoorensis *N. Ravi in Blumea* 41(1): 253 (1996)—India (Kerala).
mahendragiriensis *N. Ravi, H.O. Saxena & M. Brahmam in Rheedea,* 5(2): 142 (1995)—India (Orissa).
namboodiriana *N. Ravi & N. Mohanan in Rheedea,* 7(1): 1 (1997)—India (Kerala).
ornithopoda Trin.
subsp. subrobusta *(Hack.) S.L. Chen & G.Y. Sheng in Fl. Reipubl. Popul. Sin.,* 10(2): 172 (1997): D. ornithopoda var. subrobusta.
parva *(Keng & Y.L. Yang) S.L. Chen & G.Y. Sheng in Fl. Reipubl. Popul. Sin.,* 10(2): 175 (1997): D. ornithopoda var. parva.
sivarajanii *N. Mohanan & N. Ravi in Rheedea,* 6(2): 47 (1996)—India (Kerala).

DIMORPHOSTEMON (Crucif.).
sergievskiana *(Polozhij) S.V. Ovchinnikova in Fl. Sibir. (Berberidac.-Grossulariac.),* 7: 100 (1994): Torularia sergievskiana.

DINOCHLOA (Gramin.).
acutiflora *(Munro) S. Dransfield in Kew Bull.,* 53(4): 877 (1998): Schizostachyum acutiflorum.
albociliata *E.A. Widjaja in Reinwardtia,* 11(2): 74 (1997)—Sulawesi.
barbata *S. Dransfield in Kew Bull.,* 51(1): 115 (1996)—Sulawesi.
cordata *S. Dransfield in Kew Bull.,* 51(1): 108 (1996)—Sulawesi.
erecta *E.A. Widjaja in Reinwardtia,* 11(2): 74 (1997)—Sulawesi.
glabrescens *E.A. Widjaja in Reinwardtia,* 11(2): 77 (1997)—Sumatera.
hirsuta *S. Dransfield in Kew Bull.,* 51(1): 115 (1996)—Sulawesi.
indica *(R.B. Majumdar) U. Shukla & Das in U. Shukla, Grasses N.-East. India:* 208 (1996), isonym: Melocalamus indicus.
kostermansiana *S. Dransfield in Kew Bull.,* 51(1): 108 (1996)—Lesser Sunda Is.
malayana *S. Dransfield in Kew Bull.,* 51(1): 110 (1996)—Malaya (Peninsular Malaysia), Thailand.

DINOCHLOA:-
matmat *S. Dransfield & E.A. Widjaja in Kew Bull.,* 55(2): 495 (2000)—Jawa.
oblonga *S. Dransfield in Kew Bull.,* 51(1): 113 (1996)—Philippines.
palawanensis *(Gamble) S. Dransfield in Kew Bull.,* 51(1): 106 (1996): Schizostachyum palawanense.
truncata *E.A. Widjaja in Reinwardtia,* 11(2): 78 (1997)—Sulawesi.

DINOSPERMA *T.G. Hartley in Adansonia,* 19(2): 190 (1997). *RUTACEAE*
erythrococca *(F. Muell.) T.G. Hartley in Adansonia,* 19(2): 196 (1997): Euodia erythrococca.
longifolia *T.G. Hartley in Adansonia,* 19(2): 195 (1997)—Queensland.
melanophloia *(C.T. White) T.G. Hartley in Adansonia,* 19(2): 192 (1997): Melicope melanophloia.
stipitata *(C.T. White & W.D. Francis) T.G. Hartley in Adansonia,* 19(2): 192 (1997): Melicope stipitata.

DIOCLEA (Legumin.).
decandra *Amshoff ex F. Adema in Blumea,* 43(1): 234 (1998), nom. nov.: Macropsychanthus ferrugineus *Merr.*
grandistipula *L.P. de Queiroz in Novon,* 8(4): 433 (1998)—Brazil Southeast (São Paulo).
rosea *(Benth.) N. Zamora in Novon,* 10(2): 179 (2000): Cymbosema roseum.

DIONYSIA (Primulac.).
aubrietioides *Z. Jamzad & Mozaffarian in Iranian J. Bot.,* 7(1): 19 (1996)—Iran.
bazoftica *Z. Jamzad in Iranian J. Bot.,* 7(1): 20 (1996)—Iran.
iranica *Z. Jamzad in Iranian J. Bot.,* 7(1): 23 (1996)—Iran.
khuzistanica *Z. Jamzad in Iranian J. Bot.,* 7(1): 25 (1996)—Iran.
mozaffarianii *M. Lidén in Iranian J. Bot.,* 8(2): 304 (2000)—Iran.

DIOON (Zamiac.).
sonorense *(De Luca, Sabato & Vázq. Torres) J. Chemnick, T.J. Gregory & S. Salas-Morales in Phytologia,* 83(1): 3 (1997 publ. 1998), without exact basionym page: D. tomasellii var. sonorense.

DIOSCOREA (Dioscoreac.).
amazonum Mart. ex Griseb.
var. klugii *(R. Knuth) F. Ayala Flores, Dioscoreaceae del Perú:* 16 (1998): D. klugii.
andromedusae *O. Téllez-Valdés in Brittonia,* 48(1): 103 (1996)—Peru.
cachipuertensis *F. Ayala Flores, Dioscoreaceae del Perú:* 19 (1998)—Peru.
changjiangensis *F.W. Xing & Z.X. Li in Acta Bot. Austro Sinica,* 10: 19 (1995)—Hainan.
chaponensis R. Knuth
var. chazutensis *F. Ayala Flores, Dioscoreaceae del Perú:* 22 (1998)—Peru.
davidsei *O. Téllez-Valdés in Novon,* 7(2): 208 (1997)—Panama.
fasciculocongesta *(Sosa & B.G. Schub.) O. Téllez V. in Contrib. Univ. Mich. Herb.,* 21: 310 (1997): D. spiculiflora var. fasciculocongesta.
ferreyrae *F. Ayala Flores, Dioscoreaceae del Perú:* 27 (1998)—Peru.
gentryi *O. Téllez-Valdés in Brittonia,* 48(1): 100 (1996)—Peru.
glandulosa Klotzsch ex Kunth
var. calcencis *(R. Knuth) F. Ayala Flores, Dioscoreaceae del Perú:* 29 (1998): D. calcencis.
gomez-pompae *O. Téllez V. in Contrib. Univ. Mich. Herb.,* 21: 309 (1997), nom. nov.: D. spiculiflora var. chiapasana *Gómez Pompa.*
javariensis *F. Ayala Flores, Dioscoreaceae del Perú:* 33 (1998)—Peru.
karatana *P. Wilkin in Kew Bull.,* 55(2): 427 (2000)—Madagascar.
lijiangensis *C.L. Long & Y.H. Li in Ann. Bot. Fenn.,* 37(3): 193 (2000)—China South-central (Yunnan).
monadelphoides
var. longirachis *(R. Knuth) F. Ayala Flores, Dioscoreaceae del Perú:* 38 (1998): D. longirachis.

DIOSCOREA:-
natalia *B.E. Hammel in Novon,* 10(4): 378 (2000)—Costa Rica.
samydea Mart. ex Griseb.
var. poeppigii *(Kunth) F. Ayala Flores, Dioscoreaceae del Perú:* 47 (1998): D. poeppigii.
septemloba Thunb.
var. platyphylla *M. Mizush. ex T. Shimizu, Fl. Nagano Pref.:* 1511 (1997)—Japan.
sinoparviflora *C.T. Ting, M.G. Gilbert & N.J. Turland in Novon,* 10(1): 13 (2000), nom. nov.: D. parviflora *C.T. Ting.*

DIOSPYROS (Ebenac.).
arupaj *B. Wallnöfer in Ann. Naturhist. Mus. Wien,* 102B: 418 (2000)—Brazil North (Amazonas).
cavalcantei *C. Sothers in Kew Bull.,* 55(2): 471 (2000)—Brazil North (Amazonas, Amapá, Pará), French Guiana.
dinhensis *N.T. Hiêp in Konsp. Sosud. Rast. Fl. V'etnama,* 2: 65 (1996), nom. nov.: Maba pierrei *Lecomte.*
domarkind *B. Wallnöfer in Ann. Naturhist. Mus. Wien,* 101B: 566 (1999)—Colombia.
ekodul *B. Wallnöfer in Ann. Naturhist. Mus. Wien,* 101B: 567 (1999)—Ecuador.
esmereg *B. Wallnöfer in Ann. Naturhist. Mus. Wien,* 101B: 568 (1999)—Ecuador.
fenal *B. Wallnöfer in Ann. Naturhist. Mus. Wien,* 101B: 570 (1999)—Venezuela.
froesii *P.B. Cavalcante in Acta Amazonica,* 7(2): 191 (1977)—Brazil North (Amazonas).
gallo *B. Wallnöfer in Ann. Naturhist. Mus. Wien,* 102B: 419 (2000)—Venezuela.
guianensis (Aubl.) Gürke
subsp. coriacea *F. White in Mem. New York Bot. Gard.,* 32: 328 (1981)—Venezuela.
haivanensis *N.T. Hiêp in Konsp. Sosud. Rast. Fl. V'etnama,* 2: 67 (1996), nom. nov.: Maba nebulorum *Lecomte.*
hartmanniana *S. Knapp in Novon,* 7(3): 256 (1997)—Panama, Costa Rica.
kanizur *B. Wallnöfer in Ann. Naturhist. Mus. Wien,* 101B: 572 (1999)—Bolivia.
katendei *B. Verdcourt in Fl. Trop. E. Afr., Ebenac.:* 42 (1996)—Uganda.
ketun *B. Wallnöfer in Ann. Naturhist. Mus. Wien,* 101B: 573 (1999)—Brazil Southeast (Minas Gerais).
kolom *B. Wallnöfer in Ann. Naturhist. Mus. Wien,* 101B: 575 (1999)—Colombia.
kondor *B. Wallnöfer in Ann. Naturhist. Mus. Wien,* 102B: 421 (2000)—Brazil North (Amazonas, Pará).
kupensis *G. Gosline in Kew Bull.,* 53(2): 461 (1998)—Cameroon.
landii *P.B. Cavalcante in Acta Amazonica,* 7(2): 191 (1977)—Brazil North (Acre).
loureiriana G. Don
subsp. rufescens *(Caveney) B. Verdcourt in Fl. Trop. E. Afr., Ebenac.:* 26 (1996): D. usambarensis subsp. rufescens.
manu *B. Wallnöfer in Ann. Naturhist. Mus. Wien,* 101B: 576 (1999)—Peru.
miltonii *P.B. Cavalcante in Acta Amazonica,* 7(2): 194 (1977)—Brazil North (Pará).
morenoi *A. Pool in Novon,* 7(2): 189 (1997)—Nicaragua.
nanay *B. Wallnöfer in Ann. Naturhist. Mus. Wien,* 101B: 577 (1999)—Peru, Brazil North (Amazonas).
nenab *B. Wallnöfer in Ann. Naturhist. Mus. Wien,* 101B: 579 (1999)—Colombia.
nur *B. Wallnöfer in Ann. Naturhist. Mus. Wien,* 101B: 581 (1999)—Bolivia.
ottohuberi *B. Wallnöfer in Ann. Naturhist. Mus. Wien,* 102B: 425 (2000)—Venezuela.
panamensis *S. Knapp in Novon,* 9(4): 524 (1999), nom. nov., as 'panamense': D. whitei *S. Knapp.*
panguana *B. Wallnöfer in Ann. Naturhist. Mus. Wien,* 101B: 583 (1999)—Peru.
pemadasae *A.H.M. Jayasuriya in J. S. Asian Nat. Hist.,* 3(1): 55 (1998), as 'pemadasai'—Sri Lanka.
pilosula Wall. ex Hiern
var. andamanensis *U. Jayaraman & M.P. Nayar in Bull. Bot. Surv. India,* 34(1–4): 180 (1992 publ. 1997)—Andaman Is.
piresii *P.B. Cavalcante in Acta Amazonica,* 7(2): 189 (1977)—Brazil North (Acre).

DIOSPYROS:-
pseudoharmandii *N.T. Hiêp in Konsp. Sosud. Rast. Fl. V'etnama,* 2: 71 (1996), nom. nov.: Maba harmandii *Lecomte.*
riedelii *(Hiern) B. Wallnöfer in Ann. Naturhist. Mus. Wien,* 102B: 427 (2000): Diospyros peruviana var. riedelii.
ropourea *B. Wallnöfer in Ann. Naturhist. Mus. Wien,* 102B: 427 (2000), nom. nov.: Ropourea guianensis *Aubl.*
rotok *B. Wallnöfer in Ann. Naturhist. Mus. Wien,* 101B: 584 (1999)—Peru.
saxatilis *S.K. Lee in Guihaia,* 18(1): 1 (1998)—China Southeast (Guangxi).
tarim *B. Wallnöfer in Ann. Naturhist. Mus. Wien,* 102B: 429 (2000)—Brazil North (Amazonas).
tepu *B. Wallnöfer in Ann. Naturhist. Mus. Wien,* 102B: 432 (2000), nom. nov.: Diospyros guianensis subsp. coriacea *F. White.*
tero *B. Wallnöfer in Ann. Naturhist. Mus. Wien,* 101B: 586 (1999)—Ecuador.
thorelii *(Lecomte) N.T. Hiêp in Konsp. Sosud. Rast. Fl. V'etnama,* 2: 73 (1996): Maba thorelii.
uaupensis *P.B. Cavalcante in Acta Amazonica,* 7(2): 197 (1977)—Brazil North (Amazonas).
urep *B. Wallnöfer in Ann. Naturhist. Mus. Wien,* 101B: 587 (1999)—Peru.
uzungwaensis *C. Frimodt-Møller & H.J. Ndangalasi in Kew Bull.,* 55(2): 465 (2000)—Tanzania.
whitei *F. Dowsett-Lemaire & C.M. Pannell in Bull. Jard. Bot. Nation. Belg.,* 65(3–4): 399 (1996)—Congo.
whitei *Sandra Knapp in Novon,* 7(3): 258 (1997), homonym; *see Dowsett-Lemaire & Pannell in Bull. Jard. Bot. Nation. Belg.,* 65(3–4): 399 (1996)—Panama.
yomomo *B. Wallnöfer in Ann. Naturhist. Mus. Wien,* 101B: 589 (1999)—Bolivia.

DIPCADI (Hyacinthac.).
biflorum *S.A. Ghazanfar in Kew Bull.,* 51(4): 805 (1996)—Oman, Gulf States (United Arab Emirates).

DIPLACTIS (Compos.).
alpina *(L.) J.C. Semple in Univ. Waterloo Biol. Ser.,* 38: 17 (1996): Aster alpinus.
subsp. vierhapperi *(Onno) J.C. Semple in Univ. Waterloo Biol. Ser.,* 38: 17 (1996): Aster alpinus subsp. vierhapperi.
tongolensis *(Franch.) J.C. Semple in Univ. Waterloo Biol. Ser.,* 38: 17 (1996): Aster tongolensis.

DIPLASPIS (Umbellif.).
nivis *A. Van den Borre & M.J. Henwood in Austral. Syst. Bot.,* 11(1): 10 (1998), contrary to Art. 37.5 ICBN (1994); *et in Austral. Syst. Bot.,* 11(5–6): 739 (1999), with type herbarium specified—New South Wales, Victoria.

DIPLOCAULOBIUM (Orchidac.).
obyrnei *W.K. Harris in Lasianthera,* 1(3): 148 (1997 publ. 1998)—New Guinea (Papua New Guinea).

DIPLOLAENA (Rutac.).
cinerea *Paul G. Wilson in Nuytsia,* 12(1): 112 (1998)—Western Australia.
eneabbensis *Paul G. Wilson in Nuytsia,* 12(1): 113 (1998)—Western Australia.
geraldtonensis *Paul G. Wilson in Nuytsia,* 12(1): 113 (1998)—Western Australia.
graniticola *Paul G. Wilson in Nuytsia,* 12(1): 114 (1998)—Western Australia.
leemaniana *Paul G. Wilson in Nuytsia,* 12(1): 115 (1998)—Western Australia.
mollis *Paul G. Wilson in Nuytsia,* 12(1): 116 (1998)—Western Australia.
obovata *Paul G. Wilson in Nuytsia,* 12(1): 117 (1998)—Western Australia.
velutina *(Paul G. Wilson) Paul G. Wilson in Nuytsia,* 12(1): 118 (1998): D. microcephala var. velutina.

DIPLOPELTIS (Sapindac.).
huegelii Endl.
subsp. lehmannii *(Miq.) G.J. Keighery in Nuytsia,* 12(2): 290 (1998): D. lehmannii.
subsp. subintegra *(A.S. George) G.J. Keighery in Nuytsia,* 12(2): 290 (1998): D. huegelii var. subintegra.

DIPLOPRORA (Orchidac.).
championii (Lindl.) Hook. f.
var. uraiensis *(Hayata) S.S. Ying in Mem. Coll. Agric. Nation. Taiwan Univ.,* 28(2): 36 (1988): D. uraiensis.

DIPLOSTEPHIUM (Compos.).
fernandez-alonsoi *S. Díaz Piedrahita in Rev. Acad. Colomb. Cienc. Exact. Fis. Nat.,* 81: 405 (1997)—Colombia.
perijaense *S. Díaz Piedrahita & G.P. Méndez Ramírez in Rev. Acad. Colomb. Cienc. Exact. Fis. Nat.,* 81: 406 (1997)—Colombia.

DIPLOTAXIS (Crucif.).
sect. Heterocarpum *J. Martínez Laborde in An. Jard. Bot. Madrid,* 56(2): 380 (1998).
sect. Heteropetalum *J. Martínez Laborde in An. Jard. Bot. Madrid,* 56(2): 380 (1998).
subgen. Rhynchocarpum *(Prantl) J. Martínez Laborde in An. Jard. Bot. Madrid,* 56(2): 380 (1998): D. sect. Rhynchocarpum.
antoniensis *Ø.H. Rustan in Nordic J. Bot.,* 16(1): 35 (1996)—Cape Verde.
decipiens *M. Gandoger, Dec. Pl. Nov.,* 2: 7 (1876)—Corse.
gorgadensis *Ø.H. Rustan in Nordic J. Bot.,* 16(1): 38 (1996)—Cape Verde.
subsp. brochmannii *Ø.H. Rustan in Nordic J. Bot.,* 16(1): 41 (1996)—Cape Verde.
leiocarpa *M. Gandoger, Dec. Pl. Nov.,* 2: 6 (1876)—France.
longisiliqua *M. Gandoger, Dec. Pl. Nov.,* 2: 7 (1876)—France.
ruscinonensis *M. Gandoger, Dec. Pl. Nov.,* 2: 6 (1876)—France.
sagotii *M. Gandoger, Dec. Pl. Nov.,* 2: 6 (1876)—Spain.
sundingii *Ø.H. Rustan in Nordic J. Bot.,* 16(1): 47 (1996)—Cape Verde.
tenuifolia (L.) DC.
subsp. cretacea *(Kotov) E. Sobrino Vesperinas in An. Jard. Bot. Madrid,* 54(1): 188 (1996): D. cretacea.
varia *Ø.H. Rustan in Nordic J. Bot.,* 16(1): 47 (1996)—Cape Verde.
viminea (L.) DC.
subsp. cretacea *(Kotov) N.N. Tzvelev in Novosti Sist. Vyssh. Rast.,* 32: 184 (2000): D. cretacea.
subsp. intermedia *(Schur) N.N. Tzvelev in Novosti Sist. Vyssh. Rast.,* 32: 184 (2000): D. intermedia.

DIPLUSODON (Lythrac.).
aggregatifolius *T.B. Cavalcanti in Novon,* 8(4): 337 (1998)—Brazil Southeast (Minas Gerais).
alatus *T.B. Cavalcanti in Novon,* 8(4): 339 (1998)—Brazil West-central (Goiás).
argyrophyllus *T.B. Cavalcanti in Novon,* 8(4): 339 (1998)—Brazil Northeast (Bahia).
bolivianus *T.B. Cavalcanti & S.A. Graham in Novon,* 6(3): 253 (1996)—Bolivia.
ciliatiflorus *T.B. Cavalcanti in Novon,* 8(4): 342 (1998)—Brazil West-central (Goiás).
mattogrossensis *T.B. Cavalcanti in Novon,* 8(4): 342 (1998)—Brazil West-central (Mato Grosso).
oblongus Pohl
var. angustifolius *T.B. Cavalcanti in Novon,* 8(4): 345 (1998)—Brazil West-central (Goiás).
plumbeus *T.B. Cavalcanti in Novon,* 8(4): 345 (1998)—Brazil West-central (Goiás).
pygmaeus *T.B. Cavalcanti in Novon,* 8(4): 347 (1998)—Brazil West-central (Goiás).
trigintus *T.B. Cavalcanti in Novon,* 8(4): 347 (1998)—Brazil North (Tocantins).
ulei Koehne
var. ciliatus *T.B. Cavalcanti in Novon,* 8(4): 349 (1998)—Brazil Northeast (Bahia).

DIPODIUM (Orchidac.).
pardalinum *D.L. Jones in Muelleria,* 9: 105 (1996)—Victoria, South Australia.

DIPTERANTHUS (Orchidac.).
densiflorus *K. Senghas in Caesiana,* 8: 21 (1997)—Bolivia.
multiflorus *(Rolfe) I. Bock & Seehawer in Orchidee,* 47(6): 282 (1996): Ornithocephalus multiflorus.

DIPTEROSTELE (Orchidac.).
boylei *(J.T. Atwood) L.A. Garay & G.A. Romero-González in Harvard Pap. Bot.,* 3(1): 57 (1998): Stellilabium boylei.
distantiflorum *(Ames & C. Schweinf.) L.A. Garay & G.A. Romero-González in Harvard Pap. Bot.,* 3(1): 57 (1998): Stellilabium distantiflorum.
frymirei *(Dodson) L.A. Garay & G.A. Romero-González in Harvard Pap. Bot.,* 3(1): 57 (1998): Stellilabium frymirei.
helleri *(L.O. Williams) L.A. Garay & G.A. Romero-González in Harvard Pap. Bot.,* 3(1): 57 (1998): Stellilabium helleri.
hystrix *(Dodson) L.A. Garay & G.A. Romero-González in Harvard Pap. Bot.,* 3(1): 57 (1998): Stellilabium hystrix.
lankesteri *(Ames) L.A. Garay & G.A. Romero-González in Harvard Pap. Bot.,* 3(1): 57 (1998): Telipogon lankesteri.
lueri *(Dodson & R. Vásquez) L.A. Garay & G.A. Romero-González in Harvard Pap. Bot.,* 3(1): 57 (1998): Stellilabium lueri.
standleyi *(Ames) L.A. Garay & G.A. Romero-González in Harvard Pap. Bot.,* 3(1): 58 (1998): Telipogon standleyi.
tanii *(Dodson) L.A. Garay & G.A. Romero-González in Harvard Pap. Bot.,* 3(1): 58 (1998): Stellilabium tanii.
vasquezii *(Dodson) L.A. Garay & G.A. Romero-González in Harvard Pap. Bot.,* 3(1): 58 (1998): Stellilabium vasquezii.

DIPTERYX (Legumin.).
casiquiarensis *(Pittier) G.P. Lewis & P.E. Gasson in Kew Bull.,* 55(1): 248 (2000): Taralea casiquiarensis.

DISA (Orchidac.).
sect. Herschelianthe *(Rauschert) H.P. Linder in H.P. Linder & H. Kurzweil, Orchids of Southern Africa:* 255 (1999): gen. Herschelianthe.
chimanimaniensis *(H.P. Linder) H.P. Linder in H.P. Linder & H. Kurzweil, Orchids of Southern Africa:* 470 (1999): Herschelia chimanimaniensis.
cochlearis *S.D. Johnson & W.R. Liltved in S. Afr. J. Bot.,* 63(5): 291 (1997)—Cape Province.
ecalcarata *(G.J. Lewis) H.P. Linder in H.P. Linder & H. Kurzweil, Orchids of Southern Africa:* 224 (1999): Monadenia ecalcarata.
introrsa *H. Kurzweil, W.R. Liltved & H.P. Linder in Nordic J. Bot.,* 17(4): 353 (1997)—Cape Province.
lugens Bolus
var. nigrescens *(H.P. Linder) H.P. Linder in H.P. Linder & H. Kurzweil, Orchids of Southern Africa:* 263 (1999): Herschelia lugens var. nigrescens.
nubigena *H.P. Linder in Flow. Pl. Afr.,* 56: t. 2148 (1999)—Cape Province.
praecox *(H.P. Linder) H.P. Linder in H.P. Linder & H. Kurzweil, Orchids of Southern Africa:* 470 (1999): Herschelia praecox.
procera *H.P. Linder in H.P. Linder & H. Kurzweil, Orchids of Southern Africa:* 265 (1999), as 'nom. nov.', without type—Cape Province.
spathulata (L.f.) Sw.
subsp. tripartita *(Lindl.) H.P. Linder in H.P. Linder & H. Kurzweil, Orchids of Southern Africa:* 259 (1999): D. tripartita.
virginalis *H.P. Linder in Novon,* 8(4): 405 (1998)—Cape Province.

DISCHIDIA (Asclepiadac.).
hoyella *R. Omlor, Gen. Revis. Marsdenieae (Asclepiadac.):* 143 (1998), nom. nov.: Hoyella rosea *Ridl.*
polilloensis *D. Kloppenburg in Fraterna,* 10(4): 5 (1997)—Philippines.

DISCHISTOCALYX (Acanthac.).
champluvieranus *J. Lejoly & S. Lisowski in Bull. Jard. Bot. Nation. Belg.,* 67(1–4): 107 (1999)—Equatorial Guinea.

DISCIPHANIA (Menispermac.).
moriorum *R.C. Barneby in Brittonia,* 48(1): 20 (1996)—French Guiana.

DISCOCACTUS (Cactac.).
crassispinus P.J. Braun & Esteves
subsp. araguaiensis *P.J. Braun & E. Esteves Pereira in Kakt. And. Sukk.*, 47(9): 194 (1996)—Brazil West-central (Goiás).
hartmannii (K. Schum.) Britton & Rose
subsp. giganteus *P.J. Braun & E. Esteves Pereira in Succulenta*, 75(6): 260 (1996)—Brazil West-central (Mato Grosso do Sul).
heptacanthus
subsp. catingicola *(Buining & Brederoo) N.P. Taylor & D.C. Zappi in Cactaceae Consensus Initiatives*, 3: 7 (1997): D. catingicola.
zehntneri
var. albispinus *(Buining & Brederoo) P.J. Braun in Succulenta*, 69(10): 215 (1990): D. albispinus.
var. araneispinus *(Buining & Brederoo ex J. Theun.) P.J. Braun in Succulenta*, 69(10): 215 (1990): D. araneispinus.
var. boomianus *(Buining & Brederoo) P.J. Braun in Succulenta*, 69(10): 218 (1990): D boomianus.
var. horstiorum *P.J. Braun in Succulenta*, 69(10): 218 (1990)—Brazil Northeast (Bahia).
gentryi *S.M. Hayden in Ann. Missouri Bot. Gard.*, 83(2): 159 (1996)—Peru, Brazil North (Amazonas).

DISCRETITHECA *P.D. Cantino in Syst. Bot.*, 23(3): 379 (1998 publ. 1999). *VERBENACEAE*
nepalensis *(Moldenke) P.D. Cantino in Syst. Bot.*, 23(3): 381 (1998 publ. 1999): Caryopteris nepalensis.

DISKYPHOGYNE *D.L. Szlachetko & R. González Tamayo in Fragm. Flor. Geobot.*, 41(1): 494 (1996). *ORCHIDACEAE*
arechavaletae *(Kraenzl.) D.L. Szlachetko & R. González Tamayo in Fragm. Flor. Geobot.*, 41(1): 495 (1996): Spiranthes arechavaletae.
scabrilingua *(Szlach.) D.L. Szlachetko & R. González Tamayo in Fragm. Flor. Geobot.*, 41(1): 495 (1996): Brachystele scabrilingua.
widgrenii *(Rchb. f.) D.L. Szlachetko & R. González Tamayo in Fragm. Flor. Geobot.*, 41(1): 495 (1996): Acraea widgrenii.

DISOCACTUS (Cactac.).
aurantiacus (Kimnach) Barthlott
var. blomianus *(Kimnach) E. Meier in Kakt. And. Sukk.*, 48(1): 7 (1997): Heliocereus aurantiacus var. blomianus.

DISPERIS (Orchidac.).
aphylla Kraenzl.
var. bifolia *(Verdc.) D.L. Szlachetko & T.S. Olszewski in Fl. Cameroun*, 34: 42 (1998): D. aphylla subsp. bifolia.

DISPOROPSIS (Convallariac.).
undulata *M.N. Tamura & M. Ogisu in Acta Phytotax. Geobot.*, 49(1): 34 (1998)—China South-central (Sichuan).

DISPORUM (Convallariac.).
nantouense *S.S. Ying in Mem. Coll. Agric. Nation. Taiwan Univ.*, 30(2): 59 (1990)—Taiwan.

DISSANTHELIUM (Gramin.).
amplivaginatum *O. Tovar in Publ. Mus. Hist. Nat. Javier Prado, B*, 33: 7 (1985)—Peru.
atropidiforme *(Hack.) R.J. Soreng in Novon*, 8(2): 200 (1998): Poa atropidiformis.
var. patagonicum *(Parodi) R.J. Soreng in Novon*, 8(2): 200 (1998): D. patagonicum.
calycinum
subsp. mathewsii *(Ball) R.J. Soreng in Novon*, 8(2): 201 (1998): Deschampsia mathewsii.
giganteum *O. Tovar in Publ. Mus. Hist. Nat. Javier Prado, B*, 33: 8 (1985)—Peru.
longifolium *O. Tovar in Publ. Mus. Hist. Nat. Javier Prado, B*, 33: 9 (1985)—Peru.

DISSILIARIA (Euphorbiac.).
indistincta *P.I. Forster in Austrobaileya*, 5(1): 17 (1997)—Queensland.
surculosa *P.I. Forster in Austrobaileya*, 5(1): 22 (1997)—Queensland.
tuckeri *P.I. Forster in Austrobaileya*, 5(1): 19 (1997)—Queensland.

DISTEPHANUS (Compos.).
henryi *(Dunn) H. Robinson in Proc. Biol. Soc. Wash.*, 112(1): 238 (1999): Vernonia henryi.
qazmi *N. Kilian & A.G. Miller in Willdenowia*, 30(1): 84 (2000)—Socotra.

DISTERIGMA (Ericac.).
ecuadorense *J.L. Luteyn in Fl. Ecuador*, 54(147): 280 (1996)—Ecuador.
microphyllum *(G. Don) J.L. Luteyn in Fl. Ecuador*, 54(147): 261 (1996): Gaylussacia microphylla.
noyesiae *J.L. Luteyn in Fl. Ecuador*, 54(147): 272 (1996)—Ecuador.
pseudokillipiella *J.L. Luteyn in Fl. Ecuador*, 54(147): 285 (1996)—Colombia, Ecuador.
rimbachii *(A.C. Sm.) J.L. Luteyn in Fl. Ecuador*, 54(147): 258 (1996): Ceratostema rimbachii.
stereophyllum *(A.C. Sm.) J.L. Luteyn in Fl. Ecuador*, 54(147): 276 (1996): Killipiella stereophylla.

DISTICHIRHOPS *R.M.A.P. Haegens in Blumea*, Suppl. 12: 193 (2000). *EUPHORBIACEAE*
megale *R.M.A.P. Haegens in Blumea*, Suppl. 12: 194 (2000)—New Guinea (Papua New Guinea).
minor *R.M.A.P. Haegens in Blumea*, Suppl. 12: 195 (2000)—Borneo (Sabah), New Guinea (Irian Jaya).
mitsemosik *R.M.A.P. Haegens in Blumea*, Suppl. 12: 196 (2000)—New Guinea (Papua New Guinea).

DITASSA (Asclepiadac.).
buntingii *(Morillo) S. Liede in Novon*, 7(1): 44 (1997): Cynanchum buntingii.
farneyi *N. Marquete Ferreira da Silva & M. da C. Valente in Rodriguésia*, 42–44(68–70): 22 (1990–92 publ. 1996)—Brazil Southeast (Rio de Janeiro).
fontellae *N. Marquete Ferreira da Silva & M. da C. Valente in Rodriguésia*, 42–44(68–70): 21 (1990–92 publ. 1996)—Brazil Southeast (Rio de Janeiro).
franciscoi *(Morillo) S. Liede in Novon*, 7(1): 44 (1997): Cynanchum franciscoi.
jahnii *(Morillo) S. Liede in Novon*, 7(1): 44 (1997): Cynanchum jahnii.
oberdanii *J. Fontella Pereira & M.C. Alvarez Pereira in Novon*, 8(3): 239 (1998)—Brazil Southeast (Espírito Santo).
sobradoi *(Morillo) S. Liede in Novon*, 7(1): 44 (1997): Cynanchum sobradoi.

DITAXIS (Euphorbiac.).
argothamnoides *(Bertero ex Spreng.) A. Radcliffe-Smith & R. Govaerts in Kew Bull.*, 52(2): 479 (1997): Croton argothamnoides.
arlynniana *(J.W. Ingram) A. Radcliffe-Smith & R. Govaerts in Kew Bull.*, 52(2): 479 (1997): Argythamnia arlynniana.
clariana *(Jeps.) G.L. Webster in Aliso*, 16(1): 25 (1997): Argythamnia clariana.
dressleriana *(J.W. Ingram) A. Radcliffe-Smith & R. Govaerts in Kew Bull.*, 52(2): 479 (1997): Argythamnia dressleriana.
gentryi *(J.W. Ingram) A. Radcliffe-Smith & R. Govaerts in Kew Bull.*, 52(2): 479 (1997): Argythamnia gentryi.
guatemalensis (Müll. Arg.) Pax & K. Hoffm.
var. barrancana *(McVaugh) A. Radcliffe-Smith & R. Govaerts in Kew Bull.*, 52(2): 479 (1997): Argythamnia guatemalensis var. barrancana.
humilis (Engelm. & A. Gray) Pax
var. leiosperma *(Waterf.) A. Radcliffe-Smith & R. Govaerts in Kew Bull.*, 52(2): 479 (1997): Argythamnia humilis var. leiosperma.
micrandra *(Croizat) A. Radcliffe-Smith & R. Govaerts in Kew Bull.*, 52(2): 479 (1997): Argythamnia micrandra.
serrata (Torr.) A. Heller
var. californica *(Brandegee) V.W. Steinmann & R.S. Felger in Madroño*, 42(4): 456 (1995 publ. 1996): Argythamnia californica.
simulans *(J.W. Ingram) A. Radcliffe-Smith & R. Govaerts in Kew Bull.*, 52(2): 479 (1997): Argythamnia simulans.

DITRYSINIA (Euphorbiac.).
fruticosa *(W. Bartram) R. Govaerts & D.G. Frodin in R. Govaerts, D.G. Frodin & A. Radcliffe-Smith, World Checklist & Bibliogr.* Euphorbiaceae, 2: 586 (2000): Stillingia fruticosa.

DIURIS (Orchidac.).
arenaria *D.L. Jones in Orchadian*, 12(12): 567 (1999)—New South Wales.
chryseopsis *D.L. Jones in Austral. Orchid Res.*, 3: 74 (1998)—Tasmania, New South Wales, Victoria.
monticola *D.L. Jones in Austral. Orchid Res.*, 3: 76 (1998)—New South Wales, Victoria, Tasmania.
orientis *D.L. Jones in Austral. Orchid Res.*, 3: 77 (1998)—Tasmania, New South Wales, Victoria, South Australia.

DOCKRILLIA (Orchidac.).
bowmanii *(Benth.) M.A. Clements & D.L. Jones in Lasianthera*, 1(1): 17 (1996): Dendrobium bowmanii.
brevicauda *(D.L. Jones & M.A. Clem.) M.A. Clements & D.L. Jones in Lasianthera*, 1(1): 17 (1996): Dendrobium brevicaudum.
calamiformis *(G. Lodd.) M.A. Clements & D.L. Jones in Lasianthera*, 1(1): 17 (1996): Dendrobium calamiforme.
casuarinae *(Schltr.) M.A. Clements & D.L. Jones in Lasianthera*, 1(1): 17 (1996): Dendrobium casuarinae.
caudiculata *M.A. Clements & D.L. Jones in Lasianthera*, 1(1): 10 (1996)—New Guinea (Papua New Guinea).
convoluta *M.A. Clements & D.L. Jones in Lasianthera*, 1(1): 11 (1996)—New Guinea (Papua New Guinea).
delicata *M.A. Clements & D.L. Jones in Lasianthera*, 1(1): 12 (1996)—New Guinea (Papua New Guinea).
dolichophylla *(D.L. Jones & M.A. Clem.) M.A. Clements & D.L. Jones in Lasianthera*, 1(1): 17 (1996): Dendrobium dolichophyllum.
fuliginosa *M.A. Clements & D.L. Jones in Lasianthera*, 1(1): 13 (1996)—New Guinea (Papua New Guinea).
hepatica *M.A. Clements & D.L. Jones in Lasianthera*, 1(1): 14 (1996)—New Guinea (Papua New Guinea).
nothofageti *M.A. Clements & D.L. Jones in Lasianthera*, 1(1): 16 (1996)—New Guinea (Papua New Guinea).
nugentii *(F.M. Bailey) M.A. Clements & D.L. Jones in Lasianthera*, 1(1): 18 (1996): Dendrobium linguiforme var. nugentii.
schoenina *(Lindl.) M.A. Clements & D.L. Jones in Lasianthera*, 1(1): 18 (1996): Dendrobium schoeninum.
striolata (Rchb. f.) Rauschert
subsp. chrysantha *D.L. Jones in Austral. Orchid Res.*, 3: 9 (1998)—Tasmania.

DOLICHOS (Legumin.).
brevidentatus *B. Mackinder in Kew Bull.*, 54(2): 421 (1999), nom. nov.: D. pseudocajanus var. paniculatus *R. Wilczek*.
ichthyophone Verdc.
var. sunzuensis *B. Mackinder in Kew Bull.*, 54(2): 415 (1999)—Zambia.
subcapitatus R. Wilczek
var. angustifolius *B. Mackinder in Kew Bull.*, 54(2): 418 (1999)—Zambia.
trilobus L.
var. richardsiae *B. Mackinder in Kew Bull.*, 54(2): 416 (1999)—Zambia.
ufiomiensis *W. Heering in Mitt. Geogr. Ges. Hamburg*, 29: 211 (1915)—Tanzania.

DOLICHOTHELE (Cactac.).
carretii *(Rebut ex K. Schum.) A.B. Doweld in Sukkulenty (Moscow)*, 3(1–2): 37 (2000): Mammillaria carretii.

DOLIOCARPUS (Dilleniac.).
aracaensis *G. Aymard C. in Kew Bull.*, 53(3): 713 (1998)—Brazil North (Amazonas).
aureobaccus *G. Aymard C. in Kew Bull.*, 53(3): 715 (1998)—Brazil North (Roraima).
foreroi *G.A. Aymard C. in An. Jard. Bot. Madrid*, 55(1): 22 (1997)—Colombia.
kubitzkii *G. Aymard C. in Kew Bull.*, 53(3): 717 (1998)—Brazil West-central (Mato Grosso).
lopez-palacii *G.A. Aymard C. in An. Jard. Bot. Madrid*, 55(1): 26 (1997)—Colombia.

DOLIOCARPUS:-
pruskii *G. Aymard C. in Kew Bull.*, 53(3): 719 (1998)—Brazil North (Amapá).
subandinus *G.A. Aymard C. in Brittonia*, 52(2): 196 (2000)—Bolivia, Ecuador, Peru.

DOMBEYA (Sterculiac.).
kefaensis *I. Friis & S. Bidgood in Nordic J. Bot.*, 18(2): 217 (1998)—Ethiopia.
reginae *Jacques in Ann. Fl. Pomone*, 1832–33: 57 (1832)—Madagascar.

DONIOPHYTON (Compos.).
weddellii *L. Katinas & T.F. Stuessy in Pl. Syst. Evol.*, 206(1–4): 38 (1997)—Chile North (Coquimbo, Antofagasta, Atacama), Argentina Northwest (Catamarca, Jujuy, Mendoza, Salta, San Juan).

DONTOSTEMON (Crucif.).
pinnatifidus *(Willd.) I.A. Al-Shehbaz & H. Ohba in Novon*, 10(2): 96 (2000): Cheiranthus pinnatifidus.
subsp. linearifolius *(Maxim.) I.A. Al-Shehbaz & H. Ohba in Novon*, 10(2): 97 (2000): Sisymbrium glandulosum var. linearifolium.
tibeticus *(Maxim.) I.A. Al-Shehbaz in Novon*, 10(4): 334 (2000): Nasturtium tibeticum.

DOPATRIUM (Scrophulariac.).
tenerum *(Hiern) Eb. Fischer in Nordic J. Bot.*, 17(5): 538 (1997): Stemodiacra tenera.

DORATOXYLON (Sapindac.).
apetalum (Poir.) Radlk.
var. diphyllum *(Cordem.) F. Friedmann in Fl. Mascar.*, 76: 19 (1997), without basionym date: Hippobromus apetalus var. diphyllus.

× **DOREENHUNTARA** *hort. in Orchid Rev.*, 107(1225): 62 (1999). *ORCHIDACEAE* (Bollea × Cochleanthes × Kefersteinia × Pescatorea).

DOREMA (Umbellif.).
aitchisonii *Korovin ex M.G. Pimenov in Opred. Rast. Sred. Azii*, 7: 377 (1983)—Tadzhikistan.
badhysi *M.G. Pimenov in Opred. Rast. Sred. Azii*, 7: 379 (1983)—Afghanistan.
balchanorum *M.G. Pimenov in Opred. Rast. Sred. Azii*, 7: 378 (1983)—Turkmenistan.
kopetdaghense *M.G. Pimenov in Opred. Rast. Sred. Azii*, 7: 378 (1983)—Turkmenistan.

DORITIS (Orchidac.).
pulcherrima Wall. ex Lindl.
f. alba *O. Gruss & J. Roeth in Orchidee*, 50(6): 671 (1999)—Thailand.
f. albiflora *(Rchb. f.) J. Roeth & O. Gruss in Orchidee*, 50(6): 671 (1999): Phalaenopsis esmeralda var. albiflora.
f. coerulea *(Fowlie) O. Gruss & J. Roeth in Orchidee*, 50(6): 671 (1999): D. pulcherrima var. coerulea.

DOROBAEA (Compos.).
callacallensis *(Cuatrec.) B. Nordenstam & J.F. Pruski in Compositae Newsl.*, 27: 33 (1995): Senecio callacallensis.
laciniata *(Humb., Bonpl. & Kunth) B. Nordenstam & J.F. Pruski in Compositae Newsl.*, 27: 33 (1995): Senecio laciniatus.

DORONICUM (Compos.).
bargusinense Serg.
var. pilosum *C.H. An in Fl. Xinjiangensis*, 5: 477 (1999)—Xinjiang.
cavillieri *I. Álvarez Fernández & G. Nieto Feliner in Ann. Bot. Fenn.*, 37(4): 250 (2000)—China North-Central (Gansu), China South-Central (Sichuan).
conaense *Y.L. Chen in Acta Phytotax. Sin.*, 36(1): 75 (1998)—Xizang.
diazii *C. Pérez Morales & A. Penas Merino in Lagascalia*, 15(2): 155 (1990)—Spain.
gansuense *Y.L. Chen in Acta Phytotax. Sin.*, 36(1): 73 (1998)—China North-central (Gansu).

DORONICUM:-
grandiflorum Lam.
var. pyrenaicum *(Gay ex Gren. & Godr.) O. de Bolòs & J. Vigo, Fl. Països Catalans*, 3: 837 (1995): Aronicum scorpioides β pyrenaica.
tianshanicum *C.H. An in Fl. Xinjiangensis*, 5: 477 (1999)—Xinjiang.

DOROTHEANTHUS (Aizoac.).
clavatus *(Haw.) M. Struck in Feddes Repert.*, 109(5–6): 382 (1998): Mesembryanthemum clavatum.
ulularis *F.A. Brusse in Bradleya*, 14: 89 (1996)—Cape Province.

DORSTENIA (Morac.).
sect. Acauloma *M.E.E. Hijman in Ilicifolia*, 2: 95 (1999).
sect. Bazzemia *M.E.E. Hijman in Ilicifolia*, 2: 85 (1999).
sect. Lomatophora *M.E.E. Hijman in Ilicifolia*, 2: 42 (1999).
sect. Xylodorstenia *M.E.E. Hijman in Ilicifolia*, 2: 28 (1999).
barnimiana Schweinf.
subsp. tropaeolifolia *(Schweinf.) K.A. Lye in Lidia*, 4(3): 91 (1998): Kosaria tropaeolifolia.
ramosa (Desv.) Carauta et al.
var. ficus *(Vell.) J.P.P. Carauta in Albertoa*, 4(13): 169 (1996): D. ficus.
socotrana *A.G. Miller in Edinburgh J. Bot.*, 53(1): 145 (1996)—Socotra.

DORTMANNIA (Campanulac.).
ocimoides *(Kunze) Kuntze, Revis. Gen. Pl.*: 973 (1891), as 'ocimodes': Lobelia ocimoides.†

DORYCNIUM (Legumin.).
crantzii *(Vis.) A. Brand in Bot. Jahrb. Syst.*, 25: 226 (1898): Lotus crantzii.

DORYCNOPSIS (Legumin.).
abyssinica *(A. Rich.) V.N. Tikhomirov & D.D. Sokoloff in Feddes Repert.*, 108(5–6): 342 (1997): Helminthocarpon abyssinicum.

DRABA (Crucif.).
bajtenovii *P.V. Veselova in Izv. Min. Nauki Akad. Nauk Resp. Kazakhstan, Ser. Biol. Med.*, 1997(2): 60 (1997)—Kazakhstan.
czuensis *A.S. Revuschkin & A.L. Ebel in Sist. Zam. Mater. Gerb. Tomsk. Gos. Univ.*, 90: 4 (1998), as 'czuense'—Altay.
daurica DC.
var. hebecarpa *(Lindblom) T. Karlsson in Svensk Bot. Tidskr.*, 91(5): 249 (1997 publ. 1998): D. hirta β hebecarpa.
fuhaiensis *Z.X. An in Fl. Xinjiangensis*, 2(2): 375 (1995)—Xinjiang.
hirta L.
subsp. barguzinensis *O.D. Nikiforova in Fl. Sibir. (Berberidac.-Grossulariac.)*, 7: 120 (1994)—Buryatiya.
var. ciliata *Yu.P. Kozhevnikov in Novosti Sist. Vyssh. Rast.*, 30: 95 (1996)—Krasnoyarsk.
var. prozorovskii *(Tolm.) Yu.P. Kozhevnikov in Novosti Sist. Vyssh. Rast.*, 30: 94 (1996): D. prozorovskii.
kodarica *O.D. Nikiforova in Fl. Sibir. (Berberidac.-Grossulariac.)*, 7: 121 (1994)—Buryatiya.
kuemmerlei *V. Stevanović & D. Lakušić in Bot. Chronika*, 13: 89 (2000), nom. nov.: D. lacaitae var. albanica *Kümmerle & Jáv.*
lactea Adams
subsp. pseudopilosa *(Pohle) Yu.P. Kozhevnikov in Novosti Sist. Vyssh. Rast.*, 30: 89 (1996): D. pseudopilosa.
pilosa DC.
subsp. barbata *(Pohle) Yu.P. Kozhevnikov in Novosti Sist. Vyssh. Rast.*, 30: 85 (1996): D. barbata.
pingwuensis *Z.M. Tan & S.C. Zhou in J. Sichuan Univ., Nat. Sci. Ed.*, 33(5): 602 (1996)—China South-central (Sichuan).
pseudopilosa Pohle
f. subglabra *V.B. Kuvaev in V.B. Kuvaev et al., Rast. pokrov ostrova Sibiryakova*: 69 (1994), contrary to Art. 37.4 ICBN (1994)—Krasnoyarsk.

DRABA:-
sapozhnikovii *A.L. Ebel in Sist. Zam. Mater. Gerb. Tomsk. Gos. Univ.*, 90: 5 (1998)—Altay.
sarycheleki *P.V. Veselova in Izv. Min. Nauki Akad. Nauk Resp. Kazakhstan, Ser. Biol. Med.*, 1997(5–6): 80, 79 (1997 publ. 1998)—Kirgizstan.

DRACAENA (Dracaenac.).
chiniana *I.M. Turner in Gard. Bull. Singapore*, 47(2): 533 (1995 publ. July 1997); *et in J. Singapore Nation. Acad. Sci.*, 22–24: 43 (Aug. 1997), nom. nov.: D. congesta *Ridl.*
draco L.
subsp. ajgal *A. Benabid & F. Cuzin in Compte R. Acad. Sci., Sér. 3, (Paris)*, 320(3): 270 (1997)—Morocco.
tamaranae *A. Marrero, R.S. Almeida & M. González-Martín in Bot. J. Linn. Soc.*, 128(3): 294 (1998)—Canary Is.

DRACOCEPHALUM (Labiat.).
charkeviczii *N.S. Probatova in Sosud. Rast. Sovet. Dal'nego Vostoka*, 7: 329 (1995)—Primorye.
fragile
subsp. crenatum *N.S. Vodopianova in L.I. Malyschev (ed.), Vysokogornaya Flora Stanovogo Nagor'ya*: 131 (1972)—Irkutsk.
hypopolium *(Kharkev.) N.S. Probatova in Sosud. Rast. Sovet. Dal'nego Vostoka*, 7: 326 (1995): D. stelleranum var. hypopolium.
jacutense *G.A. Peschkova in Fl. Sibir. (Pyrolac.-Lamiac.)*, 11: 177 (1997)—Yakutskiya.
olchonense *G.A. Peschkova in Fl. Sibir. (Pyrolac.-Lamiac.)*, 11: 180 (1997)—Irkutsk.
rigidulum *Hand.-Mazz. in Oesterr. Bot. Z.*, 88(4): 306 (1939)—Inner Mongolia.

DRACONANTHES *(Luer) C.A. Luer, Ic. Pleurothallid. XIV (Monogr. Syst. Bot. Missouri Bot. Gard., 61(1))*: 2 (1996): Lepanthes subgen. Draconanthes. (Orchidac.).
aberrans *(Schltr.) C.A. Luer, Ic. Pleurothallid. XIV (Monogr. Syst. Bot. Missouri Bot. Gard., 61(1))*: 3 (1996): Lepanthes aberrans.
bufonis *(Luer & Hirtz) C.A. Luer, Ic. Pleurothallid. XIV (Monogr. Syst. Bot. Missouri Bot. Gard., 61(1))*: 4 (1996): Lepanthes bufonis.

DRACONTIUM (Arac.).
spruceanum *(Schott) G.H. Zhu in Novon*, 6(3): 308 (1996): Echidnium spruceanum.

DRACOPHYLLUM (Epacridac.).
macranthum *E.A. Brown & N. Streiber in Telopea*, 8(3): 393 (1999)—New South Wales.
oceanicum *E.A. Brown & N. Streiber in Telopea*, 8(3): 396 (1999)—New South Wales.

DRACULA (Orchidac.).
adrianae *C.A. Luer, Ic. Pleurothallid. XVII (Monogr. Syst. Bot. Missouri Bot. Gard., 72)*: 103 (1998)—Colombia.
benedictii (Rchb. f.) Luer
var. hubeinii *(Luer) J. Hermans in Orchid Dig.*, 61(4): Dracula 10 (1997): D. hubeinii.
carlueri *J. Hermans & P.J. Cribb in Orchid Rev.*, 105(1217): 301 (1997)—Costa Rica.
erythrochaete (Rchb. f.) Luer
subsp. astuta *(Rchb. f.) J. Hermans in Orchid Dig.*, 61(4): Dracula 14 (1997): Masdevallia astuta.
kareniae *C.A. Luer & Dalström in C.A. Luer, Ic. Pleurothallid. XV (Monogr. Syst. Bot. Missouri Bot. Gard.*, 64): 123 (1997)—Ecuador.
olmosii *C.A. Luer & Maduro in C.A. Luer, Ic. Pleurothallid. XVIII (Monogr. Syst. Bot. Missouri Bot.* Gard., 76): 136 (1999)—Panama.
schudelii *C.A. Luer & A.C. Hirtz in C.A. Luer, Ic. Pleurothallid. XVIII (Monogr. Syst. Bot. Missouri Bot.* Gard., 76): 136 (1999)—Ecuador.
sodiroi (Schltr.) Luer
subsp. erythrocodon *C.A. Luer & Dalström in C.A. Luer, Ic. Pleurothallid. XV (Monogr. Syst. Bot. Missouri Bot. Gard.*, 64): 125 (1997)—Ecuador.
terborchii *C.A. Luer & A.C. Hirtz in C.A. Luer, Ic. Pleurothallid. XVIII (Monogr. Syst. Bot. Missouri Bot.* Gard., 76): 137 (1999)—Ecuador.

DRACULA:-
trichroma (Schltr.) J. Hermans in Orchid Rev., 105(1217): 298 (1997): Masdevallia trichroma.
villegasii W. Königer in Arcula, 8: 218 (1999)—Colombia.

DRACUNCULUS (Arac.).
muscivorus (L.f.) Parl.
var. caprariensis A.M. Romo, Fl. Silvestres Baleares: 393 (1994), as 'D. muscivorum var. capraricnse'—Baleares.

DREPANOSTACHYUM (Gramin.).
gongshanense (T.P. Yi) T.P. Yi, Bamboos Fl. Sichuan: 180 (1997): Patellocalamus gongshanensis.
gracile (Blanch.) S. Kumar in Fl. Sikkim, 1: 239 (1996), as 'gracilis': Arundinaria gracilis.
hirsutissimum W.D. Li & Yuan C. Zhong in J. Bamboo Res., 16(1): 52 (1997)—China South-central (Guizhou).
kurzii (Gamble) D. Ohrnberger, Bamboos of the World Intro. 3: 14 (1996), isonym; see D.N. Tewari, Monogr. Bamboo: 90 (?1993): Arundinaria kurzii.
mianningense (Q. Li & Xin Jiang) T.P. Yi, Bamboos Fl. Sichuan: 191 (1997); et in Fl. Sichuanica, 12: 168 (1997 publ. 1998): Dendrocalamus mianningensis.
patellare (Gamble) J.R. Xue & T.P. Yi in T.P. Yi, Bamboos Fl. Sichuan: 180 (1997): Dendrocalamus patellaris.
polystachyum (Kurz ex Gamble) D. Ohrnberger, Bamboos of the World Intro. 2: 10 (1996), isonym; see R.B. Majumdar in Karth. et al., Fl. Ind. Enumerat. - Monocot.: 277 (1989): Arundinaria polystachya.

DRESSLERELLA (Orchidac.).
archilae C.A. Luer & M. Behar in C.A. Luer, Ic. Pleurothallid. XII (Monogr. Syst. Bot. Missouri Bot. Gard., 57): 140 (1995)—Guatemala.

DRESSLERIA (Orchidac.).
allenii H.G. Hills in Lindleyana, 15(3): 171 (2000)—Panama.
fragrans C.H. Dodson in Orquideologia, 21(1): 3 (1998)—Ecuador.
kalbreyeri H.G. Hills in Lindleyana, 15(3): 173 (2000)—Colombia.
kerryae H.G. Hills in Lindleyana, 15(3): 173 (2000)—Panama.

DRIMIA (Hyacinthac.).
albiflora (B. Nord.) J.C. Manning & P. Goldblatt in Strelitzia, 9: 711 (2000): Rhadamanthus albiflorus.
arenicola (B. Nord.) J.C. Manning & P. Goldblatt in Strelitzia, 9: 711 (2000): Rhadamanthus arenicola.
convallarioides (L.f.) J.C. Manning & P. Goldblatt in Strelitzia, 9: 711 (2000): Hyacinthus convallarioides.
cuscutoides (Burch. ex Baker) J.C. Manning & P. Goldblatt in Strelitzia, 9: 711 (2000): Asparagus cuscutoides.
cyanelloides (Baker) J.C. Manning & P. Goldblatt in Strelitzia, 9: 711 (2000): Rhadamanthus cyanelloides.
dregei (Baker) J.C. Manning & P. Goldblatt in Strelitzia, 9: 711 (2000): Urginea dregei.
fasciata (B. Nord.) J.C. Manning & P. Goldblatt in Strelitzia, 9: 711 (2000): Rhadamanthus fasciatus.
filifolia (Jacq.) J.C. Manning & P. Goldblatt in Strelitzia, 9: 711 (2000): Anthericum filifolium.
fragrans (Jacq.) J.C. Manning & P. Goldblatt in Strelitzia, 9: 711 (2000): Anthericum fragrans.
intricata (Baker) J.C. Manning & P. Goldblatt in Strelitzia, 9: 712 (2000): Anthericum intricatum.
involuta (J.C. Manning & Snijman) J.C. Manning & P. Goldblatt in Strelitzia, 9: 712 (2000): Rhadamanthus involutus.
karooica (Oberm.) J.C. Manning & P. Goldblatt in Strelitzia, 9: 712 (2000): Rhadamanthus karooicus.
namibensis (Oberm.) J.C. Manning & P. Goldblatt in Strelitzia, 9: 712 (2000): Rhadamanthus namibensis.

DRIMIA:-
obstii W. Heering in Mitt. Geogr. Ges. Hamburg, 29: 209 (1915)—Tanzania.
platyphylla (B. Nord.) J.C. Manning & P. Goldblatt in Strelitzia, 9: 712 (2000): Rhadamanthus platyphyllus.
revoluta (A.V. Duthie) J.C. Manning & P. Goldblatt in Strelitzia, 9: 712 (2000): Urginea revoluta.
rupicola (Trimen) M.D. Dassanayake in M.D. Dassanayake & W.D. Clayton, eds., Revised Handb. Fl. Ceylon, 14: 220 (2000): Urginea congesta var. rupicola.
salteri (Compton) J.C. Manning & P. Goldblatt in Strelitzia, 9: 712 (2000): Urgineopsis salteri.
sclerophylla J.C. Manning & P. Goldblatt in Strelitzia, 9: 712 (2000), nom. nov.: Urginea rigidifolia Baker.
secunda (B. Nord.) J.C. Manning & P. Goldblatt in Strelitzia, 9: 712 (2000): Rhadamanthus secundus.
simensis (Hochst. ex A. Rich.) B. Stedje in Nordic J. Bot., 15(6): 597 (1995 publ. 1996): Scilla simensis.
sudanica I. Friis & K. Vollesen in Nordic J. Bot., 19(2): 210 (1999), nom. nov.: Urginea pauciflora Baker.
undata Stearn
subsp. caeculi (Pau) G. Mateo Sanz & M.B. Crespo Villalba, Fl. Abrev. Comun. Valenciana: 430 (1995): Urginea undulata var. caeculi.
uniflora J.C. Manning & P. Goldblatt in Strelitzia, 9: 712 (2000), nom. nov.: Litanthus pusillus Harv.
uranthera (R.A. Dyer) J.C. Manning & P. Goldblatt in Strelitzia, 9: 712 (2000): Rhadamanthus urantherus.
virens (Schltr.) J.C. Manning & P. Goldblatt in Strelitzia, 9: 712 (2000): Urginea virens.

DRIMIOPSIS (Hyacinthac.).
burkei Baker
subsp. stolonissima U. Müller-Doblies & D. Müller-Doblies in Feddes Repert., 108(1–2): 63 (1997)—Transvaal.
comptonii U. Müller-Doblies & D. Müller-Doblies in Feddes Repert., 108(1–2): 64 (1997)—Swaziland.
davidsoniae U. Müller-Doblies & D. Müller-Doblies in Feddes Repert., 108(1–2): 64 (1997)—Transvaal.
fischeri (Engl.) B. Stedje in Fl. Trop. E. Afr., Hyacinthac.: 6 (25 March 1996); et in Nordic J. Bot., 15(6): 593 (1995 publ. 15 April 1996): Scilla fischeri.
fischeri (Engl.) U. Müller-Doblies & D. Müller-Doblies in Feddes Repert., 108(1–2): 64 (1997), isonym; see B. Stedje in Fl. Trop. E. Afr., Hyacinthac.: 6 (1996): Scilla fischeri.
pusilla U. Müller-Doblies & D. Müller-Doblies in Feddes Repert., 108(1–2): 64 (1997)—Swaziland.
reilleyana U. Müller-Doblies & D. Müller-Doblies in Feddes Repert., 108(1–2): 64 (1997)—Swaziland.

DRIMYCARPUS (Anacardiac.).
maximus K.M. Kochummen in Sandakania, 7: 81 (1996)—Borneo (Sarawak).

DROOGMANSIA (Legumin.).
pteropus (Baker) De Wild.
var. giorgii (De Wild.) B. Verdcourt in Fl. Zambes., 3(6): 33 (2000): D. giorgii.

DROSANTHEMUM (Aizoac.).
deciduum H.E.K. Hartmann & C. Bruckmann in Bradleya, 18: 106 (2000)—Cape Provinces (Western Cape Province).
subspinosum (Kuntze) H.E.K. Hartmann in Bradleya, 18: 84 (2000): Mesembryanthemum subspinosum.

DROSERA (Droserac.).
subgen. Arcturia (Planch.) J. Schlauer in Carniv. Pl. Newsl., 25(3): 68 (1996): D. sect. Arcturia.
subgen. Bryastrum (Planch.) J. Schlauer in Carniv. Pl. Newsl., 25(3): 71 (1996): D. sect. Bryastrum.
subgen. Coelophylla (Planch.) J. Schlauer in Carniv. Pl. Newsl., 25(3): 69 (1996): D. sect. Coelophylla.
subgen. Lasiocephala (Planch.) J. Schlauer in Carniv. Pl. Newsl., 25(3): 69 (1996): D. sect. Lasiocephala.

DROSERA:-
subgen. Meristocaulis (Maguire & Wurdack) J. Schlauer in Carniv. Pl. Newsl., 25(3): 68 (1996): D. sect. Meristocaulis.
sect. Oosperma J. Schlauer in Carniv. Pl. Newsl., 25(3): 70 (1996).
subgen. Phycopsis (Planch.) J. Schlauer in Carniv. Pl. Newsl., 25(3): 71 (1996): D. sect. Phycopsis.
subgen. Stelogyne (Diels) J. Schlauer in Carniv. Pl. Newsl., 25(3): 68 (1996): D. sect. Stelogyne.
barbigera
subsp. silvicola (Lowrie & Carlquist) J. Schlauer in Carniv. Pl. Newsl., 25(3): 72 (1996): D. silvicola.
brevicornis A. Lowrie in Nuytsia, 11(1): 56 (1996)—Northern Territory.
broomensis A. Lowrie in Nuytsia, 11(1): 57 (1996)—Western Australia.
caduca A. Lowrie in Nuytsia, 11(1): 59 (1996)—Western Australia.
citrina
var. nivea (Lowrie & Carlquist) J. Schlauer in Carniv. Pl. Newsl., 25(3): 72 (1996): D. nivea.
× corinthiaca R. Gibson & E. Green in Carniv. Pl. Newsl., 28(3): 81 (1999)—Cape Province. (D. aliciae × glabripes).
darwinensis A. Lowrie in Nuytsia, 11(1): 61 (1996)—Northern Territory.
derbyensis A. Lowrie in Nuytsia, 11(1): 63 (1996)—Northern Territory.
dichrosepala
subsp. enodes (N.G. Marchant & Lowrie) J. Schlauer in Carniv. Pl. Newsl., 25(3): 72 (1996): D. enodes.
gigantea
var. geniculata (N.G. Marchant & Lowrie) J. Schlauer in Carniv. Pl. Newsl., 25(3): 73 (1996): D. gigantea subsp. geniculata.
graomogolensis T.R.S. Silva in Novon, 7(1): 85 (1997)—Brazil Southeast (Minas Gerais).
kansaiensis P. Debbert in Sendtnera, 3: 101 (1996)—Japan.
kenneallyi A. Lowrie in Nuytsia, 10(3): 420 (1996)—Western Australia.
moorei (Diels) A. Lowrie in Nuytsia, 13(1): 80 (1999): D. subhirtella var. moorei.
nitidula
var. allantostigma (N.G. Marchant & Lowrie) J. Schlauer in Carniv. Pl. Newsl., 25(3): 73 (1996): D. nitidula subsp. allantostigma.
var. leucostigma (N.G. Marchant & Lowrie) J. Schlauer in Carniv. Pl. Newsl., 25(3): 73 (1996): D. nitidula subsp. leucostigma.
occidentalis
var. microscapa (Debbert) J. Schlauer in Carniv. Pl. Newsl., 25(3): 73 (1996): D. microscapa.
paleacea
subsp. leioblastus (N.G. Marchant & Lowrie) J. Schlauer in Carniv. Pl. Newsl., 25(3): 73 (1996): D. leioblastus.
subsp. roseana (N.G. Marchant & Lowrie) J. Schlauer in Carniv. Pl. Newsl., 25(3): 73 (1996): D. roseana.
subsp. stelliflora (Lowrie & Carlquist) J. Schlauer in Carniv. Pl. Newsl., 25(3): 72 (1996): D. stelliflora.
paradoxa A. Lowrie in Nuytsia, 11(3): 347 (1997)—Western Australia, Northern Territory.
parvula
subsp. sargentii (Lowrie & N.G. Marchant) J. Schlauer in Carniv. Pl. Newsl., 25(3): 72 (1996): D. sargentii.
stricticaulis
subsp. eremaea (N.G. Marchant & Lowrie) J. Schlauer in Carniv. Pl. Newsl., 25(3): 73 (1996): D. macrantha subsp. eremaea.
zigzagia A. Lowrie in Nuytsia, 13(1): 85 (1999)—Western Australia.

DRUMMONDITA (Rutac.).
longifolia (Paul G. Wilson) Paul G. Wilson in Nuytsia, 12(1): 86 (1998): D. hassellii var. longifolia.
microphylla Paul G. Wilson in Nuytsia, 12(1): 86 (1998)—Western Australia.

DRYADELLA (Orchidac.).
acuminata C.A. Luer & A.C. Hirtz in C.A. Luer, Ic. Pleurothallid. XVIII (Monogr. Syst. Bot. Missouri Bot. Gard., 76): 159 (1999)—Ecuador.

DRYADELLA:-
aurea *C.A. Luer & A.C. Hirtz in C.A. Luer, Ic. Pleurothallid. XVIII (Monogr. Syst. Bot. Missouri Bot.* Gard., 76): 159 (1999)—Ecuador, Peru.
butcheri *C.A. Luer, Ic. Pleurothallid. XVIII (Monogr. Syst. Bot. Missouri Bot. Gard., 76)*: 160 (1999)—Panama.
cuspidata *C.A. Luer & A.C. Hirtz in C.A. Luer, Ic. Pleurothallid. XVIII (Monogr. Syst. Bot. Missouri Bot.* Gard., 76): 160 (1999)—Ecuador, Colombia.
dodsonii *C.A. Luer, Ic. Pleurothallid. XVIII (Monogr. Syst. Bot. Missouri Bot. Gard., 76)*: 161 (1999)—Ecuador, Colombia.
dressleri *C.A. Luer, Ic. Pleurothallid. XVIII (Monogr. Syst. Bot. Missouri Bot. Gard., 76)*: 162 (1999)—Panama.
fuchsii *C.A. Luer, Ic. Pleurothallid. XVIII (Monogr. Syst. Bot. Missouri Bot. Gard., 76)*: 162 (1999)—Honduras.
odontostele *C.A. Luer in Lindleyana*, 11(2): 54 (1996)—Panama, Colombia.
pachyrhiza *C.A. Luer & A.C. Hirtz in C.A. Luer, Ic. Pleurothallid. XVIII (Monogr. Syst. Bot. Missouri Bot.* Gard., 76): 163 (1999)—Ecuador.
rodrigoi *C.A. Luer, Ic. Pleurothallid. XVIII (Monogr. Syst. Bot. Missouri Bot. Gard., 76)*: 163 (1999)—Colombia.
silvana *F. Barros in Rev. Brasil. Bot.*, 18(1): 35 (1995)—Brazil Northeast (Bahia).
sororcula *C.A. Luer in Lindleyana*, 11(2): 56 (1996)—Panama, Costa Rica, Nicaragua, El Salvador, Honduras.
verrucosa *C.A. Luer & R. Escobar in C.A. Luer, Ic. Pleurothallid. XVIII (Monogr. Syst. Bot. Missouri Bot.* Gard., 76): 164 (1999)—Colombia.

DRYANDRA (Proteac.).
ser. Acrodontae *(Meisn.) A.S. George in Nuytsia*, 10(3): 347 (1996): D. § Acrodontae.
ser. Acuminatae *A.S. George in Nuytsia*, 10(3): 391 (1996).
ser. Aphragma *(R.Br.) A.S. George in Nuytsia*, 10(3): 370 (1996): D. sect. Aphragma.
ser. Capitellatae *A.S. George in Nuytsia*, 10(3): 349 (1996).
ser. Decurrentes *(Meisn.) A.S. George in Nuytsia*, 10(3): 357 (1996): D. § Decurrentes.
subgen. Diplophragma *(R. Br.) A.S. George in Nuytsia*, 10(3): 401 (1996): D. sect. Diplophragma.
ser. Foliosae *A.S. George in Nuytsia*, 10(3): 356 (1996).
ser. Folliculosae *A.S. George in Nuytsia*, 10(3): 345 (1996).
subgen. Hemiclidia *(R. Br.) A.S. George in Nuytsia*, 10(3): 400 (1996): gen. Hemiclidia.
ser. Ilicinae *(Meisn.) A.S. George in Nuytsia*, 10(3): 352 (1996): D. § Ilicinae.
ser. Inusitatae *A.S. George in Nuytsia*, 10(3): 377 (1996).
ser. Ionthocarpae *A.S. George in Nuytsia*, 10(3): 375 (1996).
ser. Marginatae *(Meisn.) A.S. George in Nuytsia*, 10(3): 345 (1996): D. § Marginatae.
ser. Pectinatae *(Meisn.) A.S. George in Nuytsia*, 10(3): 391 (1996): D. § Pectinatae.
ser. Plumosae *A.S. George in Nuytsia*, 10(3): 382 (1996).
ser. Runcinatae *(Meisn.) A.S. George in Nuytsia*, 10(3): 360 (1996): D. § Runcinatae.
ser. Subulatae *A.S. George in Nuytsia*, 10(3): 378 (1996).
ser. Tenuifoliae *A.S. George in Nuytsia*, 10(3): 358 (1996).
ser. Triangulares *A.S. George in Nuytsia*, 10(3): 366 (1996).
acanthopoda *A.S. George in Nuytsia*, 10(3): 335 (1996)—Western Australia.
anatona *A.S. George in Nuytsia*, 10(3): 353 (1996)—Western Australia.
armata R. Br.
var. ignicida *A.S. George in Nuytsia*, 10(3): 331 (1996)—Western Australia.
aurantia *A.S. George in Nuytsia*, 10(3): 373 (1996)—Western Australia.
bipinnatifida R. Br.
subsp. multifida *A.S. George in Nuytsia*, 10(3): 402 (1996)—Western Australia.

DRYANDRA:-
borealis *A.S. George in Nuytsia*, 10(3): 343 (1996)—Western Australia.
subsp. elatior *A.S. George in Nuytsia*, 10(3): 344 (1996)—Western Australia.
catoglypta *A.S. George in Nuytsia*, 10(3): 369 (1996)—Western Australia.
columnaris *A.S. George in Nuytsia*, 10(3): 388 (1996)—Western Australia.
conferta Benth.
var. parva *A.S. George in Nuytsia*, 10(3): 388 (1996)—Western Australia.
corvijuga *A.S. George in Nuytsia*, 10(3): 365 (1996)—Western Australia.
cypholoba *A.S. George in Nuytsia*, 10(3): 392 (1996)—Western Australia.
drummondii Meisn.
subsp. hiemalis *A.S. George in Nuytsia*, 10(3): 367 (1996)—Western Australia.
subsp. macrorufa *A.S. George in Nuytsia*, 10(3): 368 (1996)—Western Australia.
echinata *A.S. George in Nuytsia*, 10(3): 339 (1996)—Western Australia.
epimicta *A.S. George in Nuytsia*, 10(3): 365 (1996)—Western Australia.
erythrocephala C.A. Gardner
var. inopinata *A.S. George in Nuytsia*, 10(3): 379 (1996)—Western Australia.
fasciculata *A.S. George in Nuytsia*, 10(3): 386 (1996)—Western Australia.
ferruginea Kippist ex Meisn.
subsp. chelomacarpa *A.S. George in Nuytsia*, 10(3): 363 (1996)—Western Australia.
subsp. flavescens *A.S. George in Nuytsia*, 10(3): 364 (1996)—Western Australia.
subsp. obliquiloba *A.S. George in Nuytsia*, 10(3): 363 (1996)—Western Australia.
subsp. pumila *A.S. George in Nuytsia*, 10(3): 362 (1996)—Western Australia.
subsp. tutanningensis *A.S. George in Nuytsia*, 10(3): 361 (1996)—Western Australia.
fililoba *A.S. George in Nuytsia*, 10(3): 372 (1996)—Western Australia.
fraseri R. Br.
var. ashbyi *(B.L. Burtt) A.S. George in Nuytsia*, 10(3): 346 (1996): D. ashbyi.
var. oxycedra *A.S. George in Nuytsia*, 10(3): 347 (1996)—Western Australia.
fuscobractea *A.S. George in Nuytsia*, 10(3): 330 (1996)—Western Australia.
glauca *A.S. George in Nuytsia*, 10(3): 400 (1996)—Western Australia.
hirsuta *A.S. George in Nuytsia*, 10(3): 332 (1996)—Western Australia.
idiogenes *A.S. George in Nuytsia*, 10(3): 377 (1996)—Western Australia.
insulanemorecincta *A.S. George in Fl. Australia*, 17B: 397 (1999)—Western Australia.
ionthocarpa *A.S. George in Nuytsia*, 10(3): 376 (1996)—Western Australia.
kippistiana Meisn.
var. paenepeccata *A.S. George in Nuytsia*, 10(3): 348 (1996)—Western Australia.
lepidorhiza *A.S. George in Nuytsia*, 10(3): 374 (1996)—Western Australia.
lindleyana Meisn.
subsp. agricola *A.S. George in Nuytsia*, 10(3): 397 (1996)—Western Australia.
subsp. media *A.S. George in Nuytsia*, 10(3): 395 (1996)—Western Australia.
var. mellicula *A.S. George in Nuytsia*, 10(3): 394 (1996)—Western Australia.
subsp. pollosta *A.S. George in Nuytsia*, 10(3): 396 (1996)—Western Australia.
subsp. sylvestris *A.S. George in Nuytsia*, 10(3): 396 (1996)—Western Australia.
longifolia R. Br.
subsp. archeos *A.S. George in Nuytsia*, 10(3): 342 (1996)—Western Australia.
subsp. calcicola *A.S. George in Nuytsia*, 10(3): 343 (1996)—Western Australia.
meganotia *A.S. George in Nuytsia*, 10(3): 351 (1996)—Western Australia.
montana *C.A. Gardner ex A.S. George in Nuytsia*, 10(3): 384 (1996)—Western Australia.
mucronulata R. Br.
subsp. retrorsa *A.S. George in Nuytsia*, 10(3): 357 (1996)—Western Australia.

DRYANDRA:-
nivea (Labill.) R. Br.
subsp. uliginosa *A.S. George in Nuytsia*, 10(3): 399 (1996)—Western Australia.
nobilis Lindl.
subsp. fragrans *A.S. George in Nuytsia*, 10(3): 355 (1996)—Western Australia.
octotriginta *A.S. George in Nuytsia*, 10(3): 369 (1996)—Western Australia.
pallida *A.S. George in Nuytsia*, 10(3): 333 (1996)—Western Australia.
platycarpa *A.S. George in Nuytsia*, 10(3): 389 (1996)—Western Australia.
plumosa R. Br.
subsp. denticulata *A.S. George in Nuytsia*, 10(3): 383 (1996)—Western Australia.
porrecta *A.S. George in Nuytsia*, 10(3): 373 (1996)—Western Australia.
praemorsa Meisn.
var. splendens *A.S. George in Nuytsia*, 10(3): 353 (1996)—Western Australia.
pseudoplumosa *A.S. George in Nuytsia*, 10(3): 383 (1996)—Western Australia.
pteridifolia R. Br.
subsp. vernalis *A.S. George in Nuytsia*, 10(3): 371 (1996)—Western Australia.
rufistylis *A.S. George in Nuytsia*, 10(3): 390 (1996)—Western Australia.
serratuloides Meisn.
subsp. perissa *A.S. George in Nuytsia*, 10(3): 350 (1996)—Western Australia.
sessilis (Knight) Domin
var. cordata *(Meisn.) A.S. George in Nuytsia*, 10(3): 329 (1996): D. floribunda var. cordata.
var. cygnorum *(Gand.) A.S. George in Nuytsia*, 10(3): 329 (1996): D. cygnorum.
var. flabellifolia *A.S. George in Nuytsia*, 10(3): 328 (1996)—Western Australia.
speciosa Meisn.
subsp. macrocarpa *A.S. George in Nuytsia*, 10(3): 381 (1996)—Western Australia.
squarrosa R. Br.
subsp. argillacea *A.S. George in Nuytsia*, 10(3): 336 (1996)—Western Australia.
stricta *A.S. George in Nuytsia*, 10(3): 338 (1996)—Western Australia.
subpinnatifida C.A. Gardner
var. imberbis *A.S. George in Nuytsia*, 10(3): 340 (1996)—Western Australia.
tenuifolia R. Br.
var. reptans *A.S. George in Nuytsia*, 10(3): 359 (1996)—Western Australia.
trifontinalis *A.S. George in Nuytsia*, 10(3): 337 (1996)—Western Australia.
viscida *A.S. George in Nuytsia*, 10(3): 380 (1996)—Western Australia.
wonganensis *A.S. George in Nuytsia*, 10(3): 337 (1996)—Western Australia.
xylothemelia *A.S. George in Nuytsia*, 10(3): 334 (1996)—Western Australia.

DRYAS (Rosac.).
ajanensis Juz.
subsp. beringensis *B.A. Jurtzev in Arktich. Fl. SSSR*, 9(1): 322 (1984)—Magadan.
subsp. ochotensis *B.A. Jurtzev in Arktich. Fl. SSSR*, 9(1): 322 (1984)—Magadan.
var. viridis *B.A. Jurtzev in Arktich. Fl. SSSR*, 9(1): 322 (1984)—Magadan.
chamissonis *B.A. Jurtzev in Arktich. Fl. SSSR*, 9(1): 322 (1984), homonym; *see Juz. in Izv. Glavn. Bot. Sada SSSR*, 28: 326 (1929)—Exact locality not stated.
var. incana *B.A. Jurtzev in Arktich. Fl. SSSR*, 9(1): 323 (1984)—Magadan.
× grandiformis *B.A. Jurtzev in Arktich. Fl. SSSR*, 9(1): 321 (1984)—Magadan. (D. grandis × punctata).
grandis Juz.
var. tomentella *B.A. Jurtzev in Arktich. Fl. SSSR*, 9(1): 321 (1984)—Yakutiya.
incisa Juz.
var. cana *B.A. Jurtzev in Arktich. Fl. SSSR*, 9(1): 322 (1984)—Magadan.
octopetala L.
subsp. subincisa *B.A. Jurtzev in Arktich. Fl. SSSR*, 9(1): 322 (1984)—West Siberia.

DRYAS:-
octopetala L.
subsp. sumneviczii *(Serg.) Y.N. Petrochenko in L.I. Malyschev (ed.), Výsokogornaya Flora Stanovogo Nagor'ya:* 110 (1972): D. sumneviczii.
var. sumneviczii *(Serg.) Yu.P. Kozhevnikov in Bot. Zhurn.*, 81(7): 18 (1996): D. sumneviczii.
punctata Juz.
var. cinerea *B.A. Jurtzev in Arktica. Fl. SSSR*, 9(1): 322 (1984)—Magadan.

DRYMOANTHUS (Orchidac.).
flavus *I.M. St. George & B.P.J. Molloy in New Zealand J. Bot.*, 32(4): 416 (1994)—New Zealand South I., New Zealand North I.

DRYMOCHLOA (Gramin.).
asthenica *(Hook. f.) J. Holub in Preslia*, 70(2): 104 (1998): Festuca aasthenica.
grandis *(Coss. & Durieu) J. Holub in Preslia*, 70(2): 104 (1998): Festuca drymeja var. grandis.
lasto *(Boiss.) J. Holub in Preslia*, 70(2): 104 (1998): Festuca lasto.
modesta *(Steud.) J. Holub in Preslia*, 70(2): 104 (1998): Festuca modesta.

DRYMONIA (Gesneriac.).
brochidodroma *H. Wiehler in Gesneriana*, 1(1): 53 (1995)—Colombia, Ecuador.
microcalyx *H. Wiehler in Gesneriana*, 1(1): 57 (1995)—Colombia.
submarginalis *J. Gómez-Laurito & M.M. Chavarría in Gesneriana*, 1(1): 15 (1995)—Costa Rica.

DRYMOPHLOEUS (Palmae).
hentyi *(Essig) S. Zona in Blumea*, 44(1): 13 (1999): Ptychosperma hentyi.

DRYPETES (Euphorbiac.).
acuminata *P.I. Forster in Austrobaileya*, 4(4): 483 (1997)—Queensland.
darcyana *G. McPherson in Adansonia*, 22(2): 205 (2000)—Comoros (Mayotte).
iodoformis *L.S. Sm. ex P.I. Forster in Austrobaileya*, 4(4): 489 (1997)—Queensland.
moliwensis *M. Cheek & A. Radcliffe-Smith in Kew Bull.*, 55(4): 895 (2000)—Cameroon.
sherffii *R. Govaerts & A. Radcliffe-Smith in Kew Bull.*, 51(1): 175 (1996), nom. nov.: D. forbesii *Sherff.*
vernicosa *P.I. Forster in Austrobaileya*, 4(4): 486 (1997)—Queensland.

DUBAUTIA (Compos.).
kenwoodii *G.D. Carr in Novon*, 8(1): 8 (1998)—Hawaiian Is.
syndetica *G.D. Carr & D.H. Lorence in Novon*, 8(1): 4 (1998)—Hawaiian Is.

DUBOUZETIA (Elaeocarpac.).
saxatilis *A.R. Bean & L.W. Jessup in Austrobaileya*, 4(4): 673 (1997)—Queensland.

DUDLEYA (Crassulac.).
cymosa (Lem.) Britton & Rose
subsp. costafolia *J.A. Bartel & J.R. Shevock in Aliso*, 12(4): 701 (1990)—California.
gnoma *S.W. McCabe in Madroño*, 44(1): 49 (1997)—California.
virens (Rose) Moran
subsp. extima *Reid Moran in Haseltonia*, 3: 8 (1995)—Mexican Pacific Is. (Guadalupe I.).
subsp. hassei *(Rose) Reid Moran in Haseltonia*, 3: 4 (1995): Stylophyllum hassei.
subsp. insularis *(Rose) Reid Moran in Haseltonia*, 3: 2 (1995): Stylophyllum insulare.

DUGUETIA (Annonac.).
aberrans *P.J.M. Maas in Bot. Jahrb.*, 118(2): 187 (1996)—Venezuela.
arenicola *P.J.M. Maas in Bot. Jahrb.*, 121(4): 466 (1999)—Brazil North (Amazonas).
aripuanae *P.J.M. Maas in Bot. Jahrb.*, 121(4): 469 (1999)—Brazil West-central (Mato Grosso).
barteri *(Benth.) L.W. Chatrou, Changing Genera:* 66 (1998): Annona barteri.

DUGUETIA:-
chrysea *P.J.M. Maas in Bot. Jahrb.*, 118(2): 190 (1996)—Brazil North (Amazonas).
chrysocarpa *P.J.M. Maas in Bot. Jahrb.*, 121(4): 471 (1999)—Brazil Southeast (Minas Gerais, Espírito Santo), Brazil Northeast (Bahia).
colombiana *P.J.M. Maas in Bot. Jahrb.*, 118(2): 192 (1996)—Colombia.
confinis *(Engl. & Diels) L.W. Chatrou, Changing Genera:* 67 (1998): Pachypodanthium confine.
confusa *P.J.M. Maas in Bot. Jahrb.*, 118(2): 195 (1996)—Panama, Costa Rica, Colombia.
dilabens *L.W. Chatrou & C.P. Repetur in L.W. Chatrou, Changing Genera:* 69 (1998)—Gabon, Cameroon.
gentryi *P.J.M. Maas in Bot. Jahrb.*, 121(4): 474 (1999)—Colombia, Panama.
granvilleana *P.J.M. Maas in Bot. Jahrb.*, 121(4): 476 (1999)—French Guiana, Brazil North (Amapá).
magnolioidea *P.J.M. Maas in Bot. Jahrb.*, 118(2): 198 (1996)—Brazil Northeast (Bahia).
manausensis *P.J.M. Maas & E.J.M.C. Boon in Bot. Jahrb.*, 118(2): 201 (1996)—Brazil North (Amazonas, Pará).
megalocarpa *P.J.M. Maas in Bot. Jahrb.*, 118(2): 203 (1996)—Brazil North (Amazonas, Pará).
nitida *P.J.M. Maas in Bot. Jahrb.*, 121(4): 479 (1999)—Peru.
oligocarpa *P.J.M. Maas & J.A.C. van Dam in Bot. Jahrb.*, 118(2): 207 (1996)—French Guiana.
phaeoclados *(Mart.) P.J.M. Maas & Rainer in Bot. Jahrb.*, 121(4): 481 (1999)—Annona phaeoclados.
restingae *P.J.M. Maas in Bot. Jahrb.*, 118(2): 210 (1996)—Brazil Northeast (Bahia).
reticulata *P.J.M. Maas in Bot. Jahrb.*, 118(2): 212 (1996)—Brazil Northeast (Bahia).
riberensis *Aristeg. ex P.J.M. Maas & E.J.M.C. Boon in Bot. Jahrb.*, 118(2): 214 (1996)—Venezuela.
ruboides *P.J.M. Maas & He in Bot. Jahrb.*, 121(4): 484 (1999)—Brazil Northeast (Maranhão).
sancticarolii *P.J.M. Maas in Bot. Jahrb.*, 118(2): 217 (1996)—Venezuela, Brazil North (Amazonas).
scottmorii *P.J.M. Maas in Bot. Jahrb.*, 118(2): 221 (1996)—Northeast (Bahia).
sessilis *(Vell.) P.J.M. Maas in Bol. Mus. Paraense Emílio Goeldi, Bot.*, 9(1): 38 (1993 publ. 1994): Uvaria sessilis.
sooretamae *P.J.M. Maas in Bot. Jahrb.*, 121(4): 486 (1999)—Brazil Southeast (Espírito Santo).
staudtii *(Engl. & Diels) L.W. Chatrou, Changing Genera:* 70 (1998): Uvaria staudtii.
subcordata *P.J.M. Maas & J.A.C. van Dam in Bot. Jahrb.*, 118(2): 222 (1996)—Brazil North (Amazonas, Pará).
trunciflora *P.J.M. Maas & A.H. Gentry in Feddes Repert.*, 106(5–8): 341 (1995 publ. 1996)—Brazil North (Amazonas, Acre, Pará), Brazil West-central (Mato Grosso), Peru, French Guiana.
tuberculata *P.J.M. Maas in Bot. Jahrb.*, 118(2): 224 (1996)—Colombia, Panama.

DUHALDEA (Compos.).
latifolia *(DC.) R. Dawar & M. Qaiser in Adansonia*, 21(2): 326 (1999): Inula latifolia.

DUNBARIA (Legumin.).
crinita *(Dunn) L.J.G. van der Maesen in Wageningen Agric. Univ. Pap.*, 98(1): 26 (1998): Atylosia crinita.
floresiana *L.J.G. van der Maesen in Wageningen Agric. Univ. Pap.*, 98(1): 44 (1998)—Lesser Sunda Is.
fusca *(Wall.) S. Kurz in J. Asiat. Soc. Bengal, Pt. 2, Nat. Hist.*, 45(4): 255 (1876 publ. 1877): Phaseolus fuscus.
fusca *(Wall.) P. Satyanarayana & K. Thothathri in Bull. Bot. Surv. India*, 36(1–4): 237 (1994 publ. 1997), isonym; *see Kurz in J. Asiat. Soc. Bengal, Pt. 2, Nat. Hist.*, 45(4): 255 (1876 publ. 1877): Phaseolus fuscus.
incana *(Zoll. & Moritzi) L.J.G. van der Maesen in Wageningen Agric. Univ. Pap.*, 98(1): 61 (1998): Phaseolus incanus.
longicarpa *(Thuan) L.J.G. van der Maesen in Wageningen Agric. Univ. Pap.*, 98(1): 69 (1998): D. fusca var. longicarpa.
trichodon *(Dunn) L.J.G. van der Maesen in Wageningen Agric. Univ. Pap.*, 98(1): 87 (1998): Atylosia trichodon.

DUNBARIA:-
truncata *(Miq.) L.J.G. van der Maesen in Wageningen Agric. Univ. Pap.*, 98(1): 89 (1998): Dolichos truncatus.

× **DUPOA** *J. Cayouette & S.J. Darbyshire in Nordic J. Bot.*, 13(6): 616 (1993). *GRAMINEAE* (Dupontia × Poa).
labradorica *(Steud.) J. Cayouette & S.J. Darbyshire in Nordic J. Bot.*, 13(6): 616 (1993): Poa labradorica.

DURANTA (Verbenac.).
repens L.
var. ellisia *(Jacq.) R.R. Fernandez, Trees of Mumbai (Bombay):* 209 (1999): D. ellisia.

DUROIA (Rubiac.).
laevis *W. Devia, C. Persson & C.M. Taylor in Novon*, 9(3): 433 (1999)—Colombia, Ecuador.
sancti-ciprianii *W. Devia, C. Persson & C.M. Taylor in Novon*, 9(3): 436 (1999)—Colombia.

DUSCHEKIA (Betulac.).
alnobetula *(Ehrh.) Z. Pouzar in Čas. Nár. Muz. Řada Přír.*, 151(1): 20 (1982): Betula alnobetula.

DUSSIA (Legumin.).
atropurpurea *N. Zamora, R.T. Pennington & C.H. Stirton in Edinburgh J. Bot.*, 56(2): 176 (1999)—Costa Rica, Panama.

DUTAILLIOPSIS *T.G. Hartley in Adansonia*, 19(2): 210 (1997). *RUTACEAE*
gordonii *T.G. Hartley in Adansonia*, 19(2): 210 (1997)—New Caledonia.

DUVALIA (Asclepiadac.).
caespitosa (Masson) Haw.
var. compacta *(Haw.) U. Meve, Gen. Duvalia (Stapelieae): stem-succulents between Cape & Arabia (Pl. Syst.* Evol. Suppl. 10): 59 (1997): D. compacta.
gracilis *U. Meve, Gen. Duvalia (Stapelieae): stem-succulents between Cape & Arabia (Pl. Syst.* Evol. Suppl. 10): 75 (1997)—Cape Province.
sulcata N.E. Br.
subsp. seminuda *(Lavranos) U. Meve, Gen. Duvalia (Stapelieae): stem-succulents between Cape & Arabia (Pl. Syst.* Evol. Suppl. 10): 116 (1997): D. sulcata var. seminuda.
subsp. somalensis *(Lavranos) U. Meve, Gen. Duvalia (Stapelieae): stem-succulents between Cape & Arabia (Pl. Syst.* Evol. Suppl. 10): 118 (1997): D. somalensis.

DUVIGNEAUDIA (Euphorbiac.).
leonardii-crispi *(J. Léonard) R.C. Kruijt & G.J. Roebers in Biblioth. Bot.*, 146: 14 (1996): Sapium leonardii-crispi.

× **DYCKCOHNIA** *G.H. Anderson ex J.R. Grant in Selbyana*, 19(1): 116 (1998). *BROMELIACEAE* (Deuterocohnia × Dyckia).

DYCKIA (Bromeliac.).
hohenbergioides *E.M.C. Leme & E. Esteves Pereira in J. Bromeliad Soc.*, 48(3): 124 (1998)—Brazil Northeast (Bahia).
ibicuiensis *T. Strehl in Bromélia*, 4(4): 14 (1997 publ. 1998)—Brazil South (Rio Grande do Sul).
milagrensis *E.M.C. Leme in Harvard Pap. Bot.*, 4(1): 164 (1999)—Brazil Northeast (Bahia).
tenebrosa *E.M.C. Leme & H.E. Luther in Selbyana*, 19(2): 183 (1998 publ. 1999)—Brazil Southeast (Minas Gerais).

× **DYCKTIA** *D.A. Beadle, Prelim. Listing Cultivar Grex Names Bromel.:* 82 (1991). *BROMELIACEAE* (Dyckia × Hechtia).

DYSCHORISTE (Acanthac.).
linearis (Torr. & A. Gray) Kuntze
var. sanpatriciensis *J. Henrickson in Lundellia*, 2: 91 (1999)—Texas, Mexico Northeast (Coahuila).
novogaliciana *T.F. Daniel in Polibotanica*, 2: 1 (1996)—Mexico Southwest (Nayarit, Jalisco), Mexico Northwest (Sinaloa), Mexico Northeast (Durango).

DYSCHORISTE:-
poliodes Leonard & Gentry
var. glabra *J. Henrickson in Lundellia,* 2: 97 (1999)—Mexico Northeast (Nuevo León).
var. obispoensis *J. Henrickson in Lundellia,* 2: 95 (1999)—Mexico Northeast (Nuevo León).
schiedeana (Nees) Kuntze
var. cinerascens *J. Henrickson & R.A. Hilsenbeck in Lundellia,* 2: 82 (1999): Texas, Mexico Northeast (Chihuahua, Coahuila, Durango, San Luis Potosí).
var. decumbens *(A. Gray) J. Henrickson in Lundellia,* 2: 84 (1999): Calophanes decumbens.
var. prostrata *J. Henrickson in Lundellia,* 2: 86 (1999)—Mexico Northeast (San Luis Potosí, Durango, Zacatecas, Aguascalientes, Guanajuato, Hidalgo), Mexico Central (Puebla), Mexico Southwest (Oaxaca).

DYSOSMA (Berberidac.).
lichuanensis *Z. Zheng & Y.J. Su in Acta Sci. Nat. Univ. Sunyatseni,* 36(2): 125 (1997)—China South-central (Hubei).

DYSOXYLUM (Meliac.).
mollissimum Blume
var. glaberrimum *(F.C. How & T.C. Chen) P.Y. Chen in Fl. Reipubl. Popul. Sin.,* 43(3): 95 (1997): D. hainanense var. glaberrimum.

E

EARINA (Orchidac.).
santoensis *T. Hashimoto in Ann. Tsukuba Bot. Gard.,* 16: 5 (1997), without type—Vanuatu.
sigmoidea *T. Hashimoto in Ann. Tsukuba Bot. Gard.,* 16: 1 (1997)—Vanuatu.

EBENOPSIS (Legumin.).
confinis *(Standl.) R.C. Barneby & J.W. Grimes, Silk Tree, Guanacaste, Monkey's Earring (Mem. New York Bot. Gard., 74(1)):* 177 (1996), isonym; *see Britton & Rose in N. Amer. Fl.,* 23: 33 (1928): Pithecellobium confine.
ebano *(Berland.) R.C. Barneby & J.W. Grimes, Silk Tree, Guanacaste, Monkey's Earring (Mem. New York Bot. Gard., 74(1)):* 175 (1996): Mimosa ebano.

EBERLANZIA (Aizoac.).
cyathiformis *(L. Bolus) H.E.K. Hartmann in Bradleya,* 16: 60 (1998):Ruschia cyathiformis.
dichotoma *(L. Bolus) H.E.K. Hartmann in Bradleya,* 16: 60 (1998): Ruschia dichotoma.
ebracteata *(L. Bolus) H.E.K. Hartmann in Bradleya,* 16: 60 (1998): Ruschia ebracteata.
gravida *(L. Bolus) H.E.K. Hartmann in Bradleya,* 16: 61 (1998): Ruschia gravida.
parvibracteata *(L. Bolus) H.E.K. Hartmann in Bradleya,* 16: 61 (1998): Ruschia parvibracteata.
schneideriana *(A. Berger) H.E.K. Hartmann in Bradleya,* 16: 61 (1998): Mesembryanthemum schneiderianum.

EBERTIA *F. Speta in Phyton (Austria),* 38(1): 65 (1998). *HYACINTHACEAE*
nana *(Oyewole) F. Speta in Phyton (Austria),* 38(1): 68 (1998): Urginea nana.
pauciflora *(Baker) F. Speta in Phyton (Austria),* 38(1): 68 (1998): Urginea pauciflora.

ECHEANDIA (Anthericac.).
subgen. Mscavea *R.W. Cruden in Novon,* 9(3): 326 (1999).
atoyacana *R.W. Cruden in Novon,* 9(3): 334 (1999)—Mexico Southwest (Guerrero), Mexico Central (México State).
attenuata *R.W. Cruden in Novon,* 9(3): 333 (1999)—Mexico Northwest (Sinaloa), Mexico Northeast (Durango).

ECHEANDIA:-
elegans *R.W. Cruden in Novon,* 9(3): 334 (1999)—Mexico Southwest (Guerrero), Mexico Central (Morelos).
falcata *R.W. Cruden in Novon,* 9(3): 332 (1999)—Mexico Northeast (Guanajuato, Querétaro).
hintonii *R.W. Cruden in Novon,* 9(3): 335 (1999)—Mexico Southwest (Guerrero).
hirticaulis *R.W. Cruden in Novon,* 9(3): 336 (1999)—Mexico Central (México State), Mexico Southwest (Guerrero, Michoacan).
mexiae *R.W. Cruden in Novon,* 9(3): 337 (1999)—Mexico Southwest (Guerrero), Mexico Central (Morelos).
nayaritensis *R.W. Cruden in Novon,* 9(3): 337 (1999)—Mexico Southwest (Nayarit), Mexico Northwest (Sinaloa).
palmeri *R.W. Cruden in Novon,* 9(3): 331 (1999)—Mexico Northeast (Durango, Chihuahua), Mexico Northwest (Sonora).
pseudopetiolata *R.W. Cruden in Novon,* 9(3): 337 (1999)—Mexico Southwest (Guerrero).
pseudoreflexa *R.W. Cruden in Novon,* 9(3): 330 (1999)—Mexico Southeast (Chiapas).
sanmiguelensis *R.W. Cruden in Novon,* 9(3): 333 (1999)—Mexico Northeast (Guanajuato).
tamaulipensis *R.W. Cruden in Novon,* 9(3): 334 (1999)—Mexico Northeast (Tamaulipas).
taxacana *R.W. Cruden in Novon,* 9(3): 332 (1999)—Mexico Central (México State, Morelos), Mexico Southwest (Guerrero, Oaxaca).
texensis *R.W. Cruden in Novon,* 9(3): 331 (1999)—Texas.

ECHEVERIA (Crassulac.).
bella
f. major *(E. Walther) M. Kimnach in Haseltonia,* 5: 51 (1997 publ. 1998): E. bella var. major.
calderoniae *E. Pérez-Calix in Acta Bot. Mex.,* 38: 9 (1997)—Mexico Northeast (Guanajuato).
cante *C. Glass & M. Mendoza-García in Cact. Succ. J. (U.S.A.),* 69(5): 241 (1997)—Mexico Northeast (Zacatecas).
chiclensis
f. backebergii *(Poelln.) M. Kimnach in Haseltonia,* 5: 51 (1997 publ. 1998): E. backebergii.
colorata
f. brandtii *(Kimnach) M. Kimnach in Haseltonia,* 5: 51 (1997 publ. 1998): E. colorata var. brandtii.
fulgens
var. obtusifolia *(Rose) M. Kimnach in Haseltonia,* 5: 51 (1997 publ. 1998): E. obtusifolia.
halbingeri
var. goldiana *(E. Walther) M. Kimnach in Haseltonia,* 5: 51 (1997 publ. 1998): E. goldiana.
var. sanchez-mejoradae *(E. Walther) M. Kimnach in Haseltonia,* 5: 51 (1997 publ. 1998): E. sanchez-mejoradae.
kimnachii *J. Meyrán G. & R. Vega A. in Cact. Suc. Mex.,* 43(1): 7 (1998)—Mexico Northwest (Sinaloa).
paniculata
var. maculata *(Rose) M. Kimnach in Haseltonia,* 5: 51 (1997 publ. 1998): E. maculata.
pringlei
var. longisepala *M. Kimnach in Haseltonia,* 5: 51 (1997 publ. 1998)—Mexico Southwest (Jalisco).
var. parva *M. Kimnach in Haseltonia,* 5: 52 (1997 publ. 1998)—Mexico Northeast (Durango).
rodolfoi *J.G. Martínez-Ávalos & A. Mora-Olivo in Acta Bot. Mex.,* 52: 43 (2000), as 'rodolfi'—Mexico Northeast (Tamaulipas).
secunda
f. byrnesii *(Rose) M. Kimnach in Haseltonia,* 5: 52 (1997 publ. 1998): E. byrnesii.
xichuensis *L.G. López Chávez & J. Reyes in Cact. Suc. Mex.,* 43(3): 65 (1998)—Mexico Northeast (Guanajuato).

ECHINOCACTUS (Cactac.).
buekii *Klein in Gartenflora,* 8: 257, t. 266 (1859), as 'buckii' but see p. 380 & t. 266—Mexico.
horizonthalonius Lem.
subsp. jarmilae *J.J. Halda & L. Horáček in Acta Mus. Richnov. Sect. Nat.,* 7(1): 34 (2000)—Mexico Northeast (Tamaulipas).

ECHINOCACTUS:-
horizonthalonius Lem.
var. subikii *Staník & Dráb in Cactaceae etc.,* 8(1): 8–9 (1998); *cf. Repert. Pl. Succ. (I.O.S.),* 49: 15 (1998 publ. 1999)—Locality not stated.
mexicanus *(Glass & W.A. Fitz Maur.) J.J. Halda in Acta Mus. Richnov. Sect. Nat.,* 7(2): 72 (2000): Geohintonia mexicana.
polycephalus
subsp. xeranthemoides *(J.M. Coult.) N.P. Taylor in Cactaceae Consensus Initiatives,* 5: 12 (1998): E. polycephalus [var.] xeranthemoides.

ECHINOCEREUS (Cactac.).
subgen. Morangaya *(G.D. Rowley) Michael Lange in W. Blum et al., Echinocereus (preprint):* [1] (1998): gen. Morangaya.
sect. Subinermes *(K. Schum.) Michael Lange in Echinocereenfreund,* 8(1): 16 (1995); *cf. Repert. Pl. Succ. (I.O.S.),* 46: 10 (1995 publ. 1996): E. ser. Subinermes.
subgen. Triglochidiati *(Bravo) W. Blum, Michael Lange & J. Rutow in W. Blum et al., Echinocereus (preprint):* [1] (1998): E. sect. Triglochidiata.
acifer (Otto ex Salm-Dyck) Jacobi
subsp. tubiflorus *W. Rischer in W. Blum et al., Echinocereus (preprint):* [1] (1998)—Mexico Northeast (Durango, Zacatecas).
adustus
subsp. bonatzii *(R.C. Römer) N.P. Taylor in Cactaceae Consensus Initiatives,* 6: 15 (1998): E. bonatzii.
subsp. schwarzii *(A.B. Lau) N.P. Taylor in Cactaceae Consensus Initiatives,* 3: 8 (1997): E. schwarzii.
apachensis *W. Blum & J. Rutow in W. Blum et al., Echinocereus (preprint):* [1] (1998)—Arizona.
arizonicus Rose ex Orcutt
subsp. nigrihorridispinus *W. Blum & J. Rutow in W. Blum et al., Echinocereus (preprint):* [3] (1998)—New Mexico.
bonatzii *R.C. Römer in Echinocereenfreund,* 8(4): 98–113 (1995); *cf. Repert. Pl. Succ. (I.O.S.),* 46: 10 (1996)—Locality not stated.
bristolii W.T. Marshall
subsp. floresii *(Backeb.) W. Blum & Michael Lange in W. Blum et al., Echinocereus (preprint):* [3] (1998): E. floresii.
carmenensis *W. Blum, Michael Lange & Scherer in W. Blum et al., Echinocereus (preprint):* [3] (1998)—Mexico Northeast (Coahuila).
chloranthus (Engelm.) Haage
subsp. cylindricus *(Engelm.) W. Blum & Michael Lange in W. Blum et al., Echinocereus (preprint):* [3] (1998): Cereus viridiflorus var. cylindricus.
subsp. rhyolithensis *W. Blum & Michael Lange in W. Blum et al., Echinocereus (preprint):* [4] (1998)—New Mexico.
cinerascens (DC.) Haage
subsp. ehrenbergii *(Pfeiffer) W. Blum & J. Rutow in W. Blum et al., Echinocereus (preprint):* [4] (1998): Cereus ehrenbergii.
subsp. septentrionalis *(N.P. Taylor) N.P. Taylor in Cactaceae Consensus Initiatives,* 3: 8 (1997): E. cinerascens var. septentrionalis.
subsp. tulensis *(Bravo) N.P. Taylor in Cactaceae Consensus Initiatives,* 3: 8 (1997): E. tulensis.
coccineus Engelm.
subsp. aggregatus *(Engelm. ex S. Watson) W. Blum, Michael Lange & J. Rutow in W. Blum et al., Echinocereus (preprint):* [4] (1998): Mammillaria aggregata.
subsp. mojavensis *(Engelm. & J.M. Bigelow) N.P. Taylor in Cactaceae Consensus Initiatives,* 3: 8 (1997): Cereus mojavensis.
subsp. paucispinus *(Engelm.) W. Blum, Michael Lange & J. Rutow in W. Blum et al., Echinocereus (preprint):* [4] (1998): Cereus paucispinus.
subsp. roemeri *(Muehlenpf.) W. Blum, Michael Lange & J. Rutow in W. Blum et al., Echinocereus (preprint):* [4] (1998): Cereus roemeri.
subsp. rosei *(Wooton & Standl.) W. Blum & J. Rutow in W. Blum et al., Echinocereus (preprint):* [4] (1998): E. rosei.

ECHINOCEREUS:-

dasyacanthus Engelm.

subsp. rectispinus *(Trocha & Fethke) W. Blum, W. Rischer & J. Rutow in W. Blum et al., Echinocereus (preprint):* [2] (1998): E. dasyacanthus var. rectispinus.

engelmannii

subsp. decumbens *(Clover & Jotter) W. Blum & Michael Lange in W. Blum et al., Echinocereus (preprint):* [2] (1998): E. decumbens.

subsp. fasciculatus *(Engelm. ex Watson) W. Blum & Michael Lange in W. Blum et al., Echinocereus (preprint):* [2] (1998): Mammillaria fasciculata.

subsp. llanuraensis *(Rutow) R.S. Felger in Proc. San Diego Soc. Nat. Hist.,* 35: 20 (1999): E. nicholii subsp. llanuraensis.

enneacanthus

subsp. brevispinus *(W.O. Moore) N.P. Taylor in Cactaceae Consensus Initiatives,* 3: 8 (1997): E. enneacanthus f. brevispinus.

fasciculatus

subsp. bonkerae *(Thornber & Bonker) N.P. Taylor in Cactaceae Consensus Initiatives,* 3: 8 (1997): E. bonkerae.

subsp. boyce-thompsonii *(Orcutt) N.P. Taylor in Cactaceae Consensus Initiatives,* 3: 8 (1997): E. boyce-thompsonii.

fendleri (Engelm.) Seitz

subsp. hempelii *(Fobe) W. Blum in W. Blum et al., Echinocereus (preprint):* [2] (1998): E. hempelii.

subsp. rectispinus *(Peebles) N.P. Taylor in Cactaceae Consensus Initiatives,* 3: 9 (1997): E. rectispinus.

ferreirianus

subsp. lindsayi *(J. Meyrán) N.P. Taylor in Cactaceae Consensus Initiatives,* 3: 9 (1997): E. lindsayi.

fitchii Britton & Rose

subsp. albertii *(L.D. Benson) W. Blum & Michael Lange in W. Blum et al., Echinocereus (preprint):* [5] (1998): E. reichenbachii var. albertii.

subsp. armatus *(Poselg.) W. Blum, Michael Lange & J. Rutow in W. Blum et al., Echinocereus (preprint):* [5] (1998): Cereus pectinatus var. armatus.

fobeanus Oehme

subsp. metornii *(G. Frank) W. Blum & Michael Lange in W. Blum et al., Echinocereus (preprint):* [5] (1998): E. metornii.

klapperi *W. Blum in W. Blum et al., Echinocereus (preprint):* [5] (1998)—Mexico Northwest (Sonora).

knippelianus

subsp. kaplani *J.J. Halda & J. Sladkovský in Acta Mus. Richnov. Sect. Nat.,* 7(1): 34 (2000)—Mexico Northeast (Nuevo León).

subsp. kruegeri *(Glass & R.A. Foster) C.E. Glass, Guía Identific. Cact. Amenazadas México,* 1: Echinoce/kn ssp. kruegeri (1998 publ. 1997): E. knippelianus var. kruegeri.

subsp. kruegeri *(Glass & R.A. Foster) W. Blum & Michael Lange in W. Blum et al., Echinocereus (preprint):* [5] (1998), isonym; *see C.E. Glass, Guía Identific. Cact. Amenazadas México,* 1: Echinoce/kn ssp. kruegeri (1998 publ. 1997): E. knippelianus var. kruegeri.

subsp. reyesii *(A.B. Lau) W. Blum & Michael Lange in W. Blum et al., Echinocereus (preprint):* [6] (1998): E. knippelianus var. reyesii.

longisetus (Engelm.) Lem.

subsp. delaetii *(Gürke) N.P. Taylor in Cactaceae Consensus Initiatives,* 3: 9 (1997): Cephalocereus delaetii.

subsp. freudenbergeri *(G. Frank) W. Blum in W. Blum et al., Echinocereus (preprint):* [6] (1998): E. freudenbergeri.

mapimiensis *E.F. Anderson, Wendy Hodgson & P. Quirk in Cact. Succ. J. (U.S.A.),* 70(6): 285 (1998)—Mexico Northeast (Coahuila).

maritimus (M.E. Jones) K. Schum.

subsp. hancockii *(E.Y. Dawson) W. Blum & J. Rutow in W. Blum et al., Echinocereus (preprint):* [6] (1998): E. hancockii.

ECHINOCEREUS:-

milleri *W. Blum et al. in Echinocereenfreund,* 12(3): 69–76 (1999); *see also Echinocereenfreund,* 12(4): 94 (1999); *cf. Repert. Pl. Succ. (I.O.S.),* 50: 12 (1999 publ. 2000)—Locality not stated.

neocapillus *(Weniger) W. Blum & Michael Lange in W. Blum et al., Echinocereus (preprint):* [6] (1998): E. chloranthus var. neocapillus.

nicholii

subsp. llanuraensis *Rutow in Echinocereenfreund,* 1995(3): 61–71 (1995); *cf. Repert. Pl. Succ. (I.O.S.),* 46: 10 (1995 publ. 1996)—Locality not stated.

ortegae

subsp. koehresianus *(G. Frank) W. Rischer & G.R.W. Frank in Echinocereenfreund,* 9(4): 100 (1996); *cf. Repert. Pl. Succ. (I.O.S.),* 47: 9 (1996 publ. 1997): E. scheeri var. koehresianus.

pacificus (Engelm.) J.N. Haage

subsp. mombergerianus *(G. Frank) W. Blum, W. Rischer & J. Rutow in W. Blum et al., Echinocereus (preprint):* [6] (1998): E. mombergerianus.

pamanesiorum

subsp. bonatzii *(R.C. Römer) R.C. Römer in Echinocereenfreund,* 10(2): 45 (1997); *cf. Repert. Pl. Succ. (I.O.S.),* 48: 8 (1997 publ. 1998): E. bonatzii.

parkeri N.P. Taylor

subsp. arteagensis *W. Blum & Michael Lange in W. Blum et al., Echinocereus (preprint):* [7] (1998)—Mexico Northeast (Coahuila, Nuevo León).

f. arteagensis *(W. Blum & M. Lange) G.R.W. Frank & E. Scherer in Echinocereenfreund,* 11 (Sonderausgabe 1998): 61 (1998), contrary to Art. 33.2 ICBN (1994); *cf. Repert. Pl. Succ. (I.O.S.),* 49: 16 (1998 publ. 1999): E. parkeri subsp. arteagensis.

subsp. gonzalezii *(N.P. Taylor) N.P. Taylor in Cactaceae Consensus Initiatives,* 3: 9 (1997): E. parkeri var. gonzalezii.

subsp. mazapilensis *W. Blum & Michael Lange in W. Blum et al., Echinocereus (preprint):* [6] (1998)—Mexico Northeast (Zacatecas).

pectinatus (Scheidw.) Engelm.

subsp. ctenoides *(Engelm.) G.R.W. Frank, Echinocereus pectinatus-E. dasyacanthus Gruppe (Echinocereenfreund, 10* Jubiläums-sonderausgabe): 135 (1997): Cereus ctenoides.

subsp. wenigeri *(L.D. Benson) W. Blum & J. Rutow in W. Blum et al., Echinocereus (preprint):* [7] (1998): E. pectinatus var. wenigeri.

pentalophus (DC.) Haage

subsp. leonensis *(Mathsson) N.P. Taylor in Cactaceae Consensus Initiatives,* 3: 9 (1997): E. leonensis.

subsp. procumbens *(Engelm.) W. Blum & Michael Lange in W. Blum et al., Echinocereus (preprint):* [7] (1998): Cereus procumbens.

polyacanthus

subsp. acifer *(Otto ex Salm-Dyck) N.P. Taylor in Cactaceae Consensus Initiatives,* 3: 9 (1997), without basionym ref.: Basionym not stated.

subsp. huitcholensis *(F.A.C. Weber) N.P. Taylor in Cactaceae Consensus Initiatives,* 3: 9 (1997): Cereus huitcholensis.

subsp. mombergerianus *(G. Frank) Breckwoldt in Echinocereenfreund,* 9(Sonderheft): 15 (1996); *cf. Repert. Pl. Succ. (I.O.S.),* 47: 9 (1996 publ. 1997): E. mombergerianus.

subsp. pacificus *(Engelm.) Breckwoldt in Echinocereenfreund,* 9(Sonderheft): 15 (1996); *cf. Repert. Pl. Succ. (I.O.S.),* 47: 9 (1996 publ. 1997): Cereus phoeniceus var. pacificus.

pulchellus (Mart.) Seitz

subsp. acanthosetus *(S. Arias & U. Guzmán) W. Blum in W. Blum et al., Echinocereus (preprint):* [8] (1998): E. pulchellus var. acanthosetus.

subsp. acanthosetus *(S. Arias & L.U. Guzmán) N.P. Taylor in Cactaceae Consensus Initiatives,* 5: 12 (1998): E. pulchellus var. acanthosetus.

var. acanthosetus *S. Arias Montes & L.U. Guzmán Cruz in Fl. Valle Tehuacán-Cuicatlán,* 14(Cactac.): 63 (1997)—Mexico Southwest (Oaxaca).

ECHINOCEREUS:-

pulchellus (Mart.) Seitz

var. costatus *J.J. Halda, P. Kupčák & J. Sladkovský in Acta Mus. Richnov. Sect. Nat.,* 7(1): 34 (2000)—Mexico Northeast (San Luis Potosí/Zacatecas border).

subsp. sharpii *(N.P. Taylor) N.P. Taylor in Cactaceae Consensus Initiatives,* 3: 9 (1997): E. pulchellus var. sharpii.

subsp. sladkovskyi *J.J. Halda & P. Kupčák in Acta Mus. Richnov. Sect. Nat.,* 7(1): 34 (2000)—Mexico Northeast (Hidalgo).

subsp. venustus *W. Blum & W. Rischer in W. Blum et al., Echinocereus (preprint):* [7] (1998)—Mexico Northeast (San Luis Potosí).

subsp. weinbergii *(Weing.) N.P. Taylor in Cactaceae Consensus Initiatives,* 3: 9 (1997): E. weinbergii.

reichenbachii (Terscheck ex Fennel) Linke

subsp. armatus *(Poselg.) N.P. Taylor in Cactaceae Consensus Initiatives,* 3: 9 (1997): Cereus pectinatus [var.] armatus.

subsp. baileyi *(Rose) N.P. Taylor in Cactaceae Consensus Initiatives,* 3: 9 (1997): E. baileyi.

subsp. caespitosus *(Engelm.) W. Blum & Michael Lange in W. Blum et al., Echinocereus (preprint):* [8] (1998): Cereus caespitosus.

subsp. fitchii *(Britton & Rose) N.P. Taylor in Cactaceae Consensus Initiatives,* 3: 9 (1997): E. fitchii.

subsp. perbellus *(Britton & Rose) N.P. Taylor in Cactaceae Consensus Initiatives,* 3: 9 (1997): E. perbellus.

rigidissimus

subsp. rubispinus *(G. Frank & A.B. Lau) N.P. Taylor in Cactaceae Consensus Initiatives,* 3: 9 (1997): E. pectinatus var. rubispinus.

russanthus Weniger

subsp. fiehnii *(Trocha) W. Blum & Michael Lange in W. Blum et al., Echinocereus (preprint):* [8] (1998): E. russanthus var. fiehnii.

var. fiehnii *W. Trocha in Echinocereenfreund,* 10(3): 69 (1997); *cf. Repert. Pl. Succ. (I.O.S.),* 48: 8 (1997 publ. 1998)—Locality not stated.

subsp. weedinii *Leuck ex W. Blum & Michael Lange in W. Blum et al., Echinocereus (preprint):* [8] (1998)—Texas.

salm-dyckianus

subsp. bacanorensis *Rischer & Trocha in Echinocereenfreund,* 11(1): 3–10 (1998); *et in Echinocereenfreund,* 11(2): 31 (1998); *cf. Repert. Pl. Succ. (I.O.S.),* 49: 17 (1998 publ. 1999)—Locality not stated.

f. obscuriensis *(A.B. Lau) W. Rischer & W. Trocha in Echinocereenfreund,* Sonderausgabe 1999: 10 (1999); *cf. Repert. Pl. Succ. (I.O.S.),* 50: 12 (1999 publ. 2000): E. scheeri var. obscuriensis.

sanpedroensis *Raudonat & Rischer in Echinocereenfreund,* 8(4): 88–97 (1995); *cf. Repert. Pl. Succ. (I.O.S.),* 46: 10 (1995 publ. 1996)—Locality not stated.

santaritensis *W. Blum & J. Rutow in W. Blum et al., Echinocereus (preprint):* [9] (1998)—Arizona.

scheeri

subsp. gentryi *(Clover) N.P. Taylor in Cactaceae Consensus Initiatives,* 3: 9 (1997): E. gentryi.

subsp. paridensis *Rischer & Trocha in Echinocereenfreund,* 11(3): 58–65 (1998); *cf. Repert. Pl. Succ. (I.O.S.),* 49: 17 (1998 publ. 1999)—Locality not stated.

var. salm-dyckianus *(Scheer) Lodé in Cact. Aventures,* 27: 29 (1995), contrary to Art. 33.2 ICBN (1994); *cf. Repert. Pl. Succ. (I.O.S.),* 46: 10 (1995 publ. 1996): E. salm-dyckianus.

sciurus

subsp. floresii *(Backeb.) N.P. Taylor in Cactaceae Consensus Initiatives,* 3: 10 (1997): E. floresii.

scopulorum Britton & Rose

subsp. pseudopectinatus *(N.P. Taylor) W. Blum & Michael Lange in W. Blum et al., Echinocereus (preprint):* [9] (1998): E. bristolii var. pseudopectinatus.

stoloniferus

subsp. tayopensis *(W.T. Marshall) N.P. Taylor in Cactaceae Consensus Initiatives,* 3: 10 (1997): E. tayopensis.

ECHINOCEREUS:-
stramineus
subsp. occidentalis *(N.P. Taylor) N.P. Taylor in Cactaceae Consensus Initiatives,* 3: 10 (1997): E. stramineus var. occidentalis.
subinermis
subsp. ochoterenae *(J.G. Ortega) N.P. Taylor in Cactaceae Consensus Initiatives,* 3: 10 (1997): E. ochoterenae.
topiensis *W. Rischer & W. Trocha in Echinocereenfreund,* 12(1): 6–11 (1999); *cf. Repert. Pl. Succ. (I.O.S.),* 50: 12 (1999 publ. 2000)—Locality not stated.
triglochidiatus Engelm.
f. inermis *(K. Schum.) W. Blum in W. Blum et al., Echinocereus (preprint):* [9] (1998): E. phoeniceus var. inermis.
subsp. mojavensis *(Engelm. & J.M. Bigelow) W. Blum & Michael Lange in W. Blum et al., Echinocereus (preprint):* [9] (1998): Cereus mojavensis.
viereckii Werderm.
subsp. huastecensis *W. Blum, Michael Lange & J. Rutow in W. Blum et al., Echinocereus (preprint):* [9] (1998)—Mexico Northeast (Nuevo León).
subsp. morricalii *(Ríha) N.P. Taylor in Cactaceae Consensus Initiatives,* 3: 10 (1997): E. morricalii.
viridiflorus Engelm.
subsp. chloranthus *(Engelm.) N.P. Taylor in Cactaceae Consensus Initiatives,* 3: 10 (1997): Cereus chloranthus.
subsp. correllii *(L.D. Benson) W. Blum & Michael Lange in W. Blum et al., Echinocereus (preprint):* [10] (1998): E. viridiflorus var. correllii.
subsp. cylindricus *(Engelm.) N.P. Taylor in Cactaceae Consensus Initiatives,* 3: 10 (1997): Cereus viridiflorus var. cylindricus.
subsp. davisii *(Houghton) N.P. Taylor in Cactaceae Consensus Initiatives,* 3: 10 (1997): E. davisii.

ECHINOCHLOA (Gramin.).
dietrichiana *P.W. Michael in Telopea,* 8(3): 403 (1999)—Queensland, Northern Territory.

ECHINODORUS (Alismatac.).
barthii *H. Mühlberg in Aquar. Terr. (Halle),* 33(11): 369 (1986)—Southern South America.
floridanus *R.R. Haynes & J.R. Burkhalter in Castanea,* 63(2): 180 (1998)—Florida.
gabrieli *K. Rataj in Akvár. Terár.,* 33: 15 (1990)—Brazil.

ECHINOFOSSULOCACTUS (Cactac.).
multicostatus
var. zacatecasensis *(Britton & Rose) Lodé in Cact. Aventures,* 27: 29 (1995); *cf. Repert. Pl. Succ. (I.O.S.),* 46: 10 (1995 publ. 1996): E. zacatecasensis.

ECHINOMASTUS (Cactac.).
gautii *(L.D. Benson) A. Mosco & C. Zanovello in Bradleya,* 15: 81 (1997): Neolloydia gautii.
mariposensis
subsp. fischeri *J.J. Halda & J. Sladkovský in Acta Mus. Richnov. Sect. Nat.,* 7(1): 35 (2000)—Mexico Northeast (Coahuila).
unguispinus
*subsp.*laui *(G. Frank & Zecher) C.E. Glass, Guía Identific. Cact. Amenazadas México,* 1: Echinom/un ssp. laui (1998 publ. 1997): E. laui.

ECHINOPEPON (Cucurbitac.).
arachoideus *(Dieterle) A.K. Monro & P.J. Stafford in Ann. Missouri Bot. Gard.,* 85(2): 268 (1998), as 'arachoides': Apatzingania arachoidea.
belizensis *A.K. Monro & P.J. Stafford in Ann. Missouri Bot. Gard.,* 85(2): 269 (1998)—Belize.
bigelovii *(S. Watson) S. Watson in Proc. Amer. Acad. Arts,* 24: 52 (1889): Elaterium bigelovii.
gemellus *(DC.) A.K. Monro & P.J. Stafford in Ann. Missouri Bot. Gard.,* 85(2): 267 (1998): Elaterium gemellum.
glutinosus *(Cogn.) A.K. Monro & P.J. Stafford in Ann. Missouri Bot. Gard.,* 85(2): 268 (1998): Echinocystis glutinosa.

ECHINOPEPON:-
micropaniculatus *A.K. Monro & P.J. Stafford in Ann. Missouri Bot. Gard.,* 85(2): 269 (1998)—Costa Rica.
minimus *(S. Watson) S. Watson in Proc. Amer. Acad. Arts,* 24: 52 (1889): Elaterium minimum.
parviflorus *(S. Watson) S. Watson in Proc. Amer. Acad. Arts,* 24: 52 (1889): Echinocystis parviflora.
tultitlanapaensis *A.K. Monro & P.J. Stafford in Ann. Missouri Bot. Gard.,* 85(2): 265 (1998)—Mexico Central (Puebla).

ECHINOPHORA (Umbellif.).
lamondiana *B. Yıldız & Z. Bahçecioğlu in Karaca Arbor. Mag.,* 4(1): 7 (1997)—Turkey.

ECHINOPS (Compos.).
abuzinadianus *S.A. Chaudhary in Fl. Kingdom Saudi Arabia,* 2(3): 198, 418 (2000), contrary to Art. 37.6 ICBN (2000)—Saudi Arabia.
latifolius Tausch
var. manshuricus *(Kitag.) C.Y. Li in Clavis Pl. Chinae Bor.-Or. ed. 2 (ed. P.Y. Fu):* 694 (1995): E. manshuricus.
rajasthanensis *R.P. Pandey & V. Singh in J. Econ. Taxon. Bot.,* 21(2): 467 (1997), as 'rajasthanense'—India (Rajasthan).
spinosus L.
subsp. fontqueri *(Pau) B. Valdés in Lagascalia,* 18(2): 306 (1996): E. fontqueri.

ECHINOPSIS (Cactac.).
arboricola *(Kimnach) R. Mottram in Cactaceae Consensus Initiatives,* 2: 8 (1997): Trichocereus arboricola.
atacamensis (Phil.) H. Friedrich & G.D. Rowley
subsp. pasacana *(F.A.C. Weber) G. Navarro in Lazaroa,* 17: 54 (1996): Pilocereus pasacanus.
clavatus *(F. Ritter) D.R. Hunt in Cactaceae Consensus Initiatives,* 4: 5 (1997): Trichocereus clavatus.
densispina Werderm.
var. amblayensis *(Rausch) J.G. Lambert, Cactus d'Argentine, ed. 2:* 50 (1998): Lobivia amblayensis.
var. pectinifera *(Wessner) J.G. Lambert, Cactus d'Argentine, ed. 2:* 51 (1998): Lobivia pectinifera.
dobeana *(Dölz) J.G. Lambert, Cactus d'Argentine, ed. 2:* 51 (1998): Lobivia dobeana.
fallax (Oehme) H. Friedrich
var. cylindrica *(Backeb.) J.G. Lambert, Cactus d'Argentine, ed. 2:* 55 (1998): Lobivia cylindrica.
haematantha *(Speg.) J.G. Lambert, Cactus d'Argentine, ed. 2:* 57 (1998), isonym; *see D.R. Hunt in Bradleya,* 9: 88 (1991): Echinocactus haematanthus.
var. chorrillosensis *(Rausch) J.G. Lambert, Cactus d'Argentine, ed. 2:* 57 (1998): Lobivia chorrillosensis.
var. elongata *(Backeb.) J.G. Lambert, Cactus d'Argentine, ed. 2:* 57 (1998): Lobivia elongata.
var. hualfinensis *(Rausch) J.G. Lambert, Cactus d'Argentine, ed. 2:* 57 (1998): Lobivia hualfinensis.
var. jasimanensis *(Rausch) J.G. Lambert, Cactus d'Argentine, ed. 2:* 57 (1998): Lobivia haematantha var. jasimanensis.
var. kuehnrichii *(Frič) J.G. Lambert, Cactus d'Argentine, ed. 2:* 59 (1998): Lobivia kuehnrichii.
hahniana *(Backeb.) R.S. Wallace in Cactaceae Consensus Initiatives,* 4: 12 (1997): Mediocactus hahnianus.
huotii (Cels) Labour.
subsp. vallegrandensis *(Cárdenas) G. Navarro in Lazaroa,* 17: 50 (1996): E. vallegrandensis.
obrepanda (Salm-Dyck) K. Schum.
subsp. calorubra *(Cárdenas) G. Navarro in Lazaroa,* 17: 50 (1996): E. calorubra.
subsp. tapecuana *(F. Ritter) G. Navarro in Lazaroa,* 17: 51 (1996): E. tapecuana.
omasensis *(Ostolaza & Mischler) C. Ostolaza Nano in Cactaceae Consensus Initiatives,* 6: 8 (1998): Islaya omasensis.
oxygona (Link) Zucc.
f. grandiflora *(Rud. Meyer) J. Ullmann in Kaktusy (Brno),* 35(4): 110 (1999): Echinopsis eyriesii var. grandiflora.

ECHINOPSIS:-
pentlandii (Hook.) Salm-Dyck
subsp. hardeniana *(Boed.) G. Navarro in Lazaroa,* 17: 53 (1996): Lobivia hardeniana.
subsp. larae *(Cárdenas) G. Navarro in Lazaroa,* 17: 53 (1996): Lobivia larae.
peruviana
subsp. puquiensis *(Rauh & Backeb.) C. Ostolaza Nano in Cactaceae Consensus Initiatives,* 6: 8 (1998): Trichocereus puquiensis.
pugionacantha Rose & Boed.
subsp. rossii *(Boed.) G. Navarro in Lazaroa,* 17: 53 (1996): Lobivia rossii.
quadratiumbonatus *(F. Ritter) D.R. Hunt in Cactaceae Consensus Initiatives,* 4: 5 (1997): Trichocereus quadratiumbonatus.
saltensis Speg.
var. nealeana *(Backeb.) J.G. Lambert, Cactus d'Argentine, ed. 2:* 65 (1998): Lobivia nealeana.
var. pseudocachensis *(Backeb.) J.G. Lambert, Cactus d'Argentine, ed. 2:* 66 (1998): Lobivia pseudocachensis.
scopulicola *(F. Ritter) R. Mottram in Cactaceae Consensus Initiatives,* 2: 8 (1997): Trichocereus scopulicola.
stilowiana *(Backeb.) J.G. Lambert, Cactus d'Argentine, ed. 2:* 70 (1998): Lobivia stilowiana.
tacaquirensis (Vaupel) H. Friedrich & G.D. Rowley
subsp. taquimbalensis *(Cárdenas) G. Navarro in Lazaroa,* 17: 55 (1996): Trichocereus taquimbalensis.
tarijensis (Vaupel) H. Friedrich & G.D. Rowley
subsp. herzogianus *(Cárdenas) G. Navarro in Lazaroa,* 17: 56 (1996): Trichocereus herzogianus.
subsp. totorensis *(Cárdenas) G. Navarro in Lazaroa,* 17: 56 (1996): Trichocereus herzogianus var. totorensis.

ECHINOSPARTUM (Legumin.).
algibicum *S. Talavera & A. Aparicio in Acta Bot. Malacitana,* 20: 297 (1995)—Spain.

ECHIOCHILON (Boraginac.).
baricum *E. Lönn in Bot. J. Linn. Soc.,* 130(3): 229 (1999)—Somalia.
callianthum *E. Lönn in Bot. J. Linn. Soc.,* 130(3): 246 (1999)—South Yemen, Oman.
cyananthum *E. Lönn in Bot. J. Linn. Soc.,* 130(3): 231 (1999)—Somalia.

ECHIOGLOSSUM (Orchidac.).
chantaburiense *(Seidenf.) D.L. Szlachetko in Fragm. Flor. Geobot.,* Supp. 3: 137 (1995): Cleisostoma chantaburiense.
complicatum *(Seidenf.) D.L. Szlachetko in Fragm. Flor. Geobot.,* Supp. 3: 137 (1995): Sarcanthus complicatus.
sacculatum *(Ridl.) D.L. Szlachetko in Fragm. Flor. Geobot.,* Supp. 3: 137 (1995): Sarcanthus sacculatus.
seidenfadenii *(Garay) D.L. Szlachetko in Fragm. Flor. Geobot.,* Supp. 3: 137 (1995): Cleisostoma seidenfadenii.
simondii *(Gagnep.) D.L. Szlachetko in Fragm. Flor. Geobot.,* Supp. 3: 137 (1995): Vanda simondii.
teretifolium *(Teijsm. & Binn.) D.L. Szlachetko in Fragm. Flor. Geobot.,* Supp. 3: 137 (1995): Cleisostoma teretifolium.
williamsonii *(Rchb. f.) D.L. Szlachetko in Fragm. Flor. Geobot.,* Supp. 3: 137 (1995): Sarcanthus williamsonii.

ECHITES (Apocynac.).
blumii *J.O. Voigt, Hort. Suburb. Calcutt.:* 528 (1845), nom. nov.: E. caryophyllata *Blume.*
darienensis *J.F. Morales in Brittonia,* 49(3): 330 (1997)—Panama.
grandiflora *W. Roxburgh, Pl. Coromandel,* 3: 78 (1820)—Bangladesh.†
puntarenensis *J.F. Morales in Brittonia,* 49(3): 332 (1997), nom. nov.: Prestonia caudata *Woodson.*

ECHIUM (Boraginac.).
acutifolium *J.G.C. Lehmann in Gött. Gel. Anz.,* 152: 1514 (1817)—Iran?
anchusoides *G. Bacchetta, S. Brullo & F. Selvi in Nordic J. Bot.,* 20(3): 273 (2000)—Sardegna.

ECHIUM:-
strigosum *J.G.C. Lehmann in Gött. Gel. Anz.*, 152: 1514 (1817)—Cape Province.
swartzii *J.G.C. Lehmann in Gött. Gel. Anz.*, 152: 1513 (1817)—Cape Province.
verrucosum *J.G.C. Lehmann in Gött. Gel. Anz.*, 152: 1514 (1817)—Cape Province.

ECLIPTA (Compos.).
prostrata (L.) L.
f. aureoreticulata *Y.T. Chang in Wuyi Sci. J.*, 5: 235 (1985), as 'aureo reticulata'—China Southeast (Fujian).
var. dixitii *Anand Kumar & K.K. Khanna in J. Econ. Taxon. Bot.*, 23(3): 713 (1999)—India (Madhya Pradesh).

ECUA *D.J. Middleton in Blumea*, 41(1): 33 (1996). *APOCYNACEAE*
moluccensis *D.J. Middleton in Blumea*, 41(1): 34 (1996)—Moluccas.

ECUADENDRON *D.A. Neill in Novon*, 8(1): 45 (1998). *LEGUMINOSAE*
acosta-solisianum *D.A. Neill in Novon*, 8(1): 45 (1998)—Ecuador.

EDMUNDOA *E.M.C. Leme, Canistrum - Bromeliads Atlantic Forest:* 42 (1997). *BROMELIACEAE*
ambigua *(Wand. & Leme) E.M.C. Leme, Canistrum - Bromeliads Atlantic Forest:* 42, 52 (1997): Nidularium ambiguum.
lindenii *(Regel) E.M.C. Leme, Canistrum - Bromeliads Atlantic Forest:* 46 (1997): Nidularium lindenii.
var. rosea *(E. Morren) E.M.C. Leme, Canistrum - Bromeliads Atlantic Forest:* 50 (1997): Canistrum roseum.
perplexa *(L.B. Sm.) E.M.C. Leme, Canistrum - Bromeliads Atlantic Forest:* 54 (1997): Canistrum perplexum.

EHRENDORFERIA *T. Fukuhara & M. Lidén in Pl. Syst. Evol.*, 206(1–4): 415 (1997), nom. nov.: Dicentra subgen. Chrysocapnos *Engelm.* (Papaverac.)
chrysantha *(Hook. & Arn.) J. Rylander in Pl. Syst. Evol.*, 206(1–4): 415 (1997): Dielytra chrysantha.
ochroleuca *(Engelm.) T. Fukuhara in Pl. Syst. Evol.*, 206(1–4): 415 (1997): Dicentra ochroleuca.

EHRETIA (Boraginac.).
dolichandra *R.R. Mill in Edinburgh J. Bot.*, 53(2): 265 (1996)—Assam (Arunachal Pradesh).
exsoluta *R.R. Mill in Edinburgh J. Bot.*, 53(2): 268 (1996)—Cultivation.
psilosiphon *R.R. Mill in Edinburgh J. Bot.*, 53(2): 260 (1996)—Bhutan, India (West Bengal), Nepal.
silvana *R.R. Mill in Edinburgh J. Bot.*, 53(2): 266 (1996), nom. nov.: Cordia acuminata *Wall.*

EIDOTHEA *A.W. Douglas & B.P.M. Hyland in Fl. Australia*, 16: 472 (1995). *PROTEACEAE*
zoexylocarya *A.W. Douglas & B.P.M. Hyland in Fl. Australia*, 16: 473 (1995)—Queensland.

EKIMIA *H. Duman & M.F. Watson in Edinburgh J. Bot.*, 56(2): 200 (1999). *UMBELLIFERAE*
bornmuelleri *(Hub.-Mor. & Reese) H. Duman & M.F. Watson in Edinburgh J. Bot.*, 56(2): 200 (1999): Prangos bornmuelleri.

ELAEAGNUS (Elaeagnac.).
epitricha *Momiy. ex H. Ohba in Fl. Japan*, 2c: 157 (1999)—Japan.
geniculata *D. Fang in Acta Phytotax. Sin.*, 38(3): 291 (2000)—China Southeast (Guangxi).
griffithii Servett.
var. multiflora *C.Y. Chang in J. Sichuan Univ., Nat. Sci. Ed.*, 1984(4): 93 (1984)—China Southeast (Guangxi).
heterophylla *D. Fang & D.R. Liang in Acta Phytotax. Sin.*, 38(3): 292 (2000)—China Southeast (Guangxi).
obovatifolia *D. Fang in Acta Phytotax. Sin.*, 38(3): 289 (2000)—China Southeast (Guangxi).
obtusa *C.Y. Chang in J. Sichuan Univ., Nat. Sci. Ed.*, 1984(4): 92 (1984)—China Southeast (Hunan).

ELAEAGNUS:-
pingnanensis *C.Y. Chang in J. Sichuan Univ., Nat. Sci. Ed.*, 1984(4): 91 (1984)—China Southeast (Guangxi).
umbellata Thunb.
f. nakaiana *(Araki) H. Ohba in Fl. Japan*, 2c: 154 (1999): E. umbellata var. nakaiana.

ELAEOCARPUS (Elaeocarpac.).
acronodia Mast.
subsp. sumatranus *M.J.E. Coode in Kew Bull.*, 51(2): 274 (1996)—Sumatera, Malaya (Peninsular Malaysia).
subsp. sundarum *M.J.E. Coode in Kew Bull.*, 51(2): 278 (1996)—Lesser Sunda Is.
chewii *M.J.E. Coode in Kew Bull.*, 53(1): 89 (1998)—Borneo (Sabah).
clementis Merr.
var. borneensis *(Ridl.) M.J.E. Coode in Kew Bull.*, 51(4): 653 (1996): E. polystachyus var. borneensis.
var. clemensiae *(R. Knuth) M.J.E. Coode in Kew Bull.*, 51(4): 654 (1996); *et in Checklist Flow. Pl. & Gymn. Brunei Darussalam (ed. M.J.E. Coode et al.)*: 435 (1996): E. clemensiae.
var. kostermansii *M.J.E. Coode in Kew Bull.*, 51(4): 657 (1996)—Borneo (Kalimantan).
cristatus *M.J.E. Coode in Kew Bull.*, 53(1): 88 (1998), nom. nov.: E. magnifolius *R. Knuth*.
dulongensis *H. Li in Fl. Dulongjian Reg.*: 117 (1993), without latin descr. or type—China South-central (Yunnan).
ferrugineus (Jack) Steud.
subsp. elliptifolius *(Merr.) M.J.E. Coode in Kew Bull.*, 51(2): 282 (1996): E. elliptifolius.
subsp. glabrescens *(Mast.) M.J.E. Coode in Kew Bull.*, 51(2): 282 (1996): E. glabrescens.
subsp. parviflorus *M.J.E. Coode in Kew Bull.*, 51(2): 283 (1996)—Borneo (Sabah).
jacobsii *M.J.E. Coode in Kew Bull.*, 51(2): 283 (1996)—Borneo (Sarawak).
jugahanus *M.J.E. Coode in Kew Bull.*, 53(1): 118 (1998)—Borneo (Brunei, Sawawak, Sabah, Kalimantan).
kjellbergii *M.J.E. Coode in Kew Bull.*, 51(1): 83 (1996)—Sulawesi.
knuthii Merr.
subsp. cuspidatus *(Ridl.) M.J.E. Coode in Kew Bull.*, 51(2): 287 (1996): E. reticulatus var. cuspidatus.
nanus Corner
subsp. congestifolius *(R. Knuth) M.J.E. Coode in Kew Bull.*, 51(2): 293 (1996): E. congestifolius.
subsp. sumatrensis *M.J.E. Coode in Kew Bull.*, 51(2): 294 (1996)—Sumatera.
nooteboomii *M.J.E. Coode in Kew Bull.*, 53(1): 95 (1998)—Borneo (Kalimantan).
obtusus Blume
subsp. apiculatus *(Mast.) M.J.E. Coode in Checklist Flow. Pl. & Gymn. Brunei Darussalam (ed. M.J.E. Coode et al.)*: 435 (1996): E. apiculatus.
pagonensis *M.J.E. Coode in Checklist Flow. Pl. & Gymn. Brunei Darussalam (ed. M.J.E. Coode et al.)*: 435 (1996)—Borneo (Brunei).
renae *M.J.E. Coode in Kew Bull.*, 53(1): 126 (1998)—Borneo (Sabah, Sarawak, Brunei, Kalimantan).
retakensis *M.J.E. Coode in Checklist Flow. Pl. & Gymn. Brunei Darussalam (ed. M.J.E. Coode et al.)*: 436 (1996)—Borneo (Brunei).
robustus Roxb.
var. collinus *M.J.E. Coode in Kew Bull.*, 51(1): 90 (1996)—Malaya (Peninsular Malaysia).
var. megacarpus *M.J.E. Coode in Kew Bull.*, 51(1): 91 (1996)—Malaya (Peninsular Malaysia).
roslii *M.J.E. Coode in Kew Bull.*, 51(1): 92 (1996)—Borneo (Sarawak).
subsp. bracteolatus *M.J.E. Coode in Kew Bull.*, 51(1): 96 (1996)—Borneo (Brunei).
subsp. opacus *M.J.E. Coode in Kew Bull.*, 51(1): 96 (1996)—Borneo (Sabah, Kalimantan).
subsp. terajanus *M.J.E. Coode in Kew Bull.*, 51(1): 98 (1996)—Borneo (Brunei).
subisensis *M.J.E. Coode in Kew Bull.*, 53(1): 103 (1998)—Borneo (Sarawak).
sylvestris (Lour.) Poir.
var. photiniifolius *(Hook. & Arn.) Yasuhiro Endo in Fl. Japan*, 2c: 127 (1999): E. photiniifolius.

ELAEOCARPUS:-
valetonii Hochr.
var. longibarbatus (Warb.) M.J.E. Coode in Kew Bull., 51(1): 100 (1996): E. longibarbatus.

ELAEODENDRON (Celastrac.).
transvaalense *(Burtt Davy) R.H. Archer in S. Afr. J. Bot.*, 64(2): 105 (1998): Salacia transvaalensis.

ELAPHANDRA (Compos.).
lehmannii *(Hieron.) J.F. Pruski in Novon*, 6(4): 405 (1996)—Aspilia lehmannii.
moriana *J.F. Pruski in Novon*, 6(4): 405 (1996)—French Guiana.
sucrensis *(Aristeg.) V.M. Badillo, Enum. Compuestas (Asterac.) Venezuela (Rev. Fac. Agron. (Maracay), 45)*: 95 (1994): Aspilia sucrensis.

ELATOSTEMA (Urticac.).
acuteserratum *B.L. Shih & Yuen P. Yang in Bot. Bull. Acad. Sin. (Taipei)*, 36(4): 260 (1995)—Taiwan.
corniculatum *(W.T. Wang) H.W. Li in Fl. Yunnanica*, 7: 263 (1997): E. subtrichotomum var. corniculatum.
cyrtandrifolium (Zoll. & Moritzi) Miq.
var. brevicaudatum *(W.T. Wang) W.T. Wang in Fl. Reipubl. Popul. Sin.*, 23(2): 267 (1995): E. herbaceifolium var. brevicaudatum.
japonicum Wedd.
var. majus *(Maxim.) H. Nakai & Hiroyoshi Ohashi in J. Jap. Bot.*, 71(2): 81 (1996): E. umbellata var. majus.
luchunense *W.T. Wang in Acta Phytotax. Sin.*, 35(5): 459 (1997)—China South-central (Yunnan).
monandrum (Buch.-Ham. ex D. Don) H. Hara
var. ciliatum *(Hook. f.) S.K. Murti in Kew Bull.*, 52(1): 193 (1997): E. surculosum var. ciliatum.
var. crassum *(Hook. f.) S.K. Murti in Kew Bull.*, 52(1): 194 (1997): E. surculosum var. crassum.
var. pedunculosum *(Hook. f.) S.K. Murti in Kew Bull.*, 52(1): 194 (1997): E. surculosum var. pedunculosum.
var. pinnatifidum *(Hook. f.) S.K. Murti in Kew Bull.*, 52(1): 194 (1997): E. surculosum var. pinnatifidum.
var. rigidiusculum *(Hook. f.) S.K. Murti in Kew Bull.*, 52(1): 195 (1997): E. surculosum var. rigidiusculum.
var. serpens *(Hook. f.) S.K. Murti in Kew Bull.*, 52(1): 195 (1997): E. surculosum var. serpens.
var. subincisum *(Hook. f.) S.K. Murti in Kew Bull.*, 52(1): 195 (1997): E. surculosum var. subincisum.
var. zeylanicum *(Hook. f.) S.K. Murti in Kew Bull.*, 52(1): 195 (1997): E. surculosum var. zeylanicum.
multicanaliculatum *B.L. Shih & Yuen P. Yang in Bot. Bull. Acad. Sin. (Taipei)*, 36(4): 268 (1995)—Taiwan.
rivulare *B.L. Shih & Yuen P. Yang in Bot. Bull. Acad. Sin. (Taipei)*, 36(4): 270 (1995)—Taiwan.
strigillosum *B.L. Shih & Yuen P. Yang in Bot. Bull. Acad. Sin. (Taipei)*, 36(4): 272 (1995)—Taiwan.
subcoriaceum *B.L. Shih & Yuen P. Yang in Bot. Bull. Acad. Sin. (Taipei)*, 36(4): 272 (1995)—Taiwan.
sukungianum *W.T. Wang in Acta Phytotax. Sin.*, 35(5): 457 (1997)—China South-central (Yunnan).
villosum *B.L. Shih & Yuen P. Yang in Bot. Bull. Acad. Sin. (Taipei)*, 36(4): 277 (1995)—Taiwan.

× **ELECALTHUSA** *hort. in Orchid Rev.*, 104(1207): 37 (1996). *ORCHIDACEAE* (Arethusa × Calopogon × Eleorchis).

ELEOCHARIS (Cyperac.).
subser. Acutae *(S.T. Blake & J. Kern) M.S. González-Elizondo & P.M. Peterson in Taxon*, 46(3): 439 (1997): E. ser. Acutae.
ser. Albidae *M.S. González-Elizondo & P.M. Peterson in Taxon*, 46(3): 440 (1997).
sect. Baeothryon *(Gray) M.S. González-Elizondo & P.M. Peterson in Taxon*, 46(3): 437 (1997): Scirpus sect. Baeothryon.

ELEOCHARIS:-
sect. Palustriformes *(Svenson) C.A. Zavaro Pérez & G. Pabón Garcés in Fontqueria*, 48: 45 (1997 publ. 1998), without exact basionym page: E. ser. Palustriformes.
sect. Pauciflorae *(Svenson) C.A. Zavaro Pérez & G. Pabón Garcés in Fontqueria*, 48: 57 (1997 publ. 1998), without exact basionym page: E. ser. Pauciflorae.
ser. Rostellatae *M.S. González-Elizondo & P.M. Peterson in Taxon*, 46(3): 440 (1997).
subser. Sulcatae *(C.B. Clarke) M.S. González-Elizondo & P.M. Peterson in Taxon*, 46(3): 441 (1997): E. b. Sulcatae.
sect. Sulcatae *(Svenson) C.A. Zavaro Pérez & G. Pabón Garcés in Fontqueria*, 48: 58 (1997 publ. 1998), without exact basionym page: E. ser. Sulcatae.
albibracteata Nees & Meyen ex Kunth
var. boliviana *(Palla ex Svenson) M.S. González-Elizondo & P.M. Peterson in Sida*, 17(2): 414 (1996): E. boliviana.
alveolatoides *M.S. González-Elizondo & A.A. Reznicek in Novon*, 6(4): 356 (1996)—Venezuela.
atrospiculata *M.S. González-Elizondo & A.A. Reznicek in Novon*, 6(4): 357 (1996)—Venezuela.
ayacuchensis *M.S. González-Elizondo & A.A. Reznicek in Novon*, 6(4): 358 (1996)—Venezuela.
debilis
f. anomala *M.S. González-Elizondo & A.A. Reznicek in Novon*, 6(4): 358 (1996)—Venezuela.
eglerioides *M.S. González-Elizondo & A.A. Reznicek in Novon*, 6(4): 359 (1996)—Venezuela, Bolivia.
geniculata *(L.) C.A. Zavaro Pérez & G. Pabón Garcés in Fontqueria*, 48: 45 (1997 publ. 1998), isonym; *see Roem. & Schult. in Syst. Veg.*, 2: 150 (1817): Scirpus geniculatus.
geniculata (L.) Roem. & Schult.
f. brunnea *M.S. González-Elizondo & A.A. Reznicek in Novon*, 6(4): 361 (1996)—Venezuela.
kamtschatica (C.A. Mey.) Kom.
var. komarovii *(Zinserl.) V.N. Voroschilov in A.K. Skvortsov (ed.), Florist. issl. v razn. raĭonakh SSSR:* 153 (1985): E. komarovii.
keigheryi *K.L. Wilson in Nuytsia*, 11(2): 273 (1997)—Western Australia.
lankana T. Koyama
subsp. mohamadii *M.A. Wadood Khan in J. Econ. Taxon. Bot.*, 22(3): 560 (1998 publ. 1999)—India (Maharashtra).
liesneri *M.S. González-Elizondo & A.A. Reznicek in Novon*, 6(4): 361 (1996)—Venezuela.
lindbergii *(Strandh.) N.N. Tzvelev in Novosti Sist. Vyssh. Rast.*, 30: 163 (1996): E. palustris var. lindbergii.
moorei *M.T. Strong & M.S. González-Elizondo in Sida*, 19(1): 118 (2000)—Mexico Northeast (Hidalgo), Mexico Gulf (Veracruz).
palustris (L.) Roem. & Schult.
subsp. intersita *(Zinserl.) N.N. Tzvelev in Novosti Sist. Vyssh. Rast.*, 32: 183 (2000): E. intersita.
subsp. lindbergii *(Strandh.) N.N. Tzvelev in Novosti Sist. Vyssh. Rast.*, 32: 183 (2000): E. palustris var. lindbergii.
parvula *(Roem. & Schult.) Palla in Bot. Jahrb. Syst.*, 10(4): 299 (4 Dec. 1888), as 'Heleocharis': Scirpus parvulus.†
quinqueflora (Hartm.) Schwarz
var. bernardina *(Munz & I.M. Johnst.) M.S. González-Elizondo & P.M. Peterson in Sida*, 17(2): 414 (1996): Scirpus bernardinus.
squamigera Svenson
subsp. straminea *M.S. González-Elizondo & A.A. Reznicek in Novon*, 6(4): 362 (1996)—Venezuela.
steyermarkii *M.S. González-Elizondo & A.A. Reznicek in Novon*, 6(4): 363 (1996)—Venezuela.
venezuelensis *M.S. González-Elizondo & A.A. Reznicek in Novon*, 6(4): 363 (1996)—Venezuela.
yecorensis *E.H. Roalson in Aliso*, 18(1): 57 (1999)—Mexico Northwest (Sonora), Mexico Central (México State).

ELEPHANTOPSIS *(Sch. Bip.) C.F. Baker in Trans. Acad. Sci. St. Louis*, 12(5): 45, 55 (1902): Elephantopus sect. Elephantopsis. (Compos.).
biflora *(Less.) C.F. Baker in Trans. Acad. Sci. St. Louis*, 12(5): 45, 55 (1902): Elephantosis biflora.

ELETTARIA (Zingiberac.).
brachycalyx *S. Sakai & H. Nagamasu in Edinburgh J. Bot.*, 57(2): 240 (2000)—Borneo (Sarawak).
kapitensis *S. Sakai & H. Nagamasu in Edinburgh J. Bot.*, 57(2): 233 (2000)—Borneo (Sarawak).
linearicrista *S. Sakai & H. Nagamasu in Edinburgh J. Bot.*, 57(2): 236 (2000)—Borneo (Sarawak, Brunei).
longipilosa *S. Sakai & H. Nagamasu in Edinburgh J. Bot.*, 57(2): 238 (2000)—Borneo (Sarawak).
stoloniflora *(K. Schum.) S. Sakai & H. Nagamasu in Edinburgh J. Bot.*, 57(2): 231 (2000), as 'stolonifera': Amomum stoloniflorum.

ELETTARIOPSIS (Zingiberac.).
stoloniflora *(K. Schum.) H.N. Ridley in J. Straits Branch Roy. Asiat. Soc.*, 46: 242 (1906), as 'stoloniferum': Amomum stoloniflorum.†
unifolia *(Gagnep.) M.F. Newman in Edinburgh J. Bot.*, 54(1): 111 (1997): Amomum unifolium.

ELEUSINE (Gramin.).
indica (L.) Gaertn.
subsp. coracana *(L.) K.A. Lye in Lidia*, 4(5): 150 (1999): Cynosurus coracanus.

ELEUTHEROCOCCUS (Araliac.).
brachypus (Harms) Nakai
var. omeiensis *C.H. Kim & B.Y. Sun in Novon*, 10(3): 209 (2000)—China South-central (Sichuan).
divaricatus (Siebold & Zucc.) S.Y. Hu
var. chiisanensis *(Nakai) C.H. Kim & B.Y. Sun in Novon*, 10(3): 209 (2000): Acanthopanax chiisanensis.
higoensis *(Hatus.) H. Ohba in Fl. Japan*, 2c: 266 (1999): Acanthopanax higoensis.
huangshanensis *C.H. Kim & B.Y. Sun in Novon*, 10(3): 210 (2000)—China Southeast (Anhui).
innovans *(Siebold & Zucc.) H. Ohba in Fl. Japan*, 2c: 264 (1999): Panax innovans.
pilosulus *(Rehder) C.H. Kim & B.Y. Sun in Novon*, 10(3): 210 (2000): Acanthopanax giraldii var. pilosulus.
pseudosetulosus *C.H. Kim & B.Y. Sun in Novon*, 10(3): 213 (2000)—China South-central (Sichuan).
pubescens *(Pamp.) C.H. Kim & B.Y. Sun in Novon*, 10(3): 213 (2000): Acanthopanax spinosus var. pubescens.
spinosus (L.f.) S.Y. Hu
f. ionanthus *(Nakai) H. Ohba in Fl. Japan*, 2c: 264 (1999): Acanthopanax japonicus f. ionanthus.
var. japonicus *(Franch. & Sav.) H. Ohba in Fl. Japan*, 2c: 263 (1999): Acanthopanax japonicus.
f. kiusianus *(Nakai) H. Ohba in Fl. Japan*, 2c: 264 (1999): Acanthopanax kiusianus.
var. nikaianus *(Nakai) H. Ohba in Fl. Japan*, 2c: 264 (1999): Acanthopanax nikaianus.

× **ELIARA** *hort. in Orchid Rev.*, 105(1215): 190 (1997). *ORCHIDACEAE* (Brassia × Epidendrum × Laelia × Sophronitis).

ELIZALDIA (Boraginac.).
calycina (Roem. & Schult.) Maire
subsp. embergeri *(Sauvage & Vindt) A. Dobignard in Candollea*, 52(1): 132 (1997): E. violacea var. embergeri.

ELLEANTHUS (Orchidac.).
carnevalii *C.H. Dodson in Orquideologia*, 20(3): 274 (1997)—Ecuador.
caveroi *D.E. Bennett & E.A. Christenson in Lindleyana*, 13(1): 33 (1998)—Peru.
escobarii *C.H. Dodson in Orquideologia*, 20(1): 4 (1996)—Colombia.
malpighiiflorus *G. Carnevali & G.A. Romero in G.A. Romero & G. Carnevali, Orchids of Venezuela, ed. 2:* 1131 (2000)—Brazil/Venezuela border.
wallnoeferi *D.L. Szlachetko in Orchidee*, Suppl. 3: 23 (1996)—Peru.

ELLISIOPHYLLUM (Scrophulariac.).
pinnatum (Wall.) Makino
subsp. bhutanense *R.R. Mill in Edinburgh J. Bot.*, 57(3): 414 (2000)—Bhutan.

ELOYELLA (Orchidac.).
thivii *K. Senghas in Caesiana*, 13: 27 (1999)—Bolivia.

ELSHOLTZIA (Labiat.).
amurensis *N.S. Probatova in Sosud. Rast. Sovet. Dal'nego Vostoka*, 7: 374 (1995)—Amur.
splendens Nakai ex F. Maek.
f. albiflora *Y.N. Lee in Korean J. Pl. Taxon.*, 28(1): 31 (1998)—South Korea.

ELTROPLECTRIS (Orchidac.).
brachycentron *D.L. Szlachetko in Novon*, 5(4): 375 (1995)—Bolivia.

ELVASIA (Ochnac.).
bisepala *C. Sastre & C. Whitefoord in Novon*, 9(2): 254 (1999)—Belize, Honduras.
canescens *(Tiegh.) E.F. Gilg in Engl. & Prantl, Nat. Pflanzenfam. ed. 2*, 21: 77 (1925): Trichovaselia canescens.†
macrostipularis *C. Sastre & J.-P. Lescure in Caldasia*, 12(57): 134 (1978)—Brazil North (Pará).

× **ELYHORDEUM** (Gramin.).
californicum *(Bowden) M.E. Barkworth in Phytologia*, 83(4): 308 (1997 publ. 1998): Sitordeum californicum. (Elymus multisetus × Hordeum jubatum).
pilosilemma *(W.W. Mitch. & H.J. Hodgs.) M.E. Barkworth in Phytologia*, 83(4): 308 (1997 publ. 1998): × Agrohordeum pilosilemma. (Elymus macrourus × Hordeum jubatum).

× **ELYLEYMUS** (Gramin.).
bergrothii *(H. Lindb.) H.J. Conert in G. Hegi, Illustr. Fl. Mitteleur., ed. 3*, 1(3: Lief. 10): 802 (1997): × Tritordeum bergrothii.
strictus *(Dethard.) H.J. Conert in G. Hegi, Illustr. Fl. Mitteleur., ed. 3*, 1(3: Lief. 10): 802 (1997): Triticum strictum.
turneri *(Lepage) M.E. Barkworth & D.R. Dewey in Amer. J. Bot.*, 72(5): 772 (1985): × Agroelymus turneri.

× **ELYMOTRIGIA** (Gramin.).
altaica *Yu.A. Kotukhov in Turczaninowia*, 1(1): 16 (1998)—Altay. (Elymus fedtschenkoi × Elytrigia repens).
bobrovica *Yu.A. Kotukhov in Turczaninowia*, 1(1): 17 (1998)—Altay. (Elymus nevski × Elytrigia repens).
gigantea *Yu.A. Kotukhov in Turczaninowia*, 1(1): 18 (1998)—Altay. (Elymus mutabilis × Elytrigia repens).

ELYMUS (Gramin.).
aegilopoides *(Drobow) V.N. Voroschilov in A.K. Skvortsov (ed.), Florist. issl. v razn. Raĭonakh SSSR:* 151 (1985): Agropyron aegilopoides.
aliena *(Keng) S.L. Chen in Novon*, 7(3): 227 (1997): Roegneria aliena.
arenosus *(Spenn.) H.J. Conert in G. Hegi, Illustr. Fl. Mitteleur., ed. 3*, 1(3: Lief. 10): 793 (1997): Triticum repens var. arenosum.
aristiglumis (Keng & S.L. Chen) S.L. Chen
var. hirsutus *(H.L. Yang) S.L. Chen in Novon*, 7(3): 227 (1997): Roegneria aristiglumis var. hirsuta.
var. leianthus *(H.L. Yang) S.L. Chen in Novon*, 7(3): 227 (1997): Roegneria aristiglumis var. leiantha.
barbicallus (Ohwi) S.L. Chen
var. pubifolius *(Keng) S.L. Chen in Novon*, 7(3): 227 (1997): Roegneria barbicalla var. pubifolia.
var. pubinodis *(Keng) S.L. Chen in Novon*, 7(3): 227 (1997): Roegneria barbicalla var. pubinodis.
besczetnovae *Yu.A. Kotukhov in Turczaninowia*, 2(4): 9 (1999)—Altay.
borealis *(Turcz.) D.F. Cui in Fl. Xinjiangensis*, 6: 184 (1996), as 'borealus', homonym; *see Scribn. in Circ. Div. Agrostol. U.S.D.A.*, 27: 9 (1900): Triticum boreale.
brachypodioides *(Nevski) G.A. Peschkova, Stepnaya Fl. Baĭkal'skoĭ Sibiri:* 45 (1972): Roegneria brachypodioides.

ELYMUS:-
calderi *M.E. Barkworth in Syst. Bot.*, 21(3): 353 (1996 publ. 1997)—Yukon, British Columbia, Alaska.
caninus (L.) L.
var. behmii *(Melderis) T. Karlsson in Svensk Bot. Tidskr.*, 91(5): 249 (1997 publ. 1998): Roegneria behmii.
subsp. donianus *(F.B. White) P.D. Sell in P.D. Sell & G. Murrell, Fl. Great Britain & Ireland*, 5: 363 (1996): Agropyron donianum.
ciliaris (Trin.) Tzvelev
var. amurensis *(Drobow) S.L. Chen in Novon*, 7(3): 228 (1997): Agropyron amurense.
var. hirtiflorus *(C.P. Wang & H.L. Yang) S.L. Chen in Novon*, 7(3): 228 (1997): Roegneria hirtiflora.
var. japonensis *(Honda) S.L. Chen in Novon*, 7(3): 228 (1997): Agropyron japonense.
var. submuticus *(Honda) S.L. Chen in Novon*, 7(3): 228 (1997): Agropyron ciliare var. submuticum.
confusus (Roshev.) Tzvelev
var. breviaristatus *(Keng) S.L. Chen in Novon*, 7(3): 228 (1997): Roegneria confusa var. breviaristata.
curvatus *(Nevski) D.F. Cui in Fl. Xinjiangensis*, 6: 197 (1996), homonym; *see Piper in Bull. Torrey Bot. Club*, 30: 233 (1903): Agropyron curvatus.
dahuricus Turcz.
var. xiningensis *(L.B. Cai) S.L. Chen in Novon*, 7(3): 228 (1997): E. xiningensis.
× dorei *(Bowden) M.E. Barkworth & D.R. Dewey in Amer. J. Bot.*, 72(5): 772 (1985): × Agroelymus dorei.
× ebingeri *G.C. Tucker in Harvard Pap. Bot.*, 9: 83 (1996)—New York. (E. hystrix × virginicus).
elongatiformis *(Drobow) M. Assadi in Willdenowia*, 26(1–2): 268 (1996): Agropyron elongatiforme.
elymoides (Raf.) Swezey
subsp. brevifolius *(J.G. Sm.) M.E. Barkworth in Phytologia*, 83(4): 305 (1997 publ. 1998): Sitanion brevifolium.
subsp. californicus *(J.G. Sm.) M.E. Barkworth in Phytologia*, 83(4): 306 (1997 publ. 1998): Sitanion californicum.
subsp. hordeoides *(Suksd.) M.E. Barkworth in Phytologia*, 83(4): 306 (1997 publ. 1998): Sitanion hordeoides.
embergeri *(Maire) M. Ibn Tattou in Bocconea*, 8: 217 (1998): Agropyron embergeri.
falcis *H.E. Connor in New Zealand J. Bot.*, 32(2): 132 (1994)—New Zealand South I.
festucoides *(Maire) M. Ibn Tattou in Bocconea*, 8: 217 (1998): Agropyron festucoides.
formosanus *(Honda) S.S. Ying in Mem. Coll. Agric. Nation. Taiwan Univ.*, 31(1): 33 (1991), as 'formosanum', isonym; *see Á. Löve in Feddes Repert.*, 95(7–8): 449 (1984): Agropyron formosanum.
formosanus (Honda) Á. Löve
var. pubigerus *(Keng) S.L. Chen in Novon*, 7(3): 228 (1997): Roegneria formosana var. pubigera.
gentryi (Melderis) Melderis
var. ciliatiglumis *M. Assadi in Willdenowia*, 26(1–2): 261 (1996)—Iran.
hispidus (Opiz) Melderis
var. podperae *(Nábělek) M. Assadi in Willdenowia*, 26(1–2): 265 (1996): Agropyron podperae.
var. villosus *(Hack.) M. Assadi in Willdenowia*, 26(1–2): 266 (1996): Triticum intermedium var. villosum.
hordeoides *(Suksd.) M.E. Barkworth & D.R. Dewey in Amer. J. Bot.*, 72(5): 772 (1985): Sitanion hordeoides.
jufinshanicus *(C.P. Wang & H.L. Yang) S.L. Chen in Novon*, 7(3): 228 (1997): Roegneria alashanica var. jufinshanica.
kengii *(Tzvelev) D.F. Cui in Fl. Xinjiangensis*, 6: 183 (1996): Agropyron kengii.
kronokensis (Kom.) Tzvelev
subsp. scandicus *(Nevski) N.N. Tzvelev in Novosti Sist. Vyssh. Rast.*, 32: 182 (2000): Roegneria scandica.
leianthus *(Keng) S.L. Chen in Novon*, 7(3): 229 (1997): Roegneria leiantha.

ELYMUS:-
× weariglumis *O. Seberg & G. Petersen in Bot. Jahrb.*, 120(4): 528 (1998), nom. nov.: Agropyron × elymoides *Hack.*
lineicus *Yu.A. Kotukhov in Turczaninowia*, 2(4): 8 (1999)—Altay.
longespicatus *Yu.A. Kotukhov in Turczaninowia*, 2(4): 7 (1999)—Altay.
× mossii *(Lepage) M.E. Barkworth & D.R. Dewey in Amer. J. Bot.*, 72(5): 772 (1985): × Agroelymus mossii.
× mucronatus *(Opiz) H.J. Conert in G. Hegi, Illustr. Fl. Mitteleur., ed. 3*, 1(3: Lief. 10): 802 (1997): Agropyron × mucronatum.
mutabilis (Drobow) Tzvelev
var. praecaespitosus *(Nevski) S.L. Chen in Novon*, 7(3): 229 (1997): Agropyron praecaespitosum.
nodosus (Nevski) Melderis
subsp. dorudicus *M. Assadi in Willdenowia*, 26(1–2): 258 (1996)—Iran.
obtusiflorus *(DC.) H.J. Conert in G. Hegi, Illustr. Fl. Mitteleur., ed. 3*, 1(3: Lief. 10): 787 (1997): Triticum obtusiflorum.
× palmerensis *(Lepage) M.E. Barkworth & D.R. Dewey in Amer. J. Bot.*, 72(5): 772 (1985): × Agroelymus palmerensis.
panormitanus (Parl.) Tzvelev
subsp. sirouanus *(Quézel) M. Ibn Tattou in Bocconea*, 8: 218 (1998): Agropyron panormitanum subsp. sirouanum.
parodii *O. Seberg & G. Petersen in Bot. Jahrb.*, 120(4): 530 (1998), nom. nov.: Agropyron condensatum *J. Presl.*
pertenuis *(C.A. Mey.) M. Assadi in Willdenowia*, 26(1–2): 256 (1996): Triticum intermedium var. pertenue.
× pinaloensis *(Pyrah) M.E. Barkworth & D.R. Dewey in Amer. J. Bot.*, 72(5): 772 (1985), without basionym date: × Agrositanion pinaloensis.
repens (L.) Gould
subsp. atlantis *(Maire) M. Ibn Tattou in Bocconea*, 8: 218 (1998): Agropyron repens var. atlantis.
subsp. littoreus *(Schumach.) H.J. Conert in G. Hegi, Illustr. Fl. Mitteleur., ed. 3*, 1:(3: Lief. 10): 798 (1997): Triticum littoreum.
sacandros *H.E. Connor in New Zealand J. Bot.*, 32(2): 138 (1994)—New Zealand South I.
sajanensis (Nevski) Tzvelev
subsp. coeruleus *(Jurtzev) N.N. Tzvelev in Novosti Sist. Vyssh. Rast.*, 32: 182 (2000): Roegneria villosa subsp. coerulea.
sarymsactensis *Yu.A. Kotukhov in Turczaninowia*, 2(4): 6 (1999)—Altay.
sauricus *Yu.A. Kotukhov in Turczaninowia*, 1(1): 19 (1998)—Kazakhstan.
scabrifolius *(Döll) J.H. Hunziker in Darwiniana*, 35(1–4): 167 (1998): Triticum repens var. scabrifolium.
sibiricus L.
var. pilosifolius *(A.P. Khokhr.) V.N. Voroschilov in A.K. Skvortsov (ed.), Florist. issl. v razn. Raĭonakh SSSR*: 152 (1985): E. confusus subsp. pilosifolius.
sinicus *(Keng) S.L. Chen in Novon*, 7(3): 229 (1997): Roegneria sinica.
solandri *(Steud.) H.E. Connor in New Zealand J. Bot.*, 32(2): 140 (1994): Triticum solandri.
stebbinsii Gould
subsp. septentrionalis *M.E. Barkworth in Phytologia*, 83(5): 360 (1997 publ. 1998)—California.
tarbagataicus *Yu.A. Kotukhov in Turczaninowia*, 1(1): 20 (1998)—Altay.
tauri (Boiss. & Balansa) Melderis
var. kosaninii *(Nábělek) M. Assadi in Willdenowia*, 26(1–2): 258 (1996): Agropyron kosaninii.
tenuispicus *(J.L. Yang & Y.H. Zhou) S.L. Chen in Novon*, 7(3): 229 (1997): Roegneria tenuispica.
tridentatus *(Yen & J.L. Yang) S.L. Chen in Novon*, 7(3): 229 (1997): Roegneria tridentata.
ubinica *Yu.A. Kotukhov in Turczaninowia*, 2(4): 5 (1999)—Altay.
violaceus (Hornem.) Feilberg
subsp. andinus *(Scribn. & J.G. Sm.) N.N. Tzvelev in Novosti Sist. Vyssh. Rast.*, 32: 182 (2000): Agropyron violaceum var. andinum.

ELYMUS:-
violaceus (Hornem.) Feilberg
subsp. latiglumis *(Scribn. & J.G. Sm.) N.N. Tzvelev in Novosti Sist. Vyssh. Rast.*, 32: 181 (2000): Agropyron violaceum var. latiglume.
wawawaiensis *J.R. Carlson & M.E. Barkworth in Phytologia*, 83(4): 327 (1997 publ. 1998)—Washington, Idaho, Oregon.

× ELYTESION *M.E. Barkworth & D.R. Dewey in Amer. J. Bot.*, 72(5): 772 (1985). *GRAMINEAE* (Critesion × Elymus).
californicum *(Bowden) M.E. Barkworth & D.R. Dewey in Amer. J. Bot.*, 72(5): 772 (1985): × Sitordeum californicum.
dakotense *(Bowden) M.E. Barkworth & D.R. Dewey in Amer. J. Bot.*, 72(5): 772 (1985): × Elymordeum dakotense.
iowense *(R.W. Pohl) M.E. Barkworth & D.R. Dewey in Amer. J. Bot.*, 72(5): 772 (1985): × Elyhordeum iowense.
macounii *(Vasey) M.E. Barkworth & D.R. Dewey in Amer. J. Bot.*, 72(5): 772 (1985): Elymus macounii.
montanense *(Scribn.) M.E. Barkworth & D.R. Dewey in Amer. J. Bot.*, 72(5): 772 (1985): Hordeum montanense.
pilosilemma *(W.W. Mitch. & H.J. Hodgs.) M.E. Barkworth & D.R. Dewey in Amer. J. Bot.*, 72(5): 772 (1985): × Agrohordeum pilosilemma.
schaakianum *(Bowden) M.E. Barkworth & D.R. Dewey in Amer. J. Bot.*, 72(5): 772 (1985): × Elymordeum schaakianum.
stebbinsianum *(Bowden) M.E. Barkworth & D.R. Dewey in Amer. J. Bot.*, 72(5): 772 (1985): × Elymordeum stebbinsianum.

× ELYTHODIA *hort. in Orchid Rev.*, 108(1232): new orchid hybrids 9 (2000). *ORCHIDACEAE* (Elythranthera × Glossodia).

ELYTRIGIA (Gramin.).
atherica *(Link) M.A. Carreras ex M. Kerguélen in Lejeunia*, 120: 86 (1987): Triticum athericum.
campestris *(Godr. & Gren.) M.A. Carreras ex M. Kerguélen in Lejeunia*, 120: 86 (1987): Agropyron campestre.
subsp. maritima *(Tzvelev) H. Scholz in Soc. Éch. Pl. Vasc. Eur. Bassin Médit. Bull.*, 27: 102 (1998): E. maritima.
czindogatuica *Yu.A. Kotukhov in Turczaninowia*, 1(1): 13 (1998)—Altay.
obtusiflora
subsp. graeca *(Melderis) H. Scholz in Willdenowia*, 28(1–2): 171 (1998): Elymus hispidus subsp. graecus.
repens
var. aristata *(Döll) P.D. Sell in P.D. Sell & G. Murrell, Fl. Great Britain & Ireland*, 5: 363 (1996): Agropyron repens var. aristatum.
subsp. longearistata *N.R. Cui in Fl. Xinjiangensis*, 6: 602 (1996); *et in Acta Bot. Bor.-Occid. Sin.*, 18(2): 286 (1998)—Xinjiang.
var. subulatum *(Roem. & Schult.) O. Seberg & G. Petersen in Bot. Jahrb.*, 120(4): 538 (1998): Agropyron repens var. subulatum.

EMARHENDIA *R. Kiew, A. Weber & B.L. Burtt in Beitr. Biol. Pfl.*, 70(2–3): 398 (1997–98 publ. 1998). *GESNERIACEAE*
bettiana *(M.R. Hend.) R. Kiew, A. Weber & B.L. Burtt in Beitr. Biol. Pfl.*, 70(2–3): 398 (1997–98 publ. 1998): Paraboea bettiana.

EMBELIA (Myrsinac.).
aurantiaca *(Wall.) B.M. Wadhwa in M.D. Dassanayake & W.D. Clayton, eds., Revised Handb. Fl. Ceylon* 13: 156 (1999): Myrsine aurantiaca.
cuneata *C.M. Hu & J.E. Vidal in J. Trop. Subtrop. Bot.*, 5(1): 14 (1997)—Vietnam.
deivanuae *R.O. Gardner in New Zealand J. Bot.*, 35(4): 487 (1997)—Fiji.
frondosa *(King ex Gamble) D.G. Long in Edinburgh J. Bot.*, 56(2): 307 (1999): Samara frondosa.
laeta (L.) Mez
subsp. papilligera *(Nakai) J.J. Pipoly & C. Chen in Novon*, 5(4): 358 (1995): Samara laeta var. papilligera.

EMBELIA:-
ribes Burm. f.
subsp. pachyphylla *(C.Y. Wu & C. Chen) J.J. Pipoly & C. Chen in Novon,* 5(4): 358 (1995): E. ribes var. pachyphylla.
tsjeriamcottam (Roem. & Schult.) A.DC.
var. ferruginea *(C.B. Clarke) K. Larsen & C.M. Hu in Fl. Thailand,* 6(2): 158 (1996): E. ferruginea.

EMELIANTHE (Loranthac.).
panganensis (Engl.) Danser
subsp. commiphorae *D. Wiens & R.M. Polhill, Mistletoes of Africa:* 97 (1998)—Kenya.

EMILIA (Compos.).
djalonensis *S. Lisowski in Bull. Jard. Bot. Nation. Belg.,* 66(3–4): 201 (1997)—Guinea.
lisowskiana *C. Jeffrey in Kew Bull.,* 52(1): 208 (1997)—Nigeria.

EMORYA (Buddlejac.).
rinconensis *M.H. Mayfield in Sida,* 18(3): 694 (1999)—Mexico Northeast (Coahuila).

EMORYCACTUS *A.B. Doweld in Succulenta,* 75(6): 270 (1996). *CACTACEAE*
parryi *(Engelm.) A.B. Doweld in Succulenta,* 75(6): 270 (1996): Echinocactus parryi.
polycephalus *(Engelm. & J.M. Bigelow) A.B. Doweld in Succulenta,* 75(6): 270 (1996): Echinocactus polycephalus.
xeranthemoides *(J.M. Coult.) A.B. Doweld in Succulenta,* 75(6): 270 (1996): Echinocactus polycephalus var. xeranthemoides.

EMPETRUM (Empetrac.).
nigrum L.
subsp. albidum *(V.N. Vassil.) V.B. Kuvaev in Bot. Zhurn.,* 81(10): 113 (1996): E. albidum.
subsp. androgynum *(V.N. Vassil.) V.B. Kuvaev in Bot. Zhurn.,* 81(10): 113 (1996): E. androgynum.
subsp. asiaticum *(Nakai) V.B. Kuvaev in Bot. Zhurn.,* 81(10): 113 (1996): E. nigrum var. asiaticum.
subsp. caucasicum *(Juz.) V.B. Kuvaev in Bot. Zhurn.,* 81(10): 113 (1996): E. caucasicum.
subsp. kardakovii *(V.N. Vassil.) V.B. Kuvaev in Bot. Zhurn.,* 81(10): 113 (1996): E. kardakovii.
subsp. sibiricum *(V.N. Vassil.) V.B. Kuvaev in Bot. Zhurn.,* 81(10): 113 (1996): E. sibiricum.

EMPODIUM (Hypoxidac.).
flexile *(Nel) M.F. Thomps. ex D.A. Snijman in Strelitzia,* 9: 714 (2000): Forbesia flexilis.

ENCELIA (Compos.).
cedrosensis *Rose in Contrib. U.S. Nat. Herb.,* 1: 17 (1890), as 'cedrosensic' in error in I.K. Supp. 1—California.†
frutescens (A. Gray) A. Gray
subsp. glandulosa *C. Clark in Aliso,* 17(2): 201 (1998)—Mexico Northwest (Baja California).
resinifera *C. Clark in Aliso,* 17(2): 201 (1998), nom. nov.: E. frutescens var. resinosa *M.E. Jones ex S.F. Blake.*
subsp. tenuifolia *C. Clark in Aliso,* 17(2): 202 (1998)—Arizona.

ENCEPHALARTOS (Zamiac.).
aplanatus *P. Vorster in S. Afr. J. Bot.,* 62(1): 57 (1996)—Swaziland.
brevifoliolatus *P. Vorster in S. Afr. J. Bot.,* 62(1): 61 (1996)—Transvaal.
hirsutus *P.J.H. Hurter in S. Afr. J. Bot.,* 62(1): 46 (1996)—Transvaal.
imbricans *P. Vorster in Novon,* 5(4): 388 (1995)—Uganda.
macrostrobilus *Scott Jones & J. Wynants in Encephalartos, J. Cycad Soc. South Africa,* 50: 13 (1997)—Uganda.
msinganus *P. Vorster in S. Afr. J. Bot.,* 62(2): 67 (1996)—Natal.
nubimontanus *P.J.H. Hurter in Phytologia,* 78(6): 409 (1995)—Transvaal.
senticosus *P. Vorster in S. Afr. J. Bot.,* 62(2): 76 (1996)—Swaziland, Natal.

ENCEPHALARTOS:-
successibus *P. Vorster in S. Afr. J. Bot.,* 61(6): 347 (1995)—Uganda.
tegulaneus Melville
subsp. powysii *B.W. Miringu & H. Beentje in J.E. Afr. Nat. Hist.,* 88(1–2): 35 (1999 publ. 2000)—Kenya.
venetus *P. Vorster in S. Afr. J. Bot.,* 62(2): 71 (1996)—Transvaal.
whitelockii *P.J.H. Hurter in Phytologia,* 78(6): 410 (1995)—Uganda.

ENCEPHALOSPHAERA (Acanthac.).
puberula *(Leonard) D.C. Wasshausen in Cat. Vasc. Pl. Ecuador (Monogr. Syst. Bot. Missouri Bot. Gard., 75):* 953 (1999): Aphelandra puberula.

ENCHOLIRIUM (Bromeliac.).
bahianum *L.B. Smith & R.W. Read in Bradea,* 5(27): 300 (1989)—Brazil Northeast (Bahia).
belemii *L.B. Smith & R.W. Read in Bradea,* 5(27): 301 (1989)—Brazil Southeast (Minas Gerais).
brachypodum *L.B. Smith & R.W. Read in Bradea,* 5(27): 302 (1989)—Brazil Northeast (Bahia).
crassiscapum *E. Gross in Trop. Subtrop. Pflanzenwelt,* 95: 7 (1997)—Brazil Southeast (Minas Gerais).
harleyi *L.B. Smith & R.W. Read in Bradea,* 5(27): 299 (1989)—Brazil Northeast (Bahia).
heloisae *(L.B. Sm.) R.C. Forzza & M.G.L. Wanderley in Bol. Bot. (São Paulo),* 17: 264 (1998): Dyckia heloisae.
luxor *L.B. Smith & R.W. Read in Bradea,* 5(27): 299 (1989)—Brazil Southeast (Minas Gerais).
paraibae *L.B. Smith & R.W. Read in Bradea,* 5(27): 302 (1989)—Brazil Northeast (Paraíba).
pernambucanum *L.B. Smith & R.W. Read in Bradea,* 5(27): 302 (1989)—Brazil Northeast (Pernambuco).
piresianum *L.B. Smith & R.W. Read in Bradea,* 5(27): 301 (1989)—Brazil West-central (Brasília Distrito Federal).

ENCYCLIA (Orchidac.).
boothiana (Lindl.) Dressler
subsp. erythronioides *(Small) Hágsater ex E.A. Christenson in Lindleyana,* 11(1): 15 (1996): Epidendrum erythronioides.
caximboensis *L.C. Menezes in Bol. CAOB,* 4(1): 15 (1992)—Brazil North (Pará).
contrerasii *R. Gonzalez Tamayo in Orchidophile,* 127: 93 (1997)—Mexico Southwest (Colima).
cordigera (Humb., Bonpl. & Kunth) Dressler
f. leucantha *(Schltr.) C.L. Withner, Cattleyas & Relatives,* 5: 104 (1998): E. atropurpurea var. leucantha.
dickinsoniana *(Withner) C.L. Withner, Cattleyas & Relatives,* 5: 105 (1998): Epidendrum dickinsonianum.
edithiana *L.C. Menezes in Bol. CAOB,* 26: 23 (1996)—Brazil North (Amazonas).
garzonensis *C.L. Withner, Cattleyas & Relatives,* 6: 75 (2000)—Colombia.
guadalupeae *R. Gonzalez Tamayo & S. Alvarado Casillas in Orchidophile,* 135: 5 (1999)—Mexcio Southwest (Nayarit, Colima).
kingsii *(C.D. Adams) G.R. Proctor in Kew Bull.,* 51(3): 488 (1996), isonym; *see M.A. Nir in Lindleyana,* 9(3): 147 (1994): Epidendrum kingsii.
kundergraberi *V.P. Castro Neto & M.A. Campacci in Orquidário,* 12(1): 7 (1998), as 'kundergraberii'—Brazil Northeast (Bahia).
magdalenae *C.L. Withner, Cattleyas & Relatives,* 6: 91 (2000)—Colombia?
meneziana *J. Gonzalez in Bol. CAOB,* 4(1): 39 (1992)—Brazil West-central (Goiás).
ochrantha *(A. Rich.) C.L. Withner, Cattleyas & Relatives,* 4: 64 (1996): Epidendrum ochranthum.
parviflora *(Regel) C.L. Withner, Cattleyas & Relatives,* 5: 121 (1998): Epidendrum alatum var. parviflorum.
peraltensis *(Ames) C.L. Withner, Cattleyas & Relatives,* 5: 122 (1998), isonym; *see Dressler in Novon,* 7(2): 124 (1997): Epidendrum peraltense.
peraltensis *(Ames) R.L. Dressler in Novon,* 7(2): 124 (1997): Epidendrum peraltense.
perazolliana *J. Gonzalez in Bol. CAOB,* 4(1): 21 (1992), without type—Brazil West-central (Goiás).

ENCYCLIA:-
perplexa *(Ames, F.T. Hubb. & C. Schweinf.) C.L. Withner, Cattleyas & Relatives,* 5: 123 (1998), isonym; *see Dressler & G.E. Pollard in Phytologia,* 21(7): 437 (1971): Epidendrum oncidioides var. perplexum.
quesneliana *(Henshall) C.L. Withner, Cattleyas & Relatives,* 6: 119 (2000), basionym not validly publ.: Epidendrum quesnelianum.
racemifera *R.L. Dressler in Lindleyana,* 11(1): 37 (1996)—Costa Rica.
randiana *(Barb. Rodr.) C.L. Withner, Cattleyas & Relatives,* 6: 122 (2000), basionym not validly publ.: Epidendrum randianum.
santos-dumontii *L.C. Menezes in Bol. CAOB,* 4(1): 12 (1992)—Brazil West-central (Brasília Distrito Federal).
tocantinensis *V.P. Castro Neto & M.A. Campacci in Orquidário,* 10(3): 74 (1996)—Brazil North (Tocantins).
trulla *(Rchb. f.) E.A. Christenson in Lindleyana,* 11(4): 222 (1996): Epidendrum trulla.
uxpanapensis *G.A. Salazar in An. Inst. Biol. Univ. Nac. Autón. Méx., Bot.,* 70(1): 3 (1999)—Mexico Gulf (Veracruz).

ENDLICHERIA (Laurac.).
canescens *A. Chanderbali in Novon,* 6(4): 328 (1996)—Surinam, Ecuador, Guyana, Peru, Venezuela.
chalisea *A. Chanderbali in Novon,* 6(4): 329 (1996)—Guyana, Bolivia, Peru, French Guiana, Venezuela, Brazil North (Amazonas, Pará).

ENDOPAPPUS (Compos.).
macrocarpus Sch. Bip.
subsp. maroccanus *(Jahand., Maire & Weiller) M. Ibn Tattou in Bocconea,* 8: 34 (1998): Chrysanthemum macrocarpum subsp. maroccanum.

ENDOSTEMON (Labiat.).
kelleri *(Briq.) Ryding ex A.J. Paton in Kew Bull.,* 49(4): 712 (1994): Orthosiphon kelleri.

ENEMION (Ranunculac.).
savilei *(Calder & R.L. Taylor) M. Tamura in Acta Phytotax. Geobot.,* 46(2): 210 (1995 publ. 1996), isonym; *see Keener in Castanea,* 45(4): 278 (1980): Isopyrum savilei.

ENGELHARDTIA (Juglandac.).
apoensis *Elmer ex K. Nagel in Bot. Jahrb.,* 50: 477 (1914)—Locality not stated.†

ENGELMANNIA (Compos.).
texana *Scheele in Linnaea,* 22: 155 (1849)—Texas.

ENGLERINA (Loranthac.).
concinna *R.M. Polhill & D. Wiens, Mistletoes of Africa:* 134 (1998)—Zambia.
cordata *R.M. Polhill & D. Wiens in Novon,* 7(3): 276 (1997)—Malawi, Tanzania.
drummondii *Balle ex R.M. Polhill & D. Wiens, Mistletoes of Africa:* 134 (1998)—Tanzania.
heckmanniana (Engl.) Polhill & Wiens
subsp. polytricha *R.M. Polhill & D. Wiens, Mistletoes of Africa:* 136 (1998)—Kenya.
longiflora *R.M. Polhill & D. Wiens, Mistletoes of Africa:* 123 (1998)—Tanzania.
macilenta *R.M. Polhill & D. Wiens, Mistletoes of Africa:* 133 (1998)—Tanzania.

ENGLEROPHYTUM (Sapotac.).
iturense *(Engl.) L. Gautier in J.-P. Lebrun & A.L. Stork, Énum. Pl. Fleurs Afr. Trop.,* 4: 29 (1997): Chrysophyllum iturense.

ENGOMEGOMA *F.J. Breteler in Bot. Jahrb.,* 118(1): 114 (1996). *OLACACEAE*
gordonii *F.J. Breteler in Bot. Jahrb.,* 118(1): 114 (1996)—Gabon.

ENKIANTHUS (Ericac.).
campanulatus (Miq.) G. Nicholson
f. lutescens *F. Yokouchi ex T. Shimizu, Fl. Nagano Pref.:* 1508 (1997)—Japan.

ENKIANTHUS:-
deflexus (Griff.) C.K. Schneid.
var. glabrescens *R.C. Fang in Novon,* 9(2): 162 (1999)—China North-central (Gansu).
serrulatus (E.H. Wilson) C.K. Schneid.
var. sichuanensis *(T.Z. Hsu) R.C. Fang in Novon,* 9(2): 162 (1999): E. sichuanensis.

ENTADA (Legumin.).
polystachya (L.) DC.
var. polyphylla *(Benth.) R.C. Barneby in Brittonia,* 48(2): 175 (1996): E. polyphylla.
var. simplicata *R.C. Barneby in Brittonia,* 48(2): 175 (1996)—Brazil North (Roraima).

ENTEROLOBIUM (Legumin.).
sect. Robrichia *R.C. Barneby & J.W. Grimes, Silk Tree, Guanacaste, Monkey's Earring (Mem. New York Bot. Gard., 74(1)):* 249 (1996).
glaziovii *(Benth.) A.L. Mesquita in Acta Bot. Brasil.,* 7(2): 18 (1993): E. schomburgkii var. glaziovii.
oldemanii *R.C. Barneby & J.W. Grimes, Silk Tree, Guanacaste, Monkey's Earring (Mem. New York Bot. Gard., 74(1)):* 249 (1996)—French Guiana.

EPACRIS (Epacridac.).
crassifolia R. Br.
subsp. macroflora *R.K. Crowden & Y. Menadue in Ann. Bot. (U.K.),* 77(4): 339 (1996)—New South Wales.
lithophila *R.K. Crowden & Y. Menadue in Ann. Bot. (U.K.),* 77(4): 338 (1996)—New South Wales.
pinoidea *R.K. Crowden & Y. Menadue in Ann. Bot. (U.K.),* 77(4): 337 (1996)—New South Wales.

EPHEDRA (Ephedrac.).
equisetina Bunge
var. monoica *Y. Yang in Acta Phytotax. Sin.,* 38(4): 385 (2000)—China North-central (Hebei).
fragilis Desf.
f. disperma *Á.M. Hernández Cardona in Blancoana,* 13: 8 (1996)—Spain Gibraltar.
laristanica *M. Assadi in Iranian J. Bot.,* 7(1): 2 (1996)—Iran.
likiangensis Florin
var. mairei *(Florin) L.K. Fu & Y.F. Yu in Novon,* 7(4): 444 (1997 publ. 1998): E. saxatilis var. mairei.
monosperma *J.G. Gmel. ex C.A. Meyer, Vers. Monogr. Ephedra:* 89 (1846)—Siberia, Mongolia.

EPHIPPIANDRA (Monimiac.).
masoalensis *D.H. Lorence in Adansonia,* 21(1): 134 (1999)—Madagascar.

EPICYCAS *D.J. de Laubenfels in Blumea,* 43(2): 388 (1998). *CYCADACEAE*
elongata *(Leandri) S.L. Yang ex D.J. de Laubenfels in Blumea,* 43(2): 393 (1998): Cycas pectinata var. elongata.
lindstromii *(S.L. Yang, K.D. Hill & Hiep) D.J. de Laubenfels in Blumea,* 43(2): 395 (1998): Cycas lindstromii.
micholitzii *(Dyer) D.J. de Laubenfels in Blumea,* 43(2): 389 (1998): Cycas micholitzii.
miquelii *(Warb.) D.J. de Laubenfels in Blumea,* 43(2): 393 (1998): Cycas miquelii.
multipinnata *(C.J. Chen & S.Y. Yang) D.J. de Laubenfels in Blumea,* 43(2): 391 (1998): Cycas multipinnata.
siamensis *(Miq.) D.J. de Laubenfels in Blumea,* 43(2): 393 (1998): Cycas siamensis.
tonkinensis *(L. Linden & Rodigas) D.J. de Laubenfels in Blumea,* 43(2): 391 (1998): Zamia tonkinensis.

EPIDENDRUM (Orchidac.).
acrostigma *E. Hágsater & J. García-Cruz in Ic. Orchidac.,* Fasc. 3: t. 301 (1999)—Panama, Costa Rica.
aguirrei *E. Hágsater in Ic. Orchidac.,* Fasc. 3: t. 302 (1999)—Colombia.
amayense *E. Hágsater in Ic. Orchidac.,* Fasc. 3: t. 303 (1999)—Colombia.
amplexigastrium *E. Hágsater & C.H. Dodson in Ic. Orchidac.,* Fasc. 3: t. 304 (1999)—Colombia, Ecuador.

EPIDENDRUM:-
amplum *D.E. Bennett & E.A. Christenson in Lindleyana,* 13(1): 36 (1998)—Peru.
anchinocturnum *E. Hágsater in Ic. Orchidac.,* Fasc. 3: t. 305 (1999)—Colombia.
andinum *G. Carnevali & G.A. Romero in G.A. Romero & G. Carnevali, Orchids of Venezuela, ed. 2:* 1131 (2000)—Venezuela.
andrei *E. Hágsater & L. Sánchez Saldaña in Ic. Orchidac.,* Fasc. 3: t. 306 (1999)—Colombia.
angaritae *E. Hágsater in Ic. Orchidac.,* Fasc. 3: t. 307 (1999)—Colombia.
angustisegmentum *(L.O. Williams) E. Hágsater in Ic. Orchidac.,* Fasc. 3: t. 308 (1999): E. rigidum var. angustisegmentum.
atrorugosum *E. Hágsater in Ic. Orchidac.,* Fasc. 3: t. 309 (1999)—Costa Rica.
atwoodchlamys *E. Hágsater in Ic. Orchidac.,* Fasc. 3: 310 (1999)—Costa Rica.
atwoodii *E. Hágsater & L. Sánchez Saldaña in Ic. Orchidac.,* Fasc. 3: t. 311 (1999)—Costa Rica.
bakrense *E. Hágsater & G. Cremers in Ic. Orchidac.,* Fasc. 3: t. 312 (1999)—French Guiana.
begotii *P. Castro & J.B.F. da Silva in Bol. CAOB,* [6](2–4): 9 (1995), as 'begotti'—Brazil North (Pará).
belloi *E. Hágsater in Ic. Orchidac.,* Fasc. 3: t. 313 (1999)—Costa Rica, Panama.
brachyclinium *E. Hágsater & J. García-Cruz in Ic. Orchidac.,* Fasc. 3: t. 314 (1999)—Costa Rica.
brachyrepens *E. Hágsater in Ic. Orchidac.,* Fasc. 3: t. 315 (1999)—Costa Rica, Nicaragua, Panama.
bracteostigma *E. Hágsater & J. García-Cruz in Ic. Orchidac.,* Fasc. 3: t. 316 (1999)—Costa Rica.
bractiacuminatum *E. Hágsater & C.H. Dodson in Ic. Orchidac.,* Fasc. 3: t. 317 (1999)—Ecuador.
campbellstigma *E. Hágsater & J. García-Cruz in Ic. Orchidac.,* Fasc. 3: t. 318 (1999)—Costa Rica.
campyloglossum *P. Ortiz Valdivieso & E. Hágsater in Ic. Orchidac.,* Fasc. 3: 319 (1999)—Colombia.
cardenasii *E. Hágsater in Ic. Orchidac.,* Fasc. 3: t. 320 (1999)—Bolivia.
carnevalii *E. Hágsater & L. Sánchez Saldaña in Ic. Orchidac.,* Fasc. 3: 321 (1999)—Venezuela.
carvalhoi *A.L.V. Toscano de Brito in Lindleyana,* 15(3): 186 (2000)—Brazil Southeast (Espírito Santo).
caveroi *D.E. Bennett & E.A. Christenson in Lindleyana,* 13(1): 36 (1998)—Peru.
chanchamayodifforme *E. Hágsater & L. Sánchez Saldaña in Ic. Orchidac.,* Fasc. 3: 322 (1999)—Peru.
chauvetii *E. Hágsater & L. Sánchez Saldaña in Ic. Orchidac.,* Fasc. 3: t. 323 (1999)—French Guiana.
chimalapense *E. Hágsater & G.A. Salazar in Ic. Orchidac.,* Fasc. 3: t. 324 (1999)—Mexico Southwest (Oaxaca), Mexico Southeast (Chiapas).
chioneoides *G. Carnevali & G.A. Romero in G.A. Romero & G. Carnevali, Orchids of Venezuela, ed. 2:* 1131 (2000)—Venezuela.
cocoense *E. Hágsater in Ic. Orchidac.,* Fasc. 3: t. 325 (1999)—C. American Pacific Is (Cocos I).
cocornocturnum *E. Hágsater in Ic. Orchidac.,* Fasc. 3: t. 326 (1999)—Colombia, Ecuador.
commelinispathum *G. Carnevali & I. Ramírez in G.A. Romero & G. Carnevali, Orchids of Venezuela, ed. 2:* 1132 (2000)—Venezuela.
compressibulbum *D.E. Bennett & E.A. Christenson in Lindleyana,* 13(1): 36 (1998)—Peru.
cremersii *E. Hágsater & L. Sánchez Saldaña in Ic. Orchidac.,* Fasc. 3: t. 327 (1999)—French Guiana.
croatii *E. Hágsater in Ic. Orchidac.,* Fasc. 3: t. 328 (1999)—Panama.
delcastilloi *D.E. Bennett & E.A. Christenson in Lindleyana,* 13(1): 39 (1998)—Peru.
dialychilum Hágsater & Dodson
subsp. peruvianum *D.E. Bennett & E.A. Christenson in Lindleyana,* 13(1): 42 (1998)—Peru.
elcimeyae *E. Hágsater & J. García-Cruz in Ic. Orchidac.,* Fasc. 3: t. 329 (1999)—Costa Rica.
envigadoense *E. Hágsater in Ic. Orchidac.,* Fasc. 3: t. 330 (1999)—Colombia.
erythrostigma *E. Hágsater in Ic. Orchidac.,* Fasc. 3: t. 331 (1999)—Panama, Costa Rica.
farallonense *E. Hágsater in Ic. Orchidac.,* Fasc. 3: t. 332 (1999)—Colombia.

EPIDENDRUM:-
grand-ansense *M.A. Nir, Orchidaceae Antillanae:* 129 (2000)—Windward Is. (Martinique).
herrenhusanum *E. Hágsater in Ic. Orchidac.,* Fasc. 3: t. 333 (1999)—Colombia.
horichii *E. Hágsater in Ic. Orchidac.,* Fasc. 3: t. 334 (1999)—Costa Rica.
humeadorense *E. Hágsater & C.H. Dodson in Ic. Orchidac.,* Fasc. 3: 335 (1999)—Ecuador.
hutchisonii *E. Hágsater in Ic. Orchidac.,* Fasc. 3: t. 336 (1999)—Peru.
ibarrae *R. González Tamayo in Bol. Inst. Bot. (Guadalajara),* 1(6): 441 (1993)—Mexico Southwest (Jalisco).
ilinizae *E. Hágsater & C.H. Dodson in Ic. Orchidac.,* Fasc. 3: t. 337 (1999)—Ecuador.
ingramii *E. Hágsater & J. García-Cruz in Ic. Orchidac.,* Fasc. 3: t. 338 (1999)—Costa Rica.
ionophyllum *P. Ortiz V. in Orquideologia,* 20(3): 316 (1997)—Colombia.
jefeallenii *E. Hágsater & J. García-Cruz in Ic. Orchidac.,* Fasc. 3: t. 339 (1999)—Panama.
jefestigma *E. Hágsater & J. García-Cruz in Ic. Orchidac.,* Fasc. 3: t. 340 (1999)—Panama.
jimenezii *E. Hágsater in Ic. Orchidac.,* Fasc. 3: t. 341 (1999)—C. American Pacific Is (Cocos I).
killipii *E. Hágsater & L. Sánchez Saldaña in Ic. Orchidac.,* Fasc. 3: t. 342 (1999)—Colombia.
kockii *E. Hágsater & C.H. Dodson in Ic. Orchidac.,* Fasc. 3: t. 343 (1999)—Ecuador.
lamprochilum *E. Hágsater in Ic. Orchidac.,* Fasc. 3: t. 344 (1999)—Colombia.
laxicaule *D.E. Bennett & E.A. Christenson in Lindleyana,* 13(1): 42 (1998)—Peru.
leonii *D.E. Bennett & E.A. Christenson in Lindleyana,* 13(1): 42 (1998)—Peru.
lindae *E. Hágsater & C.H. Dodson in Ic. Orchidac.,* Fasc. 3: t. 345 (1999)—Colombia, Peru, Ecuador.
linearidifforme *E. Hágsater & L. Sánchez Saldaña in Ic. Orchidac.,* Fasc.3: t. 346 (1999)—French Guiana, Surinam.
longibracteatum *E. Hágsater in Ic. Orchidac.,* Fasc. 3: t. 347 (1999)—Panama.
lopezii *E. Hágsater in Ic. Orchidac.,* Fasc. 3: t. 348 (1999)—Colombia.
lumbaquiense *E. Hágsater & C.H. Dodson in Ic. Orchidac.,* Fasc. 3: t. 349 (1999)—Ecuador.
macroophorum *E. Hágsater & C.H. Dodson in Ic. Orchidac.,* Fasc. 3: t. 350 (1999)—Ecuador, Colombia.
madsenii *E. Hágsater & C.H. Dodson in Ic. Orchidac.,* Fasc. 3: t. 351 (1999)—Ecuador.
maduroi *E. Hágsater & J. García-Cruz in Ic. Orchidac.,* Fasc. 3: t. 352 (1999)—Panama.
melanotrichoides *E. Hágsater & C.H. Dodson in Ic. Orchidac.,* Fasc. 3: t. 353 (1999)—Ecuador.
melanoxeros *E. Hágsater & C.H. Dodson in Ic. Orchidac.,* Fasc. 3: t. 354 (1999)—Ecuador, Colombia.
mesocarpum *E. Hágsater in Ic. Orchidac.,* Fasc. 3: t. 355 (1999)—Mexico Southeast (Chiapas), Mexico Southwest (Oaxaca), El Salvador, Guatemala, Honduras, Nicaragua.
microcephalum *E. Hágsater & L. Sánchez Saldaña in Ic. Orchidac.,* Fasc. 3: t. 356 (1999)—Venezuela.
micronocturnum *G. Carnevali & G.A. Romero in Lindleyana,* 11(4): 241 (1996)—Venezuela, Brazil North (Amazonas).
mimeticum *G. Carnevali & G.A. Romero in G.A. Romero & G. Carnevali, Orchids of Venezuela, ed. 2:* 1132 (2000)—Venezuela.
minus *(Cogn.) E. Hágsater in Cat. Vasc. Pl. Ecuador (Monogr. Syst. Bot. Missouri Bot. Gard., 75):* 956 (1999): E. nocturnum var. minus.
mocinoi *E. Hágsater in Ic. Orchidac.,* Fasc. 3: t. 357 (1999)—Mexico Southwest (Guerrero), Mexico Central (México State).
monophlebium *E. Hágsater in Ic. Orchidac.,* Fasc. 3: t. 358 (1999)—Costa Rica.
montserratense *M.A. Nir, Orchidaceae Antillanae:* 132 (2000)—Leeward Is. (Montserrat).
mutisii *E. Hágsater in Ic. Orchidac.,* Fasc. 3: t. 359 (1999)—Colombia.
nanosimplex *E. Hágsater & C.H. Dodson in Ic. Orchidac.,* Fasc. 3: t. 360 (1999)—Ecuador.
norae *G. Carnevali & G.A. Romero in Lindleyana,* 11(4): 245 (1996)—Venezuela.

EPIDENDRUM:-
orthodontum *E. Hágsater & L. Sánchez Saldaña in Ic. Orchidac.*, Fasc. 3: t. 361 (1999)—Panama.
oxapampense *E. Hágsater in Ic. Orchidac.*, Fasc. 3: t. 362 (1999)—Peru, Ecuador.
oxynanodes *E. Hágsater in Ic. Orchidac.*, Fasc. 3: t. 363 (1999)—Colombia.
palmidium *E. Hágsater in Ic. Orchidac.*, Fasc. 3: t. 364 (1999)—Costa Rica.
panegyricum *E. Hágsater & L. Sánchez Saldaña in Ic. Orchidac.*, Fasc. 3: t. 365 (1999)—Windward Is (Dominica).
paradisicola *E. Hágsater & J. García-Cruz in Ic. Orchidac.*, Fasc. 3: t. 366 (1999), as 'paradisicolum'—Costa Rica.
paraguastigma *E. Hágsater & J. García-Cruz in Ic. Orchidac.*, Fasc. 3: t. 367 (1999)—Costa Rica.
penneystigma *E. Hágsater & J. García-Cruz in Ic. Orchidac.*, Fasc. 3: t. 368 (1999)—Costa Rica.
pentacarinatum *E. Hágsater & C.H. Dodson in Ic. Orchidac.*, Fasc. 3: t. 369 (1999)—Ecuador.
perijaense *G. Carnevali & G.A. Romero in G.A. Romero & G. Carnevali, Orchids of Venezuela, ed. 2:* 1133 (2000)—Venezuela.
plagiophyllum *E. Hágsater in Ic. Orchidac.*, Fasc. 3: t. 370 (1999)—Panama.
platyclinium *E. Hágsater & C.H. Dodson in Ic. Orchidac.*, Fasc. 3: t. 371 (1999)—Ecuador.
platyphyllostigma *E. Hágsater & J. García-Cruz in Ic. Orchidac.*, Fasc. 3: t. 372 (1999)—Panama.
pogonochilum *G. Carnevali & G.A. Romero in G.A. Romero & G. Carnevali, Orchids of Venezuela, ed. 2:* 1133 (2000)—Venezuela.
polyanthogastrium *E. Hágsater & C.H. Dodson in Ic. Orchidac.*, Fasc. 3: t. 373 (1999)—Ecuador.
pomacochense *E. Hágsater in Ic. Orchidac.*, Fasc. 3: t. 374 (1999)—Peru.
pomecense *E. Hágsater in Ic. Orchidac.*, Fasc. 3: t. 375 (1999)—Colombia.
portoricense *E. Hágsater & J.D. Ackerman in Ic. Orchidac.*, Fasc. 3: t. 376 (1999)—Puerto Rico, Cuba.
pseudoanceps *D.E. Bennett & E.A. Christenson in Lindleyana,* 13(1): 46 (1998)—Peru.
pseudocernuum *G. Carnevali & I.M. Ramírez M.in Harvard Pap. Bot.*, 3(2): 241 (1998)—Venezuela.
pseudolankesteri *G. Carnevali & G.A. Romero in G.A. Romero & G. Carnevali, Orchids of Venezuela, ed. 2:* 1134 (2000)—Venezuela.
putumayoense *E. Hágsater & L. Sánchez Saldaña in Ic. Orchidac.*, Fasc. 3: t. 377 (1999)—Colombia.
rhodanthum *E. Hágsater & C.H. Dodson in Ic. Orchidac.*, Fasc. 3: t. 378 (1999)—Ecuador.
rocalderianum *P. Ortiz Valdivieso & E. Hágsater in Ic. Orchidac.*, Fasc. 3: t. 379 (1999)—Colombia.
schunkei *D.E. Bennett & E.A. Christenson in Lindleyana,* 13(1): 48 (1998)—Peru.
secundum Jacq.
var. albescens *(Pabst) F. de Barros in Acta Bot. Brasil.*, 10(1): 142 (1996): E. crassifolium var. albescens.
shigenobui *E. Hágsater in Ic. Orchidac.*, Fasc. 3: t. 380 (1999)—Colombia.
sigmoideum *E. Hágsater in Ic. Orchidac.*, Fasc. 3: t. 381 (1999)—Costa Rica.
silvae *E. Hágsater & V.P. Castro Neto in Ic. Orchidac.*, Fasc. 3: t. 382 (1999)—Brazil Northeast (Bahia).
silverstonei *E. Hágsater in Ic. Orchidac.*, Fasc. 3: t. 383 (1999)—Colombia.
stalkyi *G. Carnevali & G.A. Romero in G.A. Romero & G. Carnevali, Orchids of Venezuela, ed. 2:* 1134 (2000)—Venezuela.
tacarcunense *E. Hágsater in Ic. Orchidac.*, Fasc. 3: t. 384 (1999)—Panama.
thompsonii *E. Hágsater & C.H. Dodson in Ic. Orchidac.*, Fasc. 3: t. 385 (1999)—Ecuador.
thurstoniorum *E. Hágsater in Ic. Orchidac.*, Fasc. 3: t. 386 (1999), as 'thurstonorum'—Costa Rica.
tigriphyllum *E. Hágsater in Ic. Orchidac.*, Fasc. 3: t. 387 (1999)—Colombia.
tingo-mariae *E. Hágsater in Ic. Orchidac.*, Fasc. 3: t. 388 (1999)—Peru.
tridens Poepp. & Endl.
var. briegeri *I. Bock in Orchidee,* 33(4): 157 (1982)—Colombia.
trullichilum *E. Hágsater & C.H. Dodson in Ic. Orchidac.*, Fasc. 3: t. 389 (1999)—Ecuador.

EPIDENDRUM:-
tumuc-humaciense *(Veyret) G. Carnevali & G.A. Romero in Lindleyana,* 11(4): 246 (1996): E. nocturnum var. tumuc-humaciense.
tuxtlense *E. Hágsater, J. García-Cruz & L. Sánchez Saldaña in Ic. Orchidac.*, Fasc. 3: t. 390 (1999)—Mexico Gulf (Veracruz), Mexico Southwest (Oaxaca).
tziscaoense *E. Hágsater in Ic. Orchidac.*, Fasc. 3: t. 391 (1999)—Mexico Southeast (Chiapas).
uleinanodes *E. Hágsater in Ic. Orchidac.*, Fasc. 3: t. 392 (1999)—Brazil North (Amazonas).
upanodifforme *E. Hágsater & C.H. Dodson in Ic. Orchidac.*, Fasc. 3: t. 393 (1999)—Ecuador.
urraoense *E. Hágsater in Ic. Orchidac.*, Fasc. 3: t. 394 (1999)—Colombia.
vexillium *E. Hágsater in Ic. Orchidac.*, Fasc. 3: t. 395 (1999)—Colombia, Panama.
vidal-senegei *E. Hágsater in Ic. Orchidac.*, Fasc. 3: t. 396 (1999)—Ecuador or Peru.
villegastigma *E. Hágsater & J. García-Cruz in Ic. Orchidac.*, Fasc. 3: t. 397 (1999)—Costa Rica.
villotae *E. Hágsater & C.H. Dodson in Ic. Orchidac.*, Fasc. 3: t. 398 (1999)—Ecuador, Colombia.
wendtii *E. Hágsater & G.A. Salazar in Ic. Orchidac.*, Fasc. 3: t. 399 (1999)—Mexico Southwest (Oaxaca), Honduras.
whittenii *E. Hágsater & C.H. Dodson in Ic. Orchidac.*, Fasc. 3: t. 400 (1999)—Ecuador, Peru.
yaracuyense *G. Carnevali & G.A. Romero in G.A. Romero & G. Carnevali, Orchids of Venezuela, ed. 2:* 1135 (2000)—Venezuela.

EPIGENEIUM (Orchidac.).
sect. Katherinea *(A.D. Hawkes) L.A. Garay & G.A. Romero-Gonzalez in Harvard Pap. Bot.*, 4(2): 479 (1999): gen. Katherinea.
sect. Racemosum *L.A. Garay & G.A. Romero-Gonzalez in Harvard Pap. Bot.*, 4(2): 479 (1999).
arjunoense *J.J. Wood & J.B. Comber in Orchid Rev.*, 104(1211): 286 (1996)—Jawa.
banghamii *(Ames & C. Schweinf.) L.A. Garay & G.A. Romero-Gonzalez in Harvard Pap. Bot.*, 4(2): 476 (1999): Dendrobium banghamii.
exilifolium *(Ames & C. Schweinf.) L.A. Garay & G.A. Romero-Gonzalez in Harvard Pap. Bot.*, 4(2): 476 (1999): Dendrobium exilifolium.
gracilipes *(Burkill) L.A. Garay & G.A. Romero-Gonzalez in Harvard Pap. Bot.*, 4(2): 476 (1999): Dendrobium gracilipes.
treutleri *(Hook. f.) P. Ormerod in Oasis,* 1(3): 3 (2000): Coelogyne treutleri.

EPILOBIUM (Onagrac.).
bergianum *A.K. Skvortsov in Byull. Mosk. Obshch. Ispўt. Prir., Biol.*, 100(1): 76 (1995)—Sweden, Norway, Finland.
× brunnatum *G.D. Kitchener & D.R. McKean in Watsonia,* 22(1): 51 (1998)—Great Britain. (E. brunnescens × ciliatum).
canum (Greene) P.H. Raven
var. garrettii *(A. Nelson) N.H. Holmgren & P.K. Holmgren in A. Cronquist et al., Intermountain Fl., Vasc. Pl. Intermountain West,* U.S.A., 3A: 241 (1997): Zauschneria garrettii.
var. latifolium *(Hook.) N.H. Holmgren & P.K. Holmgren in A. Cronquist et al., Intermountain Fl., Vasc. Pl. Intermountain West,* U.S.A., 3A: 240 (1997): Zauschneria californica var. latifolia.
cephalostigma Hausskn.
var. angulatum *(Kom.) V.N. Voroschilov in A.K. Skvortsov (ed.), Florist. issl. v razn. raĭonakh SSSR:* 183 (1985): E. angulatum.
× chateri *G.D. Kitchener & D.R. McKean in Watsonia,* 22(1): 55 (1998)—Great Britain. (E. brunnescens × palustre).
× confusilobum *G.D. Kitchener & D.R. McKean in Watsonia,* 22(1): 52 (1998)—Great Britain. (E. brunnescens × montanum).
× cornubiense *G.D. Kitchener & D.R. McKean in Watsonia,* 22(1): 52 (1998)—Great Britain. (E. brunnescens × lanceolatum).
× floridulum *M. Smejkal in Čas. Morav. Muz., Vědy Přír.*, 79(1–2): 81 (1994 publ. 1995)—Czechoslovakia. (E. ciliatum × parviflorum).
× fossicola *M. Smejkal in Čas. Morav. Muz., Vědy Přír.*, 79(1–2): 83 (1994 publ. 1995)—Czechoslovakia. (E. ciliatum × palustre).

EPILOBIUM:-
glanduligerum *M.G. Pachomova in Opred. Rast. Sred. Azii,* 7: 370 (1983), homonym; *see Čelak., Prodr. Fl. Böhmen,* 4: 882 (1881)—Kirgizstan.
glandulosum Lehm.
subsp. maximowiczii *(Hausskn.) V.N. Voroschilov in A.K. Skvortsov (ed.), Florist. issl. v razn. raĭonakh SSSR:* 183 (1985): E. maximowiczii.
× iglaviense *M. Smejkal in Čas. Morav. Muz., Vědy Přír.*, 79(1–2): 82 (1994 publ. 1995)—Czechoslovakia. (E. ciliatum × lamyi).
× interjectum *M. Smejkal in Čas. Morav. Muz., Vědy Přír.*, 79(1–2): 82 (1994 publ. 1995)—Czechoslovakia. (E. ciliatum × montanum).
× josefi-holubii *F. Krahulec in Preslia,* 71(3): 241 (1999)—Czechoslovakia. (E. alsinifolium × ciliatum).
× kitcheneri *D.R. McKean in Watsonia,* 22(4): 417 (1999)—Great Britain. (E. montanum × pedunculare).
lipschitzii *M.G. Pachomova in Opred. Rast. Sred. Azii,* 7: 369 (1983)—Tadzhikistan.
× mentiens *M. Smejkal in Čas. Morav. Muz., Vědy Přír.*, 79(1–2): 82 (1994 publ. 1995)—Czechoslovakia. (E. ciliatum × tetragonum).
× nutantiflorum *M. Smejkal in Čas. Morav. Muz., Vědy Přír.*, 79(1–2): 83 (1994 publ. 1995)—Czechoslovakia. (E. ciliatum × roseum).
× obscurescens *G.D. Kitchener & D.R. McKean in Watsonia,* 22(1): 54 (1998)—Ireland (Northern Ireland). (E. brunnescens × obscurum).
petraeum *P.B. Heenan in New Zealand J. Bot.*, 34(1): 41 (1996)—New Zealand South I.
× prochazkae *F. Krahulec in Preslia,* 71(3): 242 (1999)—Czechoslovakia. (E. alpestre × ciliatum).
pseudorubescens *A.K. Skvortsov in Byull. Mosk. Obshch. Ispўt. Prir., Biol.*, 100(1): 75 (1995)—Finland, Sweden.
verticillatum *W.X. Wang, W.Y. Guo & Y.S. Fu in Abstr. XV Internat. Bot. Congr., Yokohama, Japan 1993:* 250 (1993)—China South-central (Hubei).
× vicinum *M. Smejkal in Čas. Morav. Muz., Vědy Přír.*, 79(1–2): 83 (1994 publ. 1995)—Czechoslovakia. (E. ciliatum × obscurum).

EPIMEDIUM (Berberidac.).
borealiguizhouense *S.Z. He & Y.K. Yang in J. Pl. Resourc. Environ.*, 2(4): 51 (1993), as 'baieali-guizhouense'—China South-central (Guizhou).†
brachyrrhizum *W.T. Stearn in Kew Bull.*, 52(3): 659 (1997)—China South-central (Guizhou).
campanulatum *M. Ogisu in Kew Bull.*, 51(2): 401 (1996)—China South-central (Sichuan).
chlorandrum *W.T. Stearn in Kew Bull.*, 52(3): 660 (1997)—China South-central (Sichuan).
epsteinii *W.T. Stearn in Kew Bull.*, 52(3): 662 (1997)—China Southeast (Hunan).
franchetii *W.T. Stearn in Kew Bull.*, 51(2): 396 (1996)—China South-central (Hubei).
ilicifolium *W.T. Stearn in Kew Bull.*, 53(1): 213 (1998)—China North-central (Shanxi).
lishihchenii *W.T. Stearn in Kew Bull.*, 52(3): 664 (1997)—China Southeast (Jiangxi).
lobophyllum *L.H. Liu & B.G. Li in Acta Phytotax. Sin.*, 37(3): 288 (1999)—China Southeast (Hunan).
mikinorii *W.T. Stearn in Kew Bull.*, 53(1): 214 (1998)—China South-central (Hubei).
myrianthum *W.T. Stearn in Kew Bull.*, 53(1): 218 (1998)—China Southeast (Hunan).
rhizomatosum *W.T. Stearn in Kew Bull.*, 53(1): 220 (1998)—China South-central (Sichuan).
shuichengense *S.Z. He in Acta Bot. Yunnanica,* 18(2): 209 (1996)—China South-central (Guizhou).

EPIPACTIS (Orchidac.).
albensis H. Nováková & Rydlo
var. fibri *(Scappat. & Robatsch) P. Delforge in Nat. Belg.*, 78(3): 246 (1997): E. fibri.
aspromontana *G. Bartolo, S. Pulvirenti & K. Robatsch in Caesiana,* 6: 41 (1996)—Italy.
atrorubens (Bernh.) Besser
f. sirneensis *R. Neirynck in Caesiana,* 5: 44 (1995)—Romania.
× breinerorum *P. Batousek in J. Eur. Orchideen,* 29(4): 645 (1997)—Czechoslovakia. (E. helleborine subsp. helleborine × greuteri).

EPIPACTIS:-
× bruxellensis *P. Delforge in Nat. Belg.*, 77(4): 221 (1996)—Belgium. (E. helleborine × phyllanthes var. degenera).
campeadorii *P. Delforge in Nat. Belg.*, 76(3): 90 (1995)—Spain.
cardina *J. Benito Ayuso & C.E. Hermosilla in Estud. Mus. Cienc. Nat. Alava*, 13: 109 (1998)—Spain.
× conquensis *J. Benito Ayuso & C.E. Hermosilla in Estud. Mus. Cienc. Nat. Alava*, 13: 112 (1998)—Spain. (E. cardina × parviflora).
degenii *L.J. Szentpéteri & F. Mónus in J. Eur. Orchideen*, 31(3): 647, 646 (1999)—Greece.
fageticola *(C.E. Hermos.) J. Devillers-Terschuren & P. Devillers in Nat. Belg.*, 80(3): 302 (1999): E. phyllanthes var. fageticola.
fibri *G. Scappaticci & K. Robatsch in Orchidophile*, 116: 87 (1995)—France.
futakii *P. Mered'a & O. Potůček in Preslia*, 70(3): 253 (1998)—Czechoslovakia.
× gerbaudiorum *P. Delforge in Nat. Belg.*, 78(3): 178 (1997)—France. (E. provincialis × tremolsii).
× gevaudanii *P. Delforge in Nat. Belg.*, 78(3): 179 (1997)—France. (E. helleborine × rhodanensis).
guegelii *K. Robatsch in J. Eur. Orchideen*, 28(4): 767 (1996 publ. 1997)—Romania.
helleborine (L.) Crantz
subsp. distans *(Arv.-Touv.) R. Engel & P. Quentin in Orchidophile*, 124: 205 (1996): E. distans.
f. luteola *P.M. Brown in North Amer. Native Orchid J.*, 2(4): 318 (1996)—New Hampshire.
subsp. orbicularis *(K. Richt.) E. Klein in Phyton (Austria)*, 37(1): 74 (1997): E. orbicularis.
var. papillosa *(Franch. & Sav.) T. Hashimoto in Proc. 12th World Orchid Conf., 1987 (eds. K. Saito & R. Tanaka)*: 120 (1987): E. papillosa.
subsp. renzii *(Robatsch) B. Løjtnant in Flora Fauna (Århus)*, 101(2): 71 (1995): E. renzii.
var. renzii *(Robatsch) J. Claessens, J. Kleynen & R. Wielinga in Eurorchis*, 10: 63 (1998): E. renzii.
var. sayekiana *(Makino) T. Hashimoto in Proc. 12th World Orchid Conf., 1987 (eds. K. Saito & R. Tanaka)*: 121 (1987): E. sayekiana.
hispanica *J. Benito Ayuso & C.E. Hermosilla in Estud. Mus. Cienc. Nat. Alava*, 13: 106 (1998)—Spain.
var. viridis *J. Benito Ayuso & C.E. Hermosilla in Estud. Mus. Cienc. Nat. Alava*, 13: 107 (1998)—Spain.
komoricensis *P. Mered'a in Preslia*, 68(2): 131 (1996)—Czechoslovakia.
leptochila (Godfery) Godfery
var. neglecta *(Kümpel) A. Gévcudan in Nat. Belg.*, 80(3): 367 (1999): E. leptochila subsp. neglecta.
mecsekensis *A. Molnár & K. Robatsch in J. Eur. Orchideen*, 28(4): 782 (1996 publ. 1997)—Hungary.
muelleri Godfery
var. cleistogama *(C.A. Thomas) P.D. Sell in P.D. Sell & G. Murrell, Fl. Great Britain & Ireland*, 5: 364 (1996)—E. cleistogama.
var. dunensis *(T. Stephenson & T.A. Stephenson) P.D. Sell in P.D. Sell & G. Murrell, Fl. Great Britain & Ireland*, 5: 364 (1996)—E. viridiflora f. dunensis.
var. leptochila *(Godfery) P.D. Sell in P.D. Sell & G. Murrell, Fl. Great Britain & Ireland*, 5: 364 (1996)—E. viridiflora var. leptochila.
var. peitzii *(H. Neumann & Wucherpf.) P. Delforge in Nat. Belg.*, 78(3): 246 (1997): E. peitzii.
neglecta *(Kümpel) H. Kümpel, Die wildwachsenden Orchideen der Rhön*: 67 (1996): E. leptochila subsp. neglecta.
peitzii *H. Neumann & W. Wucherpfennig in J. Eur. Orchideen*, 28(4): 748 (1996 publ. 1997)—Germany.
var. sancta *P. Delforge in Nat. Belg.*, 81(4): 397 (2000)—Great Britain.
phyllanthes G.E. Sm.
var. cambrensis *(C.A. Thomas) P.D. Sell in P.D. Sell & G. Murrell, Fl. Great Britain & Ireland*, 5: 364 (1996): E. cambrensis.
subsp. confusa *(D.P. Young) B. Løjtnant in Flora Fauna (Århus)*, 101(2): 72 (1996): E. confusa.
var. fageticola *C.E. Hermosilla in Estud. Mus. Cienc. Nat. Alava*, 13: 138 (1998)—Spain.

EPIPACTIS:-
phyllanthes G.E. Sm.
var. olarionensis *P. Delforge in Nat. Belg.*, 78(3): 242 (1997)—France.
pollinensis *B. Baumann & H. Baumann in J. Eur. Orchideen*, 32(1): 94 (2000)—Italy.
× populetorum *J. Benito Ayuso & C.E. Hermosilla in Estud. Mus. Cienc. Nat. Alava*, 13: 108 (1998)—Spain. (E. helleborine × hispanica).
provincialis *A. Aubenas & K. Robatsch in Orchidophile*, 122: 109 (1996)—France.
pseudopurpurata *P. Mered'a in Preslia*, 68(1): 27 (1996)—Czechoslovakia.
schubertiorum *G. Bartolo, S. Pulvirenti & K. Robatsch in J. Eur. Orchideen*, 28(4): 775 (1996 publ. 1997)—Italy.
× soguksuensis *C.A.J. Kreutz in Eurorchis*, 9: 71 (1997)—Turkey. (E. helleborine × turcica).
spiridonovii *J. Devillers-Terschuren & P. Devillers in Nat. Belg.*, 76(3): 87 (1995)—Bulgaria.
tallosii *A. Molnár & K. Robatsch in J. Eur. Orchideen*, 28(4): 789 (1996 publ. 1997)—Hungary.
turcica *C.A.J. Kreutz in Eurorchis*, 9: 61 (1997)—Turkey.

EPITHELANTHA (Cactac.).
micromeris
subsp. greggii *(Engelm.) N.P. Taylor in Cactaceae Consensus Initiatives*, 5: 12 (1998): Mammillaria micromeris [var.] greggii.
subsp. pachyrhiza *(W.T. Marshall) N.P. Taylor in Cactaceae Consensus Initiatives*, 5: 13 (1998): E. micromeris var. pachyrhiza.
subsp. polycephala *(Backeb.) C.E. Glass, Guia Identific. Cact. Amenazadas México*, 1: Ep/mi ssp. polycephala (1998 publ. 1997): E. polycephala.
subsp. unguispina *(Boed.) N.P. Taylor in Cactaceae Consensus Initiatives*, 5: 12 (1998): Mammillaria micromeris var. unguispina.
petri *J.J. Halda & L. Horáček in Acta Mus. Richnov. Sect. Nat.*, 7(2): 72 (2000)—Mexico Northeast (Coahuila).

EPITHEMA (Gesneriac.).
dentatum *(C.B. Clarke) O.M. Hilliard & B.L. Burtt in Edinburgh J. Bot.*, 54(1): 111 (1997): E. carnosum var. dentatum.
subsp. hispidum *(C.B. Clarke) O.M. Hilliard & B.L. Burtt in Edinburgh J. Bot.*, 54(1): 112 (1997): E. carnosum var. hispidum.
var. pusillum *(C.B. Clarke) O.M. Hilliard & B.L. Burtt in Edinburgh J. Bot.*, 54(1): 112 (1997): E. carnosum var. pusillum.
dolichopodum *O.M. Hilliard & B.L. Burtt in Edinburgh J. Bot.*, 54(1): 112 (1997): Borneo (Sabah).
horsfieldii (R. Br.) DC.
var. epiphyllum *O.M. Hilliard & B.L. Burtt in Edinburgh J. Bot.*, 54(1): 112 (1997)—Lesser Sunda Is.
longipetiolatum *(Merr.) O.M. Hilliard & B.L. Burtt in Edinburgh J. Bot.*, 54(1): 112 (1997): E. brunonis var. longipetiolatum.
longitubum *O.M. Hilliard & B.L. Burtt in Edinburgh J. Bot.*, 54(1): 112 (1997): East Timor.
madulidii *O.M. Hilliard & B.L. Burtt in Edinburgh J. Bot.*, 54(1): 112 (1997): Philippines.
parvibracteatum *O.M. Hilliard & B.L. Burtt in Edinburgh J. Bot.*, 54(1): 112 (1997): Malaya (Peninsular Malaysia).
rennellense *O.M. Hilliard & B.L. Burtt in Edinburgh J. Bot.*, 54(1): 112 (1997): Solomon Is. (South Solomons).
sarawakense *O.M. Hilliard & B.L. Burtt in Edinburgh J. Bot.*, 54(1): 113 (1997): Borneo (Sarawak).
steenisii *O.M. Hilliard & B.L. Burtt in Edinburgh J. Bot.*, 54(1): 113 (1997): Sumatera.
strigosum *(C.B. Clarke) O.M. Hilliard & B.L. Burtt in Edinburgh J. Bot.*, 54(1): 113 (1997): E. brunonis var. strigosum.
subsp. philippinum *O.M. Hilliard & B.L. Burtt in Edinburgh J. Bot.*, 54(1): 113 (1997)—Philippines.
taiwanense S.S. Ying
var. fasciculatum *(C.B. Clarke) Z.Yu Li & M.T. Kao in Fl. Taiwan, ed. 2*, 4: 697 (1998), as 'fasciculata': E. brunonis var. fasciculatum.

EPITHEMA:-
tenerum *(C.B. Clarke) O.M. Hilliard & B.L. Burtt in Edinburgh J. Bot.*, 54(1): 113 (1997): E. brunonis var. tenerum.

ERAGROSTIS (Gramin.).
acamptoclada *T.A. Cope in Kew Bull.*, 53(1): 136 (1998)—Zambia, Angola.
aegyptiaca (Willd.) Delile
subsp. humifusa *H. Scholz in Willdenowia*, 26(1–2): 231 (1996)—Sudan.
albensis *H. Scholz in Verh. Bot. Ver. Berlin Brandenburg*, 128(2): 74 (1995)—Germany.
alveiformis *M. Lazarides in Austral. Syst. Bot.*, 10(1): 96 (1997)—Queensland, New South Wales.
amabilis (L.) Wight & Arn.
var. insularis *(C.E. Hubb.) P. Umamaheswari & P. Daniel in J. Econ. Taxon. Bot.*, 22(1): 216 (1998): E. tenella var. insularis.
var. peramangalamensis *P. Umamaheswari & P. Daniel in J. Econ. Taxon. Bot.*, 22(1): 216 (1998)—India (Tamil Nadu).
anacrantha *T.A. Cope in Kew Bull.*, 53(1): 138 (1998)—Zambia.
anacranthoides *T.A. Cope in Kew Bull.*, 53(1): 139 (1998)—Zambia.
ancashensis *P.M. Peterson, N. Refulio Rodriguez & O. Tovar in Sida*, 19(1): 66 (2000)—Peru.
astreptoclada *T.A. Cope in Kew Bull.*, 53(1): 135 (1998)—Zambia.
capitula *M. Lazarides in Austral. Syst. Bot.*, 10(1): 103 (1997)—Queensland.
cassa *M. Lazarides in Austral. Syst. Bot.*, 10(1): 104 (1997)—Queensland.
crateriformis *M. Lazarides in Austral. Syst. Bot.*, 10(1): 108 (1997)—Northern Territory, Western Australia.
dayanandanii *P. Ravichandran, S. Krishnan & N.P. Samson in Kew Bull.*, 51(1): 155 (1996)—India (Tamil Nadu).
divaricata *T.A. Cope in Kew Bull.*, 53(1): 132 (1998)—Zambia.
ecarinata *M. Lazarides in Austral. Syst. Bot.*, 10(1): 114 (1997)—Northern Territory.
exigua *M. Lazarides in Austral. Syst. Bot.*, 10(1): 118 (1997)—Northern Territory, Western Australia, Queensland, New South Wales, Victoria.
fallax *M. Lazarides in Austral. Syst. Bot.*, 10(1): 120 (1997)—Northern Territory, Western Australia, Queensland.
fastigiata *T.A. Cope in Kew Bull.*, 53(1): 152 (1998)—Malawi.
filicaulis *M. Lazarides in Austral. Syst. Bot.*, 10(1): 121 (1997)—Queensland, Western Australia, Northern Territory.
fimbrillata *T.A. Cope in Kew Bull.*, 53(1): 146 (1998)—Zambia.
hirticaulis *M. Lazarides in Austral. Syst. Bot.*, 10(1): 122 (1997)—Northern Territory.
lanicaulis *M. Lazarides in Austral. Syst. Bot.*, 10(1): 126 (1997)—Northern Territory, Western Australia, Queensland, New South Wales.
latifolia *T.A. Cope in Kew Bull.*, 53(1): 144 (1998)—Zimbabwe.
lepidobasis *T.A. Cope in Kew Bull.*, 53(1): 153 (1998)—Zambia.
leptotricha *T.A. Cope in Kew Bull.*, 53(1): 147 (1998)—Botswana.
milnei *Launert ex T.A. Cope in Kew Bull.*, 53(1): 133 (1998)—Zambia.
nigricans (Kunth) Steud.
var. punensis *E.G. Nicora in Boissiera*, 54: 65 (1998)—Argentina Northwest (Jujuy, Catamarca, La Rioja, Salta, San Juan), Chile North (Antofagasta, Tarapacá).
olida *M. Lazarides in Austral. Syst. Bot.*, 10(1): 137 (1997)—Queensland, Western Australia, Northern Territory.
oligostachya *Launert ex T.A. Cope in Kew Bull.*, 53(1): 151 (1998)—Zambia.
pectinacea (Michx.) Nees
var. tracyi *(Hitchc.) P.M. Peterson in Sida*, 17(1): 106 (1996): E. tracyi.
petraea *M. Lazarides in Austral. Syst. Bot.*, 10(1): 141 (1997)—Northern Territory.
phyllacantha *T.A. Cope in Kew Bull.*, 53(1): 162 (1998)—Botswana.

ERAGROSTIS:-
pilosa (L.) P. Beauv.
var. amurensis *(Prob.) V.N. Voroschilov in A.K. Skvortsov (ed.), Florist. issl. v razn. Raĭonakh SSSR:* 149 (1985): E. amurensis.
poculiformis *T.A. Cope in Kew Bull.,* 55(2): 487 (2000)—Tanzania.
potamophila *M. Lazarides in Austral. Syst. Bot.,* 10(1): 143 (1997)—Western Australia, Northern Territory.
punctiglandulosa *T.A. Cope in Kew Bull.,* 53(1): 143 (1998)—Zambia.
rivalis *H. Scholz in Willdenowia,* 26(1–2): 229 (1996)—Belorussiya, Ukraina.
sericata *T.A. Cope in Kew Bull.,* 53(1): 163 (1998)—Mozambique.
spicigera *T.A. Cope in Kew Bull.,* 53(1): 155 (1998)—Zambia.
stagnalis *M. Lazarides in Austral. Syst. Bot.,* 10(1): 152 (1997)—Northern Territory, Queensland.
subglandulosa *T.A. Cope in Kew Bull.,* 53(1): 151 (1998)—Botswana.
subtilis *M. Lazarides in Austral. Syst. Bot.,* 10(1): 155 (1997)—Northern Territory.
sylviae *T.A. Cope in Kew Bull.,* 53(1): 163 (1998)—Malawi.
terecaulis *S.A. Renvoize, Gramineas de Bolivia:* 324 (1998)—Bolivia.
triquetra *M. Lazarides in Austral. Syst. Bot.,* 10(1): 162 (1997)—Queensland.
uvida *M. Lazarides in Austral. Syst. Bot.,* 10(1): 164 (1997)—Queensland, Northern Territory.
venustula *Launert ex T.A. Cope in Kew Bull.,* 53(1): 160 (1998)—Zambia, Zimbabwe, Angola.

EREMANTHUS (Compos.).
harleyi *H. Robinson in Rhodora,* 98(893): 90 (1996 publ. 1997)—Brazil Northeast (Bahia).
santosii *H. Robinson in Rhodora,* 98(893): 88 (1996 publ. 1997)—Brazil Northeast (Bahia).

EREMIUM *O. Seberg & I. Linde-Laursen in Syst. Bot.,* 21(1): 10 (1996). *GRAMINEAE*
erianthum *(Phil.) O. Seberg & I. Linde-Laursen in Syst. Bot.,* 21(1): 11 (1996): Elymus erianthus.

EREMOGONE (Caryophyllac.).
jakutorum *(A.P. Khokhr.) N.S. Pavlova in Sosud. Rast. Sovet. Dal'nego Vostoka,* 8: 55 (1996): Arenaria jakutorum.

EREMOPAPPUS (Compos.).
tadshicorum *(Tzvelev) G. Panigrahi in Fl. U.S.S.R. (English translation),* 28: 459 (1998): Centaurea tadshicorum.

EREMURUS (Asphodelac.).
zangezuricus *A.D. Mikheev in Bot. Zhurn.,* 81(5): 79 (1996)—Transcaucasus (Armeniya).

EREPSIA (Aizoac.).
aristata *(L. Bolus) S. Liede & H.E.K. Hartmann in Bradleya,* 16: 64 (1998): Ruschia aristata.
dunensis *(Sond.) C. Klak in Bothalia,* 30(1): 38 (2000): Mesembryanthemum dunense.

ERIA (Orchidac.).
arunachalensis *A. Nageswara Rao in J. Econ. Taxon. Bot.,* 21(3): 711 (1997 publ. 1998)—Assam (Arunachal Pradesh).
corneri Rchb. f.
var. clausa *(King & Pantl.) A.N. Rao in J. Econ. Taxon. Bot.,* 20(3): 708 (1996): E. clausa.
geboana *P. Ormerod in Austral. Orchid Rev.,* 60(6): 37 (1995)—New Guinea (Papua New Guinea).
imperatifolia *P. Ormerod in Austral. Orchid Rev.,* 60(6): 38 (1995)—New Guinea (Papua New Guinea).
jengingensis *S.N. Hegde in J. Orchid Soc. India,* 7(1–2): 13 (1993)—Assam (Arunachal Pradesh).
lactiflora *L.V. Averyanov in Lindleyana,* 14(4): 219 (1999)—Vietnam.
sabasaroe *P. Ormerod in Austral. Orchid Rev.,* 60(6): 39 (1995)—New Guinea (Papua New Guinea).
sharmae *H.J. Chowdhery, G.S. Giri & G.D. Pal in Indian J. Forest.,* 16(1): 91 (1993)—Assam (Arunachal Pradesh).

ERIANTHEMUM (Loranthac.).
lanatum *D. Wiens & R.M. Polhill, Mistletoes of Africa:* 241 (1998)—Zambia.
rotundifolium *D. Wiens & R.M. Polhill, Mistletoes of Africa:* 242 (1998)—Tanzania.
schmitzii *Balle ex D. Wiens & R.M. Polhill, Mistletoes of Africa:* 241 (1998)—Zaire.
viticola *Balle ex D. Wiens & R.M. Polhill, Mistletoes of Africa:* 239 (1998), as 'viticolum'—Central African Republic.

ERIANTHUS (Gramin.).
giganteus *(Walter) P. Beauv., Ess. Agrostogr.:* 14, 151, 177 (1812): Anthoxanthum giganteum.†
lancangensis *Y.Y. Qian in Guihaia,* 19(4): 339 (1999)—China South-central (Yunnan).

ERICA (Ericac.).
× afroeuropaea *D. McClintock in New Plantsman,* 6(4): 207 (1999)—Cultivation. (E. arborea f. alpina × baccans).
agglutinans *E.G.H. Oliver in Contrib. Bolus Herb.,* 19: 203 (2000), nom. nov.: Syndesmanthus schlechteri *N.E. Br.*
albertyniae *E.G.H. Oliver in Contrib. Bolus Herb.,* 19: 132 (2000)—Cape Province.
andringitrensis *(H. Perrier) L.J. Dorr & E.G.H. Oliver in Adansonia,* 21(1): 77 (1999): Philippia andringitrensis.
anguliger *(N.E. Br.) E.G.H. Oliver in Contrib. Bolus Herb.,* 19: 271 (2000): Anomalanthus anguliger.
angulosa *E.G.H. Oliver in Contrib. Bolus Herb.,* 19: 177 (2000), nom. nov.: Acrostemon fourcadei *L. Guthrie.*
arachnocalyx *E.G.H. Oliver in Contrib. Bolus Herb.,* 19: 325 (2000), nom. nov.: Acrostemon viscidus *N.E. Br.*
areolata *(N.E. Br.) E.G.H. Oliver in Contrib. Bolus Herb.,* 19: 378 (2000): Coccosperma areolatum.
armandiana *L.J. Dorr & E.G.H. Oliver in Adansonia,* 21(1): 77 (1999), nom. nov.: Philippia capitata *Baker.*
artemisioides *(Klotzsch) E.G.H. Oliver in Contrib. Bolus Herb.,* 19: 381 (2000): Salaxis artemisioides.
atromontana *E.G.H. Oliver in Contrib. Bolus Herb.,* 19: 251 (2000)—Cape Province.
barnettiana *L.J. Dorr & E.G.H. Oliver in Adansonia,* 21(1): 78 (1999), nom. nov.: Philippia oppositifolia *H. Perrier.*
baroniana *L.J. Dorr & E.G.H. Oliver in Adansonia,* 21(1): 78 (1999), nom. nov.: Philippia floribunda *Benth.*
benthamiana *E.G.H. Oliver in Contrib. Bolus Herb.,* 19: 191 (2000), nom. nov.: Blaeria gracilis *Bartl.*
betsileana *(H. Perrier) L.J. Dorr & E.G.H. Oliver in Adansonia,* 21(1): 78 (1999): Philippia betsileana.
binaria *E.G.H. Oliver in Contrib. Bolus Herb.,* 19: 403 (2000)—Cape Province.
bojeri *L.J. Dorr & E.G.H. Oliver in Adansonia,* 21(1): 78 (1999), nom. nov.: Philippia aristata *Benth.*
bokkeveldia *E.G.H. Oliver in Contrib. Bolus Herb.,* 19: 288 (2000), nom. nov.: Eremia calycina *Compton.*
bolusanthus *E.G.H. Oliver in Contrib. Bolus Herb.,* 19: 255 (2000), nom. nov.: Thoracosperma nanum *N.E. Br.*
bosseri *L.J. Dorr in Adansonia,* 21(1): 79 (1999)—Madagascar.
boucheri *E.G.H. Oliver in Contrib. Bolus Herb.,* 19: 394 (2000)—Cape Province.
boutonii *L.J. Dorr & E.G.H. Oliver in Adansonia,* 21(1): 80 (1999), nom. nov.: Philippia ciliata *Benth.*
bredasiana *E.G.H. Oliver in Contrib. Bolus Herb.,* 19: 396 (2000), nom. nov.: Salaxis pumila *N.E. Br.*
brownii *E.G.H. Oliver in Contrib. Bolus Herb.,* 19: 159 (2000), nom. nov.: Syndesmanthus breviflorus *N.E. Br.*
burchelliana *E.G.H. Oliver in Contrib. Bolus Herb.,* 19: 241 (2000), nom. nov.: Coilostigma glabrum *Benth.*
cabernetea *E.G.H. Oliver in Novon,* 8(3): 267 (1998)—Cape Province.
calcicola *(E.G.H. Oliv.) E.G.H. Oliver in Contrib. Bolus Herb.,* 19: 384 (2000): Scyphogyne calcicola.

ERICA:-
caprina *E.G.H. Oliver in Contrib. Bolus Herb.,* 19: 320 (2000), nom. nov.: Grisebachia minutiflora *N.E. Br.*
cedromontana *E.G.H. Oliver in Novon,* 8(3): 269 (1998)—Cape Province.
cereris *(Compton) E.G.H. Oliver in Contrib. Bolus Herb.,* 19: 327 (2000): Arachnocalyx cereris.
cetrata *E.G.H. Oliver in Contrib. Bolus Herb.,* 19: 285 (2000), nom. nov.: Eremia peltata *Compton.*
columnaris *E.G.H. Oliver in Bothalia,* 30(1): 50 (2000)—Cape Province.
comorensis *(Engl.) L.J. Dorr & E.G.H. Oliver in Adansonia,* 21(1): 90 (1999): Philippia comorensis.
subsp. anjurensis *(Alm & T.C.E. Fr.) L.J. Dorr & E.G.H. Oliver in Adansonia,* 21(1): 90 (1999): Philippia anjurensis.
cryptoclada *(Baker) L.J. Dorr & E.G.H. Oliver in Adansonia,* 21(1): 80 (1999): Philippia cryptoclada.
cubitans *E.G.H. Oliver in Bothalia,* 29(1): 112 (1999)—Cape Province.
curvistyla *(N.E. Br.) E.G.H. Oliver in Contrib. Bolus Herb.,* 19: 291 (2000): Eremiopsis curvistyla.
danguyana *(H. Perrier) L.J. Dorr & E.G.H. Oliver in Adansonia,* 21(1): 81 (1999): Philippia danguyana.
densata *L.J. Dorr & E.G.H. Oliver in Adansonia,* 21(1): 81 (1999), nom. nov.: Philippia densa *Benth.*
dispar *(N.E. Br.) E.G.H. Oliver in Contrib. Bolus Herb.,* 19: 130 (2000): Simocheilus dispar.
dregei *E.G.H. Oliver in Contrib. Bolus Herb.,* 19: 128 (2000), nom. nov.: Simocheilus oblongus *Benth.*
ecklonii *E.G.H. Oliver in Contrib. Bolus Herb.,* 19: 188 (2000), nom. nov.: Sympieza eckloniana *Klotzsch.*
eglandulosa *(Klotzsch) E.G.H. Oliver in Contrib. Bolus Herb.,* 19: 421 (2000): Omphalocaryon glandulosum.
eremioides *(MacOwan) E.G.H. Oliver in Contrib. Bolus Herb.,* 19: 306 (2000): Grisebachia eremioides.
subsp. eglandula *(N.E. Br.) E.G.H. Oliver in Contrib. Bolus Herb.,* 19: 309 (2000): Grisebachia eremioides var. eglandula.
subsp. pubescens *(E.G.H. Oliv.) E.G.H. Oliver in Contrib. Bolus Herb.,* 19: 309 (2000): Grisebachia parviflora subsp. pubescens.
erinus *(Klotzsch ex Benth.) E.G.H. Oliver in Contrib. Bolus Herb.,* 19: 170 (2000), as 'erina': Codonostigma erinus.
feminarum *E.G.H. Oliver in Bothalia,* 29(1): 97 (1999)—Cape Province.
flexistyla *E.G.H. Oliver in Feddes Repert.,* 106(5–8): 347 (1995 publ. 1996)—Cape Province.
florifera *(Compton) E.G.H. Oliver in Contrib. Bolus Herb.,* 19: 442 (2000): Eremia florifera.
garciae *E.G.H. Oliver in Bothalia,* 29(1): 115 (1999)—Cape Province.
glabella Thunb.
subsp. laevis *E.G.H. Oliver in Contrib. Bolus Herb.,* 19: 121 (2000), nom. nov., as subsp. nov.: Octogonia hirta *Klotzsch.*
globiceps *(N.E. Br.) E.G.H. Oliver in Contrib. Bolus Herb.,* 19: 143 (2000): Syndesmanthus globiceps.
subsp. consors *(N.E. Br.) E.G.H. Oliver in Contrib. Bolus Herb.,* 19: 146 (2000): Simocheilus consors.
subsp. gracilis *(Benth.) E.G.H. Oliver in Contrib. Bolus Herb.,* 19: 147 (2000): Syndesmanthus scaber var. gracilis.
goudotiana *(Klotzsch) L.J. Dorr & E.G.H. Oliver in Adansonia,* 21(1): 81 (1999): Philippia goudotiana.
× griffithsii *D.C. McClintock in New Plantsman,* 4(2): 84 (1997)—Cultivation. (E. manipuliflora × vagans).
hansfordii *E.G.H. Oliver in Bothalia,* 29(1): 95 (1999)—Cape Province.
hebdomadalis *E.G.H. Oliver & I.M. Oliver in Novon,* 8(4): 430 (1998)—Cape Province.
hebeclada *L.J. Dorr & E.G.H. Oliver in Adansonia,* 21(1): 82 (1999), nom. nov.: Philippia trichoclada *Baker.*

ERICA:-
humbertii *(H. Perrier) L.J. Dorr & E.G.H. Oliver in Adansonia,* 21(1): 82 (1999): Philippia humbertii.
humidicola *E.G.H. Oliver in Bothalia,* 30(2): 149 (2000)—Cape Province.
ibityensis *(H. Perrier) L.J. Dorr & E.G.H. Oliver in Adansonia,* 21(1): 82 (1999): Philippia ibityensis.
imerinensis *(H. Perrier) L.J. Dorr & E.G.H. Oliver in Adansonia,* 21(1): 82 (1999): Philippia imerinensis.
inaequalis *(Klotzsch) E.G.H. Oliver in Contrib. Bolus Herb.,* 19: 261 (2000)—Pachycalyx inaequalis.
inflaticalyx *E.G.H. Oliver in Contrib. Bolus Herb.,* 19: 280 (2000)—Cape Province.
innovans *E.G.H. Oliver in Contrib. Bolus Herb.,* 19: 155 (2000), nom. nov.: Syndesmanthus pumilus *N.E. Br.*
interrupta *(N.E. Br.) E.G.H. Oliver in Contrib. Bolus Herb.,* 19: 140 (2000): Thoracosperma interruptum.
isaloensis *(H. Perrier) L.J. Dorr & E.G.H. Oliver in Adansonia,* 21(1): 83 (1999): Philippia isaloensis.
jonasiana *E.G.H. Oliver in Contrib. Bolus Herb.,* 19: 334 (2000)—Cape Province.
jumellei *(H. Perrier) L.J. Dorr & E.G.H. Oliver in Adansonia,* 21(1): 83 (1999): jumellei.
juniperina *E.G.H. Oliver in Feddes Repert.,* 106(5–8): 350 (1995 publ. 1996)—Cape Province.
kammanassieae *E.G.H. Oliver in Contrib. Bolus Herb.,* 19: 282 (2000)—Cape Province.
karooica *E.G.H. Oliver in Bothalia,* 25(2): 242 (1995)—Cape Province.
karwyderi *E.G.H. Oliver in Contrib. Bolus Herb.,* 19: 337 (2000)—Cape Province.
kogelbergensis *E.G.H. Oliver in Yearbook Heather Soc.,* 1996: 3 (1996), nom. nov.: E. serratifolia var. subnuda *Bolus.*
× krameri *D.C. McClintock in Der Heidegarten,* 41: 19 (1997)—Cultivation. (E. carnea × spiculifolia).
lateriflora *E.G.H. Oliver in Contrib. Bolus Herb.,* 19: 303 (2000), nom. nov.: Grisebachia secundiflora *E.G.H. Oliv.*
lecomtei *(H. Perrier) L.J. Dorr & E.G.H. Oliver in Adansonia,* 21(1): 83 (1999): Philippia lecomtei.
subsp. ravinakely *L.J. Dorr in Adansonia,* 21(1): 84 (1999)—Madagascar.
leucoclada *(Baker) L.J. Dorr & E.G.H. Oliver in Adansonia,* 21(1): 84 (1999): Philippia leucoclada.
lyallii *L.J. Dorr & E.G.H. Oliver in Adansonia,* 21(1): 84 (1999), nom. nov.: Philippia parviflora *Benth.*
mackaiana Bab.
f. multiplicata *E.C. Nelson in Yearbook Heather Soc.,* 1995: 40 (1995)—Ireland.
macrocalyx *(Baker) L.J. Dorr & E.G.H. Oliver in Adansonia,* 21(1): 85 (1999): Philippia macrocalyx.
madagascariensis *(H. Perrier) L.J. Dorr & E.G.H. Oliver in Adansonia,* 21(1): 86 (1999): Philippia madagascariensis.
magnisylvae *E.G.H. Oliver in Bothalia,* 27(2): 142 (1997)—Cape Province.
malmesburiensis *E.G.H. Oliver in Contrib. Bolus Herb.,* 19: 160 (2000)—Cape Province.
marojejyensis *L.J. Dorr in Adansonia,* 21(1): 86 (1999)—Madagascar.
melanomontana *E.G.H. Oliver in Contrib. Bolus Herb.,* 19: 429 (2000), nom. nov. Scyphogyne orientalis *E.G.H. Oliv.*
miniscula *E.G.H. Oliver in Contrib. Bolus Herb.,* 19: 432 (2000), nom. nov.: Lagenocarpus tenuis *Benth.*
minutifolia *(Baker) L.J. Dorr & E.G.H. Oliver in Adansonia,* 21(1): 87 (1999): Philippia minutifolia.
montis-hominis *E.G.H. Oliver in Contrib. Bolus Herb.,* 19: 260 (2000)—Cape Province.
muscosa *(Aiton) E.G.H. Oliver in Contrib. Bolus Herb.,* 19: 410 (2000): Blaeria muscosa.
myriadenia *(Baker) L.J. Dorr & E.G.H. Oliver in Adansonia,* 21(1): 88 (1999): Philippia myriadenia.
navigatoris *E.G.H. Oliver in Novon,* 8(3): 272 (1998)—Cape Province.
niveniana *E.G.H. Oliver in Contrib. Bolus Herb.,* 19: 225 (2000), nom. nov.: Syndesmanthus nivenii *N.E. Br.*

ERICA:-
oakesiorum *E.G.H. Oliver in Yearbook Heather Soc.,* 1997: 18 (1997)—Cape Province.
× oldenburgensis *D.C. McClintock in Der Heidegarten,* 39: 35 (1996)—Cultivation. (E. arborea × carnea).
orthiocola *E.G.H. Oliver in Bothalia,* 30(1): 49 (2000)—Cape Province.
outeniquae *(Compton) E.G.H. Oliver in Contrib. Bolus Herb.,* 19: 347 (2000): Eremiella outeniquae.
parkeri *(Baker) L.J. Dorr & E.G.H. Oliver in Adansonia,* 21(1): 88 (1999): Philippia parkeri.
parviporandra *E.G.H. Oliver in Contrib. Bolus Herb.,* 19: 371 (2000)—Cape Province.
paucifolia *(J.C. Wendl.) E.G.H. Oliver in Contrib. Bolus Herb.,* 19: 220 (2000): Blaeria paucifolia.
subsp. ciliata *(Klotzsch) E.G.H. Oliver in Contrib. Bolus Herb.,* 19: 223 (2000): Macrolinum ciliatum.
subsp. squarrosa *(Benth.) E.G.H. Oliver in Contrib. Bolus Herb.,* 19: 223 (2000): Syndesmanthus squarrosus.
perhispida *L.J. Dorr & E.G.H. Oliver in Adansonia,* 21(1): 88 (1999), nom. nov.: Philippia hispida *Baker.*
perplexa *E.G.H. Oliver in Contrib. Bolus Herb.,* 19: 361 (2000)—Cape Province.
perrieri *L.J. Dorr & E.G.H. Oliver in Adansonia,* 21(1): 88 (1999), nom. nov.: Philippia latifolia *H. Perrier.*
phacelanthera *E.G.H. Oliver in Contrib. Bolus Herb.,* 19: 419 (2000), nom. nov.: Omphalocaryon capitatum *Klotzsch.*
phaeocarpa *E.G.H. Oliver in Contrib. Bolus Herb.,* 19: 267 (2000), nom. nov.: Simocheilus quadrisulcus *N.E. Br.*
pilosiflora *E.G.H. Oliver in Contrib. Bolus Herb.,* 19: 213 (2000), nom. nov.: Finckea eriocephala *Klotzsch.*
subsp. purpurea *E.G.H. Oliver in Contrib. Bolus Herb.,* 19: 216 (2000)—Cape Province.
piquetbergensis *(N.E. Br.) E.G.H. Oliver in Contrib. Bolus Herb.,* 19: 125 (2000): Simocheilus piquetbergensis.
platycalyx *E.G.H. Oliver in Contrib. Bolus Herb.,* 19: 340 (2000), nom. nov.: Platycalyx pumila *N.E. Br.*
puberuliflora *E.G.H. Oliver in Contrib. Bolus Herb.,* 19: 234 (2000), nom. nov.: Thamnium puberulum *Klotzsch.*
pulchelliflora *E.G.H. Oliver in Contrib. Bolus Herb.,* 19: 149 (2000), nom. nov.: Syndesmanthus pulchellus *N.E. Br.*
quadratiflora *(H. Perrier) L.J. Dorr & E.G.H. Oliver in Adansonia,* 21(1): 89 (1999): Philippia quadratiflora.
quadrifida *(Benth.) E.G.H. Oliver in Contrib. Bolus Herb.,* 19: 238 (2000): Simocheilus quadrifidus.
radicans *(L. Guthrie) E.G.H. Oliver in Contrib. Bolus Herb.,* 19: 135 (2000): Thoracosperma radicans.
subsp. schlechteri *(N.E. Br.) E.G.H. Oliver in Contrib. Bolus Herb.,* 19: 138 (2000): Acrostemon schlechteri
rakotozafyana *L.J. Dorr & E.G.H. Oliver in Adansonia,* 21(1): 89 (1999), nom. nov.: Philippia tenuissima *Klotzsch.*
recurvifolia *E.G.H. Oliver in Contrib. Bolus Herb.,* 19: 295 (2000), nom. nov.: Eremia recurvata *Klotzsch.*
remota *(N.E. Br.) E.G.H. Oliver in Contrib. Bolus Herb.,* 19: 375 (2000): Scyphogyne remota.
rigidula *(N.E. Br.) E.G.H. Oliver in Contrib. Bolus Herb.,* 19: 437 (2000): Scyphogyne rigidula.
rimarum *E.G.H. Oliver in Bothalia,* 30(2): 152 (2000)—Cape Province.
rosacea *(L. Guthrie) E.G.H. Oliver in Contrib. Bolus Herb.,* 19: 228 (2000): Thoracosperma rosaceum.
subsp. glabrata *E.G.H. Oliver in Contrib. Bolus Herb.,* 19: 231 (2000), nom. nov., as subsp. nov.: Thoracosperma marlothii *N.E. Br.*
roseoloba *E.G.H. Oliver in Feddes Repert.,* 107(5–6): 305 (1996)—Cape Province.
rugata *E.G.H. Oliver in Contrib. Bolus Herb.,* 19: 368 (2000), nom. nov.: Coccosperma rugosum *Klotzsch.*
rusticula *E.G.H. Oliver in Bothalia,* 30(2): 147 (2000)—Cape Province.

ERICA:-
saptouensis *E.G.H. Oliver in Feddes Repert.,* 107(5–6): 306 (1996)—Cape Province.
schumannii *E.G.H. Oliver in Yearbook Heather Soc.,* 1998: 35 (1998)—Cape Province.
similis *(N.E. Br.) E.G.H. Oliver in Contrib. Bolus Herb.,* 19: 163 (2000): Syndesmanthus similis.
sperata *E.G.H. Oliver in Bothalia,* 29(1): 115 (1999)—Cape Province.
spiculifolia
f. albiflora *(D.C. McClint.) E.C. Nelson & D.C. McClintock in Yearbook Heather Soc.,* 1997: 37 (1997): Bruckenthalia spiculifolia f. albiflora.
spinifera *(H. Perrier) L.J. Dorr & E.G.H. Oliver in Adansonia,* 21(1): 89 (1999): Philippia spinifera.
stokoeanthus *E.G.H. Oliver in Contrib. Bolus Herb.,* 19: 356 (2000), nom. nov.: Stokoeanthus chionophilus *E.G.H. Oliv.*
subcapitata *(N.E. Br.) E.G.H. Oliver in Contrib. Bolus Herb.,* 19: 366 (2000): Coccosperma subcapitatum.
sylvainiana *L.J. Dorr & E.G.H. Oliver in Adansonia,* 21(1): 89 (1999), nom. nov.: Philippia heterophylla *H. Perrier.*
tenuipilosa *(Engl. ex Alm & T.C.E. Fr.) M.R. Cheek in Kew Bull.,* 52(3): 753 (1997): Blaeria tenuipilosa.
subsp. spicata *(Hochst. ex A. Rich.) M.R. Cheek in Kew Bull.,* 52(3): 754 (1997): Blaeria spicata.
terniflora *E.G.H. Oliver in Contrib. Bolus Herb.,* 19: 399 (2000), nom. nov.: Salaxis triflora *Compton.*
thamnoides *E.G.H. Oliver in Contrib. Bolus Herb.,* 19: 330 (2000), nom. nov.: Thamnus multiflorus *Klotzsch.*
thomensis *(Henriq.) L.J. Dorr & E.G.H. Oliver in Kew Bull.,* 54(1): 235 (1999): Philippia thomensis.
uberiflora *E.G.H. Oliver in Contrib. Bolus Herb.,* 19: 171 (2000), nom. nov.: Blaeria multiflora *Klotzsch.*
urceolata *(Klotzsch) E.G.H. Oliver in Contrib. Bolus Herb.,* 19: 424 (2000): Tristemon urceolatus.
vallis-fluminis *E.G.H. Oliver in Contrib. Bolus Herb.,* 19: 248 (2000)—Cape Province.
velatiflora *E.G.H. Oliver in Contrib. Bolus Herb.,* 19: 350 (2000), nom. nov.: Eremia brevifolia *Benth.*
venustiflora *E.G.H. Oliver in Contrib. Bolus Herb.,* 19: 153 (2000), nom. nov.: Syndesmanthus venustus *N.E. Br.*
subsp. glandulosa *E.G.H. Oliver in Contrib. Bolus Herb.,* 19: 154 (2000)—Cape Province.
vernicosa *E.G.H. Oliver in Contrib. Bolus Herb.,* 19: 206 (2000)—Cape Province.
viguieri *(H. Perrier) L.J. Dorr & E.G.H. Oliver in Adansonia,* 21(1): 90 (1999): Philippia viguieri.
viscosissima *E.G.H. Oliver in Contrib. Bolus Herb.,* 19: 200 (2000), nom. nov.: Simocheilus viscosus *Bolus.*
vlokii *E.G.H. Oliver in Contrib. Bolus Herb.,* 19: 345 (2000)—Cape Province.
wangfatiana *L.J. Dorr & E.G.H. Oliver in Adansonia,* 21(1): 90 (1999), nom. nov.: Philippia pilulifera *H. Perrier.*
williamsiorum *E.G.H. Oliver in Contrib. Bolus Herb.,* 19: 195 (2000)—Cape Province.
zeyheriana *(Klotzsch) E.G.H. Oliver in Contrib. Bolus Herb.,* 19: 244 (2000): Coilostigma zeyherianum.

ERIGERON (Compos.).
sect. Geronpternix *G.L. Nesom & R.D. Noyes in Sida,* 18(4): 1163 (1999).
ser. Macrophylli *O. Šída in Preslia,* 70(3): 262 (1998), as 'Macrophyllae'.
ser. Politi *O. Šída in Preslia,* 70(3): 262 (1998), as 'Politae'.
ser. Trimorpha *(Cass.) O. Šída in Preslia,* 70(3): 261 (1998): gen. Trimorpha.
acer L.
var. khasianus *(Hook. f.) P.K. Hajra in Fl. India,* 12: 118 (1995), as 'khasiana': E. alpinus var. khasianus.
acris L.
subsp. manshuricus *(Kom.) V.N. Voroschilov in A.K. Skvortsov (ed.), Florist. issl. v razn. raionakh SSSR:* 194 (1985): E. acris var. manshuricus.

ERIGERON:-
bellidioides *(Hook. f.) S.J. Forbes & D.I. Morris in Muelleria,* 9: 183 (1996), homonym; *see Spenn., Fl. Friburg.,* 2: 536 (1826): Haplopappus bellidioides.
canariensis *O. Šída in Preslia,* 70(3): 265 (1998), nom. nov.: E. cabrerae *Dittrich.*
elongatus Ledeb.
var. glandulosus *Y. Wei & C.H. An in Fl. Xinjiangensis,* 5: 476 (1999)—Xinjiang.
lalehzaricus *(Rech. f.) O. Šída in Preslia,* 70(3): 265 (1998): E. acris subsp. lalehzaricus.
nitidus *S.J. Forbes in Muelleria,* 9: 181 (1996)—Victoria, New South Wales.
nubigenus *Phil. in Anales Univ. Chile,* 87: 423 (1894), as 'nubgenus' in error in I.K. Supp. 1—Chile North.
orithales *Phil. in Anales Univ. Chile,* 87: 412 (1894), in I.K. Supp. 1 as 'aristhales' in error—Chile North.†
paludicola *S.J. Forbes in Muelleria,* 9: 178 (1996)—Victoria, New South Wales.
songoricus *Y. Wei & C.H. An in Fl. Xinjiangensis,* 5: 475 (1999)—Xinjiang.
thunbergii A. Gray
f. albiflorus *Y.N. Lee, Fl. Korea:* 1162 (1996), without type—Locality not stated.
trigonus *S.J. Forbes & D.I. Morris in Muelleria,* 9: 187 (1996)—Tasmania.

ERIOCAULON (Eriocaulac.).
sect. Apoda *(Satake) Z.X. Zhang, Monogr. Gatt. Eriocaulon Ostasien (Diss. Bot., 313):* 201 (1999), without basionym ref.: Basionym not stated.
sect. Disepala *(Satake) Z.X. Zhang, Monogr. Gatt. Eriocaulon Ostasien (Diss. Bot., 313):* 164 (1999), without basionym ref. Basionym not stated.
sect. Macropoda *(Satake) Z.X. Zhang, Monogr. Gatt. Eriocaulon Ostasien (Diss. Bot., 313):* 244 (1999), without basionym ref.: Basionym not stated.
sect. Nudicuspa *Z.X. Zhang, Monogr. Gatt. Eriocaulon Ostasien (Diss. Bot., 313):* 246 (1999), without latin descr.
acutifolium *S.M. Phillips in Kew Bull.,* 52(1): 59 (1997)—Zambia.
aethiopicum *S.M. Phillips in Kew Bull.,* 52(1): 64 (1997)—Ethiopia.
asteroides *S.M. Phillips in Kew Bull.,* 53(4): 943 (1998)—Nigeria, Cameroon.
athertonense *G.J. Leach in Austral. Syst. Bot.,* 13(5): 756 (2000)—Queensland.
bamendae *S.M. Phillips in Kew Bull.,* 55(1): 195 (2000)—Cameroon, Nigeria.
burttii *S.M. Phillips in Kew Bull.,* 51(4): 636 (1996)—Tanzania.
carpentariae *G.J. Leach in Austral. Syst. Bot.,* 13(5): 757 (2000)—Northern Territory, Queensland.
chloanthe *S.M. Phillips in Kew Bull.,* 51(4): 637 (1996)—Zambia.
clarksonii *G.J. Leach in Austral. Syst. Bot.,* 13(5): 758 (2000)—Queensland.
crassiusculum *K.A. Lye in Nordic J. Bot.,* 16(1): 63 (1996)—Ethiopia.
eglandulatum *Z.X. Zhang, Monogr. Gatt. Eriocaulon Ostasien (Diss. Bot., 313):* 84 (1999)—Thailand.
ermeiense *W.L. Ma ex Z.X. Zhang, Monogr. Gatt. Eriocaulon Ostasien (Diss. Bot., 313):* 87 (1999)—China South-central (Sichuan).
fuscum *S.M. Phillips in Kew Bull.,* 51(4): 629 (1996)—Zimbabwe.
inapertum *G.J. Leach in Austral. Syst. Bot.,* 13(5): 762 (2000)—Northern Territory.
iringense *S.M. Phillips in Kew Bull.,* 51(2): 337 (1996)—Tanzania.
kunmingense *Z.X. Zhang, Monogr. Gatt. Eriocaulon Ostasien (Diss. Bot., 313):* 91 (1999), nom. nov., without exact replaced synonym page; *et in Novon,* 10(1): 92 (2000): E. bilobatum *W.L. Ma.*
laniceps *S.M. Phillips in Kew Bull.,* 51(4): 625 (1996)—Tanzania.
liberisepalum *Z.X. Zhang, Monogr. Gatt. Eriocaulon Ostasien (Diss. Bot., 313):* 221 (1999)—Japan.
maronderanum *S.M. Phillips in Kew Bull.,* 51(4): 646 (1996)—Zimbabwe, Malawi.
mbalensis *S.M. Phillips in Kew Bull.,* 52(1): 71 (1997)—Zambia.

ERIOCAULON:-
miquelianum Körn.
var. matsumurae *(Nakai) Z.X. Zhang, Monogr. Gatt. Eriocaulon Ostasien (Diss. Bot., 313):* 228 (1999): E. matsumurae.
modicum *S.M. Phillips in Kew Bull.,* 52(1): 68 (1997)—Tanzania.
mulanjeanum *S.M. Phillips in Kew Bull.,* 51(4): 638 (1996)—Malawi.
mutatum N.E. Br.
var. angustisepalum *(H.E. Hess) S.M. Phillips in Kew Bull.,* 51(4): 644 (1996): E. angustisepalum.
var. majus *S.M. Phillips in Kew Bull.,* 51(4): 644 (1996)—Zambia.
nadjae *S.M. Phillips in Kew Bull.,* 55(1): 199 (2000)—Equatorial Guinea.
nematophyllum *G.J. Leach in Austral. Syst. Bot.,* 13(5): 763 (2000)—Northern Territory, Queensland.
odontospermum *G.J. Leach in Austral. Syst. Bot.,* 13(5): 764 (2000)—Northern Territory, Queensland, Western Australia.
parvulum *S.M. Phillips in Kew Bull.,* 53(4): 944 (1998)—Cameroon.
patericola *G.J. Leach in Austral. Syst. Bot.,* 13(5): 767 (2000)—Western Australia, Northern Territory.
polhillii *S.M. Phillips in Kew Bull.,* 51(4): 626 (1996)—Tanzania.
pseudotruncatum *Z.X. Zhang, Monogr. Gatt. Eriocaulon Ostasien (Diss. Bot., 313):* 174 (1999)—India (West Bengal).
raipurense *K.K. Khanna, V. Mudgal & An. Kumar in Nordic J. Bot.,* 20(4): 413 (2000)—India (Madhya Pradesh).
ratnagiricum *S.R. Yadav, S.P. Gaikwad & M.M. Sardesai in Rheedea,* 8(2): 145 (1998), as 'ratnagiricum'—India (Maharashtra).
rosenii *(Pax) K.A. Lye in Lidia,* 3(5): 141 (1995): Mesanthemum rosenii.
scullionii *G.J. Leach in Austral. Syst. Bot.,* 13(5): 768 (2000)—Northern Territory, Western Australia.
selousii *S.M. Phillips in Kew Bull.,* 52(1): 60 (1997)—Tanzania, Malawi.
siamense Moldenke
var. leiophyllum *Z.X. Zhang, Monogr. Gatt. Eriocaulon Ostasien (Diss. Bot., 313):* 152 (1999)—Thailand.
taeniophyllum *S.M. Phillips in Kew Bull.,* 51(4): 632 (1996)—Zambia.
transvaalicum N.E. Br.
subsp. dembianense *(Chiov.) S.M. Phillips in Kew Bull.,* 52(1): 57 (1997): E. dembianense.
subsp. hanningtonii *(N.E. Br.) S.M. Phillips in Kew Bull.,* 52(1): 58 (1997): E. hanningtonii.
subsp. tofieldifolium *(Schinz) S.M. Phillips in Kew Bull.,* 52(1): 58 (1997): E. tofieldifolium.
tricornum *G.J. Leach in Austral. Syst. Bot.,* 13(5): 769 (2000)—Northern Territory.
truncatum Buch.-Ham. ex Mart.
var. florensense *Z.X. Zhang, Monogr. Gatt. Eriocaulon Ostasien (Diss. Bot., 313):* 183 (1999)—Lesser Sunda Is.
wildii *S.M. Phillips in Kew Bull.,* 52(1): 64 (1997)—Zimbabwe.
zollingerianoides *Z.X. Zhang, Monogr. Gatt. Eriocaulon Ostasien (Diss. Bot., 313):* 183 (1999)—Thailand.

ERIOCHLOA (Gramin.).
nana *M.O. Arriaga in Hickenia,* 2(32): 139 (1995)—Argentina Northeast (Corrientes, Formosa).

ERIOCHRYSIS (Gramin.).
filiformis *(Hack.) T.S. Filgueiras in Novon,* 7(3): 231 (1997): Saccharum filiforme.

ERIOCYCLA (Umbellif.).
ghafooriana *H. Akhani in Edinburgh J. Bot.,* 56(1): 21 (1999)—Iran.

ERIOGONUM (Polygonac.).
codium *J.L. Reveal, F. Caplow & K. Beck in Rhodora,* 97(892): 350 (1995 publ. 1997)—Washington.
ericifolium Torr. & A. Gray
subsp. pulchrum *(Eastw.) L.M. Shultz in Harvard Pap. Bot.,* 3(1): 51 (1998): E. pulchrum.

ERIOGONUM:-
phoeniceum *L.M. Shultz in Harvard Pap. Bot.,* 3(1): 49 (1998)—Utah, Nevada.
thornei *(Reveal & Henr.) L.M. Shultz in Harvard Pap. Bot.,* 3(1): 51 (1998): E. ericifolium var. thornei.

ERIOLARYNX *(Hunz.) A.T. Hunziker in Kurtziana,* 28(1): 66 (2000): Vassobia sect. Eriolarynx. (Solanac.).
fasciculata *(Miers) A.T. Hunziker in Kurtziana,* 28(1): 68 (2000): Lycioplesium fasciculatum.
iochromoides *(Hunz.) A.T. Hunziker in Kurtziana,* 28(1): 68 (2000): Vassobia iochromoides.
lorentzii *(Dammer) A.T. Hunziker in Kurtziana,* 28(1): 66 (2000): Acnistus lorentzii.

ERIOPEXIS (Orchidac.).
acanthophippiiflora *(J.J. Sm.) M.A. Clements & D.L. Jones in Lasianthera,* 1(2): 123 (1997): Dendrobium acanthophippiiflorum.

ERIOPHORUM (Cyperac.).
angustifolium Honck.
subsp. komarovii *(V.N. Vassil.) V.N. Voroschilov in A.K. Skvortsov (ed.), Florist. issl. v razn. Raĭonakh SSSR:* 152 (1985): E. komarovii.

ERIOPHYLLUM (Compos.).
idoneum *Jeps., Fl. W. Calif.:* 524 (1901), as 'idoreum' in error in I.K. Supp. 3—California.†

ERIOSEMA (Legumin.).
betsileense *D.J. Du Puy & J.-N. Labat in Bull. Mus. Nation. Hist. Nat., B, Adansonia,* Ser. 4, 18(1–2): 94 (1996)—Madagascar.
campestre Benth.
var. delicatulum *R.H. Fortunato in Kurtziana,* 27(2): 374 (1999), as 'delicatula'—Argentina Northeast (Corrientes), Brazil West-central (Mato Grosso do Sul), Brazil South (Rio Grande do Sul), Paraguay.
var. macrophyllum *(Grear) R.H. Fortunato in Kurtziana,* 27(2): 375 (1999): E. crinitum var. macrophyllum.
crinitum
var. discolor *R.H. Fortunato in Kurtziana,* 27(2): 377 (1999)—Brazil South (Santa Catarina, Paraná), Brazil West-central (Goiás), Brazil Southeast (Minas Gerais, São Paulo).

ERIOSPERMUM (Eriospermac.).
flagelliforme *(Baker) J.C. Manning in Bothalia,* 30(2): 157 (2000): Anthericum flagelliforme.

ERIOSYCE (Cactac.).
multicolorispina *Z. Janeba & R. Slaba in Kaktusy (Brno),* 36(3): 69 (2000)—Argentina Northwest (Mendoza).
vertongenii *(J.G. Lamb.) D.R. Hunt in Cactaceae Consensus Initiatives,* 3: 25 (1997): Pyrrhocactus vertongenii.

ERIOTHECA (Bombacac.).
squamigera *(Cuatrec.) J.L. Fernández-Alonso in An. Jard. Bot. Madrid,* 57(1): 162 (1999): Pseudobombax squamigerum.

ERISMA (Vochysiac.).
arietinum *M.L. Kawasaki in Mem. New York Bot. Gard.,* 81: 21 (1998)—Brazil Southeast (Espírito Santo).

ERITHALIS (Rubiac.).
insularis *(Ridl.) D. Zappi & T.S. Nunes in Kew Bull.,* 55(3): 655 (2000): Palicourea insularis.

ERITRICHIUM (Boraginac.).
aldanense *S.V. Ovczinnikova in Fl. Sibir. (Pyrolac.-Lamiac.),* 11: 225 (1997)—Yakutskiya.
alpinum *S.V. Ovczinnikova in Turczaninowia,* 2(4): 21 (1999)—Tuva.
armeriifolium *R.A. Philippi in Anales Univ. Chile,* 90: 551 (1895), as 'armeriifoiium' in error in I.K. Supp. 1—Chile.†
kamelinii *S.V. Ovczinnikova in Turczaninowia,* 2(4): 17 (1999)—Altay.

ERITRICHIUM:-
uralense Serg.
subsp. krascheninnikovii *M.S. Knjasev & S.V. Balandin in Bot. Zhurn.*, 84(1): 87 (1999)—Russia East.
villosum (Ledeb.) Bunge
var. micranthum *V.B. Kuvaev in V.B. Kuvaev et al., Rast. pokrov ostrova Sibiryakova:* 78 (1994)—Krasnoyarsk.

ERODIUM (Geraniac.).
aguilellae *S. López Udias, C. Fabregat & G. Mateo in Flora Montiberica*, 8: 71 (1998)—Spain.
carolinianum *J.J. Aldasoro, C. Aedo, C. Navarro & L. Sáez in An. Jard. Bot. Madrid*, 57(2): 407 (1999 publ. 2000), nom. nov.: E. cygnorum subsp. glandulosum *Carolin.*
ciconium (L.) L'Hér.
subsp. sennenii *(Bianor) A.M. Romo, Fl. Silvestres Baleares:* 183 (1994): E. sennenii.
macrocalyx *(G. López) S. López Udias, C. Fabregat & G. Mateo in Flora Montiberica*, 3: 75 (1998): E. daucoides subsp. macrocalyx.

EROPHILA (Crucif.).
verna (L.) Chevall.
subsp. macrosperma *(Sebald) B. Jonsell in Nordic J. Bot.*, 20(2): 201 (2000): E. verna var. macrosperma.

ERUCASTRUM (Crucif.).
deflexum *(Boiss.) V.I. Dorofeev in Bot. Zhurn.*, 85(7): 184 (2000): Brassica deflexa.
deserti *(Danin & Hedge) V.I. Dorofeev in Bot. Zhurn.*, 85(7): 184 (2000): Brassica deserti.
rostratum *(Balf. f.) C. Gómez Campo in An. Jard. Bot. Madrid*, 56(2): 379 (1998): Brassica rostrata.
takhtajanii *V.I. Dorofeev in Bot. Zhurn.*, 85(7): 183 (2000)—Transcaucasus (Armenia).

ERYCIBE (Convolvulac.).
tixieri *T. Deroin in Bull. Mus. Nation. Hist. Nat., B, Adansonia*, Ser. 4, 17(3–4): 184 (1995)—Vietnam.
villosa *L.L. Forman in Checklist Flow. Pl. & Gymn. Brunei Darussalam (ed. M.J.E. Coode et al.):* 435 (1996)—Borneo (Brunei).

ERYNGIUM (Umbellif.).
antiatlanticum *S.L. Jury in Lagascalia*, 18(2): 274 (1996)—Morocco.
mesopotamicum *T.M. Pedersen in Bonplandia*, 9(3–4): 252 (1997)—Argentina Northeast (Corrientes, Buenos Aires, Chaco, Entre Ríos, Santa Fé).
paludosum *(C. Moore & Betche) P.W. Michael in Telopea*, 7(2): 173 (1997): E. rostratum var. paludosum.
pendletonense *K.L. Marsden & M G. Simpson in Madroño*, 46(1): 61 (1999), as 'pendletonensis'—California.
tricuspidatum L.
subsp. bocconei *(Lam.) A. Wörz in Stuttgarter Beitr. Naturk., A*, 596: 35 (1999): E. bocconei.

ERYSIMUM (Crucif.).
altum *(Ahti) N.N. Tzvelev in Bot. Zhurn.*, 82(12): 94 (1997): E. cheiranthoides subsp. altum.
brevifolium *Z.X. An in Fl. Xinjiangensis*, 2(2): 379 (1995)—Xinjiang.
eseptatum *Z.X. An in Fl. Xinjiangensis*, 2(2): 379 (1995)—Xinjiang.
humile Pers.
subsp. duriaei *(Boiss.) G. López González in An. Jard. Bot. Madrid*, 56(2): 376 (1998): E. duriaei.
subsp. gorbeanum *(Polatschek) G. López González in An. Jard. Bot. Madrid*, 56(2): 377 (1998): E. gorbeanum.
subsp. neumannii *(Polatschek) G. López González in An. Jard. Bot. Madrid*, 56(2): 377 (1998): E. neumannii.
subsp. penyalarense *(Pau) G. López González in An. Jard. Bot. Madrid*, 56(2): 375 (1998): E. ochroleucum var. penyalarense.
subsp. pyrenaicum *(Nyman) G. López González in An. Jard. Bot. Madrid*, 56(2): 376 (1998): E. ochroleucum subsp. pyrenaicum.

ERYSIMUM:-
incanum Kunze
subsp. matritense *(Pau) G. López González in An. Jard. Bot. Madrid*, 56(2): 377 (1998): E. matritense.
javalambrense *G. Mateo Sanz, M.B. Crespo & S. López Udias in Flora Montiberica*, 9: 42 (1998)—Spain.
kykkoticum *G. Hadjikyriakou & G. Alziar in Biocosme Mésogéen*, 15(4): 244 (1999)—Cyprus.
linifolium (Pers.) J. Gay
subsp. lagascae *(Rivas Goday & Bellot) G. López González in An. Jard. Bot. Madrid*, 56(2): 371 (1998): E. lagascae.
pirinicum *M. Ančev & A. Polatschek in Ann. Naturhist. Mus. Wien*, 100B: 733 (1998)—Bulgaria, Greece.
pseudoatticum *M. Ančev & A. Polatschek in Ann. Naturhist. Mus. Wien*, 100B: 726 (1998)—Bulgaria.
slavjankae *M. Ančev & A. Polatschek in Ann. Naturhist. Mus. Wien*, 100B: 729 (1998)—Bulgaria.

ERYTHRINA (Legumin.).
ankaranensis *D.J. Du Puy & J.-N. Labat in Bull. Mus. Nation. Hist. Nat., B, Adansonia*, Ser. 4, 18(3–4): 226 (1996)—Madagascar.
arborescens *W. Roxburgh, Pl. Coromandel*, 3: 14 (1811)—Nepal.†
hazomboay *D.J. Du Puy & J.-N. Labat in Bull. Mus. Nation. Hist. Nat., B, Adansonia*, Ser. 4, 18(3–4): 228 (1996)—Madagascar.
madagascariensis *D.J. Du Puy & J.-N. Labat in Bull. Mus. Nation. Hist. Nat., B, Adansonia*, Ser. 4, 18(3–4): 230 (1996)—Madagascar, Comoros, Mayotte.
× malottiorum *D.J. Mabberley & D.H. Lorence in D.J. Mabberley, Paradisus:* 144 (1998), as 'malottorum'—Cultivation. (E. crista-galli × fusca).
× neillii *D.J. Mabberley & D.H. Lorence in D.J. Mabberley, Paradisus:* 144 (1998)—Cultivation. (E. herbacea × humeana).
resupinata *W. Roxburgh, Pl. Coromandel*, 3: 15 (1811)—India.†
variegata L.
var. alba *(Blatt. & Mill.) R.R. Fernandez & H. Santapau in R.R. Fernandez, Trees of Mumbai (Bombay):* 125 (1999): E. indica var. alba.

ERYTHROCEPHALUM (Compos.).
jeffreyanum *S. Ortiz & J. Rodríguez-Oubiña in Nordic J. Bot.*, 18(2): 191 (1998)—Tanzania.
sallyae *H.J. Beentje in Fl. Trop. E. Afr., Compos. (Part 1):* 30 (2000)—Tanzania.

ERYTHRODES (Orchidac.).
cornejoi *C.H. Dodson in Orquideologia*, 20(3): 275 (1997)—Ecuador.
hirsuta *(Griff.) P. Ormerod in G. Seidenfaden, Contrib. Orchid Fl. Thailand XIII:* 12 (1997): Goodyera hirsuta.
hirsuta *(Griff.) R.C. Srivastava in Novon*, 8(2): 203 (1998), isonym; *see Ormerod in Seidenf., Contrib. Orchid Fl. Thailand XIII:* 12 (1997): Goodyera hirsuta.
latifolia Blume
var. johorense *P. O'Byrne in Malayan Orchid Rev.*, 31: 26 (1997)—Malaya (Peninsular Malaysia).
seshagiriana *A. Nageswara Rao in Indian Forester*, 123(7): 643 (1997)—Assam (Arunachal Pradesh).

ERYTHRONIUM (Liliac.).
dens-canis L.
subsp. albiflorum *V.V. Kricsfalusý in Bot. Zhurn.*, 80(9): 48 (1995)—Ukraina.
taylorii *J.R. Shevock & G.A. Allen in Madroño*, 44(4): 360 (1997 publ. 1998), as 'taylori'—California.

ERYTHROSPERMUM (Flacourtiac.).
sifarii *S. Hul, J.-N. Labat & O. Pascal in Adansonia*, 20(2): 264 (1998)—Comoros (Mayotte).

ERYTHROXYLUM (Erythroxylac.).
cuneatum (Miq.) Kurz
f. sumatranum *(Miq.) R.C.K. Chung in Sandakania*, 7: 71 (1996): E. sumatranum.

ERYTHROXYLUM:-
iwahigense Elmer
f. calcicola *R.C.K. Chung in Sandakania*, 7: 74 (1996)—Borneo (Sarawak).
sarawakanum *R.C.K. Chung in Sandakania*, 7: 78 (1996)—Borneo (Sarawak).
socotranum *M. Thulin in Nordic J. Bot.*, 16(3): 301 (1996 publ.1997), as 'Erythroxylon'—Socotra.

ESCHWEILERA (Lecythidac.).
complanata *S.A. Mori in Bol. Bot. (São Paulo)*, 14: 16 (1995)—Brazil Northeast (Bahia).
mattos-silvae *S.A. Mori in Bol. Bot. (São Paulo)*, 14: 22 (1995)—Brazil Northeast (Bahia).
rankiniae *S.A. Mori in Mem. New York Bot. Gard.*, 75: 39 (1995)—Brazil North (Amazonas).
romeu-cardosoi *S.A. Mori in Mem. New York Bot. Gard.*, 75: 39 (1995)—Brazil North (Amazonas).

ESCOBARIA (Cactac.).
alversonii *(J.M. Coult.) N.P. Taylor in Cactaceae Consensus Initiatives*, 3: 10 (1997): Cactus radiosus var. alversonii.
chihuahuensis
subsp. henricksonii *(Glass & R.A. Foster) N.P. Taylor in Cactaceae Consensus Initiatives*, 5: 13 (1998): E. henricksonii.
dasyacantha
subsp. chaffeyi *(Britton & Rose) N.P. Taylor in Cactaceae Consensus Initiatives*, 5: 13 (1998): E. chaffeyi.
missouriensis (Sweet) D.R. Hunt
subsp. asperispina *(Boed.) N.P. Taylor in Cactaceae Consensus Initiatives*, 5: 13 (1998): Coryphantha asperispina.
subsp. navajoensis *F. Hochstätter in Succulenta*, 75(6): 257, 256 (1996)—Arizona.
sneedii
subsp. albicolumnaria *(Hester) J.M. Lüthy in Kakt. And. Sukk.*, 50(11): 278 (1999): E. albicolumnaria.
subsp. leei *(Rose ex Boed.) D.R. Hunt in Cactaceae Consensus Initiatives*, 4: 5 (1997): E. leei.
subsp. orcuttii *(Boed.) J.M. Lüthy in Kakt. And. Sukk.*, 50(11): 278 (1999): E. orcuttii.
subsp. organensis *(Zimmerman) J.M. Lüthy in Kakt. And. Sukk.*, 50(11): 278 (1999): Coryphantha organensis.
subsp. sandbergii *(Castetter, P. Pierce & K.H. Schwer.) J.M. Lüthy in Kakt. And. Sukk.*, 50(11): 278 (1999): E. sandbergii.
subsp. villardii *(Castetter, P. Pierce & K.H. Schwer.) J.M. Lüthy in Kakt. And. Sukk.*, 50(11): 278 (1999): E. villardii.
sniceri *J.J. Halda, P. Kupčák & J. Sladkovský in Acta Mus. Richnov. Sect. Nat.*, 7(1): 35 (2000)—Mexico Northeast (Coahuila).
strobiliformis *(Poselg.) F. Boedeker, Mammillarien-Vergleichs-Schlüssel:* 16 (1933): Echinocactus strobiliformis.
subsp. sisperai *J.J. Halda & J. Sladkovský in Acta Mus. Richnov. Sect. Nat.*, 7(1): 35 (2000)—Mexico Northeast (Nuevo León).
tuberculosa (Engelm.) Britton & Rose
subsp. varicolor *(Tiegel) J.M. Lüthy in Kakt. And. Sukk.*, 50(10): 257 (1999): Coryphantha varicolor.
zilziana
subsp. fricii *J.J. Halda & J. Sladkovský in Acta Mus. Richnov. Sect. Nat.*, 7(1): 35 (2000)—Mexico Northeast (Coahuila).

ESCOBARIOPSIS *A.B. Doweld in Sukkulentȳ (Moscow)*, 3(1–2): 23 (2000). *CACTACEAE*
sect. Acentracantha *(Buxb.) A.B. Doweld in Sukkulentȳ (Moscow)*, 3(1–2): 42 (2000): Chilita subgen. Acentracantha.
ser. Herrerae *(Lüthy) A.B. Doweld in Sukkulentȳ (Moscow)*, 3(1–2): 42 (2000): Mammillaria ser. Herrerae.
ser. Lasiacanthae *(D.R. Hunt) A.B. Doweld in Sukkulentȳ (Moscow)*, 3(1–2): 42 (2000): Mammillaria ser. Lasiacanthae.
ser. Pennispinosae *A.B. Doweld in Sukkulentȳ (Moscow)*, 3(1–2): 42 (2000).
sect. Polycentracantha *A.B. Doweld in Sukkulentȳ (Moscow)*, 3(1–2): 42 (2000).

ESCOBARIOPSIS:-
ser. Sphacelatae (D.R. Hunt) A.B. Doweld in Sukkulentÿ (Moscow), 3(1–2): 42 (2000): Mammillaria ser. Sphacelatae.
albicoma (Boed.) A.B. Doweld in Sukkulentÿ (Moscow), 3(1–2): 39 (2000): Mammillaria albicoma.
albiflora (Werderm.) A.B. Doweld in Sukkulentÿ (Moscow), 3(1–2): 40 (2000): Mammillaria herrerae var. albiflora.
aureilanata (Backeb.) A.B. Doweld in Sukkulentÿ (Moscow), 3(1–2): 40 (2000): Mammillaria aureilanata.
bombycina (Quehl) A.B. Doweld in Sukkulentÿ (Moscow), 3(1–2): 40 (2000): Mammillaria bombycina.
carmenae (Castañeda) A.B. Doweld in Sukkulentÿ (Moscow), 3(1–2): 40 (2000): Mammillaria carmenae.
fittkaui (Glass & R.A. Foster) A.B. Doweld in Sukkulentÿ (Moscow), 3(1–2): 40 (2000): Mammillaria fittkaui.
subsp. limonensis (Repp.) A.B. Doweld in Sukkulentÿ (Moscow), 3(1–2): 40 (2000): Mammillaria limonensis.
subsp. mathildae (Kraehenb. & Krainz) A.B. Doweld in Sukkulentÿ (Moscow), 3(1–2): 40 (2000): Mammillaria mathildae.
gracilis (Pfeiff.) A.B. Doweld in Sukkulentÿ (Moscow), 3(1–2): 40 (2000): Mammillaria gracilis.
herrerae (Werderm.) A.B. Doweld in Sukkulentÿ (Moscow), 3(1–2): 40 (2000): Mammillaria herrerae.
humboldtii (Ehrenb.) A.B. Doweld in Sukkulentÿ (Moscow), 3(1–2): 40 (2000): Mammillaria humboldtii.
jaliscana (Britton & Rose) A.B. Doweld in Sukkulentÿ Moscow), 3(1–2): 40 (2000), basionym as 'Neomammillaria bombycina': Neomammillaria jaliscana.
subsp. zacatecasensis (Shurly) A.B. Doweld in Sukkulentÿ (Moscow), 3(1–2): 40 (2000): Mammillaria zacatecasensis.
kraehenbuehlii (Krainz) A.B. Doweld in Sukkulentÿ (Moscow), 3(1–2): 40 (2000): Pseudomammillaria kraehenbuehlii.
lasiacantha (Engelm.) A.B. Doweld in Sukkulentÿ (Moscow), 3(1–2): 40 (2000): Mammillaria lasiacantha.
subsp. egregia (Backeb. ex Rogoz. & Appenz.) A.B. Doweld in Sukkulentÿ (Moscow), 3(1–2): 40 (2000): Mammillaria egregia.
subsp. magallanii (R.T. Craig) A.B. Doweld in Sukkulentÿ (Moscow), 3(1–2): 40 (2000): Mammillaria magallanii.
subsp. wohlschlageri (Repp.) A.B. Doweld in Sukkulentÿ (Moscow), 3(1–2): 40 (2000): Mammillaria wohlschlageri.
laui (D.R. Hunt) A.B. Doweld in Sukkulentÿ (Moscow), 3(1–2): 40 (2000): Mammillaria laui.
subsp. dasyacantha (D.R. Hunt) A.B. Doweld in Sukkulentÿ (Moscow), 3(1–2): 40 (2000): Mammillaria laui f. dasyacantha.
subsp. subducta (D.R. Hunt) A.B. Doweld in Sukkulentÿ (Moscow), 3(1–2): 40 (2000): Mammillaria laui f. subducta.
lenta (K. Brandegee) A.B. Doweld in Sukkulentÿ (Moscow), 3(1–2): 40 (2000): Mammillaria lenta.
mazatlanensis (K. Schum. ex Gürke) A.B. Doweld in Sukkulentÿ (Moscow), 3(1–2): 40 (2000): Mammillaria mazatlanensis.
subsp. patonii (Bravo) A.B. Doweld in Sukkulentÿ (Moscow), 3(1–2): 41 (2000): Neomammillaria patonii.
mercadensis (Patoni) A.B. Doweld in Sukkulentÿ (Moscow), 3(1–2): 41 (2000): Mammillaria mercadensis.
moelleriana (Boed.) A.B. Doweld in Sukkulentÿ (Moscow), 3(1–2): 41 (2000): Mammillaria moelleriana.
oteroi (Glass & R.A. Foster) A.B. Doweld in Sukkulentÿ (Moscow), 3(1–2): 41 (2000): Mammillaria oteroi.
pennispinosa (Krainz) A.B. Doweld in Sukkulentÿ (Moscow), 3(1–2): 41 (2000): Mammillaria pennispinosa.

ESCOBARIOPSIS:-
pennispinosa
subsp. nazasensis (Glass & R.A. Foster) A.B. Doweld in Sukkulentÿ (Moscow), 3(1–2): 41 (2000): Mammillaria pennispinosa var. nazasensis.
perezdelarosae (Bravo & Scheinvar) A.B. Doweld in Sukkulentÿ (Moscow), 3(1–2): 41 (2000): Neomammillaria perezdelarosae.
picta (Meinsh.) A.B. Doweld in Sukkulentÿ (Moscow), 3(1–2): 41 (2000): Mammillaria picta.
pilispina (J.A. Purpus) A.B. Doweld in Sukkulentÿ (Moscow), 3(1–2): 41 (2000): Mammillaria pilispina.
plumosa (F.A.C. Weber) A.B. Doweld in Sukkulentÿ (Moscow), 3(1–2): 41 (2000): Mammillaria plumosa.
prolifera (Mill.) A.B. Doweld in Sukkulentÿ (Moscow), 3(1–2): 41 (2000): Cactus proliferus.
subsp. arachnoidea (D.R. Hunt) A.B. Doweld in Sukkulentÿ (Moscow), 3(1–2): 41 (2000): Mammillaria prolifera var. arachnoidea.
subsp. haitiensis (K. Schum.) A.B. Doweld in Sukkulentÿ (Moscow), 3(1–2): 41 (2000): Mammillaria pusilla var. haitiensis.
subsp. texana (Engelm.) A.B. Doweld in Sukkulentÿ (Moscow), 3(1–2): 41 (2000): Mammillaria pusilla var. texana.
subsp. zublerae (Repp.) A.B. Doweld in Sukkulentÿ (Moscow), 3(1–2): 41 (2000): Mammillaria zublerae.
rayonesensis (Repp.) A.B. Doweld in Sukkulentÿ (Moscow), 3(1–2): 41 (2000): Mammillaria rayonesensis.
rettigiana (Boed.) A.B. Doweld in Sukkulentÿ (Moscow), 3(1–2): 41 (2000): Mammillaria rettigiana.
sanchez-mejoradae (Rodr. González) A.B. Doweld in Sukkulentÿ (Moscow), 3(1–2): 41 (2000): Mammillaria sanchez-mejoradae.
schiedeana (Ehrenb.) A.B. Doweld in Sukkulentÿ (Moscow), 3(1–2): 41 (2000): Mammillaria schiedeana.
sinistrohamata (Boed.) A.B. Doweld in Sukkulentÿ (Moscow), 3(1–2): 41 (2000): Mammillaria sinistrohamata.
sphacelata (Mart.) A.B. Doweld in Sukkulentÿ (Moscow), 3(1–2): 41 (2000), basionym as 'Mammillaria vetula': Mammillaria sphacelata.
tonalensis (D.R. Hunt) A.B. Doweld in Sukkulentÿ (Moscow), 3(1–2): 41 (2000): Mammillaria tonalensis.
vetula (Mart.) A.B. Doweld in Sukkulentÿ (Moscow), 3(1–2): 41 (2000): Mammillaria vetula.
viereckii (Boed.) A.B. Doweld in Sukkulentÿ (Moscow), 3(1–2): 42 (2000): Mammillaria viereckii.
viperina (J.A. Purpus) A.B. Doweld in Sukkulentÿ (Moscow), 3(1–2): 42 (2000): Mammillaria viperina.
weingartiana (Boed.) A.B. Doweld in Sukkulentÿ (Moscow), 3(1–2): 42 (2000): Mammillaria weingartiana.

ESCOBRITTONIA A.B. Doweld in Sukkulentÿ (Moscow), 3(1–2): 17 (2000). *CACTACEAE*
gracilis (L. Bremer & A.B. Lau) A.B. Doweld in Sukkulentÿ (Moscow), 3(1–2): 17 (2000): Coryphantha gracilis.

ESCOCORYPHANTHA A.B. Doweld in Sukkulentÿ (Moscow), 1999(1): 10 (1999); et in Novosti Sist. Vÿssh. Rast., 32: 122 (2000). *CACTACEAE*
chihuahuensis (Britton & Rose) A.B. Doweld in Sukkulentÿ (Moscow), 1999(1): 10 (1999); et in Novosti Sist. Vÿssh. Rast., 32: 123 (2000): Escobaria chihuahuensis.
henricksonii (Glass & R.A. Foster) A.B. Doweld in Sukkulentÿ (Moscow), 1999(1): 10 (1999); et in Novosti Sist. Vÿssh. Rast., 32: 122 (2000): Escobaria henricksonii.

ESENBECKIA (Rutac.).
decidua J.R. Pirani in Bot. J. Linn. Soc., 129(4): 306 (1999)—Brazil Southeast (Minas Gerais).
kallunkiae J.R. Pirani in Bot. J. Linn. Soc., 129(4): 307 (1999)—Brazil North (Rondônia), Bolivia.

ESPELETIOPSIS (Compos.).
jimenez-quesadae (Cuatrec.) Cuatrec.
var. guacharaca (S. Díaz) J. Cuatrecasas in An. Jard. Bot. Madrid, 54(1): 374 (1996): Espeletia guacharaca.

× **ESTELARA** hort. in Orchid Rev., 104(1209): 166 (1996). *ORCHIDACEAE* (Brassavola × Cattleya × Epidendrum × Tetramicra).

ESTERHAZYA (Scrophulariac.).
triflora R.B. de Moura & R.J.V. Alves in Bradea, 8(26): 145 (1999)—Brazil South (Paraná).

ESTERHUYSENIA (Aizoac.).
drepanophylla (Schltr. & A. Berger) H.E.K. Hartmann in Bradleya, 16: 65 (1998): Mesembryanthemum drepanophyllum.
inclaudens (L. Bolus) H.E.K. Hartmann in Bradleya, 16: 66 (1998): Ruschia inclaudens.
stokoei (L. Bolus) H.E.K. Hartmann in Bradleya, 16: 66 (1998): Mesembryanthemum stokoei.

ETLINGERA (Zingiberac.).
belalongensis A.D. Poulsen in Nordic J. Bot., 19(2): 141 (1999)—Borneo (Brunei).
corneri J. Mood & H. Ibrahim in Nordic J. Bot., 20(3): 282 (2000)—Thailand, Malaya (Peninsular Malaysia).
corrugata A.D. Poulsen & J. Mood in Nordic J. Bot., 19(2): 143 (1999)—Borneo (Sabah).
fulgens (Ridl.) C.K. Lim in Folia Malaysiana, 1: 6 (2000): Hornstedtia fulgens.
maingayi (Baker) R.M. Sm.
var. longibracteata (Holttum) I.M. Turner in Novon, 6(2): 223 (1996): Phaeomeria maingayi var. longibracteata.
var. ovata C.K. Lim in Folia Malaysiana, 1: 6 (2000)—Malaya (Peninsular Malaysia).
metriochilos (Griff.) R.M. Sm.
var. grandiflora (Holttum) I.M. Turner in Novon, 6(2): 223 (1996): Achasma sphaerocephalum var. grandiflorum
var. major (Holttum) I.M. Turner in Novon, 6(2): 223 (1996): Achasma sphaerocephalum var. majus.
var. petiolata (Holttum) I.M. Turner in Novon, 6(2): 223 (1996): Achasma sphaerocephalum var. petiolatum.
var. rubrostriata (Holttum) I.M. Turner in Novon, 6(2): 223 (1996): Achasma sphaerocephalum var. rubrostriatum.
rubromarginata A.D. Poulsen & J. Mood in Nordic J. Bot., 19(2): 139 (1999)—Borneo (Brunei, Sabah, Sarawak).
terengganuensis C.K. Lim in Folia Malaysiana, 1: 9 (2000), as 'terengannuensis'; see also Folia Malaysiana, 2: 83 (2000)—Malaya (Peninsular Malaysia).

EUCALYPTUS (Myrtac.).
sect. Abbreviatae M.I.H. Brooker in Austral. Syst. Bot., 13(1): 90 (2000).
subser. Abundae M.I.H. Brooker in Austral. Syst. Bot., 13(1): 100 (2000).
ser. Acaciiformes L.A.S. Johnson ex M.I.H. Brooker & A.V. Slee in Muelleria, 9: 76 (1996).
subgen. Acerosae M.I.H. Brooker in Austral. Syst. Bot., 13(1): 93 (2000), nom. nov.: E. ser. Curtisianae *Chippend.*
subser. Acetabuliformes M.I.H. Brooker in Austral. Syst. Bot., 13(1): 92 (2000).
sect. Adnataria L.D. Pryor & L.A.S. Johnson ex M.I.H. Brooker in Austral. Syst. Bot., 13(1): 117 (2000).
subsect. Alatae M.I.H. Brooker in Austral. Syst. Bot., 13(1): 87 (2000).
subgen. Alveolata (Maiden) M.I.H. Brooker in Austral. Syst. Bot., 13(1): 123 (2000): E. ser. Alveolatae.
subser. Ambitae M.I.H. Brooker in Austral. Syst. Bot., 13(1): 115 (2000).
sect. Amentum M.I.H. Brooker in Austral. Syst. Bot., 13(1): 124 (2000), nom. nov., replaced synonym not validly publ.: E. ser White-mahoganies *Blakely.*
subser. Amissae M.I.H. Brooker in Austral. Syst. Bot., 13(1): 118 (2000).

EUCALYPTUS:-

subser. Angophora *(Cav.) M.I.H. Brooker in Austral. Syst. Bot.,* 13(1): 85 (2000): gen. Angophora.
ser. Angophora *(Cav.) M.I.H. Brooker in Austral. Syst. Bot.,* 13(1): 85 (2000): gen. Angophora.
subgen. Angophora *(Cav.) M.I.H. Brooker in Austral. Syst. Bot.,* 13(1): 84 (2000): gen. Angophora.
ser. Angulares *M.I.H. Brooker in Austral. Syst. Bot.,* 13(1): 132 (2000).
suprasp. Angustae *M.I.H. Brooker in Austral. Syst. Bot.,* 13(1): 100 (2000).
ser. Angustatae *M.I.H. Brooker in Austral. Syst. Bot.,* 13(1): 135 (2000), nom. nov.: Angophora ser. Bakerinae *K.R. Thiele & Ladiges.*
ser. Angustissimae *M.I.H. Brooker in Austral. Syst. Bot.,* 13(1): 105 (2000).
subser. Annulatae *L.A.S. Johnson & K.D. Hill ex M.I.H. Brooker in Austral. Syst. Bot.,* 13(1): 101 (2000).
subsect. Apicales *M.I.H. Brooker in Austral. Syst. Bot.,* 13(1): 117 (2000).
subser. Applanatae *M.I.H. Brooker in Austral. Syst. Bot.,* 13(1): 97 (2000).
subsect. Apterae *M.I.H. Brooker in Austral. Syst. Bot.,* 13(1): 88 (2000).
ser. Aquilonares *M.I.H. Brooker in Austral. Syst. Bot.,* 13(1): 117 (2000).
subsect. Arboreae *M.I.H. Brooker in Austral. Syst. Bot.,* 13(1): 130 (2000).
sect. Aromatica *M.I.H. Brooker in Austral. Syst. Bot.,* 13(1): 125 (2000).
subser. Arrectae *M.I.H. Brooker in Austral. Syst. Bot.,* 13(1): 115 (2000).
subser. Bakerianae *(Chippend.) M.I.H. Brooker in Austral. Syst. Bot.,* 13(1): 105 (2000), isonym; *see Brooker & Hopper in Nuytsia,* 9(1) 25 (1993): E. ser. Bakerianae.
ser. Benthamianae *M.I.H. Brooker in Austral. Syst. Bot.,* 13(1): 113 (2000).
sect. Bisectae *Maiden ex M.I.H. Brooker in Austral. Syst. Bot.,* 13(1): 98 (2000).
*subgen.*Blakella *L.D. Pryor & L.A.S. Johnson ex M.I.H. Brooker in Austral. Syst. Bot.,* 13(1): 89 (2000), nom. nov.: E. subser. Clavigerae *Maiden.*
sect. Bolites *M.I.H. Brooker in Austral. Syst. Bot.,* 13(1): 94 (2000).
ser. Bridgesianae *M.I.H. Brooker & A.V. Slee in Muelleria,* 9: 80 (1996).
ser. Calcicolae *M.I.H. Brooker in Austral. Syst. Bot.,* 13(1): 132 (2000).
suprasp. Cambageanae *(Chippend.) M.I.H. Brooker in Austral. Syst. Bot.,* 13(1): 119 (2000): E. ser. Cambageanae.
sect. Capillulus *M.I.H. Brooker in Austral. Syst. Bot.,* 13(1): 125 (2000).
sect. Cineraceae *M.I.H. Brooker in Austral. Syst. Bot.,* 13(1): 128 (2000).
subser. Circulares *M.I.H. Brooker in Austral. Syst. Bot.,* 13(1): 116 (2000).
ser. Clinatae *M.I.H. Brooker in Austral. Syst. Bot.,* 13(1): 99 (2000), type species name not validly publ.
ser. Coalitae *M.I.H. Brooker in Austral. Syst. Bot.,* 13(1): 120 (2000).
subser. Cochleatae *(Maiden) M.I.H. Brooker in Austral. Syst. Bot.,* 13(1): 132 (2000): E. ser. Cochleatae.
ser. Compactae *M.I.H. Brooker in Austral. Syst. Bot.,* 13(1): 114 (2000).
sect. Complanatae *M.I.H. Brooker in Austral. Syst. Bot.,* 13(1): 92 (2000), nom. nov.: E. ser. Tetrodontae *Chippend.*
subser. Concolores *M.I.H. Brooker in Austral. Syst. Bot.,* 13(1): 122 (2000).
ser. Confines *M.I.H. Brooker in Austral. Syst. Bot.,* 13(1): 115 (2000).
subser. Conjunctae *M.I.H. Brooker in Austral. Syst. Bot.,* 13(1): 99 (2000).
ser. Connexentes *M.I.H. Brooker in Austral. Syst. Bot.,* 13(1): 97 (2000).
subser. Considenianae *M.I.H. Brooker & A.V. Slee in Muelleria,* 9: 84 (1996).
ser. Contiguae *M.I.H. Brooker & A.V. Slee in Muelleria,* 9: 84 (1996).
subser. Continentes *M.I.H. Brooker in Austral. Syst. Bot.,* 13(1): 119 (2000).

EUCALYPTUS:-

suprasp. Contrariae *M.I.H. Brooker in Austral. Syst. Bot.,* 13(1): 118 (2000), nom. nov.: E. ser. Rummeryanae *Chippend.*
ser. Cooperianae *L.A.S. Johnson ex M.I.H. Brooker in Austral. Syst. Bot.,* 13(1): 105 (2000).
ser. Cornutae *(Benth.) M.I.H. Brooker in Austral. Syst. Bot.,* 13(1): 99 (2000): E. subser. Cornutae.
ser. Corrugatae *(Blakely) M.I.H. Brooker in Austral. Syst. Bot.,* 13(1): 110 (2000): E. subser. Corrugatae.
subgen. Corymbia *(K.D. Hill & L.A.S. Johnson) M.I.H. Brooker in Austral. Syst. Bot.,* 13(1): 85 (2000): gen. Corymbia.
ser. Costatitae *(K.R. Thiele & Ladiges) M.I.H. Brooker in Austral. Syst. Bot.,* 13(1): 85 (2000): Angophora subser. Costatitae.
ser. Crenulatae *M.I.H. Brooker & A.V. Slee in Muelleria,* 9: 81 (1996).
subgen. Cruciformes *M.I.H. Brooker in Austral. Syst. Bot.,* 13(1): 93 (2000), nom. nov.: E. subsect. Papilionantherae *Blakely.*
subser. Cubiformes *M.I.H. Brooker in Austral. Syst. Bot.,* 13(1): 102 (2000).
subgen. Cuboidea *M.I.H. Brooker in Austral. Syst. Bot.,* 13(1): 123 (2000).
ser. Cymbiformes *M.I.H. Brooker in Austral. Syst. Bot.,* 13(1): 86 (2000).
subser. Decurrentes *M.I.H. Brooker in Austral. Syst. Bot.,* 13(1): 107 (2000).
subser. Decussatae *M.I.H. Brooker in Austral. Syst. Bot.,* 13(1): 106 (2000).
ser. Deflexae *M.I.H. Brooker in Austral. Syst. Bot.,* 13(1): 111 (2000).
subsect. Destitutae *M.I.H. Brooker in Austral. Syst. Bot.,* 13(1): 103 (2000).
subser. Discolores *M.I.H. Brooker in Austral. Syst. Bot.,* 13(1): 121 (2000).
ser. Disjunctae *M.I.H. Brooker in Austral. Syst. Bot.,* 13(1): 86 (2000).
ser. Dissonae *M.I.H. Brooker in Austral. Syst. Bot.,* 13(1): 109 (2000), nom. nov.: E. subser. Macrocarpae *Blakely.*
*sect.*Domesticae *M.I.H. Brooker in Austral. Syst. Bot.,* 13(1): 123 (2000), nom. nov.: E. subsect. Brachyandrae *Maiden.*
ser. Dorsiventrales *M.I.H. Brooker in Austral. Syst. Bot.,* 13(1): 87 (2000).
sect. Dumaria *L.D. Pryor & L.A.S. Johnson ex M.I.H. Brooker in Austral. Syst. Bot.,* 13(1): 109 (2000).
sect. Ebbanoenses *(Chippend.) M.I.H. Brooker in Austral. Syst. Bot.,* 13(1): 92 (2000): E. ser. Ebbanoenses.
ser. Edentatae *M.I.H. Brooker in Austral. Syst. Bot.,* 13(1): 91 (2000).
sect. Equatoria *L.D. Pryor & L.A.S. Johnson ex M.I.H. Brooker in Austral. Syst. Bot.,* 13(1): 123 (2000), nom. nov.: E. ser. Myrtiformes *Blakely.*
ser. Erectae *M.I.H. Brooker in Austral. Syst. Bot.,* 13(1): 100 (2000).
ser. Erythroxylon *(Blakely) M.I.H. Brooker in Austral. Syst. Bot.,* 13(1): 96 (2000): E. subser. Erythroxylon.
subgen. Eudesmia *(R. Br.) L.A.S. Johnson & K.D. Hill in Telopea,* 7(4): 379 (1998): gen. Eudesmia.
subsect. Euryotae *M.I.H. Brooker in Austral. Syst. Bot.,* 13(1): 113 (2000).
sect. Exsertaria *L.D. Pryor & L.A.S. Johnson ex M.I.H. Brooker in Austral. Syst. Bot.,* 13(1): 96 (2000), nom. nov.: E. ser. Exsertae *Blakely.*
sect. Extensae *M.I.H. Brooker in Austral. Syst. Bot.,* 13(1): 90 (2000), nom. nov.: E. subser. Tessellatae *Blakely.*
subser. Finales *M.I.H. Brooker in Austral. Syst. Bot.,* 13(1): 132 (2000).
ser. Floribundinae *(K.R. Thiele & Ladiges) M.I.H. Brooker in Austral. Syst. Bot.,* 13(1): 85 (2000): Angophora ser. Floribundinae.
subser. Fortes *M.I.H. Brooker in Austral. Syst. Bot.,* 13(1): 117 (2000).
subsect. Frutices *M.I.H. Brooker in Austral. Syst. Bot.,* 13(1): 131 (2000).
ser. Furfuraceae *M.I.H. Brooker in Austral. Syst. Bot.,* 13(1): 111 (2000).
subser. Fusiformes *M.I.H. Brooker in Austral. Syst. Bot.,* 13(1): 105 (2000).
suprasp. Glanduliferae *M.I.H. Brooker in Austral. Syst. Bot.,* 13(1): 120 (2000).

EUCALYPTUS:-

subsect. Glandulosae *M.I.H. Brooker in Austral. Syst. Bot.,* 13(1): 99 (2000).
subser. Glomerosae *M.I.H. Brooker in Austral. Syst. Bot.,* 13(1): 108 (2000).
subsect. Hadrotes *M.I.H. Brooker in Austral. Syst. Bot.,* 13(1): 98 (2000).
subser. Haemastomae *M.I.H. Brooker in Austral. Syst. Bot.,* 13(1): 129 (2000).
ser. Halophilae *M.I.H. Brooker in Austral. Syst. Bot.,* 13(1): 109 (2000).
subgen. Idiogenes *L.D. Pryor & L.A.S. Johnson ex M.I.H. Brooker in Austral. Syst. Bot.,* 13(1): 124 (2000), nom. nov.: E. subser. Megalanthae *Blakely.*
subser. Inclinatae *M.I.H. Brooker in Austral. Syst. Bot.,* 13(1): 93 (2000).
sect. Incognitae *(D. Nicolle) M.I.H. Brooker in Austral. Syst. Bot.,* 13(1): 96 (2000): E. ser. Incognitae.
ser. Incrassatae *L.D. Pryor & L.A.S. Johnson ex M.I.H. Brooker in Austral. Syst. Bot.,* 13(1): 110 (2000).
sect. Insolitae *M.I.H. Brooker in Austral. Syst. Bot.,* 13(1): 130 (2000), nom. nov.: E. ser. Planchonianae *Chippend.*
ser. Insulanae *M.I.H. Brooker in Austral. Syst. Bot.,* 13(1): 125 (2000).
suprasp. Intercessae *M.I.H. Brooker in Austral. Syst. Bot.,* 13(1): 120 (2000).
subser. Irregulares *M.I.H. Brooker in Austral. Syst. Bot.,* 13(1): 127 (2000).
ser. Isobilaterales *M.I.H. Brooker in Austral. Syst. Bot.,* 13(1): 87 (2000).
ser. Jacksoniae *M.I.H. Brooker in Austral. Syst. Bot.,* 13(1): 131 (2000).
subser. Jucundae *(Chippend.) M.I.H. Brooker in Austral. Syst. Bot.,* 13(1): 91 (2000): E. ser. Jucundae.
ser. Kitsonianae *L.A.S. Johnson ex M.I.H. Brooker & A.V. Slee in Muelleria,* 9: 81 (1996).
subser. Lanceolatae *M.I.H. Brooker in Austral. Syst. Bot.,* 13(1): 116 (2000).
suprasp. Latae *M.I.H. Brooker in Austral. Syst. Bot.,* 13(1): 100 (2000).
sect. Latoangulatae *M.I.H. Brooker in Austral. Syst. Bot.,* 13(1): 95 (2000).
subser. Leptopodae *(Blakely) M.I.H. Brooker in Austral. Syst. Bot.,* 13(1): 107 (2000): E. ser. Leptopodae.
subser. Levispermae *M.I.H. Brooker in Austral. Syst. Bot.,* 13(1): 103 (2000).
subser. Liberae *M.I.H. Brooker in Austral. Syst. Bot.,* 13(1): 99 (2000).
sect. Liberivalvae *(Blakely) M.I.H. Brooker in Austral. Syst. Bot.,* 13(1): 96 (2000): E. subser. Liberivalvae.
sect. Limbatae *M.I.H. Brooker in Austral. Syst. Bot.,* 13(1): 91 (2000).
subser. Limitaneae *M.I.H. Brooker in Austral. Syst. Bot.,* 13(1): 88 (2000).
ser. Limitares *M.I.H. Brooker in Austral. Syst. Bot.,* 13(1): 126 (2000).
suprasp. Longae *M.I.H. Brooker in Austral. Syst. Bot.,* 13(1): 100 (2000).
subser. Longiores *M.I.H. Brooker in Austral. Syst. Bot.,* 13(1): 108 (2000).
sect. Longistylus *M.I.H. Brooker in Austral. Syst. Bot.,* 13(1): 130 (2000).
sect. Longitudinales *(Blakely) M.I.H. Brooker in Austral. Syst. Bot.,* 13(1): 128 (2000): E. ser. Longitudinales.
suprasp. Lucasianae *(Chippend.) M.I.H. Brooker in Austral. Syst. Bot.,* 13(1): 119 (2000): E. ser. Lucasianae.
sect. Maidenaria *L.D. Pryor & L.A.S. Johnson ex M.I.H. Brooker in Austral. Syst. Bot.,* 13(1): 112 (2000).
subser. Melanoxylon *M.I.H. Brooker in Austral. Syst. Bot.,* 13(1): 85 (2000).
subser. Micrantherae *M.I.H. Brooker in Austral. Syst. Bot.,* 13(1): 105 (2000).
subgen. Minutifructus *M.I.H. Brooker in Austral. Syst. Bot.,* 13(1): 122 (2000).
suprasp. Moluccanae *(Chippend.) M.I.H. Brooker in Austral. Syst. Bot.,* 13(1): 119 (2000): E. ser. Moluccanae.
sect. Nebulosa *M.I.H. Brooker in Austral. Syst. Bot.,* 13(1): 126 (2000), nom. nov.: E. ser. Olsenianae *Chippend.*

EUCALYPTUS:-
ser. Neglectae *L.A.S. Johnson ex M.I.H. Brooker & A.V. Slee in Muelleria,* 9: 81 (1996).
subser. Neuropterae *(Maiden) M.I.H. Brooker in Austral. Syst. Bot.,* 13(1): 131 (2000): E. ser. Neuropterae.
sect. Notiales *M.I.H. Brooker in Austral. Syst. Bot.,* 13(1): 86 (2000).
subsect. Obscurae *M.I.H. Brooker in Austral. Syst. Bot.,* 13(1): 113 (2000).
subser. Oleaginae *M.I.H. Brooker in Austral. Syst. Bot.,* 13(1): 106 (2000).
suprasp. Opacae *M.I.H. Brooker in Austral. Syst. Bot.,* 13(1): 118 (2000), without latin descr.
ser. Orbiculares *M.I.H. Brooker & A.V. Slee in Muelleria,* 9: 80 (1996).
subser. Ovatae *M.I.H. Brooker in Austral. Syst. Bot.,* 13(1): 108 (2000).
subser. Oviformes *M.I.H. Brooker in Austral. Syst. Bot.,* 13(1): 105 (2000).
subser. Patelliformes *M.I.H. Brooker in Austral. Syst. Bot.,* 13(1): 91 (2000).
ser. Patentes *L.A.S. Johnson ex M.I.H. Brooker in Austral. Syst. Bot.,* 13(1): 131 (2000).
ser. Pauciflorae *L.A.S. Johnson ex M.I.H. Brooker & A.V. Slee in Muelleria,* 9: 84 (1996).
sect. Pedaria *L.A.S. Johnson ex M.I.H. Brooker in Austral. Syst. Bot.,* 13(1): 130 (2000).
subser. Peltiformes *M.I.H. Brooker in Austral. Syst. Bot.,* 13(1): 87 (2000).
subser. Phaenophylla *M.I.H. Brooker in Austral. Syst. Bot.,* 13(1): 103 (2000).
ser. Phaeoxylon *(Blakely) M.I.H. Brooker in Austral. Syst. Bot.,* 13(1): 97 (2000): E. subser. Phaeoxylon.
sect. Platysperma *M.I.H. Brooker in Austral. Syst. Bot.,* 13(1): 98 (2000), nom. nov.: E. ser. Brevifoliae *Brooker & Slee.*
suprasp. Plautae *M.I.H. Brooker in Austral. Syst. Bot.,* 13(1): 88 (2000).
subgen. Primitiva *M.I.H. Brooker in Austral. Syst. Bot.,* 13(1): 124 (2000), nom. nov.: E. ser. Rubiginosae *Chippend.*
ser. Proximae *M.I.H. Brooker in Austral. Syst. Bot.,* 13(1): 133 (2000).
sect. Pseudophloius *M.I.H. Brooker in Austral. Syst. Bot.,* 13(1): 124 (2000), nom. nov., replaced synonym not validly publ.: E. ser. Pseudostringybarks *Blakely.*
sect. Pumilio *M.I.H. Brooker in Austral. Syst. Bot.,* 13(1): 98 (2000), nom. nov.: E. ser. Pumilae *Chippend.*
subser. Pyramidales *M.I.H. Brooker in Austral. Syst. Bot.,* 13(1): 98 (2000).
sect. Racemus *M.I.H. Brooker in Austral. Syst. Bot.,* 13(1): 94 (2000), nom. nov.: E. ser. Michaelianae *Chippend.*
subser. Regulares *M.I.H. Brooker in Austral. Syst. Bot.,* 13(1): 127 (2000).
subser. Remanentes *M.I.H. Brooker in Austral. Syst. Bot.,* 13(1): 115 (2000).
sect. Reticulatae *M.I.H. Brooker in Austral. Syst. Bot.,* 13(1): 92 (2000).
ser. Rostratae *(Blakely) M.I.H. Brooker in Austral. Syst. Bot.,* 13(1): 96 (2000): E. subser. Rostratae.
suprasp. Saxatiles *M.I.H. Brooker in Austral. Syst. Bot.,* 13(1): 119 (2000).
ser. Saxicola *M.I.H. Brooker in Austral. Syst. Bot.,* 13(1): 113 (2000).
sect. Sejunctae *M.I.H. Brooker in Austral. Syst. Bot.,* 13(1): 94 (2000), nom. nov.: E. ser. Microcorythae *Blakely.*
sect. Septentrionales *M.I.H. Brooker in Austral. Syst. Bot.,* 13(1): 86 (2000).
sect. Similares *M.I.H. Brooker in Austral. Syst. Bot.,* 13(1): 95 (2000).
ser. Singulares *M.I.H. Brooker in Austral. Syst. Bot.,* 13(1): 97 (2000).
subser. Solidae *M.I.H. Brooker in Austral. Syst. Bot.,* 13(1): 122 (2000).
ser. Sphaerocarpae *M.I.H. Brooker in Austral. Syst. Bot.,* 13(1): 129 (2000).
subser. Spirales *M.I.H. Brooker in Austral. Syst. Bot.,* 13(1): 106 (2000).
suprasp. Stenophyllae *M.I.H. Brooker in Austral. Syst. Bot.,* 13(1): 88 (2000).
ser. Stenostomae *M.I.H. Brooker in Austral. Syst. Bot.,* 13(1): 129 (2000).

EUCALYPTUS:-
ser. Stricklandianae *M.I.H. Brooker in Austral. Syst. Bot.,* 13(1): 102 (2000).
ser. Strictae *L.A.S. Johnson ex M.I.H. Brooker in Austral. Syst. Bot.,* 13(1): 127 (2000).
ser. Sturgissianae *L.A.S. Johnson ex M.I.H. Brooker in Austral. Syst. Bot.,* 13(1): 117 (2000).
ser. Submelliodorae *(Blakely) M.I.H. Brooker in Austral. Syst. Bot.,* 13(1): 120 (2000): E. subser. Submelliodorae.
subgen. Symphyomyrtus *(Schauer) M.I.H. Brooker in Austral. Syst. Bot.,* 13(1): 93 (2000): gen. Symphyomyrtus.
subser. Synandrae *M.I.H. Brooker in Austral. Syst. Bot.,* 13(1): 107 (2000).
subsect. Terminales *(Maiden) M.I.H. Brooker in Austral. Syst. Bot.,* 13(1): 121 (2000): E. group Terminales.
subser. Tetraedrae *M.I.H. Brooker in Austral. Syst. Bot.,* 13(1): 91 (2000).
ser. Torquatae *L.D. Pryor & L.A.S. Johnson ex M.I.H. Brooker in Austral. Syst. Bot.,* 13(1): 110 (2000), isonym; *see Chippend. in Fl. Australia,* 19: 506 (1988).
*subsect.*Triangulares *M.I.H. Brooker in Austral. Syst. Bot.,* 13(1): 112 (2000).
subsect. Unicae *M.I.H. Brooker in Austral. Syst. Bot.,* 13(1): 133 (2000), nom. nov.: E. ser. Insulares *Chippend.*
ser. Vagitae *D. Nicolle in J. Adelaide Bot. Gard.,* 19: 91 (2000).
subser. Variabiles *M.I.H. Brooker in Austral. Syst. Bot.,* 13(1): 93 (2000).
aenea *K.D. Hill in Telopea,* 7(2): 101 (1997)—New South Wales.
albopurpurea *(Boomsma) D. Nicolle in J. Adelaide Bot. Gard.,* 19: 90 (2000): E. lansdowneana subsp. albopurpurea.
angustata *M.I.H. Brooker in Austral. Syst. Bot.,* 13(1): 135 (2000), nom. nov.: Angophora bakeri *E.C. Hall.*
aparrerinja *(K.D. Hill & L.A.S. Johnson) M.I.H. Brooker in Austral. Syst. Bot.,* 13(1): 137 (2000): Corymbia aparrerinja.
apocynifolia *(Salisb.) M.I.H. Brooker in Austral. Syst. Bot.,* 13(1): 135 (2000): Metrosideros apocynifolia.
bella *(K.D. Hill & L.A.S. Johnson) M.I.H. Brooker in Austral. Syst. Bot.,* 13(1): 137 (2000): Corymbia bella.
biterranea *L.A.S. Johnson & K.D. Hill in Telopea,* 8(4): 505 (2000)—Queensland, New Guinea (Papua New Guinea, Irian Jaya).
blakei *(K.D. Hill & L.A.S. Johnson) M.I.H. Brooker in Austral. Syst. Bot.,* 13(1): 137 (2000): Corymbia blakei.
cadophora *(K.D. Hill & L.A.S. Johnson) M.I.H. Brooker in Austral. Syst. Bot.,* 13(1): 138 (2000): Corymbia cadophora.
calycogona Turcz.
subsp. spaffordii *D. Nicolle in Nuytsia,* 13(2): 310 (2000)—South Australia.
subsp. trachybasis *D. Nicolle in Nuytsia,* 13(2): 307 (2000)—South Australia, Victoria, New South Wales.
candida *(K.D. Hill & L.A.S. Johnson) M.I.H. Brooker in Austral. Syst. Bot.,* 13(1): 138 (2000): Corymbia candida.
canescens *D. Nicolle in Nuytsia,* 11(3): 377 (1997)—South Australia.
subsp. beadellii *D. Nicolle in Nuytsia,* 11(3): 381 (1997)—South Australia.
canobolensis *(L.A.S. Johnson & K.D. Hill) J.T. Hunter in Telopea,* 8(1): 157 (1998): E. rubida subsp. canobolensis.
chartaboma *D. Nicolle in Nuytsia,* 13(2): 324 (2000)—Queensland.
chartacea *(K.D. Hill & L.A.S. Johnson) M.I.H. Brooker in Austral. Syst. Bot.,* 13(1): 138 (2000): Corymbia chartacea.
contracta *L.A.S. Johnson & K.D. Hill in Telopea,* 8(2): 228 (1999)—Queensland.
conveniens *L.A.S. Johnson & K.D. Hill in Telopea,* 7(4): 393 (1998)—Western Australia.
costuligera *L.A.S. Johnson & K.D. Hill in Telopea,* 8(4): 527 (2000)—Western Australia.
crassifolia *(G.J. Leach) M.I.H. Brooker in Austral. Syst. Bot.,* 13(1): 136 (2000): Angophora bakeri subsp. crassifolia.

EUCALYPTUS:-
delicata *L.A.S. Johnson & K.D. Hill in Telopea,* 8(2): 182 (1999)—Western Australia.
dissita *K.D. Hill in Telopea,* 7(2): 104 (1997)—New South Wales.
diversifolia Bonpl.
subsp. hesperia *I.J. Wright & P.Y. Ladiges in Austral. Syst. Bot.,* 10(5): 677 (1997)—Western Australia.
subsp. megacarpa *I.J. Wright & P.Y. Ladiges in Austral. Syst. Bot.,* 10(5): 678 (1997)—Victoria.
dolichocera *L.A.S. Johnson & K.D. Hill in Telopea,* 8(2): 196 (1999)—Western Australia.
dunlopiana *(K.D. Hill & L.A.S. Johnson) M.I.H. Brooker in Austral. Syst. Bot.,* 13(1): 138 (2000): Corymbia dunlopiana.
ebbanoensis Maiden
subsp. glauciramula *L.A.S. Johnson & K.D. Hill in Telopea,* 7(4): 402 (1998)—Western Australia.
epruinata *L.A.S. Johnson & K.D. Hill in Telopea,* 8(4): 530 (2000)—Queensland.
euryphylla *(L.A.S. Johnson & G.J. Leach) M.I.H. Brooker in Austral. Syst. Bot.,* 13(1): 136 (2000): Angophora costata subsp. euryphylla.
extrica *D. Nicolle in Nuytsia,* 13(2): 322 (2000)—Western Australia.
exul *(K.D. Hill) M.I.H. Brooker in Austral. Syst. Bot.,* 13(1): 136 (2000): Angophora exul.
farinosa *K.D. Hill in Telopea,* 7(3): 191 (1997)—Queensland.
flavescens *(K.D. Hill & L.A.S. Johnson) M.I.H. Brooker in Austral. Syst. Bot.,* 13(1): 138 (2000): Corymbia flavescens.
flocktoniae (Maiden) Maiden
subsp. hebes *D. Nicolle in Austral. Syst. Bot.,* 12(2): 225 (1999)—Western Australia.
florida *M.I.H. Brooker in Austral. Syst. Bot.,* 13(1): 136 (2000), nom. nov.: Metrosideros floribunda *Sm.*
fracta *K.D. Hill in Telopea,* 7(2): 106 (1997)—New South Wales.
fulgens *K. Rule in Muelleria,* 9: 136 (1996)—Victoria.
gittinsii Brooker & Blaxell
subsp. illucida *D. Nicolle in Nuytsia,* 13(2): 319 (2000)—Western Australia.
glomericassis *L.A.S. Johnson & K.D. Hill in Telopea,* 8(4): 516 (2000)—Northern Territory.
goniocalyx F. Muell. ex Miq.
subsp. exposa *D. Nicolle in J. Adelaide Bot. Gard.,* 19: 87 (2000)—South Australia.
gregoriensis *N.G. Walsh & D.E. Albrecht in Muelleria,* 11: 41 (1998)—Northern Territory.
grisea *L.A.S. Johnson & K.D. Hill in Telopea,* 8(4): 509 (2000)—Queensland.
gypsophila *D. Nicolle in Nuytsia,* 11(3): 373 (1997)—South Australia, Western Australia.
helidonica *K.D. Hill in Telopea,* 8(2): 225 (1999)—Queensland.
hendersonii *(K.D. Hill & L.A.S. Johnson) M.I.H. Brooker in Austral. Syst. Bot.,* 13(1): 138 (2000): Corymbia hendersonii.
hispida *(Sm.) M.I.H. Brooker in Austral. Syst. Bot.,* 13(1): 136 (2000): Metrosideros hispida.
hypolaena *L.A.S. Johnson & K.D. Hill in Telopea,* 8(2): 203 (1999)—Western Australia.
inopina *(K.D. Hill) M.I.H. Brooker in Austral. Syst. Bot.,* 13(1): 136 (2000): Angophora inopina.
intrasilvatica *L.A.S. Johnson & K.D. Hill in Telopea,* 8(2): 214 (1999)—Western Australia.
irritans *L.A.S. Johnson & K.D. Hill in Telopea,* 8(2): 233 (1999)—Queensland.
kenneallyi *K.D. Hill & L.A.S. Johnson in Telopea,* 8(4): 518 (2000)—Western Australia.
latisinensis *K.D. Hill in Telopea,* 8(2): 244 (1999)—Queensland.
leiocarpa *(L.A.S. Johnson ex G.J. Leach) M.I.H. Brooker in Austral. Syst. Bot.,* 13(1): 137 (2000): Angophora costata subsp. leiocarpa.
leucophloia Brooker
subsp. euroa *L.A.S. Johnson & K.D. Hill in Telopea,* 8(4): 520 (2000)—Northern Territory, Queensland.
leucoxylon F. Muell.
subsp. bellarinensis *K. Rule in Muelleria,* 11: 133 (1998)—Victoria.
limitaris *L.A.S. Johnson & K.D. Hill in Telopea,* 8(4): 530 (2000)—Western Australia, Northern Territory.

EUCALYPTUS:-
luculenta *L.A.S. Johnson & K.D. Hill in Telopea,* 8(2): 208 (1999)—Western Australia.
macmahonii *K. Rule in Muelleria,* 10: 13 (1997)—Victoria.
macta *L.A.S. Johnson & K.D. Hill in Telopea,* 8(4): 507 (2000)—Queensland.
mediocris *L.A.S. Johnson & K.D. Hill in Telopea,* 8(2): 237 (1999)—Queensland.
melana *M.I.H. Brooker in Austral. Syst. Bot.,* 13(1): 137 (2000), nom. nov.: Angophora melanoxylon *R.T. Baker.*
moderata *L.A.S. Johnson & K.D. Hill in Telopea,* 8(2): 198 (1999)—Western Australia.
molyneuxii *K. Rule in Muelleria,* 12(2): 163 (1999)—Victoria.
montivaga *A.R. Bean in Austrobaileya,* 5(1): 130 (1997)—Queensland.
neutra *D. Nicolle in Austral. Syst. Bot.,* 12(2): 232 (1999)—Western Australia.
oleosa F. Muell. ex Miq.
subsp. ampliata *L.A.S. Johnson & K.D. Hill in Telopea,* 8(2): 179 (1999)—South Australia.
subsp. corvina *L.A.S. Johnson & K.D. Hill in Telopea,* 8(2): 178 (1999)—Western Australia.
subsp. cylindroidea *L.A.S. Johnson & K.D. Hill in Telopea,* 8(2): 178 (1999)—Western Australia.
subsp. repleta *L.A.S. Johnson & K.D. Hill in Telopea,* 8(2): 176 (1999)—South Australia, Western Australia.
subsp. victima *L.A.S. Johnson & K.D. Hill in Telopea,* 8(2): 179 (1999)—South Australia.
subsp. wylieana *L.A.S. Johnson & K.D. Hill in Telopea,* 8(2): 180 (1999)—Western Australia, South Australia.
oligantha Schauer
subsp. modica *L.A.S. Johnson & K.D. Hill in Telopea,* 8(4): 526 (2000)—Western Australia.
optima *L.A.S. Johnson & K.D. Hill in Telopea,* 8(2): 203 (1999)—Western Australia.
oresbia *J.T. Hunter & J.J. Bruhl in Telopea,* 8(2): 260 (1999)—New South Wales.
pachycalyx Maiden & Blakely
subsp. banyabba *K.D. Hill in Telopea,* 7(3): 187 (1997)—New South Wales.
pachycarpa *(K.D. Hill & L.A.S. Johnson) M.I.H. Brooker in Austral. Syst. Bot.,* 13(1): 138 (2000): Corymbia pachycarpa.
pallida *L.A.S. Johnson & K.D. Hill in Telopea,* 7(4): 389 (1998)—Western Australia.
palustris *M.I.H. Brooker in Austral. Syst. Bot.,* 13(1): 137 (2000), nom. nov.: Angophora bakeri subsp. paludosa *G.J. Leach.*
pantoleuca *L.A.S. Johnson & K.D. Hill in Telopea,* 8(4): 513 (2000)—Western Australia.
papillosa *(K.D. Hill & L.A.S. Johnson) M.I.H. Brooker in Austral. Syst. Bot.,* 13(1): 138 (2000): Corymbia papillosa.
peninsularis *D. Nicolle in Austral. Syst. Bot.,* 12(2): 230 (1999)—South Australia.
phenax *M.I.H. Brooker & A.V. Slee in Muelleria,* 9: 77 (1996)—South Australia, Western Australia, Victoria.
subsp. compressa *D. Nicolle in J. Adelaide Bot. Gard.,* 19: 84 (2000)—South Australia.
plena *(K.D. Hill & L.A.S. Johnson) M.I.H. Brooker in Austral. Syst. Bot.,* 13(1): 138 (2000): Corymbia plena.
polyanthemos
subsp. longior *M.I.H. Brooker & A.V. Slee in Muelleria,* 9: 82 (1996)—Victoria New South Wales.
portuensis *K.D. Hill in Telopea,* 8(2): 234 (1999)—Queensland.
prolixa *D. Nicolle in Nuytsia,* 13(2): 312 (2000)—Western Australia.
provecta *A.R. Bean in Austrobaileya,* 5(4): 681 (2000)—Queensland.
pruinosa Schauer
subsp. tenuata *L.A.S. Johnson & K.D. Hill in Telopea,* 8(4): 538 (2000)—Northern Territory, Western Australia.
quinniorum *J.T. Hunter & J.J. Bruhl in Telopea,* 8(2): 257 (1999)—New South Wales.
repullulans *D. Nicolle in Nuytsia,* 11(3): 376 (1997)—Western Australia.
robur *(L.A.S. Johnson & K.D. Hill) M.I.H. Brooker in Austral. Syst. Bot.,* 13(1): 137 (2000): Angophora robur.

EUCALYPTUS:-
sabulosa *K. Rule in Muelleria,* 9: 138 (1996)—Victoria.
scopulorum *K.D. Hill in Telopea,* 7(3): 192 (1997)—New South Wales.
selachiana *L.A.S. Johnson & K.D. Hill in Telopea,* 7(4): 390 (1998)—Western Australia.
semota *C.J. Macpherson & P.M. Grayling in Nuytsia,* 10(3): 437 (1996)—Western Australia.
splendens *K. Rule in Muelleria,* 9: 140 (1996)—Victoria.
subsp. arcana *D. Nicolle & M.I.H. Brooker in J. Adelaide Bot. Gard.,* 18(2): 103 (1998)—South Australia.
striaticalyx W. Fitzg.
subsp. delicata *D. Nicolle & P.J. Lang in Nuytsia,* 11(3): 371 (1997)—Western Australia.
subcaerulea *K.D. Hill in Telopea,* 7(3): 195 (1997)—New South Wales.
sublucida *L.A.S. Johnson & K.D. Hill in Telopea,* 8(2): 186 (1999)—Western Australia, South Australia.
subvelutina *(F. Muell.) M.I.H. Brooker in Austral. Syst. Bot.,* 13(1): 137 (2000): Angophora subvelutina.
tardecidens *(L.A.S. Johnson & K.D. Hill) A.R. Bean in Austrobaileya,* 5(4): 683 (2000): Eucalyptus persistens subsp. tardecidens.
tephrodes *L.A.S. Johnson & K.D. Hill in Telopea,* 8(4): 533 (2000)—Western Australia.
transcontinentalis Maiden
subsp. semivestita *L.A.S. Johnson & K.D. Hill in Telopea,* 8(2): 201 (1999)—Western Australia.
ultima *L.A.S. Johnson & K.D. Hill in Telopea,* 8(2): 188 (1999)—Western Australia.
urna *D. Nicolle in Austral. Syst. Bot.,* 12(2): 227 (1999)—Western Australia.
uvida *K.D. Hill in Telopea,* 8(2): 231 (1999)—Queensland.
viminalis
subsp. pryoriana *(L.A.S. Johnson) M.I.H. Brooker & A.V. Slee in Muelleria,* 9: 80 (1996): E. pryoriana.
viridis R.T. Baker
subsp. wimmerensis *(Rule) M.I.H. Brooker & A.V. Slee in Muelleria,* 9: 81 (1996): E. wimmerensis.
vokesensis *D. Nicolle & L.A.S. Johnson in Telopea,* 8(2): 194 (1999)—South Australia.
woodsiana *(F.M. Bailey) M.I.H. Brooker in Austral. Syst. Bot.,* 13(1): 137 (2000): Angophora woodsiana.
xerothermica *L.A.S. Johnson & K.D. Hill in Telopea,* 8(4): 534 (2000)—Western Australia.

EUCERAEA (Flacourtiac.).
rheophytica *P.E. Berry & M.E. Olson in Brittonia,* 50(4): 493 (1998)—Venezuela.

EUCHILE *(Dressler & G.E. Pollard) C.L. Withner, Cattleyas & Relatives,* 5: 137 (1998): Encyclia sect. Euchile. (Orchidac.).
citrina *(La Llave & Lex.) C.L. Withner, Cattleyas & Relatives,* 5: 138 (1998): Sobralia citrina.
mariae *(Ames) C.L. Withner, Cattleyas & Relatives,* 5: 139 (1998): Epidendrum mariae.

EUCHITON (Compos.).
lateralis *(C.J. Webb) I. Breitwieser & J.M. Ward in New Zealand J. Bot.,* 36(2): 303 (1998): Gnaphalium laterale.
litticola *A.M. Buchanan in Pap. Proc. Roy. Soc. Tasmania,* 133(1): 115 (1999)—Tasmania.
poliochlorus *N.G. Walsh in Muelleria,* 12(2): 225 (1999)—Victoria, New South Wales, Tasmania.
polylepis *(D.G. Drury) I. Breitwieser & J.M. Ward in New Zealand J. Bot.,* 36(2): 303 (1998): Gnaphalium paludosum var. polylepis.
ruahinicus *(D.G. Drury) I. Breitwieser & J.M. Ward in New Zealand J. Bot.,* 36(2): 303 (1998): Gnaphalium audax subsp. ruahinicum.

× **EUCLADES** *hort. in Orchid Rev.,* 104(1207): 37 (1996). *ORCHIDACEAE* (Eulophia × Oeceoclades).

EUCLEA (Ebenac.).
subgen. Rymia *(Endl.) B. Verdcourt in Fl. Trop. E. Afr., Ebenac.:* 44 (1996): gen. Rymia.

EUCROSIA (Amaryllidac.).
calendulina *A.W. Meerow & A. Sagástegui Alva in Sida,* 17(4): 761 (1997)—Peru.

EUCRYPHIA (Eucryphiac.).
jinksii *P.I. Forster in Austrobaileya,* 4(4): 592 (1997)—Queensland.
milliganii Hook. f.
subsp. pubescens *R.W. Barnes, G.J. Jordan, R.S. Hill & C.J. McCoull in Austral. J. Bot.,* 48(4): 488 (2000)—Tasmania.
wilkiei *B.P.M. Hyland in Austrobaileya,* 4(4): 590 (1997)—Queensland.

EUGENIA (Myrtac.).
subgen. Hexachlamys *(O. Berg) J.R. de Mattos in Loefgrenia,* 105: 2 (1995): gen. Hexachlamys.
aloysii *C.J. Saldanha, Fl. Karnataka,* 2: 23 (1996)—India (Karnataka).
argyrophylla *B.K. Holst & M.L. Kawasaki in Brittonia,* 52(1): 20 (2000)—French Guiana.
boliviana *(D. Legrand) J.R. de Mattos in Loefgrenia,* 105: 2 (1995): Hexachlamys boliviana.
calycina Cambess.
var. herbacea *(O. Berg) J.R. de Mattos in Loefgrenia,* 113: 2 (1999): Phyllocalyx herbaceus.
capensis (Eckl. & Zeyh.) Sond.
subsp. multiflora *B. Verdcourt in Kew Bull.,* 54(1): 43 (1999)—Kenya, Tanzania.
decussata *(Vell.) J.R. de Mattos in Loefgrenia,* 102: 3 (1993): Myrtus decussatus.
flavifolia *P.C. Standley in Publ. Field Mus. Nat. Hist., Chicago, Bot. Ser.,* 8: 320 (1931), in IK Supp. 9 as 'flaviflora' in error—Belize.†
geraensis *(D. Legrand & Mattos) J.R. de Mattos in Loefgrenia,* 105: 2 (1995): Hexachlamys geraensis.
grisea *J.R. de Mattos in Loefgrenia,* 107: 2 (1995), homonym; *see C.B. Rob. In Philipp. J. Sci., C,* 4: 395 (1909)—Brazil West-central (Goiás).
hamiltonii *(Mattos) J.R. de Mattos in Loefgrenia,* 105: 2 (1995): Hexachlamys hamiltonii.
handroi *(Mattos) J.R. de Mattos in Loefgrenia,* 105: 2 (1995): Hexachlamys handroi.
indica *(Wight) V. Chithra in N.C. Nair & A.N. Henry, Fl. Tamil Nadu, India, Ser. 1,* 1: 153 (1983): Jossinia indica.
itararensis *(Mattos) J.R. de Mattos in Loefgrenia,* 105: 3 (1995): Hexachlamys itararensis.
itatiaiensis *(Mattos) J.R. de Mattos in Loefgrenia,* 105: 2 (1995), homonym; *see Kiaersk., Enum. Myrt. Bras.:* 166 (1893): Hexachlamys itatiaiensis.
legrandii *(Mattos) J.R. de Mattos in Loefgrenia,* 105: 2 (1995): Hexachlamys legrandii.
lineatifolia *J.R. de Mattos in Loefgrenia,* 102: 3 (1993), nom. nov.: E. uniflora *O. Berg*
montevidensis *J.R. de Mattos in Loefgrenia,* 105: 2 (1995), nom. nov.: Myrtus excelsa *Cambess.*
morii *B.K. Holst & M.L. Kawasaki in Brittonia,* 52(1): 21 (2000)—French Guiana.
mufindiensis *B. Verdcourt in Kew Bull.,* 54(1): 50 (1999)—Tanzania.
philippinorum *C. Nelson Sutherland in An. Jard. Bot. Madrid,* 56(1): 162 (1998), nom. nov.: E. conglobata *C.B. Rob.*
randrianasoloi *J.S. Miller in Adansonia,* 22(1): 112 (2000)—Madagascar.
robertoana *J.R. de Mattos in Loefgrenia,* 105: 2 (1995), nom. nov.: Hexachlamys itatiaiensis var. kleinii *D. Legrand.*
rojasiana *(D. Legrand) J.R. de Mattos in Loefgrenia,* 105: 2 (1995): Hexachlamys anomala var. rojasiana.
rufoflavescens *J.R. de Mattos in Loefgrenia,* 114: 1 (2000), as 'rufo-flavescens'—Brazil North (Pará).
schatzii *J.S. Miller in Adansonia,* 22(1): 114 (2000)—Madagascar.
sehnemiana *(Mattos) J.R. de Mattos in Loefgrenia,* 105: 2 (1995): Hexachlamys hamiltonii var. sehnemiana.
tanaensis *B. Verdcourt in Kew Bull.,* 54(1): 55 (1999)—Kenya, Tanzania.
teresae *J.F. Morales in Phytologia,* 81(5): 361 (1996 publ. 1997)—Costa Rica.
thikaensis *B. Verdcourt in Kew Bull.,* 54(1): 56 (1999)—Kenya.
toledoi *(Mattos) J.R. de Mattos in Loefgrenia,* 105: 3 (1995): Hexachlamys toledoi.

EUGENIA:-
toxanatolica *B. Verdcourt in Kew Bull.*, 54(1): 49 (1999)—Tanzania.
vallsiana *J.R. de Mattos in Loefgrenia*, 105: 2 (1995), nom. nov.: Hexachlamys humilis *O. Berg*

EULALIA (Gramin.).
contorta *(Brongn.) G. Panigrahi in S.K. Murti & G. Panigrahi, Fl. Bilaspur Distr.*, 2: 773 (1999), isonym; *see Kuntze, Revis. Gen. Pl.*, 2: 775 (1891): Pogonatherum contortum.
alta Fawc. & Rendle
f. pelchatii *P.M. Brown in North Amer. Native Orchid J.*, 4(1): 46 (1998)—Florida.
anjoanensis *(Rchb. f.) P.J. Cribb in Lindleyana*, 13(3): 174 (1998); *et in D. Du Puy, P. Cribb, J. Bosser, J. Hermans & C. Hermans, Orchids of* Madagascar: 4, 137 (1999): Galeandra anjoanensis.
brenanii *P.J. Cribb & I. la Croix in Fl. Zambes.*, 11(2): 536 (1998)—Zambia.
hians Spreng.
var. inaequalis *(Schltr.) S. Thomas in Lindleyana*, 13(3): 183 (Sept. 1998); *et in Kew Bull.*, 53(4): 1007 (Dec. 1998); *et in Fl. Zambes.*, 11(2): 475 (Dec. 1998): E. inaequalis.
var. nutans *(Sond.) S. Thomas in Lindleyana*, 13(3): 183 (Sept. 1998); *et in Kew Bull.*, 53(4): 1006 (Dec. 1998); *et in Fl. Zambes.*, 11(2): 476 (Dec. 1998): E. nutans.
kamarupa *Sud. Chowdhury in J. Orchid Soc. India*, 7(1–2): 49 (1993)—Assam (Assam).
obstipa *P.J. Cribb & I. la Croix in Fl. Zambes.*, 11(2): 534 (1998)—Angola.
ovalis Lindl.
var. bainesii *(Rolfe) P.J. Cribb & I. la Croix in Fl. Zambes.*, 11(2): 473 (1998): E. bainesii.
richardsiae *P.J. Cribb & I. la Croix in Kew Bull.*, 52(4): 1001 (1997)—Zambia.
sooi *Chun & T. Tang ex S.C. Chen in Fl. Reipubl. Popul. Sin.*, 18: 412 (1999)—China Southeast (Guangxi).
zollingeri (Rchb. f.) J.J. Sm.
f. viride *M. Yokota in J. Phytogeogr. Taxon.*, 46(2): 184 (1998)—Nansei-Shoto.

EUODIA (Rutac.).
daniellii (Benn.) Hemsl.
var. laxiflora *S.C. Li & X.M. Liu in S.C. Li (ed.) et al., Dendrol. Zhongtiaoshan.*: 479 (1995)—China North-central (Shanxi).

EUONYMUS (Celastrac.).
aculeolus *C.Y. Cheng ex J.S. Ma in Harvard Pap. Bot.*, 10: 94 (1997), as 'aculeola'—China South-central (Yunnan).
alatus (Thunb.) Siebold
subsp. sacrosanctus *(Koidz.) V.N. Voroschilov in A.K. Skvortsov (ed.), Florist. issl. v razn. raionakh SSSR*: 181 (1985): E. sacrosanctus.
chengiae *J.S. Ma in Harvard Pap. Bot.*, 10: 95 (1997), as 'chengii'—China Southeast (Guangdong), Hainan.
dolichopus *Merr. ex J.S. Ma in Harvard Pap. Bot.*, 10: 95 (1997), as 'dolichopa'—China Southeast (Guangxi).
echinatus Wall.
var. vagans *(Wall.) R.C. Srivastava, Fl. Sikkim (Ranunculac.-Moringac.)*: 252 (1998): E. vagans.
ficoides *C.Y. Cheng ex J.S. Ma in Harvard Pap. Bot.*, 10: 94 (1997)—China South-central (Yunnan).
furfuraceus *Q.H. Chen in Acta Bot. Yunnanica*, 21(2): 167 (1999)—China South-central (Guizhou).
huae *J.S. Ma in Harvard Pap. Bot.*, 10: 96 (1997), as 'hui'—China South-central (Sichuan).
hukuangensis *C.Y. Cheng ex J.S. Ma in Harvard Pap. Bot.*, 10: 94 (1997)—China Southeast (Guangdong, Hunan).
kengmaensis *C.Y. Cheng ex J.S. Ma in Harvard Pap. Bot.*, 10: 93 (1997)—China South-central (Yunnan).
laxicymosus *C.Y. Cheng ex J.S. Ma in Harvard Pap. Bot.*, 10: 96 (1997), as 'laxicymosa'—China Southeast (Guangxi), China South-central (Yunnan), Vietnam.
leishanensis *Q.H. Chen in Acta Bot. Yunnanica*, 21(2): 168 (1999)—China South-central (Guizhou).

EUONYMUS:-
parasimilis *C.Y. Cheng ex J.S. Ma in Harvard Pap. Bot.*, 10: 96 (1997)—China South-central (Yunnan).
percoriaceus *C.Y. Wu ex J.S. Ma in Harvard Pap. Bot.*, 10: 97 (1997), as 'pereoriacea'—China South-central (Yunnan).
pittosporoides *C.Y. Cheng ex J.S. Ma in Harvard Pap. Bot.*, 3(2): 232 (1998)—China Southeast (Guangxi, Guangdong), China South-central (Guizhou, Sichuan, Yunnan), Vietnam.
potingensis *Chun & F.C. How ex J.S. Ma in Harvard Pap. Bot.*, 10: 94 (1997)—Hainan.
prismatomerioides *C.Y. Wu ex J.S. Ma in Harvard Pap. Bot.*, 3(2): 232 (1998)—China South-central (Yunnan).
subcordatus *J.S. Ma in Harvard Pap. Bot.*, 10: 95 (1997), as 'subcordata'—China Southeast (Guangxi).
tenuiserratus *C.Y. Cheng ex J.S. Ma in Harvard Pap. Bot.*, 3(2): 232 (1998), as 'tenuiserrata'—China South-central (Yunnan).
vaganoides *C.Y. Cheng ex J.S. Ma in Harvard Pap. Bot.*, 10: 93 (1997)—China South-central (Yunnan), China Southeast (Guangxi, Hunan).
verrucocarpus *C.Y. Cheng ex J.S. Ma in Harvard Pap. Bot.*, 3(2): 231 (1998), as 'verrucocarpa'—China South-central (Yunnan).
wui *J.S. Ma in Harvard Pap. Bot.*, 10: 97 (1997)—China South-central (Yunnan).

EUPATORIADELPHUS (Compos.).
steelei *(E.E. Lamont) G.J. Schmidt & E.E. Schilling in Amer. J. Bot.*, 87(5): 726 (2000): Eupatorium steelei.

EUPATORIUM (Compos.).
angustilobum *(Y. Ling) C. Shih, S.Y. Jin & S.R. Chen in Acta Phytotax. Sin.*, 38(5): 471 (2000): E. fortunei var. angustilobum.
atrorubens *(Lem.) N.E. Brown, Suppl. Johnson's Gard. Dict.*: 927 (1882): Hebeclinium atrorubens.
bohlmannianum *(R.M. King & H. Rob.) B.L. Turner in Phytologia Mem.*, 11: 116 (1997), isonym; *see C. Nelson in Ceiba*, 27(2): 224 (1986): Fleischmannia bohlmanniana.
catarium *J.F. Veldkamp in Gard. Bull. Singapore*, 51(1): 121 (1999), nom. nov.: E. clematideum *Griseb.*
clematideum (Wall. ex DC.) Sch. Bip.
var. gracillimum *(Hayata) C.I Peng & S.W. Chung in Fl. Taiwan, ed. 2*, 4: 956 (1998): E. gracillimum.
crenifolium *Hand.-Mazz.in Oesterr. Bot. Z.*, 88(4): 307 (1939)—China South-Central (Hubei, Guizhou), China Southeast (Guangdong, Hunan).
hualienense *C.H. Ou, S.W. Chung & C.I Peng in Fl. Taiwan, ed. 2*, 4: 956 (1998)—Taiwan.
iltisii *(R.M. King & H. Rob.) Whittem. ex B.L. Turner in Phytologia Mem.*, 11: 123 (1997): Critonia iltisii.
kiirunense *(Kitam.) C.H. Ou & S.W. Chung in Fl. Taiwan, ed. 2*, 4: 960 (1998): E. luchuense var. kiirunense.
megalophyllum *(Lem.) N.E. Brown, Suppl. Johnson's Gard. Dict.*: 927 (1882): Hebeclinium megalophyllum.
naiguatensis *(V.M. Badillo) J.A. Steyermark in Pittieria*, 7: 23 (1978): Steyermarkina naiguatensis.
ornatum *(R.M. King & H. Rob.) B.L. Turner in Phytologia Mem.*, 11: 130 (1997): Bartlettina ornata.
panamense *(Carrière) N.E. Brown, Suppl. Johnson's Gard. Dict.*: 927 (1882): Hebeclinium panamense.
paneroi *(B.L. Turner) B.L. Turner in Phytologia Mem.*, 11: 131 (1997): Critonia paneroi.
urolepis *(DC.) N.E. Brown, Suppl. Johnson's Gard. Dict.*: 927 (1882): Hebeclinium urolepis.

EUPHORBIA (Euphorbiac.).
subsect. Altaicae *K.S. Baikov in Byull. Mosk. Obshch. Ispyt. Prir., Biol.*, 99(6): 124 (1994).
subsect. Arvales *D.V. Geltman in Novosti Sist. Vyssh. Rast.*, 32: 102 (2000).
subsect. Exiguae *D.V. Geltman in Novosti Sist. Vyssh. Rast.*, 32: 101 (2000).

EUPHORBIA:-
subsect. Falcatae *D.V. Geltman in Novosti Sist. Vyssh. Rast.*, 32: 106 (2000).
ser. Leptocaula *D.V. Geltman in Novosti Sist. Vyssh. Rast.*, 32: 109 (2000).
ser. Mongolicae *K.S. Baikov in Byull. Mosk. Obshch. Ispyt. Prir., Biol.*, 99(6): 122 (1994).
subsect. Phymatospermae *D.V. Geltman in Novosti Sist. Vyssh. Rast.*, 32: 107 (2000).
subsect. Potaniniae *K.S. Baikov in Byull. Mosk. Obshch. Ispyt. Prir., Biol.*, 99(6): 127 (1994).
subsect. Taurinenses *D.V. Geltman in Novosti Sist. Vyssh. Rast.*, 32: 104 (2000).
acervata *S. Carter in Kew Bull.*, 54(4): 962 (1999 publ. 2000—Zimbabwe.
acutifolia *(Boiss.) K.H. Rechinger, Fl. Aegaea*: 114 (1943): E. cybirensis var. acutifolia.
alatocaulis *V.W. Steinmann & R.S. Felger in Aliso*, 16(1): 37 (1997)—Mexico Northwest (Sonora).
ambarivatoensis *W. Rauh & M. Bardot-Vaucoulon in Succulentes*, 1999(3): 7 (1999)—Madagascar.
ambukensis *N.V. Stepanov, Florogenet. Analiz*, 1: 83 (1994)—Krasnoyarsk.
analavelonensis *W. Rauh & R.D. Mangelsdorff in Succulentes*, 2000(1): 6,3 (2000)—Madagascar.
ankazobensis *W. Rauh & S. Hofstätter in Succulentes*, 2000(3): 6 (2000)—Madagascar.
arabica *Hochst. & Steud. ex T. Anderson, Fl. Aden.*: 34 (1860)—South Yemen.†
arvalis Boiss. & Heldr.
subsp. longistyla *(Litard. & Maire) J. Molero, A.M. Rovira & J. Vicens in An. Jard. Bot. Madrid*, 54(1): 218 (1996): E. arvalis var. longistyla.
astrophora *J.G. Marx in Cact. Succ. J. (U.S.A.)*, 68(6): 311 (1996)—Cape Province.
babensis *P. Kitaibel in Linnaea*, 32: 562 (1863)—Hungary.
balakrishnanii *M.S. Binojkumar & R. Gopalan in Rheedea*, 8(1): 67 (1998)—India (Tamil Nadu).
balearica *(Willk.) A.M. Romo, Fl. Silvestres Baleares*: 205 (1994): E. gayii var. balearica.
balsamifera Aiton
subsp. adenensis *(Deflers) R. Govaerts in R. Govaerts, D.G. Frodin & A. Radcliffe-Smith, World Checklist & Bibliogr.* Euphorbiaceae, 2: 679 (2000), isonym; *see P.R.O. Bally in Candollea*, 20: 31, 34 (1965): E. adenensis.
beharensis Leandri
var. adpressifolia *W. Rauh in Trop. Subtrop. Pflanzenwelt*, 100: 13 (1999)—Madagascar.
var. guillemetii *(Ursch & Leandri) W. Rauh in Trop. Subtrop. Pflanzenwelt*, 100: 10 (1999): E. guillemetii.
var. squarrosa *W. Rauh in Trop. Subtrop. Pflanzenwelt*, 100: 14 (1999)—Madagascar.
var. truncata *W. Rauh in Trop. Subtrop. Pflanzenwelt*, 100: 12 (1999)—Madagascar.
bemarahaensis *W. Rauh & R. Mangelsdorff in Succulentes*, 1999(1): 12 (1999)—Madagascar.
bolosii *(Molero & Rovira) J. Molero & A.M. Rovira in Fontqueria*, 44: 9 (1996): E. nevadensis subsp. bolosii.
bonae *P.A.V. Mutel, Fl. Franç.*, 3: 151 (1836), nom. provis.—France.
× bothae *J.P. Lotsy & W.A. Goddijn in Genetica*, 10: 47 (1928)—Cape Province.
caloderma *S. Carter in Cact. Succ. J. (U.S.A.)*, 72(4): 189 (2000)—Tanzania.
chamaepeplus Boiss. & Gaill.
var. angustifolia *A. Danin in Bocconea*, 3: 104 (1992)—Palestine (Israel).
charleswilsoniana *V. Vlk in Kaktusy (Brno)*, 33(2): 47 (1997), nom. nov.: E. wilsonii *V. Vlk.*
cheirolepis *Fisch. & C.A. Mey. ex C.F. von Ledebour, Fl. Ross.*, 3: 558 (1850)—Siberia.†
chiribensis *V.W. Steinmann & R.S. Felger in Aliso*, 16(1): 52 (1997)—Mexico Northwest (Sonora).
clementei Boiss.
subsp. faurei *(Maire) J. Vicens, J. Molero & C. Blanché in Candollea*, 51(1): 66 (1996): E. squamigera var. faurei.
subsp. villosa *(Faure & Maire) J. Vicens, J. Molero & C. Blanché in Candollea*, 51(1): 67 (1996): E. atlantica var. villosa.
concanensis *M.K. Janarthanam & S.R. Yadav in Rheedea*, 5(2): 148 (1995)—India (Maharashtra).

EUPHORBIA:-
cremersii Rauh & Razaf.
var. rakatozafyi *(Cremers) W. Rauh, Succ. & Xeroph. Pl. Madagascar,* 1: 172 (1995): E. francoisii var. rakatozafyi.
crepuscula *(L.C. Wheeler) V.W. Steinmann & R.S. Felger in Madroño,* 42(4): 455 (1995 publ. 1996): E. villifera var. crepuscula.
czerepanovii *D.V. Geltman in Bot. Zhurn.,* 82(3): 122 (1997)—North Caucasus (Dagestan).
daghestanica *D.V. Geltman in Bot. Zhurn.,* 82(3): 122 (1997)—North Caucasus (Dagestan).
davyi
subsp. maleolens *(E. Phillips) Hargreaves in Euphorbiaceae Study Group Bull.,* 8(3): 100 (1995), without basionym ref.; *cf. Repert. Pl. Succ. (I.O.S.),* 46: 19 (1995 publ. 1996): E. maleolens.
subsp. tlapanensis *Hargreaves in Euphorbiaceae Study Group Bull.,* 8(3): 100–103 (1995), without latin descr. or type; *cf. Repert. Pl. Succ. (I.O.S.),* 46: 19 (1995 publ. 1996)—Locality not stated.
dioscoreoides Boiss.
subsp. attenuata *V.W. Steinmann in Aliso,* 14(3): 222 (1995 publ. 1996)—Mexico Northeast (Chihuahua, Durango), Mexico Northwest (Sinaloa, Sonora), Mexico Southwest (Colima, Jalisco, Michoacan, Nayarit).
ebracteolata Hayata
f. magna *Y.N. Lee in Korean J. Pl. Taxon.,* 28(1): 28 (1998)—South Korea.
eglandulosa *V.W. Steinmann in Aliso,* 14(3): 223 (1995 publ. 1996)—Mexico Southwest (Guerrero), Mexico Southeast (Chiapas), Mexico Central (Morelos).
eleanoriae *(M.E. Lawr. & W.L. Wagner) R. Govaerts in R. Govaerts, D.G. Frodin & A. Radcliffe-Smith, World Checklist & Bibliogr.* Euphorbiaceae, 2: 717 (2000): Chamaesyce eleanoriae.
enterophora *E. Drake del Castillo in Bull. Mus. Hist. Nat. (Paris),* 5: 307 (1899), in Supp. 2 as 'enterophona'—Madagascar.†
estevesii *N.F.A. Zimmermann & P.J. Braun in Cact. Succ. J. (U.S.A.),* 72(6): 318 (2000)—Brazil West-Central (Goiás).
esula L.
subsp. chankoana *(Vorosch.) V.N. Voroschilov in A.K. Skvortsov (ed.), Florist. issl. v razn. raionakh SSSR:* 180 (1985): E. chankoana.
nothosubsp. pseudovirgata *(Schur) R. Govaerts in R. Govaerts, D.G. Frodin & A. Radcliffe-Smith, World Checklist & Bibliogr.* Euphorbiaceae, 2: 723 (2000): E. virgata nothovar. pseudovirgata.
exserta *(Small) W.C. Coker, Pl. Life Hartsville:* 88 (1912): Tithymalopsis exserta.
fallax *C.F. Hagenbach, Tent. Fl. Basil.,* 1: 435 (1821)—Switzerland.
flavicoma DC.
subsp. giselae *J. Simon Pallisé in An. Jard. Bot. Madrid,* 55(1): 199 (1997)—France, Spain.
forolensis *L.E. Newton in Cact. Succ. J. (U.S.A.),* 67(6): 358 (1995)—Kenya.
fosbergii *(J. Florence) R. Govaerts in R. Govaerts, D.G. Frodin & A. Radcliffe-Smith, World Checklist & Bibliogr.* Euphorbiaceae, 2: 730 (2000): Chamaesyce fosbergii.
francoisii Leandri
var. crassicaulis *W. Rauh in Euphorbia J.,* 10: 183 (1996)—Madagascar.
gamkensis *J.G. Marx in Cact. Succ. J. (U.S.A.),* 71(1): 38 (1999)—Cape Province.
gebelica *S. Brullo in Boll. Accad. Gioenia Sci. Nat. Catania,* 27(347): 258 (1994 publ. 1996)—Libya.
gentryi *V.W. Steinmann & T.F. Daniel in Madroño,* 42(4): 450 (1995 publ. 1996)—Mexico Northwest (Sonora, Sinaloa).
gossypina Pax
subsp. mangulensis *S. Carter in Kew Bull.,* 54(4): 959 (1999 publ. 2000)—Zimbabwe.
gradyi *V.W. Steinmann & A. Ramírez-Roa in Haseltonia,* 6: 102 (1998 publ. 1999)—Mexico Southwest (Oaxaca), Mexico Central (Puebla).
gumaroi *J. Meyrán García in Cact. Suc. Mex.,* 45(3): 68 (2000)—Mexico Northeast (Hidalgo).
hsinchuensis *(S.C. Lin & Chaw) C.Y. Wu & J.S. Ma in Collect. Bot.,* 21: 106 (1992): Chamaesyce hsinchuensis.†

EUPHORBIA:-
illirica Lam.
subsp. semivillosa *(Prokh.) R. Govaerts in R. Govaerts, D.G. Frodin & A. Radcliffe-Smith, World Checklist & Bibliogr.* Euphorbiaceae, 2: 755 (2000): Tithymalus semivillosus.
jenisseiensis *K.S. Baikov in Fl. Sibir. (Geraniac.-Cornac.),* 10: 47 (1996)—Krasnoyarsk.
kamponii *W. Rauh & H. Petignat in Brit. Cact. Succ. J.,* 13(4): 131 (1995)—Madagascar.
kimmerica *Lipsky ex A.A. Grossheim, Fl. Kavkaza,* 3: 39 (1932)—Caucasus.
kirimzjulica *N.V. Stepanov, Florogenet. Analiz,* 1: 84 (1994)—Krasnoyarsk.
korshinskyi *D.V. Geltman in Fl. Vostochnoĭ Evropy̆,* 9: 281 (1996)—Russia East.
labatii *W. Rauh & M. Bardot-Vaucoulon in Succulentes,* 1999(4): 3 (1999)—Madagascar.
lancasteriana *A. Radcliffe-Smith in Kew Bull.,* 54(1): 227 (1999)—China South-central (Sichuan).
leistneri *R.H. Archer in S. Afr. J. Bot.,* 64(4): 258 (1998)—Namibia.
lenensis *K.S. Baikov in Novosti Sist. Vyssh. Rast.,* 30: 116 (1996); *et in Fl. Sibir. (Geraniac.-Cornac.),* 10: 50 (1996)—Irkutsk.
lenewtonii *S. Carter in Cact. Succ. J. (U.S.A.),* 72(4): 186 (2000)—Tanzania.
leucodendron Drake
subsp. oncoclada *(Drake) W. Rauh & Koutnik in W. Rauh, Succ. & Xeroph. Pl. Madagascar,* 2: 106 (1998): E. oncoclada.
leucophylla Benth.
subsp. comcaacorum *V.W. Steinmann & R.S. Felger in Aliso,* 16(1): 43 (1997)—Mexico Northwest (Sonora).
limpopoana *L.C. Leach ex S. Carter in Kew Bull.,* 54(4): 960 (1999 publ. 2000)—Zimbabwe, Botswana, Mozambique, Transvaal.
linguiformis McVaugh
var. actinadenia *(McVaugh) A. Radcliffe-Smith & R. Govaerts in Kew Bull.,* 51(3): 542 (1996), isonym; *see Oudejans, World cat. sp. names publ. in tribe Euphorbieae,* Cum. supp. 1: 12 (1993): Chamaesyce linguiformis var. actinadenia.
lukoseana *S. Carter in Cact. Succ. J. (U.S.A.),* 72(4): 188 (2000)—Tanzania.
mangelsdorffii *W. Rauh in Kakt. And. Sukk.,* 49(2): 44 (1998)—Madagascar.
maritae *W. Rauh in Brit. Cact. Succ. J.,* 17(4): 217 (1999)—Tanzania.
maritima *T. Nuttall in Trans. Amer. Philos. Soc., n.s.,* 5: 171 (1835)—New Jersey.
martinae *W. Rauh in Succulentes,* 1999(3): 12 (1999)—Madagascar.
meenae *S. Carter in Cact. Succ. J. (U.S.A.),* 72(4): 210 (2000)—India (Tamil Nadu, Andhra Pradesh, Karnataka).
meloformis Aiton
f. falsa *(N.E. Br.) G. Marx in Euphorbiaceae Study Group Bull.,* 12(1): 32 (1999): E. falsa.
f. magna *R.A. Dyer ex G. Marx in Euphorbiaceae Study Group Bull.,* 12(1): 13 (1999)—Cape Province.
*subsp.*valida *(N.E. Br.) G.D. Rowley in Euphorbiaceae Study Group Bull.,* 11(3): 97 (1998): E. valida.
minuta Loscos & J. Pardo
subsp. moleroi *P. Montserrat Recoder & J.V. Ferrández Palacio in An. Jard. Bot. Madrid,* 56(2): 382 (1998)—Spain.
montereyana *C.F. Millspaugh in Bull. Torrey Bot. Club,* 16(3): 66 (1889)—Mexico.
neobosseri (Rauh) Rauh
var. itampolensis *W. Rauh in Trop. Subtrop. Pflanzenwelt,* 100: 8 (1999)—Madagascar.
nicaeensis All.
var. lasiocarpa *(Boiss.) A. Radcliffe-Smith & R. Govaerts in Kew Bull.,* 51(3): 542 (1996): E. glareosa var. lasiocarpa.
obesa Hook. f.
subsp. symmetrica *(A.C. White, R.A. Dyer & B. Sloane) G.D. Rowley in Euphorbiaceae Study Group Bull.,* 11(3): 97 (1998): E. symmetrica.
obtusifolia Poir.
var. wildpretii *J. Molero & A.M. Rovira in Taxon,* 47(2): 325 (1998)—Canary Is.
occidentaustralica *A. Radcliffe-Smith & R. Govaerts in Kew Bull.,* 51(3): 542 (1996), nom. nov.: E. clementii *Domin.*

EUPHORBIA:-
pallasii Turcz.
var. komaroviana *(Prokh.) S.Z. Liu in Clavis Pl. Chinae Bor.-Or. ed. 2 (ed. P.Y. Fu):* 385 (1995), isonym; *see Y.C. Zhu in Pl. Medic. Chinae Bor.-Orient. (ed. Y.C. Zhu):* 682 (1989): E. komaroviana.
paniculata Desf.
subsp. monchiquensis *(Franco & P. Silva) J. Vicens, J. Molero & C. Blanché in Candollea,* 51(1): 82 (1996): E. monchiquensis.
subsp. welwitschii *(Boiss. & Reut.) J. Vicens, J. Molero & C. Blanché in Candollea,* 51(1): 81 (1996): E. welwitschii.
papillosa A. St.-Hil.
var. cabrerae *R. Subils in Zuloaga & Morrone (eds.), Cat. Pl. Vasc. Rep. Arg. 2 (Monogr. Syst. Bot.* Missouri Bot. Gard. 74): 1246 (1999), nom. nov.: Tithymalus erythrorrhizus *Klotzsch & Garcke.*
× paradoxa *(Schur) Simonk. in A. Kern., Sched. Fl. Exs. Austro-Hung.,* 2: 49 (1882 publ. 1883): E. esula e. paradoxa.†
nothosubsp. gusuleacii *(Prodán & Soran) R. Govaerts in R. Govaerts, D.G. Frodin & A. Radcliffe-Smith, World Checklist & Bibliogr.* Euphorbiaceae, 2: 797 (2000): E. × gusuleacii.
nothovar. podolica *(Błocki ex Borza) R. Govaerts in R. Govaerts, D.G. Frodin & A. Radcliffe-Smith, World Checklist & Bibliogr.* Euphorbiaceae, 2: 797 (2000), basionym not validly publ.: E. × podolica.
pedroi *J. Molero & A. Rovira in An. Jard. Bot. Madrid,* 55(1): 198 (1997)—Portugal.
pervittata *S. Carter in Cact. Succ. J. (U.S.A.),* 72(4): 193 (2000)—Kenya.
pilosa L.
subsp. ojensis *N.V. Stepanov, Florogenet. Analiz,* 1: 98 (1994)—Krasnoyarsk.
pionosperma *V.W. Steinmann & R.S. Felger in Aliso,* 16(1): 46 (1997)—Mexico Northwest (Sonora), Mexico Northeast (Chihuahua).
plenispina *S. Carter in Kew Bull.,* 54(4): 964 (1999 publ. 2000)—Mozambique.
polyphylla *Engelm. ex J.M. Holzinger in Contrib. U.S. Natl. Herb.,* 1(6): 215 (1892)—Florida.†
proballyana L.C. Leach
var. multangula *S. Carter in Cact. Succ. J. (U.S.A.),* 72(4): 190 (2000)—Tanzania.
prostrata Aiton
var. caudirhiza *F.R. Fosberg in Micronesica,* 25(2): 189 (1992)—Niue, Fiji.
rablensis *Wulfen ex L.C. von Vest in Syll. Pl. Nov.,* 2: 223 (1828)—Austria.
retrospina *W. Rauh & R. Gérold in Succulentes,* 2000(4): 6 (2000)—Madagascar.
rosularis *Al.A. Fedorov in Bot. Mater. Gerb. Bot. Inst. Komarova Akad. Nauk SSSR,* 9: 50 (1941)—Uzbekistan.
rumelica *E. Formánek in Verh. Naturf. Vereins Brünn,* 38: 186 (1900), nom. nov.: E. esuloides *Velen.*
sachetiana *(J. Florence) R. Govaerts in R. Govaerts, D.G. Frodin & A. Radcliffe-Smith, World Checklist & Bibliogr.* Euphorbiaceae, 2: 823 (2000): Chamaesyce sachetiana.
salsuginosa *(McVaugh) A. Radcliffe-Smith & R. Govaerts in Kew Bull.,* 51(3): 542 (1996): Chamaesyce salsuginosa.
saurica *K.S. Baikov in Bot. Zhurn.,* 84(3): 108 (1999)—Kazakhstan.
seibanica *J.J. Lavranos & A.N. Al-Gifri in Cact. Succ. J. (U.S.A.),* 71(3): 136 (1999)—South Yemen.
smirnovii *D.V. Geltman in Bot. Zhurn.,* 81(11): 102 (1996), nom. nov.: E. petrophila var. armena *Boiss.*
songweana *S. Carter in Cact. Succ. J. (U.S.A.),* 72(4): 191 (2000)—Tanzania.
specksii *W. Rauh in Cact. Succ. J. (U.S.A.),* 72(5): 247 (2000)—Tanzania.
sulcata Lens ex Loisel.
var. maroccana *J. Molero, A.M. Rovira & J. Vicens in An. Jard. Bot. Madrid,* 54(1): 227 (1996)—Morocco.
sumati *S. Carter in Cact. Succ. J. (U.S.A.),* 72(4): 185 (2000)—Kenya.

EUPHORBIA:-
suppressa *J.G. Marx in Cact. Succ. J. (U.S.A.)*, 71(1): 33 (1999)—Cape Province.
suzannae-marnierae *W. Rauh & H. Pétignat in Succulentes*, 1996(4): 10 (1996), as 'suzannae-marnieriae—Madagascar.
tenax *W.J. Burchell, Trav. S. Africa*, 1: 219 (1822)—Cape Province.
tlapanensis *Hargreaves in Euphorbiaceae Study Group Bull.*, 8(3): 100–103 (1995), without latin descr. or type; *cf. Repert. Pl. Succ. (I.O.S.)*, 46: 20 (1995 publ. 1996)—Locality not stated.
truncata *(Pers.) J.C. Loudon, Hort. Brit., ed. 2:* 192 (1832): E. serrata var. truncata.
umfoloziensis *R.G. Peckover in Aloe*, 28(2): 37 (1991)—Natal.
volhynica *Besser ex M. Raciborski in Szafer, Fl. Polska*, 2: 107 (1921)—Poland.
waringiae *W. Rauh & R. Gerold in Succulentes*, 1998(2): 7, 6 (1998)—Madagascar.
watanabei Makino
subsp. minamitanii *T. Kurosawa, S. Serizawa & H. Ohashi in Acta Phytotax. Geobot.*, 47(1): 14 (1996)—Japan.
wilsonii *V. Vlk in Kaktusy (Brno)*, 33(1): 23 (1997), homonym; *see Correll in Sida*, 8(4): 319 (1980)—Ethiopia.
yaroslavii *P.P. Poljakov in Bot. Mater. Gerb. Bot. Inst. Komarova Akad. Nauk SSSR*, 21: 484 (1961)—Kazakhstan.

EUPHRASIA (Scrophulariac.).
chumbica *R.R. Mill in Edinburgh J. Bot.*, 57(3): 415 (2000)—Xizang, Sikkim, Bhutan.
corcontica *(Smejkal) Smejkal & M. Dvořáková in Preslia*, 71(1–2): 33 (1999): E. micrantha var. corcontica.
fragosa *W.R. Barker in J. Adelaide Bot. Gard.*, 17: 217 (1996)—Tasmania.
melanosticta *R.R. Mill in Edinburgh J. Bot.*, 57(3): 417 (2000)—Bhutan, Sikkim, Xizang.
officinalis L.
subsp. kerneri *(Wettst.) Eb. Fischer in Feddes Repert.*, 108(1–2): 112 (1997): E. kerneri.
petiolaris Wettst.
var. thuschethica *Ts.N. Gviniaschvili in Fl. Gruzii, ed. 2*, 12: 331 (1999)—Transcaucasus (Gruziya).
ruptura *W.R. Barker in Telopea*, 7(3): 227 (1997)—New South Wales.
stricta D. Wolff ex J.F. Lehm.
var. pectinata *(Ten.) O. de Bolòs & J. Vigo, Fl. Països Catalans*, 3: 477 (1995): E. pectinata.
var. puberula *(Jord. ex Wettst.) O. de Bolòs & J. Vigo, Fl. Països Catalans*, 3: 477 (1995): E. pectinata var. puberula.
transmorrisonensis Hayata
var. durietziana *(Ohwi) T.C. Huang & M.J. Wu in Fl. Taiwan, ed. 2*, 4: 595 (1998): E. durietziana.
tricuspidata L.
subsp. cuspidata *(Host) D. Hartl in G. Hegi, Illustr. Fl. Mitteleur., ed. 2*, 6(1): 373 (1972): E. cuspidata.

EUPLASSA (Proteac.).
semicostata *V. Plana in Kew Bull.*, 53(4): 969 (1998)—Brazil Southeast (Minas Gerais).

EURYA (Theac.).
glaberrima Hayata
var. taitungensis *(C.E. Chang) Yuen P. Yang & Sheng Z. Yang in Taiwania*, 42(2): 87 (1997): E. taitungensis.
groffii Merr.
var. zhengkangensis *T.L. Ming in Acta Bot. Yunnanica*, 19(3): 223 (1997)—China South-central (Yunnan).
loquaiana Dunn
var. gnaphalocarpa *(Hayata) S.S. Ying in Mem. Coll. Agric. Nation. Taiwan Univ.*, 29(2): 62 (1989): E. gnaphalocarpa.
nanjenshanensis *(C.F. Hsieh, L.K. Ling & Sheng Z. Yang) Sheng Z. Yang & S.Y. Lu in Taiwania*, 42(2): 86 (1997): E. nitida var. nanjenshanensis.
nitida Korth.
var. nanjenshanensis *C.F. Hsieh, L.K. Ling & Sheng Z. Yang in Fl. Taiwan, ed. 2*, 2: 682 (1996)—Taiwan.

EURYA:-
obtusifolia Hung T. Chang
var. aurea *(H. Lév.) T.L. Ming in Fl. Yunnanica*, 8: 371 (1997): Rapanea aurea.
stenophylla Merr.
var. pubescens *(Hung T. Chang) T.L. Ming in Fl. Yunnanica*, 8: 372 (1997): E. stenophylla f. pubescens.
wuliangshanensis *T.L. Ming in Acta Bot. Yunnanica*, 19(3): 223 (1997)—China South-central (Yunnan).

EURYCARPUS (Crucif.).
marinellii *(Pamp.) I.A. Al-Shehbaz & G. Yang in Novon*, 10(4): 347 (2000): Braya marinellii.

EURYCHORDA *B.G. Briggs & L.A.S. Johnson in Telopea*, 7(4): 357 (1998). *RESTIONACEAE*
complanata *(R. Br.) B.G. Briggs & L.A.S. Johnson in Telopea*, 7(4): 359 (1998): Restio complanatus.

EURYDOCHUS (Compos.).
cortesii *S. Díaz Piedrahita in Revis. tribus Barnadesieae y Mutisieae Fl. Colombia (Monogr. Jard. Bot.* J.C. Mutis, 1): 68 (1993)—Colombia.

EURYOPS (Compos.).
jaberiana *S. Abedin & S.A. Chaudhary in Fl. Kingdom Saudi Arabia*, 2(3): 191, 418 (2000)—Saudi Arabia.

EUSCAPHIS (Staphyleac.).
japonica (Thunb.) Kanitz
f. lanata *(Masam.) K. Iwatsuki & H. Ohba in Fl. Japan*, 2c: 104 (1999): E. japonica var. lanata.

EUSTACHYS (Gramin.).
bahiensis (Steud.) Herter
var. robusta *A.M. Molina in Candollea*, 51(1): 260 (1996)—Brazil South (Paraná, Rio Grande do Sul).
paranensis *A.M. Molina in Candollea*, 51(1): 234 (1996)—Brazil South (Paraná).
swalleniana *A.M. Molina in Candollea*, 51(1): 232 (1996)—Brazil South (Rio Grande do Sul, Paraná), Uruguay.

EUSTERALIS (Labiat.).
helferi *(Hook. f.) G. Murata in Southeast Asian Stud.*, 14(2): 185 (Sept. 1976), as 'halferi', isonym; *see Panigrahi in Phytologia*, 32(6): 477 (Jan. 1976): Dysophylla helferi.
linearis *(Benth.) G. Murata in Southeast Asian Stud.*, 14(2): 185 (Sept. 1976), isonym; *see Panigrahi in Phytologia*, 32(6): 476 (Jan. 1976): Dysophylla linearis.
peguana *(Prain) G. Murata in Southeast Asian Stud.*, 14(2): 185 (Sept. 1976), isonym; *see Panigrahi in Phytologia*, 32(6): 477 (Jan. 1976): Dysophylla peguana.
pentagona *(C.B. Clarke ex Hook. f.) G. Murata in Southeast Asian Stud.*, 14(2): 185 (Sept. 1976), isonym; *see Panigrahi in Phytologia*, 32(6): 477 (Jan. 1976): Dysophylla pentagona.
quadrifolia *(D. Don) G. Murata in Southeast Asian Stud.*, 14(2): 185 (1976): Mentha quadrifolia.
stellata *(Lour.) G. Murata in Southeast Asian Stud.*, 14(2): 186 (Sept. 1976), isonym; *see Panigrahi in Phytologia*, 32(6): 474 (Jan. 1976): Mentha stellata.

EUTERPE (Palmae).
catinga Wallace
var. roraimae *(Dammer) A. Henderson & G. Galeano in Fl. Neotrop.*, 72: 28 (1996): E. roraimae.

EUTHAMIA *(Nutt.) Cassini in Dict. Sci. Nat.*, 37: 459 (1825): Solidago subgen. Euthamia. (Compos.).
graminifolia *(L.) T. Nuttall in Trans. Amer. Philos. Soc.*, n.s., 7: 325 (1840): Chrysocoma graminifolia.

EVAX (Compos.).
petro-ianii *(Rita & Dittrich) J. Holub in Preslia*, 70(2): 104 (1998): Filago petro-ianii.

EVERISTIA *S.T. Reynolds & R.J.F. Henderson in Austrobaileya*, 5(2): 354 (1999). *RUBIACEAE*

EVERISTIA:-
vacciniifolia *(F. Muell.) S.T. Reynolds & R.J.F. Henderson in Austrobaileya*, 5(2): 355 (1999): Canthium vacciniifolium.
f. crassa *S.T. Reynolds & R.J.F. Henderson in Austrobaileya*, 5(2): 357 (1999)—Queensland.
var. nervosa *S.T. Reynolds & R.J.F. Henderson in Austrobaileya*, 5(2): 357 (1999), nom. nov.: Canthium microphyllum *F. Muell.*

EVOLVULUS (Convolvulac.).
lithospermoides Mart.
var. martii *(Meisn.) R. Simão-Bianchini in Bol. Bot. (São Paulo)*, 16: 145 (1997): E. martii.

EVRARDIANTHE (Orchidac.).
asraoa *(J. Joseph & Abbar.) C. Sathish Kumar in C. Sathish Kumar & K.S. Manilal, Cat. Indian Orchids:* 94 (1994): Evrardia asraoa.

EXALARIA *L.A. Garay & G.A. Romero-Gonzalez in Harvard Pap. Bot.*, 4(2): 479 (1999). *ORCHIDACEAE*
parviflora *(C. Presl) L.A. Garay & G.A. Romero-Gonzalez in Harvard Pap. Bot.*, 4(2): 480 (1999): Ophrys parviflora.

EXARRHENA *(A.DC.) O.D. Nikiforova in Turczaninowia*, 3(1): 6 (2000), with incorrect basionym, name already publ.; *see R.Br., Prodr.:* 495 (1810): Basionym not stated. (Boraginac.).
albiflora *(Banks & Sol. ex Hook. f.) O.D. Nikiforova in Turczaninowia*, 3(1): 12 (2000): Myosotis albiflora.

EXCENTRADENIA *W.R. Anderson in Contrib. Univ. Mich. Herb.*, 21: 29 (1997). *MALPIGHIACEAE*
adenophora *(Sandwith) W.R. Anderson in Contrib. Univ. Mich. Herb.*, 21: 31 (1997): Hiraea adenophora.
boliviana *W.R. Anderson in Contrib. Univ. Mich. Herb.*, 21: 34 (1997)—Bolivia.
primaeva *(W.R. Anderson) W.R. Anderson in Contrib. Univ. Mich. Herb.*, 21: 34 (1997): Hiraea primaeva.
propinqua *(W.R. Anderson) W.R. Anderson in Contrib. Univ. Mich. Herb.*, 21: 35 (1997): Hiraea propinqua.

EXCOECARIA (Euphorbiac.).
formosana Hayata
var. daitoinsularis *(Hatus.) Hatus. ex K. Simabuku in Island Stud. Okinawa*, 7: 56 (1989): E. orientalis var. daitoinsularis.
laotica *(Gagnep.) H.-J. Esser in Nordic J. Bot.*, 16(6): 580 (1996 publ. 1997): Glyphostylus laoticus.
yunnanensis *Y.H. Li & J.C. Xu in Acta Phytotax. Sin.*, 34(3): 336 (1996)—China South-central (Yunnan).

EXOCHORDA (Rosac.).
racemosa (Lindl.) Rehder
subsp. giraldii *(Hesse) F.Y. Gao & L.J.G. van der Maesen in F.Y. Gao, Exochorda: five species or one?:* 93 (1998): E. giraldii.
subsp. serratifolia *(S. Moore) F.Y. Gao & L.J.G. van der Maesen in F.Y. Gao, Exochorda: five species or one?:* 96 (1998): E. serratifolia.

EXOGONIUM (Convolvulac.).
steudelii *(Millsp.) J. Fournet, Fl. Illustr. Phan. Guadeloupe & Martinique:* 1273 (1978), without basionym ref.: Ipomoea steudelii.

EYSENHARDTIA (Legumin.).
orthocarpa (A. Gray) S. Watson
var. tenuifolia *J.M. Lang in Iowa State J. Res.*, 56(4): 400 (1982)—Mexico Northwest (Sonora).

F

FABERIA (Compos.).
tsiangii *(C.C. Chang) C. Shih in Fl. Reipubl. Popul. Sin.*, 80(1): 166 (1997): Hieracium tsiangii.

FABERIOPSIS *C. Shih & Y.L. Chen in Acta Phytotax. Sin.,* 34(4): 438 (1996). *COMPOSITAE*
nanchuanensis *(C. Shih) C. Shih & Y.L. Chen in Acta Phytotax. Sin.,* 34(4): 439 (1996): Faberia nanchuanensis.

FACHEIROA (Cactac.).
cephaliomelana
subsp. estevesii *(P.J. Braun) N.P. Taylor & D.C. Zappi in Cactaceae Consensus Iritiatives,* 3: 7 (1997): F. estevesii.

FADOGIA (Rubiac.).
chlorantha K. Schum.
var. thamnus *(K. Schum.) B. Verdcourt in Fl. Zambes.,* 5(2): 278 (1998): F. thamnus.
luangwae *B. Verdcourt in Fl. Zambes.,* 5(2): 274 (1998)—Zambia.
schmitzii *B. Verdcourt in Fl. Zambes.,* 5(2): 279 (1998)—Zambia, Zaire.
tomentosa De Wild.
var. calvescens *(Verdc.) B. Verdcourt in Fl. Zambes.,* 5(2): 272 (1998): F. flaviflora var. calvescens.
var. flaviflora *(Robyns) B. Verdcourt in Fl. Zambes.,* 5(2): 273 (1998): F. flaviflora.

FADOGIELLA (Rubiac.).
rogersii *(Wernham) D.M. Bridson in Kew Bull.,* 51(2): 351 (1996): Fadogia rogersii.

FAGERLINDIA (Rubiac.).
canthioides *(Champ. ex Benth.) Tirveng. ex C.E. Ridsdale in Blumea,* 41(1): 176 (1996): Randia canthioides.

FAGONIA (Zygophyllac.).
californica Benth.
var. laevis *(Standl.) L.D. Benson in L.D. Benson & R.A. Darrow, Trees & Shrubs of Southwest. Deserts (ed. 3):* 119 (1981): F. laevis.
var. longipes *(Standl.) L.D. Benson in L.D. Benson & R.A. Darrow, Trees & Shrubs of Southwest. Deserts (ed. 3):* 120 (1981): F. longipes.

FAGOPYRUM (Polygonac.).
caudatum *(Sam.) A.J. Li in Fl. Reipubl. Popul. Sin.,* 25(1): 117 (1998): Polygonum caudatum.
gracilipes *(Hemsl.) Dammer in Bot. Jahrb. Syst.,* 29: 315 (1900): Polygonum gracilipes.
megaspartanium *Q.F. Chen in Bot. J. Linn. Soc.,* 130(1): 62 (1999)—China South-central (Sichuan).
pilus *Q.F. Chen in Bot. J. Linn. Soc.,* 130(1): 62 (1999)—Xizang.
zuogongense *Q.F. Chen in Bot. J. Linn. Soc.,* 130(1): 62 (1999)—Xizang.

FAGRAEA (Loganiac.).
acutibracteata *K.M. Wong & J.B Sugau in Sandakania,* 8: 49 (1996)—Borneo (Sarawak).
belukar *K.M. Wong & J.B. Sugau in Sandakania,* 8: 15 (1996)—Borneo (Sabah, Sarawak, Kalimantan).
collina *K.M. Wong & J.B. Sugau in Sandakania,* 8: 19 (1996)—Borneo (Sabah, Sarawak).
dulitensis *K.M. Wong & J.B. Sugau in Sandakania,* 8: 59 (1996)—Borneo (Sarawak).
floribunda *K.M. Wong & J.B. Sugau in Sandakania,* 8: 62 (1996)—Borneo (Sarawak).
havilandii *K.M. Wong & J.B. Sugau in Sandakania,* 8: 64 (1996)—Borneo (Sarawak).
iliasii *K.M. Wong & J.B. Sugau in Sandakania,* 8: 66 (1996)—Borneo (Sarawak).
kalimantanensis *(Leenh.) K.M. Wong & J.B. Sugau in Sandakania,* 8: 69 (1996): F. tacapala subsp. kalimantanensis.
kinabaluensis *K.M. Wong & J.B. Sugau in Sandakania,* 8: 69 (1996)—Borneo (Sabah, Brunei).
kuminii *K.M. Wong & J.B. Sugau in Sandakania,* 8: 71 (1996)—Borneo (Sabah).
littoralis Blume
var. borneensis *K.M. Wong & J.B. Sugau in Sandakania,* 8: 73 (1996)—Borneo (Sabah, Sarawak, Brunei, Kalimantan).
longipetiolata *K.M. Wong & J.B. Sugau in Sandakania,* 8: 75 (1996)—Borneo (Sarawak).

FAGRAEA:-
megalantha *K.M. Wong & J.B. Sugau in Sandakania,* 8: 77 (1996)—Borneo (Sarawak).
montana *K.M. Wong & J.B. Sugau in Sandakania,* 8: 31 (1996)—Borneo (Sabah, Sarawak).
oreophila *K.M. Wong & J.B. Sugau in Sandakania,* 8: 78 (1996)—Borneo (Sabah, Sarawak).
peninsularis *K.M. Wong & J.B. Sugau in Sandakania,* 8: 33 (1996)—Malaya (Peninsular Malaysia, Singapore), Sumatera.
philippinensis *K.M. Wong & J.B. Sugau in Sandakania,* 8: 35 (1996)—Philippines, Borneo (Sarawak, Kalimantan).
rarissima *K.M. Wong & J.B. Sugau in Sandakania,* 8: 82 (1996)—Borneo (Sarawak).
renae *K.M. Wong & J.B. Sugau in Sandakania,* 8: 82 (1996)—Borneo (Sarawak, Sabah).
rugulosa *K.M. Wong & J.B. Sugau in Sandakania,* 8: 22 (1996)—Borneo (Sarawak, Sabah, Brunei).
stonei *K.M. Wong & J.B. Sugau in Sandakania,* 8: 88 (1996)—Borneo (Sabah, Sarawak).
tuyukii *K.M. Wong & J.B. Sugau in Sandakania,* 8: 90 (1996)—Borneo (Sabah).
volubilis Wall.
var. microcalyx *K.M. Wong & J.B. Sugau in Sandakania,* 8: 43 (1996)—Borneo (Kalimantan, Sarawak, Brunei).

FAGUS (Fagac.).
hayatae Palib. ex Hayata
subsp. pashanica *(C.C. Yang) R. Peter ex J.Q. Li in J. Wuhan Bot. Res.,* 16(3): 243 (1998): F. pashanica.
japonica Maxim.
var. multinervis *(Nakai) Y.N. Lee ex R. Govaerts & D.G. Frodin, World Checklist & Bibliogr. Fagales:* 144 (1998): F. multinervis.

FALCATARIA *(I.C. Nielsen) R.C. Barneby & J.W. Grimes, Silk Tree, Guanacaste, Monkey's Earring (Mem. New York Bot. Gard., 74(1)):* 254 (1996): Paraserianthes sect. Falcataria.
moluccana *(Miq.) R.C. Barneby & J.W. Grimes, Silk Tree, Guanacaste, Monkey's Earring (Mem. New York Bot. Gard., 74(1)):* 255 (1996): Albizia moluccana.

FALCATIFOLIUM (Podocarpac.).
falcatiforme (Parl.) de Laub.
var. kinabaluensis *J. Silba in J. Int. Conifer Preserv. Soc.,* 7(1): 24 (2000)—Borneo (Sabah).

FALLOPIA (Polygonac.).
cynanchoides (Hemsl.) Haraldson
var. glabriuscula *(A.J. Li) A.J. Li in Fl. Reipubl. Popul. Sin.,* 25(1): 104 (1998): Polygonum cynanchoides var. glabriusculum.
denticulata *(C.C. Huang) J. Holub in Preslia,* 70(2): 104 (21 July 1998): Polygonum denticulatum.
denticulata *(C.C. Huang) A.J. Li in Fl. Reipubl. Popul. Sin.,* 25(1): 103 (August 1998), isonym; *see Holub in Preslia,* 70(2): 104 (21 July 1998): Polygonum denticulatum.
dumetorum (L.) Holub
var. pauciflora *(Maxim.) A.J. Li in Fl. Reipubl. Popul. Sin.,* 25(1): 100 (1998): Polygonum pauciflorum.
forbesii *(Hance) K. Yonekura & Hiroyoshi Ohashi in J. Jap. Bot.,* 72(3): 158 (1997): Polygonum forbesii.
japonica (Houtt.) Ronse Decr.
var. hachidyoensis *(Makino) K. Yonekura & Hiroyoshi Ohashi in J. Jap. Bot.,* 72(3): 158 (1997): Polygonum hachidyoense.
var. uzenensis *(Honda) K. Yonekura & Hiroyoshi Ohashi in J. Jap. Bot.,* 72(3): 158 (1997): Reynoutria japonica var. uzenensis.
koreana *B.U. Oh & J.G. Kim in Korean J. Pl. Taxon.,* 26(4): 258 (1996)—South Korea.
multiflora (Thunb.) Haraldson
var. ciliinervis *(Nakai) K. Yonekura & Hiroyoshi Ohashi in J. Jap. Bot.,* 72(3): 158 (1997): Pleuropterus ciliinervis.
var. ciliinervis *(Nakai) A.J. Li in Fl. Reipubl. Popul. Sin.,* 25(1): 103 (1998), as 'ciliinerve', isonym; *see Yonek. & H. Ohashi in J. Jap. Bot.,* 72(3): 158 (1997): Pleuropterus ciliinervis.

FALLOPIA:-
multiflora
var. hypoleuca *(Ohwi) K. Yonekura & Hiroyoshi Ohashi in J. Jap. Bot.,* 72(3): 158 (1997): Polygonum hypoleucum.
sachalinensis (F. Schmidt) Ronse Decr.
var. intermedia *(Tatew.) K. Yonekura & Hiroyoshi Ohashi in J. Jap. Bot.,* 72(3): 159 (1997): Polygonum sachalinense var. intermedium.

FARAMEA (Rubiac.).
ampla *C.M. Taylor in Novon,* 9(3): 437 (1999)—Colombia.
angusta *C.M. Taylor in Fl. Ecuador,* 62(162:3): 276 (1999)—Ecuador.
calimana *C.M. Taylor in Novon,* 9(3): 437 (1999)—Colombia.
calyptrata *C.M. Taylor in Fl. Ecuador,* 62(162:3): 280 (1999)—Ecuador, Colombia.
coffeoides *C.M. Taylor in Fl. Ecuador,* 62(162:3): 284 (1999)—Ecuador, Colombia.
correae *C.M. Taylor in Novon,* 6(3): 298 (1996)—Panama.
cupheoides *C.M. Taylor in Fl. Ecuador,* 62(162:3): 285 (1999)—Ecuador, Peru.
dichotoma *K. Schum. ex M. Gomes in Reserva Ecol. Macaé de Cima: Aspectos Flor. Espéc. Vasc.,* 2: 413 (1996)—Brazil Southeast (Rio de Janeiro).
monsalveae *C.M. Taylor in Novon,* 9(3): 439 (1999)—Colombia.
papirifolia *(Standl. ex Steyerm.) C.M. Taylor in Novon,* 6(2): 210 (1996): Cephaelis papirifolia.
permagnifolia *Dwyer ex C.M. Taylor in Novon,* 6(3): 298 (1996)—Costa Rica.
sanblasensis *C.M. Taylor in Novon,* 6(3): 300 (1996)—Panama.
uncinata *C.M. Taylor in Fl. Ecuador,* 62(162:3): 309 (1999)—Ecuador, Peru.
verticillata *C.M. Taylor in Fl. Ecuador,* 62(162:3): 313 (1999)—Colombia, Ecuador.

FARGESIA (Gramin.).
alpina *J.R. Xue & C.M. Hui in Bull. Bot. Res. North-East. Forest. Univ.,* 18(3): 271 (1998)—China South-central (Yunnan).
aurita *J.R. Xue & C.M. Hui in Bull. Bot. Res. North-East. Forest. Univ.,* 18(3): 258 (1998), homonym; *see T.P. Yi in J. Bamboo Res.,* 4(2): 23 (1985)—China South-central (Yunnan).
brevistipedis *T.P. Yi in J. Bamboo Res.,* 19(1): 14 (2000)—China South-central (Sichuan).
contracta
f. fugonensis *J.R. Xue & J.K. Duan in Bull. Bot. Res. North-East. Forest. Univ.,* 18(3): 271 (1998)—China South-central (Yunnan).
hackelii *D. Ohrnberger, Bamboos of the World Intro. 2:* 10 (1996), nom. nov.: Arundinaria parvifolia *Hack. ex Keng.*
incrassata *T.P. Yi in J. Bamboo Res.,* 19(1): 16 (2000)—China South-central (Sichuan).
longiuscula *(J.R. Xue & Y.Y. Dai) D. Ohrnberger, Bamboos of the World Intro. 3:* 14 (1996): Sinarundinaria longiuscula.
macrophylla *J.R. Xue & C.M. Hui in Bull. Bot. Res. North-East. Forest. Univ.,* 18(3): 258 (1998)—China South-central (Yunnan).
nujiangensis *J.R. Xue & C.M. Hui in Bull. Bot. Res. North-East. Forest. Univ.,* 18(3): 261 (1998)—China South-central (Yunnan).
f. lanpingensis *J.R. Xue & H.R. Zhang in Bull. Bot. Res. North-East. Forest. Univ.,* 18(3): 263 (1998)—China South-central (Yunnan).
f. striata *J.R. Xue & C.M. Hui in Bull. Bot. Res. North-East. Forest. Univ.,* 18(3): 263 (1998)—China South-central (Yunnan).
ostrina *T.P. Yi in Acta Bot. Yunnanica,* 22(3): 251 (2000)—China South-central (Sichuan).
pachyclada *J.R. Xue & C.M. Hui in Bull. Bot. Res. North-East. Forest. Univ.,* 18(3): 266 (1998)—China South-central (Yunnan).
pallens *J.R. Xue & C.M. Hui in Bull. Bot. Res. North-East. Forest. Univ.,* 18(3): 269 (1998)—China South-central (Yunnan).
sparsiflora *(Rendle) D. Ohrnberger, Bamboos of the World Intro. 3:* 14 (1996): Arundinaria sparsiflora.
stricta *J.R. Xue & C.M. Hui in Bull. Bot. Res. North-East. Forest. Univ.,* 18(3): 266 (1998)—China South-central (Yunnan).

FASCICULARIA (Bromeliac.).
bicolor (Ruiz & Pav.) Mez
subsp. canaliculata *E.C. Nelson & G. Zizka in New Plantsman,* 4(4): 238 (1997)—Chile South (Los Lagos).

FATSIA (Araliac.).
japonica (Thunb.) Decne. & Planch.
var. liukiuensis *Hatus. ex H. Ohba in Fl. Japan,* 2c: 262 (1999)—Nansei-Shoto.

FAUCARIA (Aizoac.).
felina (L.) Schwantes
subsp. britteniae *(L. Bolus) L.E. Groen in Bothalia,* 29(1): 41 (1999): F. britteniae.
subsp. tuberculosa *(Rolfe) L.E. Groen in Bothalia,* 29(1): 42 (1999): Mesembryanthemum tuberculosum.

FELICIA (Compos.).
abyssinica Sch. Bip. ex A. Rich.
var. neghelliensis *(Cufod.) Mesfin Tadesse in Compositae Newsl.,* 33: 25 (1999): F. neghelliensis.
var. schimperi *(Steud. & Hochst. ex Jaub. & Spach) Mesfin Tadesse in Compositae Newsl.,* 33: 24 (1999): F. schimperi.
flava *H.J. Beentje in Kew Bull.,* 54(1): 98 (1999)—Kenya, Tanzania.

FEMENIASIA (Compos.).
fruticosa *(Maire) D.P. Petit in Pl. Syst. Evol.,* 207(3–4): 197 (1997): Carthamus fruticosus.

FERNANDEZIA (Orchidac.).
robusta *(Schltr.) K. Senghas in Schlechter Orchideen,* I/B(31): 1926 (1995), homonym; *see Bateman in Bot. Mag.,* 92: t. 5592 (1866): Nasonia robusta.

FEROCACTUS (Cactac.).
subgen. Ferocarpus *A.B. Doweld in Sukkulentÿ (Moscow),* 1999(2): 38 (1999).
alamosanus
subsp. reppenhagenii *(G. Unger) N.P. Taylor in Cactaceae Consensus Initiatives,* 6: 15 (1998): F. reppenhagenii.
chrysacanthus
subsp. grandiflorus *(G.E. Linds.) N.P. Taylor in Cactaceae Consensus Initiatives,* 6: 15 (1998): F. fordii var. grandiflorus.
cylindraceus
subsp. lecontei *(Engelm.) N.P. Taylor in Cactaceae Consensus Initiatives,* 6: 16 (1998): Echinocactus lecontei.
subsp. tortulispinus *(H.E. Gates) N.P. Taylor in Cactaceae Consensus Initiatives,* 6: 16 (1998): F. tortulispinus.
emoryi
subsp. rectispinus *(Engelm.) N.P. Taylor in Cactaceae Consensus Initiatives,* 6: 16 (1998): Echinocactus emoryi var. rectispinus.
gracilis
subsp. coloratus *(H.E. Gates) N.P. Taylor in Cactaceae Consensus Initiatives,* 6: 16 (1998): F. coloratus.
subsp. gatesii *(G.E. Linds.) N.P. Taylor in Cactaceae Consensus Initiatives,* 6: 16 (1998): F. gatesii.
hamatacanthus
subsp. sinuatus *(A. Dietr.) N.P. Taylor in Cactaceae Consensus Initiatives,* 5: 13 (1998): Echinocactus sinuatus.
latispinus
var. recurvus *(Mill.) Lodé in Cact. Aventures,* 27: 29 (1995), without basionym ref; *cf. Repert. Pl. Succ. (I.O.S.),* 46: 10 (1995 publ. 1996): Cactus recurvus.
subsp. spiralis *(Karw. ex Pfeiff.) N.P. Taylor in Cactaceae Consensus Initiatives,* 5: 13 (1998): Echinocactus spiralis.
macrodiscus
subsp. septentrionalis *(J. Meyrán) N.P. Taylor in Cactaceae Consensus Initiatives,* 5: 13 (1998): F. macrodiscus var. septentrionalis.

FERRUMINARIA (Orchidac.).
lohokii *(J.J. Verm.) L.A. Garay & G.A. Romero-Gonzalez in Harvard Pap. Bot.,* 4(2): 481 (1999): Bulbophyllum lohokii.

FERULA (Umbellif.).
communis L.
subsp. cardonae *A. Sánchez-Cuxart & M. Bernal Cid in Acta Bot. Barcin.,* 45: 237 (1998)—Baleares.
subsp. catalaunica *(Pau) A. Sánchez-Cuxart & M. Bernal Cid in Acta Bot. Barcin.,* 45: 236 (1998): F. communis var. catalaunica.
var. microcarpa *Cauwet ex A. Sánchez-Cuxart & M. Bernal Cid in Acta Bot. Barcin.,* 45: 236 (1998), as comb. nov., basionym not validly publ., without latin descr. or type—Locality not stated.
czatkalensis *M.G. Pimenov in Opred. Rast. Sred. Azii,* 7: 376 (1983), with two types cited; *et in Bot. Zhurn.,* 85(1): 90 (2000)—Kirgizstan.
f. latisecta *M.G. Pimenov in Opred. Rast. Sred. Azii,* 7: 376 (1983)—Kirgizstan.
fedoroviorum *M.G. Pimenov in Bot. Zhurn.,* 85(1): 91 (2000)—Kirgizstan.
fontqueri *S.L. Jury in Lagascalia,* 18(2): 279 (1996)—Morocco.
hedgeana *M.G. Pimenov & E.V. Kljuykov in Edinburgh J. Bot.,* 53(2): 182 (1996)—Afghanistan.
loscosii (Lange) Willk.
subsp. cossoniana *(Batt. & Trab.) A.M. Cauwet-Marc & F.Z. Elalaoui-Faris in Acta Bot. Barcin.,* 45: 197 (1998), without basionym ref.: F. cossoniana.
ovczinnikovii *M.G. Pimenov in Opred. Rast. Sred. Azii,* 7: 377 (1983)—Uzbekistan.
seravschanica *M.G. Pimenov & J.V. Baranova in Opred. Rast. Sred. Azii,* 7: 376 (1983)—Tadzhikistan.

FERULAGO (Umbellif.).
idaea *N. Özhatay & E. Akalın in Bot. J. Linn. Soc.,* 133(4): 536 (2000)—Turkey.
macedonica *K. Micevski & E. Mayer in Acta Bot. Croat.,* 44: 77 (1985)—Yugoslavia.
ternatifolia *J.L. Solanas, M.B. Crespo & F. García-Martín in An. Jard. Bot. Madrid,* 58(1): 103 (2000)—Spain.

FESSIA *F. Speta in Phyton (Austria),* 38(1): 100 (1998). *HYACINTHACEAE*
bisotunensis *(Speta) F. Speta in Phyton (Austria),* 38(1): 101 (1998): Scilla bisotunensis.
furseorum *(Meikle) F. Speta in Phyton (Austria),* 38(1): 101 (1998): Scilla furseorum.
gorganica *(Speta) F. Speta in Phyton (Austria),* 38(1): 101 (1998): Scilla gorganica.
greilhuberi *(Speta) F. Speta in Phyton (Austria),* 38(1): 101 (1998): Scilla greilhuberi.
hohenackeri *(Fisch. & C.A. Mey.) F. Speta in Phyton (Austria),* 38(1): 101 (1998): Scilla hohenackeri.
khorassanica *(Meikle) F. Speta in Phyton (Austria),* 38(1): 101 (1998): Scilla khorassanica.
parwanica *(Speta) F. Speta in Phyton (Austria),* 38(1): 101 (1998): Scilla parwanica.
purpurea *(Griff.) F. Speta in Phyton (Austria),* 38(1): 101 (1998): Hyacinthus purpureus.
puschkinioides *(Regel) F. Speta in Phyton (Austria),* 38(1): 101 (1998): Scilla puschkinioides.
raewskiana *(Regel) F. Speta in Phyton (Austria),* 38(1): 102 (1998): Scilla raewskiana.
vvedenskyi *(Pazij) F. Speta in Phyton (Austria),* 38(1): 102 (1998): Scilla vvedenskyi.

FESTUCA (Gramin.).
sect. Subuliflorae *(E.B. Alexeev) S.J. Darbyshire in Phytologia,* 82(2): 77 (1997): F. subgen. Subuliflorae.
actae *H.E. Connor in New Zealand J. Bot.,* 36(3): 335 (1998), nom. nov.: F. ovina var. grandiflora *Howarth.*
akhanii *N.N. Tzvelev in Bot. Zhurn.,* 82(4): 118 (1997)—Iran.
alfrediana *B. Foggi & M.A. Signorini in Parlatorea,* 2: 130 (1997), nom. nov.: F. ovina var. briquetii *St.-Yves.*
ammobia *L.E. Pavlick in Phytologia,* 82(2): 77 (1997), nom. nov.: F. rubra subsp. arenicola *E.B. Alexeev.*
aragonensis *(Willk.) V. de la Fuente García & E. Ortúñez Rubio in Itinera Geobot.,* 10: 347 (1997): F. indigesta var. aragonensis.

FESTUCA:-
austrouralensis *P.V. Kulikov in Bot. Zhurn.,* 83(8): 111 (1998)—West Siberia.
baetica (Hack.) K. Richt.
subsp. moleroi *(Cebolla & Rivas Ponce) C. Cebolla Lozano & M.A. Rivas Ponce in Lagascalia,* 15(2): 163 (1990): F. paniculata var. moleroi.
beckeri *(Hack.) E.R. von Trautvetter in Trudy Imp. S.-Peterburgsk. Bot. Sada,* 9(1): 325 (1884): F. ovina subsp. beckeri.†
subsp. laeviuscula *(Klokov) N.N. Tzvelev in Novosti Sist. Vÿssh. Rast.,* 32: 183 (2000): F. laeviuscula.
brevipila R. Tracey
var. multinervis *(Stohr) J. Dengler in Kieler Not. Pflanzenk. Schleswig-Holstein Hamburg,* 24: 5 (1996): F. duvalii var. multinervis.
brigantina (Markgr.-Dann.) Markgr.-Dann.
subsp. actiophyta *M.I. Gutiérrez Villarías in Bot. J. Linn. Soc.,* 123(3): 252 (1997)—Spain.
centroapenninica *(Markgr.-Dann.) B. Foggi, F. Conti & Pignatti in Bocconea,* 10: 14, 212 (1998), as 'centro-apenninica': F. ovina var. centroapenninica.
copei *S.A. Renvoize, Gramineas de Bolivia:* 117 (1998)—Bolivia.
daghestanica *(Tzvelev) E.B. Alekseev, Ovsyanitsÿ Kavkaza:* 77 (1980): F. rubra subsp. daghestanica.
deflexa *H.E. Connor in New Zealand J. Bot.,* 36(3): 341 (1998)—New Zealand South I.
duriotagana Franco & Rocha Afonso
var. barbata *J. do Amaral Franco & M.L. Rocha Afonso in Silva Lusitana,* 5(1): 141 (1997)—Portugal.
eugenii *P.V. Kulikov in Bot. Zhurn.,* 85(3): 124 (2000)—North Caucasus (Krasnodar).
gamisansii Kerguélen
subsp. aethaliae *M.A. Signorini & B. Foggi in Pl. Biosystems,* 132(2): 106 (1998)—Italy.
hackelii *(St.-Yves) V. de la Fuente García & E. Ortúñez Rubio, Biosist. secc. Festuca del gén. Festuca en la Penins. Ibér.:* 81 (1998), homonym; *see K. Richt., Pl. Eur.,* 1: 101 (1890): F. ovina subsp. hackelii.
heterophylla Lam.
subsp. braun-blanquetii *V. de la Fuente García, E. Ortúñez Rubio & L.M. Ferrero Lomas in Itinera Geobot.,* 10: 321 (1997)—Spain.
hypsophila *R.A. Philippi in Anales Mus. Nac. Chile, sect. 2, Bot.,* 8: 89 (1891), as 'hyprophila' in error in I.K. Supp. 1—Chile.†
idahoensis Elmer
subsp. roemeri *(Pavlick) S.G. Aiken in Canad. J. Bot.,* 75(9): 1551 (1997), without basionym ref.; *et in Canad. J. Bot.,* 76(1): 179 (1998), with basionym ref.: F. idahoensis var. roemeri.
indigesta Boiss.
subsp. lagascae *C. Cebolla Lozano & M.A. Rivas Ponce in Flora Medit.,* 9: 141 (1999)—Spain.
krivotulenkoae *E.B. Alekseev, Ovsyanitsÿ Kavkaza:* 64 (1980)—North Caucasus (Kabardino-Balkariya).
kuprijanovii *D.V. Chusovljanov in Bot. Zhurn.,* 83(11): 113 (1998)—Altay.
longiauriculata *V. de la Fuente, E. Ortúñez & L.M. Ferrero in Parlatorea,* 3: 67 (1999)—Spain.
longifolia Thuill.
subsp. rivas-martinezii *(Fuente & Ortúñez) C. Cebolla Lozano, J.A. López Rodríguez & M.A. Rivas Ponce in Fontqueria,* 49: 34 (1997): F. rivas-martinezii.
luciarum *H.E. Connor in New Zealand J. Bot.,* 36(3): 343 (1998)—New Zealand North I.
madida *H.E. Connor in New Zealand J. Bot.,* 36(3): 345 (1998)—New Zealand North I., New Zealand South I., Antipodean Is.
marcopetri *J.F. Veldkamp in Blumea,* 41(1): 219 (1996)—New Guinea (Papua New Guinea).
markgrafiae *J.F. Veldkamp in Blumea,* 41(1): 220 (1996)—New Guinea (Papua New Guinea).
matthewsii (Hack.) Cheeseman
subsp. aquilonia *H.E. Connor in New Zealand J. Bot.,* 36(3): 351 (1998)—New Zealand South I.
subsp. latifundii *H.E. Connor in New Zealand J. Bot.,* 36(3): 352 (1998)—New Zealand South I.
subsp. pisamontis *H.E. Connor in New Zealand J. Bot.,* 36(3): 354 (1998)—New Zealand South I.

FESTUCA:-
melanopsis *B. Foggi, G. Rossi & M.A. Signorini in Canad. J. Bot.*, 77(7): 1001 (1999), nom. nov., with incorrect replaced synonym ref.: F. rubra var. nigricans *Hack.*
mucronata *Arv-Touv. & Dupuy in Bull. Soc. Dauphin. Échange Pl.*, 9: 384-385 (1882)—France.†
nigriflora *(Hitchc.) M.A. Negritto & A.M. Anton in Novon*, 9(4): 542 (1999): Poa nigriflora.
ovina L.
subsp. hirtula *(Hack. ex Travis) M.J. Wilkinson in Soc. Ech. Pl. Vasc. Eur. Bassin Médit. Bull.*, 20: 72 (1985): F. ovina subvar. hirtula.
subsp. kamtschatica *(St.-Yves) V.N. Voroschilov in A.K. Skvortsov (ed.), Florist. issl. v razn. Raïonakh SSSR:* 151 (1985): F. ovina var. kamtschatica.
subsp. ophioliticola *(Kerguélen) M.J. Wilkinson in Soc. Ech. Pl. Vasc. Eur. Bassin Médit. Bull.*, 20: 72 (1985): F. ophioliticola.
parodiana *(St.-Yves ex Parodi) E.G. Nicora in Hickenia*, 2(57): 267 (1998): F. subulata var. parodiana.
petersonii *S.A. Renvoize, Gramineas de Bolivia:* 122 (1998)—Bolivia.
potosiana *S.A. Renvoize, Gramineas de Bolivia:* 122 (1998)—Bolivia.
pseudovivipara *(Pavlick) L.E. Pavlick in Phytologia*, 82(2): 77 (1997): F. rubra f. pseudovivipara.
riccerii *B. Foggi & G. Rossi in Willdenowia*, 26(1–2): 206 (1996)—Italy.
rivas-martinezii Fuente & Ortúñez
subsp. rectifolia *V. de la Fuente García, E. Ortúñez Rubio & L.M. Ferrero Lomas in Itinera Geobot.*, 10: 320 (1997)—Spain.
roblensis *M. González-Ledesma in Novon*, 8(2): 147 (1998)—Mexico Northeast (Zacatecas, Guanajuato).
rubra L.
subsp. mediana *(Pavlick) L.E. Pavlick in Phytologia*, 82(2): 77 (1997): F. rubra var. mediana.
scabrifolia *S.A. Renvoize, Gramineas de Bolivia:* 109 (1998), nom. nov.: F. dissitiflora var. trachyphylla *St.-Yves.*
segimonensis *V. de la Fuente, J. Müller & E. Ortúñez in An. Jard. Bot. Madrid*, 57(1): 178 (1999), nom. nov.: F. ovina subsp. hackelii *St.-Yves.*
serana *I. Markgraf-Dannenberg in Blumea*, 41(1): 221 (1996)—Moluccas.
stebeckii *S.A. Renvoize, Gramineas de Bolivia:* 121 (1998)—Bolivia.
subverticillata (Pers.) E.B. Alexeev
f. pilosifolia *(Dore) S.J. Darbyshire in Phytologia*, 82(2): 77 (1997): F. obtusa f. pilosifolia.
ultramafica *H.E. Connor in New Zealand J. Bot.*, 36(3): 363 (1998)—New Zealand South I.
vaginalis *(Benth.) S. Laegaard in Novon*, 8(1): 30 (1998): Poa vaginalis.
vagravarica *E.B. Alekseev, Ovsyanitsÿ Kavkaza:* 88 (1980)—Transcaucasus (Armeniya).
valdesii *M. González-Ledesma & S.D. Koch in Novon*, 8(2): 149 (1998)—Mexico Northeast (Coahuila, Nuevo León).
varia
var. winnebachensis *C. Wallossek & Markgr.-Dann. in Folia Geobot.*, 34(1): 73 (1999)—Italy/Austria border.
veneris *G. Rossi, B. Foggi & M.A. Signorini in Webbia*, 51(1): 22 (1996)—Italy.
vettonica *V. de la Fuente, E. Ortúñez & L.M. Ferrero in Lazaroa*, 20: 5 (1999)—Spain.
violacea Gaudin
subsp. italica *B. Foggi, Graziano Rossi & M.A. Signorini in Flora Medit.*, 8: 34 (1998)—Italy.
subsp. puccinellii *(Parl.) B. Foggi, G. Rossi & M.A. Signorini in Canad. J. Bot.*, 77(7): 998 (1999): F. puccinellii.
woronowii Hack.
subsp. caucasica *(St.-Yves) E.B. Alekseev, Ovsyanitsÿ Kavkaza:* 74 (1980): F. varia var. caucasica.
yarochenkoi *(St.-Yves) E.B. Alekseev, Ovsyanitsÿ Kavkaza:* 82 (1980): F. ovina var. yarochenkoi.

FICINIA (Cyperac.).
borealis *K.A. Lye in Lidia*, 3(6): 205 (1996)—Ethiopia.
nodosa *(Rottb.) P. Goetghebeur, A.M. Muasya & D.A. Simpson in Novon*, 10(2): 133 (2000): Scirpus nodosus.
trollii *(Kük.) A.M. Muasya & D.A. Simpson in Novon*, 10(2): 133 (2000): Scirpus trollii.

FICUS (Morac.).
subsect. Bergianae *S. Carvajal & L.K. Shabes in Bol. Inst. Bot. (Guadalajara)*, 6(2–3): 215 (1998 publ. 2000).
subsect. Petenenses *S. Carvajal & L.K. Shabes in Bol. Inst. Bot. (Guadalajara)*, 6(2–3): 215 (1998 publ. 2000).
ashtonii *K.M. Kochummen in Gard. Bull. Singapore*, 50(2): 201 (1998)—Borneo (Sarawak).
benjamina L.
subsp. comosa *(Roxb.) G. Panigrahi & S.K. Murti, Fl. Bilaspur Distr.*, 2: 566 (1999): F. comosa.
binnendykii Miq.
var. pallescens *G.D. Weiblen in Trop. Biodivers.*, 5(3): 266 (1998)—Borneo (Kalimantan).
boliviana *C.C. Berg in Brittonia*, 51(4): 395 (1999)—Bolivia, Peru.
borneensis *K.M. Kochummen in Gard. Bull. Singapore*, 50(2): 202 (1998)—Borneo (Sabah, Sarawak).
carchiana *C.C. Berg in Novon*, 6(3): 232 (1996)—Ecuador, Panama, Colombia.
cereicarpa Corner
var. ashtonii *K.M. Kochummen in Gard. Bull. Singapore*, 50(2): 213 (1998)—Borneo (Sarawak).
chaii *K.M. Kochummen in Gard. Bull. Singapore*, 50(2): 203 (1998)—Borneo (Sarawak).
chewii *K.M. Kochummen in Gard. Bull. Singapore*, 50(2): 203 (1998)—Borneo (Sabah).
corneri *K.M. Kochummen in Gard. Bull. Singapore*, 50(2): 204 (1998)—Borneo (Sabah, Sarawak, Brunei).
corneriana *C.C. Berg in Blumea*, 45(2): 397 (2000)—Solomon Is. (South Solomons).
cotinifolia Kunth
var. hondurensis *(Standl. & L.O. Williams) C.C. Berg in Novon*, 6(3): 232 (1996): F. hondurensis.
deltoidea Jack
var. recurvata *K.M. Kochummen in Gard. Bull. Singapore*, 50(2): 213 (1998)—Borneo (Sarawak).
var. subhirsuta *K.M. Kochummen in Gard. Bull. Singapore*, 50(2): 214 (1998)—Borneo (Sarawak).
drupacea Thunb.
var. subrepanda *(Wall. ex King) D. Basu in J. Econ. Taxon. Bot.*, 20(3): 653 (1996): F. mysorensis var. subrepanda.
dulitensis *K.M. Kochummen in Gard. Bull. Singapore*, 50(2): 205 (1998)—Borneo (Sarawak).
gamostyla *K.M. Kochummen in Gard. Bull. Singapore*, 50(2): 206 (1998)—Borneo (Sabah).
ilias-paiei *K.M. Kochummen in Gard. Bull. Singapore*, 50(2): 207 (1998)—Borneo (Sarawak).
jacquelineae *S. Carvajal & C. Peña-Pinela in Bol. Inst. Bot. (Guadalajara)*, 4(1–3): 58 (1996 publ. 1997), nom. nov.: Urostigma longipes *Liebm.*
katendei *B. Verdcourt in Kew Bull.*, 53(1): 233 (1998)—Uganda.
kerangasensis *K.M. Kochummen in Gard. Bull. Singapore*, 50(2): 207 (1998)—Borneo (Sarawak).
lacunata *T.A. Kvitvik in Brittonia*, 49(2): 270 (1997)—Ecuador.
longistipulata *K.M. Kochummen in Gard. Bull. Singapore*, 50(2): 208 (1998)—Borneo (Sabah).
mariae *C.C. Berg, L. Emygdio de Mello Filho & J.P.P. Carauta in Bradea*, 8(20): 112 (1999)—Brazil Southeast (Espírito Santo, Minas Gerais), Peru, Bolivia.
microcarpa L.f.
var. nitida *(King) R.R. Fernandez, Trees of Mumbai (Bombay):* 245 (1999), isonym; *see* F.C. Ho, *Trop. Pl. Taiwan in colour*, 2: 107 (1979): F. retusa var. nitida.
ngii *K.M. Kochummen in Gard. Bull. Singapore*, 50(2): 209 (1998)—Malaya (Peninsular Malaysia).

FICUS:-
obscura Blume
var. lanata *K.M. Kochummen in Gard. Bull. Singapore*, 50(2): 215 (1998)—Borneo (Sabah, Sarawak).
oleifolia King
var. calcicola *K.M. Kochummen in Gard. Bull. Singapore*, 50(2): 215 (1998)—Borneo (Sarawak).
var. impressicostata *K.M. Kochummen in Gard. Bull. Singapore*, 50(2): 216 (1998)—Borneo (Sarawak).
osensis *C.C. Berg in Novon*, 6(3): 232 (1996)—Costa Rica.
palamoides *J.O. Voigt, Hort. Suburb. Calcutt.:* 292 (1845), nom. nov.: F. palmata *Roxb.*
palungensis *G.D. Weiblen in Trop. Biodivers.*, 5(3): 279 (1998)—Borneo (Kalimantan).
pseudotarennifolia *K.M. Kochummen in Gard. Bull. Singapore*, 50(2): 210 (1998)—Borneo (Sarawak).
rheedii *J. Graham, Cat. Pl. Bombay:* 192 (1839)—India (Maharashtra).
rieberiana *C.C. Berg in Novon*, 6(3): 234 (1996)—Ecuador.
sabahana *K.M. Kochummen in Gard. Bull. Singapore*, 50(2): 211 (1998)—Borneo (Sabah).
semicordata Buch.-Ham. ex Sm.
var. montana *S.M. Amatya in Novon*, 6(4): 323 (1996)—Nepal.
soepadmoi *K.M. Kochummen in Gard. Bull. Singapore*, 50(2): 212 (1998)—Borneo (Sarawak).
sundaica Blume
var. impressicostata *K.M. Kochummen in Gard. Bull. Singapore*, 50(2): 217 (1998)—Borneo (Sabah, Brunei, Sarawak).
triloba *Buch.-Ham. ex J.O. Voigt, Hort. Suburb. Calcutt.:* 284 (1845), nom. nov.: F. hirsuta *Roxb.*
virens Dryand.
var. lambertiana *(King) R.R. Fernandez, Trees of Mumbai (Bombay):* 241 (1999), with incorrect basionym ref.: F. infectoria var. lambertiana.
var. wightiana *(King) R.R. Fernandez, Trees of Mumbai (Bombay):* 239 (1999), with incorrect basionym ref.: F. infectoria var. wightiana.

FIELDIA (Gesneriac.).
australiana *(C.T. White) B.L. Burtt in Edinburgh J. Bot.*, 56(3): 372 (1999): Coronanthera australiana.

× **FILFIA** *J. Holub in Preslia*, 70(2): 104 (1998). *COMPOSITAE* (Filago × Logfia).
mixta *(Holuby) J. Holub in Preslia*, 70(2): 104 (1998): Filago × mixta.

FILIPENDULA (Rosac.).
subsect. Ulmaria *(Mill.) I.A. Shantser in Sosud. Rast. Sovet. Dal'nego Vostoka*, 8: 221 (1996): gen. Ulmaria.
formosa Nakai
f. albiflora *Y.N. Lee, Fl. Korea:* 1158 (1996), without type—Locality not stated.
yezoensis H. Hara
f. alba *(Nakai) Y.N. Lee, Fl. Korea:* 1158 (1996), without type—Locality not stated.

FIMBRIBAMBUSA *E.A. Widjaja in Reinwardtia*, 11(2): 80 (1997). *GRAMINEAE*
horsfieldii *(Munro) E.A. Widjaja in Reinwardtia*, 11(2): 81 (1997): Bambusa horsfieldii.
microcephala *(Pilg.) E.A. Widjaja in Reinwardtia*, 11(2): 82 (1997): Dendrocalamus microcephalus.

FIMBRIELLA (Orchidac.).
praeclara *(Sheviak & M.L. Bowles) D.L. Szlachetko & P. Rutkowski in Acta Bot. Fenn.*, 169: 380 (2000): Platanthera praeclara.

FIMBRIPETALUM (Caryophyllac.).
radians (L.) Ikonn.
var. brevipetalum *N.S. Pavlova in Sosud. Rast. Sovet. Dal'nego Vostoka*, 8: 65 (1996), without type—Locality not stated.

FIMBRISTYLIS (Cyperac.).
ambavanensis *V.P. Prasad & N.P. Singh in J. Bombay Nat. Hist. Soc.*, 96(3): 454 (1999)—India (Maharashtra).

FIMBRISTYLIS:-
benthamiana *E. Govindarajalu in Rheedea,* 7(2): 115 (1997)—India (West Bengal).
bispicula *E. Govindarajalu in Rheedea,* 6(2): 59 (1996)—India (Karnataka).
bivalvis *(Lam.) K.A. Lye in Lidia,* 3(5): 144 (1995): Scirpus bivalvis.
breviculma *E. Govindarajalu in Rheedea,* 6(2): 61 (1996), homonym (Art. 53.3 ICBN 1994); *see Boeck. in Flora,* 41: 643 (1858)—India (Karnataka).
cuneata *E. Govindarajalu & S.K. Varma in J. Econ. Taxon. Bot.,* 21(2): 377 (1997)—India (Bihar).
dichotoma (L.) Vahl
subsp. bisumbellata *(Forssk.) M. Luceño in An. Jard. Bot. Madrid,* 57(1): 176 (1999): Scirpus bisumbellatus.
diglumoides *E. Govindarajalu & S.K. Varma in J. Econ. Taxon. Bot.,* 21(2): 380, 377 (1997)—India (Bihar).
dimorphonucifera *E. Govindarajalu in Rheedea,* 7(2): 122 (1997)—India (Karnataka).
doliiformis *E. Govindarajalu in Rheedea,* 8(1): 89 (1998)—India (Tamil Nadu).
eichleriana *E. Govindarajalu in Beitr. Biol. Pfl.,* 70(1): 79 (1998)—India (Tamil Nadu, Kerala).
fuchsiana *E. Govindarajalu in Beitr. Biol. Pfl.,* 70(1): 81 (1998)—India (Uttar Pradesh).
gentarea *E. Govindarajalu in Beitr. Biol. Pfl.,* 70(1): 83 (1998)—India (Tamil Nadu).
humerosa *E. Govindarajalu in Rheedea,* 7(2): 117 (1997)—India (Kerala).
hyalina *E. Govindarajalu & N. Sasidharan in J. Econ. Taxon. Bot.,* 21(2): 373 (1997)—India (Kerala).
lithophila *E. Govindarajalu in Beitr. Biol. Pfl.,* 70(1): 87 (1998)—India (Tamil Nadu).
manilaliana *E. Govindarajalu in Rheedea,* 8(1): 87 (1998)—India (Kerala).
mycosa *E. Govindarajalu in Rheedea,* 7(2): 124 (1997)—India (Tamil Nadu).
nagpurensis *V.P. Prasad & N.P. Singh in J. Econ. Taxon. Bot.,* 21(3): 671 (1997 publ. 1998)—India (Maharashtra).
pandurata *E. Govindarajalu in Rheedea,* 7(2): 124 (1997)—India (Tamil Nadu).
perspicua *E. Govindarajalu & N. Sasidharan in J. Econ. Taxon. Bot.,* 21(2): 376 (1997)—India (Kerala).
prabatensis *D.A. Simpson in Kew Bull.,* 53(1): 225 (1998)—Thailand.
ratnagirica *V.P. Prasad & N.P. Singh in J. Econ. Taxon. Bot.,* 21(3): 673 (1997 publ. 1998)—India (Maharashtra).
simaoensis *Y.Y. Qian in Bull. Bot. Res. North-East. Forest. Univ.,* 17(2): 128 (1997)—China South-central (Yunnan).
simpsonii *V.P. Prasad & N.P. Singh in J. Bombay Nat. Hist. Soc.,* 96(3): 456 (1999)—India (Karnataka).
singularis *E. Govindarajalu in Beitr. Biol. Pfl.,* 70(1): 88 (1998)—India (Karnataka).
stigmatotecta *E. Govindarajalu in Rheedea,* 6(2): 61 (1996)—India (Kerala).
tumida *E. Govindarajalu in Rheedea,* 8(1): 91 (1998)—India (Kerala, Tamil Nadu).
ultragluma *E. Govindarajalu in Rheedea,* 7(2): 119 (1997)—India (Maharashtra).
virella *E. Govindarajalu in Rheedea,* 8(1): 93 (1998)—India (Kerala, Karnataka).
willdenowiana *E. Govindarajalu in Beitr. Biol. Pfl.,* 70(1): 90 (1998)—India (Andhra Pradesh).

FINGARDIA *D.L. Szlachetko in Fragm. Flor. Geobot.,* Supp. 3: 134 (1995). *ORCHIDACEAE*
alagensis *(Ames) D.L. Szlachetko in Fragm. Flor. Geobot.,* Supp. 3: 134 (1995): Microstylis alagensis.
carrii *(Seidenf. & J.J. Wood) D.L. Szlachetko in Fragm. Flor. Geobot.,* Supp. 3: 134 (1995): Malaxis carrii.
cordiglottis *(J.J. Sm.) D.L. Szlachetko in Fragm. Flor. Geobot.,* Supp. 3: 134 (1995): Microstylis cordiglottis.
kobi *(J.J. Sm.) D.L. Szlachetko in Fragm. Flor. Geobot.,* Supp. 3: 134 (1995): Microstylis kobi.
latilabris *(Schltr.) D.L. Szlachetko in Fragm. Flor. Geobot.,* Supp. 3: 134 (1995): Microstylis latilabris.
lokonensis *(Schltr.) D.L. Szlachetko in Fragm. Flor. Geobot.,* Supp. 3: 134 (1995): Microstylis lokonensis.

FINGARDIA:-
lunata *(Schltr.) D.L. Szlachetko in Fragm. Flor. Geobot.,* Supp. 3: 134 (1995): Microstylis lunata.
nephroglossa *(Schltr.) D.L. Szlachetko in Fragm. Flor. Geobot.,* Supp. 3: 134 (1995): Microstylis nephroglossa.
tenggerensis *(J.J. Sm.) D.L. Szlachetko in Fragm. Flor. Geobot.,* Supp. 3: 134 (1995): Microstylis tenggerensis.
yamapensis *H.B. Margońska, D.L. Szlachetko & P. Rutkowski in Fragm. Flor. Geobot.,* 43(1): 5 (1998)—New Guinea (Papua New Guinea).

FISSISTIGMA (Annonac.).
ellipticum *(King) Debika Mitra in Bull. Bot. Surv. India,* 35(1–4): 117 (1993 publ. 1997): Melodorum polyanthum var. ellipticum.

FLACOURTIA (Flacourtiac.).
indica (Burm. f.) Merr.
var. innocua *(Haines) H.O. Saxena & M. Brahmam, Fl. Orissa,* 1: 79 (1994): F. sepiaria var. innocua.
inermis *W. Roxburgh, Pl. Coromandel,* 3: 16 (1811)—Moluccas.†

FLEISCHMANNIA (Compos.).
acuminata *V.M. Badillo in Ernstia,* 10(1): 13 (2000)—Venezuela.
coibensis *S. Díaz Piedrahita in Fl. & Fauna Parque Nac. Coiba (Panamá), Invent. Prelim. (ed. S.* Castroviejo): 278 (1997)—Panama.
diversifolia *(Schrad.) H. Robinson in Phytologia,* 84(5): 348 (1998 publ. 1999): Eupatorium diversifolium.

FLEMINGIA (Legumin.).
praecox C.B. Clarke ex Prain
var. angusta *(Craib) Anand Kumar in Nation. Acad. Sci. Lett.,* 5(8): 249 (1982): F. angusta.
var. robusta *(Mukerjee) Anand Kumar in Nation. Acad. Sci. Lett.,* 5(8): 249 (1982): Moghania praecox var. robusta.
semialata *W. Roxburgh, Pl. Coromandel,* 3: 45 (1820)—Nepal.†
stricta *W. Roxburgh, Pl. Coromandel,* 3: 44 (1815)—India (Tamil Nadu).†

FLICKINGERIA (Orchidac.).
candoonensis *(Ames) H. Fessel & E. Lückel in Orchidee,* 49(5): 252 (1998), without basionym ref.; *et in Orchidee,* 50(1): 46 (1999), with basionym ref.: Dendrobium candoonense.
junctiloba *H. Fessel & E. Lückel in Orchidee,* 49(5): 254 (1998)—Philippines.

FLINDERSIA (Rutac.).
pimenteliana F. Muell.
f. oppositifolia *(F. Muell.) K.D. Scott, W.K. Harris & J. Playford in Austrobaileya,* 5(4): 668 (2000): Hypsophila oppositifolia.

FLOSCALDASIA (Compos.).
azorelloides *P. Sklenář & H. Robinson in Novon,* 10(2): 146 (2000)—Ecuador.

FLUEGGEA (Euphorbiac.).
retusa *(Roxb.) J.O. Voigt, Hort. Suburb. Calcutt.:* 152 (1845): Phyllanthus retusus.†
trigonoclada *(Ohwi) T. Kurosawa in Fl. Japan,* 2c: 8 (1999): Phyllanthus trigonocladus.
virosa *(Willd.) Royle, Ill. Bot. Himal. Mts.:* 328 (1836): Phyllanthus virosus.

FOCKEA (Asclepiadac.).
angustifolia *K. Schum. in Bot. Jahrb. Syst.,* 17: 146 (1893), in Supp. 1 as 'augustifolia'—Cape Province.†
edulis (Thunb.) K. Schum.
var. capensis *(Endl.) G.D. Rowley in Asklepios,* 75: 17 (1998): F. capensis.

FOENICULUM (Umbellif.).
vulgare Mill.
subsp. subinodorum *(Maire, Weiller & Wilczek) M. Ibn Tattou in Bocconea,* 8: 184 (1998): F. subinodorum.

FOKIENIA (Cupressac.).
hodginsii (Dunn) Henry & H.H. Thomas
var. kawaii *(Hayata) J. Silba in J. Int. Conifer Preserv. Soc.,* 7(1): 24 (2000): F. kawaii.

FOLOTSIA (Asclepiadac.).
ambovombense *S. Liede in Bull. Mus. Nation. Hist. Nat., B, Adansonia,* Ser. 4, 18(1–2): 130 (1996)—Madagascar.
humbertii *S. Liede in Bull. Mus. Nation. Hist. Nat., B, Adansonia,* Ser. 4, 18(1–2): 132 (1996)—Madagascar.

FONTAINEA (Euphorbiac.).
borealis *P.I. Forster in Austrobaileya,* 5(1): 29 (1997)—New Guinea (Papua New Guinea).
fugax *P.I. Forster in Austrobaileya,* 5(1): 34 (1997)—Queensland.
subpapuana *P.I. Forster in Austrobaileya,* 5(1): 32 (1997)—New Guinea (Papua New Guinea).
longicarpa *K.-J. Kim in J. Pl. Biol.,* 41(2): 142 (1998)—China Southeast (Zhejiang).

FORNICIUM (Compos.).
carthamoides *(Willd.) R.V. Kamelin, Mater. Istorii Fl. Azii (Altaisk. Gorn. Strana):* 139 (1998): Cnicus carthamoides.
subsp. orientale *(Serg.) R.V. Kamelin, Mater. Istorii Fl. Azii (Altaisk. Gorn. Strana):* 139 (1998): Leuzea carthamoides subsp. orientalis.
longifolium *(Hoffmanns. & Link) R.V. Kamelin, Mater. Istorii Fl. Azii (Altaisk. Gorn. Strana):* 139 (1998): Leuzea longifolia.
lyratum *(C. Winkl. ex Iljin) R.V. Kamelin, Mater. Istorii Fl. Azii (Altaisk. Gorn. Strana):* 139 (1998): Rhaponticum lyratum.
serratuloides *(Georgi) R.V. Kamelin, Mater. Istorii Fl. Azii (Altaisk. Gorn. Strana):* 139 (1998): Centaurea serratuloides.

FORSYTHIA (Oleac.).
sect. Giraldianae *X.K. Qin in Acta Bot. Yunnanica,* 18(2): 164 (1996).
sect. Suspensa *X.K. Qin in Acta Bot. Yunnanica,* 18(2): 164 (1996).

FORTUNATIA (Hyacinthac.).
acaulis *(Baker) E.R. Guaglianone & S. Arroyo-Leuenberger in Hickenia,* 2(31): 137 (1995): Scilla acaulis.

FOSBERGIA *D.D. Tirvengadum & C. Sastre in Biogeographica,* 73(2): 88 (1997). *RUBIACEAE*
alleizettii *D.D. Tirvengadum & C. Sastre in Biogeographica,* 73(2): 89 (1997)—Vietnam.
petelotii *Merr. ex D.D. Tirvengadum & C. Sastre in Biogeographica,* 73(2): 89 (1997)—Vietnam.
pubescens *D.D. Tirvengadum & C. Sastre in Biogeographica,* 73(2): 90 (1997)—Burma.
shweliensis *(J. Anthony) D.D. Tirvengadum & C. Sastre in Biogeographica,* 73(2): 88 (1997): Randia shweliensis.
thailandica *D.D. Tirvengadum & C. Sastre in Biogeographica,* 73(2): 89 (1997)—Thailand.

FOSTERELLA (Bromeliac.).
chaparensis *P.L. Ibisch, R. Vásquez & E. Gross in Rev. Soc. Boliv. Bot.,* 2(2): 118 (1999)—Bolivia.
chiquitana *P.L. Ibisch, R. Vásquez & E. Gross in Rev. Soc. Boliv. Bot.,* 2(2): 118 (1999)—Bolivia.
floridensis *P.L. Ibisch, R. Vásquez & E. Gross in Rev. Soc. Boliv. Bot.,* 2(2): 120 (1999)—Bolivia.
latifolia *P.L. Ibisch, R. Vásquez & E. Gross in Rev. Soc. Boliv. Bot.,* 2(2): 123 (1999)—Bolivia.
spectabilis *H.E. Luther in J. Bromeliad Soc.,* 47(3): 118 (1997)—Bolivia.
vasquezii *E. Gross & P.L. Ibisch in J. Bromeliad Soc.,* 47(5): 212 (1997)—Bolivia.

FRAGARIA (Rosac.).
× ananassa Duchesne
subsp. cuneifolia *(Nutt. ex Howell) G. Staudt, Syst. & Geogr. Distrib. Amer. Strawberry Sp. (Univ. Calif. Publ. Bot., 81):* 129 (1999): F. cuneifolia.
× bringhurstii *G. Staudt, Syst. & Geogr. Distrib. Amer. Strawberry Sp. (Univ. Calif. Publ. Bot., 81):* 135 (1999)—California. (F. chiloensis × vesca subsp. californica).

FRAGARIA:-
chiloensis (L.) Mill.
subsp. sandwicensis *(O. Deg. & I. Deg.) G. Staudt, Syst. & Geogr. Distrib. Amer. Strawberry Sp. (Univ. Calif. Publ. Bot., 81):* 122 (1999), homonym; *see Staudt in Canad J. Bot.*, 40: 883 (1962): F. chiloensis var. sandwicensis.
nilgerrensis Schltdl. ex J. Gay
subsp. hayatae *(Makino) G. Staudt in Bot. Jahrb.*, 121(3): 305 (1999), as 'hayatai': F. hayatae.
sandwicensis *J. Decaisne, Jard. Fruit.*, 9: 49 (1862–75)—Hawaiian Is.
vesca L.
f. albida *G. Staudt, Syst. & Geogr. Distrib. Amer. Strawberry Sp. (Univ. Calif. Publ. Bot., 81):* 62 (1999)—Idaho.
f. bracteata *(A. Heller) G. Staudt, Syst. & Geogr. Distrib. Amer. Strawberry Sp. (Univ. Calif. Publ. Bot., 81):* 57 (1999): F. bracteata.
f. helleri *(Holz.) G. Staudt, Syst. & Geogr. Distrib. Amer. Strawberry Sp. (Univ. Calif. Publ. Bot., 81):* 63 (1999), isonym; *see Scoggan, Fl. Canada*, 1: 51 (1978): F. helleri.
virginiana Duchesne
f. multicipita *(Fernald) P.M. Catling & J. Cayouette in Rhodora*, 97(891): 252 (1995 publ. 1996): F. multicipita.

FRAILEA (Cactac.).
subsect. Alacriportana *K.H. Prestlé, Gattung Frailea (Br. & R.) Prestlé:* 113 (1997), without latin descr. or type.
sect. Asterioides *K.H. Prestlé, Gattung Frailea (Br. & R.) Prestlé:* 216 (1997), without latin descr. or type.
subsect. Aurea *K.H. Prestlé, Gattung Frailea (Br. & R.) Prestlé:* 157 (1997), without latin descr. or type.
subsect. Aureaspinosior *K.H. Prestlé, Gattung Frailea (Br. & R.) Prestlé:* 120 (1997), without latin descr. or type.
subsect. Aureiflora *K.H. Prestlé, Gattung Frailea (Br. & R.) Prestlé:* 83 (1997), without latin descr. or type.
subsect. Aureoflora *K.H. Prestlé, Gattung Frailea (Br. & R.) Prestlé:* 74 (1997), without latin descr. or type.
subsect. Carminatiflora *K.H. Prestlé, Gattung Frailea (Br. & R.) Prestlé:* 98 (1997), without latin descr. or type.
subsect. Carminifilamentosa *K.H. Prestlé, Gattung Frailea (Br. & R.) Prestlé:* 174 (1997), without latin descr. or type.
sect. Castanea *K.H. Prestlé, Gattung Frailea (Br. & R.) Prestlé:* 211 (1997), without latin descr. or type.
subgen. Ceroidea *K.H. Prestlé, Gattung Frailea (Br. & R.) Prestlé:* 107 (1997), without type.
subsect. Chacoensis *K.H. Prestlé, Gattung Frailea (Br. & R.) Prestlé:* 190 (1997), without latin descr. or type.
sect. Chiquitana *K.H. Prestlé, Gattung Frailea (Br. & R.) Prestlé:* 224 (1997), without latin descr. or type.
subsect. Corrientensis *K.H. Prestlé, Gattung Frailea (Br. & R.) Prestlé:* 181 (1997), without latin descr. or type.
sect. Gracilis *K.H. Prestlé, Gattung Frailea (Br. & R.) Prestlé:* 108 (1997), without latin descr. or type.
sect. Grahliana *K.H. Prestlé, Gattung Frailea (Br. & R.) Prestlé:* 188 (1997), without latin descr. or type.
subsect. Horstii *K.H. Prestlé, Gattung Frailea (Br. & R.) Prestlé:* 128 (1997), without latin descr. or type.
subsect. Iguaraica *K.H. Prestlé, Gattung Frailea (Br. & R.) Prestlé:* 179 (1997), without latin descr. or type.
subsect. Intermedia *K.H. Prestlé, Gattung Frailea (Br. & R.) Prestlé:* 221 (1997), without latin descr. or type.
subsect. Jura *K.H. Prestlé, Gattung Frailea (Br. & R.) Prestlé:* 92 (1997), without latin descr. or type.
subsect. Jurassica *K.H. Prestlé, Gattung Frailea (Br. & R.) Prestlé:* 139 (1997), without latin descr. or type.

FRAILEA:-
subsect. Lilalunula *K.H. Prestlé, Gattung Frailea (Br. & R.) Prestlé:* 88 (1997), without latin descr. or type.
ser. Oreopygmaea *K.H. Prestlé, Gattung Frailea (Br. & R.) Prestlé:* 67 (1997), without type.
subsect. Oreopygmaea *K.H. Prestlé, Gattung Frailea (Br. & R.) Prestlé:* 67 (1997), without type.
subsect. Paraguariensis *K.H. Prestlé, Gattung Frailea (Br. & R.) Prestlé:* 194 (1997), without latin descr. or type.
sect. Phaeodiscaoides *K.H. Prestlé, Gattung Frailea (Br. & R.) Prestlé:* 29 (1997), without type.
subsect. Pseudopumila *K.H. Prestlé, Gattung Frailea (Br. & R.) Prestlé:* 163 (1997), without latin descr. or type.
sect. Pumila *K.H. Prestlé, Gattung Frailea (Br. & R.) Prestlé:* 147 (1997), without latin descr. or type.
subgen. Pumilaoides *K.H. Prestlé, Gattung Frailea (Br. & R.) Prestlé:* 146 (1997), without type.
sect. Pygmaeoides *K.H. Prestlé, Gattung Frailea (Br. & R.) Prestlé:* 65 (1997), without type.
subgen. Typicus *K.H. Prestlé, Gattung Frailea (Br. & R.) Prestlé:* 204 (1997), without type.
subsect. Uruguaya *K.H. Prestlé, Gattung Frailea (Br. & R.) Prestlé:* 166 (1997), without latin descr. or type.
alacriportana
subsp. feeburgii *K.H. Prestlé, Gattung Frailea (Br. & R.) Prestlé:* 116 (1997), without latin descr. or type—Brazil South (Rio Grande do Sul).
subsp. itapuensis *K.H. Prestlé, Gattung Frailea (Br. & R.) Prestlé:* 119 (1997), without latin descr. or type—Brazil South (Rio Grande do Sul).
subsp. longicylindrica *K.H. Prestlé, Gattung Frailea (Br. & R.) Prestlé:* 118 (1997), without latin descr. or type—Brazil South (Rio Grande do Sul).
altasensis *K.H. Prestlé, Gattung Frailea (Br. & R.) Prestlé:* 126 (1997)—Brazil South (Rio Grande do Sul).
amerhauseri *K.H. Prestlé, Gattung Frailea (Br. & R.) Prestlé:* 227 (1997), without latin descr. or type—Bolivia.
subsp. boliviensis *K.H. Prestlé, Gattung Frailea (Br. & R.) Prestlé:* 228 (1997), without latin descr. or type—Bolivia.
angelesii *R. Kiesling ex K.H. Prestlé, Gattung Frailea (Br. & R.) Prestlé:* 52 (1997), without latin descr. or type—Argentina Northeast (Entre Ríos).
buenekeri W.R. Abraham
subsp. densispina *A. Hofacker & K. Herm in Kakt. And. Sukk.*, 48(3): 61 (1997)—Brazil South (Rio Grande do Sul).
subsp. gemmrichii *K.H. Prestlé, Gattung Frailea (Br. & R.) Prestlé:* 54 (1997), without latin descr. or type—Brazil South (Rio Grande do Sul).
subsp. pygmaeoidea *K.H. Prestlé, Gattung Frailea (Br. & R.) Prestlé:* 55 (1997), without latin descr. or type—Brazil South (Rio Grande do Sul).
buiningiana *K.H. Prestlé, Gattung Frailea (Br. & R.) Prestlé:* 133 (1997)—Brazil South (Rio Grande do Sul).
f. gracilispina *K.H. Prestlé, Gattung Frailea (Br. & R.) Prestlé:* 135 (1997)—Brazil South (Rio Grande do Sul).
carminifilamentosa
subsp. dyntariana *K.H. Prestlé, Gattung Frailea (Br. & R.) Prestlé:* 175 (1997), without latin descr. or type—Brazil South (Rio Grande do Sul).
subsp. ibicuiensis *K.H. Prestlé, Gattung Frailea (Br. & R.) Prestlé:* 178 (1997), without latin descr. or type—Brazil South (Rio Grande do Sul).
subsp. itaquiensis *K.H. Prestlé, Gattung Frailea (Br. & R.) Prestlé:* 176 (1997), without latin descr. or type—Brazil South (Rio Grande do Sul).
subsp. ustinaldensis *K.H. Prestlé, Gattung Frailea (Br. & R.) Prestlé:* 177 (1997), without latin descr. or type—Brazil South (Rio Grande do Sul).

FRAILEA:-
carminifilamentosa
f. winkelmanniana *(Kilian) K.H. Prestlé, Gattung Frailea (Br. & R.) Prestlé:* 175 (1997), as 'winkelmann', basionym not validly publ.: F. carminifilamentosa var. winkelmanniana.
castanea
f. albiseta *K.H. Prestlé, Gattung Frailea (Br. & R.) Prestlé:* 216 (1997), without latin descr. or type—Brazil South (Rio Grande do Sul).
f. auricantha *K.H. Prestlé, Gattung Frailea (Br. & R.) Prestlé:* 216 (1997), without latin descr. or type—Uruguay.
f. ferruginea *K.H. Prestlé, Gattung Frailea (Br. & R.) Prestlé:* 215 (1997), without latin descr. or type—Brazil South (Rio Grande do Sul).
var. perumbilicata *(F. Ritter) K.H. Prestlé, Gattung Frailea (Br. & R.) Prestlé:* 213 (1997): F. perumbilicata.
subvar. spinosior *(F. Ritter) K.H. Prestlé, Gattung Frailea (Br. & R.) Prestlé:* 214 (1997): F. perumbilicata var. spinosior.
cataphracta (Dams) Britton & Rose
subsp. melitae *(Buining & Brederoo) K.H. Prestlé, Gattung Frailea (Br. & R.) Prestlé:* 63 (1997), isonym; *see P.J. Braun & Esteves in Succulenta*, 74(3): 130 (1995): F. melitae.
caveraensis *K.H. Prestlé, Gattung Frailea (Br. & R.) Prestlé:* 96 (1997), without latin descr. or type—Brazil South (Rio Grande do Sul).
crucifica *K.H. Prestlé, Gattung Frailea (Br. & R.) Prestlé:* 143 (1997), without latin descr. or type—Brazil South (Rio Grande do Sul).
fecotriga *K.H. Prestlé, Gattung Frailea (Br. & R.) Prestlé:* 48 (1997), without latin descr. or type—Brazil South (Rio Grande do Sul).
gigantea *K.H. Prestlé, Gattung Frailea (Br. & R.) Prestlé:* 142 (1997), without latin descr. or type—Brazil South (Rio Grande do Sul).
gracillima
subsp. alacriportana *(Backeb. & Voll) N. Gerloff in Mitteilungsbl. Frailea*, 2(1): 15 (1997); *cf. Repert. Pl. Succ. (I.O.S.)*, 48: 9 (1997 publ. 1998): F. alacriportana.
subsp. luteiflora *K.H. Prestlé, Gattung Frailea (Br. & R.) Prestlé:* 111 (1997), without latin descr. or type—Brazil South (Rio Grande do Sul).
grahliana (F. Haage) Britton & Rose
subsp. concepcionensis *(Buining & G. Moser) K.H. Prestlé, Gattung Frailea (Br. & R.) Prestlé:* 190 (1997), as 'conceptionensis', with incorrect basionym ref.: F. concepcionensis.
subsp. moseriana *(Buining & Brederoo) K.H. Prestlé, Gattung Frailea (Br. & R.) Prestlé:* 196 (1997): F. moseriana.
subsp. ybatensis *(Buining & G. Moser) K.H. Prestlé, Gattung Frailea (Br. & R.) Prestlé:* 195 (1997), as 'ybatense', with incorrect basionym ref.: F. ybatensis.
hlineckyana *L. Červinka in Frailea*, 1971(2): 13 (1971)—Locality not stated.
horstii
subsp. camaquaensis *K.H. Prestlé, Gattung Frailea (Br. & R.) Prestlé:* 132 (1997), without latin descr. or type—Brazil South (Rio Grande do Sul).
subsp. fecotrigensis *K.H. Prestlé, Gattung Frailea (Br. & R.) Prestlé:* 131 (1997)—Brazil South (Rio Grande do Sul).
subsp. nanus *K.H. Prestlé, Gattung Frailea (Br. & R.) Prestlé:* 130 (1997), without latin descr. or type—Brazil South (Rio Grande do Sul).
jajoiana *L. Červinka in Frailea*, 1971(2): 13 (1971)—Locality not stated.
klingleriana *K.H. Prestlé, Gattung Frailea (Br. & R.) Prestlé:* 191 (1997), as 'klinglerana', without latin descr. or type—Bolivia.
lepta *K.H. Prestlé, Gattung Frailea (Br. & R.) Prestlé:* 139 (1997), without latin descr. or type—Brazil South (Rio Grande do Sul).
machadoa *K.H. Prestlé, Gattung Frailea (Br. & R.) Prestlé:* 120 (1997), without type—Brazil South (Rio Grande do Sul).
subsp. gerloffii *K.H. Prestlé, Gattung Frailea (Br. & R.) Prestlé:* 123 (1997), without latin descr. or type—Brazil South (Rio Grande do Sul).

FRAILEA:-
machadoa
subsp. piratiniensis *K.H. Prestlé, Gattung Frailea (Br. & R.) Prestlé:* 122 (1997), species name not validly publ.—Brazil South (Rio Grande do Sul).
subsp. torinhaensis *K.H. Prestlé, Gattung Frailea (Br. & R.) Prestlé:* 123 (1997), without latin descr. or type—Brazil South (Rio Grande do Sul).
magnifica *Buining ex K.H. Prestlé, Gattung Frailea (Br. & R.) Prestlé:* 56 (1997), without latin descr. or type—Brazil South (Rio Grande do Sul).
mammifera Buining & Brederoo
subsp. robusta *K.H. Prestlé, Gattung Frailea (Br. & R.) Prestlé:* 51 (1997), without latin descr. or type—Brazil South (Rio Grande do Sul).
neoaureispina *K.H. Prestlé, Gattung Frailea (Br. & R.) Prestlé:* 95 (1997), as 'neo-aureispina', without latin descr. or type—Brazil South (Rio Grande do Sul).
f. albispina *K.H. Prestlé, Gattung Frailea (Br. & R.) Prestlé:* 96 (1997), without latin descr. or type—Brazil South (Rio Grande do Sul).
neogracillima *K.H. Prestlé, Gattung Frailea (Br. & R.) Prestlé:* 112 (1997), without latin descr. or type—Brazil South (Rio Grande do Sul).
oreopygmaea
subsp. afilarensis *K.H. Prestlé, Gattung Frailea (Br. & R.) Prestlé:* 70 (1997), without latin descr. or type—Uruguay.
var. aurea *(Backeb.) K.H. Prestlé, Gattung Frailea (Br. & R.) Prestlé:* 71 (1997), species name not validly publ.: F. aurea.
var. bretanhaensis *K.H. Prestlé, Gattung Frailea (Br. & R.) Prestlé:* 73 (1997), without latin descr. or type—Brazil South (Rio Grande do Sul).
var. dadakii *(Frič) K.H. Prestlé, Gattung Frailea (Br. & R.) Prestlé:* 67 (1997), without basionym ref.: Frailea dadakii.
var. feeburgii *K.H. Prestlé, Gattung Frailea (Br. & R.) Prestlé:* 71 (1997), without latin descr. or type—Uruguay.
subsp. gualeguaychuensis *K.H. Prestlé, Gattung Frailea (Br. & R.) Prestlé:* 70 (1997), without latin descr. or type—Argentina Northeast (Entre Ríos).
var. miguela *K.H. Prestlé, Gattung Frailea (Br. & R.) Prestlé:* 73 (1997), without latin descr. or type—Uruguay.
subsp. pichinangaensis *K.H. Prestlé, Gattung Frailea (Br. & R.) Prestlé:* 69 (1997), without latin descr. or type—Uruguay.
var. portoalegretensis *K.H. Prestlé, Gattung Frailea (Br. & R.) Prestlé:* 70 (1997), without latin descr. or type—Brazil South (Rio Grande do Sul).
phaeacantha *K.H. Prestlé, Gattung Frailea (Br. & R.) Prestlé:* 41 (1997), without latin descr. or type—Uruguay.
subsp. cupularia *K.H. Prestlé, Gattung Frailea (Br. & R.) Prestlé:* 43 (1997), without latin descr. or type—Uruguay.
f. glauca *K.H. Prestlé, Gattung Frailea (Br. & R.) Prestlé:* 45 (1997), without latin descr. or type—Uruguay.
subsp. heliosa *K.H. Prestlé, Gattung Frailea (Br. & R.) Prestlé:* 42 (1997), without latin descr. or type—Uruguay.
f. rubrispinus *K.H. Prestlé, Gattung Frailea (Br. & R.) Prestlé:* 44 (1997), without latin descr. or type—Uruguay.
subsp. treinta *K.H. Prestlé, Gattung Frailea (Br. & R.) Prestlé:* 45 (1997), without latin descr. or type—Uruguay.
subsp. zapicanensis *K.H. Prestlé, Gattung Frailea (Br. & R.) Prestlé:* 44 (1997), without latin descr. or type—Uruguay.
phaeoareolata *K.H. Prestlé, Gattung Frailea (Br. & R.) Prestlé:* 46 (1997), without latin descr. or type—Brazil South (Rio Grande do Sul).
subsp. ampliata *K.H. Prestlé, Gattung Frailea (Br. & R.) Prestlé:* 48 (1997), without latin descr. or type—Brazil South (Rio Grande do Sul).
subsp. santaritensis *K.H. Prestlé, Gattung Frailea (Br. & R.) Prestlé:* 47 (1997), without latin descr. or type—Brazil South (Rio Grande do Sul).

FRAILEA:-
pilzii *K.H. Prestlé, Gattung Frailea (Br. & R.) Prestlé:* 192 (1997), without latin descr. or type—Paraguay.
pseudocataphracta *K.H. Prestlé, Gattung Frailea (Br. & R.) Prestlé:* 49 (1997), without latin descr. or type—Brazil South (Rio Grande do Sul).
pumila (Lem.) Britton & Rose
subsp. albiareolata *(Buining & Brederoo) K.H. Prestlé, Gattung Frailea (Br. & R.) Prestlé:* 166 (1997): F. albiareolata.
subsp. alegretensis *K.H. Prestlé, Gattung Frailea (Br. & R.) Prestlé:* 164 (1997), without latin descr. or type—Brazil South (Rio Grande do Sul).
subsp. aquatica *K.H. Prestlé, Gattung Frailea (Br. & R.) Prestlé:* 170 (1997), without latin descr. or type—Uruguay.
subsp. aurantiaca *K.H. Prestlé, Gattung Frailea (Br. & R.) Prestlé:* 182 (1997), without latin descr. or type—Argentina Northeast (Corrientes).
subsp. boosiana *K.H. Prestlé, Gattung Frailea (Br. & R.) Prestlé:* 183 (1997), without latin descr. or type—Argentina Northeast (Misiones).
subsp. chrysacantha *(Hrabě) K.H. Prestlé, Gattung Frailea (Br. & R.) Prestlé:* 158 (1997), without full basionym ref.: F. chrysacantha.
subsp. colombiana *(Werderm.) K.H. Prestlé, Gattung Frailea (Br. & R.) Prestlé:* 156 (1997), as 'columbiana', basionym 'Frailea columbiana' in error: Echinocactus colombianus.
subsp. concordiaensis *K.H. Prestlé, Gattung Frailea (Br. & R.) Prestlé:* 151 (1997), without latin descr. or type—Argentina Northeast (Entre Ríos).
subsp. corrienta *K.H. Prestlé, Gattung Frailea (Br. & R.) Prestlé:* 181 (1997), without latin descr. or type—Argentina Northeast (Corrientes).
var. curvispina *K.H. Prestlé, Gattung Frailea (Br. & R.) Prestlé:* 180 (1997), without latin descr. or type—Brazil South (Rio Grande do Sul).
subsp. deminuta *(Buining & Brederoo) K.H. Prestlé, Gattung Frailea (Br. & R.) Prestlé:* 170 (1997): F. deminuta.
subsp. editha *K.H. Prestlé, Gattung Frailea (Br. & R.) Prestlé:* 153 (1997), without latin descr. or type—Brazil South (Rio Grande do Sul).
subsp. hlineckiana *(Červinka) K.H. Prestlé, Gattung Frailea (Br. & R.) Prestlé:* 159 (1997): F. hlineckiana.
subsp. iguaraica *K.H. Prestlé, Gattung Frailea (Br. & R.) Prestlé:* 179 (1997), without latin descr. or type—Brazil South (Rio Grande do Sul).
subsp. jajoiana *(Červinka) K.H. Prestlé, Gattung Frailea (Br. & R.) Prestlé:* 160 (1997): F. jajoiana.
subsp. meseta *K.H. Prestlé, Gattung Frailea (Br. & R.) Prestlé:* 151 (1997), without latin descr. or type—Uruguay.
subsp. multiceps *K.H. Prestlé, Gattung Frailea (Br. & R.) Prestlé:* 167 (1997), without latin descr. or type—Uruguay.
subsp. necopina *K.H. Prestlé, Gattung Frailea (Br. & R.) Prestlé:* 173 (1997), without latin descr. or type—Brazil South (Rio Grande do Sul).
subsp. neoeditha *K.H. Prestlé, Gattung Frailea (Br. & R.) Prestlé:* 154 (1997), without latin descr. or type—Brazil South (Rio Grande do Sul).
subsp. pubispina *K.H. Prestlé, Gattung Frailea (Br. & R.) Prestlé:* 165 (1997), without latin descr. or type—Brazil South (Rio Grande do Sul).
*subsp.*tenuissima *K.H. Prestlé, Gattung Frailea (Br. & R.) Prestlé:* 171 (1997), without latin descr. or type—Locality not stated.
subsp. uruguayana *K.H. Prestlé, Gattung Frailea (Br. & R.) Prestlé:* 152 (1997), without latin descr. or type—Brazil South (Rio Grande do Sul).
pygmaea
var. aceguaensis *K.H. Prestlé, Gattung Frailea (Br. & R.) Prestlé:* 103 (1997), without latin descr. or type—Uruguay.
subsp. albicolumnaris *(F. Ritter) A. Hofacker in Cactaceae Consensus Initiatives,* 6: 11 (1998): F. albicolumnaris.

FRAILEA:-
pygmaea
var. applanata *K.H. Prestlé, Gattung Frailea (Br. & R.) Prestlé:* 81 (1997), without latin descr. or type—Uruguay.
var. argenta *K.H. Prestlé, Gattung Frailea (Br. & R.) Prestlé:* 87 (1997), without latin descr. or type—Uruguay.
var. auricantha *K.H. Prestlé, Gattung Frailea (Br. & R.) Prestlé:* 81 (1997), without latin descr. or type—Uruguay.
var. bagensis *K.H. Prestlé, Gattung Frailea (Br. & R.) Prestlé:* 102 (1997), without latin descr. or type—Brazil South (Rio Grande do Sul).
subvar. cabucaensis *K.H. Prestlé, Gattung Frailea (Br. & R.) Prestlé:* 103 (1997), without latin descr. or type—Brazil South (Rio Grande do Sul).
var. carpinteria *K.H. Prestlé, Gattung Frailea (Br. & R.) Prestlé:* 78 (1997), without latin descr. or type—Uruguay.
var. castaneoidea *K.H. Prestlé, Gattung Frailea (Br. & R.) Prestlé:* 79 (1997), without latin descr. or type. Uruguay.
var. corallina *K.H. Prestlé, Gattung Frailea (Br. & R.) Prestlé:* 82 (1997), without latin descr. or type—Uruguay.
*var.*gemmrichii *K.H. Prestlé, Gattung Frailea (Br. & R.) Prestlé:* 98 (1997), without latin descr. or type—Brazil South (Rio Grande do Sul).
var. gloriosa *K.H. Prestlé, Gattung Frailea (Br. & R.) Prestlé:* 79 (1997), without latin descr. or type. Uruguay.
subsp. hyssooniana *K.H. Prestlé, Gattung Frailea (Br. & R.) Prestlé:* 83 (1997), without latin descr. or type—Uruguay.
var. insignia *K.H. Prestlé, Gattung Frailea (Br. & R.) Prestlé:* 91 (1997), without latin descr. or type—Brazil South (Rio Grande do Sul).
var. krausiana *K.H. Prestlé, Gattung Frailea (Br. & R.) Prestlé:* 80 (1997), without latin descr. or type—Uruguay.
var. largoensis *K.H. Prestlé, Gattung Frailea (Br. & R.) Prestlé:* 77 (1997), without latin descr. or type—Uruguay.
subvar. maldonadensis *K.H. Prestlé, Gattung Frailea (Br. & R.) Prestlé:* 76 (1997), without latin descr. or type—Uruguay.
subvar. mammilata *K.H. Prestlé, Gattung Frailea (Br. & R.) Prestlé:* 88 (1997), without latin descr. or type—Brazil South (Rio Grande do Sul).
var. mollispina-brunnea *K.H. Prestlé, Gattung Frailea (Br. & R.) Prestlé:* 99 (1997), without latin descr. or type—Brazil South (Rio Grande do Sul).
var. montana *K.H. Prestlé, Gattung Frailea (Br. & R.) Prestlé:* 81 (1997), without latin descr. or type—Uruguay.
var. olimarensis *K.H. Prestlé, Gattung Frailea (Br. & R.) Prestlé:* 82 (1997), without latin descr. or type—Uruguay.
var. pulcherrima *(Arechav.) K.H. Prestlé, Gattung Frailea (Br. & R.) Prestlé:* 84 (1997): Echinocactus pulcherrimus.
var. santanaensis *K.H. Prestlé, Gattung Frailea (Br. & R.) Prestlé:* 87 (1997), without latin descr. or type—Brazil South (Rio Grande do Sul).
subvar. trainta *K.H. Prestlé, Gattung Frailea (Br. & R.) Prestlé:* 77 (1997), without latin descr. or type—Uruguay.
subvar. transqueraensis *K.H. Prestlé, Gattung Frailea (Br. & R.) Prestlé:* 78 (1997), without latin descr. or type—Uruguay.
var. viridilunula *K.H. Prestlé, Gattung Frailea (Br. & R.) Prestlé:* 97 (1997), without latin descr. or type—Brazil South (Rio Grande do Sul).
ritteriana *K.H. Prestlé, Gattung Frailea (Br. & R.) Prestlé:* 142 (1997), as 'ritterana', without latin descr. or type—Brazil South (Rio Grande do Sul).
rufida *K.H. Prestlé, Gattung Frailea (Br. & R.) Prestlé:* 42 (1997), without latin descr. or type—Uruguay.
schlosseri *K.H. Prestlé, Gattung Frailea (Br. & R.) Prestlé:* 39 (1997), without latin descr. or type—Uruguay.

FRAILEA:-
stockingeri *K.H. Prestlé, Gattung Frailea (Br. & R.) Prestlé:* 136 (1997)—Brazil South (Rio Grande do Sul).
f. griseospina *K.H. Prestlé, Gattung Frailea (Br. & R.) Prestlé:* 138 (1997), without latin descr. or type—Brazil South (Rio Grande do Sul).
f. viridicastanea *K.H. Prestlé, Gattung Frailea (Br. & R.) Prestlé:* 138 (1997), without latin descr. or type—Brazil South (Rio Grande do Sul).
tunilla *K.H. Prestlé, Gattung Frailea (Br. & R.) Prestlé:* 202 (1997), without latin descr. or type—Paraguay.

FRANCOA (Saxifragac.).
alba *H. Jacq. in Ann. Fl. Pomone,* 1834–35: 112 (1835)—Chile.

FRANKENIA (Frankeniac.).
ericifolia C. Sm. ex DC.
subsp. caboverdeana *C. Brochmann, W. Lobin & P. Sunding in Nordic J. Bot.,* 15(6): 620 (1995 publ. 1996)—Cape Verde.
subsp. latifolia *(Webb & Berthel.) C. Brochmann, W. Lobin & P. Sunding in Nordic J. Bot.,* 15(6): 622 (1995 publ. 1996): F. ericifolia var. latifolia.
subsp. montana *C. Brochmann, W. Lobin & P. Sunding in Nordic J. Bot.,* 15(6): 622 (1995 publ. 1996)—Cape Verde.

FRASERA (Gentianac.).
bimaculata *(Siebold & Zucc.) H. Toyokuni in Symbolae Asahikawenses,* 1: 156 (1965): Ophelia bimaculata.
diluta *(Turcz.) H. Toyokuni in Symbolae Asahikawenses,* 1: 155 (1965): Gentiana diluta.
japonica *(Schult.) H. Toyokuni in Symbolae Asahikawenses,* 1: 156 (1965): Gentiana japonica.
pseudochinensis *(H. Hara) H. Toyokuni in Symbolae Asahikawenses,* 1: 156 (1965): Swertia pseudochinensis.
tashiroi *(Maxim.) H. Toyokuni in Symbolae Asahikawenses,* 1: 156 (1965): Ophelia tashiroi.
tetrapetala *(Pall.) H. Toyokuni in Symbolae Asahikawenses,* 1: 155 (1965): Swertia tetrapetala.
f. leucantha *(Hid. Takah.) H. Toyokuni & Y. Toyokuni in J. Asahikawa Univ.,* 4: 182 (1976): Swertia micrantha f. leucantha.
umbellata *(Makino) H. Toyokuni in Symbolae Asahikawenses,* 1: 156 (1965): Swertopsis umbellata.

FRAXINUS (Oleac.).
rhynchophylla Hance
var. densata *(Nakai) Y.N. Lee, Fl. Korea:* 1159 (1996): F. densata.

FREATULINA *J. Chrtek & Z. Slavíková in Čas. Nár. Muz. Řada Přír.,* 165(1–4): 140 (1996). *DROSERACEAE*
regia *(Stephens) J. Chrtek & Z. Slavíková in Čas. Nár. Muz. Řada Přír.,* 165(1–4): 140 (1996): Drosera regia.

FREYCINETIA (Pandanac.).
acicularis *K.-L. Huynh in Bot. Jahrb.,* 121(2): 161 (1999)—New Guinea (Papua New Guinea).
acuta *K.-L. Huynh in Candollea,* 55(2): 291 (2000)—New Guinea (Papua New Guinea).
ancistrosperma *K.-L. Huynh in Bot. Jahrb.,* 118(4): 540 (1996)—Bismarck Archipelago.
brachyclada *K.-L. Huynh in Bot. Jahrb.,* 121(2): 151 (1999)—Bismarck Archipelago.
brevipedunculata *K.-L. Huynh in Candollea,* 52(2): 369 (1997)—Solomon Is. (South Solomons, North Solomons).
brevis *K.-L. Huynh in Candollea,* 55(2): 284 (2000)—New Guinea (Papua New Guinea).
carnosa *K.-L. Huynh in Bot. Jahrb.,* 121(2): 159 (1999)—New Guinea (Papua New Guinea).
chartacea *K.-L. Huynh in Candollea,* 55(2): 285 (2000)—New Guinea (Papua New Guinea).
coagmentata *K.-L. Huynh in Candollea,* 52(2): 379 (1997)—Solomon Is. (South Solomons).
cochleatisperma *K.-L. Huynh in Bot. Jahrb.,* 118(4): 541 (1996)—Bismarck Archipelago.
concolor *K.-L. Huynh in Candollea,* 55(2): 287 (2000)—New Guinea (Papua New Guinea).

FREYCINETIA:-
craterensis *K.-L. Huynh in Candollea,* 55(2): 291 (2000)—New Guinea (Papua New Guinea).
dissita *K.-L. Huynh in Bot. Jahrb.,* 121(2): 167 (1999)—New Guinea (Papua New Guinea).
ellipticifolia *K.-L. Huynh in Bot. Jahrb.,* 121(2): 163 (1999)—New Guinea (Papua New Guinea).
falcata *K.-L. Huynh in Candollea,* 55(2): 286 (2000)—Bismarck Archipelago.
filifolia *K.-L. Huynh in Bot. Jahrb.,* 121(2): 155 (1999)—New Guinea.
filiformis *K.-L. Huynh in Bot. Jahrb.,* 118(4): 537 (1996)—Bismarck Archipelago.
fusifolia *K.-L. Huynh in Bot. Jahrb.,* 118(4): 532 (1996)—Bismarck Archipelago.
glaucifolia *K.-L. Huynh in Bot. Jahrb.,* 121(2): 172 (1999)—New Guinea (Papua New Guinea).
glomerosa *K.-L. Huynh in Bot. Jahrb.,* 121(2): 165 (1999)—New Guinea (Papua New Guinea).
insulana *K.-L. Huynh in Candollea,* 52(2): 374 (1997)—Solomon Is. (North Solomons).
lepida *K.-L. Huynh in Candollea,* 52(2): 378 (1997)—Solomon Is. (South Solomons).
lolokiensis *K.-L. Huynh in Bot. Jahrb.,* 121(2): 158 (1999)—New Guinea.
longifolia *K.-L. Huynh in Candollea,* 52(2): 374 (1997)—Solomon Is. (South Solomons).
louisiadensis *K.-L. Huynh in Bot. Jahrb.,* 121(2): 175 (1999)—New Guinea (Papua New Guinea).
lunata *K.-L. Huynh in Bot. Jahrb.,* 121(2): 170 (1999)—New Guinea (Papua New Guinea).
madangensis *K.-L. Huynh in Candollea,* 55(2): 286 (2000)—New Guinea (Papua New Guinea).
manusensis *K.-L. Huynh in Bot. Jahrb.,* 121(2): 166 (1999)—Bismarck Archipelago.
misimica *K.-L. Huynh in Bot. Jahrb.,* 121(2): 169 (1999)—New Guinea (Papua New Guinea).
morobeensis *K.-L. Huynh in Candollea,* 55(2): 292 (2000)—New Guinea (Papua New Guinea).
obtusiacuminata *K.-L. Huynh in Candollea,* 55(2): 290 (2000)—New Guinea (Papua New Guinea).
pallida *K.-L. Huynh in Bot. Jahrb.,* 118(4): 534 (1996)—Bismarck Archipelago.
pauperata *K.-L. Huynh in Bot. Jahrb.,* 121(2): 160 (1999)—New Guinea.
plana *K.-L. Huynh in Candollea,* 52(2): 376 (1997)—Solomon Is. (North Solomons).
pseudohombronii *K.-L. Huynh in Candollea,* 52(2): 371 (1997)—Solomon Is. (South Solomons).
scabrida *K.-L. Huynh in Bot. Jahrb.,* 121(2): 173 (1999)—New Guinea (Papua New Guinea).
sepikensis *K.-L. Huynh in Bot. Jahrb.,* 121(2): 177 (1999)—New Guinea (Papua New Guinea).
setosa *K.-L. Huynh in Bot. Jahrb.,* 121(2): 171 (1999)—New Guinea (Papua New Guinea).
takeuchii *K.-L. Huynh in Candollea,* 55(2): 289 (2000)—New Guinea (Papua New Guinea).
tenella *K.-L. Huynh in Candollea,* 52(2): 363 (1997)—Solomon Is. (South Solomons), Bismarck Archipelago.

FREZIERA (Theac.).
tomentosa *(Ruiz & Pav.) L.R. Tulasne in Ann. Sci. Nat., Bot.,* Sér. 3, 8: 327 (1847): Lettsomia tomentosa.†

FRITHIA (Aizoac.).
humilis *P.M. Burgoyne in Bothalia,* 30(1): 1 (2000)—Transvaal.

FRITILLARIA (Liliac.).
atrolineata *G. Bakhshi Khaniki in Edinburgh J. Bot.,* 54(2): 171 (1997)—Iran.
chitralensis *(hort.) B. Mathew in Curtis's Bot. Mag.,* 13(1): 30 (1996): F. imperialis var. chitralensis.
chlororhabdota *G. Bakhshi Khaniki in Herbertia,* 52: 140 (1997)—Iran.
lusitanica Wikstr.
subsp. macrocarpa *(Coss. ex Batt.) B. Valdés in Lagascalia,* 18(2): 327 (1996): F. macrocarpa.
subsp. oranensis *(Pomel) B. Valdés in Lagascalia,* 18(2): 327 (1996): F. oranensis.
pallidiflora Schrenk
var. halabulanica *(X.Z. Duan & X.J. Zheng) G.J. Liu in Fl. Xinjiangensis,* 6: 507 (1996): F. halabulanica.
pelinaea *G. Kamari in Bocconea,* 5(1): 227 (1996)—East Aegean Is.

FRITILLARIA:-
poluninii *(Rix) G. Bakhshi Khaniki & K. Persson in Nordic J. Bot.,* 17(6): 608 (1997 publ. 1998): F. crassifolia subsp. poluninii.
sororum *Jimmy Persson & K. Persson in New Plantsman,* 5(2): 114 (1998)—Turkey.
verticillata Willd.
var. albidiflora *(X.Z. Duan & X.J. Zheng) G.J. Liu in Fl. Xinjiangensis,* 6: 513 (1996): F. albidiflora.
var. purpurea *(X.Z. Duan & X.J. Zheng) G.J. Liu in Fl. Xinjiangensis,* 6: 513 (1996), basionym as 'F. purpurea', without exact basionym page: F. albidiflora var. purpurea.
var. rhodanthera *(X.Z. Duan & X.J. Zheng) G.J. Liu in Fl. Xinjiangensis,* 6: 515 (1996), basionym as 'F. rhodanthera', without exact basionym page: F. albidiflora var. rhodanthera.
walujewii Regel
var. xinyanensis *(Y.K. Yang & J.K. Wu) G.J. Liu in Fl. Xinjiangensis,* 6: 510 (1996), without basionym ref.: F. xinyanensis.

FROESIA (Quiinac.).
gereauana *J. Schneider & G. Zizka in Novon,* 7(4): 408 (1997 publ. 1998)—Venezuela.

FUCHSIA (Onagrac.).
barclayana *A. Cels & J.F. Cels in Ann. Fl. Pomone,* 1836–37: 154 (1837)—Locality not stated.
campii *P.E. Berry in Novon,* 5(4): 318 (1995)—Ecuador.
encliandra Steud.
subsp. microphylloides *P.E. Berry & D.E. Breedlove in Novon,* 6(2): 139 (1996)—Mexico Southwest (Oaxaca, Guerrero, Michoacan).
globifera *Jacques in Ann. Fl. Pomone,* 1834–35: 246 (1835)—Locality not stated.
mezae *P.E. Berry & E. Hermsen in Novon,* 9(4): 480 (1999)—Peru.
microphylla Humb. Bonpl. & Kunth
subsp. chiapensis *(Brandegee) P.E. Berry & D.E. Breedlove in Novon,* 6(2): 137 (1996): F. chiapensis.
paniculata Lindl.
subsp. mixensis *P.E. Berry & D.E. Breedlove in Novon,* 6(2): 135 (1996)—Mexico Southwest (Oaxaca).
rivularis J.F. Macbr.
subsp. pubescens *P.E. Berry & E. Hermsen in Novon,* 9(4): 479 (1999)—Peru.
summa *P.E. Berry in Novon,* 5(4): 320 (1995)—Ecuador.
virgata *Sweet ex Jacques in Ann. Fl. Pomone,* 1834–35: 85 (1834)—Mexico.

FUERTESIMALVA *P.A. Fryxell in Sida,* 17(1): 70 (1996). *MALVACEAE*
chilensis *(A. Braun & C.D. Bouché) P.A. Fryxell in Sida,* 17(1): 72 (1996): Malva chilensis.
corniculata *(Krapov.) P.A. Fryxell in Sida,* 17(1): 72 (1996): Urocarpidium corniculatum.
echinata *(C. Presl) P.A. Fryxell in Sida,* 17(1): 72 (1996): Malva echinata.
insularis *(Kearney) P.A. Fryxell in Sida,* 17(1): 73 (1996): Malvastrum insulare.
jacens *(S. Watson) P.A. Fryxell in Sida,* 17(1): 73 (1996): Malvastrum jacens.
killipii *(Krapov.) P.A. Fryxell in Sida,* 17(1): 73 (1996): Urocarpidium killipii.
leptocalyx *(Krapov.) P.A. Fryxell in Sida,* 17(1): 73 (1996): Urocarpidium leptocalyx.
limensis *(L.) P.A. Fryxell in Sida,* 17(1): 74 (1996): Malva limensis.
pennellii *(Ulbr.) P.A. Fryxell in Sida,* 17(1): 74 (1996): Malvastrum pennellii.
pentacocca *(Krapov.) P.A. Fryxell in Sida,* 17(1): 74 (1996): Urocarpidium pentacoccum.
pentandra *(K. Schum.) P.A. Fryxell in Sida,* 17(1): 74 (1996): Malvastrum pentandrum.
peruviana *(L.) P.A. Fryxell in Sida,* 17(1): 74 (1996): Malva peruviana.
sanambrosiana *(D.M. Bates) P.A. Fryxell in Sida,* 17(1): 75 (1996): Urocarpidium sanambrosianum.
stipulata *(Fryxell) P.A. Fryxell in Sida,* 17(1): 75 (1996): Urocarpidium stipulatum.

FUIRENA (Cyperac.).
boreocoerulescens *K.A. Lye in Candollea,* 51(1): 1 (1996)—Somalia.
lainzii *M. Luceño & M. Alves in An. Jard. Bot. Madrid,* 54(1): 415 (1996)—Brazil Northeast (Pernambuco).
striatella *K.A. Lye in Candollea,* 51(1): 4 (1996)—Somalia.

FUMANA (Cistac.).
fontqueri *J. Güemes in Folia Geobot.,* 34(3): 365 (1999)—Morocco.
thymifolia (L.) Spach ex Webb
subsp. barrelieri *(Ten.) J. Duvigneaud in Soc. Ech. Pl. Vasc. Eur. Bassin Médit. Bull.,* 20: 50 (1985): Helianthemum barrelieri.

FUMANOPSIS (Cistac.).
laevis *(Cav.) N.N. Tzvelev in Fl. Vostochnoĭ Evropy̆,* 9: 215 (1996): Cistus laevis.

FUNASTRUM (Asclepiadac.).
arenarium *(Decne. ex Benth.) S. Liede in Syst. Bot.,* 21(1): 43 (1996): Sarcostemma arenarium.
cynanchoides (Decne.) Schltr.
var. hartwegii *(Vail) A. Krings in Sida,* 19(1): 137 (2000): Philibertella hartwegii.
antillana *A. Alvarez de Zayas in An. Inst. Biol. Univ. Nac. Autón. Méx., Bot.,* 67(2): 331 (1996)—Cuba, Dominican Republic, Puerto Rico, Haiti.

FURCRAEA (Agavac.).
niquivilensis *Matuda ex A. García-Mendoza in Novon,* 9(1): 42 (1999)—Mexico Southeast (Chiapas).

FUSIFILUM (Hyacinthac.).
capitatum *(Hook.) F. Speta in Phyton (Austria),* 38(1): 69 (1998): Ornithogalum capitatum.
dregei *(Baker) F. Speta in Phyton (Austria),* 38(1): 69 (1998): Urginea dregei.
gracilis *(A.V. Duthie) F. Speta in Phyton (Austria),* 38(1): 69 (1998): Urginea gracilis.
minor *(A.V. Duthie) F. Speta in Phyton (Austria),* 38(1): 69 (1998): Urginea minor.
physodes *(Jacq.) F. Speta in Phyton (Austria),* 38(1): 69 (1998): Anthericum physodes.
pusillum *(Jacq.) F. Speta in Phyton (Austria),* 38(1): 70 (1998): Anthericum pusillum.
pygmaeum *(A.V. Duthie) F. Speta in Phyton (Austria),* 38(1): 69 (1998): Urginea pygmaea.

G

GAGEA (Liliac.).
africana *(A. Terracc.) I.G. Levichev in Komarovia,* 1: 49 (1999): G. reticulata subsp. africana.
aipetriensis *I.G. Levichev in Bot. Zhurn.,* 83(2): 111 (1998)—Krym.
ancestralis *I.G. Levichev in Bot. Zhurn.,* 83(2): 110 (1998)—Altay.
bonnassieui *M. Gandoger, Dec. Pl. Nov.,* 2: 17 (1876)—France.
brevistyla *M. Gandoger, Dec. Pl. Nov.,* 2: 18 (1876)—France.
gadzhievii *R.K. Askerova in Bot. Zhurn.,* 82(11): 84 (1997)—Transcaucasus (Azerbaydzhan).
glabriflora *M. Gandoger, Dec. Pl. Nov.,* 2: 16 (1876)—France.
longifolia *M. Gandoger, Dec. Pl. Nov.,* 2: 19 (1876)—France.
× luberonensis *J.-M. Tison in Monde Pl.,* 462: 7 (1998)—France. (G. bohemica × granatelli).
lugdunensis *M. Gandoger, Dec. Pl. Nov.,* 2: 19 (1876)—France.
mercklinii *(Pascher) I.G. Levichev in Komarovia,* 1: 53 (1999): G. stipitata var. mercklinii.
novoascanica *M.V. Klokov in Novosti Sist. Vy̆ssh. Nizsh. Rast. (Kiev),* 1974: 98 (1974)—Ukraina.
oligocephala *M. Gandoger, Dec. Pl. Nov.,* 2: 17 (1876)—France.

GAGEA:-
recondita *M. Gandoger, Dec. Pl. Nov.,* 2: 19 (1876)—France.
sarysuensis *G.Zh. Murzalieva in Izv. Natsion. Akad. Nauk Resp. Kazakhstan, Ser. Biol.,* 1993(6): 93 (1993 publ. 1994)—Kazakhstan.
schugnanica *I.G. Levichev & D. Navruzshoev in Bot. Zhurn.,* 82(9): 91 (1997)—Tadzhikistan.
subcongesta *M. Gandoger, Dec. Pl. Nov.,* 2: 20 (1876)—France.
takhtajanii *I.G. Levichev in Bot. Zhurn.,* 85(7): 186 (2000)—Uzbekistan.
versannei *M. Gandoger, Dec. Pl. Nov.,* 2: 18 (1876)—France.
villarsii *M. Gandoger, Dec. Pl. Nov.,* 2: 16 (1876)—Italy.
villosa (M. Bieb.) Duby
var. stenophylla *D. Delipavlov in Thaiszia,* 8(2): 126 (1998)—Bulgaria.

GAGNEBINA (Legumin.).
bakoliae *M. Luckow & D. Du Puy in Novon,* 10(3): 220 (2000)—Madagascar.

GAHNIA (Cyperac.).
howeana *R.O. Gardner in New Zealand J. Bot.,* 35(2): 155 (1997)—Lord Howe I.
sclerioides *K.L. Wilson in Nuytsia,* 11(2): 275 (1997)—Western Australia.

GAIADENDRON (Loranthac.).
revolutum *C.T. Rizzini in Revista Fac. Agron. (Maracay),* 8(3): 85 (1975)—Venezuela.

GAILLONIA (Rubiac.).
calcicola *(Puff) M. Thulin in Nordic J. Bot.,* 18(1): 36 (1998): Neogaillonia calcicola.
crocyllis *(Sond.) M. Thulin in Nordic J. Bot.,* 18(1): 36 (1998): Anthospermum crocyllis.
jolana *M. Thulin in Nordic J. Bot.,* 18(1): 34 (1998)—South Yemen.
somaliensis *(Puff) M. Thulin in Nordic J. Bot.,* 18(1): 36 (1998): Neogaillonia somaliensis.
yemenensis *M. Thulin in Nordic J. Bot.,* 18(1): 32 (1998)—North Yemen, South Yemen, Oman.

GALAGANIA (Umbellif.).
ferganensis *(Korovin) M.G. Vassiljeva & M.G. Pimenov in Opred. Rast. Sred. Azii,* 7: 222 (1983): Korovinia ferganensis.
platypoda *(Aitch. & Hemsl.) M.G. Vassiljeva & M.G. Pimenov in Novosti Sist. Vy̆ssh. Rast.,* 20: 156 (1983), as 'platypodae' in Supp. 18; *et in Opred. Rast. Sred. Azii,* 7: 223 (1983): Johrenia platypoda.†
tenuisecta *(Regel & Schmalh.) M.G. Vassiljeva & M.G. Pimenov in Opred. Rast. Sred. Azii,* 7: 223 (1983): Peucedanum tenuisectum.

GALANTHUS (Amaryllidac.).
subser. Glaucaefolii *(Kem.-Nath.) A.P. Davis, Genus Galanthus:* 115 (1999): G. sect. Glaucaefolii.
subser. Viridifolii *(Kem.-Nath.) A.P. Davis, Genus Galanthus:* 155 (1999): G. sect. Viridifolii.
alpinus Sosn.
var. bortkewitschianus *(Koss) A.P. Davis in Kew Bull.,* 51(4): 750 (1996): G. bortkewitschianus.
artjuschenkoae *E.Tz. Gabrieljan in Fl. Rastitel'nost' Rast. Res. Arm.,* 12: 13 (1999)—Transcaucasus (Armeniya).
byzantinus Baker
subsp. brauneri *N. Zeybek in Dogă Tu J. Bot.,* 12(1): 101 (1988), as 'braunerii'—Turkey.
subsp. saueri *N. Zeybek in Dogă Tu J. Bot.,* 12(1): 101 (1988), as 'sauerii'—Turkey.
subsp. tughrulii *N. Zeybek in Dogă Tu J. Bot.,* 12(1): 100 (1988)—Turkey.
elwesii Hook.f.
subsp. akmanii *N. Zeybek in Dogă Tu J. Bot.,* 12(1): 99 (1988)—Turkey.
subsp. baytopii *(Zeybek) N. Zeybek & E. Sauer, Türk. Kardelenleri/Beitr. Kenntn. Türk. Schneeglöckchen,* 1: 60 (1995): C. gracilis subsp. baytopii.
subsp. melihae *N. Zeybek in Dogă Tu J. Bot.,* 12(1): 98 (1988)—Turkey.
var. monostictus *P.D. Sell in P.D. Sell & G. Murrell, Fl. Great Britain & Ireland,* 5: 363 (1996)—Great Britain.

GALANTHUS:-
elwesii
subsp. tuebitaki *N. Zeybek in Dogă Tu J. Bot.,* 12(1): 98 (1988)—Turkey.
subsp. wagenitzii *N. Zeybek in Dogă Tu J. Bot.,* 12(1): 98 (1988)—Turkey.
subsp. yayintaschii *N. Zeybek in Dogă Tu J. Bot.,* 12(1): 98 (1988)—Turkey.
fosteri Baker
var. antepensis *N. Zeybek & E. Sauer, Türk. Kardelenleri/Beitr. Kenntn. Türk. Schneeglöckchen,* 1: 72 (1995), without latin descr.—Turkey.
gracilis Čelak.
subsp. baytopii *N. Zeybek in Dogă Tu J. Bot.,* 12(1): 96 (1988)—Turkey.
melihae *(Zeybek) E. Sauer & N. Zeybek, Türk. Kardelenleri/Beitr. Kenntn Türk. Schneeglöckchen,* 1: 54 (1995): G. elwesii subsp. melihae.
nivalis L.
subsp. humboldtii *N. Zeybek in Dogă Tu J. Bot.,* 12(1): 97 (1988)—Turkey.
f. pleniflorus *P.D. Sell in P.D. Sell & G. Murrell, Fl. Great Britain & Ireland,* 5: 363 (1996)—Great Britain.
subsp. subplicatus *(Zeybek) N. Zeybek & E. Sauer, Türk. Kardelenleri/Beitr. Kenntn. Türk. Schneeglöckchen,* 1: 47 (1995): G. plicatus subsp. subplicatus.
plicatus M. Bieb.
subsp. gueneri *N. Zeybek in Dogă Tu J. Bot.,* 12(1): 99 (1988), as 'guenerii'—Turkey.
subsp. karamanoghluensis *N. Zeybek in Dogă Tu J. Bot.,* 12(1): 100 (1988)—Turkey.
subsp. subplicatus *N. Zeybek in Dogă Tu J. Bot.,* 12(1): 99 (1988)—Turkey.
subsp. vardarii *N. Zeybek in Dogă Tu J. Bot.,* 12(1): 100 (1988)—Turkey.
var. viridifolius *P.D. Sell in P.D. Sell & G. Murrell, Fl. Great Britain & Ireland,* 5: 363 (1996)—Great Britain.

GALARHOEUS (Euphorbiac.).
alatavicus *(Boiss.) J.I. Prokhanov in Trudy̆ Kuĭby̆shevsk. Bot. Sada,* 1: 26 (1941): Euphorbia alatavica.
altotibeticus *(Paulsen) J.I. Prokhanov in Trudy̆ Kuĭby̆shevsk. Bot. Sada,* 1: 26 (1941): Euphorbia altotibetica.
andrachnoides *(Schrenk) J.I. Prokhanov in Trudy̆ Kuĭby̆shevsk. Bot. Sada,* 1: 63 (1941): Euphorbia andrachnoides.
borodinii *(Sambuk) J.I. Prokhanov in Trudy̆ Kuĭby̆shevsk. Bot. Sada,* 1: 71 (1941): Euphorbia borodinii.
buhsei *(Boiss.) J.I. Prokhanov in Trudy̆ Kuĭby̆shevsk. Bot. Sada,* 1: 46 (1941): Euphorbia buhsei.
falcatus *(L.) J.I. Prokhanov in Trudy̆ Kuĭby̆shevsk. Bot. Sada,* 1: 35 (1941): Euphorbia falcata.
glareosus *(M. Bieb.) J.I. Prokhanov in Trudy̆ Kuĭby̆shevsk. Bot. Sada,* 1: 40 (1941): Euphorbia glareosa.
gmelinii *(Steud.) J.I. Prokhanov in Trudy̆ Kuĭby̆shevsk. Bot. Sada,* 1: 50 (1941): Euphorbia gmelinii.
ibericus *(Boiss.) J.I. Prokhanov in Trudy̆ Kuĭby̆shevsk. Bot. Sada,* 1: 71 (1941): Euphorbia iberica.
irgisensis *(Litv.) J.I. Prokhanov in Trudy̆ Kuĭby̆shevsk. Bot. Sada,* 1: 46 (1941): Euphorbia irgisensis.
issykkulensis *(Prokh.) J.I. Prokhanov in Trudy̆ Kuĭby̆shevsk. Bot. Sada,* 1: 26 (1941): Tithymalus issykkulensis.
lamprocarpus *(Prokh.) J.I. Prokhanov in Trudy̆ Kuĭby̆shevsk. Bot. Sada,* 1: 27 (1941): Tithymalus lamprocarpus.
latifolius *(C.A. Mey.) J.I. Prokhanov in Trudy̆ Kuĭby̆shevsk. Bot. Sada,* 1: 73 (1941): Euphorbia latifolia.
leptocaulus *(Boiss.) J.I. Prokhanov in Trudy̆ Kuĭby̆shevsk. Bot. Sada,* 1: 46 (1941): Euphorbia leptocaula.
nudus *(Velen.) J.I. Prokhanov in Trudy̆ Kuĭby̆shevsk. Bot. Sada,* 1: 27 (1941): Euphorbia nuda.

GALARHOEUS:-
peplus *(L.) J.I. Prokhanov in Trudy Kuibyshevsk. Bot. Sada,* 1: 34 (1941): Euphorbia peplus.
petrophilus *(C.A. Mey.) J.I. Prokhanov in Trudy Kuibyshevsk. Bot. Sada,* 1: 40 (1941): Euphorbia petrophila.
praecox *(Fisch.) J.I. Prokhanov in Trudy Kuibyshevsk. Bot. Sada,* 1: 48 (1941), basionym not validly publ.: Euphorbia praecox.
procerus *(M. Bieb.) J.I. Prokhanov in Trudy Kuiby Kuibyshevsk. Bot. Sada,* 1: 33 (1941): Euphorbia procera.
sareptanus *(A.K. Becker) J.I. Prokhanov in Trudy Kuibyshevsk. Bot. Sada,* 1: 46 (1941): Euphorbia sareptana.
seguierianus *(Neck.) J.I. Prokhanov in Trudy Kuibyshevsk. Bot. Sada,* 1: 41 (1941): Euphorbia seguieriana.
semivillosus *(Prokh.) J.I. Prokhanov in Trudy Kuibyshevsk. Bot. Sada,* 1: 31 (1941): Tithymalus semivillosus.
soongaricus *(Boiss.) J.I. Prokhanov in Trudy Kuibyshevsk. Bot. Sada,* 1: 21 (1941): Euphorbia soongarica.
subcordatus *(C.A. Mey.) J.I. Prokhanov in Trudy Kuibyshevsk. Bot. Sada,* 1: 60 (1941): Euphorbia subcordata.
subtilis *J.I. Prokhanov in Trudy Kuibyshevsk. Bot. Sada,* 1: 48 (1941), nom. nov.: Euphorbia gracilis *M. Bieb.*
tanguticus *(Prokh.) J.I. Prokhanov in Trudy Kuibyshevsk. Bot. Sada,* 1: 26 (1941): Euphorbia tangutica.
tibeticus *(Boiss.) J.I. Prokhanov in Trudy Kuibyshevsk. Bot. Sada,* 1: 26 (1941): Euphorbia tibetica.
tranzschelii *(Prokh.) J.I. Prokhanov in Trudy Kuibyshevsk. Bot. Sada,* 1: 26 (1941): Tithymalus tranzschelii.
undulatus *(M. Bieb.) J.I. Prokhanov in Trudy Kuibyshevsk. Bot. Sada,* 1: 43 (1941): Euphorbia undulata.
uralensis *(Fisch. ex Link) J.I. Prokhanov in Trudy Kuibyshevsk. Bot. Sada,* 1: 54 (1941): Euphorbia uralensis.
villosus *(Waldst. & Kit.) J.I. Prokhanov in Trudy Kuibyshevsk. Bot. Sada,* 1: 33 (1941): Euphorbia villosa.
virgatus *(Waldst. & Kit.) J.I. Prokhanov in Trudy Kuibyshevsk. Bot. Sada,* 1: 58 (1941): Euphorbia virgata.
volgensis *(Krysht.) J.I. Prokhanov in Trudy Kuibyshevsk. Bot. Sada,* 1: 38 (1941): Euphorbia volgensis.

GALATELLA (Compos.).
sect. Chromopappae *Y. Wei & C.H. An in Fl. Xinjiangensis,* 5: 475 (1999), as 'Chrompappae'.
bogdaica *Y. Wei & C.H. An in Fl. Xinjiangensis,* 5: 475 (1999)—Xinjiang.

GALEANDRA (Orchidac.).
badia *L.A. Garay & G.A. Romero-González in Harvard Pap. Bot.,* 3(1): 58 (1998)—Venezuela.
biloba *L.A. Garay in Harvard Pap. Bot.,* 4(1): 309 (1999)—Peru.
duidensis *L.A. Garay & G.A. Romero-González in Harvard Pap. Bot.,* 3(1): 58 (1998)—Venezuela.
levyae *L.A. Garay in Harvard Pap. Bot.,* 4(1): 310 (1999)—Locality not stated.

GALEGA (Legumin.).
cirujanoi *P. García-Murillo & S. Talavera in An. Jard. Bot. Madrid,* 57(1): 218 (1999)—Spain.

GALEOBDOLON (Labiat.).
endtmannii *(G.H. Loos) J. Holub in Preslia,* 70(2): 104 (1998): Lamium endtmannii.
endtmannii *(G.H. Loos) K. Senghas in K. Senghas & S. Seybold, Schmeil & Fitschen Fl. Deutschland u. angrenz.* Länd., 91. Aufl.: 552 (2000), isonym; *see Holub in Preslia,* 70(2): 104 (1998): Lamium endtmannii.

GALEOPSIS (Labiat.).
ladanum L.
var. glandulosissima *N.N. Tzvelev in Bot. Zhurn.,* 82(12): 95 (1997)—Russia Northwest.

GALEOTTIA (Orchidac.).
peruviana *D.E. Bennett & E.A. Christenson in Lindleyana,* 13(1): 48 (1998)—Peru.

GALIANTHE (Rubiac.).
subgen. Ebelia *(Rchb.) E.L. Cabral & N.M. Bacigalupo in Ann. Missouri Bot. Gard.,* 84(4): 859 (1997 publ. 1998): gen. Ebelia.
andersonii *E.L. Cabral in Bonplandia,* 10(1–4): 119 (2000)—Brazil Southeast (Minas Gerais).
bogotensis *(Kunth) E.L. Cabral & N.M. Bacigalupo in Ann. Missouri Bot. Gard.,* 84(4): 859 (1997 publ. 1998): Spermacoce bogotensis.
brasiliensis *(Spreng.) E.L. Cabral & N.M. Bacigalupo in Ann. Missouri Bot. Gard.,* 84(4): 861 (1997 publ. 1998): Diodia brasiliensis.
subsp. angulata *(Benth.) E.L. Cabral & N.M. Bacigalupo in Ann. Missouri Bot. Gard.,* 84(4): 863 (1997 publ. 1998): Triodon angulatum.
cymosa *(Cham.) E.L. Cabral & N.M. Bacigalupo in Ann. Missouri Bot. Gard.,* 84(4): 865 (1997 publ. 1998): Diodia cymosa.
dichotoma *(Willd. ex Roem. & Schult.) E.L. Cabral & N.M. Bacigalupo in Ann. Missouri Bot. Gard.,* 84(4): 869 (1997 publ. 1998): Knoxia dichotoma.
hispidula *(A. Rich. ex DC.) E.L. Cabral & N.M. Bacigalupo in Ann. Missouri Bot. Gard.,* 84(4): 870 (1997 publ. 1998): Diodia hispidula.
humilis *E.L. Cabral & N.M. Bacigalupo in Ann. Missouri Bot. Gard.,* 84(4): 871 (1997 publ. 1998)—Brazil Southeast (Minas Gerais).
macedoi *E.L. Cabral in Bonplandia,* 10(1–4): 121 (2000)—Brazil West-Central (Goiás).
peruviana *(Pers.) E.L. Cabral in Bonplandia,* 10(1–4): 123 (2000):Spermacoce peruviana.
polygonoides *E.L. Cabral & N.M. Bacigalupo in Ann. Missouri Bot. Gard.,* 84(4): 875 (1997 publ. 1998)—Brazil Southeast (Rio de Janeiro).
reitzii *E.L. Cabral in Bonplandia,* 10(1–4): 124 (2000)—Brazil South (Santa Catarina).
souzae *E.L. Cabral & N.M. Bacigalupo in Bol. Soc. Argent. Bot.,* 34(3–4): 153 (2000)—Brazil Southeast (São Paulo).
vaginata *E.L. Cabral & N.M. Bacigalupo in Ann. Missouri Bot. Gard.,* 84(4): 875 (1997 publ. 1998)—Brazil Southeast (Minas Gerais, São Paulo).

GALINSOGA (Compos.).
parviflora Cav.
subsp. hispida *(DC.) O. de Bolòs & J. Vigo, Fl. Països Catalans,* 3: 775 (1995): G. parviflora var. hispida.

GALIUM (Rubiac.).
acutum Edgew.
var. himalayense *(Klotzsch & Garcke) R.R. Mill in Edinburgh J. Bot.,* 53(2): 195 (1996): G. himalayense.
armenum *I.A. Schanzer in Bot. Zhurn.,* 82(3): 126 (1997)—Transcaucasus (Armeniya).
attenuatum *M.V. Klokov & B.V. Zaverucha in Novosti Sist. Vyssh. Nizsh. Rast. (Kiev),* 1974: 82 (1974)—Ukraina.
boreale L.
subsp. ussuriense *(Pobed.) V.N. Voroschilov in A.K. Skvortsov (ed.), Florist. issl. v razn. raionakh SSSR:* 192 (1985): G. ussuriense.
caprarium *A. Natali in Candollea,* 53(2): 484 (1998)—Italy, Corse.
congestum *M.V. Klokov & B.V. Zaverucha in Novosti Sist. Vyssh. Nizsh. Rast. (Kiev),* 1974: 87 (1974)—Ukraina.
corsicum Spreng.
var. soleirolii *(Loisel.) A. Natali & D. Jeanmonod, Rubiaceae: Compl. Prodr. Fl. Corse (ser. ed. D. Jeanmonod):* 74 (2000): G. soleirolii.
craticulatum *R.R. Mill in Edinburgh J. Bot.,* 53(2): 202 (1996)—Bhutan.
dulongense *H. Li in Fl. Dulongjian Reg.:* 246 (1993), without latin descr. or type—China South-central (Yunnan).
fatrense *F. Ehrendorfer & H. Šípošová in Acta Bot. Slov. Acad. Sci. Slov., A, Taxon. Geobot.,* 10: 140 (1987), without type—Czechoslovakia.
glaucum L.
subsp. murcicum *(Boiss. & Reut.) O. de Bolòs & J. Vigo, Fl. Països Catalans,* 3: 569 (1995): G. murcicum.

GALIUM:-
hintoniorum *B.L. Turner in Phytologia,* 80(4): 285 (1996)—Mexico Northeast (Tamaulipas).
hoffmeisteri *(Klotzsch) Ehrend. & Schönb.-Tem. ex R.R. Mill in Edinburgh J. Bot.,* 53(1): 95 (1996): Asperula hoffmeisteri.
lucidum All.
subsp. bernardii *(Gren. & Godr.) A. Natali in Candollea,* 53(2): 496 (1998): G. bernardii.
subsp. krendlii *A. Natali in Candollea,* 53(2): 500 (1998)—Corse.
subsp. venustum *(Jord.) A. Natali in Candollea,* 53(2): 503 (1998): G. venustum.
macilentum *M.V. Klokov & B.V. Zaverucha in Novosti Sist. Vyssh. Nizsh. Rast. (Kiev),* 1974: 85 (1974)—Ukraina.
megacyttarion *R.R. Mill in Edinburgh J. Bot.,* 53(2): 200 (1996)—India (Uttar Pradesh), Sikkim, Bhutan, Nepal, Xizang.
mollugo L.
subsp. album *(Mill.) N.N. Tzvelev in Novosti Sist. Vyssh. Rast.,* 32: 185 (2000): G. album.
olgae *M.V. Klokov in Novosti Sist. Vyssh. Nizsh. Rast. (Kiev),* 1974: 107 (1974)—Ukraina.
rebae *R.R. Mill in Edinburgh J. Bot.,* 53(2): 195 (1996)—Bhutan, Sikkim, Nepal, Xizang.
rubioides L.
subsp. dasypodum *(Klokov) N.N. Tzvelev in Novosti Sist. Vyssh. Rast.,* 32: 185 (2000): G. dasypodum.
subnemorale *M.V. Klokov & B.V. Zaverucha in Novosti Sist. Vyssh. Nizsh. Rast. (Kiev),* 1974: 81 (1974)—Ukraina.
verrucosum Huds.
subsp. halophilum *(Ponzo) J. Lambinon in Soc. Ech. Pl. Vasc. Eur. Bassin Médit. Bull.,* 26: 83 (1997): G. valantia var. halophilum.

GALVEZIA (Scrophulariac.).
leucantha Wiggins
subsp. porphyrantha *A. Tye & H. Jäger in Novon,* 10(2): 165 (2000)—Galápagos.

GAMBLEA (Araliac.).
ciliata C.B. Clarke
var. evodiifolia *(Franch.) C.B. Shang, P.P. Lowry & D.G. Frodin in Adansonia,* 22(1): 51 (2000), as 'evodiaefolia': Acanthopanax evodiifolius.
innovans *(Siebold & Zucc.) C.B. Shang, P.P. Lowry & D.G. Frodin in Adansonia,* 22(1): 52 (2000): Panax innovans.
malayana *(M.R. Hend.) C.B. Shang, P.P. Lowry & D.G. Frodin in Adansonia,* 22(1): 53 (2000): Acanthopanax malayanus.
pseudoevodiifolia *(Feng) C.B. Shang, P.P. Lowry & D.G. Frodin in Adansonia,* 22(1): 54 (2000), as 'pseudoevodiaefolia': Acanthopanax evodiifolius var. pseudoevodiifolius.

GAMOCHAETA (Compos.).
lulioana *S.E. Freire & L. Iharlegui in Novon,* 7(1): 32 (1997)—Bolivia, Peru.

GANGUELIA *E. Robbrecht in S. Afr. J. Bot.,* 62(1): 21 (1996). *RUBIACEAE*
gossweileri *(S. Moore) E. Robbrecht in S. Afr. J. Bot.,* 62(1): 21 (1996): Oxyanthus gossweileri.

GAOLIGONGSHANIA *D.Z. Li, C.J. Hsueh & N.H. Xia in Acta Phytotax. Sin.,* 33(6): 598 (1995). *GRAMINEAE*
megalothyrsa *(Hand.-Mazz.) D.Z. Li, C.J. Hsueh & N.H. Xia in Acta Phytotax. Sin.,* 33(6): 600 (1995), as 'megathyrsa': Arundinaria megalothyrsa.

GARAYANTHUS *D.L. Szlachetko in Fragm. Flor. Geobot.,* Supp. 3: 136 (1995). *ORCHIDACEAE*
carinatus *(Rolfe ex Downie) D.L. Szlachetko in Fragm. Flor. Geobot.,* Supp. 3: 136 (1995): Sarcanthus carinatus.
duplicilobus *(J.J. Sm.) D.L. Szlachetko in Fragm. Flor. Geobot.,* Supp. 3: 136 (1995): Sarcanthus duplicilobus.
flexus *(Rchb. f.) D.L. Szlachetko in Fragm. Flor. Geobot.,* Supp. 3: 136 (1995): Sarcanthus flexus.
inflatus *(Rolfe) D.L. Szlachetko in Fragm. Flor. Geobot.,* Supp. 3: 136 (1995): Sarcanthus inflatus.

GARAYANTHUS:-
paniculatus *(Ker Gawl.) D.L. Szlachetko in Fragm. Flor. Geobot.*, Supp. 3: 136 (1995): Aerides paniculatum.
rostellatus *(Ridl.) D.L. Szlachetko in Fragm. Flor. Geobot.*, Supp. 3: 136 (1995): Sarcanthus rostellatus.

GARCIBARRIGOA (Compos.).
sinbundoya *S. Díaz-Piedrahita & P. Pedraza in Rev. Acad. Colomb. Cienc. Exact. Fis. Nat.*, 24(92): 344 (2000)—Colombia.

GARCINIA (Guttif.).
cymosa *(K. Schum.) I.M. Turner & P.F. Stevens in Gard. Bull. Singapore,* 51(2): 176 (1999): Tripetalum cymosum.
f. pendula *(Lauterb.) I.M. Turner & P.F. Stevens in Gard. Bull. Singapore,* 51(2): 177 (1999): Tripetalum cymosum f. pendulum.

GARDENIA (Rubiac.).
dacryoides *A. Cunn. ex C.F. Puttock in Nuytsia,* 11(2): 228 (1997)—Western Australia, Northern Territory.
ewartii *C.F. Puttock in Nuytsia,* 11(2): 231 (1997), nom. nov.: G. megasperma var. arborea *Ewart & Cookson.*
subsp. fitzgeraldii *C.F. Puttock in Nuytsia,* 11(2): 233 (1997)—Western Australia.
faucicola *C.F. Puttock in Nuytsia,* 11(2): 235 (1997)—Northern Territory.
gardneri *C.F. Puttock in Nuytsia,* 11(2): 240 (1997)—Western Australia.
jabiluka *C.F. Puttock in Nuytsia,* 11(2): 241 (1997)—Northern Territory.
kakaduensis *C.F. Puttock in Nuytsia,* 11(2): 242 (1997)—Northern Territory.
pyriformis A. Cunn. ex Benth.
subsp. keartlandii *(Tate) C.F. Puttock in Nuytsia,* 11(2): 249 (1997): G. keartlandii.
subsp. orientalis *C.F. Puttock in Nuytsia,* 11(2): 251 (1997)—Queensland, Northern Territory.
resinosa F. Muell.
subsp. kimberleyensis *C.F. Puttock in Nuytsia,* 11(2): 255 (1997)—Western Australia.
schwarzii *C.F. Puttock in Nuytsia,* 11(2): 258 (1997), nom. nov.: G. petiolata *O. Schwarz.*
sericea *C.F. Puttock in Nuytsia,* 11(2): 259 (1997)—Western Australia.

× **GASTERALOE** (Aloac.).
radlii *L.E. Newton in Brit. Cact. Succ. J.,* 15(1): 34 (1997), nom. nov.: Aloe × imbricata *A. Berger.*
sculptilis *G.D. Rowley ex L.E. Newton in Haseltonia,* 5: 94 (1997 publ. 1998)—Cultivation. (Aloe variegata × Gasteria cheilophylla).

GASTERANTHUS (Gesneriac.).
acuticarinatus *M. Freiberg in Brittonia,* 52(2): 208 (2000)—Ecuador.
adenocalyx *L.E. Skog & L.P. Kvist, Revis. Gasteranthus (Gesneriac.) (Syst. Bot. Monog., 59):* 37 (2000)—Colombia.
atrolimbus *M. Freiberg in Brittonia,* 52(2): 205 (2000)—Ecuador.
aurantiacus *M. Freiberg in Brittonia,* 52(2): 203 (2000)—Ecuador.
bilsaensis *L.E. Skog & L.P. Kvist, Revis. Gasteranthus (Gesneriac.) (Syst. Bot. Monog., 59):* 41 (2000)—Ecuador.
calcaratus (Kunth) Wiehler
subsp. calceolus *(Fritsch) L.E. Skog & L.P. Kvist, Revis. Gasteranthus (Gesneriac.) (Syst. Bot. Monog., 59):* 47 (2000): Besleria calceolus.
subsp. oncogastrus *(Hanst.) L.E. Skog & L.P. Kvist, Revis. Gasteranthus (Gesneriac.) (Syst. Bot. Monog., 59):* 48 (2000): Besleria oncogastra.
epedunculatus *L.E. Skog & L.P. Kvist, Revis. Gasteranthus (Gesneriac.) (Syst. Bot. Monog., 59):* 63 (2000)—Colombia.
extinctus *L.E. Skog & L.P. Kvist, Revis. Gasteranthus (Gesneriac.) (Syst. Bot. Monog., 59):* 65 (2000)—Ecuador.
giganteus *M. Freiberg in Phyton (Austria),* 36(2): 304 (1996)—Ecuador.
glaber *L.E. Skog & L.P. Kvist, Revis. Gasteranthus (Gesneriac.) (Syst. Bot. Monog., 59):* 67 (2000)—Ecuador, Colombia.

GASTERANTHUS:-
imbaburensis *M. Freiberg in Phyton (Austria),* 36(2): 307 (1996)—Ecuador.
leopardus *M. Freiberg in Phyton (Austria),* 36(2): 307 (1996)—Ecuador.
magentatus *M. Freiberg in Phyton (Austria),* 38(1): 168 (1998)—Ecuador.
mutabilis *L.E. Skog & L.P. Kvist, Revis. Gasteranthus (Gesneriac.) (Syst. Bot. Monog., 59):* 80 (2000)—Ecuador.
orientandinus *L.E. Skog & L.P. Kvist, Revis. Gasteranthus (Gesneriac.) (Syst. Bot. Monog., 59):* 82 (2000)—Ecuador.
osaensis *L.E. Skog & L.P. Kvist, Revis. Gasteranthus (Gesneriac.) (Syst. Bot. Monog., 59):* 84 (2000)—Costa Rica.
otongensis *M. Freiberg in Phyton (Austria),* 38(1): 170 (1998)—Ecuador.
tenellus *L.E. Skog & L.P. Kvist, Revis. Gasteranthus (Gesneriac.) (Syst. Bot. Monog., 59):* 96 (2000)—Ecuador.
ternatus *L.E. Skog & L.P. Kvist, Revis. Gasteranthus (Gesneriac.) (Syst. Bot. Monog., 59):* 98 (2000)—Ecuador.
trifoliatus *M. Freiberg in Phyton (Austria),* 36(2): 308 (1996)—Ecuador.
villosus *L.E. Skog & L.P. Kvist, Revis. Gasteranthus (Gesneriac.) (Syst. Bot. Monog., 59):* 101 (2000)—Ecuador, Colombia.

GASTERIA (Aloac.).
batesiana
var. dolomitica *E. van Jaarsveld & A.E. van Wyk in Aloe,* 36(4): 74 (1999)—Transvaal.
carinata (Mill.) Duval
var. glabra *(Salm-Dyck) E. van Jaarsveld in Cact. Succ. J. (U.S.A.),* 70(2): 70 (1998): Aloe glabra.
var. thunbergii *(N.E. Br.) E. van Jaarsveld in Cact. Succ. J. (U.S.A.),* 70(2): 70 (1998): G. thunbergii.
glauca *E. van Jaarsveld in Cact. Succ. J. (U.S.A.),* 70(2): 65 (1998)—Cape Province.

GASTRIDIUM (Gramin.).
phleoides (Nees & Meyen) C.E. Hubb.
subsp. lainzii *A.T. Romero García in An. Jard. Bot. Madrid,* 54(1): 404 (1996)—Spain.

GASTROCHILUS (Orchidac.).
sect. Caespitosi *Z.H. Tsi in Guihaia,* 16(2): 143 (1996), as 'Gaespitosi'.
carnosus *Z.H. Tsi in Guihaia,* 16(2): 137 (1996)—Assam.
garhwalensis *Z.H. Tsi in Guihaia,* 16(2): 138 (1996)—India (Uttar Pradesh).
gongshanensis *Z.H. Tsi in Guihaia,* 16(2): 149 (1996)—China South-central (Yunnan).
guangtungensis *Z.H. Tsi in Guihaia,* 16(2): 139 (1996)—China Southeast (Guangdong).
linearifolius *Z.H. Tsi & L.A. Garay in Guihaia,* 16(2): 138 (1996)—Sikkim, Xizang.
minutiflorus *L.V. Averyanov in Bot. Zhurn.,* 82(3): 146, 143 (1997)—Vietnam.
nanchuanensis *Z.H. Tsi in Guihaia,* 16(2): 149 (1996)—China South-central (Sichuan).
obliquus (Lindl.) Kuntze
var. suavis *(Seidenf.) Z.H. Tsi in Guihaia,* 16(2): 141 (1996): G. suavis.
puncticulatus *W. Cavestro in Orchidophile,* 141: 53 (2000)—Sulawesi.
saccatus *Z.H. Tsi in Guihaia,* 16(2): 149 (1996)—China South-central (Yunnan).
sessanicus *A. Nageswara Rao in J. Orchid Soc. India,* 11(1–2): 1 (1997)—Assam (Arunachal Pradesh).
simplicilabius *L.V. Averyanov in Bot. Zhurn.,* 82(3): 146 (1997)—Vietnam.
subpapillosus *Z.H. Tsi in Guihaia,* 16(2): 142 (1996)—China South-central (Yunnan).

GASTRODIA (Orchidac.).
falconeri *D.L. Jones & M.A. Clements in Orchadian,* 12(8): 350 (1998), nom. nov.: Gamoplexis orobanchoides *Falc.*
javanica (Blume) Lindl.
f. thalassina *M. Yokota in Acta Phytotax. Geobot.,* 49(2): 147 (1998 publ. 1999)—Nansei-Shoto.

GASTROGLOTTIS (Orchidac.).
latifolia *(Sm.) D.L. Szlachetko in Fragm. Flor. Geobot.*, Supp. 3: 123 (1995): Malaxis latifolia.

GASTROLYCHNIS (Caryophyllac.).
subsect. Pauciflora *N.N. Tzvelev in Bot. Zhurn.,* 85(11): 100 (2000).
pauciflora *(Ledeb.) N.N. Tzvelev in Bot. Zhurn.,* 85(11): 100 (2000): Lychnis pauciflora.
saxatilis *(Turcz. ex Fisch. & C.A. Mey.) G.A. Peschkova in Fl. Tsentral'noi Sibiri,* 1: 330 (1979): Lychnis saxatilis.
tundricola *V.B. Kuvaev in Byull. Mosk. Obshch. Ispŷt. Prir., Biol.,* 102(5): 70 (1997)—Krasnoyarsk.
uniflora *(Ledeb.) N.N. Tzvelev in Bot. Zhurn.,* 85(11): 101 (2000): Lychnis uniflora.

GASTRORCHIS (Orchidac.).
geffrayi *(Bosser) K. Senghas in Schlechter Orchideen,* I(15): 887 (1984): Phaius geffrayi.
humblotii
var. rubra *(Bosser) J. Bosser & P.J. Cribb in D. Du Puy, P. Cribb, J. Bosser, J. Hermans & C. Hermans, Orchids of* Madagascar: 4, 145 (1999), as 'ruber': Phaius humblotii var. ruber.
var. schlechteri *(H. Perrier) Senghas ex J. Bosser & P.J. Cribb in D. Du Puy, P. Cribb, J. Bosser, J. Hermans & C. Hermans, Orchids of* Madagascar: 4, 145 (1999): G. schlechteri.
lutea *(Ursch & Toill.-Gen. ex Bosser) K. Senghas in Schlechter Orchideen,* I(15): 887 (1984): Phaius luteus.
peyrotii *(Bosser) K. Senghas in Schlechter Orchideen,* I(15): 887 (1984): Phaius peyrotii.
pulchra
var. perrieri *(Bosser) J. Bosser & P.J. Cribb in D. Du Puy, P. Cribb, J. Bosser, J. Hermans & C. Hermans, Orchids of* Madagascar: 4, 146 (1999): Phaius pulcher var. perrieri.
steinhardtiana *K. Senghas in J. Orchideenfreund,* 4(3): 133 (1997)—Madagascar.

GAUDINIA (Gramin.).
valdesii *C. Romero Zarco in Lagascalia,* 18(2): 310 (1996)—Morocco.

GAULTHERIA (Ericac.).
brevistipes *(C.Y. Wu & T.Z. Hsu) R.C. Fang in Novon,* 9(2): 164 (1999): Leucothoe brevistipes.
dumicola W.W. Sm.
var. hirticaulis *R.C. Fang in Novon,* 9(2): 164 (1999)—China South-central (Yunnan).
griffithiana Wight
var. insignis *R.C. Fang in Novon,* 9(2): 164 (1999)—Xizang.
heteromera *R.C. Fang in Novon,* 9(2): 164 (1999)—Xizang.
jingdongensis *R.C. Fang in Novon,* 9(2): 166 (1999)—China South-central (Yunnan).
leucocarpa Blume
var. psilocarpa *(H.F. Copel.) R.C. Fang in Novon,* 9(2): 166 (1999): G. psilocarpa.
var. yunnanensis *(Franch.) T.Z. Hsu & R.C. Fang in Novon,* 9(2): 166 (1999): Vaccinium yunnanense.
longibracteolata *R.C. Fang in Novon,* 9(2): 166 (1999)—China South-central (Yunnan), Thailand.
pseudonotabilis *H. Li in Fl. Dulongjian Reg.:* 216 (1993), without latin descr. or type—China South-central (Yunnan).
pseudonotabilis *H. Li ex R.C. Fang in Novon,* 9(2): 169 (1999)—China South-central (Yunnan).
purpurea *R.C. Fang in Novon,* 9(2): 171 (1999)—Xizang.
straminea *R.C. Fang in Novon,* 9(2): 171 (1999), nom. nov.: G. wardii var. serrulata *C.Y. Wu & T.Z. Hsu.*
trigonoclada *R.C. Fang in Novon,* 9(2): 174 (1999)—Xizang.
wardii C. Marquand & Airy Shaw
var. elongata *R.C. Fang in Novon,* 9(2): 174 (1999)—China South-central (Yunnan).

GAURA (Onagrac.).
fruticosa *P. Loefling, Iter Hispan.:* 248 (1758)—Venezuela.

GAURA:-
glabra *J.G.C. Lehmann, Nov. stirp. pug.*, 2: 16 (1830)—Locality not stated.†
parviflora *J.G.C. Lehmann, Nov. stirp. pug.*, 2: 16 (1830)—Locality not stated.†

GAUSSENIA *A.V.F.Ch. Bobrov & A.P. Melikian in Bot. Zhurn.*, 85(7): 62 (2000). *PODOCARPACEAE*
guillauminii *(J. Buchholz) A.V.F.Ch. Bobrov & A.P. Melikian in Bot. Zhurn.*, 85(7): 62 (2000): Dacrydium guillauminii.
lycopodioides *(Brongn. & Gris) A.V.F.Ch. Bobrov & A.P. Melikian in Bot. Zhurn.*, 85(7): 62 (2000): Dacrydium lycopodioides.

GAYA (Malvac.).
albiflora *A. Krapovickas in Bonplandia*, 9(1–2): 67 (1996)—Bolivia.
atiquipana *A. Krapovickas in Bonplandia*, 9(1–2): 72 (1996)—Peru.
bordasii *A. Krapovickas in Bonplandia*, 9(1–2): 71 (1996)—Paraguay, Brazil West-central (Mato Grosso do Sul).
cruziana *A. Krapovickas in Bonplandia*, 9(1–2): 66 (1996)—Bolivia.
dentata *A. Krapovickas in Bonplandia*, 9(1–2): 74 (1996)—Brazil Northeast (Bahia).
matutina *A. Krapovickas in Bonplandia*, 9(1–2): 81 (1996)—Paraguay.
meridensis *A. Krapovickas in Bonplandia*, 9(1–2): 61 (1996)—Venezuela.
mollendoensis *A. Krapovickas in Bonplandia*, 9(1–2): 77 (1996)—Peru.
mutisiana *A. Krapovickas in Bonplandia*, 9(1–2): 69 (1996)—Colombia.
parviflora *(Phil.) A. Krapovickas in Bonplandia*, 9(1–2): 82 (1996): Tetraptera parviflora.
scopulorum *A. Krapovickas in Bonplandia*, 9(1–2): 63 (1996)—Brazil North (Pará).

GAYLUSSACIA (Ericac.).
arassatubensis *R.R. Silva & A.C. Cervi in Fontqueria*, 54(1): 4 (1999)—Brazil South (Paraná, Santa Catarina).
caratuvensis *R.R. Silva & A.C. Cervi in Fontqueria*, 54(1): 1 (1999)—Brazil South (Paraná).

GEIGERIA (Compos.).
aspalathoides S. Moore
subsp. chorleyana *S. Ortiz & J. Rodríguez Oubiña in Nordic J. Bot.*, 20(2): 195 (2000)—Angola.
decurrens *S. Ortiz, J. Rodríguez Oubiña & M. Buján in Nordic J. Bot.*, 20(2): 198 (2000)—Angola.

GEISSANTHUS (Myrsinac.).
challuayacus *J.J. Pipoly in Sida*, 17(2): 466 (1996)—Ecuador.
spectabilis *J.J. Pipoly in Sida*, 17(2): 463 (1996)—Peru.
vanderwerffii *J.J. Pipoly in Sida*, 17(2): 460 (1996)—Ecuador.

GEISSOSPERMUM (Apocynac.).
sericeum *J. Miers, Apocyn. S. Amer.*: 86 (1878)—French Guiana.†

GEMMARIA (Amaryllidac.).
discifera *(Marloth ex Snijman) D. Müller-Doblies & U. Müller-Doblies in Feddes Repert.*, 107(5–6): S.c.4 (1996). Strumaria discifera.

GENIANTHUS (Asclepiadac.).
sect. Brachyblastus *J. Klackenberg in Bot. Jahrb.*, 117(4): 447 (1995).
aurantiacus *(C.Y. Wu ex Tsiang & P.T. Li) J. Klackenberg in Bot. Jahrb.*, 117(4): 431 (1995): Toxocarpus aurantiacus.
nicobarensis *J. Klackenberg in Bot. Jahrb.*, 117(4): 460 (1995)—Nicobar Is.
subpeltatus *J. Klackenberg in Bot. Jahrb.*, 117(4): 464 (1995)—Malaya (Peninsular Malaysia).
trullatus *J. Klackenberg in Bot. Jahrb.*, 117(4): 453 (1995)—Sumatera.
valvatus *J. Klackenberg in Bot. Jahrb.*, 117(4): 442 (1995)—Borneo (Kalimantan).

GENIOSPORUM (Labiat.).
lanceolatum *Chermsir. ex G. Murata in Southeast Asian Stud.*, 14(2): 186 (1976)—Thailand.

GENISTA (Legumin.).
sect. Asterocytisus *(W.D.J. Koch) S. Talavera in Lagascalia*, 21(1): 168 (1999): Cytisus sect. Asterocytisus.
subsect. Chamaespartum *(Spach) P. Cantó, S. Rivas-Martínez, R. Greinwald & I. van Rensen in Lazaroa*, 18: 33 (1997): G. sect. Chamaespartum.
ser. Floridae *P. Cantó, S. Rivas-Martínez, R. Greinwald & I. van Rensen in Lazaroa*, 18: 23 (1997).
sect. Phyllodioides *S. Talavera & P.E. Gibbs in An. Jard. Bot. Madrid*, 57(1): 207 (1999).
× arizagae *J. Elorza, S. Patino, P. Urrutia & J. Valencia in Estud. Mus. Cienc. Nat. Alava*, 12: 69 (1997)—Spain. (G. hispanica subsp. occidentalis × micrantha).
ausetana *(O. Bolòs & Vigo) S. Talavera in An. Jard. Bot. Madrid*, 57(1): 206 (1999): G. cinerea subsp. ausetana.
florida L.
subsp. maroccana *(Ball) P. Cantó, S. Rivas-Martínez, R. Greinwald & I. van Rensen in Lazaroa*, 18: 24 (1997): G. florida var. maroccana.
longipes Pau
subsp. viciosoi *S. Talavera & B. Cabezudo in Acta Bot. Malacitana*, 23: 277 (1998)—Spain.
madoniensis *F.M. Raimondo in Flora Medit.*, 9: 319 (1999)—Sicilia.
moleroi *S. Talavera & P.E. Gibbs in Lagascalia*, 18(2): 268 (1996)—Morocco.
pumila (Debeaux & E. Rev. ex Hervier) Vierh.
subsp. rigidissima *(Vierh.) S. Talavera & L. Sáez in Acta Bot. Malacitana*, 23: 275 (1998): G. rigidissima.
tournefortii Spach
subsp. jahandiezii *(Batt.) S. Talavera & P.E. Gibbs in Lagascalia*, 18(2): 271 (1996): G. jahandiezii.
tridens (Cav.) DC.
subsp. juniperina *(Spach) S. Talavera & P.E. Gibbs in Lagascalia*, 18(2): 271 (1996): G. juniperina.
valdes-bermejoi *S. Talavera & L. Sáez in An. Jard. Bot. Madrid*, 57(1): 208 (1999), nom. nov.: G. acanthoclada var. fasciculata *Knoche.*
valentina (Willd. ex Spreng.) Steud.
subsp. murcica *(Coss.) G. Mateo Sanz & M.B. Crespo Villalba, Claves Fl. Valenciana*: 219 (1990): G. murcica.
valsecchiae *S. Brullo & G. De Marco in Pl. Syst. Evol.*, 200(3–4): 275 (1996)—Sardegna.

GENLISEA (Lentibulariac.).
subgen. Tayloria *(Fromm) Eb. Fischer, S. Porembski & W. Barthlott in Nordic J. Bot.*, 20(3): 293 (2000): G. sect. Tayloria.
barthlottii *S. Porembski, Eb. Fischer & B. Gemmel in Bull. Mus. Nation. Hist. Nat., B, Adansonia*, Ser. 4, 18(1–2): 152 (1996)—Guinea.
taylorii *Eb. Fischer, S. Porembski & W. Barthlott in Nordic J. Bot.*, 20(3): 311 (2000), as 'taylori'—Angola.

GENOPLESIUM (Orchidac.).
bishopii *D.L. Jones in Orchadian*, 13(5): 227 (2000)—New South Wales.
cranei *D.L. Jones in Orchadian*, 13(3): 102 (2000)—Queensland.
ectopum *D.L. Jones in Orchadian*, 12(12): 570 (1999)—New South Wales (Australian Capital Territory).
firthii *(Cady) D.L. Jones in Austral. Orchid Res.*, 3: 89 (1998): Prasophyllum firthii.
tasmanicum *D.L. Jones in Austral. Orchid Res.*, 3: 90 (1998)—Tasmania.

GENTIANA (Gentianac.).
ser. Angustifoliae *J.J. Halda in Acta Mus. Richnov. Sect. Nat.*, 3(1): 17 (1995).
sect. Annuae *(C. Marquand) J.J. Halda in Acta Mus. Richnov. Sect. Nat.*, 3(1): 29 (1995): G. ser. Annuae.
sect. Apteroideae *(Harry Sm.) J.J. Halda in Acta Mus. Richnov. Sect. Nat.*, 3(1): 27 (1995): G. subser. Apteroideae.

GENTIANA :-
subgen. Calathianae *(Froel.) J.J. Halda in Acta Mus. Richnov. Sect. Nat.*, 3(1): 32 (1995): G. sect. Calathianae.
sect. Caliculatae *J.J. Halda in Acta Mus. Richnov. Sect. Nat.*, 3(1): 17 (1995).
subgen. Chondrophyllae *(Bunge) J.J. Halda in Acta Mus. Richnov. Sect. Nat.*, 3(1): 34 (1995): G. sect. Chondrophyllae.
subgen. Ciminalis *(Adans.) J.J. Halda in Acta Mus. Richnov. Sect. Nat.*, 3(1): 21 (1995), isonym; *see Tzvelev in Bot. Zhurn.*, 78(6): 135 (1993): gen. Ciminalis.
ser. Coilantha *J.J. Halda in Acta Mus. Richnov. Sect. Nat.*, 3(1): 14 (1995), contrary to Art. 53.5 ICBN (1994).
sect. Coilantha *(Borkh.) J.J. Halda in Acta Mus. Richnov. Sect. Nat.*, 3(1): 13 (1995): gen. Coilantha.
subgen. Cruciata *(Gaudin) J.J. Halda in Acta Mus. Richnov. Sect. Nat.*, 3(1): 30 (1995): G. sect. Cruciata.
ser. Cruciata *J.J. Halda in Acta Mus. Richnov. Sect. Nat.*, 3(1): 31 (1995).
sect. Decoratae *(C. Marquand) J.J. Halda in Acta Mus. Richnov. Sect. Nat.*, 3(1): 16 (1995); *et Genus Gentiana*: 30 (1996 publ. 1997): G. ser. Decoratae.
sect. Decumbentes *J.J. Halda in Acta Mus. Richnov. Sect. Nat.*, 3(1): 31 (1995).
ser. Decumbentes *J.J. Halda in Acta Mus. Richnov. Sect. Nat.*, 3(1): 31 (1995).
sect. Depressae *(Harry Sm.) J.J. Halda in Acta Mus. Richnov. Sect. Nat.*, 3(1): 24 (1995): G. ser. Depressae.
subgen. Dolichocarpa *(T.N. Ho) J.J. Halda in Acta Mus. Richnov. Sect. Nat.*, 3(1): 33 (1995): G. sect. Dolichocarpa.
subgen. Frigida *(Kusn.) J.J. Halda in Acta Mus. Richnov. Sect. Nat.*, 3(1): 22 (1995): G. sect. Frigida.
sect. Glaucae *J.J. Halda in Acta Mus. Richnov. Sect. Nat.*, 3(1): 22 (1995).
subgen. Isomeria *(Kusn.) J.J. Halda in Acta Mus. Richnov. Sect. Nat.*, 3(1): 23 (1995): G. sect. Isomeria.
ser. Kaufmannianae *J.J. Halda in Acta Mus. Richnov. Sect. Nat.*, 3(1): 31 (1995).
subgen. Microsperma *(T.N. Ho) J.J. Halda in Acta Mus. Richnov. Sect. Nat.*, 3(1): 28 (1995): G. sect. Microsperma.
subgen. Monopodiae *(Harry Sm.) J.J. Halda in Acta Mus. Richnov. Sect. Nat.*, 3(1): 25 (1995): G. subsect. Monopodiae.
sect. Napuliferae *(T.N. Ho) J.J. Halda in Acta Mus. Richnov. Sect. Nat.*, 3(1): 48 (1995): G. ser. Napuliferae.
subgen. Newberryi *J.J. Halda in Acta Mus. Richnov. Sect. Nat.*, 3(1): 20 (1995).
ser. Nipponicae *J.J. Halda in Acta Mus. Richnov. Sect. Nat.*, 3(1): 45 (1995).
ser. Olivierianae *J.J. Halda in Acta Mus. Richnov. Sect. Nat.*, 3(1): 31 (1995).
sect. Orbiculatae *(C. Marquand) J.J. Halda in Acta Mus. Richnov. Sect. Nat.*, 3(1): 41 (1995): G. ser. Orbiculatae.
sect. Ornatae *(C. Marquand) J.J. Halda in Acta Mus. Richnov. Sect. Nat.*, 3(1): 26 (1995): G. ser. Ornatae.
subgen. Otophora *(Kusn.) J.J. Halda in Acta Mus. Richnov. Sect. Nat.*, 3(1): 15 (1995): G. sect. Otophora.
sect. Paradoxae *J.J. Halda in Acta Mus. Richnov. Sect. Nat.*, 3(1): 18 (1995).
subgen. Phyllocalyx *(T.N. Ho) J.J. Halda in Acta Mus. Richnov. Sect. Nat.*, 3(1): 21 (1995): G. sect. Phyllocalyx.
ser. Pneumonanthe *J.J. Halda in Acta Mus. Richnov. Sect. Nat.*, 3(1): 17 (1995).
sect. Pyrenaicae *J.J. Halda in Acta Mus. Richnov. Sect. Nat.*, 3(1): 45 (1995).
ser. Pyrenaicae *J.J. Halda in Acta Mus. Richnov. Sect. Nat.*, 3(1): 45 (1995).
sect. Septemfidae *J.J. Halda in Acta Mus. Richnov. Sect. Nat.*, 3(1): 17 (1995).
sect. Sikkimenses *(C. Marquand) J.J. Halda in Acta Mus. Richnov. Sect. Nat.*, 3(1): 24 (1995): G. ser. Sikkimenses.

GENTIANA:-

ser. Siphonanthae *J.J. Halda in Acta Mus. Richnov. Sect. Nat.*, 3(1): 32 (1995).
ser. Spathacea *J.J. Halda in Acta Mus. Richnov. Sect. Nat.*, 3(1): 17 (1995).
*ser.*Spathaceae *J.J. Halda in Acta Mus. Richnov. Sect. Nat.*, 3(1): 14 (1995).
subgen. Stenogyne *(Franch.) J.J. Halda in Acta Mus. Richnov. Sect. Nat.*, 3(1): 29 (1995): G. sect. Stenogyne.
sect. Stragulatae *(T.N. Ho) J.J. Halda in Acta Mus. Richnov. Sect. Nat.*, 3(1): 24 (1995): G. ser. Stragulatae.
sect. Suborbisepalae *(C. Marquand) J.J. Halda in Acta Mus. Richnov. Sect. Nat.*, 3(1): 28 (1995): G. ser. Suborbisepalae.
sect. Tetramerae *(C. Marquand) J.J. Halda in Acta Mus. Richnov. Sect. Nat.*, 3(1): 29 (1995): G. subser. Tetramerae.
ser. Triflorae *J.J. Halda in Acta Mus. Richnov. Sect. Nat.*, 3(1): 17 (1995), isonym; *see Gagnidze in Zam. Sist. Geogr. Rast.*, 41: 50 (1986).
sect. Uniflorae *(C. Marquand) J.J. Halda in Acta Mus. Richnov. Sect. Nat.*, 3(1): 24 (1995): G. ser. Uniflorae.
sect. Vernae *J.J. Halda in Acta Mus. Richnov. Sect. Nat.*, 3(1): 33 (1995).
sect. Verticillatae *(C. Marquand) J.J. Halda in Acta Mus. Richnov. Sect. Nat.*, 3(1): 26 (1995): G. ser. Verticillatae.

affinis Griseb.
subsp. rusbyi *(Greene ex Kusn.) J.J. Halda in Acta Mus. Richnov. Sect. Nat.*, 3(1): 18 (1995): G. rusbyi.

albicalyx Burkill
subsp. globosa *(T.N. Ho) J.J. Halda in Acta Mus. Richnov. Sect. Nat.*, 3(1): 35 (1995): G. globosa.

albomarginata C. Marquand
subsp. scytophylla *(T.N. Ho) J.J. Halda in Acta Mus. Richnov. Sect. Nat.*, 3(1): 35 (1995): G. scytophylla.

alsinoides Franch.
subsp. scabrifilamenta *(T.N. Ho) J.J. Halda in Acta Mus. Richnov. Sect. Nat.*, 3(1): 37 (1995): G. scabrifilamenta.

aquatica L.
var. baltistanica *(Omer) J.J. Halda in Acta Mus. Richnov. Sect. Nat.*, 3(1): 39 (1995): Ciminalis baltistanica.
var. bomiensis *(T.N. Ho) J.J. Halda in Acta Mus. Richnov. Sect. Nat.*, 3(1): 40 (1995): G. bomiensis.

arethusae Burkill
subsp. delicatula *(C. Marquand) J.J. Halda in Acta Mus. Richnov. Sect. Nat.*, 3(1): 26 (1995): G. arethusae var. delicatula.

argentea (Royle ex D. Don) Griseb.
subsp. albescens *(Franch.) J.J. Halda in Acta Mus. Richnov. Sect. Nat.*, 3(1): 35 (1995): G. argentea var. albescens.
subsp. alii *(Omer & Qaiser) J.J. Halda in Acta Mus. Richnov. Sect. Nat.*, 3(1): 36 (1995): Gentianodes alii.

aristata Maxim.
subsp. asparagoides *(T.N. Ho) J.J. Halda in Acta Mus. Richnov. Sect. Nat.*, 3(1): 43 (1995): G. asparagoides.

aurantiaca *(M. Laínz) F.J. Silva Pando, Valdés Berm. & Rodr. Gracia in Nova Acta Cient. Compostel. (Biol.)*, 10: 26 (2000), as 'auranthiaca': G. lutea subsp. aurantiaca.

autumnalis L.
subsp. pennelliana *(Fernald) J.J. Halda in Acta Mus. Richnov. Sect. Nat.*, 3(1): 18 (1995): G. pennelliana.

axillariflora H. Lév. & Vaniot
f. alba *Y.N. Lee, Fl. Korea:* 1160 (1996), without type—Locality not stated.

baeuerlenii L.G. Adams
subsp. bredboensis *(L.G. Adams) J.J. Halda, Genus Gentiana:* 160 (1996 publ. 1997): G. bredboensis

bokorensis *S. Hul in Adansonia*, 21(2): 246 (1999)—Cambodia.

borneensis Hook. f.
subsp. apoensis *(Merr.) J.J. Halda in Acta Mus. Richnov. Sect. Nat.*, 3(1): 45 (1995): G. apoensis.

GENTIANA:-

borneensis
subsp. malayana *(Ridl.) J.J. Halda in Acta Mus. Richnov. Sect. Nat.*, 3(1): 45 (1995): G. malayana.

capitata Buch.-Ham. ex D. Don
subsp. harwanensis *(G. Singh) J.J. Halda in Acta Mus. Richnov. Sect. Nat.*, 3(1): 36 (1995): G. harwanensis.

carinicostata Wernham
subsp. sclerophylla *(P. Royen) J.J. Halda in Acta Mus. Richnov. Sect. Nat.*, 3(1): 46 (1995): G. sclerophylla.

cephalantha Franch. ex Hemsl.
subsp. vaniotii *(H. Lév.) J.J. Halda in Acta Mus. Richnov. Sect. Nat.*, 3(1): 27 (1995): G. vaniotii.

choanantha C. Marquand
subsp. curvianthera *(T.N. Ho) J.J. Halda in Acta Mus. Richnov. Sect. Nat.*, 3(1): 43 (1995): G. curvianthera.

chungtienensis C. Marquand
subsp. macraucena *(C. Marquand) J.J. Halda in Acta Mus. Richnov. Sect. Nat.*, 3(1): 43 (1995): G. macraucena.
subsp. subuniflora *(C. Marquand) J.J. Halda in Acta Mus. Richnov. Sect. Nat.*, 3(1): 43 (1995): G. subuniflora.

clarkei Kusn.
subsp. conduplicata *(T.N. Ho) J.J. Halda in Acta Mus. Richnov. Sect. Nat.*, 3(1): 40 (1995): G. conduplicata.
var. lutescens *T.N. Ho & J.Q. Liu in Novon*, 6(2): 188 (1996)—Qinghai.

clusii E.P.Perrier & Songeon
subsp. rochelii *(A.Kern.) J.J. Halda in Acta Mus. Richnov. Sect. Nat.*, 3(1): 21 (1995): G. clusii var. rochelii.

crassa Kurz
subsp. rigescens *(Franch. ex Hemsl.) J.J. Halda in Acta Mus. Richnov. Sect. Nat.*, 3(1): 27 (1995): G. rigescens.

crassuloides Bureau & Franch.
subsp. curviphylla *(T.N. Ho) J.J. Halda in Acta Mus. Richnov. Sect. Nat.*, 3(1): 41 (1995): G. curviphylla.
subsp. mailingensis *(T.N. Ho) J.J. Halda in Acta Mus. Richnov. Sect. Nat.*, 3(1): 41 (1995): G. mailingensis.
subsp. nyingchiensis *(T.N. Ho) J.J. Halda in Acta Mus. Richnov. Sect. Nat.*, 3(1): 41 (1995): G. nyingchiensis.

crenulatotruncata (C. Marqand) T.N. Ho
var. flava *S.K. Wu & R.F. Huang in S.G. Wu (S.K. Wu) & Z.J. Feng (eds.), Biol. & Human Physiol. Hoh Xil* Region: 74 (1996)—Qinghai.

damyonensis C. Marquand
subsp. hicksii *(Harry Sm.) J.J. Halda in Acta Mus. Richnov. Sect. Nat.*, 3(1): 15 (1995): G. hicksii.

decorata Diels
subsp. leucantha *(Harry Sm.) J.J. Halda in Acta Mus. Richnov. Sect. Nat.*, 3(1): 16 (1995); *et Genus Gentiana:* 31 (1996 publ. 1997): G. leucantha.

durangensis *J.A. Villarreal Q. in Acta Bot. Mex.*, 34: 49 (1996)—Mexico Northeast (Durango).

elmeriana *J.J. Halda in Acta Mus. Richnov. Sect. Nat.*, 3(1): 46 (1995), nom. nov.: G. copelandii *Elmer.*

eonae *J.J. Halda in Acta Mus. Richnov. Sect. Nat.*, 3(1): 18 (1995), nom. nov.: G. pneumonanthe var. depressa *Boiss.*

flavomaculata Hayata
subsp. tatakensis *(Masam.) J.J. Halda in Acta Mus. Richnov. Sect. Nat.*, 3(1): 38, 42 (1995): G. tatakensis.
var. yuanyanghuensis *Chih H. Chen & J.C. Wang in Bot. Bull. Acad. Sin. (Taipei)*, 40(1): 21 (1999)—Taiwan.

flexicaulis Harry Sm. ex C. Marquand
var. complexa *(T.N. Ho) J.J. Halda in Acta Mus. Richnov. Sect. Nat.*, 3(1): 40 (1995): G. complexa.

forrestii C. Marquand
subsp. yakumontana *(Masam.) J.J. Halda in Acta Mus. Richnov. Sect. Nat.*, 3(1): 40 (1995): G. yakumontana.

GENTIANA:-

gentilis Franch.
subsp. eurycolpa *(C. Marquand) J.J. Halda in Acta Mus. Richnov. Sect. Nat.*, 3(1): 30 (1995): G. eurycolpa.

grata Harry Sm.
var. bryophylla *(Harry Sm.) J.J. Halda in Acta Mus. Richnov. Sect. Nat.*, 3(1): 35 (1995): G. bryophylla.
subsp. burmensis *(C. Marquand) J.J. Halda in Acta Mus. Richnov. Sect. Nat.*, 3(1): 34 (1995): G. burmensis.

handeliana Harry Sm.
subsp. erectosepala *(T.N. Ho) J.J. Halda in Acta Mus. Richnov. Sect. Nat.*, 3(1): 22 (1995): G. erectosepala.

haraldi-smithii *J.J. Halda in Acta Mus. Richnov. Sect. Nat.*, 3(1): 46 (1995)—Sulawesi.

helophila Balf. f. & Forrest ex C. Marquand
subsp. dolichocalyx *(T.N. Ho) J.J. Halda, Genus Gentiana:* 93 (1997): G. dolichocalyx.

hintoniorum *B.L. Turner in Phytologia*, 80(2): 118 (1996)—Mexico Northeast (Nuevo León).

hohoxiliensis *S.K. Wu & R.F. Huang in S.G. Wu (S.K. Wu) & Z.J. Feng (eds.), Biol. & Human Physiol. Hoh Xil* Region: 74 (1996)—Qinghai.

intricata C. Marquand
var. subintricata *(T.N. Ho) J.J. Halda in Acta Mus. Richnov. Sect. Nat.*, 3(1): 38 (1995): G. subintricata.

jarmilae *J.J. Halda in Acta Mus. Richnov. Sect. Nat.*, 3(1): 28 (1995), nom. nov.: G. clementis *Merr.*

kaohsiungensis *Chih H. Chen & J.C. Wang in Bot. Bull. Acad. Sin. (Taipei)*, 40(1): 24 (1999)—Taiwan.

× laengstii Hausm.
nothosubsp. komnensis *(E. Mayer) W. Starmühler in Hladnikia*, 6: 45 (1996): G. × komnensis.

langbianensis *A. Chev. ex S. Hul in Adansonia*, 21(2): 248 (1999)—Vietnam.

lateriflora Hemsl.
var. uncifolia *(H.J. Lam) T.N. Ho in Acta Phytotax. Sin.*, 38(1): 10 (2000): G. uncifolia.

leptoclada Balf. f. & Forrest
subsp. australis *(Craib) J.J. Halda in Acta Mus. Richnov. Sect. Nat.*, 3(1): 30 (1995): G. australis.

leroyana *S. Hul in Adansonia*, 21(2): 246 (1999)—Laos.

ligustica *R. de Vilmorin & R. Chopinet in VIII Congrès Internat. Bot., Paris 1954, Sect. 21–27:* 166 (1956)—France.

loerzingii Ridl.
var. timida *(Kerr) T.N. Ho in Acta Phytotax. Sin.*, 38(1): 10 (2000): G. timida.

loureirii (G. Don) Griseb.
subsp. napulifera *(Franch.) J.J. Halda in Acta Mus. Richnov. Sect. Nat.*, 3(1): 48 (1995): G. napulifera.

macgregoryi Hemsl.
subsp. piundensis *(P. Royen) J.J. Halda in Acta Mus. Richnov. Sect. Nat.*, 3(1): 47 (1995): G. piundensis.

macrophylla Pall.
subsp. fetisowii *(Regel & C. Winkl.) J.J. Halda in Acta Mus. Richnov. Sect. Nat.*, 3(1): 32 (1995): G. fetisowii.

maeulchanensis Franch.
subsp. kunmingensis *(S.W. Liu) J.J. Halda in Acta Mus. Richnov. Sect. Nat.*, 3(1): 38 (1995): G. kunmingensis.

× marceli-jouseaui *J.J. Halda in Acta Mus. Richnov. Sect. Nat.*, 7(2): 72 (2000)—Kirgizistan. (G. kaufmanniana × olgae).

marginata (G. Don) Griseb.
subsp. nasirii *(Omer, Ali & Qaiser) J.J. Halda in Acta Mus. Richnov. Sect. Nat.*, 3(1): 36 (1995): Gentianodes nasirii.

membranulifera *T.N. Ho in Acta Phytotax. Sin.*, 38(1): 10 (2000), nom. nov.: Ericala marginata *D. Don.*
var. recurvata *(Kusn.) T.N. Ho in Acta Phytotax. Sin.*, 38(1): 11 (2000): G. marginata β recurvata.

micans C.B. Clarke
subsp. kwansiensis *(T.N. Ho) J.J. Halda in Acta Mus. Richnov. Sect. Nat.*, 3(1): 44 (1995): G. kwansiensis.

GENTIANA:-
microdonta Franch.
subsp. omeiensis (T.N. Ho) J.J. Halda in Acta Mus. Richnov. Sect. Nat., 3(1): 23 (1995): G. omeiensis.
subsp. phyllopoda (H. Lév.) J.J. Halda in Acta Mus. Richnov. Sect. Nat., 3(1): 23 (1995): G. phyllopoda.
muscicola C. Marquand
subsp. lacinulata (T.N. Ho) J.J. Halda in Acta Mus. Richnov. Sect. Nat., 3(1): 35 (1995): G. lacinulata.
newberryi A. Gray
subsp. tiogana (A. Heller) J.J. Halda in Acta Mus. Richnov. Sect. Nat., 3(1): 21 (1995): G. tiogana.
nudicaulis Kurz
var. assamensis T.N. Ho in Acta Phytotax. Sin., 38(1): 11 (2000), nom. nov.: G. parryae C. Marquand.
subsp. lakshnakarae (Kerr) J.J. Halda in Acta Mus. Richnov. Sect. Nat., 3(1): 36 (1995): G. lakshnakarae.
otophora Franch. ex Hemsl.
subsp. sichitoensis (C. Marquand) J.J. Halda in Acta Mus. Richnov. Sect. Nat., 3(1): 15 (1995): G. sichitoensis.
ovatiloba Kusn.
subsp. michoacana J.A. Villarreal Quintanilla in Acta Bot. Mex., 41: 27 (1997)—Mexico Southwest (Michoacan).
pedata Harry Sm.
subsp. xingrenensis (T.N. Ho) J.J. Halda in Acta Mus. Richnov. Sect. Nat., 3(1): 36 (1995): G. xingrenensis.
pedicellata (D. Don) Wall. ex Griseb.
subsp. glabriuscula (Harry Sm. ex T.N. Ho) J.J. Halda in Acta Mus. Richnov. Sect. Nat., 3(1): 38 (1995): G. glabriuscula.
subsp. zeylanica (Griseb.) J.J. Halda in Acta Mus. Richnov. Sect. Nat., 3(1): 38 (1995): G. zeylanica.
piasezkii Maxim.
subsp. pubicaulis (Harry Sm.) J.J. Halda in Acta Mus. Richnov. Sect. Nat., 3(1): 41 (1995): G. pubicaulis.
praticola Franch.
subsp. greenwayae (Merr.) J.J. Halda in Acta Mus. Richnov. Sect. Nat., 3(1): 48 (1995): G. greenwayae.
primuliflora Franch.
subsp. melvillei (S. Moore) J.J. Halda in Acta Mus. Richnov. Sect. Nat., 3(1): 30 (1995) G. melvillei.
prostrata L.
subsp. nutans (Bunge) J.J. Halda in Acta Mus. Richnov. Sect. Nat., 3(1): 33 (1995): G. nutans.
subsp. podocarpa (Phil.) J.J. Halda in Acta Mus. Richnov. Sect. Nat., 3(1): 34 (1995): Varasia podocarpa.
pseudosquarrosa Harry Sm.
subsp. ludingensis (T.N. Ho) J.J. Halda in Acta Mus. Richnov. Sect. Nat., 3(1): 41 (1995): G. ludingensis.
pubigera C. Marquand
subsp. ninglangensis (T.N. Ho) J.J. Halda in Acta Mus. Richnov. Sect. Nat., 3(1): 40, 44 (1995): G. ninglangensis.
subsp. pubiflora (T.N. Ho) J.J. Halda in Acta Mus. Richnov. Sect. Nat., 3(1): 40, 44 (1995): G. pubiflora.
quadrifaria Blume
var. wightii (Kusn.) T.N. Ho in Acta Phytotax. Sin., 38(1): 11 (2000): G. pedicellata χ wightii.
recurvata C.B. Clarke
subsp. prainii (Burkill) J.J. Halda in Acta Mus. Richnov. Sect. Nat., 3(1): 40 (1995): G. prainii.
rickii J.J. Halda in Acta Mus. Richnov. Sect. Nat., 6(3): 234 (1999), as 'ricki'—Washington.
riparia Kar. & Kir.
subsp. daochengensis (T.N. Ho) J.J. Halda in Acta Mus. Richnov. Sect. Nat., 3(1): 39 (1995): G. daochengensis.
rubicunda Franch.
subsp. biloba (T.N. Ho) J.J. Halda in Acta Mus. Richnov. Sect. Nat., 3(1): 42 (1995): G. rubicunda var. biloba.
subsp. delicata (Hance) J.J. Halda in Acta Mus. Richnov. Sect. Nat., 3(1): 42 (1995): G. delicata.

GENTIANA:-
rubicunda
subsp. filisepala (T.N. Ho) J.J. Halda in Acta Mus. Richnov. Sect. Nat., 3(1): 42 (1995): G. filisepala.
subsp. samolifolia (Franch.) J.J. Halda in Acta Mus. Richnov. Sect. Nat., 3(1): 42 (1995): G. samolifolia.
sagarmathae F. Miyamoto & H. Ohba in J. Jap. Bot., 71(1): 45 (1996)—Nepal.
saginifolia Wernham
var. montiswilhelmi (P. Royen) J.J. Halda in Acta Mus. Richnov. Sect. Nat., 3(1): 47 (1995): G. montiswilhelmi.
scabra Bunge
f. alba Y.N. Lee, Fl. Korea: 1160 (1996), without type—Locality not stated.
var. kitadakensis (N. Yonez.) T. Shimizu, Fl. Nagano Pref.: 1509 (1997), isonym; see Halda in Acta Mus. Richnov. Sect. Nat., 3(1): 20 (1995): G. kitadakensis.
var. kitadakensis (N. Yonez.) J.J. Halda in Acta Mus. Richnov. Sect. Nat., 3(1): 20 (1995): G. kitadakensis.
f. stenophylla (Hara) W. Paik & W. Lee in Korean J. Pl. Taxon., 25(3): 21 (1995): G. scabra var. stenophylla.
scabrida Hayata
subsp. horaimontana (Masam.) J.J. Halda in Acta Mus. Richnov. Sect. Nat., 3(1): 39, 42 (1995): G. horaimontana.
subsp. itzershanensis (T.S. Liu & C.C. Kuo) J.J. Halda in Acta Mus. Richnov. Sect. Nat., 3(1): 39, 43 (1995): G. itzershanensis.
subsp. luzoniensis (Merr.) J.J. Halda in Acta Mus. Richnov. Sect. Nat., 3(1): 39 (1995): G. luzoniensis.
septemfida Pall.
subsp. grossheimii (Doluch.) J.J. Halda in Acta Mus. Richnov. Sect. Nat., 3(1): 20 (1995): G. grossheimii.
subsp. kolakovskyi (Doluch.) J.J. Halda in Acta Mus. Richnov. Sect. Nat., 3(1): 20 (1995): G. kolakovskyi.
subsp. overinii (Grossh.) J.J. Halda in Acta Mus. Richnov. Sect. Nat., 3(1): 20 (1995): G. overinii.
spathulifolia Maxim. ex Kusn.
var. abaensis (T.N. Ho) J.J. Halda in Acta Mus. Richnov. Sect. Nat., 3(1): 40 (1995): G. abaensis.
stylophora C.B. Clarke
subsp. yunnanensis J.J. Halda in Acta Mus. Richnov. Sect. Nat., 3(1): 14 (1995)—China South-central (Yunnan).
sumatrana Ridl.
var. humifusa (S. Moore) T.N. Ho in Acta Phytotax. Sin., 38(1): 12 (2000): G. humifusa.
takushii T. Yamazaki in J. Jap. Bot., 75(5): 280 (2000)—Nansei-shoto.
tarokoensis Chih H. Chen & J.C. Wang in Bot. Bull. Acad. Sin. (Taipei), 40(1): 29 (1999)—Taiwan.
ting-nung-hoae J.J. Halda in Acta Mus. Richnov. Sect. Nat., 3(1): 36 (1995)—Cambodia.
tonkinensis S. Hul in Adansonia, 21(2): 251 (1999)—Vietnam.
trichotoma Kusn.
subsp. chingii (C. Marquand) J.J. Halda in Acta Mus. Richnov. Sect. Nat., 3(1): 23 (1995): G. chingii.
tricolor Diels & Gilg
subsp. syringea (T.N. Ho) J.J. Halda in Acta Mus. Richnov. Sect. Nat., 3(1): 44 (1995): G. syringea.
triflora Pall.
f. albiflora T. Shimizu, Fl. Nagano Pref.: 1509 (1997)—Japan.
f. alboviolacea W. Paik & W. Lee in Korean J. Pl. Taxon., 25(3): 19 (1995)—South Korea.
subsp. japonica (Kusn.) V.N. Voroschilov in A.K. Skvortsov (ed.), Florist. issl. v razn. raĭonakh SSSR: 187 (1985): G. rigescens var. japonica.
f. semiglobularis H. Toyokuni in J. Fac. Liberal Arts Shinsu Univ., Nat. Sci., 13: 130 (1979)—Japan.
vandellioides Hemsl.
subsp. baoxingensis (T.N. Ho) J.J. Halda in Acta Mus. Richnov. Sect. Nat., 3(1): 43 (1995): G. baoxingensis.

GENTIANA:-
vandellioides
subsp. yiliangensis (T.N. Ho) J.J. Halda in Acta Mus. Richnov. Sect. Nat., 3(1): 43 (1995): G. yiliangensis.
verna L.
subsp. oschtenica (Kusn.) J.J. Halda in Acta Mus. Richnov. Sect. Nat., 3(1): 33 (1995): G. verna var. oschtenica.
subsp. schleicheri (Vacc.) O. de Bolòs & J. Vigo, Fl. Països Catalans, 3: 127 (1995): G. verna f. schleicheri.
vernayi C. Marquand
subsp. atropurpurea (T.N. Ho) J.J. Halda in Acta Mus. Richnov. Sect. Nat., 3(1): 28 (1995): G. atropurpurea.
walujewii Regel & Schmalh.
subsp. tenuicaulis (Ling) J.J. Halda in Acta Mus. Richnov. Sect. Nat., 3(1): 32 (1995): G. tenuicaulis.
wangchukii E. Aitken & D.G. Long in Edinburgh J. Bot., 53(1): 82 (1996)—Bhutan.
wardii W.W. Sm.
subsp. emergens (C. Marquand) J.J. Halda in Acta Mus. Richnov. Sect. Nat., 3(1): 25 (1995): G. emergens.
subsp. micrantha (C. Marquand) J.J. Halda in Acta Mus. Richnov. Sect. Nat., 3(1): 25 (1995): G. wardii var. micrantha.
wilsonii
subsp. striolata (T.N. Ho) J.J. Halda in Acta Mus. Richnov. Sect. Nat., 3(1): 23 (1995): G. striolata.
wingecarribiensis L.G. Adams
var. wissmannii (J.B. Williams) J.J. Halda, Genus Gentiana: 167 (1996 publ. 1997): G. wissmannii.
yunnanensis Franch.
subsp. kialensis (C. Marquand) J.J. Halda in Acta Mus. Richnov. Sect. Nat., 3(1): 29 (1995): G. yunnanensis var. kialensis.
zollingeri Fawc.
subsp. tentyoensis (Masam.) J.J. Halda in Acta Mus. Richnov. Sect. Nat., 3(1): 39 (1995): G. tentyoensis.
f. yaichiana H. Toyokuni & Y. Toyokuni in J. Asahikawa Univ., 4: 182 (1976)—Japan.

GENTIANELLA (Gentianac.).
amarella (L.) Börner
var. lingulata (C. Agardh) T. Karlsson in Svensk Bot. Tidskr., 91(5): 249 (1997 publ. 1998): Gentiana lingulata.
subsp. takedae (Kitag.) H. Toyokuni in Papers pl. ecol. & taxon. mem. Dr. Satoshi Nakanishi: 581 (1987): Gentiana takedae.
subsp. yuparensis (Takeda) H. Toyokuni in Papers pl. ecol. & taxon. mem. Dr. Satoshi Nakanishi: 579 (1987): Gentiana yuparensis.
androsacea J.S. Pringle in Fl. Ecuador, 53(159A): 62 (1995)—Ecuador.
× davidiana T.C.G. Rich in Watsonia, 21(4): 322 (1997)—Great Britain. (G. amarella × anglica).
heterosepala (Engelm.) Holub
subsp. durangensis J.A. Villarreal Q. in Sida, 17(1): 111 (1996)—Mexico Northeast (Durango).
oellgaardii J.S. Pringle in Fl. Ecuador, 53(159A): 49 (1995)—Ecuador.
polyantha J.S. Pringle in Fl. Ecuador, 53(159A): 64 (1995)—Ecuador.
praecox (A. Kern. & Jos. Kern.) Dostál ex E. Mayer in Biol. Vestn., 16: 26 (1968): Gentiana praecox.
subsp. bohemica (Skalický) J. Holub in Preslia, 70(2): 105 (1998): G. bohemica.
profusa J.S. Pringle in Fl. Ecuador, 53(159A): 127 (1995)—Ecuador.
pseudazurea (Grubov) J. Holub in Preslia, 70(2): 105 (1998): Gentiana pseudazurea.
tischkovii (Grubov) J. Holub in Preslia, 70(2): 105 (1998): Gentiana tischkovii.
urnigera E. Aitken & D.G. Long in Edinburgh J. Bot., 53(1): 82 (1996)—Nepal, Sikkim, Bhutan, China South-central (Yunnan).

GENTIANODES (Gentianac.).
clarkei (Kusn.) S. Omer in Fl. Pakistan, 197: 43 (1995): Gentiana clarkei.

GENTIANODES:-
eumarginata *S. Omer in Fl. Pakistan,* 197: 56 (1995), nom. nov.: G. marginata *(G. Don) Omer, Ali & Qaiser.*
var. harrissii *(Omer, Ali & Qaiser) S. Omer in Fl. Pakistan,* 197: 59 (1995): G. marginata var. harrissii.
var. scabromarginata *S. Omer in Fl. Pakistan,* 197: 59 (1995)—Pakistan.

GENTIANOPSIS (Gentianac.).
komarovii *(Grossh.) H. Toyokuni in Symbolae Asahikawenses,* 2: 64 (1967): Gentiana komarovii.
smithii *H. Toyokuni in Symbolae Asahikawenses,* 2: 63 (1967), nom. nov.: Gentiana grandis *Harry Sm.*

GENTLEA (Myrsinac.).
penduliflora *(A.DC.) J.J. Pipoly & J.M. Ricketson in Sida,* 17(4): 700 (1997): Myrsine penduliflora.

GEODORUM (Orchidac.).
terrestre *(L.) L.A. Garay in Harvard Pap. Bot.,* 2(1): 47 (1997), as No. 11 on inside pages: Epidendrum terrestre.

GEONOMA (Palmae).
brenesii *M.H. Grayum in Phytologia,* 84(4): 322 (1998 publ. 1999)—Costa Rica.
hugonis *M.H. Grayum & G. de Nevers in Principes,* 42(2): 94 (1998)—Panama.
irena *F. Borchsenius in Nordic J. Bot.,* 16(6): 605 (1996 publ. 1997)—Ecuador.
macrostachys Mart.
var. acaulis *(Mart.) Skov ex Andrew Henderson, Palms of Amazon:* 274 (1995): G. acaulis.
monospatha *G. de Nevers in Principes,* 42(2): 98 (1998)—Panama, Costa Rica.
mooreana *G.C. de Nevers & M.H. Grayum in Novon,* 5(4): 354 (1995)—Panama.
spinescens H. Wendl. & Burret
var. braunii *F.W. Stauffer in Acta Bot. Venezuel.,* 20(2): 5 (1997)—Venezuela.
talamancana *M.H. Grayum in Phytologia,* 84(4): 324 (1998 publ. 1999)—Costa Rica.

GEORGEANTHA *B.G. Briggs & L.A.S. Johnson in Telopea,* 7(4): 307 (1998). *ECDEIOCOLEACEAE*
hexandra *B.G. Briggs & L.A.S. Johnson in Telopea,* 7(4): 308 (1998)—Western Australia.

GERANIUM (Geraniac.).
sect. Azorelloida *C. Aedo, F. Muñoz Garmendia & F. Pando in An. Jard. Bot. Madrid,* 56(2): 243 (1998), nom. nov.: G. sect. Petraea *R. Knuth.*
sect. Eriantha *M.S. Novoselova in Bot. Zhurn.,* 84(5): 129 (1999).
sect. Neoandina *C. Aedo in An. Jard. Bot. Madrid,* 58(1): 183 (2000).
sect. Orientalia *(Tsyren.) M.S. Novoselova in Bot. Zhurn.,* 84(5): 132 (1999): G. subsect. Orientalia.
sect. Recurvata *(R. Knuth) M.S. Novoselova in Bot. Zhurn.,* 84(5): 130 (1999): G. subsect. Recurvata.
sect. Sobolifera *M.S. Novoselova in Bot. Zhurn.,* 84(5): 133 (1999).
sect. Strigosa *C.C. Huang & L.R. Xu in Fl. Reipubl. Popul. Sin.,* 43(1): 47 (1998).
aequale *(Bab.) C. Aedo in An. Jard. Bot. Madrid,* 55(2): 466 (1997): G. molle var. aequale.
aequatoriale *B. Halfdan-Nielsen in Nordic J. Bot.,* 16(3): 267 (1996 publ. 1997)—Ecuador, Colombia.
angelense *B. Halfdan-Nielsen in Nordic J. Bot.,* 16(3): 269 (1996 publ. 1997)—Ecuador, Colombia.
chilense *C. Aedo & F. Muñoz Garmendia in Kew Bull.,* 52(3): 725 (1997), nom. nov.: G. caespitosum *Phil.*
cinereum Cav.
var. lazicum *(Woronow) A. Güner in Turkish J. Bot.,* 12(3): 240 (1988): G. subcaulescens var. lazicum.
subsp. lazicum *(Woronow) A. Güner in Turkish J. Bot.,* 12(3): 240 (1988): G. subcaulescens var. lazicum.

GERANIUM:-
collae *C. Aedo, F. Muñoz Garmendia & F. Pando in An. Jard. Bot. Madrid,* 56(2): 224 (1998), nom. nov.: G. intermedium *Colla.*
collinum Stephan
var. londesii *(Fisch. ex Link) N.N. Tzvelev in Fl. Vostochnoĭ Evropy,* 9: 378 (1996): G. londesii.
dahuricum DC.
var. paishanense *(Y.L. Chang) C.C. Huang & L.R. Xu in Fl. Reipubl. Popul. Sin.,* 43(1): 64 (1998): G. paishanense.
foreroi *C. Aedo in An. Jard. Bot. Madrid,* 58(1): 183 (2000)—Colombia.
gorbizense *C. Aedo & F. Muñoz Garmendia in Kew Bull.,* 52(3): 725 (1997), nom. nov.: G. eriostemon var. orientale *Maxim.*
guamanense *B. Halfdan-Nielsen in Nordic J. Bot.,* 16(3): 271 (1996 publ. 1997)—Ecuador.
hepaticifolium *H.C. Andrews, Geraniums,* 2: t. 89 (1805)—Locality not stated.
hillebrandii *C. Aedo & F. Muñoz Garmendia in Kew Bull.,* 52(3): 725 (1997), nom. nov.: G. humile *Hillebr.*
holm-nielsenii *B. Halfdan-Nielsen in Nordic J. Bot.,* 16(3): 272 (1996 publ. 1997)—Ecuador.
lainzii *C. Aedo in An. Jard. Bot. Madrid,* 57(1): 162 (1999)—Colombia.
loxense *B. Halfdan-Nielsen in Nordic J. Bot.,* 16(3): 273 (1996 publ. 1997)—Ecuador.
miahuatlanum *B.L. Turner in Phytologia,* 81(1): 1 (1996 publ. 1997)—Mexico Southwest (Oaxaca).
neohispidum *C. Aedo & F. Muñoz Garmendia in Kew Bull.,* 52(3): 725 (1997), nom. nov.: G. hispidum *Phil.*
neopumilum *C. Aedo & F. Muñoz Garmendia in Kew Bull.,* 52(3): 726 (1997), nom. nov.: G. pumilum *R. Knuth.*
nepalense Sweet
var. oliganthum *(C.C. Huang) C.C. Huang & L.R. Xu in Fl. Reipubl. Popul. Sin.,* 43(1): 35 (1998): G. oliganthum.
pichinchense *C. Aedo & F. Muñoz Garmendia in Kew Bull.,* 52(3): 726 (1997), nom. nov.: G. quinquelobum *Wedd.*
planum *S.R.P. Halloy in Brittonia,* 50(4): 467 (1998)—Argentina Northwest (Tucumán).
popovii *(Tzvelev) N.N. Tzvelev, Opred. Sosud. Rast. Severo-Zap. Rossii:* 680 (2000): Geranium sibiricum subsp. popovii.
pratense L.
var. affine *(Ledeb.) C.C. Huang & L.R. Xu in Fl. Reipubl. Popul. Sin.,* 43(1): 59 (1998): G. affine.
subsp. sergievskajae *G.A. Peschkova in Fl. Sibir. (Geraniac.-Cornac.),* 10: 16 (1996), nom. nov.: G. pratense f. molle *Serg.*
subsp. transbaicalicum *(Serg.) I.A. Gubanov in Fl. Vostoch. Khangaya (MNR):* 141 (1983): G. transbaicalicum.
pseudosibiricum J. Mayer
var. subuschkanense *(Popov) G.A. Peschkova in Fl. Sibir. (Geraniac.-Cornac.),* 10: 17 (1996): G. coeruleum var. subuschkanense.
pyrenaicum
subsp. depilatum *(Sommier & Levier) M.S. Novoselova in Bot. Zhurn.,* 83(1): 121 (1998): G. pyrenaicum var. depilatum.
reuteri *C. Aedo & F. Muñoz Garmendia in Kew Bull.,* 52(3): 726 (1997), nom. nov.: G. canariense *Reut.*
selloi *C. Aedo & F. Muñoz Garmendia in Kew Bull.,* 52(3): 726 (1997), nom. nov.: G. senecioides *R. Knuth.*
shikokianum Matsum.
f. albiflorum *Y.N. Lee, Fl. Korea:* 1159 (1996), without type—Locality not stated.
sibbaldioides
subsp. beckianum *C. Aedo in An. Jard. Bot. Madrid,* 58(1): 184 (2000)—Bolivia.
subsp. elongatum *(Wedd.) C. Aedo in An. Jard. Bot. Madrid,* 58(1): 184 (2000): C. cucullatum var. elongatum.
solitarium *Z.M. Tan in Acta Phytotax. Sin.,* 33(6): 608 (1995)—China South-central (Sichuan).
sylvaticum L.
subsp. uralense *(Kuvaev) N.N. Tzvelev in Novosti Sist. Vyssh. Rast.,* 32: 185 (2000): G. uralense.
taebaek *S.J. Park & Y.S. Kim in Korean J. Pl. Taxon.,* 27(2): 189 (1997)—South Korea.

GERANIUM:-
tafiense *C. Aedo & F. Muñoz Garmendia in Kew Bull.,* 52(3): 726 (1997), nom. nov.: G. delphinifolium *R. Knuth.*
tanii *C. Aedo & F. Muñoz Garmendia in Novon,* 6(3): 229 (1996), nom. nov.: G. trifoliatum *Z.M. Tan.*
tiguense *C. Aedo & F. Muñoz Garmendia in Kew Bull.,* 52(3): 726 (1997), nom. nov.: G. decumbens *R. Knuth.*
tovarii *C. Aedo in An. Jard. Bot. Madrid,* 58(1): 184 (2000)—Peru.
transbaicalicum Serg.
subsp. turczaninovii *(Serg.) G.A. Peschkova in Fl. Sibir. (Geraniac.-Cornac.),* 10: 20 (1996): G. transbaicalicum var. turczaninovii.
trianae *C. Aedo & F. Muñoz Garmendia in Kew Bull.,* 52(3): 726 (1997), as 'nom. nov.'—Colombia.
viscosissimum Fisch. & C.A. Mey.
var. incisum *(Torr. & A. Gray) N.H. Holmgren in A. Cronquist et al., Intermountain Fl., Vasc. Pl. Intermountain West,* U.S.A., 3A: 332 (1997): G. albiflorum β incisum.
wlassovianum Fisch. ex Link
var. maximowiczii *(Regel & Maack) S.J. Park in Korean J. Pl. Taxon.,* 27(2): 213 (1997): G. maximowiczii.
yaanense *Z.M. Tan in Acta Phytotax. Sin.,* 33(6): 611 (1995)—China South-central (Sichuan).
yeoi *C. Aedo & F. Muñoz Garmendia in Kew Bull.,* 52(3): 727 (1997), nom. nov.: G. rubescens *Yeo.*

GERBERA (Compos.).
hintonii *(Bullock) L. Katinas in Novon,* 8(4): 382 (1998): Chaptalia hintonii.

× **GERBERARA** *hort. in Orchid Rev.,* 104(1212): 336 (1996). *ORCHIDACEAE* (Brassavola × Diacrium × Laelia).

GESNERIA (Gesneriac.).
coccinea *H. Jacquin in Ann. Fl. Pomone,* 1837–38: 154 (1838)—Chile.

GETHYUM (Alliac.).
cuspidatum *(Harv. ex Baker) M. Muñoz-Schick in Not. Mens. Mus. Nacion. Hist. Nat. (Chile),* 343: 22 (2000): Ancrumia cuspidata.

GEUM (Rosac.).
aleppicum Jacq.
var. glabricaule *(Juz.) V.N. Voroschilov in A.K. Skvortsov (ed.), Florist. issl. v razn. raĭonakh SSSR:* 177 (1985): G. glabricaule.
cockaynei *(F. Bolle) B.P.J. Molloy & C.J. Webb in New Zealand J. Bot.,* 32(4): 424 (1994): Oncostylus cockaynei.
japonicum Thunb.
f. pleniflorum *Okuhara ex T. Shimizu, Fl. Nagano Pref.:* 1508 (1997)—Japan.

GIBBAEUM (Aizoac.).
subgen. Imitaria *(N.E. Br.) E. van Jaarsveld & S. Hammer in Bradleya,* 14: 16 (1996): gen. Imitaria.
hortenseae *(N.E.Br.) Thiede & C. Klak in Strelitzia,* 9: 705 (2000): Muiria hortenseae.
johnstonii *E. van Jaarsveld & S. Hammer in Bradleya,* 14: 14 (1996)—Cape Province.
nebrownii
subsp. johnstonii *vVan Jaarsv. & S.A. Hammer) P. Pavelka in Astridia,* 3: 39 (1997 publ. 1998); *cf. Repert. Pl. Succ. (I.O.S.),* 49: 7 (1998 publ. 1999): G. johnstonii.

GIBSONIOTHAMNUS (Scrophulariac.).
parvifolius *K. Barringer in Novon,* 9(4): 476 (1999)—Panama, Costa Rica.

GIGANTOCHLOA (Gramin.).
albopilosa *K.M. Wong, Bamboos Penins. Malaysia (Malayan Forest Rec., 41):* 124 (1995)—Malaya (Peninsular Malaysia).
albovestita *(Holttum) K.M. Wong, Bamboos Penins. Malaysia (Malayan Forest Rec., 41):* 125 (1995): G. scortechinii var. albovestita.
calcicola *E.A. Widjaja in Reinwardtia,* 11(2): 84 (1997)—Sumatera.

GIGANTOCHLOA:-
hirtinoda *E.A. Widjaja in Reinwardtia,* 11(2): 86 (1997)—Sumatera.
kuring *E.A. Widjaja in Reinwardtia,* 11(2): 86 (1997)—Sumatera.
longiprophylla *E.A. Widjaja in Reinwardtia,* 11(2): 90 (1997)—Sumatera.
luteostriata *E.A. Widjaja in Reinwardtia,* 11(2): 92 (1997)—Borneo (Kalimantan).
magentea *E.A. Widjaja in Reinwardtia,* 11(2): 94 (1997)—Sumatera.
membranoidea *E.A. Widjaja in Reinwardtia,* 11(2): 94 (1997)—Sumatera.
papyracea *E.A. Widjaja in Reinwardtia,* 11(2): 96 (1997)—Sumatera.
pubinervis *E.A. Widjaja in Reinwardtia,* 11(2): 98 (1997)—Sumatera.
pubipetiolata *E.A. Widjaja in Reinwardtia,* 11(2): 100 (1997)—Sumatera.
serik *E.A. Widjaja in Reinwardtia,* 11(2): 102 (1997)—Sumatera.
tomentosa *E.A. Widjaja in Reinwardtia,* 11(2): 104 (1997)—Sumatera.
velutina *E.A. Widjaja in Reinwardtia,* 11(2): 106 (1997)—Sumatera.

× **GIGARA** *hort. in Orchid Rev.,* 107(1225): 62 (1999). *ORCHIDACEAE* (Baptistonia× Comparettia × Rodriguezia).

GILIA (Polemoniac.).
yorkii *J.R. Shevock & A.G. Day in Madroño,* 45(2): 137 (1998 publ. 1999)—California

GILIASTRUM (Polemoniac.).
castellanosii *J.M. Porter in Aliso,* 19(1): 75 (2000)—Argentina Northwest (Catamarca).
foetidum *(Gillies ex Benth.) J.M. Porter in Aliso,* 17(1): 83 (1998): Gilia foetida.
gypsophilum *(B.L. Turner) J.M. Porter in Aliso,* 17(1): 84 (1998): Gilia gypsophila.
incisum *(Benth.) J.M. Porter in Aliso,* 17(1): 84 (1998): Gilia incisa.
insigne *(Brand) J.M. Porter in Aliso,* 17(1): 84 (1998): Gilia rigidula subsp. insignis.
ludens *(Shinners) J.M. Porter in Aliso,* 17(1): 84 (1998): Gilia ludens.
purpusii *(Brandegee) J.M. Porter in Aliso,* 17(1): 84 (1998): Gilia purpusii.
subsp. platylobum *(I.M. Johnst.) J.M. Porter in Aliso,* 17(1): 84 (1998): Gilia platyloba.
subsp. stewartii *(I.M. Johnst.) J.M. Porter in Aliso,* 17(1): 84 (1998): Gilia stewartii.

GILLBEEA (Cunoniac.).
whypallana *A.C. Rozefelds & B. Pellow in Nordic J. Bot.,* 20(4): 435 (2000)—Queensland.

GILLIESIA (Alliac.).
nahuelbutae *P. Ravenna in Onira, Bot. Leafl.,* 4(6): 23 (2000)—Chile South (Bíobío).

GINALLOA (Viscac.).
flagellaris *B.A. Barlow in Blumea,* 41(2): 339 (1996)—New Guinea (Papua New Guinea, Irian Jaya), Bismarck Archipelalgo.

GIRARDINIA (Urticac.).
diversifolia (Link) Friis
subsp. ciliata *(C.J. Chen) H.W. Li in Fl. Yunnanica,* 7: 193 (1997): G. palmata subsp. ciliata.

GLADIOLUS (Iridac.).
sect. Acidanthera *(Hochst.) P. Goldblatt, Gladiolus in Tropical Africa:* 286 (1996): gen. Acidanthera.
sect. Blandus *(Baker) P. Goldblatt, Gladiolus in Tropical Africa:* 176 (1996): G. group Blandi.
sect. Decoratus *P. Goldblatt, Gladiolus in Tropical Africa:* 268 (1996).
sect. Densiflorus *P. Goldblatt & J.C. Manning, Gladiolus South. Africa:* 70 (1998).
sect. Homoglossum *(Salisb.) P. Goldblatt & J.C. Manning, Gladiolus South. Africa:* 234 (1998): gen. Homoglossum.
sect. Linearifolius *(M.P. de Vos) P. Goldblatt & J.C. Manning, Gladiolus South. Africa:* 153 (1998): Homoglossum sect. Linearifolius.

GLADIOLUS:-
sect. Ophiolyza *(Klatt) P. Goldblatt & J.C. Manning, Gladiolus South. Africa:* 99 (1998): G. subgen. Ophiolyza.
sect. Tenuibracteus *P. Goldblatt, Gladiolus in Tropical Africa:* 194 (1996).
albens *P. Goldblatt & J.C. Manning, Gladiolus South. Africa:* 250 (1998)—Cape Province.
amplifolius *P. Goldblatt, Gladiolus in Tropical Africa:* 124 (1996)—Angola.
atropictus *P. Goldblatt & J.C. Manning, Gladiolus South. Africa:* 235 (1998)—Cape Province.
balensis *P. Goldblatt, Gladiolus in Tropical Africa:* 181 (1996)—Ethiopia.
boranensis *P. Goldblatt, Gladiolus in Tropical Africa:* 186 (1996)—Ethiopia, Sudan, Kenya.
caeruleus *P. Goldblatt. & J.C. Manning in Novon,* 6(2): 179 (1996), nom. nov.: Gladiolus gracilis var. latifolius *G.J. Lewis.*
calcicola *P. Goldblatt, Gladiolus in Tropical Africa:* 202 (1996)—Ethiopia.
canaliculatus *P. Goldblatt, Gladiolus in Tropical Africa:* 135 (1996)—Tanzania.
chelamontanus *P. Goldblatt, Gladiolus in Tropical Africa:* 179 (1996)—Angola.
fenestratus *P. Goldblatt, Gladiolus in Tropical Africa:* 172 (1996)—Angola.
ferrugineus *P. Goldblatt & J.C. Manning, Gladiolus South. Africa:* 81 (1998), nom. nov.: G. micranthus *Baker (1904).*
filiformis *P. Goldblatt & J.C. Manning, Gladiolus South. Africa:* 182 (1998)—Transvaal.
griseus *P. Goldblatt & J.C. Manning, Gladiolus South. Africa:* 242 (1998)—Cape Province.
inflexus *P. Goldblatt & J.C. Manning, Gladiolus South. Africa:* 266 (1998)—Cape Province.
insolens *P. Goldblatt & J.C. Manning, Gladiolus South. Africa:* 139 (1998)—Cape Province.
lithicola *P. Goldblatt & J.C. Manning, Gladiolus South. Africa:* 82 (1998), homonym; *see Goldblatt, Gladiolus in Tropical Africa:* 284 (1996)—Transvaal.
lithicola *P. Goldblatt, Gladiolus in Tropical Africa:* 284 (1996)—Ethiopia.
longicollis Baker
subsp. platypetalus *(Baker) P. Goldblatt & J.C. Manning, Gladiolus South. Africa:* 297 (1998): Acidanthera platypetala.
lundaensis *P. Goldblatt, Gladiolus in Tropical Africa:* 118 (1996)—Angola, Zaire.
macrospathus *P. Goldblatt, Gladiolus in Tropical Africa:* 145 (1996)—Burundi.
malvinus *P. Goldblatt & J.C. Manning, Gladiolus South. Africa:* 155 (1998)—Transvaal.
manikaensis *P. Goldblatt, Gladiolus in Tropical Africa:* 238 (1996)—Zaire.
meliusculus *(G.J. Lewis) P. Goldblatt & J.C. Manning, Gladiolus South. Africa:* 232 (1998): G. alatus var. meliusculus.
monticola *P. Goldblatt & J.C. Manning, Gladiolus South. Africa:* 173 (1998)—Cape Province.
negeliensis *P. Goldblatt, Gladiolus in Tropical Africa:* 183 (1996)—Ethiopia.
pardalinus *P. Goldblatt & J.C. Manning, Gladiolus South. Africa:* 157 (1998)—Transvaal.
pavonia *P. Goldblatt & J.C. Manning in Novon,* 6(2): 174 (1996)—Transvaal.
phoenix *P. Goldblatt & J.C. Manning, Gladiolus South. Africa:* 126 (1998)—Cape Province.
pulcherrimus *(G.J. Lewis) P. Goldblatt & J.C. Manning, Gladiolus South. Africa:* 229 (1998): G. alatus var. pulcherrimus.
rhodanthus *J.C. Manning & P. Goldblatt in Bothalia,* 29(2): 217 (1999)—Cape Province.
richardsiae *P. Goldblatt, Gladiolus in Tropical Africa:* 253 (1996)—Tanzania.
roseovenosus *P. Goldblatt & J.C. Manning, Gladiolus South. Africa:* 238 (1998)—Cape Province.
saxatilis *P. Goldblatt & J.C. Manning in Bothalia,* 29(1): 63 (1999), nom. nov.: G. lithicola *Goldblatt & J.C. Manning.*
scabridus *P. Goldblatt & J.C. Manning, Gladiolus South. Africa:* 94 (1998)—Natal.
sekukuniensis *P.J.D. Winter in Bothalia,* 29(2): 220 (1999)—Transvaal.
serpenticola *P. Goldblatt & J.C. Manning in Novon,* 6(2): 173 (1996)—Transvaal.

GLADIOLUS:-
somalensis *P. Goldblatt & M. Thulin in Novon,* 5(4): 325 (1995)—Somalia.
stenolobus *P. Goldblatt, Gladiolus in Tropical Africa:* 282 (1996)—Tanzania.
stenosiphon *P. Goldblatt, Gladiolus in Tropical Africa:* 201 (1996)—Angola.
sudanicus *P. Goldblatt, Gladiolus in Tropical Africa:* 277 (1996)—Ethiopia, Sudan.
sufflavus *(G.J. Lewis) P. Goldblatt & J.C. Manning, Gladiolus South. Africa:* 287 (1998): G. pritzelii var. sufflavus.
sulcatus *P. Goldblatt, Gladiolus in Tropical Africa:* 113 (1996), homonym; *see Lam., Tabl. Encycl.,* 1: 119 (1791)—Tanzania.
usambarensis *Marais ex P. Goldblatt, Gladiolus in Tropical Africa:* 189 (1996)—Tanzania, Kenya.
variegatus *(G.J. Lewis) P. Goldblatt & J.C. Manning in Novon,* 6(2): 178 (1996): G. debilis var. variegatus.
virgatus *P. Goldblatt & J.C. Manning, Gladiolus South. Africa:* 260 (1998)—Cape Province.
wilsonii *(Baker) P. Goldblatt & J.C. Manning, Gladiolus South. Africa:* 206 (1998): Tritonia wilsonii.

GLANDULARIA (Verbenac.).
alejandrana *B.L. Turner in Lundellia,* 2: 57 (1999)—Mexico Northeast (Nuevo León, Zacatecas).
atacamensis *(Reiche) J.M. Watson & A.E. Hoffmann in A. Hoffmann, M.K. Arroyo, F. Liberona, M. Muñoz & J. Watson, Pl.* Altoandinas Fl. Silv. Chile: 150 (1998): Verbena atacamensis.
bipinnatifida (Nutt.) Nutt.
var. ciliata *(Benth.) B.L. Turner in Lundellia,* 1: 7 (1998): Verbena ciliata.
var. latilobata *(L.M. Perry) B.L. Turner in Lundellia,* 2: 63 (1999): Verbena bipinnatifida var. latilobata.
ciliata *(Benth.) S.M. Botta in Hickenia,* 2(28): 128 (1995): Verbena ciliata.
clavata *(Ruiz & Pav.) S.M. Botta in Hickenia,* 2(28): 128 (1995): Verbena clavata.
humifusa *(Cham.) S.M. Botta in Hickenia,* 2(28): 128 (1995): Verbena humifusa.
lucanensis *(Moldenke) S.M. Botta in Hickenia,* 2(28): 128 (1995): Verbena lucanensis.
porrigens *(Phil.) J.M. Watson & A.E. Hoffmann in A. Hoffmann, M.K. Arroyo, F. Liberona, M. Muñoz & J. Watson, Pl.* Altoandinas Fl. Silv. Chile: 150 (1998): Verbena porrigens.
pulchra *(Moldenke) S.M. Botta in Hickenia,* 2(28): 128 (1995): Verbena pulchra.
radicata *(Moldenke) Tronc. ex M.E. Múlgura in Zuloaga & Morrone (eds.), Cat. Pl. Vasc. Rep. Arg. 2 (Monogr. Syst. Bot.* Missouri Bot. Gard. 74): 1246 (1999): Verbena radicata.
spectabilis *(Moldenke) S.M. Botta in Hickenia,* 2(28): 128 (1995): Verbena spectabilis.

GLAUCIUM (Papaverac.).
elegans Fisch. & C.A. Mey.
subsp. integerrima *S. Mobayen, Fl. Iran Vasc. Pl.,* 3: 155 (1985), without latin descr. or type—Iran.
fimbrilligerum Boiss.
subsp. ophyocarpum *S. Mobayen, Fl. Iran Vasc. Pl.,* 3: 108 (1985), without latin descr. or type—Iran.
flavum Crantz
subsp. vitellinum *(Boiss. & Buhse) S. Mobayen, Fl. Iran Vasc. Pl.,* 3: 168 (1985), without basionym ref.: G. vitellinum.
var. vitellinum *(Boiss. & Buhse) S. Mobayen, Fl. Iran Vasc. Pl.,* 3: 158 (1985), without basionym ref.: G. vitellinum.

GLEBIONIS (Compos.).
carinata *(Schousb.) N.N. Tzvelev in Bot. Zhurn.,* 84(7): 117 (1999): Chrysanthemum carinatum.
coronaria *(L.) N.N. Tzvelev in Bot. Zhurn.,* 84(7): 117 (1999): Chrysanthemum coronarium.
roxburghii *(Cass.) N.N. Tzvelev in Bot. Zhurn.,* 84(7): 117 (1999): Chrysanthemum roxburghii.

GLECHOMA (Labiat.).
hederacea
f. albiflorens *S.S. Ying in Mem. Coll. Agric. Nation. Taiwan Univ.,* 31(1): 24 (1991), as 'alboflorens'—Taiwan.

GLOBBA (Zingiberac.).
acehensis *A. Takano & H. Okada in Nordic J. Bot.*, 20(1): 65 (2000)—Sumatera.
fecunda *A. Takano & H. Okada in Nordic J. Bot.*, 20(1): 61 (2000)—Sumatera.
flagellaris *K. Larsen in Nordic J. Bot.*, 17(2): 119 (1997)—Thailand.
flavibracteata *A. Takano & H. Okada in Nordic J. Bot.*, 20(1): 64 (2000)—Sumatera.
racemosa Sm.
var. hookeri *(C.B. Clarke ex Baker) S. Kumar in Fl. Sikkim*, 1: 127 (1996): G. hookeri.
simaoensis *Y.Y. Qian in Acta Bot. Austro Sinica*, 10: 68 (1995)—China South-central (Yunnan).
talangensis *A. Takano & H. Okada in Nordic J. Bot.*, 20(1): 62 (2000)—Sumatera.
virginea *I.M. Turner in Asian J. Trop. Biol.*, 4(1): 29 (2000), nom. nov.: Globba candida *Ridl.*

GLOBIMETULA (Loranthac.).
cornutibracteata *Balle ex D. Wiens & R.M. Polhill, Mistletoes of Africa:* 217 (1998)—Angola.

GLOBULARIA (Globulariac.).
borjae *(G. López) S. López Udias, C. Fabregat Llueca & G. Mateo Sanz in Flora Montiberica*, 16: 35 (2000): G. repens subsp. borjae.
× losae *L. Villar, J.A. Sesé & J.V. Ferrández in Collect. Bot.*, 23: 139 (1997 publ. 1998)—Spain. (G. gracilis × repens subsp. repens).

GLOBULARIOPSIS (Scrophulariac.).
adpressa *(Choisy) O.M. Hilliard, Tribe Selagineae (Scrophulariac.):* 20 (1999): Selago adpressa.
montana *O.M. Hilliard, Tribe Selagineae (Scrophulariac.):* 18 (1999), nom. nov.: Selago wittebergensis *Compton.*
obtusiloba *O.M. Hilliard, Tribe Selagineae (Scrophulariac.):* 21 (1999)—Cape Province.
pumila *O.M. Hilliard, Tribe Selagineae (Scrophulariac.):* 22 (1999)—Cape Province.
stricta *(P.J. Bergius) O.M. Hilliard, Tribe Selagineae (Scrophulariac.):* 21 (1999): Selago stricta.
tephrodes *(E. Mey.) O.M. Hilliard, Tribe Selagineae (Scrophulariac.):* 19 (1999): Selago tephrodes.

GLOCHIDION (Euphorbiac.).
andamanicum Kurz
var. desmogyne *(Hook. f.) T. Chakrabarty & M. Gangopadhyay in J. Econ. Taxon. Bot.*, 19(1): 186 (1995): G. desmogyne.
bachmaense *Nguyen Nghia Thin, Euphorbiac. Vietnam:* 49 (1995 publ. 1996)—Vietnam.
beehleri *W. Takeuchi in Sida*, 18(4): 945 (1999), as 'beehlerii'—New Guinea (Papua New Guinea).
bourdillonii Gamble
var. bhutanicum *(D.G. Long) T. Chakrabarty & M. Gangopadhyay in J. Econ. Taxon. Bot.*, 19(1): 190 (1995): G. bhutanicum.
brothersonii *J. Florence in Fl. Polynésie française*, 1: 68 (1997)—Society Is.
candolleanum *(Wight & Arn.) T. Chakrabarty & M. Gangopadhyay in J. Econ. Taxon. Bot.*, 19(1): 191 (1995): Lobocarpus candolleanus.
comitum *J. Florence in Novon*, 7(1): 29 (1997)—Pitcairn Is.
glomerulatum (Miq.) Boerl.
var. wallichianum *(Müll. Arg.) T. Chakrabarty & M. Gangopadhyay in J. Econ. Taxon. Bot.*, 19(1): 204 (1995): G. wallichianum.
grantii *J. Florence in Bull. Mus. Nation. Hist. Nat., B, Adansonia*, Ser. 4, 18(3–4): 250 (1996)—Society Is.
grayanum *(Müll. Arg.) J. Florence in Bull. Mus. Nation. Hist. Nat., B, Adansonia*, Ser. 4, 18(3–4): 253 (1996): Phyllanthus grayanus.
hivaoaense *J. Florence in Fl. Polynésie française*, 1: 74 (1997)—Marquesas.
hohenackeri (Müll. Arg.) Bedd.
var. johnstonei *(Hook. f.) T. Chakrabarty & M. Gangopadhyay in J. Econ. Taxon. Bot.*, 19(1): 210 (1995): G. johnstonei.
huahineense *J. Florence in Fl. Polynésie française*, 1: 75 (1997)—Society Is.
impuber *(Roxb.) R. Govaerts in R. Govaerts, D.G. Frodin & A. Radcliffe-Smith, World Checklist & Bibliogr.* Euphorbiaceae, 3: 951 (2000): Bradleia impuber.

GLOCHIDION:-
karnaticum *T. Chakrabarty & M. Gangopadhyay in J. Econ. Taxon. Bot.*, 19(1): 211 (1995)—India (Karnataka).
khasicum (Müll. Arg.) Hook. f.
var. bilobulatum *(Vasudeva Rao & Chakrab.) T. Chakrabarty & M. Gangopadhyay in J. Econ. Taxon. Bot.*, 19(1): 213 (1995): G. bilobulatum.
macphersonii *R. Govaerts & A. Radcliffe-Smith in Kew Bull.*, 51(1): 175 (1996), nom. nov.: Bradleia macrophylla *Labill.*
nadeaudii *J. Florence in Bull. Mus. Nation. Hist. Nat., B, Adansonia*, Ser. 4, 18(3–4): 253 (1996)—Society Is.
papenooense *J. Florence in Bull. Mus. Nation. Hist. Nat., B, Adansonia*, Ser. 4, 18(3–4): 254 (1996)—Society Is.
rapaense *J. Florence in Bull. Mus. Nation. Hist. Nat., B, Adansonia*, Ser. 4, 18(3–4): 258 (1996)—Tubuai Is.
rubrum Blume
var. brevistylum *(Chakrab. & M.G. Gangop.) T. Chakrabarty & M. Gangopadhyay in J. Econ. Taxon. Bot.*, 19(1): 223 (1995): G. insulare var. brevistylum.
societatis *J. Florence in Fl. Polynésie française*, 1: 90 (1997)—Society Is., Tubuai Is.
societatis *(Müll. Arg.) Florence ex S.L. Welsh, Fl. Societensis:* 108 (1998), homonym; *see Florence in Fl. Polynésie française*, 1: 90 (1997): Phyllanthus societatis.
stilpnophyllum *R. Govaerts & A. Radcliffe-Smith in Kew Bull.*, 51(1): 175 (1996), nom. nov.: G. nitidum *Merr.*
subfalcatum Elmer
var. nitidum *(Merr.) T. Chakrabarty & M. Gangopadhyay in J. Econ. Taxon. Bot.*, 19(1): 232 (1995): G. nitidum.
tooviianum *J. Florence in Bull. Mus. Nation. Hist. Nat., B, Adansonia*, Ser. 4, 18(3–4): 260 (1996)—Marquesas.
tuamotuense *J. Florence in Fl. Polynésie française*, 1: 98 (1997)—Tuamotu Is.
wilderi *J. Florence in Fl. Polynésie française*, 1: 99 (1997)—Tuamotu Is.
zeylanicum (Gaertn.) A. Juss.
var. arborescens *(Blume) T. Chakrabarty & M. Gangopadhyay in J. Econ. Taxon. Bot.*, 19(1): 228 (1995): G. arborescens.
var. arunachalense *T. Chakrabarty & M. Gangopadhyay in J. Econ. Taxon. Bot.*, 19(1): 229 (1995)—Assam (Arunachal Pradesh).
var. tomentosum *(Dalzell) T. Chakrabarty & M. Gangopadhyay in J. Econ. Taxon. Bot.*, 19(1): 229 (1995): G. tomentosum var. tomentosum.

GLOMERA (Orchidac.).
distichifolia *P. Ormerod in Austral. Orchid Rev.*, 61(6): 37 (1996)—Solomon Is. (South Solomons).

GLORIOSA (Colchicac.).
sessiliflora *I. Nordal & M.G. Bingham in Kew Bull.*, 53(2): 479 (1998)—Zambia.

GLOSSOCHILOPSIS *D.L. Szlachetko in Fragm. Flor. Geobot.*, Supp. 3: 122 (1995), nom. nov.: Microstylis sect. Glossochilus *Schltr.*
chamaeorchis *(Schltr.) D.L. Szlachetko in Fragm. Flor. Geobot.*, Supp. 3: 123 (1995): Microstylis chamaeorchis.
finetii *(Gagnep.) D.L. Szlachetko in Fragm. Flor. Geobot.*, Supp. 3: 123 (1995): Microstylis finetii.

GLOSSOSTIPULA (Rubiac.).
strigosa *D.H. Lorence in Novon*, 7(1): 46 (1997)—Honduras.

GLOVERIA *M. Jordaan in S. Afr. J. Bot.*, 64(5): 299 (1998). *CELASTRACEAE*
integrifolia *(L.f.) M. Jordaan in S. Afr. J. Bot.*, 64(5): 299 (1998): Celastrus integrifolius.

GLYCERIA (Gramin.).
acutiuscula *H. Scholz in Byull. Mosk. Obshch. Ispўt. Prir., Biol.*, 100(6): 80 (1995)—North Caucasus (Krasnodar).
pulchella *(Nash) K. Schumann in Bot. Jahresber. (Just)*, 29(1): 513 (1901 publ. 1903): Panicularia pulchella.

GLYCERIA:-
saltensis *A.A. Sulekic & Z.E. Rúgolo de Agrasar in Darwiniana*, 35(1–4): 155 (1998)—Argentina Northwest (Salta).

GLYCINE (Legumin.).
max (L.) Merr.
var. okuharae *H. Nakam. ex T. Shimizu, Fl. Nagano Pref.:* 1509 (1997)—Japan.
stenophita *B.E. Pfeil & Tindale in Syst. Bot.*, 25(3): 446 (2000)—New South Wales, Queensland.

GLYCOSMIS (Rutac.).
cyanocarpa (Blume) Spreng.
var. fuscescens *(Kurz) K. Narayanan Nair in Fl. India*, 4: 336 (1997): G. trifoliata var. fuscescens.
xizangensis *(C.Y. Wu & H. Li) D.D. Tao in Rep. & Abstr. 60th Anniv. Bot. Soc. China:* 115 (1993): Walsura xizangensis.

GMELINA (Verbenac.).
evoluta *(Däniker) D.J. Mabberley in Telopea*, 7(4): 316 (1998): Vitex evoluta.

GNAPHALIUM (Compos.).
audax *D.G. Drury in New Zealand J. Bot.*, 10(1): 136 (1972)—New Zealand South I., New Zealand North I., Chatham Is.
subsp. ruahinicum *D.G. Drury in New Zealand J. Bot.*, 10(1): 140 (1972)—New Zealand North I., New Zealand South I., Chatham Is.
cabrerae *S.E. Freire in Zuloaga & Morrone (eds.), Cat. Pl. Vasc. Rep. Arg. 2 (Monogr. Syst. Bot.* Missouri Bot. Gard. 74): 1245 (1999), nom. nov., without replaced synonym ref.: Gnaphalium fastigiatum *Phil.*
delicatum *D.G. Drury in New Zealand J. Bot.*, 10(1): 148 (1972)—New Zealand North I., New Zealand South I.
ensifer *D.G. Drury in New Zealand J. Bot.*, 10(1): 150 (1972)—New Zealand South I.
limosum *D.G. Drury in New Zealand J. Bot.*, 10(1): 133 (1972)—New Zealand North I., New Zealand South I.
paludosum Petrie
var. polylepis *D.G. Drury in New Zealand J. Bot.*, 10(1): 155 (1972), nom. nov.: G. collinum var. monocephalum *Kirk.*
teydeum *R. Knapp in Oberhess. Naturwiss. Zeitschr.*, 42: 87 (1975–76 publ. 1977), without type—Canary Is.

GNETUM (Gnetac.).
hainanense *C.Y. Cheng ex L.K. Fu, Y.F. Yu & M.G. Gilbert in Novon*, 9(2): 187 (1999)—Hainan, China Southeast (Guangxi).
latifolium Blume
var. longipes *(Markgr.) N.T. Hiêp in Fl. Cambodge, Laos & Vietnam*, 28: 142 (1996): G. latifolium f. longipes.

GNIDIA (Thymelaeac.).
eriocephala *Wall. ex J. Graham, Cat. Pl. Bombay:* 176 (1839)—India (Maharashtra).

GOCHNATIA (Compos.).
densicephala *(Cabrera) G. Sancho in Novon*, 9(4): 559 (1999): G. paniculata var. densicephala.

GOERKEMIA *Ş. Yıldırımlı in Ot Sist. Bot. Dergisi*, 6(2): 2 (1999 publ. 2000). *CRUCIFERAE*
turcica *Ş. Yıldırımlı in Ot Sist. Bot. Dergisi*, 6(2): 2 (1999 publ. 2000)—Turkey.

GOETHEA (Malvac.).
multiflora *(Juss.) N.E. Brown, Suppl. Johnson's Gard. Dict.:* 932 (1882): Pavonia multiflora.

GOMIDESIA (Myrtac.).
cordiifolia *(DC.) E. Nic Lughadha in Kew Bull.*, 53(1): 243 (1998): Myrcia cordiifolia.

GOMPHIA (Ochnac.).
coriacea *(Tiegh.) K.A. Lye in Lidia*, 4(3): 91 (1998): Monelasmum coriaceum.
florida *K.A. Lye in Lidia*, 4(3): 92 (1998), nom. nov.: Ouratea floribunda *De Wild.*

GOMPHIA:-
hiernii *(Tiegh.) K.A. Lye in Lidia,* 4(3): 92 (1998): Monelasmum hiernii.
mildbraedii *(Gilg) K.A. Lye in Lidia,* 4(3): 92 (1998): Ouratea mildbraedii.

GOMPHOCARPUS (Asclepiadac.).
cancellatus *(Burm. f.) P.V. Bruyns in Bothalia,* 25(2): 165 (1995): Asclepias cancellata.
cancellatus *(Burm. f.) A. Nicholas & P.I. Forster in Fl. Australia,* 28: 319 (1996), isonym; *see Bruyns in Bothalia,* 25(2): 165 (1995): Asclepias cancellata.
fruticosus (L.) W.T. Aiton
subsp. flavidus *(N.E. Br.) D.J. Goyder in Kew Bull.,* 51(4): 798 (1996): Asclepias flavida.
phillipsiae *(N.E. Br.) D.J. Goyder in Kew Bull.,* 51(4): 798 (1996): Asclepias phillipsiae.

GOMPHOSTEMMA (Labiat.).
javanicum (Blume) Benth.
var. helferi *(Hook. f.) G. Murata in Southeast Asian Stud.,* 14(2): 189 (1976) G. oblongum var. helferi.
subsp. lucidum *(Wall. ex Benth.) G. Murata in Southeast Asian Stud.,* 14(2): 189 (1976): G. lucidum.
f. phlomoides *(Reinw. ex Blume) G. Murata in Southeast Asian Stud.,* 14(2): 189 (1976): Prasium phlomoides.
var. rugosum *(Prain) G. Murata in Southeast Asian Stud.,* 14(2): 190 (1976): G. rugosum.
strobilinum Wall. ex Benth.
subsp. acaule *(Kurz ex Hook. f.) G. Murata in Southeast Asian Stud.,* 14(2): 192 (1976): G. acaule.
subsp. wallichii *(Prain) G. Murata in Southeast Asian Stud.,* 14(2): 191 (1976): G. wallichii.

GOMPHRENA (Amaranthac.).
affinis F. Muell. ex Benth.
subsp. pilbarensis *Kanis ex J. Palmer in Austral. Syst. Bot.,* 11(1): 87 (1998)—Western Australia.
arida *J. Palmer in Austral. Syst. Bot.,* 11(1): 106 (1998)—Queensland.
atrorubra *J. Palmer in Austral. Syst. Bot.,* 11(1): 96 (1998)—Northern Territory.
boliviana Moq.
var. tarijensis *(R.E. Fr.) T.M. Pedersen in Adansonia,* 19(2): 227 (1997): G. tarijensis.
brachystylis F. Muell.
subsp. pindanensis *J. Palmer in Austral. Syst. Bot.,* 11(1): 100 (1998)—Western Australia.
canescens R. Br.
subsp. erythrina *J. Palmer in Austral. Syst. Bot.,* 11(1): 92 (1998)—Northern Territory.
celosioides Mart.
var. fallax *(Seub.) T.M. Pedersen in Adansonia,* 19(2): 228 (1997): G. fallax.
var. hygrophila *(Mart.) T.M. Pedersen in Adansonia,* 19(2): 228 (1997): G. hygrophila.
connata *J. Palmer in Austral. Syst. Bot.,* 11(1): 97 (1998)—Western Australia, Northern Territory.
cucullata *J. Palmer in Austral. Syst. Bot.,* 11(1): 107 (1998)—Western Australia.
diffusa (R. Br.) Spreng.
subsp. arenicola *J. Palmer in Austral. Syst. Bot.,* 11(1): 113 (1998)—Northern Territory, Western Australia.
eichleri *J. Palmer in Austral. Syst. Bot.,* 11(1): 114 (1998)—Western Australia.
elegans Mart.
var. orientalis *T.M. Pedersen in Adansonia,* 19(2): 229 (1997)—Uruguay, Brazil South (Rio Grande do Sul).
floribunda *J. Palmer in Austral. Syst. Bot.,* 11(1): 102 (1998)—Northern Territory.
hatschbachiana *T.M. Pedersen in Bonplandia,* 10(1–4): 90 (2000)—Brazil Northeast (Bahia).
humifusa *J. Palmer in Austral. Syst. Bot.,* 11(1): 115 (1998)—Western Australia.
kanisii *J. Palmer in Austral. Syst. Bot.,* 11(1): 93 (1998)—Western Australia.
lacinulata *J. Palmer in Austral. Syst. Bot.,* 11(1): 94 (1998)—Northern Territory.
leptoclada Benth.
subsp. saxosa *J. Palmer in Austral. Syst. Bot.,* 11(1): 120 (1998)—Western Australia, Northern Territory.

GOMPHRENA:-
leptophylla *(Benth.) J. Palmer in Austral. Syst. Bot.,* 11(1): 103 (1998): Alternanthera leptophylla.
magentitepala *J. Palmer in Austral. Syst. Bot.,* 11(1): 94 (1998)—Northern Territory.
martiana Gillies ex Moq.
var. glutinosa *(R.E. Fr.) T.M. Pedersen in Adansonia,* 19(2): 230 (1997): G. glutinosa.
occulta *J. Palmer in Austral. Syst. Bot.,* 11(1): 108 (1998)—Western Australia.
paraguayensis Chodat
subsp. chacoensis *T.M. Pedersen in Adansonia,* 19(2): 232 (1997)—Paraguay.
pohlii Moq.
var. hassleri *(Chodat) T.M. Pedersen in Adansonia,* 19(2): 232 (1997): G. hassleri.
riparia *T.M. Pedersen in Bonplandia,* 10(1–4): 90 (2000)—Brazil South (Paraná).
rosula *J. Palmer in Austral. Syst. Bot.,* 11(1): 85 (1998)—Northern Territory.
spissa *T.M. Pedersen in Adansonia,* 19(2): 232 (1997)—Bolivia.
tomentosa (Griseb.) R.E. Fr.
var. monticola *T.M. Pedersen in Adansonia,* 19(2): 234 (1997)—Argentina Northwest (Tucumán).
triceps *T.M. Pedersen in Adansonia,* 19(2): 234 (1997)—Uruguay.

GONGORA (Orchidac.).
sect. Aceras *G. Gerlach in Schlechter Orchideen, ed. 3,* I/C(37–38): 2384 (1999 publ. 2000).
alfieana *R. Rice in Austral. Orchid Rev.,* 63(5): 14 (1998), without diagnostic latin descr.—Locality not stated.
erecta *W.M. Whitten & D.E. Bennett in Brittonia,* 50(2): 186 (1998)—Peru.
garayana *R. Rice in Oasis,* 1(4): 3 (2000)—Colombia.

GONGRODISCUS (Sapindac.).
bilocularis *H. Turner in Bull. Mus. Nation. Hist. Nat., B, Adansonia,* Ser. 4, 18(3–4): 342 (1996)—New Caledonia.

GONGROSTYLUS (Compos.).
pipolyi *H. Robinson in Phytologia,* 84(5): 349 (1998 publ. 1999)—Colombia.

GONGYLOLEPIS (Compos.).
cortesii *(S. Díaz) J.F. Pruski & S. Díaz in Fl. Venezuelan Guayana (gen. eds. J.A. Steyermark et al.),* 3: 285 (1997): Eurydochus cortesii.

GONGYLOTAXIS *M.G. Pimenov & E.V. Kljuykov in Edinburgh J. Bot.,* 53(2): 188 (1996). *UMBELLIFERAE*
rechingeri *M.G. Pimenov & E.V. Kljuykov in Edinburgh J. Bot.,* 53(2): 188 (1996), nom. nov.: Scaligeria gongylotaxis *Rech. f.*

GONIOLIMON (Plumbaginac.).
tataricum (L.) Boiss.
var. tauricum *(Klokov) N.N. Tzvelev in Fl. Vostochnoĭ Evropȳ,* 9: 161 (1996): G. tauricum.

GONIOTHALAMUS (Annonac.).
macranthus (Kurz) Boerl.
var. brevipetalus *Debika Mitra in Bull. Bot. Surv. India,* 35(1–4): 117 (1993 publ. 1997)—Andaman Is.
majestatis *Paul Kessler in Blumea,* 41(1): 27 (1996)—Sulawesi.
shraddhae *S.R. Dutta & S.M. Almeida in J. Bombay Nat. Hist. Soc.,* 95(3): 488 (1998)—India (West Bengal).

GONOCARYUM (Icacinac.).
poilanei *Gagnep. ex J.-F. Villiers in Bull. Jard. Bot. Nation. Belg.,* 66(1–2): 26 (1997)—Vietnam.
sleumeri *J.-F. Villiers in Bull. Jard. Bot. Nation. Belg.,* 66(1–2): 25 (1997)—Vietnam.

GONOLOBUS (Asclepiadac.).
ecuadorensis *P.T. Li in J. S. China Agric. Univ.,* 14(1): 59 (1993), nom. nov., homonym; *see Schltr. in Bot. Jahrb. Syst.,* 37: 624 (1906): G. macranthus *K. Schum.*

GONOLOBUS:-
fimbriatiflorus *(Morillo) W.D. Stevens in Novon,* 10(3): 243 (2000): Matelea fimbriatiflora.
haitiensis *P.T. Li in J. S. China Agric. Univ.,* 14(1): 58 (1993), nom. nov.: G. stipitatus *Alain.*

GONYSTYLUS (Thymelaeac.).
othmanii *C.S. Tawan in Bot. J. Linn. Soc.,* 130(1): 66 (1999)—Borneo (Sarawak).

GONZALAGUNIA (Rubiac.).
bifida *B. Ståhl in Fl. Ecuador,* 62(162:3): 94 (1999)—Ecuador.
exigua *B. Ståhl in Fl. Ecuador,* 62(162:3): 89 (1999)—Ecuador, Peru.
pauciflora *B. Ståhl in Fl. Ecuador,* 62(162:3): 82 (1999)—Ecuador.

GOODENIA (Goodeniac.).
drummondii Carolin
subsp. megaphylla *L.W. Sage in Nuytsia,* 12(2): 234 (1998)—Western Australia.
gypsicola *D.E. Symon in J. Adelaide Bot. Gard.,* 19: 97 (2000)—Western Australia, South Australia.
heatheriana *L.W. Sage in Nuytsia,* 13(2): 368 (2000)—Western Australia.
katabudjar *R.J. Cranfield & L.W. Sage in Nuytsia,* 11(3): 297 (1997)—Western Australia.
laevis Benth.
subsp. humifusa *L.W. Sage in Nuytsia,* 12(2): 236 (1998)—Western Australia.
lancifolia *L.W. Sage & R. Cranfield in Nuytsia,* 13(2): 369 (2000)—Western Australia.
scapigera R. Br.
subsp. graniticola *L.W. Sage in Nuytsia,* 13(2): 373 (2000)—Western Australia.

GOODYERA (Orchidac.).
biflora (Lindl.) Hook. f.
var. macrantha *(Maxim.) T. Hashimoto in Proc. 12th World Orchid Conf., 1987 (eds. K. Saito & R. Tanaka):* 122 (1987): G. macrantha.
brachyceras *(A. Rich. & Galeotti) L.A. Garay & G.A. Romero-González in Harvard Pap. Bot.,* 3(1): 61 (1998): Physurus brachyceras.
crocodiliceps *P. Ormerod in Austral. Orchid Rev.,* 61(5): 41 (1996)—New Guinea (Papua New Guinea).
dongchenii *S.Z. Lucksom in J. Ind. Bot. Soc.,* 72(1–2): 191 (1993)—Sikkim.
hachijoensis Yatabe
var. boninensis *(Nakai) T. Hashimoto in Proc. 12th World Orchid Conf., 1987 (eds. K. Saito & R. Tanaka):* 122 (1987): G. boninensis.
inmeghema *P. Ormerod in Austral. Orchid Rev.,* 61(5): 42 (1996)—Vanuatu, Santa Cruz Is.
novembrilis *(Rchb.f.) P. Ormerod in Austral. Orchid Rev.,* 61(5): 42 (1996): Salacistis novembrilis.
shixingensis *K.Y. Lang in Acta Phytotax. Sin.,* 34(6): 636 (1996)—China Southeast (Guangdong).

GORDONIA (Theac.).
dassanayakei *B.M. Wadhwa & A. Weerasooriya in M.D. Dassanayake & W.D. Clayton, eds., Revised Handb. Fl. Ceylon,* 10: 392 (1996)—Sri Lanka.
tiantangensis *L.L. Deng & G.S. Fan in J. Trop. Subtrop. Bot.,* 7(3): 193 (1999)—China South-central (Yunnan).

GOTHOFREDA (Asclepiadac.).
gracilis *Morong in Ann. New York Acad. Sci.,* 7: 162 (1893)—Paraguay.†
oblongifolia *Morong in Ann. New York Acad. Sci.,* 7: 161 (1893)—Paraguay.†

GOUANIA (Rhamnac.).
obtusifolia *Vent. ex A.T. Brongniart, Mém. Fam. Rhamnées:* 72 (1826)—Jawa.†

GOVENIA (Orchidac.).
latifolia *(Humb., Bonpl. & Kunth) L.A. Garay & G.A. Romero-Gonzalez in Harvard Pap. Bot.,* 4(2): 482 (1999): Dendrobium latifolium.

GRACIELANTHUS *R. González Tamayo & D.L. Szlachetko in Fragm. Flor. Geobot.,* 40(2): 760 (1995). *ORCHIDACEAE*

GRACIELANTHUS:-
pyramidalis *(Lindl.) R. González Tamayo & D.L. Szlachetko in Fragm. Flor. Geobot.*, 40(2): 761 (1995): Spiranthes pyramidalis.
riodelayensis *(Burns-Bal.) R. González Tamayo & D.L. Szlachetko in Fragm. Flor. Geobot.*, 40(2): 763 (1995): Deiregyne riodelayensis.

GRAMMATOPHYLLUM (Orchidac.).
finlaysonianum *Lindl., Gen. Spec. Orch.*: 173 (1833)—Malaya (Singapore).†

GRANITITES *B.L. Rye in Nuytsia*, 10(3): 451 (1996). *RHAMNACEAE*
intangendus *(F. Muell.) B.L. Rye in Nuytsia*, 10(3): 452 (1996): Pomaderris intangenda.

GRAPTOPETALUM (Crassulac.).
mendozae *C. Glass & M. Cházaro Basáñez in Cact. Suc. Mex.*, 42(4): 80 (1997)—Mexico Gulf (Veracruz).

GRASTIDIUM (Orchidac.).
subgen. Dichopus *(Blume) M.A. Clements & D.L. Jones in Lasianthera*, 1(2): 116 (1997): gen. Dichopus.
agusanense *(Ames) M.A. Clements & D.L. Jones in Lasianthera*, 1(2): 57 (1997): Dendrobium agusanense.
alagense *(Ames) M.A. Clements & D.L. Jones in Lasianthera*, 1(2): 58 (1997): Dendrobium alagense.
aliciae *(Ames & Quisumb.) M.A. Clements & D.L. Jones in Lasianthera*, 1(2): 58 (1997): Dendrobium aliciae.
angustispathum *(J.J. Sm.) M.A. Clements & D.L. Jones in Lasianthera*, 1(2): 58 (1997): Dendrobium angustispathum.
armitiae *(F.M. Bailey) M.A. Clements & D.L. Jones in Lasianthera*, 1(2): 63 (1997): Dendrobium armitiae.
biflorum *(G. Forst.) M.A. Clements & D.L. Jones in Lasianthera*, 1(2): 64 (1997): Epidendrum biflorum.
blanche-amesiae *(A.D. Hawkes & A.H. Heller) M.A. Clements & D.L. Jones in Lasianthera*, 1(2): 64 (1997), as 'blanche-amesii': Dendrobium blanche-amesiae.
campbellii *(P.J. Cribb & B.A. Lewis) M.A. Clements & D.L. Jones in Lasianthera*, 1(2): 67 (1997): Dendrobium campbellii.
carnicarinum *(Kores) M.A. Clements & D.L. Jones in Lasianthera*, 1(2): 67 (1997): Dendrobium carnicarinum.
carolinense *(Schltr.) M.A. Clements & D.L. Jones in Lasianthera*, 1(2): 67 (1997): Dendrobium carolinense.
cathcartii *(Hook. f.) M.A. Clements & D.L. Jones in Lasianthera*, 1(2): 67 (1997): Dendrobium cathcartii.
correllianum *(A.D. Hawkes & A.H. Heller) M.A. Clements & D.L. Jones in Lasianthera*, 1(2): 69 (1997): Dendrobium correllianum.
dactylodes *(Rchb. f.) M.A. Clements & D.L. Jones in Lasianthera*, 1(2): 70 (1997): Dendrobium dactylodes.
emarginatum *(J.W. Moore) M.A. Clements & D.L. Jones in Lasianthera*, 1(2): 72 (1997): Dendrobium emarginatum.
everardii *(Rolfe) M.A. Clements & D.L. Jones in Lasianthera*, 1(2): 73 (1997): Dendrobium everardii.
falcatum *(J.J. Sm.) M.A. Clements & D.L. Jones in Lasianthera*, 1(2): 75 (1997): Dendrobium falcatum.
fulgescens *(J.J. Sm.) M.A. Clements & D.L. Jones in Lasianthera*, 1(2): 76 (1997): Dendrobium fulgescens.
gibbosum *(Gilli) M.A. Clements & D.L. Jones in Lasianthera*, 1(2): 76 (1997): Dendrobium gibbosum.
glaucophyllum *(Teijsm. & Binn.) M.A. Clements & D.L. Jones in Lasianthera*, 1(2): 77 (1997): Dendrobium glaucophyllum.
greenianum *(P.J. Cribb & B.A. Lewis) M.A. Clements & D.L. Jones in Lasianthera*, 1(2): 77 (1997): Dendrobium greenianum.

GRASTIDIUM:-
guamense *(Ames) M.A. Clements & D.L. Jones in Lasianthera*, 1(2): 78 (1997): Dendrobium guamense.
haemoglossum *(Thwaites) M.A. Clements & D.L. Jones in Lasianthera*, 1(2): 78 (1997): Dendrobium haemoglossum.
implicatum *(Fukuy.) M.A. Clements & D.L. Jones in Lasianthera*, 1(2): 84 (1997): Dendrobium implicatum.
inconspicuiflorum *(J.J. Sm.) M.A. Clements & D.L. Jones in Lasianthera*, 1(2): 84 (1997): Dendrobium inconspicuiflorum.
insigne *(Blume) M.A. Clements & D.L. Jones in Lasianthera*, 1(2): 116 (1997): Dichopus insignis.
involutum *(Lindl.) M.A. Clements & D.L. Jones in Lasianthera*, 1(2): 85 (1997): Dendrobium involutum.
kandarianum *(Kraenzl.) M.A. Clements & D.L. Jones in Lasianthera*, 1(2): 85 (1997): Dendrobium kandarianum.
kraenzlinii *(L.O. Williams) M.A. Clements & D.L. Jones in Lasianthera*, 1(2): 88 (1997): Dendrobium kraenzlinii.
lepidochilum *(Kraenzl.) M.A. Clements & D.L. Jones in Lasianthera*, 1(2): 88 (1997): Dendrobium lepidochilum.
leytense *(Ames) M.A. Clements & D.L. Jones in Lasianthera*, 1(2): 89 (1997): Dendrobium leytense.
luzonense *(Lindl.) M.A. Clements & D.L. Jones in Lasianthera*, 1(2): 89 (1997): Dendrobium luzonense.
malaitense *(Rolfe) M.A. Clements & D.L. Jones in Lasianthera*, 1(2): 90 (1997): Dendrobium malaitense.
marginatum *(J.J. Sm.) M.A. Clements & D.L. Jones in Lasianthera*, 1(2): 90 (1997): Dendrobium polyschistum var. marginatum.
muriciferum *(J.J. Sm.) M.A. Clements & D.L. Jones in Lasianthera*, 1(2): 92 (1997): Dendrobium pictum var. muriciferum.
nimium *(J.J. Sm.) M.A. Clements & D.L. Jones in Lasianthera*, 1(2): 93 (1997): Dendrobium nimium.
niveopurpureum *(J.J. Sm.) M.A. Clements & D.L. Jones in Lasianthera*, 1(2): 93 (1997): Dendrobium niveopurpureum.
obcuneatum *(F.M. Bailey) M.A. Clements & D.L. Jones in Lasianthera*, 1(2): 93 (1997): Dendrobium obcuneatum.
obscure-auriculatum *(Gilli) M.A. Clements & D.L. Jones in Lasianthera*, 1(2): 94 (1997): Dendrobium obscure-auriculatum.
ochroleucum *(J.J. Sm.) M.A. Clements & D.L. Jones in Lasianthera*, 1(2): 94 (1997): Dendrobium ostrinum var. ochroleucum.
orientale *(J.J. Sm.) M.A. Clements & D.L. Jones in Lasianthera*, 1(2): 94 (1997): Dendrobium orientale.
ornithoflorum *(Ames) M.A. Clements & D.L. Jones in Lasianthera*, 1(2): 95 (1997): Dendrobium ornithoflorum.
ostrinum *(J.J. Sm.) M.A. Clements & D.L. Jones in Lasianthera*, 1(2): 95 (1997): Dendrobium ostrinum.
pergracile *(Ames) M.A. Clements & D.L. Jones in Lasianthera*, 1(2): 97 (1997): Dendrobium pergracile.
plumilobum *(J.J. Sm.) M.A. Clements & D.L. Jones in Lasianthera*, 1(2): 98 (1997): Dendrobium plumilobum.
ponapense *(Schltr.) M.A. Clements & D.L. Jones in Lasianthera*, 1(2): 103 (1997): Dendrobium ponapense.
quinquecaudatum *(J.J. Sm.) M.A. Clements & D.L. Jones in Lasianthera*, 1(2): 104 (1997): Dendrobium quinquecaudatum.
reineckei *(Schltr.) M.A. Clements & D.L. Jones in Lasianthera*, 1(2): 105 (1997): Dendrobium reineckei.
salicifolium *(J.J. Sm.) M.A. Clements & D.L. Jones in Lasianthera*, 1(2): 107 (1997): Dendrobium salicifolium.
sladei *(J.J. Wood & P.J. Cribb) M.A. Clements & D.L. Jones in Lasianthera*, 1(2): 108 (1997): Dendrobium sladei.

GRASTIDIUM:-
straussianum *(Schltr.) M.A. Clements & D.L. Jones in Lasianthera*, 1(2): 112 (1997): Dendrobium straussianum.
suaveolens *(Kraenzl.) M.A. Clements & D.L. Jones in Lasianthera*, 1(2): 112 (1997): Dendrobium suaveolens.
subelobatum *(J.J. Sm.) M.A. Clements & D.L. Jones in Lasianthera*, 1(2): 112 (1997): Dendrobium subelobatum.
subtricostatum *(J.J. Sm.) M.A. Clements & D.L. Jones in Lasianthera*, 1(2): 112 (1997): Dendrobium subtricostatum.
torquisepalum *(Kraenzl.) M.A. Clements & D.L. Jones in Lasianthera*, 1(2): 113 (1997): Dendrobium torquisepalum.
tozerense *(Lavarack) M.A. Clements & D.L. Jones in Lasianthera*, 1(1): 18 (1996): Dendrobium tozerense.
trichosepalum *(Gilli) M.A. Clements & D.L. Jones in Lasianthera*, 1(2): 114 (1997): Dendrobium trichosepalum.
trigonellodorum *(Kraenzl.) M.A. Clements & D.L. Jones in Lasianthera*, 1(2): 114 (1997): Dendrobium trigonellodorum.
trilobulatum *(Kores) M.A. Clements & D.L. Jones in Lasianthera*, 1(2): 114 (1997): Dendrobium trilobulatum.
xylophyllum *(Kraenzl.) M.A. Clements & D.L. Jones in Lasianthera*, 1(2): 116 (1997): Dendrobium xylophyllum.

GREIGIA (Bromeliac.).
atrocastanea *H.E. Luther in Selbyana*, 19(2): 218 (1998 publ. 1999)—Bolivia.
cochabambae *H.E. Luther in Selbyana*, 19(1): 83 (1998)—Bolivia.
kessleri *H.E. Luther in Selbyana*, 19(1): 83 (1998)—Bolivia.
raporum *H.E. Luther in Selbyana*, 19(2): 224 (1998 publ. 1999)—Peru.
vilcabambae *H.E. Luther in Selbyana*, 19(2): 225 (1998 publ. 1999)—Peru.

GREVEA (Montiniac.).
eggelingii Milne-Redh.
var. keniensis *(Verdc.) B. Verdcourt in Kew Bull.*, 51(4): 772 (1996): G. madagascariensis subsp. keniensis.

GREVILLEA (Proteac.).
acropogon *R.O. Makinson in Fl. Australia*, 17A: 496, 160 (2000)—Western Australia.
agrifolia A. Cunn. ex R. Br.
subsp. microcarpa *(Olde & Marriott) R.O. Makinson in Fl. Australia*, 17A: 494, 129 (2000): G. microcarpa.
amplexans F. Muell. ex Benth.
subsp. adpressa *(Olde & Marriott) R.O. Makinson in Fl. Australia*, 17A: 506, 426 (2000): G. adpressa.
subsp. semivestita *R.O. Makinson in Fl. Australia*, 17A: 505, 426 (2000)—Western Australia.
subsp. polychroma *W. Molyneux & V. Stajsic in Fl. Australia*, 17A: 502, 251 (2000)—Victoria.
cheilocarpa *R.O. Makinson in Fl. Australia*, 17A: 505, 384 (2000)—Western Australia.
commutata F. Muell.
subsp. pinnatisecta *(F. Muell.) R.O. Makinson in Fl. Australia*, 17A: 496, 178 (2000): G. commutata var. pinnatisecta.
cravenii *R.O. Makinson in Fl. Australia*, 17A: 494, 88 (2000)—Western Australia.
decora Domin
subsp. telfordii *R.O. Makinson in Fl. Australia*, 17A: 494, 114 (2000)—Queensland.
diffusa Sieber ex Spreng.
subsp. constablei *R.O. Makinson in Fl. Australia*, 17A: 499, 216 (2000)—New South Wales.
dunlopii *R.O. Makinson in Fl. Australia*, 17A: 494, 111 (2000)—Northern Territory.
epicroca *V. Stajsic & W. Molyneux in Fl. Australia*, 17A: 502, 253 (2000)—New South Wales.
florida *(McGill.) R.O. Makinson in Fl. Australia*, 17A: 504, 337 (2000): G. uncinulata subsp. florida.
gariwerdensis *R.O. Makinson in Fl. Australia*, 17A: 501, 239 (2000)—Victoria.

GREVILLEA:-
halmaturina Tate
subsp. laevis *R.O. Makinson in Fl. Australia,* 17A: 501, 243 (2000)—South Australia.
hookeriana Meisn.
subsp. apiciloba *(F. Muell.) R.O. Makinson in Fl. Australia,* 17A: 493, 82 (2000): G. apiciloba.
subsp. digitata *(F. Muell.) R.O. Makinson in Fl. Australia,* 17A: 493, 82 (2000): G. apiciloba var. digitata.
humilis *R.O. Makinson in Fl. Australia,* 17A: 499, 224 (2000)—New South Wales.
subsp. lucens *R.O. Makinson in Fl. Australia,* 17A: 500, 225 (2000)—Queensland.
subsp. maritima *R.O. Makinson in Fl. Australia,* 17A: 500, 227 (2000)—New South Wales.
imberbis *R.O. Makinson in Fl. Australia,* 17A: 501, 242 (2000)—New South Wales.
irrasa *R.O. Makinson in Fl. Australia,* 17A: 503, 256 (2000)—New South Wales.
subsp. didymochiton *R.O. Makinson in Fl. Australia,* 17A: 503, 257 (2000)—New South Wales.
juniperina R. Br.
subsp. allojohnsonii *R.O. Makinson in Fl. Australia,* 17A: 497, 211 (2000)—New South Wales, Queensland.
subsp. amphitricha *R.O. Makinson in Fl. Australia,* 17A: 498, 212 (2000)—New South Wales.
subsp. fortis *R.O. Makinson in Fl. Australia,* 17A: 498, 213 (2000)—New South Wales (Australian Capital Territory, New South Wales).
subsp. sulphurea *(A. Cunn.) R.O. Makinson in Fl. Australia,* 17A: 498, 214 (2000): G. sulphurea.
subsp. trinervis *(R. Br.) R.O. Makinson in Fl. Australia,* 17A: 497, 211 (2000): G. trinervis.
subsp. villosa *R.O. Makinson in Fl. Australia,* 17A: 498, 213 (2000)—New South Wales.
lavandulacea Schltdl.
subsp. rogersii *(Maiden) R.O. Makinson in Fl. Australia,* 17A: 503, 309 (2000): G. rogersii.
maccutcheonii *G.J. Keighery & R.J. Cranfield in Nuytsia,* 11(1): 33 (1996)—Western Australia.
maherae *R.O. Makinson & M.D. Barrett in Fl. Australia,* 17A: 493, 88 (2000)—Western Australia.
manglesioides Meisn.
subsp. ferricola *G.J. Keighery in Fl. Australia,* 17A: 497, 183 (2000)—Western Australia.
subsp. metaxa *R.O. Makinson in Fl. Australia,* 17A: 497, 183 (2000)—Western Australia.
metamorpha *R.O. Makinson in Fl. Australia,* 17A: 506, 428 (2000)—Western Australia.
microstyla *M.D. Barrett & R.O. Makinson in Fl. Australia,* 17A: 495, 142 (2000)—Western Australia.
miqueliana F. Muell.
subsp. moroka *W. Molyneux & V. Stajsic in Fl. Australia,* 17A: 502, 255 (2000)—Victoria.
monslacana *W. Molyneux & V. Stajsic in Fl. Australia,* 17A: 502, 254 (2000)—Victoria.
nematophylla F. Muell.
*subsp.*planicosta *R.O. Makinson in Fl. Australia,* 17A: 505, 371 (2000)—Western Australia.
subsp. supraplana *R.O. Makinson in Fl. Australia,* 17A: 504, 371 (2000)—Western Australia, South Australia.
neurophylla Gand.
subsp. fluviatilis *R.O. Makinson in Fl. Australia,* 17A: 501, 237 (2000)—Victoria.
obtusiflora R. Br.
subsp. fecunda *R.O. Makinson in Telopea,* 7(2): 144 (1997)—New South Wales.
oxyantha *R.O. Makinson in Telopea,* 7(2): 130 (1997)—New South Wales (New South Wales, Australian Capital Territory).
subsp. ecarinata *R.O. Makinson in Telopea,* 7(2): 133 (1997)—New South Wales.
parviflora R. Br.
subsp. supplicans *R.O. Makinson in Fl. Australia,* 17A: 501, 234 (2000)—New South Wales.
parvula *W. Molyneux & V. Stajsic in Fl. Australia,* 17A: 502, 252 (2000), nom. nov.: G. victoriae var. leptoneura *Benth.*
pauciflora R. Br.
subsp. leptophylla *W.R. Barker in J. Adelaide Bot. Gard.,* 17: 211 (1996)—South Australia.

GREVILLEA:-
petrophiloides Meisn.
subsp. remota *(Olde & Marriott) R.O. Makinson in Fl. Australia,* 17A: 504, 358 (2000): G. magnifica subsp. remota.
plurijuga F. Muell.
subsp. superba *(Olde & Marriott) R.O. Makinson in Fl. Australia,* 17A: 496, 170 (2000): G. superba.
pyramidalis A. Cunn. ex R. Br.
subsp. leucadendron *(A. Cunn. ex R. Br.) R.O. Makinson in Fl. Australia,* 17A: 505, 373 (2000): G. leucadendron.
reptans *R.O. Makinson in Fl. Australia,* 17A: 500, 231 (2000)—Queensland.
rhyolitica *R.O. Makinson in Telopea,* 7(2): 134 (1997)—New South Wales.
subsp. semivestita *R.O. Makinson in Telopea,* 7(2): 137 (1997)—New South Wales.
rosmarinifolia A. Cunn.
subsp. glabella *(R. Br.) R.O. Makinson in Fl. Australia,* 17A: 503, 301 (2000): G. glabella.
stenogyne *(Benth.) R.O. Makinson in Fl. Australia,* 17A: 506, 430 (2000): G. vestita var. stenogyne.
synapheae R. Br.
subsp. latiloba *(Meisn.) R.O. Makinson in Fl. Australia,* 17A: 505, 393 (2000): G. synapheae var. latiloba.
subsp. minyulo *R.O. Makinson in Fl. Australia,* 17A: 505, 395 (2000)—Western Australia.
tripartita Meisn.
subsp. macrostylis *(F. Muell.) R.O. Makinson in Fl. Australia,* 17A: 496, 166 (2000): G. macrostylis.
variifolia C.A. Gardner & A.S. George
subsp. bundera *G.J. Keighery in Nuytsia,* 12(2): 294 (1998)—Western Australia
victoriae F. Muell.
subsp. nivalis *V. Stajsic & W. Molyneux in Fl. Australia,* 17A: 502, 249 (2000)—New South Wales.
virgata *R.O. Makinson in Fl. Australia,* 17A: 500, 229 (2000)—New South Wales.
viridiflava *R.O. Makinson in Fl. Australia,* 17A: 500, 228 (2000)—New South Wales, Queensland.
wickhamii Meisn.
subsp. cratista *R.O. Makinson in Fl. Australia,* 17A: 495, 137 (2000)—Western Australia.
subsp. hispidula *R.O. Makinson in Fl. Australia,* 17A: 495, 135 (2000)—Western Australia.
subsp. macrodonta *R.O. Makinson in Fl. Australia,* 17A: 494, 135 (2000)—Western Australia.
subsp. pallida *R.O. Makinson in Fl. Australia,* 17A: 495, 136 (2000)—Western Australia.
wiradjuri *R.O. Makinson in Fl. Australia,* 17A: 501, 240 (2000)—New South Wales.

GREWIA (Tiliac.).
subgen. Burretia *(Hochr.) R. Capuron in Adansonia,* 21(1): 9 (1999): G. sect. Burretia.
subgen. Vincentia *(Benth.) R. Capuron in Adansonia,* 21(1): 9 (1999): gen. Vincentia.
balensis *D. Kirkup & Sebsebe Demissew in Kew Bull.,* 54(1): 231 (1999)—Ethiopia, Somalia, Kenya.
bojeri *D.J. Mabberley in Adansonia,* 21(1): 18 (1999), nom. nov.: Vincentia lancifolia *Bojer.*
cerasifera *(Chiov.) M. Thulin in Kew Bull.,* 54(4): 1007 (1999 publ. 2000): Microcos cerasifera.
erythroxyloides *R. Capuron in Adansonia,* 21(2): 290 (1999)—Madagascar.
flavicans Boivin ex Baill.
var. boinensis *R. Capuron in Adansonia,* 21(2): 294 (1999)—Madagascar.
huluperakensis *I.M. Turner in Malayan Nat. J.,* 50(4): 281 (1997)—Malaya (Peninsular Malaysia).
jinghongensis *Y.Y. Qian in Guihaia,* 17(4): 297 (1997)—China South-central (Yunnan).
lapiazicola *R. Capuron in Adansonia,* 21(1): 16 (1999)—Madagascar.
luteiflora *R. Capuron in Adansonia,* 21(1): 18 (1999)—Madagascar.
meridionalis *R. Capuron in Adansonia,* 21(1): 14 (1999)—Madagascar.
var. antandroy *R. Capuron in Adansonia,* 21(1): 16 (1999)—Madagascar.

GREWIA:-
microcyclea *(Burret) R. Capuron & D.J. Mabberley in Adansonia,* 21(2): 288 (1999): Vincentia microcyclea.
f. albida *R. Capuron in Adansonia,* 21(2): 288 (1999)—Madagascar.
monantha *Capuron ex D.J. Mabberley in Adansonia,* 21(2): 287 (1999)—Madagascar.
ribesioides *R. Capuron & D.J. Mabberley in Adansonia,* 21(2): 287 (1999)—Madagascar.
sahafariensis *R. Capuron & D.J. Mabberley in Adansonia,* 21(2): 285 (1999)—Madagascar.
sambiranensis *R. Capuron in Adansonia,* 21(1): 12 (1999)—Madagascar.
simaoensis *Y.Y. Qian in Guihaia,* 17(4): 295 (1997)—China South-central (Yunnan).
suarezensis *R. Capuron in Adansonia,* 21(1): 20 (1999)—Madagascar.
trichodes *J.O. Voigt, Hort. Suburb. Calcutt.:* 127 (1845), nom. nov.: G. pilosa *Roxb.*
tsiandrensis *R. Capuron in Adansonia,* 21(2): 292 (1999)—Madagascar.

GRIAS (Lecythidac.).
longirachis *S.A. Mori & J.L. Clark in Brittonia,* 52(2): 145 (2000)—Ecuador.

GRIFFINIA (Amaryllidac.).
alba *K.D. Preuss & A.W. Meerow in Novon,* 10(3): 230 (2000)—Brazil Northeast (Pernambuco).
arifolia *P. Ravenna in Onira, Bot. Leafl.,* 4(5): 19 (2000)—Brazil Northeast (Bahia).
colatinensis *P. Ravenna in Onira, Bot. Leafl.,* 4(5): 22 (2000)—Brazil Southeast (Espírito Santo).
ilheusiana *P. Ravenna in Onira, Bot. Leafl.,* 4(5): 21 (2000)—Brazil Northeast (Bahia).
mucurina *P. Ravenna in Onira, Bot. Leafl.,* 4(5): 20 (2000)—Brazil Northeast (Bahia).
paubrasilica *P. Ravenna in Onira, Bot. Leafl.,* 4(5): 20 (2000)—Brazil Northeast (Bahia).

GRINDELIA (Compos.).
anethifolia *(Phil.) A. Bartoli & R.D. Tortosa in Taxon,* 47(2): 345 (1998): Haplopappus anethifolius.
chacoensis *A. Bartoli & R.D. Tortosa in Hickenia,* 3(1): 1 (1999)—Argentina Northeast (Corrientes, Santa Fé).
linearifolia *A. Bartoli, R.D. Tortosa & E. Marchesi in Brittonia,* 48(1): 75 (1996)—Uruguay.
mendocina *Adr. Bartoli & R.D. Tortosa in Brittonia,* 50(4): 497 (1998)—Argentina Northwest (Mendoza).
var. radiata *A. Bartoli & R.D. Tortosa in Kurtziana,* 27(2): 346 (1999)—Argentina Northwest (Mendoza).
orientalis *A. Bartoli, R.D. Tortosa & G.H. Rua in Hickenia,* 2(26): 120 (1995)—Uruguay.
prostrata *A. Bartoli & R.D. Tortosa in Kurtziana,* 27(2): 347 (1999)—Argentina Northwest (Catamarca, La Rioja, San Luis).
prunelloides *(Poepp. ex Less.) A. Bartoli & R.D. Tortosa in Hickenia,* 3(1): 4 (1999): Diplopappus prunelloides.
pulchella Dunal
var. discoidea *(Hook. & Arn.) A. Bartoli & R.D. Tortosa in Kurtziana,* 27(2): 352 (1999): G. discoidea.
rupestris *A. Bartoli, R.D. Tortosa & E. Marchesi in Brittonia,* 48(1): 78 (1996)—Uruguay.
tarapacana Phil.
var. verrucosa *A. Granda Paucar, Adr. Bartoli & R.D. Tortosa in Bol. Soc. Argent. Bot.,* 35(1–2): 157 (2000)—Peru.

GRISEOCEREUS *(P.V. Heath) P.V. Heath in Calyx,* 6(1): 1 (1998): Stenocereus subgen. Griseocereus.

× **GRISULATOCEREUS** *P.V. Heath in Calyx,* 6(1): 1 (1998). *CACTACEAE* (Griseocereus × Insulatocereus).

GRONOPHYLLUM (Palmae).
cariosum *J.L. Dowe & M.D. Ferrero in Palms,* 44(4): 161 (2000)—New Guinea (Papua New Guinea).

GROSSHEIMIA (Compos.).
ahverdovii *(Gabrieljan) E.Tz. Gabrieljan in Fl. Armenii,* 9: 332 (1995): Centaurea ahverdovii.

GROSSHEIMIA:-
caroli-henrici *(Gabrieljan & Dittrich) E.Tz. Gabrieljan in Fl. Armenii,* 9: 332 (1995): Centaurea caroli-henrici.

GROSSULARIA (Grossulariac.).
formosana *(Hayata) S.S. Ying in Mem. Coll. Agric. Nation. Taiwan Univ.,* 25(2): 102 (1985), isonym; *see A. Berger in New York Agric. Exp. Sta. Techn. Bull.,* 109: 111 (1924): Ribes formosanum.

GRUSONIA (Cactac.).
aggeria *(Ralston & Hilsenb.) E.F. Anderson in Cact. Succ. J. (U.S.A.),* 71(6): 325 (1999): Opuntia aggeria.
agglomerata *(A. Berger) E.F. Anderson in Cact. Succ. J. (U.S.A.),* 71(6): 325 (1999): Opuntia agglomerata.
dumetorum *(A. Berger) E.F. Anderson in Cact. Succ. J. (U.S.A.),* 71(6): 325 (1999): Opuntia dumetorum.
emoryi *(Engelm.) D.J. Pinkava in J. Ariz.-Nev. Acad. Sci.,* 32(1): 50 (1999): Opuntia emoryi.
invicta *(Brandegee) E.F. Anderson in Cact. Succ. J. (U.S.A.),* 71(6): 325 (1999): Opuntia invicta.
kunzei *(Rose) D.J. Pinkava in J. Ariz.-Nev. Acad. Sci.,* 32(1): 50 (1999): Opuntia kunzei.
marenae *(S.H. Parsons) E.F. Anderson in Cact. Succ. J. (U.S.A.),* 71(6): 325 (1999): Opuntia marenae.
moelleri *(A. Berger) E.F. Anderson in Cact. Succ. J. (U.S.A.),* 71(6): 325 (1999): Opuntia moelleri.
parishii *(Orcutt) D.J. Pinkava in J. Ariz.-Nev. Acad. Sci.,* 32(1): 50 (1999): Opuntia parishii.
reflexispina *(Wiggins & Rollins) E.F. Anderson in Cact. Succ. J. (U.S.A.),* 71(6): 325 (1999): Opuntia reflexispina.

GUADUA (Gramin.).
angustifolia Kunth
var. nigra *X. Londoño in Novon,* 8(4): 423 (1998)—Colombia.
barbata *(Trin.) D. Ohrnberger, Bamboos of the World:* 353 (1999): Bambusa barbata.

GUAJAVA (Myrtac.).
ooidea *(O. Berg) Kuntze, Revis. Gen. Pl.,* 1: 239 (1891), in I.K. Supp. 1 as 'ovoidea' in error: Psidium ooideum.†

GUANCHEZIA *G.A. Romero & G. Carnevali, Orchids of Venezuela, ed. 2:* 1135 (2000). *ORCHIDACEAE*
maguirei *(C. Schweinf.) G.A. Romero & G. Carnevali, Orchids of Venezuela, ed. 2:* 1136 (2000): Bifrenaria maguirei.

GUAREA (Meliac.).
lozanii *M.E. Morales-Puentes in Rev. Acad. Colomb. Cienc. Exact. Fis. Nat.,* 84: 336 (1998)—Colombia.
penningtoniana *M.E. Morales-Puentes in Rev. Acad. Colomb. Cienc. Exact. Fis. Nat.,* 84: 339 (1998)—Colombia.
quadrangularis *M.E. Morales-Puentes in Rev. Acad. Colomb. Cienc. Exact. Fis. Nat.,* 84: 336 (1998)—Colombia.

GUATTERIA (Annonac.).
pudica *N. Zamora & P.J.M. Maas in Bot. Jahrb.,* 122(2): 244 (2000)—Costa Rica.
talamancana *N. Zamora & P.J.M. Maas in Bot. Jahrb.,* 122(2): 241 (2000)—Costa Rica.

GUELDENSTAEDTIA (Legumin.).
verna (Georgi) Boriss.
f. alba *(F.Z. Li) H.P. Tsui in Fl. Reipubl. Popul. Sin.,* 42(2): 152 (1998), homonym; *see P.C. Li in Vasc. Pl. Hengduan Mount.,* 1: 940 (1993): G. multiflora f. alba.
subsp. multiflora *(Bunge) H.P. Tsui in Fl. Reipubl. Popul. Sin.,* 42(2): 150 (1998): G. multiflora.
f. multiflora *(Bunge) H.P. Tsui in Fl. Reipubl. Popul. Sin.,* 42(2): 150 (1998): G. multiflora.

GUETTARDA (Rubiac.).
grazielae *M.R.V. Barbosa in Bradea,* 8(11): 61 (1997)—Brazil Northeast (Paraíba).
wayaensis *R.O. Gardner in New Zealand J. Bot.,* 35(4): 488 (1997)—Fiji.

GUEVARIA (Compos.).
micranthera *H. Robinson in Phytologia,* 84(5): 348 (1998 publ. 1999)—Peru.

GUIHAIOTHAMNUS *H.S. Lo in Bull. Bot. Res. North-East. Forest. Univ.,* 18(3): 279 (1998). *RUBIACEAE*
acaulis *H.S. Lo in Bull. Bot. Res. North-East. Forest. Univ.,* 18(3): 280 (1998)—China Southeast (Guangxi).

GUILLEMINEA (Amaranthac.).
chacoensis *T.M. Pedersen in Bonplandia,* 10(1–4): 94 (2000)—Argentina Northeast (Formosa), Paraguay.
fragilis *T.M. Pedersen in Bonplandia,* 10(1–4): 97 (2000)—Paraguay, Brazil West-Central (Mato Grosso do Sul).
hirsuta *T.M. Pedersen in Bonplandia,* 10(1–4): 99 (2000)—Paraguay, Argentina Northwest (Salta), Bolivia.

GUINETIA *L. Rico & M. Sousa S. in Kew Bull.,* 54(4): 977 (1999 publ. 2000). *LEGUMINOSAE*
tehuantepecensis *L. Rico & M. Sousa S.in Kew Bull.,* 54(4): 977 (1999 publ. 2000)—Mexico Southwest (Oaxaca).

GUIOA (Sapindac.).
sarcopterifructa *P.C. van Welzen in Austrobaileya,* 5(1): 104 (1997)—Queensland.

GUNDLACHIA (Compos.).
corymbosa (Urb.) Britton ex Bold.
var. apiculata *(Britton & S.F. Blake) M.A. Lane in Brittonia,* 48(4): 538 (1996 publ. 1997): G. apiculata.
var. compacta *(Urb. & Ekman) M.A. Lane in Brittonia,* 48(4): 538 (1996 publ. 1997): G. compacta.
var. cubana *(Britton & S.F. Blake) M.A. Lane in Brittonia,* 48(4): 539 (1996 publ. 1997): G. cubana.
var. foliosa *(Britton & S.F. Blake) M.A. Lane in Brittonia,* 48(4): 539 (1996 publ. 1997): G. foliosa.
var. ocoana *(Urb. & Ekman) M.A. Lane in Brittonia,* 48(4): 540 (1996 publ. 1997): G. ocoana.

GURANIA (Cucurbitac.).
lobata *(L.) J.F. Pruski in Brittonia,* 51(3): 329 (1999): Anguria lobata.

GURINGALIA *B.G. Briggs & L.A.S. Johnson in Telopea,* 7(4): 353 (1998). *RESTIONACEAE*
dimorpha *(R. Br.) B.G. Briggs & L.A.S. Johnson in Telopea,* 7(4): 353 (1998): Restio dimorphus.

GUTIERREZIA (Compos.).
amoena *(Shinners) G.M. Diggs, B.L. Lipscomb & R.J. O'Kennon, Shinners & Mahler's Illustr. Fl. N. Central Texas (Sida, Bot. Misc. 16):* 364 (1999): Xanthocephalum amoenum.
dunalii *(Spreng.) G.L. Nesom in Novon,* 10(1): 69 (2000): Grindelia dunalii.
megalocephala *(Fernald) G.L. Nesom in Novon,* 10(1): 70 (2000): Xanthocephalum megalocephalum.

GUZMANIA (Bromeliac.).
albescens *H.E. Luther & R.O. Determann in J. Bromeliad Soc.,* 43(4): 152 (1993)—Ecuador.
alcantareoides *H.E. Luther in J. Bromeliad Soc.,* 43(5): 195 (1993)—Ecuador.
alliodora *E. Gross in Trop. Subtrop. Pflanzenwelt,* 95: 18 (1997)—Peru.
bergii *H.E. Luther in Selbyana,* 18(1): 95 (1997)—Ecuador.
betancurii *H.E. Luther in Selbyana,* 20(1): 10 (1999)—Colombia.
breviscapa *H.E. Luther in Selbyana,* 19(2): 220 (1998 publ. 1999)—Colombia.
corniculata *H.E. Luther in J. Bromeliad Soc.,* 46(2): 81 (1996)—Ecuador.
darienensis *H.E. Luther in Selbyana,* 19(1): 90 (1998)—Panama.
elvallensis *H.E. Luther in Bromélia,* 3(1): 8 (1996)—Panama.

GUZMANIA:-
formosa *H.E. Luther in Selbyana,* 21(1–2): 125 (2000)—Colombia.
herrerae *H.E. Luther & W.J. Kress in Brittonia,* 48(1): 91 (1996)—Costa Rica.
izkoi *J.M. Manzanares & W. Till in J. Bromeliad Soc.,* 50(1): 17 (2000)—Ecuador.
laeta *H.E. Luther in Selbyana,* 21(1–2): 126 (2000)—Colombia.
manzanaresiorum *H.E. Luther in Selbyana,* 21(1–2): 127 (2000)—Ecuador.
marantoidea *(Rusby) H.E. Luther in Selbyana,* 19(1): 86 (1998): Tillandsia marantoidea.
pseudospectabilis *H.E. Luther in Selbyana,* 16(2): 235 (1995)—Ecuador.
remediosensis *E. Gross in Trop. Subtrop. Pflanzenwelt,* 95: 20 (1997)—Bolivia.
scandens *H.E. Luther & W.J. Kress in Brittonia,* 48(1): 93 (1996)—Costa Rica, Panama.
sieffiana *H.E. Luther in J. Bromeliad Soc.,* 48(5): 201 (1998)—Ecuador.
squarrosa
f. lutea *F. Oliva-Esteve in J. Bromeliad Soc.,* 50(6): 249 (2000)—Venezuela.
viridiflora *E. Gross in Trop. Subtrop. Pflanzenwelt,* 95: 21 (1997)—Peru.

GYMNADENIA (Orchidac.).
subgen. nigritella *(Rich.) O. Gerbaud & P. Quentin in Nat. Belg.,* 80(3): 374 (1999): gen. Nigritella.
archiducis-joannis *(Teppner & E. Klein) H. Teppner & E. Klein in Phyton (Austria),* 38(1): 222 (1998): Nigritella archiducis-joannis.
austriaca *(Teppner & E. Klein) P. Delforge in Nat. Belg.,* 79(4): 254 (1998): Nigritella nigra subsp. austriaca.
var. gallica *(E. Breiner & R. Breiner) P. Delforge in Nat. Belg.,* 79(4): 255 (1998): Nigritella nigra subsp. gallica.
var. iberica *(Teppner & E. Klein) G. Foelsche & W. Foelsche in J. Eur. Orchideen,* 31(4): 803 (1999), without exact basionym page: Nigritella nigra subsp. iberica.
borealis *(Druce) R.M. Bateman, A.M. Pridgeon & M.W. Chase in Lindleyana,* 12(2): 130 (1997): Habenaria gymnadenia var. borealis.
× breinerorum *O. Gerbaud & W. Foelsche in Nat. Belg.,* 80(3): 381 (1999)—France.(G. cenisia × corneliana).
buschmanniae *(Teppner & Ster) H. Teppner & E. Klein in Phyton (Austria),* 38(1): 222 (1998): Nigritella buschmanniae.
carpatica *(Zapał.) H. Teppner & E. Klein in Phyton (Austria),* 38(1): 221 (1998): Nigritella angustifolia var. carpatica.
cenisia *(G. Foelsche, W. Foelsche, M. Gerbaud & O. Gerbaud) G. Foelsche, W. Foelsche, M. Gerbaud & O. Gerbaud in Nat. Belg.,* 80(3): 374 (15 Oct. 1999); *et in Orchidophile,* 139: 236 (Nov. 1999): Nigritella cenisia.
× chanousiana *G. Foelsche & W. Foelsche in J. Eur. Orchideen,* 31(4): 809 (1999)—France.(G. cenisia × conopsea).
conopsea (L.) R.Br.
var. densiflora *(Wahlenb.) L.V. Averyanov in Turczaninowia,* 3(1): 47 (2000): Orchis densiflora.
corneliana *(Beauverd) H. Teppner & E. Klein in Phyton (Austria),* 38(1): 221 (1998): Nigritella nigra subsp. corneliana.corneliana (Beauverd) Teppner & E. Klein
var. vesubiana *(G. Keller) G. Foelsche & W. Foelsche in J. Eur. Orchideen,* 31(4): 803 (1999), basionym ref. not clearly indicated: Nigritella rubra lus. vesubiana.
cucullata (L.) Rich.
var. variegata *Y.N. Lee, Fl. Korea:* 1164 (1996), without type—Locality not stated.
× delphineae *(M. Gerbaud & O. Gerbaud) M. Gerbaud & O. Gerbaud in Nat. Belg.,* 80(3): 378 (1999): Nigritella × delphineae.(G. corneliana × rhellicani).
densiflora (Wahlenb.) A. Dietr.
var. candida *G. Foelsche & W. Foelsche in J. Eur. Orchideen,* 31(4): 804 (1999)—France.
dolomitensis *H. Teppner & E. Klein in Phyton (Austria),* 38(1): 223 (1998)—Italy.

GYMNADENIA:-

× eggeriana *O. Gerbaud in Nat. Belg.*, 80(3): 378 (1999), as 'eggerana'—France.(G. austriaca var. gallica × rhellicani).
gabasiana *(Teppner & E. Klein) H. Teppner & E. Klein in Phyton (Austria)*, 38(1): 221 (1998): Nigritella gabasiana.
× godferyana *(G. Keller) W. Foelsche in J. Eur. Orchideen*, 31(4): 812 (1999): × Gymnigritella godferyana.(G. conopsea × rubra).
× intermedia Peterm.
nothosubsp. proxima *C.E. Hermosilla in Estud. Mus. Cienc. Nat. Alava*, 13 143 (1998)—Spain.(G. conopsea × odoratissima subsp. longicalcarata).
lithopolitanica *(Ravnik) H. Teppner & E. Klein in Phyton (Austria)*, 38(1): 221 (1998): Nigritella lithopolitanica.
nigra (L.) Rchb. f.
subsp. austriaca *(Teppner & E. Klein) H. Teppner & E. Klein in Phyton (Austria)*, 38(1): 221 (1998): Nigritella nigra subsp. austriaca.
subsp. iberica *(Teppner & E. Klein) H. Teppner & E. Klein in Phyton (Austria)*, 38(1): 222 (1998): Nigritella nigra subsp. iberica.
odoratissima (L.) Rich.
subsp. longicalcarata *C.E. Hermosilla & J. Sabando in Estud. Mus. Cienc. Nat. Alava*, 10–11: 123, 128 (1995–96 publ. 1996)—Spain.
var. pyrenaica *(Philippe) P. Delforge in Nat. Belg.*, 81(4): 397 (2000): Orchis pyrenaica.
× pyrenaeensis *W. Foelsche in J. Eur. Orchideen*, 31(4): 812 (1999), nom. nov.: × Gymnigritella pyrenaica *C.E. Hermos. & Sabando.*
rhellicani *(Teppner & E. Klein) H. Teppner & E. Klein in Phyton (Austria)*, 38(1): 221 (1998): Nigritella rhellicani.
× robatschiana *O. Gerbaud & W. Foelsche in Nat. Belg.*, 80(3): 380 (1999)—France.(G. cenisia × rhellicani).
stiriaca *(Rech.) H. Teppner & E. Klein in Phyton (Austria)*, 38(1): 222 (1998): G. rubra var. stiriaca.
straminea *(Fernald) P. Delforge in Nat. Belg.*, 81(4): 397 (2000): Habenaria straminea.
× truongiae *(Demares) W. Foelsche in J. Eur. Orchideen*, 31(4): 812 (1999): × Gymnigritella truongiae.(G. conopsea × corneliana).
× turnowskyi *(W. Foelsche) W. Foelsche in J. Eur. Orchideen*, 31(4): 812 (1999): × Gymnigritella turnowskyi.(G. conopsea × lithopolitanica).
widderi *(Teppner & E. Klein) H. Teppner & E. Klein in Phyton (Austria)*, 38(1): 222 (1998): Nigritella widderi.

GYMNANTHEMUM (Compos.).
anceps *(C.B. Clarke ex Hook. f.) H. Robinson in Proc. Biol. Soc. Wash.*, 112(1): 239 (1999): Vernonia anceps.
andersonii *(C.B. Clarke) H. Robinson in Proc. Biol. Soc. Wash.*, 112(1): 239 (1999): Vernonia andersonii.
andrangovalense *(Humbert) H. Robinson in Proc. Biol. Soc. Wash.*, 112(1): 240 (1999): Vernonia andrangovalensis.
antanalus *(Humbert) H. Robinson in Proc. Biol. Soc. Wash.*, 112(1): 240 (1999): Vernonia antanala.
appendiculatum *(Less.) H. Robinson in Proc. Biol. Soc. Wash.*, 112(1): 240 (1999): Vernonia appendiculata.
arboreum *(Buch.-Ham.) H. Robinson in Proc. Biol. Soc. Wash.*, 112(1): 240 (1999): Vernonia arborea.
baronii *(Baker) H. Robinson in Proc. Biol. Soc. Wash.*, 112(1): 240 (1999): Vernonia baronii.
bellinghamii *(S. Moore) H. Robinson in Proc. Biol. Soc. Wash.*, 112(1): 240 (1999): Vernonia bellinghamii.
bockianum *(Diels) H. Robinson in Proc. Biol. Soc. Wash.*, 112(1): 240 (1999): Vernonia bockiana.
chapelieri *(Drake) H. Robinson in Proc. Biol. Soc. Wash.*, 112(1): 240 (1999): Vernonia chapelieri.
corymbosum *(Thunb.) H. Robinson in Proc. Biol. Soc. Wash.*, 112(1): 241 (1999): Staehelina corymbosa.
coursii *(Humbert) H. Robinson in Proc. Biol. Soc. Wash.*, 112(1): 241 (1999): Vernonia coursii.
crataegifolium *(Hutch.) H. Robinson in Proc. Biol. Soc. Wash.*, 112(1): 241 (1999): Vernonia crataegifolia.

GYMNANTHEMUM:-
cumingianum *(Benth.) H. Robinson in Proc. Biol. Soc. Wash.*, 112(1): 241 (1999): Vernonia cumingiana.
cylindriceps *(C.B. Clarke) H. Robinson in Proc. Biol. Soc. Wash.*, 112(1): 241 (1999): Vernonia cylindriceps.
dissolutum *(Baker) H. Robinson in Proc. Biol. Soc. Wash.*, 112(1): 241 (1999): Vernonia dissoluta.
esculentum *(Hemsl.) H. Robinson in Proc. Biol. Soc. Wash.*, 112(1): 241 (1999): Vernonia esculenta.
exsertiflorum *(Baker) H. Robinson in Proc. Biol. Soc. Wash.*, 112(1): 241 (1999): Vernonia exsertiflora.
exsertum *(Baker) H. Robinson in Proc. Biol. Soc. Wash.*, 112(1): 241 (1999): Vernonia exserta.
glaberrimum *(Welw. ex O. Hoffm.) H. Robinson in Proc. Biol. Soc. Wash.*, 112(1): 241 (1999): Vernonia glaberrima.
hildebrandtii *(Vatke) H. Robinson in Proc. Biol. Soc. Wash.*, 112(1): 241 (1999): Vernonia hildebrandtii.
humblotii *(Drake) H. Robinson in Proc. Biol. Soc. Wash.*, 112(1): 242 (1999): Vernonia humblotii.
louvelii *(Humbert) H. Robinson in Proc. Biol. Soc. Wash.*, 112(1): 242 (1999): Vernonia louvelii.
mespilifolium *(Less.) H. Robinson in Proc. Biol. Soc. Wash.*, 112(1): 242 (1999): Vernonia mespilifolia.
myrianthum *(Hook. f.) H. Robinson in Proc. Biol. Soc. Wash.*, 112(1): 242 (1999): Vernonia myriantha.
pectiniforme *(DC.) H. Robinson in Proc. Biol. Soc. Wash.*, 112(1): 242 (1999): Vernonia pectiniformis.
pectorale *(Baker) H. Robinson in Proc. Biol. Soc. Wash.*, 112(1): 242 (1999): Vernonia pectoralis.
pleistanthum *(Humbert) H. Robinson in Proc. Biol. Soc. Wash.*, 112(1): 242 (1999): Vernonia secundifolia subsp. pleistantha.
rueppellii *(Sch. Bip. ex Walp.) H. Robinson in Proc. Biol. Soc. Wash.*, 112(1): 242 (1999): Vernonia rueppellii.
secundifolium *(Bojer ex DC.) H. Robinson in Proc. Biol. Soc. Wash.*, 112(1): 243 (1999): Vernonia secundifolia.
solanifolium *(Benth.) H. Robinson in Proc. Biol. Soc. Wash.*, 112(1): 243 (1999): Vernonia solanifolia.
subcrassulescens *(Humbert) H. Robinson in Proc. Biol. Soc. Wash.*, 112(1): 243 (1999): Vernonia subcrassulescens.
theophrastifolium *(Schweinf. ex Oliv. & Hiern) H. Robinson in Proc. Biol. Soc. Wash.*, 112(1): 243 (1999): Vernonia theophrastifolia.
thomsonianum *(Oliv. & Hiern) H. Robinson in Proc. Biol. Soc. Wash.*, 112(1): 243 (1999): Vernonia thomsoniana.
urticifolium *(A. Rich.) H. Robinson in Proc. Biol. Soc. Wash.*, 112(1): 243 (1999): Vernonia urticifolia.
vidalii *(G. Merr.) H. Robinson in Proc. Biol. Soc. Wash.*, 112(1): 243 (1999): Vernonia vidalii.
volkameriifolium *(DC.) H. Robinson in Proc. Biol. Soc. Wash.*, 112(1): 243 (1999): Vernonia volkameriifolia.
wightianum *(Arn.) H. Robinson in Proc. Biol. Soc. Wash.*, 112(1): 243 (1999): Vernonia wightiana.
zanzibarense *(Less.) H. Robinson in Proc. Biol. Soc. Wash.*, 112(1): 243 (1999): Vernonia zanzibarensis.
zeylanicum *(L.) H. Robinson in Proc. Biol. Soc. Wash.*, 112(1): 243 (1999): Eupatorium zeylanicum.

GYMNANTHES (Euphorbiac.).
borneensis *(Pax & K. Hoffm.) H.-J. Esser in Blumea*, 44(1): 170 (1999): Sebastiania borneensis.
glabrata *(Mart.) R. Govaerts in R. Govaerts, D.G. Frodin & A. Radcliffe-Smith, World Checklist & Bibliogr.* Euphorbiaceae, 3: 979 (2000): Cnemidostachys glabrata.
remota *(Steenis) H.-J. Esser in Blumea*, 44(1): 172 (1999): Sebastiania remota.

× **GYMNIGRITELLA** (Orchidac.).
pyrenaica *C.E. Hermosilla & J. Sabando in Estud. Mus. Cienc. Nat. Alava*, 10–11: 123 (1995–96 publ. 1996)—Spain, France. (Gymnadenia conopsea × Nigritella gabasiana).

GYMNOCACTUS (Cactac.).
horripilus (Lem.) Backeb.
subsp. kvetae *J. Chvastek & J.J. Halda in Acta Mus. Richnov. Sect. Nat.*, 7(1): 35 (2000)—Mexico Northeast (Hidalgo).
subterraneus
var. zaragozae *C. Glass & R.A. Foster in Cact. Succ. J. (U.S.A.)*, 50(6): 283 (1978), as 'zaragosae'; *see Glass, Guía Identific. Cact. Amenazadas México*, 1: Tu/za (1998 publ. 1997)—Mexico Northeast (Nuevo León).†
viereckii (Werderm.) Backeb.
subsp. major *(Glass & R.A. Foster) A.B. Doweld in Sukkulenty (Moscow)*, 3(1–2): 37 (2000): Gymnocactus viereckii var. major.

GYMNOCALYCIUM *Pfeiff. ex L. Mittler, Taschenb. Cactusliebhaber*, 2: 124 (1844), as 'Gymnocalicium'. *CACTACEAE*†
andreae (Boed.) Backeb.
var. fechseri *H. Till in Gymnocalycium*, 13(4): 378 (2000)—Argentina Northeast (Córdoba).
var. longispinum *W. Rausch in Gymnocalycium*, 13(4): 378 (2000)—Argentina Northeast (Córdoba).
subsp. matznetteri *W. Rausch in Gymnocalycium*, 13(4): 379 (2000)—Argentina Northeast (Córdoba).
angelae *M. Meregalli in Kakt. And. Sukk.*, 49(12): 287 (1998)—Argentina Northeast (Corrientes).
berchtii *G.J.A. Neuhuber in Gymnocalycium*, 10(3): 219 (1997)—Argentina Northwest (San Luis).
cardenasianum F. Ritter
f. armatum *(F. Ritter) H. Till & H. Amerhauser in Gymnocalycium*, 12(4): 308 (1999): G. armatum.
carminanthum
var. montanum *R. Slaba in Kaktusy (Brno)*, 35(1): 8 (1999)—Argentina Northwest (Catamarca).
catamarcense *H. Till & W. Till in gymnocalycium*, 8(1): 142 (1995)—Argentina Northwest (Catamarca).
subsp. acinacispinum *H. Till & W. Till in Gymnocalycium*, 8(1): 146 (1995)—Argentina Northwest (Catamarca).
f. belense *H. Till & W. Till in Gymnocalycium*, 8(1): 144 (1995)—Argentina Northwest (Catamarca).
f. ensispinum *H. Till & W. Till in Gymnocalycium*, 8(1): 143 (1995)—Argentina Northwest (Catamarca).
f. montanum *H. Till & W. Till in Gymnocalycium*, 8(1): 143 (1995)—Argentina Northwest (Catamarca).
subsp. schmidianum *H. Till & W. Till in Gymnocalycium*, 8(1): 145 (1995)—Argentina Northwest (Catamarca).
chacoense *H. Amerhauser in Gymnocalycium*, 12(4): 302 (1999)—Bolivia.
denudatum *(Link & Otto) Pfeiff. ex L. Mittler, Taschenb. Cactusliebhaber*, 2: 124 (1844): Echinocactus denudatus.†
gibbosum *(Haw.) Pfeiff. ex L. Mittler, Taschenb. Cactusliebhaber*, 2: 124 (1844): Cactus gibbosus.†
var. brachypetalum *(Speg.) W. Papsch in Gymnocalycium*, 9(4): 201 (1996): G. brachypetalum.
f. cerebriforme *(Speg.) W. Papsch in Gymnocalycium*, 9(4): 201 (1996), as 'cerebriformis': Echinocactus gibbosus var. cerebriformis.
var. chubutense *(Speg.) W. Papsch in Gymnocalycium*, 9(4): 201 (1996): Echinocactus gibbosus var. chubutensis.
subsp. ferox *(Labour. ex Rümpler) W. Papsch in Gymnocalycium*, 9(4): 201 (1996): Echinocactus gibbosus var. ferox.
× heidiae *G.J.A. Neuhuber in Gymnocalycium*, 12(1): 279 (1999)—Argentina Northwest (Catamarca). (G. rosae × ?).
horridispinum
var. achirasense *(H. Till & Schatzl) Lodé in Cact. Aventures*, 27: 29 (1995); *cf. Repert. Pl. Succ. (I.O.S.)*, 46: 11 (1995 publ. 1996): G. achirasense.
kroenleinii *R. Kiesling, W. Rausch & O. Ferrari in Kakt. And. Sukk.*, 51(12): 316 (2000)—Argentina Northwest (La Rioja).
lukasikii *J.J. Halda & P. Kupčák in Acta Mus. Richnov. Sect. Nat.*, 7(2): 72 (2000), as 'lukasiki'—Argentina South.

GYMNOCALYCIUM:-
mackieanum *(Hook.) D. Metzing, M. Meregalli & R. Kiesling in Allionia,* 33: 209 (1995): Echinocactus mackieanus.
monvillei (Lem.) Pfeiff. ex Britton & Rose
var. coloratum *G.J.A. Neuhuber in Gymnocalycium,* 12(2): 285 (1999)—Argentina Northeast (Córdoba).
var. confusa *G.J.A. Neuhuber in Gymnocalycium,* 12(2): 286 (1999)—Argentina Northwest (San Luis).
subsp. gertrudae *(Neuhuber) G.J.A. Neuhuber in Gymnocalycium,* 12(2): 286 (1999): G. monvillei var. gertrudae.
nigriareolatum Backeb.
f. carmineum *H. Till in Gymnocalycium,* 11(3): 255 (1998)—Argentina Northwest (Catamarca).
var. densispinum *Backeb. ex H. Till in Gymnocalycium,* 11(3): 255 (1998)—Argentina.
var. simoi *H. Till in Gymnocalycium,* 11(3): 258 (1998)—Argentina Northwest (Catamarca).
× pazoutianum *J.J. Halda in Acta Mus. Richnov. Sect. Nat.,* 5(4): 164 (1998)—Cultivation.(G. baldianum × denudatum).
pflanzii (Vaupel) Werderm.
subsp. dorisiae *H. Amerhauser in Gymnocalycium,* 11(4): 267 (1998)—Bolivia.
var. paraguayense *H. Till & H. Amerhauser in Gymnocalycium,* 13(1): 347 (2000)—Paraguay.
poeschlii *G.J.A. Neuhuber in Gymnocalycium,* 12(3): 297 (1999)—Argentina Northwest (San Luis).
prochazkianum *Sorma in Gymnofil,* 28(1–2): 2–6 (1999); *cf. Repert. Pl. Succ. (I.O.S.),* 50: 13 (1999 publ. 2000)—Locality not stated.
reductum *(Link) Pfeiff. ex L. Mittler, Taschenb. Cactusliebhaber,* 2: 124 (1844): Cactus reductus†
var. leeanum *(Hook.) W. Papsch in Gymnocalycium,* 13(3): 371 (2000): Echinocactus leeanus.
subsp. leeanum *(Hook.) W. Papsch in Gymnocalycium,* 13(3): 371 (2000): Echinocactus leeanus.
var. leucodictyon *(Rümpler) W. Papsch in Gymnocalycium,* 10(4): 230 (1997): Echinocactus gibbosus var. leucodictyus.
var. schatzlianum *(Strigl & W. Till) W. Papsch in Gymnocalycium,* 13(3): 371 (2000): Gymnocalycium schatzlianum.
saglione (J.F.Cels) Britton & Rose
subsp. tilcarense *(Backeb.) H. Till & W. Till in Gymnocalycium,* 10(2): 215 (1997): Brachycalycium tilcarense.
spegazzinii Britton & Rose
subsp. bayrianum *(H. Till) J.J. Halda in Cactaceae etc.,* 9(3): 84 (1999); *cf. Repert. Pl. Succ. (I.O.S.),* 50: 13 (1999 publ. 2000): G. bayrianum.
subsp. cardenasianum *(F. Ritter) R. Kiesling & D. Metzing in Darwiniana,* 34(1–4): 404 (1996): G. cardenasianum.
stellatum Speg.
var. flavispinum *Bozsing ex H. Till & W. Till in Gymnocalycium,* 9(1): 172 (1996)—Argentina Northeast (Córdoba).
var. kleinianum *Rausch ex H. Till & W. Till in Gymnocalycium,* 9(1): 171 (1996)—Argentina Northeast (Córdoba).
var. obductum *(Piltz) H. Till & W. Till in Gymnocalycium,* 9(1): 174 (1996): G. obductum.
subsp. occultum *Frič ex H. Till & W. Till in Gymnocalycium,* 9(2): 176 (1996)—Argentina Northwest (Catamarca).
var. zantnerianum *(H. Schick) H. Till & W. Till in Gymnocalycium,* 9(1): 169 (1996): G. quehlianum var. zantnerianum.
terweemeanum *(Teucq ex Duursma) W. Borgmann & J. Piltz in Gymnos,* 13(25–26): 47 (1997); *cf. Repert. Pl. Succ. (I.O.S.),* 48: 9 (1997 publ. 1998): Echinocactus platensis var. terweemeanus.
zegarrae Cárdenas
subsp. millaresii *(Cárdenas) H. Till & W. Till in Gymnocalycium,* 13(2): 355 (2000): G. millaresii.
var. riograndense *(Cárdenas) H. Till & H. Amerhauser in Gymnocalycium,* 13(2): 353 (2000): G. riograndense.

GYMNOCARPOS (Illecebrac.).
algeriensis *M. Gandoger, Dec. Pl. Nov.,* 1: 43 (1875)—Algeria.

GYMNOCARPOS:-
argenteus *L. Petrusson & M. Thulin in Edinburgh J. Bot.,* 53(1): 13 (1996)—South Yemen.
bracteatus *(Balf. f.) L. Petrusson & M. Thulin in Edinburgh J. Bot.,* 53(1): 17 (1996): Lochia bracteata.
debeauxii *M. Gandoger, Dec. Pl. Nov.,* 1: 43 (1875)—Algeria.
dhofarensis *L. Petrusson & M. Thulin in Edinburgh J. Bot.,* 53(1): 15 (1996)—Oman, South Yemen.
kuriensis *(Radcl.-Sm.) L. Petrusson & M. Thulin in Edinburgh J. Bot.,* 53(1): 20 (1996): Lochia kuriensis.
mahranus *L. Petrusson & M. Thulin in Edinburgh J. Bot.,* 53(1): 22 (1996)—South Yemen.
parvibractus *(M.G. Gilbert) L. Petrusson & M. Thulin in Edinburgh J. Bot.,* 53(1): 24 (1996): Lochia parvibracta.
rotundifolius *L. Petrusson & M. Thulin in Edinburgh J. Bot.,* 53(1): 19 (1996)—Oman.

GYMNOMYOSOTIS *(A.DC.) O.D. Nikiforova in Turczaninowia,* 3(1): 6 (2000): Myosotis sect. Gymnomyosotis. (Boraginac.).
spatulata *(G. Forst.) O.D. Nikiforova in Turczaninowia,* 3(1): 12 (2000): Myosotis spatulata.

GYMNOPOGON (Gramin.).
spicatus *(Spreng.) Kuntze, Revis. Gen. Pl.,* 3(2): 354 (1898): Polypogon spicatus.

GYMNOSPORIA (Celastrac.).
arenicola *M. Jordaan in S. Afr. J. Bot.,* 65(5–6): 315 (1999)—Natal, Mozambique.
devenishii *M. Jordaan in S. Afr. J. Bot.,* 66(1): 10 (2000)—Natal, Cape Province.
haberiana *B.E. Hammel in Novon,* 7(2): 147 (1997)—Costa Rica, Mexico Gulf (Veracruz).
macrocarpa *M. Jordaan in S. Afr. J. Bot.,* 66(1): 12 (2000)—Natal.
markwardii *M. Jordaan in S. Afr. J. Bot.,* 65(5–6): 317 (1999)—Natal, Mozambique.

GYNATRIX (Malvac.).
macrophylla *N.G. Walsh in Muelleria,* 9: 191 (1996)—Victoria.

GYNOSTEMMA (Cucurbitac.).
caulopterum *S.Z. He in Acta Phytotax. Sin.,* 34(2): 207 (1996)—China South-central (Guizhou).

GYNOXYS (Compos.).
frontinoensis *S. Díaz-Piedrahita & A. Correa in Rev. Acad. Colomb. Cienc. Exact. Fis. Nat.,* 88: 333 (1999)—Colombia.

GYNURA (Compos.).
taiwanensis *S.S. Ying in Mem. Coll. Agric. Nation. Taiwan Univ.,* 30(2): 57 (1990), as 'taiwanense'—Taiwan.

GYPSOPHILA (Caryophyllac.).
fastigiata L.
subsp. belorossica *(Barkoudah) N.N. Tzvelev in Novosti Sist. Vyssh. Rast.,* 32: 183 (2000): G. belorossica.
perfoliata L.
var. glabra *(Fenzl) O.M. Tzarenko in Ukr. Bot. Zhurn.,* 55(6): 636 (1998): G. trichotoma var. glabra.
venusta Fenzl
subsp. staminea *H. Özçelik & F. Özgökçe in M.A. Öztürk, Ö. Seçmen & G. Görk (eds.), Pl. Life S.W. &* Central Asia, 1: 205 (1996), contrary to Art. 37.5 ICBN (1994)—Turkey.

H

HAAGEOCEREUS (Cactac.).
acranthus
subsp. olowinskianus *(Backeb.) C. Ostolaza Nano in Cactaceae Consensus Initiatives,* 6: 8 (1998): H. olowinskianus.

HAAGEOCEREUS:-
f. rubriflorior *(Rauh & Backeb.) C. Ostolaza Nano in Brit. Cact. Succ. J.,* 14(4): 167 (1996): H. olwinskianus f. rubriflorior.
pacalaensis
subsp. repens *(Rauh & Backeb.) C. Ostolaza in Brit. Cact. Succ. J.,* 18(3): 130 (2000): H. repens.
pseudomelanostele (Werderm. & Backeb.) Backeb.
var. aureispinus *(Rauh & Backeb.) C. Ostolaza Nano in Brit. Cact. Succ. J.,* 14(4): 170 (1996): H. aureispinus.
subsp. aureispinus *(Rauh & Backeb.) C. Ostolaza Nano in Cactaceae Consensus Initiatives,* 6: 9 (1998): H. aureispinus.
subsp. carminiflorus *(Rauh & Backeb.) C. Ostolaza Nano in Cactaceae Consensus Initiatives,* 6: 9 (1998): H. pseudomelanostele var. carminiflorus.
var. turbidus *(Rauh & Backeb.) C. Ostolaza Nano in Brit. Cact. Succ. J.,* 16(3): 132 (1998): H. turbidus.
subsp. turbidus *(Rauh & Backeb.) C. Ostolaza Nano in Cactaceae Consensus Initiatives,* 6: 9 (1998): H. turbidus.

HABENARIA (Orchidac.).
aethiopica *S. Thomas & P.J. Cribb in Kew Bull.,* 51(1): 149 (1996)—Ethiopia.
agapitae *R. González T. & J.J. Reynoso D.in Bol. Inst. Bot. (Guadalajara),* 1(5): 371 (1993)—Mexico Southwest (Jalisco).
alinae *D.L. Szlachetko in Fl. Cameroun,* 34: 116 (1998)—Cameroon.
bacata *R.L. Dressler in Bol. Inst. Bot. (Guadalajara),* 7(1–3): 94 (1999 publ. 2000)—Belize.
batesii *I.F. la Croix in Kew Bull.,* 51(2): 364 (1996), nom. nov.: H. praetermissa *la Croix.*
bertauxiana *D.L. Szlachetko & T.S. Olszewski in Fl. Cameroun,* 34: 178 (1998)—Zaire.
bosseriana *D.L. Szlachetko & T.S. Olszewski in Fl. Cameroun,* 34: 170 (1998)—Cameroon.
calicis *R. González Tamayo in Bol. Inst. Bot. (Guadalajara),* 3(1–3): 67 (1995 publ. 1997)—Mexico Southwest (Michoacan).
carlotae *R.L. Dressler in Bol. Inst. Bot. (Guadalajara),* 7(1–3): 94 (1999 publ. 2000)—Costa Rica.
chejuensis *Y.N. Lee & K. Lee in Korean J. Pl. Taxon.,* 28(1): 34 (1998)—South Korea.
chlorosepala *D.L. Jones in Orchadian,* 12(10): 461 (1998)—Queensland.
cribbiana *D.L. Szlachetko & T.S. Olszewski in Fl. Cameroun,* 34: 184 (1998)—Congo.
decumbens *S. Thomas & P.J. Cribb in Kew Bull.,* 51(1): 151 (1996)—Ethiopia.
defleurii *D. Geerinck in Belgian J. Bot.,* 129(1): 11 (1996 publ. 1997)—Rwanda.
dentata (Sw.) Schltr.
subsp. ecalcarata *(King & Pantl.) G. Panigrahi & S.K. Murti, Fl. Bilaspur Distr.,* 2: 589 (1999): H. geniculata var. ecalcarata.
excelsa *S. Thomas & P.J. Cribb in Kew Bull.,* 51(1): 147 (1996)—Ethiopia.
exilis *D.L. Jones in Orchadian,* 12(10): 464 (1998)—Queensland.
garayana *D.L. Szlachetko & T.S. Olszewski in Fl. Cameroun,* 34: 143 (1998)—Zaire.
gilbertii *S. Thomas & P.J. Cribb in Kew Bull.,* 51(1): 145 (1996)—Ethiopia.
gonzaleztamayoi *J. García-Cruz, R. Jiménez Machorro & L. Sánchez Saldaña in Acta Bot. Mex.,* 50: 35 (2000)—Mexico Central (Morelos).
greenwoodiana *R. González Tamayo in Bol. Inst. Bot. (Guadalajara),* 6(2–3): 204 (1998 publ. 2000)—Mexico Southwest (Jalisco, Guerrero).
hannae *D.L. Szlachetko in Fragm. Flor. Geobot.,* 40(2): 721 (1995)—Zaire.
harroldii *D.L. Jones in Orchadian,* 12(9): 414 (1998)—Queensland.
horaliae *R. González Tamayo in Bol. Inst. Bot. (Guadalajara),* 3(1–3): 60 (1995 publ. 1997)—Mexico Southwest (Michoacan).
humbertii *D.L. Szlachetko & T.S. Olszewski in Fl. Cameroun,* 34: 166 (1998)—Zaire.
josephi *Rchb. f. in Trans. Linn. Soc. London, Bot.,* Ser. 2, 3: 114 (1888)—East Himalaya (Sikkim).

HABENARIA:-
kornasiana *D.L. Szlachetko in Fragm. Flor. Geobot.*, 40(2): 719 (1995)—Zaire.
kornasiorum *D.L. Szlachetko & T.S. Olszewski in Fl. Cameroun*, 34: 180 (1998)—Central African Republic.
letestuana *D.L. Szlachetko & T.S. Olszewski in Fl. Cameroun*, 34: 159 (1998)—Central African Republic.
lisowskii *D.L. Szlachetko in Fragm. Flor. Geobot.*, 40(2): 721 (1995)—Congo.
luentensis *D.L. Szlachetko & T.S. Olszewski in Fl. Cameroun*, 34: 136 (1998)—Zaire.
luzmariana *R. González Tamayo in Bol. Inst. Bot. (Guadalajara)*, 6(2–3): 209 (1998 publ. 2000)—Mexico Southwest (Jalisco).
macvaughiana *R. González Tamayo in Bol. Inst. Bot. (Guadalajara)*, 3(1–3): 64 (1995 publ. 1997)—Mexcio Southwest (Michoacan).
matudae *G.A. Salazar in An. Inst. Biol. Univ. Nac. Autón. Méx., Bot.*, 70(1): 6 (1999)—Mexico Southeast (Chiapas), Mexico Southwest (Oaxaca).
mediocris *R.L. Dressler in Bol. Inst. Bot. (Guadalajara)*, 7(1–3): 96 (1999 publ. 2000)—Panama.
moratii *D.L. Szlachetko & T.S. Olszewski in Fl. Cameroun*, 34: 130 (1998), as 'morati'—Sierra Leone.
ochyrae *D.L. Szlachetko & T.S. Olszewski in Fl. Cameroun*, 34: 126 (1998)—Congo.
ortiziana *R. González Tamayo in Bol. Inst. Bot. (Guadalajara)*, 3(1–3): 56 (1995 publ. 1997)—Mexico Central (México State).
pectinata
var. arietina *(Hook.f.) R.C. Srivastava in Nation. Acad. Sci. Letters*, 18(3–4): 61 (1995): H. arietina.
periyarensis *N. Sasidharan, K.P. Rajesh & J. Augustine in Rheedea*, 8(2): 167 (1998)—India (Kerala).
praecox *P.S. Lavarack & A.W. Dockrill in Austrobaileya*, 5(2): 331 (1999)—Queensland.
radiata Spreng.
var. dianthoides *(Nevski) I.B. Vyschin in Sosud. Rast. Sovet. Dal'nego Vostoka*, 8: 320 (1996): H. dianthoides.
renziana *D.L. Szlachetko & T.S. Olszewski in Fl. Cameroun*, 34: 182 (1998)—Zaire.
rosilloana *R. González Tamayo in Bol. Inst. Bot. (Guadalajara)*, 3(1–3): 81 (1995 publ. 1997)—Mexico Southwest (Michoacan).
rosulifolia *A. Espejo Serna & A.R. López-Ferrari in Acta Bot. Mex.*, 50: 32 (2000)—Mexico Central (Morelos).
rzedowskii *R. González Tamayo in Bol. Inst. Bot. (Guadalajara)*, 3(1–3): 76 (1995 publ. 1997)—Mexico Southwest (Michoacan).
sanfordiana *D.L. Szlachetko & T.S. Olszewski in Fl. Cameroun*, 34: 156 (1998)—Cameroon.
saprophytica *J. Bosser & P.J. Cribb in Bull. Mus. Nation. Hist. Nat., B, Adansonia*, Ser. 4, 18(3–4): 337 (1996)—Madagascar.
similis *R.L. Dressler in Bol. Inst. Bot. (Guadalajara)*, 7(1–3): 98 (1999 publ. 2000), homonym; *see Schltr. in Bot. Jahrb. Syst.*, 38: 147 (1906)—Mexico Southeast (Chiapas), Guatemala.
tisserantii *D.L. Szlachetko & T.S. Olszewski in Fl. Cameroun*, 34: 174 (1998), as 'tisseranti'—Central African Republic.
uncinata *D.L. Szlachetko & T.S. Olszewski in Fl. Cameroun*, 34: 160 (1998)—Zaire.
vollesenii *S. Thomas & P.J. Cribb in Kew Bull.*, 51(1): 149 (1996)—Ethiopia.
zamudioana *R. González Tamayo in Bol. Inst. Bot. (Guadalajara)*, 3(1–3): 71 (1995 publ. 1997)—Mexico Southwest (Michoacan).

HABERLEA (Gesneriac.).
rhodopensis Friv.
var. ferdinandi-coburgii *(Urum.) M. Markova in Fl. Republ. Bŭlgariya*, 10: 290 (1995): H. ferdinandi-coburgii.

HABRANTHUS (Amaryllidac.).
concinnus *P. Ravenna in Onira, Bot. Leafl.*, 3(17): 63 (1999)—Brazil South (Paraná).

HABRANTHUS:-
conterminus *P. Ravenna in Onira, Bot. Leafl.*, 3(16): 57 (1999)—Brazil Northeast (Bahia).
crassibulbus *P. Ravenna in Onira, Bot. Leafl.*, 3(17): 63 (1999)—Argentina Northwest (Santiago del Estero).
datensis *P. Ravenna in Onira, Bot. Leafl.*, 3(16): 58 (1999)—Brazil Southeast (Minas Gerais).
gameleirensis *P. Ravenna in Onira, Bot. Leafl.*, 3(16): 57 (1999)—Brazil North (Tocantins).
ischihualastus *P. Ravenna in Onira, Bot. Leafl.*, 3(17): 65 (1999)—Argentina Northwest (San Juan).
itaobinus *P. Ravenna in Onira, Bot. Leafl.*, 3(16): 56 (1999)—Brazil Southeast (Minas Gerais).
leonensis *P. Ravenna in Onira, Bot. Leafl.*, 3(16): 56 (1999)—Paraguay.
matacus *P. Ravenna in Onira, Bot. Leafl.*, 3(17): 66 (1999)—Argentina Northeast (Formosa).
mexicanus *T.M. Howard in Herbertia*, 51: 42 (1996)—Mexico Northeast (Hidalgo, Guanajuato, San Luis Potosí, Querétaro).
oaxacanus *T.M. Howard in Herbertia*, 51: 43 (1996)—Mexico Southwest (Oaxaca).
oranensis *P. Ravenna in Onira, Bot. Leafl.*, 3(17): 64 (1999)—Argentina Northwest (Salta).
pantanalensis *P. Ravenna in Onira, Bot. Leafl.*, 3(16): 59 (1999)—Brazil West-central (Mato Grosso do Sul).
pictus *P. Ravenna in Onira, Bot. Leafl.*, 3(16): 59 (1999)—Argentina Northwest (Catamarca).
venturianus *P. Ravenna in Onira, Bot. Leafl.*, 3(17): 65 (1999)—Argentina Northwest (Santiago del Estero).

HACKELIA (Boraginac.).
bhutanica *R.R. Mill in Edinburgh J. Bot.*, 53(1): 117 (1996)—Bhutan.
obtusifolia *R.R. Mill in Edinburgh J. Bot.*, 53(1): 118 (1996)—Bhutan.

HAEMATOSPERMUM (Euphorbiac.).
neriifolium *(B. Heyne ex Roth) Wall. ex J.O. Voigt, Hort. Suburb. Calcutt.*: 157 (1845): Adelia neriifolia.†

HAKEA (Proteac.).
actites *W.R. Barker in J. Adelaide Bot. Gard.*, 17: 183 (1996)—New South Wales, Queensland.
acuminata *L. Haegi in Fl. Australia*, 17B: 396 (1999)—Western Australia.
anadenia *L. Haegi in Fl. Australia*, 17B: 394 (1999)—Western Australia.
archaeoides *W.R. Barker in Fl. Australia*, 17B: 393 (1999)—New South Wales.
decurrens R. Br.
subsp. physocarpa *W.R. Barker in J. Adelaide Bot. Gard.*, 17: 193 (1996)—Victoria.
subsp. platytaenia *W.R. Barker in J. Adelaide Bot. Gard.*, 17: 196 (1996)—New South Wales, Victoria, Tasmania.
dohertyi *L. Haegi in Fl. Australia*, 17B: 395 (1999)—New South Wales.
eneabba *L. Haegi in Fl. Australia*, 17B: 396 (1999)—Western Australia.
epiglottis Labill.
subsp. milliganii *(Meisn.) R.M. Barker in Fl. Australia*, 17B: 394 (1999): H. milliganii.
hastata *L. Haegi in Fl. Australia*, 17B: 394 (1999)—Western Australia.
laevipes Gand.
subsp. graniticola *L. Haegi in Fl. Australia*, 17B: 395 (1999)—New South Wales, Queensland.
leucoptera R. Br.
subsp. sericipes *W.R. Barker in J. Adelaide Bot. Gard.*, 17: 181 (1996)—Queensland, Western Australia, New South Wales.
lorea (R. Br.) R. Br.
subsp. borealis *W.R. Barker in Fl. Australia*, 17B: 393 (1999)—Northern Territory, Western Australia.
maconochieana *L. Haegi in Fl. Australia*, 17B: 396 (1999)—Queensland.
macrorrhyncha *W.R. Barker in J. Adelaide Bot. Gard.*, 17: 185 (1996)—Queensland, New South Wales.
ochroptera *W.R. Barker in J. Adelaide Bot. Gard.*, 17: 187 (1996)—New South Wales.

HAKEA:-
petiolaris Meisn.
subsp. angusta *L. Haegi in Fl. Australia*, 17B: 395 (1999)—Western Australia.
subsp. trichophylla *L. Haegi in Fl. Australia*, 17B: 395 (1999)—Western Australia.
recurva Meisn.
subsp. arida *(Diels) W.R. Barker & R.M. Barker in Fl. Australia*, 17B: 394 (1999): H. arida.
rigida *C.A. Gardner ex L. Haegi in Fl. Australia*, 17B: 395 (1999)—Western Australia.
salicifolia (Vent.) B.L. Burtt
subsp. angustifolia *(A.A. Ham.) W.R. Barker in Fl. Australia*, 17B: 393 (1999): H. saligna var. angustifolia.
scoparia Meisn.
subsp. trycherica *L. Haegi in Fl. Australia*, 17B: 395 (1999)—Western Australia.
stenophylla A. Cunn. ex R. Br.
subsp. notialis *R.M. Barker in Nuytsia*, 12(1): 7 (1998)—Western Australia.

HALENIA (Gentianac.).
corniculata (L.) Cornaz
f. purpurea *T. Shimizu, Fl. Nagano Pref.*: 1509 (1997)—Japan.
longicaulis *J.S. Pringle in Fl. Ecuador*, 53(159A): 80 (1995)—Ecuador.
serpyllifolia *J.S. Pringle in Fl. Ecuador*, 53(159A): 77 (1995)—Ecuador.

HALERPESTES (Ranunculac.).
tricuspis (Maxim.) Hand.-Mazz.
var. linearisecta *L.H. Zhou in Fl. Qinghaiica*, 1: 507 (1997)—Qinghai.

HALESIA (Styracac.).
hispida *(Siebold & Zucc.) N.E. Brown, Suppl. Johnson's Gard. Dict.*: 937 (1882): Pterostyrax hispida.

HALIMIUM (Cistac.).
voldii *Kit Tan, D. Perdetzoglou & T. Raus in Willdenowia*, 30(2): 232 (2000)—Greece.

HALIMOCNEMIS (Chenopodiac.).
occulta *(Bunge) I.C. Hedge in Fl. Iran.*, 172: 341 (1997): Halotis occulta.

HALLEORCHIS *D.L. Szlachetko & T.S. Olszewski in Fl. Cameroun*, 34: 246 (1998). *ORCHIDACEAE*
aspidogynoides *D.L. Szlachetko & T.S. Olszewski in Fl. Cameroun*, 34: 246 (1998)—Gabon, Cameroon.

HALORAGIS (Haloragac.).
microphylla *R.D. Hoogland in Blumea*, Suppl. 4: 228 (1958), in I.K., Supp 13 as Haumania microphylla in error—New Guinea (Papua New Guinea).

HALOXYLON (Chenopodiac.).
gracile *(Aellen) I.C. Hedge in Fl. Iran.*, 172: 322 (1997): Arthrophytum gracile.
griffithii (Moq.) Boiss.
subsp. wakhanicum *(Paulsen) I.C. Hedge in Fl. Iran.*, 172: 325 (1997): Anabasis wakhanica.

HAMATOCACTUS (Cactac.).
uncinatus
subsp. crassihamatus *(F.A.C. Weber) C.E. Glass, Guía Identific. Cact. Amenazadas México*, 1: Ha/un ssp. crassihamatus (1998 publ. 1997): Echinocactus crassihamatus.

HAMMATOLOBIUM (Legumin.).
subgen. Ludovicia *(Coss.) V.N. Tikhomirov & D.D. Sokoloff in Feddes Repert.*, 107(3–4): 214 (1996): gen. Ludovicia.

HAMMERIA *P.M. Burgoyne in Cact. Succ. J. (U.S.A.)*, 70(4): 204 (1998). *AIZOACEAE*
gracilis *P.M. Burgoyne in Cact. Succ. J. (U.S.A.)*, 70(4): 207 (1998)—Cape Province.
meleagris *(L. Bolus) C. Klak in Bothalia*, 30(1): 39 (2000): Mesembryanthemum meleagris.
salteri *(L. Bolus) P.M. Burgoyne in Cact. Succ. J. (U.S.A.)*, 70(4): 206 (1998): Ruschia salteri.

HAMPEA (Malvac.).
sect. Eohampea *R.W. Jones, P.A. Fryxell & D.M. Baro in An. Inst. Biol. Univ. Nac. Autón. Méx., Bot.*, 68(1): 34 (1997).

HANGUANA (Hanguanac.).
bogneri *H.-J. Tillich & E. Sill in Sendtnera*, 6: 216 (1999)—Borneo (Sarawak).

HANIFFIA (Zingiberac.).
albiflora *K. Larsen & J. Mood in Nordic J. Bot.*, 20(3): 287 (2000)—Thailand.
cyanescens (Ridl.) Holttum
var. penangiana *C.K. Lim in Folia Malaysiana*, 2: 36 (2000)—Malaya (Peninsular Malaysia).

HAPALINE (Arac.).
celatrix *P.C. Boyce in Kew Bull.*, 51(1): 70 (1996)—Borneo (Brunei).

HAPALOCHILUS (Orchidac.).
trichromus *(Schltr.) L.A. Garay & Sieder in Lindleyana*, 11(4): 233 (1996): Bulbophyllum trichromum.

HAPALORCHIS (Orchidac.).
cymbirostris *D.L. Szlachetko in Orchidee*, Suppl. 3: 29 (1996)—Brazil Southeast (Rio de Janeiro).
pandurata *D.L. Szlachetko in Orchidee*, Suppl. 3: 36 (1996)—Brazil Southeast (Rio de Janeiro).
stellaris *D.L. Szlachetko in Orchidee*, Suppl. 3: 33 (1996)—Brazil South (Rio Grande do Sul).

HAPLOCARPHA (Compos.).
oocephala *(DC.) J.P.B. Beyers in Strelitzia*, 9: 705 (2000): Arctotis oocephala.

HAPLOCOELOPSIS *F.G. Davies in Kew Bull.*, 52(1): 231 (1997). *SAPINDACEAE*
africana *F.G. Davies in Kew Bull.*, 52(1): 232 (1997)—Kenya, Tanzania, Angola.

HAPLOCOELUM (Sapindac.).
foliolosum (Hiern) Bullock
subsp. mombasense *(Bullock) B. Verdcourt in Kew Bull.*, 51(4): 802 (1996): H. mombasense.
subsp. strongylocarpum *(Bullock) B. Verdcourt in Fl. Trop. E. Afr., Sapindac.*: 46 (1998): H. strongylocarpum.

HAPLOPAPPUS (Compos.).
nahuelbutae *L. Klingenberg in Sendtnera*, 4: 175 (1997)—Chile South (La Araucania).

HAPLOSTICHANTHUS (Annonac.).
novoguineensis *H. Okada in Acta Phytotax. Geobot.*, 47(1): 6 (1996)—New Guinea (Papua New Guinea).

× **HARBOTRYA** *S.A. Hammer in Cact. Succ. J. (U.S.A.)*, 68(1): 8 (1996). *AIZOACEAE* (Hartmanthus × Jensenobotrya).

HARDWICKIA (Legumin.).
mopane *(J. Kirk ex Benth.) F.J. Breteler in Adansonia*, 19(2): 289 (1997): Copaifera mopane.

HARMONIA *B.G. Baldwin in Novon*, 9(4): 463 (1999). *COMPOSITAE*
doris-nilesiae *(T.W. Nelson & J.P. Nelson) B.G. Baldwin in Novon*, 9(4): 464 (1999): Madia doris-nilesiae.
hallii *(D.D. Keck) B.G. Baldwin in Novon*, 9(4): 464 (1999): Madia hallii.
nutans *(Greene) B.G. Baldwin in Novon*, 9(4): 464 (1999): Callichroa nutans.
stebbinsii *(T.W. Nelson & J.P. Nelson) B.G. Baldwin in Novon*, 9(4): 464 (1999): Madia stebbinsii.

HARMSIA (Sterculiac.).
lepidota *(Vollesen) M. Jenny in Taxon*, 7(1): 5 (1999): Aethiocarpa lepidota.

HARNIERIA (Acanthac.).
hygrophiloides *(F. Muell.) R.M. Barker in J. Adelaide Bot. Gard.*, 17: 149 (1996): Justicia hygrophiloides.

HARNIERIA:-
kempeana *(F. Muell.) R.M. Barker in J. Adelaide Bot. Gard.*, 17: 150 (1996): Justicia kempeana.
subsp. muelleri *(R.M. Barker) R.M. Barker in J. Adelaide Bot. Gard.*, 17: 150 (1996): Sarojusticia kempeana subsp. muelleri.
subsp. rhadinophylla *B.J. Lepschi in J. Adelaide Bot. Gard.*, 17: 153 (1996)—Western Australia.

HAROLDIELLA *J. Florence in Fl. Polynésie française*, 1: 218 (1997). *URTICACEAE*
rapaensis *J. Florence in Fl. Polynésie française*, 1: 220 (1997)—Tubuai Is.
sykesii *J. Florence in Fl. Polynésie française*, 1: 221 (1997)—Tubuai Is.

HARPALYCE (Legumin.).
lanata *L.P. de Queiroz in Kew Bull.*, 53(4): 985 (1998)—Brazil Northeast (Bahia).

HARPERIA (Restionac.).
confertospicata *(Steud.) B.G. Briggs & L.A.S. Johnson in Telopea*, 8(1): 28 (1998), as 'confertospicatus': Restio confertospicatus.
ferruginipes *K.A. Meney & J.S. Pate in Telopea*, 6(4): 651 (1996)—Western Australia.

HARRISIA (Cactac.).
balansae *(K. Schum.) N.P. Taylor & D.C. Zappi in Cactaceae Consensus Initiatives*, 3: 7 (1997): Cereus balansae.
pomanensis (F.A.C. Weber ex K. Schum.) Britton & Rose
subsp. regelii *(Weing.) R. Kiesling in Darwiniana*, 34(1–4): 395 (1996): Cereus regelii.

HARTIA (Theac.).
sect. Inflorescentia *C.X. Ye in Acta Sci. Nat. Univ. Sunyatseni*, 1982(4): 112 (1982).
sect. Rotundisepala *C.X. Ye in Acta Sci. Nat. Univ. Sunyatseni*, 1982(4): 112 (1982).
gracilis *(S.Z. Yan) Hung T. Chang & C.X. Ye in Fl. Reipubl. Popul. Sin.*, 49(3): 231 (1998): H. yunnanensis var. gracilis.
villosa (Merr.) Merr.
var. kwangtungensis *(Chun) S.Z. Yan ex C.X. Ye in Acta Sci. Nat. Univ. Sunyatseni*, 1982(4): 113 (1982): H. kwangtungensis.

HARTMANTHUS *S.A. Hammer in Haseltonia*, 3: 79 (1995). *AIZOACEAE*
hallii *(L. Bolus) S.A. Hammer in Haseltonia*, 3: 80 (1995): Delosperma hallii.
pergamentaceus *(L. Bolus) S.A. Hammer in Haseltonia*, 3: 80 (1995): Delosperma pergamentaceum.

HARVEYA (Scrophulariac.).
wysockiana *R. Mielcarek in Fragm. Flor. Geobot.*, 41(1): 222 (1996)—Burundi.

HASTINGSIA (Hyacinthac.).
bracteosa S. Watson
var. atropurpurea *(Becking) F.A. Lang & P.F. Zika in Madroño*, 44(2): 191 (1997 publ. 1998), without basionym date: H. atropurpurea.

HATIORA (Cactac.).
epiphylloides (Porto & Werderm.) Buxb.
subsp. bradei *(Porto & A. Cast.) W. Barthlott & N.P. Taylor in Bradleya*, 13: 74 (1995): Hariota epiphylloides var. bradei.

HAUMANIASTRUM (Labiat.).
sect. Arborescentia *A.J. Paton in Kew Bull.*, 52(2): 302 (1997).
sect. Botridiocephalum *(Briq.) A.J. Paton in Kew Bull.*, 52(2): 355 (1997): Acrocephalus sect. Botridiocephalus.
sect. Caudata *A.J. Paton in Kew Bull.*, 52(2): 340 (1997).
sect. Dubia *A.J. Paton in Kew Bull.*, 52(2): 323 (1997).
sect. Revoluta *A.J. Paton in Kew Bull.*, 52(2): 349 (1997).
bianense *A.J. Paton in Kew Bull.*, 52(2): 314 (1997)—Zaire.
chartaceum *A.J. Paton in Kew Bull.*, 52(2): 318 (1997)—Zaire.

HAUMANIASTRUM:-
coriaceum *(Robyns & Lebrun) A.J. Paton in Kew Bull.*, 52(2): 366 (1997): Acrocephalus coriaceus.
subsp. attenuifolium *A.J. Paton in Kew Bull.*, 52(2): 369 (1997)—Tanzania, Burundi.
subsp. mitwabense *A.J. Paton in Kew Bull.*, 52(2): 369 (1997)—Zaire.
cubanquense *(R.D. Good) A.J. Paton in Kew Bull.*, 52(2): 336 (1997): A. cubanquensis.
dissitifolium *(Baker) A.J. Paton in Kew Bull.*, 52(2): 326 (1997): Acrocephalus dissitifolius.
glabrifolium *A.J. Paton in Kew Bull.*, 52(2): 353 (1997)—Zambia.
graminifolium *(Robyns) A.J. Paton in Kew Bull.*, 52(2): 322 (1997): Acrocephalus graminifolius.
membranaceum *A.J. Paton in Kew Bull.*, 52(2): 361 (1997)—Angola.
minor *(Briq.) A.J. Paton in Kew Bull.*, 52(2): 330 (1997): Acrocephalus minor.
morumbense *(De Wild.) A.J. Paton in Kew Bull.*, 52(2): 354 (1997): Acrocephalus morumbensis.
paniculatum *(Briq.) A.J. Paton in Kew Bull.*, 52(2): 363 (1997): Acrocephalus paniculatus.
prealtum (Briq.) P.A. Duvign. & Plancke
var. homblei *(De Wild.) A.J. Paton in Kew Bull.*, 52(2): 359 (1997): Acrocephalus homblei.
var. succisifolium *(Baker) A.J. Paton in Kew Bull.*, 52(2): 361 (1997): Acrocephalus succisifolius.
rupestre *(R.E. Fr.) A.J. Paton in Kew Bull.*, 52(2): 323 (1997): Acrocephalus rupestris.
sericeum *(Briq.) A.J. Paton in Kew Bull.*, 52(2): 332 (1997): A. sericeus.
speciosum *(E.A. Bruce) A.J. Paton in Kew Bull.*, 52(2): 369 (1997): Acrocephalus speciosus.
stanneum *A.J. Paton in Kew Bull.*, 52(2): 355 (1997)—Zaire.
triramosum *(N.E. Br.) A.J. Paton in Kew Bull.*, 52(2): 332 (1997): Acrocephalus triramosus.
uniflorum *A.J. Paton in Kew Bull.*, 52(2): 339 (1997)—Tanzania.
villosum *(Benth.) A.J. Paton in Kew Bull.*, 52(2): 370 (1997): Acrocephalus villosus.

HAVARDIA (Legumin.).
campylacanthus *(L. Rico & M. Sousa) R.C. Barneby & J.W. Grimes, Silk Tree, Guanacaste, Monkey's Earring (Mem. New York Bot. Gard., 74(1))*: 167 (1996): Pithecellobium campylacanthus.

HAWORTHIA (Aloac.).
subgen. Fusiformes *J.J. Halda in Acta Mus. Richnov. Sect. Nat.*, 4(2): 34 (1997).
subgen. Longifolia *M. Hayashi in Haworthiad*, 13(1): 20 (1999); *cf. Repert. Pl. Succ. (I.O.S.)*, 50: 7 (1999 publ. 2000).
subgen. Oligonodes *M. Hayashi in Haworthiad*, 13(1): 20 (1999); *cf. Repert. Pl. Succ. (I.O.S.)*, 50: 7 (1999 publ. 2000).
angustifolia Haw.
var. altissima *M.B. Bayer, Haworthia Revisited*: 26 (1999)—Cape Province.
var. baylissii *(C.L. Scott) M.B. Bayer, Haworthia Revisited*: 27 (1999): H. baylissii.
arachnoidea (L.) Duval
var. aranea *(A. Berger) M.B. Bayer, Haworthia Revisited*: 30 (1999): H. bolusii var. aranea.
var. bolusii *(Baker) J.J. Halda in Acta Mus. Richnov. Sect. Nat.*, 4(2): 43 (1997): H. bolusii.
var. gracilis *(Poelln.) J.J. Halda in Acta Mus. Richnov. Sect. Nat.*, 4(2): 44 (1997): H. gracilis.
var. marumiana *(Uitewaal) J.J. Halda in Acta Mus. Richnov. Sect. Nat.*, 4(2): 44 (1997): H. marumiana.
var. namaquensis *M.B. Bayer, Haworthia Revisited*: 31 (1999)—Cape Province.
var. nigricans *(Haw.) M.B. Bayer, Haworthia Revisited*: 32 (1999): H. setata var. nigricans.
subsp. pearsonii *(C.H. Wright) J.J. Halda in Acta Mus. Richnov. Sect. Nat.*, 4(2): 43 (1997): H. pearsonii.
var. pearsonii *(C.H. Wright) J.J. Halda in Acta Mus. Richnov. Sect. Nat.*, 4(2): 44 (1997): H. pearsonii.
var. pehlemanniae *(C.L. Scott) J.J. Halda in Acta Mus. Richnov. Sect. Nat.*, 4(2): 44 (1997): H. pehlemanniae.

HAWORTHIA:-
arachnoidea (L.) Duval
var. scabrispina *M.B. Bayer, Haworthia Revisited:* 34 (1999)—Cape Province.
var. semiviva *(Poelln.) J.J. Halda in Acta Mus. Richnov. Sect. Nat.*, 4(2): 43 (1997): H. bolusii var. semiviva.
subsp. setata *(Haw.) J.J. Halda in Acta Mus. Richnov. Sect. Nat.*, 4(2): 43 (1997): H. setata.
var. setata *(Haw.) M.B. Bayer, Haworthia Revisited:* 34 (1999): H. setata.
subsp. tenera *(Poelln.) J.J. Halda in Acta Mus. Richnov. Sect. Nat.*, 4(2): 44 (1997): H. tenera.
var. tenera *(Poelln.) J.J. Halda in Acta Mus. Richnov. Sect. Nat.*, 4(2): 44 (1997): H. tenera.
var. translucens *(Haw.) J.J. Halda in Acta Mus. Richnov. Sect. Nat.*, 4(2): 43 (1997): Aloe translucens.
var. xiphiophylla *(Baker) J.J. Halda in Acta Mus. Richnov. Sect. Nat.*, 4(2): 43 (1997): H. xiphiophylla.
var. xiphiophylla *(Baker) M.B. Bayer, Haworthia Revisited:* 36 (1999), isonym; *see Halda in Acta Mus. Richnov. Sect. Nat.*, 4(2): 43 (1997): H. xiphiophylla.
aristata Haw.
subsp. helmiae *(Poelln.) J.J. Halda in Acta Mus. Richnov. Sect. Nat.*, 4(2): 48 (1997): H. helmiae.
attenuata Haw.
var. radula *(Jacq.) M.B. Bayer, Haworthia Revisited:* 167 (1999): Aloe radula.
bayeri *J.D. Venter & S.A. Hammer in Cact. Succ. J. (U.S.A.)*, 69(2): 75 (1997)—Cape Province.
blackburniae W.F. Barker
var. derustensis *M.B. Bayer, Haworthia Revisited:* 41 (1999)—Cape Province.
var. graminifolia *(G.G. Sm.) M.B. Bayer, Haworthia Revisited:* 42 (1999): H. graminifolia.
subsp. graminifolia *(G.G. Sm.) J.J. Halda in Acta Mus. Richnov. Sect. Nat.*, 4(2): 34 (1997): H. graminifolia.
chloracantha Haw.
var. angustifolia *(Haw.) J.J. Halda in Acta Mus. Richnov. Sect. Nat.*, 4(2): 42 (1997): H. angustifolia.
subsp. angustifolia *(Haw.) J.J. Halda in Acta Mus. Richnov. Sect. Nat.*, 4(2): 42 (1997): H. angustifolia.
subsp. baylissii *(C.L. Scott) J.J. Halda in Acta Mus. Richnov. Sect. Nat.*, 4(2): 41 (1997): H. baylissii.
var. floribunda *(Poelln.) J.J. Halda in Acta Mus. Richnov. Sect. Nat.*, 4(2): 42 (1997): H. floribunda.
var. mclarenii *(Poelln.) J.J. Halda in Acta Mus. Richnov. Sect. Nat.*, 4(2): 42 (1997): H. mclarenii.
var. monticola *(Fourc.) J.J. Halda in Acta Mus. Richnov. Sect. Nat.*, 4(2): 42 (1997): H. monticola.
var. pulchella *(M.B. Bayer) J.J. Halda in Acta Mus. Richnov. Sect. Nat.*, 4(2): 42 (1997): H. pulchella.
var. serrata *(M.B. Bayer) J.J. Halda in Acta Mus. Richnov. Sect. Nat.*, 4(2): 42 (1997): H. serrata.
subsp. variegata *(L. Bolus) J.J. Halda in Acta Mus. Richnov. Sect. Nat.*, 4(2): 42 (1997): H. variegata.
var. variegata *(L. Bolus) J.J. Halda in Acta Mus. Richnov. Sect. Nat.*, 4(2): 42 (1997): H. variegata.
var. zantneriana *(Poelln.) J.J. Halda in Acta Mus. Richnov. Sect. Nat.*, 4(2): 42 (1997): H. zantneriana.
coarctata Haw.
var. adelaidensis *(Poelln.) M.B. Bayer, Haworthia Revisited:* 172 (1999): H. reinwardtii var. adelaidensis.
f. greenii *(Baker) M.B. Bayer, Haworthia Revisited:* 172 (1999): H. greenii
var. tenuis *(G.G. Sm.) M.B. Bayer, Haworthia Revisited:* 173 (1999): H. reinwardtii var. tenuis.
cooperi Baker
subsp. batteniae *(C.L. Scott) J.J. Halda in Acta Mus. Richnov. Sect. Nat.*, 4(2): 48 (1997): H. batteniae.

HAWORTHIA:-
cooperi Baker
var. dielsiana *(Poelln.) M.B. Bayer, Haworthia Revisited:* 51 (1999): H. dielsiana.
var. gordoniana *(Poelln.) M.B. Bayer, Haworthia Revisited:* 52 (1999): H. gordoniana.
var. pilifera *(Baker) M.B. Bayer, Haworthia Revisited:* 54 (1999): H. pilifera.
var. truncata *(H. Jacobsen) M.B. Bayer, Haworthia Revisited:* 55 (1999): H. obtusa f. truncata.
var. venusta *(C.L. Scott) M.B. Bayer, Haworthia Revisited:* 56 (1999): H. venusta.
cymbiformis (Haw.) Duval
var. ramosa *(G.G. Sm.) M.B. Bayer, Haworthia Revisited:* 60 (1999): H. ramosa.
var. reddii *(C.L. Scott) M.B. Bayer, Haworthia Revisited:* 61 (1999): H. reddii.
var. setulifera *(Poelln.) M.B. Bayer, Haworthia Revisited:* 62 (1999): H. planifolia var. setulifera.
decipiens Poelln.
var. cyanea *M.B. Bayer, Haworthia Revisited:* 65 (1999)—Cape Province.
var. minor *M.B. Bayer, Haworthia Revisited:* 66 (1999)—Cape Province.
var. pringlei *(C.L. Scott) M.B. Bayer, Haworthia Revisited:* 67 (1999): H. pringlei.
emelyae Poelln.
var. comptoniana *(G.G. Sm.) J.D. Venter & S.A. Hammer in Cact. Succ. J. (U.S.A.)*, 69(2): 77 (1997): H. comptoniana.
var. major *(G.G. Sm.) M.B. Bayer in Aloe*, 34(1–2): 6 (1997): H. schuldtiana var. major.
floribunda Poelln.
var. dentata *M.B. Bayer, Haworthia Revisited:* 73 (1999)—Cape Province.
var. major *M.B. Bayer, Haworthia Revisited:* 74 (1999)—Cape Province.
var. isabellae *(Poelln.) M.B. Bayer, Haworthia Revisited:* 77 (1999): H. isabellae.
gracilis Poelln.
var. picturata *M.B. Bayer, Haworthia Revisited:* 78 (1999)—Cape Province.
var. tenera *(Poelln.) M.B. Bayer, Haworthia Revisited:* 77 (1999): H. tenera.
var. viridis *M.B. Bayer, Haworthia Revisited:* 79 (1999)—Cape Province.
heidelbergensis G.G. Sm.
var. minor *M.B. Bayer, Haworthia Revisited:* 82 (1999)—Cape Province.
var. scabra *M.B. Bayer, Haworthia Revisited:* 82 (1999)—Cape Province.
var. toonensis *M.B. Bayer, Haworthia Revisited:* 83 (1999)—Cape Province.
herbacea (Mill.) Stearn
var. flaccida *M.B. Bayer, Haworthia Revisited:* 86 (1999)—Cape Province.
var. lupula *M.B. Bayer, Haworthia Revisited:* 86 (1999)—Cape Province.
var. paynei *(Poelln.) M.B. Bayer, Haworthia Revisited:* 87 (1999): H. paynei.
joeyae *C.L. Scott in Bradleya*, 13: 80 (1995)—Cape Province.
koelmaniorum Oberm. & D.S. Hardy
var. mcmurtryi *(C.L. Scott) M.B. Bayer, Haworthia Revisited:* 181 (1999): H. mcmurtryi.
limifolia Marloth
var. koelmaniorum *(Oberm. & D.S. Hardy) J.J. Halda in Acta Mus. Richnov. Sect. Nat.*, 4(2): 38 (1997): H. koelmaniorum.
maculata (Poelln.) M.B. Bayer
var. intermedia *(Poelln.) M.B. Bayer, Haworthia Revisited:* 91 (1999): H. intermedia.
magnifica
var. acuminata *(M.B. Bayer) M.B. Bayer in Aloe*, 34(1–2): 6 (1997): H. retusa f. acuminata.
var. dekenahii *(G.G. Sm.) M.B. Bayer in Aloe*, 34(1–2): 6 (1997): H. dekenahii.
var. splendens *J.D. Venter & S.A. Hammer in Cact. Succ. J. (U.S.A.)*, 70(4): 180 (1998)—Cape Province.
marumiana Uitewaal
var. archeri *(W.F. Barker ex M.B. Bayer) M.B. Bayer, Haworthia Revisited:* 104 (1999): H. archeri.
var. batesiana *(Uitewaal) M.B. Bayer, Haworthia Revisited:* 105 (1999): H. batesiana.

HAWORTHIA:-
marumiana Uitewaal
var. dimorpha *(M.B. Bayer) M.B. Bayer, Haworthia Revisited:* 106 (1999): H. archeri var. dimorpha.
var. viridis *M.B. Bayer, Haworthia Revisited:* 107 (1999)—Cape Province.
minima (Aiton) Haw.
var. poellnitziana *(Uitewaal) M.B. Bayer, Haworthia Revisited:* 213 (1999): H. poellnitziana.
mirabilis Haw.
var. badia *(Poelln.) M.B. Bayer, Haworthia Revisited:* 109 (1999): H. badia.
var. beukmannii *(Poelln.) M.B. Bayer, Haworthia Revisited:* 110 (1999): H. emelyae var. beukmannii.
var. calcarea *M.B. Bayer, Haworthia Revisited:* 110 (1999)—Cape Province.
var. consanguinea *M.B. Bayer, Haworthia Revisited:* 111 (1999)—Cape Province.
var. paradoxa *(Poelln.) M.B. Bayer in Aloe*, 34(1–2): 6 (1997): H. paradoxa.
var. sublineata *(Poelln.) M.B. Bayer, Haworthia Revisited:* 113 (1999): H. triebneriana var. sublineata.
var. triebneriana *(Poelln.) M.B. Bayer, Haworthia Revisited:* 113 (1999): H. triebneriana.
monticola Fourc.
var. asema *M.B. Bayer, Haworthia Revisited:* 117 (1999)—Cape Province.
mucronata Haw.
var. altilinea *(Haw.) J.J. Halda in Acta Mus. Richnov. Sect. Nat.*, 4(2): 45 (1997): H. altilinea.
var. habdomadis *(Poelln.) M.B. Bayer, Haworthia Revisited:* 120 (1999): H. habdomadis.
var. inconfluens *(Poelln.) M.B. Bayer, Haworthia Revisited:* 121 (1999): H. altilinea f. inconfluens.
subsp. lockwoodii *(Archibald) J.J. Halda in Acta Mus. Richnov. Sect. Nat.*, 4(2): 45 (1997): H. lockwoodii.
var. nortieri *(G.G. Sm.) J.J. Halda in Acta Mus. Richnov. Sect. Nat.*, 4(2): 45 (1997): H. nortieri.
var. rooibergensis *J. Esterhuizen & Battista in Haworthiad*, 13(1): 5–7 (1999); *cf. Repert. Pl. Succ. (I.O.S.)*, 50: 9 (1999 publ. 2000)—Locality not stated.
mutica Haw.
var. nigra *M.B. Bayer, Haworthia Revisited:* 126 (1999)—Cape Province.
nortieri G.G. Sm.
var. pehlemanniae *(C.L. Scott) M.B. Bayer, Haworthia Revisited:* 130 (1999): H. pehlemanniae.
outeniquensis *M.B. Bayer, Haworthia Revisited:* 130 (1999)—Cape Province.
pubescens M.B. Bayer
var. livida *M.B. Bayer, Haworthia Revisited:* 134 (1999)—Cape Province.
pulchella M.B. Bayer
var. globifera *M.B. Bayer, Haworthia Revisited:* 136 (1999)—Cape Province.
pumila *(L.) M.B. Bayer, Haworthia Revisited:* 214 (1999), comb. already made; *see Duval, Pl. Succ. Horto Alencon.:* 7 (1809): Basionym not stated.
subsp. attenuata *(Haw.) J.J. Halda in Acta Mus. Richnov. Sect. Nat.*, 4(2): 36 (1997): Aloe attenuata.
subsp. fasciata *(Willd.) J.J. Halda in Acta Mus. Richnov. Sect. Nat.*, 4(2): 37 (1997): Apicra fasciata.
var. kingiana *(Poelln.) J.J. Halda in Acta Mus. Richnov. Sect. Nat.*, 4(2): 36 (1997): H. kingiana.
subsp. longiana *(Poelln.) J.J. Halda in Acta Mus. Richnov. Sect. Nat.*, 4(2): 37 (1997): H. longiana.
subsp. minima *(Aiton) J.J. Halda in Acta Mus. Richnov. Sect. Nat.*, 4(2): 36 (1997): Aloe margaritifera var. minima.
subsp. radula *(Jacq.) J.J. Halda in Acta Mus. Richnov. Sect. Nat.*, 4(2): 37 (1997): Aloe radula.

HAWORTHIA:-
pumila
var. smitii *(Poelln.) J.J. Halda in Acta Mus. Richnov. Sect. Nat.*, 4(2): 36 (1997): H. smitii.
pungens *M.B. Bayer, Haworthia Revisited:* 188 (1999)—Cape Province.
pygmaea
var. argenteomaculosa *(G.G. Sm.) M.B. Bayer in Aloe,* 34(1–2): 6 (1997): H. dekenahii var. argenteomaculosa.
reinwardtii (Salm-Dyck) Haw.
subsp. coarctata *(Haw.) J.J. Halda in Acta Mus. Richnov. Sect. Nat.*, 4(2): 40 (1997): H. coarctata.
var. coarctata *(Haw.) J.J. Halda in Acta Mus. Richnov. Sect. Nat.*, 4(2): 40 (1997): H. coarctata.
subsp. glauca *(Baker) J.J. Halda in Acta Mus. Richnov. Sect. Nat.*, 4(2): 41 (1997): H. glauca.
var. glauca *(Baker) J.J. Halda in Acta Mus. Richnov. Sect. Nat.*, 4(2): 41 (1997): H. glauca.
var. greenii *(Baker) J.J. Halda in Acta Mus. Richnov. Sect. Nat.*, 4(2): 41 (1997): H. greenii.
var. herrei *(Poelln.) J.J. Halda in Acta Mus. Richnov. Sect. Nat.*, 4(2): 41 (1997): H. herrei.
reticulata (Haw.) Haw.
var. attenuata *M.B. Bayer, Haworthia Revisited:* 140 (1999)—Cape Province.
subsp. batesiana *(Uitewaal) J.J. Halda in Acta Mus. Richnov. Sect. Nat.*, 4(2): 47 (1997): H. batesiana.
var. integra *(Poelln.) J.J. Halda in Acta Mus. Richnov. Sect. Nat.*, 4(2): 47 (1997): H. integra.
var. subregularis *(Baker) M.B. Bayer, Haworthia Revisited:* 141 (1999): H. subregularis.
retusa (L.) Duval
var. asperula *(Haw.) J.J. Halda in Acta Mus. Richnov. Sect. Nat.*, 4(2): 50 (1997): H. asperula.
subsp. asperula *(Haw.) J.J. Halda in Acta Mus. Richnov. Sect. Nat.*, 4(2): 50 (1997): H. asperula.
var. bruynsii *(M.B. Bayer) J.J. Halda in Acta Mus. Richnov. Sect. Nat.*, 4(2): 49 (1997): H. bruynsii.
var. comptoniana *(G.G. Sm.) J.J. Halda in Acta Mus. Richnov. Sect. Nat.*, 4(2): 50 (1997): H. comptoniana.
subsp. emelyae *(Poelln.) J.J. Halda in Acta Mus. Richnov. Sect. Nat.*, 4(2): 49 (1997): H. emelyae.
var. emelyae *(Poelln.) J.J. Halda in Acta Mus. Richnov. Sect. Nat.*, 4(2): 49 (1997): H. emelyae.
var. heidelbergensis *(G.G. Sm.) J.J. Halda in Acta Mus. Richnov. Sect. Nat.*, 4(2): 49 (1997): H. heidelbergensis.
var. magnifica *(Poelln.) J.J. Halda in Acta Mus. Richnov. Sect. Nat.*, 4(2): 50 (1997): H. magnifica.
var. mirabilis *(Haw.) J.J. Halda in Acta Mus. Richnov. Sect. Nat.*, 4(2): 50 (1997): Aloe mirabilis.
var. mutica *(Haw.) J.J. Halda in Acta Mus. Richnov. Sect. Nat.*, 4(2): 49 (1997): H. mutica.
var. springbokvlakensis *(C.L. Scott) J.J. Halda in Acta Mus. Richnov. Sect. Nat.*, 4(2): 50 (1997): H. springbokvlakensis.
scabra Haw.
subsp. granulata *(Marloth) J.J. Halda in Acta Mus. Richnov. Sect. Nat.*, 4(2): 35 (1997): H. granulata.
var. lateganiae *(Poelln.) M.B. Bayer, Haworthia Revisited:* 195 (1999): H. lateganiae.
var. lavrani *(C.L. Scott) J.J. Halda in Acta Mus. Richnov. Sect. Nat.*, 4(2): 35 (1997): H. sordida var. lavrani.
subsp. sordida *(Haw.) J.J. Halda in Acta Mus. Richnov. Sect. Nat.*, 4(2): 34 (1997): H. sordida.
var. sordida *(Haw.) J.J. Halda in Acta Mus. Richnov. Sect. Nat.*, 4(2): 34 (1997): H. sordida.
subsp. starkiana *(Poelln.) J.J. Halda in Acta Mus. Richnov. Sect. Nat.*, 4(2): 35 (1997): H. starkiana.
var. starkiana *(Poelln.) M.B. Bayer, Haworthia Revisited:* 197 (1999): H. starkiana.

HAWORTHIA:-
scabra Haw
var. tuberculata *(Poelln.) J.J. Halda in Acta Mus. Richnov. Sect. Nat.*, 4(2): 35 (1997): H. tuberculata.
truncata Schönland
var. maughanii *(Poelln.) J.J. Halda in Acta Mus. Richnov. Sect. Nat.*, 4(2): 51 (1997): H. maughanii.
turgida Haw.
var. longibracteata *(G.G. Sm.) M.B. Bayer, Haworthia Revisited:* 154 (1999): H. longibracteata.
variegata L. Bolus
var. hemicrypta *M.B. Bayer, Haworthia Revisited:* 158 (1999)—Cape Province.
var. modesta *M.B. Bayer, Haworthia Revisited:* 159 (1999)—Cape Province.
var. petrophila *M.B. Bayer, Haworthia Revisited:* 159 (1999)—Cape Province.
venosa (Lam.) Haw.
subsp. nigra *(Haw.) J.J. Halda in Acta Mus. Richnov. Sect. Nat.*, 4(2): 39 (1997): Apicra nigra.
var. tessellata *(Haw.) J.J. Halda in Acta Mus. Richnov. Sect. Nat.*, 4(2): 39 (1997): H. tessellata.
subsp. woolleyi *(Poelln.) J.J. Halda in Acta Mus. Richnov. Sect. Nat.*, 4(2): 40 (1997): H. woolleyi.
subsp. woolleyi *(Poelln.) M.B. Bayer, Haworthia Revisited:* 204 (1999), isonym; *see Halda in Acta Mus. Richnov. Sect. Nat.*, 4(2): 40 (1997): H. woolleyi.
venusta *C.L. Scott in Bradleya,* 14: 87 (1996)—Cape Province.
viscosa (L.) Haw.
subsp. derekii-clarkii *J.J. Halda in Acta Mus. Richnov. Sect. Nat.*, 5(1): 43 (1998), as 'dereki-clarki'—Cape Province.
subsp. nigra *(Haw.) J.J. Halda in Acta Mus. Richnov. Sect. Nat.*, 5(1): 44 (1998): Apicra nigra.
vlokii *M.B. Bayer, Haworthia Revisited:* 160 (1999)—Cape Province.
zantneriana Poelln.
var. minor *M.B. Bayer, Haworthia Revisited:* 164 (1999)—Cape Province.

HEBANTHE (Amaranthac.).
eriantha *(Poir.) T.M. Pedersen in Bonplandia,* 10(1–4): 101 (2000): Iresine erianthos.
f. ovatifolia *(Heimerl) T.M. Pedersen in Bonplandia,* 10(1–4): 102 (2000): Gomphrena paniculata f. ovatifolia.
grandiflora *(Hook.) T. Borsch & T.M. Pedersen in Sendtnera,* 4: 20 (1997): Iresine grandiflora.
occidentalis *(R.E. Fr.) T. Borsch & T.M. Pedersen in Sendtnera,* 4: 21 (1997): Pfaffia occidentalis.
var. bangii *(R.E. Fr.) T. Borsch & T.M. Pedersen in Sendtnera,* 4: 21 (1997): Pfaffia bangii.
paniculata Mart.
f. ovatifolia *(Heimerl) T. Borsch & T.M. Pedersen in Sendtnera,* 4: 19 (1997): Gomphrena paniculata f. ovatifolia.
reticulata *(Seub.) T. Borsch & T.M. Pedersen in Sendtnera,* 4: 20 (1997): Gomphrena reticulata.

HEBANTHODES *T.M. Pedersen in Bonplandia,* 10(1–4): 102 (2000). *AMARANTHACEAE*
peruviana *T.M. Pedersen in Bonplandia,* 10(1–4): 102 (2000)—Peru.

HEBE (Scrophulariac.).
arganthera *P.J. Garnock-Jones, M.J. Bayly, W.G. Lee & B.D. Rance in New Zealand J. Bot.*, 38(3): 380 (2000)—New Zealand South I.
× bishopiana *(Petrie) E.D. Hutch in Newsl. Auckland Bot. Soc.*, 23(1): 1 (1966): Veronica X bishopiana
brevifolia *(Cheeseman) P.J. de Lange in New Zealand J. Bot.*, 35(1): 1 (1997): Veronica speciosa var. brevifolia.
hectorii (Hook. f.) Cockayne & Allan
subsp. coarctata *(Cheeseman) S.J. Wagstaff & P. Wardle in New Zealand J. Bot.*, 37(1): 33 (1999): Veronica coarctata.
subsp. demissa *(G. Simpson) S.J. Wagstaff & P. Wardle in New Zealand J. Bot.*, 37(1): 33 (1999): H. demissa.

HEBE:-
hectorii
subsp. laingii *(Cockayne) S.J. Wagstaff & P. Wardle in New Zealand J. Bot.*, 37(1): 33 (1999): Veronica laingii.
subsp. subsimilis *(Colenso) S.J. Wagstaff & P. Wardle in New Zealand J. Bot.*, 37(1): 33 (1999): Veronica subsimilis.
subsp. subulata *(G. Simpson) S.J. Wagstaff & P. Wardle in New Zealand J. Bot.*, 37(1): 33 (1999): H. subulata.
imbricata Cockayne & Allan
subsp. poppelwellii *(Cockayne) S.J. Wagstaff & P. Wardle in New Zealand J. Bot.*, 37(1): 33 (1999): Veronica poppelwellii.
lycopodioides (Hook. f.) Cockayne & Allan
subsp. patula *(G. Simpson & J.S. Thomson) S.J. Wagstaff & P. Wardle in New Zealand J. Bot.*, 37(1): 34 (1999): H. lycopodioides var. patula.
paludosa *(Cockayne) D.A. Norton & P.J. de Lange in New Zealand J. Bot.*, 36(4): 532 (1998): Veronica salicifolia var. paludosa.
perbella *P.J. de Lange in New Zealand J. Bot.*, 36(3): 399 (1998)—New Zealand North I.
stenophylla *(Steud.) M.J. Bayly & P.J. Garnock-Jones in New Zealand J. Bot.*, 39(2): 173 (2000): Veronica stenophylla.
var. hesperia *M.J. Bayly & P.J. Garnock-Jones in New Zealand J. Bot.*, 38(2): 180 (2000)—New Zealand South I.
var. oliveri *M.J. Bayly & P.J. Garnock-Jones in New Zealand J. Bot.*, 38(2): 182 (2000)—New Zealand South I.
tairawhiti *B.D. Clarkson & P.J. Garnock-Jones in New Zealand J. Bot.*, 34(1): 51 (1996)—New Zealand North I.

HEBECLINIUM (Compos.).
escobariae *H. Robinson in Biollania,* ed. esp. 6 (J.J. Wurdack Festschr.): 512 (1997)—Colombia.
flabellatum *S. Díaz-Piedrahita & G.P. Méndez-Ramírez in Rev. Acad. Colomb. Cienc. Exact. Fis. Nat.*, 90: 39 (2000)—Colombia.
frontinoense *S. Díaz-Piedrahita & G.P. Méndez-Ramírez in Rev. Acad. Colomb. Cienc. Exact. Fis. Nat.*, 90: 37 (2000)—Colombia.
longicuspidatum *S. Díaz-Piedrahita in Rev. Acad. Colomb. Cienc. Exact. Fis. Nat.*, 90: 42 (2000)—Colombia.
palaciosii *H. Robinson in Phytologia,* 84(5): 350 (1998 publ. 1999)—Ecuador.
wurdackianum *H. Robinson in Biollania,* ed. esp. 6 (J.J. Wurdack Festschr.): 513 (1997)—Colombia.

HEBEPETALUM (Linac.).
neblinae *A. Jardim & P.E. Berry in Novon,* 9(4): 522 (1999)—Venezuela, Brazil North (Amazonas).

× **HECHCOHNIA** *G.H. Anderson ex J.R. Grant in Selbyana,* 19(1): 117 (1998). *BROMELIACEAE* (Deuterocohnia × Hechtia).

HEDEOMA (Labiat.).
matomianum *Reid Moran in Madroño,* 46(4): 212 (1999 publ. 2000)—Mexico Northwest (Baja California).
palmeri Hemsl.
var. santiagoanum *B.L. Turner in Phytologia,* 79(1): 48 (1995 publ. 1996)—Mexico Northeast (Nuevo León, Coahuila).
var. zaragozanum *B.L. Turner in Phytologia,* 79(1): 48 (1995 publ. 1996)—Mexico Northeast (Nuevo León, Tamaulipas).

HEDERA (Araliac.).
palmata *(DC.) J.O. Voigt, Hort. Suburb. Calcutt.:* 25 (1845): Gilibertia palmata.
paniculata *(DC.) Wight & Arn. ex J.O. Voigt, Hort. Suburb. Calcutt.:* 25 (1845): Gilibertia paniculata.
repanda *(DC.) Wight & Arn. ex J.O. Voigt, Hort. Suburb. Calcutt.:* 25 (1845): Gilibertia repanda.

HEDINIA (Crucif.).
lata *X.L. He & Z.X. An in Acta Phytotax. Sin.*, 34(2): 205 (1996)—Xizang.

HEDYCHIUM (Zingiberac.).
fastigiatum *J.O. Voigt, Hort. Suburb. Calcutt.:* 570 (1845), nom. nov.: H. ellipticum *Roscoe.*

HEDYCHIUM:-

larsenii *M. Dan & C. Sathish Kumar in Nordic J. Bot.*, 19(3): 333 (1999)—Assam (Manipur).

menglianense *Y.Y. Qian in J. Trop. Subtrop. Bot.*, 8(1): 17 (2000)—China South-central (Yunnan).

neocarneum *T.L. Wu, K. Larsen & N.J. Turland in Novon*, 10(1): 91 (2000), nom. nov.: H. carneum *Y.Y. Qian*.

nutantiflorum *H. Dong & G.J. Xu in Acta Phytotax. Sin.*, 35(6): 566 (1997)—China South-central (Yunnan).

puerense *Y.Y. Qian in Acta Phytotax. Sin.*, 34(4): 444 (1996)—China South-central (Yunnan).

raoi *G.D. Pal & G.S. Giri in J. Bombay Nat. Hist. Soc.*, 95(1): 102 (1998), as 'raoii'—Assam (Arunachal Pradesh).

satyanarayanum *S.C. Srivastava in Indian J. Forest.*, 22(1–2): 85 (1999)—Assam (Meghalaya).

simaoense *Y.Y. Qian in Acta Phytotax. Sin.*, 34(4): 443 (1996)—China South-central (Yunnan).

sirirugsae *M. Dan & C. Sathish Kumar in Nordic J. Bot.*, 19(3): 335 (1999)—Assam (Manipur).

HEDYOTIS (Rubiac.).

acerosa A. Gray

var. potosina *B.L. Turner in Phytologia*, 79(2): 87 (1995 publ. 1996)—Mexico Northeast (San Luis Potosí, Coahuila, Nuevo León, Tamaulipas).

var. tamaulipana *B.L. Turner in Phytologia*, 79(2): 87 (1995 publ. 1996)—Mexico Northeast (Tamaulipas).

baotingensis *W.C. Ko in J. S. China Agric. Univ.*, 16(4): 43 (1995)—Hainan.

cantoniensis *F.C.How ex W.C. Ko in J. S. China Agric. Univ.*, 16(4): 42 (1995)—China Southeast (Guangdong).

capitellata Wall. ex G. Don

var. mollis *(Pierre ex Pit.) W.C. Ko in Fl. Reipubl. Popul. Sin.*, 71(1): 66 (1999), isonym; *see N.N. Tran in Nguyen Tien Ban (ed.), Fl. Taynguyen. Enum.*: 150 (1984): Oldenlandia capitellata var. mollis.

var. mollissima *(Pit.) W.C. Ko in Fl. Reipubl. Popul. Sin.*, 71(1): 66 (1999): Oldenlandia capitellata var. mollissima.

coreana H. Lév.

var. parvifolia *(Hook. & Arn.) S. Hatusima in S. Hatusima & T. Amano, Fl. Ryukyus, S. of Amami Is., ed. 2*: 211 (1994), without basionym ref.: Basionym not stated.

dianxiensis *W.C. Ko in J. S. China Agric. Univ.*, 16(4): 44 (1995)—China South-central (Yunnan).

flynnii *W.L. Wagner & D.H. Lorence in Novon*, 8(3): 311 (1998)—Hawaiian Is.

gartmorensis *C.E. Ridsdale in Blumea*, 41(2): 455 (1996)—Sri Lanka.

hermanniana *R. Dutta in J. Econ. Taxon. Bot.*, 23(3): 734 (1999), nom. nov.: Hedyotis biflora *R. Br. ex Wight & Arn.*

lianshanensis *W.C. Ko in J. S. China Agric. Univ.*, 16(4): 41 (1995)—China Southeast (Guangdong).

lucei *D.H. Lorence & J. Florence in Adansonia*, 22(2): 225 (2000)—Marquesas.

macvaughii *E.E. Terrell in Novon*, 6(1): 128 (1996)—Mexico Southwest (Jalisco).

neolessertiana *C.E. Ridsdale in Blumea*, 41(2): 456 (1996)—Sri Lanka.

nigricans (Lam.) Fosberg

var. austrotexana *B.L. Turner in Phytologia*, 79(1): 15 (1995 publ. 1996)—Texas.

var. gypsophila *B.L. Turner in Phytologia*, 79(1): 15 (1995 publ. 1996)—Mexico Northeast (Nuevo León).

var. papillacea *B.L. Turner in Phytologia*, 79(1): 15 (1995 publ. 1996)—New Mexico.

nukuhivensis *J. Florence & D.H. Lorence in Adansonia*, 22(2): 224 (2000)—Marquesas.

palmeri (A. Gray) W.H. Lewis

var. muzquizana *B.L. Turner in Phytologia*, 79(2): 91 (1995 publ. 1996)—Mexico Northeast (Coahuila).

pooleana *B.L. Turner in Phytologia*, 79(2): 93 (1995 publ. 1996)—Texas.

schlechtendahliana Steud.

subsp. waimeae *(Wawra) W.L. Wagner & D.H. Lorence in Novon*, 8(3): 316 (1998): Kadua waimeae.

HEDYOTIS :-

shettyi *K. Ravikumar & V. Lakshmanan in Rheedea*, 9(1): 69 (1999)—India (Tamil Nadu).

tahuatensis *D.H. Lorence & J. Florence in Adansonia*, 22(2): 227 (2000)—Marquesas.

terrellii *D.H. Lorence in Novon*, 7(1): 51 (1997)—Mexico Southwest (Oaxaca).

tridentata *C.E. Ridsdale in Blumea*, 41(2): 456 (1996)—Sri Lanka.

trimenii Deb & R.M. Dutta

var. orbicularifolia *C.E. Ridsdale in Blumea*, 41(2): 457 (1996)—Sri Lanka.

yangchunensis *W.C. Ko & Zhang in J. S. China Agric. Univ.*, 16(4): 45 (1995)—China Southeast (Guangdong).

HEDYSARUM (Legumin.).

subgen. Gamotion *(Basiner) B.H. Choi & Hiroyoshi Ohashi in J. Jap. Bot.*, 71(4): 210 (1996): H. sect. Gamotion.

algidum *L.Z. Shue in Fl. Reipubl. Popul. Sin.*, 42(2): 198 (1998), nom. nov.: H. tuberosum *B. Fedtsch.*

var. speciosum *(Hand.-Mazz.) Y.H. Wu in Fl. Qinghaiica*, 2: 264 (1999): Hedysarum tuberosum var. speciosum.

var. thyrsum *Y.H. Wu in Bull. Bot. Res. North-East. Forest. Univ.*, 19(1): 8 (1999)—Qinghai.

boveanum Bunge ex Basiner

subsp. costaetalentii *(López Bernal, S. Ríos, Alcaraz & D. Rivera) B. Valdés in Lagascalia*, 21(1): 253 (1999): Hedysarum costaetalentii.

subsp. palentinum *B. Valdés in Lagascalia*, 21(1): 253 (1999)—Spain.

cisbaicalense *L.I. Malyschev in L.I. Malyschev (ed.), Vysokogornaya Flora Stanovogo Nagor'ya*: 115 (1972)—Irkutsk.

costaetalentii *J. López Bernal, S. Ríos, F. Alcaraz & D. Rivera in Israel J. Pl. Sci.*, 46(3): 225 (1998)—Spain.

ferganense Korsh.

var. minjanense *(Rech.f.) L.Z. Shue in Fl. Reipubl. Popul. Sin.*, 42(2): 217 (1998): H. minjanense.

var. poncinsii *(Franch.) L.Z. Shue in Fl. Reipubl. Popul. Sin.*, 42(2): 215 (1998): H. poncinsii.

fruticosum Pall.

var. gobicum *Y.Z. Zhao, R. Sha & R. Cao in Acta Sci. Nat. Univ. Neimonggol*, 27(5): 681 (1996)—Inner Mongolia (Nei Mongol).

gmelinii Ledeb.

var. tongtianhense *Y.H. Wu in Bull. Bot. Res. North-East. Forest. Univ.*, 19(1): 8 (1999)—Qinghai.

kungeicum *S.K. Muchtubaeva in Izv. Min. Nauki Akad. Nauk Resp. Kazakhstan, Ser. Biol. Med.*, 1998(3): 106 (1998)—Kirgizstan.

nonnae *Yu.R. Roskov in Kew Bull.*, 51(2): 376 (1996), nom. nov.: H. sachalinense subsp. confertum *N.S. Pavlova*.

× polychromum *P.V. Kulikov in Bot. Zhurn.*, 83(12): 141 (1998)—West Siberia. (H. argyrophyllum × grandiflorum).

setigerum Turcz. ex Fisch. & C.A. Mey.

var. montanum *(Malyschev) G.A. Peschkova in Fl. Tsentral'noi Sibiri*, 2: 629 (1979): H. baicalense var. montanum.

vicioides Turcz.

var. alaschanicum *(B. Fedtsch.) Y.Z. Zhao, R. Sha & R. Cao in Acta Sci. Nat. Univ. Neimonggol*, 27(5): 675 (1996): H. semenovii var. alaschanicum.

zundukii *G.A. Peschkova, Stepnaya Fl. Baĭkal'skoĭ Sibiri*: 75 (1972)—Irkutsk.

HEISTERIA (Olacac.).

povedae *Q. Jiménez & S. Knapp in Bull. Nat. Hist. Mus. (London), Bot.*, 30(1): 1 (2000)—Costa Rica, Mexico Southeast (Chiapas), Guatemala, El Salvador, Nicaragua, Panama.

HELIAMPHORA (Sarraceniac.).

hispida *A. Wistuba & J. Nerz in Carniv. Pl. Newsl.*, 29(2): 37 (2000)—Venezuela/Brazil North (Amazonas) border.

HELIANTHEMUM (Cistac.).

ser. Cana *Juz. & Pozdeeva ex M.V. Klokov & D.N. Dobroczajeva in Novosti Sist. Vyssh. Nizsh. Rast. (Kiev)*, 1974: 67 (1974).

HELIANTHEMUM:-

ser. Italica *Juz. & Pozdeeva ex M.V. Klokov & D.N. Dobroczajeva in Novosti Sist. Vyssh. Nizsh. Rast. (Kiev)*, 1974: 67 (1974).

almeriense Pau

subsp. scopulorum *(Rouy) F. Alcaraz, T.E. Díaz, S. Rivas-Martínez & P. Sánchez-Gómez in Itinera Geobot.*, 2: 110 (1989): H. lineare var. scopulorum.

apenninum (L.) Mill.

subsp. rothmaleri *(Villar ex Rothm.) M. Mayor López & M. Fernández Benito in Fontqueria*, 48: 90 (1997 publ. 1998): H. rothmaleri.

asperum Lag.

subsp. willkommii *G. Mateo Sanz & M.B. Crespo Villalba in Flora Montiberica*, 4: 15 (1996)—Spain.

baschkirorum *(Juz. ex Kupat.) N.N. Tzvelev in Fl. Vostochnoĭ Evropy*, 9: 214 (1996), isonym; *see Juz. in Fl. SSSR*, 15: 681 (1949): H. canum subsp. baschkirorum.

conquense *(Borja & Rivas Goday ex G. López) G. Mateo Sanz & V.J. Arán Resó in Flora Montiberica*, 3: 95 (1996): H. marifolium subsp. conquense.

creticola *M.V. Klokov & D.N. Dobroczajeva in Novosti Sist. Vyssh. Nizsh. Rast. (Kiev)*, 1974: 76 (1974)—Krym.

cretophilum *M.V. Klokov & D.N. Dobroczajeva in Novosti Sist. Vyssh. Nizsh. Rast. (Kiev)*, 1974: 70 (1974)—Ukraina.

glaucescens *(Murb.) N.N. Tzvelev in Fl. Vostochnoĭ Evropy*, 9: 211 (1996): H. chamaecistus var. glaucescens.

inaguae *A. Marrero, M. González-Martín & F. González-Artiles in Bot. Macaronésica*, 22: 4 (1995)—Canary Is.

× mariano-salvatoris *F. Alcaraz, T.E. Díaz, S. Rivas-Martínez & P. Sánchez-Gómez in Itinera Geobot.*, 2: 111 (1989)—Spain. (H. alypoides × almeriense subsp. almeriense).

oelandicum (L.) Dum. Cours.

subsp. levigatum *(M. Proctor) D.H. Kent, List Vasc. Pl. Brit. Isles*, Supp. 1: 36 (1996 publ. 19 Jan. 1997): H. canum subsp. levigatum.

ordosicum *Y.Z. Zhao, Zong Y. Zhu & R. Cao in Acta Phytotax. Sin.*, 38(3): 294 (2000)—Inner Mongolia (Nei Mongol).

pinegense *Juz. ex N.N. Tzvelev in Fl. Vostochnoĭ Evropy*, 9: 214 (1996), isonym; *see Juz. in Fl. SSSR*, 15: 681 (1949)—Russia North.

scopulicola *L. Sáez, J.A. Rosselló & G. Alomar in Nordic J. Bot.*, 19(4): 414 (1999), as 'scopulicolum'—Baleares.

tinetense *M. Mayor López & M. Fernández Benito in Fontqueria*, 48: 90 (1997 publ. 1998)—Spain.

violaceum (Cav.) Pers.

var. arabinianum *J.L. Solanas Ferrándiz & M.B. Crespo Villalba in Sarrià*, 1: 35 (1998–99 publ. 1999)—Spain.

violaceum (Cav.) Pers.

subsp. subobtusatum *(Maire) I. Soriano in Lagascalia*, 18(2): 240 (1996): H. pilosum var. subobtusatum.

zheguliense *Juz. ex N.N. Tzvelev in Fl. Vostochnoĭ Evropy*, 9: 214 (1996)—Russia East.

HELIANTHOPSIS (Compos.).

mollis *(Griseb.) C.A. Sala & A.M. Echarte in Fragm. Flor. Geobot.*, 41(2): 1030 (1996): Viquiera mollis.

HELIANTHUS (Compos.).

subgen. Harpalium *(Cass.) N.N. Tzvelev, Opred. Sosud. Rast. Severo-Zap. Rossii*: 681 (2000): gen. Harpalium.

sect. Onegensia *Sennikov ex N.N. Tzvelev, Opred. Sosud. Rast. Severo-Zap. Rossii*: 681 (2000).

sect. Perennes *N.N. Tzvelev, Opred. Sosud. Rast. Severo-Zap. Rossii*: 681 (2000).

porteri *(A. Gray) J.F. Pruski in Castanea*, 63(1): 75 (1998): Rudbeckia porteri.

HELICHRYSUM (Compos.).

cataractarum *H. Beentje in Kew Bull.*, 55(2): 350 (2000)—Tanzania.

kashgaricum *C.H. An in Fl. Xinjiangensis*, 5: 476 (1999)—Xinjiang.

HELICHRYSUM:-
kirkii Oliv. & Hiern
var. petersii *(Oliv. & Hiern) H. Beentje in Kew Bull.*, 55(2): 364 (2000): H. petersii.
korongoni *H. Beentje in Kew Bull.*, 55(2): 352 (2000)—Tanzania, Uganda.
mechowianum Klatt
var. ceres *(S. Moore) H. Beentje in Kew Bull.*, 55(2): 361 (2000): H. ceres.
var. vernonioides *(Wild) H. Beentje in Kew Bull.*, 55(2): 362 (2000): H. vernonioides.
nudifolium (L.) Less.
var. oxyphyllum *(DC.) H. Beentje in Kew Bull.*, 55(2): 357 (2000): H. oxyphyllum.
var. pilosellum *(L.f.) H. Beentje in Kew Bull.*, 55(2): 359 (2000): Gnaphalium pilosellum.
tenderiense *O.Yu. Umanets in Bot. Zhurn.*, 85(8): 112 (2000)—Ukraine.

HELICIA (Proteac.).
celata *D.B. Foreman in Kew Bull.*, 53(3): 672 (1998), as 'celatus'—New Guinea (Irian Jaya).
fragilis *D.B. Foreman in Kew Bull.*, 53(3): 669 (1998)—New Guinea (Irian Jaya).
mahmudii *P.P.K. Chai in Sandakania*, 7: 59 (1996)—Borneo (Sarawak).
sessilifolia *R.C.K. Chung in Gard. Bull. Singapore*, 50(2): 151 (1998)—Borneo (Sarawak, Sabah).
sleumeri *D.B. Foreman in Kew Bull.*, 53(3): 670 (1998)—New Guinea (Irian Jaya).
symplocoides *R.C.K. Chung in Gard. Bull. Singapore*, 50(2): 154 (1998)—Borneo (Sabah).
vanroyenii *D.B. Foreman in Kew Bull.*, 53(3): 675 (1998)—New Guinea (Irian Jaya).
versteeghii *D.B. Foreman in Kew Bull.*, 53(3): 674 (1998)—New Guinea (Irian Jaya).

HELICIOPSIS (Proteac.).
litseifolia *R.C.K. Chung in Gard. Bull. Singapore*, 50(2): 154 (1998)—Borneo (Sarawak, Sabah, Kalimantan), Sumatera, Malaya (Peninsular Malaysia).
percoriacea *R.C.K. Chung in Gard. Bull. Singapore*, 50(2): 158 (1998)—Borneo (Sarawak).

HELICONIA (Heliconiac.).
chartacea Lane ex Barreiros
var. meeana *W.J. Kress in Biollania*, ed. esp. 6 (J.J. Wurdack Festschr.): 414 (1997)—Venezuela.
spatho-circinada
subsp. panamensis *L. Emygdio de Mello Filho & E. Santos in Bol. Mus. Nac. Rio de Janeiro, Bot.*, 56: 1 (1979)—Panama.

HELICTERES (Sterculiac.).
lanceolata DC.
var. gagnepainiana *(Craib) C. Phengklai in Thai Forest Bull., Bot.*, 23: 89 (1995): H. gagnepainiana.
prostrata *S.Y. Liu in Acta Phytotax. Sin.*, 37(6): 601 (1999)—China Southeast (Guangxi).
urupaensis *N. Leane M. da Costa in Bol. Mus. Nac. Rio de Janeiro, Bot.*, 55: 1 (1979)—Brazil West-central (Mato Grosso).

HELICTOTRICHON (Gramin.).
subgen. Amphibromus *(Nees) G.C. Tucker in Harvard Pap. Bot.*, 9: 44 (1996): gen. Amphibromus.
austroibericum *J. Holub in Preslia*, 70(2): 105 (1998), nom. nov., as 'austro-ibericum': H. filifolium subsp. arundanum *Romero Zarco.*
cantabricum *(Lag.) C. Gervais in Denkschr. Schweiz. Naturforsch. Ges.*, 88: 69 (1973): Avena cantabrica.
convolutum (C. Presl) Henrard
subsp. heldreichii *(Parl.) C. Gervais in Denkschr. Schweiz. Naturforsch. Ges.*, 88: 58 (1973): Avena heldreichii.
f. pubescens *(St.-Yves) C. Gervais in Denkschr. Schweiz. Naturforsch. Ges.*, 88: 58 (1973): Avena filifolia f. pubescens.
hackelii (Henriq.) Henrard
var. algarbiensis *(Romero Zarco) R. Govaerts in Bol. Soc. Brot.*, 66: 294 (1993–94 publ. 1994): Avenula hackelii var. algarbiensis.
lusitanicum *(Romero Zarco) M. Röser in Flora (Germany)*, 193(4): 438 (1998): Avenula pratensis subsp. lusitanica.

HELICTOTRICHON:-
montanum (Vill.) Henrard
f. glaucophyllum *(Litard. & Maire) C. Gervais in Denkschr. Schweiz. Naturforsch. Ges.*, 88: 36 (1973): Avena montana subvar. glaucophylla.
subsp. planifolium *(Willk. & Lange) C. Gervais in Denkschr. Schweiz. Naturforsch. Ges.*, 88: 37 (1973): Avena montana var. planifolia.
parlatorei (Woods) Pilg.
f. pubescens *(St.-Yves) C. Gervais in Denkschr. Schweiz. Naturforsch. Ges.*, 88: 78 (1973): Avena parlatorei f. pubescens.
praetutianum *(Parl. ex Arcang.) M. Röser in Pl. Syst. Evol.*, 203(3–4): 234 (1996): Avena scheuchzeri subsp. praetutiana.
pratense (L.) Besser
subsp. lusitanicum *(Romero Zarco) M. Röser in Pl. Syst. Evol.*, 203(3–4): 234 (1996): Avenula pratensis subsp. lusitanica.
recurvatum *(Swallen) T.A. Cope & T.B. Ryves in Watsonia*, 21(1): 135 (1996): Amphibromus recurvatus.
romeroi-zarcoi *J. Holub in Preslia*, 70(2): 105 (1998), nom. nov.: H. sedenense subsp. gervaisii *Romero Zarco.*
rubrotinctum *(Pau) J. Holub in Preslia*, 70(2): 105 (1998): Avena rubrotincta.
scabrivalve *(Trin.) G.C. Tucker in Harvard Pap. Bot.*, 9: 44 (30 Sept. 1996): Avena scabrivalvis.
scabrivalve *(Trin.) R. Govaerts, World Checklist Seed Plants*, 2(1–2): 14 (8 Nov. 1996), as 'scabrivalvis', isonym; *see G.C. Tucker in Harvard Pap. Bot.*, 9: 44 (30 Sept. 1996): Avena scabrivalvis.
scabrivalve *(Trin.) S.A. Renvoize, Gramineas de Bolivia:* 155 (1998), as 'scabrivalvis', isonym; *see G.C. Tucker in Harvard Pap. Bot.*, 9: 44 (1996): Avena scabrivalvis.
sempervirens (Vill.) Pilg.
var. pubescens *(Husn.) C. Gervais in Denkschr. Schweiz. Naturforsch. Ges.*, 88: 47 (1973): Avena sempervirens var. pubescens.
setaceum (Vill.) Henrard
subsp. petzense *(H. Melzer) M. Röser in Pl. Syst. Evol.*, 203(3–4): 234 (1996): H. petzense.

HELIETTA (Rutac.).
glaziovii *(Engl.) J.R. Pirani in Brittonia*, 50(3): 369 (1998): Esenbeckia glaziovii.

HELIOCEREUS (Cactac.).
schrankii
var. helenae *(Scheinvar) M. Kimnach in Cact. Succ. J. (U.S.A.)*, 68(4): 217 (1996): H. elegantissimus var. helenae.
var. stenopetalus *(Bravo) M. Kimnach in Cact. Succ. J. (U.S.A.)*, 68(4): 217 (1996): H. elegantissimus var. stenopetalus.

HELIOHEBE *P.J. Garnock-Jones in New Zealand J. Bot.*, 31(4): 323 (1993). *SCROPHULARIACEAE*
acuta *P.J. Garnock-Jones in New Zealand J. Bot.*, 31(4): 326 (1993)—New Zealand South I.
× airfieldii *(Hook. f.) P.J. Garnock-Jones in New Zealand J. Bot.*, 31(4): 337 (1993): Veronica × fairfieldii.
hulkeana *(F. Muell.) P.J. Garnock-Jones in New Zealand J. Bot.*, 31(4): 328 (1993): Veronica hulkeana.
subsp. evestita *P.J. Garnock-Jones in New Zealand J. Bot.*, 31(4): 331 (1993)—New Zealand South I.
lavaudiana *(Raoul) P.J. Garnock-Jones in New Zealand J. Bot.*, 31(4): 331 (1993), without basionym ref.; *et in New Zealand J. Bot.*, 34(3): 427 (1996), with basionym ref.: Veronica lavaudiana.
pentasepala *(L.B. Moore) P.J. Garnock-Jones in New Zealand J. Bot.*, 31(4): 332 (1993): Hebe raoulii var. pentasepala.
raoulii *(Hook. f.) P.J. Garnock-Jones in New Zealand J. Bot.*, 31(4): 333 (1993): Veronica raoulii.
subsp. maccaskillii *(Allan) P.J. Garnock-Jones in New Zealand J. Bot.*, 31(4): 335 (1993): Hebe raoulii var. maccaskillii.

HELIOTROPIUM (Boraginac.).
sect. Gottliebia *B. Verdcourt in Fl. Trop. E. Afr., Boraginac.:* 50 (1991), as nom. nov. but excluding type of repl. syn.†

HELIOTROPIUM:-
sect. Pseudocoeloma *H. Förther in Sendtnera*, 5: 111 (1998).
sect. Zeylanica *H. Förther in Sendtnera*, 5: 125 (1998), nom. superfl.
aenigmatum *L.A. Craven in Austral. Syst. Bot.*, 9(4): 583 (1996)—Western Australia.
aequoreum *L.A. Craven in Austral. Syst. Bot.*, 12(6): 929 (1999), nom. nov.: H. applanatum *Craven.*
alcyonium *L.A. Craven in Austral. Syst. Bot.*, 9(4): 626 (1996)—Northern Territory, Western Australia.
ammophilum *L.A. Craven in Austral. Syst. Bot.*, 9(4): 547 (1996)—Northern Territory, Western Australia, South Australia, Queensland.
amnis-edith *L.A. Craven in Austral. Syst. Bot.*, 9(4): 584 (1996)—Northern Territory.
apertum *L.A. Craven in Austral. Syst. Bot.*, 9(4): 627 (1996)—Western Australia.
applanatum *L.A. Craven in Austral. Syst. Bot.*, 9(4): 564 (1996), homonym; *see Thulin & Verdc. in Fl. Trop. E. Afr., Boraginac.:* 119 (1991)—Queensland.
brachythrix *L.A. Craven in Austral. Syst. Bot.*, 9(4): 585 (1996)—Northern Territory, Western Australia.
calvariavis *L.A. Craven in Austral. Syst. Bot.*, 9(4): 586 (1996)—Western Australia.
chalcedonium *L.A. Craven in Austral. Syst. Bot.*, 9(4): 587 (1996)—Queensland.
chrysocarpum *L.A. Craven in Austral. Syst. Bot.*, 9(4): 554 (1996)—Western Australia.
citrifolium *J.G.C. Lehmann in Gött. Gel. Anz.*, 152: 1515 (1817)—South America.
collinum *L.A. Craven in Austral. Syst. Bot.*, 9(4): 616 (1996)—Queensland.
consimile *L.A. Craven in Austral. Syst. Bot.*, 9(4): 629 (1996)—Northern Territory.
cracens *L.A. Craven in Austral. Syst. Bot.*, 9(4): 589 (1996)—Northern Territory.
cupressinum *L.A. Craven in Austral. Syst. Bot.*, 9(4): 570 (1996)—Western Australia.
delestangii *L.A. Craven in Austral. Syst. Bot.*, 9(4): 571 (1996)—Queensland.
dichotomum *L.A. Craven in Austral. Syst. Bot.*, 9(4): 591 (1996)—Western Australia, Northern Territory.
discorde *L.A. Craven in Austral. Syst. Bot.*, 9(4): 593 (1996)—Western Australia, Northern Territory.
elongatum *(Lehm.) Gürke in Engl. & Prantl, Nat. Pflanzenfam.*, IV, 3a: 96, fig. 39c (1893): Triaridium elongatum.†
esfahanicum *M. Khatamsaz in Iranian J. Bot.*, 8(2): 184 (2000)—Iran.
euodes *L.A. Craven in Austral. Syst. Bot.*, 9(4): 594 (1996)—Western Australia.
filiforme *J.G.C. Lehmann in Gött. Gel. Anz.*, 152: 1515 (1817)—Venezuela?
foveolatum *L.A. Craven in Austral. Syst. Bot.*, 9(4): 595 (1996)—Northern Territory, Western Australia.
frohlichii *L.A. Craven in Austral. Syst. Bot.*, 9(4): 574 (1996)—Queensland.
galioides *L.A. Craven in Austral. Syst. Bot.*, 9(4): 597 (1996)—Western Australia.
glanduliferum *L.A. Craven in Austral. Syst. Bot.*, 9(4): 557 (1996)—Western Australia, Northern Territory.
haesum *L.A. Craven in Austral. Syst. Bot.*, 9(4): 600 (1996)—Western Australia, Northern Territory, Queensland.
inexplicitum *L.A. Craven in Austral. Syst. Bot.*, 9(4): 617 (1996)—Western Australia, Northern Territory, South Australia.
leptaleum *L.A. Craven in Austral. Syst. Bot.*, 9(4): 601 (1996)—Northern Territory, Western Australia, Queensland.
maroccanum *J.G.C. Lehmann in Gött. Gel. Anz.*, 152: 1514 (1817), as 'marocanum'—Morocco.
melanopedii *L.A. Craven in Austral. Syst. Bot.*, 9(4): 603 (1996)—Queensland.
moorei *L.A. Craven in Austral. Syst. Bot.*, 9(4): 604 (1996)—New South Wales, Western Australia, Northern Territory, South Australia, Queensland.
murinum *L.A. Craven in Austral. Syst. Bot.*, 9(4): 552 (1996)—Western Australia.

HELIOTROPIUM:-

nesopelydum *L.A. Craven in Austral. Syst. Bot.*, 9(4): 623 (1996)—Western Australia.
nexosum *L.A. Craven in Austral. Syst. Bot.*, 9(4): 576 (1996)—Western Australia.
pachyphyllum *L.A. Craven in Austral. Syst. Bot.*, 9(4): 560 (1996), nom. nov.: H. ovalifolium var. crassifolium *Domin.*
parviantrum *L.A. Craven in Austral. Syst. Bot.*, 9(4): 562 (1996)—Northern Territory, Western Australia.
peckhamii *L.A. Craven in Austral. Syst. Bot.*, 9(4): 624 (1996)—Western Australia.
peninsularis *L.A. Craven in Austral. Syst. Bot.*, 9(4): 619 (1996)—Queensland.
plumosum *L.A. Craven in Austral. Syst. Bot.*, 9(4): 605 (1996)—Northern Territory, Queensland.
protensum *L.A. Craven in Austral. Syst. Bot.*, 9(4): 608 (1996)—Western Australia.
pulvinum *L.A. Craven in Austral. Syst. Bot.*, 9(4): 577 (1996)—Northern Territory.
ramulipatens *L.A. Craven in Austral. Syst. Bot.*, 9(4): 621 (1996)—Northern Territory, Western Australia, Queensland.
rhadinostachyum *L.A. Craven in Austral. Syst. Bot.*, 9(4): 632 (1996)—Northern Territory, Queensland.
rottleri *J.G.C. Lehmann in Gött. Gel. Anz.*, 152: 1515 (1817)—India.
skeleton *L.A. Craven in Austral. Syst. Bot.*, 9(4): 578 (1996)—Western Australia, Northern Territory.
sphaericum *L.A. Craven in Austral. Syst. Bot.*, 9(4): 609 (1996)—Western Australia, Northern Territory.
styotrichum *L.A. Craven in Austral. Syst. Bot.*, 9(4): 580 (1996)—Queensland.
subreniforme *L.A. Craven in Austral. Syst. Bot.*, 9(4): 610 (1996)—Northern Territory.
synaimon *L.A. Craven in Austral. Syst. Bot.*, 9(4): 625 (1996)—Northern Territory, Western Australia.
tabuliplagae *L.A. Craven in Austral Syst. Bot.*, 9(4): 567 (1996)—Queensland.
tachyglossoides *L.A. Craven in Austral. Syst. Bot.*, 9(4): 563 (1996)—Western Australia.
tanythrix *L.A. Craven in Austral. Syst. Bot.*, 9(4): 611 (1996)—Northern Territory, Western Australia, South Australia, Queensland.
transforme *L.A. Craven in Austral. Syst. Bot.*, 9(4): 581 (1996)—Western Australia.
tytoides *L.A. Craven in Austral. Syst. Bot.*, 9(4): 568 (1996)—Queensland.
uniflorum *L.A. Craven in Austral. Syst. Bot.*, 9(4): 569 (1996)—Western Australia, Northern Territory.
vagum *L.A. Craven in Austral. Syst. Bot.*, 9(4): 614 (1996)—Queensland.

HELIXANTHERA (Loranthac.).
flabellifolia *Balle ex D. Wiens & R.M. Polhill, Mistletoes of Africa:* 86 (1998)—Somalia.
spathulata *D. Wiens & R.M. Polhill, Mistletoes of Africa:* 80 (1998)—Nigeria.
verruculosa *D. Wiens & R.M. Polhill, Mistletoes of Africa:* 83 (1998)—Tanzania.

HELLEBORUS (Ranunculac.).
foetidus L.
subsp. balearicus *(L. Chodat) A.M. Romo, Fl. Silvestres Baleares:* 100 (1994): H. foetidus var. balearicus.

HELONIAS (Melanthiac.).
alpina *(F.T. Wang & T. Tang) N. Tanaka in J. Jap. Bot.*, 73(2): 106 (1998): Ypsilandra alpina.
breviscapa *(Maxim.) N. Tanaka in J. Jap. Bot.*, 73(2): 111 (1998): Heloniopsis breviscapa.
kawanoi *(Koidz.) N. Tanaka in J. Jap. Bot.*, 73(2): 109 (1998): Sugerokia kawanoi.
leucantha *(Koidz.) N. Tanaka in J. Jap. Bot.*, 73(2): 107 (1998): Sugerokia leucantha.
orientalis *(Thunb.) N. Tanaka in J. Jap. Bot.*, 73(2): 109 (1998): Scilla orientalis.
thibetica *(Franch.) N. Tanaka in J. Jap. Bot.*, 73(2): 106 (1998): Ypsilandra thibetica.
umbellata *(Baker) N. Tanaka in J. Jap. Bot.*, 73(2): 108 (1998): Heloniopsis umbellata.

HELONIAS:-
yunnanensis *(W.W. Sm. & Jeffrey) N. Tanaka in J. Jap. Bot.*, 73(2): 105 (1998): Ypsilandra yunnanensis.

HELONOMA (Orchidac.).
chiropterae *(Szlach.) G. Carnevali & G.A. Romero in G.A. Romero & G. Carnevali, Orchids of Venezuela, ed. 2:* 1136 (2000): Beloglottis chiropterae.

HEMARTHRIA (Gramin.).
depressa *E. van den Heuvel in Blumea*, 45(2): 456 (2000)—Laos, Vietnam.

HEMEROCALLIS (Hemerocallidac.).
dumortieri C. Morren
subsp. middendorffii *(Trautv. & C.A. Mey.) V.N. Voroschilov in A.K. Skvortsov (ed.), Florist. issl. v razn. raionakh SSSR:* 158 (1985): H. middendorffii.
fulva (L.) L.
var. oppositibracteata *H. Kong & Ching J. Wang in Guihaia*, 16(4): 303 (1996)—China North-central (Gansu).
taeanensis *S.S. Kang & M.G. Chung in Syst. Bot.*, 22(3): 428 (1997 publ. 1998)—South Korea.

HEMIBOEA (Gesneriac.).
subcapitata C.B. Clarke
var. guangdongensis *(Z.Yu Li) Z.Yu Li in Novon*, 7(4): 429 (1997 publ. 1998): H. henryi var. guangdongensis.
wangiana *Z.Yu Li in Novon*, 7(4): 429 (1997 publ. 1998)—China South-central (Yunnan).

HEMILOPHIA (Crucif.).
franchetii *I.A. Al-Shehbaz in Adansonia*, 21(2): 241 (1999)—China South-central (Yunnan).
sessilifolia *I.A. Al-Shehbaz, Katsutoshi Arai & H. Ohba in Novon*, 9(1): 8 (1999)—China South-central (Yunnan).

HEMIPHYLACUS (Asphodelac.).
alatostylus *L. Hernández S. in Syst. Bot.*, 20(4): 548 (1995)—Mexico Northeast (San Luis Potosí, Guanajuato, Querétaro).
hintoniorum *L. Hernández S. in Syst. Bot.*, 20(4): 549 (1995)—Mexico Northeast (Nuevo León).
mahindae *L. Hernández S. in Syst. Bot.*, 20(4): 552 (1995)—Mexico Central (Puebla).
novogalicianus *L. Hernández S. in Syst. Bot.*, 20(4): 552 (1995)—Mexico Northeast (Aguascalientes).

HEMIPILIA (Orchidac.).
bidupensis *L.V. Averyanov in Lindleyana*, 14(4): 222 (1999), as 'bidupense'—Vietnam.
kwangsiensis *T. Tang & F.T. Wang ex K.Y. Lang in Guihaia*, 18(1): 7 (1998)—China Southeast (Guangxi).

HEMIPOGON (Asclepiadac.).
irwinii *J. Fontella Pereira & R.J. Paixão in Bradea*, 6(48): 424 (1996)—Brazil West-central (Goiás).

HEMITOMUS (Scrophulariac.).
acutifolius *(Ruiz & Pav.) Jacques in Ann. Fl. Pomone*, 1832–33: 249 (1833): Alonsoa acutifolia.

HEMIZYGIA (Labiat.).
comosa *(Wright ex Benth.) A.J. Paton in Kew Bull.*, 53(2): 484 (1998): Orthosiphon comosus.
madagascariensis *A.J. Paton & I.C. Hedge in Fl. Madagascar*, fam. 175: 110 (1998)—Madagascar.

HENCKELIA (Gesneriac.).
sect. Didymanthus *(C.B. Clarke) A. Weber & B.L. Burtt in Beitr. Biol. Pfl.*, 70(2–3): 334 (1997–98 publ. 1998): Didymocarpus sect. Didymanthus.
sect. Glossadenia *A. Weber & B.L. Burtt in Beitr. Biol. Pfl.*, 70(2–3): 335 (1997–98 publ. 1998).
sect. Heteroboea *(Benth.) A. Weber & B.L. Burtt in Beitr. Biol. Pfl.*, 70(2–3): 334 (1997–98 publ. 1998): Didymocarpus sect. Heteroboea.
sect. Loxocarpus *(R.Br.) A. Weber & B.L. Burtt in Beitr. Biol. Pfl.*, 70(2–3): 334 (1997–98 publ. 1998): gen. Loxocarpus.
alba *(Ridl.) A. Weber in Beitr. Biol. Pfl.*, 70(2–3): 338 (1997–98 publ. 1998): Didymocarpus albus.

HENCKELIA:-
albomarginata *(Hemsl.) A. Weber in Beitr. Biol. Pfl.*, 70(2–3): 338 (1997–98 publ. 1998): Didymocarpus albomarginatus.
alternans *(Ridl.) A. Weber in Beitr. Biol. Pfl.*, 70(2–3): 338 (1997–98 publ. 1998): Didymocarpus alternans.
alternifolia *(C.B. Clarke) B.L. Burtt in Beitr. Biol. Pfl.*, 70(2–3): 339 (1997–98 publ. 1998): Didymocarpus alternifolius.
amoena *(C.B. Clarke) B.L. Burtt in Checklist Flow. Pl. & Gymn. Brunei Darussalam (ed. M.J.E. Coode et al.):* 438 (1996): Didymocarpus amoenus.
angustifolia *(C.B. Clarke) B.L. Burtt in Beitr. Biol. Pfl.*, 70(2–3): 339 (1997–98 publ. 1998): Didymocarpus angustifolius.
anthonyi *(Kiew) A. Weber in Beitr. Biol. Pfl.*, 70(2–3): 339 (1997–98 publ. 1998): Didymocarpus anthonyi.
argentea *(B.L. Burtt) B.L. Burtt in Beitr. Biol. Pfl.*, 70(2–3): 339 (1997–98 publ. 1998): Loxocarpus argenteus.
ascendens *(Ridl.) A. Weber in Beitr. Biol. Pfl.*, 70(2–3): 339 (1997–98 publ. 1998): Didymocarpus ascendens.
atrosanguinea *(Ridl.) A. Weber in Beitr. Biol. Pfl.*, 70(2–3): 340 (1997–98 publ. 1998): Didymocarpus atrosanguineus.
bakoensis *(B.L. Burtt) B.L. Burtt in Beitr. Biol. Pfl.*, 70(2–3): 340 (1997–98 publ. 1998): Didymocarpus bakoensis.
var. tenuior *(C.B. Clarke) B.L. Burtt in Beitr. Biol. Pfl.*, 70(2–3): 340 (1997–98 publ. 1998): Didymocarpus leptocalyx var. tenuior.
battamensis *(Ridl.) B.L. Burtt in Beitr. Biol. Pfl.*, 70(2–3): 340 (1997–98 publ. 1998): Didymocarpus battamensis.
beccarii *(C.B. Clarke) B.L. Burtt in Beitr. Biol. Pfl.*, 70(2–3): 340 (1997–98 publ. 1998): Didymocarpus beccarii.
bombycina *(Ridl.) A. Weber in Beitr. Biol. Pfl.*, 70(2–3): 340 (1997–98 publ. 1998): Didymocarpus bombycinus.
breviflora *(Ridl.) A. Weber in Beitr. Biol. Pfl.*, 70(2–3): 340 (1997–98 publ. 1998): Didissandra breviflora.
browniana *A. Weber in Beitr. Biol. Pfl.*, 70(2–3): 341 (1997–98 publ. 1998), nom. nov.: Loxocarpus incanus *R.Br.*
bullata *(C.B. Clarke) B.L. Burtt in Beitr. Biol. Pfl.*, 70(2–3): 341 (1997–98 publ. 1998): Didymocarpus bullatus.
caelestis *(Ridl.) A. Weber in Beitr. Biol. Pfl.*, 70(2–3): 341 (1997–98 publ. 1998): Codonoboea caelestis.
caerulea *(Ridl.) A. Weber in Beitr. Biol. Pfl.*, 70(2–3): 341 (1997–98 publ. 1998): Didymocarpus caeruleus.
calcarea *(Ridl.) A. Weber in Beitr. Biol. Pfl.*, 70(2–3): 341 (1997–98 publ. 1998): Didymocarpus calcareus.
castaneifolia *(Ridl.) A. Weber in Beitr. Biol. Pfl.*, 70(2–3): 341 (1997–98 publ. 1998): Didymocarpus castaneifolius.
caulescens *((B.L. Burtt) A. Weber & B.L. Burtt in Beitr. Biol. Pfl.*, 70(2–3): 342 (1997–98 publ. 1998): Loxocarpus caulescens.
codonion *(Kiew) A. Weber in Beitr. Biol. Pfl.*, 70(2–3): 342 (1997–98 publ. 1998): Didymocarpus codonion.
conicapsularis *(C.B. Clarke) B.L. Burtt in Beitr. Biol. Pfl.*, 70(2–3): 342 (1997–98 publ. 1998): Didymocarpus conicapsularis.
coodei *B.L. Burtt in Checklist Flow. Pl. & Gymn. Brunei Darussalam (ed. M.J.E. Coode et al.):* 437 (1996)—Borneo (Brunei).
corneri *(Kiew) A. Weber in Beitr. Biol. Pfl.*, 70(2–3): 342 (1997–98 publ. 1998): Didymocarpus corneri.
craspedodroma *(Kiew) A. Weber in Beitr. Biol. Pfl.*, 70(2–3): 342 (1997–98 publ. 1998): Didymocarpus craspedodromus.
crenata *(Baker) B.L. Burtt in Checklist Flow. Pl. & Gymn. Brunei Darussalam (ed. M.J.E. Coode et al.):* 438 (1996): Didymocarpus crenatus.
crocea *(Ridl.) A. Weber in Beitr. Biol. Pfl.*, 70(2–3): 343 (1997–98 publ. 1998): Didymocarpus croceus.

HENCKELIA:-

curtisii *(Ridl.) A. Weber in Beitr. Biol. Pfl.*, 70(2–3): 343 (1997–98 publ. 1998): Didymocarpus crinitus var. curtisii.
davisonii *(Kiew) A. Weber in Beitr. Biol. Pfl.*, 70(2–3): 343 (1997–98 publ. 1998): Didymocarpus davisonii.
dawnii *(Kiew) A. Weber in Beitr. Biol. Pfl.*, 70(2–3): 343 (1997–98 publ. 1998): Didymocarpus dawnii.
densifolia *(Ridl.) A. Weber in Beitr. Biol. Pfl.*, 70(2–3): 343 (1997–98 publ. 1998): Didymocarpus densifolius.
dentata *(Ridl.) B.L. Burtt in Beitr. Biol. Pfl.*, 70(2–3): 343 (1997–98 publ. 1998): Didymocarpus dentatus.
descendens *(B.L. Burtt) B.L. Burtt in Beitr. Biol. Pfl.*, 70(2–3): 343 (1997–98 publ. 1998): Platyadenia descendens.
diffusa *B.L. Burtt in Checklist Flow. Pl. & Gymn. Brunei Darussalam (ed. M.J.E. Coode et al.)*: 437 (1996)—Borneo (Brunei).
doryphylla *(B.L. Burtt) A. Weber in Beitr. Biol. Pfl.*, 70(2–3): 344 (1997–98 publ. 1998): Didymocarpus doryphyllus.
elegans *(C.B. Clarke) B.L. Burtt in Beitr. Biol. Pfl.*, 70(2–3): 344 (1997–98 publ. 1998): Didymocarpus elegans.
ericiflora *(Ridl.) A. Weber in Beitr. Biol. Pfl.*, 70(2–3): 344 (1997–98 publ. 1998): Didymocarpus ericiflorus.
ericii *A. Weber in Beitr. Biol. Pfl.*, 70(2–3): 344 (1997–98 publ. 1998), nom. nov.: Loxocarpus holttumii *M.R. Hend.*
falcata *(Kiew) A. Weber in Beitr. Biol. Pfl.*, 70(2–3): 344 (1997–98 publ. 1998): Didymocarpus falcatus.
fasciata *(Ridl.) A. Weber in Beitr. Biol. Pfl.*, 70(2–3): 344 (1997–98 publ. 1998): Didymocarpus fasciatus.
fischeri *(Gamble) A. Weber & B.L. Burtt in Beitr. Biol. Pfl.*, 70(2–3): 344 (1997–98 publ. 1998): Didymocarpus fischeri.
flava *(Ridl.) A. Weber in Beitr. Biol. Pfl.*, 70(2–3): 344 (1997–98 publ. 1998): Didymocarpus flavus.
flavescens *(Ridl.) A. Weber in Beitr. Biol. Pfl.*, 70(2–3): 344 (1997–98 publ. 1998): Didymocarpus flavescens.
flavobrunnea *(Ridl.) A. Weber in Beitr. Biol. Pfl.*, 70(2–3): 344 (1997–98 publ. 1998): Didymocarpus flavobrunneus.
floccosa *(Thwaites) A. Weber & B.L. Burtt in Beitr. Biol. Pfl.*, 70(2–3): 344 (1997–98 publ. 1998): Didymocarpus floccosus.
floribunda *(M.R. Hend.) A. Weber in Beitr. Biol. Pfl.*, 70(2–3): 345 (1997–98 publ. 1998): Paraboea floribunda.
follicularis *(C.B. Clarke) B.L. Burtt in Beitr. Biol. Pfl.*, 70(2–3): 345 (1997–98 publ. 1998): Didymocarpus follicularis.
gambleana *(C.E.C. Fisch.) A. Weber & B.L. Burtt in Beitr. Biol. Pfl.*, 70(2–3): 345 (1997–98 publ. 1998): Didymocarpus gambleanus.
gardneri *B.L. Burtt in Checklist Flow. Pl. & Gymn. Brunei Darussalam (ed. M.J.E. Coode et al.)*: 437 (1996)—Borneo (Sabah).
geitleri *(A. Weber) A. Weber in Beitr. Biol. Pfl.*, 70(2–3): 345 (1997–98 publ. 1998): Didymocarpus geitleri.
glabrata *(Ridl.) A. Weber in Beitr. Biol. Pfl.*, 70(2–3): 345 (1997–98 publ. 1998): Didymocarpus glabratus.
gracilipes *(C.B. Clarke) B.L. Burtt in Checklist Flow. Pl. & Gymn. Brunei Darussalam (ed. M.J.E. Coode et al.)*: 438 (1996): Didymocarpus gracilipes.
heterophylla *(Ridl.) A. Weber in Beitr. Biol. Pfl.*, 70(2–3): 346 (1997–98 publ. 1998): Didymocarpus heterophyllus.
hirsuta *(Ridl.) A. Weber in Beitr. Biol. Pfl.*, 70(2–3): 346 (1997–98 publ. 1998): Didymocarpus hirsutus.
hirta *(Ridl.) A. Weber in Beitr. Biol. Pfl.*, 70(2–3): 346 (1997–98 publ. 1998): Didymocarpus hirtus.
hispida *(Ridl.) A. Weber in Beitr. Biol. Pfl.*, 70(2–3): 346 (1997–98 publ. 1998): Didymocarpus hispidus.
holttumii *(M.R. Hend.) A. Weber in Beitr. Biol. Pfl.*, 70(2–3): 346 (1997–98 publ. 1998): Paraboea holttumii.

HENCKELIA:-

humboldtiana *(Gardner) A. Weber & B.L. Burtt in Beitr. Biol. Pfl.*, 70(2–3): 346 (1997–98 publ. 1998): Didymocarpus humboldtianus.
humilis *(Miq.) A. Weber & B.L. Burtt in Beitr. Biol. Pfl.*, 70(2–3): 347 (1997–98 publ. 1998): Chirita humilis.
inaequalis *(Ridl.) A. Weber in Beitr. Biol. Pfl.*, 70(2–3): 347 (1997–98 publ. 1998): Didymocarpus inaequalis.
innominata *(B.L. Burtt) A. Weber & B.L. Burtt in Beitr. Biol. Pfl.*, 70(2–3): 347 (1997–98 publ. 1998): Didymocarpus innominatus.
johorica *(Ridl.) A. Weber in Beitr. Biol. Pfl.*, 70(2–3): 347 (1997–98 publ. 1998): Didissandra johorica.
kjellbergii *B.L. Burtt in Beitr. Biol. Pfl.*, 70(2–3): 378 (1997–98 publ. 1998)—Sulawesi, New Guinea (Irian Jaya).
koerperi *(B.L. Burtt) A. Weber & B.L. Burtt in Beitr. Biol. Pfl.*, 70(2–3): 347 (1997–98 publ. 1998): Didymocarpus koerperi.
kompsoboea *(C.B. Clarke) B.L. Burtt in Beitr. Biol. Pfl.*, 70(2–3): 347 (1997–98 publ. 1998): Didymocarpus kompsoboea.
lanceolata *(C.B. Clarke) B.L. Burtt in Beitr. Biol. Pfl.*, 70(2–3): 348 (1997–98 publ. 1998): Didymocarpus lanceolatus.
lancifolia *(M.R. Hend.) A. Weber in Beitr. Biol. Pfl.*, 70(2–3): 348 (1997–98 publ. 1998): Didymocarpus lancifolius.
leiophylla *(Kiew) A. Weber in Beitr. Biol. Pfl.*, 70(2–3): 348 (1997–98 publ. 1998): Didymocarpus leiophyllus.
leptocalyx *(C.B. Clarke) B.L. Burtt in Beitr. Biol. Pfl.*, 70(2–3): 348 (1997–98 publ. 1998): Didymocarpus leptocalyx.
leucantha *(Kiew) A. Weber in Beitr. Biol. Pfl.*, 70(2–3): 348 (1997–98 publ. 1998): Didymocarpus leucanthus.
leucocodon *(Ridl.) A. Weber in Beitr. Biol. Pfl.*, 70(2–3): 348 (1997–98 publ. 1998): Paraboea leucocodon.
lilacina *(Ridl.) A. Weber in Beitr. Biol. Pfl.*, 70(2–3): 348 (1997–98 publ. 1998): Didymocarpus lilacinus.
longipes *(C.B. Clarke) A. Weber in Beitr. Biol. Pfl.*, 70(2–3): 349 (1997–98 publ. 1998): Didymocarpus longipes.
longipetiolata *(B.L. Burtt) B.L. Burtt in Beitr. Biol. Pfl.*, 70(2–3): 349 (1997–98 publ. 1998): Loxocarpus longipetiolatus.
lyrata *(Wight) A. Weber & B.L. Burtt in Beitr. Biol. Pfl.*, 70(2–3): 349 (1997–98 publ. 1998): Didymocarpus lyratus.
macrostachya *(E. Barnes) A. Weber & B.L. Burtt in Beitr. Biol. Pfl.*, 70(2–3): 349 (1997–98 publ. 1998): Didymocarpus macrostachyus.
malayana *(Hook. f.) A. Weber in Beitr. Biol. Pfl.*, 70(2–3): 349 (1997–98 publ. 1998): Didymocarpus malayanus.
marginata *(C.B. Clarke) A. Weber in Beitr. Biol. Pfl.*, 70(2–3): 349 (1997–98 publ. 1998): Didymocarpus marginatus.
meeboldii *(W.W. Sm. & Ramaswami) A. Weber & B.L. Burtt in Beitr. Biol. Pfl.*, 70(2–3): 349 (1997–98 publ. 1998): Didymocarpus meeboldii.
meijeri *(B.L. Burtt) A. Weber & B.L. Burtt in Beitr. Biol. Pfl.*, 70(2–3): 350 (1997–98 publ. 1998): Loxocarpus meijeri.
miniata *(Kiew) A. Weber in Beitr. Biol. Pfl.*, 70(2–3): 350 (1997–98 publ. 1998): Didymocarpus miniatus.
minima *(Ridl.) A. Weber in Beitr. Biol. Pfl.*, 70(2–3): 350 (1997–98 publ. 1998): Loxocarpus minimus.
missionis *(R.Br.) A. Weber & B.L. Burtt in Beitr. Biol. Pfl.*, 70(2–3): 350 (1997–98 publ. 1998): Didymocarpus missionis.
modesta *(Ridl.) A. Weber in Beitr. Biol. Pfl.*, 70(2–3): 350 (1997–98 publ. 1998): Didymocarpus modestus.
murutorum *(B.L. Burtt) B.L. Burtt in Beitr. Biol. Pfl.*, 70(2–3): 350 (1997–98 publ. 1998): Didymocarpus murutorum.
myricifolia *(Ridl.) B.L. Burtt in Checklist Flow. Pl. & Gymn. Brunei Darussalam (ed. M.J.E. Coode et al.)*: 438 (1996): Didymocarpus myricifolius.

HENCKELIA:-

nana *A. Weber in Beitr. Biol. Pfl.*, 70(2–3): 350 (1997–98 publ. 1998), nom. nov.: Didymocarpus pumilus *Ridl.*
nervosa *(C.B. Clarke) B.L. Burtt in Beitr. Biol. Pfl.*, 70(2–3): 350 (1997–98 publ. 1998): Didymocarpus nervosus.
nitida *(Kiew & A. Weber) A. Weber in Beitr. Biol. Pfl.*, 70(2–3): 351 (1997–98 publ. 1998): Didymocarpus nitidus.
nivea *(Kiew) A. Weber in Beitr. Biol. Pfl.*, 70(2–3): 351 (1997–98 publ. 1998): Codonoboea nivea.
ovalifolia *(Wight) A. Weber & B.L. Burtt in Beitr. Biol. Pfl.*, 70(2–3): 351 (1997–98 publ. 1998): Didymocarpus ovalifolius.
pagonensis *B.L. Burtt in Checklist Flow. Pl. & Gymn. Brunei Darussalam (ed. M.J.E. Coode et al.)*: 437 (1996)—Borneo (Brunei, Sarawak).
papillosa *(M.R. Hend.) A. Weber in Beitr. Biol. Pfl.*, 70(2–3): 351 (1997–98 publ. 1998): Loxocarpus papillosus.
parviflora *(Ridl.) A. Weber in Beitr. Biol. Pfl.*, 70(2–3): 352 (1997–98 publ. 1998): Didymocarpus parviflorus.
pectinata *(Oliv.) A. Weber in Beitr. Biol. Pfl.*, 70(2–3): 352 (1997–98 publ. 1998): Didymocarpus pectinatus.
petiolaris *(C.B. Clarke) B.L. Burtt in Checklist Flow. Pl. & Gymn. Brunei Darussalam (ed. M.J.E. Coode et al.)*: 438 (1996): Didymocarpus petiolaris.
platypus *(C.B. Clarke) A. Weber in Beitr. Biol. Pfl.*, 70(2–3): 352 (1997–98 publ. 1998): Didymocarpus platypus.
pleuropogon *(B.L. Burtt) B.L. Burtt in Checklist Flow. Pl. & Gymn. Brunei Darussalam (ed. M.J.E. Coode et al.)*: 438 (1996): Didymocarpus pleuropogon.
polyanthoides *(Kiew) A. Weber in Beitr. Biol. Pfl.*, 70(2–3): 352 (1997–98 publ. 1998): Didymocarpus polyanthoides.
primulina *(Ridl.) A. Weber in Beitr. Biol. Pfl.*, 70(2–3): 352 (1997–98 publ. 1998): Didymocarpus primulinus.
procumbens *B.L. Burtt in Beitr. Biol. Pfl.*, 70(2–3): 353 (1997–98 publ. 1998), nom. nov.: Loxocarpus repens *B.L. Burtt.*
pulchella *(Ridl.) A. Weber in Beitr. Biol. Pfl.*, 70(2–3): 353 (1997–98 publ. 1998): Didymocarpus pulchellus.
punctata *(C.B. Clarke) B.L. Burtt in Beitr. Biol. Pfl.*, 70(2–3): 353 (1997–98 publ. 1998): Didymocarpus punctatus.
puncticulata *(Ridl.) A. Weber in Beitr. Biol. Pfl.*, 70(2–3): 353 (1997–98 publ. 1998): Didymocarpus puncticulatus.
pyroliflora *(Ridl.) A. Weber in Beitr. Biol. Pfl.*, 70(2–3): 353 (1997–98 publ. 1998): Didymocarpus pyroliflorus.
quinquevulnera *(Ridl.) A. Weber in Beitr. Biol. Pfl.*, 70(2–3): 353 (1997–98 publ. 1998): Didymocarpus quinquevulnerus.
ramosa *(Ridl.) A. Weber in Beitr. Biol. Pfl.*, 70(2–3): 353 (1997–98 publ. 1998): Didymocarpus ramosus.
repens *(Bedd.) A. Weber & B.L. Burtt in Beitr. Biol. Pfl.*, 70(2–3): 354 (1997–98 publ. 1998): Didymocarpus repens.
reptans (Jack) Spreng.
var. monticola *(Ridl.) A. Weber in Beitr. Biol. Pfl.*, 70(2–3): 354 (1997–98 publ. 1998): Didymocarpus reptans var. monticola.
reticulosa *(C.B. Clarke) A. Weber in Beitr. Biol. Pfl.*, 70(2–3): 354 (1997–98 publ. 1998): Didymocarpus reticulosus.
ridleyana *A. Weber in Beitr. Biol. Pfl.*, 70(2–3): 354 (1997–98 publ. 1998), nom. nov.: Didymocarpus grandiflorus *Ridl.*
robinsonii *(Ridl.) A. Weber in Beitr. Biol. Pfl.*, 70(2–3): 354 (1997–98 publ. 1998): Didymocarpus robinsonii.
rubiginosa *(Ridl.) A. Weber in Beitr. Biol. Pfl.*, 70(2–3): 355 (1997–98 publ. 1998): Paraboea rubiginosa.
rufescens *(C.B. Clarke) B.L. Burtt in Beitr. Biol. Pfl.*, 70(2–3): 355 (1997–98 publ. 1998): Didymocarpus rufescens.
rugosa *(Ridl.) A. Weber in Beitr. Biol. Pfl.*, 70(2–3): 355 (1997–98 publ. 1998): Didymocarpus rugosus.

HENCKELIA:-
salicina *(Ridl.) A. Weber in Beitr. Biol Pfl.*, 70(2–3): 355 (1997–98 publ. 1998): Didymocarpus salicinus.
salicinoides *(Kiew) A. Weber in Beitr. Biol. Pfl.*, 70(2–3): 355 (1997–98 publ. 1998): Didymocarpus salicinoides.
scabrinervia *(C.B. Clarke) B.L. Burtt in Beitr. Biol. Pfl.*, 70(2–3): 355 (1997–98 publ. 1998): Didymocarpus scabrinervius.
scortechinii *(Ridl.) A. Weber in Beitr. Biol. Pfl.*, 70(2–3): 356 (1997–98 publ. 1998): Paraboea scortechinii.
semitorta *(C.B. Clarke) A. Weber in Beitr. Biol. Pfl.*, 70(2–3): 356 (1997–98 publ. 1998): Didymocarpus semitortus.
sericea *(Ridl.) A. Weber in Beitr. Biol. Pfl.*, 70(2–3): 356 (1997–98 publ. 1998): Didymocarpus sericeus.
serrata *(R.Br.) A. Weber & B.L. Burtt in Beitr. Biol. Pfl.*, 70(2–3): 356 (1997–98 publ. 1998): Didymocarpus serratus.
serratifolia *(Ridl.) A. Weber in Beitr. Biol. Pfl.*, 70(2–3): 356 (1997–98 publ. 1998): Didymocarpus serratifolius.
simplex *(Kraenzl.) B.L. Burtt in Beitr. Biol. Pfl.*, 70(2–3): 356 (1997–98 publ. 1998): Didymocarpus simplex.
soldanella *(Ridl.) A. Weber in Beitr. Biol. Pfl.*, 70(2–3): 356 (1997–98 publ. 1998): Didymocarpus soldanella.
stapfii *(Kraenzl.) B.L. Burtt in Beitr. Biol. Pfl.*, 70(2–3): 356 (1997–98 publ. 1998): Didymocarpus stapfii.
stenophylla *A. Weber in Beitr. Biol. Pfl.*, 70(2–3): 357 (1997–98 publ. 1998), nom. nov.: Loxocarpus angustifolius *Ridl.*
stolonifera *(Kiew) A. Weber in Beitr. Biol. Pfl.*, 70(2–3): 357 (1997–98 publ. 1998): Didymocarpus stoloniferus.
taeniophylla *B.L. Burtt in Checklist Flow. Pl. & Gymn. Brunei Darussalam (ed. M.J.E. Coode et al.)*: 438 (1996)—Borneo (Brunei).
tahanica *(B.L. Burtt) A. Weber in Beitr. Biol. Pfl.*, 70(2–3): 357 (1997–98 publ. 1998): Didymocarpus tahanicus.
teres *(C.B. Clarke) B.L. Burtt in Beitr. Biol. Pfl.*, 70(2–3): 357 (1997–98 publ. 1998): Didymocarpus teres.
tiumanica *(Ridl.) A. Weber in Beitr. Biol. Pfl.*, 70(2–3): 357 (1997–98 publ. 1998): Paraboea tiumanica.
tunkui *(Kiew) A. Weber in Beitr. Biol. Pfl.*, 70(2–3): 357 (1997–98 publ. 1998): Loxocarpus tunkui.
urticoides *A. Weber in Beitr. Biol. Pfl.*, 70(2–3): 358 (1997–98 publ. 1998), nom. nov.: Didymocarpus urticifolius *Ridl.*
venusta *(Ridl.) A. Weber in Beitr. Biol. Pfl.*, 70(2–3): 358 (1997–98 publ. 1998): Didymocarpus venustus.
verbeniflos *(C.B. Clarke) B.L. Burtt in Beitr. Biol. Pfl.*, 70(2–3): 358 (1997–98 publ. 1998): Didymocarpus verbeniflos.
violoides *(C.B. Clarke) B.L. Burtt in Checklist Flow. Pl. & Gymn. Brunei Darussalam (ed. M.J.E. Coode et al.)*: 438 (1996): Didymocarpus violoides.
virginea *(B.L. Burtt) B.L. Burtt in Beitr. Biol. Pfl.*, 70(2–3): 358 (1997–98 publ. 1998): Didymocarpus virgineus.
viscida *(Ridl.) A. Weber in Beitr. Biol. Pfl.*, 70(2–3): 359 (1997–98 publ. 1998): Didymocarpus viscidus.
vulcanica *(Ridl.) A. Weber & B.L. Burtt in Beitr. Biol. Pfl.*, 70(2–3): 359 (1997–98 publ. 1998): Didymocarpus vulcanicus.
wightii *(C.B. Clarke) A. Weber & B.L. Burtt in Beitr. Biol. Pfl.*, 70(2–3): 359 (1997–98 publ. 1998): Didymocarpus rottlerianus var. wightii.
woodii *(Merr.) A. Weber & B.L. Burtt in Beitr. Biol. Pfl.*, 70(2–3): 359 (1997–98 publ. 1998): Didymocarpus woodii.
yongii *(Kiew) A. Weber in Beitr. Biol. Pfl.*, 70(2–3): 359 (1997–98 publ. 1998): Didymocarpus yongii.
zeylanica *(R.Br.) A. Weber & B.L. Burtt in Beitr. Biol. Pfl.*, 70(2–3): 359 (1997–98 publ. 1998): Didymocarpus zeylanicus.

HENOPHYTON (Crucif.).
zygarrhenum *(Maire) C. Gómez Campo in An. Jard. Bot. Madrid*, 56(2): 381 (1998): Oudneya zygarrhena.

HENRIETTEA (Melastomatac.).
hotteana *(Urb. & Ekman) Alain in Sida*, 18(4): 1026 (1999): Ossaea hotteana.

HERACLEUM (Umbellif.).
sect. Plurivittata *F.T. Pu & X.J. He in Acta Bot. Yunnanica*, 20(3): 299 (1998).
bhutanicum *M.F. Watson in Edinburgh J. Bot.*, 53(1): 131 (1996)—Bhutan.
deqenense *R.H. Shan & Z.H. Pan in Fl. Yunnanica*, 7: 629 (1997)—China South-central (Yunnan).
lanatum Michx.
subsp. woroschilowii *(Gorovoj) V.N. Voroschilov in A.K. Skvortsov (ed.), Florist. issl. v razn. raïonakh SSSR*: 185 (1985): H. woroschilowii.
millefolium
var. glabrum *J.Q. Liu ex J.T. Pan in Fl. Qinghaiica*, 2: 434 (1999)—Qinghai.
sphondylium
subsp. glabrum *(Huth) J. Holub in Preslia*, 68(3): 285 (1996 publ. 1997): H. sphondylium var. glabrum.
woodii *M.F. Watson in Edinburgh J. Bot.*, 53(1): 133 (1996)—Bhutan, Sikkim, Nepal.

HERNANDIA (Hernandiac.).
ovigera L.
subsp. stokesii *(F. Br.) J. Florence in Adansonia*, 21(1): 48 (1999): H. ovigera var. stokesii.
sonora
var. nymphaeifolia *(K. Presl) F.R. Fosberg in Micronesica*, 23(2): 142 (1990): Biasolettia nymphaeifolia.

HERNIARIA (Illecebrac.).
maroccana *(Font Quer) A.M. Romo in Soc. Éch. Pl. Vasc. Eur. Bassin Médit. Bull.*, 28: 114 (2000): Herniaria frigida var. maroccana.
× montenegrina *J.M. Pérez Dacosta & G. Mateo Sanz in Flora Montiberica*, 9: 49 (1998)—Spain. (H. glabra var. glabra × latifolia).
× urrutiae *J.M. Pérez Dacosta & P.M. Uribe-Echebarria Díaz in Estud. Mus. Cienc. Nat. Alava*, 12: 78 (1997)—Spain. (H. cinerea × scabrida subsp. scabrida).

HERSCHELIANTHE (Orchidac.).
chimanimaniensis *(H.P. Linder) K. Senghas in Schlechter Orchideen*, ed. 3, I/C(40): 2525 (2000), isonym; *see H.P. Linder in Kew Bull.*, 40(1): 129 (1985): Herschelia chimanimaniensis.
praecox *(H.P. Linder) K. Senghas in Schlechter Orchideen*, ed. 3, I/C(40): 2525 (2000), isonym; *see H.P. Linder in Kew Bull.*, 40(1): 127 (2000): Herschelia praecox.

HESPERALBIZIA *R.C. Barneby & J.W. Grimes, Silk Tree, Guanacaste, Monkey'sEarring (Mem. New York Bot. Gard., 74(1))*: 112 (1996). *LEGUMINOSAE*
occidentalis *(Brandegee) R.C. Barneby & J.W. Grimes, Silk Tree, Guanacaste, Monkey's Earring (Mem. New York Bot. Gard., 74(1))*: 112 (1996): Albizia occidentalis.

HESPERALOE (Agavac.).
campanulata *G.D. Starr in Madroño*, 44(3): 285 (1997 publ. 1998)—Mexico Northeast (Nuevo León).
funifera (K. Koch) Trel.
subsp. chiangii *G.D. Starr in Madroño*, 44(3): 289 (1997 publ. 1998)—Mexico Northeast (San Luis Potosí).
tenuifolia *G.D. Starr in Madroño*, 44(3): 293 (1997 publ. 1998)—Mexico Northwest (Sonora).

HESPERANTHA (Iridac.).
coccinea *(Backh. & Harv.) P. Goldblatt & J.C. Manning in Novon*, 6(3): 263 (1996): Schizostylis coccinea.

HESPERASTER (Loasac.).
nudus *(Pursh) Cockerell in Torreya*, 1: 143 (1901), in I.K. Supp. 3 as 'midus' in error: Bartonia nuda.†

HESPEROMELES (Rosac.).
obtusifolia (Pers.) Lindl.
var. microphylla *(Wedd.) K. Romoleroux in Fl. Ecuador*, 56(79): 150 (1996): H. pernettyoides var. microphylla.

HESSEA (Amaryllidac.).
tenuipedicellata *D.A. Snijman in Novon*, 9(1): 107 (1999)—Cape Province.

HETAERIA (Orchidac.).
finlaysoniana *G. Seidenfaden, Contrib. Orchid Fl. Thailand XIII*: 10 (1997), nom. nov.: Goodyera elongata *Lindl.*
pelota *N. Pearce & P.J. Cribb in Edinburgh J. Bot.*, 56(2): 278 (1999)—Bhutan.

HETEROCENTRON (Melastomatac.).
chimalapanum *C.A. Todzia in Lundellia*, 2: 142 (1999)—Mexico Southwest (Oaxaca).
evansii *F. Almeda in Novon*, 9(2): 127 (1999)—Honduras.

HETEROMORPHA (Umbellif.).
arborescens (Spreng.) Cham. & Schltdl.
var. frutescens *P.J.D. Winter in Kew Bull.*, 51(2): 246 (1996)—Transvaal, Natal, Swaziland, Namibia, Zimbabwe, Mozambique, Malawi.
var. montana *P.J.D. Winter in Kew Bull.*, 51(2): 246 (1996)—Malawi, Tanzania, Mozambique, Zimbabwe.
occidentalis *P.J.D. Winter in Kew Bull.*, 51(2): 251 (1996)—Cameroon, Nigeria, Angola, Zaire, Zambia.
stenophylla Welw. ex Schinz
var. transvaalensis *(Schltr. & H. Wolff) P.J.D. Winter in Kew Bull.*, 51(2): 250 (1996): H. transvaalensis.

HETEROPANAX (Araliac.).
balanseana *C.B. Shang in Adansonia*, 19(1): 78 (1997)—Vietnam.
hainanensis *C.B. Shang in Adansonia*, 19(1): 80 (1997)—Hainan.
phanrangensis *C.B. Shang in Adansonia*, 19(1): 78 (1997)—Vietnam.

HETEROPAPPUS (Compos.).
altaicus (Willd.) Novopokr.
subsp. appressifolius *E.A. Koroljuk in Fl. Sibir. (Asterac.)*, 13: 23 (1997 publ. 1998)—Chita.
var. canescens *(Nees) E.A. Koroljuk in Fl. Sibir. (Asterac.)*, 13: 23 (1997 publ. 1998): Calimeris canescens.
hispidus (Thunb.) Less.
f. albiflora *Y.N. Lee, Fl. Korea*: 1162 (1996), without type—Locality not stated.

HETEROPOLYGONATUM *M.N. Tamura & M. Ogisu in Kew Bull.*, 52(4): 950 (1997). *CONVALLARIACEAE*
ginfushanicum *(F.T. Wang & T. Tang) M.N. Tamura, S.C. Chen & N.J. Turland in Novon*, 10(2): 157 (2000): Smilacina ginfushanica.
pendulum *(Z.G. Liu & X.H. Hu) M.N. Tamura & M. Ogisu in Kew Bull.*, 52(4): 951 (1997): Polygonatum pendulum.
roseolum *M.N. Tamura & M. Ogisu in Kew Bull.*, 52(4): 951 (1997)—China Southeast (Guangxi).
xui *W.K. Bao & M.N. Tamura in Acta Phytotax. Geobot.*, 49(2): 143 (1998 publ. 1999)—China South-central (Sichuan).

HETEROPTERYS (Malpighiac.).
guianensis *W.R. Anderson in Contrib. Univ. Mich. Herb.*, 22: 10 (1999)—Guyana.
hatschbachii *W.R. Anderson in Contrib. Univ. Mich. Herb.*, 22: 12 (1999)—Brazil West-central (Mato Grosso do Sul, Mato Grosso).
hoffmanii *W.R. Anderson in Contrib. Univ. Mich. Herb.*, 21: 57 (1997)—Guyana.
laevifolia *(A. Juss.) R.C. Srivastava in Nation. Acad. Sci. Lett.*, 7(11): 325 (1984), as 'Heteropteris': Banisteria laevifolia.
marginata *W.R. Anderson in Contrib. Univ. Mich. Herb.*, 22: 14 (1999)—Brazil West-central (Mato Grosso).

HETEROPTERYS:-
molesta *W.R. Anderson in Contrib. Univ. Mich. Herb.*, 21: 59 (1997)—Venezuela.
mulgurae *W.R. Anderson in Novon*, 8(3): 215 (1998)—Argentina Northeast (Misiones), Brazil South (Paraná).
schulziana *W.R. Anderson in Novon*, 8(3): 216 (1998)—Argentina Northwest (Salta).
velutina *W.R. Anderson in Contrib. Univ. Mich. Herb.*, 21: 61 (1997)—Guyana.

HETEROSAMARA (Polygalac.).
bennae *(Jacq.-Fél.) J.A.R. Paiva in Fontqueria*, 50: 127 (1998): Polygala bennae.
cabrae *(Chodat) J.A.R. Paiva in Fontqueria*, 50: 128 (1998): Polygala cabrae.
cardiocarpa *(Kurz) J.A.R. Paiva in Fontqueria*, 50: 133 (1998): Polygala cardiocarpa.
carrissoana *(Exell & Mendonça) J.A.R. Paiva in Fontqueria*, 50: 129 (1998): Polygala carrissoana.
engleri *(Chodat) J.A.R. Paiva in Fontqueria*, 50: 125 (1998): Polygala engleri.
furcata *(Royle) J.A.R. Paiva in Fontqueria*, 50: 131 (1998): Polygala furcata.
galpinii *(Hook. f.) J.A.R. Paiva in Fontqueria*, 50: 125 (1998): Polygala galpinii.
isocarpa *(Chodat) J.A.R. Paiva in Fontqueria*, 50: 134 (1998): Polygala isocarpa.
malesiana *(Adema) J.A.R. Paiva in Fontqueria*, 50: 135 (1998): Polygala malesiana.
mannii *(Oliv.) J.A.R. Paiva in Fontqueria*, 50: 126 (1998): Polygala mannii.
pellucida *(Lace) J.A.R. Paiva in Fontqueria*, 50: 129 (1998): Polygala pellucida.
saxicola *(Dunn) J.A.R. Paiva in Fontqueria*, 50: 127 (1998): Polygala saxicola.
tatarinowii *(Regel) J.A.R. Paiva in Fontqueria*, 50: 130 (1998): Polygala tatarinowii.
umbonata *(Craib) J.A.R. Paiva in Fontqueria*, 50: 136 (1998): Polygala umbonata.

HETEROSPATHE (Palmae).
phillipsii *D. Fuller & J.L. Dowe in Principes*, 41(2): 66 (1997)—Fiji.
uniformis *J.L. Dowe in Austral. Syst. Bot.*, 9(1): 40 (1996)—Vanuatu.

HETEROTHECA (Compos.).
echioides (Benth.) Shinners
var. bolanderi *(A. Gray) G.L. Nesom in Phytologia*, 83(1): 17 (1997 publ. 1998): Chrysopsis bolanderi.
var. bolanderioides *(Semple) G.L. Nesom in Phytologia*, 83(1): 17 (1997 publ. 1998): H. sessiliflora var. bolanderioides.
fulcrata(Greene) Shinners
var. amplifolia *(Rydb.) J.C. Semple in Univ. Waterloo Biol. Ser.*, 37: 74 (1996): Chrysopsis amplifolia.
marginata *J.C. Semple in Univ. Waterloo Biol. Ser.*, 37: 83 (1996)—Arizona.
monarchensis *D.A. York, Shevock & J.C. Semple in Univ. Waterloo Biol. Ser.*, 37: 52 (1996)—California.
mucronata V.L. Harms ex B.L. Turner
var. harmsiana *J.C. Semple in Univ. Waterloo Biol. Ser.*, 37: 60 (1996)—Mexico Northeast (Zacatecas, Coahuila, Nuevo León).
sessiliflora (Nutt.) Shinners
var. thiniicola *(Rzed. & E. Ezcurra) G.L. Nesom in Phytologia*, 83(1): 17 (1997 publ. 1998): Haplopappus thiniicola.
shevockii *(Semple) J.C. Semple in Univ. Waterloo Biol. Ser.*, 37: 148 (1996): H. villosa var. shevockii.
villosa (Pursh) Shinners
var. sierrablancensis *J.C. Semple in Univ. Waterloo Biol. Ser.*, 37: 146 (1996)—New Mexico.

HETEROTIS (Melastomatac.).
obamae *J. Lejoly & S. Lisowski in Syst. Geogr. Pl.*, 69(2): 185 (1999)—Equatorial Guinea.

HETEROZEUXINE (Orchidac.).
amboinensis *(J.J. Sm.) T. Hashimoto in Proc. 12th World Orchid Conf., 1987 (eds. K. Saito & R. Tanaka)*: 125 (1987): Haplochilus amboinensis.

HETEROZEUXINE:-
argentea *(Schltr.) T. Hashimoto in Proc. 12th World Orchid Conf., 1987 (eds. K. Saito & R. Tanaka)*: 125 (1987): Zeuxine argentea.
elatior *(Schltr.) T. Hashimoto in Proc. 12th World Orchid Conf., 1987 (eds. K. Saito & R. Tanaka)*: 125 (1987): Zeuxine elatior.
glandulosa *(King & Pantl.) T. Hashimoto in Proc. 12th World Orchid Conf., 1987 (eds. K. Saito & R. Tanaka)*: 125 (1987): Zeuxine glandulosa.
viridiflora *(J.J. Sm.) T. Hashimoto in Proc. 12th World Orchid Conf., 1987 (eds. K. Saito & R. Tanaka)*: 125 (1987): Haplochilus viridiflorum.

HEUCHERA (Saxifragac.).
micrantha Douglas ex Lindl.
var. macropetala *B.G. Shipes & E.F. Wells in Rhodora*, 98(896): 366 (1996 publ. 1997)—Oregon, California.

HEXACHLAMYS (Myrtac.).
tapiraguaiensis *(Barb. Rodr. ex Chodat & Hassl.) J.R. de Mattos in Loefgrenia*, 102: 5 (1993), basionym not validly publ.: Eugenia tapiraguaiensis.

HIBBERTIA (Dilleniac.).
appressa *H.R. Toelken in J. Adelaide Bot. Gard.*, 18(2): 120 (1998)—Tasmania, Victoria.
appressa *H.R. Toelken in J. Adelaide Bot. Gard.*, 19: 54 (2000), homonym; *see Toelken in J. Adelaide Bot. Gard.*, 18(2): 120 (1998)—Victoria.
aspera DC.
subsp. pilosifolia *H.R. Toelken in J. Adelaide Bot. Gard.*, 18(2): 132 (1998)—Queensland, New South Wales.
axillibarba *J.R. Wheeler in Nuytsia*, 13(2): 392 (2000)—Western Australia.
carinata *J.R. Wheeler in Nuytsia*, 13(2): 387 (2000)—Western Australia.
charlesii *J.R. Wheeler in Nuytsia*, 13(2): 389 (2000)—Western Australia.
cinerea *(R. Br. ex DC.) H.R. Toelken in J. Adelaide Bot. Gard.*, 18(2): 133 (1998): Pleurandra cinerea.
crinita *H.R. Toelken in J. Adelaide Bot. Gard.*, 19: 9 (2000)—South Australia, New South Wales, Victoria.
decumbens *H.R. Toelken in J. Adelaide Bot. Gard.*, 18(2): 135 (1998)—New South Wales.
empetrifolia (DC.) Hoogland
subsp. radians *H.R. Toelken in J. Adelaide Bot. Gard.*, 18(2): 143 (1998)—South Australia.
subsp. uncinata *H.R. Toelken in J. Adelaide Bot. Gard.*, 18(2): 142 (1998)—New South Wales.
hamulosa *J.R. Wheeler in Nuytsia*, 13(2): 382 (2000)—Western Australia.
hirta *H.R. Toelken in J. Adelaide Bot. Gard.*, 18(2): 144 (1998)—New South Wales.
hirticalyx *H.R. Toelken in J. Adelaide Bot. Gard.*, 18(2): 146 (1998)—Tasmania, Victoria.
notabilis *H.R. Toelken in J. Adelaide Bot. Gard.*, 18(2): 150 (1998)—New South Wales, Victoria.
patens *H.R. Toelken in J. Adelaide Bot. Gard.*, 19: 16 (2000)—Queensland.
platyphylla *H.R. Toelken in J. Adelaide Bot. Gard.*, 19: 18 (2000)—South Australia.
subsp. halmaturina *H.R. Toelken in J. Adelaide Bot. Gard.*, 19: 24 (2000)—South Australia.
*subsp.*major *(J.M. Black) H.R. Toelken in J. Adelaide Bot. Gard.*, 19: 23 (2000): H. sericea var. major.
porongurupensis *J.R. Wheeler & R.D. Hoogland in Nuytsia*, 13(2): 395 (2000)—Western Australia.
praemorsa *H.R. Toelken in J. Adelaide Bot. Gard.*, 19: 25 (2000)—New South Wales.
puberula *H.R. Toelken in J. Adelaide Bot. Gard.*, 19: 27 (2000)—New South Wales.
rhynchocalyx *H.R. Toelken in J. Adelaide Bot. Gard.*, 18(2): 155 (1998)—New South Wales.
simulans *H.R. Toelken in J. Adelaide Bot. Gard.*, 19: 42 (2000)—New South Wales.
superans *H.R. Toelken in J. Adelaide Bot. Gard.*, 19: 44 (2000)—New South Wales.
tenuifolia *H.R. Toelken in J. Adelaide Bot. Gard.*, 19: 47 (2000)—Queensland, New South Wales.
truncata *H.R. Toelken in J. Adelaide Bot. Gard.*, 18(2): 157 (1998)—Victoria.
ulicifolia *(Benth.) J.R. Wheeler in Nuytsia*, 13(2): 385 (2000): H. acerosa var. ulicifolia.

HIBBERTIA:-
villifera *Tepper ex H.R. Toelken in J. Adelaide Bot. Gard.*, 19: 49 (2000)—South Australia.

HIBISCUS (Malvac.).
ceratophorus *M. Thulin in Nordic J. Bot.*, 19(3): 337 (1999)—Somalia.
erlangeri *(Gürke) M. Thulin in Kew Bull.*, 53(4): 1013 (1998): Symphyochlamys erlangeri.
gilletii De Wild.
subsp. hiernianus *(Exell & Mendonça) F.D. Wilson in Bull. Nat. Hist. Mus. (London), Bot.*, 29(1): 71 (1999): H. hiernianus.
subsp. lundaensis *(Baker f.) F.D. Wilson in Bull. Nat. Hist. Mus. (London), Bot.*, 29(1): 71 (1999): H. lundaensis.
goossensii *(Hauman) F.D. Wilson in Bull. Nat. Hist. Mus. (London), Bot.*, 29(1): 56 (1999): H. rostellatus var. goossensii.
partitus *(Hochr.) F.D. Wilson in Bull. Nat. Hist. Mus. (London), Bot.*, 29(1): 63 (1999): H. diversifolius var. partitus.
parvilobus *F.D. Wilson in Bull. Nat. Hist. Mus. (London), Bot.*, 29(1): 67 (1999)—Kenya.
reekmansii *F.D. Wilson in Bull. Nat. Hist. Mus. (London), Bot.*, 29(1): 64 (1999)—Burundi, Rwanda, Tanzania, Zaire.
sineaculeatus *F.D. Wilson in Bull. Nat. Hist. Mus. (London), Bot.*, 29(1): 72 (1999)—Nigeria.
syriacus L.
var. micranthus *Y.N. Lee & K.B. Yim in Korean J. Pl. Taxon.*, 22(1): 8 (1992)—South Korea.
trilobus Aubl.
subsp. hirsutus *O.J. Blanchard & P.A. Fryxell in Novon*, 10(3): 190 (2000)—Guatemala, Belize.
subsp. ingratus *(Miq.) O.J. Blanchard & P.A. Fryxell in Novon*, 10(3): 191 (2000): H. ingratus.

HIERACIUM (Compos.).
sect. Aestiva *(Üksip ex Schljakov) A.N. Sennikov in Bot. Zhurn.*, 84(12): 129 (1999): H. subsect. Aestiva.
sect. Cydoniifolia *(Arv.-Touv.) C.A. Stace in Edinburgh J. Bot.*, 55(3): 438 (1998): H. subsect. Cydoniifolia.
sect. Lucida *C.A. Stace & P.D. Sell in Edinburgh J. Bot.*, 55(3): 439 (1998).
sect. Mixta *C.A. Stace & P.D. Sell in Edinburgh J. Bot.*, 55(3): 435 (1998).
sect. Picta *(Peter) C.A. Stace in Edinburgh J. Bot.*, 55(3): 436 (1998): H. taxon Picta.
sect. Pilosissima *C.A. Stace & P.D. Sell in Edinburgh J. Bot.*, 55(3): 436 (1998).
sect. Prenanthella *A.N. Sennikov in Bot. Zhurn.*, 84(12): 128 (1999).
sect. Robusta *A.N. Sennikov in Bot. Zhurn.*, 84(12): 127 (1999).
sect. Thapsoidea *(Arv.-Touv.) C.A. Stace in Edinburgh J. Bot.*, 55(3): 436 (1998): H. taxon Thapsoidea.
sect. Umbrosa *C.A. Stace & P.D. Sell in Edinburgh J. Bot.*, 55(3): 438 (1998).
acrogymnon *(Malme) Üksip in Fl. SSSR*, 30: 389 (1960): H. sylvaticum subsp. acrogymnon.
acrosciadium *(Nägeli & Peter) Üksip in Fl. SSSR*, 30: 512 (1960): H. umbelliferum subsp. acrosciadium.
acroxanthum *(Sosn. & Zahn) Üksip in Fl. SSSR*, 30: 53 (1960): H. tschamkorijense subsp. acroxanthum.
acroxanthum *(Sosn. & Zahn) Üksip ex A.N. Sennikov in Bot. Zhurn.*, 83(2): 70 (1998), isonym; *see Üksip in Fl. SSSR*, 30: 53 (1960): H. tschamkorijense subsp. acroxanthum.
acuminatifolium *(Litv. & Zahn) Üksip in Fl. SSSR*, 30: 259 (1960): H. vulgatum subsp. acuminatifolium.
acutangulum *(Koslovsky & Zahn) Üksip in Fl. SSSR*, 30: 48 (1960): H. biebersteinii subsp. acutangulum.
acutisquamum *(Nägeli & Peter) Üksip in Fl. SSSR*, 30: 669 (1960): H. auricula subsp. acutisquamum.
akinfiewii *(Woronow & Zahn) Üksip in Fl. SSSR*, 30: 401 (1960): H. incanum subsp. akinfiewii.
alatavicum *(Zahn) Üksip in Fl. SSSR*, 30: 52 (1960): H. tschamkorijense subsp. alatavicum.

HIERACIUM:-

albellipes (Schelk. & Zahn) Üksip in Fl. SSSR, 30: 75 (1960): H. erythrocarpum subsp. albellipes.
albellipes (Schelk. & Zahn) Üksip ex A.N. Sennikov in Bot. Zhurn., 83(2): 70 (1998), isonym; see Üksip in Fl. SSSR, 30: 75 (1960): H. erythrocarpum subsp. albellipes.
albipes (Dahlst.) Üksip in Fl. SSSR, 30: 347 (1960): H. caesium subsp. albipes.
alejandrei G. Mateo Sanz in Estud. Mus. Cienc. Nat. Alava, 9: 35 (1994 publ. 1995)—Spain.
altaicum (Nägeli & Peter) Üksip in Fl. SSSR, 30: 598 (1960): H. collinum subsp. altaicum.
alticaule (Litv. & Zahn) Üksip in Fl. SSSR, 30: 536 (1960): H. brachiatum subsp. alticaule.
alupkanum (Zahn) Üksip in Fl. SSSR, 30: 488 (1960): H. auriculoides subsp. alupkanum.
amauranthum (Peter) Üksip in Fl. SSSR, 30: 510 (1960): H. zizianum subsp. amauranthum.
amaureilema (Nägeli & Peter) Üksip in Fl. SSSR, 30: 675 (1960): H. auricula subsp. amaureilema.
amaurobasis (Litv. & Zahn) Üksip in Fl. SSSR, 30: 495 (1960): H. auriculoides subsp. amaurobasis.
amnoon (Nägeli & Peter) Üksip in Fl. SSSR, 30: 466 (1960): H. magyaricum subsp. amnoon.
amoeniceps (Zahn) Üksip in Fl. SSSR, 30: 648 (1960): H. cernuiforme subsp. amoeniceps.
amphileion (Pohle & Zahn) Üksip in Fl. SSSR, 30: 125 (1960): H. crocatum subsp. amphileion.
amphitephrodes (Sosn. & Zahn) Üksip in Fl. SSSR, 30: 69 (1960): H. erythrocarpum subsp. amphitephrodes.
amphitephrodes (Sosn. & Zahn) Üksip ex A.N. Sennikov in Bot. Zhurn., 83(2): 70 (1998), isonym; see Üksip in Fl. SSSR, 30: 69 (1960): H. erythrocarpum subsp. amphitephrodes.
annae-toutoniae Zahn
subsp. vilpianense G. Gottschlich in Linz. Biol. Beitr., 32(1): 374 (2000)—Italy.
anocladum (Nägeli & Peter) Üksip in Fl. SSSR, 30: 531 (1960): H. leptophyton subsp. anocladum.
antennarioidiforme (Zahn) Üksip in Fl. SSSR, 30: 683 (1960): H. hoppeanum subsp. antennarioidiforme.
apatorium (Nägeli & Peter) Üksip in Fl. SSSR, 30: 519 (1960): H. arvicola subsp. apatorium.
aquilonare (Nägeli & Peter) Üksip in Fl. SSSR, 30: 438 (1960): H. florentinum β aquilonare.
arctogeton (Zahn) Üksip in Fl. SSSR, 30: 126 (1960): H. aestivum subsp. arctogeton.
arcuatidens (Zahn) Üksip in Fl. SSSR, 30: 263 (1960): H. vulgatum subsp. arcuatidens.
argillaceoides (Litv. & Zahn) Üksip in Fl. SSSR, 30: 269 (1960): H. vulgatum subsp. argillaceoides.
armeniacum (Nägeli & Peter) Üksip in Fl. SSSR, 30: 465 (1960): H. magyaricum subsp. armeniacum.
artabirense (Zahn) Üksip in Fl. SSSR, 30: 74 (1960): H. erythrocarpum subsp. artabirense.
artvinense (Woronow & Zahn) Üksip in Fl. SSSR, 30: 33 (1960): H. pelagae subsp. artvinense.
arvense (Nägeli & Peter) Üksip in Fl SSSR, 30: 492 (1960), homonym; see Scop., Fl. Carniol., ed. 2, 2: 110 (1772): H. pannonicum subsp. arvense.
aryslynense (Zahn) Üksip in Fl. SSSR, 30: 68 (1960): H. erythrocarpum subsp. aryslynense.
aryslynense (Zahn) Üksip ex A.N. Sennikov in Bot. Zhurn., 83(2): 71 (1998), isonym; see Üksip in Fl. SSSR, 30: 68 (1960): H. erythrocarpum subsp. aryslynense.
asiaticum (Nägeli & Peter) Üksip in Fl. SSSR, 30: 419 (1960): H. echioides subsp. asiaticum.
asterodermum (Woronow & Zahn) Üksip in Fl. SSSR, 30: 133 (1960): H. crocatum subsp. asterodermum.
atrellum (Zahn) Üksip in Fl. SSSR, 30: 189 (1960): H. atratum subsp. atrellum.
augusti-bayeri (Zlatník) J. Chrtek f.in Preslia, 67(3–4): 301 (1995 publ. 1996): H. alpinum subsp. augusti-bayeri.
baenitzii (Nägeli & Peter) Üksip in Fl. SSSR, 30: 605 (1960): H. floribundum subsp. baenitzii.
bakurianense (Fomin & Zahn) Üksip in Fl. SSSR, 30: 53 (1960): H. tschamkorijense subsp. bakurianense.

HIERACIUM:-

bakurianense (Fomin & Zahn) Üksip ex A.N. Sennikov in Bot. Zhurn., 83(2): 71 (1998), isonym; see Üksip in Fl. SSSR, 30: 53 (1960): H. tschamkorijense subsp. bakurianense.
basileucum (Litv. & Zahn) Üksip in Fl. SSSR, 30: 493 (1960): H. auriculoides subsp. basileucum.
bauhiniflorum (Nägeli & Peter) Üksip in Fl. SSSR, 30: 532 (1960): H. leptophyton subsp. bauhiniflorum.
beschtavicum (Litv. & Zahn) Üksip in Fl. SSSR, 30: 47 (1960): H. sparsiflorum subsp. beschtavicum.
beschtavicum (Litv. & Zahn) Üksip ex A.N. Sennikov in Bot. Zhurn., 83(2): 71 (1998), isonym; see Üksip in Fl. SSSR, 30: 47 (1960): H. sparsiflorum subsp. beschtavicum.
bichloricolor (Ganesch. & Zahn) Üksip in Fl. SSSR, 30: 117 (1960): H. laevigatum subsp. bichloricolor.
bichloricolor (Ganesch. & Zahn) Üksip ex A.N. Sennikov in Bot. Zhurn., 83(2): 71 (1998), isonym; see Üksip in Fl. SSSR, 30: 117 (1960): H. laevigatum subsp. bichloricolor.
bifidum Kit. ex Hornem.
subsp. hollei G. Gottschlich in Braunschweig. Naturk. Schriften, 5(4): 815 (1999), nom. nov.: H. consociatum Holle.
billyanum De Retz
var. parvulum B. De Retz in Soc. Ech. Pl. Vasc. Eur. Bassin Médit. Bull., 20: 56 (1985)—France.
borbasii R. Uechtr. ex V. Borbás in Magyar Bot. Lapok, 3: 48 (1904)—Locality not stated.
bourgaei Boiss.
subsp. bicolor (Scheele) O. de Bolòs & J. Vigo, Fl. Països Catalans, 3: 1092 (1995): H. bicolor.
subsp. catolanum (Arv.-Touv.) O. de Bolòs & J. Vigo, Fl. Països Catalans, 3: 1092 (1995): H. catolanum.
subsp. lusitanicum (Arv.-Touv.) O. de Bolòs & J. Vigo, Fl. Països Catalans, 3: 1092 (1995): H. lusitanicum.
subsp. pogonatoides (Arv.-Touv. & Gaut.) O. de Bolòs & J. Vigo, Fl. Països Catalans, 3: 1092 (1995): H. pogonatoides.
subsp. pogonatum (Arv.-Touv. & Gaut.) O. de Bolòs & J. Vigo, Fl. Països Catalans, 3: 1092 (1995): H. pogonatum.
subsp. sclerophyllum (Arv.-Touv. & Gaut.) O. de Bolòs & J. Vigo, Fl. Països Catalans, 3: 1092 (1995): H. sclerophyllum.
subsp. terianum (Arv.-Touv. & Sennen) O. de Bolòs & J. Vigo, Fl. Països Catalans, 3: 1092 (1995): H. terianum.
subsp. trichodermum (Arv.-Touv. & Gaut.) O. de Bolòs & J. Vigo, Fl. Països Catalans, 3: 1092 (1995): H. trichodermum.
subsp. vestitiforme (Zahn) O. de Bolòs & J. Vigo, Fl. Països Catalans, 3: 1092 (1995): H. bicolor subsp. vestitiforme.
branae (Nägeli & Peter) Üksip in Fl. SSSR, 30: 464 (1960): H. magyaricum subsp. branae.
buhsei (Nägeli & Peter) Üksip in Fl. SSSR, 30: 411 (1960): H. procerum subsp. buhsei.
bupleurifolioides (Zahn) Üksip in Fl. SSSR, 30: 143 (1960): H. prenanthoides subsp. bupleurifolioides.
bupleurifolium (Wimm. & Grab.) Tausch ex Üksip in Fl. SSSR, 30: 142 (1960): H. prenanthoides var. bupleurifolium.
caesiomurorum Lindeb.
var. chlorelliceps (Norrl. ex Üksip) A.N. Sennikov in Bot. Zhurn., 84(3): 124 (1999): H. chlorelliceps.
caesium (Fr.) Fr.
subsp. zabelianum G. Gottschlich in Bot. Rundbr. Mecklenburg-Vorpommern, 31: 17 (1998)—Germany.
caespitosum Dumort.
subsp. madarum (Nägeli & Peter) S. Bräutigam in Abh. Ber. Naturkundemus. Görlitz, 68(8): 3 (1995): H. collinum subsp. madarum.
callimorphopsis (Zahn) Üksip in Fl. SSSR, 30: 629 (1960): H. flagellariforme subsp. callimorphopsis.
calolepideum (Norrl.) Üksip in Fl. SSSR, 30: 657 (1960): Pilosella decolorans * calolepidea.
caloprasinum (Zahn) Üksip in Fl. SSSR, 30: 73 (1960): H. erythrocarpum subsp. caloprasinum.

HIERACIUM:-

cantabrimontanum (De Retz) G. Mateo Sanz in An. Jard. Bot. Madrid, 54(1): 367 (1996): H. gastonianum subsp. cantabrimontanum.
cardiobasis (Zahn) Üksip in Fl. SSSR, 30: 372 (1960): H. bifidum subsp. cardiobasis.
casparyanum (Nägeli & Peter) Üksip in Fl. SSSR, 30: 637 (1960): H. prussicum subsp. casparyanum.
chaetothyrsoides (Zahn) Üksip in Fl. SSSR, 30: 32 (1960): H. velenovskyi subsp. chaetothyrsoides.
chaetothyrsoides (Zahn) Üksip ex A.N. Sennikov in Bot. Zhurn., 83(2): 72 (1998), isonym; see Üksip in Fl. SSSR, 30: 32 (1960): H. velenovskyi subsp. chaetothyrsoides.
chaetothyrsum (Litv. & Zahn) Üksip in Fl. SSSR, 30: 30 (1960): H. velenovskyi subsp. chaetothyrsum.
chaetothyrsum (Litv. & Zahn) Üksip ex A.N. Sennikov in Bot. Zhurn., 83(2): 72 (1998), isonym; see Üksip in Fl. SSSR, 30: 30 (1960): H. velenovskyi subsp. chaetothyrsum.
chlorinum A.N. Sennikov in Bot. Zhurn., 84(12): 130 (1999)—West Siberia.
chloroleucolepium (Koslovsky & Zahn) Üksip in Fl. SSSR, 30: 54 (1960): H. tschamkorijense subsp. chloroleucolepium.
chloroleucolepium (Koslovsky & Zahn) Üksip ex A.N. Sennikov in Bot. Zhurn., 83(2): 72 (1998), isonym; see Üksip in Fl. SSSR, 30: 54 (1960): H. tschamkorijense subsp. chloroleucolepium.
chlorophilum (Koslovsky & Zahn) Üksip in Fl. SSSR, 30: 51 (1960): H. medschedschense subsp. chlorophilum.
chlorophilum (Koslovsky & Zahn) Üksip ex A.N. Sennikov in Bot. Zhurn., 83(2): 72 (1998), isonym; see Üksip in Fl. SSSR, 30: 51 (1960): H. medschedschense subsp. chlorophilum.
christoglossum (Zahn) Üksip in Fl. SSSR, 30: 539 (1960): H. tephrocephalum subsp. christoglossum.
chrysostyloides (Zahn) J. Chrtek f. in Preslia, 67(2): 103 (1995): H. nigrescens subsp. chrysostyloides.
cinereostriatum (Woronow & Zahn) Üksip in Fl. SSSR, 30: 321 (1960): H. sylvaticum subsp. cinereostriatum.
circumstellatum (Zahn) A.N. Sennikov in Bot. Zhurn., 83(2): 78 (1998): H. silvaticum subsp. circumstellatum.
concinnidens (Zahn) Üksip in Fl. SSSR, 30: 58 (1960): H. sparsum subsp. concinnidens.
concinnidens (Zahn) Üksip ex A.N. Sennikov in Bot. Zhurn., 83(2): 72 (1998), isonym; see Üksip in Fl. SSSR, 30: 58 (1960): H. sparsum subsp. concinnidens.
coniciforme (Litv. & Zahn) Üksip in Fl. SSSR, 30: 130 (1960): H. crocatum subsp. coniciforme.
coniciforme (Litv. & Zahn) Üksip ex A.N. Sennikov in Bot. Zhurn., 83(2): 72 (1998), isonym; see Üksip in Fl. SSSR, 30: 130 (1960): H. crocatum subsp. coniciforme.
× consociatum G. von Holle, Beobacht. Hohensteine Weserkette hybr. Form. Gatt. Hieracium: 8 (1892), homonym; see Jord. in Boreau, Fl. Centre France, ed. 3, 2: 400 (1857)—Germany. (H. caesium × subcaesioides).
cordatum Scheele ex Costa
subsp. citharocerinthe (Arv.-Touv. & Sennen) O. de Bolòs & J. Vigo, Fl. Països Catalans, 3: 1117 (1995): H. citharocerinthe.
subsp. conaticum (Arv.-Touv.) O. de Bolòs & J. Vigo, Fl. Països Catalans, 3: 1117 (1995): H. conaticum.
subsp. heteradenum (Arv.-Touv. & Cadevall) O. de Bolòs & J. Vigo, Fl. Països Catalans, 3: 1117 (1995): H. heteradenum.
subsp. leptocoleum (Arv.-Touv.) O. de Bolòs & J. Vigo, Fl. Països Catalans, 3: 1117 (1995): H. leptocoleum.
crassipedipilum (Pawł. & Zahn) J. Chrtek f.in Phyton (Austria), 37(2): 206 (1998): H. alpinum subsp. crassipedipilum.
crocatum Fr.
subsp. brachybrachion (Zahn) O. de Bolòs & J. Vigo, Fl. Països Catalans, 3: 1129 (1995): H. aestivum subsp. brachybrachion.
subsp. glareivagum (Zahn) O. de Bolòs & J. Vigo, Fl. Països Catalans, 3: 1129 (1995): H. aestivum subsp. glareivagum.

HIERACIUM:-

crocatum

subsp. phrissoides *(Arv.-Touv.) O. de Bolòs & J. Vigo, Fl. Països Catalans,* 3: 1129 (1995): H. phrissoides.

cymanthodes *(Koslovsky & Zahn) Üksip in Fl. SSSR,* 30: 457 (1960): H. bauhinii subsp. cymanthodes.

cymanthum *(Nägeli & Peter) Üksip in Fl. SSSR,* 30: 456 (1960): H. magyaricum subsp. cymanthum.

cymiramum *(Schelk. & Zahn) Üksip in Fl. SSSR,* 30: 479 (1960): H. pannoniciforme subsp. cymiramum.

cymosiforme *(Nägeli & Peter) Üksip in Fl. SSSR,* 30: 512 (1960), homonym; *see Froel. in DC., Prodr.,* 7(1): 207 (1838): H. umbelliferum subsp. cymosiforme.

debilescens *(Woronow & Zahn) Üksip in Fl. SSSR,* 30: 290 (1960): H. diaphanoides subsp. debilescens.

dechyi *(Koslovsky & Zahn) Üksip in Fl. SSSR,* 30: 104 (1960): H. laevigatum subsp. dechyi.

dechyi *(Koslovsky & Zahn) Üksip ex A.N. Sennikov in Bot. Zhurn.,* 83(2): 72 (1998), isonym; *see Üksip in Fl. SSSR,* 30: 104 (1960): H. laevigatum subsp. dechyi.

densiflorum Tausch

subsp. acrosciadium *(Nägeli & Peter) G. Gottschlich in Farn- u. Blütenpfl. Baden-Württembergs (Valerianac.-Asterac.) (ed. O.* Sebald et al.), 6 (Spez. Teil): 461 (1996): H. umbelliferum subsp. acrosciadium.

subsp. umbelliferum *(Nägeli & Peter) G. Gottschlich in Mitt. Pollich.,* 76: 75 (1989): H. umbelliferum.

derubellum *G. Gottschlich & F. Schuhwerk in Ber. Bayer. Bot. Ges.,* 69–70: 148 (2000)—Germany, New York.

× desolatum *G. von Holle, Beobacht. Hohensteine Weserkette hybr. Form. Gatt. Hieracium:* 11 (1892)—Germany.(H. murorum × subcaesioides).

diaphanoides Lindeb.

subsp. neoornatum *G. Gottschlich in Bot. Rundbr. Mecklenburg-Vorpommern,* 31: 52 (1998), nom. nov.: H. subornatum *Wiinst.*

diaphanoidiceps *(Woronow & Zahn) Üksip in Fl. SSSR,* 30: 54 (1960): H. tschamkorijense subsp. diaphanoidiceps.

discolor *(Nägeli & Peter) Üksip in Fl. SSSR,* 30: 532 (1960): H. leptophyton subsp. discolor.

dissolutum *(Nägeli & Peter) Üksip in Fl. SSSR,* 30: 600 (1960): H. collinum subsp. dissolutum.

diversifolium *L.J. Čelakovský in Sitzungsber. Königl. Böhm. Ges. Wiss. Prag, Math.-Naturwiss. Cl.,* 1887: 190 (1888)—Germany.

dmitrovense *(Peter) Üksip in Fl. SSSR,* 30: 537 (1960): H. brachiatum subsp. dmitrovense.

durisetum *(Nägeli & Peter) Üksip in Fl. SSSR,* 30: 422 (1960): H. fallax subsp. durisetum.

echiocephalum *(Nägeli & Peter) Üksip in Fl. SSSR,* 30: 494 (1960): H. pannonicum subsp. echiocephalum.

echiogenes *(Nägeli & Peter) Üksip in Fl. SSSR,* 30: 487 (1960): H. pannonicum subsp. echiogenes.

epichlorum *(Litv. & Zahn) Üksip in Fl. SSSR,* 30: 264 (1960): H. vulgatum var. epichlorum.

epichlorum *(Litv. & Zahn) Üksip ex A.N. Sennikov in Bot. Zhurn.,* 83(2): 72 (1998), isonym; *see Üksip in Fl. SSSR,* 30: 264 (1960): H. vulgatum var. epichlorum.

ericetorum *(Nägeli & Peter) Üksip in Fl. SSSR,* 30: 440 (1960): H. florentinum subsp. ericetorum.

erythrochristum *(Nägeli & Peter) Üksip in Fl. SSSR,* 30: 520 (1960): H. arvicola subsp. erythrochristum.

fariniramum *(Ganesch. & Zahn) Üksip in Fl. SSSR,* 30: 264 (1960): H. vulgatum subsp. fariniramum.

fariniramum *(Ganesch. & Zahn) Üksip ex A.N. Sennikov in Bot. Zhurn.,* 83(2): 72 (1998), isonym; *see Üksip in Fl. SSSR,* 30: 264 (1960): H. vulgatum subsp. fariniramum.

farinodermum *(Litv. & Zahn) Üksip in Fl. SSSR,* 30: 400 (1960): H. incanum subsp. farinodermum.

fastigiatum *(Tausch ex Nägeli & Peter) Üksip in Fl. SSSR,* 30: 453 (1960): H. magyaricum subsp. fastigiatum.

HIERACIUM:-

floccicomatum *(Woronow & Zahn) Üksip in Fl. SSSR,* 30: 325 (1960): H. murorum subsp. floccicomatum.

flocciparum *(Schelk. & Zahn) Üksip in Fl. SSSR,* 30: 106 (1960): H. laevigatum subsp. flocciparum.

flocciparum *(Schelk. & Zahn) Üksip ex A.N. Sennikov in Bot. Zhurn.,* 83(2): 73 (1998), isonym; *see Üksip in Fl. SSSR,* 30: 106 (1960): H. laevigatum subsp. flocciparum.

floccipedunculum *(Nägeli & Peter) Üksip in Fl. SSSR,* 30: 437 (1960): H. florentinum subsp. floccipedunculum.

floribundiforme *(Nägeli & Peter) Üksip in Fl. SSSR,* 30: 584 (1960): H. acrocomum subsp. floribundiforme.

floribundoides *(Zahn) Üksip in Fl. SSSR,* 30: 585 (1960): H. acrocomum subsp. floribundoides.

floridum *(Nägeli & Peter) Üksip in Fl. SSSR,* 30: 632 (1960): H. nigriceps subsp. floridum.

foliosissimum *(Woronow & Zahn) Üksip in Fl. SSSR,* 30: 34 (1960): H. sparsiflorum subsp. foliosissimum.

fominianum *(Woronow & Zahn) Üksip in Fl. SSSR,* 30: 480 (1960): H. pannoniciforme subsp. fominianum.

freynii *(Nägeli & Peter) Üksip in Fl. SSSR,* 30: 417 (1960): H. echioides subsp. freynii.

frickii *(Zahn) Üksip in Fl. SSSR,* 30: 431 (1960): H. bifurcum subsp. frickii.

frigidellum *(Pohle & Zahn) Üksip in Fl. SSSR,* 30: 319 (1960): H. silvaticum subsp. frigidellum.

frondiferum *M. Elfstrand in Bot. Utflygter:* 38 (1890); *cf. Hieracium Alp. Mittl. Skand.:* 21 (1893), in IK Supp. 1 as 'fronditerum' in error—Norway, Sweden.†

frondosum *(Nägeli & Peter) Üksip in Fl. SSSR,* 30: 676 (1960), homonym; *see A. Peter in Bot. Jahrb. Syst.,* 5: 452 (1884): H. auriculiforme subsp. frondosum.

fulvescens *(Nägeli & Peter) Üksip in Fl. SSSR,* 30: 618 (1960), homonym; *see Norrl. in Acta Soc. Faun. Fl. Fenn.,* 3(4): 101 (1888): H. spathophyllum subsp. fulvescens.

furfuraceoides *(Zahn) Üksip in Fl. SSSR,* 30: 318 (1960): H. murorum subsp. furfuraceoides.

fuscocinereum Norrl.

subsp. pycnodon *(Dahlst.) G. Gottschlich in Bot. Rundbr. Mecklenburg-Vorpommern,* 31: 47 (1998): H. pycnodon.

giganticaule *(Zahn) Üksip in Fl. SSSR,* 30: 569 (1960): H. glomeratum subsp. giganticaule.

glaucinum Jord.

subsp. aranicum *(Arv.-Touv.) O. de Bolòs & J. Vigo, Fl. Països Catalans,* 3: 1053 (1995): H. aranicum.

var. arnicoides *(Gren. & Godr.) O. de Bolòs & J. Vigo, Fl. Països Catalans,* 3: 1054 (1995): H. arnicoides.

subsp. arnicoides *(Gren. & Godr.) O. de Bolòs & J. Vigo, Fl. Països Catalans,* 3: 1054 (1995): H. arnicoides.

var. basalticum *(Sch. Bip.) O. de Bolòs & J. Vigo, Fl. Països Catalans,* 3: 1053 (1995): H. praecox var. basalticum.

subsp. basalticum *(Sch. Bip.) O. de Bolòs & J. Vigo, Fl. Països Catalans,* 3: 1053 (1995), isonym; *see J. Duvign. in Bull. Jard. Bot. Nation. Belg.,* 57(3–4): 459 (1987): H. praecox var. basalticum.

subsp. basalticum *(Sch. Bip.) G. Gottschlich in Mitt. Pollich.,* 76: 76 (1989), isonym; *see J. Duvign. in Bull. Jard. Bot. Nation. Belg.,* 57(3–4): 459 (1987): H. praecox var. basalticum.

subsp. bounophilum *(Jord.) O. de Bolòs & J. Vigo, Fl. Països Catalans,* 3: 1053 (1995): H. bounophilum.

subsp. brachypus *(Freyn ex Zahn) O. de Bolòs & J. Vigo, Fl. Països Catalans,* 3: 1054 (1995): H. praecox subsp. brachypus.

subsp. cinerascens *(Jord.) O. de Bolòs & J. Vigo, Fl. Països Catalans,* 3: 1054 (1995): H. cinerascens.

subsp. conjugatum *(Jord.) O. de Bolòs & J. Vigo, Fl. Països Catalans,* 3: 1053 (1995): H. conjugatum.

HIERACIUM:-

glaucinum

subsp. fallens *(Jord.) O. de Bolòs & J. Vigo, Fl. Països Catalans,* 3: 1053 (1995), isonym; *see Gottschlich in Soc. Ech. Pl. Vasc. Eur. Bassin Médit. Bull.,* 24: 69 (1993): H. fallens.

subsp. fragile *(Jord.) O. de Bolòs & J. Vigo, Fl. Països Catalans,* 3: 1053 (1995): H. fragile.

subsp. fragile *(Jord.) G. Gottschlich in Soc. Ech. Pl. Vasc. Eur. Bassin Médit. Bull.,* 26: 88 (1997), isonym; *see O. Bolòs & Vigo, Fl. Països Catalans,* 3: 1053 (1995): H. fragile.

subsp. fraternum *(Sudre) O. de Bolòs & J. Vigo, Fl. Països Catalans,* 3: 1054 (1995): H. praecox subsp. fraternum.

subsp. furcillatum *(Jord.) O. de Bolòs & J. Vigo, Fl. Països Catalans,* 3: 1054 (1995): H. furcillatum.

subsp. gougetianum *(Gren. & Godr.) O. de Bolòs & J. Vigo, Fl. Països Catalans,* 3: 1053 (1995): H. gougetianum.

subsp. guaranum *(Arv.-Touv. & Gaut.) O. de Bolòs & J. Vigo, Fl. Països Catalans,* 3: 1053 (1995): H. guaranum.

subsp. heteroschistum *(Zahn) O. de Bolòs & J. Vigo, Fl. Països Catalans,* 3: 1054 (1995): H. praecox subsp. heteroschistum.

subsp. horripilum *(Arv.-Touv. & Gaut.) O. de Bolòs & J. Vigo, Fl. Països Catalans,* 3: 1053 (1995): H. horripilum.

subsp. incisoides *(Arv.-Touv. & Gaut.) O. de Bolòs & J. Vigo, Fl. Països Catalans,* 3: 1054 (1995): H. praecox subsp. incisoides.

subsp. jaubertianum *(Loret & Timb.-Lagr.) O. de Bolòs & J. Vigo, Fl. Països Catalans,* 3: 1053 (1995): H. praecox subsp. jaubertianum.

subsp. leptocauloides *(Zahn) O. de Bolòs & J. Vigo, Fl. Països Catalans,* 3: 1054 (1995): H. leptocauloides.

subsp. medium *(Jord.) O. de Bolòs & J. Vigo, Fl. Països Catalans,* 3: 1054 (1995): H. medium.

subsp. montolearense *(Jeanb. & Timb.-Lagr.) O. de Bolòs & J. Vigo, Fl. Països Catalans,* 3: 1053 (1995): H. montolearense.

subsp. ovalifolium *(Jord.) O. de Bolòs & J. Vigo, Fl. Països Catalans,* 3: 1053 (1995): H. ovalifolium.

subsp. pallidifrons *(Sudre) O. de Bolòs & J. Vigo, Fl. Països Catalans,* 3: 1053 (1995): H. praecox subsp. pallidifrons.

subsp. pallidulum *(Jord.) O. de Bolòs & J. Vigo, Fl. Països Catalans,* 3: 1054 (1995): H. pallidulum.

subsp. pinicola *(Sudre) G. Gottschlich in Soc. Ech. Pl. Vasc. Eur. Bassin Médit. Bull.,* 25: 69 (1995): H. praecox b. pinicola.

subsp. praecox *(Sch. Bip.) O. de Bolòs & J. Vigo, Fl. Països Catalans,* 3: 1053 (1995): H. praecox.

subsp. prasinifolium *(Jord.) O. de Bolòs & J. Vigo, Fl. Països Catalans,* 3: 1053 (1995): H. prasinifolium.

*subsp.*pseudocyaneum *(Zahn) O. de Bolòs & J. Vigo, Fl. Països Catalans,* 3: 1053 (1995): H. praecox subsp. pseudocyaneum.

subsp. recensitum *(Jord.) O. de Bolòs & J. Vigo, Fl. Països Catalans,* 3: 1054 (1995), isonym; *see Gottschlich in Soc. Ech. Pl. Vasc. Eur. Bassin Médit. Bull.,* 24: 69 (1993): H. recensitum.

subsp. sagittibifidum *(Arv.-Touv. & Gaut.) O. de Bolòs & J. Vigo, Fl. Països Catalans,* 3: 1053 (1995): H. sagittibifidum.

var. sericopodioides *(Zahn) O. de Bolòs & J. Vigo, Fl. Països Catalans,* 3: 1054 (1995): H. praecox var. sericopodioides.

var. serratense *(Arv.-Touv. & Gaut.) O. de Bolòs & J. Vigo, Fl. Països Catalans,* 3: 1053 (1995): H. praecox var. serratense.

subsp. similatum *(Jord.) O. de Bolòs & J. Vigo, Fl. Països Catalans,* 3: 1053 (1995), isonym; *see Gottschlich in Soc. Ech. Pl. Vasc. Eur. Bassin Médit. Bull.,* 23: 72 (1991): H. similatum.

subsp. subalatiforme *(Zahn) O. de Bolòs & J. Vigo, Fl. Països Catalans,* 3: 1054 (1995): H. praecox subsp. subalatiforme.

subsp. subalatum *(Arv.-Touv. & Gaut.) O. de Bolòs & J. Vigo, Fl. Països Catalans,* 3: 1054 (1995): H. subalatum.

subsp. subpinicola *(Zahn) O. de Bolòs & J. Vigo, Fl. Països Catalans,* 3: 1053 (1995): H. praecox subsp. subpinicola.

HIERACIUM:-

glaucinum
  subsp. suntaliense *(Holle) G. Gottschlich in Braunschweig. Naturk. Schriften,* 5(4): 813 (1999): H. caesium subsp. suntalense.
  subsp. verlotii *(Jord.) O. de Bolòs & J. Vigo, Fl. Països Catalans,* 3: 1053 (1995): H. verlotii.
  subsp. vernum *(Sauzé & Maillard) O. de Bolòs & J. Vigo, Fl. Països Catalans,* 3: 1053 (1995): H. vernum.
glomeratiforme *(Zahn) Üksip in Fl. SSSR,* 30: 612 (1960): H. floribundum subsp. glomeratiforme.
glomerellum *(Zahn) Üksip in Fl. SSSR,* 30: 70 (1960): H. erythrocarpum subsp. glomerellum.
glomerellum *(Zahn) Üksip ex A.N. Sennikov in Bot. Zhurn.,* 83(2): 73 (1998), isonym; *see Üksip in Fl. SSSR,* 30: 70 (1960): H. erythrocarpum subsp. glomerellum.
gnaphalium *(Nägeli & Peter) Üksip in Fl. SSSR,* 30: 634 (1960): H. prussicum subsp. gnaphalium.
goriense *(Koslovsky & Zahn) Üksip in Fl. SSSR,* 30: 95 (1960): H. laevigatum subsp. goriense.
goriense *(Koslovsky & Zahn) Üksip ex A.N. Sennikov in Bot. Zhurn.,* 83(2): 73 (1998), isonym; *see Üksip in Fl. SSSR,* 30: 95 (1960): H. laevigatum subsp. goriense.
gothicifrons *(Zahn) Üksip in Fl. SSSR,* 30: 39 (1960): H. litwinowianum subsp. gothicifrons.
guttenfeldense *(Zahn) Üksip in Fl. SSSR,* 30: 635 (1960): H. prussicum subsp. guttenfeldense.
haematoglossum *(Koslovsky & Zahn) Üksip in Fl. SSSR,* 30: 487 (1960): H. auriculoides subsp. haematoglossum.
harpago *M. Gandoger, Dec. Pl. Nov.,* 1 28 (1875)—Hungary.
heothinum *(Nägeli & Peter) Üksip in Fl. SSSR,* 30: 462 (1960): H. magyaricum subsp. heothinum.
heterodontoides *(Litv. & Zahn) Üksip in Fl. SSSR,* 30: 72 (1960): H. erythrocarpum subsp. heterodontoides.
heterodontoides *(Litv. & Zahn) Üksip ex A.N. Sennikov in Bot. Zhurn.,* 83(2): 73 (1998), isonym; *see Üksip in Fl. SSSR,* 30: 72 (1960): H. erythrocarpum subsp. heterodontoides.
hispidissimum *(Rehmann ex Nägeli & Peter) Üksip in Fl. SSSR,* 30: 459 (1960): H. magyaricum subsp. hispidissimum.
hohenackeri *(Sch. Bip. ex Nägeli & Peter) Üksip in Fl. SSSR,* 30: 403 (1960): H. caucasicum subsp. hohenackeri.
holmiense *(Nägeli & Peter) K. Johansson in Kongl. Svenska Vetenskapsakad. Handl.,* 29(1): 127 (1897): H. cymosum γ holmiense.
hylogeton *(Koslovsky & Zahn) Üksip in Fl. SSSR,* 30: 300 (1960): H. murorum subsp. hylogeton.
hylogeton *(Koslovsky & Zahn) Üksip ex A.N. Sennikov in Bot. Zhurn.,* 83(2): 73 (1998), isonym; *see Üksip in Fl. SSSR,* 30: 300 (1960): H. murorum subsp. hylogeton.
hypoglaucum *(Litv. & Zahn) Üksip in Fl. SSSR,* 30: 141 (1960): H. prenanthoides subsp. hypoglaucum.
hypoglaucum *(Litv. & Zahn) Üksip ex A.N. Sennikov in Bot. Zhurn.,* 83(2): 73 (1998), isonym; *see Üksip in Fl. SSSR,* 30: 141 (1960): H. prenanthoides subsp. hypoglaucum.
hypopitys *(Litv. & Zahn) Üksip ex A.N. Sennikov in Bot. Zhurn.,* 83(2): 73 (1998), isonym; *see Üksip in Fl. SSSR,* 30: 261 (1960): H. divisum subsp. hypopitys.
hypopogon *(Litv. & Zahn) Üksip in Fl SSSR,* 30: 50 (1960): H. biebersteinii subsp. hypopogon.
ilyassowoense *(Zahn) Üksip in Fl. SSSR,* 30: 535 (1960): H. brachiatum subsp. ilyassowoense.
ingricum *(Nägeli & Peter) Üksip in Fl. SSSR,* 30: 454 (1960): H. magyaricum subsp. ingricum.
insigne *Backh. f. ex P.D. Sell, C. West & D.J. Tennant in Watsonia,* 20(4): 354 (1995)—Great Britain.
insolitum *(Zahn) Üksip in Fl. SSSR,* 30: 73 (1960): H. erythrocarpum subsp. insolitum.
jailanum *(Zahn) Üksip in Fl. SSSR,* 30: 543 (1960): H. ruprechtii subsp. jailanum.
karpinskyanum *(Nägeli & Peter) Üksip in Fl. SSSR,* 30: 400 (1960): H. incanum subsp. karpinskyanum.
kebeshense *N.V. Stepanov, Florogenet. Analiz,* 1: 103 (1994)—Krasnoyarsk.

HIERACIUM:-

kentii *P.D. Sell in Watsonia,* 23(1): 216 (2000)—Great Britain.
kiderense *(Zahn) Üksip in Fl. SSSR,* 30: 57 (1960): H. sparsiflorum subsp. kiderense.
kiderense *(Zahn) Üksip ex A.N. Sennikov in Bot. Zhurn.,* 83(2): 73 (1998), isonym; *see Üksip in Fl. SSSR,* 30: 57 (1960): H. sparsiflorum subsp. kiderense.
kihlmanii *(Norrl.) Üksip in Fl. SSSR,* 30: 661 (1960): Pilosella kihlmanii.
kochtanum *(Koslovsky & Zahn) Üksip in Fl. SSSR,* 30: 49 (1960): H. callichlorum subsp. kochtanum.
koenigianum *(Zahn) Üksip in Fl. SSSR,* 30: 200 (1960): H. pilosissimum subsp. koenigianum.
kofelicum *G. Gottschlich in Linz. Biol. Beitr.,* 32(1): 368 (2000)—Italy.
kopsicum *G. Gottschlich in Linz. Biol. Beitr.,* 32(1): 370 (2000)—Austria.
kubanicum *(Litv. & Zahn) Üksip in Fl. SSSR,* 30: 104 (1960): H. laevigatum subsp. kubanicum.
kubanicum *(Litv. & Zahn) Üksip ex A.N. Sennikov in Bot. Zhurn.,* 83(2): 74 (1998), isonym; *see Üksip in Fl. SSSR,* 30: 104 (1960): H. laevigatum subsp. kubanicum.
kulkowianum *(Zahn) Üksip in Fl. SSSR,* 30: 112 (1960): H. knafii subsp. kulkowianum.
lackschewitzii *(Dahlst.) Üksip in Fl. SSSR,* 30: 386 (1960): H. silvaticum subsp. lackschewitzii.
lactucella Wallr.
  subsp. amaureilema *(Nägeli & Peter) O. de Bolòs & J. Vigo, Fl. Països Catalans,* 3: 1153 (1995): H. auricula subsp. amaureilema.
  subsp. tricheilema *(Nägeli & Peter) O. de Bolòs & J. Vigo, Fl. Països Catalans,* 3: 1153 (1995): H. auricula subsp. tricheilema.
laevigans *(Zahn) Üksip in Fl. SSSR,* 30: 102 (1960): H. laevigatum subsp. laevigans.
lailanum *(Schelk. & Zahn) Üksip in Fl. SSSR,* 30: 59 (1960): H. sparsiflorum subsp. lailanum.
lailanum *(Schelk. & Zahn) Üksip ex A.N. Sennikov in Bot. Zhurn.,* 83(2): 74 (1998), isonym; *see Üksip in Fl. SSSR,* 30: 59 (1960): H. sparsiflorum subsp. lailanum.
lamprocomoides *(Woronow & Zahn) Üksip in Fl. SSSR,* 30: 688 (1960): H. hypeuryum subsp. lamprocomoides.
lamprocomum *(Nägeli & Peter) Üksip in Fl. SSSR,* 30: 688 (1960): H. hypeuryum subsp. lamprocomum.
lancidens *(Zahn) Üksip in Fl. SSSR,* 30: 101 (1960): H. laevigatum subsp. lancidens.
lasiophorum *(Nägeli & Peter) Üksip in Fl. SSSR,* 30: 493 (1960): H. pannonicum subsp. lasiophorum.
lasiothrix *(Nägeli & Peter) Üksip in Fl. SSSR,* 30: 687 (1960): H. hypeuryum subsp. lasiothrix.
latifolium Froel. ex Link
  subsp. brevifolium *(Tausch) O. de Bolòs & J. Vigo, Fl. Països Catalans,* 3: 1123 (1995): H. brevifolium
  subsp. halimifolium *(Froel. ex Fr.) O. de Bolòs & J. Vigo, Fl. Països Catalans,* 3: 1123 (1995): H. halimifolium.
leptocaulon *(Nägeli & Peter) Üksip in Fl. SSSR,* 30: 598 (1960), homonym; *see Arv.-Touv. & Sennen in Arv.-Touv., Hier. Gall. Hisp. Cat.:* 234 (1913): H. collinum subsp. leptocaulon.
leptophyes *(Peter) Üksip in Fl. SSSR,* 30: 587 (1960), homonym; *see Omang in Nyt Mag. Naturvidensk.,* 48: 248 (1910): H. sciadophorum subsp. leptophyes.
leptophyllum *(Nägeli & Peter) Üksip in Fl. SSSR,* 30: 510 (1960): H. zizianum subsp. leptophyllum.
leptophytomorphum *(Litv. & Zahn) Üksip in Fl. SSSR,* 30: 505 (1960): H. euchaetium subsp. leptophytomorphum.
lespinassei *(Koslovsky & Zahn) Üksip in Fl. SSSR,* 30: 134 (1960): H. conicum subsp. lespinassei.
lespinassei *(Koslovsky & Zahn) Üksip ex A.N. Sennikov in Bot. Zhurn.,* 83(2): 74 (1998), isonym; *see Üksip in Fl. SSSR,* 30: 134 (1960): H. conicum subsp. lespinassei.
leucothyrsogenes *(Koslovsky & Zahn) Üksip in Fl. SSSR,* 30: 314 (1960): H. murorum subsp. leucothyrsogenes.

HIERACIUM:-

leucothyrsogenes *(Koslovsky & Zahn) Üksip ex A.N. Sennikov in Bot. Zhurn.,* 83(2): 74 (1998), isonym; *see Üksip in Fl. SSSR,* 30: 314 (1960): H. murorum subsp. leucothyrsogenes.
leucothyrsoides *(Koslovsky & Zahn) Üksip in Fl. SSSR,* 30: 223 (1960): H. knafii subsp. leucothyrsoides.
leucothyrsoides *(Koslovsky & Zahn) Üksip ex A.N. Sennikov in Bot. Zhurn.,* 83(2): 74 (1998), isonym; *see Üksip in Fl. SSSR,* 30: 223 (1960): H. knafii subsp. leucothyrsoides.
leucothyrsum *(Litv. & Zahn) Üksip in Fl. SSSR,* 30: 106 (1960): H. knafii subsp. leucothyrsum.
leucothyrsum *(Litv. & Zahn) Üksip ex A.N. Sennikov in Bot. Zhurn.,* 83(2): 74 (1998), isonym; *see Üksip in Fl. SSSR,* 30: 106 (1960): H. knafii subsp. leucothyrsum.
littuanicum *(Nägeli & Peter) Üksip in Fl. SSSR,* 30: 673 (1960): H. auricula subsp. littuanicum.
longatum *(Peter) Üksip in Fl. SSSR,* 30: 620 (1960): H. spathophyllum subsp. longatum.
longiradiatum *(Zahn) Üksip in Fl. SSSR,* 30: 511 (1960): H. umbelliferum subsp. longiradiatum.
longum *(Nägeli & Peter) Üksip in Fl. SSSR,* 30: 502 (1960): H. euchaetium subsp. longum.
loscosianum Scheele
  subsp. baeticum *(Arv.-Touv. & Reverch.) O. de Bolòs & J. Vigo, Fl. Països Catalans,* 3: 1090 (1995): H. baeticum.
  subsp. palomerae *(Zahn) O. de Bolòs & J. Vigo, Fl. Països Catalans,* 3: 1090 (1995): H. baeticum subsp. palomerae.
  subsp. pseudobaeticum *(Zahn) O. de Bolòs & J. Vigo, Fl. Països Catalans,* 3: 1090 (1995): H. baeticum subsp. pseudobaeticum.
lugdunense *(Rouy) Üksip in Fl. SSSR,* 30: 86 (1960), homonym; *see Gand., Fl. Lyon.:* 142 (1875): H. virgultorum δ lugdunense.
lyccense *(Nägeli & Peter) Üksip in Fl. SSSR,* 30: 444 (1960): H. florentinum subsp. lyccense.
lychnaeum *(Norrl.) Üksip in Fl. SSSR,* 30: 655 (1960): Pilosella lychnaea.
macranthum (Ten.) Ten.
  subsp. testimoniale *(Peter) G. Gottschlich in Willdenowia,* 25(2): 507 (1996): H. hoppeanum subsp. testimoniale.
macrochaetium *(Nägeli & Peter) Üksip in Fl. SSSR,* 30: 406 (1960): H. procerum subsp. macrochaetium.
macrocymum *(Nägeli & Peter) Üksip in Fl. SSSR,* 30: 416 (1960): H. echioides subsp. macrocymum.
macrolepidiforme *(Zahn) Üksip in Fl. SSSR,* 30: 69 (1960): H. erythrocarpum subsp. macrolepidiforme.
macrolepidiforme *(Zahn) Üksip ex A.N. Sennikov in Bot. Zhurn.,* 83(2): 74 (1998), isonym; *see Üksip in Fl. SSSR,* 30: 69 (1960): H. erythrocarpum subsp. macrolepidiforme.
macrolepioides *(Zahn) Üksip in Fl. SSSR,* 30: 69 (1960): H. erythrocarpum subsp. macrolepioides.
macrolepioides *(Zahn) Üksip ex A.N. Sennikov in Bot. Zhurn.,* 83(2): 74 (1998), isonym; *see Üksip in Fl. SSSR,* 30: 69 (1960): H. erythrocarpum subsp. macrolepioides.
macrolepium *(Nägeli & Peter) Üksip in Fl. SSSR,* 30: 683 (1960): H. hoppeanum subsp. macrolepium.
macrolygodes *(Zahn) Üksip in Fl. SSSR,* 30: 101 (1960): H. laevigatum subsp. macrolygodes.
macrophyllopodum *(Zahn) Üksip in Fl. SSSR,* 30: 255 (1960): H. vulgatum subsp. macrophyllopodum.
macroradium *(Zahn) Üksip in Fl. SSSR,* 30: 500 (1960): H. auriculoides subsp. macroradium.
magnauricula *(Nägeli & Peter) Üksip in Fl. SSSR,* 30: 674 (1960): H. auricula subsp. magnauricula.
marginale *(Nägeli & Peter) Üksip in Fl. SSSR,* 30: 468 (1960): H. magyaricum subsp. marginale.
matrense *(Nägeli & Peter) Üksip in Fl. SSSR,* 30: 537 (1960), homonym; *see Schur in Verh. Naturf. Vereins Brünn,* 36: 230 (1897 publ. 1898): H. brachiatum subsp. matrense.
medianiforme *(Litv. & Zahn) Üksip in Fl. SSSR,* 30: 325 (1960): H. silvaticum subsp. medianiforme.
mediopositum *G. Gottschlich in Linz. Biol. Beitr.,* 32(1): 367 (2000)—Italy.

HIERACIUM:-

megalomastix (Nägeli & Peter) Üksip in Fl. SSSR, 30: 465 (1960): H. magyaricum subsp. megalomastix.
membranulatum (Litv. & Zahn) Üksip in Fl. SSSR, 30: 229 (1960): H. vulgatum subsp. membranulatum.
membranulatum (Litv. & Zahn) Üksip ex A.N. Sennikov in Bot. Zhurn., 83(2): 75 (1998), isonym; see Üksip in Fl. SSSR, 30: 229 (1960): H. vulgatum subsp. membranulatum.
miansarofii (Koslovsky & Zahn) Üksip in Fl. SSSR, 30: 62 (1960): H. pseudosvaneticum subsp. miansarofii.
miansarofii (Koslovsky & Zahn) Üksip ex A.N. Sennikov in Bot. Zhurn., 83(2): 75 (1998), isonym; see Üksip in Fl. SSSR, 30: 62 (1960): H. pseudosvaneticum subsp. miansarofii.
micrastrum (Zahn) Üksip in Fl. SSSR, 30: 583 (1960): H. dubium subsp. micrastrum.
microbauhinii (Zahn) Üksip in Fl. SSSR, 30: 530 (1960), as 'miro-bauhini': H. leptophyton subsp. microbauhinii.
microsphaericum (Zahn) Üksip in Fl. SSSR, 30: 649 (1960): H. piloselliflorum subsp. microsphaericum.
mirum (Nägeli & Peter) Üksip in Fl. SSSR, 30: 495 (1960): H. pannonicum subsp. mirum.
mohrungenense (Zahn) Üksip in Fl. SSSR, 30: 635 (1960): H. prussicum subsp. mohrungenense.
moscoviticum (Peter) Üksip in Fl. SSSR, 30: 647 (1960): H. flagellare subsp. moscoviticum.
multiceps (Nägeli & Peter) Üksip in Fl. SSSR, 30: 484 (1960): H. calodon subsp. multiceps.
murlainzii G. Mateo Sanz in An. Jard. Bot. Madrid, 54(1): 367 (1996)—Spain.
murorum L.
    subsp. rugianum G. Gottschlich in Bot. Rundbr. Mecklenburg-Vorpommern, 31: 15 (1998)—Germany.
nasimovae N.V. Stepanov in Turczaninowia, 1(1): 5 (1998)—Krasnoyarsk.
nassovicum (Zahn) G. Gottschlich in Feddes Repert., 108(1–2): 105 (1997): H. heterodoxiforme subsp. nassovicum.
neogelmii G. Gottschlich in Linz. Biol. Beitr., 32(1): 365 (2000)—Italy.
neorupicola P.D. Sell & C. West
    subsp. riofriori (Pau & Font Quer) M. Ibn Tattou in Bocconea, 8: 38 (1998): H. riofriori.
nigrisetum (Nägeli & Peter) Üksip in Fl. SSSR, 30: 463 (1960): H. magyaricum subsp. nigrisetum.
nigrocephalum G. Gottschlich in Linz. Biol. Beitr., 32(1): 374 (2000)—Italy.
niphocladum (Schelk. & Zahn) Üksip in Fl. SSSR, 30: 52 (1960): H. pseudosvaneticum subsp. niphocladum.
niphocladum (Schelk. & Zahn) Üksip ex A.N. Sennikov in Bot. Zhurn., 83(2): 75 (1998), isonym; see Üksip in Fl. SSSR, 30: 52 (1960): H. pseudosvaneticum subsp. niphocladum.
nivimontis (Oborný & Zahn) J. Chrtek f.in Preslia, 67(2): 104 (1995): H. nigrescens subsp. nivimontis.
nobile Gren. & Godr.
    subsp. bartlingii (Scheele) O. de Bolòs & J. Vigo, Fl. Països Catalans, 3: 1137 (1995): H. bartlingii.
    subsp. burserianum (Arv.-Touv.) O. de Bolòs & J. Vigo, Fl. Països Catalans, 3: 1137 (1995): H. burserianum.
    subsp. ibericum (Zahn) O. de Bolòs & J. Vigo, Fl. Països Catalans, 3: 1137 (1995): H. pyrenaicum subsp. ibericum.
    subsp. perclusum (Arv.-Touv.) O. de Bolòs & J. Vigo, Fl. Països Catalans, 3: 1137 (1995): H. perclusum.
    subsp. tarnense (Arv.-Touv. & Sudre) O. de Bolòs & J. Vigo, Fl. Països Catalans, 3: 1137 (1995): H. tarnense.
nydeggerorum G. Gottschlich in Bauhinia, 11(4): 243 (1996)—Turkey.
obscuribracteum (Nägeli & Peter) Üksip in Fl. SSSR, 30: 453 (1960): H. magyaricum subsp. obscuribracteum.
obscuricaule (Litv. & Zahn) Üksip in Fl. SSSR, 30: 30 (1960): H. chloroprenanthes subsp. obscuricaule.

HIERACIUM:-

obscuricaule (Litv. & Zahn) Üksip ex A.N. Sennikov in Bot. Zhurn., 83(2): 75 (1998), isonym; see Üksip in Fl. SSSR, 30: 30 (1960): H. chloroprenanthes subsp. obscuricaule.
obscuriceps (Dahlst.) R.N. Schljakov in Fl. Evropeĭskoĭ Chasti SSSR, 8: 254 (1989): H. irriguum subsp. obscuriceps.
obscuriceps (Dahlst.) Schljakov ex A.N. Sennikov in Bot. Zhurn., 83(3): 71 (1998), isonym; see Schljakov in Fl. Evropeĭskoĭ Chasti SSSR, 8: 254 (1989): H. irriguum subsp. obscuriceps.
ochrophyllum (Nägeli & Peter) Üksip in Fl. SSSR, 30: 485 (1960): H. calodon subsp. ochrophyllum.
oioense (Dahlst.) Üksip in Fl. SSSR, 30: 374 (1960): H. silvaticum subsp. oioense.
onosmaceum (Zahn) Üksip in Fl. SSSR, 30: 55 (1960): H. tschamkorijense subsp. onosmaceum.
osiliae (Dahlst.) Üksip in Fl. SSSR, 30: 352 (1960): H. caesium subsp. osiliae.
ovalifrons (Woronow & Zahn) Üksip in Fl. SSSR, 30: 315 (1960): H. sylvaticum subsp. ovalifrons.
pachycymigerum G. Gottschlich in Linz. Biol. Beitr., 32(1): 366 (2000)—Italy.
panaeoliforme (Pohle & Zahn) Üksip in Fl. SSSR, 30: 315 (1960): H. silvaticum subsp. panaeoliforme.
parreyssianum (Nägeli & Peter) Üksip in Fl. SSSR, 30: 498 (1960): H. pannonicum subsp. parreyssianum.
perasperum (Zahn) Üksip in Fl. SSSR, 30: 482 (1960): H. calodon subsp. perasperum.
pergrandidens (Zahn) Üksip in Fl. SSSR, 30: 177 (1960): H. atratum subsp. pergrandidens.
perileucum (Schelk. & Zahn) Üksip in Fl. SSSR, 30: 683 (1960): H. hoppeanum subsp. perileucum.
permense (Zahn) Üksip in Fl. SSSR, 30: 422 (1960): H. fallax subsp. permense.
petunnikovii (Peter) Üksip in Fl. SSSR, 30: 638 (1960): H. flagellare subsp. petunnikovii.
phrygicum (Zahn) Üksip in Fl. SSSR, 30: 410 (1960), as 'phrygium': H. procerum subsp. phrygicum.
pilisquamum (Nägeli & Peter) Üksip in Fl. SSSR, 30: 682 (1960): H. hoppeanum subsp. pilisquamum.
pinegense A.N. Sennikov in Bot. Zhurn., 84(12): 131 (1999)—Russia North.
pinetophilum (Degen & Zahn) J. Chrtek f.in Phyton (Austria), 37(2): 209 (1998): H. fritzei subsp. pinetophilum.
pleuroleucum (Dahlst.) Üksip in Fl. SSSR, 30: 326 (1960): H. serratifrons subsp. pleuroleucum.
plicatulum (Zahn) Üksip in Fl. SSSR, 30: 454 (1960): H. bauhinii subsp. plicatulum.
polatschekii G. Gottschlich in Linz. Biol. Beitr., 32(1): 371 (2000)—Austria.
polygonifolium G. Gottschlich & K. Cos,kunçelebi in Willdenowia, 30(2): 281 (2000)—Turkey.
prolixiforme Norrl.
    var. nemorum (Fr. ex Lindeb.) A.N. Sennikov in Bot. Zhurn., 84(3): 123 (1999): H. caesium var. nemorum.
prolongatum (Nägeli & Peter) Üksip in Fl. SSSR, 30: 568 (1960): H. glomeratum subsp. prolongatum.
psammophilum (Nägeli & Peter) Üksip in Fl. SSSR, 30: 483 (1960): H. calodon subsp. psammophilum.
pseudauricula (Nägeli & Peter) Üksip in Fl. SSSR, 30: 627 (1960): H. floribundum subsp. pseudauricula.
pseudauriculoides (Nägeli & Peter) Üksip in Fl. SSSR, 30: 471 (1960): H. magyaricum subsp. pseudauriculoides.
pseudodulacianum (De Retz) G. Mateo Sanz in An. Jard. Bot. Madrid, 54(1): 367 (1996): H. gastonianum subsp. pseudodulacianum.
pseudophyllodes (Zahn) Üksip in Fl. SSSR, 30: 173 (1960): H. nigrescens subsp. pseudophyllodes.
psilobrachion (Woronow & Zahn) Üksip in Fl. SSSR, 30: 534 (1960): H. brachiatum subsp. psilobrachion.
pujattii G. Gottschlich in Linz. Biol. Beitr., 32(1): 369 (2000)—Italy.
purpurascens Scheele ex Willk.
    subsp. berganum (Arv.-Touv.) O. de Bolòs & J. Vigo, Fl. Països Catalans, 3: 1080 (1995): H. berganum.

HIERACIUM:-

purpurascens
    subsp. lanuzae (Zahn) O. de Bolòs & J. Vigo, Fl. Països Catalans, 3: 1080 (1995): H. tephrocerinthe subsp. lanuzae.
    subsp. serdanyolae (Zahn) O. de Bolòs & J. Vigo, Fl. Països Catalans, 3: 1080 (1995): H. tephrocerinthe subsp. serdanyolae.
purpureibracteum (Zahn) Üksip in Fl. SSSR, 30: 540 (1960): H. leptocephalum subsp. purpureibracteum.
purpureovittatum (Zahn) Üksip in Fl. SSSR, 30: 540 (1960): H. leptophyton subsp. purpureovittatum.
puschlachtae (Pohle & Zahn) Üksip in Fl. SSSR, 30: 110 (1960): H. laevigatum subsp. puschlachtae.
pycnothyrsum (Peter) Üksip in Fl. SSSR, 30: 569 (1960): H. glomeratum subsp. pycnothyrsum.
radiatellum (Woronow & Zahn) Üksip in Fl. SSSR, 30: 313 (1960): H. silvaticum subsp. radiatellum.
regiomontanum (Nägeli & Peter) Üksip in Fl. SSSR, 30: 609 (1960): H. floribundum subsp. regiomontanum.
rehmannii (Nägeli & Peter) Üksip in Fl. SSSR, 30: 566 (1960), homonym; see Woł. in Spraw. Komis. Fizjogr., 21: 128 (1888): H. cruentum subsp. rehmannii.
retroversilobatum (Schelk. & Zahn) Üksip in Fl. SSSR, 30: 319 (1960): H. murorum subsp. retroversilobatum.
rizense G. Gottschlich & K. Coşkunçelebi in Willdenowia, 30(2): 279 (2000)—Turkey.
rojowskii (Rehmann) Üksip in Fl. SSSR, 30: 452 (1960): H. magyaricum subsp. rojowskii.
rostanii Nägeli & Peter
    subsp. centroalpinum G. Gottschlich & G. Brandstätter in Linz. Biol. Beitr., 32(1): 376 (2000)—Austria, Italy.
rubricymigerum (Nägeli & Peter) Üksip in Fl. SSSR, 30: 565 (1960): H. cruentum subsp. rubricymigerum.
rubrobauhinii (Schelk. & Zahn) Üksip in Fl. SSSR, 30: 460 (1960): H. bauhinii subsp. rubrobauhinii.
rubropannonicum (Litv. & Zahn) Üksip in Fl. SSSR, 30: 486 (1960): H. auriculoides subsp. rubropannonicum.
rupicola Jord.
    subsp. aldeanum (Arv.-Touv.) O. de Bolòs & J. Vigo, Fl. Països Catalans, 3: 1118 (1995): H. aldeanum.
    subsp. sonchifolium (Scheele) O. de Bolòs & J. Vigo, Fl. Països Catalans, 3: 1118 (1995): H. sonchifolium.
rupicoliforme Zahn
    subsp. curunense G. Gottschlich in Linz. Biol. Beitr., 32(1): 375 (2000)—Italy.
rusanum (Zahn) Üksip in Fl. SSSR, 30: 570 (1960): H. ambiguum subsp. rusanum.
sabiniceps (Litv. & Zahn) Üksip in Fl. SSSR, 30: 499 (1960): H. auriculoides subsp. sabiniceps.
sabinopsis (Ganesch. & Zahn) Üksip in Fl. SSSR, 30: 421 (1960): H. fallax subsp. sabinopsis.
sabulosorum (Dahlst.) K. Johansson in Kongl. Svenska Vetenskapsakad. Handl., 29(1): 123 (1897): H. peleterianum subsp. sabulosorum.
samurense (Zahn) Üksip in Fl. SSSR, 30: 68 (1960): H. erythrocarpum subsp. samurense.
samurense (Zahn) Üksip ex A.N. Sennikov in Bot. Zhurn., 83(2): 76 (1998), isonym; see Üksip in Fl. SSSR, 30: 68 (1960): H. erythrocarpum subsp. samurense.
scabiosum H. Sudre in Bull. Assoc. Pyrén. Échange Pl., 9: 10 (1898–99 publ. 1899)—France.
schmidtii Tausch
    subsp. arcicola (Arv.-Touv.) O. de Bolòs & J. Vigo, Fl. Països Catalans, 3: 1044 (1995): H. arcicola.
    subsp. barbulatiforme (Zahn) O. de Bolòs & J. Vigo, Fl. Països Catalans, 3: 1044 (1995): H. pallidum subsp. barbulatiforme.
    subsp. brunelliforme (Arv.-Touv.) O. de Bolòs & J. Vigo, Fl. Països Catalans, 3: 1044 (1995): H. brunelliforme.
    subsp. ceratodon (Arv.-Touv.) O. de Bolòs & J. Vigo, Fl. Països Catalans, 3: 1045 (1995): H. ceratodon.

HIERACIUM:-
schmidtii
subsp. comatulum *(Jord.) O. de Bolòs & J. Vigo, Fl. Països Catalans,* 3: 1044 (1995): H. comatulum.
subsp. comosulum *(Arv.-Touv. & Gaut.) O. de Bolòs & J. Vigo, Fl. Països Catalans,* 3: 1045 (1995): H. comosulum.
subsp. cyaneum *(Arv.-Touv.) O. de Bolòs & J. Vigo, Fl. Països Catalans,* 3: 1044 (1995): H. cyaneum.
subsp. echinanthum *(Arv.-Touv. & Gaut.) O. de Bolòs & J. Vigo, Fl. Països Catalans,* 3: 1044 (1995): H. echinanthum.
subsp. glaucopallens *(Zahn) O. de Bolòs & J. Vigo, Fl. Països Catalans,* 3: 1044 (1995): H. pallidum subsp. glaucopallens.
subsp. graniticum *(Sch. Bip.) O. de Bolòs & J. Vigo, Fl. Països Catalans,* 3: 1044 (1995), with incorrect basionym ref.: H. graniticum.
subsp. graniticum *(Sch. Bip.) G. Gottschlich in Mitt. Pollich.,* 76: 79 (1989): H. graniticum.
subsp. lasiophyllum *(W.D.J. Koch) O. de Bolòs & J. Vigo, Fl. Països Catalans,* 3: 1044 (1995): H. lasiophyllum.
subsp. pallidiglaucinum *(Zahn) G. Gottschlich in Soc. Éch. Pl. Vasc. Eur. Bassin Médit. Bull.,* 27: 86 (1998): H. pallidum subsp. pallidiglaucinum.
subsp. pallidum *(Biv.) O. de Bolòs & J. Vigo, Fl. Països Catalans,* 3: 1044 (1995): H. pallidum.
subsp. rupicola *(Fr.) O. de Bolòs & J. Vigo, Fl. Països Catalans,* 3: 1045 (1995): H. bifidum var. rupicola.
subsp. schmidtiiforme *(Zahn) O. de Bolòs & J. Vigo, Fl. Països Catalans,* 3: 1044 (1995): H. pallidum subsp. schmidtiiforme.
subsp. semirupicola *(Zahn) O. de Bolòs & J. Vigo, Fl. Països Catalans,* 3: 1045 (1995): H. pallidum subsp. semirupicola.
subsp. subcaesioides *(Holle) G. Gottschlich in Braunschweig. Naturk. Schriften,* 5(4): 813 (1999): H. subcaesioides.
subsp. subcomatulum *(Zahn) O. de Bolòs & J. Vigo, Fl. Països Catalans,* 3: 1044 (1995): H. pallidum subsp. subcomatulum.
subsp. trichellum *(Arv.-Touv.) O. de Bolòs & J. Vigo, Fl. Països Catalans,* 3: 1045 (1995): H. trichellum.
subsp. vestitum *(Gren. & Godr.) O. de Bolòs & J. Vigo, Fl. Països Catalans,* 3: 1044 (1995): H. vestitum.
semicymigerum *(Zahn) Üksip in Fl. SSSR,* 30: 513 (1960): H. tauschii subsp. semicymigerum.
semilimbatum *A.N. Sennikov in Bot. Zhurn.,* 84(7): 125 (1999)—Belorussiya.
sericicaule *(Schelk. & Zahn) Üksip in Fl. SSSR,* 30: 55 (1960): H. tschamkorijense subsp. sericicaule.
sericicaule *(Schelk. & Zahn) Üksip ex A.N. Sennikov in Bot. Zhurn.,* 83(2): 76 (1998), isonym; *see Üksip in Fl. SSSR,* 30: 55 (1960): H. tschamkorijense subsp. sericicaule.
signiferum *(Norrl.) Üksip in Fl. SSSR,* 30: 548 (1960): Pilosella signifera.
slovacum *J. Chrtek f. in Phyton (Austria),* 37(2): 203 (1998)—Czechoslovakia.
sobrinatum *(Litv. & Zahn) Üksip in Fl SSSR,* 30: 66 (1960): H. pseudosvaneticum subsp. sobrinatum.
sparsum Friv.
subsp. paulii *Z. Szeląg in Feddes Repert.,* 111(5–6): 257 (2000)—Italy.
spodocephalum *G. Gottschlich & K. Coşkunçelebi in Willdenowia,* 30(2): 281 (2000)—Turkey.
stellatum *(Tausch ex Zahn) Üksip in Fl. SSSR,* 30: 442 (1960): H. florentinum subsp. stellatum.
stenozon *(Zahn) Üksip in Fl. SSSR,* 30: 649 (1960): H. piloselliflorum subsp. stenozon.
sterromastix *(Nägeli & Peter) Üksip in Fl. SSSR,* 30: 432 (1960): H. bifurcum subsp. sterromastix.
stranigense *G. Gottschlich in Linz. Biol. Beitr.,* 32(1): 373 (2000)—Austria.
strictiramum *(Nägeli & Peter) Üksip in Fl. SSSR,* 30: 483 (1960): H. calodon subsp. strictiramum.
stupposipilum *(Woronow & Zahn) Üksip in Fl. SSSR,* 30: 445 (1960): H. florentinum subsp. stupposipilum.

HIERACIUM:-
subambiguum *(Nägeli & Peter) Üksip in Fl. SSSR,* 30: 570 (1960): H. glomeratum subsp. subambiguum.
subasperellum *(Zahn) Üksip in Fl. SSSR,* 30: 247 (1960): H. vulgatum subsp. subasperellum.
subatriceps *(Zahn) Üksip in Fl. SSSR,* 30: 677 (1960): H. schultesii subsp. subatriceps.
subauricula *(Nägeli & Peter) Üksip in Fl. SSSR,* 30: 627 (1960), homonym; *see Schur in Verh. Naturf. Vereins Brünn,* 36: 232 (1897 publ. 1898): H. floribundum subsp. subauricula.
subbauhiniiflorum *(Woronow & Zahn) Üksip in Fl. SSSR,* 30: 531 (1960): H. leptophyton subsp. subbauhiniiflorum.
subcaesioides *G. von Holle, Beobacht. Hohensteine Weserkette hybr. Form. Gatt. Hieracium:* 5 (1892) Germany.
subcrassifolium *(Zahn) Üksip in Fl. SSSR,* 30: 307 (1960): H. murorum subsp. subcrassifolium.
subfallaciforme *(Zahn) Üksip in Fl. SSSR,* 30: 425 (1960): H. fuckelianum subsp. subfallaciforme.
subfariniramum *(Ganesch. & Zahn) Üksip in Fl. SSSR,* 30: 268 (1960): H. vulgatum subsp. subfariniramum.
subfiliferum *(Zahn) Üksip in Fl. SSSR,* 30: 466 (1960): H. bauhinii subsp. subfiliferum.
subfloribundum *(Nägeli & Peter) Üksip in Fl. SSSR,* 30: 582 (1960), homonym; *see Schur in Verh. Naturf. Vereins Brünn,* 36: 248 (1897 publ. 1898): H. floribundum β subfloribundum.
subhastulatum *(Zahn) Üksip in Fl. SSSR,* 30: 257 (1960): H. vulgatum subsp. subhastulatum.
subincomptum *(Zahn) Üksip in Fl. SSSR,* 30: 175 (1960): H. nigrescens subsp. subincomptum.
sublasiophorum *(Litv. & Zahn) Üksip in Fl. SSSR,* 30: 496 (1960): H. auriculoides subsp. sublasiophorum.
submaculosum *(Dahlst.) Üksip in Fl. SSSR,* 30: 361 (1960): H. silvaticum subsp. submaculosum.
submarginellum *(Zahn) Üksip in Fl. SSSR,* 30: 307 (1960): H. silvaticum subsp. submarginellum.
submedianum *(Zahn) Üksip in Fl. SSSR,* 30: 282 (1960): H. diaphanoides subsp. submedianum.
submirum *(Litv. & Zahn) Üksip in Fl. SSSR,* 30: 497 (1960): H. auriculoides subsp. submirum.
subnigriceps *(Zahn) Üksip in Fl. SSSR,* 30: 631 (1960): H. iseranum subsp. subnigriceps.
subobscuriceps *(Zahn) Üksip in Fl. SSSR,* 30: 251 (1960): H. vulgatum subsp. subobscuriceps.
subramosum Lönnr.
subsp. basilare *(Wiinst.) G. Gottschlich in Bot. Rundbr. Mecklenburg-Vorpommern,* 31: 65 (1998): H. basilare.
subrubellum *(Schelk. & Zahn) Üksip in Fl. SSSR,* 30: 539 (1960): H. tephrocephalum subsp. subrubellum.
substoloniferum *(Nägeli & Peter) Üksip in Fl. SSSR,* 30: 467 (1960): H. magyaricum subsp. substoloniferum.
subsvaneticum *(Litv. & Zahn) Üksip in Fl. SSSR,* 30: 67 (1960): H. pseudosvaneticum subsp. subsvaneticum.
subsvaneticum *(Litv. & Zahn) Üksip ex A.N. Sennikov in Bot. Zhurn.,* 83(2): 77 (1998), isonym; *see Üksip in Fl. SSSR,* 30: 67 (1960): H. pseudosvaneticum subsp. subsvaneticum.
subumbelliforme *(Zahn) Üksip in Fl. SSSR,* 30: 501 (1960): H. auriculoides subsp. subumbelliforme.
subviolascentiforme *(Pohle & Zahn) Üksip in Fl. SSSR,* 30: 229 (1960): H. vulgatum subsp. subviolascentiforme.
sudavicum *(Nägeli & Peter) Üksip in Fl. SSSR,* 30: 609 (1960): H. floribundum subsp. sudavicum.
sulphurelliforme *(Koslovsky & Zahn) Üksip in Fl. SSSR,* 30: 41 (1960): H. litwinowianum subsp. sulphurelliforme.
sulphurellum *(Koslovsky & Zahn) Üksip in Fl. SSSR,* 30: 41 (1960): H. litwinowianum subsp. sulphurellum.
syrjaenorum *(Norrl.) Üksip in Fl. SSSR,* 30: 547 (1960): Pilosella syrjaenorum.
sysolskiense *(Zahn) Üksip in Fl. SSSR,* 30: 581 (1960): H. dubium subsp. sysolskiense.
szovitsii *(Nägeli & Peter) Üksip in Fl. SSSR,* 30: 433 (1960): H. bifurcum subsp. szovitsii.

HIERACIUM:-
tabergense *(Dahlst.) Üksip in Fl. SSSR,* 30: 551 (1960): H. cymosum subsp. tabergense.
tanythrix *(Nägeli & Peter) Üksip in Fl. SSSR,* 30: 500 (1960): H. pannonicum subsp. tanythrix.
tatrense *(Nägeli & Peter) Üksip in Fl. SSSR,* 30: 643 (1960): H. flagellare subsp. tatrense.
taurinense Jord.
subsp. elocatum *Huter ex G. Gottschlich in Linz. Biol. Beitr.,* 32(1): 377 (2000)—Austria.
teberdense *(Litv. & Zahn) Üksip in Fl. SSSR,* 30: 133 (1960): H. inuloides subsp. teberdense.
teberdifontis *(Litv. & Zahn) Üksip in Fl. SSSR,* 30: 491 (1960): H. auriculoides subsp. teberdifontis.
tenuiceps *(Nägeli & Peter) Üksip in Fl. SSSR,* 30: 482 (1960), homonym; *see Omang in Nyt Mag. Naturvidensk.,* 56: 75 (1919): H. calodon subsp. tenuiceps.
tephrantheloides *(Zahn) Üksip in Fl. SSSR,* 30: 634 (1960): H. prussicum subsp. tephrantheloides.
tephranthelum *(Zahn) Üksip in Fl. SSSR,* 30: 566 (1960): H. norrliniiforme subsp. tephranthelum.
tephrochlamys *M. Gandoger, Dec. Pl. Nov.,* 1: 28 (1875)—Hungary.
tephrochlorellum *(Ganesch. & Zahn) Üksip in Fl. SSSR,* 30: 426 (1960): H. heterodoxum subsp. tephrochlorellum.
tephrophilum *(Koslovsky & Zahn) Üksip in Fl. SSSR,* 30: 223 (1960): H. knafii subsp. tephrophilum.
tephrophilum *(Koslovsky & Zahn) Üksip ex A.N. Sennikov in Bot. Zhurn.,* 83(2): 77 (1998), isonym; *see Üksip in Fl. SSSR,* 30: 223 (1960): H. knafii subsp. tephrophilum.
tephropodum *(Zahn) Üksip in Fl. SSSR,* 30: 543 (1960): H. ruprechtii subsp. tephropodum.
terekianum *(Litv. & Zahn) Üksip in Fl. SSSR,* 30: 33 (1960): H. dacicum subsp. terekianum.
tersundagense *G. Gottschlich & K. Coşkunçelebi in Willdenowia,* 30(2): 280 (2000)—Turkey.
thracicum *(Nägeli & Peter) Üksip in Fl. SSSR,* 30: 498 (1960): H. pannonicum subsp. thracicum.
tonalense *G. Gottschlich in Linz. Biol. Beitr.,* 32(1): 372 (2000)—Italy.
transbalticum *(Dahlst.) Üksip in Fl. SSSR,* 30: 585 (1960): H. poliodermum subsp. transbalticum.
tricheilema *(Nägeli & Peter) Üksip in Fl. SSSR,* 30: 674 (1960): H. auricula subsp. tricheilema.
trichocymosum *(Zahn) Üksip in Fl. SSSR,* 30: 548 (1960): H. cymosum subsp. trichocymosum.
tridentaticeps *(Zahn) Üksip in Fl. SSSR,* 30: 113 (1960): H. knafii subsp. tridentaticeps.
tridentaticeps *(Zahn) Üksip ex A.N. Sennikov in Bot. Zhurn.,* 83(2): 77 (1998), isonym; *see Üksip in Fl. SSSR,* 30: 113 (1960): H. knafii subsp. tridentaticeps.
tubuliflorum *(Nägeli & Peter) Üksip in Fl. SSSR,* 30: 534 (1960): H. brachiatum subsp. tubuliflorum.
turkestanicum *(Zahn) Üksip in Fl. SSSR,* 30: 217 (1960): H. pohlei subsp. turkestanicum.
tuscheticum *(Zahn) Üksip in Fl. SSSR,* 30: 543 (1960): H. ruprechtii subsp. tuscheticum.
tzagwericum *(Koslovsky & Zahn) Üksip in Fl. SSSR,* 30: 48 (1960), as 'tzagwerianum': H. callichlorum subsp. tzagwericum.
umbellaticeps *(Pohle & Zahn) Üksip in Fl. SSSR,* 30: 96 (1960): H. laevigatum subsp. umbellaticeps.
umbellosum *(Nägeli & Peter) Üksip in Fl. SSSR,* 30: 499 (1960): H. pannonicum subsp. umbellosum.
ussense *(Pohle & Zahn) Üksip in Fl. SSSR,* 30: 192 (1960): H. atratum subsp. ussense.
variegatisquamum *(Zahn) Üksip in Fl. SSSR,* 30: 75 (1960): H. erythrocarpum subsp. variegatisquamum.
venostorum *(Zahn) G. Gottschlich in Veröff. Tiroler Landesmus. Ferdinandeum,* 79: 201 (1999 publ. 2000): H. illyricum subsp. venostorum.
villosellipes *(Zahn) Üksip in Fl. SSSR,* 30: 62 (1960): H. pseudosvaneticum subsp. villosellipes.
villosellipes *(Zahn) Üksip ex A.N. Sennikov in Bot. Zhurn.,* 83(2): 77 (1998), isonym; *see Üksip in Fl. SSSR,* 30: 62 (1960): H. pseudosvaneticum subsp. villosellipes.

HIERACIUM:-
vindobonae *(Zahn) Üksip in Fl. SSSR,* 30: 431 (1960): H. bifurcum subsp. vindobonae.
violaceipes *(Zahn) Üksip in Fl. SSSR,* 30: 588 (1960): H. sciadophorum subsp. violaceipes.
violascentiforme *(Pohle & Zahn) Üksip in Fl. SSSR,* 30: 269 (1960): H. vulgatum subsp. violascentiforme.
virenticeps *(Dahlst.) Üksip in Fl. SSSR,* 30: 265 (1960), homonym; *see Norrl. in Acta Soc. Fauna Fl. Fenn.,* 36(4): 37 (1912): H. porrigens subsp. virenticeps.
virentisquamum *(Nägeli & Peter) Üksip in Fl. SSSR,* 30: 681 (1960): H. hoppeanum subsp. virentisquamum.
viscidulum *(Tausch ex Nägeli & Peter) Üksip in Fl. SSSR,* 30: 461 (1960): H. magyaricum subsp. viscidulum.
vogesiacum (Kirschl.) Fr.
subsp. angliciforme *(Zahn) O. de Bolòs & J. Vigo, Fl. Països Catalans,* 3: 1106 (1995): H. mougeotii subsp. angliciforme.
subsp. asturicum *(Arv.-Touv.) O. de Bolòs & J. Vigo, Fl. Països Catalans,* 3: 1106 (1995): H. asturicum.
subsp. attractum *(Arv.-Touv.) O. de Bolòs & J. Vigo, Fl. Països Catalans,* 3: 1106 (1995): H. attractum.
subsp. benthamianum *(Arv.-Touv. & Gaut.) O. de Bolòs & J. Vigo, Fl. Països Catalans,* 3: 1106 (1995): H. benthamianum.
subsp. chrysanthemoides *(Arv.-Touv. & Gaut.) O. de Bolòs & J. Vigo, Fl. Països Catalans,* 3: 1106 (1995): H. chrysanthemoides.
subsp. doronicoides *(Arv.-Touv.) O. de Bolòs & J. Vigo, Fl. Països Catalans,* 3: 1106 (1995): H. doronicoides.
subsp. lagopinum *(Arv.-Touv. & Gaut.) O. de Bolòs & J. Vigo, Fl. Països Catalans,* 3: 1106 (1995): H. lagopinum.
subsp. stelliceps *(Zahn) O. de Bolòs & J. Vigo, Fl. Països Catalans,* 3: 1106 (1995): H. mougeotii subsp. stelliceps.
subsp. tournefortianum *(Arv.-Touv. & Gaut.) O. de Bolòs & J. Vigo, Fl. Països Catalans,* 3: 1106 (1995): H. tournefortianum.
volhynicum *(Nägeli & Peter) Üksip in Fl. SSSR,* 30: 468 (1960): H. magyaricum subsp. volhynicum.
walteri-langii *G. Gottschlich in Mitt. Pollich.,* 77: 183 (1990)—Germany.
wjazowoense *(Zahn) Üksip in Fl. SSSR,* 30: 511 (1960), as 'wjasowoense': H. tauschii subsp. wjazowoense.
wologdense *(Pohle & Zahn) Üksip in Fl. SSSR,* 30: 219 (1960): H. vulgatum subsp. wologdense.

HIERONYMIELLA (Amaryllidac.).
sect. Synexa *P. Ravenna in Onira, Bot. Leafl.,* 4(7): 27 (2000).
angacoana *P. Ravenna in Onira, Bot. Leafl.,* 4(7): 26 (2000)—Argentina Northwest (San Juan).
ferruginea *P. Ravenna in Onira, Bot. Leafl.,* 4(7): 25 (2000)—Argentina Northwest (Salta).
tarijensis *(Fern. Casas & R. Lara) F.J. Fernández Casas & R. Lara in Fontqueria,* 44: 244 (1996): Androstephanos tarijensis.

HILDEBRANDTIA (Convolvulac.).
africana Vatke
subsp. arabica *Sebsebe Demissew in Kew Bull.,* 51(3): 529 (1996)—Oman, Saudi Arabia, North Yemen.
aloysii *(Chiov.) Sebsebe Demissew in Kew Bull.,* 51(3): 539 (1996): Sabaudiella aloysii.
diredawaensis *Sebsebe Demissew in Kew Bull.,* 51(3): 531 (1996)—Ethiopia, Somalia.
obcordata S. Moore
var. puberula *Sebsebe Demissew in Kew Bull.,* 51(3): 531 (1996)—Ethiopia.
promontorii *T. Deroin in Candollea,* 51(1): 147 (1996)—Madagascar.
valo *T. Deroin in Candollea,* 51(1): 149 (1996)—Madagascar.

HILGERIA *H. Förther in Sendtnera,* 5: 132 (1998). *BORAGINACEAE*
hypogaea *(Urb. & Ekman) H. Förther in Sendtnera,* 5: 133 (1998): Heliotropium hypogaeum.

HILGERIA:-
pedicellaris *(Urb. & Ekman) H. Förther in Sendtnera,* 5: 133 (1998): Heliotropium pedicellare.
serpylloides *(Griseb.) H. Förther in Sendtnera,* 5: 133 (1998): Heliotropium serpylloides.

HILLIARDIELLA *H. Robinson in Proc. Biol. Soc. Wash.,* 112(1): 229 (1999). *COMPOSITAE*
aristata *(DC.) H. Robinson in Proc. Biol. Soc. Wash.,* 112(1): 230 (1999): Webbia aristata.
calyculata *(S. Moore) H. Robinson in Proc. Biol. Soc. Wash.,* 112(1): 230 (1999): Vernonia calyculata.
hirsuta *(Sch. Bip. ex Walp.) H. Robinson in Proc. Biol. Soc. Wash.,* 112(1): 230 (1999): Vernonia hirsuta.
leopoldii *(Vatke) H. Robinson in Proc. Biol. Soc. Wash.,* 112(1): 230 (1999): Vernonia leopoldii.
nudicaulis *(DC.) H. Robinson in Proc. Biol. Soc. Wash.,* 112(1): 230 (1999): Webbia nudicaulis.
oligocephala *(DC.) H. Robinson in Proc. Biol. Soc. Wash.,* 112(1): 230 (1999): Webbia oligocephala.
pinifolia *(Less.) H. Robinson in Proc. Biol. Soc. Wash.,* 112(1): 230 (1999): Vernonia pinifolia.
smithiana *(Less.) H. Robinson in Proc. Biol. Soc. Wash.,* 112(1): 230 (1999): Vernonia smithiana.

HILLIELLA (Crucif.).
xiangguiensis *Y.H. Zhang in Acta Bot. Yunnanica,* 19(2): 139 (1997)—China Southeast (Hunan, Guangxi).

HIMALAYACALAMUS (Gramin.).
collaris *(T.P. Yi) D. Ohrnberger, Bamboos of the World Intro. 3:* 14 (1996): Fargesia collaris.
gyirongensis *(T.P. Yi) D. Ohrnberger, Bamboos of the World Intro. 3:* 14 (1996): Fargesia gyirongensis.

HIMANTOGLOSSUM (Orchidac.).
comperianum *(Steven) P. Delforge in Nat. Belg.,* 80(3): 401 (1999): Orchis comperiana.
metlesicsianum *(W.P. Teschner) P. Delforge in Nat. Belg.,* 80(3): 401 (1999): Barlia metlesicsiana.
montis-tauri *C.A.J. Kreutz & W. Lüders in J. Eur. Orchideen,* 29(4): 655 (1997)—Turkey.
robertianum *(Loisel.) P. Delforge in Nat. Belg.,* 80(3): 401 (1999): Orchis robertiana.

HINDSIA (Rubiac.).
arianeae *F.R. Di Maio in Bradea,* 8(9): 46 (1997)—Brazil Southeast (Rio de Janeiro).
cucullata *F.R. Di Maio in Bradea,* 8(9): 46 (1997)—Brazil Southeast (Rio de Janeiro).
ibitipocensis *F.R. Di Maio in Bradea,* 8(9): 47 (1997)—Brazil Southeast (Minas Gerais).
longiflora (Cham.) Benth.
subsp. colorata *F.R. Di Maio in Bradea,* 8(9): 48 (1997)—Brazil Southeast (Rio de Janeiro).
republicana *F.R. Di Maio in Bradea,* 8(9): 49 (1997)—Brazil Southeast (Rio de Janeiro).
sessilifolia *F.R. Di Maio in Bradea,* 8(9): 50 (1997)—Brazil Northeast (Bahia).

HIPPEASTRUM (Amaryllidac.).
amaru *(Vargas) A.W. Meerow in Taxon,* 46(1): 16 (1997): Amaryllis amaru.
anzaldoi *(Cárdenas) J. Van Scheepen in Taxon,* 46(1): 16 (1997): Amaryllis anzaldoi.
araripinum *(Ravenna) A.W. Meerow in Taxon,* 46(1): 16 (1997): Amaryllis araripina.
arboricola *(Ravenna) A.W. Meerow in Taxon,* 46(1): 16 (1997): Amaryllis arboricola.
argilagae *(Traub) J.H.A. Dutilh in Taxon,* 46(1): 16 (1997): Amaryllis argilagae.
aviflorum *(Ravenna) J.H.A. Dutilh in Taxon,* 46(1): 17 (1997): Amaryllis aviflora.
blossfeldiae *(Traub & J.L. Doran) J. Van Scheepen in Taxon,* 46(1): 17 (1997): Amaryllis blossfeldiae.
brasilianum *(Traub & J.L. Doran) J.H.A. Dutilh in Taxon,* 46(1): 17 (1997): Amaryllis brasiliana.
caiaponicum *(Ravenna) J.H.A. Dutilh in Taxon,* 46(1): 17 (1997): Amaryllis caiaponica.
caupolicanense *(Cárdenas) J. Van Scheepen in Taxon,* 46(1): 17 (1997): Amaryllis caupolicanensis.
chionedyanthum *(Cárdenas) J. Van Scheepen in Taxon,* 46(1): 17 (1997): Amaryllis chionedyantha.
condemaitae *(Vargas & F. Peréz) A.W. Meerow in Taxon,* 46(1): 17 (1997): Amaryllis condemaitae.

HIPPEASTRUM:-
curitibanum *(Ravenna) J.H.A. Dutilh in Taxon,* 46(1): 17 (1997): Amaryllis curitibana.
divijulianum *(Cárdenas) A.W. Meerow in Taxon,* 46(1): 17 (1997), as 'divijuliani': Amaryllis divijulianus.
doraniae *(Traub) A.W. Meerow in Taxon,* 46(1): 17 (1997): Amaryllis doraniae.
escobaruriae *(Cárdenas) J. Van Scheepen in Taxon,* 46(1): 17 (1997): Amaryllis escobaruriae.
fragrantissimum *(Cárdenas) A.W. Meerow in Taxon,* 46(1): 17 (1997): Amaryllis fragrantissima.
gertianum *(Ravenna) J.H.A. Dutilh in Taxon,* 46(1): 17 (1997): Amaryllis gertiana.
goianum *(Ravenna) A.W. Meerow in Taxon,* 46(1): 17 (1997): Amaryllis goiana.
guarapuavicum *(Ravenna) J. Van Scheepen in Taxon,* 46(1): 17 (1997): Amaryllis guarapuavica.
hemographes *(Ravenna) J.H.A. Dutilh in Taxon,* 46(1): 17 (1997): Amaryllis hemographes.
illustre *(Vell.) J.H.A. Dutilh in Taxon,* 46(1): 17 (1997): Amaryllis illustris.
incachacanum *(Cárdenas) J. Van Scheepen in Taxon,* 46(1): 17 (1997): Amaryllis incachacana.
kromeri *(Worsley) A.W. Meerow in Taxon,* 46(1): 17 (1997): Amaryllis kromeri.
lapacense *(Cárdenas) J. Van Scheepen in Taxon,* 46(1): 17 (1997): Amaryllis lapacensis.
leucobasis *(Ravenna) J.H.A. Dutilh in Taxon,* 46(1): 18 (1997): Amaryllis leucobasis.
marumbiense *(Ravenna) J. Van Scheepen in Taxon,* 46(1): 18 (1997): Amaryllis marumbiensis.
minasgerais *(Traub) A.W. Meerow in Taxon,* 46(1): 18 (1997): Amaryllis minasgerais.
mollevillquense *(Cárdenas) J. Van Scheepen in Taxon,* 46(1): 18 (1997): Amaryllis mollevillquensis.
monanthum *(Ravenna) A.W. Meerow in Taxon,* 46(1): 18 (1997): Amaryllis monantha.
nelsonii *(Cárdenas) J. Van Scheepen in Taxon,* 46(1): 18 (1997): Amaryllis nelsonii.
papilio *(Ravenna) J. Van Scheepen in Taxon,* 46(1): 18 (1997): Amaryllis papilio.
paquichanum *(Cárdenas) J.H.A. Dutilh in Taxon,* 46(1): 18 (1997): Amaryllis paquichana.
paradisiacum *(Ravenna) A.W. Meerow in Taxon,* 46(1): 18 (1997): Amaryllis paradisiaca.
paranaense *(Traub) A.W. Meerow in Taxon,* 46(1): 18 (1997): Amaryllis paranaensis.
pilcomaicum *(Ravenna) A.W. Meerow in Taxon,* 46(1): 18 (1997): Amaryllis pilcomaica.
rubropictum *(Ravenna) A.W. Meerow in Taxon,* 46(1): 18 (1997): Amaryllis rubropicta.
santacatarina *(Traub) J.H.A. Dutilh in Taxon,* 46(1): 18 (1997): Amaryllis santacatarina.
starkiorum *(I.S. Nelson & Traub) J. Van Scheepen in Taxon,* 46(1): 18 (1997): Amaryllis starkiorum
teyucuarense *(Ravenna) J. Van Scheepen in Taxon,* 46(1): 18 (1997): Amaryllis teyucuarensis.
umabisanum *(Cárdenas) A.W. Meerow in Taxon,* 46(1): 18 (1997): Amaryllis umabisana.
yungacense *(Cárdenas & I.S. Nelson) A.W. Meerow in Taxon,* 46(1): 18 (1997): Amaryllis yungacensis.

HIPPEOPHYLLUM (Orchidac.).
pumilum *Fukuy. ex S.C. Chen & K.Y. Lang in Acta Phytotax. Sin.,* 36(1): 72 (1998)—Taiwan.
sinicum *S.C. Chen & K.Y. Lang in Acta Phytotax. Sin.,* 36(1): 70 (1998)—China North-central (Gansu).

HIPPOCREPIS (Legumin.).
subsect. Comosae *(Hrabětova) S. Talavera & E. Domínguez in An. Jard. Bot. Madrid,* 57(2): 456 (1999 publ. 2000): H. ser. Comosae.
subsect. Eriocarpae *(Hrabětova) S. Talavera & E. Domínguez in An. Jard. Bot. Madrid,* 57(2): 460 (1999 publ. 2000): H. ser. Eriocarpae.
subsect. Occidentales *(Hrabětova) S. Talavera & E. Domínguez in An. Jard. Bot. Madrid,* 57(2): 462 (1999 publ. 2000): H. sect. Occidentales.
subsect. Scorpioideae *S. Talavera & E. Domínguez in An. Jard. Bot. Madrid,* 57(2): 459 (1999 publ. 2000).
castroviejoi *S. Talavera & E. Domínguez in An. Jard. Bot. Madrid,* 57(2): 461 (1999 publ. 2000)—Spain.

HIPPOCREPIS:-
grosii *(Pau) H. Boira, L. Gil & L. Llorens in Flora Medit.*, 5: 170 (1995 publ. 1996): H balearica var. grosii.
grosii *(Pau) S. Talavera & E. Domínguez in An. Jard. Bot. Madrid*, 57(2): 459 (1999 publ. 2000), isonym; *see Boira, Gil & L. Llorens in Flora Medit.*, 5: 170 (1995 publ. 1996): H. balearica var. grosii.
nevadensis *(Hrabětova) S. Talavera & E. Domínguez in An. Jard. Bot. Madrid*, 57(2): 460 (1999 publ. 2000): H. comosa f. nevadensis.
tavera-mendozae *S. Talavera & E. Domínguez in An. Jard. Bot. Madrid*, 57(2): 457 (1999 publ. 2000)—Spain.

HIPPOPHAE (Elaeagnac.).
rhamnoides L.
subsp. neurocarpa *(S.W. Liu & T.N. He) J. Hyvönen in Nordic J. Bot.*, 16(1): 59 (1996): H. neurocarpa.
salicifolia D. Don
subsp. sinensis *(Rousi) J. Hyvönen in Nordic J. Bot.*, 16(1): 60 (1996): H. rhamnoides subsp. sinensis.
subsp. yunnanensis *(Rousi) J. Hyvönen in Nordic J. Bot.*, 16(1): 60 (1996): H. rhamnoides subsp. yunnanensis.

HIPTAGE (Malpighiac.).
benghalensis
var. rothinii *R.C. Srivastava in Nation. Acad. Sci. Lett.*, 6(7): 223 (1983)—Assam (Manipur).
var. tonkinensis *(Dop) S.K. Chen in Fl. Reipubl. Popul. Sin.*, 43(3): 120 (March 1997); *et in Fl. Yunnanica*, 8: 398 (Nov. 1997): H. madablota var. tonkinensis.
luodianensis *S.K. Chen in Acta Bot. Yunnanica*, 18(4): 409 (1996)—China South-central (Guizhou).
yunnanensis *Huang ex S.K. Chen in Acta Bot. Yunnanica*, 18(4): 405 (1996)—China South-central (Yunnan).

HIRAEA (Malpighiac.).
buntingii *W.R. Anderson in Contrib. Univ. Mich. Herb.*, 21: 63 (1997)—Venezuela.
hypoleuca *W.R. Anderson in Contrib. Univ. Mich. Herb.*, 21: 64 (1997)—Venezuela, Colombia.
neblinensis *W.R. Anderson in Contrib. Univ. Mich. Herb.*, 21: 66 (1997)—Venezuela.
perplexa *W.R. Anderson in Contrib. Univ. Mich. Herb.*, 21: 67 (1997)—Ecuador.
steyermarkii *W.R. Anderson in Contrib. Univ. Mich. Herb.*, 21: 69 (1997)—Venezuela.
valida *W.R. Anderson in Contrib. Univ. Mich. Herb.*, 21: 70 (1997)—Ecuador, Peru.

HIRSCHFELDIA (Crucif.).
incana (L.) Lagr.-Foss.
subsp. incrassata *(Thell. ex Hegi) C. Gómez Campo in An. FJard. Bot. Madrid*, 56(2): 379 (1998): Erucastrum thellungii var. incrassatum.

HIRTELLA (Chrysobalanac.).
aequatoriensis *G.T. Prance in Kew Bull.*, 54(4): 995 (1999 publ. 2000)—Ecuador.
beckii *G.T. Prance in Kew Bull.*, 54(1): 112 (1999)—Bolivia.
castillana *G.T. Prance in Kew Bull.*, 54(1): 110 (1999)—Venezuela, Colombia.

HIRTELLINA (Compos.).
fruticosa *(L.) M. Dittrich in Boissiera*, 51: 75 (1996): Centaurea fruticosa.
kurdica *(Merxm. & Rech. f.) M. Dittrich in Boissiera*, 51: 75 (1996): Staehelina kurdica.
lobelii *(DC.) M. Dittrich in Boissiera*, 51: 75 (1996): Staehelina lobelii.

HIRTZIA (Orchidac.).
escobarii *C.H. Dodson in Orquideología*, 20(1): 8 (1996)—Colombia.

HOFFMANNIA (Rubiac.).
dwyeri *W.C. Burger in Novon*, 9(1): 15 (1999)—Panama.
fortunensis *Dwyer ex W.C. Burger in Novon*, 9(1): 17 (1999)—Panama.

HOFFMANNSEGGIA (Legumin.).
oxycarpa Benth.
subsp. arida *(Rose) B.B. Simpson in Lundellia*, 2: 40 (1999): H. arida.

HOHENBERGIA (Bromeliac.).
burle-marxii *E.M.C. Leme & W. Till in Bromélia*, 3(1): 28 (1996)—Brazil Northeast (Bahia).
hatschbachii *E.M.C. Leme in Harvard Pap. Bot.*, 4(1): 141 (1999)—Brazil Northeast (Bahia).
tetaensis *G.R. Proctor & J.A. Cedeño-Maldonado in Harvard Pap. Bot.*, 4(1): 111 (1999)—Puerto Rico.
undulatifolia *E.M.C. Leme & H.E. Luther in J. Bromeliad Soc.*, 48(4): 153 (1998)—Brazil Northeast (Bahia).

× **HOHENTEA** *D.A. Beadle, Prelim. Listing Cultivar Grex Names Bromel.*: 89 (1991). *BROMELIACEAE* (Hohenbergia × Portea).

HOHERIA (Malvac.).
equitum *M. Heads in New Zealand J. Bot.*, 38(3): 374 (2000)—New Zealand North I.

HOLBOELLIA (Lardizabalac.).
sect. Racemosa *H.N. Qin in Cathaya*, 8–9: 88 (1997).
angustifolia Wall.
subsp. linearifolia *T. Chen & H.N. Qin in Cathaya*, 8–9: 112 (1997)—China South-central (Sichuan, Guizhou, Hubei, Yunnan).
subsp. obtusa *(Gagnep.) H.N. Qin in Cathaya*, 8–9: 116 (1997): H. latifolia var. obtusa.
subsp. trifoliata *H.N. Qin in Cathaya*, 8–9: 114 (1997)—China South-central (Sichuan, Hubei).
brachyandra *H.N. Qin in Cathaya*, 8–9: 126 (1997)—China South-central (Yunnan).
crassifolia *H.N. Qin in Cathaya*, 8–9: 88 (1997)—Burma.
filamentosa *(Griff.) H.N. Qin in Cathaya*, 8–9: 90 (1997): Stauntonia filamentosa.
latifolia Wall.
subsp. chartacea *C.Y. Wu & S.H. Huang ex H.N. Qin in Cathaya*, 8–9: 124 (1997)—Xizang, China South-central (Yunnan), Bhutan, Sikkim, India (West Bengal).
medogensis *H.N. Qin in Cathaya*, 8–9: 93 (1997)—Xizang.
rotundifolia *C.Y. Wu in Fl. Dulongjian Reg.*: 68 (1993), without latin descr. or type—China South-central (Yunnan).

HOLCOGLOSSUM (Orchidac.).
tangii *E.A. Christenson in Lindleyana*, 13(2): 121 (1998)—China South-central (Yunnan).
tsii *T. Yukawa in Ann. Tsukuba Bot. Gard.*, 19: 1 (2000)—China South-Central (Yunnan).
wangii *E.A. Christenson in Lindleyana*, 13(2): 123 (1998)—China South-central (Yunnan).

HOLCUS (Gramin.).
annuus C.A. Mey.
subsp. duriensis *(P. Silva) J. do Amaral Franco & M.L. Rocha Afonso in Silva Lusitana*, 5(1): 141 (1997): H. setiglumis subsp. duriensis.
lanatus L.
var. soboliferus *H.A. Duwensee in Phyton (Austria)*, 35(2): 292 (1995)—Germany, Great Britain, Norway, Austria, Italy, Portugal, Madeira, California, Argentina South (Neuquén).

HOLLANDAEA (Proteac.).
riparia *B.P.M. Hyland in Fl. Australia*, 16: 499 (1995)—Queensland.

HOLOGRAPHIS (Acanthac.).
velutifolia *(House) T.F. Daniel in Bol. Inst. Bot. (Guadalajara)*, 5(1–3): 117 (1998): Exogonium velutifolium.

HOLOPOGON (Orchidac.).
gaudissartii *(Hand.-Mazz.) S.C. Chen in Acta Phytotax. Sin.*, 35(2): 179 (1997): Neottia gaudissartii.
japonicus *(M. Furuse) S.C. Chen in Acta Phytotax. Sin.*, 35(2): 180 (1997): Archineottia japonica.

HOLOPOGON:-
microglottis *(Duthie) S.C. Chen in Acta Phytotax. Sin.*, 35(2): 179 (1997): Listera microglottis.
pantlingii *(W.W. Sm.) S.C. Chen in Acta Phytotax. Sin.*, 35(2): 179 (1997): Aphyllorchis pantlingii.
smithianus *(Schltr.) S.C. Chen in Acta Phytotax. Sin.*, 35(2): 179 (1997): Neottia smithiana.

HOLOSTEUM (Caryophyllac.).
filiforme *A. Risso, Hist. nat. prod. Eur. Mérid.*, 2: 412 (1826)—France.†
subglutinosum *M.V. Klokov in Novosti Sist. Vÿssh. Nizsh. Rast. (Kiev)*, 1974: 38 (1974)—Krym.

HOLOTHRIX (Orchidac.).
majubensis *C. Archer & R.H. Archer in S. Afr. J. Bot.*, 62(4): 209 (1996)—Natal.

HOMALANTHUS (Euphorbiac.).
remotus *H.-J. Esser in Blumea*, 42(2): 457 (1997)—New Guinea (Irian Jaya).

HOMALIUM (Flacourtiac.).
palawanense *J.C. Regalado, D.D. Soejarto & D.A. Madulid in Edinburgh J. Bot.*, 57(3): 317 (2000)—Philippines.

HOMALOMENA (Arac.).
adiensis *A. Hay in Blumea*, 44(1): 66 (1999)—New Guinea (Irian Jaya).
davidiana *A. Hay in Blumea*, 44(1): 48 (1999)—New Guinea (Papua New Guinea).
hammelii *T.B. Croat & M.H. Grayum in Phytologia*, 82(1): 37 (1997)—Costa Rica.
hooglandii *A. Hay in Blumea*, 44(1): 53 (1999)—New Guinea (Papua New Guinea).
jacobsiana *A. Hay in Blumea*, 44(1): 55 (1999)—New Guinea (Papua New Guinea).
kalkmanii *A. Hay in Blumea*, 44(1): 62 (1999)—New Guinea (Irian Jaya).
magna *A. Hay in Blumea*, 44(1): 51 (1999)—New Guinea (Papua New Guinea).
melanesica *A. Hay in Blumea*, 44(1): 50 (1999)—Bismarck Archipelago.
producta *A. Hay in Blumea*, 44(1): 54 (1999)—New Guinea (Papua New Guinea).
soniae *A. Hay in Blumea*, 44(1): 54 (1999)—New Guinea (Papua New Guinea).
steenisiana *A. Hay in Blumea*, 44(1): 63 (1999)—New Guinea (Papua New Guinea).

HOMALOPETALUM (Orchidac.).
subgen. Pachyphyllum *L.O. Williams ex C.L. Withner, Cattleyas & Relatives*, 5: 151 (1998).
kienastii *(Rchb. f.) C.L. Withner, Cattleyas & Relatives*, 5: 158 (1998): Ponera kienastii.

HOMERIA (Iridac.).
hantamensis *P. Goldblatt & J.C. Manning in Flow. Pl. Afr.*, 55: t.2128 (1997)—Cape Province.

HOMORANTHUS (Myrtac.).
bornhardtiensis *J.T. Hunter in Telopea*, 8(1): 37 (1998)—New South Wales.
coracinus *A.R. Bean in Austrobaileya*, 5(4): 687 (2000)—Queensland.
croftianus *J.T. Hunter in Telopea*, 8(1): 39 (1998)—New South Wales.

HOODIA (Asclepiadac.).
sect. Cactoidea *J.J. Halda in Acta Mus. Richnov. Sect. Nat.*, 5(1): 31 (1998), as comb. nov.
sect. Lavrania *(Plowes) J.J. Halda in Acta Mus. Richnov. Sect. Nat.*, 5(1): 32 (1998): gen. Lavrania.
subgen. Lavrania *(Plowes) J.J. Halda in Acta Mus. Richnov. Sect. Nat.*, 5(1): 30 (1998): gen. Lavrania.
cactiformis *(Hook.) J.J. Halda in Acta Mus. Richnov. Sect. Nat.*, 5(1): 31 (1998): Stapelia cactiformis.
dinteri *(A. Berger) J.J. Halda in Acta Mus. Richnov. Sect. Nat.*, 5(1): 31 (1998): Trichocaulon dinteri.
felina *(D.T. Cole) J.J. Halda in Acta Mus. Richnov. Sect. Nat.*, 5(1): 31 (1998): Trichocaulon felinum.
haagnerae *(Plowes) J.J. Halda in Acta Mus. Richnov. Sect. Nat.*, 5(1): 33 (1998): Lavrania haagnerae.

HOODIA:-
marlothii *(N.E. Br.) J.J. Halda in Acta Mus. Richnov. Sect. Nat.,* 5(1): 31 (1998): Trichocaulon marlothii.
meloformis *(Marloth) J.J. Halda in Acta Mus. Richnov. Sect. Nat.,* 5(1): 32 (1998): Trichocaulon meloforme.
perlata *(Dinter) J.J. Halda in Acta Mus. Richnov. Sect. Nat.,* 5(1): 32 (1998): Trichocaulon perlatum.
picta *(N.E. Br.) J.J. Halda in Acta Mus. Richnov. Sect. Nat.,* 5(1): 32 (1998): Trichocaulon pictum.
similis *(N.E. Br.) J.J. Halda in Acta Mus. Richnov. Sect. Nat.,* 5(1): 32 (1998): Trichocaulon simile.
sociarum *(A.C. White & B. Sloane) J.J. Halda in Acta Mus. Richnov. Sect. Nat.,* 5(1): 32 (1998): Trichocaulon sociarum.
tirasmontana *(Plowes) J.J. Halda in Acta Mus. Richnov. Sect. Nat.,* 5(1): 32 (1998), without exact basionym page: Leachiella tirasmontana.

HOPKINSIA (Restionac.).
adscendens *B.G. Briggs & L.A.S. Johnson in Telopea,* 8(4): 487 (2000)—Western Australia.

HORDEUM (Gramin.).
murinum L.
subsp. montanum *(Hack.) H. Scholz & T. Raus in Feddes Repert.,* 108(7–8): 530 (1997): H. murinum f. montanum.
subsp. setariurum *H. Scholz & T. Raus in Feddes Repert.,* 108(7–8): 528 (1997)—Greece.
pubiflorum Hook. f.
subsp. breviaristatum *(Parodi & Nicora) C. Baden in Nordic J. Bot.,* 16(3): 236 (1996 publ. 1997): H. halophilum var. breviaristatum.

HORMIDIUM (Orchidac.).
caetense *H.D. Bicalho in Bol. Soc. Campineira Orq.,* 2(4): 26 (1972)—Brazil Southeast (Minas Gerais).

HORNSTEDTIA (Zingiberac.).
arunachalensis *Sunil Tripathi & Ved Prakash in Nordic J. Bot.,* 19(3): 329 (1999)—Assam (Arunachal Pradesh).

HORNUNGIA (Crucif.).
alpina *(L.) O. Appel in Novon,* 7(4): 339 (1997 publ. 1998): Lepidium alpinum.
subsp. auerswaldii *(Willk.) O. Appel in Novon,* 7(4): 339 (1997 publ. 1998): Hutchinsia auerswaldii.
subsp. austroalpina *(Trpin) O. Appel in Novon,* 7(4): 339 (1997 publ. 1998): Hutchinsia alpina subsp. austroalpina.
subsp. brevicaulis *(Spreng.) O. Appel in Novon,* 7(4): 340 (1997 publ. 1998): Hutchinsia brevicaulis.
subsp. font-queri *(Sauvage) O. Appel in Novon,* 7(4): 340 (1997 publ. 1998): Hutchinsia font-queri.
subsp. polatschekii *(M. Laínz) O. Appel in Novon,* 7(4): 340 (1997 publ. 1998): Hutchinsia alpina subsp. polatschekii.

HOROVITZIA *V.M. Badillo, Caricaceae Seg. Esquema (Rev. Fac. Agron. (Maracay), 43):* 102 (1993), nom. nov.: Carica sect. Holostigma *Lorence & R. Torres.* (Caricac.).
cnidoscoloides *(Lorence & R. Torres) V.M. Badillo, Caricaceae Seg. Esquema (Rev. Fac. Agron. (Maracay), 43):* 104 (1993): Carica cnidoscoloides.

HORSFIELDIA (Myristicac.).
penangiana J. Sinclair
subsp. obtusifolia *W.J.J.O. de Wilde in Blumea,* 41(2): 379 (1996)—Borneo (Sarawak, Kalimantan).
perangusta *W.J.J.O. de Wilde in Blumea,* 41(2): 379 (1996)—Malaya (Peninsular Malaysia).

HOSACKIA (Legumin.).
confinis *(Greene) A. Brand in Bot. Jahrb. Syst.,* 25: 226 (1898): Lotus confinis.
guadalupensis *(Greene) A. Brand in Bot. Jahrb. Syst.,* 25: 228 (1898): Lotus guadalupensis.

HOSACKIA:-
helleri *(Britton) A. Brand in Bot. Jahrb. Syst.,* 25: 228 (1898): Lotus helleri.
hintoniorum *(B.L. Turner) D.D. Sokoloff in Kew Bull.,* 55(4): 1009 (2000): Lotus hintoniorum.
hirtella *(Greene) A. Brand in Bot. Jahrb. Syst.,* 25: 228 (1898): Lotus hirtellus.
neomexicana *(Greene) A. Brand in Bot. Jahrb. Syst.,* 25: 229 (1898): Lotus neomexicanus.
tomentella *(Greene) A. Brand in Bot. Jahrb. Syst.,* 25: 231 (1898): Lotus tomentellus.
trisperma *(Greene) A. Brand in Bot. Jahrb. Syst.,* 25: 232 (1898): Lotus trispermus.
watsonii *G. Vasey & J.N. Rose in Proc. U.S. Natl. Mus.,* 11: 528 (1889), in Supp. 1 as 'waltoni'—Mexico Northwest (Baja California).†
yollaboliiensis *(Munz) D.D. Sokoloff in Kew Bull.,* 55(4): 1010 (2000): Lotus yollabolliensis.

HOSTA (Hostac.).
capitata (Koidz.) Nakai
f. albiflora *Y.N. Lee, Fl. Korea:* 1163 (1996), without type—Locality not stated.

HOTTARUM (Arac.).
brevipedunculatum *H. Okada & Y. Mori in Acta Phytotax. Geobot.,* 51(1): 7 (2000)—Borneo (Kalimantan).

HOULLETIA (Orchidac.).
subgen. Neohoulletia *G. Gerlach & Whitten in Schlechter Orchideen, ed. 3,* I/C(37–38): 2409 (1999 publ. 2000).

× **HOULLORA** *hort. in Orchid Rev.,* 108(1234): new orchid hybrids 15 (2000). *ORCHIDACEAE* (Gongora × Houlletia).

HOUSTONIA (Rubiac.).
sect. Amphiotis *(DC.) E.E. Terrell, Revis. Houstonia (Syst. Bot. Monog., 48):* 49 (1996): Anotis sect. Amphiotis.
sect. Ericotis *(Terrell) E.E. Terrell, Revis. Houstonia (Syst. Bot. Monog., 48):* 73 (1996): H. subgen. Ericotis.
sect. Mullera *E.E. Terrell, Revis. Houstonia (Syst. Bot. Monog., 48):* 45 (1996).

HOVANELLA *A. Weber & B.L. Burtt in Beitr. Biol. Pfl.,* 70(2–3): 333 (1997–98 publ. 1998), nom. nov.: Didymocarpus sect. Hova *C.B. Clarke.* (Gesneriac.)
madagascarica *(C.B. Clarke) A. Weber & B.L. Burtt in Beitr. Biol. Pfl.,* 70(2–3): 349 (1997–98 publ. 1998): Didymocarpus madagascaricus.
vestita *(Baker) A. Weber & B.L. Burtt in Beitr. Biol. Pfl.,* 70(2–3): 358 (1997–98 publ. 1998): Didymocarpus vestitus.

HOYA (Asclepiadac.).
subsect. Acersuccus *C.M. Burton in Hoyan,* 17(3:2): 14 (1996), contrary to Art. 22.2 ICBN (1994).
sect. Angustialatus *C.M. Burton in Hoyan,* 17(3:2): 17 (1996).
subsect. Externatae *D. Kloppenburg, Hoya section Acanthostemma:* 10 (1995).
subsect. Lactisuccus *C.M. Burton in Hoyan,* 17(3:2): 14 (1996).
sect. Latiretinacula *C.M. Burton in Hoyan,* 17(3:2): 17 (1996).
sect. Physostelma *(Wight) C.M. Burton in Hoyan,* 18(1:2): 5 (1996): gen. Physostelma.
sect. Pseudohoya *C.M. Burton in Hoyan,* 17(2:2): 11 (1995).
sect. Pseudopterostelma *C.M. Burton in Hoyan,* 17(3:2): 17 (1996).
bicknellii *D. Kloppenburg in Fraterna,* 12(4): 12 (1999)—Philippines.
blashernaezii *D. Kloppenburg in Fraterna,* 12(1): 9 (1999)—Philippines.
crassicaulis *Elmer ex D. Kloppenburg in Fraterna,* 1995(3): 10 (1995)—Philippines.
deykei *T. Green in Fraterna,* 13(1): 15 (2000)—Sumatera.
dickasoniana *P.T. Li in J. S. China Agric. Univ.,* 15(2): 74 (1994)—Burma.
gongshanica *P.T. Li in J. S. China Agric. Univ.,* 12(3): 40 (1991), nom. nov.: Centrostemma yunnanense *P.T. Li.*

HOYA:-
kingdonwardii *P.T. Li in J. S. China Agric. Univ.,* 15(2): 74 (1994)—Burma.
myanmarica *P.T. Li in J. S. China Agric. Univ.,* 15(2): 75 (1994)—Burma.
odetteae *D. Kloppenburg in Fraterna,* 11(2): 8 (1998)—Philippines.
pimenteliana *D. Kloppenburg in Fraterna,* 12(3): 7 (1999)—Philippines.
telosmoides *R. Omlor in Novon,* 6(3): 290 (1996)—Borneo (Sabah, Sarawak).
verticillata (Vahl) G. Don
var. citrina *(Ridl.) D. Kloppenburg in Fraterna,* 1996(2): 9 (1996), without basionym ref.; *et in Fraterna,* 1996(3): 10 (1996), without basionym ref.: Basionym not stated.
var. citrina *(Ridl.) J.F. Veldkamp in Blumea,* 41(2): 440 (1996): H. citrina.
var. hendersonii *(Kiew) D. Kloppenburg in Fraterna,* 1996(2): 9 (1996), contrary to Art. 26.1 ICBN (1994); *et in Fraterna,* 1996(3): 10 (1996): H. parasitica var. hendersonii.
var. hendersonii *(Kiew) J.F. Veldkamp in Blumea,* 41(2): 440 (27 Nov. 1996), isonym; *see Kloppenburg in Fraterna,* 1996(3): 10 (Sept. 1996): H. parasitica var. hendersonii.
wallichii *(Wight) C.M. Burton in Hoyan,* 18(1:2): 5 (1996): Physostelma wallichii.

HUBERIA (Melastomatac.).
carvalhoi *J.F.A. Baumgratz in Brittonia,* 52(1): 25 (2000)—Brazil Northeast (Bahia).
cogniauxii *J.F.A. Baumgratz in Novon,* 9(2): 143 (1999)—Peru.
consimilis *J.F.A. Baumgratz in Brittonia,* 52(1): 29 (2000)—Brazil Southeast (Rio de Janeiro), Brazil Northeast (Bahia, Pernambuco).
espirito-santensis *J.F.A. Baumgratz in Rodriguésia,* 50(76–77): 39 (1998)—Brazil Southeast (Espírito Santo).
piranii *J.F.A. Baumgratz in Rodriguésia,* 50(76–77): 40 (1998)—Brazil Southeast (Minas Gerais).
staminodia *J.F.A. Baumgratz in Novon,* 9(2): 139 (1999)—Peru.
weberbaueriana *J.F.A. Baumgratz in Novon,* 9(2): 140 (1999)—Peru.

HUEGELIA *Rchb. in Mitt. Geb. Fl. Pomona,* 1829(13): 50 (1829). *UMBELLIFERAE*
coerulea *Rchb. in Mitt. Geb. Fl. Pomona,* 1829(13): 50 (1829)—Australia.

HUERNIA (Asclepiadac.).
keniensis R.E. Fr.
var. globosa *L.E. Newton in Asklepios,* 74: 23 (1998)—Kenya.
macrocarpa (A. Rich.) Sprenger
f. schimperi *(A. Berger) S. Brodie in Curtis's Bot. Mag.,* 15(1): 6 (1998): H. penzigii var. schimperi.

HUILAEA (Melastomatac.).
minor *(L. Uribe) G. Lozano-C. & N. Ruiz-R. in Rev. Acad. Colomb. Cienc. Exact. Fis. Nat.,* 77: 240 (1996), without basionym ref.; *et in Rev. Acad. Colomb. Cienc. Exact. Fis. Nat.,* 81: 419 (1997), with basionym ref.: H. macrocarpa subsp. minor.
occidentalis *G. Lozano-C. & N. Ruiz-R. in Rev. Acad. Colomb. Cienc. Exact. Fis. Nat.,* 77: 238 (1996)—Colombia.

HUNTERIA (Apocynac.).
hexaloba *(Pichon) E. Omino in Wageningen Agric. Univ. Pap.,* 96(1): 104 (1996); *et in Belmontia,* 29: 104 (1996): Tetradoa hexaloba.
macrosiphon *E. Omino in Wageningen Agric. Univ. Pap.,* 96(1): 107 (1996); *et in Belmontia,* 29: 107 (1996)—Gabon, Congo.
myriantha *E. Omino in Wageningen Agric. Univ. Pap.,* 96(1): 109 (1996); *et in Belmontia,* 29: 109 (1996)—Zaire.
oxyantha *E. Omino in Wageningen Agric. Univ. Pap.,* 96(1): 112 (1996); *et in Belmontia,* 29: 112 (1996)—Gabon, Congo, Zaire.

HUTCHINSIA (Crucif.).
brevicaulis *Spreng., Syst. Veg.,* 2: 863 (1825)—Austria.†

HYACINTHELLA (Hyacinthac.).
leucophaea (K. Koch) Schur
subsp. atchleyi *(A.K. Jacks. & Turrill) K. Persson & Jim. Persson in Candollea,* 55(2): 222 (2000): Bellevalia atchleyi.
venusta *K. Persson in New Plantsman,* 7(4): 204 (2000)—Turkey.

HYACINTHOIDES (Hyacinthac.).
sect. Somera *(Salisb.) S. Ortiz & J. Rodríguez-Oubiña in Pl. Syst. Evol.,* 202(1–2): 118 (1996): gen. Somera.
× massartiana *D. Geerinck in Belgian J. Bot.,* 129(1): 83 (1996 publ. 1997)—Belgium. (H. hispanica × non-scripta).
mauritanica (Schousb.) Speta
subsp. vincentina *(Link & Hoffmanns.) S. Ortiz, M. Buján & J. Rodríguez-Oubiña in Pl. Syst. Evol.,* 217(1–2): 172 (1999): Scilla vincentina.
paivae *S. Ortiz & J. Rodríguez-Oubiña in Pl. Syst. Evol.,* 202(1–2): 113 (1996)—Spain.

HYBANTHUS (Violac.).
denticulatus *H.E. Ballard, M.A. Wetter & N. Zamora in Novon,* 7(3): 222 (1997)—Costa Rica, Nicaragua, Panama.
floribundus
subsp. chloroxanthus *E.M. Bennett in Nuytsia,* 13(1): 4 (1999)—Western Australia.
hespericlivus *H.E. Ballard, M.A. Wetter & N. Zamora in Novon,* 7(3): 225 (1997)—Costa Rica.

HYDRANGEA (Hydrangeac.).
chinensis Maxim.
var. yaeyamensis *(Koidz.) T. Yamazaki in J. Jap. Bot.,* 75(5): 321 (2000): Hydrangea yaeyamensis.
involucrata Siebold
var. tokarensis *M. Hotta & T. Shiuchi in J. Jap. Bot.,* 71(4): 184 (1996)—Nansei-Shoto.
macrophylla (Thunb.) Ser.
f. buergeri *(Nakai) Y.N. Lee, Fl. Korea:* 1158 (1996): H. serrata f. buergeri.
f. fertilis *(Nakai) Y.N. Lee, Fl. Korea:* 1158 (1996): H. serrata f. fertilis.

HYDRIASTELE (Palmae).
rheophytica *J.L. Dowe & M.D. Ferrero in Palms,* 44(4): 195 (2000)—New Guinea (Irian Jaya, Papua New Guinea).

HYDROBRYUM (Podostemac.).
koribanum *Imamura ex S. Nakayama & T. Minamitani in J. Jap. Bot.,* 74(5): 314 (1999)—Japan.

HYDROCHOREA *R.C. Barneby & J.W. Grimes, Silk Tree, Guanacaste, Monkey's Earring (Mem. New York Bot. Gard., 74(1)):* 23 (1996). *LEGUMINOSAE*
acreana *(J.F. Macbr.) R.C. Barneby & J.W. Grimes, Silk Tree, Guanacaste, Monkey's Earring (Mem. New York Bot. Gard., 74(1)):* 33 (1996): Pithecellobium acreanum.
corymbosa *(Rich.) R.C. Barneby & J.W. Grimes, Silk Tree, Guanacaste, Monkey's Earring (Mem. New York Bot. Gard., 74(1)):* 27 (1996): Mimosa corymbosa.
gonggrijpii *(Kleinhoonte) R.C. Barneby & J.W. Grimes, Silk Tree, Guanacaste, Monkey's Earring (Mem. New York Bot. Gard., 74(1)):* 25 (1996): Pithecellobium gonggrijpii.
marginata *(Benth.) R.C. Barneby & J.W. Grimes, Silk Tree, Guanacaste, Monkey's Earring (Mem. New York Bot. Gard., 74(1)):* 29 (1996): Pithecellobium marginatum.
var. panurensis *(Benth.) R.C. Barneby & J.W. Grimes, Silk Tree, Guanacaste, Monkey's Earring (Mem. New York Bot. Gard., 74(1)):* 32 (1996): Pithecellobium panurense.
var. scheryi *R.C. Barneby & J.W. Grimes, Silk Tree, Guanacaste, Monkey's Earring (Mem. New York Bot. Gard., 74(1)):* 32 (1996), nom. nov.: Pithecellobium bijugatum *Schery.*

HYDROCOTYLE (Umbellif.).
burmanica Kurz
subsp. craibii *(H. Eichler) C.Y. Wu & F.T. Pu in Novon,* 8(1): 70 (1998): H. craibii.

HYDROCOTYLE:-
dulongensis *H. Li in Fl. Dulongjian Reg.:* 209 (1993), without latin descr. or type—China South-central (Yunnan).
sibthorpioides Lam.
var. calcicola *(Y.H. Li) S.L. Liou in Fl. Yunnanica,* 7: 363 (1997): H. calcicola.

HYDROLEA (Hydrophyllac.).
spinosa L.
var. cervantesii *(Brand) L.J. Davenport & A. Pool in Novon,* 10(1): 12 (2000): H. cervantesii.

HYDROSME (Arac.).
gigantiflora *(Hayata) S.S. Ying in Mem. Coll. Agric. Nation. Taiwan Univ.,* 31(1): 31 (1991), as 'Hydroseme gigantiflorus': Amorphophallus gigantiflorus.
niimurae *(Yamam.) S.S. Ying in Mem. Coll. Agric. Nation. Taiwan Univ.,* 31(1): 31 (1991), as 'Hydroseme niimurai': Amorphophallus niimurae.

HYGROPHILA (Acanthac.).
bengalensis *S.K. Mandal, A. Bhattacharjee & A.K. Nayek in J. Bombay Nat. Hist. Soc.,* 94(3): 546 (1997)—India (West Bengal).
salicifolia (Vahl) Nees
var. longihirsuta *H.S. Lo & D. Fang in Guihaia,* 17(1): 41 (1997)—China Southeast (Guangxi).
var. megalantha *(Merr.) H.S. Lo & L.D. Chou in Guihaia,* 16(4): 374 (1996): H. megalantha.

HYLAEA *J.F. Morales in Novon,* 9(1): 83 (1999). *APOCYNACEAE*
arborescens *(Monach.) J.F. Morales in Novon,* 9(1): 83 (1999): Prestonia arborescens.
leptoloba *(Monach.) J.F. Morales in Novon,* 9(1): 84 (1999): Prestonia leptoloba.

HYLAEORCHIS *G. Carnevali & G.A. Romero in G.A. Romero & G. Carnevali, Orchids of Venezuela, ed. 2:* 1136 (2000).*ORCHIDACEAE*
petiolaris *(Schltr.) G. Carnevali & G.A. Romero in G.A. Romero & G. Carnevali, Orchids of Venezuela, ed. 2:* 1136 (2000): Maxillaria petiolaris.

HYLEBIA (Caryophyllac.).
bungeana *(Fenzl) N.N. Tzvelev in Novosti Sist. Vyssh. Rast.,* 32: 36 (2000): Stellaria bungeana.

HYLODESMUM *H. Ohashi & R.R. Mill in Edinburgh J. Bot.,* 57(2): 173 (2000), nom. nov.: Desmodium sect. Podocarpium *Benth.* (Legumin.).
densum *(C. Chen & X.J. Cui) H. Ohashi & R.R. Mill in Edinburgh J. Bot.,* 57(2): 176 (2000): Podocarpium fallax var. densum.
dolabriforme *(Benth.) H. Ohashi & R.R. Mill in Edinburgh J. Bot.,* 57(2): 176 (2000): Desmodium dolabriforme.
glutinosum *(Muhl. ex Willd.) H. Ohashi & R.R. Mill in Edinburgh J. Bot.,* 57(2): 177 (2000): Hedysarum glutinosum.
laterale *(Schindl.) H. Ohashi & R.R. Mill in Edinburgh J. Bot.,* 57(2): 177 (2000): Desmodium laterale.
laxum *(DC.) H. Ohashi & R.R. Mill in Edinburgh J. Bot.,* 57(2): 178 (2000): Desmodium laxum.
subsp. falfolium *(H. Ohashi) H. Ohashi & R.R. Mill in Edinburgh J. Bot.,* 57(2): 179 (2000): Desmodium laxum subsp. falfolium.
subsp. lateraxum *(H. Ohashi) H. Ohashi & R.R. Mill in Edinburgh J. Bot.,* 57(2): 179 (2000): Desmodium laxum subsp. lateraxum.
leptopus *(A. Gray ex Benth.) H. Ohashi & R.R. Mill in Edinburgh J. Bot.,* 57(2): 179 (2000): Desmodium leptopus.
longipes *(Franch.) H. Ohashi & R.R. Mill in Edinburgh J. Bot.,* 57(2): 179 (2000): Shuteria longipes.
menglaense *(C. Chen & X.J. Cui) H. Ohashi & R.R. Mill in Edinburgh J. Bot.,* 57(2): 180 (2000): Podocarpium menglaense.
nudiflorum *(L.) H. Ohashi & R.R. Mill in Edinburgh J. Bot.,* 57(2): 180 (2000): Hedysarum nudiflorum.

HYLODESMUM:-
oldhamii *(Oliv.) H. Ohashi & R.R. Mill in Edinburgh J. Bot.,* 57(2): 180 (2000): Desmodium oldhamii.
pauciflorum *(Nutt.) H. Ohashi & R.R. Mill in Edinburgh J. Bot.,* 57(2): 181 (2000): Hedysarum pauciflorum.
podocarpum *(DC.) H. Ohashi & R.R. Mill in Edinburgh J. Bot.,* 57(2): 181 (2000): Desmodium podocarpum.
subsp. fallax *(Schindl.) H. Ohashi & R.R. Mill in Edinburgh J. Bot.,* 57(2): 182 (2000): Desmodium fallax.
var. mandshuricum *(Maxim.) H. Ohashi & R.R. Mill in Edinburgh J. Bot.,* 57(2): 183 (2000): Desmodium podocarpum var. mandshuricum.
var. oxyphyllum *(DC.) H. Ohashi & R.R. Mill in Edinburgh J. Bot.,* 57(2): 183 (2000): Desmodium oxyphyllum.
subsp. oxyphyllum *(DC.) H. Ohashi & R.R. Mill in Edinburgh J. Bot.,* 57(2): 182 (2000): Desmodium oxyphyllum.
subsp. szechuenense *(Craib) H. Ohashi & R.R. Mill in Edinburgh J. Bot.,* 57(2): 185 (2000): Desmodium podocarpum var. szechuenense.
repandum *(Vahl) H. Ohashi & R.R. Mill in Edinburgh J. Bot.,* 57(2): 185 (2000): Hedysarum repandum.
williamsii *(H. Ohashi) H. Ohashi & R.R. Mill in Edinburgh J. Bot.,* 57(2): 186 (2000):Desmodium williamsii.

HYLOTELEPHIUM (Crassulac.).
sect. Sieboldiana *(H. Ohba) H. Ohba in H. 't Hart & U. Eggli (eds.), Evol. & Syst. Crassulac.:* 155 (1995): H. ser. Sieboldiana.
angustum (Maxim.) H. Ohba
var. longipedunculum *J.M. Zhang & K.T. Fu in Guihaia,* 19(3): 207 (1999)—China North-central (Shaanxi, Gansu).
decumbens *(Lucé) V.V. Byalt in Bot. Zhurn.,* 80(11): 35 (1995): Sedum decumbens.
pakistanicum *(G.R. Sarwar) J. Holub in Preslia,* 70(2): 105 (1998): Sedum pakistanicum.
sanguineum *(Ortega) S. Castroviejo & M. Velayos in An. Jard. Bot. Madrid,* 53(2): 280 (1995 publ. 1996): Sedum sanguineum.
triphyllum (Haw.) Holub
var. pluricaule *(Maxim.) V.V. Byalt in Bot. Zhurn.,* 84(9): 47 (1999): Sedum telephium γ pluricaule.
uralense *(Rupr.) V.V. Byalt in Bot. Zhurn.,* 83(1): 128 (1998): Sedum uralense.

HYMENANDRA (Myrsinac.).
acutissima *(Cuatrec.) J.J. Pipoly & J.M. Ricketson in Sida,* 18(3): 738 (1999): Parathesis acutissima.
callejasii *(Pipoly) J.J. Pipoly & J.M. Ricketson in Sida,* 18(3): 736 (1999): Ardisia callejasii.
calycosa *(Hemsl.) J.J. Pipoly & J.M. Ricketson in Sida,* 18(3): 718 (1999): Ardisia calycosa.
crosbyi *(Lundell) J.J. Pipoly & J.M. Ricketson in Sida,* 18(3): 739 (1999): Ardisia crosbyi.
pittieri *(Mez) J.J. Pipoly & J.M. Ricketson in Sida,* 18(3): 731 (1999): Ardisia pittieri.
sordida *(Lundell) J.J. Pipoly & J.M. Ricketson in Sida,* 18(3): 722 (1999): Auriculardisia sordida.
squamata *(Lundell) J.J. Pipoly & J.M. Ricketson in Sida,* 18(3): 725 (1999): Auriculardisia squamata.
stenophylla *(Donn. Sm.) J.J. Pipoly & J.M. Ricketson in Sida,* 18(3): 715 (1999): Ardisia stenophylla.
wilburiana *(Lundell) J.J. Pipoly & J.M. Ricketson in Sida,* 18(3): 728 (1999): Auriculardisia wilburiana.

HYMENIDIUM (Umbellif.).
album *(C.B. Clarke ex H. Wolff) M.G. Pimenov & E.V. Kljuykov in Feddes Repert.,* 111(7–8): 549 (2000): Pleurospermum album.
amabile *(Craib & W.W. Sm.) M.G. Pimenov & E.V. Kljuykov in Feddes Repert.,* 111(7–8): 545 (2000): Pleurospermum amabile.
apiolens *(C.B. Clarke) M.G. Pimenov & E.V. Kljuykov in Feddes Repert.,* 111(7–8): 546 (2000): Pleurospermum apiolens.
astrantioideum *(H. Boissieu) M.G. Pimenov & E.V. Kljuykov in Feddes Repert.,* 111(7–8): 545 (2000): Trachydium astrantioideum.

HYMENIDIUM:-
benthamii *(DC.) M.G. Pimenov & E.V. Kljuykov in Feddes Repert.*, 111(7–8): 543 (2000): Hymenolaena benthamii.
bicolor *(Franch.) M.G. Pimenov & E.V. Kljuykov in Feddes Repert.*, 111(7–8): 545 (2000): Pleurospermum govanianum var. bicolor.
chloroleucum *(Diels) M.G. Pimenov & E.V. Kljuykov in Feddes Repert.*, 111(7–8): 548 (2000): Trachydium chloroleucum.
corydalifolium *(Aitch. & Hemsl.) M.G. Pimenov & E.V. Kljuykov in Feddes Repert.*, 111(7–8): 547 (2000): Pleurospermum corydalifolium.
cristatum *(H. Boissieu) M.G. Pimenov & E.V. Kljuykov in Feddes Repert.*, 111(7–8): 542 (2000): Pleurospermum cristatum.
davidii *(Franch.) M.G. Pimenov & E.V. Kljuykov in Feddes Repert.*, 111(7–8): 542 (2000): Pleurospermum davidii.
decurrens *(Franch.) M.G. Pimenov & E.V. Kljuykov in Feddes Repert.*, 111(7–8): 542 (2000): Pleurospermum decurrens.
delavayi *(Franch.) M.G. Pimenov & E.V. Kljuykov in Feddes Repert.*, 110(7–8): 484 (1999): Ligusticum delavayi.
dentatum *(DC.) M.G. Pimenov & E.V. Kljuykov in Feddes Repert.*, 111(7–8): 546 (2000): Hymenolaena dentata.
foetens *(Franch.) M.G. Pimenov & E.V. Kljuykov in Feddes Repert.*, 111(7–8): 544 (2000): Pleurospermum foetens.
giraldii *(Diels) M.G. Pimenov & E.V. Kljuykov in Feddes Repert.*, 111(7–8): 548 (2000): Pleurospermum giraldii.
hedinii *(Diels) M.G. Pimenov & E.V. Kljuykov in Feddes Repert.*, 111(7–8): 550 (2000): Pleurospermum hedinii.
heracleifolium *(Franch. ex H. Boissieu) M.G. Pimenov & E.V. Kljuykov in Feddes Repert.*, 111(7–8): 543 (2000): Pleurospermum heracleifolium.
heterosciadium *(H. Wolff) M.G. Pimenov & E.V. Kljuykov in Feddes Repert.*, 111(7–8): 548 (2000): Pleurospermum heterosciadium.
hookeri *(C.B. Clarke) M.G. Pimenov & E.V. Kljuykov in Feddes Repert.*, 111(7–8): 547 (2000): Pleurospermum hookeri.
kansuense *(H. Wolff) M.G. Pimenov & E.V. Kljuykov in Feddes Repert.*, 111(7–8): 547 (2000): Pleurospermum kansuense.
linearilobum *(W.W. Sm.) M.G. Pimenov & E.V. Kljuykov in Feddes Repert.*, 111(7–8): 543 (2000): Pleurospermum linearilobum.
macrochlaenum *(K.T. Fu & Y.C. Ho) M.G. Pimenov & E.V. Kljuykov in Feddes Repert.*, 111(7–8): 543 (2000): Pleurospermum macrochlaenum.
nanum *(Rupr.) M.G. Pimenov & E.V. Kljuykov in Feddes Repert.*, 111(7–8): 549 (2000): Hymenolaena nana.
nubigenum *(H. Wolff.) M.G. Pimenov & E.V. Kljuykov in Feddes Repert.*, 111(7–8): 549 (2000): Pleurospermum nubigenum.
pachycaule *M.G. Pimenov & E.V. Kljuykov in Edinburgh J. Bot.*, 53(2): 275 (1996)—China North-central (Gansu).
pilosum *(C.B. Clarke ex H. Wolff) M.G. Pimenov & E.V. Kljuykov in Feddes Repert.*, 111(7–8): 546 (2000): Pleurospermum pilosum.
pulszkyi *(Kanitz) M.G. Pimenov & E.V. Kljuykov in Feddes Repert.*, 111(7–8): 544 (2000): Pleurospermum pulszkyi.
sikkimense *(C.B. Clarke) M.G. Pimenov & E.V. Kljuykov in Feddes Repert.*, 111(7–8): 545 (2000): Pleurospermum sikkimense.
stellatum *(D. Don) M.G. Pimenov & E.V. Kljuykov in Feddes Repert.*, 111(7–8): 549 (2000): Selinum stellatum.
szechenyi *(Kanitz) M.G. Pimenov & E.V. Kljuykov in Feddes Repert.*, 111(7–8): 544 (2000): Pleurospermum szechenyi.
tsekuense *(R.H. Shan) M.G. Pimenov & E.V. Kljuykov in Feddes Repert.*, 111(7–8): 545 (2000): Pleurospermum tsekuense.
wilsonii *(H. Boissieu) M.G. Pimenov & E.V. Kljuykov in Feddes Repert.*, 111(7–8): 549 (2000): Pleurospermum wilsonii.
wrightianum *(H. Boissieu) M.G. Pimenov & E.V. Kljuykov in Feddes Repert.*, 111(7–8): 544 (2000): Pleurospermum wrightianum.

HYMENIDIUM:-
yunnanense *(Franch.) M.G. Pimenov & E.V. Kljuykov in Feddes Repert.*, 111(7–8): 550 (2000): Pleurospermum yunnanense.

HYMENOCALLIS (Amaryllidac.).
azteciana *Traub ex J.E. Laferriere in Herbertia,* 51: 66 (1996)—Mexico Northeast (Zacatecas).
choctawensis *Traub ex J.E. Laferriere in Herbertia,* 51: 66 (1996)—Florida.
clivorum *J.E. Laferrière in Novon,* 8(3): 242 (1998)—Mexico Northwest (Sonora).
duvalensis *Traub ex J.E. Laferriere in Herbertia,* 51: 66 (1996)—Florida.
floridana
subsp. amplifolia *Traub ex J.E. Laferriere in Herbertia,* 51: 66 (1996)—Florida.
henryae *Traub ex J.E. Laferriere in Herbertia,* 51: 66 (1996)—Florida.
var. glaucifolia *J. de N. Henry & Gerald L. Smith in Herbertia,* 54: 114 (1999 publ. 2000)—Florida.
incaica *Ravenna ex J.E. Laferriere in Herbertia,* 51: 66 (1996)—Locality not stated.
maximilianii *T.M. Howard in Herbertia,* 51: 67 (1996)—Mexico Southwest (Guerrero).
moldenkiana *Traub ex J.E. Laferriere in Herbertia,* 51: 67 (1996)—Georgia.
palusvirensis *Traub ex J.E. Laferriere in Herbertia,* 51: 67 (1996)—North Carolina.
puntagordensis *Traub ex J.E. Laferriere in Herbertia,* 51: 67 (1996)—Florida.

HYMENOXYS (Compos.).
ambigens (S.F. Blake) Bierner
var. floribunda *(A. Gray) W.L. Wagner in Brittonia,* 51(1): 84 (1999): Plummera floribunda.
var. neomexicana *W.L. Wagner in Brittonia,* 51(1): 85 (1999)—New Mexico.

HYOSCYAMUS (Solanac.).
subgen. Parahyoscyamus *M. Khatamsaz in Fl. Iran (ed. M. Assadi et al.)*, 24: 70 (1998).
bornmulleri *M. Khatamsaz in Iranian J. Bot.*, 7(1): 32 (1996)—Iran.

HYPECOUM (Papaverac.).
alpinum *Z.X. An in Fl. Xinjiangensis,* 2(2): 373 (1995)—Xinjiang.
ferrugineomaculatum *Z.X. An in Fl. Xinjiangensis,* 2(2): 373 (1995), as 'ferrugineo-maculatum'—Xinjiang.

HYPERICOPHYLLUM (Compos.).
altissimum *(Klatt) J.-P. Lebrun & A.L. Stork, Enum. Pl. Fleurs Afr. Trop.*, 4: 300 (1997): Jaumea altissima.

HYPERICUM (Guttif.).
subsect. Ascyrum *(L.) N.K.B. Robson in Bull. Nat. Hist. Mus. (London), Bot.*, 26(2): 124 (1996): gen. Ascyrum.
choji-suzukii *H. Koidzumi in J. Pl. Iwateken,* 2(2): 91 (1937)—Japan.
× desetangsii Lamotte
nothosubsp. carinthiacum *(A. Fröhl.) N.K.B. Robson in R. Wisskirchen & H. Haeupler, Standardliste Farn- und Blütenpfl.* Deutschl.: 268 (1998): H. × carinthiacum.
nothof. maculatiforme *(A. Fröhl.) N.K.B. Robson in R. Wisskirchen & H. Haeupler, Standardliste Farn- und Blütenpfl.* Deutschl.: 269 (1998): H. × carinthiacum f. maculatiforme.
nothof. perforatiforme *(A. Fröhl.) N.K.B. Robson in R. Wisskirchen & H. Haeupler, Standardliste Farn- und Blütenpfl.* Deutschl.: 269 (1998): H. × carinthiacum f. perforatiforme.
fukudae *H. Koidzumi in J. Pl. Iwateken,* 2(2): 88 (1937)—Japan.
ganjuense *H. Koidzumi in J. Pl. Iwateken,* 2(2): 86 (1937)—Japan.
hayachinense *H. Koidzumi in J. Pl. Iwateken,* 2(2): 90 (1937)—Japan.
iwabuchii *Makino & H. Koidzumi in J. Pl. Iwateken,* 2(2): 82 (1937)—Japan.

HYPERICUM:-
iwakiense *H. Koidzumi in J. Pl. Iwateken,* 2(2): 83 (1937)—Japan.
iwate-littorale *H. Koidzumi in J. Pl. Iwateken,* 2(2): 89 (1937)—Japan.
kirilowii *A.N. Sennikov in Novosti Sist. Vyssh. Rast.*, 30: 120 (1996)—Kazakhstan.
kurikomense *H. Koidzumi in J. Pl. Iwateken,* 2(2): 94 (1937)—Japan.
lygophyllum *H. Koidzumi in J. Pl. Iwateken,* 2(2): 85 (1937)—Japan.
maleevii *A.S. Zernov & A.G. Jelenevski in Bot. Zhurn.*, 85(10): 54 (2000), nom. nov.: Hypericum strictum *Maleev.*
nakaii *H. Koidzumi in J. Pl. Iwateken,* 2(2): 99 (1937)—Japan.
nambuense *H. Koidzumi in J. Pl. Iwateken,* 2(2): 85 (1937)—Japan.
novi-ramosissimum *H. Koidzumi in J. Pl. Iwateken,* 2(2): 87, 105 (1937)—Japan.
peshmenii *Ş. Yıldırımlı in Ot Sist. Bot. Dergisi,* 2(2): 12 (1995)—Turkey.
pseudonikkoense *H. Koidzumi in J. Pl. Iwateken,* 2(2): 87 (1937), as 'pseudo-nikkoense'—Japan.
silenoides
subsp. minus *N.K.B. Robson in Bull. Nat. Hist. Mus. (London), Bot.*, 26(2): 76 (1996)—Galápagos.
subquadrangulare *H. Koidzumi in J. Pl. Iwateken,* 2(2): 84 (1937)—Japan.
subrotundifolium *H. Koidzumi in J. Pl. Iwateken,* 2(2): 92 (1937)—Japan.
swertioides *H. Koidzumi in J. Pl. Iwateken,* 2(2): 89 (1937)—Japan.
yakeisidakense *H. Koidzumi in J. Pl. Iwateken,* 2(2): 92 (1937)—Japan.
yamamotoanum *H. Koidzumi in J. Pl. Iwateken,* 2(2): 102 (1937)—Japan.
yoitiense *H. Koidzumi in J. Pl. Iwateken,* 2(2): 101 (1937)—Japan.

HYPOCHAERIS (Compos.).
alliatae *(Biv.) A. Galán de Mera, E. de Castro & J.A. Vicente Orellana in Nordic J. Bot.*, 19(5): 588 (1999): Seriola alliatae.
chubutensis *E. Bortiri in Hickenia,* 2(46): 223 (1997), as 'Hypochoeris'—Argentina (Chubut, Tierra del Fuego).
leontodontoides Ball
var. atlantica *A. Galán de Mera, E. de Castro & J.A. Vicente Orellana in Nordic J. Bot.*, 19(5): 591 (1999)—Morocco.
montana (Phil.) Reiche
var. hookeri *(Phil.) E. Bortiri in Hickenia,* 2(46): 229 (1997), as 'Hypochoeris': H. hookeri.
tenuifolia (Hook. & Arn.) Griseb.
var. clarionoides *(J. Rémy) E. Bortiri in Hickenia,* 2(46): 229 (1997), as 'Hypochoeris': Achyrophorus clarionoides.
var. linearifolia *(DC.) E. Bortiri in Hickenia,* 2(46): 230 (1997), as 'Hypochoeris': Achyrophorus tenuifolius var. linearifolius.

HYPOLAENA (Restionac.).
humilis *(Gilg) B.G. Briggs & L.A.S. Johnson in Telopea,* 8(1): 28 (1998): Leptocarpus humilis.
robusta *K.A. Meney & J.S. Pate in Telopea,* 6(4): 653 (1996)—Western Australia.

HYPOPITYS (Ericac.).
monotropa Crantz
subsp. hypophegea *(Wallr.) N.N. Tzvelev in Novosti Sist. Vyssh. Rast.*, 32: 184 (2000): Monotropa hypophegea.

HYPOXIS (Hypoxidac.).
bampsiana *J. Wiland in Bull. Jard. Bot. Nation. Belg.*, 66(3–4): 207 (1997)—Zaire, Zambia.
lejolyana *J. Wiland in Fragm. Flor. Geobot.*, 42(2): 418 (1997)—Zaire.
lusalensis *J. Wiland in Fragm. Flor. Geobot.*, 42(2): 421 (1997)—Zaire.
malaissei *J. Wiland in Fragm. Flor. Geobot.*, 42(2): 418 (1997)—Zaire.
muhilensis Wiland
subsp. kansimbensis *J. Wiland in Fragm. Flor. Geobot.*, 42(2): 414 (1997)—Zaire.

HYPOXIS:-
symoensiana *J. Wiland in Fragm. Flor. Geobot.*, 42(2): 421 (1997)—Zaire.
upembensis *J. Wiland in Fragm. Flor. Geobot.*, 42(2): 414 (1997)—Zaire.

HYPSERPA (Menispermac.).
mackeei *L.L. Forman in Kew Bull.*, 51(3): 550 (1996)—New Caledonia.

HYPTIS (Labiat.).
pyriformis Epling & Játiva
var. yutajeensis *R.M. Harley in Kew Bull.*, 53(4): 975 (1998)—Venezuela.

HYSTERIA *Reinw.in Syll. Pl. Nov.*, 2: 5 (1825). *ORCHIDACEAE*†
veratrifolia *Reinw.in Syll. Pl. Nov.*, 2: 5 (1825)—Jawa.†

HYSTERIONICA (Compos.).
pinnatiloba *N.I. Matzenbacher & M. Sobral in Comun. Mus. Ciênc. PUCRS, Sér. Bot.*, 2(1): 17 (1996)—Brazil South (Rio Grande do Sul).
pinnatisecta *N.I. Matzenbacher & M. Sobral in Comun. Mus. Ciênc. PUCRS, Sér. Bot.*, 2(1): 16 (1996)—Brazil South (Santa Catarina).

HYSTRIX (Gramin.).
duthiei (Stapf) Bor
subsp. japonica *(Hack.) C. Baden, S. Frederiksen & O. Seberg in Nordic J. Bot.*, 17(5): 461 (1997): Asprella japonica.
subsp. longearistata *(Hack.) C. Baden, S. Frederiksen & O. Seberg in Nordic J. Bot.*, 17(5): 461 (1997): Asprella sibirica var. longearistata.

I

× **IACOVIELLOARA** *hort. in Orchid Rev*, 108(1234): new orchid hybrids 15 (2000). *ORCHIDACEAE* (Brassavola × Cattleya × Diacrium × Epidendrum × Laelia).

IANHEDGEA *I.A. Al-Shehbaz & S.L. O'Kane in Edinburgh J. Bot.*, 56(3): 322 (1999). *CRUCIFERAE*
minutiflora *(Hook. f. & Thomson) I.A. Al-Shehbaz & S.L. O'Kane in Edinburgh J. Bot.*, 56(3): 322 (1999): Sisymbrium minutiflorum.
subsp. brevipedicellata *(Hedge) I.A. Al-Shehbaz & S.L. O'Kane in Edinburgh J. Bot.*, 56(3): 324 (1999): Microsisymbrium minutiflorum var. brevipedicellatum.

IBERIS (Crucif.).
ciliata All.
subsp. vinetorum *(Pau) G. Mateo Sanz & M.B. Crespo Villalba, Fl. Abrev. Comun. Valenciana:* 430 (1995): I. vinetorum.
grosmiquelii Pau & Font Quer
subsp. senneniana *(Pau) A. Dobignard in Candollea*, 52(1): 130 (1997): I. grosmiquelii var. senneniana.
lagascana DC.
subsp. rhomarensis *J.M. Montserrat in Lagascalia*, 18(2): 244 (1996)—Morocco.
linifolia L.
subsp. boppardensis *(Jord.) D. Korneck in R. Wisskirchen & H. Haeupler, Standardliste Farn- und Blütenpfl.* Deutschl.: 271 (1998): I. boppardensis.
nazarita Moreno
subsp. maroccana *(Pau) J.M. Montserrat in Lagascalia*, 18(2): 245 (1996): I. linifolia var. maroccana.

ICHTHYOSTOMA *M. Hedrén & K. Vollesen in Nordic J. Bot.*, 16(4): 441 (1996 publ. 1997). *ACANTHACEAE*
thulinii *M. Hedrén & K. Vollesen in Nordic J. Bot.*, 16(4): 441 (1996 publ. 1997)—Somalia.

ICHTHYOTHERE (Compos.).
macdanielii *H. Robinson in Sida*, 16(4): 731 (1995)—Peru.
matogrossensis *R.C.A. Pereira & J. Semir in Bradea*, 8(28): 155 (2000)—Brazil West-central (Mato Grosso do Sul).
pastazensis *H. Robinson in Sida*, 16(4): 734 (1995)—Ecuador.

ICHTYOSELMIS *M. Lidén & T. Fukuhara in Pl. Syst. Evol.*, 206(1–4): 415 (1997), nom. nov.: Dicentra sect. Macranthos *K.R. Stern.* (Papaverac.)
macrantha *(Oliv.) M. Lidén in Pl. Syst. Evol.*, 206(1–4): 415 (1997): Dicentra macrantha.

ICURIA *J.J. Wieringa in Wageningen Agric. Univ. Pap.*, 99(4): 241 (1999). *LEGUMINOSAE*
dunensis *J.J. Wieringa in Wageningen Agric. Univ. Pap.*, 99(4): 242 (1999)—Mozambique.

IDESIA (Flacourtiac.).
polycarpa Maxim.
var. fujianensis *(G.S. Fan) S.S. Lai in Fl. Reipubl. Popul. Sin.*, 52(1): 58 (1999): Idesia fujianensis.

IDIOPAPPUS (Compos.).
saloyensis *(Domke) H. Robinson & J.L. Panero in Cat. Vasc. Pl. Ecuador (Monogr. Syst. Bot. Missouri Bot. Gard., 75)*: 953 (1999): Verbesina saloyensis.

IGIDIA *F. Speta in Phyton (Austria)*, 38(1): 70 (1998). *HYACINTHACEAE*
volubilis *(H. Perrier) F. Speta in Phyton (Austria)*, 38(1): 70 (1998): Urginea volubilis.

IGUANURA (Palmae).
asli *C.K. Lim in Gard. Bull. Singapore*, 48(1–2): 28 (1996 publ. 1998)—Malaya (Peninsular Malaysia).
belumensis *C.K. Lim in Gard. Bull. Singapore*, 48(1–2): 52 (1996 publ. 1998)—Malaya (Peninsular Malaysia).
divergens *D.R. Hodel in Palm J.*, 136: 7 (1997)—Thailand.
humilis *(Kiew) C.K. Lim in Gard. Bull. Singapore*, 48(1–2): 37 (1996 publ. 1998): I. wallichiana var. humilis.
kelantanensis *C.K. Lim in Gard. Bull. Singapore*, 48(1–2): 30 (1996 publ. 1998)—Malaya (Peninsular Malaysia).
mirabilis *C.K. Lim in Gard. Bull. Singapore*, 48(1–2): 57 (1996 publ. 1998)—Malaya (Peninsular Malaysia).
multifida *D.R. Hodel in Palm J.*, 136: 8 (1997)—Thailand.
perdana *C.K. Lim in Gard. Bull. Singapore*, 48(1–2): 55 (1996 publ. 1998)—Malaya (Peninsular Malaysia).
piahensis *C.K. Lim in Gard. Bull. Singapore*, 48(1–2): 31 (1996 publ. 1998)—Malaya (Peninsular Malaysia).
polymorpha Becc.
var. integra *C.K. Lim in Principes*, 42(2): 112 (1998)—Thailand.
speciosa *D.R. Hodel in Palm J.*, 134: 29 (1997)—Thailand.
tenuis *D.R. Hodel in Palm J.*, 136: 11 (1997)—Thailand.
var. khaosokensis *C.K. Lim in Principes*, 42(2): 112 (1998)—Thailand.
thalangensis *C.K. Lim in Principes*, 42(2): 114 (1998)—Thailand.
wallichiana (Wall. ex Mart.) Hook. f.
var. rosea *C.K. Lim in Gard. Bull. Singapore*, 48(1–2): 23 (1996 publ. 1998)—Malaya (Peninsular Malaysia).

ILEX (Aquifoliac.).
ser. Spicatae *T.R. Dudley in F.C. Galle, Hollies, The Genus Ilex:* 100 (1997), without latin descr. or type.
alainii *T.R. Dudley in F.C. Galle, Hollies, The Genus Ilex:* 154 (1997), nom. nov.: I. victorinii *Alain.*
apiensis *S. Andrews in Sandakania*, 11: 1 (1998)—Borneo (Sarawak).

ILEX:-
atabapoensis *T.R. Dudley in F.C. Galle, Hollies, The Genus Ilex:* 159 (1997), nom. nov.: I. spathulata *Steyerm.*
× attenuata
f. buxtonica *B.M. Bauers in Holly Soc. J.*, 16(3): 37 (1998), without type—North Carolina.
buswellii Small
f. channellii *(Edwin) T.R. Dudley in F.C. Galle, Hollies, The Genus Ilex:* 134 (1997): I. ambigua f. channellii.
celebensis Capit.
var. ranauana *S. Andrews in Sandakania*, 11: 2 (1998)—Borneo (Sabah), Sumatera.
charrascosensis *T.R. Dudley in F.C. Galle, Hollies, The Genus Ilex:* 168 (1997), nom. nov., without replaced synonym ref.: I. paucinervis *Alain.*
chimantaensis *T.R. Dudley in F.C. Galle, Hollies, The Genus Ilex:* 168 (1997), nom. nov., without replaced synonym page: I. sulcata *Edwin.*
var. pygmaea *(Edwin) T.R. Dudley in F.C. Galle, Hollies, The Genus Ilex:* 169 (1997), specific name not validly publ.: I. sulcata var. pygmaea.
clethriflora *S. Andrews in Sandakania*, 11: 3 (1998)—Borneo (Sabah).
crenata Thunb.
f. multicrenata *(C.J. Tseng) S.K. Chen in Fl. Reipubl. Popul. Sin.*, 45(2): 62 (1999): I. crenata var. multicrenata.
decidua Walter
var. mulleri *Edwin ex T.R. Dudley in F.C. Galle, Hollies, The Genus Ilex:* 136 (1997)—Mexico Northeast (Nuevo León).
dehongensis *S.K. Chen & Y.X. Feng in Bot. Bull. Acad. Sin. (Taipei)*, 40(2): 173 (1999)—China South-central (Yunnan).
discolor Hemsl.
var. tolucana *(Hemsl.) Edwin ex T.R. Dudley in F.C. Galle, Hollies, The Genus Ilex:* 182 (1997), without basionym page: I. tolucana.
excelsa *(Wall.) J.O. Voigt, Hort. Suburb. Calcutt.:* 347 (1845): Cassine excelsa.
ficifolia *C.J. Tseng ex S.K. Chen & Y.X. Feng in Acta Phytotax. Sin.*, 37(2): 143 (1999)—China Southeast (Fujian).
fruticosa *S. Andrews in Sandakania*, 11: 4 (1998)—Borneo (Sarawak).
fuertensiana *(Loes.) T.R. Dudley in F.C. Galle, Hollies, The Genus Ilex:* 189 (1997): I. caroliniana var. fuertensiana.
haberi *(Lundell) W.J. Hahn in Novon*, 6(2): 183 (1996) Euonymus haberi.
hemiepiphytica *W.J. Hahn in Novon*, 6(2): 183 (1996)—Costa Rica.
hirsuta *C.J. Tseng ex S.K. Chen & Y.X. Feng in Acta Phytotax. Sin.*, 37(2): 143 (1999)—China South-central (Hubei).
huiana *C.J. Tseng ex S.K. Chen & Y.X. Feng in Acta Phytotax. Sin.*, 37(2): 143 (1999)—Hainan.
ijuensis *S. Andrews in Sandakania*, 11: 4 (1998)—Borneo (Sarawak).
integra Thunb.
f. lutea *H. Ohba & S. Akiyama in Fl. Japan*, 2c: 87 (1999)—Japan.
var. undulata *S. Hatusima in S. Hatusima & T. Amano, Fl. Ryukyus, S. of Amami Is., ed. 2:* 116 (1994), without latin descr. or type—Locality not stated.
jacobsii *S. Andrews in Sandakania*, 11: 5 (1998)—Borneo (Sarawak, Brunei).
kelabitana *S. Andrews in Sandakania*, 11: 6 (1998)—Borneo (Sarawak, Kalimantan).
keranjiensis *S. Andrews in Sandakania*, 11: 7 (1998)—Borneo (Sarawak).
kinabaluensis *S. Andrews in Sandakania*, 11: 8 (1998)—Borneo (Sabah).
knucklesensis *D. Philcox in M.D. Dassanayake & W.D. Clayton, eds., Revised Handb. Fl. Ceylon*, 13: 6 (1999)—Sri Lanka.
laurifolia *Zipp. ex L.E.T. Loesener, Monogr. Aquifol.*, 1: 116 (1901)—Moluccas, Philippines.
limii *C.J. Tseng in Acta Sci. Nat. Univ. Amoiensis*, 9(4): 305 (1962)—China Southeast (Fujian).
longipes Chapm. ex Trel.
f. hirsuta *(Lundell) T.R. Dudley in F.C. Galle, Hollies, The Genus Ilex:* 139 (1997): I. longipes var. hirsuta.

ILEX:-
megalophylla *Edwin ex T.R. Dudley in F.C. Galle, Hollies, The Genus Ilex:* 213 (1997), nom. nov., illegit. superfl.: I. grandis *Reissek*.
megaphylla *S. Andrews in Sandakania,* 11: 9 (1998)—Borneo (Sarawak).
mesilauensis *S. Andrews in Sandakania,* 11: 9 (1998)—Borneo (Sabah).
micrococca Maxim.
f. luteocarpa *H. Ohba & S. Akiyama in Fl. Japan,* 2c: 90 (1999)—Japan.
mucronata *(L.) M. Powell, V. Savolainen & S. Andrews in Kew Bull.,* 55(2): 345 (2000): Vaccinium mucronatum.
nervulosa *(Loes.) S. Andrews in Sandakania,* 11: 14 (1998): I. venulosa var. nervulosa.
nuculicava S.Y. Hu
f. brevipedicellata *(S.Y. Hu) T.R. Dudley in F.C. Galle, Hollies, The Genus Ilex:* 219 (1997): I. nuculicava var. brevipedicellata.
× owariensis *S. Hatusima & M. Kobayashi in J. Phytogeogr. Taxon.,* 44(1–2): 85 (1996)—Japan. (I. integra × latifolia).
pentagona *S.K. Chen, Y.X. Feng & C.F. Liang in Acta Phytotax. Sin.,* 36(4): 357 (1998)—China Southeast (Guangxi, Hunan), China South-central (Guizhou, Yunnan).
poiensis *S. Andrews in Sandakania,* 11: 10 (1998)—Borneo (Sarawak).
promecophylla *S. Andrews in Sandakania,* 11: 11 (1998)—Borneo (Sabah, Sarawak, Brunei).
pseudoumbelliformis *T.R. Dudley in F.C. Galle, Hollies, The Genus Ilex:* 228 (1997), nom. nov.: I. attenuata *Steyerm*.
pubigera *(C.Y. Wu ex Y.R. Li) S.K. Chen & Y.X. Feng in Fl. Reipubl. Popul. Sin.,* 45(2): 34 (1999): I. purpurea var. pubigera.
reisseckiana *T.R. Dudley in F.C. Galle, Hollies, The Genus Ilex:* 231 (1997), nom. nov., replaced synonym not validly publ.: I. andarensis *Rusby ex Loes*.
renae *S. Andrews in Sandakania,* 11: 12 (1998)—Borneo (Sarawak).
repanda Griseb.
var. hypaneura *(Loes.) Edwin ex T.R. Dudley in F.C. Galle, Hollies, The Genus Ilex:* 231 (1997), with incorrect basionym ref.: I. hypaneura.
robustinervosa *C.J. Tseng ex S.K. Chen & Y.X. Feng in Acta Phytotax. Sin.,* 37(2): 144 (1999)—China Southeast (Guangdong).
skutchii *Edwin ex T.R. Dudley & W.J. Hahn in Novon,* 6(2): 181 (1996)—Costa Rica, Nicaragua.
spicata Blume
var. harmsiana *(Loes.) S. Andrews in Sandakania,* 11: 13 (1998): I. harmsiana.
sulcata *Wall. ex J.D. Hooker, Fl. Brit. India,* 1: 604 (1875)—Burma.
syzygiophylla *C.J. Tseng ex S.K. Chen & Y.X. Feng in Acta Phytotax. Sin.,* 37(2): 144 (1999)—China Southeast (Guangdong).
tamii *T.R. Dudley in F.C. Galle, Hollies, The Genus Ilex:* 244 (1997), nom. nov.: I. revoluta *P.C. Tam*.
urbaniana Loes.
var. riedlaei *(Loes.) Edwin ex T.R. Dudley in F.C. Galle, Hollies, The Genus Ilex:* 251 (1997): I. riedlaei.
verisimilis *C.J. Tseng ex S.K. Chen & Y.X. Feng in Acta Phytotax. Sin.,* 37(2): 144 (1999)—China Southeast (Guangxi).
vietnamensis *T.R. Dudley in F.C. Galle, Hollies, The Genus Ilex:* 252 (1997), nom. nov.: I. merrillii *Tardieu*.
wugongshanensis *C.J. Tseng ex S.K. Chen & Y.X. Feng in Acta Phytotax. Sin.,* 37(2): 144 (1999)—China Southeast (Jiangxi).
zhejiangensis *C.J. Tseng ex S.K. Chen & Y.X. Feng in Acta Phytotax. Sin.,* 37(2): 144 (1999)—China Southeast (Zhejiang).
zippeliana *T.R. Dudley in F.C. Galle, Hollies, The Genus Ilex:* 259 (1997), nom. nov., with incorrect replaced synonym ref.: I. laurifolia *Zipp. ex Loes.*

ILLICIUM (Illiciac.).
subgen. Cymbostemon *(Spach) Y.W. Law & Q. Lin in Fl. Reipubl. Popul. Sin.,* 30(1): 271 (1996): gen. Cymbostemon.
sect. Impressicosta *Y.W. Law & Q. Lin in Fl. Reipubl. Popul. Sin.,* 30(1): 271 (1996).

ILLICIUM:-
fargesii Finet & Gagnep.
var. szechuanense *(C.Y. Cheng) Q. Lin in Acta Bot. Yunnanica,* 17(3): 297 (1995), as 'szechuanensis': I. szechuanense.
subsp. szechuanense *(C.Y. Cheng) Q. Lin in Acta Phytotax. Sin.,* 38(2): 172 (2000), as 'szechuanensis': I. szechuanense.
jiadifengpi B.N. Chang
var. szechuanense *(C.Y. Cheng) Y.W. Law & B.N. Chang in Fl. Reipubl. Popul. Sin.,* 30(1): 205 (1996): I. szechuanense.
micranthum Dunn
subsp. tsangii *(A.C. Sm.) Q. Lin in Acta Phytotax. Sin.,* 38(2): 176 (2000): I. tsangii.
parvifolium Merr.
subsp. oligandrum *(Merr. & Chun) Q. Lin in Acta Phytotax. Sin.,* 38(2): 176 (2000): I. oligandrum.
pseudosimonsii *Q. Lin in Acta Bot. Austro Sinica,* 10: 12 (1995)—China South-central (Yunnan).

IMPATIENS (Balsaminac.).
anhuiensis *Y.L. Chen in Acta Phytotax. Sin.,* 37(1): 90 (1999)—China Southeast (Anhui).
chishuiensis *Y.X. Xiong in Acta Phytotax. Sin.,* 34(1): 99 (1996)—China South-central (Guizhou).
chumphonensis *T. Shimizu in Bull. Nation. Sci. Mus., B (Tokyo),* 26(2): 35 (2000)—Thailand.
clavicalcar *Eb. Fischer in Bull. Jard. Bot. Nation. Belg.,* 66(1–2): 64 (1997)—Zaire.
clavigeroides *S. Akiyama, H. Ohba & S.K. Wu in Bull. Nation. Sci. Mus., B (Tokyo),* 22(4): 141 (1996)—China South-central (Yunnan).
etindensis *M. Cheek & Eb. Fischer in Kew Bull.,* 54(2): 472 (1999)—Cameroon.
fanjingshanica *Y.L. Chen in Acta Phytotax. Sin.,* 37(1): 93 (1999)—China South-central (Guizhou).
gongshanensis *Y.L. Chen in Acta Phytotax. Sin.,* 38(6): 560 (2000)—China South-Central (Yunnan).
guizhouensis *Y.L. Chen in Acta Phytotax. Sin.,* 37(1): 97 (1999)—China South-central (Guizhou).
halongensis *R. Kiew & T.H. Nguyen in Gard. Bull. Singapore,* 52(1): 196 (2000)—Vietnam.
henanensis *Y.L. Chen in Acta Phytotax. Sin.,* 37(1): 88 (1999)—China Southeast (Henan), China North-central (Shanxi).
irangiensis *Eb. Fischer in Bull. Jard. Bot. Nation. Belg.,* 66(1–2): 64 (1997)—Zaire.
jiulongshanica *Y.L. Xu & Y.L. Chen in Acta Phytotax. Sin.,* 37(2): 194 (1999)—China Southeast (Zhejiang).
× kaskazini *J.M. Grimshaw & C. Grey-Wilson in Curtis's Bot. Mag.,* 14(1): 29 (1997)—Tanzania. (I. kilimanjari subsp. pocsii × I. pseudoviola).
kilimanjari Oliv.
subsp. pocsii *J.M. Grimshaw & C. Grey-Wilson in Curtis's Bot. Mag.,* 14(1): 28 (1997)—Tanzania.
kojeensis *Y.N. Lee in Korean J. Pl. Taxon.,* 28(1): 28 (1998)—South Korea.
laumonieri *T. Shimizu in Kew Bull.,* 52(2): 435 (1997)—Sumatera.
linearisepala *S. Akiyama, H. Ohba & S.K. Wu in Bull. Nation. Sci. Mus., B (Tokyo),* 22(4): 142 (1996)—China South-central (Yunnan).
longicornuta *Y.L. Chen in Acta Phytotax. Sin.,* 37(1): 93 (1999)—China Southeast (Hunan).
luchunensis *S. Akiyama, H. Ohba & S.K. Wu in Bull. Nation. Sci. Mus., B (Tokyo),* 22(4): 138 (1996)—China South-central (Yunnan).
macrovexilla *Y.L. Chen in Acta Phytotax. Sin.,* 38(6): 560 (2000)—China Southeast (Guangxi).
maguanensis *S. Akiyama, H. Ohba & S.K. Wu in Bull. Nation. Sci. Mus., B (Tokyo),* 22(4): 138 (1996)—China South-central (Yunnan).
namkatensis *T. Shimizu in Bull. Nation. Sci. Mus., B (Tokyo),* 26(2): 37 (2000)—Thailand.
napoensis *Y.L. Chen in Acta Phytotax. Sin.,* 38(6): 557 (2000)—China Southeast (Guangxi).
neglecta *Y.L. Xu & Y.L. Chen in Acta Phytotax. Sin.,* 37(2): 196 (1999)—China Southeast (Zhejiang, Anhui).
noli-tangere L.
f. hemileuca *Koji Ito & Hinoma in J. Jap. Bot.,* 71(5): 301 (1996)—Japan.

IMPATIENS:-
parviflora DC.
subsp. nevskii *(Pobed.) N.N. Tzvelev in Novosti Sist. Vyssh Rast.,* 32: 185 (2000): I. nevskii.
polyneura *K.M. Liu in Acta Phytotax. Sin.,* 37(2): 201 (1999)—China Southeast (Hunan).
santisukii *T. Shimizu in Bull. Nation. Sci. Mus., B (Tokyo),* 26(2): 39 (2000)—Thailand.
sarawakensis *T. Shimizu in Kew Bull.,* 52(2): 437 (1997)—Borneo (Sarawak).
sivarajanii *M. Kumar & S. Sequiera in Rheedea,* 6(2): 51 (1996)—India (Kerala).
spathulata *Y.X. Xiong in Acta Phytotax. Sin.,* 34(1): 98 (1996)—China South-central (Guizhou).
suichangensis *Y.L. Xu & Y.L. Chen in Acta Phytotax. Sin.,* 37(2): 198 (1999)—China Southeast (Zhejiang).
tienmushanica Y.L. Chen
var. longicalcarata *Y.L. Xu & Y.L. Chen in Acta Phytotax. Sin.,* 37(2): 200 (1999)—China Southeast (Zhejiang).
trichosepala *Y.L. Chen in Acta Phytotax. Sin.,* 37(1): 95 (1999)—China South-central (Guizhou), China Southeast (Guangxi).
tujuhensis *N. Utami & T. Shimizu in Kew Bull.,* 52(2): 439 (1997)—Sumatera.
uzungwaensis *C. Grey-Wilson & C. Frimodt-Møller in Kew Bull.,* 54(1): 179 (1999)—Tanzania.
volkensii Warb.
subsp. macrosepala *C. Grey-Wilson & C. Frimodt-Møller in Kew Bull.,* 54(1): 180 (1999)—Tanzania.
wuchengyihii *S. Akiyama, H. Ohba & S.K. Wu in Bull. Nation. Sci. Mus., B (Tokyo),* 22(4): 137 (1996)—China South-central (Yunnan).

INCARVILLEA (Bignoniac.).
compacta Maxim.
var. qinghaiensis *C. Grey-Wilson in New Plantsman,* 5(2): 84 (1998)—Qinghai.
himalayensis *C. Grey-Wilson in New Plantsman,* 5(2): 88 (1998)—Bhutan, Nepal, Xizang.
parasitica *W. Roxburgh, Pl. Coromandel,* 3: 88 (1820)—Assam (Meghalaya).†
zhongdianensis *C. Grey-Wilson in New Plantsman,* 5(2): 96 (1998)—China South-central (Yunnan).

INDIA *A. Nageswara Rao in J. Econ. Taxon. Bot.,* 22(3): 701 (1998 publ. 1999). *ORCHIDACEAE*
arunachalensis *A. Nageswara Rao in J. Econ. Taxon. Bot.,* 22(3): 701 (1998 publ. 1999)—Assam (Arunachal Pradesh).

INDIGOFERA (Legumin.).
emarginella Steud. ex A. Rich.
var. marrupaensis *B.D. Schrire in Kew Bull.,* 53(3): 661 (1998)—Mozambique.
grandiflora *B.H. Choi & S.K. Cho in J. Pl. Biol.,* 39(2): 147 (1996)—South Korea.
jucunda *B. Schrire in Curtis's Bot. Mag.,* 14(4): 216 (1997)—Natal, Cape Province.
longistaminata *B.D. Schrire in Kew Bull.,* 53(3): 655 (1998), nom. nov.: I. homblei subsp. longiflora *J.B. Gillett*.
patula Baker
subsp. okuensis *B.D. Schrire & J.-M. Onana in Kew Bull.,* 55(1): 223 (2000)—Cameroon.
pseudomoniliformis *B.D. Schrire in Kew Bull.,* 53(3): 662 (1998)—Mozambique.
rubromarginata *M. Thulin in Nordic J. Bot.,* 15(5): 519 (1995 publ. 1996)—South Yemen, Oman.
rumphiensis *B.D. Schrire in Kew Bull.,* 52(1): 157 (1997)—Malawi, Zambia.
silvestrii Pamp.
var. alii *M. Sanjappa in Fasc. Fl. India,* 21: 123 (1995)—Assam, Bhutan, China South-central (Yunnan).
sinuspersica *V. Mozaffarian in Iranian J. Bot.,* 7(1): 133 (1996)—Iran.
subcorymbosa Baker
var. grandis *B.D. Schrire in Kew Bull.,* 53(3): 665 (1998)—Zambia, Tanzania.
trita L.f.
var. nubica *(J.B. Gillett) L. Boulos & B.D. Schrire in L. Boulos, Fl. Egypt,* 1: xiii, 313 (1999): I. subulata var. nubica.

INDOCALAMUS (Gramin.).
chebalingensis *W.T. Lin in J. Bamboo Res.*, 19(1): 6 (2000)—Southeast (Guangdong).
confertus *C.H. Hu in J. Bamboo Res.*, 15(1): 1 (1996)—China South-central (Sichuan).
multinervis *(W.T. Lin & Z.M. Wu) W.T. Lin in J. Bamboo Res.*, 19(4): 17 (2000): Arundinaria multinervis.
petelotii *(A. Camus) D. Ohrnberger, Bamboos of the World Intro. 3:* 14 (1996): Arundinaria petelotii.
semifalcatus *(H.R. Zhao & Y.L. Yang) T.P. Yi in J. Bamboo Res.*, 19(1): 26 (2000): I. longiauritus var. semifalcatus.
f. luteolivittatus *T.P. Yi in J. Bamboo Res.*, 19(1): 26 (2000)—China South-central (Sichuan).
tessellatus
f. hamadae *(Hatus.) Rifat ex D. Ohrnberger, Bamboos of the World Intro. 2:* 10 (1996): Pseudosasa hamadae.

INDOSASA (Gramin.).
glabrata C.D. Chu & C.S. Chao
var. albo-hispidula *(Q.H. Dai & C.F. Huang) C.S. Chao & C.D. Chu in Fl. Reipubl. Popul. Sin.*, 9(1): 212 (1996): I. albo-hispidula.
longiligula T.H. Wen
var. amara *T.H. Wen in J. Bamboo Res.*, 2(1): 70 (1983)—China Southeast (Fujian).

INGA (Legumin.).
sect. Affonsea *(A. St.-Hil.) T.D. Pennington, Genus Inga. Botany:* 678, 103 (1997): gen. Affonsea.
sect. Complanatae *T.D. Pennington, Genus Inga. Botany:* 603 (1997).
sect. Grandiflorae *T.D. Pennington, Genus Inga. Botany:* 584 (1997).
sect. Longiflorae *(Benth.) T.D. Pennington, Genus Inga. Botany:* 629 (1997): I. ser. Longiflorae.
sect. Multijugae *(J. León) T.D. Pennington, Genus Inga. Botany:* 473 (1997): I. ser. Multijugae.
sect. Pilosulae *(J. León) T.D. Pennington, Genus Inga. Botany:* 493 (1997): I. ser. Pilosulae.
sect. Spectabiles *(J. León) T.D. Pennington, Genus Inga. Botany:* 466 (1997): I. ser. Spectabiles.
sect. Tetragonae *(Pittier) T.D. Pennington, Genus Inga. Botany:* 756 (1997): I. ser. Tetragonae.
sect. Urceolatae *T.D. Pennington, Genus Inga. Botany:* 772 (1997).
sect. Vulpinae *(Benth.) T.D. Pennington, Genus Inga. Botany:* 518 (1997): I. ser. Vulpinae.
acicularis *T.D. Pennington, Genus Inga. Botany:* 261 (1997)—Peru, Brazil North (Amazonas).
albicoria *O. Poncy in Bull. Mus. Nation. Hist. Nat., B, Adansonia*, Ser. 4, 18(1–2): 57 (1996)—Guyana, French Guiana, Brazil North (Amapá, Pará).
amboroensis *T.D. Pennington, Genus Inga. Botany:* 387 (1997)—Bolivia.
approximata *T.D. Pennington, Genus Inga. Botany:* 552 (1997)—Bolivia.
aptera *(Vinha) T.D. Pennington, Genus Inga. Botany:* 678 (1997): Affonsea aptera
arenicola *T.D. Pennington, Genus Inga. Botany:* 296 (1997)—Brazil Southeast (Rio de Janeiro).
balsapambensis *T.D. Pennington, Genus Inga. Botany:* 698 (1997)—Ecuador.
bracteifera *N. Zamora & T.D. Pennington in T.D. Pennington, Genus Inga. Botany:* 538 (1997)—Costa Rica.
cabelo *T.D. Pennington, Genus Inga. Botany:* 525 (1997)—Brazil Southeast (Espírito Santo, Rio de Janeiro).
canonegrensis *N. Zamora & T.D. Pennington in T.D. Pennington, Genus Inga. Botany:* 593 (1997)—Costa Rica.
carinata *T.D. Pennington, Genus Inga. Botany:* 731 (1997)—Ecuador.
ciliata C. Presl
subsp. subcapitata *T.D. Pennington, Genus Inga. Botany:* 522 (1997)—Ecuador, Colombia, Peru.
congesta *T.D. Pennington, Genus Inga. Botany:* 681 (1997), nom. nov.: Affonsea bullata *Benth.*
edwallii *(Harms) T.D. Pennington, Genus Inga. Botany:* 687 (1997): Affonsea edwallii.
enterolobioides *T.D. Pennington, Genus Inga. Botany:* 294 (1997)—Brazil Southeast (Rio de Janeiro).

INGA:-
exalata T.S. Elias
subsp. umbilicata *N. Zamora & T.D. Pennington in T.D. Pennington, Genus Inga. Botany:* 620 (1997)—Costa Rica.
exfoliata *T.D. Pennington & F.C.P. García in T.D. Pennington, Genus Inga. Botany:* 245 (1997)—Brazil Southeast (Espírito Santo).
exilis *T.D. Pennington, Genus Inga. Botany:* 320 (1997)—Brazil North (Amazonas).
extra-nodis *T.D. Pennington, Genus Inga. Botany:* 620 (1997)—Ecuador.
fosteriana *T.D. Pennington, Genus Inga. Botany:* 582 (1997)—Peru.
gereauana *(Pipoly & R. Vásquez) T.D. Pennington, Genus Inga. Botany:* 781 (1997): Affonsea gereauana.
globularis *T.D. Pennington, Genus Inga. Botany:* 690 (1997), nom. nov.: Affonsea juglandifolia *A. St.-Hil.*
grandis *T.D. Pennington, Genus Inga. Botany:* 584 (1997)—Brazil North (Acre, Amazonas), Peru.
grazielae *(Vinha) T.D. Pennington, Genus Inga. Botany:* 685 (1997); *see Kew Bull.*, 54(4): 982 (1999 publ. 2000): Affonsea grazielae.
guadeloupensis *C. Sastre in Biogeographica*, 72(4): 184 (1996), homonym; *see Desv. in J. Bot. Agric.*, 3: 70 (1814)—Leeward Is. (Guadeloupe).
ilta *T.D. Pennington, Genus Inga. Botany:* 437 (1997)—Ecuador, Peru.
japurensis *T.D. Pennington, Genus Inga. Botany:* 560 (1997)—Peru, Bolivia, Brazil North (Amazonas).
leptocarpa *T.D. Pennington, Genus Inga. Botany:* 510 (1997)—Brazil North (Amazonas).
manabiensis *T.D. Pennington, Genus Inga. Botany:* 400 (1997)—Ecuador, Colombia.
megalobotrys *T.D. Pennington, Genus Inga. Botany:* 580 (1997)—Peru.
megaphylla *O. Poncy & H. Vester in Novon*, 5(4): 384 (1995)—Colombia, Peru.
mitaraka *O. Poncy in Bull. Mus. Nation. Hist. Nat., B, Adansonia*, Ser. 4, 18(1–2): 70 (1996)—Guyana, French Guiana, Brazil North (Amapá).
multijuga Benth.
subsp. aestuariorum *(Pittier) T.D. Pennington, Genus Inga. Botany:* 483 (1997): I. aestuariorum.
subsp. mexicana *T.D. Pennington, Genus Inga. Botany:* 483 (1997)—Mexico Southwest (Oaxaca), Mexico Southeast (Chiapas), Mexico Gulf (Veracruz), Guatemala, Belize.
multinervis *T.D. Pennington, Genus Inga. Botany:* 353 (1997)—Ecuador, Peru.
nobilis Willd.
subsp. quaternata *(Poepp. & Endl.) T.D. Pennington, Genus Inga. Botany:* 383 (1997): I. quaternata.
nouragensis *O. Poncy in Bull. Mus. Nation. Hist. Nat., B, Adansonia*, Ser. 4, 18(1–2): 72 (1996)—French Guiana.
pedunculata *(Vinha) T.D. Pennington, Genus Inga. Botany:* 692 (1997): Affonsea pedunculata.
pleiogyna *T.D. Pennington, Genus Inga. Botany:* 683 (1997), nom. nov.: Affonsea densiflora *Benth.*
pluricarpellata *T.D. Pennington, Genus Inga. Botany:* 586 (1997)—Peru.
porcata *T.D. Pennington, Genus Inga. Botany:* 243 (1997)—Peru.
praegnans *T.D. Pennington, Genus Inga. Botany:* 777 (1997)—Brazil Southeast (São Paulo).
salicifolia *T.D. Pennington, Genus Inga. Botany:* 188 (1997)—Brazil North (Amazonas).
sarayacuensis *T.D. Pennington, Genus Inga. Botany:* 783 (1997)—Ecuador.
sertulifera DC.
subsp. leptopus *(Benth.) T.D. Pennington, Genus Inga. Botany:* 234 (1997): I. leptopus.
silanchensis *T.D. Pennington, Genus Inga. Botany:* 364 (1997)—Ecuador.
striolata *T.D. Pennington, Genus Inga. Botany:* 402 (1997)—Bolivia, Ecuador, Peru, Brazil North (Acre).
suberosa *T.D. Pennington, Genus Inga. Botany:* 651 (1997)—Brazil North (Amazonas).
subnuda Salzm. ex Benth.
subsp. luschnathiana *(Benth.) T.D. Pennington, Genus Inga. Botany:* 753 (1997): Inga luschnathiana.

INGA:-
suborbicularis *T.D. Pennington, Genus Inga. Botany:* 542 (1997)—Brazil Southeast (Minas Gerais), Brazil Northeast (Bahia).
tayronaensis *T.D. Pennington, Genus Inga. Botany:* 424 (1997)—Colombia.
tenuicalyx *T.D. Pennington, Genus Inga. Botany:* 426 (1997)—Peru.
tenuiloba *N. Zamora & T.D. Pennington in T.D. Pennington, Genus Inga. Botany:* 540 (1997)—Costa Rica.
thibaudiana DC.
subsp. peltadenia *(Harms) T.D. Pennington, Genus Inga. Botany:* 489 (1997): I. peltadenia.
subsp. russotomentella *(Malme) T.D. Pennington, Genus Inga. Botany:* 491 (1997): I. russotomentella.
vera Willd.
subsp. affinis *(DC.) T.D. Pennington, Genus Inga. Botany:* 716 (1997): I. affinis.
yasuniana *T.D. Pennington, Genus Inga. Botany:* 286 (1997)—Ecuador.

INOBULBUM (Orchidac.).
layardii *(F. Muell. & Kraenzl.) M.A. Clements & D.L. Jones in Orchadian*, 12(7): 308 (1998): Cirrhopetalum layardii.

INULA (Compos.).
caspica Blume
var. paniculata *C.H. An in Fl. Xinjiangensis*, 5: 476 (1999)—Xinjiang.
koelzii *R. Dawar & M. Qaiser in Candollea*, 52(2): 283 (1997)—Jammu-Kashmir.
rungwensis *H.J. Beentje in Kew Bull.*, 54(1): 100 (1999)—Tanzania.
salicina L.
var. minipetala *Y.N. Lee, Fl. Korea:* 1161 (1996), contrary to Arts. 37.4, 37.5 ICBN (1994)—South Korea.
stenocalathia *(Rech. f.) A. Soldano in Compositae Newsl.*, 32: 19 (1998): Codonocephalum stenocalathium.

IOCHROMA (Solanac.).
edule *S. Leiva Gonzalez in Arnaldoa*, 3(1): 41 (1995)—Peru.
gesnerioides (Humb., Bonpl. & Kunth) Miers
var. flavum *(André) J.M.H. Shaw in New Plantsman*, 5(3): 176 (1998): I. flavum.
nitidum *S. Leiva González & V. Quipuscoa Silvestre in Arnaldoa*, 5(2): 172 (1998)—Peru.
schjellerupii *S. Leiva González & V. Quipuscoa Silvestre in Arnaldoa*, 5(2): 175 (1998)—Peru.
stenanthum *S. Leiva Gonzalez, V. Quipuscoa Silvestre & N.W. Sawyer in Arnaldoa*, 5(1): 78 (1998)—Peru.

IONE (Orchidac.).
arunachalense *A. Nageswara Rao in Rheedea*, 7(1): 48 (1997)—Assam (Arunachal Pradesh).
pallida *L.V. Averyanov in Bot. Zhurn.*, 82(3): 143 (1997)—Vietnam.

IONOPSIS (Orchidac.).
minutiflora *(Dodson & N.H. Williams) F. Pupulin in Harvard Pap. Bot.*, 3(2): 227 (1998): Konantzia minutiflora.
papillosa *F. Pupulin in Harvard Pap. Bot.*, 3(2): 227 (1998)—Ecuador.
utricularioides (Sw.) Lindl.
f. virginalis *(L.C. Menezes) E.A. Christenson in Lindleyana*, 11(1): 17 (1996): I. utricularioides var. virginalis.

IPHIGENIA (Colchicac.).
socotrana *M. Thulin in Nordic J. Bot.*, 15(4): 403 (1995 publ. 1996)—Socotra.

IPHIONOPSIS (Compos.).
oblanceolata *N. Kilian in Willdenowia*, 29(1–2): 119 (1999)—Somalia.

IPOMOEA (Convolvulac.).
ser. Batatas *(Choisy) D.F. Austin in Taxon*, 45(1): 13 (1996): gen. Batatas.
sect. Tricolores *J.A. McDonald ex D.F. Austin in J. Torrey Bot. Soc.*, 124(2): 159 (1997).

IPOMOEA:-
eremnobrocha *D.F. Austin in J. Torrey Bot. Soc.*, 124(2): 145 (1997)—Panama.
involucrata P. Beauv.
f. bicolor *K.O. Ogunwenmo in Feddes Repert.*, 110(7–8): 510 (1999)—Nigeria.
katangensis *S. Lisowski & J. Wiland-Szymańska in Syst. Geogr. Pl.*, 69(1): 135 (1999)—Zaire.
mcphersonii *D.F. Austin in Taxon*, 45(1): 12 (1996), nom. nov.: I. praecox *McPherson & Meacham.*
pes-caprae (L.) R. Br.
var. perunkulamensis *P. Umamaheswari & P. Daniel in J. Econ. Taxon. Bot.*, 23(3): 691 (1999)—India (Tamil Nadu).
rzedowskii *E. Carranza Gonzalez, S. Zamudio Ruiz & G. Murguia Sanchez in Acta Bot. Mex.*, 45: 32 (1998)—Mexico Northeast (Hidalgo, Guanajuato, Querétaro).
sofomarensis *Sebsebe Demissew in Kew Bull.*, 53(2): 487 (1998); *see also Kew Bull.*, 54(2): 424 (1999)—Ethiopia.
ugborea *K.O. Ogunwenmo in Feddes Repert.*, 109(5–6): 365 (1998)—Nigeria.

IPOMOPSIS (Polemoniac.).
congesta
var. nevadensis *(Tidestr.) A. Tiehm in Mem. New York Bot. Gard.*, 77: 91 (1996): Gilia nevadensis.
monticola *J.M. Porter & L.A. Johnson in Aliso*, 19(1): 77 (2000)—Mexico Northwest (Sinaloa).

IPSEA (Orchidac.).
thailandica *G. Seidenfaden, Contrib. Orchid Fl. Thailand XIII:* 14 (1997)—Thailand.

IRESINE (Amaranthac.).
diffusa Willd.
f. herbstii *(Hook.) T.M. Pedersen in Bonplandia*, 10(1–4): 104 (2000): Iresine herbstii.

IRIARTEA (Palmae).
nobilis *(H. Wendl. ex Seem.) N.E. Brown, Suppl. Johnson's Gard. Dict.*: 946 (1882): Deckeria nobilis.

IRIDODICTYUM (Iridac.).
kopetdaghense *D. Kurbanov in Bot. Zhurn.*, 83(6): 110 (1998)—Turkmenistan.

IRIS (Iridac.).
bicapitata *M. Colasante in Flora Medit.*, 6: 214 (1996)—Italy.
cengialti *Ambrosi ex A. Kerner in Oesterr. Bot. Z.*, 21(9): 227 (1871)—Italy.
daliensis *X.D. Dong & Y.T. Zhao in Acta Phytotax. Sin.*, 35(1): 81 (1997)—China South-central (Yunnan).
decora Wall.
var. leucantha *X.D. Dong & Y.T. Zhao in Bull. Bot. Res. North-East. Forest. Univ.*, 18(2): 150 (1998), holotype not indicated—China South-central (Yunnan).
ensata Thunb.
f. alba *Y.N. Lee, Fl. Korea:* 1163 (1996), without type—Locality not stated.
odaesanensis Y.N. Lee
f. albiflora *Y.N. Lee, Fl. Korea:* 1163 (1996), without type—Locality not stated.
polychroa *M. Gandoger, Dec. Pl. Nov.*, 1: 14 (1875)—France.
pseudopumila Tineo
subsp. gozoensis *N. Service in New Plantsman*, 6(3): 186 (1999)—Sicilia (Malta).
relicta *M. Colasante in Flora Medit.*, 6: 214 (1996), nom. nov.: I. germanica var. suaveolens *N. Terracc.*
rossii
var. latifolia *J.K. Sim & Y.S. Kim in Korean J. Pl. Taxon.*, 22(1): 1 (1992)—South Korea.
sanguinea Hornem.
f. albiflora *Y.N. Lee, Fl. Korea:* 1163 (1996), without type—Locality not stated.
× setosothungbergii *H. Koidz. ex T. Shimizu, Fl. Nagano Pref.:* 1510 (1997)—Japan. (I. sanguinea × setosa).
uniflora Pall. ex Link
f. caricina *(Kitag.) P.Y. Fu & Y.A. Chen in Fl. Plant. Herb. Chinae Bor.-Or.*, 12: 213 (1998): I. uniflora var. caricina.

IRLBACHIA (Gentianac.).
sect. Elegans *B. Maguire in Mem. New York Bot. Gard.*, 32: 368 (1981).
sect. Tatei *B. Maguire in Mem. New York Bot. Gard.*, 32: 377 (1981).
cardonae *(Gleason) B. Maguire in Mem. New York Bot. Gard.*, 32: 381 (1981): Chelonanthus cardonae.
paruana *B. Maguire in Mem. New York Bot. Gard.*, 32: 380 (1981)—Venezuela.
phelpsiana *B. Maguire in Mem. New York Bot. Gard.*, 32: 371 (1981)—Venezuela.
plantaginifolia *B. Maguire in Mem. New York Bot. Gard.*, 32: 372 (1981)—Venezuela.
pumila *(Benth.) B. Maguire in Mem. New York Bot. Gard.*, 32: 372 (1981): Lisianthus pumilus.
ramosissima *(Benth.) B. Maguire in Mem. New York Bot. Gard.*, 32: 370 (1981): Lisianthus ramosissimus.
tatei *(Gleason) B. Maguire in Mem. New York Bot. Gard.*, 32: 377 (1981): Calolisianthus tatei.

ISACHNE (Gramin.).
jayachandranii *R. Gopalan & V. Chandrasekaran in Kew Bull.*, 55(4): 1005 (2000)—India (Kerala).

ISATIS (Crucif.).
cappadocica Desv.
subsp. gudrunensis *(Boiss.) Ş. Yıldırımlı in Turkish J. Bot.*, 12(3): 352 (1988): I. latisiliqua var. gudrunensis.
ermenekensis *Ş. Yıldırımlı in Turkish J. Bot.*, 12(3): 381 (1988)—Turkey.
glauca Aucher ex Boiss.
subsp. galatica *Ş. Yıldırımlı in Turkish J. Bot.*, 12(3): 379 (1988)—Turkey.
subsp. sivasica *(P.H. Davis) Ş. Yıldırımlı in Turkish J. Bot.*, 12(3): 379 (1988): I. sivasica.
pavlii *A.P. Khokhrjakov in Byull. Glavn. Bot. Sada (Moscow)*, 172: 23 (1995)—Transcaucasus (Adzhariya).
tinctoria L.
subsp. parchalensis *A.P. Khokhrjakov in Byull. Glavn. Bot. Sada (Moscow)*, 175: 49 (1997)—Turkey.

ISCHAEMUM (Gramin.).
kumarakodiense *N. Ravi, N. Mohanan & M.S. Kiran Raj in Rheedea*, 8(2): 149 (1998), as 'kumarakodiensis'—India (Kerala).
pappinisseriense *N. Ravi, N. Mohanan & R. Rajesh in Rheedea*, 8(2): 155 (1998), as 'pappinisseriensis'—India (Kerala).
pushpangadanii *N. Ravi, N. Mohanan & M.S. Kiranraj in Rheedea*, 10(1): 49 (2000)—India (Kerala).
quilonense *N. Ravi & T. Shaju in Rheedea*, 8(2): 152 (1998), as 'quilonensis'—India (Kerala).

ISCHNOCARPUS (Crucif.).
exilis *P.B. Heenan in New Zealand J. Bot.*, 37(1): 46 (1999)—New Zealand South I.

ISCHNOLEPIS (Asclepiadac.).
graminifolia *(Costantin & Gallaud) J. Klackenberg in Candollea*, 54(2): 332 (1999): Pentopetia graminifolia.

ISCHNOSIPHON (Marantac.).
morlae *H. Eggers in Bot. Centralbl.*, 53(10): 307 (1893), in Supp. 1 as 'mariae' in error—Ecuador.†

ISIDODENDRON *J.L. Fernández-Alonso, J.A. Pérez-Zabala & A. Idarraga-Piedrahita in Rev. Acad. Colomb. Cienc. Exact. Fis. Nat.*, 24(92): 348 (2000). *TRIGONIACEAE*
tripterocarpum *J.L. Fernández-Alonso, J.A. Pérez-Zabala & A. Idarraga-Piedrahita in Rev. Acad. Colomb. Cienc. Exact. Fis. Nat.*, 24(92): 348 (2000)—Colombia.

ISOCARPHA (Compos.).
oppositifolia *(L.) Cass. in Dict. Sci. Nat.*, 24: 19 (1822): Calea oppositifolia†

ISOCHILUS (Orchidac.).
bracteatus *(Lex.) Salazar & Soto Arenas ex A. Espejo Serna & A.R. López-Ferrari in Acta Bot. Mex.*, 51: 65 (2000): Dendrobium bracteatum.

ISODON *(Schrad. ex Benth.) Spach, Hist. Nat. Vég.*, 9: 162 (1840): Plectranthus sect. Isodon. (Labiat.).
dakglayensis *(Phuong) A.L. Budantzev in Komarovia*, 1: 25 (1999): Rabdosia dakglayensis.
lophanthoides (Buch.-Ham. ex D. Don) H. Hara
var. kerrii *(Doan) A.L. Budantzev in Komarovia*, 1: 26 (1999): Plectranthus striatus var. kerrii.
oreophilus *(Diels) A.J. Paton & O. Ryding in Kew Bull.*, 53(3): 730 (1998): Plectranthus oreophilus.
var. elongatus *(Hand.-Mazz.) A.J. Paton & O. Ryding in Kew Bull.*, 53(3): 730 (1998): Plectranthus oreophilus var. elongatus.
schimperi *(Vatke) J.K. Morton in Novon*, 8(3): 265 (1998): Plectranthus schimperi.
taiwanensis *(Masam.) S.S. Ying in Mem. Coll. Agric. Nation. Taiwan Univ.*, 31(1): 25 (1991): Amethystanthus taiwanensis.

ISOLEPIS (Cyperac.).
cernua (Vahl) Roem. & Schult.
f. aphylla *(Boeck.) A. Vegetti in Hickenia*, 2(31): 138 (1995): Scirpus aphyllus.
leucoloma *(Nees) C. Archer in Bothalia*, 28(1): 42 (1998): Cyperus leucoloma.
ranko *(Steud.) A. Vegetti in Hickenia*, 2(31): 138 (1995): Cyperus ranko.

ISOLOMA (Gesneriac.).
ceciliae *(André) N.E. Brown, Suppl. Johnson's Gard. Dict.*: 947 (1882): Tydaea ceciliae.
concinnum *(Hook. f.) N.E. Brown, Suppl. Johnson's Gard. Dict.*: 947 (1882): Stenogastra concinna.
digitaliflorum *(Linden & André) N.E. Brown, Suppl. Johnson's Gard. Dict.*: 947 (1882): Sciadocalyx digitaliflora.
hypocyrtiflorum *(Hook. f.) N.E. Brown, Suppl. Johnson's Gard. Dict.*: 947 (1882): Gloxinia hypocyrtiflora.
lindenianum *(Regel) N.E. Brown, Suppl. Johnson's Gard. Dict.*: 947 (1882): Tydaea lindeniana.

ISOPOGON (Proteac.).
formosus R. Br.
subsp. dasylepis *(Meisn.) D.B. Foreman in Fl. Australia*, 16: 479 (1995): I. formosus var. dasylepis.
gardneri *D.B. Foreman in Fl. Australia*, 16: 479 (1995)—Western Australia.
inconspicuus *(Meisn.) D.B. Foreman in Fl. Australia*, 16: 479 (1995): Petrophile inconspicua.
scabriusculus Meisn.
subsp. pubifloris *D.B. Foreman in Fl. Australia*, 16: 480 (1995)—Western Australia.
subsp. stenophyllus *D.B. Foreman in Fl. Australia*, 16: 480 (1995)—Western Australia.
teretifolius R. Br.
subsp. petrophiloides *(R. Br.) D.B. Foreman in Fl. Australia*, 16: 481 (1995): I. petrophiloides.

ISOTOMA (Campanulac.).
rivalis *(E. Wimm.) T.G. Lammers in Novon*, 9(1): 75 (1999): Hypsela rivalis.
sessiliflora *(E. Wimm.) T.G. Lammers in Novon*, 9(1): 75 (1999): Hypsela sessiliflora.
tridens *(E. Wimm.) T.G. Lammers in Novon*, 9(1): 75 (1999): Hypsela tridens.

ITEA (Escalloniac.).
chingiana *S.Y. Jin in Acta Phytotax. Sin.*, 33(5): 497 (1995)—China Southeast (Guangxi).
yangchunensis *S.Y. Jin in Acta Phytotax. Sin.*, 33(5): 497 (1995)—China Southeast (Guangdong).

IXERIDIUM (Compos.).
beauverdianum *(H. Lév.) L.S. Springate in Edinburgh J. Bot.*, 57(3): 402 (2000): Lactuca beauverdiana.
oldhamii *(Maxim.) A.N. Sennikov in Bot. Zhurn.*, 82(5): 116 (1997): Lactuca oldhamii.

IXIA (Iridac.).
sect. Dichone *(Salisb. ex Baker) P. Goldblatt & J.C. Manning in Bothalia*, 29(1): 63 (1999): Tritonia subgen. Dichone.
subgen. Morphixia *(Ker Gawl.) P. Goldblatt & J.C. Manning in Bothalia*, 29(1): 63 (1999): gen. Morphixia.
aurea *J.C. Manning & P. Goldblatt in Bothalia*, 29(1): 61 (1999)—Cape Province.

IXIA:-
longituba N.E. Br.
var. bellendenii *(R.C. Foster) M.P. de Vos in Fl. S. Afr.*, 7(2:1): 63 (1999): I. bellendenii.
polystachya L.
var. longistylis *M.P. de Vos in Fl. S. Afr.*, 7(2:1): 39 (1999)—Cape Province.
viridiflora Lam.
var. minor *M.P. de Vos in Fl. S. Afr.*, 7(2:1): 45 (1999)—Cape Province.

IXIOLIRION (Ixioliriac.).
songaricum *P. Yan in Fl. Xinjiangensis*, 6: 605 (1996)—Xinjiang.
tataricum (Pall.) Herb.
var. album *A. Ghahreman, Fl. Iran*, 16: No. 1941 (1997), without type—Iran.

IXORA (Rubiac.).
amplexicaulis *C.Y. Wu & W.C. Ko in Guihaia*, 19(2): 98 (1999), homonym; *see Gillespie in Bernice P. Bishop Mus. Bull.*, 74: 30 (1930)—China South-central (Yunnan).
aneimenodesma K. Schum.
subsp. kizuensis *P. De Block, Afr. Sp. Ixora (Rubiac.-Pavett.) (Op. Bot. Belg., 9)*: 82 (1998)—Zaire, Congo, Cabinda, Gabon.
auricularis *Chun & How ex W.C. Ko in Guihaia*, 19(2): 99, 100 (1999)—China South-central (Yunnan).
foetida *(L.f.) F.R. Fosberg in Allertonia*, 7(2): 78 (1993): Daphne foetida.
foonchewii *W.C. Ko in Guihaia*, 19(2) 102 (1999)—China South-central (Yunnan).
gracilis *W.C. Ko in Guihaia*, 19(2) 103 (1999), homonym; *see Fawc. in H.O. Forbes, Naturalist's Wanderings E. Archipelago*: 508 (1885)—China South-central (Yunnan).
hartiana *P. De Block, Afr. Sp. Ixora (Rubiac.-Pavett.) (Op. Bot. Belg., 9)*: 106 (1998)—Zaire, Tanzania, Angola.
insignis *Chun & How ex W.C. Ko in Guihaia*, 19(2): 97 (1999)—China South-central (Yunnan).
laxiflora Sm.
var. linderi *(Hutch. & Dalziel) P. De Block, Afr. Sp. Ixora (Rubiac.-Pavett.) (Op. Bot. Belg., 9)*: 118 (1998): I. linderi.
liberiensis *P. De Block, Afr. Sp. Ixora (Rubiac.-Pavett.) (Op. Bot. Belg., 9)*: 122 (1998)—Liberia.
macilenta *P. De Block, Afr. Sp. Ixora (Rubiac.-Pavett.) (Op. Bot. Belg., 9)*: 126 (1998)—Cameroon, Congo.
minutiflora Hiern
subsp. chasalliensis *P. De Block, Afr. Sp. Ixora (Rubiac.-Pavett.) (Op. Bot. Belg., 9)*: 132 (1998)—Cameroon, Congo, Gabon, Nigeria.
nigerica Keay
subsp. occidentalis *P. De Block, Afr. Sp. Ixora (Rubiac.-Pavett.) (Op. Bot. Belg., 9)*: 142 (1998)—Ivory Coast, Ghana, Sierra Leone.
paraopara *W.C. Ko in Guihaia*, 19(2): 101 (1999)—China South-central (Yunnan).
praetermissa *P. De Block, Afr. Sp. Ixora (Rubiac.-Pavett.) (Op. Bot. Belg., 9)*: 148 (1998)—Gabon, Cameroon, Congo, Zaire.
sivarajiana *A.K. Pradeep in Nordic J. Bot.*, 17(3): 315 (1997)—India (Kerala).
synactica *P. De Block, Afr. Sp. Ixora (Rubiac.-Pavett.) (Op. Bot. Belg., 9)*: 156 (1998)—Cameroon.
tenuis *P. De Block, Afr. Sp. Ixora (Rubiac.-Pavett.) (Op. Bot. Belg., 9)*: 160 (1998)—Ghana.

IZOZOGIA *G. Navarro in Novon*, 7(1): 1 (1997). *ZYGOPHYLLACEAE*
nellii *G. Navarro in Novon*, 7(1): 1 (1997)—Bolivia.

J

JACARANDA (Bignoniac.).
caucana Pittier
subsp. glabrata *A.H. Gentry in Fl. Venezuela*, 8(4): 201 (1982)—Venezuela.

JACOBSENIA (Aizoac.).
vaginata *(L. Bolus) H.-D. Ihlenfeldt in Mitt. Inst. Allg. Bot. Hamburg*, 27: 124 (1997): Drosanthemum vaginatum.

× **JACONECIO** *P.V. Heath in Calyx*, 6(2): 55 (1999). *COMPOSITAE* (Jacobaea × Senecio).
cambrensis *(Rosser) P.V. Heath in Calyx*, 6(2): 55 (1999): Senecio cambrensis.

× **JACORARIA** *P.V. Heath in Calyx*, 6(2): 55 (1999). *COMPOSITAE* (Cineraria × Jacobaea).
albescens *(Burb. & Colgan) P.V. Heath in Calyx*, 6(2): 55 (1999): Senecio × albescens.

JACQUEMONTIA (Convolvulac.).
revoluta *R. Simão-Bianchini in Novon*, 9(1): 104 (1999)—Brazil Southeast (Minas Gerais).

JACQUESHUBERIA (Legumin.).
pustulata *B. Stergios & P.E. Berry in Novon*, 6(4): 429 (1996)—Venezuela.
splendens *B. Stergios & P.E. Berry in Novon*, 6(4): 431 (1996)—Venezuela.

JACQUINIA (Theophrastac.).
frutescens *(Mill.) B. Ståhl in Nordic J. Bot.*, 15(5): 507 (1995 publ. 1996): Ruscus frutescens.
morenoana *G. Castillo-Campos & M.E. Medina Abreo in Novon*, 8(2): 129 (1998)—Mexico Gulf (Veracruz).
pauciflora *B. Ståhl & F.S. Axelrod in Brittonia*, 50(1): 1 (1998)—Puerto Rico.

JALTOMATA (Solanac.).
auriculata *(Miers) T. Mione in Brittonia*, 51(1): 31 (1999): Saracha auriculata.
aypatensis *S. Leiva González, T. Mione & V. Quipuscoa Silvestre in Arnaldoa*, 5(2): 180 (1998)—Peru.
bernardelloana *S. Leiva González & T. Mione in Arnaldoa*, 5(2): 183 (1998)—Peru.
cajacayensis *S. Leiva G. & T. Mione in Novon*, 10(1): 53 (2000)—Peru.
cajamarca *T. Mione in Novon*, 6(3): 281 (1996)—Peru.
chihuahuensis *(Bitter) T. Mione & R. Bye in Novon*, 6(1): 79 (1996): Saracha chihuahuensis.
contorta *(Ruiz & Pav.) T. Mione in Brittonia*, 51(1): 32 (1999): Saracha contorta.
cuyasensis *S. Leiva González, V. Quipuscoa Silvestre & Sawyer in Arnaldoa*, 5(2): 186 (1998)—Peru.
diversa *(J.F. Macbr.) T. Mione in Brittonia*, 51(1): 32 (1999): Saracha dentata var. diversa.
herrerae *(C.V. Morton) T. Mione in Brittonia*, 51(1): 32 (1999): Saracha herrerae.
lezamae *S. Leiva González & T. Mione in Arnaldoa*, 6(1): 66 (1999 publ. 2000)—Peru.
lojae *T. Mione in Rhodora*, 101(906): 137 (1999)—Ecuador, Peru.
lomana *T. Mione & S. Leiva G.in Novon*, 10(1): 54 (2000)—Peru.
mionei *S. Leiva González & V. Quipuscoa Silvestre in Arnaldoa*, 5(2): 188 (1998)—Peru.
nitida *(Bitter) T. Mione in Brittonia*, 51(1): 33 (1999): Saracha nitida.
paneroi *T. Mione & S. Leiva G. in Rhodora*, 99(900): 283 (1997 publ. 1998)—Peru.
sagastegui *T. Mione in Novon*, 6(3): 281 (1996)—Peru.
salpoensis *S. Leiva González & T. Mione in Arnaldoa*, 6(1): 69 (1999 publ. 2000)—Peru.
yungayensis *T. Mione & S. Leiva G. in Novon*, 10(1): 58 (2000)—Peru.

JAMBOSA (Myrtac.).
polypetala *N. Wallich, Report on Calcutta Bot. Gard. to G.A. Bushby*: 27 (1840), nom. nov.: Eugenia angustifolia *Roxb.*

JARAVA (Gramin.).
plumosa *(Spreng.) S.W.L. Jacobs & J. Everett in Telopea*, 7(3): 301 (1997): Calamagrostis plumosa.

JASIONE (Campanulac.).
crispa (Pourr.) Samp.
subsp. cedretorum *(Pau & Font Quer) A. Dobignard in Candollea*, 52(1): 139 (1997): J. perennis var. cedretorum.
subsp. mesatlantica *(Emb. & Maire) A. Dobignard in Candollea*, 52(1): 140 (1997): J. humilis subvar. mesatlantica.

JASIONE:-
crispa
var. tazzekana *(Emb. & Maire) A. Dobignard in Candollea*, 52(1): 139 (1997): J. humilis subvar. tazzekana.
subsp. tristis *(Bory) G. López González in An. Jard. Bot. Madrid*, 56(1): 166 (1998): J. tristis.
foliosa Cav.
subsp. xauenensis *A. Dobignard in Candollea*, 52(1): 141 (1997)—Morocco.
mansanetiana *R. Roselló & J.B. Peris in Bol. Soc. Castellonense Cult.*, 68(1–2): 210 (1992)—Spain.

JASMINUM (Oleac.).
dispermum Wall.
subsp. forrestianum *(Kobuski) P.S. Green in Kew Bull.*, 54(2): 936 (1997): J. forrestianum.
domatiigerum Lingelsh.
subsp. australis *W.K. Harris & W.J. McDonald in Austrobaileya*, 5(4): 699 (2000)—Queensland.
jenniae *W.K. Harris & G. Holmes in Austrobaileya*, 5(2): 341 (1999)—Queensland.
leptophyllum *R.A. Rafiq in Novon*, 6(3): 295 (1996)—Pakistan.
pipolyi *W. Takeuchi in Sida*, 18(4): 948 (1999)—New Guinea (Papua New Guinea).
simplicifolium G. Forst.
subsp. leratii *(Schltr.) P.S. Green in Fl. N. Caléd.*, 22: 54 (1998): J. leratii.
vidalii *P.S. Green in Kew Bull.*, 54(2): 394 (1999)—Laos.

JASONIA (Compos.).
antiatlantica *(Emb. & Maire) F. Gómiz in An. Jard. Bot. Madrid*, 57(1): 191 (1999): J. glutinosa var. antiatlantica.

JATROPHA (Euphorbiac.).
subgen. Manihotoides *A. Radcliffe-Smith in Kew Bull.*, 52(1): 178 (1997).
afrotuberosa *A. Radcliffe-Smith & R. Govaerts in Kew Bull.*, 52(1): 188 (1997), nom. nov.: J. tuberosa *Pax.*
krusei *J. Jiménez Ramírez & M. Martínez Gordillo in An. Inst. Biol. Univ. Nac. Autón. Méx., Bot.*, 71(1): 35 (2000)—Mexico Southwest (Guerrero).
paraguayensis *A. Radcliffe-Smith & R. Govaerts in Kew Bull.*, 52(1): 189 (1997), nom. nov.: J. ricinifolia *Pax.*

JAVORKAEA (Rubiac.).
acuminata *(Oerst. ex Standl.) A. Borhidi in Acta Bot. Hung.*, 40(1–4): 18 (1996–97 publ. 1997): Sommera acuminata.
megalantha *(Lorence) A. Borhidi in Acta Bot. Hung.*, 40(1–4): 16 (1996–97 publ. 1997): Rondeletia megalantha.
uxpanapensis *(Lorence & Cast.-Campos) A. Borhidi in Acta Bot. Hung.*, 40(1–4): 19 (1996–97 publ. 1997): Rondeletia uxpanapensis.

JEJEWOODIA *D.L. Szlachetko in Fragm. Flor. Geobot.*, Supp. 3: 135 (1995). *ORCHIDACEAE*
jiewhoei *(J.J. Wood & Shim) D.L. Szlachetko in Fragm. Flor. Geobot.*, Supp. 3: 135 (1995): Ceratochilus jiewhoei.

JENNYELLA *E. Lückel & H. Fessel in Caesiana*, 13: 3 (1999). *ORCHIDACEAE*
clarae *(Schltr.) E. Lückel & H. Fessel in Caesiana*, 13: 6 (1999): Houlletia clarae.
kalbreyeriana *(Kraenzl.) E. Lückel & H. Fessel in Caesiana*, 13: 5 (1999): Houlletia kalbreyeriana.
lowiana *(Rchb. f.) E. Lückel & H. Fessel in Caesiana*, 13: 6 (1999): Houlletia lowiana.
sanderi *(Rolfe) E. Lückel & H. Fessel in Caesiana*, 13: 4 (1999): Houlletia sanderi.

JENSIA *B.G. Baldwin in Novon*, 9(4): 464 (1999). *COMPOSITAE*
rammii *(Greene) B.G. Baldwin in Novón*, 9(4): 465 (1999): Madia rammii.
yosemitana *(Parry ex A. Gray) B.G. Baldwin in Novon*, 9(4): 465 (1999): Madia yosemitana.

JESSEA (Compos.).
gunillae *B. Nordenstam in Bot. Jahrb.*, 118(2): 148 (1996)—Costa Rica.

JOBINIA (Asclepiadac.).
balslevii *(Morillo) W.D. Stevens in Novon,* 10(3): 244 (2000): Cynanchum balslevii.
campii *(Morillo) W.D. Stevens in Novon,* 10(3): 244 (2000): Cynanchum campii.
eulaxiflora *(Lundell) W.D. Stevens in Novon,* 10(3): 244 (2000): Cynanchum eulaxiflorum.

JOOSIA (Rubiac.).
longisepala *L. Andersson in Brittonia,* 49(1): 37 (1997)—Ecuador.
multiflora *L. Andersson in Brittonia,* 49(1): 31 (1997)—Peru.
obtusa *L. Andersson in Brittonia,* 49(1): 37 (1997)—Ecuador.

JOSTIA *C.A. Luer, Ic. Pleurothallid. XX (Monogr. Syst. Bot. Missouri Bot. Gard., 79):* 2 (2000). *ORCHIDACEAE*
teaguei *(Luer) C.A. Luer, Ic. Pleurothallid. XX (Monogr. Syst. Bot. Missouri Bot. Gard., 79):* 2 (2000): Masdevallia teaguei.

JOVIBARBA (Crassulac.).
globifera (L.) J. Parn.
subsp. glabrescens *(Sabr.) J. Holub in Preslia,* 70(2): 106 (1998): Sempervivum hirtum f. glabrescens.
subsp. preissiana *(Domin) J. Holub in Preslia,* 70(2): 106 (1998): Sempervivum preissianum.
subsp. pseudohirta *(Leute) R. Letz in Thaiszia,* 8(1): 15 (1998): Dipogon arenarius subsp. pseudohirtus.

JOYCEA *H.P. Linder in Telopea,* 6(4): 611 (1996). *GRAMINEAE*
clelandii *(Vickery) H.P. Linder in Telopea,* 6(4): 612 (1996): Danthonia clelandii.
lepidopoda *(N.G. Walsh) H.P. Linder in Telopea,* 6(4): 612 (1996): Danthonia lepidopoda.
pallida *(R. Br.) H.P. Linder in Telopea,* 6(4): 611 (1996): Danthonia pallida.

JUGLANS (Juglandac.).
× intermedia *Jacques in Ann. Fl. Pomone,* 1833–34: 214 (1834)—Cultivation.(J. nigra × regia).
petiolata *P. Custis in Philad. Med. Phys. J.,* 2(2): 45 (1806)—Louisiana.

JUMELLEA (Orchidac.).
walleri *(Rolfe) I. la Croix in Curtis's Bot. Mag.,* 17(4): 214 (2000): Mystacidium walleri.

JUNCELLUS (Cyperac.).
serotinus (Rottb.) C.B. Clarke
var. capitatus *D.Z. Ma in Fl. Ningxiaensis,* 2: 522 (1988 publ. 1990)—Inner Mongolia (Ningxia).

JUNCUS (Juncac.).
sect. Iridifolii *S. Snogerup & J. Kirschner in Folia Geobot.,* 34(3): 382 (1999).
alpinoarticulatus Chaix
var. marshallii *(Pugsley) P.D. Sell in P.D. Sell & G. Murrell, Fl. Great Britain & Ireland,* 5: 363 (1996): J. marshallii.
alpinus Vill.
subsp. turczaninovii *(Buchenau) V.N. Voroschilov in A.K. Skvortsov (ed.),. issl. v razn. raĭonakh SSSR:* 157 (1985): J. lamprocarpus var. turczaninovii.
amuricus (Maxim.) V.I. Krecz. & Gontsch.
var. wui *V.S. Novikov in Byull. Mosk. Obshch. Ispўt. Prir., Biol.,* 103(4): 66 (1998), nom. nov.: J. subglobosus *K.F. Wu.*
anthelatus *(Wiegand) R.E. Brooks in Novon,* 9(1): 11 (1999): J. tenuis var. anthelatus.
articulatus L.
subsp. limosus *(Vorosch.) V.N. Voroschilov in A.K. Skvortsov (ed.), Florist. issl. v razn. raĭonakh SSSR:* 157 (1985): J. limosus.
subsp. tatewakii *(Satake) V.N. Voroschilov in A.K. Skvortsov (ed.), Florist. issl. v razn. raĭonakh SSSR:* 157 (1985): J. tatewakii.
bufonius L.
var. juzepczukii *(V.I. Krecz. & Gontsch.) V.N. Voroschilov in A.K. Skvortsov (ed.), Florist. issl. v razn. raĭonakh SSSR:* 157 (1985): J. juzepczukii.

JUNCUS:-
bufonius
subsp. rhiphaenus *(Pau & Font Quer) M. Ibn Tattou in Bocconea,* 8: 206 (1998), with incorrect basionym ref.: Basionym not stated.
concinnus D. Don
var. gracilicaulis *(A. Camus) R.C. Srivastava in Novon,* 8(2): 203 (1998): J. gracilicaulis.
deosaicus *H.J. Noltie in Edinburgh J. Bot.,* 55(1): 41 (1998)—Pakistan.
dulongjiongensis *V.S. Novikov in Byull. Mosk. Obshch. Ispўt. Prir., Biol.,* 103(4): 65 (1998)—China South-central (Yunnan).
fimbristyloides *H.J. Noltie in Edinburgh J. Bot.,* 55(1): 39 (1998)—China South-central (Yunnan).
fuscocapitatus *H. Balslev in Fl. Neotrop.,* 68: 111 (1996)—Bolivia, Peru.
gonggae *F. Miyamoto & H. Ohba in J. Jap. Bot.,* 72(3): 162 (1997)—China South-central (Sichuan).
inflexus L.
subsp. austro-occidentalis *K.F. Wu in Fl. Reipubl. Popul. Sin.,* 13(3): 261 (1997)—China South-central (Sichuan).
lanpinguensis *V.S. Novikov in Byull. Mosk. Obshch. Ispўt. Prir., Biol.,* 103(4): 67 (1998), nom. nov.: J. glomeratus *K.F. Wu.*
novikovii *A.P. Khokhrjakov in Byull. Glavn. Bot. Sada (Moscow),* 172: 25 (1995)—Transcaucasus (Adzhariya).
pallescens Lam.
var. achalensis *(Barros) L. Novara in Zuloaga & Morrone (eds.), Cat. Pl. Vasc. Rep. Arg. 1 (Monogr. Syst. Bot.* Missouri Bot. Gard. 60): 221 (1996): J. achalensis.
papillosus Franch. & Sav.
var. virens *(Buchenau) V.N. Voroschilov in A.K. Skvortsov (ed.), Florist. issl. v razn. raĭonakh SSSR:* 157 (1985): J. virens.
sherei *F. Miyamoto & H. Ohba in J. Jap. Bot.,* 72(5): 293 (1997)—Nepal.
subulitepalus *H. Balslev in Fl. Neotrop.,* 68: 136 (1996)—Ecuador, Peru, Bolivia.
tobdeniorum *H.J. Noltie in Edinburgh J. Bot.,* 55(1): 42 (1998), as 'tobdenii'—Sikkim.
trachyphyllus *F. Miyamoto & H. Ohba in J. Jap. Bot.,* 72(3): 164 (1997)—China South-central (Sichuan).
valvatus Link
subsp. conimbrigensis *A. Fernandes & R.B. Fernandes, Flórula Vasc. Mata Bufarda:* 154 (1991)—Portugal.
yanshanuensis *V.S. Novikov in Byull. Mosk. Obshch. Ispўt. Prir., Biol.,* 103(4): 67 (1998), nom. nov.: J. sphaerocephalus *K.F. Wu.*

JUNELLIA (Verbenac.).
bisulcata (Hayek) Moldenke
var. campestris *(Griseb.) S.M. Botta in Hickenia,* 2(28): 127 (1995): Verbena juniperina var. campestris.
digitata (Phil.) Moldenke
var. integerrima *(Botta) S.M. Botta in Hickenia,* 2(28): 128 (1995): Verbena digitata var. integerrima.
longidentata Moldenke
var. glandulosa *S.M. Botta in Hickenia,* 2(28): 128 (1995)—Argentina Northwest (La Rioja).
scoparia (Gillies ex Hook.) Botta
var. puberula *(Tronc.) S.M. Botta in Hickenia,* 2(28): 128 (1995): Verbena scoparia var. puberula.
selaginoides (Kunth ex Walp.) Moldenke
var. illapelina *(Phil.) S.M. Botta in Hickenia,* 2(28): 128 (1995): Verbena illapelina.
spathulata (Gillies & Hook.) Moldenke
var. glauca *(Gillies & Hook.) S.M. Botta in Hickenia,* 2(28): 128 (1995): Verbena glauca.
var. grandiflora *(Schauer) S.M. Botta in Hickenia,* 2(28): 128 (1995): Verbena spathulata var. grandiflora.

JUNGIA (Compos.).
beckii *G. Harling in Novon,* 7(3): 246 (1997)—Bolivia.

JUNIPERUS (Cupressac.).
baimashanensis *Y.F. Yu & L.K. Fu in Novon,* 7(4): 443 (1997 publ. 1998)—China South-central (Yunnan).

JUNIPERUS:-
barbadensis L.
var. jamaicensis *J. Silba in J. Int. Conifer Preserv. Soc.,* 7(1): 24 (2000)—Jamaica.
var. lucayana *(Britton) J. Silba in J. Int. Conifer Preserv. Soc.,* 7(1): 25 (2000), isonym; *see R.P. Adams in Phytologia,* 78(2): 145 (1995): J. lucayana.
var. saxicola *(Britton & P. Wilson) J. Silba in J. Int. Conifer Preserv. Soc.,* 7(1): 25 (2000): J. saxicola.
californica Carrière
var. osteosperma *(Torr.) L.D. Benson in L.D. Benson & R.A. Darrow, Trees & Shrubs of Southwest. Deserts (ed. 3):* 377 (1981): J. tetragona var. osteosperma.
carinata *(Y.F. Yu & L.K. Fu) R.P. Adams in Biochem. Syst. Ecol.,* 28(6): 541 (2000): J. pingii var. carinata.
cedrus Webb & Berthel.
f. fastigiata *R. Vieira in Bol. Mus. Mun. Funchal,* 49: 143 (1997)—Madeira.
chengii *L.K. Fu & Y.F. Yu in Novon,* 7(4): 443 (1997 publ. 1998)—China South-central (Yunnan).
erectopatens *(Cheng & L.K. Fu) R.P. Adams in Biochem. Syst. Ecol.,* 27(7): 723 (1999): Sabina vulgaris var. erectopatens.
microsperma *(W.C. Cheng & L.K. Fu) R.P. Adams in Biochem. Syst. Ecol.,* 28(6): 540 (2000): Sabina convallium var. microsperma.
mucronata *R.P. Adams in Biochem. Syst. Ecol.,* 28(2): 158 (2000)—Mexico Northwest (Sonora).
oxycedrus L.
var. macrocarpa *(Sibth. & Sm.) J. Silba in J. Int. Conifer Preserv. Soc.,* 7(1): 25 (2000): J. macrocarpa.
phoenicea L.
f. prostrata *Z. Debreczy & I. Rácz in Stud. Bot. Hung.,* 29 : 87 (1998 publ. 1999)—Portugal.
pingii Cheng ex Ferré
var. carinata *Y.F. Yu & L.K. Fu in Novon,* 7(4): 443 (1997 publ. 1998)—China South-central (Sichuan, Yunnan).
sabina L.
var. erectopatens *(W.C. Cheng & L.K. Fu) Y.F. Yu & L.K. Fu in Novon,* 7(4): 444 (1997 publ. 1998): Sabina vulgaris var. erectopatens.
var. yulinensis *(T.C. Chang & C.G. Chen) Y.F. Yu & L.K. Fu in Novon,* 7(4): 444 (1997 publ. 1998): Sabina vulgaris var. yulinensis.
squamata Buch.-Ham. ex D. Don
var. hongxiensis *Y.F. Yu & L.K. Fu in Novon,* 7(4): 444 (1997 publ. 1998)—China South-central (Sichuan).
var. parvifolia *Y.F. Yu & L.K. Fu in Novon,* 7(4): 444 (1997 publ. 1998)—China South-central (Sichuan).
turcomanica B. Fedtsch.
var. seravschanica *(Kom.) R.P. Adams in Biochem. Syst. Ecol.,* 27(7): 721 (1999), without basionym ref.: J. seravschanica.

JURINEA (Compos.).
subsect. Auriculatae *O.V. Tscherneva in Bot. Zhurn.,* 83(9): 104 (1998).
subsect. Cataonicae *O.V. Tscherneva in Bot. Zhurn.,* 83(9): 105 (1998).
subsect. Dumulosae *O.V. Tscherneva in Bot. Zhurn.,* 83(9): 100 (1998).
subsect. Eriobases *O.V. Tscherneva in Bot. Zhurn.,* 83(9): 101 (1998).
subsect. Integrae *O.V. Tscherneva in Bot. Zhurn.,* 84(8): 115 (1999).
subsect. Macrocephalae *O.V. Tscherneva in Bot. Zhurn.,* 83(9): 103 (1998).
subsect. Multiflorae *O.V. Tscherneva in Novosti Sist. Vўssh. Rast.,* 31: 252 (1998).
subsect. Oblongifoliae *O.V. Tscherneva in Bot. Zhurn.,* 83(9): 102 (1998).
sect. Pendentiflorae *O.V. Tscherneva in Bot. Zhurn.,* 84(8): 115 (1999).
subsect. Pendulifoliae *O.V. Tscherneva in Bot. Zhurn.,* 83(9): 105 (1998).
subsect. Pinnatae *O.V. Tscherneva in Novosti Sist. Vўssh. Rast.,* 31: 254 (1998).
subsect. Pinnatae *O.V. Tscherneva in Bot. Zhurn.,* 84(8): 114 (1999), homonym; *see Tscherneva in Novosti Sist. Vўssh Rast.,* 31: 254 (1998).

JURINEA:-
subsect. Pulvinatae *O.V. Tscherneva in Bot. Zhurn.*, 83(9): 99 (1998).
subsect. Ramosissimae *O.V. Tscherneva in Bot. Zhurn.*, 83(9): 100 (1998).
cynaroides *(DC.) Mutel, Fl. Franç.*, 2: 184 (1835): Serratula cynaroides.
elbursensis *(Wagenitz) O.V. Tscherneva in Bot. Zhurn.*, 83(9): 103 (1998): J. macrocephala subsp. elbursensis.
kyzylkyrensis *R.V. Kamelin & O.V. Tscherneva in Bot. Zhurn.*, 85(1): 126 (2000)—Kirgizstan.
mobayenii *A. Ghahreman & S.M. Mirtadzadini in Iranian J. Bot.*, 8(2): 248 (2000)—Iran.
mollis (L.) Rchb.
f. erectobracteata *F. Conti in Willdenowia*, 28(1–2): 51 (1998)—Italy.
narynensis *R.V. Kamelin & O.V. Tscherneva in Bot. Zhurn.*, 85(1): 127 (2000)—Kirgizstan.

JUSTICIA (Acanthac.).
acutangula *H.S. Lo & D. Fang in Guihaia*, 17(1): 56 (1997)—China Southeast (Guangxi).
amblyosepala *D. Fang & H.S. Lo in Guihaia*, 17(1): 55 (1997)—China Southeast (Guangxi).
arborescens *L.H. Durkee & L.A. McDade in Novon*, 6(1): 13 (1996)—Costa Rica.
asclepiadea *(Nees) D.C. Wasshausen & C. Ezcurra in Candollea*, 52(1): 172 (1997): Simonisia asclepiadea.
austrocapensis *T.F. Daniel in Proc. Calif. Acad. Sci.*, 49(10): 353 (1997), nom. nov.: Dianthera incerta *Brandegee*.
austroguangxiensis *H.S. Lo & D. Fang in Guihaia*, 17(1): 54 (1997)—China Southeast (Guangxi).
f. albinervia *D. Fang & H.S. Lo in Guihaia*, 17(1): 55 (1997)—China Southeast (Guangxi).
austrosinensis *H.S. Lo & D. Fang in Guihaia*, 17(1): 52 (1997)—China Southeast (Guangdong, Guangxi, Jiangxi), China South-central (Guizhou).
aymardii *D.C. Wasshausen in Novon*, 10(3): 260 (2000)—Venezuela.
bizuneshiae *Ensermu Kelbessa in Kew Bull.*, 54(2): 445 (1999)—Ethiopia.
buxifolia *H.S. Lo & D. Fang in Guihaia*, 17(1): 51 (1997)—China Southeast (Guangxi).
candicans *(Nees) L.D. Benson in L.D. Benson & R.A. Darrow, Trees & Shrubs of Southwest. Deserts (ed. 3)*: 218 (1981): Adhatoda candicans.
var. subglabra *(S. Watson) L.D. Benson in L.D. Benson & R.A. Darrow, Trees & Shrubs of Southwest. Deserts (ed. 3)*: 218 (1981): Jacobinia ovata var. subglabra.
capitata *(T. Anderson ex Hook. f.) L.H. Cramer in M.D. Dassanayake & W.D. Clayton, eds., Revised Handb. Fl. Ceylon*, 12: 114 (1998): J. ceylanica var. capitata.
cardiophylla *D. Fang & H.S. Lo in Guihaia*, 17(1): 57 (1997)—China Southeast (Guangxi).
cauliflora *L.H. Durkee in Novon*, 9(4): 505 (1999)—Panama.
chacoensis *D.C. Wasshausen & C. Ezcurra in Candollea*, 52(1): 178 (1997), nom. nov.: Beloperone riparia *S. Moore*.
chapana *D.C. Wasshausen in Novon*, 9(3): 450 (1999)—Venezuela.
chiriquiensis *L.H. Durkee in Novon*, 9(4): 507 (1999)—Panama.
circulibracteata *L.H. Durkee & L.A McDade in Novon*, 6(1): 16 (1996)—Costa Rica.
clausseniana *(Nees) S.R. Profice in Reserva Ecol. Macaé de Cima: Aspectos Flor. Espéc. Vasc.*, 2: 31 (1996): Sericographis clausseniana.
consanguinea *(Lindau) D.C. Wasshausen & C. Ezcurra in Candollea*, 52(1): 174 (1997): Beloperone consanguinea.
var. pubescens *(Lindau) D.C. Wasshausen & C. Ezcurra in Candollea*, 52(1): 174 (1997): Beloperone consanguinea var. pubescens.
coppenamensis *D.C. Wasshausen in Brittonia*, 47(4): 424 (1995), nom. nov.: Beloperone modesta *Bremek*.
corumbensis *(Lindau) D.C. Wasshausen & C. Ezcurra in Candollea*, 52(1): 178 (1997): Beloperone corumbensis.
cuicatlana *T.F. Daniel in Bol. Inst. Bot. (Guadalajara)*, 5(1–3): 120 (1998)—Mexico Southwest (Oaxaca).

JUSTICIA:-
cyrtantheriformis *(Rizzini) S.R. Profice in M. do Carmo Mendes Marques et al. (eds.), Flórula Apa Cairuçu, Parati*, RJ: Esp. Vasc.: 20 (1997): Sericographis cyrtantheriformis.
damingensis *(H.S. Lo) H.S. Lo in Guihaia*, 17(1): 58 (1997): Mananthes damingensis.
danielii *L.H. Durkee in Novon*, 9(4): 508 (1999), nom. nov.: Chaetothylax cuspidatus *D.N. Gibson*.
densibracteata *L.H. Durkee & L.A. McDade in Novon*, 6(1): 16 (1996)—Costa Rica.
ferruginea *H.S. Lo & D. Fang in Guihaia*, 17(1): 58 (1997)—China Southeast (Guangxi).
flaviflora *(Turrill) D.C. Wasshausen in Brittonia*, 47(4): 424 (1995): Beloperone flaviflora.
floribunda *(K. Koch) D.C. Wasshausen in Darwiniana*, 35(1–4): 151 (1998): Libonia floribunda.
galeata *M. Hedrén in Nordic J. Bot.*, 13(6): 647 (1993)—Kenya.
insolita Brandegee
subsp. tastensis *(Brandegee) T.F. Daniel in Proc. Calif. Acad. Sci.*, 49(10): 369 (1997): J. insolita var. tastensis.
kucharii *M. Hedrén in Nordic J. Bot.*, 13(6): 649 (1993), as 'kuchari'—Somalia, Ethiopia.
kwangsiensis *(H.S. Lo) H.S. Lo in Guihaia*, 17(1): 51 (1997): Calophanoides kwangsiensis.
lianshanica *(H.S. Lo) H.S. Lo in Guihaia*, 17(1): 55 (1997): Mananthes lianshanica.
lilloi *(Lotti) C. Ezcurra in A.L. Cabrera, Fl. Prov. Jujuy (Inst. Nac. Tecn. Agropec., 13)*, 9: 349 (1993): Chaetochlamys lilloi.
linearifolia *(Bremek.) H.S. Lo in Guihaia*, 17(1): 58 (1997): Rostellularia linearifolia.
subsp. liangkwangensis *(H.S. Lo) H.S. Lo in Guihaia*, 17(1): 59 (1997): Rostellularia linearifolia subsp. liangkwangensis.
mandonii *(Lindau) D.C. Wasshausen & C. Ezcurra in Candollea*, 52(1): 174 (1997): Beloperone mandonii.
manserichensis *D.C. Wasshausen in Biollania*, ed. esp. 6 (J.J. Wurdack Festschr.): 545 (1997)—Peru.
masiaca *T.F. Daniel in Brittonia*, 47(4): 408 (1995)—Mexico Northwest (Sonora, Sinaloa).
nervata *(Lindau) S.R. Profice in Reserva Ecol. Macaé de Cima: Aspectos Flor. Espéc. Vasc.*, 2: 32 (1996): Jacobina nervata.
nicaraguensis *L.H. Durkee in Novon*, 9(4): 508 (1999)—Nicaragua.
oaxacana *(Greenm.) T.F. Daniel in Bol. Inst. Bot. (Guadalajara)*, 5(1–3): 120 (1998): Jacobinia oaxacana.
oncodes *(Lindau) D.C. Wasshausen & C. Ezcurra in Candollea*, 52(1): 175 (1997): Poikilacanthus oncodes.
paucifolia *T.F. Daniel in Bol. Inst. Bot. (Guadalajara)*, 5(1–3): 122 (1998)—Mexico Central (Puebla).
phyllocalyx *(Lindau) D.C. Wasshausen & C. Ezcurra in Candollea*, 52(1): 175 (1997): Poikilacanthus phyllocalyx.
polita *(Nees) S.R. Profice in Reserva Ecol. Macaé de Cima: Aspectos Flor. Espéc. Vasc.*, 2: 30 (1996): Sericographis polita.
potarensis *(Bremek.) D.C. Wasshausen in Brittonia*, 47(4): 424 (1995): Duvernoia potarensis.
procumbens L.
subsp. latispica *(C.B. Clarke) L.H. Cramer in M.D. Dassanayake & W.D. Clayton, eds., Revised Handb. Fl. Ceylon*, 12: 120 (1998): J. procumbens var. latispica.
pseudospicata *H.S. Lo & D. Fang in Guihaia*, 17(1): 52 (1997)—China Southeast (Guangxi).
puberula *K.L. Immelman in Fl. S. Afr.*, 30(3:1): 33 (1995), nom. nov.: J. parvibracteata *Immelman*.
riparia *C. Kameyama in Bol. Bot. (São Paulo)*, 14: 204 (1995), nom. nov.: Beloperone mollis *Nees*.
rothschuhii *(Lindau) L.H. Durkee in Novon*, 9(4): 510 (1999): Chaetothylax rothschuhii.
serrana *C. Kameyama in Bol. Bot. (São Paulo)*, 14: 201 (1995), nom. nov.: Ortothactus montanus *Nees*.
thyrsiformis *T. Hardwicke in Asiat. Res.*, 6: 349 (1799)—India.
tobagensis *(Urb.) D.C. Wasshausen in Brittonia*, 47(4): 425 (1995): Drejerella tobagensis.
violaceotincta *D. Champluvier in Belgian J. Bot.*, 130(2): 222 (1998)—Burundi, Rwanda, Zaire.

K

KABUYEA *R.K. Brummitt in Kew Bull.*, 53(4): 771 (1998). *TECOPHILAEACEAE*
hostifolia *(Engl.) R.K. Brummitt in Kew Bull.*, 53(4): 774 (1998): Cyanastrum hostifolium.

KADENICARPUS *A.B. Doweld in Sukkulentÿ (Moscow)* 1998(1): 22 (1998). *CACTACEAE*
pseudomacrochele *(Backeb.) A.B. Doweld in Sukkulentÿ (Moscow)*, 1998(1): 22 (1998): Strombocactus pseudomacrochele.
var. lausseri *(Diers & Gerhart Frank) A.B. Doweld in Sukkulentÿ (Moscow)*, 1998(1): 22 (1998): Turbinicarpus pseudomacrochele var. lausseri.
subsp. lausseri *(Diers & Gerhart Frank) A.B. Doweld in Sukkulentÿ (Moscow)*, 3(1–2): 37 (2000): Turbinicarpus pseudomacrochele var. lausseri.
var. sphacellatus *(Diers & Gerhart Frank) A.B. Doweld in Sukkulentÿ (Moscow)*, 1998(1): 22 (1998): Turbinicarpus pseudomacrochele var. sphacellatus.

KADSURA (Schisandrac.).
subgen. Cosbaea *(Lem.) Y.W. Law in Fl. Reipubl. Popul. Sin.*, 30(1): 272 (1996), as 'Cosbaca': gen. Cosbaea.
acsmithii *R.M.K. Saunders in Blumea*, 42(1): 109 (1997)—Borneo (Sarawak, Kalimantan).
coccinea (Lem.) A.C. Sm.
var. sichuanensis *Y.W. Law in Fl. Reipubl. Popul. Sin.*, 30(1): 272 (1996): China South-central (Sichuan).

KAEMPFERIA (Zingiberac.).
rubromarginata *(S.Q. Tong) R.J. Searle in Telopea*, 8(3): 375 (1999): Stahlianthus rubromarginatus.
simaoensis *Y.Y. Qian in Acta Bot. Austro Sinica*, 10: 65 (1995)—China South-central (Yunnan).

KALANCHOE (Crassulac.).
aromatica H. Perrier
var. brevicorolla *P. Boiteau in P. Boiteau & L. Allorge-Boiteau, Kalanchoe de Madagascar*: 192 (1995)—Madagascar.
beauverdii Raym.-Hamet
var. juelii *(Raym.-Hamet & H. Perrier) W. Rauh & Hebding in Succulentes*, 1995(1): 15 (1995), as 'jueli', without basionym ref.: K. juelii.
bracteata Scott-Elliot
var. aurantiaca *W. Rauh & R. Hebding in Bradleya*, 15: 8 (1997)—Madagascar.
var. glabra *W. Rauh & R. Hebding in Bradleya*, 15: 5 (1997)—Madagascar.
subsp. glabra *W. Rauh & R. Hebding in Bradleya*, 15: 5 (1997)—Madagascar.
var. longisepala *Boiteau ex L. Allorge-Boiteau in P. Boiteau & L. Allorge-Boiteau, Kalanchoe de Madagascar*: 144 (1995)—Madagascar.
var. pubescens *W. Rauh & R. Hebding in Bradleya*, 15: 8 (1997)—Madagascar.
brevicalyx *(Raym.-Hamet & H. Perrier) P. Boiteau in P. Boiteau & L. Allorge-Boiteau, Kalanchoe de Madagascar*: 133 (1995), without basionym ref.: K. pinnata var. brevicalyx.
calcicola *(H. Perrier) P. Boiteau in P. Boiteau & L. Allorge-Boiteau, Kalanchoe de Madagascar*: 132 (1995): K. pinnata var. calcicola.
coccinea *(H. Perrier) P. Boiteau in P. Boiteau & L. Allorge-Boiteau, Kalanchoe de Madagascar*: 188 (1995), not validly publ., see Art. 34.2 ICBN (1994): K. globulifera var. coccinea.
var. blossfeldiana *(Poelln.) P. Boiteau in P. Boiteau & L. Allorge-Boiteau, Kalanchoe de Madagascar*: 190 (1995), not validly publ., see Art. 34.2 ICBN (1994): K. blossfeldiana.
gastonis-bonnieri Raym.-Hamet & H. Perrier
var. ankaizinensis *Boiteau ex L. Allorge-Boiteau in P. Boiteau & L. Allorge-Boiteau, Kalanchoe de Madagascar*: 119 (1995)—Madagascar.
gracilipes (Baker) Baill.
var. microphylla *Humbert ex L. Allorge-Boiteau in P. Boiteau & L. Allorge-Boiteau, Kalanchoe de Madagascar*: 46 (1995)—Madagascar.

KALANCHOE:-
hildebrandtii Baill.
var. glabra *W. Rauh & R. Hebding in Bradleya,* 15: 12 (1997)—Madagascar.
integrifolia Baker
var. flava *P. Boiteau in P. Boiteau & L. Allorge-Boiteau, Kalanchoe de Madagascar:* 184 (1995)—Madagascar.
lanceolata (Forssk.) Pers.
var. glabra *(C.B. Clarke) S.R. Srinivasan in N.C. Nair & A.N. Henry, Fl. Tamil Nadu, India, Ser. 1,* 1: 145 (1983): K. floribunda var. glabra.
ndotoensis *L.E. Newton in Cact. Succ. J. (U.S.A.),* 70(2): 85 (1998)—Kenya.
poincarei Raym.-Hamet & H. Perrier
var. mortagei *(Raym.-Hamet & H. Perrier) P. Boiteau in P. Boiteau & L. Allorge-Boiteau, Kalanchoe de Madagascar:* 122 (1995): K. mortagei.
var. suarezensis *L. Allorge-Boiteau in P. Boiteau & L. Allorge-Boiteau, Kalanchoe de Madagascar:* 122 (1995): K. suarezensis.
pubescens Baker
f. anjirensis *(H. Perrier) L. Allorge-Boiteau in P. Boiteau & L. Allorge-Boiteau, Kalanchoe de Madagascar:* 65 (1995): K. miniata var. anjirensis.
pumila Baker
f. venustior *P. Boiteau in P. Boiteau & L. Allorge-Boiteau, Kalanchoe de Madagascar:* 175 (1995)—Madagascar.
rechingeri *Raym.-Hamet ex W. Rauh & Hebding in Succulentes,* 1995(1): 17 (1995)—Madagascar.
rhombopilosa Mannoni & Boiteau
var. argentea *W. Rauh in Bradleya,* 15: 9 (1997)—Madagascar.
var. viridifolia *W. Rauh in Bradleya,* 15: 9 (1997)—Madagascar.

KALIMERIS (Compos.).
incisa (Fisch.) DC.
subsp. macrodon *(Vaniot & H. Lév.) H.Y. Gu in Ann. Missouri Bot. Gard.,* 84(4): 778 (1997 publ. 1998): Aster macrodon.
indica (L.) Sch. Bip.
subsp. collina *(Hance) H.Y. Gu in Ann. Missouri Bot. Gard.,* 84(4): 789 (1997 publ. 1998): Boltonia indica var. collina.
subsp. stenolepis *(Hand.-Mazz.) H.Y. Gu in Ann. Missouri Bot. Gard.,* 84(4): 791 (1997 publ. 1998): Asteromoea indica var. stenolepis.
lautureana (Debeaux) Kitam.
subsp. mangtaoensis *(Kitag.) H.Y. Gu in Ann. Missouri Bot. Gard.,* 84(4): 807 (1997 publ. 1998): Aster mangtaoensis.
yomena (Kitam.) Kitam.
subsp. angustifolia *(Nakai) H.Y. Gu in Ann. Missouri Bot. Gard.,* 84(4): 799 (1997 publ. 1998): K. pinnatifida var. angustifolia.
subsp. dentata *(Kitam.) H.Y. Gu in Ann. Missouri Bot. Gard.,* 84(4): 801 (1997 publ. 1998): K. pinnatifida var. dentata.

KALOPANAX (Araliac.).
septemlobus (Thunb.) Koidz.
var. lutchuensis *(Nakai) Ohwi ex H. Ohba in Fl. Japan,* 2c: 267 (1999): K. ricinifolius var. lutchuensis.

KARIMBOLEA (Asclepiadac.).
macrantha *(Jum. & H. Perrier) S. Liede & U. Meve in Brittonia,* 48(4): 502 (1996 publ. 1997): Cynanchum macranthum.
mariensis *U. Meve & S. Liede in Bradleya,* 16: 10 (1998)—Madagascar.

KAYEA (Guttif.).
borneensis *P.F. Stevens in Checklist Flow. Pl. & Gymn. Brunei Darussalam (ed. M.J.E. Coode et al.):* 439 (1996)—Borneo (Sabah).
elmeri Merr.
subsp. tenuis *P.F. Stevens in Checklist Flow. Pl. & Gymn. Brunei Darussalam (ed. M.J.E. Coode et al.):* 439 (1996)—Borneo (Brunei).
scalarinervosa *P.F. Stevens in Checklist Flow. Pl. & Gymn. Brunei Darussalam (ed. M.J.E. Coode et al.):* 439 (1996)—Borneo (Brunei).

KEENANIA (Rubiac.).
flava *H.S. Lo in Bull. Bot. Res. North-East. Forest. Univ.,* 18(3): 282 (1998), as 'Koenania', without type—China Southeast (Guangxi).

KEFERSTEINIA (Orchidac.).
koechliniorum *E.A. Christenson in Orchid Dig.,* 64(3): 139 (2000), as 'koechlinorum'—Peru.
lehmannii *P. Ortiz V.in Orquideologia,* 20(2): 234 (1996)—Colombia.
niesseniae *P. Ortiz V. in Orquideologia,* 20(2): 239 (1996)—Colombia.
orbicularis *F. Pupulin in Lindleyana,* 15(1): 25 (2000)—Costa Rica.
oscarii *P. Ortiz V. in Orquideologia,* 20(2): 240 (1996)—Colombia.
retanae *G. Gerlach in Brenesia,* 41–42: 100 (1994 publ. 1995), contrary to Art. 37.3 ICBN (1994)—Costa Rica.
retanae *G. Gerlach ex C.O. Morales in Brenesia,* 52: 75 (1999)—Costa Rica.
ricii *R. Vásquez Ch. & C.H. Dodson in Rev. Soc. Boliv. Bot.,* 2(1): 4 (1998)—Bolivia.
villosa *D.E. Bennett & E.A. Christenson in Lindleyana,* 13(1): 51 (1998)—Peru.

× **KEFOREA** *hort. in Orchid Rev.,* 105(1213): 62 (1997). *ORCHIDACEAE* (Kefersteinia × Pescatoria).

KENGYILIA (Gramin.).
sect. Hyalolepis *(Nevski) L.B. Cai in Acta Phytotax. Sin.,* 37(5): 462 (1999): Agropyron sect. Hyalolepis.
sect. Stenachyra *L.B. Cai in Acta Phytotax. Sin.,* 37(5): 456 (1999).
alatavica (Drobow) J.L. Yang, C. Yen & B.R. Baum
var. longiglumis *(Keng & S.L. Chen) C. Yen, J.L. Yang & B.R. Baum in Novon,* 8(1): 94 (1998): Roegneria longiglumis.
batalinii (Krasn.) J.L. Yang, C. Yen & B.R. Baum
var. nana *(J.L. Yang, C. Yen & B.R. Baum) C. Yen, J.L. Yang & B.R. Baum in Novon,* 8(1): 95 (1998), without exact basionym page: K. nana.
var. villosissima *Roshev. ex C. Yen, J.L. Yang & B.R. Baum in Novon,* 8(1): 94 (1998)—Turkmenistan.
carinata *(Ovcz. & Sidorenko) C. Yen, J.L. Yang & B.R. Baum in Novon,* 8(1): 95 (1998): Roegneria carinata.
eremopyroides *Nevski ex C. Yen, J.L. Yang & B.R. Baum in Novon,* 8(1): 96 (1998)—China.
grandiglumis (Keng & S.L. Chen) J.L. Yang, C. Yen & B.R. Baum
var. laxiuscula *(Melderis) L.B. Cai in Acta Phytotax. Sin.,* 37(5): 458 (1999): Agropyron thoroldianum var. laxiusculum.
guidenensis *C. Yen, J.L. Yang & B.R. Baum in Novon,* 5(4): 395 (1995)—Qinghai.
hirsuta (Keng & S.L. Chen) J.L. Yang, C. Yen & B.R. Baum
var. tahopaica *(Keng & S.L. Chen) L.B. Cai in Acta Phytotax. Sin.,* 37(5): 466 (1999): Roegneria melanthera var. tahopaica.
var. variabilis *(Keng & S.L. Chen) L.B. Cai in Acta Phytotax. Sin.,* 37(5): 466 (Oct. 1999); *et in Fl. Qinghaiica,* 4: 93 (Dec. 1999): Roegneria hirsuta var. variabilis.
kaschgarica *(D.F. Cui) L.B. Cai in Novon,* 6(2): 142 (1996): Elymus kaschgaricus.
kryloviana *(Schischk.) C. Yen, J.L. Yang & B.R. Baum in Novon,* 8(1): 100 (1998): Agropyron krylovianum.
leiantha *(Keng) L.B. Cai in Novon,* 6(2): 142 (1996): Roegneria leiantha.
obviaristata *(L.B. Cai) L.B. Cai in Acta Phytotax. Sin.,* 37(5): 456 (1999): K. hirsuta. var. obviaristata.
pendula *L.B. Cai in Acta Phytotax. Sin.,* 37(5): 460 (1999)—Qinghai.
pulcherrima *(Grossh.) C. Yen, J.L. Yang & B.R. Baum in Novon,* 8(1): 100 (1998): Agropyron pulcherrimum.
rigidula (Keng & S.L. Chen) J.L. Yang, C. Yen & B.R. Baum
var. intermedia *(Keng & S.L. Chen) L.B. Cai in Acta Phytotax. Sin.,* 37(5): 459 (1999), isonym; *see S.L. Chen in Bull. Bot. Res. North-East. Forest. Univ.,* 14(2): 141 (1994): Roegneria rigidula var. intermedia.

KENGYILIA:-
shawanensis *L.B. Cai in Guihaia,* 16(3): 202 (1996)—Xinjiang.
thoroldiana (Oliv.) J.L. Yang, C. Yen & Baum
var. laxiuscula *(Melderis) S.L. Chen in Novon,* 7(3): 229 (1997): Agropyron thoroldianum var. laxiusculum.
var. melanthera *(Keng) L.B. Cai in Acta Phytotax. Sin.,* 37(5): 463 (1999): Agropyron melantherum.

KENTIA (Palmae).
macarthurii *H. Wendl. ex H.J. Veitch, Catalogue of Plants:* 26 (1879)—New Guinea.

KENTIOPSIS (Palmae).
magnifica *(H.E. Moore) J.-C. Pintaud & D.R. Hodel in Principes,* 42(1): 42 (1998): Mackeea magnifica.
piersoniorum *J.-C. Pintaud & D.R. Hodel in Principes,* 42(1): 45 (1998)—New Caledonia.
pyriformis *J.-C. Pintaud & D.R. Hodel in Principes,* 42(1): 49 (1998)—New Caledonia.

KERAUDRENIA (Sterculiac.).
exastia *C.F. Wilkins in Nuytsia,* 13(1): 235 (1999)—Western Australia.
katatona *C.F. Wilkins in Nuytsia,* 13(1): 239 (1999)—Western Australia.

KERAYMONIA (Umbellif.).
pinnatifolia *M.F. Watson in Edinburgh J. Bot.,* 53(1): 134 (1996)—Bhutan.

KERGELIELLA (Orchidac.).
orientalis *G. Gerlach in G.A. Romero & G. Carnevali, Orchids of Venezuela, ed. 2:* 1137 (2000)—Venezuela.

KETELEERIA (Pinac.).
evelyniana Mast.
var. dopiana *(Flous) J. Silba in J. Int. Conifer Preserv. Soc.,* 7(1): 26 (2000): K. dopiana.
fortunei (A. Murray) Carrière
var. oblonga *(W.C. Cheng & L.K. Fu) L.K. Fu & Nan Li in Novon,* 7(3): 261 (1997): K. oblonga.

KHADIA (Aizoac.).
alticola *P. Chesselet & H.E.K. Hartmann in Aloe,* 32(2): 48 (1995)—Transvaal, Natal.
media *P.J.D. Winter & N. Hahn in Aloe,* 36(1): 16 (1999)—Transvaal.

KICKXIA (Scrophulariac.).
elegans (G. Forst.) D.A. Sutton
subsp. dichondrifolia *(Benth.) Ø.H. Rustan & C. Brochmann in Sommerfeltia,* 24: 284 (1997): Linaria dichondrifolia.
subsp. webbiana *(Sunding) Ø.H. Rustan & C. Brochmann in Sommerfeltia,* 24: 286 (1997): K. webbiana.
papillosa *R.R. Mill in Edinburgh J. Bot.,* 57(3): 419 (2000)—Bhutan.

KIELMEYERA (Guttif.).
coriacea Mart.
var. pseudocoriacea *N. Saddi, Gen. Kielmeyera Fl. Mato Grosso (Brasil) (Publ. Avuls. Herb. Centr., 10):* 54 (1996)—Brazil West-central (Mato Grosso).
var. tomentosa *(Cambess.) N. Saddi, Gen. Kielmeyera Fl. Mato Grosso (Brasil) (Publ. Avuls. Herb. Centr., 10):* 57 (1996): K. tomentosa.

KILLIPIA (Melastomatac.).
verticalis *N. Ruiz-Rodgers in Caldasia,* 19(1–2): 175 (1997)—Colombia.

KINGIDIUM (Orchidac.).
sect. Conspicuum *O. Gruss & L. Roellke in Orchidee,* 48(6): 267 (1997).
chibae *(T. Yukawa) O. Gruss & L. Roellke in Orchidee,* 48(6): 261 (1997): Phalaenopsis chibae.
mysorensis *(C.J. Saldanha) C. Sathish Kumar in C. Sathish Kumar & K.S. Manilal, Cat. Indian Orchids:* 95 (1994): Phalaenopsis mysorensis.

KINGIDIUM:-
niveum *C. Sathish Kumar in C. Sathish Kumar & K.S. Manilal, Cat. Indian Orchids:* 53 (1994)—India (Kerala).
wilsonii *(Rolfe) O. Gruss & L. Röllke in Orchidee,* 47(3): 149 (1996): Phalaenopsis wilsonii.

× **KIPPENARA** *hort. in Orchid Rev.,* 106(1223): 318 (1998). *ORCHIDACEAE* (Ascocentrum × Doritis × Rhynchostylis × Vanda).

KLAROBELIA *L.W. Chatrou, Changing Genera:* 121 (1998). *ANNONACEAE*
anomala *(R.E. Fr.) L.W. Chatrou, Changing Genera:* 123 (1998): Crematosperma anomalum.
candida *L.W. Chatrou, Changing Genera:* 127 (1998)—Peru.
cauliflora *L.W. Chatrou, Changing Genera:* 128 (1998)—Peru, Colombia, Brazil North (Amazonas).
inundata *L.W. Chatrou, Changing Genera:* 131 (1998)—Peru, Brazil North (Amazonas).
lucida *(Diels) L.W. Chatrou, Changing Genera:* 133 (1998): Malmea lucida.
megalocarpa *L.W. Chatrou, Changing Genera:* 136 (1998)—Ecuador.
napoensis *L.W. Chatrou, Changing Genera:* 138 (1998)—Peru, Ecuador.
peruviana *(R.E. Fr.) L.W. Chatrou, Changing Genera:* 141 (1998): Malmea peruviana.
pumila *L.W. Chatrou, Changing Genera:* 141 (1998)—Peru, Brazil North (Acre).
stipitata *L.W. Chatrou, Changing Genera:* 143 (1998)—Costa Rica.

KLASEA (Compos.).
donetzica *(Dubovik) J. Holub in Preslia,* 70(2): 106 (1998): Serratula donetzica.
insularis *(Iljin) J. Holub in Preslia,* 70(2): 106 (1998): Serratula insularis.
lusitanica *(Cantó) J. Holub in Preslia,* 70(2): 106 (1998): Serratula baetica subsp. lusitanica.
pallida *(DC.) J. Holub in Preslia,* 70(2): 106 (1998): Serratula pallida.
tanaitica *(P.A. Smirn.) J. Holub in Preslia,* 70(2): 106 (1998): Serratula tanaitica.

KLEINIA (Compos.).
isabellae *M. Dioli & Mesfin Tadesse in Cact. Succ. J. (U.S.A.),* 72(6): 303 (2000), with two types cited—Ethiopia.

KNAUTIA (Dipsacac.).
subgen. Trichera *(Schrad. ex Roem. & Schult.) N.N. Tzvelev, Opred. Sosud. Rast. Severo-Zap. Rossii:* 680 (2000): gen. Trichera.
alboi *Sennen in Ann. Junta Ciènc. Nat.,* 2: 644 (1917)—Spain.
arvernensis (Briq.) Szabó
subsp. catalaunica *(Sennen ex Szabó) O. de Bolòs & J. Vigo, Fl. Països Catalans,* 3: 635 (1995): K. catalaunica.
godetii Reut.
subsp. lebrunii *(J. Prudhomme) O. de Bolòs & J. Vigo, Fl. Països Catalans,* 3: 636 (1995): K. lebrunii.
× hungarica *V. Borbás, Revis. Knaut.:* 64 (1904)—Hungary. (K. kitaibelii × turocensis).
× kohleri *J. Briquet in Annuaire Conserv. Jard. Bot. Genève,* 6: 134 (1902)—Switzerland. (K. arvensis × godetii).
rupicola (Willk.) Font Quer
var. cardonica *(O. Bolòs & Vigo) J. Molero, L. Sáez & J. Vallverdú in But. Inst. Catalana Hist. Nat.,* 64: 65 (1996 publ. 1997): K. arvensis subvar. cardonica.
× sambucifolia *(Godet) J. Briquet in Annuaire Conserv. Jard. Bot. Genève,* 6: 131 (1902): K. sylvatica var. sambucifolia.
subscaposa Boiss. & Reut.
subsp. alboi *(Sennen) J. Molero, L. Sáez & J. Vallverdú in But. Inst. Catalana Hist. Nat.,* 64: 66 (1996 publ. 1997): K. alboi.

KNEMA (Myristicac.).
emmae *W.J.J.O. de Wilde in Blumea,* 41(2): 384 (1996)—Borneo (Kalimantan, Sabah).

KNEMA:-
korthalsii Warb.
subsp. rimosa *W.J.J.O. de Wilde in Blumea,* 43(1): 242 (1998)—Borneo (Sabah, Sarawak).
krusemaniana *W.J.J.O. de Wilde in Blumea,* 41(2): 385 (1996)—Borneo (Kalimantan).
kunstleri (King) Warb.
subsp. leptophylla *W.J.J.O. de Wilde in Blumea,* 41(2): 386 (1996)—Borneo (Kalimantan).
subsp. pseudostellata *W.J.J.O. de Wilde in Blumea,* 43(1): 244 (1998)—Borneo (Kalimantan).
latericia Elmer
f. nana *W.J.J.O. de Wilde in Blumea,* 43(1): 246 (1998)—Borneo (Sarawak, Brunei, Sabah).
f. olivacea *W.J.J.O. de Wilde in Blumea,* 43(1): 248 (1998)—Borneo (Kalimantan, Sabah).
longepilosa *(W.J. de Wilde) W.J.J.O. de Wilde in Blumea,* 41(2): 389 (1996): K. percoriacea f. longepilosa.
minima *W.J.J.O. de Wilde in Blumea,* 43(1): 249 (1998)—Borneo (Brunei).
ridsdaleana *W.J.J.O. de Wilde in Blumea,* 41(2): 392 (1996)—Philippines.
tridactyla Airy Shaw
subsp. pachydactyla *W.J.J.O. de Wilde in Blumea,* 43(1): 251 (1998)—Borneo (Sabah).
subsp. salicifolia *W.J.J.O. de Wilde in Blumea,* 43(1): 251 (1998)—Borneo (Sabah).

KNOXIA (Rubiac.).
spicata *(Thwaites ex Trimen) C.E. Ridsdale in Blumea,* 41(2): 457 (1996): K. platycarpa var. spicata.

KOANOPHYLLON (Compos.).
constanzae *(Urb.) H. Robinson in Phytologia,* 84(5): 348 (1998 publ. 1999): Eupatorium constanzae.
galeanum *(B.L. Turner) B.L. Turner in Phytologia,* 82(6): 386 (1997 publ. 1998): Eupatorium galeanum.
peninsulare *(Brandegee) B.L. Turner in Phytologia,* 82(6): 386 (1997 publ. 1998): Eupatorium peninsulare.

KOBRESIA (Cyperac.).
*subgen.*Compositae *(C.B. Clarke) I. Kukkonen in Ann. Naturhist. Mus. Wien,* 98B(Suppl.): 91 (1996): Hemicarex sect. Compositae.
subgen. Hemicarex *(Benth.) I. Kukkonen in Ann. Naturhist. Mus. Wien,* 98B(Suppl.): 92 (1996): gen. Hemicarex.
sect. Nitentes *(N.A. Ivanova) I. Kukkonen in Ann. Naturhist. Mus. Wien,* 98B(Suppl.): 92 (1996): K. ser. Nitentes.
sect. Sibiricae *(N.A. Ivanova) I. Kukkonen in Ann. Naturhist. Mus. Wien,* 98B(Suppl.): 92 (1996): K. ser. Sibiricae.
filicina (C.B. Clarke) C.B. Clarke
var. subfilicinoides *P.C. Li in Acta Phytotax. Sin.,* 37(2): 155 (1999)—China South-central (Yunnan).
glaucifolia *Wang & Tang ex P.C. Li in Acta Phytotax. Sin.,* 37(2): 153 (1999)—Xizang.
hohxilensis *R.F. Huang in S.G. Wu (S.K. Wu) & Z.J. Feng (eds.), Biol. & Human Physiol. Hoh Xil Region:* 101 (1996)—Qinghai.
laxa *C.G.D. Nees von Esenbeck in R. Wight, Contrib. Bot. India:* 119 (1834), as 'Cobresia'—Nepal.†
var. curvirostris *(C.B. Clarke) R.C. Srivastava in Fl. Sikkim,* 1: 224 (1996): K. curvirostris.
subsp. hissarica *(Pissjauk.) I. Kukkonen in Ann. Naturhist. Mus. Wien,* 98B(Suppl.): 91 (1996): Schoenoxiphium hissaricum.
myosuroides (Vill.) Fiori & Paol.
subsp. bistaminata *(W.Z. Di & M.J. Zhong) S.R. Zhang in Novon,* 9(3): 453 (1999): K. bistaminata.
nepalensis
var. stiebritziana *(Hand.-Mazz.) R.C. Srivastava in Novon,* 8(2): 203 (1998): K. stiebritziana.
var. vaginosa *(C.B. Clarke) R.C. Srivastava in Fl. Sikkim,* 1: 225 (1996): K. vaginosa.
royleana (Nees) Boeck.
subsp. minshanica *(Y.C. Yang) S.R. Zhang in Novon,* 9(3): 453 (1999): K. minshanica.
setschwanensis Hand.-Mazz.
subsp. squamiformis *(Y.C. Yang) S.R. Zhang in Novon,* 9(3): 453 (1999): K. squamiformis.

KOBRESIA:-
stolonifera *Y.C. Tang ex P.C. Li in Acta Phytotax. Sin.,* 37(2): 154 (1999)—Xizang.

KOCHIA (Chenopodiac.).
angustifolia *(Turcz.) G.A. Peschkova, Stepnaya Fl. Baikal'skoĭ Sibiri:* 53 (1972): K. scoparia var. angustifolia.

KOELERIA (Gramin.).
macrantha (Ledeb.) Schult.
subsp. glauca *(Schrad.) P.D. Sell in P.D. Sell & G. Murrell, Fl. Great Britain & Ireland,* 5: 363 (1996): Aira glauca.
riguorum *E. Edgar & E.S. Gibb in New Zealand J. Bot.,* 37(1): 59 (1999)—New Zealand South I.

KOELLENSTEINIA (Orchidac.).
spiralis *A.B. Gomes-Ferreira & L.C. Menezes in Bol. CAOB,* 29: 87 (1997); *et in Orchidee,* 48(6): 258 (1997); *et in Orchidophile,* 132: 114 (1998)—Brazil Northeast (Bahia).

KOENIGIA (Polygonac.).
binsarii *(Silas & R.D. Gaur) R.D. Gaur, Fl. Distr. Garhwal N.W. Himalaya:* 135 (1999): Polygonum binsarii.

KOGELBERGIA *J.P. Rourke in Bothalia,* 30(1): 12 (2000). *STILBACEAE*
phylicoides *(A.DC.) J.P. Rourke in Bothalia,* 30(1): 13 (2000): Stilbe phylicoides.
verticillata *(Eckl. & Zeyh.) J.P. Rourke in Bothalia,* 30(1): 12 (2000): Trichocephalus verticillatus.

KOHLRAUSCHIA (Caryophyllac.).
dubia *(Raf.) J. Holub in Preslia,* 70(2): 107 (1998): Dianthus dubius.

KOPSIA (Apocynac.).
terengganensis *L. Allorge & C. Wiart in Acta Bot. Gallica,* 142(5): 434 (1995 publ. 1996)—Malaya (Peninsular Malaysia).

KOPSIOPSIS (Scrophulariac.).
hookeri *(Walp.) R. Govaerts, World Checklist Seed Plants,* 2(1–2): 14 (1996): Boschniakia hookeri.

KORNASIA *D.L. Szlachetko in Fragm. Flor. Geobot.,* Supp. 3: 120 (1995). *ORCHIDACEAE*
chevalieri *(Summerh.) D.L. Szlachetko in Fragm. Flor. Geobot.,* Supp. 3: 121 (1995): Malaxis chevalieri.
maclaudii *(Finet) D.L. Szlachetko in Fragm. Flor. Geobot.,* Supp. 3: 121 (1995): Microstylis maclaudii.
schliebenii *(Mansf.) D.L. Szlachetko in Fragm. Flor. Geobot.,* Supp. 3: 121 (1995): Microstylis schliebenii.

KORTHALSELLA (Viscac.).
japonica (Thunb.) Engl.
f. grayi *(Barlow) M. Molvray in Novon,* 7(3): 272 (1997): K. grayi.
f. rubra *(Tiegh.) M. Molvray in Novon,* 7(3): 272 (1997): Bifaria rubra.
taenioides (Comm. ex DC.) Engl.
f. disticha *(Endl.) M. Molvray in Novon,* 7(3): 271 (1997): Viscum distichum.
f. emersa *(Barlow) M. Molvray in Novon,* 7(3): 271 (1997): K. emersa.
f. horneana *(Tiegh.) M. Molvray in Novon,* 7(3): 271 (1997): K. horneana.
f. pendula *(Wawra) M. Molvray in Novon,* 7(3): 271 (1997): Viscum moniliforme f. pendula.
f. remyana *(Tiegh.) M. Molvray in Novon,* 7(3): 271 (1997): K. remyana.
vitifolia *(L.) M.R. Almeida & N. Patil in J. Bombay Nat. Hist. Soc.,* 93(1): 111 (1996), without basionym ref. Hibiscus vitifolius.

KOYAMASIA *H. Robinson in Proc. Biol. Soc. Wash.,* 112(1): 234 (1999). *COMPOSITAE*
calcarea *(Kitam.) H. Robinson in Proc. Biol. Soc. Wash.,* 112(1): 235 (1999): Camchaya calcarea.

KRAINZIA (Cactac.).
sect. Decipientes *(D.R. Hunt) A.B. Doweld in Sukkulentÿ (Moscow),* 3(1–2): 43 (2000): Mammillaria ser. Decipientes.

KRAINZIA:-
sect. Leptocladodae (Lem.) A.B. Doweld in Sukkulentÿ (Moscow), 3(1–2): 43 (2000): Mammillaria § Leptocladodae.
sect. Stylothelae (Pfeiff.) A.B. Doweld in Sukkulentÿ (Moscow), 3(1–2): 43 (2000): Mammillaria § Stylothelae.
bocasana (Poselg.) A.B. Doweld in Sukkulentÿ (Moscow), 3(1–2): 42 (2000): Mammillaria bocasana.
subsp. eschauzieri (J.M. Coult.) A.B. Doweld in Sukkulentÿ (Moscow), 3(1–2): 42 (2000): Cactus eschauzieri.
crinita (DC.) A.B. Doweld in Sukkulentÿ (Moscow), 3(1–2): 42 (2000): Mammillaria crinita.
subsp. leucantha (Boed.) A.B. Doweld in Sukkulentÿ (Moscow), 3(1–2): 42 (2000): Mammillaria leucantha.
subsp. wildii (A. Dietr.) A.B. Doweld in Sukkulentÿ (Moscow), 3(1–2): 42 (2000): Mammillaria wildii.
decipiens (Scheidw.) A.B. Doweld in Sukkulentÿ (Moscow), 3(1–2): 42 (2000): Mammillaria decipiens.
subsp. albescens (Tiegel) A.B. Doweld in Sukkulentÿ (Moscow), 3(1–2): 42 (2000): Mammillaria albescens.
subsp. camptotricha (Dams) A.B. Doweld in Sukkulentÿ (Moscow), 3(1–2): 43 (2000): Mammillaria camptotricha.
densispina (J.M. Coult.) A.B. Doweld in Sukkulentÿ (Moscow), 3(1–2): 43 (2000): Cactus densispinus.
elongata (DC.) A.B. Doweld in Sukkulentÿ (Moscow), 3(1–2): 43 (2000): Mammillaria elongata.
erythrosperma (Boed.) A.B. Doweld in Sukkulentÿ (Moscow), 3(1–2): 43 (2000): Mammillaria erythrosperma.
longiflora (Britton & Rose) Backeb.
subsp. stampferi (Repp.) A.B. Doweld in Sukkulentÿ (Moscow), 3(1–2): 43 (2000): Mammillaria stampferi.
microhelia (Werderm.) A.B. Doweld in Sukkulentÿ (Moscow), 3(1–2): 43 (2000): Mammillaria microhelia.
pottsii (Scheer ex Salm-Dyck) A.B. Doweld in Sukkulentÿ (Moscow), 3(1–2): 43 (2000): Mammillaria pottsii.
schwarzii (Shurly) A.B. Doweld in Sukkulentÿ (Moscow), 3(1–2): 43 (2000): Mammillaria schwarzii.
tepexicensis (J. Meyrán) A.B. Doweld in Sukkulentÿ (Moscow), 3(1–2): 43 (2000): Mammillaria tepexicensis.

KRAPOVICKASIA (Malvac.).
araujoana M. Brandão in Daphne, 4(2): 6 (1994), contrary to Art. 37.5 ICBN (1994)—Brazil Southeast (Minas Gerais).

KRASCHENINIKOVIA (Caryophyllac.).
rhaphanorrhiza (Hemsl.) Korsh. in Izv. Imp. Akad. Nauk, Ser. 5, 9(1): 39 (June 1898), as 'raphanorhiza': Stellaria rhaphanorrhiza.†
rupestris Turcz.in Bull. Soc. Imp. Naturalistes Moscou, 15(3): 610 (May 1842)—Siberia.†

KRASCHENINNIKOVIA (Chenopodiac.).
longipilosa (C.P. Tsien & C.G. Ma) G.L. Chu in Higher Plants China, 4: 335 (2000), with incorrect basionym ref.: Ceratoides compacta var. longipilosa.

KRECZETOVICZIA N.N. Tzvelev in Bot. Zhurn., 84(7): 112 (1999). CYPERACEAE
caespitosa (L.) N.N. Tzvelev in Bot. Zhurn., 84(7): 112 (1999): Scirpus caespitosus.
pumila (Vahl) N.N. Tzvelev in Bot. Zhurn., 84(7): 112 (1999): Scirpus pumilus.
uniflora (Trautv.) N.N. Tzvelev in Bot. Zhurn., 84(7): 112 (1999): Scirpus uniflorus.

KRYPTOSTOMA (Orchidac.).
sect. Multipartita (Kraenzl.) D.L. Szlachetko & T.S. Olszewski in Fl. Cameroun, 34: 230 (1998), as 'Multipartitae': Habenaria sect. Multipartitae.

KRYPTOSTOMA:-
sect. Pseudocoeloglossum D.L. Szlachetko & T.S. Olszewski in Fl. Cameroun, 34: 228 (1998).
cavatibrachia (Summerh.) T.S. Olszewski & D.L. Szlachetko in Ann. Bot. Fenn., 37(4): 299 (2000): Habenaria cavatibrachia.
coeloglossoides (Summerh.) D.L. Szlachetko in Fragm. Flor. Geobot., Supp. 3: 114 (1995): Habenaria coeloglossoides.
davidii (Franch.) D.L. Szlachetko & T.S. Olszewski in Ann. Bot. Fenn., 37(4): 299 (2000): Habenaria davidii.
egregium (Summerh.) D.L. Szlachetko in Fragm. Flor. Geobot., Supp. 3: 114 (1995): Habenaria egregia.
excelsum (S. Thomas & P.J. Cribb) D.L. Szlachetko & T.S. Olszewski in Fl. Cameroun, 34: 230 (1998), as 'excelsa': Habenaria excelsa.
falcilobum (Summerh.) D.L. Szlachetko & T.S. Olszewski in Ann. Bot. Fenn., 37(4): 297 (2000), as 'falciloba': Habenaria falciloba.
gilbertii (S. Thomas & P.J. Cribb) D.L. Szlachetko & T.S. Olszewski in Fl. Cameroun, 34: 230 (1998): Habenaria gilbertii.
godefroyi (Rchb. f.) T.S. Olszewski & D.L. Szlachetko in Ann. Bot. Fenn., 37(4): 299 (2000): Habenaria godefroyi.
intermedium (D. Don) T.S. Olszewski & D.L. Szlachetko in Ann. Bot. Fenn., 37(4): 299 (2000), as 'intermedia': Habenaria intermedia.
keniense (Summerh.) D.L. Szlachetko in Fragm. Flor. Geobot., Supp. 3: 114 (1995): Habenaria keniensis.
limprichtii (Schltr.) D.L. Szlachetko & T.S. Olszewski in Fl. Cameroun, 34: 231 (1998): Habenaria limprichtii.
macranthum (A. Rich.) D.L. Szlachetko in Fragm. Flor. Geobot., Supp. 3: 114 (1995): Habenaria macrantha.
mannii (Hook. f.) D.L. Szlachetko in Fragm. Flor. Geobot., Supp. 3: 114 (1995): Habenaria mannii.
multipartitum (Blume ex Kraenzl.) D.L. Szlachetko & T.S. Olszewski in Fl. Cameroun, 34: 231 (1998), as 'multipartita': Habenaria multipartita.
occlusum (Summerh.) D.L. Szlachetko in Fragm. Flor. Geobot., Supp. 3: 114 (1995): Habenaria occlusa.
oligoschistum (Schltr.) D.L. Szlachetko & T.S. Olszewski in Fl. Cameroun, 34: 231 (1998), as 'oligoschista': Habenaria oligoschista.
pectinatum (D. Don) T.S. Olszewski & D.L. Szlachetko in Ann. Bot. Fenn., 37(4): 299 (2000), as 'pectinata': Habenaria pectinata.
praestans (Rendle) D.L. Szlachetko in Fragm. Flor. Geobot., Supp. 3: 114 (1995): Habenaria praestans.
var. umbrosum (G. Will.) D.L. Szlachetko & T.S. Olszewski in Fl. Cameroun, 34: 231 (1998), as 'umbrosa': Habenaria praestans var. umbrosa.
quartinianum (A. Rich.) D.L. Szlachetko in Fragm. Flor. Geobot., Supp. 3: 114 (1995): Habenaria quartiniana.
schaijesii (Geerinck) D.L. Szlachetko in Fragm. Flor. Geobot., Supp. 3: 114 (1995): Habenaria schaijesii.
splendens (Rendle) D.L. Szlachetko in Fragm. Flor. Geobot., Supp. 3: 114 (1995): Habenaria splendens.
splendentius (Summerh.) D.L. Szlachetko & T.S. Olszewski in Fl. Cameroun, 34: 231 (1998), as 'splendentior': Habenaria splendentior.
tosariense (J.J. Sm.) D.L. Szlachetko & T.S. Olszewski in Fl. Cameroun, 34: 231 (1998), as 'tosariensis': Habenaria tosariensis.

KULINIA B.G. Briggs & L.A.S. Johnson in Telopea, 7(4): 349 (1998). RESTIONACEAE
eludens B.G. Briggs & L.A.S. Johnson in Telopea, 7(4): 349 (1998)—Western Australia.

KUNGIA K.T. Fu in J. Northwest. Teachers Coll. (Nat. Sci.), 1988(1): 3 (1988). CRASSULACEAE
aliciae (Raym.-Hamet) K.T. Fu in J. Northwest. Teachers Coll. (Nat. Sci.), 1988(1): 4 (1988): Crassula aliciae.
var. komarovii (Raym.-Hamet) K.T. Fu in J. Northwest. Teachers Coll. (Nat. Sci.), 1988(1): 5 (1988): Sedum aliciae var. komarovii.

KUNGIA:-
schoenlandii (Raym.-Hamet) K.T. Fu in J. Northwest. Teachers Coll. (Nat. Sci.), 1988(1): 5 (1988): Sedum schoenlandii.
var. lepidotricha (K.T. Fu) K.T. Fu in J. Northwest. Teachers Coll. (Nat. Sci.), 1988(1): 5 (1988): Sinocrassula stenostachya var. lepidotricha.
var. stenostachya (Fröd.) K.T. Fu in J. Northwest. Teachers Coll. (Nat. Sci.), 1988(1): 5 (1988): Sedum stenostachyum.

KUNZEA (Myrtac.).
subsect. Arborescentes H.R. Toelken in J. Adelaide Bot. Gard., 17: 32 (1996).
subsect. Floridae H.R. Toelken in J. Adelaide Bot. Gard., 17: 82 (1996).
subsect. Globosae H.R. Toelken in J. Adelaide Bot. Gard., 17: 53 (1996).
sect. Zeanuk H.R. Toelken in J. Adelaide Bot. Gard., 17: 32 (1996).
acuminata H.R. Toelken in J. Adelaide Bot. Gard., 17: 88 (1996)—Western Australia.
capitata (Sm.) G. Heynhold, Alph. Aufz. Gew.: 337 (1846): Metrosideros capitata.
ciliata H.R. Toelken in J. Adelaide Bot. Gard., 17: 64 (1996)—Western Australia.
cincinnata H.R. Toelken in J. Adelaide Bot. Gard., 17: 101 (1996)—Western Australia.
clavata H.R. Toelken in J. Adelaide Bot. Gard., 17: 50 (1996)—Western Australia.
corifolia (Vent.) J.C. Schauer in Lehm., Pl. Preiss., 1: 124 (1844): Metrosideros corifolia.
ericifolia (Sm.) G. Heynhold, Alph. Aufz. Gew.: 338 (1846): Metrosideros ericifolia.
subsp. subulata H.R. Toelken in J. Adelaide Bot. Gard., 17: 47 (1996)—Western Australia.
glabrescens H.R. Toelken in J. Adelaide Bot. Gard., 17: 41 (1996)—Western Australia.
micrantha Schauer
subsp. hirtiflora H.R. Toelken in J. Adelaide Bot. Gard., 17: 77 (1996)—Western Australia.
subsp. oligandra (Turcz.) H.R. Toelken in J. Adelaide Bot. Gard., 17: 75 (1996): K. oligandra.
subsp. petiolata H.R. Toelken in J. Adelaide Bot. Gard., 17: 73 (1996)—Western Australia.
newbeyi H.R. Toelken in J. Adelaide Bot. Gard., 17: 60 (1996)—Western Australia.
rostrata H.R. Toelken in J. Adelaide Bot. Gard., 17: 62 (1996)—Western Australia.
similis H.R. Toelken in J. Adelaide Bot. Gard., 17: 86 (1996)—Western Australia.
spathulata H.R. Toelken in J. Adelaide Bot. Gard., 17: 48 (1996)—Western Australia.

KYHOSIA B.G. Baldwin in Novon, 9(4): 465 (1999). COMPOSITAE
bolanderi (A. Gray) B.G. Baldwin in Novon, 9(4): 466 (1999): Anisocarpus bolanderi.

KYLLINGA (Cyperac.).
brevifolia Rottb.
subsp. kamtschatica (Meinsh.) V.N. Voroschilov in A.K. Skvortsov (ed.), Florist. issl. v razn. Raĭonakh SSSR: 152 (1985): K. kamtschatica.
kilianii A.M. Muasya & D.A. Simpson in Kew Bull., 51(1): 183 (1996)—Kenya.
nervosa Steud.
subsp. jubensis (Mtot.) J.-P. Lebrun & A.L. Stork, Enum. Pl. Fleurs Afr. Trop., 4: 602 (1997): K. flava subsp. jubensis.

KYRSTENIOPSIS (Compos.).
eriocarpa (B.L. Rob. & Greenm.) B.L. Turner in Phytologia, 82(6): 387 (1997 publ. 1998): Eupatorium eriocarpum.
heathiae (B.L. Turner) B.L. Turner in Phytologia, 82(6): 387 (1997 publ. 1998): Eupatorium heathiae.
iltisii (R.M. King & H. Rob.) B.L. Turner in Phytologia, 82(6): 387 (1997 publ. 1998): Critonia iltisii.
perpetiolata (R.M. King & H. Rob.) B.L. Turner in Phytologia, 82(6): 387 (1997 publ. 1998): Pseudokyrsteniopsis perpetiolata.
spinaciifolia (DC.) B.L. Turner in Phytologia, 82(6): 387 (1997 publ. 1998): Bulbostylis spinaciifolia.

L

LABLAB (Legumin.).
purpureus (L.) Sweet
var. lignosus *(Prain) G.R. Kumari in N.C. Nair & A.N. Henry, Fl. Tamil Nadu, India, Ser. 1:* 1: 114 (1983): Dolichos lablab var. lignosus.

LABRAMIA (Sapotac.).
mayottensis *J.-N. Labat, M. Pignal & O. Pascal in Adansonia,* 19(2): 214 (1997)—Comoros (Mayotte).

LACHEMILLA (Rosac.).
angustata *K. Romoleroux in Fl. Ecuador,* 56(79): 107 (1996)—Ecuador.
involucrata *J. Gaviria in Plantula,* 1(1): 76 (1996)—Venezuela.
loki-schmidtiae *J. Gaviria in Plantula,* 1(1): 78 (1996), as 'loki-schmidtii'—Venezuela.
pelusae *J. Gaviria in Plantula,* 1(1): 78 (1996)—Venezuela.

LACHENALIA (Hyacinthac.).
alba *W.F. Barker ex G.D. Duncan in Bothalia,* 26(1): 6 (1996)—Cape Province.
angelica *W.F. Barker in S. Afr. J. Bot.,* 55(6): 631 (1989)—Cape Province.
attenuata *W.F. Barker ex G.D. Duncan in Bothalia,* 28(2): 132 (1998)—Cape Province.
aurioliae *G.D. Duncan in Bothalia,* 27(1): 7 (1997)—Cape Province.
doleritica *G.D. Duncan in Bothalia,* 28(2): 134 (1998)—Cape Province.
duncanii *W.F. Barker in S. Afr. J. Bot.,* 55(6): 630 (1989)—Cape Province.
elegans W.F. Barker
var. flava *W.F. Barker in S. Afr. J. Bot.,* 55(6): 634 (1989)—Cape Province.
var. membranacea *W.F. Barker in S. Afr. J. Bot.,* 55(6): 636 (1989)—Cape Province.
var. suaveolens *W.F. Barker in S. Afr. J. Bot.,* 55(6): 635 (1989)—Cape Province.
hirta (Thunb.) Thunb.
var. exserta *W.F. Barker in S. Afr. J. Bot.,* 55(6): 638 (1989)—Cape Province.
inconspicua *G.D. Duncan in Bothalia,* 27(1): 11 (1997)—Cape Province.
juncifolia Baker
var. campanulata *W.F. Barker in S. Afr. J. Bot.,* 55(6): 640 (1989)—Cape Province.
karooica *W.F. Barker ex G.D. Duncan in Bothalia,* 26(1): 1 (1996)—Orange Free State, Cape Province.
lactosa *G.D. Duncan in Bothalia,* 28(2): 135 (1998)—Cape Province.
leipoldtii *G.D. Duncan in Bothalia,* 28(2): 137 (1998)—Cape Province.
marginata W.F. Barker
subsp. neglecta *Schltr. ex G.D. Duncan in Bothalia,* 26(1): 7 (1996)—Cape Province.
marlothii *W.F. Barker ex G.D. Duncan in Bothalia,* 27(1): 12 (1997)—Cape Province.
mediana Jacq.
var. rogersii *(Baker) W.F. Barker in S. Afr. J. Bot.,* 55(6): 642 (1989): L. unifolia var. rogersii.
neilii *W.F. Barker ex G.D. Duncan in Bothalia,* 26(1): 5 (1996)—Cape Province.
nutans *G.D. Duncan in Bothalia,* 28(2) 131 (1998)—Namibia.
obscura *Schltr. ex G.D. Duncan in Bothalia,* 27(1): 9 (1997)—Cape Province.
orchioides (L.) Aiton
var. glaucina *(Jacq.) W.F. Barker in S. Afr. J. Bot.,* 55(6): 640 (1989): L. glaucina.
perryae *G.D. Duncan in Bothalia,* 26(1): 3 (1996)—Cape Province.
unifolia Jacq.
var. schlechteri *(Baker) W.F. Barker in S. Afr. J. Bot.,* 55(6): 643 (1989): L. schlechteri.
violacea Jacq.
var. glauca *W.F. Barker in S. Afr. J. Bot.,* 55(6): 639 (1989)—Cape Province.
xerophila *Schltr. ex G.D. Duncan in Bothalia,* 27(1): 14 (1997)—Cape Province.

LACHNAEA (Thymelaeac.).
filicaulis *(Meisn.) J.B.P. Beyers in Bothalia,* 27(1): 45 (1997): Cryptadenia filicaulis.
laxa *(C.H. Wright) J.B.P. Beyers in Bothalia,* 27(1): 45 (1997): Cryptadenia laxa.
pudens *J.B.P. Beyers in S. Afr. J. Bot.,* 64(1): 66 (1998)—Cape Province.
sociorum *J.B.P. Beyers in Bothalia,* 28(1): 49 (1998)—Cape Province.
uniflora *(L.) J.B.P. Beyers in Bothalia,* 27(1): 45 (1997), homonym; *see Crantz, Inst. Rei Herb.,* 1: 130 (1766): Passerina uniflora.

LACHNAGROSTIS (Gramin.).
ammobia *E. Edgar in New Zealand J. Bot.,* 33(1): 14 (1995)—New Zealand South I.
elata *E. Edgar in New Zealand J. Bot.,* 33(1): 18 (1995)—New Zealand South I., New Zealand North I.
glabra *(Petrie) E. Edgar in New Zealand J. Bot.,* 33(1): 20 (1995): Deyeuxia glabra.
littoralis *(Hack.) E. Edgar in New Zealand J. Bot.,* 33(1): 23 (1995): Deyeuxia forsteri var. littoralis.
subsp. salaria *E. Edgar in New Zealand J. Bot.,* 33(1): 24 (1995)—New Zealand South I.
pilosa *(Buchanan) E. Edgar in New Zealand J. Bot.,* 33(1): 27 (1995): Deyeuxia pilosa.
subsp. nubifera *E. Edgar in New Zealand J. Bot.,* 33(1): 28 (1995)—Antipodean Is.
tenuis *(Cheeseman) E. Edgar in New Zealand J. Bot.,* 33(1): 30 (1995): Deyeuxia billardierei var. tenuis.
uda *E. Edgar in New Zealand J. Bot.,* 33(1): 32 (1995)—New Zealand South I.

LACKEYA *R.H. Fortunato, L.P. de Queiroz & G.P. Lewis in Kew Bull.,* 51(2): 365 (1996). *LEGUMINOSAE*
multiflora *(Torr. & A. Gray) R.H. Fortunato, L.P. de Queiroz & G.P. Lewis in Kew Bull.,* 51(2): 366 (1996): Dolichos multiflorus.

LACMELLEA (Apocynac.).
zamorae *J.F. Morales in Novon,* 8(3): 262 (1998)—Costa Rica.

LACTUCA (Compos.).
sativa L.
var. yemensis *X.S. Shen & Run L. Lu in Bull. Bot. Res. North-East. Forest. Univ.,* 18(1): 32 (1998)—China Southeast (Anhui).
undulata Ledeb.
var. albicaulis *C.H. An in Fl. Xinjiangensis,* 5: 480 (1999)—Xinjiang.

LADENBERGIA (Rubiac.).
dwyeri *L. Andersson in Nordic J. Bot.,* 17(3): 290 (1997)—Panama.
klugii *L. Andersson in Nordic J. Bot.,* 17(3): 273 (1997)—Peru.
lehmanniana *L. Andersson in Nordic J. Bot.,* 17(3): 295 (1997)—Colombia, Ecuador.
magdalenae *L. Andersson in Nordic J. Bot.,* 17(3): 279 (1997)—Colombia.
nubigena *L. Andersson in Nordic J. Bot.,* 17(3): 287 (1997)—Panama.
obovata *L. Andersson in Nordic J. Bot.,* 17(3): 292 (1997)—Colombia.
pauciflora *L. Andersson in Nordic J. Bot.,* 17(3): 282 (1997)—Colombia.

LAELIA (Orchidac.).
sect. Esalqueanae *C.L. Withner, Cattleyas & Relatives,* 2: 18 (1990).
sect. Harpophyllae *C.L. Withner, Cattleyas & Relatives,* 2: 18 (1990).
sect. Liliputanae *C.L. Withner, Cattleyas & Relatives,* 2: 18 (1990).
sect. Perriniae *C.L. Withner, Cattleyas & Relatives,* 2: 18 (1990).
sect. Rupestres *C.L. Withner, Cattleyas & Relatives,* 2: 18 (1990).
sect. Sinocoranae *C.L. Withner, Cattleyas & Relatives,* 2: 18 (1990).
alaorii Brieger & Bicalho
f. dietliana *O. Gruss in Orchidee,* Suppl. 3: 12 (1996)—Brazil Northeast (Bahia).
alvaroana *F.E. Miranda in Bradea,* 8(22): 124 (1999)—Brazil Southeast (Rio de Janeiro).

LAELIA:-
autumnalis (La Llave & Lex.) Lindl.
f. xanthotrophis *(Rchb. f.) F. Halbinger & M. Soto Arenas, Laelias ofMexico (Orquídea (México), 15):* 95 (1997): L. autumnalis var. xanthotrophis.
munchowiana *F.E. Miranda in Bradea,* 8(22): 126 (1999)—Brazil Southeast (Espírito Santo).
× zaslawskii *L.C. Menezes in Bol. CAOB,* 24: 11 (1996)—Brazil Southeast (Espírito Santo). (L. harpophylla × spectabilis).

LAGOTIS (Scrophulariac.).
chumbica *R.R. Mill in Edinburgh J. Bot.,* 57(3): 420 (2000)—Xizang.

LAGYNIAS (Rubiac.).
monteiroi *(Oliv.) D.M. Bridson in Kew Bull.,* 51(2): 351 (1996): Ancylanthos monteiroi.

× **LAIPENCHIHARA** *hort. in Orchid Rev.,* 105(1218): 382 (1997). *ORCHIDACEAE* (Ascocentrum × Doritis × Neofinetia × Rhynchostylis × Vanda).

LALLDHWOJIA (Umbellif.).
acronemifolia *(H. Wolff) M.F. Watson ex D.G. Long in Edinburgh J. Bot.,* 56(2): 307 (1999): Peucedanum acronemifolium.

LAMBERTIA (Proteac.).
echinata R. Br.
subsp. citrina *R.J. Hnatiuk in Fl. Australia,* 16: 500 (1995)—Western Australia.
subsp. occidentalis *G.J. Keighery in Nuytsia,* 11(2): 283 (1997)—Western Australia.
inermis R. Br.
var. drummondii *(Fielding & Gardner) R.J. Hnatiuk in Fl. Australia,* 16: 500 (1995): L. drummondii.
multiflora Lindl.
var. darlingensis *R.J. Hnatiuk in Fl. Australia,* 16: 500 (1995)—Western Australia.
rariflora Meisn.
subsp. lutea *R.J. Hnatiuk in Fl. Australia,* 16: 500 (1995)—Western Australia.

LAMIASTRUM (Labiat.).
argentatum *(Smejkal) J. Soják in Zprávy Krajsk. Vlastiv. Muz. Olomouci,* 215: 4 (1982): Galeobdolon argentatum.

LAMIUM (Labiat.).
album L.
f. barbatum *(Siebold & Zucc.) Y.S. Kim & S.J. Park in Korean J. Pl. Taxon.,* 25(2): 114 (1995): L. barbatum.
argentatum *(Smejkal) Henker ex G.H. Loos in Flor. Rundbr.,* 31(1): 43 (1997): Galeobdolon argentatum.
demirizii *A.P. Khokhrjakov in Byull. Mosk. Obshch. Ispÿt. Prir., Biol.,* 102(4): 47 (1997)—Turkey.
endtmannii *G.H. Loos in Flor. Rundbr.,* 31(1): 43 (1997)—Germany.
taiwanense *S.S. Ying in Mem. Coll. Agric. Nation. Taiwan Univ.,* 31(1): 22 (1991)—Taiwan.
tschorochense *A.P. Khokhrjakov in Byull. Mosk. Obshch. Ispÿt. Prir., Biol.,* 100(6): 82 (1995)—Turkey.
villosifolium *(R.R. Mill) A.P. Khokhrjakov in Byull. Mosk. Obshch. Ispÿt. Prir., Biol.,* 100(6): 83 (1995): L. maculatum subsp. villosifolium.
violaceovelutinum *A.P. Khokhrjakov in Byull. Mosk. Obshch. Ispÿt. Prir., Biol.,* 102(4): 47 (1997), as 'violaceo-velutinum'—Turkey.
vreemanii *A.P. Khokhrjakov in Byull. Mosk. Obshch. Ispÿt. Prir., Biol.,* 100(6): 82 (1995)—Turkey.

LAMPRANTHUS (Aizoac.).
amphibolius *(G.D. Rowley) H.E.K. Hartmann in Bradleya,* 16: 67 (1998): Ruschia amphibolia.
arenarius *H.E.K. Hartmann in Bradleya,* 16: 67 (1998), nom. nov.: Ruschia arenosa *L. Bolus.*
curviflorus *(Haw.) H.E.K. Hartmann in Bradleya,* 16: 67 (1998): Mesembryanthemum curviflorum.
dissimilis *(G.D. Rowley) H.E.K. Hartmann in Bradleya,* 16: 67 (1998): Ruschia dissimilis.
francesiae *H.E.K. Hartmann in Bradleya,* 16: 68 (1998), nom. nov.: Ruschia leightoniae *L. Bolus.*

LAMPRANTHUS:-
marginatus *(L. Bolus) H.E.K. Hartmann in Bradleya*, 16: 68 (1998): Ruschia marginata.
mutatus *(G.D. Rowley) H.E.K. Hartmann in Bradleya*, 16: 68 (1998): Ruschia mutata.
pakpassensis *H.E.K. Hartmann in Bradleya*, 16: 70 (1998), nom. nov.: Ruschia pakhuisensis *L. Bolus*.
profundus *(L. Bolus) H.E.K. Hartmann in Bradleya*, 16: 70 (1998): Ruschia profunda.
tanquanus *H.E.K. Hartmann in Bradleya*, 16: 70 (1998), nom. nov.: Ruschia salteri *L. Bolus*.

LAMPROCAPNOS (Papaverac.).
spectabilis *(L.) T. Fukuhara in Pl. Syst. Evol.*, 206(1–4): 415 (1997): Fumaria spectabilis.

LAMPROPAPPUS *(O. Hoffm.) H. Robinson in Proc. Biol. Soc. Wash.*, 112(1): 245 (1999): Vernonia sect. Lampropappus. (Compos.).
eremanthifolia *(O. Hoffm.) H. Robinson in Proc. Biol. Soc. Wash.*, 112(1): 245 (1999): Vernonia eremanthifolia.
hoffmannii *H. Robinson in Proc. Biol. Soc. Wash.*, 112(1): 245 (1999), nom. nov.: Vernonia lampropappus *O. Hoffm.*
turbinellus *(S. Moore) H. Robinson in Proc. Biol. Soc. Wash.*, 112(1): 245 (1999): Vernonia turbinella.

LANDIOPSIS *Capuron ex J. Bosser in Adansonia*, 20(1): 132 (1998). *RUBIACEAE*
capuronii *J. Bosser in Adansonia*, 20(1): 132 (1998)—Madagascar.

LANDOLTIA *D.H. Les & D.J. Crawford in Novon*, 9(4): 532 (1999). *LEMNACEAE*
punctata *(G. Mey.) D.H. Les & D.J. Crawford in Novon*, 9(4): 532 (1999): Lemna punctata.

LANKESTERELLA (Orchidac.).
alainii *M.A. Nir, Orchidaceae Antillanae:* 170 (2000)—Dominican Republic.

× **LANTANA** (Verbenac.).
elenievskii *I.E. Méndez Santos in Kew Bull.*, 54(2): 487 (1999)—Cuba.

LAPIDERMA *S.A. Hammer in Mesemb Study Group Bull.*, 14(4): 92 (1999); *cf. Repert. Pl. Succ. (I.O.S.)*, 50: 7 (1999 publ. 2000). *AIZOACEAE* (Argyroderma × Lapidaria).

LAPORTEA (Urticac.).
sect. Sceptrocnide *(Maxim.) C.J. Chen in Fl. Reipubl. Popul. Sin.*, 23(2): 37 (1995): gen. Sceptrocnide.
bulbifera (Siebold & Zucc.) Wedd.
subsp. dielsii *(Pamp.) C.J. Chen in Fl. Reipubl. Popul. Sin.*, 23(2): 34 (1995): L. dielsii.
cuspidata (Wedd.) Friis
f. bulbifera *(Kitam.) N. Fukuoka & N. Kurosaki in Mem. Nation. Sci. Mus. (Tokyo)*, 28: 89 (1995)—L. macrostachya f. bulbifera.

LAPPULA (Boraginac.).
heteracantha (Ledeb.) Borbás
subsp. heterocarpa *(Klokov & Artemczuk) J. Holub in Fl. Slovenska*, 5(1): 153 (1993): L. heterocarpa.
lenensis *Popov ex S.V. Ovczinnikova in Fl. Sibir. (Pyrolac.-Lamiac.)*, 11: 137 (1997)—Yakutskiya.
rechingeri *H. Riedl in Ann. Naturhist. Mus. Wien*, 98B(Suppl.): 80 (1996)—Pakistan.
saissanica *N.K. Aralbaev in Izv. Akad. Nauk Resp. Kazakhstan, Ser. Biol. Med.*, 1997(2): 58 (1997), as comb. nov.—Kazakhstan.
saphronovae *R.V. Kamelin in Turczaninowia*, 2(3): 5 (1999), as 'saphronoviae'—Kazakhstan.
tuvinica *S.V. Ovczinnikova in Fl. Sibir. (Pyrolac.-Lamiac.)*, 11: 142 (1997)—Tuva.

LARIX (Pinac.).
gmelinii (Rupr.) Rupr.
f. genhensis *(S.Y. Li & K.T. Adair) L.K. Fu & Nan Li in Novon*, 7(3): 262 (1997): L. gmelinii var. genhensis.
f. pendula *(D.S. Zhang & Y.M. Chen) L.K. Fu & Nan Li in Novon*, 7(3): 261 (1997): L. principis-ruprechtii var. pendula.

LARIX:-
griffithiana Carrière
var. mastersiana *(Rehder & E.H. Wilson) J. Silba in J. Int. Conifer Preserv. Soc.*, 7(1): 26 (2000): L. mastersiana.
griffithii Hook. f.
var. speciosa *(W.C. Cheng & Y.W. Law) A. Farjon, World Checklist & Bibliogr. Conifers:* 139 (1998): L. speciosa.
kongboensis *R.R. Mill in Novon*, 9(1): 79 (1999)—Xizang.
potaninii Batalin
var. chinensis *(Beissn.) J. Silba in J. Int. Conifer Preserv. Soc.*, 7(1): 26 (2000), isonym; *see L.K. Fu & Nan Li in Novon*, 7(3): 262 (1997): L. chinensis.
var. chinensis *(Beissn.) L.K. Fu & Nan Li in Novon*, 7(3): 262 (1997): L. chinensis.

LARNAX (Solanac.).
andersonii *N.W. Sawyer in Novon*, 8(1): 72 (1998)—Ecuador.
dilloniana *S. Leiva Gonzalez, V. Quipuscoa Silvestre & N.W. Sawyer in Arnaldoa*, 5(1): 85 (1998)—Peru.
longipedunculata *S. Leiva González, E. Rodríguez Rodríguez & J. Campos de la Cruz in Arnaldoa*, 5(2): 194 (1998)—Peru.
lutea *S. Leiva Gonzalez in Arnaldoa*, 4(1): 19 (1996)—Peru.
macrocalyx *S. Leiva González, E. Rodríguez Rodríguez & J. Campos de la Cruz in Arnaldoa*, 5(2): 197 (1998)—Peru.
peruviana *(Zahlbr.) A.T. Hunziker in Darwiniana*, 33(1–4): 371 (1995): Athenaea peruviana.
pilosa *S. Leiva González, E. Rodríguez Rodríguez & J. Campos de la Cruz in Arnaldoa*, 5(2): 200 (1998)—Peru.
psilophyta *N.W. Sawyer in Novon*, 8(1): 74 (1998)—Ecuador.
purpurea *S. Leiva Gonzalez in Arnaldoa*, 4(1): 16 (1996)—Peru.
sagasteguii *S. Leiva Gonzalez, V. Quipuscoa Silvestre & N.W. Sawyer in Arnaldoa*, 5(1): 86 (1998)—Peru.
sawyeriana *S. Leiva González, E. Rodríguez Rodríguez & J. Campos de la Cruz in Arnaldoa*, 5(2): 203 (1998)—Peru.
vasquezii *S. Leiva González, E. Rodríguez Rodríguez & J. Campos de la Cruz in Arnaldoa*, 5(2): 206 (1998)—Peru.

LARREA (Zygophyllac.).
tridentata (DC.) Coville
var. arenaria *L.D. Benson in L.D. Benson & R.A. Darrow, Trees & Shrubs of Southwest. Deserts (ed. 3):* 123 (1981)—California.

LARRYLEACHIA *D.C.H. Plowes in Excelsa*, 17: 5 (1996), nom. nov.: gen. Leachia *Plowes*. (Asplepiadac.).
cactiformis *(Hook.) D.C.H. Plowes in Excelsa*, 17: 5 (1996): Stapelia cactiformis.
dinteri *(A. Berger) D.C.H. Plowes in Excelsa*, 17: 5 (1996): Trichocaulon dinteri.
felina *(D.T. Cole) D.C.H. Plowes in Excelsa*, 17: 7 (1996): Trichocaulon felinum.
marlothii *(N.E. Br.) D.C.H. Plowes in Excelsa*, 17: 7 (1996): Trichocaulon marlothii.
meloformis *(Marloth) D.C.H. Plowes in Excelsa*, 17: 7 (1996): Trichocaulon meloforme.
perlata *(Dinter) D.C.H. Plowes in Excelsa*, 17: 9 (1996): Trichocaulon perlatum.
picta *(N.E. Br.) D.C.H. Plowes in Excelsa*, 17: 9 (1996): Trichocaulon pictum.
similis *(N.E. Br.) D.C.H. Plowes in Excelsa*, 17: 12 (1996): Trichocaulon simile.
sociarum *(A.C. White & B. Sloane) D.C.H. Plowes in Excelsa*, 17: 15 (1996): Trichocaulon sociarum.
tirasmontana *(Plowes) D.C.H. Plowes in Excelsa*, 17: 15 (1996): Leachiella tirasmontana.

LASER (Umbellif.).
rechingeri *H. Akhani in Ann. Naturhist. Mus. Wien*, 98B(Suppl.): 99 (1996)—Iran.

LASERPITIUM (Umbellif.).
gallicum
subsp. majoricum *A.M. Romo, Fl. Silvestres Baleares:* 196 (1994)—Baleares.

LASIANTHAEA (Compos.).
machucana *B.L. Turner in Phytologia*, 82(4): 265 (1997 publ. 1998)—Mexico Southwest (Jalisco).

LASIANTHUS (Rubiac.).
capitatus Blume
subsp. pendulus *(Ridl.) H. Zhu in Acta Bot. Yunnanica*, 22(4): 398 (2000): Lasianthus pendulus.
subsp. vietnamensis *H. Zhu in Acta Bot. Yunnanica*, 22(4): 397 (2000)—Vietnam.
chrysocaulis *C.E. Ridsdale in Blumea*, 41(2): 457 (1996)—Sri Lanka.
foetulentus *C.E. Ridsdale in Blumea*, 41(2): 458 (1996)—Sri Lanka.
var. pseudovarians *C.E. Ridsdale in Blumea*, 41(2): 459 (1996)—Sri Lanka.
japonicus Miq.
var. taiheizanensis *(Masam. & Suzuki) S. Hatusima in S. Hatusima & T. Amano, Fl. Ryukyus, S. of Amami Is., ed. 2:* 213 (1994), without basionym ref.: Basionym not stated.
kurzii Hook. f.
var. microphyllus *(Pit.) H. Zhu in Acta Bot. Yunnanica*, 20(2): 153 (1998): L. hoaensis var. microphyllus.
larsenii *H. Zhu in Acta Bot. Yunnanica*, 22(4): 395 (2000)—Vietnam.
lucidus Blume
var. inconspicuus *(Hook. f.) H. Zhu in Acta Bot. Yunnanica*, 20(2): 154 (1998): L. inconspicuus.
malaiensis *H. Zhu in Gard. Bull. Singapore*, 52(2): 257 (2000), as 'malaiense'—Malaya (Peninsular Malaysia).
nebulosus *(Simizu) S.S. Ying in Mem. Coll. Agric. Nation. Taiwan Univ.*, 28(2): 45 (1988): Mephitidia nebulosa.
neolanceolatus *C.E. Ridsdale in Blumea*, 41(2): 459 (1996), nom. nov.: L. walkerianus var. lanceolatus *Hook. f.*
rhinocerotis Blume
subsp. pedunculatus *(Pit.) H. Zhu in Acta Bot. Yunnanica*, 20(2): 154 (1998): L. rhinocerotis var. pedunculatus.
subsp. xishuangbannaensis *H. Zhu & H. Wang in Acta Phytotax. Sin.*, 38(3): 282 (2000)—China South-central (Yunnan), Thailand.
tomentosus Blume
var. hiiranensis *(Hayata) H. Zhu in Acta Bot. Yunnanica*, 20(2): 156 (1998): L. hiiranensis.
wallichii (Wight & Arn.) Wight
var. hispidocostatus *H. Zhu in Acta Bot. Yunnanica*, 20(2): 158 (1998)—China South-central (Yunnan), Xizang, Nepal, Bangladesh, Assam (Meghalaya, Nagaland), Burma.

LASIOCARYUM (Boraginac.).
ludlowii *R.R. Mill in Edinburgh J. Bot.*, 53(1): 121 (1996)—Bhutan, Nepal, Xizang.

LASIOLAENA (Compos.).
carvalhoi *D.J.N. Hind in Kew Bull.*, 54(4): 917 (1999 publ. 2000)—Brazil Northeast (Bahia).

LATHROCASIS *L.A. Johnson in Aliso*, 19(1): 67 (2000). *POLEMONIACEAE*
tenerrima *(A. Gray) L.A. Johnson in Aliso*, 19(1): 67 (2000): Gilia tenerrima.

LATHYRUS (Legumin.).
sect. Verni *(Fritsch ex Czefr.) N.N. Tzvelev, Opred. Sosud. Rast. Severo-Zap. Rossii:* 680 (2000): Lathyrus ser. Verni.
frenchii *Sennen in Ann. Junta Ciènc. Nat.*, 2: 642 (1917)—Spain.
japonicus Willd.
var. pubescens *(Hartm.) T. Karlsson in Svensk Bot. Tidskr.*, 91(5): 249 (1997 publ. 1998): Pisum maritimum β pubescens.
latifolius L.
subsp. heterophyllus *(L.) C. Asmussen in Svensk Bot. Tidskr.*, 91(5): 249 (1997 publ. 1998): L. heterophyllus.
polymorphus Nutt.
var. incanus *(J.G. Sm. & Rydb.) D. Isely, Native & Natural. Legumin. (Fabac.) U.S. (excl. Alaska & Hawaii):* 615 (1998), with incorrect basionym ref.: Basionym not stated.

LAUBENFELSIA *A.V.F.Ch. Bobrov & A.P. Melikian in Bot. Zhurn.*, 85(7): 60 (2000). *PODOCARPACEAE*
vieillardii *(Parl.) A.V.F.Ch. Bobrov & A.P. Melikian in Bot. Zhurn.*, 85(7): 60 (2000), without full basionym ref.: Podocarpus vieillardii.

LAUNAEA (Compos.).
sect. Castanospermae *N. Kilian in Englera*, 17: 306 (1997).
sect. Cervicornes *N. Kilian in Englera* 17: 186 (1997).
sect. Cornutae *N. Kilian in Englera*, 17: 169 (1997).
sect. Pseudosonchus *(Amin ex Rech. f.) N. Kilian in Englera*, 17: 94 (1997): L. subgen. Pseudosonchus.
acanthodes (Boiss.) Kuntze
subsp. tomentella *(Rech. f.) N. Kilian in Englera*, 17: 163 (1997): L. tomentella.
almahrahensis *N. Kilian in Englera*, 17: 339 (1997)—South Yemen, Oman.
amal-aminiae *N. Kilian in Englera*, 17: 346 (1997), as 'amal-aminae'—Algeria, Morocco, Western Sahara, Mauritania, Libya, Egypt, Saudi Arabia, North Yemen.
angolensis *N. Kilian in Englera*, 17: 128 (1997)—Angola.
cabrae *(De Wild.) N. Kilian in Englera*, 17: 122 (1997): Lactuca cabrae.
subsp. nanella *(R.E. Fr.) N. Kilian in Englera*, 17: 127 (1997): Sonchus nanellus.
fragilis (Asso) Pau
subsp. asirensis *N. Kilian in Englera*, 17: 390 (1997)—North Yemen, Saudi Arabia.
lackii *N. Kilian in Englera*, 17: 337 (1997)—Somalia.
omanensis *N. Kilian in Englera*, 17 343 (1997)—Oman.
petitiana *(A. Rich.) N. Kilian in Englera*, 17: 231 (1997): Lactuca petitiana.
polydichotoma *(Ostenf.) Amin ex N. Kilian in Englera*, 17: 166 (1997): Chondrilla polydichotoma.
pseudoabyssinica *(Chiov.) N. Kilian in Englera*, 17: 259 (1997): Lactuca pseudoabyssinica.
quettaensis *N. Kilian in Englera*, 17 350 (1997)—Pakistan.
rhynchocarpa *(Balf. f.) B. Mies in Senckenberg. Biol.*, 74: 231 (1994): Lactuca rhynchocarpa.
socotrana *N. Kilian in Englera*, 17: 265 (1997)—Socotra.

LAUREMBERGIA (Haloragac.).
minor *(C.B.. Clarke) D. Philcox in M.D. Dassanayake & W.D. Clayton, eds., Revised Handb. Fl. Ceylon*, 13: 110 (1999): Serpicula zeylanica var. minor.

LAURIDIA (Celastrac.).
tetragona *(L.f.) R.H. Archer in S. Afr. J. Bot.*, 63(4): 230 (1997): Rhamnus tetragona.

LAURUS (Laurac.).
rumphii *J. Graham, Cat. Pl. Bombay* 174 (1839)—India (Maharashtra).

LAVANDULA (Labiat.).
approximata *M. Gandoger, Dec. Pl Nov.*, 1: 19 (1875)—France.
cladophora *M. Gandoger, Dec. Pl. Nov.*, 1: 38 (1875)—France.
corsica *M. Gandoger, Dec. Pl. Nov.*, 1: 18 (1875)—Corse.
debeauxii *M. Gandoger, Dec. Pl. Nov.*, 1: 18 (1875)—Corse.
decipiens *M. Gandoger, Dec. Pl. Nov.*, 1: 38 (1875)—France.
fascicularis *M. Gandoger, Dec. Pl. Nov.*, 1: 17 (1875)—Italy.
guinandii *M. Gandoger, Dec. Pl. Nov.*, 1: 38 (1875)—France.
olbiensis *M. Gandoger, Dec. Pl. Nov.*, 1: 18 (1875)—France.

LAVATERA (Malvac.).
× clementii *M. Cheek in Kew Bull.*, 55(4): 1013 (2000)—Cultivation. (L. olbia × thuringiaca).
plazzae *A.D. Atzei in Boll. Soc. Sarda Sci. Nat.*, 30: 151 (1994–95 publ. 1995)—Sardegna.
thuringiaca L.
var. cachemiriana *(Cambess.) S. Mobayen, Fl. Iran Vasc. Pl.*, 3: 210 (1985), without basionym ref.: L. cachemiriana.

LAVOISIERA (Melastomatac.).
fragilis *Cogn. ex C. Munhoz & C. Proença in Novon*, 10(1): 60 (2000)—Brazil West-central (Goiás).

LAVRANIA (Asclepiadac.).
sect. Cactoidea *P.V. Bruyns in S. Afr. J. Bot.*, 65(4): 305 (1999).

LEANDRA (Melastomatac.).
humilis *(Urb. & Ekman) Alain in Sida*, 18(4): 1026 (1999), homonym; *see Wurdack in Sellowia*, 14: 202 (1962): Ossaea humilis.
hybophylla *(Urb.) Alain in Sida*, 18(4): 1026 (1999): Ossaea hybophylla.
marigotiana *(Urb. & Ekman) Alain in Sida*, 18(4): 1026 (1999): Ossaea marigotiana.
polychaeta *(Urb. & Ekman) Alain in Sida*, 18(4): 1026 (1999), homonym; *see Cogn. in A.DC., Monogr. Phan.*, 7: 1186 (1891): Ossaea polychaeta.

LECANANTHUS (Rubiac.).
peduncularis *C. Puff in Blumea*, 43(2): 343 (1998)—Borneo (Sarawak).
pentander *(Merr.) C. Puff in Blumea*, 43(2): 343 (1998): Lucinaea pentandra.

LECANORCHIS (Orchidac.).
seidenfadenii *D.L. Szlachetko & J. Mytnik in Ann. Bot. Fenn.*, 37(3): 228 (2000), as 'seidenfadeni'—Malaya (Peninsular Malaysia).
sikkimensis *N. Pearce & P.J. Cribb in Edinburgh J. Bot.*, 56(2): 280 (1999)—Sikkim.

LECANTHUS (Urticac.).
corniculatus *(C.J. Chen) H.W. Li in Fl. Yunnanica*, 7: 237 (1997): L. petelotii var. corniculatus.
var. yunnanensis *(C.J. Chen) H.W. Li in Fl. Yunnanica*, 7: 237 (1997): L. petelotii var. yunnanensis.
taiwanensis *S.S. Ying in Mem. Coll. Agric. Nation. Taiwan Univ.*, 31(1): 28 (1991)—Taiwan.

LECOSIA *T.M. Pedersen in Bonplandia*, 10(1–4): 107 (2000). *AMARANTHACEAE*
formicarum *T.M. Pedersen in Bonplandia*, 10(1–4): 107 (2000)—Brazil Northeast (Bahia).
oppositifolia *T.M. Pedersen in Bonplandia*, 10(1–4): 107 (2000)—Brazil Southeast (Espírito Santo).

LECTICULA *J.H. Barnhart in J.K. Small, Fl. Miami*: 170 (26 April 1913). *LENTIBULARIACEAE*†
resupinata *(B.D. Greene ex Bigelow) J.H. Barnhart in J.K. Small, Fl. Miami*: 170 (26 April 1913): Utricularia resupinata†

LECYTHIS (Lecythidac.).
gracieana *S.A. Mori in Mem. New York Bot. Gard.*, 75: 47 (1995)—Brazil (Amazonas).

LEDEBOURIA (Hyacinthac.).
atrobrunnea *S. Venter in Bothalia*, 28(1): 15 (1998)—Transvaal.
camerooniana *(Baker) F. Speta in Phyton (Austria)*, 38(1): 105 (1998): Scilla camerooniana.
confertiflora *(Dammer) U. Müller-Doblies & D. Müller-Doblies in Feddes Repert.*, 108(1–2): 56 (1997): Drimia confertiflora.
cordifolia *(Baker) B. Stedje & M. Thulin in Fl. Trop. E. Afr., Hyacinthac.*: 14 (25 March 1996); *et in Nordic J. Bot.*, 15(6): 595 (1995 publ. 15 April 1996): Scilla cordifolia.
crispa *S. Venter in Bothalia*, 28(2): 179 (1998)—Transvaal.
dolomiticola *S. Venter in Bothalia*, 28(1): 16 (1998)—Transvaal.
edulis *(Engl.) B. Stedje in Fl. Trop. E. Afr., Hyacinthac.*: 14 (25 March 1996); *et in Nordic J. Bot.*, 15(6): 595 (1995 publ. 15 April 1996): Scilla edulis.
grandifolia *(Balf. f.) A.G. Miller & D. Alexander in Edinburgh J. Bot.*, 53(1): 44 (1996): Haemanthus grandifolius.
hildebrandtii *(Baker) U. Müller-Doblies & D. Müller-Doblies in Feddes Repert.*, 108(1–2): 58 (1997): Drimia hildebrandtii.
kirkii *(Baker) B. Stedje & M. Thulin in Fl. Trop. E. Afr., Hyacinthac.*: 12 (25 March 1996); *et in Nordic J. Bot.*, 15(6): 595 (1995 publ. 15 April 1996): Scilla kirkii.

LEDEBOURIA:-
lilacina *(Fenzl ex Kunth) F. Speta in Phyton (Austria)*, 38(1): 106 (1998): Eratobotrys lilacina.
macowanii *(Baker) S. Venter in Bothalia*, 28(2): 181 (1998): Scilla macowanii.
maesta *(Baker) F. Speta in Phyton (Austria)*, 38(1): 106 (1998): Scilla maesta.
parvifolia *S. Venter in Bothalia*, 28(2): 180 (1998)—Transvaal.
sickenbergeri *(Deflers) F. Speta in Phyton (Austria)*, 38(1): 106 (1998): Scilla sickenbergeri.
somaliensis *(Baker) B. Stedje & M. Thulin in Fl. Trop. E. Afr., Hyacinthac.*: 12 (25 March 1996); *et in Nordic J. Bot.*, 15(6): 595 (1995 publ. 15 April 1996): Scilla somaliensis.
urceolata *B. Stedje in Nordic J. Bot.*, 15(6): 595 (1995 publ. 1996)—Ethiopia.
yemenensis *(Deflers) F. Speta in Phyton (Austria)*, 38(1): 106 (1998): Scilla yemenensis.
zambesiaca *(Baker) F. Speta in Phyton (Austria)*, 38(1): 106 (1998): Scilla zambesiaca.

LEEA (Leeac.).
glabra *C.L. Li in Chinese J. Appl. Environ. Biol.*, 2(1): 43 (1996)—China South-central (Yunnan), China Southeast (Guangxi).

LEGOUSIA (Campanulac.).
scabra (Lowe) Gamisans
var. grandiflora *(Willk.) O. de Bolòs & J. Vigo, Fl. Països Catalans*, 3: 666 (1995): Specularia castellana var. grandiflora.

LEIONEMA *(F. Muell.) Paul G. Wilson in Nuytsia*, 12(2): 270 (1998): Eriostemon sect. Leionema. (Rutac.).
ambiens *(F. Muell.) Paul G. Wilson in Nuytsia*, 12(2): 271 (1998): Eriostemon ambiens.
bilobum *(Lindl.) Paul G. Wilson in Nuytsia*, 12(2): 271 (1998): Phebalium bilobum.
carruthersii *(F. Muell.) Paul G. Wilson in Nuytsia*, 12(2): 272 (1998): Eriostemon carruthersii.
coxii *(F. Muell.) Paul G. Wilson in Nuytsia*, 12(2): 272 (1998): Eriostemon coxii.
dentatum *(Sm.) Paul G. Wilson in Nuytsia*, 12(2): 272 (1998): Phebalium dentatum.
diosmeum *(A. Juss.) Paul G. Wilson in Nuytsia*, 12(2): 272 (1998): Phebalium diosmeum.
elatius *(F. Muell.) Paul G. Wilson in Nuytsia*, 12(2): 273 (1998): Eriostemon elatior.
subsp. beckleri *(F. Muell.) Paul G. Wilson in Nuytsia*, 12(2): 273 (1998): Eriostemon beckleri.
ellipticum *Paul G. Wilson in Nuytsia*, 12(2): 273 (1998)—Queensland.
equestre *(D.A. Cooke) Paul G. Wilson in Nuytsia*, 12(2): 274 (1998): Phebalium equestre.
gracile *(C.T. White) Paul G. Wilson in Nuytsia*, 12(2): 274 (1998): Phebalium gracile.
hillebrandii *(J.H. Willis) Paul G. Wilson in Nuytsia*, 12(2): 274 (1998): Phebalium hillebrandii.
lachnaeoides *(A. Cunn.) Paul G. Wilson in Nuytsia*, 12(2): 275 (1998): Phebalium lachnaeoides.
lamprophyllum *(F. Muell.) Paul G. Wilson in Nuytsia*, 12(2): 275 (1998): Eriostemon lamprophyllus.
subsp. obovatum *F.M. Anderson in Muelleria*, 12(2): 233 (1999)—New South Wales (Australian Capital Territory, New South Wales), Victoria.
subsp. orbiculare *F.M. Anderson in Muelleria*, 12(2): 232 (1999)—New South Wales.
microphyllum *(F. Muell.) Paul G. Wilson in Nuytsia*, 12(2): 275 (1998): Eriostemon microphyllus.
montanum *(Hook.) Paul G. Wilson in Nuytsia*, 12(2): 275 (1998): Phebalium montanum.
nudum *(Hook.) Paul G. Wilson in Nuytsia*, 12(2): 275 (1998): Phebalium nudum.
obtusifolium *(Paul G. Wilson) Paul G. Wilson in Nuytsia*, 12(2): 276 (1998): Phebalium obtusifolium.
oldfieldii *(F. Muell.) Paul G. Wilson in Nuytsia*, 12(2): 276 (1998): Eriostemon oldfieldii.
phylicifolium *(F. Muell.) Paul G. Wilson in Nuytsia*, 12(2): 276 (1998): Phebalium phylicifolium.
ralstonii *(F. Muell.) Paul G. Wilson in Nuytsia*, 12(2): 276 (1998): Eriostemon ralstonii.
rotundifolium *(Endl.) Paul G. Wilson in Nuytsia*, 12(2): 276 (1998): Eriostemon rotundifolius.

LEIONEMA:-
sympetalum *(Paul G. Wilson) Paul G. Wilson in Nuytsia*, 12(2): 277 (1998): Phebalium sympetalum.
viridiflorum *(Paul G. Wilson) Paul G. Wilson in Nuytsia*, 12(2): 277 (1998): Phebalium viridiflorum.

LEIOSPORA (Crucif.).
sect. Pseudoleiospora *R.V. Kamelin, Mater. Istorii Fl. Azii (Altaisk. Gorn. Strana)*: 21 (1998).
bellidifolia (Danguy) Botsch. & Pachom.
subsp. pamirica *(Botsch. & Vved.) R.V. Kamelin, Mater. Istorii Fl. Azii (Altaisk. Gorn. Strana)*: 21 (1998): Parrya pamirica.
maidantalica *(Popov & P.A. Baranov) R.V. Kamelin, Mater. Istorii Fl. Azii (Altaisk. Gorn. Strana)*: 22 (1998): Parrya maidantalica.

LEIPOLDTIA (Aizoac.).
*subgen.*Cephalophylloides *H.E.K. Hartmann & S. Rust in Verh. Naturwiss. Ver. Hamburg*, 34: 348 (1994).
subgen. Spinosae *H.E.K. Hartmann & D. Stüber in Verh. Naturwiss. Ver. Hamburg*, 34: 370 (1994).
lunata *H.E.K. Hartmann & S. Rust in Verh. Naturwiss. Ver. Hamburg*, 34: 342 (1994)—Cape Province.
weigangiana (Dinter) Dinter & Schwantes
subsp. grandifolia *(L. Bolus) H.E.K. Hartmann & S. Rust in Verh. Naturwiss. Ver. Hamburg*, 34: 346 (1994): L. grandifolia.
subsp. littlewoodii *(L. Bolus) H.E.K. Hartmann & S. Rust in Verh. Naturwiss. Ver. Hamburg*, 34: 347 (1994): L. littlewoodii.

LEMBOGLOSSUM (Orchidac.).
× humeanum *(Rchb. f.) Bockemühl ex R. Diekmann in J. Orchideenfreund*, 7(2): 105 (2000): Odontoglossum × humeanum. (L. cordatum × rossii).

LEMNA (Lemnac.).
yungensis *E. Landolt in Bull. Geobot. Inst. ETH*, 64: 18 (1998)—Bolivia.

LENS (Legumin.).
culinaris Medik.
subsp. odemensis *(Ladiz.) M.E. Ferguson, N. Maxted, M.W. van Slageren & L.D. Robertson in Bot. J. Linn. Soc.*, 133(1): 57 (2000): L. odemensis.
subsp. tomentosus *(Ladiz.) M.E. Ferguson, N. Maxted, M.W. van Slageren & L.D. Robertson in Bot. J. Linn. Soc.*, 133(1): 57 (2000): L. tomentosus.
tomentosus *G. Ladizinsky in Bot. J. Linn. Soc.*, 123(3): 260 (1997)—Turkey.

LEONARDOXA (Legumin.).
africana (Baill.) Aubrév.
subsp. gracilicaulis *D.B. McKey in Adansonia*, 22(1): 102 (2000)—Cameroon, Equatorial Guinea, Congo.
subsp. letouzeyi *D.B. McKey in Adansonia*, 22(1): 100 (2000)—Cameroon.
subsp. rumpiensis *D.B. McKey in Adansonia*, 22(1): 101 (2000)—Cameroon.

LEONTOCHIR (Alstroemeriac.).
ovallei Phil.
var. luteus *M. Muñoz Schick in Gayana, Bot.*, 57(1): 58 (2000), without type—Chile North (Atacama).

LEONTODON (Compos.).
muelleri (Sch. Bip.) Ball
subsp. austromaroccanus *(Maire) A. Izuzquiza in Lagascalia*, 20(2): 309 (1998): L. hispidulus var. austromaroccanus.
taraxacoides (Vill.) Mérat
subsp. mesorrhynchus *(Maire) B. Valdés in Lagascalia*, 18(2): 308 (1996), as 'mesorrhyncha': L. nudicaule subsp. mesorrhynchus.
subsp. mesorrhynchus *(Maire) M. Ibn Tattou in Bocconea*, 8: 41 (1998), homonym: L. saxatilis subsp. mesorrhynchus.

LEONTOPODIUM (Compos.).
jacotianum P. Beauv.
var. haastioides *(Hand.-Mazz.) R.C. Srivastava in Novon*, 8(2): 203 (1998): L. haastioides.

LEONTOPODIUM:-
japonicum Miq.
f. happoense *Hid. Takah. ex T. Shimizu, Fl. Nagano Pref.*: 1510 (1997)—Japan.
montisganeshii *S. Akiyama in Bull. Nation. Sci. Mus., B (Tokyo)*, 25(1): 1 (1999)—Nepal.
ochroleucum Beauverd
subsp. campestre *(Ledeb.) V.M. Khanminchun in Fl. Sibir. (Asterac.)*, 13: 48 (1997 publ. 1998): L. alpinum var. campestre.
subsp. conglobatum *(Turcz.) V.M. Khanminchun in Fl. Sibir. (Asterac.)*, 13: 48 (1997 publ. 1998): L. sibiricum var. conglobatum.

LEONURUS (Labiat.).
cardiaca L.
subsp. intermedius *(Holub) N.N. Tzvelev in Novosti Sist. Vyssh. Rast.*, 32: 185 (2000): L. intermedius.

LEPANTHES (Orchidac.).
subsect. Bilabiatae *C.A. Luer, Ic. Pleurothallid. XIV (Monogr. Syst. Bot. Missouri Bot. Gard., 61(3))*: 7 (1996).
subsect. Breves *(Luer) C.A. Luer, Ic. Pleurothallid. XIV (Monogr. Syst. Bot. Missouri Bot. Gard., 61(3))*: 7 (1996): L. ser. Breves.
ser. Mucronatae *C.A. Luer, Ic. Pleurothallid. XIV (Monogr. Syst. Bot. Missouri Bot. Gard., 61(3))*: 7 (1996).
abitaguae *C.A. Luer & L. Jost in C.A. Luer, Ic. Pleurothallid. XVIII (Monogr. Syst. Bot. Missouri Bot.* Gard., 76): 139 (1999)—Ecuador.
absens *C.A. Luer & A.C. Hirtz in C.A. Luer, Ic. Pleurothallid. XIV (Monogr. Syst. Bot. Missouri Bot. Gard.*, 61(3)): 28 (1996)—Ecuador.
aciculifolia *C.A. Luer in Orquideologia*, 20(2): 194 (1996)—Colombia.
acuminata Schltr.
subsp. ernesti *G.A. Salazar Chávez & M.A. Soto Arenas in Orquídea (México)*, 14: 51 (1996)—Mexico Southeast (Chiapas).
adelphe *C.A. Luer & A.C. Hirtz in C.A. Luer, Ic. Pleurothallid. XIV (Monogr. Syst. Bot. Missouri Bot. Gard.*, 61(3)): 31 (1996)—Ecuador.
aithalos *G. Carnevali & I. Ramírez in G.A. Romero & G. Carnevali, Orchids of Venezuela, ed. 2*: 1137 (2000)—Venezuela.
alopex *C.A. Luer & A.C. Hirtz in C.A. Luer, Ic. Pleurothallid. XIV (Monogr. Syst. Bot. Missouri Bot. Gard.*, 61(3)): 33 (1996)—Ecuador.
amphioxa *C.A. Luer & A.C. Hirtz in C.A. Luer, Ic. Pleurothallid. XIV (Monogr. Syst. Bot. Missouri Bot. Gard.*, 61(3)): 35 (1996)—Ecuador.
anatina *C.A. Luer & R. Escobar R. in Orquideologia*, 20(3): 281 (1997)—Colombia.
angulata *C.A. Luer & A.C. Hirtz in C.A. Luer, Ic. Pleurothallid. XIV (Monogr. Syst. Bot. Missouri Bot. Gard.*, 61(3)): 36 (1996)—Ecuador.
aprina *C.A. Luer & L. Jost in C.A. Luer, Ic. Pleurothallid. XVIII (Monogr. Syst. Bot. Missouri Bot.* Gard., 76): 139 (1999)—Ecuador.
arethusa *C.A. Luer & R. Escobar R. in Orquideologia*, 21(2): 135 (1999)—Colombia.
ariasiana *C.A. Luer & L. Jost in C.A. Luer, Ic. Pleurothallid. XVII (Monogr. Syst. Bot. Missouri Bot.* Gard., 72): 104 (1998)—Ecuador.
athena *C.A. Luer, Ic. Pleurothallid. XIV (Monogr. Syst. Bot. Missouri Bot. Gard., 61(3))*: 38 (1996)—Ecuador, Colombia.
attenuata *G.A. Salazar Chávez, M.A. Soto Arenas & O. Suárez in Orquídea (México)*, 14: 65 (1996)—Mexico Southwest (Oaxaca), Mexico Gulf (Veracruz).
auspicata *C.A. Luer & R. Escobar R. in Orquideologia*, 21(2): 138 (1999)—Colombia.
barbelifera *C.A. Luer & A.C. Hirtz in C.A. Luer, Ic. Pleurothallid. XIV (Monogr. Syst. Bot. Missouri Bot. Gard.*, 61(3)): 41 (1996)—Ecuador.
barbigera *C.A. Luer & L. Jost in C.A. Luer, Ic. Pleurothallid. XVIII (Monogr. Syst. Bot. Missouri Bot.* Gard., 76): 140 (1999)—Ecuador.
beatriziae *C.A. Luer & R. Escobar R. in Orquideologia*, 20(3): 283 (1997), as 'beatrizae'—Colombia.
bifida *C.A. Luer & Behar in Lindleyana*, 12(1): 34 (1997)—Guatemala.
bifurcata *C.A. Luer & R. Escobar R. in Orquideologia*, 21(2): 140 (1999)—Colombia.

LEPANTHES:-
bipinnatula *C.A. Luer & R. Escobar R. in Orquideologia*, 20(3): 284 (1997)—Colombia.
brachystele *G.A. Salazar Chávez & M.A. Soto Arenas in Orquídea (México)*, 14: 71 (1996)—Mexico Southwest (Guerrero, Oaxaca).
breedlovei *G.A. Salazar Chávez & M.A. Soto Arenas in Orquídea (México)*, 14: 75 (1996)—Mexico Southeast (Chiapas), Mexico Southwest (Oaxaca).
cacique-tone *C.A. Luer & R. Escobar R. in Orquideologia*, 20(3): 286 (1997)—Colombia.
calimae *P. Ortiz V. in Orquideologia*, 21(1): 69 (1998)—Colombia.
calliope *C.A. Luer & A.C. Hirtz in C.A. Luer, Ic. Pleurothallid. XIV (Monogr. Syst. Bot. Missouri Bot. Gard.*, 61(3)): 46 (1996)—Ecuador.
calopetala *G.A. Salazar Chávez & M.A. Soto Arenas in Orquídea (México)*, 14: 77 (1996)—Mexico Southwest (Oaxaca), Mexico Gulf (Veracruz).
calypso *C.A. Luer & A.C. Hirtz in C.A. Luer, Ic. Pleurothallid. XIV (Monogr. Syst. Bot. Missouri Bot. Gard.*, 61(3)): 49 (1996)—Ecuador.
camposii *G.A. Salazar Chávez & M.A. Soto Arenas in Orquídea (México)*, 14: 79 (1996)—Mexico Southwest (Oaxaca, Guerrero).
camptica *C.A. Luer & A.C. Hirtz in C.A. Luer, Ic. Pleurothallid. XIV (Monogr. Syst. Bot. Missouri Bot. Gard.*, 61(3)): 50 (1996)—Ecuador.
canaliculata *C.A. Luer & R. Escobar R. in Orquideologia*, 20(3): 287 (1997)—Colombia.
catlingii *G.A. Salazar Chávez, M.A. Soto Arenas & O. Suárez in Orquídea (México)*, 14: 81 (1996)—Mexico Southwest (Oaxaca).
cauda-avis *C.A. Luer, Ic. Pleurothallid. XIV (Monogr. Syst. Bot. Missouri Bot. Gard., 61(3))*: 53 (1996), holotype not indicated; *et Ic. Pleurothallid. XV (Monogr. Syst. Bot. Missouri Bot. Gard., 64)*: 122 (1997)—Ecuador.
caudigera *C.A. Luer & Hirtz in C.A. Luer, Ic. Pleurothallid. XVI (Monogr. Syst. Bot. Missouri Bot. Gard.*, 65): 101 (1998)—Ecuador.
chiangii *G.A. Salazar Chávez, M.A. Soto Arenas & O. Suárez in Orquídea (México)*, 14: 85 (1996)—Mexico Southwest (Oaxaca).
ciliicampa *C.A. Luer & A.C. Hirtz in C.A. Luer, Ic. Pleurothallid. XIV (Monogr. Syst. Bot. Missouri Bot. Gard.*, 61(3)): 57 (1996)—Ecuador.
clarkii *C.A. Luer, Ic. Pleurothallid. XIV (Monogr. Syst. Bot. Missouri Bot. Gard., 61(3))*: 58 (1996)—Ecuador.
cloesii *C.A. Luer, Ic. Pleurothallid. XVIII (Monogr. Syst. Bot. Missouri Bot. Gard., 76)*: 165 (1999)—Peru.
comadresina *C.A. Luer in Lindleyana*, 13(3): 138 (1998)—Cuba.
conchilabia *C.A. Luer & Hirtz in C.A. Luer, Ic. Pleurothallid. XVI (Monogr. Syst. Bot. Missouri Bot. Gard.*, 65): 102 (1998)—Ecuador.
confusoides *C.A. Luer, Ic. Pleurothallid. XVIII (Monogr. Syst. Bot. Missouri Bot. Gard., 76)*: 165 (1999)—Panama.
corrugata *C.A. Luer & Dalström in C.A. Luer, Ic. Pleurothallid. XIV (Monogr. Syst. Bot. Missouri Bot. Gard.*, 61(3)): 63 (1996), contrary to Art. 37.4 ICBN (1994)—Ecuador.
cotylisca *C.A. Luer & A.C. Hirtz in C.A. Luer, Ic. Pleurothallid. XIV (Monogr. Syst. Bot. Missouri Bot. Gard.*, 61(3)): 64 (1996)—Ecuador.
cremasta *C.A. Luer & A.C. Hirtz in C.A. Luer, Ic. Pleurothallid. XIV (Monogr. Syst. Bot. Missouri Bot. Gard.*, 61(3)): 65 (1996)—Ecuador.
cremersii *C.A. Luer in Adansonia*, 21(1): 63 (1999)—French Guiana.
cryptostele *G.A. Salazar Chávez & M.A. Soto Arenas in Orquídea (México)*, 14: 87 (1996)—Mexico Southwest (Guerrero).
cunicularis *C.A. Luer & R. Escobar R. in Orquideologia*, 20(3): 289 (1997)—Colombia.
curiosa *C.A. Luer, Ic. Pleurothallid. XX (Monogr. Syst. Bot. Missouri Bot. Gard., 79)*: 117 (2000)—Ecuador.
cyrtostele *C.A. Luer & Hirtz in C.A. Luer, Ic. Pleurothallid. XVI (Monogr. Syst. Bot. Missouri Bot. Gard.*, 65): 102 (1998)—Ecuador.
dactylina *C.A. Luer, Ic. Pleurothallid. XVIII (Monogr. Syst. Bot. Missouri Bot. Gard., 76)*: 166 (1999)—Peru.

LEPANTHES:-
darioi *C.A. Luer & R. Escobar R.in Orquideologia,* 20(3): 291 (1997)—Colombia.
davidsei *C.A. Luer in Lindleyana,* 11(2): 56 (1996)—Costa Rica.
declivis *C.A. Luer & R. Escobar R. in Orquideologia,* 21(2): 142 (1999)—Colombia.
dewildei *C.A. Luer & R. Escobar R. in Orquideologia,* 20(3): 292 (1997)—Colombia.
diaziae *C.A. Luer in Lindleyana,* 13(3): 138 (1998)—Cuba.
dichroma *C.A. Luer in Lindleyana,* 11(2): 59 (1996)—Costa Rica.
didyma *C.A. Luer & A.C. Hirtz in C.A. Luer, Ic. Pleurothallid. XIV (Monogr. Syst. Bot. Missouri Bot. Gard.,* 61(3)): 71 (1996)—Ecuador.
dolabrata *C.A. Luer & R. Escobar R. in Orquideologia,* 20(3): 294 (1997)—Colombia.
dolabriformis *C.A. Luer in Lindleyana,* 11(2): 59 (1996)—Costa Rica.
elytrifera *C.A. Luer & L. Jost in C.A. Luer, Ic. Pleurothallid. XVIII (Monogr. Syst. Bot. Missouri Bot.* Gard., 76): 140 (1999)—Ecuador.
empticia *C.A. Luer & Béhar in C.A. Luer, Ic. Pleurothallid. XVIII (Monogr. Syst. Bot. Missouri Bot.* Gard., 76): 166 (1999)—Costa Rica.
eriocampa *C.A. Luer & A.C. Hirtz in C.A. Luer, Ic. Pleurothallid. XIV (Monogr. Syst Bot. Missouri Bot. Gard.,* 61(3)): 79 (1996); *see also C.A. Luer, Ic. Pleurothallid. XV (Monogr. Syst. Bot. Missouri Bot.* Gard., 64): 122 (1997)—Ecuador.
erythroxantha *G.A. Salazar Chávez & M.A. Soto Arenas in Orquídea (México),* 14: 95 (1996)—Mexico Southwest (Oaxaca).
evansiae *C.A. Luer & A.C. Hirtz in C.A. Luer, Ic. Pleurothallid. XIV (Monogr. Syst. Bot. Missouri Bot. Gard.,* 61(3)): 81 (1996)—Ecuador.
expansa *C.A. Luer & A.C. Hirtz in C.A. Luer, Ic. Pleurothallid. XIV (Monogr. Syst. Bot. Missouri Bot. Gard.,* 61(3)): 82 (1996)—Ecuador.
exserta *C.A. Luer & Hirtz in C.A. Luer, Ic. Pleurothallid. XVI (Monogr. Syst. Bot. Missouri Bot. Gard.,* 65): 103 (1998)—Ecuador.
flaccida *C.A. Luer & A.C. Hirtz in C.A. Luer, Ic. Pleurothallid. XIV (Monogr. Syst. Bot. Missouri Bot. Gard.,* 61(3)): 84 (1996)—Ecuador.
gabriellae *G.A. Salazar Chávez & M.A. Soto Arenas in Orquídea (México),* 14: 97 (1996)—Mexico Southwest (Oaxaca).
gaileana *C.A. Luer & A.C. Hirtz in C.A. Luer, Ic. Pleurothallid. XIV (Monogr. Syst. Bot. Missouri Bot. Gard.,* 61(3)): 87 (1996)—Ecuador.
galeottiana *G.A. Salazar Chávez & M.A. Soto Arenas in Orquídea (México),* 14: 101 (1996)—Mexico Southwest (Oaxaca).
gelata *C.A. Luer & R. Escobar R. in Orquideologia,* 20(3): 295 (1997)—Colombia.
generi *C.A. Luer & Hirtz in C.A. Luer, Ic. Pleurothallid. XVI (Monogr. Syst. Bot. Missouri Bot. Gard.,* 65): 103 (1998)—Ecuador.
glabella *C.A. Luer & A.C. Hirtz in C.A. Luer, Ic. Pleurothallid. XIV (Monogr. Syst. Bot. Missouri Bot. Gard.,* 61(3)): 90 (1996)—Ecuador.
glomerulosa *C.A. Luer & A.C. Hirtz in C.A. Luer, Ic. Pleurothallid. XIV (Monogr. Syst. Bot. Missouri Bot. Gard.,* 61(3)): 90 (1996)—Ecuador.
gloris *C.A. Luer & Hirtz in C.A. Luer, Ic. Pleurothallid. XVI (Monogr. Syst. Bot. Missouri Bot. Gard.,* 65): 104 (1998)—Ecuador.
glossites *C.A. Luer, Ic. Pleurothallid. XVIII (Monogr. Syst. Bot. Missouri Bot Gard., 76):* 167 (1999)—Peru.
gnoma *C.A. Luer & A.C. Hirtz in C.A. Luer, Ic. Pleurothallid. XIV (Monogr. Syst. Bot. Missouri Bot. Gard.,* 61(3)): 91 (1996), contrary to Art. 37.4 ICBN (1994)—Ecuador.
greenwoodii *G.A. Salazar Chávez & M.A. Soto Arenas in Orquídea (México),* 14: 103 (1996)—Mexico Southwest (Oaxaca).
guaduasensis *C.A. Luer & R. Escobar R. in Orquideologia,* 21(2): 145 (1999)—Colombia.
guerrerensis *G.A. Salazar Chávez & M.A. Soto Arenas in Orquídea (México),* 14: 111 (1996)—Mexico Southwest (Guerrero, Oaxaca).
gustavoi *C.A. Luer & R. Escobar R. in Orquideologia,* 20(3): 298 (1997)—Colombia.
hamiltonii *C.A. Luer, Ic. Pleurothallid. XVIII (Monogr. Syst. Bot. Missouri Bot. Gard., 76):* 167 (1999)—Peru.

LEPANTHES:-
herzogii *C.A. Luer in Rev. Soc. Boliv. Bot.,* 2(1): 23 (1998)—Bolivia.
hispidosa *C.A. Luer & L. Jost in C.A. Luer, Ic. Pleurothallid. XVIII (Monogr. Syst. Bot. Missouri Bot.* Gard., 76): 141 (1999)—Ecuador.
horribilis *C.A. Luer & A.C. Hirtz in C.A. Luer, Ic. Pleurothallid. XIV (Monogr. Syst. Bot. Missouri Bot. Gard.,* 61(3)): 97 (1996)—Ecuador.
hydrae *C.A. Luer & L. Jost in C.A. Luer, Ic. Pleurothallid. XVIII (Monogr. Syst. Bot. Missouri Bot.* Gard., 76): 141 (1999)—Ecuador.
imbricans *C.A. Luer & R. Escobar R.in Orquideologia,* 21(2): 147 (1999)—Colombia.
imitator *C.A. Luer & A.C. Hirtz in C.A. Luer, Ic. Pleurothallid. XIV (Monogr. Syst. Bot. Missouri Bot. Gard.,* 61(3)): 101 (1996)—Ecuador.
impotens *C.A. Luer & R. Escobar R. in Orquideologia,* 20(3): 299 (1997)—Colombia.
incredibilis *C.A. Luer & R. Vásquez Ch. in Rev. Soc. Boliv. Bot.,* 2(1): 19 (1998)—Bolivia.
ingridiana *C.A. Luer, Ic. Pleurothallid. XX (Monogr. Syst. Bot. Missouri Bot. Gard., 79):* 118 (2000)—Ecuador.
isochila *C.A. Luer in Orquideologia,* 20(2): 196 (1996)—Colombia.
jamboeensis *C.A. Luer & A.C. Hirtz in C.A. Luer, Ic. Pleurothallid. XIV (Monogr. Syst. Bot. Missouri Bot. Gard.,* 61(3)): 105 (1996)—Ecuador.
jayandella *C.A. Luer, Ic. Pleurothallid. XIV (Monogr. Syst. Bot. Missouri Bot. Gard., 61(3)):* 107 (1996)—Ecuador.
jostii *C.A. Luer, Ic. Pleurothallid. XVIII (Monogr. Syst. Bot. Missouri Bot. Gard., 76):* 142 (1999)—Ecuador.
laevis *C.A. Luer, Ic. Pleurothallid. XVIII (Monogr. Syst. Bot. Missouri Bot. Gard., 76):* 168 (1999)—Panama.
lappacea *C.A. Luer in Lindleyana,* 11(2): 61 (1996)—Costa Rica.
lingulosa *C.A. Luer & R. Escobar R. in Orquideologia,* 21(2): 149 (1999)—Colombia.
llamachoi *C.A. Luer in Lindleyana,* 13(3): 141 (1998)—Cuba.
longiacuminata *C.A. Luer & Hirtz in C.A. Luer, Ic. Pleurothallid. XVI (Monogr. Syst. Bot. Missouri Bot. Gard.,* 65): 104 (1998)—Ecuador.
lymphosa *C.A. Luer & J. Hermans in Orchid Rev.,* 105(1216): 206 (1997)—Venezuela.
maccolmiana *C.A. Luer, Ic. Pleurothallid. XIV (Monogr. Syst. Bot. Missouri Bot. Gard., 61(3)):* 112 (1996), contrary to Art. 37.4 ICBN (1994); *et Ic. Pleurothallid. XV (Monogr. Syst. Bot. Missouri Bot. Gard., 64):* 122 (1997)—Ecuador.
machorroi *G.A. Salazar Chávez & M.A. Soto Arenas in Orquídea (México),* 14: 121 (1996)—Mexico Southwest (Oaxaca).
macrostylis *C.A. Luer & R. Escobar R. in Orquideologia,* 21(2): 151 (1999)—Colombia.
macrotica *C.A. Luer & Dalström in C.A. Luer, Ic. Pleurothallid. XIV (Monogr. Syst. Bot. Missouri Bot. Gard.,* 61(4)): 3 (1996)—Ecuador.
mammillata *C.A. Luer, Ic. Pleurothallid. XIV (Monogr. Syst. Bot. Missouri Bot. Gard., 61(3)):* 113 (1996)—Ecuador.
mariae *G.A. Salazar Chávez, M.A. Soto Arenas & O. Suárez in Orquídea (México),* 14: 123 (1996)—Mexico Southwest (Oaxaca).
marshana *C.A. Luer & L. Jost in C.A. Luer, Ic. Pleurothallid. XVIII (Monogr. Syst. Bot. Missouri Bot.* Gard., 76): 142 (1999)—Ecuador.
martinezii *G.A. Salazar Chávez & M.A. Soto Arenas in Orquídea (México),* 14: 125 (1996)—Mexico Southeast (Chiapas), Mexico Southwest (Oaxaca).
matudana *G.A. Salazar Chávez & M.A. Soto Arenas in Orquídea (México),* 14: 129 (1996)—Mexico Southeast (Chiapas).
maxillaris *C.A. Luer & A.C. Hirtz in C.A. Luer, Ic. Pleurothallid. XIV (Monogr. Syst. Bot. Missouri Bot. Gard.,* 61(3)): 115 (1996)—Ecuador.
maxima *G.A. Salazar Chávez & M.A. Soto Arenas in Orquídea (México),* 14: 131 (1996)—Mexico Southeast (Chiapas).
melpomene *C.A. Luer & A.C. Hirtz in C.A. Luer, Ic. Pleurothallid. XIV (Monogr. Syst. Bot. Missouri Bot. Gard.,* 61(3)): 117 (1996)—Ecuador.

LEPANTHES:-
mendozae *C.A. Luer & D'Alessandro in C.A. Luer, Ic. Pleurothallid. XVII (Monogr. Syst. Bot. Missouri Bot.* Gard., 72): 104 (1998)—Ecuador.
meniscophora *C.A. Luer & A.C. Hirtz in C.A. Luer, Ic. Pleurothallid. XIV (Monogr. Syst. Bot. Missouri Bot. Gard.,* 61(3)): 118 (1996); *see also C.A. Luer, Ic. Pleurothallid. XV (Monogr. Syst. Bot. Missouri Bot.* Gard., 64): 122 (1997)—Ecuador.
micronyx *C.A. Luer & R. Escobar R.in Orquideologia,* 20(3): 302 (1997)—Colombia.
minima *G.A. Salazar Chávez & M.A. Soto Arenas in Orquídea (México),* 14: 133 (1996)—Mexico Southeast (Chiapas).
minima-mundana *C.A. Luer in Lindleyana,* 13(3): 142 (1998)—Cuba.
mirador *C.A. Luer & A.C. Hirtz in C.A. Luer, Ic. Pleurothallid. XIV (Monogr. Syst. Bot. Missouri Bot. Gard.,* 61(3)): 119 (1996)—Ecuador.
mixe *G.A. Salazar Chávez & M.A. Soto Arenas in Orquídea (México),* 14: 135 (1996)—Mexico Southwest (Oaxaca).
mononeura *C.A. Luer & J. Hermans in Orchid Rev.,* 105(1216): 206 (1997)—Venezuela.
montis-rotundi *P. Ortiz V. in Orquideologia,* 20(3): 318 (1997)—Colombia.
mooreana *C.A. Luer & L. Jost in C.A. Luer, Ic. Pleurothallid. XVII (Monogr. Syst. Bot. Missouri Bot.* Gard., 72): 106 (1998)—Ecuador.
morleyi *C.A. Luer & Dalström in C.A. Luer, Ic. Pleurothallid. XIV (Monogr. Syst. Bot. Missouri Bot. Gard.,* 61(3)): 122 (1996)—Ecuador.
motozintlensis *G.A. Salazar Chávez & M.A. Soto Arenas in Orquídea (México),* 14: 143 (1996)—Mexico Southeast (Chiapas).
mulleriana *C.A. Luer, Ic. Pleurothallid. XVII (Monogr. Syst. Bot. Missouri Bot. Gard., 72):* 107 (1998)—Ecuador.
nagelii *G.A. Salazar Chávez & M.A. Soto Arenas in Orquídea (México),* 14: 145 (1996)—Mexico Southwest (Guerrero, Michoacan, Oaxaca), Mexico Central (México State, Morelos).
narcissus *C.A. Luer & A.C. Hirtz in C.A. Luer, Ic. Pleurothallid. XIV (Monogr. Syst. Bot. Missouri Bot. Gard.,* 61(3)): 125 (1996)—Ecuador.
nicolasii *C.A. Luer & R. Escobar R. in Orquideologia,* 20(3): 304 (1997)—Colombia.
noelii *C.A. Luer & Behar in Orquideologia,* 21(3): 318 (2000)—Guatemala.
oaxacana *G.A. Salazar Chávez, M.A. Soto Arenas & O. Suárez in Orquídea (México),* 14: 153 (1996)—Mexico Southwest (Oaxaca).
octopus *C.A. Luer & R. Escobar R. in Orquideologia,* 20(3): 305 (1997)—Colombia.
odontocera *C.A. Luer & A.C. Hirtz in C.A. Luer, Ic. Pleurothallid. XIV (Monogr. Syst. Bot. Missouri Bot. Gard.,* 61(3)): 129 (1996)—Ecuador.
omnifera *C.A. Luer & Hirtz in C.A. Luer, Ic. Pleurothallid. XVI (Monogr. Syst. Bot. Missouri Bot. Gard.,* 65): 105 (1998)—Ecuador.
orbella *C.A. Luer in Lindleyana,* 11(2): 64 (1996)—Costa Rica.
ordonezii *C.A. Luer & Behar in Lindleyana,* 12(1): 34 (1997)—Guatemala.
ortegae *C.A. Luer & A.C. Hirtz in C.A. Luer, Ic. Pleurothallid. XIV (Monogr. Syst. Bot. Missouri Bot. Gard.,* 61(3)): 132 (1996)—Ecuador.
pabloi *C.A. Luer & Behar in Orquideologia,* 21(3): 321 (2000)—Guatemala.
palpebralis *C.A. Luer in Lindleyana,* 13(3): 143 (1998)—Cuba.
pan *C.A. Luer & Dalström in C.A. Luer, Ic. Pleurothallid. XIV (Monogr. Syst. Bot. Missouri Bot. Gard.,* 61(3)): 133 (1996)—Ecuador.
papilionacea *G.A. Salazar Chávez, M.A. Soto Arenas & O. Suárez in Orquídea (México),* 14: 161 (1996)—Mexico Southwest (Oaxaca).
pelyx *C.A. Luer & A.C. Hirtz in C.A. Luer, Ic. Pleurothallid. XIV (Monogr. Syst. Bot. Missouri Bot. Gard.,* 61(3)): 136 (1996)—Ecuador.
pendula *C.A. Luer & L. Jost in C.A. Luer, Ic. Pleurothallid. XX (Monogr. Syst. Bot. Missouri Bot. Gard.,* 79): 118 (2000)—Ecuador.
perdita *C.A. Luer & A.C. Hirtz in C.A. Luer, Ic. Pleurothallid. XIV (Monogr. Syst. Bot. Missouri Bot. Gard.,* 61(3)): 138 (1996)—Locality not stated.

LEPANTHES:-

phalloides *C.A. Luer & R. Escobar R. in Orquideologia*, 21(2): 154 (1999)—Colombia.
pilosa *C.A. Luer & R. Vásquez Ch. in Rev. Soc. Boliv. Bot.*, 2(1): 21 (1998)—Bolivia.
pinnatula *C.A. Luer & R. Escobar R. in Orquideologia*, 20(3): 307 (1997)—Colombia.
popayanensis *C.A. Luer & R. Escobar R. in Orquideologia*, 21(2): 156 (1999)—Colombia.
porracea *C.A. Luer & R. Escobar R. in Orquideologia*, 20(3): 309 (1997)—Colombia.
portillae *C.A. Luer, Ic. Pleurothallid. XVIII (Monogr. Syst. Bot. Missouri Bot. Gard., 76)*: 143 (1999)—Ecuador.
privigna *C.A. Luer & L. Jost in C.A. Luer, Ic. Pleurothallid. XVIII (Monogr. Syst. Bot. Missouri Bot.* Gard., 76): 143 (1999)—Ecuador.
proboscidis *C.A. Luer & A.C. Hirtz in C.A. Luer, Ic. Pleurothallid. XIV (Monogr. Syst. Bot. Missouri Bot. Gard.*, 61(3)): 141 (1996), contrary to Art. 37.4 ICBN (1994); *et in C.A. Luer, Ic. Pleurothallid. XV (Monogr. Syst. Bot. Missouri Bot.* Gard., 64): 122 (1997)—Ecuador.
protuberans *C.A. Luer & P. Jesup in C.A. Luer, Ic. Pleurothallid. XIV (Monogr. Syst. Bot. Missouri Bot. Gard.*, 61(3)): 142 (1996)—Ecuador.
psomion *C.A. Luer & A.C. Hirtz in C.A. Luer, Ic. Pleurothallid. XIV (Monogr. Syst. Bot. Missouri Bot. Gard.*, 61(3)): 143 (1996)—Ecuador.
retusa *C.A. Luer & J. Hermans in Orchid Rev.*, 105(1216): 209 (1997)—Venezuela.
ricina *C.A. Luer & Dalström in C.A. Luer, Ic. Pleurothallid. XIV (Monogr. Syst. Bot. Missouri Bot. Gard.*, 61(3)): 148 (1996)—Ecuador.
ridicula *C.A. Luer, Ic. Pleurothallid. XVIII (Monogr. Syst. Bot. Missouri Bot. Gard., 76)*: 168 (1999)—Panama.
rodrigoi *C.A. Luer in Orquideologia*, 20(2): 198 (1996)—Colombia.
rosoria *C.A. Luer & Hirtz in C.A. Luer, Ic. Pleurothallid. XX (Monogr. Syst. Bot. Missouri Bot. Gard.*, 79): 119 (2000)—Ecuador.
rudicula *C.A. Luer & A.C. Hirtz in C.A. Luer, Ic. Pleurothallid. XIV (Monogr. Syst. Bot. Missouri Bot. Gard.*, 61(3)): 149 (1996)—Ecuador.
ruthiana *C.A. Luer & L. Jost in C.A. Luer, Ic. Pleurothallid. XVIII (Monogr. Syst. Bot. Missouri Bot.* Gard., 76): 147 (1999)—Ecuador.
saltatrix *C.A. Luer & A.C. Hirtz in C.A. Luer, Ic. Pleurothallid. XIV (Monogr. Syst. Bot. Missouri Bot. Gard.*, 61(3)): 150 (1996)—Ecuador.
scapha *C.A. Luer & A.C. Hirtz in C.A. Luer, Ic. Pleurothallid. XIV (Monogr. Syst. Bot. Missouri Bot. Gard.*, 61(3)): 152 (1996)—Ecuador.
schultesii *G.A. Salazar Chávez & M.A. Soto Arenas in Orquídea (México)*, 14: 183 (1996)—Mexico Southwest (Oaxaca).
scopulifera *C.A. Luer & R. Escobar R.in Orquideologia*, 20(3): 310 (1997)—Colombia.
scrotifera *C.A. Luer & A.C. Hirtz in C.A. Luer, Ic. Pleurothallid. XIV (Monogr. Syst. Bot. Missouri Bot. Gard.*, 61(3)): 154 (1996)—Ecuador.
semilaminata *C.A. Luer & A.C. Hirtz in C.A. Luer, Ic. Pleurothallid. XIV (Monogr. Syst. Bot. Missouri Bot. Gard.*, 61(3)): 154 (1996)—Ecuador.
serialina *C.A. Luer & L. Jost in C.A. Luer, Ic. Pleurothallid. XVII (Monogr. Syst. Bot. Missouri Bot.* Gard., 72): 107 (1998)—Ecuador.
silenus *C.A. Luer & A.C. Hirtz in C.A. Luer, Ic. Pleurothallid. XIV (Monogr. Syst. Bot. Missouri Bot. Gard.*, 61(3)): 156 (1996)—Ecuador.
similis *C.A. Luer in Lindleyana*, 11(2): 64 (1996)—Costa Rica.
sousae *G.A. Salazar Chávez & M.A. Soto Arenas in Orquídea (México)*, 14: 189 (1996)—Mexico Southwest (Oaxaca).
staatsiana *C.A. Luer & L. Jost in C.A. Luer, Ic. Pleurothallid. XVIII (Monogr. Syst. Bot. Missouri Bot.* Gard., 76): 147 (1999)—Ecuador.
stelidipetala *C.A. Luer, Ic. Pleurothallid. XIV (Monogr. Syst. Bot. Missouri Bot. Gard., 61(3))*: 160 (1996)—Ecuador.
stenosepala *C.A. Luer & Behar in Lindleyana*, 12(1): 36 (1997)—Guatemala, El Salvador.
suarezii *G.A. Salazar Chávez & M.A. Soto Arenas in Orquídea (México)*, 14: 195 (1996)—Mexico Southwest (Oaxaca), Mexico Northeast (Guanajuato).

LEPANTHES:-

suavium *C.A. Luer & A.C. Hirtz in C.A. Luer, Ic. Pleurothallid. XIV (Monogr. Syst. Bot. Missouri Bot. Gard.*, 61(3)): 162 (1996)—Ecuador.
sulcata *C.A. Luer & A.C. Hirtz in C.A. Luer, Ic. Pleurothallid. XIV (Monogr. Syst. Bot. Missouri Bot. Gard.*, 61(3)): 162 (1996)—Ecuador.
sybax *C.A. Luer & A.C. Hirtz in C.A. Luer, Ic. Pleurothallid. XIV (Monogr. Syst. Bot. Missouri Bot. Gard.*, 61(3)): 164 (1996)—Ecuador.
teaguei *C.A. Luer, Ic. Pleurothallid. XIV (Monogr. Syst. Bot. Missouri Bot. Gard., 61(3))*: 165 (1996)—Ecuador.
telipogoniflora *A. Schuiteman & A. de Wilde in Orquideologia*, 20(1): 28 (1996)—Colombia.
tentaculata *C.A. Luer & A.C. Hirtz in C.A. Luer, Ic. Pleurothallid. XIV (Monogr. Syst. Bot. Missouri Bot. Gard.*, 61(3)): 167 (1996)—Ecuador.
terpsichore *C.A. Luer & A.C. Hirtz in C.A. Luer, Ic. Pleurothallid. XIV (Monogr. Syst. Bot. Missouri Bot. Gard.*, 61(3)): 168 (1996)—Ecuador.
tetrachaeta *C.A. Luer & L. Jost in C.A. Luer, Ic. Pleurothallid. XX (Monogr. Syst. Bot. Missouri Bot. Gard.*, 79): 119 (2000)—Ecuador.
thalia *C.A. Luer & A.C. Hirtz in C.A. Luer, Ic. Pleurothallid. XIV (Monogr. Syst. Bot. Missouri Bot. Gard.*, 61(3)): 168 (1996)—Ecuador.
thoerleae *C.A. Luer in Lindleyana*, 12(1): 39 (1997)—Surinam.
thurstoniorum *G.A. Salazar Chávez, M.A. Soto Arenas & O. Suárez in Orquídea (México)*, 14: 203 (1996)—Mexico Southwest (Oaxaca), Mexico Southeast (Chiapas).
tortilis *C.A. Luer & A.C. Hirtz in C.A. Luer, Ic. Pleurothallid. XIV (Monogr. Syst. Bot. Missouri Bot. Gard.*, 61(3)): 170 (1996)—Ecuador.
totontepecensis *G.A. Salazar Chávez & M.A. Soto Arenas in Orquídea (México)*, 14: 206 (1996)—Mexico Southwest (Oaxaca).
trinaria *C.A. Luer & R. Escobar R.in Orquideologia*, 20(3): 311 (1997)—Colombia.
unijuga *C.A. Luer & Dalström in C.A. Luer, Ic. Pleurothallid. XIV (Monogr. Syst. Bot. Missouri Bot. Gard.*, 61(3)): 173 (1996)—Ecuador.
urania *C.A. Luer & A.C. Hirtz in C.A. Luer, Ic. Pleurothallid. XIV (Monogr. Syst. Bot. Missouri Bot. Gard.*, 61(3)): 174 (1996)—Ecuador.
uxoria *C.A. Luer & A.C. Hirtz in C.A. Luer, Ic. Pleurothallid. XIV (Monogr. Syst. Bot. Missouri Bot. Gard.*, 61(3)): 175 (1996)—Ecuador.
valerioi *C.A. Luer in Lindleyana*, 11(2): 67 (1996)—Costa Rica.
vellicata *C.A. Luer & A.C. Hirtz in C.A. Luer, Ic. Pleurothallid. XIV (Monogr. Syst. Bot. Missouri Bot. Gard.*, 61(3)): 177 (1996)—Ecuador, Colombia.
viahoensis *C.A. Luer & R. Escobar R. in Orquideologia*, 20(3): 311 (1997)—Colombia.
viebrockiana *C.A. Luer & L. Jost in C.A. Luer, Ic. Pleurothallid. XVII (Monogr. Syst. Bot. Missouri Bot.* Gard., 72): 108 (1998)—Ecuador.
vivipara *G.A. Salazar Chávez & M.A. Soto Arenas in Orquídea (México)*, 14: 207 (1996)—Mexico Southeast (Chiapas).
volador *C.A. Luer & A.C. Hirtz in C.A. Luer, Ic. Pleurothallid. XIV (Monogr. Syst. Bot. Missouri Bot. Gard.*, 61(3)): 180 (1996)—Ecuador.
wendtii *G.A. Salazar Chávez & M.A. Soto Arenas in Orquídea (México)*, 14: 211 (1996)—Mexico Southwest (Oaxaca).
woodfredensis *C.A. Luer in Lindleyana*, 13(3): 145 (1998)—Cuba.
zamorensis *C.A. Luer & A.C. Hirtz in C.A. Luer, Ic. Pleurothallid. XIV (Monogr. Syst. Bot. Missouri Bot. Gard.*, 61(3)): 183 (1996)—Ecuador.
zunagensis *C.A. Luer & A.C. Hirtz in C.A. Luer, Ic. Pleurothallid. XIV (Monogr. Syst. Bot. Missouri Bot. Gard.*, 61(3)): 184 (1996)—Ecuador.

LEPIDAGATHIS (Acanthac.).

floribunda *(Pohl) C. Kameyama in Bol. Bot. (São Paulo)*, 14: 197 (1995): Lophostachys floribunda.
gracilis *(Bremek.) D.C. Wasshausen in Fl. Venezuelan Guayana (gen. eds. J.A. Steyermark et al.)*, 2: 358 (1995): Teliostachya gracilis.

LEPIDAGATHIS:-

incurva Buch.-Ham. ex D. Don
var. lophostachyoides *(Nees) P. Mathew in V.V. Sivarajan & P. Mathew, Fl. Nilambur:* 512 (1997), isonym; *see B.D. Sharma et al., Fl. Karnataka:* 209 (1984): L. hyalina var. lophostachyoides.
lanceolata *(Nees) D.C. Wasshausen in Biollania*, ed. esp. 6 (J.J. Wurdack Festschr.): 542 (1997); *et in Cat. Vasc. Pl. Ecuador (Monogr. Syst. Bot. Missouri Bot. Gard., 75)*: 953 (1999): Teliostachya lanceolata.
medicaginea *(Bremek.) D.C. Wasshausen in Fl. Venezuelan Guayana (gen. eds. J.A. Steyermark et al.)*, 2: 359 (1995): Teliostachya medicaginea.
nickeriensis *D.C. Wasshausen in Fl. Venezuelan Guayana (gen. eds. J.A. Steyermark et al.)*, 2: 359 (1995), nom. nov.: Teliostachya glandulosa *Bremek.*
surinamensis *(Bremek.) D.C. Wasshausen in Fl. Venezuelan Guayana (gen. eds. J.A. Steyermark et al.)*, 2: 359 (1995): Teliostachya surinamensis.

LEPIDAPLOA (Compos.).

amambaia *H. Robinson in Phytologia*, 84(1): 40 (1998)—Brazil West-central (Mato Grosso do Sul, Goiás), Paraguay.
boquerona *(B.L. Turner) H. Robinson in Smithsonian Contrib. Bot.*, 89: 70 (1999): Vernonia boquerona.
chinchipensis *H. Robinson in Phytologia*, 84(1): 41 (1998)—Ecuador.
diazlunana *(B.L. Turner) H. Robinson in Smithsonian Contrib. Bot.*, 89: 71 (1999): Vernonia diazlunana.
imeriensis *(V.M. Badillo) J.F. Pruski in Fl. Venezuelan Guayana (gen. eds. J.A. Steyermark et al.)*, 3: 305 (1997): Vernonia imeriensis.
koelzii *(McVaugh) H. Robinson in Smithsonian Contrib. Bot.*, 89: 71 (1999): Vernonia koelzii.
polypleura *(S.F. Blake) H. Robinson in Smithsonian Contrib. Bot.*, 89: 72 (1999): Vernonia polypleura.
ruiz-teranii *(V.M. Badillo) V.M. Badillo in Ernstia*, 10(1): 4 (2000), without basionym ref.: Basionym not stated.
sprengeliana *(Sch. Bip.) H. Robinson in Smithsonian Contrib. Bot.*, 89: 72 (1999): Vernonia sprengeliana.
vauthieriana *(DC.) H. Robinson in Smithsonian Contrib. Bot.*, 89: 72 (1999): Vernonia vauthieriana.
yuruanensis *V.M. Badillo in Ernstia*, 10(1): 3 (2000)—Venezuela.
zumbadorensis *(V.M. Badillo) V.M. Badillo in Ernstia*, 10(1): 5 (2000): Vernonia zumbadorensis.

LEPIDIUM (Crucif.).

amelum *B.J. Lepschi in Nuytsia*, 12(2): 191 (1998)—Western Australia.
beckii *I.A. Al-Shehbaz in Novon*, 9(1): 5 (1999)—Bolivia.
heterophyllum Benth.
subsp. atlanticum *(Ball) J.M. Montserrat in Lagascalia*, 18(2): 241 (1996): L. nebrodense var. atlanticum.
subsp. rifanum *(Emb. & Maire) J.M. Montserrat in Lagascalia*, 18(2): 241 (1996): L. hirtum subsp. rifanum.
hirtum L.
subsp. afrum *(Pau & Font Quer) J.M. Montserrat in Lagascalia*, 18(2): 243 (1996): L. petrophilum var. afrum.
loulanicum *Z.X. An & G.L. Zhou in Fl. Xinjiangensis*, 2(2): 374 (1995)—Xinjiang.
naufragorum *P.J. Garnock-Jones & D.A. Norton in New Zealand J. Bot.*, 33(1): 43 (1995)—New Zealand South I.
strictum *(S. Watson) Rattan ex B.L. Robinson in A. Gray, Syn. Fl. N. Amer.*, 1(1): 129 (1895): L. oxycarpum var. strictum.

LEPIDOBOLUS (Restionac.).

basiflorus *J.S. Pate & K.A. Meney in Telopea*, 6(4): 655 (1996)—Western Australia.
preissianus Nees
subsp. volubilis *(Nees) B.G. Briggs & L.A.S. Johnson in Telopea*, 8(1): 28 (1998): L. preissianus var. volubilis.
spiralis *K.A. Meney & K.W. Dixon in Telopea*, 6(4): 656 (1996)—Western Australia.

LEPIDOCORYPHANTHA (Cactac.).
macromeris
subsp. runyonii *(Britton & Rose) A.B. Doweld in Sukkulentÿ (Moscow),* 1999(1): 28 (1999): Coryphantha runyonii.

LEPIDOSTEMON (Crucif.).
everestianus *I.A. Al-Shehbaz in Novon,* 10(4): 331 (2000)—Tibet.
glaricola *(H. Hara) I.A. Al-Shehbaz in Novon,* 10(4): 333 (2000): Chrysobraya glaricola.
gouldii *I.A. Al-Shehbaz in Novon,* 10(4): 331 (2000)—East Himalaya (Bhutan), Nepal.
rosularis *(K.C. Kuan & C.H. An) I.A. Al-Shehbaz in Novon,* 10(4): 332 (2000): Christolea rosularis.

LEPIDOTRICHILIA *(Harms) T.D. Pennington & B.T. Styles in Blumea,* 22(3): 473 (1975): Trichilia sect. Lepidotrichilia. (Meliac.).†
ambrensis *J.-F. Leroy in Bull. Mus. Nation. Hist. Nat., B, Adansonia,* Ser. 4, 18(1–2): 10 (1996)—Madagascar.
convallariiodora *(Baill.) J.-F. Leroy in Bull. Mus. Nation. Hist. Nat., B, Adansonia,* Ser. 4, 18(1–2): 7 (1996): Ekebergia ? convallariiodora.
sambiranensis *J.-F. Leroy in Bull. Mus. Nation. Hist. Nat., B, Adansonia,* Ser. 4, 18(1–2): 9 (1996)—Madagascar.
volkensii *(Gürke) J.-F. Leroy ex B.T. Styles & F. White in Fl. Trop. E. Afr., Meliac.:* 37 (1991): Trichilia volkensii.

LEPILAENA (Zannichelliac.).
marina *E.L. Robertson in H.B.S. Womersley, Marine Benthic Fl. South. Australia,* 1: 80 (1984)—South Australia, Victoria, Tasmania.

LEPINIA (Apocynac.).
marquisensis *D.H. Lorence & W.L. Wagner in Allertonia,* 7(4): 263 (1997)—Marquesas.

LEPISANTHES (Sapindac.).
tetraphylla (Vahl) Radlk.
var. trichocarpa *(Thwaites) B.M. Wadhwa & W. Meijer in M.D. Dassanayake & W.D. Clayton, eds., Revised Handb. Fl. Ceylon,* 12: 369 (1998): Hemigyrosa trichocarpa.

LEPISMIUM (Cactac.).
subgen. Houlletia *W. Barthlott & N.P. Taylor in Bradleya,* 13: 46 (1995).
asuntapatense *M. Kessler, P.L. Ibisch & W. Barthlott in Bradleya,* 18: 13 (2000)—Bolivia.
houlletianum
f. regnellii *(G. Lindb.) F. Süpplie, Rhipsalidinae, ed. 4:* [100] (1994): Rhipsalis regnellii.
f. regnellii *(G. Lindb.) W. Barthlott & N.P. Taylor in Bradleya,* 13: 47 (1995): Rhipsalis regnellii.
lumbricoides (Lem.) Barthlott
f. aculeatum *(F.A.C. Weber) W. Barthlott & N.P. Taylor in Bradleya,* 13: 48 (1995): Rhipsalis aculeata.

LEPTACTINA (Rubiac.).
liebrechtsiana De Wild. & T. Durand
var. mollis *E. Robbrecht in Syst. Geogr. Pl.,* 69(1): 131 (1999)—Congo.
platyphylla *(Hiern) Wernham in J. Bot.,* 51: 278 (1913): Mussaenda platyphylla.

LEPTECOPHYLLA *C.M. Weiller in Muelleria,* 12(2): 196 (1999). *EPACRIDACEAE*
abietina *(Labill.) C.M. Weiller in Muelleria,* 12(2): 211 (1999): Styphelia abietina.
brassii *(Sleumer) C.M. Weiller in Muelleria,* 12(2): 213 (1999): Styphelia brassii.
brevistyla *(J.W. Moore) C.M. Weiller in Muelleria,* 12(2): 213 (1999): Styphelia brevistyla.
divaricata *(Hook. f.) C.M. Weiller in Muelleria,* 12(2): 207 (1999): Lissanthe divaricata.
imbricata *(Stschegl.) C.M. Weiller in Muelleria,* 12(2): 213 (1999): Cyathodes imbricata.
juniperina *(J.R. Forst. & G. Forst.) C.M. Weiller in Muelleria,* 12(2): 200 (1999): Epacris juniperina.
subsp. oxycedrus *(Labill.) C.M. Weiller in Muelleria,* 12(2): 205 (1999): Styphelia oxycedrus.
subsp. parvifolia *(R. Br.) C.M. Weiller in Muelleria,* 12(2): 204 (1999): Cyathodes parvifolia.

LEPTECOPHYLLA:-
mariannensis *(Kaneh.) C.M. Weiller in Muelleria,* 12(2): 213 (1999): Cyathodes mariannensis.
pendulosa *(Jarman) C.M. Weiller in Muelleria,* 12(2): 209 (1999): Cyathodes pendulosa.
pogonocalyx *C.M. Weiller in Muelleria,* 12(2): 206 (1999)—Tasmania.
pomarae *(A. Gray) C.M. Weiller in Muelleria,* 12(2): 213 (1999): Cyathodes pomarae.
rapae *(Sleumer) C.M. Weiller in Muelleria,* 12(2): 213 (1999): Styphelia rapae.
robusta *(Hook. f.) C.M. Weiller in Muelleria,* 12(2): 210 (1999): Cyathodes robusta.
tameiameiae *(Cham. & Schltdl.) C.M. Weiller in Muelleria,* 12(2): 213 (1999): Cyathodes tameiameiae.

LEPTOCARPUS (Restionac.).
crassipes *J.S. Pate & K.A. Meney in Telopea,* 6(4): 658 (1996)—Western Australia.
diffusus *(Spreng.) B.G. Briggs & L.A.S. Johnson in Telopea,* 8(1): 29 (1998): Restio diffusus.

LEPTOCHLOA (Gramin.).
caudata *(K. Schum.) N. Snow in Novon,* 8(1): 77 (1998): Diplachne caudata.
decipiens (R. Br.) Stapf ex Maiden
subsp. asthenes *(Roem. & Schult.) N. Snow in Novon,* 8(1): 77 (1998): Poa asthenes.
subsp. peacockii *(Maiden & Betche) N. Snow in Novon,* 8(1): 78 (1998): Diplachne peacockii.
eleusine *(Nees) T.A. Cope & N. Snow in Novon,* 8(1): 78 (1998): Diplachne eleusine.
fusca (L.) Kunth
subsp. fascicularis *(Lam.) N. Snow in Novon,* 8(1): 78 (1998): Festuca fascicularis.
subsp. muelleri *(Benth.) N. Snow in Novon,* 8(1): 78 (1998): Diplachne muelleri.
subsp. uninervia *(J. Presl) N. Snow in Novon,* 8(1): 79 (1998): Megastachya uninervia.
gigantea *(Launert) T.A. Cope & N. Snow in Novon,* 8(1): 79 (1998): Diplachne gigantea.
panicea (Retz.) Ohwi
subsp. brachiata *(Steud.) N. Snow in Novon,* 8(1): 79 (1998): Leptochloa brachiata.
simoniana *N. Snow in Novon,* 10(3): 238 (2000)—Queensland, New Guinea (Papua New Guinea).
southwoodii *N. Snow & B.K. Simon in Austrobaileya,* 5(1): 138 (1997)—Queensland.
srilankensis *N. Snow in Novon,* 8(2): 183 (1998)—Sri Lanka.

LEPTODERMIS (Rubiac.).
amoena *L.S. Springate in Edinburgh J. Bot.,* 53(1): 99 (1996)—Bhutan, Assam (Arunachal Pradesh).
beichuanensis *H.S. Lo in J. Trop. Subtrop. Bot.,* 7(1): 19 (1999)—China South-central (Sichuan).
brevisepala *H.S. Lo in J. Trop. Subtrop. Bot.,* 7(1): 21 (1999)—China South-central (Sichuan).
buxifolia *H.S. Lo in J. Trop. Subtrop. Bot.,* 7(1): 20 (1999)—China South-central (Sichuan), China North-central (Gansu).
f. strigosa *H.S. Lo in J. Trop. Subtrop. Bot.,* 7(1): 21 (1999)—China South-central (Sichuan).
gracilis C.E.C. Fisch.
var. longiflora *H.S. Lo in Fl. Reipubl. Popul. Sin.,* 71(2): 338 (1999), without type—China South-central (Sichuan).
hirsutiflora *H.S. Lo in J. Trop. Subtrop. Bot.,* 7(1): 19 (1999)—Xizang.
var. ciliata *H.S. Lo in J. Trop. Subtrop. Bot.,* 7(1): 20 (1999)—Xizang.
huashanica *H.S. Lo in J. Trop. Subtrop. Bot.,* 7(1): 16 (1999)—China North-central (Shaanxi).
lanata *H.S. Lo in J. Trop. Subtrop. Bot.,* 7(1): 19 (1999)—China South-central (Yunnan).
ludlowii *L.S. Springate in Edinburgh J. Bot.,* 53(1): 102 (1996)—Bhutan, Xizang, ?Sikkim.
pilosa Diels
var. acanthoclada *H.S. Lo in Fl. Reipubl. Popul. Sin.,* 71(2): 338 (1999), without type—China South-central (Sichuan), Xizang.
var. spicatiformis *H.S. Lo in Fl. Reipubl. Popul. Sin.,* 71(2): 338 (1999), without type—China North-central (Shanxi, Gansu).
potanini Batalin
var. angustifolia *H.S. Lo in Fl. Reipubl. Popul. Sin.,* 71(2): 338 (1999), without type—China South-central (Yunnan).

LEPTODERMIS:-
pumila *H.S. Lo in J. Trop. Subtrop. Bot.,* 7(1): 18 (1999)—China South-central (Yunnan).
tubicalyx *H.S. Lo in J. Trop. Subtrop. Bot.,* 7(1): 17 (1999)—Xizang.
velutiniflora *H.S. Lo in J. Trop. Subtrop. Bot.,* 7(1): 17 (1999)—China South-central (Sichuan).
var. tenera *H.S. Lo in J. Trop. Subtrop. Bot.,* 7(1): 17 (1999)—Xizang, China South-central (Sichuan, Yunnan).
xizangensis *H.S. Lo in J. Trop. Subtrop. Bot.,* 7(1): 16 (1999)—Xizang.
yui *H.S. Lo in J. Trop. Subtrop. Bot.,* 7(1): 15 (1999)—China South-central (Sichuan).

LEPTOMERIA (Santalac.).
ellytes *B.J. Lepschi in Austral. Syst. Bot.,* 12(1): 73 (1999)—Western Australia.
furtiva *B.J. Lepschi in Austral. Syst. Bot.,* 12(1): 76 (1999)—Western Australia.

LEPTOMISCHUS (Rubiac.).
erianthus *H.S. Lo in Bull. Bot. Res. North-East. Forest. Univ.,* 18(3): 277 (1998)—China South-central (Yunnan).
funingensis *H.S. Lo in Bull. Bot. Res. North-East. Forest. Univ.,* 18(3): 278 (1998)—China South-central (Yunnan).
guangxiensis *H.S. Lo in Bull. Bot. Res. North-East. Forest. Univ.,* 18(3): 278 (1998)—China South-east (Guangxi).

LEPTOSEMA (Legumin.).
sect. Burgesia *(F. Muell.) M.D. Crisp in Austral. Syst. Bot.,* 12(1): 14 (1999): gen. Burgesia.
sect. Oolobium *M.D. Crisp in Austral. Syst. Bot.,* 12(1): 11 (1999).
ser. Pereostemon *M.D. Crisp in Austral. Syst. Bot.,* 12(1): 34 (1999).
aphyllum *(Hook.) M.D. Crisp in Austral. Syst. Bot.,* 12(1): 18 (1999)—Brachysema aphyllum.
cervicorne *M.D. Crisp in Austral. Syst. Bot.,* 12(1): 38 (1999)—Western Australia.
chapmanii *M.D. Crisp in Austral. Syst. Bot.,* 12(1): 34 (1999)—Queensland.
macrocarpum *(Benth.) M.D. Crisp in Austral. Syst. Bot.,* 12(1): 22 (1999): Brachysema macrocarpum.
tomentosum *(Benth.) M.D. Crisp in Austral. Syst. Bot.,* 12(1): 41 (1999): Brachysema tomentosum.
uniflorum *(R.Br. ex Benth.) M.D. Crisp in Austral. Syst. Bot.,* 12(1): 15 (1999): Brachysema uniflorum.
villosum *M.D. Crisp in Austral. Syst. Bot.,* 12(1): 31 (1999)—Northern Territory.

LEPTOSIPHON (Polemoniac.).
ambiguus *(Rattan) J.M. Porter & L.A. Johnson in Aliso,* 19(1): 80 (2000): Gilia ambigua.
aureus *(Nutt.) J.M. Porter & L.A. Johnson in Aliso,* 19(1): 80 (2000): Gilia aurea.
subsp. decorus *(A. Gray) J.M. Porter & L.A. Johnson in Aliso,* 19(1): 80 (2000): Gilia aurea var. decora.
bolanderi *(A. Gray) J.M. Porter & L.A. Johnson in Aliso,* 19(1): 80 (2000): Gilia bolanderi.
breviculus *(A. Gray) J.M. Porter & L.A. Johnson in Aliso,* 19(1): 80 (2000): Gilia brevicula.
croceus *(Eastw.) J.M. Porter & L.A. Johnson in Aliso,* 19(1): 80 (2000), with incorrect basionym ref.: Linanthus croceus.
filipes *(Benth.) J.M. Porter & L.A. Johnson in Aliso,* 19(1): 80 (2000): Gilia filipes.
floribundus *(A. Gray) J.M. Porter & L.A. Johnson in Aliso,* 19(1): 80 (2000): Gilia floribunda.
subsp. glaber *(R. Patt.) J.M. Porter & L.A. Johnson in Aliso,* 19(1): 80 (2000): Linanthus floribundus subsp. glaber.
subsp. hallii *(Jeps.) J.M. Porter & L.A. Johnson in Aliso,* 19(1): 80 (2000): Siphonella floribunda var. hallii.
harknessii *(Curran) J.M. Porter & L.A. Johnson in Aliso,* 19(1): 80 (2000): Gilia harknessii.
jamauensis *(Moran) J.M. Porter & L.A. Johnson in Aliso,* 19(1): 80 (2000): Linanthus jamauensis.
jepsonii *(Schemske & Goodw.) J.M. Porter & L.A. Johnson in Aliso,* 19(1): 81 (2000): Linanthus jepsonii.
latisectus *(E.G. Buxton) J.M. Porter & L.A. Johnson in Aliso,* 19(1): 81 (2000): Linanthus latisectus.

LEPTOSIPHON:-
laxus *(Vasey & Rose) J.M. Porter & L.A. Johnson in Aliso,* 19(1): 81 (2000): Gilia laxa.
lemmonii *(A. Gray) J.M. Porter & L.A. Johnson in Aliso,* 19(1): 81 (2000): Gilia lemmonii.
liniflorus *(Benth.) J.M. Porter & L.A. Johnson in Aliso,* 19(1): 81 (2000): Gilia liniflora.
melingii *(Wiggins) J.M. Porter & L.A. Johnson in Aliso,* 19(1): 81 (2000): Leptodactylon melingii.
montanus *(Greene) J.M. Porter & L.A. Johnson in Aliso,* 19(1): 81 (2000): Linanthus montanus.
nudatus *(Greene) J.M. Porter & L.A. Johnson in Aliso,* 19(1): 81 (2000): Linanthus nudatus.
nuttallii *(A. Gray) J.M. Porter & L.A. Johnson in Aliso,* 19(1): 81 (2000): Gilia nuttallii.
subsp. howellii *(T.W. Nelson & R. Patt.) J.M. Porter & L.A. Johnson in Aliso,* 19(1): 81 (2000): Linanthus nuttallii subsp. howellii.
subsp. pubescens *(R. Patt.) J.M. Porter & L.A. Johnson in Aliso,* 19(1): 81 (2000): Linanthus nuttallii subsp. pubescens.
oblanceolatus *(Brand) J.M. Porter & L.A. Johnson in Aliso,* 19(1): 81 (2000): Gilia oblanceolata.
pachyphyllus *(R. Patt.) J.M. Porter & L.A. Johnson in Aliso,* 19(1): 81 (2000): Linanthus pachyphyllus.
pusillus *(Benth.) J.M. Porter & L.A. Johnson in Aliso,* 19(1): 81 (2000): Gilia pusilla.
pygmaeus *(Brand) J.M. Porter & L.A. Johnson in Aliso,* 19(1): 81 (2000): Gilia pygmaea.
subsp. continentalis *(P.H. Raven) J.M. Porter & L.A. Johnson in Aliso,* 19(1): 81 (2000): Linanthus pygmaeus subsp. continentalis.
rattanii *(A. Gray) J.M. Porter & L.A. Johnson in Aliso,* 19(1): 81 (2000): Gilia rattanii.
septentrionalis *(H. Mason) J.M. Porter & L.A. Johnson in Aliso,* 19(1): 81 (2000): Linanthus septentrionalis.
serrulatus *(Greene) J.M. Porter & L.A. Johnson in Aliso,* 19(1): 81 (2000): Linanthus serrulatus.

LEPTOSPERMUM (Myrtac.).
jingera *A.M. Lyne & M.D. Crisp in Austral. Syst. Bot.,* 9(3): 301 (1996)—Victoria.

LEPTOSTACHYA (Acanthac.).
caudatifolia *H.S. Lo & D. Fang in Guihaia,* 17(1): 45 (1997)—China Southeast (Guangxi).

LERCHEA (Rubiac.).
micrantha *(Drake) H.S. Lo in Bull. Bot. Res. North-East. Forest. Univ.,* 18(3): 275 (1998): Ophiorrhiza micrantha.
sinica *(H.S. Lo) H.S. Lo in Bull. Bot. Res. North-East. Forest. Univ.,* 18(3): 275 (1998): Xanthophytum sinicum.

LESPEDEZA (Legumin.).
bicolor Turcz.
var. melanantha *(Nakai) Y.N. Lee, Fl. Korea:* 1159 (1996), isonym; *see Akiyama & H. Ohba in J. Jap. Bot.,* 59(6): 191 (1984): L. melanantha.
grandiflora *H. Koidzumi in J. Pl. Iwateken,* 2(2): 75 (1937)—Japan.
hisauchii *T. Nemoto & Hiroyoshi Ohashi in J. Jap. Bot.,* 74(5): 273 (1999)—Japan.
tobae *H. Koidzumi in J. Pl. Iwateken,* 2(2): 75 (1937)—Japan.

LESQUERELLA (Crucif.).
navajoensis *S.L. O'Kane in Madroño,* 46(2): 88 (1999)—New Mexico.
tuplashensis *R.C. Rollins, K.A. Beck & F.E. Caplow in Rhodora,* 97(891): 203 (1995 publ. 1996)—Washington.
vicina *J.L. Anderson, J.L. Reveal & R.C. Rollins in Novon,* 7(1): 9 (1997)—Colorado.

LESSERTIA (Legumin.).
canescens *P. Goldblatt & J.C. Manning in Strelitzia,* 9: 708 (2000), nom. nov.: Sutherlandia tomentosa *Eckl. & Zeyh.*
contracta *M.-J. Balkwill in S. Afr. J. Bot.,* 65(5–6): 352 (1999)—Natal.
frutescens *(L.) P. Goldblatt & J.C. Manning in Strelitzia,* 9: 708 (2000): Colutea frutescens.
humilis *(E. Phillips & R.A. Dyer) P. Goldblatt & J.C. Manning in Strelitzia,* 9: 708 (2000): Sutherlandia humilis.

LESSERTIA:-
ingeliensis *M.-J. Balkwill in S. Afr. J. Bot.,* 65(5–6): 347 (1999)—Natal.
macroflora *M.-J. Balkwill in S. Afr. J. Bot.,* 65(5–6): 352 (1999)—Natal.
microphylla *(Burch. ex DC.) P. Goldblatt & J.C. Manning in Strelitzia,* 9: 708 (2000): Sutherlandia microphylla.
montana *(E. Phillips & R.A. Dyer) P. Goldblatt & J.C. Manning in Strelitzia,* 9: 708 (2000): Sutherlandia montana.
speciosa *(E. Phillips & R.A. Dyer) P. Goldblatt & J.C. Manning in Strelitzia,* 9: 708 (2000): Sutherlandia speciosa.

LESSINGIANTHUS (Compos.).
parvifolius *(Chodat) H. Robinson in Smithsonian Contrib. Bot.,* 89: 74 (1999): Vernonia glabrata var. parvifolia.
ramellae *(Cabrera) H. Robinson in Smithsonian Contrib. Bot.,* 89: 74 (1999): Vernonia ramellae.

LETESTUDOXA (Annonac.).
glabrifolia *L.W. Chatrou & C.P. Repetur in L.W. Chatrou, Changing Genera:* 76 (1998)—Gabon.

LEUCADENDRON (Proteac.).
lanigerum *H. Buek ex Meisn. in DC. Prodr.,* 14: 222 (1856), as 'lanigenum'—Cape Province.†
osbornei *J.P. Rourke in Bothalia,* 27(1): 52 (1997)—Cape Province.

LEUCAENA (Legumin.).
confertiflora Zárate
var. adenotheloidea *(Zárate) C.E. Hughes in Contrib. Univ. Mich. Herb.,* 21: 287 (1997): L. confertiflora subsp. adenotheloidea.
lanceolata S. Watson
var. sousae *(Zárate) C.E. Hughes in Contrib. Univ. Mich. Herb.,* 21: 288 (1997): L. lanceolata subsp. sousae.
lempirana *C.E. Hughes in Contrib. Univ. Mich. Herb.,* 21: 279 (1997)—Honduras.
leucocephala (Lam.) de Wit
subsp. ixtahuacana *C.E. Hughes in Contrib. Univ. Mich. Herb.,* 21: 285 (1997)—Guatemala, Mexico Southeast (Chiapas).
macrophylla Benth.
subsp. istmensis *C.E. Hughes in Contrib. Univ. Mich. Herb.,* 21: 283 (1997)—Mexico Southwest (Oaxaca), Mexico Gulf (Veracruz).
magnifica *(C.E. Hughes) C.E. Hughes in Contrib. Univ. Mich. Herb.,* 21: 286 (1997): L. shannonii subsp. magnifica.
matudae *(Zárate) C.E. Hughes in Contrib. Univ. Mich. Herb.,* 21: 286 (1997): L. esculenta subsp. matudae.
× mixtec *C.E. Hughes & S.A. Harris in C.E. Hughes, Monogr. Leucaena (Legumin.-Mimosoid.) (Syst. Bot. Monog.,* 55): 209 (1998)—Mexico Central (Puebla, Morelos), Mexico Northeast (Querétaro), Mexico Southwest (Guerrero, Oaxaca), Mexico Southeast (Chiapas). (L. esculenta × leucocephala subsp. glabrata).
× spontanea *C.E. Hughes & S.A. Harris in Pl. Syst. Evol.,* 212(1–2): 70 (1998)—Mexico Gulf (Veracruz), Dominican Republic, Guatemala, New Guinea (Papua New Guinea). (L. diversifolia × leucocephala subsp. glabrata).

LEUCANTHEMUM (Compos.).
vulgare Lam.
subsp. aligulatum *(Vogt) O. de Bolòs & J. Vigo, Fl. Països Catalans,* 3: 813 (1995): L. aligulatum.
subsp. barrelieri *(Dufour ex DC.) O. de Bolòs & J. Vigo, Fl. Països Catalans,* 3: 816 (1995): Pyrethrum halleri var. barrelieri.
subsp. catalaunicum *(Vogt) O. de Bolòs & J. Vigo, Fl. Països Catalans,* 3: 813 (1995): L. catalaunicum.
subsp. montserratianum *(Vogt) O. de Bolòs & J. Vigo, Fl. Països Catalans,* 3: 814 (1995): L. montserratianum.

LEUCAS (Labiat.).
anandaraoana *P. Umamaheswari & P. Daniel in Kew Bull.,* 54(4): 1003 (1999 publ. 2000)—India (Tamil Nadu).

LEUCAS:-
decemdentata (Willd.) R. Br.
var. angustifolia *(Benth.) V. Singh in J. Econ. Taxon. Bot.,* 22(2): 388 (1998): L. angustifolia.
var. sebastiana *(Subba Rao & Kumari) V. Singh in J. Econ. Taxon. Bot.,* 22(2): 388 (1998): L. mollissima var. sebastiana.
manipurensis *V. Singh in J. Econ. Taxon. Bot.,* 21(3): 743 (1997 publ. 1998)—Assam (Manipur, Meghalaya).
mollissima Wall. ex Benth.
var. riukiuensis *(Ohwi) S. Hatusima in S. Hatusima & T. Amano, Fl. Ryukyus, S. of Amami Is., ed. 2:* 188 (1994), without basionym ref.: Basionym not stated.
subsp. scaberula *(Hook. f.) S.K. Murti & G. Panigrahi, Fl. Bilaspur Distr.,* 2: 487 (1999): L. mollissima var. scaberula.

LEUCOCHLORON *R.C. Barneby & J.W. Grimes, Silk Tree, Guanacaste, Monkey's Earring (Mem. New York Bot. Gard., 74(1)):* 130 (1996). *LEGUMINOSAE*
foederale *(Barneby & J.W. Grimes) R.C. Barneby & J.W. Grimes, Silk Tree, Guanacaste, Monkey's Earring (Mem. New York Bot. Gard., 74(1)):* 135 (1996): Pithecellobium foederale.
incuriale *(Vell.) R.C. Barneby & J.W. Grimes, Silk Tree, Guanacaste, Monkey's Earring (Mem. New York Bot. Gard., 74(1)):* 134 (1996): Mimosa incurialis.
limae *R.C. Barneby & J.W. Grimes, Silk Tree, Guanacaste, Monkey's Earring (Mem. New York Bot. Gard., 74(1)):* 132 (1996)—Brazil Northeast (Bahia).
minarum *(Harms) R.C. Barneby & J.W. Grimes, Silk Tree, Guanacaste, Monkey's Earring (Mem. New York Bot. Gard., 74(1)):* 135 (1996): Pithecellobium minarum.

LEUCOCORYNE (Alliac.).
ser. Subulatae *P. Ravenna in Onira, Bot. Leafl.,* 5(6): 28 (2000).
conconensis *P. Ravenna in Onira, Bot. Leafl.,* 5(6): 24 (2000), contrary to Art. 37.6 ICBN (2000)—Chile Central (Valparaíso).
coronata *P. Ravenna in Onira, Bot. Leafl.,* 5(6): 29 (2000), nom. nov.: Stemmatium narcissoides *Phil.*
fragrantissima *P. Ravenna in Onira, Bot. Leafl.,* 4(2): 9 (2000)—Chile North (Santiago).
fuscostriata *P. Ravenna in Onira, Bot. Leafl.,* 4(2): 8 (2000)—Chile North (Coquimbo).
gilliesioides *(Phil.) P. Ravenna in Onira, Bot. Leafl.,* 4(2): 10 (2000): Erinna gilliesioides.
inclinata *P. Ravenna in Onira, Bot. Leafl.,* 4(2): 7 (2000)—Chile North (Coquimbo).
lilacea *P. Ravenna in Onira, Bot. Leafl.,* 4(2): 4 (2000)—Chile North (Valparaíso).
lurida *P. Ravenna in Onira, Bot. Leafl.,* 4(2): 6 (2000)—Chile North (Coquimbo).
modesta *P. Ravenna in Onira, Bot. Leafl.,* 5(6): 26 (2000)—Chile Central (Valparaíso).
mollensis *P. Ravenna in Onira, Bot. Leafl.,* 5(6): 26 (2000)—Chile Central (Valparaíso).
porphyrea *P. Ravenna in Onira, Bot. Leafl.,* 4(2): 6 (2000)—Chile North (Coquimbo).
praealta *P. Ravenna in Onira, Bot. Leafl.,* 4(2): 4 (2000)—Chile North (Valparaíso).
quilimarina *P. Ravenna in Onira, Bot. Leafl.,* 5(6): 25 (2000)—Chile Central (Coquimbo).
simulans *P. Ravenna in Onira, Bot. Leafl.,* 4(2): 3 (2000)—Chile North (Coquimbo).
subulata *P. Ravenna in Onira, Bot. Leafl.,* 5(6): 27 (2000)—Chile Central (Santiago).
tricornis *P. Ravenna in Onira, Bot. Leafl.,* 5(6): 28 (2000)—Chile Central (Santiago).
valparadisea *P. Ravenna in Onira, Bot. Leafl.,* 4(2): 8 (2000)—Chile North (Valparaíso).
vittata *P. Ravenna in Onira, Bot. Leafl.,* 4(2): 5 (2000)—Chile North (Valparaíso).

LEUCOGENES (Compos.).
neglecta *B.P.J. Molloy in New Zealand J. Bot.,* 33(1): 54 (1995)—New Zealand South I.
tarahaoa *B.P.J. Molloy in New Zealand J. Bot.,* 33(1): 56 (1995)—New Zealand South I.

LEUCOMPHALOS (Legumin.).
brachycarpus *(Harms) F.J. Breteler in Wageningen Agric. Univ. Pap.*, 94(4): 13 (1994): Baphiastrum brachycarpum.
callicarpus *(Champ. ex Benth.) F.J. Breteler in Wageningen Agric. Univ. Pap.*, 94(4): 18 (1994): Bowringia callicarpa.
discolor *(J.B. Hall) F.J. Breteler in Wageningen Agric. Univ. Pap.*, 94(4): 25 (1994): Bowringia discolor.
libericus *F.J. Breteler in Wageningen Agric. Univ. Pap.*, 94(4): 27 (1994)—Liberia.
mildbraedii *(Harms) F.J. Breteler in Wageningen Agric. Univ. Pap.*, 94(4): 31 (1994): Bowringia mildbraedii.

LEUCOTHOE (Ericac.).
grayana Maxim.
var. pruinosa *(H. Hara) F. Yokouchi ex T. Shimizu, Fl. Nagano Pref.*: 1508 (1997): Eubotryoides grayana var. pruinosa.

LEWISIA (Portulacac.).
kelloggii K. Brandegee
subsp. hutchisonii *L.T. Dempster in Madroño*, 43(3): 415 (1996)—California.

LEWISIOPSIS *R. Govaerts, World Checklist Seed Plants*, 3(1): 21 (1999), nom. nov.: Lewisia subgen. Strophiolum *B. Mathew*. (Portulacac.).
tweedyi *(A. Gray) R. Govaerts, World Checklist Seed Plants*, 3(1): 21 (1999): Calandrinia tweedyi.

× **LEYDEUM** *M.E. Barkworth in Phytologia*, 83(4): 308 (1997 publ. 1998). *GRAMINEAE* (Hordeum × Leymus).
dutillyanum *(Lepage) M.E. Barkworth in Phytologia*, 83(4): 309 (1997 publ. 1998): Elymordeum dutillyanum. (Hordeum jubatum × Leymus mollis).
piperi *(Bowden) M.E. Barkworth in Phytologia*, 83(4): 309 (1997 publ. 1998): Elymordeum piperi. (Hordeum jubatum × Leymus triticoides).

LEYMUS (Gramin.).
altus *D.F. Cui in Fl. Xinjiangensis*, 6: 604 (1996); *et in Bull. Bot. Res. North-East. Forest. Univ.*, 18(2): 144 (1998)—Xinjiang.
angustus (Trin.) Pilg.
subsp. macrantherus *D.F. Cui in Fl. Xinjiangensis*, 6: 604 (1996); *et in Bull. Bot. Res. North-East. Forest. Univ.*, 18(2): 148 (1998), as 'macroantherus'—Xinjiang.
aristiglumus *L.B. Cai in Bull. Bot. Res. North-East. Forest. Univ.*, 17(1): 28 (1997)—Qinghai.
arjinshanicus *D.F. Cui in Fl. Xinjiangensis*, 6: 602 (1996); *et in Bull. Bot. Res. North-East. Forest. Univ.*, 18(2): 146 (1998)—Xinjiang.
subsp. ruoqiangensis *(S.L. Lu & Y.H. Wu) D.F. Cui in Fl. Xinjiangensis*, 6: 218 (1996), as 'aergishanicus subsp. ruqiangensis': L. ruoqiangensis.
bruneostachyus *N.R. Cui & D.F. Cui in Fl. Xinjiangensis*, 6: 603 (1996), holotype not indicated—Xinjiang.
coreanus *(Honda) K.B. Jensen & R.R.-C. Wang in Int. J. Pl. Sci.*, 158(6): 872 (1997): Elymus coreanus.
crassiusculus *L.B. Cai in Acta Phytotax. Sin.*, 33(5): 494 (1995)—Qinghai, China North-central (Shanxi).
erianthus *(Phil.) J. Dubcovsky in Genome*, 40(4): 518 (1997): Elymus erianthus.
flexus *L.B. Cai in Acta Phytotax. Sin.*, 33(5): 491 (1995)—Qinghai.
innovatus (Beal) Pilg.
subsp. velutinus *(Bowden) N.N. Tzvelev in Novosti Sist. Vyssh. Rast.*, 32: 182 (2000): Elymus inovatus subsp. velutinus.
latiglumis *L.B. Cai in Acta Phytotax. Sin.*, 33(5): 493 (1995), as 'latiglumus', homonym; *see Tzvelev in Novosti Sist. Vyssh. Rast.*, 9: 62 (1972)—China North-central (Shanxi).
mendocinus *(Parodi) J. Dubcovsky in Genome*, 40(4): 518 (1997): Agropyron mendocinus.
obvipodus *L.B. Cai in Novon*, 10(1): 9 (2000)—Qinghai.

LEYMUS:-
pendulus *L.B. Cai in Novon*, 10(1): 7 (2000)—Qinghai.
secalinus (Georgi) Tzvelev
var. laxinodis *L.B. Cai in Bull. Bot. Res. North-East. Forest. Univ.*, 17(1): 32 (1997)—Qinghai.
var. tenuis *L.B. Cai in Acta Phytotax. Sin.*, 33(5): 496 (1995)—Xizang.
yiunensis *N.R. Cui & D.F. Cui in Fl. Xinjiangensis*, 6: 603 (1996)—Xinjiang.
yiwuensis *L.B. Cai in Bull. Bot. Res. North-East. Forest. Univ.*, 17(1): 29 (1997), orth. var. of L. yiwuensis N.R. Cui & D.F. Cui; *see Fl. Xinjiangensis*, 6: 603 (1996)—Xinjiang.

LIABUM (Compos.).
crispum *Sch. Bip. in J. Bot.*, 1: 237 (1863)—Cuba.
cubense *Sch. Bip. in J. Bot.*, 1: 236 (1863)—Cuba.

LIATRIS (Compos.).
scariosa
var. nieuwlandii *(Lunell) E.G. Voss in Mich. Bot.*, 34(4): 139 (1995 publ. 1996): Laciniaria scariosa var. nieuwlandii.

LICANIA (Chrysobalanac.).
cuyabenensis *G.T. Prance in Kew Bull.*, 54(1): 104 (1999)—Ecuador.
ferreirae *G.T. Prance in Kew Bull.*, 54(1): 104 (1999)—Brazil North (Amazonas).

LICARIA (Laurac.).
aureosericea *H. van der Werff in Novon*, 10(3): 264 (2000)—Brazil North (Amazonas), Guyana.
caribaea *J. Gómez-Laurito & A. Cascante in Novon*, 9(2): 199 (1999)—Costa Rica.
hirsuta *H. van der Werff in Novon*, 10(3): 266 (2000)—Brazil North (Amazonas).

LICUALA (Palmae).
ahlidurii *L.G. Saw in Sandakania*, 10: 30 (1997)—Malaya (Peninsular Malaysia).
aurantiaca *D.R. Hodel in Palm J.*, 134: 30 (1997)—Thailand.
bayana *L.G. Saw in Sandakania*, 10: 21 (1997)—Malaya (Peninsular Malaysia).
cameronensis *L.G. Saw in Sandakania*, 10: 76 (1997)—Malaya (Peninsular Malaysia).
crassiflora *A.S. Barfod in Palms*, 44(4): 199 (2000)—New Guinea (Papua New Guinea).
delicata *D.R. Hodel in Palm J.*, 136: 11 (1997)—Thailand.
egregia *L.G. Saw in Sandakania*, 10: 16 (1997)—Malaya (Peninsular Malaysia).
filiformis *D.R. Hodel in Palm J.*, 134: 32 (1997)—Thailand.
fractiflexa *L.G. Saw in Sandakania*, 10: 32 (1997)—Malaya (Peninsular Malaysia).
hirta *D.R. Hodel in Palm J.*, 136: 12 (1997)—Thailand.
kamarudinii *L.G. Saw in Sandakania*, 10: 24 (1997)—Malaya (Peninsular Malaysia).
khoonmengii *L.G. Saw in Sandakania*, 10: 43 (1997)—Malaya (Peninsular Malaysia).
malajana Becc.
var. humilis *L.G. Saw in Sandakania*, 10: 90 (1997)—Malaya (Peninsular Malaysia).
mustapana *L.G. Saw in Sandakania*, 10: 55 (1997)—Malaya (Peninsular Malaysia).
palas *L.G. Saw in Sandakania*, 10: 59 (1997)—Malaya (Peninsular Malaysia).
peltata Roxb. ex Buch.-Ham.
var. sumawongii *L.G. Saw in Sandakania*, 10: 10 (1997)—Thailand, Malaya (Peninsular Malaysia).
poonsakii *D.R. Hodel in Palm J.*, 134: 32 (1997)—Thailand.
ruthiae *L.G. Saw in Sandakania*, 10: 41 (1997)—Malaya (Peninsular Malaysia).
sallehana *L.G. Saw in Sandakania*, 10: 11 (1997)—Malaya (Peninsular Malaysia).
var. incisifolia *L.G. Saw in Sandakania*, 10: 13 (1997)—Malaya (Peninsular Malaysia).
stenophylla *D.R. Hodel in Palm J.*, 136: 16 (1997)—Thailand.
stongensis *L.G. Saw in Sandakania*, 10: 90 (1997)—Malaya (Peninsular Malaysia).
tansachana *D.R. Hodel in Palm J.*, 134: 34 (1997)—Thailand.

LICUALA:-
tenuissima *L.G. Saw in Sandakania*, 10: 37 (1997)—Malaya (Peninsular Malaysia).
terengganuensis *L.G. Saw in Sandakania*, 10: 61 (1997)—Malaya (Peninsular Malaysia).
whitmorei *L.G. Saw in Sandakania*, 10: 78 (1997)—Malaya (Peninsular Malaysia).

LIGNARIELLA (Crucif.).
hinkuensis *Kats. Arai, H. Ohba & I.A. Al-Shehbaz in Harvard Pap. Bot.*, 5(1): 117 (2000)—Nepal.
ohbana *I.A. Al-Shehbaz & Kats. Arai in Harvard Pap. Bot.*, 5(1): 120 (2000)—Nepal, China South-central (Yunnan).
serpens *(W.W. Sm.) I.A. Al-Shehbaz, Kats. Arai & H. Ohba in Harvard Pap. Bot.*, 5(1): 119 (2000): Cochlearia serpens.

LIGULARIA (Compos.).
cuneata *S.W. Liu & T.N. Ho in Acta Phytotax. Sin.*, 38(3): 286 (2000)—Xizang.
eriocaulis *M. Zhang & L.S. Xu in Acta Bot. Yunnanica*, 19(3): 241 (1997)—China South-central (Yunnan).
hamiica *C.H. An in Fl. Xinjiangensis*, 5: 478 (1999)—Xinjiang.
kunlunshanica *C.H. An in Fl. Xinjiangensis*, 5: 477 (1999)—Xinjiang.
lancifera *(J.R. Drumm.) A.J.C. Grierson in Edinburgh J. Bot.*, 57(3): 402 (2000): Senecio lancifer.
latiligulata *(R.D. Good) L.S. Springate in Edinburgh J. Bot.*, 57(3): 402 (2000): Cremanthodium hookeri f. latiligulatum.
shifangensis *G.H. Chen & Wen Jin Zhang in Acta Phytotax. Sin.*, 35(2): 181 (1997)—China South-central (Sichuan).
sibirica (L.) Cass.
subsp. abakanica *(Pojark.) E.I. Wiebe in Fl. Sibir. (Asterac.)*, 13: 173 (1997 publ. 1998): L. abakanica.
subsp. lydiae *(Minderova) N.N. Tzvelev in Novosti Sist. Vyssh. Rast.*, 32: 185 (2000): L. lydiae.
tulupanica *C.H. An in Fl. Xinjiangensis*, 5: 478 (1999)—Xinjiang.
virgaurea (Maxim.) Mattf.
var. oligocephala *(R.D. Good) S.W. Liu in Fl. Qinghaiica*, 3: 427 (1996): Cremanthodium plantaginifolium subsp. oligocephalum.
var. pilosa *S.W. Liu & T.N. Ho in Acta Phytotax. Sin.*, 38(3): 288 (2000)—Xizang, China South-central (Sichuan).
yakla *(C.B. Clarke) V. Singh & P. Singh in J. Bombay Nat. Hist. Soc.*, 92(2): 290 (1995): Senecio yakla.
zhouquensis *W.D. Peng & Z.X. Peng in Acta Phytotax. Sin.*, 33(6): 612 (1995)—China North-central (Gansu).

LIGULARIOPSIS *Y.L. Chen in Acta Phytotax. Sin.*, 34(6): 631 (1996). *COMPOSITAE*
shichuana *Y.L. Chen in Acta Phytotax. Sin.*, 34(6): 632 (1996)—China North-central (Shaanxi, Gansu).

LIGUSTICOPSIS (Umbellif.).
dielsiana *(H. Wolff) M.G. Pimenov & E.V. Kljuykov in Feddes Repert.*, 110(7–8): 484 (1999): Ligusticum dielsianum.
longicalycia *(M.L. Sheh) M.G. Pimenov & E.V. Kljuykov in Feddes Repert.*, 110(7–8): 484 (1999): Selinum longicalycium.
pseudodaucoides *(H. Peng & Yin Z. Wang) M.G. Pimenov & E.V. Kljuykov in Feddes Repert.*, 110(7–8): 485 (1999): Ligusticum pseudodaucoides.

LIGUSTICUM (Umbellif.).
pseudodaucoides *H. Peng & Yin Z. Wang in Novon*, 8(1): 50 (1998)—China South-central (Yunnan).
yushuense *J.T. Pan in Fl. Qinghaiica*, 2: 433 (1999)—Qinghai.

LILAEOPSIS (Umbellif.).
brisbanica *A.R. Bean in Austrobaileya*, 5(1): 145 (1997)—Queensland.
mauritiana *G. Petersen & J. Affolter in Novon*, 9(1): 92 (1999)—Mauritius.

LILIUM (Liliac.).
akkusianum *R. Gämperle in Quart. Bull. Alp. Gard. Soc.*, 66(3): 380 (1998)—Turkey.
floridum *J.L. Ma & Yan Jun Li in J. Wuhan Bot. Res.*, 18(2): 115 (2000)—Manchuria (Liaoning).
hansonii Leichtlin ex Baker
f. mutatum *Y.N. Lee, Fl. Korea:* 1163 (1996), without type—Locality not stated.
tsingtauense Gilg
f. carneum *(Nakai) Y.N. Lee, Fl. Korea:* 1163 (1996): L. carneum.

LIMACIA (Menispermac.).
selwynii *F.M. Bailey, Queensl. Fl.*, 1: 30 (1899); see also *F.M. Bailey, Queensl. Fl.*, 6: 1996 (1902)—Queensland.

LIMNOPHILA (Scrophulariac.).
aromaticoides *Y.P. Yang & S.H. Yen in Bot. Bull. Acad. Sin. (Taipei)*, 38(4): 285 (1997)—Taiwan.
taoyuanensis *Y.P. Yang & S.H. Yen in Bot. Bull. Acad. Sin. (Taipei)*, 38(4): 287 (1997)—Taiwan.

LIMODORUM (Orchidac.).
abortivum
var. rubrum *H. Sund. ex C.A.J. Kreutz in Eurorchis*, 9: 72 (1997)—Turkey.
rubriflorum *G. Bartolo & S. Pulvirenti in Caesiana*, 12: 3 (1999)—Turkey.
trabutianum L.
var. brulloi *(Bartolo & Pulv.) P. Delforge in Nat. Belg.*, 81(4): 397 (2000): Limodorum brulloi.

LIMONIASTRUM (Plumbaginac.).
rechingeri *J.R. Edmondson in Ann. Naturhist. Mus. Wien*, 98B(Suppl.): 400 (1996)—South Yemen, Somalia.

LIMONIUM (Plumbaginac.).
alicunense *F. Gómiz García in An. Jard. Bot. Madrid*, 53(2): 255 (1995 publ. 1996)—Spain.
aphroditae *R. Artelari & O. Georgiou in Bot. J. Linn. Soc.*, 131(4): 401 (1999)—Greece.
articulatum (Loisel.) Kuntze
subsp. pseudarticulatum *(Erben) O. de Bolòs & J. Vigo, Fl. Països Catalans*, 3: 94 (1995): L. pseudarticulatum.
caprariae *M. Rizzotto in Webbia*, 53(2): 255 (1999)—Italy.
carvalhoi *J.A. Rosselló & L. Sáez in An. Jard. Bot. Madrid*, 56(1): 24 (1998)—Baleares.
cordovillense *G. Stübing & S. Cirujano in An. Jard. Bot. Madrid*, 55(2): 471 (1997)—Spain.
cornarianum *Z. Kypriotakis & R. Artelari in Phyton (Austria)*, 38(1): 144 (1998)—Kriti.
cythereum *R. Artelari & O. Georgiou in Bot. J. Linn. Soc.*, 131(4): 404 (1999)—Greece.
dianiae *(Pau) A. Barber i Vallès, M.B. Crespo & M.D. Lledó in A. Barber i Vallès, Contrib. Coneix. Fl. Fitogeogr. Lit. Comarca Marina* Alta: 126 (1999): Statice dianiae.
furfuraceum (Lag.) Kuntze
subsp. lucentinum *(Pignatti & Freitag) O. de Bolòs & J. Vigo, Fl. Països Catalans*, 3: 75 (1995): L. lucentinum.
inexpectans *L. Sáez & J.A. Rosselló in An. Jard. Bot. Madrid*, 54(1): 286 (1996)—Baleares.
interjectum *J.X. Soler & J.A. Rosselló in An. Jard. Bot. Madrid*, 55(1): 11 (1997)—Spain.
leonardi-llorensii *L. Sáez, A.C. Carvalho & J.A. Rosselló in An. Jard. Bot. Madrid*, 56(1): 34 (1998)—Baleares.
melancholicum *S. Brullo, C. Marcenò & S. Romano in Candollea*, 51(1): 100 (1996)—Sicilia.
messeniacum *R. Artelari & G. Kamari in Bot. Chronika*, 13: 45 (2000)—Greece.
minutum (L.) Chaz.
var. pseudebusitanum *(Erben) O. de Bolòs & J. Vigo, Fl. Països Catalans*, 3: 94 (1995): L. pseudebusitanum.
subsp. pseudominutum *(Erben) O. de Bolòs & J. Vigo, Fl. Països Catalans*, 3: 77 (1995): L. pseudominutum.
montis-christi *M. Rizzotto in Webbia*, 53(2): 269 (1999)—Italy.
nydeggeri *M. Erben in Sendtnera*, 6: 103 (1999)—Portugal.
perplexum *L. Sáez & J.A. Rosselló in An. Jard. Bot. Madrid*, 57(1): 48 (1999)—Spain.

LIMONIUM:-
pinillense *R. Roselló & J.B. Peris in An. Jard. Bot. Madrid*, 55(2): 475 (1997)—Spain.
sinense (Girard) Kuntze
var. spinulosum *Y. Huang in Bull. Bot. Res. North-East. Forest. Univ.*, 17(4): 361 (1997)—China North-central (Shandong).
sommieranum *(Fiori) P.V. Arrigoni in P.V. Arrigoni & P.L. Di Tommaso, Carta Vegetaz. Isola Giannutri:* 7 (1981): Statice minuta var. sommieriana.
vigoi *L. Sáez, A. Curcó & J.A. Rosselló in An. Jard. Bot. Madrid*, 56(2): 270 (1998)—Spain.

LINANTHUS (Polemoniac.).
caespitosus *(Nutt.) J.M. Porter & L.A. Johnson in Aliso*, 19(1): 82 (2000): Leptodactylon caespitosum.
californicus *(Hook. & Arn.) J.M. Porter & L.A. Johnson in Aliso*, 19(1): 82 (2000): Leptodactylon californicum.
campanulatus *(A. Gray) J.M. Porter & L.A. Johnson in Aliso*, 19(1): 82 (2000): Gilia campanulata.
filiformis *(Parry ex A. Gray) J.M. Porter & L.A. Johnson in Aliso*, 19(1): 82 (2000): Gilia filiformis.
glaber *(R. Patt. & Yoder-Will.) J.M. Porter & L.A. Johnson in Aliso*, 19(1): 82 (2000), as 'glabrum': Leptodactylon glabrum.
inyoensis *(I.M. Johnst.) J.M. Porter & L.A. Johnson in Aliso*, 19(1): 82 (2000): Gilia inyoensis.
jaegeri *(Munz) J.M. Porter & L.A. Johnson in Aliso*, 19(1): 82 (2000): Gilia jaegeri.
jepsonii *D.W. Schemske & C. Goodwillie in Madroño*, 43(4): 453 (1996 publ. 1997)—California.
pungens *(Torr.) J.M. Porter & L.A. Johnson in Aliso*, 19(1): 82 (2000): Cantua pungens.
veachii *(Parry ex Greene) J.M. Porter & L.A. Johnson in Aliso*, 19(1): 82 (2000): Gilia veachii.

LINARIA (Scrophulariac.).
ser. Dalmaticeae *Klokov ex M.B. Kapanadze in Fl. Gruzii, ed. 2*, 12: 329 (1999).
ser. Kurdicae *Kuprian. ex M.B. Kapanadze in Fl. Gruzii, ed. 2*, 12: 329 (1999).
subsect. Macrocentrum *(D.A. Sutton) Ju.L. Menitsky in Bot. Zhurn.*, 84(1): 130 (1999): L. sect. Macrocentrum.
*ser.*Macrourae *Kuprian. ex M.B. Kapanadze in Fl. Gruzii, ed. 2*, 12: 329 (1999).
subsect. Pyramidales *Ju.L. Menitsky in Bot. Zhurn.*, 83(12): 130 (1998).
ser. Vulgares *Klokov ex M.B. Kapanadze in Fl. Gruzii, ed. 2*, 12: 329 (1999).
alpina (L.) Mill.
var. aciculifolia *(Braun-Blanq.) O. de Bolòs & J. Vigo, Fl. Països Catalans*, 3: 427 (1995): L. alpina subsp. aciculifolia.
arenosa *(Tzvelev) N.N. Tzvelev, Opred. Sosud. Rast. Severo-Zap. Rossii:* 680 (2000): Linaria vulgaris subsp. arenosa.
dalmatica (L.) Mill.
var. tessala *(Formánek) D. Delipavlov & M. Popova in Fl. Republ. Bŭlgariya*, 10: 116 (1995): L. dalmatica subsp. tessala.
depauperata Leresche ex Lange
subsp. hegelmaieri *(Lange) A. De la Torre, F. Alcaraz & M.B. Crespo in Lazaroa*, 16: 155 (1995 publ. 1996): L. hegelmaieri.
ikonnikovii *J. Stasiak in Fragm. Flor. Geobot.*, 41(2): 777 (1996)—Tadzhikistan.
f. dolomitica *J. Stasiak in Fragm. Flor. Geobot.*, 41(2): 778 (1996)—Tadzhikistan.
f. tanymasii *J. Stasiak in Fragm. Flor. Geobot.*, 41(2): 780 (1996)—Tadzhikistan.
ilergabona *M.B. Crespo & V.J. Arán in Flora Montiberica*, 14: 24 (2000)—Spain.
multicaulis (L.) Mill.
subsp. gigantea *(Dobignard & D. Jord.) A. Dobignard in Candollea*, 52(1): 134 (1997): L. heterophylla subsp. gigantea.
subsp. humilis *(Guss.) W. De Leonardis, G. Giardina & A. Zizza in Flora Medit.*, 9: 101 (1999), without basionym page: L. stricta var. humilis.
subsp. pseudosupina *(Ball) A. Dobignard in Candollea*, 52(1): 136 (1997): L. galioides var. pseudosupina.
pamirica *(Junussov) J. Stasiak in Fragm. Flor. Geobot.*, 41(2): 780 (1996): L. sessilis f. pamirica.

LINCONIA (Bruniac.).
ericoides *E.G.H. Oliver in Bothalia*, 29(2): 256 (1999)—Cape Province.

LINDENBERGIA (Scrophulariac.).
awashensis *M.L. Hjertson in Bot. J. Linn. Soc.*, 119(4): 313 (1995 publ. 1996)—Ethiopia.
fengkaiensis *R.H. Miau & Q.Y. Cen in Acta Sci. Nat. Univ. Sunyatseni*, 35(1): 104 (1996)—China Southeast (Guangdong).
serpyllifolia *M.L. Hjertson in Bot. J. Linn. Soc.*, 119(4): 317 (1995 publ. 1996)—Somalia.

LINDERA (Laurac.).
aggregata (Sims) Kosterm.
var. hemsleyana *(Diels) S.S. Ying in Mem. Coll. Agric. Nation. Taiwan Univ.*, 29(2): 65 (1989): L. strychnifolia var. hemsleyana.
gracilipes H.W. Li
var. macrocarpa *H. Zhu & H. Wang in Acta Bot. Yunnanica*, 20(1): 32 (1998)—China South-central (Yunnan).

LINDERNIA (Scrophulariac.).
andringitrae *Eb. Fischer in Bull. Mus. Nation. Hist. Nat., B, Adansonia*, Ser. 4, 17(3–4): 236 (1995), nom. nov.: Craterostigma cerastioides *Bonati*.
bonatii *Eb. Fischer in Bull. Mus. Nation. Hist. Nat., B, Adansonia*, Ser. 4, 17(3–4): 242 (1995), nom. nov., homonym; *see Philcox in Taxon*, 19: 649 (1970): Ilysanthes micrantha *Bonati*.
bryoides *Eb. Fischer in Bull. Mus. Nation. Hist. Nat., B, Adansonia*, Ser. 4, 17(3–4): 248 (1995)—Madagascar.
cowiei *W.R. Barker in J. Adelaide Bot. Gard.*, 18(2): 162 (1998)—Northern Territory.
dubia (L.) Pennell
var. mexicana *(S. Watson) D.Q. Lewis in Castanea*, 65(2): 112 (2000): Gratiola mexicana.
var. rhizomatosa *Pennell ex D.Q. Lewis in Castanea*, 65(2): 112 (2000)—Mexico Northwest (Sonora).
horombensis *Eb. Fischer in Bull. Mus. Nation. Hist. Nat., B, Adansonia*, Ser. 4, 17(3–4): 238 (1995)—Madagascar.
kigomensis *Eb. Fischer in Bull. Jard. Bot. Nation. Belg.*, 67(1–4): 362 (1999)—Tanzania.
mbalaensis *Eb. Fischer in Bull. Jard. Bot. Nation. Belg.*, 67(1–4): 361 (1999)—Zambia.
natans *Eb. Fischer in Bull. Mus. Nation. Hist. Nat., B, Adansonia*, Ser. 4, 17(3–4): 250 (1995)—Madagascar.
paludosa *(Bonati) Eb. Fischer in Bull. Mus. Nation. Hist. Nat., B, Adansonia*, Ser. 4, 17(3–4): 245 (1995): Ilysanthes paludosa.
pygmaea *(Bonati) Eb. Fischer in Bull. Mus. Nation. Hist. Nat., B, Adansonia*, Ser. 4, 17(3–4): 234 (1995): Craterostigma pygmaea.
rotundata *(Pilg.) Eb. Fischer in Bull. Jard. Bot. Nation. Belg.*, 67(1–4): 366 (1999): Ilysanthes rotundata.
tridentata *(Small) D.Q. Lewis in Castanea*, 65(2): 105 (2000): Ilysanthes tridentata.
viguieri *(Bonati) Eb. Fischer in Bull. Mus. Nation. Hist. Nat., B, Adansonia*, Ser. 4, 17(3–4): 240 (1995): Ilysanthes viguieri.

LINDMANIA (Bromeliac.).
candelabriforma *B.K. Holst in J. Bromeliad Soc.*, 46(4): 156 (1996)—Venezuela.
oliva-estevae *Steyerm. & L.B. Sm. ex B.K. Holst in J. Bromeliad Soc.*, 46(4): 158 (1996)—Venezuela.

LINGELSHEIMIA (Euphorbiac.).
abbayesii *(Leandri) A. Radcliffe-Smith in Kew Bull.*, 52(1): 172 (1997): Danguyodrypetes abbayesii.
ambigua *(Leandri) A. Radcliffe-Smith in Kew Bull.*, 52(1): 172 (1997): Danguyodrypetes ambigua.
fiherenensis *(Leandri) A. Radcliffe-Smith in Kew Bull.*, 52(1): 172 (1997): Danguyodrypetes fiherenensis.
sylvestris *(Radcl.-Sm.) A. Radcliffe-Smith in Kew Bull.*, 52(1): 172 (1997): Aerisilvaea sylvestris.

LINGNANIA (Gramin.).
distegia (Keng & Keng f.) Keng f.
f. flavidostriata *X. Jiang & Q. Li in J. Sichuan Agric. Coll.*, 2(2): 129 (1984)—China South-central (Sichuan).

LINGNANIA:-
intermedia *(J.R. Xue & T.P. Yi) T.P. Yi, Bamboos Fl. Sichuan:* 74 (1997), with basionym page; *et in Fl. Sichuanica,* 12: 51 (1997 publ. 1998), without basionym page: Bambusa intermedia.

LINNAEOPSIS (Gesneriac.).
alba (E.A. Bruce) B.L. Burtt
subsp. edwardsii *M. Weigend in Flora (Germany),* 195(1): 49 (2000)—Tanzania.
heckmanniana Engl.
subsp. gracilis *(E.A. Bruce) M. Weigend in Flora (Germany),* 195(1): 49 (2000): L gracilis.

LINOSPADIX (Palmae).
apetiolata *J.L. Dowe & A.K. Irvine in Principes,* 41(4): 215 (1997)—Queensland.

LINUM (Linac.).
sect. Heleolinum *T.V. Egorova in Novosti Sist. Vyssh. Rast.,* 31: 160 (1998), nom. nov.: L. sect. Halolinum *T.V. Egorova.*
bellidifolium *M. Gandoger, Dec. Pl. Nov.,* 1: 48 (1875)—France.
hirsutum L.
subsp. lanuginosum *(Juz.) T.V. Egorova in Fl. Vostochnoĭ Evropy,* 9: 358 (1996): L. lanuginosum.
komarovii Juz.
subsp. boreale *(Juz.) T.V. Egorova in Fl. Vostochnoĭ Evropy,* 9: 354 (1996): L. boreale.
nervosum Waldst. & Kit.
subsp. jailicola *(Juz.) T.V. Egorova in Bot. Zhurn.,* 85(7): 170 (2000): Linum jailicola.
rochelianum *M. Gandoger, Dec. Pl. Nov.,* 1: 48 (1875)—France.
seljukorum P.H. Davis
subsp. barsegjanii *(Gabrieljan & Dittrich) T.V. Egorova in Bot. Zhurn.,* 85(7): 172 (2000): Linum barsegjanii.
taymirense *G.A. Peschkova in Fl. Sibir. (Geraniac.-Cornac.),* 10: 28 (1996)—Krasnoyarsk.
ucranicum (Griseb. ex Planch.) Czern.
subsp. uralense *(Juz.) T.V. Egorova in Fl. Vostochnoĭ Evropy,* 9: 353 (1996): L. uralense.

LINZIA (Compos.).
gerberiformis *(Oliv. & Hiern) H. Robinson in Proc. Biol. Soc. Wash.,* 112(1): 237 (1999): Vernonia gerberiformis.
infundibulariformis *(Oliv. & Hiern) H. Robinson in Proc. Biol. Soc. Wash.,* 112(1): 237 (1999): Vernonia infundibulariformis.
ituriensis *(Muschl.) H. Robinson in Proc. Biol. Soc. Wash.,* 112(1): 237 (1999): Vernonia ituriensis.
melleri *(Oliv. & Hiern) H. Robinson in Proc. Biol. Soc. Wash.,* 112(1): 238 (1999): Vernonia melleri.
usafuensis *(O. Hoffm.) H. Robinson in Proc. Biol. Soc. Wash.,* 112(1): 238 (1999): Vernonia usafuensis.

LIPARIA (Legumin.).
sect. Decussatae *A.L. Schutte in Nordic J. Bot.,* 17(1): 26 (1997).
racemosa *A.L. Schutte in Nordic J. Bot.,* 17(1): 22 (1997)—Cape Province.

LIPARIS (Orchidac.).
sect. Aphyllum *G.A. Romero-Gonzalez & L.A. Garay in Harvard Pap. Bot.,* 4(2): 485 (1999).
sect. Decumbentes *L.A. Garay & G.A. Romero-Gonzalez in Harvard Pap. Bot.,* 4(2): 486 (1999).
sect. Tipuloidea *(Ridl.) L.A. Garay & G.A. Romero-Gonzalez in Harvard Pap. Bot.,* 4(2): 486 (1999): Microstylis sect. Tipuloidea.
abbreviata *(Blume) N.E. Brown, Suppl. Johnson's Gard. Dict.:* 954 (1882): Dendrochilum abbreviatum.
aphylla *G.A. Romero-Gonzalez & L.A. Garay in Harvard Pap. Bot.,* 4(2): 483 (1999)—Colombia, Ecuador.
ascendens *P.J. Cribb in Kew Bull.,* 51(2): 357 (1996)—Gabon, Congo.
bituberculata (Hook.) Lindl.
var. hachijoensis *(Nakai) T. Hashimoto in Proc. 12th World Orchid Conf., 1987 (eds. K. Saito & R. Tanaka):* 123 (1987): L. hachijoensis.
conopea *L.V. Averyanov in Bot. Zhurn.,* 82(3): 132, 131 (1997)—Vietnam.

LIPARIS:-
derchiensis *S.S. Ying in Mem. Coll. Agric. Nation. Taiwan Univ.,* 31(1): 21 (1991)—Taiwan.
emarginata *L.V. Averyanov in Bot. Zhurn.,* 82(3): 133 (1997)—Vietnam.
fujisanensis *F. Maek. ex F. Konta & S. Matsumoto in Ann. Tsukuba Bot. Gard.,* 16: 11 (1997)—Japan.
hallei *D.L. Szlachetko in Orchidee,* Suppl. 3: 43 (1996)—Gabon.
layardii F. Muell.
var. santoensis *T. Hashimoto in Ann. Tsukuba Bot. Gard.,* 17: 56 (1998)—Vanuatu.
nervosa (Thunb.) Lindl.
f. kappleri *(Rchb. f.) E.A. Christenson & Carnevali in Lindleyana,* 11(1): 17 (1996): Sturmia kappleri.
nigra Seidenf.
var. flava *L.V. Averyanov in Bot. Zhurn.,* 84(10): 128 (1999)—Vietnam.
pentagonalis *D.L. Szlachetko in Orchidee,* Suppl. 3: 40 (1996)—Cameroon.
rosseelii *T. Stévart in Syst. Geogr. Pl.,* 70(1): 141 (2000)—Gulf of Guinea Is. (São Tomé).
verticillata *G.A. Romero-Gonzalez & L.A. Garay in Harvard Pap. Bot.,* 4(2): 484 (1999)—Venezuela.
vexillifera (La Llave & Lex.) Cogn.
subsp. lindeniana *(A.Rich. & Galeotti) R.L. Dressler in Orquideologia,* 20(3): 264 (1997): Malaxis lindeniana.

LIPPIA (Verbenac.).
gratissima *(Gillies & Hook.) L.D. Benson in L.D. Benson & R.A. Darrow, Trees & Shrubs of Southwest. Deserts (ed. 3):* 202 (1981): Verbena gratissima.
var. schulziae *(Standl.) L.D. Benson in L.D. Benson & R.A. Darrow, Trees & Shrubs of Southwest. Deserts (ed. 3):* 203 (1981): L. ligustrina var. schulziae.
turnerifolia Cham.
var. polytricha *(Briq.) M.E. Múlgura de Romero in Candollea,* 55(2): 241 (2000): Lippia polytricha.

LIQUIDAMBAR (Hamamelidac.).
cerasifolia *(Wall. & Griff.) J.O. Voigt, Hort. Suburb. Calcutt.:* 301 (1845): Sedgwickia cerasifolia.

LIRIOPE (Convallariac.).
muscari (Decne.) L.H. Bailey
var. stolonifera *S. Hatusima in S. Hatusima & T. Amano, Fl. Ryukyus, S. of Amami Is., ed. 2:* 305 (1994), without latin descr. or type—Locality not stated.
yingdeensis *R.H. Miau in Acta Sci. Nat. Univ. Sunyatseni,* 1982(3): 75 (1982)—China Southeast (Guangdong).

LISOWSKIA *D.L. Szlachetko in Fragm. Flor. Geobot.,* Supp. 3: 121 (1995). *ORCHIDACEAE*
katangensis *(Summerh.) D.L. Szlachetko in Fragm. Flor. Geobot.,* Supp. 3: 121 (1995): Malaxis katangensis.
melanotoessa *(Summerh.) D.L. Szlachetko in Fragm. Flor. Geobot.,* Supp. 3: 121 (1995): Malaxis melanotoessa.
physuroides *(Schltr.) D.L. Szlachetko in Fragm. Flor. Geobot.,* Supp. 3: 121 (1995): Microstylis physuroides.
prorepens *(Kraenzl.) D.L. Szlachetko in Fragm. Flor. Geobot.,* Supp. 3: 121 (1995): Microstylis prorepens.
weberbaueriana *(Kraenzl.) D.L. Szlachetko in Fragm. Flor. Geobot.,* Supp. 3: 121 (1995), basionym as 'Microstylis': Liparis weberbaueriana.

LISSOCARPA (Lissocarpac.).
tetramera *(Rusby) P.E. Berry in Brittonia,* 51(2): 214 (1999): Diospyros tetramera.

LISTERA (Orchidac.).
puberula Maxim.
var. maculata *(T. Tang & F.T. Wang) S.C. Chen & Y.B. Luo in Fl. Reipubl. Popul. Sin.,* 17: 109 (1999): Listera savatieri var. maculata.

LISTERA:-
smallii Wiegand
var. nipponica *(Makino) T. Hashimoto in Proc. 12th World Orchid Conf., 1987 (eds. K. Saito & R. Tanaka):* 121 (1987): L. nipponica.

LITHOCARPUS (Fagac.).
brochidodromus *S. Julia & E. Soepadmo in Gard. Bull. Singapore,* 50(2): 125 (1998)—Borneo (Sabah, Sarawak).
corneri *S. Julia & E. Soepadmo in Gard. Bull. Singapore,* 50(2): 128 (1998)—Borneo (Sabah).
damiaoshanicus *C.C. Huang & Y.T. Chang in Guihaia,* 16(4): 301 (1996)—China Southeast (Guangxi).
kalkmanii *S. Julia & E. Soepadmo in Gard. Bull. Singapore,* 50(2): 130 (1998)—Borneo (Sabah).
keningauensis *S. Julia & E. Soepadmo in Gard. Bull. Singapore,* 50(2): 132 (1998)—Borneo (Sabah).
kochummenii *S. Julia & E. Soepadmo in Gard. Bull. Singapore,* 50(2): 134 (1998)—Borneo (Sarawak, Brunei).
melataiensis *S. Julia & E. Soepadmo in Gard. Bull. Singapore,* 50(2): 137 (1998)—Borneo (Sarawak).
muluensis *S. Julia & E. Soepadmo in Gard. Bull. Singapore,* 50(2): 139 (1998)—Borneo (Sarawak).
oblancifolius *S. Julia & E. Soepadmo in Gard. Bull. Singapore,* 50(2): 141 (1998)—Borneo (Sabah).
palungensis *C.H. Cannon & P.S. Manos in Bot. J. Linn. Soc.,* 133(3): 346 (2000)—Borneo (Sarawak).
pseudoxizangensis *Z.K. Zhou & H. Sun in Acta Bot. Yunnanica,* 18(2): 216 (1996)—Xizang.
sandakanensis *S. Julia & E. Soepadmo in Gard. Bull. Singapore,* 50(2): 143 (1998)—Borneo (Sabah).
stonei *S. Julia & E. Soepadmo in Gard. Bull. Singapore,* 50(2): 145 (1998)—Borneo (Sabah, Sarawak).
tawaiensis *S. Julia & E. Soepadmo in Gard. Bull. Singapore,* 50(2): 148 (1998)—Borneo (Sabah).

LITHOMYRTUS *F. Muell. in Hooker's J. Bot. Kew Gard. Misc.,* 9: 228 (1857). *MYRTACEAE*
cordata *(A.J. Scott) N. Snow & G.P. Guymer in Austrobaileya,* 5(2): 184 (1999): Myrtella cordata.
densifolia *N. Snow & G.P. Guymer in Austrobaileya,* 5(2): 185 (1999)—Northern Territory.
dunlopii *N. Snow & G.P. Guymer in Austrobaileya,* 5(2): 186 (1999)—Northern Territory.
grandifolia *N. Snow & G.P. Guymer in Austrobaileya,* 5(2): 189 (1999)—Northern Territory.
hypoleuca *F. Muell. ex N. Snow & G.P. Guymer in Austrobaileya,* 5(2): 191 (1999)—Northern Territory, Queensland.
kakaduensis *N. Snow & G.P. Guymer in Austrobaileya,* 5(2): 192 (1999)—Northern Territory.
linariifolia *N. Snow & G.P. Guymer in Austrobaileya,* 5(2): 193 (1999)—Northern Territory.
microphylla *(Benth.) N. Snow & G.P. Guymer in Austrobaileya,* 5(2): 193 (1999): Fenzlia obtusa var. microphylla.
obtusa *(Endl.) N. Snow & G.P. Guymer in Austrobaileya,* 5(2): 194 (1999): Fenzlia obtusa.
repens *N. Snow & G.P. Guymer in Austrobaileya,* 5(2): 197 (1999)—Northern Territory.
retusa *(Endl.) N. Snow & G.P. Guymer in Austrobaileya,* 5(2): 198 (1999): Fenzlia retusa.

LITHOPS (Aizoac.).
deboeri Schwantes
var. kennedyi *(de Boer) R. Heine in Mesemb Stud. Group Bull.,* 15(2): 30 (2000): L. fulleri var. kennedyi.
hermetica *D.T. Cole in Cactus & Co.,* 4(4): 158 (2000)—Namibia.

LITHOSCIADIUM (Umbellif.).
kamelinii *(V.M. Vinogr.) Pimenov ex I.A. Gubanov, Konsp. Fl. Vneshnei Mongolii (Sosud. Rast.):* 79 (1996): Cnidium kamelinii.

LITHOSPERMUM (Boraginac.).
multiflorum *Torr. ex A. Gray in Proc. Amer. Acad. Arts,* 10: 51 (1874)—Colorado.†

LITOTHAMNUS (Compos.).
nitidus *(DC.) W.C. Holmes in Phytologia,* 81(6): 386 (1996 publ. 1997): Eupatorium nitidum.

LITSEA (Laurac.).
akoensis Hayata
f. hayatae *(Kaneh.) S.S. Ying in Mem. Coll. Agric. Nation. Taiwan Univ.*, 25(1): 95 (1985): L. hayatae.
elongata (Wall. ex Nees) Benth.
var. acutivena *(Hayata) S.S. Ying, Coloured Illustr. Fl. Taiwan*, 1: 285 (March 1985); *et in Mem. Coll. Agric. Nation. Taiwan Univ.*, 25(1): 99 (April 1985): L. acutivena.†
glutinosa (Lour.) C.B. Rob.
var. glabraria *(Juss.) V. Mudgal & K.K. Khanna in Fl. Madhya Pradesh*, 2: 501 (1997): L. glabraria.
morrisonensis Hayata
var. lii *(C.E. Chang) S.S. Ying, Coloured Illustr. Fl. Taiwan*, 1: 280 (March 1985); *et in Mem. Coll. Agric. Nation. Taiwan Univ.*, 25(1): 97 (April 1985): L. lii.†
napoensis *D. Fang in Acta Phytotax. Sin.*, 37(6): 595 (1999)—China Southeast (Guangxi).
ochracea (Blume) Boerl.
var. oblanceolata *(Gamble) Kochummen ex I.M. Turner in Novon*, 6(2): 222 (1996): L. oblanceolata.
umbellata (Lour.) Merrill
var. fuscotomentosa *(Meisn.) I.M. Turner in Novon*, 6(2): 222 (1996): Tetranthera amara var. fuscotomentosa.

LIVISTONA (Palmae).
fulva *A.N. Rodd in Telopea*, 8(1): 103 (1998)—Queensland.
halongensis *T.H. Nguyen & R. Kiew in Gard. Bull. Singapore*, 52(1): 198 (2000)—Vietnam.
kimberleyana *A.N. Rodd in Telopea*, 8(1): 121 (1998)—Western Australia.
lanuginosa *A.N. Rodd in Telopea*, 8(1): 82 (1998)—Queensland.
mariae F. Muell.
subsp. occidentalis *A.N. Rodd in Telopea*, 8(1): 81 (1998)—Western Australia.
subsp. rigida *(Becc.) A.N. Rodd in Telopea*, 8(1): 80 (1998): L. rigida.
nitida *A.N. Rodd in Telopea*, 8(1): 96 (1998)—Queensland.
victoriae *A.N. Rodd in Telopea*, 8(1): 123 (1998)—Northern Territory, Western Australia.

LOASA (Loasac.).
aequatoriana *(Urb. & Gilg) M. Weigend in Sendtnera*, 3: 228 (1996): L. triphylla var. aequatoriana.
amaluzensis *M. Weigend in Bot. Jahrb.*, 118(2): 264 (1996)—Ecuador.
auca *M. Weigend in Bot. Jahrb.*, 118(2): 266 (1996)—Ecuador.
bicornuta *M. Weigend in Sendtnera*, 3: 229 (1996)—Peru, Ecuador.
bryoniifolia *H.A. Schrader in Index Seminum Horti Academici Gottingensis:* 4 (1831)—Locality not stated.†
campaniflora *Triana & Planch. ex I. Urban & E.F. Gilg, Monogr. Loas. (Nova Acta Leop., 76(1)):* 218 (1900)—Colombia, Ecuador.
glabra *M. Weigend in Bot. Jahrb.*, 118(2): 268 (1996)—Ecuador.
hornii *M. Weigend in Bot. Jahrb.*, 118(2): 254 (1996)—Ecuador.
jungifolia *M. Weigend in Bot. Jahrb.*, 118(2): 258 (1996)—Ecuador.
peltiphylla *M. Weigend in Sendtnera*, 3: 241 (1996)—Colombia, Ecuador.
perijensis *M. Weigend in Sendtnera*, 3: 235 (1996)—Venezuela, Colombia.
puma-chini *M. Weigend in Bot. Jahrb.*, 188(2): 270 (1996)—Ecuador.
ramirezii *M. Weigend in Sendtnera*, 3: 231 (1996)—Colombia.
roseoalba *M. Weigend in Sendtnera*, 3: 230 (1996)—Ecuador.
rubrastra *M. Weigend in Bot. Jahrb.*, 118(2): 272 (1996)—Ecuador.
santa-martae *M. Weigend in Sendtnera*, 3: 236 (1996)—Colombia.
triphylla Juss.
subsp. dyeri *(Urb. & Gilg) M. Weigend in Bot. Jahrb.*, 118(2): 288 (1996): L. dyeri.

LOASA:-
triphylla
subsp. papaverifolia *(Humb., Bonpl. & Kunth) M. Weigend in Sendtnera*, 3: 226 (1996): L. papaverifolia.
subsp. rudis *(Benth.) M. Weigend in Sendtnera*, 3: 227 (1996): L. rudis.
venezuelensis *(Steyerm.) M. Weigend in Sendtnera*, 3: 234 (1996): Cajophora venezuelensis.

LOBELIA (Campanulac.).
australiensis *T.G. Lammers in Novon*, 8(1): 34 (1998), nom. nov.: Pratia prostrata *E. Wimm.*
ayavacensis *J.A. Schultes in J.J. Roemer & J.A. Schultes, Syst. Veg.*, 5: 57 (1819)—South America.†
brasiliensis *A.O.S. Vieira & G.J. Shepherd in Novon*, 8(4): 457 (1998)—Brazil West-central (Brasília Distrito Federal).
brevisepala *(Y.S. Lian) T.G. Lammers in Novon*, 8(1): 34 (1998): Pratia brevisepala.
chinensis Lour.
var. tetrapetala *Y.N. Lee, Fl. Korea:* 1161 (1996), without type—Locality not stated.
diazlunae *J. Rzedowski & G. Calderón de Rzedowski in Acta Bot. Mex.*, 40: 59 (1997)—Mexico Southwest (Nayarit).
guerrerensis *M.J. Eakes & T.G. Lammers in Novon*, 9(3): 381 (1999)—Mexico Southwest (Guerrero).
hintoniorum *B.L. Turner in Phytologia*, 79(4): 293 (1995 publ. 1996)—Mexico Southwest (Oaxaca).
laxiflora Kunth
subsp. angustifolia *(A.DC.) M.J. Eakes & T.G. Lammers in Novon*, 9(3): 384 (1999): L. laxiflora var. angustifolia.
leucotos *D. Albrecht in Austrobaileya*, 5(4): 706 (2000)—Queensland.
macrodon *(Hook. f.) T.G. Lammers in Novon*, 8(1): 34 (1998): Pratia macrodon.
oligophylla *(Wedd.) T.G. Lammers in Novon*, 9(1): 74 (1999): Pratia oligophylla.
orientalis *J. Rzedowski & G. Calderón de Rzedowski in Acta Bot. Mex.*, 40: 62 (1997)—Mexico Northeast (Querétaro).
praetervisa *A. Borhidi in Acta Bot. Hung.*, 38(1–4): 192 (1993–94 publ. 1995)—Haiti.
pratiana *Gaudich. ex T.G. Lammers in Novon*, 8(1): 34 (1998), nom. nov.: Pratia repens *Gaudich.*
racemosa Sims
subsp. ryanii *(Rendle) C. Sastre in Biogeographica*, 73(1): 33 (1997): L. ryanii.
reflexisepala *T.G. Lammers in Novon*, 8(1): 34 (1998), nom. nov.: Pratia reflexa *Y.S. Lian.*
sonderiana *(Kuntze) T.G. Lammers in Novon*, 9(3): 386 (1999): Dortmanna sonderiana.
sylvicola *J. Lejoly & S. Lisowski in Acta Bot. Gallica*, 147(2): 120 (2000)—Equatorial Guinea, Gabon, Congo.
tibetica *W.L. Cheng in Acta Phytotax. Sin.*, 36(6): 549 (1998)—Xizang.

LOBIVIA (Cactac.).
bonnieae *J.J. Halda, Hogan & Janeba in Acta Mus. Richnov. Sect. Nat.*, 6(3): 235 (1999)—Argentina Northwest (Catamarca).
gonjianii *(R. Kiesling) R. Kiesling in Zuloaga & Morrone (eds.), Cat. Pl. Vasc. Rep. Arg. 2 (Monogr. Syst. Bot.* Missouri Bot. Gard. 74): 1245 (1999): Rebutia gonjianii.

LOBOSTEMON (Boraginac.).
belliformis *M.H. Buys in S. Afr. J. Bot.*, 62(1): 31 (1996)—Cape Province.
daltonii *M.H. Buys in S. Afr. J. Bot.*, 65(2): 144 (1999), vol. 65(1) at top of page in error—Cape Province.

LOCHEMIA (Sterculiac.).
supina *(DC.) Arn. ex J.O. Voigt, Hort. Suburb. Calcutt.:* 109 (1845): Riedlea supina.

LOCKHARTIA (Orchidac.).
subgen. Pseudobulbosa *D.E. Bennett & E.A. Christenson in Lindleyana*, 13(1): 51 (1998).
biserra *(Rich.) E.A. Christenson & Garay in Lindleyana*, 11(1): 17 (1996): Epidendrum biserrum.
genegeorgei *D.E. Bennett & E.A. Christenson in Lindleyana*, 13(1): 53 (1998)—Peru.

LOCKHARTIA:-
oblongicallosa *G. Carnevali & G.A. Romero in G.A. Romero & G. Carnevali, Orchids of Venezuela, ed. 2:* 1138 (2000)—Venezuela.
pandurata *F. Pupulin in Rev. Biol. Trop.*, 46(4): 998 (1998)—Costa Rica.

LODIA *A. Mosco & C. Zanovello in Bradleya*, 18: 44 (2000). *CACTACEAE*
mandragora *(Frič ex A. Berger) A. Mosco & C. Zanovello in Bradleya*, 18: 44 (2000): Echinocactus mandragora.

LOESELIA (Polemoniac.).
ciliata L.
subsp. echinophylla *(Brand) J.M. Porter & L.A. Johnson in Aliso*, 19(1): 78 (2000): L. ciliata var. echinophylla.
glandulosa (Cav.) G. Don
subsp. nepetifolia *(Cham. & Schltdl.) J.M. Porter & L.A. Johnson in Aliso*, 19(1): 78 (2000): Hoitzia nepetifolia.

LOESELIASTRUM (Polemoniac.).
depressum *(M.E. Jones ex A. Gray) J.M. Porter & L.A. Johnson in Aliso*, 19(1): 79 (2000): Gilia depressa.

LOGFIA (Compos.).
arizonica *(A. Gray) J. Holub in Preslia*, 70(2): 107 (1998): Filago arizonica.
californica *(Nutt.) J. Holub in Preslia*, 70(2): 107 (1998): Filago californica.
depressa *(A. Gray) J. Holub in Preslia*, 70(2): 107 (1998): Filago depressa.

LOLIUM (Gramin.).
edwardii *H. Scholz, C. Stierstorfer & M. von Gaisberg in Feddes Repert.*, 111(7–8): 561 (2000)—Canary Is.
grandispicum *Y.J. Fei in Guihaia*, 19(3): 205 (1999)—China South-central (Hubei).

LOMATIA (Proteac.).
bidwillii *R. Hogg in Gard. Year Book*, 1860: 75 (1860)—Australia.

LOMATIUM (Umbellif.).
hooveri *(Mathias & Constance) L. Constance & B. Ertter in Madroño*, 43(4): 519 (1996 publ. 1997): L. ciliolatum var. hooveri.
observatorium *L. Constance & B. Ertter in Madroño*, 43(4): 515 (1996 publ. 1997)—California.

LOMATOGONIUM (Gentianac.).
chilaiensis *Chih H. Chen & J.C. Wang in Bot. Bull. Acad. Sin. (Taipei)*, 41(4): 324 (2000)—Taiwan.
himalayense *(Klotzsch) E. Aitken in Edinburgh J. Bot.*, 53(1): 87 (1996): Pleurogyne himalayensis.
sichuanense *Z.Y. Zhu in Guihaia*, 20(4): 323 (2000)—China South-Central (Sichuan).

LOMATOPHYLLUM (Aloac.).
anivoranoense *W. Rauh & Hebding in Bradleya*, 16: 93 (1998)—Madagascar.
namorokaense *W. Rauh in Bradleya*, 16: 94 (1998)—Madagascar.
pembanum *(L.E. Newton) W. Rauh in Bradleya*, 16: 96 (1998): Aloe pembana.
peyrierasii *(Cremers) W. Rauh in Bradleya*, 16: 95 (1998): Aloe peyrierasii.
propaguliferum *W. Rauh & Razafindratsira in Bradleya*, 16: 93 (1998)—Madagascar.

LOMELOSIA (Dipsacac.).
sect. Callistemma *(Mert. & W.D.J. Koch) V. Mayer & F. Ehrendorfer in Pl. Syst. Evol.*, 216(1–2): 161 (1999): Scabiosa sect. Callistemma.
sect. Olivierianae *(Rech. f.) V. Mayer & F. Ehrendorfer in Pl. Syst. Evol.*, 216(1–2): 162 (1999): Scabiosa sect. Olivierianae.
× auriaultii *F. Lagarde in Monde Pl.*, 463: 15 (1998), as 'Lomeliosa', without type—Locality not stated. (L. cretica × graminifolia).
flavida (Boiss. & Hausskn.) Soják
f. enigmatica *(Rech. f.) V. Mayer & F. Ehrendorfer in Pl. Syst. Evol.*, 216(1–2): 162 (1999): Scabiosiopsis enigmatica.

LONCHOCARPUS (Legumin.).
chiangii *M. Sousa S. in An. Inst. Biol. Univ. Nac. Autón. Méx., Bot.*, 70(2): 137 (1999 publ. 2000)—Costa Rica, Nicaragua.
fluvialis *(Lindm.) R.H. Fortunato & R. Palese in Candollea*, 52(2): 510 (14 Nov. 1997): Coublandia fluvialis.
fluvialis *(Lindm.) E. Gómez-Sosa in Hickenia*, 2(51): 247 (15 Nov. 1997), isonym; *see Fortunato & Palese in Candollea*, 52(2): 510 (14 Nov. 1997): Coublandia fluvialis.
seorsus *(J.F. Macbr.) M. Sousa ex D.A. Neill, B.B. Klitgaard & G.P. Lewis in Cat. Vasc. Pl. Ecuador (Monogr. Syst. Bot. Missouri Bot Gard., 75)*: 954 (1999): Derris seorsa.

LONICERA (Caprifoliac.).
sect. Melanospermae *(Zaitsev ex Nedol.) R.V. Kamelin, Mater. Istorii Fl. Azii (Altaisk. Gorn. Strana)*: 27 (1998): L. subsect. Melanospermae.
alpigena L.
subsp. glutinosa *(Vis.) Kit Tan & J. Zieliński in Willdenowia*, 30(2): 232 (2000): Lonicera glutinosa.
subsp. hellenica *(Boiss.) Kit Tan & J. Zieliński in Willdenowia*, 30(2): 232 (2000): Lonicera hellenica.
caerulea L.
subsp. baltica *(Pojark.) N.N. Tzvelev in Novosti Sist. Vyssh. Rast.*, 32: 185 (2000): L. baltica.
var. pallasii *(Ledeb.) R. Cinovskis in Fl. Baltiĭskikh Respublik*, 2: 353 (1990): L. pallasii.
subsp. venulosa *(Maxim.) V.N. Voroschilov in A.K. Skvortsov (ed.), Florist. issl. v razn. raĭonakh SSSR*: 192 (1985): L. venulosa.
chrysantha
var. linearifolia *S.W. Liu & T.N. Ho in Fl. Qinghaiica*, 3: 511 (1996)—Qinghai.
gongshanensis *H. Li in Fl. Dulongjian Reg.*: 252 (1993), without latin descr. or type—China South-central (Yunnan).
guatemalensis *M.E. Véliz-Pérez & E. Carrillo in An. Inst. Biol. Univ. Nac. Autón. Méx., Bot.*, 71(1): 31 (2000)—Guatemala.
hispida Pall. ex Roem. & Schult.
var. anisocalyx *(Rehder) B.K. Zhou in Fl. Sichuanica*, 11: 136 (1994): L. anisocalyx.
jarmilae *J.J. Halda in Acta Mus. Richnov. Sect. Nat.*, 7(2): 73 (2000)—Kirgizistan.
macranthoides Hand.-Mazz.
var. heterotricha *(P.S. Hsu & H.J. Wang) B.K. Zhou in Fl. Sichuanica*, 11: 159 (1994): L. macrantha var. heterotricha.
omeiensis *(P.S. Hsu & H.J. Wang) B.K. Zhou in Fl. Sichuanica*, 11: 156 (1994): L. similis var. omeiensis.
taronensis *C.Y. Wu in Fl. Dulongjian Reg.*: 253 (1993), without latin descr. or type—China South-central (Yunnan).

LOPHIARIS (Orchidac.).
maculata *(Aubl.) J.D. Ackerman in Lindleyana*, 15(2): 92 (2000): Epidendrum maculatum.
microchila *(Lindl.) K. Senghas in Schlechter Orchideen, ed. 3*, I/C(33–36): 2130 (1997): Oncidium microchilum.
schwambachiae *(V.P. Castro & Toscano) K. Senghas in Schlechter Orchideen, ed. 3*, I/C(33–36): 2128 (1997): Oncidium schwambachiae.

LOPHOMYRTUS (Myrtac.).
aotearoana *E.C. Nelson in New Zealand J. Bot.*, 33(4): 558 (1995), nom. nov. inval. (Art. 34.2 ICBN, 1994), see Myrtus aotearoana (E.C. Nelson) E.C. Nelson: Myrtus bullata *Sol. ex A. Cunn.*

LOPHOPHORA (Cactac.).
diffusa
subsp. fricii *(Haberm.) J.J. Halda in Cactaceae etc.*, 7(2): 61 (1997); *cf. Repert. Pl. Succ. (I.O.S.)*, 48: 10 (1997 publ. 1998): L. fricii.
var. koehresii *J. Říha in Kaktusy (Brno)*, 32(3): 70 (1996)—Mexico Northeast (San Luis Potosí).
subsp. viridescens *J.J. Halda in Cactaceae etc.*, 7(2): 61 (1997), nom. nov.; *cf. Repert. Pl. Succ. (I.O.S.)*, 48: 10 (1997 publ. 1998): L. diffusa var. koehresii *Říha*.

LOPHOPHORA:-
viridescens *(Halda) J.J. Halda in Acta Mus. Richnov. Sect. Nat.*, 4(2): 71 (1997): L. diffusa subsp. viridescens.
williamsii
var. fricii *(Haberm.) R. Grym, Rod Lophophora*: 56 (1997): L. fricii.
subsp. grymi *J.J. Halda, J. Kupčák & J. Sladkovský in Acta Mus. Richnov. Sect. Nat.*, 7(1): 36 (2000)—Mexico Northeast (San Luis Potosí).
var. koehresii *(Říha) R. Grym, Rod Lophophora*: 55 (1997): L. diffusa var. koehresii.

LOPHOSTACHYS (Acanthac.).
sessilifolia *Pohl, Pl. Bras. icon. descr.*, 2: 96, t. 163 (1831), I.K. Orig. as 'sessiliflora' in error—Brazil west-central.

LORANTHUS (Loranthac.).
lageniferus *Wight ex J. Graham, Cat. Pl. Bombay*: 86 (1839)—India (Maharashtra).

LOTONONIS (Legumin.).
carnea *B.-E. van Wyk in S. Afr. J. Bot.*, 55(6): 653 (1989)—Cape Province.
magnifica *B.-E. van Wyk in S. Afr. J. Bot.*, 55(6): 647 (1989)—Cape Province.
plicata *B.-E. van Wyk in S. Afr. J. Bot.*, 55(6): 649 (1989)—Cape Province.
rostrata Benth.
subsp. brachybotrys *B.-E. van Wyk in S. Afr. J. Bot.*, 55(6): 658 (1989)—Cape Province.
subsp. namaquensis *(Bolus) B.-E. van Wyk in S. Afr. J. Bot.*, 55(6): 658 (1989): L. namaquensis.
sparsiflora *(E. Mey.) B.-E. van Wyk in S. Afr. J. Bot.*, 55(6): 660 (1989): Crotalaria sparsiflora.
stenophylla *(Eckl. & Zeyh.) B.-E. van Wyk in S. Afr. J. Bot.*, 55(6): 652 (1989): Crotalaria stenophylla.

LOTUS (Legumin.).
sect. Pedrosia *(Lowe) B. Valdés in Lagascalia*, 21(1): 255 (1999): gen. Pedrosia.
sect. Pseudosimpeteria *T.E. Kramina & D.D. Sokoloff in Adansonia*, 19(2): 326 (1997).
arinagensis *D. Bramwell in Bot. Macaronésica*, 22: 114 (1995)—Canary Is.
corniculatus L.
subsp. ambiguus *(Besser ex Spreng.) N.N. Tzvelev in Novosti Sist. Vyssh. Rast.*, 32: 184 (2000): L. ambiguus.
subsp. callunetorum *(Üksip) N.N. Tzvelev in Novosti Sist. Vyssh. Rast.*, 32: 184 (2000): L. corniculatus f. callunetorum.
subsp. glacialis *(Boiss.) B. Valdés Castrillón in An. Jard. Bot. Madrid*, 57(2): 454 (1999 publ. 2000): L. corniculatus var. glacialis.
subsp. komarovii *(Miniaev) N.N. Tzvelev in Novosti Sist. Vyssh. Rast.*, 32: 185 (2000): L. komarovii.
subsp. ruprechtii *(Miniaev) N.N. Tzvelev in Novosti Sist. Vyssh. Rast.*, 32: 184 (2000): L. ruprechtii.
subsp. ucrainicus *(Klokov) N.N. Tzvelev in Novosti Sist. Vyssh. Rast.*, 32: 185 (2000): L. ucrainicus.
mearnsii (Britton) Greene
var. equisolensis *J.L. Anderson in Madroño*, 43(2): 261 (1996)—Arizona.

LOURTEIGIA (Compos.).
morenoi *V.M. Badillo in Ernstia*, 10(1): 14 (2000)—Venezuela.

LOXOCARYA (Restionac.).
albipes *J.S. Pate & K.A. Meney in Telopea*, 6(4): 660 (1996)—Western Australia.
magna *K.A. Meney & K.W. Dixon in Telopea*, 6(4): 662 (1996)—Western Australia.
striata *(F. Muell.) B.G. Briggs & L.A.S. Johnson in Telopea*, 8(1): 29 (1998): Megalotheca striata.

LUCILIA (Compos.).
paramora *(S.F. Blake) V.M. Badillo in Biollania*, ed. esp. 6 (J.J. Wurdack Festschr.): 236 (1997): Gnaphalium paramorum.
pinceana Hook.
var. pubescens *(W.C. Chen) W.C. Chen in Fl. Reipubl. Popul. Sin.*, 71(1): 241 (1999): L. intermedia var. pubescens.

LUDWIGIA (Onagrac.).
grandiflora (Michx.) Greuter & Burdet
subsp. hexapetala *(Hook. & Arn.) G.L. Nesom & J.T. Kartesz in Castanea*, 65(2): 125 (2000): Jussiaea hexapetala.
× kentiana *E.J. Clement in Watsonia*, 23(1): 168 (2000)—Great Britain. (L. palustris × repens).

LUECKELIA *R. Jenny in Austral. Orchid Rev.*, 64(4): 15 (1999). *ORCHIDACEAE*
breviloba *(Summerh. ex E.W. Cooper) R. Jenny in Austral. Orchid Rev.*, 64(4): 15 (1999): Polycycnis breviloba.

LUISIA (Orchidac.).
appressifolia *L.V. Averyanov in Lindleyana*, 15(2): 79 (2000)—Vietnam.
curtisii *G. Seidenfaden, Contrib. Orchid Fl. Thailand XIII*: 49 (1997)—Malaya (Peninsular Malaysia).

LUPINASTER (Legumin.).
pentaphyllus Moench
subsp. albus *(Link) N.N. Tzvelev in Novosti Sist. Vyssh. Rast.*, 32: 185 (2000): L. albus.

LUPINUS (Legumin.).
sect. Polyphylli *N.N. Tzvelev, Opred. Sosud. Rast. Severo-Zap. Rossii*: 680 (2000).
albifrons Benth. ex Lindl.
var. hallii *(Abrams) D. Isely, Native & Natural. Legumin. (Fabac.) U.S. (excl. Alaska & Hawaii)*: 675 (1998): L. hallii.
var. longifolius *(S. Watson) D. Isely, Native & Natural. Legumin. (Fabac.) U.S. (excl. Alaska & Hawaii)*: 675 (1998): L. chamissonis var. longifolius.
ammophilus *E.L. Greene in Pittonia*, 4: 136 (1900), in Supp. 2 as 'anemophilus' in error—New Mexico, Colorado.†
var. crassus *(Payson) D. Isely, Native & Natural. Legumin. (Fabac.) U.S. (excl. Alaska & Hawaii)*: 676 (1998): L. crassus.
anatolicus *W. Święcicki & W.K. Święcicki in Genet. Resourc. Crop Evol.*, 43(2): 116 (1996)—Turkey.
arbustus Douglas ex Lindl.
var. calcaratus *(Kellogg) D. Isely, Native & Natural. Legumin. (Fabac.) U.S. (excl. Alaska & Hawaii)*: 680 (1998), isonym; *see S.L. Welsh in Great Basin Nat.*, 38(3): 324 (1978): L. calcaratus.
argenteus Pursh
var. lemmonii *(C.P. Sm.) D. Isely, Native & Natural. Legumin. (Fabac.) U.S. (excl. Alaska & Hawaii)*: 684 (1998): L. lemmonii.
argenteus Pursh
var. myrianthus *(Greene) D. Isely, Native & Natural. Legumin. (Fabac.) U.S. (excl. Alaska & Hawaii)*: 687 (1998): L. myrianthus.
× hispanicoluteus *W. Święcicki & W.K. Święcicki in Acta Soc. Bot. Pol.*, 68(3): 220 (1999)—Cultivation. (L. hispanicus × luteus).
jaimehintonianus *B.L. Turner in Phytologia*, 79(2): 102 (1995 publ. 1996), as 'jaimehintoniana'—Mexico Southwest (Oaxaca).
latifolius J. Agardh
var. leucanthus *(Rydb.) D. Isely, Native & Natural. Legumin. (Fabac.) U.S. (excl. Alaska & Hawaii)*: 703 (1998): L. leucanthus.
var. simplex *(C.P. Sm.) D. Isely, Native & Natural. Legumin. (Fabac.) U.S. (excl. Alaska & Hawaii)*: 703 (1998): L. columbianus var. simplex.
lepidus Douglas ex Lindl.
var. ashlandensis *(B.J. Cox) D. Isely, Native & Natural. Legumin. (Fabac.) U.S. (excl. Alaska & Hawaii)*: 705 (1998): L. aridus subsp. ashlandensis.
var. constancei *(T.W. Nelson & J.P. Nelson) D. Isely, Native & Natural. Legumin. (Fabac.) U.S. (excl. Alaska & Hawaii)*: 705 (1998): L. constancei.
littoralis Douglas ex Lindl.
var. layneae *(Eastw.) D. Isely, Native & Natural. Legumin. (Fabac.) U.S. (excl. Alaska & Hawaii)*: 711 (1998): L. layneae.
var. variicolor *(Steud.) D. Isely, Native & Natural. Legumin. (Fabac.) U.S. (excl. Alaska & Hawaii)*: 711 (1998): L. variicolor.

LUPINUS:-
rubriflorus *A.M. Planchuelo in Brittonia,* 48(2): 263 (1996)—Brazil South (Rio Grande do Sul).
sericeus Pursh
var. jonesii *(Rydb.) S.L. Welsh ex D. Isely, Native & Natural. Legumin. (Fabac.) U.S. (excl. Alaska & Hawaii):* 726 (1998): L. jonesii.

LUVUNGA (Rutac.).
monophylla *(DC.) D.J. Mabberley in Telopea,* 7(4): 334 (1998): Triphasia monophylla.

LUZULA (Juncac.).
sect. Effusae *K.F. Wu in J. E. China Norm. Univ., Nat. Sci. Ed.,* 1992(3): 91 (1992).
ser. Inaequales *K.F. Wu in J. E. China Norm. Univ., Nat Sci. Ed.,* 1992(3): 97 (1992).
sect. Spicatae *K.F. Wu in J. E. China Norm. Univ., Nat. Sci. Ed.,* 1992(3): 101 (1992).
badia *K.F. Wu in J. E. China Norm. Univ., Nat. Sci. Ed.,* 1992(3): 99 (1992)—Xinjiang.
bomiensis *K.F. Wu in J. E. China Norm. Univ., Nat. Sci. Ed.,* 1992(3): 100 (1992)—Xizang.
confusa Lindeb.
f. minor *V.B. Kuvaev in V.B. Kuvaev et al., Rast. pokrov ostrova Sibiryakova:* 55 (1994), without type—Krasnoyarsk.
effusa Buchenau
var. chinensis *(N.E. Br.) K.F. Wu in J. E. China Norm. Univ., Nat. Sci. Ed.,* 1992(3): 92 (1992): L. chinensis.
hieronymii Buchenau & Griseb.
f. pusilla *(Castillón) L. Novara in Hickenia,* 2(31): 138 (1995): L. hieronymii var. pusilla.
inaequalis *K.F. Wu in J. E. China Norm. Univ., Nat. Sci. Ed.,* 1992(3): 97 (1992)—China Southeast (Jiangxi).
jilongensis *K.F. Wu in J. E. China Norm. Univ., Nat. Sci. Ed.,* 1992(3): 95 (1992)—Xizang.
mannii *(Buchenau) J. Kirschner & M. Cheek in Kew Bull.,* 55(4): 900 (2000): L. campestris var. mannii.
subsp. gracilis *(S. Carter) J. Kirschner & M. Cheek in Kew Bull.,* 55(4): 901 (2000): L. campestris var. gracilis.
multiflora (Ehrh.) Lej.
subsp. hibernica *J. Kirschner & T.C.G. Rich in Watsonia,* 21(1): 93 (1996)—Ireland.
subsp. monticola *J. Kirschner in Pl. Syst. Evol.,* 200(1–2): 6 (1996)—Spain.
odaesanensis *Y.N. Lee & Y.O. Chae in Y.N. Lee, Fl. Korea:* 1163 (1996), contrary to Arts. 37.4, 37.5 ICBN (1994)—South Korea.
pallescens (Wahlenb.) Sw.
var. castanescens *K.F. Wu in J. E. China Norm. Univ., Nat. Sci. Ed.,* 1992(3): 99 (1992)—Manchuria (Jilin).
sichuanensis *K.F. Wu in J. E. China Norm. Univ., Nat. Sci. Ed.,* 1992(3): 92 (1992)—China South-central (Sichuan).

LYCASTE (Orchidac.).
jarae *D.E. Bennett & E.A. Christenson in Orchid Dig.,* 60(1): 14 (1996)—Peru.
skinneri (Bateman ex Lindl.) Lindl.
f. virginalis *(Scheidw.) E.A. Christenson in Lindleyana,* 11(1): 18 (1996): Maxillaria virginalis.

LYCHNIS (Caryophyllac.).
atropurpurea (Griseb.) Nyman
subsp. sartorii *(Boiss.) K. Micevski in Prilozi Oddel. Biol. Med. Nauki Makedonska Akad. Nauk. Umet.,* 9(1–2): 48 (1988 publ. 1991): Viscaria sartorii.
fulgens Fisch.
var. glabra *T.A. Bezdeleva in Sosud. Rast. Sovet. Dal'nego Vostoka,* 8: 113 (1996)—Primorye.
sibirica L.
var. arenosa *G.A. Peschkova, Stepnaya Fl. Baikal'skoĭ Sibiri:* 56 (1972), without type—Buryatiya.
f. glabra *G.A. Peschkova, Stepnaya Fl. Baikal'skoĭ Sibiri:* 56 (1972), without type—Buryatiya.
viscaria L.
subsp. subnivalis *P.P. Panov in Fitologiya,* 48: 8 (1996)—Bulgaria.

LYCHNOPHORA (Compos.).
sericea *D.J.N. Hind in Kew Bull.,* 55(2): 394 (2000)—Brazil Northeast (Bahia).

LYCIANTHES (Solanac.).
biflorum (Lour.) Bitter
var. glabrum *(Koidz. ex Hats.) S. Hatusima in S. Hatusima & T. Amano, Fl. Ryukyus, S. of Amami Is., ed. 2:* 193 (1994), without basionym ref.: Basionym not stated.
crassipetala *(Wall.) R.R. Mill in Edinburgh J. Bot.,* 57(3): 465 (2000), as 'crassipetalum': Solanum crassipetalum.
gongshanensis *H. Li in Fl. Dulongjian Reg.:* 288 (1993), without latin descr. or type—China South-central (Yunnan).
jalicensis *E. Dean in Novon,* 8(2): 133 (1998)—Mexico Southwest (Jalisco).

LYCIUM (Solanac.).
ser. Ruthenica *Pojark. ex L.I. Ivanina in Bot. Zhurn.,* 85(6): 149 (2000).
mascarenense *A.M. Venter & A.J. Scott in S. Afr. J. Bot.,* 65(5–6): 428 (1999)—Rodrigues, Mauritius, Réunion, Madagascar, Mozambique, Natal.

LYCOPERSICON (Solanac.).
sect. Juglandifolium *(Rydb.) J.M.H. Shaw in New Plantsman,* 5(2): 109 (1998): Solanum [no rank indicated] Juglandifolium.
sect. Lycopersicoides *(A. Child) A. Child ex J.M.H. Shaw in New Plantsman,* 5(2): 109 (1998): Solanum subsect. Lycopersicoides.
juglandifolium *(Dunal) J.M.H. Shaw in New Plantsman,* 5(2): 109 (1998): Solanum juglandifolium.
lycopersicoides *(Dunal) A. Child ex J.M.H. Shaw in New Plantsman,* 5(2): 109 (1998): Solanum lycopersicoides.
ochranthum *(Dunal) J.M.H. Shaw in New Plantsman,* 5(2): 109 (1998): Solanum ochranthum.
sitiens *(I.M. Johnst.) J.M.H. Shaw in New Plantsman,* 5(2): 109 (1998): Solanum sitiens.

LYCOPUS (Labiat.).
alissoviae *N.S. Probatova in Sosud. Rast. Sovet. Dal'nego Vostoka,* 7: 347 (1995)—Primorye.
charkeviczii *N.S. Probatova in Sosud. Rast. Sovet. Dal'nego Vostoka,* 7: 351 (1995)—Primorye.
cokeri *H.E. Ahles ex B.A. Sorrie in Castanea,* 62(2): 124 (1997)—North Carolina, South Carolina.
kurilensis *N.S. Probatova in Sosud. Rast. Sovet. Dal'nego Vostoka,* 7: 350 (1995)—Kuril Is.
sichotensis *N.S. Probatova in Sosud. Rast. Sovet. Dal'nego Vostoka,* 7: 351 (1995)—Primorye.

LYCORIS (Amaryllidac.).
× chejuensis *S. Kurita & P.S. Hsu D.E. Boufford & H. Ohba (eds.), Sino-Japan. Fl. Its Char. & Diversific.* (Univ. Mus. Univ. Tokyo Bull. 37): 178 (1998), homonym; *see K. Tae & S. Ko in Korean J. Pl. Taxon.,* 23(4): 234 (1993)—South Korea. (L. chinensis × flavescens × sanguinea var. koreana).
flavescens *M. Kim & S. Lee in Korean J. Pl. Taxon.,* 21(2): 127 (1991), with two types cited—South Korea.
var. uydoensis *M. Kim in Korean J. Pl. Taxon.,* 26(4): 263 (1996), species name not validly publ.—South Korea.

LYGINIA (Restionac.).
excelsa *B.G. Briggs & L.A.S. Johnson in Telopea,* 8(4): 497 (2000)—Western Australia.

LYSILOMA (Legumin.).
latisiliquum (L.) Benth.
var. latifoliolatum *(A. Barreto & Yakovlev) M. Bässler in Fl. Repúbl. Cuba,* Ser. A, 2: 148 (1998): L. latisiliquum subsp. latifoliolatum.

LYSIMACHIA (Primulac.).
chingshuiensis *C.I Peng & C.M. Hu in Bot. Bull. Acad. Sin. (Taipei),* 40(1): 49 (1999)—Taiwan.
christinae Hance
subsp. tanakae *(Maxim.) C.M. Hu in J. Trop. Subtrop. Bot.,* 4(3): 15 (1996): L. tanakae.

LYSIMACHIA:-
iniki *K.L. Marr in Pac. Sci.,* 51(3): 270 (1997)—Hawaiian Is.
longshengensis *G.Z. Li & S.C. Tang in Guihaia,* 19(4): 342 (1999)—China Southeast (Guangxi).
pendens *K.L. Marr in Pac. Sci.,* 51(3): 275 (1997)—Hawaiian Is.
perfoliata *Hand.-Mazz.in Oesterr. Bot. Z.,* 88(4): 305 (1939)—China Southeast (Anhui).
remyi Hillebr.
subsp. caliginis *(H. St. John) K.L. Marr in Pac. Sci.,* 51(3): 277 (1997): L. caliginis.
subsp. kipahuluensis *(H. St. John) K.L. Marr in Pac. Sci.,* 51(3): 278 (1997): L. kipahuluensis.
subsp. subherbacea *(Hillebr.) K.L. Marr in Pac. Sci.,* 51(3): 281 (1997): L. hillebrandii var. subherbacea.
scopulensis *K.L. Marr in Pac. Sci.,* 51(3): 282 (1997)—Hawaiian Is.
stenosepala Hemsl.
var. lutea *Z.E. Zhao & D.X. Li in J. Wuhan Bot. Res.,* 18(4): 284 (2000)—China South-central (Hubei).
violascens Franch.
var. brevistamina *Y.Y. Qian in Guihaia,* 18(4): 354 (1998)—China South-central (Yunnan).

LYSIONOTUS (Gesneriac.).
microphyllus W.T. Wang
var. omeiensis *(W.T. Wang) W.T. Wang in Novon,* 7(4): 429 (1997 publ. 1998): L. omeiensis.
palinensis *G.D. Pal in J. Bombay Nat. Hist. Soc.,* 97(1): 131 (2000)—Assam (Arunachal Pradesh).
pauciflorus Maxim.
var. ikedae *(Hatus.) W.T. Wang in Novon,* 7(4): 430 (1997 publ. 1998): L. ikedae.

LYSIPOMIA (Campanulac.).
caespitosa *T.J. Ayers in Brittonia,* 49(4): 434 (1997)—Ecuador.
cylindrocarpa *T.J. Ayers in Brittonia,* 49(4): 435 (1997)—Ecuador.
speciosa *T.J. Ayers in Brittonia,* 49(4): 437 (1997)—Ecuador.

LYTHRUM (Lythrac.).
anatolicum *E. Leblebici & Ö. Seçmen in Turkish J. Bot.,* 19(5): 559 (1995)—Turkey.
lydiae *A.K. Sytin in Novosti Sist. Vÿssh. Rast.,* 32: 124 (2000)—Russia East.

M

MABEA (Euphorbiac.).
bahiensis *M. Emmerich in Bradea,* 5(26): 289 (1989)—Brazil Northeast (Bahia).
juinensis *M. Emmerich in Bradea,* 5(26): 285 (1989)—Brazil North (Rondonia), Brazil West-central (Mato Grosso).
prancei *M. Emmerich in Bol. Mus. Nac. Rio de Janeiro, Bot.,* 62: 4 (1981)—Brazil North (Amazonas).
tenorioi *M. Martínez Gordillo, J. Jiménez Ramírez & R. Cruz Durán in An. Inst. Biol. Univ. Nac. Autón. Méx., Bot.,* 71(2): 93 (2000)—Mexico Southwest (Oaxaca).

MACADAMIA (Proteac.).
erecta *J.A. McDonald & Ismail R. in Harvard Pap. Bot.,* 7: 7 (1995)—Sulawesi.

MACARANGA (Euphorbiac.).
angulata *S.J. Davies in Kew Bull.,* 54(1): 150 (1999)—Borneo (Sabah, Brunei, Sarawak).
glandibracteolata *S.J. Davies in Kew Bull.,* 54(1): 147 (1999)—Borneo (Sabah).
grallata *G. McPherson in Bull. Mus. Nation. Hist. Nat., B, Adansonia,* Ser. 4, 18(3–4): 275 (1996)—Madagascar.
henricorum *W.B. Hemsley in J. Linn. Soc., Bot.,* 26: 442 (1894)—Hainan.
huahineensis *J. Florence in Bull. Mus. Nation. Hist. Nat., B, Adansonia,* Ser. 4, 18(3–4): 263 (1996)—Society Is.

MACARANGA:-
moluccana *A. Radcliffe-Smith & R. Govaerts in Kew Bull.*, 52(2): 480 (1997), nom. nov.: Adelia cordifolia *Roxb.*
truncata *J. Florence in Fl. Polynésie française*, 1: 111 (1997)—Society Is.
velutiniflora *S.J. Davies in Harvard Pap. Bot.*, 4(2): 433 (1999)—Borneo (Sarawak).

MACARTHURIA (Molluginac.).
keigheryi *B.J. Lepschi in Nuytsia*, 11(1): 48 (1996)—Western Australia.
vertex *B.J. Lepschi in Nuytsia*, 11(1) 51 (1996)—Northern Territory, Western Australia.

× **MACCOLMARA** *hort. in Orchid Rev.*, 107(1226): 126 (1999). *ORCHIDACEAE* (Cochleanthes × Colax × Promenaea × Zygopetalum).

MACDONALDIA (Orchidac.).
apiculata *(A.S. George) D.L. Szlachetko in Fragm. Flor. Geobot.*, Supp. 3: 112 (1995): Thelymitra variegata var. apiculata.
carnea *(R. Br.) D.L. Szlachetko in Fragm. Flor. Geobot.*, Supp. 3: 112 (1995): Thelymitra carnea.
cucullata *(Rupp) D.L. Szlachetko in Fragm. Flor. Geobot.*, Supp. 3: 113 (1995): Thelymitra cucullata.
flexuosa *(Endl.) D.L. Szlachetko in Fragm. Flor. Geobot.*, Supp. 3: 113 (1995): Thelymitra flexuosa.
mackibbinii *(F. Muell.) D.L. Szlachetko in Fragm. Flor. Geobot.*, Supp. 3: 113 (1995) Thelymitra mackibbinii.
macmillanii *(F. Muell.) D.L. Szlachetko in Fragm. Flor. Geobot.*, Supp. 3: 113 (1995) Thelymitra macmillanii.
matthewsii *(Cheeseman) D.L. Szlachetko in Fragm. Flor. Geobot.*, Supp. 3: 113 (1995): Thelymitra matthewsii.
psammophila *(C.R.P. Andrews) D.L. Szlachetko in Fragm. Flor. Geobot.*, Supp. 3: 113 (1995): Thelymitra psammophila.
venosa *(R. Br.) D.L. Szlachetko in Fragm. Flor. Geobot.*, Supp. 3: 113 (1995), isonym; *see Lindl., Sketch Veg. Swan R.*: 50 (1840): Thelymitra venosa.

MACHAERANTHERA (Compos.).
carnosa (A. Gray) G.L. Nesom
var. intricata *(A. Gray) G.L. Nesom in Phytologia*, 82(2): 106 (1997): Bigelovia intricata.

MACHAERINA (Cyperac.).
austrobrasiliensis *M.T. Strong in Novon*, 7(3): 317 (1997)—Brazil South (Paraná, Rio Grande do Sul, Santa Catarina).
ayangannensis *M.T. Strong in Novon*, 7(3): 313 (1997)—Guyana.
raiateensis *(J.W. Moore) S.L. Welsh, Fl. Societensis*: 320 (1998): Mariscus raiateensis.

MACHILUS (Laurac.).
pachyclada *D. Fang in Acta Phytotax. Sin.*, 37(6): 597 (1999)—China Southeast (Guangxi).

MACLEANIA (Ericac.).
subgen. Aponema *J.L. Luteyn in Biollania*, ed. esp. 6 (J.J. Wurdack Festschr.): 459 (1997).
alata *J.L. Luteyn in Fl. Ecuador*, 54(147): 137 (1996)—Ecuador.
dodsonii *J.L. Luteyn in Fl. Ecuador*, 54(147): 138 (1996)—Ecuador.
maldonadensis *J.L. Luteyn in Fl. Ecuador*, 54(147): 133 (1996)—Ecuador.
smithiana *J.L. Luteyn in Fl. Ecuador*, 54(147): 131 (1996)—Ecuador.
subsessilis *J.L. Luteyn in Fl. Ecuador*, 54(147): 142 (1996)—Ecuador.

MACLEAYA (Papaverac.).
cordata (Willd.) R.Br.
f. glabra *H. Ohba in J. Jap. Bot.*, 75(3): 186 (2000)—Japan.

MACOUBEA (Apocynac.).
mesoamericana *J.F. Morales in Novon*, 9(1): 86 (1999)—Costa Rica, Panama.

MACRADENIA (Orchidac.).
brassavolae Rchb. f.
var. albiflora *K. Senghas & Seeger in Orchidee*, 47(4): centre page pull-out, Orchideenkartei 832 (1996)—Colombia.

MACROCARPAEA (Gentianac.).
autanae *Weaver ex B. Maguire in Mem. New York Bot. Gard.*, 32: 336 (1981)—Venezuela.
harlingii *J.S. Pringle in Fl. Ecuador*, 53(159A): 101 (1995)—Ecuador.
marahuacae *L. Struwe & V.A. Albert in Harvard Pap. Bot.*, 3(2): 182 (1998)—Venezuela.
neblinae *B. Maguire & J.A. Steyermark in Mem. New York Bot. Gard.*, 32: 332 (1981)—Venezuela.
piresii *B. Maguire in Mem. New York Bot. Gard.*, 32: 338 (1981)—Brazil North (Amazonas).

MACROCLINIUM (Orchidac.).
alleniorum *R.L. Dressler & F. Pupulin in Lindleyana*, 11(1): 34 (1996), as 'allenorum'—Costa Rica.
confertum *F. Pupulin in Lindleyana*, 11(3): 138 (1996)—Costa Rica.
doderoi *D.E. Mora & F. Pupulin in Selbyana*, 18(1): 7 (1997)—Costa Rica.
escobarianum *Dodson ex F. Pupulin in Orquideologia*, 21(3): 272 (2000)—Colombia.
generalense *F. Pupulin in Lindleyana*, 11(3): 136 (1996)—Costa Rica.
robustum *F. Pupulin & D.E. Mora in Selbyana*, 18(1): 7 (1997)—Costa Rica.

MACRODITASSA (Asclepiadac.).
lagoensis
var. cucullata *(E. Fourn.) J. Fontella Pereira & M.V. Ferreira in Bradea*, 8(18): 102 (1998): Ditassa cucullata.
marianae *J. Fontella Pereira & M.V. Ferreira in Bradea*, 8(18): 102 (1998)—Brazil Southeast (São Paulo).
morilloana *J. Fontella Pereira & M.V. Ferreira in Bradea*, 8(18): 101 (1998)—Brazil Southeast (Minas Gerais).

MACROLOBIUM (Legumin.).
savannarum R.S. Cowan
var. foldatsii *(R.S. Cowan) P.E. Berry in Fl. Venezuelan Guayana (gen. eds. J.A. Steyermark et al.)*, 4: 80 (1998): M. foldatsii.

MACROPHARYNX (Apocynac.).
steyermarkii *(Markgr.) J.F. Morales in Rhodora*, 99(899): 258 (1997 publ. 1998): Asketanthera steyermarkii.

MACROPIPER (Piperac.).
excelsum
f. delangei *R.O. Gardner in New Zealand J. Bot.*, 35(3): 302 (1997)—New Zealand North I.
subsp. peltatum *R.O. Gardner in New Zealand J. Bot.*, 35(3): 301 (1997)—New Zealand North I.
f. peltatum *R.O. Gardner in New Zealand J. Bot.*, 35(3): 302 (1997)—New Zealand North I.

MACROPSYCHANTHUS (Legumin.).
lauterbachii Harms
var. glabricalyx *(Verdc.) F. Adema in Blumea*, 43(1): 236 (1998): M. lauterbachii subsp. glabricalyx.
var. parviflorus *(Verdc.) F. Adema in Blumea*, 43(1): 237 (1998): M. lauterbachii subsp. parviflorus.

MACROPTILIUM (Legumin.).
arenarium *(Bacigalupo) S.I. Drewes & R.A. Palacios in Bol. Soc. Argent. Bot.*, 30(3–4): 187 (1995): M. prostratum var. arenarium.
supinum *(Wiggins & Rollins) A. Delgado-Salinas & L. Torres-Colín in An. Inst. Biol. Univ. Nac. Autón. Méx., Bot.*, 66(2): 189 (1995 publ. 1996): Phaseolus supinus.

MACROSAMANEA (Legumin.).
amplissima *(Ducke) R.C. Barneby & J.W. Grimes, Silk Tree, Guanacaste, Monkey's Earring (Mem. New York Bot. Gard., 74(1))*: 198 (1996): Pithecellobium amplissimum.

MACROSAMANEA:-
consanguinea *(R.S. Cowan) R.C. Barneby & J.W. Grimes, Silk Tree, Guanacaste, Monkey's Earring (Mem. New York Bot. Gard., 74(1))*: 189 (1996): Pithecellobium consanguineum.
discolor (Willd.) Britton & Rose
var. arenicola *(R.S. Cowan) R.C. Barneby & J.W. Grimes, Silk Tree, Guanacaste, Monkey's Earring (Mem. New York Bot. Gard., 74(1))*: 193 (1996): Pithecellobium arenicola.
duckei *(Huber) R.C. Barneby & J.W. Grimes, Silk Tree, Guanacaste, Monkey's Earring (Mem. New York Bot. Gard., 74(1))*: 193 (1996): Pithecellobium duckei.
froesii *R.C. Barneby & J.W. Grimes, Silk Tree, Guanacaste, Monkey's Earring (Mem. New York Bot. Gard., 74(1))*: 191 (1996)—Brazil North (Amazonas).
macrocalyx *(Ducke) R.C. Barneby & J.W. Grimes, Silk Tree, Guanacaste, Monkey's Earring (Mem. New York Bot. Gard., 74(1))*: 186 (1996): Pithecellobium macrocalyx.
prancei *(Barneby) R.C. Barneby & J.W. Grimes, Silk Tree, Guanacaste, Monkey's Earring (Mem. New York Bot. Gard., 74(1))*: 200 (1996): Pithecellobium prancei.
pubiramea *(Steud.) R.C. Barneby & J.W. Grimes, Silk Tree, Guanacaste, Monkey's Earring (Mem. New York Bot. Gard., 74(1))*: 194 (1996): Inga pubiramea.
var. lindsaeifolia *(Benth.) R.C. Barneby & J.W. Grimes, Silk Tree, Guanacaste, Monkey's Earring (Mem. New York Bot. Gard., 74(1))*: 197 (1996): Pithecellobium lindsaeifolium.

MACROZAMIA (Zamiac.).
cardiacensis *P.I. Forster & D.L. Jones in Fl. Australia*, 48: 717 (1998)—Queensland.
concinna *D.L. Jones in Fl. Australia*, 48: 718 (1998)—New South Wales.
elegans *K.D. Hill & D.L. Jones in Fl. Australia*, 48: 718 (1998)—New South Wales.
glaucophylla *D.L. Jones in Fl. Australia*, 48: 718 (1998)—New South Wales.
humilis *D.L. Jones in Fl. Australia*, 48: 719 (1998)—New South Wales.
longispina *P.I. Forster & D.L. Jones in Fl. Australia*, 48: 717 (1998)—Queensland.
montana *K.D. Hill in Fl. Australia*, 48: 717 (1998)—New South Wales.
polymorpha *D.L. Jones in Fl. Australia*, 48: 718 (1998)—New South Wales.
reducta *K.D. Hill & D.L. Jones in Fl. Australia*, 48: 718 (1998)—New South Wales.

MACULIGILIA *V. Grant in Phytologia*, 84(2): 78 (1998 publ. 1999). *POLEMONIACEAE*
maculata *(Parish) V. Grant in Phytologia*, 84(2): 79 (1998 publ. 1999): Gilia maculata.

MADANGIA *P.I. Forster, D.J. Liddle & I.M. Liddle in Austrobaileya*, 5(1): 53 (1997). *ASCLEPIADACEAE*
inflata *P.I. Forster, D.J. Liddle & I.M. Liddle in Austrobaileya*, 5(1): 54 (1997)—New Guinea (Papua New Guinea).

MADHUCA (Sapotac.).
chai-ananii *P. Chantaranothai in Nordic J. Bot.*, 18(4): 494 (1998)—Thailand.
chiangmaiensis *P. Chantaranothai in Nordic J. Bot.*, 18(4): 493 (1998)—Thailand.
fusca *(Engl.) L.L. Forman in Checklist Flow. Pl. & Gymn. Brunei Darussalam (ed. M.J.E. Coode et al.)*: 439 (1996): Illipe fusca.
klackenbergii *P. Chantaranothai in Nordic J. Bot.*, 18(4): 494 (1998)—Thailand.
krabiensis *(Aubrév.) P. Chantaranothai in Novon*, 8(3): 230 (1998): Aesandra krabiensis.
lecomtei *(Aubrév.) P. Chantaranothai in Novon*, 8(3): 230 (1998): Aesandra lecomtei.
palembanica *(Miq.) L.L. Forman in Checklist Flow. Pl. & Gymn. Brunei Darussalam (ed. M.J.E. Coode et al.)*: 439 (1996): Podocarpus palembanicus.
smitinandii *P. Chantaranothai in Nordic J. Bot.*, 18(4): 496 (1998)—Thailand.

MADIA (Compos.).
exigua *(Sm.) A. Gray in Proc. Amer. Acad. Arts,* 8: 391 (1872): Sclerocarpus exiguus.†

MADLABIUM *I.C. Hedge in Fl. Madagascar,* fam. 175: 260 (1998). *LABIATAE*
magenteum *I.C. Hedge in Fl. Madagascar,* fam. 175: 261 (1998)—Madagascar.

MAESA (Myrsinac.).
beamanii *T.M.A. Utteridge in Kew Bull.,* 55(2): 443 (2000)—New Guinea (Irian Jaya).
brevipaniculata *(C.Y. Wu & C. Chen) J.J. Pipoly & C. Chen in Novon,* 5(4): 357 (1995): M. parvifolia var. brevipaniculata.
cambodiana *C.M. Hu & J.E. Vidal in J. Trop. Subtrop. Bot.,* 5(1): 15 (1997)—Cambodia, Vietnam.
confusa *(C.M. Hu) J.J. Pipoly & C. Chen in Novon,* 5(4): 357 (1995): M. consanguinea var. confusa.
kerrii *C.M. Hu & J.E. Vidal in J. Trop. Subtrop. Bot.,* 5(1): 16 (1997)—Laos.
laevis *C.M. Hu & J.E. Vidal in J. Trop. Subtrop. Bot.,* 5(1): 16 (1997)—Vietnam.
lanyuensis *Yuen P. Yang in Bot. Bull. Acad. Sin. (Taipei),* 40(1): 44 (1999)—Taiwan.
perlaria (Lour.) Merr.
var. formosana *(Mez) Yuen P. Yang in Fl. Taiwan, ed. 2,* 4: 52 (1998): M. formosana.
procumbens *T.M.A. Utteridge in Kew Bull.,* 55(2): 446 (2000)—New Guinea (Irian Jaya).
reflexa *T.M.A. Utteridge & R.M.K. Saunders in Kew Bull.,* 55(4): 981 (2000)—Philippines.

MAGNOLIA (Magnoliac.).
sect. Cylindricae *S.A. Spongberg in D.R. Hunt (ed.), Magnolias & their allies:* 115 (1998).
amazonica *(Ducke) R. Govaerts in D.G. Frodin & R. Govaerts, World Checklist & Bibliogr. Magnoliac.:* 70 (1996): Talauma amazonica.
arcabucoana *(Lozano) R. Govaerts in D.G. Frodin & R. Govaerts, World Checklist & Bibliogr. Magnoliac.:* 70 (1996): Talauma arcabucoana.
argyrotricha *(Lozano) R. Govaerts in D.G. Frodin & R. Govaerts, World Checklist & Bibliogr. Magnoliac.:* 70 (1996): Dugandiodendron argyrotrichum.
boliviana *(M. Nee) R. Govaerts in D.G. Frodin & R. Govaerts, World Checklist & Bibliogr. Magnoliac.:* 26 (1996): Talauma boliviana.
× brooklynensis *G. Kalmbacher in Newslett. Amer. Magnolia Soc.,* 8(2): 7 (1972), without diagnostic latin descr.—Cultivation. (M. acuminata × liliflora).
calimaensis *(Lozano) R. Govaerts in D.G. Frodin & R. Govaerts, World Checklist & Bibliogr. Magnoliac.:* 70 (1996): Dugandiodendron calimaense.
calophylla *(Lozano) R. Govaerts in D.G. Frodin & R. Govaerts, World Checklist & Bibliogr. Magnoliac.:* 70 (1996): Dugandiodendron calophyllum.
cararensis *(Lozano) R. Govaerts in D.G. Frodin & R. Govaerts, World Checklist & Bibliogr. Magnoliac.:* 70 (1996): Dugandiodendron cararense.
caricifragrans *(Lozano) R. Govaerts in D.G. Frodin & R. Govaerts, World Checklist & Bibliogr. Magnoliac.:* 70 (1996): Talauma caricifragrans.
cespedesii *(Triana & Planch.) R. Govaerts in D.G. Frodin & R. Govaerts, World Checklist & Bibliogr. Magnoliac.:* 70 (1996): Talauma cespedesii.
championii Benth.
subsp. fistulosa *(Finet & Gagnep.) J. Li in Acta Bot. Yunnanica,* 19(2): 133 (1997): Talauma fistulosa.
chocoensis *(Lozano) R. Govaerts in D.G. Frodin & R. Govaerts, World Checklist & Bibliogr. Magnoliac.:* 70 (1996): Talauma chocoensis.
colombiana *(Little) R. Govaerts in D.G. Frodin & R. Govaerts, World Checklist & Bibliogr. Magnoliac.:* 70 (1996): Talauma colombiana.
dixonii *(Little) R. Govaerts in D.G. Frodin & R. Govaerts, World Checklist & Bibliogr. Magnoliac.:* 70 (1996): Talauma dixonii.
dodecapetala *(Lam.) R. Govaerts in D.G. Frodin & R. Govaerts, World Checklist & Bibliogr. Magnoliac.:* 70 (1996): Annona dodecapetala.

MAGNOLIA:-
espinalii *(Lozano) R. Govaerts in D.G. Frodin & R. Govaerts, World Checklist & Bibliogr. Magnoliac.:* 70 (1996): Talauma espinalii.
georgii *(Lozano) R. Govaerts in D.G. Frodin & R. Govaerts, World Checklist & Bibliogr. Magnoliac.:* 70 (1996): Talauma georgii.
gilbertoi *(Lozano) R. Govaerts in D.G. Frodin & R. Govaerts, World Checklist & Bibliogr. Magnoliac.:* 70 (1996): Talauma gilbertoi.
globosa Hook. f. & Thomson
subsp. wilsonii *(Finet & Gagnep.) J. Li in Acta Bot. Yunnanica,* 19(2): 134 (1997): M. parviflora var. wilsonii.
gloriensis *(Pittier) R. Govaerts in D.G. Frodin & R. Govaerts, World Checklist & Bibliogr. Magnoliac.:* 71 (1996): Talauma gloriensis.
× gotoburgensis *T.G. Nitzelius in Magnolia,* 28(1): 5, 3 (1992), contrary to Art. 37.3 ICBN (2000)—Cultivation.(M. hypoleuca × wilsonii).
guatapensis *(Lozano) R. Govaerts in D.G. Frodin & R. Govaerts, World Checklist & Bibliogr. Magnoliac.:* 71 (1996): Dugandiodendron guatapense.
henaoi *(Lozano) R. Govaerts in D.G. Frodin & R. Govaerts, World Checklist & Bibliogr. Magnoliac.:* 71 (1996): Talauma henaoi.
hernandezii *(Lozano) R. Govaerts in D.G. Frodin & R. Govaerts, World Checklist & Bibliogr. Magnoliac.:* 71 (1996): Talauma hernandezii.
irwiniana *(Lozano) R. Govaerts in D.G. Frodin & R. Govaerts, World Checklist & Bibliogr. Magnoliac.:* 71 (1996): Talauma irwiniana.
katiorum *(Lozano) R. Govaerts in D.G. Frodin & R. Govaerts, World Checklist & Bibliogr. Magnoliac.:* 71 (1996): Talauma katiorum.
lenticellata *(Lozano) R. Govaerts in D.G. Frodin & R. Govaerts, World Checklist & Bibliogr. Magnoliac.:* 71 (1996): Dugandiodendron lenticellatum.
liliifera (L.) Baill.
var. angatensis *(Blanco) R. Govaerts in D.G. Frodin & R. Govaerts, World Checklist & Bibliogr. Magnoliac.:* 71 (1996): M. angatensis.
var. beccarii *(Ridl.) R. Govaerts in D.G. Frodin & R. Govaerts, World Checklist & Bibliogr. Magnoliac.:* 71 (1996): Talauma beccarii.
var. obovata *(Korth.) R. Govaerts in D.G. Frodin & R. Govaerts, World Checklist & Bibliogr. Magnoliac.:* 71 (1996): Talauma obovata.
var. singapurensis *(Ridl.) R. Govaerts in D.G. Frodin & R. Govaerts, World Checklist & Bibliogr. Magnoliac.:* 71 (1996): Talauma singapurensis.
× loebneri *P. Kache in Gartenschönheit,* 1: 20 (1920)—Cultivation.(M. kobus × stellata).
macrophylla
subsp. dealbata *(Zucc.) J.D. Tobe in D.R. Hunt (ed.), Magnolias & their allies:* 170 (1998): M. dealbata.
magnifolia *(Lozano) R. Govaerts in D.G. Frodin & R. Govaerts, World Checklist & Bibliogr. Magnoliac.:* 71 (1996): Dugandiodendron magnifolium.
mahechae *(Lozano) R. Govaerts in D.G. Frodin & R. Govaerts, World Checklist & Bibliogr. Magnoliac.:* 71 (1996): Dugandiodendron mahechae.
minor *(Urb.) R. Govaerts in D.G. Frodin & R. Govaerts, World Checklist & Bibliogr. Magnoliac.:* 71 (1996): Talauma minor.
morii *(Lozano) R. Govaerts in D.G. Frodin & R. Govaerts, World Checklist & Bibliogr. Magnoliac.:* 71 (1996): Talauma morii.
narinensis *(Lozano) R. Govaerts in D.G. Frodin & R. Govaerts, World Checklist & Bibliogr. Magnoliac.:* 71 (1996): Talauma narinensis.
neillii *(Lozano) R. Govaerts in D.G. Frodin & R. Govaerts, World Checklist & Bibliogr. Magnoliac.:* 71 (1996): Talauma neillii.
officinalis Rehder & E.H. Wilson
subsp. biloba *(Rehder & E.H. Wilson) Y.W. Law in Acta Phytotax. Sin.,* 34(1): 91 (Feb. 1996); *et in Fl. Reipubl. Popul. Sin.,* 30(1): 271 (May 1996): M. officinalis var. biloba.
polyhypsophylla *(Lozano) R. Govaerts in D.G. Frodin & R. Govaerts, World Checklist & Bibliogr. Magnoliac.:* 71 (1996): Talauma polyhypsophylla.

MAGNOLIA:-
rimachii *(Lozano) R. Govaerts in D.G. Frodin & R. Govaerts, World Checklist & Bibliogr. Magnoliac.:* 71 (1996): Talauma rimachii.
sambuensis *(Pittier) R. Govaerts in D.G. Frodin & R. Govaerts, World Checklist & Bibliogr. Magnoliac.:* 72 (1996): Talauma sambuensis.
santanderiana *(Lozano) R. Govaerts in D.G. Frodin & R. Govaerts, World Checklist & Bibliogr. Magnoliac.:* 72 (1996): Talauma santanderiana.
sellowiana *(A. St.-Hil.) R. Govaerts in D.G. Frodin & R. Govaerts, World Checklist & Bibliogr. Magnoliac.:* 72 (1996): Talauma sellowiana.
silvioi *(Lozano) R. Govaerts in D.G. Frodin & R. Govaerts, World Checklist & Bibliogr. Magnoliac.:* 72 (1996): Talauma silvioi.
sirindhorniae *H.P. Nooteboom & P. Chalermglin in Blumea,* 45(1): 245 (2000)—Thailand.
urraoensis *(Lozano) R. Govaerts in D.G. Frodin & R. Govaerts, World Checklist & Bibliogr. Magnoliac.:* 72 (1996): Dugandiodendron urraoense.
venezuelensis *(Lozano) R. Govaerts in D.G. Frodin & R. Govaerts, World Checklist & Bibliogr. Magnoliac.:* 72 (1996): Talauma venezuelensis.
virolinensis *(Lozano) R. Govaerts in D.G. Frodin & R. Govaerts, World Checklist & Bibliogr. Magnoliac.:* 72 (1996): Talauma virolinensis.
wolfii *(Lozano) R. Govaerts in D.G. Frodin & R. Govaerts, World Checklist & Bibliogr. Magnoliac.:* 72 (1996): Talauma wolfii.
wugangensis *T.B. Zhao, W.B. Sun & Zhi X. Chen in Acta Bot. Yunnanica,* 21(2): 171 (1999)—China Southeast (Henan).
yarumalensis *(Lozano) R. Govaerts in D.G. Frodin & R. Govaerts, World Checklist & Bibliogr. Magnoliac.:* 72 (1996): Dugandiodendron yarumalense.

MAHONIA (Berberidac.).
dulongensis *H. Li in Fl. Dulongjian Reg.:* 67 (1993), without latin descr. or type—China South-central (Yunnan).
microphylla *T.S. Ying & G.R. Long in Acta Phytotax. Sin.,* 37(3): 282 (1999)—China Southeast (Guangxi).

MAIANTHEMUM (Convallariac.).
comaltepecense *A. Espejo Serna, A.R. López-Ferrari & J. Ceja Romero in Acta Bot. Mex.,* 36: 22 (1996)—Mexico Southwest (Oaxaca).
cordatum *H. Li in Fl. Dulongjian Reg.:* 319 (1993), without latin descr. or type—China South-central (Yunnan).
fusciduliflorum *(Kawano) S.C. Chen & S. Kawano in Novon,* 10(2): 113 (2000): Smilacina fusciduliflora.
henryi (Baker) LaFrankie
var. szechuanicum *(F.T. Wang & T. Tang) H. Li in Fl. Yunnanica,* 7: 747 (1997): Oligobotrya szechuanica.
stenolobum *(Franch.) S.C. Chen & S. Kawano in Novon,* 10(2): 113 (2000): Tovaria stenoloba.
tatsienense (Franch.) LaFrankie
var. stenolobum *(Franch.) H. Li in Fl. Yunnanica,* 7: 744 (1997): Tovaria stenoloba.

MAIHUENIOPSIS (Cactac.).
bonnieae *(D.J. Ferguson & R. Kiesling) E.F. Anderson in Cact. Succ. J. (U.S.A.),* 71(6): 325 (1999): Puna bonnieae.
clavarioides *(Otto ex Pfeiff.) E.F. Anderson in Cact. Succ. J. (U.S.A.),* 71(6): 325 (1999): Opuntia clavarioides.
subterranea *(R.E. Fr.) E.F. Anderson in Cact. Succ. J. (U.S.A.),* 71(6): 325 (1999): Opuntia subterranea.

MALACOTHRIX (Compos.).
foliosa A. Gray
subsp. crispifolia *W.S. Davis in Madroño,* 44(3): 234 (1997 publ. 1998)—California.
subsp. philbrickii *W.S. Davis in Madroño,* 44(3): 236 (1997 publ. 1998)—California.
subsp. polycephala *W.S. Davis in Madroño,* 44(3): 237 (1997 publ. 1998)—California.
junakii *W.S. Davis in Madroño,* 44(3): 241 (1997 publ. 1998)—California.

MALANEA (Rubiac.).
sect. Obtusae *J.H. Kirkbride in Brittonia,* 49(3): 366 (1997).
harleyi *J.H. Kirkbride in Brittonia,* 49(3): 366 (1997)—Brazil Northeast (Bahia).

MALAXIS (Orchidac.).
sect. Umbellulatae *(Ridl.) F. Barros in Bol. Bot. (São Paulo),* 15: 33 (1996): Microstylis sect. Umbellulatae.
alvaroi *J. García Cruz, R. Jiménez Machorro & L. Sánchez Saldaña in Sida,* 18(2): 414 (1998)—Mexico Central (Morelos), Mexico Northeast (Guanajuato).
buchtienii *(Schltr.) E.A. Christenson in Lindleyana,* 11(1): 19 (1996): Microstylis buchtienii.
casillasii *R. González Tamayo in Bol. Inst. Bot. (Guadalajara),* 3(1–3): 123 (1995 publ. 1997)—Mexico Southwest (Jalisco).
cipoensis *F. Barros in Bol. Bot. (São Paulo),* 15: 31 (1996)—Brazil Southeast (Minas Gerais).
contrerasii *R. Gonzalez Tamayo in Orchidophile,* 138: 180 (1999)—Mexico Southwest (Jalisco).
insularis *(H. Dietr. & M.A. Díaz) M.A. Nir, Orchidaceae Antillanae:* 234 (2000): M. domingensis subsp. insularis.
iwashinae *T. Yukawa & T. Hashimoto in Ann. Tsukuba Bot. Gard.,* 17: 64 (1998)—Vanuatu.
latifolia Sm.
var. nana *S.S. Ying in Mem. Coll. Agric. Nation. Taiwan Univ.,* 25(2): 101 (1985)—Taiwan.
licatae *G. Carnevali & I.M. Ramírez M.in Harvard Pap. Bot.,* 3(2): 244 (1998), as 'licatiae'—Venezuela.
lyonnetii *G.A. Salazar in Brittonia* 49(4): 449 (1997)—Mexico Central (Morelos).
monophyllos (L.) Sw.
var. obtusa *H. Tsukaya & H. Ohba in J. Jap. Bot.,* 75(5): 314 (2000)—Nepal.
nidiae *G. Carnevali & I.M. Ramírez M.in Harvard Pap. Bot.,* 3(2): 246 (1998)—Venezuela.
ophrydis *(J. König) P. Ormerod in G. Seidenfaden, Descr. Epidendrorum J.G. König 1791:* 18 (1995): Epidendrum ophrydis.
ottonis *(Schltr.) G. Carnevali & I.M. Ramírez M.in Harvard Pap. Bot.,* 3(2): 246 (1998) Microstylis ottonis.
palustris *A. Espejo Serna & A.R. López-Ferrari in Sida,* 18(2): 412 (1998)—Mexico Central (Morelos).
perezii *R. González Tamayo in Bol. Inst. Bot. (Guadalajara),* 4(1–3): 50 (1996 publ. 1997)—Mexico Southwest (Jalisco).
roblesgiliana *R. González Tamayo in Bol. Inst. Bot. (Guadalajara),* 6(2–3): 219 (1998 publ. 2000), as 'roblesgilana'—Mexico Southwest (Nayarit).
rodrigueziana *R. González Tamayo in Bol. Inst. Bot. (Guadalajara),* 2(3–4): 97 (1994), as 'rodriguezana'—Mexico Southwest (Jalisco, Nayarit).
tridentula *(Schltr.) E.A. Christenson in Lindleyana,* 11(1): 19 (1996): Microstylis tridentula.

× **MALCOLMCAMPBELLARA** *hort. in Orchid Rev.,* 105(1218): 382 (1997). *ORCHIDACEAE* (Drymoanthus × Plectorrhiza × Sarcochilus).

MALCOLMIA (Crucif.).
humilis *Z.X. An in Fl. Xinjiangensis,* 2(2): 378 (1995)—Xinjiang.

MALLEASTRUM (Meliac.).
contractum *J.-F. Leroy in Bull. Mus. Nation. Hist. Nat., B, Adansonia,* Ser. 4, 18(1–2): 16 (1996)—Madagascar.
horokoke *J.-F. Leroy in Bull. Mus. Nation. Hist. Nat., B, Adansonia,* Ser. 4, 18(1–2): 17 (1996)—Madagascar.
minutifoliolatum *J.-F. Leroy in Bull. Mus. Nation. Hist. Nat., B, Adansonia,* Ser. 4, 18(1–2): 14 (1996)—Madagascar.
orientale *J.-F. Leroy in Bull. Mus. Nation. Hist. Nat., B, Adansonia,* Ser. 4, 18(1–2): 18 (1996)—Madagascar.
schatzii *J.-F. Leroy & M. Lescot in Bull. Mus. Nation. Hist. Nat., B, Adansonia,* Ser 4, 18(1–2): 18 (1996)—Madagascar.
tampolense *J.-F. Leroy in Bull. Mus. Nation. Hist. Nat., B, Adansonia,* Ser. 4, 18(1–2): 13 (1996)—Madagascar.

MALLEOLA (Orchidac.).
seidenfadenii *E.A. Christenson in Lindleyana,* 13(4): 232 (1998)—Thailand.

MALLOTUS (Euphorbiac.).
canii *Nguyen Nghia Thin, Euphorbiac. Vietnam:* 46 (1995 publ. 1996)—Vietnam.
capuronii *(Leandri) G. McPherson in Bull. Mus. Nation. Hist. Nat., B, Adansonia,* Ser. 4, 17(3–4): 170 (1995): Deuteromallotus capuronii.
cordatifolius *J.W.F. Slik in Blumea,* 43(1): 225 (1998)—Philippines.
dispersus *P.I. Forster in Austrobaileya,* 5(3): 465 (1999)—Queensland, Western Australia, Northern Territory.
grandistipularis *J.W.F. Slik in Blumea,* 43(1): 227 (1998)—Sumatera.
hanheoensis *Nguyen Nghia Thin, Euphorbiac. Vietnam:* 46 (1995 publ. 1996)—Vietnam.
hellwigianus *K. Schumann in K. Schumann & M. Hollrung, Fl. Kais. Wilh. Land.:* 79 (1889), in Supp. 1 as 'hollrungianus'—New Guinea (Papua New Guinea).†
megadontus *P.I. Forster in Austrobaileya,* 5(3): 470 (1999), nom. nov.: M. claoxyloides var. angustifolia *F.M. Bailey.*
puber *S.M. Bollendorff in Blumea,* 45(2): 334 (2000), nom. nov.: Mallotus leucodermis var. puberulus *Airy Shaw.*
sathayensis *Nguyen Nghia Thin, Euphorbiac. Vietnam:* 46 (1995 publ. 1996)—Vietnam.
spinulosus *G. McPherson in Bull. Mus. Nation. Hist. Nat., B, Adansonia,* Ser. 4, 17(3–4): 170 (1995)—Madagascar.
surculosus *P.I. Forster in Austrobaileya,* 5(3): 488 (1999)—Queensland.
wenzelianus *J.W.F. Slik in Blumea,* 43(1): 229 (1998)—Philippines.

MALMEA (Annonac.).
depressa (Baill.) R.E. Fr.
subsp. abscondita *L.W. Chatrou in Amer. J. Bot.,* 84(6): 868 (1997)—Mexico Gulf (Veracruz).
dimera *L.W. Chatrou in Novon,* 7(4): 346 (1997 publ. 1998)—Panama, Colombia.
surinamensis *L.W. Chatrou in Feddes Repert.,* 106(5–8): 344 (1995 publ. 1996)—Surinam.

MALPIGHIA (Malpighiac.).
ser. Artiambe *(Nied.) F.K. Meyer, Revis. Gatt. Malpighia L. (Phanerogam. Monogr., 23):* 253 (2000): Malpighia subsect. Artiambe.
subsect. Bahamenses *F.K. Meyer, Revis. Gatt. Malpighia L. (Phanerogam. Monogr., 23):* 405 (2000).
ser. Calcaratae *F.K. Meyer, Revis. Gatt. Malpighia L. (Phanerogam. Monogr., 23):* 184 (2000).
ser. Caribeae *F.K. Meyer, Revis. Gatt. Malpighia L. (Phanerogam. Monogr., 23):* 298 (2000).
ser. Cnideae *F.K. Meyer, Revis. Gatt. Malpighia L. (Phanerogam. Monogr., 23):* 338 (2000).
ser. Furcatae *F.K. Meyer, Revis. Gatt. Malpighia L. (Phanerogam. Monogr., 23):* 324 (2000).
ser. Lineares *F.K. Meyer, Revis. Gatt. Malpighia L. (Phanerogam. Monogr., 23):* 248 (2000).
sect. Lingulatae *F.K. Meyer, Revis. Gatt. Malpighia L. (Phanerogam. Monogr., 23):* 367 (2000).
ser. Lingulatae *F.K. Meyer, Revis. Gatt. Malpighia L. (Phanerogam. Monogr., 23):* 367 (2000).
ser. Obovatae *F.K. Meyer, Revis. Gatt. Malpighia L. (Phanerogam. Monogr., 23):* 377 (2000).
ser. Polytrichae *F.K. Meyer, Revis. Gatt. Malpighia L. (Phanerogam. Monogr., 23):* 393 (2000).
ser. Pubescentes *F.K. Meyer, Revis. Gatt. Malpighia L. (Phanerogam. Monogr., 23):* 415 (2000).
ser. Racemosae *F.K. Meyer, Revis. Gatt. Malpighia L. (Phanerogam. Monogr., 23):* 293 (2000).
subgen. Romeroa *F.K. Meyer, Revis. Gatt. Malpighia L. (Phanerogam. Monogr., 23):* 137 (2000).
sect. Schizocarpae *F.K. Meyer, Revis. Gatt. Malpighia L. (Phanerogam. Monogr., 23):* 144 (2000).
ser. Spathulifoliae *F.K. Meyer, Revis. Gatt. Malpighia L. (Phanerogam. Monogr., 23):* 217 (2000).
subsect. Tergacanthae *F.K. Meyer, Revis. Gatt. Malpighia L. (Phanerogam. Monogr., 23):* 240 (2000).

MALPIGHIA:-
ser. Tergacanthae *F.K. Meyer, Revis. Gatt. Malpighia L. (Phanerogam. Monogr., 23):* 240 (2000).
subsect. Urentes *(DC.) F.K. Meyer, Revis. Gatt. Malpighia L. (Phanerogam. Monogr., 23):* 253 (2000): Malpighia sect. Urentes.
acutifolia *F.K. Meyer, Revis. Gatt. Malpighia L. (Phanerogam. Monogr., 23):* 299 (2000)—Dominican Republic.
arborescens *F.K. Meyer, Revis. Gatt. Malpighia L. (Phanerogam. Monogr., 23):* 380 (2000)—Cuba.
articulata *F.K. Meyer, Revis. Gatt. Malpighia L. (Phanerogam. Monogr., 23):* 317 (2000)—Cuba.
atlantica *F.K. Meyer, Revis. Gatt. Malpighia L. (Phanerogam. Monogr., 23):* 469 (2000)—Turks-Caicos Is., Bahamas.
aurea *F.K. Meyer, Revis. Gatt. Malpighia L. (Phanerogam. Monogr., 23):* 377 (2000)—Cuba.
avilensis *F.K. Meyer, Revis. Gatt. Malpighia L. (Phanerogam. Monogr., 23):* 365 (2000)—Cuba.
azucarensis *F.K. Meyer, Revis. Gatt. Malpighia L. (Phanerogam. Monogr., 23):* 290 (2000)—Dominican Republic.
bahamensis *F.K. Meyer, Revis. Gatt. Malpighia L. (Phanerogam. Monogr., 23):* 411 (2000)—Bahamas.
subsp. androsana *F.K. Meyer, Revis. Gatt. Malpighia L. (Phanerogam. Monogr., 23):* 414 (2000)—Bahamas, Cuba.
baracoensis *F.K. Meyer, Revis. Gatt. Malpighia L. (Phanerogam. Monogr., 23):* 338 (2000)—Cuba.
bissei *F.K. Meyer, Revis. Gatt. Malpighia L. (Phanerogam. Monogr., 23):* 346 (2000)—Cuba.
cajalbanensis *F.K. Meyer, Revis. Gatt. Malpighia L. (Phanerogam. Monogr., 23):* 516 (2000)—Cuba.
capitis-crucis *F.K. Meyer, Revis. Gatt. Malpighia L. (Phanerogam. Monogr., 23):* 428 (2000)—Cuba.
caribaea *F.K. Meyer, Revis. Gatt. Malpighia L. (Phanerogam. Monogr., 23):* 301 (2000)—Cuba.
cornistipulata *F.K. Meyer, Revis. Gatt. Malpighia L. (Phanerogam. Monogr., 23):* 315 (2000)—Cuba.
cristalensis *(F.K. Mey.) F.K. Meyer, Revis. Gatt. Malpighia L. (Phanerogam. Monogr., 23):* 463 (2000): Malpighia nummulariifolia subsp. cristalensis.
cuneiformis *F.K. Meyer, Revis. Gatt. Malpighia L. (Phanerogam. Monogr., 23):* 511 (2000)—Cuba.
dentata *F.K. Meyer, Revis. Gatt. Malpighia L. (Phanerogam. Monogr., 23):* 480 (2000)—Dominican Republic.
dura *F.K. Meyer, Revis. Gatt. Malpighia L. (Phanerogam. Monogr., 23):* 471 (2000)—Cuba.
epedunculata *F.K. Meyer, Revis. Gatt. Malpighia L. (Phanerogam. Monogr., 23):* 505 (2000)—Cuba.
subsp. arenaria *F.K. Meyer, Revis. Gatt. Malpighia L. (Phanerogam. Monogr., 23):* 508 (2000)—Cuba.
erinacea *F.K. Meyer, Revis. Gatt. Malpighia L. (Phanerogam. Monogr., 23):* 406 (2000)—Cuba.
flavescens *F.K. Meyer, Revis. Gatt. Malpighia L. (Phanerogam. Monogr., 23):* 426 (2000)—Cuba.
fucata Ker Gawl.
subsp. longifolia *(Nied.) F.K. Meyer, Revis. Gatt. Malpighia L. (Phanerogam. Monogr., 23):* 279 (2000): Malpighia fucata var. longifolia.
fuertesii *F.K. Meyer, Revis. Gatt. Malpighia L. (Phanerogam. Monogr., 23):* 498 (2000)—Dominican Republic.
subsp. minor *F.K. Meyer, Revis. Gatt. Malpighia L. (Phanerogam. Monogr., 23):* 500 (2000)—Dominican Republic.
gonavensis *F.K. Meyer, Revis. Gatt. Malpighia L. (Phanerogam. Monogr., 23):* 350 (2000)—Haiti.
granitica *F.K. Meyer, Revis. Gatt. Malpighia L. (Phanerogam. Monogr., 23):* 190 (2000)—Mexico Southwest (Guerrero).
guantanamensis *F.K. Meyer, Revis. Gatt. Malpighia L. (Phanerogam. Monogr., 23):* 430 (2000)—Cuba.
habanensis *F.K. Meyer, Revis. Gatt. Malpighia L. (Phanerogam. Monogr., 23):* 519 (2000)—Cuba.
harrisii Small
subsp. vivaldii *F.K. Meyer, Revis. Gatt. Malpighia L. (Phanerogam. Monogr., 23):* 244 (2000)—Jamaica.
higueyensis *F.K. Meyer, Revis. Gatt. Malpighia L. (Phanerogam. Monogr., 23):* 248 (2000)—Dominican Republic.

MALPIGHIA:-
hispaniolica *F.K. Meyer, Revis. Gatt. Malpighia L. (Phanerogam. Monogr., 23):* 484 (2000), nom. nov.: Malpighia ilicifolia f. eglandulosa *Ekman & Nied.*
subsp. cuneata *F.K. Meyer, Revis. Gatt. Malpighia L. (Phanerogam. Monogr., 23):* 487 (2000)—Haiti.
hondurensis *F.K. Meyer, Revis. Gatt. Malpighia L. (Phanerogam. Monogr., 23):* 213 (2000)—Honduras.
humilis *F.K. Meyer, Revis. Gatt. Malpighia L. (Phanerogam. Monogr., 23):* 419 (2000)—Dominican Republic.
imiensis *F.K. Meyer, Revis. Gatt. Malpighia L. (Phanerogam. Monogr., 23):* 423 (2000)—Cuba.
jaguensis *F.K. Meyer, Revis. Gatt. Malpighia L. (Phanerogam. Monogr., 23):* 353 (2000)—Cuba.
lancifolia *(Nied.) F.K. Meyer, Revis. Gatt. Malpighia L. (Phanerogam. Monogr., 23):* 222 (2000): Malpighia punicifolia var. lancifolia.
latifolia *F.K. Meyer, Revis. Gatt. Malpighia L. (Phanerogam. Monogr., 23):* 197 (2000)—Mexico Gulf (Veracruz).
linearifolia *F.K. Meyer, Revis. Gatt. Malpighia L. (Phanerogam. Monogr., 23):* 438 (2000)—Cuba.
longifolia *F.K. Meyer, Revis. Gatt. Malpighia L. (Phanerogam. Monogr., 23):* 319 (2000)—Cuba.
macracantha Urb. & Nied.
subsp. centralis *F.K. Meyer, Revis. Gatt. Malpighia L. (Phanerogam. Monogr., 23):* 397 (2000)—Dominican Republic, Haiti.
macrocarpa *F.K. Meyer, Revis. Gatt. Malpighia L. (Phanerogam. Monogr., 23):* 191 (2000)—Mexico Gulf (Veracruz), Mexico Southeast (Chiapas).
manacensis *F.K. Meyer, Revis. Gatt. Malpighia L. (Phanerogam. Monogr., 23):* 312 (2000)—Cuba.
maxima *F.K. Meyer, Revis. Gatt. Malpighia L. (Phanerogam. Monogr., 23):* 240 (2000)—Jamaica.
melbensis *F.K. Meyer, Revis. Gatt. Malpighia L. (Phanerogam. Monogr., 23):* 355 (2000)—Cuba.
mexicana A. Juss.
subsp. guadalajarensis *(S. Watson) F.K. Meyer, Revis. Gatt. Malpighia L. (Phanerogam. Monogr., 23):* 203 (2000): Bunchosia guadalajarensis.
mirabilis *F.K. Meyer, Revis. Gatt. Malpighia L. (Phanerogam. Monogr., 23):* 432 (2000)—Cuba.
moncionensis *F.K. Meyer, Revis. Gatt. Malpighia L. (Phanerogam. Monogr., 23):* 502 (2000)—Dominican Republic.
montecristensis *F.K. Meyer, Revis. Gatt. Malpighia L. (Phanerogam. Monogr., 23):* 326 (2000)—Cuba.
subsp. naranjensis *F.K. Meyer, Revis. Gatt. Malpighia L. (Phanerogam. Monogr., 23):* 328 (2000)—Cuba.
mucronata *F.K. Meyer, Revis. Gatt. Malpighia L. (Phanerogam. Monogr., 23):* 385 (2000)—Cuba.
subsp. insulae-pinorum *F.K. Meyer, Revis. Gatt. Malpighia L. (Phanerogam. Monogr., 23):* 390 (2000)—Cuba.
nothosubsp. intermedia *F.K. Meyer, Revis. Gatt. Malpighia L. (Phanerogam. Monogr., 23):* 392 (2000)—Cuba.(M. mucronata subsp. insulae-pinorum × subsp. mucronata).
multiflora *F.K. Meyer, Revis. Gatt. Malpighia L. (Phanerogam. Monogr., 23):* 269 (2000)—Jamaica.
mutabilis *F.K. Meyer, Revis. Gatt. Malpighia L. (Phanerogam. Monogr., 23):* 362 (2000)—Cuba.
nayaritensis *(Vivaldi) F.K. Meyer, Revis. Gatt. Malpighia L. (Phanerogam. Monogr., 23):* 142 (2000): Malpighia romeroana var. nayaritensis.
neglecta *F.K. Meyer, Revis. Gatt. Malpighia L. (Phanerogam. Monogr., 23):* 449 (2000)—Cuba.
ophiticola *F.K. Meyer, Revis. Gatt. Malpighia L. (Phanerogam. Monogr., 23):* 451 (2000)—Cuba.
ovalis *(Ekman & Nied.) F.K. Meyer, Revis. Gatt. Malpighia L. (Phanerogam. Monogr., 23):* 284 (2000): Malpighia cnide var. ovalis.
oviedensis *F.K. Meyer, Revis. Gatt. Malpighia L. (Phanerogam. Monogr., 23):* 344 (2000)—Dominican Republic.
palenquensis *F.K. Meyer, Revis. Gatt. Malpighia L. (Phanerogam. Monogr., 23):* 324 (2000)—Cuba.
pallidior *F.K. Meyer, Revis. Gatt. Malpighia L. (Phanerogam. Monogr., 23):* 287 (2000)—Cuba.

MALPIGHIA:-
pasorealensis *F.K. Meyer, Revis. Gatt. Malpighia L. (Phanerogam. Monogr., 23):* 367 (2000)—Cuba.
phillyreifolia *F.K. Meyer, Revis. Gatt. Malpighia L. (Phanerogam. Monogr., 23):* 360 (2000)—Cuba.
polytricha A. Juss.
subsp. parvifolia *F.K. Meyer, Revis. Gatt. Malpighia L. (Phanerogam. Monogr., 23):* 404 (2000)—Bahamas.
subsp. vivaldiana *F.K. Meyer, Revis. Gatt. Malpighia L. (Phanerogam. Monogr., 23):* 402 (2000)—Bahamas.
racemiflora *F.K. Meyer, Revis. Gatt. Malpighia L. (Phanerogam. Monogr., 23):* 295 (2000)—Cuba.
racemosa *F.K. Meyer, Revis. Gatt. Malpighia L. (Phanerogam. Monogr., 23):* 293 (2000)—Cuba.
raunkiaeri *F.K. Meyer, Revis. Gatt. Malpighia L. (Phanerogam. Monogr., 23):* 181 (2000)—Leeward Is. (Virgin Is.).
revoluta *F.K. Meyer, Revis. Gatt. Malpighia L. (Phanerogam. Monogr., 23):* 321 (2000)—Cuba.
reyensis *F.K. Meyer, Revis. Gatt. Malpighia L. (Phanerogam. Monogr., 23):* 465 (2000)—Cuba.
salicifolia *F.K. Meyer, Revis. Gatt. Malpighia L. (Phanerogam. Monogr., 23):* 291 (2000)—Dominican Republic.
samanensis *F.K. Meyer, Revis. Gatt. Malpighia L. (Phanerogam. Monogr., 23):* 282 (2000)—Dominican Republic.
serpentinicola *F.K. Meyer, Revis. Gatt. Malpighia L. (Phanerogam. Monogr., 23):* 408 (2000)—Cuba.
spathulifolia *F.K. Meyer, Revis. Gatt. Malpighia L. (Phanerogam. Monogr., 23):* 219 (2000)—Belize, Guatemala, Mexico Southeast (Quintana Roo, Yucatán).
squarrosa *F.K. Meyer, Revis. Gatt. Malpighia L. (Phanerogam. Monogr., 23):* 513 (2000)—Cuba.
subpilosa *F.K. Meyer, Revis. Gatt. Malpighia L. (Phanerogam. Monogr., 23):* 334 (2000)—Cuba.
substrigosa *F.K. Meyer, Revis. Gatt. Malpighia L. (Phanerogam. Monogr., 23):* 436 (2000), nom. nov.: Malpighia nummulariifolia subsp. oblongifolia *Vivaldi.*
tortolensis *F.K. Meyer, Revis. Gatt. Malpighia L. (Phanerogam. Monogr., 23):* 503 (2000)—Leeward Is. (British Virgin Is.).
torulosa *F.K. Meyer, Revis. Gatt. Malpighia L. (Phanerogam. Monogr., 23):* 331 (2000)—Cuba.
tunensis *F.K. Meyer, Revis. Gatt. Malpighia L. (Phanerogam. Monogr., 23):* 357 (2000)—Cuba.
urens L.
subsp. ovatifolia *(Nied.) F.K. Meyer, Revis. Gatt. Malpighia L. (Phanerogam. Monogr., 23):* 275 (2000): M. urens f. ovatifolia.
velutina Ekman & Nied.
subsp. glabrescens *F.K. Meyer, Revis. Gatt. Malpighia L. (Phanerogam. Monogr., 23):* 447 (2000)—Haiti.
subsp. intermedia *(Ekman & Nied.) F.K. Meyer, Revis. Gatt. Malpighia L. (Phanerogam. Monogr., 23):* 446 (2000): Malpighia velutina var. intermedia.
verruculosa W.R. Anderson
subsp. antillana *(Vivaldi) F.K. Meyer, Revis. Gatt. Malpighia L. (Phanerogam. Monogr., 23):* 150 (2000): Malpighia albiflora subsp. antillana.
vertientensis *F.K. Meyer, Revis. Gatt. Malpighia L. (Phanerogam. Monogr., 23):* 441 (2000)—Cuba.
yucatanaea *F.K. Meyer, Revis. Gatt. Malpighia L. (Phanerogam. Monogr., 23):* 225 (2000)—Mexico Southeast (Campeche).

MALUS (Rosac.).
baccata (L.) Borkh.
var. jinxianensis *(J.Q. Deng & J.Y. Hong) C.Y. Li in Clavis Pl. Chinae Bor.-Or. ed. 2 (ed. P.Y. Fu):* 318 (1995): M. jinxianensis.
doumeri (Bois) A. Chev.
var. formosana *(Kawak. & Koidz.) S.S. Ying in Mem. Coll. Agric. Nation. Taiwan Univ.,* 31(1): 33 (1991): M. formosana.

MALVA (Malvac.).
assurgentiflora *(Kellogg) M.F. Ray in Novon,* 8(3): 290 (1998): Lavatera assurgentiflora.
australiana *M.F. Ray in Novon,* 8(3): 291 (1998), nom. nov.: Lavatera plebeia *Sims.*

MALVA:-
canariensis *M.F. Ray in Novon,* 8(3): 292 (1998), nom. nov.: Lavatera acerifolia *Cav.*
dendromorpha *M.F. Ray in Novon,* 8(3): 292 (1998), nom. nov.: Lavatera arborea *L.*
lindsayi *(Moran) M.F. Ray in Novon,* 8(3): 292 (1998): Lavatera lindsayi.
linnaei *M.F. Ray in Novon,* 8(3): 292 (1998), nom. nov.: Lavatera cretica *L.*
occidentalis *(S. Watson) M.F. Ray in Novon,* 8(3): 293 (1998): Lavatera occidentalis.
pacifica *M.F. Ray in Novon,* 8(3): 293 (1998), nom. nov.: Lavatera venosa *S. Watson.*
verticillata L.
subsp. chinensis *(Mill.) N.N. Tzvelev in Novosti Sist. Vyssh. Rast.,* 32: 184 (2000): M. chinensis.
subsp. crispa *(L.) N.N. Tzvelev in Novosti Sist. Vyssh. Rast.,* 32: 184 (2000): M. verticillata var. crispa.
subsp. pulchella *(Bernh.) N.N. Tzvelev in Novosti Sist. Vyssh. Rast.,* 32: 184 (2000): M. pulchella.
wigandii *(Alef.) M.F. Ray in Novon,* 8(3): 293 (1998): Axolopha wigandii.

MALVASTRUM (Malvac.).
hillii *P.A. Fryxell, J.L. León de la Luz & M. Domínguez L. in Contrib. Univ. Mich. Herb.,* 21: 198 (1997)—Mexico Northwest (Baja California Sur).

MAMMEA (Guttif.).
calciphila Kosterm.
var. fasciculata *P.F. Stevens in Checklist Flow. Pl. & Gymn. Brunei Darussalam (ed. M.J.E. Coode et al.):* 438 (1996)—Borneo (Sarawak, Brunei).

MAMMILLARIA (Cactac.).
ser. Bartschella *(Britton & Rose) J.M. Lüthy, Taxon. Unters. Gatt. Mammillaria Haw. (Cactac.):* 136 (1995): gen. Bartschella.
ser. Bombycinae *J.M. Lüthy, Taxon. Unters. Gatt. Mammillaria Haw. (Cactac.):* 154 (1995).
sect. Cochemiea *(K. Brandegee) J.M. Lüthy, Taxon. Unters. Gatt. Mammillaria Haw. (Cactac.):* 132 (1995): M. subgen. Cochemiea.
ser. Cochemiea *(K. Brandegee) J.M. Lüthy, Taxon. Unters. Gatt. Mammillaria Haw. (Cactac.):* 136 (1995): M. subgen. Cochemiea.
ser. Herrerae *J.M. Lüthy, Taxon. Unters. Gatt. Mammillaria Haw. (Cactac.):* 141 (1995).
ser. Phellosperma *(Britton & Rose) J.M. Lüthy, Taxon. Unters. Gatt. Mammillaria Haw. (Cactac.):* 128 (1995): gen. Phellosperma.
ser. Rhodanthae *J.M. Lüthy, Taxon. Unters. Gatt. Mammillaria Haw. (Cactac.):* 176 (1995).
albicans
subsp. fraileana *(Britton & Rose) D.R. Hunt in Mammillaria Postscripts,* 6: 5 (1997): Neomammillaria fraileana.
f. slevinii *(Britton & Rose) T. Neutelings in Succulenta,* 65(5): 120 (1986): M. slevinii.
albilanata
subsp. oaxacana *D.R. Hunt in Mammillaria Postscripts,* 6: 9 (1997), nom. nov.: M. ignota *Repp.*
subsp. reppenhagenii *(D.R. Hunt) D.R. Hunt in Mammillaria Postscripts,* 6: 9 (1997): M. reppenhagenii.
subsp. tegelbergiana *(H.E. Gates ex G.E. Linds.) D.R. Hunt in Mammillaria Postscripts,* 6: 9 (1997): M. tegelbergiana.
amajacensis *C. Brachet & M. Lacoste in Cact. Succ. J. (U.S.A.),* 69(4): 185 (1997)—Mexico Northeast (Hidalgo).
armillata
subsp. cerralboa *(Britton & Rose) D.R. Hunt in Mammillaria Postscripts,* 6: 5 (1997): Neomammillaria cerralboa.
aureilanata Backeb.
f. alba *(Backeb.) C.E. Glass, Guía Identific. Cact. Amenazadas México,* 1: Ma/au f. alba (1998 publ. 1997): M. aureilanata var. alba.
backebergiana
subsp. ernestii *(Fittkau) D.R. Hunt in Mammillaria Postscripts,* 6: 8 (1997): M. ernestii.

MAMMILLARIA:-
bombycina
subsp. perezdelarosae *(Bravo & Scheinvar) D.R. Hunt in Mammillaria Postscripts,* 6: 5 (1997): M. perezdelarosae.
brandegeei
subsp. gabbii *(J.M. Coult.) D.R. Hunt in Mammillaria Postscripts,* 6: 11 (1997): Cactus gabbii.
subsp. glareosa *(Boed.) D.R. Hunt in Mammillaria Postscripts,* 6: 11 (1997): M. glareosa.
subsp. lewisiana *(H.E. Gates) D.R. Hunt in Mammillaria Postscripts,* 7: 3 (1998): M. lewisiana.
coahuilensis
subsp. albiarmata *(Boed.) D.R. Hunt in Mammillaria Postscripts,* 7: 3 (1998): M. albiarmata.
columbiana
subsp. yucatanensis *(Britton & Rose) D.R. Hunt in Mammillaria Postscripts,* 6: 9 (1997): Neomammillaria yucatanensis.
compressa
subsp. centralifera *(Repp.) D.R. Hunt in Mammillaria Postscripts,* 6: 11 (1997): M. centralifera.
crinita
subsp. leucantha *(Boed.) D.R. Hunt in Mammillaria Postscripts,* 6: 6 (1997): M. leucantha.
subsp. scheinvariana *(R. Ortega V. & Glass) W.A. Fitz Maurice & B. Fitz Maurice in Cact. Succ. J. (U.S.A.),* 70(2): 94 (1998): M. scheinvariana.
subsp. wildii *(A. Dietr.) D.R. Hunt in Mammillaria Postscripts,* 6: 6 (1997): M. wildii.
f. zeilmanniana *(Boed.) C.E. Glass, Guía Identific. Cact. Amenazadas México,* 1: Ma/cr f. zeilmanniana (1998 publ. 1997): Mammillaria zeilmanniana.
crucigera
var. grandinosa *T. Linzen et al. in Mitteilungsbl. AfM,* 21(4): 203–207 (1997 publ. 1998); *cf. Repert. Pl. Succ. (I.O.S.),* 49: 19 (1998 publ. 1999)—Locality not stated.
var. tlalocii *(Repp.) T. Linzen et al. in Mitteilungsbl. AfM,* 21(4): 202 (1997 publ. 1998); *cf. Repert. Pl. Succ. (I.O.S.),* 49: 19 (1998 publ. 1999): M. tlalocii.
subsp. tlalocii *(Repp.) D.R. Hunt in Mammillaria Postscripts,* 6: 9 (1997): M. tlalocii
decipiens
subsp. albescens *(Tiegel) D.R. Hunt in Mammillaria Postscripts,* 6: 7 (1997): M. albescens.
subsp. camptotricha *(Dams) D.R. Hunt in Mammillaria Postscripts,* 6: 7 (1997): M. camptotricha.
deherdtiana
subsp. dodsonii *(Bravo) D.R. Hunt in Mammillaria Postscripts,* 6: 5 (1997): M. dodsonii.
dioica
f. angelensis *(R.T. Craig) T. Neutelings in Succulenta,* 65(5): 119 (1986): M. angelensis.
subsp. angelensis *(R.T. Craig) D.R. Hunt in Mammillaria Postscripts,* 7: 3 (1998): M. angelensis.
var. armillata *(K. Brandegee) T. Neutelings in Succulenta,* 65(5): 119 (1986): M. armillata.
var. capensis *(H.E. Gates) T. Neutelings in Succulenta,* 65(5): 119 (1986): Neomammillaria capensis.
var. cerralboa *(Britton & Rose) T. Neutelings in Succulenta,* 65(5): 119 (1986): Neomammillaria cerralboa.
f. estebanensis *(G.E. Linds.) T. Neutelings in Succulenta,* 65(5): 119 (1986): M. estebanensis.
subsp. estebanensis *(G.E. Linds.) D.R. Hunt in Mammillaria Postscripts,* 7: 3 (1998): M. estebanensis.
f. incerta *(Parish) T. Neutelings in Succulenta,* 65(5): 119 (1986): M. incerta.
var. multidigitata *(Radley ex G.E. Linds.) T. Neutelings in Succulenta,* 65(5): 119 (1986): M. multidigitata.
var. neopalmeri *(R.T. Craig) T. Neutelings in Succulenta,* 65(5): 119 (1986): M. neopalmeri.
f. phitauiana *(E.M. Baxter) T. Neutelings in Succulenta,* 65(5): 119 (1986): Neomammillaria phitauiana.

MAMMILLARIA:-
dioica
f. verhaertiana *(Boed.) T. Neutelings in Succulenta,* 65(5): 119 (1986): M. verhaertiana.
discolor
subsp. esperanzaensis *(Boed.) D.R. Hunt in Mammillaria Postscripts,* 6: 7 (1997): M. esperanzaensis.
elongata
subsp. echinaria *(DC.) D.R. Hunt in Mammillaria Postscripts,* 6: 7 (1997): M. echinaria.
fittkaui Glass & R.A. Foster
subsp. limonensis *(Repp.) J.M. Lüthy, Taxon. Unters. Gatt. Mammillaria Haw. (Cactac.):* 162 (1995): M. limonensis
subsp. mathildae *(Kraehenb. & Krainz) J.M. Lüthy, Taxon. Unters. Gatt. Mammillaria Haw. (Cactac.):* 162 (1995): M. mathildae.
formosa
subsp. chionocephala *(J.A. Purpus) D.R. Hunt in Mammillaria Postscripts,* 7: 3 (1998): M. chionocephala.
subsp. microthele *(Muehlenpf.) D.R. Hunt in Mammillaria Postscripts,* 6: 9 (1997): M. microthele.
subsp. pseudocrucigera *(R.T. Craig) D.R. Hunt in Mammillaria Postscripts,* 6: 10 (1997): M. pseudocrucigera.
× gaji *J. Chvastek & J.J. Halda in Acta Mus. Richnov. Sect. Nat.,* 7(1): 36 (2000)—Mexico Southwest (Oaxaca). (M. deherdtiana var. dodsonii × rekoi).
geminispina
subsp. leucocentra *(A. Berg) D.R. Hunt in Mammillaria Postscripts,* 6: 10 (1997): M. leucocentra.
giselae *J.G. Martínez-Avalos & C. Glass in Bradleya,* 15: 73 (1997)—Mexico Northeast (Tamaulipas).
glassii
subsp. ascensionis *(Repp.) D.R. Hunt in Mammillaria Postscripts,* 6: 6 (1997): M. ascensionis.
goodridgii
var. blossfeldiana *(Boed.) T. Neutelings in Succulenta,* 65(5): 120 (1986): M. blossfeldiana.
var. bullardiana *(H.E. Gates) T. Neutelings in Succulenta,* 65(5): 120 (1986): Neomammillaria bullardiana.
var. hutchisoniana *(H.E. Gates) T. Neutelings in Succulenta,* 65(5): 119 (1986): Neomammillaria hutchisoniana.
var. louisae *(G.E. Linds.) T. Neutelings in Succulenta,* 65(5): 119 (1986): M. louisae.
f. shurliana *(H.E. Gates) T. Neutelings in Succulenta,* 65(5): 120 (1986): M. shurliana.
haageana
subsp. acultzingensis *(Linzen et al.) D.R. Hunt in Mammillaria Postscripts,* 6: 9 (1997): M. acultzingensis.
subsp. conspicua *(J.A. Purpus) D.R. Hunt in Mammillaria Postscripts,* 6: 9 (1997): M. conspicua.
subsp. elegans *D.R. Hunt in Mammillaria Postscripts,* 6: 9 (1997), nom. nov.: M. collina *J.A. Purpus.*
subsp. san-angelensis *(Sánchez-Mej.) D.R. Hunt in Mammillaria Postscripts,* 6: 9 (1997): M. san-angelensis.
subsp. schmollii *(R.T. Craig) D.R. Hunt in Mammillaria Postscripts,* 6: 9 (1997): M. elegans var. schmollii.
hahniana
subsp. bravoae *(R.T. Craig) D.R. Hunt in Mammillaria Postscripts,* 6: 10 (1997): M. bravoae.
subsp. mendeliana *(Bravo) D.R. Hunt in Mammillaria Postscripts,* 6: 10 (1997): Neomammillaria mendeliana.
subsp. woodsii *(R.T. Craig) D.R. Hunt in Mammillaria Postscripts,* 6: 10 (1997): M. woodsii.
heyderi Muehlenpf.
subsp. coahuilensis *(Boed.) J.M. Lüthy, Taxon. Unters. Gatt. Mammillaria Haw. (Cactac.):* 192 (1995): Porfiria coahuilensis.
subsp. gaumeri *(Britton & Rose) D.R. Hunt in Mammillaria Postscripts,* 6: 10 (1997): Neomammillaria gaumeri.

MAMMILLARIA:-
heyderi Muehlenpf.
subsp. gummifera *(Engelm.) D.R. Hunt in Mammillaria Postscripts,* 6: 10 (1997): M. gummifera.
subsp. hemisphaerica *(Engelm.) D.R. Hunt in Mammillaria Postscripts,* 6: 10 (1997): M. hemisphaerica.
subsp. macdougalii *(Rose) D.R. Hunt in Mammillaria Postscripts,* 6: 10 (1997): M. macdougalii.
subsp. meiacantha *(Engelm.) D.R. Hunt in Mammillaria Postscripts,* 6: 10 (1997): M. meiacantha.
huitzilopochtli D.R. Hunt
subsp. niduliformis *(A.B. Lau) J. Pilbeam, Mammillaria (Cactus File Handbook, 6):* 143 (1999): M. huitzilopochtli var. niduliformis.
hutchisoniana
subsp. louisae *(G.E. Linds.) D.R. Hunt in Mammillaria Postscripts,* 6: 5 (1997): M. louisae.
jaliscana
subsp. zacatecasensis *(Shurly) D.R. Hunt in Mammillaria Postscripts,* 6: 6 (1997): M. zacatecasensis.
karwinskiana
subsp. beiselii *(Diers) D.R. Hunt in Mammillaria Postscripts,* 6: 11 (1997): M. beiselii.
subsp. collinsii *(Britton & Rose) D.R. Hunt in Mammillaria Postscripts,* 6: 11 (1997): Neomammillaria collinsii.
subsp. nejapensis *(R.T. Craig & E.Y. Dawson) D.R. Hunt in Mammillaria Postscripts,* 6: 11 (1997): M. nejapensis.
kupcaki *J.J. Halda & J. Sladkovský in Acta Mus. Richnov. Sect. Nat.,* 7(1): 36 (2000)—Mexico Northeast (Coahuila).
lasiacantha
subsp. egregia *(Backeb. ex Rogoz. & Appenz.) D.R. Hunt in Mammillaria Postscripts,* 7: 3 (1998): M. egregia.
subsp. hyalina *D.R. Hunt in Mammillaria Postscripts,* 6: 6 (1997), nom. nov.: M. wohlschlageri *Repp.*
subsp. magallanii *(F. Schmoll ex R.T. Craig) D.R. Hunt in Mammillaria Postscripts,* 6: 7 (1997): M. magallanii.
laui
subsp. dasyacantha *(D.R. Hunt) D.R. Hunt in Mammillaria Postscripts,* 6: 7 (1997): M. laui f. dasyacantha.
subsp. subducta *(D.R. Hunt) D.R. Hunt in Mammillaria Postscripts,* 6: 7 (1997): M. laui f. subducta.
longiflora
subsp. stampferi *(Repp.) D.R. Hunt in Mammillaria Postscripts,* 6: 5 (1997): M. stampferi.
subsp. tepexicensis *(J. Meyrán) J.M. Lüthy in Kakt. And. Sukk.,* 51(7): 171 (2000): M. tepexicensis.
luethyi *G.S. Hinton in Phytologia,* 80(1): 58 (1996)—Mexico Northeast (Coahuila).
luthieniae *J.E. Laferrière in J. Mammillaria Soc.,* 38(2): 18 (1998)—Mexico Northeast (Chihuahua).
machucae *W.A. Fitz Maurice & B. Fitz Maurice in Cact. Succ. J. (U.S.A.),* 72(2): 66 (2000)—Mexico Southwest (Jalisco).
marcosii *W.A. Fitz Maurice, B. Fitz Maurice & C. Glass in Cact. Succ. J. (U.S.A.),* 69(1): 10 (1997)—Mexico Northeast (Guanajuato).
mazatlanensis
var. occidentalis *(Britton & Rose) T. Neutelings in Succulenta,* 65(5): 119 (1986): Neomammillaria occidentalis.
f. patonii *(Bravo) T. Neutelings in Succulenta,* 65(5): 119 (1986): Neomammillaria patonii.
subsp. patonii *(Bravo) D.R. Hunt in Mammillaria Postscripts,* 7: 3 (1998): Neomammillaria patonii.
f. sinalensis *(R.T. Craig) T. Neutelings in Succulenta,* 65(5): 119 (1986): M. occidentalis var. sinalensis.
melanocentra
subsp. linaresensis *(R. Wolf & F. Wolf) D.R. Hunt in Mammillaria Postscripts,* 6: 10 (1997): M. linaresensis.

MAMMILLARIA:-
melanocentra
subsp. rubrograndis *(Repp. & A.B. Lau) D.R. Hunt in Mammillaria Postscripts,* 6: 10 (1997): M. rubrograndis.
melilotiae *J.E. Laferrière in J. Mammillaria Soc.,* 38(2): 18 (1998)—Mexico Northeast (Chihuahua).
milleri (Britton & Rose) Boed.
f. auricarpa *(W.T. Marshall) T. Neutelings in Succulenta,* 65(5): 119 (1986): M. microcarpa var. auricarpa.
var. grahamii *(Engelm.) T. Neutelings in Succulenta,* 65(5): 119 (1986): M. grahamii.
f. gueldemanniana *(Backeb.) T. Neutelings in Succulenta,* 65(5): 119 (1986): M. gueldemanniana.
f. oliviae *(Orcutt) T. Neutelings in Succulenta,* 65(5): 119 (1986): M. oliviae.
var. sheldonii *(Britton & Rose) T. Neutelings in Succulenta,* 65(5): 119 (1986): Neomammillaria sheldonii.
f. swinglei *(Britton & Rose) T. Neutelings in Succulenta,* 65(5): 119 (1986): Neomammillaria swinglei.
nana Backeb. ex Mottram
subsp. duwei *(Rogoz. & P.J. Braun) J. Pilbeam, Mammillaria (Cactus File Handbook, 6):* 203 (1999): M. duwei.
nunezii
subsp. bella *(Backeb.) D.R. Hunt in Mammillaria Postscripts,* 6: 8 (1997): M. bella.
odehnali *J.J. Halda, P. Kupčák & J. Sladkovský in Acta Mus. Richnov. Sect. Nat.,* 7(1): 36 (2000)—Mexico Northeast (Nuevo León).
pennispinosa
subsp. nazasensis *(Glass & R.A. Foster) D.R. Hunt in Mammillaria Postscripts,* 6: 6 (1997): M. pennispinosa var. nazasensis.
petrophila
subsp. arida *(Rose ex Quehl) D.R. Hunt in Mammillaria Postscripts,* 7: 3 (1998): M. arida.
subsp. baxteriana *(H.E. Gates) D.R. Hunt in Mammillaria Postscripts,* 7: 3 (1998): Neomammillaria baxteriana.
phantasma *J.M. Sotomayor M., A. Arredondo Gómez & M. Martínez Méndez in Cact. Suc. Mex.,* 44(2): 34 (1999)—Mexico Northeast (San Luis Potosí).
picta
subsp. viereckii *(Boed.) D.R. Hunt in Mammillaria Postscripts,* 6: 6 (1997): M. viereckii.polythele
polythele
subsp. durispina *(Boed.) D.R. Hunt in Mammillaria Postscripts,* 6: 7 (1997): M. durispina.
subsp. obconella *(Scheidw.) D.R. Hunt in Mammillaria Postscripts,* 6: 7 (1997): M. obconella.
pondii
subsp. maritima *(G.E. Linds.) D.R. Hunt in Mammillaria Postscripts,* 6: 5 (1997): Cochemiea maritima.
subsp. setispina *(J.M. Coult.) D.R. Hunt in Mammillaria Postscripts,* 6: 5 (1997): Cactus setispinus.
prolifera
subsp. arachnoidea *(D.R. Hunt) D.R. Hunt in Mammillaria Postscripts,* 6: 6 (1997): M. prolifera var. arachnoidea.
subsp. haitiensis *(K. Schum.) D.R. Hunt in Mammillaria Postscripts,* 6: 6 (1997): M. pusilla var. haitiensis.
subsp. texana *(Engelm.) D.R. Hunt in Mammillaria Postscripts,* 6: 6 (1997): M. pusilla var. texana.
subsp. zublerae *(Repp.) D.R. Hunt in Mammillaria Postscripts,* 6: 6 (1997): M. zublerae.
rekoi
subsp. aureispina *(A.B. Lau) D.R. Hunt in Mammillaria Postscripts,* 6: 8 (1997): M. rekoi var. aureispina.
subsp. leptacantha *(A.B. Lau) D.R. Hunt in Mammillaria Postscripts,* 6: 8 (1997): M. rekoi var. leptacantha.
rhodantha
subsp. aureiceps *(Lem.) D.R. Hunt in Mammillaria Postscripts,* 6: 7 (1997): M. aureiceps.
subsp. fera-rubra *(Schmoll ex R.T. Craig) D.R. Hunt in Mammillaria Postscripts,* 6: 8 (1997): M. fera-rubra.

MAMMILLARIA:-
rhodantha
subsp. mccartenii *D.R. Hunt in Mammillaria Postscripts,* 6: 8 (1997), nom. nov.: M. verticealba *Repp.*
subsp. mollendorffiana *(Shurly) D.R. Hunt in Mammillaria Postscripts,* 6: 8 (1997): M. mollendorffiana.
subsp. pringlei *(J.M. Coult.) D.R. Hunt in Mammillaria Postscripts,* 6: 8 (1997): Cactus pringlei.
saboae
subsp. goldii *(Glass & R.A. Foster) D.R. Hunt in Mammillaria Postscripts,* 6: 5 (1997): M. goldii.
subsp. haudeana *(A.B. Lau & K. Wagner) D.R. Hunt in Mammillaria Postscripts,* 7: 3 (1998): M. haudeana.
scheinvariana *R. Ortega-Varela & C.E. Glass in C.E. Glass, Guía Identific. Cact. Amenazadas México,* 1: Ma/sche (1998 publ. 1997)—Mexico Northeast (Querétaro).
schiedeana
subsp. dumetorum *(J.A. Purpus) D.R. Hunt in Mammillaria Postscripts,* 6: 7 (1997): M. dumetorum.
subsp. giselae *(Mart.-Aval. & Glass) J.M. Lüthy in Kakt. And. Sukk.,* 49(8): 184 (1998): M. giselae.
sphacelata
subsp. viperina *(J.A. Purpus) D.R. Hunt in Mammillaria Postscripts,* 6: 7 (1997): M. viperina.
f. viperina *(J.A. Purpus) L. Battaia & C. Zanovello in Piante Grasse,* 16(1): 18 (1996): M. viperina.
spinosissima
subsp. pilcayensis *(Bravo) D.R. Hunt in Mammillaria Postscripts,* 6: 8 (1997): M. pilcayensis.
subsp. tepoxtlana *D.R. Hunt in Mammillaria Postscripts,* 6: 8 (1997), nom. nov.: M. crassior *Repp.*
thornberi
var. yaquensis *(R.T. Craig) T. Neutelings in Succulenta,* 65(5): 119 (1986): M. yaquensis.
subsp. yaquensis *(R.T. Craig) D.R. Hunt in Mammillaria Postscripts,* 6: 5 (1997): M. yaquensis.
tinuvieliae *J.E. Laferrière in J. Mammillaria Soc.,* 38(2): 21 (1998)—Mexico Northeast (Chihuahua).
trichacantha K. Schum.*subsp.*duwei *(Rogoz. & P.J. Braun) R. Mottram in J. Mammillaria Soc.,* 40(1): 8 (2000): M. duwei.
vetula
subsp. gracilis *(Pfeiff.) D.R. Hunt in Mammillaria Postscripts,* 6: 6 (1997): M. gracilis.
voburnensis
subsp. eichlamii *(Quehl) D.R. Hunt in Mammillaria Postscripts,* 6: 11 (1997): M. eichlamii.
wiesingeri
subsp. apamensis *(Repp.) D.R. Hunt in Mammillaria Postscripts,* 6: 8 (1997): M. apamensis.
winterae
subsp. aramberri *D.R. Hunt in Mammillaria Postscripts,* 6: 10 (1997): M. crassimammillis *Repp.*
wrightii
subsp. wilcoxii *(Toumey ex K. Schum.) D.R. Hunt in Mammillaria Postscripts,* 6: 5 (1997): M. wilcoxii.
xaltianguensis
subsp. bambusiphila *(Repp.) D.R. Hunt in Mammillaria Postscripts,* 6: 8 (1997): M. bambusiphila.
zephyranthoides Scheidw.
subsp. heidiae *(Krainz) J.M. Lüthy in Kakt. And. Sukk.,* 51(3): 63 (2000): M. heidiae

MANDEVILLA (Apocynac.).
holosericea *(Sessé & Moç.) J.K. Williams in Sida,* 18(1): 237 (1998): Echites holosericea.
kalmiifolia *(Woodson) J.F. Morales in Novon,* 8(4): 429 (1998), as 'kalmiaefolia': Salpinctes kalmiifolius.
pringlei *J.K. Williams in Sida,* 18(1): 231 (1998)—Mexico Southwest (Jalisco).
rigidifolia *J.F. Morales in Brittonia,* 50(2): 224 (1998)—Panama.
rugellosa *(Rich.) L. Allorge-Boiteau in Phytologia,* 84(4): 305 (1998 publ. 1999): Echites rugellosa.

MANFREDA (Agavac.).
littoralis *A. García-Mendoza, A. Castañeda Rojas & I.S. Franco Martínez in Acta Bot. Mex.,* 50: 39 (2000)—Mexico Southwest (Guerrero, Oaxaca).
virginica (L.) Rose
subsp. lata *(Shinners) R.J. O'Kennon, G.M. Diggs & B.L. Lipscomb in G.M. Diggs et al., Shinners & Mahler's Illustr. Fl. N. Central Texas* (Sida, Bot. Misc. 16): 1079 (1999): Agave lata.

MANGIFERA (Anacardiac.).
khoonmengiana *K.M. Kochummen in Sandakania,* 7: 83 (1996)—Borneo (Brunei, Sarawak).

MANGLIETIA (Magnoliac.).
conifera Dandy
subsp. chingii *(Dandy) J. Li in Acta Bot. Yunnanica,* 19(2): 131 (1997): M. chingii.
hebecarpa *C.Y. Wu & Y.W. Law in Acta Phytotax. Sin.,* 34(1): 88 (Feb. 1996); *et in Fl. Reipubl. Popul. Sin.,* 30(1): 270 (May 1996)—China South-central (Yunnan).
obovatifolia *C.Y. Wu & Y.W. Law in Acta Phytotax. Sin.,* 34(1): 89 (Feb. 1996); *et in Fl. Reipubl. Popul. Sin.,* 30(1): 270 (May 1996), as 'obovalifolia'—China South-central (Yunnan, Guizhou).

MANIHOT (Euphorbiac.).
baccata *A.C. Allem in Int. J. Pl. Sci.,* 160(1): 181 (1999)—Brazil Northeast (Maranhão, Ceará), Brazil North (Amazonas, Pará), French Guiana.
var. ferruginea *A.C. Allem in Int. J. Pl. Sci.,* 160(1): 184 (1999)—Brazil North (Pará, Amapá), Brazil Northeast (Pernambuco).

MANOAO *B.P.J. Molloy in New Zealand J. Bot.,* 33(2): 196 (1995). *PODOCARPACEAE*
colensoi *(Hook.) B.P.J. Molloy in New Zealand J. Bot.,* 33(2): 196 (1995): Dacrydium colensoi.

MANSOA (Bignoniac.).
montecillensis *(A. Molina) C.H. Nelson Sutherland in Fontqueria,* 48: 28 (1997 publ. 1998): Chodanthus montecillensis.

MANTISALCA (Compos.).
duriaei (Spach) Briq. & Cavill.
var. iserniana *(Gay & Webb ex Graells) O. de Bolòs & J. Vigo, Fl. Països Catalans,* 3: 912 (1995): Microlonchus isernianus.

MANTISIA (Zingiberac.).
radicalis *(Roxb.) D.P. Dam & N. Dam in Bull. Bot. Surv. India,* 34(1–4): 190 (1992 publ. 1997): Globba radicalis.

MANYONIA *H. Robinson in Proc. Biol. Soc. Wash.,* 112(1): 224 (1999). *COMPOSITAE*
peculiaris *(Verdc.) H. Robinson in Proc. Biol. Soc. Wash.,* 112(1): 225 (1999): Vernonia peculiaris.

MAPANIA (Cyperac.).
sect. Mapaniopsis *(C.B. Clarke) D.A. Simpson in Kew Bull.,* 51(4): 734 (1996): gen. Mapaniopsis.
arunachalensis *G.D. Pal in J. Bombay Nat. Hist. Soc.,* 91(3): 438 (1994 publ. 1995)—Assam (Arunachal Pradesh).
aturensis *D.A. Simpson in Kew Bull.,* 51(4): 730 (1996)—Venezuela.
coodei *D.A. Simpson in Kew Bull.,* 55(4): 997 (2000)—New Guinea (Irian Jaya).
micrococca *(T. Koyama) D.A. Simpson in Kew Bull.,* 51(4): 737 (1996): Mapaniopsis micrococca.
pedunculata *D.A. Simpson in Kew Bull.,* 51(4): 732 (1996)—Venezuela.
rionegrensis *D.A. Simpson in Kew Bull.,* 51(4): 732 (1996)—Venezuela.
tamdaoensis *N.K. Khoi in J. Biol. (Vietnam),* 18(4): 4 (1996)—Vietnam.
theobromina *D.A. Simpson in Kew Bull.,* 51(4): 738 (1996)—Venezuela.

MAPROUNEA (Euphorbiac.).
amazonica *H.-J. Esser in Novon,* 9(1): 32 (1999), nom. nov.: M. guianensis f. obtusata *Müll. Arg.*

MAQUIRA (Morac.).
guianensis Aubl.
subsp. costaricana *(Standl.) C.C. Berg in Novon,* 6(3): 235 (1996): Perebea costaricana.

MARANTA (Marantac.).
lietzei *(E. Morren ex Petersen) C.H. Nelson Sutherland & F.J. Fernández Casas in Fontqueria,* 51: 3 (1998): Calathea lietzei.

MARANTOCHLOA (Marantac.).
incertifolia *M.M. Dhetchuvi in Bull. Jard. Bot. Nation. Belg.,* 65(3–4): 374 (1996)—Gabon.
microphylla *(Koechlin) M.M. Dhetchuvi in Bull. Jard. Bot. Nation. Belg.,* 65(3–4): 375 (1996): M. congensis var. microphylla.

MARCETIA (Melastomatac.).
bahiana *(Ule) A.B. Martins in Acta Bot. Brasil.,* 9(1): 149 (1995): Fritzschia bahiana.
eimeariana *A.B. Martins & E.M. Woodgyer in Kew Bull.,* 55(1): 189 (2000)—Brazil Northeast (Bahia).
hatschbachii *A.B. Martins in Bol. Bot. (São Paulo),* 14: 44 (1995)—Brazil Southeast (Minas Gerais).
lychnophoroides *A.B. Martins in Novon,* 10(3): 226 (2000)—Brazil Northeast (Bahia).
semiriana *A.B. Martins in Novon,* 10(3): 229 (2000)—Brazil Southeast (Minas Gerais).
shepherdii *A.B. Martins in Novon,* 10(3): 224 (2000)—Brazil Northeast (Bahia).

MARCGRAVIA (Marcgraviac.).
evenia Krug & Urb.
subsp. calcicola *(Britton) S. Dressler in Fl. Repúbl. Cuba,* ser. A, 5(4): 7 (2000): Marcgravia calcicola.
panamensis *S. Dressler in Novon,* 6(3): 256 (1996)—Panama.
punctifolia *S. Dressler in Novon,* 8(2): 137 (1998)—Venezuela, Brazil North (Roraima).
roonii *S. Dressler in Novon,* 6(3): 258 (1996)—Colombia, Costa Rica, Ecuador, Panama.
yukunarum *P. Picca & D. Giraldo-Cañas in Hickenia,* 2(63): 294 (1999)—Colombia.
zonopunctata *S. Dressler in Willdenowia,* 30(2): 369 (2000)—Brazil North (Amazonas), Colombia.

MARCGRAVIASTRUM *(Wittm. ex Szyszył.) A.C. de Roon & S. Dressler in Bot. Jahrb.,* 119(3): 332 (1997): Norantea subsect. Marcgraviastrum. (Marcgraviac.).
cuneifolium *(Gardner) H.G. Bedell in Bot. Jahrb.,* 119(3): 332 (1997): Marcgravia cuneifolia.
gigantophyllum *(Gilg) Bedell ex S. Dressler in Cat. Vasc. Pl. Ecuador (Monogr. Syst. Bot. Missouri Bot. Gard., 75):* 955 (1999): Norantea gigantophylla.
macrocarpum *(G. Don) Bedell ex S. Dressler in Cat. Vasc. Pl. Ecuador (Monogr. Syst. Bot. Missouri Bot. Gard., 75):* 955 (1999): Norantea macrocarpa.
mixtum *(Triana & Planch.) H.G. Bedell in Bot. Jahrb.,* 119(3): 332 (1997): Norantea mixta.
pauciflorum *A.C. de Roon & H.G. Bedell in Bot. Jahrb.,* 119(3): 333 (1997)—Panama.
pendulum *(Lanj. & Heerdt) H.G. Bedell in Novon,* 9(2): 252 (1999): Norantea pendula.
sodiroi *(Gilg) Bedell ex S. Dressler in Cat. Vasc. Pl. Ecuador (Monogr. Syst. Bot. Missouri Bot. Gard., 75):* 955 (1999): Norantea sodiroi.
subsessile *(Benth.) H.G. Bedell in Bot. Jahrb.,* 119(3): 333 (1997): Ruyschia subsessilis.

MAREYA (Euphorbiac.).
congolensis *(J. Léonard) J. Léonard in Bull. Jard. Bot. Nation. Belg.,* 65(1–2): 12 (1996): M. micrantha subsp. congolensis.

MAREYOPSIS (Euphorbiac.).
oligogyna *F.J. Breteler in Bull. Jard. Bot. Nation. Belg.,* 66(1–2): 134 (1997)—Gabon.

MARGARETTA (Asclepiadac.).
rosea Oliv.
subsp. whytei *(K. Schum.) M.L. Mwanyambo in Kew Bull.,* 51(4): 726 (1996): M. whytei.

MARGARITARIA (Euphorbiac.).
indica (Dalzell) Airy Shaw
f. vestita *(J.J. Sm.) C. Barker in Kew Bull.,* 53(3): 755 (1998): Phyllanthus indicus f. vestitus.

MARGBENSONIA *A.V. Bobrov & A.P. Melikyan in Byull. Mosk. Obshch. Ispȳt. Prir., Biol.,* 103(1): 59 (1998). *PODOCARPACEAE*
archboldii *(N.E. Gray) A.V. Bobrov & A.P. Melikyan in Byull. Mosk. Obshch. Ispȳt. Prir., Biol.,* 103(1): 59 (1998): Podocarpus archboldii.
atjehensis *(Wasscher) A.V. Bobrov & A.P. Melikyan in Byull. Mosk. Obshch. Ispȳt. Prir., Biol.,* 103(1): 59 (1998): Podocarpus neriifolius var. atjehensis.
chinensis *A.V. Bobrov & A.P. Melikyan in Byull. Mosk. Obshch. Ispȳt. Prir., Biol.,* 103(1): 59 (1998), nom. nov., as 'comb. nov.': Juniperus chinensis *Roxb.*
chingiana *(S.Y. Hu) A.V. Bobrov & A.P. Melikyan in Byull. Mosk. Obshch. Ispȳt. Prir., Biol.,* 103(1): 59 (1998): Podocarpus chingianus.
degeneri *(N.E. Gray) A.V. Bobrov & A.P. Melikyan in Byull. Mosk. Obshch. Ispȳt. Prir., Biol.,* 103(1): 60 (1998): Podocarpus neriifolius var. degeneri.
disperma *(C.T.White) A.V. Bobrov & A.P. Melikyan in Byull. Mosk. Obshch. Ispȳt. Prir., Biol.,* 103(1): 60 (1998): Podocarpus dispermus.
drouyniana *(F. Muell.) A.V. Bobrov & A.P. Melikyan in Byull. Mosk. Obshch. Ispȳt. Prir., Biol.,* 103(1): 60 (1998): Podocarpus drouynianus.
elata *(R. Br. ex Endl.) A.V. Bobrov & A.P. Melikyan in Byull. Mosk. Obshch. Ispȳt. Prir., Biol.,* 103(1): 60 (1998): Podocarpus elatus.
forrestii *(Craib & W.W. Sm.) A.V. Bobrov & A.P. Melikyan in Byull. Mosk. Obshch. Ispȳt. Prir., Biol.,* 103(1): 60 (1998): Podocarpus forrestii.
koordersii *(Pilg. ex Koord. & Valeton) A.V. Bobrov & A.P. Melikyan in Byull. Mosk. Obshch. Ispȳt. Prir., Biol.,* 103(1): 60 (1998): Podocarpus koordersii.
macrophylla *(Thunb.) A.V. Bobrov & A.P. Melikyan in Byull. Mosk. Obshch. Ispȳt. Prir., Biol.,* 103(1): 59 (1998): Taxus macrophylla.
maki *(Siebold & Zucc. ex Endl.) A.V. Bobrov & A.P. Melikyan in Byull. Mosk. Obshch. Ispȳt. Prir., Biol.,* 103(1): 60 (1998): Podocarpus macrophyllus var. maki.
neriifolia *(D. Don) A.V. Bobrov & A.P. Melikyan in Byull. Mosk. Obshch. Ispȳt. Prir., Biol.,* 103(1): 60 (1998): Podocarpus neriifolius.
philippinensis *(Foxw.) A.V. Bobrov & A.P. Melikyan in Byull. Mosk. Obshch. Ispȳt. Prir., Biol.,* 103(1): 60 (1998): Podocarpus philippinensis.
polystachya *(R. Br. ex Endl.) A.V. Bobrov & A.P. Melikyan in Byull. Mosk. Obshch. Ispȳt. Prir., Biol.,* 103(1): 60 (1998): Podocarpus polystachyus.
ridleyi *(Wasscher) A.V. Bobrov & A.P. Melikyan in Byull. Mosk. Obshch. Ispȳt. Prir., Biol.,* 103(1): 60 (1998): Podocarpus neriifolius var. ridleyi.
rumphii *(Blume) A.V. Bobrov & A.P. Melikyan in Byull. Mosk. Obshch. Ispȳt. Prir., Biol.,* 103(1): 60 (1998): Podocarpus rumphii.
spinulosa *(Sm.) A.V. Bobrov & A.P. Melikyan in Byull. Mosk. Obshch. Ispȳt. Prir., Biol.,* 103(1): 60 (1998): Taxus spinulosa.
sweetii *(C. Presl) A.V. Bobrov & A.P. Melikyan in Byull. Mosk. Obshch. Ispȳt. Prir., Biol.,* 103(1): 60 (1998): Podocarpus sweetii.
thevetiifolia *(Blume) A.V. Bobrov & A.P. Melikyan in Byull. Mosk. Obshch. Ispȳt. Prir., Biol.,* 103(1): 60 (1998): Podocarpus thevetiifolius.

MARILA (Guttif.).
subgen. Poranthera *P.F. Stevens in Biollania,* ed. esp. 6 (J.J. Wurdack Festschr.): 560 (1997).

MARISCUS (Cyperac.).
concinnus Nees
var. evolutior *(C.B. Clarke) G. Panigrahi in S.K. Murti & G. Panigrahi, Fl. Bilaspur Distr.,* 2: 699 (1999): M. sieberianus var. evolutior.
var. khasianus *(C.B. Clarke) G. Panigrahi in S.K. Murti & G. Panigrahi, Fl. Bilaspur Distr.,* 2: 699 (1999), as 'khasiana': M. sieberianus var. khasianus.
var. subcompositus *(C.B. Clarke) G. Panigrahi in S.K. Murti & G. Panigrahi, Fl. Bilaspur Distr.,* 2: 699 (1999): M. sieberianus var. subcompositus.

MARKEA (Solanac.).
antioquiensis *S. Knapp in Novon,* 8(2): 153 (1998)—Colombia.

MARKEA:-
epifita *S. Knapp in Novon,* 8(2): 155 (1998)—Ecuador.
fosbergii *A.T. Hunziker in Kurtziana,* 25: 85 (1997)—Ecuador.
harlingiana *A.T. Hunziker in Kurtziana,* 25: 96 (1997)—Ecuador, Colombia.
pilosa *S. Knapp in Novon,* 8(2): 157 (1998)—Ecuador, Colombia.
plowmanii *A.T. Hunziker in Kurtziana,* 25: 102 (1997)—Colombia.
spruceana *A.T. Hunziker in Kurtziana,* 25: 99 (1997)—Ecuador.

MARLIEREA (Myrtac.).
leal-costae *G.M. Barroso & A.L. Peixoto in Rev. Brasil. Bot.,* 18(1): 105 (1995)—Brazil Northeast (Bahia).
silvestris *(Vell.) J.R. de Mattos in Loefgrenia,* 102: 4 (1993): Myrtus silvestris.

MARLOTHISTELLA (Aizoac.).
stenophylla *(L. Bolus) S.A. Hammer in Bradleya,* 13: 87 (1995): Mesembryanthemum stenophyllum.

MARSDENIA (Asclepiadac.).
sect. Dregea *(E. Mey.) R. Omlor, Gen. Revis. Marsdenieae (Asclepiadac.):* 76 (1998): gen. Dregea.
sect. Obscurae *R. Omlor, Gen. Revis. Marsdenieae (Asclepiadac.):* 74 (1998).
sect. Pterygocarpus *(Hochst.) R. Omlor, Gen. Revis. Marsdenieae (Asclepiadac.):* 80 (1998): gen. Pterygocarpus.
acuminata *(Roxb.) I.M. Turner in Asian J. Trop. Biol.,* 1(2): 25 (1995): Asclepias acuminata.
var. glabra *(M.A. Rahman & Wilcock) I.M. Turner in Asian J. Trop. Biol.,* 1(2): 25 (1995): Gymnema acuminatum var. glabrum.
amorimii *G. Morillo in Ernstia,* 8(1): 4 (1998)—Brazil Northeast (Bahia).
arabica *(Decne.) R. Omlor, Gen. Revis. Marsdenieae (Asclepiadac.):* 79 (1998): Dregea arabica.
callosa *V. Juárez-Jaimes & W.D. Stevens in Novon,* 7(3): 253 (1997)—Mexico Southwest (Guerrero, Michoacan).
carterae *W.D. Stevens & V. Juárez-Jaimes in Novon,* 9(4): 565 (1999)—Mexico Northwest (Baja California Sur).
faulknerae *(Bullock) R. Omlor, Gen. Revis. Marsdenieae (Asclepiadac.):* 79 (1998): Dregea faulknerae.
forsteri *I.M. Turner in Asian J. Trop. Biol.,* 1(2): 25 (1995), nom. nov.: Gymnema maingayi *Hook. f.*
gallardoae *L. Lozada-Pérez in Novon,* 10(2): 128 (2000)—Mexico Southwest (Oaxaca), Mexico Southeast (Chiapas).
gymnemopsis *R. Omlor, Gen. Revis. Marsdenieae (Asclepiadac.):* 117 (1998), nom. nov.: Gymnemopsis pierrei *Costantin.*
harmandiella *R. Omlor, Gen. Revis. Marsdenieae (Asclepiadac.):* 118 (1998), nom. nov.: Harmandiella cordifolia *Costantin.*
hilariana E. Fourn.
var. macieliana *J. Fontella Pereira & R.J. Paixão in Pabstia,* 7(2): 1 (1996)—Brazil Southeast (Rio de Janeiro).
jensenii *P.I. Forster in Austrobaileya,* 5(1): 59 (1997)—Queensland.
lactifera *(L.) I.M. Turner in Asian J. Trop. Biol.,* 1(2): 25 (1995): Asclepias lactifera.
var. thwaitesii *(Hook. f.) I.M. Turner in Asian J. Trop. Biol.,* 1(2): 26 (1995): Gymnema lactiferum var. thwaitesii.
nicaraguensis *W.D. Stevens in Novon,* 10(3): 244 (2000)—Nicaragua.
normaniana *R. Omlor, Gen. Revis. Marsdenieae (Asclepiadac.):* 81 (1998), nom. nov.: M. grandiflora *C. Norman.*
oaxacana *G. Morillo in Ernstia,* 8(1): 5 (1998)—Mexico Southwest (Oaxaca).
oblanceolata *(Turrill) R. Omlor, Gen. Revis. Marsdenieae (Asclepiadac.):* 81 (1998): Dalzielia oblanceolata.
obscura *(Bullock) R. Omlor, Gen. Revis. Marsdenieae (Asclepiadac.):* 75 (1998): Gongronema obscurum.
olgamarthae *W.D. Stevens in Novon,* 10(3): 246 (2000)—Nicaragua.

MARSDENIA:-
pickelii *J. Fontella Pereira & G. Morillo in Pabstia,* 7(4): 2 (1996)—Brazil Northeast (Pernambuco).
purpusiana *W.D. Stevens in Novon,* 10(3): 244 (2000), nom. nov.: Ecliptostelma molle *Brandegee.*
rzedowskiana *V. Juárez-Jaimes & W.D. Stevens in Novon,* 5(4): 337 (1995)—Mexico Central (Puebla), Mexico Gulf (Veracruz).
secamonoides *(Schltr.) R. Omlor, Gen. Revis. Marsdenieae (Asclepiadac.):* 134 (1998): Astelma secamonoides
squamulosa *R. Omlor, Gen. Revis. Marsdenieae (Asclepiadac.):* 117 (1998), nom. nov.: Gymnemopsis calcicola *Kerr.*
thomasii *G. Morillo in Ernstia,* 8(1): 6 (1998)—Brazil West-central (Mato Grosso).
tolimensis *G. Morillo in Ernstia,* 10(2): 80 (2000)—Colombia.
veronicae *W.D. Stevens in Novon,* 10(3): 246 (2000)—Nicaragua, Costa Rica, El Salvador.
woodburyana *P. Acevedo-Rodríguez in Brittonia,* 51(2): 167 (1999)—Puerto Rico.

MASCAGNIA (Malpighiac.).
anderssonii *W.R. Anderson in Contrib. Univ. Mich. Herb.,* 22: 15 (1999)—Ecuador, Peru.
haenkeana *W.R. Anderson in Contrib. Univ. Mich. Herb.,* 21: 73 (1997)—Ecuador.
lugoi *W.R. Anderson in Contrib. Univ. Mich. Herb.,* 21: 74 (1997)—Ecuador, Peru, Brazil North (Amazonas).

MASDEVALLIA (Orchidac.).
subsect. Amaluzae *(Luer) C.A. Luer, Ic. Pleurothallid. XIX (Monogr. Syst. Bot. Missouri Bot. Gard., 77):* 10 (2000): M. sect. Amaluzae.
subsect. Coaetaneae *C.A. Luer, Ic. Pleurothallid. XIX (Monogr. Syst. Bot. Missouri Bot. Gard., 77):* 13 (2000).
subgen. Cucullatia *C.A. Luer, Ic. Pleurothallid. XIX (Monogr. Syst. Bot. Missouri Bot. Gard., 77):* 10 (2000).
sect. Dentatae *(Luer) C.A. Luer, Ic. Pleurothallid. XIX (Monogr. Syst. Bot. Missouri Bot. Gard., 77):* 10 (2000): M. subsect. Dentatae.
sect. Durae *(Luer) C.A. Luer, Ic. Pleurothallid. XIX (Monogr. Syst. Bot. Missouri Bot. Gard., 77):* 10 (2000): M. subsect. Durae.
subgen. Fissia *C.A. Luer, Ic. Pleurothallid. XIX (Monogr. Syst. Bot. Missouri Bot. Gard., 77):* 10 (2000).
subgen. Nidificia *C.A. Luer, Ic. Pleurothallid. XIX (Monogr. Syst. Bot. Missouri Bot. Gard., 77):* 10 (2000).
subgen. Polyantha *C.A. Luer, Ic. Pleurothallid. XIX (Monogr. Syst. Bot. Missouri Bot. Gard., 77):* 11 (2000): M. sect. Polyanthae.
subsect. Pterygiophorae *C.A. Luer, Ic. Pleurothallid. XIX (Monogr. Syst. Bot. Missouri Bot. Gard., 77):* 10 (2000).
subgen. Pygmaeia *C.A. Luer, Ic. Pleurothallid. XIX (Monogr. Syst. Bot. Missouri Bot. Gard., 77):* 10 (2000).
subgen. Scabripes *C.A. Luer, Ic. Pleurothallid. XIX (Monogr. Syst. Bot. Missouri Bot. Gard., 77):* 10 (2000).
subsect. Successiviflorae *C.A. Luer, Ic. Pleurothallid. XIX (Monogr. Syst. Bot. Missouri Bot. Gard., 77):* 223 (2000).
subgen. Volvula *C.A. Luer, Ic. Pleurothallid. XIX (Monogr. Syst. Bot. Missouri Bot. Gard., 77):* 10 (2000).
subsect. Zahlbrucknerae *C.A. Luer, Ic. Pleurothallid. XIX (Monogr. Syst. Bot. Missouri Bot. Gard., 77):* 10 (2000).
adrianae *C.A. Luer, Ic. Pleurothallid. XVII (Monogr. Syst. Bot. Missouri Bot. Gard., 72):* 110 (1998)—Ecuador.
aguirrei *C.A. Luer & R. Escobar in Harvard Pap. Bot.,* 9: 1 (1996)—Colombia.
amoena *C.A. Luer, Ic. Pleurothallid. XVIII (Monogr. Syst. Bot. Missouri Bot. Gard., 76):* 148 (1999)—Ecuador.
amplexa *C.A. Luer, Ic. Pleurothallid. XV (Monogr. Syst. Bot. Missouri Bot. Gard., 64):* 125 (1997)—Peru.
antonii *W. Königer in Arcula,* 8: 221 (1999)—Peru.

MASDEVALLIA:-
aphanes *W. Königer in Orchidee,* 30(5): 196 (1979), contrary to Art. 37.3 ICBN (1994)—Peru.
bicornis *C.A. Luer, Ic. Pleurothallid. XV (Monogr. Syst. Bot. Missouri Bot. Gard., 64):* 128 (1997), nom. nov.: Portillia popowiana *Königer & J. Portilla.*
blanda *W. Königer & J. Portilla in Arcula,* 10: 266 (2000)—Ecuador.
bryophila *C.A. Luer in Harvard Pap. Bot.,* 9: 1 (1996)—Peru.
burianii *C.A. Luer & Dalström in C.A. Luer, Ic. Pleurothallid. XX (Monogr. Syst. Bot. Missouri Bot. Gard.,* 79): 120 (2000)—Bolivia.
calosiphon *C.A. Luer, Ic. Pleurothallid. XX (Monogr. Syst. Bot. Missouri Bot. Gard., 79):* 120 (2000)—Peru.
carpishica *C.A. Luer & P. Cloes in C.A. Luer, Ic. Pleurothallid. XVIII (Monogr. Syst. Bot. Missouri Bot.* Gard., 76): 148 (1999)—Peru.
cocapatae *C.A. Luer, Teague & R. Vásquez in Harvard Pap. Bot.,* 2(1): 55 (1997), as No. 11 on inside pages—Bolivia.
collantesii *D.E. Bennett & E.A. Christenson in Lindleyana,* 13(1): 53 (1998)—Peru.
cosmia *W. Königer in Arcula,* 8: 223 (1999)—Peru.
crassicaulis *C.A. Luer & J. Portilla in C.A. Luer, Ic. Pleurothallid. XVIII (Monogr. Syst. Bot. Missouri Bot.* Gard., 76): 149 (1999)—Ecuador.
dorisiae *C.A. Luer, Ic. Pleurothallid. XVIII (Monogr. Syst. Bot. Missouri Bot. Gard., 76):* 149 (1999)—Ecuador.
ejiriana *C.A. Luer & J. Portilla in C.A. Luer, Ic. Pleurothallid. XVIII (Monogr. Syst. Bot. Missouri Bot.* Gard., 76): 150 (1999)—Ecuador.
formosa *C.A. Luer & Cloes in C.A. Luer, Ic. Pleurothallid. XX (Monogr. Syst. Bot. Missouri Bot. Gard.,* 79): 121 (2000)—Peru.
fosterae *C.A. Luer, Ic. Pleurothallid. XVIII (Monogr. Syst. Bot. Missouri Bot. Gard., 76):* 150 (1999)—Locality not stated.
harlequina *C.A. Luer, Ic. Pleurothallid. XV (Monogr. Syst. Bot. Missouri Bot. Gard., 64):* 130 (1997)—Peru.
helgae *W. Königer & J. Portilla in Arcula,* 7: 194 (1997)—Ecuador.
immensa *C.A. Luer, Ic. Pleurothallid. XX (Monogr. Syst. Bot. Missouri Bot. Gard., 79):* 122 (2000)—Peru.
jarae *C.A. Luer in Harvard Pap. Bot.,* 9: 4 (1996)—Peru.
jimenezii *W. Königer in Arcula,* 5: 122 (1996)—Ecuador.
josei *C.A. Luer in Harvard Pap. Bot.,* 9: 6 (1996)—Ecuador.
juan-albertoi *C.A. Luer & M. Arias in C.A. Luer, Ic. Pleurothallid. XX (Monogr. Syst. Bot. Missouri Bot. Gard.,* 79): 122 (2000)—Peru.
karineae *Nauray ex C.A. Luer, Ic. Pleurothallid. XX (Monogr. Syst. Bot. Missouri Bot. Gard., 79):* 123 (2000)—Peru.
maduroi *C.A. Luer, Ic. Pleurothallid. XVII (Monogr. Syst. Bot. Missouri Bot. Gard., 72):* 110 (1998)—Panama.
manoloi *C.A. Luer & M. Arias in C.A. Luer, Ic. Pleurothallid. XVI (Monogr. Syst. Bot. Missouri Bot. Gard.,* 65): 108 (1998)—Peru.
manta *W. Königer & Sijm in Arcula,* 10: 268 (2000)—Ecuador.
marizae *C.A. Luer & Rolando in C.A. Luer, Ic. Pleurothallid. XX (Monogr. Syst. Bot. Missouri Bot. Gard.,* 79): 124 (2000)—Peru.
martineae *C.A. Luer in Harvard Pap. Bot.,* 2(1): 55 (1997), as No. 11 on inside pages—Bolivia.
martiniana *C.A. Luer, Ic. Pleurothallid. XVI (Monogr. Syst. Bot. Missouri Bot. Gard., 65):* 108 (1998)—Ecuador.
melina *W. Königer & J. Meza in Arcula,* 5: 125 (1996)—Peru.
merinoi *C.A. Luer & J. Portilla in C.A. Luer, Ic. Pleurothallid. XVIII (Monogr. Syst. Bot. Missouri Bot.* Gard., 76): 151 (1999)—Ecuador.
misasii *L.A. Braas in Orchidee,* 33(4): 148 (1982), nom. nov.: M. reflexa *Misas.*
monicana *C.A. Luer, Ic. Pleurothallid. XVII (Monogr. Syst. Bot. Missouri Bot. Gard., 72):* 111 (1998)—Ecuador.

MASDEVALLIA:-
newmaniana *C.A. Luer & Teague in C.A. Luer, Ic. Pleurothallid. XVII (Monogr. Syst. Bot. Missouri Bot.* Gard., 72): 111 (1998)—Ecuador.
niesseniae *C.A. Luer, Ic. Pleurothallid. XVI (Monogr. Syst. Bot. Missouri Bot. Gard., 65):* 110 (1998)—Colombia.
nikoleana *C.A. Luer & J. Portilla in C.A. Luer, Ic. Pleurothallid. XX (Monogr. Syst. Bot. Missouri Bot. Gard.,* 79): 124 (2000)—Ecuador, Peru.
obscurans *(Luer) C.A. Luer, Ic. Pleurothallid. XVII (Monogr. Syst. Bot. Missouri Bot. Gard., 72):* 112 (1998): M. infracta subsp. obscurans.
omorenoi *C.A. Luer & R. Vásquez in Harvard Pap. Bot.,* 2(1): 58 (1997), as No. 11 on inside pages—Bolivia.
oscitans *(Luer) C.A. Luer, Ic. Pleurothallid. XVII (Monogr. Syst. Bot. Missouri Bot. Gard., 72):* 112 (1998): M. infracta subsp. oscitans.
papillosa *C.A. Luer in Lindleyana,* 10(4): 232 (1995)—Ecuador.
pastinata *C.A. Luer in Harvard Pap. Bot.,* 2(1): 60 (1997), as No. 11 on inside pages—Colombia.
phasmatodes *W. Königer in Orchidee,* 30(5): 197 (1979), contrary to Art. 37.3 ICBN (1994)—Peru.
polychroma *C.A. Luer, Ic. Pleurothallid. XVII (Monogr. Syst. Bot. Missouri Bot. Gard., 72):* 113 (1998)—Ecuador.
popowiana *W. Königer & J.G. Weinmann in Arcula,* 5: 127 (1996)—Peru.
prodigiosa *W. Königer in Orchidee,* 30(5): 197 (1979), contrary to Art. 37.3 ICBN (1994)—Peru.
prolixa *C.A. Luer, Ic. Pleurothallid. XX (Monogr. Syst. Bot. Missouri Bot. Gard., 79):* 125 (2000)—Peru.
rana-aurea *C.A. Luer, Ic. Pleurothallid. XVIII (Monogr. Syst. Bot. Missouri Bot. Gard., 76):* 151 (1999)—Peru.
rauhii *K. Senghas & L.A. Braas in Orchidee,* 30(5): 194 (1979)—Peru.
recurvata *C.A. Luer & Dalström in C.A. Luer, Ic. Pleurothallid. XV (Monogr. Syst. Bot. Missouri Bot. Gard.,* 64): 132 (1997)—Peru.
regina *C.A. Luer, Ic. Pleurothallid. XVI (Monogr. Syst. Bot. Missouri Bot. Gard., 65):* 113 (1998)—Peru.
renzii *C.A. Luer in Lindleyana,* 10(4): 232 (1995)—Colombia.
robusta *C.A. Luer, Ic. Pleurothallid. XVII (Monogr. Syst. Bot. Missouri Bot. Gard., 72):* 113 (1998)—Ecuador.
santiagoorum *W. Königer in Arcula,* 7: 196 (1997)—Ecuador.
saulii *W. Königer in Arcula,* 8: 226 (1999)—Peru.
scalpellifera *C.A. Luer in Harvard Pap. Bot.,* 2(1): 60 (1997), as No. 11 on inside cover—Ecuador.
scapha *L.A. Braas in Orchidee,* 33(4): 148 (1982), nom. nov.: M. navicularis *Garay & Dunst.*
schizostigma *C.A. Luer, Ic. Pleurothallid. XVII (Monogr. Syst. Bot. Missouri Bot. Gard., 72):* 114 (1998)—Peru.
schudelii *C.A. Luer, Ic. Pleurothallid. XVIII (Monogr. Syst. Bot. Missouri Bot. Gard., 76):* 152 (1999)—Ecuador.
shiraishii *W. Königer & M. Arias in Arcula,* 7: 199 (1997)—Peru.
titan *C.A. Luer in Harvard Pap. Bot.,* 9: 6 (1996)—Peru.
trautmanniana *C.A. Luer & J. Portilla in C.A. Luer, Ic. Pleurothallid. XVIII (Monogr. Syst. Bot. Missouri Bot.* Gard., 76): 153 (1999)—Ecuador.
trechsliniana *W. Königer & J. Meza in Arcula,* 7: 202 (1997)—Peru.
trivenia *W. Königer in Arcula,* 7: 205 (1997)—Peru.
venus *C.A. Luer & Hirtz in C.A. Luer, Ic. Pleurothallid. XX (Monogr. Syst. Bot. Missouri Bot. Gard.,* 79): 125 (2000)—Ecuador.
villegasii *W. Königer in Arcula,* 9: 253 (1999)—Colombia.
vittatula *C.A. Luer & R. Escobar in Harvard Pap. Bot.,* 9: 9 (1996)—Colombia.
vomeris *C.A. Luer in Harvard Pap. Bot.,* 2(1): 64 (1997), as No. 11 on inside cover—Peru.
wageneriana Linden ex Lindl. & Paxton
var. pteroglossa *(Schltr.) L.A. Braas in Orchidee,* 33(4): 148 (1982): M. pteroglossa.

MASDEVALLIA:-
wuelfinghoffiana *C.A. Luer & J. Portilla in C.A. Luer, Ic. Pleurothallid. XVIII (Monogr. Syst. Bot. Missouri Bot.* Gard., 76): 153 (1999)—Ecuador.
wuerstlei *C.A. Luer, Ic. Pleurothallid. XVIII (Monogr. Syst. Bot. Missouri Bot Gard., 76):* 154 (1999)—Colombia.
zongoensis *C.A. Luer & A. Hirtz in Lindleyana,* 10(4): 235 (1995)—Bolivia.

MASSONIA (Hyacinthac.).
pygmaea Kunth
subsp. kamiesbergensis *U. Müller-Doblies & D. Müller-Doblies in Feddes Repert.,* 108(1–2): 69 (1997)—Cape Province.
sessiliflora *(Dinter) U. Müller-Doblies & D. Müller-Doblies in Feddes Repert.,* 108(1–2): 70 (1997), without basionym ref.: Haemanthus sessiliflorus.

MATELEA (Asclepiadac.).
carmenaemiliae *G. Morillo in Acta Bot. Venezuel.,* 20(2): 16 (1997)—Venezuela.
elachyantha *W.D. Stevens in Novon,* 10(3): 249 (2000)—Nicaragua.
fulvida *(F. Ballard) W.D. Stevens in Novon,* 9(4): 564 (1999): Gonolobus fulvidus.
furvescens *W.D. Stevens in Novon,* 10(3): 251 (2000)—Costa Rica, Nicaragua.
gracieae *G. Morillo in Brittonia,* 50(3): 298 (1998)—French Guiana.
jenmanii *(Morillo) G. Morillo in Ernstia,* 10(2): 82 (2000): Cynanchum jenmanii.
marcoassisii *J. Fontella Pereira in Bradea,* 8(18): 103 (1998)—Brazil Southeast (São Paulo).
maritima
subsp. ganglinosa *(Vell.) J. Fontella Pereira in Bradea,* 5(23): 263 (1989): Cynanchum ganglinosum.
ocellata *W.D. Stevens in Novon,* 10(3): 252 (2000)—Mexico Southwest (Oaxaca, Michoacan), Nicaragua.
oxypetaloides
var. grazielae *J. Fontella Pereira in Pabstia,* 7(4): 2 (1996)—Brazil West-central (Brasília Distrito Federal).
punensis *G. Morillo in Pittieria,* 23: 50 (1995)—Peru.
roulinioides *M.F. Agra & W.D. Stevens in Novon,* 6(1): 6 (1996)—Brazil Northeast (Paraíba).
sugillata *W.D. Stevens in Novon,* 10(3): 252 (2000)—Mexico Gulf (Veracruz), Mexico Southwest (Oaxaca), Mexico Northeast (Querétaro), Nicaragua.

MATISIA (Bombacac.).
tinamastiana *A. Estrada & A. Cascante in Brenesia,* 49–50: 80 (1998)—Costa Rica.

MATRICARIA (Compos.).
sect. Tetragonospermae *N.M. Boldÿreva & O.S. Zhirova in Fl. Sibir. (Asterac.),* 13: 84 (1997 publ. 1998), as 'Tetragonosperma'.

MATUCANA (Cactac.).
aurantiaca
subsp. currundayensis *(F. Ritter) R. Mottram in Cactaceae Consensus Initiatives,* 3: 11 (1997): M. currundayensis.
haynei
subsp. herzogiana *(Backeb.) R. Mottram in Cactaceae Consensus Initiatives,* 3: 11 (1997): M. herzogiana.
subsp. hystrix *(Rauh & Backeb.) R. Mottram in Cactaceae Consensus Initiatives,* 3: 11 (1997): M. hystrix.
subsp. myriacantha *(Vaupel) R. Mottram in Cactaceae Consensus Initiatives,* 3: 11 (1997): Echinocactus myriacanthus.
f. purpureoalba *(F. Ritter) Lodé in Cact. Aventures,* 27: 29 (1995); *cf. Repert. Pl. Succ. (I.O.S.),* 46: 12 (1995 publ. 1996): M. purpureoalba.

MATUDAEA (Hamamelidac.).
colombiana *G. Lozano-Contreras in Rev. Acad. Colomb. Cienc. Exact. Fis. Nat.,* 78: 444 (1996)—Colombia.

MAULOUTCHIA *(Baill.) O. Warburg in Ber. Deutsch. Bot. Ges.,* 13(G.V.1): (83), (94) (1895 publ. 1896): Myristica sect. Mauloutchia. (Myristicac.).†

MAXILLARIA (Orchidac.).
admonens *I. Bock in Orchidee,* 48(3): 105 (1997)—Costa Rica.
cacaoensis *J.T. Atwood in Selbyana,* 19(2): 254 (1998 publ. 1999)—Costa Rica.
caveroi *D.E. Bennett & E.A. Christenson in Lindleyana,* 13(2): 64 (1998)—Peru.
cedralensis *J.T. Atwood & D.E. Mora in Selbyana,* 18(1): 31 (1997)—Costa Rica.
chionantha *J.T. Atwood in Selbyana,* 16(2): 242 (1995)—Panama, Costa Rica.
christensonii *D.E. Bennett in Brittonia,* 50(2): 187 (1998)—Peru.
colemanii *G. Carnevali & W. Fritz in Selbyana,* 21(1–2): 145 (2000)—Venezuela.
cymbidioides *C.H. Dodson, J.T. Atwood & Carnevali in Orquideologia,* 20(3): 267 (1997)—Ecuador.
darienensis *J.T. Atwood in Selbyana,* 19(2): 254 (1998 publ. 1999)—Panama.
dillonii *D.E. Bennett & E.A. Christenson in Lindleyana,* 13(2): 66 (1998)—Peru.
dressleriana *G. Carnevali & J.T. Atwood in Lindleyana,* 11(1): 29 (1996)—Panama, Costa Rica.
edwardsii *D.E. Bennett & E.A. Christenson in Lindleyana,* 13(2): 68 (1998)—Peru.
encyclioides *J.T. Atwood & C.H. Dodson in Orquideologia,* 20(3): 268 (1997)—Ecuador.
fimbriatiloba *G. Carnevali & G.A. Romero in G.A. Romero & G. Carnevali, Orchids of Venezuela, ed. 2:* 1138 (2000)—Venezuela.
formosa *G. Carnevali & G.A. Romero in G.A. Romero & G. Carnevali, Orchids of Venezuela, ed. 2:* 1139 (2000)—Venezuela.
gigantea (Lindl.) Dodson
subsp. condorensis *D.E. Bennett & E.A. Christenson in Lindleyana,* 13(2): 68 (1998)—Peru.
gomeziana *J.T. Atwood in Lindleyana,* 11(3): 202 (1996)—Costa Rica, Panama.
granditenuis *D.E. Bennett & E.A. Christenson in Lindleyana,* 13(2): 71 (1998)—Peru.
grisebachiana *M.A. Nir & Dod in M.A. Nir, Orchidaceae Antillanae:* 243 (2000)—Haiti, Cuba, Puerto Rico.
guiardiana *G. Chiron in Orchidées Cult. Protect.,* 39(3): 28 (1999)—Peru.
haberi *J.T. Atwood in Selbyana,* 16(2): 245 (1995)—Costa Rica.
lutheri *J.T. Atwood in Selbyana,* 19(2): 257 (1998 publ. 1999)—Panama.
machinazensis *D.E. Bennett & E.A. Christenson in Lindleyana,* 13(2): 71 (1998)—Peru.
mejiae *G. Carnevali & G.A. Romero in G.A. Romero & G. Carnevali, Orchids of Venezuela, ed. 2:* 1139 (2000)—Venezuela.
moralesii *G. Carnevali & J.T. Atwood in Lindleyana,* 11(1): 31 (1996)—Costa Rica.
pastorellii *D.E. Bennett & E.A. Christenson in Lindleyana,* 13(2): 74 (1998), nom. nov.: M. unguiculata *D.E. Benn. & Christenson.*
scalariformis *J.T. Atwood in Selbyana,* 19(2): 257 (1998 publ. 1999)—Panama.
synsepala *J.T. Atwood in Selbyana,* 19(2): 260 (1998 publ. 1999)—Panama.
tiaraensis *G. Carnevali & G.A. Romero in G.A. Romero & G. Carnevali, Orchids of Venezuela, ed. 2:* 1140 (2000)—Venezuela.
tonsbergii *E.A. Christenson in Orchid Dig.,* 64(1): 13 (2000)—Venezuela?
tricarinata *J.T. Atwood in Selbyana,* 19(2): 260 (1998 publ. 1999)—Panama.
turkeliae *E.A. Christenson in Orchid Dig.,* 64(2): 65 (2000)—Ecuador.
tutae *J.T. Atwood in Selbyana,* 19(2): 262 (1998 publ. 1999)—Panama.
wojii *E.A. Christenson in Orchids,* 69(8): 765 (2000)—Colombia.

× **MAXIMBIGNYA** *S.F. Glassman, Taxon. Treatment Palm Subtribe Attaleinae (Tribe Cocoeae) (Illinois Biol.* Monogr., 59): 199 (1999). *PALMAE* (Maximiliana × Orbignya).

× **MAXIMBIGNYA:-**
dahlgreniana *(Bondar) S.F. Glassman, Taxon. Treatment Palm Subtribe Attaleinae (Tribe Cocoeae) (Illinois Biol.* Monogr., 59): 199 (1999): Markleya dahlgreniana.

× **MAYSARA** *D.M. Cumming in Haworthiad,* 13(3): 115 (1999), nom. nov.; *cf. Repert. Pl. Succ. (I.O.S.),* 50: 9 (1999 publ. 2000): × Rowleyara *D.M. Cumming.* (Aloac.). (Astroloba × Gasteria × Haworthia).

MAYTENUS (Celastrac.).
fournieri (Pancher & Sebert) Loes.
subsp. drakeana *(Loes.) I.H. Müller in Fl. N. Caléd.,* 20: 73 (1996): Gymnosporia drakeana.
recondita *B.E. Hammel in Novon,* 7(2): 150 (1997)—Costa Rica, Panama.
rupestris *J.R. Pirani & R.M. Carvalho-Okano in Novon,* 9(1): 95 (1999)—Brazil Southeast (Minas Gerais).
urquiolae *B. Mory in Feddes Repert.,* 104(7–8): 432 (1993)—Cuba.

MAZAEA (Rubiac.).
shaferi *(Standl.) P.G. Delprete in Brittonia,* 51(2): 221 (1999): Neomazaea shaferi.

MAZUS (Scrophulariac.).
arenarius *P.B. Heenan, P.N. Johnson & C.J. Webb in New Zealand J. Bot.,* 34(1): 33 (1996)—New Zealand South I.
novaezeelandiae W.R. Barker
f. hirtus *P.B. Heenan in New Zealand J. Bot.,* 36(3): 410 (1998)—New Zealand North I.
subsp. impolitus *P.B. Heenan in New Zealand J. Bot.,* 36(3): 409 (1998)—New Zealand North I., New Zealand South I.
pulchellus Hemsl.
var. wangii *(H.L. Li) C.Y. Wu in Index Fl. Yunnan.,* 2: 1596 (1984): Mazus wangii.
pumilus (Burm. f.) Steenis
var. delavayi *(Bonati) C.Y. Wu in Index Fl. Yunnan.,* 2: 1597 (1984), without basionym date: Mazus delavayi.
var. delavayi *(Bonati) Chin ex D.Y. Hong in Novon,* 6(4): 374 (1996): M. delavayi.
var. macrocalyx *(Bonati) Chin ex D.Y. Hong in Novon,* 6(4): 374 (1996), isonym; *see T. Yamaz. in J. Jap. Bot.,* 55(1): 11 (1980): M. macrocalyx.
var. wangii *(H.L. Li) Chin ex D.Y. Hong in Novon,* 6(4): 374 (1996): M. wangii.
quadriprotuberans *N. Yonezawa in J. Phytogeogr. Taxon.,* 48(1): 1 (2000)—Japan.
wanmuliensis *M. Qian & L.B. Geng in Wuyi Sci. J.,* 7: 263 (1987)—China Southeast (Fujian).

MECARDONIA (Scrophulariac.).
procumbens (Mill.) Small
var. caespitosa *(Cham.) V.C. Souza in Acta Bot. Brasil.,* 11(2): 188 (1997): Herpestis caespitosa.
var. flagellaris *(Cham. & Schltdl.) V.C. Souza in Acta Bot. Brasil.,* 11(2): 186 (1997): Herpestis flagellaris.
var. herniarioides *(Cham.) V.C. Souza in Acta Bot. Brasil.,* 11(2): 187 (1997): Herpestis herniarioides.
var. tenella *(Cham. & Schltdl.) V.C. Souza in Acta Bot. Brasil.,* 11(2): 185 (1997): Herpestis tenella.

MECONOPSIS (Papaverac.).
integrifolia (Maxim.) Franch.
subsp. lijiangensis *C. Grey-Wilson in New Plantsman,* 3(1): 33 (1996)—China South-central (Yunnan).
pseudointegrifolia Prain
subsp. daliensis *C. Grey-Wilson in New Plantsman,* 3(1): 36 (1996)—China South-central (Yunnan).
subsp. robusta *C. Grey-Wilson in New Plantsman,* 3(1): 35 (1996)—China South-central (Yunnan).

MECRANIUM (Melastomatac.).
juddii *J.D. Skean in Brittonia,* 52(2): 191 (2000)—Haiti.

MEDICAGO (Legumin.).
anguina *(Delile) C. Romero Zarco in Lagascalia,* 18(2): 265 (1996): Trigonella anguina.

MEDICAGO:-
laxispira *C.C. Heyn in Fl. Iran.*, 157: 266 (1984)—Iraq.
makranica *C.C. Heyn in Fl. Iran.*, 157: 259 (1984)—Pakistan.
polymorpha L.
subsp. reticulata *(Benth.) J.L. Solanas Ferrándiz & M.B. Crespo Villalba in Sarrià*, 1: 35 (1998–99 publ. 1999): M. reticulata.

MEDINILLA (Melastomatac.).
arunachalica *G.D. Pal in J. Bombay Nat. Hist. Soc.*, 92(2): 240 (1995)—Assam (Arunachal Pradesh).
bismarck-ramuensis *W. Takeuchi in Sida,* 18(4): 1071 (1999)—New Guinea (Papua New Guinea).
sapoi-riverensis *W. Takeuchi in Sida,* 18(4): 1074 (1999)—New Guinea (Papua New Guinea).

MEDIOCALCAR (Orchidac.).
congestum *A. Schuiteman in Orchid Monogr.*, 8: 49 (1997)—New Guinea (Papua New Guinea).
paradoxum (Kraenzl.) Schltr.
subsp. latifolium *(Schltr.) A. Schuiteman in Orchid Monogr.*, 8: 56 (1997): M. latifolium.
subsp. robustum *(Schltr.) A. Schuiteman in Orchid Monogr.*, 8: 57 (1997): M. robustum.
umboiense *A. Schuiteman in Orchid Monogr.*, 8: 63 (1997)—Bismarck Archipelago.
versteegii J.J. Sm.
subsp. amphigeneum *A. Schuiteman in Orchid Monogr.*, 8: 68 (1997)—New Guinea (Papua New Guinea).
subsp. intermedium *A. Schuiteman in Orchid Monogr.*, 8: 69 (1997)—Moluccas, New Guinea (Irian Jaya, Papua New Guinea), Bismarck Archipelago, Solomon Is., Vanuatu.
subsp. vulcanicum *A. Schuiteman in Orchid Monogr.*, 8: 70 (1997)—Solomon Is. (North Solomons).

MEEBOLDIA (Umbellif.).
digitata *(Kljuykov) M.F. Watson in Edinburgh J. Bot.*, 53(1): 137 (1996): Sinodielsia digitata.
yunnanensis *(H. Wolff) Constance & F.T. Pu ex S.L. Liou in Fl. Yunnanica*, 7: 405 (1997); *et Constance & F.T. Pu in Novon*, 8(1): 70 (1998): Sinodielsia yunnanensis.

MEEBOLDINA (Restionac.).
cana *(Nees) B.G. Briggs & L.A.S. Johnson in Telopea*, 8(1): 29 (1998): Leptocarpus canus.
coangustata *(Nees) B.G. Briggs & L.A.S. Johnson in Telopea*, 8(1): 30 (1998): Leptocarpus coangustatus.
crassipes *(Pate & Meney) B.G. Briggs & L.A.S. Johnson in Telopea*, 8(1): 30 (1998): Leptocarpus crassipes.
scariosa *(R. Br.) B.G. Briggs & L.A.S. Johnson in Telopea*, 8(1): 30 (1998): Leptocarpus scariosus.

× **MEECHAIARA** *hort.in Orchid Rev.*, 104(1210): 220 (1996). *ORCHIDACEAE* (Ascocentrum × Doritis × Phalaenopsis × Rhynchostylis × Vanda).

MEGAHERTZIA *A.S. George & B.P.M. Hyland in Fl. Australia*, 16: 497 (1995). *PROTEACEAE*
amplexicaulis *A.S. George & B.P.M. Hyland in Fl. Australia*, 16: 497 (1995)—Queensland.

MEGASTYLIS (Orchidac.).
elliptica *(R. Br.) D.L. Szlachetko & P. Rutkowski in Acta Bot. Fenn.*, 169: 380 (2000): Lyperanthus ellipticus.

MEGISTOSTIGMA (Euphorbiac.).
glabratum *(Kurz) R. Govaerts in R. Govaerts, D.G. Frodin & A. Radcliffe-Smith, World Checklist & Bibliogr.* Euphorbiaceae, 3: 1166 (2000): Cnesmone glabrata.

MEINECKIA (Euphorbiac.).
sect. Decaryana *Jean F. Brunel ex A. Radcliffe-Smith in Kew Bull.*, 52(1): 173 (1997).
sect. Zimmermannia *(Pax) Jean F. Brunel ex A. Radcliffe-Smith in Kew Bull.*, 52(1): 174 (1997): gen. Zimmermannia.
sect. Zimmermanniopsis *(Radcl.-Sm.) A. Radcliffe-Smith in Kew Bull.*, 52(1): 175 (1997): gen. Zimmermanniopsis.

MEINECKIA:-
acuminata *(Verdc.) Jean F. Brunel ex A. Radcliffe-Smith in Kew Bull.*, 52(1): 174 (1997): Zimmermannia acuminata.
decaryi *(Leandri) Jean F. Brunel ex A. Radcliffe-Smith in Kew Bull.*, 52(1): 173 (1997): Cluytiandra decaryi.
var. occidentalis *(Leandri) A. Radcliffe-Smith in Kew Bull.*, 52(1): 174 (1997): Cluytiandra decaryi var. occidentalis.
grandiflora *(Verdc.) Jean F. Brunel ex A. Radcliffe-Smith in Kew Bull.*, 52(1): 174 (1997): Zimmermannia grandiflora.
nguruensis *(Radcl.-Sm.) Jean F. Brunel ex A. Radcliffe-Smith in Kew Bull.*, 52(1): 174 (1997): Zimmermannia nguruensis.
ovata *(E.A. Bruce) Jean F. Brunel ex A. Radcliffe-Smith in Kew Bull.*, 52(1): 174 (1997): Zimmermannia ovata.
paxii *Jean F. Brunel ex A. Radcliffe-Smith in Kew Bull.*, 52(1): 174 (1997), nom. nov.: Zimmermannia capillipes *Pax*.
stipularis *(Radcl.-Sm.) Jean F. Brunel ex A. Radcliffe-Smith in Kew Bull.*, 52(1): 175 (1997): Zimmermannia stipularis.
uzungwaensis *(Radcl.-Sm.) A. Radcliffe-Smith in Kew Bull.*, 52(1): 175 (1997): Zimmermanniopsis uzungwaensis.

MEIOGYNE (Annonac.).
baillonii *(Guillaumin) E.C.H. van Heusden in Bull. Mus. Nation. Hist. Nat., B, Adansonia*, Ser. 4, 18(1–2): 77 (1996): Uvaria baillonii.
dumetosa *(Guillaumin) E.C.H. van Heusden in Bull. Mus. Nation. Hist. Nat., B, Adansonia*, Ser. 4, 18(1–2): 79 (1996): Uvaria dumetosa.
lecardii *(Guillaumin) E.C.H. van Heusden in Bull. Mus. Nation. Hist. Nat., B, Adansonia*, Ser. 4, 18(1–2): 80 (1996): Unona lecardii.
tiebaghiensis *(Däniker) E.C.H. van Heusden in Bull. Mus. Nation. Hist. Nat., B, Adansonia*, Ser. 4, 18(1–2): 81 (1996): Unona tiebaghiensis.

MELALEUCA (Myrtac.).
acuminata F. Muell.
subsp. websteri *(S. Moore) Barlow ex L.A. Craven in Austral. Syst. Bot.*, 12(6): 858 (1999): M. websteri.
amydra *L.A. Craven in Austral. Syst. Bot.*, 12(6): 859 (1999)—Western Australia.
apodocephala Turcz.
subsp. calcicola *Barlow ex L.A. Craven in Austral. Syst. Bot.*, 12(6): 860 (1999)—Western Australia.
barlowii *L.A. Craven in Austral. Syst. Bot.*, 12(6): 861 (1999)—Western Australia.
beardii *L.A. Craven in Austral. Syst. Bot.*, 12(6): 862 (1999)—Western Australia.
boeophylla *L.A. Craven in Austral. Syst. Bot.*, 12(6): 863 (1999)—Western Australia.
borealis *L.A. Craven in Austral. Syst. Bot.*, 12(6): 863 (1999)—Queensland.
brevisepala *(J.W. Dawson) L.A. Craven & J.W. Dawson in Adansonia*, 20(1): 193 (1998): Callistemon brevisepalus.
brophyi *L.A. Craven in Austral. Syst. Bot.*, 12(6): 864 (1999)—Western Australia.
buseana *(Guillaumin) L.A. Craven & J.W. Dawson in Adansonia*, 20(1): 192 (1998): Callistemon buseanus.
caeca *L.A. Craven in Austral. Syst. Bot.*, 12(6): 865 (1999)—Western Australia.
cajuputi Powell
subsp. cumingiana *(Turcz.) B.A. Barlow in Novon*, 7(2): 113 (1997): M. cumingiana.
subsp. platyphylla *B.A. Barlow in Novon*, 7(2): 113 (1997)—New Guinea (Papua New Guinea, Irian Jaya), Queensland.
calyptroides *L.A. Craven in Austral. Syst. Bot.*, 12(6): 866 (1999)—Western Australia.
campanae *L.A. Craven in Austral. Syst. Bot.*, 12(6): 866 (1999)—Western Australia.
carrii *L.A. Craven in Austral. Syst. Bot.*, 12(6): 867 (1999)—Western Australia.
clarksonii *B.A. Barlow in Novon*, 7(2): 114 (1997)—Queensland.
clavifolia *L.A. Craven in Austral. Syst. Bot.*, 12(6): 868 (1999)—Western Australia.

MELALEUCA:-
condylosa *L.A. Craven in Austral. Syst. Bot.*, 12(6): 869 (1999)—Western Australia.
croxfordiae *L.A. Craven in Austral. Syst. Bot.*, 12(6): 870 (1999)—Western Australia.
dawsonii *L.A. Craven in Adansonia*, 20(1): 192 (1998), nom. nov.: Callistemon suberosus *Pancher ex Brongn. & Gris*.
delta *L.A. Craven in Austral. Syst. Bot.*, 12(6): 872 (1999)—Western Australia.
dempta *(Barlow) L.A. Craven in Austral. Syst. Bot.*, 12(6): 872 (1999): M. calycina subsp. dempta.
eulobata *L.A. Craven in Austral. Syst. Bot.*, 12(6): 875 (1999)—Western Australia.
eurystoma *Barlow ex L.A. Craven in Austral. Syst. Bot.*, 12(6): 875 (1999)—Western Australia.
eximia *(K.J. Cowley) L.A. Craven in Austral. Syst. Bot.*, 12(6): 876 (1999): M. coccinea subsp. eximia.
fabri *L.A. Craven in Austral. Syst. Bot.*, 12(6): 876 (1999)—Western Australia.
fluviatilis *B.A. Barlow in Novon*, 7(2): 116 (1997)—Queensland.
glena *L.A. Craven in Austral. Syst. Bot.*, 12(6): 878 (1999)—Western Australia.
grieveana *L.A. Craven in Austral. Syst. Bot.*, 12(6): 879 (1999)—Western Australia.
halophila *L.A. Craven in Austral. Syst. Bot.*, 12(6): 880 (1999)—Western Australia.
hnatiukii *L.A. Craven in Austral. Syst. Bot.*, 12(6): 880 (1999)—Western Australia.
hollidayi *L.A. Craven in Austral. Syst. Bot.*, 12(6): 881 (1999)—Western Australia.
huttensis *L.A. Craven in Austral. Syst. Bot.*, 12(6): 881 (1999)—Western Australia.
idana *L.A. Craven in Austral. Syst. Bot.*, 12(6): 882 (1999)—Western Australia.
johnsonii *L.A. Craven in Austral. Syst. Bot.*, 12(6): 883 (1999)—Western Australia.
keigheryi *L.A. Craven in Austral. Syst. Bot.*, 12(6): 883 (1999)—Western Australia.
laetifica *L.A. Craven in Austral. Syst. Bot.*, 12(6): 884 (1999)—Western Australia.
lara *L.A. Craven in Austral. Syst. Bot.*, 12(6): 885 (1999)—Western Australia.
lateriflora Benth.
subsp. acutifolia *(Benth.) Barlow ex L.A. Craven in Austral. Syst. Bot.*, 12(6): 885 (1999): M. lateriflora var. acutifolia.
leuropoma *L.A. Craven in Austral. Syst. Bot.*, 12(6): 888 (1999)—Western Australia.
linguiformis *L.A. Craven in Austral. Syst. Bot.*, 12(6): 888 (1999)—Western Australia.
longistaminea *(F. Muell.) Barlow ex L.A. Craven in Austral. Syst. Bot.*, 12(6): 889 (1999): M. cardiophylla var. longistaminea.
subsp. spectabilis *Barlow ex L.A. Craven in Austral. Syst. Bot.*, 12(6): 889 (1999)—Western Australia.
monantha *(Barlow) L.A. Craven in Austral. Syst. Bot.*, 12(6): 891 (1999): M. minutifolia subsp. monantha.
nematophylla *F. Muell. ex L.A. Craven in Austral. Syst. Bot.*, 12(6): 892 (1999)—Western Australia.
nervosa (Lindl.) Cheel
subsp. crosslandiana *(W. Fitzg.) Barlow ex L.A. Craven in I. Southwell & R. Lowe (eds.), Tea Tree:* 23 (1999), without basionym page: M. crosslandiana.
orbicularis *L.A. Craven in Austral. Syst. Bot.*, 12(6): 893 (1999)—Western Australia.
pancheri *(Brongn. & Gris) L.A. Craven & J.W. Dawson in Adansonia*, 20(1): 192 (1998): Callistemon pancheri.
papillosa *Turcz. ex L.A. Craven in Austral. Syst. Bot.*, 12(6): 894 (1999)—Western Australia.
penicula *(K.J. Cowley) L.A. Craven in Austral. Syst. Bot.*, 12(6): 896 (1999): M. coccinea subsp. penicula.
pentagona Labill.
var. raggedensis *L.A. Craven in Austral. Syst. Bot.*, 12(6): 897 (1999)—Western Australia.
phoidophylla *Barlow ex L.A. Craven in Austral. Syst. Bot.*, 12(6): 897 (1999)—Western Australia.
plumea *L.A. Craven in Austral. Syst. Bot.*, 12(6): 897 (1999)—Western Australia.
podiocarpa *Barlow ex L.A. Craven in Austral. Syst. Bot.*, 12(6): 898 (1999)—Western Australia.

MELALEUCA:-
procera *L.A. Craven in Austral. Syst. Bot.*, 12(6): 899 (1999)—Western Australia.
ryeae *L.A. Craven in Austral. Syst. Bot.*, 12(6): 901 (1999)—Western Australia.
sapientes *L.A. Craven in Austral. Syst. Bot.*, 12(6): 901 (1999)—Western Australia.
similis *L.A. Craven in Austral. Syst. Bot.*, 12(6): 903 (1999)—Western Australia.
societatis *L.A. Craven in Austral. Syst. Bot.*, 12(6): 904 (1999)—Western Australia.
sphaerodendra *L.A. Craven & J.W. Dawson in Adansonia*, 20(1): 192 (1998), nom. nov.: Callistemon gnidioides *Guillaumin*.
var. microphylla *(Virot) L.A. Craven & J.W. Dawson in Adansonia*, 20(1): 193 (1998): Callistemon gnidioides var. microphyllus.
squamophloia *(Byrnes) L.A. Craven in Novon*, 7(2): 118 (1997): M. styphelioides var. squamophloia
stereophloia *L.A. Craven in Austral. Syst. Bot.*, 12(6): 905 (1999)—Western Australia.
stipitata *L.A. Craven in Novon*, 7(2): 118 (1997)—Northern Territory.
stramentosa *L.A. Craven in Austral. Syst. Bot.*, 12(6): 906 (1999)—Western Australia.
systena *L.A. Craven in Austral. Syst. Bot.*, 12(6): 907 (1999), nom. nov.: M. acerosa *Schauer*.
thapsina *L.A. Craven in Austral. Syst. Bot.*, 12(6): 908 (1999)—Western Australia.
tinkeri *L.A. Craven in Austral. Syst. Bot.*, 12(6): 909 (1999)—Western Australia.
triumphalis *L.A. Craven in Muelleria*, 11: 1 (1998)—Northern Territory.
tuberculata Schauer
var. arenaria *(C.A. Gardner) L.A. Craven in Austral. Syst. Bot.*, 12(6): 911 (1999): M. arenaria.
var. macrophylla *L.A. Craven in Austral. Syst. Bot.*, 12(6): 911 (1999)—Western Australia.
venusta *L.A. Craven in Austral. Syst. Bot.*, 12(6): 912 (1999)—Western Australia.
villosisepala *L.A. Craven in Austral. Syst. Bot.*, 12(6): 913 (1999)—Western Australia.
viminea Lindl.
subsp. demissa *F.C. Quinn ex L.A. Craven in Austral. Syst. Bot.*, 12(6): 913 (1999)—Western Australia.
virgata *(Benth.) L.A. Craven in Austral. Syst. Bot.*, 12(6): 914 (1999), homonym; *see L.f., Suppl. Pl.*: 343 (1781 publ. 1782): M. urceolaris var. virgata.
wonganensis *L.A. Craven in Austral. Syst. Bot.*, 12(6): 915 (1999)—Western Australia.
zonalis *L.A. Craven in Austral. Syst. Bot.*, 12(6): 916 (1999)—Western Australia.

MELAMPYRUM (Scrophulariac.).
hians *(Druce) N.N. Tzvelev, Opred. Sosud. Rast. Severo-Zap. Rossii*: 680 (2000), isonym; *see Soó in Feddes Repert.*, 83(3): 182 (1972): Melampyrum pratense var. hians.
nemorosum L.
subsp. polonicum *(Beauverd) N.N. Tzvelev in Novosti Sist. Vÿssh. Rast.*, 32: 185 (2000): M. nemorosum var. polonicum.
var. praecox *M. Štech in Preslia*, 72(2–4): 363 (2000)—Czechoslovakia.
setaceum (Maxim.) Nakai
f. albiflorum *Y.N. Lee, Fl. Korea*: 1161 (1996), without type—Locality not stated.

MELANDRIUM (Caryophyllac.).
album (Mill.) Garcke
var. advenum *N.N. Tzvelev in Novosti Sist. Vÿssh. Rast.*, 30: 165 (1996)—Russia Northwest.
angustiflorum
subsp. tenellum *(Tolm.) Yu.P. Kozhevnikov in Novosti Sist. Vÿssh. Rast.*, 18: 233 (1981), as 'angustifolium' in Supp. 18: M. tenellum.†
multifurcatum *(C.L. Tang) Yu.P. Kozhevnikov in Rast. Tsentral. Azii*, 11: 88 (1994 publ. 1995): Silene multifurcata.
nangqenense *(C.L. Tang) L.H. Zhou in Fl. Qinghaiica*, 1: 277 (1997): Silene nangqenensis.
obscurum *(Vorosch.) T.A. Bezdeleva in Sosud. Rast. Sovet. Dal'nego Vostoka*, 8: 121 (1996): Silene obscura.

MELANOCHYLA (Anacardiac.).
condensata *K.M. Kochummen in Sandakania*, 7: 83 (1996)—Borneo (Sarawak, Brunei).
montana *K.M. Kochummen in Sandakania*, 7: 86 (1996)—Borneo (Sabah).
scalarinervis *K.M. Kochummen in Sandakania*, 7: 88 (1996)—Borneo (Sarawak).
woodiana *K.M. Kochummen in Sandakania*, 7: 88 (1996)—Borneo (Sabah).

MELANOPHYLLA (Melanophyllac.).
modestei *G.E. Schatz, P.P. Lowry & A.-E. Wolf in Adansonia*, 20(2): 239 (1998)—Madagascar.

MELANOSTACHYA *B.G. Briggs & L.A.S. Johnson in Telopea*, 7(4): 361 (1998). *RESTIONACEAE*
ustulata *(F. Muell. ex Ewart & Sharman) B.G. Briggs & L.A.S. Johnson in Telopea*, 7(4): 362 (1998): Restio ustulatus.

MELANTHIUM (Melanthiac.).
woodii *(J.W. Robbins ex A.W. Wood) N.L. Bodkin in Novon*, 8(4): 332 (1998): Veratrum woodii.

MELASTOMA (Melastomatac.).
malabathricum L.
var. normale *(D. Don) R.C. Srivastava in Novon*, 8(2): 203 (1998): M. normale.
vimineum *(D. Don) J.O. Voigt, Hort. Suburb. Calcutt.*: 44 (1845): Rhexia viminea.

MELHANIA (Sterculiac.).
annua *M. Thulin in Nordic J. Bot.*, 19(2): 193 (1999)—Somalia, Kenya.
hiranensis *M. Thulin in Nordic J. Bot.*, 19(2): 195 (1999)—Somalia.
quercifolia *M. Thulin in Nordic J. Bot.*, 19(2): 196 (1999)—Somalia.

MELICA (Gramin.).
hunzikeri *E.G. Nicora in Kurtziana*, 27(2): 361 (1999)—Argentina Northeast (Misiones).

MELICOCCUS (Sapindac.).
jimenezii *(Alain) P. Acevedo-Rodríguez in Moscosoa*, 9: 60 (1997): Talisia jimenezii.

MELICOPE (Rutac.).
brachiata *(Nadeaud) S.L. Welsh, Fl. Societensis*: 254 (1998): E. tahitensis var. brachiata.
bracteata *(Nadeaud) S.L. Welsh, Fl. Societensis*: 255 (1998): Euodia bracteata.
emarginata *(Drake) S.L. Welsh, Fl. Societensis*: 255 (1998): Euodia emarginata.
hivaoaensis *J. Florence in Allertonia*, 7(4): 243 (1997)—Marquesas.
nishimurae *(Koidz.) T. Yamazaki in J. Jap. Bot.*, 72(4): 249 (1997): Euodia nishimurae.
perlmanii *J. Florence in Allertonia*, 7(4): 246 (1997)—Marquesas.
tiarapuensis *S.L. Welsh, Fl. Societensis*: 255 (1998), nom. nov.: Euodia tahitensis *Nadeaud*.

MELICYTUS (Violac.).
dentatus *(R. Br. ex DC.) Molloy & D.J. Mabberley in Curtis's Bot. Mag.*, 17(4): 234 (2000): Hymenanthera dentata.
drucei *B.P.J. Molloy & B.D. Clarkson in New Zealand J. Bot.*, 34(4): 431 (1996)—New Zealand North I.
flexuosus *B.P.J. Molloy & A.P. Druce in New Zealand J. Bot.*, 32(2): 114 (1994)—New Zealand South I., New Zealand North I.

MELINIA (Asclepiadac.).
corymbosa *(Malme) J. Fontella Pereira & M.A. Farinaccio in Bradea*, 8(12): 67 (1997): Widgrenia corymbosa.

MELIOSMA (Meliosmac.).
arnottiana (Wight) Walp.
var. hachijoensis *(Nakai) H. Ohba in Fl. Japan*, 2c: 79 (1999): M. hachijoensis.
subsp. oldhamii *(Maxim.) H. Ohba in Fl. Japan*, 2c: 78 (1999): M. oldhamii.
var. oldhamii *(Maxim.) H. Ohba in Fl. Japan*, 2c: 78 (1999): M. oldhamii.
costata *G. Cufodontis in Oesterr. Bot. Z.*, 88(4): 266 (1939)—China Southeast (Guangxi), China South-Central (Yunnan).

MELIOSMA:-
glandulosa *G. Cufodontis in Oesterr. Bot. Z.*, 88(4): 252 (1939)—China South-Central (Guizhou), China Southeast (Guangxi).
mairei *G. Cufodontis in Oesterr. Bot. Z.*, 88(4): 257 (1939)—China South-Central (Yunnan).
pseudopaupera *G. Cufodontis in Oesterr. Bot. Z.*, 88(4): 264 (1939)—Hainan.

MELOCACTUS (Cactac.).
inclinatus *H. Antesberger in Bradleya*, 13: 14 (1995)—Netherlands Antilles.
intortus (Mill.) Urb.
subsp. domingensis *A.E. Areces-Mallea in Cact. Succ. J. (U.S.A.)*, 69(5): 246 (1997)—Dominican Republic.
pedernalensis *M. Mejía & Ricardo García in Moscosoa*, 9: 12 (1997)—Dominican Republic.
praerupticola *A.E. Areces-Mallea in Cact. Succ. J. (U.S.A.)*, 72(1): 27 (2000)—Dominican Republic.
violaceus Pfeiff.
subsp. natalensis *P.J. Braun & E. Esteves Pereira in Cact. Succ. J. (U.S.A.)*, 69(2): 71 (1997)—Brazil Northeast (Rio Grande do Norte).

MELOCALAMUS (Gramin.).
gracilis *W.T. Lin in J. S. China Agric. Univ.*, 14(3): 110 (1993), homonym; *see R.B. Majumdar in S. Karthikeyan et al., Fl. Ind. Enumerat.-Monocot.*: 278 (1989)—China Southeast (Guangxi).
maclellandii *(Munro) H.B. Naithani in Indian Forester*, 125(11): 1135 (1999): Bambusa maclellandii.
ningmingensis *D. Ohrnberger, Bamboos of the World Intro. 4*: 19 (1997), nom. nov.M. gracilis *W.T. Lin*.

MELOCHIA (Sterculiac.).
canescens *C.L. Cristóbal in Bonplandia*, 9(1–2): 43 (1996)—Paraguay, Argentina Northeast (Formosa), Argentina Northwest (Salta), Bolivia.
ministella *C.L. Cristóbal in Bonplandia*, 9(1–2): 46 (1996)—Argentina Northeast (Formosa, Chaco), Paraguay.
villosissima (K. Presl) Merr.
var. compacta *(Hochr.) F.R. Fosberg in Micronesica*, 23(2): 149 (1990): M. compacta.

MELODINUS (Apocynac.).
insularis *(Markgr.) F.R. Fosberg in Micronesica*, 23(2): 150 (1990): Clitandropsis insularis.

MELOTHRIA (Cucurbitac.).
heterophylla (Lour.) Cogn.
var. nepalensis *(C.B. Clarke) D.S. Dhaliwal & M. Sharma, Fl. Kullu Distr. (Himachal Pradesh)*: 302 (1999): Zehneria umbellata var. nepalensis.
javanica *(Miq.) Panigrahi & S.C. Mishra in J. Econ. Taxon. Bot.*, 5(2): 416 (1984), isonym see Cogn.*in A.DC. & C.DC, Monogr. Phan.*, 3: 625 (1881): Karivia javanica.

MEMECYLON (Melastomatac.).
manickamii *C. Murugan, V. Sundaresan & G.J. Jothi in Kew Bull.*, 55(4): 1001 (2000)—India (Tamil Nadu).
talbotianum *D. Brandis in W.A. Talbot, Trees, Shrubs, Woody Climbers Bombay*: appendix (1902)—India.†

MEMORA (Bignoniac.).
cidii *A.H. Gentry ex W.D. Hauk in Novon*, 9(1): 48 (1999)—Brazil North (Amazonas).
contracta *A.H. Gentry ex W.D. Hauk in Novon*, 9(1): 50 (1999)—Brazil North (Rondônia, Amazonas, Pará), Venezuela.
sastrei *A.H. Gentry & W.D. Hauk in Novon*, 9(1): 52 (1999)—Colombia.
velutina *A.H. Gentry ex W.D. Hauk in Novon*, 9(1): 52 (1999)—Brazil North (Pará).

MENDONCIA (Acanthac.).
bahiensis *S.R. Profice in Brittonia*, 49(1): 67 (1997)—Brazil Northeast (Bahia).
blanchetiana *S.R. Profice in Brittonia*, 49(1): 69 (1997)—Brazil Northeast (Bahia).

MENEPETALUM (Celastrac.).
schlechteri Loes.
subsp. crassiusculum *I.H. Müller in Fl. N. Caléd.*, 20: 48 (1996)—New Caledonia.

MENODORA (Oleac.).
gypsophila *B.L. Turner in Phytologia*, 79(1): 8 (1995 publ. 1996)—Mexico Northeast (Nuevo León).
yecorana *T.R. Van Devender & B.L. Turner in Phytologia*, 82(4): 269 (1997 publ. 1998)—Mexico Northwest (Sonora).

MENTHA (Labiat.).
canadensis L.
var. retrorsa *J.L. Liu in Acta Bot. Yunnanica*, 18(4): 410 (1996)—China South-central (Sichuan).

MENTZELIA (Loasac.).
collomiae *C.M. Christy in Novon*, 7(1): 25 (1997)—Arizona.
conspicua *T.K. Todsen in Sida*, 18(3): 819 (1999)—New Mexico.
decapetala *(Pursh) I. Urban in Ber. Deutsch. Bot. Ges.*, 10(5): 263 (1892): Bartonia decapetala.†
heterosepala *M. Weigend & E. Rodríguez Rodríguez in Arnaldoa*, 5(1): 52 (1998)—Peru.
hintoniorum *B.L. Turner & A.L. Hempel in Phytologia*, 79(4): 298 (1995 publ. 1996)—Mexico Northeast (Nuevo León).
humilis (A. Gray) J. Darl.
var. guadalupensis *R. Spellenberg in Sida*, 18(4): 995 (1999)—New Mexico.

MERCURIALIS (Euphorbiac.).
sect. Annuae *N.N. Tzvelev, Opred. Sosud. Rast. Severo-Zap. Rossii:* 679 (2000).

MERENDERA (Colchicac.).
greuteri *E.Tz. Gabrieljan in Fl. Rastitel'nost' Rast. Res. Arm.*, 12: 15 (1999)—Transcaucasus (Armeniya).

MERIANIA (Melastomatac.).
sanchezii *R. Goldenberg in Kew Bull.*, 54(2): 465 (1999)—Brazil Southeast (São Paulo).

MEROSTACHYS (Gramin.).
annulifera *T. Sendulsky in Novon*, 7(3): 286 (1997)—Brazil Northeast (Bahia).
argentea *T. Sendulsky in Novon*, 7(3): 287 (1997)—Brazil Northeast (Bahia).
bifurcata *T. Sendulsky in Novon*, 7(3): 290 (1997)—Brazil Northeast (Bahia).
calderoniana *T. Sendulsky in Novon*, 7(3): 290 (1997)—Brazil Northeast (Bahia).
fimbriata *T. Sendulsky in Novon*, 7(3): 302 (1997)—Brazil North (Rondônia).
lanata *T. Sendulsky in Novon*, 7(3): 292 (1997)—Brazil Northeast (Bahia).
leptophylla *T. Sendulsky in Novon*, 7(3): 295 (1997)—Brazil Northeast (Bahia).
magnispicula *T. Sendulsky in Novon*, 7(3): 296 (1997)—Brazil Northeast (Bahia).
medullosa *T. Sendulsky in Novon*, 7(3): 298 (1997)—Brazil Northeast (Bahia).
procerrima *T. Sendulsky in Novon*, 7(3): 300 (1997)—Brazil Northeast (Bahia), Brazil Southeast (Espírito Santo).
ramosissima *T. Sendulsky in Novon*, 7(3): 300 (1997)—Brazil Northeast (Bahia).
rondoniensis *T. Sendulsky in Novon*, 7(3): 305 (1997)—Brazil North (Rondônia).

MERREMIA (Convolvulac.).
macdonaldii *S. Valencia Ávalos & M. Martínez Gordillo in An. Inst. Biol. Univ. Nac. Autón. Méx., Bot.*, 66(1): 108 (1995)—Mexico Southwest (Guerrero).
umbellata (L.) Hallier f.
var. cochleata *(Haines) H.O. Saxena & M. Brahmam, Fl. Orissa*, 2: 1196 (1995): Ipomoea cymosa var. cochleata.

MERWILLA *F. Speta in Phyton (Austria)*, 38(1): 107 (1998). *HYACINTHACEAE*
dracomontana *(Hilliard & B.L. Burtt) F. Speta in Phyton (Austria)*, 38(1): 109 (1998): Scilla dracomontana.

MERWILLA:-
kraussii *(Baker) F. Speta in Phyton (Austria)*, 38(1): 109 (1998): Scilla kraussii.
lazulina *(Wild) F. Speta in Phyton (Austria)*, 38(1): 109 (1998): Scilla lazulina.
natalensis *(Planch.) F. Speta in Phyton (Austria)*, 38(1): 109 (1998): Scilla natalensis.
plumbea *(Lindl.) F. Speta in Phyton (Austria)*, 38(1): 109 (1998): Scilla plumbea.

MERXMUELLERA (Gramin.).
cincta (Nees) Conert
subsp. sericea *N.P. Barker in S. Afr. J. Bot.*, 65(1): 105 (1999)—Cape Province.

MESANTHOPHORA (Compos.).
rojasii *(Cabrera) H. Robinson in Smithsonian Contrib. Bot.*, 89: 76 (1999): Vernonia rojasii.

METABOLUS (Rubiac.).
decipiens *(Thwaites) C.E. Ridsdale in Blumea*, 41(2): 460 (1996): Allaeophania decipiens.

METADACRYDIUM *Baum.-Bod. ex A.P. Melikian & A.V.F.Ch. Bobrov in Bot. Zhurn.*, 85(7): 63 (2000). *PODOCARPACEAE*
araucarioides *(Brongn. & Gris) Baum.-Bod. ex A.P. Melikian & A.V.F.Ch. Bobrov in Bot. Zhurn.*, 85(7): 63 (2000): Dacrydium araucarioides.
balansae *(Brongn. & Gris) Baum.-Bod. ex A.P. Melikian & A.V.F.Ch. Bobrov in Bot. Zhurn.*, 85(7): 63 (2000): Dacrydium balansae.

METAPORANA (Convolvulac.).
verdcourtii *T. Deroin in Bull. Jard. Bot. Nation. Belg.*, 66(1–2): 31 (1997)—Madagascar.

METASEQUOIA (Taxodiac.).
honshuenensis *J. Silba & F.T. Callahan in J. Int. Conifer Preserv. Soc.*, 7(1): 2 (2000)—Japan?

METASTELMA (Asclepiadac.).
abietoides *(E. Fourn.) J. Fontella Pereira in Bradea*, 5(23): 261 (1989): Hemipogon abietoides.
chimantense *(Morillo) S. Liede in Novon*, 7(1): 42 (1997): Cynanchum chimantense.
eulaxiflorum *(Lundell) S. Liede in Novon*, 7(1): 42 (1997): Cynanchum eulaxiflorum.
exasperatum *(R.W. Holm) S. Liede in Novon*, 7(1): 42 (1997): Cynanchum exasperatum.
hirtellum *(Oliv.) S. Liede in Novon*, 7(1): 42 (1997): Vincetoxicum hirtellum.
huberi *(Morillo) S. Liede in Novon*, 7(1): 42 (1997): Cynanchum huberi.
liesnerianum *(L.O. Williams) S. Liede in Novon*, 7(1): 42 (1997): Cynanchum liesneriana.
longicoronatum *(L.O. Williams) S. Liede in Novon*, 7(1): 42 (1997): Cynanchum longicoronatum.
mexicanum *(Brandegee) M. Fishbein & R. Levin in Madroño*, 44(3): 270 (1997 publ. 1998): Melinia mexicana.
nubicola *(Morillo) S. Liede in Novon*, 7(1): 43 (1997): Cynanchum nubicola.
ovalifolium *(A. Rich.) S. Liede in Novon*, 7(1): 43 (1997): Astephanus ovalifolius.
paraquense *(Morillo) S. Liede in Novon*, 7(1): 43 (1997): Cynanchum paraquense.
pubipetalum *(Alain) S. Liede in Novon*, 7(1): 43 (1997): Cynanchum pubipetalum.
quitense *(K. Schum.) S. Liede in Novon*, 7(1): 43 (1997): Cynanchum quitense.
spruceanum *(Morillo) S. Liede in Novon*, 7(1): 43 (1997): Cynanchum spruceanum.
stipitatum *(Correll) S. Liede in Novon*, 7(1): 43 (1997): Cynanchum stipitatum.

METROSIDEROS (Myrtac.).
rotundifolia *J.W. Dawson in Blumea*, 45(2): 437 (2000)—New Caledonia.
whitakeri *J.W. Dawson in Blumea*, 45(2): 435 (2000)—New Caledonia.

METROXYLON (Palmae).
paulcoxii *W.C. McClatchey, Revis. Gen. Metroxylon sect. Coelococcus (Ph.D. Thesis)*: 158 (1996); *et in Novon*, 8(3): 255 (1998)—Samoa (Western Samoa).

MEYEROCACTUS *A.B. Doweld in Succulenta*, 75(6): 271 (1996). *CACTACEAE*
horizonthalonius *(Lem.) A.B. Doweld in Succulenta*, 75(6): 271 (1996): Echinocactus horizonthalonius.
subsp. nicholii *(L.D. Benson) A.B. Doweld in Sukkulentÿ (Moscow)*, 1999(1): 27 (1999): Echinocactus horizonthalonius var. nicholii.

MEZIA (Malpighiac.).
angelica *W.R. Anderson in Contrib. Univ. Mich. Herb.*, 21: 76 (1997)—French Guiana, Guyana, Brazil North (Amazonas, Pará).
beckii *W.R. Anderson in Contrib. Univ. Mich. Herb.*, 22: 17 (1999)—Bolivia.
mariposa *W.R. Anderson in Contrib. Univ. Mich. Herb.*, 21: 80 (1997)—Brazil North (Acre, Amazonas, Pará, Rondônia), Peru.
tomentosa *W.R. Anderson in Contrib. Univ. Mich. Herb.*, 21: 82 (1997)—Ecuador.

MEZONEURON (Legumin.).
enneaphyllum *(Roxb.) Wight & Arn. ex J.O. Voigt, Hort. Suburb. Calcutt.*: 245 (1845): Caesalpinia enneaphylla.
sumatranum *(Roxb.) Wight & Arn. ex J.O. Voigt, Hort. Suburb. Calcutt.*: 245 (1845): Caesalpinia sumatrana.

MIBORA (Gramin.).
minima (L.) Desv.
subsp. littorea *(Samp.) S. Ortiz, J. Rodríguez-Oubiña & P. Guitián in Nordic J. Bot.*, 19(5): 585 (1999): Chamagrostis minima raça littorea.

MICHELIA (Magnoliac.).
subgen. Metamichelia *Y.W. Law & Y.F. Wu in Acta Phytotax. Sin.*, 34(1): 87 (Feb. 1996); *et in Fl. Reipubl. Popul. Sin.*, 30(1): 270 (May 1996).
compressa (Maxim.) Sarg.
var. cumingii *(Merr. & Rolfe) S. Hatusima in S. Hatusima & T. Amano, Fl. Ryukyus, S. of Amami Is., ed. 2*: 60 (1994), without basionym ref.: Basionym not stated.
f. macrantha *(Hats.) S. Hatusima in S. Hatusima & T. Amano, Fl. Ryukyus, S. of Amami Is., ed. 2*: 60 (1994), without basionym ref.: Basionym not stated.
dolsopa Buch.-Ham. ex DC.
subsp. kisopa *(Buch.-Ham. ex DC.) J. Li in Acta Bot. Yunnanica*, 19(2): 135 (1997): M. kisopa.
guangxiensis *Y.W. Law & R.Z. Zhou in J. Trop. Subtrop. Bot.*, 7(3): 191 (1999)—China Southeast (Guangxi).
wilsonii Finet & Gagnep.
subsp. szechuanica *(Dandy) J. Li in Acta Bot. Yunnanica*, 19(2): 137 (1997): M. szechuanica.

MICONIA (Melastomatac.).
angelana *Rosana Romero & R. Goldenberg in Novon*, 9(1): 98 (1999)—Brazil Southeast (Minas Gerais).
bordoncilloana *G. Lozano-Contreras & M.E. Morales-Puentes in Rev. Acad. Colomb. Cienc. Exact. Fis. Nat.*, 23(Suplemento Especial): 85 (1999 publ. 2000)—Colombia.
capixaba *R. Goldenberg in Novon*, 9(4): 514 (1999)—Brazil Southeast (Espírito Santo).
celaquensis *F. Almeda in Novon*, 6(4): 319 (1996)—Honduras.
cipoensis *R. Goldenberg in Bol. Bot. (São Paulo)*, 18: 29 (1999)—Brazil Southeast (Minas Gerais).
colliculosa *F. Almeda in Proc. Calif. Acad. Sci.*, 52(4): 33 (2000)—Costa Rica, Panama.
correae *F. Almeda in Proc. Calif. Acad. Sci.*, 52(4): 37 (2000)—Panama.
crocata *F. Almeda in Proc. Calif. Acad. Sci.*, 52(4): 39 (2000)—Panama.
howardiana *W.S. Judd, V.T. Salzman & J.D. Skean in Brittonia*, 47(4): 416 (1995)—Dominican Republic.
jefensis *F. Almeda in Proc. Calif. Acad. Sci.*, 52(4): 43 (2000)—Panama.
morii *F. Almeda in Proc. Calif. Acad. Sci.*, 52(4): 46 (2000)—Panama.
oellgaardii *E. Cotton in Nordic J. Bot.*, 20(2): 181 (2000)—Ecuador.
picinguabensis *R. Goldenberg & A.B. Martins in Kew Bull.*, 54(2): 466 (1999)—Brazil Southeast (São Paulo).

MICONIA:-
rennerae E. Cotton in Nordic J. Bot., 20(2): 179 (2000), as 'renneri'—Ecuador.
talamancensis F. Almeda in Proc. Calif. Acad. Sci., 52(4): 48 (2000)—Costa Rica.
teotepecensis J.R. de Santiago Gómez in Acta Bot. Mex., 50: 22 (2000)—Mexico Southwest (Guerrero, Oaxaca).
vestita F. Almeda in Proc. Calif. Acad. Sci., 52(4): 51 (2000)—Costa Rica.

MICRAIRA (Gramin.).
lazaridis L.G. Clark, J.F. Wendel & L.A. Craven in Beagle, 12: 2 (1995)—Western Australia.

MICROCARPAEA (Scrophulariac.).
agonis A.R. Bean in Austrobaileya, 5(1): 149 (1997)—Queensland.

MICROCEPHALA (Compos.).
discoidea (Ledeb.) K. Bremer, H. Eklund, G. Medhanie, S. Heidmarsson, N. Laurent, J. Maad, J. Niklasson & A. Nordin in Pl. Syst. Evol., 200(3–4): 269 (1996): Pyrethrum discoideum.

MICROCHARIS (Legumin.).
omissa M. Thulin in Nordic J. Bot., 16(3): 295 (1996 publ. 1997)—Somalia.

MICROCHLOA (Gramin.).
annua (Kupicha & Cope) T.A. Cope in Fl. Zambes., 10(2): 231 (1999): Rendlia annua.

MICROCOCCA (Euphorbiac.).
wightii (Hook. f.) Prain
var. angustata (Susila & N.P. Balakr.) A. Radcliffe-Smith & R. Govaerts in Kew Bull., 52(2): 480 (1997): Claoxylon wightii var. angustatum.
var. glabrata (Susila & N.P. Balakr.) A. Radcliffe-Smith & R. Govaerts in Kew Bull., 52(2): 480 (1997): Claoxylon wightii var. glabratum.

MICROCYBE (Rutac.).
multiflora Turcz.
subsp. baccharoides (F. Muell.) Paul G. Wilson in Nuytsia, 12(1): 87 (1998): Eriostemon capitatus var. baccharoides.
pauciflora Turcz.
subsp. grandis Paul G. Wilson in Nuytsia, 12(1): 88 (1998)—Western Australia.

MICRODON (Scrophulariac.).
dubius (L.) O.M. Hilliard, Tribe Selagineae (Scrophulariac.): 12 (1999): Selago dubia.
nitidus (E. Mey.) O.M. Hilliard, Tribe Selagineae (Scrophulariac.): 12 (1999): Agathelpis nitida.
parviflorus (P.J. Bergius) O.M. Hilliard, Tribe Selagineae (Scrophulariac.): 12 (1999): Eranthemum parviflorum.

MICROGILIA J.M. Porter & L.A. Johnson in Aliso, 19(1): 79 (2000). POLEMONIACEAE
minutiflora (Benth.) J.M. Porter & L.A. Johnson in Aliso, 19(1): 79 (2000): Gilia minutiflora.

MICROLICIA (Melastomatac.).
flava R. Romero in Brittonia, 52(2): 142 (2000)—Brazil Southeast (Minas Gerais).

MICROMERIA (Labiat.).
albanica (Griseb. ex K. Malý) Č. Šilić, Monografija Satureja, Calamintha, Micromeria, Acinos i Clinopodium fl. Jugoslavije: 202 (1979): Satureja albanica.
danaensis A. Danin in Willdenowia, 27(1–2): 161 (1997)—Palestine (Jordan).
kosaninii Č. Šilić, Monografija Satureja, Calamintha, Micromeria, Acinos i Clinopodium fl. Jugoslavije: 234 (1979)—Yugoslavia.
microphylla (d'Urv.) Benth.
subsp. rodriguezii (Freyn & Janka) A.M. Romo, Fl. Silvestres Baleares: 261 (1994): M. rodriguezii.
parviflora (Vis.) Rchb.
f. effusa Rohlena ex Č. Šilić, Monografija Satureja, Calamintha, Micromeria, Acinos i Clinopodium fl. Jugoslavije: 262 (1979)—Yugoslavia.

MICROMERIA:-
parviflora
f. monantha Latzel ex Č. Šilić, Monografija Satureja, Calamintha, Micromeria, Acinos i Clinopodium fl. Jugoslavije: 262 (1979)—Yugoslavia.
pseudocroatica Č. Šilić, Monografija Satureja, Calamintha, Micromeria, Acinos i Clinopodium fl. Jugoslavije: 250 (1979)—Yugoslavia.

MICROMYRTUS (Myrtac.).
albicans A.R. Bean in Austrobaileya, 4(4): 463 (1997)—Queensland.
capricornia A.R. Bean in Austrobaileya, 4(4): 464 (1997)—Queensland.
carinata A.R. Bean in Austrobaileya, 4(4): 464 (1997)—Queensland.
delicata A.R. Bean in Austrobaileya, 4(4): 457 (1997)—Queensland.
forsteri A.R. Bean in Austrobaileya, 4(4): 471 (1997)—Queensland.
gracilis A.R. Bean in Austrobaileya, 4(4): 466 (1997)—Queensland.
grandis J.T. Hunter in Telopea, 7(1): 77 (1996)—New South Wales.
littoralis A.R. Bean in Austrobaileya, 4(4): 467 (1997)—Queensland.
patula A.R. Bean in Austrobaileya, 4(4): 458 (1997)—Queensland.
rotundifolia A.R. Bean in Austrobaileya, 4(4): 460 (1997)—Queensland.
vernicosa A.R. Bean in Austrobaileya, 4(4): 469 (1997)—Queensland.

MICRONYCHIA (Anacardiac.).
acuminata A. Randrianasolo in Adansonia, 22(1): 147 (2000)—Madagascar.

MICROPAPPUS (Sch. Bip.) C.F. Baker in Trans. Acad. Sci. St. Louis, 12(5): 45, 55 (1902): Elephantopus sect. Micropappus. (Compos.).

MICROPEPLIS (Chenopodiac.).
foliosa (L.) G.L. Chu in Higher Plants China, 4: 350 (2000): Anabasis foliosa.

MICROPERA (Orchidac.).
draco (Tuyama) P.J. Cribb & P. Ormerod in Kew Bull., 55(1): 231 (2000): Sarcanthus draco.

MICRORPHIUM (Gentianac.).
elmerianum J.C. Regalado & D.D. Soejarto in Novon, 7(1): 77 (1997)—Philippines.

MICROSISYMBRIUM (Crucif.).
taxkorganicum Z.X. An in Fl. Xinjiangensis, 2(2): 380 (1995)—Xinjiang.

MICROSTACHYS (Euphorbiac.).
acetosella (Milne-Redh.) H.-J. Esser in Kew Bull., 53(4): 958 (1998): Sapium acetosella.
bidentata (Mart. & Zucc.) H.-J. Esser in Kew Bull., 53(4): 958 (1998): Cnemidostachys bidentata.
dalzielii (Hutch.) H.-J. Esser in Kew Bull., 53(4): 958 (1998): Sapium dalzielii.
ditassoides (Didr.) H.-J. Esser in Kew Bull., 53(4): 958 (1998): Cnemidostachys ditassoides.
faradianensis (Beille) H.-J. Esser in Kew Bull., 53(4): 959 (1998): Excoecaria faradianensis.
heterodoxa (Müll. Arg.) H.-J. Esser in Kew Bull., 53(4): 959 (1998): Stillingia heterodoxa.
hispida (Mart.) R. Govaerts in R. Govaerts, D.G. Frodin & A. Radcliffe-Smith, World Checklist & Bibliogr. Euphorbiaceae, 3: 1190 (2000): Cnemidostachys hispida.
nummularifolia (Cordeiro) H.-J. Esser in Kew Bull., 53(4): 959 (1998): Sebastiania nummularifolia.
revoluta (Ule) H.-J. Esser in Kew Bull., 53(4): 959 (1998): Sebastiania revoluta.
uleana (Pax & K. Hoffm.) H.-J. Esser in Kew Bull., 53(4): 959 (1998): Sebastiania uleana.

MICROSTEGIUM (Gramin.).
reticulatum B.S. Sun ex H. Peng & X. Yang in Acta Phytotax. Sin., 34(2): 213 (1996)—China South-central (Yunnan).
vimineum (Trin.) A. Camus
f. willdenowianum (Nees) T. Osada, Illustr. Grasses Japan: 756 (1993): M. willdenowianum.

MICROTATORCHIS (Orchidac.).
taenioides P. O'Byrne in Malayan Orchid Rev., 32: 47 (1998)—Sulawesi.

MICROTEA (Phytolaccac.).
bahiensis M.S. Marchioretto & J.C. de Siqueira in Pesquisas, Bot., Bras., 48: 11 (1998)—Brazil Northeast (Bahia).
papillosa M.S. Marchioretto & J.C. de Siqueira in Pesquisas, Bot., Bras., 48: 30 (1998), as 'papilosa'—Brazil Southeast (Minas Gerais).

MICROTHELYS (Orchidac.).
sect. Erectifloreae (Szlach.) D.L. Szlachetko in Fragm. Flor. Geobot., 41(2): 852 (1996): Galeottiella sect. Erectifloreae.
constricta (Szlach.) D.L. Szlachetko in Fragm. Flor. Geobot., 41(2): 853 (1996): Galeottiella constricta.
intagana (Dodson & Dressler) D.L. Szlachetko in Fragm. Flor. Geobot., 41(2): 853 (1996): Ecuadoria intagana.
markowskiana (Szlach.) D.L. Szlachetko in Fragm. Flor. Geobot., 41(2): 853 (1996): Galeottiella markowskiana.
santa-elenae (Szlach.) D.L. Szlachetko in Fragm. Flor. Geobot., 41(2): 853 (1996): Galeottiella rubrocalosa var. santa-elenae.

MICROTIS (Orchidac.).
angusii D.L. Jones in Orchadian, 12(1): 10 (1996)—New South Wales.
eremaea R.J. Bates in J. Adelaide Bot. Gard., 17: 119 (1996)—Western Australia.
graniticola R.J. Bates in J. Adelaide Bot. Gard., 17: 120 (1996)—Western Australia.
media R. Br.
subsp. eremicola R.J. Bates in J. Adelaide Bot. Gard., 17: 123 (1996)—Western Australia.

MIEGIA (Gramin.).
pumila T. Nuttall in Trans. Amer. Philos. Soc., n.s., 5: 149 (1837)—Arkansas.†

MIERSIA (Alliac.).
leporina P. Ravenna in Onira, Bot. Leafl., 5(7): 31 (2000)—Chile Central (Coquimbo).
tenuiseta P. Ravenna in Onira, Bot. Leafl., 5(7): 32 (2000)—Chile Central (Santiago).
f. castanea P. Ravenna in Onira, Bot. Leafl., 5(7): 33 (2000)—Chile Central (Santiago).

MIKANIA (Compos.).
sect. Summikania W.C. Holmes in D.J.N. Hind & H.J. Beentje (eds.), Compos.: Syst. (Proc. Internat. Compos. Conf., Kew, 1994, 1): 624 (1996).
glandulosissima W.C. Holmes & D.J.N. Hind in Kew Bull., 55(2): 399 (2000)—Brazil Northeast (Bahia).
palmata J.F. Pruski & W.C. Holmes in Syst. Bot., 25(4): 571 (2000)—Ecuador.
wurdackii J.F. Pruski & W.C. Holmes in Syst. Bot., 25(4): 573 (2000)—Peru.

MIKANIOPSIS (Compos.).
camarae S. Lisowski in Fragm. Flor. Geobot., 40(2): 905 (1995)—Guinea.

MILLA (Alliac.).
conspicua (Baker) N.E. Brown, Suppl. Johnson's Gard. Dict.: 963 (1882): Triteleia conspicua.
filifolia T.M. Howard in Herbertia, 54: 235 (1999 publ. 2000)—Mexico Central.
mexicana T.M. Howard in Herbertia, 54: 233 (1999 publ. 2000)—Mexico Central, Mexico Southwest.
potosina T.M. Howard in Herbertia, 54: 234 (1999 publ. 2000)—Mexico Northeast.

MILLERIA (Compos.).
perfoliata B.L. Turner in Phytologia, 81(5): 359 (1996 publ. 1997)—Mexico Southwest (Jalisco).

MILLETTIA (Legumin.).
austroyunnanensis Y.Y. Qian in Acta Phytotax. Sin., 35(2): 183 (1997), as 'austro-yunnanensis'—China South-central (Yunnan).

MILLETTIA:-
borneensis *F. Adema in Blumea,* 45(2): 407 (2000)—Borneo (Sarawak).
chrysamaryssa *F. Adema in Novon,* 9(3): 289 (1999)—Malaya (Peninsular Malaysia).
glabra *F. Adema in Blumea,* 45(2): 413 (2000)—Maluku.
japonica (Siebold & Zucc.) A. Gray
f. microphylla *(Makino) T. Yamanaka & H. Ohashi in J. Jap. Bot.,* 72(5): 310 (1997): M. japonica var. microphylla.
pachyloba
var. lasiopetala *(Hayata) P.K. Lôc in Bot. Zhurn.,* 81(10): 99 (1996), without basionym ref.: Basionym not stated.
pseudoracemosa *K. Thothathri & S. Ravikumar in J. Econ. Taxon. Bot.,* 21(1): 239 (1997), as 'pseudo-racemosa'—India (Tamil Nadu).
velvetina *F. Adema in Blumea,* 45(2): 414 (2000), nom. nov.: Pongamia pinnata var. velutina *C.T. White.*

MILTONIA (Orchidac.).
bismarckii *(Dodson & D.E. Benn.) P.F. Hunt in Orchid Rev.,* 108(1233): new orchid hybrids 15 (2000): Miltoniopsis bismarckii.

MILTONIOIDES (Orchidac.).
oviedomotae *(Hágsater) K. Senghas in Schlechter Orchideen, ed. 3,* I/C(33–36): 2283 (1997): Oncidium oviedomotae.

MIMOSA (Legumin.).
ser. Lactifluae *(Barneby) R. Grether in Novon,* 10(1): 31 (2000): M. subser. Lactifluae.
ser. Teledactylae *(Barneby) R. Grether in Novon,* 10(1): 35 (2000): M. subser. Teledactylae.
acantholoba (Willd.) Poir.
var. seticuspis *(Barneby) R. Grether in Novon,* 10(1): 29 (2000): M. seticuspis.
bispiculata *R.C. Barneby in Brittonia,* 49(4): 454 (1997)—Brazil Southeast (Minas Gerais).
candollei *R. Grether in Novon,* 10(1): 34 (2000), nom. nov.: Schrankia leptocarpa *DC.*
craspedisetosa *R.H. Fortunato & R. Palese in Candollea,* 54(1): 84 (1999)—Bolivia.
crumenarioides *L.P. Queiroz & G.P. Lewis in Kew Bull.,* 54(4): 983 (1999 publ. 2000)—Brazil Northeast (Bahia).
cuiabensis *L. Rico & R. Grether in Kew Bull.,* 55(1): 224 (2000), nom. nov.: M. diptera *Barneby.*
dasilvae *A.S.L. da Silva & R. de S. Secco in Acta Amazonica,* 30(3): 449 (2000)—Brazil North (Pará).
demissa *R.C. Barneby in Brittonia,* 49(4): 456 (1997)—Brazil West-central (Goiás).
epitropica *R.C. Barneby & León de la Luz in Brittonia,* 49(4): 452 (1997)—Mexico Northwest (Baja California Sur).
pigra L.
var. dehiscens *(Barneby) D. Glazier & B.A. Mackinder in Kew Bull.,* 52(2): 462 (1997): M. pellita var. dehiscens.
var. hispida *(Barneby) D. Glazier & B.A. Mackinder in Kew Bull.,* 52(2): 463 (1997), without basionym ref.: Basionym not stated.
platycarpa Benth.
var. liesneri *(Barneby) R. Grether in Novon,* 10(1): 30 (2000): M. acantholoba var. liesneri.
robusta *R. Grether in Novon,* 10(1): 34 (2000), nom. nov.: Schrankia distachya *DC.*
sundra *W. Roxburgh, Pl. Coromandel,* 3: 19 (1811)—India (Tamil Nadu).†
tejupilcana *R. Grether & A. Martínez-Bernal in Syst. Bot.,* 21(4): 617 (1996 publ. 1997)—Mexico Central (México State).
tricephala Cham. & Schltdl.
var. lignosa *(Micheli) T. Chehaibar & R. Grether in Novon,* 10(1): 31 (2000): M. lignosa.
var. nelsonii *(B.L. Rob.) T. Chehaibar & R. Grether in Novon,* 10(1): 31 (2000): M. nelsonii.
var. xanti *(A. Gray) T. Chehaibar & R. Grether in Novon,* 10(1): 32 (2000): M. xanti.
tweedieana *Barneby ex D. Glazier & B.A. Mackinder in Kew Bull.,* 52(2): 460 (1997)—Argentina Northeast (Buenos Aires), Uruguay, Paraguay, Brazil South (Rio Grande do Sul).
velloziana Mart.
var. maxonii *(Standl.) R. Grether in Novon,* 10(1): 33 (2000): M. maxonii.

MIMULUS (Scrophulariac.).
minutiflorus *R.K. Vickery in Madroño,* 44(4): 392 (1997 publ. 1998)—Mexico Northeast (Durango).
naiandinus *J.M. Watson & C. von Bohlen in Curtis's Bot. Mag.,* 17(4): 199 (2000)—Chile North (Maule).
yecorensis *R.K. Vickery in Madroño,* 44(4): 391 (1997 publ. 1998)—Mexico Northwest (Sonora).

MINASIA (Compos.).
cabralensis *H. Robinson in Phytologia,* 80(5): 350 (1996 publ. 1997)—Brazil Southeast (Minas Gerais).

MINUARTIA (Caryophyllac.).
sect. Tricostatae *N.S. Pavlova in Sosud. Rast. Sovet. Dal'nego Vostoka,* 8: 40 (1996).
armena *Schischk. ex S.S. Ikonnikov in Bot. Zhurn.,* 81(6): 102 (1996)—Transcaucasus (Armeniya).
asiyeae *H. Duman in Karaca Arbor. Mag.,* 4(4): 159 (1998)—Turkey.
athoa *(Griseb.) G. Kamari in A. Strid & Kit Tan (eds.), Fl. Hellenica,* 1: 180 (1997): Alsine setacea var. athoa.
subsp. neoiraklitsa *G. Kamari in A. Strid & Kit Tan (eds.), Fl. Hellenica,* 1: 180 (1997)—Greece.
barkalovii *N.S. Pavlova in Sosud. Rast. Sovet. Dal'nego Vostoka,* 8: 39 (1996)—Kuril Is.
bosniaca *(Beck) A. von Degen in Mitt. Naturwiss. Vereines Steiermark,* 46: 319 (1910): Alsine bosniaca.
corcontica *M. Dvořáková in Preslia,* 70(4): 335 (1998 publ. 1999)—Czechoslovakia.
daralagexica *Schischk. ex S.S. Ikonnikov in Bot. Zhurn.,* 81(6): 102 (1996)—Transcaucasus (Armeniya).
eglandulosa *(Fenzl) M.V. Klokov in Novosti Sist. Vÿssh. Nizsh. Rast. (Kiev),* 1974: 47 (1974): Alsine pinifolia var. eglandulosa.
euxina *M.V. Klokov in Novosti Sist. Vÿssh. Nizsh. Rast. (Kiev),* 1974: 44 (1974)—Krym.
glomerata *(M. Bieb.) A. von Degen in Mitt. Naturwiss. Vereines Steiermark,* 46: 319 (1910): Arenaria glomerata.
subsp. burnatii *(Rouy & Foucaud) C. Favarger & F. Conti in Candollea,* 55(1): 209 (2000): Alsine rostrata f. burnatii.
subsp. cymifera *(Rouy & Foucaud) C. Favarger & F. Conti in Candollea,* 55(1): 209 (2000): Alsine cymifera.
subsp. trichocalycina *(Ten. & Guss.) F. Conti in Willdenowia,* 27(1–2): 75 (1997): Arenaria trichocalycina.
graminifolia (Ard.) Jáv.
subsp. brachypetala *G. Kamari in A. Strid & Kit Tan (eds.), Fl. Hellenica,* 1: 190 (1997)—Greece.
hirsuta (M. Bieb.) Hand.-Mazz.
subsp. asiatica *(McNeill) C. Favarger & T. Çelebioglu in Ann. Naturhist. Mus. Wien,* 98B(Suppl.) 169 (1996): M. juressi subsp. asiatica.
subsp. laxa *P.P. Panov in Fitologiya,* 48: 8 (1996)—Bulgaria.
intermedia *P.P. Panov in Fitologiya,* 48: 5 (1996), homonym; *see Hand.-Mazz. in Ann. K.K. Naturhist. Hofmus.,* 26: 148 (1912)—Bulgaria.
janevii *P.P. Panov in Fitologiya,* 48: 6 (1996)—Bulgaria.
juniperina (L.) Maire & Petitm.
var. glandulifera *(Halácsy) G. Kamari in Phyton (Austria),* 36(1): 98 (1996): Alsine juniperina var. glandulifera.
subsp. glandulifera *(Halácsy) G. Kamari in Phyton (Austria),* 36(1): 97 (1996): Alsine juniperina var. glandulifera.
subsp. kosaninii *V. Stevanović & G. Kamari in Phyton (Austria),* 36(1): 103 (1996)—Greece.
var. micropetala *G. Kamari in Phyton (Austria),* 36(1): 101 (1996)—Greece.
leptophylla *(Rchb.) M. Dvořáková in Preslia,* 70(4): 338 (1998 publ. 1999): Tryphane verna b. leptophylla.
macrocarpa (Pursh) Ostenf.
subsp. koreana *(Nakai) N.S. Pavlova in Sosud. Rast. Sovet. Dal'nego Vostoka,* 8: 37 (1996): Alsine macrocarpa var. koreana.
subsp. kurilensis *(Ikonn. & Barkalov) N.S. Pavlova in Sosud. Rast. Sovet. Dal'nego Vostoka,* 8: 37 (1996): M. kurilensis.

MINUARTIA:-
macrocarpa
subsp. minutiflora *(Hultén) V.N. Voroschilov in A.K. Skvortsov (ed.), Florist. issl. v razn. raĭonakh SSSR:* 166 (1985): M. macrocarpa var. minutiflora.
mutabilis *(Lapeyr.) Schinz & Thell. ex A. Becherer in Ber. Schweiz. Bot. Ges.,* 48: 296 (1938): Arenaria mutabilis.
subsp. lesurina *(Braun-Blanq.) C. Favarger & F. Conti in Candollea,* 55(1): 209 (2000): M. rostrata subsp. lesurina.
perdita *M. Dvořáková in Preslia,* 70(4): 337 (1998 publ. 1999)—Russia North.
pseudohybrida *M.V. Klokov in Novosti Sist. Vÿssh. Nizsh. Rast. (Kiev),* 1974: 40 (1974)—Krym.
rubella (Wahlenb.) Hiern
var. glabrata *(Cham. & Schltdl.) G.A. Peschkova in Fl. Tsentral'noi Sibiri,* 1: 321 (1979), basionym as 'Minuartia': Arenaria hirta var. glabrata.
var. glabrata *(Cham. & Schltdl.) N.S. Pavlova in Sosud. Rast. Sovet. Dal'nego Vostoka,* 8: 43 (1996), isonym; *see Peschkova in Fl. Tsentral'noi Sibiri,* 1: 321 (1979): Arenaria hirta var. glabrata.
setacea (Thuill.) Hayek
subsp. olympica *G. Kamari in A. Strid & Kit Tan (eds.), Fl. Hellenica,* 1: 179 (1997)—Greece.
subsp. subsetacea *P.P. Panov in Fitologiya,* 48: 6 (1996)—Bulgaria.
sibirica *(Regel & Tiling) N.S. Pavlova in Sosud. Rast. Sovet. Dal'nego Vostoka,* 8: 39 (1996): Cherleria sibirica.
strandjensis *P.P. Panov in Fitologiya,* 48: 7 (1996)—Bulgaria.
wettsteinii Mattf.
subsp. parnonia *G. Kamari in A. Strid & Kit Tan (eds.), Fl. Hellenica,* 1: 182 (1997)—Greece.

MINURIA (Compos.).
scoparia *P.S. Short & J.R. Hosking in Telopea,* 8(4): 407 (2000)—New South Wales.

MIQUELIOPUNTIA (Cactac.).
miquelii
f. flava *P.V. Heath in Calyx,* 6(2): 41 (1999), nom. nov., as 'Michaelopuntia michaelis': Austrocylindropuntia miquelii var. jilesii *Backeb.*

MIRABILIS (Nyctaginac.).
donahooana *A. Le Duc in Sida,* 16(4): 636 (1995), as 'donahooiana'—Mexico Southwest (Michoacan).
gracilis *(Standl.) A. Le Duc in Sida,* 16(4): 638 (1995): M. jalapa var. gracilis.
latifolia *(A. Gray) G.M. Diggs, B.L. Lipscomb & R.J. O'Kennon, Shinners & Mahler's Illustr. Fl. N. Central Texas (Sida, Bot. Misc. 16):* 840 (1999): Oxybaphus nyctagineus var. latifolius.
polonii *A. Le Duc in Sida,* 16(4): 627 (1995)—Mexico Northwest (Sinaloa), Mexico Southwest (Nayarit, Oaxaca), Mexico Northeast (Nuevo León, San Luis Potosí, Tamaulipas).
sanguinea Heimerl
var. breviflora *A. Le Duc in Sida,* 16(4): 634 (1995)—Mexico Southwest (Jalisco), Mexico Northeast (Nuevo León).

MISBROOKEA *V.A. Funk in Brittonia,* 49(1): 111 (1997). *COMPOSITAE*
strigosissima *(A. Gray) V.A. Funk in Brittonia,* 49(1): 111 (1997): Werneria strigosissima.

MISCANTHUS (Gramin.).
ridleyi *(Hack.) I.M. Turner in Asian J. Trop. Biol.,* 1(2): 26 (1995): Saccharum ridleyi.
sinensis Andersson
var. chejuensis *(Y.N. Lee) Y.N. Lee, Fl. Korea:* 1164 (1996): M. chejuensis.

MISOPATES (Scrophulariac.).
fontqueri *(Emb.) M. Ibn Tattou in Bocconea,* 8: 171 (1998): Antirrhinum fontqueri.
orontium (L.) Raf.
subsp. pusillus *(Molero) G. Mateo Sanz & M.B. Crespo Villalba, Claves Fl. Valenciana:* 307 (1990): M. orontium var. pusillus.

MITRACARPUS (Rubiac.).
steyermarkii *E.L. Cabral & N.M. Bacigalupo in Acta Bot. Brasil.*, 11(1): 50 (1997)—Brazil Northeast (Bahia).

MITRASACME (Loganiac.).
subgen. Plecocalyx *(G. Don) C.R. Dunlop in Fl. Australia*, 28: 308 (1996): M. sect. Plecocalyx.
subgen. Schizogyne *C.R. Dunlop in Fl. Australia*, 28: 308 (1996).
aggregata *C.R. Dunlop in Fl. Australia*, 28: 312 (1996)—Northern Territory.
clarksonii *C.R. Dunlop in Fl. Australia*, 28: 312 (1996)—Queensland.
epigaea *C.R. Dunlop in Fl. Australia*, 28: 311 (1996)—Western Australia, Northern Territory.
floribunda *C.R. Dunlop in Fl. Australia*, 28: 311 (1996)—Northern Territory.
galbina *C.R. Dunlop in Fl. Australia*, 28: 313 (1996)—Western Australia.
geniculosa *C.R. Dunlop in Fl. Australia*, 28: 313 (1996)—Northern Territory.
glaucescens *C.R. Dunlop in Fl. Australia*, 28: 309 (1996)—Northern Territory.
graminea *C.R. Dunlop in Fl. Australia*, 28: 309 (1996)—Western Australia.
inornata *C.R. Dunlop in Fl. Australia*, 28: 312 (1996)—Northern Territory, Queensland.
kenneallyi *C.R. Dunlop in Fl. Australia*, 28: 313 (1996)—Western Australia.
laxiceps *C.R. Dunlop in Fl. Australia*, 28: 314 (1996)—Western Australia.
maritima *C.R. Dunlop in Fl. Australia*, 28: 312 (1996)—Queensland.
neldneri *C.R. Dunlop in Fl. Australia*, 28: 311 (1996)—Queensland.
nidulifera *C.R. Dunlop in Fl. Australia*, 28: 310 (1996)—Northern Territory, Western Australia.
nudicaulis Reinw. ex Blume
var. citrina *C.R. Dunlop in Fl. Australia*, 28: 311 (1996)—Western Australia, Queensland.
oasena *C.R. Dunlop in Fl. Australia*, 28: 308 (1996)—Queensland.
patens *C.R. Dunlop in Fl. Australia*, 28: 311 (1996)—Queensland, Northern Territory.
retroloba *C.R. Dunlop in Fl. Australia*, 28: 309 (1996)—Northern Territory, Western Australia.
scopata *C.R. Dunlop in Fl. Australia*, 28: 310 (1996)—Western Australia.
scrithicola *C.R. Dunlop in Fl. Australia*, 28: 309 (1996)—Western Australia, Northern Territory.
secedens *C.R. Dunlop in Fl. Australia*, 28: 313 (1996)—Northern Territory.
squamigera *C.R. Dunlop in Fl. Australia*, 28: 309 (1996), nom. nov.: M. arnhemica var. pygmaea *Specht*.
troglodytica *C.R. Dunlop in Fl. Australia*, 28: 309 (1996)—Northern Territory.

MITREOLA (Loganiac.).
minima *B.J. Conn in Kew Bull.*, 51(1): 169 (1996)—Western Australia.

MODECCA (Passiflorac.).
trilobata *W. Roxburgh, Pl. Coromandel*, 3: 93 (1820)—Bangladesh.†

MOEHRINGIA (Caryophyllac.).
lateriflora (L.) Fenzl
var. elongata *(Schischk.) V.N. Voroschilov in A.K. Skvortsov (ed.), Florist. issl. v razn. raĭonakh SSSR*: 166 (1985): M. elongata.

MOERENHOUTIA (Orchidac.).
bombus *(J.J. Sm.) A. Schuiteman in Blumea*, 41(2): 400 (1996): Platylepis bombus.
tidorensis *(J.J. Sm.) A. Schuiteman in Blumea*, 41(2): 400 (1996): Platylepis tidorensis.

MOLDENHAWERA (Legumin.).
sect. Acuminatae *L.P. Queiroz, G.P. Lewis & R. Allkin in Kew Bull.*, 54(4): 837 (1999 publ. 2000).
sect. Brasilianae *L.P. Queiroz, G.P. Lewis & R. Allkin in Kew Bull.*, 54(4): 838 (1999 publ. 2000).
blanchetiana Tul.
var. multijuga *L.P. Queiroz, G.P. Lewis & R. Allkin in Kew Bull.*, 54(4): 840 (1999 publ. 2000)—Brazil Northeast (Bahia).

MOLDENHAWERA:-
emarginata *(Spreng.) L.P. de Queiroz & R. Allkin in Bol. Bot. (São Paulo)*, 16: 107 (1997): Cubaea emarginata.
nutans *L.P. Queiroz, G.P. Lewis & R. Allkin in Kew Bull.*, 54(4): 844 (1999 publ. 2000)—Brazil Northeast (Bahia).
papillanthera *L.P. Queiroz, G.P. Lewis & R. Allkin in Kew Bull.*, 54(4): 833 (1999 publ. 2000)—Brazil Southeast (Espírito Santo).

MOLINERIA (Hypoxidac.).
latifolia (Dryand.) Herb. ex Kurz
var. megacarpa *(Ridl.) I.M. Turner in Novon*, 6(2): 222 (1996): Curculigo megacarpa.

MOLLINEDIA (Monimiac.).
macrophylla *J.F. Morales & Q. Jiménez M. in Novon*, 6(4): 395 (1996)—Costa Rica.
marliae *A.L. Peixoto & M.V.L. Pereira in Reserva Ecol. Macaé de Cima: Aspectos Flor. Espéc. Vasc.*, 2: 316 (1996), holotype not indicated—Brazil Southeast (Rio de Janeiro).
oaxacana *D.H. Lorence in Candollea*, 54(2): 433 (1999)—Mexico Southwest (Oaxaca).
torresiorum *D.H. Lorence in Candollea*, 54(2): 436 (1999)—Mexico Southwest (Oaxaca).

MONADENIUM (Euphorbiac.).
kimberleyanum *G. Williamson in Excelsa*, 19: 56 (1999), as 'kimberleyana'—Zimbabwe.
pudibundum P.R.O. Bally
var. lanatum *S. Carter in Kew Bull.*, 55(2): 436 (2000)—Zambia.
rugosum *S. Carter in Kew Bull.*, 55(2): 437 (2000)—Tanzania, Mozambique.
spinulosum *S. Carter in Kew Bull.*, 55(2): 440 (2000)—Zimbabwe, Malawi, Mozambique.

MONANTHES (Crassulac.).
sect. Petrophyllea *Svent. ex H. 't Hart in Pl. Syst. Evol.*, 193(1–4): 215 (1994).
× isabellae Bañares
nothosubsp. gomerensis *(Bañares) Á. Bañares in Vieraea*, 25: 192 (1996): M. × gomerensis.

MONARDELLA (Labiat.).
stebbinsii *C.B. Hardham & J.A. Bartel in Aliso*, 12(4): 693 (1990)—California.

MONECHMA (Acanthac.).
genistifolium (Engl.) C.B. Clarke
subsp. australe *(P.G. Mey.) J. Munday in Fl. S. Afr.*, 30(3:1): 54 (1995): M. australe.

× **MONKHOUSEARA** *hort. in Orchid Rev.*, 105(1213): 62 (1997). *ORCHIDACEAE* (Aganisia × Batemannia × Colax × Otostylis × Zygopetalum × Zygosepalum).

MONOCARPIA (Annonac.).
borneensis *J.B. Mols & P.J.A. Kessler in Bot. Jahrb.*, 122(2): 235 (2000)—Borneo (Brunei).

MONOCHILUS (Verbenac.).
obovatus *P.D. Cantino in Novon*, 9(3): 323 (1999)—Brazil West-central (Goiás).

MONOCLADUS (Gramin.).
saxatilis L.C. Chia, H.L. Fung & Y.L. Yang
var. solidus *(C.D. Chu & C.S. Chao) L.C. Chia in Fl. Reipubl. Popul. Sin.*, 9(1): 40 (1996): Indocalamus solidus.

MONOGYNELLA (Convolvulac.).
tiricensis *J. Chrtek in Čas. Nár. Muz. Řada Přír.*, 166(1–4): 2 (1997)—Pakistan.

MONOLLUMA *D.C.H. Plowes in Haseltonia*, 3: 64 (1995). *ASCLEPIADACEAE*
cicatricosa *(Deflers) D.C.H. Plowes in Haseltonia*, 3: 65 (1995): Boucerosia cicatricosa.
quadrangula *(Forssk.) D.C.H. Plowes in Haseltonia*, 3: 65 (1995): Stapelia quadrangula.

MONOLOPIA (Compos.).
congdonii *(A. Gray) B.G. Baldwin in Novon*, 9(4): 460 (1999): Eatonella congdonii.

MONOPHYLLAEA (Gesneriac.).
albicalyx *A. Weber in Beitr. Biol. Pfl.*, 70(2–3): 405 (1997–98 publ. 1998)—Malaya (Peninsular Malaysia).
hendersonii *(B.L. Burtt) A. Weber in Beitr. Biol. Pfl.*, 70(2–3): 406 (1997–98 publ. 1998): M. horsfieldii var. hendersonii.
hirticalyx Franch.
var. furcata *A. Weber in Beitr. Biol. Pfl.*, 70(2–3): 406 (1997–98 publ. 1998)—Malaya (Peninsular Malaysia).
musangensis *A. Weber in Beitr. Biol. Pfl.*, 70(2–3): 406 (1997–98 publ. 1998)—Malaya (Peninsular Malaysia).

MONSONIA (Geraniac.).
camdeboensis *(Moffett) F. Albers in S. Afr. J. Bot.*, 62(6): 346 (1996): Sarcocaulon camdeboense.
ciliata *(Moffett) F. Albers in S. Afr. J. Bot.*, 62(6): 346 (1996): Sarcocaulon ciliatum.
crassicaulis *(S.E.A. Rehm) F. Albers in S. Afr. J. Bot.*, 62(6): 346 (1996): Sarcocaulon crassicaule.
flavescens *(S.E.A. Rehm) F. Albers in S. Afr. J. Bot.*, 62(6): 346 (1996): Sarcocaulon flavescens.
herrei *(L. Bolus) F. Albers in S. Afr. J. Bot.*, 62(6): 346 (1996): Sarcocaulon herrei.
inermis *(S.E.A. Rehm) F. Albers in S. Afr. J. Bot.*, 62(6): 346 (1996): Sarcocaulon inerme.
marlothii *(Engl.) F. Albers in S. Afr. J. Bot.*, 62(6): 346 (1996): Sarcocaulon marlothii.
peniculina *(Moffett) F. Albers in S. Afr. J. Bot.*, 62(6): 346 (1996): Sarcocaulon peniculinum.
salmoniflora *(Moffett) F. Albers in S. Afr. J. Bot.*, 62(6): 346 (1996): Sarcocaulon salmoniflorum.
vanderietiae *(L. Bolus) F. Albers in S. Afr. J. Bot.*, 62(6): 346 (1996): Sarcocaulon vanderietiae.

MONSTERA (Arac.).
buseyi *T.B. Croat & M.H. Grayum in Phytologia*, 82(1): 38 (1997)—Costa Rica, Panama.
filamentosa *T.B. Croat & M.H. Grayum in Phytologia*, 82(1): 43 (1997)—Costa Rica, Panama, Colombia.
glaucescens *T.B. Croat & M.H. Grayum in Phytologia*, 82(1): 44 (1997)—Costa Rica, Nicaragua, Panama.
lentii *T.B. Croat & M.H. Grayum in Phytologia*, 82(1): 46 (1997)—Costa Rica, Panama.
molinae *M.H. Grayum in Phytologia*, 82(1): 48 (1997)—Costa Rica, Panama.

MONTICALIA *C. Jeffrey in Kew Bull.*, 47(1): 69 (1992), nom. nov.: gen. Microchaete *Benth.* (Compos.).
abietina (Willd. ex Wedd.) C. Jeffrey
var. aciculata *(Cuatrec.) B. Nordenstam in Compositae Newsl.*, 34: 31 (1999): Senecio aciculatus.
albi-panquei *(S. Díaz & Cuatrec.) B. Nordenstam in Compositae Newsl.*, 34: 34 (1999): Pentacalia albi-panquei.
almorzana *(S. Díaz & Cuatrec.) B. Nordenstam in Compositae Newsl.*, 34: 33 (1999): Pentacalia almorzana.
andicola (Turcz.) C. Jeffrey
var. pseudonitida *(Cuatrec.) B. Nordenstam in Compositae Newsl.*, 29: 49 (1996): Senecio andicola var. pseudonitidus.
angustifolia *(Kunth) B. Nordenstam in Compositae Newsl.*, 30: 48 (1997): Cacalia angustifolia.
befarioides *(Cuatrec.) B. Nordenstam in Compositae Newsl.*, 29: 49 (1996): Senecio befarioides.
cacaosensis *(Cuatrec.) B. Nordenstam in Compositae Newsl.*, 34: 34 (1999): Senecio cacaosensis.
carrikeri (Cuatrec.) C. Jeffrey
var. macrophylla *(S. Díaz) B. Nordenstam in Compositae Newsl.*, 34: 32 (1999): Pentacalia carrikeri var. macrophylla.
carupana *(S. Díaz) B. Nordenstam in Compositae Newsl.*, 34: 32 (1999), without exact basionym page: Pentacalia carupana.
elatoides *(Wedd.) B. Nordenstam in Compositae Newsl.*, 34: 31 (1999): Senecio elatoides.
fimbriifera *(S. Díaz & G.P. Méndez) B. Nordenstam in Compositae Newsl.*, 34: 35 (1999): Pentacalia fimbriifera.

MONTICALIA:-
guadalupe (Cuatrec.) C. Jeffrey
subsp. caucana *(Cuatrec.) B. Nordenstam in Compositae Newsl.*, 34: 35 (1999): Pentacalia guadalupe subsp. caucana.
hammenii *(S. Díaz & Cuatrec.) B. Nordenstam in Compositae Newsl.*, 34: 34 (1999): Pentacalia hammenii.
harrietiae *(Cuatrec.) B. Nordenstam in Compositae Newsl.*, 34: 31 (1999), as 'harrietae': Senecio harrietiae.
iguaquensis *(S. Díaz & Cuatrec.) B. Nordenstam in Compositae Newsl.*, 34: 33 (1999): Pentacalia iguaquensis.
juajibioyi *(S. Díaz & Cuatrec.) B. Nordenstam in Compositae Newsl.*, 34: 30 (1999): Pentacalia juajibioyi.
ledifolia (Humb., Bonpl. & Kunth) C. Jeffrey
var. glabrata *(S. Díaz & G.P. Méndez) B. Nordenstam in Compositae Newsl.*, 34: 34 (1999): Pentacalia ledifolia var. glabrata.
var. lanata *(Humb., Bonpl. & Kunth) B. Nordenstam in Compositae Newsl.*, 34: 33 (1999): Cacalia lanata.
microdon *(Wedd.) B. Nordenstam in Compositae Newsl.*, 29: 49 (1996): Senecio microdon.
pulchella (Humb., Bonpl. & Kunth) C. Jeffrey
subsp. guantivana *(Cuatrec.) B. Nordenstam in Compositae Newsl.*, 34: 32 (1999), isonym; *see C. Jeffrey in Kew Bull.*, 47(1): 72 (1992): Senecio guantivanus.
romeroana *(S. Díaz & M.L. Bueno) B. Nordenstam in Compositae Newsl.*, 34: 30 (1999): Pentacalia romeroana.
schultzei *(Cuatrec.) B. Nordenstam in Compositae Newsl.*, 34: 30 (1999): Pentacalia schultzei.
subsp. sevillana *(Cuatrec.) B. Nordenstam in Compositae Newsl.*, 34: 31 (1999): Pentacalia schultzei subsp. sevillana.
subarachnoidea (Wedd.) C. Jeffrey
var. pauciflora *(S. Díaz & M.L. Bueno) B. Nordenstam in Compositae Newsl.*, 34: 30 (1999): Pentacalia subarachnoidea var. pauciflora.
summapacis *(Cuatrec.) B. Nordenstam in Compositae Newsl.*, 34: 33 (1999): Senecio summapacis.
taironae *(S. Díaz & Cuatrec.) B. Nordenstam in Compositae Newsl.*, 34: 34 (1999): Pentacalia taironae.
teretifolia *(Kunth) B. Nordenstam in Compositae Newsl.*, 29: 49 (1996): Cacalia teretifolia.
tolimensis (Cuatrec.) C. Jeffrey
var. colombiana *(Cuatrec.) B. Nordenstam in Compositae Newsl.*, 34: 35 (1999): Senecio colombianus.
vaccinioides (Humb., Bonpl. & Kunth) C. Jeffrey
var. microdentata *(Cuatrec.) B. Nordenstam in Compositae Newsl.*, 34: 32 (1999): Senecio vaccinioides var. microdentatus.

MONTIGENA *P.B. Heenan in New Zealand J. Bot.*, 36(1): 42 (1998). *LEGUMINOSAE*
novae-zelandiae *(Hook. f.) P.B. Heenan in New Zealand J. Bot.*, 36(1): 43 (1998): Swainsona novae-zelandiae.

× **MOOREARA** *hort.in Orchid Rev.*, 106(1219): 62 (1998). *ORCHIDACEAE* (Brassavola × Broughtonia × Cattleya × Laelia × Schomburgkia × Sophronitis).

MORAEA (Iridac.).
ser. Eurystigma *(Goldblatt) P. Goldblatt in Novon*, 8(4): 377 (1998): Galaxia subgen. Eurystigma.
sect. Galaxia *(Thunb.) P. Goldblatt in Novon*, 8(4): 376 (1998): gen. Galaxia.
sect. Gynandriris *(Parl.) P. Goldblatt in Novon*, 8(4): 376 (1998): gen. Gynandriris.
sect. Hexaglottis *(Vent.) P. Goldblatt in Novon*, 8(4): 376 (1998): gen. Hexaglottis.
sect. Homeria *(Vent.) P. Goldblatt in Novon*, 8(4): 374 (1998): gen. Homeria.
albiflora *(G.J. Lewis) P. Goldblatt in Novon*, 8(4): 376 (1998): Galaxia albiflora.
angulata *P. Goldblatt in Novon*, 8(4): 376 (1998), nom. nov.: Galaxia alata *Goldblatt*.
aspera *P. Goldblatt in Novon*, 8(4): 374 (1998), nom. nov.: Homeria spiralis *L. Bolus*.

MORAEA:-
australis *(Goldblatt) P. Goldblatt in Novon*, 8(4): 376 (1998): Gynandris australis.
autumnalis *(Goldblatt) P. Goldblatt in Novon*, 8(4): 374 (1998): Homeria autumnalis.
barnardiella *P. Goldblatt in Novon*, 8(4): 377 (1998), nom. nov.: Galaxia barnardii *Goldblatt*.
bifida *(L. Bolus) P. Goldblatt in Novon*, 8(4): 374 (1998): Homeria bifida.
brachygyne *(Schltr.) P. Goldblatt in Novon*, 8(4): 374 (1998): Homeria brachygyne.
brevituba *(Goldblatt) P. Goldblatt in Novon*, 8(4): 376 (1998): Hexaglottis brevituba.
britteniae *(L. Bolus) P. Goldblatt in Novon*, 8(4): 374 (1998): Homeria britteniae.
bulbillifera *(G.J. Lewis) P. Goldblatt in Novon*, 8(4): 374 (1998): Homeria bulbillifera.
cedarmontana *(Goldblatt) P. Goldblatt in Novon*, 8(4): 376 (1998): Gynandriris cedarmontana.
cedarmonticola *P. Goldblatt in Novon*, 8(4): 374 (1998), nom. nov.: Homeria cedarmontana *Goldblatt*.
citrina *(G.J. Lewis) P. Goldblatt in Novon*, 8(4): 377 (1998): Galaxia citrina.
comptonii *(L. Bolus) P. Goldblatt in Novon*, 8(4): 375 (1998): Homeria comptonii.
contorta *P. Goldblatt in Novon*, 8(4): 376 (1998), nom. nov.: Gynandriris anomala *Goldblatt*.
cookii *(L. Bolus) P. Goldblatt in Novon*, 8(4): 375 (1998): Homeria cookii.
deltoidea *P. Goldblatt & J.C. Manning in Novon*, 10(1): 16 (2000)—Cape Province.
demissa *P. Goldblatt in Novon*, 8(4): 375 (1998), nom. nov.: Homeria tenuis *Schltr.*
fenestralis *(Goldblatt & E.G.H. Oliv.) P. Goldblatt in Novon*, 8(4): 377 (1998): Galaxia fenestralis.
fenestrata *(Goldblatt) P. Goldblatt in Novon*, 8(4): 375 (1998): Homeria fenestrata.
fistulosa *(Goldblatt) P. Goldblatt in Novon*, 8(4): 374 (1998): Roggeveldia fistulosa.
flavescens *(Goldblatt) P. Goldblatt in Novon*, 8(4): 375 (1998): Homeria flavescens.
fragrans *P. Goldblatt in Novon*, 8(4): 375 (1998), nom. nov.: Homeria odorata *L. Bolus*.
fugacissima *(L.f.) P. Goldblatt in Novon*, 8(4): 376 (1998): Ixia fugacissima.
fuscomontana *(Goldblatt) P. Goldblatt in Novon*, 8(4): 375 (1998): Homeria fuscomontana.
galaxia *(L.f.) P. Goldblatt & J.C. Manning in Strelitzia*, 9: 714 (2000): Ixia galaxia.
herrei *(L. Bolus) P. Goldblatt in Novon*, 8(4): 374 (1998): Homeria herrei.
hesperantha *(Goldblatt) P. Goldblatt in Novon*, 8(4): 376 (1998): Gynandriris hesperantha.
kamiesensis *P. Goldblatt in Novon*, 8(4): 377 (1998), nom. nov.: Galaxia grandiflora *Andrews*.
kamiesmontana *(Goldblatt) P. Goldblatt in Novon*, 8(4): 377 (1998): Galaxia kamiesmontana.
karroica *P. Goldblatt in Novon*, 8(4): 375 (1998), nom. nov.: Homeria tricolor *G.J. Lewis*.
knersvlaktensis *P. Goldblatt in Novon*, 8(4): 375 (1998), nom. nov.: Homeria ramosissima *Schltr.*
lewisiae *(Goldblatt) P. Goldblatt in Novon*, 8(4): 376 (1998): Hexaglottis lewisiae.
longistyla *(Goldblatt) P. Goldblatt in Novon*, 8(4): 375 (1998): Homeria longistyla.
louisabolusiae *P. Goldblatt in Novon*, 8(4): 375 (1998), nom. nov.: Homeria bolusiae *Goldblatt*.
luteoalba *(Goldblatt) P. Goldblatt in Novon*, 8(4): 377 (1998): Galaxia luteoalba.
marlothii *(L. Bolus) P. Goldblatt in Novon*, 8(4): 375 (1998): Homeria marlothii.
mediterranea *P. Goldblatt in Novon*, 8(4): 376 (1998), nom. nov.: Gynandriris monophylla *Boiss. & Heldr. ex Klatt*.
melanops *P. Goldblatt & J.C. Manning in Novon*, 10(1): 19 (2000)—Cape Province.
minima *P. Goldblatt in Novon*, 8(4): 377 (1998), nom. nov.: Galaxia parva *Goldblatt*.
monticola *P. Goldblatt in Novon*, 8(4): 374 (1998), nom. nov.: Roggeveldia montana *Goldblatt*.
namaquana *(Goldblatt) P. Goldblatt in Novon*, 8(4): 376 (1998): Hexaglottis namaquana.
ovalifolia *P. Goldblatt in Novon*, 8(4): 377 (1998), nom. nov.: Galaxia ovata *Thunb.*
pallida *(Baker) P. Goldblatt in Novon*, 8(4): 375 (1998): Homeria pallida.
patens *(Goldblatt) P. Goldblatt in Novon*, 8(4): 375 (1998): Homeria patens.

MORAEA:-
pendula *(Goldblatt) P. Goldblatt in Novon*, 8(4): 375 (1998): Homeria pendula.
pilifolia *P. Goldblatt in Novon*, 8(4): 377 (1998), nom. nov.: Galaxia ciliata *Pers.*
pyrophila *P. Goldblatt in Novon*, 8(4): 375 (1998), nom. nov.: Homeria galpinii *L. Bolus*.
radians *(Goldblatt) P. Goldblatt in Novon*, 8(4): 375 (1998): Sessilistigma radians.
reflexa *P. Goldblatt in Novon*, 8(4): 375 (1998), nom. nov.: Homeria hantamensis *Goldblatt & J.C. Manning*.
riparia *(Goldblatt) P. Goldblatt in Novon*, 8(4): 376 (1998): Hexaglottis riparia.
schlechteri *(L. Bolus) P. Goldblatt in Novon*, 8(4): 375 (1998): Homeria schlechteri.
serratostyla *(Goldblatt) P. Goldblatt in Novon*, 8(4): 376 (1998): Homeria serratostyla.
stagnalis *(Goldblatt) P. Goldblatt in Novon*, 8(4): 377 (1998): Galaxia stagnalis.
vallisbelli *(Goldblatt) P. Goldblatt in Novon*, 8(4): 376 (1998): Homeria vallisbelli.
variabilis *(G.J. Lewis) P. Goldblatt in Novon*, 8(4): 377 (1998): Galaxia variabilis.
versicolor *(Salisb. ex Klatt) P. Goldblatt in Novon*, 8(4): 377 (1998): Galaxia versicolor.
vespertina *P. Goldblatt & J.C. Manning in Novon*, 10(1): 14 (2000)—Cape Province.
vigilans *P. Goldblatt & J.C. Manning in Novon*, 10(1): 17 (2000)—Orange Free State.

MORELLA (Myricac.).
arborea *(Hutch.) M. Cheek in M. Cheek et al., Pl. Mount Oku & Ijim Ridge, Cameroon, Conserv. Checklist*: 149 (2000): Myrica arborea.
brevifolia *(E. Mey. ex C.DC.) D.J.B. Killick in Kew Bull.*, 53(4): 994 (1998): Myrica brevifolia.
cordifolia *(L.) D.J.B. Killick in Kew Bull.*, 53(4): 994 (1998): Myrica cordifolia.
diversifolia *(Adamson) D.J.B. Killick in Kew Bull.*, 53(4): 994 (1998): Myrica diversifolia.
humilis *(Cham. & Schltdl.) D.J.B. Killick in Kew Bull.*, 53(4): 994 (1998): Myrica humilis.
integra *(A. Chev.) D.J.B. Killick in Kew Bull.*, 53(4): 993 (1998): Myrica conifera var. integra.
kandtiana *(Engl.) B. Verdcourt & R.M. Polhill in Kew Bull.*, 53(4): 994 (1998): Myrica kandtiana.
kraussiana *(Buchinger ex Meisn.) D.J.B. Killick in Kew Bull.*, 53(4): 994 (1998): Myrica kraussiana.
microbracteata *(Weim.) B. Verdcourt & R.M. Polhill in Kew Bull.*, 53(4): 995 (1998): Myrica microbracteata.
pilulifera *(Rendle) D.J.B. Killick in Kew Bull.*, 53(4): 994 (1998): Myrica pilulifera.
quercifolia *(L.) D.J.B. Killick in Kew Bull.*, 53(4): 994 (1998): Myrica quercifolia.
salicifolia *(A. Rich.) B. Verdcourt & R.M. Polhill in Kew Bull.*, 53(4): 995 (1998): Myrica salicifolia.
var. goetzei *(Engl.) B. Verdcourt & R.M. Polhill in Kew Bull.*, 53(4): 995 (1998), without basionym ref.; *et in Fl. Trop. E. Afr., Myricac.*: 10 (2000), with basionym ref.: Myrica goetzei.
subsp. kilimandscharica *(Engl.) B. Verdcourt & R.M. Polhill in Kew Bull.*, 53(4): 995 (1998): Myrica kilimandscharica.
var. kilimandscharica *(Engl.) B. Verdcourt & R.M. Polhill in Kew Bull.*, 53(4): 995 (1998): Myrica kilimandscharica.
subsp. meyeri-johannis *(Engl.) B. Verdcourt & R.M. Polhill in Kew Bull.*, 53(4): 995 (1998): Myrica meyeri-johannis.
subsp. mildbraedii *(Engl.) B. Verdcourt & R.M. Polhill in Kew Bull.*, 53(4): 995 (1998): Myrica mildbraedii.
serrata *(Lam.) D.J.B. Killick in Kew Bull.*, 53(4): 993 (1998): Myrica serrata.
spathulata *(Mirb.) B. Verdcourt & R.M. Polhill in Kew Bull.*, 53(4): 994 (1998): Myrica spathulata.

MORINDA (Rubiac.).
sect. Bajitian *Y.Z. Ruan in Fl. Reipubl. Popul. Sin.*, 71(2): 337 (1999).
angustifolia *W. Roxburgh, Pl. Coromandel*, 3: 32 (1815)—Bangladesh.†
badia *Y.Z. Ruan in Fl. Reipubl. Popul. Sin.*, 71(2): 327 (1999)—China Southeast (Hunan, Fujian, Guangdong, Guangxi).

MORINDA:-
callicarpifolia *Y.Z. Ruan in Fl. Reipubl. Popul. Sin.*, 71(2): 325 (1999), as 'callicarpaefolia'—China South-central (Yunnan, Guizhou Sichuan).
canthoides *(F. Muell.) D.A. Halford & R.J.F. Henderson in Austrobaileya*, 5(4): 731 (2000): Coprosma canthoides.
cinnamomifoliata *Y.Z. Ruan in Fl. Reipubl. Popul. Sin.*, 71(2): 336 (1999)—China Southeast (Guangxi).
citrina *Y.Z. Ruan in Fl. Reipubl. Popul. Sin.*, 71(2): 331 (1999)—China Southeast (Guangdong, Guangxi, Hunan).
var. chlorina *Y.Z. Ruan in Fl. Reipubl. Popul. Sin.*, 71(2): 332 (1999)—China Southeast (Zhejiang, Hunan, Fujian, Guangxi).
coreia Buch.-Ham.
var. tomentosa *(Hook. f.) R.R. Fernandez, Trees of Mumbai (Bombay)*: 174 (1999): M. tinctoria var. tomentosa.
litseifolia *Y.Z. Ruan in Fl. Reipubl. Popul. Sin.*, 71(2): 335 (1999)—China Southeast (Jiangxi, Hunan), China South-central (Sichuan).
longissima *Y.Z. Ruan in Fl. Reipubl. Popul. Sin.*, 71(2): 323 (1999)—China South-central (Yunnan).
nanlingensis *Y.Z. Ruan in Fl. Reipubl. Popul. Sin.*, 71(2): 329 (1999)—China Southeast (Hunan, Fujian, Guangdong, Guangxi).
var. pauciflora *Y.Z. Ruan in Fl. Reipubl. Popul. Sin.*, 71(2): 331 (1999), contrary to Art. 37.6 ICBN (2000)—China Southeast (Zhejiang).
var. pilophora *Y.Z. Ruan in Fl. Reipubl. Popul. Sin.*, 71(2): 331 (1999)—China Southeast (Hunan).
pubescens Sm.
var. stenophylla *(Spreng.) G.R. Kumari in A.N. Henry, G.R. Kumari & V. Chithra, Fl. Tamil Nadu, India, Ser. 1*, 2: 255 (1987): M. stenophylla.
pubiofficinalis *Y.Z. Ruan in Fl. Reipubl. Popul. Sin.*, 71(2): 334 (1999)—China South-central (Guizhou), China Southeast (Guangdong).
rosiflora *Y.Z. Ruan in Fl. Reipubl. Popul. Sin.*, 71(2): 324 (1999)—China South-central (Yunnan).
rugulosa *Y.Z. Ruan in Fl. Reipubl. Popul. Sin.*, 71(2): 328 (1999)—China Southeast (Guangxi, Hunan).
scabrifolia *Y.Z. Ruan in Fl. Reipubl. Popul. Sin.*, 71(2): 332 (1999)—China South-central (Yunnan), China Southeast (Guangxi).
umbellata L.
subsp. obovata *Y.Z. Ruan in Fl. Reipubl. Popul. Sin.*, 71(2): 325 (1999)—China Southeast (Guangdong).
undulata *Y.Z. Ruan in Fl. Reipubl. Popul. Sin.*, 71(2): 321 (1999)—China South-central (Yunnan).

MORMODES (Orchidac.).
aureum *L.C. Menezes & K. Tadaiesky in Orchidee*, 48(6): 256 (1997); *et in Bol. CAOB*, 29: 94 (1997)—Brazil North (Pará).
morenoi *R. Vásquez Ch. & C.H. Dodson in Rev. Soc. Boliv. Bot.*, 2(1): 6 (1998)—Bolivia.

MORMOLYCA (Orchidac.).
aurorae *D.E. Bennett & E.A. Christenson in Lindleyana*, 13(2): 74 (1998)—Peru

MORUS (Morac.).
australis Poir.
var. hastifolia *(Z.Y. Cao) Z.Y. Cao in Acta Bot. Yunnanica*, 17(2): 158 (1995): M. hastifolia.
var. linearipartita *Z.Y. Cao in Acta Bot. Yunnanica*, 17(2): 158 (1995)—China South-central (Sichuan, Hubei), China North-central (Shaanxi), China Southeast (Jiangxi).
var. oblongifolia *Z.Y. Cao in Acta Bot. Yunnanica*, 17(2): 154 (1995)—China South-central (Hubei), China North-central (Shaanxi).
cathayana Hemsl.
var. gongshanensis *(Z.Y. Cao) Z.Y. Cao in Acta Bot. Yunnanica*, 17(2): 154 (1995): M. gongshanensis.
macroura Miq.
var. mawu *(Koidz.) C.Y. Wu & Z.Y. Cao in Acta Bot. Yunnanica*, 17(2): 153 (1995): M. wittiorum var. mawu.
mongolica (Bureau) C.K. Schneid.
var. barkamensis *(S.S. Chang) C.Y. Wu & Z.Y. Cao in Acta Bot. Yunnanica*, 17(2): 154 (1995): M. barkamensis.

MORUS:-
mongolica
var. pubescens *S.C. Li & X.M. Liu in S.C. Li (ed.) et al., Dendrol. Zhongtiaoshan.*: 366 (1995)—China North-central (Shanxi).
var. yunnanensis *(Koidz.) C.Y. Wu & Z.Y. Cao in Acta Bot. Yunnanica*, 17(2): 154 (1995): M. yunnanensis.

MOSANNONA *L.W. Chatrou, Changing Genera*: 155 (1998). *ANNONACEAE*
costaricensis *(R.E. Fr.) L.W. Chatrou, Changing Genera*: 157 (1998): Malmea costaricensis.
depressa *(Baill.) L.W. Chatrou, Changing Genera*: 159 (1998): Annona depressa.
subsp. abscondita *(Chatrou) L.W. Chatrou, Changing Genera*: 162 (1998): Malmea depressa subsp. abscondita.
discolor *(R.E. Fr.) L.W. Chatrou, Changing Genera*: 163 (1998): Malmea discolor.
garwoodiae *L.W. Chatrou & Welzenis in L.W. Chatrou, Changing Genera*: 164 (1998), as 'garwoodii'—Panama.
guatemalensis *(Lundell) L.W. Chatrou, Changing Genera*: 166 (1998): Malmea guatemalensis.
hypoglauca *(Standl.) L.W. Chatrou, Changing Genera*: 166 (1998): Guatteria hypoglauca.
maculata *L.W. Chatrou & Welzenis in L.W. Chatrou, Changing Genera*: 168 (1998)—Colombia, Panama.
pachiteae *(D.R. Simpson) L.W. Chatrou, Changing Genera*: 169 (1998): Malmea pachiteae.
pacifica *L.W. Chatrou, Changing Genera*: 170 (1998)—Ecuador.
papillosa *L.W. Chatrou, Changing Genera*: 171 (1998)—Ecuador, Colombia.
parva *L.W. Chatrou, Changing Genera*: 175 (1998)—Peru.
raimondii *(Diels) L.W. Chatrou, Changing Genera*: 176 (1998): Guatteria raimondii.
vasquezii *L.W. Chatrou, Changing Genera*: 178 (1998)—Peru.
xanthochlora *(Diels) L.W. Chatrou, Changing Genera*: 179 (1998): Guatteria xanthochlora.

MOSIERA (Myrtac.).
prismatica *(D. Legrand) L.R. Landrum in Brittonia*, 49(4): 522 (1997): Eugenia prismatica.

MOURETIA (Rubiac.).
guangdongensis *H.S. Lo in Bull. Bot. Res. North-East. Forest. Univ.*, 18(3): 282 (1998), holotype not indicated—China Southeast (Guangdong, Guangxi).
inaequalis *(H.S. Lo) C. Tange in Nordic J. Bot.*, 17(2): 127 (1997): M. tonkinensis var. inaequalis.
larsenii *C. Tange in Nordic J. Bot.*, 17(2): 128 (1997)—Thailand, Laos.
vietnamensis *C. Tange in Nordic J. Bot.*, 17(2): 124 (1997)—Vietnam.

MOURIRI (Melastomatac.).
laxiflora *T. Morley in Novon*, 8(1): 41 (1998)—Ecuador.
tuberculata *T. Morley & K. Thomsen in Biollania*, ed. esp. 6 (J.J. Wurdack Festschr.): 467 (1997)—Costa Rica.

MOUTABEA (Polygalac.).
gentryi *T. Wendt in Lundellia*, 3: 6 (2000)—Panama, Costa Rica, Colombia.

MTONIA *H.J. Beentje in Kew Bull.*, 54(1): 97 (1999). *COMPOSITAE*
glandulifera *H.J. Beentje in Kew Bull.*, 54(1): 97 (1999)—Tanzania.

MUCUNA (Legumin.).
hexandra *T.S. Ralph, Icon. Carpolog.*: 30 (1849)—India.
manongarivensis *D.J. Du Puy & J.-N. Labat in Bull. Mus. Nation. Hist. Nat., B, Adansonia*, Ser. 4, 18(3–4): 232 (1996)—Madagascar.
pruriens (L.) DC.
var. thekkadiensis *K. Thothathri & S. Ravikumar in J. Econ. Taxon. Bot.*, 21(3): 703 (1997 publ. 1998)—India (Kerala).

MUEHLENBECKIA (Polygonac.).
horrida H. Gross
subsp. abdita *K.L. Wilson in Nuytsia*, 11(1): 136 (1996)—Western Australia.
injucunda *(Lindl.) N.E. Brown, Suppl. Johnson's Gard. Dict.*: 965 (1882): Polygonum injucundum.
tuggeranong *D.J. Mallinson in Telopea*, 7(3): 215 (1997)—New South Wales (Australian Capital Territory).

MUHLENBERGIA (Gramin.).
jaime-hintonii *P.M. Peterson & J. Valdés-Reyna in Sida*, 18(3): 686 (1999)—Mexico Northeast (Nuevo León).
villiflora Hitchc.
var. villosa *(Swallen) C.W. Morden in Phytologia*, 79(1): 29 (1995 publ. 1996): M. villosa.

MULGEDIUM (Compos.).
sect. Lagedium *(Soják) A.N. Sennikov in Bot. Zhurn.*, 82(5): 116 (1997): gen. Lagedium.
alatoicum *C.H. An in Fl. Xinjiangensis*, 5: 480 (1999)—Xinjiang.
quercinum (L.) C. Jeffrey
var. integrifolium *(Bogenh.) T. Karlsson in Svensk Bot. Tidskr.*, 91(5): 249 (1997 publ. 1998): Lactuca stricta δ integrifolia.

MURALTIA (Polygalac.).
levynsiana *J.A.R. Paiva in Fontqueria*, 50: 321 (1998), nom. nov. superfl.: Polygala tenuifolia *Poir.*

MURDANNIA (Commelinac.).
audreyae *R.B. Faden in M.D. Dassanayake & W.D. Clayton, eds., Revised Handb. Fl. Ceylon*, 14: 170 (2000), without latin descr.—Sri Lanka.
dimorphoides *R.B. Faden in M.D. Dassanayake & W.D. Clayton, eds., Revised Handb. Fl. Ceylon*, 14: 154 (2000), without latin descr.—Sri Lanka.
subsp. perennis *R.B. Faden in M.D. Dassanayake & W.D. Clayton, eds., Revised Handb. Fl. Ceylon*, 14: 156 (2000), without latin descr.—Sri Lanka.
spirata (L.) G. Brückn.*var.*parviflora *R.B. Faden in M.D. Dassanayake & W.D. Clayton, eds., Revised Handb. Fl. Ceylon*, 14: 161 (2000), without latin descr.—Sri Lanka.
striatipetala *R.B. Faden in M.D. Dassanayake & W.D. Clayton, eds., Revised Handb. Fl. Ceylon*, 14: 163 (2000), without latin descr.—Sri Lanka.
vaginata (L.) G. Brückn.
var. glabrisepala *R.B. Faden in M.D. Dassanayake & W.D. Clayton, eds., Revised Handb. Fl. Ceylon*, 14: 167 (2000), without latin descr.—Sri Lanka.

MUSA (Musac.).
balbisiana Colla
var. andamanica *D.B. Singh, P.V. Sreekumar, T.V.R.S. Sharma & A.K. Bandyopadhyay in Malayan Nat. J.*, 52(3–4): 157 (1998)—Andaman Is.
basjoo Siebold
var. formosana *(Warb.) S.S. Ying in Mem. Coll. Agric. Nation. Taiwan Univ.*, 25(2): 100 (1985): M. paradisiaca var. formosana.
glauca *W. Roxburgh, Pl. Coromandel*, 3: 96 (1820)—Burma.†
lawitiensis *R.E. Nasution & Supardiyono in Bul. Kebun Raya*, 8(4): 128 (1998)—Borneo (Kalimantan).
monticola *M. Hotta ex G. Argent in Gard. Bull. Singapore*, 52(1): 206 (2000)—Borneo (Sabah).
superba *W. Roxburgh, Pl. Coromandel*, 3: 17 (1811)—India (Tamil Nadu).†
suratii *G. Argent in Gard. Bull. Singapore*, 52(1): 203 (2000)—Borneo (Sabah, Sarawak).

MUSCARI (Hyacinthac.).
adilii *M.B. Güner & H. Duman in Karaca Arbor. Mag.*, 5(2): 59 (1999)—Turkey.
botryoides (L.) Mill.
subsp. longifolium *(Rigo) F. Garbari in Inform. Bot. Ital.*, 28(1): 59 (1996): M. longifolium.
breviscapum *(Tod.) N.E. Brown, Suppl. Johnson's Gard. Dict.*: 965 (1882): Botryanthus breviscapus.

MUSCARI:-
cazorlanum *C. Soriano, M.A. Rivas Ponce, R. Lozano & M. Ruíz Rejón in Lagascalia*, 15(2): 145 (1990)—Spain.
creticum *(Baker) N.E. Brown, Suppl. Johnson's Gard. Dict.*: 965 (1882): M. moschatum var. creticum.
cycladicum P.H. Davis & D.C. Stuart
subsp. subsessilis *(Bentzer) T. Raus in Ann. Naturhist. Mus. Wien*, 98B(Suppl.): 272 (1996): Leopoldia cycladica subsp. subsessilis.

MUTISIA (Compos.).
castellanosii Cabrera
var. comechingoniana *L. Ariza Espinar in Bol. Soc. Argent. Bot.*, 35(1–2): 173 (2000)—Argentina Northeast (Córdoba).
var. hieronymi *(Kuntze) L. Ariza Espinar in Bol. Soc. Argent. Bot.*, 35(1–2): 173 (2000): M. retrorsa var. hieronymi.
magnifica *C. Ulloa Ulloa & Peter M. Jørgensen in Novon*, 6(1): 131 (1996)—Ecuador.

MYANMARIA *H. Robinson in Proc. Biol. Soc. Wash.*, 112(1): 244 (1999). *COMPOSITAE*
calycina *(DC.) H. Robinson in Proc. Biol. Soc. Wash.*, 112(1): 245 (1999): Vernonia calycina.

MYCELIS (Compos.).
sect. Multicostaria *(Kirp.) A.N. Sennikov in Bot. Zhurn.*, 82(2): 111 (1997): Cicerbita sect. Multicostaria.
deltoidea *(M. Bieb.) A.N. Sennikov in Bot. Zhurn.*, 82(2): 111 (1997): Prenanthes deltoidea.
sagittarioides *(C.B. Clarke) A.N. Sennikov in Bot. Zhurn.*, 82(5): 112 (1997): Lactuca sagittarioides.

MYOSOTIS (Boraginac.).
sect. Alpestres *(T.N. Popova) O.D. Nikiforova in Bot. Zhurn.*, 85(1): 146 (2000): M. ser. Alpestres.
sect. Alpestres *(T.N. Popova) O.D. Nikiforova in Turczaninowia*, 3(1): 8 (2000): Myosotis ser. Alpestres.
sect. Arvenses *(Popov ex Riedl) N.N. Tzvelev, Opred. Sosud. Rast. Severo-Zap. Rossii*: 680 (2000): Myosotis ser. Arvenses.
sect. Sylvaticae *(Popov ex Riedl) O.D. Nikiforova in Turczaninowia*, 3(1): 8 (2000): Myosotis ser. Sylvaticae.
sect. Sylvaticae *(Popov ex Riedl) N.N. Tzvelev, Opred. Sosud. Rast. Severo-Zap. Rossii*: 680 (1 March 2000): Myosotis ser. Sylvaticae.
asiatica (Vestergr.) Schischk. & Serg.
subsp. kolymensis *O.D. Nikiforova in Fl. Sibir. (Pyrolac.-Lamiac.)*, 11: 122 (1997)—Yakutskiya.
austrobaicalensis *O.D. Nikiforova in Fl. Sibir. (Pyrolac.-Lamiac.)*, 11: 122 (1997)—Irkutsk.
austrosibirica *O.D. Nikiforova in Fl. Sibir. (Pyrolac.-Lamiac.)*, 11: 123 (1997)—Irkutsk.
baicalensis *O.D. Nikiforova in Fl. Sibir. (Pyrolac.-Lamiac.)*, 11: 125 (1997)—Buryatiya.
chakassica *O.D. Nikiforova in Fl. Sibir. (Pyrolac.-Lamiac.)*, 11: 126 (1997)—Krasnoyarsk.
guneri *A.P. Khokhrjakov in Byull. Mosk. Obshch. Ispŷt. Prir., Biol.*, 102(4): 46 (1997), as 'gunnerii'—Turkey.
kolakovskyi *A.P. Khokhrjakov in Byull. Glavn. Bot. Sada (Moscow)*, 175: 55 (1997), nom. nov.: M. deflexa *A.P. Khokhr.*
ochotensis *O.D. Nikiforova in Bot. Zhurn.*, 83(10): 118 (1998)—Magadan.
palustris (L.) L.
subsp. lithuanica *(Schmalh.) N.N. Tzvelev in Novosti Sist. Vŷssh. Rast.*, 32: 185 (2000): M. palustris var. lithuanica.
sajanensis *O.D. Nikiforova in Fl. Sibir. (Pyrolac.-Lamiac.)*, 11: 129 (1997)—Buryatiya.
scorpioides L.
var. nemorosa *(Besser) M.M. Ivanova in Fl. Tsentral'noi Sibiri*, 2: 736 (1979): M. nemorosa.
subsp. praecox *(Hülph.) W.B. Dickoré in Feddes Repert.*, 108(1–2): 106 (1997): M. praecox.

MYOXANTHUS (Orchidac.).
montanus *P. Ortiz V. in Orquideologia*, 20(3): 320 (1997)—Colombia.

MYRCIA (Myrtac.).
majaguitana *Alain & M. Mejía in Moscosoa*, 9: 18 (1997)—Dominican Republic.

MYRCIANTHES (Myrtac.).
cavalcantei *J.R. de Mattos in Loefgrenia*, 115: 1 (2000)—Brazil North (Roraima).
cymosa *(O. Berg) J.R. de Mattos in Loefgrenia*, 115: 1 (2000): Calyptromyrcia cymosa.
montana *(Aubl.) C.H. Nelson Sutherland in An. Jard. Bot. Madrid*, 57(2): 406 (1999 publ. 2000): Eugenia montana.

MYRICA (Myricac.).
singularis *C.A. Parra-Osorio in Caldasia*, 21(1): 5 (1999), nom. nov.: Alchornea cerifera *Croizat.*

MYRIOCEPHALUS (Compos.).
gascoynensis *P.S. Short in Austral. Syst. Bot.*, 13(5): 730 (2000)—Western Australia.
walcottii *P.S. Short in Austral. Syst. Bot.*, 13(5): 731 (2000)—Western Australia.

MYRISTICA (Myristicac.).
agusanensis Elmer
subsp. squamulosa *W.J.J.O. de Wilde in Blumea*, 42(1): 146 (1997)—Philippines.
basilanica *W.J.J.O. de Wilde in Blumea*, 42(1): 148 (1997)—Philippines.
beddomei King
subsp. sphaerocarpa *W.J.J.O. de Wilde in Blumea*, 42(1): 152 (1997)—India (Kerala).
subsp. ustulata *W.J.J.O. de Wilde in Blumea*, 42(1): 152 (1997)—India (Kerala).
colinridsdalei *W.J.J.O. de Wilde in Blumea*, 42(1): 156 (1997)—Philippines.
corticata *W.J.J.O. de Wilde in Blumea*, 42(1): 157 (1997)—Borneo (Sabah, Brunei).
dasycarpa *W.J.J.O. de Wilde in Blumea*, 43(1): 171 (1998)—New Guinea (Papua New Guinea).
depressa *W.J.J.O. de Wilde in Blumea*, 42(1): 161 (1997)—Borneo (Sarawak, Sabah, Kalimantan), Sumatera, Malaya (Peninsular Malaysia).
devogelii *W.J.J.O. de Wilde in Blumea*, 42(1): 163 (1997)—Sulawesi.
extensa *W.J.J.O. de Wilde in Blumea*, 42(1): 164 (1997)—Borneo (Sarawak, Kalimantan, Brunei).
fatua Houtt.
subsp. affinis *(Warb.) W.J.J.O. de Wilde in Blumea*, 42(1): 166 (1997): M. affinis.
frugifera *W.J.J.O. de Wilde in Blumea*, 42(1): 166 (1997)—Philippines.
inaequalis *W.J.J.O. de Wilde in Blumea*, 43(1): 252 (1998)—New Guinea (Irian Jaya).
kjellbergii *W.J.J.O. de Wilde in Blumea*, 42(1): 169 (1997)—Sulawesi.
laevis *W.J.J.O. de Wilde in Blumea*, 42(1): 170 (1997)—Philippines.
subsp. badia *W.J.J.O. de Wilde in Blumea*, 42(1): 172 (1997)—Philippines.
longepetiolata *W.J.J.O. de Wilde in Blumea*, 42(1): 173 (1997)—Philippines.
malaccensis Hook. f.
subsp. papillosa *W.J.J.O. de Wilde in Blumea*, 42(1): 177 (1997)—Borneo (Sarawak).
perlaevis *W.J.J.O. de Wilde in Blumea*, 42(1): 178 (1997)—Moluccas.
pilosigemma *W.J.J.O. de Wilde in Blumea*, 42(1): 179 (1997)—Philippines.
rosselensis J. Sinclair *subsp.*minutiflora *W.J.J.O. de Wilde in Blumea*, 43(1): 254 (1998)—New Guinea (Papua New Guinea).
rubrinervis *W.J.J.O. de Wilde in Blumea*, 42(1): 180 (1997)—Philippines.
var. duplex *W.J.J.O. de Wilde in Blumea*, 42(1): 182 (1997)—Philippines.
rumphii (Blume) Kosterm.
var. florentis *W.J.J.O. de Wilde in Blumea*, 42(1): 183 (1997)—Lesser Sunda Is.
sarcantha *W.J.J.O. de Wilde in Blumea*, 43(1): 175 (1998)—New Guinea (Irian Jaya).
simiarum A.DC.
subsp. calcarea *W.J.J.O. de Wilde in Blumea*, 42(1): 184 (1997)—Borneo (Kalimantan, Sarawak).
subcordata Blume
var. rimosa *W.J.J.O. de Wilde in Blumea*, 43(1): 181 (1998)—New Guinea (Irian Jaya).

MYRISTICA:-
tristis Warb.
subsp. moluccana *W.J.J.O. de Wilde in Blumea*, 42(1): 186 (1997)—Moluccas.
subsp. sessilifructa *W.J.J.O. de Wilde in Blumea*, 42(1): 186 (1997)—Moluccas.
ultrabasica *W.J.J.O. de Wilde in Blumea*, 42(1): 187 (1997)—Sulawesi.
verruculosa *W.J.J.O. de Wilde in Blumea*, 43(1): 177 (1998)—New Guinea (Irian Jaya).

MYRMECHIS (Orchidac.).
aurea *(J.J. Sm.) A. Schuiteman in Blumea*, 41(2): 401 (1996): Tubilabium aureum.
bilobulifera *(J.J. Sm.) A. Schuiteman in Blumea*, 41(2): 401 (1996): Tubilabium bilobuliferum.
chalmersii *(Schltr.) A. Schuiteman in Blumea*, 41(2): 401 (1996): Zeuxine chalmersii.
grandiflora *(Lindl.) P. Ormerod in Austral. Orchid Rev.*, 63(4): 10 (1998): Rhamphidia grandiflora.
quadrilobata *(Schltr.) A. Schuiteman in Blumea*, 41(2): 403 (1996): Cheirostylis quadrilobata.
urceolata *Tang & K.Y. Lang in Acta Phytotax. Sin.*, 34(6): 638 (1996)—China Southeast (Guangdong), China South-central (Yunnan), Hainan.

MYRMECOPHILA (Orchidac.).
albopurpurea *(H. Strachan ex Fawc.) M.A. Nir, Orchidaceae Antillanae*: 250 (2000): Schomburgkia thomsoniana var. albopurpurea.

MYRSINE (Myrsinac.).
amorosoana *J.J. Pipoly in Sida*, 17(1): 126 (1996), nom. nov.: Rapanea retusa *Merr.*
apoensis *(Elmer) J.J. Pipoly in Sida*, 17(1): 125 (1996): Rapanea apoensis.
aralioides *(Philipson) J.J. Pipoly in Sida*, 17(1): 152 (1996): Rapanea aralioides.
argentea *P.B. Heenan & P.J. de Lange in New Zealand J. Bot.*, 36(3): 382 (1998)—New Zealand South I.
balansae *(Mez) M. Otegui in Candollea*, 53(1): 155 (1998): Rapanea balansae.
calcarata *(Lundell) J.M. Ricketson & J.J. Pipoly in Sida*, 18(2): 506 (1998): Rapanea calcarata.
ceylanica *(Mez) B.M. Wadhwa in M.D. Dassanayake & W.D. Clayton, eds., Revised Handb. Fl. Ceylon*, 13: 161 (1999): Rapanea ceylanica.
cicatricosa *(C. Chen) J.J. Pipoly & C. Chen in Novon*, 5(4): 360 (1995): Rapanea cicatricosa.
collina Nadeaud
var. falcata *(Nadeaud) S.L. Welsh, Fl. Societensis*: 192 (1998): M. falcata.
coriacea (Sw.) R. Br. ex Roem. & Schult.
subsp. nigrescens *(Lundell) J.M. Ricketson & J.J. Pipoly in Sida*, 17(3): 585 (1997): M. nigrescens.
cruciata *(Philipson) J.J. Pipoly in Sida*, 17(1): 130 (1996): Rapanea cruciata.
faberi *(Mez) J.J. Pipoly & C. Chen in Novon*, 5(4): 360 (1995): Rapanea faberi.
fastigiata *(Elmer) J.J. Pipoly in Sida*, 17(1): 149 (1996): Rapanea fastigiata.
fusca Nadeaud
var. fasciculata *(J.W. Moore) S.L. Welsh, Fl. Societensis*: 192 (1998): Rapanea fasciculata.
glandulosa *(Elmer) J.J. Pipoly in Sida*, 17(1): 154 (1996): Rapanea glandulosa.
juergensenii *(Mez) J.M. Ricketson & J.J. Pipoly in Sida*, 18(2): 506 (1998): Rapanea juergensenii.
kwangsiensis *(E. Walker) J.J. Pipoly & C. Chen in Novon*, 5(4): 361 (1995): Rapanea kwangsiensis.
ligustrina *(Mez) M. Otegui in Candollea*, 53(1): 149 (1998): Rapanea ligustrina.
lineata *(Mez) N.N. Imkhanitskaya in Bot. Zhurn.*, 81(10): 28 (1996): Rapanea lineata.
loefgrenii *(Mez) N.N. Imkhanitskaya in Bot. Zhurn.*, 81(10): 28 (1996): Rapanea loefgrenii.
loefgrenii *(Mez) M. Otegui in Candollea*, 53(1): 150 (1998), isonym; *see Imkhan. in Bot. Zhurn.*, 81(10): 28 (1996): Rapanea loefgrenii.
luae *J.M. Ricketson & J.J. Pipoly in Sida*, 18(3): 747 (1999)—Brazil North (Roraima).
matensis *(Mez) M. Otegui in Candollea*, 53(1): 136 (1998): Rapanea matensis.
medeciloae *J.J. Pipoly in Sida*, 17(1): 122 (1996), nom. nov.: Rapanea angustifolia *Merr.*

MYRSINE:-
mindanaensis *(Elmer) J.J. Pipoly in Sida,* 17(1): 140 (1996): Rapanea mindanaensis.
miquelii *N.N. Imkhanitskaya in Bot. Zhurn.,* 81(10): 26 (1996), nom. nov.: M. ovalifolia *Miq.*
multibracteata *(Merr.) J.J. Pipoly in Sida,* 17(1): 148 (1996): Rapanea multibracteata.
oblongibacca *(Merr.) J.J. Pipoly in Sida,* 17(1): 129 (1996): Rapanea oblongibacca.
parvula *(Mez) M. Otegui in Candollea,* 53(1): 152 (1998): Rapanea parvula.
penibukana *(Philipson) J.J. Pipoly in Sida,* 17(1): 147 (1996): Rapanea penibukana.
peregrina *(Mez) J.J. Pipoly in Sida,* 17(1): 136 (1996): Rapanea peregrina.
robusta *(Mez) B.M. Wadhwa in M.D. Dassanayake & W.D. Clayton, eds., Revised Handb. Fl. Ceylon,* 13: 163 (1999): Rapanea robusta.
thwaitesii *(Mez) B.M. Wadhwa in M.D. Dassanayake & W.D. Clayton, eds., Revised Handb. Fl. Ceylon,* 13: 159 (1999): Rapanea thwaitesii.
verruculosa *(C. Chen) J.J. Pipoly & C. Chen in Novon,* 5(4): 360 (1995): Rapanea verruculosa.
villicaulis *(Mez) N.N. Imkhanitskaya in Bot. Zhurn.,* 81(10): 29 (1996): Rapanea villicaulis.
wettsteinii *(Mez) M. Otegui in Candollea,* 53(1): 146 (1998): Rapanea wettsteinii.

MYRTUS (Myrtac.).
aotearoana *(E.C. Nelson) E.C. Nelson in New Zealand J. Bot.,* 33(4): 558 (1995), nom. inval. (Art. 34.2 ICBN, 1994), see Lophomyrtus aotearoana E.C. Nelson: Lophomyrtus aotearoana.

MYSTROXYLON (Celastrac.).
aethiopicum (Thunb.) Loes.
subsp. schlechteri *(Loes.) R.H. Archer in Flow. Pl. Afr.,* 55: t.2134 (1997): M. schlechteri.

N

NABALUS (Compos.).
sect. Acerifolii *A.N. Sennikov in Novosti Sist. Vyssh. Rast.,* 32: 180 (2000).
sect. Altissimi *A.N. Sennikov in Novosti Sist. Vyssh. Rast.,* 32: 179 (2000).
sect. Asperi *A.N. Sennikov in Novosti Sist. Vyssh. Rast.,* 32: 179 (2000).
sect. Ochroleuci *A.N. Sennikov in Novosti Sist. Vyssh. Rast.,* 32: 179 (2000).
sect. Racemosi *A.N. Sennikov in Novosti Sist. Vyssh. Rast.,* 32: 178 (2000).
sect. Sinoseris *A.N. Sennikov in Novosti Sist. Vyssh. Rast.,* 32: 180 (2000).
angustilobus *(C. Shih) A.N. Sennikov in Novosti Sist. Vyssh. Rast.,* 32: 180 (2000): Prenanthes angustiloba.
faberi *(Hemsl.) A.N. Sennikov in Novosti Sist. Vyssh. Rast.,* 32: 180 (2000): Prenanthes faberi.
leptanthus *(C. Shih) A.N. Sennikov in Novosti Sist. Vyssh. Rast.,* 32: 180 (2000): Prenanthes leptantha.
pyramidalis *(C. Shih) A.N. Sennikov in Novosti Sist. Vyssh. Rast.,* 32: 180 (2000): Prenanthes pyramidalis.
racemiformis *(C. Shih) A.N. Sennikov in Novosti Sist. Vyssh. Rast.,* 32: 180 (2000): Prenanthes racemiformis.

NAGEIA (Podocarpac.).
falcata (Thunb.) Kuntze
var. gaussenii *(Woltz) J. Silba in J. Int. Conifer Preserv. Soc.,* 7(1): 27 (2000), species name illegit.: Podocarpus gaussenii.
var. gracilior *(Pilg.) J. Silba in J. Int. Conifer Preserv. Soc.,* 7(1): 27 (2000), species name illegit.: Podocarpus gracilior.
mannii (Hook. f.) Kuntze
var. dawei *(Stapf) J. Silba in J. Int. Conifer Preserv. Soc.,* 7(1): 27 (2000): Podocarpus dawei.

NAGEIA:-
mannii
var. usambarensis *(Pilg.) J. Silba in J. Int. Conifer Preserv. Soc.,* 7(1): 28 (2000): Podocarpus usambarensis.
nankoensis *(Hayata) R.R. Mill in Novon,* 9(1): 77 (1999): Podocarpus nankoensis.

NAJAS (Hydrocharitac.).
guadalupensis (Spreng.) Magnus
subsp. floridana *(R.R. Haynes & Wentz) R.R. Haynes & C.B. Hellquist in Novon,* 6(4): 371 (1996): N. guadalupensis var. floridana.
subsp. muenscheri *(R.T. Clausen) R.R. Haynes & C.B. Hellquist in Novon,* 6(4): 371 (1996): N. muenscheri.
subsp. olivacea *(Rosend. & Butters) R.R. Haynes & C.B. Hellquist in Novon,* 6(4): 371 (1996): N. olivacea.

NAMATAEA *D.W. Thomas & D.J. Harris in Kew Bull.,* 54(4): 951 (1999 publ. 2000). *SAPINDACEAE*
simplicifolia *D.W. Thomas & D.J. Harris in Kew Bull.,* 54(4): 952 (1999 publ. 2000)—Cameroon, Nigeria.

NAMOPHILA *U. Müller-Doblies & D. Müller-Doblies in Feddes Repert.,* 108(1–2): 77 (1997). *HYACINTHACEAE*
urotepala *U. Müller-Doblies & D. Müller-Doblies in Feddes Repert.,* 108(1–2): 77 (1997)—Namibia.

NANAREPENTA (Dioscoreac.).
juxtlahuacensis *O. Téllez-Valdés & P. Dávila-Aranda in Novon,* 8(2): 210 (1998)—Mexico Southwest (Oaxaca).

NANODES (Orchidac.).
congesta *(Rolfe) C.H. Nelson Sutherland & F.J. Fernández Casas in Fontqueria,* 51: 4 (1998): Epidendrum congestum.

NARAVELIA (Ranunculac.).
pilulifera Hance
var. yunnanensis *Y. Fei in Acta Bot. Yunnanica,* 19(4): 406 (1997)—China South-central (Yunnan).

NARCISSUS (Amaryllidac.).
sect. Angustifolii *F.J. Fernández Casas in An. Jard. Bot. Madrid,* 55(1): 174 (1997).
nothosect. Angustizettae *F.J. Fernández Casas in An. Jard. Bot. Madrid,* 55(1): 174 (1997). (N. sect. Angustifolii × sect. Tazettae).
nothosect. Apozettae *I. Sánchez García & C. Martínez Ortega in Lagascalia,* 18(1): 105 (1995). (N. sect. Apodanthi × sect. Tazettae).
nothosect. Chlorifolii *F.J. Fernández Casas in An. Jard. Bot. Madrid,* 55(1): 174 (1997). (N. sect. Angustifolii × sect. Chloraster).
nothosect. Cyclacissi *F.J. Fernández Casas in An. Jard. Bot. Madrid,* 55(1): 174 (1997). (N. sect. Cyclaminei × sect. Pseudonarcissi).
sect. Perezlarae *Fern. Casas ex J.X. Soler in An. Jard. Bot. Madrid,* 56(1): 166 (1998).
nothosect. Seroangustifolii *F.J. Fernández Casas in An. Jard. Bot. Madrid,* 55(1): 174 (1997). (N. sect. Angustifolii × sect. Serotini).
alcaracensis *S. Ríos Ruiz, D. Rivera Nuñez, F. Alcaraz Ariza & C. Obón de Castro in Bot. J. Linn. Soc.,* 131(2): 160 (1999)—Spain.
× alleniae *D. Donnison-Morgan in New Plantsman,* 7(1): 42 (2000)—Spain. (N. serotinus × viridiflorus).
bicolor L.
var. concolor *A. Barra in An. Jard. Bot. Madrid,* 57(1): 178 (1999)—Spain.
cantabricus DC.
subsp. kesticus *(Maire & Wilczek) M. Ibn Tattou in Bocconea,* 8: 193 (1998): N. bulbocodium subsp. kesticus.
× christopheri *J. Blanchard in Daff. Tulip Yearbook,* 1997–8: 11 (1997)—Spain. (N. assoanus × panizzianus).
× dichromus *P.D. Sell in P.D. Sell & G. Murrell, Fl. Great Britain & Ireland,* 5: 364 (1996)—Cultivation. (N. cyclamineus × moschatus).
× dordae *F.J. Fernández Casas in Fontqueria,* 53: 8 (1999), nom. nov.: N. × pyrenaicus *Dorda, Rivas Ponce & Fern. Casas.*

NARCISSUS:-
enemeritoi *(Sánchez-Gómez, A.F. Carillo, A. Hern., M.A. Carrión & Güemes) P. Sánchez-Gómez, A.F. Carrillo López, A. Hernández González,* M.Á. Carrión Vilches & J. Güemes Heras *in An. Jard. Bot. Madrid,* 57(2): 430 (1999 publ. 2000): N. nevadensis subsp. enemeritoi.
fernandesii Pedro
var. rivas-martinezii *(Fern. Casas) F.J. Fernández Casas in Fontqueria,* 48: 151 (1997 publ. 1998): N. rivas-martinezii.
× galdoanus *F.J. Fernández Casas in An. Jard. Bot. Madrid,* 55(1): 174 (1997)—Spain. (N. nobilis × triandrus).
× georgemawii *F.J. Fernández Casas in An. Jard. Bot. Madrid,* 55(1): 174 (1997)—Morocco. (N. elegans × viridiflorus).
hesperidis *F.J. Fernández Casas in Fontqueria,* 53: 6 (1999)—Morocco.
hispanicus Gouan
subsp. bujei *(Fern. Casas) F.J. Fernández Casas in Fontqueria,* 54(1–6): 126 (1999–2000 publ. 2000): N. longispathus var. bujei.
subsp. longispathus *(Degen & Hervier ex Pugsley) F.J. Fernández Casas in Fontqueria,* 54(1–6): 126 (1999–2000 publ. 2000): N. longispathus.
× koshinomurae *F.J. Fernández Casas in Fontqueria,* 44: 254 (1996)—Spain. (N. fernandesii × cf. panizzianus).
× libarensis *I. Sánchez García & C. Martínez Ortega in Lagascalia,* 18(1): 105 (1995)—Spain. (N. cuatrecasasii × papyraceus).
× monochromus *P.D. Sell in P.D. Sell & G. Murrell, Fl. Great Britain & Ireland,* 5: 364 (1996)—Cultivation. (N. cyclamineus × pseudonarcissus).
nevadensis Pugsley
subsp. enemeritoi *P. Sánchez Gómez, A.F. Carrillo López, A. Hernández González, M.A.* Carrión Vilches & J. Güemes Heras *in P. Sánchez Gómez et al., Fl. Murciana Inter. Nac. Eur. Protecc.* Legislac.: 63 (1998)—Spain.
× pacensis *F.J. Fernández Casas in Fontqueria,* 44: 255 (1996)—Spain. (N. confusus × pallidulus).
peroccidentalis *F.J. Fernández Casas in Fontqueria,* 44: 256 (1996)—Mauritania.
poeticus L.
subsp. majalis *(Curtis) P.D. Sell in P.D. Sell & G. Murrell, Fl. Great Britain & Ireland,* 5: 363 (1996): N. majalis.
subsp. recurvus *(Haw.) P.D. Sell in P.D. Sell & G. Murrell, Fl. Great Britain & Ireland,* 5: 363 (1996): N. recurvus.
subsp. verbanensis *(Herb.) P.D. Sell in P.D. Sell & G. Murrell, Fl. Great Britain & Ireland,* 5: 363 (1996): N. poeticus var. verbanensis.
× praviani Fern. Casas
nothovar. bergidensis *(Fern. Casas & Rivas Ponce) F.J. Fernández Casas in Fontqueria,* 48: 156 (1997 publ. 1998): N. bergidensis.
pseudonarcissus L.
f. pleniflorus *P.D. Sell in P.D. Sell & G. Murrell, Fl. Great Britain & Ireland,* 5: 364 (1996): Great Britain.
radiiflorus Salisb.
subsp. exertus *(Haw.) P.D. Sell in P.D. Sell & G. Murrell, Fl. Great Britain & Ireland,* 5: 364 (1996): N. majalis var. exertus.
subsp. poetarum *(Haw.) P.D. Sell in P.D. Sell & G. Murrell, Fl. Great Britain & Ireland,* 5: 364 (1996): N. poetarum.
subsp. stellaris *(Haw.) P.D. Sell in P.D. Sell & G. Murrell, Fl. Great Britain & Ireland,* 5: 364 (1996): N. stellaris.
rivasmartinezii *F.J. Fernández Casas in Fontqueria,* 44: 256 (1996)—Spain.
segurensis *S. Ríos Ruiz, D. Rivera Nuñez, F. Alcaraz Ariza & C. Obón de Castro in Bot. J. Linn. Soc.,* 131(2): 155 (1999)—Spain.
× somedanus *M.Á. Fernández Casado, H.S. Nava & F.J. Suárez Pérez in Fontqueria,* 48: 30 (1997 publ. 1998)—Spain. (N. asturiensis var. villarvildensis × triandrus).
tananicus *(Maire) M. Ibn Tattou in Bocconea,* 8: 194 (1998): N. bulbocodium subsp. tananicus.
tingitanus *F.J. Fernández Casas in Fontqueria,* 44: 262 (1996)—Mauritania.
triandrus L.
var. alejandrei *A. Barra Lázaro in An. Jard. Bot. Madrid,* 58(1): 185 (2000)—Spain.

NARCISSUS:-
triandrus
subsp. lusitanicus *(Dorda & Fern. Casas) A. Barra Lázaro in An. Jard. Bot. Madrid,* 58(1): 186 (2000): N. lusitanicus.
var. paivae *A. Barra Lázaro in An. Jard. Bot. Madrid,* 58(1): 185 (2000)—Portugal.
× vallrutae *A. Barra in An. Jard. Bot. Madrid,* 57(1): 179 (1999)—Spain. (N. bicolor var. concolor × poeticus).
yepesii *S. Ríos Ruiz, D. Rivera Nuñez, F. Alcaraz Ariza & C. Obón de Castro in Bot. J. Linn. Soc.,* 131(2): 161 (1999)—Spain.

NARENGA (Gramin.).
fallax (Balansa) Bor
var. aristata *(Balansa) L. Liou in Fl. Reipubl. Popul. Sin.,* 10(2): 37 (1997): Saccharum fallax var. aristata.

NASA *M. Weigend, Nasa & Conquest S. America - Syst. Rearrang. in Loasac.:* 214 (1997). *LOASACEAE*
aequatoriana *(Urb. & Gilg) M. Weigend in Cat. Vasc. Pl. Ecuador (Monogr. Syst. Bot. Missouri Bot. Gard., 75):* 954 (1999): Loasa triphylla var. aequatoriana.
amaluzensis *(Weigend) M. Weigend in Cat. Vasc. Pl. Ecuador (Monogr. Syst. Bot. Missouri Bot. Gard., 75):* 954 (1999): Loasa amaluzensis.
anderssonii *M. Weigend in Nordic J. Bot.,* 20(1): 17 (2000)—Ecuador, Peru.
aspiazui *(J.F. Macbr.) M. Weigend in Arnaldoa,* 5(2): 162 (1998): Loasa aspiazui.
auca *(Weigend) M. Weigend in Cat. Vasc. Pl. Ecuador (Monogr. Syst. Bot. Missouri Bot. Gard., 75):* 954 (1999): Loasa auca.
bicornuta *(Weigend) M. Weigend in Arnaldoa,* 5(2): 162 (1998); *et in Cat. Vasc. Pl. Ecuador (Monogr. Syst. Bot. Missouri Bot. Gard., 75):* 954 (1999): Loasa bicornuta.
carnea *(Urb. & Gilg) M. Weigend in Arnaldoa,* 5(2): 162 (1998): Loasa carnea.
carunculata *(Urb. & Gilg) M. Weigend in Arnaldoa,* 5(2): 162 (1998): Loasa carunculata.
chenopodiifolia *(Desr.) M. Weigend in Arnaldoa,* 5(2): 163 (1998): Loasa chenopodiifolia.
colanii *N. Dostert & M. Weigend in Harvard Pap. Bot.,* 4(2): 461 (1999)—Peru.
cuatrecasasii *M. Weigend, Nasa & Conquest S. America - Syst. Rearrang. in Loasac.:* 218 (1997)—Colombia.
cymbopetala *(Urb. & Gilg) M. Weigend in Arnaldoa,* 5(2): 163 (1998): Loasa cymbopetala.
dillonii *M. Weigend in Nordic J. Bot.,* 20(1): 18 (2000)—Peru.
driessleae *M. Weigend in Nordic J. Bot.,* 20(1): 20 (2000), as 'driesslei'—Peru.
dyeri *(Urb. & Gilg) M. Weigend in Cat. Vasc. Pl. Ecuador (Monogr. Syst. Bot. Missouri Bot. Gard., 75):* 954 (1999): Loasa dyeri.
dyeri (Urb. & Gilg) Dostert & Weigend
subsp. australis *N. Dostert & M. Weigend in Harvard Pap. Bot.,* 4(2): 467 (1999)—Peru.
ferruginea *(Urb. & Gilg) M. Weigend in Arnaldoa,* 5(2): 164 (1998): Loasa ferruginea.
glabra *(Weigend) M. Weigend in Cat. Vasc. Pl. Ecuador (Monogr. Syst. Bot. Missouri Bot. Gard., 75):* 955 (1999): Loasa glabra.
glandulosissima *M. Weigend in Arnaldoa,* 5(2): 152 (1998)—Peru.
grandiflora *(Desr.) M. Weigend in Cat. Vasc. Pl. Ecuador (Monogr. Syst. Bot. Missouri Bot. Gard., 75):* 955 (1999): Loasa grandiflora.
hastata *(Killip) M. Weigend in Arnaldoa,* 5(2): 164 (1998): Loasa hastata.
hornii *(Weigend) M. Weigend in Cat. Vasc. Pl. Ecuador (Monogr. Syst. Bot. Missouri Bot. Gard., 75):* 955 (1999): Loasa hornii.
humboldtiana *(Urb. & Gilg) M. Weigend in Cat. Vasc. Pl. Ecuador (Monogr. Syst. Bot. Missouri Bot. Gard., 75):* 955 (1999): Loasa humboldtiana.
subsp. obliqua *N. Dostert & M. Weigend in Harvard Pap. Bot.,* 4(2): 464 (1999)—Peru.
subsp. roseoalba *(Weigend) Dostert ex M. Weigend in Cat. Vasc. Pl. Ecuador (Monogr. Syst. Bot. Missouri Bot. Gard., 75):* 955 (1999): Loasa roseoalba.

NASA:-
humboldtiana
subsp. tricolor *N. Dostert & M. Weigend in Harvard Pap. Bot.,* 4(2): 464 (1999)—Peru.
insignis *M. Weigend & E. Rodríguez Rodríguez in Arnaldoa,* 5(2): 154 (1998)—Peru.
jungifolia *(Weigend) M. Weigend in Cat. Vasc. Pl. Ecuador (Monogr. Syst. Bot. Missouri Bot. Gard., 75):* 955 (1999): Loasa jungifolia.
laxa *(J.F. Macbr.) M. Weigend in Arnaldoa,* 5(2): 164 (1998): Loasa laxa.
lenta *(J.F. Macbr.) M. Weigend in Arnaldoa,* 5(2): 164 (1998): Loasa lenta.
limata *(J.F. Macbr.) M. Weigend in Arnaldoa,* 5(2): 164 (1998): Loasa limata.
loxensis *(Kunth) M. Weigend in Cat. Vasc. Pl. Ecuador (Monogr. Syst. Bot. Missouri Bot. Gard., 75):* 955 (1999): Loasa loxensis.
macrantha *(Urb. & Gilg) M. Weigend in Arnaldoa,* 5(2): 164 (1998): Loasa macrantha.
macrorrhiza *(Urb. & Gilg) M. Weigend in Arnaldoa,* 5(2): 164 (1998): Loasa macrorrhiza.
macrothyrsa *(Urb. & Gilg) M. Weigend in Arnaldoa,* 5(2): 165 (1998): Loasa macrothyrsa.
magnifica *(Urb. & Gilg) M. Weigend in Arnaldoa,* 5(2): 165 (1998): Loasa magnifica.
nubicolorum *M. Weigend in Nordic J. Bot.,* 20(1): 22 (2000)—Peru.
olmosiana *(Gilg ex J.F. Macbr.) M. Weigend in Arnaldoa,* 5(2): 165 (1998); *et in Cat. Vasc. Pl. Ecuador (Monogr. Syst. Bot. Missouri Bot. Gard., 75):* 955 (1999): Loasa olmosiana.
orbicularis *M. Weigend, Nasa & Conquest S. America - Syst. Rearrang. in Loasac.:* 218 (1997)—Colombia.
pascoensis *M. Weigend in Nordic J. Bot.,* 20(1): 25 (2000)—Peru.
peltata *(Spruce ex Urb. & Gilg) M. Weigend in Cat. Vasc. Pl. Ecuador (Monogr. Syst. Bot. Missouri Bot. Gard., 75):* 955 (1999): Loasa peltata.
peltiphylla *(Weigend) M. Weigend in Cat. Vasc. Pl. Ecuador (Monogr. Syst. Bot. Missouri Bot. Gard., 75):* 955 (1999): Loasa peltiphylla.
picta *(Hook.) M. Weigend in Arnaldoa,* 5(2): 165 (1998), without basionym date: Loasa picta.
pilovena *M. Weigend, Nasa & Conquest S. America - Syst. Rearrang. in Loasac.:* 219 (1997)—Colombia.
poissoniana *(Urb. & Gilg) M. Weigend in Arnaldoa,* 5(2): 165 (1998): Loasa poissoniana.
profundilobata *(Werderm.) M. Weigend in Cat. Vasc. Pl. Ecuador (Monogr. Syst. Bot. Missouri Bot. Gard., 75):* 955 (1999): Loasa profundilobata.
profundiserrata *M. Weigend, Nasa & Conquest S. America - Syst. Rearrang. in Loasac.:* 219 (1997)—Colombia.
pteridophylla *M. Weigend & N. Dostert in Harvard Pap. Bot.,* 4(2): 457 (1999)—Peru.
subsp. geniculata *M. Weigend & N. Dostert in Harvard Pap. Bot.,* 4(2): 459 (1999)—Peru.
puma-chini *(Weigend) M. Weigend in Cat. Vasc. Pl. Ecuador (Monogr. Syst. Bot. Missouri Bot. Gard., 75):* 955 (1999): Loasa puma-chini.
raimondii *(Standl. & F.A. Barkley) M. Weigend in Arnaldoa,* 5(2): 166 (1998): Loasa raimondii.
ramirezii *(Weigend) M. Weigend in Cat. Vasc. Pl. Ecuador (Monogr. Syst. Bot. Missouri Bot. Gard., 75):* 955 (1999): Loasa ramirezii.
ranunculifolia *(Kunth) M. Weigend in Arnaldoa,* 5(2): 166 (1998): Loasa ranunculifolia.
rubrastra *(Weigend) M. Weigend, Nasa & Conquest S. America - Syst. Rearrang. in Loasac.:* 214 (1997): Loasa rubrastra.
rufipila *M. Weigend, Nasa & Conquest S. America - Syst. Rearrang. in Loasac.:* 219 (1997)—Ecuador.
rugosa *(Killip) M. Weigend in Arnaldoa,* 5(2): 166 (1998): Loasa rugosa.
sagasteguii *M. Weigend in Nordic J. Bot.,* 20(1): 26 (2000)—Peru.
solaria *(J.F. Macbr.) M. Weigend in Arnaldoa,* 5(2): 166 (1998): Loasa solaria.
solata *(J.F. Macbr.) M. Weigend in Arnaldoa,* 5(2): 166 (1998): Loasa solata.
stuebeliana *(Urb. & Gilg) M. Weigend in Arnaldoa,* 5(2): 166 (1998): Loasa stuebeliana.
tingomariensis *(J.F. Macbr.) M. Weigend in Arnaldoa,* 5(2): 166 (1998): Loasa tingomariensis.

NASA:-
triphylla *(Juss.) M. Weigend in Arnaldoa,* 5(2): 167 (1998); *et in Cat. Vasc. Pl. Ecuador (Monogr. Syst. Bot. Missouri Bot. Gard., 75):* 955 (1999): Loasa triphylla.
subsp. colonchensis *M. Weigend in Harvard Pap. Bot.,* 4(2): 454 (1999)—Ecuador.
subsp. elegans *N. Dostert & M. Weigend in Harvard Pap. Bot.,* 4(2): 456 (1999)—Peru.
subsp. flavipes *M. Weigend & N. Dostert in Harvard Pap. Bot.,* 4(2): 456 (1999)—Peru.
subsp. loxensis *N. Dostert & M. Weigend in Harvard Pap. Bot.,* 4(2): 454 (1999)—Ecuador.
subsp. lutescens *M. Weigend & N. Dostert in Harvard Pap. Bot.,* 4(2): 453 (1999)—Ecuador.
subsp. papaverifolia *(Kunth) M. Weigend in Cat. Vasc. Pl. Ecuador (Monogr. Syst. Bot. Missouri Bot. Gard., 75):* 955 (1999): Loasa papaverifolia.
subsp. rudis *(Benth.) M. Weigend in Harvard Pap. Bot.,* 4(2): 453 (1999): Loasa rudis.
umbraculifera *E. Rodríguez Rodríguez & M. Weigend in Arnaldoa,* 6(1): 51 (1999 publ. 2000)—Peru.
urens *(Jacq.) M. Weigend in Arnaldoa,* 5(2): 167 (1998): Loasa urens.
urentivelutina *M. Weigend in Nordic J. Bot.,* 20(1): 29 (2000)—Peru.
vargasii *(J.F. Macbr.) M. Weigend in Arnaldoa,* 5(2): 167 (1998): Loasa vargasii.
weberbaueri *(Urb. & Gilg) M. Weigend in Arnaldoa,* 5(2): 167 (1998): Loasa weberbaueri.

NASSELLA (Gramin.).
arcaensis *(Speg.) M.A. Torres in Monogr. Comis. Invest. Ci. Prov. Buenos Aires,* 13: 13 (1997): Stipa arcaensis.
arcuata *(R.E. Fr.) M.A. Torres in Monogr. Comis. Invest. Ci. Prov. Buenos Aires,* 13: 13 (1997): Stipa arcuata.
argentinensis *(Speg.) P. Peñailillo in Gayana, Bot.,* 55(2): 86 (1998 publ. 1999): Stipa argentinensis.
brasiliensis *(A. Zanin & Longhi-Wagner) P. Peñailillo in Gayana, Bot.,* 55(2): 86 (1998 publ. 1999): Stipa brasiliensis.
cabrerae *M.A. Torres in Monogr. Comis. Invest. Ci. Prov. Buenos Aires,* 13: 14 (1997), as 'cabrerai'—Argentina Northwest (Jujuy), Bolivia.
carettei *(Hauman) M.A. Torres in Monogr. Comis. Invest. Ci. Prov. Buenos Aires,* 13: 16 (1997): Stipa carettei.
catamarcensis *M.A. Torres in Monogr. Comis. Invest. Ci. Prov. Buenos Aires,* 13: 17 (1997)—Argentina Northwest (Catamarca).
coquimbensis *(Matthei) P. Peñailillo in Gayana, Bot.,* 55(2): 86 (1998 publ. 1999): Stipa coquimbensis.
dasycarpa *(Hitchc.) M.A. Torres in Monogr. Comis. Invest. Ci. Prov. Buenos Aires,* 13: 19 (1997): Stipa dasycarpa.
depauperata (Pilg.) Barkworth
var. glabrata *(Pilg.) M.A. Torres in Monogr. Comis. Invest. Ci. Prov. Buenos Aires,* 13: 20 (1997): Stipa depauperata var. glabrata.
elata *(Speg.) M.A. Torres in Monogr. Comis. Invest. Ci. Prov. Buenos Aires,* 13: 20 (1997): Stipa caespitosa var. elata.
entrerriensis *(Burkart) P. Peñailillo in Gayana, Bot.,* 55(2): 86 (1998 publ. 1999): Stipa entrerriensis.
fabrisii *M.A. Torres in Monogr. Comis. Invest. Ci. Prov. Buenos Aires,* 13: 20 (1997)—Argentina Northwest (Tucumán).
famatinensis *M.A. Torres in Monogr. Comis. Invest. Ci. Prov. Buenos Aires,* 13: 21 (1997)—Argentina Northwest (La Rioja).
glabripoda *M.A. Torres in Monogr. Comis. Invest. Ci. Prov. Buenos Aires,* 13: 22 (1997)—Argentina Northwest (Jujuy, La Rioja).
var. planifolia *M.A. Torres in Monogr. Comis. Invest. Ci. Prov. Buenos Aires,* 13: 23 (1997)—Argentina Northwest (La Rioja, Jujuy).
ibarrensis *(Kunth) S. Laegaard in Novon,* 8(1): 30 (1998): Stipa ibarrensis.
ibarrensis *(Kunth) P. Peñailillo in Gayana, Bot.,* 55(2): 86 (1998 publ. 1999), isonym; *see Laegaard in Novon,* 8(1): 30 (1998): Stipa ibarrensis.
karstenii *(Hitchc.) P. Peñailillo in Gayana, Bot.,* 55(2): 86 (1998 publ. 1999): Stipa karstenii.

NASSELLA:-
leptothera *(Speg.) M.A. Torres in Monogr. Comis. Invest. Ci. Prov. Buenos Aires,* 13: 25 (1997): Stipa leptothera.
meyeri *M.A. Torres in Monogr. Comis. Invest. Ci. Prov. Buenos Aires,* 13: 26 (1997)—Argentina Northwest (Jujuy, Catamarca, La Rioja, Salta, Tucumán).
novari *M.A. Torres in Monogr. Comis. Invest. Ci. Prov. Buenos Aires,* 13: 29 (1997)—Argentina Northwest (Jujuy, Salta).
nubicola *(Speg.) M.A. Torres in Monogr. Comis. Invest. Ci. Prov. Buenos Aires,* 13: 30 (1997): Stipa nubicola.
paramilloensis *(Speg.) M.A. Torres in Monogr. Comis. Invest. Ci. Prov. Buenos Aires,* 13: 31 (1997): Stipa paramilloensis.
parva *M.A. Torres in Monogr. Comis. Invest. Ci. Prov. Buenos Aires,* 13: 32 (1997)—Argentina Northwest (Catamarca, Jujuy, Tucumán).
pittieri *(Hitchc.) P. Peñailillo in Gayana, Bot.,* 55(2): 86 (1998 publ. 1999): Stipa pittieri.
planaltina *(A. Zanin & Longhi-Wagner) P. Peñailillo in Gayana, Bot.,* 55(2): 87 (1998 publ. 1999): Stipa planaltina.
psittacorum *(Speg.) P. Peñailillo in Gayana, Bot.,* 55(2): 86 (1998 publ. 1999): Stipa psittacorum.
pubiflora (Trin. & Rupr.) E. Desv.
var. humilior *(Hack.) M.A. Torres in Monogr. Comis. Invest. Ci. Prov. Buenos Aires,* 13: 33 (1997): N. flaccidula var. humilior.
var. levipoda *M.A. Torres in Monogr. Comis. Invest. Ci. Prov. Buenos Aires,* 13: 34 (1997)—Argentina Northwest (Jujuy).
punensis *M.A. Torres in Monogr. Comis. Invest. Ci. Prov. Buenos Aires,* 13: 34 (1997)—Argentina Northwest (Jujuy, Salta).
ragonesei *M.A. Torres in Monogr. Comis. Invest. Ci. Prov. Buenos Aires,* 13: 35 (1997)—Argentina Northwest (Catamarca).
rhizomata *(A. Zanin & Longhi-Wagner) P. Peñailillo in Gayana, Bot.,* 55(2): 87 (1998 publ. 1999): Stipa rhizomata.
rupestris *(Phil.) M.A. Torres in Monogr. Comis. Invest. Ci. Prov. Buenos Aires,* 13: 36 (1997): Stipa rupestris.
sellowiana *(Nees ex Trin. & Rupr.) P. Peñailillo in Gayana, Bot.,* 55(2): 87 (1998 publ. 1999): Stipa sellowiana.
tenuiculmis *(Hack.) P. Peñailillo in Gayana, Bot.,* 55(2): 87 (1998 publ. 1999): Stipa tenuiculmis.
tucumana *(Parodi) M.A. Torres in Monogr. Comis. Invest. Ci. Prov. Buenos Aires,* 13: 38 (1997): Stipa tucumana.
uspallatensis *(Speg.) M.A. Torres in Monogr. Comis. Invest. Ci. Prov. Buenos Aires,* 13: 38 (1997): Stipa uspallatensis.
vallsii *(A. Zanin & Longhi-Wagner) P. Peñailillo in Gayana, Bot.,* 55(2): 87 (1998 publ. 1999): Stipa vallsii.
vargasii *(Tovar) P. Peñailillo in Gayana, Bot.,* 55(2): 87 (1998 publ. 1999): Stipa vargasii
yaviensis *M.A. Torres in Monogr. Comis. Invest. Ci. Prov. Buenos Aires,* 13: 39 (1997)—Argentina Northwest (Jujuy).

NASTURTIUM (Crucif.).
floridanum *(Al-Shehbaz & Rollins) I.A. Al-Shehbaz & R.A. Price in Novon,* 8(2): 125 (1998): Rorippa floridana.

NASTUS (Gramin.).
elatoides *E.A. Widjaja in Reinwardtia,* 11(2): 108 (1997)—New Guinea (Irian Jaya).
glaucus *E.A. Widjaja in Reinwardtia,* 11(2): 110 (1997)—New Guinea (Irian Jaya).

NAUCLEOPSIS (Morac.).
capirensis *C.C. Berg in Novon,* 6(3): 235 (1996)—Panama, Costa Rica.
herrerensis *C.C. Berg in Novon,* 6(3): 237 (1996)—Peru, Colombia, Ecuador.
humilis *C.C. Berg in Novon,* 6(3): 238 (1996)—Ecuador, Colombia, Peru.
naga Pittier
subsp. meridionalis *C.C. Berg in Novon,* 6(3): 238 (1996)—Colombia, Ecuador.
straminea *C.C. Berg in Novon,* 6(3): 240 (1996)—Colombia, Panama.

NAUCLEOPSIS:-
ulei (Warb.) Ducke
subsp. puberula *C.C. Berg in Novon,* 6(3): 240 (1996)—Panama, Costa Rica, Colombia, Ecuador.
velutina *C.C. Berg in Novon,* 6(3): 241 (1996)—Peru.

NAUTILOCALYX (Gesneriac.).
antioquensis *H. Wiehler in Gesneriana,* 1(1): 63 (1995)—Colombia.
melittifolius (L.) Wiehler
var. guadalupensis *(C.DC.) J. Fournet, Fl. Illustr. Phan. Guadeloupe & Martinique:* 1325 (1978): Besleria guadalupensis.

NAVAJOA (Cactac.).
peeblesiana
f. maia *F. Hochstätter in Cactaceae Rev.,* 2(2): 11 (1999); *cf. Repert. Pl. Succ. (I.O.S.),* 50: 13 (1999 publ. 2000)—Locality not stated.
f. menzelii *F. Hochstätter in Cactaceae Rev.,* 2(2): 10–11 (1999); *cf. Repert. Pl. Succ. (I.O.S.),* 50: 13 (1999 publ. 2000)—Locality not stated.

NAVARRETIA (Polemoniac.).
leptalea *(A. Gray) L.A. Johnson in Aliso,* 19(1): 68 (2000): Collomia leptalea.
subsp. bicolor *(H. Mason & A.D. Grant) L.A. Johnson in Aliso,* 19(1): 68 (2000): Gilia leptalea subsp. bicolor.
sinistra *(M.E. Jones) L.A. Johnson in Aliso,* 19(1): 68 (2000): Gilia sinistra.
subsp. pinnatisecta *(H. Mason & A.D. Grant) L.A. Johnson in Aliso,* 19(1): 68 (2000): Gilia leptalea subsp. pinnatisecta.

NAVIA (Bromeliac.).
ebracteata *J. Betancur & M.V. Arbeláez in Mutisia,* 74: 2 (1995)—Colombia.
glandulifera *B.K. Holst in J. Bromeliad Soc.,* 46(4): 160 (1996)—Venezuela.
paruana *B.K. Holst in J. Bromeliad Soc.,* 46(4): 161 (1996)—Venezuela.
tentaculata *B.K. Holst in J. Bromeliad Soc.,* 46(4): 165 (1996)—Venezuela.

NEANOTIS (Rubiac.).
ingrata (Wall. ex Hook. f.) W.H. Lewis
f. parvifolia *F.C. How ex W.C. Ko in J. S. China Agric. Univ.,* 16(4): 46 (1995)—China South-central (Sichuan).

NELUMBO (Nelumbonac.).
nucifera Gaertn.
subsp. lutea *(Willd.) T. Borsch & W. Barthlott in Beitr. Biol. Pfl.,* 68(3): 443 (1994 publ. 1996): Nymphaea lutea.

NEMATOLEPIS (Rutac.).
elliptica *(Paul G. Wilson) Paul G. Wilson in Nuytsia,* 12(2): 278 (1998): Phebalium ellipticum
frondosa *(N.G. Walsh & Albr.) Paul G. Wilson in Nuytsia,* 12(2): 278 (1998): Phebalium frondosum.
ovatifolia *(F. Muell.) Paul G. Wilson in Nuytsia,* 12(2): 278 (1998): Phebalium ovatifolium.
rhytidophylla *(Albr. & N.G. Walsh) Paul G. Wilson in Nuytsia,* 12(2): 279 (1998): Phebalium rhytidophyllum.
squamea *(Labill.) Paul G. Wilson in Nuytsia,* 12(2): 279 (1998): Eriostemon squameus.
subsp. coriacea *(Paul G. Wilson) Paul G. Wilson in Nuytsia,* 12(2): 279 (1998): Phebalium squameum subsp. coriaceum.
subsp. retusa *(Hook.) Paul G. Wilson in Nuytsia,* 12(2): 279 (1998): Phebalium retusum.
wilsonii *(N.G. Walsh & Albr.) Paul G. Wilson in Nuytsia,* 12(2): 280 (1998): Phebalium wilsonii.

× **NEOAERISTYLIS** *hort. in Orchid Rev.,* 106(1220): 126 (1998). *ORCHIDACEAE* (Aerides × Neofinetia × Rhynchostylis).

× **NEOBERGIA** *E.L. Smith in J. Bromeliad Soc.,* 33(2): 73 (1983). *BROMELIACEAE* (Billbergia × Neoregelia).

NEOBESSEYA (Cactac.).
missouriensis (Sweet) Britton & Rose
subsp. similis *(Engelm.) A.B. Doweld in Sukkulentÿ (Moscow),* 3(1–2): 37 (2000): Mammillaria similis.
robbinsiorum *(W.H. Earle) A.B. Doweld in Sukkulentÿ (Moscow),* 3(1–2): 37 (2000), as 'robbinsorum': Cochiseia robbinsiorum.

NEOBOUTELOUA (Gramin.).
pauciracemosa *M.G. López & F. Biurrun in Hickenia,* 2(67): 317 (1999)—Argentina Northwest (La Rioja, San Luis).

NEOBUXBAUMIA (Cactac.).
laui *(P.V. Heath) D.R. Hunt in Cactaceae Consensus Initiatives,* 4: 5 (1997): Carnegiea laui.

NEOCRYPTODISCUS (Umbellif.).
papillaris *(Boiss.) I. Herrnstadt & C.C. Heyn in Candollea,* 52(1): 184 (1997): Cachrys papillaris.

NEODILLENIA *G.A. Aymard C. in Harvard Pap. Bot.,* 10: 121 (1997). *DILLENIACEAE*
coussapoana *G.A. Aymard C. in Harvard Pap. Bot.,* 10: 124 (1997)—Ecuador, Colombia.
peruviana *G.A. Aymard C. in Harvard Pap. Bot.,* 10: 126 (1997)—Peru.
venezuelana *G.A. Aymard C. in Harvard Pap. Bot.,* 10: 128 (1997)—Venezuela.

NEODRYAS (Orchidac.).
schildhaueri *W. Königer in Arcula,* 5: 130 (1996)—Peru.

× **NEODSTROPSIS** *D. Butcher in J. Bromeliad Soc.,* 49(1): 14 (1999), contrary to Art. H.6.2 ICBN (1994). *BROMELIACEAE* (Canistropsis × Neoregelia).

NEOFINETIA (Orchidac.).
richardsiana *E.A. Christenson in Lindleyana,* 11(4): 220 (1996)—China.

× **NEOISTRUM** *D.A. Beadle, Prelim. Listing Cultivar Grex Names Bromel.:* 89 (1991). *BROMELIACEAE* (Canistrum × Neoregelia).

NEOKOEHLERIA (Orchidac.).
langkastii *K. Senghas in Caesiana,* 10: 28 (1998)—Ecuador.
neudeckeri *W. Königer in Arcula,* 7: 208 (1997)—Bolivia.
penduliflora *K. Senghas & M. Thiv in Caesiana,* 10: 25 (1998)—Bolivia.
thivii *K. Senghas in Caesiana,* 10: 25 (1998)—Bolivia.

× **NEOLARIUM** *Racine C. Foster & M.B. Foster in J. Bromeliad Soc.,* 23(5): 175 (1973). *BROMELIACEAE* (Neoregelia × Nidularium).

NEOLLOYDIA (Cactac.).
subgen. Echinothele *A.B. Doweld in Cactaceae Syst. Initiatives,* 9: 21 (2000).
subgen. Rapicactus *(Buxb. & Oehme) A.B. Doweld in Cactaceae Syst. Initiatives,* 9: 21 (2000): gen. Rapicactus.
booleana *(G.S. Hinton) A.B. Doweld in Cactaceae Syst. Initiatives,* 9: 19 (2000): Turbinicarpus booleanus.
glassii *A.B. Doweld in Cactaceae Syst. Initiatives,* 9: 20 (2000), nom. nov.: Thelocactus beguinii *N.P. Taylor.*
hoferi *(Lüthy & A.B. Lau) A.T. Powell in Brit. Cact. Succ. J.,* 13(4): 155 (1995), without basionym ref.; *cf. Repert. Pl. Succ. (I.O.S.),* 46: 12 (1995 publ. 1996): Turbinicarpus hoferi.
knuthiana
var. ysabelae *(Schlange) A.T. Powell in Brit. Cact. Succ. J.,* 13(4): 154 (1995), without basionym ref.; *cf. Repert. Pl. Succ. (I.O.S.),* 46: 12 (1995 publ. 1996): Thelocactus ysabelae.
krainziana *(G. Frank) A.T. Powell in Brit. Cact. Succ. J.,* 13(4): 156 (1995), without basionym ref.; *cf. Repert. Pl. Succ. (I.O.S.),* 46: 13 (1995 publ. 1996): Toumeya krainziana.

NEOLLOYDIA
krainziana
var. minima *(G. Frank) A.T. Powell in Brit. Cact. Succ. J.*, 13(4): 156 (1995), without basionym ref.; *cf. Repert. Pl. Succ. (I.O.S.)*, 46: 13 (1995 publ. 1996): Turbinicarpus krainzianus f. minimus.
pailana *(Halda & Panar.) A.B. Doweld in Sukkulentÿ (Moscow)*, 3(1–2): 70 (2000): Turbinicarpus pailanus.
pseudopectinata
var. rubriflora *A.T. Powell in Brit. Cact. Succ. J.*, 13(4): 155 (1995), without latin descr. or type; *cf. Repert. Pl. Succ. (I.O.S.)*, 46: 13 (1995 publ. 1996)—Locality not stated.
roseiflora *A.T. Powell in Brit. Cact. Succ. J.*, 13(4): 156 (1995), without latin descr. or type; *cf. Repert. Pl. Succ. (I.O.S.)*, 46: 13 (1995 publ. 1996)—Locality not stated.
subterranea (Backeb.) H.E. Moore
subsp. zaragozae *(Glass & R.A. Foster) A.B. Doweld in Cactaceae Consensus Initiatives*, 7: 12 (1999): Gymnocactus subterraneus var. zaragozae.
swobodae *(Diers) A.T. Powell in Brit. Cact. Succ. J.*, 13(4): 156 (1995), without basionym ref.; *cf. Repert. Pl. Succ. (I.O.S.)*, 46: 13 (1995 publ. 1996): Turbinicarpus swobodae.
valdeziana
var. albiflora *A.T. Powell in Brit. Cact. Succ. J.*, 13(4): 155 (1995), without latin descr. or type; *cf. Repert. Pl. Succ. (I.O.S.)*, 46: 13 (1995 publ. 1996)—Locality not stated.
zaragozae *(Glass & R.A. Foster) A.B. Doweld in Cactaceae Syst. Initiatives*, 9: 19 (2000): Gymnocactus subterraneus var. zaragozae.

NEOLOLEBA *E.A. Widjaja in Reinwardtia*, 11(2): 112 (1997). *GRAMINEAE*
amahussana *(Lindl.) E.A. Widjaja in Reinwardtia*, 11(2): 113 (1997): Bambusa amahussana.
atra *(Lindl.) E.A. Widjaja in Reinwardtia*, 11(2): 114 (1997): Bambusa atra.
glabra *E.A. Widjaja in Reinwardtia*, 11(2): 116 (1997)—New Guinea (Irian Jaya).
hirsuta *(Holttum) E.A. Widjaja in Reinwardtia*, 11(2): 118 (1997): Bambusa hirsuta.
inaurita *E.A. Widjaja in Reinwardtia*, 11(2): 119 (1997)—New Guinea (Irian Jaya).

NEOMARTINELLA (Crucif.).
grandiflora *I.A. Al-Shehbaz in Novon*, 10(4): 339 (2000)—China Southeast (Hunan), China South-Central (Sichuan).
guizhouensis *S.Z. He & Y.Z. Lan in Acta Phytotax. Sin.*, 35(1): 73 (1997)—China South-central (Guizhou).
yungshunensis *(W.T. Wang) I.A. Al-Shehbaz in Novon*, 10(4): 338 (2000): Cardamine yungshunensis.

NEOMEZIA (Theophrastac.).
cubensis (Radlk.) Votsch
subsp. oligospinosa *(Lepper) A. Borhidi in Acta Bot. Hung.*, 38(1–4): 45 (1993–94 publ. 1995): Deherainia cubensis subsp. oligospinosa.

NEOMICROCALAMUS (Gramin.).
ciliatus *(A. Camus) J.-P. Demoly in Bambou*, 21: 14 (1995): Arundinaria ciliata.
mannii *(Gamble) D. Ohrnberger, Bamboos of the World Intro. 2:* 10 (1996), isonym; *see R.B. Majumdar in Karth. et al., Fl. Ind. Enumerat. - Monocot.*: 279 (1989): Arundinaria mannii.
pusillus *(A. Chev. & A. Camus) J.-P. Demoly in Bambou*, 21: 14 (1995): Arundinaria pusilla.
yunnanensis *(T.H. Wen) D. Ohrnberger, Bamboos of the World Intro. 4:* 19 (1997): Racemobambos yunnanensis.

NEOMITRANTHES (Myrtac.).
nitida *J.R. de Mattos in Loefgrenia*, 110: 2 (1997)—Brazil Southeast (São Paulo).

NEOMORTONIA (Gesneriac.).
alba *H. Wiehler in Gesneriana*, 1(1): 63 (1995)—Panama.

NEONAVAJOA *A.B. Doweld in Sukkulentÿ (Moscow)*, 1999(2): 43 (1999), nom. nov.: gen. Navajoa *Croizat*. (Cactac.).
peeblesiana *(Croizat) A.B. Doweld in Sukkulentÿ (Moscow)*, 1999(2): 43 (1999): Navajoa peeblesiana.
subsp. fickeisenii *(Hochstätter) A.B. Doweld in Sukkulentÿ (Moscow)*, 1999(2): 43 (1999), as 'fickeiseniae': navajoa peeblesiana var. fickeisenii.

NEOPAXIA (Portulacac.).
calycina *(Colenso) P.B. Heenan in New Zealand J. Bot.*, 37(2): 220 (1999): Claytonia calycina.
campylostigma *P.B. Heenan in New Zealand J. Bot.*, 37(2): 222 (1999)—New Zealand North I, New Zealand South I.
drucei *P.B. Heenan in New Zealand J. Bot.*, 37(2): 224 (1999)—New Zealand South I.
erythrophylla *P.B. Heenan in New Zealand J. Bot.*, 37(2): 226 (1999)—New Zealand South I.
linearifolia *P.B. Heenan in New Zealand J. Bot.*, 37(2): 227 (1999), as 'lineariifolia'—New Zealand South I.
racemosa *(Buchanan) P.B. Heenan in New Zealand J. Bot.*, 37(2): 229 (1999): Claytonia australasica var. racemosa.
sessiliflora *(G. Simpson) P.B. Heenan in New Zealand J. Bot.*, 37(2): 231 (1999): Claytonia australasica var. sessiliflora.

NEOPICRORHIZA (Scrophulariac.).
minima *R.R. Mill in Edinburgh J. Bot.*, 57(3): 423 (2000)—Bhutan.

NEORAIMONDIA (Cactac.).
arequipensis
subsp. roseiflora *(Werderm. & Backeb.) C. Ostolaza Nano in Cactaceae Consensus Initiatives*, 6: 9 (1998): Cereus macrostibas var. roseiflorus.

NEOREGELIA (Bromeliac.).
subgen. Longipetalopsis *E.M.C. Leme, Canistropsis - Bromeliads of Atlantic Forest:* 60 (1998).
subgen. Protoregelia *W. Till & E.M.C. Leme in E.M.C. Leme, Canistropsis - Bromeliads of Atlantic Forest:* 96 (1998).
azevedoi *E.M.C. Leme, Canistropsis - Bromeliads of Atlantic Forest:* 80 (1998)—Brazil Northeast (Bahia).
bragarum *(E. Pereira & L.B. Sm.) E.M.C. Leme, Canistropsis - Bromeliads of Atlantic Forest:* 78 (1998): Wittrockia bragarum.
brigadeirensis *C.C. de Paula & E.M.C. Leme in Pabstia*, 8(1): 2 (1997)—Brazil Southeast (Minas Gerais).
brownii *E.M.C. Leme in Pabstia*, 8(1): 3 (1997)—Brazil Southeast (Minas Gerais).
burlemarxii *R.W. Read in J. Bromeliad Soc.*, 46(6): 261 (1996)—Brazil Southeast (São Paulo).
subsp. meeana *R.W. Read in J. Bromeliad Soc.*, 46(6): 263 (1996)—Brazil Southeast (São Paulo).
carinata *E.M.C. Leme in Harvard Pap. Bot.*, 4(1): 138 (1999)—Brazil Southeast (Rio de Janeiro).
crispata *E.M.C. Leme in J. Bromeliad Soc.*, 46(1): 22 (1996)—Brazil Northeast (Bahia).
ibitipocensis *(Leme) E.M.C. Leme, Canistropsis - Bromeliads of Atlantic Forest:* 67 (1998): Wittrockia ibitipocensis.
inexspectata *E.M.C. Leme, Canistropsis - Bromeliads of Atlantic Forest:* 75 (1998)—Brazil Southeast (Espírito Santo).
kerryi *E.M.C. Leme, Canistropsis - Bromeliads of Atlantic Forest:* 82 (1998)—Brazil Northeast (Bahia).
lactea *H.E. Luther & E.M.C. Leme in Bromélia*, 2(2): 8 (1995)—Brazil Southeast (Rio de Janeiro).
longipedicellata *E.M.C. Leme, Canistropsis - Bromeliads of Atlantic Forest:* 74 (1998)—Brazil Southeast (Rio de Janeiro).
menescalii *E.M.C. Leme, Canistropsis - Bromeliads of Atlantic Forest:* 73 (1998)—Brazil Southeast (Espírito Santo).
mucugensis *E.M.C. Leme, Canistropsis - Bromeliads of Atlantic Forest:* 83 (1998)—Brazil Northeast (Bahia).

NEOREGELIA:-
nevaresii *E.M.C. Leme & H.E. Luther in J. Bromeliad Soc.*, 48(4): 155 (1998)—Brazil Southeast (Rio de Janeiro).
nivea *E.M.C. Leme in J. Bromeliad Soc.*, 46(5): 219 (1996)—Brazil Southeast (São Paulo).
odorata *E.M.C. Leme in Harvard Pap. Bot.*, 4(1): 140 (1999)—Brazil Southeast (São Paulo).
pernambucana *E.M.C. Leme & J.A. Siqueira in E.M.C. Leme, Nidularium - Bromeliads of Atlantic Forest:* 232 (2000)—Brazil Northeast (Pernambuco).
petropolitana *E.M.C. Leme in J. Bromeliad Soc.*, 45(5): 199 (1995)—Brazil Southeast (Rio de Janeiro).
rubrovittata *E.M.C. Leme in J. Bromeliad Soc.*, 46(5): 217 (1996)—Brazil Southeast (Rio de Janeiro).
tenebrosa *E.M.C. Leme, Canistropsis - Bromeliads of Atlantic Forest:* 70 (1998)—Brazil Southeast (Rio de Janeiro).

× **NEOROCKIA** *D. Butcher in J. Bromeliad Soc.*, 49(1): 14 (1999). *BROMELIACEAE* (Neoregelia × Wittrockia).

NEOSHIRAKIA *H.-J. Esser in Blumea*, 43(1): 129 (1998), nom. nov.: gen. Shirakia *Hurus.* (Euphorbiac.).
japonica *(Siebold & Zucc.) H.-J. Esser in Blumea*, 43(1): 129 (1998): Stillingia japonica.

NEOSINOCALAMUS (Gramin.).
farinosus
f. flavostriatus *(T.P. Yi) D. Ohrnberger, Bamboos of the World Intro.*, 3: 14 (1996): Dendrocalamus farinosus f. flavostriatus.

× **NEOSTYLOPSIS** *hort. in Orchid Rev.*, 108 (1236): new orchid hybrids 12 (2000). *ORCHIDACEAE* (Neofinetia × Phalaenopsis × Rhynchostylis).

NEOSTYRAX *G.S. Fan in Guihaia*, 16(4): 305 (1996), without latin descr. or type. *STYRACACEAE*
polysperma *(Clarke) G.S. Fan in Guihaia*, 16(4): 306 (1996), generic name not validly publ.: Basionym not stated.

NEOTATEA (Guttif.).
duidae *(Kobuski & Steyerm.) P.F. Stevens & A.L. Weitzman in Biollania*, ed. esp. 6 (J.J. Wurdack Festschr.): 559 (1997): Bonnetia duidae.

NEOTINEA (Orchidac.).
lactea *(Poir.) R.M. Bateman, A.M. Pridgeon & M.W. Chase in Lindleyana*, 12(3): 122 (1997): Orchis lactea.
tridentata *(Scop.) R.M. Bateman, A.M. Pridgeon & M.W. Chase in Lindleyana*, 12(3): 122 (1997): Orchis tridentata.
subsp. commutata *(Tod.) R.M. Bateman, A.M. Pridgeon & M.W. Chase in Lindleyana*, 12(3): 122 (1997): Orchis commutata.
subsp. conica *(Willd.) R.M. Bateman, A.M. Pridgeon & M.W. Chase in Lindleyana*, 12(3): 122 (1997): Orchis conica.
ustulata *(L.) R.M. Bateman, A.M. Pridgeon & M.W. Chase in Lindleyana*, 12(3): 122 (1997): Orchis ustulata.

NEOTORULARIA (Crucif.).
conferta *R.F. Huang in Acta Phytotax. Sin.*, 35(6): 558 (1997)—Qinghai.
conferta *(Steven ex Turcz.) V.I. Dorofeev in Turczaninowia*, 1(3): 25 (1998), without basionym ref.: Basionym not stated.
gamosepala *(Hedge) I.A. Al-Shehbaz & S.L. O'Kane in Novon*, 7(2): 93 (1997): Arabidopsis gamosepala.
pamirica *Z.X. An in Fl. Xinjiangensis*, 2(2): 380 (1995)—Xinjiang.
polyceratia *(L.) V.I. Dorofeev in Turczaninowia*, 1(3): 25 (1998), without basionym ref.: Basionym not stated.
qingshuiheense *(Ma & Zong Y. Zhu) I.A. Al-Shehbaz, S.L. O'Kane & G. Yang in Edinburgh J. Bot.*, 56(3): 326 (1999): Microsisymbrium qingshuiheense.

NEOTORULARIA:-
runcinata *(Lag. ex DC.) V.I. Dorofeev in Turczaninowia,* 1(3): 25 (1998), without basionym ref.: Basionym not stated.

NEOTTIA (Orchidac.).
subgen. Archineottia *(S.C. Chen) D.L. Szlachetko in Fragm. Flor. Geobot.,* Supp. 3: 116 (1995): gen. Archineottia.
subgen. Listera *(R. Br.) D.L. Szlachetko in Fragm. Flor. Geobot.,* Supp. 3: 116 (1995): gen. Listera.
alternifolia *(King & Pantl.) D.L. Szlachetko in Fragm. Flor. Geobot.,* Supp. 3: 117 (1995): Listera alternifolia.
auriculata *(Wiegand) D.L. Szlachetko in Fragm. Flor. Geobot.,* Supp. 3: 117 (1995): Listera auriculata.
australis *(Lindl.) D.L. Szlachetko in Fragm. Flor. Geobot.,* Supp. 3: 117 (1995), homonym; *see R. Br., Prodr.:* 319 (1810): Listera australis.
bambusetorum *(Hand.-Mazz.) D.L Szlachetko in Fragm. Flor. Geobot.,* Supp. 3: 117 (1995): Listera bambusetorum.
biflora *(Schltr.) D.L. Szlachetko in Fragm. Flor. Geobot.,* Supp. 3: 117 (1995): Listera biflora.
borealis *(Morong) D.L. Szlachetko in Fragm. Flor. Geobot.,* Supp. 3: 117 (1995): Listera borealis.
brachybotryosa *(T. Tang & F.T. Wang) D.L. Szlachetko in Fragm. Flor. Geobot.* Supp. 3: 117 (1995): Listera brachybotryosa.
brevicaulis *(King & Pantl.) D.L. Szlachetko in Fragm. Flor. Geobot.,* Supp. 3: 117 (1995): Listera brevicaulis.
bungeana *(Y. Yabe) D.L. Szlachetko in Fragm. Flor. Geobot.,* Supp. 3: 117 (1995): Listera bungeana.
caurina *(Piper) D.L. Szlachetko in Fragm. Flor. Geobot.,* Supp. 3: 117 (1995): Listera caurina.
deltoidea *(Fukuy.) D.L. Szlachetko in Fragm. Flor. Geobot.,* Supp. 3: 117 (1995): Listera deltoidea.
dentata *(King & Pantl.) D.L. Szlachetko in Fragm. Flor. Geobot.,* Supp. 3: 117 (1995): Listera dentata.
divaricata *(Panigrahi & P. Taylor) D.L. Szlachetko in Fragm. Flor. Geobot.,* Supp. 3: 117 (1995): Listera divaricata.
flabellata *(W.W. Sm.) D.L. Szlachetko in Fragm. Flor. Geobot.,* Supp. 3: 117 (1995): Listera flabellata.
grandiflora *(Rolfe) D.L. Szlachetko in Fragm. Flor. Geobot.,* Supp. 3: 117 (1995), homonym; *see Hook. in Bot. Mag.,* 54: t. 2730 (1827): Listera grandiflora.
japonica *(Blume) D.L. Szlachetko in Fragm. Flor. Geobot.,* Supp. 3: 117 (1995): Listera japonica.
japonica *(M. Furuse) K. Inoue in Acta Phytotax. Geobot.,* 49(2): 199 (1998 publ. 1999), homonym; *see Szlach. in Fragm. Flor. Geobot.,* Supp. 3: 117 (1995): Archineottia japonica.
karoana *D.L. Szlachetko in Fragm. Flor. Geobot.,* Supp. 3: 117 (1995), nom. nov.: Listera micrantha *Lindl.*
longicaulis *(King & Pantl.) D.L. Szlachetko in Fragm. Flor. Geobot.,* Supp. 3: 117 (1995): Listera longicaulis.
macrantha *(Fukuy.) D.L. Szlachetko in Fragm. Flor. Geobot.,* Supp. 3: 117 (1995), homonym; *see Sweet, Hort. Brit.,* ed. 2: 485 (1830): Listera macrantha.
major *(Nakai) D.L. Szlachetko in Fragm. Flor. Geobot.,* Supp. 3: 118 (1995): Listera major.
makinoana *(Ohwi) D.L. Szlachetko in Fragm. Flor. Geobot.,* Supp. 3: 118 (1995): Listera makinoana.
morrisonicola *(Hayata) D.L. Szlachetko in Fragm. Flor. Geobot.,* Supp. 3: 118 (1995): Listera morrisonicola.
mucronata *(Panigrahi & J.J. Wood) D.L. Szlachetko in Fragm. Flor. Geobot.,* Supp. 3: 118 (1995): Listera mucronata.
nanchuanica *(S.C. Chen) D.L. Szlachetko in Fragm. Flor. Geobot.,* Supp. 3: 118 (1995): Listera nanchuanica.
nandadeviensis *(Hajra) D.L. Szlachetko in Fragm. Flor. Geobot.,* Supp. 3: 118 (1995): Listera nandadeviensis.
nankomontana *(Fukuy.) D.L. Szlachetko in Fragm. Flor. Geobot.,* Supp. 3: 118 (1995): Listera nankomontana.
nepalensis *(N.P. Balakr.) D.L. Szlachetko in Fragm. Flor. Geobot.,* Supp. 3: 118 (1995): Listera nepalensis.

NEOTTIA:-
nephrophylla *(Rydb.) D.L. Szlachetko in Fragm. Flor. Geobot.,* Supp. 3: 118 (1995): Listera nephrophylla.
nipponica *(Makino) D.L. Szlachetko in Fragm. Flor. Geobot.,* Supp. 3: 118 (1995): Listera nipponica.
oblata *(S.C. Chen) D.L. Szlachetko in Fragm. Flor. Geobot.,* Supp. 3: 118 (1995): Listera oblata.
pinetorum *(Lindl.) D.L. Szlachetko in Fragm. Flor. Geobot.,* Supp. 3: 118 (1995): Listera pinetorum.
pseudonipponica *(Fukuy.) D.L. Szlachetko in Fragm. Flor. Geobot.,* Supp. 3: 118 (1995): Listera pseudonipponica.
puberula *(Maxim.) D.L. Szlachetko in Fragm. Flor. Geobot.,* Supp. 3: 118 (1995): Listera puberula.
schlechteriana *D.L. Szlachetko in Fragm. Flor. Geobot.,* Supp. 3: 118 (1995), nom. nov., as 'schlechterana': N. grandiflora *Schltr.*
shaoi *(S.S. Ying) D.L. Szlachetko in Fragm. Flor. Geobot.,* Supp. 3: 118 (1995): Listera shaoi.
smallii *(Wiegand) D.L. Szlachetko in Fragm. Flor. Geobot.,* Supp. 3: 118 (1995): Listera smallii.
smithii *(Schltr.) D.L. Szlachetko in Fragm. Flor. Geobot.,* Supp. 3: 118 (1995): Listera smithii.
suzuki *(Masam.) D.L. Szlachetko in Fragm. Flor. Geobot.,* Supp. 3: 119 (1995): Listera suzuki.
taiwaniana *(S.S. Ying) D.L. Szlachetko in Fragm. Flor. Geobot.,* Supp. 3: 119 (1995): Listera taiwaniana.
taizanensis *(Fukuy.) D.L. Szlachetko in Fragm. Flor. Geobot.,* Supp. 3: 119 (1995): Listera taizanensis.
tenuis *(Lindl.) D.L. Szlachetko in Fragm. Flor. Geobot.,* Supp. 3: 119 (1995): Listera tenuis.
tianschanica *(Grubov) D.L. Szlachetko in Fragm. Flor. Geobot.,* Supp. 3: 119 (1995): Listera tianschanica.
unguiculata *(W.W. Sm.) D.L. Szlachetko in Fragm. Flor. Geobot.,* Supp. 3: 119 (1995): Listera unguiculata.
uraiensis *(S.S. Ying) D.L. Szlachetko in Fragm. Flor. Geobot.,* Supp. 3: 119 (1995): Listera uraiensis.
wardii *(Rolfe) D.L. Szlachetko in Fragm. Flor. Geobot.,* Supp. 3: 119 (1995): Listera wardii.
yatabei *(Makino) D.L. Szlachetko in Fragm. Flor. Geobot.,* Supp. 3: 119 (1995): Listera yatabei.
yueana *(T. Tang & F.T. Wang) D.L. Szlachetko in Fragm. Flor. Geobot.,* Supp. 3: 119 (1995): Listera yueana.
yunnanensis *(S.C. Chen) D.L. Szlachetko in Fragm. Flor. Geobot.,* Supp. 3: 119 (1995): Listera yunnanensis.

NEOTTIANTHE (Orchidac.).
angustifolia *K.Y. Lang in Acta Phytotax. Sin.,* 35(6): 538 (1997)—China South-central (Sichuan).
gymnadenioides *(Hand.-Mazz.) K.Y. Lang & S.C. Chen in Acta Phytotax. Sin.,* 35(6): 541 (1997): Gymphyosepalum gymnadenioides.
luteola *K.Y. Lang & S.C. Chen in Acta Phytotax. Sin.,* 35(6): 545 (1996)—China South-central (Yunnan).
oblonga *K.Y. Lang in Acta Phytotax. Sin.,* 35(6): 544 (1997)—China South-central (Yunnan).
ovata *K.Y. Lang in Acta Phytotax. Sin.,* 35(6): 542 (1997), with two types cited—China South-central (Sichuan).

NEOVEITCHIA (Palmae).
brunnea *J.L. Dowe in Austral. Syst. Bot.,* 9(1): 36 (1996)—Vanuatu.

NEOWERDERMANNIA (Cactac.).
chilensis
subsp. peruviana *(F. Ritter) C. Ostolaza Nano in Cactaceae Consensus Initiatives,* 6: 9 (1998): N. peruviana.

NEPENTHES (Nepenthac.).
angasanensis *R. Maulder, D. Schubert, B. Salmon & B. Quinn in Carniv. Pl. Newsl.,* 28(1): 15 (1999)—Sumatera.
argentii *M. Jebb & M. Cheek in Blumea,* 42(1): 19 (1997)—Philippines.
aristolochioides *M. Jebb & M. Cheek in Blumea,* 42(1): 22 (1997)—Sumatera.

NEPENTHES:-
benstonei *C. Clarke in Sandakania,* 13: 80 (1999)—Malaya (Peninsular Malaysia).
curtisii Mast.
subsp. zakriana *J.H. Adam & C.C. Wilcock in Sarawak Mus. J.,* 71: 151 (1996)—Borneo (Sabah).
danseri *M. Jebb & M. Cheek in Blumea,* 42(1): 30 (1997)—New Guinea (Irian Jaya), Moluccas.
diatas *M. Jebb & M. Cheek in Blumea,* 42(1): 33 (1997)—Sumatera.
eymae *Shigeo Kurata in J. Insectiv. Pl. Soc. (Japan),* 35(2): 41 (1984)—Sulawesi.
kinabaluensis *Sh. Kurata ex J.H. Adam & C.C. Wilcock in Sarawak Mus. J.,* 71: 152 (1996)—Borneo (Sabah).
lamii *M. Jebb & M. Cheek in Blumea,* 42(1): 54 (1997)—New Guinea (Irian Jaya).
lavicola *A. Wistuba & H. Rischer in Carniv. Pl. Newsl.,* 25(4): 106 (1996)—Sumatera.
macrophylla *(Marabini) M. Jebb & M. Cheek in Blumea,* 42(1): 58 (1997): N. edwardsiana subsp. macrophylla.
mira *M. Jebb & M. Cheek in Kew Bull.,* 53(4): 966 (1998)—Philippines.
murudensis *Culham ex M. Jebb & M. Cheek in Blumea,* 42(1): 66 (1997)—Borneo (Sarawak).
rubromaculata *Shigeo Kurata in J. Insectiv. Pl. Soc. (Japan),* 35(2): 42 (1984), as 'rubro-maculata'—Sulawesi.
sandakanensis *J.H. Adam & C.C. Wilcock in Sarawak Mus. J.,* 71: 156 (1996)—Borneo (Sabah).
var. eglandulosa *J.H. Adam & C.C. Wilcock in Sarawak Mus. J.,* 71: 158 (1996)—Borneo (Sabah).
var. ferruginea *J.H. Adam & C.C. Wilcock in Sarawak Mus. J.,* 71: 158 (1996)—Borneo (Sabah).
sibuyanensis *J. Nerz in Carniv. Pl. Newsl.,* 27(1): 18 (1998)—Philippines.
wilkiei *M. Jebb & M. Cheek in Kew Bull.,* 53(4): 966 (1998)—Philippines.

NEPETA (Labiat.).
apulei Ucria ex Guss.
subsp. oulmesiana *F. Gómiz García in An. Jard. Bot. Madrid,* 55(2): 476 (1997)—Morocco.
subsp. pallescens *(Maire) F. Gómiz García in An. Jard. Bot. Madrid,* 55(2): 476 (1997)—N. apulei var. pallescens.
bokhonica *Z. Jamzad in Iranian J. Bot.,* 8(1): 44 (1999)—Iran.
cataria L.
subsp. verticillata *K. Micevski in Prilozi Oddel. Biol. Med. Nauki Makedonska Akad. Nauk. Umet.,* 9(1–2): 33 (1988 publ. 1991)—Yugoslavia.
macedonica *K. Micevski in Prilozi Oddel. Biol. Med. Nauki Makedonska Akad. Nauk. Umet.,* 9(1–2): 30 (1988 publ. 1991)—Yugoslavia.
makuensis *Z. Jamzad & Mozaffarian in Iranian J. Bot.,* 7(2): 250 (1998)—Iran.
minuticephala *Z. Jamzad in Iranian J. Bot.,* 8(1): 46 (1999)—Iran.
nuda L.
subsp. coerulea *(Aiton) O. de Bolòs & J. Vigo, Fl. Països Catalans,* 3: 286 (1995): N. coerulea.
subcaespitosa *A. Jehan in Willdenowia,* 25(2): 647 (1996)—Pakistan.

NEPHTHYTIS (Arac.).
afzelii Schott
var. graboensis *J. Bogner & M. Knecht in Sendtnera,* 4: 11 (1997)—Ivory Coast.

NERTERA (Rubiac.).
villosa *B.H. Macmillan & R. Mason in New Zealand J. Bot.,* 33(4): 435 (1995)—New Zealand South I., New Zealand North I.
yamashitae *T. Yamazaki in J. Jap. Bot.,* 73(2): 89 (1998)—Nansei-Shoto.

NERVILIA (Orchidac.).
jacksoniae *A.F. Rinehart & F.R. Fosberg in Micronesica,* 24(1): 81 (1991)—Marianas (Northern Marianas, Guam).
lanyuensis *S.S. Ying in Mem. Coll. Agric. Nation. Taiwan Univ.,* 29(2): 55 (1989), as 'lanyuense'—Taiwan.

NESOGENES (Nesogenac.).
euphrasioides (Hook. & Arn.) A.DC.
var. lineata *F.R. Fosberg in Micronesica,* 23(1): 2 (1990)—Cook Is.

NESOGORDONIA (Sterculiac.).
holtzii *(Engl.) Capuron ex L.C. Barnett & L.J. Dorr in Kew Bull.,* 55(4): 985 (2000): Cistanthera holtzii.
suzannae *J.-N. Labat, J. Munzinger & O. Pascal in Candollea,* 55(2): 278 (2000)—Comoros (Mayotte).

NESPHOSTYLIS (Legumin.).
junodii *(Harms) M.P. Munyenyembe & F.A. Bisby in Kew Bull.,* 52(3): 690 (1997): Vigna junodii.

NEURACANTHUS (Acanthac.).
sect. Didymosperma *S. Bidgood & R.K. Brummitt in Kew Bull.,* 53(1): 24 (1998).
africanus T. Anderson ex S. Moore
var. limpopoensis *S. Bidgood & R.K. Brummitt in Kew Bull.,* 53(1): 36 (1998)—Zimbabwe, Mozambique, Transvaal.
var. masaicus *S. Bidgood & R.K. Brummitt in Kew Bull.,* 53(1): 37 (1998)—Tanzania, Kenya.
var. ruahae *S. Bidgood & R.K. Brummitt in Kew Bull.,* 53(1): 38 (1998)—Tanzania.
decorus S. Moore
subsp. strobilinus *(C.B. Clarke) S. Bidgood & R.K. Brummitt in Kew Bull.,* 53(1): 40 (1998): N. strobilinus.
mahajangensis *S. Bidgood & R.K. Brummitt in Kew Bull.,* 53(1): 27 (1998)—Madagascar.
matsabdianus *S. Bidgood & R.K. Brummitt in Kew Bull.,* 53(1): 61 (1998)—Somalia.
migiurtinus *S. Bidgood & R.K. Brummitt in Kew Bull.,* 53(1): 59 (1998)—Somalia.
ovalifolius *(Fiori) S. Bidgood & R.K. Brummitt in Kew Bull.,* 53(1): 54 (1998): N. stolonosus var. ovalifolius.
pedalis *S. Bidgood & R.K. Brummitt in Kew Bull.,* 53(1): 29 (1998)—Madagascar.
robecchii (Lindau) C.B. Clarke
subsp. atratus *S. Bidgood & R.K. Brummitt in Kew Bull.,* 53(1): 67 (1998)—Somalia, Oman, South Yemen.
tephrophyllus *S. Bidgood & R.K. Brummitt in Kew Bull.,* 53(1): 44 (1998)—Somalia, Kenya, Tanzania.
subsp. conifer *S. Bidgood & R.K. Brummitt in Kew Bull.,* 53(1): 48 (1998)—Kenya, Tanzania.
subsp. tsavoensis *S. Bidgood & R.K. Brummitt in Kew Bull.,* 53(1): 50 (1998)—Kenya.
tetragonostachyus Nees
subsp. trinervius *(Wight) S. Bidgood & R.K. Brummitt in Kew Bull.,* 53(1): 16 (1998): N. trinervius.
umbraticus *S. Bidgood & R.K. Brummitt in Kew Bull.,* 53(1): 30 (1998)—Madagascar.

NEURADA (Neuradac.).
al-eisawii *G. Barsotti, A. Borzatti de Loewenstern & F. Garbari in Bot. Chronika,* 13: 113 (2000)—Palestine (Jordan).

NEUROLAENA (Compos.).
jannaweissiana *B.L. Turner in Phytologia,* 84(2): 91 (1998 publ. 1999), as 'jannaweissana'—Mexico Southwest (Oaxaca).

NEUROLEPIS (Gramin.).
asymmetrica *L.G. Clark in Novon,* 6(4): 336 (1996)—Ecuador.
fimbriligulata *L.G. Clark in Novon,* 6(4): 338 (1996)—Ecuador.
subsp. peruviana *L.G. Clark in Novon,* 6(4): 342 (1996)—Peru.
laegaardii *L.G. Clark in Novon,* 6(4): 342 (1996)—Ecuador.
nana *L.G. Clark in Novon,* 6(4): 344 (1996)—Ecuador.
petiolata *G. Davidse & L.G. Clark in Novon,* 6(2): 153 (1996)—Colombia.
rigida *L.G. Clark in Novon,* 6(4): 347 (1996)—Ecuador.
silverstonei *G. Davidse & L.G. Clark in Novon,* 6(2): 150 (1996)—Colombia.
villosa *L.G. Clark in Novon,* 6(4): 349 (1996)—Ecuador.

NEWCASTELIA (Verbenac.).
roseoazurea *B.L. Rye in Nuytsia,* 11(1): 98 (1996)—Western Australia.

NEZAHUALCOYOTLIA *R. González Tamayo in Bol. Inst. Bot. (Guadalajara),* 4(1–3): 67 (1996 publ. 1997). *ORCHIDACEAE*
gracilis *(L.O. Williams) R. González Tamayo in Bol. Inst. Bot. (Guadalajara),* 4(1–3): 69 (1996 publ. 1997): Cranichis gracilis.

NICORAELLA *M.A. Torres in Monogr. Comis. Invest. Ci. Prov. Buenos Aires,* 13: 70 (1997). *GRAMINEAE*
subgen. Mucronella *M.A. Torres in Monogr. Comis. Invest. Ci. Prov. Buenos Aires,* 13: 72 (1997).
bomanii *(Hauman) M.A. Torres in Monogr. Comis. Invest. Ci. Prov. Buenos Aires,* 13: 73 (1997): Stipa bomanii.
brevis *M.A. Torres in Monogr. Comis. Invest. Ci. Prov. Buenos Aires,* 13: 73 (1997)—Argentina Northwest (Jujuy).
henrardiana *(Parodi) M.A. Torres in Monogr. Comis. Invest. Ci. Prov. Buenos Aires,* 13: 74 (1997): Stipa henrardiana.
mucronata *(Griseb.) M.A. Torres in Monogr. Comis. Invest. Ci. Prov. Buenos Aires,* 13: 72 (1997): Piptochaetium mucronatum.
obtusa *(Nees & Meyen) M.A. Torres in Monogr. Comis. Invest. Ci. Prov. Buenos Aires,* 13: 74 (1997): Piptatherum obtusum.
rigidiseta *(Pilg.) M.A. Torres in Monogr. Comis. Invest. Ci. Prov. Buenos Aires,* 13: 75 (1997): Oryzopsis rigidiseta.
venusta *(Phil.) M.A. Torres in Monogr. Comis. Invest. Ci. Prov. Buenos Aires,* 13: 75 (1997): Stipa venusta.

NICOTIANA (Solanac.).
truncata *D.E. Symon in J. Adelaide Bot. Gard.,* 18(1): 1 (1998)—South Australia.

NIDEMA (Orchidac.).
ottonis
var. triandra *(Schltr.) M.A. Nir, Orchidaceae Antillanae:* 407 (2000): Nidema boothii var. triandrum.

NIDULARIUM (Bromeliac.).
albiflorum *(L.B. Sm.) E.M.C. Leme, Nidularium - Bromeliads of Atlantic Forest:* 140 (2000): N. rubrum var. albiflorum.
amorimii *E.M.C. Leme, Nidularium - Bromeliads of Atlantic Forest:* 118 (2000)—Brazil Northeast (Bahia).
azureum *(L.B. Sm.) E.M.C. Leme, Nidularium - Bromeliads of Atlantic Forest:* 94 (2000): Wittrockia azurea.
bicolor *(E. Pereira) E.M.C. Leme, Nidularium - Bromeliads of Atlantic Forest:* 49 (2000): N. innocentii var. bicolor.
campos-portoi *(L.B. Sm.) E.M.C. Leme, Nidularium - Bromeliads of Atlantic Forest:* 164 (2000): Wittrockia campos-portoi.
var. robustum *(E. Pereira & I.A. Penna) E.M.C. Leme, Nidularium - Bromeliads of Atlantic Forest:* 166 (2000): Wittrockia campos-portoi var. robusta.
cariacicaense *(W. Weber) E.M.C. Leme, Nidularium - Bromeliads of Atlantic Forest:* 113 (2000): N. procerum var. cariacicaense.
catarinense *E.M.C. Leme, Nidularium - Bromeliads of Atlantic Forest:* 93 (2000)—Brazil South (Santa Catarina).
corallinum *(Leme) E.M.C. Leme, Nidularium - Bromeliads of Atlantic Forest:* 121 (2000): Wittrockia corallina.
exostigmum *R.C. Tardivo in Bromélia,* 2(2): 26 (1995)—Brazil South (Paraná).
ferrugineum *E.M.C. Leme, Nidularium - Bromeliads of Atlantic Forest:* 48 (2000)—Brazil Southeast (Espírito Santo).
gracile *R.C. Tardivo in Bromélia,* 2(2): 29 (1995)—Brazil South (Paraná).
krisgreeniae *E.M.C. Leme, Nidularium - Bromeliads of Atlantic Forest:* 171 (2000)—Brazil Southeast (São Paulo).
linehamii *E.M.C. Leme in J. Bromeliad Soc.,* 43(5): 199 (1993)—Brazil Southeast (Minas Gerais).

NIDULARIUM:-
mangaratibense *E.M.C. Leme, Nidularium - Bromeliads of Atlantic Forest:* 73 (2000)—Brazil Southeast (Rio de Janeiro).
organense *E.M.C. Leme, Nidularium - Bromeliads of Atlantic Forest:* 142 (2000)—Brazil Southeast (Rio de Janeiro).
serratum *E.M.C. Leme, Nidularium - Bromeliads of Atlantic Forest:* 111 (2000), nom. nov.: N. purpureum var. coeruleum *E. Pereira & Moutinho.*
viridipetalum *E.M.C. Leme, Nidularium - Bromeliads of Atlantic Forest:* 109 (2000)—Brazil Southeast (Rio de Janeiro).

× **NIDUREGELIA** (Bromeliac.).
edmundoi *(Leme) E.M.C. Leme, Nidularium - Bromeliads of Atlantic Forest:* 178 (2000): Nidularium edmundoi.
fraudulenta *(Leme) E.M.C. Leme, Nidularium - Bromeliads of Atlantic Forest:* 179 (2000): Nidularium fraudulentum.
lyman-smithii *(Leme) E.M.C. Leme, Nidularium - Bromeliads of Atlantic Forest:* 180 (2000): Nidularium lyman-smithii.

NIEBUHRIA (Capparac.).
linearifolia *J. Graham, Cat. Pl. Bombay:* 8 (1839)—India (Maharashtra).

NIEREMBERGIA (Solanac.).
linariifolia Graham
var. glabriuscula *(Dunal) A.A. Cocucci & A.T. Hunziker in Darwiniana,* 33(1–4): 38 (1995), as 'linariaefolia': N. hippomanica var. glabriuscula.
var. pampeana *(Millán) A.A. Cocucci & A.T. Hunziker in Darwiniana,* 33(1–4): 38 (1995), as 'linariaefolia': N. hippomanica var. pampeana.
var. pinifolioides *(Millán) A.A. Cocucci & A.T. Hunziker in Darwiniana,* 33(1–4): 38 (1995), as 'linariaefolia': N. hippomanica var pinifolioides.
pulchella Miers
var. longistyla *(Millán) A.A. Cocucci & A.T. Hunziker in Darwiniana,* 33(1–4): 40 (1995): N. longistyla.
var. macrocalyx *(Millán) A.A. Cocucci & A.T. Hunziker in Darwiniana,* 33(1–4): 40 (1995): N. macrocalyx.

NIGELLICEREUS *(P.V. Heath) P.V. Heath in Calyx,* 6(1): 1 (1998): Stenocereus subgen. Nigellicereus. (Cactac.).

NIGRITELLA (Orchidac.).
buschmanniae *H. Teppner & T. Ster in Phyton (Austria),* 36(2): 278 (1996)—Italy.
cenisia *G. Foelsche, W. Foelsche, M. Gerbaud & O. Gerbaud in Orchidophile,* 134: 248 (1998)—France.
corneliana (Beauverd) Gölz & H.R. Reinhard
var. bourneriasii *(E. Breiner & R. Breiner) E. Klein in Phyton (Austria),* 36(1): 59 (1996): N. corneliana subsp. bourneriasii.
× delphineae *M. Gerbaud & O. Gerbaud in Orchidophile,* 120: 33, 30 (1996)—France. (N. corneliana × rhellicani).
dolomitensis *(Teppner & E. Klein) M. Hedrén, E. Klein & H. Teppner in Phyton (Austria),* 40(2): 243 (2000): Gymnadenia dolomitensis.

NIPHAEA (Gesneriac.).
peruviana *H. Wiehler in Gesneriana,* 1(1): 65 (1995)—Peru.

NITRARIA (Zygophyllac.).
sect. Primaria *P. Khalkuziev, Rodstv. Svyaz. Nekotor. Semeĭstv Rast. Pust. Oblast.:* 55 (1990).
schoberi L.
var. roborowskii *(Kom.) V.I. Grubov in Novosti Sist. Vyssh. Rast.,* 31: 184 (1998), isonym; *see Hadidi in Fl. Iran.,* 98: 11 (1972): N. roboroskii.

NIVENIA (Iridac.).
parviflora *P. Goldblatt in Bothalia,* 27(2): 101 (1997)—Cape Province.

NOAEA (Chenopodiac.).
mucronata (Forssk.) Asch. & Schweinf.
subsp. leptoclada *(Woronow) M. Assadi in Iranian J. Bot.*, 8(1): 28 (1999): N. tournefortii var. leptoclados.

NOCCAEA (Crucif.).
arctica *(A.E. Porsild) J. Holub in Preslia*, 70(2): 107 (1998): Thlaspi arcticum.
caerulescens (J. Presl & C. Presl) F.K. Mey.
subsp. brachypetala *(Jord.) N.N. Tzvelev in Novosti Sist. Vyssh. Rast.*, 32: 184 (2000): Thlaspi brachypetalum.
subsp. calaminaris *(Lej.) J. Holub in Preslia*, 70(2): 107 (1998): Thlaspi alpestre var. calaminare.
coloradensis *(Rydb.) J. Holub in Preslia*, 70(2): 107 (1998): Thlaspi coloradense.
cuneifolia *(Pant.) J. Holub in Preslia*, 70(2): 108 (1998): Thlaspi cuneifolium.
fendleri *(A. Gray) J. Holub in Preslia*, 70(2): 108 (1998): Thlaspi fendleri.
glauca *(A. Nelson) J. Holub in Preslia*, 70(2): 108 (1998): Thlaspi alpestre var. glaucum.
magellanica *(Pers.) J. Holub in Preslia*, 70(2): 108 (1998): Thlaspi magellanicum.
mexicana *(Standl.) J. Holub in Preslia*, 70(2): 108 (1998): Thlaspi mexicanum.
oligosperma *(Merino) J. Holub in Preslia*, 70(2): 108 (1998): Thlaspi sylvestre var. oligospermum.
parviflora *(A. Nelson) J. Holub in Preslia*, 70(2): 108 (1998): Thlaspi parviflorum.
pawlowskii *(Dvořáková) J. Holub in Preslia*, 70(2): 108 (1998): Thlaspi pawlowskii.
pindica *(Hausskn.) J. Holub in Preslia*, 70(2): 108 (1998): Thlaspi pindicum.
suffruticosa *(Loscos & J. Pardo) J. Holub in Preslia*, 70(2): 108 (1998): Thlaspi suffruticosum.

NORANTEA (Marcgraviac.).
aurantiaca *Spruce ex Gilg in Bot. Jahrb. Syst.*, 25, Beibl. 60: 32 (1898)—Locality not stated.
aurantiaca *Spruce ex G.L. Ferreira in Arq. Jard. Bot. Rio de Janeiro*, 33(2): 29 (1995), isonym; *see Gilg in Bot. Jahrb. Syst.*, 25, Beibl. 60: 32 (1898)—Brazil North (Amazonas).
eoetvoesorum *V.A. Richter in Math. Term. Ertes.*, 34: 580 (1916)—Peru.†
gracilis *(Wittm.) V.A. Richter in Math. Term. Ertes.*, 34: 801 (1916): N. guianensis var. gracilis.
guianensis Aubl.
var. goyazensis *(Cambess.) G.L. Ferreira in Arq. Jard. Bot. Rio de Janeiro*, 33(2): 41 (1995): N. goyazensis.
var. japurensis *(Mart.) G.L. Ferreira in Arq. Jard. Bot. Rio de Janeiro*, 33(2): 39 (1995): N. japurensis.
loczyi *V.A. Richter in Math. Term. Ertes.*, 34: 816 (1916)—Guatemala.†
zahlbruckneri *V.A. Richter in Math. Term. Ertes.*, 34: 828 (1916)—Peru.

NORONHIA (Oleac.).
cochleata *J.-N. Labat, Marc Pignal & O. Pascal in Novon*, 9(1): 66 (1999)—Comoros (Mayotte).

NOTHERIA *P. O'Byrne & J.J. Vermeulen in Gard. Bull. Singapore*, 52(2): 286 (2000). *ORCHIDACEAE*
diaphana *P. O'Byrne & J.J. Vermeulen in Gard. Bull. Singapore*, 55(2): 286 (2000)—Sulawesi.

NOTHOBACCAUREA *R.M.A.P. Haegens in Blumea*, Suppl. 12: 198 (2000). *EUPHORBIACEAE*
pulvinata *(A.C. Sm.) R.M.A.P. Haegens in Blumea*, Suppl. 12: 199 (2000): Baccaurea pulvinata.
stylaris *(Müll. Arg.) R.M.A.P. Haegens in Blumea*, Suppl. 12: 201 (2000): Baccaurea stylaris.

NOTHOFAGUS (Fagac.).
× apiculata (Colenso) Cockayne
nothovar. blairii *(Kirk) R. Govaerts in R. Govaerts & D.G. Frodin, World Checklist & Bibliogr. Fagales:* 193 (1998): Fagus × blairii.
nothovar. dubia *(Kirk) R. Govaerts in R. Govaerts & D.G. Frodin, World Checklist & Bibliogr. Fagales:* 193 (1998): Fagus × apiculata var. dubia.
macrocarpa *(A.DC.) F.M. Vázquez & R.A. Rodríguez in Bot. J. Linn. Soc.*, 129(1): 81 (1999): F. obliqua var. macrocarpa.

NOTHOFAGUS:-
obliqua (Mirb.) Oerst.
subsp. andina *F.M. Vázquez & R.A. Rodríguez in Bot. J. Linn. Soc.*, 129(1): 80 (1999)—Chile South (Bíobío, La Araucania, Los Lagos), Chile North (Maule).
subsp. valdiviana *(Phil.) F.M. Vázquez & R.A. Rodríguez in Bot. J. Linn. Soc.*, 129(1): 79 (1999): Fagus valdiviana.
rutila *P. Ravenna in Onira, Bot. Leafl.*, 5(1): 1 (2000)—Chile North (Santiago).

NOTHOPEGIA (Anacardiac.).
monadelpha *(Roxb.) L.L. Forman in Kew Bull.*, 53(3): 754 (1998): Blackburnia monadelpha.
racemosa (Dalzell) Ramamoorthy
var. angustifolia *(Gamble) V. Chithra in N.C. Nair & A.N. Henry, Fl. Tamil Nadu, India, Ser. 1*, 1: 88 (1983): N. dalzellii var. angustifolia.
vajravelui *K. Ravikumar & V. Lakshmanan in Rheedea*, 9(1): 56 (1999)—India (Tamil Nadu).

NOTHOSCORDUM (Alliac.).
andinum (Poepp.) Fuentes
f. luteum *P. Ravenna in Onira, Bot. Leafl.*, 4(4): 17 (2000)—Chile North (Valparaíso).
aparadense *P. Ravenna in Onira, Bot. Leafl.*, 5(9): 36 (2001)—Brazil South (Santa Catarina).
gracile (Dryand. ex Aiton) Stearn
var. macrostemon *(Kunth) E.R. Guaglianone in Hickenia*, 2(31): 137 (1995): N. macrostemon.
jaibanum *P. Ravenna in Onira, Bot. Leafl.*, 4(14): 55 (2000)—Brazil Southeast (Minas Gerais).
saltense *P. Ravenna in Onira, Bot. Leafl.*, 4(14): 54 (2000)—Argentina Northwest (Salta, Catamarca).

NOTOCACTUS (Cactac.).
sect. Brasilicactus *(Backeb.) N. Gerloff, J. Neduchal & S. Stuchlik, Notokakteen:* 145 (1995), without basionym ref.; *cf. Repert. Pl. Succ. (I.O.S.)*, 46: 13 (1995 publ 1996): Basionym not stated.
subgen. Eonotocactus *A.B. Doweld in Sukkulenty (Moscow)*, 1999(2): 23 (1999).
sect. Eriocactus *(Backeb.) N. Gerloff, J. Neduchal & S. Stuchlik, Notokakteen:* 136 (1995), without basionym ref.; *cf. Repert. Pl. Succ. (I.O.S.)*, 46: 13 (1995 publ. 1996): Basionym not stated.
ser. Holkospermae *Prestlé in Minimus*, 27(4): 39 (1996), without latin descr. or type; *cf. Repert. Pl. Succ. (I.O.S.)*, 47: 10 (1996 publ. 1997).
ser. Leiospermae *Prestlé in Minimus*, 27(4): 39 (1996), without latin descr. or type; *cf. Repert. Pl. Succ. (I.O.S.)*, 47: 10 (1996 publ. 1997).
ser. Linkianae *A.B. Doweld in Sukkulenty (Moscow)*, 3(1–2): 55 (2000).
sect. Neonotocactus *(Backeb.) N. Gerloff, J. Neduchal & S. Stuchlik, Notokakteen:* 112 (1995), without basionym ref.; *cf. Repert. Pl. Succ. (I.O.S.)*, 46: 13 (1995 publ. 1996): Basionym not stated.
ser. Pampocactus *A.B. Doweld in Sukkulenty (Moscow)*, 1999(2): 23 (1999).
ser. Praeparodia *A.B. Doweld in Sukkulenty (Moscow)*, 3(1–2): 55 (2000).
subgen. Scopacactus *A.B. Doweld in Sukkulenty (Moscow)*, 3(1–2): 55 (2000).
sect. Wigginsia *(D.M. Porter) N. Gerloff, J. Neduchal & S. Stuchlik, Notokakteen:* 149 (1995), without basionym ref.; *cf. Repert. Pl. Succ. (I.O.S.)*, 46: 13 (1995 publ. 1996): Basionym not stated.
alacriportanus
f. hamatacanthus *K.H. Prestlé in Internoto*, 17(4): 124 (1996), without type—Brazil South (Rio Grande do Sul).
arachnitis
f. darilhoensis *N. Gerloff in Internoto*, 18(4): 119 (1997)—Brazil South (Rio Grande do Sul).
brevihamatus
f. albispinosus *K.H. Prestlé in Internoto*, 17(4): 132 (1996), without type—Brazil South (Rio Grande do Sul).
concinnus (Monv.) A. Berger
subsp. agnetae *(Vliet) A.B. Doweld in Sukkulenty (Moscow)*, 3(1–2): 55 (2000): Notocactus agnetae.

NOTOCACTUS:-
concinnus
subsp. multicostatus *(Buining & Brederoo) A.B. Doweld in Sukkulenty (Moscow)*, 3(1–2): 55 (2000): Notocactus multicostatus.
var. pororensis *M. Dumon in Internoto*, 17(3): 100 (1996), without latin descr. or type—Uruguay.
erythracanthus *H. Schlosser & A.J. Brederoo in Kakt. And. Sukk.*, 36(9): 189 (1985)—Uruguay.
gracilis *W. Gemmrich in Internoto*, 17(3): 98 (1996), nom. provis. without latin descr. or type—Brazil South (Rio Grande do Sul).
langsdorfii
f. eddiewarasii *Gerloff in Minimus*, 25(2): 5–9 (1996); *cf. Repert. Pl. Succ. (I.O.S.)*, 47: 10 (1996 publ. 1997)—Locality not stated.
f. multiceps *(Hofacker & K.Herm) J. Neduchal in Minimus*, 30(4): 92 (1999); *cf. Repert. Pl. Succ. (I.O.S.)*, 50: 13 (1999 publ. 2000): Parodia langsdorfii subsp. multiceps.
var. pulvinatus *(Vliet) J. Neduchal in Minimus*, 30(2): 28 (1999); *cf. Repert. Pl. Succ. (I.O.S.)*, 50: 13 (1999 publ. 2000): N. pulvinatus.
minusculus *A. Hofacker & K. Herm in Internoto*, 17(3): 75 (1996)—Brazil South (Rio Grande do Sul).
neoarechavaletae
f. kovarikii *(Frič) J. Neduchal in Minimus*, 30(2): 29 (1999); *cf. Repert. Pl. Succ. (I.O.S.)*, 50: 13 (1999 publ. 2000): Malacocarpus kovarikii.
ottonis (Lehm.) A. Berger
subsp. horstii *(F. Ritter) A.B. Doweld in Sukkulenty (Moscow)*, 3(1–2): 55 (2000): Notocactus arechavaletae var. horstii.
subvar. rubrispinus *Prestlé in Minimus*, 27(2): 25–27 (1996), without latin descr. or type; *cf. Repert. Pl. Succ. (I.O.S.)*, 47: 10 (1996 publ. 1997)—Locality not stated.
oxycostatus Buining & Brederoo
subsp. gracilis *(F. Ritter) A.B. Doweld in Sukkulenty (Moscow)*, 3(1–2): 54 (2000): Notocactus glaucinus var. gracilis.
rudibuenekeri W.R. Abraham
subsp. glomeratus *(N. Gerloff) A.B. Doweld in Sukkulenty (Moscow)*, 3(1–2): 55 (2000): Notocactus glomeratus.
schumannianus (Nicolai) Frič
f. ampliocostatus *(F. Ritter) J. Neduchal in Internoto*, 18(3): 111 (1997): Eriocactus ampliocostatus.
subsp. claviceps *(F. Ritter) A.B. Doweld in Sukkulenty (Moscow)*, 1999(2): 23 (1999): Eriocactus claviceps.
var. grossei *(K. Schum.) J. Neduchal in Internoto*, 18(3): 111 (1997): Echinocactus grossei.
f. nigrispinus *(K. Schum.) J. Neduchal in Internoto*, 18(3): 111 (1997): Echinocactus nigrispinus.
scopa (Spreng.) A. Berger
subsp. marchesii *(W.R. Abraham) A.B. Doweld in Sukkulenty (Moscow)*, 3(1–2): 55 (2000): Notocactus scopa var. marchesii.
subsp. neobuenekeri *(F. Ritter) A.B. Doweld in Sukkulenty (Moscow)*, 3(1–2): 55 (2000): Notocactus neobuenekeri.
subsp. succineus *(F. Ritter) A.B. Doweld in Sukkulenty (Moscow)*, 3(1–2): 56 (2000): Notocactus succineus.
sellowii
var. beltranii *(Frič) J. Neduchal in Minimus*, 30(2): 30 (1999), basionym not validly publ.; *cf. Repert. Pl. Succ. (I.O.S.)*, 50: 14 (1999 publ. 2000): Malacocarpus beltranii.
var. bezrucii *(Frič) J. Neduchal in Minimus*, 30(2): 30 (1999); *cf. Repert. Pl. Succ. (I.O.S.)*, 50: 14 (1999 publ. 2000): Malacocarpus bezrucii.
f. centrispinus *(Frič) J. Neduchal in Minimus*, 30(2): 31 (1999); *cf. Repert. Pl. Succ. (I.O.S.)*, 50: 13 (1999 publ. 2000): Malacocarpus bezrucii var. centrispinus.
f. cornifer *(Frič) J. Neduchal in Minimus*, 30(2): 32 (1999); *cf. Repert. Pl. Succ. (I.O.S.)*, 50: 13 (1999 publ. 2000): Malacocarpus bezrucii var. cornifer.
f. tephracanthus *(Link & Otto) J. Neduchal in Minimus*, 30(2): 32 (1999); *cf. Repert. Pl. Succ. (I.O.S.)*, 50: 14 (1999 publ. 2000): Echinocactus tephracanthus.

NOTOCACTUS:-
sessiliflorus
f. tetracanthus *(Lem.) J. Neduchal in Minimus,* 30(2): 33 (1999); *cf. Repert. Pl. Succ. (I.O.S.),* 50: 14 (1999 publ. 2000): Echinocactus tetracanthus.
f. vorwerkianus *(Werderm.) J. Neduchal in Minimus,* 30(2): 34 (1999); *cf. Repert. Pl. Succ. (I.O.S.),* 50: 14 (1999 publ. 2000): Echinocactus vorwerkianus.
tabularis (Cels ex Rümpler) A. Berger
subsp. muricatus *(Doweld) A.B. Doweld in Sukkulentý (Moscow),* 3(1–2): 55 (2000): Peronocactus tabularis subsp. muricatus.
tetracanthus *(Lem.) N. Gerloff, J. Neduchal & S. Stuchlik, Notokakteen:* 167 (1995), contrary to Art. 34.1a ICBN (1994); *cf. Repert. Pl. Succ. (I.O.S.),* 46: 13 (1995 publ. 1996): Echinocactus tetracanthus.
turbinatus
f. schaeferianus *(W.R. Abraham & J. Theun.) J. Neduchal in Minimus,* 30(2): 35 (1999); *cf. Repert. Pl. Succ. (I.O.S.),* 50: 14 (1999 publ. 2000): Wigginsia schaeferiana.

NOTODANTHONIA (Gramin.).
bonthainica *(Jansen) H.P. Linder in Telopea,* 6(4): 615 (1996): Danthonia pilosa var. bonthainica.
diemenica *(D.I. Morris) H.P. Linder in Telopea,* 6(4): 616 (1996): Danthonia diemenica.
fulva *(Vickery) H.P. Linder in Telopea,* 6(4): 616 (1996): Danthonia linkii var. fulva.
linkii *(Kunth) H.P. Linder in Telopea,* 6(4): 616 (1996): Danthonia linkii.
mera *(Connor & Edgar) H.P. Linder in Telopea,* 6(4): 616 (1996): Rytidosperma merum.
oreophila *(H.P. Linder & N.G. Walsh) H.P. Linder in Telopea,* 6(4): 616 (1996): Rytidosperma oreophilum.
popinensis *(D.I. Morris) H.P. Linder in Telopea,* 6(4): 616 (1996): Danthonia popinensis.
remota *(D.I. Morris) H.P. Linder in Telopea,* 6(4): 617 (1996): Danthonia remota.

NOTOPTERYGIUM (Umbellif.).
sect. Tenuifolium *F.T. Pu in Acta Phytotax. Sin.,* 38(5): 434 (2000).
forbesii H. Boissieu
subsp. oviforme *(R.H. Shan) F.T. Pu in Acta Phytotax. Sin.,* 38(5): 433 (2000): N. oviforme.
tenuifolium *M.L. Sheh & F.T. Pu in J. Pl. Resourc. Environ.,* 6(2): 41 (1997), with two types cited; *et in Acta Phytotax. Sin.,* 38(5): 435 (2000), with holotype as 'lectotype' indicated—China South-central (Sichuan).
weberbauerianum *(Fedde ex H. Wolff) M.G. Pimenov & E.V. Kljuykov in Feddes Repert.,* 110(7–8): 485 (1999): Ligusticum weberbauerianum.

NOTOTRICHIUM (Amaranthac.).
divaricatum *D.H. Lorence in Novon,* 6(1): 64 (1996)—Hawaiian Is.

NOTYLIA (Orchidac.).
leucantha *G.A. Salazar in An. Inst. Biol. Univ. Nac. Autón. Méx., Bot.,* 70(1): 9 (1999), nom. nov.: N. multiflora *Hook.*

NOTYLIOPSIS *P. Ortiz V. in Orquideologia,* 20(2): 184 (1996). *ORCHIDACEAE*
beatricis *P. Ortiz V.in Orquideologia,* 20(2): 184 (1996)—Colombia.

NUPHAR (Nymphaeac.).
sect. Astylus *D.J. Padgett in Sida,* 18(3): 824 (1999).
advena (Aiton) W.T. Aiton
subsp. orbiculata *(Small) D.J. Padgett in Sida,* 18(3): 824 (1999): Nymphaea orbiculata.
subsp. ozarkana *(G.S. Mill. & Standl.) D.J. Padgett in Sida,* 18(3): 825 (1999): Nymphaea ozarkana.
subsp. ulvacea *(G.S. Mill. & Standl.) D.J. Padgett in Sida,* 18(3): 825 (1999): Nymphaea ulvacea.
pumila (Timm) DC.
subsp. oguraensis *(Miki) D.J. Padgett in Sida,* 18(3): 825 (1999): Nuphar oguraensis.
subsp. sinensis *(Hand.-Mazz.) D.J. Padgett in Sida,* 18(3): 825 (1999): Nuphar sinensis.

NYLANDTIA (Polygalac.).
scoparia *(Eckl. & Zeyh.) P. Goldblatt & J.C. Manning in Strelitzia,* 9: 715 (2000): Mundtia scoparia.

NYMPHAEA (Nymphaeac.).
sect. Chamaenymphaea *(Planch.) J.H. Wiersema in Brittonia,* 48(4): 524 (1996 publ. 1997): N. b. Chamaenymphaea.
sect. Xanthantha *(Casp.) J.H. Wiersema in Brittonia,* 48(4): 523 (1996 publ. 1997): N. c. Xanthantha.
tetragona Georgi
var. wenzelii *(Maack) V.N. Voroschilov in A.K. Skvortsov (ed.), Florist. issl. v razn. raĭonakh SSSR:* 167 (1985): N. wenzelii.

NYMPHOIDES (Menyanthac.).
spinulosperma *H.I. Aston in Muelleria,* 10: 21 (1997)—Victoria, Queensland, New South Wales.

O

OBERNA (Caryophyllac.).
aeoniopsis *(Bornm.) N.M. Fedoronchuk in Ukr. Bot. Zhurn.,* 54(1): 74 (1997); *et in Ukr. Bot. Zhurn.,* 54(2): 179 (1997): Silene aeoniopsis.
anatolica *(Melzh. & A. Baytop) N.M. Fedoronchuk in Ukr. Bot. Zhurn.,* 54(1): 74 (1997); *et in Bot. Zhurn.,* 54(2): 179 (1997): Silene anatolica.
behen (L.) Ikonn.
var. carpatica *(Zapał.) N.M. Fedoronczuk in Ukr. Bot. Zhurn.,* 57(1): 26 (2000): Silene venosa var. carpatica.
subsp. littoralis *(Rupr.) N.N. Tzvelev in Novosti Sist. Vyssh. Rast.,* 32: 183 (2000): Silene inflata var. littoralis.

OBERONIA (Orchidac.).
agastyamalayana *C. Sathish Kumar in C. Sathish Kumar & K.S. Manilal, Cat. Indian Orchids:* 57 (1994)—India (Kerala).
anthropophora Lindl.
var. arisanensis *(Hayata) T. Hashimoto in Proc. 12th World Orchid Conf., 1987 (eds. K. Saito & R. Tanaka):* 122 (1987): O. arisanensis.
arunachalensis *A. Nageswara Rao in Rheedea,* 7(2): 130 (1997)—Assam (Arunachal Pradesh).
brevifolia *(Hook. f.) G. Panigrahi in S.K. Murti & G. Panigrahi, Fl. Bilaspur Distr.,* 2: 591 (1999), homonym; *see Lindl., Gen. Sp. Orchid. Pl.:* 16 (1830): O. iridifolia var. brevifolia.
katakiana *A. Nageswara Rao in J. Econ. Taxon. Bot.,* 20(3): 711 (1996)—Assam (Arunachal Pradesh).
longifolia *M. Kumar & S. Sequiera in J. Orchid Soc. India,* 12(1–2): 29 (1998), homonym; *see Ridl. in J. Straits Branch Roy. Asiat. Soc.,* 50: 127 (1908)—India (Kerala).
mucronata *(D. Don) P. Ormerod & G. Seidenfaden in G. Seidenfaden, Contrib. Orchid Fl. Thailand XIII:* 20 (1997): Stelis mucronata.
nepalensis *L.R. Shakya & R.P. Chaudhary in Harvard Pap. Bot.,* 4(1): 359 (1999)—Nepal.
pakshipadalensis *M. Kumar & S. Sequiera in J. Orchid Soc. India,* 12(1–2): 31 (1998)—India (Kerala).
pantlingiana *L.R. Shakya & R.P. Chaudhary in Harvard Pap. Bot.,* 4(1): 360 (1999)—Sikkim.
raoi *L.R. Shakya & R.P. Chaudhary in Rheedea,* 10(1): 57 (2000), as 'raoii'—Assam (Meghalaya).
smisrae *G. Panigrahi in S.K. Murti & G. Panigrahi, Fl. Bilaspur Distr.,* 2: 593 (1999), nom. nov.: Cymbidium iridifolium *Roxb.*
var. angustifolia *(Hook. f.) G. Panigrahi in S.K. Murti & G. Panigrahi, Fl. Bilaspur Distr.,* 2: 593 (1999): O. iridifolia var. angustifolia.
vieillardii *(Rchb. f.) M.A. Clements & D.L. Jones in Orchadian,* 13(1): 35 (1999): Malaxis vieillardii.

OBERONIOIDES *D.L. Szlachetko in Fragm. Flor. Geobot.,* Supp. 3: 134 (1995). *ORCHIDACEAE*
microtatantha *(Schltr.) D.L. Szlachetko in Fragm. Flor. Geobot.,* Supp. 3: 135 (1995): Microstylis microtatantha.

OBERONIOIDES:-
oberoniiflora *(Seidenf.) D.L. Szlachetko in Fragm. Flor. Geobot.,* Supp. 3: 135 (1995): Malaxis oberoniiflora.

OBTEGOMERIA *A. Doroszenko & P.D. Cantino in Novon,* 8(1): 2 (1998). *LABIATAE*
caerulescens *(Benth.) A. Doroszenko & P.D. Cantino in Novon,* 8(1): 2 (1998): Hedeoma caerulescens.

OCHLANDRA (Gramin.).
soderstromiana *M. Kumar & S. Sequiera in Rheedea,* 9(1): 33 (1999)—India (Kerala).
spirostylis *M. Kumar, K.K. Seethalakshmi & S. Sequiera in Rheedea,* 9(1): 31 (1999)—India (Kerala).

OCHROSPERMA (Myrtac.).
obovatum *A.R. Bean in Austrobaileya,* 5(3): 499 (1999)—Queensland.
sulcatum *A.R. Bean in Austrobaileya,* 4(4): 648 (1997)—Northern Territory.

OCHYRELLA *D.L. Szlachetko & R. Gonzalez Tamayo in Fragm. Flor. Geobot.,* 41(2): 698 (1996). *ORCHIDACEAE*
brachycentron *(Szlach.) D.L. Szlachetko & R. Gonzalez Tamayo in Fragm. Flor. Geobot.,* 41(2): 700 (1996): Eltroplectris brachycentron.
dalessandroi *(Dodson) D.L. Szlachetko & R. Gonzalez Tamayo in Fragm. Flor. Geobot.,* 41(2): 700 (1996): Eltroplectris dalessandroi.
lurida *(M.N. Correa) D.L. Szlachetko & R. Gonzalez Tamayo in Fragm. Flor. Geobot.,* 41(2): 700 (1996): Centrogenium luridum.
misera *(Kraenzl.) D.L. Szlachetko & R. Gonzalez Tamayo in Fragm. Flor. Geobot.,* 41(2): 700 (1996): Spiranthes misera.
triloba *(Lindl.) D.L. Szlachetko & R. Gonzalez Tamayo in Fragm. Flor. Geobot.,* 41(2): 700 (1996): Pelexia triloba.

OCIMUM (Labiat.).
subsect. Erythrochlamys *(Gürke) A.J. Paton in R. Hiltunen & Y. Holm (eds.), Basil, Genus Ocimum:* 15 (1999): gen. Erythrochlamys.
sect. Gratissima *(Benth.) A.J. Paton in R. Hiltunen & Y. Holm (eds.), Basil, Genus Ocimum:* 15 (1999): O. subsect. Gratissima.
subgen. Gymnocimum *(Benth.) A.J. Paton in R. Hiltunen & Y. Holm (eds.), Basil, Genus Ocimum:* 17 (1999): O. sect. Gymnocimum.
sect. Hiantia *(Benth.) A.J. Paton in R. Hiltunen & Y. Holm (eds.), Basil, Genus Ocimum:* 15 (1999): O. subsect. Hiantia.
ser. Hiantia *(Benth.) A.J. Paton in R. Hiltunen & Y. Holm (eds.), Basil, Genus Ocimum:* 16 (1999): O. subsect. Hiantia.
ser. Monobecium *(Sebald) A.J. Paton in R. Hiltunen & Y. Holm (eds.), Basil, Genus Ocimum:* 16 (1999): Becium subgen. Monobecium.
*subgen.*Nautochilus *(Bremek.) A.J. Paton in R. Hiltunen & Y. Holm (eds.), Basil, Genus Ocimum:* 16 (1999): gen. Nautochilus.
ser. Serpyllifolia *(Sebald) A.J. Paton in R. Hiltunen & Y. Holm (eds.), Basil, Genus Ocimum:* 16 (1999), as 'Serpyllifolium': Becium sect. Serpyllifolia.
albostellatum *(Verdc.) A.J. Paton in R. Hiltunen & Y. Holm (eds.), Basil, Genus Ocimum:* 30 (1999): Becium albostellatum.
amicorum *A.J. Paton in R. Hiltunen & Y. Holm (eds.), Basil, Genus Ocimum:* 29 (1999), nom. nov.: Becium fastigiatum *A.J. Paton.*
canescens *A.J. Paton in R. Hiltunen & Y. Holm (eds.), Basil, Genus Ocimum:* 34 (1999), nom. nov.: Becium virgatum *A.J. Paton.*
coddii *(S.D. Will. & K. Balkwill) A.J. Paton in R. Hiltunen & Y. Holm (eds.), Basil, Genus Ocimum:* 33 (1999): Becium coddii.
dambicola *A.J. Paton in R. Hiltunen & Y. Holm (eds.), Basil, Genus Ocimum:* 33 (1999), nom. nov.: O. punctatum *Baker.*
dhofarense *(Sebald) A.J. Paton in R. H35* (1999): Becium dhofarense.
dolomiticola *A.J. Paton in R. Hiltunen & Y. Holm (eds.), Basil, Genus Ocimum:* 32 (1999), nom. nov.: Becium citriodorum *S.D. Will. & K. Balkwill.*

OCIMUM:-
ericoides (P.A. Duvign. & Plancke) A.J. Paton in R. Hiltunen & Y. Holm (eds.), Basil, Genus Ocimum: 31 (1999): Becium ericoides.
fimbriatum Briq.
var. angustilanceolatum (De Wild.) A.J. Paton in R. Hiltunen & Y. Holm (eds.), Basil, Genus Ocimum: 28 (1999): Ocimum angustilanceolatum.
var. bequaertii (De Wild.) A.J. Paton in R. Hiltunen & Y. Holm (eds.), Basil, Genus Ocimum: 28 (1999): Becium bequaertii.
var. ctenodon (Gilli) A.J. Paton in R. Hiltunen & Y. Holm (eds.), Basil, Genus Ocimum: 28 (1999): Becium ctenodon.
var. microphyllum (Sebald) A.J. Paton in R. Hiltunen & Y. Holm (eds.), Basil, Genus Ocimum: 29 (1999): Becium fimbriatum var. microphyllum.
fruticosum (Ryding) A.J. Paton in R. Hiltunen & Y. Holm (eds.), Basil, Genus Ocimum: 27 (1999): Erythrochlamys fruticosa.
grandiflorum Lam.
subsp. densiflorum (A.J. Paton) A.J. Paton in R. Hiltunen & Y. Holm (eds.), Basil, Genus Ocimum: 31 (1999): Becium grandiflorum subsp. densiflorum.
subsp. turkanaense (Sebald) A.J. Paton in R. Hiltunen & Y. Holm (eds.), Basil, Genus Ocimum: 31 (1999), without basionym ref.: Basionym not stated.
hirsutissimum (P.A. Duvign.) A.J. Paton in R. Hiltunen & Y. Holm (eds.), Basil, Genus Ocimum: 32 (1999): Becium hirsutissimum.
labiatum (N.E. Br.) A.J. Paton in R. Hiltunen & Y. Holm (eds.), Basil, Genus Ocimum: 35 (1999): Orthosiphon labiatus.
mearnsii (Ayob. ex Sebald) A.J. Paton in R. Hiltunen & Y. Holm (eds.), Basil, Genus Ocimum: 32 (1999): Becium grandiflorum var. mearnsii.
metallorum (P.A. Duvign.) A.J. Paton in R. Hiltunen & Y. Holm (eds.), Basil, Genus Ocimum: 30 (1999): Becium metallorum.
minutiflorum (Sebald) A.J. Paton in R. Hiltunen & Y. Holm (eds.), Basil, Genus Ocimum: 30 (1999): Becium minutiflorum.
mitwabense (Ayob.) A.J. Paton in R. Hiltunen & Y. Holm (eds.), Basil, Genus Ocimum: 32 (1999): Becium mitwabense.
monocotyloides (Ayob.) A.J. Paton in R. Hiltunen & Y. Holm (eds.), Basil, Genus Ocimum: 31 (1999): Becium monocotyloides.
obovatum E. Mey. ex Benth.
subsp. cordatum (A.J. Paton) A.J. Paton in R. Hiltunen & Y. Holm (eds.), Basil, Genus Ocimum: 33 (1999): Becium obovatum subsp. cordatum.
subsp. crystallinum (A.J. Paton) A.J. Paton in R. Hiltunen & Y. Holm (eds.), Basil, Genus Ocimum: 34 (1999): Becium obovatum subsp. crystallinum.
var. galpinii (Gürke) A.J. Paton in R. Hiltunen & Y. Holm (eds.), Basil, Genus Ocimum: 33 (1999): Ocimum galpinii.
pseudoserratum (M. Ashby) A.J. Paton in R. Hiltunen & Y. Holm (eds.), Basil, Genus Ocimum: 35 (1999): Orthosiphon pseudoserratus.
pyramidatum (A.J. Paton) A.J. Paton in R. Hiltunen & Y. Holm (eds.), Basil, Genus Ocimum: 34 (1999): Becium pyramidatum.
reclinatum (S.D. Will. & K. Balkwill) A.J. Paton in R. Hiltunen & Y. Holm (eds.), Basil, Genus Ocimum: 33 (1999): Becium reclinatum.
serratum (Schltr.) A.J. Paton in R. Hiltunen & Y. Holm (eds.), Basil, Genus Ocimum: 35 (1999): Orthosiphon serratus.
spectabile (Gürke) A.J. Paton in R. Hiltunen & Y. Holm (eds.), Basil, Genus Ocimum: 27 (1999): Erythrochlamys spectabilis.
tubiforme (R.D. Good) A.J. Paton in R. Hiltunen & Y. Holm (eds.), Basil, Genus Ocimum: 35 (1999): Orthosiphon tubiformis.
vandenbrandei (Ayob.) A.J. Paton in R. Hiltunen & Y. Holm (eds.), Basil, Genus Ocimum: 34 (1999): Becium vandenbrandei.
vanderystii (De Wild.) A.J. Paton in R. Hiltunen & Y. Holm (eds.), Basil, Genus Ocimum: 30 (1999): Becium vanderystii.

OCIMUM:-
vihpyense A.J. Paton in R. Hiltunen & Y. Holm (eds.), Basil, Genus Ocimum: 31 (1999), nom. nov.: Becium grandiflorum var. frutescens Sebald.
waterbergense (S.D. Will. & K. Balkwill) A.J. Paton in R. Hiltunen & Y. Holm (eds.), Basil, Genus Ocimum: 32 (1999): Becium waterbergense.

OCOTEA (Laurac.).
betazensis (Mez) H. van der Werff in Novon, 9(4): 572 (1999): Phoebe betazensis.
bourgeauviana (Mez) H. van der Werff in Novon, 9(4): 574 (1999): Phoebe bourgeauviana.
cinerea H. van der Werff in Novon, 10(3): 268 (2000)—Brazil North (Amazonas, Amapá), French Guiana, Peru.
congregata H. van der Werff in Novon, 9(4): 574 (1999)—Mexico Southeast (Chiapas).
corrugata H. van der Werff in Novon, 9(4): 574 (1999)—Mexico Southwest (Oaxaca).
delicata A. Vicentini in Novon, 10(3): 269 (2000)—Brazil North (Amazonas, Pará).
diffusa H. van der Werff in Brittonia, 49(2): 189 (1997)—French Guiana, Brazil North (Pará, Amapá).
gordonii H. van der Werff in Novon, 9(4): 575 (1999)—Panama.
grayi H. van der Werff in Novon, 6(4): 465 (1996)—Madagascar.
heribertoi T. Wendt in Lundellia, 1: 40 (1998)—Mexico Southwest (Oaxaca), Mexico Gulf (Veracruz).
ikonyokpe H. van der Werff in Novon, 6(4): 460 (1996)—Cameroon.
immersa H. van der Werff in Novon, 10(3): 272 (2000)—Peru, Brazil North (Amazonas).
ligulata H. van der Werff in Novon, 10(3): 273 (2000)—Brazil North (Amazonas).
longipedicellata H. van der Werff in Novon, 6(4): 465 (1996)—Madagascar.
malcomberi H. van der Werff in Novon, 6(4): 467 (1996)—Madagascar.
minor A. Vicentini in Novon, 10(3): 275 (2000)—Brazil North (Amazonas), Brazil Northeast (Maranhão).
morae J. Gómez-Laurito in Novon, 7(2): 145 (1997)—Costa Rica.
multiflora H. van der Werff in Novon, 6(4): 481 (1996)—Costa Rica.
nigrescens A. Vicentini in Novon, 10(3): 277 (2000)—Brazil North (Amazonas, Pará, Rondônia), Brazil West-central (Mato Grosso).
obliqua A. Vicentini in Novon, 10(3): 279 (2000)—Brazil North (Amazonas).
oligantha H. van der Werff in Novon, 6(4): 469 (1996)—Madagascar.
oppositifolia S. Yasuda in Novon, 6(4): 484 (1996)—Brazil Northeast (Bahia), Brazil Southeast (Minas Gerais).
patula H. van der Werff in Novon, 9(4): 577 (1999)—Costa Rica.
percurrens A. Vicentini in Novon, 10(3): 281 (2000)—Brazil North (Amazonas, Amapá, Pará), Brazil Northeast (Bahia).
praetermissa H. van der Werff in Novon, 6(4): 482 (1996)—Costa Rica, Panama.
purpurea (Mez) H. van der Werff in Novon, 9(4): 579 (1999): Phoebe purpurea.
rhodophylla A. Vicentini in Novon, 10(3): 283 (2000)—Brazil North (Amazonas, Roraima), Colombia, Peru, Venezuela.
rufescens H. van der Werff in Novon, 6(4): 479 (1996)—Costa Rica.
salvadorensis (Lundell) H. van der Werff in Novon, 6(4): 481 (1996): Nectandra salvadorensis.
sambiranensis H. van der Werff in Novon, 6(4): 470 (1996)—Madagascar.
scabrella H. van der Werff in Novon, 10(3): 287 (2000)—Brazil North (Pará, Amazonas, Rondónia), French Guiana.
subterminalis H. van der Werff in Novon, 10(3): 289 (2000)—Brazil North (Amapá, Amazonas, Pará), Brazil Northeast (Maranhão), French Guiana.
tonii (Lundell) H. van der Werff in Novon, 9(4): 579 (1999): Nectandra tonii.
tsaratananensis H. van der Werff in Novon, 6(4): 472 (1996)—Madagascar.

OCTOMERIA (Orchidac.).
excentrica C.A. Luer in Lindleyana, 14(2): 106 (1999)—Cuba.
ffrenchiana P. Feldmann & N. Barré in Lindleyana, 11(3): 199 (1996)—Leeward Is. (Guadeloupe).

OCTOPOMA (Aizoac.).
quadrisepalum (L. Bolus) H.E.K. Hartmann in Bradleya, 16: 73 (1998): Ruschia quadrisepala.
tetrasepalum (L. Bolus) H.E.K. Hartmann in Bradleya, 16: 74 (1998): Ruschia tetrasepala.

× **ODDYARA** hort. in Orchid Rev., 105(1215): 190 (1997). ORCHIDACEAE (Cochleanthes × Kefersteinia × Pescatorea).

ODONTADENIA (Apocynac.).
sect. Gracilipae J.F. Morales in Bull. Jard. Bot. Nation. Belg., 67(1–4): 424 (1999).
campanulata J.F. Morales in Bull. Jard. Bot. Nation. Belg., 67(1–4): 442 (1999)—Colombia.
markgrafiana J.F. Morales in Bull. Jard. Bot. Nation. Belg., 67(1–4): 455 (1999)—Brazil North (Amazonas), French Guiana.
matogrossana J.F. Morales in Bull. Jard. Bot. Nation. Belg., 67(1–4): 450 (1999)—Brazil West-central (Mato Grosso, Goiás).

ODONTITES (Scrophulariac.).
citrinus M. Bolliger in Willdenowia, 26(1–2): 157 (1996)—Tunisia.
discolor Pomel
subsp. ciliatus (Pomel) M. Bolliger in Willdenowia, 26(1–2): 147 (1996): O. ciliatus.
hispidulus (Boiss.) M. Bolliger in Willdenowia, 26(1–2): 104 (1996): O. luteus var. hispidulus.
linkii Boiss.
subsp. cyprius (Boiss.) M. Bolliger in Willdenowia, 26(1–2): 135 (1996): O. cyprius.
longiflora (Vahl) Webb
subsp. lateritia (Charpin & Fern. Casas) P. Sánchez-Gómez in P. Sánchez-Gómez et al., Fl. Murcia: Claves ident. pl. vasc.: 347 (1996): O. longiflora var. lateritia.
maroccanus M. Bolliger in Willdenowia, 26(1–2): 135 (1996), nom. nov.: O. violaceus var. brevifolius Emb. & Maire.
squarrosus (Salzm. ex Rchb.) M. Bolliger in Willdenowia, 26(1–2): 148 (1996): Euphrasia squarrosa.
subsp. foliosus (Pérez Lara) M. Bolliger in Willdenowia, 26(1–2): 151 (1996): O. foliosus.
verna (Bellardi) Dumort.
subsp. salina (Kotov) N. Andreev in Fl. Republ. Bŭlgariya, 10: 244 (1995): O. serotina var. salina.
viscosus (L.) Clairv.
subsp. lusitanicus M. Bolliger in Willdenowia, 26(1–2): 97 (1996)—Portugal.
vulgaris Moench
subsp. himalayicus (Pennell) M. Bolliger in Willdenowia, 26(1–2): 113 (1996): O. himalayicus.
subsp. mesatlanticus (Emb. & Maire) M. Bolliger in Willdenowia, 26(1–2): 111 (1996): O. mesatlanticus.
subsp. siculus (Guss.) M. Bolliger in Willdenowia, 26(1–2): 113 (1996): Euphrasia serotina var. sicula.

ODONTOCHILUS (Orchidac.).
pomrangianus (Seidenf.) D.L. Szlachetko in Fragm. Flor. Geobot., Supp. 3: 115 (1995): Anoectochilus pomrangianus.
reniformis (Hook. f.) P. Ormerod in Austral. Orchid Rev., 63(4): 10 (1998): Zeuxine reniformis.

ODONTOGLOSSUM (Orchidac.).
bachmannii I. Bock in Orchidee, 49(6): 265 (1998)—Colombia.
dracoceps S. Dalström in Lindleyana, 14(2): 87 (1999)—Bolivia.
fidicularium S. Dalström in Lindleyana, 14(3): 168 (1999)—Ecuador.
helgae W. Königer in Arcula, 7: 211 (1997)—Ecuador.
hoeijeri S. Dalström in Lindleyana, 14(3): 171 (1999)—Ecuador.

ODONTOGLOSSUM :-
lapacense *R. Vásquez Ch. & Dalström in Rev. Soc. Boliv. Bot.*, 2(1): 10 (1998)—Bolivia.
posadarum *W. Königer in Arcula*, 6: 158 (1996)—Colombia.
povedanum *P. Ortiz V. in Orquideologia*, 20(3): 321 (1997)—Colombia.
quadridentatum *D.E. Bennett & E.A. Christenson in Lindleyana*, 13(2): 76 (1998)—Peru.
tenuifolium *S. Dalström in Lindleyana*, 11(2): 114 (1996)—Bolivia.
vierlingii *K. Senghas in J. Orchideenfreund*, 7(2): 136 (2000)—Bolivia.

ODONTOPHORUS (Aizoac.).
pusillus *S.A. Hammer in Piante Grasse*, 15(4 Suppl.): 86 (1995 publ. 1996)—Cape Province.

ODONTORRHYNCHUS (Orchidac.).
domeykoanus *D.L. Szlachetko in Fragm. Flor. Geobot.*, 41(2): 853 (1996)—Chile North (Valparaíso).
erosus *D.L. Szlachetko in Fragm. Flor. Geobot.*, 41(2): 854 (1996)—Chile.

OECEOCLADES (Orchidac.).
ampullacea *(Roxb.) Lindl. ex J.O. Voigt, Hort. Suburb. Calcutt.*: 630 (1845): Aerides ampullaceum.
flexuosa *J. Lindley, Gen. Sp. Orchid. Pl.*: 236 (1833)—India.†

OEDINA (Loranthac.).
brevispicata *R.M. Polhill & D. Wiens, Mistletoes of Africa*: 228 (1998)—Tanzania.
congdoniana *R.M. Polhill & D. Wiens in Novon*, 7(3): 277 (1997)—Malawi, Tanzania.

OENANTHE (Umbellif.).
cyclocarpa *M.G. Pimenov & E.V. Kljuykov in Bot. Zhurn.*, 83(1): 142 (1998)—Turkey.
javanica (Blume) DC.
subsp. rosthornii *(Diels) F.T. Pu in Novon*, 8(1): 70 (1998): O. rosthornii.
thomsonii C.B. Clarke
subsp. stenophyllum *(Boissieu) F.T. Pu in Novon*, 8(1): 71 (1998): O. dielsii var. stenophylla.

OENOTHERA (Onagrac.).
arizonica *(Munz) W.L. Wagner in Novon*, 8(3): 308 (1998): O. deltoides var. arizonica.
bulgarica *D. Delipavlov in Thaiszia*, 8(2): 122 (1998)—Bulgaria.
elata Kunth
var. hirsutissima *(A. Gray ex S. Watson) A. Cronquist in A. Cronquist et al., Intermountain Fl., Vasc. Pl. Intermountain West*, U.S.A., 3A: 202 (1997): O. biennis var. hirsutissima.
moravica *V. Jehlík & K. Rostański in Folia Geobot. Phytotax.*, 30(4): 440 (1995)—Czechoslovakia.
nivertiana *(Goujon) N.E. Brown, Suppl. Johnson's Gard. Dict.*: 971 (1882): Godetia nivertiana.
renneri *H. Scholz in Wiss. Zeitschr. Pädag. Hochsch. Potsdam, Math.-Naturwiss. Reihe*, 2(2): 206 (1956)—Germany.
victorini R.R. Gates & Catches.
f. rostanskii *(V. Jehlík) V. Jehlík & K. Rostański in Folia Geobot. Phytotax.*, 30(4): 437 (1995): O. rostanskii.

OEONIA (Orchidac.).
curvata *J. Bosser in Adansonia*, 22(2): 233 (2000)—Madagascar.

OERSTEDELLA (Orchidac.).
hansenii *(C.D. Adams) M.A. Nir, Orchidaceae Antillanae*: 259 (2000): Epidendrum verrucosum var. hansenii.

OHBAEA *V.V. Byalt & I.V. Sokolova in Kew Bull.*, 54(2): 476 (1999), nom. nov.: Sedum subgen. Balfouria *H. Ohba*. (Crassulac.).
balfourii *(Raym.-Hamet) V.V. Byalt & I.V. Sokolova in Kew Bull.*, 54(2): 476 (1999), as 'balfouri': Sedum balfourii.

OHWIA *Hiroyoshi Ohashi in Sci. Rep. Tohoku Univ., Ser. 4*, 40(3): 243 (1999), nom. nov.: gen. Catenaria *Benth.* (Legumin.).

OHWIA:-
caudata *(Thunb.) Hiroyoshi Ohashi in Sci. Rep. Tohoku Univ., Ser. 4*, 40(3): 243 (1999): Hedysarum caudatum.
luteola *(H. Ohashi & T. Nemoto) Hiroyoshi Ohashi in Sci. Rep. Tohoku Univ., Ser. 4*, 40(3): 244 (1999): Desmodium luteolum.

OLDENLANDIA (Rubiac.).
erecta *(Manilal & Sivar.) R.R. Mill in Edinburgh J. Bot.*, 53(1): 104 (Feb. 1996): Hedyotis erecta.
erecta *(Manilal & Sivar.) C.E. Ridsdale in Blumea*, 41(2): 460 (27 Nov. 1996), isonym; *see R.R. Mill in Edinburgh J. Bot.*, 53(1): 104 (Feb. 1996): Hedyotis erecta.
gibsonii *D.A. Halford in Austrobaileya*, 5(2): 337 (1999)—Queensland.

OLEA (Oleac.).
borneensis *J.G. Boerlage, Handl. Fl. Ned. Ind.*, 2(2): 332 (1899)—Locality not stated.
gitingensis *(Elmer) R. Kiew in Gard. Bull. Singapore*, 51(1): 90 (1999): Linociera gitingensis.
moluccensis *R. Kiew in Gard. Bull. Singapore*, 51(1): 92 (1999)—Moluccas.
obovata *(Merr.) R. Kiew in Gard. Bull. Singapore*, 51(1): 94 (1999), homonym; *see Baker, Fl. Mauritius*: 219 (1877): Linociera obovata.
rubrovenia *(Elmer) R. Kiew in Gard. Bull. Singapore*, 51(1): 95 (1999): Linociera rubrovenia.

OLEARIA (Compos.).
sect. Divaricaster *M. Heads in Bot. J. Linn. Soc.*, 127(3): 254 (1998).
fimbriata *M. Heads in Bot. J. Linn. Soc.*, 127(3): 271 (1998)—New Zealand South I.
gardneri *M. Heads in Bot. J. Linn. Soc.*, 127(3): 265 (1998)—New Zealand North I.
stenophylla *N.G. Walsh in Muelleria*, 12(2): 223 (1999)—New South Wales.
virgata (Hook. f.) Hook. f.
subsp. centralis *M. Heads in Bot. J. Linn. Soc.*, 127(3): 279 (1998)—New Zealand North I.
subsp. implicita *(G. Simpson) M. Heads in Bot. J. Linn. Soc.*, 127(3): 275 (1998): O. virgata var. implicita.
subsp. serpentina *(G. Simpson) M. Heads in Bot. J. Linn. Soc.*, 127(3): 277 (1998): O. serpentina.

OLIGOSTACHYUM (Gramin.).
orthotropoides *W.T. Lin in J. S. China Agric. Univ.*, 14(1): 49 (1993)—China Southeast (Guangdong).

OLIMARABIDOPSIS *I.A. Al-Shehbaz, S.L. O'Kane & R.A. Price in Novon*, 9(3): 302 (1999). *CRUCIFERAE*
cabulica *(Hook. f. & Thomson) I.A. Al-Shehbaz, S.L. O'Kane & R.A. Price in Novon*, 9(3): 303 (1999): Sisymbrium cabulicum.
pumila *(Stephan) I.A. Al-Shehbaz, S.L. O'Kane & R.A. Price in Novon*, 9(3): 303 (1999): Sisymbrium pumilum.
umbrosa *(Botsch. & Vved.) I.A. Al-Shehbaz, S.L. O'Kane & R.A. Price in Novon*, 9(3): 303 (1999): Trichochiton umbrosum.

OLSYNIUM (Iridac.).
junceum (E. Mey. ex C. Presl) Goldblatt
subsp. colchaguense *(Phil.) J.M. Watson & A.R. Flores in A. Hoffmann, M.K. Arroyo, F. Liberona, M. Muñoz & J. Watson, Pl.* Altoandinas Fl. Silv. Chile: 240 (1998): Sisyrinchium colchaguense.
philippii (Klatt) Goldblatt
subsp. illapelinum *(Phil.) J.M. Watson & A.R. Flores in A. Hoffmann, M.K. Arroyo, F. Liberona, M. Muñoz & J. Watson, Pl.* Altoandinas Fl. Silv. Chile: 242 (1998): Sisyrinchium illapelinum.

OLYRA (Gramin.).
paniculata *O. Swartz, Prodr.*: 21 (1788)—Locality not stated.†

OMALOTHECA (Compos.).
hoppeana
subsp. magellensis *(Fiori & Paol.) J. Holub in Preslia*, 70(2): 108 (1998): Gnaphalium supinum f. magellense.

OMALOTHECA:-
nanchuensis *(Y. Ling & Y.Q. Tseng) J. Holub in Preslia*, 70(2): 108 (1998): Gnaphalium nanchuense.

OMPHALEA (Euphorbiac.).
ankaranensis *L.J. Gillespie in Novon*, 7(2): 128 (1997)—Madagascar.
oppositifolia *(Willd.) L.J. Gillespie in Novon*, 7(2): 133 (1997): Hecatea oppositifolia.

OMPHALOCARPUM (Sapotac.).
strombocarpum *Y.B. Harvey & J.C. Lovett in Kew Bull.*, 54(1): 198 (1999)—Tanzania.

ONCELLA (Loranthac.).
gracilis *Balle ex R.M. Polhill & D. Wiens, Mistletoes of Africa*: 233 (1998)—Tanzania.
schliebeniana *Balle ex R.M. Polhill & D. Wiens, Mistletoes of Africa*: 230 (1998)—Tanzania.

ONCIDIUM (Orchidac.).
andreae *W. Königer in Arcula*, 6: 162 (1996)—Colombia.
aspecum *W. Königer in Arcula*, 5: 133 (1996)—Peru.
aurorae *D.E. Bennett & E.A. Christenson in Lindleyana*, 13(2): 76 (1998)—Peru.
ayabacanum *D.E. Bennett & E.A. Christenson in Lindleyana*, 13(2): 78 (1998)—Peru.
bismarckii *D.E. Bennett & E.A. Christenson in Lindleyana*, 13(2): 78 (1998)—Peru.
chrysopteranthum *E. Lückel in Orchidee*, 49(1): 43 (1998), nom. nov.: O. micropogon var. chrysopterum *Rchb. f.*
erucatum *W. Königer in Arcula*, 5: 136 (1996)—Ecuador.
fragae *W. Königer in Arcula*, 9: 256 (1999)—Brazil Southeast (Rio de Janeiro).
hannelorae *M.A. Nir, Orchidaceae Antillanae*: 264 (2000)—Windward Is. (Dominica).
helgae *W. Königer in Arcula*, 8: 229 (1999)—Ecuador.
hirtzii *C.H. Dodson in Orquideologia*, 20(3): 276 (1997)—Ecuador.
hydrophilum Barb. Rodr.
f. immaculatum *(L.C. Menezes) E.A. Christenson in Lindleyana*, 11(1): 20 (1996): O. hydrophilum var. immaculatum.
imitans *R.L. Dressler in Novon*, 7(2): 120 (1997)—Costa Rica.
luerorum *C.H. Dodson in Orquideologia*, 20(1): 91 (1996)—Ecuador.
macasense *C.H. Dodson in Orquideologia*, 20(1): 92 (1996)—Ecuador.
maduroi *R.L. Dressler in Orchids*, 69(8): 762 (2000)—Panama.
miradorense *D.E. Bennett & E.A. Christenson in Lindleyana*, 13(2): 80 (1998)—Peru.
× monacranthum *Andreetta ex C.H. Dodson in Orquideologia*, 20(1): 93 (1996)—Ecuador. (O. macranthum × monachicum).
niesseniae *W. Königer in Arcula*, 6: 165 (1996)—Colombia.
ouricanense *V.P. Castro Neto & M.A. Campacci in Bol. CAOB*, 4(1): 17 (1992)—Brazil Northeast (Bahia).
palaciosii *C.H. Dodson in Orquideologia*, 20(1): 97 (1996)—Ecuador.
peltiforme *W. Königer in Arcula*, 8: 232 (1999)—Ecuador.
portillae *W. Königer in Arcula*, 10: 274 (2000)—Ecuador.
riopalenqueanum *C.H. Dodson in Orquideologia*, 20(1): 99 (1996)—Ecuador.
rodrigoi *W. Königer in Arcula*, 8: 235 (1999)—Colombia.
rutkisii *E. Foldats in Acta Bot. Venezuel.*, 19(1): 39 (1996)—Venezuela.
santiagoorum *W. Königer in Arcula*, 8: 238 (1999)—Ecuador.
sprucei *(Lindl.) W. Königer & D. Pongratz in Arcula*, 9: 265 (1999), intended combination not made: Oncidium sprucei.
subcruciforme *(A.H. Heller) G.A. Salazar in An. Inst. Biol. Univ. Nac. Autón. Méx., Bot.*, 70(1): 10 (1999), isonym; *see Garay & Stacy in Bradea*, 1(40): 408 (1974): Odontoglossum subcruciforme.
tsubotae *W. Königer in Arcula*, 6: 168 (1996)—Colombia.

ONCIDIUM:-
vasquezii *E.A. Christenson in Lindleyana*, 11(1): 21 (1996)—Bolivia.
williamsianum *C.H. Dodson in Orquideologia*, 20(1): 102 (1996)—Ecuador.

ONCOBA (Flacourtiac.).
bukobensis *(Gilg) S. Hul & F.J. Breteler in Adansonia*, 19(2): 256 (1997): Lindackeria bukobensis.
cuneatoacuminata *(De Wild.) S. Hul & F.J. Breteler in Adansonia*, 19(2): 257 (1997): O. dentata var. cuneatoacuminata.
kivuensis *(Bamps) S. Hul & F.J. Breteler in Adansonia*, 19(2): 258 (1997): Camptostylus kivuensis.
paludosa *(Benth.) S. Hul & F.J. Breteler in Adansonia*, 19(2): 259 (1997): Mayna paludosa.
paraensis *(Kuhlm.) S. Hul & F.J. Breteler in Adansonia*, 19(2): 259 (1997): Lindackeria paraensis.
schweinfurthii *(Gilg) S. Hul & F.J. Breteler in Adansonia*, 19(2): 260 (1997): Lindackeria schweinfurthii.
somalensis *(Chiov.) S. Hul & F.J. Breteler in Adansonia*, 19(2): 260 (1997): Lindackeria somalensis.
subtomentosa *(Gilg) S. Hul & F.J. Breteler in Adansonia*, 19(2): 260 (1997): Caloncoba subtomentosa.
suffruticosa *(Milne-Redh.) S. Hul & F.J. Breteler in Adansonia*, 19(2): 260 (1997): Paraphyadanthe suffruticosa.
tettensis (Klotzsch) Harv.
var. fissistyla *(Warb.) S. Hul & F.J. Breteler in Adansonia*, 19(2): 261 (1997): O. fissistyla.
var. kirkii *(Oliv.) S. Hul & F.J. Breteler in Adansonia*, 19(2): 261 (1997): O. kirkii.
var. macrophylla *(Klotzsch) S. Hul & F.J. Breteler in Adansonia*, 19(2): 260 (1997): Chlanis macrophylla.

ONCOCALYX (Loranthac.).
sect. Longicalyculati *(Engl.) D. Wiens & R.M. Polhill, Mistletoes of Africa:* 102 (1998): L. group. Longicalyculati.
cordifolius *D. Wiens & R.M. Polhill, Mistletoes of Africa:* 110 (1998)—Kenya.
doberae *(Schweinf.) D. Wiens & R.M. Polhill, Mistletoes of Africa:* 108 (1998), isonym; *see A.G. Mill. & J.A. Nyberg in Edinburgh J. Bot.*, 51(1): 39 (1994): Loranthus doberae.
quinquenervius *(Hochst.) D. Wiens & R.M. Polhill, Mistletoes of Africa:* 105 (1998): Loranthus quinquenervius.

ONCODOSTIGMA (Annonac.).
microflorum *H. Okada in Acta Phytotax. Geobot.*, 47(1): 8 (1996)—Sumatera.

ONCOSTEMA (Hyacinthac.).
cupanii *(Guss.) F. Speta in Phyton (Austria)*, 38(1): 110 (1998): Scilla cupanii.

ONOBRYCHIS (Legumin.).
sect. Insignes *(Širj.) K.H. Rechinger in Fl. Iran.*, 157: 460 (1984): O. subsect. Insignes.
sect. Laxiflorae *(Širj.) K.H. Rechinger in Fl. Iran.*, 157: 415 (1984): O. subsect. Laxiflorae.
afghanica Širj. & Rech. f.
subsp. brachycalyx *K.H. Rechinger in Fl. Iran.*, 157: 397 (1984)—Afghanistan.
subsp. codringtonii *(Širj. & Rech f.) K.H. Rechinger in Fl. Iran.*, 157: 398 (1984): O. afghanica var. codringtonii.
aliacmonia Rech. f.
subsp. peloponnesiaca *G. Iatroú & Kit Tan in Ann. Naturhist. Mus. Wien*, 98B(Suppl.): 308 (1996)—Greece.
aucheri Boiss.
subsp. psammophila *(Bornm.) K.H. Rechinger in Fl. Iran.*, 157: 447 (1984): O. psammophila.
subsp. teheranica *(Bornm.) K.H. Rechinger in Fl. Iran.*, 157: 445 (1984): O. teheranica.
cornuta (L.) Desv.
subsp. leptacantha *K.H. Rechinger in Fl. Iran.*, 157: 394 (1984), with two types cited—Afghanistan, Pakistan.

ONOBRYCHIS:-
freitagii *K.H. Rechinger in Fl. Iran.*, 157: 463 (1984)—Afghanistan.
gypsicola *K.H. Rechinger in Fl. Iran.*, 157: 436 (1984)—Iran.
iranshahrii *K.H. Rechinger in Fl. Iran.*, 157: 442 (1984)—Iran.
kermanensis *(Širj. & Rech. f.) K.H. Rechinger in Fl. Iran.*, 157: 441 (1984): O. melanotricha var. kermanensis.
laxiflora Baker
subsp. kabulica *K.H. Rechinger in Fl. Iran.*, 157: 419 (1984)—Afghanistan.
subsp. macrodonta *K.H. Rechinger in Fl. Iran.*, 157: 418 (1984)—Pakistan, Afghanistan.
subsp. shahrestanica *K.H. Rechinger in Fl. Iran.*, 157: 418 (1984)—Afghanistan.
subsp. taftanica *K.H. Rechinger in Fl. Iran.*, 157: 418 (1984)—Iran.
luristanica *K.H. Rechinger in Fl. Iran.*, 157: 409 (1984)—Iran.
mazanderanica *K.H. Rechinger in Fl. Iran.*, 157: 457 (1984)—Iran.
peloponnesiaca *(Iatroú & Kit Tan) G. Iatroú & Kit Tan in Willdenowia*, 29(1–2): 59 (1999): O. aliacmonia subsp. peloponnesiaca.
poikilantha *K.H. Rechinger in Fl. Iran.*, 157: 419 (1984)—Afghanistan.
samanganica *K.H. Rechinger in Fl. Iran.*, 157: 396 (1984)—Afghanistan.
shahpurensis *K.H. Rechinger in Fl. Iran.*, 157: 414 (1984)—Iran, Turkey.
sojakii *K.H. Rechinger in Fl. Iran.*, 157: 443 (1984)—Iran.
splendida *K.H. Rechinger & D. Podlech in Fl. Iran.*, 157: 462 (1984)—Afghanistan.
takhtajanii *A.K. Sytin in Bot. Zhurn.*, 85(7): 177 (2000)—Transcaucasus (Armenia).
talagonica *K.H. Rechinger in Fl. Iran.*, 157: 457 (1984)—Iran.

ONONIS (Legumin.).
cuatrecasasii *J.A. Devesa Alcaraz in Acta Bot. Malacitana*, 21: 192 (1996)—Morocco.
gines-lopezii *J.A. Devesa Alcaraz in Acta Bot. Malacitana*, 21: 192 (1996)—Morocco.
natrix L.
var. font-queri *Širj. ex O. Bolòs & J. Vigo in But. Inst. Catalana Hist. Nat.*, 38 Bot., 1: 72 (1974)—Baleares.
subsp. font-queri *(Širj. ex O. Bolòs & Vigo) A.M. Romo, Fl. Silvestres Baleares:* 145 (1994): O. natrix var. font-queri.
pusilla L.
var. calycina *(Rouy) J.A. Devesa Alcaraz & G. López González in An. Jard. Bot. Madrid*, 55(2): 259 (1997): O. columnae var. calycina.
talaverae *J.A. Devesa Alcaraz & G. López González in An. Jard. Bot. Madrid*, 55(2): 252 (1997)—Spain.
tridentata L.
subsp. angustifolia *(Lange) J.A. Devesa Alcaraz & G. López González in An. Jard. Bot. Madrid*, 55(2): 258 (1997): O. tridentata f. angustifolia.

ONOPORDUM (Compos.).
acanthophorum *M. Gandoger, Dec. Pl. Nov.*, 2: 27 (1876)—Corse.
macracanthum Schousb.
subsp. micropterum *(Pau) G. Mateo Sanz & M.B. Crespo Villalba, Fl. Abrev. Comun. Valenciana:* 430 (1995), as 'Onopordon': O. micropterum.
nivescens *M. Gandoger, Dec. Pl. Nov.*, 2: 27 (1876)—France.
transiens *M. Gandoger, Dec. Pl. Nov.*, 2: 28 (1876)—France.

ONOSMA (Boraginac.).
emodi Wall.
var. stelligera *(I.M. Johnst.) R.R. Mill in Edinburgh J. Bot.*, 53(1): 122 (1996): Maharanga emodi var. stelligera.
iranshahrii *A. Ghahreman & F. Attar in Iranian J. Bot.*, 7(1): 52 (1996)—Iran.
mirabilis *A.P. Khokhrjakov in Byull. Mosk. Obshch. Ispÿt. Prir., Biol.*, 102(4): 47 (1997)—Turkey.
stojanoffii *(Turrill) H. Teppner in Feddes Repert.*, 106(5–8): 528 (1995 publ. 1996): O. echioides var. stojanoffii.

ONYCHOSEPALUM (Restionac.).
microcarpum *K.A. Meney & J.S. Pate in Telopea*, 6(4): 664 (1996)—Western Australia.

OOCEPHALA *(S.B. Jones) H. Robinson in Proc. Biol. Soc. Wash.*, 112(1): 231 (1999): Vernonia subsect. Oocephalae. (Compos.).
agrianthoides *(C. Jeffrey) H. Robinson in Proc. Biol. Soc. Wash.*, 112(1): 232 (1999): Vernonia agrianthoides.
stenocephala *(Oliv.) H. Robinson in Proc. Biol. Soc. Wash.*, 112(1): 232 (1999): Vernonia stenocephala.

OPARANTHUS (Compos.).
sect. Albiflori *R.K. Shannon & W.L. Wagner in Allertonia*, 7(4): 282 (1997).
hivoanus *(O. Deg. & Sherff) R.K. Shannon & W.L. Wagner in Allertonia*, 7(4): 282 (1997): Bidens hivoana.
teikiteetinii *(Florence & Stuessy) R.K. Shannon & W.L. Wagner in Allertonia*, 7(4): 286 (1997): Bidens teikiteetinii.

OPERCULARIA (Rubiac.).
loganioides *G.J. Keighery in Nuytsia*, 13(1): 29 (1999)—Western Australia.

OPERCULICARYA (Anacardiac.).
borealis *U. Eggli in Bull. Mus. Nation. Hist. Nat., B, Adansonia*, Ser. 4, 17(3–4): 152 (1995)—Madagascar.
hirsutissima *U. Eggli in Bull. Mus. Nation. Hist. Nat., B, Adansonia*, Ser. 4, 17(3–4): 154 (1995)—Madagascar.
pachypus *U. Eggli in Bull. Mus. Nation. Hist. Nat., B, Adansonia*, Ser. 4, 17(3–4): 156 (1995)—Madagascar.

OPERCULINA (Convolvulac.).
aurea *(Kellogg) H.D. House in Muhlenbergia*, 5: 68 (1909): Aniseia aurea.†

OPHELIA (Gentianac.).
japonica (Schult.) Griseb.
f. littoralis *(Hid. Takah.) H. Toyokuni & Y. Toyokuni in J. Asahikawa Univ.*, 4: 182 (1976): Swertia japonica var. littoralis.
noguchiana *(Hatus.) H. Toyokuni & Y. Toyokuni in J. Asahikawa Univ.*, 4: 182 (1976): Swertia noguchiana.

OPHIOGLOSSELLA *A. Schuiteman & P. Ormerod in Kew Bull.*, 53(3): 742 (1998). *ORCHIDACEAE*
chrysostoma *A. Schuiteman & P. Ormerod in Kew Bull.*, 53(3): 743 (1998)—New Guinea (Papua New Guinea).

OPHIONELLA (Asclepiadac.).
arcuata (N.E. Br.) Bruyns
subsp. mirkinii *(Pillans) P.V. Bruyns in Bot. J. Linn. Soc.*, 131(4): 395 (1999): Pectinaria mirkinii.
willowmorensis *P.V. Bruyns in Bot. J. Linn. Soc.*, 131(4): 396 (1999)—Cape Province.

OPHIOPOGON (Convallariac.).
albimarginatus *D. Fang in J. Trop. Subtrop. Bot.*, 6(2): 97 (1998)—China Southeast (Guangxi).
caulescens (Blume) Backer
var. prolifer *(Lindl.) N. Tanaka in J. Jap. Bot.*, 75(6): 363 (2000): Ophiopogon prolifer.
filipes *D. Fang in J. Trop. Subtrop. Bot.*, 6(2): 98 (1998)—China Southeast (Guangxi).
longifolius Decne.
f. albostriatus *Noriyuki Tanaka in J. Jap. Bot.*, 75(5): 269 (2000)—Vietnam, Nansei-shoto, Lesser Sunda Is.
pseudotonkinensis *D. Fang in J. Trop. Subtrop. Bot.*, 6(2): 100 (1998)—China Southeast (Guangxi).
sar-garhwalensis *R.D. Gaur & D.S. Rawat in R.D. Gaur, Fl. Distr. Garhwal N.W. Himalaya:* 714 (1999)—India (Uttar Pradesh).
siamensis *M.N. Tamura in Acta Phytotax. Geobot.*, 49(1): 27 (1998)—Thailand.
vietnamensis *Noriyuki Tanaka in J. Jap. Bot.*, 75(5): 265 (2000)—Vietnam.

OPHIORRHIZA (Rubiac.).
howii *H.S. Lo in Bull. Bot. Res. North-East. Forest. Univ.*, 18(3): 277 (1998), nom. nov.: O. longiflora *How ex H.S. Lo*.
japonica Blume
var. cantoniensis *(Hance) S. Hatusima in S. Hatusima & T. Amano, Fl. Ryukyus, S. of Amami Is., ed. 2:* 214 (1994), without basionym ref.: Basionym not stated.
longii *J.R.I. Wood in Edinburgh J. Bot.*, 53(1): 104 (1996)—Bhutan.
michelloides *(Masam.) H.S. Lo in Bull. Bot. Res. North-East. Forest. Univ.*, 18(3): 277 (1998): Hayataella michelloides.
rugosa Wall.
var. angustifolia *(Thwaites) C.E. Ridsdale in Blumea*, 41(2): 461 (1996): O. angustifolia.
shendurunii *A.E. Shanavas Khan, E.S. Santhosh Kumar & P. Pushpangadan in J. Bombay Nat. Hist. Soc.*, 95(2): 317 (1998)—India (Kerala).

OPHRESTIA (Legumin.).
breviracemosa *B. Verdcourt in Kew Bull.*, 52(2): 495 (1997)—Zambia.

OPHRYOSPORUS (Compos.).
carchiensis *H. Robinson in Phytologia*, 84(5): 350 (1998 publ. 1999)—Ecuador.
ferreyrae *H. Robinson in Phytologia*, 84(5): 351 (1998 publ. 1999)—Peru.
sagasteguii *H. Robinson in Phytologia*, 84(5): 352 (1998 publ. 1999)—Peru.

OPHRYS (Orchidac.).
× abdita *C.E. Hermosilla in J. Eur. Orchideen*, 31(4): 883 (1999)—Spain. (O. riojana × sphegodes).
× acina *C.E. Hermosilla in J. Eur. Orchideen*, 31(4): 883 (1999)—Spain. (O. passionis × riojana).
aegirtica *P. Delforge in Nat. Belg.*, 77(4): 205 (1996)—France.
aeoli *P. Delforge in Nat. Belg.*, 78(3): 163 (1997)—Greece.
andria P. Delforge
var. halkionis *G. Kretzschmar & H. Kretzschmar in Ber. Arbeitskr. Heim. Orch.*, 13(1): 9 (1996)—Greece.
antalyensis *C.A.J. Kreutz & B.J. Seckel in C.A.J. Kreutz, Orchideen Türkei:* 752 (1998)—Turkey.
apifera Huds.
var. belgarum *D.M. Turner Ettlinger in Watsonia*, 22(1): 107 (1998)—Great Britain.
f. bicolor *(O. Nägeli) P.D. Sell in P.D. Sell & G. Murrell, Fl. Great Britain & Ireland*, 5: 365 (1996): O. bicolor.
f. botteronii *(Chodat) P.D. Sell in P.D. Sell & G. Murrell, Fl. Great Britain & Ireland*, 5: 365 (1996): O. botteronii.
f. flavescens *(Rosbach) P.D. Sell in P.D. Sell & G. Murrell, Fl. Great Britain & Ireland*, 5: 365 (1996): O. apifera var. flavescens.
var. tilaventina *U. Nonis & P. Liverani in Caesiana*, 9: 57 (1997), contrary to Art. 37.3 ICBN (1994)—Italy.
f. trollii *(Hegetschw.) P.D. Sell in P.D. Sell & G. Murrell, Fl. Great Britain & Ireland*, 5: 365 (1996): O. trollii.
aramaeorum *P. Delforge in Nat. Belg.*, 81(3): 228 (2000)—Turkey.
archimedea *P. Delforge & M. Walravens in Nat. Belg.*, 81(3): 256 (2000)—Sicilia.
× arciensis *G. Bartolo & S. Pulvirenti in J. Eur. Orchideen*, 30(3): 647 (1998)—Sicilia. (O. oxyrrhynchos × tenthredinifera).
argensonensis *J.-C. Guérin & A. Merlet in Orchidophile*, 133: 173 (1998)—France.
arnoldii *P. Delforge in Nat. Belg.*, 80(3): 247 (1999)—Spain.
aspea *J. Devillers-Terschuren & P. Devillers in Nat. Belg.*, 81(3): 297 (2000)—Tunisia.
astypalaeica *P. Delforge in Nat. Belg.*, 78(3): 154 (1997)—Greece.
× ayusoi *C.E. Hermosilla & R. Soca in Caesiana*, 13: 35 (1999)—Spain. (O. aveyronensis × castellana).
× barauensis *J.-M. Lewin in Orchidophile*, 140: 16 (2000)—France. (O. lutea × sulcata).
× bibbiensis *R. Soca in Monde Pl.*, 460: 21 (1997)—Sicilia.
blitopertha *H.F. Paulus & C. Gack in J. Eur. Orchideen*, 30(1): 181 (1998)—Greece, East Aegean Is.

OPHRYS:-
× bonnyorum *R. Soca in Monde Pl.*, 91(458): 15 (1997), as 'bonnyiorum'—Italy. (O. sipotensis × tenthredinifera).
caesiella *P. Delforge in Nat. Belg.*, 81(3): 233 (2000)—Sicilia (Malta).
calliantha *G. Bartolo & S. Pulvirenti in Caesiana*, 9: 41 (1997)—Sicilia.
calocaerina *J. Devillers-Terschuren & P. Devillers in Nat. Belg.*, 75("Orchidées" no. 7, suppl.): 374 (1994)—Greece.
× campoi *R. Soca in Monde Pl.*, 95(470): 13 (2000)—Sicilia. (O. lacaitae × oxyrrhynchos).
× carnazzae *R. Galesi in J. Eur. Orchideen*, 31(2): 309 (1999)—Sicilia. (O. calliantha × incubacea).
× cascalesii *R. Soca in Monde Pl.*, 95(470): 12 (2000)—France. (O. araneola × passionis).
celiensis *(O. Danesch & E. Danesch) P. Delforge in Nat. Belg.*, 81(4): 397 (2000): Ophrys fuciflora subsp. celiensis.
cilentana *J. Devillers-Terschuren & P. Devillers in Nat. Belg.*, 81(3): 351 (2000)—Italy.
cinereophila *H.F. Paulus & C. Gack in J. Eur. Orchideen*, 30(1): 170 (1998)—Kriti.
classica *J. Devillers-Terschuren & P. Devillers in Nat. Belg.*, 81(3): 351 (2000)—Italy.
climacis *H. Heimeier & T. Perschke in J. Eur. Orchideen*, 30(1): 205 (1998)—Turkey.
× corvey-bironii *J.-M. Lewin in Orchidophile*, 140: 18 (2000)—France. (O. apifera × aveyronensis).
× coulotii *R. Soca in Monde Pl.*, 91(458): 15 (1997)—Italy. (O. garganica × parvimaculata).
creberrima *H.F. Paulus in J. Eur. Orchideen*, 30(1): 168 (1998)—Kriti.
cressa *H.F. Paulus in J. Eur. Orchideen*, 30(1): 176 (1998)—Kriti.
cretica
subsp. beloniae *G. Kretzschmar & H. Kretzschmar in Ber. Arbeitskr. Heim. Orch.*, 13(1): 10 (1996)—Greece.
creticola *H.F. Paulus in J. Eur. Orchideen*, 30(1): 177 (1998)—Kriti.
× cugniensis *R. Soca in Monde Pl.*, 460: 20 (1997)—Sicilia. (O. bertolonii × lutea).
× daissiorum *(H. Baumann et al.) P. Delforge in Nat. Belg.*, 78(3): 188 (1997): O. × maladroxiensis nothosubsp. daissiorum.
× deheyniana *P. Delforge in Nat. Belg.*, 78(3): 180 (1997)—Greece. (O. bremifera × thesei).
× dekegheliana *P. Delforge in Nat. Belg.*, 79(3): 105 (1998)—Italy. (O. bertolonii × majellensis).
× denglerensis *A. Riechelmann in Ber. Arbeitskr. Heim. Orch.*, 15(1): 59 (1998)—Cyprus. (O. kotschyi × lapethica).
× dessartiana *P. Delforge in Nat. Belg.*, 78(3): 180 (1997)—Greece. (O. astypalaeica × parosica).
elatior *Gumpr. ex H.F. Paulus in Ber. Arbeitskr. Heim. Orch.*, 13(2): 10 (1996)—Germany.
× estacensis *R. Soca in Monde Pl.*, 91(458): 17 (1997)—France. (O. fuciflora × splendida).
× ettlingeriana *P. Delforge in Nat. Belg.*, 78(3): 182 (1997), as 'ettlingerana'—Greece. (O. argolica × attica).
× evvia *A. Riechelmann in Ber. Arbeitskr. Heim. Orch.*, 17(2): 45, 44 (2000)—Greece. (O. aesculapii × mammosa).
exaltata Ten.
subsp. mateolana *(Medagli, D'Emerico, Bianco & Ruggiero) H.F. Paulus & C. Gack in J. Eur. Orchideen*, 31(2): 374 (1999): O. mateolana.
× ezcaraiensis *C.E. Hermosilla & R. Soca in Caesiana*, 13: 36 (1999)—Spain. (O. aveyronensis × sphegodes).
× ferliana *G. Bartolo & S. Pulvirenti in J. Eur. Orchideen*, 30(3): 645 (1998)—Sicilia. (O. calliantha × oxyrrhynchos).
flammeola *P. Delforge in Nat. Belg.*, 81(3): 256 (2000)—Sicilia.
× fonsaudiensis *R. Soca in Monde Pl.*, 95(470): 12 (2000)—France. (O. insectifera × passionis).
× fontechensis *C.E. Hermosilla in J. Eur. Orchideen*, 31(4): 884 (1999)—Spain. (O. lutea × riojana).
× frascae *R. Galesi in J. Eur. Orchideen*, 31(2): 311 (1999)—Sicilia. (O. lunulata × sphegodes subsp. sphegodes).
fuciflora (F.W. Schmidt) Moench
subsp. annae *(Devillers-Tersch. & Devillers) R. Engel & P. Quentin in Orchidophile*, 124: 205 (1996): O. annae.

OPHRYS:-
fuciflora
subsp. elatior *Gumpr. ex R. Engel & P. Quentin in Orchidophile*, 123: 168 (1996)—France.
subsp. elatior *(Paulus) R. Engel & P. Quentin in Orchidophile*, 127: 133 (1997), homonym; *see* R. Engel & Quentin in *Orchidophile*, 123: 168 (1996): O. elatior.
× galesii *R. Soca in Monde Pl.*, 91(458): 16 (1997)—Sicilia. (O. oxyrrhynchos × sphegodes).
gallica *H. van Looken in Liparis*, 6.2(extra nummer): 66, 65 (2000)—France.
× garganensis *R. Soca in Caesiana*, 9: 49 (1997), contrary to Art. 37.5 ICBN (1994)—Italy. (O. archipelagi × biscutella).
garganica
subsp. passionis *(Sennen ex Devillers & Devillers-Tersch.) H.F. Paulus & C. Gack in J. Eur. Orchideen*, 31(2): 378 (1999): O. passionis.
gazella *J. Devillers-Terschuren & P. Devillers in Nat. Belg.*, 81(3): 322 (2000)—Tunisia.
× gelana H. Baumann & Künkele *[unranked]* murgettensis *R. Soca in Monde Pl.*, 95(470): 12 (2000), contrary to Art. 35.1 ICBN (2000)—Italy. (O. incubacea × oxyrrhynchos subsp. celiensis).
× geniezii *R. Soca in Monde Pl.*, 91(458): 14 (1997)—Italy. (O. archipelagi × incubacea).
× glanensis *R. Soca in Monde Pl.*, 95(470): 12 (2000)—France. (O. fuciflora × provincialis).
gracilis *(Büel, O. Danesch & E. Danesch) H.F. Paulus in Ber. Arbeitskr. Heim. Orch.*, 13(2): 9 (1996): O. fuciflora subsp. gracilis.
heldreichii Schltr.
var. calypsus *(M. Hirth & H. Spaeth) P. Delforge in Nat. Belg.*, 76(3): 176 (1995): O. calypsus.
var. pseudoapulica *P. Delforge in Nat. Belg.*, 76(3): 176 (1995)—Greece.
var. scolopaxoides *P. Delforge in Nat. Belg.*, 76(3): 176 (1995)—Greece.
× hermosillae *R. Soca & J. Benito Ayuso in Monde Pl.*, 95(470): 14 (2000)—France. (O. passionis × scolopax).
hespera *J. Devillers-Terschuren & P. Devillers in Nat. Belg.*, 81(3): 322 (2000)—Italy.
hittitica *C.A.J. Kreutz & Ruedi Peter in J. Eur. Orchideen*, 30(1): 128 (prepr. 30 Sept. 1997): Turkey.
holosericea (Burm. f.) Greuter
subsp. lacaitae *(Lojac.) W. Rossi in Inform. Bot. Ital.*, 13(2–3): 201 (1981 publ. 1983): O. lacaitae.
homeri *M. Hirth & H. Spaeth in J. Eur. Orchideen*, 30(1): 9, 7 (prepr. 30 Sept. 1997)—East Aegean Is., Turkey.
hygrophila *E. Gügel, C.A.J. Kreutz, D. Rückbrodt & U. Rückbrodt in J. Eur. Orchideen*, 29(4): 671 (1997)—Turkey.
hystera *C.A.J. Kreutz & Ruedi Peter in J. Eur. Orchideen*, 30(1): 128 (prepr. 30 Sept.1997), nom. nov.: O. mammosa subsp. serotina *B. Willing & E. Willing*.
iceliensis *C.A.J. Kreutz in J. Eur. Orchideen*, 32(3–4): 529 (2000)—Turkey.
× ingaranensis *R. Soca in Monde Pl.*, 460: 18 (1997)—Italy. (O. archipelagi × parvimaculata).
iricolor Desf.
subsp. eleonorae *(Devillers-Tersch. & Devillers) H.F. Paulus & C. Gack in Jahresber. Naturwiss. Ver. Wuppertal*, 48: 199 (1995): O. eleonorae.
subsp. maxima *(A. Terracc.) H.F. Paulus & C. Gack in Jahresber. Naturwiss. Ver. Wuppertal*, 48: 199 (1995); *et in J. Eur. Orchideen*, 31(2): 361 (1999): O. fusca f. maxima.
subsp. vallesiana *(Devillers & Devillers-Tersch.) H.F. Paulus & C. Gack in Jahresber. Naturwiss. Ver. Wuppertal*, 48: 199 (1995): O. vallesiana.
karadenizensis *M. Schönfelder & H. Schönfelder in Ber. Arbeitskr. Heim. Orch.*, 14(2): 18, 17 (1997 publ. 1998)—Turkey.
× kohlmuellerorum *R. Soca in Monde Pl.*, 460: 21 (1997)—France. (O. scolopax × sulcata).
× kolymbiae *G. Kretzschmar & H. Kretzschmar in Ber. Arbeitskr. Heim. Orch.*, 15(1): 49 (1998)—East Aegean Is. (O. lucis × tenthredinifera).
labiosa *C.A.J. Kreutz in J. Eur. Orchideen*, 29(4): 683 (1997)—Turkey.
× laureantis *R. Galesi in J. Eur. Orchideen*, 31(4): 913 (1999)—Sicilia. (O. exaltata × fusca).

OPHRYS:-

× lebeaultii *J.-F. Perrin & R. Soca in Orchidophile,* 138: 176 (1999)—France. (O. passionis × splendida).
leptomera *P. Delforge in Nat. Belg.,* 81(3): 192 (2000)—Greece.
leucophthalma *J. Devillers-Terschuren & P. Devillers in Nat. Belg.,* 75 ("Orchidées" no. 7, suppl.): 380 (1994)—Greece.
× lewinii *R. Soca in Monde Pl.,* 91(458): 14 (1997)—Italy. (O. archipelagi × oestrifera).
linearis *(Moggr.) P. Delforge, P. Devillers & J. Devillers-Terschuren in Nat. Belg.,* 81(3): 153 (2000): O. insectifera subvar. linearis.
lojaconoi *P. Delforge in Nat. Belg.,* 76(3): 289 (1995)—Italy.
× lorenzii *R. Soca in Monde Pl.,* 460: 20 (1997)—Italy. (O. bertoloniiformis × incubacea).
lucana *P. Delforge, J. Devillers-Terschuren & P. Devillers in Nat. Belg.,* 81(3): 264 (2000)—Italy.
lucentina *P. Delforge in Nat. Belg.,* 80(3): 253 (1999)—Spain.
lucifera *J. Devillers-Terschuren & P. Devillers in Nat. Belg.,* 81(3): 322 (2000)—Italy.
× lumenii *R. Soca in Monde Pl.,* 460: 20 (1997)—Sicilia. (O. bertolonii × ciliata).
lupercalis *J. Devillers-Terschuren & P. Devillers in Nat. Belg.,* 75("Orchidées" no. 7, suppl.): 373 (1994)—France.
macedonica *(H. Fleischm. ex Soó) J. Devillers-Terschuren & P. Devillers in Nat. Belg.,* 75("Orchidées" no. 7, suppl.): 381 (1994): O. aranifera nothosubsp. macedonica.
× mahilloniana *P. Delforge in Nat. Belg.,* 78(3): 182 (1997)—Greece. (O. aeoli × ferrum-equinum).
majellensis *(Helga Daiss & Herm. Daiss) P. Delforge in Nat. Belg.,* 79(3): 104 (1998): O. sphegodes subsp. majellensis.
marmorata *G. Foelsche & W. Foelsche in Orchidophile,* 133: 178 (1998)—Corse.
× martorinae *R. Galesi in J. Eur. Orchideen,* 31(2): 313 (1999)—Sicilia. (O. discors × incubacea).
× marzensis *R. Soca in Monde Pl.,* 460: 20 (1997)—Sicilia. (O. bertolonii × explanata).
massiliensis *J. Viglione & E. Véla in Orchidophile,* 135: 13 (1999)—France.
melitensis *(Salk.) J. Devillers-Terschuren & P. Devillers in Nat. Belg.,* 75("Orchidées" no. 7, suppl.): 380 (1994): O. sphegodes subsp. melitensis.
× melkii *R. Soca in Monde Pl.,* 91(458): 15 (1997)—Sicilia. (O. explanata × incubacea).
× minuticauda Duffort
nothosubsp. donorensis *V. Gulli & M.P. Grasso in Orchidophile,* 128: 177 (1997)—Sardegna. (O. apifera × scolopax subsp. conradiae).
× mirtiae *G. Kretzschmar & H. Kretzschmar in Ber. Arbeitskr. Heim. Orch.,* 15(1): 53 (1998)—Greece. (O. sicula × speculum).
× motolesensis *R. Soca in Monde Pl.,* 460: 21 (1997)—Italy. (O. bertoloniiformis × tarentina).
numida *J. Devillers-Terschuren & P. Devillers in Nat. Belg.,* 81(3): 297 (2000)—Tunisia.
omegaifera
subsp. israelitica *(H. Baumann & Künkele) G. Morschek & K. Morschek, Orchids of Cyprus-Zyperns Orchideen:* 126 (1996), without exact basionym page: O. israelitica.
panormitana
subsp. praecox *(Corrias) H.F. Paulus & C. Gack in Jahresber. Naturwiss. Ver. Wuppertal,* 48: 204 (1995): O. sphegodes subsp. praecox.
var. praecox *(Corrias) P. Delforge in Nat. Belg.,* 81(4): 398 (2000): Ophrys sphegodes subsp. praecox.
× pantalicae *G. Bartolo & S. Pulvirenti in J. Eur. Orchideen,* 30(3): 646 (1998)—Sicilia. (O. calliantha × lacaitae).
parosica *P. Delforge in Nat. Belg.,* 76(3): 160 (1995)—Greece.
passionis *Sennen ex J. Devillers-Terschuren & P. Devillers in Nat. Belg.,* 75("Orchidées" no. 7, suppl.): 379 (1994)—France.
subsp. majellensis *(Helga Daiss & Herm. Daiss) R. Romolini & R. Soca in Caesiana,* 15: 23 (2000): Ophrys sphegodes subsp. majellensis.
peraiolae *G. Foelsche, W. Foelsche, M. Gerbaud & O. Gerbaud in J. Eur. Orchideen,* 32(3–4): 421 (2000)—Corse.

OPHRYS:-

peraiolae
var. rubra *G. Foelsche, W. Foelsche, M. Gerbaud & O. Gerbaud in J. Eur. Orchideen,* 32(3–4): 422 (2000)—Corse.
× perrinii *R. Soca in Monde Pl.,* 95(470): 14 (2000)—France. (O. passionis × provincialis).
× pesseiae *P. Delforge in Nat. Belg.,* 78(3): 186 (1997)—Sardegna. (O. morisii × panattensis).
phaseliana *D. Rückbrodt & U. Rückbrodt in J. Eur. Orchideen,* 28(2): 395 (1996)—Turkey.
× pizzulensis *R. Soca in Monde Pl.,* 460: 21 (1997)—Italy. (O. biscutella × parvimaculata).
× poisneliae *J.L. Menos in Orchidophile,* 143: 197 (2000), as 'poisnelae'—Spain. (O. catalaunica × subinsectifera).
pollinensis *E. Nelson ex J. Devillers-Terschuren & P. Devillers in Nat. Belg.,* 81(3): 352 (2000)—Italy.
posidonia *P. Delforge in Nat. Belg.,* 81(3): 173 (2000)—Italy.
× pourteiniae *J.M. Mathé in Orchidophile,* 140: 25 (2000)—France. (O. araneola × lupercalis).
pseudoscolopax *(Moggr.) H.F. Paulus & C. Gack in J. Eur. Orchideen,* 31(2): 394 (1999): O. insectifera var. pseudoscolopax.
× quintartiana *P. Delforge in Nat. Belg.,* 78(3): 183 (1997)—Greece. (O. ariadnae × ferrum-equinum).
riojana *C.E. Hermosilla in J. Eur. Orchideen,* 31(4): 881 (1999)—Spain.
× saintenoy-simoniana *P. Delforge in Nat. Belg.,* 78(3): 183 (1997)—Greece. (O. ferrum-equinum × thesei).
× samuelii *R. Soca in Monde Pl.,* 91(458): 14 (1997)—France. (O. drumana × scolopax).
× sancticyrensis *R. Soca in Monde Pl.,* 95(470): 14 (2000)—France. (O. lupercalis × passionis).
sardoa *(H. Baumann, Giotta, R. Lorenz, Künkele & Piccitto) H.F. Paulus & C. Gack in J. Eur. Orchideen,* 31(2): 386 (1999): O. scolopax subsp. sardoa.
schlechteriana *(Soó) J. Devillers-Terschuren & P. Devillers in Nat. Belg.,* 75("Orchidées" no. 7, suppl.): 378 (1994), isonym of suggested basionym see also IK. Supp. 8.: O. heldreichii subsp. schlechteriana.
scolopax Cav.
subsp. bremifera *(Steven) B. Biel in Ber. Arbeitskr. Heim. Orch.,* 16(1): 54 (1999): O. bremifera.
var. grandiflora *(B. Willing & E. Willing) B. Biel in Ber. Arbeitskr. Heim. Orch.,* 17(2): 61 (2000), without exact basionym page: O. oestrifera var. grandiflora.
subsp. lapethica *(Gölz & H.R. Reinhard) G. Morschek & K. Morschek, Orchids of Cyprus-Zyperns Orchideen:* 128 (1996), without exact basionym page: O. lapethica.
var. minutula *(Gölz & H.R. Reinhard) H.A. Pedersen & N. Faurholdt in Orchidee,* 48(6): 235 (1997): O. minutula.
subsp. minutula *(Gölz & H.R. Reinhard) B. Biel in Ber. Arbeitskr. Heim. Orch.,* 16(1): 54 (1999): O. minutula.
subsp. rhodia *(H. Baumann & Künkele) H.A. Pedersen & N. Faurholdt in Orchidee,* 48(6): 235 (1997): O. umbilicata subsp. rhodia.
subsp. santonica *(J.M. Mathé & Melki) R. Engel & P. Quentin in Orchidophile,* 124: 205 (1996): O. santonica.
serotina *Rolli ex H.F. Paulus in Ber. Arbeitskr. Heim. Orch.,* 17(2): 23, 24 (2000)—Italy.
× socae *J.-M. Lewin in Orchidophile,* 140: 17 (2000)—France. (O. lupercalis × vasconica).
× soller *M. Henkel in Ber. Arbeitskr. Heim. Orch.,* 17(2): 121 (2000)—Baleares. (O. apifera × speculum).
× spanui *P. Delforge in Nat. Belg.,* 78(3): 187 (1997)—Sardegna. (O. annae × tenthredinifera).
sphegodes Mill.
nothosubsp. jeanpertii *(E.G. Camus) C. Del Prete & R. Conte in Webbia,* 34(2): 560 (1980), as 'O. speculum' in Supp. 17: O. × jeanpertii.†
subsp. majellensis *Helga Daiss & Hermann Daiss in J. Eur. Orchideen,* 28(4): 614 (1996 publ. 1997)—Italy.
subsp. passionis *(Sennen) H. Sanz i Gonel & J. Nuet i Badia, Guia Camp Orquid. Catalunya:* 176 (1995): O. passionis.

OPHRYS:-

× stefaniae *G. Kretzschmar & H. Kretzschmar in Ber. Arbeitskr. Heim. Orch.,* 15(1): 54 (1998)—Greece. (O. ferrum-equinum × iricolor).
subinsectifera *C.E. Hermosilla & J. Sabando in Estud. Mus. Cienc. Nat. Alava,* 10–11: 133, 135 (1995–96 publ. 1996)—Spain.
sulcata *J. Devillers-Terschuren & P. Devillers in Nat. Belg.,* 75("Orchidées" no. 7, suppl.): 375 (1994)—France.
tarquinia *P. Delforge in Nat. Belg.,* 81(3): 84 (2000)—Italy.
× tavignanensis *H. Mathé, J.-M. Mathé & M. Pena in Orchidophile,* 125: 12 (1997), without type—Corse. (O. incubacea × iricolor subsp. maxima).
× telchinensis *A. Riechelmann in Ber. Arbeitskr. Heim. Orch.,* 14(1): 67 (1997)—East Aegean Is. (O. ferrum-equinum × lucis).
tenthredinifera Willd.
subsp. ficalhoana *(J.A. Guim.) M.R. Lowe & D. Tyteca in J. Eur. Orchideen,* 32(2): 297 (2000): O. ficalhoana.
subsp. guimaraesii *D. Tyteca in J. Eur. Orchideen,* 32(2): 297 (2000), nom. nov.: O. tenthredinifera subsp. praecox *D. Tyteca.*
× teresae *R. Soca in Monde Pl.,* 460: 18 (1997)—Italy. (O. argentaria × tenthredinifera).
thesei *P. Delforge in Nat. Belg.,* 78(3): 167 (1997)—Greece.
thriptiensis *H.F. Paulus in J. Eur. Orchideen,* 30(1): 173 (1998)—Kriti.
× torrensis *H. Dekker in Eurorchis,* 11: 24 (1999)—Italy. (O. argentaria × pseudoscolopax).
× triadensis *G. Kretzschmar & H. Kretzschmar in Ber. Arbeitskr. Heim. Orch.,* 15(1): 52 (1998)—Greece. (O. sicula × tenthredinifera).
× trombettensis *R. Soca in Monde Pl.,* 460: 18 (1997)—Italy. (O. archipelagi × sphegodes).
umbilicata Desf.
subsp. bucephala *(Gölz & H.R. Reinhard) B. Biel in Ber. Arbeitskr. Heim. Orch.,* 16(1): 58 (1999): O. bucephala.
untchjii *(M. Schulze) P. Delforge in Nat. Belg.,* 81(4): 398 (2000): Ophrys fuciflora subsp. untchjii.
vallesiana *J. Devillers-Terschuren & P. Devillers in Nat. Belg.,* 75("Orchidées" no. 7, suppl.): 373 (1994)—Tunisia.
× vespertilio *W. Rossi & M. Contorni in Orchid Rev.,* 104(1209): 184 (1996)—Italy. (O. apifera × bertolonii).
× zamba *C.E. Hermosilla in Estud. Mus. Cienc. Nat. Alava,* 13: 118 (1998)—Spain. (O. passionis × sphegodes).
ziyaretiana *C.A.J. Kreutz & Ruedi Peter in J. Eur. Orchideen,* 30(1): 104 (prepr. 30 Sept. 1997)—Turkey, Palestine (Israel).
zonata *J. Devillers-Terschuren & P. Devillers in Nat. Belg.,* 75("Orchidées" no. 7, suppl.): 374 (1994)—Sardegna.

OPHTHALMOBLAPTON (Euphorbiac.).
parviflorum *M. Emmerich in Bol. Mus. Nac. Rio de Janeiro, Bot.,* 62: 1 (1981)—Brazil Northeast (Bahia).

OPITHANDRA (Gesneriac.).
burttii *W.T. Wang in Novon,* 7(4): 430 (1997 publ. 1998)—China Southeast (Jiangxi).

OPLISMENUS (Gramin.).
hirtellus *(L.) P. Beauv., Ess. Agrostogr.:* 54, 168, 170 (1812): Panicum hirtellum.

OPUNTIA (Cactac.).
albicarpa *L. Scheinvar in Agro-ecol., cult. & uses of cactus pear, ed. G. Barbera et al. (FAO Pl.* Prod. & Prot. Paper 132): 22 (1995), without latin descr., contrary to Art. 37.5 ICBN (1994); *et in Rev. Jard. Bot. Nacion. Univ. Habana,* 20: 167 (1999), contrary to Art. 37.5 ICBN (1994)—Mexico Central (Puebla).
alcerrecensis *J. Iliff in Cactaceae Consensus Initiatives,* 2: 9 (1997), nom. nov.: Platyopuntia chilensis *F. Ritter.*
anacantha Speg.
var. kiska-loro *(Speg.) R. Kiesling in Candollea,* 53(2): 475 (1998): O. kiska-loro.

OPUNTIA:-
anacantha
var. retrorsa *(Speg.) R. Kiesling in Candollea,* 53(2): 475 (1998): O. retrorsa.
var. utkilio *(Speg.) R. Kiesling in Candollea,* 53(2): 476 (1998): O. utkilio.
apurimacensis *(F. Ritter) R. Crook & R. Mottram in Bradleya,* 13: 100 (1995): Platyopuntia apurimacensis.
araucana *(A. Cast.) M. Kiessling, Tephrocactus u. and. Feigenkakt.:* 38 (2000), as 'araucanus': Pterocactus araucanus.
archiconoidea *(F. Ritter) D.R. Hunt in Cactaceae Consensus Initiatives,* 4: 5 (1997): Maihueniopsis archiconoidea.
atroglobosa *(F. Ritter) R. Crook & R. Mottram in Bradleya,* 13: 106 (1995): Platyopuntia atroglobosa.
bonnieae *(D.J. Ferguson & R. Kiesling) Halda & Janeba in Cactaceae etc.,* 9(4): 160 (1999); *cf. Repert. Pl. Succ. (I.O.S.),* 50: 14 (1999 publ. 2000): Puna bonnieae.
brachyacantha *(F. Ritter) R. Crook & R. Mottram in Bradleya,* 13: 115 (1995): Platyopuntia brachyacantha.
brunneogemmia *(F. Ritter) C. Schlindwein in Bradleya,* 13: 29 (1995): Platyopuntia brunneogemmia.
californica (Torr. & A. Gray) Coville
var. parkeri *(J.M. Coult.) D.J. Pinkava in Haseltonia,* 4: 103 (1996): O. echinocarpa var. parkeri.
× campii *M.A. Baker & D.J. Pinkava in Cact. Succ. J. (U.S.A.),* 71(6): 320 (1999)—Arizona. (O. acanthocarpa × bigelovii).
colorea *(F. Ritter) D.R. Hunt in Cactaceae Consensus Initiatives,* 4: 5 (1997): Tephrocactus coloreus.
crassispina *(F. Ritter) D.R. Hunt in Cactaceae Consensus Initiatives,* 4: 5 (1997): Maihueniopsis crassispina.
darwinii
var. hickenii *(Britton & Rose) M. Kiessling, Tephrocactus u. and. Feigenkakt.:* 100 (2000): Opuntia hickenii.
domeykoensis *(F. Ritter) D.R. Hunt in Cactaceae Consensus Initiatives,* 4: 5 (1997): Maihueniopsis domeykoensis.
frigida *(F. Ritter) G. Navarro in Lazaroa,* 17: 66 (1996): Cumulopuntia frigida.
galerasensis *(F. Ritter) D.R. Hunt in Cactaceae Consensus Initiatives,* 4: 5 (1997): Cumulopuntia galerasensis.
ganderi *(C.B. Wolf) J. Rebman & D.J. Pinkava in Haseltonia,* 4: 103 (1996): O. acanthocarpa subsp. ganderi.
guatinensis *D.R. Hunt in Cactaceae Consensus Initiatives,* 4: 5 (1997), nom. nov.: Cumulopuntia tortispina *F. Ritter.*
humifusa (Raf.) Raf.
subsp. minor *(Engelm.) R. Crook & R. Mottram in Bradleya,* 16: 135 (1998): O. rafinesquei var. minor.
ianthinantha *(F. Ritter) J. Iliff in Cactaceae Consensus Initiatives,* 2: 8 (1997): Platyopuntia ianthinantha.
inaequilateralis A. Berger
var. angustior *(F. Ritter) R. Crook & R. Mottram in Bradleya,* 17: 113 (1999): Platyopuntia inaequilateralis var. angustior.
inaperta *(Schott ex Griffiths) D.R. Hunt in Cactaceae Consensus Initiatives,* 4: 5 (1997): Nopalea inaperta.
infesta *(F. Ritter) J. Iliff in Cactaceae Consensus Initiatives,* 2: 8 (1997): Platyopuntia infesta.
kuntzei *(K. Schum.) M. Kiessling, Tephrocactus u. and. Feigenkakt.:* 156 (2000): Pterocactus kuntzei.
lindsayi *J.P. Rebman in Cact. Succ. J. (U.S.A.),* 69(2): 67 (1997)—Mexico Northwest (Baja California, Baja California Sur).
llanos-de-huanta *D.R. Hunt in Cactaceae Consensus Initiatives,* 4: 6 (1997), nom. nov.: Maihueniopsis grandiflora *F. Ritter.*
lutea *(Rose) D.R. Hunt in Cactaceae Consensus Initiatives,* 4: 6 (1997): Nopalea lutea.
lutea *(Rose) J.C. Solomon in Novon,* 9(3): 422 (1999), isonym: Nopalea lutea.

OPUNTIA:-
megliolii *(R. Kiesling) M. Kiessling, Tephrocactus u. and. Feigenkakt.:* 182 (2000): Pterocactus megliolii.
moniliformis
f. guantanamana *(Areces) R. Govaerts, World Checklist Seed Plants,* 3(1): 21 (1999): Consolea moniliformis subsp. guantanamana.
nashii
subsp. gibarensis *(Areces) R. Govaerts, World Checklist Seed Plants,* 3(1): 21 (1999): Consolea nashii subsp. gibarensis.
parviclada *S. Arias Montes & S. Gama López in Fl. Valle Tehuacán-Cuicatlán,* 14(Cactac.): 125 (1997)—Mexico Southwest (Oaxaca).
pinkavae *B.D. Parfitt in Rhodora,* 99(899): 223 (1997 publ. 1998)—Arizona, Utah.
pituitosa *(F. Ritter) J. Iliff in Cactaceae Consensus Initiatives,* 2: 9 (1997): Platyopuntia pituitosa.
polyacantha Haw.
var. arenaria *(Engelm.) B.D. Parfitt in Cact. Succ. J. (U.S.A.),* 70(4): 188 (1998): O. arenaria.
var. erinacea *(Engelm. & J.M. Bigelow) B.D. Parfitt in Cact. Succ. J. (U.S.A.),* 70(4): 188 (1998): O. erinacea.
var. hystricina *(Engelm. & J.M. Bigelow) B.D. Parfitt in Cact. Succ. J. (U.S.A.),* 70(4): 188 (1998): O. hystricina.
var. nicholii *(L.D. Benson) B.D. Parfitt in Cact. Succ. J. (U.S.A.),* 70(4): 188 (1998): O. nicholii.
rossiana *(Heinrich & Backeb.) D.R. Hunt in Cactaceae Consensus Initiatives,* 4: 6 (1997): Tephrocactus pentlandii var. rossianus.
var. fuauxiana *(Backeb.) M. Kiessling, Tephrocactus u. and. Feigenkakt.:* 244 (2000): Tephrocactus pentlandii var. fuauxianus.
sanctae-barbarae *D.R. Hunt in Cactaceae Consensus Initiatives,* 4: 6 (1997), nom. nov.: Cumulopuntia hystrix *F. Ritter.*
sanfelipensis *J.P. Rebman in Haseltonia,* 6: 17 (1998 publ. 1999)—Mexico Northwest (Baja California).
tehuacana *S. Arias Montes & L.U. Guzmán Cruz in Fl. Valle Tehuacán-Cuicatlán,* 14(Cactac.): 131 (1997)—Mexico Central (Puebla), Mexico Southwest (Oaxaca).
ticnamarensis *(F. Ritter) D.R. Hunt in Cactaceae Consensus Initiatives,* 4: 6 (1997): Cumulopuntia ticnamarensis.
tumida *(F. Ritter) D.R. Hunt in Cactaceae Consensus Initiatives,* 4: 6 (1997): Cumulopuntia tumida.
valentinii *(Speg.) M. Kiessling, Tephrocactus u. and. Feigenkakt.:* 272 (2000): Pterocactus valentinii.
violacea
var. azurea *(Rose) Lodé, Fichier Encycl. Cact. Autres Succ.,* Sér. 17: no. 1576 (1996), without basionym ref., species name not validly publ.; *cf. Repert. Pl. Succ. (I.O.S.),* 47: 11 (1996 publ. 1997): O. azurea.
viridirubra *(F. Ritter) C. Schlindwein in Bradleya,* 13: 29 (Dec. 1995), isonym; *see P.J. Braun & Esteves in Succulenta,* 74(3): 133 (June 1995): Platyopuntia viridirubra.
wagenknechtii *(F. Ritter) D.R. Hunt in Cactaceae Consensus Initiatives,* 4: 6 (1997): Maihueniopsis wagenknechtii.
zamudioi *L. Scheinvar in Cact. Suc. Mex.,* 44(4): 88 (1999)—Mexico Northeast (Querétaro).

ORBEA (Asclepiadac.).
abayensis *(M.G. Gilbert) P.V. Bruyns in Aloe,* 37(4): 73 (2000): Caralluma abayensis.
albocastanea *(Marloth) P.V. Bruyns in Aloe,* 37(4): 73 (2000): Stapelia albocastanea.
araysiana *(Lavranos & Bilaidi) P.V. Bruyns in Aloe,* 37(4): 73 (2000): Stultitia araysiana.
baldratii *(A.C. White & B. Sloane) P.V. Bruyns in Aloe,* 37(4): 73 (2000): Caralluma baldratii.
subsp. somalensis *P.V. Bruyns in Aloe,* 37(4): 73 (2000)—Somalia.
carnosa *(Stent) P.V. Bruyns in Aloe,* 37(4): 73 (2000): Caralluma carnosa.
caudata *(N.E. Br.) P.V. Bruyns in Aloe,* 37(4): 73 (2000): Caralluma caudata.
subsp. rhodesiaca *(L.C. Leach) P.V. Bruyns in Aloe,* 37(4): 74 (2000): Caralluma caudata subsp. rhodesiaca.
chrysostephana *(Deflers) P.V. Bruyns in Aloe,* 37(4): 74 (2000): Stapelia chrysostephana.

ORBEA:-
commutata *(A. Berger) P.V. Bruyns in Aloe,* 37(4): 74 (2000): Caralluma commutata.
conjuncta *(A.C. White & B. Sloane) P.V. Bruyns in Aloe,* 37(4): 74 (2000): Stultitia conjuncta.
decaisneana *(Lem.) P.V. Bruyns in Aloe,* 37(4): 74 (2000): Boucerosia decaisneana.
deflersiana *(Lavranos) P.V. Bruyns in Aloe,* 37(4): 74 (2000): Caralluma deflersiana.
denboefii *(Lavranos) P.V. Bruyns in Aloe,* 37(4): 74 (2000): Caralluma denboefii.
distincta *(E.A. Bruce) P.V. Bruyns in Aloe,* 37(4): 74 (2000): Caralluma distincta.
dummeri *(N.E. Br.) P.V. Bruyns in Aloe,* 37(4): 74 (2000): Stapelia dummeri.
subsp. circes *(M.G. Gilbert) P.V. Bruyns in Aloe,* 37(4): 74 (2000): Caralluma circes.
gemugofana *(M.G. Gilbert) P.V. Bruyns in Aloe,* 37(4): 74 (2000): Caralluma gemugofana.
gerstneri *(Letty) P.V. Bruyns in Aloe,* 37(4): 74 (2000): Caralluma gerstneri.
subsp. elongata *(R.A. Dyer) P.V. Bruyns in Aloe,* 37(4): 74 (2000): Caralluma gerstneri subsp. elongata.
hardyi *(R.A. Dyer) P.V. Bruyns in Aloe,* 37(4): 74 (2000): Stultitia hardyi.
huernioides *(P.R.O. Bally) P.V. Bruyns in Aloe,* 37(4): 74 (2000): Caralluma huernioides.
huillensis *(Hiern) P.V. Bruyns in Aloe,* 37(4): 74 (2000): Caralluma huillensis.
subsp. flava *P.V. Bruyns in Aloe,* 37(4): 75 (2000)—Namibia.
knobelii *(E. Phillips) P.V. Bruyns in Aloe,* 37(4): 75 (2000): Stapelia knobelii.
laikipiensis *(M.G. Gilbert) P.V. Bruyns in Aloe,* 37(4): 75 (2000): Pachycymbium laikipiense.
laticorona *(M.G. Gilbert) P.V. Bruyns in Aloe,* 37(4): 75 (2000): Caralluma sprengeri subsp. laticorona.
lugardii *(N.E. Br.) P.V. Bruyns in Aloe,* 37(4): 75 (2000): Caralluma lugardii.
luntii *(N.E. Br.) P.V. Bruyns in Aloe,* 37(4): 75 (2000): Caralluma luntii.
lutea *(N.E. Br.) P.V. Bruyns in Aloe,* 37(4): 75 (2000): Caralluma lutea.
subsp. vaga *(N.E. Br.) P.V. Bruyns in Aloe,* 37(4): 75 (2000): Stapelia vaga.
maculata (N.E. Br.) L.C. Leach
subsp. kaokoensis *P.V. Bruyns in Aloe,* 37(4): 75 (2000)—Namibia.
subsp. rangeana *(Dinter & A. Berger) P.V. Bruyns in Aloe,* 37(4): 76 (2000): Caralluma rangeana.
melanantha *(Schltr.) P.V. Bruyns in Aloe,* 37(4): 76 (2000): Stapelia melanantha.
miscella *(N.E. Br.) U. Meve in Kakt. And. Sukk.,* 51(7): 186 (2000): Stapelia miscella.
rogersii *(L. Bolus) P.V. Bruyns in Aloe,* 37(4): 76 (2000): Stapelia rogersii.
sacculata *(N.E. Br.) P.V. Bruyns in Aloe,* 37(4): 76 (2000): Caralluma sacculata.
schweinfurthii *(A. Berger) P.V. Bruyns in Aloe,* 37(4): 76 (2000): Caralluma schweinfurthii.
semitubiflora *(L.E. Newton) P.V. Bruyns in Aloe,* 37(4): 76 (2000): Angolluma semitubiflora.
sprengeri *(Schweinf.) P.V. Bruyns in Aloe,* 37(4): 76 (2000): Huernia sprengeri.
subsp. foetida *(M.G. Gilbert) P.V. Bruyns in Aloe,* 37(4): 76 (2000): Caralluma sprengeri subsp. foetida.
subsp. ogadensis *(M.G. Gilbert) P.V. Bruyns in Aloe,* 37(4): 76 (2000): Caralluma sprengeri subsp. ogadensis.
subterranea *(E.A. Bruce & P.R.O. Bally) P.V. Bruyns in Aloe,* 37(4): 76 (2000): Caralluma subterranea.
tubiformis *(E.A. Bruce & P.R.O. Bally) P.V. Bruyns in Aloe,* 37(4): 76 (2000): Caralluma tubiformis.
ubomboensis *(I. Verd.) P.V. Bruyns in Aloe,* 37(4): 76 (2000): Caralluma ubomboensis.
valida *(N.E. Br.) P.V. Bruyns in Aloe,* 37(4): 76 (2000): Caralluma valida.
subsp. occidentalis *P.V. Bruyns in Aloe,* 37(4): 76 (2000)—Botswana.
vibratilis *(E.A. Bruce & P.R.O. Bally) P.V. Bruyns in Aloe,* 37(4): 76 (2000): Caralluma vibratilis.
wilsonii *(P.R.O. Bally) P.V. Bruyns in Aloe,* 37(4): 76 (2000): Caralluma wilsonii.

ORBEA:-
wissmannii *(O. Schwartz) P.V. Bruyns in Aloe,* 37(4): 76 (2000): Caralluma wissmannii.
var. eremastrum *(O. Schwartz) P.V. Bruyns in Aloe,* 37(4): 76 (2000): Caralluma eremastrum.
var. parviloba *P.V. Bruyns in Aloe,* 37(4): 76 (2000)—Yemen.

ORBIGNYA (Palmae).
brejinhoensis *S.F. Glassman, Taxon. Treatment Palm Subtribe Attaleinae (Tribe Cocoeae) (Illinois Biol.* Monogr., 59): 84 (1999)—Brazil Northeast (Bahia).
crassispatha *(Mart.) S.F. Glassman, Taxon. Treatment Palm Subtribe Attaleinae (Tribe Cocoeae) (Illinois Biol.* Monogr., 59): 94 (1999): Maximiliana crassispatha.

ORBIVESTUS *H. Robinson in Proc. Biol. Soc. Wash.*, 112(1): 230 (1999). *COMPOSITAE*
cinerascens *(Sch. Bip.) H. Robinson in Proc. Biol. Soc. Wash.*, 112(1): 231 (1999): Vernonia cinerascens.
homilanthus *(S. Moore) H. Robinson in Proc. Biol. Soc. Wash.*, 112(1): 231 (1999): Vernonia homilantha.
karaguensis *(Oliv. & Hiern) H. Robinson in Proc. Biol. Soc. Wash.*, 112(1): 231 (1999): Vernonia karaguensis.
undulatus *(Oliv. & Hiern) H. Robinson in Proc. Biol. Soc. Wash.*, 112(1): 231 (1999): Vernonia undulata.

× **ORCHIDACTYLORHIZA** (Orchidac.).
atacina *(P. Delforge) P. Delforge in Nat. Belg.*, 81(4): 398 (2000): × Orchidactyla atacina.

ORCHIDANTHA (Lowiac.).
chinensis T.L. Wu
var. longisepala *(D. Fang) T.L. Wu in Novon,* 7(4): 440 (1997 publ. 1998): O. longisepala.
inouei *H. Nagamasu & S. Sakai in Nordic J. Bot.*, 19(2): 149 (1999)—Borneo (Sarawak).
longisepala *D. Fang in Guihaia*, 16(1) 4 (1996)—China Southeast (Guangxi).

ORCHIDOTYPUS (Orchidac.).
bryophytus *(Schltr.) K. Senghas in Schlechter Orchideen*, I/B(31): 1923 (1995): Pachyphyllum bryophytum.
gracillimus *(C. Schweinf.) K. Senghas in Schlechter Orchideen*, I/B(31): 1923 (1995): Pachyphyllum gracillimum.
hispidulus *(Rchb.f.) K. Senghas in Schlechter Orchideen*, I/B(31): 1923 (1995): Aeranthus hispidulus.
mexicanus *(Dressler & Hágsater) K. Senghas in Schlechter Orchideen*, I/B(31): 1923 (1995): Pachyphyllum mexicanum.
schultesii *(L.O. Williams) K. Senghas in Schlechter Orchideen*, I/B(31): 1923 (1995): Pachyphyllum schultesii.
tenuis *(Schltr.) K. Senghas in Schlechter Orchideen*, I/B(31): 1923 (1995): Pachyphyllum tenue.
vaginatus *(Schltr.) K. Senghas in Schlechter Orchideen*, I/B(31): 1923 (1995): Pachyphyllum vaginatum.

× **ORCHIOPHRYS** *hort.in Orchid Rev.*, 107(1230): 391 (1999). *ORCHIDACEAE* (Ophrys × Orchis).

ORCHIS (Orchidac.).
subgen. Anacamptis *(Rich.) P. Quentin in Orchidophile*, 143: 181 (2000): gen. Anacamptis.
subgen. Neotinea *(Rchb. f.) P. Quentin in Orchidophile*, 143: 181 (2000): gen. Neotinea.
× bugarachensis *J. Claessens & J.M. Lewin in Orchidophile*, 117: 109 (1995)—France. (O. conica × ustulata).
chingshuishania *S.S. Ying in Mem. Coll. Agric. Nation. Taiwan Univ.*, 28(2): 34 (1988)—Taiwan.
× haussknechtii M. Schulze
nothof. alboflavida *H. Kümpel, Die wildwachsenden Orchideen der Rhön:* 105 (1996)—Germany. (O. mascula lus. alba × pallens).
holoserica *N.L. Burman in Nova Acta Phys.-Med. Acad. Caes. Leop.-Carol. Nat. Cur.*, 4(Appendix): 237 (1770)—Corse.

ORCHIS:-
mascula (L.) L.
subsp. tenera *(Landwehr) C. Del Prete in Caesiana*, 12: 37 (1999): O. mascula var. tenera.
militaris L.
var. tenuifrons *P.D. Sell in P.D. Sell & G. Murrell, Fl. Great Britain & Ireland*, 5: 365 (1996): Great Britain.
papilionacea L.
var. cyrenaica *(E.A. Durand & Baratte) P. Delforge in Nat. Belg.*, 81(4): 398 (2000): Orchis cyrenaica.
prosteveniella *P. Delforge in Nat. Belg.*, 81(3): 187 (2000), nom. nov.: O. satyrioides *Steven.*
× richardiorum *R. Soca in Caesiana*, 11: 59 (1998)—Sardegna. (O. ichnusae × papilionacea var. grandiflora).

× **ORCHISERAPIAS** (Orchidac.).
myrtoa *J. Kalopissis & M. Aperghis in J. Eur. Orchideen*, 28(4): 738 (1996 publ. 1997)—Greece. (Orchis papilionacea × Serapias vomeracea subsp. laxiflora).

OREACANTHUS (Acanthac.).
coeruleus *(S. Moore) D. Champluvier & E. Figueiredo in Bull. Jard. Bot. Nation. Belg.*, 65(3–4): 416 (1996): Brachystephanus coeruleus.
cristalensis *D. Champluvier & E. Figueiredo in Bull. Jard. Bot. Nation. Belg.*, 65(3–4): 416 (1996), nom. nov.: Brachystephanus mannii *C.B. Clarke.*

OREANTHES (Ericac.).
fragilis *(A.C. Sm.) J.L. Luteyn in Fl. Ecuador*, 54(147): 334 (1996): Semiramisia fragilis.
hypogaeus *(A.C. Sm.) J.L. Luteyn in Fl. Ecuador*, 54(147): 335 (1996): Semiramisia hypogaea.
sperlingii *J.L. Luteyn in Fl. Ecuador*, 54(147): 336 (1996)—Ecuador.

OREOBATUS (Rosac.).
deliciosus (E. James ex Torr.) Rydb.
subsp. neomexicanus *(A. Gray) W.A. Weber in Phytologia*, 79(2): 65 (1995 publ. 1996): Rubus neomexicanus.

× **OREOBIVIA** *M. Lowry in Brit. Cact. Succ. J.*, 18(4): 216 (2000). *CACTACEAE* (Lobivia × Oreocereus).

OREOCEREUS (Cactac.).
doelzianus
var. sericatus *(F. Ritter) J. Lodé, Fichier Encycl. Cact. Autres Succ.*, Sér. 17: no. 1583 (1996); *cf. Repert. Pl. Succ. (I.O.S.)*, 47: 11 (1996 publ. 1997): Morawetzia sericata.

OREOCHARIS (Gesneriac.).
aurea Dunn
var. cordato-ovata *(C.Y. Wu ex H.W. Li) K.Y. Pan, A.L. Weitzman & L.E. Skog in Novon*, 7(4): 430 (1997 publ. 1998): O. cordato-ovata.
dentata *A.L. Weitzman & L.E. Skog in Novon*, 7(4): 431 (1997 publ. 1998)—China South-central (Sichuan).

OREOCNIDE (Urticac.).
integrifolia (Gaudich.) Miq.
var. sylvatica *(Miq.) B.M. Wadhwa in M.D. Dassanayake & W.D. Clayton, eds., Revised Handb. Fl. Ceylon*, 13: 283 (1999), with incorrect basionym & basionym ref.: O. sylvatica.
obovata (C.H. Wright) Merr.
var. paradoxa *(Gagnep.) C.J. Chen in Fl. Reipubl. Popul. Sin.*, 23(2): 384 (1995): Villebrunea paradoxa.

OREOCOMOPSIS *M.G. Pimenov & E.V. Kljuykov in Acta Phytotax. Sin.*, 34(1): 2 (1996). *UMBELLIFERAE*
aromatica *(W.W. Sm.) M.G. Pimenov & E.V. Kljuykov in Feddes Repert.*, 111(7–8): 519 (2000): Pleurospermum aromaticum.
stelliphora *(Cauwet & Farille) M.G. Pimenov & E.V. Kljuykov in Acta Phytotax. Sin.*, 34(1): 3 (1996): Oreocome stelliphora.
xizangensis *M.G. Pimenov & E.V. Kljuykov in Acta Phytotax. Sin.*, 34(1): 3 (1996)—Xizang.

OREOPANAX (Araliac.).
anomalus *M.J. Cannon & J.F.M. Cannon in Novon*, 7(2): 102 (1997)—Costa Rica.
grandifolius *F. Borchsenius in Nordic J. Bot.*, 17(4): 385 (1997)—Ecuador.
impolitus *F. Borchsenius in Nordic J. Bot.*, 17(4): 387 (1997)—Ecuador.
microflorus *F. Borchsenius in Nordic J. Bot.*, 17(4): 387 (1997)—Ecuador, Peru.
obscurus *F. Borchsenius in Nordic J. Bot.*, 17(4): 389 (1997)—Ecuador.
robustus *F. Borchsenius in Nordic J. Bot.*, 17(4): 391 (1997)—Ecuador.
trifidus *F. Borchsenius in Nordic J. Bot.*, 17(4): 393 (1997)—Ecuador, Peru.

OREORCHIS (Orchidac.).
angustata *L.O. Williams ex N. Pearce & P.J. Cribb in Edinburgh J. Bot.*, 54(3): 294 (1997)—China South-central (Yunnan, Sichuan).
aurantiaca *P.J. Cribb & N. Pearce in Edinburgh J. Bot.*, 54(3): 296 (1997)—Burma.
foliosa (Lindl.) Lindl.
var. indica *(Lindl.) N. Pearce & P.J. Cribb in J. Orchid Soc. India*, 10(1–2): 5 (1996); *et in Edinburgh J. Bot.*, 54(3): 307 (1997): Corallorhiza indica.
nepalensis *N. Pearce & P.J. Cribb in J. Orchid Soc. India*, 10(1–2): 8 (1996); *et in Edinburgh J. Bot.*, 54(3): 315 (1997)—Nepal, Xizang.
patens Lindl.
subsp. coreana *(Finet) Y.N. Lee, Fl. Korea:* 1164 (1996): O. coreana.

ORFILEA (Euphorbiac.).
ankafinensis *(Baill.) A. Radcliffe-Smith & R. Govaerts in Kew Bull.*, 52(2): 480 (1997): Macaranga ankafinensis.

ORIGANUM (Labiat.).
× adanense *K.H.C. Başer & H. Duman in Turkish J. Bot.*, 22: 55 (1998)—Turkey. (O. bargyli × laevigatum).
husnucan-baseri *H. Duman, Z. Aytaç & A. Duran in Flora Medit.*, 5: 226 (1995 publ. 1996), as 'husnucan-baserii'—Turkey.
jordanicum *A. Danin & I. Künne in Willdenowia*, 25(2): 603 (1996)—Palestine (Jordan).

ORITES (Proteac.).
sect. Excelsae *A.S. George & B.P.M. Hyland in Fl. Australia*, 16: 496 (1995).
megacarpa *A.S. George & B.P.M. Hyland in Fl. Australia*, 16: 497 (1995)—Queensland.

ORITROPHIUM (Compos.).
durangense *G.L. Nesom in Sida*, 18(2): 523 (1998)—Mexico Northeast (Durango).
llanganatense *P. Sklenář & H. Robinson in Novon*, 10(2): 144 (2000)—Ecuador.

ORLEANESIA (Orchidac.).
ecuadorana *C.H. Dodson in Orquideologia*, 20(3): 277 (1997)—Ecuador.

ORMOCARPOPSIS (Legumin.).
itremoensis *D.J. Du Puy & J.-N. Labat in Novon*, 6(1): 54 (1996)—Madagascar.
tulearensis *D.J. Du Puy & J.-N. Labat in Novon*, 6(1): 56 (1996)—Madagascar.

ORMOCARPUM (Legumin.).
bernierianum *(Baill.) D.J. Du Puy & J.-N. Labat in Novon*, 6(1): 57 (1996): Diphaca bernieriana.
zambesianum *B. Verdcourt in Fl. Zambes.*, 3(6): 57 (2000)—Mozambique, Zimbabwe.

ORMOSIA (Legumin.).
bolivarensis *(Rudd) C.H. Stirton in Fl. Venezuelan Guayana (gen. eds. J.A. Steyermark et al.)*, 5: 366 (1999): O. nobilis var. bolivarensis.
robusta *(Wight) J.O. Voigt, Hort. Suburb. Calcutt.:* 205 (1845): Sophora robusta.

ORNITHOCEPHALUS (Orchidac.).
aurorae *D.E. Bennett & E.A. Christenson in Lindleyana*, 13(2): 80 (1998)—Peru.
brachyceras *G.A. Romero & G. Carnevali, Orchids of Venezuela, ed. 2:* 1140 (2000)—Venezuela.

ORNITHOCEPHALUS:-
castelfrancoi *F. Pupulin in Lindleyana,* 15(1): 27 (2000)—Costa Rica.
cryptanthus *(C. Schweinf. & P.H. Allen) A.L.V. Toscano de Brito & R.L. Dressler in Lindleyana,* 15(4): 255 (2000): Oakes-amesia cryptantha.
dalstroemii *(Dodson) A.L.V. Toscano de Brito & R.L. Dressler in Lindleyana,* 15(4): 255 (2000), as 'dalströmii': Sphyrastylis dalstroemii.
dressleri *(Toscano) A.L.V. Toscano de Brito & R.L. Dressler in Lindleyana,* 15(4): 255 (2000): Sphyrastylis dressleri.
ecuadorensis *(Garay) A.L.V. Toscano de Brito & R.L. Dressler in Lindleyana,* 15(4): 255 (2000): Sphyrastylis ecuadorensis.
escobarianus *(Garay) A.L.V. Toscano de Brito & R.L. Dressler in Lindleyana,* 15(4): 255 (2000): Sphyrastylis escobariana.
garayi *(D.E. Benn. & Christenson) A.L.V. Toscano de Brito & R.L. Dressler in Lindleyana,* 15(4): 255 (2000): Sphyrastylis garayi.
grex-anserinus *R.L. Dressler & D.E. Mora-Retana in Novon,* 7(2): 120 (1997)—Costa Rica.
hoppii *(Schltr.) A.L.V. Toscano de Brito & R.L. Dressler in Lindleyana,* 15(4): 255 (2000): Sphyrastylis hoppii.
numenius *A.L.V. Toscano de Brito & R.L. Dressler in Lindleyana,* 15(4): 252 (2000)—Honduras, Nicaragua.
oberonioides *(Schltr.) A.L.V. Toscano de Brito & R.L. Dressler in Lindleyana,* 15(4): 255 (2000): Sphyrastylis oberonioides.
torresii *G.A. Salazar & M.A. Soto Arenas in Brittonia,* 48(2): 209 (1996)—Mexico Southwest (Oaxaca).
tsubotae *(P. Ortiz) A.L.V. Toscano de Brito & R.L. Dressler in Lindleyana,* 15(4): 255 (2000): Sphyrastylis tsubotae.
urceilabris *(P. Ortiz & R. Escobar) A.L.V. Toscano de Brito & R.L. Dressler in Lindleyana,* 15(4): 256 (2000): Sphyrastylis urceilabris.

ORNITHOGALUM (Hyacinthac.).
subser. Acuminata *U. Müller-Doblies & D. Müller-Doblies in Feddes Repert.,* 107(5–6): 495 (1996).
sect. Aspasia *U. Müller-Doblies & D. Müller-Doblies in Feddes Repert.,* 107(5–6): 384 (1996).
subsect. Aspasia *U. Müller-Doblies & D. Müller-Doblies in Feddes Repert.,* 107(5–6): 384 (1996).
ser. Aspasia *U. Müller-Doblies & D. Müller-Doblies in Feddes Repert.,* 107(5–6): 384 (1996).
ser. Ciliifera *U. Müller-Doblies & D. Müller-Doblies in Feddes Repert.,* 107(5–6): 414 (1996).
ser. Clavifolia *U. Müller-Doblies & D. Müller-Doblies in Feddes Repert.,* 107(5–6): 516 (1996).
ser. Coriacea *U. Müller-Doblies & D. Müller-Doblies in Feddes Repert.,* 107(5–6): 518 (1996).
ser. Discifera *U. Müller-Doblies & D. Müller-Doblies in Feddes Repert.,* 107(5–6): 487 (1996).
ser. Drakensbergensia *U. Müller-Doblies & D. Müller-Doblies in Feddes Repert.,* 107(5–6): 398 (1996).
ser. Erecta *U. Müller-Doblies & D. Müller-Doblies in Feddes Repert.,* 107(5–6): 495 (1996).
ser. Flexuosa *U. Müller-Doblies & D. Müller-Doblies in Feddes Repert.,* 107(5–6): 484 (1996).
sect. Helogalum *U. Müller-Doblies & D. Müller-Doblies in Feddes Repert.,* 107(5–6): 483 (1996).
ser. Hispida *U. Müller-Doblies & D. Müller-Doblies in Feddes Repert.,* 107(5–6): 404 (1996).
sect. Hispidaspasia *U. Müller-Doblies & D. Müller-Doblies in Feddes Repert.,* 107(5–6): 404 (1996).
ser. Hispidula *U. Müller-Doblies & D. Müller-Doblies in Feddes Repert.,* 107(5–6): 408 (1996).
ser. Juncifolia *U. Müller-Doblies & D. Müller-Doblies in Feddes Repert.,* 107(5–6): 436 (1996).
ser. Leeupoortensia *U. Müller-Doblies & D. Müller-Doblies in Feddes Repert.,* 107(5–6): 396 (1996).
sect. Linaspasia *U. Müller-Doblies & D. Müller-Doblies in Feddes Repert.,* 107(5–6): 436 (1996).
subser. Linearia *U. Müller-Doblies & D. Müller-Doblies in Feddes Repert.,* 107(5–6): 504 (1996).
ser. Maculata *U. Müller-Doblies & D. Müller-Doblies in Feddes Repert.,* 107(5–6): 390 (1996).
sect. Monarchos *U. Müller-Doblies & D. Müller-Doblies in Feddes Repert.,* 107(5–6): 470 (1996).
sect. Monosmyne *U. Müller-Doblies & D. Müller-Doblies in Feddes Repert.,* 107(5–6): 515 (1996).

ORNITHOGALUM:-
sect. Namibiogalum *U. Müller-Doblies & D. Müller-Doblies in Feddes Repert.,* 107(5–6): 487 (1996).
sect. Nanaspasia *U. Müller-Doblies & D. Müller-Doblies in Feddes Repert.,* 107(5–6): 459 (1996).
ser. Nanula *U. Müller-Doblies & D. Müller-Doblies in Feddes Repert.,* 107(5–6): 459 (1996).
ser. Noctiflora *U. Müller-Doblies & D. Müller-Doblies in Feddes Repert.,* 107(5–6): 412 (1996).
ser. Notopterygia *U. Müller-Doblies & D. Müller-Doblies in Feddes Repert.,* 107(5–6): 397 (1996).
sect. Osmyne *U. Müller-Doblies & D. Müller-Doblies in Feddes Repert.,* 107(5–6): 495 (1996).
ser. Osmyne *U. Müller-Doblies & D. Müller-Doblies in Feddes Repert.,* 107(5–6): 505 (1996).
ser. Prasinocorymbosa *U. Müller-Doblies & D. Müller-Doblies in Feddes Repert.,* 107(5–6): 490 (1996).
ser. Rupestria *U. Müller-Doblies & D. Müller-Doblies in Feddes Repert.,* 107(5–6): 400 (1996).
subgen. Spetagalum *U. Müller-Doblies & D. Müller-Doblies in Feddes Repert.,* 107(5–6): 520 (1996), without latin descr.
ser. Spirifolia *U. Müller-Doblies & D. Müller-Doblies in Feddes Repert.,* 107(5–6): 509 (1996).
subsect. Teretaspasia *U. Müller-Doblies & D. Müller-Doblies in Feddes Repert.,* 107(5–6): 398 (1996).
ser. Unifolia *U. Müller-Doblies & D. Müller-Doblies in Feddes Repert.,* 107(5–6): 470 (1996).
sect. Urophyllon *U. Müller-Doblies & D. Müller-Doblies in Feddes Repert.,* 107(5–6): 488 (1996).
ser. Urophyllon *U. Müller-Doblies & D. Müller-Doblies in Feddes Repert.,* 107(5–6): 488 (1996).
sect. Vaginaspasia *U. Müller-Doblies & D. Müller-Doblies in Feddes Repert.,* 107(5–6): 428 (1996).
ser. Xanthochlora *U. Müller-Doblies & D. Müller-Doblies in Feddes Repert.,* 107(5–6): 492 (1996).
ser. Zebrina *U. Müller-Doblies & D. Müller-Doblies in Feddes Repert.,* 107(5–6): 515 (1996).
ser. Zebrinella *U. Müller-Doblies & D. Müller-Doblies in Feddes Repert.,* 107(5–6): 429 (1996).
adseptentrionesvergentulum *U. Müller-Doblies & D. Müller-Doblies in Feddes Repert.,* 107(5–6): 446 (1996)—Cape Province.
aetfatense *U. Müller-Doblies & D. Müller-Doblies in Feddes Repert.,* 107(5–6): 459 (1996)—Transvaal.
anamurense *F. Speta in Preslia,* 72(2–4): 373 (2000)—Turkey.
autumnulum *U. Müller-Doblies & D. Müller-Doblies in Feddes Repert.,* 107(5–6): 511 (1996)—Cape Province.
bruce-bayeri *U. Müller-Doblies & D. Müller-Doblies in Feddes Repert.,* 107(5–6): 505 (1996)—Cape Province.
campanulatum *U. Müller-Doblies & D. Müller-Doblies in Feddes Repert.,* 107(5–6): 450 (1996)—Cape Province.
ciliiferum *U. Müller-Doblies & D. Müller-Doblies in Feddes Repert.,* 107(5–6): 414 (1996)—Cape Province.
collinum Guss.
subsp. rhodium *F. Speta in Preslia,* 72(2–4): 376 (2000)—East Aegean Is.
concordianum *(Baker) U. Müller-Doblies & D. Müller-Doblies in Feddes Repert.,* 107(5–6): 509 (1996): Albuca concordiana.
costatulum *U. Müller-Doblies & D. Müller-Doblies in Feddes Repert.,* 107(5–6): 482 (1996)—Namibia.
decus-montium *G. Williamson in Bothalia,* 28(1): 63 (1998)—Cape Province.
demirizianum *H. Malyer & M. Koyuncu in Turkish J. Bot.,* 13(2): 301 (1989)—Turkey.
dolichopharynx *U. Müller-Doblies & D. Müller-Doblies in Feddes Repert.,* 107(5–6): 466 (1996)—Cape Province.
dyeri Poelln.
var. leistneri *U. Müller-Doblies & D. Müller-Doblies in Feddes Repert.,* 107(5–6): 476 (1996)—Cape Province.
etesiogaripense *U. Müller-Doblies & D. Müller-Doblies in Feddes Repert.,* 107(5–6): 476 (1996)—Namibia, Cape Province.
subsp. longipilosum *U. Müller-Doblies & D. Müller-Doblies in Feddes Repert.,* 107(5–6): 478 (1996)—Namibia.
fissurisedulum *U. Müller-Doblies & D. Müller-Doblies in Feddes Repert.,* 107(5–6): 466 (1996)—Namibia.

ORNITHOGALUM:-
flexuosum *(Thunb.) U. Müller-Doblies & D. Müller-Doblies in Feddes Repert.,* 107(5–6): 484 (1996): Hyacinthus flexuosus.
gabrielianiae *N. Agapova in Willdenowia,* 27(1–2): 199 (1997), as 'gabrielianae'—Transcaucasus (Armeniya).
gethylloides *U. Müller-Doblies & D. Müller-Doblies in Feddes Repert.,* 107(5–6): 513 (1996)—Cape Province.
gifbergense *U. Müller-Doblies & D. Müller-Doblies in Feddes Repert.,* 107(5–6): 418 (1996)—Cape Province.
glaucifolium *U. Müller-Doblies & D. Müller-Doblies in Feddes Repert.,* 107(5–6): 503 (1996)—Cape Province.
gregorianum *U. Müller-Doblies & D. Müller-Doblies in Feddes Repert.,* 107(5–6): 462 (1996)—Cape Province.
haalenbergense *U. Müller-Doblies & D. Müller-Doblies in Feddes Repert.,* 107(5–6): 492 (1996)—Namibia.
hesperanthum *U. Müller-Doblies & D. Müller-Doblies in Feddes Repert.,* 107(5–6): 443 (1996)—Cape Province.
hispidulum *U. Müller-Doblies & D. Müller-Doblies in Feddes Repert.,* 107(5–6): 411 (1996)—Namibia.
immaculatum *F. Speta in Preslia,* 72(2–4): 384 (2000)—Greece.
karachabpoortense *U. Müller-Doblies & D. Müller-Doblies in Feddes Repert.,* 107(5–6): 517 (1996)—Cape Province.
knersvlaktense *U. Müller-Doblies & D. Müller-Doblies in Feddes Repert.,* 107(5–6): 497 (1996)—Cape Province.
leeupoortense *U. Müller-Doblies & D. Müller-Doblies in Feddes Repert.,* 107(5–6): 396 (1996)—Cape Province.
longicollum *U. Müller-Doblies & D. Müller-Doblies in Feddes Repert.,* 107(5–6): 462 (1996)—Cape Province.
mater-familias *U. Müller-Doblies & D. Müller-Doblies in Feddes Repert.,* 107(5–6): 444 (1996)—Cape Province.
microcarpum *F. Speta in Phyton (Austria),* 40(1): 128 (2000)—Turkey.
monarchos *U. Müller-Doblies & D. Müller-Doblies in Feddes Repert.,* 107(5–6): 476 (1996)—Cape Province.
monophyllum Baker
subsp. eckardtianum *U. Müller-Doblies & D. Müller-Doblies in Feddes Repert.,* 107(5–6): 434 (1996)—Cape Province.
munzurense *F. Speta in Preslia,* 72(2–4): 392 (2000)—Turkey.
namaquanulum *U. Müller-Doblies & D. Müller-Doblies in Feddes Repert.,* 107(5–6): 410 (1996)—Cape Province.
namaquanum *U. Müller-Doblies & D. Müller-Doblies in Feddes Repert.,* 107(5–6): 500 (1996)—Cape Province.
nathoanum *U. Müller-Doblies & D. Müller-Doblies in Feddes Repert.,* 107(5–6): 488 (1996)—Cape Province.
navaschinii *N.D. Agapova in Novosti Sist. Vyssh. Rast.,* 31: 31 (1998)—Transcaucasus (Armeniya), Krym.
osmynellum *U. Müller-Doblies & D. Müller-Doblies in Feddes Repert.,* 107(5–6): 511 (1996)—Cape Province.
ovatum Thunb.
subsp. oliverorum *U. Müller-Doblies & D. Müller-Doblies in Feddes Repert.,* 107(5–6): 474 (1996)—Cape Province.
paucifolium *U. Müller-Doblies & D. Müller-Doblies in Feddes Repert.,* 107(5–6): 514 (1996)—Cape Province.
subsp. karooparkense *U. Müller-Doblies & D. Müller-Doblies in Feddes Repert.,* 107(5–6): 515 (1996)—Cape Province.
pendulinum *U. Müller-Doblies & D. Müller-Doblies in Feddes Repert.,* 107(5–6): 480 (1996)—Namibia.
polyodontulum *U. Müller-Doblies & D. Müller-Doblies in Feddes Repert.,* 107(5–6): 498 (1996)—Namibia.
psammophorum *U. Müller-Doblies & D. Müller-Doblies in Feddes Repert.,* 107(5–6): 482 (1996)—Cape Province.

ORNITHOGALUM:-
puberulum Oberm.
subsp. chris-bayeri *U. Müller-Doblies & D. Müller-Doblies in Feddes Repert.,* 107(5–6): 390 (1996)—Namibia.
rossouwii *U. Müller-Doblies & D. Müller-Doblies in Feddes Repert.,* 107(5–6): 394 (1996)—Cape Province.
rotatum *U. Müller-Doblies & D. Müller-Doblies in Feddes Repert.,* 107(5–6): 482 (1996)—Cape Province.
sabulosum *U. Müller-Doblies & D. Müller-Doblies in Feddes Repert.,* 107(5–6): 507 (1996)—Cape Province.
scabrocostatum *U. Müller-Doblies & D. Müller-Doblies in Feddes Repert.,* 107(5–6): 516 (1996)—Cape Province.
semipedale *(Baker) U. Müller-Doblies & D. Müller-Doblies in Feddes Repert.,* 107(5–6): 504 (1996): Albuca semipedalis.
strigosulum *U. Müller-Doblies & D. Müller-Doblies in Feddes Repert.,* 107(5–6): 480 (1996)—Namibia.
stuetzelianum *U. Müller-Doblies & D. Müller-Doblies in Feddes Repert.,* 107(5–6): 503 (1996)—Namibia, Cape Province.
subglandulosum *U. Müller-Doblies & D. Müller-Doblies in Feddes Repert.,* 107(5–6): 513 (1996)—Cape Province.
synadelphicum *U. Müller-Doblies & D. Müller-Doblies in Feddes Repert.,* 107(5–6): 452 (1996)—Cape Province.
thunbergianulum *U. Müller-Doblies & D. Müller-Doblies in Feddes Repert.,* 107(5–6): 440 (1996), nom. nov.: Anthericum filifolium *Thunb.*
toxicarium *C. Archer & R.H. Archer in S. Afr. J. Bot.,* 65(5–6): 431 (1999)—Cape Province, Namibia.
umbellatum L.
subsp. angustifolium *(Boreau) P.D. Sell in P.D. Sell & G. Murrell, Fl. Great Britain & Ireland,* 5: 363 (1996): O. angustifolium.
unifolium Retz.
var. vestitum *U. Müller-Doblies & D. Müller-Doblies in Feddes Repert.,* 107(5–6): 471 (1996)—Cape Province.
vasakii *F. Speta in Linz. Biol. Beitr.,* 31(1): 438 (1999)—Turkey.
verae *U. Müller-Doblies & D. Müller-Doblies in Feddes Repert.,* 107(5–6): 432 (1996)—Cape Province.

ORNITHOPUS (Legumin.).
glaber *M. Gandoger, Dec. Pl. Nov.,* 1: 45 (1875)—France.
littoralis *M. Gandoger, Dec. Pl. Nov.,* 1: 45 (1875)—France.
microphyllus *M. Gandoger, Dec. Pl. Nov.,* 1: 45 (1875)—France.
sativus Brot.
nothosubsp. macrorrhynchus *(Willk.) S. Talavera, M. Arista & P.L. Ortiz in An. Jard. Bot. Madrid,* 57(1): 227 (1999): O. roseus β macrorrhynchus.

OROBANCHE (Scrophulariac.).
ser. Alsaticae *E.S. Teryokhin in Opred. Zarazikhovŷkh Fl. SSSR (ed. L.Yu. Budantsev):* 39 (1993).
subsect. Amoenae *(Beck) E.S. Teryokhin in Opred. Zarazikhovŷkh Fl. SSSR (ed. L.Yu. Budantsev):* 37 (1993): O. ser. Amoenae.
subsect. Arcuatae *(Beck) E.S. Teryokhin in Opred. Zarazikhovŷkh Fl. SSSR (ed. L.Yu. Budantsev):* 40 (1993): O. ser. Arcuatae.
subsect. Camptolepides *E.S. Teryokhin in Opred. Zarazikhovŷkh Fl. SSSR (ed. L.Yu. Budantsev):* 38 (1993).
ser. Clarkei *E.S. Teryokhin in Opred. Zarazikhovŷkh Fl. SSSR (ed. L.Yu. Budantsev):* 37 (1993).
subsect. Coerulescentes *(Beck) E.S. Teryokhin in Opred. Zarazikhovŷkh Fl. SSSR (ed. L.Yu. Budantsev):* 37 (1993): O. ser. Coerulescentes.
subsect. Cruentae *(Beck) E.S. Teryokhin in Opred. Zarazikhovŷkh Fl. SSSR (ed. L.Yu. Budantsev):* 40 (1993): O. ser. Cruentae.
subsect. Galeatae *(Beck) E.S. Teryokhin in Opred. Zarazikhovŷkh Fl. SSSR (ed. L.Yu. Budantsev):* 39 (1993): O. ser. Galeatae.

OROBANCHE:-
sect. Gamosepalae *E.S. Teryokhin in Opred. Zarazikhovŷkh Fl. SSSR (ed. L.Yu. Budantsev):* 40 (1993).
subsect. Glandulosae *(Beck) E.S. Teryokhin in Opred. Zarazikhovŷkh Fl. SSSR (ed. L.Yu. Budantsev):* 38 (1993): O. ser. Glandulosae.
subsect. Hederae *E.S. Teryokhin in Opred. Zarazikhovŷkh Fl. SSSR (ed. L.Yu. Budantsev):* 38 (1993).
ser. Kotschyanae *E.S. Teryokhin in Opred. Zarazikhovŷkh Fl. SSSR (ed. L.Yu. Budantsev):* 41 (1993).
sect. Kotschyinae *E.S. Teryokhin in Opred. Zarazikhovŷkh Fl. SSSR (ed. L.Yu. Budantsev):* 40 (1993).
subsect. Minores *(Beck) E.S. Teryokhin in Opred. Zarazikhovŷkh Fl. SSSR (ed. L.Yu. Budantsev):* 39 (1993): O. ser. Minores.
ser. Sordidae *E.S. Teryokhin in Opred. Zarazikhovŷkh Fl. SSSR (ed. L.Yu. Budantsev):* 41 (1993).
subsect. Speciosae *(Beck) E.S. Teryokhin in Opred. Zarazikhovŷkh Fl. SSSR (ed. L.Yu. Budantsev):* 38 (1993): O. ser. Speciosae.
subsect. Vitellinae *E.S. Teryokhin in Opred. Zarazikhovŷkh Fl. SSSR (ed. L.Yu. Budantsev):* 41 (1993).
acaulis *W. Roxburgh, Pl. Coromandel,* 3: 89 (1820)—India (West Bengal).†
alba Stephan ex Willd.
subsp. major *(Čelak.) J. Zázvorka in Preslia,* 72(1): 88 (2000): O. epithymum β major.
arenaria Borkh.
var. euxina *(Velen.) D. Delipavlov in Fl. Republ. Bŭlgariya,* 10: 302 (1995): Phelipaea arenaria var. euxina.
artemisiae-campestris Vaucher ex Gaudin
subsp. picridis *(F.W. Schultz ex W.D.J. Koch) O. de Bolòs & J. Vigo, Fl. Països Catalans,* 3: 516 (1995): O. picridis.
ballotae *A.J. Pujadas Salvá in Acta Bot. Malacitana,* 22: 30 (1997)—Spain.
cernua Loefl.
subsp. rajahmundrica *E.S. Teryokhin, Weed Broomrapes:* 15 (1997), without type—Locality not stated.
clausonis Pomel
subsp. hesperina *(J.A. Guim.) M.J.Y. Foley in An. Jard. Bot. Madrid,* 54(1): 324 (1996): O. lucorum var. hesperina.
dhofarensis *M.J.Y. Foley in Edinburgh J. Bot.,* 55(2): 230 (1998)—Oman.
hypertomentosa *M.J.Y. Foley in Edinburgh J. Bot.,* 55(2): 232 (1998)—Saudi Arabia, Gulf States (Bahrain).
major L.
subsp. icterica *(Pau) A.J. Pujadas Salvà in Flora Montiberica,* 11: 16 (1999): O. icterica.
montserratii *A.J. Pujadas Salvà & D. Gómez García in An. Jard. Bot. Madrid,* 57(2): 269 (1999 publ. 2000)—Spain.
ramosa L.
var. monoclonos *(Wallr.) D. Delipavlov in Fl. Republ. Bŭlgariya,* 10: 295 (1995): Phelipaea ramosa var. monoclonos.

OROBUS (Legumin.).
baicalensis *(Turcz.) A.K. Stankevich & Yu.R. Roskov in Kew Bull.,* 53(3): 732 (1998): O. venosus var. baicalensis.
davidii *(Hance) A.K. Stankevich & Yu.R. Roskov in Kew Bull.,* 53(3): 732 (1998): Lathyrus davidii.
komarovii *(Ohwi) A.K. Stankevich & Yu.R. Roskov in Kew Bull.,* 53(3): 732 (1998): Lathyrus komarovii.
krylovii *(Serg.) A.K. Stankevich & Yu.R. Roskov in Kew Bull.,* 53(3): 732 (1998): Lathyrus krylovii.
lacaitae *(Czefr.) A.K. Stankevich & Yu.R. Roskov in Kew Bull.,* 53(3): 732 (1998): Lathyrus lacaitae.
multijugus *(Ledeb.) A.K. Stankevich & Yu.R. Roskov in Kew Bull.,* 53(3): 732 (1998): O. lacteus var. multijugus.
nipponicus *(Matsum.) A.K. Stankevich & Yu.R. Roskov in Kew Bull.,* 53(3): 732 (1998): Vicia nipponica.
ohwianus *(Hosok.) A.K. Stankevich & Yu.R. Roskov in Kew Bull.,* 53(3): 732 (1998): Vicia ohwiana.

OROBUS:-
subrotundus *(Maxim.) A.K. Stankevich & Yu.R. Roskov in Kew Bull.,* 53(3): 732 (1998): Lathyrus subrotundus.

OROPHACA (Legumin.).
triphylla
var. purpurea *(Dorn) D. Isely, Native & Natural. Legumin. (Fabac.) U.S. (excl. Alaska & Hawaii):* 757 (1998): Astragalus gilviflorus var. purpureus.

OROPHEA (Annonac.).
sivarajanii *N. Sasidharan in Nordic J. Bot.,* 19(3): 301 (1999)—India (Kerala).

OROSTACHYS (Crassulac.).
ser. Aggregatae *V.V. Byalt in Novosti Sist. Vŷssh. Rast.,* 32: 42 (2000).
sect. Caespitosae *T.A. Bezdeleva in Sosud. Rast. Sovet. Dal'nego Vostoka,* 7: 231 (1995), without latin descr. or type.
subgen. Schoenlandia *(H. Ohba) V.V. Byalt in Novosti Sist. Vŷssh. Rast.,* 32: 48 (2000): O. sect. Schoenlandia.
sect. Woroschilovia *V.V. Byalt in Novosti Sist. Vŷssh. Rast.,* 32: 41 (2000).
aliciae (Raym.-Hamet) H. Ohba
var. komarovii *(Raym.-Hamet) H. Ohba in J. Jap. Bot.,* 74(1): 61 (1999): Sedum aliciae var. komarovii.
komarovii *(Raym.-Hamet) V.V. Byalt in Novosti Sist. Vŷssh. Rast.,* 32: 49 (2000): Sedum aliciae var. komarovii.
maximowiczii *V.V. Byalt in Novosti Sist. Vŷssh. Rast.,* 32: 44 (2000)—Manchuria.
minutus (Kom.) A. Berger
f. albus *Y.N. Lee, Fl. Korea:* 1158 (1996), without type—Locality not stated.
ramosissimus *(Maxim.) V.V. Byalt in Problemi Botaniki n rubeshe 20/21 vekov:* 167 (1998); *cf. Repert. Pl. Succ. (I.O.S.),* 50: 18 (1999 publ. 2000): Umbilicus ramosissimus.
stenostachya (Fröd.) H. Ohba
var. integrifolia *(S.H. Fu) V.V. Byalt in Novosti Sist. Vŷssh. Rast.,* 32: 49 (2000): Sinocrassula stenostachya var. integrifolia.
var. lepidotricha *(K.T. Fu) V.V. Byalt in Novosti Sist. Vŷssh. Rast.,* 32: 50 (2000): Sinocrassula stenostachya var. lepidotricha.
thyrsiflora Fisch.
var. rosea *(Less.) V.V. Byalt in Novosti Sist. Vŷssh. Rast.,* 32: 46 (2000): Cotyledon rosea.
vyschinii *T.A. Bezdeleva in Sosud. Rast. Sovet. Dal'nego Vostoka,* 7: 232 (1995)—Primorye.

ORTHAEA (Ericac.).
ecuadorensis *J.L. Luteyn in Fl. Ecuador,* 54(147): 117 (1996)—Ecuador.
oriens *J.L. Luteyn in Fl. Ecuador,* 54(147): 114 (1996)—Ecuador.

× **ORTHOLARIUM** *Racine C. Foster & M.B. Foster in J. Bromeliad Soc.,* 23(5): 175 (1973). *BROMELIACEAE* (Nidularium × Orthophytum).

× **ORTHOMEA** *E.L. Smith in J. Bromeliad Soc.,* 33(2): 75 (1983). *BROMELIACEAE* (Aechmea × Orthophytum).

ORTHOPHYTUM (Bromeliac.).
benzingii *E.M.C. Leme & H.E. Luther in J. Bromeliad Soc.,* 48(4): 150 (1998)—Brazil Southeast (Minas Gerais).
eddie-estevesii *E.M.C. Leme in J. Bromeliad Soc.,* 50(2): 55 (2000)—Brazil Southeast (Minas Gerais).
lucidum *E.M.C. Leme & H.E. Luther in Selbyana,* 19(2): 189 (1998 publ. 1999)—Brazil Southeast (Minas Gerais).
sucrei *H.E. Luther in Selbyana,* 18(1): 95 (1997)—Brazil Northeast (Bahia).

ORTHOSIPHON (Labiat.).
adenocaulis *A.J. Paton & I.C. Hedge in Fl. Madagascar,* fam. 175: 132 (1998)—Madagascar.
argenteus *A.J. Paton & I.C. Hedge in Fl. Madagascar,* fam. 175: 122 (1998)—Madagascar.

ORTHOSIPHON:-
biflorus *A.J. Paton & I.C. Hedge in Fl. Madagascar,* fam. 175: 126 (1998)—Madagascar.
discolor *A.J. Paton & I.C. Hedge in Fl. Madagascar,* fam. 175: 129 (1998)—Madagascar.
ellipticus *A.J. Paton & I.C. Hedge in Fl. Madagascar,* fam. 175: 127 (1998)—Madagascar.
exilis *A.J. Paton & I.C. Hedge in Fl. Madagascar,* fam. 175: 131 (1998)—Madagascar.
miserabilis *A.J. Paton & I.C. Hedge in Fl. Madagascar,* fam. 175: 133 (1998)—Madagascar.
ruber *A.J. Paton & I.C. Hedge in Fl. Madagascar,* fam. 175: 120 (1998)—Madagascar.
sarmentosus *A.J. Paton & I.C. Hedge in Fl. Madagascar,* fam. 175: 124 (1998)—Madagascar.

ORYCHOPHRAGMUS (Crucif.).
diffusus *Z.M. Tan & J.M. Xu in Acta Phytotax. Sin.,* 36(6): 547 (1998)—China Southeast (Zhejiang, Jiangsu).
hupehensis *(Pamp.) Z.M. Tan & X.Liang Zhang in Acta Phytotax. Sin.,* 36(6): 546 (1998): O. sonchifolius var. hupehensis.
limprichtianus *(Pax) I.A. Al-Shehbaz & G. Yang in Novon,* 10(4): 351 (2000): Cardamine limprichtiana.
taibaiensis *Z.M. Tan & B.X. Zhao in Acta Phytotax. Sin.,* 36(6): 544 (1998)—China North-central (Shaanxi).

ORYCTINA (Loranthac.).
eubrachioides *J. Kuijt in Novon,* 10(4): 391 (2000)—Brazil Southeast (Minas Gerais).
quadrangularis *J. Kuijt in Novon,* 10(4): 393 (2000)—Brazil Southeast (Minas Gerais).

ORYXIS *A. Delgado Salinas & G.P. Lewis in Kew Bull.,* 52(1): 221 (1997). *LEGUMINOSAE*
monticola *(Mart. ex Benth.) A. Delgado Salinas & G.P. Lewis in Kew Bull.,* 52(1): 221 (1997): Dolichos monticola.

ORYZA (Gramin.).
meyeriana (Zoll. & Moritzi) Baill.
subsp. tuberculata *W.C. Wu, Y.G. Lu & G.C. Wang in Chinese J. Rice Sci.,* 4(1): 35 (1990)—Hainan, China South-central (Yunnan).

OSCULARIA (Aizoac.).
alba *(L. Bolus) H.E.K. Hartmann in Bradleya,* 16: 75 (1998): Mesembryanthemum album.
cedarbergensis *(L. Bolus) H.E.K. Hartmann in Bradleya,* 16: 76 (1998): Mesembryanthemum cedarbergense.
compressa *(L. Bolus) H.E.K. Hartmann in Bradleya,* 16: 76 (1998): Lampranthus compressus.
comptonii *(L. Bolus) H.E.K. Hartmann in Bradleya,* 16: 76 (1998): M. comptonii.
copiosa *(L. Bolus) H.E.K. Hartmann in Bradleya,* 16: 76 (1998): Mesembryanthemum copiosum.
excedens *(L. Bolus) H.E.K. Hartmann in Bradleya,* 16: 77 (1998): Mesembryanthemum excedens.
falciformis *(Haw.) H.E.K. Hartmann in Bradleya,* 16: 77 (1998): Mesembryanthemum falciforme.
guthrieae *(L. Bolus) H.E.K. Hartmann in Bradleya,* 16: 77 (1998), as 'guthriae': Mesembryanthemum guthrieae.
lunata *(Willd.) H.E.K. Hartmann in Bradleya,* 16: 77 (1998): M. lunatum.
ornata *(L. Bolus) H.E.K. Hartmann in Bradleya,* 16: 78 (1998): Lampranthus ornatus.
paardebergensis *(L. Bolus) H.E.K. Hartmann in Bradleya,* 16: 78 (1998): Mesembryanthemum paardebergense.
piquetbergensis *(L. Bolus) H.E.K. Hartmann in Bradleya,* 16: 78 (1998): Mesembryanthemum piquetbergense.
prasina *(L. Bolus) H.E.K. Hartmann in Bradleya,* 16: 78 (1998): Lampranthus prasinus.
primiverna *(L. Bolus) H.E.K. Hartmann in Bradleya,* 16: 79 (1998): Mesembryanthemum primivernum.
steenbergensis *(L. Bolus) H.E.K. Hartmann in Bradleya,* 16: 79 (1998): Mesembryanthemum steenbergense.
superans *(L. Bolus) H.E.K. Hartmann in Bradleya,* 16: 79 (1998): Mesembryanthemum superans.

OSCULARIA:-
thermarum *(L. Bolus) H.E.K. Hartmann in Bradleya,* 16: 79 (1998): Mesembryanthemum thermarum.
vernicolor *(L. Bolus) H.E.K. Hartmann in Bradleya,* 16: 79 (1998): Mesembryanthemum vernicolor.
vredenburgensis *(L. Bolus) H.E.K. Hartmann in Bradleya,* 16: 79 (1998): Lampranthus vredenburgensis.

OSMANTHUS (Oleac.).
austrocaledonicus (Vieill.) Knobl.
subsp. badula *(Vieill. ex Pancher & Sebert) P.S. Green in Fl. N. Caléd.,* 22: 40 (1998): Notelaea badula.
subsp. collinus *(Schltr.) P.S. Green in Fl. N. Caléd.,* 22: 38 (1998): Notelaea collina.
var. crassifolius *(Guillaumin) P.S. Green in Fl. N. Caléd.,* 22: 37 (1998): O. crassifolius.
fragrans Lour.
f. aurantiacus *(Makino) S. Hatusima in S. Hatusima & T. Amano, Fl. Ryukyus, S. of Amami Is., ed. 2:* 170 (1994), without basionym ref.: Basionym not stated.
heterophyllus (G. Don) P.S. Green
var. iriomotensis *S. Hatusima in S. Hatusima & T. Amano, Fl. Ryukyus, S. of Amami Is., ed. 2:* 170 (1994), without latin descr. or type—Locality not stated.

OSMITES *L., Sp. Pl., ed. 2:* 1285 (1763). *COMPOSITAE†*
asteriscoides *L., Sp. Pl., ed. 2:* 1285 (1763)—South Africa (Cape Province).†

OSMITOPSIS (Compos.).
asteriscoides *(L.) C.F. Lessing Syn. Comp.:* 386 (1832): Osmites asteriscoides†

OSMOXYLON (Araliac.).
arrhenicum *B.J. Conn & D.G. Frodin in Handbooks Fl. Papua New Guinea (ed. B.J. Conn),* 3: 273 (1995)—Solomon Is. (South Solomons).
chrysanthum *B.J. Conn & D.G. Frodin in Handbooks Fl. Papua New Guinea (ed. B.J. Conn),* 3: 273 (1995)—Solomon Is. (South Solomons).
confertiflorum *B.J. Conn & D.G. Frodin in Handbooks Fl. Papua New Guinea (ed. B.J. Conn),* 3: 274 (1995)—Solomon Is. (South Solomons).
corneri *B.J. Conn & D.G. Frodin in Handbooks Fl. Papua New Guinea (ed. B.J. Conn),* 3: 275 (1995)—Solomon Is. (South Solomons).
ellipsoideum *B.J. Conn & D.G. Frodin in Handbooks Fl. Papua New Guinea (ed. B.J. Conn),* 3: 276 (1995)—New Guinea (Papua New Guinea).
pseudofoliatum *B.J. Conn & D.G. Frodin in Handbooks Fl. Papua New Guinea (ed. B.J. Conn),* 3: 278 (1995)—Solomon Is. (South Solomons).
rectibrachiatum *B.J. Conn & D.G. Frodin in Handbooks Fl. Papua New Guinea (ed. B.J. Conn),* 3: 280 (1995)—New Guinea (Papua New Guinea).
striatifructum *B.J. Conn & D.G. Frodin in Handbooks Fl. Papua New Guinea (ed. B.J. Conn),* 3: 282 (1995)—Solomon Is. (South Solomons).
superantiflorum *B.J. Conn & D.G. Frodin in Handbooks Fl. Papua New Guinea (ed. B.J. Conn),* 3: 283 (1995)—Solomon Is. (South Solomons).
whitmorei *B.J. Conn & D.G. Frodin in Handbooks Fl. Papua New Guinea (ed. B.J. Conn),* 3: 284 (1995)—Solomon Is. (South Solomons).

OSTEOCARPUM (Chenopodiac.).
acropterum *(F. Muell. & Tate) Volkens in Engl. & Prantl, Nat. Pflanzenfam.,* III(1a): 72 (1893): Babbagia acroptera.
dipterocarpum *(F. Muell.) Volkens in Engl. & Prantl, Nat. Pflanzenfam.,* III(1a): 72 (1893): Babbagia dipterocarpa.
pentapterum *(F. Muell. & Tate) Volkens in Engl. & Prantl, Nat. Pflanzenfam.,* III(1a): 72 (1893): Babbagia pentaptera.

OSTEOCARPUM:-
scleropterum *(F. Muell.) Volkens in Engl. & Prantl, Nat. Pflanzenfam.,* III(1a): 72 (1893): Babbagia scleroptera.

OSTEOPHLOEUM (Myristicac.).
platyspermum (A.DC.) Warb.
var. sulcatum *(Little) T.S. Jaramillo & H. Balslev in Nordic J. Bot.,* 20(4): 445 (2000): Osteophloeum sulcatum.

OSTERICUM (Umbellif.).
huadongense *Z.H. Pan & X.H. Li in J. Pl. Resourc. Environ.,* 5(2): 48 (1996), as 'huadongensis'—China Southeast (Anhui, Hunan, Jiangsu, Jiangxi).
scaberulum (Franch.) R.H. Shan & C.Q. Yuan
var. longiinvolucellatum *C.Y. Wu & F.T. Pu in Novon,* 8(1): 70 (1998)—China South-central (Yunnan).
tenuifolium *(Pall. ex Spreng.) Y.C. Chu in Clavis Pl. Chinae Bor.-Or. ed. 2 (ed. P.Y. Fu):* 478 (1995), as 'tenuifolia': Athamanta tenuifolia.

OSYRIS (Santalac.).
speciosa *(A.W. Hill) J.C. Manning & P. Goldblatt in Strelitzia,* 9: 716 (2000): O. abyssinica var. speciosa.
wightiana *J. Graham, Cat. Pl. Bombay:* 177 (1839)—India (Maharashtra).

OTACHYRIUM (Gramin.).
boliviensis *S.A. Renvoize, Gramineas de Bolivia:* 415 (1998)—Bolivia.

OTHOCALLIS (Hyacinthac.).
armena *(Grossh.) F. Speta in Phyton (Austria),* 38(1): 112 (1998): Scilla armena.
caucasica *(Miscz.) F. Speta in Phyton (Austria),* 38(1): 112 (1998): Scilla caucasica.
cilicica *(Siehe) F. Speta in Phyton (Austria),* 38(1): 112 (1998): Scilla cilicica.
ingridiae *(Speta) F. Speta in Phyton (Austria),* 38(1): 112 (1998): Scilla ingridiae.
koenigii *(Fomin) F. Speta in Phyton (Austria),* 38(1): 112 (1998): Scilla koenigii.
kurdistanica *(Speta) F. Speta in Phyton (Austria),* 38(1): 112 (1998): Scilla kurdistanica.
leepii *(Speta) F. Speta in Phyton (Austria),* 38(1): 112 (1998): Scilla leepii.
libanotica *(Speta) F. Speta in Phyton (Austria),* 38(1): 112 (1998): Scilla libanotica.
melaina *(Speta) F. Speta in Phyton (Austria),* 38(1): 112 (1998): Scilla melaina.
mesopotamica *(Speta) F. Speta in Phyton (Austria),* 38(1): 112 (1998): Scilla mesopotamica.
mischtschenkoana *(Grossh.) F. Speta in Phyton (Austria),* 38(1): 112 (1998): Scilla mischtschenkoana.
monanthos *(K. Koch) F. Speta in Phyton (Austria),* 38(1): 112 (1998): Scilla monanthos.
mordakiae *(Speta) F. Speta in Phyton (Austria),* 38(1): 112 (1998): Scilla mordakiae.
morrisii *(Meikle) F. Speta in Phyton (Austria),* 38(1): 112 (1998): Scilla morrisii.
otschiauriae *(Mordak) F. Speta in Phyton (Austria),* 38(1): 112 (1998): Scilla otschiauriae.
poluninii *(Speta) F. Speta in Phyton (Austria),* 38(1): 112 (1998): Scilla poluninii.
rosenii *(K. Koch) F. Speta in Phyton (Austria),* 38(1): 113 (1998): Scilla rosenii.
siberica *(Haw.) F. Speta in Phyton (Austria),* 38(1): 113 (1998): Scilla siberica.
winogradowii *(Sosn.) F. Speta in Phyton (Austria),* 38(1): 113 (1998): Scilla winogradowii.

OTHOLOBIUM (Legumin.).
foliosum (Oliv.) C.H. Stirt.
subsp. gazense *(Baker f.) B. Verdcourt in Fl. Zambes.,* 3(6): 48 (2000): Psoralea foliosa var. gazensis.

OTHONNA (Compos.).
herrei
subsp. armiana *(van Jaarsv.) J.J. Halda in Cactaceae etc.,* 8(1): 4 (1998); *cf. Repert. Pl. Succ. (I.O.S.),* 49: 13 (1998 publ. 1999): O. armiana.

OTHONNA (Compos.).
wrinkleana *J.J. Lavranos in Piante Grasse*, 20(1): 35 (2000)—Cape Province.

OTITES (Caryophyllac.).
sect. Albopetalae *(Panov) Rabeler ex J. Holub in Preslia*, 70(2): 108 (1998): Silene sect. Albopetalae.
sect. Ciliatae *M.V. Klokov in Novosti Sist. Vÿssh. Nizsh. Rast. (Kiev)*, 1974: 59 (1974).
baldaccii *(Rohlena) J. Holub in Preslia*, 70(2): 109 (1998): Silene otites var. baldaccii.
cuneifolia
subsp. hungarica *(Wrigley) J. Holub in Preslia*, 70(2): 109 (1998): Silene otites subsp. hungarica.
eugeniae *(Kleopow) M.V. Klokov in Novosti Sist. Vÿssh. Nizsh. Rast. (Kiev)*, 1974: 61 (1974): Silene eugeniae.
jenissensis *M.V. Klokov in Novosti Sist. Vÿssh. Nizsh. Rast. (Kiev)*, 1974: 59 (1974)—Krasnoyarsk.
krymensis *(Kleopow) M.V. Klokov in Novosti Sist. Vÿssh. Nizsh. Rast. (Kiev)*, 1974: 58 (1974): Silene krymensis.
michaelae *J. Holub in Preslia*, 70(2): 117 (1998), nom. superfl.—Bulgaria.
staminea *(Bertol.) J. Holub in Preslia*, 70(2): 109 (1998): Silene staminea.

OTOBA (Myristicac.).
glycycarpa *(Ducke) W.A. Rodrigues & T.S. Jaramillo in Nordic J. Bot.*, 20(4): 446 (2000): Virola glycycarpa.

OTTELIA (Hydrocharitac.).
javanica *(Blume) J.O. Voigt, Hort. Suburb. Calcutt.*: 618 (1845): Damasonium javanicum.

OTTLEYA *D.D. Sokoloff in Feddes Repert.*, 110(1–2): 91 (1999). *LEGUMINOSAE*
argyraea *(Greene) D.D. Sokoloff in Feddes Repert.*, 110(1–2): 94 (1999): Hosackia argyraea.
flexuosa *(Greene) D.D. Sokoloff in Feddes Repert.*, 110(1–2): 94 (1999): Hosackia flexuosa.
grandiflora *(Benth.) D.D. Sokoloff in Feddes Repert.*, 110(1–2): 94 (1999): Hosackia grandiflora.
greenei *(Wooton & Standl.) D.D. Sokoloff in Feddes Repert.*, 110(1–2): 94 (1999), contrary to Art. 11.4 ICBN (1994): Anisolotus greenei.
intricata *(Eastw.) D.D. Sokoloff in Feddes Repert.*, 110(1–2): 95 (1999): Lotus intricatus.
mearnsii *(Britton) D.D. Sokoloff in Feddes Repert.*, 110(1–2): 95 (1999): Hosackia mearnsii.
nivea *(S. Watson) D.D. Sokoloff in Feddes Repert.*, 110(1–2): 95 (1999): Hosackia nivea.
oroboides *(Humb., Bonpl. & Kunth) D.D. Sokoloff in Feddes Repert.*, 110(1–2): 95 (1999): Tephrosia oroboides.
rigida *(Benth.) D.D. Sokoloff in Feddes Repert.*, 110(1–2): 95 (1999): Hosackia rigida.
strigosa *(Nutt.) D.D. Sokoloff in Feddes Repert.*, 110(1–2): 95 (1999): Hosackia strigosa.
utahensis *(Ottley) D.D. Sokoloff in Feddes Repert.*, 110(1–2): 96 (1999): Lotus utahensis.
wrightii *(A. Gray) D.D. Sokoloff in Feddes Repert.*, 110(1–2): 92, 96 (1999): Hosackia wrightii.

OURATEA (Ochnac.).
dybovskii *(Tiegh.) L. Aké Assi in Candollea*, 55(2): 281 (2000): Campylospermum dybovskii.
hatschbachii *K. Yamamoto in Bol. Bot. (São Paulo)*, 14: 34 (1995)—Brazil Southeast (Minas Gerais).
megaphylla *C. Sastre in Ann. Naturhist. Mus. Wien*, 98B: 573 (1996)—Peru.
wallnoeferiana *C. Sastre in Ann. Naturhist. Mus. Wien*, 98B: 575 (1996)—Peru.

OURISIA (Scrophulariac.).
microphylla Poepp. & Endl.
f. alba *R. Rolfe in Quart. Bull. Alp. Gard. Soc.*, 68(4): 504 (2000), without type—Chile Central (Maule).

OXALIS (Oxalidac.).
subsect. Dimidiatae *(R. Knuth) A. Lourteig in Bradea*, 7(2): 520 (2000): O. ser. Dimidiatae.

OXALIS:-
sect. Giganteae *A. Lourteig in Bradea*, 7(2): 311 (2000).
sect. Lotoideae *A. Lourteig in Bradea*, 7(2): 210 (2000).
sect. Rhombifoliae *(R. Knuth) A. Lourteig in Bradea*, 7(2): 205 (2000): O. ser. Rhombifoliae.
sect. Ripariae *A. Lourteig in Bradea*, 7(2): 297 (2000).
subgen. Trifidus *A. Lourteig in Bradea*, 6(45): 389 (1995).
acetosella L.
subsp. leucolepis *(Diels) C.C. Huang & L.R. Xu in Fl. Reipubl. Popul. Sin.*, 43(1): 9 (1998): O. leucolepis.
andrieuxii *Lémon in Ann. Fl. Pomone*, 1836–37: 30 (1836)—Mexico.
apurimacensis *A. Lourteig in Bradea*, 7(2): 280 (2000)—Peru.
benjaminii *A. Lourteig in Bradea*, 7(2): 306 (2000)—Paraguay.
bulbocastanum Phil.
subsp. hirta *(R. Knuth) A. Lourteig in Bradea*, 7(2): 356 (2000): O. carnosa var. hirta.
campanensis *A. Lourteig in Bradea*, 7(2): 464 (2000)—Chile North (Valparaíso).
cardenasiana *A. Lourteig in Bradea*, 7(2): 581 (2000)—Bolivia.
colchaguensis *A. Lourteig in Bradea*, 7(2): 482 (2000)—Chile North (O'Higgins).
conventionensis *A. Lourteig in Bradea*, 7(2): 281 (2000)—Peru.
corniculata L.
var. stricta *(L.) C.C. Huang & L.R. Xu in Fl. Reipubl. Popul. Sin.*, 43(1): 13 (1998): O. stricta.
cuatrecasasii *A. Lourteig in Bradea*, 7(2): 582 (2000)—Colombia.
dudleyi *A. Lourteig in Bradea*, 7(2): 326 (2000), as 'dudleii'—Peru.
ecuadorensis R. Knuth
subsp. gentryi *A. Lourteig in Bradea*, 7(2): 318 (2000), as 'gentrii'—Ecuador.
famatinae R. Knuth
subsp. sanjuaninae *A. Lourteig in Bradea*, 7(2): 454 (2000)—Argentina Northwest (San Juan).
fendleri *A. Lourteig in Bradea*, 7(2): 238 (2000)—Venezuela, Colombia.
flagellata *(Rusby) A. Lourteig in Bradea*, 7(2): 231 (2000): Xanthoxalis flagellata.
humbertii *A. Lourteig in Bradea*, 7(2): 278 (2000)—Peru.
kalbreyeri *A. Lourteig in Bradea*, 7(2): 276 (2000)—Colombia, Venezuela.
latemucronata *A. Lourteig in Bradea*, 7(2): 247 (2000)—Peru, Ecuador.
latifolia Kunth
subsp. galeottii *(Turcz.) A. Lourteig in Bradea*, 7(2): 591 (2000): O. galeottii.
subsp. schraderiana *(Kunth) A. Lourteig in Bradea*, 7(2): 586 (2000): O. schraderiana.
subsp. vespertilionis *(Zucc.) A. Lourteig in Bradea*, 7(2): 591 (2000): O. vespertilionis.
leucolepis Diels
var. griffithii *(Edgew. & Hook. f.) R.C. Srivastava, Fl. Sikkim (Ranunculac.-Moringac.)*: 214 (April 1998); *et in Novon*, 8(2): 203 (15 July 1998): O. griffithii.
libertatis *A. Lourteig in Bradea*, 7(2): 215 (2000)—Peru.
lucumayensis R. Knuth
subsp. subiens *A. Lourteig in Bradea*, 7(2): 273 (2000)—Bolivia.
subsp. woytkowskii *A. Lourteig in Bradea*, 7(2): 274 (2000)—Peru.
matancillae *A. Lourteig in Bradea*, 7(2): 351 (2000)—Chile North (Antofagasta).
melindae *A. Lourteig in Bradea*, 7(2): 544 (2000)—Mexico Southwest (Guerrero).
mira *A. Lourteig in Bradea*, 7(2): 481 (2000)—Chile North (Coquimbo).
morenoensis *A. Lourteig in Bradea*, 7(2): 359 (2000)—Chile North (Antofagasta).
ortigiesii Regel
subsp. tingoensis *A. Lourteig in Bradea*, 7(2): 326 (2000)—Peru.
oulophora *A. Lourteig in Bradea*, 7(2): 245 (2000)—Bolivia.

OXALIS:-
paranaensis *A. Lourteig in Bradea*, 7(2): 310 (2000)—Brazil South (Paraná).
pedunculata *(Chodat & Wilczek) A. Lourteig in Bradea*, 7(2): 488 (2000): O. sericea f. pedunculata.
phaeotricha Diels
var. glandulosa *(R. Knuth) A. Lourteig in Bradea*, 7(2): 237 (2000): O. mollis var. glandulosa.
subsp. sonsonensis *A. Lourteig in Bradea*, 7(2): 237 (2000)—Colombia.
pretoensis *A. Lourteig in Bradea*, 7(2): 509 (2000)—Brazil West-central (Goiás).
psammophila *G. Williamson in Aloe*, 36(4): 68 (1999)—Cape Province.
ptychoclada Diels
var. trichocarpa *A. Lourteig in Bradea*, 7(2): 338 (2000)—Peru.
ricardii *A. Lourteig in Bradea*, 7(2): 352 (2000)—Chile North (Antofagasta).
rusbyi *A. Lourteig in Bradea*, 7(2): 251 (2000), as 'rusbii'—Bolivia.
salticola *A. Lourteig in Bradea*, 7(2): 314 (2000)—Peru.
san-miguelii R. Knuth
subsp. urubambensis *(R. Knuth) A. Lourteig in Bradea*, 7(2): 340 (2000): O. urubambensis.
sandemanii *A. Lourteig in Bradea*, 7(2): 230 (2000)—Colombia.
semitruncata *A. Lourteig in Bradea*, 7(2): 322 (2000)—Peru.
sleumeri *A. Lourteig in Bradea*, 6(45): 390 (1995)—Argentina Northwest (Catamarca, Salta).
solomonii *A. Lourteig in Bradea*, 7(2): 246 (2000)—Bolivia.
spiralis Ruiz & Pav. ex G. Don
subsp. membranifolia *(R. Knuth) A. Lourteig in Bradea*, 7(2): 267 (2000): O. membranifolia.
subsp. trichophora *A. Lourteig in Bradea*, 7(2): 270 (2000)—Costa Rica.
tuberosa Molina
subsp. unduavensis *(Rusby) A. Lourteig in Bradea*, 7(2): 256 (2000): Xanthoxalis unduavensis.
vargasii *A. Lourteig in Bradea*, 7(2): 271 (2000)—Peru.
violacella *A.P. Khokhrjakov in Byull. Glavn. Bot. Sada (Moscow)*, 175: 53 (1997)—Turkey.
westii *A. Lourteig in Bradea*, 7(2): 260 (2000)—Peru.
wurdackii *A. Lourteig in Bradea*, 7(2): 250 (2000)—Peru.

OXANDRA (Annonac.).
espintana *(Spruce ex Benth.) Baill., Hist. pl.*, 1: 207 (1868): Bocagea espintana.

OXERA (Verbenac.).
baladica Vieill.
subsp. nuda *(Virot) R.P.J. de Kok in Kew Bull.*, 54(2): 270 (1999): O. nuda.
brevicalyx *(Moldenke) R.P.J. de Kok in Kew Bull.*, 54(2): 282 (1999): O. pulchella var. brevicalyx.
coronata *R.P.J. de Kok in Kew Bull.*, 54(2): 275 (1999)—New Caledonia.
inodora *R.P.J. de Kok in Kew Bull.*, 54(2): 293 (1999)—New Caledonia.
macrocalyx Dubard
subsp. sororia *(Däniker) R.P.J. de Kok in Kew Bull.*, 54(2): 293 (1999): O. sororia.
pulchella Labill.
subsp. grandiflora *(Dubard) R.P.J. de Kok in Kew Bull.*, 54(2): 284 (1999): O. pulchella var. grandiflora.
vanuatuensis *R.P.J. de Kok in Kew Bull.*, 54(2): 271 (1999)—Vanuatu.

OXYANTHUS (Rubiac.).
okuensis *M. Cheek & B. Sonké in Kew Bull.*, 55(4): 890 (2000)—Cameroon.

OXYBAPHUS (Nyctaginac.).
himalaicus Edgew.
var. chinensis *(Heimerl) D.Q. Lu in Rep. & Abstr. 60th Anniv. Bot. Soc. China*: 102 (1993): Mirabilis himalaica var. chinensis.

OXYCARYUM (Cyperac.).
cubense (Poepp. & Kunth) Lye
f. paraguayense *(Maury) T.M. Pedersen in Hickenia,* 2(31): 138 (1995): Anosporum paraguayense.

OXYCEROS (Rubiac.).
griffithii *(Hook. f.) W.C. Chen in Fl. Reipubl. Popul. Sin.,* 71(1): 346 (1999): Randia griffithii.

OXYGONUM (Polygonac.).
annuum *S. Ortiz & J.A.R. Paiva in Bot. J. Linn. Soc.,* 131(2): 172 (1999)—Zambia.

OXYGRAPHIS (Ranunculac.).
delavayi Franch.
var. nyingchiensis *W.L. Cheng in Acta Phytotax. Sin.,* 37(3): 304 (1999)—Xizang.

OXYPETALUM (Asclepiadac.).
capitatum Mart.
subsp. mirabile *(Malme) J. Fontella Pereira & M.A. Farinaccio in Bradea,* 8(12): 67 (1997): O. hasslerianum var. mirabile.
insigne (Decne.) Malme
var. nitidum *(Malme) J. Fontella Pereira & E.A. Schwarz in Bradea,* 8(12): 67 (1997): O. nitidum.
leonii *J. Fontella Pereira in Pabstia,* 7(4): 1 (1996)—Brazil Southeast (Minas Gerais).
streptanthum *(Malme) S. Liede & G.H. Rua in Novon,* 7(1): 41 (1997): Cynanchum streptanthum.

OXYTROPIS (Legumin.).
ser. Campanulatae *M.S. Knjasev in Bot. Zhurn.,* 84(9): 121 (1999).
ser. Chiliophyllae *Schischk. ex C.W. Chang in Fl. Reipubl. Popul. Sin.,* 42(2): 145 (1998).
ser. Hallerae *M.S. Knjasev in Bot. Zhurn.,* 84(9): 114 (1999).
sect. Neimonggolica *X.Y. Zhu & Hiroyoshi Ohashi in Cathaya,* 11–12: 161 (1999–2000 publ. 2000).
sect. Ochrocea *G.Z. Qian in Bull. Bot. Res. North-East. Forest. Univ.,* 19(1): 2, 1 (1999).
subgen. Orobia *(Bunge) C.W. Chang in Fl. Reipubl. Popul. Sin.,* 42(2): 81 (1998): O. sect. Orobia.
sect. Rechingeria *I.T. Vassilczenko in Fl. Iran.,* 157: 139 (1984).
ser. Vesicatae *C.W. Chang in Fl. Reipubl. Popul. Sin.,* 42(2): 143 (1998), nom. nov.: O. ser. Physocarpae *Gontsch. ex N. Ulziykh.*
alavae *K.H. Rechinger in Fl. Iran.,* 157: 162 (1984)—Iran.
anertii Nakai
f. alba *Y.N. Lee, Fl. Korea:* 1159 (1996), without type—Locality not stated.
f. albiflora *(Z.J. Zong & X.R. He) X.Y. Zhu & Hiroyoshi Ohashi in Cathaya,* 11–12: 123 (1999–2000 publ. 2000): Oxytropis anertii var. albiflora.
assadliensis *I.T. Vassilczenko in Fl. Iran.,* 157: 121 (1984)—Iran.
asterocarpa *I.T. Vassilczenko in Fl. Iran.,* 157: 125 (1984)—Afghanistan.
azerbaijanica *D. Podlech in Sendtnera,* 6: 171 (1999)—Iran.
barkolensis *X.Y. Zhu, Hiroyoshi Ohashi & Y.B. Deng in J. Jap. Bot.,* 74(2): 65 (1999)—Xinjiang.
bicolor Bunge
f. luteola *(C.W. Chang) X.Y. Zhu & Hiroyoshi Ohashi in Cathaya,* 11–12: 66 (1999–2000 publ. 2000): Oxytropis bicolor var. luteola.
bicornis *I.T. Vassilczenko in Fl. Iran.,* 157: 123 (1984)—Iran.
binaludensis *I.T. Vassilczenko in Fl. Iran.,* 157: 135 (1984)—Iran.
campestris (L.) DC.
subsp. johannensis *(Fernald) M. Blondeau & C. Gervais in Bull. Soc. Neuchâtel. Sci. Nat.,* 122: 49 (1999), without exact basionym page: O. campestris var. johannensis.
cinerea *I.T. Vassilczenko in Fl. Iran.,* 157: 138 (1984)—Iran.
coerulea (Pall.) DC.
f. albiflora *(H.C. Fu) X.Y. Zhu & Hiroyoshi Ohashi in Cathaya,* 11–12: 137 (1999–2000 publ. 2000), as 'caerulea': Oxytropis mandshurica f. albiflora.

OXYTROPIS:-
coerulea
subsp. monticola *(Malyschev) G.A. Peschkova in Fl. Tsentral'noi Sibiri,* 2: 613 (1979): O. filiformis var. monticola.
danorum *K.H. Rechinger in Fl. Iran.,* 157: 163 (1984)—Afghanistan.
daqingshanica *Y.Z. Zhao & Zong Y. Zhu in Acta Sci. Nat. Univ. Neimonggol,* 27(1): 83 (1996)—Inner Mongolia (Nei Mongol).
darwasica *B.A. Fedtschenko in Fl. Tadzhikistana,* 5: 685 (1937)—Tadzhikistan.
demidovii *M.S. Knjasev in Bot. Zhurn.,* 84(9): 118 (1999)—Russia East.
densiflora P.C. Li
var. multiramosa *(P.C. Li) X.Y. Zhu & Hiroyoshi Ohashi in Cathaya,* 11–12: 91 (1999–2000 publ. 2000): Oxytropis multiramosa.
flavovirens *H. Ohba, Shinobu Akiyama & S.K. Wu in Bull. Nation. Sci. Mus., B (Tokyo),* 26(2): 44 (2000)—Xinjiang.
friabilis *H. Ohba, Shinobu Akiyama & S.K. Wu in Bull. Nation. Sci. Mus., B (Tokyo),* 26(2): 47 (2000)—Xinjiang.
gracillima *I.T. Vassilczenko in Fl. Iran.,* 157: 131 (1984), homonym; *see Bunge in Linnaea,* 17: 5 (1843)—Iran.
guinanensis *Y.H. Wu in Acta Phytotax. Sin.,* 35(4): 348 (1997)—Qinghai.
hissarica *B.A. Fedtschenko in Fl. Tadzhikistana,* 5: 686 (1937)—Tadzhikistan.
iskanderica *B.A. Fedtschenko in Fl. Tadzhikistana,* 5: 685 (1937)—Tadzhikistan.
ivdelensis *M.S. Knjasev in Bot. Zhurn.,* 84(9): 119 (1999)—West Siberia.
var. microphylla *M.S. Knjasev in Bot. Zhurn.,* 84(9): 121 (1999)—West Siberia.
var. rectirostra *M.S. Knjasev in Bot. Zhurn.,* 84(9): 121 (1999)—Russia North.
jabalambrensis *(Pau) D. Podlech in Sendtnera,* 3: 147 (1996): Astragalus jabalambrensis.
junatovii *Ch. Sanchir in Trudy̆ Inst. Bot. Akad. Nauk MNR,* 1985(7): 90 (1985)—Mongolia.
khorasanica *M. Ranjbar in Sendtnera,* 6: 194 (1999), nom. nov., without exact replaced synonym page: O, gracillima *Vassilcz.*
konlonica *H. Ohba, Shinobu Akiyama & S.K. Wu in Bull. Nation. Sci. Mus., B (Tokyo),* 26(2): 49 (2000)—Qinghai, Xinjiang.
kozhuharovii *D. Pavlova, D. Dimitrov & M. Nikolova in Willdenowia,* 29(1–2): 69 (1999)—Bulgaria.
kukkonenii *I.T. Vassilczenko in Fl. Iran.,* 157: 141 (1984)—Afghanistan.
kungurensis *M.S. Knjasev in Bot. Zhurn.,* 84(9): 114 (1999)—Russia East.
lanceatifoliola *H. Ohba, Shinobu Akiyama & S.K. Wu in Bull. Nation. Sci. Mus., B (Tokyo),* 26(2): 53 (2000)—Xinjiang.
lasiocarpa *N.F. Gontscharow in Bot. Mater. Gerb. Bot. Inst. Komarova Akad. Nauk SSSR,* 9: 86 (1941)—Tadzhikistan.
latibracteata Jurtzev
var. longibracteata *Y.H. Wu in Acta Phytotax. Sin.,* 35(4): 352 (1997)—Qinghai.
liliputa *I.T. Vassilczenko in Fl. Iran.,* 157: 119 (1984)—Afghanistan.
maduoensis *Y.H. Wu in Acta Bot. Yunnanica,* 19(1): 33 (1997)—Qinghai.
maqinensis *Y.H. Wu in Acta Bot. Yunnanica,* 19(1): 34 (1997)—Qinghai.
margacea *I.T. Vassilczenko in Fl. Iran.,* 157: 116 (1984)—Afghanistan.
michelsonii *B.A. Fedtschenko in Fl. Tadzhikistana,* 5: 684 (1937), as 'michelsoni'—Tadzhikistan.
nangqianensis *X.Y. Zhu in Cathaya,* 11–12: 154 (1999–2000 publ. 2000), nom. nov.: Oxytropis qinghaiensis *Y.H. Wu.*
neorechingeriana *I.T. Vassilczenko in Fl. Iran.,* 157: 134 (1984), as 'neo-rechingeriana'—Iran.
ochrantha Turcz.
var. chrysotricha *(Franch.) G.Z. Qian in Bull. Bot. Res. North-East. Forest. Univ.,* 19(1): 2 (1999): Oxytropis chrysotricha.
var. diversicolor *(H.C. Fu & Ma) G.Z. Qian in Bull. Bot. Res. North-East. Forest. Univ.,* 19(1): 3 (1999), with incorrect basionym ref.: Oxytropis ochrantha f. diversicolor.

OXYTROPIS:-
ochrolongibracteata *X.Y. Zhu & Hiroyoshi Ohashi in Cathaya,* 11–12: 75 (1999–2000 publ. 2000), nom. nov.: Oxytropis ochrocephala var. longibracteata *P.C. Li.*
oreophila *I.T. Vassilczenko in Fl. Iran.,* 157: 119 (1984), homonym; *see A. Gray in Proc. Amer. Acad. Arts,* 20: 3 (1884)—Afghanistan.
oxyphylla (Pall.) DC.
var. psilocarpa *(Kitag.) G.Z. Qian in Bull. Bot. Res. North-East. Forest. Univ.,* 17(1): 59 (1997): O. lanata var. psilocarpa.
popoviana *G.A. Peschkova, Stepnaya Fl. Baĭkal'skoĭ Sibiri:* 73 (1972), nom. nov.: O. popovii *Peschkova.*
pseudomyriophylla *Cheng f. ex X.Y. Zhu, H. Ohashi & Y.B. Deng in J. Jap. Bot.,* 74(3): 127 (1999)—Inner Mongolia (Ningxia), China North-central (Gansu, Shanxi).
qiemoensis *H. Ohba, Shinobu Akiyama & S.K. Wu in Bull. Nation. Sci. Mus., B (Tokyo),* 26(2): 56 (2000)—Xinjiang.
qinghaiensis *Y.H. Wu in Novon,* 6(2): 187 (1996)—Qinghai.
qinghaiensis *Y.H. Wu in Acta Bot. Bor.-Occid. Sin.,* 17(1): 109 (1997), homonym; *see Y.H. Wu in Novon,* 6(2): 187 (1996)—Qinghai.
qingnanensis *Y.H. Wu in Acta Bot. Yunnanica,* 19(1): 36 (1997), as 'qinghaiensis'; *see Acta Bot. Yunnanica,* 19(3): 223 (1997)—Qinghai.
qitaiensis *X.Y. Zhu, H. Ohashi & Y.B. Deng in J. Jap. Bot.,* 74(2): 66 (1999)—Xinjiang.
racemosa Turcz.
f. albiflora *(P.Y. Fu & Y.A. Chen) C.W. Chang in Fl. Reipubl. Popul. Sin.,* 42(2): 139 (1998): O. psammocharis f. albiflora.
scabrida *N.F. Gontscharow in Bot. Mater. Gerb. Bot. Inst. Komarova Akad. Nauk SSSR,* 9: 89 (1941)—Uzbekistan.
schachimardanica *Z.N. Filimonova in Opred. Rast. Sred. Azii,* 7: 379 (1983)—Kirgizstan.
sitaipaiensis T.P. Wang ex C.W. Chang
var. brevidentata *(C.W. Chang) X.Y. Zhu & Hiroyoshi Ohashi in Cathaya,* 11–12: 112 (1999–2000 publ. 2000): Oxytropis melanocalyx var. brevidentata.
sojakii *I.T. Vassilczenko in Fl. Iran.,* 157: 119 (1984)—Iran, Afghanistan.
subcapitata *N.F. Gontscharow in Bot. Mater. Gerb. Bot. Inst. Komarova Akad. Nauk SSSR,* 9: 88 (1941)—Turkmenistan.
talassica *N.F. Gontscharow in Bot. Mater. Gerb. Bot. Inst. Komarova Akad. Nauk SSSR,* 9: 85 (1941)—Kirgizstan.
tashkurensis *S.H. Cheng ex X.Y. Zhu, Y.F. Du & Hiroyoshi Ohashi in J. Jap. Bot.,* 75(5): 289 (2000)—Xinjiang.
thaumasiomorpha *K.H. Rechinger in Fl. Iran.,* 157: 163 (1984)—Iran.
tudanensis *X.Y. Zhu, H. Ohashi & Si Feng Li in J. Jap. Bot.,* 74(3): 130 (1999)—Xizang.
tukemansuensis *X.Y. Zhu, H. Ohashi & Y.B. Deng in J. Jap. Bot.,* 74(2): 68 (1999)—Xinjiang.
tyttantha *N.F. Gontscharow in Bot. Mater. Gerb. Bot. Inst. Komarova Akad. Nauk SSSR,* 9: 90 (1941)—Tadzhikistan.
ulzijchutagii *Ch. Sanchir in Trudy̆ Inst. Bot. Akad. Nauk MNR,* 1985(7): 93 (1985)—Mongolia.
yanchiensis *X.Y. Zhu, Hiroyoshi Ohashi & L.R. Xu in J. Jap. Bot.,* 74(3): 133 (1999)—Xinjiang.
yekenensis *X.Y. Zhu, H. Ohashi & Y.B. Deng in J. Jap. Bot.,* 74(2): 68 (1999)—Xinjiang.
zekogensis *Y.H. Wu in Acta Phytotax. Sin.,* 35(4): 350 (1997)—Qinghai.

OYEDAEA (Compos.).
camargoana *(S. Díaz) S. Díaz Piedrahita in Rev. Acad. Colomb. Cienc. Exact. Fis. Nat.,* 81: 408 (1997): Calea camargoana.
cuatrecasasii *J.F. Pruski in Compositae Newsl.,* 34: 3 (1999)—Colombia.
neei *J.F. Pruski in Compositae Newsl.,* 34: 5 (1999)—Bolivia.
tepuiana *(V.M. Badillo) J.F. Pruski in Novon,* 6(4): 407 (1996): Zexmenia tepuiana.
wurdackii *J.F. Pruski in Novon,* 6(4): 407 (1996)—Venezuela.

OZIROE (Hyacinthac.).
sect. Nidia *P. Ravenna in Onira, Bot. Leafl.*, 4(11): 48 (2000).
acaulis *(Baker) F. Speta in Phyton (Austria)*, 38(1): 56 (14 Aug. 1998): Scilla acaulis.
acaulis *(Baker) P. Ravenna in Onira, Bot. Leafl.*, 3(14): 41 (8 Nov. 1998), isonym; *see Speta in Phyton (Austria)*, 38(1): 56 (14 Aug. 1998): Scilla acaulis.
argentinensis *(Lillo & Hauman) F. Speta in Phyton (Austria)*, 38(1): 56 (14 Aug. 1998): Scilla argentinensis.
argentinensis *(Lillo & Hauman) P. Ravenna in Onira, Bot. Leafl.*, 3(14): 40 (8 Nov. 1998), isonym; *see Speta in Phyton (Austria)*, 38(1): 56 (14 Aug. 1998): Scilla argentinensis.
arida *(Poepp.) F. Speta in Phyton (Austria)*, 38(1): 56 (14 Aug. 1998): Ornithogalum aridum.
arida *(Poepp.) P. Ravenna in Onira, Bot. Leafl.*, 3(14): 40 (8 Nov. 1998), isonym; *see Speta in Phyton (Austria)*, 38(1): 56 (14 Aug. 1998): Ornithogalum aridum.
biflora *(Ruiz & Pav.) F. Speta in Phyton (Austria)*, 38(1): 56 (14 Aug. 1998): Scilla biflora.
biflora *(Ruiz & Pav.) P. Ravenna in Onira, Bot. Leafl.*, 3(14): 40 (8 Nov. 1998), isonym; *see Speta in Phyton (Austria)*, 38(1): 56 (14 Aug. 1998): Scilla biflora.
chloroleucum *(Lindl.) F. Speta in Phyton (Austria)*, 38(1): 56 (1998): Ornithogalum chloroleucum.
correntina *P. Ravenna in Onira, Bot. Leafl.*, 3(18): 68 (1999)—Argentina Northeast (Corrientes).
pomensis *P. Ravenna in Onira, Bot. Leafl.*, 3(14): 41 (1998)—Argentina Northwest (Salta).
sessilis *(R.E. Fr.) P. Ravenna in Onira, Bot. Leafl.*, 3(14): 41 (1998): Allium sessile.
totorensis *P. Ravenna in Onira, Bot. Leafl.*, 3(18): 69 (1999)—Bolivia.

OZOTHAMNUS (Compos.).
leptophyllus *(G. Forst.) I. Breitwieser & J.M. Ward in New Zealand J. Bot.*, 35(1): 127 (1997): Calea leptophylla.

P

PACHIRA (Bombacac.).
calophylla *(K. Schum.) J.L. Fernández-Alonso in An. Jard. Bot. Madrid*, 56(2): 309 (1998): Bombax calophyllum.
duckei *(A. Robyns) J.L. Fernández-Alonso in An. Jard. Bot. Madrid*, 56(2): 310 (1998): Rhodognaphalopsis duckei.
dugandeana *(A. Robyns) J.L. Fernández-Alonso in An. Jard. Bot. Madrid*, 56(2): 310 (1998): Bombacopsis dugandeana.
flaviflora *(Pulle) J.L. Fernández-Alonso in An. Jard. Bot. Madrid*, 56(2): 310 (1998): Bombax flaviflorum.
macrocalyx *(Ducke) J.L. Fernández-Alonso in An. Jard. Bot. Madrid*, 56(2): 310 (1998): Bombax macrocalyx.
nervosa *(Uittien) J.L. Fernández-Alonso in An. Jard. Bot. Madrid*, 56(2): 310 (1998): Bombax nervosum.
nukakica *J.L. Fernández-Alonso in An. Jard. Bot. Madrid*, 56(2): 306 (1998)—Colombia.
patinoi *(Dugand & A. Robyns) J.L. Fernández-Alonso in Rev. Acad. Colomb. Cienc. Exact. Fis. Nat.*, 82: 8 (1998): Bombacopsis patinoi.
subandina *(Dugand) J.L. Fernández-Alonso in An. Jard. Bot. Madrid*, 56(2): 310 (1998): Pseudobombax subandinum.
tocantina *(Ducke) J.L. Fernández-Alonso in An. Jard. Bot. Madrid*, 56(2): 310 (1998): Bombax tocantinum.

PACHYCARPUS (Asclepiadac.).
bisacculatus *(Oliv.) D.J. Goyder in Kew Bull.*, 51(4): 798 (1996): Gomphocarpus bisacculatus.
chirindensis *(S. Moore) D.J. Goyder in Kew Bull.*, 53(2): 356 (1998): Schizoglossum chirindense.

PACHYCARPUS:-
firmus *(N.E. Br.) D.J. Goyder in Kew Bull.*, 53(2): 367 (1998): Schizoglossum firmum.
grandiflorus (L.f.) E. Mey.
subsp. tomentosus *(Schltr.) D.J. Goyder in Kew Bull.*, 53(2): 370 (1998): Gomphocarpus grandiflorus var. tomentosus.
grantii (Oliv.) Bullock
subsp. marroninus *D.J. Goyder in Kew Bull.*, 53(2): 359 (1998)—Tanzania.
pachyglossus *D.J. Goyder in Kew Bull.*, 53(2): 368 (1998)—Tanzania.
petherickianus *(Oliv.) D.J. Goyder in Kew Bull.*, 51(4): 798 (1996): Schizoglossum petherickianum.
richardsiae *D.J. Goyder in Kew Bull.*, 53(2): 369 (1998)—Tanzania.

PACHYCENTRIA (Melastomatac.).
glauca Triana
subsp. maingayi *(C.B. Clarke) G. Clausing in Blumea*, 45(2): 356 (2000): Medinilla maingayi.
hanseniana *G. Clausing in Blumea*, 45(2): 356 (2000)—Borneo (Kalimantan).
pulverulenta *(Jack) G. Clausing in Blumea*, 45(2): 362 (2000): Melastoma pulverulentum.
vogelkopensis *G. Clausing in Blumea*, 45(2): 367 (2000)—New Guinea (Irian Jaya).

PACHYCEREUS (Cactac.).
tepamo *S. Gama-López & S. Arias in Novon*, 8(4): 359 (1998)—Mexico Southwest (Michoacan).

PACHYPHYLLUM (Orchidac.).
dalstroemii *C.H. Dodson in Orquideologia*, 20(3): 278 (1997)—Ecuador.
hagsateri *C.H. Dodson in Orquideologia*, 20(1): 104 (1996)—Ecuador.

PACHYPHYTUM (Crassulac.).
caesium *M. Kimnach & R. Moran in Cact. Succ. J. (U.S.A.)*, 70(1): 41 (1998)—Mexico Northeast (Aguascalientes).
garciae *E. Pérez-Calix & C. Glass in Acta Bot. Mex.*, 48: 4 (1999)—Mexico Northeast (Querétaro).
machucae *I. García Ruiz, C. Glass & M. Cházaro Basañez in Acta Bot. Mex.*, 47: 10 (1999)—Mexico Southwest (Michoacan).

PACHYPLECTRON (Orchidac.).
aphyllum *T. Hashimoto in Ann. Tsukuba Bot. Gard.*, 16: 7 (1997)—New Caledonia.

PACHYPODIUM (Apocynac.).
bicolor *J.J. Lavranos & S.H.J.V. Rapanarivo in Cact. Succ. J. (U.S.A.)*, 69(5): 232 (1997)—Madagascar.
eburneum *J.J. Lavranos & S.H.J.V. Rapanarivo in Cact. Succ. J. (U.S.A.)*, 69(5): 227 (1997)—Madagascar.
gracilius *(H. Perrier) S.H.J.V. Rapanarivo in S.H.J.V. Rapanarivo et al., Pachypodium (Apocynac.) Taxon., habitats & cultiv.*: 35 (1999): P. rosulatum var. gracilius.
× hojnyi *J.J. Halda in Cactaceae etc.*, 8(3): 83–85 (1998); *cf. Repert. Pl. Succ. (I.O.S.)*, 49: 12 (1998 publ. 1999)—Locality not stated. (P. densiflorum subsp. densiflorum × rosulatum subsp. rosulatum).
inopinatum *J.J. Lavranos in Cact. Succ. J. (U.S.A.)*, 68(4): 175 (1996)—Madagascar.
× rauhii *J.J. Halda in Cactaceae etc.*, 7(3): 108–111 (1997); *cf. Repert. Pl. Succ. (I.O.S.)*, 48: 6 (1997 publ. 1998)—Locality not stated. (P. brevicaule × densiflorum).

PACKERA (Compos.).
bolanderi
var. harfordii *(Greenm.) D.K. Trock & T.M. Barkley in Sida*, 18(2): 386 (1998): Senecio harfordii.
buekii *D.K. Trock & T.M. Barkley in Sida*, 18(2): 387 (1998), nom. nov.: Senecio subnudus *DC.*
contermina *(Greenm.) J.F. Bain in Novon*, 9(4): 457 (1999): Senecio conterminus.
cymbalarioides *W.A. Weber & Á. Löve in Phytologia*, 49(1): 46 (1981), nom. nov., as 'comb. nov.': Senecio cymbalarioides *Nutt.*†

PACKERA:-
cymbalarioides *(H. Buek) W.A. Weber & Á. Löve in Phytologia*, 55(1): 9 (1984), homonym; *see W.A. Weber & Á. Löve in Phytologia*, 49(1): 46 (1981): Senecio cymbalarioides.†
dimorphophylla
var. intermedia *(T.M. Barkley) D.K. Trock & T.M. Barkley in Sida*, 18(2): 386 (1998): Senecio dimorphophyllus var. intermedius.
var. paysonii *(T.M. Barkley) D.K. Trock & T.M. Barkley in Sida*, 18(2): 386 (1998): Senecio dimorphophyllus var. paysonii.
eurycephala
var. lewisrosei *(J.T. Howell) J.F. Bain in Novon*, 9(4): 457 (1999): Senecio lewisrosei.
heterophylla *(Fisch.) E.I. Wiebe in Fl. Sibir. (Asterac.)*, 13: 162 (1997 publ. 1998): Cineraria heterophylla.
moresbiensis *(Calder & R.L. Taylor) J.F. Bain in Novon*, 9(4): 457 (1999): Senecio cymbalarioides subsp. moresbiensis.
neomexicana
var. metcalfei *(Greene ex Wooton & Standl.) D.K. Trock & T.M. Barkley in Sida*, 18(2): 386 (1998): Senecio metcalfei.
var. toumeyi *(Greene) D.K. Trock & T.M. Barkley in Sida*, 18(2): 386 (1998): Senecio toumeyi.
ovina *(Greene) J.F. Bain in Novon*, 9(4): 457 (1999): Senecio ovinus.
pseudaurea
var. flavula *(Greene) D.K. Trock & T.M. Barkley in Sida*, 18(2): 386 (1998): Senecio flavulus.
var. semicordata *(Mack. & Bush) D.K. Trock & T.M. Barkley in Sida*, 18(2): 386 (1998): Senecio semicordatus.
scalaris (Greene) C. Jeffrey
var. carmenensis *(C.C. Freeman) C.C. Freeman & T.M. Barkley in Sida*, 16(4): 706 (1995): Senecio scalaris var. carmenensis.
var. parrasiana *(Greenm.) C.C. Freeman & T.M. Barkley in Sida*, 16(4): 706 (1995): Senecio parrasianus.
subnuda *(DC.) D.K. Trock & T.M. Barkley in Sida*, 18(3): 635 (1999): Senecio subnudus.
zimapanica *(Hemsl.) C.C. Freeman & T.M. Barkley in Sida*, 16(4): 708 (1995): Senecio zimpanicus.

PADUS (Rosac.).
sect. Phyllomahaleb *(Koehne) V.A. Nedoluzhko in Sosud. Rast. Sovet. Dal'nego Vostoka*, 8: 238 (1996): Prunus subsect. Phyllomahaleb.

PAEDERIA (Rubiac.).
scandens (Lour.) Merr.
var. angustifolia *S.S. Ying in Mem. Coll. Agric. Nation. Taiwan Univ.*, 29(2): 51 (1989)—Taiwan.
spectatissima *H. Li in Novon*, 9(2): 220 (1999)—China South-central (Yunnan).
taolagnarensis *S.G. Razafimandimbison & C.M. Taylor in Novon*, 10(1): 71 (2000)—Madagascar.

PAEDEROTELLA (Scrophulariac.).
pontica (Rupr. ex Boiss.) Kem.-Nath.
subsp. teberdensis *(Kem.-Nath.) R.I. Gagnidze in Fl. Gruzii, ed. 2*, 12: 122 (1999): Paederotella teberdensis.

PAEONIA (Paeoniac.).
subgen. Albiflorae *(Salm-Dyck) J.J. Halda in Acta Mus. Richnov. Sect. Nat.*, 5(1): 4 (1998), as 'Albiflora': P. sect. Albiflorae.
subsect. Anomalae *(Kem.-Nath. ex M.S. Uspensk.) J.J. Halda in Acta Mus. Richnov. Sect. Nat.*, 4(2): 26 (1997): P. ser. Anomalae.
sect. Delavayanae *(Stern) J.J. Halda in Acta Mus. Richnov. Sect. Nat.*, 4(2): 29 (1997): P. subsect. Delavayanae.
sect. Emodi *J.J. Halda in Acta Mus. Richnov. Sect. Nat.*, 5(1): 4 (1998).
subsect. Masculae *(Stern ex M.S. Uspensk.) J.J. Halda in Acta Mus. Richnov. Sect. Nat.*, 5(1): 4 (1998): P. ser. Masculae.
subsect. Tenuifoliae *(Stern ex M.S. Uspensk.) J.J. Halda in Acta Mus. Richnov. Sect. Nat.*, 4(2): 27 (1997): P. ser. Tenuifoliae.
sect. Tenuifoliae *(Stern ex M.S. Uspensk.) J.J. Halda in Acta Mus. Richnov. Sect. Nat.*, 5(1): 4 (1998): P. ser. Tenuifoliae.

PAEONIA:-
anomala L.
subsp. hybrida *(Pall.) J.J. Halda in Acta Mus. Richnov. Sect. Nat.,* 4(2): 29 (1997): P. hybrida.
baokangensis *Z.L. Dai & T. Hong in Bull. Bot. Res. North-East. Forest. Univ.,* 17(1): 2 (1997)—China South-central (Hubei).
brownii Douglas
subsp. californica *(Nutt.) J.J. Halda in Acta Mus. Richnov. Sect. Nat.,* 4(2): 32 (1997): P. californica.
daurica Andrews
subsp. lagodechiana *(Kem.-Nath.) J.J. Halda in Acta Mus. Richnov. Sect. Nat.,* 4(2): 29 (1997): P. lagodechiana.
decomposita Hand.-Mazz.
subsp. rotundiloba *D.Y. Hong in Kew Bull.,* 52(4): 961 (1997)—China South-central (Sichuan).
emodi Wall. ex Royle
subsp. sterniana *(H.R. Fletcher) J.J. Halda in Acta Mus. Richnov. Sect. Nat.,* 4(2): 30 (1997): P. sterniana.
× franchetii *J.J. Halda in Acta Mus. Richnov. Sect. Nat.,* 4(2): 31 (1997), as 'francheti'—China South-central (Sichuan). (P. lutea × potaninii).
× handel-mazzettii *J.J. Halda in Acta Mus. Richnov. Sect. Nat.,* 4(2): 31 (1997)—China South-central (Yunnan). (P. delavayi × lutea).
ludlowii *(Stern & G. Taylor) D.Y. Hong in Novon,* 7(2): 157 (1997): P. lutea var. ludlowii.
ludlowii *(Stern & G. Taylor) J.J. Li & D.Z. Chen in Bull. Bot. Res. North-East. Forest. Univ.,* 18(2): 154 (1998), isonym; *see D.Y. Hong in Novon,* 7(2): 157 (1997): P. lutea var. ludlowii.
lutea Delavay ex Franch.
subsp. ludlowii *(Stern & G. Taylor) J.J. Halda in Acta Mus. Richnov. Sect. Nat.,* 6(3): 234 (1999): P. lutea var. ludlowii.
mascula (L.) Mill.
subsp. kesrouanensis *(Thiébaut) J.J. Halda in Acta Mus. Richnov. Sect. Nat.,* 4(2): 29 (1997): P. kesrouanensis.
subsp. orientalis *(Thiébaut) D.Y. Hong in Acta Phytotax. Sin.,* 38(4): 382 (2000): P. corallina var. orientalis.
obovata Maxim.
f. albiflora *M. Mizush. ex T. Shimizu, Fl. Nagano Pref.:* 1505 (1997)—Japan.
subsp. japonica *(Makino) J.J. Halda in Acta Mus. Richnov. Sect. Nat.,* 4(2): 30 (1997): P. obovata var. japonica.
subsp. willmottiae *(Stapf) D.Y. Hong & K.Y. Pan in Higher Plants China,* 4: 560 (2000): Paeonia willmottiae.
ostii T. Hong & J.X. Zhang
var. lishizhenii *B.A. Shen in Acta Phytotax. Sin.,* 35(4): 360 (1997)—China Southeast (Anhui).
ridleyi *Z.L. Dai & T. Hong in Bull. Bot. Res. North-East. Forest. Univ.,* 17(1): 1 (1997)—China South-central (Hubei).
rockii *(S.G. Haw & Lauener) T. Hong & J.J. Li ex D.Y. Hong in Acta Phytotax. Sin.,* 36(6): 539 (1998): P. suffruticosa subsp. rockii.
subsp. taibaishanica *D.Y. Hong in Acta Phytotax. Sin.,* 36(6): 542 (1998)—China North-central (Shaanxi, Gansu).
suffruticosa Andrews
var. jishanensis *(T. Hong & W.Z. Zhao) J.J. Halda in Acta Mus. Richnov. Sect. Nat.,* 4(2): 30 (1997): P. jishanensis.
var. linyanshanii *(T. Hong & Osti) J.J. Halda in Acta Mus. Richnov. Sect. Nat.,* 4(2): 30 (1997), basionym not validly publ. at ref. stated: P. rockii subsp. linyanshanii.
subsp. ostii *(T. Hong & J.X. Zhang) J.J. Halda in Acta Mus. Richnov. Sect. Nat.,* 4(2): 30 (1997): P. ostii.
var. qiui *(Y.L. Pei & D.Y. Hong) J.J. Halda in Acta Mus. Richnov. Sect. Nat.,* 4(2): 31 (1997): P. qiui.
var. yananensis *(T. Hong & M.R. Li) J.J. Halda in Acta Mus. Richnov. Sect. Nat.,* 4(2): 30 (1997): P. yananensis.
subsp. yinpingmudan *D.Y. Hong, K.Y. Pan & Zhang W. Xie in Acta Phytotax. Sin.,* 36(6): 519 (1998)—China Southeast (Anhui, Henan).
tenuifolia L.
subsp. biebersteiniana *(Rupr.) J.J. Halda in Acta Mus. Richnov. Sect. Nat.,* 4(2): 29 (1997): P. biebersteiniana.

PAEONIA:-
veitchii Lynch
subsp. altaica *(K.M. Dai & T.H. Ying) J.J. Halda in Acta Mus. Richnov. Sect. Nat.,* 4(2): 29 (1997): P. altaica.
wittmanniana Hartwiss ex Lindl.
subsp. macrophylla *(Albov) J.J. Halda in Acta Mus. Richnov. Sect. Nat.,* 4(2): 30 (1997): P. corallina f. macrophylla.

PAEPALANTHUS *Mart. in Ann. Sci. Nat., Bot.,* Sér. 2, 2: 28 (1834). *ERIOCAULACEAE†*
albovaginatus Silveira
var. albobracteatus *M.L. Tissot-Squalli H., Monogr. Bearbeit. Paepalanthus subgen. Platycaulon (Diss. Bot., 280):* 164 (1997), contrary to Art. 37.5 ICBN (1994); *et in Feddes Repert.,* 108(5–6): 372 (Aug. 1997)—Brazil South (Santa Catarina, Paraná).
var. fuscobracteatus *M.L. Tissot-Squalli H., Monogr. Bearbeit. Paepalanthus subgen. Platycaulon (Diss. Bot., 280):* 167 (1997), contrary to Art. 37.5 ICBN (1994); *et in Feddes Repert.,* 108(5–6): 375 (Aug. 1997)—Brazil South (Paraná).
alpestris *(Körn.) M.L. Tissot-Squalli H., Monogr. Bearbeit. Paepalanthus subgen. Platycaulon (Diss. Bot., 280):* 91 (1997); *et in Feddes Repert.,* 108(5–6): 361 (Aug. 1997): P. planifolius var. alpestris.
celsus *M.L. Tissot-Squalli H., Monogr. Bearbeit. Paepalanthus subgen. Platycaulon (Diss. Bot., 280):* 103 (1997); *et in Feddes Repert.,* 108(5–6): 364 (Aug. 1997)Ecuador.
crinitus *M.L. Tissot-Squalli H., Monogr. Bearbeit. Paepalanthus subgen. Platycaulon (Diss. Bot., 280):* 175 (1997); *et in Feddes Repert.,* 108(5–6): 377 (Aug. 1997)—Brazil Southeast (Minas Gerais).
humilis *P.T. Sano in Novon,* 8(3): 298 (1998)—Brazil Northeast (Bahia).
longivaginatus *M.L. Tissot-Squalli H., Monogr. Bearbeit. Paepalanthus subgen. Platycaulon (Diss. Bot., 280):* 136 (1997), contrary to Art. 37.5 ICBN (1994); *et in Feddes Repert.,* 108(5–6): 367 (Aug. 1997)—Peru.
obnatus *M.L. Tissot-Squalli H., Monogr. Bearbeit. Paepalanthus subgen. Platycaulon (Diss. Bot., 280):* 142 (1997); *et in Feddes Repert.,* 108(5–6): 369 (Aug. 1997)—Ecuador, Peru, Venezuela.
parviflorus *(Hensold) N. Hensold in Phytologia,* 81(1): 25 (1996 publ. 1997): P. chlorocephalus f. parviflorus.
rigidulus *Mart.in Ann. Sci. Nat., Bot.*Sér. 2, 2: 29 (1834)—Locality not stated.†
spixianus *Mart. in Ann. Sci. Nat., Bot.,* Sér. 2, 2: 28 (1834)—Brazil.†
stuetzelii *N. Hensold in Phytologia,* 81(1): 25 (1996 publ. 1997), nom. nov.: P. lanuginosus *Hensold.*

PALAQUIUM (Sapotac.).
hansenii *P. Chantaranothai in Nordic J. Bot.,* 18(3): 365 (1998)—Thailand.

PALAUA (Malvac.).
sandemanii *(Sandwith) P.A. Fryxell in Brittonia,* 49(2): 246 (1997): Malvastrum sandemanii.

× **PALERMOARA** *hort. in Orchid Rev.,* 106(1220): 126 (1998). *ORCHIDACEAE* (Ada × Comparettia × Gomesa).

PALICOUREA (Rubiac.).
sect. Cephaeloides *C.M. Taylor in Ann. Missouri Bot. Gard.,* 84(2): 235 (1997).
sect. Corymbiferae *(Müll. Arg.) C.M. Taylor in Ann. Missouri Bot. Gard.,* 84(2): 232 (1997): Psychotria ser. Corymbiferae.
sect. Grandiflorae *C.M. Taylor in Ann. Missouri Bot. Gard.,* 84(2): 231 (1997).
subgen. Montanae *C.M. Taylor in Ann. Missouri Bot. Gard.,* 84(2): 232 (1997).
sect. Montanae *C.M. Taylor in Ann. Missouri Bot. Gard.,* 84(2): 232 (1997).
sect. Obovoideae *C.M. Taylor in Ann. Missouri Bot. Gard.,* 84(2): 234 (1997).
sect. Pseudoamethystinae *C.M. Taylor in Ann. Missouri Bot. Gard.,* 84(2): 234 (1997).

PALICOUREA:-
sect. Psychotrioides *C.M. Taylor in Ann. Missouri Bot. Gard.,* 84(2): 235 (1997).
subser. Subcymosae *(Müll. Arg.) C.M. Taylor in Ann. Missouri Bot. Gard.,* 84(2): 231 (1997): Psychotria ser. Subcymosae.
acanthacea *Standl. ex C.M. Taylor in Novon,* 6(2): 211 (1996)—Colombia.
amplissima *(Standl. ex Steyerm.) C.M. Taylor in Novon,* 6(2): 211 (1996): Psychotria amplissima.
anderssoniana *C.M. Taylor in Ann. Missouri Bot. Gard.,* 84(2): 236 (1997)—Ecuador.
anianguana *C.M. Taylor in Ann. Missouri Bot. Gard.,* 84(2): 237 (1997)—Ecuador.
asplundii *C.M. Taylor in Ann. Missouri Bot. Gard.,* 84(2): 239 (1997)—Ecuador, Colombia.
australis *C.M. Taylor in Novon,* 10(2): 161 (2000)—Brazil South (Paraná, Santa Catarina).
awa *C.M. Taylor in Novon,* 10(1): 78 (2000)—Colombia, Ecuador.
azurea *C.M. Taylor in Ann. Missouri Bot. Gard.,* 84(2): 240 (1997)—Ecuador.
canarina *C.M. Taylor in Ann. Missouri Bot. Gard.,* 84(2): 242 (1997)—Ecuador.
candida *C.M. Taylor in Ann. Missouri Bot. Gard.,* 84(2): 243 (1997)—Ecuador.
chignul *C.M. Taylor in Ann. Missouri Bot. Gard.,* 84(2): 244 (1997)—Ecuador, Colombia.
clerodendroides *C.M. Taylor in Novon,* 10(1): 80 (2000)—Ecuador.
cogolloi *C.M. Taylor in Novon,* 10(1): 82 (2000)—Colombia.
condorica *C.M. Taylor in Ann. Missouri Bot. Gard.,* 84(2): 246 (1997)—Ecuador.
corniculata *C.M. Taylor in Ann. Missouri Bot. Gard.,* 84(2): 246 (1997)—Ecuador.
cornigera *C.M. Taylor in Ann. Missouri Bot. Gard.,* 84(2): 249 (1997)—Ecuador.
deviae *C.M. Taylor in Ann. Missouri Bot. Gard.,* 84(2): 250 (1997)—Ecuador.
dimorphandrioides *(Dwyer) C.M. Taylor in Novon,* 6(2): 211 (1996): dimorphandrioides.
eburnea *C.M. Taylor in Novon,* 10(1): 84 (2000)—Colombia.
frontinoensis *A. Cogollo P. & C.M. Taylor in Caldasia,* 19(1–2): 91 (1997)—Colombia.
fuchsioides *C.M. Taylor in Fl. Ecuador,* 62(162:3): 228 (1999), nom. nov.: Cephaelis jamesonii *Standl.*
garcioides *C.M. Taylor in Novon,* 10(1): 85 (2000)—Colombia.
gentryi *C.M. Taylor in Ann. Missouri Bot. Gard.,* 84(2): 250 (1997)—Ecuador.
glandulifera *C.M. Taylor in Novon,* 10(1): 86 (2000)—Colombia.
grandistipula *(Standl. ex Steyerm.) C.M. Taylor in Novon,* 6(2): 211 (1996): Cephaelis grandistipula.
harlingii *C.M. Taylor in Ann. Missouri Bot. Gard.,* 84(2): 251 (1997)—Ecuador.
jaramilloi *C.M. Taylor in Ann. Missouri Bot. Gard.,* 84(2): 252 (1997)—Ecuador.
kirkbrideae *C.M. Taylor in Novon,* 10(1): 86 (2000), nom. nov.: P. justicioides var. macrocalyx *M.C.G. Kirkbr.*
lugoana *C.M. Taylor in Ann. Missouri Bot. Gard.,* 84(2): 253 (1997)—Ecuador, Colombia.
luteonivea *C.M. Taylor in Ann. Missouri Bot. Gard.,* 84(2): 253 (1997)—Ecuador, Colombia.
macarthuriorum *C.M. Taylor in Novon,* 5(4): 380 (1995), as 'macarthurorum'—Colombia, Ecuador.
melheana *S.L. Jung-Mendaçolli in Hoehnea,* 18(2): 143 (1991 publ. 1992)—Brazil Southeast (São Paulo).
microcarpa *(Ruiz & Pav.) D.C. Zappi in Kew Bull.,* 55(1): 168 (2000): Coffea microcarpa.
orquidea *C.M. Taylor in Novon,* 7(2): 191 (1997)—Colombia.
ottohuberi *J.H. Kirkbride in Biollania,* ed. esp. 6 (J.J. Wurdack Festschr.): 400 (1997)—Venezuela.
pensilis *J.H. Kirkbride in Biollania,* ed. esp. 6 (J.J. Wurdack Festschr.): 402 (1997)—Venezuela.
prodiga *Standl. ex C.M. Taylor in Ann. Missouri Bot. Gard.,* 84(2): 254 (1997)—Ecuador.
quadrifolia (Rudge) DC.
subsp. leticiana *C.M. Taylor in Novon,* 5(4): 382 (1995)—Colombia, Peru.
quadrilateralis *C.M. Taylor in Novon,* 7(2): 193 (1997)—Colombia.

PALICOUREA:-
remyana *(Baill.) C.M. Taylor in Novon,* 6(2): 212 (1996): Uragoga remyana.
sopkinii *C.M. Taylor in Novon,* 7(2): 194 (1997)—Colombia.
subalatoides *C.M. Taylor in Ann. Missouri Bot. Gard.,* 84(2): 255 (1997)—Ecuador.
subtomentosa *(Ruiz & Pav.) C.M. Taylor in Ann. Missouri Bot. Gard.,* 84(2): 256 (1997): Psychotria subtomentosa.
subsp. lojana *C.M. Taylor in Ann. Missouri Bot. Gard.,* 84(2): 257 (1997)—Ecuador.
tunjaensis *C.M. Taylor in Novon,* 6(2): 212 (1996), nom. nov.: Psychotria boyacana *Standl.*
ulloana *C.M. Taylor in Ann. Missouri Bot. Gard.,* 84(2): 257 (1997)—Ecuador, Peru.
vulcanalis *Standl. ex C.M. Taylor in Ann. Missouri Bot. Gard.,* 84(2): 259 (1997)—Ecuador.
wurdackiana *J.H. Kirkbride in Biollania,* ed. esp. 6 (J.J. Wurdack Festschr.): 403 (1997)—Venezuela.
zarucchii *C.M. Taylor in Novon,* 7(2): 195 (1997)—Colombia.

PALLENIS (Compos.).
subgen. Saulcya *(Michon) W. Greuter in Flora Medit.,* 7: 47 (1997): gen. Saulcya.
cuspidata Pomel
subsp. canescens *(Maire) W. Greuter in Flora Medit.,* 7: 47 (1997): P. spinosa var. canescens.
hierochuntica *(Michon) W. Greuter in Flora Medit.,* 7: 47 (1997): Saulcya hierochuntica.
maritima *(L.) W. Greuter in Flora Medit.,* 7: 47 (1997): Buphthalmum maritimum.
spinosa (L.) Cass.
subsp. asteroidea *(Viv.) W. Greuter in Flora Medit.,* 7: 47 (1997): Buphthalmum asteroideum.
subsp. maroccana *(Aurich & Podlech) W. Greuter in Flora Medit.,* 7: 47 (1997): Asteriscus spinosus subsp. maroccanus.

PALMORCHIS (Orchidac.).
deceptorius *Y. Veyret & D.L. Szlachetko in Fragm. Flor. Geobot.,* 40(2): 889 (1995)—Colombia.
paludicola *R.L. Dressler in Orquideologia,* 20(3): 261 (1997)—Costa Rica.
sordida *R.L. Dressler in Orquideologia,* 20(3): 262 (1997)—Costa Rica.
trinotata *R.L. Dressler in Orquideologia,* 20(3): 263 (1997)—Panama.

PANCOVIA (Sapindac.).
holtzii Radlk.
subsp. faulknerae *B. Verdcourt in Fl. Trop. E. Afr., Sapindac.:* 58 (1998)—Tanzania.

PANCRATIUM (Amaryllidac.).
longifolium *Jacques in Ann. Fl. Pomone,* 1835–36: 252 (1836)—Locality not known.

× **PANCZAKARA** *hort. in Orchid Rev.,* 105(1214): 126 (1997). *ORCHIDACEAE* (Brassavola × Epidendrum × Laelia × Sophronitis).

PANDANUS (Pandanac.).
subsect. Barbellati *K.-L. Huynh in Bull. Soc. Neuchâtel. Sci. Nat.,* 121: 60 (1998).
sect. Pseudophalanx *K.-L. Huynh in Bull. Soc. Neuchâtel. Sci. Nat.,* 118: 38 (1995).
alveolatus *K.-L. Huynh in Bull. Soc. Neuchâtel. Sci. Nat.,* 122: 36 (1999)—Madagascar.
ambalavaoensis *K.-L. Huynh in Bot. Jahrb.,* 122(2): 208 (2000)—Madagascar.
ambohitantelensis *K.-L. Huynh in Candollea,* 54(1): 156 (1999)—Madagascar.
amissus *K.-L. Huynh in Gard. Bull. Singapore,* 51(2): 163 (1999)—Caroline Is. (Micronesia Federated States).
analamerensis *K.-L. Huynh in Bot. Jahrb.,* 122(2): 213 (2000)—Madagascar.
andringitrensis *K.-L. Huynh in Bull. Soc. Neuchâtel. Sci. Nat.,* 122: 38 (1999)—Madagascar.
arenicola *K.-L. Huynh in Candollea,* 54(1): 151 (1999)—Madagascar.
barbellatus *K.-L. Huynh in Bull. Soc. Neuchâtel. Sci. Nat.,* 121: 57 (1998)—Madagascar.
bemarahensis *K.-L. Huynh in Candollea,* 54(1): 152 (1999)—Madagascar.

PANDANUS:-
cissei *K.-L. Huynh in Bull. Soc. Neuchâtel. Sci. Nat.,* 119: 27 (1996)—Mali.
coriaceus *K.-L. Huynh in Bot. Jahrb.,* 122(2): 215 (2000)—Madagascar.
cuneatus *K.-L. Huynh in Candollea,* 54(1): 157 (1999)—Madagascar.
echinops *K.-L. Huynh in Bull. Soc. Neuchâtel. Sci. Nat.,* 118: 31 (1995)—Mali.
farakoensis *K.-L. Huynh in Bull. Soc. Neuchâtel. Sci. Nat.,* 119: 32 (1996)—Mali.
ferox *K.-L. Huynh in Bot. Jahrb.,* 122(2): 219 (2000)—Madagascar.
fetosus *K.-L. Huynh in Bull. Soc. Neuchâtel. Sci. Nat.,* 123: 30 (2000)—Madagascar.
flagellibracteatus *K.-L. Huynh in Candollea,* 54(1): 150 (1999)—Madagascar.
globulatus *K.-L. Huynh in Bot. Helvetica,* 107(1): 94 (1997)—Mozambique.
insuetus *K.-L. Huynh in Bot. Jahrb.,* 122(2): 217 (2000)—Madagascar.
isalicus *K.-L. Huynh in Candollea,* 54(1): 155 (1999)—Madagascar.
latistigmaticus *K.-L. Huynh in Candollea,* 54(1): 149 (1999)—Madagascar.
lorencei *K.-L. Huynh in Gard. Bull. Singapore,* 51(2): 166 (1999)—Caroline Is. (Palau).
majungensis *K.-L. Huynh in Candollea,* 54(1): 154 (1999)—Madagascar.
malgrasii *K.-L. Huynh in Bull. Soc. Neuchâtel. Sci. Nat.,* 118: 28 (1995)—Mali.
manongarivensis *K.-L. Huynh in Candollea,* 54(1): 153 (1999)—Madagascar.
masoalensis *M.O. Laivao & M.W. Callmander in Bot. Helvetica,* 110(1): 43 (2000)—Madagascar.
membranaceus *K.-L. Huynh in Candollea,* 54(1): 151 (1999)—Madagascar.
mosambicius *H. St. John ex K.-L. Huynh in Bot. Helvetica,* 107(1): 94 (1997)—Mozambique.
muralis *K.-L. Huynh in Bull. Soc. Neuchâtel. Sci. Nat.,* 118: 32 (1995)—Mali.
nosibicus *K.-L. Huynh in Bull. Soc. Neuchâtel. Sci. Nat.,* 120: 37 (1997)—Madagascar.
odoratissimus L.f.
f. ferreus *(Y. Kimura) S. Hatusima in S. Hatusima & T. Amano, Fl. Ryukyus, S. of Amami Is., ed. 2:* 237 (1994), without basionym ref.: Basionym not stated.
f. laevis *S. Hatusima in S. Hatusima & T. Amano, Fl. Ryukyus, S. of Amami Is., ed. 2:* 237 (1994), without latin descr. or type—Locality not stated.
pluriaculeatus *K.-L. Huynh in Bull. Soc. Neuchâtel. Sci. Nat.,* 122: 39 (1999)—Madagascar.
serrimarginalis *H. St. John ex K.-L. Huynh in Bot. Helvetica,* 107(1): 99 (1997)—Mozambique.
sikassoensis *K.-L. Huynh in Bull. Soc. Neuchâtel. Sci. Nat.,* 119: 29 (1996)—Mali.
silvanus *K.-L. Huynh in Bot. Jahrb.,* 122(2): 206 (2000)—Madagascar.
sylvicola *K.-L. Huynh in Bull. Soc. Neuchâtel. Sci. Nat.,* 121: 54 (1998)—Madagascar.
tabellarius *K.-L. Huynh in Bull. Soc. Neuchâtel. Sci. Nat.,* 122: 36 (1999)—Madagascar.
tenuiflagellatus *K.-L. Huynh in Bot. Jahrb.,* 122(2): 210 (2000)—Madagascar.
tolanarensis *K.-L. Huynh in Candollea,* 54(1): 158 (1999)—Madagascar.
tubulatus *K.-L. Huynh in Candollea,* 54(1): 159 (1999)—Madagascar.

PANICUM (Gramin.).
sect. Agrostoides *(Nash) F.O. Zuloaga in T.R. Soderstrom et al. (eds.), Grass Syst. & Evol.:* 294 (1987 publ. March. 1988): P. group Agrostoidea.
sect. Dichanthelium *(Hitchc. & Chase) F.O. Zuloaga in T.R. Soderstrom et al. (eds.), Grass Syst. & Evol.:* 299 (1987 publ. March 1988): P. subgen. Dichanthelium.
acuminatum Sw.
var. wrightianum *(Scribn.) C.F. Reed in Phytologia,* 80(4): 284 (1996): P. wrightianum.
alatum *F.O. Zuloaga & O. Morrone in Ann. Missouri Bot. Gard.,* 83(2): 224 (1996)—Mexico Northwest (Baja California Sur, Sinaloa, Sonora), Mexico Northeast (Durango, San Luis Potosí, Tamaulipas), Mexico Southwest (Guerrero, Michoacan, Nayarit), Arizona, California.

PANICUM:-
alatum
var. longiflorum *F.O. Zuloaga & O. Morrone in Ann. Missouri Bot. Gard.,* 83(2): 226 (1996)—Mexico Northwest (Baja California Sur, Sonora), California.
var. minus *(Andersson) F.O. Zuloaga & O. Morrone in Ann. Missouri Bot. Gard.,* 83(2): 226 (1996): P. hirticaule var. minus.
aztecanum *F.O. Zuloaga & O. Morrone in Ann. Missouri Bot. Gard.,* 83(2): 229 (1996)—Mexico Central (México State, Mexico Distrito Federal), Mexico Northwest (Sinaloa), Mexico Southwest (Jalisco).
bivonianum *S. Brullo, P. Minissale, F. Scelsi & G. Spampinato in Giorn. Bot. Ital.,* 129(2): 173 (1995), nom. nov.: P. compressum *Biv.*
brunonianum *Wall. & Griff.in J. Asiat. Soc. Bengal,* 5: 574 (1836)—Bangladesh†
ephemeroides *F.O. Zuloaga & O. Morrone in Ann. Missouri Bot. Gard.,* 83(2): 239 (1996), nom. nov.: P. ephemerum *Zuloaga, Morrone & Valls.*
ephemerum *F.O. Zuloaga, O. Morrone & J.F.M. Valls in Iheringia, Bot.,* 42: 4 (1992 publ. 1993), homonym; *see Renvoize in Kew Bull.,* 34(3): 551 (1980)—Brazil West-central (Goiás).
hirticaule J. Presl
var. verrucosum *F.O. Zuloaga & O. Morrone in Ann. Missouri Bot. Gard.,* 83(2): 253 (1996)—Mexico Northeast (Chihuahua), Mexico Northwest (Sinaloa, Sonora), Mexico Southwest (Guerrero, Jalisco, Nayarit), Arizona.
implicatum *F.L. Scribner in Britton & A. Br., Ill. Fl. N. U. S.,* 3: 498 (20 June 1898)—Northeastern U.S.A.†
jaliscana *(Santana Mich.) A. Espejo Serna & A.R. López-Ferrari in Acta Bot. Mex.,* 51: 62 (2000): Brachiaria jaliscana.
laterale *Nees in Mart., Fl. Bras. Enum. Pl.,* 2(1): 213 (1829)—Brazil.†
linearifolium *F.L. Scribner in Britton & A. Br., Ill. Fl. N. U. S.,* 3: 500 (20 June 1898)—Northeastern U.S.A., North-central U.S.A.†
littorale *M.S.M. Sosef in Bull. Jard. Bot. Nation. Belg.,* 67(1–4): 6 (1999), homonym; *see Vasey in Bot. Gaz.,* 4(1): 106 (1879)—Gabon.
longiglume *H. Peng & L.H. Zhou in Acta Phytotax. Sin.,* 36(1): 77 (1998)—China South-central (Yunnan).
macrospermum *(S. Gould) A. Espejo Serna & A.R. López-Ferrari in Acta Bot. Mex.,* 51: 62 (2000): Dichanthelium macrospermum.
magnispicula *F.O. Zuloaga, O. Morrone & J.F.M. Valls in Iheringia, Bot.,* 42: 5 (1992 publ. 1993)—Brazil South (Santa Catarina).
mystasipum *F.O. Zuloaga & O. Morrone in Iheringia, Bot.,* 42: 14 (1992 publ. 1993)—Brazil West-central (Brasília Distrito Federal).
nutabundum *F.O. Zuloaga & O. Morrone in Novon,* 6(3): 312 (1996)—Brazil North (Pará).
paludosum Roxb.
var. parvispiculosum *P.C. Keng & L.C. Li in Pl. Shanghai,* 1: 740 (1999)—China Southeast (Jiangsu).
sendulskyi *F.O. Zuloaga & O. Morrone in Novon,* 6(3): 310 (1996), as 'sendulskyii'—Brazil Southeast (Minas Gerais).
sphaerocarpon Elliott
var. isophyllum *(Scribn.) C.F. Reed in Phytologia,* 80(4): 284 (1996), isonym; *see R. Angelo in Phytologia,* 71(2): 85 (1991): P. microcarpon var. isophyllum.
werneri *F.L. Scribner in Britton & A. Br., Ill. Fl. N. U. S.,* 3: 501 (20 June 1898)—Northeastern U.S.A.

PANZERINA (Labiat.).
lanata (L.) Soják
subsp. argyracea *(Kuprian.) T.V. Krestovskaja in Fl. Sibir. (Pyrolac.-Lamiac.),* 11: 196 (1997): Panzeria argyracea.
parviflora *(C.Y. Wu & H.W. Li) Y.Z. Zhao in Acta Phytotax. Sin.,* 36(3): 204 (1998): Panzeria parviflora.

PAPAVER (Papaverac.).
subgen. Feddia *T.V. Egorova in Novosti Sist. Vyssh. Rast.,* 31: 118 (1998).

PAPAVER:-

subgen. Tolmatchevia *T.V. Egorova in Novosti Sist. Vÿssh. Rast.*, 31: 96 (1998).
ammophilum *(Turcz.) G.A. Peschkova in Fl. Sibir. (Berberidac.-Grossulariac.)*, 7: 16 (1994): P. nudicaule var. ammophilum.
atlanticum (Ball) Coss.
subsp. mesatlanticum *(Maire) J.W. Kadereit in Edinburgh J. Bot.*, 53(3): 303 (1996): P. rupifragum var. mesatlanticum.
bivalve (DC.) K.-F. Günther
subsp. hybridum *(L.) T. Karlsson in Svensk Bot. Tidskr.*, 91(5): 249 (1997 publ. 1998): Chelidonium hybridum.
chakassicum *G.A. Peschkova in Fl. Sibir. (Berberidac.-Grossulariac.)*, 7: 18 (1994)—Krasnoyarsk, Altay, Tuva.
croceum Ledeb.
subsp. stanovense *Y.N. Petrochenko in L.I. Malyschev (ed.), Vÿsokogornaya Flora Stanovogo Nagor'ya*: 96 (1972)—Yakutiya.
jacuticum *G.A. Peschkova in Fl. Sibir. (Berberidac.-Grossulariac.)*, 7: 19 (1994), nom. nov.: P. nudicaule subsp. gracile *Tolm.*
var. atrofuscum *G.A. Peschkova in Fl. Sibir. (Berberidac.-Grossulariac.)*, 7: 20 (1994), contrary to Art. 37.5 ICBN (1994)—Yakutiya.
lateritium K. Koch
subsp. monanthum *(Trautv.) J.W. Kadereit in Edinburgh J. Bot.*, 53(3): 305 (1996): P. monanthum.
ledebourianum C.E. Lundstr.
var. ammophilum *(Turcz.) G.A. Peschkova in Fl. Tsentral'noi Sibiri*, 1: 378 (1979): P. nudicaule var. ammophilum.
var. leucanthum *(Trautv.) G.A. Peschkova, Stepnaya Fl. Baĭkal'skoĭ Sibiri*: 59 (1972): P. nudicaule var. leucanthum.
maschukense *A.D. Mikheev in Bot. Zhurn.*, 84(8): 105 (1999)—North Caucasus (Stavropol).
nudicaule L.
f. amurense *(N. Busch) H. Chuang in Fl. Reipubl. Popul. Sin.*, 32: 58 (1999): P. nudicaule subsp. amurense.
subsp. anomalum *(Fedde) V.N. Voroschilov in A.K. Skvortsov (ed.), Florist. issl. v razn. raĭonakh SSSR*: 171 (1985): P. anomalum.
var. calcareum *G.A. Peschkova in Fl. Tsentral'noi Sibiri*, 1: 379 (1979)—Tuva.
f. seticarpum *(P.Y. Fu) H. Chuang in Fl. Reipubl. Popul. Sin.*, 32: 58 (1999): Papaver nudicaule var. seticarpum.
olchonense *G.A. Peschkova in Fl. Sibir. (Berberidac.-Grossulariac.)*, 7: 23 (1994)—Irkutsk, Buryatiya.
pilosum Sibth. & Sm.
subsp. glabrisepalum *J.W. Kadereit in Edinburgh J. Bot.*, 53(3): 292 (1996)—Turkey.
subsp. sparsipilosum *(Boiss.) J.W. Kadereit in Edinburgh J. Bot.*, 53(3): 295 (1996): P. heldreichii var. sparsipilosum.
subsp. spicatum *(Boiss. & Balansa) N. Wendt ex J.W. Kadereit in Edinburgh J. Bot.*, 53(3): 295 (1996): P. spicatum.
subsp. strictum *(Boiss. & Balansa) N. Wendt ex J.W. Kadereit in Edinburgh J. Bot.*, 53(3): 292 (1996): P. strictum.
pseudoradicatum *M. Kitagawa in Rep. Inst. Sci. Res. Manchoukuo*, 6(4): 122 (1942), as 'pseudo-radicatum'—Manchuria.
radicatum Rottb.
f. albiflorum *Y.N. Lee, Fl. Korea*: 1158 (1996), without type—Locality not stated.
refractum (DC.) K.-F. Günther
subsp. occidentale *(Kadereit) T. Karlsson in Svensk Bot. Tidskr.*, 91(5): 249 (1997 publ. 1998): Roemeria refracta subsp. occidentalis.
rubroaurantiacum (Fisch.) Sweet
subsp. changaicum *(Kamelin) R.V. Kamelin & I.A. Gubanov in I.A. Gubanov, Konsp. Fl. Vneshnei Mongolii (Sosud. Rast.)*: 52 (1996): P. changaicum.
subsp. saichanense *(Grubov) R.V. Kamelin & I.A. Gubanov in I.A. Gubanov, Konsp. Fl. Vneshnei Mongolii (Sosud. Rast.)*: 52 (1996): P. saichanense.
subsp. smirnovii *(Peschkova) R.V. Kamelin & I.A. Gubanov in I.A. Gubanov, Konsp. Fl. Vneshnei Mongolii (Sosud. Rast.)*: 52 (1996): P. smirnovii.

PAPAVER:-

setosum *(Tolm.) G.A. Peschkova, Stepnaya Fl. Baĭkal'skoĭ Sibiri*: 59 (1972); *et in Fl. Sibir. (Berberidac.-Grossulariac.)*, 7: 27 (1994): P. rubroaurantiacum subsp. setosum.
stanovense *(Petroch.) G.A. Peschkova in Fl. Sibir. (Berberidac.-Grossulariac.)*, 7: 28 (1994): P. croceum subsp. stanovense.
stevenianum *A.D. Mikheev in Bot. Zhurn.*, 78(5): 123 (1993), 1981 publ. invalid because 2 types cited—North Caucasus (Stavropol).†
tolmatschevianum *N.S. Pavlova in Bot. Zhurn.*, 84(2): 112 (1999)—Sakhalin.
turczaninovii *G.A. Peschkova in Fl. Sibir. (Berberidac.-Grossulariac.)*, 7: 30 (1994), nom. nov.: P. nudicaule var. calcareum *Peschkova*.
udocanicum *(Peschkova) G.A. Peschkova in Fl. Sibir. (Berberidac.-Grossulariac.)*, 7: 30 (1994): P. pseudocanescens subsp. udocanicum.

PAPHINIA (Orchidac.).

levyae *L.A. Garay in Harvard Pap. Bot.*, 4(1): 311 (1999)—Ecuador.
var. angustisegmenta *L.A. Garay in Harvard Pap. Bot.*, 4(1): 311 (1999)—Ecuador.
subclausa *R.L. Dressler in Novon*, 7(2): 121 (1997)—Costa Rica.
zamorae *L.A. Garay in Harvard Pap. Bot.*, 4(1): 310 (1999)—Ecuador.

PAPHIOPEDILUM (Orchidac.).

adductum Asher
var. anitum *(Golamco) H. Koopowitz in Orchid Dig.*, 64(4): 157 (2000): P. anitum.
angustatum *Z.J. Liu & S.C. Chen in Acta Phytotax. Sin.*, 38(5): 464 (2000)—China South-central (Yunnan).
anitum *A.S. Golamco in Waling-Waling Rev.*, 6(2): 10 (1998)—Philippines.
appletonianum (Gower) Rolfe
f. album *Asher ex O. Gruss in Orchidee*, 51(6): 746 (2000)—Thailand.
var. hainanense *(Fowlie) G.J. Braem, C.O. Baker & M.L. Baker, Gen. Paphiopedilum Nat. Hist. & Cult.*, 2: 295 (1999), as 'hainanensis': P. hainanense.
f. immaculatum *(Braem) G.J. Braem in Orchidées Cult. Protect.*, 36: 35, 37 (1998): P. appletonianum var. immaculatum.
armeniacum S.C. Chen & F.Y. Liu
f. markii *(O. Gruss) G.J. Braem in G.J. Braem, C.O. Baker & M.L. Baker, Gen. Paphiopedilum Nat. Hist. &* Cult., 1: 69 (1998): P. armeniacum var. markii.
var. markii *O. Gruss in Orchidee*, 48(5): 215 (1997)—China South-central (Yunnan).
barbigerum T. Tang & F.T. Wang
var. aureum *H.S. Hua in Orchids*, 68(3): 242 (1999)—China South-central (Yunnan).
f. aureum *(H.S. Hua) O. Gruss & J. Roeth in Caesiana*, 12: 59 (1999): P. barbigerum var. aureum.
f. aureum *(H.S. Hua) G.J. Braem, C.O. Baker & M.L. Baker, Gen. Paphiopedilum Nat. Hist. & Cult.*, 2: 217 (1999): P. barbigerum var. aureum.
bellatulum (Rchb. f.) Stein
f. album *(O'Brien) G.J. Braem in G.J. Braem, C.O. Baker & M.L. Baker, Gen. Paphiopedilum Nat. Hist. &* Cult., 1: 97 (1998): P. bellatulum var. album.
bougainvilleanum Fowlie
var. saskianum *O. Gruss & J. Roeth in Orchidee*, 50(5): 534 (1999)—Solomon Is (South Solomons).
bullenianum (Rchb. f.) Pfitzer
var. johorense *(Fowlie & Yap) G.J. Braem, C.O. Baker & M.L. Baker, Gen. Paphiopedilum Nat. Hist. & Cult.*, 2: 302 (1999), as 'johorensis': P. johorense.
callosum (Rchb. f.) Stein
var. potentianum *(O. Gruss & J. Roeth) P.J. Cribb, Genus Paphiopedilum ed. 2*: 337 (1998): P. potentianum.
f. viridiflorum *(Hort.) O. Gruss & J. Roeth in Caesiana*, 12: 60 (1999), with incorrect basionym ref.: Cypripedium callosum var. viridiflorum.

PAPHIOPEDILUM:-

cerveranum *G.J. Braem in Orchidées Cult. Protect.*, 38: 28 (1999)—Vietnam.
f. viride *(Braem) G.J. Braem in Orchidées Cult. Protect.*, 38: 29 (1999), without exact basionym page; *et in G.J. Braem, C.O. Baker & M.L. Baker, Gen. Paphiopedilum Nat. Hist. &* Cult., 2: 309 (1999), without exact basionym page: P. robinsonii f. viride.
chaoi *H.S. Hua in Orchidee*, 50(5): 495 (1999)—China Southeast (Guangxi).
charlesworthii
f. crawshawiae *(O'Brien) O. Gruss in Caesiana*, 12: 63 (1999): Cypripedium crawshawiae.
f. sandowiae *G.J. Braem in Orchidées Cult. Protect.*, 36: 35, 37 (1998)—Burma.
coccineum *H. Perner & R. Herrmann in Orchidee*, 51(5): 623 (2000)—Vietnam.
concolor (Bateman) Pfitzer
f. album *(Braem) G.J. Braem in G.J. Braem, C.O. Baker & M.L. Baker, Gen. Paphiopedilum Nat. Hist. &* Cult., 1: 101 (1998), basionym not validly publ.: P. concolor var. album.
var. dahuaense *Z.J. Liu & J.Yong Zhang in Acta Bot. Yunnanica*, 22(4): 393 (2000)—China Southeast (Guangxi).
var. immaculatum *Z.J. Liu & J.Yong Zhang in Acta Bot. Yunnanica*, 22(4): 393 (2000)—China Southeast (Guangxi).
crossii *(E. Morren) G.J. Braem & K. Senghas in Sida*, 19(2): 251 (2000): Cypripedium crossii.
var. potentianum *(O. Gruss & Roeth) G.J. Braem & K. Senghas in Sida*, 19(2): 254 (2000): Paphiopedilum potentianum.
var. sublaeve *(Rchb. f.) G.J. Braem & K. Senghas in Sida*, 19(2): 254 (2000): Cypripedium callosum var. sublaeve.
delenatii Guillaumin
f. albinum *G.J. Braem in G.J. Braem, C.O. Baker & M.L. Baker, Gen. Paphiopedilum Nat. Hist. &* Cult., 1: 73 (1998)—Vietnam.
fairieanum
f. bohlmannianum *(Matho) G.J. Braem in Orchidées Cult. Protect.*, 36: 35, 37 (1998): P. fairieanum var. bohlmannianum.
fowliei
f. christianae *G.J. Braem in Orchidées Cult. Protect.*, 36: 36, 38 (1998)—Philippines.
f. sangianum *(Braem) O. Gruss & J. Roeth in Caesiana*, 12: 60 (1999): P. fowliei var. sangianum.
gigantifolium *G.J. Braem, M.L. Baker & C.O. Baker in Orchidées Cult. Protect.*, 30: 5 (May 1997), without diagnostic latin descr.; *et in Orchidées Cult. Protect.*, 31: 34 (Sept. 1997), with latin descr.—Sulawesi.
× grussianum *H.S. Hua in Orchidee*, 49(4): 180 (1998)—China Southeast (Guangxi). (P. hirsutissimum var. esquirolei × parishii var. dianthum).
hangianum *H. Perner & O. Gruss in Orchidee*, Suppl. 6: 5 (1999); *et in Oasis*, 1(3): 14 (2000)—Vietnam.
haynaldianum (Rchb. f.) Stein
f. album *(Asher) G.J. Braem in G.J. Braem, C.O. Baker & M.L. Baker, Gen. Paphiopedilum Nat. Hist. &* Cult., 1: 112 (1998), basionym not validly publ.: P. haynaldianum var. album.
var. album *J.H. Asher in Orchid Dig.*, 44(6): 214 (1980), without latin descr. or type—Locality not stated.
f. album *Asher ex O. Gruss & J. Roeth in Orchidee*, 51(6): 749 (2000)—Philippines.
helenae *L.V. Averyanov in Orchids*, 65(10): 1064 (1996); *et in Bot. Zhurn.*, 81(9): 108 (1996)—Vietnam.
f. aureum *O. Gruss & J. Roeth in Orchidee*, 50(1): 3 (1999)—Vietnam.
hennisianum (Schoser ex M.W. Wood) Fowlie
var. christiansenii *O. Gruss & J. Röth in Orchidee*, 47(5): 236 (1996)—Philippines.
f. christiansenii *(O. Gruss & Roeth) O. Gruss & J. Roeth in Caesiana*, 12: 59 (1999): P. hennisianum var. christiansenii.
henryanum
f. christae *(Braem) O. Gruss & J. Roeth in Caesiana*, 12: 62 (1999): P. henryanum var. christae.

PAPHIOPEDILUM:-
henryanum
f. christae *(Braem) G.J. Braem in G.J. Braem, C.O. Baker & M.L. Baker, Gen. Paphiopedilum Nat. Hist.* & Cult., 2: 206 (1999): P. henryanum var. christae.
herrmannii *F. Fuchs & H. Reisinger in Linz. Biol. Beitr.*, 27(2): 1213 (1995)—Vietnam.
hiepii *L.V. Averyanov in Orchids*, 67(3): 261 (1998)—Vietnam.
hilmarii *K. Senghas & R. Schettler in J. Orchideenfreund*, 6(1): 4 (1999), as 'hilmari'—Vietnam.
hirsutissimum (Lindl. ex Hook.) Stein
f. viride *O. Gruss & J. Roeth in Caesiana*, 12: 57 (1999)—Cultivation.
hookerae
f. sandowiae *(Braem) O. Gruss in Caesiana*, 12: 62 (1999): P. volonteanum f. sandowiae.
insigne
f. sanderae *(Rchb. f.) O. Gruss & J. Roeth in Caesiana*, 12: 62 (1999): Cypripedium insigne var. sanderae.
f. sanderae *(Rchb. f.) G.J. Braem in G.J. Braem, C.O. Baker & M.L. Baker, Gen. Paphiopedilum Nat. Hist.* & Cult., 2: 233 (1999): Cypripedium insigne var. sanderae.
f. sanderianum *(hort.) O. Gruss & J. Roeth in Caesiana*, 12: 60 (1999): Cypripedium insigne var. sanderianum.
f. sanderianum *(Rolfe) G.J. Braem in G.J. Braem, C.O. Baker & M.L. Baker, Gen. Paphiopedilum Nat. Hist.* & Cult., 2: 233 (1999), basionym as 'Paphiopedilum': Cypripedium insigne var. sanderianum.
intaniae *W. Cavestro in Rhône-Alpes Orchidées*, 25: 5 (15 April 2000), preprint—Sulawesi.
jackii H.S. Hua
var. hiepii *(Aver.) H. Koopowitz in Orchid Dig.*, 64(4): 168 (2000): P. hiepii.
javanicum (Reinw. ex Lindl.) Pfitzer
var. nymphenburgianum *J. Röth & O. Gruss in Orchidee*, 48(1): 10 (1997)—Cultivation.
f. nymphenburgianum *(J. Roeth & O. Gruss) P.J. Cribb, Genus Paphiopedilum ed. 2*: 371 (1998): P. javanicum var. nymphenburgianum.
lawrenceanum
f. hyeanum *(L. Linden & Rodigas) O. Gruss & J. Roeth in Caesiana*, 12: 62 (1999): Cypripedium lawrenceanum var. hyeanum.
lowii (Lindl.) Stein
f. aureum *(P.J. Cribb) P.J. Cribb, Slipper Orchids of Borneo*: 73 (1997); *et in J.J. Wood, Orchids of Borneo*, 3: 199 (1997): P. lowii var. aureum.
var. lynniae *(Garay) O. Gruss & J. Roeth in Orchidee*, 48(2): 72 (1997): P. lynniae.
var. richardianum *(Asher & Beaman) O. Gruss in Orchidee*, 45(5): U-Seite 4 (1994): P. richardianum.
lynniae *L.A. Garay in Lindleyana*, 11(4): 233 (1996)—Borneo.
malipoense Chen & Tsi
f. concolor *G.J. Braem in G.J. Braem, C.O. Baker & M.L. Baker, Gen. Paphiopedilum Nat. Hist.* & Cult., 1: 82 (1998)—Vietnam.
var. hiepii *(Aver.) P.J. Cribb, Genus Paphiopedilum ed. 2*: 88 (1998): P. hiepii.
var. jackii *(H.S. Hua) L.V. Averyanov in Orchids*, 66(2): 153 (1997): P. jackii.
f. tonnianum *J. Roeth in Orchidee*, 51(5): 561 (2000)—China/Vietnam border.
mastersianum (Rchb. f.) Stein
var. mohrianum *(Braem) H. Koopowitz in Orchid Dig.*, 64(4): 169 (2000): P. mohrianum.
× mattesii *Pittenauer ex J. Röth & O. Gruss in Orchidee*, 47(3): 131 (1996)—Malaya (Peninsular Malaysia). (P. barbatum × bullenianum).
micranthum T. Tang & F.T. Wang
f. alboflavum *(Braem) G.J. Braem in G.J. Braem, C.O. Baker & M.L. Baker, Gen. Paphiopedilum Nat. Hist.* & Cult., 1: 91 (1998): P. micranthum var. alboflavum.
f. glanzeanum *(O. Gruss & Roeth) O. Gruss & J. Roeth in Caesiana*, 12: 59 (1999): P. micranthum var. glanzeanum.
mirabile *W. Cavestro & G. Chiron in Orchidées Cult. Protect.*, 38: 32 (preprint, 19 Feb. 1999)—Vietnam.

PAPHIOPEDILUM:-
niveum
f. album *(hort.) O. Gruss in Caesiana*, 12: 63 (1999): Cypripedium niveum var. album.
ooii *H. Koopowitz in Orchid Dig.*, 63(2): 106 (1999)—Borneo.
parnatanum *W. Cavestro in Orchidées Cult. Protect.*, 38: 30 (preprint, 19 Feb. 1999)—Philippines.
philippinense (Rchb. f.) Stein
f. alboflavum *O. Gruss in Orchidee*, 51(3): 354 (2000)—Philippines.
potentianum *O. Gruss & J. Röth in Caesiana*, 5: 39 (1995)—Thailand.
× powellii *E.A. Christenson in Orchidee*, 47(4): 220 (1996)—Thailand. (P. callosum × exul).
praestans (Rchb. f.) Pfitzer
var. wilhelminiae *(L.O. Williams) G.J. Braem in G.J. Braem, C.O. Baker & M.L. Baker, Gen. Paphiopedilum Nat. Hist.* & Cult., 1: 144 (1998): P. wilhelminiae.
robinsonii
f. viride *G.J. Braem in Orchidées Cult. Protect.*, 36: 36, 38 (1998)—Malaya (Peninsular Malaysia).
saccopetalum *H.S. Hua in Orchidee*, 49(1): 38 (1998)—China Southeast (Guangxi).
singchii *Z.J. Liu & J.Y. Zhang in Acta Phytotax. Sin.*, 38(5): 468 (2000)—China South-central (Yunnan).
spicerianum (Mast. & T. Moore) Pfitzer
f. immaculatum *G.J. Braem in G.J. Braem, C.O. Baker & M.L. Baker, Gen. Paphiopedilum Nat. Hist.* & Cult., 2: 270 (1999)—India.
striatum *M.A. Clements & D.L. Jones in Lasianthera*, 1(1): 3 (1996)—New Guinea (Papua New Guinea).
sukhakulii
f. aureum *Van Delden ex O. Gruss in Orchidee*, Suppl. 6: 16 (1999)—Thailand.
tonsum
f. alboviride *G.J. Braem in Orchidées Cult. Protect.*, 36: 36, 38 (1998)—Sumatera.
tranlienianum *O. Gruss & H. Perner in Caesiana*, 11: 66 (1998)—Vietnam.
urbanianum
f. alboviride *G.J. Braem in Orchidées Cult. Protect.*, 36: 36, 38 (1998)—Philippines.
usitanum *O. Gruss & J. Roeth in Orchidee*, 50(1): 4 (1999)—Philippines.
venustum
f. measuresianum *(hort.) G.J. Braem in Orchidées Cult. Protect.*, 36: 35, 37 (1998); *et in G.J. Braem, C.O. Baker & M.L. Baker, Gen. Paphiopedilum Nat. Hist.* & Cult., 2: 274 (1999): Cypripedium venustum var. measuresianum.
f. pardinum *(Rchb. f.) G.J. Braem in G.J. Braem, C.O. Baker & M.L. Baker, Gen. Paphiopedilum Nat. Hist.* & Cult., 2: 274 (1999): Cypripedium pardinum.
vietnamense *O. Gruss & H. Perner in Orchidee*, Suppl. 5: 3 (1999)—Vietnam.
villosum (Lindl.) Stein
f. annamense *(Rolfe) G.J. Braem in G.J. Braem, C.O. Baker & M.L. Baker, Gen. Paphiopedilum Nat. Hist.* & Cult., 2: 245 (1999): P. villosum var. annamense.
f. aureum *G.J. Braem in Orchidées Cult. Protect.*, 36: 36, 37 (1998)—Burma.
viniferum *H. Koopowitz & N. Hasegawa in Orchid Dig.*, 64(4): 150 (2000)—Burma.
violascens Schltr.
var. saskianum *(O. Gruss & Roeth) H. Koopowitz in Orchid Dig.*, 64(4): 176 (2000), without exact basionym page: P. bougainvilleanum var. saskianum.
volonteanum
f. sandowii *G.J. Braem in Orchidées Cult. Protect.*, 36: 36, 38 (1998)—Borneo (Sabah).
wardii Summerh.
var. alboviride *O. Gruss & J. Roeth in Orchidee*, 49(3): 143 (1998)—Burma.
f. alboviride *(O. Gruss & Roeth) G.J. Braem in Orchidées Cult. Protect.*, 36: 35, 37 (1998): P. wardii var. alboviride.
wenshanense *Z.J. Liu & J. Yong Zhang in Acta Bot. Yunnanica*, 22(4): 392, 391 (2000)—China South-Central (Yunnan).

PAPILIONANTHE (Orchidac.).
cylindrica *(Lindl.) G. Seidenfaden, Descr. Epidendrorum J.G. König 1791*: 33 (1995): Aerides cylindrica.

PARABAMBUSA *E.A. Widjaja in Reinwardtia*, 11(2): 121 (1997). *GRAMINEAE*
kaini *E.A. Widjaja in Reinwardtia*, 11(2): 121 (1997)—New Guinea (Irian Jaya).

PARABOEA (Gesneriac.).
clavisepala *D. Fang & D.H. Qin in Acta Phytotax. Sin.*, 33(6): 602 (1995)—China Southeast (Guangxi).
glutinosa *(Hand.-Mazz.) K.Y. Pan in Novon*, 7(4): 431 (1997 publ. 1998): Boea glutinosa.
halongensis *R. Kiew & T.H. Nguyen in Gard. Bull. Singapore*, 52(1): 195 (2000)—Vietnam.
leopoldii *K.M. Wong, J.T. Pereira, J.B. Sugau & S.P. Lim in Sandakania*, 13: 28 (1999)—Borneo (Sabah).
nutans *D. Fang & D.H. Qin in Acta Phytotax. Sin.*, 33(6): 604 (1995)—China Southeast (Guangxi).
peltifolia *D. Fang & L. Zeng in Acta Phytotax. Sin.*, 33(6): 606 (1995)—China Southeast (Guangxi).
tribracteata *D. Fang & W.Y. Rao in Acta Phytotax. Sin.*, 33(6): 607 (1995)—China South-central (Guizhou).

PARACARYUM (Boraginac.).
khorassanicum *M. Khatamsaz in Iranian J. Bot.*, 8(1): 4 (1999)—Iran.

PARACROTON (Euphorbiac.).
sterrhopodus *(Airy Shaw) A. Radcliffe-Smith & R. Govaerts in Kew Bull.*, 52(1): 189 (1997): Fahrenheitia sterrhopoda.

PARADRYMONIA (Gesneriac.).
longipetiolata *(Donn. Sm.) H. Wiehler in Gesneriana*, 1(1): 122 (1995): Episcia longipetiolata.
sastrei *H. Wiehler in Gesneriana*, 1(1): 71 (1995)—Colombia.
splendens *M. Freiberg in Phyton (Austria)*, 37(1): 136 (1997)—Ecuador.
ulei *H. Wiehler in Gesneriana*, 1(1): 71 (1995)—Peru.

PARAISOMETRUM *W.T. Wang in Novon*, 7(4): 431 (1997 publ. 1998). *GESNERIACEAE*
mileense *W.T. Wang in Novon*, 7(4): 434 (1997 publ. 1998)—China South-central (Yunnan).

PARAJUBAEA (Palmae).
sunkha *M. Moraes R. in Novon*, 6(1): 85 (1996)—Bolivia.
torallyi (Mart.) Burret
var. microcarpa *M. Moraes R. in Novon*, 6(1): 87 (1996)—Bolivia.

PARAKEELYA *M.A. Hershkovitz in Phytologia*, 84(2): 101 (1998 publ. 1999). *PORTULACACEAE*
arenicola *(Syeda) M.A. Hershkovitz in Phytologia*, 84(2): 101 (1998 publ. 1999): Calandrinia arenicola.
balonensis *(Lindl.) M.A. Hershkovitz in Phytologia*, 84(2): 101 (1998 publ. 1999): Calandrinia balonensis.
brevipedata *(F. Muell.) M.A. Hershkovitz in Phytologia*, 84(2): 101 (1998 publ. 1999): Calandrinia brevipedata.
calyptrata *(Hook. f.) M.A. Hershkovitz in Phytologia*, 84(2): 101 (1998 publ. 1999): Calandrinia calyptrata.
composita *(Nees) M.A. Hershkovitz in Phytologia*, 84(2): 101 (1998 publ. 1999): Calandrinia polypetala var. composita.
corrigioloides *(F. Muell. ex Benth.) M.A. Hershkovitz in Phytologia*, 84(2): 101 (1998 publ. 1999): Calandrinia corrigioloides.
creethiae *(Tratman ex Morrison) M.A. Hershkovitz in Phytologia*, 84(2): 101 (1998 publ. 1999), as 'creethae': Calandrinia creethiae.
disperma *(J.M. Black) M.A. Hershkovitz in Phytologia*, 84(2): 101 (1998 publ. 1999): Calandrinia disperma.
eremaea *(Ewart) M.A. Hershkovitz in Phytologia*, 84(2): 101 (1998 publ. 1999): Calandrinia eremaea.
gracilis *(Benth.) M.A. Hershkovitz in Phytologia*, 84(2): 102 (1999): Calandrinia gracilis.
granulifera *(Benth.) M.A. Hershkovitz in Phytologia*, 84(2): 102 (1999): Calandrinia granulifera.

PARAKEELYA:-
lehmannii *(Endl.) M.A. Hershkovitz in Phytologia,* 84(2): 102 (1998 publ. 1999): Calandrinia lehmannii.
liniflora *(Fenzl) M.A. Hershkovitz in Phytologia,* 84(2): 102 (1998 publ. 1999): Calandrinia liniflora.
nana *(Nees) M.A. Hershkovitz in Phytologia,* 84(2): 102 (1998 publ. 1999): Talinum nanum.
papillata *(Syeda) M.A. Hershkovitz in Phytologia,* 84(2): 102 (1998 publ. 1999): Calandrinia papillata.
pickeringii *(A. Gray) M.A. Hershkovitz in Phytologia,* 84(2): 102 (1998 publ. 1999): Calandrinia pickeringii.
pleiopetala *(F. Muell.) M.A. Hershkovitz in Phytologia,* 84(2): 102 (1998 publ. 1999): Calandrinia pleiopetala.
polyandra *(Benth.) M.A. Hershkovitz in Phytologia,* 84(2): 102 (1998 publ. 1999): polyandra.
polypetala *(Fenzl) M.A. Hershkovitz in Phytologia,* 84(2): 102 (1998 publ. 1999): Calandrinia polypetala.
porifera *(Syeda) M.A. Hershkovitz in Phytologia,* 84(2): 103 (1998 publ. 1999): Calandrinia porifera.
primuliflora *(Diels) M.A. Hershkovitz in Phytologia,* 84(2): 103 (1998 publ. 1999): Calandrinia primuliflora.
pumila *(Benth.) M.A. Hershkovitz in Phytologia,* 84(2): 103 (1998 publ 1999): Calandrinia calyptrata var. pumila.
pytchosperma *(F. Muell.) M.A. Hershkovitz in Phytologia,* 84(2): 103 (1998 publ. 1999): Calandrinia ptychosperma.
quadrivalvis *(F. Muell.) M.A. Hershkovitz in Phytologia,* 84(2): 103 (1998 publ. 1999): Calandrinia quadrivalvis.
remota *(J.M. Black) M.A. Hershkovitz in Phytologia,* 84(2): 103 (1998 publ. 1999): Calandrinia remota.
reticulata *(Syeda) M.A. Hershkovitz in Phytologia,* 84(2): 103 (1998 publ. 1999): Calandrinia reticulata.
schistorhiza *(Morrison) M.A. Hershkovitz in Phytologia,* 84(2): 103 (1998 publ. 1999): Calandrinia schistorhiza.
spergularina *(F. Muell.) M.A. Hershkovitz in Phytologia,* 84(2): 103 (1998 publ. 1999): Calandrinia spergularina.
sphaerophylla *(J.M. Black) M.A. Hershkovitz in Phytologia,* 84(2): 103 (1998 publ. 1999): Calandrinia sphaerophylla.
stagnensis *(J.M. Black) M.A. Hershkovitz in Phytologia,* 84(2): 103 (1998 publ. 1999): Calandrinia stagnensis.
stenogyna *(Domin) M.A. Hershkovitz in Phytologia,* 84(2): 103 (1998 publ. 1999): Calandrinia stenogyna.
strophiolata *(F. Muell.) M.A. Hershkovitz in Phytologia,* 84(2): 103 (1998 publ. 1999): Claytonia strophiolata.
tumida *(Syeda) M.A. Hershkovitz in Phytologia,* 84(2): 103 (1998 publ. 1999): Calandrinia tumida.
uniflora *(F. Muell.) M.A. Hershkovitz in Phytologia,* 84(2): 104 (1998 publ. 1999): Calandrinia uniflora.
volubilis *(Benth.) M.A. Hershkovitz in Phytologia,* 84(2): 104 (1998 publ. 1999): Calandrinia volubilis.

PARAKMERIA (Magnoliac.).
kachirachirai *(Kaneh. & Yamam.) Y.W. Law in Acta Phytotax. Sin.,* 34(1): 91 (Feb. 1996), with basionym date; *et in Fl. Reipubl. Popul. Sin.,* 30(1): 271 (May 1996), without basionym date: Michelia kachirachirai.
lotungensis *(Chun & C.H. Tsoong) Y.W. Law in Acta Phytotax. Sin.,* 34(1): 91 (Feb. 1996); *et in Fl. Reipubl. Popul. Sin.,* 30(1): 271 (May 1996): Magnolia lotungensis.
nitida *(W.W. Sm.) Y.W. Law in Acta Phytotax. Sin.,* 34(1): 91 (Feb. 1996); *et in Fl. Reipubl. Popul. Sin.,* 30(1): 271 (May 1996): Magnolia nitida.

PARALYCHNOPHORA (Compos.).
atkinsiae *D.J.N. Hind in Kew Bull.,* 55(2): 375 (2000)—Brazil Northeast (Bahia).

PARALYCHNOPHORA:-
harleyi *(H. Rob.) D.J.N. Hind in Kew Bull.,* 55(2): 370 (2000): Eremanthus harleyi.
patriciana *D.J.N. Hind in Kew Bull.,* 55(2): 371 (2000)—Brazil Northeast (Bahia).
santosii *(H. Rob.) D.J.N. Hind in Kew Bull.,* 55(2): 370 (2000): Eremanthus santosii.

PARAMIFLOS *J. Cuatrecasas in Proc. Biol. Soc. Wash.,* 108(4): 748 (1995). *COMPOSITAE*
glandulosus *(Cuatrec.) J. Cuatrecasas in Proc. Biol. Soc. Wash.,* 108(4): 749 (1995): Espeletia glandulosa.

PARAPRENANTHES (Compos.).
glandulosissima *(C.C. Chang) C. Shih in Fl. Reipubl. Popul. Sin.,* 80(1): 182 (1997): Lactuca glandulosissima.
meridionalis *(C. Shih) A.N. Sennikov in Bot. Zhurn.,* 82(5): 111 (1997): Mulgedium meridionale.
polypodifolia *(Franch.) C.C. Chang ex C. Shih in Fl. Reipubl. Popul. Sin.,* 80(1): 181 (1997): Lactuca polypodifolia.
polypodiifolia *(Franch.) A.N. Sennikov in Bot. Zhurn.,* 82(5): 111 (1997): Lactuca polypodiifolia.
umbrosa *(Dunn) A.N. Sennikov in Bot. Zhurn.,* 82(5): 111 (1997): Lactuca umbrosa.

PARARISTOLOCHIA (Aristolochiac.).
alexandriana *Michael J. Parsons in Bot. J. Linn. Soc.,* 120(3): 218 (1996)—New Guinea (Papua New Guinea).
australopithecurus *Michael J. Parsons in Bot. J. Linn. Soc.,* 120(3): 227 (1996)—Queensland.
biakensis *Michael J. Parsons in Bot. J. Linn. Soc.,* 120(3): 232 (1996)—New Guinea (Irian Jaya).
decandra *(Ding Hou) M.J. Parsons in Bot. J. Linn. Soc.,* 120(3): 206 (1996): Aristolochia decandra.
deltantha *(F. Muell.) M.J. Parsons in Bot. J. Linn. Soc.,* 120(3): 206 (1996): Aristolochia deltantha.
dictyophlebia *(Merr. & L.M. Perry) M.J. Parsons in Bot. J. Linn. Soc.,* 120(3): 206 (1996): Aristolochia dictyophlebia.
dielsiana *(O.C. Schmidt) M.J. Parsons in Bot. J. Linn. Soc.,* 120(3): 206 (1996): Aristolochia dielsiana.
engleriana *(O.C. Schmidt) M.J. Parsons in Bot. J. Linn. Soc.,* 120(3): 206 (1996): Aristolochia engleriana.
gracilifolia *(O.C. Schmidt) M.J. Parsons in Bot. J. Linn. Soc.,* 120(3): 206 (1996): Aristolochia gracilifolia.
kepara *Michael J. Parsons in Bot. J. Linn. Soc.,* 120(3): 219 (1996)—New Guinea (Papua New Guinea).
laheyana *(F.M. Bailey) Michael J. Parsons in Bot. J. Linn. Soc.,* 120(3): 206 (1996): Aristolochia deltantha var. laheyana.
linearifolia *Michael J. Parsons in Bot. J. Linn. Soc.,* 120(3): 235 (1996)—Queensland.
manokwariensis *Michael J. Parsons in Bot. J. Linn. Soc.,* 120(3): 220 (1996)—New Guinea (Irian Jaya).
meridionaliana *Michael J. Parsons in Bot. J. Linn. Soc.,* 120(3): 214 (1996)—New Guinea (Papua New Guinea).
subsp. milnensis *Michael J. Parsons in Bot. J. Linn. Soc.,* 120(3): 218 (1996)—New Guinea (Papua New Guinea).
subsp. popondettensis *Michael J. Parsons in Bot. J. Linn. Soc.,* 120(3): 216 (1996)—New Guinea (Papua New Guinea).
momandul *(K. Schum.) Michael J. Parsons in Bot. J. Linn. Soc.,* 120(3): 207 (1996): Aristolochia momandul.
paradisiana *Michael J. Parsons in Bot. J. Linn. Soc.,* 120(3): 223 (1996)—New Guinea (Papua New Guinea).
peninsulensis *Michael J. Parsons in Bot. J. Linn. Soc.,* 120(3): 234 (1996)—Queensland.
pithecurus *(Ridl.) Michael J. Parsons in Bot. J. Linn. Soc.,* 120(3): 207 (1996): Aristolochia pithecurus.
praevenosa *(F. Muell.) Michael J. Parsons in Bot. J. Linn. Soc.,* 120(3): 207 (1996): Aristolochia praevenosa.
schlechteri *(Lauterb.) Michael J. Parsons in Bot. J. Linn. Soc.,* 120(3): 207 (1996): Aristolochia schlechteri.

PARARISTOLOCHIA:-
sepikensis *Michael J. Parsons in Bot. J. Linn. Soc.,* 120(3): 231 (1996)—New Guinea (Papua New Guinea).
sparusifolia *Michael J. Parsons in Bot. J. Linn. Soc.,* 120(3): 228 (1996)—Queensland.
tithonusiana *Michael J. Parsons in Bot. J. Linn. Soc.,* 120(3): 222 (1996)—New Guinea (Irian Jaya).

PARARTOCARPUS (Morac.).
spinulosus *R. Go in Sandakania,* 12: 2 (1998)—Borneo (Sabah).

PARASENECIO (Compos.).
ser. Ainsliiflorae *Y.L. Chen in Fl. Reipubl. Popul. Sin.,* 77(1): 329 (1999).
ser. Candidae *(Kitam.) Y.L. Chen in Fl. Reipubl. Popul. Sin.,* 77(1): 30 (1999): Cacalia ser. Candidae.
sect. Koyamacalia *(H. Rob. & Brettell) Y.L. Chen in Fl. Reipubl. Popul. Sin.,* 77(1): 28 (1999): gen. Koyamacalia.
sect. Monanthae *(Kitam.) Y.L. Chen in Fl. Reipubl. Popul. Sin.,* 77(1): 52 (1999): Cacalia ser. Monanthae.
ser. Rubescentes *Y.L. Chen in Fl. Reipubl. Popul. Sin.,* 77(1): 329 (1999).
ser. Trilobae *(H. Koyama) Y.L. Chen in Fl. Reipubl. Popul. Sin.,* 77(1): 29 (1999): Cacalia ser. Trilobae.
× abukumensis *H. Koyama in Mem. Nation. Sci. Mus. (Tokyo),* 28: 85 (1995)—Japan. (P. farfarifolia × nikomontana).
ainsliiflorus *(Franch.) Y.L. Chen in Fl. Reipubl. Popul. Sin.,* 77(1): 47 (1999): Senecio ainsliiflorus.
ambiguus *(Ling) Y.L. Chen in Fl. Reipubl. Popul. Sin.,* 77(1): 45 (1999): Cacalia ambigua.
var. wangianus *(Ling) Y.L. Chen in Fl. Reipubl. Popul. Sin.,* 77(1): 47 (1999): Cacalia wangiana.
begoniifolius *(Franch.) Y.L. Chen in Fl. Reipubl. Popul. Sin.,* 77(1): 28 (1999), as 'begoniaefolius': Senecio begoniifolius.
bulbiferoides *(Hand.-Mazz.) Y.L. Chen in Fl. Reipubl. Popul. Sin.,* 77(1): 73 (1999): Cacalia bulbiferoides.
chenopodifolius *(DC.) A.J.C. Grierson in Edinburgh J. Bot.,* 57(3): 405 (2000): Senecio chenopodifolius.
chola *(W.W. Sm.) Y.L. Chen in Fl. Reipubl. Popul. Sin.,* 77(1): 29 (1999): Senecio chola.
cyclotus *(Bureau & Franch.) Y.L. Chen in Fl. Reipubl. Popul. Sin.,* 77(1): 64 (1999): Senecio cyclotus.
dasythyrsus *(Hand.-Mazz.) Y.L. Chen in Fl. Reipubl. Popul. Sin.,* 77(1): 78 (1999): Cacalia dasythyrsa.
delphiniphyllus *(H. Lév.) Y.L. Chen in Fl. Reipubl. Popul. Sin.,* 77(1): 79 (1999): Senecio delphiniphyllus.
deltophyllus *(Maxim.) Y.L. Chen in Fl. Reipubl. Popul. Sin.,* 77(1): 30 (1999): Senecio deltophyllus.
firmus *(Kom.) Y.L. Chen in Fl. Reipubl. Popul. Sin.,* 77(1): 26 (1999): Cacalia firma.
gansuensis *Y.L. Chen in Acta Phytotax. Sin.,* 34(6): 643 (1996)—China North-central (Gansu, Shaanxi).
hastatus (L.) H. Koyama
var. glaber *(Ledeb.) Y.L. Chen in Fl. Reipubl. Popul. Sin.,* 77(1): 33 (1999): Cacalia hastata var. glabra.
hastiformis *Y.L. Chen in Acta Phytotax. Sin.,* 34(6): 641 (1996)—China South-central (Yunnan).
hwangshanicus *(Ling) C.I Peng & S.W. Chung in Fl. Taiwan, ed. 2,* 4: 1024 (1998), 'hwangshanica': Cacalia hwangshanica.
hwangshanicus *(Ling) Y.L. Chen in Fl. Reipubl. Popul. Sin.,* 77(1): 75 (1999), isonym; *see C.I Peng & S.W. Chung in Fl. Taiwan, ed. 2,* 4: 1024 (1998): Cacalia hwangshanica.
ianthophyllus *(Franch.) Y.L. Chen in Fl. Reipubl. Popul. Sin.,* 77(1): 71 (1999): Senecio ianthophyllus.
jiulongensis *Y.L. Chen in Acta Phytotax. Sin.,* 34(6): 645 (1996)—China South-central (Sichuan).
kangxianensis *(Z. Ying Zhang & Y.H. Guo) Y.L. Chen in Fl. Reipubl. Popul. Sin.,* 77(1): 49 (1999): Cacalia kangxianensis.
komarovianus *(Pojark.) Y.L. Chen in Fl. Reipubl. Popul. Sin.,* 77(1): 34 (1999): Hasteola komaroviana.

PARASENECIO:-
koualapensis *(Franch.) Y.L. Chen in Fl. Reipubl. Popul. Sin.*, 77(1): 59 (1999): Senecio koualapensis.
lancifolius *(Franch.) Y.L. Chen in Fl. Reipubl. Popul. Sin.*, 77(1): 33 (1999): Senecio sagittatus var. lancifolius.
latipes *(Franch.) Y.L. Chen in Fl. Reipubl. Popul. Sin.*, 77(1): 52 (1999): Senecio latipes.
leucocephalus *(Franch.) Y.L. Chen in Fl. Reipubl. Popul. Sin.*, 77(1): 71 (1999): Senecio leucocephalus.
lidjiangensis *(Hand.-Mazz.) Y.L. Chen in Fl. Reipubl. Popul. Sin.*, 77(1): 53 (1999): Cacalia lidjiangensis.
longispicus *(Hand.-Mazz.) Y.L. Chen in Fl. Reipubl. Popul. Sin.*, 77(1): 72 (1999): Cacalia longispica.
maowenensis *Y.L. Chen in Acta Phytotax. Sin.*, 34(6): 645 (1996)—China South-central (Sichuan).
matsudae *(Kitam.) Y.L. Chen in Fl. Reipubl. Popul. Sin.*, 77(1): 50 (1999), as 'matsudai': Cacalia matsudae.
monanthus *(Diels) C.I Peng & S.W. Chung in Fl. Taiwan, ed. 2*, 4: 1024 (1998), as 'monantha': Senecio monanthus.
morrisonensis *Y.L. Chen in Fl. Reipubl. Popul. Sin.*, 77(1): 67 (1999), as nom. nov., without latin descr. or type—Locality not stated.
nokoensis *(Masam. & Suzuki) Y.L. Chen in Fl. Reipubl. Popul. Sin.*, 77(1): 42 (1999), isonym; *see C.I Peng & S.W. Chung in Fl. Taiwan, ed. 2*, 4: 1026 (1998): Cacalia nokoensis.
nokoensis *(Masam. & Suzuki) C.I Peng & S.W. Chung in Fl. Taiwan, ed. 2*, 4: 1026 (1998): Cacalia nokoensis.
otopteryx *(Hand.-Mazz.) Y.L. Chen in Fl. Reipubl. Popul. Sin.*, 77(1): 40 (1999): Cacalia otopteryx.
palmatisectus *(Jeffrey) Y.L. Chen in Fl. Reipubl. Popul. Sin.*, 77(1): 82 (1999): Senecio palmatisectus.
var. moupinensis *(Franch.) Y.L. Chen in Fl. Reipubl. Popul. Sin.*, 77(1): 82 (1999): Senecio quinquelobus var. moupinensis.
petasitoides *(H. Lév.) Y.L. Chen in Fl. Reipubl. Popul. Sin.*, 77(1): 75 (1999): Senecio petasitoides.
phyllolepis *(Franch.) Y.L. Chen in Fl. Reipubl. Popul. Sin.*, 77(1): 69 (1999): Senecio phyllolepis.
praetermissus *(Pojark.) Y.L. Chen in Fl. Reipubl. Popul. Sin.*, 77(1): 38 (1999): Hasteola praetermissa.
profundorum *(Dunn) Y.L. Chen in Fl. Reipubl. Popul. Sin.*, 77(1): 68 (1999): Senecio profundorum.
quinquelobus *(Wall. ex DC.) Y.L. Chen in Fl. Reipubl. Popul. Sin.*, 77(1): 59 (1999): Prenanthes quinqueloba.
var. sinuatus *(H. Koyama) Y.L. Chen in Fl. Reipubl. Popul. Sin.*, 77(1): 63 (1999): Cacalia pentaloba var. sinuata.
roborowskii *(Maxim.) Y.L. Chen in Fl. Reipubl. Popul. Sin.*, 77(1): 65 (1999)—Senecio roborowskii.
rockianus *(Hand.-Mazz.) Y.L. Chen in Fl. Reipubl. Popul. Sin.*, 77(1): 63 (1999): Cacalia rockiana.
rubescens *(S. Moore) Y.L. Chen in Fl. Reipubl. Popul. Sin.*, 77(1): 49 (1999): Senecio rubescens.
rufipilis *(Franch.) Y.L. Chen in Fl. Reipubl. Popul. Sin.*, 77(1): 76 (1999): Senecio rufipilis.
sinicus *(Ling) Y.L. Chen in Fl. Reipubl. Popul. Sin.*, 77(1): 81 (1999): Cacalia sinica.
souliei *(Franch.) Y.L. Chen in Fl. Reipubl. Popul. Sin.*, 77(1): 63 (1999): Senecio souliei.
subglaber *(C.C. Chang) Y.L. Chen in Fl. Reipubl. Popul. Sin.*, 77(1): 43 (1999): Cacalia subglabra.
taliensis *(Franch.) Y.L. Chen in Fl. Reipubl. Popul. Sin.*, 77(1): 55 (1999): Senecio taliensis.
tenianus *(Hand.-Mazz.) Y.L. Chen in Fl. Reipubl. Popul. Sin.*, 77(1): 57 (1999): Cacalia teniana.
tripteris *(Hand.-Mazz.) Y.L. Chen in Fl. Reipubl. Popul. Sin.*, 77(1): 85 (1999): Cacalia tripteris.
tsinlingensis *(Hand.-Mazz.) Y.L. Chen in Fl. Reipubl. Popul. Sin.*, 77(1): 77 (1999): Cacalia tsinlingensis.
vespertilo *(Franch.) Y.L. Chen in Fl. Reipubl. Popul. Sin.*, 77(1): 44 (1999): Senecio vespertilo.

PARASENECIO:-
xinjiashanensis *(Z. Ying Zhang & Y.H. Guo) Y.L. Chen in Fl. Reipubl. Popul. Sin.*, 77(1): 66 (1999): Cacalia xinjiashanensis.

PARATHESIS (Myrsinac.).
acostensis *J.F. Morales in Phytologia*, 81(5): 362 (1996 publ. 1997)—Costa Rica.
costaricensis *J.M. Ricketson in Novon*, 7(4): 398 (1997 publ. 1998)—Costa Rica.
glendae *J.M. Ricketson in Novon*, 7(4): 400 (1997 publ. 1998)—Costa Rica.
longipedicellata *J.M. Ricketson in Novon*, 7(4): 402 (1997 publ. 1998)—Costa Rica.
sinuata *(Lundell) J.M. Ricketson & J.J. Pipoly in Novon*, 7(4): 405 (1997 publ. 1998): P. reticulata var. sinuata.

PARIETARIA (Urticac.).
erronea *P.P. Panov in Fitologiya*, 48: 4 (1996)—Bulgaria.
judaica L.
subsp. speluncarum *J. Chrtek & J. Sádlo in Thaiszia*, 5(2): 106 (1995 publ. 1996)—Syria.
rhodopaea *P.P. Panov in Fitologiya*, 48: 3 (1996)—Bulgaria.

PARIS (Trilliac.).
birmanica *(Takht.) H. Li & H.J. Noltie in Edinburgh J. Bot.*, 54(3): 351 (1997): Daiswa birmanica.
delavayi Franch.
var. petiolata *(Baker ex C.H. Wright) H. Li in H. Li (ed.), Gen. Paris (Trilliac.)*: 30 (1998), isonym; *see S.F. Wang in Fl. Sichuanica*, 7: 264 (1991): P. petiolata.
fargesii Franch.
var. brevipetalata *(T.C. Huang & K.C. Yang) H. Li in H. Li (ed.), Gen. Paris (Trilliac.)*: 50 (1998), isonym; *see T.C. Huang & K.C. Yang in Taiwania*, 34(1): 52 (1989): Daiswa fargesii var. brevipetalata.
kwantungensis *R.H. Miau in Acta Sci. Nat. Univ. Sunyatseni*, 1982(3): 74 (1982)—China Southeast (Guangdong).
polyphylla Sm.
var. kwantungensis *(R.H. Miao) S.C. Chen & S.Yun Liang in Acta Phytotax. Sin.*, 33(5): 490 (1995): P. kwantungensis.
f. velutina *H. Li & H.J. Noltie in Edinburgh J. Bot.*, 54(3): 352 (1997)—China South-central (Yunnan, Sichuan).
quadrifolia L.
var. angustiovata *D.Z. Ma & H.L. Liu in Fl. Ningxiaensis*, 2: 523 (1988 publ. 1990)—Inner Mongolia (Ningxia).
verticillata M. Bieb.
var. nigra *Y.N. Lee, Fl. Korea*: 1163 (1996), without type—Locality not stated.

PARISHIA (Anacardiac.).
dinghouiana *K.M. Kochummen in Sandakania*, 7: 91 (1996)—Borneo (Sarawak).
maingayi Hook. f.
var. minor *(Ridl.) K.M. Kochummen in Sandakania*, 7: 94 (1996): P. minor.
trifoliolata *K.M. Kochummen in Sandakania*, 7: 91 (1996)—Borneo (Sarawak).

PARITI (Malvac.).
tortuosum *(Roxb.) J.O. Voigt, Hort. Suburb. Calcutt.*: 120 (1845): Hibiscus tortuosus.

PARKIA (Legumin.).
barnebyana *H.C. Hopkins in Kew Bull.*, 55(1): 133 (2000)—Venezuela.
lutea *H.C. Hopkins in Adansonia*, 22(1): 140 (2000)—Brazil North (Amazonas).
paya *H.C. Hopkins in Kew Bull.*, 55(1): 123 (2000)—Borneo (Sarawak), Sumatera.

PARKINSONIA (Legumin.).
× carterae *J.A. Hawkins in Pl. Syst. Evol.*, 216(1–2): 63 (1999)—Mexico Southwest (Guerrero, Michoacan, Oaxaca), Mexico Northwest (Baja California Sur, Sonora), Mexico Central (Puebla), Honduras, Ecuador. (P. aculeata × praecox).

PARKINSONIA:-
praecox *(Ruiz & Pav.) J.A. Hawkins in Pl. Syst. Evol.*, 216(1–2): 63 (1999), basionym as 'Cercidium': Caesalpinia praecox.

PARNASSIA (Parnassiac.).
cirrata Piper
var. intermedia *(Rydb.) P.K. Holmgren & N.H. Holmgren in A. Cronquist et al., Intermountain Fl., Vasc. Pl. Intermountain West, U.S.A.*, 3A: 61 (1997): P. intermedia.
guilinensis *G.Z. Li & S.C. Tang in Guihaia*, 19(4): 341 (1999)—China Southeast (Guangxi).
neogaea *(Fernald) S.S. Kharkevich in Sosud. Rast. Sovet. Dal'nego Vostoka*, 7: 236 (1995): P. palustris var. neogaea.
palustris L.
subsp. multiseta *(Ledeb.) V.N. Voroschilov in A.K. Skvortsov (ed.), Florist. issl. v razn. raĭonakh SSSR*: 175 (1985): P. palustris var. multiseta.
qinghaiensis *J.T. Pan in Novon*, 6(2): 188 (1996)—Qinhai.
simaoensis *Y.Y. Qian in Bull. Bot. Res. North-East. Forest. Univ.*, 17(3): 305 (1997)—China South-central (Yunnan).

PARODIA (Cactac.).
alacriportana
subsp. brevihamata *(W. Haage) A. Hofacker & P.J. Braun in Cactaceae Consensus Initiatives*, 6: 10 (1998): P. brevihamata.
subsp. buenekeri *(Buining) A. Hofacker & P.J. Braun in Cactaceae Consensus Initiatives*, 6: 10 (1998): P. buenekeri.
subsp. catarinensis *(F. Ritter) A. Hofacker & P.J. Braun in Cactaceae Consensus Initiatives*, 6: 10 (1998): Brasiliparodia catarinensis.
arnostiana *(Lisal & Kolarik) A. Hofacker in Cactaceae Consensus Initiatives*, 6: 11 (1998): Notocactus arnostianus.
augustinii *W. Weskamp, Gattung Parodia*, 3: 196 (1997)—Bolivia.
aureicentra Backeb.
var. muhrii *(F.H. Brandt) J.G. Lambert, Cactus d'Argentine, ed. 2*: 207 (1998): P. muhrii.
var. rauschii *(Backeb.) J.G. Lambert, Cactus d'Argentine, ed. 2*: 209 (1998): P. rauschii.
var. variicolor *(F. Ritter) J.G. Lambert, Cactus d'Argentine, ed. 2*: 210 (1998): P. variicolor.
carambeiensis *(Buining & Brederoo) A. Hofacker in Cactaceae Consensus Initiatives*, 6: 11 (1998): Notocactus carambeiensis.
catamarcensis Backeb.
var. riojensis *(F. Ritter & Weskamp) J.G. Lambert, Cactus d'Argentine, ed. 2*: 226 (1998): P. riojensis.
concinna
subsp. agnetae *(Vliet) A. Hofacker in Cactaceae Consensus Initiatives*, 6: 11 (1998): Notocactus agnetae.
subsp. blaauwiana *(Vliet) A. Hofacker in Cactaceae Consensus Initiatives*, 6: 11 (1998): Notocactus blaauwianus.
copavilquensis *W. Weskamp, Gattung Parodia*, 3: 194, 193 (1997)—Bolivia.
curvispina *(F. Ritter) D.R. Hunt in Cactaceae Consensus Initiatives*, 4: 6 (1997): Notocactus curvispinus.
erubescens *(Osten) D.R. Hunt in Cactaceae Consensus Initiatives*, 4: 6 (1997): Echinocactus erubescens.
fusca *(F. Ritter) A. Hofacker & P.J. Braun in Cactaceae Consensus Initiatives*, 6: 10 (1998): Notocactus fuscus.
haselbergii
subsp. graessneri *(K. Schum.) A. Hofacker & P.J. Braun in Cactaceae Consensus Initiatives*, 6: 10 (1998): Echinocactus graessneri.
ladae *J.J. Halda & L. Horáček in Acta Mus. Richnov. Sect. Nat.*, 7(2): 73 (2000)—Bolivia.
langsdorfii *(Lehm.) D.R. Hunt in Cactaceae Consensus Initiatives*, 4: 6 (1997): Cactus langsdorfii.
subsp. multiceps *A. Hofacker & K. Herm in Kakt. And. Sukk.*, 50(3): 62 (1999)—Uruguay.
subsp. pulvinata *(Vliet) A. Hofacker & K. Herm in Kakt. And. Sukk.*, 50(3): 62 (1999): Notocactus pulvinatus.

PARODIA:-
mammulosa
subsp. brasiliensis *(Havlicek) A. Hofacker in Cactaceae Consensus Initiatives*, 6: 11 (1998): Notocactus mammulosus var. brasiliensis.
subsp. erythracantha *(H. Schloss. & Brederoo) A. Hofacker in Cactaceae Consensus Initiatives*, 6: 11 (1998): Notocactus erythracanthus.
subsp. eugeniae *(Vliet) A. Hofacker in Cactaceae Consensus Initiatives*, 6: 11 (1998): Notocactus eugeniae.
subsp. submammulosa *(Lem.) A. Hofacker in Cactaceae Consensus Initiatives*, 6: 11 (1998): Echinocactus submammulosus.
meonacantha *(Prestlé) A. Hofacker in Cactaceae Consensus Initiatives*, 6: 11 (1998): Notocactus meonacanthus.
microsperma (F.A.C. Weber) Speg.
var. weberiana *(F.H. Brandt) J.G. Lambert, Cactus d'Argentine, ed. 2:* 240 (1998): P. weberiana.
mueller-melchersii
subsp. gutierrezii *(W.R. Abraham) A. Hofacker in Cactaceae Consensus Initiatives*, 6: 11 (1998): Notocactus gutierrezii.
subsp. winkleri *(Vliet) A. Hofacker in Cactaceae Consensus Initiatives*, 6: 11 (1998): Notocactus winkleri.
muricata *(Otto ex Pfeiff.) A. Hofacker in Cactaceae Consensus Initiatives*, 6: 12 (1998): Echinocactus muricatus.
neoarechavaletae *(Havlicek) D.R. Hunt in Cactaceae Consensus Initiatives*, 4: 6 (1997): Notocactus neoarechavaletae.
nothominuscula *A. Hofacker in Cactaceae Consensus Initiatives*, 6: 11 (1998), nom. nov.: Notocactus minusculus *Hofacker & K. Herm.*
subsp. gravior *A. Hofacker in Kakt. And. Sukk.*, 51(9): 234 (2000)—Brazil South (Rio Grande do Sul).
nothorauschii *D.R. Hunt in Cactaceae Consensus Initiatives*, 4: 6 (1997), nom. nov.: Notocactus rauschii *Vliet*.
ottonis
subsp. horstii *(F. Ritter) A. Hofacker in Cactaceae Consensus Initiatives*, 6: 12 (1998): Notocactus arechavaletae var. horstii.
oxycostata *(Buining & Brederoo) A. Hofacker in Cactaceae Consensus Initiatives*, 6: 12 (1998): Notocactus oxycostatus.
subsp. gracilis *(F. Ritter) A. Hofacker in Cactaceae Consensus Initiatives*, 6: 12 (1998): Notocactus glaucinus var. gracilis.
permutata *(F. Ritter) A. Hofacker in Cactaceae Consensus Initiatives*, 6: 12 (1998): Notocactus permutatus.
prolifera *(F. Ritter) W. Weskamp, Gattung Parodia*, 3: 159 (1997): P. camargensis var. prolifera.
rauschii *Backeb. ex D.R. Hunt in Cactaceae Consensus Initiatives*, 4: 4 (1997)—Argentina Northwest (Salta).
rudibuenekeri *(W.R. Abraham) A. Hofacker & P.J. Braun in Cactaceae Consensus Initiatives*, 6: 10 (1998): Notocactus rudibuenekeri.
subsp. glomerata *(N. Gerloff) A. Hofacker in Cactaceae Consensus Initiatives*, 6: 12 (1998): Notocactus glomeratus.
rutilans
subsp. veeniana *(Vliet) A. Hofacker in Cactaceae Consensus Initiatives*, 6: 12 (1998): Notocactus veenianus.
schumanniana
subsp. claviceps *(F. Ritter) A. Hofacker & P.J. Braun in Cactaceae Consensus Initiatives*, 6: 12 (1998): Eriocactus claviceps.
scopa
subsp. marchesii *(W.R.Abraham) A. Hofacker in Cactaceae Consensus Initiatives*, 6: 12 (1998): Notocactus scopa var. marchesii.
subsp. neobuenekeri *(F. Ritter) A. Hofacker & P.J. Braun in CactaceaeConsensus Initiatives*, 6: 10 (1998): Notocactus neobuenekeri.
subsp. succinea *(F. Ritter) A. Hofacker & P.J. Braun in Cactaceae Consensus Initiatives*, 6: 10 (1998): Notocactus succineus.
sellowii *(Link & Otto) D.R. Hunt in Cactaceae Consensus Initiatives*, 4: 6 (1997): Echinocactus sellowii.

PARODIA:-
stockingeri *(Prestlé) A. Hofacker & P.J. Braun in Cactaceae Consensus Initiatives*, 6: 10 (1998): Notocactus stockingeri.
tabularis *(Cels ex Rümpler) D.R. Hunt in Cactaceae Consensus Initiatives*, 4: 6 (1997): Echinocactus concinnus var. tabularis.
subsp. bommeljei *(Vliet) A. Hofacker in Cactaceae Consensus Initiatives*, 6: 12 (1998): Notocactus bommeljei.
tenuicylindrica *(F. Ritter) D.R. Hunt in Cactaceae Consensus Initiatives*, 4: 6 (1997): Notocactus tenuicylindricus.
turbinata *(Arechav.) A. Hofacker in Cactaceae Consensus Initiatives*, 6: 12 (1998): Echinocactus sellowii var. turbinatus.
werneri *A. Hofacker in Cactaceae Consensus Initiatives*, 6: 12 (1998), nom. nov.: Notocactus uebelmannianus *Buining*.
subsp. pleiocephala *(N. Gerloff & Königs) A. Hofacker in Cactaceae Consensus Initiatives*, 6: 12 (1998): Notocactus uebelmannianus var. pleiocephalus.
winbergii *W. Weskamp in Kakt. And. Sukk.*, 47(7): 158 (1996)—Argentina Northwest (Jujuy).

PARODIOLYRA (Gramin.).
colombiensis *G. Davidse & F. Zuloaga in Novon*, 9(4): 587 (1999)—Colombia.
micrantha *(Kunth) G. Davidse & F. Zuloaga in Novon*, 9(4): 590 (1999): Olyra micrantha.

PARONYCHIA (Illecebrac.).
cephalotes (M. Bieb.) Besser
subsp. taurica *(Borhidi & Sikura) N.N. Tzvelev in Novosti Sist. Vÿssh. Rast.*, 32: 32 (2000): P. taurica.
hintoniorum *B.L. Turner in Phytologia*, 79(1): 38 (1995 publ. 1996)—Mexico Northeast (Nuevo León), Mexico Gulf (Veracruz).
velata (Maire) Chaudhri
var. subvelata *(Litard. & Maire) A. Dobignard in Candollea*, 52(1): 128 (1997): P. argentea var. subvelata.

PARROTIA (Hamamelidac.).
subaequalis *(Hung T. Chang) R.M. Hao & H.T. Wei in Acta Phytotax. Sin.*, 36(1): 80 (1998): Hamamelis subaequalis.

PARRYCACTUS *A.B. Doweld in Novosti Sist. Vÿssh. Rast.*, 32: 117 (2000). *CACTACEAE*
alamosanus *(Britton & Rose) A.B. Doweld in Sukkulentÿ (Moscow)*, 3(1–2): 36 (2000): Echinocactus alamosanus.
subsp. reppenhagenii *(G. Unger) A.B. Doweld in Sukkulentÿ (Moscow)*, 3(1–2): 36 (2000): Ferocactus reppenhagenii.
echidne *(DC.) A.B. Doweld in Novosti Sist. Vÿssh. Rast.*, 32: 118 (2000): Echinocactus echidne.
flavovirens *(Scheidw.) A.B. Doweld in Novosti Sist. Vÿssh. Rast.*, 32: 118 (2000): Echinocactus flavovirens.
glaucescens *(DC.) A.B. Doweld in Novosti Sist. Vÿssh. Rast.*, 32: 118 (2000): Echinocactus glaucescens.
reppenhagenii *(G. Unger) A.B. Doweld in Novosti Sist. Vÿssh. Rast.*, 32: 118 (2000): Ferocactus reppenhagenii.
schwarzii *(G.E. Linds.) A.B. Doweld in Novosti Sist. Vÿssh. Rast.*, 32: 118 (2000): Ferocactus schwarzii.

PARSONSIA (Apocynac.).
apiculata *(Bakh. f.) D.J. Middleton in Blumea*, 42(1): 200 (1997): Grisseea apiculata.
appressa *D.J. Middleton in Blumea*, 42(1): 203 (1997)—New Guinea (Papua New Guinea).
bartlensis *J.B. Williams in Fl. Australia*, 28: 319 (1996)—Queensland.
blakeana *J.B. Williams in Fl. Australia*, 28: 316 (1996)—Queensland.
brisbanensis *J.B. Williams in Fl. Australia*, 28: 316 (1996)—Queensland.
constricta *D.J. Middleton in Blumea*, 42(1): 207 (1997)—New Guinea (Irian Jaya).
densiflora *D.J. Middleton in Blumea*, 42(1): 211 (1997)—New Guinea (Papua New Guinea).

PARSONSIA:-
dorrigoensis *J.B. Williams in Fl. Australia*, 28: 317 (1996)—New South Wales.
ferruginea *J.B. Williams in Fl. Australia*, 28: 318 (1996)—Queensland.
grandiflora *D.J. Middleton in Blumea*, 42(1): 213 (1997)—Moluccas.
grayana *J.B. Williams in Fl. Australia*, 28: 316 (1996)—Queensland.
kimberleyensis *J.B. Williams in Fl. Australia*, 28: 317 (1996)—Western Australia.
kroombitensis *J.B. Williams in Fl. Australia*, 28: 319 (1996)—Queensland.
larcomensis *J.B. Williams in Fl. Australia*, 28: 316 (1996)—Queensland.
longiloba *D.J. Middleton in Blumea*, 42(1): 217 (1997)—Bismarck Archipelago.
longipetiolata *J.B. Williams in Fl. Australia*, 28: 317 (1996)—Queensland, New South Wales.
novoguineensis *D.J. Middleton in Blumea*, 42(1): 218 (1997)—New Guinea (Papua New Guinea, Irian Jaya).
oligantha *(K. Schum.) D.J. Middleton in Blumea*, 42(1): 220 (1997): Delphyodon oliganthus.
paulforsteri *J.B. Williams in Fl. Australia*, 28: 318 (1996)—Queensland.
praeruptis *M.J. Heads & P.J. de Lange in New Zealand J. Bot.*, 37(1): 1 (1999)—New Zealand North I.
purpurascens *J.B. Williams in Fl. Australia*, 28: 318 (1996)—New South Wales.
sanguinea (Wernham) Markgr.
var. albiflora *(Gibbs) D.J. Middleton in Blumea*, 42(1): 229 (1997): Lyonsia albiflora.
var. brassii *(Markgr.) D.J. Middleton in Blumea*, 42(1): 228 (1997): Parsonsia brassii.
var. cruttwellii *D.J. Middleton in Blumea*, 42(1): 230 (1997)—New Guinea (Papua New Guinea).
var. millikenii *D.J. Middleton in Blumea*, 42(1): 230 (1997)—New Guinea (Irian Jaya).
var. sepikensis *D.J. Middleton in Blumea*, 42(1): 233 (1997)—New Guinea (Papua New Guinea).
var. vinkii *D.J. Middleton in Blumea*, 42(1): 231 (1997)—New Guinea (Papua New Guinea).
sankowskyana *J.B. Williams in Fl. Australia*, 28: 316 (1996)—Queensland.
schoddei *D.J. Middleton in Blumea*, 42(1): 233 (1997)—New Guinea (Papua New Guinea).
sundensis *D.J. Middleton in Blumea*, 42(1): 235 (1997)—Lesser Sunda Is. (East Timor, Lesser Sunda Is.).
tenuiflora *D.J. Middleton in Blumea*, 42(1): 237 (1997)—New Guinea (Irian Jaya).
wildensis *J.B. Williams in Fl. Australia*, 28: 319 (1996)—Queensland.
wongabelensis *J.B. Williams in Fl. Australia*, 28: 318 (1996)—Queensland.

PARTHENOCISSUS (Vitac.).
ser. Heterophyllae *C.L. Li in Chinese J. Appl. Environ. Biol.*, 2(1): 44 (1996), without type.
sect. Margaritaceae *C.L. Li in Chinese J. Appl. Environ. Biol.*, 2(1): 44 (1996), without type.
sect. Tuberculiformes *C.L. Li in Chinese J. Appl. Environ. Biol.*, 2(1): 44 (1996), without type.
chinensis *C.L. Li in Chinese J. Appl. Environ. Biol.*, 2(1): 45 (1996)—China South-central (Sichuan).
cuspidifera (Miq.) Planch.
var. pubifolia *C.L. Li in Chinese J. Appl. Environ. Biol.*, 2(1): 44 (1996)—China South-central (Yunnan).
feddei *(H. Lév.) C.L. Li in Chinese J. Appl. Environ. Biol.*, 2(1): 45 (1996): Vitis feddei.
var. pubescens *C.L. Li in Chinese J. Appl. Environ. Biol.*, 2(1): 45 (1996), without type—China Southeast (Guangdong, Guangxi).
semicordata (Wall.) Planch.
var. rubrifolia *(H. Lév. & Vaniot) C.L. Li in Chinese J. Appl. Environ. Biol.*, 2(1): 45 (1996), as 'rubifolia': Vitis rubrifolia.

PARVATIA (Lardizabalac.).
brunoniana (Wall.) Decne.
subsp. elliptica *(Hemsl.) H.N. Qin in Cathaya*, 8–9: 81 (1997): Stauntonia elliptica.

PASCOPYRUM (Gramin.).
smithii *(Rydb.) M.E. Barkworth & D.R. Dewey in Amer. J. Bot.*, 72(5): 772 (1985): Agropyron smithii.

PASPALUM (Gramin.).
sect. Parviflora *H. Rodríguez R. in Ernstia*, 8(4): 100 (1999).
barretoi *T.S. do Canto-Dorow, J.F.M. Valls & H.M. Longhi-Wagner in Bradea*, 6(40): 332 (1995)—Brazil South (Santa Catarina, Rio Grande do Sul).
cardenasianum *H. Rodríguez-R. & F. Camacho de Torres in Ernstia*, 8(1): 9 (1998)—Venezuela.
conduplicatum *T.S. do Canto-Dorow, J.F.M. Valls & H.M. Longhi-Wagner in Bradea*, 6(40): 333 (1995)—Brazil South (Rio Grande do Sul, Santa Catarina, Paraná).
dilatatum Poir.
var. lorirhachis *(Bor) P.K. Hajra & S.K. Jain, Botany Kaziranga & Manas:* 290 (1996): P. longifolium var. lorirhachis.
quarinii *O. Morrone & F.O. Zuloaga in Candollea*, 55(2): 311 (2000)—Paraguay, Argentina Northeast (Chaco, Corrientes, Misiones), Brazil South (Rio Grande do Sul).
ramianum *E.F. Camacho de Torres, H Rodríguez R. & L. Guevara in Ernstia*, 8(4): 105 (1999)—Venezuela.
volcanensis *F.O. Zuloaga, O. Morrone & S. Denham in Novon*, 10(2): 183 (2000)—Argentina Northwest (Jujuy), Bolivia.

PASSERINA (Thymelaeac.).
choulettei *M. Gandoger, Dec. Pl. Nov.*, 1: 6 (1875)—Algeria.
clementii *M. Gandoger, Dec. Pl. Nov.*, 1: 6 (1875)—France.
telonensis *M. Gandoger, Dec. Pl. Nov.*, 1: 6 (1875)—France.

PASSIFLORA (Passiflorac.).
ser. Digitatae *Killip ex A.C. Cervi in Fontqueria*, 45: 20 (1997).
sect. Elkea *C. Feuillet & J.M. MacDougal in Biollania*, ed. esp. 6 (J.J. Wurdack Festschr.): 337 (1997).
ser. Imbricatae *Killip ex A.C. Cervi in Fontqueria*, 45: 50 (1997).
nothosect. Inkea *C. Feuillet & J.M. MacDougal in Biollania*, ed. esp. 6 (J.J. Wurdack Festschr.): 339 (1997). (P. sect. Elkea × sect. Insignes).
sect. Insignes *(Harms) C. Feuillet & J.M. MacDougal in Biollania*, ed. esp. 6 (J.J. Wurdack Festschr.): 339 (1997): P. § Insignes.
ser. Kermesinae *Killip ex A.C. Cervi in Fontqueria*, 45: 45 (1997).
ser. Laurifoliae *Killip ex A.C. Cervi in Fontqueria*, 45: 22 (1997).
ser. Marginatae *Killip ex A.C. Cervi in Fontqueria*, 45: 22 (1997).
ser. Menispermifoliae *Killip ex A.C. Cervi in Fontqueria*, 45: 84 (1997).
ser. Pedatae *Killip ex A.C. Cervi in Fontqueria*, 45: 37 (1997).
ser. Serratifoliae *Killip ex A.C. Cervi in Fontqueria*, 45: 30 (1997).
ser. Setaceae *Killip ex A.C. Cervi in Fontqueria*, 45: 34 (1997).
aimae *H. Annonay & C. Feuillet in Adansonia*, 20(2): 297 (1998)—French Guiana.
anadenia *I. Urban, Symb. Antill.*, 3: 323 (1902)—Haiti.
cerasina *H. Annonay & C. Feuillet in Sida*, 17(3): 551 (1997)—French Guiana.
chaparensis *R. Vásquez Ch. in Rev. Soc Boliv. Bot.*, 2(1): 38 (1998)—Bolivia.
chrysosepala *M. Schwerdtfeger in Haussknechtia, Mitt. Thüring. Bot. Ges.*, 6: 41 (1997)—Ecuador.
formosa *T. Ulmer in Edinburgh J. Bot*, 56(2): 196 (1999)—Colombia.
jatunsachensis *M. Schwerdtfeger in Haussknechtia, Mitt. Thüring. Bot. Ges.*, 6: 43 (1997)—Ecuador.
linda *J.L. Panero in Brittonia*, 48(2): 192 (1996)—Ecuador.
loefgrenii *F.A. Vitta in Novon*, 7(2): 210 (1997)—Brazil Southeast (São Paulo).
luzmarina *Peter M. Jørgensen in Novon*, 7(4): 385 (1997 publ. 1998)—Ecuador.

PASSIFLORA:-
microstipula *L.E. Gilbert & J.M. MacDougal in Lundellia*, 3: 1 (2000)—Mexico Gulf (Veracruz), Mexico Southwest (Oaxaca).
neumannii *A. Cels & J.F. Cels in Ann. Fl. Pomone*, 1837–38: 251 (1838)—Brazil.
quadrifaria *R.J.R. Vanderplank in Curtis's Bot. Mag.*, 13(2): 66 (1996)—Brazil North (Pará).
smilacifolia *J.M. MacDougal in Novon*, 7(4): 382 (1997 publ. 1998)—Ecuador.
suberosa L.
var. perhastata *F.R. Fosberg in Micronesica*, 25(2): 193 (1992)—Society Is., Hawaiian Is., Fiji.
telesiphe *S. Knapp & J. Mallet in Novon*, 8(2): 162 (1998)—Ecuador.
trialata *C. Feuillet & J.M. MacDougal in Novon*, 6(4): 351 (1996)—French Guiana.
trochlearis *Peter M. Jørgensen in Novon*, 7(4): 379 (1997 publ. 1998)—Ecuador.
ulmeri *M. Schwerdtfeger in Haussknechtia, Mitt. Thüring. Bot. Ges.*, 6: 46 (1997)—Ecuador.
weigendii *T. Ulmer & M. Schwerdtfeger in Nordic J. Bot.*, 20(1): 47 (2000)—Peru.

PASTINACA (Umbellif.).
clausii *(Ledeb.) M.G. Pimenov in Opred. Rast. Sred. Azii*, 7: 315 (1983), isonym; *see Calest. in Webbia*, 1: 245 (1905): Heracleum clausii.

PATELLOCALAMUS *W.T. Lin in J. S. China Agric. Univ.*, 10(2): 45 (1989). *GRAMINEAE*
patellaris *(Gamble) W.T. Lin in J. S. China Agric. Univ.*, 10(2): 46 (1989): Dendrocalamus patellaris.

PAULLINIA (Sapindac.).
cearensis *G.V. Somner & M.S. Ferrucci in Bonplandia*, 9(3–4): 241 (1997)—Brazil Northeast (Ceará, Piauí).
lingulata *P. Acevedo-Rodríguez in Brittonia*, 50(4): 516 (1998)—French Guiana.
talamancensis *J.F. Morales in Phytologia*, 81(5): 363 (1996 publ. 1997)—Costa Rica.

PAULOWNIA (Scrophulariac.).
catalpifolia *T. Gong ex D.Y. Hong in Novon*, 7(4): 366 (1997 publ. 1998)—China North-central (Shandong).
× henanensis *C.Y. Zhang & Y.H. Zhao in Acta Phytotax. Sin.*, 33(5): 503 (1995)—China Southeast (Henan). (P. catalpifolia × tomentosa).

PAUSINYSTALIA (Rubiac.).
lane-poolei (Hutch.) Hutch. ex Lane-Poole
subsp. ituriense *(De Wild.) P. Stoffelen & E. Robbrecht in Bot. J. Linn. Soc.*, 120(4): 310 (1996): P. ituriense.

PAVETTA (Rubiac.).
badullensis *C.E. Ridsdale in Blumea*, 41(2): 461 (1996), nom. nov.: P. indica var. montana *Thwaites ex Hook. f.*
bidentata Hiern
var. sessilifolia *S.D. Manning in Ann. Missouri Bot. Gard.*, 83(1): 105 (1996)—Cameroon.
bourdillonii *M. Sivadasan & N. Mohanan in Bot. Bull. Acad. Sin. (Taipei)*, 40(1): 61 (1999)—India (Kerala).
camerounensis *S.D. Manning in Ann. Missouri Bot. Gard.*, 83(1): 107 (1996)—Cameroon.
subsp. brevirama *S.D. Manning in Ann. Missouri Bot. Gard.*, 83(1): 108 (1996)—Cameroon.
grossissima *S.D. Manning in Ann. Missouri Bot. Gard.*, 83(1): 114 (1996)—Cameroon.
hookeriana Hiern
var. pubinervata *S.D. Manning in Ann. Missouri Bot. Gard.*, 83(1): 117 (1996)—Cameroon.
kribiensis *S.D. Manning in Ann. Missouri Bot. Gard.*, 83(1): 118 (1996)—Cameroon.
kupensis *S.D. Manning in Ann. Missouri Bot. Gard.*, 83(1): 118 (1996)—Cameroon.
laxa *S.D. Manning in Ann. Missouri Bot. Gard.*, 83(1): 123 (1996)—Cameroon.
longistyla *S.D. Manning in Ann. Missouri Bot. Gard.*, 83(1): 125 (1996)—Cameroon.
mpomii *S.D. Manning in Ann. Missouri Bot. Gard.*, 83(1): 128 (1996)—Cameroon.
muiriana *S.D. Manning in Ann. Missouri Bot. Gard.*, 83(1): 128 (1996)—Cameroon.

PAVETTA:-
namatae *S.D. Manning in Ann. Missouri Bot. Gard.*, 83(1): 131 (1996)—Cameroon.
owariensis P. Beauv.
var. glaucescens *(Hiern) S.D. Manning in Ann. Missouri Bot. Gard.*, 83(1): 136 (1996): P. glaucescens.
var. opaca *S.D. Manning in Ann. Missouri Bot. Gard.*, 83(1): 134 (1996)—Cameroon.
var. satabiei *S.D. Manning in Ann. Missouri Bot. Gard.*, 83(1): 137 (1996)—Cameroon.
rubentifolia *S.D. Manning in Ann. Missouri Bot. Gard.*, 83(1): 139 (1996)—Cameroon.
tenuissima *S.D. Manning in Ann. Missouri Bot. Gard.*, 83(1): 142 (1996)—Cameroon.
tshikondeni *N. Hahn in Bothalia*, 29(1): 107 (1999)—Transvaal.
urophylla Bremek.
subsp. bosii *S.D. Manning in Ann. Missouri Bot. Gard.*, 83(1): 144 (1996)—Cameroon.
viridiloba K. Krause
var. meurillonii *S.D. Manning in Ann. Missouri Bot. Gard.*, 83(1): 146 (1996)—Cameroon.

PAVONIA (Malvac.).
sect. Albae *P.A. Fryxell in Fl. Neotrop.*, 76: 162 (1999).
subgen. Asterochlamys *P.A. Fryxell in Fl. Neotrop.*, 76: 11, 132 (1999), nom. nov.: gen. Asterochlaena *Garcke.*
sect. Asterochlamys *P.A. Fryxell in Fl. Neotrop.*, 76: 132 (1999), nom. nov.: gen Asterochlaena *Garcke.*
sect. Carcerariae *P.A. Fryxell in Fl. Neotrop.*, 76: 231 (1999).
sect. Collicolae *P.A. Fryxell in Fl. Neotrop.*, 76: 171 (1999).
subsect. Collicolae *P.A. Fryxell in Fl. Neotrop.*, 76: 171 (1999).
subsect. Conditiflorae *P.A. Fryxell in Fl. Neotrop.*, 76: 180 (1999).
subsect. Cordifoliae *P.A. Fryxell in Fl. Neotrop.*, 76: 95 (1999).
sect. Diathericae *P.A. Fryxell in Fl. Neotrop.*, 76: 228 (1999).
subsect. Hastifoliae *P.A. Fryxell in Fl. Neotrop.*, 76: 102 (1999).
sect. Laminares *P.A. Fryxell in Fl. Neotrop.*, 76: 182 (1999).
subgen. Malache *(B. Vogel) P.A. Fryxell in Fl. Neotrop.*, 76: 11, 164 (1999): gen. Malache.
sect. Malache *(B. Vogel) P.A. Fryxell in Fl. Neotrop.*, 76: 164 (1999): gen. Malache.
subsect. Rubriflorae *P.A. Fryxell in Fl. Neotrop.*, 76: 177 (1999).
sect. Varians *P.A. Fryxell in Fl. Neotrop.*, 76: 234 (1999).
aschersonioides *P.A. Fryxell in Fl. Neotrop.*, 76: 37 (1999)—Brazil West-central (Goiás, Mato Grosso).
atlantica *P.A. Fryxell in Fl. Neotrop.*, 76: 239 (1999)—Brazil Northeast (Bahia).
aurelii *P.A. Fryxell in Fl. Neotrop.*, 76: 137 (1999)—Paraguay.
cauliflora *(Nees) Fryxell ex G.L. Esteves in Bol. Inst. Bot. (São Paulo)*, 11: 200 (1998): Goethea cauliflora.
cauliflora *(Nees) P.A. Fryxell in Fl. Neotrop.*, 76: 240 (1999), isonym; *see G.L. Esteves in Bol. Inst. Bot. (São Paulo)*, 11: 200 (1998): Goethea cauliflora.
ciliata *G.L. Esteves & A. Krapovickas in Hoehnea*, 22(1–2): 41 (1995)—Brazil Northeast (Bahia).
clarkii *P.A. Fryxell in Fl. Neotrop.*, 76: 181 (1999)—Mexico Northeast (San Luis Potosí).
cracens *P.A. Fryxell & G.L. Esteves in Bol. Bot. (São Paulo)*, 17: 42 (1998)—Brazil Northeast (Bahia).
cristobaliae *P.A. Fryxell in Fl. Neotrop.*, 76: 16 (1999)—Brazil West-central (Mato Grosso, Mato Grosso do Sul).
decora *P.A. Fryxell in Fl. Neotrop.*, 76: 67 (1999)—Brazil Northeast (Bahia).
fonsecana *(Standl.) P.A. Fryxell in Cuscatlania*, 1(10: 173): 8 (1997): Malache fonsecana.
formosa *P.A. Fryxell in Fl. Neotrop.*, 76: 209 (1999)—Bolivia, Argentina Northwest (Salta).
friisii *M. Thulin & K. Vollesen in Nordic J. Bot.*, 19(2): 199 (1999)—Ethiopia, Somalia.

PAVONIA:-
froesii *P.A. Fryxell in Fl. Neotrop.*, 76: 243 (1999)—Brazil Northeast (Bahia).
glaziovii *(Ulbr.) P.A. Fryxell in Fl. Neotrop.*, 76: 71 (1999): Codonochlamys glaziovii.
goetheoides *(Hassl.) Fryxell ex G.L. Esteves in Bol. Inst. Bot. (São Paulo)*, 11: 197 (1998): Blanchetiastrum goetheoides.
goetheoides *(Hassl.) P.A. Fryxell in Fl. Neotrop.*, 76: 245 (1999), isonym; *see G.L. Esteves in Bol. Inst. Bot. (São Paulo)*, 11: 197 (1998): Blanchetiastrum goetheoides.
heterotricha *P.A. Fryxell in Fl. Neotrop.*, 76: 21 (1999)—Bolivia.
hirticalyx *P.A. Fryxell in Fl. Neotrop.*, 76: 43 (1999)—Brazil Southeast (Minas Gerais).
immitis *P.A. Fryxell in Fl. Neotrop.*, 76: 46 (1999)—Brazil West-central (Goiás).
krapovickasii *P.A. Fryxell in Fl. Neotrop.*, 76: 148 (1999)—Brazil West-central (Mato Grosso, Goiás), Brazil North (Tocantins), Brazil Northeast (Bahia).
longipilosa *M. Thulin in Nordic J. Bot.*, 19(2): 203 (1999)—Ethiopia.
longitricha *P.A. Fryxell in Fl. Neotrop.*, 76: 46 (1999)—Brazil West-central (Goiás).
longitudinalis *P.A. Fryxell in Fl. Neotrop.*, 76: 196 (1999)—Peru.
marginata *M. Thulin in Nordic J. Bot.*, 19(2): 206 (1999)—Somalia.
matteiana *M. Thulin in Nordic J. Bot.*, 19(2): 202 (1999)—Somalia.
mogotensis *(Carabia) P.A. Fryxell in Fl. Neotrop.*, 76: 93 (1999): Hibiscus mogotensis.
nayarensis *P.A. Fryxell in Fl. Neotrop.*, 76: 178 (1999)—Mexico Southwest (Nayarit, Jalisco).
neei *P.A. Fryxell in Fl. Neotrop.*, 76: 153 (1999)—Bolivia, Paraguay, Argentina Northeast (Formosa).
nigrescens *M. Thulin in Nordic J. Bot.*, 19(2): 201 (1999)—Somalia.
ovaliphylla *G.L. Esteves & A. Krapovickas in Hoehnea*, 22(1–2): 43 (1995)—Brazil Northeast (Bahia).
paneroi *P.A. Fryxell in Lundellia*, 1: 75 (1998)—Mexico Central (Puebla).
paraensis *P.A. Fryxell in Fl. Neotrop.*, 76: 73 (1999)—Brazil North (Pará).
paucibracteata *M. Thulin in Nordic J. Bot.*, 19(2): 206 (1999)—Somalia.
pedersenii *P.A. Fryxell in Fl. Neotrop.*, 76: 123 (1999)—Brazil South (Rio Grande do Sul).
pseudolaxifolia *P.A. Fryxell in Fl. Neotrop.*, 76: 27 (1999)—Brazil West-central (Brasília Distrito Federal).
quadrifida *P.A. Fryxell in Fl. Neotrop.*, 76: 154 (1999)—Brazil West-central (Goiás).
ramosissima *(Arechav.) P.A. Fryxell & A. Krapovickas in Fl. Neotrop.*, 76: 218 (1999): P. sepium var. ramosissimum.
rotundifolia *M. Thulin & K. Vollesen in Nordic J. Bot.*, 19(2): 204 (1999)—Ethiopia, Somalia.
rubriphylla *G.L. Esteves in Bol. Bot. (São Paulo)*, 17: 43 (1998)—Brazil Northeast (Bahia).
rudis *P.A. Fryxell in Fl. Neotrop.*, 76: 218 (1999)—Peru.
sepioides *P.A. Fryxell & A. Krapovickas in Fl. Neotrop.*, 76: 221 (1999)—Colombia, Venezuela, Ecuador, Peru, Bolivia.
setifer *(C. Presl) P.A. Fryxell in Fl. Neotrop.*, 76: 57 (1999): Hibiscus setifer.
spuria *P.A. Fryxell in Fl. Neotrop.*, 76: 158 (1999)—Mexico Southwest (Oaxaca).
strictiflora *(Hook.) G.L. Esteves in Bol. Bot. (São Paulo)*, 17: 45 (1998): Goethea strictiflora.
strictiflora *(Hook.) P.A. Fryxell in Fl. Neotrop.*, 76: 252 (1999), isonym; *see G.L. Esteves in Bol. Bot. (São Paulo)*, 17: 45 (1998): Goethea strictiflora.
tiliifolia *(Ulbr.) P.A. Fryxell in Fl. Neotrop.*, 76: 75 (1999): Codonochlamys tiliifolia.
vannii *P.A. Fryxell in Fl. Neotrop.*, 76: 160 (1999)—Paraguay, Bolivia.
vinosa *G.L. Esteves in Bol. Inst. Bot. (São Paulo)*, 11: 176 (1998)—Brazil Northeast (Ceará).

PAYENA (Sapotac.).
ferruginea *J.T. Pereira in Kew Bull.*, 52(4): 905 (1997)—Borneo (Sabah, Brunei, Kalimantan, Sarawak).

PAYENA:-
grandistipula *J.T. Pereira in Kew Bull.*, 52(4): 908 (1997)—Borneo (Sarawak).
kapitensis *J.T. Pereira in Kew Bull.*, 52(4): 910 (1997)—Borneo (Sarawak).
khoonmengiana *J.T. Pereira in Kew Bull.*, 52(4): 912 (1997)—Borneo (Sabah).
kinabaluensis *J.T. Pereira in Kew Bull.*, 52(4): 914 (1997)—Borneo (Sabah).
obscura Burck
var. havilandii *(King & Gamble) J.T. Pereira in Kew Bull.*, 52(4): 916 (1997): P. havilandii.

PEARCEA (Gesneriac.).
abunda *(Wiehler) L.P. Kvist & L.E. Skog in Smithsonian Contrib. Bot.*, 84: 20 (1996): Parakohleria abunda.
bella *L.P. Kvist & L.E. Skog in Smithsonian Contrib. Bot.*, 84: 20 (1996)—Colombia.
bilabiata *L.P. Kvist & L.E. Skog in Smithsonian Contrib. Bot.*, 84: 22 (1996)—Ecuador.
cordata *L.P. Kvist & L.E. Skog in Smithsonian Contrib. Bot.*, 84: 22 (1996)—Ecuador.
fuscicalyx *L.P. Kvist & L.E. Skog in Smithsonian Contrib. Bot.*, 84: 25 (1996)—Peru.
glabrata *L.P. Kvist & L.E. Skog in Smithsonian Contrib. Bot.*, 84: 25 (1996)—Ecuador.
gracilis *L.P. Kvist & L.E. Skog in Smithsonian Contrib. Bot.*, 84: 27 (1996)—Ecuador.
grandifolia *L.P. Kvist & L.E. Skog in Smithsonian Contrib. Bot.*, 84: 29 (1996)—Peru.
hispidissima *(Wiehler) L.P. Kvist & L.E. Skog in Smithsonian Contrib. Bot.*, 84: 29 (1996): Parakohleria hispidissima.
intermedia *L.P. Kvist & L.E. Skog in Smithsonian Contrib. Bot.*, 84: 31 (1996)—Ecuador.
purpurea *(Poepp.) L.P. Kvist & L.E. Skog in Smithsonian Contrib. Bot.*, 84: 32 (1996): Rhytidophyllum purpureum.
reticulata *(Fritsch) L.P. Kvist & L.E. Skog in Smithsonian Contrib. Bot.*, 84: 33 (1996): Kohleria reticulata.
rhodotricha *(Cuatrec.) L.P. Kvist & L.E. Skog in Smithsonian Contrib. Bot.*, 84: 35 (1996): Diastema rhodotrichum.
sprucei *(Britton) L.P. Kvist & L.E. Skog in Smithsonian Contrib. Bot.*, 84: 37 (1996): Isoloma sprucei.
var. parviflora *(Rusby) L.P. Kvist & L.E. Skog in Smithsonian Contrib. Bot.*, 84: 39 (1996): Isoloma parviflorum.
strigosa *L.P. Kvist & L.E. Skog in Smithsonian Contrib. Bot.*, 84: 40 (1996)—Peru.

PEARSONIA (Legumin.).
hirsuta *G. Germishuizen in Bothalia*, 28(1): 57 (1998)—Transvaal.

PECTEILIS (Orchidac.).
dianthoides *(Nevski) L.A. Garay & G.A. Romero-González in Harvard Pap. Bot.*, 3(1): 61 (1998): Habenaria dianthoides.

PECTIANTIA (Saxifragac.).
formosana *(Hayata) S.S. Ying in Mem. Coll. Agric. Nation. Taiwan Univ.*, 25(2): 102 (1985): Mitella japonica var. formosana.

PECTIS (Compos.).
vandevenderi *B.L. Turner in Phytologia*, 79(3): 133 (1995 publ. 1996)—Mexico Northwest (Sonora).

PEDERSENIA *J. Holub in Preslia*, 70(2): 181 (1998), nom. nov.: gen. Trommsdorffia *Mart.* (Amaranthac.).
argentata *(Mart.) J. Holub in Preslia*, 70(2): 181 (1998): Trommsdorffia argentata.
canescens *(Willd.) J. Holub in Preslia*, 70(2): 181 (1998): Iresine canescens.
cardenasii *(Standl.) J. Holub in Preslia*, 70(2): 181 (1998): Iresine cardenasii.
costaricensis *(Standl.) J. Holub in Preslia*, 70(2): 181 (1998): Iresine costaricensis.
hassleriana *(Chodat) T.M. Pedersen in Bonplandia*, 10(1–4): 108 (2000): Iresine hassleriana.
macrophylla *(R.E. Fr.) J. Holub in Preslia*, 70(2): 181 (1998): Iresine macrophylla.

PEDERSENIA :-
pulverulenta *(Mart.) J. Holub in Preslia*, 70(2): 181 (1998): Trommsdorffia pulverulenta.
weberbaueri *(Suess.) J. Holub in Preslia*, 70(2): 181 (1998): Trommsdorffia weberbaueri.

PEDICULARIS (Scrophulariac.).
ser. Acmodontae *T.N. Popova ex Sh.I. Kuthatheladze & M.T. Davlianidze in Fl. Gruzii, ed. 2*, 12: 330 (1999).
ser. Albertianae *Vved. ex L.I. Ivanina & T.N. Popova in Novon*, 7(4): 374 (1997 publ. 1998).
ser. Caucasicae *Vved. ex Sh.I. Kuthatheladze & M.T. Davlianidze in Fl. Gruzii, ed. 2*, 12: 330 (1999).
ser. Chamissonianae *Vved. ex L.I. Ivanina & T.N. Popova in Bot. Zhurn.*, 83(10): 97 (1998).
ser. Condensatae *Galushko & T.N. Popova ex Sh.I. Kuthatheladze & M.T. Davlianidze in Fl. Gruzii, ed. 2*, 12: 330 (1999).
ser. Crassirostres *Vved. ex Sh.I. Kuthatheladze & M.T. Davlianidze in Fl. Gruzii, ed. 2*, 12: 330 (1999).
subsect. Crassirostres *L.I. Ivanina & T.N. Popova in Bot. Zhurn.*, 83(10): 96 (1998).
ser. Crassirostres *Vved. ex L.I. Ivanina & T.N. Popova in Bot. Zhurn.*, 83(10): 97 (1998).
ser. Czuilienses *L.I. Ivanina & T.N. Popova in Bot. Zhurn.*, 83(10): 96 (1998).
ser. Maximoviczianae *L.I. Ivanina & T.N. Popova in Novon*, 7(4): 373 (1997 publ. 1998).
ser. Palustres *Vved. ex Sh.I. Kuthatheladze & M.T. Davlianidze in Fl. Gruzii, ed. 2*, 12: 330 (1999).
ser. Rostratae *Vved. ex Sh.I. Kuthatheladze & M.T. Davlianidze in Fl. Gruzii, ed. 2*, 12: 330 (1999).
ser. Sarawschanicae *L.I. Ivanina & T.N. Popova in Bot. Zhurn.*, 83(10): 95 (1998).
ser. Semenowianae *Vved ex L.I. Ivanina & T.N. Popova in Novon*, 7(4): 375 (1997 publ. 1998).
ser. Sibthorpianae *T.N. Popova ex Sh.I. Kuthatheladze & M.T. Davlianidze in Fl. Gruzii, ed. 2*, 12: 330 (1999).
ser. Spicatae *Vved. ex L.I. Ivanina & T.N. Popova in Bot. Zhurn.*, 83(10): 98 (1998).
sect. Spicatae *L.I. Ivanina & T.N. Popova in Bot. Zhurn.*, 83(10): 98 (1998).
ser. Veraeanae *L.I. Ivanina & T.N. Popova in Bot. Zhurn.*, 83(10): 95 (1998).
subsect. Verticillatae *(Benth.) L.I. Ivanina & T.N. Popova in Bot. Zhurn.*, 83(10): 94 (1998): P. sect. Verticillatae.
ser. Wilhelmsianae *Galushko & T.N. Popova ex Sh.I. Kuthatheladze & M.T. Davlianidze in Fl. Gruzii, ed. 2*, 12: 330 (1999).
arctoeuropaea *(Hultén) U. Molau & D.F. Murray in Symb. Bot. Upsal.*, 31(3): 38 (1996): P. sudetica subsp. arctoeuropaea.
armata Maxim.
var. trimaculata *X.F. Lu in Novon*, 6(2): 190 (1996)—Qinghai, China North-central (Gansu).
cholashanensis *T. Yamazaki in J. Jap. Bot.*, 75(4): 220 (2000)—China South-central (Sichuan).
columigera *T. Yamazaki in J. Jap. Bot.*, 75(4): 216 (2000)—Xizang.
comosa L.
subsp. nevadensis *(Pau) À.M. Romo in Soc. Éch. Pl. Vasc. Eur. Bassin Médit. Bull.*, 27: 95 (1998): P. comosa var. nevadensis.
elsholtzioides *T. Yamazaki in J. Jap. Bot.*, 75(4): 213 (2000)—Xizang.
gruiflora *T. Yamazaki in J. Jap. Bot.*, 75(5): 285 (2000)—Nepal.
incisopetala *E. Menzel, H. Menzel & H. Zoller in Schlern*, 53: 419 (1979)—Italy.
interior *(Hultén) U. Molau & D.F. Murray in Symb. Bot. Upsal.*, 31(3): 39 (1996): P. sudetica var. interior.
koshiensis *T. Yamazaki in J. Jap. Bot.*, 75(5): 287 (2000)—Nepal.
kuruchuensis *T. Yamazaki in J. Jap. Bot.*, 75(4): 218 (2000)—Xizang.
lhasana *T. Yamazaki in J. Jap. Bot.*, 75(4): 214 (2000)—Xizang.
liguliflora *T. Yamazaki in J. Jap. Bot.*, 75(4): 215 (2000)—Xizang.
longiflora Rudolph
var. hongyuanensis *Y. Tang in Novon*, 8(4): 455 (1998)—China South-central (Sichuan).

PEDICULARIS:-
ningjuingensis *T. Yamazaki in J. Jap. Bot.*, 75(4): 216 (2000)—Xizang.
resupinata L.
f. alba *Y.N. Lee, Fl. Korea:* 1161 (1996), without type—Locality not stated.
f. albiflora *Y.N. Lee, Fl. Korea.* 1161 (1996), without type—Locality not stated.
var. albiflora *(Nakai) S.X. Li in Clavis Pl. Chinae Bor.-Or. ed. 2 (ed. P.Y. Fu):* 600 (1995), homonym; *see Y.Z. Zhao in Fl. Intramongolica*, 5: 413 (1980 publ. 1981): P. resupinata f. albiflora.
verticillata L.
var. hallaisanensis *(Hurus.) Y.N. Lee, Fl. Korea:* 1161 (1996): P. hallaisanensis.
xiqingshanensis *H.Y. Feng & J.Z. Sun in Acta Bot. Yunnanica*, 21(1): 28 (1999)—China North-central (Gansu).

PEDIOCACTUS (Cactac.).
sect. Gymnocactus *J.J. Halda in Acta Mus. Richnov. Sect. Nat.*, 5(1): 17 (1998).
sect. Neolloydia *(Britton & Rose) J.J. Halda in Acta Mus. Richnov. Sect. Nat.*, 5(1): 23 (1998): gen. Neolloydia.
subgen. Neolloydia *(Britton & Rose) J.J. Halda in Acta Mus. Richnov. Sect. Nat.*, 5(1): 23 (1998): gen. Neolloydia.
subsect. Pectinati *J.J. Halda in Acta Mus. Richnov. Sect. Nat.*, 5(1): 22 (1998), as 'Pectinatae'.
sect. Rhytidospermae *F. Hochstätter in Cactaceae Rev.*, 2(2): 14 (1999); *cf. Repert. Pl. Succ. (I.O.S.)*, 50: 14 (1999 publ. 2000).
sect. Sclerocactus *(Britton & Rose) J.J. Halda in Acta Mus. Richnov. Sect. Nat.*, 5(1): 14 (1998): gen. Sclerocactus.
subgen. Strombocactus *(Britton & Rose) J.J. Halda in Acta Mus. Richnov. Sect. Nat.*, 5(5): 185 (1998): gen. Strombocactus.
subgen. Toumeya *(Britton & Rose) J.J. Halda in Acta Mus. Richnov. Sect. Nat.*, 5(1): 17 (1998): gen. Toumeya.
sect. Turbinicarpus *J.J. Halda in Acta Mus. Richnov. Sect. Nat.*, 5(1): 19 (1998).
subsect. Turbinicarpus *J.J. Halda in Acta Mus. Richnov. Sect. Nat.*, 5(1): 19 (1998).
alonsoi *(Glass & S. Arias) J.J. Halda in Acta Mus. Richnov. Sect. Nat.*, 5(1): 22 (1998): Turbinicarpus alonsoi.
bonatzii *(Gerhart Frank) J.J. Halda in Acta Mus. Richnov. Sect. Nat.*, 5(1): 22 (1998): Turbinicarpus bonatzii.
brevihamatus *(Engelm.) J.J. Halda in Acta Mus. Richnov. Sect. Nat.*, 5(1): 14 (1998): Echinocactus brevihamatus.
subsp. tobuschii *(W.T. Marshall) J.J. Halda in Acta Mus. Richnov. Sect. Nat.*, 5(1): 15 (1998): Mammillaria tobuschii.
cloverae *(K.D. Heil & J.M. Porter) J.J. Halda in Acta Mus. Richnov. Sect. Nat.*, 5(1): 15 (1998): Sclerocactus cloverae.
subsp. brackii *(K.D. Heil & J.M. Porter) J.J. Halda in Acta Mus. Richnov. Sect. Nat.*, 5(1): 15 (1998): Sclerocactus cloverae subsp. brackii.
conoideus *(DC.) J.J. Halda in Acta Mus. Richnov. Sect. Nat.*, 5(1): 23 (1998): Mammillaria conoidea.
disciformis *(DC.) J.J. Halda in Acta Mus. Richnov. Sect. Nat.*, 5(5): 186 (1998): Mammillaria disciformis.
erectocentrus *(J.M. Coult.) J.J. Halda in Acta Mus. Richnov. Sect. Nat.*, 5(1): 24 (1998): Echinocactus erectocentrus.
gautii *(L.D. Benson) J.J. Halda in Acta Mus. Richnov. Sect. Nat.*, 5(1): 19 (1998): Neolloydia gautii.
gielsdorfianus *(Werderm.) J.J. Halda in Acta Mus. Richnov. Sect. Nat.*, 5(1): 18 (1998): Echinocactus gielsdorfianus.
hoferi *(Lüthy & A.B. Lau) J.J. Halda in Acta Mus. Richnov. Sect. Nat.*, 5(1): 22 (1998): Turbinicarpus hoferi.
horripilus *(Lem.) J.J. Halda in Acta Mus. Richnov. Sect. Nat.*, 5(1): 18 (1998): Mammillaria horripila.
intertextus *(Engelm.) J.J. Halda in Acta Mus. Richnov. Sect. Nat.*, 5(1): 24 (1998): Echinocactus intertextus.

PEDIOCACTUS:-
intertextus
var. dasyacanthus *(Engelm.) J.J. Halda in Acta Mus. Richnov. Sect. Nat.*, 5(1): 24 (1998): Echinocactus intertextus var. dasyacanthus.
jarmilae *(Halda) J.J. Halda in Acta Mus. Richnov. Sect. Nat.*, 5(5): 186 (1998): Strombocactus jarmilae.
johnsonii *(Engelm.) J.J. Halda in Acta Mus. Richnov. Sect. Nat.*, 5(1): 24 (1998): Echinocactus johnsonii.
knuthianus *(Boed.) J.J. Halda in Acta Mus. Richnov. Sect. Nat.*, 5(1): 18 (1998): Echinocactus knuthianus.
laui *(Glass & R.A. Foster) J.J. Halda in Acta Mus. Richnov. Sect. Nat.*, 5(1): 21 (1998): Turbinicarpus laui.
lophophoroides *(Werderm.) J.J. Halda in Acta Mus. Richnov. Sect. Nat.*, 5(1): 21 (1998): Thelocactus lophophoroides.
var. jauernigii *(Gerhart Frank) J.J. Halda in Acta Mus. Richnov. Sect. Nat.*, 5(1): 21 (1998): Turbinicarpus jauernigii.
mandragora *(Frič ex A. Berger) J.J. Halda in Acta Mus. Richnov. Sect. Nat.*, 5(1): 18 (1998): Echinocactus mandragora.
mariposensis *(Hester) J.J. Halda in Acta Mus. Richnov. Sect. Nat.*, 5(1): 24 (1998): Echinomastus mariposensis.
nyensis *(Hochstätter) J.J. Halda in Acta Mus. Richnov. Sect. Nat.*, 5(1): 15 (1998): Sclerocactus nyensis.
parviflorus *(Clover & Jotter) J.J. Halda in Acta Mus. Richnov. Sect. Nat.*, 5(1): 15 (1998): Sclerocactus parviflorus.
subsp. havasupaiensis *(Clover) J.J. Halda in Acta Mus. Richnov. Sect. Nat.*, 5(1): 15 (1998): Sclerocactus havasupaiensis.
subsp. intermedius *(Peebles) J.J. Halda in Acta Mus. Richnov. Sect. Nat.*, 5(1): 16 (1998): Sclerocactus intermedius.
subsp. terrae-canyonae *(K.D. Heil) J.J. Halda in Acta Mus. Richnov. Sect. Nat.*, 5(1): 16 (1998): Sclerocactus terrae-canyonae.
peeblesianus
subsp. fickeisenii *(Hochstätter) J.M. Lüthy in Kakt. And. Sukk.*, 50(11): 278 (1999): Navajoa peeblesiana var. fickeisenii.
f. maius *(Hochstätter) M. Hájek in Kaktusy (Brno)*, 36(3): 73 (2000), as 'maianus': Navajoa peeblesiana f. maia.
f. menzelii *(Hochstätter) M. Hájek in Kaktusy (Brno)*, 36(3): 73 (2000): Navajoa peeblesiana f. menzelii.
pseudomacrochele *(Backeb.) J.J. Halda in Acta Mus. Richnov. Sect. Nat.*, 5(1): 21 (1998): Strombocactus pseudomacrochele.
var. krainzianus *(Gerhart Frank) J.J. Halda in Acta Mus. Richnov. Sect. Nat.*, 5(1): 21 (1998): Toumeya krainziana.
var. lausseri *(Diers & Gerhart Frank) J.J. Halda in Acta Mus. Richnov. Sect. Nat.*, 5(1): 22 (1998): Turbinicarpus pseudomacrochele var. lausseri.
var. sphacellatus *(Diers & Gerhart Frank) J.J. Halda in Acta Mus. Richnov. Sect. Nat.*, 5(1): 22 (1998): Turbinicarpus pseudomacrochele var. sphacellatus.
pseudopectinatus *(Backeb.) J.J. Halda in Acta Mus. Richnov. Sect. Nat.*, 5(1): 22 (1998): Pelecyphora pseudopectinata.
pubispinus (Engelm.) Arp
subsp. sileri *(L.D. Benson) J.J. Halda in Acta Mus. Richnov. Sect. Nat.*, 5(1): 16 (1998): Sclerocactus pubispinus var. sileri.
pulcherrimus *(Halda) J.J. Halda in Acta Mus. Richnov. Sect. Nat.*, 5(5): 186 (1998): Strombocactus pulcherrimus.
rioverdensis *(Gerhart Frank) J.J. Halda in Acta Mus. Richnov. Sect. Nat.*, 5(1): 22 (1998): Turbinicarpus rioverdensis.
saueri *(Boed.) J.J. Halda in Acta Mus. Richnov. Sect. Nat.*, 5(1): 18 (1998): Echinocactus saueri.
scheeri *(Salm-Dyck) J.J. Halda in Acta Mus. Richnov. Sect. Nat.*, 5(1): 25 (1998): Echinocactus scheeri.
schmiedickeanus *(Boed.) J.J. Halda in Acta Mus. Richnov. Sect. Nat.*, 5(1): 20 (1998): Echinocactus schmiedickeanus.

PEDIOCACTUS :-
schmiedickeanus
var. dickisoniae *(Glass & R.A. Foster) J.J. Halda in Acta Mus. Richnov. Sect. Nat.*, 5(1): 20 (1998): Turbinicarpus schmiedickeanus var. dickisoniae.
var. flaviflorus *(Gerhart Frank & A.B. Lau) J.J. Halda in Acta Mus. Richnov. Sect. Nat.*, 5(1): 21 (1998): Turbinicarpus flaviflorus.
var. gracilis *(Glass & R.A. Foster) J.J. Halda in Acta Mus. Richnov. Sect. Nat.*, 5(1): 20 (1998): Turbinicarpus gracilis.
var. klinkerianus *(Backeb. & H. Jacobsen) J.J. Halda in Acta Mus. Richnov. Sect. Nat.*, 5(1): 20 (1998): Turbinicarpus klinkerianus.
var. macrochele *(Werderm.) J.J. Halda in Acta Mus. Richnov. Sect. Nat.*, 5(1): 20 (1998): Echinocactus macrochele.
var. rubriflorus *(Gerhart Frank) J.J. Halda in Acta Mus. Richnov. Sect. Nat.*, 5(1): 21 (1998): Turbinicarpus schwarzii var. rubriflorus.
var. schwarzii *(Shurly) J.J. Halda in Acta Mus. Richnov. Sect. Nat.*, 5(1): 21 (1998): Strombocactus schwarzii.
simpsonii
f. acklinii *F. Hochstätter in Succulenta*, 76(6): 264 (1997)—Utah.
subsp. bensonii *F. Hochstätter, Genera Pediocactus, Navajoa, Toumeya Revised:* 46 (1995), nom. nov., as comb. nov.: Echinocactus simpsonii var. minor *Engelm.*†
subsp. bradyi *(L.D. Benson) J.J. Halda in Acta Mus. Richnov. Sect. Nat.*, 5(1): 13 (1998): P. bradyi.
f. campestris *F. Hochstätter in Succulenta*, 76(6): 268 (1997)—Wyoming.
var. despainii *(S.L. Welsh & Goodrich) J.J. Halda in Acta Mus. Richnov. Sect. Nat.*, 5(1): 14 (1998): P. despainii.
f. dinosauriensis *F. Hochstätter in Succulenta*, 76(6): 266 (1997)—Utah.
f. flaminggorgensis *F. Hochstätter in Succulenta*, 76(6): 269 (1997)—Utah/ Wyoming.
f. flinspachii *F. Hochstätter in Succulenta*, 76(6): 263 (1997)—Utah.
subsp. idahoensis *F. Hochstätter in Succulenta*, 76(3): 124 (1997)—Idaho, Nevada, Utah.
var. knowltonii *(L.D. Benson) J.J. Halda in Acta Mus. Richnov. Sect. Nat.*, 5(1): 13 (1998): P. knowltonii.
f. kuenzleri *F. Hochstätter in Succulenta*, 76(6): 264 (1997)—Utah.
f. montanensis *F. Hochstätter in Succulenta*, 76(6): 267 (1997)—Montana.
f. muehlii *F. Hochstätter in Succulenta*, 76(6): 264 (1997)—Utah.
var. paradinei *(B.W. Benson) J.J. Halda in Acta Mus. Richnov. Sect. Nat.*, 5(1): 13 (1998): P. paradinei.
f. prairie *F. Hochstätter, Genera Pediocactus, Navajoa, Toumeya Revised:* 92 (1995), without latin descr. or type; *cf. Repert. Pl. Succ. (I.O.S.)*, 46: 14 (1995 publ. 1996)—Wyoming.
f. sandiamontanus *F. Hochstätter in Succulenta*, 76(6): 266 (1997), as 'sandiamontana'—New Mexico.
var. winkleri *(K.D. Heil) J.J. Halda in Acta Mus. Richnov. Sect. Nat.*, 5(1): 14 (1998), without exact basionym page: P. winkleri.
smithii *(Muehlenpf.) J.J. Halda in Acta Mus. Richnov. Sect. Nat.*, 5(1): 19 (1998): Echinocactus smithii.
spinosior *(Engelm.) J.J. Halda in Acta Mus. Richnov. Sect. Nat.*, 5(1): 16 (1998): Echinocactus whipplei var. spinosior.
subsp. blainei *(S.L. Welsh & K.H. Thorne) J.J. Halda in Acta Mus. Richnov. Sect. Nat.*, 5(1): 16 (1998), without basionym ref.: Sclerocactus blainei.
var. schlesseri *(K.D. Heil & S.L. Welsh) J.J. Halda in Acta Mus. Richnov. Sect. Nat.*, 5(1): 16 (1998): Sclerocactus schlesseri.
subterraneus *(Backeb.) J.J. Halda in Acta Mus. Richnov. Sect. Nat.*, 5(1): 17 (1998)—Echinocactus subterraneus.
var. zaragosae *(Glass & R.A. Foster) J.J. Halda in Acta Mus. Richnov. Sect. Nat.*, 5(1): 18 (1998): Gymnocactus subterraneus var. zaragosae.

PEDIOCACTUS:-
swobodae *(Diers) J.J. Halda in Acta Mus. Richnov. Sect. Nat.*, 5(1): 22 (1998): Turbinicarpus swobodae.
uncinatus *(Galeotti) J.J. Halda in Acta Mus. Richnov. Sect. Nat.*, 5(1): 25 (1998): Echinocactus uncinatus.
var. crassihamatus *(F.A.C. Weber) J.J. Halda in Acta Mus. Richnov. Sect. Nat.*, 5(1): 25 (1998): Echinocactus crassihamatus.
var. wrightii *(Engelm.) J.J. Halda in Acta Mus. Richnov. Sect. Nat.*, 5(1): 25 (1998): Echinocactus uncinatus var. wrightii.
unguispinus *(Engelm.) J.J. Halda in Acta Mus. Richnov. Sect. Nat.*, 5(1): 25 (1998): Echinocactus unguispinus.
var. durangensis *(Runge) J.J. Halda in Acta Mus. Richnov. Sect. Nat.*, 5(1): 26 (1998): Echinocactus durangensis.
valdezianus *(A.F. Moeller) J.J. Halda in Acta Mus. Richnov. Sect. Nat.*, 5(1): 22 (1998): Pelecyphora valdeziana.
viereckii *(Werderm.) J.J. Halda in Acta Mus. Richnov. Sect. Nat.*, 5(1): 19 (1998): Echinocactus viereckii.
var. major *(Glass & R.A. Foster) J.J. Halda in Acta Mus. Richnov. Sect. Nat.*, 5(1): 19 (1998): Gymnocactus viereckii var. major.
warnockii *(L.D. Benson) J.J. Halda in Acta Mus. Richnov. Sect. Nat.*, 5(1): 26 (1998): Neolloydia warnockii.
wetlandicus *(Hochstätter) J.J. Halda in Acta Mus. Richnov. Sect. Nat.*, 5(1): 16 (1998): Sclerocactus wetlandicus.
f. ilseae *(Hochstätter) J.J. Halda in Acta Mus. Richnov. Sect. Nat.*, 5(1): 16 (1998): Sclerocactus wetlandicus var. ilseae.
whipplei (Engelm. & J.M. Bigelow) Arp
subsp. busekii *(Hochstätter) J.J. Halda in Acta Mus. Richnov. Sect. Nat.*, 5(1): 17 (1998): Sclerocactus whipplei subsp. busekii.
ysabelae *(Schlange) J.J. Halda in Acta Mus. Richnov. Sect. Nat.*, 5(1): 18 (1998): Thelocactus ysabelae.

PEGAEOPHYTON (Crucif.).
angustiseptatum *I.A. Al-Shehbaz, T.Y. Cheo, L.L. Lu & G. Yang in Edinburgh J. Bot.*, 57(2): 167 (2000)—China South-central (Yunnan).
nepalense *I. Al-Shehbaz, Katsutoshi Arai & H. Ohba in Novon*, 8(4): 327 (1998)—Nepal, Sikkim.
scapiflorum (Hook. f. & Thomson) C. Marquand & Airy Shaw
subsp. robustum *(O.E. Schulz) I.A. Al-Shehbaz, T.Y. Cheo, L.L. Lu & G. Yang in Edinburgh J. Bot.*, 57(2): 164 (2000): P. sinense var. robustum.
sulphureum *I.A. Al-Shehbaz in Edinburgh J. Bot.*, 57(2): 169 (2000)—Bhutan.
watsonii *I.A. Al-Shehbaz in Edinburgh J. Bot.*, 57(2): 168 (2000)—Sikkim.

PELARGONIUM (Geraniac.).
sect. Reniformia *L.L. Dreyer in S. Afr. J. Bot.*, 66(1): 44 (2000).
sect. Subsucculentia *J.J.A. van der Walt in S. Afr. J. Bot.*, 61(6): 335 (1995).
angustipetalum *E.M. Marais in S. Afr. J. Bot.*, 65(1): 52 (1999)—Cape Province.
aridicola *E.M. Marais in S. Afr. J. Bot.*, 63(2): 86 (1997)—Cape Province.
buysii *D. Hellbrügge in S. Afr. J. Bot.*, 62(3): 143 (1996)—Cape Province.
githagineum *E.M. Marais in S. Afr. J. Bot.*, 64(5): 308 (1998)—Cape Province.
glabriphyllum *E.M. Marais in S. Afr. J. Bot.*, 62(5): 228 (1996)—Cape Province.
hirtipetalum *E.M. Marais in S. Afr. J. Bot.*, 63(2): 82 (1997)—Cape Province.
isidorianum *Lémon in Ann. Fl. Pomone*, 1833–34: 281 (1834)—Locality not stated.
jarmilae *J.J. Halda in Acta Mus. Richnov. Sect. Nat.*, 5(1): 41 (1998)—Cape Province.
laevigatum (L.f.) Willd.
subsp. diversifolium *(J.C. Wendl.) C.M. Schonken in Bothalia*, 25(2): 137 (1995): P. diversifolium.
subsp. oxyphyllum *(DC.) C.M. Schonken in Bothalia*, 25(2): 138 (1995): P. oxyphyllum.

PELARGONIUM:-
parvipetalum *E.M. Marais in S. Afr. J. Bot.*, 65(1): 54 (1999)—Cape Province.
pubipetalum *E.M. Marais in S. Afr. J. Bot.*, 63(2): 84 (1997)—Cape Province.
quarciticola *U. Meve & E.M. Marais in S. Afr. J. Bot.*, 66(2): 96 (2000)—Cape Province.
redactum *P. Vorster in S. Afr. J. Bot.*, 62(1): 54 (1996)—Namibia, Cape Province.
reflexipetalum *E.M. Marais in S. Afr. J. Bot.*, 63(2): 68 (1997), nom. nov.: P. pulchellum *Salisb.*
reniforme Curtis
subsp. velutinum *(Eckl. & Zeyh.) L.L. Dreyer in S. Afr. J. Bot.*, 61(6): 328 (1995): Cortusina velutina.
rubiginosum *E.M. Marais in S. Afr. J. Bot.*, 65(1): 54 (1999)—Cape Province.
tripalmatum *E.M. Marais in S. Afr. J. Bot.*, 62(5): 225 (1996)—Cape Province.

PELEXIA (Orchidac.).
ochyrae *D.L. Szlachetko in Fragm. Flor. Geobot.*, 41(2): 858 (1996)—Panama.
palmorchidis *D.L. Szlachetko in Fragm. Flor. Geobot.*, 41(2): 859 (1996)—Colombia.

PELIOSANTHES (Convallariac.).
dehongensis *H. Li in Fl. Yunnanica*, 7: 720 (1997)—China South-central (Yunnan).
gracilipes *(Craib) N. Tanaka in Acta Phytotax. Geobot.*, 50(2): 148 (1999 publ. 2000): Ophiopogon gracilipes.
sessilis *H. Li in Fl. Yunnanica*, 7: 721 (1997)—China South-central (Yunnan).
subcoronata *N. Tanaka in Acta Phytotax. Geobot.*, 50(2): 149 (1999 publ. 2000)—Laos.

PELLIONIA (Urticac.).
menglianensis *Y.Y. Qian in Acta Phytotax. Sin.*, 37(5): 523 (1999)—China South-central (Yunnan).
tsoongii (Merr.) Merr.
subsp. subglabra *H.W. Li in Fl. Yunnanica*, 7: 242 (1997)—China South-central (Yunnan).

PELTAEA (Malvac.).
obsita *(Mart. ex Colla) A. Krapovickas & C.L. Cristóbal in Bonplandia*, 9(1–2): 148 (1996): Sida obsita.

PELTIERA *D.J. Du Puy & J.-N. Labat in Adansonia*, 19(1): 88 (1997). *LEGUMINOSAE*
alaotrensis *D.J. Du Puy & J.-N. Labat in Adansonia*, 19(1): 88 (1997)—Madagascar.
nitida *D.J. Du Puy & J.-N. Labat in Adansonia*, 19(1): 90 (1997)—Madagascar.

PELTOPHORUM (Legumin.).
dubium (Spreng.) Taub.
var. adnatum *(Griseb.) R.C. Barneby in Brittonia*, 48(2): 180 (1996): P. adnatum.
var. berteroanum *(Urb.) R.C. Barneby in Brittonia*, 48(2): 180 (1996): P. berteroanum.

PENNISETUM (Gramin.).
advena *J.K. Wipff & J.F. Veldkamp in Sida*, 18(4): 1033 (1999)—Texas.

PENSTEMON (Scrophulariac.).
rhizomatosus *N.H. Holmgren in Brittonia*, 50(2): 162 (1998)—Nevada.
tiehmii *N.H. Holmgren in Brittonia*, 50(2): 159 (1998)—Nevada.

PENTACALIA (Compos.).
arborea *(Kunth) H. Robinson & J. Cuatrecasas in Phytologia*, 40(1): 39 (1978): Cacalia arborea.†
carrikeri (Cuatrec.) Cuatrec.
var. macrophylla *S. Díaz-Piedrahita in Rev. Acad. Colomb. Cienc. Exact. Fis. Nat.*, 83: 192 (1998)—Colombia.
carupana *S. Díaz-Piedrahita in Rev. Acad. Colomb. Cienc. Exact. Fis. Nat.*, 83: 191 (1998)—Colombia.
cuatrecasana *S. Díaz-Piedrahita in Rev. Acad. Colomb. Cienc. Exact. Fis. Nat.*, 78: 435 (1996)—Colombia.
encanoana *S. Díaz-Piedrahita & G.P. Méndez in Rev. Acad. Colomb. Cienc. Exact. Fis. Nat.*, 83: 190 (1998)—Colombia.

PENTACALIA:-
fimbriifera *S. Díaz Piedrahita & G.P. Méndez Ramírez in Rev. Acad. Colomb. Cienc. Exact. Fis. Nat.*, 81: 404 (1997)—Colombia.
guambiana *S. Díaz Piedrahita in Rev. Acad. Colomb. Cienc. Exact. Fis. Nat.*, 81: 401 (1997)—Colombia.
ledifolia (Humb., Bonpl. & Kunth) Cuatrec.
var. glabrata *S. Díaz Piedrahita & G.P. Méndez in S. Díaz Piedrahita & J. Cuatrecasas, Asterac. Fl. Colombia, Senecioneae*-1: 313 (1999)—Colombia.
var. lanata *(Humb., Bonpl. & Kunth) S. Díaz Piedrahita & J. Cuatrecasas, Asterac. Fl. Colombia, Senecioneae-1*: 311 (1999): Cacalia lanata.
odorata *S. Díaz-Piedrahita & G.P. Méndez in Rev. Acad. Colomb. Cienc. Exact. Fis. Nat.*, 83: 188 (1998)—Colombia.
petiolincrassata *(Cabrera & Zardini) H. Beltran in Compositae Newsl.*, 34: 51 (1999): Senecio petiolincrassatus.
retroflexa *S. Díaz-Piedrahita in Rev. Acad. Colomb. Cienc. Exact. Fis. Nat.*, 83: 188 (1998)—Colombia.
romeroana *S. Díaz-Piedrahita & M.L. Bueno in Rev. Acad. Colomb. Cienc. Exact. Fis. Nat.*, 80: 202 (1997)—Colombia.
sinforosi *S. Díaz-Piedrahita & G.P. Méndez Ramírez in Rev. Acad. Colomb. Cienc. Exact. Fis. Nat.*, 80: 206 (1997)—Colombia.
subarachnoidea (Cuatrec.) Cuatrec.
var. pauciflora *S. Díaz-Piedrahita & M.L. Bueno in Rev. Acad. Colomb. Cienc. Exact. Fis. Nat.*, 80: 204 (1997)—Colombia.
tolimensis (Sch. Bip. ex Wedd.) Cuatrec.
var. colombiana *(Cuatrec.) S. Díaz Piedrahita in S. Díaz Piedrahita & J. Cuatrecasas, Asterac. Fl. Colombia, Senecioneae*-1: 343 (1999): Senecio colombianus.
ucumariana *S. Díaz Piedrahita & G.P. Méndez in S. Díaz Piedrahita & J. Cuatrecasas, Asterac. Fl. Colombia, Senecioneae*-1: 165 (1999)—Colombia.
ullucosana *(Hieron.) S. Díaz Piedrahita & J. Cuatrecasas, Asterac. Fl. Colombia, Senecioneae-1*: 149 (1999): Senecio ullucosanus.
vaccinioides (Humb., Bonpl. & Kunth) Cuatrec.
var. microdentata *(Cuatrec.) J. Cuatrecasas & S. Díaz Piedrahita in S. Díaz Piedrahita & J. Cuatrecasas, Asterac. Fl. Colombia, Senecioneae*-1: 266 (1999): Senecio vaccinioides var. microdentatus.

PENTACHLAENA (Sarcolaenac.).
betamponensis *P.P. Lowry, T. Haevermans, J.-N. Labat & G.E. Schatz in Adansonia*, 22(1): 16 (2000)—Madagascar.

PENTACYPHUS (Asclepiadac.).
andinus *(Ball) S. Liede in Syst. Bot.*, 21(1): 43 (1996), as 'andinum': Lugonia andina.
lehmannii *(Schltr.) S. Liede in Syst. Bot.*, 21(1): 43 (1996): Tetraphysa lehmannii.
tamanus *(Morillo) S. Liede in Taxon*, 46(2): 235 (1997): Tetraphysa tamana.

PENTADENIA (Gesneriac.).
inconspicua *(L.P. Kvist & L.E. Skog) H. Wiehler in Gesneriana*, 1(1): 122 (1995): Columnea inconspicua.
lutea *M. Freiberg in Phyton (Austria)*, 37(1): 138 (1997)—Ecuador.
ovatifolia *(L.P. Kvist & L.E. Skog) H. Wiehler in Gesneriana*, 1(1): 122 (1995): Columnea ovatifolia.

PENTAGONIA (Rubiac.).
involucrata *C.M. Taylor in Novon*, 6(2): 217 (1996)—Ecuador.

PENTANEMA (Compos.).
alanyense *H. Duman & A.A. Anderberg in Bot. J. Linn. Soc.*, 129(4): 335 (1999)—Turkey.
indicum (L.) Ling
var. sivarajanianum *R.R. Venkata Raju & C. Prabhakar Raju in Rheedea*, 6(2): 55 (1996)—India (Andhra Pradesh).

PENTAPLARIS (Tiliac.).
davidsmithii *L.J. Dorr & C. Bayer in Brittonia*, 51(2): 139 (1999)—Bolivia, Peru.
huaoranica *L.J. Dorr & C. Bayer in Brittonia*, 51(2): 143 (1999)—Ecuador.

PENTARRHINUM (Asclepiadac.).
abyssinicum Decne.
subsp. ijimense *D.J. Goyder in M. Cheek et al., Pl. Mount Oku & Ijim Ridge, Cameroon, Conserv. Checklist*: 92 (2000)—Cameroon.
balense *(Liede) S. Liede in Syst. Bot.*, 22(2): 368 (1997): Cynanchum balense.
gonoloboides *(Schltr.) S. Liede in Syst. Bot.*, 22(2): 368 (1997): Cynanchum gonoloboides.
somaliense *(N.E. Br.) S. Liede in Syst. Bot.*, 22(2): 368 (1997): Schizostephanus somaliensis.

PENTOPETIA (Asclepiadac.).
albicans *(Jum. & H. Perrier) J. Klackenberg in Candollea*, 54(2): 308 (1999): Cryptolepis albicans.
bosseri *J. Klackenberg in Candollea*, 54(2): 290 (1999)—Madagascar.
calycina *J. Klackenberg in Candollea*, 54(2): 325 (1999)—Madagascar.
dolichopodia *J. Klackenberg in Candollea*, 54(2): 329 (1999)—Madagascar.
ecoronata *J. Klackenberg in Candollea*, 54(2): 307 (1999)—Madagascar.
elastica *(Costantin & Gallaud) H.J.T. Venter in Taxon*, 46(4): 714 (1997), homonym; *see Jum. & H. Perrier in Ann. Mus. Col. Marseille*, Ser. 2, 6: 174 (1908): Kompitsia elastica.
grevei *(Baill.) H.J.T. Venter in Taxon*, 46(4): 714 (1997): Gonocrypta grevei.
intermedia *J. Klackenberg in Candollea*, 54(2): 302 (1999)—Madagascar.
longipetala *J. Klackenberg in Candollea*, 54(2): 304 (1999)—Madagascar.
lutea *J. Klackenberg & L. Civeyrel in Adansonia*, 19(2): 267 (1997)—Madagascar.
ovalifolia *(Costantin & Gallaud) J. Klackenberg in Candollea*, 54(2): 312 (1999): Fentopetiopsis ovalifolia.
subsp. glabrata *J. Klackenberg in Candollea*, 54(2): 314 (1999)—Madagascar.
urceolata *J. Klackenberg in Candollea*, 54(2): 297 (1999)—Madagascar.

PEPEROMIA (Piperac.).
alwynii *R. Callejas & J. Betancur in Novon*, 7(1): 17 (1997)—Colombia.
angularis C.DC.
var. bethaniana *(Trel. & Yunck.) J.A. Steyermark in Fl. Venezuela*, 2(2): 43 (1984): P. bethaniana.
aroensis *J.A. Steyermark in Fl. Venezuela*, 2(2): 50 (1984)—Venezuela.
atabapoensis *J.A. Steyermark in Fl. Venezuela*, 2(2): 52 (1984)—Venezuela.
beckeri *E.F. Guimarães & R.J.V. Alves in Bradea*, 8(17): 98 (1998)—Brazil Southeast (Trinidade).
berryi *J.A. Steyermark in Fl. Venezuela*, 2(2): 53 (1984)—Venezuela.
blanda (Jacq.) Humb., Bonpl. & Kunth
var. microphyllophora *(Trel. & Yunck.) J.A. Steyermark in Fl. Venezuela*, 2(2): 57 (1984): P. microphyllophora.
var. porriginifera *(Trel. & Yunck.) J.A. Steyermark in Fl. Venezuela*, 2(2) 58 (1984): P. porriginifera.
borburatensis *J.A. Steyermark in Fl. Venezuela*, 2(2): 61 (1984)—Venezuela.
carnevalii *J.A. Steyermark in Fl. Venezuela*, 2(2): 67 (1984)—Venezuela.
chapensis *J.A. Steyermark in Fl. Venezuela*, 2(2): 70 (1984)—Venezuela.
choroniana C.DC.
var. heterodoxa *J.A. Steyermark in Fl. Venezuela*, 2(2): 74 (1984)—Venezuela.
cyclophylla Miq.
f. zulianensis *J.A. Steyermark in Fl. Venezuela*, 2(2): 80 (1984)—Venezuela.
distachya (L.) A. Dietr.
f. geminispica *(Trel. & Yunck.) J.A. Steyermark in Fl. Venezuela*, 2(2): 86 (1984): P. geminispica.

PEPEROMIA:-
falconensis *J.A. Steyermark in Fl. Venezuela*, 2(2): 95 (1984)—Venezuela.
foveolata *J.A. Steyermark in Fl. Venezuela*, 2(2): 99 (1984)—Venezuela.
fundacionensis *J.A. Steyermark in Fl. Venezuela*, 2(2): 101 (1984)—Venezuela.
glabella (Sw.) A. Dietr.
var. obtusa *J.A. Steyermark in Fl. Venezuela*, 2(2): 112 (1984)—Venezuela.
gracieana *A.R.A. Görts-van Rijn in Brittonia*, 50(1): 56 (1998)—French Guiana.
hammelii *M.H. Grayum in Phytologia*, 79(2): 108 (1995 publ. 1996)—Costa Rica.
honigii *J.A. Steyermark in Fl. Venezuela*, 2(2): 125 (1984)—Venezuela.
hunteriana *P.I. Forster in Austrobaileya*, 5(3): 573 (1999)—Queensland.
japonica Makino
f. okinawensis *(T. Yamaz.) S. Hatusima in S. Hatusima & T. Amano, Fl. Ryukyus, S. of Amami Is., ed. 2*: 29 (1994), without basionym ref.: Basionym not stated.
kimnachii *W. Rauh in Cact. Succ. J. (U.S.A.)*, 72(5): 264, 260 (2000)—Bolivia.
lancifolia Hook.
subsp. erasmiiformis *(Trel.) J.A. Steyermark in Fl. Venezuela*, 2(2): 131 (1984), without basionym page: P. erasmiiformis.
liesneri *J.A. Steyermark in Fl. Venezuela*, 2(2): 131 (1984)—Venezuela.
longemucronata C.DC.
f. longirostrata *(Yunck.) J.A. Steyermark in Fl. Venezuela*, 2(2): 133 (1984): P. longirostrata.
macrostachya (Vahl) A. Dietr.
var. rupestris *(Humb., Bonpl. & Kunth) J.A. Steyermark in Fl. Venezuela*, 2(2): 140 (1984): P. rupestris.
manarae *J.A. Steyermark in Fl. Venezuela*, 2(2): 148 (1984)—Venezuela.
minarum *J.A. Steyermark in Fl. Venezuela*, 2(2): 158 (1984), homonym; *see Standl. & Steyerm. in Fieldiana, Bot.*, 24(3): 258 (1952)—Venezuela.
molleri C.DC.
subsp. ukagurensis *B. Verdcourt in Fl. Trop. E. Afr., Piperac.*: 17 (1996)—Tanzania.
mosenii *Dahlst. in Kongl. Svenska Vetenskapsakad. Handl.*, 33(2): 115 (1900), in I.K. Supp. 2 as 'mosemi' in error—Brazil Southeast.
oliveri *J. Florence & W.L. Wagner in Bull. Mus. Nation. Hist. Nat., B, Adansonia*, Ser. 4, 18(3–4): 265 (1996)—Marquesas.
ouabianae C.DC.
var. sanluisensis *J.A. Steyermark in Fl. Venezuela*, 2(2): 173 (1984)—Venezuela.
pariensis *J.A. Steyermark in Fl. Venezuela*, 2(2): 176 (1984)—Venezuela.
portuguesensis *J.A. Steyermark in Fl. Venezuela*, 2(2): 195 (1984)—Venezuela.
pseudoglabella *J.A. Steyermark in Fl. Venezuela*, 2(2): 198 (1984)—Venezuela.
pseudojamesoniana *J.A. Steyermark in Fl. Venezuela*, 2(2): 199 (1984)—Venezuela.
rotundata Humb., Bonpl. & Kunth
f. glabrilimba *(Yunck.) J.A. Steyermark in Fl. Venezuela*, 2(2): 216 (1984): P. epilobioides var. glabrilimba.
var. tequendamana *(Trel.) J.A. Steyermark in Fl. Venezuela*, 2(2): 217 (1984): P. tequendamana.
var. trinervula *(C.DC.) J.A. Steyermark in Fl. Venezuela*, 2(2): 217 (1984): P. trinervula.
saintpauliella *M.H. Grayum in Phytologia*, 79(2): 110 (1995 publ. 1996)—Costa Rica.
sui *T.T. Lin & S.Y. Lu in Taiwania*, 40(4): 353 (1995)—Taiwan.
tenella (Sw.) A. Dietr.
var. tyleri *(Trel.) J.A. Steyermark in Fl. Venezuela*, 2(2): 246 (1984): P. tyleri.
ternata C.DC.
var. abrupte-acutata *(Trel. & Yunck.) J.A. Steyermark in Fl. Venezuela*, 2(2): 249 (1984): P. abrupte-acutata.
tillettii *J.A. Steyermark in Fl. Venezuela*, 2(2): 257 (1984)—Venezuela.
tooviiana *J. Florence in Bull. Mus. Nation. Hist. Nat., B, Adansonia*, Ser. 4, 18(3–4): 267 (1996)—Marquesas.
trichomanoides *M.H. Grayum in Phytologia*, 79(2): 111 (1995 publ. 1996)—Costa Rica.

PEPEROMIA:-
trujilloi *J.A. Steyermark in Fl. Venezuela*, 2(2): 261 (1984)—Venezuela.
ursina *M.H. Grayum in Phytologia*, 79(2): 112 (1995 publ. 1996)—Costa Rica.
vestita
var. lindenii *W. Rauh in Cact. Succ. J. (U.S.A.)*, 72(5): 258 (2000)—Bolivia.
yatuensis *J.A. Steyermark in Fl. Venezuela*, 2(2): 275 (1984)—Venezuela.
yutajensis *J.A. Steyermark in Fl. Venezuela*, 2(2): 276 (1984)—Venezuela.

PEPINIA (Bromeliac.).
minicorallina *H.E. Luther in Selbyana*, 21(1–2): 130 (2000)—Peru.
sanguinea *H.E. Luther in Bromélia*, 4(3): 6 (1997)—Colombia.
tatzyanae *H.E. Luther in Bromélia*, 2(4): 38 (1995)—Peru.
verrucosa *E. Gross in Trop. Subtrop. Pflanzenwelt*, 95: 9 (1997)—Ecuador.

PEREBEA (Morac.).
mollis *(Poepp. & Endl.) J.E. Huber in Bol. Mus. Goeldi Paraense Hist. Nat. Ethnogr.*, 5: 334 (1909): Olmedia mollis.
subsp. lecithogalacta *(R.E. Schult.) C.C. Berg in Fl. Ecuador*, 60(27B): 86 (1998): Castilla ulei f. lecithogalacta.
rubra *(Trécul) C.C. Berg in Fl. Ecuador*, 60(27B): 87 (1998): Noyera rubra.
subsp. glabrifolia *(Ducke) C.C. Berg in Fl. Ecuador*, 60(27B): 87 (1998): Noyera glabrifolia.

PERESKIA (Cactac.).
grandifolia
subsp. violacea *(Leuenb.) N.P. Taylor & D.C. Zappi in Cactaceae Consensus Initiatives*, 3: 7 (1997): P. grandifolia var. violacea.
horrida
subsp. rauhii *(Backeb.) C. Ostolaza Nano in Cactaceae Consensus Initiatives*, 6: 9 (1998): P. vargasii var. rauhii.

PERIANDRA (Legumin.).
subgen. Coccinea *L.S. Funch & G.M. Barroso in Rev. Brasil. Bot.*, 22(3): 352 (1999).

PERIBOEA (Hyacinthac.).
oliveri *U. Müller-Doblies & D. Müller-Doblies in Feddes Repert.*, 108(1–2): 84 (1997)—Cape Province.
paucifolia *(W.F. Barker) U. Müller-Doblies & D. Müller-Doblies in Feddes Repert.*, 108(1–2): 84 (1997): Hyacinthus paucifolius.

PERICALYMMA (Myrtac.).
ellipticum (Endl.) Schauer
var. floridum *(Schauer) R.J. Cranfield in Nuytsia*, 13(1): 15 (1999): P. floridum.
megaphyllum *R.J. Cranfield in Nuytsia*, 13(1): 18 (1999)—Western Australia.
spongiocaule *R.J. Cranfield in Nuytsia*, 13(1): 20 (1999)—Western Australia.

PERILLA (Labiat.).
setoyensis *G. Honda in J. Jap. Bot.*, 71(1): 39 (1996)—Japan.

PERIPLOCA (Asclepiadac.).
ser. Calophyllae *H.J.T. Venter in S. Afr. J. Bot.*, 63(3): 126 (1997), contrary to Arts. 37.2, 37.4 ICBN (1994).
sect. Monocoronata *H.J.T. Venter in S. Afr. J. Bot.*, 63(3): 127 (1997).
mexicoensis *P.T. Li in J. S. China Agric. Univ.*, 14(1): 54 (1993), nom. nov.: P. viridiflora *Sessé & Moç.*
sesseiana *P.T. Li in J. S. China Agric. Univ.*, 14(1): 54 (1993), nom. nov.: P. parviflora *Sessé & Moç.*

PERISSOCARPA (Ochnac.).
ondox *B. Wallnöfer in Ann. Naturhist. Mus. Wien*, 100B: 694 (1998)—Peru.

× **PERISTERCHILUS** *hort. in Orchid Rev.*, 105(1214): 122 (1997). *ORCHIDACEAE* (Peristeranthus × Sarcochilus).

PERISTROPHE (Acanthac.).
guangxiensis *H.S. Lo & D. Fang in Guihaia,* 17(1): 44 (1997)—China Southeast (Guangxi).

PERISTYLUS (Orchidac.).
albomarginatus *(King & Pantl.) K.Y. Lang in Acta Phytotax. Sin.,* 34(6): 640 (1996): Habenaria albomarginata.
alpinus *(Hand.-Mazz.) K.Y. Lang in Acta Phytotax. Sin.,* 34(6): 639 (1996): Habenaria alpina.
humidicolus *K.Y. Lang & D.S. Deng in Novon,* 6(2): 190 (1996)—Qinghai.
timorensis *(Ridl.) J.J. Wood & P. Ormerod in Orchid Rev.,* 106(1222): 239 (1998): Habenaria timorensis.
whistleri *P.J. Cribb in P.J. Cribb & W.A. Whistler, Orchids Samoa:* 42 (1996), nom. nov.: Habenaria vaupelii *Schltr.*

PERITASSA (Celastrac.).
hatschbachii *J.A. Lombardi in Novon,* 9(2): 222 (1999)—Brazil South (Paraná, Rio Grande do Sul, Santa Catarina), Brazil Southeast (Rio de Janeiro, São Paulo).

PERITYLE (Compos.).
gilensis (M.E. Jones) J.F. Macbr.
var. longiloba *(W.E. Niles) A.M. Powell & S.C. Yarborough in Phytologia,* 82(4): 328 (1997 publ. 1998), as 'longilobus': Laphamia gilensis subsp. longiloba.
pseudociliata *A.M. Powell & S.C. Yarborough in Phytologia,* 82(4): 326 (1997 publ. 1998)—Mexico Northeast (Chihuahua).

PERONOCACTUS *A.B. Doweld in Sukkulentÿ (Moscow),* 1999(2): 20 (1999). *CACTACEAE*
sect. Eowigginsia *A.B. Doweld in Sukkulentÿ (Moscow),* 1999(2): 24 (1999).
carambeiensis *(Buining & Brederoo) A.B. Doweld in Sukkulentÿ (Moscow),* 1999(2): 20 (1999): Notocactus carambeiensis
concinnus *(Monv.) A.B. Doweld in Sukkulentÿ (Moscow),* 1999(2): 20 (1999): Echinocactus concinnus.
subsp. agnetae *(Vliet) A.B. Doweld in Sukkulentÿ (Moscow),* 1999(2): 21 (1999): Notocactus agnetae.
subsp. multicostatus *(Buining & Brederoo) A.B. Doweld in Sukkulentÿ (Moscow),* 1999(2): 21 (1999): Notocactus multicostatus.
fuscus *(F. Ritter) A.B. Doweld in Sukkulentÿ (Moscow),* 1999(2): 21 (1999): Notocactus fuscus.
horstii *(F. Ritter) A.B. Doweld in Sukkulentÿ (Moscow),* 1999(2): 21 (1999): Notocactus horstii.
linkii *(Lehm.) A.B. Doweld in Sukkulentÿ (Moscow),* 1999(2): 21 (1999): Cactus linkii.
minimus *(Frič & Kreuz.) A.B. Doweld in Sukkulentÿ (Moscow),* 1999(2): 21 (1999): Notocactus minimus.
minusculus *(Hofacker & K. Herm) A.B. Doweld in Sukkulentÿ (Moscow),* 1999(2): 21 (1999): Notocactus minusculus.
muricatus *(Otto ex Pfeiff.) A.B. Doweld in Sukkulentÿ (Moscow),* 1999(2): 21 (1999): Echinocactus muricatus.
neohorstii *(S. Theun.) A.B. Doweld in Sukkulentÿ (Moscow),* 1999(2): 21 (1999): Notocactus neohorstii.
ottonis *(Lehm.) A.B. Doweld in Sukkulentÿ (Moscow),* 1999(2): 20 (1999): Cactus ottonis.
subsp. horstii *(F. Ritter) A.B. Doweld in Sukkulentÿ (Moscow),* 1999(2): 21 (1999): Notocactus arechavaletae var. horstii.
oxycostatus *(Buining & Brederoo) A.B. Doweld in Sukkulentÿ (Moscow),* 1999(2): 21 (1999): Notocactus oxycostatus.
subsp. gracilis *(F. Ritter) A.B. Doweld in Sukkulentÿ (Moscow),* 1999(2): 21 (1999): Notocactus glaucinus var. gracilis.
rudibuenekeri *(W.R. Abraham) A.B. Doweld in Sukkulentÿ (Moscow),* 1999(2): 21 (1999): Notocactus rudibuenekeri.
subsp. glomeratus *(Menges ex N. Gerloff) A.B. Doweld in Sukkulentÿ (Moscow),* 1999(2): 21 (1999): Notocactus glomeratus.

PERONOCACTUS:-
scopa *(Spreng.) A.B. Doweld in Sukkulentÿ (Moscow),* 1999(2): 21 (1999): Cactus scopa.
subsp. marchesii *(W.R. Abraham) A.B. Doweld in Sukkulentÿ (Moscow),* 1999(2): 21 (1999): Notocactus scopa var. marchesii.
subsp. neobuenekeri *(F. Ritter) A.B. Doweld in Sukkulentÿ (Moscow),* 1999(2): 21 (1999): Notocactus neobuenekeri.
subsp. succineus *(F. Ritter) A.B. Doweld in Sukkulentÿ (Moscow),* 1999(2): 21 (1999): Notocactus succineus.
stockingeri *(Prestlé) A.B. Doweld in Sukkulentÿ (Moscow),* 1999(2): 21 (1999): Notocactus stockingeri.
tabularis *(Cels ex Rümpler) A.B. Doweld in Sukkulentÿ (Moscow),* 1999(2): 21 (1999): Echinocactus concinnus var. tabularis.
subsp. muricatus *A.B. Doweld in Sukkulentÿ (Moscow),* 1999(2): 21 (1999), nom. nov.: Echinocactus muricatus *K. Schum.*
werdermannianus *(Herter) A.B. Doweld in Sukkulentÿ (Moscow),* 1999(2): 22 (1999): Notocactus werdermannianus.

PEROTIS (Gramin.).
indica (L.) Kuntze
var. keelakaraiensis *P. Umamaheswari & P. Daniel in J. Econ. Taxon. Bot.,* 23(3): 691 (1999)—India (Tamil Nadu).

PERRIERODENDRON (Sarcolaenac.).
capuronii *J.-F. Leroy, P.P. Lowry, T. Haevermans, J.-N. Labat & G.E. Schatz in Adansonia,* 22(1): 21 (2000)—Madagascar.
occidentale *J.-F. Leroy, P.P. Lowry, T. Haevermans, J.-N. Labat & G.E. Schatz in Adansonia,* 22(1): 23 (2000), as 'occidentalis'—Madagascar.
quartzitorum *J.-F. Leroy, P.P. Lowry, T. Haevermans, J.-N. Labat & G.E. Schatz in Adansonia,* 22(1): 25 (2000)—Madagascar.
rodoense *J.-F. Leroy, P.P. Lowry, T. Haevermans, J.-N. Labat & G.E. Schatz in Adansonia,* 22(1): 28 (2000)—Madagascar.

PERROTTETIA (Celastrac.).
colorata *R. Sánchez Montaño & J.L. Fernández Alonso in An. Jard. Bot. Madrid,* 58(1): 110 (2000)—Colombia.
guacharana *R. Sánchez Montaño & J.L. Fernández Alonso in An. Jard. Bot. Madrid,* 58(1): 113 (2000)—Colombia.
simplicissima *R. Sánchez Montaño & J.L. Fernández Alonso in An. Jard. Bot. Madrid,* 58(1): 111 (2000)—Colombia.

PERRYODENDRON *T.G. Hartley in Adansonia,* 19(2): 198 (1997). *RUTACEAE*
parviflorum *(C.T. White) T.G. Hartley in Adansonia,* 19(2): 198 (1997): Melicope parviflora.

PERSICA (Rosac.).
vulgaris Mill.
subsp. atropurpurea *(C.K. Schneid.) V.A. Zaiats in Ukr. Bot. Zhurn.,* 57(1): 54 (2000): Prunus persica f. atropurpurea.
var. compressa *V.A. Zaiats in Ukr. Bot. Zhurn.,* 57(1): 54 (2000), homonym; *see Loudon, Arbor. Frutic. Brit.,* 2: 680 (1838)—Ukraine.
var. densa *V.A. Zaiats in Ukr. Bot. Zhurn.,* 57(1): 54 (2000)—Ukraine.
var. densa *(Makino) V.A. Zaiats in Ukr. Bot. Zhurn.,* 57(1): 55 (2000), with incorrect basionym ref.: Prunus persica var. densa.
subsp. erythrocarpa *V.A. Zaiats in Ukr. Bot. Zhurn.,* 57(1): 55 (2000)—Ukraine.
var. magnifica *(Bean) V.A. Zaiats in Ukr. Bot. Zhurn.,* 57(1): 55 (2000), isonym; *see Holub in Vedecke Prace Ovocnarske,* 6: 308 (1977): Prunus persica var. magnifica.
var. pendula *V.A. Zaiats in Ukr. Bot. Zhurn.,* 57(1): 54 (2000)—Ukraine.

PERSICARIA (Polygonac.).
sect. Hydropiper *M.A. Hassan in Bangladesh J. Pl. Taxon.,* 4(1): 99 (1997).
sect. Planocarpon *M.A. Hassan in Bangladesh J. Pl. Taxon.,* 4(1): 99 (1997).

PERSICARIA:-
barbata (L.) H. Hara
var. hirtella *(Meisn.) M.A. Hassan in Bangladesh J. Pl. Taxon.,* 3(1): 81 (1996): Polygonum barbatum var. hirtellum.
cliopolitana *(Hance) K. Yonekura & Hiroyoshi Ohashi in J. Jap. Bot.,* 72(3): 159 (1997): Polygonum cliopolitanum.
eciliata *M.A. Hassan in Bangladesh J. Pl. Taxon.,* 3(2): 87 (1996)—Bangladesh.
glabra *(Willd.) A.E. Borodina-Grabovskaya in Konsp. Sosud. Rast. Fl. V'etnama,* 2: 148 (1996), isonym; *see M. Gómez in Anales Inst. Segunda Enseñ.,* 2: 278 (1896): Polygonum glabrum.
glomerata *(Dammer) S. Ortiz & J.A.R. Paiva in Bot. J. Linn. Soc.,* 131(2): 171 (1999): Polygonum glomeratum.
jucunda *(Meisn.) K. Yonekura & Hiroyoshi Ohashi in J. Jap. Bot.,* 72(3): 159 (1997), isonym; *see Migo in J. Shanghai Sci. Inst.,* Sect. 3, 4: 142 (1939): Polygonum jucundum.
lanata *(Roxb.) M.A. Hassan in Bangladesh J. Pl. Taxon.,* 3(1): 81 (1996), isonym; *see Tzvelev in Novosti Sist. Vÿssh. Rast.,* 25: 81 (1988): Polygonum lanatum.
lapathifolia (L.) Gray
f. alba *Y.N. Lee, Fl. Korea:* 1156 (1996), without type—Locality not stated.
subsp. leptoclada *(Danser) R. Wisskirchen in R. Wisskirchen & H. Haeupler, Standardliste Farn- und Blütenpfl.* Deutschl.: 357 (1998): Polygonum lapathifolium subsp. leptocladum.
subsp. linicola *(Sutulov) N.N. Tzvelev in Novosti Sist. Vÿssh. Rast.,* 32: 184 (2000): Polygonum linicola.
limicola *(Sam.) K. Yonekura & Hiroyoshi Ohashi in J. Jap. Bot.,* 72(3): 159 (1997): Polygonum limicola.
longiflora *(Courchet) A.E. Borodina-Grabovskaya in Konsp. Sosud. Rast. Fl. V'etnama,* 2: 149 (1996): Polygonum longiflorum.
maculosa Gray
subsp. hirticaulis *(Danser) S. Ekman & Knutsson in Nordic J. Bot.,* 16(1): 5 (1996): Polygonum persicaria subsp. hirticaule.
madagascariensis *(Meisn.) S. Ortiz & J.A.R. Paiva in Bot. J. Linn. Soc.,* 131(2): 172 (1999): Polygonum poiretii var. madagascariense.
manshuricola *M. Kitagawa in Rep. Inst. Sci. Res. Manchoukuo,* 6(4): 121 (1942)—Manchuria (Jilin).
nogueirae *S. Ortiz & J.A.R. Paiva in Bot. J. Linn. Soc.,* 131(2): 168 (1999)—Zambia.
palmata *(Dunn) K. Yonekura & Hiroyoshi Ohashi in J. Jap. Bot.,* 72(3): 159 (1997): Polygonum palmatum.
posumbu (Buch.-Ham. ex D. Don) H. Gross
var. stenophylla *(Makino) K. Yonekura & Hiroyoshi Ohashi in J. Jap. Bot.,* 72(3): 159 (1997): Polygonum yokusaianum var. stenophyllum.
× pseudolapathum *(Schur) D.H. Kent, List Vasc. Pl. Brit. Isles,* Supp. 1: 36 (1996 publ. 19 Jan. 1997): Polygonum × pseudolapathum.
senticosa (Meisn.) H. Gross
var. sagittifolia *(H. Lév. & Vaniot) K. Yonekura & Hiroyoshi Ohashi in J. Jap. Bot.,* 72(3): 160 (1997): Polygonum sagittifolium.
serrulata *(Lag.) M.A. Hassan in Bangladesh J. Pl. Taxon.,* 4(1): 99 (1997), isonym; *see Webb & Moq. in Webb & Berthel., Hist. Nat. Iles Canaries,* 3: 219 (1846): Polygonum serrulatum.
sinuata *(Royle ex Bab.) K.R. Rajbhandari & R. Joshi in J. Jap. Bot.,* 70(6): 332 (1995): Polygonum sinuatum.
stagnina *(Buch.-Ham. ex Meisn.) M.A. Hassan in Bangladesh J. Pl. Taxon.,* 3(1): 82 (1996): Polygonum stagninum.

PERSOONIA (Proteac.).
hindii *P.H. Weston & L.A.S. Johnson in Telopea,* 7(3): 199 (1997)—New South Wales.
muelleri (P. Parm.) Orchard
subsp. angustifolia *(Benth.) L.A.S. Johnson & P.H. Weston in Fl. Australia,* 16: 472 (1995): P. gunnii var. angustifolia.
subsp. densifolia *(Orchard) L.A.S. Johnson & P.H. Weston in Fl. Australia,* 16: 472 (1995): P. muelleri var. densifolia.

PERSOONIA:-
pauciflora *P.H. Weston in Telopea,* 8(2): 159 (1999)—New South Wales.

PERVILLEA (Asclepiadac.).
phillipsonii *J. Klackenberg in Nordic J. Bot.,* 16(2): 181 (1996)—Madagascar.
venenata (Baill.) Klack.
subsp. macrantha *J. Klackenberg in Nordic J. Bot.,* 16(2): 180 (1996)—Madagascar.

PERYMENIUM (Compos.).
basaseachicanum *B.L. Turner in Phytologia,* 81(2): 104 (1996 publ. 1997)—Mexico Northeast (Chihuahua).
hintoniorum B.L. Turner
var. gypsophilum *B.L. Turner in Phytologia,* 79(5): 343 (1995 publ. 1996)—Mexico Northeast (Nuevo León).

PESCATORIA (Orchidac.).
aprilensis *Y. Norich' & A. Tangor in Orchidées Cult. Protect.,* 34: 12 (1998), as 'Piscatoria', contrary to Arts. 37.3, 37.5 ICBN (1994)—Mexico.

PETALONYX (Loasac.).
thurberi A. Gray
var. gilmanii *(Munz) L.D. Benson in L.D. Benson & R.A. Darrow, Trees & Shrubs of Southwest. Deserts (ed. 3):* 284 (1981): P. gilmanii.

PETASITES (Compos.).
frigidus (L.) Fr.
var. sagittatus *(Banks ex Pursh) D.M. Cherniawsky in Canad. J. Bot.,* 76(12): 2070 (1998 publ. 1999): Tussilago sagittata.
nothovar. vitifolius *(Greene) D.M. Cherniawsky in Canad. J. Bot.,* 76(12): 2072 (1998 publ. 1999): P. vitifolius.

PETCHIA (Apocynac.).
africana *A.J.M. Leeuwenberg in Wageningen Agric. Univ. Pap.,* 97(2): 56 (1997)—Cameroon.
cryptophlebia *(Baker) A.J.M. Leeuwenberg in Wageningen Agric. Univ. Pap.,* 97(2): 60 (1997): Carissa cryptophlebia.
erythrocarpa *(Vatke) A.J.M. Leeuwenberg in Wageningen Agric. Univ. Pap.,* 97(2): 64 (1997): Ellertonia madagascariensis.
humbertii *(Markgr.) A.J.M. Leeuwenberg in Wageningen Agric. Univ. Pap.,* 97(2): 69 (1997): Cabucala humbertii.
madagascariensis *(A.DC.) A.J.M. Leeuwenberg in Wageningen Agric. Univ. Pap.,* 97(2): 71 (1997): Alyxia madagascariensis.
montana *(Pichon) A.J.M. Leeuwenberg in Wageningen Agric. Univ. Pap.,* 97(2): 78 (1997): Cabucala montana.
plectaneiifolia *(Pichon) A.J.M. Leeuwenberg in Wageningen Agric. Univ. Pap.,* 97(2): 79 (1997): Cabucala plectaneiifolia.

PETRAEOMYRTUS *L.A. Craven in Austral. Syst. Bot.,* 12(5): 678 (1999). *MYRTACEAE*
punicea *(Byrnes) L.A. Craven in Austral. Syst. Bot.,* 12(5): 678 (1999): Melaleuca punicea.

PETROPHILE (Proteac.).
aculeata *D.B. Foreman in Fl. Australia,* 16: 474 (1995)—Western Australia.
arcuata *D.B. Foreman in Fl. Australia,* 16: 474 (1995)—Western Australia.
cyathiforma *D.B. Foreman in Fl. Australia,* 16: 474 (1995)—Western Australia.
ericifolia R. Br.
subsp. subpubescens *(Domin) D.B. Foreman in Fl. Australia,* 16: 475 (1995): P. ericifolia f. subpubescens.
glauca *D.B. Foreman in Fl. Australia,* 16: 475 (1995)—Western Australia.
imbricata *D.B. Foreman in Fl. Australia,* 16: 475 (1995)—Western Australia.
merrallii *D.B. Foreman in Fl. Australia,* 16: 476 (1995)—Western Australia.
misturata *D.B. Foreman in Fl. Australia,* 16: 476 (1995)—Western Australia.
pauciflora *D.B. Foreman in Fl. Australia,* 16: 476 (1995)—Western Australia.

PETROPHILE:-
recurva *D.B. Foreman in Fl. Australia,* 16: 477 (1995)—Western Australia.
trifurcata *D.B. Foreman in Fl. Australia,* 16: 479 (1995)—Western Australia.
wonganensis *D.B. Foreman in Fl. Australia,* 16: 479 (1995)—Western Australia.

PETROSIMONIA (Chenopodiac.).
sect. Triandra *N.N. Tzvelev in Fl. Vostochnoĭ Evropy̆,* 9: 77 (1996).

PETUNIA (Solanac.).
axillaris (Lam.) Britton, Sterns & Poggenb.
subsp. subandina *T. Ando in Acta Phytotax. Geobot.,* 47(1): 21 (1996)—Argentina Northwest (Jujuy, Catamarca, La Rioja, Mendoza, Salta, Santiago del Estero, San Juan, San Luis, Tucumán), Argentina Northeast (Córdoba, La Pampa), Bolivia.
bajeensis *T. Ando & G. Hashimoto in Brittonia,* 50(4): 483 (1998)—Brazil South (Rio Grande do Sul).
interior *T. Ando & G. Hashimoto in Brittonia,* 48(2): 217 (1996)—Brazil South (Santa Catarina, Rio Grande do Sul).
phoenicea *H. Jacquin in Ann. Fl. Pomone,* 1832–33: 380 (1833)—Locality not stated.
riograndensis *T. Ando & G. Hashimoto in Brittonia,* 50(4): 485 (1998)—Brazil South (Rio Grande do Sul).

PEUCEDANUM (Umbellif.).
japonicum Thunb.
var. australe *M. Hotta & A. Seo in Acta Phytotax. Geobot.,* 51(1): 113 (2000)—Nansei-Shoto, Japan, Taiwan, Philippines.
var. latifolium *M. Hotta & T. Shiuchi in J. Jap. Bot.,* 71(4): 183 (1996)—Nansei-Shoto.
paucipartitum *M.L. Sheh in Fl. Yunnanica,* 7: 617 (1997), nom. nov.: P. heterophyllum *Franch.*
polyactinum *B.L. Burtt in Bothalia,* 27(1): 52 (1997), nom. nov.: Oreoselinum capense *Eckl. & Zeyh.*
terebinthaceum (Fisch. ex Trevir.) Turcz.
f. aciculare *(T. Inoue) H. Ohba in Fl. Japan,* 2c: 301 (1999): P. terebinthaceum var. aciculare.
subsp. deltoideum *(Makino ex Y. Yabe) V.N. Voroschilov in A.K. Skvortsov (ed.), Florist. issl. v razn. rinakh SSSR:* 184 (1985): P. deltoideum.
townsendii *A. Charpin & J. Fernández Casas in An. Jard. Bot. Madrid,* 54(1): 276 (1996)—Gulf of Guinea Is. (Bioko), Cameroon.

PFAFFIA (Amaranthac.).
argyrea *T.M. Pedersen in Bonplandia,* 10(1–4): 108 (2000)—Brazil Southeast (Minas Gerais).
fruticulosa Suess.
var. diffusa *T.M. Pedersen in Bonplandia,* 10(1–4): 111 (2000)—Paraguay, Bolivia.
minarum *T.M. Pedersen in Adansonia,* 19(2): 237 (1997)—Brazil Southeast (Minas Gerais).
ninae *T.M. Pedersen in Adansonia,* 19(2): 237 (1997)—Uruguay.
rotundifolia *T.M. Pedersen in Adansonia,* 19(2): 239 (1997)—Bolivia.
sarcophylla *T.M. Pedersen in Adansonia,* 19(2): 239 (1997)—Brazil West-central (Goiás).
tuberculosa *T.M. Pedersen in Adansonia,* 19(2): 241 (1997)—Brazil Northeast (Bahia).
tuberosa (Spreng.) Hicken
subsp. goiana *T.M. Pedersen in Adansonia,* 19(2): 243 (1997)—Brazil West-central (Brasília Distrito Federal).
f. rubella *T.M. Pedersen in Adansonia,* 19(2): 243 (1997)—Argentina Northeast (Corrientes), Uruguay.

PFITZERIA *K. Senghas in J. Orchideenfreund,* 5(1): 30 (1998). *ORCHIDACEAE*
schaeferi *K. Senghas in J. Orchideenfreund,* 5(1): 30 (1998)—Peru.

PFOSSERIA *F. Speta in Phyton (Austria),* 38(1): 113 (1998). *HYACINTHACEAE*
bithynica *(Boiss.) F. Speta in Phyton (Austria),* 38(1): 113 (1998): Scilla bithynica.

PHACELURUS (Gramin.).
latifolius (Steud.) Ohwi
var. trichophyllus *(S.L. Zhong) B.S. Sun & Z.H. Hu in Fl. Reipubl. Popul. Sin.,* 10(2): 261 (1997): P. trichophyllus.

PHAEONYCHIUM (Crucif.).
fengii *I.A. Al-Shehbaz in Novon,* 10(4): 335 (2000)—China South-Central (Yunnan).
jafrii *I.A. Al-Shehbaz in Nordic J. Bot.,* 20(2): 160 (2000)—Xizang, Bhutan, Nepal.
kashgaricum *(Botsch.) I.A. Al-Shehbaz in Nordic J. Bot.,* 20(2): 162 (2000): Vvedenskyella kashgarica.
villosum *(Maxim.) I.A. Al-Shehbaz in Nordic J. Bot.,* 20(2): 161 (2000): Parrya villosa.

PHAGNALON (Compos.).
helichrysoides (Ball) Coss. ex Maire
subsp. nitidum *(Emb.) M. Ibn Tattou in Bocconea,* 8: 46 (1998): P. helichrysoides var. nitidum.
sordidum (L.) Rchb.
var. corsicum *(Rouy) J. Gamisans in J. Gamisans & D. Jeanmonod, Asteraceae I: Compl. Prodr. Fl. Corse (ser.* eds. Jeanm. & Burdet): 158 (1998): P. sordidum subsp. corsicum.
stenolepis Chiov.
subsp. abdulazizianum *S.A. Chaudhary in Fl. Kingdom Saudi Arabia,* 2(3): 143, 418, 427 (2000), rank not clearly indicated, see Art. 35.1 ICBN (2000)—Saudi Arabia.

PHAIUS (Orchidac.).
subgen. Limatodes *(Blume) K. Senghas in Schlechter Orchideen,* I(15): 881 (1984), as 'Phajus': gen. Limatodes.
mishmensis (Lindl. & Paxton) Rchb. f.
var. tonkinensis *L.V. Averyanov in Bot. Zhurn.,* 82(3): 133 (1997)—Vietnam.

PHALAENOPSIS (Orchidac.).
chibae *T. Yukawa in Ann. Tsukuba Bot. Gard.,* 15: 19 (1996)—Vietnam.
chuxiongensis *F.Y. Liu in Acta Bot. Yunnanica,* 18(4): 411 (1996)—China South-central (Yunnan).

PHALARIS (Gramin.).
coerulescens Desf.
subsp. lusitanica *M.L. Rocha Afonso & J. do Amaral Franco in Silva Lusitana,* 5(1): 142 (1997)—Portugal.
occidentalis *T. Nuttall in Trans. Amer. Philos. Soc., n.s.,* 5: 144 (1835)—Arkansas.

PHARNACEUM (Molluginac.).
lanuginosum *J.C. Manning & P. Goldblatt in Strelitzia,* 9: 715 (2000), nom. nov.: P. microphyllum var. albens *Adamson.*

PHASEOLUS (Legumin.).
albescens *McVaugh ex R. Ramírez-Delgadillo & A. Delgado-Salinas in Sida,* 18(3): 638 (1999)—Mexico Southwest (Jalisco, Colima, Guerrero, Michoacan).
costaricensis *G.F. Freytag & D.G. Debouck in Novon,* 6(2): 157 (1996)—Costa Rica, Panamá.
hintonii *A. Delgado-Salinas in Syst. Bot.,* 25(3): 431 (2000)—Mexico Central (México State), Mexico Southwest (Guerrero, Jalisco, Michoacan, Nayarit, Oaxaca).
juquilensis *A. Delgado-Salinas in Syst. Bot.,* 25(3): 421 (2000)—Mexico Southwest (Oaxaca).
macvaughii *A. Delgado-Salinas in Syst. Bot.,* 25(3): 414 (2000)—Mexico Southwest (Jalisco, Guerrero, Michoacan), Mexico Northwest (Sinaloa).
marechalii *A. Delgado-Salinas in Syst. Bot.,* 25(3): 427 (2000)—Mexico Central (Morelos).
pedicellatus Benth.
var. grayanus *(Wooton & Standl.) A. Delgado ex D. Isely, Native & Natural. Legumin. (Fabac.) U.S. (excl. Alaska & Hawaii):* 798 (1998): P. grayanus.
perplexus *A. Delgado-Salinas in Syst. Bot.,* 25(3): 418 (2000)—Mexico Southwest (Jalisco, Guerrero, Michoacan, Nayarit, Oaxaca), Mexico Northwest (Sinaloa), Mexico Northeast (Durango), Mexico Central (México State).

PHASEOLUS:-
xolocotzii *A. Delgado-Salinas in Syst. Bot.*, 25(3): 429 (2000)—Mexico Southwest (Guerrero, Jalisco), Mexico Central (México State, Morelos).
zimapanensis *A. Delgado-Salinas in Syst. Bot.*, 25(3): 432 (2000)—Mexico Northeast (Hidalgo, Aguascalientes, Coahuila, Durango, San Luis Potosí, Tamaulipas).

PHAULOPSIS (Acanthac.).
aequivoca *M. Manktelow in Symb. Bot. Upsal.*, 31(2): 107 (1996)—Zambia, Zaire.
barteri T. Anderson
var. pauciglandula *M. Manktelow in Symb. Bot. Upsal.*, 31(2): 97 (1996)—Ghana, Senegal, Mali, Guinea, Sierra Leone, Ivory Coast, Nigeria, Cameroon, Sudan.
gediensis *M. Manktelow in Symb. Bot. Upsal.*, 31(2): 152 (1996)—Kenya, Tanzania.
grandiflora *M. Manktelow in Symb. Bot. Upsal.*, 31(2): 84 (1996)—Zaire.
imbricata (Forssk.) Sweet
subsp. madagascariensis *M. Manktelow in Symb. Bot. Upsal.*, 31(2): 134 (1996)—Madagascar, Comoros (Comoros, Mayotte).
subsp. pallidifolia *M. Manktelow in Symb. Bot. Upsal.*, 31(2): 144 (1996)—Zambia, Zaire.
subsp. poggei *(Lindau) M. Manktelow in Symb. Bot. Upsal.*, 31(2): 138 (1996): Micranthus poggei.
latiloba *M. Manktelow in Symb. Bot. Upsal.*, 31(2): 89 (1996)—Zaire.
marcelinoi *M. Manktelow in Symb. Bot. Upsal.*, 31(2): 82 (1996)—Angola.
pulchella *M. Manktelow in Symb. Bot. Upsal.*, 31(2): 150 (1996)—Tanzania.
savannicola *M. Manktelow in Symb. Bot. Upsal.*, 31(2): 86 (1996)—Central African Republic, Mali, Nigeria, Cameroon, Chad, Sudan, Zaire.
symmetrica *M. Manktelow in Symb. Bot. Upsal.*, 31(2): 121 (1996)—Cameroon, Nigeria.
verticillaris *(Nees) M. Manktelow in Symb. Bot. Upsal.*, 31(2): 154 (1996): Aetheilema imbricatum var. verticillare.

PHEBALIUM (Rutac.).
appressum *Paul G. Wilson in Nuytsia*, 12(2): 280 (1998)—Western Australia.
brevifolium *Paul G. Wilson in Nuytsia*, 12(2): 281 (1998), nom. nov.: P. tuberculosum subsp. brachyphyllum *Paul G. Wilson*.
elegans *Paul G. Wilson in Nuytsia*, 12(2): 282 (1998)—Western Australia.
festivum *Paul G. Wilson in Nuytsia*, 12(2): 283 (1998)—Victoria.
glandulosum Hook.
subsp. nitidum *Paul G. Wilson in Nuytsia*, 12(2): 284 (1998)—New South Wales.
laevigatum *Paul G. Wilson in Nuytsia*, 12(2): 284 (1998)—Western Australia.
megaphyllum *(Ewart) Paul G. Wilson in Nuytsia*, 12(2): 285 (1998): Eriostemon tuberculosus var. megaphyllus.
obovatum *(Paul G. Wilson) Paul G. Wilson in Nuytsia*, 12(2): 285 (1998): P. lepidotum var. obovatum.

PHEDIMUS (Crassulac.).
subgen. Aizoon *(L.K.A. Koch ex Schönland) H. Ohba & N.J. Turland in Novon*, 10(4): 401 (2000): Sedum sect. Aizoon.
subgen. Aizoon *(W.D.J. Koch) H. 't Hart in H. 't Hart & U. Eggli (eds.), Evol. & Syst. Crassulac.*: 168 (1995): Sedum sect. Aizoon.
sect. Spathulata *(Boriss.) H. Ohba in H. 't Hart & U. Eggli (eds.), Evol. & Syst. Crassulac.*: 156 (1995): Sedum subsect. Spathulata.
aizoon *(L.) H. 't Hart in H. 't Hart & U. Eggli (eds.), Evol. & Syst. Crassulac.*: 168 (1995): Sedum aizoon.
var. latifolius *(Maxim.) H. Ohba, K.T. Fu & B.M. Bartholomew in Novon*, 10(4): 401 (2000): Sedum aizoon var. latifolium.
var. scabrus *(Maxim.) H. Ohba, K.T. Fu & B.M. Bartholomew in Novon*, 10(4): 401 (2000): Sedum aizoon var. scabrum.
var. yamatutae *(Kitag.) H. Ohba, K.T. Fu & B.M. Bartholomew in Novon*, 10(4): 401 (2000): Sedum aizoon var. yamatutae.

PHEDIMUS:-
ellacombianus *(Praeger) H. 't Hart in H. 't Hart & U. Eggli (eds.), Evol. & Syst. Crassulac.*: 168 (1995): Sedum ellacombianum.
floriferus *(Praeger) H. 't Hart in H. 't Hart & U. Eggli (eds.), Evol. & Syst. Crassulac.*: 168 (1995): Sedum floriferum.
hsinganicus *(Y.C. Chu ex S.H. Fu & Y.H. Huang) H. Ohba, K.T. Fu & B.M. Bartholomew in Novon*, 10(4): 402 (2000): Sedum hsinganicum.
hybridus *(L.) H. 't Hart in H. 't Hart & U. Eggli (eds.), Evol. & Syst. Crassulac.*: 168 (1995): Sedum hybridum.
kamtschaticus *(Fisch. & C.A. Mey.) H. 't Hart in H. 't Hart & U. Eggli (eds.), Evol. & Syst. Crassulac.*: 168 (1995): Sedum kamtschaticum.
litoralis *(Kom.) H. 't Hart in H. 't Hart & U. Eggli (eds.), Evol. & Syst. Crassulac.*: 168 (1995): Sedum litorale.
maximowiczii *(Regel) H. 't Hart in H. 't Hart & U. Eggli (eds.), Evol. & Syst. Crassulac.*: 168 (1995): Sedum maximowiczii.
middendorfianus *(Maxim.) H. 't Hart in H. 't Hart & U. Eggli (eds.), Evol. & Syst. Crassulac.*: 169 (1995): Sedum middendorfianum.
obtusifolius *(C.A. Mey.) H. 't Hart in H. 't Hart & U. Eggli (eds.), Evol. & Syst. Crassulac.*: 168 (1995): Sedum obtusifolium.
odontophyllus *(Fröd.) H. 't Hart in H. 't Hart & U. Eggli (eds.), Evol. & Syst. Crassulac.*: 169 (1995): Sedum odontophyllum.
selskanianus *(Regel & Maack) H. 't Hart in H. 't Hart & U. Eggli (eds.), Evol. & Syst. Crassulac.*: 169 (1995): Sedum selskanianum.
sichotensis *(Vorosch.) H. 't Hart in H. 't Hart & U. Eggli (eds.), Evol. & Syst. Crassulac.*: 169 (1995): Sedum sichotense.
sikokianus *(Maxim.) H. 't Hart in H. 't Hart & U. Eggli (eds.), Evol. & Syst. Crassulac.*: 169 (1995): Sedum sikokianum.
spurius *(M. Bieb.) H. 't Hart in H. 't Hart & U. Eggli (eds.), Evol. & Syst. Crassulac.*: 168 (1995): Sedum spurium.
stevenianus *(Rouy & E.G. Camus) H. 't Hart in H. 't Hart & U. Eggli (eds.), Evol. & Syst. Crassulac.*: 168 (1995): Sedum stevenianum.
stoloniferus *(S.G. Gmel.) H. 't Hart in H. 't Hart & U. Eggli (eds.), Evol. & Syst. Crassulac.*: 168 (1995): Sedum stoloniferum.
takesimensis *(Nakai) H. 't Hart in H. 't Hart & U. Eggli (eds.), Evol. & Syst. Crassulac.*: 169 (1995): Sedum takesimense.
zokuriensis *(Nakai) H. 't Hart in H. 't Hart & U. Eggli (eds.), Evol. & Syst. Crassulac.*: 169 (1995): Sedum zokuriense.

PHELIPANCHE (Scrophulariac.).
ser. Arenariae *E.S. Teryokhin in Opred. Zarazikhovÿkh Fl. SSSR (ed. L.Yu. Budantsev)*: 43 (1993).
sect. Arenariae *E.S. Teryokhin in Opred. Zarazikhovÿkh Fl. SSSR (ed. L.Yu. Budantsev)*: 43 (1993).
ser. Borissovae *E.S. Teryokhin in Opred. Zarazikhovÿkh Fl. SSSR (ed. L.Yu. Budantsev)*: 42 (1993).
ser. Cilicicae *E.S. Teryokhin in Opred. Zarazikhovÿkh Fl. SSSR (ed. L.Yu. Budantsev)*: 42 (1993).
sect. Ibericae *E.S. Teryokhin in Opred. Zarazikhovÿkh Fl. SSSR (ed. L.Yu. Budantsev)*: 43 (1993).
sect. Nikitae *E.S. Teryokhin in Opred. Zarazikhovÿkh Fl. SSSR (ed. L.Yu. Budantsev)*: 43 (1993).
ser. Purpureae *E.S. Teryokhin in Opred. Zarazikhovÿkh Fl. SSSR (ed. L.Yu. Budantsev)*: 43 (1993).
purpurea (Jacq.) Soják
subsp. bohemica *(Čelak.) J. Zázvorka in Preslia*, 72(1): 88 (2000): Orobanche bohemica.

PHEMERANTHUS (Portulacac.).
brevifolius *(Torr.) M.A. Hershkovitz in Taxon*, 46(2): 222 (1997): Talinum brevifolium.
confertiflorus *(Greene) M.A. Hershkovitz in Taxon*, 46(2): 222 (1997): Talinum confertiflorum.
spinescens *(Torr.) M.A. Hershkovitz in Taxon*, 46(2): 222 (1997): Talinum spinescens.

PHIALANTHUS (Rubiac.).
peduncularis *A. Borhidi in Acta Bot. Hung.*, 38(1–4): 167 (1993–94 publ. 1995)—Cuba.
rigidus Griseb.
subsp. bissei *A. Borhidi in Acta Bot. Hung.*, 38(1–4): 168 (1993–94 publ. 1995)—Cuba.

PHILADELPHUS (Hydrangeac.).
kunmingensis *S.M. Hwang in J. Trop. Subtrop. Bot.*, 8(3): 263 (2000), with type—China South-central (Yunnan).
var. parvifolius *S.M. Hwang in J. Trop. Subtrop. Bot.*, 8(3): 263 (2000), with type—China South-central (Yunnan).
seoulensis *Y.H. Chung & H. Shin in Korean J. Pl. Taxon.*, 21(3): 212 (1991)—South Korea.

PHILCOXIA *P. Taylor & V.C. Souza in Kew Bull.*, 55(1): 159 (2000). *SCROPHULARIACEAE*
bahiensis *V.C. Souza & R.M. Harley in Kew Bull.*, 55(1): 160 (2000)—Brazil Northeast (Bahia).
goiasensis *P. Taylor in Kew Bull.*, 55(1): 160 (2000)—Brazil West-central (Goiás).
minensis *V.C. Souza & A.M. Giulietti in Kew Bull.*, 55(1): 161 (2000)—Brazil Southeast (Minas Gerais).

PHILENOPTERA (Legumin.).
bussei *(Harms) B.D. Schrire in Kew Bull.*, 55(1): 87 (2000): Lonchocarpus bussei.
eriocalyx *(Harms) B.D. Schrire in Kew Bull.*, 55(1): 86 (2000): Lonchocarpus eriocalyx.
kanurii *(Brenan & J.B. Gillett) B.D. Schrire in Kew Bull.*, 55(1): 87 (2000): Lonchocarpus kanurii.
katangensis *(De Wild.) B.D. Schrire in Kew Bull.*, 55(1): 90 (2000): Lonchocarpus katangensis.
madagascariensis *(Vatke) B.D. Schrire in Kew Bull.*, 55(1): 93 (2000): Millettia madagascariensis.
nelsii *(Schinz) B.D. Schrire in Kew Bull.*, 55(1): 91 (2000): Dalbergia nelsii.
pallescens *(Welw. ex Baker) B.D. Schrire in Kew Bull.*, 55(1): 92 (2000): Lonchocarpus pallescens.
sutherlandii *(Harv.) B.D. Schrire in Kew Bull.*, 55(1): 92 (2000): Millettia sutherlandii.
violacea *(Klotzsch) B.D. Schrire in Kew Bull.*, 55(1): 89 (2000): Capassa violacea.
wankieensis *(Mendonça & E.P. Sousa) J.M. Lock in Kew Bull.*, 55(1): 95 (2000): Lonchocarpus eriocalyx subsp. wankieensis.

PHILIBERTIA (Asclepiadac.).
fiebrigii *(Schltr.) S. Liede in Novon*, 7(1): 44 (1997): Stelmatodocon fiebrigii.
vaileae *(Rusby) S. Liede in Syst. Bot.*, 21(1): 43 (1996): Oxystelma vaileae.

PHILLYREA (Oleac.).
latifolia L.
subsp. daninii *B. Valdés in Bocconea*, 3: 161 (1992)—Palestine (Israel).
subsp. rodriguezii *(P. Monts.) A.M. Romo, Fl. Silvestres Baleares*: 235 (1994): P. media var. rodriguezii.

PHILODENDRON (Arac.).
ser. Albisuccosa *T.B. Croat in Ann. Missouri Bot. Gard.*, 84(3): 331 (1997).
ser. Ecordata *T.B. Croat in Ann. Missouri Bot. Gard.*, 84(3): 333 (1997).
ser. Fibrosa *T.B. Croat in Ann. Missouri Bot. Gard.*, 84(3): 331 (1997).
sect. Fruticosa *M.H. Grayum, Revis. Philodendron subgen. Pteromischum Pacific & Caribbean Trop. Amer.* (Syst. Bot. Monog., 47): 117 (1996).
ser. Glossophyllum *(Schott) T.B. Croat in Ann. Missouri Bot. Gard.*, 84(3): 334 (1997): P. grex Glossophyllum.
subsect. Glossophyllum *(Schott) T.B. Croat in Ann. Missouri Bot. Gard.*, 84(3): 334 (1997): P. grex Glossophyllum.
ser. Impolita *T.B. Croat in Ann. Missouri Bot. Gard.*, 84(3): 330 (1997).
*ser.*Macrobelium *(Schott) T.B. Croat in Ann. Missouri Bot. Gard.*, 84(3): 333 (1997): P. grex Macrobelium.
ser. Ovata *T.B. Croat in Ann. Missouri Bot. Gard.*, 84(3): 335 (1997).

PHILODENDRON:-
ser. Pachycaulia *T.B. Croat in Ann. Missouri Bot. Gard.*, 84(3): 333 (1997).
ser. Reticulata *T.B. Croat in Ann. Missouri Bot. Gard.*, 84(3): 333 (1997).
ser. Velvetina *T.B. Croat in Ann. Missouri Bot. Gard.*, 84(3): 330 (1997).
albisuccus *T.B. Croat in Ann. Missouri Bot. Gard.*, 84(3): 393 (1997)—Panama.
alticola *T.B. Croat & M.H. Grayum in Ann. Missouri Bot. Gard.*, 84(3): 394 (1997)—Panama, Costa Rica.
altomacaense *M.A. Nadruz Coelho & S.J. Mayo in Bol. Bot. (São Paulo)*, 17: 48 (1998)—Brazil Southeast (Rio de Janeiro).
angustilobum *T.B. Croat & M.H. Grayum in Ann. Missouri Bot. Gard.*, 84(3): 396 (1997)—Costa Rica, Honduras, Nicaragua, Panama.
annulatum *T.B. Croat in Ann. Missouri Bot. Gard.*, 84(3): 400 (1997)—Panama.
antonioanum *T.B. Croat in Ann. Missouri Bot. Gard.*, 84(3): 402 (1997)—Panama.
appendiculatum *M.A. Nadruz Coelho & S.J. Mayo in Bol. Bot. (São Paulo)*, 17: 50 (1998)—Brazil Southeast (Rio de Janeiro).
aromaticum *T.B. Croat & M.H. Grayum in Ann. Missouri Bot. Gard.*, 84(3): 403 (1997)—Costa Rica.
aurantiifolium Schott
*subsp.*calderense *(K. Krause) M.H. Grayum, Revis. Philodendron subgen. Pteromischum Pacific & Caribbean Trop. Amer.* (Syst. Bot. Monog., 47): 199 (1996): P. calderense.
bakeri *T.B. Croat & M.H. Grayum in Ann. Missouri Bot. Gard.*, 84(3): 406 (1997)—Costa Rica, Panama.
biribiriense *C.M. Sakuragui & S.J. Mayo in Kew Bull.*, 52(3): 673 (1997)—Brazil Southeast (Minas Gerais).
brandii *M.H. Grayum, Revis. Philodendron subgen. Pteromischum Pacific & Caribbean Trop. Amer.* (Syst. Bot. Monog., 47): 86 (1996)—Colombia.
breedlovei *T.B. Croat in Ann. Missouri Bot. Gard.*, 84(3): 409 (1997)—Mexico Southeast (Chiapas).
brewsterense *T.B. Croat in Ann. Missouri Bot. Gard.*, 84(3): 413 (1997)—Panama.
brunneicaule *T.B. Croat & M.H. Grayum in Ann. Missouri Bot. Gard.*, 84(3): 414 (1997)—Panama, Costa Rica, Ecuador, Colombia.
burgeri *M.H. Grayum, Revis. Philodendron subgen. Pteromischum Pacific & Caribbean Trop. Amer.* (Syst. Bot. Monog., 47): 188 (1996)—Costa Rica, Panama.
chiriquense *T.B. Croat in Ann. Missouri Bot. Gard.*, 84(3): 416 (1997)—Panama.
chirripoense *T.B. Croat & M.H. Grayum in Ann. Missouri Bot. Gard.*, 84(3): 417 (1997)—Costa Rica.
cipoense *C.M. Sakuragui & S.J. Mayo in Kew Bull.*, 52(3): 674 (1997)—Brazil Southeast (Minas Gerais).
clewellii *T.B. Croat in Ann. Missouri Bot. Gard.*, 84(3): 418 (1997)—Panama.
coloradense *T.B. Croat in Ann. Missouri Bot. Gard.*, 84(3): 419 (1997)—Panama.
copense *T.B. Croat in Ann. Missouri Bot. Gard.*, 84(3): 420 (1997)—Panama.
correae *T.B. Croat in Ann. Missouri Bot. Gard.*, 84(3): 421 (1997)—Panama.
cotobrusense *T.B. Croat & M.H. Grayum in Ann. Missouri Bot. Gard.*, 84(3): 422 (1997)—Costa Rica.
cotonense *T.B. Croat & M.H. Grayum in Ann. Missouri Bot. Gard.*, 84(3): 423 (1997)—Panama, Costa Rica.
crassispathum *T.B. Croat & M.H. Grayum in Ann. Missouri Bot. Gard.*, 84(3): 425 (1997)—Panama, Costa Rica.
cretosum *T.B. Croat & M.H. Grayum in Ann. Missouri Bot. Gard.*, 84(3): 427 (1997)—Panama, Costa Rica.
croatii *M.H. Grayum, Revis. Philodendron subgen. Pteromischum Pacific & Caribbean Trop. Amer.* (Syst. Bot. Monog., 47): 168 (1996)—Panama.
davidsonii Croat
subsp. bocatoranum *T.B. Croat in Ann. Missouri Bot. Gard.*, 84(3): 430 (1997)—Panama.

PHILODENDRON:-
dodsonii *T.B. Croat & M.H. Grayum in Ann. Missouri Bot. Gard.*, 84(3): 430 (1997)—Ecuador, Costa Rica, Colombia.
dolichophyllum *T.B. Croat in Ann. Missouri Bot. Gard.*, 84(3): 432 (1997)—Panama.
dominicalense *T.B. Croat & M.H. Grayum in Ann. Missouri Bot. Gard.*, 84(3): 434 (1997)—Costa Rica.
dwyeri *T.B. Croat in Ann. Missouri Bot. Gard.*, 84(3): 436 (1997)—Belize.
edenudatum *T.B. Croat in Ann. Missouri Bot. Gard.*, 84(3): 437 (1997)—Panama.
elegantulum *T.B. Croat & M.H. Grayum in Revis. Philodendron subgen. Pteromischum Pacific & Caribbean Trop. Amer.* (Syst. Bot. Monog., 47): 135 (1996)—Panama.
ensifolium Croat & Grayum
subsp. campanense *M.H. Grayum, Revis. Philodendron subgen. Pteromischum Pacific & Caribbean Trop. Amer.* (Syst. Bot. Monog., 47): 95 (1996)—Panama.
subsp. colonense *M.H. Grayum, Revis. Philodendron subgen. Pteromischum Pacific & Caribbean Trop. Amer.* (Syst. Bot. Monog., 47): 97 (1996)—Panama.
ferrugineum *T.B. Croat in Ann. Missouri Bot. Gard.*, 84(3): 438 (1997)—Panama.
findens *T.B. Croat & M.H. Grayum in Ann. Missouri Bot. Gard.*, 84(3): 440 (1997)—Panama, Costa Rica, Colombia.
flumineum *E.G. Gonçalves in Kew Bull.*, 55(1): 175 (2000)—Brazil West-central (Goiás, Brasília Distrito Federal).
folsomii *T.B. Croat in Ann. Missouri Bot. Gard.*, 84(3): 442 (1997)—Panama.
fortunense *T.B. Croat in Ann. Missouri Bot. Gard.*, 84(3): 443 (1997)—Panama.
fragile *M.A. Nadruz Coelho & S.J. Mayo in Bol. Bot. (São Paulo)*, 17: 52 (1998)—Brazil Southeast (Rio de Janeiro).
gigas *T.B. Croat in Ann. Missouri Bot. Gard.*, 84(3): 446 (1997)—Panama.
gonzalezii *M.H. Grayum, Revis. Philodendron subgen. Pteromischum Pacific & Caribbean Trop. Amer.* (Syst. Bot. Monog., 47): 129 (1996)—Venezuela.
granulare *T.B. Croat in Ann. Missouri Bot. Gard.*, 84(3): 451 (1997)—Panama.
grayumii *T.B. Croat in Ann. Missouri Bot. Gard.*, 84(3): 452 (1997)—Panama, Costa Rica.
guaraense *E.G. Gonçalves in Kew Bull.*, 52(2): 499 (1997)—Brazil West-central (Brasília Distrito Federal).
hammelii *T.B. Croat in Ann. Missouri Bot. Gard.*, 84(3): 454 (1997)—Panama.
hatschbachii *M.A. Nadruz Coelho & S.J. Mayo in Bol. Bot. (São Paulo)*, 17: 52 (1998)—Brazil Southeast (Rio de Janeiro, Espírito Santo).
hebetatum *T.B. Croat in Ann. Missouri Bot. Gard.*, 84(3): 455 (1997)—Panama, Colombia, Ecuador.
hederaceum (Jacq.) Schott
var. kirkbridei *T.B. Croat in Ann. Missouri Bot. Gard.*, 84(3): 463 (1997)—Costa Rica, Panama, Ecuador.
var. oxycardium *(Schott) T.B. Croat in Ann. Missouri Bot. Gard.*, 84(3): 464 (1997), without full basionym ref.: P. oxycardium.
heleniae *T.B. Croat in Ann. Missouri Bot. Gard.*, 84(3): 465 (1997)—Panama, Colombia, Ecuador.
hooveri *T.B. Croat & M.H. Grayum in Revis. Philodendron subgen. Pteromischum Pacific & Caribbean Trop. Amer.* (Syst. Bot. Monog., 47): 111 (1996)—Ecuador.
ichthyoderma *T.B. Croat & M.H. Grayum in Revis. Philodendron subgen. Pteromischum Pacific & Caribbean Trop. Amer.* (Syst. Bot. Monog., 47): 147 (1996)—Panama, Colombia.
immixtum *T.B. Croat in Ann. Missouri Bot. Gard.*, 84(3): 467 (1997)—Panama, Colombia.
jefense *T.B. Croat in Ann. Missouri Bot. Gard.*, 84(3): 471 (1997)—Panama.
knappiae *T.B. Croat in Ann. Missouri Bot. Gard.*, 84(3): 475 (1997)—Panama.
lazorii *T.B. Croat in Ann. Missouri Bot. Gard.*, 84(3): 476 (1997)—Panama.
lentii *T.B. Croat & M.H. Grayum in Ann. Missouri Bot. Gard.*, 84(3): 477 (1997)—Panama, Costa Rica.

PHILODENDRON:-
ligulatum Schott
var. heraclioanum *T.B. Croat in Ann. Missouri Bot. Gard.*, 84(3): 482 (1997)—Panama.
var. ovatum *T.B. Croat in Ann. Missouri Bot. Gard.*, 84(3): 483 (1997)—Panama.
llanense *T.B. Croat in Ann. Missouri Bot. Gard.*, 84(3): 484 (1997)—Panama.
madronense *T.B. Croat in Ann. Missouri Bot. Gard.*, 84(3): 486 (1997)—Panama.
malesevichiae *T.B. Croat in Ann. Missouri Bot. Gard.*, 84(3): 487 (1997)—Panama, Colombia.
mayoi *E.G. Gonçalves in Kew Bull.*, 55(1): 178 (2000)—Brazil West-central (Brasília Distrito Federal, Goiás).
morii *T.B. Croat in Ann. Missouri Bot. Gard.*, 84(3): 492 (1997)—Panama.
niqueanum *T.B. Croat in Ann. Missouri Bot. Gard.*, 84(3): 493 (1997)—Panama.
pirrense *T.B. Croat in Ann. Missouri Bot. Gard.*, 84(3): 495 (1997)—Panama.
pseudauriculatum *T.B. Croat in Ann. Missouri Bot. Gard.*, 84(3): 498 (1997)—Panama, Colombia.
purulhense *T.B. Croat in Ann. Missouri Bot. Gard.*, 84(3): 504 (1997)—Guatemala, Honduras, Mexico Southeast (Chiapas).
radiatum Schott
var. pseudoradiatum *(Matuda) T.B. Croat in Ann. Missouri Bot. Gard.*, 84(3): 510 (1997): P. pseudoradiatum.
rayanum *T.B. Croat & M.H. Grayum in Revis. Philodendron subgen. Pteromischum Pacific & Caribbean Trop. Amer.* (Syst. Bot. Monog., 47): 131 (1996)—Panama, Costa Rica.
rhizomatosum *C.M. Sakuragui & S.J. Mayo in Kew Bull.*, 52(3): 676 (1997)—Brazil Southeast (Minas Gerais).
rhodoaxis G.S. Bunting
subsp. lewisii *T.B. Croat & M.H. Grayum in Revis. Philodendron subgen. Pteromischum Pacific & Caribbean Trop. Amer.* (Syst. Bot. Monog., 47): 106 (1996)—Costa Rica, Panama, Colombia, Ecuador.
rigidifolium K. Krause
subsp. sanctae-ritae *M.H. Grayum, Revis. Philodendron subgen. Pteromischum Pacific & Caribbean Trop. Amer.* (Syst. Bot. Monog., 47): 167 (1996)—Panama.
rodrigueziae *T.B. Croat & M.H. Grayum in Revis. Philodendron subgen. Pteromischum Pacific & Caribbean Trop. Amer.* (Syst. Bot. Monog., 47): 152 (1996)—Ecuador.
romeroi *M.H. Grayum, Revis. Philodendron subgen. Pteromischum Pacific & Caribbean Trop. Amer.* (Syst. Bot. Monog., 47): 70 (1996)—Colombia.
roseopetiolatum *M.A. Nadruz Coelho & S.J. Mayo in Bol. Bot. (São Paulo)*, 17: 55 (1998)—Brazil Southeast (Rio de Janeiro).
roseospathum *T.B. Croat in Ann. Missouri Bot. Gard.*, 84(3): 511 (1997)—Panama, Colombia.
var. angustilaminatum *T.B. Croat in Ann. Missouri Bot. Gard.*, 84(3): 513 (1997)—Panama.
scalarinerve *T.B. Croat & M.H. Grayum in Ann. Missouri Bot. Gard.*, 84(3): 521 (1997)—Panama, Costa Rica, Colombia, Ecuador.
scandens K. Koch & Sello
subsp. cubense *(Engl.) I. Arias Granda in Feddes Repert.*, 109(1–2): 37 (1998): P. scandens var. cubense.
schottii K. Koch
subsp. talamancae *(Engl.) M.H. Grayum, Revis. Philodendron subgen. Pteromischum Pacific & Caribbean Trop. Amer.* (Syst. Bot. Monog., 47): 142 (1996): P. talamancae.
seguine Schott
subsp. lingua-bovis *M.H. Grayum, Revis. Philodendron subgen. Pteromischum Pacific & Caribbean Trop. Amer.* (Syst. Bot. Monog., 47): 186 (1996)—Mexico Gulf (Veracruz), Mexico Southeast (Chiapas), Mexico Southwest (Oaxaca).
sousae *T.B. Croat in Ann. Missouri Bot. Gard.*, 84(3): 528 (1997)—Mexico Southeast (Chiapas).
squamicaule *T.B. Croat & M.H. Grayum in Ann. Missouri Bot. Gard.*, 84(3): 529 (1997)—Panama, Costa Rica, Colombia, Ecuador.
squamipetiolatum *T.B. Croat in Ann. Missouri Bot. Gard.*, 84(3): 531 (1997)—Panama, Colombia, Ecuador.

PHILODENDRON:-
straminicaule *T.B. Croat in Ann. Missouri Bot. Gard.*, 84(3): 532 (1997)—Panama, Costa Rica.
sulcicaule *T.B. Croat & M.H. Grayum in Ann. Missouri Bot. Gard.*, 84(3): 537 (1997)—Costa Rica, Panama.
thalassicum *T.B. Croat & M.H. Grayum in Ann. Missouri Bot. Gard.*, 84(3): 541 (1997)—Panama, Costa Rica.
tuerckheimii *M.H. Grayum, Revis. Philodendron subgen. Pteromischum Pacific & Caribbean Trop. Amer.* (Syst. Bot. Monog., 47): 174 (1996)—Rica, Guatemala, Honduras, Nicaragua, Panama, Mexico Southwest (Oaxaca), Mexico Southeast (Tabasco), Mexico Gulf (Veracruz), Colombia, Ecuador, Venezuela.
tysonii *T.B. Croat in Ann. Missouri Bot. Gard.*, 84(3): 549 (1997)—Panama.
ubigantupense *T.B. Croat in Ann. Missouri Bot. Gard.*, 84(3): 551 (1997)—Panama.
utleyanum *T.B. Croat in Ann. Missouri Bot. Gard.*, 84(3): 551 (1997)—Panama.
venustifoliatum *E.G. Gonçalves & S.J. Mayo in Kew Bull.*, 55(2): 483 (2000)—Brazil West-central (Goiás, Brasília Distrito Federal).
verapazense *T.B. Croat in Ann. Missouri Bot. Gard.*, 84(3): 552 (1997)—Guatemala, Mexico Southeast (Chiapas).
wilburii *T.B. Croat & M.H. Grayum in Ann. Missouri Bot. Gard.*, 84(3): 559 (1997)—Costa Rica, Panama.
var. longipedunculatum *T.B. Croat & M.H. Grayum in Ann. Missouri Bot. Gard.*, 84(3): 563 (1997)—Costa Rica, Panama.
zhuanum *T.B. Croat in Ann. Missouri Bot. Gard.*, 84(3): 564 (1997)—Panama.

PHILODICE *Mart. in Ann. Sci. Nat., Bot.*, Sér. 2, 2: 26 (1834). *ERIOCAULACEAE*†
hoffmannseggii *Mart. in Ann. Sci. Nat., Bot.*, Sér. 2, 2: 29 (1834), as 'hoffmanseggii'—Brazil.†

PHILOTHECA (Rutac.).
sect. Corynonema *(Paul G. Wilson) Paul G. Wilson in Nuytsia*, 12(2): 260 (1998): Eriostemon sect. Corynonema.
sect. Cyanochlamys *(F. Muell.) Paul G. Wilson in Nuytsia*, 12(2): 261 (1998): Eriostemon subgen. Cyanochlamys.
sect. Erionema *(F. Muell.) Paul G. Wilson in Nuytsia*, 12(2): 256 (1998): Eriostemon sect. Erionema.
acrolopha *Paul G. Wilson in Nuytsia*, 12(2): 243 (1998)—Queensland.
angustifolia *(Paul G. Wilson) Paul G. Wilson in Nuytsia*, 12(2): 246 (1998): Eriostemon angustifolius.
subsp. montana *(Paul G. Wilson) Paul G. Wilson in Nuytsia*, 12(2): 246 (1998): Eriostemon angustifolius subsp. montanus.
apiculata *(Paul G. Wilson) Paul G. Wilson in Nuytsia*, 12(2): 246 (1998): Eriostemon apiculatus.
brevifolia *(Endl.) Paul G. Wilson in Nuytsia*, 12(2): 246 (1998): Eriostemon brevifolius.
brucei *(F. Muell.) Paul G. Wilson in Nuytsia*, 12(2): 257 (1998): Eriostemon brucei.
subsp. brevifolia *(Paul G. Wilson) Paul G. Wilson in Nuytsia*, 12(2): 257 (1998): Eriostemon brucei subsp. brevifolius.
subsp. cinerea *(Paul G. Wilson) Paul G. Wilson in Nuytsia*, 12(2): 258 (1998): Eriostemon brucei subsp. cinereus.
buxifolia *(Sm.) Paul G. Wilson in Nuytsia*, 12(2): 258 (1998): Eriostemon buxifolius.
subsp. falcata *Paul G. Wilson in Nuytsia*, 12(2): 258 (1998): New South Wales (Australian Capital Territory, New South Wales).
subsp. obovata *(G. Don) Paul G. Wilson in Nuytsia*, 12(2): 259 (1998): Eriostemon buxifolius var. obovatus.
coateana *Paul G. Wilson in Nuytsia*, 12(2): 247 (1998)—Western Australia.
coccinea *(C.A. Gardner) Paul G. Wilson in Nuytsia*, 12(2): 247 (1998): Eriostemon coccineus.
cuticularis *Paul G. Wilson in Nuytsia*, 12(2): 248 (1998)—Queensland.
cymbiformis *(Paul G. Wilson) Paul G. Wilson in Nuytsia*, 12(2): 248 (1998): Eriostemon cymbiformis.

PHILOTHECA:-
deserti *(E. Pritz.) Paul G. Wilson in Nuytsia*, 12(2): 248 (1998): Eriostemon deserti.
subsp. brevifolia *Paul G. Wilson in Nuytsia*, 12(2): 249 (1998)—Western Australia.
difformis *(A. Cunn. ex Endl.) Paul G. Wilson in Nuytsia*, 12(2): 249 (1998): Eriostemon difformis.
subsp. smithiana *(Benth.) Paul G. Wilson in Nuytsia*, 12(2): 251 (1998): Eriostemon difformis var. smithianus.
eremicola *Paul G. Wilson in Nuytsia*, 12(2): 251 (1998)—Western Australia.
ericifolia *(A. Cunn. ex Benth.) Paul G. Wilson in Nuytsia*, 12(2): 251 (1998): Eriostemon ericifolius.
falcata *(Paul G. Wilson) Paul G. Wilson in Nuytsia*, 12(2): 251 (1998): Eriostemon falcatus.
fitzgeraldii *(C.R.P. Andrews) Paul G. Wilson in Nuytsia*, 12(2): 261 (1998): Eriostemon fitzgeraldii.
gardneri *(Paul G. Wilson) Paul G. Wilson in Nuytsia*, 12(2): 252 (1998): Eriostemon gardneri.
subsp. globosa *Paul G. Wilson in Nuytsia*, 12(2): 252 (1998)—Western Australia.
glabra *(Paul G. Wilson) Paul G. Wilson in Nuytsia*, 12(2): 252 (1998): Eriostemon glaber.
hispidula *(Spreng.) Paul G. Wilson in Nuytsia*, 12(2): 259 (1998): Eriostemon hispidulus.
kalbarriensis *Paul G. Wilson in Nuytsia*, 12(2): 252 (1998)—Western Australia.
linearis *(A. Cunn. ex Endl.) Paul G. Wilson in Nuytsia*, 12(2): 253 (1998): Eriostemon linearis.
myoporoides *(DC.) M.J. Bayly in Muelleria*, 11: 118 (1998): Eriostemon myoporoides.
subsp. acuta *(Blakely) M.J. Bayly in Muelleria*, 11: 120 (1998): Eriostemon myoporoides var. acutus.
subsp. brevipedunculata *M.J. Bayly in Muelleria*, 11: 121 (1998)—New South Wales.
subsp. conduplicata *(Paul G. Wilson) M.J. Bayly in Muelleria*, 11: 124 (1998): Eriostemon myoporoides subsp. conduplicatus.
subsp. epilosa *(Paul G. Wilson) M.J. Bayly in Muelleria*, 11: 120 (1998): Eriostemon myoporoides subsp. epilosus.
subsp. euroensis *M.J. Bayly in Muelleria*, 11: 122 (1998)—Victoria.
subsp. leichhardtii *(Benth.) M.J. Bayly in Muelleria*, 11: 124 (1998): Eriostemon trachyphyllus var. leichhardtii.
subsp. obovatifolia *M.J. Bayly in Muelleria*, 11: 123 (1998)—Queensland.
subsp. queenslandica *(C.T. White) M.J. Bayly in Muelleria*, 11: 125 (1998): Eriostemon queenslandicus.
nodiflora *(Lindl.) Paul G. Wilson in Nuytsia*, 12(2): 263 (1998): Eriostemon nodiflorus.
subsp. calycina *(Turcz.) Paul G. Wilson in Nuytsia*, 12(2): 263 (1998): Eriostemon calycinus.
subsp. lasiocalyx *(Domin) Paul G. Wilson in Nuytsia*, 12(2): 263 (1998): Eriostemon nodiflorus var. lasiocalyx.
subsp. latericola *Paul G. Wilson in Nuytsia*, 12(2): 263 (1998)—Western Australia.
nutans *(Paul G. Wilson) Paul G. Wilson in Nuytsia*, 12(2): 253 (1998): Eriostemon nutans.
obovalis *(A. Cunn.) Paul G. Wilson in Nuytsia*, 12(2): 259 (1998): Eriostemon obovalis.
pachyphylla *(Paul G. Wilson) Paul G. Wilson in Nuytsia*, 12(2): 254 (1998): Eriostemon pachyphyllus.
pinoides *(Paul G. Wilson) Paul G. Wilson in Nuytsia*, 12(2): 261 (1998): Eriostemon pinoides.
pungens *(Lindl.) Paul G. Wilson in Nuytsia*, 12(2): 261 (1998): Eriostemon pungens.
rhomboidea *(Paul G. Wilson) Paul G. Wilson in Nuytsia*, 12(2): 254 (1998): Eriostemon rhomboideus.
salsolifolia (Sm.) Druce
subsp. pedicellata *Paul G. Wilson in Nuytsia*, 12(2): 255 (1998)—New South Wales.
scabra *(Paxton) Paul G. Wilson in Nuytsia*, 12(2): 259 (1998): Eriostemon scaber.
subsp. latifolia *(Paul G. Wilson) Paul G. Wilson in Nuytsia*, 12(2): 259 (1998): Eriostemon scaber subsp. latifolius.
sericea *(Paul G. Wilson) Paul G. Wilson in Nuytsia*, 12(2): 255 (1998): Eriostemon sericeus.

PHILOTHECA:-
spicata *(A. Rich.) Paul G. Wilson in Nuytsia*, 12(2): 264 (1998): Eriostemon spicatus.
sporadica *(Bayly) Paul G. Wilson in Nuytsia*, 12(2): 255 (1998): Eriostemon sporadicus.
thryptomenoides *(S. Moore) Paul G. Wilson in Nuytsia*, 12(2): 255 (1998): Eriostemon thryptomenoides.
tomentella *(Diels) Paul G. Wilson in Nuytsia*, 12(2): 256 (1998): Eriostemon tomentellus.
trachyphylla *(F. Muell.) Paul G. Wilson in Nuytsia*, 12(2): 260 (1998): Eriostemon trachyphyllus.
verrucosa *(A. Rich.) Paul G. Wilson in Nuytsia*, 12(2): 260 (1998): Eriostemon verrucosus.
virgata *(Hook. f.) Paul G. Wilson in Nuytsia*, 12(2): 260 (1998): Eriostemon virgatus.
wonganensis *(Paul G. Wilson) Paul G. Wilson in Nuytsia*, 12(2): 256 (1998): Eriostemon wonganensis.

PHINAEA (Gesneriac.).
ecuadorana *H. Wiehler in Gesneriana*, 1(1): 84 (1995)—Ecuador.
macrophylla *H. Wiehler in Gesneriana*, 1(1): 86 (1995)—Colombia.

PHITOSIA *G. Kamari & W. Greuter in Bot. Chronika*, 13: 14 (2000). *COMPOSITAE*
crocifolia *(Boiss. & Heldr.) G. Kamari & W. Greuter in Bot. Chronika*, 13: 14 (2000): Crepis crocifolia.

PHLEUM (Gramin.).
sect. Echinata *M. Doğan & J. Us in M.A. Öztürk, Ö. Seçmen & G. Görk (eds.), Pl. Life S.W. & Central Asia*, 1: 164 (1996).

PHLOJODICARPUS (Umbellif.).
baicalensis Popov
var. arenarius *G.A. Peschkova, Stepnaya Fl. Baĭkal'skoĭ Sibiri:* 83 (1972), without type—Irkutsk.
sibiricus (Fisch. ex Spreng.) Koso-Pol.
subsp. komarovii *(Gorovoj) V.N. Voroschilov in A.K. Skvortsov (ed.), Florist. issl. v razn. raĭonakh SSSR:* 184 (1985): P. komarovii.
subsp. villosus *(Turcz. ex Fisch. & C.A. Mey.) V.N. Voroschilov in A.K. Skvortsov (ed.), Florist. issl. v razn. raĭonakh SSSR:* 184 (1985): Libanotis villosa.

PHLOMIS (Labiat.).
iberica *Pépin in Ann. Fl. Pomone*, 1836–37: 19 (1836)—Spain.

PHLOMOIDES (Labiat.).
ajdarovae *G.A. Lazkov in Bot. Zhurn.*, 85(8): 114 (2000)—Kirgizistan.
woroschilowii *(Makarov) N.S. Probatova in Sosud. Rast. Sovet. Dal'nego Vostoka*, 7: 335 (1995): Phlomis woroschilowii.

PHOENIX (Palmae).
andamanensis *S.C. Barrow in Kew Bull.*, 53(3): 538 (1998)—Andaman Is.
loureiri Kunth
var. humilis *S.C. Barrow in Kew Bull.*, 53(3): 563 (1998), nom. nov. illegit., see Art. 52.1 ICBN (1994): P. humilis *Becc.*

PHOLIDOTA (Orchidac.).
katakiana *S. Phukan in Orchid Rev.*, 104(1210): 238 (1996)—Assam (Arunachal Pradesh).
pachyglossa *L.V. Averyanov in Lindleyana*, 14(4): 224 (1999)—Vietnam.
pygmaea *H.J. Chowdhery & G.D. Pal in Nordic J. Bot.*, 15(4): 411 (1995 publ. 1996)—Assam (Arunachal Pradesh).

PHORADENDRON (Viscac.).
alpestre *C.T. Rizzini in Revista Fac. Agron. (Maracay)*, 8(3): 85 (1975)—Venezuela.
apiciflorum *C.T. Rizzini in Revista Fac. Agron. (Maracay)*, 8(3): 86 (1975)—Brazil Southeast (Minas Gerais).
crispulum *C.T. Rizzini in Revista Fac. Agron. (Maracay)*, 8(3): 86 (1975)—Venezuela.
dendrophthora *C.T. Rizzini in Revista Fac. Agron. (Maracay)*, 8(3): 87 (1975)—Venezuela.

PHORADENDRON:-
diminutivum *E.A. Kellogg in Novon,* 6(1): 41 (1996)—Venezuela.
distans *C.T. Rizzini in Revista Fac. Agron. (Maracay),* 8(3): 87 (1975)—Brazil Southeast (Minas Gerais).
kelloggii *J. Kuijt in Novon,* 6(1): 42 (1996)—Venezuela.
kingii *J. Kuijt in Novon,* 6(1): 43 (1996)—Mexico Southeast (Chiapas).
longiarticulatum *C.T. Rizzini in Revista Fac. Agron. (Maracay),* 8(3): 88 (1975)—Venezuela.
olae *J. Kuijt in Brittonia,* 49(2): 185 (1997)—Mexico Central (Puebla).
orbiculare *C.T. Rizzini in Revista Fac. Agron. (Maracay),* 8(3): 88 (1975)—Brazil West-central (Mato Grosso), Brazil Southeast (São Paulo).
prancei *C.T. Rizzini in Revista Fac. Agron. (Maracay),* 8(3): 89 (1975)—Brazil North (Amazonas).
roldanii *J. Kuijt in Novon,* 6(1): 43 (1996)—Colombia.
rubrum *(L.) Griseb., Fl. Brit. W.I.:* 314 (1860): Viscum rubrum.
sulfuratum *C.T. Rizzini in Revista Fac. Agron. (Maracay),* 8(3): 89 (1975)—Venezuela.
tomentosum (DC.) Engelm.
var. macrophyllum *(Engelm.) L.D. Benson in L.D. Benson & R.A. Darrow, Trees & Shrubs of Southwest. Deserts (ed. 3):* 286 (1981): P. flavescens var. macrophyllum.
triflorum *E.A. Kellogg in Novon,* 6(1): 44 (1996)—Venezuela.
triplinervium *C.T. Rizzini in Revista Fac. Agron. (Maracay),* 8(3): 90 (1975)—Brazil Northeast (Bahia).
websteri *J. Kuijt in Novon,* 6(1): 47 (1996)—Panama.

PHOTINIA (Rosac.).
chingiana Hand.-Mazz.
var. lipingensis *(Y.K. Li & M.Z. Yang) L.T. Lu & C.L. Li in Acta Phytotax. Sin.,* 38(3): 277 (2000): P. lipingensis.
dulongensis *H. Li in Fl. Dulongjien Reg.:* 130 (1993), without latin descr. or type—China South-central (Yunnan).
komarovii *(H. Lév. & Vaniot) L.T. Lu & C.L. Li in Acta Phytotax. Sin.,* 38(3): 278 (2000): Viburnum komarovii.
lasiogyna *(Franch.) C.K. Schneider in Repert. Nov. Spec. Regni Veg.,* 3: 153 (1906): Eriobotrya lasiogyna.
var. glabrescens *L.T. Lu & C.L. Li in Acta Phytotax. Sin.,* 38(3): 278 (2000)—China Southeast (Jiangxi, Hunan, Zhejiang, Fujian, Guangdong, Guangxi), China South-central (Sichuan, Yunnan).
lucida *(Decne.) C.K. Schneider, Ill. Handb. Laubholzk.,* 1: 710 (1906): Pourthiaea lucida.
parvifolia (E. Pritz.) C.K. Schneid.
var. subparvifolia *(Y.K. Li & X.M. Wang) L.T. Lu & C.L. Li in Acta Phytotax Sin.,* 38(3): 278 (2000): P. subparvifolia.
salicifolia *(Decne.) C.K. Schneider, Ill. Handb. Laubholzk.,* 1: 709 (1906), homonym; *see C. Presl, Epimel. Bot.:* 204 (1851): Pourthiaea salicifolia.
schneideriana Rehder & E.H. Wilson
var. parviflora *(Cardot) L.T. Lu & C.L. Li in Acta Phytotax. Sin.,* 38(3): 277 (2000): P. parviflora.
semiserrata *H. Li in Fl. Dulongjian Reg.:* 131 (1993), without latin descr. or type—China South-central (Yunnan).
serratifolia (Desf.) Kalkman
var. daphniphylloides *(Hayata) L.T. Lu in Acta Phytotax. Sin.,* 38(3): 277 (2000): P. daphniphylloides.
villosa (Thunb.) DC.
var. glabricalycina *L.T. Lu & C.L. Li in Acta Phytotax. Sin.,* 38(3): 279 (2000)—China Southeast (Jiangxi, Jiangsu, Zhejiang, Hunan, Guangxi), China South-central (Guizhou).

PHRAGMANTHERA (Loranthac.).
sect. Eubracteatae *(Engl.) R.M. Polhill & D. Wiens, Mistletoes of Africa:* 251 (1998): Loranthus group Eubracteati.

PHRAGMANTHERA:-
sect. Lepidotae *(Engl.) R.M. Polhill & D. Wiens, Mistletoes of Africa:* 247 (1998): Loranthus group Lepidoti.
sect. Rufescentes *(Engl.) R.M. Polhill & D. Wiens, Mistletoes of Africa:* 247 (1998): Loranthus group Rufescentes.
exellii *Balle ex R.M. Polhill & D. Wiens, Mistletoes of Africa:* 255 (1998)—Angola.
polycrypta (Didr.) Balle
subsp. subglabriflora *Balle ex R.M. Polhill & D. Wiens, Mistletoes of Africa:* 265 (1998)—Uganda.

PHRAGMIPEDIUM (Orchidac.).
besseae Dodson & J. Kuhn
var. dalessandroi *(Dodson & O. Gruss) Alan Moon & P.J. Cribb in Orchid Rev.,* 105(1216): 229 (1997): P. dalessandroi.
f. flavum *(Braem) O. Gruss & J. Roeth in Caesiana,* 12: 65 (1999): P. besseae var. flavum.
caricinum *(Lindl. & Paxton) R.A. Rolfe in Orchid Rev.,* 4: 332 (1896), in I.K. Supp. 3 as 'Phragmopedilum': Cypripedium caricinum.†
caudatum *(Lindl.) R.A. Rolfe in Orchid Rev.,* 4: 332 (1896), in Supp. 3 as 'Phragmopedilum': Cypripedium caudatum.†
chapadense *M.A. Campacci & R. Takase in J. Hokkaido Orchid Soc.,* 28: Supp. 1 (2000)—Brazil West-central (Goiás).
czerwiakowianum *(Rchb.f.) R.A. Rolfe in Orchid Rev.,* 4: 332 (1896), in I.K. Supp. 3 as 'Phragmopedilum': Selenipedium czerwiakowianum.†
dalessandroi *C.H. Dodson & O. Gruss in Orchidee,* 47(4): 217 (1996)—Ecuador.
fischeri *G.J. Braem & H. Mohr in Leafl. Schlechter Inst.,* 3: 30 (1996)—Ecuador.
hartwegii (Rchb. f.) Pfitzer
var. baderi *J. Roeth & O. Gruss in Orchidee,* 48(3): 95 (1997)—Ecuador.
humboldtii *(Warsz. ex Rchb. f.) J.T. Atwood & R.L. Dressler in Selbyana,* 19(2): 246 (1998 publ. 1999): Cypripedium humboldtii.
subsp. exstaminodium *(Castaño, Hágsater & E. Aguirre) J.T. Atwood & R.L. Dressler in Selbyana,* 19(2): 247 (1998 publ. 1999), without basionym date: P. exstaminodium.
klotzschianum *(Rchb.f.) R.A. Rolfe in Orchid Rev.,* 4: 332 (1896), in Supp. 3 as 'Phragmopedilum': Cypripedium klotzschianum.†
lindleyanum *(R.H.Schomb. ex Lindl.) R.A. Rolfe in Orchid Rev.,* 4: 332 (1896), in Supp. 3 as 'Phragmopedilum': Cypripedium lindleyanum.†
longifolium *(Warsz. ex Rchb. f.) R.A. Rolfe in Orchid Rev.,* 4: 332 (1896), in Supp. 3 as 'Phragmopedilum': Cypripedium longifolium.†
× roethianum *O. Gruss & T. Kalina in Orchidee,* 49(5): 245 (1998)—Ecuador. (P. hirtzii × longifolium).
sargentianum *(Rolfe) R.A. Rolfe in Orchid Rev.,* 4: 332 (1896), in Supp. 3 as 'Phragmopedilum': Selenipedium sargentianum.†
schlimii *(Rchb.f.) R.A. Rolfe in Orchid Rev.,* 4: 332 (1896), in Supp. 3 as 'Phragmopedilum': Selenipedium schlimii.†
f. albiflorum *(Linden) O. Gruss in Orchidee,* 47(1): 22 (1996): Cypripedium schlimii var. albiflorum.
var. albiflorum *(Linden) G.J. Braem in Orchids,* 65(2): 128 (1996): Cypripedium schlimii var. albiflorum.
tetzlaffianum *O. Gruss in Caesiana,* 15: 37 (2000)—Venezuela?
vittatum *(Vell.) R.A. Rolfe in Orchid Rev.,* 4: 332 (1896), in Supp. 3 as 'Phragmopedilum': Cypripedium vittatum.†

PHRAGMITES (Gramin.).
frutescens *H. Scholz in Taxon,* 45(3): 522 (1996)—Kriti.

PHRAGMOTHECA (Bombacac.).
amazonica *(W.S. Alverson) J.L. Fernández-Alonso in Caldasia,* 18(3): 274 (1996): P. mammosa subsp. amazonica.
var. leticiana *J.L. Fernández-Alonso in Caldasia,* 18(3): 274 (1996)—Colombia.
hydra *J.L. Fernández-Alonso in Caldasia,* 18(3): 270 (1996)—Colombia.

PHRAGMOTHECA:-
lemniscata *J.L. Fernández-Alonso in Caldasia,* 18(3): 262 (1996)—Colombia.
mambitana *J.L. Fernández-Alonso & R. Jaramillo-Mejía in Caldasia,* 21(2): 125 (1999)—Colombia.
rubriflora *J.L. Fernández-Alonso in Caldasia,* 18(3): 272 (1996)—Colombia.
sidereotricha *J.L. Fernández-Alonso in Caldasia,* 18(3): 279 (1996)—Peru.
siderosa Cuatrec.
subsp. megacarpa *J.L. Fernández-Alonso in Caldasia,* 18(3): 267 (1996)—Colombia.

PHRYNIUM (Marantac.).
venustum *I.M. Turner in Gard. Bull. Singapore,* 50(1): 39 (1998), nom. nov.: P. gracile *Holttum.*

PHTHEIROSPERMUM (Scrophulariac.).
muliense *C.Y. Wu & D.D. Tao in Acta Bot. Yunnanica,* 18(3): 307 (1996)—China South-central (Sichuan).

PHTHIRUSA (Loranthac.).
caatingae *C.T. Rizzini in Revista Fac. Agron. (Maracay),* 8(3): 90 (1975)—Brazil Northeast (Bahia).
maritima *C.T. Rizzini in Revista Fac. Agron. (Maracay),* 8(3): 92 (1975)—Venezuela.
perforata *C.T. Rizzini in Revista Fac. Agron. (Maracay),* 8(3): 91 (1975)—Venezuela.
trichodes *C.T. Rizzini in Revista Fac. Agron. (Maracay),* 8(3): 92 (1975)—Venezuela, Brazil Northeast (Paraíba).

PHYLLAGATHIS (Melastomatac.).
siamensis *N. Cellinese & S.S. Renner in Novon,* 7(2): 109 (1997), nom. nov.: Tylanthera cordata *C. Hansen.*
tuberosa *(C. Hansen) N. Cellinese & S.S. Renner in Novon,* 7(2): 109 (1997): Tylanthera tuberosa.

PHYLLANGIUM *C.R. Dunlop in Fl. Australia,* 28: 315 (1996). *LOGANIACEAE*
distylis *(F. Muell.) C.R. Dunlop in Fl. Australia,* 28: 315 (1996): Mitrasacme distylis.
divergens *(Hook. f.) C.R. Dunlop in Fl. Australia,* 28: 315 (1996): Mitrasacme divergens.
palustre *(W. Fitzg.) C.R. Dunlop in Fl. Australia,* 28: 315 (1996): Mitrasacme palustris.
paradoxum *(R. Br.) C.R. Dunlop in Fl. Australia,* 28: 315 (1996): Mitrasacme paradoxa.
sulcatum *C.R. Dunlop in Fl. Australia,* 28: 315 (1996)—Western Australia, South Australia, New South Wales, Victoria.

PHYLLANTHUS (Euphorbiac.).
amarus Schumach. & Thonn.
subsp. sanyaensis *P.T. Li & Y.T. Zhu in J. S. China Agric. Univ.,* 17(3): 118 (1996)—Hainan.
arachnodes *R. Govaerts & A. Radcliffe-Smith in Kew Bull.,* 51(1): 175 (1996), nom. nov.: Arachnodes chevalieri *Gagnep.*
baeckeoides *J.T. Hunter & J.J. Bruhl in Nuytsia,* 11(2): 149 (1997)—Western Australia.
balakrishnairii *R. Govaerts & A. Radcliffe-Smith in Kew Bull.,* 51(1): 176 (1996), nom. nov.: P. andamanicus *N.P. Balakr. & N.G. Nair.*
bernierianus Baill. ex Müll. Arg.
var. glaber *A. Radcliffe-Smith in Kew Bull.,* 51(2): 305 (1996)—Mozambique.
binhii *Nguyen Nghia Thin, Euphorbiac. Vietnam:* 48 (1995 publ. 1996)—Vietnam.
breynioides *(P.T. Li) R. Govaerts & A. Radcliffe-Smith in Kew Bull.,* 51(1): 176 (1996): Phyllanthodendron breynioides.
casticum P. Willemet
var. kirganelia *(Willd.) R. Govaerts in R. Govaerts, D.G. Frodin & A. Radcliffe-Smith, World Checklist & Bibliogr.* Euphorbiaceae, 4: 1276 (2000): P. kirganelia.
cauticola *J.T. Hunter & J.J. Bruhl in Nuytsia,* 11(2): 151 (1997)—Northern Territory.
chandrabosei *R. Govaerts & A. Radcliffe-Smith in Kew Bull.,* 51(1): 176 (1996), nom. nov.: Reidia stipulacea *Gamble.*
distichus Hook. & Arn.
var. degeneri *(Sherff) R. Govaerts & A. Radcliffe-Smith in Kew Bull.,* 51(1): 176 (1996): P. sandwicensis var. degeneri.

PHYLLANTHUS:-
distichus
var. ellipticus *(Müll. Arg.) R. Govaerts & A. Radcliffe-Smith in Kew Bull.*, 51(1): 176 (1996): P. sandwicensis var. ellipticus.
eliae *(Brunel & J.P. Roux) Brunel ex R. Govaerts & A. Radcliffe-Smith in Kew Bull.*, 51(1): 176 (1996): P. sublanatus subsp. eliae.
erwinii *J.T. Hunter & J.J. Bruhl in J. Adelaide Bot. Gard.*, 17: 130 (1996)—Northern Territory.
× fluminis-sabi *A. Radcliffe-Smith in Kew Bull.*, 51(2): 306 (1996)—Mozambique. (P. muelleranus × reticulatus).
× fluminis-zambesi *A. Radcliffe-Smith in Kew Bull.*, 51(2): 306 (1996)—Zambia. (P. engleri × polyanthus).
gillettianus *Jean F. Brunel ex A. Radcliffe-Smith in Kew Bull.*, 51(2): 306 (1996)—Kenya.
goniostemon *A. Radcliffe-Smith in Kew Bull.*, 51(2): 307 (1996)—Uganda.
guangxiensis *R. Govaerts & A. Radcliffe-Smith in Kew Bull.*, 51(1): 176 (1996), nom. nov.: Phyllanthodendron petraeum *P.T. Li*.
hesperonotos *R. Govaerts & A. Radcliffe-Smith in Kew Bull.*, 51(1): 177 (1996), nom. nov.: P. polycladus *W. Fitzg.*
hortensis *R. Govaerts & A. Radcliffe-Smith in Kew Bull.*, 51(1): 177 (1996), nom. nov.: P. ovatus *Desf. ex Poir.*
involutus *J.T. Hunter & J.J. Bruhl in Telopea*, 7(2): 155 (1997)—New South Wales, Queensland.
lativenius *(Croizat) R. Govaerts & A. Radcliffe-Smith in Kew Bull.*, 51(1): 177 (1996): Phyllanthodendron lativenium.
lii *R. Govaerts & A. Radcliffe-Smith in Kew Bull.*, 51(1): 177 (1996), nom. nov.: Phyllanthodendron caudatifolium *P.T. Li*.
macranthus Pax
var. gilletii *(De Wild.) Jean F. Brunel ex A. Radcliffe-Smith in Kew Bull.*, 51(2): 308 (1996): P. gilletii.
mafingensis *A. Radcliffe-Smith in Kew Bull.*, 51(2): 308 (1996)—Malawi.
manicaensis *Jean F. Brunel ex A. Radcliffe-Smith in Kew Bull.*, 51(2): 309 (1996)—Mozambique.
martinii *A. Radcliffe-Smith in Kew Bull.*, 51(2): 311 (1996)—Zambia.
mendesii *Jean F. Brunel ex A. Radcliffe-Smith in Kew Bull.*, 51(2): 312 (1996)—Angola, Zambia, Zimbabwe, Botswana, Namibia.
mendoncae *Jean F. Brunel ex A. Radcliffe-Smith in Kew Bull.*, 51(2): 313 (1996)—Mozambique.
microdendron Welw. ex Müll. Arg.
var. asper *A. Radcliffe-Smith in Kew Bull.*, 51(2): 315 (1996)—Zambia.
micromeris Radcl.-Sm.
var. mughessensis *A. Radcliffe-Smith in Kew Bull.*, 51(2): 316 (1996)—Malawi.
var. sesbanioides *A. Radcliffe-Smith in Kew Bull.*, 51(2): 316 (1996)—Malawi.
myrsinites Humb., Bonpl. & Kunth
subsp. francavillanus *(Müll. Arg.) G.L. Webster in Fl. Venezuelan Guayana (gen. eds. J.A. Steyermark et al.)*, 5: 198 (1999): P. francavillanus.
nummulariifolius Poir.
var. capillaris *(Schumach. & Thonn.) A. Radcliffe-Smith in Kew Bull.*, 51(2): 316 (1996): P. capillaris.
nyikae *A. Radcliffe-Smith in Kew Bull.*, 51(2): 317 (1996)—Malawi.
oblanceolatus *J.T. Hunter & J.J. Bruhl in J. Adelaide Bot. Gard.*, 17: 128 (1996)—South Australia.
occidentalis *J.T. Hunter & J.J. Bruhl in Telopea*, 7(2): 157 (1997)—New South Wales, Queensland.
orbicularifolius *(P.T. Li) R. Govaerts & A. Radcliffe-Smith in Kew Bull.*, 51(1): 177 (1996): Phyllanthodendron orbicularifolium.
parvulus Sond.
var. garipensis *(E. Mey. ex Drège) A. Radcliffe-Smith in Kew Bull.*, 51(2): 317 (1996): P. garipensis.
pinaiensis *S.L. Welsh, Fl. Societensis:* 112 (1998), nom. nov.: P. urceolatus *Baill.*
prominulatus *J.T. Hunter & J.J. Bruhl in Nuytsia*, 11(2): 153 (1997)—Northern Territory.
pseudocarunculatus *A. Radcliffe-Smith in Kew Bull.*, 51(2): 318 (1996)—Zaire, Zambia.

PHYLLANTHUS:-
reticulatus Poir.
var. orae-solis *A. Radcliffe-Smith in Kew Bull.*, 51(2): 319 (1996)—Mozambique.
rubristipulus *R. Govaerts & A. Radcliffe-Smith in Kew Bull.*, 51(1): 177 (1996), nom. nov.: P. rubriflorus *Beille*.
sarothamnoides *R. Govaerts & A. Radcliffe-Smith in Kew Bull.*, 51(1): 177 (1996), nom. nov.: P. scoparius *Müll. Arg.*
serpentinicola *A. Radcliffe-Smith in Kew Bull.*, 51(2): 320 (1996)—Zimbabwe.
striaticaulis *J.T. Hunter & J.J. Bruhl in J. Adelaide Bot. Gard.*, 17: 133 (1996)—South Australia.
subapicalis Jabl.
subsp. sequoiifolius *(Jabl.) G.L. Webster in Fl. Venezuelan Guayana (gen. eds. J.A. Steyermark et al.)*, 5: 200 (1999): P. sequoiifolius.
sulcatus *J.T. Hunter & J.J. Bruhl in Nuytsia*, 11(2): 155 (1997)—Northern Territory, Queensland.
taylorianus *Jean F. Brunel ex A. Radcliffe-Smith in Kew Bull.*, 51(2): 322 (1996)—Tanzania, Central African Republic, Cameroon, Ethiopia, Uganda, Malawi, Zambia, Zimbabwe.
tener *A. Radcliffe-Smith in Kew Bull.*, 51(2): 323 (1996)—Zambia.
tenuis *A. Radcliffe-Smith in Kew Bull.*, 51(2): 323 (1996)—Zambia.
tsetserrae *Jean F. Brunel ex A. Radcliffe-Smith in Kew Bull.*, 51(2): 325 (1996)—Mozambique.
tui *Nguyen Nghia Thin, Euphorbiac. Vietnam:* 48 (1995 publ. 1996)—Vietnam.
udoricola *A. Radcliffe-Smith in Kew Bull.*, 51(2): 326 (1996)—Zambia, Malawi, Zimbabwe.
virgatus
var. gardnerianus *(Wight) R. Govaerts & A. Radcliffe-Smith in Kew Bull.*, 51(1): 177 (1996): Macraea gardneriana.
xiphephorus *Jean F. Brunel ex A. Radcliffe-Smith in Kew Bull.*, 51(2): 327 (1996)—Zambia.
yunnanensis *(Croizat) R. Govaerts & A. Radcliffe-Smith in Kew Bull.*, 51(1): 178 (1996): Phyllanthodendron yunnanense.
zambicus *A. Radcliffe-Smith in Kew Bull.*, 51(2): 328 (1996)—Zambia.
zornioides *A. Radcliffe-Smith in Kew Bull.*, 51(2): 328 (1996)—Zambia, Malawi, Zimbabwe.

PHYLLOBOLUS (Aizoac.).
abbreviatus *(L. Bolus) M. Gerbaulet in Bot. Jahrb.*, 119(2): 170 (1997): Aridaria abbreviata.
amabilis *M. Gerbaulet & M. Struck in Haseltonia*, 4: 25 (1996)—Cape Province.
caudatus *(L. Bolus) M. Gerbaulet in Bot. Jahrb.*, 119(2): 171 (1997): Mesembryanthemum caudatum.
chrysophthalmus *M. Gerbaulet & M. Struck in Haseltonia*, 4: 26 (1996)—Cape Province.
congestus *(L. Bolus) M. Gerbaulet in Bot. Jahrb.*, 119(2): 172 (1997): Aridaria congesta.
deciduus *(L. Bolus) M. Gerbaulet in Bot. Jahrb.*, 119(2): 173 (1997): Aridaria decidua.
decurvatus *(L. Bolus) M. Gerbaulet in Bot. Jahrb.*, 119(2): 173 (1997): Aridaria decurvata.
delus *(L. Bolus) M. Gerbaulet in Bot. Jahrb.*, 119(2): 174 (1997): Mesembryanthemum delum.
digitatus *(Aiton) M. Gerbaulet in Bot. Jahrb.*, 119(2): 174 (1997): Mesembryanthemum digitatum.
subsp. littlewoodii *(L. Bolus) M. Gerbaulet in Bot. Jahrb.*, 119(2): 176 (1997): Dactylopsis littlewoodii.
gariepensis *M. Gerbaulet & M. Struck in Haseltonia*, 4: 26 (1996)—Cape Province.
grossus *(Aiton) M. Gerbaulet in Bot. Jahrb.*, 119(2): 177 (1997): Mesembryanthemum grossum.
herbertii *(N.E. Br.) M. Gerbaulet in Bot. Jahrb.*, 119(2): 178 (1997): Sphalmanthus herbertii.
latipetalus *(L. Bolus) M. Gerbaulet in Bot. Jahrb.*, 119(2): 178 (1997): Aridaria latipetala.
lignescens *(L. Bolus) M. Gerbaulet in Bot. Jahrb.*, 119(2): 179 (1997): Sphalmanthus lignescens.
melanospermus *(Dinter & Schwantes) M. Gerbaulet in Bot. Jahrb.*, 119(2): 179 (1997): Sphalmanthus melanospermus.
nitidus *(Haw.) M. Gerbaulet in Bot. Jahrb.*, 119(2): 180 (1997): Mesembryanthemum nitidum.
oculatus *(N.E. Br.) M. Gerbaulet in Bot. Jahrb.*, 119(2): 181 (1997): Mesembryanthemum oculatum.

PHYLLOBOLUS :-
prasinus *(L. Bolus) M. Gerbaulet in Bot. Jahrb.*, 119(2): 182 (1997): Aridaria prasina.
pumilus *(L. Bolus) M. Gerbaulet in Bot. Jahrb.*, 119(2): 183 (1997): Aridaria pumila.
quartziticus *(L. Bolus) M. Gerbaulet in Bot. Jahrb.*, 119(2): 184 (1997): Aridaria quartzitica.
rabiei *(L. Bolus) M. Gerbaulet in Bot. Jahrb.*, 119(2): 184 (1997): Aridaria rabiei.
roseus *(L. Bolus) M. Gerbaulet in Bot. Jahrb.*, 119(2): 186 (1997): Amoebophyllum roseum.
saturatus *(L. Bolus) M. Gerbaulet in Bot. Jahrb.*, 119(2): 186 (1997): Aridaria saturata.
sinuosus *(L. Bolus) M. Gerbaulet in Bot. Jahrb.*, 119(2): 187 (1997): Mesembryanthemum sinuosum.
spinuliferus *(Haw.) M. Gerbaulet in Bot. Jahrb.*, 119(2): 188 (1997): Mesembryanthemum spinuliferum.
splendens *(L.) M. Gerbaulet in Bot. Jahrb.*, 119(2): 189 (1997): Mesembryanthemum splendens.
subsp. pentagonus *(L. Bolus) M. Gerbaulet in Bot. Jahrb.*, 119(2): 193 (1997): Aridaria pentagona.
suffruticosus *(L. Bolus) M. Gerbaulet in Bot. Jahrb.*, 119(2): 194 (1997): Cryophytum suffruticosum.
tenuiflorus *(Jacq.) M. Gerbaulet in Bot. Jahrb.*, 119(2): 194 (1997): Mesembryanthemum tenuiflorum.
trichotomus *(Thunb.) M. Gerbaulet in Bot. Jahrb.*, 119(2): 195 (1997): Mesembryanthemum trichotomum.
viridiflorus *(Aiton) M. Gerbaulet in Bot. Jahrb.*, 119(2): 196 (1997): Mesembryanthemum viridiflorum.

PHYLLOCEPHALUM (Compos.).
hookeri *(C.B. Clarke) B.P. Uniyal in Fl. India*, 13: 339 (1995): Centratherum hookeri.
microcephalum *(Dalzell) H. Robinson in Proc. Biol. Soc. Wash.*, 112(1): 235 (1999): Decaneurum microcephalum.

PHYLLOCLADUS (Phyllocladac.).
toatoa *B.P.J. Molloy in New Zealand J. Bot.*, 34(3): 290 (1996)—New Zealand North I.

PHYLLODIUM (Legumin.).
hackeri *L. Pedley in Austrobaileya*, 5(2): 256 (1999)—Queensland.
pulchellum (L.) Desv.
var. glabrius *L. Pedley in Austrobaileya*, 5(2): 257 (1999)—Northern Territory.

× **PHYLLOSASA** *J.-P. Demoly in Bambou*, 21: 14 (1995). *GRAMINEAE* (Phyllostachys × Sasa).
tranquillans *(Koidz.) J.-P. Demoly in Bambou*, 21: 14 (1995): Semiarundinaria tranquillans.

PHYLLOSTACHYS (Gramin.).
cantoniensis *W.T. Lin in J. S. China Agric. Univ.*, 14(3): 112 (1993)—China Southeast (Guangdong).
heteroclada Oliv.
var. funhuaensis *X.G. Wang & Z.M. Lu in J. Bamboo Res.*, 16(4): 15 (1997)—China Southeast (Zhejiang).
makinoi Hayata
f. wuyishanensis *S.S. You & H.L. Yu ex G.H. Lai in J. Wuhan Bot. Res.*, 17(4): 320 (1999)—China Southeast (Zhejiang).
mazelii *Carrière ex V. Nohl in Mitt. Deutsch. Dendrol. Ges.*, 24: 99 (1915)—Japan.
rigida *X. Jiang & Q. Li in J. Sichuan Agric. Coll.*, 2(2): 127 (1984)—China South-central (Sichuan).

PHYLLOSTEGIA (Labiat.).
parviflora (Gaudich.) Benth.
var. lydgatei *(Sherff) W.L. Wagner in Novon*, 9(2): 282 (1999): P. mollis var. lydgatei.
renovans *W.L. Wagner in Novon*, 9(2): 280 (1999)—Hawaiian Is.

PHYMATIDIUM (Orchidac.).
seehaweri *I. Bock in Orchidee*, 49(3): 102 (1998)—Brazil Southeast (Rio de Janeiro).

PHYMATOCARPUS (Myrtac.).
interioris *L.A. Craven in Muelleria*, 12(2): 133 (1999)—Western Australia.

PHYSALIS (Solanac.).
sect. Campanulae *M. Martínez in M. Nee et al. (eds.), Solanaceae IV - Advances Biol. & Utiliz.*: 280 (1999 publ. 2000).
sect. Coztomatae *M. Martínez in M. Nee et al. (eds.), Solanaceae IV - Advances Biol. & Utiliz.*: 280 (1999 publ. 2000).
subgen. Physalodendron *(G. Don) M. Martínez in M. Nee et al. (eds.), Solanaceae IV - Advances Biol. & Utiliz.*: 279 (1999 publ. 2000): P. sect. Physalodendron.
subgen. Quincula *(Raf.) M. Martínez in M. Nee et al. (eds.), Solanaceae IV - Advances Biol. & Utiliz.*: 280 (1999 publ. 2000): gen. Quincula.
sect. Rydbergae *M. Martínez in M. Nee et al. (eds.), Solanaceae IV - Advances Biol. & Utiliz.*: 281 (1999 publ. 2000).
sect. Tehuacanae *M. Martínez in M. Nee et al. (eds.), Solanaceae IV - Advances Biol. & Utiliz.*: 281 (1999 publ. 2000).
hunzikeriana *M. Martínez in Kurtziana*, 27(2): 383 (1999)—Mexico Northeast (Nuevo León).
queretaroensis *Mahinda Martínez & L. Hernández in Acta Bot. Mex.*, 46: 73 (1999)—Mexico Northeast (Querétaro).
quillabambensis *D. Medina Castro in Arnaldoa*, 5(2): 211 (1998)—Peru.
solanaceous *(Schtdl.) B. Axelius in Phytologia*, 79(1): 11 (1995 publ. 1996): Margaranthus solanaceous.
waterfallii *O. Vargas Ponce, M. Martínez Diaz & P.A. Dávila Aranda in Acta Bot. Mex.*, 48: 22 (1999)—Mexico Southwest (Jalisco, Michoacan).

PHYSARIA (Crucif.).
newberryi A. Gray
var. yesicola *R.C. Sivinski in Sida*, 18(3): 673 (1999)—New Mexico.

PHYSODIA (Anthericac.).
capitata *(Hook.) U. Müller-Doblies, J.S. Tang & D. Müller-Doblies in Feddes Repert.*, 107(5–6): 520 (1996): Ornithogalum capitatum.
minor *(Duthie) U. Müller-Doblies, J.S Tang & D. Müller-Doblies in Feddes Repert.*, 107(5–6): 520 (1996): Urginea minor.
physodes *(Jacq.) U. Müller-Doblies, J.S. Tang & D. Müller-Doblies in Feddes Repert.*, 107(5–6): 519 (1996): Anthericum physodes.
pusilla *(Jacq.) U. Müller-Doblies, J.S. Tang & D. Müller-Doblies in Feddes Repert.*, 107(5–6): 520 (1996): Anthericum pusillum.

PHYSOGYNE (Orchidac.).
garayana *R. González Tamayo & D.L. Szlachetko in Fragm. Flor. Geobot.*, 40(2): 767 (1995)—Mexico Southwest (Colima).

PHYSOKENTIA (Palmae).
petiolata *(Burret) D. Fuller in A. Henderson & Borchsenius (eds.), Evol., Variat. & Classif. Palms (Mem.* New York Bot. Gard. 83): 208 (1999), as 'petiolatus': Goniocladus petiolatus.

PHYSOPSIS (Verbenac.).
chrysophylla *(C.A. Gardner) B.L. Rye in Nuytsia*, 11(1): 102 (1996): Newcastelia chrysophylla.
chrysotricha *(F. Muell.) B.L. Rye in Nuytsia*, 11(1): 104 (1996): Newcastelia chrysotricha
viscida *(E. Pritz.) B.L. Rye in Nuytsia*, 11(1): 106 (1996): Newcastelia viscida.

PHYSOSPERMOPSIS (Umbellif.).
fuscopurpurea *(Hand.-Mazz.) M.G. Pimenov & E.V. Kljuykov in Feddes Repert.*, 111(7–8): 538 (2000): Trachydium fuscopurpureum.
handelii *(H. Wolff ex Hand.-Mazz.) M G. Pimenov & E.V. Kljuykov in Feddes Repert.*, 111(7–8): 538 (2000): Pleurospermum handelii.
nana *(Franch.) M.G. Pimenov & E.V. Kljuykov in Feddes Repert.*, 111(7–8): 538 (2000): Pleurospermum nanum.
stricta *(C.B. Clarke) M.G. Pimenov & E.V. Kljuykov in Feddes Repert.*, 111(7–8): 539 (2000): Trachydium obtusiusculum var. strictum.

PIARANTHUS (Asclepiadac.).
atrosanguineus *(N.E. Br.) P.V. Bruyns in Syst. Bot.*, 24(3): 396 (1999): Stapelia atrosanguinea.
decipiens *(N.E. Br.) P.V. Bruyns in Syst. Bot.*, 24(3): 396 (1999): Huerniopsis decipiens.

PICEA (Pinac.).
asperata Mast.
var. heterolepis *(Rehder & E.H. Wilson) L.K. Fu & Nan Li in Novon*, 7(3): 262 (1997): P. heterolepis.
austromandshurica *J. Silba in J. Int. Conifer Preserv. Soc.*, 6(2): 35 (1999)—Cultivation.
austropanlanica *J. Silba in J. Int. Conifer Preserv. Soc.*, 6(2): 33 (1999)—China South-central (Sichuan).
brachytyla (Franch.) E. Pritz.
var. ascendens *(Patschke) J. Silba in J. Int. Conifer Preserv. Soc.*, 7(1): 28 (2000): P. ascendens.
engelmannii Parry ex Engelm.
var. mohinorsensis *J. Silba in J. Int. Conifer Preserv. Soc.*, 7(1): 28 (2000)—Mexico Northeast (Chihuahua).
fricksii *J. Silba in J. Int. Conifer Preserv. Soc.*, 6(2): 32 (1999)—China North-central (Hebei).
likiangensis (Franch.) Pritz.
subsp. balfouriana *(Rehder & E.H. Wilson) K. Rushforth in Int. Dendrol. Soc. Yearbook*, 1998: 61 (1999): P. balfouriana.
meyeri Rehder & E.H. Wilson
f. pyramidalis *(H.W. Jen & C.G. Bai) L.K. Fu & Nan Li in Novon*, 7(3): 262 (1997): P. meyeri var. pyramidalis.
shennongjianensis *J. Silba in J. Int. Conifer Preserv. Soc.*, 6(2): 34 (1999)—China South-central (Hubei).

PICRADENIA (Compos.).
richardsonii
subsp. floribunda *(A. Gray) W.A. Weber in Phytologia*, 79(2): 65 (1995 publ. 1996): Actinella richardsonii var. floribunda.

PICRAMNIA (Simaroubac.).
coccinea *W.W. Thomas in Brittonia*, 49(3): 380 (1997)—Brazil Northeast (Bahia).

PICRIS (Compos.).
hieracioides L.
subsp. koreana *(Kitam.) V.N. Voroschilov in A.K. Skvortsov (ed.), Florist. issl. v razn. raĭonakh SSSR*: 198 (1985): P. japonica var. koreana.
f. laevicaulis *M. Mizush. ex T. Shimizu, Fl. Nagano Pref.*: 1510 (1997)—Japan.
scabra Forssk.
subsp. abyssinica *(Sch. Bip.) M. Smalla in Willdenowia*, 30(2): 319 (2000): Picris abyssinica.

PICTETIA (Legumin.).
mucronata *(Griseb.) A. Beyra M. & M. Lavin, Monogr. Pictetia (Legumin.-Papilion.) & Rev. Aeschynomeneae (Syst. Bot.* Monog., 56): 56 (1999): Belairia mucronata.
nipensis *(Urb.) A. Beyra M. & M. Lavin, Monogr. Pictetia (Legumin.-Papilion.) & Rev. Aeschynomeneae (Syst. Bot.* Monog., 56): 65 (1999): Belairia nipensis.
spinosa *(A. Rich.) A. Beyra M. & M. Lavin, Monogr. Pictetia (Legumin.-Papilion.) & Rev. Aeschynomeneae (Syst. Bot.* Monog., 56): 63 (1999): Belairia spinosa.
sulcata *(P. Beauv.) A. Beyra M. & M. Lavin, Monogr. Pictetia (Legumin.-Papilion.) & Rev. Aeschynomeneae (Syst. Bot.* Monog., 56): 45 (1999): Ormocarpum sulcatum.

PIERREBRAUNIA *E. Esteves Pereira in Cact. Succ. J. (U.S.A.)*, 69(6): 296 (1997). *CACTACEAE*
bahiensis *(P.J. Braun & Esteves) E. Esteves Pereira in Cact. Succ. J. (U.S.A.)*, 69(6): 296 (1997): Floribunda bahiensis.
brauniorum *E. Esteves Pereira in Kakt. And. Sukk.*, 50(12): 312 (1999)—Brazil Southeast (Minas Gerais).

PIETROSIA (Compos.).
sect. Andryalopsis *A.N. Sennikov in Komarovia*, 1: 78 (1999).

PIETROSIA:-
agardhii *(Haens. ex DC.) A.N. Sennikov in Komarovia*, 1: 78 (1999): Andryala agardhii.
laevitomentosa *Nyár. ex A.N. Sennikov in Komarovia*, 1: 78 (1999)—Romania.

PILEA (Urticac.).
adamsiana *A.K. Monro in Bull. Nat. Hist. Mus. (London), Bot.*, 30(1): 9 (2000)—Panama.
cangyuanensis *H.W. Li in Fl. Yunnanica*, 7: 215 (1997)—China South-central (Yunnan).
conjugalis *A.K. Monro in Bull. Nat. Hist. Mus. (London), Bot.*, 30(1): 7 (2000)—Panama, Costa Rica.
elliptifolia *B.L. Shih & Yuen P. Yang in Taiwania*, 40(3): 262 (1995)—Taiwan.
hedemarkii *W. Takeuchi in Sida*, 18(3): 770 (1999)—New Guinea (Papua New Guinea).
hilliana Hand.-Mazz.
var. corniculata *H.W. Li in Fl. Yunnanica*, 7: 201 (1997)—China South-central (Yunnan).
howelliana Hand.-Mazz.
var. longipedunculata *(J.J. Chien & C.J. Chen) H.W. Li in Fl. Yunnanica*, 7: 222 (1997): P. longipedunculata.
magnicarpa *A.K. Monro in Novon*, 9(3): 398 (1999)—Panama.
microphylla (L.) Liebm.
var. nanophylla *M.-L. Groult in Compte R. Acad. Sci., Sér. 3 (Paris)*, 322(9): 819 (1999)—Windward Is (Martinique).
occulta *J. Florence in Fl. Polynésie française*, 1: 236 (1997)—Tubuai Is.
plumulosa *A.K. Monro in Novon*, 9(3): 392 (1999)—Panama.
pteridophylla *A.K. Monro in Novon*, 9(3): 390 (1999)—Mexico Southeast (Tabasco).
quadrata *A.K. Monro in Novon*, 9(3): 395 (1999)—Panama.
rostulata *A.K. Monro in Novon*, 9(3): 395 (1999)—Panama.
sancti-johannis *J. Florence in Fl. Polynésie française*, 1: 237 (1997)—Tuamotu Is.
solandri *(Seem.) J. Florence in Bull. Mus. Nation. Hist. Nat., B, Adansonia*, Ser. 4, 18(3–4): 270 (1996): Lecanthus solandri.
spathulifolia *M.-L. Groult in Adansonia*, 21(2): 270 (1999)—Dominican Republic.
succulenta *(Griseb.) M.-L. Groult in Compte R. Acad. Sci., Sér. 3 (Paris)*, 322(9): 822 (1999), homonym; *see Wedd. in Arch. Mus. Hist. Nat. Paris*, 9: 256 (1856): P. microphylla var. succulenta.
var. domingensis *M.-L. Groult in Adansonia*, 21(2): 272 (1999)—Dominican Republic.
trichomanophylla *A.K. Monro in Bull. Nat. Hist. Mus. (London), Bot.*, 30(1): 9 (2000)—Panama.
tripartita *A.K. Monro in Novon*, 9(3): 393 (1999)—Costa Rica.
tutensis *A.K. Monro in Novon*, 9(3): 397 (1999)—Panama.

PILOCARPUS (Rutac.).
alatus *C.J. Joseph ex L.A. Skorupa in Novon*, 8(4): 447 (1998)—Brazil North (Pará).
carajaensis *L.A. Skorupa in Novon*, 8(4): 449 (1998)—Brazil North (Pará).
manuensis *L.A. Skorupa in Novon*, 9(2): 257 (1999)—Peru.
sulcatus *L.A. Skorupa in Novon*, 8(4): 453 (1998)—Brazil Northeast (Bahia), Brazil Southeast (Minas Gerais).

PILOSELLA (Compos.).
akinfiewii *(Woronow & Zahn) A.N. Sennikov in Bot. Zhurn.*, 83(3): 62 (1998): Hieracium incanum subsp. akinfiewii.
antennarioidiformis *(Zahn) A.N. Sennikov in Bot. Zhurn.*, 83(3): 62 (1998): Hieracium hoppeanum subsp. antennarioidiforme.
branae *(Nägeli & Peter) A.N. Sennikov in Bot. Zhurn.*, 83(3): 63 (1998): Hieracium magyaricum subsp. branae.
buhsei *(Nägeli & Peter) A.N. Sennikov in Bot. Zhurn.*, 83(3): 63 (1998): Hieracium procerum subsp. buhsei.
calolepidea *(Norrl.) R.N. Schljakov in Fl. Evropeĭskoĭ Chasti SSSR*, 8: 354 (1989): P. decolorans * calolepidea.

PILOSELLA:-
caucasica *(Nägeli & Peter) A.N. Sennikov in Bot. Zhurn.*, 83(2): 78 (1998): Hieracium caucasicum.
× ciesielskii *(Błocki) Val.N. Tikhomirov in Bot. Zhurn.*, 85(11): 112 (2000): Hieracium ciesielskii.
cymanthodes *(Koslovsky & Zahn) A.N. Sennikov in Bot. Zhurn.*, 83(3): 63 (1998): Hieracium bauhinii subsp. cymanthodes.
czerepninii *N.N. Tupitsȳna in Bot. Zhurn.*, 81(3): 128 (1996)—Krasnoyarsk.
dublitzkii *(B. Fedtsch. & Nevski) N.N. Tupitzina in Fl. Sibir. (Asterac.)*, 13: 348 (1997 publ. 1998): Hieracium dublitzkii.
dublitzkii *(B. Fedtsch. & Nevski) A.N. Sennikov in Bot. Zhurn.*, 83(2): 78 (1998): Hieracium dublitzkii.
farinoderma *(Litv. & Zahn) A.N. Sennikov in Bot. Zhurn.*, 83(3): 64 (1998): Hieracium incanum subsp. farinodermum.
fastigiatiformis *(Zahn) A.N. Sennikov in Bot. Zhurn.*, 83(3): 64 (1998): Hieracium bauhinii subsp. fastigiatiforme.
freynii *(Nägeli & Peter) A.N. Sennikov in Bot. Zhurn.*, 83(3): 64 (1998): Hieracium echioides subsp. freynii.
katunensis *N.N. Tupitsȳna in Bot. Zhurn.*, 81(3): 124 (1996)—Altay.
kebeshensis *(Stepanov) N.N. Tupitzina in Fl. Sibir. (Asterac.)*, 13: 345 (1997 publ. 1998): Hieracium kebeshense.
kumbelica *(B. Fedtsch. & Nevski) A.N. Sennikov in Bot. Zhurn.*, 83(2): 78 (1998): Hieracium kumbelicum.
lamprocomoides *(Woronow & Zahn) A.N. Sennikov in Bot. Zhurn.*, 83(3): 66 (1998): Hieracium hypeuryum subsp. lamprocomoides.
lasiothrix *(Nägeli & Peter) A.N. Sennikov in Bot. Zhurn.*, 83(3): 66 (1998): Hieracium hypeuryum subsp. lasiothrix.
× leptadeniiformis *(Üksip) N.N. Tupitzina in Fl. Sibir. (Asterac.)*, 13: 352 (1997 publ. 1998): Hieracium leptadeniiforme.
leptadeniiformis *(Üksip) A.N. Sennikov in Bot. Zhurn.*, 83(2): 78 (1998): Hieracium leptadeniiforme.
longiscapa *(Nägeli & Peter) A.N. Sennikov in Bot. Zhurn.*, 83(2): 78 (1998): Hieracium spathophyllum subsp. longiscapum.
lydiae *(Schischk. & Steinb.) N.N. Tupitzina in Fl. Sibir. (Asterac.)*, 13: 343 (1997 publ. 1998)—Hieracium lydiae.
massagetovii *(Gamajun. ex Kamelin & Zuckerw.) A.N. Sennikov in Bot. Zhurn.*, 83(2): 79 (1998): Hieracium massagetovii.
novosibirskensis *N.N. Tupitsȳna in Bot. Zhurn.*, 81(3): 126 (1996)—West Siberia.
perileuca *(Schelk. & Zahn) A.N. Sennikov in Bot. Zhurn.*, 83(3): 68 (1998): Hieracium hoppeanum subsp. perileucum.
pinea *(Schischk. & Serg.) N.N. Tupitzina in Fl. Sibir. (Asterac.)*, 13: 342 (1997 publ. 1998): Hieracium pineum.
procerigena *(Litv. & Zahn) A.N. Sennikov in Bot. Zhurn.*, 83(3): 69 (1998): Hieracium auriculoides subsp. procerigenum.
rubrobauhinii *(Schelk. & Zahn) A.N. Sennikov in Bot. Zhurn.*, 83(3): 69 (1998): Hieracium bauhinii subsp. rubrobauhinii.
sabinopsis *(Ganesch. & Zahn) N.N. Tupitzina in Fl. Sibir. (Asterac.)*, 13: 352 (1997 publ. 1998): Hieracium fallax subsp. sabinopsis.
× solacolui *(Prodán ex Nyár.) Val.N. Tikhomirov in Bot. Zhurn.*, 85(11): 113 (2000): Hieracium solacolui.
sosnowskyi *(Zahn) A.N. Sennikov in Bot. Zhurn.*, 83(2): 79 (1998): Hieracium incanum subsp. sosnowskyi.
stupposipila *(Woronow & Zahn) A.N. Sennikov in Bot. Zhurn.*, 83(3): 70 (1998): Hieracium florentinum subsp. stupposipilum.
subrubella *(Schelk. & Zahn) A.N. Sennikov in Bot. Zhurn.*, 83(3): 71 (1998): Hieracium tephrocephalum subsp. subrubellum.
× tephrochlorella *(Ganesch. & Zahn) N.N. Tupitzina in Fl. Sibir. (Asterac.)*, 13: 353 (1997 publ. 1998): Hieracium heterodoxum subsp. tephrochlorellum.
tjumentzevii *(Serg. & Üksip) N.N. Tupitzina in Fl. Sibir. (Asterac.)*, 13: 342 (1997 publ. 1998): Hieracium tjumentzevii.

PILOSELLA:-
tremedalis Mateo
subsp. caballeroi *G. Mateo Sanz in Flora Montiberica*, 3: 31 (1996)—Spain.
vahlii (Froel.) F.W. Schultz & Sch. Bip.
subsp. pseudovahlii *(De Retz) G. Mateo Sanz in Flora Montiberica*, 2: 36 (1996): H. pseudovahlii.
virentisquama *(Nägeli & Peter) A.N. Sennikov in Bot. Zhurn.*, 83(3): 73 (1998): Hieracium hoppeanum subsp. virentisquamum.

PILOSOCEREUS (Cactac.).
azulensis *N.P. Taylor & D.C. Zappi in Cactaceae Consensus Initiatives*, 3: 8 (1997)—Brazil Southeast (Minas Gerais).
chrysostele (Vaupel) Byles & G.D. Rowley
subsp. cearensis *P.J. Braun & E. Esteves Pereira in Brit. Cact. Succ. J.*, 17(1): 26 (1999)—Brazil Northeast (Ceará).
densiareolatus F. Ritter
subsp. brunneolanatus *P.J. Braun & E. Esteves Pereira in Kakt. And. Sukk.*, 50(11): 286 (1999)—Brazil Northeast (Bahia).
estevesii *P.J. Braun in Cact. Succ. J. (U.S.A.)*, 71(2): 74 (1999)—Brazil Northeast (Bahia).
occultiflorus *P.J. Braun & E. Esteves Pereira in Cact. Succ. J. (U.S.A.)*, 71(6): 311 (1999)—Brazil Southeast (Minas Gerais).
palmeri
var. sartorianus *(Rose) Lodé in Fichier Encycl. Cact. autres Succ.*, 19: 1776 (1997); *cf. Repert. Pl. Succ. (I.O.S.)*, 48: 14 (1997 publ. 1998): Cephalocereus sartorianus.

PILOURATEA (Ochnac.).
glabrifolia *Tiegh. in Ann. Sci. Nat., Bot., ser. 8*, 16: 239 (1902), in I.K. Supp. 3 with incorrect ref.—Brazil.†

PIMELEA (Thymelaeac.).
neokyrea *B.L. Rye in Nuytsia*, 13(1): 183 (1999)—Western Australia.
rosea R. Br.
subsp. annelsii *B.L. Rye in Nuytsia*, 13(1): 186 (1999)—Western Australia.

PIMPINELLA (Umbellif.).
subsect. Corymbosae *(H. Wolff) O.V. Yurtseva in Byull. Mosk. Obshch. Ispȳt. Prir., Biol.*, 100(3): 66 (1995): P. subser. Corymbosae.
brachycarpa (Kom.) Nakai
var. uchiyamana *(Nakai) G. Jang, W. Paik & W. Lee in Korean J. Pl. Taxon.*, 29(2): 161 (1999): P. koreana var. uchiyamana.
candolleana Wight & Arn.
var. rotundifolia *R.H. Shan & Z.H. Pan in Fl. Yunnanica*, 7: 495 (1997)—China South-central (Yunnan).
hallaisanensis *(W. Lee & G. Jang) G. Jang, W. Paik & W. Lee in Korean J. Pl. Taxon.*, 29(2): 163 (1999): P. brachycarpa var. hallaisanensis.
saxifraga L.
subsp. nigra *(Mill.) N.N. Tzvelev in Novosti Sist. Vȳssh. Rast.*, 32: 185 (2000): P. nigra.
thellungiana H. Wolff
var. gustavohegiana *(Koidz.) Kitam. ex H. Ohba in Fl. Japan*, 2c: 281 (1999): P. gustavohegiana.
tragium Vill.
subsp. balearica *(Knoche) A.M. Romo, Fl. Silvestres Baleares:* 192 (1994), isonym; *see Malagarriga, Las Subesp. y Variac. Geogr.:* 13 (1973): P. tragium var. balearica.

PINANGA (Palmae).
auriculata Becc.
var. leucocarpa *C.K. Lim in Gard. Bull. Singapore*, 50(1): 93 (1998)—Malaya (Peninsular Malaysia, Singapore), Thailand.
var. merguensis *(Becc.) C.K. Lim in Gard. Bull. Singapore*, 50(1): 89 (1998): P. patula var. merguensis.
badia *D.R. Hodel in Palm J.*, 136: 16 (1997)—Thailand.
bowiana *D.R. Hodel in Palm J.*, 134: 35 (1997)—Thailand.
fractiflexa *D.R. Hodel in Palm J.*, 136: 19 (1997)—Thailand.

PINANGA:-
jamariensis *C.K. Lim in Gard. Bull. Singapore*, 50(1): 100 (1998)—Malaya (Peninsular Malaysia).
johorensis *C.K. Lim & L.G. Saw in Gard. Bull. Singapore*, 50(1): 104 (1998)—Malaya (Peninsular Malaysia, Singapore).
palustris *R. Kiew in Gard. Bull. Singapore*, 50(1): 106 (1998)—Malaya (Peninsular Malaysia).
pantiensis *J. Dransfield in Gard. Bull. Singapore*, 50(1): 110 (1998)—Malaya (Peninsular Malaysia).
sylvestris *(Lour.) D.R. Hodel in Palm J.*, 139: 55 (1998): Areca sylvestris.
watanaiana *C.K. Lim in Principes*, 42(2): 116 (1998)—Thailand.

PINAROPAPPUS (Compos.).
mojadanus *B.L. Turner in Phytologia*, 80(2): 100 (1996)—Mexico Northeast (Coahuila).
powellii *B.L. Turner in Phytologia*, 80(2): 101 (1996)—Mexico Northeast (Coahuila).

PINGA *E.A. Widjaja in Reinwardtia*, 11(2): 123 (1997). *GRAMINEAE*
marginata *E.A. Widjaja in Reinwardtia*, 11(2): 124 (1997)—New Guinea (Irian Jaya).

PINGUICULA (Lentibulariac.).
sect. Longifoliae *(Casper) G. Blanca, M. Ruiz-Rejón & Reg. Zamora in Folia Geobot.*, 34(3): 347 (1999), as 'Longifolia': P. ser. Longifoliae.
*sect.*Orchidioides *H. Luhrs in Phytologia*, 79(2): 118 (1995 publ. 1996).
dertosensis *(Cañig.) G. Mateo Sanz & M.B. Crespo Villalba, Fl. Abrev. Comun. Valenciana:* 430 (1995): P. grandiflora var. dertosensis.
elizabethiae *S. Zamudio in Acta Bot. Mex.*, 47: 16 (1999)—Mexico Northeast (Hidalgo, Querétaro).
fontiqueriana *A. Romo, J.B. Peris & G. Stübing in Ann. Bot. Fenn.*, 33(1): 31 (1996)—Morocco.
gigantea *H. Luhrs in Phytologia*, 79(6): 389 (1995 publ. 1996)—Mexico Southwest (Oaxaca).
laxifolia *H. Luhrs in Phytologia*, 79(2): 116 (1995 publ. 1996)—Mexico Northeast (Tamaulipas).
medusina *S. Zamudio Ruiz & M. Studnicka in Acta Bot. Mex.*, 53: 68 (2000)—Mexico Southwest (Oaxaca).
mesophytica *S. Zamudio in Acta Bot. Mex.*, 40: 65 (1997)—El Salvador, Guatemala, Honduras.
mirandae *S. Zamudio Ruiz & A. Salinas T.in Acta Bot. Mex.*, 37: 40 (1996)—Mexico Southwest (Oaxaca).
moranensis Humb., Bonpl. & Kunth
var. neovolcanica *S. Zamudio in Acta Bot. Mex.*, 49: 27 (1999)—Mexico Northeast (Hidalgo), Mexico Central (Mexico Distrito Federal, México State, Morelos, Puebla, Tlaxcala), Mexico Gulf (Veracruz), Mexico Southwest (Michoacan).
mundi *G. Blanca, M. Jamilena, M. Ruiz-Rejón & R. Zamora in Pl. Syst. Evol.*, 200(1–2): 58 (1996)—Spain.
stolonifera *H. Luhrs in Phytologia*, 79(2): 114 (1995 publ. 1996)—Mexico Southwest (Oaxaca).
submediterranea *G. Blanca, M. Jamilena, M. Ruiz-Rejón & R. Zamora in Pl. Syst. Evol.*, 200(1–2): 57 (1996)—Spain.
vulgaris L.
var. floribunda *S. Watanabe & A. Takeda in J. Jap. Bot.*, 72(4): 237 (1997)—Japan.

PINUS (Pinac.).
arizonica Engelm.
var. cooperi *(C.E. Blanco) A. Farjon in Fl. Neotrop.*, 75: 107 (1997): P. cooperi.
armandii Franch.
var. farjonii *J. Silba in J. Int. Conifer Preserv. Soc.*, 7(1): 29 (2000)—Burma.
subsp. mastersiana *(Hayata) R. Businský in Acta Průhoniciana*, 68: 13 (1999): P. mastersiana.
bhutanica Grierson, D.G. Long & C.N. Page
var. ludlowii *J. Silba in J. Int. Conifer Preserv. Soc.*, 7(1): 29 (2000)—Xizang.
brutia Ten.
var. densifolia *F. Yaltırık & M. Boydak in Karaca Arbor. Mag.*, 5(4): 175 (2000)—Turkey.
f. kruepericola *M.P. Frankis in Papers, Int. Symp. Pinus brutia Ten., Marmaris:* 14 (1993)—Turkey.

PINUS:-

brutia
var. pendulifolia *M.P. Frankis in Papers, Int. Symp. Pinus brutia Ten., Marmaris:* 14 (1993)—Turkey.
var. stankewiczii *(Sukaczev) M.P. Frankis in Papers, Int. Symp. Pinus brutia Ten., Marmaris:* 14 (1993): P. pityusa var. stankewiczii.
cembroides Zucc.
var. quadrifolia *(Parl. ex Sudw.) J. Silba in J. Int. Conifer Preserv. Soc.,* 7(1): 29 (2000): P. quadrifolia.
dalatensis Ferré
var. bidoupensis *R. Businský in Candollea,* 54(1): 127 (1999)—Vietnam.
subsp. procera *R. Businský in Candollea,* 54(1): 133 (1999)—Vietnam.
fenzeliana Hand.-Mazz.
var. annamiensis *J. Silba in J. Int. Conifer Preserv. Soc.,* 7(1): 30 (2000)—Vietnam.
var. dabeshanensis *(W.C. Cheng & Y.W. Law) L.K. Fu & Nan Li in Novon,* 7(3): 262 (1997): P. dabeshanensis.
flexilis E. James
var. alpina *J. Silba in J. Int. Conifer Preserv. Soc.,* 7(1): 30 (2000)—California.
var. callahanii *J. Silba in J. Int. Conifer Preserv. Soc.,* 7(1): 30 (2000)—Montana, South Dakota.
greggii Engelm. ex Parl.
var. australis *J.K. Donahue & J. Lopez Upton in Sida,* 18(4): 1092 (1999)—Mexico Northeast (Querétaro).
kesiya Royle ex Gordon
subsp. insularis *(Endl.) D.Z. Li in Edinburgh J. Bot.,* 54(3): 346 (1997): P. insularis.
subsp. yunnanensis *(Franch.) R. Businský in Acta Průhoniciana,* 68: 24 (1999): P. yunnanensis.
kwangtungensis Chun ex Tsiang
var. varifolia *Nan Li & Y.C. Zhong in Novon,* 7(3): 262 (1997)—China Southeast (Guangxi).
luchuensis Mayr
subsp. hwangshanensis *(W.Y. Hsia) D.Z. Li in Edinburgh J. Bot.,* 54(3): 341 (1997): P. hwangshanensis.
var. shenkanensis *J. Silba in J. Int. Conifer Preserv. Soc.,* 6(1): 26 (1999)—China.
subsp. taiwanensis *(Hayata) D.Z. Li in Edinburgh J. Bot.,* 54(3): 342 (1997): P. taiwanensis.
luzmariae *J.A. Pérez de la Rosa in Bol. Inst. Bot. (Guadalajara),* 5(1–3): 127 (1998)—Mexico Southwest (Jalisco), Mexico Northeast (Durango).
massoniana
var. hainanensis *W.C. Cheng & L.K. Fu in Acta Phytotax. Sin.,* 13(4): 85 (1975), in IK Supp. 16 as 'haianensis' in error—China.†
merkusii Jungh. & de Vriese
subsp. latteri *(Mason) D.Z. Li in Edinburgh J. Bot.,* 54(3): 346 (1997): P. latteri.
nigra
var. angustisquama *(Willk.) E. Laguna Lumbreras in Flora Montiberica,* 15: 28 (2000): P. laricio var. angustisquama.
subsp. caramanica *(Loudon) R. Businský in Acta Průhoniciana,* 68: 22 (1999): P. laricio var. caramanica.
var. latisquama *(Willk.) E. Laguna Lumbreras in Flora Montiberica,* 15: 28 (2000): P. laricio var. latisquama.
var. salzmannii *(Dunal) E. Laguna Lumbreras in Flora Montiberica,* 15: 28 (2000): P. salzmannii.
parviflora Siebold & Zucc.
subsp. pentaphylla *(Mayr) R. Businský in Acta Průhoniciana,* 68: 12 (1999): P. pentaphylla.
quadrifolia *Parl. ex G.B. Sudworth, Nomencl. Arb. Fl. U.S.:* 17 (1897), nom. nov.: P. llaveana *Torr.*
sylvestris L.
subsp. densiflora *(Siebold & Zucc.) V.N. Voroschilov in A.K. Skvortsov (ed.), Florist. issl. v razn. raĭonakh SSSR:* 146 (1985): P. densiflora.
subsp. krylovii *(Serg. & Kondr.) V.V. Busik in Fl. Tsentral'noi Sibiri,* 1: 50 (1979): P. krylovii.
subsp. sibirica *(Ledeb.) R. Businský in Acta Průhoniciana,* 68: 20 (1999): P. sylvestris var. sibirica.

PINUS:-

tabuliformis Carrière
subsp. henryi *(Mast.) R. Businský in Acta Průhoniciana,* 68: 26 (1999): P. henryi.
subsp. mukdensis *(Nakai) R. Businský in Acta Průhoniciana,* 68: 26 (1999): P. mukdensis.
wallichiana A.B. Jacks.
subsp. bhutanica *(Grierson, D.G. Long & C.N. Page) R. Businský in Acta Průhoniciana,* 68: 10 (1999): P. bhutanica.
wangii Hu & W.C. Cheng
subsp. kwangtungensis *(Chun ex Tsiang) R. Businský in Acta Průhoniciana,* 68: 11 (1999): P. kwangtungensis.
subsp. varifolia *(Nan Li & Y.C. Zhong) R. Businský in Acta Průhoniciana,* 68: 11 (1999): P. kwangtungensis var. varifolia.

PIPER (Piperac.).
sect. Radula *(Miq.) Tebbs ex B. Verdcourt in Fl. Trop. E. Afr., Piperac.:* 3 (1996): Artanthe sect. Radula.
aequale Vahl
var. glanduligerum *(C.DC.) J.A. Steyermark in Fl. Venezuela,* 2(2): 314 (1984): P. glanduligerum.
agostiniorum *J.A. Steyermark in Fl. Venezuela,* 2(2): 317 (1984)—Venezuela.
amalago L.
f. ceanothifolium *(Humb., Bonpl. & Kunth) J.A. Steyermark in Fl. Venezuela,* 2(2): 322 (1984): P. ceanothifolium.
anonifolium *(Kunth) E.G. Steudel, Nomencl. Bot., ed. 2,* 2: 339 (1841): Steffensia anonifolia.
var. parkerianum *(C.DC.) J.A. Steyermark in Fl. Venezuela,* 2(2): 328 (1984): P. parkerianum.
arieianum C.DC.
f. barbulaense *J.A. Steyermark in Fl. Venezuela,* 2(2): 343 (1984)—Venezuela.
var. chimantanum *(Yunck.) J.A. Steyermark in Fl. Venezuela,* 2(2): 344 (1984): P. chimantanum.
var. duidaense *(Trel.) J.A. Steyermark in Fl. Venezuela,* 2(2): 344 (1984): P. duidaense.
f. falconense *J.A. Steyermark in Fl. Venezuela,* 2(2): 341 (1984)—Venezuela.
f. nubicola *J.A. Steyermark in Fl. Venezuela,* 2(2): 342 (1984), as 'nubicolum'—Venezuela.
f. puberulum *J.A. Steyermark in Fl. Venezuela,* 2(2): 341 (1984)—Venezuela.
var. yaracuyense *(Yunck.) J.A. Steyermark in Fl. Venezuela,* 2(2): 340 (1984): P. yaracuyense.
baccans (Miq.) C.DC.
var. sancarlosianum *(C.DC.) J.A. Steyermark in Fl. Venezuela,* 2(2): 349 (1984): P. sancarlosianum.
bartlingianum (Miq.) C.DC.
var. parimanum *(Trel.) J.A. Steyermark in Fl. Venezuela,* 2(2): 352 (1984): P. parimanum.
berryi *J.A. Steyermark in Fl. Venezuela,* 2(2): 354 (1984)—Venezuela.
boehmeriifolium (Miq.) C.DC.
var. glabricaule *(C.DC.) M.G. Gilbert & N.H. Xia in Novon,* 9(2): 191 (1999): P. glabricaule.
bredemeyeri Jacq.
f. escabridum *J.A. Steyermark in Fl. Venezuela,* 2(2): 361 (1984)—Venezuela.
canovillosum *J.A. Steyermark in Fl. Venezuela,* 2(2): 362 (1984)—Venezuela.
capense L.f.
var. brachyrhachis *(C.H. Wright) B. Verdcourt in Fl. Trop. E. Afr., Piperac.:* 7 (1996): P. brachyrhachis.
cathayanum *M.G. Gilbert & N.H. Xia in Novon,* 9(2): 191 (1999), nom. nov.: Chavica sinensis *Champ. ex Benth.*
cernuum Vell.
var. araguense *(Trel. & Yunck.) J.A. Steyermark in Fl. Venezuela,* 2(2): 372 (1984): P. araguense.
f. papillatum *J.A. Steyermark in Fl. Venezuela,* 2(2): 369 (1984)—Venezuela.
var. perlongispicum *(Yunck.) J.A. Steyermark in Fl. Venezuela,* 2(2): 372 (1984): P. perlongispicum.
f. pubens *J.A. Steyermark in Fl. Venezuela,* 2(2): 374 (1984)—Venezuela.
f. subglabrirameum *(Yunck.) J.A. Steyermark in Fl. Venezuela,* 2(2): 370 (1984): P. arbelaezii var. subglabrirameum.
cerronianum *J.A. Steyermark in Fl. Venezuela,* 2(2): 375 (1984)—Venezuela.

PIPER:-

cordiforme *J.A. Steyermark in Fl. Venezuela,* 2(2): 380 (1984)—Venezuela.
coruscans Humb., Bonpl. & Kunth
var. membranaceum *(C.DC.) J.A. Steyermark in Fl. Venezuela,* 2(2): 383 (1984): P. pseudochurumayu var. membranaceum.
crenulatum *J.A. Steyermark in Fl. Venezuela,* 2(2): 385 (1984)—Venezuela.
curvinervium *R. Callejas & J. Betancur in Novon,* 7(1): 21 (1997)—Colombia.
cuyunianum *J.A. Steyermark in Fl. Venezuela,* 2(2): 391 (1984)—Venezuela.
deliciasanum *J.A. Steyermark in Fl. Venezuela,* 2(2): 394 (1984)—Venezuela.
diffamatum Trel. & Yunck.
var. angustius *J.A. Steyermark in Fl. Venezuela,* 2(2): 402 (1984)—Venezuela.
dilatatum Rich.
f. dilatatifolium *(Trel. & Yunck.) J.A. Steyermark in Fl. Venezuela,* 2(2): 406 (1984): P. dilatatifolium.
f. guayranum *(C.DC.) J.A. Steyermark in Fl. Venezuela,* 2(2): 407 (1984): P. guayranum.
f. longipilum *J.A. Steyermark in Fl. Venezuela,* 2(2): 412 (1984)—Venezuela.
f. regresivanum *(Trel. & Yunck.) J.A. Steyermark in Fl. Venezuela,* 2(2): 411 (1984): P. regresivanum.
f. taboganum *(C.DC.) J.A. Steyermark in Fl. Venezuela,* 2(2): 409 (1984): P. taboganum.
dolichostachyum *M.G. Gilbert & N.H. Xia in Novon,* 9(2): 192 (1999)—China South-central (Yunnan).
dolichostylum *R. Callejas & J. Betancur in Novon,* 7(1): 20 (1997)—Colombia.
dunstervilleorum *J.A. Steyermark in Fl. Venezuela,* 2(2): 418 (1984)—Venezuela.
fundacionense *J.A. Steyermark in Fl. Venezuela,* 2(2): 426 (1984)—Venezuela.
galicianum *J.A. Steyermark in Fl. Venezuela,* 2(2): 427 (1984)—Venezuela.
f. hirsutopetiolatum *J.A. Steyermark in Fl. Venezuela,* 2(2): 429 (1984)—Venezuela.
heterobracteum *J.A. Steyermark in Fl. Venezuela,* 2(2): 439 (1984)—Venezuela.
hippocrepiforme *J.A. Steyermark in Fl. Venezuela,* 2(2): 440 (1984)—Venezuela.
hispidum Sw.
f. lanceolatum *(Trel. & Yunck.) J.A. Steyermark in Fl. Venezuela,* 2(2): 449 (1984): P. hispidum var. lanceolatum.
f. patulipilum *(Yunck.) J.A. Steyermark in Fl. Venezuela,* 2(2): 451 (1984): P. patulipilum.
f. surinamense *(Miq.) J.A. Steyermark in Fl. Venezuela,* 2(2): 448 (1984): Artanthe asperifolia f. surinamensis.
liesneri *J.A. Steyermark in Fl. Venezuela,* 2(2): 467 (1984)—Venezuela.
linguliforme *J.A. Steyermark in Fl. Venezuela,* 2(2): 468 (1984)—Venezuela.
var. adpressipilum *J.A. Steyermark in Fl. Venezuela,* 2(2): 469 (1984)—Venezuela.
longiappendiculatum *J.A. Steyermark in Fl. Venezuela,* 2(2): 469 (1984)—Venezuela.
longispicum C.DC.
var. glabratum *(Yunck.) J.A. Steyermark in Fl. Venezuela,* 2(2): 471 (1984): P. nobile var. glabratum.
f. tomentosum *J.A. Steyermark in Fl. Venezuela,* 2(2): 474 (1984)—Venezuela.
marginatum Jacq.
f. catalpifolium *(Humb., Bonpl. & Kunth) J.A. Steyermark in Fl. Venezuela,* 2(2): 480 (1984): P. marginatum var. catalpifolium.
var. niceforoi *(Trel. & Yunck.) J.A. Steyermark in Fl. Venezuela,* 2(2): 482 (1984): P. niceforoi.
marturetense Trel. & Yunck.
f. pilosum *J.A. Steyermark in Fl. Venezuela,* 2(2): 485 (1984)—Venezuela.
methysticum G. Forst.
var. wichmannii *(C.DC.) V. Lebot in Biochem. Syst. Ecol.,* 24(7–8): 782 (1996), without basionym ref.: P. wichmannii.
minasarum *J.A. Steyermark in Fl. Venezuela,* 2(2): 486 (1984)—Venezuela.
morilloi *J.A. Steyermark in Fl. Venezuela,* 2(2): 489 (1984)—Venezuela.

PIPER:-
nubigenum Kunth
var. venezuelense *J.A. Steyermark in Fl. Venezuela,* 2(2): 497 (1984)—Venezuela.
otto-huberi *J.A. Steyermark in Fl. Venezuela,* 2(2): 503 (1984)—Venezuela.
var. eciliatum *J.A. Steyermark in Fl. Venezuela,* 2(2): 504 (1984)—Venezuela.
papilliferum *J.A. Steyermark in Fl. Venezuela,* 2(2): 507 (1984)—Venezuela.
pavasense *J.A. Steyermark in Fl. Venezuela,* 2(2): 510 (1984), homonym; *see Trel. in Contr. U.S. Natl. Herb.,* 26: 173 (1929)—Venezuela.
pendentispicum Trel. & Yunck.
var. furcatipilosum *(Yunck.) J.A. Steyermark in Fl. Venezuela,* 2(2): 515 (1984): P. furcatipilosum.
perijaense *J.A. Steyermark in Fl. Venezuela,* 2(2): 517 (1984)—Venezuela.
perstipulare *J.A. Steyermark in Fl. Venezuela,* 2(2): 519 (1984)—Guyana, Venezuela.
phytolaccifolium Opiz
var. carabobense *(C.DC.) J.A. Steyermark in Fl. Venezuela,* 2(2): 527 (1984): P. carabobense.
f. peltobryon *(C.DC.) J.A. Steyermark in Fl. Venezuela,* 2(2): 526 (1984): P. peltobryon.
politii Yunck.
subsp. sipapoense *J.A. Steyermark in Fl. Venezuela,* 2(2): 532 (1984)—Venezuela.
subsp. toronoense *J.A. Steyermark in Fl. Venezuela,* 2(2): 530 (1984)—Venezuela.
portobellense *C. DC.in Smithsonian Misc. Collect.,* 71(6): 6 (1920), in I.K. Supp. 6 as 'porlobellense' in error—Panama.†
prunifolium Jacq.
var. carboneranum *(Yunck.) J.A. Steyermark in Fl. Venezuela,* 2(2): 535 (1984): P. carboneranum.
f. strigosum *(Trel. & Yunck.) J.A. Steyermark in Fl. Venezuela,* 2(2): 537 (1984): P. artanthe var. strigosum.
var. tachiranum *(Yunck.) J.A. Steyermark in Fl. Venezuela,* 2(2): 538 (1984): P. tachiranum.
pseudoacreanum *J.A. Steyermark in Fl. Venezuela,* 2(2): 539 (1984)—Venezuela.
pseudoaequale *J.A. Steyermark in Fl. Venezuela,* 2(2): 540 (1984)—Venezuela.
pseudobredemeyeri *J.A. Steyermark in Fl. Venezuela,* 2(2): 542 (1984)—Venezuela.
pseudodivulgatum *J.A. Steyermark in Fl. Venezuela,* 2(2): 543 (1984)—Venezuela.
pseudohastularum *J.A. Steyermark in Fl. Venezuela,* 2(2): 548 (1984)—Venezuela.
pubivaginatum *J.A. Steyermark in Fl. Venezuela,* 2(2): 549 (1984)—Venezuela.
ronaldii *J.A. Steyermark in Fl. Venezuela,* 2(2): 556 (1984)—Venezuela.
setebarraense *E.F. Guimarães & L.H.P. da Costa in Bradea,* 8(27): 150 (1999)—Brazil Southeast (São Paulo).
var. pilosum *E.F. Guimarães & L.H.P. da Costa in Bradea,* 8(27): 150 (1999)—Brazil Southeast (São Paulo).
sierra-aroense *J.A. Steyermark in Fl. Venezuela,* 2(2): 563 (1984)—Venezuela.
stenophyllum *J.O. Voigt, Hort. Suburb. Calcutt.:* 300 (1845), nom. nov.: P. angustifolium *Roxb.*
subsessilifolium C.DC.
var. morii *J.A. Steyermark in Fl. Venezuela,* 2(2): 567 (1984)—Venezuela.
tacariguense *J.A. Steyermark in Fl. Venezuela,* 2(2): 570 (1984)—Venezuela.
taiwanense *T.T. Lin & S.Y. Lu in Taiwania,* 40(4): 356 (1995)—Taiwan.
tamayoanum *J.A. Steyermark in Fl. Venezuela,* 2(2): 572 (1984)—Venezuela.
tepuiense *J.A. Steyermark in Fl. Venezuela,* 2(2): 575 (1984)—Venezuela.
toronotepuiense *J.A. Steyermark in Fl. Venezuela,* 2(2): 576 (1984)—Venezuela.
tortivenulosum Yunck.
f. glabrum *(Yunck.) J.A. Steyermark in Fl. Venezuela,* 2(2): 578 (1984): P. tortivenulosum var. glabrum.
tsengianum *M.G. Gilbert & N.H. Xia in Novon,* 9(2): 196 (1999)—China South-central (Yunnan).
venamoense *J.A. Steyermark in Fl. Venezuela,* 2(2): 589 (1984)—Venezuela.

PIPER:-
veraguense C.DC.
var. mutisii *(Trel. & Yunck.) J.A. Steyermark in Fl. Venezuela,* 2(2): 590 (1984): P. mutisii.
var. venezuelense *J.A. Steyermark in Fl. Venezuela,* 2(2): 590 (1984)—Venezuela.
vitaceum Yunck.
var. venezuelense *J.A. Steyermark in Fl. Venezuela,* 2(2): 594 (1984)—Venezuela.
wangii *M.G. Gilbert & N.H. Xia in Novon,* 9(2): 197 (1999)—China South-central (Yunnan).
wingfieldii *J.A. Steyermark in Fl. Venezuela,* 2(2): 595 (1984)—Venezuela.
yui *M.G. Gilbert & N.H. Xia in Novon,* 9(2): 197 (1999)—China South-central (Yunnan).

PIPERIA (Orchidac.).
dilatata *(Pursh) D.L. Szlachetko & P. Rutkowski in Acta Bot. Fenn.,* 169: 380 (2000): Orchis dilatata.
var. albiflora *(Cham.) D.L. Szlachetko & P. Rutkowski in Acta Bot. Fenn.,* 169: 380 (2000): Habenaria borealis var. albiflora.
var. leucostachys *(Lindl.) D.L. Szlachetko & P. Rutkowski in Acta Bot. Fenn.,* 169: 380 (2000): Platanthera leucostachys.

PIPTANTHUS (Legumin.).
nepalensis (Hook.) D. Don
f. leiocarpus *(Stapf) S.Q. Wei in Fl. Reipubl. Popul. Sin.,* 42(2): 392 (1998): P. leiocarpus.
f. sericopetalus *(P.C. Li) S.Q. Wei in Fl. Reipubl. Popul. Sin.,* 42(2): 392 (1998): P. leiocarpus var. sericopetalus.

PIPTOCARPHA (Compos.).
verticillata *(Vell.) G.L. Sm. ex H. Robinson in Smithsonian Contrib. Bot.,* 89: 79 (1999): Chrysocoma verticillata.

PIPTOCOMA (Compos.).
sect. Dialesta *(Humb., Bonpl. & Kunth) J.F. Pruski in Novon,* 6(1): 97 (1996): gen. Dialesta.
sect. Odontoloma *(Humb., Bonpl. & Kunth) J.F. Pruski in Novon,* 6(1): 97 (1996): gen. Odontoloma.
sect. Pollalesta *(Humb., Bonpl. & Kunth) J.F. Pruski in Novon,* 6(1): 100 (1996): gen. Pollalesta.
acevedoi *J.F. Pruski in Novon,* 6(1): 98 (1996)—Puerto Rico.
acuminata *(Humb., Bonpl. & Kunth) J.F. Pruski in Novon,* 6(1): 98 (1996): Odontoloma acuminatum.
areolata *(Wurdack) J.F. Pruski in Novon,* 6(1): 100 (1996): Oliganthes areolata.
barinensis *(Aristeg.) J.F. Pruski in Novon,* 6(1): 98 (1996): Pollalesta barinensis.
discolor *(Humb., Bonpl. & Kunth) J.F. Pruski in Novon,* 6(1): 97 (1996): Dialesta discolor.
hypochlora *(S.F. Blake) J.F. Pruski in Novon,* 6(1): 98 (1996): Oliganthes hypochlora.
macrophylla *(Sch. Bip.) J.F. Pruski in Novon,* 6(1): 100 (1996): Oliganthes macrophylla.
milleri *(J.R. Johnst.) J.F. Pruski in Novon,* 6(1): 100 (1996): Vernonia milleri.
neglecta *(Stutts) J.F. Pruski in Novon,* 6(1): 101 (1996): Pollalesta neglecta.
niceforoi *(Cuatrec.) J.F. Pruski in Novon,* 6(1): 101 (1996): Oliganthes niceforoi.
roraimensis *(Steyerm.) J.F. Pruski in Novon,* 6(1): 101 (1996): Oliganthes roraimensis.
schomburgkii *(Sch. Bip.) J.F. Pruski in Novon,* 6(1): 101 (1996): Oliganthes schomburgkii.
spruceana *(Benth.) J.F. Pruski in Novon,* 6(1): 101 (1996): Oliganthes spruceana.
trujillensis *(Aristeg.) J.F. Pruski in Novon,* 6(1): 98 (1996): Pollalesta trujillensis.
vernonioides *(Humb., Bonpl. & Kunth) J.F. Pruski in Novon,* 6(1): 101 (1996): Pollalesta vernonioides.

PIPTOSPATHA (Arac.).
brevipedunculata *(H. Okada & Y. Mori) J. Bogner & A. Hay in Telopea,* 9(1): 203 (2000): Hottarum brevipedunculatum.
kinabaluensis *(Bogner) J. Bogner & A. Hay in Telopea,* 9(1): 216 (2000): Hottarum kinabaluense.
lucens *(Bogner) J. Bogner & A. Hay in Telopea,* 9(1): 217 (2000): Hottarum lucens.
manduensis *A. Hay & J. Bogner in Telopea,* 9(1): 207 (2000)—Borneo (Kalimantan).
truncata *(M. Hotta) J. Bogner & A. Hay in Telopea,* 9(1): 213 (2000): Microcasia truncata.

PIPTURUS (Urticac.).
argenteus (G. Forst.) Wedd.
var. tuamotensis *(F. Br.) J. Florence in Bull. Mus. Nation. Hist. Nat., B, Adansonia,* Ser. 4, 18(3–4): 270 (1996): P. incanus var. tuamotensis.
australium *J. Florence in Fl. Polynésie française,* 1: 248 (1997)—Tubuai Is.
grantii *J. Florence in Fl. Polynésie française,* 1: 250 (1997)—Society Is.
oreophilus *J. Florence in Fl. Polynésie française,* 1: 252 (1997)—Society Is.
schaeferi *J. Florence in Bull. Mus. Nation. Hist. Nat., B, Adansonia,* Ser. 4, 18(3–4): 271 (1996)—Marquesas.
tooviianus *J. Florence in Fl. Polynésie française,* 1: 257 (1997)—Marquesas.

PIRANHEA (Euphorbiac.).
mexicana *(Standl.) A. Radcliffe-Smith in Kew Bull.,* 51(3): 546 (1996): Celaenodendron mexicanum.
securinega *A. Radcliffe-Smith & J.A. Ratter in Kew Bull.,* 51(3): 543 (1996)—Brazil Southeast (Minas Gerais).
trifoliolata Baill.
var. pubescens *A. Radcliffe-Smith in Kew Bull.,* 51(3): 547 (1996)—Venezuela.

PIRIQUETA (Turnerac.).
abairana *M.M. Arbo in Kew Bull.,* 54(2): 459 (1999)—Brazil Northeast (Bahia).
revoluta *M.M. Arbo in Kew Bull.,* 54(2): 462 (1999)—Brazil Northeast (Bahia).

PISONIA (Nyctaginac.).
amplifolia *(Heimerl) J. Florence in Adansonia,* 21(1): 49 (1999): Ceodes umbellifera f. amplifolia.
coronata *(Heimerl) J. Florence in Adansonia,* 21(1): 49 (1999): Ceodes umbellifera var. coronata.
tahitensis *(Heimerl) F. Friedmann ex J. Florence in Adansonia,* 21(1): 49 (1999): Calpidia tahitensis.

PISTACIA (Anacardiac.).
brevifolia *M. Gandoger, Dec. Pl. Nov.,* 1: 44 (1875)—France.
choulettei *M. Gandoger, Dec. Pl. Nov.,* 1: 44 (1875)—Algeria.
multiflora *M. Gandoger, Dec. Pl. Nov.,* 1: 44 (1875)—Italy.
subfalcata *M. Gandoger, Dec. Pl. Nov.,* 1: 44 (1875)—France.

PISUM (Legumin.).
sativum L.
subsp. arvense *(L.) N.N. Tzvelev in Novosti Sist. Vyssh. Rast.,* 32: 185 (2000): P. arvense.
syriacum *(A. Berger) C.O. Lehmann in Züchter,* 24(11–12): 318 (1954): P. sativum subsp. syriacum.

PITAVIASTER *T.G. Hartley in Adansonia,* 19(2): 200 (1997). *RUTACEAE*
haplophyllus *(F. Muell.) T.G. Hartley in Adansonia,* 19(2): 202 (1997): Euodia haplophylla.

PITCAIRNIA (Bromeliac.).
alexanderi *(H. Luther) D.C. Taylor & H. Robinson in Harvard Pap. Bot.,* 4(1): 206 (1999): Pepinia alexanderi.
amboroensis *P.L. Ibisch, R. Vásquez, E. Gross & M. Kessler in J. Bromeliad Soc.,* 49(3): 129 (1999)—Bolivia.
calcicola *J.R. Grant & J.F. Morales in Novon,* 6(4): 366 (1996)—Costa Rica.
clarkii *H.E. Luther in Selbyana,* 18(1): 98 (1997)—Ecuador.
cristalinensis *(Leme) D.C. Taylor & H. Robinson in Harvard Pap. Bot.,* 4(1): 207 (1999): Pepinia cristalinensis.
croatii *H.E. Luther in Selbyana,* 18(1): 101 (1997)—Panama.
elvirae *D.C. Taylor & H. Robinson in Harvard Pap. Bot.,* 4(1): 207 (1999), nom. nov.: Pepinia verrucosa *E. Gross.*
ferrell-ingramiae *Harry E. Luther & Dalstrom in J. Bromeliad Soc.,* 47(5): 231 (1997)—Ecuador.
geotropa *J.R. Grant in J. Bromeliad Soc.,* 48(3): 99 (1998)—Panama.

PITCAIRNIA:-
harrylutheri *D.C. Taylor & H. Robinson in Harvard Pap. Bot.*, 4(1): 207 (1999), nom. nov.: Pepinia fulgens *H. Luther.*
heydlauffii *R. Vásquez & P.L. Ibisch in Bromelie*, 2000(1): 14 (2000)—Bolivia.
hooveri *(H. Luther) D.C. Taylor & H. Robinson in Harvard Pap. Bot.*, 4(1): 207 (1999): Pepinia hooveri.
koeneniana *E. Gross & Barthlott in Trop. Subtrop. Pflanzenwelt*, 95: 12 (1997)—Mexico Southwest (Nayarit).
kroemeri *H.E. Luther in Selbyana*, 21(1–2): 125 (2000)—Bolivia.
leopoldii *(W. Till & S. Till) B.K. Holst in Fl. Venezuelan Guayana (gen. eds. J.A. Steyermark et al.)*, 3: 644 (1997): Pepinia leopoldii.
longissimiflora *P.L. Ibisch, R. Vásquez & E. Gross in J. Bromeliad Soc.*, 49(3): 127 (1999)—Bolivia.
mirandae *J.F. Utley & K. Burt-Utley in Novon*, 8(3): 304 (1998)—Mexico Southeast (Chiapas).
neglecta *(H. Luther) D.C. Taylor & H. Robinson in Harvard Pap. Bot.*, 4(1): 208 (1999): Pepinia neglecta.
roseoalba *E. Gross & W. Rauh in Trop. Subtrop. Pflanzenwelt*, 95: 14 (1997)—Peru.
var. rubra *E. Gross & W. Rauh in Trop. Subtrop. Pflanzenwelt*, 95: 16 (1997)—Peru.
rundelliana *J.R. Grant in J. Bromeliad Soc.*, 48(3): 103 (1998)—Panama.
sanguinea *(H. Luther) D.C. Taylor & H. Robinson in Harvard Pap. Bot.*, 4(1): 208 (1999): Pepinia sanguinea.
tatzyanae *(H. Luther) D.C. Taylor & H. Robinson in Harvard Pap. Bot.*, 4(1): 208 (1999): Pepinia tatzyanae.

PITHECELLOBIUM (Legumin.).
cynodonticum *R.C. Barneby & J.W. Grimes, Silk Tree, Guanacaste, Monkey's Earring (Mem. New York Bot. Gard., 74(2))*: 13 (1997)—Haiti.
roseum *(Vahl) R.C. Barneby & J.W. Grimes, Silk Tree, Guanacaste, Monkey's Earring (Mem. New York Bot. Gard., 74(2))*: 21 (1997): Mimosa rosea.
var. glaucescens *(Pittier) R.C. Barneby & J.W. Grimes, Silk Tree, Guanacaste, Monkey's Earring (Mem. New York Bot. Gard., 74(2))*: 22 (1997): P. glaucescens.

PITHOCARPA (Compos.).
pulchella Lindl.
var. melanostigma *(P. Lewis & Summerh.) B.J. Lepschi in Nuytsia*, 13(1): 71 (1999): P. melanostigma.

× **PITINIA** *J. Irvin ex E. Baskerville in J. Bromeliad Soc.*, 48(2): 64 (1998). *BROMELIACEAE* (Pepinia × Pitcairnia).

PITTOCAULON (Compos.).
calzadanum *B.L. Turner in Phytologia*, 79(1): 43 (1995 publ. 1996)—Mexico Southwest (Oaxaca).
hintonii H. Rob. & Brettell
var. cerrograndensis *B.L. Clark in Sida*, 19(2): 235 (2000)—Mexico Southwest (Jalisco).
velatum (Greenm.) H. Rob. & Brettell
var. tzimolensis *(T.M. Barkley) B.L. Clark in Sida*, 19(2): 235 (2000): Senecio praecox var. tzimolensis.

PITTOSPORUM (Pittosporac.).
bernardii *C. Tirel & J.-M. Veillon in Adansonia*, 19(1): 40 (1997)—New Caledonia.
bouletii *J.-M. Veillon & C. Tirel in Adansonia*, 21(2): 218 (1999)—New Caledonia.
ellipticum Kirk
subsp. serpentinum *P.J. de Lange in New Zealand J. Bot.*, 36(3): 390 (1998)—New Zealand North I.
ferrugineum Dryand. ex W.T. Aiton
subsp. linifolium *(A. Cunn.) L.W. Cayzer, M.D. Crisp & I.R.H. Telford in Austral. Syst. Bot.*, 13(6): 892 (2000): Pittosporum linifolium.
lancifolium *(F.M. Bailey) L.W. Cayzer, M.D. Crisp & I.R.H. Telford in Austral. Syst. Bot.*, 13(6): 869 (2000): Citriobatus lancifolius.
lanipetalum *C. Tirel & J.-M. Veillon in Adansonia*, 20(2): 244 (1998)—New Caledonia.

PITTOSPORUM:-
leroyanum *C. Tirel & J.-M. Veillon in Adansonia*, 21(2): 214 (1999)—New Caledonia.
letocartiorum *J.-M. Veillon & C. Tirel in Adansonia*, 21(2): 216 (1999)—New Caledonia.
mackeei *C. Tirel & J.-M. Veillon in Adansonia*, 19(1): 41 (1997)—New Caledonia.
multiflorum *(A. Cunn. ex Loudon) L.W. Cayzer, M.D. Crisp & I.R.H. Telford in Austral. Syst. Bot.*, 13(6): 864 (2000): Citriobatus multiflorus.
pentandrum (Blanco) Merr.
var. formosanum *(Hayata) S. Hatusima in S. Hatusima & T. Amano, Fl. Ryukyus, S. of Amami Is., ed. 2*: 74 (1994), without basionym ref.: Basionym not stated.
spinescens *(F. Muell.) L.W. Cayzer, M.D. Crisp & I.R.H. Telford in Austral. Syst. Bot.*, 13(6): 865 (2000): Ixiosporum spinescens.
spissescens *T.M.A. Utteridge in Kew Bull.*, 55(3): 706 (2000)—New Guinea (Irian Jaya).
trilobum *L.W. Cayzer, M.D. Crisp & I.R.H. Telford in Austral. Syst. Bot.*, 13(6): 897 (2000)—Queensland.
verticillatum *N. Wallich, Report on Calcutta Bot. Gard. to G.A. Bushby*: 29 (1840), nom. nov.: Celastrus verticillatus *Roxb.*
viscidum *L.W. Cayzer, M.D. Crisp & I.R.H. Telford in Austral. Syst. Bot.*, 13(6): 868 (2000), nom. nov.: Citriobatus multiflorus var. linearis *F.M. Bailey.*
wenxianense *G.H. Wang & Y.S. Lian in Acta Phytotax. Sin.*, 34(2): 210 (1996)—China North-central (Gansu).

PITYPHYLLUM (Orchidac.).
hirtzii *C.H. Dodson in Orquideologia*, 20(1): 106 (1996)—Ecuador.

PLACODISCUS (Sapindac.).
pedicellatus *F.G. Davies in Fl. Trop. E. Afr., Sapindac.*: 55 (1998)—Tanzania.

PLAGIOSTACHYS (Zingiberac.).
breviramosa *J. Cowley in Kew Bull.*, 54(1): 142 (1999)—Borneo (Brunei, Sabah, Kalimantan).
parva *J. Cowley in Kew Bull.*, 54(1): 140 (1999)—Borneo (Sabah, Brunei).

PLAGIUS (Compos.).
fontanesii *(Boiss. & Reut.) R. Vogt in Lagascalia*, 18(2): 303 (1996): Leucanthemum fontanesii.

PLAKOTHIRA (Loasac.).
parviflora *J. Florence in Allertonia*, 7(4): 239 (1997)—Marquesas.
perlmanii *J. Florence in Allertonia*, 7(4): 240 (1997)—Marquesas.

PLANOCARPA *C.M. Weiller in Austral. Syst. Bot.*, 9(4): 510 (1996). *EPACRIDACEAE*
nitida *(Jarman) C.M. Weiller in Austral. Syst. Bot.*, 9(4): 517 (1996): Cyathodes nitida.
petiolaris *(DC.) C.M. Weiller in Austral. Syst. Bot.*, 9(4): 514 (1996): Leucopogon petiolaris.
sulcata *(Mihaich) C.M. Weiller in Austral. Syst. Bot.*, 9(4): 516 (1996): Cyathodes sulcata.

PLANTAGO (Plantaginac.).
subgen. Albicans *K. Rahn in Bot. J. Linn. Soc.*, 120(2): 197 (1996).
subgen. Bougeria *(Decne.) K. Rahn in Bot. J. Linn. Soc.*, 120(2): 197 (1996): gen. Bougeria.
ser. Carpophorae *(Rahn) K. Rahn in Bot. J. Linn. Soc.*, 120(2): 196 (1996): P. sect. Carpophorae.
subgen. Littorella *(P.J. Bergius) K. Rahn in Bot. J. Linn. Soc.*, 120(2): 197 (1996): gen. Littorella.
ser. Microcalyx *(Pilg.) K. Rahn in Bot. J. Linn. Soc.*, 120(2): 196 (1996): P. sect. Microcalyx.
ser. Oliganthos *K. Rahn in Bot. J. Linn. Soc.*, 120(2): 196 (1996).
sect. Uralenses *A.B. Shipunov in Novosti Sist. Vÿssh. Rast.*, 32: 145 (2000).
alpina L.
subsp. incana *(Ramond) J. Chrtek in Čas. Nár. Muz. Řada Přír.*, 168(1–4): 60 (1999): P. incana.
americana *(Fernald) K. Rahn in Bot. J. Linn. Soc.*, 120(2): 197 (1996), without basionym ref.: Littorella americana.

PLANTAGO:-
araucana *K. Rahn in Bot. J. Linn. Soc.*, 120(2): 197 (1996), nom. nov.: Littorella australis *Griseb.*
coronopus L.
subsp. humilis *(Guss.) J. Gamisans, Plantaginaceae: Compl. Prodr. Fl. Corse (ser. eds. D. Jeanmonod & H.M.* Burdet): 18 (1988): P. macrorhiza var. humilis.
holosteum Scop.
var. gallica *J. Chrtek in Čas. Nár. Muz. Řada Přír.*, 168(1–4): 20 (1999)—France.
subsp. libanotica *(Pilg.) J. Chrtek in Čas. Nár. Muz. Řada Přír.*, 168(1–4): 20 (1999): P. holosteum var. libanotica.
subsp. littoralis *(Rouy) J. Chrtek in Čas. Nár. Muz. Řada Přír.*, 168(1–4): 19 (1999): P. recurvata β littoralis.
subsp. macrantha *J. Chrtek in Čas. Nár. Muz. Řada Přír*, 168(1–4): 20 (1999)—France.
jehohlensis *Koidz. in Acta Phytotax. Geobot.*, 10(2): 142 (1941), nom. nov., in I.K. Supp. 11 as 'jepohlensis': P. villifera *Kitag.*†
jurtzevii *(Tzvelev) N.N. Tzvelev in Novosti Sist. Vÿssh. Rast.*, 30: 144 (1996); *et in Sosud. Rast. Sovet. Dal'nego Vostoka*, 8: 258 (1996): P. canescens subsp. jurtzevii.
major L.
var. jehohlensis *(Koidz.) S.X. Li in Clavis Pl. Chinae Bor.-Or. ed. 2 (ed. P.Y. Fu)*: 615 (1995): P. jehohlensis.
nubicola *(Decne.) K. Rahn in Bot. J. Linn. Soc.*, 120(2): 197 (1996): Bougeria nubicola.
podlechii *H. Akhani in Edinburgh J. Bot.*, 56(1): 16 (1999)—Iran.
popovii *N.N. Tzvelev in Novosti Sist. Vÿssh. Rast.*, 30: 146 (1996); *et in Sosud. Rast. Sovet. Dal'nego Vostoka*, 8: 257 (1996)—Sakhalin.
sarda C. Presl
var. alpinoides *J. Gamisans, Plantaginaceae: Compl. Prodr. Fl. Corse (ser. eds. D. Jeanmonod & H.M.* Burdet): 44 (1988)—Corse.
uliginosa F.W. Schmidt
var. winteri *(Wirtg.) A.B. Shipunov in Novosti Sist. Vÿssh. Rast.*, 32: 146 (2000): P. winteri.
unibracteata *K. Rahn in Bot. J. Linn. Soc.*, 120(2): 196 (1996), nom. nov.: P. uniflora *Hook. f.*
weldenii Rchb.
var. commutata *(Guss.) J. Gamisans, Plantaginaceae: Compl. Prodr. Fl. Corse (ser. eds. D. Jeanmonod & H.M.* Burdet): 29 (1988): P. commutata.
var. purpurascens *(Nyman) J. Gamisans, Plantaginaceae: Compl. Prodr. Fl. Corse (ser. eds. D. Jeanmonod & H.M.* Burdet): 29 (1988): P. macrorhiza subsp. purpurascens.

PLATANTHERA (Orchidac.).
subgen. Stigmatosae *K.Y. Lang in Acta Phytotax. Sin.*, 36(5): 450 (1998).
aquilonis *C.J. Sheviak in Lindleyana*, 14(4): 193 (1999)—Illinois.
bifolia L.
var. kuenkelei *(H. Baumann) P. Delforge in Nat. Belg.*, 81(4): 398 (2000): Platanthera kuenkelei.
chingshuishania *S.S. Ying in Mem. Coll. Agric. Nation. Taiwan Univ.*, 28(2): 36 (1988)—Taiwan.
edgeworthii *(Hook. f. ex Collett) K.Y. Lang in Acta Phytotax. Sin.*, 36(5): 457 (1998), isonym; *see R.K. Gupta, Fl. Nainital.*, 349 (1968): Habenaria edgeworthii.
× hollandiae *P.M. Catling & V.R. Brownell in Canad. J. Bot.*, 77(8): 1147 (1999)—Ontario. (P. lacera × leucophaea).
× keenanii *P.M. Brown, Field & Study Guide Orchids New Engl. & New York*: 189 (1994), without diagnostic latin descr.—Maine. (P. grandiflora × lacera).
kwangsiensis *K.Y. Lang in Guihaia*, 18(1): 5 (1998)—China Southeast (Guangxi).
mandarinorum Rchb. f.
var. oreades *(Franch. & Sav.) T. Hashimoto in Proc. 12th World Orchid Conf., 1987 (eds. K. Saito & R. Tanaka)*: 119 (1987): P. oreades.
purpurascens *(Rydb.) C.J. Sheviak & W.F. Jennings in North Amer. Native Orchid J.*, 3(4): 445 (1997): Limnorchis purpurascens.
replicata *(A. Rich.) J.D. Ackerman in Lindleyana*, 12(3): 151 (1997): Habenaria replicata.

PLATANTHERA:-
× reznicekii *P.M. Catling, V.R. Brownell & G. Allen in Lindleyana*, 14(2): 81 (1999)—Ontario. (P. leucophaea × psycodes).

PLATANUS (Platanac.).
occidentalis L.
var. palmeri *(Kuntze) K. Nixon & J. Poole ex D. Geerinck in Belgian J. Bot.*, 130(2): 127 (1998): P. orientalis var. palmeri.

PLATOSTOMA (Labiat.).
subgen. Acrocephalus *(Benth.) A.J. Paton in Kew Bull.*, 52(2): 272 (1997): gen. Acrocephalus.
sect. Acrocephalus *(Benth.) A.J. Paton in Kew Bull.*, 52(2): 273 (1997): gen. Acrocephalus.
sect. Ceratanthus *(G. Taylor) A.J. Paton in Kew Bull.*, 52(2): 279 (1997): gen. Ceratanthus.
sect. Heterodonta *(Briq.) A.J. Paton in Kew Bull.*, 52(2): 274 (1997): Geniosporum sect. Heterodonta.
subsect. Limniboza *(R.E. Fr.) A.J. Paton in Kew Bull.*, 52(2): 286 (1997): gen. Limniboza.
sect. Limniboza *(R.E. Fr.) A.J. Paton in Kew Bull.*, 52(2): 286 (1997): gen. Limniboza.
sect. Mesona *(Blume) A.J. Paton in Kew Bull.*, 52(2): 281 (1997): gen. Mesona.
subgen. Octomeron *(Robyns) A.J. Paton in Kew Bull.*, 52(2): 276 (1997): gen. Octomeron.
subsect. Rotundifolia *A.J. Paton in Kew Bull.*, 52(2): 286 (1997).
annamense *(G. Taylor) A.J. Paton in Kew Bull.*, 52(2): 280 (1997): Ceratanthus annamensis.
axillare *(Benth.) A.J. Paton in Kew Bull.*, 52(2): 274 (1997): Acrocephalus axillaris.
becquerellii *(Doan) A.J. Paton in Kew Bull.*, 52(2): 280 (1997): Ceratanthus becquerellii.
calcaratum *(Hemsl.) A.J. Paton in Kew Bull.*, 52(2): 281 (1997): Plectranthus calcaratus.
cambodgense *(Doan) A.J. Paton in Kew Bull.*, 52(2): 285 (1997): Mesona cambodgensis.
chinense *(Benth.) A.J. Paton in Kew Bull.*, 52(2): 282 (1997): Mesona chinensis.
clausum *(Merr.) A.J. Paton in Kew Bull.*, 52(2): 285 (1997): Mesona clausa.
cochinchinense *(Lour.) A.J. Paton in Kew Bull.*, 52(2): 285 (1997): Dracocephalum cochinchinense.
coeruleum *(R.E. Fr.) A.J. Paton in Kew Bull.*, 52(2): 286 (1997): Limniboza coerulea.
coloratum *(D. Don) A.J. Paton in Kew Bull.*, 52(2): 274 (1997): Plectranthus coloratum.
dilungense *(Lisowski & Mielcarek) A.J. Paton in Kew Bull.*, 52(2): 286 (1997): Limniboza dilungensis.
elongatum *(Benth.) A.J. Paton in Kew Bull.*, 52(2): 274 (1997): Geniosporum elongatum.
fastigiatum *A.J. Paton & I.C. Hedge in Kew Bull.*, 52(2): 275 (1997)—Madagascar.
fimbriatum *A.J. Paton in Kew Bull.*, 52(2): 284 (1997)—Thailand.
gabonense *A.J. Paton in Kew Bull.*, 52(2): 277 (1997)—Gabon.
garrettii *(Craib) A.J. Paton in Kew Bull.*, 52(2): 281 (1997): Plectranthus garrettii.
glomerulatum *A.J. Paton & I.C. Hedge in Kew Bull.*, 52(2): 276 (1997)—Madagascar.
grandiflorum *(Doan) A.J. Paton in Kew Bull.*, 52(2): 286 (1997): Mesona grandiflora.
hildebrandtii *(Vatke) A.J. Paton in Kew Bull.*, 52(2): 275 (1997): Ocimum hildebrandtii.
hispidum *(L.) A.J. Paton in Kew Bull.*, 52(2): 273 (1997): Gomphrena hispida.
intermedium *A.J. Paton in Kew Bull.*, 52(2): 279 (1997)—Thailand.
kerrii *(Doan) A.J. Paton in Kew Bull.*, 52(2): 281 (1997): Ceratanthus kerrii.
lanceolatum *(Chermsir. ex Murata) A.J. Paton in Kew Bull.*, 52(2): 275 (1997): Geniosporum lanceolatum.
laxiflorum *A.J. Paton & I.C. Hedge in Kew Bull.*, 52(2): 275 (1997)—Madagascar.
lisowskianum *(Bamps) A.J. Paton in Kew Bull.*, 52(2): 286 (1997): Limniboza lisowskiana.
longicorne *(F. Muell.) A.J. Paton in Kew Bull.*, 52(2): 280 (1997): Plectranthus longicornis.
madagascariense *(Benth.) A.J. Paton & I.C. Hedge in Kew Bull.*, 52(2): 275 (1997): Geniosporum madagascariense.

PLATOSTOMA:-
menthoides *(L.) A.J. Paton in Kew Bull.*, 52(2): 274 (1997): Ocimum menthoides.
montanum *(Robyns) A.J. Paton in Kew Bull.*, 52(2): 276 (1997): Octomeron montanum.
ocimoides *(G. Taylor) A.J. Paton in Kew Bull.*, 52(2): 281 (1997): Ceratanthus ocimoides.
palniense *(Mukerjee) A.J. Paton in Kew Bull.*, 52(2): 273 (1997): Acrocephalus palniensis.
palustre *(Blume) A.J. Paton in Kew Bull.*, 52(2): 281 (1997): Mesona palustris.
rotundifolium *(Briq.) A.J. Paton in Kew Bull.*, 52(2): 287 (1997): Geniosporum rotundifolium.
rubrum *(Doan) A.J. Paton in Kew Bull.*, 52(2): 285 (1997): Mesona rubra.
siamense *(Murata) A.J. Paton in Kew Bull.*, 52(2): 275 (1997): Geniosporum siamense.
stoloniferum *(G. Taylor) A.J. Paton in Kew Bull.*, 52(2): 280 (1997): Ceratanthus stolonifer.
strictum *(Hiern) A.J. Paton in Kew Bull.*, 52(2): 287 (1997): Geniosporum strictum.
taylorii *(Doan) A.J. Paton in Kew Bull.*, 52(2): 282 (1997): Geniosporum taylorii.
tectum *A.J. Paton in Kew Bull.*, 52(2): 282 (1997)—Thailand.
tenellum *(Benth.) A.J. Paton & I.C. Hedge in Kew Bull.*, 52(2): 274 (1997): Ocimum tenellum.
thymiflorum *(Benth.) A.J. Paton & I.C. Hedge in Kew Bull.*, 52(2): 287 (1997): Geniosporum thymiflorum.
verbenifolium *(Mukerjee) A.J. Paton in Kew Bull.*, 52(2): 273 (1997): Acrocephalus verbenifolius.

PLATYCARYA (Juglandac.).
longzhouensis *S.Y. Liang & G.J. Liang in Guihaia*, 19(2): 119 (1999)—China Southeast (Guangxi).

PLATYCHORDA *B.G. Briggs & L.A.S. Johnson in Telopea*, 7(4): 359 (1998). *RESTIONACEAE*
applanata *(Spreng.) B.G. Briggs & L.A.S. Johnson in Telopea*, 7(4): 361 (1998): Restio applanatus.

PLATYCLADUS (Cupressac.).
chengii *(Bordères & Gaussen) A.V.F.Ch. Bobrov in 14. Symp. Biodivers. & Evolutionsbiol. Jena 1999 (ed. H. Manitz & F.H.* Hellwig): 18 (1999): Thuja chengii.

PLATYCORYNE (Orchidac.).
alinae *D.L. Szlachetko in Fl. Cameroun*, 34: 202 (1998)—Cameroon.
rautanenii *(Kraenzl.) D.L. Szlachetko & T.S. Olszewski in Fl. Cameroun*, 34: 200 (1998): Habenaria rautanenii.

PLATYCRASPEDUM (Crucif.).
wuchengyii *I. Al-Shehbaz, T.Y. Cheo, L.L. Lu & G. Yang in Novon*, 10(1): 3 (2000)—Xizang, China South-central (Sichuan).

PLATYMISCIUM (Legumin.).
curuense *N. Zamora & B.B. Klitgaard in Novon*, 7(2): 216 (1997)—Costa Rica.
floribundum Vogel
var. nitens *(Vogel) B.B. Klitgaard in Kew Bull.*, 54(4): 971 (1999 publ. 2000): P. nitens.
var. obtusifolium *(Harms.) B.B. Klitgaard in Kew Bull.*, 54(4): 971 (1999 publ. 2000): P. obtusifolium.
jejunum *B.B. Klitgaard in Kew Bull.*, 54(4): 968 (1999 publ. 2000)—Mexico Southwest (Guerrero), Mexico Southeast (Chiapas), Mexico Gulf (Veracruz).
pinnatum (Jacq.) Dugand
var. diadelphum *(S.F. Blake) B.B. Klitgaard in Kew Bull.*, 54(4): 972 (1999 publ. 2000): P. diadelphum.
subsp. polystachyum *(Benth.) B.B. Klitgaard in Kew Bull.*, 54(4): 973 (1999 publ. 2000): P. polystachyum.
var. ulei *(Harms) B.B. Klitgaard in Kew Bull.*, 54(4): 972 (1999 publ. 2000): P. ulei.
pubescens Micheli
subsp. fragrans *(Rusby) B.B. Klitgaard in Kew Bull.*, 54(4): 970 (1999 publ. 2000): P. fragrans.
subsp. zehntneri *(Harms) B.B. Klitgaard in Kew Bull.*, 54(4): 970 (1999 publ. 2000): P. zehntneri.

PLATYMISCIUM:-
trinitatis Benth.
var. duckei *(Huber) B.B. Klitgaard in Kew Bull.*, 54(4): 970 (1999 publ. 2000): P. duckei.
var. nigrum *(Ducke) B.B. Klitgaard in Kew Bull.*, 54(4): 971 (1999 publ. 2000): P. nigrum.

PLATYPODANTHERA (Compos.).
melissifolia (DC.) R.M. King & H. Rob.
subsp. riocontensis *D.J.N. Hind in Kew Bull.*, 54(4): 928 (1999 publ. 2000)—Brazil Northeast (Bahia).

PLATYPODIUM (Legumin.).
elegans Vogel
subsp. maxonianum *(Pittier) H.C. Lima in Fl. Venezuelan Guayana (gen. eds. J.A. Steyermark et al.)*, 5: 377 (1999): P. maxonianum.

PLATYSACE (Umbellif.).
saxatilis *G.J. Keighery in Nordic J. Bot.*, 16(2): 135 (1996)—Northern Territory, Western Australia.

PLATYSTELE (Orchidac.).
bernoullii *C.A. Luer, Ic. Pleurothallid. XX (Monogr. Syst. Bot. Missouri Bot. Gard., 79)*: 126 (2000)—Guatemala.
bovilinguis *C.A. Luer, Ic. Pleurothallid. XII (Monogr. Syst. Bot. Missouri Bot. Gard., 57)*: 142 (1995)—Bolivia.
hampshireae *C.A. Luer, Ic. Pleurothallid. XVI (Monogr. Syst. Bot. Missouri Bot. Gard., 65)*: 113 (1998)—Panama.
ingramii *C.A. Luer & Dalström in C.A. Luer, Ic. Pleurothallid. XIV (Monogr. Syst. Bot. Missouri Bot. Gard., 61(4))*: 3 (1996)—Ecuador.

PLATYTHELYS (Orchidac.).
pedicellata *(Cogn.) D.L. Szlachetko in Fragm. Flor. Geobot.*, Supp. 3: 115 (1995): Spiranthes pedicellata.
roseoalba *(Dressler) D.L. Szlachetko in Fragm. Flor. Geobot.*, Supp. 3: 115 (1995): Erythrodes roseoalba.

PLECONAX (Caryophyllac.).
grisebachii *(Davidov) J. Holub in Preslia*, 70(2): 109 (1998): Silene subconica var. grisebachii.

PLECTOCEPHALUS (Compos.).
rothrockii *(Greenm.) D.J.N. Hind in Curtis's Bot. Mag.*, 13(1): 5 (1996): Centaurea rothrockii.

PLECTOCOMIA (Palmae).
barthiana *D.R. Hodel in Palm J.*, 139: 54 (1998), as 'barthana'—Thailand.

PLECTRANTHUS (Labiat.).
agnewii *C. Lukhoba & A.J. Paton in Kew Bull.*, 55(4): 957 (2000)—Kenya, Uganda, Tanzania.
aliciae *(Codd) E.J. Van Jaarsveld & T.J. Edwards in Bothalia*, 27(1): 5 (1997): P. madagascariensis var. aliciae.
alpinus *(Vatke) O. Ryding in Bot. Jahrb.*, 121(2): 147 (1999): Coleus alpinus.
amoenus *P.I. Forster in Austrobaileya*, 4(4): 656 (1997)—Queensland.
amplexicaulis *I.C. Hedge in Fl. Madagascar*, fam. 175: 203 (1998)—Madagascar.
angulatus *I.C. Hedge in Fl. Madagascar*, fam. 175: 211 (1998)—Madagascar.
antongilicus *I.C. Hedge in Fl. Madagascar*, fam. 175: 160 (1998)—Madagascar.
argentifolius *O. Ryding in Kew Bull.*, 54(1): 122 (1999)—Somalia, North Yemen.
atroviolaceus *I.C. Hedge in Fl. Madagascar*, fam. 175: 156 (1998)—Madagascar.
bariensis *O. Ryding in Kew Bull.*, 54(1): 123 (1999)—Somalia.
betamponus *I.C. Hedge in Fl. Madagascar*, fam. 175: 175 (1998)—Madagascar.
bojeri *(Benth.) I.C. Hedge in Fl. Madagascar*, fam. 175: 218 (1998): Coleus bojeri.
bosseri *I.C. Hedge in Fl. Madagascar*, fam. 175: 214 (1998)—Madagascar.
brevicaulis *(Baker) I.C. Hedge in Fl. Madagascar*, fam. 175: 154 (1998): Orthosiphon brevicaulis.
burorum *(Chiov.) J.K. Morton in Novon*, 8(3): 265 (1998): Coleus burorum.

PLECTRANTHUS:-

capuronii I.C. Hedge in Fl. Madagascar, fam. 175: 164 (1998)—Madagascar.

cardiophyllus I.C. Hedge in Fl. Madagascar, fam. 175: 172 (1998)—Madagascar.

cicatricosus (Hutch. & E.A. Bruce) J.R.I. Wood, Handbook Yemen Fl.: 249 (1997): Coleus cicatricosus.

circinnatus I.C. Hedge in Fl. Madagascar, fam. 175: 228 (1998)—Madagascar.

clementiae I.C. Hedge in Fl. Madagascar, fam. 175: 168 (1998)—Madagascar.

cuneatus (Baker f.) O. Ryding in Kew Bull., 54(1): 124 (1999): Coleus cuneatus.

decaryi I.C. Hedge in Fl. Madagascar, fam. 175: 182 (1998)—Madagascar.

delicatissimus I.C. Hedge in Fl. Madagascar, fam. 175: 197 (1998)—Madagascar.

ellipticus I.C. Hedge in Fl. Madagascar, fam. 175: 186 (1998)—Madagascar.

emirnensis (Baker) I.C. Hedge in Fl. Madagascar, fam. 175: 187 (1998): Orthosiphon emirnensis.

erubescens I.C. Hedge in Fl. Madagascar, fam. 175: 216 (1998)—Madagascar.

fasciculatus P.I. Forster in Haseltonia, 6: 14 (1998 publ. 1999)—Queensland.

forsythii I.C. Hedge in Fl. Madagascar, fam. 175: 177 (1998)—Madagascar.

garckeanus (Vatke) J.K. Morton in Novon, 8(3): 265 (1998): Coleus garckeanus.

gibbosus I.C. Hedge in Fl. Madagascar, fam. 175: 200 (1998)—Madagascar.

gillettii J.K. Morton in Kew Bull., 53(4): 998 (1998)—Kenya, Ethiopia, Somalia.

globosus O. Ryding in Bull. Jard. Bot. Nation. Belg., 66(1–2): 101 (1997)—Zaire.

grandibracteatus I.C. Hedge in Fl. Madagascar, fam. 175: 152 (1998)—Madagascar.

grandicalyx (E.A. Bruce) J.K. Morton in Novon, 8(3): 265 (1998): Coleus grandicalyx.

hirsutus I.C. Hedge in Fl. Madagascar, fam. 175: 158 (1998)—Madagascar.

humbertii I.C. Hedge in Fl. Madagascar, fam. 175: 194 (1998)—Madagascar.

kapatensis (R.E. Fr.) J.K. Morton in Novon, 8(3): 265 (1998): Coleus kapatensis.

laurifolius I.C. Hedge in Fl. Madagascar, fam. 175: 170 (1998)—Madagascar.

linearis I.C. Hedge in Fl. Madagascar, fam. 175: 191 (1998)—Madagascar.

longipetiolatus I.C. Hedge in Fl. Madagascar, fam. 175: 147 (1998)—Madagascar.

lucidus (Benth.) E.J. Van Jaarsveld & T.J. Edwards in Bothalia, 27(1): 5 (1997): P. strigosus var. lucidus.

macilentus I.C. Hedge in Fl. Madagascar, fam. 175: 170 (1998)—Madagascar.

malvinus E.J. Van Jaarsveld & T.J. Edwards in Bothalia, 27(1): 1 (1997)—Cape Province.

marrubatus J.K. Morton in Kew Bull., 53(4): 997 (1998)—Ethiopia, Kenya.

membranaceus (Scott-Elliott) I.C. Hedge in Fl. Madagascar, fam. 175: 145 (1998): P. canescens var. membranaceus.

minutiflorus O. Ryding in Nordic J. Bot., 20(1): 43 (2000)—Ethiopia, Kenya.

oblanceolatus I.C. Hedge in Fl. Madagascar, fam. 175: 161 (1998)—Madagascar.

ombrophilus I.C. Hedge in Fl. Madagascar, fam. 175: 166 (1998)—Madagascar.

otostegioides (Gürke) O. Ryding in Kew Bull., 54(1): 126 (1999): Capitanya otostegioides.

pachyphyllus Gürke ex T. Cooke in Dyer, Fl. Cap., 5(1): 285 (1910)—Natal.

pachyphyllus I.C. Hedge in Fl. Madagascar, fam. 175: 176 (1998), homonym; see T. Cooke in Dyer, Fl. Cap., 5(1): 285 (1910)—Madagascar.

pentheri (Gürke) E.J. Van Jaarsveld & T.J. Edwards in Bothalia, 27(1): 5 (1997): Coleus pentheri.

perrieri I.C. Hedge in Fl. Madagascar, fam. 175: 199 (1998)—Madagascar.

persoonii (Benth.) I.C. Hedge in Fl. Madagascar, fam. 175: 222 (1998): Coleus persoonii.

pichompae I.C. Hedge in Fl. Madagascar, fam. 175: 163 (1998)—Madagascar.

porphyranthus T.J. Edwards & N.R. Crouch in Kew Bull., 55(2): 459 (2000)—Zimbabwe.

pseudobarbatus J.K. Morton in Novon, 8(3): 265 (1998), nom. nov.: Coleus vestitus Baker.

puberulentus J.K. Morton in Novon, 8(3): 265 (1998), nom. nov.: Coleus gracilis Gürke.

purpuratus Harv.
- subsp. montanus E.J. Van Jaarsveld & T.J. Edwards in Bothalia, 27(1): 4 (1997)—Swaziland.
- subsp. tongaensis E.J. Van Jaarsveld & T.J. Edwards in Bothalia, 27(1): 2 (1997)—Natal.

rubroviolaceus I.C. Hedge in Fl. Madagascar, fam. 175: 151 (1998)—Madagascar.

saccatus Benth.
- subsp. pondoensis E.J. Van Jaarsveld & S. Milstein in Bothalia, 27(1): 4 (1997)—Natal.

scaposus I.C. Hedge in Fl. Madagascar, fam. 175: 157 (1998)—Madagascar.

scebeli (Chiov.) O. Ryding in Kew Bull., 54(1): 125 (1999): Coleus scebeli.

secundiflorus (Baker) I.C. Hedge in Fl. Madagascar, fam. 175: 180 (1998): Orthosiphon secundiflorus.

thalassoscopicus P.I. Forster in Austrobaileya, 4(4): 653 (1997)—Queensland.

trilobus I.C. Hedge in Fl. Madagascar, fam. 175: 205 (1998)—Madagascar.

venteri E. Van Jaarsveld & A. Hankey in Aloe, 34(1–2): 40 (1997)—Transvaal.

veyretiae (Guillaumet & Cornet) I.C. Hedge in Fl. Madagascar, fam. 175: 226 (1998): Solenostemon veyretiae.

vinaceus I.C. Hedge in Fl. Madagascar, fam. 175: 148 (1998)—Madagascar.

xylopodus C. Lukhoba & A.J. Paton in Kew Bull., 55(4): 960 (2000)—Kenya, Ethiopia, Uganda.

PLECTROPHORA (Orchidac.).

schmidtii R. Jenny & F. Pupulin in Orquidário, 11(3): 79 (1997)—Brazil West-central (Mato Grosso).

PLEIARINA (Salicac.).

subgen. Lygus (Dumort.) N. Chao & G.T. Gong in J. Sichuan Forest. Sci. Techn., 19(4): 14 (1998): Salix subgen. Lygus.

acmophylla (Boiss.) N. Chao & G.T. Gong in J. Sichuan Forest. Sci. Techn., 17(2): 4 (1996): Salix acmophylla.

amygdaloides (Andersson) N. Chao & G.T. Gong in J. Sichuan Forest. Sci. Techn., 17(2): 4 (1996): Salix amygdaloides.

balansae (Seemen) N. Chao & G.T. Gong in J. Sichuan Forest. Sci. Techn., 17(2): 4 (1996), as 'balansaei': Salix balansae.

bonplandiana (Kunth) N. Chao & G.T. Gong in J. Sichuan Forest. Sci. Techn., 17(2): 4 (1996): Salix bonplandiana.

boseensis (N. Chao) N. Chao & G.T. Gong in J. Sichuan Forest. Sci. Techn., 17(2): 4 (1996): Salix boseensis.

capensis (Thunb.) N. Chao & G.T. Gong in J. Sichuan Forest. Sci. Techn., 17(2): 4 (1996): Salix capensis.

caroliniana (Michx.) N. Chao & G.T. Gong in J. Sichuan Forest. Sci. Techn., 17(2): 4 (1996): Salix caroliniana.

cavaleriei (H. Lév.) N. Chao & G.T. Gong in J. Sichuan Forest. Sci. Techn., 17(2): 4 (1996): Salix cavaleriei.

chevalieri (Seemen) N. Chao & G.T. Gong in J. Sichuan Forest. Sci. Techn., 19(4): 14 (1998): Salix chevalieri.

coluteoides (Mirb.) N. Chao & G.T. Gong in J. Sichuan Forest. Sci. Techn., 19(4): 14 (1998): Salix coluteoides.

crateradenia (Seemen) N. Chao & G.T. Gong in J. Sichuan Forest. Sci. Techn., 19(4): 14 (1998): Salix crateradenia.

dealbata (Andersson) N. Chao & G.T. Gong in J. Sichuan Forest. Sci. Techn., 17(2): 4 (1996): Salix dealbata.

dictyoneura (Seemen) N. Chao & G.T. Gong in J. Sichuan Forest. Sci. Techn., 17(2): 4 (1996): Salix dictyoneura.

dunnii (C.K. Schneid.) N. Chao & G.T. Gong in J. Sichuan Forest. Sci. Techn., 17(2): 4 (1996): Salix dunnii.

fangiana (N. Chao & Z.Y. Wang) N. Chao & G.T. Gong in J. Sichuan Forest. Sci. Techn., 17(2): 4 (1996): Salix fangiana.

PLEIARINA:-

floridana (Chapm.) N. Chao & G.T. Gong in J. Sichuan Forest. Sci. Techn., 17(2): 4 (1996): Salix floridana.

glandulosa (Seemen) N. Chao & G.T. Gong in J. Sichuan Forest. Sci. Techn., 17(2): 4 (1996): Salix glandulosa.

hirsuta (Thunb.) N. Chao & G.T. Gong in J. Sichuan Forest. Sci. Techn., 17(2): 4 (1996): Salix hirsuta.

huillensis (Seemen) N. Chao & G.T. Gong in J. Sichuan Forest. Sci. Techn., 17(2): 5 (1996): Salix huillensis.

humaensis (Y.L. Chou & R.C. Chou) N. Chao & G.T. Gong in J. Sichuan Forest. Sci. Techn., 17(2): 5 (1996): Salix humaensis.

hutchinsii (Skan) N. Chao & G.T. Gong in J. Sichuan Forest. Sci. Techn., 17(2): 5 (1996): Salix hutchinsii.

ichnostachya (Lindl. ex Andersson) N. Chao & G.T. Gong in J. Sichuan Forest. Sci. Techn., 17(2): 5 (1996): Salix ichnostachya.

jaliscana (M.E. Jones) N. Chao & G.T. Gong in J. Sichuan Forest. Sci. Techn., 17(2): 5 (1996): Salix jaliscana.

kusanoi (Hayata) N. Chao & G.T. Gong in J. Sichuan Forest. Sci. Techn., 17(2): 5 (1996): Salix tetrasperma var. kusanoi.

lasiandra (Benth.) N. Chao & G.T. Gong in J. Sichuan Forest. Sci. Techn., 17(2): 5 (1996): Salix lasiandra.

ledermannii (Seemen) N. Chao & G.T. Gong in J. Sichuan Forest. Sci. Techn., 19(4): 14 (1998): Salix ledermannii.

lucida (Muhl.) N. Chao & G.T. Gong in J. Sichuan Forest. Sci. Techn., 17(2): 5 (1996): Salix lucida.

madagascariensis (Bojer ex Andersson) N. Chao & G.T. Gong in J. Sichuan Forest. Sci. Techn., 17(2): 5 (1996): Salix madagascariensis.

mesnyi (Hance) N. Chao & G.T. Gong in J. Sichuan Forest. Sci. Techn., 17(2): 5 (1996): Salix mesnyi.

mucronata (Thunb.) N. Chao & G.T. Gong in J. Sichuan Forest. Sci. Techn., 17(2): 5 (1996): Salix mucronata.

murielii (Skan) N. Chao & G.T. Gong in J. Sichuan Forest. Sci. Techn., 19(4): 14 (1998): Salix murielii.

neowilsonii (W.P. Fang) N. Chao & G.T. Gong in J. Sichuan Forest. Sci. Techn., 17(2): 5 (1996): Salix neowilsonii.

nigra (Marshall) N. Chao & G.T. Gong in J. Sichuan Forest. Sci. Techn., 17(2): 5 (1996): Salix nigra.

paraplesia (C.K. Schneid.) N. Chao & G.T. Gong in J. Sichuan Forest. Sci. Techn., 17(2): 5 (1996): Salix paraplesia.

pentandra (L.) N. Chao & G.T. Gong in J. Sichuan Forest. Sci. Techn., 17(2): 5 (1996): Salix pentandra.

plectilis (Kimura) N. Chao & G.T. Gong in J. Sichuan Forest. Sci. Techn., 17(2): 6 (1996): Salix plectilis.

safsaf (Forssk.) N. Chao & G.T. Gong in J. Sichuan Forest. Sci. Techn., 17(2): 6 (1996): Salix safsaf.

serissima (L.H. Bailey ex Arthur) N. Chao & G.T. Gong in J. Sichuan Forest. Sci. Techn., 17(2): 6 (1996): Salix lucida var. serissima.

songarica (Andersson) N. Chao & G.T. Gong in J. Sichuan Forest. Sci. Techn., 17(2): 6 (1996): Salix songarica.

subserrata (Willd.) N. Chao & G.T. Gong in J. Sichuan Forest. Sci. Techn., 19(4): 14 (1998): Salix subserrata.

tetrasperma (Roxb.) N. Chao & G.T. Gong in J. Sichuan Forest. Sci. Techn., 17(2): 6 (1996): Salix tetrasperma.

thorelii (Dode) N. Chao & G.T. Gong in J. Sichuan Forest. Sci. Techn., 17(2): 6 (1996): Salix thorelii.

tonkinensis (Seemen) N. Chao & G.T. Gong in J. Sichuan Forest. Sci. Techn., 17(2): 6 (1996): Salix tonkinensis.

triandra (L.) N. Chao & G.T. Gong in J. Sichuan Forest. Sci. Techn., 17(2): 6 (1996): Salix triandra.

tsoongii (Cheng) N. Chao & G.T. Gong in J. Sichuan Forest. Sci. Techn., 17(2): 6 (1996): Salix tsoongii.

warburgii (Seemen) N. Chao & G.T. Gong in J. Sichuan Forest. Sci. Techn., 17(2): 6 (1996): Salix warburgii.

PLEIARINA:-
woodii *(Seemen) N. Chao & G.T. Gong in J. Sichuan Forest. Sci. Techn.*, 17(2): 6 (1996): Salix woodii.

PLEIOBLASTUS (Gramin.).
albosericeus *W.T. Lin in J. S. China Agric. Univ.*, 11(3): 47 (1990), as 'albo-sericeus'—China Southeast (Guangdong).
altiligulatus S.L. Chen & S.Y. Chen
var. spongiosus *B.M. Yang in J. Hunan Sci. Technol. Univ.*, 1(1): 113 (1985)—China Southeast (Hunan).
amarus (Keng) Keng f.
f. huangshanensis *C.L. Huang in J. Bamboo Res.*, 15(3): 14 (1996)—China Southeast (Anhui).
chino (Franch. & Sav.) Makino
f. humilis *(Makino) Sadao Suzuki, Index Japanese Bambusac.*: 369 (1978): Arundinaria variegata f. humilis.
hispidulus *W.T. Lin & Z.J. Feng in J. S. China Agric. Univ.*, 14(1): 50 (1993)—China Southeast (Guangdong).
patellaris *W.T. Lin & Z.M. Wu in J. S. China Agric. Univ.*, 14(3): 113 (1993)—China Southeast (Guangdong).
wuyishanensis *Q.F. Zheng & K.F. Huang in Wuyi Sci. J.*, 2: 18 (1982)—China Southeast (Fujian).
yingdeensis *W.T. Lin & Z.M. Wu in J. S. China Agric. Univ.*, 11(3): 47 (1990)—China Southeast (Guangdong).

PLEIOCARPA (Apocynac.).
brevistyla *E. Omino in Wageningen Agric. Univ. Pap.*, 96(1): 139 (1996); *et in Belmontia*, 29: 139 (1996)—Gabon.
picralimoides *(Pichon) E. Omino in Wageningen Agric. Univ. Pap.*, 96(1): 156 (1996); *et in Belmontia*, 29: 156 (1996): Carpodinopsis (?) picralimoides.

PLEIOCORYNE (Rubiac.).
fernandensis (Hiern) Rauschert
var. pobeguinii *(N. Hallé) J.-P. Lebrun & A.L. Stork, Enum. Pl. Fleurs Afr. Trop.*, 4: 197 (1997): Polycoryne fernandensis var. pobeguinii.

PLEIONE (Orchidac.).
sect. Bulbocodioides *G. Torelli & M. Riccaboni in Caesiana*, 10: 6 (1998).
subgen. Dictyopleione *G. Torelli & M. Riccaboni in Caesiana*, 10: 6 (1998), as 'comb. nov.'.
sect. Hookerianae *G. Torelli & M. Riccaboni in Caesiana*, 10: 6 (1998).
sect. Humiles *G.H. Zhu & S.C. Chen in Novon*, 8(4): 462 (1998).
subgen. Humiles *(G.H. Zhu & S.C. Chen) G. Torelli & M. Riccaboni in Caesiana*, 14: 23 (2000), without basionym ref: Pleione sect. Humiles.
subgen. Saxicola *G. Torelli & M. Riccaboni in Caesiana*, 10: 6 (1998).
subgen. Scopulorum *G. Torelli & M. Riccaboni in Caesiana*, 10: 6 (1998).
autumnalis *S.C. Chen & G.H. Zhu in Harvard Pap. Bot.*, 4(2): 429 (1999)—China South-central (Yunnan).
barbarae *G.J. Braem in Orchidées Cult. Protect.*, 38: 20 (1999)—China South-central (Yunnan).
braemii *H. Pinkepank in Orchidées Cult. Protect.*, 38: 27 (1999)—China South-central (Yunnan).
× christianii *H. Perner in Orchidee*, Suppl. 6: 12 (1999)—China South-central (Yunnan?). (P. forrestii × yunnanensis).
formosana Hayata
f. alba *G. Torelli & M. Riccaboni in Caesiana*, 14: 55 (2000)—Taiwan.
forrestii Schltr.
f. alba *(H. Li & G.H. Feng) G. Torelli & M. Riccaboni in Caesiana*, 14: 60 (2000): Pleione alba.
harberdii *G.J. Braem in Orchidées Cult. Protect.*, 38: 21 (1999)—China South-central (Yunnan).
hookeriana (Lindl.) B.S. Williams
f. alba *G. Torelli & M. Riccaboni in Caesiana*, 14: 72 (2000), without type—Locality not stated.
var. milanii *(Braem) G. Torelli & M. Riccaboni in Caesiana*, 14: 73 (2000), without exact basionym page: Pleione milanii.

PLEIONE:-
hookeriana
var. sinensis *G. Kleinh. ex G. Torelli & M. Riccaboni in Caesiana*, 14: 72 (2000)—China South-Central (Yunnan).
hubeiensis *G. Torelli & M. Riccaboni in Caesiana*, 10: 31 (1998)—China South-central (Hubei).
microphylla *S.C. Chen & Z.H. Tsi in Acta Phytotax. Sin.*, 38(2): 182 (2000)—China Southeast (Guangdong).
milanii *G.J. Braem in Orchidées Cult. Protect.*, 38: 23 (1999)—China South-central (Yunnan).
moelleri *G.J. Braem in Orchidées Cult. Protect.*, 38: 23 (1999)—China South-central (Yunnan).
mohrii *G.J. Braem in Orchidées Cult. Protect.*, 38: 21 (1999)—China South-central (Yunnan).
praecox (Sm.) D. Don
var. reichenbachiana *(T. Moore & Veitch) G. Torelli & M. Riccaboni in Caesiana*, 14: 101 (2000): Coelogyne reichenbachiana.
× taliensis *Butterf. ex G. Torelli & M. Riccaboni in Caesiana*, 14: 137 (2000)—China South-Central (Yunnan). (P. bulbocodioides × yunnanensis).
voltolinii *G. Torelli & M. Riccaboni in Caesiana*, 12: 51 (1999)—China South-central (Hubei).
yunnanensis (Rolfe) Rolfe
var. chiwuana *(T. Tang & F.T. Wang) G. Kleinh. ex G. Torelli & M. Riccaboni in Caesiana*, 14: 127 (2000): Pleione chiwuana.

PLEIOSORBUS *L.H. Zhou & C.Y. Wu in Acta Bot. Yunnanica*, 22(4): 383 (2000). *ROSACEAE*
megacarpa *L.H. Zhou & C.Y. Wu in Acta Bot. Yunnanica*, 22(4): 384 (2000), as 'megacarpus'—Tibet.

PLEIOTAXIS (Compos.).
angolensis *J. Rodríguez Oubiña & S. Ortiz in Nordic J. Bot.*, 16(5): 477 (1996 publ. 1997)—Angola.
huillensis O. Hoffm.
subsp. argentea *(M. Taylor) S. Ortiz & J. Rodríguez Oubiña in Nordic J. Bot.*, 19(1): 25 (1999): P. argentea.
subsp. axillaris *S. Ortiz & J. Rodríguez Oubiña in Nordic J. Bot.*, 19(1): 25 (1999)—Angola.
var. macrocephala *S. Ortiz & J. Rodríguez Oubiña in Nordic J. Bot.*, 19(1): 25 (1999)—Angola.

PLEUROSPERMUM (Umbellif.).
hookeri
var. haidongense *J.T. Pan in Fl. Qinghaiica*, 2: 434 (1999)—Qinghai.

PLEUROTHALLIS (Orchidac.).
ser. Acronia *(C. Presl) C.A. Luer, Ic. Pleurothallid. XVII (Monogr. Syst. Bot. Missouri Bot. Gard., 72)*: 5 (1998), as 'Acroniae': gen. Acronia.
subgen. Acuminatia *C.A. Luer, Ic. Pleurothallid. XVIII (Monogr. Syst. Bot. Missouri Bot. Gard., 76)*: 98 (1999).
sect. Alatae *C.A. Luer, Ic. Pleurothallid. XVIII (Monogr. Syst. Bot. Missouri Bot. Gard., 76)*: 99 (1999).
ser. Amphigyae *C.A. Luer, Ic. Pleurothallid. XVII (Monogr. Syst. Bot. Missouri Bot. Gard., 72)*: 5 (1998).
subsect. Antenniferae *C.A. Luer, Ic. Pleurothallid. XVII (Monogr. Syst. Bot. Missouri Bot. Gard., 72)*: 4 (1998).
subgen. Antilla *C.A. Luer, Ic. Pleurothallid. XX (Monogr. Syst. Bot. Missouri Bot. Gard., 79)*: 39 (2000).
sect. Brobdingnagia *C.A. Luer, Ic. Pleurothallid. XVII (Monogr. Syst. Bot. Missouri Bot. Gard., 72)*: 66 (1998).
sect. Cylindria *C.A. Luer, Ic. Pleurothallid. XVII (Monogr. Syst. Bot. Missouri Bot. Gard., 72)*: 66 (1998).
subgen. Effusia *C.A. Luer, Ic. Pleurothallid. XX (Monogr. Syst. Bot. Missouri Bot. Gard., 79)*: 54 (2000).
subsect. Longiracemosae *C.A. Luer, Ic. Pleurothallid. XVII (Monogr. Syst. Bot. Missouri Bot. Gard., 72)*: 4 (1998).
subsect. Perplexae *C.A. Luer, Ic. Pleurothallid. XVIII (Monogr. Syst. Bot. Missouri Bot. Gard., 76)*: 83 (1999).

PLEUROTHALLIS:-
subgen. Pseudostelis *(Schltr.) C.A. Luer, Ic. Pleurothallid. XVIII (Monogr. Syst. Bot. Missouri Bot. Gard., 76)*: 87 (1999): gen. Pseudostelis.
sect. Racemosae *C.A. Luer, Ic. Pleurothallid. XX (Monogr. Syst. Bot. Missouri Bot. Gard., 79)*: 106 (2000).
sect. Satyria *(Luer) C.A. Luer, Ic. Pleurothallid. XII (Monogr. Syst. Bot. Missouri Bot. Gard., 57)*: 146 (1995), without basionym ref.; *et in Lindleyana*, 11(2): 118 (1996), with basionym ref.: Myoxanthus subgen. Satyria.
sect. Silenia *(Luer) C.A. Luer, Ic. Pleurothallid. XII (Monogr. Syst. Bot. Missouri Bot. Gard., 57)*: 146 (1995), without basionym ref.; *et in Lindleyana*, 11(2): 118 (1996), with basionym ref.: Myoxanthus subgen. Silenia.
subgen. Unciferae *(Luer) C.A. Luer, Ic. Pleurothallid. XVII (Monogr. Syst. Bot. Missouri Bot. Gard., 72)*: 89 (1998): P. sect. Unciferae.
aculeata *C.A. Luer & Hirtz in Lindleyana*, 11(3): 141 (1996)—Ecuador.
acutilabia *C.A. Luer in Orquideologia*, 21(2): 116 (1999)—Colombia.
adelphe *C.A. Luer & Hirtz in Lindleyana*, 11(3): 142 (1996)—Ecuador.
aggeris *C.A. Luer in Lindleyana*, 11(3): 143 (1996)—Ecuador, Peru.
alta *C.A. Luer in Lindleyana*, 11(3): 143 (1996)—Ecuador.
ancistra *C.A. Luer & Hirtz in Lindleyana*, 11(3): 144 (1996)—Ecuador.
anderssonii *C.A. Luer in Lindleyana*, 11(3): 145 (1996)—Ecuador.
andrei *C.A. Luer & R. Escobar R. in Orquideologia*, 20(1): 34 (1996)—Colombia.
angulosa *C.A. Luer & Hirtz in Lindleyana*, 11(3): 146 (1996)—Ecuador.
ankyloglossa *C.A. Luer & Hirtz in C.A. Luer, Ic. Pleurothallid. XX (Monogr. Syst. Bot. Missouri Bot. Gard.*, 79): 127 (2000)—Ecuador.
aondae *G. Carnevali & G.A. Romero in G.A. Romero & G. Carnevali, Orchids of Venezuela, ed. 2*: 1141 (2000)—Venezuela.
applanata *C.A. Luer & Dalström in Lindleyana*, 11(3): 147 (1996)—Ecuador.
apposita *C.A. Luer, Ic. Pleurothallid. XVI (Monogr. Syst. Bot. Missouri Bot. Gard., 65)*: 12 (1998)—Ecuador.
archidonopsis *C.A. Luer & Hirtz in C.A. Luer, Ic. Pleurothallid. XVIII (Monogr. Syst. Bot. Missouri Bot.* Gard., 76): 42 (1999)—Ecuador.
arctata *C.A. Luer in Lindleyana*, 12(1): 39 (1997)—Peru.
arenicola *(Carnevali & I. Ramírez) G. Carnevali & I. Ramírez in G.A. Romero & G. Carnevali, Orchids of Venezuela, ed. 2*: 1141 (2000): Myoxanthus aspasicensis subsp. arenicola.
ariasii *C.A. Luer & Hirtz in Lindleyana*, 12(1): 42 (1997)—Peru.
ascera *C.A. Luer & R. Escobar R.in Orquideologia*, 20(1): 36 (1996)—Colombia.
asplundii *C.A. Luer in Lindleyana*, 11(3): 147 (1996)—Ecuador.
atrorubra *C.A. Luer & Hirtz in Lindleyana*, 11(3): 148 (1996)—Ecuador.
atwoodii *C.A. Luer in Lindleyana*, 11(2): 67 (1996)—Costa Rica.
avirostris *C.A. Luer & Hirtz in C.A. Luer, Ic. Pleurothallid. XVI (Monogr. Syst. Bot. Missouri Bot. Gard.*, 65): 13 (1998)—Ecuador.
bahorucensis *C.A. Luer in Lindleyana*, 14(2): 108 (1999)—Dominican Republic.
batracha *C.A. Luer & Hirtz in Lindleyana*, 11(3): 149 (1996)—Ecuador.
baudoensis *C.A. Luer & R. Escobar R. in Orquideologia*, 21(1): 73 (1998)—Colombia.
bevilacquana *G. Carnevali & I. Ramírez in G.A. Romero & G. Carnevali, Orchids of Venezuela, ed. 2*: 1141 (2000)—Venezuela.
bicallosa *C.A. Luer & Hirtz in Lindleyana*, 11(3): 150 (1996)—Ecuador.
biceps *C.A. Luer & Hirtz in Lindleyana*, 11(3): 150 (1996)—Ecuador.
bicochlearis *C.A. Luer in Orquideologia*, 21(2): 118 (1999)—Colombia.
bissei *C.A. Luer in Lindleyana*, 14(2): 108 (1999)—Cuba.

PLEUROTHALLIS:-

bitumida *C.A. Luer in Lindleyana,* 11(2): 70 (1996)—Costa Rica.

blancoi *F. Pupulin in Caesiana,* 15: 1 (2000)—Costa Rica.

blepharoglossa *C.A. Luer in Lindleyana,* 14(2): 111 (1999)—Cuba.

bothros *C.A. Luer in Lindleyana,* 11(2): 70 (1996)—Costa Rica.

bovilingua *C.A. Luer & R. Escobar R. in Orquideologia,* 21(1): 75 (1998)—Colombia.

brighamella *C.A. Luer, Ic. Pleurothallid. XVIII (Monogr. Syst. Bot. Missouri Bot. Gard., 76):* 171 (1999); *et in Orquideologia,* 21(3): 323 (2000)—Panama.

bucaramangae *C.A. Luer & R. Escobar R. in Orquideologia,* 20(1): 38 (1996)—Colombia.

calamifolia *C.A. Luer & R. Escobar R. in Orquideologia,* 20(1): 40 (1996)—Colombia.

calolalax *C.A. Luer & R. Escobar R. in Orquideologia,* 21(1): 78 (1998)—Colombia.

candida *C.A. Luer & Hirtz in Lindleyana,* 12(1): 42 (1997), contrary to Art. 37.5 ICBN (1994)—Peru.

canidentis *C.A. Luer & R. Escobar R. in Orquideologia,* 21(1): 80 (1998)—Colombia.

capitonis *C.A. Luer & R. Escobar R. in Orquideologia,* 20(1): 44 (1996)—Colombia.

cassidata *C.A. Luer & R. Escobar R. in Orquideologia,* 21(1): 82 (1998)—Colombia.

catoxys *C.A. Luer & Hirtz in Lindleyana,* 11(3): 151 (1996)—Ecuador.

cedrinorum *C.A. Luer & Dalström in Lindleyana,* 11(3): 152 (1996)—Ecuador.

chamelopoda *C.A. Luer in Orquideologia,* 20(2): 201 (1996)—Colombia.

chavezii *C.A. Luer in Lindleyana,* 11(2): 72 (1996)—Costa Rica.

choerorhyncha *C.A. Luer in Orquideologia,* 20(2): 203 (1996)—Colombia.

chuscalica *C.A. Luer in Orquideologia,* 21(2): 120 (1999)—Colombia.

ciliifera *C.A. Luer in Lindleyana,* 14(2): 111 (1999)—Dominican Republic.

circinata *C.A. Luer in Orquideologia* 20(2): 206 (1996)—Colombia.

claviculata *C.A. Luer & Hirtz in Lindleyana,* 11(3): 153 (1996)—Ecuador.

cleistogama *C.A. Luer, Ic. Pleurothallid. XVII (Monogr. Syst. Bot. Missouri Bot. Gard., 72):* 41 (1998)—Panama.

cocornaensis *C.A. Luer & R. Escobar R. in Orquideologia,* 20(1): 45 (1996)—Colombia.

coeloglossa *C.A. Luer & Hirtz in Lindleyana,* 11(3): 154 (1996)—Ecuador.

complanata *C.A. Luer & Hirtz in Lindleyana,* 11(3): 154 (1996)—Ecuador.

compressa *C.A. Luer in Lindleyana,* 11(2): 75 (1996)—Costa Rica.

conochila *C.A. Luer in Lindleyana,* 11(2): 75 (1996)—Costa Rica.

constricta *C.A. Luer & R. Escobar R.in Orquideologia,* 21(1): 84 (1998)—Colombia.

convexa *C.A. Luer & Hirtz in Lindleyana,* 11(3): 155 (1996)—Ecuador.

coronula *C.A. Luer, Ic. Pleurothallid. XVIII (Monogr. Syst. Bot. Missouri Bot. Gard., 76):* 171 (1999); *et in Orquideologia,* 21(3): 325 (2000)—Colombia.

cosangae *C.A. Luer & Hirtz in C.A. Luer, Ic. Pleurothallid. XVI (Monogr. Syst. Bot. Missouri Bot. Gard.,* 65): 18 (1998)—Ecuador.

cotyligera *C.A. Luer & R. Escobar R.in Orquideologia,* 20(1): 46 (1996)—Colombia.

croatii *C.A. Luer, Ic. Pleurothallid. XVI (Monogr. Syst. Bot. Missouri Bot. Gard., 65):* 18 (1998)—Panama.

crossota *C.A. Luer & Dalström in Lindleyana,* 11(3): 156 (1996)—Ecuador.

crucifera *C.A. Luer & Hirtz in Lindleyana,* 11(3): 157 (1996)—Ecuador.

cuatrecasasii *C.A. Luer in Orquideologia,* 20(2): 208 (1996)—Colombia.

cunabularis *C.A. Luer, Ic. Pleurothallid. XVII (Monogr. Syst. Bot. Missouri Bot. Gard., 72):* 42 (1998)—Colombia.

cycesis *C.A. Luer & R. Escobar R. in Orquideologia,* 20(1): 49 (1996)—Colombia.

PLEUROTHALLIS:-

dasypetala *C.A. Luer & Hirtz in Lindleyana,* 11(3): 158 (1996)—Ecuador.

depressa *C.A. Luer & R. Escobar R. in Orquideologia,* 20(1): 50 (1996)—Colombia.

dewildei *C.A. Luer & R. Escobar R.in Orquideologia,* 21(1): 86 (1998)—Colombia.

dimidia *C.A. Luer, Ic. Pleurothallid. XVIII (Monogr. Syst. Bot. Missouri Bot. Gard., 76):* 109 (1999)—Bolivia, Peru, Colombia, Venezuela.

displosa *C.A. Luer, Ic. Pleurothallid. XVIII (Monogr. Syst. Bot. Missouri Bot. Gard., 76):* 172 (1999); *et in Orquideologia,* 21(3): 329 (2000)—Panama.

dorrii *C.A. Luer in Lindleyana,* 12(1): 45 (1997)—Venezuela.

dracula *H. Seehawer in Orchidee,* 50(6): 635 (1999)—Brazil Southeast (Rio de Janeiro).

drewii *C.A. Luer, Ic. Pleurothallid. XX (Monogr. Syst. Bot. Missouri Bot. Gard., 79):* 127 (2000)—Ecuador.

driessenii *C.A. Luer, Ic. Pleurothallid. XX (Monogr. Syst. Bot. Missouri Bot. Gard., 79):* 128 (2000)—Ecuador.

dukei *C.A. Luer in Lindleyana,* 12(1): 45 (1997)—Panama.

endresii *C.A. Luer, Ic. Pleurothallid. XVIII (Monogr. Syst. Bot. Missouri Bot. Gard., 76):* 172 (1999); *et in Orquideologia,* 21(3): 331 (2000)—Costa Rica.

erucosa *C.A. Luer & R. Escobar R. in Orquideologia,* 21(1): 88 (1998)—Colombia.

esmeraldae *C.A. Luer & Hirtz in Lindleyana,* 11(3): 158 (1996)—Ecuador.

espinalii *C.A. Luer in Orquideologia,* 21(2): 122 (1999)—Colombia.

fabiobarrosii *E.L. Borba & J. Semir in Lindleyana,* 15(1): 3 (2000)—Brazil Southeast (Minas Gerais).

flaveola *C.A. Luer & Hirtz in Lindleyana,* 11(3): 159 (1996)—Ecuador.

flosculifera *C.A. Luer in Lindleyana,* 14(2): 113 (1999)—Dominican Republic.

fornicata *C.A. Luer in Lindleyana,* 11(3): 160 (1996)—Ecuador, Peru.

fossulata *C.A. Luer & R. Escobar R. in Orquideologia,* 21(1): 91 (1998)—Colombia.

fritzii *G. Carnevali & G.A. Romero in G.A. Romero & G. Carnevali, Orchids of Venezuela, ed. 2:* 1142 (2000)—Venezuela.

fustifera *C.A. Luer, Ic. Pleurothallid. XVIII (Monogr. Syst. Bot. Missouri Bot. Gard., 76):* 173 (1999)—Peru.

galerasensis *C.A. Luer, Ic. Pleurothallid. XVI (Monogr. Syst. Bot. Missouri Bot. Gard., 65):* 28 (1998)—Colombia.

garciae *C.A. Luer, Ic. Pleurothallid. XVIII (Monogr. Syst. Bot. Missouri Bot. Gard., 76):* 173 (1999)—Venezuela.

gargantua *C.A. Luer in Lindleyana,* 11(3): 161 (1996)—Ecuador.

gigas *C.A. Luer & R. Escobar R. in Orquideologia,* 20(1): 52 (1996)—Colombia.

giraldoi *C.A. Luer in Orquideologia,* 21(2): 125 (1999)—Colombia.

glabra *C.A. Luer & R. Escobar R. in Orquideologia,* 20(1): 56 (1996)—Colombia.

globosa *C.A. Luer & R. Escobar R. in Orquideologia,* 20(1): 60 (1996)—Colombia.

gomezii *C.A. Luer & R. Escobar R. in Orquideologia,* 20(1): 62 (1996)—Colombia.

granitica *C.A. Luer & G.A. Romero in G.A. Romero & G. Carnevali, Orchids of Venezuela, ed. 2:* 1142 (2000)—Venezuela.

grayumii *C.A. Luer in Lindleyana,* 11(2): 78 (1996)—Costa Rica.

hammelii *C.A. Luer in Lindleyana,* 12(1): 48 (1997)—Panama.

henniae *C.A. Luer & Dalström in Lindleyana,* 11(3): 162 (1996)—Ecuador.

hilariana *G. Carnevali & G.A. Romero in G.A. Romero & G. Carnevali, Orchids of Venezuela, ed. 2:* 1143 (2000)—Venezuela.

hjertingii *C.A. Luer in Lindleyana,* 12(1): 48 (1997)—Peru.

imbaburae *C.A. Luer & Hirtz in Lindleyana,* 11(3): 163 (1996)—Ecuador.

PLEUROTHALLIS:-

imberbis *C.A. Luer & Hirtz in Lindleyana,* 11(3): 163 (1996)—Ecuador, Bolivia, Brazil North (Amazonas).

imitor *C.A. Luer in Lindleyana,* 11(2): 78 (1996)—Costa Rica.

incongrua *C.A. Luer in Orquideologia,* 21(2): 127 (1999)—Colombia.

infundibulosa *C.A. Luer in Orquideologia,* 20(2): 210 (1996)—Colombia.

ingramii *C.A. Luer in Lindleyana,* 11(2): 81 (1996)—Costa Rica.

jaculifera *C.A. Luer & R. Escobar R. in Orquideologia,* 21(1): 93 (1998)—Colombia.

jesupii *C.A. Luer in Lindleyana,* 14(2): 116 (1999)—Dominican Republic.

jesupiorum *C.A. Luer & Hirtz in Lindleyana,* 11(3): 164 (1996)—Ecuador.

jurisdixii *C.A. Luer & R. Escobar R. in Orquideologia,* 20(1): 64 (1996)—Colombia.

kareniae *C.A. Luer in Lindleyana,* 11(2): 83 (1996), as 'karenae'—Costa Rica.

knappiae *C.A. Luer, Ic. Pleurothallid. XX (Monogr. Syst. Bot. Missouri Bot. Gard., 79):* 128 (2000), as 'knappii'—Panama.

laevis *C.A. Luer & Hirtz in C.A. Luer, Ic. Pleurothallid. XVI (Monogr. Syst. Bot. Missouri Bot. Gard.,* 65): 33 (1998)—Ecuador.

lagarophyta *C.A. Luer, Ic. Pleurothallid. XVIII (Monogr. Syst. Bot. Missouri Bot. Gard., 76):* 112 (1999), nom. nov.: Pleurothallis tenuifolia var. longisepala *C. Schweinf.*

laminata *C.A. Luer, Ic. Pleurothallid. XVI (Monogr. Syst. Bot. Missouri Bot. Gard., 65):* 33 (1998)—Ecuador.

languida *C.A. Luer & R. Escobar R. in Orquideologia,* 21(1): 95 (1998)—Colombia.

lappacea *C.A. Luer, Ic. Pleurothallid. XX (Monogr. Syst. Bot. Missouri Bot. Gard., 79):* 129 (2000)—Ecuador.

lehmanneptis *C.A. Luer & R. Escobar R. in Orquideologia,* 21(1): 100 (1998)—Colombia.

lenae *C.A. Luer & Dalström in Lindleyana,* 11(3): 165 (1996)—Ecuador.

llamachoi *C.A. Luer in Lindleyana,* 13(3): 146 (1998)—Cuba.

lobata *C.A. Luer in Orquideologia,* 20(2): 214 (1996)—Colombia.

loejtnantii *C.A. Luer in Lindleyana,* 11(3): 166 (1996)—Ecuador.

londonoi *C.A. Luer, Ic. Pleurothallid. XIV (Monogr. Syst. Bot. Missouri Bot. Gard., 61(4)):* 6 (1996)—Colombia.

lopezii *C.A. Luer & R. Escobar R. in Orquideologia,* 21(1): 102 (1998)—Colombia.

loxensis *C.A. Luer & Hirtz in Lindleyana,* 11(3): 167 (1996)—Ecuador.

madsenii *C.A. Luer, Ic. Pleurothallid. XVIII (Monogr. Syst. Bot. Missouri Bot. Gard., 76):* 16 (1999)—Ecuador.

maduroi *C.A. Luer in Lindleyana,* 12(1): 51 (1997)—Panama.

maguirei *C.A. Luer, Ic. Pleurothallid. XVIII (Monogr. Syst. Bot. Missouri Bot. Gard., 76):* 113 (1999)—Guyana.

manicosa *C.A. Luer & R. Escobar R. in Orquideologia,* 20(1): 66 (1996)—Colombia.

marthae *C.A. Luer & R. Escobar R. in Orquideologia,* 20(1): 68 (1996)—Colombia.

mastodon *C.A. Luer & Hirtz in Lindleyana,* 11(3): 167 (1996)—Ecuador.

medinae *C.A. Luer & J. Portilla in C.A. Luer, Ic. Pleurothallid. XX (Monogr. Syst. Bot. Missouri Bot. Gard.,* 79): 110 (2000)—Ecuador.

medusa *C.A. Luer in Lindleyana,* 12(1): 51 (1997)—?Costa Rica.

megaglossa *C.A. Luer & Dalström in Lindleyana,* 11(3): 168 (1996)—Ecuador.

megalorhina *C.A. Luer & R. Escobar R. in Orquideologia,* 20(1): 71 (1996)—Colombia.

melanochthoda *C.A. Luer & Hirtz in Lindleyana,* 11(3): 169 (1996)—Ecuador.

melanoglossa *C.A. Luer & R. Escobar R. in Orquideologia,* 20(1): 74 (1996)—Colombia.

melanosticta *C.A. Luer in Lindleyana,* 11(3): 170 (1996)—Ecuador.

PLEUROTHALLIS:-

mexiae *C.A. Luer, Ic. Pleurothallid. XVIII (Monogr. Syst. Bot. Missouri Bot. Gard., 76):* 174 (1999); *et in Orquideologia,* 21(3): 333 (2000), nom. nov.: Pleurothallis chamensis var. tenuis *C. Schweinf.*

millipeda *C.A. Luer in Orquideologia,* 20(2): 216 (1996)—Colombia.

minutissima *C.A. Luer in Orquideologia,* 20(2): 218 (1996)—Colombia.

molleturoi *C.A. Luer & Dodson in C.A. Luer, Ic. Pleurothallid. XVI (Monogr. Syst. Bot. Missouri Bot. Gard.,* 65): 38 (1998)—Ecuador.

montezumae *C.A. Luer in Lindleyana,* 11(2): 83 (1996)—Costa Rica.

morganii *C.A. Luer in Lindleyana,* 11(3): 171 (1996)—Ecuador.

moronae *C.A. Luer & Hirtz in Lindleyana,* 11(3): 171 (1996)—Ecuador.

muriculata *C.A. Luer & Hirtz in Lindleyana,* 11(3): 172 (1996)—Ecuador.

napintzae *C.A. Luer & Hirtz in Lindleyana,* 11(3): 173 (1996)—Ecuador.

nellyae *P. Ortiz V.in Orquideologia,* 20(3): 325 (1997)—Colombia.

nortonii *C.A. Luer, Ic. Pleurothallid. XVIII (Monogr. Syst. Bot. Missouri Bot. Gard., 76):* 174 (1999)—Venezuela.

nox-media *C.A. Luer & R. Escobar R. in Orquideologia,* 21(1): 104 (1998)—Colombia.

oblonga *C.A. Luer & Hirtz in Lindleyana,* 11(3): 174 (1996)—Ecuador.

omissa *C.A. Luer, Ic. Pleurothallid. XVIII (Monogr. Syst. Bot. Missouri Bot. Gard., 76):* 175 (1999)—Venezuela.

oncoglossa *C.A. Luer in Lindleyana,* 11(2): 86 (1996)—Costa Rica.

orecta *C.A. Luer & R. Escobar R. in Orquideologia,* 20(1): 75 (1996)—Colombia.

ortegae *C.A. Luer & Hirtz in Lindleyana,* 11(3): 174 (1996)—Ecuador.

pandurata *C.A. Luer & Hirtz in Lindleyana,* 11(3): 175 (1996)—Ecuador.

panica *C.A. Luer & Dalström in C.A. Luer, Ic. Pleurothallid. XIV (Monogr. Syst. Bot. Missouri Bot. Gard.,* 61(4)): 6 (1996); *see also C.A. Luer, Ic. Pleurothallid. XV (Monogr. Syst. Bot. Missouri Bot.* Gard., 64): 122 (1997)—Ecuador.

pantherina *H. Seehawer in Orchidee,* 49(3): 130 (1998)—Brazil Southeast (Rio de Janeiro).

papulifolia *C.A. Luer in Lindleyana,* 14(2): 116 (1999)—Cuba.

paquishae *C.A. Luer in Lindleyana,* 11(3): 176 (1996)—Ecuador.

paradoxa *C.A. Luer & Dalström in Lindleyana,* 11(3): 177 (1996)—Ecuador.

pariaensis *G. Carnevali & G.A. Romero in G.A. Romero & G. Carnevali, Orchids of Venezuela, ed. 2:* 1144 (2000)—Venezuela.

patateensis *C.A. Luer, Ic. Pleurothallid. XVI (Monogr. Syst. Bot. Missouri Bot. Gard., 65):* 42 (1998)—Ecuador.

pellucida *C.A. Luer & Hirtz in C.A. Luer, Ic. Pleurothallid. XVI (Monogr. Syst. Bot. Missouri Bot. Gard.,* 65): 42 (1998)—Ecuador.

pennelliana *C.A. Luer, Ic. Pleurothallid. XVI (Monogr. Syst. Bot. Missouri Bot. Gard., 65):* 43 (1998)—Colombia.

perennis *C.A. Luer in Lindleyana,* 11(2): 86 (1996)—Costa Rica.

perforata *C.A. Luer & Hirtz in Lindleyana,* 11(3): 178 (1996)—Ecuador.

petiolaris *C.A. Luer in Orquideologia,* 20(2): 220 (1996)—Colombia.

phratria *C.A. Luer & Hirtz in Lindleyana,* 11(3): 179 (1996)—Ecuador.

phrynoglossa *C.A. Luer & Hirtz in C.A. Luer, Ic. Pleurothallid. XVI (Monogr. Syst. Bot. Missouri Bot. Gard.,* 65): 116 (1998)—Ecuador.

pilostoma *C.A. Luer in Lindleyana,* 11(2): 89 (1996)—Costa Rica.

platypetala *C.A. Luer & R. Escobar R.in Orquideologia,* 20(1): 78 (1996)—Colombia.

possoae *C.A. Luer, Ic. Pleurothallid. XX (Monogr. Syst. Bot. Missouri Bot. Gard., 79):* 129 (2000)—Colombia.

praealta *C.A. Luer & Hirtz in C.A. Luer, Ic. Pleurothallid. XVI (Monogr. Syst. Bot. Missouri Bot. Gard.,* 65): 45 (1998)—Ecuador.

privigna *C.A. Luer in Lindleyana,* 14(2): 118 (1999)—Dominican Republic.

prolificans *C.A. Luer & Hirtz in C.A. Luer, Ic. Pleurothallid. XVI (Monogr. Syst. Bot. Missouri Bot. Gard.,* 65): 45 (1998)—Ecuador.

prolixa *C.A. Luer & Hirtz in Lindleyana,* 11(3): 179 (1996)—Ecuador.

pseudopogon *C.A. Luer & Hirtz in C.A. Luer, Ic. Pleurothallid. XVII (Monogr. Syst. Bot. Missouri Bot.* Gard., 72): 30 (1998)—Ecuador.

psichion *C.A. Luer in Lindleyana,* 11(2): 89 (1996)—Costa Rica.

psilantha *C.A. Luer, Ic. Pleurothallid. XVII (Monogr. Syst. Bot. Missouri Bot. Gard., 72):* 95 (1998)—Costa Rica.

ptychophora *C.A. Luer & Hirtz in Lindleyana,* 11(3): 180 (1996)—Ecuador.

pulcherrima *C.A. Luer, Ic. Pleurothallid. XVIII (Monogr. Syst. Bot. Missouri Bot. Gard., 76):* 58 (1999)—Ecuador.

purpurascens *C.A. Luer & Hirtz in Lindleyana,* 11(3): 181 (1996)—Ecuador.

purpurella *C.A. Luer, Ic. Pleurothallid. XVIII (Monogr. Syst. Bot. Missouri Bot. Gard., 76):* 176 (1999); et *in Orquideologia,* 21(3): 335 (2000)—Colombia.

quaternaria *C.A. Luer, Ic. Pleurothallid. XVII (Monogr. Syst. Bot. Missouri Bot. Gard., 72):* 31 (1998)—Colombia.

questionis *C.A. Luer & R. Escobar R. in Orquideologia,* 20(1): 79 (1996)—Colombia.

ramosii *C.A. Luer, Ic. Pleurothallid. XVII (Monogr. Syst. Bot. Missouri Bot. Gard., 72):* 32 (1998)—Colombia.

recula *C.A. Luer in Lindleyana,* 11(2): 92 (1996)—Ecuador, Costa Rica, Panama.

recurvata *C.A. Luer & Hirtz in Lindleyana,* 11(3): 182 (1996)—Ecuador.

reptilis *C.A. Luer & Dalström in Lindleyana,* 11(3): 183 (1996)—Ecuador.

roseola *C.A. Luer & Hirtz in Lindleyana,* 11(3): 183 (1996)—Ecuador.

sagittilabia *C.A. Luer, Ic. Pleurothallid. XVII (Monogr. Syst. Bot. Missouri Bot. Gard., 72):* 32 (1998); *et in Orquideologia,* 21(2): 130 (1999)—Colombia.

salpingantha *C.A. Luer & Hirtz in C.A. Luer, Ic. Pleurothallid. XVI (Monogr. Syst. Bot. Missouri Bot. Gard.,* 65): 50 (1998)—Ecuador, Colombia, Venezuela, Peru.

sandaliorum *G.A. Romero & G. Carnevali in Harvard Pap. Bot.,* 5(1): 182 (2000)—Venezuela.

sandemanii *C.A. Luer in Orquideologia,* 20(2): 222 (1996)—Colombia.

sannio *C.A. Luer & R. Escobar R. in Orquideologia,* 21(1): 106 (1998)—Colombia.

sarcosepala *G. Carnevali & I. Ramírez in G.A. Romero & G. Carnevali, Orchids of Venezuela, ed. 2:* 1144 (2000)—Venezuela.

scaphipetala *C.A. Luer in Lindleyana,* 11(2): 94 (1996)—Costa Rica.

schlimii *C.A. Luer, Ic. Pleurothallid. XVIII (Monogr. Syst. Bot. Missouri Bot. Gard., 76):* 120 (1999)—Colombia.

silvestris *C.A. Luer & Hirtz in Lindleyana,* 11(3): 184 (1996)—Ecuador.

siphoglossa *C.A. Luer & R. Escobar R. in Orquideologia,* 20(1): 82 (1996)—Colombia, Venezuela.

siphonantha *C.A. Luer, Ic. Pleurothallid. XVI (Monogr. Syst. Bot. Missouri Bot. Gard., 65):* 52 (1998)—Peru.

sobrina *C.A. Luer & Hirtz in Lindleyana,* 11(3): 185 (1996)—Ecuador.

somnolenta *C.A. Luer in Orquideologia,* 20(2): 225 (1996)—Colombia.

spathosa *C.A. Luer & R. Escobar R. in Orquideologia,* 20(1): 86 (1996)—Colombia.

spathuliformis *C.A. Luer & R. Vásquez in C.A. Luer, Ic. Pleurothallid. XVIII (Monogr. Syst. Bot. Missouri Bot.* Gard., 76): 123 (1999)—Bolivia.

spathulipetala *C.A. Luer in Orquideologia,* 20(2): 227 (1996)—Colombia.

stergiosii *G. Carnevali & I.M. Ramírez M. in Harvard Pap. Bot.,* 3(2): 247 (1998)—Venezuela.

sugdenii *C.A. Luer in Orquideologia,* 20(2): 229 (1996)—Colombia.

syringodes *C.A. Luer, Ic. Pleurothallid. XVIII (Monogr. Syst. Bot. Missouri Bot. Gard., 76):* 176 (1999)—Colombia.

tarantula *C.A. Luer & Hirtz in Lindleyana,* 11(3): 186 (1996)—Ecuador, Colombia.

teaguei *C.A. Luer in Lindleyana,* 11(3): 187 (1996)—Ecuador.

tectosa *C.A. Luer & Hirtz in Lindleyana,* 11(3): 187 (1996)—Ecuador.

tenebrosa *C.A. Luer, Ic. Pleurothallid. XVIII (Monogr. Syst. Bot. Missouri Bot. Gard., 76):* 25 (1999)—Venezuela.

tetrachaeta *C.A. Luer & Hirtz in C.A. Luer, Ic. Pleurothallid. XVII (Monogr. Syst. Bot. Missouri Bot.* Gard., 72): 35 (1998)—Ecuador.

thomasii *C.A. Luer, Ic. Pleurothallid. XX (Monogr. Syst. Bot. Missouri Bot. Gard., 79):* 84, 130 (2000)—Honduras.

tintinnabula *C.A. Luer in Lindleyana,* 11(2): 94 (1996)—Costa Rica.

toachica *C.A. Luer & Dodson in Lindleyana,* 11(3): 188 (1996)—Ecuador.

tokachii *C.A. Luer in Lindleyana,* 12(1): 55 (1997)—Peru.

tomentosa *C.A. Luer, Ic. Pleurothallid. XVIII (Monogr. Syst. Bot. Missouri Bot. Gard., 76):* 177 (1999); *et in Orquideologia,* 21(3): 337 (2000)—Costa Rica.

torrana *C.A. Luer in Orquideologia,* 20(2): 231 (1996); *see also Ic. Pleurothallid. XV (Monogr. Syst. Bot. Missouri Bot. Gard., 64):* 122 (1997)—Colombia.

tragulosa *C.A. Luer & Hirtz in Lindleyana,* 11(3): 189 (1996)—Ecuador.

transversilabia *G. Carnevali & I. Ramírez in G.A. Romero & G. Carnevali, Orchids of Venezuela, ed. 2:* 1145 (2000)—Venezuela.

trifurcata *C.A. Luer & Hirtz in Lindleyana,* 11(3): 190 (1996)—Ecuador.

trinitensis *(Griseb.) G. Carnevali & G.A. Romero in G.A. Romero & G. Carnevali, Orchids of Venezuela, ed. 2:* 1145 (2000): Pleurothallis sertularioides var. trinitensis.

trullifera *C.A. Luer & Hirtz in Lindleyana,* 11(3): 190 (1996)—Ecuador.

tsubotae *C.A. Luer & R. Escobar R. in Orquideologia,* 20(1): 88 (1996)—Colombia.

tuberculosa *C.A. Luer & Hirtz in Lindleyana,* 11(3): 191 (1996)—Ecuador.

tuzae *C.A. Luer in Orquideologia,* 21(2): 132 (1999)—Colombia.

tyria *C.A. Luer & Hirtz in Lindleyana,* 11(3): 192 (1996)—Ecuador.

uninervia *C.A. Luer & Dodson in C.A. Luer, Ic. Pleurothallid. XVIII (Monogr. Syst. Bot. Missouri Bot.* Gard., 76): 35 (1999)—Ecuador.

ursula *C.A. Luer & Hirtz in Lindleyana,* 11(3): 193 (1996)—Ecuador.

vegrandis *C.A. Luer & Dodson in C.A. Luer, Ic. Pleurothallid. XVI (Monogr. Syst. Bot. Missouri Bot. Gard.,* 65): 56 (1998)—Ecuador.

veliformis *C.A. Luer & Dalström in Lindleyana,* 11(3): 194 (1996)—Ecuador.

virgata *C.A. Luer, Ic. Pleurothallid. XVI (Monogr. Syst. Bot. Missouri Bot. Gard., 65):* 59 (1998)—Peru.

viridiflora *H. Seehawer in Orchidee,* 50(6): 637 (1999)—Brazil Southeast (Rio de Janeiro).

viridis *C.A. Luer & Hirtz in Lindleyana,* 11(3): 194 (1996)—Ecuador.

volans *C.A. Luer & Hirtz in Lindleyana,* 11(3): 195 (1996)—Ecuador.

zarumae *C.A. Luer & Hirtz in Lindleyana,* 11(3): 196 (1996)—Ecuador.

zunagensis *C.A. Luer & Hirtz in C.A. Luer, Ic. Pleurothallid. XVI (Monogr. Syst. Bot. Missouri Bot. Gard.,* 65): 61 (1998)—Ecuador.

PLEUROTHYRIUM (Laurac.).

guindonii *H. van der Werff in Novon,* 6(4): 476 (1996)—Costa Rica.

oblongum *H. van der Werff in Novon,* 6(4): 476 (1996)—Costa Rica, Panama.

PLICOSEPALUS (Loranthac.).

foliosus *D. Wiens & R.M. Polhill, Mistletoes of Africa:* 91 (1998)—Zambia, Malawi.

PLICOSEPALUS:-
somalensis *D. Wiens & R.M. Polhill, Mistletoes of Africa:* 93 (1998)—Somalia.

PLINIA (Myrtac.).
aureana *(Mattos) J.R. de Mattos in Loefgrenia,* 112: 5 (1998): Myrciaria aureana.
brachybotrya *(D. Legrand) M. Sobral in Hoehnea,* 21(1–2): 202 (1994 publ. 1995): Marlierea parviflora var. brachybotrya.
cordifolia *(D. Legrand) M. Sobral in Hoehnea,* 21(1–2): 202 (1994 publ. 1995): Myrciaria cordifolia.
coronata *(Mattos) J.R. de Mattos in Loefgrenia,* 112: 5 (1998): Myrciaria coronata.
grandifolia *(Mattos) M. Sobral in Hoehnea,* 21(1–2): 202 (1994 publ. 1995): Myrciaria grandifolia.
hatschbachii *(Mattos) M. Sobral in Hoehnea,* 21(1–2): 202 (1994 publ. 1995): Myrciaria hatschbachii.
muricata *M. Sobral in Hoehnea,* 21(1–2): 200 (1994 publ. 1995)—Brazil Northeast (Bahia).
oblongata *(Mattos) J.R. de Mattos in Loefgrenia,* 112: 5 (1998): Myrciaria oblongata.
phitrantha *(Kiaersk.) M. Sobral in Hoehnea,* 21(1–2): 202 (1994 publ. 1995): Eugenia phitrantha.
rara *M. Sobral in Hoehnea,* 21(1–2): 200 (1994 publ. 1995)—Brazil Northeast (Bahia).
rogersiana *J.R. Mattos in Loefgrenia,* 106: 3 (1995)—Brazil Southeast (São Paulo).
spiciflora *(Nees & Mart.) M. Sobral in Hoehnea,* 21(1–2): 202 (1994 publ. 1995): Eugenia spiciflora.
spirito-santensis *(Mattos) J.R. de Mattos in Loefgrenia,* 112: 5 (1998): Myrciaria spirito-santensis.

PLUCHEA (Compos.).
dunlopii *S. Hunger in Willdenowia,* 26(1–2): 276 (1996)—Western Australia, Northern Territory, South Australia, New South Wales, Queensland.
indica (L.) Less.
subsp. yemenensis *S. King-Jones in Willdenowia,* 29(1–2): 209 (1999)—South Yemen.

PLUTONOPUNTIA *P.V. Heath in Calyx,* 6(2): 41 (1999), nom. nov.: Opuntia ser. Chaffeyanae *Britton & Rose.* (Cactac.).
chaffeyi *(Britton & Rose) P.V. Heath in Calyx,* 6(2): 41 (1999): Opuntia chaffeyi.

POA (Gramin.).
sect. Alpinae *(Hegetschw. ex Nyman) R.J. Soreng in Novon,* 8(2): 193 (1998): P. [unranked] Alpinae.
sect. Dasypoa *(Pilg.) R.J. Soreng in Novon,* 8(2): 187 (1998): gen. Dasypoa.
subsect. Glaucopoa *(Asch. & Graebn.) M.V. Olonova in Turczaninowia,* 1(4): 18 (1998): Poa sect. Glaucopoa.
sect. Malacanthae *(Roshev.) M.V. Olonova in Turczaninowia,* 1(4): 7 (1998): Poa ser. Malacanthae.
sect. Sylvestres *V.L. Marsh ex R.J. Soreng in Novon,* 8(2): 188 (1998).
akmanii *R.J. Soreng, P. Hein & H. Scholz in Willdenowia,* 27(1–2): 196 (1997)—Turkey.
alascana *(Soreng) N.N. Tzvelev in Novosti Sist. Vÿssh. Rast.,* 32: 182 (2000): P. hartzii subsp. alascana.
anae *O. Tovar in Publ. Mus. Hist. Nat. Javier Prado, B,* 33: 6 (1985)—Peru.
andicola *S.A. Renvoize, Gramineas de Bolivia:* 138 (1998)—Bolivia.
anfamensis *M.A. Negritto & A.M. Anton in Darwiniana,* 35(1–4): 159 (1998)—Argentina Northwest (Tucumán).
angustifolia L.
var. alpigena *(Blytt) V.N. Voroschilov in A.K. Skvortsov (ed.), Florist. issl. v razn. Raĭonakh SSSR:* 149 (1985): P. pratensis var. alpigena.
var. sublanata *(Reverd.) V.N. Voroschilov in A.K. Skvortsov (ed.), Florist. issl. v razn. Raĭonakh SSSR:* 150 (1985): P. sublanata.
annua L.
subsp. raniglumis *(S.E. Fröhner) J. Chrtek in Čas. Nár. Muz. Řada Přír.,* 165(1–4): 128 (1996): P. annua var. raniglumis.

POA:-
arnowiae *R.J. Soreng in Novon,* 8(2): 197 (1998)—Utah, Idaho, Wyoming.
asiae-minoris *H. Scholz & A.J. Byfield in Turkish J. Bot.,* 24(4): 265 (2000)—Turkey.
attenuata Trin.
subsp. dahurica *(Trin.) M.V. Olonova in Turczaninowia,* 1(4): 17 (1998), isonym; *see Gubanov, Konsp. Fl. Vneshnei Mongolii (Sosud. Rast.):* 22 (1996): Poa dahurica.
subsp. dahurica *(Trin.) I.A. Gubanov, Konsp. Fl. Vneshnei Mongolii (Sosud. Rast.):* 22 (1996): P. dahurica.
subsp. tczuensis *(Serg.) M.V. Olonova in Turczaninowia,* 1(4): 17 (1998): Poa argunensis var. tczuensis.
babiogorensis *D. Bernátová, J. Májovský & J. Obuch in Biologia, Bot. (Slovakia),* 54(1): 38 (1999)—Poland.
botryoides Trin. ex Griseb.
subsp. argunensis *(Roshev.) M.V. Olonova in Turczaninowia,* 1(4): 17 (1998): Poa argunensis.
bracteosa Kom.
var. penicillata *(Kom.) V.N. Voroschilov in A.K. Skvortsov (ed.), Florist. issl. v razn. Raĭonakh SSSR:* 150 (1985): P. penicillata.
var. platyantha *(Kom.) V.N. Voroschilov in A.K. Skvortsov (ed.), Florist. issl. v razn. Raĭonakh SSSR:* 150 (1985): P. platyantha.
chambersii *R.J. Soreng in Novon,* 8(2): 195 (1998)—Oregon.
chumbiensis *H.J. Noltie in Edinburgh J. Bot.,* 57(2): 282 (2000)—Xizang.
compressa L.
var. expansa *N. Böhling & H. Scholz in Carolinea,* 55: 7 (1997)—Austria, Germany.
cooperi *H.J. Noltie in Edinburgh J. Bot.,* 57(2): 283 (2000)—Sikkim.
dentigluma *O. Tovar in Publ. Mus. Hist. Nat. Javier Prado, B,* 33: 5 (1985)—Peru.
dolichachyra Keng
var. longiflora *S.L. Chen ex D.Z. Ma in Fl. Ningxiaensis,* 2: 522 (1988 publ. 1990)—Inner Mongolia (Ningxia).
durifolia *L.M. Giussani, E.G. Nicora & F.A. Roig in Darwiniana,* 38(1–2): 53 (2000), nom. nov.: P. ligularis var. stricta *Parodi ex Nicora & F.A. Roig.*
dzongicola *H.J. Noltie in Edinburgh J. Bot.,* 57(2): 283 (2000)—Bhutan, Sikkim.
garhwalensis *D.C. Nautiyal & R.D. Gaur in J. Bombay Nat. Hist. Soc.,* 96(2): 285 (1999)—Jammu-Kashmir.
golestanensis *H. Scholz & H. Akhani in Edinburgh J. Bot.,* 55(3): 449 (1998)—Iran.
harveyi *J. Chrtek in Čas. Nár. Muz. Řada Přír.,* 165(1–4): 128 (1996)—Yugoslavia.
× inconspicua *H. Scholz in Bot. Chronika,* 12: 16 (1996), homonym; *see Veldkamp in P. Royen, Alp. Fl. New Guinea,* 2: 1062 (1979)—Kriti. (P. infirma × maroccana).
krasnoborovii *N.V. Stepanov, Florogenet. Analiz,* 1: 104 (1994)—Krasnoyarsk.
lachenensis *H.J. Noltie in Edinburgh J. Bot.,* 57(2): 286 (2000)—Sikkim.
lanata Scribn. & Merr.
var. trivialiformis *(Kom.) V.N. Voroschilov in A.K. Skvortsov (ed.), Florist. issl. v razn. Raĭonakh SSSR:* 150 (1985): P. trivialiformis.
ligularis Nees
var. stricta *Parodi ex E.G. Nicora & F.A. Roig in Hickenia,* 2(58): 275 (1998)—Argentina Northwest (Mendoza).
longii *H.J. Noltie in Edinburgh J. Bot.,* 57(2): 285 (2000)—Sikkim.
macrocalyx Trautv. & C.A. Mey.
var. beringiana *(Prob.) V.N. Voroschilov in A.K. Skvortsov (ed.), Florist. issl. v razn. Raĭonakh SSSR:* 150 (1985): P. beringiana.
var. sugawarae *(Ohwi) V.N. Voroschilov in A.K. Skvortsov (ed.), Florist. issl. v razn. Raĭonakh SSSR:* 150 (1985): P. sugawarae.
margilicola *D. Bernátová & J. Májovský in Biologia, Bot. (Slovakia),* 52(1): 28 (1997)—Czechoslovakia.
mendocina *E.G. Nicora & F.A. Roig in Hickenia,* 2(58): 273 (1998)—Argentina Northwest (Mendoza).

POA:-
nemoralis L.
var. sichotensis *(Prob.) V.N. Voroschilov in A.K. Skvortsov (ed.), Florist. issl. v razn. Raĭonakh SSSR:* 150 (1985): P. sichotensis.
subsp. subpolaris *(Kuvaev) N.N. Tzvelev in Novosti Sist. Vÿssh. Rast.,* 32: 182 (2000): P. subpolaris.
paramoensis *S. Laegaard in Novon,* 8(1): 28 (1998)—Ecuador.
parviceps Hack.
var. jujuyensis *Parodi ex E.G. Nicora in Hickenia,* 2(33): 143 (1995)—Argentina Northwest (Jujuy, Salta).
× perinconspicua *H. Scholz in Willdenowia,* 28(1–2): 172 (1998), nom. nov.: P. × inconspicua *H. Scholz.*
pratensis L.
subsp. turfosa *(Litv.) M.V. Olonova in Turczaninowia,* 1(4): 9 (1998), isonym; *see Vorosch. in Byull. Mosk. Obshch. Ispÿt. Prir., Biol.,* 95(2): 91 (1990): Poa turfosa.
pseudotibetica *H.J. Noltie in Edinburgh J. Bot.,* 57(2): 279 (2000)—Sikkim, Xizang.
ragonesei *E.G. Nicora in Hickenia,* 2(33): 145 (1995)—Argentina Northwest (Salta, Jujuy).
rajbhandarii *H.J. Noltie in Edinburgh J. Bot.,* 57(2): 288 (2000)—Sikkim.
rigidifolia Steud.
var. ibari *(Phil.) L.M. Giussani in Ann. Missouri Bot. Gard.,* 87(2): 221 (2000): P. ibari.
rivas-martinezii *O. Tovar in Publ. Mus. Hist. Nat. Javier Prado, B,* 33: 3 (1985)—Peru.
rohmooiana *H.J. Noltie in Edinburgh J. Bot.,* 57(2): 281 (2000)—Sikkim.
schistacea *E. Edgar & H.E. Connor in New Zealand J. Bot.,* 37(1): 63 (1999)—New Zealand South I.
× scotica *H.A. McAllister in Bot. J. Scotland,* 49(2): 273 (1997), without type—Great Britain.
sejuncta *D. Bernátová, J. Májovský & J. Obuch in Biologia, Bot. (Slovakia),* 54(1): 37 (1999)—Czechoslovakia.
sibirica Roshev.
subsp. insignis *(Litv. ex Roshev.) M.V. Olonova in Turczaninowia,* 1(4): 13 (1998): Poa insignis.
smirnovii Roshev.
subsp. polozhiae *(Revjakina) M.V. Olonova in Turczaninowia,* 1(4): 7 (1998): Poa polozhiae.
var. vivipara *L.I. Malyschev in L.I. Malyschev (ed.), Vÿsokogornaya Flora Stanovogo Nagor'ya:* 53 (1972)—Yakutiya.
spania *E. Edgar & B.P.J. Molloy in New Zealand J. Bot.,* 37(2): 43 (1999)—New Zealand South I.
spiciformis *D.F. Cui in Fl. Xinjiangensis,* 6: 601 (1996), holotype not indicated—Xinjiang.
sublanata Reverd.
subsp. vivipara *(Tzvelev) M.V. Olonova in Turczaninowia,* 1(4): 11 (1998): Poa sublanata var. vivipara.
tovarii *R.J. Soreng in Novon,* 8(2): 200 (1998), nom. nov., as 'tovari': P. geniculata *Tovar.*
tristis Trin.
subsp. bryophila *(Trin.) M.V. Olonova in Turczaninowia,* 1(4): 19 (1998): Poa bryophila.
subsp. litvinoviana *(Ovcz.) M.V. Olonova in Turczaninowia,* 1(4): 18 (1998): Poa litvinoviana.
unilateralis
subsp. pachypholis *(Piper) D.D. Keck ex R.J. Soreng in Novon,* 8(2): 199 (1998): P. pachypholis.
urssulensis Trin.
subsp. krylovii *(Reverd.) M.V. Olonova in Turczaninowia,* 1(4): 16 (1998): Poa krylovii.
versicolor Besser
subsp. filiculmis *(Roshev.) M.V. Olonova in Turczaninowia,* 1(4): 16 (1998): Poa filiculmis.
xenica *E. Edgar & H.E. Connor in New Zealand J. Bot.,* 37(1): 65 (1999)—New Zealand South I.

PODACHENIUM (Compos.).
chimalapanum *B.L. Turner in Phytologia,* 84(2): 61 (1998 publ. 1999)—Mexico Southwest (Oaxaca).

PODALYRIA (Legumin.).
bracteata *W. Roxburgh, Pl. Coromandel,* 3: 55 (1820)—Bangladesh.†

PODANDRIELLA *D.L. Szlachetko in Fl. Cameroun,* 34: 192 (1998), nom. nov.: gen. Podandria *Rolfe.* (Orchidac.).
batesii *(la Croix) D.L. Szlachetko & T.S. Olszewski in Fl. Cameroun,* 34: 196 (1998): Habenaria batesii.
letouzeyana *D.L. Szlachetko & T.S. Olszewski in Fl. Cameroun,* 34: 196 (1998)—Gulf of Guinea Is. (Principe).
macrandra *(Lindl.) D.L. Szlachetko & T.S. Olszewski in Fl. Cameroun,* 34: 194 (1998): Habenaria macrandra.
phantasma *(la Croix) D.L. Szlachetko & T.S. Olszewski in Fl. Cameroun,* 34: 198 (1998): Habenaria phantasma.

PODANDROGYNE (Capparac.).
hispidula *(DC.) T.S. Cochrane in Novon,* 7(4): 353 (1997 publ. 1998): Gynandropsis hispidula.
jamesonii *(Briq.) T.S. Cochrane in Novon,* 7(4): 353 (1997 publ. 1998): Gynandropsis jamesonii.
mathewsii *(Briq.) T.S. Cochrane in Novon,* 7(4): 354 (1997 publ. 1998): Gynandropsis mathewsii.
subsp. ulei *(Briq.) T.S. Cochrane in Novon,* 7(4): 354 (1997 publ. 1998): Gynandropsis ulei.

PODOCARPIUM (Legumin.).
densum *(C. Chen & X.J. Cui) P.H. Huang in Bull. Bot. Res. North-East. Forest. Univ.,* 18(1): 6 (1998): P. fallax var. densum.
podocarpum
var. japonicum *(Matsum.) P.H. Huang in Bull. Bot. Res. North-East. Forest. Univ.,* 18(1): 6 (1998): Desmodium oxyphyllum var. japonicum.
var. mandshuricum *(Maxim.) P.H. Huang in Bull. Bot. Res. North-East. Forest. Univ.,* 18(1): 6 (1998): Desmodium podocarpum var. mandshuricum.

PODOCARPUS (Podocarpac.).
aristulatus Parl.
var. buchii *(Urb.) J. Silba in J. Int. Conifer Preserv. Soc.,* 7(1): 31 (2000): P. buchii.
atjehensis *(Wasscher) D.J. de Laubenfels in Phytologia Mem.,* 7: 60 (1984): P. neriifolius var. atjehensis.†
ballivianensis *J. Silba in J. Int. Conifer Preserv. Soc.,* 7(1): 32 (2000), without latin descr.—Bolivia.
capuronii de Laub.
var. woltzii *(Gaussen) J. Silba in J. Int. Conifer Preserv. Soc.,* 7(1): 31 (2000): P. woltzii.
costalis
var. muroii *J. Silba in J. Int. Conifer Preserv. Soc.,* 7(1): 33 (2000)—Japan.
degeneri *(N.E. Gray) D.J. de Laubenfels in Phytologia Mem.,* 7: 61 (1984): P. neriifolius var. degeneri.†
expansus *(de Laub.) T.C. Whitmore in T.C. Whitmore, I.G.M. Tantra & U. Sutisna (eds.), Tree Flora Indonesia,* Check List Irian Jaya: 234 (1997), with incorrect basionym ref.: Dacrycarpus expansus.
ingensis *D.J. de Laubenfels in Bol. Lima,* 13(73): 58 (1991); *et in Bol. Lima,* 16(91–96): 36 (1994)—Peru, Bolivia.
lambertii Klotzsch. ex Endl.
var. horsmanii *J. Silba in J. Int. Conifer Preserv. Soc.,* 7(1): 34 (2000)—Brazil Southeast (Minas Gerais).
var. tigreensis *J. Silba in J. Int. Conifer Preserv. Soc.,* 7(1): 34 (2000)—Brazil South (Rio Grande do Sul).
neriifolius D. Don
var. decipiens *(N.E. Gray) J. Silba in J. Int. Conifer Preserv. Soc.,* 7(1): 35 (2000): P. decipiens.
var. penibukanensis *J. Silba in J. Int. Conifer Preserv. Soc.,* 7(1): 34 (2000)—Borneo (Sabah).
var. staintonii *J. Silba in J. Int. Conifer Preserv. Soc.,* 7(1): 36 (2000)—Bhutan-Sikkim.
rostratus L. Laurent
var. perrieri *(Gaussen & Woltz) J. Silba in J. Int. Conifer Preserv. Soc.,* 7(1): 39 (2000): P. perrieri.
rubens de Laub.
var. pabinamaensis *J. Silba in J. Int. Conifer Preserv. Soc.,* 7(1): 37 (2000)—New Guinea (Papua New Guinea).

PODOCARPUS:-
rubens
var. sumatrana *J. Silba in J. Int. Conifer Preserv. Soc.,* 7(1): 37 (2000)—Sumatera.
rumphii Blume
var. arbainii *J. Silba in J. Int. Conifer Preserv. Soc.,* 7(1): 38 (2000)—Sumatera.
var. aruensis *J. Silba in J. Int. Conifer Preserv. Soc.,* 7(1): 38 (2000)—New Guinea (Irian Jaya).
spathoides de Laub.
var. solomonensis *J. Silba in J. Int. Conifer Preserv. Soc.,* 7(1): 39 (2000)—Solomon Is (South Solomons).
subtropicalis de Laub.
var. medogensis *J. Silba in J. Int. Conifer Preserv. Soc.,* 7(1): 40 (2000)—Xizang.
transiens *(Pilg.) D.J. de Laubenfels in Phytologia Mem.,* 7: 68 (1984): P. lambertii var. transiens.†
var. harleyi *J. Silba in J. Int. Conifer Preserv. Soc.,* 7(1): 39 (2000)—Brazil Northeast (Bahia).

PODOCHILUS (Orchidac.).
sect. Sarganella *A. Schuiteman in Blumea,* 43(2): 492 (1998), as 'nom. nov'.
marsupialis *A. Schuiteman in Blumea,* 43(2): 489 (1998)—Borneo (Sarawak, Sabah, Kalimantan).

PODOLEPIS (Compos.).
papillosa *Brown ex Pépin in Ann. Fl. Pomone,* 1833–34: 88 (Dec. 1833)—Australia.

PODOPHYLLUM (Berberidac.).
sect. Dysosma *(Woodson) J.M.H. Shaw in New Plantsman,* 6(3): 159 (1999).gen. Dysosma.
sect. Paradysosma *J.M.H. Shaw in New Plantsman,* 6(3): 163 (1999).
aurantiocaule Hand.-Mazz.
subsp. furfuraceum *(S.Y. Bao) J.M.H. Shaw in New Plantsman,* 6(3): 164 (1999): Dysosma furfuracea.
var. uniflorum *J.M.H. Shaw in New Plantsman,* 6(3): 163 (1999)—Bhutan.
delavayi Franch.
var. longipetalum *J.L. Wu & P. Zhuang ex J.M.H. Shaw in New Plantsman,* 6(3): 163 (1999)—China South-central (Sichuan).
glaucescens *J.M.H. Shaw in New Plantsman,* 6(3): 162 (1999)—China Southeast (Guangxi).
guangxiense *(Y.S. Wang) J.M.H. Shaw in New Plantsman,* 6(3): 163 (1999): Dysosma guangxiensis.
hemsleyi *J.M.H. Shaw & W.T. Stearn in New Plantsman,* 6(3): 161 (1999)—China South-central (Hubei).
hexandrum Royle
var. sikkimense *(R. Chatterjee & Mukerjee) R.C. Srivastava, Fl. Sikkim (Ranunculac.-Moringac.):* 62 (1998), as 'sikkimensis': P. sikkimense.
× inexpectatum *J.M.H. Shaw in New Plantsman,* 6(3): 159 (1999)—Cultivation. (P. peltatum × pleianthum).
peltatum L.
var. annulare *J.M.H. Shaw in New Plantsman,* 6(3): 158 (1999)—Virginia.
trilobulus *J.M.H. Shaw in New Plantsman,* 7(3): 158 (2000)—China South-central (Sichuan).
versipelle Hance
subsp. boreale *J.M.H. Shaw in New Plantsman,* 6(3): 160 (1999)—China South-central (Hubei).
var. sichuanense *J.M.H. Shaw in New Plantsman,* 6(3): 161 (1999)—China South-central (Sichuan).

PODOSTEMUM (Podostemac.).
munnarense *(Nagendran & Arekal) C.J. Mathew & V.K. Satheesh in Aquatic Bot.,* 57(1–4): 255 (1997): Polypleurum munnarense.

POGONIA (Orchidac.).
bifaria *(Fernald) P.M. Brown & R.P. Wunderlin in North Amer. Native Orchid J.,* 3(4): 451 (1997): Cleistes divaricata var. bifaria.
japonica Rchb. f.
f. albiflora *Y.N. Lee, Fl. Korea:* 1164 (1996), without type—Locality not stated.

POGOSTEMON (Labiat.).
subgen. Allopogostemon *G.R. Bhatti & M. Ingrouille in Bull. Nat. Hist. Mus. (London), Bot.,* 27(2): 97 (1997), without type.
*subgen.*Dysophylla *(Blume) G.R. Bhatti & M. Ingrouille in Bull. Nat. Hist. Mus. (London), Bot.,* 27(2): 107 (1997), as 'Dysophyllus', without basionym ref.: gen. Dysophylla.
subsect. Glabriusculi *(Briq.) G.R. Bhatti & M. Ingrouille in Bull. Nat. Hist. Mus. (London), Bot.,* 27(2): 102 (1997), as 'Glabriusculus', without basionym ref., name same as basionym: P. subsect. Glabriusculae.
subsect. Racemosi *(Benth.) G.R. Bhatti & M. Ingrouille in Bull. Nat. Hist. Mus. (London), Bot.,* 27(2): 97 (1997), as 'Racemosus', without basionym ref.: P. sect. Racemosae.
sect. Racemosi *(Benth.) G.R. Bhatti & M. Ingrouille in Bull. Nat. Hist. Mus. (London), Bot.,* 27(2): 97 (1997), as 'Racemosus', without basionym ref., same name as basionym: P. sect. Racemosae.
sect. Verticillati *(Benth.) G.R. Bhatti & M. Ingrouille in Bull. Nat. Hist. Mus. (London), Bot.,* 27(2): 111 (1997), as 'Verticillatus', without basionym ref.: Dysophylla sect. Verticillatae.
sect. Zygocalyx *G.R. Bhatti & M. Ingrouille in Bull. Nat. Hist. Mus. (London), Bot.,* 27(2): 104 (1997), without type.
auricularius (L.) Hassk.
var. giganteus *(Doan) P.H. Hô ex A.L. Budantzev in Komarovia,* 1: 22 (1999), basionym not validly publ.: Dysophylla auricularia var. gigantea.
barbatus *G.R. Bhatti & M. Ingrouille in Bull. Nat. Hist. Mus. (London), Bot.,* 27(2): 109 (1997)—China Southeast (Hong Kong).
ciliatus *G.R. Bhatti & M. Ingrouille in Bull. Nat. Hist. Mus. (London), Bot.,* 27(2): 111 (1997)—New Guinea (Irian Jaya).
elatispicatus *G.R. Bhatti & M. Ingrouille in Bull. Nat. Hist. Mus. (London), Bot.,* 27(2): 102 (1997)—Philippines.
erectus (Dalzell) Kuntze
var. diplolobatus *G.R. Bhatti & M. Ingrouille in Bull. Nat. Hist. Mus. (London), Bot.,* 27(2): 117 (1997)—India (Kerala).
globulosus *(Doan) A.L. Budantzev in Komarovia,* 1: 23 (1999), basionym not validly publ.: Dysophylla globulosa.
peethapushpum *A.K. Pradeep in Candollea,* 53(2): 419 (1998)—India (Kerala).
petelotii *(Doan) A.L. Budantzev in Komarovia,* 1: 23 (1999), basionym not validly publ.: Dysophylla petelotii.
tisserantii *(Pellegr.) J.-P. Lebrun & A.L. Stork, Enum. Pl. Fleurs Afr. Trop.,* 4: 536 (1997): Dysophylla tisserantii.
tisserantii *(Pellegr.) G.R. Bhatti & M. Ingrouille in Bull. Nat. Hist. Mus. (London), Bot.,* 27(2): 111 (Nov. 1997), isonym; *see J.-P. Lebrun & Stork, Enum. Pl. Fleurs Afr. Trop.,* 4: 536 (Sept. 1997): Dysophylla tisserantii.
verticillatus *(Roxb.) G.R. Bhatti & M. Ingrouille in Bull. Nat. Hist. Mus. (London), Bot.,* 27(2): 115 (1997), isonym; *see Miq., Fl. Ned. Ind.,* 2: 965 (1859): Mentha verticillata.

POIKILOSPERMUM (Cecropiac.).
longifolium *A.C. Church in Kew Bull.,* 53(1): 232 (1998)—Borneo (Kalimantan).

POLEMONIUM (Polemoniac.).
sect. Borealia *N.N. Tzvelev in Novosti Sist. Vÿssh. Rast.,* 30: 141 (1996).
boreale Adams
var. hyperboreum *(Tolm.) N.N. Tzvelev in Sosud. Rast. Sovet. Dal'nego Vostoka,* 7: 291 (1995); *et in Novosti Sist. Vÿssh. Rast.,* 30: 142 (1996): P. hyperboreum.
var. macranthum *(Cham.) N.N. Tzvelev in Sosud. Rast. Sovet. Dal'nego Vostoka,* 7: 291 (1995); *et in Novosti Sist. Vÿssh. Rast.,* 30: 142 (1996): P. humile var. macranthum.
caeruleum L.
subf. albiflorum *Tatew. ex Koji Ito in Environm. Sci., Hokkaido,* 6(2): 263 (1983)—Sakhalin.
f. hidakanum *(Koji Ito) Koji Ito in Environm. Sci., Hokkaido,* 6(2): 270 (1983): P. yezoense var. hidakanum.

POLEMONIUM:-
caeruleum
f. insulare *Koji Ito in Environm. Sci., Hokkaido,* 6(2): 263 (1983)—Japan.
subsp. laxiflorum *(Regel) Koji Ito in Environm. Sci., Hokkaido,* 6(2): 261 (1983): P. caeruleum lusus laxiflorum.
var. nipponicum *(Kitam.) Koji Ito in Environm. Sci., Hokkaido,* 6(2): 268 (1983): P. nipponicum.
f. paludosum *Koji Ito in Environm. Sci., Hokkaido,* 6(2): 264 (1983), without type—Japan, Sakhalin, Kuril Is.
kiushianum Kitam.
f. albiflorum *Y.N. Lee, Fl. Korea:* 1160 (1996), without type—Locality not stated.
pauciflorum S. Watson
subsp. melindae *(Rzed., Calderón & Villarreal) J.M. Porter & L.A. Johnson in Aliso,* 19(1): 85 (2000): P. melindae.
villosum Rudolph ex Georgi
f. glabrum *(S.D. Zhao) S.Z. Liu in Clavis Pl. Chinae Bor.-Or. ed. 2 (ed. P.Y. Fu):* 538 (1995), isonym; *see Y.C. Zhu in Pl. Medic. Chinae Bor.-Orient. (ed. Y.C. Zhu):* 936 (1989): P. villosum var. glabrum.

POLIANTHES (Agavac.).
bicolor *E. Solano Camacho & A. García-Mendoza in Sida,* 18(2): 473 (1998)—Mexico Southwest (Oaxaca).

POLLIA (Commelinac.).
macrobracteata *D.Y. Hong in Novon,* 7(3): 252 (1997)—China Southeast (Guangxi).
siamensis *(Craib) Faden ex D.Y. Hong in Fl. Reipubl. Popul. Sin.,* 13(3): 89 (1997): Aneilema siamense.

POLYALTHIA (Annonac.).
endertii *D.M. Johnson in Contrib. Univ. Mich. Herb.,* 22: 103 (1999)—Borneo (Kalimantan).
igniflora *D.M. Johnson in Contrib. Univ. Mich. Herb.,* 22: 96 (1999)—Borneo (Sabah, Brunei, Kalimantan).
jambosifolia *(Ridl.) D.M. Johnson in Contrib. Univ. Mich. Herb.,* 22: 99 (1999): Unona jambosifolia.
montis-silam *D.M. Johnson in Contrib. Univ. Mich. Herb.,* 22: 101 (1999)—Borneo (Sabah).
tipuliflora *D.M. Johnson in Contrib. Univ. Mich. Herb.,* 22: 99 (1999)—Borneo (Kalimantan).

POLYCARPAEA (Caryophyllac.).
basaltica *M. Thulin in Nordic J. Bot.,* 16(3): 291 (1996 publ. 1997)—Somalia.
haufensis *A.G. Miller in Edinburgh J. Bot.,* 53(1): 146 (1996)—South Yemen.

POLYCARPON (Caryophyllac.).
polycarpoides *(Biv.) Adr. Fiori in Nuovo Giorn. Bot. Ital.,* 15(3): 347 (1908): Hagaea polycarpoides.
var. colomense *(Porta) A.M. Romo, Fl. Silvestres Baleares:* 214 (1994), isonym; *see O. Bolòs & Vigo in But. Inst. Catalana Hist. Nat.,* 38 Bot., 1: 86 (1974): P. colomense.
tetraphyllum (L.) L.
var. diphyllum *(Cav.) A.M. Romo, Fl. Silvestres Baleares:* 215 (1994): P. diphyllum.

POLYCYCNIS (Orchidac.).
subgen. Angustilabia *G. Gerlach in Schlechter Orchideen, ed. 3,* I/C(37–38): 2333 (1999 publ. 2000).
pfisteri *K. Senghas, P. Taggesell & G. Gerlach in J. Orchideenfreund,* 7(1): 19 (20 Dec. 1999), preprint—Colombia.
tortuosa *R.L. Dressler in Orquideologia,* 12(2-3): 120 (1977)—Panamá.

POLYDORA (Compos.).
angustifolia *(Steetz) H. Robinson in Proc. Biol. Soc. Wash.,* 112(1): 232 (1999): Crystallopollen angustifolium.
bainesii *(Oliv. & Hiern) H. Robinson in Proc. Biol. Soc. Wash.,* 112(1): 232 (1999): Vernonia bainesii.
chloropappa *(Baker) H. Robinson in Proc. Biol. Soc. Wash.,* 112(1): 233 (1999): Vernonia chloropappa.

POLYDORA:-
jelfiae *(S. Moore) H. Robinson in Proc. Biol. Soc. Wash.,* 112(1): 233 (1999): Vernonia jelfiae.
poskeana *(Vatke & Hildebrandt) H. Robinson in Proc. Biol. Soc. Wash.,* 112(1): 233 (1999): Vernonia poskeana.
serratuloides *(DC.) H. Robinson in Proc. Biol. Soc. Wash.,* 112(1): 233 (1999): Webbia serratuloides.
steetziana *(Oliv. & Hiern) H. Robinson in Proc. Biol. Soc. Wash.,* 112(1): 233 (1999): Vernonia steetziana.
sylvicola *(G.V. Pope) H. Robinson in Proc. Biol. Soc. Wash.,* 112(1): 233 (1999): Vernonia sylvicola.

POLYGALA (Polygalac.).
subgen. Acanthocladus *(Klotzsch ex Hassk.) J.A.R. Paiva in Fontqueria,* 50: 146 (1998): gen. Acanthocladus.
subsect. Arenariae *(Chodat) J.A.R. Paiva in Fontqueria,* 50: 185 (1998): P. IV. Arenariae.
subsect. Asiaticae *(Chodat) J.A.R. Paiva in Fontqueria,* 50: 172 (1998): P. VI. Asiaticae.
sect. Chloropterae *(Chodat) J.A.R. Paiva in Fontqueria,* 50: 240 (1998): P. β. Chloropterae.
subgen. Chodatia *J.A.R. Paiva in Fontqueria,* 50: 148 (1998).
sect. Conosperma *J.A.R. Paiva in Fontqueria,* 50: 239 (1998).
subsect. Ecristatae *J.A.R. Paiva in Fontqueria,* 50: 225 (1998).
subgen. Ecristatae *L. Bernardi, Considerac. Taxon. Fitogeogr. Acerca 101 Polygalae Americ.:* 269 (2000), without latin descr. or type.
subgen. Gymnospora *(Chodat) J.A.R. Paiva in Fontqueria,* 50: 147 (1998): P. sect. Gymnospora.
subsect. Heterolophus *J.A.R. Paiva in Fontqueria,* 50: 254 (1998).
subsect. Heterophyllae *J.A.R. Paiva in Fontqueria,* 50: 167 (1998).
subsect. Hexandrae *J.A.R. Paiva in Fontqueria,* 50: 229 (1998).
sect. Laureolae *L. Bernardi, Considerac. Taxon. Fitogeogr. Acerca 101 Polygalae Americ.:* 317 (2000).
sect. Leptaleae *(Chodat) J.A.R. Paiva in Fontqueria,* 50: 165 (1998): P. subsect. Leptaleae.
subgen. Ligustrina *(Chodat) J.A.R. Paiva in Fontqueria,* 50: 147 (1998): P. sect. Ligustrina.
sect. Megatropis *J.A.R. Paiva in Fontqueria,* 50: 247 (1998).
subgen. Procerae *L. Bernardi, Considerac. Taxon. Fitogeogr. Acerca 101 Polygalae Americ.:* 380 (2000), without type.
subgen. Rhinotropis *(S.F. Blake) J.A.R. Paiva in Fontqueria,* 50: 148 (1998): P. sect. Rhinotropis.
subsect. Sativae *J.A.R. Paiva in Fontqueria,* 50: 193 (1998).
subsect. Sphenopterae *(Chodat) J.A.R. Paiva in Fontqueria,* 50: 180 (1998): P. II. Sphenopterae.
subsect. Tetrasepalae *(Chodat) J.A.R. Paiva in Fontqueria,* 50: 212 (1998): P. α. Tetrasepalae.
sect. Tetrasepalae *(Chodat) J.A.R. Paiva in Fontqueria,* 50: 211 (1998): P. α. Tetrasepalae.
subsect. Tinctoriae *(Chodat) J.A.R. Paiva in Fontqueria,* 50: 176 (1998): P. I. Tinctoriae.
abyssinica R. Br. ex Fresen.
var. subaphylla *(O. Schwartz) J.R.I. Wood, Handbook Yemen Fl.:* 195 (1997): P. subaphylla.
afra *J.A.R. Paiva in Fontqueria,* 50: 162 (1998)—Tanzania.
albida Schinz
subsp. stanleyana *(Chodat) J.A.R. Paiva in Fontqueria,* 50: 191 (1998): P. stanleyana.
andringitrensis *J.A.R. Paiva in Fontqueria,* 50: 260 (1998), nom. nov.: P. bicornis *H. Perrier.*
ariadnae *L. Bernardi in B. Dubs, Prodr. Fl. Matogrossensis (Bot. Mato Grosso, Ser. B, 3):* 241 (1998), nom. nov.: P. filiformis *A.St.-Hil. & Moq.*
baikiei Chodat
subsp. pobeguinii *(A. Chev. & Jacq.-Fél.) J.A.R. Paiva in Fontqueria,* 50: 201 (1998): P. pobeguinii.
bellidifolia *M. Gandoger, Dec. Pl. Nov.,* 2: 5 (1876)—Germany.
butyracea Heckel
var. carunculata *J.A.R. Paiva in Fontqueria,* 50: 201 (1998)—Ghana.

POLYGALA:-
capillaris *E. Mey. ex W.H. Harvey in Harv. & Sond., Fl. Cap.,* 1: 93 (1860)—Cape Province.
subsp. perrottetiana *(Paiva) J.A.R. Paiva in Fontqueria,* 50: 161 (1998): P. perrottetiana.
chloroptera *M. Gandoger, Dec. Pl. Nov.,* 2: 5 (1876)—France.
declinata *(Harv.) J.A.R. Paiva in Fontqueria,* 50: 210 (1998): P. hispida var. declinata.
erioptera DC.
subsp. petraea *(Chodat) J.A.R. Paiva in Fontqueria,* 50: 172 (1998): P. petraea.
gillettii *J.A.R. Paiva in Fontqueria,* 50: 174 (1998)—Kenya.
glaucoides L.
var. laotica *(Gagnep.) C.A. Pendry in Novon,* 9(4): 545 (1999): P. ciliata var. laotica.
goudahensis *J.A.R. Paiva in Fontqueria,* 50: 256 (1998)—Djibouti.
grondonae *F.O. Zuloaga & O. Morrone in Zuloaga & Morrone (eds.), Cat. Pl. Vasc. Rep. Arg. 2 (Monogr. Syst. Bot.* Missouri Bot. Gard. 74): 1246 (1999), nom. nov.: Polygala reflexa *Grondona.*
humifusa *J.A.R. Paiva in Fontqueria,* 50: 174 (1998)—Tanzania.
kunzei *M. Gandoger, Dec. Pl. Nov.,* 2: 5 (1876)—Germany.
latistyla *C.A. Pendry in Blumea,* 42(2): 467 (1997)—Vietnam.
leptophylla Burch.
var. armata *(Chodat) J.A.R. Paiva in Fontqueria,* 50: 260 (1998): P. armata.
levynsiana *J.A.R. Paiva in Fontqueria,* 50: 272 (1998), nom. nov.: P. microlopha var. gracilis *Levyns.*
macowaniana *J.A.R. Paiva in Fontqueria,* 50: 185 (1998), nom. nov.: Polygala confusa *MacOwan.*
misella *L. Bernardi in B. Dubs, Prodr. Fl. Matogrossensis (Bot. Mato Grosso, Ser. B, 3):* 242 (1998); *et in Considerac. Taxon. Fitogeogr. Acerca 101 Polygalae Americ.:* 174 (2000), nom. nov.: P. exigua *A.W.Benn.*
mooneyi *M.G. Gilbert in Nordic J. Bot.,* 20(2): 199 (2000)—Ethiopia.
myrtifolia L.
var. pinifolia *(Lam. ex Poir.) J.A.R. Paiva in Fontqueria,* 50: 280 (1998): P. pinifolia.
nicaeensis Risso ex W.D.J. Koch
subsp. gerundensis *(O. Bolòs & Vigo) G. Mateo Sanz & M.B. Crespo Villalba, Fl. Abrev. Comun. Valenciana:* 430 (1995): P. vulgaris var. gerundensis.
oxyphylla DC.
var. salicina *(Chodat) M.C.M. Marques in Albertoa,* 4(Supl. 1): 147 (1997): P. salicina.
perdurans *C.A. Pendry in Blumea,* 42(2): 470 (1997)—Laos.
perrieri *J.A.R. Paiva in Fontqueria,* 50: 167 (1998)—Madagascar.
petitiana A. Rich.
var. abercornensis *J.A.R. Paiva in Fontqueria,* 50: 228 (1998)—Zambia, Tanzania.
subsp. parviflora *(Exell) J.A.R. Paiva in Fontqueria,* 50: 228 (1998): P. petitiana var. parviflora.
reinii Franch. & Sav.
f. angustifolia *(T. Itô) K. Iwatsuki & H. Ohba in Fl. Japan,* 2c: 54 (1999): P. reinii var. angustifolia.
rodrigueana *J.A.R. Paiva in Fontqueria,* 50: 247 (1998), nom. nov.: P. arcuata *Chodat.*
selloi *(Spreng.) L. Bernardi, Considerac. Taxon. Fitogeogr. Acerca 101 Polygalae Americ.:* 333 (2000): Monnina selloi.
sinaica Botsch.
var. glabrescens *(Zohary) L. Boulos, Fl. Egypt,* 2: 71 (2000): Polygala spinescens var. glabrescens.
sparsiflora Oliv.
var. ukirensis *(Gürke) J.A.R. Paiva in Fontqueria,* 50: 207 (1998): P. ukirensis.
stenopetala Klotzsch
subsp. casuarina *(Chodat) J.A.R. Paiva in Fontqueria,* 50: 220 (1998): P. viminalis var. casuarina.
subglobosa *J.A.R. Paiva in Fontqueria,* 50: 224 (1998), nom. nov.: P. pilosa *Baker.*
subvirgata *M. Gandoger, Dec. Pl. Nov.,* 2: 5 (1876)—France.

POLYGALA:-
tenuicaulis Hook. f.
var. tayloriana *J.A.R. Paiva in Fontqueria,* 50: 207 (1998)—Nigeria, Cameroon.
timoutou Aubl.
var. comata *(A.W. Benn.) L. Bernardi, Considerac. Taxon. Fitogeogr. Acerca 101 Polygalae Americ.:* 224 (2000): P. comata.
transvaalensis Chodat
subsp. kagerensis *(Lebrun & Taton) J.A.R. Paiva in Fontqueria,* 50: 246 (1998): P. kagerensis.
vittata *J.A.R. Paiva in Fontqueria,* 50: 183 (1998)—Kenya.
welwitschii Chodat
subsp. pygmaea *(Gürke) J.A.R. Paiva in Fontqueria,* 50: 189 (1998): P. pygmaea.
zambesiaca *J.A.R. Paiva in Fontqueria,* 50: 275 (1998)—Mozambique, Zimbabwe.

POLYGONATUM (Convallariac.).
chingshuishanianum *S.S. Ying in Mem. Coll. Agric. Nation. Taiwan Univ.,* 28(2): 42 (1988)—Taiwan.
grandicaule *Y.S. Kim, B.U. Oh & C.G. Jang in Korean J. Pl. Taxon.,* 28(1): 42 (1998)—South Korea.
humile Fisch.
var. humillimum *(Nakai) Y.N. Lee, Fl. Korea:* 1163 (1996): P. humillimum.
infundiflorum *Y.S. Kim, B.U. Oh & C.G. Jang in Korean J. Pl. Taxon.,* 28(2): 210 (1998)—South Korea.
jinzhaiense *D.C. Zhang & J.Z. Shao in Guihaia,* 20(1): 34 (2000)—China Southeast (Anhui).
kiotense *N. Yonezawa in Acta Phytotax. Geobot.,* 49(1): 19 (1998)—Japan.
odoratum (Mill.) Druce
subsp. glaberrimum *(K. Koch) A.G. Jelenevsky & A.S. Zernov in Novosti Sist. Vÿssh. Rast.,* 32: 25 (2000): P. glaberrimum.
f. variegatum *Y.N. Lee, Fl. Korea:* 1163 (1996), without type—Locality not stated.

POLYGONUM (Polygonac.).
ajanense (Regel & Tiling) Grig.
subsp. middendorfii *(Kongar) V.N. Voroschilov in A.K. Skvortsov (ed.), Florist. issl. v razn. raĭonakh SSSR:* 163 (1985): P. middendorfii.
alopecuroides Turcz. ex Meisn.
subsp. subauriculatum *(Petrov ex Kom.) V.N. Voroschilov in A.K. Skvortsov (ed.), Florist. issl. v razn. raĭonakh SSSR:* 163 (1985): P. subauriculatum.
arenastrum Boreau
subsp. calcatum *(Lindm.) R. Wisskirchen in R. Wisskirchen & H. Haeupler, Standardliste Farn- und Blütenpfl.* Deutschl.: 378 (1998): P. calcatum.
subsp. caspicum *(Kom.) N.N. Tzvelev in Fl. Vostochnoĭ Evropÿ,* 9: 146 (1996): P. caspicum.
var. depressum *(Meisn.) J. do Amaral Franco & M. da Luz Rocha Afonso in Stud. Bot. (Salamanca),* 14: 172 (1995 publ. 1996): P. aviculare var. depressum.
var. microspermum *(Boreau) J. do Amaral Franco & M. da Luz Rocha Afonso in Stud. Bot. (Salamanca),* 14: 174 (1995 publ. 1996): P. microspermum.
subsp. microspermum *(Jord. ex Boreau) H. Scholz in R. Wisskirchen & H. Haeupler, Standardliste Farn- und Blütenpfl.* Deutschl.: 378 (1998): P. microspermum.
argentum *M. Brandão & J.P. Laca-Buendia in Daphne,* 5(1): 5 (1995)—Brazil Southeast (Minas Gerais).
aviculare L.
subsp. boreale *(Lange) T. Karlsson in Svensk Bot. Tidskr.,* 91(5): 249 (1997 publ. 1998): P. aviculare var. boreale.
subsp. excelsius *Karlsson in Nordic J. Bot.,* 19(4): 386 (1999)—Sweden.
var. fusco-ochreatum *(Kom.) A.J. Li in Fl. Reipubl. Popul. Sin.,* 25(1): 9 (1998): P. fusco-ochreatum.
subsp. sabulosum *(Vorosch.) N.N. Tzvelev in Novosti Sist. Vÿssh. Rast.,* 32: 184 (2000): P. sabulosum.
bistorta L.
subsp. ochotense *(Petrov ex Kom.) V.N. Voroschilov in A.K. Skvortsov (ed.), Florist. issl. v razn. raĭonakh SSSR:* 163 (1985): P. ochotense.

POLYGONUM:-
bistorta
subsp. pacificum *(Petrov ex Kom.) V.N. Voroschilov in A.K. Skvortsov (ed.), Florist. issl. v razn. raĭonakh SSSR:* 163 (1985): P. pacificum.
borgoicum *N.N. Tupitsÿna in Fl. Sibir. (Salicac.-Amaranthac.),* 5: 129 (1992), contrary to Art. 37.5 ICBN (1994)—Buryatiya.
bourdinii *M. Gandoger, Dec. Pl. Nov.,* 1: 20 (1875)—France.
ciliinerve *(Nakai) S.X. Li & Y.L. Chang in Fl. Plant. Herb. Chinae bor.-or.,* 2: 66 (1959), isonym; *see Ohwi in Acta Phytotax. Geobot.,* 6: 146 (1937): Pleuropterus ciliinervis.
convolvulus L.
var. pauciflorum *(Maxim.) V.N. Voroschilov in A.K. Skvortsov (ed.), Florist. issl. v razn. raĭonakh SSSR:* 164 (1985): P. pauciflorum.
crispatum *M. Gandoger, Dec. Pl. Nov.,* 2: 11 (1876)—France.
fertile *(Maxim.) A.J. Li in Fl. Reipubl. Popul. Sin.,* 25(1): 67 (2000): Koenigia fertilis.
fusco-ochreatum Kom.
f. stans *(Kitag.) C.F. Fang in Clavis Pl. Chinae Bor.-Or. ed. 2 (ed. P.Y. Fu):* 123 (1995): P. stans.
glaciale Hook. f.
var. przewalskii *(A.K. Skvortsov & Borodina) A.J. Li in Fl. Reipubl. Popul. Sin.,* 25(1): 64 (1998): P. przewalskii.
glaucescens *N.A. Ivanova ex N.N. Tupitsÿna in Fl. Sibir. (Salicac.-Amaranthac.),* 5: 130 (1992), contrary to Art. 37.5 ICBN (1994)—Buryatiya.
hickmanii *H.R. Hinds & Randall Morgan in Novon,* 5(4): 336 (1995)—California.
hygrogenes *M. Gandoger, Dec. Pl. Nov.,* 2: 11 (1876)—Corse.
jeholense *(Kitag.) Baranov & Skvortsov ex S.X. Li & Y.L. Chang in Fl. Plant. Herb. Chinae bor.-or.,* 2: 48 (1959): Pleuropteropyrum jeholense.
kirinense *S.X. Li & Y.L. Chang in Fl. Plant. Herb. Chinae bor.-or.,* 2: 109 (1959)—Manchuria (Jilin).
ligerinum *M. Gandoger, Dec. Pl. Nov.,* 2: 13 (1876)—France.
mollifolium *M. Gandoger, Dec. Pl. Nov.,* 1: 10 (1875)—Switzerland.
nanetense *M. Gandoger, Dec. Pl. Nov.,* 2: 13 (1876)—France.
neglectum Besser
subsp. rectum *(Chrtek) N.N. Tzvelev in Fl. Vostochnoĭ Evropÿ,* 9: 143 (1996): P. aviculare subsp. rectum.
oligophyllum *M. Gandoger, Dec. Pl. Nov.,* 2: 10 (1876)—Austria.
pallens *M. Gandoger, Dec. Pl. Nov.,* 2: 14 (1876)—France.
persicaria L.
var. opacum *(Sam.) A.J. Li in Fl. Reipubl. Popul. Sin.,* 25(1): 23 (1998): P. opacum.
plebejum R. Br.
subsp. changii *(Kitag.) V.N. Voroschilov in A.K. Skvortsov (ed.), Florist. issl. v razn. raĭonakh SSSR:* 164 (1985): P. changii.
renii *L.C. Wang in Acta Bot. Bor.-Occid. Sin.,* 18(3): 457 (1998)—China North-central (Gansu).
richteri *M. Gandoger, Dec. Pl. Nov.,* 2: 12 (1876)—Hungary.
roseoviride *(Kitag.) S.X. Li & Y.L. Chang in Fl. Plant. Herb. Chinae bor.-or.,* 2: 46 (1959): Persicaria roseoviridis.
royi *M. Gandoger, Dec. Pl. Nov.,* 1: 10 (1875)—France.
ruscinonense *M. Gandoger, Dec. Pl. Nov.,* 2: 33 (1876)—France.
sagittatum L.
subsp. sieboldii *(Meisn.) V.N. Voroschilov in A.K. Skvortsov (ed.), Florist. issl. v razn. raĭonakh SSSR:* 164 (1985): P. sieboldii.
salinum *Baranov & Skvortsov ex S.X. Li & Y.L. Chang in Fl. Plant. Herb. Chinae bor.-or.,* 2: 107 (1959)—Manchuria (Heilongjiang).
sparsipilosum *A.J. Li in Fl. Reipubl. Popul. Sin.,* 25(1): 65 (1998), nom. nov.: Koenigia pilosa *Maxim.*
var. hubertii *(Lingelsh.) A.J. Li in Fl. Reipubl. Popul. Sin.,* 25(1): 65 (1998): P. hubertii.
stricticaule *M. Gandoger, Dec. Pl. Nov.,* 2: 33 (1876)—France.

POLYGONUM:-
uniflorum *Y.X. Ma & Y.T. Zhao in Acta Phytotax. Sin.,* 35(6): 569 (1997)—Manchuria (Liaoning).
vesulum *M. Gandoger, Dec. Pl. Nov.,* 1: 10 (1875)—France.
viviparum L.
f. linearifolium *J.Q. Fu in Fl. Loess-Plat. Sin.,* 1: 610 (2000), contrary to Art. 37.5 ICBN (1994)—Qinghai.

POLYMITA (Aizoac.).
steenbokensis *H.E.K. Hartmann in Bradleya,* 14: 42 (1996)—Cape Province.

POLYPOGON (Gramin.).
exasperatus *(Trin.) S.A. Renvoize, Gramineas de Bolivia:* 235 (1998): Agrostis exasperata.
hackelii *(R.E. Fr.) S.A. Renvoize, Gramineas de Bolivia:* 236 (1998): Agrostis hackelii.

POLYSCIAS (Araliac.).
aequatoguineensis *J. Lejoly & S. Lisowski in Bull. Jard. Bot. Nation. Belg.,* 67(1–4): 112 (1999)—Equatorial Guinea.
bastardianum *(Decne.) S.L. Welsh, Fl. Societensis:* 43 (1998): Panax bastardianum.
mayottensis *P.P. Lowry, O. Pascal & J.-N. Labat in Adansonia,* 21(1): 69 (1999)—Comoros (Mayotte).

POLYSTACHYA (Orchidac.).
albescens Ridl.
subsp. principensis *T. Stévart in Syst. Geogr. Pl.,* 70(1): 144 (2000)—Gulf of Guinea Is. (Principe).
anthoceros *I. la Croix & P.J. Cribb in Kew Bull.,* 51(3): 571 (1996)—Nigeria.
armeniaca *I. la Croix & P.J. Cribb in Kew Bull.,* 51(3): 573 (1996)—Zambia, Zaire.
biteaui *P.J. Cribb, I. la Croix & T. Stévart in Kew Bull.,* 54(4): 999 (1999 publ. 2000)—Gulf of Guinea Is (São Tomé).
couloniana *D. Geerinck & M. Arbonnier in Belgian J. Bot.,* 129(1): 68 (1996 publ. 1997)—Burundi.
johnstonii Rolfe
var. roseopurpurea *I. la Croix & P.J. Cribb in Kew Bull.,* 52(4): 1002 (1997)—Malawi.
peruviana *C.H. Nelson Sutherland & F.J. Fernández Casas in Fontqueria,* 51: 4 (1998), nom. nov.: Encyclia nana *Poepp. & Endl.*
phiriae *W. Fibeck in Orchids S. Afr.,* 30: 42 (1999), as 'phirii'—Zimbabwe.
valentina *I. la Croix & P.J. Cribb in Kew Bull.,* 51(3): 575 (1996)—Zimbabwe.
walravensiana *D. Geerinck & M. Arbonnier in Belgian J. Bot.,* 129(1): 69 (1996 publ. 1997)—Burundi.

POLYTRIAS (Gramin.).
amaura (Büse) Kuntze
var. nana *(Keng & S.L. Chen) S.L. Chen in Fl. Reipubl. Popul. Sin.,* 10(2): 101 (1997): Eulalia nana.

POLYXENA (Hyacinthac.).
calcicola *U. Müller-Doblies & D. Müller-Doblies in Feddes Repert.,* 108(1–2): 85 (1997)—Cape Province.
paucifolia *(W.F. Barker) A.M. van der Merwe & J.C. Manning in Strelitzia,* 9: 714 (2000): Hyacinthus paucifolius.

POMADERRIS (Rhamnac.).
sect. Annulares *N.G. Walsh in Muelleria,* 10: 29 (1997).
sect. Apetalae *N.G. Walsh in Muelleria,* 10: 28 (1997).
sect. Biloculares *N.G. Walsh in Muelleria,* 10: 30 (1997).
sect. Flabellares *N.G. Walsh in Muelleria,* 10: 28 (1997).
sect. Psilogyne *N.G. Walsh in Muelleria,* 10: 28 (1997).
sect. Umbelliflorae *N.G. Walsh in Muelleria,* 10: 30 (1997).
adnata *N.G. Walsh & F. Coates in Muelleria,* 10: 39 (1997)—New South Wales.
andromedifolia A. Cunn.
subsp. confusa *N.G. Walsh & F. Coates in Muelleria,* 10: 48 (1997)—New South Wales (Australian Capital Territory, New South Wales).

POMADERRIS:-
apetala Labill.
subsp. maritima *N.G. Walsh & F. Coates in Muelleria,* 10: 34 (1997)—Tasmania, Victoria, New Zealand North I.
argyrophylla N.A. Wakef.
subsp. graniticola *N.G. Walsh & F. Coates in Muelleria,* 10: 46 (1997)—Queensland, New South Wales.
betulina A. Cunn.
subsp. actensis *N.G. Walsh & F. Coates in Muelleria,* 10: 49 (1997)—New South Wales (Australian Capital Territory, New South Wales).
bodalla *N.G. Walsh & F. Coates in Muelleria,* 10: 42 (1997)—New South Wales.
coomingalensis *N.G. Walsh & F. Coates in Muelleria,* 10: 37 (1997)—Queensland.
crassifolia *N.G. Walsh & F. Coates in Muelleria,* 10: 35 (1997)—Queensland, New South Wales.
delicata *N.G. Walsh & F. Coates in Muelleria,* 10: 40 (1997)—New South Wales.
elliptica Labill.
var. diemenica *N.G. Walsh & F. Coates in Muelleria,* 10: 51 (1997)—Tasmania.
helianthemifolia (Reissek) N.A. Wakef.
subsp. hispida *N.G. Walsh & F. Coates in Muelleria,* 10: 31 (1997)—Victoria, New South Wales.
subsp. minor *N.G. Walsh & F. Coates in Muelleria,* 10: 33 (1997)—Victoria.
ligustrina Sieb. ex DC.
subsp. latifolia *N.G. Walsh & F. Coates in Muelleria,* 10: 50 (1997)—New South Wales, Queensland.
mediora *N.G. Walsh & F. Coates in Muelleria,* 10: 38 (1997)—New South Wales.
phylicifolia Lodd. ex Link
subsp. ericoides *(Maiden & Betche) N.G. Walsh & F. Coates in Muelleria,* 10: 52 (1997): P. phylicifolia var. ericoides.
precaria *N.G. Walsh & F. Coates in Muelleria,* 10: 43 (1997)—New South Wales.
reperta *N.G. Walsh & F. Coates in Muelleria,* 10: 45 (1997)—New South Wales.
rotundifolia *(F. Muell.) B.L. Rye in Nuytsia,* 11(1): 109 (1996): Spyridium rotundifolium.

POMARIA (Legumin.).
austrotexana *B.B. Simpson in Lundellia,* 1: 51 (1998)—Texas, Mexico Northeast (Tamaulipas).
brachycarpa *(A. Gray) B.B. Simpson in Lundellia,* 1: 54 (1998): Hoffmannseggia brachycarpa.
canescens *(Fisher) B.B. Simpson in Lundellia,* 1: 56 (1998): Hoffmannseggia canescens.
fruticosa *(S. Watson) B.B. Simpson in Lundellia,* 1: 57 (1998): Hoffmannseggia fruticosa.
multijuga *(S. Watson) B.B. Simpson in Lundellia,* 1: 68 (1998): Hoffmannseggia multijuga.
wootonii *(Britton) B.B. Simpson in Lundellia,* 1: 69 (1998): Larrea wootonii.

× **POMATOCHILUS** *hort.in Orchid Rev.,* 106(1224): 380 (1998).*ORCHIDACEAE* (Pomatocalpa × Sarcochilus).

PONERORCHIS (Orchidac.).
chusua (D. Don) Soó
var. nana *(King & Pantl.) R.C. Srivastava in Nation. Acad. Sci. Letters,* 18(3–4): 61 (1996): Orchis chusua var. nana.
graminifolia Rchb. f.
var. kurokamiana *(Ohwi & Hatus.) T. Hashimoto in Proc. 12th World Orchid Conf., 1987 (eds. K. Saito & R. Tanaka):* 119 (1987): Orchis kurokamiana.

PONGAMIA (Legumin.).
fruticosa *(Roxb.) T.S. Ralph, Icon. Carpolog.:* 23 (1849): Robinia fruticosa.
macrophylla *T.S. Ralph, Icon. Carpolog.:* 23 (1849), nom. nov.: Robinia macrophylla *Roxb.*

PONTECHIUM *U.-R. Böhle & H.H. Hilger in Taxon,* 49(4): 743 (2000).*BORAGINACEAE*
maculatum *(L.) U.-R. Böhle & H.H. Hilger in Taxon,* 49(4): 744 (2000): Echium maculatum.
var. acutifolium *(Lehm.) U.-R. Böhle & H.H. Hilger in Taxon,* 49(4): 744 (2000): Echium acutifolium.

PONTHIEVA (Orchidac.).
hameri *R.L. Dressler in Bol. Inst. Bot. (Guadalajara),* 5(1–3): 80 (1998)—El Salvador.
pilosissima *(Senghas) C.H. Dodson in Orquideologia,* 20(1): 110 (1996): Cranichis pilosissima.
poitaei *Rchb. f. ex M.A. Nir, Orchidaceae Antillanae:* 312 (2000)—Dominican Republic.

POPULUS (Salicac.).
subgen. Leucoides *(Spach) N. Chao & J. Liu in J. Sichuan Forest. Sci. Techn.,* 19(4): 12 (1998): P. sect. Leucoides.
× robusta *C.K. Schneider, Ill. Handb. Laubholzk.,* 1: 11 (1904)—Cultivation.(P. angulata × eugenei).
serotina *T. Hartig, Vollst. Naturgesch. Culturpfl. Deutschl.:* 437 (1851)—Locality not stated.
serrata T.B. Chao & J.S. Chen
f. rotundata *T.B. Chao & Zhi X. Chen in Bull. Bot. Res. North-East. Forest. Univ.,* 18(3): 287 (1998)—China Southeast (Henan).
undulata J. Zhang
f. purpurea *T.B. Chao & Zhi X. Chen in Bull. Bot. Res. North-East. Forest. Univ.,* 18(3): 287 (1998)—China Southeast (Henan).
wenxianica *Z.C. Feng & J.L. Guo ex G.H. Zhu in Novon,* 8(4): 464 (1998)—China North-central (Gansu).

POROCYSTIS (Sapindac.).
acuminata *(Radlk.) P. Acevedo-Rodríguez in Novon,* 8(2): 106 (1998): Toulicia acuminata.

POROGLOSSUM (Orchidac.).
tripollex *C.A. Luer, Ic. Pleurothallid. XVII (Monogr. Syst. Bot. Missouri Bot. Gard., 72):* 114 (1998)—Ecuador.

POROPHYLLUM (Compos.).
cabrerae *D.J.N. Hind in Kew Bull.,* 55(2): 388 (2000)—Argentina Northwest (Salta).

PORPAX (Orchidac.).
meirax
var. elwesii *(Rchb. f.) R.C. Srivastava in Nation. Acad. Sci. Letters,* 18(3–4): 61 (1995): Eria elwesii.

PORROGLOSSUM (Orchidac.).
josei *C.A. Luer, Ic. Pleurothallid. XII (Monogr. Syst. Bot. Missouri Bot. Gard., 57):* 144 (1995)—Ecuador.

× **PORTEMEA** *L. Ariza-Julia in J. Bromeliad Soc.,* 28(1): 21 (1978). *BROMELIACEAE* (Aechmea × Portea).

PORTERANDIA (Rubiac.).
sericantha *(W.C. Chen) W.C. Chen in Fl. Reipubl. Popul. Sin.,* 71(1): 384 (1999): Randia sericantha.

× **PORTERARA** *hort. in Orchid Rev.,* 106(1220): 126 (1998). *ORCHIDACEAE* (Rhynchostylis × Sarcochilus × Vanda).

PORTILLIA *W. Königer in Arcula,* 6: 154 (1996). *ORCHIDACEAE*
popowiana *W. Königer & J. Portilla in Arcula,* 6: 155 (1996)—Ecuador.

PORTULACA (Portulacac.).
conoidea *S.M. Phillips in Kew Bull.,* 55(3): 693 (2000)—Kenya.
coralloides *S.M. Phillips in Kew Bull.,* 55(3): 695 (2000)—Kenya.
dhofarica *M.G. Gilbert in Kew Bull.,* 55(4): 788 (2000)—Oman.
dodomaensis *M.G. Gilbert in Kew Bull.,* 55(4): 800 (2000)—Tanzania.
granulatostellulata *(Poelln.) C. Ricceri & P.V. Arrigoni in Parlatorea,* 4: 93 (2000), as 'granulato-stellulata': P. oleracea var. granulatostellulata.
greenwayi *M.G. Gilbert in Kew Bull.,* 55(4): 775 (2000)—Kenya.
kermesina N.E. Br.
var. lutea *(Poelln.) S.M. Phillips in Kew Bull.,* 55(3): 696 (2000): P. fischeri var. lutea.
kuriensis *M.G. Gilbert in Kew Bull.,* 55(4): 789 (2000)—Socotra.

PORTULACA:-
macrantha *(Maire) C. Ricceri & P.V. Arrigoni in Parlatorea,* 4: 93 (2000): P. oleracea var. macrantha.
massaica *S.M. Phillips in Kew Bull.,* 55(3): 689 (2000)—Kenya, Tanzania.
nitida *(Danin & H.G. Baker) C. Ricceri & P.V. Arrigoni in Parlatorea,* 4: 93 (2000): P. oleracea subsp. nitida.
olosirwa *S.M. Phillips in Kew Bull.,* 55(3): 690 (2000)—Kenya.
stellata *(Danin & H.G. Baker) C. Ricceri & P.V. Arrigoni in Parlatorea,* 4: 93 (2000): P. oleracea subsp. stellata.
yecorensis *J. Henrickson & T.R. Van Devender in Sida,* 18(4): 1002 (1999)—Mexico Northwest (Sonora).

POSOQUERIA (Rubiac.).
tarairensis *C.M. Taylor & R. Cortés-Ballén in Novon,* 9(3): 428 (1999)—Colombia.

POTALIA (Loganiac.).
elegans *L. Struwe & V.A. Albert in Harvard Pap. Bot.,* 3(2): 185 (1998)—Colombia, Venezuela.
maguireorum *L. Struwe & V.A. Albert in Harvard Pap. Bot.,* 3(2): 189 (1998)—Brazil North (Amazonas), Venezuela.

POTAMEIA (Laurac.).
micrantha *H. van der Werff in Novon,* 6(4): 472 (1996)—Madagascar.

POTAMOGETON (Potamogetonac.).
sect. Alternatiophylli *Z.S. Diao in J. Yuzhou Univ. (Nat. Sci. Ed.),* 13(4): 2 (1996), without type.
subgen. Rectiembryo *Z.S. Diao in J. Yuzhou Univ. (Nat. Sci. Ed.),* 13(4): 2 (1996), contrary to Art. 37.4 ICBN (1994).
subgen. Unciniembryo *Z.S. Diao in J. Yuzhou Univ. (Nat. Sci. Ed.),* 13(4): 2 (1996), without type.
berchtoldii Fieber
subsp. groenlandicus *(Hagstr.) J. Feilberg in Flora Fauna (Århus),* 101(2): 72 (1996): P. groenlandicus.
× biformoides *V.G. Papchenkov in Bot. Zhurn.,* 82(12): 69 (1997)—Russia East.(P. biformis × graminifolius).
distinctus A. Benn.
var. tonkinensis *(A. Camus) V.C. Vu ex L.V. Averyanov in Konsp. Sosud. Rast. Fl. V'etnama,* 2: 155 (1996): P. tonkinensis.
foliosus Raf.
subsp. fibrillosus *(Fernald) R.R. Haynes & C.B. Hellquist in Novon,* 6(4): 370 (1996): P. fibrillosus.
juncifolius *A. Kern.in Verh. K. K. Zool.-Bot. Ges. Wien,* 45(9): 364 (1895)—Austria.
× mariensis *V.G. Papchenkov in Bot. Zhurn.,* 82(12): 71 (1997)—Russia East.(P. biformis × nodosus).
octandrus Poir.
subsp. limosellifolius *(Maxim. ex Korsh.) V.N. Voroschilov in A.K. Skvortsov (ed.), Florist. issl. v razn. Raĭonakh SSSR:* 147 (1985): P. limosellifolius.
pusillus L.
subsp. gemmiparus *(J.W. Robbins) R.R. Haynes & C.B. Hellquist in Novon,* 6(4): 370 (1996): P. pusillus var. gemmiparus.
subsp. tenuissimus *(Mert. & W.D.J. Koch) R.R. Haynes & C.B. Hellquist in Novon,* 6(4): 370 (1996): P. pusillus var. tenuissimus.
sibiricus A. Benn.
subsp. subsibiricus *(Hagstr.) N.N. Tzvelev in Novosti Sist. Vÿssh. Rast.,* 32: 182 (2000): P. subsibiricus.

POTENTILLA (Rosac.).
sect. Butkovii *M.M. Mirachmedova in Uzb. Biol. Zhurn.,* 1998(3): 33 (1998), without latin descr.
sect. Fedtschenkoana *M.M. Mirachmedova in Uzb. Biol. Zhurn.,* 1998(3): 32 (1998), without latin descr.
subsect. Incisae *B.A. Jurtzev in Arktich. Fl. SSSR,* 9(1): 321 (1984).
ser. Leuconotae *(Dikshit & Panigrahi) H. Ikeda & H. Ohba in H. Ohba (ed.), Himalayan Pl., 3 (Univ. Mus. Univ. Tokyo Bull., 39):* 76 (1999): P. sect. Leuconotae.

POTENTILLA:-
ser. Lineatae *H. Ikeda & H. Ohba in H. Ohba (ed.), Himalayan Pl., 3 (Univ. Mus. Univ. Tokyo Bull., 39):* 43 (1999).
subsect. Microphyllae *(T.T. Yu & C.L. Li) H. Ikeda & H. Ohba in H. Ohba (ed.), Himalayan Pl., 3 (Univ. Mus. Univ. Tokyo Bull., 39):* 43 (1999): P. ser. Microphyllae.
ser. Oxyodontae *B.A. Jurtzev in Arktich. Fl. SSSR,* 9(1): 321 (1984).
ser. Smithianae *H. Ikeda & H. Ohba in H. Ohba (ed.), Himalayan Pl., 3 (Univ. Mus. Univ. Tokyo Bull., 39):* 85 (1999), holotype not indicated.
sect. Vvedenskyi *M.M. Mirachmedova in Uzb. Biol. Zhurn.*, 1998(3): 33 (1998), without latin descr.
anachoretica Soják
var. planiuscula *B.A. Jurtzev in Arktich. Fl. SSSR,* 9(1): 321 (1984)—Magadan.
arenosa (Turcz.) Juz.
var. pinnatisecta *B.A. Jurtzev in Arktich. Fl. SSSR,* 9(1): 179 (1984)—Magadan.
argentea L.
var. acutifida *(Markl.) T. Karlsson in Svensk Bot. Tidskr.,* 91(5): 249 (1997 publ. 1998): P. impolita subsp. acutifida.
var. decora *(Markl.) T. Karlsson in Svensk Bot. Tidskr.,* 91(5): 249 (1997 publ. 1998): P. decora.
beringensis *B.A. Jurtzev in Arktich. Fl. SSSR,* 9(1): 317 (1984)—Magadan.
beringii *B.A. Jurtzev in Arktich. Fl. SSSR,* 9(1): 321 (1984)—Magadan.
bifurca L.
subsp. semiglabra *(Juz.) V.N. Voroschilov in A.K. Skvortsov (ed.), Florist. issl. v razn. raĭonakh SSSR:* 176 (1985): P. semiglabra.
cardotiana Hand.-Mazz.
var. nepalensis *H. Ikeda & H. Ohba in H. Ohba (ed.), Himalayan Pl., 3 (Univ. Mus. Univ. Tokyo Bull., 39):* 72 (1999)—Nepal.
caulescens L.
subsp. achhalii *A. Romo in Lagascalia,* 18(2): 256 (1996)—Morocco.caulescens L.
subsp. djurjurae *(Chabert) A. Romo in Lagascalia,* 18(2): 257 (1996): P. caulescens var. djurjurae.
crebridens Juz.
subsp. hemicryophila *B.A. Jurtzev in Arktich. Fl. SSSR,* 9(1): 318 (1984)—Magadan.
czegitunica *B.A. Jurtzev in Arktich. Fl. SSSR,* 9(1): 319 (1984)—Magadan.
davisii *R.R. Mill & H. Duman in Edinburgh J. Bot.,* 56(3): 351 (1999)—Turkey.
× dezhnevii *B.A. Jurtzev in Arktich. Fl. SSSR,* 9(1): 318 (1984)—Magadan.(P. anachoretica × subvahliana or uniflora).
drummondii Lehm.
var. breweri *(S. Watson) N.H. Holmgren in A. Cronquist et al., Intermountain Fl., Vasc. Pl. Intermountain West,* U.S.A., 3A: 92 (1997): P. breweri.
var. bruceae *(Rydb.) N.H. Holmgren in A. Cronquist et al., Intermountain Fl., Vasc. Pl. Intermountain West,* U.S.A., 3A: 94 (1997): P. bruceae.
egedii Wormsk.
var. thermalis *B.A. Jurtzev in Arktich. Fl. SSSR,* 9(1): 321 (1984)—Magadan.
fruticosa L.
subsp. glabrata *(Willd. ex Schltdl.) V.N. Voroschilov in A.K. Skvortsov (ed.), Florist. issl. v razn. raĭonakh SSSR:* 176 (1985): P. glabrata.
gelida C.A. Mey.
subsp. boreoasiatica *B.A. Jurtzev & R.V. Kamelin in Arktich. Fl. SSSR,* 9(1): 320 (1984)—Krasnoyarsk.
subsp. boreojacutica *B.A. Jurtzev in Arktich. Fl. SSSR,* 9(1): 320 (1984)—Yakutiya.
× gorodkovii *B.A. Jurtzev in Arktich. Fl. SSSR,* 9(1): 319 (1984)—Magadan.(P. nivea × uniflora).
huparctica Malte
subsp. gelida *(C.A. Mey.) V.N. Voroschilov in A.K. Skvortsov (ed.), Florist. issl. v razn. raĭonakh SSSR:* 176 (1985): P. gelida.
hyparctica Malte
subsp. nivicola *B.A. Jurtzev & V.V. Petrovsky in Arktich. Fl. SSSR,* 9(1): 319 (1984)—Magadan.
intermedia L.
subsp. heidenreichii *(Zimmeter) N.N. Tzvelev in Novosti Sist. Vўssh. Rast.,* 32: 184 (2000): P. heidenreichii.

POTENTILLA:-
laevissima *R.V. Kamelin in Byull. Mosk. Obshch. Ispўt. Prir., Biol.,* 100(1): 88 (1995)—Mongolia.
leuconota D. Don
var. omeiensis *H. Ikeda & H. Ohba in H. Ohba (ed.), Himalayan Pl., 3 (Univ. Mus. Univ. Tokyo Bull., 39):* 78 (1999)—China South-central (Sichuan).
lyngei *B.A. Jurtzev & J. Soják in Arktich. Fl. SSSR,* 9(1): 317 (1984)—West Siberia.
matsumurae Th. Wolf
f. alba *Y.N. Lee, Fl. Korea:* 1158 (1996), without type—Locality not stated.
× micropeduncularis *H. Ikeda & H. Ohba in J. Jap. Bot.,* 71(5): 253 (1996)—Nepal.(P. microphylla × peduncularis var. ganeshii).
microphylla D. Don
var. luteopilosa *(T.T. Yu & C.L. Li) H. Ikeda & H. Ohba in H. Ohba (ed.), Himalayan Pl., 3 (Univ. Mus. Univ. Tokyo Bull., 39):* 48 (1999): P. luteopilosa.
var. tapetodes *(Soják) H. Ikeda & H. Ohba in H. Ohba (ed.), Himalayan Pl., 3 (Univ. Mus. Univ. Tokyo Bull., 39):* 49 (1999): P. tapetodes.
nerimaniae *H. Duman in Edinburgh J. Bot.,* 56(3): 349 (1999)—Turkey.
nivea L.
var. insularis *G.A. Peschkova in Fl. Tsentral'noi Sibiri,* 2: 568 (1979)—Buryatiya, Irkutsk.
subsp. mischkinii *(Juz.) B.A. Jurtzev in Arktich. Fl. SSSR,* 9(1): 173 (1984): P. mischkinii.
var. niveiformis *B.A. Jurtzev in Arktich. Fl. SSSR,* 9(1): 321 (1984)—West Siberia.
ozjorensis *G.A. Peschkova, Stepnaya Fl. Baĭkal'skoĭ Sibiri:* 66 (1972)—Irkutsk, Buryatiya, Mongolia.
peduncularis D. Don
var. curta *(Soják) H. Ikeda & H. Ohba in H. Ohba (ed.), Himalayan Pl., 3 (Univ. Mus. Univ. Tokyo Bull., 39):* 68 (1999): P. curta.
var. ganeshii *H. Ikeda & H. Ohba in J. Jap. Bot.,* 71(5): 252 (1996)—Nepal.
var. shweliensis *(H.R. Fletcher) H. Ikeda & H. Ohba in H. Ohba (ed.), Himalayan Pl., 3 (Univ. Mus. Univ. Tokyo Bull., 39):* 66 (1999): P. shweliensis.
var. vittata *(Soják) H. Ikeda & H. Ohba in H. Ohba (ed.), Himalayan Pl., 3 (Univ. Mus. Univ. Tokyo Bull., 39):* 67 (1999): P. vittata.
recta
subsp. afra *(Pau & Font Quer) A. Romo in Lagascalia,* 18(2): 262 (1996): P. hirta var. afra.
stenophylla (Franch.) Diels
var. cristata *(H.R. Fletcher) H. Ikeda & H. Ohba in H. Ohba (ed.), Himalayan Pl., 3 (Univ. Mus. Univ. Tokyo Bull., 39):* 52 (1999): P. cristata.
var. taliensis *(W.W. Sm.) H. Ikeda & H. Ohba in H. Ohba (ed.), Himalayan Pl., 3 (Univ. Mus. Univ. Tokyo Bull., 39):* 52 (1999): P. taliensis.
subvahliana *B.A. Jurtzev in Arktich. Fl. SSSR,* 9(1): 319 (1984)—Magadan.
tanacetifolia Willd. ex Schltdl.
var. longifolia *(Willd. ex Schltdl.) V.N. Voroschilov in A.K. Skvortsov (ed.), Florist. issl. v razn. raĭonakh SSSR:* 177 (1985): P. longifolia.
tikhomirovii *B.A. Jurtzev in Arktich. Fl. SSSR,* 9(1): 318 (1984)—Krasnoyarsk.
× tolmatchevii *B.A. Jurtzev & J. Soják in Arktich. Fl. SSSR,* 9(1): 320 (1984)—Krasnoyarsk.(P. arenosa × pulchella).
var. zubkovii *B.A. Jurtzev & J. Soják in Arktich. Fl. SSSR,* 9(1): 320 (1984)—West Siberia.
× tomentulosa *B.A. Jurtzev in Arktich. Fl. SSSR,* 9(1): 318 (1984)—Magadan.(P. arenosa × nivea).
tristis Soják
f. ciliata *H. Ikeda & H. Ohba in J. Jap. Bot.,* 71(2): 69 (1996)—Nepal.
× tschaunensis *Juz. ex B.A. Jurtzev in Arktich. Fl. SSSR,* 9(1): 317 (1984)—Magadan.(P. anachoretica × nivea).
tschukotica *B.A. Jurtzev & V.V. Petrovsky in Arktich. Fl. SSSR,* 9(1): 320 (1984)—Magadan.
tucumanensis *A. Castagnaro & M. Arias bis in Novon,* 8(4): 333 (1998)—Argentina Northwest (Tucumán, Catamarca, Salta, San Luis), Argentina Northeast (Córdoba).
turfosa Hand.-Mazz.
var. gracilescens *(Soják) H. Ikeda & H. Ohba in H. Ohba (ed.), Himalayan Pl., 3 (Univ. Mus. Univ. Tokyo Bull., 39):* 61 (1999): P. glabriuscula var. gracilescens.

POTENTILLA:-
turfosoides *H. Ikeda & H. Ohba in H. Ohba (ed.), Himalayan Pl., 3 (Univ. Mus. Univ. Tokyo Bull., 39):* 62 (1999)—Nepal.
villosula *B.A. Jurtzev in Arktich. Fl. SSSR,* 9(1): 319 (1984)—Magadan.
wenchuensis *H. Ikeda & H. Ohba in H. Ohba (ed.), Himalayan Pl., 3 (Univ. Mus. Univ. Tokyo Bull., 39):* 81 (1999)—China South-central (Sichuan).

POTHOMORPHE (Piperac.).
umbellata (L.) Miq.
f. glabra *(C.DC.) J.A. Steyermark in Fl. Venezuela,* 2(2): 604 (1984), without basionym ref.: Basionym not stated.

POTHOS (Arac.).
dzui *P.C. Boyce in Blumea,* 45(1): 160 (2000)—Vietnam.
gigantipes *Buchet ex P.C. Boyce in Blumea,* 45(1): 162 (2000)—Vietnam.
keralensis *A.G. Pandurangan & V.J. Nair in Indian J. Forest.,* 17(1): 64 (1994)—India (Kerala).
kerrii *Buchet ex P.C. Boyce in Blumea,* 45(1): 168 (2000)—Laos, Vietnam, China Southeast (Guangxi).
pilulifer *Buchet ex P.C. Boyce in Blumea,* 45(1): 175 (2000)—Vietnam, China Southeast (Guangxi).

POUPARTIA (Anacardiac.).
acuminata *(Roxb.) N. Wallich, Report on Calcutta Bot. Gard. to G.A. Bushby:* 32 (1840): Spondias acuminata.
orientalis *Capuron ex A. Randrianasolo & J.S. Miller in Novon,* 9(4): 546 (1999)—Madagascar.

POUTERIA (Sapotac.).
adolfi-friedericii (Engl.) A. Meeuse
subsp. australis *(J.H. Hemsl.) L. Gautier in J.-P. Lebrun & A.L. Stork, Énum. Pl. Fleurs Afr. Trop.,* 4: 34 (1997): Aningeria adolfi-friedericii subsp. australis.
subsp. floccosa *(J.H. Hemsl.) L. Gautier in J.-P. Lebrun & A.L. Stork, Énum. Pl. Fleurs Afr. Trop.,* 4: 34 (1997): Aningeria adolfi-friedericii subsp. floccosa.
subsp. keniensis *(R.E. Fr.) L. Gautier in J.-P. Lebrun & A.L. Stork, Énum. Pl. Fleurs Afr. Trop.,* 4: 34 (1997): Sideroxylon adolfi-friedericii subsp. keniensis.
subsp. usambarensis *(J.H. Hemsl.) L. Gautier in J.-P. Lebrun & A.L. Stork, Énum. Pl. Fleurs Afr. Trop.,* 4: 34 (1997): Aningeria adolfi-friedericii subsp. usambarensis.
alnifolia (Baker) Roberty
var. sacleuxii *(Lecomte) L. Gautier in J.-P. Lebrun & A.L. Stork, Énum. Pl. Fleurs Afr. Trop.,* 4: 34 (1997): Malacantha sacleuxii.
bulliformis *Q. Jiménez Madrigal & T.D. Pennington in Novon,* 7(2): 169 (1997)—Costa Rica, Colombia.
glabra *(Ridl.) I.M. Turner in Novon,* 6(2): 223 (1996): Sideroxylon glabrum.
glabra *(Ridl.) S.P. Teo in Sarawak Mus. J.,* 72: 215 (1997), isonym; *see I.M. Turner in Novon,* 6(2): 223 (1996): Sideroxylon glabrum.
grayana *(H. St. John) F.R. Fosberg in Micronesica,* 25(2): 196 (1992): Planchonella grayana.
var. florencei *F.R. Fosberg in Micronesica,* 25(2): 196 (1992)—Society Is.
mongaguensis *J.R. de Mattos in Loefgrenia,* 101: 1 (1992)—Brazil Southeast (São Paulo).
nitida *(Blume) S.P. Teo in Sarawak Mus. J.,* 72: 217 (1997), homonym; *see Radlk. in Sitzungsber. Math.-Phys. Cl. Königl. Bayer. Akad. Wiss.* München, 12: 333 (1882): Sideroxylon nitidum.
pseudoracemosa *(J.H. Hemsl.) L. Gautier in J.-P. Lebrun & A.L. Stork, Énum. Pl. Fleurs Afr. Trop.,* 4: 35 (Sept. 1997): Aningeria pseudoracemosa.
pseudoracemosa *(J.H. Hemsl.) Y.B. Harvey in Kew Bull.,* 52(3): 672 (Nov. 1997), isonym; *see L. Gaut. in J.-P. Lebrun & Stork, Énum. Pl. Fleurs Afr. Trop.,* 4: 35 (Sept. 1997): Aningeria pseudoracemosa.
superba *(Vermoesen) L. Gautier in J.-P. Lebrun & A.L. Stork, Énum. Pl. Fleurs Afr. Trop.,* 4: 35 (1997): Malacantha superba.

POUZOLZIA (Urticac.).
guatemalana (Blume) Wedd.
var. nivea *(S. Watson) I. Friis & C.M. Wilmot-Dear in Op. Bot.*, 129: 87 (1996): P. nivea.
latistipula *I. Friis & C.M. Wilmot-Dear in Op. Bot.*, 129: 65 (1996)—Colombia.
matsudae *(Yamam.) S.S. Ying, Coloured Illustr. Fl. Taiwan*, 3: 479 (Oct. 1988); *et in Mem. Coll. Agric. Nation. Taiwan Univ.*, 28(2): 41 (Oct. 1988): Memorialis matsudae.†
neurocarpa *(Yamam.) S.S. Ying, Coloured Illustr. Fl. Taiwan*, 3: 481 (Oct. 1988); *et in Mem. Coll. Agric. Nation. Taiwan Univ.*, 28(2): 42 (Oct. 1988): Memorialis neurocarpa.†
nudiflora *(Willd.) I. Friis & C.M. Wilmot-Dear in Op. Bot.*, 129: 71 (1996): Boehmeria nudiflora.
occidentalis (Liebm.) Wedd.
var. palmeri *(S. Watson) I. Friis & C.M. Wilmot-Dear in Op. Bot.*, 129: 83 (1996): P. palmeri.
pentandra (Roxb.) Benn.
var. akoensis *(Yamam.) S.S. Ying, Coloured Illustr. Fl. Taiwan*, 3: 477 (Oct. 1988); *et in Mem. Coll. Agric. Nation. Taiwan Univ.*, 28(2): 41 (Oct. 1988): Memorialis pentandra var. akoensis.†
var. hypericifolia *(Blume) S.S. Ying, Coloured Illustr. Fl. Taiwan*, 3: 476 (Oct 1988); *et in Mem. Coll. Agric. Nation. Taiwan Univ.*, 28(2): 41 (Oct. 1988): P. hypericifolia.†
zeylanica (L.) Benn.
var. microphylla *(Wedd.) W.T. Wang in Fl. Reipubl. Popul. Sin.*, 23(2): 365 (1995): P. indica subvar. microphylla.

PRAECEREUS (Cactac.).
euchlorus *(F.A.C. Weber) N.P. Taylor in Cactaceae Consensus Initiatives*, 3: 10 (1997): Cereus euchlorus.
subsp. amazonicus *(K. Schum. ex Vaupel) N.P. Taylor in Cactaceae Consensus Initiatives*, 3: 10 (1997): Cereus amazonicus.
subsp. diffusus *(Britton & Rose) N.P. Taylor in Cactaceae Consensus Initiatives*, 3: 10 (1997): Monvillea diffusa.
subsp. jaenensis *(Rauh & Backeb.) C. Ostolaza Nano in Cactaceae Consensus Initiatives*, 6: 9 (1998): Monvillea jaenensis.
subsp. smithianus *(Britton & Rose) N.P. Taylor in Cactaceae Consensus Initiatives*, 3: 10 (1997): Cephalocereus smithianus.
saxicola *(Morong) N.P. Taylor in Cactaceae Consensus Initiatives*, 3: 10 (1997): Cereus saxicola.

PRAECOXANTHUS *S.D. Hopper & A.P. Brown in Lindleyana*, 15(2): 124 (2000). *ORCHIDACEAE*
aphyllus *(Benth.) S.D. Hopper & A.P. Brown in Lindleyana*, 15(2): 124 (2000): Caladenia aphylla.

PRANGOS (Umbellif.).
sect. Bracteatae *M.G. Pimenov & E.V. Kljuykov in Candollea*, 52(1): 111 (1997).
dzhungarica *M.G. Pimenov in Opred. Rast. Sred. Azii*, 7: 371 (1983)—Kazakhstan.
heyniae *H. Duman & M.F. Watson in Edinburgh J. Bot.*, 56(2): 203 (1999)—Turkey.
longistylis *(Boiss.) M.G. Pimenov & E.V. Kljuykov in Candollea*, 52(1): 110 (1997): Ferulago longistylis.
platychlaena Boiss. ex Tchich.
subsp. engizekensis *H. Duman & M.F. Watson in Edinburgh J. Bot.*, 56(2): 203 (1999), as 'platychloenae'—Turkey.

PRASOPHYLLUM (Orchidac.).
alpestre *D.L. Jones in Austral. Orchid Res.*, 3: 98 (1998)—New South Wales (New South Wales, Australian Capital Territory), Victoria, Tasmania.
amoenum *D.L. Jones in Austral. Orchid Res.*, 3: 99 (1998)—Tasmania.
apoxychilum *D.L. Jones in Austral. Orchid Res.*, 3: 100 (1998)—Tasmania.
bagoense *D.L. Jones in Orchadian*, 13(4) 150 (2000), as 'bagoensis'—New South Wales.
canaliculatum *D.L. Jones in Orchadian*, 12(3): 125 (1997)—New South Wales.
caricetum *D.L. Jones in Orchadian*, 13(4): 151 (2000)—New South Wales.

PRASOPHYLLUM:-
castaneum *D.L. Jones in Austral. Orchid Res.*, 3: 102 (1998)—Tasmania.
caudiculum *D.L. Jones in Orchadian*, 13(4): 155 (2000)—New South Wales.
favonium *D.L. Jones in Austral. Orchid Res.*, 3: 104 (1998)—Tasmania.
fosteri *D.L. Jones in Orchadian*, 13(4): 156 (2000)—Victoria.
incurvum *D.L. Jones in Austral. Orchid Res.*, 3: 106 (1998)—Tasmania.
milfordense *D.L. Jones in Austral. Orchid Res.*, 3: 107 (1998)—Tasmania.
murfetii *D.L. Jones in Orchadian*, 13(4): 160 (2000)—South Australia.
niphopedium *D.L. Jones in Orchadian*, 13(4): 161 (2000)—Victoria.
odoratissimum *D.L. Jones in Orchadian*, 12(2): 53 (1996)—Western Australia.
olidum *D.L. Jones in Austral. Orchid Res.*, 3: 108 (1998)—Tasmania.
pallens *D.L. Jones in Orchadian*, 13(4): 162 (2000)—New South Wales.
paulinae *D.L. Jones & M.A. Clements in Nuytsia*, 10(3): 414 (1996)—Western Australia.
perangustum *D.L. Jones in Austral. Orchid Res.*, 3: 109 (1998)—Tasmania.
pulchellum *D.L. Jones in Austral. Orchid Res.*, 3: 110 (1998)—Tasmania.
retroflexum *D.L. Jones in Orchadian*, 13(4): 167 (2000)—New South Wales.
secutum *D.L. Jones in Austral. Orchid Res.*, 3: 113 (1998)—Tasmania.
solstitium *D.L. Jones in Orchadian*, 13(4): 169 (2000)—New South Wales.
sphacelatum *D.L. Jones in Muelleria*, 9: 59 (1996)—New South Wales, Australian Capital Territory, Victoria.
stellatum *D.L. Jones in Austral. Orchid Res.*, 3: 115 (1998)—Tasmania.
tunbridgense *D.L. Jones in Austral. Orchid Res.*, 3: 117 (1998)—Tasmania.
wilkinsoniorum *D.L. Jones in Orchadian*, 13(4): 170 (2000)—New South Wales.

PRAXELIS (Compos.).
diffusa *(Rich.) J.F. Pruski in Brittonia*, 50(4): 475 (1998): Cacalia diffusa.

PREMNA (Verbenac.).
balakrishnanii *A. Rajendran & P. Daniel in Bull. Bot. Surv. India*, 34(1–4): 174 (1992 publ. 1997)—India (Tamil Nadu).
barbata *J.O. Voigt, Hort. Suburb. Calcutt.*: 468 (1845), nom. nov.: P. serratifolia *Roxb.*
debiana *A. Rajendran & P. Daniel in Bull. Bot. Surv. India*, 34(1–4): 177 (1992 publ. 1997)—Assam (Arunachal Pradesh).
hircina *N. Wallich, Report on Calcutta Bot. Gard. to G.A. Bushby*: 34 (1840), nom. nov.: P. integrifolia *Roxb.*
nimmoniana *J. Graham, Cat. Pl. Bombay*: 155 (1839)—India (Maharashtra).

PRENANTHES (Compos.).
sect. Abietinae *A.N. Sennikov in Bot. Zhurn.*, 82(2): 113 (1997).
sect. Petiolatae *(Gagnidze) A.N. Sennikov in Bot. Zhurn.*, 82(2): 113 (1997): Cicerbita sect. Petiolatae.
petiolata *(K. Koch) A.N. Sennikov in Bot. Zhurn.*, 82(2): 113 (1997): Mulgedium petiolatum.

PRENIA (Aizoac.).
englishiae *(L. Bolus) M. Gerbaulet in Bot. Jahrb.*, 118(1): 30 (1996): Mesembryanthemum englishiae.
pallens (Aiton) N.E. Br.
subsp. lancea *(Thunb.) M. Gerbaulet in Bot. Jahrb.*, 118(1): 32 (1996): Mesembryanthemum lanceum.
subsp. namaquensis *M. Gerbaulet in Bot. Jahrb.*, 118(1): 34 (1996)—Cape Province.
radicans *(L. Bolus) M. Gerbaulet in Bot. Jahrb.*, 118(1): 34 (1996): Aridaria radicans.
tetragona *(Thunb.) M. Gerbaulet in Bot. Jahrb.*, 118(1): 36 (1996): Mesembryanthemum tetragonum.

PREONANTHUS (Ranunculac.).
scherfelii
f. lutescens *(Šourek) J. Holub in Preslia*, 70(2): 109 (1998): Pulsatilla alba var. lutescens.

PRESCOTTIA (Orchidac.).
auyantepuiensis *G. Carnevali & G.A. Romero in G.A. Romero & G. Carnevali, Orchids of Venezuela, ed. 2*: 1145 (2000)—Venezuela.
lojana *C.H. Dodson in Orquideologia*, 20(1): 108 (1996)—Ecuador.
tepuyensis *G. Carnevali & C.A. Vargas in Lindleyana*, 11(4): 236 (1996)—Venezuela.

PRESLIOPHYTUM *(Urb. & Gilg) M. Weigend, Nasa & Conquest S. America - Syst. Rearrang. in Loasac.*: 215 (1997): Loasa sect. Presliophytum. (Loasac.).
arequipensis *M. Weigend, Nasa & Conquest S. America - Syst. Rearrang. in Loasac.*: 219 (1997)—Peru.
heucherifolia *(Killip) M. Weigend, Nasa & Conquest S. America - Syst. Rearrang. in Loasac.*: 219 (1997), as 'heucheraefolia': Loasa heucherifolia.
incanum *(Graham) M. Weigend, Nasa & Conquest S. America - Syst. Rearrang. in Loasac.*: 215, 219 (1997): Loasa incana.

PRESTOEA (Palmae).
acuminata (Willd.) H.E. Moore
var. dasystachys *(Burret) A. Henderson & G. Galeano in Fl. Neotrop.*, 72: 53 (1996): Euterpe dasystachys.
var. montana *(Graham) A. Henderson & G. Galeano in Fl. Neotrop.*, 72: 53 (1996): Euterpe montana.
longepetiolata (Oerst.) H.E. Moore
var. cuatrecasasii *(H.E. Moore) A. Henderson & G. Galeano in Fl. Neotrop.*, 72: 67 (1996): P. cuatrecasasii.
var. roseospadix *(L.H. Bailey) A. Henderson & G. Galeano in Fl. Neotrop.*, 72: 67 (1996): Euterpe roseospadix.
pubens H.E. Moore
var. semispicata *(Nevers & Andr. Hend.) A. Henderson & G. Galeano in Fl. Neotrop.*, 72: 68 (1996): P. semispicata.

PRESTONIA (Apocynac.).
clandestina *J.F. Morales in Novon*, 7(1): 60 (1997)—Mexico Gulf (Veracruz), Mexico Southeast (Chiapas), Mexico Southwest (Oaxaca).
erecta *J.F. Morales in Novon*, 9(1): 90 (1999), nom. nov., as 'comb. nov.': Echites erecta *A.DC.*
folsomii *J.F. Morales in Novon*, 6(3): 285 (1996)—Colombia, Panama.
hammelii *J.F. Morales in Novon*, 7(1): 60 (1997)—Costa Rica.
longifolia *(Sessé & Moç.) J.F. Morales in Novon*, 6(3): 287 (1996): Echites longifolia.
riverae *J.F. Morales in Novon*, 7(1): 63 (1997)—Costa Rica.

PRIMULA (Primulac.).
subsect. Acaules *Ju.L. Menitsky in Bot. Zhurn.*, 85(6): 154 (2000).
adenophylla *M. Gandoger, Dec. Pl. Nov.*, 2: 25 (1876)—Austria.
apicicallosa *D. Fang in J. Trop. Subtrop. Bot.*, 6(1): 49 (1998)—China Southeast (Guangxi), China South-central (Yunnan).
arunachalensis *S.K. Basak & G.G. Maiti in Acta Phytotax. Geobot.*, 51(1): 11 (2000)—Assam (Arunachal Pradesh).
bavarica *M. Gandoger, Dec. Pl. Nov.*, 2: 26 (1876)—Germany.
chamaejasme *(Wulfen) K.K. Khanna & Anand Kumar in J. Econ. Taxon. Bot.*, 22(1): 237 (1998), with incorrect basionym ref.: Androsace chamaejasme.
cuneifolia Ledeb.
subsp. hakusanensis *(Franch.) V.N. Voroschilov in A.K. Skvortsov (ed.), Florist. issl. v razn. raĭonakh SSSR*: 186 (1985): P. hakusanensis.
delavayi *(Franch.) K.K. Khanna & Anand Kumar in J. Econ. Taxon. Bot.*, 22(1): 237 (1998), homonym; *see Franch. in Bull. Soc. Bot. France*, 32: 272 (1885): Androsace delavayi.

PRIMULA:-
× digenea A. Kern.
nothosubsp. rothmaleri *A. Kress, Primulac.-Stud.*, 16: 9 (1999), nom. nov.: P. legionensis *Rothm.*
elatior
subsp. bergidensis *A. Kress, Primulac.-Stud.*, 16: 7 (1999)—Spain.
subsp. corcontica *(Domin) M. Kovanda in Thaiszia*, 7(1): 21 (1997): P. elatior var. corcontica.
subsp. impigrorum *A. Kress, Primulac.-Stud.*, 16: 7 (1999)—France.
farinosa L.
var. pinnata *(Popov & Fed.) M.M. Ivanova in Fl. Tsentral'noi Sibiri*, 2: 702 (1979): P. pinnata.
fragilis *(Duthie) K.K. Khanna & Anand Kumar in J. Econ. Taxon. Bot.*, 22(1): 237 (1998), homonym; *see Balf. f. & Kingdon-Ward in Notes Roy. Bot. Gard. Edinburgh*, 9: 18 (1915): Androsace fragilis.
garhwalica *(Balodi & Singh) K.K. Khanna & Anand Kumar in J. Econ. Taxon. Bot.*, 22(1): 237 (1998), as 'garhwalicum': Androsace garhwalica.
grignensis *D.M. Moser in Candollea*, 53(2): 388 (1998)—Italy.
illibata *M. Gandoger, Dec. Pl. Nov.*, 2: 25 (1876)—France.
kweichouensis W.W. Sm.
var. guangxiensis *D. Fang in J. Trop. Subtrop. Bot.*, 6(1): 51 (1998)—China Southeast (Guangxi).
leiocalyx *M. Gandoger, Dec. Pl. Nov.*, 2: 27 (1876)—France.
maximowiczii Regel
f. flaviflorida *(D.Z. Lu) T.J. Feng & Y.F. Hu in J. Hebei Agric. Univ.*, 17(4): 16 (1994), as 'maximoviczii': P. maximowiczii var. flaviflorida.
munroi Lindl.
subsp. yargongensis *(Petitm.) D.G. Long in Edinburgh J. Bot.*, 56(2): 307 (1999): P. yargongensis.
nanocapitata *Balf. f. & W.W. Sm. ex S.K. Basak & G.G. Maiti in J. Econ. Taxon. Bot.*, 23(3): 699 (1999)—East Himalaya (Sikkim).
nivalis Pall.
subsp. subintegerrima *(Regel) V.N. Voroschilov in A.K. Skvortsov (ed.), Florist. issl. v razn. raionakh SSSR:* 186 (1985): P. nivalis var. subintegerrima.
recubariensis *F. Prosser & S. Scortegagna in Willdenowia*, 28(1–2): 28 (1998)—Italy.
sachalinensis Nakai
f. albida *Y.N. Lee, Fl. Korea:* 1159 (1996), without type—Locality not stated.
serratifolia *M. Gandoger, Dec. Pl. Nov.*, 2: 26 (1876)—France.
sieboldii E. Morren
f. albiflora *Y.N. Lee, Fl. Korea:* 1159 (1996), without type—Locality not stated.
sinoexscapa *C.M. Hu in J. Trop. Subtrop. Bot.*, 8(3): 214 (2000), nom. nov.: P. exscapa *F.H. Chen & C.M. Hu*.
sinoplantaginea Balf. f.
var. fengxiangiana *W.L. Zheng in Acta Bot. Yunnanica*, 20(3): 275 (1998)—Xizang.
sorachina Miyabe & Tatew.
f. candida *H. Nakai & H. Ohashi in J. Jap. Bot.*, 72(1): 62 (1997)—Japan.
× vallarsae *F. Prosser & S. Scortegagna in Willdenowia*, 28(1–2): 42 (1998)—Italy.(P. auricula × recubariensis).
villosa *(L.) K.K. Khanna & Anand Kumar in J. Econ. Taxon. Bot.*, 22(1): 237 (1998), homonym; *see Wulfen in Jacq., Fl. Austriac.*, 5: 41 (1778): Androsace villosa.

PRIONOSCIADIUM (Umbellif.).
saraviki *J.E. Laferrière in Phytologia*, 81(2): 122 (1996 publ. 1997)—Mexico Northeast (Chihuahua).

PRIORIA (Legumin.).
alternifolia *(Elmer) F.J. Breteler in Wageningen Agric. Univ. Pap.*, 99(3): 50 (1999): Cynometra alternifolia.
balsamifera *(Vermoesen) F.J. Breteler in Wageningen Agric. Univ. Pap.*, 99(3): 21 (1999): Pterygopodium balsamiferum.
buchholzii *(Harms) F.J. Breteler in Wageningen Agric. Univ. Pap.*, 99(3): 24 (1999): Oxystigma buchholzii.

PRIORIA:-
gilbertii *(J. Léonard) F.J. Breteler in Wageningen Agric. Univ. Pap.*, 99(3): 28 (1999): Oxystigma gilbertii.
joveri *(Normand ex Aubrév.) F.J. Breteler in Wageningen Agric. Univ. Pap.*, 99(3): 31 (1999): Gossweilerodendron joveri.
mannii *(Baill.) F.J. Breteler in Wageningen Agric. Univ. Pap.*, 99(3): 33 (1999): Copaifera mannii.
micrantha *(Burtt) F.J. Breteler in Wageningen Agric. Univ. Pap.*, 99(3): 50 (1999): Kingiodendron micranthum.
msoo *(Harms) F.J. Breteler in Wageningen Agric. Univ. Pap.*, 99(3): 35 (1999): Oxystigma msoo.
novoguineensis *(Verdc.) F.J. Breteler in Wageningen Agric. Univ. Pap.*, 99(3): 51 (1999): Kingiodendron novoguineense.
oxyphylla *(Harms) F.J. Breteler in Wageningen Agric. Univ. Pap.*, 99(3): 37 (1999): Pterygopodium oxyphyllum.
pinnata *(Roxb. ex DC.) F.J. Breteler in Wageningen Agric. Univ. Pap.*, 99(3): 52 (1999): Hardwickia pinnata.
platycarpa *(Burtt) F.J. Breteler in Wageningen Agric. Univ. Pap.*, 99(3): 53 (1999): Kingiodendron platycarpum.
tenuicarpa *(Verdc.) F.J. Breteler in Wageningen Agric. Univ. Pap.*, 99(3): 54 (1999): Kingiodendron tenuicarpum.

PRISTIGLOTTIS (Orchidac.).
torta *(King & Pantl.) L.V. Averyanov in Bot. Zhurn.*, 81(10): 78 (1996): Odontochilus tortus.
umbrosa *L.V. Averyanov in Bot. Zhurn.*, 81(10): 78, 75 (1996)—Vietnam.

PRITCHARDIA (Palmae).
perlmanii *C.E.C. Gemmill in Novon*, 8(1): 18 (1998)—Hawaiian Is.

PRITZELAGO (Crucif.).
alpina (L.) Kuntze
subsp. brevicaulis *(Spreng.) J. Soják in Zprávy Krajsk. Vlastiv. Muz. Olomouci*, 215: 2 (1982), basionym as 'Noccaea brevicaulis Hoppe', see Art. 33.6 ICBN (2000): Hutchinsia brevicaulis.
subsp. dubia *(Pawł.) J. Soják in Zprávy Krajsk. Vlastiv. Muz. Olomouci*, 215: 2 (1982): Hutchinsia alpina var. dubia.
auerswaldii *(Willk.) J. Soják in Zprávy Krajsk. Vlastiv. Muz. Olomouci*, 215: 2 (1982): Hutchinsia auerswaldii.

PRIVA (Verbenac.).
ibugana *J. Rzedowski & G. Calderón de Rzedowski in Bol. Inst. Bot. (Guadalajara)*, 5(1–3): 65 (1998)—Mexico Northeast (Guanajuato, Querétaro).

× **PROMELLIA** *hort. in Orchid Rev.*, 108(1236): new orchid hybrids 12 (2000). *ORCHIDACEAE* (Ansellia × Promenaea).

PROMENAEA (Orchidac.).
ovatiloba
var. robertii *L.C. Menezes in Bol. CAOB*, 6(1): 12 (1995), as 'roberti'—Brazil Southeast (Espírito Santo)
silvana *F. de Barros & E.L.M. Catharino in Hoehnea*, 21(1–2): 94 (1994 publ. 1995)—Brazil Northeast (Bahia).
sincorana *P. Castro & Campacci in Bol. CAOB*, 5(3): 10 (1993)—Brazil Northeast (Bahia).

PROMETHEUM (Crassulac.).
aizoon *(Fenzl) H. 't Hart in H. 't Hart & U. Eggli (eds.), Evol. & Syst. Crassulac.:* 170 (1995): Umbilicus aizoon.
chrysanthum *(Boiss.) H. 't Hart in H. 't Hart & U. Eggli (eds.), Evol. & Syst. Crassulac.:* 170 (1995): Umbilicus chrysantha.
muratdaghense *(Kit Tan) H. 't Hart in Succulenta*, 78(5): 236 (1999): Rosularia muratdaghensis.
rechingeri *(C.-A. Jansson) H. 't Hart in H. 't Hart & U. Eggli (eds.), Evol. & Syst. Crassulac.:* 170 (1995): Rosularia rechingeri.
serpentinicum *(Werdermann) H. 't Hart in H. 't Hart & U. Eggli (eds.), Evol. & Syst. Crassulac.:* 170 (1995): Umbilicus serpentinicus.

PROMETHEUM:-
serpentinicum
var. giganteum *(Eggli) H. 't Hart in Succulenta*, 78(5): 236 (1999): Rosularia serpentinica var. gigantea.
tymphaeum *(Quézel & Contandr.) H. 't Hart in H. 't Hart & U. Eggli (eds.), Evol. & Syst. Crassulac.:* 170 (1995): Sedum tymphaeum.

× **PROMOSEPALUM** *hort. in Orchid Rev.*, 108 (1235): new orchid hybrids 14 (2000). *ORCHIDACEAE* (Promenaea × Zygosepalum).

PROSARTES (Convallariac.).
ovalis *(Ohwi) M.N. Tamura in K. Kubitzki (ed.), Fam. & Gen. Vasc. Pl.*, 3: 171 (1998): Disporum ovale.

PROSOPIS (Legumin.).
aculeata *J.G. König ex W. Roxburgh in Asiat. Res.*, 4: 405 (1795)—India.

PROSPERO (Hyacinthac.).
battagliae *F. Speta in Linz. Biol. Beitr.*, 32(2): 1325 (2000)—Kriti.
cyrenaicum *(Pamp.) F. Speta in Phyton (Austria)*, 38(1): 116 (1998): Scilla autumnalis var. cyrenaica.
depressum *F. Speta in Linz. Biol. Beitr.*, 32(2): 1325 (2000)—Kriti.
hierapytnense *F. Speta in Linz. Biol. Beitr.*, 32(2): 1325 (2000)—Kriti.
idaeum *F. Speta in Linz. Biol. Beitr.*, 32(2): 1324 (2000)—Kriti.
minimum *F. Speta in Linz. Biol. Beitr.*, 32(2): 1324 (2000)—Kriti.
rhadamanthi *F. Speta in Linz. Biol. Beitr.*, 32(2): 1324 (2000)—Kriti.
talosii *(Tzanoud. & Kypr.) F. Speta in Phyton (Austria)*, 38(1): 116 (1998): Scilla talosii.

PROSTANTHERA (Labiat.).
askania *B.J. Conn in Telopea*, 7(3): 231 (1997)—New South Wales.
clotteniana *(F.M. Bailey) A.R. Bean in Austrobaileya*, 5(4): 733 (2000): Hemigenia clotteniana.
cryptandroides A. Cunn. ex Benth.
subsp. euphrasioides *(Benth.) B.J. Conn in Telopea*, 8(2): 269 (1999): P. euphrasioides.
galbraithiae *B.J. Conn in Telopea*, 7(4): 321 (1998)—Victoria.
hindii *B.J. Conn in Telopea*, 7(3): 234 (1997)—New South Wales.
junonis *B.J. Conn in Telopea*, 7(3): 237 (1997)—New South Wales.
palustris *B.J. Conn in Telopea*, 7(3): 241 (1997)—New South Wales.

PROSTHECHEA (Orchidac.).
abbreviata *(Schltr.) W.E. Higgins in Phytologia*, 82(5): 376 (1997 publ. 1998): Epidendrum abbreviatum.
aemula *(Lindl.) W.E. Higgins in Phytologia*, 82(5): 376 (1997 publ. 1998): Epidendrum aemulum.
alagoensis *(Pabst) W.E. Higgins in Phytologia*, 82(5): 376 (1997 publ. 1998): Epidendrum alagoense.
allemanii *(Barb. Rodr.) W.E. Higgins in Phytologia*, 82(5): 376 (1997 publ. 1998): Epidendrum allemanii.
allemanoides *(Hoehne) W.E. Higgins in Phytologia*, 82(5): 376 (1997 publ. 1998): Epidendrum allemanoides.
aloisii *(Schltr.) C.H. Dodson & E. Hágsater in Cat. Vasc. Pl. Ecuador (Monogr. Syst. Bot. Missouri Bot. Gard., 75):* 956 (1999): Epidendrum aloisii.
baculus *(Rchb. f.) W.E. Higgins in Phytologia*, 82(5): 376 (1997 publ. 1998): Epidendrum baculus.
bennettii *(Christenson) W.E. Higgins in Phytologia*, 82(5): 376 (1997 publ. 1998): Encyclia bennettii.
bicamerata *(Rchb. f.) W.E. Higgins in Phytologia*, 82(5): 376 (1997 publ. 1998): Epidendrum bicameratum.
boothiana *(Lindl.) W.E. Higgins in Phytologia*, 82(5): 376 (1997 publ. 30 March 1998): Epidendrum boothianum.
var. erythronioides *(Small) W.E. Higgins in North Amer. Native Orchid J.*, 4(1): 52 (20 March 1998), species name not validly publ.: Epidendrum erythronioides.

PROSTHECHEA:-
boothiana
var. erythronioides *(Small) M.A. Nir, Orchidaceae Antillanae:* 407 (2000): Epidendrum erythronioides.
brachiata *(A. Rich. & Galeotti) W.E. Higgins in Phytologia,* 82(5): 376 (1997 publ. 1998): Epidendrum brachiatum.
brachychila *(Lindl.) W.E. Higgins in Phytologia,* 82(5): 376 (1997 publ. 1998): Epidendrum brachychilum.
brassavolae *(Rchb. f.) W.E. Higgins in Phytologia,* 82(5): 376 (1997 publ. 1998): Epidendrum brassavolae.
bulbosa *(Vell.) W.E. Higgins in Phytologia,* 82(5): 377 (1997 publ. 1998): Epidendrum bulbosum.
caetensis *(Bicalho) W.E. Higgins in Phytologia,* 82(5): 377 (1997 publ. 1998): Hormidium caetense.
calamaria *(Lindl.) W.E. Higgins in Phytologia,* 82(5): 377 (1997 publ. 1998): Epidendrum calamarium.
campos-portoi *(Pabst) W.E. Higgins in Phytologia,* 82(5): 377 (1997 publ. 1998): Encyclia campos-portoi.
campylostalix *(Rchb. f.) W.E. Higgins in Phytologia,* 82(5): 377 (1997 publ. 1998): Epidendrum campylostalix.
chacaoensis *(Rchb. f.) W.E. Higgins in Phytologia,* 82(5): 377 (1997 publ. 1998): Epidendrum chacaoense.
chimborazoensis *(Schltr.) W.E. Higgins in Phytologia,* 82(5): 377 (1997 publ. 1998): Epidendrum chimborazoensis.
chondylobulbon *(A. Rich. & Galeotti) W.E. Higgins in Phytologia,* 82(5): 377 (1997 publ. 1998): Epidendrum chondylobulbon.
christi *(Rchb. f.) C.H. Dodson & E. Hágsater in Cat. Vasc. Pl. Ecuador (Monogr. Syst. Bot. Missouri Bot. Gard., 75):* 956 (1999): Epidendrum christi.
citrina *(La Llave & Lex.) W.E. Higgins in Phytologia,* 82(5): 377 (1997 publ. 1998): Sobralia citrina.
cochleata *(L.) W.E. Higgins in Phytologia,* 82(5): 377 (1997 publ. 30 March 1998): Epidendrum cochleatum.
f. albidoflava *(P.M. Br.) P.M. Brown in North Amer. Native Orchid J.,* 4(1): 52 (20 March 1998), species name not validly publ.: Encyclia cochleata f. albidoflava.
var. triandra *(Ames) W.E. Higgins in North Amer. Native Orchid J.,* 4(1): 52 (20 March 1998), species name not validly publ.: Epidendrum cochleatum var. triandrum.
subsp. triandra *(Ames) M.A. Nir, Orchidaceae Antillanae:* 407 (2000): Epidendrum cochleatum var. triandrum.
concolor *(La Llave & Lex.) W.E. Higgins in Phytologia,* 82(5): 377 (1997 publ. 1998): Epidendrum concolor.
cretacea *(Dressler & G.E. Pollard) W.E. Higgins in Phytologia,* 82(5): 377 (1997 publ. 1998): Encyclia cretacea.
faresiana *(Bicalho) W.E. Higgins in Phytologia,* 82(5): 377 (1997 publ. 1998): Hormidium faresianum.
fausta *(Rchb. f. ex Cogn.) W.E. Higgins in Phytologia,* 82(5): 377 (1997 publ. 1998): Epidendrum faustum.
fortunae *(Dressler) W.E. Higgins in Phytologia,* 82(5): 377 (1997 publ. 1998): Encyclia fortunae.
fragrans *(Sw.) W.E. Higgins in Phytologia,* 82(5): 377 (1997 publ. 1998): Epidendrum fragrans.
garciana *(Garay & Dunst.) W.E. Higgins in Phytologia,* 82(5): 378 (1997 publ. 1998): Epidendrum garcianum.
ghiesbreghtiana *(A. Rich. & Galeotti) W.E. Higgins in Phytologia,* 82(5): 378 (1997 publ. 1998): Epidendrum ghiesbreghtianum.
gilbertoi *(Garay) W.E. Higgins in Phytologia,* 82(5): 378 (1997 publ. 1998): Epidendrum gilbertoi.
glumacea *(Lindl.) W.E. Higgins in Phytologia,* 82(5): 378 (1997 publ. 1998): Epidendrum glumaceum.
grammatoglossa *(Rchb. f.) W.E. Higgins in Phytologia,* 82(5): 378 (1997 publ. 1998): Epidendrum grammatoglossum.

PROSTHECHEA:-
greenwoodiana *(Aguirre-Olav.) W.E. Higgins in Phytologia,* 82(5): 378 (1997 publ. 1998): Encyclia greenwoodiana.
grisebachianum *(Cogn.) W.E. Higgins in Phytologia,* 82(5): 378 (1997 publ. 1998): Epidendrum grisebachianum.
hartwegii *(Lindl.) W.E. Higgins in Phytologia,* 82(5): 378 (1997 publ. 1998): Epidendrum hartwegii.
hastata *(Lindl.) W.E. Higgins in Phytologia,* 82(5): 378 (1997 publ. 1998): Epidendrum hastatum.
hoehnei *(A.D. Hawkes) W.E. Higgins in Phytologia,* 82(5): 378 (1997 publ. 1998): Epidendrum hoehnei.
inversa *(Lindl.) W.E. Higgins in Phytologia,* 82(5): 378 (1997 publ. 1998): Epidendrum inversum.
ionocentra *(Rchb. f.) W.E. Higgins in Phytologia,* 82(5): 378 (1997 publ. 1998): Epidendrum ionocentrum.
ionophlebia *(Rchb. f.) W.E. Higgins in Phytologia,* 82(5): 378 (1997 publ. 1998): Epidendrum ionophlebium.
jauana *(Carnevali & I. Ramírez) W.E. Higgins in Phytologia,* 82(5): 378 (1997 publ. 1998): Encyclia jauana.
kautzkii *(Pabst) W.E. Higgins in Phytologia,* 82(5): 378 (1997 publ. 1998): Encyclia kautzkii.
lambda *(Linden & Rchb. f.) W.E. Higgins in Phytologia,* 82(5): 378 (1997 publ. 1998): Epidendrum lambda.
leopardina *(Rchb. f.) C.H. Dodson & E. Hágsater in Cat. Vasc. Pl. Ecuador (Monogr. Syst. Bot. Missouri Bot. Gard., 75):* 956 (1999): Epidendrum leopardinum.
lindenii *(Lindl.) W.E. Higgins in Phytologia,* 82(5): 379 (1997 publ. 1998): Epidendrum lindenii.
linkiana *(Klotzsch) W.E. Higgins in Phytologia,* 82(5): 379 (1997 publ. 1998): Epidendrum linkianum.
livida *(Lindl.) W.E. Higgins in Phytologia,* 82(5): 379 (1997 publ. 1998): Epidendrum lividum.
maculosa *(Ames, F.T. Hubb. & C. Schweinf.) W.E. Higgins in Phytologia,* 82(5): 379 (1997 publ. 1998): Epidendrum maculosum.
magnispatha *(Ames, F.T. Hubb. & C. Schweinf.) W.E. Higgins in Phytologia,* 82(5): 379 (1997 publ. 1998): Epidendrum magnispathum.
mariae *(Ames) W.E. Higgins in Phytologia,* 82(5): 379 (1997 publ. 1998): Epidendrum mariae.
megahybos *(Schltr.) C.H. Dodson & E. Hágsater in Cat. Vasc. Pl. Ecuador (Monogr. Syst. Bot. Missouri Bot. Gard., 75):* 956 (1999): Epidendrum megahybos.
michuacana *(La Llave & Lex.) W.E. Higgins in Phytologia,* 82(5): 379 (1997 publ. 1998): Epidendrum michuacanum.
moojenii *(Pabst) W.E. Higgins in Phytologia,* 82(5): 379 (1997 publ. 1998): Epidendrum moojenii.
neurosa *(Ames) W.E. Higgins in Phytologia,* 82(5): 379 (1997 publ. 1998): Epidendrum neurosum.
obpiribulbon *(Hágsater) W.E. Higgins in Phytologia,* 82(5): 379 (1997 publ. 1998): Encyclia obpiribulbon.
ochracea *(Lindl.) W.E. Higgins in Phytologia,* 82(5): 379 (1997 publ. 1998): Epidendrum ochraceum.
ortizii *(Dressler) W.E. Higgins in Phytologia,* 82(5): 379 (1997 publ. 1998): Encyclia ortizii.
pamplonensis *(Rchb. f.) W.E. Higgins in Phytologia,* 82(5): 379 (1997 publ. 1998): Epidendrum pamplonense.
panthera *(Rchb. f.) W.E. Higgins in Phytologia,* 82(5): 379 (1997 publ. 1998): Epidendrum panthera.
papilio *(Vell.) W.E. Higgins in Phytologia,* 82(5): 379 (1997 publ. 1998): Epidendrum papilio.
pastoris *(Lex.) A. Espejo Serna & A.R. López-Ferrari in Acta Bot. Mex.,* 51: 63 (2000): Epidendrum pastoris.
pipio *(Rchb. f.) W.E. Higgins in Phytologia,* 82(5): 379 (1997 publ. 1998): Epidendrum pipio.
pringlei *(Rolfe) W.E. Higgins in Phytologia,* 82(5): 380 (1997 publ. 1998): Epidendrum pringlei.
prismatocarpa *(Rchb. f.) W.E. Higgins in Phytologia,* 82(5): 380 (1997 publ. 1998): Epidendrum prismatocarpum.
pseudopygmaea *(Finet) W.E. Higgins in Phytologia,* 82(5): 380 (1997 publ. 1998): Hormidium pseudopygmaeum.

PROSTHECHEA:-
pterocarpa *(Lindl.) W.E. Higgins in Phytologia,* 82(5): 380 (1997 publ. 1998): Epidendrum pterocarpum.
pulcherrima *(Klotzsch) W.E. Higgins in Phytologia,* 82(5): 380 (1997 publ. 1998): Epidendrum pulcherrimum.
punctifera *(Rchb. f.) W.E. Higgins in Phytologia,* 82(5): 380 (1997 publ. 1998): Epidendrum punctiferum.
pygmaea *(Hook.) W.E. Higgins in Phytologia,* 82(5): 380 (1997 publ. 1998): Epidendrum pygmaeum.
racemifera *(Dressler) W.E. Higgins in Phytologia,* 82(5): 380 (1997 publ. 1998): Encyclia racemifera.
radiata *(Lindl.) W.E. Higgins in Phytologia,* 82(5): 380 (1997 publ. 1998): Epidendrum radiatum.
regnelliana *(Hoehne & Schltr.) W.E. Higgins in Phytologia,* 82(5): 380 (1997 publ. 1998): Epidendrum regnellianum.
rhombilabia *(S. Rosillo) W.E. Higgins in Phytologia,* 82(5): 380 (1997 publ. 1998): Encyclia rhombilabia.
rhynchophora *(A. Rich. & Galeotti) W.E. Higgins in Phytologia,* 82(5): 380 (1997 publ. 1998): Epidendrum rhynchophorum.
sceptra *(Lindl.) W.E. Higgins in Phytologia,* 82(5): 380 (1997 publ. 1998): Epidendrum sceptrum.
semiaptera *(Hágsater) W.E. Higgins in Phytologia,* 82(5): 380 (1997 publ. 1998): Encyclia semiaptera.
serrulata *(Sw.) W.E. Higgins in Phytologia,* 82(5): 380 (1997 publ. 1998): Epidendrum serrulatum.
sessiliflora *(Edwall) W.E. Higgins in Phytologia,* 82(5): 380 (1997 publ. 1998): Epidendrum sessiliflorum.
sima *(Dressler) W.E. Higgins in Phytologia,* 82(5): 381 (1997 publ. 1998): Encyclia sima.
spondiada *(Rchb. f.) W.E. Higgins in Phytologia,* 82(5): 381 (1997 publ. 1998): Epidendrum spondiadum.
suzanensis *(Hoehne) W.E. Higgins in Phytologia,* 82(5): 381 (1997 publ. 1998): Epidendrum suzanense.
tigrina *(Lindl.) W.E. Higgins in Phytologia,* 82(5): 381 (1997 publ. 1998): Epidendrum tigrinum.
tripunctata *(Lindl.) W.E. Higgins in Phytologia,* 82(5): 381 (1997 publ. 1998): Epidendrum tripunctatum.
trulla *(Rchb. f.) W.E. Higgins in Phytologia,* 82(5): 381 (1997 publ. 1998): Epidendrum trulla.
vagans *(Ames) W.E. Higgins in Phytologia,* 82(5): 381 (1997 publ. 1998): Epidendrum vagans.
varicosa *(Lindl.) W.E. Higgins in Phytologia,* 82(5): 381 (1997 publ. 1998): Epidendrum varicosum.
venezuelana *(Schltr.) W.E. Higgins in Phytologia,* 82(5): 381 (1997 publ. 1998): Epidendrum venezuelanum.
venosa *(Lindl.) W.E. Higgins in Phytologia,* 82(5): 381 (1997 publ. 1998): Epidendrum venosum.
vespa *(Vell.) W.E. Higgins in Phytologia,* 82(5): 381 (1997 publ. 1998): Epidendrum vespa.
subsp. triandra *(Dod) M.A. Nir, Orchidaceae Antillanae:* 407 (2000): Encyclia vespa subsp. triandra.
vitellina *(Lindl.) W.E. Higgins in Phytologia,* 82(5): 381 (1997 publ. 1998): Epidendrum vitellinum.
widgrenii *(Lindl.) W.E. Higgins in Phytologia,* 82(5): 381 (1997 publ. 1998): Epidendrum widgrenii.

PROTASPARAGUS (Asparagac.).
falcatus *(L.) S.Y. Kamble in J. Econ. Taxon. Bot.,* 20(2): 270 (1996), isonym; *see Oberm. in S. Afr. J. Bot.,* 2(3): 244 (1983): Asparagus falcatus.
gharoensis *(Blatt.) S.Y. Kamble in J. Econ. Taxon. Bot.,* 20(2): 269 (1996): Asparagus gharoensis.
karthikeyanii *S.Y. Kamble in J. Econ. Taxon. Bot.,* 19(3): 735 (1995 publ. 1996)—India (Maharashtra).
penicillatus *(H. Hara) S.Y. Kamble in J. Econ. Taxon. Bot.,* 20(2): 269 (1996): Asparagus penicillatus.
rubricaulis *(Baker) S.Y. Kamble in J. Econ. Taxon. Bot.,* 20(2): 270 (1996): Asparagus rubricaulis.
sarmentosus *(L.) S.Y. Kamble in J. Econ. Taxon. Bot.,* 20(2): 270 (1996): Asparagus sarmentosus.
virgatus *(Baker) S.Y. Kamble in J. Econ. Taxon. Bot.,* 20(2): 270 (1996), isonym; *see Oberm. in Fl. Australia,* 45: 494 (1987): Asparagus virgatus.

PROTASPARAGUS:-
zeylanicus *(Hook. f.) S.Y. Kamble in J. Econ. Taxon. Bot.*, 20(2): 269 (1996): Asparagus zeylanicus.

PROTIUM (Burserac.).
amazonicum *(Cuatrec.) D.C. Daly in Fl. Venezuelan Guayana (gen. eds. J.A. Steyermark et al.)*, 3: 713 (1997): Paraprotium amazonicum.
gallicum *D.C. Daly in Brittonia*, 50(4): 520 (1998)—French Guiana, Surinam.
heptaphyllum (Aubl.) Marchand
subsp. ulei *(Swart) D.C. Daly in Fl. Venezuelan Guayana (gen. eds. J.A. Steyermark et al.)*, 3: 716 (1997): P. almecega var. ulei.
inodorum *D.C. Daly in Brittonia*, 50(4): 517 (1998)—French Guiana, Brazil North (Pará).

PROUSTIA (Compos.).
cuneifolia D. Don
var. mendocina *(Phil.) L. Ariza Espinar in Darwiniana*, 33(1–4): 371 (1995): P. mendocina.

PRUMNOPITYS (Podocarpac.).
spicata *(Poepp.) B.P.J. Molloy & M. Muñoz-Schick in New Zealand J. Bot.*, 37(2): 190 (1999): Podocarpus spicatus.
utilior *(Pilg.) A.P. Melikian & A.V.F.Ch. Bobrov in Bot. Zhurn.*, 85(7): 58 (2000): Podocarpus utilior.

PRUNELLA (Labiat.).
gracillicaulia *A.P. Khokhrjakov in Byull. Glavn. Bot. Sada (Moscow)*, 175: 54 (1997)—Transcaucasus (Adzhariya).
vulgaris
var. taiwaniana *S.S. Ying in Mem. Coll. Agric. Nation. Taiwan Univ.*, 25(2): 98 (1985)—Taiwan.

PRUNUS (Rosac.).
accumulans *(Koehne) C.L. Li & G. Aymard in Biollania*, ed. esp. 6 (J.J. Wurdack Festschr.): 449 (1997): P. myrtifolia var. accumulans.
acutiserrata *H. Koidzumi in J. Pl. Iwateken*, 2(2): 80 (1937)—Japan.
davidiana *(Carrière) N.E. Brown, Suppl. Johnson's Gard. Dict.*: 991 (1882): Persica davidiana.
espinozana *C.L. Li in Biollania*, ed. esp. 6 (J.J. Wurdack Festschr.): 450 (1997)—Venezuela.
foliibracteata *H. Koidzumi in J. Pl. Iwateken*, 2(2): 79 (1937)—Japan.
gideonii *W. Takeuchi in Sida*, 18(3): 765 (1999)—New Guinea (Papua New Guinea).
iwatensis *H. Koidzumi in J. Pl. Iwateken*, 2(2): 82 (1937)—Japan.
jamasakura *Siebold ex G. Koidzumi in Bot. Mag. Tokyo*, 25(295): 184 (Aug. 1911)—Japan.
lichoana *G. Aymard in Biollania*, ed. esp. 6 (J.J. Wurdack Festschr.): 451 (1997)—Venezuela.
× misayamensis *Okuhara & Y. Matsuda ex T. Shimizu, Fl. Nagano Pref.*: 1508 (1997), with two types cited—Japan.(P. apetala × nipponica).
nambuana *H. Koidzumi in J. Pl. Iwateken*, 2(2): 81 (1937)—Japan.
ramiformis *H. Koidzumi in J. Pl. Iwateken*, 2(2): 80 (1937)—Japan.
rivularis Scheele
var. pubescens *M.F. Enquist in Phytologia*, 82(4): 296 (1997 publ. 1998)—Texas.
serotina Ehrh.
var. capuli *(Cav.) S. Hatusima in S. Hatusima & T. Amano, Fl. Ryukyus, S. of Amami Is., ed. 2*: 77 (1994), without basionym ref.: Basionym not stated.
wurdackii *C.L. Li in Biollania*, ed. esp. 6 (J.J. Wurdack Festschr.): 451 (1997)—Venezuela.

PSACALIOPSIS (Compos.).
paneroi (B.L. Turner) C. Jeffrey
var. juxtlahuacensis *J.L. Panero & J.L. Villaseñor in Brittonia*, 48(1): 81 (1996)—Mexico Southwest (Oaxaca).

PSACALIUM (Compos.).
hintoniorum *B.L. Turner in Phytologia*, 80(2): 92 (1996)—Mexico Southwest (Oaxaca).
schillingii *J.L. Panero & J.L. Villaseñor in Brittonia*, 48(4): 566 (1996 publ. 1997)—Mexico Southwest (Oaxaca).

PSAMMETES (Scrophulariac.).
madagascariensis *(Bonati) Eb. Fischer & F.N. Hepper in Kew Bull.*, 52(3): 750 (1997): Bryodes madagascariensis.

PSAMMISIA (Ericac.).
citrina *J.L. Luteyn & D.S. Sylva S.in Brittonia*, 51(3): 291 (1999)—Colombia.
cuyujensis *J.L. Luteyn in Fl. Ecuador*, 54(147): 171 (1996)—Ecuador.
debilis Sleumer
var. ecuadorensis *J.L. Luteyn in Fl. Ecuador*, 54(147): 187 (1996)—Ecuador.
mediobullata *J.L. Luteyn & D.S. Sylva S.in Brittonia*, 51(3): 294 (1999)—Colombia.
montana *J.L. Luteyn in Fl. Ecuador*, 54(147): 183 (1996)—Ecuador, Colombia.
orientalis *J.L. Luteyn in Fl. Ecuador*, 54(147): 188 (1996)—Ecuador.

PSAMMOPHORA (Aizoac.).
saxicola *H.E.K. Hartmann in Bradleya*, 14: 44 (1996)—Namibia.

PSAMMOSILENE *W.C. Wu & C.Y. Wu in L.P. King et al., Icones Plantarum Medicarum e Libro Tien-Nan-Pen-Tsao* Lanmaoano, 1: Table 1 (1945). *CARYOPHYLLACEAE*
tunicoides *W.C. Wu & C.Y. Wu in L.P. King et al., Icones Plantarum Medicarum e Libro Tien-Nan-Pen-Tsao* Lanmaoano, 1: Table 1 (1945), nom. nov.: Silene cryptantha *Diels*.

PSATHYROSTACHYS (Gramin.).
juncea (Fisch.) Nevski
var. hyalantha *(Rupr.) S.L. Chen in Novon*, 7(3): 229 (1997): Elymus hyalanthus.

PSECTRA *(Endl.) P. Tomšovic in Preslia*, 69(1): 33 (1997): Echinops sect. Psectra. (Compos.).
strigosa *(L.) P. Tomšovic in Preslia*, 69(1): 33 (1997): Echinops strigosus.

PSEDNOTRICHIA (Compos.).
newtonii *(O. Hoffm.) A.A. Anderberg & P.O. Karis in Nordic J. Bot.*, 15(4): 378 (1995 publ. 1996): Oligothrix newtonii.
xyridopsis *(O. Hoffm.) A.A. Anderberg & P.O. Karis in Nordic J. Bot.*, 15(4): 378 (1995 publ. 1996): Oligothrix xyridopsis.

PSEPHELLUS (Compos.).
sect. Aetheopappus *(Cass.) G. Wagenitz & F.H. Hellwig in Willdenowia*, 30(1): 36 (2000): gen. Aetheopappus.
sect. Amblyopogon *(DC.) G. Wagenitz & F.H. Hellwig in Willdenowia*, 30(1): 37 (2000): Amberboa sect. Amblyopogon.
sect. Buschiani *A.D. Mikheev in Bot. Zhurn.*, 81(7): 114 (1996).
sect. Czerniakovskya *(Czerep.) G. Wagenitz & F.H. Hellwig in Willdenowia*, 30(1): 38 (2000): Centaurea subgen. Czerniakovskya.
sect. Heterolophus *(Cass.) G. Wagenitz & F.H. Hellwig in Willdenowia*, 30(1): 37 (2000), isonym; *see Mikheev in Bot. Zhurn.*, 81(7): 112 (1996): gen. Heterolophus.
sect. Heterolophus *(Cass.) A.D. Mikheev in Bot. Zhurn.*, 81(7): 112 (1996): gen. Heterolophus.
sect. Hyalinella *(Tzvelev) G. Wagenitz & F.H. Hellwig in Willdenowia*, 30(1): 35 (2000): Centaurea sect. Hyalinella.
subsect. Leucophylli *(Sosn.) A.D. Mikheev in Bot. Zhurn.*, 81(7): 116 (1996): P. sect. Leucophylli.
sect. Odontolophoidei *(Tzvelev) G. Wagenitz & F.H. Hellwig in Willdenowia*, 30(1): 38 (2000): Centaurea sect. Odontolophoideae.
sect. Odontolophus *(Cass.) G. Wagenitz & F.H. Hellwig in Willdenowia*, 30(1): 39 (2000): gen. Odontolophus.
sect. Psephelloidei *(Boiss.) G. Wagenitz & F.H. Hellwig in Willdenowia*, 30(1): 34 (2000): Centaurea [unranked] Psephelloideae.
subsect. Schistosi *(Alieva) A.D. Mikheev in Bot. Zhurn.*, 81(7): 112 (1996): P. sect. Schistosi.
subsect. Somcheticae *E.Tz. Gabrieljan in Fl. Armenii*, 9: 345 (1995).
sect. Sosnovskya *(Takht.) G. Wagenitz & F.H. Hellwig in Willdenowia*, 30(1): 41 (2000): gen. Sosnovskya.

PSEPHELLUS:-
sect. Uralepis *(DC.) G. Wagenitz & F.H. Hellwig in Willdenowia*, 30(1): 40 (2000): Microlonchus sect. Uralepis.
sect. Xanthopsis *(DC.) G. Wagenitz & F.H. Hellwig in Willdenowia*, 30(1): 40 (2000): Amberboa sect. Xanthopsis.
alexeenkoi *A.A. Alieva in Bot. Zhurn.*, 83(1): 116 (1998)—North Caucasus (Dagestan).
amblyolepis *(Ledeb.) G. Wagenitz in Willdenowia*, 30(1): 41 (2000): Centaurea amblyolepis.
appendicigerus *(K. Koch) G. Wagenitz in Willdenowia*, 30(1): 36 (2000): Centaurea appendicigera.
arpensis *(Czerep.) G. Wagenitz in Willdenowia*, 30(1): 41 (2000): Sosnovskya arpensis.
avaricus *(Tzvelev) G. Wagenitz in Willdenowia*, 30(1): 39 (2000): Centaurea avarica.
bellus *(Trautv.) G. Wagenitz in Willdenowia*, 30(1): 36 (2000): Centaurea bella.
bornmuelleri *(Hausskn. ex Bornm.) G. Wagenitz in Willdenowia*, 30(1): 34 (2000): Centaurea bornmuelleri.
brevifimbriatus *(Hub.-Mor.) G. Wagenitz in Willdenowia*, 30(1): 34 (2000): Centaurea brevifimbriata.
congestus *(Wagenitz) G. Wagenitz in Willdenowia*, 30(1): 37 (2000): Centaurea congesta.
cronquistii *(Takht. & Gabrieljan) E.Tz. Gabrieljan in Fl. Armenii*, 9: 352 (1995): Centaurea cronquistii.
debedicus *(Gabrieljan) E.Tz. Gabrieljan in Fl. Armenii*, 9: 348 (1995): Centaurea debedica.
dittrichii *(Gabrieljan) E.Tz. Gabrieljan in Fl. Armenii*, 9: 352 (1995): Centaurea dittrichii.
eugenii *(Sosn.) G. Wagenitz in Willdenowia*, 30(1): 39 (2000): Centaurea eugenii.
gamidii *A.A. Alieva in Bot. Zhurn.*, 83(1): 117 (1998)—North Caucasus (Dagestan).
geghamensis *(Gabrieljan) E.Tz. Gabrieljan in Fl. Armenii*, 9: 345 (1995): Centaurea geghamensis.
gilanicus *(Bornm.) G. Wagenitz in Willdenowia*, 30(1): 34 (2000): Centaurea gilanica.
gracillimus *(Wagenitz) G. Wagenitz in Willdenowia*, 30(1): 34 (2000): Centaurea gracillima.
hadimensis *(Wagenitz, Ertugrul & Dural) G. Wagenitz in Willdenowia*, 30(1): 34 (2000): Centaurea hadimensis.
holtzii *(Wagenitz) G. Wagenitz in Willdenowia*, 30(1): 34 (2000): Centaurea holtzii.
huber-morathii *(Wagenitz) G. Wagenitz in Willdenowia*, 30(1): 34 (2000): Centaurea huber-morathii.
iljinii *(Czerniak.) G. Wagenitz in Willdenowia*, 30(1): 38 (2000): Centaurea iljinii.
karduchorum *(Boiss.) G. Wagenitz in Willdenowia*, 30(1): 37 (2000): Centaurea karduchorum.
kobstanicus *(Tzvelev) G. Wagenitz in Willdenowia*, 30(1): 39 (2000): Centaurea kobstanica.
kopet-daghensis *(Iljin) G. Wagenitz in Willdenowia*, 30(1): 38 (2000): Centaurea kopet-daghensis.
leuzeoides *(Jaub. & Spach) G. Wagenitz in Willdenowia*, 30(1): 34 (2000): Hyalea leuzeoides.
manakyanii *(Gabrieljan) E.Tz. Gabrieljan in Fl. Armenii*, 9: 345 (1995): Centaurea manakyanii.
marschallianus *(Spreng.) A.D. Mikheev in Bot. Zhurn.*, 81(7): 112 (1996), isonym; *see K. Koch in Linnaea*, 24: 438 (1851): Centaurea marschalliana.
meskheticus *(Sosn.) E.Tz. Gabrieljan in Fl. Armenii*, 9: 351 (1995): Centaurea meskhetica.
mucroniferus *(DC.) G. Wagenitz in Willdenowia*, 30(1): 34 (2000): Centaurea mucronifera.
oltensis *G. Wagenitz in Willdenowia*, 30(1): 34 (2000), nom. nov.: Centaurea taochia *Sosn.*
pecho *(Albov) G. Wagenitz in Willdenowia*, 30(1): 36 (2000): Centaurea pecho.
pergamaceus *(DC.) G. Wagenitz in Willdenowia*, 30(1): 34 (2000): Centaurea pergamacea.
persicus *(DC.) G. Wagenitz in Willdenowia*, 30(1): 41 (2000): Microlonchus persicus.
phaeopappoides *(Bordz.) G. Wagenitz in Willdenowia*, 30(1): 39 (2000): Centaurea phaeopappoides.
poluninii *(Wagenitz) G. Wagenitz in Willdenowia*, 30(1): 37 (2000): Centaurea poluninii.
popovae *(Gabrieljan) E.Tz. Gabrieljan in Fl. Armenii*, 9: 352 (1995): Centaurea popovae.

PSEPHELLUS:-
psephelloides *(Freyn & Sint.) G. Wagenitz in Willdenowia*, 30(1): 34 (2000): Centaurea psephelloides.
pseudoandinus *Galushko & A.A. Alieva in Bot. Zhurn.*, 83(1): 119 (1998)—North Caucasus (Checheno-Ingushetiya).
pulcherrimus *(Willd.) G. Wagenitz in Willdenowia*, 30(1): 36 (2000): Centaurea pulcherrima.
pyrroblepharus *(Boiss.) G. Wagenitz in Willdenowia*, 30(1): 34 (2000): Centaurea pyrrhoblephara.
schischkinii *(Tzvelev) G. Wagenitz in Willdenowia*, 30(1): 35 (2000): Centaurea schischkinii.
sibiricus *(L.) G. Wagenitz in Willdenowia*, 30(1): 38 (2000): Centaurea sibirica.
simplicicaulis *(Boiss. & A. Huet) G. Wagenitz in Willdenowia*, 30(1): 36 (2000): Centaurea simplicicaulis.
straminicephalus *(Hub.-Mor.) G. Wagenitz in Willdenowia*, 30(1): 40 (2000) Centaurea straminicephala.
trinervius *(Willd.) G. Wagenitz in Willdenowia*, 30(1): 39 (2000): Centaurea trinervia.
xeranthemoides *(Rech. f.) G. Wagenitz in Willdenowia*, 30(1): 35 (2000) Centaurea xeranthemoides.
zuvandicus Sosn.
subsp. gegharkunikensis *(Gabrieljan) E.Tz. Gabrieljan in Fl. Armenii*, 9: 345 (1995): Centaurea zuvandica subsp. gegharkunikensis.
subsp. jelenevskyi *(Gabrieljan) E.Tz. Gabrieljan in Fl. Armenii*, 9: 344 (1995): Centaurea zuvandica subsp. jelenevskyi.

PSEUDABUTILON (Malvac.).
benense *(Britton) P.A. Fryxell in Contrib. Univ. Mich. Herb.*, 21: 178 (1997): Sida benensis.
cowanii *P.A. Fryxell in Contrib. Univ. Mich. Herb.*, 21: 180 (1997)—Peru.
cymosum *(Triana & Planch.) P.A. Fryxell in Contrib. Univ. Mich. Herb.*, 21: 181 (1997): Abutilon cymosum.
ellipticum *(Schltdl.) P.A. Fryxell in Contrib. Univ. Mich. Herb.*, 21: 182 (1997): Abutilon ellipticum.
glomeratum *P.A. Fryxell in Contrib. Univ. Mich. Herb.*, 21: 184 (1997)—Bolivia.
leucothrix *P.A. Fryxell in Contrib. Univ. Mich. Herb.*, 21: 184 (1997)—Bolivia.
orientale *(Standl. & Steyerm.) P.A. Fryxell in Contrib. Univ. Mich. Herb.*, 21: 186 (1997): Abutilon orientale.
thurberi *(A. Gray) P.A. Fryxell in Contrib. Univ. Mich. Herb.*, 21: 189 (1997): Abutilon thurberi.
umbellatum *(L.) P.A. Fryxell in Contrib. Univ. Mich. Herb.*, 21: 190 (1997): Sida umbellata.
virgatum *(Cav.) P.A. Fryxell in Contrib. Univ. Mich. Herb.*, 21: 191 (1997): Sida virgata.

× **PSEUDADENIA** (Orchidac.).
vitosana *(H. Baumann) O. Gerbaud & W. Schmid, Hybrides Genres Nigritella et/ou Pseudorchis (Cahiers S.F.O., 5)*: 87 (1999): Leucorchis × vitosana.

PSEUDOARABIDOPSIS *I.A. Al-Shehbaz, S.L. O'Kane & R.A. Price in Novon*, 9(3) 304 (1999). *CRUCIFERAE*
toxophylla *(M. Bieb.) I.A. Al-Shehbaz, S.L. O'Kane & R.A. Price in Novon*, 9(3): 304 (1999): Arabis toxophylla.

PSEUDOBAMBUSA (Gramin.).
kurzii *(Munro) D. Ohrnberger, Bamboos of the World Intro. 4:* 19 (1997): Melocanna kurzii.

PSEUDOCALYX (Acanthac.).
ochraceus *D. Champluvier in Syst. Geogr. Pl.*, 69(2): 201 (1999)—Zaire.
pasquierorum *F.J. Breteler in Adansonia*, 20(2): 273 (1998), as 'pasquieri'—Gabon.

PSEUDOCANNABOIDES *B.-E. van Wyk in Taxon*, 48(4): 742 (1999). *UMBELLIFERAE*
andringitrensis *(Humbert) B.-E. van Wyk in Taxon*, 48(4): 742 (1999): Heteromorpha andringitrensis.

PSEUDOCARUM (Umbellif.).
laxiflorum *(Baker) B.-E. van Wyk in Taxon*, 48(4): 744 (1999): Pimpinella laxiflora.
var. alticola *(Humbert) B.-E. van Wyk in Taxon*, 48(4): 744 (1999): Pimpinella laxiflora var. alticola.

PSEUDOCARYOPTERIS *(Briq.) P.D. Cantino in Syst. Bot.*, 23(3): 380 (1998 publ. 1999): Caryopteris sect. Pseudocaryopteris. (Verbenac.).
bicolor *(Hardw.) P.D. Cantino in Syst. Bot.*, 23(3): 381 (1998 publ. 1999): Volkameria bicolor.
foetida *(D. Don) P.D. Cantino in Syst. Bot.*, 23(3): 381 (1998 publ. 1999): Clerodendrum foetidum.
paniculata *(C.B. Clarke) P.D. Cantino in Syst. Bot.*, 23(3): 381 (1998 publ. 1999): Caryopteris paniculata.

PSEUDOCERASTIUM *C.Y. Wu, X.H. Guo & X.P. Zhang in Acta Bot. Yunnanica*, 20(4): 395 (1998). *CARYOPHYLLACEAE*
stellarioides *X.H. Guo & X.P. Zhang in Acta Bot. Yunnanica*, 20(4): 396 (1998)—China Southeast (Anhui).

PSEUDOCORCHORUS (Tiliac.).
rostratus *(Danguy) D.J. Mabberley in Adansonia*, 21(1): 22 (1999): Corchorus rostratus.

PSEUDODACRYODES *R. Pierlot in Bull. Jard. Bot. Nation. Belg.*, 66(3–4): 182 (1997). *BURSERACEAE*
leonardiana *R. Pierlot in Bull. Jard. Bot. Nation. Belg.*, 66(3–4): 182 (1997)—Zaire.

PSEUDODRIMYS *A.B. Doweld in Novosti Sist. Vyssh. Rast.*, 32: 38 (2000). *WINTERACEAE*
stipitata *(Vickery) A.B. Doweld in Novosti Sist. Vyssh. Rast.*, 32: 39 (2000): Drimys stipitata.

PSEUDOEURYSTYLES (Orchidac.).
splendissima *(Szlach.) D.L. Szlachetko & Tamayo in Acta Bot. Fenn.*, 169: 380 (2000): Eurystyles splendissima.

PSEUDOGNAPHALIUM (Compos.).
dichotomum *V.M. Badillo in Biollania*, ed. esp. 6 (J.J. Wurdack Festschr.): 238 (1997), nom. nov.: Gnaphalium dichotomum *Willd. ex Spreng.*
moritzianum *(Klatt) V.M. Badillo in Biollania*, ed. esp. 6 (J.J. Wurdack Festschr.): 237 (1997): Achyrocline moritzianum.

PSEUDOGYNOXYS (Compos.).
alajuelana *B.L. Turner in Phytologia*, 80(4): 254 (1996)—Costa Rica.
durandii *(Klatt) B.L. Turner in Phytologia*, 80(4): 255 (1996): Senecio durandii.
lobata *J.F. Pruski in Syst. Bot.*, 21(1): 104 (1996)—Brazil North (Rondônia), Bolivia.
westonii *(H. Rob. & Cuatrec.) B.L. Turner in Phytologia*, 80(4): 255 (1996): Talamancalia westonii.

PSEUDOLAELIA (Orchidac.).
canaanensis *(Ruschi) F. Barros in Acta Bot. Brasil.*, 8(1): 15 (1994): Renata canaanensis.

PSEUDOLASIACIS (Gramin.).
leptolomoides (A. Camus) A. Camus
var. coursii *(A. Camus) J. Bosser & D. Florens in Adansonia*, 21(2): 233 (1999): Panicum coursii.
neoperrieri (A. Camus) A. Camus
var. leroyana *J. Bosser & D. Florens in Adansonia*, 21(2): 235 (1999)—Madagascar.

PSEUDOLIPARIS (Orchidac.).
balabacensis *(Ames) H.B. Margońska & D.L. Szlachetko in Ann. Bot. Fenn.*, 37(4): 281 (2000): Malaxis balabacensis.
brachycaulos *(Schltr.) D.L. Szlachetko & H.B. Margońska in Adansonia*, 21(2): 276 (1999): Microstylis brachycaulos.
breviscapa *(Schltr.) D.L. Szlachetko & H.B. Margońska in Adansonia*, 21(2): 276 (1999): Microstylis breviscapa.
cruciatis *H.B. Margońska & D.L. Szlachetko in Ann. Bot. Fenn.*, 37(4): 281 (2000)—New Guinea (Irian Jaya).

PSEUDOLIPARIS:-
curvatula *(Schltr.) D.L. Szlachetko & H.B. Margońska in Adansonia*, 21(2): 276 (1999): Microstylis curvatula.
curviauriculata *D.L. Szlachetko & H.B. Margońska in Adansonia*, 21(2): 279 (1999)—New Guinea (Irian Jaya).
diploceras *(Schltr.) D.L. Szlachetko & H.B. Margońska in Adansonia*, 21(2): 276 (1999): Microstylis diploceras.
graminifolia *(Schltr.) D.L. Szlachetko & H.B. Margońska in Adansonia*, 21(2): 277 (1999): Microstylis graminifolia.
gregorii *H.B. Margońska & D.L. Szlachetko in Ann. Bot. Fenn.*, 37(4): 279 (2000)—New Guinea (Irian Jaya).
incurva *(J.J. Sm.) D.L. Szlachetko & H.B. Margońska in Adansonia*, 21(2): 277 (1999): Microstylis incurva.
laevis *(Schltr.) D.L. Szlachetko & H.B. Margońska in Adansonia*, 21(2): 277 (1999): Microstylis laevis.
latipetala *(J.J. Sm.) H.B. Margońska & D.L. Szlachetko in Ann. Bot. Fenn.*, 37(4): 283 (2000): Microstylis latipetala.
maboroensis *(Schltr.) D.L. Szlachetko & H.B. Margońska in Adansonia*, 21(2): 277 (1999): Microstylis maboroensis.
macrotis *(Kraenzl.) D.L. Szlachetko & H.B. Margońska in Adansonia*, 21(2): 277 (1999): Liparis macrotis
microhybos *(Schltr.) D.L. Szlachetko & H.B. Margońska in Adansonia*, 21(2): 277 (1999): Microstylis microhybos.
raciborskii *D.L. Szlachetko & H.B. Margońska in Adansonia*, 21(2): 279 (1999)—New Guinea (Irian Jaya, Papua New Guinea).
rhinoceros *(J.J. Sm.) D.L. Szlachetko & H.B. Margońska in Adansonia*, 21(2): 277 (1999): Microstylis rhinoceros.
schumanniana *(Schltr.) D.L. Szlachetko & H.B. Margońska in Adansonia*, 21(2): 277 (1999): Microstylis schumanniana.
seleniglossa *(Schltr.) D.L. Szlachetko & H.B. Margońska in Adansonia*, 21(2): 277 (1999): Microstylis seleniglossa.
stenophylla *(Schltr.) D.L. Szlachetko & H.B. Margońska in Adansonia*, 21(2): 277 (1999): Microstylis stenophylla.
var. crispatula *(Schltr.) D.L. Szlachetko & H.B. Margońska in Adansonia*, 21(2): 277 (1999): Microstylis stenophylla var. crispatula.
torricellensis *(Schltr.) D.L. Szlachetko & H.B. Margońska in Adansonia*, 21(2): 277 (1999): Microstylis torricellensis.
tubulosa *(J.J. Sm.) D.L. Szlachetko & H.B. Margońska in Adansonia*, 21(2): 277 (1999): Microstylis tubulosa.
umbonata *(Schltr.) D.L. Szlachetko & H.B. Margońska in Adansonia*, 21(2): 277 (1999): Microstylis umbonata.
undulata *(Schltr.) D.L. Szlachetko & H.B. Margońska in Adansonia*, 21(2): 277 (1999): Microstylis undulata.
van-royenii *D.L. Szlachetko & H.B. Margońska in Adansonia*, 21(2): 282 (1999)—New Guinea (Irian Jaya).

PSEUDOLMEDIA (Morac.).
gentryi *C.C. Berg in Fl. Ecuador*, 60(27B): 93 (1998)—Peru, Ecuador.
glabrata *(Liebm.) C.C. Berg in Novon*, 6(3): 241 (1996): Trophis glabrata.
manabiensis *C.C. Berg in Fl. Ecuador*, 60(27B): 98 (1998)—Ecuador.

PSEUDOLYSIMACHION (Scrophulariac.).
sect. Longifolia *(Holub) B. Trávníček in Preslia*, 70(3): 195 (1998): P. ser. Longifolia.
bashkiriense *(Klokov ex Tzvelev) N.N. Tzvelev in Bot. Zhurn.*, 84(7): 116 (1999), as 'bashkiriensis': Veronica spicata subsp. bashkiriensis.
× blockianum *B. Trávníček in Preslia*, 70(3): 201 (1998)—Ukraina.(P. incanum subsp. incanum × spicatum subsp. spicatum).
dahuricum *(Steven) Y.N. Lee, Fl. Korea:* 1160 (1996), isonym; *see Holub in Folia Geobot. Phytotax.*, 2: 424 (1967): Veronica dahurica.

PSEUDOLYSIMACHION:-
incanum (L.) Holub
subsp. pallens (Host) B. Trávníček in Fl. Slovenska, 5(2): 294 (1997): Veronica pallens.
kiusianum (Furumi) T. Yamaz.
var. diamanticum (Nakai) Y.N. Lee, Fl. Korea: 1160 (1996): Veronica diamantica.
klokovii (Tzvelev) J. Holub in Preslia, 70(2): 109 (1998): Veronica spicata subsp. klokovii.
linariifolium (Pall. ex Link) Holub
f. album (T.B. Lee) Y.N. Lee, Fl. Korea: 1161 (1996), as 'linearifolium', without basionym ref.: Veronica linariifolia f. alba.
var. dilatatum (Nakai & Kitag.) Y.N. Lee, Fl. Korea: 1160 (1996), as 'linearifolium', with incorrect basionym ref.: Basionym not stated.
subsp. dilatatum (Nakai & Kitag.) D.Y. Hong in Novon, 6(1): 23 (1996), as 'linarifolium': Veronica angustifolia var. dilatata.
var. villosulus (Nakai) Y.N. Lee, Fl. Korea: 1161 (1996), as 'linearifolium': Veronica villosula.
longifolium
subsp. pseudolongifolium (Printz) J. Holub in Preslia, 70(2): 109 (1998): Veronica pseudolongifolia.
maeoticum (Klokov) J. Holub in Preslia, 70(2): 109 (1998): Veronica maeotica.
miyabei (Nakai & Honda) J. Holub in Preslia, 70(2): 109 (1998): Veronica miyabei.
nakaianum (Ohwi) Y.N. Lee, Fl. Korea: 1161 (1996), isonym; see T. Yamaz. in J. Jap. Bot., 43: 411 (1968): Veronica nakaiana.
× neglectum (Vahl) B. Trávníček in Preslia, 70(3): 200 (1998): Veronica neglecta.
orchideum (Crantz) Wraber
var. bulgaricum (Peev) D. Peev & M.A. Fischer in Fl. Republ. Bŭlgariya, 10: 397 (1995): Veronica spicata var. bulgarica.
var. eglandulosum D. Peev & M.A. Fischer in Fl. Republ. Bŭlgariya, 10: 397 (1995)—Bulgaria.
var. glandulopilosum D. Peev & M.A. Fischer in Fl. Republ. Bŭlgariya, 10: 397 (1995)—Bulgaria.
petschoricum (Tzvelev) N.N. Tzvelev in Bot. Zhurn., 84(7): 116 (1999): Veronica spicata subsp. petschorica.
pyrethrinum (Nakai) Y.N. Lee, Fl. Korea: 1161 (1996), isonym; see T. Yamaz. in J. Jap. Bot., 43: 410 (1968): Veronica pyrethrina.
rotundum (Nakai) T. Yamaz.
subsp. coreanum (Nakai) D.Y. Hong in Novon, 6(1): 23 (1996): Veronica coreana.
subsp. subintegrum (Nakai) D.Y. Hong in Novon, 6(1): 23 (1996): Veronica spuria var. subintegra.
septentrionale (Boriss.) N.N. Tzvelev in Bot. Zhurn., 84(7): 116 (1999): Veronica septentrionalis.
spicatum (L.) Opiz
subsp. fischeri B. Trávníček in Preslia, 70(3): 219 (1998)—Czechoslovakia, Hungary, Romania, Bulgaria, Yugoslavia, Italy.
subsp. lanisepalum B. Trávníček in Preslia, 70(3): 218 (1998)—Yugoslavia.
subsp. porphyrianum (Pavlov) B. Trávníček in Preslia, 70(3): 210 (1998): Veronica porphyriana.
var. pseudorchideum (Pacz.) N.N. Tzvelev in Bot. Zhurn., 84(7): 117 (1999): Veronica spicata var. pseudorchidea.
subsp. viscosulum (Klokov) B. Trávníček in Preslia, 70(3): 210 (1998): Veronica viscosula.
taigischense (Stepanov) J. Holub in Preslia, 70(2): 109 (1998): Veronica taigischensis.
viscosulum (Klokov) N.N. Tzvelev in Bot. Zhurn., 84(7): 116 (1999), isonym; see Soják in Čas. Nár. Muz. (Prague), 152(1): 21 (1983): Veronica viscosula.

PSEUDOMALMEA L.W. Chatrou, Changing Genera: 180 (1998). ANNONACEAE
boyacana (J.F. Macbr.) L.W. Chatrou, Changing Genera: 182 (1998): Guatteria boyacana.
diclina (R.E. Fr.) L.W. Chatrou, Changing Genera: 183 (1998): Malmea diclina.

PSEUDOMELASMA Eb. Fischer in Bull. Mus. Nation. Hist. Nat., B, Adansonia, Ser. 4, 18(1–2): 61 (1996). SCROPHULARIACEAE
pedicularioides (Baker) Eb. Fischer in Bull. Mus. Nation. Hist. Nat., B, Adansonia, Ser. 4, 18(1–2): 61 (1996): Alectra pedicularioides.

PSEUDOMERTENSIA (Boraginac.).
flavescens R.A. Rafiq in Novon, 6(3): 296 (1996)—Pakistan.

PSEUDOMISOPATES J. Güemes in An. Jard. Bot. Madrid, 55(2): 493 (1997). SCROPHULARIACEAE
rivas-martinezii (Sánchez Mata) J. Güemes in An. Jard. Bot. Madrid, 55(2): 493 (1997): Misopates rivas-martinezii.

PSEUDOPANAX (Araliac.).
laetevirens (Gay) H. Baillon, Adansonia, 12: 148 (1878): Aralia laetevirens.†
valdiviensis (Gay) H. Baillon, Adansonia, 12: 148 (1878): Aralia valdiviensis.†

PSEUDOPENTAMERIS (Gramin.).
caespitosa N.P. Barker in Bothalia, 25(2): 147 (1995)—Cape Province.

PSEUDOPERISTYLUS (P.F. Hunt) D.L. Szlachetko & T.S. Olszewski in Fl. Cameroun, 34: 210 (1998): Habenaria sect. Pseudoperistylus. (Orchidac.).
aethiopicus D.L. Szlachetko & T.S. Olszewski in Ann. Bot. Fenn., 37(4): 310 (2000)—Ethiopia.
attenuatus (Hook. f.) D.L. Szlachetko & T.S. Olszewski in Fl. Cameroun, 34: 216 (1998): Habenaria attenuata.
bequaertii (Summerh.) D.L. Szlachetko & T.S. Olszewski in Fl. Cameroun, 34: 214 (1998): Habenaria bequaertii.
bracteosus (A. Rich.) D.L. Szlachetko & T.S. Olszewski in Fl. Cameroun, 34: 220 (1998): Habenaria bracteosa.
clarencensis (Rolfe) D.L. Szlachetko & T.S. Olszewski in Fl. Cameroun, 34: 218 (1998): Habenaria clarencensis.
ituriensis D.L. Szlachetko & T.S. Olszewski in Fl. Cameroun, 34: 216 (1998)—Zaire.
lefebureanus (A. Rich.) D.L. Szlachetko & T.S. Olszewski in Ann. Bot. Fenn., 37(4): 312 (2000): Peristylus lefebureanus.
microceras (Hook. f.) D.L. Szlachetko & T.S. Olszewski in Fl. Cameroun, 34: 212 (1998): Habenaria microceras.
montolivaea (Kraenzl. ex Engl.) D.L. Szlachetko & T.S. Olszewski in Ann. Bot. Fenn., 37(4): 312 (2000): Habenaria montolivaea.
petitianus (A. Rich.) D.L. Szlachetko & T.S. Olszewski in Fl. Cameroun, 34: 213 (1998): Peristylus petitianus.

PSEUDOPROSPERO F. Speta in Phyton (Austria), 38(1): 116 (1998). HYACINTHACEAE
firmifolium (Baker) F. Speta in Phyton (Austria), 38(1): 116 (1998): Scilla firmifolia.

PSEUDORAPHIS Griff. ex R. Pilger in Notizbl. Bot. Gart. Berlin-Dahlem, 10(93): 209, 210 (1928). GRAMINEAE†
brunoniana (Wall. & Griff.) R. Pilger in Notizbl. Bot. Gart. Berlin-Dahlem, 10(93): 209, 210 (1928): Panicum brunonianum.†
simaoensis Y.Y. Qian in Guihaia, 15(4): 305 (1995)—China South-central (Yunnan).

PSEUDORCHIS (Orchidac.).
albida (L.) Á. Löve & D. Löve
subsp. tricuspis (Beck) E. Klein in Phyton (Austria), 40(1): 143 (2000): Gymnadenia albida β tricuspis.

× **PSEUDORHIZA** (Orchidac.).
bulgarica (D. Lindig) O. Gerbaud & W. Schmid, Hybrides Genres Nigritella et/ou Pseudorchis (Cahiers S.F.O., 5): 110 (1999): × Dactyleucorchis bulgarica.
foelscheorum M. Gerbaud & O. Gerbaud in Nat. Belg., 81(3): 91 (2000)—France.(Dactylorhiza lapponica × Pseudorchis albida).

PSEUDOSAMANEA (Legumin.).
cubana (Britton & Rose) R.C. Barneby & J.W. Grimes, Silk Tree, Guanacaste, Monkey's Earring (Mem. New York Bot. Gard., 74(1)): 116 (1996): Albizia cubana.

PSEUDOSASA (Gramin.).
amplexicaulis W.T. Lin & Z.J. Feng in J. S. China Agric. Univ., 14(1): 51 (1993)—China Southeast (Guangdong).
flexuosa T.P. Yi & X.M. Zhou in J. Bamboo Res., 15(3): 1 (1996)—China Southeast (Jiangxi).
hindsii (Munro) C.D. Chu & C.S. Chao in Fl. Reipubl. Popul. Sin., 9(1): 653 (1996): Arundinaria hindsii.
magilaminaris B.M. Yang in J. Hunan Sci. Technol. Univ., 1(1): 111 (1985)—China Southeast (Hunan).
multifloscula (W.T. Lin) D. Ohrnberger, Bamboos of the World Intro. 2: 10 (1996), isonym; see W.T. Lin in Guihaia, 10(1): 18 (1990): Arundinaria multifloscula.
nigrinodis G.A. Fu in J. Bamboo Res., 15(1): 4 (1996), as 'nigro-nodis'—Hainan.
victorialis (Keng f.) T.P. Yi in J. Bamboo Res., 15(3): 6 (1996): Indocalamus victorialis.

PSEUDOSCABIOSA (Dipsacac.).
africana (Font Quer) A. Romo, S. Cirujano, J.B. Peris & G. Stübing in Feddes Repert., 108(1–2): 31 (1997): Scabiosa saxatilis var. africana.
grosii (Font Quer) Devesa
var. africana (Font Quer) A.M. Romo in Lagascalia, 18(2): 337 (1996): Scabiosa saxatilis var. africana.

PSEUDOSEDUM (Crassulac.).
ferganense Boriss.
subsp. parvum R.V. Kamelin & V.V. Byalt in Komarovia, 1: 39 (1999)—Kirgizstan.

PSEUDOSTELLARIA (Caryophyllac.).
angustifolia Y.N. Lee, Fl. Korea: 1156 (1996), contrary to Arts. 37.4, 37.5 ICBN (1994)—South Korea.
dalaolingensis Z.E. Zhao & J.Q. Wu in J. Wuhan Bot. Res., 18(4): 283 (2000)—China South-central (Hubei).
ebracteata (Kom.) N.S. Pavlova in Sosud. Rast. Sovet. Dal'nego Vostoka, 8: 63 (1996): Stellaria ebracteata.
heterantha (Maxim.) Pax
var. tibetica (Ohwi) Yu.P. Kozhevnikov in Rast. Tsentral. Azii, 11: 39 (1994 publ. 1995): P. tibetica.
multiflora Y.N. Lee, Fl. Korea: 1156 (1996), contrary to Arts. 37.4, 37.5 ICBN (1994)—South Korea.
terminalis W.Z. Di & Y. Ren in J. Northwest Univ. (Nat. Sci.), 17(2): 42 (1987)—Inner Mongolia (Nei Mongol).

PSEUDOTAXUS (Taxac.).
liana J. Silba in Phytologia, 81(4): 327 (1996 publ. 1997), as 'liiana'—China Southeast (Guangxi, Hunan, Jiangxi).

PSEUDOTRACHYDIUM (Kljuykov, Pimenov & V.N. Tikhom.) M.G. Pimenov & E.V. Kljuykov in Feddes Repert., 111(7–8): 526 (2000): Aulacospermum sect. Pseudotrachydium. (Umbellif.).
depressum (Boiss.) M.G. Pimenov & E.V. Kljuykov in Feddes Repert., 111(7–8): 527 (2000): Rumia depressa.
dichotomum (Korovin) M.G. Pimenov & E.V. Kljuykov in Feddes Repert., 111(7–8): 527 (2000): Trachydium dichotomum.
kotschyi (Boiss.) M.G. Pimenov & E.V. Kljuykov in Feddes Repert., 111(7–8): 528 (2000): Rumia kotschyi.
pauciradiatum (Boiss. & Hohen.) M.G. Pimenov & E.V. Kljuykov in Feddes Repert., 111(7–8): 527 (2000): Prangos pauciradiata.
vesiculosoalatum (Rech. f.) M.G. Pimenov & E.V. Kljuykov in Feddes Repert., 111(7–8): 527 (2000): Trachydium vesiculosoalatum.

PSEUDOTRIMEZIA (Iridac.).
diamantinensis P. Ravenna in Onira, Bot. Leafl., 5(2): 6 (2000)—Brazil Southeast (Minas Gerais).
subsp. scabra P. Ravenna in Onira, Bot. Leafl., 5(2): 7 (2000)—Brazil Southeast (Minas Gerais).

PSEUDOTRIMEZIA:-
fulva *P. Ravenna in Onira, Bot. Leafl.*, 5(2): 5 (2000)—Brazil Southeast (Minas Gerais).
laevis *P. Ravenna in Onira, Bot. Leafl.*, 5(2): 7 (2000)—Brazil Southeast (Minas Gerais).
pumila *P. Ravenna in Onira, Bot. Leafl.*, 5(2): 8 (2000)—Brazil Southeast (Minas Gerais).

PSEUDOTSUGA (Pinac.).
sinensis Dode
var. wilsoniana *(Hayata) L.K. Fu & Nan Li in Novon*, 7(3): 263 (1997): P. wilsoniana.

PSEUDOXYTENANTHERA (Gramin.).
bourdillonii *(Gamble) D. Ohrnberger, Bamboos of the World Intro. 4:* 20 (1997), isonym; *see H.B. Naithani in J. Bombay Nat. Hist. Soc.*, 87(3): 440 (1990 publ. 1991): Oxytenanthera bourdillonii. *GRAMINEAE*
ritchiei *(Munro) D. Ohrnberger, Bamboos of the World Intro. 4:* 20 (1997), isonym; *see H.B. Naithani in J. Bombay Nat. Hist. Soc.*, 87(3): 440 (1990 publ. 1991): Bambusa ritchiei.

PSIDIUM (Myrtac.).
bahorucanum *Alain & Ricardo García in Moscosoa*, 9: 22 (1997)—Dominican Republic.

PSIGURIA (Cucurbitac.).
aurantiaca *(S.F. Blake) C.H. Nelson Sutherland & F.J. Fernández Casas in Fontqueria*, 52: 4 (1998): Anguria aurantiaca.
diversifolia *(Cogn.) C.H. Nelson Sutherland & F.J. Fernández Casas in Fontqueria*, 52: 4 (1998): Anguria diversifolia.

PSILOTHONNA *(E. Mey. ex DC.) E.P. Phillips in J. S. Afr. Bot.*, 16: 16 (1950): Gamolepis sect. Psilothonna (Compos.).†

PSILOTRICHUM (Amaranthac.).
ferrugineum *(Roxb.) J.O. Voigt, Hort. Suburb. Calcutt.:* 318 (1845): Achyranthes ferruginea.
yunnanense *D.D. Tao in Acta Phytotax. Sin.*, 35(1): 86 (1997)—China South-central (Yunnan).

PSITTACANTHUS (Loranthac.).
breedlovei *J. Kuijt in Novon*, 6(1): 51 (1996)—Mexico Southeast (Chiapas).
excrenulatus *C.T. Rizzini in Revista Fac. Agron. (Maracay)*, 8(3): 93 (1975)—Brazil Northeast (Bahia).
intermedius *C.T. Rizzini in Revista Fac. Agron. (Maracay)*, 8(3): 94 (1975)—Venezuela.
irwinii *C.T. Rizzini in Revista Fac. Agron. (Maracay)*, 8(3): 94 (1975)—Brazil North (Amazonas).
julianus *C.T. Rizzini in Revista Fac. Agron. (Maracay)*, 8(3): 95 (1975)—Venezuela.
stenanthus *C.T. Rizzini in Revista Fac. Agron. (Maracay)*, 8(3): 95 (1975)—Brazil North (Roraima).

PSORALEA (Legumin.).
bituminosa L.
subsp. pontica *(A.P. Khokhr.) A.S. Zernov, Rast. Severo-Zapad. Zakavkaz'ya (Pl. Transcaucas. Bor.-Occ.):* 70 (2000): P. pontica.
peratica *C.H. Stirton in S. Afr. J. Bot.*, 64(4): 244 (1998)—Cape Province.
pontica *A.P. Khokhrjakov in Byull. Glavn. Bot. Sada (Moscow)*, 175: 52 (1997)—Transcaucasus (Gruziya, Abkhasiya), North Caucasus (Krasnodar), Krym.

PSYCHOPSIS (Orchidac.).
papilio
var. albiflora *(Rchb. f. ex Lindl.) E. Lückel & Jenny in Orchidee*, 50(1): 47 (1999): Oncidium papilio var. albiflorum.
f. latourae *(Broadway) E. Lückel & Jenny in Orchidee*, 50(1): 47 (1999): Oncidium papilio var. latourae.

PSYCHOTRIA (Rubiac.).
albopedunculata *I.M. Turner in Asian J. Trop. Biol.*, 1(2): 27 (1995), nom. nov.: Cephaelis melanocarpa *Ridl.*

PSYCHOTRIA:-
alto-macahensis *M. Gomes in Reserva Ecol. Macaé de Cima: Aspectos Flor. Espéc. Vasc.*, 2: 387 (1996), nom. nov.: Cephaelis glaziovii *Standl.*
ararum *C.M. Taylor in Novon*, 9(2): 260 (1999), nom. nov.: Palicourea glaziovii *Standl.*
cauligera *C.M. Taylor in Sida*, 17(4): 709 (1997)—Ecuador.
deneversii *C.M. Taylor in Novon*, 6(3): 300 (1996)—Panama.
diguana *(Standl. ex Steyerm.) C.M. Taylor in Novon*, 6(2): 212 (1996): Cephaelis diguana.
dives *(Standl.) C.M. Taylor in Novon*, 6(2): 213 (1996): Palicourea dives.
fusiformis *C.M. Taylor in Sida*, 17(4): 712 (1997)—Ecuador.
hamiltoniana *C.M. Taylor in Novon*, 9(3): 425 (1999)—Nicaragua.
howcroftii *W. Takeuchi in Sida*, 18(3): 767 (1999)—New Guinea (Papua New Guinea).
hypochlorina *C.M. Taylor in Sida*, 17(4): 713 (1997)—Peru.
leitana *C.M. Taylor in Novon*, 9(2): 260 (1999)—Brazil Southeast (São Paulo).
malaccensis *I.M. Turner in Asian J. Trop. Biol.*, 1(2): 27 (1995), nom. nov.: Cephaelis griffithii *Hook. f.*
matambuaii *W. Takeuchi in Sida*, 18(4): 950 (1999), as 'matambuai'—New Guinea (Papua New Guinea).
mayana *W. Takeuchi in Sida*, 19(1): 17 (2000)—New Guinea (Papua New Guinea).
molinarum *D.H. Lorence in Novon*, 7(1): 51 (1997), nom. nov.: Evea chiapensis *Standl.*
monsalveae *C.M. Taylor in Novon*, 9(3): 439 (1999)—Colombia.
naviculiformis *I.M. Turner in Asian J. Trop. Biol.*, 1(2): 27 (1995), nom. nov.: Cephaelis ridleyi *King.*
ophirensis *I.M. Turner in Asian J. Trop. Biol.*, 1(2): 26 (1995), nom. nov.: Cephaelis angustifolia *Ridl.*
osiana *W. Takeuchi & J.J. Pipoly in Sida*, 18(1): 162 (1998)—Bismarck Archipelago.
ovatistipula *C.M. Taylor in Novon*, 7(2): 197 (1997)—Colombia.
paeonia *C.M. Taylor in Sida*, 17(4): 716 (1997)—Peru, Colombia, Ecuador.
porcata *I.M. Turner in Asian J. Trop. Biol.*, 1(2): 27 (1995), nom. nov.: Cephaelis elongata *Ridl.*
purpureocapitata *Dwyer ex C.M. Taylor in Novon*, 6(3): 301 (1996)—Panama.
rhombibractea *C.M. Taylor & M.T. Campos in Novon*, 9(1): 118 (1999)—Brazil North (Amazonas).
sanchezii *C.M. Taylor in Novon*, 7(2): 198 (1997)—Colombia.
setulifera *C.M. Taylor in Novon*, 9(2): 261 (1999), nom. nov.: Palicourea hispidula *Standl.*
singapurensis *(Ridl.) I.M. Turner in Asian J. Trop. Biol.*, 1(2): 27 (1995): Cephaelis singapurensis.
sohmeriana *I.M. Turner in Asian J. Trop. Biol.*, 1(2): 26 (1995), nom. nov., as 'sohmerana': P. minutiflora *Ridl.*
subcoriacea *(Ridl.) I.M. Turner in Asian J. Trop. Biol.*, 1(2): 27 (1995): Cephaelis subcoriacea.
subspathulata *(Müll. Arg.) C.M. Taylor in Novon*, 9(2): 261 (1999): Mapouria subspathulata.
tahanica *I.M. Turner in Asian J. Trop. Biol.*, 1(2): 26 (1995), nom. nov.: Cephaelis albiflora *Ridl.*
triceps *(Ridl.) I.M. Turner in Asian J. Trop. Biol.*, 1(2): 27 (1995): Cephaelis triceps.
vesciculifera *C.M. Taylor in Novon*, 7(2): 199 (1997)—Colombia.
volii *R.O. Gardner in New Zealand J. Bot.*, 35(4): 489 (1997)—Fiji.
whistleri *F.R. Fosberg in Micronesica*, 23(1): 3 (1990)—Cook Is.

PSYDRAX (Rubiac.).
dicoccos Gaertn.
var. lanceolatus *(Arn.) C.E. Ridsdale in Blumea*, 41(2): 461 (1996), as 'lanceolatum': Canthium lanceolatum.
grandifolius *(Thwaites) C.E. Ridsdale in Blumea*, 41(2): 462 (1996), as 'grandifolia': Canthium didymum var. grandifolium.
kingii *(Hook. f.) D.M. Bridson & L.S. Springate in Edinburgh J. Bot.*, 53(1): 107 (1996): Canthium didymum var. kingii.

PSYDRAX:-
montanus *(Thwaites) C.E. Ridsdale in Blumea*, 41(2): 462 (1996), as 'montanum': Canthium montanum.
pergracilis *(Bourd.) C.E. Ridsdale in Blumea*, 41(2): 462 (1996), as 'pergracile': Canthium pergracile.

PSYLLIUM (Plantaginac.).
libycum *(Bég. & A. Vacc.) J. Holub in Preslia*, 70(2): 109 (1998): Plantago libyca.

PSYLLOCARPUS (Rubiac.).
intermedius *E.L. Cabral & N.M. Bacigalupo in Acta Bot. Brasil.*, 11(1): 52 (1997)—Brazil Northeast (Bahia).

PTERALYXIA (Apocynac.).
laurifolia *(Lodd.) A.J.M. Leeuwenberg in Wageningen Agric. Univ. Pap.*, 98(3): 41 (1999): Cerbera laurifolia.

PTERANDRA (Malpighiac.).
andersonii *C. Anderson in Contrib. Univ. Mich. Herb.*, 21: 12 (1997)—Brazil Northeast (Maranhão).
colombiana *C. Anderson in Contrib. Univ. Mich. Herb.*, 21: 16 (1997)—Colombia.
hirsuta *C. Anderson in Contrib. Univ. Mich. Herb.*, 21: 22 (1997)—Bolivia, Brazil North (Amazonas, Rondônia).
mcphersonii *C. Anderson in Contrib. Univ. Mich. Herb.*, 21: 18 (1997)—Panama.

PTERNOPETALUM (Umbellif.).
trachycarpum *C.Y. Wu ex R.H. Shan & Z.H. Pan in Fl. Yunnanica*, 7: 535 (1997), contrary to Art. 37.5 ICBN (1994)—China South-central (Yunnan).

PTEROCACTUS (Cactac.).
tuberosus (Pfeiff.) Britton & Rose
f. lelongii *(Ruiz Leal ex R. Kiesling) R. Kiesling in Zuloaga & Morrone (eds.), Cat. Pl. Vasc. Rep. Arg. 2 (Monogr. Syst. Bot.* Missouri Bot. Gard. 74): 1246 (1999): Pterocactus kuntzei f. lelongii.
michelianus *N. Zamora in Novon*, 10(2): 177 (2000)—Costa Rica, Mexico Southwest (Guerrero), Guatemala, Nicaragua, Honduras.
monophyllus *B.B. Klitgaard, L.P. de Queiroz & G.P. Lewis in Kew Bull.*, 55(4): 989 (2000)—Brazil Northeast (Bahia).
rotundifolius (Sond.) Druce
subsp. martinii *(Dunkley) J.M. Lock in Kew Bull.*, 54(1): 208 (1999): P. martinii.

PTEROCEPHALODES *V. Mayer & F. Ehrendorfer in Bot. J. Linn. Soc.*, 132(1): 69, 67 (2000). *DIPSACACEAE*
bretschneideri *(Batalin) V. Mayer & F. Ehrendorfer in Bot. J. Linn. Soc.*, 132(1): 69 (2000): Scabiosa bretschneideri.
hookeri *(C.B. Clarke) V. Mayer & F. Ehrendorfer in Bot. J. Linn. Soc.*, 132(1): 69 (2000): Scabiosa hookeri.
siamensis *(Craib) V. Mayer & F. Ehrendorfer in Bot. J. Linn. Soc.*, 132(1): 69 (2000): Scabiosa siamensis.

PTEROCEPHALUS (Dipsacac.).
kunkelianus *J. Peris, A. Romo & G. Stübing in Feddes Repert.*, 110(1–2): 13 (1999)—Morocco.

PTEROCYCLUS (Umbellif.).
forrestii *(Diels) M.G. Pimenov & E.V. Kljuykov in Feddes Repert.*, 110(7–8): 485 (1999): Angelica forrestii.
rotundatus *(DC.) M.G. Pimenov & E.V. Kljuykov in Feddes Repert.*, 111(7–8): 522 (2000): Hymenolaena rotundata.

PTEROCYPSELA (Compos.).
× mansuensis *(Hayata) C.I Peng in Fl. Taiwan, ed. 2*, 4: 1045 (1998): Lactuca × mansuensis.

PTEROGLOSSA (Orchidac.).
magnifica *D.L. Szlachetko in Orchidee*, Suppl. 3: 49 (1996)—Paraguay.

PTEROMONNINA (Polygalac.).
boliviana *(Chodat) B. Eriksen in Biollania,* ed. esp. 6 (J.J. Wurdack Festschr.): 315 (1997): Monnina boliviana.
cyanea *(Chodat) B. Eriksen in Biollania,* ed. esp. 6 (J.J. Wurdack Festschr.): 315 (1997): Monnina cyanea.
herzogii *(Chodat) B. Eriksen in Biollania,* ed. esp. 6 (J.J. Wurdack Festschr.): 315 (1997): Monnina herzogii.

PTEROSPARTUM (Legumin.).
tridentatum (L.) Willk.
subsp. cantabricum *(Spach) S. Talavera & P.E. Gibbs in Lagascalia,* 18(2): 266 (1996): Genista cantabrica.
subsp. lasianthum *(Spach) S. Talavera & P.E. Gibbs in Lagascalia,* 18(2): 266 (1996): Genista lasiantha.
subsp. riphaeum *(Pau & Font Quer) S. Talavera & P.E. Gibbs in Lagascalia,* 18(2): 266 (1996): Genistella riphaea.

PTEROSPERMUM (Sterculiac.).
littorale Craib
var. venustrum *(Craib) C. Phengklai in Thai Forest Bull., Bot.,* 23: 80 (1995): P. venustrum.

PTEROSTEMON (Pterostemonac.).
bravoanus *J. Jiménez Ramirez & M. Martínez Gordillo in Acta Bot. Mex.,* 41: 21 (1997)—Mexico Southwest (Guerrero).

PTEROSTYLIS (Orchidac.).
atriola *D.L. Jones in Austral. Orchid Res.,* 3: 140 (1998)—Tasmania.
bryophila *D.L. Jones in Orchadian,* 12(4): 180 (1997)—South Australia.
hians *D.L. Jones in Orchadian,* 12(4): 185 (1997)—New South Wales.
melagramma *D.L. Jones in Austral. Orchid Res.,* 3: 145 (1998)—Tasmania, New South Wales, Victoria, South Australia.
pratensis *D.L. Jones in Austral. Orchid Res.,* 3: 149 (1998)—Tasmania.
rubenachii *D.L. Jones in Austral. Orchid Res.,* 3: 150 (1998)—Tasmania.
saxicola *D.L. Jones & M.A. Clements in Orchadian,* 12(3): 132 (1997)—New South Wales.
splendens *D.L. Jones & M.A. Clements in Orchadian,* 12(7): 322 (1998)—New Caledonia.
stenochila *D.L. Jones in Austral. Orchid Res.,* 3: 153 (1998)—Tasmania.
uliginosa *D.L. Jones in Austral. Orchid Res.,* 3: 155 (1998)—New South Wales, Victoria, Tasmania, South Australia.
wapstrarum *D.L. Jones in Austral. Orchid Res.,* 3: 156 (1998), as 'wapstreorum'—Tasmania.
williamsonii *D.L. Jones in Austral. Orchid Res.,* 3: 157 (1998)—Tasmania.
ziegeleri *D.L. Jones in Austral. Orchid Res.,* 3: 158 (1998)—Tasmania.

PTERYGIELLA (Scrophulariac.).
suffruticosa *D.Y. Hong in Novon,* 6(4): 372 (1996)—China South-central (Sichuan, Yunnan).

PUCCINELLIA (Gramin.).
arjinshanensis *D.F. Cui in Fl. Xinjiangensis,* 6: 601 (1996); *et in Bull. Bot. Res. North-East. Forest. Univ.,* 17(2): 123 (1997)—Xinjiang.
atacamensis *(Parodi) R.J. Soreng in Novon,* 8(2): 200 (1998): Poa atacamensis.
fasciculata (Torr.) E.P. Bicknell
var. pseudodistans *(Crép.) P.D. Sell in P.D. Sell & G. Murrell, Fl. Great Britain & Ireland,* 5: 363 (1996): Glyceria pseudodistans.
festuciformis (Host) Parl.
subsp. lagascana *M.A. Julià Berruezo & J.M. Montserrat Martí in Fontqueria,* 53: 4 (1999)—Spain.
florida *D.F. Cui in Fl. Xinjiangensis,* 6: 600 (1996); *et in Bull. Bot. Res. North-East. Forest. Univ.,* 17(2): 121 (1997)—Xinjiang.
hispanica *M.A. Julià Berruezo & J.M. Montserrat Martí in Fontqueria,* 53: 3 (1999)—Spain.
× hybrida Holmb.
nothosubsp. mixta *(Holmb.) P.D. Sell in P.D. Sell & G. Murrell, Fl. Great Britain & Ireland,* 5: 363 (1996): P. × mixta.

PUCCINELLIA:-
micranthera *D.F. Cui in Fl. Xinjiangensis,* 6: 600 (1996); *et in Bull. Bot. Res. North-East. Forest. Univ.,* 17(2): 125 (1997), as 'microanthera'—Xinjiang.
parodii *E.G. Nicora in Darwiniana,* 37(3–4): 311 (1999), nom. nov.: Catabrosa tenuifolia *J. Presl.*
raroflorens *E. Edgar in New Zealand J. Bot.,* 34(1): 22 (1996)—New Zealand South I.
walkeri (Kirk) Allan
subsp. antipoda *(Petrie) E. Edgar in New Zealand J. Bot.,* 34(1): 26 (1996): Atropis antipoda.
subsp. chathamica *(Cheeseman) E. Edgar in New Zealand J. Bot.,* 34(1): 26 (1996): Atropis chathamica.

× **PUCOHNIA** *D.A. Beadle, Prelim. Listing Cultivar Grex Names Bromel.:* 200 (1991). *BROMELIACEAE* (Deuterocohnia × Puya).

PUEBLOA *A.B. Doweld in Sukkulentý (Moscow),* 1999(1): 20 (1999). *CACTACEAE*
bradyi *(L.D. Benson) A.B. Doweld in Sukkulentý (Moscow),* 1999(1): 22 (1999): Pediocactus bradyi.
var. despainii *(S.L. Welsh & Goodrich) A.B. Doweld in Sukkulentý (Moscow),* 1999(1): 22 (1999): Pediocactus despainii.
var. winkleri *(K.D. Heil) A.B. Doweld in Sukkulentý (Moscow),* 1999(1): 22 (1999), without exact basionym page: Pediocactus winkleri.
subsp. winkleri *(K.D. Heil) A.B. Doweld in Sukkulentý (Moscow),* 1999(1): 22 (1999), without exact basionym page: Pediocactus winkleri.

PUERARIA (Legumin.).
garhwalensis *L.R. Dangwal & D.S. Rawat in J. Bombay Nat. Hist. Soc.,* 93(3): 570 (1996)—India (Uttar Pradesh).
montana (Lour.) Merr.
var. thomsonii *(Benth.) Wiersema ex D.B. Ward in Phytologia,* 84(6): 386 (1998 publ. 1999): P. thomsonii.

PULICARIA (Compos.).
sect. Sclerostephane *(Chiov.) N. Kilian in Willdenowia,* 29(1–2): 174 (1999): gen. Sclerostephane.
argyrophylla Franch.
subsp. oligophylla *(E. Gamal-Eldin) N. Kilian & P. Hein in Willdenowia,* 29(1–2): 193 (1999): P. argyrophylla var. oligophylla.
aromatica *(Balf. f.) S. King-Jones & N. Kilian in Willdenowia,* 29(1–2): 199 (1999): Pluchea aromatica.
collenettei *(Wagenitz & E. Gamal-Eldin) N. Kilian in Willdenowia,* 29(1–2): 176 (1999): Sclerostephane collenettei.
discoidea *(Chiov.) N. Kilian in Willdenowia,* 29(1–2): 176 (1999): Sclerostephane discoidea.
gabrielii *N. Kilian in Willdenowia,* 29(1–2): 115 (1999)—Somalia.
gamal-eldiniae *N. Kilian & P. Hein in Willdenowia,* 29(1–2): 173 (1999), as 'gamal-eldinae'—South Yemen.
longifolia *(Wagenitz & E. Gamal-Eldin) N. Kilian in Willdenowia,* 29(1–2): 176 (1999), homonym; *see Boiss., Diagn. Pl. Orient.,* Ser. 2, 3: 16 (1856): Sclerostephane longifolia.
paludosa Link
subsp. inuloides *(DC.) B. Valdés in Lagascalia,* 18(2): 305 (1996): P. inuloides.
samhanensis *N. Kilian & P. Hein in Willdenowia,* 29(1–2): 189 (1999)—Oman.

PULMONARIA (Boraginac.).
apennina *G. Cristofolini & G. Puppi in Webbia,* 51(1): 17 (1996)—Italy.
chrysoblephara *M. Gandoger, Dec. Pl. Nov.,* 2: 22 (1876)—Austria.
fissa *M. Gandoger, Dec. Pl. Nov.,* 2: 22 (1876)—Switzerland.
iodocalyx *M. Gandoger, Dec. Pl. Nov.,* 2: 21 (1876)—Hungary.
longistyla *M. Gandoger, Dec. Pl. Nov.,* 2: 20 (1876)—Switzerland.
primuliflora *M. Gandoger, Dec. Pl. Nov.,* 2: 21 (1876)—Hungary.

PULSATILLA (Ranunculac.).
ajanensis Regel & Tiling
subsp. tatewakii *(Kudô) V.N. Voroschilov in A.K. Skvortsov (ed.), Florist. issl. v razn. raĭonakh SSSR:* 169 (1985): P. tatewakii.
archarensis *S.G. Kudrin in Bot. Zhurn.,* 84(4): 112 (1999)—Amur.
bungeana C.A. Mey.
var. astragalifolia *(Pobed.) I.A. Gubanov, Konsp. Fl. Vneshnei Mongolii (Sosud. Rast.):* 50 (1996): P. astragalifolia.
var. tenuiloba *(Turcz.) I.A. Gubanov, Konsp. Fl. Vneshnei Mongolii (Sosud. Rast.):* 50 (1996): P. vulgaris var. tenuiloba.
cernua (Thunb.) Spreng.
f. flava *Y.N. Lee, Fl. Korea:* 1157 (1996), without type—Locality not stated.
var. koreana *(Nakai) Y.N. Lee, Fl. Korea:* 1157 (1996): P. koreana.
jarmilae *J.J. Halda in Acta Mus. Richnov. Sect. Nat.,* 7(2): 74 (2000)—Kirgizistan.

PULTENAEA (Legumin.).
kraehenbuehlii *P.J. Lang in J. Adelaide Bot. Gard.,* 18(1): 25 (1998)—South Australia.

PUNA (Cactac.).
bonnieae *D.J. Ferguson & R. Kiesling in Cact. Succ. J. (U.S.A.),* 69(6): 290 (1997)—Argentina Northwest (Catamarca).

PURDIAEA (Cyrillac.).
nipensis Vict.
subsp. maestrensis *(Borhidi & Catasus) R. Berazaín Iturralde & S. Rodríguez Salgueiro in Rev. Jard. Bot. Nacion. Univ. Habana,* 13: 24 (1992 publ. 1995): P. maestrensis.
ophiticola Vict.
var. oblongisepala *R. Berazaín Iturralde & S. Rodríguez Salgueiro in Rev. Jard. Bot. Nacion. Univ. Habana,* 13: 24 (1992 publ. 1995)—Cuba.
stenopetala Griseb.
var. stereosepala *(J.L. Thomas) R. Berazaín Iturralde & S. Rodríguez Salgueiro in Rev. Jard. Bot. Nacion. Univ. Habana,* 13: 23 (1992 publ. 1995): P. stereosepala.

PUTTERLICKIA (Celastrac.).
saxatilis *(Burch.) M. Jordaan in S. Afr. J. Bot.,* 64(6): 328 (1998): Celastrus saxatilis.

PUYA (Bromeliac.).
butcheriana *H.E. Luther in Selbyana,* 16(2): 237 (1995)—Locality not stated.
claudiae *P.L. Ibisch, R. Vásquez & E. Gross in J. Bromeliad Soc.,* 49(3): 132 (1999)—Bolivia.
cochabambensis *R. Vásquez & P.L. Ibisch in Bromelie,* 2000(1): 16 (2000)—Bolivia.
entre-riosensis *P.L. Ibisch & E. Gross in J. Bromeliad Soc.,* 48(5): 217 (1998)—Bolivia.
ochroleuca *J. Betancur & R. Callejas in Caldasia,* 19(1–2): 74 (1997)—Colombia.
prosanae *P. Ibisch & E. Gross in J. Bromeliad Soc.,* 43(5): 212 (1993)—Bolivia.
pumila *P. Ravenna in Onira, Bot. Leafl.,* 4(15): 57 (2000)—Chile South (Bíobío).
pusilla *H.E. Luther in Bromélia,* 4(4): 30 (1997 publ. 1998)—Bolivia.
robin-fosteri *G.S. Varadarajan & H.E. Luther in Selbyana,* 16(2): 235 (1995)—Peru.
roldanii *J. Betancur & R. Callejas in Caldasia,* 19(1–2): 78 (1997)—Colombia.
vasquezii *P.L. Ibisch & E. Gross in Rev. Soc. Boliv. Bot.,* 2(1): 26 (1998)—Bolivia.

PYCNANTHUS (Myristicac.).
angolensis (Welw.) Warb.
subsp. schweinfurthii *(Warb.) B. Verdcourt in Fl. Zambes.,* 9(2): 38 (1997): P. schweinfurthii.

PYCNOCOMA (Euphorbiac.).
bampsiana *J. Léonard in Bull. Jard. Bot. Nation. Belg.,* 65(1–2): 62 (1996)—Zaire.
devredii *J. Léonard in Bull. Jard. Bot. Nation. Belg.,* 65(1–2): 50 (1996)—Zaire.
louisii *J. Léonard in Bull. Jard. Bot. Nation. Belg.,* 65(1): 59 (1996)—Zaire.

PYCNOCOMA:-
subflava *J. Léonard in Bull. Jard. Bot. Nation. Belg.*, 65(1–2): 57 (1996)—Zaire.

PYCNOCYCLA (Umbellif.).
saxatilis *A. Danin, I.C. Hedge & J.M. Lamond in Willdenowia*, 30(1): 77 (2000)—Palestine (Jordan), Saudi Arabia.

PYCNOSTACHYS (Labiat.).
recurvata *O. Ryding in Novon*, 9(1): 101 (1999)—Ethiopia.

PYCREUS (Cyperac.).
fasciculatus *E. Govindarajalu in J. Econ. Taxon. Bot.*, 20(2): 301 (1996)—India (Kerala, Tamil Nadu).
kanarensis *V.P. Prasad & N.P. Singh in J. Econ. Taxon. Bot.*, 21(3): 667 (1997 publ. 1998)—India (Karnataka).
mahadevanii *E. Govindarajalu in J. Econ. Taxon. Bot.*, 20(2): 299 (1996)—India (Karnataka).
megapotamicus (Kunth) Nees
var. jaeggii *(Boeck.) E.R. Guaglianone in Hickenia*, 2(31): 137 (1995): Cyperus jaeggii.
pyramidalis *E. Govindarajalu in J. Econ. Taxon. Bot.*, 20(2): 301 (1996)—India (Tamil Nadu).
sanguinolentus (Vahl) Nees
subsp. cyrtostachys *(Miq.) D.A. Simpson in J.H. Beaman & R.S. Beaman, Pl. Mt. Kinabalu, 3. Gymnosperms & Non-Orchid Monocot.*: 128 (1998), with incorrect basionym ref.: Basionym not stated.

PYRACANTHA (Rosac.).
heterophylla *T.B. Chao & Z.X. Chen in Bull. Bot. Res. North-East. Forest. Univ.*, 17(3): 302 (1997)—China Southeast (Henan).
stoloniformis *T.B. Chao & Z.X. Chen in Bull. Bot. Res. North-East. Forest. Univ.*, 17(3): 301 (1997)—China Southeast (Henan).

PYRENACANTHA (Icacinac.).
gabonica *F.J. Breteler & J.-F. Villiers in Adansonia*, 22(2): 201 (2000)—Gabon.

PYRENARIA (Theac.).
grandiflora *(Y.C. Wu) T.L. Ming & S.X. Yang in Chinese J. Appl. Environ. Biol.*, 3(3): 284 (1997): Tutcheria villosa var. grandiflora.
hirta (Hand.-Mazz.) H. Keng
var. cordatula *(H.L. Li) S.X. Yang & T.L. Ming in Chinese J. Appl. Environ. Biol.*, 3(3): 286 (1997): Tutcheria hirta var. cordatula.
microcarpa (Dunn) H. Keng
var. ovalifolia *(H.L. Li) T.L. Ming & S.X. Yang in Chinese J. Appl. Environ. Biol.*, 3(3): 287 (1997): Tutcheria ovalifolia.
var. shinkoensis *(Hayata) T.L. Ming & S.X. Yang in Chinese J. Appl. Environ. Biol.*, 3(3): 286 (1997): Thea shinkoensis.
var. tenuifolia *(Hung T. Chang) T.L. Ming & S.X. Yang in Chinese J. Appl. Environ. Biol.*, 3(3): 287 (1997): Tutcheria tenuifolia.
pingpienensis *(Hung T. Chang) S.X. Yang & T.L. Ming in Chinese J. Appl. Environ. Biol.*, 3(3): 283 (Sept. 1997); *et in Fl. Yunnanica*, 8: 312 (Nov. 1997): Tutcheria pingpienensis.
rostrata *(Hung T. Chang) S.X. Yang & T.L. Ming in Chinese J. Appl. Environ. Biol.*, 3(3): 285 (1997): Tutcheria rostrata.
sophiae *(Hu) S.X. Yang & T.L. Ming in Chinese J. Appl. Environ. Biol.*, 3(3): 283 (Sept. 1997); *et in Fl. Yunnanica*, 8: 312 (Nov. 1997): Camellia sophiae.
turbinata *S.X. Yang in Chinese J. Appl. Environ. Biol.*, 3(3): 281 (1997)—China Southeast (Guangxi).

PYRETHRUM (Compos.).
alatavicum (Herder) O. Fedtsch. & B. Fedtsch.
subsp. krylovianum *(Krasch.) N.M. Boldÿreva in Fl. Sibir. (Asterac.)*, 13: 74 (1997 publ. 1998): P. krylovianum.
brachanthemoides *R.V. Kamelin & G.A. Lazkov in Bot. Zhurn.*, 82(7): 123 (1997)—Kirgizstan.
sussamyrense *G.A. Lazkov in Bot. Zhurn.*, 84(5): 125 (1999)—Kirgizstan.

PYROLA (Ericac.).
alpestris *M. Gandoger, Dec. Pl. Nov.*, 2: 30 (1876)—France.
atrovirens *M. Gandoger, Dec. Pl. Nov.*, 2: 30 (1876)—France.
longifolia *M. Gandoger, Dec. Pl. Nov.*, 2: 31 (1876)—France.
mixta *M. Gandoger, Dec. Pl. Nov.*, 2: 30 (1876)—Italy.
pedemontana *M. Gandoger, Dec. Pl. Nov.*, 2: 29 (1876)—Italy.
rossica *M. Gandoger, Dec. Pl. Nov.*, 2: 31 (1876)—Russia Northwest.
winkleriana *M. Gandoger, Dec. Pl. Nov.*, 2: 30 (1876)—Czechoslovakia.

PYROSTRIA (Rubiac.).
chapmanii *D.M. Bridson in Fl. Zambes.*, 5(2): 343 (1998)—Malawi.

PYRRHOCACTUS (Cactac.).
kattermannii *R. Kiesling in Zuloaga & Morrone (eds.), Cat. Pl. Vasc. Rep. Arg. 2 (Monogr. Syst. Bot.* Missouri Bot. Gard. 74): 1246 (1999), nom. nov.: Eriosyce andreana *Katt.*
marayesensis *(Backeb.) J.G. Lambert, Cactus d'Argentine, ed. 2*: 265 (1998): P. umadeave var. marayesensis.

PYRUS (Rosac.).
× babadagensis Prodán
nothosubsp. eminensis *J. Dostálek in Feddes Repert.*, 108(5–6): 354 (1997)—Bulgaria.(P. elaeagrifolia subsp. bulgarica × pyraster).
× dasyphylla *(Tausch) J. Dostálek in Feddes Repert.*, 108(5–6): 356 (1997): P. communis γ dasyphylla.
hopeiensis T.T. Yu
var. peninsula *D.K. Zang & W.D. Peng in Bull. Bot. Res. North-East. Forest. Univ.*, 17(1): 50 (1997)—China North-central (Shandong).
× nivalis Jacq.
nothosubsp. austriaca *(A. Kern.) J. Dostálek in Feddes Repert.*, 108(5–6): 352 (1997): P. austriaca.
pyraster (L.) Burgsd.
var. achras *(Gaertn.) R. Cinovskis in Fl. Baltiĭskikh Respublik*, 2: 97 (1996): P. achras.
ussuriensis Maxim.
var. viris *(Nakai) Y.N. Lee, Fl. Korea*: 1158 (1996): P. viris.

Q

QIONGZHUEA *(T.H. Wen & Ohrnb.) J.R. Xue & T.P. Yi in Taxon*, 45(2): 218 (1996): Chimonobambusa sect. Qiongzhuea. (Gramin.).
communis *J.R. Xue & T.P. Yi in Taxon*, 45(2): 219 (1996), as comb. nov., basionym not validly publ.—China South-central (Sichuan).
gracilis *W.T. Lin in J. Bamboo Res.*, 19(4): 1 (2000)—China Southeast (Guangdong).
intermedia *J.R. Xue & D.Z. Li in Taxon*, 45(2): 219 (1996), as comb. nov., basionym not validly publ.—China South-central (Sichuan).
luzhiensis *J.R. Xue & T.P. Yi in Taxon*, 45(2): 219 (1996), as comb. nov., basionym not validly publ.—China South-central (Guizhou).
macrophylla *J.R. Xue & T.P. Yi in Taxon*, 45(2): 219 (1996), as comb. nov., basionym not validly publ.—China South-central (Sichuan).
var. leiboensis *J.R. Xue & D.Z. Li in Taxon*, 45(2): 219 (1996), as comb. nov., basionym not validly publ.—China South-central (Sichuan).
maculata *(T.H. Wen) J.R. Xue & T.P. Yi in Taxon*, 45(2): 219 (1996): Chimonobambusa maculata.
montigena *T.P. Yi in Taxon*, 45(2): 219 (1996), as comb. nov., basionym not validly publ.—China South-central (Yunnan).
multigemmia *T.P. Yi in J. Bamboo Res.*, 19(1): 18 (2000)—China South-central (Sichuan).
opienensis *J.R. Xue & T.P. Yi in Taxon*, 45(2): 220 (1996), as comb. nov., basionym not validly publ.—China South-central (Sichuan).

QIONGZHUEA:-
puberula *J.R. Xue & T.P. Yi in Taxon*, 45(2): 220 (1996), as comb. nov., basionym not validly publ.—China South-central (Guizhou).
rigidula *J.R. Xue & T.P. Yi in Taxon*, 45(2): 220 (1996), as comb. nov., basionym not validly publ.—China South-central (Sichuan).
tumidissinoda *(Ohrnb.) J.R. Xue & T.P. Yi in Taxon*, 45(1): 220 (1996): Chimonobambusa tumidissinoda.
unifolia *(T.H. Wen) T.P. Yi in Taxon*, 45(1): 220 (1996): Chimonobambusa unifolia.
verruculosa *T.P. Yi in Taxon*, 45(2): 220 (1996), as comb. nov., basionym not validly publ.—China South-central (Sichuan).

QUAQUA (Asclepiadac.).
sect. Pauciflora *P.V. Bruyns in Bot. Jahrb.*, 121(3): 336 (1999).
sect. Pseudorhytidocaulon *P.V. Bruyns in Bot. Jahrb.*, 121(3): 336 (1999).
arenicola (N.E. Br.) Plowes
subsp. pilifera *(Bruyns) P.V. Bruyns in Bot. Jahrb.*, 121(3): 383 (1999): Q. armata subsp. pilifera.
cincta *(C.A. Lückh.) P.V. Bruyns in Asklepios*, 65: 17 (1995): Caralluma cincta.
incarnata (L.f.) Bruyns
subsp. hottentotorum *(N.E. Br.) P.V. Bruyns in Bot. Jahrb.*, 121(3): 370 (1999): Q. hottentotorum.
subsp. tentaculata *(Bruyns) P.V. Bruyns in Bot. Jahrb.*, 121(3): 369 (1999): Q. incarnata var. tentaculata.
pallens *P.V. Bruyns in Bot. Jahrb.*, 121(3): 355 (1999)—Cape Province.
parviflora (Masson) Bruyns
subsp. confusa *(Plowes) P.V. Bruyns in Bot. Jahrb.*, 121(3): 354 (1999): Q. confusa.

QUARARIBEA (Bombacac.).
ruiziana *J.L. Fernández-Alonso in Rev. Acad. Colomb. Cienc. Exact. Fis. Nat.*, 23(Suplemento Especial): 50 (1999 publ. 2000)—Colombia.
yunckeri Standl.
subsp. izabalensis *W.S. Alverson ex M.E. Véliz Pérez in An. Inst. Biol. Univ. Nac. Autón. Méx., Bot.*, 71(2): 81 (2000)—Guatemala.
subsp. sessiliflora *Miranda ex W.S. Alverson in Novon*, 8(2): 127 (1998)— Mexico Gulf (Veracruz).
subsp. veracruzana *W.S. Alverson in Novon*, 8(2): 127 (1998)—Mexico Gulf (Veracruz).

QUATERNELLA (Amaranthac.).
ephedroides *T.M. Pedersen in Adansonia*, 19(2): 245 (1997)—Brazil Northeast (Bahia), Brazil Southeast (Minas Gerais).
glabratoides *(Suess.) T.M. Pedersen in Adansonia*, 19(2): 247 (1997): Pfaffia glabratoides.

QUERCUS (Fagac.).
aliena Blume
f. alticupuliformis *(H.W. Jen & L.M. Wang) H.W. Jen & L.M. Wang in J. Beijing Forest. Univ.*, 15(4): 44 (1993): Q. aliena var. alticupuliformis.
f. jeholensis *(Liou & S.X. Li) H.W. Jen & L.M. Wang in Fl. Reipubl. Popul. Sin.*, 22: 233 (1998): Q. aliena var. jeholensis.
× andresii *R. Alonso, F. Llamas, E. Puente & A. Penas in Bot. Helvetica*, 109(1): 91 (1999)—Spain. (Q. faginea × pauciradiata).
× arrimatensis *A. Penas, F. Llamas, C. Pérez-Morales & C. Acedo in Bot. Helvetica*, 107(1): 80 (1997)—Spain. (Q. pauciradiata × petraea).
bawanglingensis *C.C. Huang, Ze X. Li & F.W. Xing in Guihaia*, 16(4): 302 (1996)—Hainan.
camusiae *Trel. ex Hickel & A. Camus in Lecomte, Fl. Indo-Chine*, 5(9): 957 (1929), nom. nov.: Q. geminata *Hickel & A. Camus.*
castanicarpa *W. Roxburgh, Pl. Coromandel*, 3: 93 (1820)—Bangladesh.†
coccifera L.
f. brevicupulata *(Batt. & Trab.) F.M. Vázquez Pardo, Semillas de Quercus: Biol., Ecol. Manejo*: 81 (1998): Q. coccifera var. brevicupulata.
f. crassicupulata *(Batt. & Trab.) F.M. Vázquez Pardo, Semillas de Quercus: Biol., Ecol. Manejo*: 81 (1998): Q. coccifera var. crassicupulata.

QUERCUS:-

coccifera

f. dolichocarpa (A. Camus) F.M. Vázquez Pardo, Semillas de Quercus: Biol., Ecol. Manejo: 81 (1998): Q. coccifera var. dolichocarpa.

f. imbricata (DC.) F.M. Vázquez Pardo, Semillas de Quercus: Biol., Ecol. Manejo: 81 (1998): Q. coccifera var. imbricata.

f. macrocarpa (Cout.) F.M. Vázquez Pardo, Semillas de Quercus: Biol., Ecol. Manejo: 81 (1998): Q. coccifera var. macrocarpa.

dolicholepis A. Camus

var. elliptica (Y.C. Hsu & H.W. Jen) Y.C. Hsu & H.W. Jen in J. Beijing Forest. Univ., 15(4): 44 (1993): Q. spathulata var. elliptica.

dongfangensis C.C. Huang, F.W. Xing & Ze X. Li in Guihaia, 16(4): 302 (1996)— Hainan.

faginea Lam.

f. brevicupulata (Cout.) F.M. Vázquez Pardo, Semillas de Quercus: Biol., Ecol. Manejo: 81 (1998): Q. lusitanica var. brevicupulata.

f. macrobalanus (A. Camus) F.M. Vázquez Pardo, Semillas de Quercus: Biol., Ecol. Manejo: 81 (1998): Q. faginea var. macrobalanus.

f. suboccultata (Cout.) F.M. Vázquez Pardo, Semillas de Quercus: Biol., Ecol. Manejo: 81 (1998): Q. lusitanica var. suboccultata.

fulviseriacus (Y.C. Hsu & D.M. Wang) R. Govaerts in R. Govaerts & D.G. Frodin, World Checklist & Bibliogr. Fagales: 247 (April 1998): Cyclobalanopsis fulviseriacus.

fulvisericea (Y.C. Hsu & D.M. Wang) Z.K. Zhou in Acta Bot. Yunnanica, 20(1): 43 (Feb. 1998); et in Int. Oaks, 8: 25 (Summer 1998), as 'fulvisericeous': Cyclobalanopsis fulvisericea.

garryana Douglas ex Hook.

var. fruticosa (Engelm.) R. Govaerts in R. Govaerts & D.G. Frodin, World Checklist & Bibliogr. Fagales: 249 (1998): Q. lobata var. fruticosa.

humilis Mill.

f. brachycarpa (Guss.) F.M. Vázquez Pardo, Semillas de Quercus: Biol., Ecol. Manejo: 81 (1998): Q. brachycarpa.

f. globulosa (Vuk.) F.M. Vázquez Pardo, Semillas de Quercus: Biol., Ecol. Manejo: 81 (1998): Q. pubescens var. globulosa.

f. leptobalanus (Guss.) F.M. Vázquez Pardo, Semillas de Quercus: Biol., Ecol. Manejo: 81 (1998): Q. leptobalanus.

f. microcarpa (Guss.) F.M. Vázquez Pardo, Semillas de Quercus: Biol., Ecol. Manejo: 82 (1998): Q. microcarpa.

f. oxycarpa (Vuk.) F.M. Vázquez Pardo, Semillas de Quercus: Biol., Ecol. Manejo: 82 (1998): Q. pubescens var. oxycarpa.

f. parviglandis (Vuk.) F.M. Vázquez Pardo, Semillas de Quercus: Biol., Ecol. Manejo: 82 (1998): Q. pinnatifida var. parviglandis.

f. rostrata (Vuk.) F.M. Vázquez Pardo, Semillas de Quercus: Biol., Ecol. Manejo: 82 (1998): Q. pubescens var. rostrata.

ichnusae L. Mossa, G. Bacchetta & S. Brullo in Israel J. Pl. Sci., 47(3): 199 (1999)—Sardegna.

ilex L.

f. macrocarpa (Albert) F.M. Vázquez Pardo, Semillas de Quercus: Biol., Ecol. Manejo: 82 (1998): Q. ilex var. macrocarpa.

f. obesa (Albert) F.M. Vázquez Pardo, Semillas de Quercus: Biol., Ecol. Manejo: 82 (1998): Q. ilex var. obesa.

f. stenocarpa (Albert) F.M. Vázquez Pardo, Semillas de Quercus: Biol., Ecol. Manejo: 82 (1998): Q. ilex var. stenocarpa.

f. urceolata (Martrin-Donos & Timb.-Lagr.) F.M. Vázquez Pardo, Semillas de Quercus: Biol., Ecol. Manejo: 82 (1998), as 'urceolada': Q. ilex var. urceolata.

kiukiangensis (Y.T. Chang ex Y.C. Hsu & H. Wei Jen) Y.T. Chang

var. xizangensis (Y.C. Hsu & H. Wei Jen) Z.K. Zhou & H. Sun in Acta Bot. Yunnanica, 18(2): 219 (1996): Cyclobalanopsis xizangensis.

lamellosa Sm.

var. nigrinervis (Hu) Z.K. Zhou & H. Sun in Acta Bot. Yunnanica, 18(2): 220 (1996): Cyclobalanopsis nigrinervis.

QUERCUS:-

liboensis Z.K. Zhou in Acta Bot. Yunnanica, 20(1): 43 (Feb. 1998); et in Int. Oaks, 8: 25 (Summer 1998), nom. nov.: Cyclobalanopsis pseudoglauca Y.K. Li & Xue M. Wang.

lusitanica Lam.

f. subinclusa (Cout.) F.M. Vázquez Pardo, Semillas de Quercus: Biol., Ecol. Manejo: 82 (1998): Q. fruticosa subvar. subinclusa.

× macdanielii T.L. Green & W.J. Hess in Int. Oaks, 8: 16 (1998)—Cultivation. (Q. macrocarpa × robur).

× macnabiana Sudw.

nothovar. mahonii (E.J. Palmer) R. Govaerts in R. Govaerts & D.G. Frodin, World Checklist & Bibliogr. Fagales: 270 (1998): Q. × mahonii.

multinervis (W.C. Cheng & T. Hong) R. Govaerts in R. Govaerts & D.G. Frodin, World Checklist & Bibliogr. Fagales: 276 (April 1998): Cyclobalanopsis multinervis.

multinervis (W.C. Cheng & T. Hong) J.Q. Li in J. Wuhan Bot. Res., 16(3): 247 (Aug. 1998), isonym; see R. Govaerts in R. Govaerts & D.G. Frodin, World Checklist & Bibliogr. Fagales: 276 (April 1998): Cyclobalanopsis multinervis.

nanchuanica C.C. Huang in Guihaia, 16(4): 302 (1996)—China South-central (Sichuan).

obconica Y.C. Hsu ex Z.K. Zhou in Acta Bot. Yunnanica, 20(1): 43 (Feb. 1998); et in Int. Oaks, 8: 26 (Summer 1998), nom. nov., as 'obconicus': Cyclobalanopsis litoralis Chun & P.C. Tam ex Y.C. Hsu & H.W. Jen

× pacensis F.M. Vázquez in An. Jard. Bot. Madrid, 53(2): 249 (1995 publ. 1996)—Spain. (Q. faginea subsp. broteroi × suber).

pachucana F. Zavala-Chávez in Int. Oaks, 10: 31 (2000)—Mexico Northeast (Hidalgo).

pauciradiata A. Penas, F. Llamas, C. Pérez-Morales & C. Acedo in Bot. Helvetica, 107(1): 75 (1997)—Spain.

petraea (Matt.) Liebl.

subsp. austrotyrrhenica S. Brullo, R. Guarino & G. Siracusa in Webbia, 54(1): 24 (1999)—Sicilia.

pubescens Willd.

subsp. subpyrenaica (Villar) R. Govaerts in R. Govaerts & D.G. Frodin, World Checklist & Bibliogr. Fagales: 294 (1998): Q. subpyrenaica.

pyrenaica Willd.

f. expansa (C. Vicioso) F.M. Vázquez Pardo, Semillas de Quercus: Biol., Ecol. Manejo: 82 (1998): Q. pyrenaica var. expansa.

f. normalis (Rouy) F.M. Vázquez Pardo, Semillas de Quercus: Biol., Ecol. Manejo: 82 (1998): Q. toza var. normalis.

f. sphaerocarpa (Rouy) F.M. Vázquez Pardo, Semillas de Quercus: Biol., Ecol. Manejo: 82 (1998): Q. toza var. sphaerocarpa.

f. stenocarpa (Rouy) F.M. Vázquez Pardo, Semillas de Quercus: Biol., Ecol. Manejo: 82 (1998): Q. toza var. stenocarpa.

robur L.

f. brachybalanus (Schur) F.M. Vázquez Pardo, Semillas de Quercus: Biol., Ecol. Manejo: 83 (1998): Q. pedunculata var. brachybalanus.

f. globulata (Schur) F.M. Vázquez Pardo, Semillas de Quercus: Biol., Ecol. Manejo: 83 (1998): Q. pedunculata var. globulata.

f. macrobalanus (Schur) F.M. Vázquez Pardo, Semillas de Quercus: Biol., Ecol. Manejo: 83 (1998): Q. pedunculata var. macrobalanus.

f. microbalanus (Heuff.) F.M. Vázquez Pardo, Semillas de Quercus: Biol., Ecol. Manejo: 83 (1998): Q. pedunculata var. microbalanus.

f. sphaerobalanus (Schur) F.M. Vázquez Pardo, Semillas de Quercus: Biol., Ecol. Manejo: 83 (1998): Q. pedunculata var. sphaerobalanus.

f. subglobosa (Schur) F.M. Vázquez Pardo, Semillas de Quercus: Biol., Ecol. Manejo: 83 (1998): Q. robur var. subglobosa.

× otensis A. Penas, F. Llamas, C. Pérez-Morales & C. Acedo in Bot. Helvetica, 107(1): 78 (1997)—Spain. (Q. pauciradiata × pyrenaica).

rotundifolia Lam.

f. avellaniformis (Colmeiro & E. Boutelou) F.M. Vázquez Pardo, Semillas de Quercus: Biol., Ecol. Manejo: 83 (1998), as 'avellanaeformis': Q. avellaniformis.

QUERCUS:-

rotundifolia

f. brevicupulata (Laguna) F.M. Vázquez Pardo, Semillas de Quercus: Biol., Ecol. Manejo: 83 (1998): Q. ilex f. brevicupulata.

f. crassicupulata (Cout.) F.M. Vázquez Pardo, Semillas de Quercus: Biol., Ecol. Manejo: 83 (1998): Q. ilex f. crassicupulata.

f. dolichocalyx (C. Vicioso) F.M. Vázquez Pardo, Semillas de Quercus: Biol., Ecol. Manejo: 83 (1998): Q. ilex var. dolichocalyx.

f. expansa (Laguna) F.M. Vázquez Pardo, Semillas de Quercus: Biol., Ecol. Manejo: 83 (1998): Q. ilex var. expansa.

f. macrocarpa (Cout.) F.M. Vázquez Pardo, Semillas de Quercus: Biol., Ecol. Manejo: 83 (1998): Q. ilex f. macrocarpa.

f. pilosella F.M. Vázquez Pardo, Semillas de Quercus: Biol., Ecol. Manejo: 84 (1998)—Spain.

senescens Hand.-Mazz.

var. muliensis (Hu) Y.C. Hsu & H.W. Jen in J. Beijing Forest. Univ., 15(4): 44 (1993): Q. muliensis.

var. muliensis (Hu) Y.C. Hsu & H.W. Jen ex R. Govaerts in R. Govaerts & D.G. Frodin, World Checklist & Bibliogr. Fagales: 308 (1998), isonym; see Y.C. Hsu & H.W. Jen in J. Beijing Forest. Univ., 15(4): 44 (1993): Q. muliensis.

× senneniana

nothosubsp. tentudaicus F.M. Vázquez in An. Jard. Bot. Madrid, 53(2): 250 (1995 publ. 1996)—Spain. (Q. faginea subsp. broteroi × rotundifolia).

serrata Thunb.

var. tomentosa (B.C. Ding & T.B. Chao) Y.C. Hsu & H.W. Jen in J. Beijing Forest. Univ., 15(4): 44 (1993): Q. glandulifera var. tomentosa.

shangxiensis Z.K. Zhou in Acta Bot. Yunnanica, 20(1): 44 (Feb. 1998); et in Int. Oaks, 8: 26 (Summer 1998), nom. nov.: Q. lanceolata S.Z. Qu & W.H. Zhang.

suber L.

f. brevicupulata (Batt. & Trab.) F.M. Vázquez Pardo, Semillas de Quercus: Biol., Ecol. Manejo: 37, 84 (1998): Q. suber var. brevicupulata.

f. clavata (Cout.) F.M. Vázquez Pardo, Semillas de Quercus: Biol., Ecol. Manejo: 37, 84 (1998): Q. suber var. clavata.

f. dolichocarpa (A. Camus) F.M. Vázquez Pardo, Semillas de Quercus: Biol., Ecol. Manejo: 37, 84 (1998): Q. suber var. dolichocarpa.

f. longicalyx (A. Camus) F.M. Vázquez Pardo, Semillas de Quercus: Biol., Ecol. Manejo: 37, 84 (1998), as 'longicalix': Q. suber var. longicalyx.

f. macrocarpa (Willk. & Lange) F.M. Vázquez Pardo, Semillas de Quercus: Biol., Ecol. Manejo: 37, 84 (1998): Q. suber var. macrocarpa.

f. microcarpa (Batt. & Trab.) F.M. Vázquez Pardo, Semillas de Quercus: Biol., Ecol. Manejo: 37, 84 (1998): Q. suber var. microcarpa.

f. racemosa (Borzí) F.M. Vázquez Pardo, Semillas de Quercus: Biol., Ecol. Manejo: 37, 84 (1998): Q. suber var. racemosa.

f. suboccultata (Cout.) F.M. Vázquez Pardo, Semillas de Quercus: Biol., Ecol. Manejo: 37, 84 (1998): Q. suber var. suboccultata.

trojana Webb

subsp. euboica (Papaioannou) K.I. Christensen in A. Strid & Kit Tan (eds.), Fl. Hellenica, 1: 45 (1997): Q. euboica.

turbinella Greene

var. ajoensis (C.H. Mull.) L.D. Benson in L.D. Benson & R.A. Darrow, Trees & Shrubs of Southwest. Deserts (ed. 3): 370 (1981), isonym; see Little in Phytologia, 42(3): 221 (1979): Q. ajoensis.

var. californica (J.M. Tucker) L.D. Benson in L.D. Benson & R.A. Darrow, Trees & Shrubs of Southwest. Deserts (ed. 3): 369 (1981): Q. turbinella subsp. californica.

variabilis Blume

var. zhongtiaoensis D.F. Ke & X.M. Liu in S.C. Li (ed.) et al., Dendrol. Zhongtiaoshan.: 324 (1995)—China North-central (Shanxi).

QUERCUS:-
× warei *T.L. Green & W.J. Hess in Int. Oaks,* 8: 19 (1998)—Cultivation. (Q. bicolor × robur).
yingjiangensis *(Y.C. Hsu & Q.Z. Dong) R. Govaerts in R. Govaerts & D.G. Frodin, World Checklist & Bibliogr. Fagales:* 322 (1998): Cyclobalanopsis yingjiangensis.
yongchuniana *Z.K. Zhou in Acta Bot. Yunnanica,* 20(1): 44 (Feb. 1998); *et in Int. Oaks,* 8: 26 (Summer 1998), nom. nov., as 'yongchunana': Cyclobalanopsis longifolia *Y.C. Hsu & Q.Z. Dong.*

× **QUESMEA** *E. Knobloch in J. Bromeliad Soc.,* 22(3): 58 (1972). *BROMELIACEAE* (Aechmea × Quesnelia).

QUESNELIA (Bromeliac.).
kautskyi *C.M. Vieira in Bradea,* 8(23): 131 (1999)—Brazil Southeast (Espírito Santo, Minas Gerais).

× **QUESREGELIA** *J.F. Carrone in J. Bromeliad Soc.,* 33(5): 207 (1983). *BROMELIACEAE* (Neoregelia × Quesnelia).

QUIINA (Quiinac.).
yatuensis *J. Schneider & G. Zizka in Novon,* 7(4): 406 (1997 publ. 1998)—Venezuela.

R

RABDOSIA (Labiat.).
glaucocalyx *(Maxim.) N.S. Probatova in Sosud. Rast. Sovet. Dal'nego Vostoka,* 7: 378 (1995): Plectranthus glaucocalyx.
inflexa (Thunb.) H. Hara
f. albiflora *Y.N. Lee, Fl. Korea:* 1160 (1996), without type—Locality not stated.
serra *(Maxim.) Y.N. Lee, Fl. Korea:* 1160 (1996), with incorrect basionym ref.: Basionym not stated.
umbrosa (Maxim.) H. Hara
var. chiisanensis *(Nakai) Y.N. Lee, Fl. Korea:* 1160 (1996): Amethystanthus excisus var. chiisanensis.
var. coreana *(T.B. Lee) Y.N. Lee, Fl. Korea:* 1160 (1996): Isodon excisus var. coreanus.

RACEMOBAMBOS (Gramin.).
rupicola *E.A. Widjaja in Reinwardtia,* 11(2): 126 (1997)—New Guinea (Irian Jaya).
sessilis *E.A. Widjaja in Reinwardtia,* 11(2): 128 (1997)—New Guinea (Irian Jaya).

RACHELIA *J.M. Ward & I. Breitwieser in New Zealand J. Bot.,* 35(2): 146 (1997). *COMPOSITAE*
glaria *J.M. Ward & I. Breitwieser in New Zealand J. Bot.,* 35(2): 146 (1997)—New Zealand South I.

RACIBORSKANTHOS *D.L. Szlachetko in Fragm. Flor. Geobot.,* Supp. 3: 135 (1995). *ORCHIDACEAE*
capricornis *(Ridl.) D.L. Szlachetko in Fragm. Flor. Geobot.,* Supp. 3: 136 (1995): Ascochilus capricornis.
halophilus *(Ridl.) D.L. Szlachetko in Fragm. Flor. Geobot.,* Supp. 3: 136 (1995), as 'halophilum': Saccolabium halophilum.
minax *(Rchb. f.) D.L. Szlachetko in Fragm. Flor. Geobot.,* Supp. 3: 136 (1995): Echioglossum minax.

RACINAEA (Bromeliac.).
kessleri *H.E. Luther in Selbyana,* 19(1): 86 (1998)—Bolivia.

RAFFENALDIA (Crucif.).
primuloides Godr.
subsp. riphaensis *J.M. Montserrat in Lagascalia,* 18(2): 248 (1996)—Morocco.

RAFFLESIA (Rafflesiac.).
arnoldii R. Br.
var. atjehensis *(Koord.) W. Meijer in Fl. Males., Ser. 1,* 13: 23 (1997): R. atjehensis.

RANDIA (Rubiac.).
armata (Sw.) DC.
subsp. panamensis *(Standl.) D.H. Lorence in Novon,* 8(3): 250 (1998): R. panamensis.
decussata *N. Wallich, Report on Calcutta Bot. Gard. to G.A. Bushby:* 30 (1840), nom. nov.: Rondeletia tetrandra *Roxb.*
genipifolia *(Standl. & Steyerm.) D.H. Lorence in Novon,* 8(3): 249 (1998): Duroia genipifolia.
longifolia *C.G.R. Gustafsson in Novon,* 10(3): 206 (2000)—Ecuador.
micracantha *(Lillo) N.M. Bacigalupo in A.L. Cabrera, Fl. Prov. Jujuy (Inst. Nac. Tecn. Agropec., 13),* 9: 388 (1993): Basanacantha micracantha.
var. glabra *N.M. Bacigalupo in A.L. Cabrera, Fl. Prov. Jujuy (Inst. Nac. Tecn. Agropec., 13),* 9: 391 (1993)—Argentina Northwest (Jujuy).
nicaraguensis *D.H. Lorence & J.D. Dwyer in Novon,* 8(3): 247 (1998)—Nicaragua.
pubistyla *C.G.R. Gustafsson in Novon,* 10(3): 204 (2000)—Ecuador, Colombia.
wigginsii *Standl. ex C.G.R. Gustafsson in Novon,* 10(3): 202 (2000)—Ecuador, Peru.

RANUNCULUS (Ranunculac.).
subsect. Cassubici *G.H. Loos in Thaiszia,* 7(1): 3 (1997), isonym; *see Tzvelev in Byull. Mosk. Obshch. Ispȳt. Prir., Biol.,* 99(5): 69 (1994).
sect. Chinenses *(Luferov) E.V. Nevidomova-Malacha in Bot. Zhurn.,* 81(11): 98 (1996): R. subsect. Chinenses.
sect. Gmelinii *(Luferov) E.V. Nevidomova-Malacha in Bot. Zhurn.,* 81(11): 94 (1996): R. subsect. Gmelinia.
sect. Hyperborei *E.V. Nevidomova-Malacha in Bot. Zhurn.,* 81(11): 94 (1996).
ser. Indecori *G.H. Loos in Thaiszia,* 7(1): 4 (1997).
sect. Nivales *(Prantl) E.V. Nevidomova-Malacha in Bot. Zhurn.,* 81(11): 93 (1996), without basionym ref.: Basionym not stated.
sect. Pedatifidi *(Tzvelev) E.V. Nevidomova-Malacha in Bot. Zhurn.,* 81(11): 92 (1996): R. subsect. Pedatifidi.
ser. Phragmitetori *G.H. Loos in Thaiszia,* 7(1): 5 (1997).
sect. Polyanthemos *(Luferov) E.V. Nevidomova-Malacha in Bot. Zhurn.,* 81(11): 96 (1996): R. subsect. Polyanthemos.
sect. Punctati *E.V. Nevidomova-Malacha in Bot. Zhurn.,* 81(11): 93 (1996).
sect. Pygmaei *(Luferov) E.V. Nevidomova-Malacha in Bot. Zhurn.,* 81(11): 92 (1996): R. subsect. Pygmaea.
subgen. Repentes *(Luferov) E.V. Nevidomova-Malacha in Bot. Zhurn.,* 81(11): 98 (1996): R. subsect. Repentes.
sect. Repentes *(Luferov) E.V. Nevidomova-Malacha in Bot. Zhurn.,* 81(11): 98 (1996): R. subsect. Repentes.
sect. Stenoglossa *(W.T. Wang) W.T. Wang in Phytologia,* 79(5): 387 (1995 publ. 1996): R. subgen. Stenoglossa.
subgen. Xanthobatrachium *E.V. Nevidomova-Malacha in Bot. Zhurn.,* 81(11): 94 (1996), nom. illegit., see Arts. 48.1, 53.5 ICBN (1994).
acris L.
subsp. hultenii *V.N. Voroschilov in Byull. Mosk. Obshch. Ispȳt. Prir., Biol.,* 99(6): 117 (1994)—Kamchatka.
subsp. novus *(H. Lév. & Vaniot) V.N. Voroschilov in A.K. Skvortsov (ed.), Florist. issl. v razn. raĭonakh SSSR:* 169 (1985): R. novus.
altaicus Laxm.
subsp. sulphureus *(Phipps) A.G. Jelenevsky & T.G. Derviz-Sokolova in Novosti Sist. Vȳssh. Rast.,* 30: 49 (1996): R. sulphureus.
ambranus *E. Hörandl & W. Gutermann in Bot. Jahrb.,* 120(1): 35 (1998), nom. nov.: R. ponticus *Borch.-Kolb.*
auricomus L.
subsp. montserratii *(Grau) A. Segura Zubizarreta & G. Mateo Sanz in Stud. Bot. (Salamanca),* 14: 198 (1995 publ. 1996): R. montserratii.
subsp. valdesii *(Grau) G. Mateo Sanz, C. Fabregat Llueca & S. López Udias in An. Biol. Fac. Biol. Univ. Murcia,* 20: 107 (1995): R. valdesii.

RANUNCULUS:-
borealis Trautv.
subsp. turneri *(Greene) V.N. Voroschilov in Byull. Mosk. Obshch. Ispȳt. Prir., Biol.,* 99(6): 119 (1994): R. turneri.
bupleuroides Brot.
subsp. cherubicus *J.A. Sánchez Rodríguez, M.J. Elías Rivas & M.A. Martín Marcos in An. Jard. Bot. Madrid,* 57(2): 402 (1999 publ. 2000)—Spain.
carpinetorum *E. Hörandl & W. Gutermann in Bot. Jahrb.,* 121(1): 104 (1999)—Austria.
chopericus *N.N. Tzvelev in Bot. Zhurn.,* 80(7): 78 (1995)—Russia Central.
crenatolobus *E. Hörandl & W. Gutermann in Phyton (Austria),* 37(2): 280 (1998)—Austria.
debreceniensis *E. Hörandl & W. Gutermann in Bot. Jahrb.,* 120(1): 35 (1998), nom. nov.: R. acriformis *Soó.*
densiciliatus W.T. Wang
var. glabrescens *W.L. Cheng in Acta Phytotax. Sin.,* 37(3): 304 (1999)—Xizang.
var. nyingchiensis *W.L. Cheng in Acta Phytotax. Sin.,* 37(3): 303 (1999)—Xizang.
deripovae *N.N. Tzvelev in Bot. Zhurn.,* 80(7): 77 (1995)—Ukraina.
elegantifrons *E. Hörandl & W. Gutermann in Bot. Jahrb.,* 121(1): 120 (1999)—Austria.
ganeschinii *N.N. Tzvelev in Bot. Zhurn.,* 80(7): 79 (1995)—Russia Northwest.
var. lugensis *N.N. Tzvelev in Novosti Sist. Vȳssh. Rast.,* 30: 70 (1996)—Russia Northwest.
gaurii *L.R. Dangwal & D.S. Rawat in J. Econ. Taxon. Bot.,* 20(3): 703 (1996)—India (Uttar Pradesh).
golitzynii *N.N. Tzvelev in Bot. Zhurn.,* 80(7): 78 (1995)—Russia Central.
graecensis *E. Hörandl & W. Gutermann in Bot. Jahrb.,* 121(1): 126 (1999)— Austria.
gratiosus *T. Brodtbeck in Bauhinia,* 12(1–2): 52 (1998)—France, Switzerland, Germany.
hirtellus Royle
var. glabrescens *W.L. Cheng in Acta Phytotax. Sin.,* 37(3): 303 (1999)—Xizang.
var. sigyilaicus *W.L. Cheng in Acta Phytotax. Sin.,* 37(3): 303 (1999)—Xizang.
istriacus *E. Hörandl & W. Gutermann in Bot. Jahrb.,* 120(4): 563 (1998)—Yugoslavia.
janischevskyi *N.N. Tzvelev in Bot. Zhurn.,* 80(7): 79 (1995)—Russia East.
japonicus Thunb.
f. albiflorus *Y.N. Lee, Fl. Korea:* 1157 (1996), without type—Locality not stated.
f. plenus *Sig. Tanaka & T. Shimizu in T. Shimizu, Fl. Nagano Pref.:* 1505 (1997), as 'pleniflorus', see amendment sheet, homonym; *see Y.C. Chu & D.C. Wu in Nat. Resources Res.,* 2: 4 (1979)—Japan.
kosinensis *E.G. Chugaynova in Bot. Zhurn.,* 84(2): 115 (1999)—Russia East.
kunlunshanicus *J.G. Liu in Fl. Xinjiangensis,* 2(1): 352 (1994)—Xinjiang.
laticrenatus *E. Hörandl & W. Gutermann in Bot. Jahrb.,* 121(1): 108 (1999)—Austria, Czechoslovakia.
lingulatus *T. Brodtbeck in Bauhinia,* 12(1–2): 62 (1998)—France.
longicaulis C.A. Mey.
subsp. pulchellus *(C.A. Mey.) I.A. Gubanov in Fl. Vostoch. Khangaya (MNR):* 130 (1983): R. pulchellus.
lunaris Brodtb.
var. plenus *T. Brodtbeck in Bauhinia,* 12(1–2): 35 (1998), without type—Switzerland.
lyratus *T. Brodtbeck in Bauhinia,* 12(1–2): 65 (1998)—Switzerland, France.
macrophyllus Desf.
subsp. barceloi *(Grau) A.M. Romo, Fl. Silvestres Baleares:* 110 (1994): R. barceloi.
macrotis Brodtb.
var. paradoxus *T. Brodtbeck in Bauhinia,* 12(1–2): 35 (1998), without type—Switzerland.
mediosectus *E. Hörandl & W. Gutermann in Phyton (Austria),* 37(2): 283 (1998)—Austria.
megalocaulis *E. Hörandl & W. Gutermann in Bot. Jahrb.,* 121(1): 100 (1999)—Austria.
melzeri *E. Hörandl & W. Gutermann in Bot. Jahrb.,* 120(4): 561 (1998)—Austria.

RANUNCULUS:-
mendosus *E. Hörandl & W. Gutermann in Bot. Jahrb.*, 120(4): 581 (1998)—Austria.
natans C.A. Mey.
var. intramongolicus *(Y.Z. Zhao) S.X. Li in Clavis Pl. Chinae Bor.-Or. ed. 2 (ed. P.Y. Fu)*: 202 (1995): R. intramongolicus.
nemorosifolius *E. Hörandl & W. Gutermann in Bot. Jahrb.*, 120(4): 568 (1998)—Austria.
nemorosus DC.
subsp. mixtus *(Jord.) N.N. Tzvelev in Novosti Sist. Vÿssh. Rast.*, 32: 183 (2000): R. mixtus.
nephelogenes Edgew.
var. pseudohirculus *(Trautv.) J.G. Liu in Fl. Xinjiangensis*, 2(1): 308 (1994): R. pulchellus var. pseudohirculus.
nivalis L.
subsp. altaicus *(Laxm.) V.N. Voroschilov in A.K. Skvortsov (ed.), Florist. issl. v razn. raĭonakh SSSR*: 170 (1985): R. altaicus.
subsp. intercedens *(Hultén) V.N. Voroschilov in A.K. Skvortsov (ed.), Florist. issl. v razn. raĭonakh SSSR*: 170 (1985): R. sulphureus var. intercedens.
subsp. sulphureus *(Phipps) V.N. Voroschilov in A.K. Skvortsov (ed.), Florist. issl. v razn. raĭonakh SSSR*: 170 (1985): R. sulphureus.
noricus *E. Hörandl & W. Gutermann in Phyton (Austria)*, 37(2): 289 (1998)—Austria.
notabilis *E. Hörandl & W. Gutermann in Phyton (Austria)*, 37(2): 268 (1998)—Austria.
ovessnovii *N.N. Tzvelev in Bot. Zhurn.*, 80(7): 81 (1995)—Russia East.
oxyodon *E. Hörandl & W. Gutermann in Phyton (Austria)*, 37(2): 292 (1998)—Austria.
pachycaulus *(Nevski) A.N. Luferov in Byull. Glavn. Bot. Sada (Moscow)*, 175: 56 (1997): Batrachium pachycaulum.
pekinensis *(L. Liou) G. Dahlgren in U. Jensen & J.W. Kadereit (eds.), Syst. Evol. Ranunculiflorae (Pl. Syst.* Evol., Supp. 9): 306 (1995): Batrachium pekinense.
pekinensis *(L. Liou) A.N. Luferov in Byull. Glavn. Bot. Sada (Moscow)*, 175: 56 (1997), isonym; *see G. Dahlgren in Jensen & Kadereit (eds.), Syst. Evol. Ranunculiflorae (Pl.* Syst. Evol., Supp. 9): 306 (1995): Batrachium pekinense.
pentadactylus *E. Hörandl & W. Gutermann in Bot. Jahrb.*, 121(1): 116 (1999)—Austria.
permiensis *E.G. Chugaynova in Bot. Zhurn.*, 84(2): 117 (1999)—Russia East.
polyanthemos L.
subsp. schennikovii *(Ovcz. ex Tzvelev) N.N. Tzvelev in Novosti Sist. Vÿssh. Rast.*, 32: 183 (2000): R. schennikovii.
praetermissus *E. Hörandl & W. Gutermann in Phyton (Austria)*, 37(2): 277 (1998)—Austria.
pseudoacris *N.N. Tzvelev in Novosti Sist. Vÿssh. Rast.*, 30: 67 (1996), nom. nov.: R. acriformis *Soó*.
pseudohirculus *Schrenk ex F.E.L. Fischer & C.A. Meyer, Enum. Pl. Nov.*, 2: 65 (1842)—Middle Asia.
puringii *N.N. Tzvelev in Bot. Zhurn.*, 80(7): 77 (1995)—Russia Northwest.
quinatus Brodtb.
var. heteranthus *T. Brodtbeck in Bauhinia*, 12(1–2): 35 (1998), without type—Switzerland.
recticaulis *E. Hörandl & W. Gutermann in Bot. Jahrb.*, 120(1): 35 (1998), nom. nov.: R. rectus *Haas*.
schmalhausenii *A.N. Luferov in Byull. Glavn. Bot. Sada (Moscow)*, 175: 57 (1997), nom. nov.: Ranunculus aquatilis var. dichotomus *Schmalh*.
shuichengensis *L. Liao in Acta Phytotax. Sin.*, 35(1): 57 (1997)—China South-central (Guizhou).
sieboldii Miq.
var. glabriusculus *J.Q. Fu in Fl. Loess-Plat. Sin.*, 1: 611 (2000)—China North-central (Shaanxi, Gansu).
silerifolius Lév.
var. dolicanthus *L. Liao in Acta Phytotax. Sin.*, 35(1): 59 (1997)—China South-central (Yunnan).
var. prostratus *(Nakai) Sig. Tanaka ex T. Shimizu, Fl. Nagano Pref.*: 1505 (1997): R. vernyi var. prostratus.
var. quelpaertensis *(H. Lév.) Sig. Tanaka ex T. Shimizu, Fl. Nagano Pref.*: 1505 (1997): R. repens var. quelpaertensis.

RANUNCULUS:-
sphinx *T. Brodtbeck in Bauhinia*, 12(1–2): 59 (1998)—France.
pryginii *N.N. Tzvelev in Bot. Zhurn.*, 80(7): 80 (1995)—Russia Central.
stellaris Brodtb.
var. supernova *T. Brodtbeck in Bauhinia*, 12(1–2): 36 (1998), without type—Switzerland.
styriacus *E. Hörandl & W. Gutermann in Phyton (Austria)*, 37(2): 309 (1998)—Austria.
sulphureus Sol.
subsp. taimyrensis *V.B. Kuvaev in Byull. Mosk. Obshch. Ispÿt. Prir., Biol.*, 102(5): 71 (1997)—Krasnoyarsk.
tenuirostrus *J.Q. Fu in Fl. Loess-Plat. Sin.*, 1: 611 (2000), homonym; *see Pomel, Nouv. Mat. Fl. Atl.*, 1: 250 (1874)—China North-central (Shaanxi, Gansu).
truniacus *E. Hörandl & W. Gutermann in Phyton (Austria)*, 37(2): 299 (1998)—Austria.
turczaninovii *(Luferov) V.N. Voroschilov in Byull. Mosk. Obshch. Ispÿt. Prir., Biol.*, 99(6): 121 (1994): R. pedatifidus subsp. turczaninovii.
udicola *E. Hörandl & W. Gutermann in Bot. Jahrb.*, 121(1): 129 (1999)—Austria.
variabilis *E. Hörandl & W. Gutermann in Phyton (Austria)*, 37(2): 301 (1998)—Austria.
veronicae *N. Böhling in Willdenowia*, 30(2): 245 (2000)—Kriti.
vindobonensis *E. Hörandl & W. Gutermann in Bot. Jahrb.*, 120(4): 584 (1998)—Austria.
vjatkensis *N.N. Tzvelev in Bot. Zhurn.*, 80(7): 80 (1995)—Russia East.
vorkutensis *N.N. Tzvelev in Bot. Zhurn.*, 80(7): 81 (1995)—Russia North.
walo-kochii *E. Hörandl & W. Gutermann in Bot. Jahrb.*, 120(1): 30 (1998), nom. nov.: R. kochii *Jasiewicz*.
wangianus *Q.E. Yang in Acta Phytotax. Sin.*, 38(6): 551 (2000)—China South-Central (Yunnan).

RAPANEA (Myrsinac.).
hermogenesii *S.L. Jung-Mendaçolli & L.C. Bernacci in Bol. Bot. (São Paulo)*, 16: 32 (1997)—Brazil Southeast (São Paulo, Rio de Janeiro), Brazil South (Santa Catarina).
kwangsiensis E. Walker
var. tibetica *W.L. Cheng & K. Yao in Acta Bot. Yunnanica*, 19(2): 176 (1997)—Xizang.

RAPHANUS (Crucif.).
acanthiformis *Morel ex L. Sisley in Rev. Hortic.*, 46: 445 (1874)—Japan.†
confusus *(Greuter & Burdet) I.A. Al-Shehbaz & S.I. Warwick in Novon*, 7(3): 219 (1997): Quidproquo confusum.

RAPHIOCARPUS (Gesneriac.).
annamensis *(Pellegr.) B.L. Burtt in Beitr. Biol. Pfl.*, 70(2–3): 172 (1997–98 publ. 1998): Didissandra annamensis.
asper *(Drake) B.L. Burtt in Beitr. Biol. Pfl.*, 70(2–3): 172 (1997–98 publ. 1998): Didissandra aspera.
aureus *(Dunn) B.L. Burtt in Beitr. Biol. Pfl.*, 70(2–3): 173 (1997–98 publ. 1998): Loxostigma aureum.
begoniifolius *(H. Lév.) B.L. Burtt in Beitr. Biol. Pfl.*, 70(2–3): 173 (1997–98 publ. 1998): Didissandra begoniifolia.
clemensiae *(Pellegr.) B.L. Burtt in Beitr. Biol. Pfl.*, 70(2–3): 173 (1997–98 publ. 1998): Didissandra clemensiae.
evrardii *(Pellegr.) B.L. Burtt in Beitr. Biol. Pfl.*, 70(2–3): 173 (1997–98 publ. 1998): Didissandra evrardii.
longipedunculatus *(H.W. Li) B.L. Burtt in Beitr. Biol. Pfl.*, 70(2–3): 174 (1997–98 publ. 1998): Didissandra longipedunculata.
macrosiphon *(Hance) B.L. Burtt in Beitr. Biol. Pfl.*, 70(2–3): 174 (1997–98 publ. 1998): Chirita macrosiphon.
petelotii *(Pellegr.) B.L. Burtt in Beitr. Biol. Pfl.*, 70(2–3): 174 (1997–98 publ. 1998): Didissandra petelotii.
sesquifolius *(C.B. Clarke) B.L. Burtt in Beitr. Biol. Pfl.*, 70(2–3): 174 (1997–98 publ. 1998): Didissandra sesquifolia.

RAPHIONACME (Asclepiadac.).
ernstiana *U. Meve in Bradleya*, 18: 71 (2000)—Tanzania.
haeneliae *H.J.T. Venter & R.L. Verhoeven in S. Afr. J. Bot.*, 62(6): 316 (1996)—Namibia.
moyalicus *H.J.T. Venter & R.L. Verhoeven in S. Afr. J. Bot.*, 63(2): 101 (1997)—Kenya.
sylvicola *H.J.T. Venter & R.L. Verhoeven in Novon*, 10(2): 170 (2000)—Zambia.

RAUHOCEREUS (Cactac.).
riosaniensis
subsp. jaenensis *(Rauh ex Backeb.) C. Ostolaza Nano in Cactaceae Consensus Initiatives*, 6: 9 (1998): R. riosaniensis var. jaenensis.

RAUIA (Rutac.).
nodosa *(Engl.) J.A. Kallunki in Kew Bull.*, 53(2): 324 (1998): Cusparia nodosa.

RAUKAUA *B.C. Seemann in J. Bot.*, 4: 352 (1866), as 'Raukana'; *see J. Bot.*, 6: 142 (1868). *ARALIACEAE*†
anomalus *(Hook.) A.D. Mitchell, D.G. Frodin & M.J. Heads in New Zealand J. Bot.*, 35(3): 312 (1997): Panax anomalus.
edgerleyi *(Hook. f.) B.C. Seemann in J. Bot.*, 4: 352 (1866), as 'Raukana'; *see J. Bot.*, 6: 142 (1868): Panax edgerleyi.†
var. serratus *(Kirk) A.D. Mitchell, D.G. Frodin & M.J. Heads in New Zealand J. Bot.*, 35(3): 312 (1997): Panax edgerleyi var. serratus.
× parvus *(Kirk) P.B. Heenan in New Zealand J. Bot.*, 36(2): 309 (1998): Panax simplex var. parvus.
× serratus *(Kirk) P.B. Heenan in New Zealand J. Bot.*, 36(2): 308 (1998): Panax edgerleyi var. serratus.
simplex *(G. Forst.) A.D. Mitchell, D.G. Frodin & M.J. Heads in New Zealand J. Bot.*, 35(3): 313 (1997): Panax simplex.
var. sinclairii *(Hook. f.) A.D. Mitchell, D.G. Frodin & M.J. Heads in New Zealand J. Bot.*, 35(3): 313 (1997): Panax sinclairii.

RAUVOLFIA (Apocynac.).
kamarora *Hendrian in Blumea*, 44(2): 454 (1999)—Sulawesi.
oligantha *Hendrian in Blumea*, 44(2): 457 (1999)—Jawa.

RAYJACKSONIA *R.L. Hartman & M.A. Lane in Amer. J. Bot.*, 83(3): 368 (1996). *COMPOSITAE*
annua *(Rydb.) R.L. Hartman & M.A. Lane in Amer. J. Bot.*, 83(3): 368 (1996): Sideranthus annuus.
aurea *(A. Gray) R.L. Hartman & M.A. Lane in Amer. J. Bot.*, 83(3): 369 (1996): Haplopappus aureus.
phyllocephala *(DC.) R.L. Hartman & M.A. Lane in Amer. J. Bot.*, 83(3): 369 (1996): Haplopappus phyllocephalus.

RAZISEA (Acanthac.).
ericae *Mildbr. ex D.C. Wasshausen in Novon*, 7(1): 88 (1997)—Ecuador, Peru.

REBUTIA (Cactac.).
subgen. Pygmaeolobivia *(Backeb.) O. Šída, Rod Rebutia*: 22 (1997): Lobivia subgen. Pygmaeolobivia.
subgen. Reicheocactus *(Backeb.) O. Šída, Rod Rebutia*: 29 (1997): gen. Reicheocactus.
sect. Reicheocactus *(Backeb.) O. Šída, Rod Rebutia*: 29 (1997): gen. Reicheocactus.
albispina *(Rausch) O. Šída, Rod Rebutia*: 29 (1997): Lobivia nigricans var. albispina.
amblypetala *(F. Ritter) S. Mosti in Cactus & Co.*, 3(4): 196 (1999): R. rosalbiflora var. amblypetala.
applanata *(Rausch) O. Šída, Rod Rebutia*: 27 (1997): Lobivia steinmannii var. applanata.
applanata *(Rausch) S. Mosti in Cactus & Co.*, 3(4): 196 (1999), isonym; *see Šída, Rod Rebutia*: 27 (1997): Lobivia steinmannii var. applanata.
atrovirens *(Backeb.) J. Pilbeam, Rebutia (Cactus File Handbook, 2)*: 20 (1997), isonym; *see O. Šída, Atlas Kakt.*, 10: 40 (1996): Lobivia atrovirens.
atrovirens *(Backeb.) O. Šída, Atlas Kakt.*, 10: 40 (1996); *cf. Repert. Pl. Succ. (I.O.S.)*, 47: 11 (1996 publ. 1997); *et in Rod Rebutia*: 28 (1997): Lobivia atrovirens.

REBUTIA:-
atrovirens
var. pseudoritteri *(Rausch) S. Mosti in Cactus & Co.*, 3(4): 198 (1999): Lobivia atrovirens var. pseudoritteri.
aureiflora Backeb.
f. boedekeriana *(Backeb.) O. Šída, Rod Rebutia:* 23 (1997): Mediolobivia boedekeriana.
f. duursmaiana *(Backeb.) O. Šída, Rod Rebutia:* 23 (1997), as 'duursmana': Mediolobivia duursmaiana.
canigueralii
subsp. crispata *(Rausch) Donald ex D.R. Hunt in Cactaceae Consensus Initiatives*, 3: 6 (1997): Sulcorebutia crispata.
subsp. pulchra *(Cárdenas) Donald ex D.R. Hunt in Cactaceae Consensus Initiatives*, 3: 6 (1997): R. pulchra.
cardenasiana *(R. Vásquez) G. Navarro in Lazaroa*, 17: 76 (1996): Sulcorebutia cardenasiana.
chrysacantha Backeb.
var. iseliniana *(Krainz) Donald ex O. Šída, Rod Rebutia:* 18 (1997): R. senilis var. iseliniana.
var. kesselringiana *(Bewer.) Donald ex O. Šída, Rod Rebutia:* 18 (1997): R. senilis var. kesselringiana.
var. schieliana *(Bewer.) O. Šída, Rod Rebutia:* 18 (1997): R. senilis var. schieliana.
crassa *(Rausch) O. Šída, Rod Rebutia:* 26 (1997): Lobivia haagei var. crassa.
crassa *(Rausch) S. Mosti in Cactus & Co.*, 3(4): 203 (1999), isonym; *see Šída, Rod Rebutia:* 26 (1997): Lobivia haagei var. crassa.
diersiana Rausch
subsp. atrovirens *(Rausch) S. Mosti in Cactus & Co.*, 3(4): 205 (1999): R. diersiana var. atrovirens.
elegantula *(Rausch) O. Šída, Rod Rebutia:* 25 (1997): Lobivia haagei var. elegantula.
famatimensis (Speg.) Speg.
var. jachalensis *(Rausch) O. Šída, Rod Rebutia:* 29 (1997): Lobivia famatimensis var jachalensis.
var. sanjuanensis *(Rausch) O. Šída, Rod Rebutia:* 29 (1997): Lobivia famatimensis var. sanjuanensis.
fidaiana
subsp. cintiensis *(Cárdenas) D.R. Hunt in Cactaceae Consensus Initiatives*, 4: 7 (1997): Weingartia cintiensis.
fuauxiana *(Backeb.) O. Šída, Rod Rebutia:* 28 (1997): Mediolobivia fuauxiana.
gavazzii *S. Mosti in Cactus & Co.*, 3(4): 207 (1999)—Bolivia.
haagei Frič & Schelle
var. elegantula *(Rausch) S. Mosti in Cactus & Co.*, 4(1): 37 (2000): Lobivia haagei var. elegantula.
f. flavovirens *(Backeb.) O. Šída, Rod Rebutia:* 25 (1997): Lobivia neohaageana var. flavovirens.
subsp. mudanensis *(Rausch) S. Mosti in Cactus & Co.*, 4(1): 37 (2000): R. mudanensis.
haefneriana *(Cullman) O. Šída, Rod Rebutia:* 28 (1997): Mediolobivia haefneriana.
haefneriana *(Cullman) S. Mosti in Cactus & Co.*, 4(1): 38 (2000), isonym; *see O. Šída, Rod Rebutia:* 28 (1997): Mediolobivia haefneriana.
knizei *(Rausch) O. Šída, Rod Rebutia:* 26 (1997): Lobivia pygmaea var. knizei.
knizei *(Rausch) S. Mosti in Cactus & Co.*, 4(1): 40 (2000), isonym; *see O. Šída, Rod Rebutia:* 26 (1997): Lobivia pygmaea var. knizei.
leucacantha *(Rausch) O. Šída, Rod Rebutia:* 27 (1997): Lobivia steinmannii var. leucacantha.
major *(Rausch) O. Šída, Rod Rebutia:* 27 (1997): Lobivia steinmannii var. major.
major *(Rausch) S. Mosti in Cactus & Co.*, 4(1): 41 (2000), isonym; *see O. Šída, Rod Rebutia:* 27 (1997): Lobivia steinmannii var. major.
melanocentra *(Rausch) O. Šída, Rod Rebutia:* 28 (1997): Lobivia steinmannii var. melanocentra.
mentosa
subsp. purpurea *(Donald & A.B. Lau) Donald ex D.R. Hunt in Cactaceae Consensus Initiatives*, 3: 6 (1997): Weingartia purpurea.
minor *(Rausch) S. Mosti in Cactus & Co.*, 4(1): 42 (2000): R. diersiana var. minor.

REBUTIA:-
minuscula
var. marsoneri *(Werderm.) Eb. Scholz in Inform.-brief Freundeskreis Echinopseen*, 20: 9 (1995); *cf. Repert. Pl. Succ. (I.O.S.)*, 46: 15 (1995 publ. 1996): R. marsoneri.
var. wessneriana *(Bewer.) Eb. Scholz in Inform.-brief Freundeskreis Echinopseen*, 20: 9 (1995); *cf. Repert. Pl. Succ. (I.O.S.)*, 46: 15 (1995 publ. 1996): R. wessneriana.
nazarenoensis *(Rausch) S. Mosti in Cactus & Co.*, 4(1): 43 (2000), isonym; *see B. Fearn & Pearcy, The Genus Rebutia 1895–1981:* 71 (1981): Digitorebutia nazarenoensis.
neocumingii
subsp. pilcomayensis *(Cárdenas) D.R. Hunt in Cactaceae Consensus Initiatives*, 4: 8 (1997): Weingartia pilcomayensis.
subsp. riograndensis *(F. Ritter) D.R. Hunt in Cactaceae Consensus Initiatives*, 4: 8 (1997): Weingartia riograndensis.
subsp. saipinensis *(F.H. Brandt) D.R. Hunt in Cactaceae Consensus Initiatives*, 4: 8 (1997): Weingartia saipinensis.
subsp. trollii *(Oeser) D.R. Hunt in Cactaceae Consensus Initiatives*, 4: 8 (1997): Weingartia trollii.
nigricans *(Wessner) J. Pilbeam, Rebutia (Cactus File Handbook, 2):* 61 (8 May 1997): Lobivia nigricans.
var. albispina *(Rausch) S. Mosti in Cactus & Co.*, 4(1): 44 (2000): Lobivia nigricans var. albispina.
subsp. carmeniana *(Rausch) S. Mosti in Cactus & Co.*, 4(1): 44 (2000): R. carmeniana.
var. peterseimii *Buining & Donald ex O. Šída, Rod Rebutia:* 29 (1997)—Locality not stated.
var. peterseimii *(Frič) S. Mosti in Cactus & Co.*, 4(1): 44 (2000): R. peterseimii.
nigricans *(Wessner) D.R. Hunt in Cactaceae Consensus Initiatives*, 4: 6 (Oct. 1997), isonym: Lobivia nigricans.
nigricans *(Wessner) O. Šída, Rod Rebutia:* 29 (June 1997), isonym: Lobivia nigricans.
oculata Werderm.
subsp. tilcarensis *(Rausch) S. Mosti in Cactus & Co.*, 4(1): 46 (2000): Lobivia euanthema var. tilcarensis.
oligacantha *(F.H. Brandt) Donald ex D.R. Hunt in Cactaceae Consensus Initiatives*, 3: 6 (1997): Weingartia oligacantha.
orurensis *(Backeb.) F. Ritter ex O. Šída, Rod Rebutia:* 25 (1997): Lobivia orurensis.
parvula *(Rausch) O. Šída, Rod Rebutia:* 28 (1997): Lobivia steinmannii var. parvula.
parvula *(Rausch) S. Mosti in Cactus & Co.*, 4(1): 48 (2000), isonym; *see O. Šída, Rod Rebutia:* 28 (1997): Lobivia steinmannii var. parvula.
pelzliana *(Rausch) O. Šída, Rod Rebutia:* 26 (1997), as 'peltzliana': Lobivia haagei var. pelzliana.
pelzliana *(Rausch) S. Mosti in Cactus & Co.*, 4(1): 49 (2000), isonym; *see O. Šída, Rod Rebutia:* 26 (1997): Lobivia haagei var. pelzliana.
polypetala *(Rausch) O. Šída, Rod Rebutia:* 27 (1997): Lobivia pygmaea var. polypetala.
polypetala *(Rausch) S. Mosti in Cactus & Co.*, 4(2): 87 (2000), isonym; *see Šída, Rod Rebutia:* 27 (1997): Lobivia pygmaea var. polypetala.
pseudodeminuta Backeb.
f. schumanniana *(Backeb.) O. Šída, Rod Rebutia:* 20 (1997): R. pseudodeminuta var. schumanniana.
pseudoritteri *(Rausch) O. Šída, Rod Rebutia:* 28 (1997): Lobivia atrovirens var. pseudoritteri.
pygmaea (R.E. Fr.) Britton & Rose
var. pectinata *(Backeb.) O. Šída, Rod Rebutia:* 26 (1997): Lobivia pectinata.
spegazziniana Backeb.
var. atroviridis *(Backeb.) O. Šída, Rod Rebutia:* 21 (1997): Aylostera spegazziniana var. atroviridis.
steinbachii
subsp. kruegerae *(Cárdenas) D.R. Hunt in Cactaceae Consensus Initiatives*, 3: 6 (1997), as 'kruegeri': Aylostera kruegerae.
subsp. tiraquensis *(Cárdenas) D.R. Hunt in Cactaceae Consensus Initiatives*, 3: 6 (1997): R. tiraquensis.

REBUTIA:-
steinbachii
subsp. verticillacantha *(F. Ritter) Donald ex D.R. Hunt in Cactaceae Consensus Initiatives*, 3: 6 (1997): Sulcorebutia verticillacantha.
steinmannii
var. brachyantha *(Wessner) Pilbeam in Cact. File*, 10(2): 18 (1996), without basionym ref.; *cf. Repert. Pl. Succ. (I.O.S.)*, 47: 11 (1996 publ. 1997): Lobivia brachyantha.
subsp. brachyantha *(Wessner) S. Mosti in Cactus & Co.*, 4(2): 93 (2000): Lobivia brachyantha.
var. camargoensis *(Rausch) Pilbeam in Cact. File*, 10(2): 18 (1996), without basionym ref.; *cf. Repert. Pl. Succ. (I.O.S.)*, 47: 11 (1996 publ. 1997): R. camargoensis.
var. christinae *(Rausch) Pilbeam in Cact. File*, 10(2): 18 (1996), without basionym ref.; *cf. Repert. Pl. Succ. (I.O.S.)*, 47: 11 (1996 publ. 1997): R. christinae.
var. costata *(Werderm.) Pilbeam in Cact. File*, 10(2): 18 (1996), without basionym ref.; *cf. Repert. Pl. Succ. (I.O.S.)*, 47: 11 (1996 publ. 1997): R. costata.
var. costata *(Werderm.) S. Mosti in Cactus & Co.*, 4(2): 91 (2000): R. costata.
var. eucaliptana *(Backeb.) O. Šída, Rod Rebutia:* 27 (1997): Lobivia eucaliptana.
var. leucacantha *(Rausch) S. Mosti in Cactus & Co.*, 4(2): 91 (2000): Lobivia steinmannii var. leucacantha.
var. melanocentra *(Rausch) S. Mosti in Cactus & Co.*, 4(2): 92 (2000): Lobivia steinmannii var. melanocentra.
var. parvula *(Rausch) Pilbeam in Cact. File*, 10(2): 18 (1996), without basionym ref.; *cf. Repert. Pl. Succ. (I.O.S.)*, 47: 11 (1996 publ. 1997): Lobivia steinmannii var. parvula.
var. rauschii *(Zecher) Pilbeam in Cact. File*, 10(2): 18 (1996), without basionym ref.; *cf. Repert. Pl. Succ. (I.O.S.)*, 47: 11 (1996 publ. 1997): R. rauschii.
var. tuberculata *(Rausch) S. Mosti in Cactus & Co.*, 4(2): 92 (2000): Lobivia steinmannii var. tuberculata.
tafnaensis *(Rausch) O. Šída, Rod Rebutia:* 27 (1997): Lobivia pygmaea var. tafnaensis.
tafnaensis *(Rausch) S. Mosti in Cactus & Co.*, 4(2): 94 (2000), isonym; *see Šída, Rod Rebutia:* 27 (1997): Lobivia pygmaea var. tafnaensis.
tilcarensis *(Rausch) O. Šída, Rod Rebutia:* 24 (1997): Lobivia euanthema var. tilcarensis.
tuberculata *(Rausch) O. Šída, Rod Rebutia:* 27 (1997): Lobivia steinmannii var. tuberculata.
violaceostaminata *(Rausch) O. Šída, Rod Rebutia:* 27 (1997), as 'violaceostaminea': Lobivia pygmaea var. violaceostaminata.
violaceostaminata *(Rausch) S. Mosti in Cactus & Co.*, 4(2): 95 (2000), isonym; *see Šída, Rod Rebutia:* 27 (1997): Lobivia pygmaea var. violaceostaminata.
xanthocarpa Backeb.
var. graciliflora *Backeb. ex O. Šída, Rod Rebutia:* 18 (1997)—Locality not stated.
yuncharasensis *(Rausch) O. Šída, Rod Rebutia:* 29 (1997): Lobivia atrovirens var. yuncharasensis.
yuncharasensis *(Rausch) S. Mosti in Cactus & Co.*, 4(2): 96 (2000), isonym; *see Šída, Rod Rebutia:* 29 (1997): Lobivia atrovirens var. yuncharasensis.

REEVESIA (Sterculiac.).
lumlingensis *H.H. Hsue ex S.J. Xu in Acta Phytotax. Sin.*, 38(6): 566 (2000)—China Southeast (Guangxi).

REHIA *F. Fijten in Blumea*, 22(3): 416 (1975), in Supp. 16 as 'nom. nov.'. *GRAMINEAE*†
nervata *F. Fijten in Blumea*, 22(3): 416 (1975), in Supp. 16 as 'comb. nov.'—Locality not stated.†

REICHENBACHANTHUS (Orchidac.).
subulatus *(Schltr.) R.L. Dressler in Novon*, 7(2): 124 (1997): Scaphyglottis subulata.

REMIJIA (Rubiac.).
uniflora *C.M. Taylor in Novon*, 6(2): 218 (1996)—Venezuela.

RENANTHERA (Orchidac.).
citrina *L.V. Averyanov in Orchids*, 66(12): 1287 (1997)—Vietnam.

RENEALMIA (Zingiberac.).
sancti-thomae *I.M. Turner in Edinburgh J. Bot.*, 56(1): 149 (1999), nom. nov.: Alpinia africana *Ridl.*

RENGGERIA (Guttif.).
comans *(Mart.) Engler in Mart., Fl. Bras.*, 12(1): 441 (1888): Schweiggera comans.

RENNERA (Compos.).
stellata *P.P.J. Herman in Bot. J. Linn. Soc.*, 129(4): 368 (1999)—Cape Province.

RENZORCHIS *D.L. Szlachetko & T.S. Olszewski in Adansonia*, 20(2): 324 (1998). *ORCHIDACEAE*
pseudoplatycoryne *D.L. Szlachetko & T.S. Olszewski in Adansonia*, 20(2): 326 (1998)—Gabon.
singularis *(Summerh.) T.S. Olszewski & D.L. Szlachetko in Ann. Bot. Fenn.*, 37(4): 299 (2000): Habenaria singularis.

RESIA (Gesneriac.).
ichthyoides Leeuwenb.
subsp. bracteata *L.E. Skog & F.F. de Jesus in Biollania*, ed. esp. 6 (J.J. Wurdack Festschr.): 520 (1997), without latin descr.—Colombia.

RESNOVA (Hyacinthac.).
humifusa *(Baker) U. Müller-Doblies & D. Müller-Doblies in Feddes Repert.*, 108(1–2): 59 (1997): Scilla humifusa.
nossibeensis *(H. Perrier) F. Speta in Phyton (Austria)*, 38(1): 119 (1998): Scilla nossibeensis.

RESTREPIA (Orchidac.).
subgen. Ecmeles *C.A. Luer, Ic. Pleurothallid. XIII (Monogr. Syst. Bot. Missouri Bot. Gard., 59)*: 12 (1996).
subgen. Pachymeles *(Garay) C.A. Luer, Ic. Pleurothallid. XIII (Monogr. Syst. Bot. Missouri Bot. Gard., 59)*: 13 (1996): R. sect. Pachymeles.
subgen. Pleurothallopsis *(Schltr.) H. Mohr in Leafl. Schlechter Inst.*, 2: 7 (1996), without basionym ref.: R. sect. Pleurothallopsis.
sect. Schlimii *H. Mohr in Leafl. Schlechter Inst.*, 2: 16 (1996), without latin descr. or type.
aberrans *C.A. Luer in Orquideologia*, 20(2): 117 (1996)—Panama.
angustilabia Schltr.
subsp. subserrata *(Schltr.) H. Mohr in Leafl. Schlechter Inst.*, 2: 14 (1996): R. subserrata.
antennifera Kunth
subsp. erythroxantha *(Rchb. f.) H. Mohr in Leafl. Schlechter Inst.*, 2: 13 (1996): R. erythroxantha.
subsp. hemsleyana *(Schltr.) H. Mohr in Leafl. Schlechter Inst.*, 2: 13 (1996): R. hemsleyana.
subsp. klabochorum *H. Mohr in Leafl. Schlechter Inst.*, 2: 11 (1996), contrary to Art. 8.1 ICBN (1994)—Locality not stated.
subsp. leontoglossa *(Schltr.) H. Mohr in Leafl. Schlechter Inst.*, 2: 13 (1996): R. leontoglossa.
subsp. striata *(Rolfe) H. Mohr in Leafl. Schlechter Inst.*, 2: 13 (1996): R. striata.
brachypus Rchb. f.
subsp. serrilabia *(Schltr.) H. Mohr in Leafl. Schlechter Inst.*, 2: 14 (1996): R. serrilabia.
chameleon *C.A. Luer & R. Escobar in Orquideologia*, 20(2): 120 (1996)—Colombia.
chrysoglossa *C.A. Luer & R. Escobar in Orquideologia*, 20(2): 123 (1996)—Colombia.
cloesii *C.A. Luer in Orquideologia*, 20(2): 125 (1996)—Peru.
condorensis *C.A. Luer & R. Escobar in Orquideologia*, 20(2): 128 (1996)—Ecuador.
contorta *(Ruiz & Pav.) C.A. Luer, Ic. Pleurothallid. XIII (Monogr. Syst. Bot. Missouri Bot. Gard., 59)*: 50 (1996): Humboldtia contorta.
cuprea *C.A. Luer & R. Escobar in Orquideologia*, 20(2): 130 (1996)—Colombia.
cymbula *C.A. Luer & R. Escobar in Orquideologia*, 20(2): 133 (1996)—Ecuador.
echinata *C.A. Luer & R. Escobar in Orquideologia*, 20(2): 135 (1996)—Colombia, Peru.

RESTREPIA:-
echo *C.A. Luer & R. Escobar in Orquideologia*, 20(2): 138 (1996)—Colombia.
ephippium *C.A. Luer & Hirtz in Orquideologia*, 20(2): 141 (1996)—Ecuador.
escobariana *C.A. Luer in Orquideologia*, 20(2): 144 (1996)—Colombia.
flosculata Luer
subsp. picta *(H. Mohr & Herzum) H. Mohr in Leafl. Schlechter Inst.*, 2: 8 (1996): R. flosculata var. picta.
jesupiana *C.A. Luer in Orquideologia*, 20(2): 146 (1996)—Venezuela.
maculata Lindl.
subsp. ecuadoriensis *(Rolfe) H. Mohr in Leafl. Schlechter Inst.*, 2: 16 (1996): R. ecuadoriensis.
subsp. robledorum *(Braas & Braem) H. Mohr in Leafl. Schlechter Inst.*, 2: 16 (1996): R. robledorum.
mendozae *C.A. Luer in Orquideologia*, 20(2): 157 (1996)—Ecuador.
metae *C.A. Luer in Orquideologia*, 20(2): 159 (1996)—Colombia.
muscifera (Lindl.) Rchb. f.
subsp. shuttleworthii *(Rolfe) H. Mohr in Leafl. Schlechter Inst.*, 2: 8 (1996): R. shuttleworthii.
piperitosa *C.A. Luer, Ic. Pleurothallid. XVI (Monogr. Syst. Bot. Missouri Bot. Gard., 65)*: 119 (1998)—Peru.
purpurea *C.A. Luer & R. Escobar in Orquideologia*, 20(2): 162 (1996)—Colombia.
radulifera *C.A. Luer & R. Escobar in Orquideologia*, 20(2): 165 (1996)—Venezuela.
renzii *C.A. Luer in Orquideologia*, 20(2): 167 (1996)—Venezuela.
robledorum *L.A. Braas & G.J. Braem in Orchidee*, 33(4): 155 (1982)—Colombia.
roseola *C.A. Luer & R. Escobar in Orquideologia*, 20(2): 170 (1996)—Venezuela.
schizosepala *C.A. Luer & Hirtz in Orquideologia*, 20(2): 172 (1996)—Ecuador.
seketii *C.A. Luer & R. Escobar in Orquideologia*, 20(2): 175 (1996)—Colombia.
tabeae *H. Mohr in Leafl. Schlechter Inst.*, 2: 10 (1996)—Colombia.
tsubotae *C.A. Luer & R. Escobar in Orquideologia*, 20(2): 178 (1996)—Colombia.
vasquezii *C.A. Luer in Orquideologia*, 20(2): 180 (1996)—Bolivia.

RESTREPIELLA (Orchidac.).
bicallosa *(Luer & R. Escobar) L.A. Braas & H. Mohr in Orchidee*, 33(4): 154 (1982): Restrepiopsis bicallosa.
clausa *(Luer & R. Escobar) L.A. Braas & H. Mohr in Orchidee*, 33(4): 154 (1982): Restrepiopsis clausa.
lenkenhoffii *L.A. Braas & H. Mohr in Orchidee*, 33(4): 154 (1982)—Costa Rica.
ophiocephala (Lindl.) Garay & Dunst.
f. clausa *I. Bock in Orchidee*, Suppl. 3: 6 (1996)—Costa Rica.
powersii *(Luer) L.A. Braas & H. Mohr in Orchidee*, 33(4): 154 (1982): Restrepiopsis powersii.

RESTREPIOPSIS (Orchidac.).
niesseniae *C.A. Luer, Ic. Pleurothallid. XX (Monogr. Syst. Bot. Missouri Bot. Gard., 79)*: 130 (2000)—Colombia.

RETAMA (Legumin.).
raetam (Forssk.) Webb
subsp. bovei *(Spach) S. Talavera & P.E. Gibbs in Lagascalia*, 18(2): 271 (1996): Spartium bovei.

RETANILLA *(DC.) A.T. Brongniart, Mém. Fam. Rhamnées*: 57 (1826): Colletia sect. Retanilla. (Rhamnac.).†
ephedra *(Vent.) A.T. Brongniart, Mém. Fam. Rhamnées*: 58 (1826): Colletia ephedra.†
obcordata *(Vent.) A.T. Brongniart, Mém. Fam. Rhamnées*: 58 (1826): Colletia obcordata.†

REYLLOPIA *J. Holub in Preslia*, 70(2): 110 (1998). *POLYGONACEAE* (Fallopia × Reynoutria).

REYNOUTRIA (Polygonac.).
ciliinervis *(Nakai) C.F. Fang in Clavis Pl. Chinae Bor.-Or. ed. 2 (ed. P.Y. Fu)*: 132 (1995), as 'ciliinerve': Pleuropterus ciliinervis.

REYNOUTRIA:-
× mizushimae *Yokouchi ex T. Shimizu, Fl. Nagano Pref.*: 1505 (1997)— Japan. (R. japonica var. uzenensis × sachalinensis).

RHADAMANTHOPSIS *(Oberm.) F. Speta in Phyton (Austria)*, 38(1): 74 (1998): Rhadamanthus subgen. Rhadamanthopsis. (Hyacinthac.).
karooicus *(Oberm.) F. Speta in Phyton (Austria)*, 38(1): 74 (1998): Rhadamanthus karooicus.
namibensis *(Oberm.) F. Speta in Phyton (Austria)*, 38(1): 74 (1998): Rhadamanthus namibensis.

RHADAMANTHUS (Hyacinthac.).
apiculatus *(H. Perrier) F. Speta in Phyton (Austria)*, 38(1): 76 (1998): Rhodocodon apiculatus.
calcicola *(H. Perrier) F. Speta in Phyton (Austria)*, 38(1): 76 (1998): Rhodocodon calcicola.
cyathiformis *(H. Perrier) F. Speta in Phyton (Austria)*, 38(1): 76 (1998): Rhodocodon cyathiformis.
floribundus *(H. Perrier) F. Speta in Phyton (Austria)*, 38(1): 76 (1998): Rhodocodon floribundus.
intermedius *(H. Perrier) F. Speta in Phyton (Austria)*, 38(1): 77 (1998): Rhodocodon intermedius.
involutus *J.C. Manning & D.A. Snijman in Novon*, 9(1): 113 (1999)—Cape Province.
madagascariensis *(Baker) F. Speta in Phyton (Austria)*, 38(1): 77 (1998): Rhodocodon madagascariensis.
mascarenensis *(Baker) F. Speta in Phyton (Austria)*, 38(1): 77 (1998): Rhodocodon mascarenensis.
rotundus *(H. Perrier) F. Speta in Phyton (Austria)*, 38(1): 77 (1998): Rhodocodon rotundus.
urgineoides *(Baker) F. Speta in Phyton (Austria)*, 38(1): 77 (1998): Rhodocodon urgineoides.

RHADINOTHAMNUS (Rutac.).
anceps *(DC.) Paul G. Wilson in Nuytsia*, 12(2): 286 (1998): Phebalium anceps.
rudis *(Bartl.) Paul G. Wilson in Nuytsia*, 12(2): 287 (1998): Phebalium rude.
subsp. amblycarpus *(F. Muell.) Paul G. Wilson in Nuytsia*, 12(2): 287 (1998): Eriostemon amblycarpus.
subsp. linearis *(C.A. Gardner) Paul G. Wilson in Nuytsia*, 12(2): 287 (1998): Phebalium lineare.

RHAMNUS (Rhamnac.).
subgen. Alaternus *(Mill.) Ju.E. Alexeev & N.N. Tzvelev in Fl. Vostochnoĭ Evropy̆*, 9: 396 (1996): gen. Alaternus.
sect. Chamaethamnos *(W. Vent) M.J. López Pacheco, E. Puente García, F. Llamas García & Á. Penas* Merino *in Stud. Bot. (Salamanca)*, 16: 55 (1997 publ. 1998): Oreoherzogia sect. Chamaethamnos.
subgen. Oreoherzogia *(W. Vent) Ju.E. Alexeev & N.N. Tzvelev in Fl. Vostochnoĭ Evropy̆*, 9: 396 (1996): gen. Oreoherzogia.
calderoniae *R. Fernández Nava in Polibotanica*, 1: 2 (1996)—Mexico Southwest (Oaxaca).
chimalapensis *R. Fernández Nava in Polibotanica*, 1: 4 (1996)—Mexico Southwest (Oaxaca).
daliensis *G.S. Fan & L.L. Deng in Sida*, 17(4): 680 (1997)—China South-central (Yunnan).
laoshanensis *D.K. Zang in Bull. Bot. Res. North-East. Forest. Univ.*, 19(4): 371 (1999)—China North-Central (Shandong).
utilis Decne.
var. multinervis *Y.Q. Zhu & D.K. Zang in Bull. Bot. Res. North-East. Forest. Univ.*, 18(1): 9 (1998)—China North-central (Shandong).

RHAPHIDOPHORA (Arac.).
araea *P.C. Boyce in Gard. Bull. Singapore*, 52(1): 116 (2000)—Sumatera.
balgooyi *P.C. Boyce in Gard. Bull. Singapore*, 52(1): 118 (2000)—Moluccas.
banosensis *P.C. Boyce in Gard. Bull. Singapore*, 52(2): 222 (2000), nom. nov.: Rhaphidophora stenophylla *Elmer.*
corneri *P.C. Boyce in Gard. Bull. Singapore*, 51(2): 205 (1999)—Malaya (Peninsular Malaysia).

RHAPHIDOPHORA:-
cretosa *P.C. Boyce in Gard. Bull. Singapore,* 52(2): 228 (2000)—Philippines.
floresensis *P.C. Boyce in Gard. Bull. Singapore,* 52(1): 126 (2000)—Lesser Sunda Is.
hayi *P.C. Boyce & J. Bogner in Gard. Bull. Singapore,* 52(1): 91 (2000)—Queensland, New Guinea (Irian Jaya, Papua New Guinea), Bismarck Archipelago.
nicolsonii *P.C. Boyce in Gard. Bull. Singapore,* 51(2): 236 (1999)—Malaya (Peninsular Malaysia).
okapensis *P.C. Boyce & J. Bogner in Gard. Bull. Singapore,* 52(1): 94 (2000)—New Guinea (Papua New Guinea).
sabit *P.C. Boyce in Gard. Bull. Singapore,* 52(1): 161 (2000)—Sulawesi.
ustulata *P.C. Boyce in Gard. Bull. Singapore,* 52(1): 176 (2000)—Sumatera.

RHAPIS (Palmae).
siamensis *D.R. Hodel in Palm J.,* 136 19 (1997)—Thailand.

RHAPONTICUM (Compos.).
carthamoides (Willd.) Iljin
var. chamarense *(Peschkova) O.S. Zhirova in Fl. Sibir. (Asterac.),* 13: 230 (1997 publ. 1998): R. chamarense.
uniflorum (L.) DC.
subsp. satzyperovii *(Soskov) V.N. Voroschilov in A.K. Skvortsov (ed.), Florist. issl. v razn. raĭonakh SSSR:* 198 (1985): R. satzyperovii.

RHEEDIA (Guttif.).
arubayensis *G. Coimbra Sanz in Rev. Soc. Boliv. Bot.,* 2(1): 31 (1998)—Bolivia.

RHEOCHLOA *T.S. Filgueiras, P.M. Peterson & Y. Herrera-Arrieta in Syst. Bot.,* 24(2) 123 (1999). *GRAMINEAE*
scabriflora *T.S. Filgueiras, P.M. Peterson & Y. Herrera-Arrieta in Syst. Bot.,* 24(2): 123 (1999)—Brazil West-central (Goiás).

RHEXIA (Melastomatac.).
× brevibracteata *D. Snyder in Bartonia,* 59: 65 (1996), without diagnostic latin descr.—New Jersey. (R. aristosa × virginica).

RHINACANTHUS (Acanthac.).
albus *(Roxb.) J.O. Voigt, Hort. Suburb. Calcutt.:* 491 (1845): Justicia alba.
gracilis Klotzsch
var. latilabiatus *K. Balkwill in Fl. S. Afr.,* 30(3:1): 13 (1995)—Cape Province, Natal.
kaokoensis *K. Balkwill & S. Williamson in Fl. S. Afr.,* 30(3:1): 14 (1995)—Namibia.
polonnaruwensis *L.H. Cramer in J. Nation. Sci. Counc. Sri Lanka,* 18(2): 168 (1990)—Sri Lanka.

RHINACTINIDIA (Compos.).
eremophila (Bunge) Novopokr. ex Botsch.
subsp. tuvinica *E.A. Koroljuk in Fl. Sibir. (Asterac.),* 13: 29 (1997 publ. 1998), contrary to Art. 37.5 ICBN (1994)—Tuva.

RHINANTHUS (Scrophulariac.).
elatior *(Schur) N.N. Tzvelev, Opred. Sosud. Rast. Severo-Zap. Rossii:* 680 (2000): Rhinanthus minor var. elatior.
pumilus (Sterneck) Pau
subsp. apuanus *(Soldano) L. Sáez in Acta Bot. Barcin.,* 46: 137 (2000): R. apuanus.

RHIPIDOCLADUM (Gramin.).
abregoense *X. Londoño & L.G. Clark in Novon,* 8(4): 419 (1998), as 'abregoensis'—Colombia.
neumannii *A.A. Sulekic, Z.E. Rúgolo de Agrasar & L.G. Clark in Darwiniana,* 37(3–4): 317 (1999)—Argentina Northwest (Salta, Tucumán), Bolivia.

RHIPSALIS (Cactac.).
baccifera (J.S. Muell.) Stearn
subsp. cleistogama *M. Kessler, P.L. Ibisch & W. Barthlott in Bradleya,* 18: 22 (2000)—Bolivia.
subsp. fasciculata *(Willd.) Süpplie in Tijdschr. Liefhebb. Cact. Vetpl. Kamerpl.,* 9(9): 136–137 (1996); *cf. Repert. Pl. Succ. (I.O.S.),* 47: 11 (1996 publ. 1997): Cactus fasciculatus.

RHIPSALIS:-
baccifera
subsp. fortdauphinensis *F. Süpplie in Brit. Cact. Succ. J.,* 14(2): 90 (1996), contrary to Art. 37.3 ICBN (1994)—Madagascar.
subsp. hileiabaiana *N.P. Taylor & W. Barthlott in Bradleya,* 13: 63 (1995)—Brazil Northeast (Bahia).
subsp. shaferi *(Britton & Rose) W. Barthlott & N.P. Taylor in Bradleya,* 13: 64 (1995): R. shaferi.
brevispina *(Barthlott) M. Kimnach in Cact. Succ. J. (U.S.A.),* 68(3): 156 (1996): Lepismium brevispinum.
clavata F.A.C. Weber
f. delicatula *(Loefgr.) W. Barthlott & N.P. Taylor in Bradleya,* 13: 72 (1995): R. clavata var. delicatula.
dissimilis
f. epiphyllanthoides *(Backeb.) F. Süpplie, Rhipsalidinae, ed. 4:* [100] (1994): Lepismium epiphyllanthoides.
f. epiphyllanthoides *(Backeb.) W. Barthlott & N.P. Taylor in Bradleya,* 13: 54 (1995): Lepismium epiphyllanthoides.
epiphylloides
var. bradei *(Porto & A. Cast.) M. Kimnach in Cact. Succ. J. (U.S.A.),* 68(3): 156 (1996): Hariota epiphylloides var. bradei.
ewaldiana *W. Barthlott & N.P. Taylor in Bradleya,* 13: 66 (1995)— Locality not stated.
floccosa Salm-Dyck ex Pfeiff.
subsp. hohenauensis *(F. Ritter) W. Barthlott & N.P. Taylor in Bradleya,* 13: 57 (1995): R. hohenauensis.
subsp. oreophila *N.P. Taylor & D. Zappi in Cactaceae Consensus Initiatives,* 6: 7 (1998), nom. nov.: R. monteazulensis *F. Ritter.*
subsp. pittieri *(Britton & Rose) W. Barthlott & N.P. Taylor in Bradleya,* 13: 55 (1995): R. pittieri.
subsp. pulvinigera *(G. Lindb.) W. Barthlott & N.P. Taylor in Bradleya,* 13: 55 (1995): R. pulvinigera.
subsp. tucumanensis *(F.A.C. Weber) W. Barthlott & N.P. Taylor in Bradleya,* 13: 57 (1995): R. tucumanensis.
herminiae *(Porto & A. Cast.) M. Kimnach in Cact. Succ. J. (U.S.A.),* 68(3): 156 (1996): Hariota herminiae.
hoelleri *W. Barthlott & N.P. Taylor in Bradleya,* 13: 50 (1995)—Brazil Southeast (Espírito Santo).
houlletiana
var. regnellii *(G. Lindb.) M. Kimnach in Cact. Succ. J. (U.S.A.),* 68(3): 156 (1996): R. regnellii.
juengeri *W. Barthlott & N.P. Taylor in Bradleya,* 13: 69 (1995)— Brazil Southeast (São Paulo).
lumbricoides
var. aculeata *(F.A.C. Weber) M. Kimnach in Cact. Succ. J. (U.S.A.),* 68(3): 156 (1996): R. aculeata.
micrantha
f. kirbergii *(Barthlott) F. Süpplie, Rhipsalidinae, ed. 4:* [100] (1994): R. kirbergii.
f. kirbergii *(Barthlott) W. Barthlott & N.P. Taylor in Bradleya,* 13: 62 (1995), isonym: R. kirbergii.
f. rauhiorum *(Barthlott) F. Süpplie, Rhipsalidinae, ed. 4:* [101] (1994): R. rauhiorum.
f. rauhiorum *(Barthlott) W. Barthlott & N.P. Taylor in Bradleya,* 13: 62 (1995), isonym: R. rauhiorum.
miyagawae *(Barthlott & Rauh) M. Kimnach in Cact. Succ. J. (U.S.A.),* 68(3): 156 (1996): Pfeiffera miyagawae.
neves-armondii K. Schum.
f. megalantha *(Loefgr.) W. Barthlott & N.P. Taylor in Bradleya,* 13: 50 (1995): R. megalantha.
olivifera *N.P. Taylor & D.C. Zappi in Cactaceae Consensus Initiatives,* 3: 8 (1997)— Brazil Southeast (Rio de Janeiro).
ormindoi *N.P. Taylor & D.C. Zappi in Cactaceae Consensus Initiatives,* 3: 8 (1997)—Brazil Southeast (Rio de Janeiro).
pacheco-leonis Loefgr.
subsp. catenulata *(Kimnach) W. Barthlott & N.P. Taylor in Bradleya,* 13: 59 (1995): R. paradoxa var. catenulata.

RHIPSALIS:-
paradoxa (Salm-Dyck ex Pfeiff.) Salm-Dyck
subsp. septentrionalis *N.P. Taylor & W. Barthlott in Bradleya,* 13: 57 (1995)—Brazil Northeast (Bahia, Pernambuco), Brazil Southeast (Espírito Santo, Minas Gerais).
salicornioides
var. cylindrica *(Britton & Rose) M. Kimnach in Cact. Succ. J. (U.S.A.),* 68(3): 156 (1996): Hatiora cylindrica.
spinescens *J.A. Lombardi in Acta Bot. Brasil.,* 5(2): 71 (1991 publ. 1992)—Brazil Southeast (São Paulo).
teres (Vell.) Steud.
f. capilliformis *(F.A.C. Weber) W. Barthlott & N.P. Taylor in Bradleya,* 13: 65 (1995): R. capilliformis.
f. heteroclada *(Britton & Rose) W. Barthlott & N.P. Taylor in Bradleya,* 13: 65 (1995): R. heteroclada.
f. prismatica *(Lem.) W. Barthlott & N.P. Taylor in Bradleya,* 13: 65 (1995): Hariota prismatica.

RHIZANTHES (Rafflesiac.).
deceptor *H. Bänziger & B. Hansen in Nat. Hist. Bull. Siam Soc.,* 48(1): 130 (2000)—Sumatera.
infanticida *H. Bänziger & B. Hansen in Nat. Hist. Bull. Siam Soc.,* 48(1): 127 (2000)—Thailand, Malaya (Peninsular Malaysia), Sumatera.

RHODALSINE (Caryophyllac.).
geniculata (Poir.) F.N. Williams
var. fontqueri *(Maire) A. Dobignard in Candollea,* 52(1): 127 (1997): Minuartia geniculata var. fontqueri.Geniculata (Poir.) F.N. Williams
var. maroccana *(Batt.) A. Dobignard in Candollea,* 52(1): 127 (1997): Alsine maroccana.

RHODAMNIA (Myrtac.).
angustifolia *N. Snow & G.P. Guymer in Austrobaileya,* 5(3): 421 (1999)—Queensland.

RHODANTHEMUM (Compos.).
gayanum (Coss. & Durieu) B.H. Wilcox, K. Bremer & Humphries
subsp. antiatlanticum *(Emb. & Maire) M. Ibn Tattou in Bocconea,* 8: 49 (1998): Leucanthemum gayanum subsp. antiatlanticum.
subsp. demnatense *(Murb.) M. Ibn Tattou in Bocconea,* 8: 49 (1998), with incorrect basionym ref.: Basionym not stated.
subsp. demnatense *(Murb.) R. Vogt in Lagascalia,* 18(2): 304 (1996): Chrysanthemum demnatense.
ifniense *(Font Quer) M. Ibn Tattou in Bocconea,* 8: 49 (1998): Leucanthemum ifniense.
redieri (Maire) B.H. Wilcox, K. Bremer & Humphries
subsp. humbertii *F. Gómiz in An. Jard. Bot. Madrid,* 58(1): 200 (2000), nom. nov.: Leucanthemum atlanticum var. sericeum *Humbert.*

RHODAX (Cistac.).
distachyos *(Roth) J. Holub in Preslia,* 70(2): 110 (1998): Cistus distachyos.
subsp. rubellus *(Fiori) J. Holub in Preslia,* 70(2): 110 (1998): Helianthemum canum var. rubellum.
subsp. stipulatus *(Willk.) J. Holub in Preslia,* 70(2): 110 (1998): Helianthemum rubellum γ stipulatum.
hieronymi *(Sennen) J. Holub in Preslia,* 70(2): 110 (1998): Helianthemum hieronymi.
raynaudii *(Ortega Oliv., Romero García & C. Morales) J. Holub in Preslia,* 70(2): 110 (1998): Helianthemum raynaudii.
viscaria *(Boiss. & Reut.) J. Holub in Preslia,* 70(2): 110 (1998): Helianthemum viscarium.

RHODIOLA (Crassulac.).
integrifolia Raf.
subsp. leedyi *(Rosend. & J.W. Moore) H. Ohba in J. Jap. Bot.,* 74(1): 61 (1999): Sedum roseum var. leedyi.

RHODIOLA:-
integrifolia
subsp. leedyi *(Rosend. & J.W. Moore) Reid Moran in Cact. Succ. J. (U.S.A.)*, 72(3): 138 (2000), isonym; *see H. Ohba in J. Jap. Bot.*, 74(1): 61 (1999): Sedum roseum var. leedyi.
subsp. neomexicana *(Britton) H. Ohba in J. Jap. Bot.*, 74(1): 61 (1999): R. neomexicana.
subsp. neomexicana *(Britton) Reid Moran in Cact. Succ. J. (U.S.A.)*, 72(3): 138 (2000), isonym; *see H. Ohba in J. Jap. Bot.*, 74(1): 61 (1999): R. neomexicana.
subsp. procera *(R.T. Clausen) H. Ohba in J. Jap. Bot.*, 74(1): 61 (1999): Sedum integrifolium subsp. procerum.
multibracteata *H. Chuang in Acta Bot. Yunnanica*, 19(3): 225 (1997)—China South-central (Yunnan).
rosea L.
subsp. integrifolia *(Raf.) V.V. Petrovsky in Arktich. Fl. SSSR*, 9(1): 14 (1984), isonym; *see Kozhevn. in Novosti Sist. Vyssh. Rast.*, 18: 240 (1981): R. integrifolia.
subsp. krivochizhinii *(Sipliv.) S.B. Gontcharova in Bull. Nation. Sci. Mus., B (Tokyo)*, 25(2): 56 (1999): R. krivochizhinii.
f. purpurascens *Y. Meng & J.J. Tian in Bull. Bot. Res. North-East. Forest. Univ.*, 17(2): 143 (1997)—China North-central (Hebei).
subsp. sachalinensis *(Boriss.) S.B. Gontcharova in Bull. Nation. Sci. Mus., B (Tokyo)*, 25(2): 56 (1999): R. sachalinensis.

RHODODENDRON (Ericac.).
subsect. Coriifolia *T. Yamazaki, Revis. Gen. Rhododendron Japan, Taiwan, Korea & Sakhalin*: 34 (1996).
ser. Coriifolia *T. Yamazaki, Revis. Gen. Rhododendron Japan, Taiwan, Korea & Sakhalin*: 34 (1996).
ser. Kiyosumensia *T. Yamazaki, Revis. Gen. Rhododendron Japan, Taiwan, Korea & Sakhalin*: 111 (1996).
subgen. Lutescentia *T. Yamazaki, Revis. Gen. Rhododendron Japan, Taiwan, Korea & Sakhalin*: 2 (1996).
ser. Maculifera *(Tagg) T. Yamazaki, Revis. Gen. Rhododendron Japan, Taiwan, Korea & Sakhalin*: 126 (1996): R. subser. Maculiferum.
ser. Mariesia *T. Yamazaki, Revis. Gen. Rhododendron Japan, Taiwan, Korea & Sakhalin*: 106 (1996).
sect. Micrantha *(Hutch.) T. Yamazaki, Revis. Gen. Rhododendron Japan, Taiwan, Korea & Sakhalin*: 20 (1996): R. ser. Micranthum.
sect. Schlippenbachia *T. Yamazaki, Revis. Gen. Rhododendron Japan, Taiwan, Korea & Sakhalin*: 78 (1996).
ser. Schlippenbachia *T. Yamazaki, Revis. Gen. Rhododendron Japan, Taiwan, Korea & Sakhalin*: 78 (1996).
adenanthum *M.Y. He in J. Sichuan Univ., Nat. Sci. Ed.*, 1984(1): 94 (1984)—China Southeast (Guangxi).
albrechtii Maxim.
f. albiflorum *T. Yamazaki, Revis. Gen. Rhododendron Japan, Taiwan, Korea & Sakhalin*: 121 (1996)—Japan.
araiophyllum Balf. f. & W.W. Sm.
subsp. lapidosum *(T.L. Ming) M.Y. Fang in Proc. Fourth Internat. Rhodod. Conf.*: 57 (1989): R. lapidosum.
arboreum Sm.
subsp. albotomentosum *(Davidian) D.F. Chamberlain in R.H.S. Rhododendron Handb. 1998*: 350 (1998), as 'albomentosum': R. delavayi var. albotomentosum.
argyrophyllum Franch.
var. glabriovarium *M.Y. He in Acta Phytotax. Sin.*, 35(1): 66 (1997)—China South-central (Sichuan).
aureum Georgi
subsp. hypopitys *(Pojark.) V.N. Voroschilov in A.K. Skvortsov (ed.), Florist. issl. v razn. raïonakh SSSR*: 185 (1985): R. hypopitys.
chilanshanense *Y. Kurashige in Edinburgh J. Bot.*, 56(1): 75 (1999)—Taiwan.
chrysocalyx H. Lév. & Vaniot
var. xiushanense *(W.P. Fang) M.Y. He in Fl. Reipubl. Popul. Sin.*, 57(2): 407 (1994): R. xiushanense.

RHODODENDRON:-
× chukirishima *T. Yamazaki, Revis. Gen. Rhododendron Japan, Taiwan, Korea & Sakhalin*: 147 (1996)—Japan. (R. indicum × scabrum).
crassistylum *M.Y. He in J. Sichuan Univ., Nat. Sci. Ed.*, 24(1): 83 (1987)—China Southeast (Jiangxi).
dauricum L.
f. albiflorum *(Turcz.) C.F. Fang in Clavis Pl. Chinae Bor.-Or. ed. 2 (ed. P.Y. Fu)*: 496 (1995), isonym; *see Y.C. Zhu in Pl. Medic. Chinae Bor.-Orient. (ed. Y.C. Zhu)*: 864 (1989): R. dauricum var. albiflorum.
subsp. mucronulatum *(Turcz.) V.N. Voroschilov in A.K. Skvortsov (ed.), Florist. issl. v razn. raïonakh SSSR*: 185 (1985): R. mucronulatum.
dayiense *M.Y. He in Acta Phytotax. Sin.*, 35(1): 63 (1997)—China South-central (Sichuan).
dilatatum Miq.
var. lasiocarpum *(H. Hara) T. Yamazaki, Revis. Gen. Rhododendron Japan, Taiwan, Korea & Sakhalin*: 93 (1996): R. decandrum f. lasiocarpum.
diversipilosum *(Nakai) H. Harmaja in Ann. Bot. Fenn.*, 35(4): 263 (1998 publ. 1999): Ledum palustre var. diversipilosum.
eastmanii *K.A. Kron & M. Creel in Novon*, 9(3): 377 (1999)—South Carolina.
fulvum Balf. f. & W.W. Sm.
subsp. fulvoides *(Balf. f. & Forrest) D.F. Chamberlain in R.H.S. Rhododendron Handb. 1998*: 350 (1998): R. fulvoides.
fuscipilum *M.Y. He in J. Sichuan Univ., Nat. Sci. Ed.*, 1984(1): 94 (1984)—China Southeast (Guangxi).
fuyuanense *Z.H. Yang in Acta Phytotax. Sin.*, 35(2): 189 (1997)—China South-central (Yunnan).
guihainianum *G.Z. Li in Guihaia*, 15(4): 299 (1995)—China Southeast (Guangxi).
guizhouense *M.Y. Fang in Proc. Fourth Internat. Rhodod. Conf.*: 58 (1989)—China South-central (Guizhou), China Southeast (Hunan).
heizhugouense *M.Y. He & L.C. Hu in Acta Bot. Yunnanica*, 18(3): 295 (1996)—China South-central (Sichuan).
henryi Hance
var. dunnii *(E.H. Wilson) M.Y. He in Fl. Reipubl. Popul. Sin.*, 57(2): 365 (1994): R. dunnii.
jasminoides *M.Y. He in J. Sichuan Univ., Nat. Sci. Ed.*, 1984(1): 93 (1984)—China Southeast (Guangxi).
jinchangense *Z.H. Yang in Acta Phytotax. Sin.*, 35(2): 186 (1997)—China South-central (Yunnan).
kaempferi Planch.
f. kozushimense *T. Yamazaki, Revis. Gen. Rhododendron Japan, Taiwan, Korea & Sakhalin*: 60 (1996)—Japan.
var. saikaiense *(T. Yamaz.) T. Yamazaki, Revis. Gen. Rhododendron Japan, Taiwan, Korea & Sakhalin*: 61 (1996): R. obtusum var. saikaiense.
× kisoanum *Okuhara ex T. Shimizu, Fl. Nagano Pref.*: 1508 (1997), as 'kisoana'—Japan. (R. dilatatum × reticulatum).
lamprophyllum Hayata
var. ovalifolium *T. Yamazaki, Revis. Gen. Rhododendron Japan, Taiwan, Korea & Sakhalin*: 28 (1996)—Taiwan.
lamrialianum *G. Argent & T.J. Barkman in New Plantsman*, 7(4): 208 (2000)—Borneo (Sabah).
subsp. gunsalamianum *G. Argent & T.J. Barkman in New Plantsman*, 7(4): 214 (2000)—Borneo (Sabah).
lapponicum (L.) Wahlenb.
var. alpinum *(Glehn) T. Yamazaki, Revis. Gen. Rhododendron Japan, Taiwan, Korea & Sakhalin*: 19 (1996): R. parvifolium f. alpinum.
subsp. parvifolium *(Adams) T. Yamazaki, Revis. Gen. Rhododendron Japan, Taiwan, Korea & Sakhalin*: 18 (1996): R. parvifolium.
subsp. parvifolium *(Adams) L.I. Malyschev in Fl. Sibir. (Pyrolac.-Lamiac.)*, 11: 18 (1997), isonym: Rhododendron parvifolium.

RHODODENDRON:-
latoucheae Franch.
var. amamiense *(Ohwi) T. Yamazaki, Revis. Gen. Rhododendron Japan, Taiwan, Korea & Sakhalin*: 25 (1996): R. amamiense.
linguiense *G.Z. Li in Guihaia*, 15(4): 295 (1995)—China Southeast (Guangxi).
loranthiflorum Sleumer
subsp. lakekamuensis *W.N. Takeuchi in Edinburgh J. Bot.*, 57(3): 333 (2000)—New Guinea (Papua New Guinea).
madulidii *G. Argent in New Plantsman*, 5(4): 204 (1998)—Philippines.
malipoense *M.Y. He in J. Sichuan Univ., Nat. Sci. Ed.*, 1984(1): 95 (1984)—China South-central (Yunnan).
morii Hayata
var. taitunense *(T. Yamaz.) D.F. Chamberlain in R.H.S. Rhododendron Handb. 1998*: 350 (1998): R. pseudochrysanthum var. taitunense.
neriiflorum Franch.
var. appropinquans *(Tagg & Forrest) W.K. Hu in Fl. Reipubl. Popul. Sin.*, 57(2): 290 (1994): R. floccigerum var. appropinquans.
notiale *L.A. Craven in Edinburgh J. Bot.*, 53(1): 33 (1996)—Queensland.
octandrum *M.Y. He in J. Sichuan Univ., Nat. Sci. Ed.*, 24(1): 84 (1987)—China South-Central (Guizhou).
orbiculare Decne.
subsp. maolingense *G.Z. Li in Guihaia*, 15(4): 294 (1995)—China Southeast (Guangxi).
ovatum Planch.
var. setuliferum *M.Y. He in J. Sichuan Univ., Nat. Sci. Ed.*, 1984(1): 96 (1984)—China Southeast (Zhejiang, Fujian).
pachytrichum Franch.
var. monosematum *(Hutch.) D.F. Chamberlain in R.H.S. Rhododendron Handb. 1998*: 350 (1998): R. monosematum.
ponticum L.
var. heterophyllum *R. Anşin in Turkish J. Agric. Forest.*, 18: 139 (1994)—Turkey.
pseudochrysanthum Hayata
var. rufovelutinum *(T. Yamaz.) T. Yamazaki, Revis. Gen. Rhododendron Japan, Taiwan, Korea & Sakhalin*: 127 (1996): R. pseudochrysanthum f. rufovelutinum.
purdomii Rehder & E.H. Wilson
var. villosum *L.H. Wu in J. Wuhan Bot. Res.*, 17(1): 21 (1999)—China Southeast (Henan).
rousei *G. Argent & D. Madulid in New Plantsman*, 5(1): 25 (1998)—Philippines.
rubropilosum Hayata
var. breviperulatum *(Hayata) T. Yamazaki, Revis. Gen. Rhododendron Japan, Taiwan, Korea & Sakhalin*: 35 (1996): R. breviperulatum.
var. grandiflorum *T. Yamazaki, Revis. Gen. Rhododendron Japan, Taiwan, Korea & Sakhalin*: 36 (1996)—Taiwan.
var. taiwanalpinum *(Ohwi) S.Y. Lu, Yuen P. Yang & Y.H. Tseng in Fl. Taiwan, ed. 2*, 4: 32 (1998): R. taiwanalpinum.
rushforthii *G. Argent & D. Chamberlain in New Plantsman*, 3(4): 195 (1996)—Vietnam.
sanctum Nakai
f. albiflorum *T. Yamazaki, Revis. Gen. Rhododendron Japan, Taiwan, Korea & Sakhalin*: 90 (1996)—Japan.
scabrum D. Don
f. roseum *S. Hatusima & Kudaka in J. Phytogeogr. Taxon.*, 44(1–2): 86 (1996)—Japan.
simiarum Hance
var. grandifolium *G.Z. Li in Guihaia*, 15(4): 295 (1995)—China Southeast (Guangxi), Hainan.
simsii Planch.
var. strigosistylum *G.Z. Li in Guihaia*, 15(4): 298 (1995), as 'strigoso-stylum'—China Southeast (Guangxi).
taichungianum *S.S. Ying in Mem. Coll. Agric. Nation. Taiwan Univ.*, 31(1): 29 (1991), contrary to Art. 37.5 ICBN (1994)—Taiwan.
tiantangense *G.Z. Li in Guihaia*, 15(4): 298 (1995)—China Southeast (Guangxi).
tschonoskii Maxim.
f. roseotinctum *Y. Kurashige, N. Kobayashi & T. Yukawa in Ann. Tsukuba Bot. Gard.*, 19: 6 (2000)—Japan (Honshu).
tuhanensis *G. Argent & T.J. Barkman in New Plantsman*, 7(4): 215 (2000)—Borneo (Sabah).

RHODODENDRON:-
ungeonticum *A.P. Khokhrjakov & M.T. Mazurenko in Byull. Glavn. Bot. Sada (Moscow),* 172: 24 (1995)—Transcaucasus (Adzhariya).
xiaoxidongense *W.K. Hu in J. Sichuan Univ., Nat. Sci. Ed.,* 27(4): 492 (1990)—China Southeast (Jiangxi).
yuefengense *G.Z. Li in Guihaia,* 15(4): 293 (1995)—China Southeast (Guangxi).

RHODODON (Labiat.).
angulatus *(Tharp) B.L. Turner in Phytologia,* 78(6): 449 (1995): Stachydeoma angulata.

RHODOLAENA (Sarcolaenac.).
coriacea *G.E. Schatz, P.P. Lowry & A.-E. Wolf in Adansonia,* 22(2): 244 (2000)—Madagascar.
leroyana *G.E. Schatz, P.P. Lowry & A.-E. Wolf in Adansonia,* 22(2): 247 (2000)—Madagascar.
macrocarpa *G.E. Schatz, P.P. Lowry & A.-E. Wolf in Adansonia,* 22(2): 248 (2000)—Madagascar.

RHODOSPATHA (Arac.).
pellucida *T.B. Croat & Grayum in Novon,* 9(4): 500 (1999)—Panama, Colombia, Costa Rica, Ecuador, Nicaragua.

RHODOSTEMONODAPHNE (Laurac.).
elephantopus *S. Madriñán in Brittonia,* 48(1): 46 (1996)—French Guiana.
leptoclada *S. Madriñán in Brittonia,* 48(1): 49 (1996)—French Guiana, Brazil North (Amapá).
morii *S. Madriñán in Brittonia,* 48(1): 52 (1996)—French Guiana, Surinam.
recurva *H. van der Werff in Novon,* 10(3): 291 (2000)—Brazil North (Amazonas).
revolutifolia *S. Madriñán in Brittonia,* 48(1): 55 (1996)—French Guiana.
rufovirgata *S. Madriñán in Brittonia,* 48(1): 58 (1996)—French Guiana.
saulensis *S. Madriñán in Brittonia,* 48(1): 60 (1996)—French Guiana.
scandens *S. Madriñán in Brittonia,* 48(1): 63 (1996)—French Guiana, Brazil North (Amapá).

RHODOTHYRSUS *H.-J. Esser in Brittonia,* 51(2): 177 (1999). *EUPHORBIACEAE*
hirsutus *H.-J. Esser in Brittonia,* 51(2): 178 (1999)—Venezuela.
macrophyllus *(Ducke) H.-J. Esser in Brittonia,* 51(2): 179 (1999): Senefeldera macrophylla.

RHOEADIA *C. Lemaire in Jard. Fleur,* 4: t. 356 (1853). *MALVACEAE*

RHOICISSUS (Vitac.).
kougabergensis *E. Retief & E.J. Van Jaarsveld in Bothalia,* 27(1): 49 (1997)—Cape Province.

RHOMBODA (Orchidac.).
arunachalensis *A. Nageswara Rao in J. Econ. Taxon. Bot.,* 22(2): 426 (1998)—Assam (Arunachal Pradesh)
blackii *(Ames) P. Ormerod in Austral. Orchid Rev.,* 63(4): 11 (1998): Hetaeria blackii.
pauciflora *(Ridl.) P. Ormerod in Austral. Orchid Rev.,* 63(4): 11 (1998): Hetaeria pauciflora.
petelottii *(Gagnep.) P. Ormerod in Austral. Orchid Rev.,* 63(4): 11 (1998): Haemaria petelottii.
tokioi *(Fukuy.) P. Ormerod in Austral. Orchid Rev.,* 63(4): 11 (1998): Hetaeria tokioi.

RHOMBOPHYLLUM (Aizoac.).
albanense *(L. Bolus) H.E.K. Hartmann in Bradleya,* 16: 81 (1998): Heteroa albanensis.
dyeri *(L. Bolus) H.E.K. Hartmann in Bradleya,* 16: 81 (1998): Hereroa dyeri.

RHOPALEPHORA (Commelinac.).
scaberrima (Blume) Faden
var. fruticosa *(Hook. f.) S.L. Vrinda & M.V.N. Panikkar in J. Econ. Taxon. Bot.,* 23(3): 670 (1999), basionym non-existent: Aneilema scaberrimum var. fruticose.

RHOPALOCARPUS (Sphaerosepalac.).
crassinervius *(Capuron) G.E. Schatz, P.P. Lowry & A.-E. Wolf in Adansonia,* 21(1): 117 (1999): R. coriaceus var. crassinervius.

RHOPALOCARPUS:-
parvifolius *(Capuron) G.E. Schatz, P.P. Lowry & A.-E. Wolf in Adansonia,* 21(1): 120 (1999): R. louvelii var. parvifolius.

RHUS (Anacardiac.).
pygmaea *R.O. Moffett in Bot. J. Linn. Soc.,* 130(1): 39 (1999)—Transvaal.
tamaulipana *B.L. Turner in Phytologia,* 80(5): 370 (1996 publ. 1997)—Mexico Northeast (Tamaulipas, Nuevo León).
virens Lindh. ex A. Gray
var. australis *B.L. Turner in Phytologia,* 80(5): 375 (1996 publ. 1997)—Mexico Southwest (Oaxaca), Mexico Central (Puebla).
var. choriophylla *(Wooton & Standl.) L.D. Benson in L.D. Benson & R.A. Darrow, Trees & Shrubs of Southwest. Deserts (ed. 3):* 143 (1981): R. choriophylla.
var. choriophylla *(Wooton & Standl.) B.L. Turner in Phytologia,* 80(5): 375 (1996 publ. 1997), isonym; *see L.D. Benson in L.D. Benson & R.A. Darrow, Trees & Shrubs of Southwest.* Deserts (ed. 3): 143 (1981): R. choriophylla.

RHYNCHOSIA (Legumin.).
androyensis *D.J. Du Puy & J.-N. Labat in Bull. Mus. Nation. Hist. Nat., B, Adansonia,* Ser. 4, 18(1–2): 86 (1996)—Madagascar.
atropurpurea *G. Germishuizen in Bothalia,* 28(1): 58 (1998)—Transvaal.
baukea *D.J. Du Puy & J.-N. Labat in Bull. Mus. Nation. Hist. Nat., B, Adansonia,* Ser. 4, 18(1–2): 86 (1996), nom. nov.: Baukea insignis *Vatke.*
chapmanii *B. Verdcourt in Kew Bull.,* 55(1): 140 (2000)—Malawi.
chimanimaniensis *B. Verdcourt in Kew Bull.,* 55(1): 144 (2000)—Mozambique, Zimbabwe.
delicatula *O. Téllez V. & M. Sousa S. in Novon,* 10(3): 257 (2000)—Mexico Southwest (Jalisco).
leandrii *D.J. Du Puy & J.-N. Labat in Bull. Mus. Nation. Hist. Nat., B, Adansonia,* Ser. 4, 18(1–2): 89 (1996)—Madagascar.
stipitata *B. Verdcourt in Kew Bull.,* 55(1): 142 (2000)—Botswana, Zambia, Zimbabwe, Angola.
variegata *(Deflers) J.R.I. Wood, Handbook Yemen Fl.:* 161 (1997): Vigna variegata.
versicolor Baker
subsp. imerinensis *D.J. Du Puy & J.-N. Labat in Bull. Mus. Nation. Hist. Nat., B, Adansonia,* Ser. 4, 18(1–2): 92 (1996)—Madagascar.
wildii *B. Verdcourt in Kew Bull.,* 55(1): 137 (2000)—Zimbabwe, Mozambique, Zambia.

RHYNCHOSPORA (Cyperac.).
sect. Rigidifoliae *W.W. Thomas in Brittonia,* 48(4): 482 (1996 publ. 1997).
capillifolia *W.W. Thomas in Brittonia,* 48(4): 484 (1996 publ. 1997), nom. nov.: R. rigidifolia subsp. capillacea *T. Koyama.*
corymbosa *(L.) N.L. Britton in Trans. New York Acad. Sci.,* 11: 84 (1892): Scirpus corymbosus.†
subsp. legrandii *(Kük. ex Barros) M. Luceño & A.L.L. Vanzela in Canad. J. Bot.,* 76(3): 448 (1998): R. legrandii.
depressirostris *M.T. Strong in Brittonia,* 52(3): 241 (2000)—Puerto Rico.
dilectissimi-patris *M. Luceño in An. Jard. Bot. Madrid,* 55(2): 309 (1997)—Brazil Northeast (Piauí).
harveyi W. Boott
var. culixa *(Gale) R. Kral in Novon,* 9(2): 206 (1999): R. culixa.
imeriensis *(Kük.) W.W. Thomas in Brittonia,* 48(4): 481 (1996 publ. 1997): R. cephalotes var. imeriensis.
marcelo-guerrae *M. Luceño & M.V. Alves in Candollea,* 52(1): 192 (1997)—Brazil North (Tocantins).
megaplumosa *E.L. Bridges & S.L. Orzell in Lundellia,* 3: 20 (2000)—Florida.
modesti-lucennoi *S. Castroviejo in Nordic J. Bot.,* 15(6): 569 (1995 publ. 1996)—Spain.
planifolia *(T. Koyama) W.W. Thomas in Brittonia,* 48(4): 485 (1996 publ. 1997): R. rigidifolia subsp. planifolia.
plusquamrobusta *M. Luceño & M.L.L. Martins in An. Jard. Bot. Madrid,* 57(1): 174 (1999)—Brazil Southeast (Espírito Santo).

RHYNCHOSPORA:-
recognita *(Gale) R. Kral in Novon,* 9(2): 205 (1999): R. globularis var. recognita.
rubra *(Lour.) T. Makino in Bot. Mag. Tokyo,* 17: 180 (1903): Schoenus ruber.†
zacualtipanensis *M.T. Strong in Sida,* 19(1): 115 (2000)—Mexico Northeast (Hidalgo).

RHYNCHOTECHUM (Gesneriac.).
ellipticum DC.
var. saurauifolia *(S.S. Ying) S.S. Ying in Mem. Coll. Agric. Nation. Taiwan Univ.,* 29(2): 47 (1989): Lysimachia saurauifolia.

RHYTACHNE (Gramin.).
anisonodis *B.S. Sun in Acta Bot. Yunnanica,* 20(1): 34 (1998)—China South-central (Yunnan).
lijiangensis *B.S. Sun in Acta Bot. Yunnanica,* 20(1): 33 (1998)—China South-central (Yunnan).

RHYTIDOCAULON (Asclepiadac.).
ciliatum *P. Hanáček & M. Říčánek in Asklepios,* 80: 19 (2000)—South Yemen.
macrolobum Lavranos
subsp. minimum *U. Meve & S. Collenette in Edinburgh J. Bot.,* 56(1): 80 (1999)—Saudi Arabia.

RHYTIDOSPORUM (Pittosporac.).
diosmoides *(Putt.) L.W. Cayzer, M.D. Crisp & I.R.H. Telford in Austral. Syst. Bot.,* 12(5): 702 (1999): Bursaria diosmoides.
inconspicuum *L.W. Cayzer, M.D. Crisp & I.R.H. Telford in Austral. Syst. Bot.,* 12(5): 700 (1999)—Victoria, New South Wales, Tasmania.

RHYTIONANTHOS (Orchidac.).
strigosus *L.A. Garay in Harvard Pap. Bot.,* 4(1): 311 (1999), as 'strigosum'—Vietnam.

RIBES (Grossulariac.).
altissimum Turcz. ex Pojark.
var. verruculosum *L.I. Malyschev in Fl. Sibir. (Berberidac.-Grossulariac.),* 7: 211 (1994)—Altay.
austroecuadorense *A. Freire-Fierro in Novon,* 8(4): 356 (1998)—Ecuador.
diacanthum Pall.
f. weichangense *J.X. Huang & J.Z. Wang in J. Hebei Agric. Univ.,* 17(4): 14 (1994), as 'weichangensis'—China North-central (Hebei).
erectum *A. Freire-Fierro in Novon,* 8(4): 354 (1998)—Ecuador, Colombia.
griffithii Hook. f. & Thomson
var. gongshanense *(T.C. Ku) L.T. Lu in Fl. Reipubl. Popul. Sin.,* 35(1): 309 (1995): R. gongshanense.
himalense Royle ex Decne.
var. verruculosum *(Rehder) L.T. Lu in Fl. Reipubl. Popul. Sin.,* 35(1): 306 (1995): R. emodense var. verruculosum.
khorasanicum *F. Saghafi & M. Assadi in Iranian J. Bot.,* 7(1): 8 (1996), as 'khorasanica'—Iran.
nigrum L.
var. eglandulosum *L.I. Malyschev in Fl. Sibir. (Berberidac.-Grossulariac.),* 7: 214 (1994)—Buryatiya.
oxyacanthoides L.
var. hendersonii *(C.L. Hitchc.) P.K. Holmgren in A. Cronquist et al., Intermountain Fl., Vasc. Pl. Intermountain West,* U.S.A., 3A: 16 (1997): R. hendersonii.
sinanense F. Maek.
f. inerme *Karayama ex T. Shimizu, Fl. Nagano Pref.:* 1508 (1997)—Japan.
spicatum E. Robson
var. pubescens *(Hartm.) R. Cinovskis in Fl. Baltiiskikh Respublik,* 2: 33 (1996): R. rubrum β pubescens.
takare D. Don
var. desmocarpum *(Hook. f. & Thomson) L.T. Lu in Fl. Reipubl. Popul. Sin.,* 35(1): 351 (1995): R. desmocarpum.
tomentosum *M. Gandoger, Dec. Pl. Nov.,* 1: 4 (1875)—France.
triste Pall.
var. repens *(A.I. Baranov) L.T. Lu in Fl. Reipubl. Popul. Sin.,* 35(1): 320 (1995): R. repens.

RIBES:-
viridiflorum *(W.C.Cheng) L.T. Lu & G. Yao in Fl. Reipubl. Popul. Sin.*, 35(1): 345 (1995): R. tenue var. viridiflorum.

× **RIBULARIA** *J. Holub in Preslia*, 70(2): 110 (1998). *GROSSULARIACEAE* (Grossularia × Ribes).
paui *(Blanca) J. Holub in Preslia*, 70(2): 110 (1998): Ribes paui.

RICHEA (Epacridac.).
alpina *Y. Menadue in Austral. Syst. Bot.*, 13(5): 798 (2000)—Tasmania.
pandanifolia Hook. f.
subsp. ramulosa *Y. Menadue in Austral. Syst. Bot.*, 13(5): 795 (2000)—Tasmania.

RICINODENDRON (Euphorbiac.).
heudelotii *(Baill.) Pierre ex Heckel in Ann. Inst. Colon. Marseille*, 5(2): 40 (1898): Jatropha heudelotii.†

RIDLEYANDRA *A. Weber & B.L. Burtt in Beitr. Biol. Pfl.*, 70(2–3): 171 (1997–98 publ. 1998). *GESNERIACEAE*
sect. Stilpnothrix *(C.B. Clarke) A. Weber & B.L. Burtt in Beitr. Biol. Pfl.*, 70(2–3): 172 (1997–98 publ. 1998): Didissandra sect. Stilpnothrix.
atrocyanea *(Ridl.) A. Weber in Beitr. Biol. Pfl.*, 70(2–3): 171 (1997–98 publ. 1998): Didissandra atrocyanea.
atropurpurea *(Ridl.) A. Weber in Beitr. Biol. Pfl.*, 70(2–3): 247 (1997–98 publ. 1998): Didissandra atropurpurea.
castaneifolia *(Ridl.) A. Weber in Beitr. Biol. Pfl.*, 70(2–3): 254 (1997–98 publ. 1998): Didissandra castaneifolia.
corneri *A. Weber in Beitr. Biol. Pfl.*, 70(2–3): 245 (1997–98 publ. 1998)—Malaya (Peninsular Malaysia).
flammea *(Ridl.) A. Weber in Beitr. Biol. Pfl.*, 70(2–3): 248 (1997–98 publ. 1998): Didissandra flammea.
glabrescens *(Ridl.) A. Weber in Beitr. Biol. Pfl.*, 70(2–3): 242 (1997–98 publ. 1998): Didissandra glabrescens.
kerrii *A. Weber in Beitr. Biol. Pfl.*, 70(2–3): 253 (1997–98 publ. 1998)—Thailand.
kiewii *(Kiew) A. Weber in Beitr. Biol. Pfl.*, 70(2–3): 243 (1997–98 publ. 1998): Didissandra kiewii.
latisepala *(Ridl.) A. Weber in Beitr. Biol. Pfl.*, 70(2–3): 250 (1997–98 publ. 1998): Didissandra latisepala.
longisepala *(Ridl.) A. Weber in Beitr. Biol. Pfl.*, 70(2–3): 259 (1997–98 publ. 1998): Didissandra longisepala.
morganii *(Franch.) A. Weber in Beitr. Biol. Pfl.*, 70(2–3): 251 (1997–98 publ. 1998): Didissandra morganii.
natunae *B.L. Burtt in Beitr. Biol. Pfl.*, 70(2–3): 266 (1997–98 publ. 1998)—Borneo (Kalimantan).
ornata *(C.B. Clarke) B.L. Burtt in Beitr. Biol. Pfl.*, 70(2–3): 172, 267 (1997–98 publ. 1998): Didissandra ornata.
petiolata *(Ridl.) A. Weber in Beitr. Biol. Pfl.*, 70(2–3): 247 (1997–98 publ. 1998): Didissandra petiolata.
porphyrantha *(A. Weber & Kiew) A. Weber in Beitr. Biol. Pfl.*, 70(2–3): 244 (1997–98 publ. 1998): Didissandra porphyrantha.
quercifolia *(Ridl.) A. Weber in Beitr. Biol. Pfl.*, 70(2–3): 255 (1997–98 publ. 1998): Didissandra quercifolia.
rufa *(C.B. Clarke) B.L. Burtt in Beitr. Biol. Pfl.*, 70(2–3): 270 (1997–98 publ. 1998): Didissandra rufa.
serratifolia *(Ridl.) A. Weber in Beitr. Biol. Pfl.*, 70(2–3): 258 (1997–98 publ. 1998): Didissandra serratifolia.
stellata *A. Weber in Beitr. Biol. Pfl.*, 70(2–3): 262 (1997–98 publ. 1998)—Malaya (Peninsular Malaysia).
synaptica *(B.L. Burtt) B.L. Burtt in Beitr. Biol. Pfl.*, 70(2–3): 265 (1997–98 publ. 1998): Didissandra synaptica.
tenella *(B.L. Burtt) B.L. Burtt in Beitr. Biol. Pfl.*, 70(2–3): 264 (1997–98 publ. 1998): Didissandra tenella.

RIDLEYANDRA:-
violacea *(Ridl.) A. Weber in Beitr. Biol. Pfl.*, 70(2–3): 261 (1997–98 publ. 1998): Didissandra violacea.
wrayi *(Ridl.) A. Weber in Beitr. Biol. Pfl.*, 70(2–3): 260 (1997–98 publ. 1998): Didissandra wrayi.

RIENCOURTIA (Compos.).
pedunculosa *(Rich.) J.F. Pruski in Brittonia*, 50(4): 478 (1998): Trixis pedunculosa.

RINDERA (Boraginac.).
alaica *G.A. Lazkov in Bot. Zhurn.*, 85(8): 116 (2000)—Kirgizistan.

RINOREA (Violac.).
abbreviata *G. Achoundong & J.J. Bos in Adansonia*, 21(1): 126 (1999)—Gabon.
apertior *G. Achoundong & J.J. Bos in Adansonia*, 21(1): 126 (1999)—Gabon.
belalongii *P.F. Stevens in Novon*, 10(2): 153 (2000)—Borneo (Brunei, Sabah).
cinerea *(King) J.K. Jarvie & P.F. Stevens in Harvard Pap. Bot.*, 3(2): 259 (1998): Alsodeia cinerea.
subsp. mahayarii *J.K. Jarvie & P.F. Stevens in Harvard Pap. Bot.*, 3(2): 260 (1998)—Borneo (Kalimantan).
congesta *L.L. Forman in Checklist Flow. Pl. & Gymn. Brunei Darussalam (ed. M.J.E. Coode et al.)*: 440 (1996)—Borneo (Brunei).
crassa *P.F. Stevens in Harvard Pap. Bot.*, 3(2): 261 (1998)—Borneo (Kalimantan).
curtirama *G. Achoundong & J.J. Bos in Adansonia*, 21(1): 128 (1999)—Gabon.

RIODOCEA *P.G. Delprete in Brittonia*, 51(1): 17 (1999). *RUBIACEAE*
pulcherrima *P.G. Delprete in Brittonia*, 51(1): 18, 17 (1999)—Brazil Southeast (Espírito Santo).

× **RITTELLICEREUS** *P.V. Heath in Calyx*, 6(1): 1 (1998). *CACTACEAE* (Nigellicereus × Ritterocereus).

RITTEROCACTUS *A.B. Doweld in Sukkulentÿ (Moscow)*, 1999(2): 22 (1999). *CACTACEAE*
sect. Herteria *A.B. Doweld in Sukkulentÿ (Moscow)*, 3(1–2): 59 (2000).
sect. Pseudoparodia *A.B. Doweld in Sukkulentÿ (Moscow)*, 3(1–2): 61 (2000).
sect. Uebelmannia *A.B. Doweld in Sukkulentÿ (Moscow)*, 3(1–2): 60 (2000).
allosiphon *(Marchesi) A.B. Doweld in Sukkulentÿ (Moscow)*, 1999(2): 22 (1999): Notocactus allosiphon.
arnostianus *(Lisal & Kolarik) A.B. Doweld in Sukkulentÿ (Moscow)*, 1999(2): 22 (1999): Notocactus arnostianus.
ayopayanus *(Cárdenas) A.B. Doweld in Sukkulentÿ (Moscow)*, 3(1–2): 61 (2000): Parodia ayopayana.
buiningii *(Buxb.) A.B. Doweld in Sukkulentÿ (Moscow)*, 1999(2): 22 (1999): Notocactus buiningii.
crassigibbus *(F. Ritter) A.B. Doweld in Sukkulentÿ (Moscow)*, 1999(2): 22 (1999): Notocactus crassigibbus.
curvispinus *(F. Ritter) A.B. Doweld in Sukkulentÿ (Moscow)*, 1999(2): 22 (1999): Notocactus curvispinus.
erubescens *(Osten) A.B. Doweld in Sukkulentÿ (Moscow)*, 1999(2): 22 (1999): Echinocactus × erubescens.
fuscus *(F. Ritter) A.B. Doweld in Sukkulentÿ (Moscow)*, 3(1–2): 60 (2000): Notocactus fuscus.
herteri *(Werderm.) A.B. Doweld in Sukkulentÿ (Moscow)*, 1999(2): 22 (1999): Echinocactus herteri.
horstii *(F. Ritter) A.B. Doweld in Sukkulentÿ (Moscow)*, 3(1–2): 60 (2000): Wigginsia horstii.
f. juvenaliformis *(F. Ritter) A.B. Doweld in Sukkulentÿ (Moscow)*, 3(1–2): 60 (2000): Wigginsia horstii var. juvenaliformis.
langsdorfii *(Lehm.) A.B. Doweld in Sukkulentÿ (Moscow)*, 3(1–2): 60 (2000): Cactus langsdorfii.
subsp. multiceps *(Hofacker & K. Herm) A.B. Doweld in Sukkulentÿ (Moscow)*, 3(1–2): 60 (2000): Parodia langsdorfii subsp. multiceps.

RITTEROCACTUS:-
langsdorfii:-
subsp. pulvinatus *(Vliet) A.B. Doweld in Sukkulentÿ (Moscow)*, 3(1–2): 60 (2000): Notocactus pulvinatus.
mammulosus *(Lem.) A.B. Doweld in Sukkulentÿ (Moscow)*, 1999(2): 22 (1999): Echinocactus mammulosus.
subsp. brasiliensis *(Havlíček) A.B. Doweld in Sukkulentÿ (Moscow)*, 1999(2): 22 (1999): Notocactus mammulosus var. brasiliensis.
subsp. erythracanthus *(H. Schloss. & Brederoo) A.B. Doweld in Sukkulentÿ (Moscow)*, 1999(2): 22 (1999): Notocactus erythracanthus.
subsp. eugeniae *(Vliet) A.B. Doweld in Sukkulentÿ (Moscow)*, 1999(2): 22 (1999): Notocactus eugeniae.
*subsp.*submammulosus *(Lem.) A.B. Doweld in Sukkulentÿ (Moscow)*, 1999(2): 22 (1999): Echinocactus submammulosus.
megalanthus *(H. Schloss. & Brederoo) A.B. Doweld in Sukkulentÿ (Moscow)*, 1999(2): 22 (1999): Notocactus megalanthus.
meonacanthus *(Prestlé) A.B. Doweld in Sukkulentÿ (Moscow)*, 1999(2): 22 (1999): Notocactus meonacanthus.
miguillensis *(Cárdenas) A.B. Doweld in Sukkulentÿ (Moscow)*, 3(1–2): 61 (2000): Parodia miguillensis.
mueller-melchersii *(Frič ex Backeb.) A.B. Doweld in Sukkulentÿ (Moscow)*, 1999(2): 22 (1999): Notocactus mueller-melchersii.
subsp. gutierrezii *(W.R. Abraham) A.B. Doweld in Sukkulentÿ (Moscow)*, 1999(2): 22 (1999): Notocactus gutierrezii.
subsp. winkleri *(Vliet) A.B. Doweld in Sukkulentÿ (Moscow)*, 1999(2): 23 (1999): Notocactus winkleri.
permutatus *(F. Ritter) A.B. Doweld in Sukkulentÿ (Moscow)*, 1999(2): 23 (1999): Notocactus permutatus.
rauschii *(Vliet) A.B. Doweld in Sukkulentÿ (Moscow)*, 1999(2): 23 (1999): Notocactus rauschii.
rutilans *(Däniker & Krainz) A.B. Doweld in Sukkulentÿ (Moscow)*, 1999(2): 23 (1999): Notocactus rutilans.
subsp. veenianus *(Vliet) A.B. Doweld in Sukkulentÿ (Moscow)*, 1999(2): 23 (1999): Notocactus veenianus.
uebelmannianus *(Buining) A.B. Doweld in Sukkulentÿ (Moscow)*, 1999(2): 23 (1999): Notocactus uebelmannianus.
subsp. pleiocephalus *(N. Gerloff & Königs) A.B. Doweld in Sukkulentÿ (Moscow)*, 1999(2): 23 (1999): Notocactus uebelmannianus var. pleiocephalus.

× **RITTERSHAUSENARA** *hort. in Orchid Rev.*, 108(1235): new orchid hybrids 14 (2000). *ORCHIDACEAE* (Colax × Galeottia × Promenaea × Zygopetalum).

RIVEA (Convolvulac.).
roxburghii *Prain ex D. Brandis, Indian Trees*: 484 (1906)—India (West Bengal).
wightiana *R.R. Mill in Edinburgh J. Bot.*, 53(2): 241 (1996)—India, Sri Lanka.

ROBINSONECIO *T.M. Barkley & J.P. Janovec in Sida*, 17(1): 79 (1996). *COMPOSITAE*
gerberifolius *(Sch. Bip.) T.M. Barkley & J.P. Janovec in Sida*, 17(1): 79 (1996): Senecio gerberifolius.
porphyresthes *(T.M. Barkley) T.M. Barkley & J.P. Janovec in Sida*, 17(1): 80 (1996): Senecio porphyresthes.

ROBSONODENDRON *R.H. Archer in S. Afr. J. Bot.*, 63(3): 116 (1997). *CELASTRACEAE*
eucleiforme *(Eckl. & Zeyh.) R.H. Archer in S. Afr. J. Bot.*, 63(3): 118 (1997): Mystroxylon eucleiforme.
maritimum *(Bolus) R.H. Archer in S. Afr. J. Bot.*, 63(3): 119 (1997): Celastrus maritimus.

RODRIGOA *L.A. Braas in Orchidee*, 30(5): 203 (1979). *ORCHIDACEAE*
alexandri *(Luer) L.A. Braas in Orchidee*, 33(4): 147 (1982): Masdevallia alexandri.

RODRIGOA:-
meleagris *(Lindl.) L.A. Braas in Orchidee,* 30(5): 203 (1979): Masdevallia meleagris.

RODRIGUEZIA (Orchidac.).
decora
var. lactea *L.C. Menezes in Bol. CAOB,* 6(1): 11 (1995)—Brazil West-central (Brasília Distrito Federal).

ROEGNERIA (Gramin.).
ser. Alashanicae *L.B. Cai in Acta Phytotax. Sin.,* 35(2): 163 (1997).
ser. Angustae *L.B. Cai in Acta Phytotax. Sin.,* 35(2): 174 (1997).
sect. Goulardia *(Husn.) L.B. Cai in Acta Phytotax. Sin.,* 35(2): 162 (1997): gen. Goulardia.
ser. Hongyuanenses *L.B. Cai in Acta Phytotax. Sin.,* 35(2): 157 (1997).
ser. Platyphyllae *L.B. Cai in Acta Phytotax. Sin.,* 35(2): 177 (1997).
ser. Schrenkianae *L.B. Cai in Acta Phytotax. Sin.,* 35(2): 161 (1997).
ser. Serotinae *L.B. Cai in Acta Phytotax. Sin.,* 35(2): 167 (1997).
ser. Tridentatae *L.B. Cai in Acta Phytotax. Sin.,* 35(2): 150 (1997).
alashanica Keng & S.L. Chen
var. elytrigioides *(C. Yen & J.L. Yang) L.B. Cai in Acta Phytotax. Sin.,* 35(2): 163 (1997): R. elytrigioides.
angusta *L.B. Cai in Acta Phytotax. Sin.,* 34(3): 332 (1996)—Qinghai.
anthosachnoides Keng & S.L. Chen
var. scabrilemmata *L.B. Cai in Acta Phytotax. Sin.,* 35(2): 165 (1997)—Qinghai, China South-central (Sichuan).
barbicalla Ohwi
var. foliosa *(Keng & S.L. Chen) L.B. Cai in Acta Phytotax. Sin.,* 35(2): 172 (1997): R. foliosa.
breviarista *(D.F. Cui) L.B. Cai in Acta Phytotax. Sin.,* 35(2): 170 (1997): Elymus glaberrimus var. breviaristus.
breviglumis Keng & S.L. Chen
var. brevipes *(Keng & S.L. Chen) L.B. Cai in Acta Phytotax. Sin.,* 35(2): 160 (1997): R. brevipes.
cacumina *(B. Rong Lu & B. Salomon) L.B. Cai in Acta Phytotax. Sin.,* 35(2): 160 (1997): Elymus cacuminus.
cheniae *L.B. Cai in Acta Phytotax. Sin.,* 34(3): 333 (1996)—Xinjiang.
ciliaris (Trin.) Nevski
var. hackeliana *(Honda) L.B. Cai in Acta Phytotax. Sin.,* 35(2): 176 (1997): Agropyron japonicum var. hackelianum.
crassa *L.B. Cai in Acta Phytotax. Sin.,* 34(3): 332 (1996)—Inner Mongolia (Ningxia), Qinghai.
curtiaristata *L.B. Cai in Guihaia,* 16(3): 200 (1996)—Xizang.
debilis *L.B. Cai in Acta Phytotax. Sin.,* 34(3): 327 (1996)—China North-central (Gansu), Qinghai.
flexuosa *L.B. Cai in Acta Phytotax. Sin.,* 34(3): 330 (1996)—China North-central (Gansu), Xinjiang.
glabrispicula *(D.F. Cui) L.B. Cai in Acta Phytotax. Sin.,* 35(2): 167 (1997): Elymus tschimganicus var. glabrispiculus.
gracilis *L.B. Cai in Acta Phytotax. Sin.,* 34(3): 328 (1996)—Xizang.
hondae *M. Kitagawa in Rep. Inst. Sci. Res. Manchoukuo,* 6(4): 118 (1942)—Manchuria (Liaoning), China North-central (Hebei).
hongyuanensis *L.B. Cai in Acta Phytotax. Sin.,* 35(2): 157 (1997)—China South-central (Sichuan).
jacquemontii (Hook. f.) Ovcz. & Sidorenko
var. pulanensis *(H.L. Yang) L.B. Cai in Acta Phytotax. Sin.,* 35(2): 169 (1997): R. pulanensis.
jufinshanica *(C.P. Wang & H.L. Yang) L.B. Cai in Acta Phytotax. Sin.,* 35(2): 170 (1997): R. alashanica var. jufinshanica.
kaschgarica *(D.F. Cui) Y.H. Wu in Y.H. Wu & Q.J. Wang, Grasses of Karakorum & Kunlun Mts.:* 77 (1999): Elymus kaschgaricus.
laxinodis *L.B. Cai in Guihaia,* 16(3): 199 (1996)—China South-central (Sichuan), Qinghai.
longearistata (Boiss.) Drobow
var. canaliculata *(Nevski) L.B. Cai in Acta Phytotax. Sin.,* 35(2): 165 (1997): Agropyron canaliculatum.

ROEGNERIA:-
macerrima *(Keng & S.L. Chen) L.B. Cai in Acta Phytotax. Sin.,* 35(2): 164 (1997): R. kamoji var. macerrima.
macrathera *(Ohwi) L.B. Cai in Acta Phytotax. Sin.,* 35(2): 176 (1997): R. turczaninovii var. macrathera.
magnipoda *L.B. Cai in Acta Phytotax. Sin.,* 35(2): 164 (1997)—Qinghai.
media *(Keng & S.L. Chen) L.B. Cai in Acta Phytotax. Sin.,* 35(2): 175 (1997): R. sinica var. media.
nudiuscula *L.B. Cai in Acta Phytotax. Sin.,* 35(2): 171 (1997)—Xinjiang.
retroflexa *(B. Rong Lu & B. Salomon) L.B. Cai in Acta Phytotax. Sin.,* 35(2): 161 (1997): Elymus retroflexus.
serpentina *L.B. Cai in Acta Phytotax. Sin.,* 35(2): 167 (1997)—North-central (Hebei).
shandongensis *(B. Salomon) J.L. Yang, Y.H. Zhou & C. Yen in Guihaia,* 17(1): 22 (Feb. 1997): Elymus shandongensis.
shandongensis *(B. Salomon) L.B. Cai in Acta Phytotax. Sin.,* 35(2): 166 (April 1997): Elymus shandongensis.
shouliangiae *L.B. Cai in Acta Phytotax. Sin.,* 35(2): 161 (1997)—Xizang.
tschimganica (Drobow) Nevski
var. glabrispicula *(D.F. Cui) Y.H. Wu in Y.H. Wu & Q.J. Wang, Grasses of Karakorum & Kunlun Mts.:* 78 (1999): Elymus tschimganicus var. glabrispiculus.
var. variiglumis *(Keng & S.L. Chen) L.B. Cai in Acta Phytotax. Sin.,* 35(2): 160 (1997): R. dura var. variiglumis.
tsukushiensis (Honda) B. Rong Lu, C. Yen & J.L. Yang
var. hybrida *(Keng & S.L. Chen) L.B. Cai in Acta Phytotax. Sin.,* 35(2): 167 (1997): R. hybrida.
yangiae *(B. Rong Lu) L.B. Cai in Acta Phytotax. Sin.,* 35(2): 158 (1997): Elymus yangiae.

ROEPEROCHARIS (Orchidac.).
bequaertii *(Summerh.) D.L. Szlachetko in Fragm. Flor. Geobot.,* Supp. 3: 114 (1995): Habenaria bequaertii.
camerunensis *D.L. Szlachetko & T.S. Olszewski in Fl. Cameroun,* 34: 224 (1998)—Cameroun.
nigrescens *(Summerh.) D.L. Szlachetko & T.S. Olszewski in Fl. Cameroun,* 34: 226 (1998): Habenaria nigrescens.
peristyloides *(A. Rich.) D.L. Szlachetko & T.S. Olszewski in Fl. Cameroun,* 34: 222 (1998), isonym; *see Schltr. in Bot. Jahrb. Syst.,* 53: 520 (1915): Habenaria peristyloides.

ROGERSONANTHUS (Gentianac.).
coccineus *Steyerm. ex L. Struwe, S. Nilsson & V.A. Albert in Harvard Pap. Bot.,* 3(2): 191 (1998)—Venezuela/Brazil North (Amazonas) border.

ROGIERA (Rubiac.).
breedlovei *(Lorence) A. Borhidi in Acta Bot. Hung.,* 40(1–4): 16 (1996–97 publ. 1997): Rondeletia breedlovei.
macdougallii *(Lorence) A. Borhidi in Acta Bot. Hung.,* 38(1–4): 140 (1993–94 publ. 1995): Rondeletia macdougallii.

ROHDEA (Convallariac.).
japonica
var. watanabei *(Hayata) S.S. Ying in Mem. Coll. Agric. Nation. Taiwan Univ.,* 31(1): 31 (1991), as 'Rhodea': R. watanabei.

× **ROHRLARA** *hort.in Orchid Rev.,* 105(1217): 318 (1997). *ORCHIDACEAE* (Ada × Aspasia × Brassia).

ROLDANA (Compos.).
calzadana *B.L. Turner in Phytologia,* 80(4): 276 (1996)—Mexico Southwest (Oaxaca).
gonzaleziae *(B.L. Turner) B.L. Turner in Phytologia,* 80(4): 278 (1996): Senecio gonzaleziae.
manantlanensis *(R.R. Kowal) B.L. Turner in Phytologia,* 80(4): 278 (1996): Senecio galicianus var. manantlanensis.

ROLDANA:-
mixtecana *J.L. Panero & J.L. Villaseñor in Brittonia,* 48(1): 83 (1996)—Mexico Southwest (Oaxaca).
neogibsonii *(B.L. Turner) B.L. Turner in Phytologia,* 80(4): 278 (1996): Senecio neogibsonii.
sundbergii *(B.L. Turner) B.L. Turner in Phytologia,* 80(4): 278 (1996): Senecio sundbergii.

ROMANZOFFIA (Hydrophyllac.).
thompsonii *V.M. Marttala in Madroño,* 43(3): 405 (1996)—Oregon.

ROMULEA (Iridac.).
arnaudii *J. Moret in Monde Pl.,* 95(468): 28 (2000)—France.
columnae Sebast. & Mauri
subsp. occidentalis *(Bég.) J. Moret in Monde Pl.,* 95(468): 29 (2000): R. columnae var. occidentalis.
subsp. subalbida *(Jord. & Fourr.) J. Moret in Monde Pl.,* 95(468): 29 (2000): R. subalbida.
florentii *J. Moret in Monde Pl.,* 95(468): 28 (2000)—France.
petraea *D.M.H. Al-Eisawi, Field Guide Wild Flow. Jordan & Neighb. Countries:* 165 (1998), without latin descr. or type— Palestine (Jordan).
villaretii *A. Dobignard in Candollea,* 52(1): 123 (1997)—Morocco.

RONDELETIA (Rubiac.).
evansii *D.H. Lorence in Novon,* 7(1): 52 (1997)—Honduras.
pringlei *D.H. Lorence in Novon,* 7(1): 54 (1997)—Mexico Gulf (Veracruz), Mexico Northeast (Hidalgo), Mexico Central (Puebla).

RONNBERGIA (Bromeliac.).
neoregelioides *E.M.C. Leme in J. Bromeliad Soc.,* 49(3): 102 (1999)—Brazil Northeast (Bahia).

RORIPPA (Crucif.).
subgen. Jonsellia *V.I. Dorofeev in Bot. Zhurn.,* 83(8): 105 (1998).
ampullicarpa *V.I. Dorofeev in Bot. Zhurn.,* 83(8): 102 (1998)—Transcaucasus (Azerbaydzhan).
× brachyceras *(Honda) Kitam. ex T. Shimizu, Fl. Nagano Pref.:* 1507 (1997): R. atrovirens f. brachyceras.
chubutica *(O.E. Schulz) J.B. Martínez-Laborde in Zuloaga & Morrone (eds.), Cat. Pl. Vasc. Rep. Arg. 2 (Monogr. Syst. Bot.* Missouri Bot. Gard. 74): 1245 (1999): Descurainia chubutica.
cochlearioides *(Roth) I.A. Al-Shehbaz & B. Jonsell in Novon,* 10(4): 344 (2000): Alyssum cochlearioides.
neocaledonica *B. Jonsell in Bull. Mus. Nation. Hist. Nat., B, Adansonia,* Ser. 4, 17(3–4): 177 (1995)—New Caledonia.
spasskajae *V.I. Dorofeev in Bot. Zhurn.,* 83(8): 102 (1998)—Transcaucasus (Nagorno-Karabakh).
vallicola *V.I. Dorofeev in Bot. Zhurn.,* 83(8): 104 (1998), nom. nov.: R. viaria *V.I. Dorof.*
villosa *R.F. Huang in Acta Phytotax. Sin.,* 35(6): 561 (1997)—Qinghai.

ROSA (Rosac.).
sect. Glaucae *N.N. Tzvelev, Opred. Sosud. Rast. Severo-Zap. Rossii:* 680 (2000), without latin descr.
affinior *M. Gandoger, Dec. Pl. Nov.,* 1: 9 (1875)—France.
amblyotis C.A. Mey.
subsp. jacutica *(Juz.) V.N. Voroschilov in A.K. Skvortsov (ed.), Florist. issl. v razn. raĭonakh SSSR:* 177 (1985): R. jacutica.
amici *M. Gandoger, Dec. Pl. Nov.,* 1: 16 (1875)—France.
apostigma *M. Gandoger, Dec. Pl. Nov.,* 1: 36 (1875)—France.
× archipelagica *E.A. Tchubar in Byull. Glavn. Bot. Sada (Moscow),* 178: 67 (1999), contrary to Art. 37.6 ICBN (2000)—Primorye. (R. maximowicziana × rugosa).
arvensis
var. pseudorepens *(Rouy) P.V. Heath in Calyx,* 6(2): 47 (1999): R. gallicoides var. pseudorepens.

ROSA:-
× biturigensis
nothovar. paui *(Cuatrec.) P.V. Heath in Calyx,* 6(2): 47 (1999), as 'pauli': R. paui.
briggsii *M. Gandoger, Dec. Pl. Nov.,* 1: 26 (1875), as 'brippii'—Great Britain.
brilonensis *G.H. Loos in Dortmunder Beitr. Landesk.,* 32: 87 (1998)—Germany.
caesia Sm.
subsp. vosagiaca *(N.H.F. Desp.) D.H. Kent, List Vasc. Pl. Brit. Isles,* Supp. 1: 36 (1996 publ. 19 Jan. 1997): R. vosagiaca.
calcarea *M. Gandoger, Dec. Pl. Nov.,* 1: 32 (1875)—France.
calyptocalyx *M. Gandoger, Dec. Pl. Nov.,* 1: 35 (1875)—France.
canina
var. montana *(Vill.) P.V. Heath in Calyx,* 6(2): 47 (1999): R. montana.
nothovar. permiscibilis *(Schmidely) P.V. Heath in Calyx,* 6(2): 47 (1999): R. × permiscibilis.
subsp. virens *(Wahlenb.) D. Šmite in Fl. Baltiiskikh Respublik,* 2: 54 (1996): R. canina γ virens.
cedrorum *M. Gandoger, Dec. Pl. Nov.,* 1: 42 (1875)—Algeria.
× centifolia
var. semiduplex *(Andrews) P.V. Heath in Calyx,* 6(2): 70 (1999): R. provincialis var. semiduplex.
commiserata *M. Gandoger, Dec. Pl. Nov.,* 1: 25 (1875)—France.
cordatifolia *M. Gandoger, Dec. Pl. Nov.,* 1: 19 (1875)—France.
cyanescens *M. Gandoger, Dec. Pl. Nov.,* 2: 23 (1876)—France.
djurdjurensis *Debeaux ex M. Gandoger, Dec. Pl. Nov.,* 1: 42 (1875)—Algeria.
eglanteria
var. cretica *(Thory) P.V. Heath in Calyx,* 6(2): 47 (1999): R. rubiginosa [var.] cretica.
var. glabra *(A. Rau) P.V. Heath in Calyx,* 6(2): 47 (1999): R. rubiginosa var. glabra.
nothovar. ladae *P.V. Heath in Calyx,* 6(2): 47 (1999), nom. nov.: R. ladanifera *Timb.-Lagr.*
nothovar. rara *P.V. Heath in Calyx,* 6(2): 47 (1999), nom. nov.: R. × muennerstadtiana *Fedde.*
geracantha *M. Gandoger, Dec. Pl. Nov.,* 1: 32 (1875)—Switzerland.
hezhangensis *T.L. Xu in Acta Phytotax. Sin.,* 38(1): 74 (2000)—China South-central (Guizhou).
indifferens *M. Gandoger, Dec. Pl. Nov.,* 1: 36 (1875)—France.
jacetana *P. Montserrat in Soc. Ech. Pl. Vasc. Eur. Bassin Médit. Bull.,* 26: 107 (1997)—Spain.
kurganica *N.V. Mironova in Bot. Zhurn.,* 85(8): 110 (2000)—South European Russia.
langyashanica *D.C. Zhang & J.Z. Shao in Acta Phytotax. Sin.,* 35(3): 265 (1997)—China Southeast (Anhui).
laxa Retz.
var. kaschgarica *(Rupr.) Y.L. Han in Fl. Xinjiangensis,* 2(2): 343 (1995): R. kaschgarica.
leptostyla *M. Gandoger, Dec. Pl. Nov.,* 1: 8 (1875)—France.
macrophylla
var. davidi *(Crép.) P.V. Heath in Calyx,* 6(2): 47 (1999), as 'davidis': R. davidi.
majalis Herrm.
var. petropolitana *(N.P. Smirn.) I.O. Buzunova in Bot. Zhurn.,* 81(11): 47 (1996): R. cinnamomea var. petropolitana.
× majorugosa *A. Palmén & L. Hämet-Ahti in Symb. Bot. Upsal.,* 31(3): 233 (1996)—Finland. (R. majalis × rugosa)
manshurica *I.O. Buzunova in Novosti Sist. Vÿssh. Rast.,* 30: 100 (1996)—Manchuria (Jilin).
var. glabriuscula *I.O. Buzunova in Novosti Sist. Vÿssh. Rast.,* 30: 102 (1996)—Manchuria (Jilin).
megalochlamys *M. Gandoger, Dec. Pl. Nov.,* 1: 27 (1875)—France.
naias *M. Gandoger, Dec. Pl. Nov.,* 1: 34 (1875)—France.
oblongicalyx *M. Gandoger, Dec. Pl. Nov.,* 2: 23 (1876)—France.
obtusispina *M. Gandoger, Dec. Pl. Nov.,* 1: 37 (1875)—France.

ROSA:-
onoei Makino
var. hakonensis *(Franch. & Sav.) H. Ohba in J. Jap. Bot.,* 75(3): 161 (2000): R. luciae var. hakonensis.
var. oligantha *(Franch. & Sav.) H. Ohba in J. Jap. Bot.,* 75(3): 157 (2000): R. luciae var. oligantha.
phalacropoda *M. Gandoger, Dec. Pl. Nov.,* 1: 24 (1875)—France.
podostigma *M. Gandoger, Dec. Pl. Nov.,* 2: 24 (1876)—France.
polyodon *M. Gandoger, Dec. Pl. Nov.,* 1: 33 (1875)—France.
pomifera Herrm.
subsp. glandulosa *(Schmalh.) D. Šmite in Fl. Baltiiskikh Respublik,* 2: 59 (1996): R. mollis var. glandulosa.
pumila
var. holosericea *(Du Roi) P.V. Heath in Calyx,* 6(2): 76 (1999): R. holosericea.
var. officinalis *(Andrews) P.V. Heath in Calyx,* 6(2): 76 (1999): R. gallica var. officinalis.
var. rosamundi *(Weston) P.V. Heath in Calyx,* 6(2): 77 (1999): R. gallica var. rosamundi.
var. subnigra *(Andrews) P.V. Heath in Calyx,* 6(2): 76 (1999): R. centifolia var. subnigra.
var. tricolor *(Andrews) P.V. Heath in Calyx,* 6(2): 76 (1999): R. centifolia var. tricolor.
sericea Lindl.
f. chrysocarpa *(Rehder) S.G. Haw in New Plantsman,* 3(3): 146 (1996), with incorrect basionym ref.: Basionym not stated.
subsp. morrisonensis *(Hayata) S.G. Haw in New Plantsman,* 3(3): 146 (1996): R. morrisonensis.
f. taronensis *(T.T. Yu) S.G. Haw in New Plantsman,* 3(3): 146 (1996): R. taronensis.
stellata Wooton
var. abyssa *(A.M. Phillips) N.H. Holmgren in A. Cronquist et al., Intermountain Fl., Vasc. Pl. Intermountain West,* U.S.A., 3A: 140 (1997): R. stellata subsp. abyssa.
stupens *M. Gandoger, Dec. Pl. Nov.,* 1: 3 (1875)—France.
tanaitica *N.V. Mironova in Bot. Zhurn.,* 85(8): 111 (2000)—South European Russia.

ROSCOEA (Zingiberac.).
bhutanica *C. Ngamriabsakul in Edinburgh J. Bot.,* 57(2): 272 (2000)—Bhutan, Xizang.
cautleoides Gagnep.
var. pubescens *(Z.Y. Zhu) T.L. Wu in Novon,* 7(4): 441 (1997 publ. 1998): R. pubescens.
ganeshensis *E.J. Cowley & W.J. Baker in Curtis's Bot. Mag.,* 13(1): 10 (1996)—Nepal.
humeana Balf. f. & W.W. Sm.
f. lutea *E.J. Cowley in Curtis's Bot. Mag.,* 17(1): 27 (2000)—China South-central (Sichuan).
f. tyria *E.J. Cowley in Curtis's Bot. Mag.,* 17(1): 27 (2000)—Cultivation.

ROSENBERGIODENDRON (Rubiac.).
formosum (Jacq.) Fagerl.
var. nitidum *(K. Schum.) C.G.R. Gustafsson in Brittonia,* 50(4): 457 (1998): Randia formosa var. nitida.

ROSMARINUS (Labiat.).
aunieri *M. Gandoger, Dec. Pl. Nov.,* 1: 21 (1875)—France.
cyanocalyx *M. Gandoger, Dec. Pl. Nov.,* 1: 22 (1875)—France.
eriocalyx Jord. & Fourr.
var. pallescens *(Maire) T.M. Upson & S.L. Jury in Lagascalia,* 18(2): 296 (1996): R. tournefortii f. pallescens.
massiliensis *M. Gandoger, Dec. Pl. Nov.,* 1: 21 (1875)—France.
officinalis L.
subsp. palaui *(O. Bolòs & Molin.) A.M. Romo, Fl. Silvestres Baleares:* 266 (1994), isonym; *see Malagarriga, Las Subesp. y Variac. Geogr.:* 23 (1973): R. officinalis var. palaui.
platyphyllus *M. Gandoger, Dec. Pl. Nov.,* 1: 21 (1875)—France.
verrietii *M. Gandoger, Dec. Pl. Nov.,* 1: 21 (1875)—France.

ROSTELLULARIA (Acanthac.).
ciliata *(Yamam.) S.S. Ying in Mem. Coll. Agric. Nation. Taiwan Univ.,* 29(2): 45 (1989): Justicia hayatae var. ciliata.
hayatae *(Yamam.) S.S. Ying in Mem. Coll. Agric. Nation. Taiwan Univ.,* 29(2): 45 (1989): Justicia hayatae.
procumbens (L.) Nees
var. hirsuta *(Yamam.) S.S. Ying in Mem. Coll. Agric. Nation. Taiwan Univ.,* 29(2): 44 (1989): Justicia procumbens var. hirsuta.
var. linearifolia *(Yamam.) S.S. Ying in Mem. Coll. Agric. Nation. Taiwan Univ.,* 29(2): 45 (1989): Justicia procumbens var. linearifolia.
var. riukiuensis *(Yamam.) S.S. Ying in Mem. Coll. Agric. Nation. Taiwan Univ.,* 29(2): 44 (1989): Justicia procumbens var. riukiuensis.
quadrifaria *(Wall.) S.S. Ying in Mem. Coll. Agric. Nation. Taiwan Univ.,* 29(2): 45 (1989), basionym not validly publ.: Justicia quadrifaria.

ROSTRARIA (Gramin.).
amblyantha *(Boiss.) J. Holub in Preslia,* 70(2): 110 (1998): Koeleria phleoides var. amblyantha.
festucoides *(Link) C. Romero Zarco in Lagascalia,* 18(2): 310 (1996): Bromus festucoides.
salzmannii
subsp. maroccana *(Domin) H. Scholz in Willdenowia,* 28(1–2): 172 (1998): Koeleria salzmannii var. maroccana.

ROSULARIA (Crassulac.).
sempervivum (M. Bieb.) A. Berger
var. glabrum *(Raym.-Hamet) M. Assadi in Fl. Iran (ed. M. Assadi et al.),* 32: 50 (2000): Sedum sempervivum var. glabrum.

ROTHECA (Verbenac.).
alata *(Gürke) B. Verdcourt in Kew Bull.,* 55(1): 148 (2000): Clerodendrum alatum.
amplifolia *(S. Moore) R.B. Fernandes in Kew Bull.,* 55(1): 148 (2000): Clerodendrum amplifolium.
aurantiaca *(Baker) R.B. Fernandes in Kew Bull.,* 55(1): 148 (2000): Clerodendrum aurantiacum.
f. faulkneri *(R. Fern.) R.B. Fernandes in Kew Bull.,* 55(1): 148 (2000): Clerodendrum aurantiacum f. faulkneri.
bukobensis *(Gürke) B. Verdcourt in Kew Bull.,* 55(1): 148 (2000): Clerodendrum bukobense.
calundensis *(R. Fern.) R.B. Fernandes in Kew Bull.,* 55(1): 148 (2000): Clerodendrum calundense.
commiphoroides *(Verdc.) D.A. Steane & D.J. Mabberley in Novon,* 8(2): 205 (1998): Clerodendrum commiphoroides.
cyanea *(R. Fern.) R.B. Fernandes in Kew Bull.,* 55(1): 148 (2000): Clerodendrum cyaneum.
farinosa *(Roxb.) R. Govaerts, World Checklist Seed Plants,* 3(1): 15 (1999): Volkameria farinosa.
hirsuta *(Hochst.) R.B. Fernandes in Kew Bull.,* 55(1): 148 (2000): Cyclonema ? hirsutum.
f. triphylla *(Harv.) R.B. Fernandes in Kew Bull.,* 55(1): 149 (2000): Cyclonema triphyllum.
incisa *(Klotzsch) D.A. Steane & D.J. Mabberley in Novon,* 8(2): 205 (1998): Clerodendrum incisum.
kissakensis *(Gürke) B. Verdcourt in Kew Bull.,* 55(1): 149 (2000): Clerodendrum kissakense.
luembensis *(De Wild.) R.B. Fernandes in Kew Bull.,* 55(1): 149 (2000): Clerodendrum luembense.
f. herbacea *(Hiern) R.B. Fernandes in Kew Bull.,* 55(1): 149 (2000): Siphonanthus myricoides var. herbaceus.
var. malawiensis *(R. Fern.) R.B. Fernandes in Kew Bull.,* 55(1): 149 (2000): Clerodendrum luembense var. malawiense.
subsp. niassensis *(R. Fern.) R.B. Fernandes in Kew Bull.,* 55(1): 149 (2000): Clerodendrum luembense subsp. niassense.
makanjana *(H.J.P. Winkler) D.A. Steane & D.J. Mabberley in Novon,* 8(2): 205 (1998): Clerodendrum makanjanum.
mendesii *(R. Fern.) R.B. Fernandes in Kew Bull.,* 55(1): 149 (2000): Clerodendrum mendesii.
myricoides *(Hochst.) D.A. Steane & D.J. Mabberley in Novon,* 8(2): 205 (1998): Spironema myricoides.
f. alatipetiolatum *(R. Fern.) R.B. Fernandes in Kew Bull.,* 55(1): 150 (2000): Clerodendrum myricoides f. alatipetiolatum.

ROTHECA:-
myricoides
f. angustilobata *(R. Fern.) R.B. Fernandes in Kew Bull.*, 55(1): 150 (2000): Clerodendrum myricoides f. angustilobatum.
subsp. austromonticola *(Verdc.) B. Verdcourt in Kew Bull.*, 55(1): 151 (2000): Clerodendrum myricoides subsp. austromonticola.
f. brevilobata *(R. Fern.) R.B. Fernandes in Kew Bull.*, 55(1): 150 (2000): Clerodendrum myricoides f. brevilobatum.
var. capiriensis *(R. Fern.) R.B. Fernandes in Kew Bull.*, 55(1): 150 (2000): Clerodendrum myricoides var. capiriense.
f. cubangensis *(R. Fern.) R.B. Fernandes in Kew Bull.*, 55(1): 151 (2000): Clerodendrum myricoides f. cubangense.
var. discolor *(Klotzsch) B. Verdcourt in Kew Bull.*, 55(1): 150 (2000): Cyclonema discolor.
var. dumalis *(Hiern) R.B. Fernandes in Kew Bull.*, 55(1): 151 (2000): Siphonanthus dumalis.
var. eleanorae *(R. Fern.) R.B. Fernandes in Kew Bull.*, 55(1): 151 (2000): Clerodendrum myricoides var. eleanorae.
var. kilimandscharensis *(Verdc.) B. Verdcourt in Kew Bull.*, 55(1): 151 (2000): Clerodendrum myricoides var. kilimandscharense.
f. lanceolatilobata *(R. Fern.) R.B. Fernandes in Kew Bull.*, 55(1): 150 (2000): Clerodendrum myricoides f. lanceolatilobatum.
f. lobulata *(R. Fern.) R.B. Fernandes in Kew Bull.*, 55(1): 150 (2000): Clerodendrum myricoides f. lobulatum.
subsp. mafiensis *(Verdc.) B. Verdcourt in Kew Bull.*, 55(1): 152 (2000): Clerodendrum myricoides subsp. mafiense.
var. moldenkei *(R. Fern.) R.B. Fernandes in Kew Bull.*, 55(1): 151 (2000): Clerodendrum myricoides var. moldenkei.
subsp. muenzneri *(B. Thomas) B. Verdcourt in Kew Bull.*, 55(1): 152 (2000): Clerodendrum muenzneri.
subsp. namibiensis *(R. Fern.) R.B. Fernandes in Kew Bull.*, 55(1): 152 (2000): Clerodendrum myricoides subsp. namibiense.
subsp. napperae *(Verdc.) B. Verdcourt in Kew Bull.*, 55(1): 152 (2000), as 'napperi': Clerodendrum myricoides subsp. napperae.
f. reflexilobata *(R. Fern.) R.B. Fernandes in Kew Bull.*, 55(1): 151 (2000): Clerodendrum myricoides f. reflexilobatum.
subsp. ussukumae *(Verdc.) B. Verdcourt in Kew Bull.*, 55(1): 152 (2000): Clerodendrum myricoides subsp. ussukumae.
var. viridiflora *(Verdc.) B. Verdcourt in Kew Bull.*, 55(1): 151 (2000): Clerodendrum myricoides var. viridiflorum.
prittwitzii *(B. Thomas) B. Verdcourt in Kew Bull.*, 55(1): 152 (2000): Clerodendrum prittwitzii.
quadrangulata *(B. Thomas) R.B. Fernandes in Kew Bull.*, 55(1): 152 (2000): Clerodendrum quadrangulatum.
reflexa *(H. Pearson) R.B. Fernandes in Kew Bull.*, 55(1): 152 (2000): Clerodendrum reflexum.
rupicola *(Verdc.) B. Verdcourt in Kew Bull.*, 55(1): 152 (2000): Clerodendrum rupicola.
sansibarensis *(Gürke) D.A. Steane & D.J. Mabberley in Novon*, 8(2): 205 (1998): Clerodendrum sansibarense.
subsp. caesia *(Gürke) D.A. Steane & D.J. Mabberley in Novon*, 8(2): 206 (1998): Clerodendrum caesium.
var. eratensis *(R. Fern.) R.B. Fernandes in Kew Bull.*, 55(1): 153 (2000): Clerodendrum sansibarense var. eratense.
subsp. occidentalis *(Verdc.) D.A. Steane & D.J. Mabberley in Novon*, 8(2): 205 (1998): Clerodendrum sansibarense subsp. occidentale.
f. tomentella *(R. Fern.) R.B. Fernandes in Kew Bull.*, 55(1): 153 (2000): Clerodendrum sansibarense f. tomentellum.
serrata *(L.) D.A. Steane & D.J. Mabberley in Novon*, 8(2): 206 (1998): Volkameria serrata.
suffruticosa *(Gürke) B. Verdcourt in Kew Bull.*, 55(1): 153 (2000): Clerodendrum suffruticosum.

ROTHECA:-
taborensis *(Verdc.) B. Verdcourt in Kew Bull.*, 55(1): 153 (2000): Clerodendrum taborense.
var. latifolia *(Verdc.) B. Verdcourt in Kew Bull.*, 55(1): 153 (2000): Clerodendrum taborense var. latifolium.
tanneri *(Verdc.) B. Verdcourt in Kew Bull.*, 55(1): 153 (2000): Clerodendrum tanneri.
teaguei *(Hutch.) R.B. Fernandes in Kew Bull.*, 55(1): 153 (2000): Clerodendrum teaguei.
verdcourtii *(R. Fern.) R.B. Fernandes in Kew Bull.*, 55(1): 153 (2000): Clerodendrum verdcourtii.
violacea *(Gürke) B. Verdcourt in Kew Bull.*, 55(1): 154 (2000): Clerodendrum violaceum.
subsp. kigeziensis *(Verdc.) B. Verdcourt in Kew Bull.*, 55(1): 154 (2000): Clerodendrum violaceum subsp. kigeziense.
wildii *(Moldenke) R.B. Fernandes in Kew Bull.*, 55(1): 154 (2000): Clerodendrum wildii.
f. glabra *(R. Fern.) R.B. Fernandes in Kew Bull.*, 55(1): 154 (2000): Clerodendrum wildii f. glabrum.

ROTHIA (Legumin.).
indica (L.) Druce
subsp. australis *A.E. Holland in Austrobaileya*, 5(1): 94 (1997)—Queensland, Northern Territory, Western Australia.

ROTHMANNIA (Rubiac.).
ebamutensis *B. Sonké in Syst. Geogr. Pl.*, 70(1): 149 (2000)—Cameroon.

ROUCHERIA (Linac.).
sipapoensis *A. Jardim & P.E. Berry in Novon*, 9(4): 520 (1999)—Venezuela.

ROUREA (Connarac.).
foreroi *G.A. Aymard & P.E. Berry in Brittonia*, 48(4): 580 (1996 publ. 1997)—Venezuela.

× **ROWLEYARA** *D.M. Cumming in Haworthiad*, 13(1): 20 (1999), nom. illegit.; *cf. Repert. Pl. Succ. (I.O.S.)*, 50: 9 (1999 publ. 2000). *ALOACEAE* (Astroloba × Gasteria × Haworthia).

ROYSTONEA (Palmae).
maisiana *(L.H. Bailey) S. Zona in Fl. Neotrop.*, 71: 22 (1996): R. regia var. maisiana.
oleracea (Jacq.) O.F. Cook
var. jenmanii *(Waby) S. Zona in Fl. Neotrop.*, 71: 24 (1996): Oreodoxa regia var. jenmanii.

RUBIA (Rubiac.).
angustifolia L.
subsp. caespitosa *(Font Quer & Marcos) A.M. Romo, Fl. Silvestres Baleares*: 284 (1994), isonym; *see Rosselló et al. in Candollea*, 48(2): 599 (1993): R. angustifolia var. caespitosa.
balearica (Willk.) Porta
subsp. caespitosa *(Font Quer & Marcos) J.A. Rosselló, L. Sáez & Mus in An. Jard. Bot. Madrid*, 55(2): 479 (1997): R. angustifolia var. caespitosa.
danaensis *A. Danin in Willdenowia*, 27(1–2): 163 (1997)—Palestine (Jordan).
falciformis *H.S. Lo in J. Trop. Subtrop. Bot.*, 7(1): 23 (1999)—China South-central (Yunnan).
filiformis *How ex H.S. Lo in J. Trop. Subtrop. Bot.*, 7(1): 24 (1999)—China South-central (Yunnan).
hispidicaulis *D.G. Long in Edinburgh J. Bot.*, 53(1): 109 (1996)—Bhutan, Sikkim, India (West Bengal).
latipetala *H.S. Lo in J. Trop. Subtrop. Bot.*, 7(1): 23 (1999)—China South-central (Sichuan).
linii *J.M. Chao in Biol. Bull. Natl. Taiwan Norm. Univ.*, 1: 48 (1966)—Taiwan.
ovatifolia Zhi Y. Zhang
var. oligantha *H.S. Lo in Fl. Reipubl. Popul. Sin.*, 71(2): 338 (1999), without type—China South-central (Hubei).
polyphlebia *H.S. Lo in J. Trop. Subtrop. Bot.*, 7(1): 22 (1999)—China South-central (Sichuan).
pterygocaulis *H.S. Lo in J. Trop. Subtrop. Bot.*, 7(1): 22 (1999)—China South-central (Sichuan).
salicifolia *H.S. Lo in J. Trop. Subtrop. Bot.*, 7(1): 24 (1999)—China South-central (Sichuan).
tenuis *H.S. Lo in J. Trop. Subtrop. Bot.*, 7(1): 24 (1999)—China South-central (Sichuan).

RUBIA:-
trichocarpa *H.S. Lo in J. Trop. Subtrop. Bot.*, 7(1): 23 (1999)—China South-central (Sichuan).

RUBIMONS *B.S. Sun in Acta Bot. Yunnanica*, 19(3): 239 (1997). *GRAMINEAE*
paniculatus *B.S. Sun in Acta Bot. Yunnanica*, 19(3): 239 (1997)—China South-central (Yunnan).

RUBITEUCRIS (Labiat.).
siccanea *(W.W. Sm.) P.D. Cantino in Syst. Bot.*, 23(3): 381 (1998 publ. 1999): Caryopteris siccanea.

RUBOVIETNAMIA *D.D. Tirvengadum in Biogeographica*, 74(4): 166 (1998). *RUBIACEAE*
aristata *D.D. Tirvengadum in Biogeographica*, 74(4): 167 (1998)—Vietnam.

RUBUS (Rosac.).
ser. Alleghenienses *(L.H. Bailey) L.S. Krassovskaja in Novosti Sist. Vÿssh. Rast.*, 32: 56 (2000): R. sect. Alleghenienses.
nothosubgen. Anobatus *J. Holub in Preslia*, 70(2): 110 (1998). (R. subgen. Anoplobatus × subgen. Idaeobatus).
ser. Canadenses *(L.H. Bailey) L.S. Krassovskaja in Novosti Sist. Vÿssh. Rast.*, 32: 56 (2000): R. sect. Canadenses.
nothosubgen. Chabatus *J. Holub in Preslia*, 70(2): 110 (1998). (R. subgen. Chamaerubus × subgen. Idaeobatus).
subgen. Diemenicus *A.R. Bean in Austrobaileya*, 5(1): 48 (1997).
ser. Idaei *J. Holub in Preslia*, 69(4): 291 (1997), isonym; *see Sinkova in Novosti Sist. Vÿssh. Rast.*, 9: 195 (1972).
acanthoblepharon *M. Gandoger in Mém. Soc. Emul. Doubs*, Sér. 5, 8: 155 (1884)—France.
acanthocaulon *M. Gandoger in Mém. Soc. Emul. Doubs*, Sér. 5, 8: 181 (1884)—France.
acanthocephalus *M. Gandoger in Mém. Soc. Emul. Doubs*, Sér. 5, 8: 222 (1884)—France.
acantholytus *A. Foerster, Fl. Excurs. Aachen*: 132 (1878)—Germany.
acanthopodus *M. Gandoger in Mém. Soc. Emul. Doubs*, Sér. 5, 8: 160 (1884)—France.
acanthothyrsus *M. Gandoger in Mém. Soc. Emul. Doubs*, Sér. 5, 8: 208 (1884)—France.
acaracis *A. Foerster, Fl. Excurs. Aachen*: 116 (1878)—Germany.
acmophorus *M. Gandoger in Mém. Soc. Emul. Doubs*, Sér. 5, 8: 157 (1884)—France.
acmophyllus *A. Foerster, Fl. Excurs. Aachen*: 103 (1878)—Germany.
acupungens *A. Foerster, Fl. Excurs. Aachen*: 142 (1878)—Germany.
acutispinus *A. Foerster, Fl. Excurs. Aachen*: 106 (1878)—Germany.
adulans *A. van de Beek in Osnabrücker Naturwiss. Mitt.*, 23: 40 (1997)—Netherlands.
aequiparus *A. Foerster, Fl. Excurs. Aachen*: 163 (1878)—Germany.
alixensis *M. Gandoger in Mém. Soc. Emul. Doubs*, Sér. 5, 8: 180 (1884)—France.
ambiguus *A. Foerster, Fl. Excurs. Aachen*: 144 (1878), homonym; *see P.J. Müll. in Jahresber. Pollichia*, 16–17: 258 (1859)—Germany.
amici *M. Gandoger in Mém. Soc. Emul. Doubs*, Sér. 5, 8: 185 (1884)—France.
amphipetalus *M. Gandoger in Mém. Soc. Emul. Doubs*, Sér. 5, 8: 158 (1884)—France.
amphistrophos *(Focke) H.E. Weber in Ber. Bayer. Bot. Ges.*, 68: 74 (1997): R. melanoxylon subsp. amphistrophos.
amplificus *A. Foerster, Fl. Excurs. Aachen*: 106 (1878)—Germany.
anglobelgicus *D.E. Allen & H. Vannerom in Watsonia*, 22(1): 85 (1998)—Belgium, Great Britain.
angustulus *M. Gandoger in Mém. Soc. Emul. Doubs*, Sér. 5, 8: 179 (1884)—France.
apetalus Poir.
var. glaber *(Cordem.) F. Friedmann in Fl. Mascar.*, 81: 6 (1997), without basionym date: R. glaber.
aphanes *A. Foerster, Fl. Excurs. Aachen*: 95 (1878)—Germany.
apiceobtusus *M. Gandoger in Mém. Soc. Emul. Doubs*, Sér. 5, 8: 213 (1884)—France.

RUBUS:-
aquisgranensis *A. Foerster, Fl. Excurs. Aachen:* 147 (1878)—Germany.
arcticus L.
var. humilis *V.N. Gladkova in Arktich. Fl. SSSR,* 9(1): 125 (1984)—Russia North.
arcturus *M. Gandoger in Mém. Soc. Emul. Doubs,* Sér. 5, 8: 175 (1884)—France.
argus *M. Gandoger in Mém. Soc. Emul. Doubs,* Sér. 5, 8: 158 (1884)—France.
× armiger *Focke ex J. Utsch in Jahres-Ber. Westfäl. Prov.-Vereins Wiss.,* 22: 207 (1893–94 publ. 1894)—Germany. (R. carpinifoluis × rosaceus).
assimulatus *A. Foerster, Fl. Excurs. Aachen:* 129 (1878)—Germany.
aunieri *M. Gandoger in Mém. Soc. Emul. Doubs,* Sér. 5, 8: 187 (1884)—France.
avellanoides *M. Gandoger in Mém. Soc. Emul. Doubs,* Sér. 5, 8: 217 (1884)—France.
azuayensis *K. Romoleroux in Fl. Ecuador,* 56(79): 9 (1996)—Ecuador.
barmondierei *M. Gandoger in Mém. Soc. Emul. Doubs,* Sér. 5, 8: 184 (1884)—France.
baruthicus *H.E. Weber in Ber. Bayer. Bot. Ges.,* 66–67: 42 (1996)—Germany, Luxembourg, France.
batos-weberi *G.H. Loos in Flor. Rundbr.,* 32(1): 49 (1998)—Germany.
beijerinckii *K. Meijer in Gorteria,* 26(5): 209 (2000)—Netherlands.
bellojocensis *M. Gandoger in Mém. Soc. Emul. Doubs,* Sér. 5, 8: 199 (1884)—France.
blepharocarpon *M. Gandoger in Mém. Soc. Emul. Doubs,* Sér. 5, 8: 197 (1884)—France.
bombycinus *G. Matzke-Hajek in Decheniana,* 149: 40 (1996)—Germany.
bonus-henricus *G. Matzke-Hajek in Decheniana,* 148: 15 (1995)—Germany, Belgium (Luxembourg).
borderi *M. Gandoger in Mém. Soc. Emul. Doubs,* Sér. 5, 8: 140 (1884)—France.
brachyphlaeus *M. Gandoger in Mém. Soc. Emul. Doubs,* Sér. 5, 8: 177 (1884)—France.
calopetalus *M. Gandoger in Mém. Soc. Emul. Doubs,* Sér. 5, 8: 180 (1884)—France.
calophyton *M. Gandoger in Mém. Soc. Emul. Doubs,* Sér. 5, 8: 193 (1884)—France.
calotemnus *A. van de Beek in Gorteria,* 24(1): 19 (1998)—Netherlands.
campaniensis *Winkel ex A. van de Beek in Gorteria,* 24(1): 21 (1998)—Netherlands.
caninitergi *H.E. Weber in Ber. Bayer. Bot. Ges.,* 66–67: 35 (1996)—Germany, Luxembourg.
carduelis *G. Matzke-Hajek in Decheniana,* 149: 46 (1996)—Germany.
cariotii *M. Gandoger in Mém. Soc. Emul. Doubs,* Sér. 5, 8: 210 (1884)—France.
carthusianorum *M. Gandoger in Mém. Soc. Emul. Doubs,* Sér. 5, 8: 141 (1884)—France.
catalpifolius *M. Gandoger in Mém. Soc. Emul. Doubs,* Sér. 5, 8: 214 (1884)—France.
caudatus *A. Foerster, Fl. Excurs. Aachen:* 146 (1878)—Germany.
cenisius *M. Gandoger in Mém. Soc. Emul. Doubs,* Sér. 5, 8: 141 (1884)—Italy.
ceratifolius *A. van de Beek in Gorteria,* 23(1–2): 8 (1997)—Netherlands.
chabertii *M. Gandoger in Mém. Soc. Emul. Doubs,* Sér. 5, 8: 222 (1884)—France.
chartaceus *A. Foerster, Fl. Excurs. Aachen:* 153 (1878)—Germany.
chavinii *M. Gandoger in Mém. Soc. Emul. Doubs,* Sér. 5, 8: 196 (1884)—France.
chingii Hu
var. suavissimus *(S.K. Lee) L.T. Lu in Acta Phytotax. Sin.,* 38(3): 280 (2000): R. suavissimus.
cineraceus *M. Gandoger in Mém. Soc. Emul. Doubs,* Sér. 5, 8: 183 (1884)—France.
cladocamptus *M. Gandoger in Mém. Soc. Emul. Doubs,* Sér. 5, 8: 156 (1884)—France.
cladotrichus *Gand. ex L.G. Genevier, Monogr. Rubus Bassin Loire,* ed. 2: 352 (1880)—France.
coccinatus *K. Meijer in Gorteria,* 26(5): 212 (2000)—Netherlands.
cochlearis *G. Matzke-Hajek in Decheniana,* 152: 43 (1999)—Germany.
colligenus *M. Gandoger in Mém. Soc. Emul. Doubs,* Sér. 5, 8: 201 (1884)—France.

RUBUS:-
comatulus *M. Gandoger in Mém. Soc. Emul. Doubs,* Sér. 5, 8: 176 (1884)—France.
congruens *M. Gandoger in Mém. Soc. Emul. Doubs,* Sér. 5, 8: 217 (1884)—France.
convivalis *A. Foerster, Fl. Excurs. Aachen:* 133 (1878)—Germany.
cosmophyllus *A. Foerster, Fl. Excurs. Aachen:* 113 (1878)—Germany.
costifolius *A. Foerster, Fl. Excurs. Aachen:* 105 (1878)—Germany.
cretatus *G. Matzke-Hajek in Decheniana,* 149: 37 (1996)—Germany, Netherlands.
crispulus *M. Gandoger in Mém. Soc. Emul. Doubs,* Sér. 5, 8: 157 (1884)—France.
cromerensis *A.L. Bull in Watsonia,* 22(1): 101 (1998)—Great Britain.
curvaciculatus *Walsemann ex H.E. Weber in Osnabrücker Naturwiss. Mitt.,* 22: 111 (1996)—Denmark, Germany, Poland.
curvispinus *A. Foerster, Fl. Excurs. Aachen:* 96 (1878)—Germany.
cycloideus *M. Gandoger in Mém. Soc. Emul. Doubs,* Sér. 5, 8: 159 (1884)—France.
cyclophyllus *A. Foerster, Fl. Excurs. Aachen:* 132 (1878)—Germany.
cymophorus *A. Foerster, Fl. Excurs. Aachen:* 103 (1878)—Germany.
davisii *D.E. Allen in Watsonia,* 22(1): 87 (1998)—Great Britain, France (Channel Is.).
debeauxii *M. Gandoger, Dec. Pl. Nov.,* 1: 7 (1875)—France.
debiliatus *A. Foerster, Fl. Excurs. Aachen:* 113 (1878)—Germany.
decipiens *A. Foerster, Fl. Excurs. Aachen:* 166 (1878), homonym; *see P.J. Müll. in Jahresber. Pollichia,* 16–17: 158 (1859)—Germany.
declinatus *A. Foerster, Fl. Excurs. Aachen:* 128 (1878)—Germany.
deloynei *M. Gandoger in Mém. Soc. Emul. Doubs,* Sér. 5, 8: 175 (1884)—France.
depromptus *M. Gandoger in Mém. Soc. Emul. Doubs,* Sér. 5, 8: 142 (1884)—France.
depulsorius *A. Foerster, Fl. Excurs. Aachen:* 166 (1878)—Germany.
deruyveri *H. Vannerom & E. Jacques in Belgian J. Bot.,* 129(1): 83 (1996 publ. 1997), nom. nov.: R. chlorothyrsos var. micradenes *Sudre.*
deweveri *A. van de Beek in Osnabrücker Naturwiss. Mitt.,* 23: 48 (1997)—Netherlands.
dichromophyllus *A. Foerster, Fl. Excurs. Aachen:* 104 (1878)—Germany.
dicthyophlaeus *M. Gandoger in Mém. Soc. Emul. Doubs,* Sér. 5, 8: 212 (1884)—France.
dinacanthus *A. Foerster, Fl. Excurs. Aachen:* 105 (1878)—Germany.
diremptus *A. Foerster, Fl. Excurs. Aachen:* 165 (1878)—Germany.
distortifolius *G. Matzke-Hajek in Decheniana,* 148: 18 (1995)—Germany.
echinosepalus *H.E. Weber in Mitt. Pollich. Pfälz. Naturk. Naturschutz,* 75: 204 (1988)—Germany.
elegantulus *A. Foerster, Fl. Excurs. Aachen:* 164 (1878)—Germany.
elongellus *M. Gandoger in Mém. Soc. Emul. Doubs,* Sér. 5, 8: 158 (1884)—France.
emarginatifolius *M. Gandoger in Mém. Soc. Emul. Doubs,* Sér. 5, 8: 153 (1884)—France.
erythrophlaeus *M. Gandoger in Mém. Soc. Emul. Doubs,* Sér. 5, 8: 183 (1884)—France.
eucalyptus *W.O. Focke in Biblioth. Bot.,* 72(2): 169 (1911)—China South-central (Hubei).
euphyllotrophus *A. Foerster, Fl. Excurs. Aachen:* 135 (1878)—Germany.
exemptus *A. Foerster, Fl. Excurs. Aachen:* 146 (1878)—Germany.
exstans *E. Walsemann & G. Stohr in Osnabrücker Naturwiss. Mitt.,* 23: 318 (1997)—Germany.
exsul *M. Gandoger in Mém. Soc. Emul. Doubs,* Sér. 5, 8: 149 (1884)—France.
fabrimontanus *F.J. Spribille in Jahresber. Schles. Ges. Vaterl. Cult.,* 83(2): 109 (1905 publ. 1906)—Germany.
festii *H.E. Weber in Ann. Mus. Civ. Rovereto,* 12: 180 (1996 publ. 1998)—Italy.
× firmus *Frid. ex J. Utsch in Jahres-Ber. Westfäl. Prov.-Vereins Wiss.,* 25: 185 (1896–97 publ. 1897)—Germany.

RUBUS:-
flexuosellus *M. Gandoger in Mém. Soc. Emul. Doubs,* Sér. 5, 8: 150 (1884)—France.
foersteri *G. Matzke-Hajek in Feddes Repert.,* 110(3–4): 162 (1999)—Germany, Netherlands.
foliolosus D. Don
var. incanus *(Sasaki ex T.S. Liu & T.Y. Yang) S.S. Ying in Mem. Coll. Agric. Nation. Taiwan Univ.,* 30(2): 59 (1990): R. incanus.
fragariiformis *M. Gandoger in Mém. Soc. Emul. Doubs,* Sér. 5, 8: 188 (1884)—France.
franchetianus *H. Lév. in Bull. Acad. Int. Géogr. Bot.,* 20(1): 71 (1909), nom. nov.: R. fragarioides var. adenophorus *Franch.*
fraudulentus *A. Foerster, Fl. Excurs. Aachen:* 125 (1878)—Germany.
frederici *A. van de Beek in Osnabrücker Naturwiss. Mitt.,* 23: 38 (1997), nom. nov.: R. laetus *A. Beek.*
friburgensis *M. Gandoger in Mém. Soc. Emul. Doubs,* Sér. 5, 8: 161 (1884)—Switzerland.
frigidus *M. Gandoger in Mém. Soc. Emul. Doubs,* Sér. 5, 8: 141 (1884)—Italy.
frondicomus *A. Foerster, Fl. Excurs. Aachen:* 152 (1878)—Germany.
fuernrohrii *H.E. Weber in Ber. Bayer. Bot. Ges.,* 66–67: 185 (1996)—Germany.
fundipartitus *A. Foerster, Fl. Excurs. Aachen:* 116 (1878)—Germany.
gariannensis *A.L. Bull in Watsonia,* 22(1): 97 (1998)—Great Britain.
geocamptus *M. Gandoger in Mém. Soc. Emul. Doubs,* Sér. 5, 8: 151 (1884)—France.
glossoides *H.E. Weber & G. Stohr in Feddes Repert.,* 109(5–6): 373 (1998)—Germany.
gymnocaulus *A. Foerster, Fl. Excurs. Aachen:* 154 (1878)—Germany.
hanryi *M. Gandoger in Mém. Soc. Emul. Doubs,* Sér. 5, 8: 194 (1884)—France.
hassicus *H.E. Weber in Ber. Bayer. Bot. Ges.,* 66–67: 28 (1996)—Germany.
henrici-weberi *A. van de Beek in Osnabrücker Naturwiss. Mitt.,* 23: 44 (1997)—Netherlands.
hermaniifolius *M. Gandoger in Mém. Soc. Emul. Doubs,* Sér. 5, 8: 191 (1884)—France.
hilsianus *H.E. Weber in Osnabrücker Naturwiss. Mitt.,* 20–21: 147 (1995)—Germany.
hindii *A.L. Bull in Watsonia,* 22(1): 102 (1998)—Great Britain.
hobroensis *A. Pedersen & J.C. Schou in Flora Fauna (Århus),* 99(2): 61 (1993)—Denmark.
holmiensis *M. Gandoger in Mém. Soc. Emul. Doubs,* Sér. 5, 8: 179 (1884)—Sweden.
holubyanus *M. Gandoger in Mém. Soc. Emul. Doubs,* Sér. 5, 8: 198 (1884)—Hungary.
humifixus *M. Gandoger in Mém. Soc. Emul. Doubs,* Sér. 5, 8: 186 (1884)—France.
hungaricus *M. Gandoger in Mém. Soc. Emul. Doubs,* Sér. 5, 8: 202 (1884), homonym; *see Holuby in Oesterr. Bot. Z.,* 25: 314 (1875)—Hungary.
hypocyclus *M. Gandoger in Mém. Soc. Emul. Doubs,* Sér. 5, 8: 150 (1884)—France.
hypodontus *M. Gandoger in Mém. Soc. Emul. Doubs,* Sér. 5, 8: 179 (1884)—France.
idaeus L.
subsp. komarovii *(Nakai) V.N. Voroschilov in A.K. Skvortsov (ed.), Florist. issl. v razn. raionakh SSSR:* 176 (1985): R. komarovii.
idanensis *M. Gandoger in Mém. Soc. Emul. Doubs,* Sér. 5, 8: 189 (1884)—France.
imbellis *G. Matzke-Hajek in Feddes Repert.,* 110(3–4): 164 (1999)—Germany, Belgium (Luxembourg).
immixtus *A. Foerster, Fl. Excurs. Aachen:* 155 (1878)—Germany.
inaccumulatus *A. Foerster, Fl. Excurs. Aachen:* 133 (1878)—Germany.
inanis *M. Gandoger in Mém. Soc. Emul. Doubs,* Sér. 5, 8: 197 (1884)—France.
infracanus *M. Gandoger in Mém. Soc. Emul. Doubs,* Sér. 5, 8: 219 (1884)—France.
inframollis *M. Gandoger in Mém. Soc. Emul. Doubs,* Sér. 5, 8: 198 (1884)—France.
innoxius *A. Foerster, Fl. Excurs. Aachen:* 162 (1878)—Germany.
insigniferus *M. Gandoger in Mém. Soc. Emul. Doubs,* Sér. 5, 8: 195 (1884)—France.
intercalaris *A. Foerster, Fl. Excurs. Aachen:* 162 (1878)—Germany.

RUBUS:-

intermedius *A. Foerster, Fl. Excurs. Aachen:* 126 (1878), homonym; *see Ortmann in Flora,* 18: 496 (1835)—Germany.

ischnophyllus *M. Gandoger in Mém. Soc. Emul. Doubs,* Sér. 5, 8: 178 (1884)—France.

ischnopodus *M. Gandoger in Mém. Soc. Emul. Doubs,* Sér. 5, 8: 182 (1884)—France.

josholubii *H.E. Weber in Preslia,* 72(2–4): 234 (2000)—Czechoslovakia.

klimmekianus *G. Matzke-Hajek in Decheniana,* 149: 49 (1996)—Germany.

kuleszae *J. Zieliński in Fragm. Flor. Geobot.,* 41(1): 249 (1996)—Poland, Czechoslovakia.

laegaardii *K. Romoleroux in Fl. Ecuador,* 56(79): 15 (1996)—Ecuador.

landoltii *H.E. Weber in Bot. Helvetica,* 107(2): 212 (1997)—Switzerland.

lasiotrichos Focke

var. blinii *(H. Lév.) L.T. Lu in Acta Phytotax. Sin.,* 38(3): 281 (2000): R. blinii.

× laurifolius *J. Utsch in Jahres-Ber. Westfäl. Prov.-Vereins Wiss.,* 24: 150 (1895–96 publ. 1896)—Germany.

laurocerasus *M. Gandoger in Mém. Soc. Emul. Doubs,* Sér. 5, 8: 221 (1884)—France.

laxissimus *M. Gandoger in Mém. Soc. Emul. Doubs,* Sér. 5, 8: 207 (1884)—France.

laxiusculus *M. Gandoger in Mém. Soc. Emul. Doubs,* Sér. 5, 8: 193 (1884)—France.

leptaleus *M. Gandoger in Mém. Soc. Emul. Doubs,* Sér. 5, 8: 185 (1884)—France.

leptargyrophyllus *A. Foerster, Fl. Excurs. Aachen:* 97 (1878)—Germany.

leptostictus *M. Gandoger in Mém. Soc. Emul. Doubs,* Sér. 5, 8: 212 (1884)—France.

liboensis *T.L. Xu in Acta Bot. Yunnanica,* 20(2): 163 (1998)—China South-central (Guizhou).

limitis *G. Matzke-Hajek & H. Grossheim in Mainzer Naturwiss. Arch.,* 34: 37 (1996)—Germany.

lobulatus *M. Gandoger in Mém. Soc. Emul. Doubs,* Sér. 5, 8: 210 (1884)—France.

longeacuminatus *M. Gandoger in Mém. Soc. Emul. Doubs,* Sér. 5, 8: 181 (1884)—France.

longebracteatus *M. Gandoger in Mém. Soc. Emul. Doubs,* Sér. 5, 8: 191 (1884)—France.

longispinus *M. Gandoger in Mém. Soc. Emul. Doubs,* Sér. 5, 8: 197 (1884)—France.

loosii *H.E. Weber in Osnabrücker Naturwiss. Mitt.,* 20–21: 151 (1995)—Germany.

lucovagus *M. Gandoger in Mém. Soc. Emul. Doubs,* Sér. 5, 8: 195 (1884)—France.

lugdunensis *M. Gandoger in Mém. Soc. Emul. Doubs,* Sér. 5, 8: 216 (1884)—France

macrobotrys *M. Gandoger in Mém. Soc. Emul. Doubs,* Sér. 5, 8: 192 (1884)—France

macrolimbos *M. Gandoger in Mém. Soc. Emul. Doubs,* Sér. 5, 8: 177 (1884)—France

macrotrichus *M. Gandoger in Mém. Soc. Emul. Doubs,* Sér. 5, 8: 183 (1884)—France

magnisepalus *K. Meijer in Gorteria,* 26(5): 214 (2000)—Netherlands.

magnolioides *M. Gandoger in Mém. Soc. Emul. Doubs,* Sér. 5, 8: 178 (1884)—France

malarbetii *M. Gandoger in Mém. Soc. Emul. Doubs,* Sér. 5, 8: 162 (1884)—France.

mathonnetii *M. Gandoger in Mém. Soc. Emul. Doubs,* Sér. 5, 8: 141 (1884)—France.

mediocris *A. Foerster, Fl. Excurs. Aachen:* 158 (1878)—Germany.

meierottii *H.E. Weber in Ber. Bayer Bot. Ges.,* 66–67: 180 (1996)—Germany.

melamporphyrus *A. van de Beek in Gorteria,* 24(1): 23 (1998)—Poland.

mesocladus *M. Gandoger in Mém. Soc. Emul. Doubs,* Sér. 5, 8: 190 (1884)—France.

molestus *A. Foerster, Fl. Excurs. Aachen:* 161 (1878)—Germany.

mollior *M. Gandoger in Mém. Soc. Emul. Doubs,* Sér. 5, 8: 148 (1884)—France.

moluccanus L.

var. trilobus *A.R. Bean in Austrobaileya,* 5(1): 45 (1997)—New South Wales, Queensland, Victoria.

montivagus *M. Gandoger in Mém. Soc. Emul. Doubs,* Sér. 5, 8: 215 (1884)—France.

× muenderleinii *J. Utsch in Jahres-Ber. Westfäl. Prov.-Vereins Wiss.,* 25: 183 (1896–97 publ. 1897)—Germany.

multicaudatus *H.E. Weber in Carolinea,* 54: 16 (1996)—Germany.

muridens *A. van de Beek in Osnabrücker Naturwiss. Mitt.,* 23: 46 (1997)—Netherlands.

mus *A. van de Beek in Gorteria,* 24(1): 25 (1998)—Netherlands.

neanias *A. van de Beek in Osnabrücker Naturwiss. Mitt.,* 23: 50 (1997)—Netherlands.

nemoralis P.J. Müll.

var. argyriophyllus *(Ranke) H.E. Weber in Kieler Not. Pflanzenk. Schleswig-Holstein,* 7(4): 90 (1975), without basionym ref.: Basionym not stated.

nessensis Hall

subsp. cubirianus *H.E. Weber in Abh. Naturwiss. Ver. Bremen,* 44(2–3): 240 (1999)—Germany.

neumannianus *H.E. Weber & H. Vannerom in Mitt. Pollich. Pfälz. Naturk. Naturschutz,* 75: 199 (1988)—Germany, Belgium, Luxembourg.

oblongellus *M. Gandoger in Mém. Soc. Emul. Doubs,* Sér. 5, 8: 186 (1884)—France.

oculus-junonis *M. Gandoger in Mém. Soc. Emul. Doubs,* Sér. 5, 8: 173 (1884)—France.

odontophyllus *M. Gandoger in Mém. Soc. Emul. Doubs,* Sér. 5, 8: 163 (1884)—Hungary.

olbiensis *M. Gandoger in Mém. Soc. Emul. Doubs,* Sér. 5, 8: 195 (1884)—France.

opiflorus *A. Foerster, Fl. Excurs. Aachen:* 161 (1878)—Germany.

oreophilus *A. Foerster, Fl. Excurs. Aachen:* 92 (1878), homonym; *see P.J. Müll. in Jahresber. Pollichia,* 16–17: 227 (1859)—Germany.

orotavensis *M. Gandoger in Mém. Soc. Emul. Doubs,* Sér. 5, 8: 154 (1884)—Canary Is.

oscillans *M. Gandoger in Mém. Soc. Emul. Doubs,* Sér. 5, 8: 187 (1884)—France.

osseus *G. Matzke-Hajek in Osnabrücker Naturwiss. Mitt.,* 23: 212 (1997)—Germany.

ostryifolius *M. Gandoger in Mém. Soc. Emul. Doubs,* Sér. 5, 8: 153 (1884)—France.

patulus *A. Foerster, Fl. Excurs. Aachen:* 152 (1878), homonym; *see P.J. Müll. & Lefèvre in Jahresber. Pollichia,* 16–17: 97 (1859)—Germany.

pauanus *E. Monasterio-Huelin in Nordic J. Bot.,* 15(4): 368 (1995 publ. 1996)—Spain.

× paxii *W.O. Focke in Abh. Naturwiss. Verein Bremen,* 19(2): 205 (1908)—Cultivation. (R. idaeus × phoenicolasius).

perperus *H.E. Weber in Ber. Bayer. Bot. Ges.,* 66–67: 176 (1996)—Germany, Belgium, Czechoslovakia.

peyronii *M. Gandoger in Mém. Soc. Emul. Doubs,* Sér. 5, 8: 219 (1884)—France.

phaeoleucus *M. Gandoger in Mém. Soc. Emul. Doubs,* Sér. 5, 8: 213 (1884)—France.

pictaviensis *M. Gandoger in Mém. Soc. Emul. Doubs,* Sér. 5, 8: 184 (1884)—France.

pilatensis *M. Gandoger in Mém. Soc. Emul. Doubs,* Sér. 5, 8: 218 (1884)—France.

pinivagus *M. Gandoger in Mém. Soc. Emul. Doubs,* Sér. 5, 8: 209 (1884)—France.

poliothyrsus *A. van de Beek in Osnabrücker Naturwiss. Mitt.,* 23: 38 (1997)—Netherlands.

polycarpiformis *F. Spribille in Verh. Bot. Vereins Prov. Brandenburg,* 55: 138 (1913)—Poland.

polyphyllus *A. Foerster, Fl. Excurs. Aachen:* 104 (1878)—Germany.

praestans *H.E. Weber in Mitt. Pollich. Pfälz. Naturk. Naturschutz,* 75: 203 (1988)—Germany.

praticolor *A. van de Beek in Gorteria,* 24(1): 26 (1998)—Netherlands.

prostratus *A. Foerster, Fl. Excurs. Aachen:* 149 (1878)—Germany.

protractus *M. Gandoger in Mém. Soc. Emul. Doubs,* Sér. 5, 8: 162 (1884)—France.

pseudoinfestus *H.E. Weber in Carolinea,* 47: 49 (1989)—Germany.

pugiunculosus *G. Matzke-Hajek in Feddes Repert.,* 110(3–4): 167 (1999)—Germany, Belgium, Netherlands.

pycnophyllus *A. Foerster, Fl. Excurs. Aachen:* 107 (1878)—Germany.

pyracanthus *M. Gandoger in Mém. Soc. Emul. Doubs,* Sér. 5, 8: 221 (1884)—France.

pyrenaeicola *M. Gandoger in Mém. Soc. Emul. Doubs,* Sér. 5, 8: 140 (1884)—France.

queenslandicus *A.R. Bean in Austrobaileya,* 4(4): 683 (1997)—Queensland.

racemulosus *M. Gandoger in Mém. Soc. Emul. Doubs,* Sér. 5, 8: 202 (1884)—France.

remotior *M. Gandoger in Mém. Soc. Emul. Doubs,* Sér. 5, 8: 151 (1884)—France.

rhombicus *H.E. Weber in Bot. Helvetica,* 107(2): 215 (1997)—Switzerland, France, Germany.

riparium *M. Gandoger in Mém. Soc. Emul. Doubs,* Sér. 5, 8: 199 (1884)—France.

ripuaricus *G. Matzke-Hajek in Decheniana,* 148: 22 (1995)—Germany.

rubercadaver *A. van de Beek in Osnabrücker Naturwiss. Mitt.,* 23: 42 (1997)—Netherlands.

rubrinaevus *M. Gandoger in Mém. Soc. Emul. Doubs,* Sér. 5, 8: 153 (1884)—France.

rugosifolius *G.H. Loos in Flor. Rundbr.,* 32(1): 46 (1998)—Germany.

rupifragus *M. Gandoger in Mém. Soc. Emul. Doubs,* Sér. 5, 8: 218 (1884)—France.

rurae *G.H. Loos in Flor. Rundbr.,* 32(1): 48 (1998)—Germany.

sachalinensis H. Lév.

var. eglandulatus *(Y.B. Chang) L.T. Lu in Acta Phytotax. Sin.,* 38(3): 280 (2000), as 'eglanduratus': R. matsumuranus var. eglandulatus.

var. przewalskii *(Prokh.) L.T. Lu in Acta Phytotax. Sin.,* 38(3): 280 (2000): R. przewalskii.

saguramicus *Juz. ex L.S. Krassovskaja in Bot. Zhurn.,* 85(10): 110 (2000)—Transcaucasus (Gruziya).

salaevensis *M. Gandoger in Mém. Soc. Emul. Doubs,* Sér. 5, 8: 174 (1884)—France.

saxivagus *M. Gandoger in Mém. Soc. Emul. Doubs,* Sér. 5, 8: 189 (1884)—France.

× scherzeri *J. Utsch in Jahres-Ber. Westfäl. Prov.-Vereins Wiss.,* 25: 151 (1896–97 publ. 1897)—Germany.

schumacheri *(Ade) G. Matzke-Hajek in Decheniana,* 149: 43 (1996): R. menkei subsp. schumacheri.

sciaphilus *A. Foerster, Fl. Excurs. Aachen:* 121 (1878)—Germany.

segregatus *A. Foerster, Fl. Excurs. Aachen:* 157 (1878)—Germany.

semitarum *M. Gandoger in Mém. Soc. Emul. Doubs,* Sér. 5, 8: 175 (1884)—France.

sempernitens *D.E. Allen & L.J. Margetts in Watsonia,* 22(1): 89 (1998)—Great Britain, France.

sepiiformis *M. Gandoger in Mém. Soc. Emul. Doubs,* Sér. 5, 8: 152 (1884)—France.

sergii *L.S. Krassovskaja in Bot. Zhurn.,* 85(10): 110 (2000)—Transcaucasus (Gruziya).

serrulatus *A. Foerster, Fl. Excurs. Aachen:* 140 (1878)—Germany.

setosus *A. Foerster, Fl. Excurs. Aachen:* 134 (1878), homonym; *see Bigelow, Fl. Boston., ed. 2:* 198 (1824)—Germany.

sierrae *J.E. Laferrière in Phytologia,* 84(3): 169 (1998 publ. 1999)—Mexico Northeast (Chihuahua).

silvivagus *M. Gandoger in Mém. Soc. Emul. Doubs,* Sér. 5, 8: 223 (1884)—France.

sparmanniifolius *M. Gandoger in Mém. Soc. Emul. Doubs,* Sér. 5, 8: 152 (1884)—France.

speculans *K. Meijer in Gorteria,* 26(5): 216 (2000)—Netherlands.

speculatus *G. Matzke-Hajek in Osnabrücker Naturwiss. Mitt.,* 23: 215 (1997)—Germany.

spiculus *K. Meijer in Gorteria,* 26(5): 221 (2000)—Netherlands.

spinirameus *M. Gandoger in Mém. Soc. Emul. Doubs,* Sér. 5, 8: 182 (1884)—France.

spiniramulis *A. Foerster, Fl. Excurs. Aachen:* 113 (1878)—Germany.

× stellulans *Beck, Fl. Nieder-Österreich,* 2(1): 727 (1893), in Supp. 1 as 'stellulatus'—Austria.†

stictochlamys *M. Gandoger in Mém. Soc. Emul. Doubs,* Sér. 5, 8: 160 (1884)—France.

stohrii *H.E. Weber & M. Ranft in Feddes Repert.,* 109(5–6): 370 (1998)—Germany.

subalpestris *M. Gandoger in Mém. Soc. Emul. Doubs,* Sér. 5, 8: 142 (1884)—France.

RUBUS:-
subcordatus *H.E. Weber in Carolinea,* 47: 50 (1989)—Germany, France.
sublanatus *M. Gandoger in Mém. Soc. Emul. Doubs,* Sér. 5, 8: 192 (1884)—France.
submitis *M. Gandoger in Mém. Soc. Emul. Doubs,* Sér. 5, 8: 190 (1884)—France.
subplumbeus *A. Foerster, Fl. Excurs. Aachen:* 128 (1878)—Germany.
suevicola *H.E. Weber in Ber. Bayer. Bot. Ges.,* 66–67: 38 (1996)—Germany.
superbus *A. Foerster, Fl. Excurs. Aachen:* 130 (1878)—Germany.
surrectus *K. Meijer in Gorteria,* 26(5): 219 (2000)—Netherlands.
× sylvulicola *J. Utsch in Jahres-Ber. Westfäl. Prov.-Vereins Wiss.,* 25: 165 (1896–97 publ. 1897)—Germany.
talenceanus *M. Gandoger in Mém. Soc. Emul. Doubs,* Sér. 5, 8: 174 (1884)—France.
tassinensis *M. Gandoger in Mém. Soc. Emul. Doubs,* Sér. 5, 8: 216 (1884)—France.
tauni *W. Schnedler & H. Grossheim in Osnabrücker Naturwiss. Mitt.,* 23: 305 (1997)—Germany.
taxandriae *Vannerom ex A. van de Beek in Gorteria,* 24(1): 27 (1998)—Netherlands.
tenuihabitus *G.H. Loos in Flor. Rundbr.,* 32(1): 47 (1998)—Germany.
tephrocalyx *M. Gandoger in Mém. Soc. Emul. Doubs,* Sér. 5, 8: 149 (1884)—France.
tephrocladus *M. Gandoger in Mém. Soc. Emul. Doubs,* Sér. 5, 8: 156 (1884)—France.
tephrodes Hance
var. holadenus *(H. Lév.) L.T. Lu in Acta Phytotax. Sin.,* 38(3): 281 (2000): R. holadenus.
thalassarctos *A. van de Beek in Gorteria,* 23(1–2): 10 (1997)—Netherlands.
tholeyronianus *M. Gandoger in Mém. Soc. Emul. Doubs,* Sér. 5, 8: 159 (1884)—France.
tomophyllus *M. Gandoger in Mém. Soc. Emul. Doubs,* Sér. 5, 8: 194 (1884)—France.
transvestitus *G. Matzke-Hajek in Decheniana,* 148: 25 (1995)—Germany, Belgium (Luxembourg).
tsangii Merr.
var. yanshanensis *(Z.X. Yu & W.T. Ji) L.T. Lu in Acta Phytotax. Sin.,* 38(3): 280 (2000): R. yanshanensis.
turnerifolius *M. Gandoger in Mém. Soc. Emul. Doubs,* Sér. 5, 8: 161 (1884)—France.
ubericus *G. Matzke-Hajek in Decheniana,* 149: 51 (1996)—Germany, Belgium.
vacillans *M. Gandoger in Mém. Soc. Emul. Doubs,* Sér. 5, 8: 208 (1884)—France.
vadalis *A. van de Beek in Gorteria,* 23(1–2): 6 (1997)—Netherlands.
vandermeijdenii *A. van de Beek in Osnabrücker Naturwiss. Mitt.,* 23: 53 (1997)—Netherlands.
vapillonii *M. Gandoger in Mém. Soc. Emul. Doubs,* Sér. 5, 8: 187 (1884)—France.
venetorum *D.E. Allen in Watsonia,* 22(1): 92 (1998)—France, Great Britain.
villosior *A.L. Bull in Watsonia,* 22(1): 100 (1998)—Great Britain.
vindobonensis *M. Gandoger in Mém. Soc. Emul. Doubs,* Sér. 5, 8: 200 (1884)—Austria, Hungary.
waddellii *D.E. Allen in Watsonia,* 22(1): 83 (1998)—Ireland (Northern Ireland), Great Britain.
wahlbergii Arrh.
subsp. nordicus *H.E. Weber & A. Pedersen in Nordic J. Bot.,* 18(1): 39 (1998)—Norway, Sweden.
walteri *H.E. Weber & H. Grossheim in Mitt. Pollich.,* 85: 223 (1998)—Germany.
wessbergii *A. Pedersen & H.O. Martensen in Flora Fauna (Århus),* 93(1–2): 3 (1987)—Denmark.
× wolfredoi-wildpretii *H.E. Weber in An. Jard. Bot. Madrid,* 57(1): 33 (1999)—Canary Is. (R. bollei × ulmifolius).
xanthoacanthus *M. Gandoger in Mém. Soc. Emul. Doubs,* Sér. 5, 8: 201 (1884)—France.

RUDBECKIA (Compos.).
speciosa *H.A. Schrader in Index Seminum Horti Academici Gottingensis:* 5 (1831)—Northern America.

RUDGEA (Rubiac.).
grandifructa *C.M. Taylor & M. Monsalve B.in Novon,* 7(2): 205 (1997)—Colombia.
hemisphaerica *Dwyer ex C.M. Taylor in Novon,* 6(3): 302 (1996)—Panama.
mandevilliifolia *Dwyer ex C.M. Taylor in Novon,* 6(3): 304 (1996)—Panama.
mcphersonii *Dwyer ex C.M. Taylor in Novon,* 6(3): 304 (1996)—Panama.
panamensis *(Dwyer) C.M. Taylor in Novon,* 6(3): 305 (1996): Cephaelis panamensis.
sanblasensis *C.M. Taylor in Novon,* 6(2): 220 (1996)—Panamá.
tayloriae *G. Aymard C., L.J. Dorr & N. Cuello in Novon,* 9(3): 315 (1999)—Venezuela.
vallis *(Standl. ex Steyerm.) C.M. Taylor in Novon,* 6(2): 213 (1996): Cephaelis vallis.

RUDOLFIELLA (Orchidac.).
peruviana *D.E. Bennett & E.A. Christenson in Lindleyana,* 13(2): 83 (1998)—Peru.

RUELLIA (Acanthac.).
baikiei *(Hook.) N.E. Brown, Suppl. Johnson's Gard. Dict.:* 999 (1882): Stephanophysum baikiei.
boranica *Ensermu Kelbessa in Kew Bull.,* 54(2): 453 (1999)—Ethiopia.
californica (Rose) I.M. Johnst.
subsp. peninsularis *(Rose) T.F. Daniel in Proc. Calif. Acad. Sci.,* 49(10): 384 (1997): Calophanes peninsularis.
chilpancingana *T.F. Daniel in Brittonia,* 51(2): 124 (1999)—Mexico Southwest (Guerrero).
comonduensis *T.F. Daniel in Brittonia,* 51(2): 124 (1999), nom. nov.: R. cordata *Brandegee.*
jimulcensis *J.A. Villarreal Q.in Sida,* 18(1): 223 (1998)—Mexico Northeast (Coahuila, Durango).
leucantha Brandegee
subsp. postinsularis *(Gentry) T.F. Daniel in Polibotanica,* 2: 7 (1996): R. leucantha var. postinsularis.
pearcei *(Hook. f.) N.E. Brown, Suppl. Johnson's Gard. Dict.:* 999 (1882): Stemonacanthus pearcei.
saulensis *D.C. Wasshausen in Brittonia,* 47(4): 422 (1995)—French Guiana.
terminalis *(Nees) D.C. Wasshausen in Cat. Vasc. Pl. Ecuador (Monogr. Syst. Bot. Missouri Bot. Gard., 75):* 953 (1999), as 'terminale': Scorodoxylum terminale.
verbasciformis *(Nees) C. Ezcurra & D. Zappi in Kew Bull.,* 51(4): 819 (1996): Eurychanes verbasciformis.
woolstonii *C. Ezcurra in Brittonia,* 50(3): 339 (1998)—Paraguay.

RUMEX (Polygonac.).
acetosa L.
subsp. islandicus *(Á. Löve & D. Löve) Ö. Nilsson in Nordic J. Bot.,* 16(1): 5 (1996): Acetosa alpestris subsp. islandica.
acetosella L.
subsp. arenicola *Y. Mäkinen ex Elven in Nordic J. Bot.,* 19(4): 387 (1999)—Greenland.
acmophorus *M. Gandoger, Dec. Pl. Nov.,* 1: 4 (1875)—France.
× akeroydii *F.J. Rumsey in Watsonia,* 22(4): 414 (1999)—Great Britain. (R. cristatus × palustris).
crispus L.
subsp. robustus *(Rech.) I.O. Pestova in Ukr. Bot. Zhurn.,* 55(1): 28 (1998): R. crispus var. robustus.
subsp. strictissimus *(Rech.) I.O. Pestova in Ukr. Bot. Zhurn.,* 55(1): 27 (1998): R. crispus var. strictissimus.
erythrocarpus *M. Gandoger, Dec. Pl. Nov.,* 1: 17 (1875)—France.
gmelinii Turcz.
var. latus *A.J. Li ex J.Q. Fu in Fl. Loess-Plat. Sin.,* 1: 610 (2000), holotype not indicated—China North-central (Gansu, Shaanxi, Shanxi), Inner Mongolia (Ningxia).
nepalensis Spreng.
var. remotiflorus *(Sam.) A.J. Li in Fl. Reipubl. Popul. Sin.,* 25(1): 161 (1998): R. remotiflorus.
× rosemurphyae *D.T. Holyoak in Watsonia,* 23(1): 84 (2000), contrary to Art. 8.2 ICBN (2000)—Great Britain. (R. conglomeratus × rupestris).

RUMEX:-
roseus L.
f. lacerus *(Boiss.) J. Lambinon in Soc. Éch. Pl. Vasc. Eur. Bassin Médit. Bull.,* 28: 128 (2000): R. tingitanus var. lacerus.
× togaensis *T. Kawahara in J. Phytogeogr. Taxon.,* 43(1–2): 48 (1995)—Japan. (R. madaio × obtusifolius).
× transiens *(Simonk.) N.N. Tzvelev, Opred. Sosud. Rast. Severo-Zap. Rossii:* 679 (2000): Rumex sylvestris var. transiens.
verrietianus *M. Gandoger, Dec. Pl. Nov.,* 1: 4 (1875)—France.

RUNGIA (Acanthac.).
bisaccata *D. Fang & H.S. Lo in Guihaia,* 17(1): 48 (1997)—China Southeast (Guangxi).
camerunensis *D. Champluvier in Syst. Geogr. Pl.,* 70(1): 140 (2000)—Cameroon.
guangxiensis *H.S. Lo & D. Fang in Guihaia,* 17(1): 46 (1997)—China Southeast (Guangxi).
longifolia Nees
subsp. latior *(Nees) L.H. Cramer in Kew Bull.,* 51(3): 556 (1996): R. latior.
longipes *D. Fang & H.S. Lo in Guihaia,* 17(1): 49 (1997)—China Southeast (Guangxi).
napoensis *D. Fang & H.S. Lo in Guihaia,* 17(1): 46 (1997)—China Southeast (Guangxi).
parviflora (Retz.) Nees
subsp. pectinata *(L.) L.H. Cramer in M.D. Dassanayake & W.D. Clayton, eds., Revised Handb. Fl. Ceylon,* 12: 105 (1998): Justicia pectinata.
pungens *D. Fang & H.S. Lo in Guihaia,* 17(1): 47 (1997)—China Southeast (Guangxi).

× **RUPPARA** *hort. in Orchid Rev.,* 107(1230): 391 (1999). *ORCHIDACEAE* (Gomesa × Odontoglossum × Oncidium).

RUPRECHTIA (Polygonac.).
howardiana *G. Aymard C. & P.E. Berry in Novon,* 9(3): 313 (1999)—Venezuela.

RUSBYELLA (Orchidac.).
quadridentata *(D.E. Benn. & Christenson) K. Senghas in J. Orchideenfreund,* 7(2): 135 (2000), as 'quadridentatum': Odontoglossum quadridentatum.

RUSCHIA (Aizoac.).
dielsiana *(A. Berger) H. Jacobsen ex H.E.K. Hartmann in Bradleya,* 16: 84 (1998): Mesembryanthemum dielsianum.
punctulata *(L. Bolus) H.E.K. Hartmann in Bradleya,* 16: 85 (1998): Mesembryanthemum punctulatum.
purpureostyla *(L. Bolus) P.V. Bruyns in S. Afr. J. Bot.,* 63(4): 241 (1997): Mesembryanthemum purpureostylum.
recurva *(Moench) H.E.K. Hartmann in Bradleya,* 16: 85 (1998): Mesembryanthemum recurvum.
scabra *H.E.K. Hartmann in Bradleya,* 14: 46 (1996)—Cape Province.
sessilis *(Thunb.) H.E.K. Hartmann in Bradleya,* 16: 85 (1998): Mesembryanthemum sessile.
vetovalida *H.E.K. Hartmann in Bradleya,* 16: 85 (1998), nom. nov.: Mesembryanthemum validum *Haw.*

RUSSELIA (Scrophulariac.).
contrerasii *B.L. Turner in Phytologia,* 81(5): 335 (1996 publ. 1997)—Guatemala, Mexico Southeast (Chiapas).
cora *I. Méndez-Larios & O. Téllez Valdés in An. Inst. Biol. Univ. Nac. Autón. Méx., Bot.,* 66(2): 184 (1995 publ. 1996)—Mexico Southwest (Nayarit).
iltisneeana *B.L. Turner in Phytologia,* 81(5): 333 (1996 publ. 1997)—Mexico Southwest (Jalisco).
manantlana *B.L. Turner in Phytologia,* 81(1): 22 (1996 publ. 1997)—Mexico Southwest (Jalisco).

RUSTIA (Rubiac.).
bilsana *P.G. Delprete in Novon,* 8(3): 234 (1998)—Ecuador.
schunkeana *P.G. Delprete in Brittonia,* 48(1): 29 (1996)—Peru, Ecuador.

RUSTIA:-
simpsonii *P.G. Delprete in Fl. Neotrop.*, 77: 89 (1999)—Brazil Southeast (Rio de Janeiro).
thibaudioides *(H. Karst.) P.G. Delprete in Fl. Neotrop.*, 77: 81 (1999): Henlea thibaudioides.

RUTIDOSIS (Compos.).
glandulosa *A.E. Holland in Austrobaileya*, 5(3): 565 (1999)—Queensland.
helichrysoides DC.
subsp. acutiglumis *(Philipson) A.E. Holland in Austrobaileya*, 5(3): 569 (1999): R acutiglumis.

RYPAROSA (Flacourtiac.).
porcata *P.F. Stevens in Harvard Pap. Bot.*, 3(2): 253 (1998)—Borneo (Kalimantan, Sarawak).

RYTIDOSPERMA (Gramin.).
australe *(Petrie) H.P. Linder in Telopea*, 6(4): 613 (1996), isonym; *see Connor & Edgar in New Zealand J. Bot.*, 25(1): 166 (1987): Triodia australis.
craigii *(Veldkamp) H.P. Linder in Telopea*, 6(4): 613 (1996): Danthonia craigii.
davyi *(C.E. Hubb.) T.A. Cope in Fl. Zambes.*, 10(2): 7 (1999): Danthonia davyi.
dendeniwae *(Veldkamp) H.P. Linder in Telopea*, 6(4): 613 (1996): Danthonia dendeniwae.
distichum *(Nees) T.A. Cope in Fl. Zambes.*, 10(2): 9 (1999), as 'disticha': Danthonia disticha.
exiguum *(Kirk) H.P. Linder in Telopea*, 6(4): 614 (1996): Triodia exigua.
grandiflorum *(Hochst. ex A. Rich.) S.M. Phillips in Fl. Ethiopia & Eritrea*, 7: 74 (1995), as 'grandiflora': Danthonia grandiflora.
irianense *(Veldkamp) H.P. Linder in Telopea*, 6(4): 614 (1996): Danthonia irianensis.
javanicum *(Ohwi ex Veldkamp) H.P. Linder in Telopea*, 6(4): 614 (1996): Danthonia javanica.
montis-wilhelmii *(Veldkamp & Fortuin) H.P. Linder in Telopea*, 6(4): 614 (1996): Danthonia montis-wilhelmii.
nardifolium *(Veldkamp) H.P. Linder in Telopea*, 6(4): 614 (1996): Danthonia nardifolia.
nitens *(D.I. Morris) H.P. Linder in Telopea*, 6(4): 614 (1996): Danthonia nitens.
oreoboloides *(F. Muell.) H.P. Linder in Telopea*, 6(4): 614 (1996): Festuca oreoboloides.
pumilum *(Kirk) H.P. Linder in Telopea*, 6(4): 615 (1996), isonym; *see Connor & Edgar in New Zealand J. Bot.*, 25(1): 166 (1987): Atropis pumila.
vickeryae *M. Gray & H.P. Linder in Austral. Syst. Bot.*, 12(5): 744 (1999)—New South Wales.

RYTIGYNIA (Rubiac.).
xanthotricha *(K. Schum.) B. Verdcourt in Kew Bull.*, 36(3): 539 (1981): Plectronia xanthotricha.

S

SABAUDIA (Labiat.).
atriplicifolia *(Benth.) E. Chiovenda in Bull. Soc. Bot. Ital.*, 4: 61 (1916); *see also Bull. Soc. Bot. Ital.*, 6–7: 56 (1917): Lavandula atriplicifolia.

SABAZIA (Compos.).
breedlovei *B.L. Turner in Phytologia*, 82(4): 278 (1997 publ. 1998), homonym; *see B.L. Turner in Wrightia*, 5(8): 303 (1976)—Mexico Southwest (Guerrero).
lapsensis *B.L. Turner in Phytologia*, 83(1): 34 (1997 publ. 1998), nom. nov.: S. breedlovei *B.L. Turner*.

SABICEA (Rubiac.).
ceylanica *C. Puff in Pl. Diversity Malesia III (eds. J. Dransfield et al.)*: 201 (1997 publ. 1998), nom. nov.: Schizostigma hirsutum *Arn. ex Meisn.*
marojejyensis *S.G. Razafimandimbison & J.S. Miller in Adansonia*, 21(1): 42 (1999)—Madagascar.
pyramidalis *L. Andersson in Fl. Ecuador*, 62(162:3): 110 (1999)—Ecuador.
thyrsiflora *L. Andersson in Fl. Ecuador*, 62(162:3): 112 (1999)—Ecuador.

SABINA (Cupressac.).
indica *(Bertol.) L.K. Fu & Y.F. Yu in L.K. Fu & T. Hong (eds.), Higher Pl. China*, 3: 90 (2000): Juniperus indica.
semiglobosa *(Regel) L.K. Fu & Y.F. Yu in L.K. Fu & T. Hong (eds.), Higher Pl. China*, 3: 88 (2000): Juniperus semiglobosa.
squamata (Buch.-Ham. ex D. Don) Antoine
var. fargesii *(Rehder & E.H. Wilson) L.K. Fu & Y.F. Yu in L.K. Fu & T. Hong (eds.), Higher Pl. China*, 3: 86 (2000): Juniperus squamata var. fargesii.

SACCHARUM (Gramin.).
longisetosum (Andersson) V. Naray. ex Bor
var. hookeri *(Hack.) U. Shukla, Grasses N.-East. India*: 125 (1996): Erianthus hookeri.
sinense *W. Roxburgh, Pl. Coromandel*, 3: 26 (1815)—China.†

SACCOLABIUM (Orchidac.).
filiforme *(Wight) Rchb. f. in Walpers, Ann. Bot. Syst.*, 6: 887 (1864): Sarcanthus filiformis.†
flexuosum *(Lindl.) Rchb. f. in Walpers, Ann. Bot. Syst.*, 6: 886 (1864): Oeceoclades flexuosa.†

SAGERAEA (Annonac.).
sarawakensis *E.C.H. van Heusden in Nordic J. Bot.*, 17(1): 48 (1997)—Borneo (Sarawak).
zeylanica *E.C.H. van Heusden in Nordic J. Bot.*, 17(1): 50 (1997)—Sri Lanka.

SAGERETIA *A.T. Brongniart, Mém. Fam. Rhamnées*: 52 (1826). *RHAMNACEAE*†
elegans *(Kunth) A.T. Brongniart, Mém. Fam. Rhamnées*: 53 (1826): Rhamnus elegans.†
gongshanensis *G.S. Fan & L.L. Deng in Sida*, 17(4): 677 (1997)—China South-central (Yunnan).
guayaquilensis *(Kunth) A.T. Brongniart, Mém. Fam. Rhamnées*: 53 (1826), as 'gayaquilensis': Rhamnus guayaquilensis.†
hamosa *(Wall. ex Roxb.) A.T. Brongniart, Mém. Fam. Rhamnées*: 53 (1826): Ziziphus hamosa.†
lijiangensis *G.S. Fan & S.K. Chen in Acta Bot. Yunnanica*, 19(1): 38 (1997)—China South-central (Yunnan).
michauxii *A.T. Brongniart, Mém. Fam. Rhamnées*: 53 (1826), nom. nov. superfl.: Rhamnus minutiflora *Michx*.†
oppositifolia *(Wall. ex Roxb.) A.T. Brongniart, Mém. Fam. Rhamnées*: 53 (1826): Ziziphus oppositifolia.†
senticosa *(Kunth) A.T. Brongniart, Mém. Fam. Rhamnées*: 53 (1826): Rhamnus senticosa.†
spicata *A.T. Brongniart, Mém. Fam. Rhamnées*: 53 (1826)—Locality not stated.†
theezans *(L.) A.T. Brongniart, Mém. Fam. Rhamnées*: 53 (1826): Rhamnus theezans.†
yilinii *G.S. Fan & S.K. Chen in Acta Bot. Yunnanica*, 19(1): 38 (1997)—China South-central (Yunnan).

SAGINA (Caryophyllac.).
diemensis *L.G. Adams in Muelleria*, 9: 64 (1996)—Tasmania.
karakorensis *(Em.Schmid) Yu.P. Kozhevnikov in Rast. Tsentral. Azii*, 11: 49 (1994 publ. 1995): Arenaria karakorensis.
maxima A. Gray
var. crassicaulis *(S. Watson) V.N. Voroschilov in A.K. Skvortsov (ed.), Florist. issl. v razn. raĭonakh SSSR*: 166 (1985): S. crassicaulis.
namadgi *L.G. Adams in Muelleria*, 9: 63 (1996)—Australian Capital Territory, New South Wales, Victoria, Tasmania.
procumbens L.
subsp. atlasica *A. Dobignard in Candollea*, 52(1): 126 (1997), nom. nov.: S. procumbens var. parviflora *Ball*.
rupestris *R.T.A. Schouten ex Kai Larsen in Nordic J. Bot.*, 18(4): 421 (1998)—New Guinea (Papua New Guinea).

SAGITTARIA (Alismatac.).
graminea Michx.
subsp. chapmanii *(J.G. Sm.) R.R. Haynes & C.B. Hellquist in Novon*, 6(4): 370 (1996): S. graminea var. chapmanii.

SAGITTARIA:-
graminea
subsp. weatherbiana *(Fernald) R.R. Haynes & C.B. Hellquist in Novon*, 6(4): 371 (1996): S. weatherbiana.

SAGRAEA (Melastomatac.).
abbottii *(Urb.) Alain in Sida*, 18(4): 1027 (1999): Ossaea abbottii.
barahonensis *(Urb. & Ekman) Alain in Sida*, 18(4): 1027 (1999): Ossaea barahonensis.
cinerea *(Cogn.) Alain in Sida*, 18(4): 1027 (1999): Ossaea cinerea.
cornuta *G. Lozano-Contreras & N. Ruiz-Rodgers in Mutisia*, 75: 2,3 (1995)—Colombia.
curvipila *(Urb. & Ekman) Alain in Sida*, 18(4): 1027 (1999): Ossaea curvipila.
discoidea *G. Lozano-Contreras & N. Ruiz-Rodgers in Mutisia*, 75: 3,4 (1995)—Colombia.
ellipsoidea *(Urb. & Ekman) Alain in Sida*, 18(4): 1027 (1999): Ossaea ellipsoidea.
fuertesii *(Cogn.) Alain in Sida*, 18(4): 1028 (1999): Clidemia fuertesii.
gracilis *(Alain) Alain in Sida*, 18(4): 1028 (1999): Ossaea gracilis.
lanceifolia *(Urb.) Alain in Sida*, 18(4): 1028 (1999): Ossaea lanceifolia.
munchicana *G. Lozano-Contreras & N. Ruiz-Rodgers in Mutisia*, 75: 3,4 (1995)—Colombia.
oligantha *(Urb.) Alain in Sida*, 18(4): 1028 (1999): Clidemia oligantha.
polychaete *(Urb. & Ekman) Alain in Sida*, 18(4): 1028 (1999): Ossaea polychaete.
pusilliflora *(Cogn.) Alain in Sida*, 18(4): 1028 (1999): Clidemia pusilliflora.
setulosa *(Urb.) Alain in Sida*, 18(4): 1028 (1999): Ossaea setulosa.
woodsii *(Judd & Skean) Alain in Sida*, 18(4): 1029 (1999): Ossaea woodsii.

SAHARANTHUS *M.B. Crespo & M.D. Lledó in Bot. J. Linn. Soc.*, 132(2): 169 (2000), nom. nov.: gen. Lerrouxia *Caball*. (Plumbaginac.).
ifniensis *(Caball.) M.B. Crespo & M.D. Lledó in Bot. J. Linn. Soc.*, 132(2): 169 (2000): Lerrouxia ifniensis.

× **SAKABAARA** *hort. in Orchid Rev.*, 108(1233): new orchid hybrids 14 (2000). *ORCHIDACEAE* (Brassavola × Broughtonia × Cattleya × Diacrium × Laelia).

SALACCA (Palmae).
stolonifera *D.R. Hodel in Palm J.*, 134: 35 (1997)—Thailand.

SALACIOPSIS (Celastrac.).
longistyla *I.H. Müller in Fl. N. Caléd.*, 20: 65 (1996)—New Caledonia.
tiwakae *I.H. Müller in Fl. N. Caléd.*, 20: 58 (1996)—New Caledonia.

SALICORNIA (Chenopodiac.).
europaea L.
subsp. brachystachya *(G. Mey.) R. Dahmen & R. Wisskirchen in R. Wisskirchen & H. Haeupler, Standardliste Farn- und Blütenpfl.* Deutschl.: 448 (1998): S. herbacea subvar. brachystachya.

SALIX (Salicac.).
sect. Allochroae *K.S. Hao ex C.F. Fang & A.K. Skvortsov in Novon*, 8(4): 469 (1998).
sect. Biondianae *K.S. Hao ex C.F. Fang & A.K. Skvortsov in Novon*, 8(4): 469 (1998).
sect. Cathayanae *K.S. Hao ex C.F. Fang & A.K. Skvortsov in Novon*, 8(4): 469 (1998).
sect. Cheilophilae *K.S. Hao ex C.F. Fang & A.K. Skvortsov in Novon*, 8(4): 470 (1998).
sect. Chingianae *K.S. Hao ex C.F. Fang & A.K. Skvortsov in Novon*, 8(4): 470 (1998).
sect. Dodecandrae *K.S. Hao ex C.F. Fang & A.K. Skvortsov in Novon*, 8(4): 470 (1998).
sect. Doliae *K.S. Hao ex C.F. Fang & A.K. Skvortsov in Novon*, 8(4): 470 (1998).
sect. Eriocladae *K.S. Hao ex C.F. Fang & A.K. Skvortsov in Novon*, 8(4): 470 (1998).
sect. Ernestia *K.S. Hao ex C.F. Fang & A.K. Skvortsov in Novon*, 8(4): 470 (1998).

SALIX:-
sect. Floccosae *K.S. Hao ex C.F. Fang & A.K. Skvortsov in Novon*, 8(4): 470 (1998).
sect. Geyerianae *G.W. Argus, Infragen. Classif. Salix New World (Syst. Bot. Monog., 52)*: 85 (1997).
subgen. Longifoliae *(Andersson) G.W. Argus, Infragen. Classif. Salix New World (Syst. Bot. Monog., 52)*: 57 (1997): S. [unranked] Longifoliae.
sect. Setchellianae *G.W. Argus, Infragen. Classif. Salix New World (Syst. Bot. Monog., 52)*: 62 (1997).
subsect. Sitchenses *(Bebb) R.D. Dorn in Brittonia*, 52(1): 4 (2000): S. group Sitchenses.
sect. Tangia *K.S. Hao ex C.F. Fang & A.K. Skvortsov in Novon*, 8(4): 470 (1998).
sect. Tetradeniae *K.S. Hao ex C.F. Fang & A.K. Skvortsov in Novon*, 8(4): 470 (1998).
subgen. Urbanianae *(Seemen) Hiroyoshi Ohashi in J. Jap. Bot.*, 75(1): 6 (2000): S. [unranked] Urbanianae.
sect. Variegatae *K.S. Hao ex C.F. Fang & A.K. Skvortsov in Novon*, 8(4): 470 (1998).
*sect.*Wilsonia *K.S. Hao ex C.F. Fang & A.K. Skvortsov in Novon*, 8(4): 470 (1998).
acridentata *M. Gandoger in Flora*, 65: 267 (1882)—France.
acuminatomicrophylla *K.S. Hao ex C.F. Fang & A.K. Skvortsov in Novon*, 8(4): 467 (1998)—China South-central (Yunnan).
acutata *M. Gandoger in Flora*, 65: 226 (1882)—Switzerland.
acutibasis *M. Gandoger in Flora*, 65: 263 (1882)—Norway.
alfredii Goerz ex Rehder & Kobuski
var. fengxianica *(N. Chao) G.H. Zhu in Novon*, 8(4): 464 (1998): S. fengxianica.
alniformis *M. Gandoger in Flora*, 65: 232 (1882)—Switzerland.
alpestrivaga *M. Gandoger in Flora*, 65: 236 (1882)—France.
alpivaga *M. Gandoger in Flora*, 65: 264 (1882)—France.
amnematchinensis *K.S. Hao ex C.F. Fang & A.K. Skvortsov in Novon*, 8(4): 467 (1998)—Qinghai.
angusta *M. Gandoger in Flora*, 64: 334 (1881)—Switzerland.
appropinquata *M. Gandoger in Flora*, 65: 266 (1882)—France.
areskutana *M. Gandoger in Flora*, 65: 259 (1882)—Sweden.
aria *M. Gandoger in Flora*, 65: 230 (1882)—France.
arvernensis *M. Gandoger in Flora*, 64: 134 (1881)—France.
atopantha C.K. Schneid.
var. glabra *K.S. Hao ex C.F. Fang & A.K. Skvortsov in Novon*, 8(4): 467 (1998)—China Southeast (Zhejiang).
atrocinerea Brot.
subsp. jahandiezii *(Chass.) M. Ibn Tattou in Bocconea*, 8: 162 (1998): S. jahandiezii.
× atroelaeagnos *L. Serra & M.B. Crespo in Thaiszia*, 5(1): 3 (1995)—Spain. (S. atrocinerea × elaeagnos).
autaretica *M. Gandoger in Flora*, 64: 317 (1881)—France.
bactiacensis *M. Gandoger in Flora*, 65: 264 (1882)—Switzerland.
bellula *M. Gandoger in Flora*, 64: 331 (1881)—France.
berberifolia Pall.
subsp. tschuktschorum *(A.K. Skvortsov) V.N. Voroschilov in A.K. Skvortsov (ed.), Florist. issl. v razn. raĭonakh SSSR*: 161 (1985): S. tschuktschorum.
bernensis *M. Gandoger in Flora*, 65: 234 (1882)—Switzerland.
bertholletii *M. Gandoger in Flora*, 65: 269 (1882)—Switzerland.
bichroophylla *M. Gandoger in Flora*, 65: 259 (1882)—France.
binata *M. Gandoger in Flora*, 65: 236 (1882)—Switzerland.
blepharostachya *M. Gandoger in Flora*, 65: 270 (1882)—France.
borderi *M. Gandoger in Flora*, 64: 333 (1881)—France.
bothnica *M. Gandoger in Flora*, 65: 263 (1882)—Sweden.
bouffordii *A.K. Skvortsov in Harvard Pap. Bot.*, 4(1): 324 (1999)—China South-central (Yunnan).

SALIX:-
bourdinii *M. Gandoger in Flora*, 65: 262 (1882)—France.
brevicaulis *M. Gandoger in Flora*, 65: 231 (1882)—Switzerland.
brigantiaca *M. Gandoger in Flora*, 65: 262 (1882)—France
cacuminis *A.K. Skvortsov in Harvard Pap. Bot.*, 4(1): 323 (1999)—North Korea.
caesia Vill.
subsp. tschujensis *N.M. Bolschakov in Fl. Sibir. (Salicac.-Amaranthac.)*, 5: 44 (1992)—Altay.
calceoloides *M. Gandoger in Flora*, 65: 272 (1882)—France
camptosperma *M. Gandoger in Flora*, 65: 262 (1882)—Sweden.
camusii H. Lév.
var. inamoena *(Hand.-Mazz.) N. Chao in Sichuan Forest. Sci. Techn.*, 15(2): 2 (1994): S. inamoena.
canifolia *M. Gandoger in Flora*, 65: 257 (1882)—France.
caprea L.
var. dentata *K.S. Hao ex C.F. Fang & A.K. Skvortsov in Novon*, 8(4): 467 (1998)—China North-central (Hebei).
var. sinica *K.S. Hao ex C.F. Fang & A.K. Skvortsov in Novon*, 8(4): 467 (1998)—China North-central (Hebei).
var. subsessilis *K.S. Hao ex C.F. Fang & A.K. Skvortsov in Novon*, 8(4): 468 (1998)—China.
cardiophyllos *M. Gandoger, Dec. Pl. Nov.*, 1: 29 (1875)—Switzerland.
cariotii *M. Gandoger in Flora*, 64: 316 (1881)—France.
cenisia *M. Gandoger in Flora*, 65: 268 (1882)—Italy.
chaenomeloides Kimura
var. pilosa *(Nakai) Y.N. Lee, Fl. Korea*: 1156 (1996): S. glandulosa var. pilosa.
chavanisii *M. Gandoger, Dec. Pl. Nov.*, 1: 29 (1875)—France.
chengfui *N. Chao in Sichuan Forest. Sci. Techn.*, 15(2): 3 (1994), nom. nov.: S. trichocarpa *C.F. Fang*.
chingiana *K.S. Hao ex C.F. Fang & A.K. Skvortsov in Novon*, 8(4): 468 (1998)—China North-central (Gansu).
chingshuishanensis *S.S. Ying in Mem. Coll. Agric. Nation. Taiwan Univ.*, 27(2): 31 (1987)—Taiwan.
clementii *M. Gandoger in Flora*, 64: 317 (1881)—France.
conformis *Schleich. ex M. Gandoger in Flora*, 65: 228 (1882)—Switzerland.
corsica *M. Gandoger, Dec. Pl. Nov.*, 1: 6 (1875)—Corse.
crenata *K.S. Hao ex C.F. Fang & A.K. Skvortsov in Novon*, 8(4): 468 (1998)—China South-central (Yunnan).
cuneifolia *M. Gandoger, Dec. Pl. Nov.*, 1: 5 (1875)—Corse.
cyclophylla *M. Gandoger in Flora*, 65: 271 (1882)—France.
danica *M. Gandoger in Flora*, 64: 135 (1881)—Denmark.
davidiana *M. Gandoger in Flora*, 65: 273 (1882)—France.
debeauxii *M. Gandoger in Flora*, 64: 136 (1881)—China.
delavayana Hand.-Mazz.
var. medogensis *(Y.L. Chou) N. Chao in Sichuan Forest. Sci. Techn.*, 15(2): 3 (1994): S. medogensis.
delphinensis *M. Gandoger in Flora*, 65: 272 (1882)—France.
dendrocharis *M. Gandoger in Flora*, 65: 264 (1882)—France.
divaricata Pall.
subsp. pulchra *(Cham.) V.N. Voroschilov in A.K. Skvortsov (ed.), Florist. issl. v razn. raĭonakh SSSR*: 161 (1985): S. pulchra.
dizhindanensis *P.I. Mao & H. Li in Fl. Dulongjian Reg.*: 153 (1993), without latin descr. or type—China South-central (Yunnan).
dolabrifolia *M. Gandoger in Flora*, 65: 237 (1882)—Switzerland.
dolia C.K. Schneid.
var. characta *(C.K. Schneid.) N. Chao in Sichuan Forest. Sci. Techn.*, 15(2): 4 (1994): S. characta.

SALIX:-
dolia
var. leveilleana *(C.K. Schneid.) N. Chao in Sichuan Forest. Sci. Techn.*, 15(2): 4 (1994): S. leveilleana.
drumensis *M. Gandoger in Flora*, 65: 229 (1882)—France.
dulongensis *P.I. Mao & H. Li in Fl. Dulongjian Reg.*: 153 (1993), without latin descr. or type—China South-central (Yunnan).
elongatula *M. Gandoger in Flora*, 65: 269 (1882)—France.
empetrifolia *M. Gandoger in Flora*, 65: 236 (1882)—Switzerland.
eriostachya Wall. ex Andersson
var. argyrotrichocarpa *(C.F. Fang) N. Chao in Sichuan Forest. Sci. Techn.*, 15(2): 4 (1994): S. argyrotrichocarpa.
var. lineariloba *(N. Chao) G.H. Zhu in Novon*, 8(4): 464 (1998): S. dolia var. lineariloba.
euthymei *M. Gandoger in Flora*, 65: 239 (1882)—France.
exigua Nutt.
var. columbiana *R.D. Dorn in Brittonia*, 50(2): 204 (1998)—Oregon, Washington.
var. hindsiana *(Benth.) R.D. Dorn in Brittonia*, 50(2): 203 (1998): S. hindsiana.
var. sessilifolia *(Nutt.) R.D. Dorn in Brittonia*, 50(2): 204 (1998): S. sessilifolia.
exilifolia *R.D. Dorn in Brittonia*, 50(2): 196 (1998)—Arizona.
fangiana *N. Chao & Z.Y. Wang in Sichuan Forest. Sci. Techn.*, 15(2): 5 (1994), nom. nov.: S. triandroides *W.P. Fang*.
fargesii Burkill
var. kansuensis *(K.S. Hao ex C.F. Fang & A.K. Skvortsov) G.H. Zhu in Novon*, 8(4): 464 (1998): S. kansuensis.
fausta *M. Gandoger in Flora*, 64: 135 (1881)—France.
forrestii *K.S. Hao ex C.F. Fang & A.K. Skvortsov in Novon*, 8(4): 468 (1998)—China South-central (Yunnan).
frayi *M. Gandoger in Flora*, 65: 265 (1882)—Italy.
fulvopubescens Hayata
var. doii *(Hayata) K.C. Yang & T.C. Huang in Taiwania*, 41(1): 1 (1996): S. doii.
var. eriostroma *(Hayata) S.S. Ying in Mem. Coll. Agric. Nation. Taiwan Univ.*, 27(2): 35 (1987): S. eriostroma.
var. tagawana *(Koidz.) K.C. Yang & T.C. Huang in Taiwania*, 41(1): 3 (1996): S. tagawana.
futura Seemen
f. rufa *(Kimura) Hiroyoshi Ohashi in J. Jap. Bot.*, 75(1): 30 (2000): S. futura var. rufa.
ganderi *M. Gandoger, Dec. Pl. Nov.*, 1: 22 (1875)—Austria.
glauca L.
var. callicarpaea *(Trautv.) G.W. Argus, Infragen. Classif. Salix New World (Syst. Bot. Monog., 52)*: 70 (1997): S. callicarpaea.
gonggashanica *C.F. Fang & A.K. Skvortsov in J. Trop. Subtrop. Bot.*, 7(1): 29 (1999)—China South-central (Sichuan).
gracilistyla Miq.
f. adscendens *(Kimura) Hiroyoshi Ohashi in J. Jap. Bot.*, 75(1): 24 (2000): S. gracilistyla var. adscendens.
var. graciliglans *(Nakai) Hiroyoshi Ohashi in J. Jap. Bot.*, 75(1): 24 (2000): S. graciliglans.
f. pendula *(Kimura) Hiroyoshi Ohashi in J. Jap. Bot.*, 75(1): 24 (2000): S. gracilistyla var. pendula.
guinandii *M. Gandoger in Flora*, 65: 269 (1882)—France.
hainanica *A.K. Skvortsov in Harvard Pap. Bot.*, 3(1): 107 (1998)—Hainan.
heteromorpha *M. Gandoger in Flora*, 64: 318 (1881)—France.
humilior *M. Gandoger in Flora*, 65: 237 (1882)—France.
hupehensis *K.S. Hao ex C.F. Fang & A.K. Skvortsov in Novon*, 8(4): 468 (1998)—China South-central (Hubei).
hylonoma C.K. Schneid.
var. liocarpa *(Goerz) G.H. Zhu in Novon*, 8(4): 465 (1998): S. hylonoma f. liocarpa.
hypargyrea *M. Gandoger in Flora*, 65: 233 (1882)—Switzerland.

SALIX:-

incompta *M. Gandoger in Flora*, 65: 272 (1882)—Austria.

indefinita *M. Gandoger in Flora*, 64: 136 (1881)—Sweden.

integra Thunb.

f. pendula *Okuhara ex T. Shimizu, Fl. Nagano Pref.*: 1507 (1997), homonym; *see Kimura in J. Jap. Bot.*, 57(6): 161 (1982)—Japan.

iodocarpa *M. Gandoger in Flora*, 65: 271 (1882)—Italy.

iodophylla *M. Gandoger in Flora*, 65: 235 (1882)—France.

ischnoclada *M. Gandoger in Flora*, 65: 264 (1882)—Italy.

islandica *M. Gandoger in Flora*, 65: 260 (1882)—Iceland.

isophylla *M. Gandoger in Flora*, 65: 233 (1882)—Switzerland.

iva *M. Gandoger in Flora*, 65: 233 (1882)—Switzerland.

jayetiana *M. Gandoger in Flora*, 64: 317 (1881)—France.

jessoensis Seemen

subsp. serissifolia *(Kimura) Hiroyoshi Ohashi in J. Jap. Bot.*, 75(1): 17 (2000), as 'serissaefolia': S. serissifolia.

jichoushanica *N. Chao & G.T. Gong in Sichuan Forest. Sci. Techn.*, 15(2): 6 (1994)—China South-central (Sichuan).

jucundissima *M. Gandoger in Flora*, 65: 261 (1882)—Sweden.

jurana *M. Gandoger in Flora*, 64: 316 (1881)—France.

kansuensis *K.S. Hao ex C.F. Fang & A.K. Skvortsov in Novon*, 8(4): 468 (1998)—China North-central (Gansu, Shanxi).

lactaris *M. Gandoger in Flora*, 65: 258 (1882)—Switzerland.

laishuiensis *N. Chao & G.T. Gong in Sichuan Forest. Sci. Techn.*, 15(2): 7 (1994)—China North-central (Hebei).

lamashanensis *K.S. Hao ex C.F. Fang & A.K. Skvortsov in Novon*, 8(4): 468 (1998)—China North-central (Gansu).

× lasiogyne Seemen

nothosubsp. yuhkii *(Kimura) Hiroyoshi Ohashi in J. Jap. Bot.*, 75(1): 34 (2000): S. × yuhkii.

lavanduloides *M. Gandoger in Flora*, 64: 318 (1881)—France.

leindanensis *P.I. Mao & H. Li in Fl. Dulongjian Reg.*: 153 (1993), without latin descr. or type—China South-central (Yunnan).

liebmannii *M. Gandoger, Dec. Pl. Nov.*, 1: 34 (1875)—Norway.

longiramea *M. Gandoger in Flora*, 64: 332 (1881)—France.

macrosperma *M. Gandoger in Flora*, 65: 226 (1882)—Switzerland.

magistri *M. Gandoger in Flora*, 65: 236 (1882)—Switzerland.

malarbetii *M. Gandoger in Flora*, 64: 134 (1881)—France.

marichalii *M. Gandoger in Flora*, 65: 237 (1882)—France.

mauriana *M. Gandoger in Flora*, 65: 235 (1882)—France.

mazzettiana *N. Chao in Sichuan Forest. Sci. Techn.*, 15(2): 8 (1994), nom. nov.: S. vaccinioides *Hand.-Mazz.*

megalostachya *M. Gandoger in Flora*, 65: 258 (1882)—Switzerland.

melachroa *M. Gandoger, Dec. Pl. Nov.*, 1: 39 (1875)—Switzerland.

micromegas *M. Gandoger in Flora*, 65: 272 (1882)—France.

micromeriifolia *M. Gandoger, Dec. Pl. Nov.*, 1: 16 (1875)—Italy.

miyabeana Seemen

subsp. gilgiana *(Seemen) Hiroyoshi Ohashi in J. Jap. Bot.*, 75(1): 22 (2000): S. gilgiana.

f. gilgiana *(Seemen) Hiroyoshi Ohashi in J. Jap. Bot.*, 75(1): 22 (2000): S. gilgiana.

f. pendula *(Kimura) Hiroyoshi Ohashi in J. Jap. Bot.*, 75(1): 22 (2000): S. gilgiana var. pendula.

montivaga *M. Gandoger in Flora*, 65: 229 (1882)—France.

morelii *M. Gandoger in Flora*, 64: 315 (1881)—France.

SALIX:-

muscoides *M. Gandoger in Flora*, 65: 274 (1882)—France.

myrioblephara *M. Gandoger, Dec. Pl. Nov.*, 1: 34 (1875)—France.

myrsinifolia Salisb.

subsp. kolaensis *(Schljakov) R. Elven in Nordic J. Bot.*, 16(1): 5 (1996): S. kolaensis.

nakamurana Koidz.

subsp. kurilensis *(Koidz.) Hiroyoshi Ohashi in J. Jap. Bot.*, 75(1): 12 (2000): S. kurilensis.

f. neoreticulata *(Nakai) Hiroyoshi Ohashi in J. Jap. Bot.*, 75(1): 12 (2000): S. neoreticulata.

f. yezoalpina *(Koidz.) Hiroyoshi Ohashi in J. Jap. Bot.*, 75(1): 12 (2000): S. yezoalpina.

subsp. yezoalpina *(Koidz.) Hiroyoshi Ohashi in J. Jap. Bot.*, 75(1): 12 (2000): S. yezoalpina.

nipponica Franch. & Sav.

var. mengshanensis *(S.B. Liang) G.H. Zhu in Novon*, 8(4): 465 (1998): S. triandra var. mengshanensis.

nubicola *M. Gandoger in Flora*, 65: 266 (1882)—France.

oblongella *M. Gandoger in Flora*, 65: 238 (1882)—France.

obovatifolia *M. Gandoger in Flora*, 65: 274 (1882)—France.

obtusisquama *M. Gandoger, Dec. Pl. Nov.*, 2: 10 (1876)—France.

obtusiuscula *M. Gandoger in Flora*, 65: 265 (1882)—Switzerland.

occitanica *M. Gandoger, Dec. Pl. Nov.*, 2: 32 (1876)—France.

odontophylla *M. Gandoger in Flora*, 65: 270 (1882)—France.

opsimantha C.K. Schneid.

var. wawashanica *(P.I. Mao & P.X. He) G.H. Zhu in Novon*, 8(4): 465 (1998): S. oreinoma var. wawashanica.

oritrepha C.K. Schneid.

var. amnematchinensis *(K.S. Hao ex C.F. Fang & A.K. Skvortsov) G.H. Zhu in Novon*, 8(4): 465 (1998): S. amnematchinensis.

orogenes *M. Gandoger in Flora*, 65: 266 (1882)—Switzerland.

oropotamica *S. Brullo, F. Scelsi & G. Spampinato in Flora Medit.*, 5: 53 (1995 publ. 1996)—Italy.

ovatomicrophylla *K.S. Hao ex C.F. Fang & A.K. Skvortsov in Novon*, 8(4): 468 (1998)—China South-central (Yunnan).

pachysperma *M. Gandoger in Flora*, 65: 261 (1882)—Sweden.

parisiensis *M. Gandoger, Dec. Pl. Nov.*, 1: 5 (1875)—France.

peraffinis *M. Gandoger in Flora*, 64: 136 (1881)—Denmark.

perrieri *M. Gandoger, Dec. Pl. Nov.*, 1: 17 (1875)—France.

petiolosa *M. Gandoger in Flora*, 65: 227 (1882)—Switzerland.

phalacrocarpa *M. Gandoger in Flora*, 65: 267 (1882)—Switzerland.

phylicifolia L.

subsp. divaricata *(Pall.) N.M. Bolschakov in Fl. Sibir. (Salicac.-Amaranthac.)*, 5: 33 (1992): S. divaricata.

platycarpa *M. Gandoger in Flora*, 65: 260 (1882)—Greenland.

platystachya *M. Gandoger in Flora*, 65: 232 (1882)—Switzerland.

populoides *M. Gandoger, Dec. Pl. Nov.*, 1: 39 (1875)—France.

pruna *M. Gandoger in Flora*, 65: 234 (1882)—Switzerland.

pseudopyrus *M. Gandoger, Dec. Pl. Nov.*, 1: 30 (1875)—France.

pseudowolohoensis *K.S. Hao ex C.F. Fang & A.K. Skvortsov in Novon*, 8(4): 469 (1998)—China South-central (Yunnan).

pugetii *M. Gandoger in Flora*, 65: 261 (1882)—France.

purpurea L.

subsp. leucodermis *F. Yaltırık in Istanbul Univ. Orman Fak. Dergisi, A*, 38(2): 95 (1988 publ. 1989)—Turkey.

pyrenaeicola *M. Gandoger in Flora*, 65: 273 (1882)—France.

pyrrha *M. Gandoger in Flora*, 65: 227 (1882)—Switzerland.

SALIX:-

reinii *Franch. & Sav. ex Seemen, Salic. Jap.*: 41 (1903)—Japan.

subsp. tontomussirensis *(Koidz.) V.A. Nedolushko in Sosud. Rast. Sovet. Dal'nego Vostoka*, 7: 173 (1995): S. tontomussirensis.

repens L.

subsp. brachypoda *(Trautv. & C.A. Mey.) V.N. Voroschilov in A.K. Skvortsov (ed.), Florist. issl. v razn. raĭonakh SSSR*: 161 (1985): S. repens var. brachypoda.

rhamnifolia Pall.

subsp. saposhnikovii *(A.K. Skvortsov) N.M. Bolschakov in Fl. Sibir. (Salicac.-Amaranthac.)*, 5: 35 (1992): S. saposhnikovii.

roffavieri *M. Gandoger in Flora*, 65: 235 (1882)—France.

rorida Laksch.

f. rorìdiformis *(Nakai) Kimura ex Hiroyoshi Ohashi in J. Jap. Bot.*, 75(1): 26 (2000), as 'roridaeformis': S. roridiformis.

rupifraga Koidz.

f. eriocarpa *(Kimura) Kimura ex Hiroyoshi Ohashi in J. Jap. Bot.*, 75(1): 19 (2000): S. rupifraga var. eriocarpa.

saxatilis Turcz. ex Ledeb.

subsp. stoloniferoides *(Kimura) V.N. Voroschilov in A.K. Skvortsov (ed.), Florist. issl. v razn. raĭonakh SSSR*: 161 (1985): S. stoloniferoides.

× schaburovii *I.V. Beljaeva in Bot. Zhurn.*, 83(11): 111 (1998)—Russia East. (S. dasyclados × triandra).

schneideriana *K.S. Hao ex C.F. Fang & A.K. Skvortsov in Novon*, 8(4): 469 (1998)—China South-central (Yunnan).

schwerinii E.L. Wolf

subsp. yezoensis *(C.K. Schneid.) V.N. Voroschilov in A.K. Skvortsov (ed.), Florist. issl. v razn. raĭonakh SSSR*: 162 (1985): S. viminalis var. yezoensis.

sclerophylla Andersson

var. sclerophylloides *(Y.L. Chou) T.Y. Ding in Fl. Qinghaiica*, 1: 102 (1997): S. sclerophylloides.

scrupea *M. Gandoger in Flora*, 65: 274 (1882)—France.

sericophylla *M. Gandoger in Flora*, 65: 228 (1882)—Switzerland.

serissifolia Kimura

f. pendula *Okuhara ex T. Shimizu, Fl. Nagano Pref.*: 1507 (1997)—Japan.

shansiensis *K.S. Hao ex C.F. Fang & A.K. Skvortsov in Novon*, 8(4): 469 (1998)—China North-central (Shanxi).

sieberi *M. Gandoger in Flora*, 65: 230 (1882)—Austria.

sinica *(K.S. Hao ex C.F. Fang & A.K. Skvortsov) G.H. Zhu in Novon*, 8(4): 465 (1998): S. caprea var. sinica.

var. dentata *(K.S. Hao ex C.F. Fang & A.K. Skvortsov) G.H. Zhu in Novon*, 8(4): 465 (1998): S. caprea var. dentata.

var. subsessilis *(K.S. Hao ex C.F. Fang & A.K. Skvortsov) G.H. Zhu in Novon*, 8(4): 466 (1998): S. caprea var. subsessilis.

sinorivularis *P.I. Mao & H. Li in Fl. Dulongjian Reg.*: 155 (1993), without latin descr. or type—China South-central (Yunnan).

siphuncula *M. Gandoger in Flora*, 65: 262 (1882)—Norway.

spathulifolia Seemen

var. glabra *Z. Wang & C.F. Fang ex T.Y. Ding in Fl. Qinghaiica*, 1: 506 (1997)—Qinghai.

spodiophylla Hand.-Mazz.

var. liocarpa *(K.S. Hao ex C.F. Fang & A.K. Skvortsov) G.H. Zhu in Novon*, 8(4): 466 (1998): S. spodiophylla f. liocarpa.

f. liocarpa *K.S. Hao ex C.F. Fang & A.K. Skvortsov in Novon*, 8(4): 469 (1998)—China South-central (Sichuan).

stilbophylla *M. Gandoger in Flora*, 65: 239 (1882)—France.

× stottii *R. Parfitt in New Plantsman*, 6(4): 244 (1999), contrary to Art. 37.5 ICBN (1994)—Cultivation. (S. burjatica × viminalis).

streptodon *M. Gandoger in Flora*, 65: 226 (1882)—Switzerland.

styligera *M. Gandoger in Flora*, 64: 333 (1881)—Switzerland.

subdentosa *M. Gandoger in Flora*, 65: 271 (1882)—France.

SALIX:-

subemarginata *M. Gandoger in Flora,* 65: 273 (1882)—Switzerland.

subnigrescens *M. Gandoger in Flora,* 65: 225 (1882)—Switzerland.

subobovata *M. Gandoger in Flora,* 65: 237 (1882)—Switzerland.

suchowensis *W.C. Cheng ex G.H. Zhu in Novon,* 8(4): 466 (1998)—China Southeast (Jiangsu).

taiwanalpina Kimura
- *var.* morrisonicola *(Kimura) K.C. Yang & T.C. Huang in Taiwania,* 41(1): 4 (1996): S. morrisonicola.
- *var.* takasagoalpina *(Koidz.) S.S. Ying in Mem. Coll. Agric. Nation. Taiwan Univ.,* 27(2): 33 (1987): S. takasagoalpina.

tangii *K.S. Hao ex C.F. Fang & A.K. Skvortsov in Novon,* 8(4): 469 (1998)—China North-central (Shanxi).

thymoides *M. Gandoger in Flora,* 65: 268 (1882)—Italy.

tomentelloidea *M. Gandoger in Flora,* 65: 229 (1882)—Switzerland.

tomophylla *M. Gandoger in Flora,* 65: 267 (1882)—Switzerland.

tractabilis *M. Gandoger in Flora,* 65: 230 (1882)—France.

turneroides *M. Gandoger in Flora,* 64: 317 (1881)—Switzerland.

vahlii *M. Gandoger in Flora,* 65: 260 (1882)—Greenland.

vendeana *M. Gandoger in Flora,* 65: 234 (1882)—France.

vesula *M. Gandoger in Flora,* 65: 238 (1882)—France.

viminalis L.
- *var.* pyramidalis *I.V. Beljaeva in Bot. Zhurn.,* 85(8): 104 (2000)—West Siberia.
- *subsp.* rossica *(Nasarow) N.N. Tzvelev in Novosti Sist. Vÿssh. Rast.,* 32: 184 (2000): S. rossica.

viticulosa *M. Gandoger in Flora,* 65: 273 (1882)—France.

wangiana *K.S. Hao ex C.F. Fang & A.K. Skvortsov in Novon,* 8(4): 469 (1998)—China North-central (Shanxi).
- *var.* minjiangensis *(N. Chao) T.Y. Ding in Fl. Qinghaiica,* 1: 91 (1997), species name not validly publ.: S. minjiangensis.

warburgii Seemen
- *var.* eucalyptifolia *S.S. Ying in Mem. Coll. Agric. Nation. Taiwan Univ.,* 27(2): 38 (1987), with two types cited—Taiwan.

wuiana *K.S. Hao ex C.F. Fang & A.K. Skvortsov in Novon,* 8(4): 469 (1998)—China North-central (Shanxi).

zhouquensis *X.G. Sun in Bull. Bot. Res. North-East. Forest. Univ.,* 17(4): 357 (1997)—China North-central (Gansu).

SALMONOPUNTIA *P.V. Heath in Calyx,* 6(2): 41 (1999), nom. nov.: Opuntia ser. Salmianae *Britton & Rose.* (Cactac.).

salmiana *(Pfeiff.) P.V. Heath in Calyx,* 6(2): 41 (1999), as 'salmonana': Opuntia salmiana.
- *f.* alba *P.V. Heath in Calyx,* 6(2): 41 (1999), nom. nov., as 'salmonana': Opuntia albiflora *K. Schum.*
- *f.* glauca *P.V. Heath in Calyx,* 6(2): 41 (1999), nom. nov., as 'salmonana': Opuntia spegazzinii *A.A. Weber.*
- *f.* rosea *P.V. Heath in Calyx,* 6(2): 41 (1999), nom. nov., as 'salmonana': Opuntia ipatiana *Cárdenas.*

SALOMONIA (Polygalac.).

edentula *(DC.) R.C. Srivastava in Novon,* 8(2): 203 (1998), nom. superfl., basionym publ. as 'Salomonia' not 'Polygala': Polygala edentula.

SALSOLA (Chenopodiac.).
- *sect.* Androssowia *S. Rilke in Biblioth. Bot.,* 149: 77 (1999).
- *subsect.* Griffithia *S. Rilke in Biblioth. Bot.,* 149: 82 (1999).
- *subsect.* Rosaceae *S. Rilke in Biblioth. Bot.,* 149: 85 (1999).
- *sect.* Sogdiana *S. Rilke in Biblioth. Bot.,* 149: 69 (1999), as 'comb. nov.'.

SALSOLA:-

androssowii Litv.
- *subsp.* deserticola *(Iljin) S. Rilke in Biblioth. Bot.,* 149: 79 (1999): S. deserticola.

canescens (Moq.) Spach
- *subsp.* serpentinicola *H. Freitag & E. Özhatay in Willdenowia,* 27(1–2): 185 (1997)—Turkey.

crassa M. Bieb.
- *subsp.* longipistillata *(Botsch.) H. Freitag in Fl. Iran.,* 172: 252 (1997): Climacoptera longipistillata.
- *subsp.* turcomanica *(Litv.) H. Freitag in Fl. Iran.,* 172: 250 (1997): S. turcomanica.

cyrenaica (Maire & Weiller) Brullo
- *subsp.* antalyensis *H. Freitag & H. Duman in Edinburgh J. Bot.,* 57(3): 339 (2000)—Turkey.

grandis *H. Freitag, M. Vural & N. Adıgüzel in Willdenowia,* 29(1–2): 131 (1999)—Turkey.

kali L.
- *subsp.* iberica *(Sennen & Pau) S. Rilke in Fl. Iran.,* 172: 183 (1997): S. tragus subsp. iberica.
- *subsp.* pontica *(Pall.) S.L. Mosyakin in Ann. Missouri Bot. Gard.,* 83(3): 389 (1996): S. kali var. pontica.

lanata Pall.
- *subsp.* longistylosa *(Iljin) H. Freitag in Fl. Iran.,* 172: 248 (1997): S. longistylosa.

maimanica *H. Freitag in Fl. Iran.,* 172: 254 (1997)—Afghanistan.

makranica *H. Freitag in Fl. Iran.,* 172: 166 (1997)—Pakistan, Syria.

paulsenii Litv.
- *subsp.* praecox *(Litv.) S. Rilke in Biblioth. Bot.,* 149: 162 (1999): S. kali var. praecox.

tragus L.
- *subsp.* grandiflora *S. Rilke in Biblioth. Bot.,* 149: 135 (1999)—Western Australia, Northern Territory, New South Wales.
- *subsp.* pontica *(Pall.) S. Rilke in Biblioth. Bot.,* 149: 133 (1999): S. kali var. pontica.
- *var.* pseudocollina *N.N. Tzvelev in Fl. Vostochnoĭ Evropÿ,* 9: 82 (1996)—Ukraina.

zehzadii *H. Akhani in Sendtnera,* 3: 6 (1996)—Iran.

SALTUGILIA *(V.E. Grant) L.A. Johnson in Aliso,* 19(1): 69 (2000): Gilia sect. Saltugilia. (Polemoniac.).

australis *(H. Mason & A.D. Grant) L.A. Johnson in Aliso,* 19(1): 69 (2000): Gilia splendens subsp. australis.

caruifolia *(Abrams) L.A. Johnson in Aliso,* 19(1): 69 (2000): Gilia caruifolia.

grinnellii *(Brand) L.A. Johnson in Aliso,* 19(1): 69 (2000): Gilia grinnellii.
- *subsp.* grantii *(Brand) L.A. Johnson in Aliso,* 19(1): 69 (2000): Gilia collina var. grantii.

SALVADORA (Salvadorac.).

persica L.
- *var.* indica *(Wight) T.A. Rao & S. Chakraborti in Fasc. Fl. India,* 22: 9 (1996): S. indica.
- *var.* tuticornica *T.A. Rao & S. Chakraborti in Fasc. Fl. India,* 22: 9 (1996)—India (Tamil Nadu).

SALVIA (Labiat.).

× accidentalis *P. Sánchez Gómez & R. Morales in An. Jard. Bot. Madrid,* 57(2): 421 (1999 publ. 2000)—Spain. (S. lavandulifolia subsp. vellerea × officinalis).

× atroviolacea *J.L. Fernández Alonso in L. Montes & P. Eguiluz, El Cerro, Frontera Abierta:* 156 (1996)—Colombia. (S. amethystina × rubescens).

axillaris Moç. & Sessé ex Benth.
- *var.* hidalgoana *B.L. Turner in Phytologia,* 81(1): 17 (1996 publ. 1997)—Mexico Northeast (Hidalgo), Mexico Gulf (Veracruz).
- *var.* potosina *B.L. Turner in Phytologia,* 81(1): 20 (1996 publ. 1997)—Mexico Northeast (Guanajuato, Aguascalientes, Durango, Hidalgo, Querétaro, San Luis Potosí, Zacatecas).

aytachii *M. Vural & N. Adıgüzel in Turkish J. Bot.,* 20: 531 (1996)—Turkey.

booleana *B.L. Turner in Phytologia,* 79(4): 289 (1995 publ. 1996)—Mexico Northeast (San Luis Potosí, Nuevo León).

SALVIA:-

dorrii (Kellogg) Abrams
- *var.* argentea *(Rydb.) L.D. Benson in L.D. Benson & R.A. Darrow, Trees & Shrubs of Southwest. Deserts (ed. 3):* 208 (1981): Audibertiella argentea.
- *var.* gilmanii *(Epling) L.D. Benson in L.D. Benson & R.A. Darrow, Trees & Shrubs of Southwest. Deserts (ed. 3):* 208 (1981): S. carnosa subsp. gilmanii.
- *var.* mearnsii *(Britton) L.D. Benson in L.D. Benson & R.A. Darrow, Trees & Shrubs of Southwest. Deserts (ed. 3):* 207 (1981): Audibertia mearnsii.

glabrescens (Franch. & Sav.) Makino
- *var.* purpureomaculata *(Makino) K. Inoue ex T. Shimizu, Fl. Nagano Pref.:* 1510 (1997): S. robusta f. purpureomaculata.

jaimehintoniana *Ramamoorthy ex B.L. Turner in Phytologia,* 79(2): 97 (1995 publ. 1996)—Mexico Northeast (Nuevo León, Durango, Hidalgo, Tamaulipas).

jorgehintoniana *Ramamoorthy ex B.L. Turner in Phytologia,* 79(2): 80 (1995 publ. 1996)—Mexico Northeast (Nuevo León).

officinalis L.
- *subvar.* adenostachys *(O. Bolòs & Vigo) O. de Bolòs & J. Vigo, Fl. Països Catalans,* 3: 341 (1995): S. officinalis f. adenostachys.
- *var.* adenostachys *(O. Bolòs & Vigo) O. de Bolòs & J. Vigo, Fl. Països Catalans,* 3: 341 (1995): S. officinalis f. adenostachys.
- *var.* approximata *(Pau) O. de Bolòs & J. Vigo, Fl. Països Catalans,* 3: 342 (1995): S. approximata.
- *var.* gallica *(W. Lippert) O. de Bolòs & J. Vigo, Fl. Països Catalans,* 3: 342 (1995): S. lavandulifolia subsp. gallica.
- *var.* lavandulifolia *(Vahl) O. de Bolòs & J. Vigo, Fl. Països Catalans,* 3: 342 (1995): S. lavandulifolia.
- *subvar.* mariolensis *(Figuerola) O. de Bolòs & J. Vigo, Fl. Països Catalans,* 3: 341 (1995): S. blancoana subsp. mariolensis.
- *var.* pyrenaeorum *(W. Lippert) O. de Bolòs & J. Vigo, Fl. Països Catalans,* 3: 342 (1995), without exact basionym page: S. lavandulifolia subsp. pyrenaeorum.
- *var.* trichostachya *(Font Quer ex O. Bolòs & Vigo) O. de Bolòs & J. Vigo, Fl. Països Catalans,* 3: 342 (1995): S. officinalis f. trichostachya.
- *f.* trichostachya *Font Quer ex O. de Bolòs & J. Vigo in Collect. Bot.,* 14: 96 (1983)—Spain.

ovalifolia Benth.
- *var.* nitidula *(Briq.) E.P. dos Santos in Mém. Soc. Biogéogr.,* Sér. 3, 4: 17 (1994): S. nitidula.

pinosiana *V.N. Gladkova in Novosti Sist. Vÿssh. Rast.,* 30: 151 (1996)—Cuba.

saccardiana *(Pamp.) F. Del Carratore & F. Garbari in Pl. Biosystems,* 133(2): 158 (1999): S. pratensis var. saccardiana.

sylvestris L.
- *var.* barcinonensis *(O. Bolòs & Vigo) O. de Bolòs & J. Vigo, Fl. Països Catalans,* 3: 347 (1995): S. nemorosa var. barcinonensis.
- *subsp.* villicaulis *(Borbás) O. de Bolòs & J. Vigo, Fl. Països Catalans,* 3: 347 (1995): S. villicaulis.

thermara *E.J. Van Jaarsveld in Bothalia,* 29(1): 100 (1999)—Cape Province.

turneri *T.P. Ramamoorthy in Phytologia,* 81(5): 330 (1996 publ. 1997)—Mexico Southwest (Guerrero).

verapazana *B.L. Turner in Phytologia,* 81(5): 340 (1996 publ. 1997)—Guatemala.

SAMADERA (Simaroubac.).

lucida *(Wall.) J.O. Voigt, Hort. Suburb. Calcutt.:* 182 (1845): Niota lucida.

SAMANEA (Legumin.).

inopinata *(Harms) R.C. Barneby & J.W. Grimes, Silk Tree, Guanacaste, Monkey's Earring (Mem. New York Bot. Gard., 74(1)):* 123 (1996): Serianthes inopinata.

tubulosa *(Benth.) R.C. Barneby & J.W. Grimes, Silk Tree, Guanacaste, Monkey's Earring (Mem. New York Bot. Gard., 74(1)):* 121 (1996): Calliandra tubulosa.

SAMBUCUS (Caprifoliac.).
racemosa L.
subsp. manshurica *(Kitag.) V.N. Voroschilov in A.K. Skvortsov (ed.), Florist. issl. v razn. raĭonakh SSSR:* 192 (1985): S. manshurica.

SANCHEZIA (Acanthac.).
macrocnemis *(Nees) D.C. Wasshausen in Cat. Vasc. Pl. Ecuador (Monogr. Syst. Bot. Missouri Bot. Gard., 75):* 953 (1999): Ancylogyne macrocnemis.
wurdackii *D.C. Wasshausen in Biollania,* ed. esp. 6 (J.J. Wurdack Festschr.): 543 (1997)—Peru.

SANGUILLUMA *D.C.H. Plowes in Haseltonia,* 3: 65 (1995). *ASCLEPIADACEAE*
socotrana *(Balf. f.) D.C.H. Plowes in Haseltonia,* 3: 66 (1995): Boucerosia socotrana.

SANGUISORBA (Rosac.).
megacarpa *(Lowe) F. Muñoz Garmendia & C. Navarro in An. Jard. Bot. Madrid,* 56(1): 176 (1998): Poterium megacarpon.
minor Scop.
subsp. balearica *(Bourg. ex Nyman) F. Muñoz Garmendia & C. Navarro in An. Jard. Bot. Madrid,* 56(1): 176 (1998): Poterium spachianum subsp. balearicum.

SANICULA (Umbellif.).
moranii *P. Vargas, L. Constance & B.G. Baldwin in Brittonia,* 50(1): 123 (1998)—Mexico Northwest (Baja California).

SANIELLA (Hypoxidac.).
occidentalis *(Nel) B.L. Burtt in Edinburgh J. Bot.,* 57(1): 69 (2000): Forbesia occidentalis.

SANRAFAELIA *B. Verdcourt in Garcia de Orta, Ser. Bot.,* 13(1): 43 (1996). *ANNONACEAE*
ruffonammari *B. Verdcourt in Garcia de Orta, Ser. Bot.,* 13(1): 43 (1996)—Tanzania.

SANSEVIERIA (Dracaenac.).
arborescens *Cornu ex Gérôme & Labroy in Bull. Mus. Hist. Nat. (Paris),* 9: 173 (1903)—Tanzania.†
bella *L.E. Newton in Cact. Succ. J. (U.S.A.),* 72(4): 224 (2000)—Kenya.
burdettii *B.J. Chahinian in Brit. Cact. Succ. J.,* 18(3): 132 (2000)—Malawi.
downsii *B.J. Chahinian in Brit. Cact. Succ. J.,* 18(3): 133 (2000)—Malawi.
frequens *B.J. Chahinian in Cact. Succ. J. (U.S.A.),* 72(3): 130 (2000)—Kenya.
hallii *B.J. Chahinian in Sansevieria J.,* 5(1): 7 (1996)—Zimbabwe.
laurentii *De Wild. in Rev. Cult. Colon.,* 14: 231 (1904)—Congo.†
masoniana *B.J. Chahinian in Cact. Succ. J. (U.S.A.),* 72(1): 31 (2000)—Zaire.
perrotii *O. Warburg in Tropenfl.,* 5(4): 191 (1901),in IK Supp. 3 as 'perrotti' in error—Tanzania.
sulcata *Bojer ex D. Morris in Kew Bull.,* 1915(5): 10, 11 (1915)—E. Africa†

SANTOLINA (Compos.).
magonica *(O. Bolòs, Molin. & P. Monts.) A.M. Romo, Fl. Silvestres Baleares:* 303 (1994): S. chamaecyparissus subsp. magonica.
subsp. vedraensis *A.M. Romo, Fl. Silvestres Baleares:* 305 (1994), as 'vedraense'—Baleares.
melidensis *(Rodr. Oubiña & S. Ortíz) J. Rodríguez-Oubiña & S. Ortíz in Nova Acta Cient. Compostel. (Biol.),* 8: 299 (1997 pub. 1998): S. rosmarinifolia subsp. melidensis.
semidentata Hoffmanns. & Link
subsp. melidensis *(Rodr. Oubiña & S. Ortíz) S. López Udias, C. Fabregat & G. Mateo in An. Jard. Bot. Madrid,* 55(2): 293 (1997): S. rosmarinifolia subsp. melidensis.

SAPRANTHUS (Annonac.).
viridiflorus *G.E. Schatz in Novon,* 8(4): 436 (1998)—Costa Rica, Nicaragua, Panama.

SAPRIA (Rafflesiac.).
ram *H. Bänziger & B. Hansen in Nat. Hist. Bull. Siam Soc.,* 45(2): 157 (1997)—Thailand.

SARCOCAULON (Geraniac.).
lavrani *J.J. Halda in Cactaceae etc.,* 7(4): 123–125 (1997); *cf. Repert. Pl. Succ. (I.O.S.),* 48: 17 (1997 publ. 1998)—Locality not stated.

SARCOCHILUS (Orchidac.).
ceciliae F. Muell.
var. eriochilus *(Fitzg.) A.W. Dockrill in Austral. Orchid Rev.,* 61(2): 5 (1996): S. eriochilus.

SARCODUM (Legumin.).
bicolor *F. Adema in Blumea,* 44(2): 407 (1999)—Lesser Sunda Is.

SARCOGLOTTIS (Orchidac.).
allemanii *J. Barbosa Rodrigues, Gen. spec. Orchid.,* 2: 288 (1881)—Brazil Northeast (Ceará).†
curvisepala *D.L. Szlachetko & P. Rutkowski in Ann. Bot. Fenn.,* 34(4): 277 (1997)—Brazil Northeast (Bahia).
depinctrix *E.A. Christenson & A.L.V. Toscano de Brito in Orchids,* 69(9): 875 (2000)—Brazil Southeast (Rio de Janeiro, Espírito Santo).
gonzalezii *L.C. Menezes in Orchidee,* 49(4): 181 (1998); *et in Bol. CAOB,* 31: 28 (1998)—Brazil Southeast (Minas Gerais).
multiflora *J. Barbosa Rodrigues, Gen. spec. Orchid.,* 2: 287 (1881)—Brazil Southeast (Minas Gerais).†
ornithocephala *J. Barbosa Rodrigues, Gen. spec. Orchid.,* 2: 288 (1881)—Brazil Southeast (Rio de Janeiro).†
viscosa *D.L. Szlachetko & P. Rutkowski in Ann. Bot. Fenn.,* 34(4): 275 (1997), as 'viscosus'—Brazil West-central (Brasília Distrito Federal).

SARCOLOBUS (Asclepiadac.).
merrillii *(Schltr.) R. Omlor, Gen. Revis. Marsdenieae (Asclepiadac.):* 135 (1998): Petalonema merrillii.

SARCOPERA *H.G. Bedell in Bot. Jahrb.,* 119(3): 328 (1997), nom. nov.: Norantea subgen. Pseudostachyum *Delpino.* (Marcgraviac.).
anomala *(Kunth) H.G. Bedell in Bot. Jahrb.,* 119(3): 329 (1997): Norantea anomala.
aurantiaca *(Spruce ex Gilg) A.C. de Roon & S. Dressler in Novon,* 9(2): 248 (1999): Norantea aurantiaca.
cordachida *(Ruiz & Pav. ex G. Don) Bedell ex S. Dressler in Cat. Vasc. Pl. Ecuador (Monogr. Syst. Bot. Missouri Bot. Gard., 75):* 955 (1999): Norantea cordachida.
flammifera *A.C. de Roon & H.G. Bedell in Novon,* 9(2): 249 (1999)—Brazil North (Roraima, Amazonas), Venezuela.
rosulata *A.C. de Roon & H.G. Bedell in Bot. Jahrb.,* 119(3): 329 (1997)—Panama, Costa Rica.
sessiliflora *(Triana & Planch.) H.G. Bedell in Bot. Jahrb.,* 119(3): 329 (1997): Norantea sessiliflora.
tepuiensis *(de Roon) H.G. Bedell in Novon,* 9(2): 251 (1999): Norantea tepuiensis.
subsp. coccinea *A.C. de Roon & S. Dressler in Novon,* 9(2): 251 (1999)—Venezuela, Brazil North (Amazonas).

SARCOSTEMMA (Asclepiadac.).
antsiranense *U. Meve & S. Liede in Kew Bull.,* 52(2): 491 (1997)—Madagascar.
viminale (L.) R. Br.
subsp. odontolepis *(Balf. f.) U. Meve & S. Liede in Bot. J. Linn. Soc.,* 120(1): 31 (1996): S. odontolepis.
subsp. stipitaceum *(Forssk.) U. Meve & S. Liede in Bot. J. Linn. Soc.,* 120(1): 32 (1996): Asclepias stipitacea.
subsp. suberosum *U. Meve & S. Liede in Bot. J. Linn. Soc.,* 120(1): 35 (1996)—Transvaal, Ethiopia, Kenya, Somalia, Zimbabwe.

SARCOSTIGMA (Icacinac.).
lucidum *N. Wallich, Report on Calcutta Bot. Gard. to G.A. Bushby:* 34 (1840), nom. nov.: Hopea eglandulosa *Roxb.*

SARCOZYGIUM (Zygophyllac.).
kaschgaricum *(Boriss.) Y.X. Liou in Fl. Reipubl. Popul. Sin.,* 43(1): 142 (1998): Zygophyllum kaschgaricum.

SARMENTICOLA *K. Senghas & L.A. Garay in Willdenowia,* 25(2): 657 (1996). *ORCHIDACEAE*
calceolaris *(Garay) K. Senghas & L.A. Garay in Willdenowia,* 25(2): 657 (1996): Pterostemma calceolare.

SAROPSIS *B.G. Briggs & L.A.S. Johnson in Telopea,* 7(4): 355 (1998). *RESTIONACEAE*
fastigiata *(R. Br.) B.G. Briggs & L.A.S. Johnson in Telopea,* 7(4): 356 (1998): Restio fastigiatus.

× **SARPARIZA** *hort. in Orchid Rev.,* 107(1226): 126 (1999). *ORCHIDACEAE* (Parasarcochilus × Plectorrhiza × Sarcochilus).

SARRACENIA (Sarraceniac.).
purpurea L.
f. luteola *R. Hanrahan & James M. Miller in Carniv. Pl. Newsl.,* 27(1): 16 (1998)—Alabama.
var. montana *D.E. Schnell & R.O. Determann in Castanea,* 62(1): 60 (1997)—North Carolina, South Carolina, Georgia.
rosea *R.F.C. Naczi, F.W. Case & R.B. Case in Sida,* 18(4): 1188 (1999)—Florida.
f. luteola *(Hanrahan & James M. Mill.) R.F.C. Naczi, F.W. Case & R.B. Case in Sida,* 18(4): 1192 (1999): S. purpurea f. luteola.
rubra Walter
subsp. jonesii *(Wherry) E.T. Wherry in Castanea,* 37(2): 146 (1972): S. jonesii.

SASA (Gramin.).
borealis (Hack.) Makino & Shibata
subsp. viridescens *(Nakai) J.-P. Demoly in Bambou,* 21: 14 (1995): Sasamorpha purpurascens var. viridescens.
chiisanensis *(Nakai) Y.N. Lee, Fl. Korea:* 1164 (1996): Sasamorpha chiisanensis.
jotanii *(Kenji Inoue & Tanim.) M. Kobayashi in J. Jap. Bot.,* 75(4): 245 (2000): S. kurilensis var. jotanii.

SASAELLA (Gramin.).
atamiana (Nakai) Sad. Suzuki
f. leucorhoda *(Koidz.) Sadao Suzuki, Index Japanese Bambusac.:* 363 (1978): Arundinaria leucorhoda.
shiobarensis (Nakai) Nakai ex Koidz.
var. yessaensis *(Koidz.) Sadao Suzuki, Index Japanese Bambusac.:* 361 (1978): Arundinaria yessaensis.

× **SASINARIA** *J.-P. Demoly in Bambou,* 21: 15 (1995). *GRAMINEAE* (Arundinaria × Sasa).
hidaensis *(Makino) J.-P. Demoly in Bambou,* 21: 15 (1995): Sasa hidaensis.
iwatekensis *(Makino & Uchida) J.-P. Demoly in Bambou,* 21: 15 (1995): Sasaella iwatekensis.
masamuneana *(Makino) J.-P. Demoly in Bambou,* 21: 15 (1995): Pleioblastus masamuneanus.
ramosa *(Makino) J.-P. Demoly in Bambou,* 21: 15 (1995): Arundinaria ramosa.
sasakiana *(Makino & Uchida) J.-P. Demoly in Bambou,* 21: 15 (1995): Sasaella sasakiana.
sawadae *(Makino) J.-P. Demoly in Bambou,* 21: 15 (1995), as 'sawadai': Pleioblastus sawadae.
suwekoana *(Makino) J.-P. Demoly in Bambou,* 21: 15 (1995): Sasa suwekoana.

SASSIA (Oxalidac.).
sect. Ionoxalis *(Small) J. Holub in Preslia,* 70(2): 111 (1998): gen. Ionoxalis.
debilis *(Humb., Bonpl. & Kunth) J. Holub in Preslia,* 68(3): 285 (1996 publ. 1997): Oxalis debilis.
latifolia *(Humb., Bonpl. & Kunth) J. Holub in Preslia,* 68(3): 285 (1996 publ. 1997): Oxalis latifolia.
tetraphylla *(Cav.) J. Holub in Preslia,* 70(2): 111 (1998): Oxalis tetraphylla.
violacea *(L.) J. Holub in Preslia,* 70(2): 111 (1998): Oxalis violacea.

SATUREJA (Labiat.).
adamovicii *Č. Šilić, Monografija Satureja, Calamintha, Micromeria, Acinos i Clinopodium fl. Jugoslavije:* 71 (1979)—Yugoslavia.

SATUREJA:-
fukarekii *Č. Šilić, Monografija Satureja, Calamintha, Micromeria, Acinos i Clinopodium fl.* Jugoslavije: 63 (1979), nom. nov.: S. tenuis *Formánek.*
microphylla (d'Urv.) Guss.
subsp. filiformis *(Aiton) O. de Bolòs & J. Vigo, Fl. Països Catalans,* 3: 301 (1995): Thymus filiformis.
subsp. rodriguezii *(Freyn & Janka) O. de Bolòs & J. Vigo, Fl. Països Catalans,* 3: 301 (1995): Micromeria rodriguezii.
nabateorum *A. Danin & I.C. Hedge in Willdenowia,* 28(1–2): 135 (1998)—Palestine (Jordan).
obovata Lag.
subsp. valentina *(G. López) M.B. Crespo in Acta Bot. Malacitana,* 24: 219 (1999): S. cuneifolia var. valentina.
rechingeri *Z. Jamzad in Ann. Naturhist. Mus. Wien,* 98B(Suppl.): 75 (1996)—Iran.
rouyana *J.I. Briquet, Lab. Alp. Mar.,* 3: 442 (1895)—Baleares.†
subspicata Bartl. ex Vis.
var. blavii *(Asch. ex K. Malý) Č. Šilić, Monografija Satureja, Calamintha, Micromeria, Acinos i Clinopodium fl.* Jugoslavije: 105 (1979), isonym; *see K. Malý in Dörfler, Herbarium Normale,* Sched. ad Cent. 49 & 50: 302 (1908): S. montana var. blavii.
f. liburnica *Č. Šilić, Monografija Satureja, Calamintha, Micromeria, Acinos i Clinopodium fl.* Jugoslavije: 109 (1979)—Italy.
var. obcordata *(Horák) Č. Šilić, Monografija Satureja, Calamintha, Micromeria, Acinos i Clinopodium fl.* Jugoslavije: 105 (1979): S. pygmaea var. obcordata.
visianii *Č. Šilić, Monografija Satureja, Calamintha, Micromeria, Acinos i Clinopodium fl.* Jugoslavije: 90 (1979)—Yugoslavia.

SATYRIUM (Orchidac.).
subgen. Bifidum *H. Kurzweil & H.P. Linder in Beitr. Biol. Pfl.,* 71(1): 167 (1999).
subgen. Brachysaccium *(Schltr.) H. Kurzweil & H.P. Linder in Beitr. Biol. Pfl.,* 71(1): 165 (1999): S. sect. Brachysaccium.
neilgherrensis *P.F. Fyson, Fl. S. Ind. Hill Stations,* 1: 595 (1932)—India (Tamil Nadu).
pallens *S.D. Johnson & H. Kurzweil in Bot. J. Linn. Soc.,* 127(3): 191 (1998)—Cape Province.
pulchrum *S.D. Johnson & H. Kurzweil in Bot. J. Linn. Soc.,* 127(3): 192 (1998)—Cape Province.

SAURAUIA (Actinidiac.).
homotricha *A. Pool in Novon,* 8(3): 280 (1998)—Nicaragua, Honduras.

SAURIA *M.S. Bajtenov in Izv. Natsion. Akad. Nauk Resp. Kazakhstan, Ser. Biol.,* 1995(6): 81 (1995 publ. 1996). *BORAGINACEAE*
akkolia *M.S. Bajtenov in Izv. Natsion. Akad. Nauk Resp. Kazakhstan, Ser. Biol.,* 1995(6): 81 (1995 publ. 1996)—Kazakhstan.

SAUROLLUMA *D.C.H. Plowes in Haseltonia,* 3: 52 (1995). *ASCLEPIADACEAE*
furta *(P.R.O. Bally) D.C.H. Plowes in Haseltonia,* 3: 53 (1995): Caralluma furta.

SAUROMATUM (Arac.).
gaoligongense *Z.L. Wang & H. Li in Acta Bot. Yunnanica,* Supp. 11: 61 (1999)—China South-central (Yunnan).

SAUROPUS (Euphorbiac.).
anemoniflorus *J.T. Hunter & J.J. Bruhl in Austrobaileya,* 4(4): 662 (1997)—Queensland.
aphyllus *J.T. Hunter & J.J. Bruhl in Austrobaileya,* 4(4): 664 (1997)—Queensland.
arenosus *J.T. Hunter & J.J. Bruhl in Nuytsia,* 11(2): 166 (1997)—Western Australia.
bishnupadae *M. Gangopadhyay & T. Chakrabarty in J. Econ. Taxon. Bot.,* 20(3): 524 (1996)—Sikkim, India (West Bengal).
compressus Müll. Arg.
var. puberulus *(Kurz) T. Chakrabarty & M. Gangopadhyay in J. Econ. Taxon. Bot.,* 20(3): 528 (1996): S. quadrangularis var. puberulus.

SAUROPUS:-
convallarioides *J.T. Hunter & J.J. Bruhl in Austrobaileya,* 4(4): 666 (1997)—Queensland.
decrescentifolia *J.T. Hunter & J.J. Bruhl in Austrobaileya,* 4(4): 668 (1997)—Queensland.
dunlopii *J.T. Hunter & J.J. Bruhl in Nuytsia,* 11(2): 169 (1997)—Northern Territory.
filicinus *J.T. Hunter & J.J. Bruhl in Nuytsia,* 11(2): 170 (1997)—Northern Territory.
gour-maitii *T. Chakrabarty & M. Gangopadhyay in J. Econ. Taxon. Bot.,* 20(3): 529 (1996)—India (Kerala).
gracilis *J.T. Hunter & J.J. Bruhl in Nuytsia,* 11(2): 171 (1997)—Northern Territory.
kitanovii *Nguyen Nghia Thin, Euphorbiac. Vietnam:* 49 (1995 publ. 1996)—Vietnam.
paucifolius *J.T. Hunter & J.J. Bruhl in Nuytsia,* 11(2): 172 (1997)—Northern Territory.
po-khantii *T. Chakrabarty & M. Gangopadhyay in J. Econ. Taxon. Bot.,* 20(3): 531 (1996)—Burma.
rimophilus *J.T. Hunter & J.J. Bruhl in Nuytsia,* 11(2): 173 (1997)—Northern Territory.
salignus *J.T. Hunter & J.J. Bruhl in Nuytsia,* 11(2): 175 (1997)—Western Australia.
siamensis *T. Chakrabarty & M. Gangopadhyay in J. Econ. Taxon. Bot.,* 19(2): 452 (1995)—Thailand.
stenocladus *(Müll. Arg.) J.T. Hunter & J.J. Bruhl in Nuytsia,* 11(2): 176 (1997): Phyllanthus stenocladus.
subsp. pinifolius *J.T. Hunter & J.J. Bruhl in Nuytsia,* 11(2): 177 (1997)—Northern Territory.
thoii *Nguyen Nghia Thin, Euphorbiac. Vietnam:* 48 (1995 publ. 1996)—Vietnam.
tiepii *Nguyen Nghia Thin, Euphorbiac. Vietnam:* 49 (1995 publ. 1996), contrary to Art. 37.5 ICBN (1994)—Vietnam.
torridus *J.T. Hunter & J.J. Bruhl in Nuytsia,* 11(2): 178 (1997)—Western Australia, Northern Territory.

SAUSSUREA (Compos.).
acuminata Turcz.
var. duiensis *(F. Schmidt) V.N. Voroschilov in A.K. Skvortsov (ed.), Florist. issl. v razn. raĭonakh SSSR:* 197 (1985): S. duiensis.
aerjingensis *K.M. Shen in Acta Phytotax. Sin.,* 36(3): 275 (1998); *et in Fl. Xinjiangensis,* 5: 478 (1999)—Xinjiang.
chabyoungsanica *H.T. Im in J. Pl. Biol.,* 40(4): 288 (1997)—South Korea.
congesta Turcz.
subsp. soczavae *(Lipsch.) V.N. Voroschilov in A.K. Skvortsov (ed.), Florist. issl. v razn. raĭonakh SSSR, as* 'soczavet': 197 (1985): S. soczavae.
diffusa *S.Ju. Lipschitz, Rod Saussurea DC.:* 140 (1979), nom. nov.: S. albescens var. brachycephala *Hook. f.*
donkiah *C.B. Clarke ex L.S. Springate in Edinburgh J. Bot.,* 57(3): 405 (2000)—Sikkim, Nepal.
elongata DC.
subsp. recurvata *(Maxim.) V.N. Voroschilov in A.K. Skvortsov (ed.), Florist. issl. v razn. raĭonakh SSSR:* 197 (1985): S. elongata var. recurvata.
eriostemon *Wall. ex C.B. Clarke, Compos. Ind.:* 229 (1876)—Nepal.†
glabrata *(DC.) C. Shih in Fl. Reipubl. Popul. Sin.,* 78(2): 84 (1999): Aplotaxis denticulata β glabrata.
gorbunovae *R.V. Kamelin in Turczaninowia,* 2(4): 25 (1999)—Kirgizistan.
inconspicua *Hand.-Mazz. in Oesterr. Bot. Z.,* 88(4): 312 (1939), nom. nov.: Saussurea modesta *Hand.-Mazz.*
involucrata Kar. & Kir.
var. axillicalathina *J.S. Li in Bull. Bot. Res. North-East. Forest. Univ.,* 17(1): 41 (1997)—Xinjiang.
ischnoides *J.S. Li in Bull. Bot. Res. North-East. Forest. Univ.,* 17(1): 40 (1997)—Xinjiang.
× iwateyamensis *M. Kikuchi in Ann. Rep. Gakugei Fac. Iwate Univ.,* 18(3): 36 (1961)—Japan. (S. brachycephala × sagitta).
macrota Franch.
var. papyracea *S.Ju. Lipschitz, Rod Saussurea DC.:* 210 (1979)—China South-central (Sichuan).
nuda Ledeb.
var. oxyodonta *(Hultén) V.N. Voroschilov in A.K. Skvortsov (ed.), Florist. issl. v razn. raĭonakh SSSR:* 197 (1985): S. oxyodonta.

SAUSSUREA:-
parvialata *H.S. Pak & Y.P. Hong in Fragm. Flor. Geobot.,* 40(2): 725 (1995)—North Korea.
parviflora (Poir.) DC.
subsp. neoserrata *(Nakai) V.N. Voroschilov in A.K. Skvortsov (ed.), Florist. issl. v razn. raĭonakh SSSR:* 197 (1985): S. neoserrata.
pinnatiphylla *A.J.C. Grierson & L.S. Springate in Edinburgh J. Bot.,* 57(3): 408 (2000)—Bhutan.
polylada *J.S. Li in Bull. Bot. Res. North-East. Forest. Univ.,* 17(1): 39 (1997)—Xinjiang.
qinghaiensis *S.W. Liu & T.N. Ho in Fl. Qinghaiica,* 3: 511 (1996)—Qinghai.
ruoqiangensis *K.M. Shen in Acta Phytotax. Sin.,* 36(3): 273 (1998)—Xinjiang.
ruoqiangensis *K.M. Shen in Fl. Xinjiangensis,* 5: 478 (1999), homonym; *see K.M. Shen in Acta Phytotax. Sin.,* 36(3): 273 (1998)—Xinjiang.
sclerolepis Nakai & Kitag.
var. parasclerolepis *(A.I. Baranov & Skvortsov) C.Y. Li in Clavis Pl. Chinae Bor.-Or. ed. 2 (ed. P.Y. Fu):* 711 (1995), with incorrect basionym ref.: S. parasclerolepis.
spodochroa *H.S. Pak & Y.P. Hong in Fragm. Flor. Geobot.,* 40(2): 726 (1995)—North Korea.
× subgracilis *K. Asano ex T. Shimizu, Fl. Nagano Pref.:* 1510 (1997)—Japan. (S. gracilis × ussuriensis).
taraxacifolia Wall. ex DC.
var. glabrescens *Kitam. ex K. Fujikawa & H. Ohba in H. Ohba & H. Ikeda (eds.), Fl. Hinku & Hunku Valleys, E Nepal (Univ. Mus.* Tokyo Nat. & Cult., 6): 221 (2000)—Nepal.
yushuensis *S.W. Liu & T.N. Ho in Fl. Qinghaiica,* 3: 512 (1996)—Qinghai.

SAUVAGESIA (Ochnac.).
lanceolata *C. Sastre in Bol. Bot. (São Paulo),* 16: 71 (1997)—Brazil West-central (Goiás).

SAXICOLELLA (Podostemac.).
marginalis *(G. Taylor) C. Cusset ex M. Cheek in M. Cheek et al., Pl. Mount Oku & Ijim Ridge, Cameroon, Conserv. Checklist:* 153 (2000): Butumia marginalis.

SAXIFRAGA (Saxifragac.).
ser. Aestivales *P.Yu. Zhmylev in Byull. Mosk. Obshch. Ispȳt. Prir., Biol.,* 100(3): 77 (1995).
subsect. Axilliflorae *(Willk.) P.Yu. Zhmylev in Byull. Mosk. Obshch. Ispȳt. Prir., Biol.,* 102(6): 46 (1997): S. group Axilliflorae.
ser. Calthophyllum *(A.M. Johnson) P.Yu. Zhmylev in Byull. Mosk. Obshch. Ispȳt. Prir., Biol.,* 101(6): 74 (1996): S. sect. Calthophyllum.
subsect. Intermediae *(Engl. & Irmsch.) P.Yu. Zhmylev in Byull. Mosk. Obshch. Ispȳt. Prir., Biol.,* 101(6): 68 (1996): S. grex Intermediae.
subsect. Juniperifoliae *(Engl. & Irmsch.) P.Yu. Zhmylev in Byull. Mosk. Obshch. Ispȳt. Prir., Biol.,* 102(6): 43 (1997): S. grex Juniperifoliae.
sect. Komarovia *P.Yu. Zhmylev in Byull. Mosk. Obshch. Ispȳt. Prir., Biol.,* 102(6): 46 (1997).
subsect. Lacteae *P.Yu. Zhmylev in Byull. Mosk. Obshch. Ispȳt. Prir., Biol.,* 102(6): 47 (1997).
subsect. Melanocentrae *(Engl. & Irmsch.) P.Yu. Zhmylev in Byull. Mosk. Obshch. Ispȳt. Prir., Biol.,* 101(6): 71 (1996): S. grex Melanocentrae.
subsect. Merkianae *(Engl. & Irmsch.) P.Yu. Zhmylev in Byull. Mosk. Obshch. Ispȳt. Prir., Biol.,* 102(3): 47 (1997): S. grex Merkianae.
subsect. Mesogyne *(Sternb.) P.Yu. Zhmylev in Byull. Mosk. Obshch. Ispȳt. Prir., Biol.,* 102(6): 45 (1997): S. sect. Mesogyne.
ser. Radioflorae *(A.M. Johnson) P.Yu. Zhmylev in Byull. Mosk. Obshch. Ispȳt. Prir., Biol.,* 101(6): 74 (1996): S. subsect. Radioflorae.
ser. Reniformes *P.Yu. Zhmylev in Byull. Mosk. Obshch. Ispȳt. Prir., Biol.,* 100(3): 77 (1995).
sect. Rosulares *(Gornall) P.Yu. Zhmylev in Byull. Mosk. Obshch. Ispȳt. Prir., Biol.,* 102(3): 50 (1997): S. subsect. Rosulares.
subsect. Rufescentes *(J.T. Pan) P.Yu. Zhmylev in Byull. Mosk. Obshch. Ispȳt. Prir., Biol.,* 102(3): 49 (1997): S. ser. Rufescentes.
ser. Staminosae *P.Yu. Zhmylev in Byull. Mosk. Obshch. Ispȳt. Prir., Biol.,* 101(6): 69 (1996).
subsect. Stoloniferae *(Haw.) P.Yu. Zhmylev in Byull. Mosk. Obshch. Ispȳt. Prir., Biol.,* 102(6): 46 (1997): S. sect. Stoloniferae.

SAXIFRAGA:-
× abingdonensis *B. Arundel & R.J. Gornall in Saxifraga Group J.*, 2: 43 (1992)—Cultivation. (S. burseriana × poluniniana).
alaskana *(Small) Fedde in Just's Bot. Jahresber.*, 33(1): 613 (1905 publ. 1906): Leptasea alaskana.
almeriensis *Willk. ex P. Vargas in An. Jard. Bot. Madrid*, 54(1): 194 (1996), nom. nov.: S. camposii var. leptophylla *Willk.*
× arizagae *J.A. Alejandre Sáenz, J.A. Arizaleta Urarte & J. Benito Ayuso in Flora Montiberica*, 12: 60 (1999)—Spain. (S. fragosoi × losae).
× berolinensis *A. Engler & E. Irmscher, Pflanzenreich IV*, 117 I (Heft 67): 182 (1916)—Cultivation. (S. cuneifolia × umbrosa).
boreo-olympica *J.J. Halda in Acta Mus. Richnov. Sect. Nat.*, 4(2): 59 (1997)—Washington.
brachypoda D. Don
var. eglandulosa *(Harry Sm.) Shinobu Akiyama, H. Ohba & S.K. Wu in Bull. Nation. Sci. Mus., B (Tokyo)*, 26(3): 82 (2000): S. gouldii var. eglandulosa.
var. gouldii *(C.E.C. Fisch.) Shinobu Akiyama, H. Ohba & S.K. Wu in Bull. Nation. Sci. Mus., B (Tokyo)*, 26(3): 85 (2000): S. gouldii.
bronchialis L.
subsp. compacta *(Adams) L.I. Malyschev in Fl. Sibir. (Berberidac.-Grossulariac.)*, 7: 177 (1994): S. compacta.
bryomorpha *J.J. Halda in Acta Mus. Richnov. Sect. Nat.*, 4(2): 63 (1997)—Tadzhikistan.
cartilaginea Willd. & Sternb.
subsp. erubescens *S. Mobayen, Fl. Iran (Tehran Univ. Publ. 2283)*, 4: 67 (1996), without latin descr. or type—Iran.
cherlerioides D. Don
subsp. ascoldica *(Sipliv.) V.N. Voroschilov in A.K. Skvortsov (ed.), Florist. issl. v razn. raĭonakh SSSR:* 174 (1985): S. ascoldica.
subsp. stellerana *(Merk. ex Ser.) V.N. Voroschilov in A.K. Skvortsov (ed.), Florist. issl. v razn. raĭonakh SSSR:* 174 (1985): S. stellerana.
corsica (Duby) Gren. & Godr.
subsp. ebusitana *A.M. Romo, Fl. Silvestres Baleares:* 123 (1994)—Baleares.
cuneata Willd.
subsp. paniculata *(Pau) G. Mateo Sanz & M.B. Crespo Villalba, Fl. Abrev. Comun. Valenciana:* 430 (1995): S. trifurcata subsp. paniculata.
eastwoodiae *(Small) Fedde in Just's Bot. Jahresber.*, 33(1): 613 (1905 publ. 1906): Heterisia eastwoodiae.
epiphylla *R.J. Gornall & H. Ohba in Novon*, 10(4): 375 (2000)—China South-Central (Yunnan).
exarata Vill.
subsp. basaltica *(Braun-Blanq.) J. Jalas in Ann. Bot. Fenn.*, 36(1): 10 (1999): S. moschata subsp. basaltica.
subsp. carniolica *(Huter) J. Jalas in Ann. Bot. Fenn.*, 36(1): 10 (1999): S. moschata var. carniolica.
felineri *P. Vargas in Fl. Iberica*, 5: 293 (1997)—Spain.
fortunei Hook. f.
f. crassa *(Nakai) Otsuka ex T. Shimizu, Fl. Nagano Pref.:* 1508 (1997): S. fortunei var. crassa.
var. mutabilis *(Koidz.) H. Nakai & Hiroyoshi Ohashi in J. Jap. Bot.*, 71(2): 78 (1996): S. mutabilis.
funstonii *(Small) Fedde in Just's Bot. Jahresber.*, 33(1): 613 (1905 publ. 1906): Leptasea funstonii.
ganeshii *H. Ohba & S. Akiyama in J. Jap. Bot.*, 74(4): 223 (1999)—Nepal.
gemmigera Engl.
var. gemmuligera *(Engl.) J.T. Pan & R.J. Gornall in Novon*, 10(4): 376 (2000): Saxifraga unguiculata var. gemmuligera.
genesiana *P. Vargas in An. Jard. Bot. Madrid*, 55(1): 192 (1997)—Spain.
× grata *A. Engler & E. Irmscher, Pflanzenreich IV*, 117 II (Heft 69): 600 (1919)—Cultivation. (S. aretioides × ferdinandi-coburgi).
heterotricha C. Marquand & Airy Shaw
var. anadena *(Harry Sm.) J.T. Pan & R.J. Gornall in Novon*, 10(4): 376 (2000): Saxifraga anadena.

SAXIFRAGA:-
× hoerhammeri *Sünd. ex A. Engler & E. Irmscher, Pflanzenreich IV*, 117 II (Heft 69): 576 (1919)—Cultivation. (S. grisebachii × marginata var. coriophylla).
hypericoides Franch.
var. aurantiascens *(Engl. & Irmsch.) J.T. Pan & R.J. Gornall in Novon*, 10(4): 376 (2000): Saxifraga macrostigma var. aurantiascens.
var. rockii *(Mattf.) J.T. Pan & R.J. Gornall in Novon*, 10(4): 377 (2000): Saxifraga rockii.
jamalensis *P.Yu. Zhmylev & V.Yu. Razzhivin in Bot. Zhurn.*, 85(5): 119 (2000)—West Siberia.
jarmilae *J.J. Halda in Acta Mus. Richnov. Sect. Nat.*, 4(2): 60 (1997)—China South-central (Yunnan).
× karacardica *J. Bürgel in Čas. Nár. Muz. Řada Přír.*, 164(1–4): 107 (1995)—Yugoslavia. (S. karadzicensis × scardica).
karadzicensis *(Degen & Košanin) J. Bürgel in Čas. Nár. Muz. Řada Přír.*, 164(1–4): 107 (1995): S. coriophylla var. karadzicensis.
× lainzii *P. Vargas in An. Jard. Bot. Madrid*, 54(1): 197 (1996)—Spain. (S. canaliculata × praetermissa).
lancangensis *Y.Y. Qian in Acta Bot. Austro Sinica*, 10: 70 (1995)—China South-central (Yunnan).
nudicaulis D. Don
subsp. soczavae *O.V. Rebristaya in Arktich. Fl. SSSR*, 9(1): 60 (1984)—Amur, Khabarovsk, Buriyatiya.
subsp. vaginalis *(Turcz. ex Ledeb.) O.V. Rebristaya in Arktich. Fl. SSSR*, 9(1): 60 (1984): S. vaginalis.
oppositifolia L.
subsp. ajanica *(Sipliv.) L.I. Malyschev in Fl. Sibir. (Berberidac.-Grossulariac.)*, 7: 192 (1994), isonym; *see Vorosch. in Byull. Mosk. Obshch. Ispўt. Prir., Biol.*, 96(1): 133 (1991): S. ajanica.
pacifica *(Hultén) P.Yu. Zhmylev & V.V. Petrovsky in Byull. Mosk. Obshch. Ispўt. Prir., Biol.*, 100(3): 78 (1995): S. punctata subsp. pacifica.
paniculata Mill.
subsp. laestadii *(Neuman) T. Karlsson in Svensk Bot. Tidskr.*, 91(5): 249 (1997 publ. 1998): S. aizoon * [subsp.] laestadii.
pedemontana All.
subsp. pedatifida *(Bonnier & Layens) J. Jalas in Ann. Bot. Fenn.*, 36(1): 10 (1999), with incorrect basionym & basionym ref.: S. geranioides subsp. pedatifida.
punctata L.
subsp. astilbeoides *(Losinsk.) V.N. Voroschilov in A.K. Skvortsov (ed.), Florist. issl. v razn. raĭonakh SSSR:* 174 (1985): S. astilbeoides.
subsp. redowskyana *(Sternb.) V.N. Voroschilov in A.K. Skvortsov (ed.), Florist. issl. v razn. raĭonakh SSSR:* 174 (1985): S. redowskyana.
sinomontana *J.T. Pan & R.J. Gornall in Novon*, 10(4): 377 (2000): Saxifraga montana *Harry Sm.*
var. amabilis *Harry Sm. ex J.T. Pan in Novon*, 10(4): 377 (2000)—China South-Central (Sichuan).
spinulosa Adams
subsp. derbekii *(Sipliv.) V.N. Voroschilov in A.K. Skvortsov (ed.), Florist. issl. v razn. raĭonakh SSSR:* 175 (1985): S. derbekii.
subsp. funstonii *(Small) V.N. Voroschilov in A.K. Skvortsov (ed.), Florist. issl. v razn. raĭonakh SSSR:* 175 (1985): Leptasea funstonii.
× tolmatchevii *P.Yu. Zhmylev in Byull. Mosk. Obshch. Ispўt. Prir., Biol.*, 100(3): 81 (1995)—Magadan. (S. nelsoniana × porsildiana).
× trifurcatoides *A. Engler & E. Irmscher, Pflanzenreich IV*, 117 I (Heft 67): 337 (1916)—Cultivation. (S. geranioides × trifurcata).
vespertina *(Small) Fedde in Just's Bot. Jahresber.*, 33(1): 613 (1905 publ. 1906): Leptasea vespertina.
× vetteri *É. Burnat, Fl. Alpes Marit.*, 3(2): 251 (1902)—Italy. (S. exarata × pedemontana).
vreelandii *(Small) Fedde in Just's Bot. Jahresber.*, 33(1): 613 (1905 publ. 1906): Spatularia vreelandii.

SCABIOSA (Dipsacac.).
atropurpurea L.
subvar. amansii *(Rouy) O. de Bolòs & J. Vigo, Fl. Països Catalans*, 3: 639 (1995): S. maritima var. amansii.

SCABIOSA:-
cyanocephala *L.C. von Vest in Syll. Pl. Nov.*, 2: 222 (1828)—Locality not stated.
opaca *M.V. Klokov in Novosti Sist. Vўssh. Nizsh. Rast. (Kiev)*, 1974: 112 (1974)—Ukraina.
turolensis Pau
subsp. maroccana *(Pau & Font Quer) A.M. Romo in Lagascalia*, 18(2): 336 (1996): S. tomentosa var. maroccana.
subsp. weyleri *(Pau & Font Quer) A.M. Romo in Lagascalia*, 18(2): 335 (1996): S. weyleri.

SCAEVOLA (Goodeniac.).
hobdyi *W.L. Wagner in Novon*, 6(2): 225 (1996)—Hawaiian Is.
kallophylla *G.J. Howell in Nuytsia*, 11(1): 28 (1996)—Western Australia.
paulayi *F.R. Fosberg in Micronesica*, 23(1): 3 (1990)—Cook Is.
temarii *S.L. Welsh, Fl. Societensis:* 122 (1998), nom. nov.: Temminckia tahitensis *Nadeaud.*

SCAPHOSEPALUM (Orchidac.).
cloesii *C.A. Luer, Ic. Pleurothallid. XVII (Monogr. Syst. Bot. Missouri Bot. Gard., 72):* 115 (1998)—Ecuador.
globosum *C.A. Luer & Hirtz in C.A. Luer, Ic. Pleurothallid. XVII (Monogr. Syst. Bot. Missouri Bot.* Gard., 72): 115 (1998)—Ecuador.
jostii *C.A. Luer, Ic. Pleurothallid. XVII (Monogr. Syst. Bot. Missouri Bot. Gard., 72):* 116 (1998)—Ecuador.
manningii *C.A. Luer, Ic. Pleurothallid. XVI (Monogr. Syst. Bot. Missouri Bot. Gard., 65):* 119 (1998)—Venezuela.
martineae *C.A. Luer, Ic. Pleurothallid. XVII (Monogr. Syst. Bot. Missouri Bot. Gard., 72):* 116 (1998)—Peru.
medinae *C.A. Luer & J. Portilla in C.A. Luer, Ic. Pleurothallid. XX (Monogr. Syst. Bot. Missouri Bot. Gard.*, 79): 131 (2000)—Ecuador.
panduratum *C.A. Luer & R. Escobar in C.A. Luer, Ic. Pleurothallid. XV (Monogr. Syst. Bot. Missouri Bot. Gard.*, 64): 134 (1997)—Colombia.

SCAPHYGLOTTIS (Orchidac.).
atwoodii *R.L. Dressler in Novon*, 7(2): 124 (1997)—Costa Rica.
bicallosa *R.L. Dressler in Novon*, 10(3): 198 (2000)—Costa Rica.
condorana *C.H. Dodson in Orquideologia*, 21(1): 7 (1998)—Ecuador.
gentryi *C.H. Dodson & M. Monsalve in Orquideologia*, 21(1): 8 (1998)—Colombia.
hirtzii *C.H. Dodson in Orquideologia*, 21(1): 11 (1998)—Ecuador.
monspirrae *R.L. Dressler in Novon*, 10(3): 199 (2000)—Panama.
pachybulbon *(Schltr.) R.L. Dressler in Novon*, 10(3): 199 (2000): Hexadesmia pachybulbon.

SCARIOLA (Compos.).
longidentata *(Moris) J. Holub in Preslia*, 70(2): 111 (1998): Lactuca longidentata.

SCELETIUM (Aizoac.).
exalatum *M. Gerbaulet in Bot. Jahrb.*, 118(1): 17 (1996)—Cape Province.
varians *(Haw.) M. Gerbaulet in Bot. Jahrb.*, 118(1): 20 (1996): Mesembryanthemum varians.

SCELOCHILOIDES (Orchidac.).
coimbrae *C.H. Dodson & R. Vásquez Ch.in Rev. Soc. Boliv. Bot.*, 2(1): 13 (1998)—Bolivia.

SCELOCHILOPSIS *C.H. Dodson & M.W. Chase in Orquideologia*, 21(1): 61 (1998). *ORCHIDACEAE*
ecalcarata *(Determann) C.H. Dodson & M.W. Chase in Orquideologia*, 21(1): 61 (1998): Scelochilus ecalcaratus.

SCELOCHILUS (Orchidac.).
amboroensis *R. Vásquez Ch. & C.H. Dodson in Rev. Soc. Boliv. Bot.*, 2(1): 15 (1998), as 'amboroense'—Bolivia.
blankei *K. Senghas in Caesiana*, 8: 23 (1997)—Colombia.

SCELOCHILUS:-
crucicornibus *K. Senghas, D.E. Bennett & E.A. Christenson in Brittonia,* 50(2): 183 (1998)—Peru.
granizoi *W. Königer in Arcula,* 5: 140 (1996)—Peru.
hauensteinii *W. Königer in Arcula,* 5: 143 (1996)—Bolivia.
portillae *W. Königer in Arcula,* 7: 214 (1997)—Ecuador.
topoanus *C.H. Dodson in Orquideologia,* 21(1): 12 (1998)—Ecuador.

× **SCHEDOLIUM** *J. Holub in Preslia,* 70(2): 111 (21 July 1998). *GRAMINEAE* (Lolium × Schedonorus).
brinkmannii *(A. Braun) J. Holub in Preslia,* 70(2): 112 (21 July 1998): Festuca × brinkmannii.
holmbergii *(Dörfl.) J. Holub in Preslia,* 70(2): 112 (21 July 1998): Festuca × holmbergii.
loliaceum *(Huds.) J. Holub in Preslia,* 70(2): 112 (21 July 1998): Festuca loliacea.

× **SCHEDOLOLIUM** *R.J. Soreng & E.E. Terrell in Phytologia,* 83(2): 86 (1997 publ. 27 July 1998). *GRAMINEAE* (Lolium × Schedonorus).
braunii *(K. Richt.) R.J. Soreng & E.E. Terrell in Phytologia,* 83(2): 87 (1997 publ. 27 July 1998): Festuca × braunii.
brinkmannii *(A. Braun) R.J. Soreng & E.E. Terrell in Phytologia,* 83(2): 87 (1997 publ. 27 July 1998): Festuca × brinkmannii.
holmbergii *(Dörfl.) R.J. Soreng & E.E. Terrell in Phytologia,* 83(2): 87 (1997 publ. 27 July 1998): Festuca × holmbergii.
loliaceum *(Huds.) R.J. Soreng & E.E. Terrell in Phytologia,* 83(2): 87 (1997 publ. 27 July 1998): Festuca loliacea.
loliaceum *(Huds.) N.N. Tzvelev in Novosti Sist. Vÿssh. Rast.,* 31: 259 (Dec. 1998), isonym; *see Soreng & Terrell in Phytologia,* 83(2): 87 (1997 publ. 27 July 1998): Festuca loliacea.
nilssonii *(Cugnac & A. Camus) R.J. Soreng & E.E. Terrell in Phytologia,* 83(2): 87 (1997 publ. 27 July 1998): × Festulolium nilssonii.

SCHEDONORUS (Gramin.).
subgen. Drymonaetes *(V.I. Krecz. & Bobrov) J. Holub in Preslia,* 70(2): 112 (1998): Festuca subgen. Drymonaetes.
apenninus *(De Not.) N.N. Tzvelev in Novosti Sist. Vÿssh. Rast.,* 31: 259 (1998): Festuca apennina.
× aschersonianus *(Dörfl.) J. Holub in Preslia,* 70(2): 113 (1998): Festuca × aschersoniana.
× fleischeri *(Rohlena) J. Holub in Preslia,* 70(2): 113 (1998): Festuca × fleischeri.
giganteus *(L.) J. Holub in Preslia,* 70(2): 113 (1998): Bromus giganteus.
giganteus *(L.) R.J. Soreng & E.E. Terrell in Phytologia,* 83(2): 86 (1997 publ. 1998): Bromus giganteus.
interruptus *(Desf.) N.N. Tzvelev in Novosti Sist. Vÿssh. Rast.,* 31: 259 (1998): Festuca interrupta.
littoreus *(Retz.) N.N. Tzvelev in Novosti Sist. Vÿssh. Rast.,* 31: 258 (1998): Bromus littoreus.
mazzettianus *(E.B. Alexeev) J. Holub in Preslia,* 70(2): 113 (1998): Festuca mazzettiana.
phoenix *(Scop.) J. Holub in Preslia,* 70(2): 113 (1998): Poa phoenix.
× schlickumii *(Grantzow) J. Holub in Preslia,* 70(2): 113 (1998): Festuca × schlickumii.
uechtritzianus *(Wiesb.) J. Holub in Preslia,* 70(2): 113 (1998): Festuca uechtritziana.

SCHEELEA (Palmae).
camopiensis *S.F. Glassman, Taxon. Treatment Palm Subtribe Attaleinae (Tribe Cocoeae) (Illinois Biol.* Monogr., 59): 138 (1999)—French Guiana.
degranvillei *S.F. Glassman, Taxon. Treatment Palm Subtribe Attaleinae (Tribe Cocoeae) (Illinois Biol.* Monogr., 59): 139 (1999)—French Guiana.
fairchildensis *S.F. Glassman, Taxon. Treatment Palm Subtribe Attaleinae (Tribe Cocoeae) (Illinois Biol.* Monogr., 59): 163 (1999)—Locality not stated.
guianensis *S.F. Glassman, Taxon. Treatment Palm Subtribe Attaleinae (Tribe Cocoeae) (Illinois Biol.* Monogr., 59): 137 (1999)—French Guiana.
maripensis *S.F. Glassman, Taxon. Treatment Palm Subtribe Attaleinae (Tribe Cocoeae) (Illinois Biol.* Monogr., 59): 140 (1999)—French Guiana.

SCHEELEA:-
moorei *S.F. Glassman, Taxon. Treatment Palm Subtribe Attaleinae (Tribe Cocoeae) (Illinois Biol.* Monogr., 59): 127 (1999)—Peru.
plowmanii *S.F. Glassman, Taxon. Treatment Palm Subtribe Attaleinae (Tribe Cocoeae) (Illinois Biol.* Monogr., 59): 144 (1999)—Peru.
salazarii *S.F. Glassman, Taxon. Treatment Palm Subtribe Attaleinae (Tribe Cocoeae) (Illinois Biol.* Monogr., 59): 146 (1999)—Peru.
wesselsboeri *S.F. Glassman, Taxon. Treatment Palm Subtribe Attaleinae (Tribe Cocoeae) (Illinois Biol.* Monogr., 59): 170 (1999), as 'wesselsboerii' —Venezuela.

SCHEFFLERA (Araliac.).
exaltata *(Thwaites) D.G. Frodin in M.D. Dassanayake & W.D. Clayton, eds., Revised Handb. Fl. Ceylon,* 10: 11 (1996): Hedera exaltata.
heterobotrya *D.G. Frodin in M.D. Dassanayake & W.D. Clayton, eds., Revised Handb. Fl. Ceylon,* 10: 8 (1996)—Sri Lanka.
maduraiensis *K. Ravikumar & V. Lakshmanan in Rheedea,* 9(1): 66 (1999)—India (Tamil Nadu).
splendens *(Kunth) Frodin ex J.C. Lindeman in Fl. Suriname,* 3(1–2, addit. & correct.): 352 (1986): Panax splendens.
venulosa (Wight & Arn.) Harms
var. roxburghii *(Gamble) H.O. Saxena & M. Brahmam, Fl. Orissa,* 2: 799 (1995): S. roxburghii.

SCHIEDEA (Caryophyllac.).
jacobii *W.L. Wagner, S.G. Weller & A.C. Medeiros in Novon,* 9(2): 284 (1999)—Hawaiian Is.
pentandra *W.L. Wagner & E.M. Harris in Amer. J. Bot.,* 87(2): 159 (2000)—Hawaiian Is.

SCHIEDEELLA (Orchidac.).
confusa *(Garay) A. Espejo Serna & A.R. López-Ferrari in Phytologia,* 82(2): 80 (1997): Deiregyne confusa.
crenulata *(L.O. Williams) A. Espejo Serna & A.R. López-Ferrari in Phytologia,* 82(2): 80 (1997): Spiranthes trilineata var. crenulata.
dressleri *D.L. Szlachetko in Fragm. Flor. Geobot.,* 41(2): 855 (1996)—Costa Rica.
llaveana (Lindl.) Schltr.
subsp. szlachetkoana *R. González Tamayo in Bol. Inst. Bot. (Guadalajara),* 7(1–3): 80 (1999 publ. 2000)—Mexico Southwest (Nayarit).
pandurata *(Garay) A. Espejo Serna & A.R. López-Ferrari in Phytologia,* 82(2): 80 (1997): Deiregyne pandurata.

SCHIEKIA (Haemodorac.).
flavescens *P. Maury in J. Bot. (Morot),* 3: 269 (1889), in IK Supp. 1 as 'Sansevieria' in error—Venezuela/Colombia border.

× **SCHILLIGERARA** *hort. in Orchid Rev.,* 107(1226): 126 (1999). *ORCHIDACEAE* (Aspasia × Gomesa × Miltonia).

SCHIMA (Theac.).
sericans *(Hand.-Mazz.) T.L. Ming in Iconogr. Arbor. Yunnan.,* 2: 64 (1991), without basionym ref.: S. khasiana var. sericans.
sericans *(Hand.-Mazz.) T.L. Ming in Fl. Yunnanica,* 8: 326 (1997): S. khasiana var. sericans.
var. paracrenata *(Hung T. Chang) T.L. Ming in Fl. Yunnanica,* 8: 327 (1997): S. paracrenata.

SCHISANDRA (Schisandrac.).
subgen. Maximowiczia *(Rupr.) Y.W. Law in Fl. Reipubl. Popul. Sin.,* 30(1): 272 (1996): gen. Maximowiczia.
subgen. Pleiostema *(A.C. Sm.) Y.W. Law in Fl. Reipubl. Popul. Sin.,* 30(1): 272 (1996): S. sect. Pleiostema.
subgen. Plena *Y.W. Law in Fl. Reipubl. Popul. Sin.,* 30(1): 272 (1996).
subgen. Sinoschisandra *Y.W. Law in Fl. Reipubl. Popul. Sin.,* 30(1): 272 (1996).
subgen. Sphaerostema *(Blume) Y.W. Law in Fl. Reipubl. Popul. Sin.,* 30(1): 272 (1996): gen. Sphaerostema.

SCHISANDRA:-
arisanensis Hayata
subsp. viridis *(A.C. Sm.) R.M.K. Saunders, Monogr. Schisandra (Schisandrac.) (Syst. Bot. Monog., 58):* 72 (2000): S. viridis.
bicolor W.C. Cheng
var. tuberculata *(Y.W. Law) Y.W. Law in Fl. Reipubl. Popul. Sin.,* 30(1): 273 (1996): S. tuberculata.
henryi C.B. Clarke
subsp. marginalis *(A.C. Sm.) R.M.K. Saunders, Monogr. Schisandra (Schisandrac.) (Syst. Bot. Monog., 58):* 90 (2000): S. henryi var. marginalis.
subsp. yunnanensis *(A.C. Sm.) R.M.K. Saunders, Monogr. Schisandra (Schisandrac.) (Syst. Bot. Monog., 58):* 89 (2000): S. henryi var. yunnanensis.
longipes *(Merr. & Chun) R.M.K. Saunders, Monogr. Schisandra (Schisandrac.) (Syst. Bot. Monog., 58):* 90 (2000): S. sphenanthera var. longipes.
propinqua (Wall.) Baill.
subsp. axillaris *(Blume) R.M.K. Saunders in Fl. Malesiana, Ser. 1,* 13: 204 (2 Oct. 1997); *et in Edinburgh J. Bot.,* 54(3): 282 (31 Oct. 1997): Sphaerostema axillare.
subsp. intermedia *(A.C. Sm.) R.M.K. Saunders in Edinburgh J. Bot.,* 54(3): 278 (1997): S. propinqua var. intermedia.
subsp. sinensis *(Oliv.) R.M.K. Saunders in Edinburgh J. Bot.,* 54(3): 280 (1997): S. propinqua var. sinensis.
pubinervis *(Rehder & E.H. Wilson) R.M.K. Saunders, Monogr. Schisandra (Schisandrac.) (Syst. Bot. Monog., 58):* 81 (2000): S. sphenanthera var. pubinervis.

SCHISMATOGLOTTIS (Arac.).
ahmadii *A. Hay in Telopea,* 9(1): 102 (2000)—Borneo (Sabah, Sarawak).
bauensis *A. Hay & C. Lee in Telopea,* 9(1): 84 (2000)—Borneo (Sarawak).
bogneri *A. Hay in Telopea,* 9(1): 105 (2000)—Philippines.
bulbifera *H. Okada, H. Tsukaya & Y. Mori in Syst. Bot.,* 24(1): 62 (1999)—Borneo (Kalimantan).
ciliata *A. Hay in Telopea,* 9(1): 60 (2000)—Borneo (Sarawak).
clarae *A. Hay in Telopea,* 9(1): 118 (2000)—Borneo (Sarawak).
clemensiorum *A. Hay in Telopea,* 9(1): 61 (2000)—Borneo (Sabah).
convolvula *P.C. Boyce in S.J. Mayo, J. Bogner & P.C. Boyce, The Genera of Araceae:* 346 (1997)—Borneo (Sarawak).
corneri *A. Hay in Telopea,* 9(1): 29 (2000)—Borneo (Sabah, Kalimantan).
crinitissima *A. Hay in Telopea,* 9(1): 64 (2000)—Borneo (Sabah, Brunei, Sarawak).
decipiens *A. Hay in Telopea,* 9(1): 120 (2000)—Borneo (Sabah).
ecaudata *A. Hay in Telopea,* 9(1): 121 (2000)—Sumatera.
edanoi *A. Hay in Telopea,* 9(1): 121 (2000)—Philippines.
elegans *A. Hay in Telopea,* 9(1): 67 (2000)—Borneo (Sarawak, Sabah, Kalimantan).
eymae *A. Hay in Telopea,* 9(1): 122 (2000)—Sulawesi.
josefii *A. Hay in Telopea,* 9(1): 89 (2000)—Borneo (Sarawak).
lingua *A. Hay in Telopea,* 9(1): 124 (2000)—Borneo (Sabah).
moodii *A. Hay in Telopea,* 9(1): 131 (2000)—Borneo (Sabah).
niahensis *A. Hay in Telopea,* 9(1): 137 (2000)—Borneo (Sarawak).
nicolsonii *A. Hay in Telopea,* 9(1): 95 (2000)—Borneo (Sarawak).
pectinervia *A. Hay in Telopea,* 9(1): 138 (2000)—Borneo (Brunei).
petri *A. Hay in Telopea,* 9(1): 162 (2000)—Borneo (Brunei).
pudenda *A. Hay in Telopea,* 9(1): 98 (2000)—Borneo (Sarawak).
pyrrhias *A. Hay in Telopea,* 9(1): 78 (2000)—Borneo (Sarawak).

SCHISMATOGLOTTIS:-
samarensis *A. Hay in Telopea*, 9(1): 143 (2000)—Philippines.
sarikeensis *(Bogner & M. Hotta) A. Hay & J. Bogner in Telopea*, 9(1): 100 (2000): Hottarum sarikeense.
sejuncta *A. Hay in Telopea*, 9(1): 83 (2000)—Borneo (Brunei).
silamensis *A. Hay in Telopea*, 9(1): 144 (2000)—Borneo (Sabah).
trusmadiensis *A. Hay & J. Mood in Telopea*, 9(1): 151 (2000)—Borneo (Sabah).
unifolia *A. Hay & P.C. Boyce in Telopea*, 9(1): 151 (2000)—Borneo (Sabah).
venusta *A. Hay in Telopea*, 9(1): 152 (2000)—Borneo (Sabah).
viridissima *A. Hay in Telopea*, 9(1): 154 (2000)—Borneo (Sarawak).
wongii *A. Hay in Telopea*, 9(1): 160 (2000)—Borneo (Sabah).

SCHISTOCARPHA (Compos.).
eupatorioides *(Fenzl) O. Kuntze, Revis. gen. pl.*, 3(3): 170 (1898): Neilreichia eupatorioides.†

SCHIVERECKIA (Crucif.).
hyperborea *(L.) A.N. Berkutenko in Linz. Biol. Beitr.*, 27(2): 1120 (1995): Alyssum hyperboreum.

SCHIZACHYRIUM (Gramin.).
scoparium (Michx.) Nash
subsp. divergens *(Hack.) K.N. Gandhi & F.E. Smeins in Harvard Pap. Bot.*, 8: 68 (1996): Andropogon scoparius var. divergens.
subsp. littorale *(Nash) K.N. Gandhi & F.E. Smeins in Harvard Pap. Bot.*, 8: 68 (1996): Andropogon littoralis.
subsp. neomexicanum *(Nash) K.N. Gandhi & F.E. Smeins in Harvard Pap. Bot.*, 8: 68 (1996): Andropogon neomexicanus.
var. stoloniferum *(Nash) J.K. Wipff in Phytologia*, 80(1): 37 (1996): S. stoloniferum.
f. virile *(Shinners) K.N. Gandhi & F.E. Smeins in Harvard Pap. Bot.*, 8: 68 (1996): Andropogon scoparius var. virilis.
spadiceum *(Swallen) J.K. Wipff in Phytologia*, 80(1): 35 (1996): Andropogon spadiceus.

SCHIZACME *C.R. Dunlop in Fl. Australia*, 28: 314 (1996). *LOGANIACEAE*
archeri *(Hook. f.) C.R. Dunlop in Fl. Australia*, 29: 314 (1996): Mitrasacme archeri.
montana *(Hook. f. ex Benth.) C.R. Dunlop in Fl. Australia*, 28: 314 (1996): Mitrasacme montana.

SCHIZANTHUS (Solanac.).
coccineus *(Phil.) J.M. Watson in A. Hoffmann, M.K. Arroyo, F. Liberona, M. Muñoz & J. Watson, Pl.* Altoandinas Fl. Silv. Chile: 140 (1998): S. grahamii var. coccineus.

SCHIZOCARPHUS (Hyacinthac.).
hispidulus *(Baker) F. Speta in Phyton (Austria)*, 38(1): 120 (1998): Scilla hispidula.
setiferus *(Baker) F. Speta in Phyton (Austria)*, 38(1): 120 (1998): Scilla setifera.

SCHIZODIUM (Orchidac.).
satyrioides *(L.) L.A. Garay in Harvard Pap. Bot.*, 2(1): 49 (1997), as No. 11 on inside pages: Orchis satyrioides.

SCHIZOGLOSSUM (Asclepiadac.).
saccatum *P.V. Bruyns in Bothalia*, 25(2): 169 (1995)—Namibia.

SCHIZOLAENA (Sarcolaenac.).
capuronii *P.P. Lowry, G.E. Schatz, J.-F. Leroy & A.-E. Wolf in Adansonia*, 21(2): 190 (1999)—Madagascar.
cavacoana *P.P. Lowry, G.E. Schatz, J.-F. Leroy & A.-E. Wolf in Adansonia*, 21(2): 193 (1999)—Madagascar.
gereaui *P.P. Lowry, G.E. Schatz, J.-F. Leroy & A.-E. Wolf in Adansonia*, 21(2): 195 (1999)—Madagascar.
manomboensis *P.P. Lowry, G.E. Schatz, J.-F. Leroy & A.-E. Wolf in Adansonia*, 21(2): 199 (1999)—Madagascar.

SCHIZOLAENA:-
masoalensis *P.P. Lowry, G.E. Schatz, J.-F. Leroy & A.-E. Wolf in Adansonia*, 21(2): 200 (1999)—Madagascar.
milleri *P.P. Lowry, G.E. Schatz, J.-F. Leroy & A.-E. Wolf in Adansonia*, 21(2): 202 (1999)—Madagascar.
tampoketsana *P.P. Lowry, G.E. Schatz, J.-F. Leroy & A.-E. Wolf in Adansonia*, 21(2): 207 (1999)—Madagascar.
turkii *P.P. Lowry, G.E. Schatz, J.-F. Leroy & A.-E. Wolf in Adansonia*, 21(2): 207 (1999)—Madagascar.

SCHIZOLOBIUM (Legumin.).
parahyba (Vell.) Blake
var. amazonicum *(Ducke) R.C. Barneby in Brittonia*, 48(2): 178 (1996): S. amazonicum.

SCHIZONEPETA (Labiat.).
multifida (L.) Briq.
var. longistyla *J.X. Huang, T.J. Feng & J.Z. Wang in J. Hebei Agric. Univ.*, 17(4): 17 (1994), as 'longostyla'—China North-central (Hebei).

SCHIZOSTACHYUM (Gramin.).
aequiramosum *E.A. Widjaja in Reinwardtia*, 11(2): 131 (1997)—Jawa.
atrocingulare *E.A. Widjaja in Reinwardtia*, 11(2): 132 (1997)—Sumatera.
auriculatum *Q.H. Dai & D.Y. Huang in J. Bamboo Res.*, 16(3): 29 (1997)—China Southeast (Guangxi).
bamban *E.A. Widjaja in Reinwardtia*, 11(2): 134 (1997)—Sumatera.
castaneum *E.A. Widjaja in Reinwardtia*, 11(2): 134 (1997)—Lesser Sunda Is. (Bali).
cornutum *E.A. Widjaja in Reinwardtia*, 11(2): 136 (1997)—Sumatera.
cuspidatum *E.A. Widjaja in Reinwardtia*, 11(2): 137 (1997)—Sumatera.
flexuosum *E.A. Widjaja in Reinwardtia*, 11(2): 138 (1997)—Borneo (Kalimantan).
glaucocladum *E.A. Widjaja in Reinwardtia*, 11(2): 140 (1997)—Borneo (Kalimantan).
khoonmengii *S. Dransfield in Kew Bull.*, 55(2): 491 (2000)—Borneo (Brunei).
lengguanii *K.M. Wong, Bamboos Penins. Malaysia (Malayan Forest Rec., 41)*: 176 (1995)—Malaya (Peninsular Malaysia).
lutescens *E.A. Widjaja in Reinwardtia*, 11(2): 142 (1997)—Sumatera.
mampouw *E.A. Widjaja in Reinwardtia*, 11(2): 144 (1997)—Sumatera.
munroi *S. Kumar & Paramjit Singh in J. Ind. Bot. Soc.*, 70(1–4): 423 (1991), nom. nov.: Cephalostachyum capitatum *Munro*.
var. decompositum *(Gamble) S. Kumar & Paramjit Singh in J. Ind. Bot. Soc.*, 70(1–4): 423 (1991): Cephalostachyum capitatum var. decompositum.
sharmae *S. Kumar & Paramjit Singh in J. Ind. Bot. Soc.*, 70(1–4): 423 (1991), nom. nov.: Cephalostachyum latifolium *Munro*.
silicatum *E.A. Widjaja in Reinwardtia*, 11(2): 145 (1997)—Sumatera, Jawa, Lesser Sunda Is. (Bali).
subvexorum *Q.H. Dai & D.Y. Huang in J. Bamboo Res.*, 16(3): 27 (1997)—China Southeast (Guangxi).

SCHKUHRIA (Compos.).
pinnata (Lam.) Kuntze ex Thell.
var. wislizeni *(A. Gray) B.L. Turner in Phytologia*, 79(5): 364 (1995 publ. 1996): S. wislizeni.

SCHLECHTERELLA (Asclepiadac.).
abyssinica *(Chiov.) H.J.T. Venter & R.L. Verhoeven in S. Afr. J. Bot.*, 64(6): 353 (1998): Raphionacme abyssinica.

SCHLUMBERGERA (Cactac.).
microsphaerica (K. Schum.) P.V. Heath
subsp. candida *(Loefgr.) D.R. Hunt in A.J.S. McMillan & J.F. Horobin, Christmas Cacti: gen. Schlumbergera &* hybr. (Succ. Pl. Res., 4): 79 (1995): Zygocactus candidus.

SCHMIDELIA (Sapindac.).
aporetica *(Roxb.) J.O. Voigt, Hort. Suburb. Calcutt.*: 93 (1845): Ornithrophe aporetica.

SCHNABELIA (Verbenac.).
aureoglandulosa *(Vaniot) P.D. Cantino in Syst. Bot.*, 23(3): 381 (1998 publ. 1999): Ocimum aureoglandulosum.
nepetifolia *(Benth.) P.D. Cantino in Syst. Bot.*, 23(3): 381 (1998 publ. 1999): Teucrium nepetifolium.
terniflora *(Maxim.) P.D. Cantino in Syst. Bot.*, 23(3): 382 (1998 publ. 1999): Caryopteris terniflora.

SCHNARFIA *F. Speta in Phyton (Austria)*, 38(1): 120 (1998). *HYACINTHACEAE*
albanica *(Turrill) F. Speta in Phyton (Austria)*, 38(1): 121 (1998): Scilla albanica.
messeniaca *(Boiss.) F. Speta in Phyton (Austria)*, 38(1): 121 (1998): Scilla messeniaca.

SCHOENOCAULON (Melanthiac.).
caricifolium (Schltdl.) A. Gray
var. oaxacense *D. Frame in Acta Bot. Mex.*, 48: 29 (1999)—Mexico Southwest (Oaxaca).
ignigenum *D. Frame in Acta Bot. Mex.*, 48: 29 (1999)—Mexico Northeast (Tamaulipas, Nuevo León).
madidorum *D. Frame in Acta Bot. Mex.*, 48: 33 (1999)—Mexico Central (Puebla), Mexico Gulf (Veracruz), Mexico Southwest (Oaxaca).
megarrhizum M.E. Jones
var. deminutum *D. Frame in Acta Bot. Mex.*, 48: 35 (1999)—Mexico Northeast (Durango), Mexico Southwest (Nayarit).
plumosum *D. Frame in Acta Bot. Mex.*, 48: 36 (1999)—Mexico Northeast (Nuevo León, Coahuila, Tamaulipas).
rzedowskii *D. Frame in Acta Bot. Mex.*, 48: 38 (1999)—Mexico Central (Puebla, México State).
tenorioi *D. Frame in Acta Bot. Mex.*, 48: 40 (1999)—Mexico Southwest (Oaxaca), Mexico Central (Puebla).

SCHOENOLIRION (Hyacinthac.).
wrightii *H.L. Sherman in Southwest. Nat.*, 24(1): 125 (1979)—Texas.

SCHOENOPLECTUS (Cyperac.).
americanus (Pers.) Volkart
var. longebracteatus *(Osten & Barros) A.C. Vegetti in Beitr. Biol. Pfl.*, 67(2): 247 (1992 publ. 1993), without basionym ref.: Basionym not stated.
subsp. monophyllus *(C. Presl. & J. Presl) A.C. Vegetti in Beitr. Biol. Pfl.*, 67(2): 247 (1992 publ. 1993), without basionym ref.: Basionym not stated.
articulatus *(L.) Palla in Bot. Jahrb. Syst.*, 10(4): 299 (4 Dec. 1888): Scirpus articulatus.
californicus (C.A. Mey.) Soják
var. spoliatus *(Barros) A.C. Vegetti in Beitr. Biol. Pfl.*, 67(2): 248 (1992 publ. 1993), without basionym ref.; *et in Hickenia*, 2(31): 138 (1995), with basionym ref.: Scirpus californicus var. spoliatus.
var. tereticulmis *(Steud.) A.C. Vegetti in Beitr. Biol. Pfl.*, 67(2): 247 (1992 publ. 1993), without basionym ref.; *et in Hickenia*, 2(31): 138 (1995), with basionym ref.: Scirpus tereticulmis.
× carinatus *(Sm.) Palla in Bot. Jahrb. Syst.*, 10(4): 299 (4 Dec. 1888): Scirpus carinatus.†
nothosubsp. kuekenthalianus *(Junge) P.D. Sell in P.D. Sell & G. Murrell, Fl. Great Britain & Ireland*, 5: 363 (1996): Scirpus kuekenthalianus.
× igaensis *(T. Koyama) E. Hayasaka & H. Ohashi in J. Jap. Bot.*, 75(4): 224 (2000): Scirpus × igaensis.
javanus *(Nees) Palla in Bot. Jahrb. Syst.*, 10(4): 299 (4 Dec. 1888): Scirpus javanus.
juncoides *(Roxb.) Palla in Bot. Jahrb. Syst.*, 10(4): 299 (4 Dec. 1888): Scirpus juncoides.†
lacustris *(L.) Palla in Bot. Jahrb. Syst.*, 10(4): 299 (4 Dec. 1888): Scirpus lacustris.†
subsp. hippolyti *(V.I. Krecz.) I. Kukkonen in Ann. Naturhist. Mus. Wien*, 98B(Suppl.): 89 (1996); *et in Fl. Iran.*, 173: 25 (1998): Scirpus hippolyti.

SCHOENOPLECTUS:-
litoralis *(Schrad.) Palla in Bot. Jahrb. Syst.*, 10(4): 299 (4 Dec. 1888), as 'littoralis': Scirpus litoralis.†
mucronatus *(L.) Palla in Bot. Jahrb. Syst.*, 10(4): 299 (4 Dec. 1888): Scirpus mucronatus.†
f. hosoiri *E. Hayasaka & Hiroyoshi Ohashi in J. Jap. Bot.*, 75(5): 319 (2000)— Japan (Honshu).
f. robustus *(Miq.) E. Hayasaka & Hiroyoshi Ohashi in J. Jap. Bot.*, 75(5): 319 (2000): Scirpus mucronatus var. robustus.
nevadensis (S. Watson) Soják
var. remiroides *(Griseb.) A.C. Vegetti in Beitr. Biol. Pfl.*, 67(2): 247 (1992 publ. 1993), without basionym ref.: Basionym not stated.
nobilis *(Ridl.) K.A. Lye in Lidia*, 3(5): 144 (1995): Scirpus nobilis.
× oguraensis *(T. Koyama) E. Hayasaka & H. Ohashi in J. Jap. Bot.*, 75(4): 224 (2000): Scirpus × oguraensis.
olneyi *(A. Gray) Palla in Bot. Jahrb. Syst.*, 10(4): 299 (4 Dec. 1888): Scirpus olneyi.
f. australis *(Osten & Barros) A.C. Vegetti in Beitr. Biol. Pfl.*, 67(2): 247 (1992 publ. 1993), without basionym ref.: Basionym not stated.
orthorhizomatus *(Kats. Arai & Miyam.) E. Hayasaka & H. Ohashi in J. Jap. Bot.*, 75(4): 224 (2000): Scirpus orthorhizomatus.
paludicola *(Kunth) Palla in Bot. Jahrb. Syst.*, 10(4): 299 (4 Dec. 1888): Scirpus paludicola.†
pungens *(Vahl) Palla in Bot. Jahrb. Syst.*, 10(4): 299 (4 Dec. 1888): Scirpus pungens.†
var. longebracteatus *(Osten & Barros) A.C. Vegetti in Beitr. Biol. Pfl.*, 67(2): 247 (1992 publ. 1993), without basionym ref.: Basionym not stated.
quinquefarius *(Buch.-Ham. ex Boeck.) Palla in Bot. Jahrb. Syst.*, 10(4): 299 (4 Dec. 1888): Scirpus quinquefarius.
rigidus *(Steud.) A.C. Vegetti in Beitr. Biol. Pfl.*, 67(2): 247 (1992 publ. 1993), without basionym ref.: Basionym not stated.
senegalensis *(Hochst. ex Steud.) Palla in Bot. Jahrb. Syst.*, 10(4): 299 (4 Dec. 1888): Isolepis senegalensis.†
supinus *(L.) Palla in Bot. Jahrb. Syst.*, 10(4): 299 (4 Dec. 1888): Scirpus supinus.†
tabernaemontani *(C.C. Gmel.) Palla in Bot. Jahrb. Syst.*, 10(4): 299 (4 Dec. 1888): Scirpus tabernaemontani.†
tatora *(Kunth) Palla in Bot. Jahrb. Syst.*, 10(4): 299 (4 Dec. 1888): Scirpus tatora.
× trapezoideus *(Koidz.) E. Hayasaka & H. Ohashi in J. Jap. Bot.*, 75(4): 224 (2000): Scirpus trapezoideus.
triqueter *(L.) Palla in Bot. Jahrb. Syst.*, 10(4): 299 (4 Dec. 1888): Scirpus triqueter.†
× uzenensis *(T. Koyama) E. Hayasaka & H. Ohashi in J. Jap. Bot.*, 75(4): 224 (2000): Scirpus × uzenensis.

SCHOENORCHIS (Orchidac.).
manilaliana *M. Kumar & S. Sequiera in Kew Bull.*, 55(1): 241 (2000)—India (Kerala).

SCHOENUS (Cyperac.).
attenuatus *M. Gandoger, Dec. Pl. Nov.*, 2: 35 (1876)—France.
badius *B.L. Rye in Nuytsia*, 11(2): 263 (1997)—Western Australia.
calcatus *K.L. Wilson in Nuytsia*, 11(2): 276 (1997)—Western Australia.
griffinianus *K.L. Wilson in Nuytsia*, 11(2): 278 (1997)—Western Australia.
insolitus *K.L. Wilson in Nuytsia*, 11(2): 280 (1997)—Western Australia.
longus *M. Gandoger, Dec. Pl. Nov.*, 2: 35 (1876)—France.
lymansmithii *M.T. Strong in Harvard Pap. Bot.*, 4(1): 275 (1999)—Brazil South (Santa Catarina).
megalocephalus *M. Gandoger, Dec. Pl. Nov.*, 2: 36 (1876)—France.
pallescens *M. Gandoger, Dec. Pl. Nov.*, 2: 35 (1876)—France.
plumosus *B.L. Rye in Nuytsia*, 11(2): 264 (1997)—Western Australia.
subcontortus *M. Gandoger, Dec. Pl. Nov.*, 2: 36 (1876)—France.

SCHOENUS:-
variicellae *B.L. Rye in Nuytsia*, 11(2): 266 (1997)—Western Australia.
vendeanus *M. Gandoger, Dec. Pl. Nov.*, 2: 34 (1876)—France.

× **SCHOMBOCATTLEYA** (Orchidac.).
felix *L.C. Menezes in Orchidee*, 50(1): 12 (1999)—Brazil Southeast (Minas Gerais). (Cattleya walkeriana × Schomburgkia crispa).

× **SCHOMBOLAELIOPSIS** *hort. in Orchid Rev.*, 104(1209): 166 (1996). *ORCHIDACEAE* (Laeliopsis × Schomburgkia).

SCHRADERA (Rubiac.).
acutifolia *(Valeton) C. Puff, R. Buchner & J. Greimler in Blumea*, 43(2): 326 (1998): Lucinaea acutifolia.
elmeri *C. Puff, R. Buchner & J. Greimler in Blumea*, 43(2): 315 (1998), nom. nov.: Lucinaea involucrata *Elmer.*
grandiflora *C. Puff, R. Buchner & J. Greimler in Blumea*, 43(2): 297 (1998)—Borneo (Sarawak, Brunei).
korthalsiana *(Miq.) C. Puff, R. Buchner & J. Greimler in Blumea*, 43(2): 312 (1998): Lucinaea korthalsiana.
subsp. robusta *C. Puff, R. Buchner & J. Greimler in Blumea*, 43(2): 315 (1998)—Borneo (Brunei, Sabah, Sarawak).
ledermannii *(Valeton) C. Puff, R. Buchner & J. Greimler in Blumea*, 43(2): 324 (1998): Lucinaea ledermannii.
membranacea *(King) C. Puff, R. Buchner & J. Greimler in Blumea*, 43(2): 300 (1998): Lucinaea membranacea.
subsp. flagellarioides *C. Puff, R. Buchner & J. Greimler in Blumea*, 43(2): 305 (1998)—Borneo (Sarawak).
subsp. parvifolia *(W.W. Sm.) C. Puff, R. Buchner & J. Greimler in Blumea*, 43(2): 305 (1998): Lucinaea parvifolia.
monantha *(Merr. & L.M. Perry) C. Puff, R. Buchner & J. Greimler in Blumea*, 43(2): 328 (1998): Lucinaea monantha.
monocephala *(Merr.) C. Puff, R. Buchner & J. Greimler in Blumea*, 43(2): 299 (1998): Lucinaea monocephala.
montana *(Korth.) C. Puff, R. Buchner & J. Greimler in Blumea*, 43(2): 308 (1998): Lucinaea montana.
nervulosa *(Stapf) C. Puff, R. Buchner & J. Greimler in Blumea*, 43(2): 316 (1998): Lucinaea nervulosa.
novoguineensis *(Valeton) C. Puff, R. Buchner & J. Greimler in Blumea*, 43(2): 306 (1998): Lucinaea novoguineensis.
pentacme *(Stapf) C. Puff, R. Buchner & J. Greimler in Blumea*, 43(2): 311 (1998): Lucinaea pentacme.
polysperma *(Jack) C. Puff, R. Buchner & J. Greimler in Blumea*, 43(2): 293 (1998): Morinda polysperma.
pseudonervulosa *C. Puff, R. Buchner & J. Greimler in Blumea*, 43(2): 321 (1998)—Borneo (Sabah, Sarawak, Kalimantan), Moluccas.
ramiflora *(Valeton) C. Puff, R. Buchner & J. Greimler in Blumea*, 43(2): 329 (1998): Lucinaea ramiflora.
schlechteri *(Valeton) C. Puff, R. Buchner & J. Greimler in Blumea*, 43(2): 327 (1998): Lucinaea schlechteri.

SCHRENKIA (Umbellif.).
pulverulenta *M.G. Pimenov in Opred. Rast. Sred. Azii*, 7: 370 (1983)—Kirgizstan.

SCHUBERTIA (Asclepiadac.).
hatschbachii *G. Morillo in Asklepios*, 74: 26 (1998)—Brazil Southeast (Minas Gerais).

SCHULZIA (Umbellif.).
bhutanica *M.F. Watson in Edinburgh J. Bot.*, 53(1): 138 (1996)—Bhutan.
crinita (Pall.) Spreng.
subsp. albiflora *(Kar. & Kir.) M.G. Pimenov in Opred. Rast. Sred. Azii*, 7: 263 (1983): Chamaesciadium albiflorum.

SCHUNKEA *K. Senghas in Palmengarten*, 58(2): 128 (1994), contrary to Art. 37.2 ICBN (1994). *ORCHIDACEAE*

SCHUNKEA:-
vierlingii *K. Senghas in Palmengarten*, 58(2): 128 (1994), contrary to Art. 37.3 ICBN (1994)—Peru.

SCHWANTESIA (Aizoac.).
constanceae *N.F.A. Zimmermann in Kakt. And. Sukk.*, 47(4): 84 (1996)—Namibia.

SCHWARTZIA (Marcgraviac.).
brenesii *(Standl.) H.G. Bedell in Bot. Jahrb.*, 119(3): 328 (1997): Norantea brenesii.
costaricensis *(Gilg) H.G. Bedell in Bot. Jahrb.*, 119(3): 328 (1997): Norantea costaricensis.
jimenezii *(Standl.) H.G. Bedell in Bot. Jahrb.*, 119(3): 328 (1997): Ruyschia jimenezii.

SCILLA (Hyacinthac.).
borealijaponica *M. Kikuchi in Ann. Rep. Gakugei Fac. Iwate Univ.*, 11(2): 67 (1957)—Japan.
cydonia *F. Speta in Linz. Biol. Beitr.*, 30(1): 432 (1998)—Kriti.
nana (Schult. & Schult. f.) Speta
subsp. albescens *(Speta) F. Speta in Linz. Biol. Beitr.*, 30(1): 435 (1998): S. albescens.
numidica Poir.
subsp. ebusitana *A.M. Romo, Fl. Silvestres Baleares:* 325 (1994)—Baleares.
talosii *D. Tzanoudakis & Z. Kypriotakis in Folia Geobot.*, 33(1): 104 (1998)—Kriti.

SCINDAPSUS (Arac.).
sumatranus *(Miq.) P.C. Boyce & A. Hay in Telopea*, 9(1): 170 (2000): Schismatoglottis sumatrana.

SCIRPODENDRON (Cyperac.).
bogneri *S.S. Hooper in Kew Bull.*, 52(3): 683 (1997)—Malaya (Peninsular Malaysia).

SCIRPOIDES (Cyperac.).
burkei *(C.B. Clarke) P. Goetghebeur, A.M. Muasya & D.A. Simpson in Novon*, 10(2): 132 (2000): Scirpus burkei.

SCIRPUS (Cyperac.).
sect. Pseudotrichophorum *T.V. Egorova in Novosti Sist. Vyssh. Rast.*, 31: 10 (1998).
kimsonensis *N.K. Khoi in J. Biol. (Vietnam)*, 18(4): 5 (1996)—Vietnam.
littoralis Schrad.
subsp. subulatus *(Vahl) M.A. Wadood Khan in Rheedea*, 10(1): 28 (2000): Scirpus subulatus.
monocephalus *J.Q. He in Acta Phytotax. Sin.*, 37(3): 291 (1999)—China Southeast (Anhui).
naikianus *M.A. Wadood Khan in Rheedea*, 8(1): 71 (1998)—India (Maharashtra).
orthorhizomatus *K. Arai & F. Miyamoto in J. Jap. Bot.*, 72(5): 299 (1997)—Japan.
× osoreyamensis *M. Kikuchi in Ann. Rep. Gakugei Fac. Iwate Univ.*, 18(3): 35 (1961)—Japan. (S. hondoensis × triangulatus).
sylvaticus L.
subsp. orientalis *(Ohwi) V.N. Voroschilov in A.K. Skvortsov (ed.), Florist. issl. v razn. Raĭonakh SSSR:* 152 (1985): S. orientalis.

SCLERIA (Cyperac.).
amazonica *K. Camelbeke, M.T. Strong & P. Goetghebeur in Novon*, 7(2): 98 (1997)—Venezuela.
jiangchengensis *Y.Y. Qian in Bull. Bot. Res. North-East. Forest. Univ.*, 19(3): 241 (1999)—China South-Central (Yunnan).
levis Retz.
var. pubescens *(Steud.) S. Tjitrosoedirdjo in Rheedea*, 5(2): 111 (1995), isonym; *see C.Z. Zheng in Fl. Zhejiang*, 7: 286 (1993): S. pubescens.
pernambucana *M. Luceño & M.V. Alves in Candollea*, 52(1): 196 (1997)—Brazil Northeast (Pernambuco).
poklei *M.A. Wadood Khan in J. Econ. Taxon. Bot.*, 22(3): 559 (1998 publ. 1999), as 'poklii'—India (Maharashtra).
robusta *K. Camelbeke & P. Goetghebeur in Novon*, 10(4): 359 (2000), nom. nov.: Scleria grandis *Core.*

SCLEROCACTUS (Cactac.).
sect. Coloradoa *(Boissev. & C. Davidson) A.B. Doweld in Sukkulentŷ (Moscow)*, 1999(1): 27 (1999): gen. Coloradoa.
sect. Mesae-verdae *F. Hochstätter in Piante Grasse*, 19(1): 18 (1999).
sect. Parviflori *F. Hochstätter in Piante Grasse*, 19(1): 17 (1999).
parviflorus Clover & Jotter
f. macrospermus *F. Hochstätter in Cactaceae Rev.*, 1(1): 11–12 (1998); *cf. Repert. Pl. Succ. (I.O.S.)*, 49: 24 (1998 publ. 1999); *et in Piante Grasse*, 19(1): 9 (1999), without latin descr.—Utah.
f. terrae-canyonae *(K.D. Heil) F. Hochstätter in Cactaceae Rev.*, 1(1): 14–15 (1998); *cf. Repert. Pl. Succ. (I.O.S.)*, 49: 24 (1998 publ. 1999); *et in Piante Grasse*, 19(1): 12 (1999), without exact basionym page: S. terrae-canyonae.
f. variiflorus *F. Hochstätter in Cactaceae Rev.*, 1(1): 8–10 (1998); *cf. Repert. Pl. Succ. (I.O.S.)*, 49: 24 (1998 publ. 1999); *et in Piante Grasse*, 19(1): 8 (1999), without latin descr.—Arizona.
uncinatus
subsp. crassihamatus *(F.A.C. Weber) N.P. Taylor in Cactaceae Consensus Initiatives*, 5: 13 (1998): Echinocactus crassihamatus.
subsp. wrightii *(Engelm.) N.P. Taylor in Cactaceae Consensus Initiatives*, 5: 13 (1998): Echinocactus uncinatus [var.] wrightii.
whipplei (Engelm. & J.M. Bigelow) Britton & Rose
subvar. aztecia *F. Hochstätter in Piante Grasse*, 17(3): 228 (1997), nom. nov.: S. whipplei var. heilii *Castetter, P. Pierce & K.H. Schwer.*

SCLEROCROTON (Euphorbiac.).
carterianus *(J. Léonard) R.C. Kruijt & G.J. Roebers in Biblioth. Bot.*, 146: 18 (1996): Sapium carterianum.
cornutus *(Pax) R.C. Kruijt & G.J. Roebers in Biblioth. Bot.*, 146: 20 (1996): Sapium cornutum.
melanostictus *(Baill.) R.C. Kruijt & G.J. Roebers in Biblioth. Bot.*, 146: 23 (1996): Stillingia melanosticta.
oblongifolius *(Müll. Arg.) R.C. Kruijt & G.J. Roebers in Biblioth. Bot.*, 146: 24 (1996): Excoecaria oblongifolia.
schmitzii *(J. Léonard) R.C. Kruijt & G.J. Roebers in Biblioth. Bot.*, 146: 26 (1996): Sapium schmitzii.

SCLEROPHYLAX (Solanac.).
cuyanus *T.E. Di Fulvio in Kurtziana*, 27(1): 244 (1999)—Argentina Northwest (San Juan, Mendoza).

SCLEROTHECA (Campanulac.).
magdalenae *J. Florence in Bull. Mus. Nation. Hist. Nat., B, Adansonia*, Ser. 4, 18(1–2): 97 (1996)—Society Is.

SCOLOPHYLLUM (Scrophulariac.).
longitubum *T. Yamazaki & W. Chuakul in J. Jap. Bot.*, 74(2): 82 (1999)—Thailand.

SCOLOPIA (Flacourtiac.).
maoulidae *S. Hul, J.-N. Labat & O. Pascal in Adansonia*, 20(2): 266 (1998)—Comoros (Mayotte).

SCOLYMUS (Compos.).
hispanicus L.
var. aggregatus *(Ruch.) F.M. Vázquez in An. Jard. Bot. Madrid*, 58(1): 89 (2000): S. aggregatus.
subsp. occidentalis *F.M. Vázquez in An. Jard. Bot. Madrid*, 58(1): 91 (2000)—Spain, Algeria, Morocco, Tunisia, France, Italy, Portugal, Baleares, Sardegna, Canary Is.

SCOPARIA (Scrophulariac.).
praedensa *(R.E. Fr.) S.M. Botta & A.L. Cabrera in A.L. Cabrera, Fl. Prov. Jujuy (Inst Nac. Tecn. Agropec., 13)*, 9: 186 (1993): S. nudicaulis subsp. praedensa.

SCOPELOGENA (Aizoac.).
bruynsii *C. Klak in Bothalia*, 30(1): 39 (2000)—Cape Province.

SCORZONERA (Compos.).
albicaulis Bunge
subsp. macrosperma *(Turcz. ex DC.) V.N. Voroschilov in A.K. Skvortsov (ed.), Florist. issl. v razn. raĭonakh SSSR*: 198 (1985): S. macrosperma.
austriaca Willd.
var. hebecarpa *C.H. An in Fl. Xinjiangensis*, 5: 479 (1999), as 'hebecarpus'—Xinjiang.
elongata *C.H. An & X.L. He in Fl. Xinjiangensis*, 5: 480 (1999), homonym; *see Willd., Sp. Pl.*, 3(3): 1508 (1803)—Xinjiang.
hotanica *C.H. An in Fl. Xinjiangensis*, 5: 479 (1999)—Xinjiang.
pseudodivaricata Lipsch.
var. leiocarpa *C.H. An in Fl. Xinjiangensis*, 5: 480 (1999)—Xinjiang.
tianshanensis *C.H. An in Fl. Xinjiangensis*, 5: 480 (1999)—Xinjiang.

SCROPHULARIA (Scrophulariac.).
bulgarica *(Stoj.) D. Peev in Fl. Republ. Bŭlgariya*, 10: 106 (1996): S. variegata var. bulgarica.
chasmophila W.W. Sm.
subsp. xizangensis *D.Y. Hong in Novon*, 6(4): 374 (1996)—Xizang.
cooperi *R.R. Mill in Edinburgh J. Bot.*, 57(3): 424 (2000)—Bhutan.
fontqueri *A. Ortega-Olivencia & J.A. Devesa in Bot. J. Linn. Soc.*, 128(2): 191 (1998)—Morocco.
jallui *(Gattef. & Weiller) M. Ibn Tattou in Bocconea*, 8: 173 (1998): S. aquatica subsp. jallui.
kermanica *A. Ghahreman & S.M. Mirtadzadini in Iranian J. Bot.*, 8(2): 246 (2000)—Iran.
lhasaensis *D.Y. Hong in Novon*, 6(4): 372 (1996)—Xizang.
macrorrhyncha *(Humbert, Litard. & Maire) M. Ibn Tattou in Bocconea*, 8: 173 (1998): S. ramosissima subsp. macrorrhyncha.
ramosissima Loisel.
subsp. minoricensis *(P. Monts.) A.M. Romo, Fl. Silvestres Baleares*: 274 (1994): S. canina var. minoricensis.
subsessilis *R.R. Mill in Edinburgh J. Bot.*, 57(3): 426 (2000)—Bhutan.
taihangshanensis *C.S. Zhu & H.W. Yang in Acta Phytotax. Sin.*, 35(1): 76 (1997)—China Southeast (Henan).

SCURRULA (Loranthac.).
buddleioides (Desr.) G. Don
var. heynei *(DC.) H.S. Kiu in Guihaia*, 17(4): 308 (1997): Loranthus heynei.
tsaii *(S.T. Chiu) Yuen P. Yang & S.Y. Lu in Taiwania*, 42(2): 87 (1997): Taxillus tsaii.

SCUTELLARIA (Labiat.).
austrotaiwanensis *T.H. Hsieh & T.C. Huang in Taiwania*, 42(2): 111 (1997)—Taiwan.
dentata H. Lév. & Vaniot
f. albida *Y.N. Lee, Fl. Korea*: 1160 (1996), without type—Locality not stated.
ebracteata *A. Pool in Novon*, 8(1): 62 (1998)—Nicaragua.
grandiflora Sims
subsp. tuvensis *(Juz.) R.V. Kamelin & I.A. Gubanov in I.A. Gubanov, Konsp. Fl. Vneshnei Mongolii (Sosud. Rast.)*: 89 (1996): S. tuvensis.
indica L.
f. albiflora *Y.N. Lee, Fl. Korea*: 1160 (1996), without type—Locality not stated.
lilungensis *S.S. Ying in Mem. Coll. Agric. Nation. Taiwan Univ.*, 31(1): 26 (1991), as 'lilungense'—Taiwan.
ochotensis *N.S. Probatova in Sosud. Rast. Sovet. Dal'nego Vostoka*, 7: 312 (1995)—Khabarovsk, Amur.
pekinensis Maxim.
var. maxima *S. Kim & S. Lee in Korean J. Pl. Taxon.*, 25(2): 98 (1995)—South Korea.
platyphylla *(Epling) B.L. Turner & P.G. Delprete in Pl. Syst. Evol.*, 199(1–2): 119 (1996): S. potosina subsp. platyphylla.
scordiifolia Fisch. ex Schrank
var. polyphylla *(Juz.) V.N. Voroschilov in A.K. Skvortsov (ed.), Florist. issl. v razn. raĭonakh SSSR*: 188 (1985): S. polyphylla.

SCUTELLARIA:-
shikokiana *T. Makino in Bot. Mag. Tokyo*, 18(206): 44 (1904)—Japan.†
strigillosa Hemsl.
f. albiflora *Y.N. Lee, Fl. Korea*: 1160 (1996), without type—Locality not stated.
taiwanensis *S.S. Ying in Mem. Coll. Agric. Nation. Taiwan Univ.*, 31(1): 24 (1991), homonym; *see C.Y. Wu in Fl. Reipubl. Popul. Sin.*, 65(2): 580 (1977)—Taiwan.
tenuipetiolata *A. Pool in Novon*, 8(1): 64 (1998)—Costa Rica.
ternejica *N.S. Probatova in Sosud. Rast. Sovet. Dal'nego Vostoka*, 7: 308 (1995)—Primorye, Khabarovsk.
tortumensis *(Kit Tan & Sorger) A.P. Khokhrjakov in Byull. Mosk. Obshch. Ispŷt. Prir., Biol.*, 102(4): 47 (1997): S. orientalis subsp. tortumensis.
uzunderensis *A.P. Khokhrjakov in Byull. Mosk. Obshch. Ispŷt. Prir., Biol.*, 102(4): 47 (1997)—Turkey.

SCUTIA *(Comm. ex DC.) A.T. Brongniart, Mém. Fam. Rhamnées*: 55 (1826): Ceanothus sect. Scutia. (Rhamnac.).†
commersonii *A.T. Brongniart, Mém. Fam. Rhamnées*: 56 (1826)—†
ferrea *(Vahl) A.T. Brongniart, Mém. Fam. Rhamnées*: 56 (1826): Rhamnus ferrea.†
indica *A.T. Brongniart, Mém. Fam. Rhamnées*: 56 (1826)—India.†

SEBAEA (Gentianac.).
brevicaulis *Sileshi Nemomissa in Kew Bull.*, 55(1): 213 (2000)—Tanzania.
dimidiata *Sileshi Nemomissa in Kew Bull.*, 55(1): 214 (2000)—Tanzania.
leariformis *Sileshi Nemomissa in Kew Bull.*, 54(1): 193 (1999)—Tanzania.
macroptera *Sileshi Nemomissa in Kew Bull.*, 54(1): 191 (1999)—Tanzania.
primuliflora *(Welw.) Sileshi Nemomissa in Kew Bull.*, 55(1): 217 (2000), as 'primulaeflora': Exochaenium primuliflorum.

SEBASTIANIA (Euphorbiac.).
acetosella *(Milne-Redh.) R.C. Kruijt in Biblioth. Bot.*, 146: 83 (1996): Sapium acetosella.
appendiculata *(Müll. Arg.) R.C. Kruijt in Biblioth. Bot.*, 146: 83 (1996): Stillingia appendiculata.
faradianensis *(Beille) R.C. Kruijt in Biblioth. Bot.*, 146: 88 (1996): Excoecaria faradianensis.

SECALE (Gramin.).
strictum (C. Presl) C. Presl
var. ciliatoglume *(Boiss.) S. Frederiksen & G. Petersen in Nordic J. Bot.*, 18(4): 409 (1998): S. montanum var. ciliatoglume.

SECAMONE (Asclepiadac.).
bemarahensis *J. Klackenberg in Candollea*, 55(1): 76 (2000)—Madagascar.
brevicoronata *J. Klackenberg in Willdenowia*, 30(1): 47 (2000)—Madagascar.
drepanoloba *J. Klackenberg in Bot. Jahrb.*, 120(1): 120 (1998)—Madagascar.
jongkindii *J. Klackenberg in Candollea*, 55(1): 80 (2000)—Madagascar.
linearis *J. Klackenberg in Novon*, 10(3): 217 (2000)—Madagascar.
marsupiata *J. Klackenberg in Candollea*, 52(2): 302 (1997)—Madagascar.
pedicellaris *J. Klackenberg in Willdenowia*, 30(1): 48 (2000)—Madagascar.
schatzii *J. Klackenberg in Novon*, 10(3): 215 (2000)—Madagascar.
sulfurea *(Jum. & H. Perrier) J. Klackenberg in Nordic J. Bot.*, 16(2): 183 (1996): Toxocarpus sulfureus.
subsp. stylosa *(Klack.) J. Klackenberg in Candollea*, 55(1): 78 (2000): S. stylosa.
tsingycola *J. Klackenberg in Candollea*, 55(1): 82 (2000)—Madagascar.

SECAMONOPSIS (Asclepiadac.).
microphylla *L. Civeyrel & J. Klackenberg in Novon*, 6(2): 144 (1996)—Madagascar.

SECHIUM (Cucurbitac.).
edule (Jacq.) Sw.
subsp. sylvestre *R. Lira & J. Castrejón in Acta Bot. Mex.*, 49: 57 (1999)—Mexico Northeast (Querétaro).
mexicanum *R. Lira & M. Nee in Brittonia*, 51(2): 204 (1999)—Mexico Gulf (Veracruz), Mexico Northeast (Hidalgo, Querétaro), Mexico Central (Puebla).

SECURIDACA (Polygalac.).
dolod *B. Wallnöfer in Ann. Naturhist. Mus. Wien*, 100B: 709 (1998)—Peru.
revoluta *(A.W. Benn.) M.C.M. Marques in Arq. Jard. Bot. Rio de Janeiro*, 34(1): 104 (1996): Bredemeyera revoluta.

SEDUM (Crassulac.).
ser. Elegans *H. 't Hart & K. Alpınar in Edinburgh J. Bot.*, 56(2): 193 (1999).
ser. Gracile *H. 't Hart & K. Alpınar in M.A. Öztürk, Ö. Seçmen & G. Görk (eds.), Pl. Life S.W. &* Central Asia, 1: 87 (1996).
ser. Inconspicua *H. 't Hart & K. Alpınar in M.A. Öztürk, Ö. Seçmen & G. Görk (eds.), Pl. Life S.W. &* Central Asia, 1: 87 (1996).
ser. Nana *H. 't Hart & K. Alpınar in M.A. Öztürk, Ö. Seçmen & G. Görk (eds.), Pl. Life S.W. &* Central Asia, 1: 88 (1996).
ser. Pubescens *T.H.M. Mes in H. 't Hart & U. Eggli (eds.), Evol. & Syst. Crassulac.*: 42 (1995).
ser. Telmissa *(Fenzl) H. 't Hart & K. Alpınar in M.A. Öztürk, Ö. Seçmen & G. Görk (eds.), Pl. Life S.W. &* Central Asia, 1: 90 (1996), without basionym page: gen. Telmissa.
ser. Tenella *H. 't Hart & K. Alpınar in M.A. Öztürk, Ö. Seçmen & G. Görk (eds.), Pl. Life S.W. &* Central Asia, 1: 90 (1996).
abchasicum *Kolak. ex V.V. Byalt in Novosti Sist. Vyssh. Rast.*, 31: 123 (1998)—North Caucasus (Krasnodar).
aizoon L.
subsp. baicalense *G.A. Peschkova in Fl. Sibir. (Berberidac.-Grossulariac.)*, 7: 161 (1994)—Buryatiya.
subsp. maximowiczii *(Regel) V.N. Voroschilov in A.K. Skvortsov (ed.), Florist. issl. v razn. raĭonakh SSSR*: 173 (1985): S. maximowiczii.
booleanum *B.L. Turner in Phytologia*, 79(1): 31 (1995 publ. 1996)—Mexico Northeast (Nuevo León).
catorce *G.L. Nesom in Phytologia*, 79(4): 264 (1995 publ. 1996), nom. nov.: S. parvum subsp. dendroides *R.T. Clausen*.
clausenii *V.V. Byalt in Novosti Sist. Vyssh. Rast.*, 31: 121 (Dec. 1998), nom. nov., homonym; *see Pérez-Calix in Acta Bot. Mex.*, 43: 1 (July 1998): S. decumbens *R.T. Clausen*.
clausenii *E. Pérez-Calix in Acta Bot. Mex.*, 43: 1 (1998)—Mexico Northeast (Guanajuato, San Luis Potosí).
copalense *M. Kimnach in Cact. Succ. J. (U.S.A.)*, 68(5): 242 (1996)—Mexico Northwest (Sinaloa).
dasyphyllum L.
f. donatianum *(Vis. & Sacc.) G. Tavormina in Piante Grasse*, 18(4): 131 (1998 publ. 1999), without basionym ref.: Basionym not stated.
var. elisae *G. Tavormina in Piante Grasse*, 18(4): 134 (1998 publ. 1999), without latin descr.—Italy.
f. glaucum *(Litard.) G. Tavormina in Piante Grasse*, 18(4): 131 (1998 publ. 1999), without basionym ref.: S. dasyphyllum var. glaucum.
subsp. granatense *(Pau) S. Castroviejo & M. Velayos in An. Jard. Bot. Madrid*, 53(2): 275 (1995 publ. 1996): S. granatense.
decipiens *(Baker) J. Thiede & H. 't Hart in Novon*, 9(1): 124 (1999): Cotyledon decipiens.
diminutum *(R.T. Clausen) G.L. Nesom in Phytologia*, 79(4): 261 (1995 publ. 1996): S. parvum subsp. diminutum.
dulcinomen *G.L. Nesom in Phytologia*, 79(4): 265 (1995 publ. 1996)—Mexico Northeast (Nuevo León).
edwardsii *(R.T. Clausen) B.L. Turner in Phytologia*, 78(6): 406 (1995): S. rhodocarpum subsp. edwardsii.
elburzense *Kh. Akhiani & M. Assadi in Iranian J. Bot.*, 8(2): 164 (2000)—Iran.

SEDUM:-
eriocarpum Sibth. & Sm.
subsp. apertiflorum *H. 't Hart in Ot Sist. Bot. Dergisi*, 2(2): 7 (1995), contrary to Art. 37.5 ICBN (1994)—Greece.
subsp. caricum *(Carlström) H. 't Hart in Ot Sist. Bot. Dergisi*, 2(2): 8 (1995): Sedum caricum.
subsp. delicum *(Vierh.) H. 't Hart in Ot Sist. Bot. Dergisi*, 2(2): 7 (1995): S. rubens subsp. delicum.
subsp. epiroticum *(Bald.) H. 't Hart in Ot Sist. Bot. Dergisi*, 2(2): 7 (1995): S. annuum subsp. epiroticum.
subsp. orientale *H. 't Hart in Ot Sist. Bot. Dergisi*, 2(2): 8 (1995), contrary to Art. 37.5 ICBN (1994)—Palestine (Israel).
subsp. porphyreum *(Kotschy) H. 't Hart in Ot Sist. Bot. Dergisi*, 2(2): 8 (1995): S. porphyreum.
subsp. spathulifolium *H. 't Hart in Ot Sist. Bot. Dergisi*, 2(2): 7 (1995), contrary to Art. 37.5 ICBN (1994)—Kriti.
goldmanii *(Rose) Reid Moran in Haseltonia*, 4: 46 (1996): Altamiranoa goldmanii.
hintoniorum *B.L. Turner in Phytologia*, 78(6): 405 (1995)—Mexico Northeast (Nuevo León, Tamaulipas).
ince *H. 't Hart & K. Alpınar in Edinburgh J. Bot.*, 56(2): 182 (1999)—Turkey.
jinianum *X.H. Guo in Acta Bot. Yunnanica*, 18(3): 297 (1996)—China Southeast (Anhui).
jurgensenii *(Hemsl.) Reid Moran in Haseltonia*, 4: 46 (1996): Cotyledon jurgensenii.
subsp. attenuatum *R.V. Moran in Haseltonia*, 5: 76 (1997 publ. 1998)—Mexico Central (Puebla), Mexico Gulf (Veracruz), Mexico Southwest (Oaxaca).
kimnachii *V.V. Byalt in Kew Bull.*, 54(2): 458 (1999), nom. nov.: S. decumbens *R.T. Clausen*.
latiovalifolium *Y.N. Lee in Korean J. Pl. Taxon.*, 22(1): 8 (1992)—South Korea.
litoreum Guss.
var. creticum *H. 't Hart in Ot Sist. Bot. Dergisi*, 2(2): 3 (1995), contrary to Art. 37.5 ICBN (1994)—Kriti.
subsp. praesidis *(Runemark & Greuter) W. Greuter in Bocconea*, 3: 187 (1992): S. praesidis.
macdonaldii *G.L. Nesom in Phytologia*, 79(4): 266 (1995 publ. 1996)—Mexico Northeast (Nuevo León, Coahuila).
meyranianum *J. Metzger in Cact. Succ. J. (U.S.A.)*, 71(6): 316 (1999)—Mexico Southwest (Jalisco).
microcarpum *(Sibth. & Sm.) S. Schönland in Engl. & Prantl, Nat. Pflanzenfam.*, 3(2a): 31 (1890): Crassula microcarpa.
middendorfianum Maxim.
subsp. sichotense *(Vorosch.) V.N. Voroschilov in A.K. Skvortsov (ed.), Florist. issl. v razn. raĭonakh SSSR*: 174 (1985): S. sichotense.
mocinianum *E. Pérez-Calix in Acta Bot. Mex.*, 45: 50 (1998)—Mexico Northeast (Guanajuato).
mucizonia (Ortega) Raym.-Hamet
subsp. abylaeum *(Font Quer & Maire) L.S. Springate in Lagascalia*, 18(2): 254, 337 (1996), see also footnote p. 221: Cotyledon mucizonia subsp. abyleae.
papillicaulum *G.L. Nesom in Phytologia*, 79(4): 265 (1995 publ. 1996)—Mexico Northeast (Nuevo León).
plicatum *J. Thiede & H. 't Hart in Novon*, 9(1): 124 (1999), nom. nov.: Cotyledon stricta *Diels*.
pseudomontanum *J. Holub in Preslia*, 70(2): 114 (1998), nom. nov.: S. rupestre subsp. erectum *'t Hart*.
reniforme *(H. Jacobsen) J. Thiede & H. 't Hart in Novon*, 9(1): 125 (1999): Villadia reniformis.
scopulinum *(Rose) Reid Moran in Haseltonia*, 4: 46 (1996): Altamiranoa scopulina.
sikokianum Maxim.
subsp. kurilense *(Vorosch.) V.N. Voroschilov in A.K. Skvortsov (ed.), Florist. issl. v razn. raĭonakh SSSR*: 174 (1985): S. kurilense.
trullipetalum Hook. f. & Thomson
var. gagei *(Raym.-Hamet) H. Ohba in H. Ohba & H. Ikeda (eds.), Fl. Hinku & Hunku Valleys, E Nepal (Univ. Mus.* Tokyo Nat. & Cult., 6): 75 (2000): S. gagei.

SEDUM:-
urvillei DC.
subsp. sartorianum *(Boiss.) V.V. Byalt in Bot. Zhurn.*, 82(1): 108 (1997): S. sartorianum.
wannanense *X.H. Guo, X.P. Zhang & X.H. Chen in Acta Bot. Yunnanica*, 21(1): 25 (1999)—China Southeast (Anhui).
var. incarnatum *X.H. Guo, X.P. Zhang & X.H. Chen in Acta Bot. Yunnanica*, 21(1): 26 (1999)—China Southeast (Anhui).
weberbaueri *(Diels) J. Thiede & H. 't Hart in Novon*, 9(1): 125 (1999): Cotyledon weberbaueri.
yildizianum *H. Sümbül in Turkish J. Bot.*, 14(1): 55 (1990)—Turkey.

SEEGERIELLA *K. Senghas in J. Orchideenfreund*, 4(4): 190 (1997). *ORCHIDACEAE*
pinifolia *K. Senghas in J. Orchideenfreund*, 4(4): 190 (1997)—Bolivia.

SEIDENFADENIELLA *C. Sathish Kumar in C. Sathish Kumar & K.S. Manilal, Cat. Indian Orchids*: 43 (1994). *ORCHIDACEAE*
chrysantha *(Alston) C. Sathish Kumar in C. Sathish Kumar & K.S. Manilal, Cat. Indian Orchids*: 47 (1994): Saccolabium chrysanthum.
filiformis *(Rchb. f.) E.A. Christenson & P. Ormerod in K.M. Matthew, Fl. Palni Hills, S. India*, 3: 1258 (1999): Saccolabium filiforme.
rosea *(Wight) C. Sathish Kumar in C. Sathish Kumar & K.S. Manilal, Cat. Indian Orchids*: 46 (1994): Sarcanthus roseus.

SEIDENFIA *D.L. Szlachetko in Fragm. Flor. Geobot.*, Supp. 3: 122 (1995). *ORCHIDACEAE*
crenulata *(Ridl.) D.L. Szlachetko in Fragm. Flor. Geobot.*, Supp. 3: 122 (1995): Microstylis crenulata.
densiflora *(A. Rich.) D.L. Szlachetko in Fragm. Flor. Geobot.*, Supp. 3: 122 (1995): Liparis densiflora.
discolor *(Lindl.) D.L. Szlachetko in Fragm. Flor. Geobot.*, Supp. 3: 122 (1995): Microstylis discolor.
intermedia *(A. Rich.) D.L. Szlachetko in Fragm. Flor. Geobot.*, Supp. 3: 122 (1995): Liparis intermedia.
lancifolia *(Thwaites) D.L. Szlachetko in Fragm. Flor. Geobot.*, Supp. 3: 122 (1995): Microstylis lancifolia.
rheedii *(Sw.) D.L. Szlachetko in Fragm. Flor. Geobot.*, Supp. 3: 122 (1995): Malaxis rheedii.
seychellarum *(Kraenzl.) D.L. Szlachetko in Fragm. Flor. Geobot.*, Supp. 3: 122 (1995): Liparis seychellarum.
stocksii *(Hook. f.) D.L. Szlachetko in Fragm. Flor. Geobot.*, Supp. 3: 122 (1995): Microstylis stocksii.

SELAGO (Scrophulariac.).
acocksii *O.M. Hilliard in Edinburgh J. Bot.*, 55(2): 221 (1998)—Cape Province.
adenodes *O.M. Hilliard, Tribe Selagineae (Scrophulariac.)*: 171 (1999)—Cape Province.
albomarginata *O.M. Hilliard in Edinburgh J. Bot.*, 54(1): 106 (1997)—Namibia.
albomontana *O.M. Hilliard, Tribe Selagineae (Scrophulariac.)*: 83 (1999)—Cape Province.
anatrichota *O.M. Hilliard in Edinburgh J. Bot.*, 55(2): 221 (1998)—Zimbabwe.
angustibractea *O.M. Hilliard in Edinburgh J. Bot.*, 55(2): 221 (1998)—Namibia.
barabei *(Mielcarek) O.M. Hilliard in Edinburgh J. Bot.*, 54(1): 107 (1997): Walafrida barabei.
beaniana *O.M. Hilliard in Edinburgh J. Bot.*, 54(1): 107 (1997)— Cape Province.
bilacunosa *O.M. Hilliard, Tribe Selagineae (Scrophulariac.)*: 87 (1999)—Cape Province.
cedrimontana *O.M. Hilliard in Edinburgh J. Bot.*, 55(2): 222 (1998)—Cape Province.
centralis *O.M. Hilliard in Edinburgh J. Bot.*, 55(2): 222 (1998)—Cape Province.
chalarantha *O.M. Hilliard, Tribe Selagineae (Scrophulariac.)*: 145 (1999)—Cape Province.
comptonii *O.M. Hilliard in Edinburgh J. Bot.*, 55(2): 222 (1998)—Cape Province.
confusa *O.M. Hilliard, Tribe Selagineae (Scrophulariac.)*: 189 (1999)—Cape Province.
crassifolia *(Rolfe) O.M. Hilliard, Tribe Selagineae (Scrophulariac.)*: 147 (1999): Walafrida crassifolia.

SELAGO:-
cryptadenia *O.M. Hilliard, Tribe Selagineae (Scrophulariac.)*: 89 (1999)—Cape Province.
cucullata *O.M. Hilliard, Tribe Selagineae (Scrophulariac.)*: 277 (1999)—Natal, Transvaal.
cupressoides *O.M. Hilliard, Tribe Selagineae (Scrophulariac.)*: 90 (1999)—Cape Province.
diabolica *O.M. Hilliard in Edinburgh J. Bot.*, 55(2): 222 (1998)—Cape Province.
dinteri Rolfe
subsp. pseudodinteri *O.M. Hilliard, Tribe Selagineae (Scrophulariac.)*: 266 (1999)—Namibia, Botwsana, Cape Province.
dolichonema *O.M. Hilliard in Edinburgh J. Bot.*, 53(2): 277 (1996)—Cape Province.
dolosa *O.M. Hilliard in Edinburgh J. Bot.*, 53(2): 277 (1996)—Cape Province.
dregeana *O.M. Hilliard, Tribe Selagineae (Scrophulariac.)*: 106 (1999)—Cape Province.
elsiae *O.M. Hilliard, Tribe Selagineae (Scrophulariac.)*: 154 (1999)—Cape Province.
esterhuyseniae *O.M. Hilliard in Edinburgh J. Bot.*, 55(2): 223 (1998)—Cape Province.
exigua *O.M. Hilliard in Edinburgh J. Bot.*, 55(2): 223 (1998)—Cape Province.
farrago *O.M. Hilliard in Edinburgh J. Bot.*, 55(2): 223 (1998)—Cape Province.
florifera *O.M. Hilliard, Tribe Selagineae (Scrophulariac.)*: 215 (1999)—Cape Province.
fourcadei *O.M. Hilliard in Edinburgh J. Bot.*, 55(2): 223 (1998)—Cape Province.
gloiodes *O.M. Hilliard, Tribe Selagineae (Scrophulariac.)*: 99 (1999)—Cape Province.
glutinosa E. Mey.
subsp. cylindriphylla *O.M. Hilliard, Tribe Selagineae (Scrophulariac.)*: 104 (1999)—Cape Province.
goetzei Rolfe
subsp. ambigua *O.M. Hilliard in Edinburgh J. Bot.*, 55(2): 224 (1998)—Zimbabwe.
gossweileri *O.M. Hilliard, Tribe Selagineae (Scrophulariac.)*: 260 (1999)—Angola.
gracilis *(Rolfe) O.M. Hilliard, Tribe Selagineae (Scrophulariac.)*: 190 (1999): Walafrida gracilis.
grandiceps *O.M. Hilliard, Tribe Selagineae (Scrophulariac.)*: 91 (1999)—Cape Province.
griquana *O.M. Hilliard in Edinburgh J. Bot.*, 53(2): 278 (1996)—Natal.
heterotricha *O.M. Hilliard, Tribe Selagineae (Scrophulariac.)*: 78 (1999)—Cape Province.
huilana *O.M. Hilliard, Tribe Selagineae (Scrophulariac.)*: 278 (1999)—Angola.
impedita *O.M. Hilliard in Edinburgh J. Bot.*, 55(2): 224 (1998)—Cape Province.
inaequifolia *O.M. Hilliard in Edinburgh J. Bot.*, 55(2): 224 (1998)—Cape Province.
inconstans *O.M. Hilliard in Edinburgh J. Bot.*, 53(2): 278 (1996)—Cape Province.
intermedia *O.M. Hilliard in Edinburgh J. Bot.*, 53(2): 278 (1996)—Cape Province.
karooica *O.M. Hilliard in Edinburgh J. Bot.*, 53(2): 278 (1996)—Cape Province.
kurtdinteri *O.M. Hilliard in Edinburgh J. Bot.*, 55(2): 224 (1998)—Namibia.
lepida *O.M. Hilliard in Edinburgh J. Bot.*, 54(1): 107 (1997)—Namibia.
leptothrix *O.M. Hilliard in Edinburgh J. Bot.*, 53(2): 278 (1996)—Cape Province.
levynsiae *O.M. Hilliard in Edinburgh J. Bot.*, 55(2): 224 (1998)—Cape Province.
lilacina *O.M. Hilliard, Tribe Selagineae (Scrophulariac.)*: 168 (1999)—Cape Province.
magnakarooica *O.M. Hilliard in Edinburgh J. Bot.*, 55(2): 225 (1998)—Cape Province.
marlothii *O.M. Hilliard, Tribe Selagineae (Scrophulariac.)*: 94 (1999)—Cape Province.
mediocris *O.M. Hilliard in Edinburgh J. Bot.*, 53(2): 278 (1996)—Cape Province.
michelliae *O.M. Hilliard, Tribe Selagineae (Scrophulariac.)*: 209 (1999)—Cape Province.
micradenia *O.M. Hilliard in Edinburgh J. Bot.*, 55(2): 225 (1998)—Cape Province.
mixta *O.M. Hilliard, Tribe Selagineae (Scrophulariac.)*: 272 (1999)—Cape Province, Transvaal, Orange Free State.
mucronata *O.M. Hilliard, Tribe Selagineae (Scrophulariac.)*: 80 (1999)—Cape Province.
multiflora *O.M. Hilliard in Edinburgh J Bot.*, 55(2): 225 (1998)—Cape Province.

SELAGO:-
multispicata *O.M. Hilliard, Tribe Selagineae (Scrophulariac.)*: 275 (1999)—Cape Province, Lesotho.
myriophylla *O.M. Hilliard in Edinburgh J. Bot.*, 53(2): 279 (1996)—Cape Province.
neglecta *O.M. Hilliard, Tribe Selagineae (Scrophulariac.)*: 82 (1999)—Cape Province.
nigromontana *O.M. Hilliard in Edinburgh J. Bot.*, 53(2): 279 (1996)—Cape Province.
oppositifolia *O.M. Hilliard, Tribe Selagineae (Scrophulariac.)*: 125 (1999)—Cape Province.
parvibractea *O.M. Hilliard in Edinburgh J. Bot.*, 53(2): 279 (1996)—Cape Province.
perplexa *O.M. Hilliard, Tribe Selagineae (Scrophulariac.)*: 95 (1999)—Cape Province.
persimilis *O.M. Hilliard in Edinburgh J. Bot.*, 55(2): 225 (1998)—Cape Province.
praetermissa *O.M. Hilliard in Edinburgh J. Bot.*, 55(2): 226 (1998)—Cape Province.
prostrata *O.M. Hilliard in Edinburgh J. Bot.*, 53(2): 279 (1996)—Cape Province.
psammophila *O.M. Hilliard in Edinburgh J. Bot.*, 55(2): 226 (1998)—Cape Province.
pulchra *O.M. Hilliard, Tribe Selagineae (Scrophulariac.)*: 169 (1999)—Cape Province.
pustulosa *O.M. Hilliard, Tribe Selagineae (Scrophulariac.)*: 109 (1999)—Cape Province.
retropilosa *O.M. Hilliard, Tribe Selagineae (Scrophulariac.)*: 185 (1999)—Cape Province.
rubromontana *O.M. Hilliard in Edinburgh J. Bot.*, 55(2): 226 (1998)—Cape Province.
scabribractea *O.M. Hilliard, Tribe Selagineae (Scrophulariac.)*: 107 (1999)—Cape Province.
serpentina *O.M. Hilliard in Edinburgh J. Bot.*, 54(1): 107 (1997)—Zimbabwe.
seticaulis *O.M. Hilliard in Edinburgh J. Bot.*, 55(2): 226 (1998)—Cape Province.
singularis *O.M. Hilliard in Edinburgh J. Bot.*, 54(1): 107 (1997)—Cape Province.
spectabilis *O.M. Hilliard in Edinburgh J. Bot.*, 54(1): 108 (1997)—Cape Province.
stenostachya *O.M. Hilliard in Edinburgh J. Bot.*, 55(2): 226 (1998)—Cape Province.
subspinosa *O.M. Hilliard in Edinburgh J. Bot.*, 54(1): 108 (1997)—Cape Province.
swynnertonii *(S. Moore) O.M. Hilliard in Edinburgh J. Bot.*, 55(2): 226 (1998), isonym; *see* Eyles *in Trans. Roy. Soc. South Africa*, 5(4): 473 (1916): Walafrida swynnertonii.
swynnertonii (S. Moore) Eyles
var. leiophylla *(Brenan) O.M. Hilliard in Edinburgh J. Bot.*, 55(2): 227 (1998): Walafrida swynnertonii var. leiophylla.
tenuifolia *(Rolfe) O.M. Hilliard, Tribe Selagineae (Scrophulariac.)*: 255 (1999): Walafrida tenuifolia.
thermalis *O.M. Hilliard in Edinburgh J. Bot.*, 55(2): 227 (1998)—Cape Province.
trichophylla *O.M. Hilliard, Tribe Selagineae (Scrophulariac.)*: 102 (1999)—Cape Province.
valliscitri *O.M. Hilliard, Tribe Selagineae (Scrophulariac.)*: 76 (1999)—Cape Province.
variicalyx *O.M. Hilliard in Edinburgh J. Bot.*, 53(2): 279 (1996)—Cape Province.
venosa *O.M. Hilliard in Edinburgh J. Bot.*, 53(2): 280 (1996)—Cape Province.
verna *O.M. Hilliard in Edinburgh J. Bot.*, 55(2): 227 (1998)—Cape Province.
welwitschii Rolfe
var. australis *O.M. Hilliard in Edinburgh J. Bot.*, 54(1): 108 (1997)—Cape Province.

SELLIERA (Goodeniac.).
rotundifolia *P.B. Heenan in New Zealand J. Bot.*, 35(2): 133 (1997)—New Zealand North I.

SELLOA (Compos.).
ligulata *(Cuatrec.) H. Robinson in Brittonia*, 49(1): 77 (1997): Aphanactis ligulata.
macdonaldii *(B.L. Turner) H. Robinson in Brittonia*, 49(1): 77 (1997): Aphanactis macdonaldii.

SELLULOCALAMUS *W.T. Lin in J. S. China Agric. Univ.*, 10(2): 43 (1989). *GRAMINEAE*
bambusoides *(C.J. Hsueh & D.Z. Li) W.T. Lin in J. S. China Agric. Univ.*, 10(2): 44 (1989): Dendrocalamus bambusoides.

SELLULOCALAMUS:-
tibeticus *(C.J. Hsueh & T.P. Yi) W.T. Lin in J. S. China Agric. Univ.*, 10(2): 45 (1989): Dendrocalamus tibeticus.

SEMECARPUS (Anacardiac.).
angulatus *K.M. Kochummen in Sandakania*, 7: 94 (1996)—Borneo (Sabah).
angustifolius *K.M. Kochummen in Sandakania*, 7: 94 (1996)—Borneo (Sabah).
calcicola *K.M. Kochummen in Sandakania*, 7: 97 (1996), as 'calcicolus'—Borneo (Sabah).
cupularis *K.M. Kochummen in Sandakania*, 7: 97 (1996)—Borneo (Sarawak).
euodiifolius *K.M. Kochummen in Sandakania*, 7: 100 (1996)—Borneo (Sabah).
impressicostatus *K.M. Kochummen in Sandakania*, 7: 102 (1996)—Borneo (Sabah).
kathalekanensis *Dasappa & M.H. Swaminath in Indian Forester*, 126(1): 78 (2000), contrary to Art. 37.6 ICBN (2000)—India (Karnataka).
kinabaluensis *K.M. Kochummen in Sandakania*, 7: 102 (1996)—Borneo (Sabah).
minutipetalus *K.M. Kochummen in Sandakania*, 7: 105 (1996)—Borneo (Sarawak).
poyaensis *M. Hoff in Bull. Mus. Nation. Hist. Nat., B, Adansonia*, Ser. 4, 17(3–4): 164 (1995)—New Caledonia.
pulvinatus *K.M. Kochummen in Sandakania*, 7: 105 (1996)—Borneo (Sabah).
sandakanus *K.M. Kochummen in Sandakania*, 7: 108 (1996)—Borneo (Sabah).
virotii *M. Hoff in Bull. Mus. Nation. Hist. Nat., B, Adansonia*, Ser. 4, 17(3–4): 165 (1995)—New Caledonia.

SEMIARUNDINARIA (Gramin.).
bicorniculata *(S.F. Li & Z.P. Wang) R. Govaerts, World Checklist Seed Plants*, 3(1): 21 (1999): Chimonocalamus bicorniculatus.

SEMIRIA *D.J.N. Hind in Kew Bull.*, 54(2): 425 (1999). *COMPOSITAE*
viscosa *D.J.N. Hind in Kew Bull.*, 54(2): 426 (1999)—Brazil Northeast (Bahia).

SEMPERVIVUM (Crassulac.).
ciliosum
subsp. octopodes *(Turrill) B.J.M. Zonneveld in Succulenta*, 78(2): 92 (1999): S. octopodes.
globiferum L.
subsp. allionii *(Jord. & Fourr.) H. 't Hart & B. Bleij in Succulenta*, 78(1): 40 (1999): Diopogon allionii.
subsp. arenarium *(W.D.J. Koch) H. 't Hart & B. Bleij in Succulenta*, 78(1): 40 (1999): S. arenarium.
subsp. hirtum *(L.) H. 't Hart & B. Bleij in Succulenta*, 78(1): 40 (1999): S. hirtum.
marmoreum Griseb.
subsp. ballsii *(Wale) B.J.M. Zonneveld in Succulenta*, 78(2): 92 (1999): S. ballsii.
subsp. erythreum *(Velen.) B.J.M. Zonneveld in Succulenta*, 78(2): 92 (1999): S. erythreum.
subsp. reginae-amaliae *(Heldr. & Sartori ex Boiss.) B.J.M. Zonneveld in Succulenta*, 78(2): 92 (1999): S. reginae-amaliae.
tectorum
var. arvernense *(Lecoq & Lamotte) B.J.M. Zonneveld in Succulenta*, 78(2): 92 (1999): S. arvernense.
transcaucasicum Muirhead
subsp. aghricum *(Kit Tan & Sorger) V.V. Byalt in Novosti Sist. Vȳssh. Rast.*, 31: 121 (1998): S. globiferum subsp. aghricum.

SENECIO (Compos.).
subser. Amplectentes *(Cabrera) Á.L. Cabrera & S.E. Freire in Fl. Fan. Argentina*, 62: 22 (1999): S. subsect. Amplectentes.
ser. Andina *(Cabrera) Á.L. Cabrera & S.E. Freire in Fl. Fan. Argentina*, 62: 18 (1999): S. sect. Andina.
sect. Bahiana *D.J.N. Hind in Kew Bull.*, 54(4): 901 (1999 publ. 2000).
subser. Brasilienses *(Cabrera) Á.L. Cabrera & S.E. Freire in Fl. Fan. Argentina*, 62: 22 (1999): S. subsect. Brasilienses.

SENECIO:-
subser. Caespitosi *(O. Hoffm.) Á.L. Cabrera & S.E. Freire in Fl. Fan. Argentina,* 62: 38 (1999): Basionym not stated.
subser. Candidi *(Cabrera) Á.L. Cabrera & S.E. Freire in Fl. Fan. Argentina,* 62: 43 (1999): S. subsect. Candidi.
ser. Columbaria *(Cabrera) Á.L. Cabrera & S.E. Freire in Fl. Fan. Argentina,* 62: 20 (1999): S. sect. Columbaria.
subser. Cuneati *(Cabrera) Á.L. Cabrera & S.E. Freire in Fl. Fan. Argentina,* 62: 26 (1999): S. subsect. Cuneati.
subser. Filaginoidei *(Cabrera) Á.L. Cabrera & S.E. Freire in Fl. Fan. Argentina,* 62: 52 (1999): S. subsect. Filaginoidei.
subser. Glabrati *(Cabrera) Á.L. Cabrera & S.E. Freire in Fl. Fan. Argentina,* 62: 27 (1999): S. subsect. Sandwithiani.
ser. Haplostichia *(Phil.) Á.L. Cabrera & S.E. Freire in Fl. Fan. Argentina,* 62: 32 (1999): gen. Haplostichia.
ser. Leucanthemifolia *(Cabrera) Á.L. Cabrera & S.E. Freire in Fl. Fan. Argentina,* 62: 34 (1999): S. sect. Leucanthemifolium.
ser. Lomincola *H. Beltrán Santiago & A. Galán de Mera in Bot. Complutensis,* 21: 107 (1996).
subser. Megacephali *(Cabrera) Á.L. Cabrera & S.E. Freire in Fl. Fan. Argentina,* 62: 46 (1999): S. subsect. Megacephali.
ser. Metazanthus *(Meyen) Á.L. Cabrera & S.E. Freire in Fl. Fan. Argentina,* 62: 35 (1999): gen. Metazanthus.
subser. Microcephali *(Cabrera) Á.L. Cabrera & S.E. Freire in Fl. Fan. Argentina,* 62: 55 (1999): S. subsect. Microcephali.
subser. Radiati *(Cabrera) Á.L. Cabrera & S.E. Freire in Fl. Fan. Argentina,* 62: 47 (1999): S. subsect. Radiati.
subser. Sandwithiani *(Cabrera) Á.L. Cabrera & S.E. Freire in Fl. Fan. Argentina,* 62: 27 (1999): S. subsect. Sandwithiani.
subser. Simplices *(Cabrera) Á.L. Cabrera & S.E. Freire in Fl. Fan. Argentina,* 62: 28 (1999): S. subsect. Simplices.
subser. Subincani *(Cabrera) Á.L. Cabrera & S.E. Freire in Fl. Fan. Argentina,* 62: 49 (1999): S. sect. Subincani.
subser. Viscosi *(Baker) Á.L. Cabrera & S.E. Freire in Fl. Fan. Argentina,* 62: 30 (1999): S. [unranked] Viscosi.
ser. Websteria *(Cabrera) Á.L. Cabrera & S.E. Freire in Fl. Fan. Argentina,* 62: 51 (1999): S. sect. Websteria.
ser. Xerosenecio *(Cabrera) Á.L. Cabrera & S.E. Freire in Fl. Fan. Argentina,* 62: 52 (1999): S. sect. Xerosenecio.
aetfatensis *B. Nordenstam in Novon,* 9(2): 245 (1999)— Zimbabwe.
aquaticus Hill
subsp. calvescens *(Moris & De Not.) M.G. Mariotti in Allionia,* 30: 97 (1990–91 publ. 1991): S. calvescens.
botijae *C. Ehrhart in Sendtnera,* 5: 30 (1998)—Chile North (Antofagasta).
candollii *H.A. Weddell, Chlor. Andina,* 1: 106 (1856)—Peru.
carbonensis *C. Ezcurra, M. Ferreyra & S. Clayton in Hickenia,* 2(34): 150 (1995)—Argentina South (Rio Negro).
chienii *Hand.-Mazz. in Oesterr. Bot. Z.,* 88(4): 311 (1939)—China South-Central (Sichuan).
cymbalarioides *H.W. Buek, Gen. Spec. Syn. Candolleana,* 2: vi (1840), nom. nov.: S. subnudus *DC.*†
cymbalarioides *T. Nuttall in Trans. Amer. Philos. Soc.,* n.s., 7: 412 (1841), homonym; *see H. Buek, Gen. Spec. Syn. Candolleana,* 2: vi (1840)—Oregon.†
dacicus *I. Hodálová & K. Marhold in Bot. J. Linn. Soc.,* 128(3): 288 (1998)—Romania.
× dominii *I. Hodálová in Folia Geobot.,* 34(3): 334 (1999)—Romania. (S. germanicus × ucranicus).
doria L.
subsp. laderoi *(Pérez Morales, M.E. García & Penas) G. Blanca in Lagascalia,* 18(2): 308 (1996): S. laderoi.

SENECIO:-
dubitabilis C. Jeffrey & Y.L. Chen
var. densicapitatus *C.H. An in Fl. Xinjiangensis,* 5: 477 (1999), as 'deniscapitata'—Xinjiang.
var. linearifolius *C.H. An in Fl. Xinjiangensis,* 5: 477 (1999)—Xinjiang.
eriopus Willk.
subsp. hosmariensis *(Ball) G. Blanca in Lagascalia,* 18(2): 309 (1996): S. doronicum var. hosmariensis.
erucifolius L.
subsp. argunensis *(Turcz.) E.I. Wiebe in Fl. Sibir. (Asterac.),* 13: 166 (1997 publ. 1998): S. argunensis.
× futakii *I. Hodálová in Flora (Germany),* 191(3): 289 (1996)—Czechoslovakia. (S. germanicus × ovatus).
harleyi *D.J.N. Hind in Kew Bull.,* 54(4): 897 (1999 publ. 2000)—Brazil Northeast (Bahia).
hebdingii *(Rauh & Buchloh) W. Rauh, Succ. & Xeroph. Pl. Madagascar,* 2: 213 (1998), isonym; *see G.D. Rowley, Succulent Compositae:* 105 (1994): Notonia hebdingii.
icaensis *H. Beltrán Santiago & A. Galán de Mera in An. Jard. Bot. Madrid,* 55(1): 168 (1997)—Peru.
kunlunshanicus *C.H. An in Fl. Xinjiangensis,* 5: 477 (1999)—Xinjiang.
larahuinensis *H. Beltrán Santiago & A. Galán de Mera in An. Jard. Bot. Madrid,* 56(1): 168 (1998)—Peru.
linaresensis *A. Soldano in Compositae Newsl.,* 32: 18 (1998), nom. nov.: S. subdentatus *Phil.*
malaissei *S. Lisowski in Fragm. Flor. Geobot.,* 41(1): 484 (1996)—Zaire.
modestus Wedd.
var. candollii *(Wedd.) H. Beltrán Santiago & A. Galán de Mera in Bot. Complutensis,* 21: 104 (1996): S. candollii.
mweroensis Baker
f. schwartzii *(L.E. Newton) G.D. Rowley in Bradleya,* 14: 83 (1996): Kleinia schwartzii.
neoviscidulus *A. Soldano in Compositae Newsl.,* 32: 18 (1998), nom. nov.: S. viscidulus *Compton.*
nublensis *A. Soldano in Compositae Newsl.,* 32: 18 (1998), nom. nov.: S. carnosus *Phil.*
ozolotepecanus *B.L. Turner in Phytologia,* 79(2): 77 (1995 publ. 1996)—Mexico Southwest (Oaxaca).
pectinatus DC.
var. major *F. Muell. ex R.O. Belcher in Muelleria,* 9: 120 (1996)—Victoria, New South Wales, Australian Capital Territory.
pemehuensis *A. Soldano in Compositae Newsl.,* 32: 19 (1998), nom. nov.: S. scoparius *Phil.*
perralderianus Coss. & Durieu
subsp. hosmariensis *(Pau & Font Quer) G. Blanca in Lagascalia,* 18(2): 309 (1996): S. perralderianus var. hosmariensis.
pinnatifolius A. Rich.
var. pleiocephalus *R.O. Belcher in Muelleria,* 9: 127 (1996), as 'comb. nov.'—Victoria, Tasmania, New South Wales, Australian Capital Territory.
promatensis *N.I. Matzenbacher in Comun. Mus. Ciênc. PUCRS, Sér. Bot.,* 2(1): 4 (1996)—Brazil South (Rio Grande do Sul).
pulcher Hook. & Arn.
f. albiflorus *N.I. Matzenbacher in Comun. Mus. Ciênc. PUCRS, Sér. Bot.,* 2(1): 8 (1996)—Brazil South (Rio Grande do Sul).
quaylei *T.M. Barkley in Sida,* 19(2): 286 (2000)—Texas.
repangae *P.J. de Lange & B.G. Murray in New Zealand J. Bot.,* 36(4): 510 (1998)—New Zealand North I.
subsp. pokohinuensis *P.J. de Lange & B.G. Murray in New Zealand J. Bot.,* 36(4): 512 (1998)—New Zealand North I.
riograndensis *N.I. Matzenbacher in Comun. Mus. Ciênc. PUCRS, Sér. Bot.,* 2(1): 6 (1996)—Brazil South (Rio Grande do Sul).
ritoveganus *B.L. Turner in Phytologia,* 80(2): 95 (1996), as 'ritovegana'—Mexico Northwest (Sinaloa), Mexico Northeast (Chihuahua).
sandersii *B.L. Turner in Phytologia,* 80(2): 96 (1996)—Mexico Northwest (Sonora), Mexico Northeast (Chihuahua).
schultzii Hochst. ex A. Rich.
var. lanatus *D.F. Otieno & Mesfin Tadesse in Compositae Newsl.,* 33: 30 (1999)—Ethiopia.

SENECIO:-
taitungensis *S.S. Ying in Mem. Coll. Agric. Nation. Taiwan Univ.,* 30(2): 62 (1990)—Taiwan.
tarokoensis *C.I Peng in Bot. Bull. Acad. Sin. (Taipei),* 40(1): 57 (1999)—Taiwan.
tuberosus Sch. Bip. ex A. Rich.
var. pubescens *(Mesfin) G.D. Rowley in Bradleya,* 14: 83 (1996): Solanecio tuberosus var. pubescens.
ucranicus *I. Hodálová in Folia Geobot.,* 34(3): 334 (1999)—Ukraina.
viejoanus *B.L. Turner in Phytologia,* 79(2): 78 (1995 publ. 1996)—Mexico Northeast (Nuevo León, Tamaulipas).

× **SENGHASARA** *hort. in Orchid Rev.,* 108(1233): new orchid hybrids 14 (2000). *ORCHIDACEAE* (Ada × Gomesa × Odontoglossum).

SENNA (Legumin.).
artemisioides *(Gaudich. ex DC.) D. Isely, Native & Natural. Legumin. (Fabac.) U.S. (excl. Alaska & Hawaii):* 44 (1998), isonym; *see Randell in J. Adelaide Bot. Gard.,* 12(2): 220 (1989): Cassia artemisioides.
cearensis *Afr. Fernandes in Albertoa, n.s.,* 7: 6 (2000), nom. nov.: S. barnebyana *Afr. Fern.*
charlesiana *(Symon) B.R. Randell in Fl. Australia,* 12: 194 (1998): Cassia charlesiana.
circinnata *(Benth.) B.R. Randell in Fl. Australia,* 12: 194 (1998): Cassia circinnata.
cuthbertsonii *(F. Muell.) B.R. Randell in Fl. Australia,* 12: 195 (1998): Cassia cuthbertsonii.
ferraria *(Symon) B.R. Randell in Fl. Australia,* 12: 194 (1998): Cassia ferraria.
flexuosa *(Randell) B.R. Randell in Fl. Australia,* 12: 195 (1998): S. cardiosperma subsp. flexuosa.
glaucifolia *(Randell) B.R. Randell in Fl. Australia,* 12: 195 (1998): S. artemisioides subsp. glaucifolia.
hamersleyensis *(Symon) B.R. Randell in Fl. Australia,* 12: 195 (1998): Cassia hamersleyensis.
manicula *(Symon) B.R. Randell in Fl. Australia,* 12: 195 (1998): Cassia manicula.
pendula (Willd.) H.S. Irwin & Barneby
var. indecora *(Kunth) M. Luckow in Baileya,* 23(4): 229 (1996): Cassia indecora.
phyllodinea *(R. Br.) D.E. Symon in J. Adelaide Bot. Gard.,* 18(2): 102 (1998): Cassia phyllodinea.
pilocarina *(Symon) B.R. Randell in Fl. Australia,* 12: 195 (1998): Cassia pilocarina.
sericea *(Symon) D. Albrecht & D.E. Symon in J. Adelaide Bot. Gard.,* 19: 96 (2000): Cassia oligophylla var. sericea.
stowardii *(S. Moore) B.R. Randell in Fl. Australia,* 12: 195 (1998): Cassia stowardii.
stricta *(Randell) B.R. Randell in Fl. Australia,* 12: 194 (1998): S. artemisioides subsp. stricta.
symonii *(Randell) B.R. Randell in Fl. Australia,* 12: 194 (1998): S. artemisioides subsp. symonii.

SENYUMIA *R. Kiew, A. Weber & B.L. Burtt in Beitr. Biol. Pfl.,* 70(2–3): 400 (1997–98 publ. 1998). *GESNERIACEAE*
minutiflora *(Ridl.) R. Kiew, A. Weber & B.L. Burtt in Beitr. Biol. Pfl.,* 70(2–3): 400 (1997–98 publ. 1998): Boea minutiflora.

SERAPIAS (Orchidac.).
× ambigua
nothosubsp. panormosana *B. Baumann & H. Baumann in J. Eur. Orchideen,* 31(2): 517 (1999)—Kriti. (S. cordigera subsp. cretica × lingua).
carica (H. Baumann & Künkele) P. Delforge
var. monantha *P. Delforge in Nat. Belg.,* 80(3): 419 (1999)—Greece.
cordigera L.
subsp. cretica *B. Baumann & H. Baumann in J. Eur. Orchideen,* 31(2): 510, 509 (1999)—Kriti.
cossyrensis *B. Baumann & H. Baumann in J. Eur. Orchideen,* 31(2): 503, 502 (1999)—Sicilia.
× halicarnassia *(H. Baumann & Künkele) P. Delforge in Nat. Belg.,* 76(3): 192 (1995): S. × wettsteinii nothosubsp. halicarnassia.
istriaca *M.L. Perko in Ber. Arbeitskr. Heim. Orch.,* 15(2): 24 (1998)—Yugoslavia.
nurrica Corrias
subsp. argensii *M. Gerbaud & O. Gerbaud in Orchidophile,* 126: 72 (1997)—France.

SERAPIAS:-
orientalis (Greuter) H. Baumann & Künkele
var. siciliensis *(Bartolo & Pulv.) P. Delforge in Nat. Belg.*, 81(4): 398 (2000): Serapias orientalis subsp. siciliensis.
× pulae *M.L. Perko in Ber. Arbeitskr. Heim. Orch.*, 15(2): 25 (1998)—Yugoslavia. (S. istriaca × lingua).
× walravensiana *P. Delforge in Nat. Belg.*, 78(3): 186 (1997)—Greece. (S. carica × lingua).

SERICOSTOMA (Boraginac.).
pauciflorum Stocks ex Wight
var. qeshmensis *A. Ghahreman & F. Attar in Iranian J. Bot.*, 7(1): 58 (1996)—Iran.

SERIPHIDIUM (Compos.).
densiflorum *(Viv.) A. Soldano in Compositae Newsl.*, 32: 19 (1998): Artemisia densiflora.
huguetii *(Caball.) A. Dobignard in Candollea*, 52(1): 150 (1997): Artemisia huguetii.
ifranensis *(J. Didier) A. Dobignard in Candollea*, 52(1): 154 (1997): Artemisia ifranensis.
maritimum (L.) Poljakov
subsp. humifusum *(Fr. ex C. Hartm.) T. Karlsson in Svensk Bot. Tidskr.*, 91(5): 2–9 (1997 publ. 1998): Artemisia maritima β humifusa.
ramosum *(C. Sm.) A. Dobignard in Candollea*, 52(1): 151 (1997): Artemisia ramosa.

SERJANIA (Sapindac.).
adenophylla *M.S. Ferrucci in Bonplandia*, 9(1–2): 21 (1996)—Brazil West-central (Mato Grosso).
brevipetiolata *M.S. Ferrucci & G.V. Somner in Kew Bull.*, 54(4): 987 (1999 publ. 2000)—Brazil Northeast (Bahia).
lachnocarpa *(Benth. ex Radlk.) P. Acevedo-Rodríguez in Novon*, 8(2): 106 (1998): Paullinia lachnocarpa.
matogrossensis *M.S. Ferrucci & P. Acevedo-Rodríguez in Brittonia*, 49(4): 442 (1997)—Brazil West-central (Mato Grosso).
neei *P. Acevedo-Rodríguez in Brittonia*, 49(4): 498 (1997)—Bolivia.
rhytidococca *P. Acevedo-Rodríguez in Brittonia*, 49(4): 500 (1997)—Mexico Southwest (Guerrero).

SERRATULA (Compos.).
centauroides L.
subsp. komarovii *(Iljin) O.S. Zhirova in Fl. Sibir. (Asterac.)*, 13: 226 (1997 publ. 1998): S. komarovii.
coronata L.
subsp. manshurica *(Kitag.) O.S. Zhirova in Fl. Sibir. (Asterac.)*, 13: 224 (1997 publ. 1998): S. manshurica.
gmelinii Tausch
subsp. tanaitica *(P.A. Smirn.) A.K. Skvortsov in Byull. Mosk. Obshch. Ispýt. Prir., Biol.*, 102(1): 63 (1997): S. tanaitica.
khuzistanica *(Mozaff.) V. Mozaffarian in Bot. J. Linn. Soc.*, 128(4): 420 (1998): Centaurea khuzistanica.

SERRURIA (Proteac.).
lacunosa *J.P. Rourke in Bothalia*, 26(2): 154 (1996)—Cape Province.
rebeloi *J.P. Rourke in Bothalia*, 29(2): 263 (1999)—Cape Province.

SESAMOIDES (Resedac.).
purpurascens (L.) G. López
subsp. pinetorum *G. Mateo Sanz & M.B. Crespo Villalba, Fl. Abrev. Comun. Valenciana:* 430 (1995)—Spain.

SESBANIA (Legumin.).
bispinosa (Jacq.) W. Wight
var. grandiflora *D.J. Du Puy & J.-N. Labat in Adansonia*, 19(1): 94 (1997)—Madagascar, Comoros (Mayotte).
laevigata *Afrânio Fernandes & E.P. Nunes in Albertoa*, 4(26): 338 (1998)—Brazil Northeast (Ceará, Maranhão, Piauí, Rio Grande do Norte).
madagascariensis *D.J. Du Puy & J.-N. Labat in Adansonia*, 19(1): 96 (1997)—Madagascar.
melanocaulis *S. Bidgood & I. Friis in Nordic J. Bot.*, 18(2): 211 (1998)—Ethiopia.

SESELI (Umbellif.).
subsect. Cordata *M.G. Pimenov & E.V. Kljuykov in Bot. Zhurn.*, 85(10): 106 (2000).
sect. Davidosciadium *M.G. Pimenov & E.V. Kljuykov in Bot. Zhurn.*, 85(10): 107 (2000).
sect. Eriocycla *(Lindl.) M.G. Pimenov & E.V. Kljuykov in Bot. Zhurn.*, 85(10): 104 (2000): gen. Eriocycla.
sect. Petrosciadium *(Edgew.) M.G. Pimenov & E.V. Kljuykov in Bot. Zhurn.*, 85(10): 107 (2000): gen. Petrosciadium.
sect. Rostroseseli *M.G. Pimenov & E.V. Kljuykov in Bot. Zhurn.*, 85(10): 106 (2000).
albescens *(Franch.) M.G. Pimenov & E.V. Kljuykov in Bot. Zhurn.*, 85(10): 107 (2000): Pimpinella albescens.
elbursense *M.G. Pimenov & E.V. Kljuykov in Bot. Zhurn.*, 85(10): 106 (2000), nom. nov.: Trachydium eriocarpum *Bornm. & Gauba*.
ghafoorianum *(Akhani) M.G. Pimenov & E.V. Kljuykov in Bot. Zhurn.*, 85(10): 106 (2000): Eriocycla ghafooriana.
jinanense *(L.C. Xu & M.D. Xu) M.G. Pimenov in Feddes Repert.*, 110(7–8): 487 (1999): Libanotis jinanensis.
lancifolium *(K.T. Fu) M.G. Pimenov in Feddes Repert.*, 110(7–8): 487 (1999): Libanotis lancifolia.
laticalycinum *(R.H. Shan & M.L. Sheh) M.G. Pimenov in Feddes Repert.*, 110(7–8): 487 (1999): Libanotis laticalycina.
libanotis (L.) K. Koch
var. alpicola *(Kitag.) H. Ohba in Fl. Japan*, 2c: 285 (1999): Libanotis coreana var. alpicola.
nudum *(Lindl.) M.G. Pimenov & E.V. Kljuykov in Bot. Zhurn.*, 85(10): 105 (2000): Eriocycla nuda.
pelliotii *(H. Boissieu) M.G. Pimenov & E.V. Kljuykov in Bot. Zhurn.*, 85(10): 105 (2000): Pituranthos pelliotii.
petrosciadium *M.G. Pimenov & E.V. Kljuykov in Bot. Zhurn.*, 85(10): 107 (2000), nom. nov.: Petrosciadium caespitosum *Edgew.*
simplicifolium *(C.Y. Wu ex R.H. Shan & M.L. Shen) M.G. Pimenov & E.V. Kljuykov in Feddes Repert.*, 110(7–8): 488 (1999): S. mairei var. simplicifolium.
spodotrichomum *(K.T. Fu) M.G. Pimenov in Feddes Repert.*, 110(7–8): 487 (1999), without basionym page: Libanotis spodotrichoma.
staintonii *(Riedl & Kuber) M.G. Pimenov & E.V. Kljuykov in Bot. Zhurn.*, 85(10): 106 (2000): Eriocycla staintonii.
thomsonii *(C.B. Clarke) M.G. Pimenov & E.V. Kljuykov in Bot. Zhurn.*, 85(10): 106 (2000): Pituranthos thomsonii.
togasii *(M. Hiroe) M.G. Pimenov & E.V. Kljuykov in Feddes Repert.*, 110(7–8): 488 (1999): Deverra togasii.
tortuosum L.
subsp. kiabii *H. Akhani in Edinburgh J. Bot.*, 56(1): 26 (1999)—Iran.
subsp. thracicum *D. Delipavlov in Bot. Chronika*, 13: 107 (2000), contrary to Art. 8.2 ICBN (2000)—Bulgaria.
wannienchun *(K.T. Fu) M.G. Pimenov in Feddes Repert.*, 110(7–8): 487 (1999): Libanotis wanniechun.

SETARIA (Gramin.).
subgen. Cernatum *J.F. Pensiero in Darwiniana*, 37(1–2): 47 (1999).
subgen. Reverchoniae *W.E. Fox in Sida*, 18(4): 1039 (1999).
cordobensis R.A.W. Herrm.
var. limensis *(Tovar) J.F. Pensiero in Darwiniana*, 37(1–2): 56 (1999), as 'limense': S. limensis.
elegantula *(Mez) P. Morat in Adansonia*, 22(2): 270 (2000): Panicum elegantulum.
inopinata *L. Toolin in Novon*, 10(4): 415 (2000)—Argentina Northwest (Jujuy, Salta, Tucumán), Argentina Northeast (Chaco, Formosa).
macrocarpa *Z.I. Luchnik in Trudy Dal'nevost. Fil. Akad. Nauk SSSR, Ser. Bot.*, 2: 880 (1937)—Amur, Primorye, Manchuria, Japan.
nicorae *J.F. Pensiero in Novon*, 7(2): 186 (1997)—Paraguay, Argentina Northeast, Argentina Northwest, Brazil West-central, Bolivia, Colombia, Peru, Venezuela, Trinidad-Tobago.

SETARIA:-
pflanzii *J.F. Pensiero in Hickenia*, 2(27): 123 (1995)—Bolivia, Paraguay, Argentina Northwest (Salta).
reverchonii (Vasey) Pilg.
subsp. firmula *(Hitchc. & Chase) W.E. Fox in Sida*, 18(4): 1041 (1999): Panicum firmulum.
subsp. ramiseta *(Scribn.) W.E. Fox in Sida*, 18(4): 1041 (1999): Panicum ramisetum.
rosengurttii Nicora
var. uruguayensis *J.F. Pensiero in Darwiniana*, 37(1–2): 100 (1999)—Uruguay.
utowanaea (Scribn.) Pilg.
var. ophiticola *(Hitchc. & Ekman) W.E. Fox in Sida*, 18(4): 1043 (1999): Panicum ophiticola.
var. subtransiens *(Hitchc. & Ekman) W.E. Fox in Sida*, 18(4): 1043 (1999), basionym as 'Panicum subtransiens': S. subtransiens.
viridis (L.) P. Beauv.
subsp. weinmanii *(Roem. & Schult.) N.N. Tzvelev in Novosti Sist. Výssh. Rast.*, 32: 183 (2000): S. weinmanii.

SETISCAPELLA *J.H. Barnhart in J.K. Small, Fl. Miami:* 170 (26 April 1913) *LENTIBULARIACEAE*†
subulata *(L.) J.H. Barnhart in J.K. Small, Fl. Miami:* 170 (26 April 1913): Utricularia subulata.†

SETULOCARYA *R.R. Mill & D.G. Long in Edinburgh J. Bot.*, 53(1): 113 (1996). *BORAGINACEAE*
diffusa *(Brand) R.R. Mill & D.G. Long in Edinburgh J. Bot.*, 53(1): 114 (1996): Microcaryum diffusum.

SEXTONIA *H. van der Werff in Novon*, 7(4): 436 (1997 publ. 1998). *LAURACEAE*
pubescens *H. van der Werff in Novon*, 7(4): 438 (1997 publ. 1998)—Peru.
rubra *(Mez) H. van der Werff in Novon*, 7(4): 438 (1997 publ. 1998): Ocotea rubra.

SEYRIGIA (Cucurbitac.).
napifera *W. Rauh, Succ. & Xeroph. Pl. Madagascar*, 2: 193 (1998)—Madagascar.

× **SHEEHANARA** *hort. in Orchid Rev.*, 104(1211): 276 (1996). *ORCHIDACEAE* (Ascoglossum × Renanthera × Trichoglottis).

SHIBATAEA *Makino ex T. Nakai in J. Jap. Bot.*, 9(2): 83 (1933). *GRAMINEAE*
kumasaca (Zoll. ex Steud.) Makino
f. aureostriata *(Regel) Sadao Suzuki, Index Japanese Bambusac.:* 339 (1978): Bambusa aureostriata.
nanpingensis *Q.F. Zheng & K.F. Huang in Wuyi Sci. J.*, 2: 17 (1982) as 'nanpinensis'—China Southeast (Fujian).†
var. fujianica *(C.D. Chu & H.Y. Zhou ex C.H. Hu, Q.F. Zheng & K.F. Huang) C.H. Hu in Fl. Reipubl. Popul. Sin.*, 9(1): 321 (1996): Shibataea fujianica.

SHIRAKIA (Euphorbiac.).
aubrevillei *(Leandri) R.C. Kruijt in Biblioth. Bot.*, 146: 92 (1996): Sapium aubrevillei.
elliptica *(Hochst.) R.C. Kruijt in Biblioth. Bot.*, 146: 93 (1996): Sclerocroton ellipticus.
sanchezii *(Merr.) R.C. Kruijt in Biblioth. Bot.*, 146: 93 (1996): Sapium sanchezii.
trilocularis *(Pax & K. Hoffm.) R.C. Kruijt in Biblioth. Bot.*, 146: 93 (1996): Sapium triloculare.
virgata *(Zoll. & Moritzi) R.C. Kruijt in Biblioth. Bot.*, 146: 93 (1996): Excoecaria virgata.

SHIRAKIOPSIS *H.-J. Esser in Blumea*, 44(1): 184 (1999). *EUPHORBIACEAE*
aubrevillei *(Leandri) H.-J. Esser in R. Govaerts, D.G. Frodin & A. Radcliffe-Smith, World Checklist & Bibliogr.* Euphorbiaceae, 4: 1470 (2000): Sapium aubrevillei.
elliptica *(Hochst.) H.-J. Esser in R. Govaerts, D.G. Frodin & A. Radcliffe-Smith, World Checklist & Bibliogr.* Euphorbiaceae, 4: 1470 (2000): Sclerocroton ellipticus.

SHIRAKIOPSIS:-
indica *(Willd.) H.-J. Esser in Blumea,* 44(1): 185 (1999): Sapium indicum.
sanchezii *(Merr.) H.-J. Esser in Blumea,* 44(1): 189 (1999): Sapium sanchezii.
trilocularis *(Pax & K. Hoffm.) H.-J. Esser in R. Govaerts, D.G. Frodin & A. Radcliffe-Smith, World Checklist & Bibliogr.* Euphorbiaceae, 4: 1471 (2000), without basionym ref.: Basionym not stated.
virgata *(Zoll. & Moritzi ex Miq.) H.-J. Esser in Blumea,* 44(1): 189 (1999): Excoecaria virgata.

SHOREA (Dipterocarpac.).
obtusa *Wall. ex Blume, Mus. Bot.,* 2: 32 t. 8 (1856)—Burma.†
stenoptera Burck
var. scabra *(Burck) M. Hotta in Tropics,* 7(1–2): 56 (1997): S. gysbertsiana var. scabra.

SIAMANTHUS *K. Larsen & J. Mood in Nordic J. Bot.,* 18(4): 393 (1998). *ZINGIBERACEAE*
siliquosus *K. Larsen & J. Mood in Nordic J. Bot.,* 18(4): 393 (1998): Thailand.

SIAPAEA *J.F. Pruski in Brittonia,* 48(2): 190 (1996). *COMPOSITAE*
liesneri *J.F. Pruski in Brittonia,* 48(2): 190 (1996)—Venezuela.

SIBAROPSIS *S. Boyd & T.S. Ross in Madroño,* 44(1): 30 (1997). *CRUCIFERAE*
hammittii *S. Boyd & T.S. Ross in Madroño,* 44(1): 30 (1997)—California.

SIBBALDIA (Rosac.).
emodi *H. Ikeda & H. Ohba in J. Jap. Bot.,* 71(4): 188 (1996)—Nepal.

SICYOS (Cucurbitac.).
dieterleae *I. Rodríguez-Arévalo & R. Lira in Acta Bot. Mex.,* 48: 15 (1999)—Mexico Southwest (Michoacan, Jalisco).
lanceoloideus *(H. St. John) W.L. Wagner & D.R. Herbst in Novon,* 9(3): 444 (1999): Sicyocarya lanceoloidea.

SIDA (Malvac.).
atherophora Domin
subsp. magnifica *(Domin) R.M. Barker in J. Adelaide Bot. Gard.,* 18(1): 39 (1998): S. magnifica.
calliantha *M. Thulin in Nordic J. Bot.,* 19(4): 423 (1999)—Somalia.
diplotricha *F.J.H. von Mueller, Enum. Pl. collected by Gregory (N.S.W. Votes & Proc. Legislative Assembly* 1858–59): 4 (1859): Abutilon diplotrichum.
floccosa *M. Thulin & K. Vollesen in Nordic J. Bot.,* 19(4): 421 (1999)—Ethiopia, Somalia.
halophila *F.J.H. von Mueller, Pl. Victoria,* 1: 165 (1862): Abutilon halophilum.
hemitropousa *S.C. Pandeya in J. Bombay Nat. Hist. Soc.,* 95(2): 323, 322 (1998)—India (Uttar Pradesh).
jamaicensis L.
subsp. hermannioides *(Humb., Bonpl. & Kunth) F.J. Fuertes Aguilar in Fl. Colombia,* 17: 70 (1995): S. hermannioides.
longipedicellata *M. Thulin in Nordic J. Bot.,* 19(4): 424 (1999)—Somalia.
ogadensis *M. Thulin & K. Vollesen in Nordic J. Bot.,* 19(4): 422 (1999)—Somalia, Ethiopia.
ravii *M. Sivadasan & N. Anil Kumar in Willdenowia,* 25(2): 651 (1996)—India (Kerala).
rohlenae Domin
subsp. occidentalis *R.M. Barker in J. Adelaide Bot. Gard.,* 18(1): 36 (1998)—Western Australia, Northern Territory.

SIDASTRUM (Malvac.).
burrerense *P.A. Fryxell, J.L. León de la Luz & M. Domínguez L. in Contrib. Univ. Mich. Herb.,* 21: 201 (1997)—Mexico Northwest (Baja California Sur).

SIDERITIS (Labiat.).
ser. Peyrei *C. Coulomb, Étude Sideritis Français,* 2: 20 (1999).

SIDERITIS:-
ser. Rossii *C. Coulomb, Étude Sideritis Français,* 2: 50 (1999).
akmanii *Z. Aytaç, M. Ekici & A. Dönmez in Flora Medit.,* 5: 221 (1995 publ. 1996)—Turkey.
arborescens Salzm. ex Benth.
subsp. maireana *(Font Quer & Pau) O. Socorro & M.L. Arrebola in Lagascalia,* 17(2): 357 (1994), as 'mairea': S. maireana.
subsp. ortenedae *(Font Quer & Pau) O. Socorro & M.L. Arrebola in Lagascalia,* 17(2): 357 (1994): S. arborescens var. ortenedae.
caesarea *H. Duman, Z. Aytaç & K.H.C. Başer in Turkish J. Bot.,* 22: 51 (1998)—Turkey.
carbonellis Socorro
var. hispanica *(Font Quer) O. Socorro & M.L. Arrebola in Lagascalia,* 17(2): 357 (1994): S. ochroleuca var. hispanica.
carolipauana *J.B. Peris & G. Stübing in Feddes Repert.,* 104(5–6): 289 (1993)—Morocco.
caureliana *(Obón & D. Rivera) C. Coulomb, Étude Sideritis Français,* 2: 31 (1999): S. hyssopifolia subsp. caureliana.
× costa-talensii *C. Obón de Castro, D. Rivera Núñez, F. Alcaraz Ariza & A. de la Torre* García *in An. Jard. Bot. Madrid,* 54(1): 299 (1996)—Spain. (S. hirsuta × sericea).
× delgadilloi *C. Obón de Castro, D. Rivera Núñez, F. Alcaraz Ariza & A. de la Torre* García *in An. Jard. Bot. Madrid,* 54(1): 297 (1996)—Spain. (S. edetana × tragoriganum).
dubia *C. Coulomb, Étude Sideritis Français,* 1: 88 (1999)—Switzerland/France/Germany border.
endressii Willk.
var. mathonii *C. Coulomb, Étude Sideritis Français,* 2: 35 (1999)—France.
subsp. provincialis *(Jord. & Fourr.) C. Coulomb, Étude Sideritis Français,* 2: 32 (1999): S. provincialis.
foetens Clemente ex Lag.
subsp. rivasgodayi *(Fern. Casas) F. Alcaraz, T.E. Díaz, S. Rivas-Martínez & P. Sánchez-Gómez in Itinera Geobot.,* 2: 116 (1989): S. foetens var. rivasgodayi.
granatensis (Pau) Font Quer
subsp. briquetiana *(Font Quer & Pau) O. Socorro & M.L. Arrebola in Lagascalia,* 17(2): 356 (1994): S. briquetiana.
grandiflora Salzm. ex Benth.
subsp. baetica *(Lange) R. Roselló, J.B. Peris, Á. Romo & G. Stübing in An. Jard. Bot. Madrid,* 56(2): 387 (1998): S. baetica.
× gudarica *G. Mateo, S. López Udias & C. Fabregat in An. Jard. Bot. Madrid,* 57(2): 419 (1999 publ. 2000)—Spain. (S. fernandez-casasii × pungens).
gulendamii *H. Duman & F.A. Karavelioğullari in Flora Medit.,* 5: 224 (1995 publ. 1996)—Turkey.
hirsuta L.
subsp. gypsicola *S. Cirujano, R. Roselló, G. Stübing & J.B. Peris in An. Jard. Bot. Madrid,* 53(2): 260 (1995 publ. 1996)—Spain.
subsp. maritima *(Font Quer) O. Socorro & M.L. Arrebola in Lagascalia,* 17(2): 356 (1994): S. hirsuta var. maritima.
subsp. vulgaris *(Willk.) C. Coulomb, Étude Sideritis Français,* 2: 46 (1999): S. hirsuta var. vulgaris.
hyssopifolia L.
var. australis *C. Coulomb, Étude Sideritis Français,* 2: 13 (1999)—France.
subsp. brachycalyx *(Pau) O. Socorro & M.L. Arrebola in Lagascalia,* 17(2): 357 (1994): S. brachycalyx.
f. corbariensis *C. Coulomb, Étude Sideritis Français,* 2: 17 (1999)—France.
subsp. javalambrensis *(Pau) J. Borja in Soc. Ech. Pl. Vasc. Eur. Bassin Médit. Bull.,* 20: 54 (1985): S. javalambrensis.
incana L.
subsp. edetana *(Pau ex Font Quer) G. Mateo Sanz & M.B. Crespo Villalba, Claves Fl. Valenciana:* 200 (1990): S. incana var. edetana.
× lainzii *C. Obón de Castro & D. Rivera Núñez in An. Jard. Bot. Madrid,* 54(1): 296 (1996)—Spain. (S. ibanyezii × murgetana).
leucantha Cav.
subsp. illicitana *F. Alcaraz, T.E. Díaz, S. Rivas-Martínez & P. Sánchez-Gómez in Itinera Geobot.,* 2: 117 (1989)—Spain.

SIDERITIS:-
leucantha
subsp. mohamedii *(Rejdali) O. Socorro & M.L. Arrebola in Lagascalia,* 17(2): 356 (1994): S. mohamedii.
× liantei *C. Obón de Castro & D. Rivera Núñez in An. Jard. Bot. Madrid,* 54(1): 297 (1996)—Spain. (S. chamaedryfolia × leucantha subsp. bourgeana).
lurida J. Gay
subsp. borgiae *(J. Andrés) M. Luceño, B. García Muñoz & A. González Canalejo in An. Jard. Bot. Madrid,* 57(2): 412 (1999 publ. 2000): S. borgiae.
mansanetiana *G. Stübing, J.B. Peris & R. Figuerola in Feddes Repert.,* 103(5–6): 309 (1992)—Spain.
× murcica (Font Quer) Romo
nothosubsp. mediovalentina *M.B. Crespo & E. Laguna in Flora Montiberica,* 6: 87 (1997)—Spain. (S. incana subsp. incana × tragoriganum subsp. tragoriganum).
nothosubsp. stricta *(Font Quer) M.B. Crespo & E. Laguna in Flora Montiberica,* 6: 87 (1997): S. × viciosoi var. stricta.
obonisriveraeque *G. Stübing, J.B. Peris, R. Roselló & S. Cirujano in Fontqueria,* 44: 41 (1996)—Portugal.
ochroleuca Noë
subsp. antiatlantica *(Maire) O. Socorro & M.L. Arrebola in Lagascalia,* 17(2): 357 (1994): S. ochroleuca var. antiatlantica.
× petriludovici *C. Obón de Castro & D. Rivera Núñez in An. Jard. Bot. Madrid,* 54(1): 296 (1996)—Spain. (S. ibanyezii × lasiantha).
peyrei Timb.-Lagr.
subsp. castellana *(Sennen & Pau) C. Coulomb, Étude Sideritis Français,* 2: 29 (1999): S. hyssopifolia var. castellana.
var. gouanii *(Timb.-Lagr.) C. Coulomb, Étude Sideritis Français,* 2: 25 (1999): S. gouanii.
subsp. guillonii *(Timb.-Lagr.) C. Coulomb, Étude Sideritis Français,* 2: 28 (1999): S. guillonii.
var. incana *(Benth.) C. Coulomb, Étude Sideritis Français,* 2: 23 (1999): S. scordioides var. incana.
var. laxispica *(López Pach.) C. Coulomb, Étude Sideritis Français,* 2: 30 (1999): S. hyssopifolia var. laxispica.
subsp. pastoris *(Sennen) C. Coulomb, Étude Sideritis Français,* 2: 21 (1999): S. hyssopifolia proles pastoris.
var. vidalii *(Sennen) C. Coulomb, Étude Sideritis Français,* 2: 27 (1999): S. hyssopifolia proles vidalii.
pungens Benth.
subsp. vigoi *J.B. Peris, R. Roselló & G. Stübing in An. Jard. Bot. Madrid,* 57(2): 416 (1999 publ. 2000)—Spain.
pusilla (Lange) Pau
subsp. carthaginensis *(Font Quer) F. Alcaraz, T.E. Díaz, S. Rivas-Martínez & P. Sánchez-Gómez in Itinera Geobot.,* 2: 117 (1989): S. pusilla var. carthaginensis.
rossii Peris, Stübing, Jury & Rejdali
subsp. moorei *(Peris, Stübing, Jury & Rejdali) C. Coulomb, Étude Sideritis Français,* 2: 50 (1999), without exact basionym page: S. moorei.
sierrarafolsiana *R. Roselló, G. Stübing, J.B. Peris & Á.M. Romo in Acta Bot. Barcin.,* 45: 208 (1998)—Morocco.
× tomas-barberanii *C. Obón de Castro, D. Rivera Núñez, F. Alcaraz Ariza & A. de la Torre* García *in An. Jard. Bot. Madrid,* 54(1): 298 (1996)—Spain. (S. edetana × hirsuta).
vuralii *H. Duman & K.H.C. Başer in Turkish J. Bot.,* 22: 53 (1998)—Turkey.

SIDEROXYLON (Sapotac.).
alachuense *L.C. Anderson in Sida,* 17(3): 565 (1997), nom. nov.: Bumelia lanuginosa var. anomala *Sarg.*
rufohirtum *B.J. Herring & W.S. Judd in Castanea,* 60(4): 358 (1995 publ. 1996), nom. nov.: Bumelia rufotomentosa *Small.*

× **SIEGERISTARA** *hort. in Orchid Rev.,* 104(1207): 37 (1996). *ORCHIDACEAE* (Bulbophyllum × Cirrhopetalum × Trias).

SIEGLINGIA (Gramin.).
decumbens (L.) Bernh.
subsp. decipiens *(O. Schwarz & Bässler) K. Senghas in K. Senghas & S. Seybold, Schmeil & Fitschen Fl. Deutschland u. angrenz. Länd.*, 91. Aufl.: 792 (2000), with incorrect basionym ref.: Danthonia decumbens subsp. decipiens.

SIEVEKINGIA (Orchidac.).
subgen. Cirrhata *R. Jenny in Schlechteriana*, 3(2): 60 (1992).
subgen. Cirrhosa *R. Jenny in Schlechteriana*, 3(2): 60 (1992).
subgen. Lamellata *R. Jenny in Schlechteriana*, 3(2): 60 (1992).

SIEVERSANDREAS *Eb. Fischer in Bull. Mus. Nation. Hist. Nat., B, Adansonia*, Ser. 4, 18(3–4): 213 (1996). *SCROPHULARIACEAE*
madagascarianus *Eb. Fischer in Bull. Mus. Nation. Hist. Nat., B, Adansonia*, Ser. 4, 18(3–4): 214 (1996)—Madagascar.

× **SIGMACIDIUM** *hort. in Orchid Rev.*, 106(1220): 126 (1998). *ORCHIDACEAE* (Oncidium × Sigmatostalix).

SIGMATOSTALIX (Orchidac.).
ariasii *W. Königer in Arcula*, 8: 241 (1999)—Peru.
brevicornis *W. Königer & J. Portilla in Arcula*, 10: 271 (2000)—Ecuador.
gentryi *C.H. Dodson in Orquideologia*, 21(1): 15 (1998)—Ecuador.
hermansiana *W. Königer in Arcula*, 8: 243 (1999)—Colombia.
hirtzii *C.H. Dodson in Orquideologia*, 21(1): 17 (1998)—Ecuador.
× luerorum *C.H. Dodson in Orquideologia*, 21(1): 18 (1998)—Ecuador. (S. morganii × picta).
oxyceras *W. Königer & J.G. Weinmann in Arcula*, 5: 146 (1996)—Ecuador.
papilio *W. Königer & R. Escobar in Arcula*, 5: 148 (1996)—Colombia.
pichinchensis *C.H. Dodson in Orquideologia*, 21(1): 21 (1998)—Ecuador.
pseudounguiculata *F. Pupulin & R.L. Dressler in Lindleyana*, 15(1): 27 (2000)—Costa Rica.
trimorion *W. Königer in Arcula*, 6: 172 (1996)—Peru.

SILENE (Caryophyllac.).
subsect. Armeriae *M.M. Fedoronchuk in Ukr. Bot. Zhurn.*, 54(2): 179 (1997).
subsect. Cretaceae *M.M. Fedoronchuk in Ukr. Bot. Zhurn.*, 54(2): 180 (1997).
sect. Dianthoidea *(Chowdhuri) G.A. Lazkov in Bot. Zhurn.*, 81(8): 103 (1996): S. subsect. Dianthoideae.
subsect. Dichotomae *(Rohrb.) M.M. Fedoronchuk in Ukr. Bot. Zhurn.*, 54(2): 179 (1997): S. ser. Dichotomae.
subsect. Erectorefractae *(Chowdhuri) M.M. Fedoronchuk in Ukr. Bot. Zhurn.*, 54(2): 179 (1997): S. sect. Erectorefractae.
sect. Graminiformes *G.A. Lazkov in Bot. Zhurn.*, 82(1): 109 (1997).
subsect. Hypanicae *M.M. Fedoronchuk in Ukr. Bot. Zhurn.*, 54(2): 179 (1997).
sect. Koreanae *G.A. Lazkov in Bot. Zhurn.*, 83(10): 117 (1998).
sect. Laciniatae *(Boiss.) C.L. Tang in Fl. Reipubl. Popul. Sin.*, 26: 292 (1996): S. § Laciniatae.
subsect. Paniculatae *(Boiss.) M.M. Fedoronchuk in Ukr. Bot. Zhurn.*, 54(2): 180 (1997): S. § Paniculatae.
subgen. Petrocoptis *(A. Braun ex Endl.) M. Mayol & J.A. Rosselló in Taxon*, 48(3): 476 (1999): gen. Petrocoptis.
subsect. Scorpioideae *(Rohrb.) M.M. Fedoronchuk in Ukr. Bot. Zhurn.*, 54(2): 179 (1997): S. ser. Scorpioideae.
subsect. Viridiflorae *(Boiss.) M.M. Fedoronchuk in Ukr. Bot. Zhurn.*, 54(2): 180 (1997): S. § Viridiflorae.
sect. Viscosae *(Boiss.) C.L. Tang in Fl. Reipubl. Popul. Sin.*, 26: 338 (1996): S. § Viscosae.

SILENE:-
ajanensis (Regel & Tiling) Vorosch.
subsp. villosula *(Trautv.) V.N. Voroschilov in A.K. Skvortsov (ed.), Florist. issl. v razn. raĭonakh SSSR*: 166 (1985): Lychnis ajanensis var. villosula.
akaisialpina *(T. Yamaz.) Hiroyoshi Ohashi, Y. Tateishi & H. Nakai in J. Jap. Bot.*, 72(4): 195 (1997): Melandryum keiskei var. akaisialpinum.
f. leucantha *(Takeda) Hiroyoshi Ohashi, Y. Tateishi & H. Nakai in J. Jap. Bot.*, 72(4): 196 (1997): S. keiskei lusus leucantha.
apetala Willd.
var. glabrata *A.I. Hosny & E. Shamso in Taeckholmia*, 14: 33 (1992–93 publ. 1993)—Egypt.
aprica *Turcz. ex F.E.L. Fischer & C.A. Meyer, Index sem. hort. petrop.*, 1: 38 (1835)—Irkutsk.†
var. oldhamiana *(Miq.) C.Y. Wu in Fl. Reipubl. Popul. Sin.*, 26: 343 (1996): S. oldhamiana.
var. ryukyuensis *T. Yamazaki in J. Jap. Bot.*, 70(6): 344 (1995)—Nansei-Shoto.
arghireica *F. Valsecchi in Boll. Soc. Sarda Sci. Nat.*, 30: 459 (1994–95 publ. 1995)— Sardegna, Sicilia, Italy, Spain, Portugal, Algeria.
asterias Griseb.
f. radikae *(Trinajstić & Zi. Pavletić) K. Micevski in Prilozi Oddel. Biol. Med. Nauki Makedonska Akad. Nauk. Umet.*, 9(1–2): 45 (1988 publ. 1991): S. asterias subsp. radikae.
baccifera (L.) Roth
var. japonica *(Miq.) H. Ohashi & H. Nakai in J. Jap. Bot.*, 71(5): 268 (1996): Cucubalus baccifer var. japonicus.
banksia *(Meerb.) D.J. Mabberley in Telopea*, 8(2): 253 (1999): Agrostemma banksia.
beguinotii *F. Valsecchi in Boll. Soc. Sarda Sci. Nat.*, 30: 462 (1994–95 publ. 1995)—Sardegna.
bungeana *(D. Don) H. Ohashi & H. Nakai in J. Jap. Bot.*, 71(5): 269 (1996): Agrostemma bungeana.
caryophylloides (Poir.) Otth
subsp. binbogaensis *H. Duman & M. Vural in Turkish J. Bot.*, 19(4): 478 (1995), as 'binbogaense'—Turkey.
chamarensis Turcz.
var. udocanica *G.A. Peschkova in Fl. Tsentral'noi Sibiri*, 1: 325 (1979)—Buryatiya.
cognata *(Maxim.) H. Ohashi & H. Nakai in J. Jap. Bot.*, 71(5): 269 (1996): Lychnis cognata.
danaensis *A. Danin in Willdenowia*, 27(1–2): 166 (1997)—Palestine (Jordan).
denizliensis *Z. Aytaç in Thaiszia*, 8(1): 8 (1998), as 'denizliense'—Turkey.
foliosa Maxim.
var. macrostyla *(Maxim.) V.N. Voroschilov in A.K. Skvortsov (ed.), Florist. issl. v razn. raĭonakh SSSR*: 167 (1985): S. macrostyla.
gazulensis *A. Galán de Mera, J.E. Cortés, J.A. Vicente Orellana & R. Morales Alonso in Acta Bot. Malacitana*, 24: 238 (1999)—Spain.
georgievskyi *G.A. Lazkov in Bot. Zhurn.*, 84(9): 123 (1999)—Syria.
glaucifolia Lag.
subsp. pseudoviscosa *(Fern. Casas) M. Mayol & J.A. Rosselló in Taxon*, 48(3): 478 (1999): Petrocoptis pseudoviscosa.
grandiflora *(Jacq.) Hiroyoshi Ohashi & H. Nakai in J. Jap. Bot.*, 71(2): 110 (1996), homonym; *see Franch. in Bull. Soc. Bot. France*, 33: 427 (1886): Lychnis grandiflora.
huochenensis *X.M. Pi & X.L. Pan in J. Xinjiang Univ. (Nat. Sci.)*, 8(1): 72 (1991)—Xinjiang.
ichnusae *S. Brullo, G. De Marco & G. De Marco f. in Nordic J. Bot.*, 17(2): 161 (1997)—Sardegna.
ikonnikovii *G.A. Lazkov in Bot. Zhurn.*, 81(7): 109 (1996)—North Caucasus (Dagestan).
intramongolica *G.A. Lazkov in Komarovia*, 1: 42 (1999)—Mongolia.
ispirensis *A.P. Khokhrjakov in Byull. Mosk. Obshch. Ispyt. Prir., Biol.*, 102(4): 46 (1997), homonym; *see Boiss., Diagn. Pl. Orient.*, Ser. 2, 5: 55 (1856)—Turkey.
jenisseensis Willd.
var. graminifolia *(Otth) Y.C. Chu in Clavis Pl. Chinae Bor.-Or. ed. 2 (ed. P.Y. Fu)*: 159 (1995): S. graminifolia.
var. kodarensis *G.A. Peschkova in Fl. Tsentral'noi Sibiri*, 1: 326 (1979)—Buryatiya.

SILENE:-
keiskei Miq.
f. albescens *(Takeda) Hiroyoshi Ohashi, Y. Tateishi & H. Nakai in J. Jap. Bot.*, 72(4): 197 (1997): S. keiskei subf. albescens.
kiusiana *(Makino) H. Ohashi & H. Nakai in J. Jap. Bot.*, 71(5): 269 (1996): Lychnis kiusiana.
laxipruinosa *M. Mayol & J.A. Rosselló in Taxon*, 48(3): 478 (1999), nom. nov.: Petrocoptis grandiflora *Rothm.*
lerchenfeldiana Baumg.
f. pilosa *K. Micevski in Prilozi Oddel. Biol. Med. Nauki Makedonska Akad. Nauk. Umet.*, 9(1–2): 45 (1988 publ. 1991)—Yugoslavia.
linnaeana *V.N. Voroschilov in A.K. Skvortsov (ed.), Florist. issl. v razn. raĭonakh SSSR*: 167 (1985), nom. nov.: Lychnis sibirica *L.*
lipskyi *G.A. Laz'kov in Bot. Zhurn.*, 81(4): 96 (1996)—Transcaucasus (Gruziya).
marschallii C.A. Mey.
subsp. propinqua *(Schischk.) A. Nersesian in Phyton (Austria)*, 36(1): 48 (1996): S. propinqua.
maximowicziana *Yu.P. Kozhevnikov in Rast. Tsentral. Azii*, 11: 78 (1994 publ. 1995), nom. nov., isonym; *see Rohrb. in Linnaea*, 36: 680 (1870): S. foliosa var. mongolica *Maxim.*
miqueliana *(Rohrb.) H. Ohashi & H. Nakai in J. Jap. Bot.*, 71(5): 269 (1996): Lychnis miqueliana.
f. argyrata *(M. Mizush.) H. Ohashi & H. Nakai in J. Jap. Bot.*, 71(5): 270 (1996): Lychnis miqueliana f. argyrata.
f. plena *(Makino) H. Ohashi & H. Nakai in J. Jap. Bot.*, 71(5): 270 (1996): Lychnis miqueliana var. plena.
× montistellensis *M. Ladero, S. Rivas-Martinez, A. Amor, M.T. Santos & M.T. Alonso in Bot. J. Linn. Soc.*, 130(1): 72 (1999)—Portugal. (S. acutifolia × foetida subsp. foetida).
montserratii *(Fern. Casas) M. Mayol & J.A. Rosselló in Taxon*, 48(3): 479 (1999): Petrocoptis montserratii.
subsp. crassifolia *(Rouy) M. Mayol & J.A. Rosselló in Taxon*, 48(3): 480 (1999): Petrocoptis crassifolia.
multiflora (Ehrh.) Pers.
var. glabra *(Kleopow) M.M. Fedoronchuk in Ukr. Bot. Zhurn.*, 54(2): 180 (1997): S. steppicola var. glabra.
var. macrothyrsa *M.M. Fedoronchuk in Ukr. Bot. Zhurn.*, 54(2): 180 (1997), without type—Ukraina.
var. pubescens *(Kleopow) M.M. Fedoronchuk in Ukr. Bot. Zhurn.*, 54(2): 180 (1997): S. steppicola var. pubescens.
nummica *F. Valsecchi in Boll. Soc. Sarda Sci. Nat.*, 30: 452 (1994–95 publ. 1995)—Sardegna, Corse, Sicilia, Italy, Libya.
oenotriae *S. Brullo in Nordic J. Bot.*, 17(6): 649 (1997 publ. 1998)—Italy.
ordossica *G.A. Lazkov in Komarovia*, 1: 42 (1999), nom. nov.: S. foliosa var. mongolica *Maxim.*
pakistanica *G.A. Lazkov in Ann. Naturhist. Mus. Wien*, 102B: 438 (2000)—Pakistan.
pardoi *(Pau) M. Mayol & J.A. Rosselló in Taxon*, 48(3): 480 (1999): Petrocoptis pardoi.
subsp. guarensis *(Fern. Casas) M. Mayol & J.A. Rosselló in Taxon*, 48(3): 480 (1999): Petrocoptis guarensis.
prilepensis *K. Micevski in Prilozi Oddel. Biol. Med. Nauki Makedonska Akad. Nauk. Umet.*, 9(1–2): 39 (1988 publ. 1991)—Yugoslavia.
pseudoholopetala *G.A. Laz'kov in Bot. Zhurn.*, 80(11): 105 (1995)—Xinjiang.
puranensis *(L.H. Zhou) C.Y. Wu & H. Chuang in Fl. Reipubl. Popul. Sin.*, 26: 321 (1996): Melandrium puranense.
pyrenaica *(Bergeret) M. Mayol & J.A. Rosselló in An. Jard. Bot. Madrid*, 57(2): 404 (1999 publ. 2000): Lychnis pyrenaica.
subsp. pseudoviscosa *(Fern. Casas) M. Mayol & J.A. Rosselló in An. Jard. Bot. Madrid*, 57(2): 404 (1999 publ. 2000): Petrocoptis pseudoviscosa.
repens Patrin
f. glabra *G.A. Peschkova, Stepnaya Fl. Baĭkal'skoĭ Sibiri*: 55 (1972)—Buryatiya.
var. sinensis *(F.N. Williams) C.L. Tang in Fl. Reipubl. Popul. Sin.*, 26: 292 (1996): S. repens f. sinensis.

SILENE:-
sieboldii *(Van Houtte) H. Ohashi & H. Nakai in J. Jap. Bot.*, 71(5): 270 (1996): Lychnis sieboldii.
sinensis *(Lour.) H. Ohashi & H. Nakai in J. Jap. Bot.*, 71(5): 270 (1996): Hedona sinensis.
f. verticillata *(Makino) H. Ohashi & H. Nakai in J. Jap. Bot.*, 71(5): 271 (1996): Lychnis coronata var. verticillata.
spergulifolia (Willd.) M. Bieb.
subsp. soskae *(Černjavski) K. Micevski in Prilozi Oddel. Biol. Med. Nauki Makedonska Akad. Nauk. Umet.*, 9(1–2): 43 (1988 publ. 1991): S. soskae.
subodhii *S.R. Kundu in Novon*, 10(1): 38 (2000), nom. nov.: S. stewartii *Ghaz.*
sytnikii *L.I. Krytzka, V.V. Novosad & V.V. Protopopova in Ukr. Bot. Zhurn.*, 53(5): 581 (1996)—Ukraina.
tamaranae *D. Bramwell in Bot. Macaronésica*, 22: 122 (1995)—Canary Is.
transalpina *(Hayata) S.S. Ying in Fl. Taiwan, ed. 2*, 2: 368 (1996): Melandrium transalpinum.
turczaninovii *G.A. Lazkov in Bot. Zhurn.*, 84(9): 125 (1999), nom. nov.: S. ambigua *Turcz.*
uniflora Roth
subsp. petraea *(C. Hartm.) B. Jonsell & H.C. Prentice in Nordic J. Bot.*, 16(1): 6 (1996): S. maritima var. petraea.
wilfordii *(Regel) H. Ohashi & H. Nakai in J. Jap. Bot.*, 71(5): 271 (1996): Lychnis fulgens var. wilfordii.
zhongbaensis *(L.H. Zhou) C.Y. Wu & C.L. Tang in Fl. Reipubl. Popul. Sin.*, 26: 314 (1996): Melandrium zhongbaense.
zhoui *C.Y. Wu in Fl. Reipubl. Popul. Sin.*, 26: 315 (1996), nom. nov.: Melandrium integripetalum *L.H. Zhou.*

SILVIANTHUS (Carlemanniac.).
bracteatus Hook. f.
subsp. tonkinensis *(Gagnep.) H.W. Li in Acta Bot. Yunnanica*, 17(4): 414 (1995): Quiducia tonkinensis.

SIMABA (Simaroubac.).
docensis *E.V. Franceschinelli & K. Yamamoto in Novon*, 9(3): 345 (1999)—Brazil Southeast (Minas Gerais).

SIMIRA (Rubiac.).
cesariana *C.M. Taylor in Novon*, 9(4): 568 (1999)—Colombia.
gardneriana *M.R.V. Barbosa & A.L. Peixoto in Novon*, 10(2): 110 (2000)—Brazil Northeast (Ceará, Bahia, Pernambuco, Piauí).
hatschbachiorum *J.H. Kirkbride in Brittonia*, 49(3): 360 (1997)—Brazil Southeast (Espírito Santo).
hirsuta *C.M. Taylor in Novon*, 9(4): 569 (1999)—Colombia.

SINARUNDINARIA (Gramin.).
papillosa *W.T. Lin in J. S. China Agric. Univ.*, 14(3): 112, 111 (1993)—China Southeast (Guangdong).

SINCLAIRIA (Compos.).
ismaelis *J.L. Panero & J.L. Villaseñor in Brittonia*, 48(1): 85 (1996)—Mexico Southwest (Oaxaca).
manriqueae *J.L. Panero & J.L. Villaseñor in Brittonia*, 48(4): 568 (1996 publ. 1997), as 'manriquei'—Mexico Southwest (Oaxaca).

SINNINGIA (Gesneriac.).
araneosa *A. Chautems in Candollea*, 52(1): 168 (1997), nom. nov., as 'areneosa': Corytholoma pusillum *Fritsch.*
brasiliensis *(Regel & E. Schmidt) H. Wiehler & A. Chautems in Gesneriana*, 1(1): 7 (1995): Lietzia brasiliensis.
carangolensis *A. Chautems in Gesneriana*, 1(1): 8 (1995)—Brazil Southeast (Minas Gerais).
hatschbachii *A. Chautems in Candollea*, 52(1): 165 (1997)—Brazil South (Paraná), Brazil Southeast (São Paulo).
iarae *A. Chautems in Gesneriana*, 1(1): 9 (1995)—Brazil Southeast (São Paulo).
mauroana *A. Chautems in Gesneriana*, 1(1): 9 (1995)—Brazil Southeast (São Paulo).
micans *(Fritsch) A. Chautems in Gesneriana*, 1(1): 10 (1995): Corytholoma micans.

SINNINGIA:-
nordestina *A. Chautems, G.S. Baracho & J.A. Siqueira Filho in Brittonia*, 52(1): 50 (2000)—Brazil Northeast (Pernambuco, Alagoas, Bahia, Ceará, Paraíba, Pernambuco, Sergipe).
striata *(Fritsch) A. Chautems in Gesneriana*, 1(1): 11 (1995): Corytholoma striatum.

SINOBAMBUSA (Gramin.).
glabrata *W.T. Lin & Z.J. Feng in J. S. China Agric. Univ.*, 14(1): 47 (1993)—China Southeast (Guangdong).
tootsik (Siebold) Makino
var. tenuifolia *(Koidz.) Sadao Suzuki, Index Japanese Bambusac.*: 339 (1978): Semiarundinaria tenuifolia.

SINOCALAMUS (Gramin.).
oldhamii Munro
f. revolutus *(W.T. Lin & J.Y. Lin) W.T. Lin in J. S. China Agric. Univ.*, 14(3): 111 (1993): Bambusa oldhamii f. revoluta.
rectocuneatus *(W.T. Lin) W.T. Lin in J. S. China Agric. Univ.*, 14(3): 111 (1993): Neosinocalamus rectocuneatus.

SINOCARUM *H. Wolff ex R.H. Shan & F.T. Pu in Acta Phytotax. Sin.*, 18(3): 374 (1980). *UMBELLIFERAE*
coloratum *(Diels) H. Wolff ex F.T. Pu in Fl. Reipubl. Popul. Sin.*, 55(2): 33 (1985): Carum coloratum.
cruciatum *(Franch.) H. Wolff ex F.T. Pu in Fl. Reipubl. Popul. Sin.*, 55(2): 33 (1985): Carum cruciatum.
dolichopodum *(Diels) H. Wolff ex F.T. Pu in Fl. Reipubl. Popul. Sin.*, 55(2): 38 (1985): Carum dolichopodum.
longii *M.F. Watson in Edinburgh J. Bot.*, 53(1): 138 (1996)—Bhutan, Sikkim.
minus *M.F. Watson in Edinburgh J. Bot.*, 53(1): 140 (1996)— Nepal, Bhutan.
pulchellum *C. Norman ex M.F. Watson in Edinburgh J. Bot.*, 53(1): 141 (1996)—Nepal, Bhutan, Sikkim.
schizopetalum *(Franch.) H. Wolff ex F.T. Pu in Fl. Reipubl. Popul. Sin.*, 55(2): 33 (1985): Carum schizopetalum.
var. bijiangense *(S.L. Liou) X.T. Liu in Fl. Yunnanica*, 7: 521 (1997): S. bijiangense.
woodii *M.F. Watson in Edinburgh J. Bot.*, 53(1): 143 (1996)—Bhutan, Sikkim, Nepal.

SINOCRASSULA (Crassulac.).
diversifolia *H. Chuang in Acta Bot. Yunnanica*, 19(3): 224 (1997)—China South-central (Yunnan).
indica (Decne.) A. Berger
var. maculosa *H. Chuang ex J. Thiede in Cact. Succ. J. (U.S.A.)*, 71(3): 160 (1999), as 'nom. nov.'—China South-central (Yunnan).
luteorubra *(Praeger) H. Chuang in Fl. Yunnanica*, 8: 142 (1997), without basionym ref.: Sedum indicum var. luteorubrum.
var. forrestii *(Raym.-Hamet) H. Chuang in Fl. Yunnanica*, 8: 142 (1997), without basionym ref.: Sedum indicum var. forrestii.
var. maculosa *H. Chuang in Acta Bot. Yunnanica*, 19(3): 224 (1997), contrary to Art. 43.1 ICBN (1994)—China South-central (Yunnan).
paoshingensis *(S.H. Fu) H. Ohba, Shinobu Akiyama & S.K. Wu in J. Jap. Bot.*, 75(5): 296 (2000): Sedum paoshingense.

SINODIELSIA (Umbellif.).
cuneata *(H. Wolff) M.G. Pimenov & E.V. Kljuykov in Feddes Repert.*, 110(7–8): 489 (1999): Physospermopsis cuneata.
delavayi *(Franch.) M.G. Pimenov & E.V. Kljuykov in Feddes Repert.*, 110(7–8): 488 (1999): Peucedanum delavayi.

SINOJACKIA (Styracac.).
microcarpa *Tao Chen & G.Y. Li in Novon*, 7(4): 350 (1997 publ. 1998)—China Southeast (Zhejiang).
oblongicarpa *C.T. Chen & T.R. Cao in Edinburgh J. Bot.*, 55(2): 235 (1998)—China Southeast (Hunan).

SINOMARSDENIA *P.T. Li & J.J. Chen in J. S. China Agric. Univ.*, 18(1): 39 (1997). *ASCLEPIADACEAE*
incisa *(P.T. Li & Y.H. Li) P.T. Li & J.J. Chen in J. S. China Agric. Univ.*, 18(1): 40 (1997): Marsdenia incisa.

SINOSENECIO (Compos.).
denticulatus *J. Quan Liu in Acta Phytotax. Sin.*, 38(2): 192 (2000)—China South-central (Sichuan).
newcombei *(Greene) J.P. Janovec & T.M. Barkley in Novon*, 6(3): 266 (1996): Senecio newcombei.

SINOSOPHIOPSIS *I.A. Al-Shehbaz in Novon*, 10(4): 340 (2000). *CRUCIFERAE*
bartholomewii *I.A. Al-Shehbaz in Novon*, 10(4): 341 (2000)—Qinghai, Tibet.
heishuiensis *(W.T. Wang) I.A. Al-Shehbaz in Novon*, 10(4): 341 (2000): Cardamine heishuiensis.

SIPANEA (Rubiac.).
palustris *(A. Rich. ex DC.) J.H. Kirkbride in Brittonia*, 49(3): 360 (1996), homonym; *see Seem., Bot. Voy. Herald*: 136 (1854): Bertiera palustris.
saxicola *J.H. Kirkbride in Brittonia*, 49(3): 362 (1997)—Brazil North (Pará).

SIPARUNA (Monimiac.).
calantha *(Perkins) S.S. Renner & G. Hausner in Novon*, 10(2): 140 (2000): S. riparia var. calantha.
campii *S.S. Renner & G. Hausner in Novon*, 6(1): 103 (1996)—Ecuador.
cascada *S.S. Renner & G. Hausner in Novon*, 6(1): 116 (1996)—Ecuador.
conica *S.S. Renner & G. Hausner in Novon*, 6(1): 108 (1996)—Ecuador, Colombia.
gentryana *S.S. Renner in Novon*, 10(2): 134 (2000)—Ecuador, Colombia.
guajalitensis *S.S. Renner & G. Hausner in Novon*, 6(1): 105 (1996)—Ecuador.
lozaniana *S.S. Renner & G. Hausner in Novon*, 10(2): 138 (2000)—Colombia.
multiflora *S.S. Renner & G. Hausner in Novon*, 6(1): 108 (1996)—Ecuador.
pubancura *S.S. Renner & G. Hausner in Novon*, 6(1): 105 (1996)—Ecuador.
stellulata Perkins
var. manarae *(Steyerm.) J.A. Steyermark in Pittieria*, 7: 17 (1978): S. manarae.
vasqueziana *S.S. Renner & G. Hausner in Novon*, 10(2): 143 (2000)—Peru.
verticillata *S.S. Renner & G. Hausner in Novon*, 6(1): 111 (1996)—Ecuador.

SIPHANTHERA (Melastomatac.).
fasciculata *(Gleason) F. Almeda & O.R. Robinson in Novon*, 9(2): 131 (1999): Farringtonia fasciculata.

SIPHOCAMPYLUS (Campanulac.).
ayersiae *T.G. Lammers in Brittonia*, 50(2): 236 (1998)—Bolivia.
fallax *T.G. Lammers in Brittonia*, 50(2): 238 (1998)—Peru.
foliosus Griseb.
var. diversifolius *(Hicken) A. Pontiroli in A.L. Cabrera, Fl. Prov. Jujuy (Inst. Nac. Tecn. Agropec., 13)*, 9: 528 (1993): S. tupaeformis var. diversifolius.
rictus *T.G. Lammers in Brittonia*, 50(2): 239 (1998)—Peru.
sanchezii *T.G. Lammers in Brittonia*, 50(2): 237 (1998)—Peru.
smilax *T.G. Lammers in Brittonia*, 50(2): 240 (1998)—Bolivia.

SIPHOKENTIA (Palmae).
dransfieldii *G.G. Hambali, R.A. Maturbongs, J. Wanggai & W.J. Baker in Palms*, 44(4): 179 (2000)—New Guinea (Irian Jaya).

SIPHONOCHILUS (Zingiberac.).
bambutiorum *A.D. Poulsen & J.M. Lock in Kew Bull.*, 54(1): 203 (1999)—Zaire.
decorus *(Druten) J.M. Lock in Kew Bull.*, 54(2): 346 (1999): Kaempferia decora.

SISYMBRIOPSIS (Crucif.).
pamirica *(Y.Z. Lan & C.H. An) I.A. Al-Shehbaz, C.H. An & G. Yang in Novon,* 9(3): 311 (1999): Arabis pamirica.
shuanghuica *(K.C. Kuan & C.H. An) I.A. Al-Shehbaz, C.H. An & G. Yang in Novon,* 9(3): 311 (1999): Torularia shuanghuica.
yechengnica *(C.H. An) I.A. Al-Shehbaz, C.H. An & G. Yang in Novon,* 9(3): 312 (1999): Microsisymbrium yechengnicum.

SISYMBRIUM (Crucif.).
subgen. Buschia *V.I. Dorofeev in Turczaninowia,* 1(3): 26 (1998).
sect. Perennia *V.I. Dorofeev in Bot. Zhurn.,* 82(5): 109 (1997).
ferreyrae *H. Förther & M. Weigend in Brittonia,* 51(2): 120 (1999)—Peru.
junceum M. Bieb.
subsp. intermedium *S.V. Ovchinnikova in Fl. Sibir. (Berberidac.-Grossulariac.),* 7: 54 (1994)—Tuva.
lactucoides *H. Förther & M. Weigend in Brittonia,* 51(2): 120 (1999)—Peru.
polymorphum (Murray) Roth
var. pinnatisectum *Vassilcz. ex V.I. Dorofeev in Bot. Zhurn.,* 82(5): 109 (1997), contrary to Art. 37.5 ICBN (1994)—Russia East.
var. pinnatisectum *Vassilcz. ex V.I. Dorofeev in Turczaninowia,* 1(3): 28 (1998), contrary to Art. 37.6 ICBN (2000)—East European Russia.

SISYRINCHIUM (Iridac.).
sect. Hydastylus *(Salisb.) P. Ravenna in Onira, Bot. Leafl.,* 5(4): 17 (2000): gen. Hydastylus.
sect. Lenitium *P. Ravenna in Onira, Bot. Leafl.,* 5(4): 18 (2000).
sect. Scirpeocharis *P. Ravenna in Onira, Bot. Leafl.,* 5(4): 17 (2000).
sect. Sipacapa *P. Ravenna in Onira, Bot. Leafl.,* 5(4): 18 (2000).
angustifolium
f. album *J.K. Sim & Y.S. Kim in Korean J. Pl. Taxon.,* 22(1): 3 (1992)—South Korea.
arguellesiae *J. Ceja Romero, A. Espejo Serna & A.R. López-Ferrari in Acta Bot. Mex.,* 44: 88 (1998)—Mexico Northeast (Querétaro).
binervatum *P. Ravenna in Onira, Bot. Leafl.,* 5(4): 15 (2000)—Brazil Southeast (Minas Gerais).
brasiliense *(Ravenna) P. Ravenna in Onira, Bot. Leafl.,* 5(4): 17 (2000): Phaiophleps brasiliensis.
cholewae *A. Espejo-Serna, A.R. López-Ferrari & J. Ceja-Romero in Acta Bot. Mex.,* 49: 20 (1999)—Mexico Northeast (Durango).
coalitum *P. Ravenna in Onira, Bot. Leafl.,* 5(4): 14 (2000)—Brazil South (Santa Catarina).
cyperellum *P. Ravenna in Onira, Bot. Leafl.,* 5(4): 14 (2000)—Brazil Southeast (Minas Gerais).
firmifolium *P. Ravenna in Onira, Bot. Leafl.,* 5(4): 16 (2000)—Peru.
maulense *P. Ravenna in Onira, Bot. Leafl.,* 5(4): 17 (2000), nom. nov.: Susarium nigricans *Phil.*
parvum *(E.P. Bicknell) A. Espejo Serna & A.R. López-Ferrari in Acta Bot. Mex.,* 41: 4 (1997): Hydastylus parvus.
serrulatum *(E.P. Bicknell) A. Espejo Serna & A.R. López-Ferrari in Acta Bot. Mex.,* 41: 4 (1997): Hydastylus serrulatus.
teleanthum *P. Ravenna in Onira, Bot. Leafl.,* 5(4): 13 (2000)—Brazil South (Paraná).
translucens *(E.P. Bicknell) A. Espejo Serna & A.R. López-Ferrari in Acta Bot. Mex.,* 41: 5 (1997): Hydastylus translucens.
zamudioi *A. Espejo Serna, A.R. López-Ferrari & J. Ceja Romero in Acta Bot. Mex.,* 45: 44 (1998)—Mexico Northeast (San Luis Potosí).

SIUM (Umbellif.).
bracteatum *(Roxb.) Q.C.B. Cronk in Bull. Nat. Hist. Mus. (London), Bot.,* 25(1): 98 (1995): Angelica bracteata.

SKEPTROSTACHYS (Orchidac.).
correana *D.L. Szlachetko in Fragm. Flor. Geobot.,* 41(2): 857 (1996)—Argentina Northeast (Corrientes).

SKIMMIA (Rutac.).
japonica Thunb.
var. lutchuensis *(Nakai) T. Yamazaki in J. Jap. Bot.,* 68(5): 303 (1993): S. lutchuensis.
f. yatabei *H. Ohba in Fl. Japan,* 2c: 42 (1999), nom. nov.: S. japonica var. rugosa *Yatabe.*

SLOANEA (Elaeocarpac.).
autanae *J.A. Steyermark in Pittieria,* 7: 13 (1978)—Venezuela.
dulongensis *H. Li in Fl. Dulongjian Reg.:* 117 (1993), without latin descr. or type—China South-central (Yunnan).
geniculata *D.A. Smith in Novon,* 6(1): 120 (1996)—Costa Rica.
laevigata *D.A. Smith in Novon,* 6(1): 123 (1996)—Costa Rica.
maroana *J.A. Steyermark in Pittieria,* 7: 14 (1978)—Venezuela.
rugosa *D.A. Smith in Novon,* 6(1): 124 (1996)—Costa Rica, Nicaragua.
yapacanae *J.A. Steyermark in Pittieria,* 7: 15 (1978)—Venezuela.

SMALLANTHUS (Compos.).
macroscyphus *(Baker ex Mart.) A. Grau in M. Hermann & J. Heller (eds.), Andean roots & tubers: Ahipa etc.:* 205 (1997), without basionym ref.: Basionym not stated.
macroscyphus *(Baker) A. Grau ex A.A. Sáenz in Zuloaga & Morrone (eds.), Cat. Pl. Vasc. Rep. Arg. 2 (Monogr. Syst. Bot.* Missouri Bot. Gard. 74): 1245 (1999): Polymnia macroscypha.

SMILACINA (Convallariac.).
bicolor
var. flavovirens *N.S. Lee & J.Y. Kim in J. Pl. Biol.,* 41(1): 56 (1998)—South Korea.
japonica A. Gray
subsp. hirta *(Maxim.) V.N. Voroschilov in A.K. Skvortsov (ed.), Florist. issl. v razn. raĭonakh SSSR:* 159 (1985): S. hirta.

SMILAX (Smilacac.).
sect. Heterosmilax *(Kunth) W.S. Judd in Harvard Pap. Bot.,* 3(2): 158 (1998)—gen. Heterosmilax.
aspera L.
subsp. balearica *(Willk.) A.M. Romo, Fl. Silvestres Baleares:* 338 (1994):S. aspera var. balearica.
banglaoensis *R.H. Miau in Acta Sci. Nat. Univ. Sunyatseni,* 1982(3): 76 (1982)—China South-central (Yunnan).
guiyangensis *C.X. Fu & C.D. Shen in Acta Phytotax. Sin.,* 35(1): 70 (1997)—China South-central (Guizhou).
hajrae *R.C. Srivastava in Fl. Sikkim,* 1: 164 (1996), nom. nov.: S. rigida *Wall. ex Kunth.*
mengmaensis *R.H. Miau in Acta Sci. Nat. Univ. Sunyatseni,* 1982(3): 75 (1982)—China South-central (Yunnan).
munita *S.C. Chen in Acta Phytotax. Sin.,* 34(4): 436 (1996), nom. nov.: S. rigida *Wall. ex Kunth.*
retroflexa *(F.T. Wang & T. Tang) S.C. Chen in Acta Phytotax. Sin.,* 34(4): 436 (1996): S. aberrans var. retroflexa.

× **SMITHARA** *D.M. Cumming in Haworthiad,* 13(1): 20 (1999), nom. illegit.; *cf. Repert. Pl. Succ. (I.O.S.),* 50: 9 (1999 publ. 2000). *ALOACEAE* (Gasteria × Haworthia × Poellnitzia).

SMITHIANTHA (Gesneriac.).
aurantiaca *H. Wiehler in Gesneriana,* 1(1): 90 (1995)—Mexico Southwest (Oaxaca).

SOBRALIA (Orchidac.).
altissima *D.E. Bennett & E.A. Christenson in Orchids,* 68(11): 1112 (1999)—Peru.
ecuadorana *C.H. Dodson in Orquideologia,* 21(1): 22 (1998)—Ecuador.
gentryi *C.H. Dodson in Orquideologia,* 21(1): 26 (1998)—Ecuador.
granitica *G.A. Romero & G. Carnevali in Harvard Pap. Bot.,* 5(1): 184 (2000)—Venezuela.
hagsateri *C.H. Dodson in Orquideologia,* 21(1): 30 (1998)—Ecuador.
kerryae *R.L. Dressler in Orchid Dig.,* 62(2): 90 (1998)—Costa Rica.

SOBRALIA:-
luerorum *C.H. Dodson in Orquideologia,* 21(1): 33 (1998)—Ecuador.
neudeckeri *C.H. Dodson in Orquideologia,* 21(1): 34 (1998)—Ecuador.
oroana *C.H. Dodson in Orquideologia,* 21(1): 37 (1998)—Ecuador.
piedadiae *C.H. Dodson in Orquideologia,* 21(1): 40 (1998), as 'piedadae'—Ecuador.
purpurea *R.L. Dressler in Novon,* 10(3): 200 (2000)—Costa Rica.
stevensonii *C.H. Dodson in Orquideologia,* 21(1): 42 (1998)—Ecuador.
tamboana *C.H. Dodson in Orquideologia,* 21(1): 44 (1998)—Ecuador.

SOLANECIO *(Sch.Bip.) W.G. Walpers, Repert. bot. syst.,* 6: 273 (1846): Senecio subgen. Solanecio. (Compos.).
lainzii *F.J. Fernández Casas in Fontqueria,* 44: 140 (1996)—Gulf of Guinea Is. (Bioko).

SOLANUM (Solanac.).
ser. Anarrhichomenum *(Bitter) A.T. Hunziker in Kurtziana,* 28(1): 63 (2000), as 'Anarrhichomenia', isonym; *see A. Child in Feddes Repert.,* 101(5–6): 222, 221 (1990): S. sect. Anarrhichomenum.
sect. Californisolanum *A. Child in Feddes Repert.,* 109(5–6): 410 (1998).
subsect. Cliocarpus *(Miers) A. Child in Feddes Repert.,* 109(5–6): 413 (1998): gen. Cliocarpus.
sect. Crinita *A. Child in Feddes Repert.,* 109(5–6): 423 (1998), as 'Crinitum'.
sect. Delitescens *Barboza & A.T. Hunziker in Kurtziana,* 28(1): 63 (2000).
sect. Diamonon *(Raf.) A. Child in Feddes Repert.,* 109(5–6): 409 (1998): gen. Diamonon.
ser. Dulcamara *(Moench) Pojark. ex L.I. Ivanina in Bot. Zhurn.,* 85(6): 142 (2000): gen. Dulcamara.
sect. Erythrotricha *A. Child in Feddes Repert.,* 109(5–6): 419 (1998), as 'Erythrotrichum'.
sect. Giganteiformia *(Bitter) A. Child in Feddes Repert.,* 109(5–6): 415 (1998): S. ser. Giganteiformia.
ser. Kieseritzkianae *Pojark. ex L.I. Ivanina in Bot. Zhurn.,* 85(6): 142 (2000).
ser. Lutea *Pojark. ex L.I. Ivanina in Bot. Zhurn.,* 85(6): 144 (2000).
subsect. Multispina *A. Child in Feddes Repert.,* 109(5–6): 418 (1998), as 'Multispinum'.
ser. Neosuaveolentia *A.T. Hunziker in Kurtziana,* 28(1): 63 (2000).
subsect. Nitida *A. Child in Feddes Repert.,* 109(5–6): 409 (1998), without latin descr., as comb. nov. 'Nitidum':
sect. Polygama *(G. Don) A. Child in Feddes Repert.,* 109(5–6): 416 (1998), as 'Polygamum': S. § Polygama.
sect. Polytricha *A. Child in Feddes Repert.,* 109(5–6): 422 (1998), as 'Polytrichum'.
sect. Stellatigeminata *A. Child in Feddes Repert.,* 109(5–6): 412 (1998), as 'Stellatigeminatum'.
sect. Taeniotricha *A. Child in Feddes Repert.,* 109(5–6): 411 (1998), as 'Taeniotrichum'.
castaneum *L. d'A. Freire de Carvalho in Pesquisas, Bot., Bras.,* 46: 19 (1996), nom. nov.: S. jubatum var. gigantifolium *Dunal.*
corcovadense *A. Gras in Bull. Soc. Bot. France,* 10: 590 (1863), nom. nov.: S. capsicoides *Mart.*
delitescens *C.V. Morton, Revis. Argent. Solanum:* 168 (1976)—Argentina Northwest (Tucumán, Jujuy, Salta).
diamantinense *M.F. Agra in Novon,* 9(3): 292 (1999)—Brazil Northeast (Bahia).
dimorphandrum *S. Knapp in Bull. Nat. Hist. Mus. (London), Bot.,* 30(1): 21 (2000)—Colombia, Panama.
glabratum Dunal
var. sepicula *(Dunal) J.R.I. Wood, Handbook Yemen Fl.:* 227 (1997): S. sepicula.
habrocaulon *S. Knapp in Novon,* 6(1): 31 (1996)—Peru.
habrochaites *S. Knapp & D.M. Spooner in Novon,* 9(3): 375 (1999), nom. nov.: Lycopersicon hirsutum *Dunal.*
jabrense *M. de F. Agra & M. Nee in Brittonia,* 49(3): 350 (1997)—Brazil Northeast (Paraíba).

SOLANUM:-
litoraneum *A.E. Gonçalves in Kew Bull.*, 52(3): 703 (1997)—Mozambique.
maternum *L. Bohs in Novon*, 7(4): 341 (1997 publ. 1998)—Bolivia.
mesopliarthrum *Bitter ex Rojas & J.A. Steyermark in Pittieria*, 7: 20 (1978)—Venezuela.
mirikense *C.R. Mukhopadhyay in J. Ind. Bot. Soc.*, 72(1–2): 185 (1993)—India (West Bengal).
monarchostemon *S. Knapp in Bull. Nat. Hist. Mus. (London), Bot.*, 30(1): 23 (2000)—Ecuador, Colombia, Peru.
quebradense *S. Knapp in Novon*, 6(1): 29 (1996)—Venezuela.
schimperianum Hochst. ex A. Rich.
var. carense *(Dunal) J.R.I. Wood, Handbook Yemen Fl.*: 228 (1997): S. carense.
swartzianum Roem. & Schult.
subsp. argyrophyllum *(Dunal) L. d'A. Freire de Carvalho in Pesquisas, Bot., Bras.*, 46: 52 (1996): S. swartzianum var. argyrophyllum.
subsp. chrysophyllum *(Dunal) L. d'A. Freire de Carvalho in Pesquisas, Bot., Bras.*, 46: 50 (1996): S. swartzianum var. chrysophyllum.
tettense Klotzsch
var. renschii *(Vatke) A.E. Gonçalves in Kirkia*, 16(1): 89 (1996 publ. 1997): S. renschii.
torreanum *A.E. Gonçalves in Kew Bull.*, 52(3): 706 (1997)—Mozambique.
verecundum *M. Nee in Kurtziana*, 28(1): 137 (2000)—Ecuador, Peru.
vespertilio Aiton
subsp. doramae *Á. Marrero & M. González Martín in An. Jard. Bot. Madrid*, 56(2): 388 (1998)—Canary Is.
youngii *S. Knapp in Novon*, 6(1): 28 (1996)—Ecuador, Peru.

SOLDANELLA (Primulac.).
alpina L.
subsp. cantabrica *A. Kress, Primulac.-Stud.*, 12: 8 (1997)—Spain.

SOLENANTHUS (Boraginac.).
bakhtiaricus *M. Khatamsaz in Iranian J. Bot.*, 8(1): 6 (1999)—Iran.

SOLENOCENTRUM (Orchidac.).
maasii *R.L. Dressler in Bol. Inst. Bot. (Guadalajara)*, 5(1–3): 83 (1998)—Costa Rica.

SOLENOPHORA (Gesneriac.).
tuxtlensis *A. Ramírez-Roa & G. Ibarra-Manríquez in Novon*, 7(3): 281 (1997)—Mexico Gulf (Veracruz).

SOLENOPSIS (Campanulac.).
bivonae *(Tineo) M.B. Crespo, L. Serra & A. Juan in Pl. Syst. Evol.*, 210(3–4): 219 (1998): Lobelia bivonae.
corsica *(Meikle) M.B. Crespo, L. Serra & A. Juan in Pl. Syst. Evol.*, 210(3–4): 221 (1998): S. minuta subsp. corsica.
minima *(Sims) M.B. Crespo, L. Serra & A. Juan in Pl. Syst. Evol.*, 210(3–4): 224 (1998): Lobelia minima.

SOLENOSTEMON (Labiat.).
autranii *(Briq.) J.K. Morton in Novon*, 8(3): 266 (1998): Coleus autranii.
giorgii *(De Wild.) D. Champluvier in Syst. Geogr. Pl.*, 69(1): 16 (1999): Coleus giorgii.
porpeodon *(Baker) J.K. Morton in Novon*, 8(3): 266 (1998): Plectranthus porpeodon.

SOLIDAGO (Compos.).
subgen. Oreochrysum *(Rydb.) J.C. Semple in Univ. Waterloo Biol. Ser.*, 39: 23 (1999): gen. Oreochrysum.
asteroides *J.C. Semple in Univ. Waterloo Biol. Ser.*, 39: 72 (1999), nom. nov.: Inula alba *Nutt.*
elongata *Pépin in Ann. Fl. Pomone*, 1834–35: 55 (1834)—Georgia.
hispida Muhl. ex Willd.
var. huronensis *J.C. Semple in Univ. Waterloo Biol. Ser.*, 39: 36 (1999)—Ontario.
ridigula *Pépin in Ann. Fl. Pomone*, 1834–35: 53 (1834)—Northern America.0

SOLIDAGO :-
villosicarpa *R.J. LeBlond in Sida*, 19(2): 292 (2000)—North Carolina.
virgaurea L.
subsp. kurilensis *(Juz.) V.N. Voroschilov in A.K. Skvortsov (ed.), Florist. issl. v razn. raĭonakh SSSR*: 193 (1985): S. kurilensis.
subsp. pinetorum *J. Holub in Preslia*, 70(2): 114 (1998), nom. nov.: S. lapponica subsp. stenophylla *G.E. Schultz*.

SOLISIA (Cactac.).
solisioides *(Backeb.) A.B. Doweld in Sukkulentÿ (Moscow)*, 3(1–2): 37 (2000): Mammillaria solisioides.

SOLMS-LAUBACHIA (Crucif.).
carnosifolia *Z.X. An in Fl. Xinjiangensis*, 2(2): 377 (1995)—Xinjiang.
eurycarpa (Maxim.) Botsch.
var. lasiophylla *R.F. Huang in Fl. Qinghaiica*, 1: 510 (1997)—Qinghai.
pamirica *Z.X. An in Fl. Xinjiangensis*, 2(2): 377 (1995)—Xinjiang.

SOMALLUMA *D.C.H. Plowes in Haseltonia*, 3: 56 (1995). *ASCLEPIADACEAE*
baradii *(Lavranos) D.C.H. Plowes in Haseltonia*, 3: 57 (1995): Caralluma baradii.

SONCHUS (Compos.).
sect. Chrysoprenanthes *(Sch. Bip.) A.N. Sennikov in Compositae Newsl.*, 34: 56 (1999): Prenanthes subgen. Chrysoprenanthes.
arvensis L.
subsp. arenicola *(Vorosch.) V.N. Voroschilov in A.K. Skvortsov (ed.), Florist. issl. v razn. raĭonakh SSSR*: 199 (1985): S. arenicola.
subsp. humilis *(N.I. Orlova) N.N. Tzvelev in Novosti Sist. Vÿssh. Rast.*, 32: 185 (2000): S. humilis.
× beltraniae *U. Reifenberger & A. Reifenberger in Vieraea*, 25: 197 (1996)—Canary Is. (S. hierrensis × wildpretii).
heterophyllus *(Boulos) U. Reifenberger & A. Reifenberger in Feddes Repert.*, 107(5–6): 312 (1996): Taeckholmia heterophylla.
microcarpus *(Boulos) U. Reifenberger & A. Reifenberger in Feddes Repert.*, 107(5–6): 312 (1996): Taeckholmia microcarpa.
pendulus *(Sch. Bip.) A.N. Sennikov in Compositae Newsl.*, 34: 56 (1999): Prenanthes pendula.
sventenii *U. Reifenberger & A. Reifenberger in Feddes Repert.*, 107(5–6): 312 (1996), nom. nov.: Taeckholmia canariensis *Boulos*.
wightianus DC.
var. glaber *(Haines) H.O. Saxena & M. Brahmam, Fl. Orissa*, 2: 958 (1995): S. arvensis var. glaber.

SONDERA (Droserac.).
auriculata *(Backh.) J. Chrtek & Z. Slavíková in Novit. Bot. Univ. Carol.*, 13: 43 (1999 publ. 2000): Drosera auriculata.
bulbosa *(Hook.) J. Chrtek & Z. Slavíková in Novit. Bot. Univ. Carol.*, 13: 43 (1999 publ. 2000): Drosera bulbosa.
collina *(N.G. Marchant & Lowrie) J. Chrtek & Z. Slavíková in Novit. Bot. Univ. Carol.*, 13: 43 (1999 publ. 2000): Drosera erythrorhiza subsp. collina.
eremaea *(N.G. Marchant & Lowrie) J. Chrtek & Z. Slavíková in Novit. Bot. Univ. Carol.*, 13: 43 (1999 publ. 2000): Drosera macrantha subsp. eremaea.
erythrogyne *(N.G. Marchant & Lowrie) J. Chrtek & Z. Slavíková in Novit. Bot. Univ. Carol.*, 13: 43 (1999 publ. 2000): Drosera erythrogyne.
erythrorhiza *(Lindl.) J. Chrtek & Z. Slavíková in Novit. Bot. Univ. Carol.*, 13: 43 (1999 publ. 2000): Drosera erythrorhiza.
fimbriata *(DeBuhr) J. Chrtek & Z. Slavíková in Novit. Bot. Univ. Carol.*, 13: 43 (1999 publ. 2000): Drosera fimbriata.
gigantea *(Lindl.) J. Chrtek & Z. Slavíková in Novit. Bot. Univ. Carol.*, 13: 43 (1999 publ. 2000): Drosera gigantea.

SONDERA:-
graniticola *(N.G. Marchant) J. Chrtek & Z. Slavíková in Novit. Bot. Univ. Carol.*, 13: 43 (1999 publ. 2000): Drosera graniticola.
heterophylla *(Lindl.) J. Chrtek & Z. Slavíková in Novit. Bot. Univ. Carol.*, 13: 43 (1999 publ. 2000): Drosera heterophylla.
huegelii *(Endl.) J. Chrtek & Z. Slavíková in Novit. Bot. Univ. Carol.*, 13: 44 (1999 publ. 2000): Drosera huegelii.
humilis *(Planch.) J. Chrtek & Z. Slavíková in Novit. Bot. Univ. Carol.*, 13: 44 (1999 publ. 2000): Drosera humilis.
lowriei *(N.G. Marchant) J. Chrtek & Z. Slavíková in Novit. Bot. Univ. Carol.*, 13: 44 (1999 publ. 2000): Drosera lowriei.
macrantha *(Endl.) J. Chrtek & Z. Slavíková in Novit. Bot. Univ. Carol.*, 13: 44 (1999 publ. 2000), homonym; *see Lehm., Nov. Stirp. Pug.*, 8: 44 (1844): Drosera macrantha.
macrophylla *(Lindl.) J. Chrtek & Z. Slavíková in Novit. Bot. Univ. Carol.*, 13: 44 (1999 publ. 2000): Drosera macrophylla.
magna *(N.G. Marchant & Lowrie) J. Chrtek & Z. Slavíková in Novit. Bot. Univ. Carol.*, 13: 44 (1999 publ. 2000): Drosera erythrorhiza subsp. magna.
marchantii *(DeBuhr) J. Chrtek & Z. Slavíková in Novit. Bot. Univ. Carol.*, 13: 44 (1999 publ. 2000): Drosera marchantii.
menziesii *(R. Br.) J. Chrtek & Z. Slavíková in Novit. Bot. Univ. Carol.*, 13: 44 (1999 publ. 2000): Drosera menziesii.
microphylla *(Endl.) J. Chrtek & Z. Slavíková in Novit. Bot. Univ. Carol.*, 13: 44 (1999 publ. 2000): Drosera microphylla.
modesta *(Diels) J. Chrtek & Z. Slavíková in Novit. Bot. Univ. Carol.*, 13: 44 (1999 publ. 2000): Drosera modesta.
myriantha *(Planch.) J. Chrtek & Z. Slavíková in Novit. Bot. Univ. Carol.*, 13: 44 (1999 publ. 2000): Drosera myriantha.
neesii *(Lehm.) J. Chrtek & Z. Slavíková in Novit. Bot. Univ. Carol.*, 13: 44 (1999 publ. 2000): Drosera neesii.
orbiculata *(N.G. Marchant & Lowrie) J. Chrtek & Z. Slavíková in Novit. Bot. Univ. Carol.*, 13: 44 (1999 publ. 2000): Drosera orbiculata.
pallida *(Lindl.) J. Chrtek & Z. Slavíková in Novit. Bot. Univ. Carol.*, 13: 44 (1999 publ. 2000): Drosera pallida.
peltata *(Sm.) J. Chrtek & Z. Slavíková in Novit. Bot. Univ. Carol.*, 13: 44 (1999 publ. 2000): Drosera peltata.
planchonii *(Hook. f.) J. Chrtek & Z. Slavíková in Novit. Bot. Univ. Carol.*, 13: 44 (1999 publ. 2000): Drosera planchonii.
platypoda *(Turcz.) J. Chrtek & Z. Slavíková in Novit. Bot. Univ. Carol.*, 13: 44 (1999 publ. 2000): Drosera platypoda.
porrecta *(Lehm.) J. Chrtek & Z. Slavíková in Novit. Bot. Univ. Carol.*, 13: 44 (1999 publ. 2000): Drosera porrecta.
ramellosa *(Lehm.) J. Chrtek & Z. Slavíková in Novit. Bot. Univ. Carol.*, 13: 45 (1999 publ. 2000): Drosera ramellosa.
rosulata *(Lehm.) J. Chrtek & Z. Slavíková in Novit. Bot. Univ. Carol.*, 13: 45 (1999 publ. 2000): Drosera rosulata.
salina *(N.G. Marchant & Lowrie) J. Chrtek & Z. Slavíková in Novit. Bot. Univ. Carol.*, 13: 45 (1999 publ. 2000): Drosera salina.
squamosa *(Benth.) J. Chrtek & Z. Slavíková in Novit. Bot. Univ. Carol.*, 13: 45 (1999 publ. 2000): Drosera squamosa.
stolonifera *(Endl.) J. Chrtek & Z. Slavíková in Novit. Bot. Univ. Carol.*, 13: 45 (1999 publ. 2000): Drosera stolonifera.
subhirtella *(Planch.) J. Chrtek & Z. Slavíková in Novit. Bot. Univ. Carol.*, 13: 45 (1999 publ. 2000): Drosera subhirtella.
sulphurea *(Lehm.) J. Chrtek & Z. Slavíková in Novit. Bot. Univ. Carol.*, 13: 45 (1999 publ. 2000): Drosera sulphurea.
thysanosepala *(Diels) J. Chrtek & Z. Slavíková in Novit. Bot. Univ. Carol.*, 13: 45 (1999 publ. 2000): Drosera thysanosepala.

SONDERA:-
tubaestylis *(N.G. Marchant & Lowrie) J. Chrtek & Z. Slavíková in Novit. Bot. Univ. Carol.*, 13: 45 (1999 publ. 2000): Drosera tubaestylis.
whittakeri *(Planch.) J. Chrtek & Z. Slavíková in Novit. Bot. Univ. Carol.*, 13: 45 (1999 publ. 2000): Drosera whittakeri.
zonaria *(Planch.) J. Chrtek & Z. Slavíková in Novit. Bot. Univ. Carol.*, 13: 45 (1999 publ. 2000): Drosera zonaria.

SONERILA (Melastomatac.).
loeiensis *N. Cellinese & S.S. Renner in Novon*, 7(2): 111 (1997)—Thailand.
nagyana *N. Cellinese in Novon*, 7(2) 104 (1997)—Borneo (Kalimantan).
parameswaranii *K. Ravikumar & V. Lakshmanan in Rheedea*, 9(1): 63 (1999)—India (Tamil Nadu).
urceolata *N. Cellinese & S.S. Renner in Novon*, 7(2): 109 (1997)—Thailand.
vatphouensis *J. Munzinger & C.V. Martin in Adansonia*, 22(2): 212 (2000)—Laos.

SOPHORA (Legumin.).
tomentosa L.
var. angustifoliolata *A. Barreto Valdés in Fontqueria*, 44: 16 (1996)—Cuba.

SOPHRONITIS (Orchidac.).
alaorii *(Brieger & Bicalho) C. Van den Berg & M.W. Chase in Lindleyana*, 15(2): 115 (2000): Laelia alaorii.
alvaroana *(F.E.L. Miranda) C. Van den Berg & M.W. Chase in Lindleyana*, 15(2): 115 (2000): Laelia alvaroana.
angereri *(Pabst) C. Van den Berg & M.W. Chase in Lindleyana*, 15(2): 115 (2000): Laelia angereri.
bahiensis *(Schltr.) C. Van den Berg & M.W. Chase in Lindleyana*, 15(2): 116 (2000): Laelia bahiensis.
blumenscheinii *(Pabst) C. Van den Berg & M.W. Chase in Lindleyana*, 15(2): 116 (2000): Laelia blumenscheinii.
bradei *(Pabst) C. Van den Berg & M.W. Chase in Lindleyana*, 15(2): 116 (2000): Laelia bradei.
brevicaulis *(H.G. Jones) C. Van den Berg & M.W. Chase in Lindleyana*, 15(2): 116 (2000): Hoffmannseggella brevicaulis.
briegeri *(Blumensch. ex Pabst) C. Van den Berg & M.W. Chase in Lindleyana*, 15(2): 116 (2000): Laelia briegeri.
× caetensis *(Pabst) C. Van den Berg & M.W. Chase in Lindleyana*, 15(2): 116 (2000): Laelia × caetensis.
× carassana *(Pabst) C. Van den Berg & M.W. Chase in Lindleyana*, 15(2): 116 (2000): Laelia × carassana.
caulescens *(Lindl.) C. Van den Berg & M.W. Chase in Lindleyana*, 15(2): 116 (2000): Laelia caulescens.
cernua Lindl.
var. alagoensis *A.B. Gomes Ferreira in Orquidário*, 12(1): 15 (1998)—Brazil Northeast (Alagoas).
cinnabarina *(Bateman ex Lindl.) C. Van den Berg & M.W. Chase in Lindleyana*, 15(2): 116 (2000): Laelia cinnabarina.
× cipoensis *(Pabst) C. Van den Berg & M.W. Chase in Lindleyana*, 15(2): 116 (2000): Laelia × cipoensis.
crispa *(Lindl.) C. Van den Berg & M.W. Chase in Lindleyana*, 15(2): 116 (2000): Cattleya crispa.
crispata *(Thunb.) C. Van den Berg & M.W. Chase in Lindleyana*, 15(2): 116 (2000): Cymbidium crispatum.
dayana *(Rchb. f.) C. Van den Berg & M.W. Chase in Lindleyana*, 15(2): 116 (2000): Laelia dayana.
duveenii *(Fowlie) C. Van den Berg & M.W. Chase in Lindleyana*, 15(2): 116 (2000): Laelia duveenii.
endsfeldzii *(Pabst) C. Van den Berg & M.W. Chase in Lindleyana*, 15(2): 116 (2000): Laelia endsfeldzii.
esalqueana *(Blumensch. ex Pabst) C. Van den Berg & M.W. Chase in Lindleyana*, 15(2): 116 (2000): Laelia esalqueana.
fidelensis *(Pabst) C. Van den Berg & M.W. Chase in Lindleyana*, 15(2): 117 (2000): Laelia fidelensis.
fournieri *(Cogn.) C. Van den Berg & M.W. Chase in Lindleyana*, 15(2): 117 (2000): Laelia longipes var. fournieri.
× gerhard-santosii *(Pabst) C. Van den Berg & M.W. Chase in Lindleyana*, 15(2): 117 (2000): Laelia × gerhard-santosii.

SOPHRONITIS:-
ghillanyi *(Pabst) C. Van den Berg & M.W. Chase in Lindleyana*, 15(2): 117 (2000): Laelia ghillanyi.
gloedeniana *(Hoehne) C. Van den Berg & M.W. Chase in Lindleyana*, 15(2): 117 (2000): Laelia gloedeniana.
gracilis *(Pabst) C. Van den Berg & M.W. Chase in Lindleyana*, 15(2): 117 (2000): Laelia gracilis.
grandis *(Lindl. & Paxton) C. Van den Berg & M.W. Chase in Lindleyana*, 15(2): 117 (2000): Laelia grandis.
harpophylla *(Rchb. f.) C. Van den Berg & M.W. Chase in Lindleyana*, 15(2): 117 (2000): Laelia harpophylla.
hispidula *(Pabst & A.F. Mello) C. Van den Berg & M.W. Chase in Lindleyana*, 15(2): 117 (2000): Laelia hispidula.
itambana *(Pabst) C. Van den Berg & M.W. Chase in Lindleyana*, 15(2): 117 (2000): Laelia itambana.
jongheana *(Rchb. f.) C. Van den Berg & M.W. Chase in Lindleyana*, 15(2): 117 (2000): Laelia jongheana.
kautskyi *(Pabst) C. Van den Berg & M.W. Chase in Lindleyana*, 15(2): 117 (2000): Laelia kautskyi.
kettieana *(Pabst) C. Van den Berg & M.W. Chase in Lindleyana*, 15(2): 117 (2000): Laelia kettieana.
liliputana *(Pabst) C. Van den Berg & M.W. Chase in Lindleyana*, 15(2): 117 (2000): Laelia liliputana.
lobata *(Lindl.) C. Van den Berg & M.W. Chase in Lindleyana*, 15(2): 117 (2000): Cattleya lobata.
longipes *(Rchb. f.) C. Van den Berg & M.W. Chase in Lindleyana*, 15(2): 117 (2000): Laelia longipes.
lundii *(Rchb. f. & Warm.) C. Van den Berg & M.W. Chase in Lindleyana*, 15(2): 117 (2000): Bletia lundii.
macrobulbosa *(Pabst) C. Van den Berg & M.W. Chase in Lindleyana*, 15(2): 117 (2000): Laelia macrobulbosa.
milleri *(Blumensch. ex Pabst) C. Van den Berg & M.W. Chase in Lindleyana*, 15(2): 118 (2000): Laelia milleri.
mirandae *C. Van den Berg & M.W. Chase in Lindleyana*, 15(2): 118 (2000), nom. nov., as 'mirandai': Laelia rupestris *Lindl.*
mixta *(Hoehne) C. Van den Berg & M.W. Chase in Lindleyana*, 15(2): 118 (2000): Laelia mixta.
× mucugense *(F.E.L. Miranda) C. Van den Berg & M.W. Chase in Lindleyana*, 15(2): 118 (2000): Laelia × mucugense.
munchowiana *(F.E.L. Miranda) C. Van den Berg & M.W. Chase in Lindleyana*, 15(2): 118 (2000): Laelia munchowiana.
perrinii *(Lindl.) C. Van den Berg & M.W. Chase in Lindleyana*, 15(2): 118 (2000): Cattleya perrinii.
pfisteri *(Pabst & Senghas) C. Van den Berg & M.W. Chase in Lindleyana*, 15(2): 118 (2000): Laelia pfisteri.
praestans *(Linden & Rchb. f.) C. Van den Berg & M.W. Chase in Lindleyana*, 15(2): 118 (2000): Laelia praestans.
pumila *(Hook.) C. Van den Berg & M.W. Chase in Lindleyana*, 15(2): 118 (2000): Cattleya pumila.
purpurata *(Lindl. & Paxton) C. Van den Berg & M.W. Chase in Lindleyana*, 15(2): 118 (2000): Laelia purpurata.
reginae *(Pabst) C. Van den Berg & M.W. Chase in Lindleyana*, 15(2): 118 (2000): Laelia reginae.
sanguiloba *(Withner) C. Van den Berg & M.W. Chase in Lindleyana*, 15(2): 118 (2000): Laelia sanguiloba.
sincorana *(Schltr.) C. Van den Berg & M.W. Chase in Lindleyana*, 15(2): 118 (2000): Laelia sincorana.
tenebrosa *(Rolfe) C. Van den Berg & M.W. Chase in Lindleyana*, 15(2): 118 (2000): Laelia grandis var. tenebrosa.
tereticaulis *(Hoehne) C. Van den Berg & M.W. Chase in Lindleyana*, 15(2): 118 (2000): Laelia tereticaulis.
verboonenii *(F.E.L. Miranda) C. Van den Berg & M.W. Chase in Lindleyana*, 15(2): 118 (2000): Laelia verboonenii.
virens *(Lindl.) C. Van den Berg & M.W. Chase in Lindleyana*, 15(2): 118 (2000): Laelia virens.
xanthina *(Lindl.) C. Van den Berg & M.W. Chase in Lindleyana*, 15(2): 119 (2000): Laelia xanthina.

SOPUBIA (Scrophulariac.).
sect. Densiflorae *R. Mielcarek in Fragm. Flor. Geobot.*, 41(1): 210 (1996).
sect. Laxiflorae *R. Mielcarek in Fragm. Flor. Geobot.*, 41(1): 199 (1996).
conferta S. Moore
var. congensis *(S. Moore) R. Mielcarek in Fragm. Flor. Geobot.*, 41(1): 218 (1996): S. congensis.
gracilis *H.-P. Hofmann & Eb. Fischer in Adansonia*, 20(2): 308 (1998)—Madagascar.
lemuriana *H.-P. Hofmann & Eb. Fischer in Adansonia*, 20(2): 308 (1998)—Madagascar.
mannii Skan
var. metallorum *(P.A. Duvign.) R. Mielcarek in Fragm. Flor. Geobot.*, 41(1): 209 (1996): S. metallorum.
matsumurae *(T. Yamaz.) C.Y. Wu in Novon*, 9(2): 288 (1999): Petitmenginia matsumurae.
parviflora Engl.
subsp. eminii *(Engl.) R. Mielcarek in Fragm. Flor. Geobot.*, 41(1): 202 (1996): S. eminii.

SORBARIA (Rosac.).
grandiflora (Sweet) Maxim.
subsp. rhoifolia *(Kom.) V.V. Jakubov in Sosud. Rast. Sovet. Dal'nego Vostoka*, 8: 139 (1996): S. rhoifolia.
pallasii (G. Don) Pojark.
subsp. rhoifolia *(Kom.) V.N. Voroschilov in A.K. Skvortsov (ed.), Florist. issl. v razn. raĭonakh SSSR*: 176 (1985): S. rhoifolia.

SORBUS (Rosac.).
× abscondita *M. Kovanda in Verh. Zool.-Bot. Ges. Österr.*, 133: 339 (1996)—Czechoslovakia. (S. aucuparia × danubialis).
acutisecta *R. Reuther & O. Schwarz in Wiss. Zeitschr. Pädag. Hochsch. Erfurt-Mühlhausen, Mat.-Naturwiss.* Reihe, 7(1): 54 (1971)—Germany.
alnifrons *M. Kovanda in Verh. Zool.-Bot. Ges. Österr.*, 133: 356 (1996)—Czechoslovakia.
alnoides *M. Gandoger, Dec. Pl. Nov.*, 1: 23 (1875)—Europe.
arvernensis *M. Gandoger, Dec. Pl. Nov.*, 1: 24 (1875)—France.
betulifolia *(Bunge) S. Schauer in Übers. Arbeiten Veränd. Schles. Ges. Vaterl. Kult.*, 1847: 292 (1848): Pyrus betulifolia.†
controversa *M. Gandoger, Dec. Pl. Nov.*, 1: 23 (1875)—Europe.
coronata (Cardot) T.T. Yu & H.T. Tsai
var. ambrozyana *(C.K. Schneid.) L.T. Lu in Acta Phytotax. Sin.*, 38(3): 280 (2000): S. ambrozyana.
crenata *(Royle) S. Schauer in Übers. Arbeiten Veränd. Schles. Ges. Vaterl. Kult.*, 1847: 292 (1848): Pyrus crenata.†
dolomiticola *V. Mikoláš in Thaiszia*, 6(1–2): 2 (1996 publ. 1997)—Czechoslovakia.
flabellifolia *(Spach) S. Schauer in Übers. Arbeiten Veränd. Schles. Ges. Vaterl. Kult.*, 1847: 292 (1848): Crataegus flabellifolia.†
gemella *M. Kovanda in Verh. Zool.-Bot. Ges. Österr.*, 133: 329 (1996)—Czechoslovakia.
graeca *(Spach) Lodd. ex S. Schauer in Übers. Arbeiten Veränd. Schles. Ges. Vaterl. Kult.*, 1847: 292 (1848): Crataegus graeca.†
hardeggensis *M. Kovanda in Verh. Zool.-Bot. Ges. Österr.*, 133: 348 (1996)—Czechoslovakia, Austria.
hupehensis C.K. Schneid.
var. paucijuga *(D.K. Zang & P.C. Huang) L.T. Lu in Acta Phytotax. Sin.*, 38(3): 279 (2000): S. discolor var. paucijuga.
isenacensis *R. Reuther in Haussknechtia, Mitt. Thüring. Bot. Ges.*, 6: 17 (1997)—Germany.
kamaonensis *(Wall. ex G. Don) S. Schauer in Übers. Arbeiten Veränd. Schles. Ges. Vaterl. Kult.*, 1847: 292 (1848): Pyrus kamaonensis.†
lanata *(D. Don) S. Schauer in Übers. Arbeiten Veränd. Schles. Ges. Vaterl. Kult.*, 1847: 292 (1848): Pyrus lanata.†
pekarovae *J. Májovský & D. Bernátová in Biologia, Bot., (Slovakia)*, 51(1): 25 (1996)—Czechoslovakia.
platyodon *M. Gandoger, Dec. Pl. Nov.*, 1: 23 (1875)—Europe.

SORBUS:-
quernea *M. Kovanda in Verh. Zool.-Bot. Ges. Österr.,* 133: 335 (1996)—Czechoslovakia.
reverchonii *M. Gandoger, Dec. Pl. Nov.,* 1: 24 (1875)—France.
rhodanthera *M. Kovanda in Verh. Zool.-Bot. Ges. Österr.,* 133: 321 (1996)—Czechoslovakia.
tianschanica Rupr.
var. tomentosa *Chang Y. Yang & Y.L. Han in Fl. Xinjiangensis,* 2(2): 381 (1995)—Xinjiang.
ursina *(Wall. ex G. Don) S. Schauer in Übers. Arbeiten Veränd. Schles. Ges. Vaterl. Kult.,* 1847: 292 (1848): Pyrus ursina.†
variolosa *(Wall. ex G. Don) S. Schauer in Übers. Arbeiten Veränd. Schles. Ges. Vaterl. Kult.,* 1847: 292 (1848): Pyrus variolosa.†
vestita *(Wall. ex G. Don) S. Schauer in Übers. Arbeiten Veränd. Schles. Ges. Vaterl. Kult.,* 1847: 292 (1848): Pyrus vestita.†
× yokouchii *M. Mizush. ex T. Shimizu, Fl. Nagano Pref.:* 1508 (1997)—Japan. (S. matsumurana × sambucifolia).
yunnanensis *L.T. Lu in Acta Phytotax. Sin.,* 38(3): 279 (2000), nom. nov.: S. carpinifolia *T.T. Yu & L.T. Lu.*

SORGHASTRUM (Gramin.).
pohlianum *P. Dávila, L.I. Cabrera & R. Lira in Brittonia,* 50(3): 305 (1998)—Mexico Southwest (Colima).

SORINDEIA (Anacardiac.).
gabonensis *H. Bourobou Bourobou & F.J. Breteler in Bull. Jard. Bot. Nation. Belg.,* 66(3–4): 342 (1997)—Gabon.
oxyandra *H. Bourobou Bourobou & F.J. Breteler in Syst. Geogr. Pl.,* 69(1): 115 (1999)—Gabon.
ripicola *D. Champluvier in Syst. Geogr. Pl.,* 69(1): 42 (1999)—Zaire.

SOROCEA (Morac.).
jaramilloi *C.C. Berg in Novon,* 6(3): 241 (1996)—Ecuador, Colombia.
jureiana *S. Romaniuc Neto in Albertoa,* 4(8): 98 (1996)—Brazil Southeast (São Paulo).
pubivena Hemsl.
subsp. hirtella *(Mildbr.) C.C. Berg in Novon,* 6(3): 243 (1996): S. hirtella.
subsp. oligotricha *(Akkermans & C.C. Berg) C.C. Berg in Novon,* 6(3): 243 (1996): S. hirtella subsp. oligotricha.
ruminata *C.C. Berg in Novon,* 6(3): 244 (1996)—Panama, Colombia.
sprucei (Baill.) J.F. Macbr.
subsp. subumbellata *C.C. Berg in Novon,* 6(3): 245 (1996)—Ecuador, Peru.

SOULANGIA *A.T. Brongniart, Mém. Fam. Rhamnées:* 70 (1826). *RHAMNACEAE*†
axillaris *(Lam.) A.T. Brongniart, Mém. Fam. Rhamnées:* 71 (1826): Phylica axillaris.†
buxifolia *(L.) A.T. Brongniart, Mém. Fam. Rhamnées:* 71 (1826): Phylica buxifolia.†
cordata *(L.) A.T. Brongniart, Mém. Fam. Rhamnées:* 71 (1826): Phylica cordata.†
oleifolia *(Vent.) A.T. Brongniart, Mém. Fam. Rhamnées:* 71 (1826): Phylica oleifolia.†
paniculata *(Willd.) A.T. Brongniart, Mém. Fam. Rhamnées:* 71 (1826): Phylica paniculata.†
thymifolia *(Vent.) A.T. Brongniart, Mém. Fam. Rhamnées:* 71 (1826): Phylica thymifolia.†

SOUROUBEA (Marcgraviac.).
gilgii *V.A. Richter in Math. Term. Ertes.,* 34: 835 (1916)—Guatemala.

SPARAXIS (Iridac.).
auriculata *P. Goldblatt & J.C. Manning in Bothalia,* 29(1): 59 (1999)—Cape Province.
metelerkampiae *(L. Bolus) P. Goldblatt & J.C. Manning in Bothalia,* 29(1): 61 (1999): Synnotia metelerkampiae.

SPARGANIUM (Typhac.).
neglectum
subsp. microcarpum *(Neuman) J. Holub in Preslia,* 70(2): 114 (1998): S. ramosum f. microcarpum.

SPATHANTHEUM (Arac.).
heterandrum *(Baker) N.E. Brown, Suppl. Johnson's Gard. Dict.:* 1007 (1882): Gamochlamys heterandra.
intermedium *J. Bogner in Sendtnera,* 4: 5 (1997)—Peru.

SPATHIPHYLLUM (Arac.).
montanum *(R.A. Baker) M.H. Grayum in Phytologia,* 82(1): 50 (1997): S. wendlandii subsp. montanum.

SPATHOGLOTTIS (Orchidac.).
umbraticola *L.A. Garay in Harvard Pap. Bot.,* 4(1): 312 (1999)—New Guinea (Papua New Guinea).

SPATHOLIRION (Commelinac.).
decumbens *N. Fukuoka & N. Kurosaki in Acta Phytotax. Geobot.,* 48(2): 147 (1997 publ. 1998)—Thailand.

SPATHULOPETALUM (Asclepiadac.).
arachnoideum *(P.R.O. Bally) D.C.H. Plowes in Haseltonia,* 3: 56 (1995): Caralluma gracilipes var. arachnoidea.
congestiflorum *(P.R.O. Bally) D.C.H. Plowes in Haseltonia,* 3: 56 (1995): Caralluma congestiflora.
edwardsiae *(M.G. Gilbert) D.C.H. Plowes in Haseltonia,* 3: 56 (1995): Caralluma gracilipes subsp. edwardsiae.
gracilipes *(K. Schum.) D.C.H. Plowes in Haseltonia,* 3: 56 (1995): Caralluma gracilipes.
longiflorum *(M.G. Gilbert) D.C.H. Plowes in Haseltonia,* 3: 56 (1995): Caralluma longiflora.
mogadoxense *(Chiov.) D.C.H. Plowes in Haseltonia,* 3: 56 (1995): Caralluma mogadoxensis.
moniliforme *(P.R.O. Bally) D.C.H. Plowes in Haseltonia,* 3: 56 (1995): Caralluma moniliformis.
peckii *(P.R.O. Bally) D.C.H. Plowes in Haseltonia,* 3: 56 (1995): Caralluma peckii.
priogonium *(K. Schum.) D.C.H. Plowes in Haseltonia,* 3: 56 (1995): Caralluma priogonium.
turneri *(E.A. Bruce) D.C.H. Plowes in Haseltonia,* 3: 56 (1995): Caralluma turneri.
vaduliae *(Lavranos) D.C.H. Plowes in Haseltonia,* 3: 56 (1995): Caralluma vaduliae.

SPELAEANTHUS *R. Kiew, A. Weber & B.L. Burtt in Beitr. Biol. Pfl.,* 70(2–3): 401 (1997–98 publ. 1998). *GESNERIACEAE*
chinii *R. Kiew, A. Weber & B.L. Burtt in Beitr. Biol. Pfl.,* 70(2–3): 401 (1997–98 publ. 1998)—Malaya (Peninsular Malaysia).

SPERGULA (Caryophyllac.).
sect. Pentandrae *N.N. Tzvelev in Novosti Sist. Výssh. Rast.,* 32: 27 (2000).

SPERGULARIA (Caryophyllac.).
subgen. Delia *(Dumort.) N.N. Tzvelev in Novosti Sist. Výssh. Rast.,* 32: 30 (2000): gen. Delia.
sect. Diandrae *N.N. Tzvelev in Novosti Sist. Výssh. Rast.,* 32: 30 (2000).
sect. Leiosperma *(Kindb.) N.N. Tzvelev in Novosti Sist. Výssh. Rast.,* 32: 27 (2000): Lepigonum sect. Leiosperma.
angustata *(Clavaud) N.N. Tzvelev in Novosti Sist. Výssh. Rast.,* 32: 27 (2000): S. marginata var. angustata.
bungeana *N.N. Tzvelev in Novosti Sist. Výssh. Rast.,* 32: 29 (2000), nom. nov.: Arenaria halophila *Bunge.*
media *(L.) C. Presl, Fl. Sicul.:* 161 (1826): Arenaria media.†
pinguis *(Fenzl) N.N. Tzvelev in Novosti Sist. Výssh. Rast.,* 32: 29 (2000): S. rubra var. pinguis.
sezer-zenginii *Ş. Yıldırımlı in Ot Sist. Bot. Dergisi,* 4(2): 4 (1997 publ. 1998)—Turkey.
syvaschica *N.N. Tzvelev in Novosti Sist. Výssh. Rast.,* 32: 28 (2000)—Krym.

SPERMACOCE (Rubiac.).
aprica *(Hiern) R. Govaerts, World Checklist Seed Plants,* 2(1–2): 14 (1996): Tardavel aprica.
arenicola *(Brandegee) R. Govaerts, World Checklist Seed Plants,* 2(1–2): 14 (1996): Borreria arenicola.
aristeguietiana *(Steyerm.) R. Govaerts, World Checklist Seed Plants,* 2(1–2): 14 (1996): Borreria aristeguietiana.

SPERMACOCE:-
asperula *(DC.) R. Govaerts, World Checklist Seed Plants,* 2(1–2): 14 (1996): Borreria asperula.
balansae *(Standl.) R. Govaerts, World Checklist Seed Plants,* 2(1–2): 15 (1996): Borreria balansae.
bolivarensis *(Steyerm.) R. Govaerts, World Checklist Seed Plants,* 2(1–2): 15 (1996): Borreria bolivarensis.
bradei *(Standl.) R. Govaerts, World Checklist Seed Plants,* 2(1–2): 15 (1996): Borreria bradei.
buchneri *(K. Schum.) R. Govaerts, World Checklist Seed Plants,* 2(1–2): 15 (1996): Borreria buchneri.
carolinensis *R. Govaerts, World Checklist Seed Plants,* 2(1–2): 15 (1996), nom. nov.: Borreria rotundifolia *Valeton.*
casmae *R. Govaerts, World Checklist Seed Plants,* 2(1–2): 15 (1996), nom. nov.: Borreria humifusa *Mart.*
cataractarum *(Steyerm.) R. Govaerts, World Checklist Seed Plants,* 2(1–2): 15 (1996): Borreria cataractarum.
chacoensis *(Hassl.) R. Govaerts, World Checklist Seed Plants,* 2(1–2): 15 (1996): Borreria chacoensis.
chartensis *(Standl.) R. Govaerts, World Checklist Seed Plants,* 2(1–2): 15 (1996): Borreria chartensis.
chodatiana *(Standl.) R. Govaerts, World Checklist Seed Plants,* 2(1–2): 15 (1996): Borreria chodatiana.
clinopodioides *(Standl.) R. Govaerts, World Checklist Seed Plants,* 2(1–2): 15 (1996): Borreria clinopodioides.
confertifolia *(Steyerm.) R. Govaerts, World Checklist Seed Plants,* 2(1–2): 15 (1996): Borreria confertifolia.
cyperoides *(Chodat) R. Govaerts, World Checklist Seed Plants,* 2(1–2): 15 (1996): Borreria cyperoides.
dichasia *(Sucre & C.G. Costa) R. Govaerts, World Checklist Seed Plants,* 2(1–2): 15 (1996): Borreria dichasia.
egleri *(Sucre & C.G. Costa) R. Govaerts, World Checklist Seed Plants,* 2(1–2): 15 (1996): Borreria egleri.
evenia *(Standl.) R. Govaerts, World Checklist Seed Plants,* 2(1–2): 15 (1996): Borreria evenia.
exigua *(S. Moore) R. Govaerts, World Checklist Seed Plants,* 2(1–2): 16 (1996): Borreria exigua.
fruticosa *(Standl.) R. Govaerts, World Checklist Seed Plants,* 2(1–2): 16 (1996): Borreria fruticosa.
garuensis *(K. Krause) R. Govaerts, World Checklist Seed Plants,* 2(1–2): 16 (1996): Borreria garuensis.
gilliesiae *(Specht) R. Govaerts, World Checklist Seed Plants,* 2(1–2): 16 (1996), isonym; *see J.R. Clarkson in Austrobaileya,* 4(1): 117 (1993): Borreria gilliesiae.
guaranitica *(Chodat) R. Govaerts, World Checklist Seed Plants,* 2(1–2): 16 (1996): Borreria guaranitica.
hassleriana *(Chodat) R. Govaerts, World Checklist Seed Plants,* 2(1–2): 16 (1996): Borreria hassleriana.
herzogii *(S. Moore) R. Govaerts, World Checklist Seed Plants,* 2(1–2): 16 (1996): Borreria herzogii.
heterophylla *(Brandegee) R. Govaerts, World Checklist Seed Plants,* 2(1–2): 16 (1996): Borreria heterophylla.
hillii *(Chippend.) R. Govaerts, World Checklist Seed Plants,* 2(1–2): 16 (1996): Borreria hillii.
hombbei *(De Wild.) R. Govaerts, World Checklist Seed Plants,* 2(1–2): 16 (1996): Borreria hombbei.
hondurensis *R. Govaerts, World Checklist Seed Plants,* 2(1–2): 16 (1996), nom. nov.: Borreria malacophylla *Standl. & L.O. Williams.*
hondurensis *C.H. Nelson in An. Jard. Bot. Madrid,* 58(1): 203 (2000), nom. nov., homonym; *see Govaerts, World Checklist Seed Plants,* 2(1–2): 16 (1996): Borreria fruticosa *Standl.*
intricata *(Steyerm.) R. Govaerts, World Checklist Seed Plants,* 2(1–2): 16 (1996): Borreria intricata.
islana *R. Govaerts, World Checklist Seed Plants,* 2(1–2): 16 (1996), nom. nov.: Spermacoce exserta *Benth.*
ivorensis *R. Govaerts, World Checklist Seed Plants,* 2(1–2): 16 (1996), nom. nov.: Borreria macrantha *Hepper.*
jangouxii *(Steyerm.) R. Govaerts, World Checklist Seed Plants,* 2(1–2): 16 (1996): Borreria jangouxii.

SPERMACOCE:-
krausei *(Suess.) R. Govaerts, World Checklist Seed Plants,* 2(1–2): 16 (1996): Borreria krausei.
lagunensis *(M.E. Jones) R. Govaerts, World Checklist Seed Plants,* 2(1–2): 16 (1996): Borreria lagunensis.
lagurus *(S. Moore) R. Govaerts, World Checklist Seed Plants,* 2(1–2): 16 (1996): Borreria lagurus.
lancea *(Hiern) R. Govaerts, World Checklist Seed Plants,* 2(1–2): 17 (1996): Tardavel lancea.
ledermannii *(K. Krause) R. Govaerts, World Checklist Seed Plants,* 2(1–2): 17 (1996): Borreria ledermannii.
liliifolia *(Standl.) R. Govaerts, World Checklist Seed Plants,* 2(1–2): 17 (1996): Borreria liliifolia.
limae *(Sucre) R. Govaerts, World Checklist Seed Plants,* 2(1–2): 17, 382 (1996): Borreria limae.
loretiana *(E.L. Cabral) R. Govaerts, World Checklist Seed Plants,* 2(1–2): 17 (1996): Borreria loretiana.
luteovirens *(Standl.) R. Govaerts, World Checklist Seed Plants,* 2(1–2): 17 (1996): Borreria luteovirens.
macrocephala *(Standl. & Steyerm.) R. Govaerts, World Checklist Seed Plants,* 2(1–2): 17 (1996): Borreria macrocephala.
malacophylla *(K. Schum.) R. Govaerts, World Checklist Seed Plants,* 2(1–2): 17 (1996): Borreria malacophylla.
marticrovettiana *(E.L. Cabral) R. Govaerts, World Checklist Seed Plants,* 2(1–2): 17 (1996): Borreria marticrovettiana.
matogrossensis *R. Govaerts, World Checklist Seed Plants,* 2(1–2): 17 (1996), nom. nov.: Borreria cristata *S. Moore.*
melochioides *(Standl.) R. Govaerts, World Checklist Seed Plants,* 2(1–2): 17 (1996): Borreria melochioides.
mexicana *R. Govaerts, World Checklist Seed Plants,* 2(1–2): 17 (1996), nom. nov.: Borreria aspera *M. Martens & Galeotti.*
mitreoloides *(Standl.) R. Govaerts, World Checklist Seed Plants,* 2(1–2): 17 (1996): Borreria mitreoloides.
molleri *(Gand.) R. Govaerts, World Checklist Seed Plants,* 2(1–2): 17 (1996): Borreria molleri.
nelidae *(E.L. Cabral) R. Govaerts, World Checklist Seed Plants,* 2(1–2): 17 (1996): Borreria nelidae.
neodiffusa *R. Govaerts, World Checklist Seed Plants,* 2(1–2): 17 (1996), nom. nov.: Borreria diffusa *DC.*
neohispida *R. Govaerts, World Checklist Seed Plants,* 2(1–2): 18 (1996), nom. nov.: Borreria hispida *Spruce ex K. Schum.*
neosuffruticosa *R. Govaerts, World Checklist Seed Plants,* 2(1–2): 18 (1996), nom. nov.: Borreria suffruticosa *R. Br. ex Benth.*
neotenera *R. Govaerts, World Checklist Seed Plants,* 2(1–2): 18 (1996), nom. nov.: Borreria tenera *DC.*
neotenuis *R. Govaerts, World Checklist Seed Plants,* 2(1–2): 18 (1996), nom. nov.: Borreria tenuis *DC.*
neoterminalis *R. Govaerts, World Checklist Seed Plants,* 2(1–2): 18 (1996), nom. nov.: Borreria terminalis *Small*
neovirgata *R. Govaerts, World Checklist Seed Plants,* 2(1–2): 18 (1996), nom. nov.: Borreria virgata *Cham. & Schltdl.*
noronhensis *(Sucre) R. Govaerts, World Checklist Seed Plants,* 2(1–2): 18 (1996): Borreria noronhensis.
oligodonta *(Steyerm.) R. Govaerts, World Checklist Seed Plants,* 2(1–2): 18 (1996): Borreria oligodonta.
paraguariensis *(Chodat) R. Govaerts, World Checklist Seed Plants,* 2(1–2): 18 (1996): Borreria paraguariensis.
parviceps *(Ridl.) R. Govaerts, World Checklist Seed Plants,* 2(1–2): 18 (8 Nov. 1996), isonym; *see I.M. Turner in Novon,* 6(2): 222 (24 June 1996): Borreria parviceps.
parviceps *(Ridl.) I.M. Turner in Novon,* 6(2): 222 (1996): Borreria parviceps.
petraea *(Pires-O'Brien) R. Govaerts, World Checklist Seed Plants,* 2(1–2): 18 (1996): Borreria petraea.
pilulifera *(Ridl.) I.M. Turner in Novon,* 6(2): 222 (24 June 1996): Borreria pilulifera.
pilulifera *(Ridl.) R. Govaerts, World Checklist Seed Plants,* 2(1–2): 18 (8 Nov. 1996), isonym; *see I.M. Turner in Novon,* 6(2): 222 (24 June 1996): Borreria pilulifera.

SPERMACOCE:-
pumilio *(Standl.) R. Govaerts, World Checklist Seed Plants,* 2(1–2): 18 (1996): Borreria pumilio.
quadrifaria *(E.L. Cabral) R. Govaerts, World Checklist Seed Plants,* 2(1–2): 18 (1996): Borreria quadrifaria.
reflexa *(J.H. Kirkbr.) R. Govaerts, World Checklist Seed Plants,* 2(1–2): 18 (1996): Borreria reflexa.
schumanniana *(Taub. ex Ule) R. Govaerts, World Checklist Seed Plants,* 2(1–2): 18 (1996): Borreria schumanniana.
shandongensis *(F.Z. Li & X.D. Chen) R. Govaerts, World Checklist Seed Plants,* 2(1–2): 18 (1996): Borreria shandongensis.
simplicicaulis *(K. Schum. ex Sucre) R. Govaerts, World Checklist Seed Plants,* 2(1–2): 19 (1996): Borreria simplicicaulis.
somalica *(K. Schum.) R. Govaerts, World Checklist Seed Plants,* 2(1–2): 19 (1996): Borreria somalica.
stachydea DC.
var. phyllocephala *(DC.) J.-P. Lebrun & A.L. Stork, Enum. Pl. Fleurs Afr. Trop.,* 4: 227 (1997): S. phyllocephala.
tonalensis *(Brandegee) R. Govaerts, World Checklist Seed Plants,* 2(1–2): 19 (1996): Borreria tonalensis.
valens *(Standl.) R. Govaerts, World Checklist Seed Plants,* 2(1–2): 19 (1996): Borreria valens.
viridiflora *(Hassl.) R. Govaerts, World Checklist Seed Plants,* 2(1–2): 19 (1996): Borreria viridiflora.
vulpina *(Standl.) R. Govaerts, World Checklist Seed Plants,* 2(1–2): 19 (1996): Borreria vulpina.
weygaertii *R. Govaerts, World Checklist Seed Plants,* 2(1–2): 19 (1996), nom. nov.: Borreria nana *Standl.*
wurdackii *(Steyerm.) R. Govaerts, World Checklist Seed Plants,* 2(1–2): 19 (1996): Borreria wurdackii.
xanthophylla *(Bremek.) R. Govaerts, World Checklist Seed Plants,* 2(1–2): 19 (1996): Borreria xanthophylla.

SPHAENOLOBIUM (Umbellif.).
coriaceum *(Korovin) M.G. Pimenov in Opred. Rast. Sred. Azii,* 7: 264 (1983): Selinum coriaceum.

SPHAERADENIA (Cyclanthac.).
alba *R. Eriksson in Op. Bot.,* 126: 87 (1995)—Panama.
brachiolata *R. Eriksson in Op. Bot.,* 126: 71 (1995)—Ecuador.
carnosa *R. Eriksson in Op. Bot.,* 126: 64 (1995)—Colombia.
columnaris *R. Eriksson in Op. Bot.,* 126: 74 (1995)—Colombia.
compacta *R. Eriksson in Op. Bot.,* 126: 54 (1995)—Colombia.
crassiceps *R. Eriksson in Op. Bot.,* 126: 54 (1995)—Peru.
distans *R. Eriksson in Op. Bot.,* 126: 72 (1995)—Colombia.
garciae Harling
subsp. cuadrosii *(Galeano & R. Bernal) R. Eriksson in Op. Bot.,* 126: 92 (1995): S. cuadrosii.
gigantea *R. Eriksson in Op. Bot.,* 126: 49 (1995)—Colombia.
laucheana (Mast.) Harling
subsp. irazuensis *(Cufod.) R. Eriksson in Op. Bot.,* 126: 40 (1995): Carludovica irazuensis.
magniglobula *R. Eriksson in Op. Bot.,* 126: 87 (1995)—Panama, Costa Rica.
meridionalis *R. Eriksson in Op. Bot.,* 126: 55 (1995)—Bolivia, Peru.
occidentalis *R. Eriksson in Op. Bot.,* 126: 80 (1995)—Costa Rica.
oxystigma *R. Eriksson in Op. Bot.,* 126: 42 (1995)—Ecuador, Peru.
pallida *R. Eriksson in Op. Bot.,* 126: 76 (1995)—Colombia.
perangusta *R. Eriksson in Op. Bot.,* 126: 47 (1995)—Peru.
praetermissa *R. Eriksson in Op. Bot.,* 126: 79 (1995)—Costa Rica.
proboscidifera *R. Eriksson in Op. Bot.,* 126: 74 (1995)—Colombia.
pulchra *R. Eriksson in Op. Bot.,* 126: 92 (1995)—Colombia.

SPHAERADENIA:-
sanctae-barbarae Harling
subsp. monostyla *R. Eriksson in Op. Bot.,* 126: 46 (1995)—Ecuador.
scandens *R. Eriksson in Op. Bot.,* 126: 82 (1995)—Colombia.
versicolor *R. Eriksson in Op. Bot.,* 126: 51 (1995)—Ecuador.
virella *R. Eriksson in Op. Bot.,* 126: 99 (1995)—Panama.

SPHAERALCEA (Malvac.).
janeae *(S.L. Welsh) S.L. Welsh in Great Basin Nat.,* 58(4): 389 (1998): S. leptophylla var. janeae.

SPHAERANTHUS (Compos.).
samburuensis *H.J. Beentje in Kew Bull.,* 54(1): 100 (1999)—Kenya.

SPHAEROLOBIUM (Legumin.).
pubescens *R. Butcher in Nuytsia,* 12(2): 172 (1998)—Western Australia.
rostratum *R. Butcher in Nuytsia,* 12(2): 174 (1998)—Western Australia.

SPHAGNETICOLA (Compos.).
brachycarpa *(Baker) J.F. Pruski in Novon,* 6(4): 411 (1996): Wedelia brachycarpa.
calendulacea *(L.) J.F. Pruski in Novon,* 6(4): 411 (1996): Verbesina calendulacea.
gracilis *(Rich.) J.F. Pruski in Novon,* 6(4): 412 (1996): Wedelia gracilis.
trilobata *(L.) J.F. Pruski in Fl. St. John (Mem. New York Bot. Gard., 78):* 114 (17 Dec. 1996); *et in Novon,* 6(4): 412 (27 Dec. 1996): Silphium trilobatum.

SPHENOSTEMON (Sphenostemonac.).
thibaudii *J. Jérémie in Bull. Mus. Nation. Hist. Nat., B, Adansonia,* Ser. 4, 18(3–4): 222 (1996)—New Caledonia.
tireliae *J. Jérémie in Bull. Mus. Nation. Hist. Nat., B, Adansonia,* Ser. 4, 18(3–4): 219 (1996), as 'tirelii'—New Caledonia.

SPHINGA *R.C. Barneby & J.W. Grimes, Silk Tree, Guanacaste, Monkey's Earring (Mem. New York Bot. Gard., 74(1)):* 160 (1996). *LEGUMINOSAE*
acatlensis *(Benth.) R.C. Barneby & J.W. Grimes, Silk Tree, Guanacaste, Monkey's Earring (Mem. New York Bot. Gard., 74(1)):* 161 (1996): Pithecellobium acatlense.
platyloba *(DC.) R.C. Barneby & J.W. Grimes, Silk Tree, Guanacaste, Monkey's Earring (Mem. New York Bot. Gard., 74(1)):* 164 (1996): Acacia platyloba.
prehensilis *(C. Wright) R.C. Barneby & J.W. Grimes, Silk Tree, Guanacaste, Monkey's Earring (Mem. New York Bot. Gard., 74(1)):* 161 (1996): Calliandra prehensilis.

SPHYRASTYLIS (Orchidac.).
dressleri *A.L.V. Toscano de Brito in Lindleyana,* 14(1): 24 (1999)—Panama.
garayi *D.E. Bennett & E.A. Christenson in Lindleyana,* 13(2): 83 (1998)—Peru.
tsubotae *P. Ortiz V.in Orquideologia,* 20(2): 242 (1996)—Colombia.

SPHYROSPERMUM (Ericac.).
lanceolatum *J.L. Luteyn in Fl. Ecuador,* 54(147): 350 (1996)—Ecuador.

SPIGELIA (Loganiac.).
chiapensis *K.R. Gould in Brittonia,* 51(4): 410 (1999)—Mexico Southeast (Chiapas).
coelostylioides *K.R. Gould in Brittonia,* 51(4): 407 (1999)—Mexico Southeast (Chiapas), Mexico Southwest (Oaxaca), Mexico Northeast (San Luis Potosí, Tamaulipas), Mexico Gulf (Veracruz), Mexico Northwest (Sonora), Guatemala.
cremnophila *D.C. Zappi & E. Lucas in Kew Bull.,* 53(4): 981 (1998)—Brazil Northeast (Bahia).
gentianoides Chapm. ex A.DC.
var. alabamensis *K. Gould in Sida,* 17(2): 418 (1996)—Alabama.

SPIGELIA:-
trispicata *H. Hurley ex K.R. Gould in Brittonia,* 51(4): 412 (1999)—Mexico Southwest (Guerrero).

SPIRADICLIS (Rubiac.).
subgen. Sinospiradiclis *H.S. Lo in Bull. Bot. Res. North-East. Forest. Univ.,* 18(3): 275 (1998).
corymbosa *W.L. Sha & X.X. Chen ex H.S. Lo in Bull. Bot. Res. North-East. Forest. Univ.,* 18(3): 276 (1998), contrary to Art. 37.4, 37.5 ICBN (1994)—China Southeast (Guangxi).
microcarpa *H.S. Lo in Bull. Bot. Res. North-East. Forest. Univ.,* 18(3): 276 (1998), contrary to Art. 37.4, 37.5 ICBN (1994)—China Southeast (Guangxi).

SPIRAEA (Rosac.).
aemiliana C.K. Schneid.
subsp. stevenii *(C.K. Schneid.) V.N. Voroschilov in A.K. Skvortsov (ed.), Florist. issl. v razn. raĭonakh SSSR:* 175 (1985): S. beauverdiana var. stevenii.
cantoniensis Lour.
var. jiangxiensis *(Z.X. Yu) L.T. Lu in Acta Phytotax. Sin.,* 38(3): 276 (2000): S. jiangxiensis.
fritschiana C.K. Schneid.
var. villosa *Y.Q. Zhu & D.K. Zang in Bull. Bot. Res. North-East. Forest. Univ.,* 18(1): 9 (1998)—China North-central (Shandong).
grefsheimii *N.N. Tzvelev, Opred. Sosud. Rast. Severo-Zap. Rossii:* 679 (2000)—Cultivation.
× hitchcockii *W.J. Hess & N.A. Stoynoff in Sida,* 18(3): 827 (1999)—Oregon.(S. douglasii var. menziesii × splendens var. rosea).
nanojaponica *N.N. Tzvelev, Opred. Sosud. Rast. Severo-Zap. Rossii:* 679 (2000), as 'nano-japonica'—Cultivation.
roxburghiana *N. Wallich, Report on Calcutta Bot. Gard. to G.A. Bushby:* 29 (1840), nom. nov.: S. corymbosa *Roxb.*
teniana Rehder
var. mairei *(H. Lév.) L.T. Lu in Acta Phytotax. Sin.,* 38(3): 276 (2000): S. mairei.
ussuriensis Pojark.
subsp. elegans *(Pojark.) V.V. Jakubov in Sosud. Rast. Sovet. Dal'nego Vostoka,* 8: 132 (1996): S. elegans.
vanhouttei *(Briot) E.A. Carrière in Rev. Hort.,* 48: 260 (1876): S. aquilegifolia var. vanhouuttei.†
xizangensis *L.T. Lu in Acta Phytotax. Sin.,* 38(3): 276 (2000), nom. nov.: S. tibetica *T.T. Yu & L.T. Lu.*

SPIRALLUMA *D.C.H. Plowes in Haseltonia,* 3: 53 (1995). *ASCLEPIADACEAE*
longidens *(N.E. Br.) D.C.H. Plowes in Haseltonia,* 3: 53 (1995): Caralluma longidens.
mouretii *(A. Chev.) D.C.H. Plowes in Haseltonia,* 3: 54 (1995): Caralluma mouretii.

SPIRANTHES (Orchidac.).
actinosophila *(Barb.Rodr.) B.D. Jackson, Index Kewensis,* vol. 2, fasc. 4: 966 (1895): Stenorrhynchos actinosophila.†
× australis *P.M. Brown in North Amer. Native Orchid J.,* 5(4): 358 (1999), homonym; *see Lindl. in Bot. Reg.,* 10: sub t. 823 (1824)—Florida. (S. praecox × vernalis).
× itchetuckneensis *P.M. Brown in North Amer. Native Orchid J.,* 5(4): 362 (1999)—Florida. (S. odorata × ovalis).
multiflora *(Barb.Rodr.) B.D. Jackson, Index Kewensis,* vol. 2 fasc. 4: 966 (1895): Sarcoglottis multiflora.†
ornithocephala *(Barb.Rodr.) B.D. Jackson, Index Kewensis,* vol. 2, fasc. 4: 967 (1895): Sarcoglottis ornithocephala.†

SPONDIAS (Anacardiac.).
testudinis *J.D. Mitchell & D.C. Daly in Brittonia,* 50(4): 448 (1998)—Brazil North (Acre), Peru, Bolivia.

SPORADANTHUS (Restionac.).
caudatus *(L.A.S. Johnson & O.D. Evans) B.G. Briggs & L.A.S. Johnson in Telopea,* 8(1): 30 (1998): Lepyrodia caudata.

SPORADANTHUS:-
ferrugineus *P.J. de Lange, P.B. Heenan & B.D. Clarkson in New Zealand J. Bot.,* 37(3): 415 (1999)—New Zealand North I.
gracilis *(R. Br.) B.G. Briggs & L.A.S. Johnson in Telopea,* 8(1): 30 (1998): Lepyrodia gracilis.
interruptus *(F. Muell.) B.G. Briggs & L.A.S. Johnson in Telopea,* 8(1): 30 (1998): Lepyrodia interrupta.
strictus *(R. Br.) B.G. Briggs & L.A.S. Johnson in Telopea,* 8(1): 30 (1998): Lepyrodia stricta.
tasmanicus *(Hook. f.) B.G. Briggs & L.A.S. Johnson in Telopea,* 8(1): 31 (1998): Lepyrodia tasmanica.

SPOROBOLUS (Gramin.).
crucensis *S.A. Renvoize, Gramineas de Bolivia:* 334 (1998)—Bolivia.
eximius
var. latifolius *S. de Castro Boechat & H.M. Longhi-Wagner in Acta Bot. Brasil.,* 9(1): 50 (1995)—Brazil South (Paraná).
fibrosus *T.A. Cope in Kew Bull.,* 53(1): 171 (1998)—Zambia.
hajrae *P. Umamaheswari & P. Daniel in Nordic J. Bot.,* 18(5): 577 (1998)—India (Tamil Nadu).
latzii *B.K. Simon in Austral. Syst. Bot.,* 12(3): 399 (1999)—Northern Territory.
metallicola *H.M. Longhi-Wagner & S.C. Boechat in Acta Bot. Brasil.,* 7(2): 150 (1993), as 'metallicolus'—Brazil Southeast (Minas Gerais), Brazil South (Paraná).
minarum *S.C. Boechat & H.M. Longhi-Wagner in Iheringia, Bot.,* 44: 34 (1994)—Brazil Southeast (Minas Gerais).
mopane *T.A. Cope in Kew Bull.,* 53(1): 168 (1998)—Zimbabwe.
multiramosus *H.M. Longhi-Wagner & S.C. Boechat in Acta Bot. Brasil.,* 7(2): 152 (1993)—Brazil North (Pará).
nudiramus *S.C. Boechat & H.M. Longhi-Wagner in Iheringia, Bot.,* 44: 36 (1994)— Brazil South (Paraná).
paucifolius *S.C. Boechat & H.M. Longhi-Wagner in Iheringia, Bot.,* 44: 37 (1994)—Brazil West-central (Goiás).
pinetorum *A.S. Weakley & P.M. Peterson in Sida,* 18(1): 258 (1998)—North Carolina, Georgia, South Carolina.
pyramidatus (Lam.) Hitchc.
subsp. domingensis *(Trin.) L. Catasús Guerra in Fontqueria,* 44: 143 (1996): Vilfa domingensis.
reflexus *S.C. Boechat & H.M. Longhi-Wagner in Iheringia, Bot.,* 44: 38 (1994)—Brazil West-central (Brasilia Distrito Federal), Brazil Southeast (Minas Gerais, Rio de Janeiro).
sessilis *B.K. Simon in Austral. Syst. Bot.,* 12(3): 408 (1999)—Jawa, Queensland, New South Wales.

SPRAGUEANELLA (Loranthac.).
curta *D. Wiens & R.M. Polhill, Mistletoes of Africa:* 113 (1998)—Tanzania.

SPREKELIA (Amaryllidac.).
howardii *D.J. Lehmiller in Herbertia,* 54: 230 (1999 publ. 2000)—Mexico Southwest (Guerrero).

STACHYPITYS *A.V.F.Ch. Bobrov & A.P. Melikian in Bot. Zhurn.,* 85(7): 58 (2000). *PODOCARPACEAE*
disticha *(J. Buchholz) A.V.F.Ch. Bobrov & A.P. Melikyan in Bot. Zhurn.,* 85(7): 58 (2000), as 'distichus': Podocarpus distichus.
ferruginea *(G. Benn. ex D. Don) A.V.F.Ch. Bobrov & A.P. Melikian in Bot. Zhurn.,* 85(7): 58 (2000), as 'ferrugineus': Podocarpus ferrugineus.
ferruginoides *(Compton) A.V.F.Ch. Bobrov & A.P. Melikian in Bot. Zhurn.,* 85(7): 58 (2000): Podocarpus ferruginoides.
ladei *(F.M. Bailey) A.V.F.Ch. Bobrov & A.P. Melikian in Bot. Zhurn.,* 85(7): 58 (2000): Podocarpus ladei.

STACHYS (Labiat.).
aculeolata Hook. f.
var. inermis *Y.B. Harvey in Kew Bull.,* 51(3): 440 (1996)—Uganda, Burundi, Zaire, Tanzania, Malawi.

STACHYS:-
anamurensis *H. Sümbül in Turkish J. Bot.,* 14(1): 57 (1990)—Turkey.
arenaria Vahl
subsp. mollis *(Willd. ex Benth.) F. Gómiz in An. Jard. Bot. Madrid,* 58(1): 199 (2000): S. mollis.
aspera Michx.
var. hispidula *(Regel) V.N. Voroschilov in A.K. Skvortsov (ed.), Florist. issl. v razn. raĭonakh SSSR:* 189 (1985): S. palustris var. hispidula.
var. hispidula *(Regel) T.V. Krestovskaja in Sosud. Rast. Sovet. Dal'nego Vostoka,* 7: 344 (1995), isonym: S. palustris var. hispidula.
var. japonica *(Miq.) V.N. Voroschilov in A.K. Skvortsov (ed.), Florist. issl. v razn. raĭonakh SSSR:* 189 (1985): S. japonica.
asperata *I.C. Hedge in Fl. Madagascar,* fam. 175: 28 (1998)—Madagascar.
cydni *Kotschy ex Y. Gemici & E. Leblebici in Turkish J. Bot.,* 22(5): 359 (1998)—Turkey.
filifolia *I.C. Hedge in Fl. Madagascar,* fam. 175: 37 (1998)—Madagascar.
forsythii *I.C. Hedge in Fl. Madagascar,* fam. 175: 30 (1998)—Madagascar.
glandulibracteata *Y.B. Harvey in Kew Bull.,* 51(3): 452 (1996)—Tanzania.
humbertii *I.C. Hedge in Fl. Madagascar,* fam. 175: 34 (1998)—Madagascar.
macedonica *K. Micevski in Prilozi Oddel. Biol. Med. Nauki Makedonska Akad. Nauk. Umet.,* 9(1–2): 20 (1988 publ. 1991)—Yugoslavia.
oreophila *I.C. Hedge in Fl. Madagascar,* fam. 175: 31 (1998)—Madagascar.
pseudohumifusa Sebsebe
subsp. minutiflora *Y.B. Harvey in Kew Bull.,* 51(3): 451 (1996)—Zambia.
subsp. saxeri *Y.B. Harvey in Kew Bull.,* 51(3): 451 (1996)—Cameroon, Nigeria.
reptans *I.C. Hedge in Fl. Madagascar,* fam. 175: 26 (1998)—Madagascar.
rubella *I.C. Hedge in Fl. Madagascar,* fam. 175: 33 (1998)—Madagascar.
scardica (Griseb.) Hayek
var. bistrae *K. Micevski in Prilozi Oddel. Biol. Med. Nauki Makedonska Akad. Nauk. Umet.,* 9(1–2): 23 (1988 publ. 1991)—Yugoslavia.
viridis Boiss. & Heldr.
var. gevgeliensis *K. Micevski in Prilozi Oddel. Biol. Med. Nauki Makedonska Akad. Nauk. Umet.,* 9(1–2): 26 (1988 publ. 1991)—Yugoslavia.

STACHYTARPHETA (Verbenac.).
fallax *A.E. Gonçalves in Garcia de Orta, Ser. Bot.,* 13(1): 70 (1996), nom. nov.: Ubochea dichotoma *Baill.*

STACHYURUS (Stachyurac.).
himalaicus Hook. f. & Thomson ex Benth.
var. lancifolius *(Koidz.) S. Hatusima in S. Hatusima & T. Amano, Fl. Ryukyus, S. of Amami Is., ed. 2:* 135 (1994), without basionym ref.: Basionym not stated.
subsp. macrocarpus *(Koidz.) S. Hatusima in S. Hatusima & T. Amano, Fl. Ryukyus, S. of Amami Is., ed. 2:* 135 (1994), without basionym ref.: Basionym not stated.
praecox Siebold & Zucc.
var. macrocarpus *(Koidz.) Tuyama ex H. Ohba in Fl. Japan,* 2c: 192 (1999), as 'macrocarpa': S. macrocarpus.

STAELIA (Rubiac.).
hatschbachii *J.H. Kirkbride in Brittonia,* 49(3): 376 (1997)—Brazil Southeast (Minas Gerais).

STAHLIANTHUS (Zingiberac.).
rubrimarginatus *S.Q. Tong in Acta Phytotax. Sin.,* 33(5): 499 (1995), as 'rubro-marginatus'—China South-central (Yunnan).

STANHOPEA (Orchidac.).
subgen. Candida *G. Gerlach & Dodson in Schlechter Orchideen, ed. 3,* I/C(37–38): 2357 (1999 publ. 2000).
sect. Connata *G. Gerlach & Dodson in Schlechter Orchideen, ed. 3,* I/C(37–38): 2357 (1999 publ. 2000).

STANHOPEA:-
sect. Saccata *G. Gerlach & Dodson in Schlechter Orchideen, ed. 3,* I/C(37–38): 2357 (1999 publ. 2000).
subgen. Stanhopeastrum *G. Gerlach & Dodson in Schlechter Orchideen, ed. 3,* I/C(37–38): 2357 (1999 publ. 2000).
sect. Tricornis *G. Gerlach & Dodson in Schlechter Orchideen, ed. 3,* I/C(37–38): 2357 (1999 publ. 2000).
sect. Wardii *G. Gerlach & Dodson in Schlechter Orchideen, ed. 3,* I/C(37–38): 2357 (1999 publ. 2000).
carchiensis *C.H. Dodson in Orquideologia,* 21(1): 46 (1998)—Ecuador.
greeri *R. Jenny in Austral. Orchid Rev.,* 65(3): 6 (2000), as 'greerii', contrary to Art. 37.5 ICBN (1994); *et in Austral. Orchid Rev.,* 65(6): 28 (2000)—Peru.
maduroi *C.H. Dodson & Dressler in Orquideologia,* 21(1): 62 (1998)—Panama.
platyceras Rchb. f.
f. gunnii *R. Jenny in Orchidee,* 48(1): 7 (1997)—Colombia.
pseudoradiosa *R. Jenny & R. Gonzalez-Tamayo in Caesiana,* 9: 38 (1997)—Mexico Southwest (Oaxaca).

STAPELIA (Asclepiadac.).
johni-lavrani *J.J. Halda in Acta Mus. Richnov. Sect. Nat.,* 5(1): 42 (1998)—Namibia.

STATHMOSTELMA (Asclepiadac.).
angustatum K. Schum.
subsp. vomeriforme *(S. Moore) D.J. Goyder in Kew Bull.,* 53(3): 603 (1998): Asclepias vomeriformis.
diversifolium *D.J. Goyder in Kew Bull.,* 53(3): 604 (1998)—Ethiopia, Kenya, Tanzania.
fornicatum (N.E. Br.) Bullock
subsp. tridentatum *D.J. Goyder in Kew Bull.,* 53(3): 593 (1998)—Tanzania.
katangense *(De Wild.) D.J. Goyder in Kew Bull.,* 53(3): 612 (1998): Asclepias katangensis.
spectabile (N.E. Br.) Schltr.
subsp. frommii *(Schltr.) D.J. Goyder in Kew Bull.,* 53(3): 585 (1998): S. frommii.
welwitschii Britten & Rendle
var. bagshawei *(S. Moore) D.J. Goyder in Kew Bull.,* 53(3): 599 (1998): Asclepias bagshawei.

STAUNTONIA (Lardizabalac.).
sect. Leuconota *H.N. Qin in Cathaya,* 8–9: 130 (1997).
libera *H.N. Qin in Cathaya,* 8–9: 136 (1997)—Xizang, Burma, China South-central (Yunnan).
obovatifoliola Hayata
subsp. urophylla *(Hand.-Mazz.) H.N. Qin in Cathaya,* 8–9: 164 (1997): S. hexaphylla var. urophylla.

STAURACANTHUS (Legumin.).
*subgen.*Nepa *(Webb) Paiva ex S. Talavera & F.J. Salgueiro in An. Jard. Bot. Madrid,* 57(1): 217 (1999): gen. Nepa.

STAUROCHILUS (Orchidac.).
agusanensis *(Ames & Quisumb.) H. Fessel & E. Lückel in Orchidee,* 49(1): 47 (1998): Trichoglottis agusanensis.
intermedius *(L.O. Williams) H. Fessel & E. Lückel in Orchidee,* 49(1): 47 (1998): Trichoglottis intermedia.
tamesii *(Quisumb. & C. Schweinf.) H. Fessel & E. Lückel in Orchidee,* 49(1): 47 (1998): Trichoglottis tamesii.

STAUROGETON (Lemnac.).
sect. Biformes *(Landolt) N.N. Tzvelev in Bot. Zhurn.,* 84(7): 114 (1999): Lemna sect. Biformes.
tener *(Kurz) N.N. Tzvelev in Bot. Zhurn.,* 84(7): 114 (1999): Lemna tenera.

STAUROGYNE (Acanthac.).
carvalhoi *S.R. Profice in Bradea,* 8(33): 203 (2000)—Brazil Northeast (Bahia), Brazil Southeast (Espirito Santo).
parvicaulis *B. Hansen in Nordic J. Bot.,* 15(6): 585 (1995 publ. 1996)—Thailand.

STEINCHISMA (Gramin.).
stenophylla *(Hack.) F.O. Zuloaga & O. Morrone in Ann. Missouri Bot. Gard.,* 85(4): 651 (1998): Panicum stenophyllum.

STELIS (Orchidac.).
asseris *O. Duque in Orquideologia,* 20(3): 329 (1997)—Colombia.
barbuda *O. Duque in Orquideologia,* 20(3): 330 (1997)—Colombia.
bella *O. Duque in Orquideologia,* 20(3): 331 (1997)—Colombia.
carcharodonta *G. Carnevali & G.A. Romero in G.A. Romero & G. Carnevali, Orchids of Venezuela, ed. 2:* 1146 (2000)—Venezuela.
clipeus *O. Duque in Orquideologia,* 20(3): 333 (1997)—Colombia.
comica *O. Duque in Orquideologia,* 20(3): 338 (1997)—Colombia.
croatii *C.A. Luer in Lindleyana,* 11(2): 97 (1996)—Costa Rica.
dentata *O. Duque in Orquideologia,* 20(3): 339 (1997)—Colombia.
erecta *O. Duque in Orquideologia,* 20(3): 341 (1997)—Colombia.
foetida *O. Duque in Orquideologia,* 20(3): 342 (1997)—Colombia.
franciscana *O. Duque in Orquideologia,* 20(3): 343 (1997)—Colombia.
frontinensis *O. Duque in Orquideologia,* 20(3): 347 (1997)—Colombia.
gustavii *O. Duque in Orquideologia,* 20(3): 348 (1997)—Colombia.
inclinata *O. Duque in Orquideologia,* 20(3): 349 (1997)—Colombia.
ingramii *C.A. Luer in Lindleyana,* 11(2): 100 (1996)—Costa Rica.
kareniae *C.A. Luer in Lindleyana,* 11(2): 100 (1996), as 'karenae'—Costa Rica.
londonnii *O. Duque in Orquideologia,* 20(3): 352 (1997)—Colombia.
longipetala *O. Duque in Orquideologia,* 20(3): 353 (1997)—Colombia.
longirepens *G. Carnevali & J.L. Tapia in G.A. Romero & G. Carnevali, Orchids of Venezuela, ed. 2:* 1146 (2000)—Venezuela.
lumbricosa *O. Duque in Orquideologia,* 20(3): 354 (1997)—Colombia.
morae *C.A. Luer in Lindleyana,* 11(2): 103 (1996)—Costa Rica.
oscitans *C.A. Luer in Lindleyana,* 11(2): 105 (1996)—Panama, Costa Rica.
papilio *O. Duque in Orquideologia,* 20(3): 358 (1997)—Colombia.
papiliopsis *O. Duque in Orquideologia,* 20(3): 361 (1997)—Colombia.
pyramidalis *O. Duque in Orquideologia,* 20(3): 364 (1997)—Colombia.
retroversa *O. Duque in Orquideologia,* 20(3): 365 (1997)—Colombia.
salazarii *R. Solano Gómez in Lindleyana,* 15(1): 48 (2000)— Mexico Southwest (Oaxaca).
scaphoides *O. Duque in Orquideologia,* 20(3): 369 (1997)—Colombia.
scutella *O. Duque in Orquideologia,* 20(3): 370 (1997)—Colombia.
stapedia *O. Duque in Orquideologia,* 20(3): 372 (1997)—Colombia.
stella *O. Duque in Orquideologia,* 20(3): 373 (1997)—Colombia.
taurina *O. Duque in Orquideologia,* 20(3): 374 (1997)—Colombia.
wendtii *R. Solano Gómez in Lindleyana,* 15(1): 50 (2000)—Mexico Southwest (Oaxaca).

STELLARIA (Caryophyllac.).
sect. Alsine *(L.) N.N. Tzvelev, Opred. Sosud. Rast. Severo-Zap. Rossii:* 679 (2000): gen. Alsine.
subgen. Alsine *(L.) N.N. Tzvelev, Opred. Sosud. Rast. Severo-Zap. Rossii:* 679 (2000): gen. Alsine.
subsect. Asterotrichae *T. Yamazaki in J. Jap. Bot.,* 70(6): 340 (1995).
alsine Grimm.
subsp. inundata *(Vorosch.) V.N. Voroschilov in A.K. Skvortsov (ed.), Florist. issl. v razn. raĭonakh SSSR:* 165 (1985): S. inundata.
subsp. undulata *(Thunb.) V.N. Voroschilov in A.K. Skvortsov (ed.), Florist. issl. v razn. raĭonakh SSSR:* 165 (1985): S. undulata.

STELLARIA:-
altimontana *N.S. Pavlova in Sosud. Rast. Sovet. Dal'nego Vostoka,* 8: 81 (1996)—Khabarovsk.
capillipes *(Franch.) C.Y. Wu in Index Fl. Yunnan.,* 1: 261 (1984): Stellaria saxatilis var. capillipes.
dichotoma L.
f. glabra *G.A. Peschkova, Stepnaya Fl. Baĭkal'skoĭ Sibiri:* 55 (1972), as comb. nov., without type—Siberia.
fragilis *M.V. Klokov in Novosti Sist. Vȳssh. Nizsh. Rast. (Kiev),* 1974: 19 (1974)—Ukraina.
gypsophylloides Fenzl
var. lanceolata *(Bunge) Yu.P. Kozhevnikov in Rast. Tsentral. Azii,* 11: 31 (1994 publ. 1995): S. dichotoma γ lanceolata.
hassiana *Loes. in Beih. Bot. Centralbl.,* 37(2): 119 (1919)—China North-Central (Shandong).
hippoctona *(Czern.) M.V. Klokov in Novosti Sist. Vȳssh. Nizsh. Rast. (Kiev),* 1974: 15 (1974): S. graminea β hippoctona.
holostea L.
subsp. hispidula *A.P. Khokhrjakov in Byull. Glavn. Bot. Sada (Moscow),* 175: 49 (1997)—Turkey.
miahuatlana *B.L. Turner in Phytologia,* 79(4): 303 (1995 publ. 1996)—Mexico Southwest (Oaxaca).
neglecta Weihe
var. prespaensis *K. Micevski in Prilozi Oddel. Biol. Med. Nauki Makedonska Akad. Nauk. Umet.,* 9(1–2): 46 (1988 publ. 1991)—Yugoslavia.
strongylosepala *Hand.-Mazz. in Oesterr. Bot. Z.,* 88(4): 301 (1939)—Inner Mongolia.
vestita Kurz
var. amplexicaulis *(Hand.-Mazz.) C.Y. Wu in Index Fl. Yunnan.,* 1: 263 (1984): Stellaria saxatilis var. amplexicaulis.
williamsiana *Yu.P. Kozhevnikov in Rast. Tsentral. Azii,* 11: 37 (1994 publ. 1995), nom. nov.: Arenaria monosperma *F.N. Williams.*

STELLEROPSIS (Thymelaeac.).
altaica (Thiéb.-Bern. ex Pers.) Pobed.
subsp. tianschanica *(Pobed.) R.V. Kamelin, Mater. Istorii Fl. Azii (Altaisk. Gorn. Strana):* 69 (1998): S. tianschanica.

STELLILABIUM (Orchidac.).
sect. Dipterostele *(Schltr.) R.L. Dressler in Harvard Pap. Bot.,* 4(2): 470 (1999): gen. Dipterostele.
sect. Pseudobulbosa *D.E. Bennett & E.A. Christenson in Lindleyana,* 13(2): 86 (1998), as 'Pseudobulbosae'.
sect. Rhamphostele *R.L. Dressler in Harvard Pap. Bot.,* 4(2): 470 (1999).
sect. Taeniorhachis *R.L. Dressler in Harvard Pap. Bot.,* 4(2): 470 (1999).
aciculare *R.L. Dressler in Harvard Pap. Bot.,* 4(2): 471 (1999)—Panama, Costa Rica.
alticola *C.H. Dodson & R. Escobar in Orquideologia,* 21(1): 48 (1998), as 'alticolum'—Ecuador.
barbozae *J.T. Atwood & R.L. Dressler in Selbyana,* 16(2): 239 (1995)—Costa Rica.
bergoldii *(Garay & Dunst.) G. Carnevali & G.A. Romero in G.A. Romero & G. Carnevali, Orchids of Venezuela, ed. 2:* 1147 (2000): Trichoceros bergoldii.
butcheri *R.L. Dressler in Harvard Pap. Bot.,* 4(2): 472 (1999)—Panama.
fortunae *R.L. Dressler in Harvard Pap. Bot.,* 4(2): 473 (1999)—Panama.
ibischii *R. Vásquez Ch. in Lindleyana,* 13(2): 61 (1998)—Bolivia.
lankesteri *(Ames) R.L. Dressler in Novon,* 7(2): 124 (1997): Telipogon lankesteri.
morii *R.L. Dressler in Harvard Pap. Bot.,* 4(2): 473 (1999)—Panama.
peruvianum *D.E. Bennett & E.A. Christenson in Lindleyana,* 13(2): 86 (1998)—Peru.
pseudobulbosum *D.E. Bennett & E.A. Christenson in Lindleyana,* 13(2): 88 (1998)— Peru.

STEMMACANTHA (Compos.).
caulescens *(Coss. & Balansa) A. Soldano in Compositae Newsl.,* 32: 19 (1998): Rhaponticum caulescens.
imatongensis *(Philipson) A. Soldano in Compositae Newsl.,* 32: 19 (1998): Centaurea imatongensis.

STEMMADENIA (Apocynac.).
abbreviata *J.F. Morales in Novon,* 9(2): 236 (1999)—Costa Rica.
simulans *J.F. Morales & Q. Jiménez in Novon,* 9(2): 236 (1999)—Costa Rica, Panama.

STEMODIOPSIS (Scrophulariac.).
ruandensis *Eb. Fischer in Bot. Jahrb.,* 119(3): 319 (1997)—Rwanda.

STEMONA (Stemonac.).
hutanguriana *W. Chuakul in Kew Bull.,* 55(4): 977 (2000)—Thailand.

STEMOTRIA (Scrophulariac.).
triandra *(Cav.) R. Govaerts, World Checklist Seed Plants,* 3(1): 21 (1999): Jovellana triandra.

STENANDRIUM (Acanthac.).
andrei *(Leonard) D.C. Wasshausen in Nordic J. Bot.,* 16(4): 387 (1996 publ. 1997): Aphelandra andrei.
arnoldii *(Mildbr.) D.C. Wasshausen in Nordic J. Bot.,* 16(4): 387 (1996 publ. 1997), homonym; *see H. Dietr. in Wiss. Zeitschr. Friedrich-Schiller Univ. Jena, Mat.* Naturwiss., 33(6): 725 (1984): Aphelandra arnoldii.
fosbergii *(Leonard) D.C. Wasshausen in Nordic J. Bot.,* 16(4): 387 (1996 publ. 1997): Aphelandra fosbergii.
grandiflorum *K. Vollesen in Kew Bull.,* 55(4): 967 (2000)—Tanzania.
harlingii *D.C. Wasshausen in Nordic J. Bot.,* 16(4): 384 (1996 publ. 1997)—Ecuador.
var. angustifolia *D.C. Wasshausen in Nordic J. Bot.,* 16(4): 387 (1996 publ. 1997)—Ecuador.
nephoica *(Wassh.) D.C. Wasshausen in Nordic J. Bot.,* 16(4): 388 (1996 publ. 1997): Aphelandra nephoica.
stenophyllum *C. Kameyama in Novon,* 6(3): 268 (1996)—Brazil Southeast (Minas Gerais).

STENANTHEMUM (Rhamnac.).
bilobum *B.L. Rye in Nuytsia,* 10(2): 281 (1995)—Western Australia.
complicatum *(F. Muell.) B.L. Rye in Nuytsia,* 10(2): 282 (1995): Spyridium complicatum.
cristatum *B.L. Rye in Nuytsia,* 10(2): 284 (1995)—Western Australia.
divaricatum *(Benth.) B.L. Rye in Nuytsia,* 10(2): 284 (1995): Spyridium divaricatum.
emarginatum *B.L. Rye in Nuytsia,* 10(2): 286 (1995)—Western Australia.
intricatum *B.L. Rye in Nuytsia,* 10(2): 287 (1995)—Western Australia.
limitatum *B.L. Rye in Nuytsia,* 10(2): 289 (1995)—Western Australia.
mediale *B.L. Rye in Nuytsia,* 10(2): 290 (1995)—Western Australia.
nanum *B.L. Rye in Nuytsia,* 10(2): 291 (1995)—Western Australia.
newbeyi *B.L. Rye in Nuytsia,* 10(2): 293 (1995)—Western Australia.
notiale *B.L. Rye in Nuytsia,* 10(2): 294 (1995)—Western Australia.
subsp. chamelum *B.L. Rye in Nuytsia,* 10(2): 294 (1995), nom. nov.: Cryptandra tridentata var. tomentosa *Reissek.*
petraeum *B.L. Rye in Nuytsia,* 10(2): 298 (1995)—Western Australia, Northern Territory.
poicilum *B.L. Rye in Nuytsia,* 10(2): 299 (1995)—Western Australia.
reissekii *B.L. Rye in Nuytsia,* 10(2): 300 (1995)—Western Australia.
stipulosum *B.L. Rye in Nuytsia,* 10(2): 302 (1995)—Western Australia.

STENIA (Orchidac.).
angustilabia *D.E. Bennett & E.A. Christenson in Lindleyana,* 13(2): 88 (1998)— Peru.
aurorae *D.E. Bennett & E.A. Christenson in Lindleyana,* 13(2): 91 (1998)—Peru.
christensonii *D.E. Bennett in Brittonia,* 50(2): 189 (1998)—Peru.
glatzii *T. Neudecker & G. Gerlach in Orquideologia,* 21(3): 259 (2000)—Ecuador.
luerorum *D.E. Bennett & E.A. Christenson in Lindleyana,* 13(2): 91 (1998), as 'lueriorum'—Peru.

× **STENIELLA** *hort. in Orchid Rev.,* 106(1219): 62 (1998). *ORCHIDACEAE* (Chaubardiella × Stenia).

× **STENILLOCACTUS** *P.V. Heath in Calyx,* 5(3): 102 (1996). *CACTACEAE* (Myrtillocactus × Stenocereus).
nothosect. Isillocactus *P.V. Heath in Calyx,* 5(3): 102 (1996), as 'Insillocactus'. (Myrtillocactus sect. Myrtillocactus × Stenocereus subgen. Isolatocereus).
glassii *(P.V. Heath) P.V. Heath in Calyx,* 5(3): 102 (1996): × Rathbunillocactus glassii.

STENOCACTUS (Cactac.).
jarmilae *J.J. Halda & L. Horáček in Acta Mus. Richnov. Sect. Nat.,* 7(1): 37 (2000)—Mexico Northeast (Nuevo León).
kaplani *J.J. Halda, J. Kupčák & J. Sladkovský in Acta Mus. Richnov. Sect. Nat.,* 7(1): 37 (2000)—Mexico Northeast (Querétaro).
magnificus *J.J. Halda & L. Horáček in Acta Mus. Richnov. Sect. Nat.,* 7(1): 37 (2000)—Mexico Northeast (Nuevo León).
pentacanthus
subsp. tulensis *J.J. Halda, P. Kupčák & J. Sladkovský in Acta Mus. Richnov. Sect. Nat.,* 7(1): 37 (2000), contrary to Art. 37.6 ICBN (2000)—Mexico Northeast (Tamaulipas).
verticis *D.B. Foreman in Fl. Australia,* 16: 498 (1995)—Northern Territory.

STENOCEREUS (Cactac.).
subgen. Griseocereus *P.V. Heath in Calyx,* 5(3): 101 (1996), nom. nov.: Cereus ser. Pruinosi *K. Schum.*
nothosubgen. Grisolatocereus *P.V. Heath in Calyx,* 5(3): 101 (1996), as 'Grisulatocereus'. (S. sect. Griseocereus × sect. Isolatocereus).
subgen. Hertrichocereus *(Backeb.) P.V. Heath in Calyx,* 5(3): 100 (1996): gen. Hertrichocereus.
subgen. Isolatocereus *(Backeb.) P.V. Heath in Calyx,* 5(3): 101 (1996), as 'Insulatocereus': Lemaireocereus subgen. Isolatocereus.
subgen. Machaerocereus *(Britton & Rose) P.V. Heath in Calyx,* 5(3): 100 (1996): gen. Machaerocereus.
subgen. Nigellicereus *P.V. Heath in Calyx,* 5(3): 100 (1996), nom. nov.: Cereus ser. Chlorotini *K. Schum.*
nothosubgen. Rittellicereus *P.V. Heath in Calyx,* 5(3): 101 (1996).(Stenocereus subgen. Nigellicereus × subgen. Ritterocereus).
*subgen.*Ritterocereus *(Backeb.) P.V. Heath in Calyx,* 5(3): 100 (1996): gen. Ritterocereus.
alamosensis (J.M. Coult.) A.C. Gibson & K.E. Horak
f. cristatus *(P.V. Heath) P.V. Heath in Calyx,* 5(3): 100 (1996): Rathbunia alamosensis f. cristata.
var. sonorensis *(K. Schum.) P.V. Heath in Calyx,* 5(3): 100 (1996): Cereus sonorensis.
beneckei (C. Ehrenb.) Buxb.
f. cristatus *(P.V. Heath) P.V. Heath in Calyx,* 5(3): 100 (1996): Rathbunia beneckei f. cristata.
griseus (Haw.) Buxb.
f. cristatus *(P.V. Heath) P.V. Heath in Calyx,* 5(3): 101 (1996): Rathbunia grisea f. cristata.
var. monstrosus *(DC.) P.V. Heath in Calyx,* 5(3): 101 (1996), as 'monstruosus': Cereus eburneus β monstrosus.
gummosus (Brandegee) A.C. Gibson & K.E. Horak
f. cristatus *(P.V. Heath) P.V. Heath in Calyx,* 5(3): 100 (1996): Rathbunia gummosa f. cristata.
f. tortuosus *(P.V. Heath) P.V. Heath in Calyx,* 5(3): 100 (1996): Rathbunia gummosa f. tortuosa.
laevigatus (Salm-Dyck) Buxb.
var. schumannii *(K. Schum.) P.V. Heath in Calyx,* 5(3): 101 (1996): Cereus schumannii.
thurberi (Engelm.) Buxb.
f. cristatus *(P.V. Heath) P.V. Heath in Calyx,* 5(3): 100 (1996): Rathbunia thurberi f. cristata.
f. dichotomus *(P.V. Heath) P.V. Heath in Calyx,* 5(3): 100 (1996): Rathbunia thurberi f. dichotoma.
subsp. littoralis *(K. Brandegee) N.P. Taylor in Cactaceae Consensus Initiatives,* 5: 13 (1998): Cereus thurberi var. littoralis.

STENOCEREUS:-
yunckeri *(Standl.) P.V. Heath in Calyx,* 5(3): 101 (1996): Cereus yunckeri.

STENOGLOTTIS (Orchidac.).
macloughlinii *(L. Bolus) G. McDonald in S. Afr. Orchid J.,* 26(4): 117 (1995), without basionym ref.: Cynorchis macloughlinii.

STENOGYNE (Labiat.).
kaalae Wawra
subsp. sherffii *(O. Deg.) W.L. Wagner & S.G. Weller in Novon,* 9(3): 449 (1999): S. sherffii.

STENOMESSON (Amaryllidac.).
rupense *P. Ravenna in Onira, Bot. Leafl.,* 4(13): 52 (2000)—Peru.

STENOPADUS (Compos.).
andicola *J.F. Pruski in Novon,* 8(1): 67 (1998)—Ecuador.

STENOPTERA (Orchidac.).
ecuadorana *C.H. Dodson & C. Vargas in Orquideologia,* 21(1): 51 (1998)—Ecuador.

STENORRHYNCHOS (Orchidac.).
actinosophila *J. Barbosa Rodrigues, Gen. spec. Orchid.,* 2: 286 (1881), as 'Stenorhynchus'—Brazil Southeast (Minas Gerais).†

STENOSEPALA *C. Persson in Novon,* 10(4): 403 (2000). *RUBIACEAE*
hirsuta *C. Persson in Novon,* 10(4): 403 (2000)—Colombia, Panama.

STENOSPERMATION (Arac.).
majus *M.H. Grayum in Phytologia,* 82(1): 50 (1997)—Costa Rica.
pteropus *M.H. Grayum in Phytologia,* 82(1): 52 (1997)—Costa Rica.

STENOSTACHYS (Gramin.).
deceptorix *H.E. Connor in New Zealand J. Bot.,* 32(2): 144 (1994)—New Zealand South I.
gracilis *(Hook. f.) H.E. Connor in New Zealand J. Bot.,* 32(2): 146 (1994): Gymnostichum gracile.
laevis *(Petrie) H.E. Connor in New Zealand J. Bot.,* 32(2): 146 (1994): Asprella laevis.

STENOSTEGIA *A.R. Bean in Muelleria,* 11: 127 (1998). *MYRTACEAE*
congesta *A.R. Bean in Muelleria,* 11: 130 (1998)—Northern Territory.

STENOSTEPHANUS (Acanthac.).
alushii *T.F. Daniel in Contrib. Univ. Mich. Herb.,* 22: 59 (1999)—Mexico Southeast (Chiapas).
asplundii *(Wassh.) D.C. Wasshausen in Cat. Vasc. Pl. Ecuador (Monogr. Syst. Bot. Missouri Bot. Gard., 75):* 953 (1999): Habracanthus asplundii.
cochabambensis *D.C. Wasshausen in Harvard Pap. Bot.,* 4(1): 282 (1999)—Bolivia.
crenulatus *(Britton ex Rusby) D.C. Wasshausen in Harvard Pap. Bot.,* 4(1): 288 (1999): Hansteinia crenulata.
davidsonii *D.C. Wasshausen in Harvard Pap. Bot.,* 4(1): 285 (1999)—Bolivia.
guerrerensis *T.F. Daniel in Contrib. Univ. Mich. Herb.,* 22: 69 (1999)—Mexico Southwest (Guerrero).
haematodes *(Schltdl.) T.F. Daniel in Contrib. Univ. Mich. Herb.,* 22: 71 (1999): Justicia haematodes.
harleyi *(Wassh.) T.F. Daniel in Contrib. Univ. Mich. Herb.,* 22: 77 (1999): Hansteinia harleyi.
harlingii *(Wassh.) D.C. Wasshausen in Cat. Vasc. Pl. Ecuador (Monogr. Syst. Bot. Missouri Bot. Gard., 75):* 953 (1999): Habracanthus harlingii.
jamesonii *(Wassh.) D.C. Wasshausen in Cat. Vasc. Pl. Ecuador (Monogr. Syst. Bot. Missouri Bot. Gard., 75):* 953 (1999): Hansteinia jamesonii.
krukoffii *D.C. Wasshausen in Harvard Pap. Bot.,* 4(1): 282 (1999)—Bolivia.
laxus *(Wassh.) D.C. Wasshausen in Cat. Vasc. Pl. Ecuador (Monogr. Syst. Bot. Missouri Bot. Gard., 75):* 953 (1999): Habracanthus laxus.
lugonis *(Wassh.) D.C. Wasshausen in Cat. Vasc. Pl. Ecuador (Monogr. Syst. Bot. Missouri Bot. Gard., 75):* 953 (1999): Habracanthus lugonis.

STENOSTEPHANUS:-
luteynii *(Wassh.) D.C. Wasshausen in Cat. Vasc. Pl. Ecuador (Monogr. Syst. Bot. Missouri Bot. Gard., 75)*: 953 (1999): Habracanthus luteynii.
lyman-smithii *D.C. Wasshausen in Harvard Pap. Bot.*, 4(1): 281 (1999)—Bolivia.
madrensis *T.F. Daniel in Contrib. Univ. Mich. Herb.*, 22: 79 (1999)—Mexico Southwest (Oaxaca).
oaxacanus *T.F. Daniel in Contrib. Univ. Mich. Herb.*, 22: 83 (1999)—Mexico Southwest (Oaxaca).
pyramidalis *(Lindau) D.C. Wasshausen in Harvard Pap. Bot.*, 4(1): 288 (1999): Habracanthus pyramidalis.
sanguineus *(Willd. ex Nees) D.C. Wasshausen in Cat. Vasc. Pl. Ecuador (Monogr. Syst. Bot. Missouri Bot. Gard., 75)*: 953 (1999): Habracanthus sanguineus.

STENOSTOMUM (Rubiac.).
sect. Neolaugeria *A. Borhidi in Acta Bot. Hung.*, 38(1–4): 158 (1993–94 publ. 1995), as 'comb. nov.' without basionym ref.:
sect. Resinanthus *A. Borhidi in Acta Bot. Hung.*, 38(1–4): 158 (1993–94 publ. 1995).
abbreviatum *(Urb.) A. Borhidi & M.Z. Fernandez in Acta Bot. Hung.*, 38(1–4): 152 (1993–94 publ. 1995): Antirhea abbreviata.
subsp. moaense *(M.Z. Fernandez) A. Borhidi & M.Z. Fernandez in Acta Bot. Hung.*, 38(1–4): 152 (1993–94 publ. 1995): Antirhea abbreviata var. moaensis.
subsp. obcordatum *(Alain) A. Borhidi & M.Z. Fernandez in Acta Bot. Hung.*, 38(1–4): 152 (1993–94 publ. 1995): Antirhea obcordata.
acutatum DC.
var. latifolium *(Urb.) A. Borhidi in Acta Bot. Hung.*, 38(1–4): 159 (1993–94 publ. 1995): Antirhea acutata var. latifolia.
albobruneum *(Urb. & Ekman) A. Borhidi in Acta Bot. Hung.*, 38(1–4): 159 (1993–94 publ. 1995): Antirhea albobrunea.
aromaticum *(Cast.-Campos & Lorence) A. Borhidi in Acta Bot. Hung.*, 38(1–4): 160 (1993–94 publ. 1995): Antirhea aromatica.
baracoense *A. Borhidi in Acta Bot. Hung.*, 38(1–4): 144 (1993–94 publ. 1995)— Cuba.
biflorum *A. Borhidi in Acta Bot. Hung.*, 38(1–4): 145 (1993–94 publ. 1995)—Cuba.
cahosianum *(Urb. & Ekman) A. Borhidi in Acta Bot. Hung.*, 38(1–4): 160 (1993–94 publ. 1995): Antirhea cahosiana.
cuspidatum *A. Borhidi in Acta Bot. Hung.*, 38(1–4): 145 (1993–94 publ. 1995)—Cuba.
ekmanii *(Borhidi) A. Borhidi in Acta Bot. Hung.*, 38(1–4): 160 (1993–94 publ. 1995): Antirhea ekmanii.
ellipticum *(Urb. & Ekman) A. Borhidi in Acta Bot. Hung.*, 38(1–4): 160 (1993–94 publ. 1995): Antirhea elliptica.
heteroneurum *(Urb. & Ekman) A. Borhidi in Acta Bot. Hung.*, 38(1–4): 160 (1993–94 publ. 1995): Antirhea heteroneura.
hotteanum *(Urb.) A. Borhidi in Acta Bot. Hung.*, 38(1–4): 161 (1993–94 publ. 1995): Laugeria hotteana.
imbricatum *A. Borhidi in Acta Bot. Hung.*, 38(1–4): 146 (1993–94 publ. 1995)—Cuba.
involucratum *(Urb. & Ekman) A. Borhidi in Acta Bot. Hung.*, 38(1–4): 161 (1993–94 publ. 1995): Antirhea involucrata.
jamaicense *(Urb.) A. Borhidi in Acta Bot. Hung.*, 38(1–4): 161 (1993–94 publ. 1995): Antirhea jamaicensis.
lineolatum *(Urb.) A. Borhidi in Acta Bot. Hung.*, 38(1–4): 161 (1993–94 publ. 1995): Laugeria lineolata.
maestrense *(Urb.) A. Borhidi & M.Z. Fernandez in Acta Bot. Hung.*, 38(1–4): 154 (1993–94 publ. 1995): Antirhea maestrensis.
minutifolium *(Borhidi & Capote) A. Borhidi & M.Z. Fernandez in Acta Bot. Hung.*, 38(1–4): 152 (1993–94 publ. 1995): Antirhea minutifolia.
montecristinum *(Urb.) A. Borhidi in Acta Bot. Hung.*, 38(1–4): 162 (1993–94 publ. 1995): Antirhea montecristina.
mucronatum *(Urb.) A. Borhidi & M.Z. Fernandez in Acta Bot. Hung.*, 38(1–4): 151 (1993–94 publ. 1995): Antirhea mucronata.

STENOSTOMUM:-
multinerve *(Urb.) A. Borhidi & M.Z. Fernandez in Acta Bot. Hung.*, 38(1–4): 154 (1993–94 publ. 1995): Antirhea multinervis.
nipense *(Borhidi & O. Muñiz) A. Borhidi & M.Z. Fernandez in Acta Bot. Hung.*, 38(1–4): 152 (1993–94 publ. 1995): Antirhea nipensis.
occidentale *(Urb.) A. Borhidi & M.Z. Fernandez in Acta Bot. Hung.*, 38(1–4): 155 (1993–94 publ. 1995): Antirhea occidentalis.
oliganthum *(Urb.) A. Borhidi in Acta Bot. Hung.*, 38(1–4): 162 (1993–94 publ. 1995): Antirhea oligantha.
ophiticola *(Alain) A. Borhidi & M.Z. Fernandez in Acta Bot. Hung.*, 38(1–4): 153 (1993–94 publ. 1995): Antirhea ophiticola.
orbiculare *(Alain) A. Borhidi & M.Z. Fernandez in Acta Bot. Hung.*, 38(1–4): 152 (1993–94 publ. 1995): Antirhea orbicularis.
pedicellare *(Borhidi & Bisse) A. Borhidi & M.Z. Fernandez in Acta Bot. Hung.*, 38(1–4): 154 (1993–94 publ. 1995): Antirhea pedicellaris.
pitonianum *(Urb. & Ekman) A. Borhidi in Acta Bot. Hung.*, 38(1–4): 163 (1993–94 publ. 1995): Antirhea pitoniana.
radiatum Griseb.
subsp. haitiense *(Borhidi) A. Borhidi in Acta Bot. Hung.*, 38(1–4): 163 (1993–94 publ. 1995): Antirhea radiata subsp. haitiensis.
reticulare *A. Borhidi & M.Z. Fernandez in Acta Bot. Hung.*, 38(1–4): 147 (1993–94 publ. 1995)—Cuba.
revolutum *A. Borhidi in Acta Bot. Hung.*, 38(1–4): 148 (1993–94 publ. 1995)—Cuba.
scrobiculatum *(Urb.) A. Borhidi & M.Z. Fernandez in Acta Bot. Hung.*, 38(1–4): 154 (1993–94 publ. 1995): Antirhea scrobiculata.
shaferi *(Urb.) A. Borhidi & M.Z. Fernandez in Acta Bot. Hung.*, 38(1–4): 155 (1993–94 publ. 1995): Antirhea shaferi.
urbanianum *(C.T. White) A. Borhidi & M.Z. Fernandez in Acta Bot. Hung.*, 38(1–4): 155 (1993–94 publ. 1995): Antirhea urbaniana.

STENOTALIS *B.G. Briggs & L.A.S. Johnson in Telopea*, 7(4): 368 (1998). *RESTIONACEAE*
ramosissima *(Gilg) B.G. Briggs & L.A.S. Johnson in Telopea*, 7(4): 369 (1998): Hypolaena ramosissima.

STEPHANIA (Menispermac.).
japonica (Thunb.) Miers
var. timoriensis *(DC.) J. Florence in Adansonia*, 21(1): 50 (1999): Cocculus japonicus var. timoriensis.

STEPHANOMERIA (Compos.).
fluminea *L.D. Gottlieb in Madroño*, 46(1): 58 (1999)—Wyoming.

STERCULIA (Sterculiac.).
guangxiensis *S.J. Xu & P.T. Li in Acta Phytotax. Sin.*, 38(6): 568 (2000)—China Southeast (Guangxi).
lanceolata Cav.
var. coccinea *(Jack) C. Phengklai in Thai Forest Bull., Bot.*, 23: 99 (1995): S. coccinea.
var. principis *(Gagnep.) C. Phengklai in Thai Forest Bull., Bot.*, 23: 99 (1995): S. principis.
simaoensis *Y.Y. Qian in Acta Phytotax. Sin.*, 35(1): 79 (1997)—China South-central (Yunnan).
urena Roxb.
var. thorelii *(Pierre) C. Phengklai in Thai Forest Bull., Bot.*, 23: 95 (1995): S. thorelii.
xolocotzii *T. Wendt & E.L. Taylor in Lundellia*, 2: 128 (1999)—Mexico Gulf (Veracruz), Mexico Southwest (Oaxaca).

STEREOCHILUS (Orchidac.).
brevirachis *E.A. Christenson in Orchid Dig.*, 62(3): 123 (1998)—China.

STERIS (Caryophyllac.).
× viburgensis *N.N. Tzvelev, Opred. Sosud. Rast. Severo-Zap. Rossii*: 679 (2000)—Northwest European Russia. (S. alpina × viscaria).

STEVENIELLA (Orchidac.).
caucasica *L.A. Garay in Harvard Pap. Bot.*, 2(1): 50 (1997), nom. nov., as No. 11 on inside pages: Orchis satyrioides *Steven*.

STEVIA (Compos.).
calzadana *B.L. Turner in Phytologia*, 79(1): 5 (1995 publ. 1996)—Mexico Southwest (Oaxaca).
pulcherrima *(B.L. Rob.) M.E. Jones in Contr. W. Bot.*, 18: 71 (1933): Brickellia pulcherrima.
quiexobra *B.L. Turner in Phytologia*, 79(4): 301 (1995 publ. 1996)—Mexico Southwest (Oaxaca).
stolonifera *T. Yahara & A. Soejima in Phytologia*, 79(1): 35 (1995 publ. 1996)—Mexico Southwest (Jalisco, Michoacan).

STEWARTIA (Theac.).
subgen. Dialystyla *(Szyszył.) J. Li & Ming in Acta Phytotax. Sin.*, 34(1): 53 (1996), without exact basionym page: S. sect. Dialystyla.
sect. Foliobractea *C.X. Ye in Acta Sci. Nat. Univ. Sunyatseni*, 1982(4): 115 (1982).
sect. Pteropetiolatae *J. Li & Ming in Acta Phytotax. Sin.*, 34(1): 54 (1996).
sect. Racemosae *(Y.C. Wu) J. Li & Ming in Acta Phytotax. Sin.*, 34(1): 53 (1996): Hartia sect. Racemosa.
calcicola *Ming & J. Li in Acta Phytotax. Sin.*, 34(1): 63 (1996), nom. nov.: Hartia yunnanensis *Hu*.
cordifolia *(H.L. Li) J. Li & Ming in Acta Phytotax. Sin.*, 34(1): 63 (1996): Hartia cordifolia.
crassifolia *(S.Z. Yan) J. Li & Ming in Acta Phytotax. Sin.*, 34(1): 57 (1996): Hartia crassifolia.
damingshanica *J. Li & Ming in Acta Phytotax. Sin.*, 34(1): 61 (1996)—China Southeast (Guangxi).
densivillosa *(Hu ex H.T. Chang & C.X. Ye) J. Li & Ming in Acta Phytotax. Sin.*, 34(1): 57 (1996): Hartia densivillosa.
laotica *(Gagnep.) J. Li & Ming in Acta Phytotax. Sin.*, 34(1): 58 (1996): Hartia laotica.
medogensis *J. Li & Ming in Acta Phytotax. Sin.*, 34(1): 64 (1996), nom. nov.: Hartia brevicalyx *Hung T. Chang*.
obovata *(Chun ex Hung T. Chang) J. Li & Ming in Acta Phytotax. Sin.*, 34(1): 57 (1996): Hartia obovata.
sichuanensis *(S.Z. Yan) J. Li & Ming in Acta Phytotax. Sin.*, 34(1): 61 (1996)—Hartia sichuanensis.
sinensis Rehder & E.H. Wilson
var. acutisepala *(P.L. Chiu & G.R. Zhong) Ming & J. Li in Acta Phytotax. Sin.*, 34(1): 60 (1996): S. acutisepala.
var. brevicalyx *(S.Z. Yan) Ming & J. Li in Acta Phytotax. Sin.*, 34(1): 60 (1996): S. brevicalyx.
var. rostrata *(Spongberg) Hung T. Chang in Fl. Reipubl. Popul. Sin.*, 49(3): 239 (1998): S. rostrata.
var. shensiensis *(H.T. Chang) Ming & J. Li in Acta Phytotax. Sin.*, 34(1): 60 (1996): S. shensiensis.
tonkinensis *(Merr.) C.Y. Wu ex J. Li in Acta Phytotax. Sin.*, 34(1): 57 (1996): Hartia tonkinensis.
villosa Merr.
var. grandifolia *(Chun) J. Li & Ming in Acta Phytotax. Sin.*, 34(1): 64 (1996): Hartia kwangtungensis var. grandifolia.
var. kwangtungensis *(Chun) J. Li & Ming in Acta Phytotax. Sin.*, 34(1): 64 (1996): Hartia kwangtungensis.

STEYERBROMELIA (Bromeliac.).
neblinae *B.K. Holst in Fl. Venezuelan Guayana (gen. eds. J.A. Steyermark et al.)*, 3: 655 (1997), nom. nov.: Navia diffusa *L.B. Sm.*
plowmanii *(L.B. Sm., Steyerm. & H. Rob.) H. Robinson & D.C. Taylor in Harvard Pap. Bot.*, 4(1): 200 (1999): Navia plowmanii.
ramosa *(L.B. Sm.) B.K. Holst in Fl. Venezuelan Guayana (gen. eds. J.A. Steyermark et al.)*, 3: 655 (1997): Navia ramosa.
thomasii *(L.B. Sm., Steyerm. & H. Rob.) B.K. Holst in Fl. Venezuelan Guayana (gen. eds. J.A. Steyermark et al.)*, 3: 655 (1997): Navia thomasii.

STICTOCARDIA (Convolvulac.).
sivarajanii *S.D. Biju, P. Pushpangadan & P. Mathew in Novon*, 9(2): 147 (1999)—India (Kerala).

STIFFTIA (Compos.).
fruticosa *(Vell.) D.J.N. Hind & J. Semir in Kew Bull.*, 53(3): 618 (1998): Aristomenia fruticosa.

STIGMAPHYLLON (Malpighiac.).
boliviense *C.E. Anderson, Monogr. Stigmaphyllon (Malpighiac.) (Syst. Bot. Monog., 51):* 178 (1997)—Bolivia.
bonariense *(Hook. & Arn.) C.E. Anderson in Brittonia,* 48(4): 543 (1996 publ. 1997): Banisteria bonariensis.
carautae *C.E. Anderson, Monogr. Stigmaphyllon (Malpighiac.) (Syst. Bot. Monog., 51):* 82 (1997)—Brazil Southeast (Rio de Janeiro).
hispidum *C.E. Anderson in Novon,* 10(2): 107 (2000)—Brazil Northeast (Bahia).

STIGMATOSEMA (Orchidac.).
draculoides *D.L. Szlachetko in Fragm. Flor. Geobot.,* 41(2): 861 (1996)—Brazil.
longibracteata *G. Carnevali & G.A. Romero in G.A. Romero & G. Carnevali, Orchids of Venezuela, ed. 2:* 1147 (2000)—Venezuela.

STILBE (Stilbac.).
gymnopharyngia *(Rourke) J.P. Rourke in Bothalia,* 30(1): 11 (2000): Xeroplana gymnopharyngia.
overbergensis *J.P. Rourke in Bothalia,* 30(1): 11 (2000), nom. nov.: Xeroplana zeyheri *Briq.*
serrulata *(Hochst.) J.P. Rourke in Bothalia,* 30(1): 11 (2000): Eurylobium serrulatum.

STILIFOLIUM *W. Königer & D. Pongratz in Arcula,* 7: 186 (1997). *ORCHIDACEAE*
aguirrei *W. Königer in Arcula,* 9: 259 (1999)—Colombia.
ascendens *(Lindl.) W. Königer & D. Pongratz in Arcula,* 7: 186 (1997): Oncidium ascendens.
cebolleta *(Jacq.) W. Königer & D. Pongratz in Arcula,* 7: 187 (1997): Dendrobium cebolleta.
jonesianum *(Rchb. f.) W. Königer & D. Pongratz in Arcula,* 7: 189 (1997): Oncidium jonesianum.
× marvraganii *E. Lückel in Orchidee,* 49(2): 90 (1998), contrary to Art. 37.4 ICBN (1994), as 'marvreganii'; *see Orchidee,* 49(3): 148 (1998)—Bolivia. (S. jonesianum × stacyi).
nudum *(Bateman ex Lindl.) W. Königer & D. Pongratz in Arcula,* 7: 189 (1997): Oncidium nudum.
ostenianum *(Schltr.) W. Königer & D. Pongratz in Arcula,* 7: 189 (1997): Oncidium ostenianum.
pongratzianum *W. Königer in Arcula,* 9: 261 (1999)—Peru.
stacyi *(Garay) E. Lückel in Orchidee,* 49(2): 90 (1998): Oncidium stacyi.
stipitatum *(Lindl.) W. Königer & D. Pongratz in Arcula,* 7: 189 (1997): Oncidium stipitatum.
teres *(Ames & Schweinf.) W. Königer & D. Pongratz in Arcula,* 7: 190 (1997): Oncidium teres.
wittii *(Oppenheim) W. Königer & D. Pongratz in Arcula,* 7: 190 (1997): Oncidium wittii.

STILLINGIA (Euphorbiac.).
oblongifolia *M.C. Leavenworth in Amer. J. Sci. Arts,* 49: 128 (1845)—Florida.

STIPA (Gramin.).
sect. Giganteae *(Martinovský) F.M. Vázquez & J.A. Devesa in Acta Bot. Malacitana,* 21: 163 (1996): S. ser. Giganteae.
sect. Inaequiglumes *(Bor) F.M. Vázquez & J.A. Devesa in Acta Bot. Malacitana,* 21: 153 (1996): S. ser. Inaequiglumes.
sect. Jarava *(Ruiz & Pav.) F.M. Vázquez & J.A. Devesa in Acta Bot. Malacitana,* 21: 158 (1996): gen. Jarava.
sect. Macrochloa *(Kunth) F.M. Vázquez & J.A. Devesa in Acta Bot. Malacitana,* 21: 161 (1996): gen. Macrochloa.
akseirica *Yu.A. Kotukhov in Turczaninowia,* 1(2): 11 (1998)—Kazakhstan.
apertifolia Martinovský
subsp. apenninicola *(Martinovský & Moraldo) F.M. Vázquez & J.A. Devesa in Lagascalia,* 18(2): 323 (1996): S. dasyvaginata subsp. apenninicola.
subsp. longiglumis *(H. Scholz) F.M. Vázquez & J.A. Devesa in Lagascalia,* 18(2): 323 (1996): S. dasyvaginata subsp. longiglumis.
var. nevadensis *F.M. Vázquez & J.A. Devesa in Acta Bot. Malacitana,* 21: 139 (1996)—Spain.

STIPA:-
arenicola *F.A. Roig in Hickenia,* 2(55): 257 (1998)—Argentina South (Santa Cruz).
argillosa *Yu.A. Kotukhov in Turczaninowia,* 1(1): 8 (1998)—Altay.
azutavica *Yu.A. Kotukhov in Turczaninowia,* 1(2): 9 (1998)—Altay.
balansae *H. Scholz in Willdenowia,* 26(1–2): 226 (1996)—Algeria.
barbata
subsp. brevipila *(Coss. & Durieu) F.M. Vázquez & J.A. Devesa in Bot. J. Linn. Soc.,* 124(2): 202 (1997): S. barbata var. brevipila.
bhutanica *H.J. Noltie in Edinburgh J. Bot.,* 56(2): 289 (1999)—Bhutan.
borysthenica Klokov ex Prokudin
var. germanica *(Endtm.) J. Dengler in Gleditschia,* 28(1–2): 20 (2000): Stipa joannis subsp. germanica.
bufensis *F.M. Vázquez, H. Scholz & M. Sonnentag in Acta Bot. Malacitana,* 24: 30 (1999)—Spain.
cazorlensis *(F.M. Vázquez & Devesa) F.M. Vázquez, H. Scholz & M. Sonnentag in Acta Bot. Malacitana,* 24: 30 (1999): S. clausa subsp. cazorlensis.
clausa Trab.
subsp. cazorlensis *F.M. Vázquez Pardo & J.A. Devesa Alcaraz in An. Jard. Bot. Madrid,* 54(1): 413 (1996)—Spain, Morocco, Algeria.
var. matritensis *F.M. Vázquez Pardo & J.A. Devesa Alcaraz in An. Jard. Bot. Madrid,* 54(1): 412 (1996)—Spain.
czerepanovii *Yu.A. Kotukhov in Turczaninowia,* 1(2): 13 (1998)—Kazakhstan.
durifolia *Parodi ex M.A. Torres in Monogr. Comis. Invest. Ci. Prov. Buenos Aires,* 13: 54 (1997)—Argentina Northwest (Jujuy, Tucumán).
filabrensis *H. Scholz, M. Sonnentag & F.M. Vázquez in Acta Bot. Malacitana,* 24: 28 (1999)—Spain.
gigantea Link
subsp. donyanae *F.M. Vázquez & J.A. Devesa in Acta Bot. Malacitana,* 21: 165 (1996)—Spain.
subsp. maroccana *(Pau & Font Quer) F.M. Vázquez & J.A. Devesa in Bot. J. Linn. Soc.,* 124(2): 202 (1997): S. gigantea var. maroccana.
iberica Martinovský
var. austroiberica *(H. Scholz) F.M. Vázquez & J.A. Devesa in Acta Bot. Malacitana,* 21: 142 (1996): S. iberica subsp. austroiberica.
subsp. bolosii *À.M. Romo, E. Sierra, Lluis de Torres & A. Cervi in Acta Bot. Barcin.,* 45: 214 (1998)—Spain.
var. pseudodasyphylla *(Martinovský) F.M. Vázquez & J.A. Devesa in Acta Bot. Malacitana,* 21: 141 (1996): S. iberica f. pseudodasyphylla.
subsp. pseudodasyphylla *(Martinovský) A.M. Romo in Soc. Éch. Pl. Vasc. Eur. Bassin Médit. Bull.,* 28: 123 (2000): Stipa iberica f. pseudodasyphylla.
jacquemontii Jaub. & Spach
subsp. chuzomica *H.J. Noltie in Edinburgh J. Bot.,* 56(2): 290 (1999)—Bhutan.
juncea L.
var. cabanasii *F.M. Vázquez & J.A. Devesa in Acta Bot. Malacitana,* 21: 149 (1996)—Spain.
kamelinii *Yu.A. Kotukhov in Turczaninowia,* 1(1): 10 (1998)—Kazakhstan.
kyzylkiensis *Yu.A. Kotukhov in Turczaninowia,* 1(2): 12 (1998)—Kazakhstan.
langshanica *(Y.Z. Zhao) Y.Z. Zhao in Acta Sci. Nat. Univ. Neimonggol,* 27(2): 211 (1996): S. glareosa var. langshanica.
letourneuxii
subsp. pellita *(Trin. & Rupr.) H. Scholz in Willdenowia,* 28(1–2): 173 (1998): S. lagascae var. pellita.
magrebensis *F.M. Vázquez & J.A. Devesa in Bot. J. Linn. Soc.,* 124(2): 206 (1997)—Morocco, Algeria.
maroccana *H. Scholz in Willdenowia,* 26(1–2): 227 (1996)—Morocco.
meridionalis *F.M. Vázquez & J.A. Devesa in Bot. J. Linn. Soc.,* 124(2): 204 (1997)—Morocco.
milleri *H.J. Noltie in Edinburgh J. Bot.,* 56(2): 288 (1999)—Sikkim, Bhutan.
monticola *Yu.A. Kotukhov in Turczaninowia,* 1(1): 11 (1998), homonym; *see H. Scholz in Willdenowia,* 23(1–2): 117 (1993)—Kazakhstan.

STIPA:-
parviflora Desf.
var. pilosa *(Chrtek & Martinovský) F.M. Vázquez & J.A. Devesa in Acta Bot. Malacitana,* 21: 155 (1996): S. parviflora f. pilosa.
pauneroana *(Martinovský) F.M. Vázquez & J.A. Devesa in Acta Bot. Malacitana,* 21: 143 (1996): S. iberica subsp. pauneroana.
pavlovii *Yu.A. Kotukhov in Turczaninowia,* 1(1): 7 (1998)—Kazakhstan.
purpurea Griseb.
subsp. arenosa *(Tzvelev) D.F. Cui in Fl. Xinjiangensis,* 6: 307 (1996): S. purpurea var. arenosa.
rohmooiana *H.J. Noltie in Edinburgh J. Bot.,* 56(2): 287 (1999)—Sikkim.
saikanica *Yu.A. Kotukhov in Turczaninowia,* 1(2): 10 (1998)—Kazakhstan.
saposhnikowii *(Roshev.) M. Kitagawa in Rep. Inst. Sci. Res. Manchoukuo,* 6(4): 118 (1942): Timouria saposhnikowii.
sareptana A.K. Becker
subsp. krylovii *(Roshev.) D.F. Cui in Fl. Xinjiangensis,* 6: 299 (1996): S. krylovii.
scabrifolia *M.A. Torres in Monogr. Comis. Invest. Ci. Prov. Buenos Aires,* 13: 61 (1997)—Argentina Northwest (Tucumán, Catamarca, La Rioja, Salta).
shoshoneana *M. Curto & Douglass M. Henderson in Madroño,* 45(1): 59 (1998 publ. 1999)—Idaho, Nevada.
subsessiliflora (Rupr.) Roshev.
subsp. basiplumosa *(Munro ex Hook. f.) D.F. Cui in Fl. Xinjiangensis,* 6: 309 (1996): S. basiplumosa.
tianschanica Roshev.
subsp. gobica *(Roshev.) D.F. Cui in Fl. Xinjiangensis,* 6: 299 (1996): S. gobica.
wulateica *(Y.Z. Zhao) Y.Z. Zhao in Acta Sci. Nat. Univ. Neimonggol,* 27(2): 211 (1996): S. gobica var. wulateica.

STIPAGROSTIS (Gramin.).
affinis *H. Scholz in Willdenowia,* 30(2): 274 (2000)—Egypt, Sudan.
fallax *H. Scholz in Willdenowia,* 30(2): 274 (2000)—Pakistan.
libyca (H. Scholz) H. Scholz
subsp. darfurensis *H. Scholz in Willdenowia,* 30(2): 276 (2000)—Sudan.

STIRLINGIA (Proteac.).
divaricatissima *A.S. George in Fl. Australia,* 16: 473 (1995)—Western Australia.

STOEBERIA (Aizoac.).
porphyrea *H.E.K. Hartmann in Bradleya,* 14: 51 (1996)—Namibia, Cape Province.

STOMOTECHIUM *J.G.C. Lehmann in Gött. Gel. Anz.,* 152: 1513 (1817). *BORAGINACEAE*
papillosum *J.G.C. Lehmann in Gött. Gel. Anz.,* 152: 1513 (1817)—Cape Province.

STONESIELLA *M.D. Crisp & P.H. Weston in Taxon,* 48(4): 711 (1999). *LEGUMINOSAE*
selaginoides *(Hook. f.) M.D. Crisp & P.H. Weston in Taxon,* 48(4): 712 (1999): Pultenaea selaginoides.

STREBLACANTHUS (Acanthac.).
amoenus *(Bremek.) T.F. Daniel in Novon,* 6(2): 147 (1996): Sciaphyllum amoenum.
roseus *(Radlk.) B.L. Burtt in Edinburgh J. Bot.,* 54(2): 253 (1997): Pseuderanthemum roseum.

STREBLUS (Morac.).
sect. Tetrastigma *T.H. Nguyên in Adansonia,* 19(2): 263 (1997).
vidalii *T.H. Nguyên in Adansonia,* 19(2): 265 (1997)—Vietnam.

STREPHIUM (Gramin.).
distichophyllum *Schrad. ex Nees in Mart., Fl. Bras. Enum. Pl.,* 2(1): 298 (1829)—Brazil Northeast (Bahia).†
guianense *A.T. Brongniart in Bull. Soc. Bot. France,* 7: 471 (1860)—French Guiana.

STREPTOCARPUS (Gesneriac.).
huamboensis *O.M. Hilliard & B.L. Burtt in Edinburgh J. Bot.*, 56(3): 378 (1999)—Angola.

× **STREPTOLARIUM** *D.A. Beadle, Prelim. Listing Cultivar Grex Names Bromel.*: 201 (1991). *BROMELIACEAE* (Nidularium × Streptocalyx).

× **STREPTOMEA** *E.L. Smith in J. Bromeliad Soc.*, 33(2): 75 (1983). *BROMELIACEAE* (Aechmea × Streptocalyx).

STREPTOPUS (Convallariac.).
ovalis (Ohwi) F.T. Wang & Y.C. Tang
f. albus *Y.N. Lee, Fl. Korea:* 1163 (1996), without type—Locality not stated.

STRIGA (Scrophulariac.).
angolensis *K.I. Mohamed & L.J. Musselman in Brittonia*, 49(1): 118 (1997)—Angola.

STROBILANTHES (Acanthac.).
alatiramosa *H.S. Lo & D. Fang in Guihaia*, 17(1): 29 (1997)—China Southeast (Guangxi), China South-central (Yunnan).
botryantha *D. Fang & H.S. Lo in Guihaia*, 17(1): 35 (1997)—China Southeast (Guangxi)
compacta *D. Fang & H.S. Lo in Guihaia*, 17(1): 31 (1997)—China Southeast (Guangxi)
ferruginea *D. Fang & H.S. Lo in Guihaia*, 17(1): 29 (1997)—China Southeast (Guangxi)
fulvihispida *D. Fang & H.S. Lo in Guihaia*, 17(1): 37 (1997)—China Southeast (Guangxi).
glandibracteata *D. Fang & H.S. Lo in Guihaia*, 17(1): 38 (1997)—China Southeast (Guangxi).
heteroclita *D. Fang & H.S. Lo in Guihaia*, 17(1): 32 (1997)—China Southeast (Guangxi).
longgangensis *D. Fang & H.S. Lo in Guihaia*, 17(1): 33 (1997)—China Southeast (Guangxi).
longzhouensis *H.S. Lo & D. Fang in Guihaia*, 17(1): 34 (1997)—China Southeast (Guangxi).
matthewiana *R.W. Scotland in Bot. J. Linn. Soc.*, 128(2): 204 (1998)—India (Tamil Nadu).
myriostachya *D. Fang & H.S. Lo in Guihaia*, 17(1): 35 (1997)—China Southeast (Guangxi).
neglectus *T. Anderson ex H.O. Saxena & M. Brahmam, Fl. Orissa*, 3: 1390 (1995), nom. nov., illegit. superfl.: S. consanguineus *C.B. Clarke*.
ningmingensis *D. Fang & H.S. Lo in Guihaia*, 17(1): 39 (1997), contrary to Art. 37.4 ICBN (1994)—China Southeast (Guangxi).
ovatibracteata *H.S. Lo & D. Fang in Guihaia*, 17(1): 40 (1997)—China Southeast (Guangxi).
refracta *D. Fang, Y.G. Wei & J. Murata in Acta Phytotax. Sin.*, 38(2): 185 (2000)—China Southeast (Guangxi).
retusa *D. Fang in Acta Phytotax. Sin.*, 38(2): 187 (2000)—China Southeast (Guangxi).
strigosa *D. Fang & H.S. Lo in Guihaia*, 17(1): 36 (1997)—China Southeast (Guangxi).
truncata *D. Fang & H.S. Lo in Guihaia*, 17(1): 31 (1997)—China Southeast (Guangxi).
willisii *M.A. Carine in Kew Bull.*, 55(4): 971 (2000)—Sri Lanka.

STROMANTHE (Marantac.).
palustris *Helen Kennedy in Novon*, 9(1): 63 (1999)—Costa Rica.
popolucana *G. Castillo-Campos, A.P. Vovides & M. Vázquez Torres in Polibotánica*, 8: 14 (1998 publ. 1999)—Mexico Gulf (Veracruz).

STROMBOCACTUS (Cactac.).
disciformis
subsp. esperanzae *C. Glass & S. Arias in Brit. Cact. Succ. J.*, 14(4): 202 (1995)—Mexico Northeast (Guanajuato).
var. esperanzae *(Glass & S. Arias) Don Pedro & J. Říha in Kaktusy (Brno)*, 33(3): 84 (1997): S. disciformis subsp. esperanzae.
subsp. jarmilae *(Halda) J.J. Halda in Acta Mus. Richnov. Sect. Nat.*, 4(2): 71 (1997): S. jarmilae.
f. minimus *Don Pedro & J. Říha in Kaktusy (Brno)*, 33(3): 84 (1997), nom. nov.: S. jarmilae *Halda*.
jarmilae *Halda in Cactaceae etc.*, 6(3): 83–86 (1996); *cf. Repert. Pl. Succ. (I.O.S.)*, 47: 12 (1996 publ. 1997)—Locality not stated.

STROMBOCACTUS:-
pulcherrimus *Halda in Cactaceae etc.*, 6(3): 88–91 (1996); *cf. Repert. Pl. Succ. (I.O.S.)*, 47: 12 (1996 publ. 1997)—Locality not stated.

STROPHIOBLACHIA (Euphorbiac.).
fimbricalyx Boerl.
var. cordifolia *(Airy Shaw) H.S. Kiu in Guihaia*, 19(3): 195 (1999): S. glandulosa var. cordifolia.

STRUTHANTHUS (Loranthac.).
concinnus Mart.
var. centralis *C.T. Rizzini in Revista Fac. Agron. (Maracay)*, 8(3): 96 (1975)—Brazil Southeast (Minas Gerais), Brazil Northeast (Bahia).
corymbifer *C.T. Rizzini in Revista Fac. Agron. (Maracay)*, 8(3): 96 (1975)—Venezuela.
granulatus *C.T. Rizzini in Revista Fac. Agron. (Maracay)*, 8(3): 97 (1975)—Venezuela.
savannae *C.T. Rizzini in Revista Fac. Agron. (Maracay)*, 8(3): 97 (1975)—Brazil West-central (Mato Grosso).

STUCKENIA (Potamogetonac.).
amblyphylla *(C.A. Mey.) J. Holub in Preslia*, 68(4): 364 (1996 publ. 1997): Potamogeton amblyphyllus.
austrosibirica *(Kaschina) N.N. Tzvelev in Bot. Zhurn.*, 84(7): 111 (1999): Potamogeton austrosibiricus.
borealis *(Raf.) J. Holub in Preslia*, 68(4): 364 (1996 publ. 1997): Potamogeton borealis.
× bottnica *(Hagstr.) J. Holub in Preslia*, 68(4): 364 (1996 publ. 1997): Potamogeton bottnicus.
× fennica *(Hagstr.) J. Holub in Preslia*, 68(4): 364 (1996 publ. 1997): Potamogeton fennicus.
filiformis (Pers.) Börner
subsp. alpina *(Blytt) R.R. Haynes, D.H. Les & M. Král in Novon*, 8(3): 241 (1998): Potamogeton marinus f. alpinus.
subsp. occidentalis *(J.W. Robbins) R.R. Haynes, D.H. Les & M. Král in Novon*, 8(3): 241 (1998): Potamogeton marinus var. occidentalis.
helvetica *(G. Fisch.) J. Holub in Preslia*, 68(4): 364 (1996 publ. 1997): Potamogeton vaginatus var. helveticus.
interior *(Rydb.) J. Holub in Preslia*, 68(4): 364 (1996 publ. 1997): Potamogeton interior.
intramongolica *(Ma) N.N. Tzvelev in Bot. Zhurn.*, 84(7): 111 (1999): Potamogeton intramongolicus.
macrocarpa *(Dobrocz.) N.N. Tzvelev in Bot. Zhurn.*, 84(7): 111 (1999): Potamogeton macrocarpus.
marina *(L.) N.N. Tzvelev in Bot. Zhurn.*, 84(7): 111 (1999): Potamogeton marinus.
× meinshausenii *(Juz.) N.N. Tzvelev in Bot. Zhurn.*, 84(7): 111 (1999): Potamogeton × meinshausenii.
punensis *(A. Galán) A. Galán de Mera in Novon*, 10(2): 115 (2000): Potamogeton punense.
striata *(Ruiz & Pav.) J. Holub in Preslia*, 68(4): 364 (1996 publ. 1997): Potamogeton striatus.
subretusa *(Hagstr.) J. Holub in Preslia*, 68(4): 364 (1996 publ. 1997): Potamogeton subretusus.
× suecica *(K. Richt.) J. Holub in Preslia*, 68(4): 364 (1996 publ. 1997): Potamogeton × suecicus.
zosteracea *(Fr.) N.N. Tzvelev in Bot. Zhurn.*, 84(7): 111 (1999): Potamogeton zosteraceus.

STYLIDIUM (Stylidiac.).
sect. Alsinoida *(Mildbr.) A.R. Bean in Austrobaileya*, 5(4): 634 (2000): Stylidium subgen. Alsinoida.
sect. Biloba *A.R. Bean in Austrobaileya*, 5(4): 631 (2000).
sect. Lanata *A.R. Bean in Austrobaileya*, 5(3): 451 (1999).
sect. Tenella *(Benth.) A.R. Bean in Austrobaileya*, 5(4): 619 (2000): Stylidium ser. Tenella.
sect. Uniflora *A.R. Bean in Austrobaileya*, 5(4): 613 (2000).
accedens *A.R. Bean in Austrobaileya*, 5(4): 605 (2000)—Northern Territory.
aceratum *A. Lowrie & K.F. Kenneally in Nuytsia*, 12(2): 198 (1998)—Western Australia.
adenophorum *A. Lowrie & K.F. Kenneally in Nuytsia*, 11(2): 199 (1997)—Western Australia, Northern Territory, Queensland.
aquaticum *A.R. Bean in Austrobaileya*, 5(4): 622 (2000)—Northern Territory.

STYLIDIUM:-
austrocapense *A.R. Bean in Austrobaileya*, 5(3): 442 (1999)—Queensland.
barrettiorum *A. Lowrie & K.F. Kenneally in Nuytsia*, 11(2): 202 (1997), as 'barrettorum'—Western Australia.
begoniifolium *(Roxb.) J.O. Voigt, Hort. Suburb. Calcutt.*: 40 (1845): Marlea begoniifolia.
burbidgeanum *A. Lowrie & K.F. Kenneally in Nuytsia*, 11(2): 185 (1997)—Western Australia.
candelabrum *A. Lowrie & K.F. Kenneally in Nuytsia*, 13(1): 251 (1999)—Northern Territory.
chiddarcoopingense *A. Lowrie, D.J. Coates & K.F. Kenneally in Nuytsia*, 13(1): 255 (1999)—Western Australia.
cilium *A. Lowrie, A.H. Burbidge & K.F. Kenneally in Nuytsia*, 13(1): 118 (1999), nom. nov.: S. bulbiferum β ciliatum *Sond.*
clarksonii *A. Lowrie & K.F. Kenneally in Nuytsia*, 11(2): 204 (1997)—Queensland.
confertum *A.R. Bean in Austrobaileya*, 5(4): 625 (2000)—Queensland.
daphne *A. Lowrie & K.F. Kenneally in Nuytsia*, 12(1): 76 (1998)—Western Australia.
delicatum *A.R. Bean in Austrobaileya*, 5(3): 434 (1999)—Queensland.
diceratum *A. Lowrie & K.F. Kenneally in Nuytsia*, 12(2): 201 (1998)—Western Australia.
diplectroglossum *(R.O. Erickson & J.H. Willis) A. Lowrie, A.H. Burbidge & K.F. Kenneally in Nuytsia*, 13(1): 132 (1999): S. repens var. diplectroglossum.
divergens *A.R. Bean in Austrobaileya*, 5(4): 604 (2000)—Northern Territory.
ensatum *A.R. Bean in Austrobaileya*, 5(4): 603 (2000)—Northern Territory.
fimbriatum *A. Lowrie & K.F. Kenneally in Nuytsia*, 10(3): 425 (1996)—Western Australia.
flagellum *A. Lowrie, A.H. Burbidge & K.F. Kenneally in Nuytsia*, 13(1): 136 (1999)—Western Australia.
foveolatum *A.R. Bean in Austrobaileya*, 5(3): 430 (1999)—Queensland.
glabrifolium *A. Lowrie & K.F. Kenneally in Nuytsia*, 11(2): 187 (1997)—Western Australia.
hortiorum *A. Lowrie & K.F. Kenneally in Nuytsia*, 13(2): 293 (2000)—Western Australia.
ireneae *A. Lowrie & K.F. Kenneally in Nuytsia*, 12(1): 78 (1998)—Western Australia.
kalbarriense *A. Lowrie & K.F. Kenneally in Nuytsia*, 11(2): 189 (1997)—Western Australia.
leeuwinense *A. Lowrie & K.F. Kenneally in Nuytsia*, 11(3): 356 (1997)—Western Australia.
leiophyllum *A.R. Bean in Austrobaileya*, 5(2): 326 (1999)—Queensland.
longissimum *A.R. Bean in Austrobaileya*, 5(4): 629 (2000)—Queensland.
maritimum *A. Lowrie, D.J. Coates & K.F. Kenneally in Nuytsia*, 12(1): 48 (1998)—Western Australia.
marradongense *A. Lowrie & K.F. Kenneally in Nuytsia*, 11(3): 358 (1997)—Western Australia.
megacarpum *A. Lowrie, A.H. Burbidge & K.F. Kenneally in Nuytsia*, 13(1): 120 (1999), nom. nov.: S. bulbiferum var. macrocarpum *Benth.*
mucronatum *A. Lowrie & K.F. Kenneally in Nuytsia*, 11(2): 206 (1997)—Western Australia.
oviflorum *A.R. Bean in Austrobaileya*, 5(4): 624 (2000)—Queensland.
paniculatum *(Maiden & Betche) A.R. Bean in Austrobaileya*, 5(3): 446 (1999): S. debile var. paniculatum.
paulineae *A. Lowrie & K.F. Kenneally in Nuytsia*, 12(1): 80 (1998)—Western Australia.
perizostera *A. Lowrie & K.F. Kenneally in Nuytsia*, 11(2): 208 (1997)—Western Australia.
pingrupense *A. Lowrie, A.H. Burbidge & K.F. Kenneally in Nuytsia*, 13(1): 138 (1999)—Western Australia.
prophyllum *A. Lowrie & K.F. Kenneally in Nuytsia*, 11(2): 210 (1997)—Western Australia.
pseudosacculatum *A. Lowrie, A.H. Burbidge & K.F. Kenneally in Nuytsia*, 13(1): 143 (1999)—Western Australia.
ramosissimum *A.R. Bean in Austrobaileya*, 5(2): 328 (1999)—Queensland.
rivulosum *A. Lowrie & K.F. Kenneally in Nuytsia*, 11(2): 213 (1997)—Western Australia.

STYLIDIUM:-
sejunctum *A. Lowrie, D.J. Coates & K.F. Kenneally in Nuytsia*, 12(1): 52 (1998)—Western Australia.
semaphorum *A. Lowrie & K.F. Kenneally in Nuytsia*, 11(3): 362 (1997)—Western Australia.
septentrionale *(Mildbr.) A. Lowrie, A.H. Burbidge & K.F. Kenneally in Nuytsia*, 13(1): 123 (1999): S. bulbiferum var. septentrionale.
sidjamesii *A. Lowrie & K.F. Kenneally in Nuytsia*, 13(2): 296 (2000)—Western Australia.
stenophyllum *A.R. Bean in Austrobaileya*, 5(4): 612 (2000)—Northern Territory.
tinkeri *A. Lowrie & K.F. Kenneally in Nuytsia*, 13(2): 299 (2000)—Western Australia.
torticarpum *A. Lowrie & K.F. Kenneally in Nuytsia*, 11(2): 191 (1997)—Western Australia.
turbinatum *A. Lowrie & K.F. Kenneally in Nuytsia*, 11(2): 215 (1997)—Western Australia, Northern Territory.
tylosum *A. Lowrie & K.F. Kenneally in Nuytsia*, 11(2): 195 (1997)—Western Australia.
udusicola *A. Lowrie & K.F. Kenneally in Nuytsia*, 11(2): 197 (1997)—Western Australia.
velleioides *A.R. Bean in Austrobaileya*, 5(3): 448 (1999)—Queensland.
warriedarense *A. Lowrie, A.H. Burbidge & K.F. Kenneally in Nuytsia*, 13(1): 131 (1999)—Western Australia.
weeliwolli *A. Lowrie & K.F. Kenneally in Nuytsia*, 12(2): 203 (1998)—Western Australia.
wilroyense *A. Lowrie, D.J. Coates & K.F. Kenneally in Nuytsia*, 12(1): 54 (1998)—Western Australia.

STYLOGYNE (Myrsinac.).
aguarunana *J.J. Pipoly & J.M. Ricketson in Sida*, 19(2): 269 (2000)—Peru.
incognita *J.J. Pipoly in Brittonia*, 51(2): 130 (1999)—French Guiana.
latipes *(Mart.) N.N. Imkhanitskaya in Bot. Zhurn.*, 81(9): 32 (1996): Ardisia latipes.
turbacensis (Kunth) Mez
subsp. laevis *(Oerst.) J.M. Ricketson & J.J. Pipoly in Sida*, 17(3): 595 (1997): Ardisia laevis.
viridis *(Lundell) J.M. Ricketson & J.J. Pipoly in Sida*, 18(4): 1171 (1999): Parathesis viridis.

STYLOSANTHES (Legumin.).
longicarpa *M. Brandão & N.M. Sousa Costa in Daphne*, 2(3): 5 (1992)—Brazil Southeast (Minas Gerais).
nunoi *M. Brandão in Daphne*, 2(1): 5 (1991)—Brazil Southeast (Minas Gerais).

STYLOTRICHIUM (Compos.).
glomeratum *H.P. Bautista, J. Rodríguez-Oubiña & S. Ortiz in Brittonia*, 52(3): 215 (2000)—Brazil Northeast (Bahia).

STYPANDRA (Phormiac.).
jamesii *S.D. Hopper in Nuytsia*, 13(1): 248 (1999)—Western Australia.

STYPHELIA (Epacridac.).
tameiameiae (Cham. & Schltdl.) F. Muell.
var. societas *(A. Gray) S.L. Welsh, Fl. Societensis:* 97 (1998): Cyathodes tameiameiae var. societas.

STYRAX (Styracac.).
ser. Andinae *B. Wallnöfer in Ann. Naturhist. Mus. Wien*, 99B: 712 (1997).
ser. Benzoin *(Hayne) P.W. Fritsch in Syst. Bot.*, 24(3): 374 (1999): gen. Benzoin.
ser. Cyrta *(Lour.) P.W. Fritsch in Syst. Bot.*, 24(3): 373 (1999): gen. Cyrta.
sect. Pamphilia *(Mart. ex A. DC.) B. Wallnöfer in Ann. Naturhist. Mus. Wien*, 99B: 696 (1997): gen. Pamphilia.
argenteus
var. grandiflorus *E. Carranza Gonzalez in Acta Bot. Mex.*, 36: 15 (1996)—Mexico Southwest (Oaxaca).
austromexicanus *P.W. Fritsch in Ann. Missouri Bot. Gard.*, 84(4): 718 (1997 publ. 1998), nom. nov.: S. argenteus var. grandiflorus *E. Carranza*.
gentryi *P.W. Fritsch in Ann. Missouri Bot. Gard.*, 84(4): 723 (1997 publ. 1998)—Mexico Northwest (Sinaloa, Baja California Sur).

STYRAX:-
incarnatus *P.W. Fritsch in Ann. Missouri Bot. Gard.*, 84(4): 730 (1997 publ. 1998)—Guatemala, Honduras.
lanceolatus *P.W. Fritsch in Ann. Missouri Bot. Gard.*, 84(4): 733 (1997 publ. 1998), nom. nov.: S. argenteus var. parvifolius *E. Carranza*.
maninul *B. Wallnöfer in Ann. Naturhist. Mus. Wien*, 99B: 701 (1997), nom. nov.: Pamphilia aurea *Mart. ex A. DC.*
nicaraguensis *P.W. Fritsch in Ann. Missouri Bot. Gard.*, 84(4): 736 (1997 publ. 1998)—Nicaragua, Costa Rica.
subsp. ellipsoidalis *P.W. Fritsch in Ann. Missouri Bot. Gard.*, 84(4): 737 (1997 publ. 1998)—Costa Rica.
nui *B. Wallnöfer in Linz. Biol. Beitr.*, 28(2): 1054 (1996)—Peru.
omuk *B. Wallnöfer in Ann. Naturhist. Mus. Wien*, 99B: 712 (1997)—Peru.
pedicellatus *(Perkins) B. Wallnöfer in Ann. Naturhist. Mus. Wien*, 99B: 706 (1997): Pamphilia pedicellata.
pefrit *B. Wallnöfer in Ann. Naturhist. Mus. Wien*, 99B: 710 (1997)—Colombia.
platanifolius Engelm. ex Torr.
subsp. mollis *P.W. Fritsch in Ann. Missouri Bot. Gard.*, 84(4): 742 (1997 publ. 1998)—Mexico Northeast (Coahuila, Nuevo León).
var. mollis *(P.W. Fritsch) B.L. Turner in Sida*, 19(2): 261 (2000): Styrax platanifolius subsp. mollis.
subsp. stellatus *(Cory) P.W. Fritsch in Ann. Missouri Bot. Gard.*, 84(4): 743 (1997 publ. 1998): S. platanifolius var. stellatus.
subsp. texanus *(Cory) P.W. Fritsch in Ann. Missouri Bot. Gard.*, 84(4): 744 (1997 publ. 1998): S. texanus.
var. texanus *(Cory) B.L. Turner in Sida*, 19(2): 261 (2000): Styrax texanus.
var. youngiae *(Cory) B.L. Turner in Sida*, 19(2): 261 (2000): Styrax youngiae.
subsp. youngiae *(Cory) P.W. Fritsch in Ann. Missouri Bot. Gard.*, 84(4): 744 (1997 publ. 1998): S. youngiae.
radians *P.W. Fritsch in Ann. Missouri Bot. Gard.*, 84(4): 745 (1997 publ. 1998)—Mexico Southwest (Jalisco, Guerrero, Nayarit, Oaxaca), Mexico Northwest (Sinaloa), Mexico Central (México State).
steyermarkii *P.W. Fritsch in Ann. Missouri Bot. Gard.*, 84(4): 751 (1997 publ. 1998)—Guatemala, Costa Rica.
tuxtlensis *P.W. Fritsch in Ann. Missouri Bot. Gard.*, 84(4): 753 (1997 publ. 1998)—Mexico Gulf (Veracruz).
vilcabambae *(D.R. Simpson) B. Wallnöfer in Ann. Naturhist. Mus. Wien*, 99B: 715 (1997): Pamphilia vilcabambae.

SUAEDA (Chenopodiac.).
anonyma *J.F. Gmelin, Onomat. Bot. Compl.*, 8: 798 (1776)—Locality not stated.†
baccata *Forssk. ex J.F. Gmelin, Onomat. Bot. Compl.*, 8: 797 (1776)—Egypt.†
baluchestanica *H. Akhani & D. Podlech in Fl. Iran.*, 172: 151 (1997)—Iran, Pakistan.
corniculata (C.A. Mey.) Bunge
subsp. erecta *(Bunge) M.N. Lomonosova in Fl. Sibir. (Salicac.-Amaranthac.)*, 5: 175 (1992), without basionym ref.: Basionym not stated.
fruticosa *Forssk. ex J.F. Gmelin, Onomat. Bot. Compl.*, 8: 798 (1776)—Saudi Arabia.
hortensis *Forssk. ex J.F. Gmelin, Onomat. Bot. Compl.*, 8: 798 (1776)— Egypt.†
monoica *Forssk. ex J.F. Gmelin, Onomat. Bot. Compl.*, 8: 798 (1776)—Egypt.†
vera *Forssk. ex J.F. Gmelin, Onomat. Bot. Compl.*, 8: 797 (1776)—Egypt.†
vermiculata *Forssk. ex J.F. Gmelin, Onomat. Bot. Compl.*, 8: 798 (1776), as 'vermucularis'—Egypt.†

SULCOLLUMA *D.C.H. Plowes in Haseltonia*, 3: 61 (1995). *ASCLEPIADACEAE*
foulcheri-delboscii *(Lavranos) D.C.H. Plowes in Haseltonia*, 3: 62 (1995): Caralluma foulcheri-delboscii.

SULCOLLUMA:-
foulcheri-delboscii
var. greenbergiana *(Lavranos) D.C.H. Plowes in Haseltonia*, 3: 62 (1995): Caralluma foulcheri-delboscii var. greenbergiana.
hexagona *(Lavranos) D.C.H. Plowes in Haseltonia*, 3: 62 (1995): Caralluma hexagona.
var. septentrionalis *(Lavranos & L.E. Newton) D.C.H. Plowes in Haseltonia*, 3: 62 (1995): Caralluma hexagona var. septentrionalis.
shadhbana *(Lavranos) D.C.H. Plowes in Haseltonia*, 3: 62 (1995): Caralluma shadhbana.
var. barhana *(Lavranos & L.E. Newton) D.C.H. Plowes in Haseltonia*, 3: 62 (1995): Caralluma shadhbana var. barhana.

SULCOREBUTIA (Cactac.).
candiae (Cárdenas) Buining & Donald
var. kamiensis *(Brederoo & Donald) K. Augustin & W. Gertel in K. Augustin, W. Gertel & G. Hentzschel, Sulcorebutia: Kakteenzw. boliv. Anden*: 69 (2000): Sulcorebutia menesesii var. kamiensis.
crispata Rausch
subsp. aureicapillata *J.J. Halda, P. Heřtus & L. Horáček in Acta Mus. Richnov. Sect. Nat.*, 7(2): 74 (2000)—Bolivia.
subsp. hertusii *J.J. Halda & L. Horáček in Acta Mus. Richnov. Sect. Nat.*, 7(2): 74 (2000), as 'hertusi'—Bolivia.
gemmae *S. Mosti & G. Rovida in Cactus & Co.*, 4(4): 174 (2000)—Bolivia.
krugeri (Cárdenas) F. Ritter
var. hoffmannii *K. Augustin & G. Hentzschel in K. Augustin, W. Gertel & G. Hentzschel, Sulcorebutia: Kakteenzw. boliv. Anden*: 89 (2000)—Bolivia.
langeri *K. Augustin & G. Hentzschel in Kakt. And. Sukk.*, 50(8): 202, 201 (1999)—Bolivia.
losenickyana Rausch
var. chatajillensis *(Oeser & Brederoo) K. Augustin & W. Gertel in K. Augustin, W. Gertel & G. Hentzschel, Sulcorebutia: Kakteenzw. boliv. Anden*: 95 (2000): Sulcorebutia verticillacantha var. chatajillensis.
var. vasqueziana *(Rausch) K. Augustin & W. Gertel in K. Augustin, W. Gertel & G. Hentzschel, Sulcorebutia: Kakteenzw. boliv. Anden*: 98 (2000): Sulcorebutia vasqueziana.
mariana Swoboda
var. laui *(Brederoo & Donald) K. Augustin & W. Gertel in K. Augustin, W. Gertel & G. Hentzschel, Sulcorebutia: Kakteenzw. boliv. Anden*: 102 (2000): Sulcorebutia vizcarrae var. laui.
markusii Rausch
subsp. tintiniensis *W. Gertel in Cactus & Co.*, 4(4): 167 (2000)—Bolivia.
mentosa F. Ritter
var. swobodae *(K. Augustin) K. Augustin in K. Augustin, W. Gertel & G. Hentzschel, Sulcorebutia: Kakteenzw. boliv. Anden*: 110 (2000): Sulcorebutia swobodae.
oenantha Rausch
var. pampagrandensis *(Rausch) K. Augustin & W. Gertel in K. Augustin, W. Gertel & G. Hentzschel, Sulcorebutia: Kakteenzw. boliv. Anden*: 113 (2000): Sulcorebutia pampagrandensis.
purpurea (Donald & A.B. Lau) Brederoo & Donald
var. unguispina *(Rausch) K. Augustin & W. Gertel in K. Augustin, W. Gertel & G. Hentzschel, Sulcorebutia: Kakteenzw. boliv. Anden*: 123 (2000): Sulcorebutia unguispina.
steinbachii (Werderm.) Backeb.
subvar. aguilari *W. Gertel in Kakt. And. Sukk.*, 47(6): 135 (1996), nom. prov., without latin descr. or type—Bolivia.
subvar. augustinii *W. Gertel in Kakt. And. Sukk.*, 47(6): 139 (1996), nom. prov., without latin descr. or type—Bolivia.
subsp. australis *W. Gertel in Kakt. And. Sukk.*, 47(6): 133 (1996), nom. prov., without latin descr. or type—Bolivia.
f. bicolorispina *W. Gertel in Kakt. And. Sukk.*, 47(6): 134 (1996), nom. prov., without latin descr. or type—Bolivia.

SULCOREBUTIA:-
steinbachii
var. krahnii *W. Gertel in Kakt. And. Sukk.*, 47(6): 135 (1996), nom. prov., without latin descr. or type—Bolivia.
subsp. kruegeri *W. Gertel in Kakt. And. Sukk.*, 47(6): 133 (1996), nom. prov., without latin descr. or type—Bolivia.
var. lepida *W. Gertel in Kakt. And. Sukk.*, 47(6): 134 (1996), nom. prov., without latin descr. or type—Bolivia.
subvar. longiseta *(Cárdenas) Gertel in Inform.-brief Freundeskreis Echinopseen*, 22: 26 (1996), without basionym ref.; *cf. Repert. Pl. Succ. (I.O.S.)*, 47: 12 (1996 publ. 1997): Rebutia tiraquensis var. longiseta.
subvar. oenantha *W. Gertel in Kakt. And. Sukk.*, 47(6): 139 (1996), nom. prov., without latin descr. or type—Bolivia.
subvar. pampagrandensis *W. Gertel in Kakt. And. Sukk.*, 47(6): 139 (1996), nom. prov., without latin descr. or type—Bolivia.
subvar. polymorpha *(Cárdenas) Gertel in Inform.-brief Freundeskreis Echinopseen*, 22: 26 (1996), without basionym ref.; *cf. Repert. Pl. Succ. (I.O.S.)*, 47: 12 (1996 publ. 1997): Rebutia polymorpha.
var. tiraquensis *W. Gertel in Kakt. And. Sukk.*, 47(6): 133 (1996), nom. prov., without latin descr. or type—Bolivia.
subsp. tiraquensis *W. Gertel in Kakt. And. Sukk.*, 47(6): 133 (1996), nom. prov., without latin descr. or type—Bolivia.
var. totorensis *W. Gertel in Kakt. And. Sukk.*, 47(6): 136 (1996), nom. prov., without latin descr. or type—Bolivia.
subvar. totorensis *W. Gertel in Kakt. And. Sukk.*, 47(6): 139 (1996), nom. prov., without latin descr. or type—Bolivia.
var. tunariensis *(Cárdenas) K. Augustin & W. Gertel in K. Augustin, W. Gertel & G. Hentzschel, Sulcorebutia: Kakteenzw. boliv. Anden*: 132 (2000): Rebutia tunariensis.
var. tunariensis *(Cárdenas) Gertel in Inform.-brief Freundeskreis Echinopseen*, 22: 26 (1996), without basionym ref.; *cf. Repert. Pl. Succ. (I.O.S.)*, 47: 13 (1996 publ. 1997): Rebutia tunariensis.
subsp. verticillacantha *W. Gertel in Kakt. And. Sukk.*, 47(6): 133 (1996), nom. prov., without latin descr. or type—Bolivia.
tarabucoensis Rausch
var. aureiflora *(Rausch) K. Augustin & W. Gertel in K. Augustin, W. Gertel & G. Hentzschel, Sulcorebutia: Kakteenzw. boliv.* Anden: 136 (2000): Sulcorebutia verticillacantha var. aureiflora.
var. callecallensis *(F.H. Brandt) K. Augustin & W. Gertel in K. Augustin, W. Gertel & G. Hentzschel, Sulcorebutia: Kakteenzw. boliv. Anden*: 136 (2000): Weingartia callecallensis.
tiraquensis (Cárdenas) F. Ritter
var. aguilarii *K. Augustin & W. Gertel in Cactus & Co.*, 3(3): 117, 120 (1999)—Bolivia.
var. lepida *(F. Ritter) K. Augustin & W. Gertel in K. Augustin, W. Gertel & G. Hentzschel, Sulcorebutia: Kakteenzw. boliv. Anden*: 143 (2000): Sulcorebutia lepida.
var. renatae *G. Hentzschel & S. Beck in Cactus & Co.*, 3(3): 119, 122 (1999)—Bolivia.
var. totorensis *(Cárdenas) K. Augustin & W. Gertel in K. Augustin, W. Gertel & G. Hentzschel, Sulcorebutia: Kakteenzw. boliv. Anden*: 147 (2000): Rebutia totorensis.
veronikae *J.J. Halda, P. Heřtus & L. Horáček in Acta Mus. Richnov. Sect. Nat.*, 7(2): 74 (2000)—Bolivia.
verticillacantha F. Ritter
var. taratensis *(Cárdenas) K. Augustin & W. Gertel in K. Augustin, W. Gertel & G. Hentzschel, Sulcorebutia: Kakteenzw. boliv. Anden*: 154 (2000): Rebutia taratensis.

SUNIPIA (Orchidac.).
dulongensis *H. Li in Fl. Dulongjian Reg.*: 357 (1993), without latin descr. or type—China South-central (Yunnan).
hainanensis *Z.H. Tsi in Acta Phytotax. Sin.*, 33(6): 590 (1995)—Hainan.

SUNIPIA:-
pallida *(Aver.) L.V. Averyanov in Bot. Zhurn.*, 84(10): 129 (1999): Ione pallida.

× **SUSANPERREIRAARA** *hort. in Orchid Rev.*, 108(1231): new orchid hybrids 15 (2000). *ORCHIDACEAE* (Broughtonia × Cattleya × Tetramicra).

SVENKOELTZIA (Orchidac.).
luzmariana *R. González Tamayo in Bol. Inst. Bot. (Guadalajara)*, 7(1–3): 40 (1999 publ. 2000)—Mexico Southwest (Jalisco).
patriciae *R. González Tamayo in Bol. Inst. Bot. (Guadalajara)*, 7(1–3): 45 (1999 publ. 2000)—Mexico Southwest (Jalisco).

SWARTZIA (Legumin.).
alternifoliolata *V. de F. Mansano in Brittonia*, 51(2): 149 (1999)—Brazil Southeast (Espírito Santo).
capixabensis *V. de F. Mansano in Brittonia*, 51(2): 152 (1999)—Brazil Southeast (Espírito Santo).
invenusta *R.C. Barneby in Brittonia*, 48(2): 186 (1996)—Guyana.

SWERTIA (Gentianac.).
crassiuscula Gilg
var. leucantha *(T.C.E. Fr.) Sileshi Nemomissa in Kew Bull.*, 53(2): 425 (1998): S. leucantha.
subsp. robusta *Sileshi Nemomissa in Kew Bull.*, 53(2): 425 (1998)—Ethiopia.
diluta *(Turcz.) V.V. Zuev in Fl. Sibir. (Pyrolac.-Lamiac.)*, 11: 77 (1997): Sczukinia diluta.
engleri Gilg
var. woodii *(J. Shah) Sileshi Nemomissa in Kew Bull.*, 53(2): 427 (1998): S. woodii.
macrosepala Gilg
subsp. microsperma *Sileshi Nemomissa in Kew Bull.*, 53(2): 432 (1998)—Ethiopia.
perennis L.
subsp. stenopetala *(Regel & Tiling) V.N. Voroschilov in A.K. Skvortsov (ed.), Florist. issl. v razn. raĭonakh SSSR*: 187 (1985): S. obtusa var. stenopetala.
pseudochinensis (Bunge) H. Hara
f. alba *Y.N. Lee, Fl. Korea*: 1159 (1996), without type—Locality not stated.
radiata (Kellogg) Kuntze
var. maderensis *J. Henrickson in Sida*, 17(1): 83 (1996)—Mexico Northeast (Coahuila).
squamigera *Sileshi Nemomissa in Kew Bull.*, 52(2): 487 (1997)—Tanzania.
tetrapetala Pall.
var. happoensis *Hid. Takah. ex T. Shimizu, Fl. Nagano Pref.*: 1509 (1997)—Japan.
usambarensis Engl.
var. curtioides *(Gilg) Sileshi Nemomissa in Kew Bull.*, 53(2): 428 (1998): S. curtioides.
volkensii Gilg
var. baleensis *Sileshi Nemomissa in Kew Bull.*, 53(2): 434 (1998)—Ethiopia.

SWIDA (Cornac.).
subgen. Umbellicrania *(Murrell) J. Holub in Preslia*, 70(2): 114 (1998): Cornus sect. Umbellicrania.
subgen. Yinquania *(Z.Y. Zhu) J. Holub in Preslia*, 70(2): 114 (1998): gen. Yinquania.
alba (L.) Opiz
subsp. stolonifera *(Michx.) N.N. Tzvelev in Novosti Sist. Vÿssh. Rast.*, 32: 185 (2000): Cornus stolonifera.
controversa (Hemsl. ex Prain) Soják
var. alpina *(Wangerin) H. Hara ex S. Noshiro in Fl. Japan*, 2c: 256 (1999): Cornus controversa var. alpina.
var. shikokumontana *(Hiyama) H. Hara ex S. Noshiro in Fl. Japan*, 2c: 256 (1999): Cornus controversa var. shikokumontana.
× friedlanderi *(W.H. Wagner) J. Holub in Preslia*, 70(2): 114 (1998): Cornus × friedlanderi.
muchuanensis *(Z.Y. Zhu) J. Holub in Preslia*, 70(2): 114 (1998): Yinquania muchuanensis.
peruviana *(J.F. Macbr.) J. Holub in Preslia*, 70(2): 115 (1998): Cornus peruviana.

SWINTONIA (Anacardiac.).
sarawakana *K.M. Kochummen in Sandakania*, 7: 111 (1996)—Borneo (Sarawak).

SYAGRUS (Palmae).
× nabonnandii *(Prosch.) J.-P. Demoly in Bull. Ass. Parcs Bot. France*, 12: 30 (1989): × Butiarecastrum nabonnandii.
yungasensis *M. Moraes R. in Novon*, 6(1): 89 (1996)—Bolivia.

SYLVICHADSIA *D.J. Du Puy & J.-N. Labat in Adansonia*, 20(1): 165 (1998). *LEGUMINOSAE*
grandidieri *(Baill.) D.J. Du Puy & J.-N. Labat in Adansonia*, 20(1): 168 (1998): Chadsia grandidieri.
grandifolia *(R. Vig.) D.J. Du Puy & J.-N. Labat in Adansonia*, 20(1): 166 (1998): Chadsia grandifolia.
macrophylla *(R. Vig.) D.J. Du Puy & J.-N. Labat in Adansonia*, 20(1): 170 (1998): Mundulea macrophylla.
perrieri *(R. Vig.) D.J. Du Puy & J.-N. Labat in Adansonia*, 20(1): 170 (1998): Strongylodon perrieri.

SYMBOLANTHUS (Gentianac.).
aureus *L. Struwe & V.A. Albert in Harvard Pap. Bot.*, 3(2): 194 (1998)—Venezuela.
rosmarinifolius *L. Struwe & V.A. Albert in Harvard Pap. Bot.*, 3(2): 194 (1998)— Venezuela.

× **SYMMONSARA** *hort. in Orchid Rev.*, 105(1217): 318 (1997). *ORCHIDACEAE* (Brassavola × Cattleya × Epidendrum × Schomburgkia).

SYMPHYOTRICHUM (Compos.).
eatonii *(A. Gray) G.L. Nesom in Phytologia*, 82(4): 283 (1997 publ. 1998): Aster foliaceus var. eatonii.
kralii *G.L. Nesom in Phytologia*, 82(4): 284 (1997 publ. 1998), nom. nov.: Aster pinifolius *Alexander.*
novi-belgii (L.) G.L. Nesom
var. crenifolium *(Fernald) J. Labrecque & L. Brouillet in Phytologia*, 82(3): 138 (1997): Aster foliaceus var. crenifolius.
var. villicaule *(A. Gray) J. Labrecque & L. Brouillet in Phytologia*, 82(3): 138 (1997): Aster longifolius var. villicaulis.
ontarionis (Wiegand) G.L. Nesom
var. glabratum *(Semple) L. Brouillet & D. Bouchard in Phytologia*, 82(3): 138 (1997): Aster ontarionis var. glabratus.
praealtum (Poir.) G.L. Nesom
var. nebraskense *(Britton) G.L. Nesom in Phytologia*, 82(4): 285 (1997 publ. 1998): Aster praealtus var. nebraskensis.
robynsianum *(J. Rousseau) L. Brouillet & J. Labrecque in Phytologia*, 82(3): 138 (1997): Aster robynsianus.

SYMPHYTUM (Boraginac.).
× mosquense *S.R. Majorov & D.D. Sokoloff in Novosti Sist. Vÿssh. Rast.*, 31: 241 (1998)—Russia Central. (S. caucasicum × officinale).

SYMPLOCARPUS (Arac.).
renifolius *Schott ex N.N. Tzvelev in Novosti Sist. Vÿssh. Rast.*, 28: 28 (1991)—Japan.

SYMPLOCOS (Symplocac.).
bombycina *B. Ståhl in Candollea*, 51(2): 328 (1996)—Colombia.
crassulacea *B. Ståhl in Candollea*, 51(2): 340 (1996)—Colombia.
cundinamarcensis *B. Ståhl in Candollea*, 51(2): 339 (1996)—Colombia.
decorticans *B. Ståhl in Candollea*, 51(2): 337 (1996)—Colombia.
leochaii *P.P.K. Chai in Sandakania*, 7: 63 (1996)—Borneo (Sarawak).
macrophylla Wall. ex DC.
subsp. namboodiriana *M. Sivadasan & N. Mohanan in Rheedea*, 7(2): 89 (1997), as 'namboodirianus'—India (Kerala).
magdalenae *B. Ståhl in Candollea*, 51(2): 362 (1996)—Colombia.
menglianensis *Y.Y. Qian in Bull. Bot. Res. North-East. Forest. Univ.*, 19(1): 5 (1999)— China South-Central (Yunnan).
migoi *H. Nagamasu in Fl. Taiwan, ed. 2*, 4: 116 (1998)—Taiwan.

SYMPLOCOS:-
phaeoneura *B. Ståhl in Candollea*, 51(2): 358 (1996)—Colombia.
ramuliflora *B. Ståhl in Candollea*, 51(2): 343 (1996)—Colombia.
rhomboidea *B. Ståhl in Candollea*, 51(2): 332 (1996)—Colombia.
rimosa *B. Ståhl in Candollea*, 51(2): 347 (1996)—Colombia.
sawafutagi Nagam.
var. terrae-nivosae *Baba ex T. Shimizu, Fl. Nagano Pref.*: 1508 (1997)—Japan.
silverstonei *B. Ståhl in Candollea*, 51(2): 340 (1996)—Colombia.

SYNAPHEA (Proteac.).
sect. Bicornis *A.S. George in Fl. Australia*, 16: 487 (1995).
sect. Oulopha *A.S. George in Fl. Australia*, 16: 487 (1995).
sect. Pinnata *A.S. George in Fl. Australia*, 16: 487 (1995).
aephynsa *A.S. George in Fl. Australia*, 16: 487 (1995)—Western Australia.
bifurcata *A.S. George in Fl. Australia*, 16: 488 (1995)—Western Australia.
boyaginensis *A.S. George in Fl. Australia*, 16: 488 (1995)—Western Australia.
brachyceras *Ryonen Butcher in Nuytsia*, 13(2): 266 (2000)—Western Australia.
canaliculata *A.S. George in Fl. Australia*, 16: 488 (1995)—Western Australia.
cervifolia *A.S. George in Fl. Australia*, 16: 488 (1995)—Western Australia.
constricta *A.S. George in Fl. Australia*, 16: 488 (1995)—Western Australia.
cuneata *A.S. George in Fl. Australia*, 16: 489 (1995)—Western Australia.
damopsis *A.S. George in Fl. Australia*, 16: 489 (1995)—Western Australia.
decumbens *A.S. George in Fl. Australia*, 16: 489 (1995)—Western Australia.
divaricata *(Benth.) A.S. George in Fl. Australia*, 16: 489 (1995): S. favosa var. divaricata.
endothrix *A.S. George in Fl. Australia*, 16: 489 (1995)—Western Australia.
flabelliformis *A.S. George in Fl. Australia*, 16: 490 (1995)—Western Australia.
flexuosa *A.S. George in Fl. Australia*, 16: 490 (1995)—Western Australia.
floribunda *A.S. George in Fl. Australia*, 16: 490 (1995)—Western Australia.
grandis *A.S. George in Fl. Australia*, 16: 490 (1995)—Western Australia.
hians *A.S. George in Fl. Australia*, 16: 490 (1995)—Western Australia.
incurva *A.S. George in Fl. Australia*, 16: 491 (1995)—Western Australia.
interioris *A.S. George in Fl. Australia*, 16: 491 (1995)—Western Australia.
intricata *A.S. George in Fl. Australia*, 16: 491 (1995)—Western Australia.
lesueurensis *A.S. George in Fl. Australia*, 16: 491 (1995)—Western Australia.
macrophylla *A.S. George in Fl. Australia*, 16: 491 (1995)—Western Australia.
media *A.S. George in Fl. Australia*, 16: 492 (1995)—Western Australia.
nexosa *A.S. George in Fl. Australia*, 16: 492 (1995)—Western Australia.
obtusata *(Meisn.) A.S. George in Fl. Australia*, 16: 492 (1995): S. preissii var. obtusata.
odocoileops *A.S. George in Fl. Australia*, 16: 492 (1995)—Western Australia.
oligantha *A.S. George in Fl. Australia*, 16: 492 (1995)—Western Australia.
otiostigma *A.S. George in Fl. Australia*, 16: 492 (1995)—Western Australia.
oulopha *A.S. George in Fl. Australia*, 16: 493 (1995)—Western Australia.
panhesya *A.S. George in Fl. Australia*, 16: 493 (1995)—Western Australia.
parviflora *A.S. George in Fl. Australia*, 16: 493 (1995)—Western Australia.
petiolaris R. Br.
subsp. simplex *A.S. George in Fl. Australia*, 16: 493 (1995)—Western Australia.
subsp. triloba *A.S. George in Fl. Australia*, 16: 493 (1995)—Western Australia.

SYNAPHEA:-
platyphylla *A.S. George in Fl. Australia*, 16: 493 (1995)—Western Australia.
quartzitica *A.S. George in Fl. Australia*, 16: 494 (1995)—Western Australia.
rangiferops *A.S. George in Fl. Australia*, 16: 494 (1995)—Western Australia.
recurva *A.S. George in Fl. Australia*, 16: 494 (1995)—Western Australia.
sparsiflora *A.S. George in Fl. Australia*, 16: 494 (1995)—Western Australia.
spinulosa (Burm. f.) Merr.
subsp. borealis *A.S. George in Fl. Australia*, 16: 494 (1995)—Western Australia.
subsp. major *A.S. George in Fl. Australia*, 16: 495 (1995)—Western Australia.
stenoloba *A.S. George in Fl. Australia*, 16: 495 (1995)—Western Australia.
tamminensis *A.S. George in Fl. Australia*, 16: 495 (1995)—Western Australia.
tripartita *A.S. George in Fl. Australia*, 16: 495 (1995)—Western Australia.
whicherensis *A.S. George in Fl. Australia*, 16: 495 (1995)—Western Australia.

SYNCALATHIUM (Compos.).
chrysocephalum *(C. Shih) S.W. Liu in Fl. Qinghaiica*, 3: 498 (1996), as 'chrysocephala': Soroseris chrysocephala.
chrysocephalum *(C. Shih) C. Shih in Fl. Reipubl. Popul. Sin.*, 80(1): 204 (1997), isonym see S.W. Liu (1996): Soroseris chrysocephala.
qinghaiense *(C. Shih) C. Shih in Fl. Reipubl. Popul. Sin.*, 80(1): 204 (1997): Soroseris qinghaiensis.

SYNCOLOSTEMON (Labiat.).
flabellifolius *(S. Moore) A.J. Paton in Kew Bull.*, 53(2): 484 (1998): Hemizygia flabellifolia.

SYNGONANTHUS (Eriocaulac.).
anthemidiflorus (Bong.) Ruhland
var. similis *(Ruhland) L.R. Parra & A.M. Giulietti in Willdenowia*, 27(1–2): 228 (1997): S. similis.
brasilianus *A.M. Giulietti in Bol. Bot. (São Paulo)*, 15: 63 (1996), as 'brasiliana'—Brazil Southeast (Minas Gerais).
exilis *S.M. Phillips in Kew Bull.*, 52(1): 85 (1997)—Zambia.
magnificus *A.M. Giulietti in Bol. Bot. (São Paulo)*, 15: 67 (1996)—Brazil Southeast (Minas Gerais).
mucugensis *A.M. Giulietti in Bol. Bot. (São Paulo)*, 15: 69 (1996)—Brazil Northeast (Bahia).
mwinilungensis *S.M. Phillips in Kew Bull.*, 52(1): 81 (1997)—Zambia.
paleaceus *S.M. Phillips in Kew Bull.*, 53(2): 491 (1998)—Zambia.
schlechteri *Ruhland in Engl., Pflanzenr.*, 13(4.30): 247 (1903)— Congo.†
suberosus *A.M. Giulietti in Bol. Bot. (São Paulo)*, 15: 65 (1996)—Brazil Southeast (Minas Gerais).
wahlbergii (Wikstr. ex Körn.) Ruhland
var. sinkabolensis *S.M. Phillips in Kew Bull.*, 52(1): 76 (1997)—Zambia, Zimbabwe.

SYNGONIUM (Arac.).
castroi *M.H. Grayum in Phytologia*, 82(1): 52 (1997)—Costa Rica.
rayi *T.B. Croat & M.H. Grayum in Phytologia*, 82(1): 53 (1997)—Costa Rica, Panama.

SYNOTIS (Compos.).
reniformis *Y.L. Chen in Acta Phytotax. Sin.*, 34(6): 648 (1996)—China South-central (Yunnan).

SYNOUM (Meliac.).
glandulosum (Sm.) A. Juss.
subsp. paniculosum *(F. Muell.) D.J. Mabberley in Telopea*, 8(1): 48 (1998): S. glandulosum var. paniculosum.

SYNSEPALUM (Sapotac.).
laurentii *(De Wild.) D.J. Harris in Kew Bull.*, 54(1): 80 (1999): Sersalisia laurentii.
nyangense *(Pellegr.) G. McPherson & L.J.T. White in Novon*, 9(4): 534 (1999): Gambeya nyangensis.
pobeguinianum *(Pierre ex Lecomte) L. Aké Assi & L. Gautier in Candollea*, 55(2): 282 (2000): Pachystela pobeguiniana.

SYNSTEMON (Crucif.).
deserticola *Y.Z. Zhao in Acta Phytotax. Sin.*, 36(4): 373 (1998), as 'deserticolus'—Inner Mongolia (Nei Mongol).
lulianlianus *I.A. Al-Shehbaz, T.Y. Cheo & G. Yang in Novon*, 10(2): 102 (2000)—China North-central (Gansu).

SYRINGA (Oleac.).
komarowii C.K. Schneid.
subsp. reflexa *(C.K. Schneid.) P.S. Green & M.C. Chang in Novon*, 5(4): 329 (1995): S. reflexa.
oblata Lindl.
subsp. dilatata *(Nakai) P.S. Green & M.C. Chang in Novon*, 5(4): 329 (1995): S. dilatata.
pubescens Turcz.
var. potaninii *(C.K. Schneid.) P.S. Green & M.C. Chang in Novon*, 5(4): 332 (1995): S. potaninii.
reticulata (Blume) H. Hara
subsp. amurensis *(Rupr.) P.S. Green & M.C. Chang in Novon*, 5(4): 329 (1995): S. amurensis.
subsp. pekinensis *(Rupr.) P.S. Green & M.C. Chang in Novon*, 5(4): 330 (1995): S. pekinensis.
spontanea *(M.C. Chang) X.K. Qin in Acta Phytotax. Sin.*, 36(4): 362 (1998): S. meyeri var. spontanea.
f. alba *(W. Wang, P.Y. Fu & T.C. Chao) X.K. Qin in Acta Phytotax. Sin.*, 36(4): 363 (1998): S. microphylla var. alba.
velutina Kom.
var. kamibayashii *(Nakai) Y.N. Lee, Fl. Korea*: 1159 (1996): S. kamibayashii.
var. lactea *(Nakai) Y.N. Lee, Fl. Korea*: 1159 (1996): S. fauriei var. lactea.

SYRINGANTHA (Rubiac.).
coulteri *(Hook. f.) T. McDowell in Novon*, 6(3): 277 (1996): Exostema coulteri.

SYRMATIUM (Legumin.).
biolettii *(Greene) A. Brand in Bot. Jahrb. Syst.*, 25: 225 (1898): Lotus biolettii.
haydonii *(Orcutt) A. Brand in Bot. Jahrb. Syst.*, 25: 228 (1898): Hosackia haydonii.
heermannii *(Durand & Hilg.) E.L. Greene in Bull. Calif. Acad. Sci.*, 2(6): 148 (1886): Hosackia heermannii.
leucophyllum *(Greene) A. Brand in Bot. Jahrb. Syst.*, 25: 229 (1898): Lotus leucophyllus.
nevadense *(S. Watson) E.L. Greene in Bull. Calif. Acad. Sci.*, 2(6): 148 (1886): Hosackia decumbens var. nevadensis.
niveum *E.L. Greene in Bull. Calif. Acad. Sci.*, 2(6): 148 (1886)—California.
nudatum *E.L. Greene in Pittonia*, 1(4): 173 (1888), in Supp. 1 with incorrect ref.—Mexico Northwest (Baja California).†
sulphureum *A. Brand in Bot. Jahrb. Syst.*, 25: 231 (1898), nom. nov.: Lotus sulphureus *Greene*.
watsonii *(Vasey & Rose) A. Brand in Bot. Jahrb. Syst.*, 25: 232 (1898): Hosackia watsonii.

SYZYGIUM (Myrtac.).
acre *(Pancher ex Guillaumin) J.W. Dawson in Fl. N. Caléd.*, 23: 44 (1999): Jambosa acris.
aggregatum *J.W. Dawson in Fl. N. Caléd.*, 23: 19 (1999)—New Caledonia.
alyxiifolium *(Ridl.) I.M. Turner in J. Singapore Nation. Acad. Sci.*, 22–24: 16 (1997): Eugenia alyxiifolia.
amieuense *(Guillaumin) J.W. Dawson in Fl. N. Caléd.*, 23: 31 (1999): Caryophyllus amieuensis.
anisosepalum *(Duthie) I.M. Turner in J. Singapore Nation. Acad. Sci.*, 22–24: 16 (1997): Eugenia anisosepala.
aoupinianum *J.W. Dawson in Fl. N. Caléd.*, 23: 127 (1999)—New Caledonia.
apetiolatum *J.W. Dawson in Fl. N. Caléd.*, 23: 28 (1999)—New Caledonia.
arboreum *(Baker f.) J.W. Dawson in Fl. N. Caléd.*, 23: 30 (1999): Eugenia arborea.
attenuatum (Miq.) Merr. & L.M. Perry
var. ophirense *(M.R. Hend.) I.M. Turner in J. Singapore Nation. Acad. Sci.*, 22–24: 16 (1997): Eugenia attenuata var.. ophirensis.
baladense *(Brongn. & Gris) J.W. Dawson in Fl. N. Caléd.*, 23: 88 (1999): Caryophyllus baladensis.
balansae *(Guillaumin) J.W. Dawson in Fl. N. Caléd.*, 23: 24 (1999): Caryophyllus balansae.

SYZYGIUM:-

beddomei (Duthie) V. Chithra in N.C. Nair & A.N. Henry, Fl. Tamil Nadu, India, Ser. 1, 1: 155 (1983): Eugenia beddomei.

borneense (Miq.) Miq.
  var. myrtillus (Stapf) P. Chantaranothai & J. Parnell in Thai Forest Bull., Bot., 21: 45 (1994): Eugenia myrtillus.

boulindaense J.W. Dawson in Fl. N. Caléd., 23: 91 (1999)—New Caledonia.

brachycalyx (Baker f.) J.W. Dawson in Fl. N. Caléd., 23: 133 (1999): Eugenia brachycalyx.

brevioperculatum J.W. Dawson in Fl. N. Caléd., 23: 18 (1999)—New Caledonia.

brevipes (Brongn. & Gris) J.W. Dawson in Fl. N. Caléd., 23: 101 (1999): S. tenuiflorum var. brevipes.

brongniartii (Merr. & L.M. Perry) J.W. Dawson in Fl. N. Caléd., 23: 16 (1999): Cleistocalyx brongniartii.

burkillianum (King) I.M. Turner in J. Singapore Nation. Acad. Sci., 22–24: 16 (1997): Eugenia burkilliana.
  var. garcinifolioides (M.R. Hend.) I.M. Turner in J. Singapore Nation. Acad. Sci., 22–24: 16 (1997): Eugenia burkilliana var. garcinifolioides.

buxifolium Hook. & Arn.
  var. cleyerifolium (Yatabe) Y. Tateishi in Fl. Japan, 2c: 212 (1999): Eugenia cleyerifolia.

cameronum I.M. Turner in J. Singapore Nation. Acad. Sci., 22–24: 16 (1997), nom. nov.: Eugenia goniocalyx Ridl.

camptophyllum (M.R. Hend.) I.M. Turner in J. Singapore Nation. Acad. Sci., 22–24: 16 (1997): Eugenia camptophylla.

capillaceum (Brongn. & Gris) J.W. Dawson in Fl. N. Caléd., 23: 121 (1999): S. tenuiflorum var. capillaceum.

cerinum (M.R. Hend.) I.M. Turner in J. Singapore Nation. Acad. Sci., 22–24: 17 (1997): Eugenia cerina.
  var. montanum (M.R. Hend.) I.M. Turner in J. Singapore Nation. Acad. Sci., 22–24: 17 (1997): Eugenia cerina var. montana.
  var. turbinatum (M.R. Hend.) I.M. Turner in J. Singapore Nation. Acad. Sci., 22–24: 17 (1997): Eugenia cerina var. turbinata.

claviflorum (Roxb.) Wall. ex A.M. Cowan & Cowan
  var. excavatum (King) I.M. Turner in J. Singapore Nation. Acad. Sci., 22–24: 17 (1997): Eugenia claviflora var. excavata.
  var. montanum (M.R. Hend.) I.M. Turner in J. Singapore Nation. Acad. Sci., 22–24: 17 (1997): Eugenia claviflora var. montana.
  var. riparium (M.R. Hend.) I.M. Turner in J. Singapore Nation. Acad. Sci., 22–24: 18 (1997): Eugania claviflora var. riparia.

clypeolatum (Ridl.) I.M. Turner in J. Singapore Nation. Acad. Sci., 22–24: 18 (1997): Eugenia clypeolata.

coccineum J.W. Dawson in Fl. N. Caléd., 23: 140 (1999)—New Caledonia.

conglomeratum (Duthie) I.M. Turner in J. Singapore Nation. Acad. Sci., 22–24: 18 (1997): Eugenia conglomerata.
  var. paniculatum (M.R. Hend.) I.M. Turner in J. Singapore Nation. Acad. Sci., 22–24: 18 (1997): Eugenia conglomerata var. paniculata.

cordifoliatum (Ridl.) I.M. Turner in J. Singapore Nation. Acad. Sci., 22–24: 18 (1997): Eugenia cordifoliata.

curtisii (King) Merr. & L.M. Perry
  var. holttumii (Ridl.) I.M. Turner in J. Singapore Nation. Acad. Sci., 22–24: 18 (1997): Eugenia holttumii.
  var. minus (King) I.M. Turner in J. Singapore Nation. Acad. Sci., 22–24: 18 (1997): Eugenia curtisii var. minor.

cyrtophylloides (Ridl.) I.M. Turner in J. Singapore Nation. Acad. Sci., 22–24: 18 (1997): Eugenia cyrtophylloides.

deplanchei (Guillaumin) J.W. Dawson in Fl. N. Caléd., 23: 83 (1999): Caryophyllus deplanchei.

elegans (Brongn. & Gris) J.W. Dawson in Fl. N. Caléd., 23: 82 (1999): Caryophyllus elegans.

filiflorum J.W. Dawson in Fl. N. Caléd., 23: 107 (1999)—New Caledonia.

SYZYGIUM:-

filiforme (Wall. ex Duthie) Chantar. & J. Parn.
  var. clavimyrtus (Koord. & Valeton) I.M. Turner in J. Singapore Nation. Acad. Sci., 22–24: 18 (1997): Eugenia clavimyrtus.
  var. constrictum (Kochummen) I.M. Turner in J. Singapore Nation. Acad. Sci., 22–24: 18 (1997): Eugenia filiformis var. constricta.

fullagarii (F. Muell.) L.A. Craven in Muelleria, 11: 95 (1998): Acicalyptus fullagarii.

gageanum (King) I.M. Turner in J. Singapore Nation. Acad. Sci., 22–24: 19 (1997): Eugenia gageana.

gambleanum Rathakr. & V. Chithra in N.C. Nair & A.N. Henry, Fl. Tamil Nadu, India, Ser. 1, 1: 156 (1983), nom. nov.: Eugenia microphylla Bedd.

gongshanense P.Y. Bai in Fl. Dulongjian Reg.: 111 (1993), without latin descr. or type—China South-central (Yunnan).

graeme-andersoniae (Ridl.) I.M. Turner in J. Singapore Nation. Acad. Sci., 22–24: 19 (1997): Eugenia graeme-andersoniae.

guehoi J. Bosser & D. Florens in Adansonia, 22(2): 183 (2000)—Mauritius.

guillauminii J.W. Dawson in Fl. N. Caléd., 23: 49 (1999), nom. nov.: Jambosa nervosa Vieill. ex Guillaumin.

haniffii (M.R. Hend.) I.M. Turner in J. Singapore Nation. Acad. Sci., 22–24: 19 (1997): Eugenia haniffii.

inasense (King) I.M. Turner in J. Singapore Nation. Acad. Sci., 22–24: 20 (1997): Eugenia inasensis.

inophyllum DC.
  var. bernardi (King) I.M. Turner in J. Singapore Nation. Acad. Sci., 22–24: 20 (1997): Eugenia bernardi.

jaffrei J.W. Dawson in Fl. N. Caléd., 23: 112 (1999)—New Caledonia.

kemamanense (M.R. Hend.) I.M. Turner in J. Singapore Nation. Acad. Sci., 22–24: 20 (1997): Eugenia kemamanensis.

kiahii (M.R. Hend.) I.M. Turner in J. Singapore Nation. Acad. Sci., 22–24: 20 (1997): Eugenia kiahii.
  var. angustifolium (M.R. Hend.) I.M. Turner in J. Singapore Nation. Acad. Sci., 22–24: 20 (1997): Eugenia kiahii var. angustifolia.

koniamboense J.W. Dawson in Fl. N. Caléd., 23: 63 (1999)—New Caledonia.

koordersianum (King) I.M. Turner in J. Singapore Nation. Acad. Sci., 22–24: 20 (1997): Eugenia koordersiana.

koumacense J.W. Dawson in Fl. N. Caléd., 23: 126 (1999)—New Caledonia.

kuebiniense J.W. Dawson in Fl. N. Caléd., 23: 13 (1999)—New Caledonia.

laxeracemosum (Guillaumin) J.W. Dawson in Fl. N. Caléd., 23: 43 (1999): Caryophyllus laxeracemosus.

linocieroideum (King) I.M. Turner in J. Singapore Nation. Acad. Sci., 22–24: 21 (1997), as 'linoceroides': Eugenia linocieroidea.

longifolium (Brongn. & Gris) J.W. Dawson in Fl. N. Caléd., 23: 46 (1999): Jambosa longifolia.

malayanum (Gagnep.) I.M. Turner in J. Singapore Nation. Acad. Sci., 22–24: 21 (1997): Eugenia malayana.

meorianum J.W. Dawson in Fl. N. Caléd., 23: 132 (1999)—New Caledonia.

micklethwaitii B. Verdcourt in Kew Bull., 52(3): 682 (1997), nom. nov.: S. sclerophyllum Brenan.

millsii (M.R. Hend.) I.M. Turner in J. Singapore Nation. Acad. Sci., 22–24: 21 (1997): Eugenia millsii.

mundagam (Bourd.) V. Chithra in N.C. Nair & A.N. Henry, Fl. Tamil Nadu, India, Ser. 1, 1: 157 (1983): Eugenia mundagam.

myrianthum (King) I.M. Turner in J. Singapore Nation. Acad. Sci., 22–24: 21 (1997): Eugenia myriantha.

nanum J.W. Dawson in Fl. N. Caléd., 23: 92 (1999)—New Caledonia.

nemestrinum (M.R. Hend.) I.M. Turner in J. Singapore Nation. Acad. Sci., 22–24: 21 (1997): Eugenia nemestrina.

neocaledonicum (Seem.) J.W. Dawson in Fl. N. Caléd., 23: 32 (1999): Cupheanthus neocaledonicus.

SYZYGIUM:-

ngadimanianum (M.R. Hend.) I.M. Turner in J. Singapore Nation. Acad. Sci., 22–24: 21 (1997): Eugenia ngadimaniana.

nitens J.W. Dawson in Fl. N. Caléd., 23: 26 (1999)—New Caledonia.

nitidulum (Ridl.) I.M. Turner in J. Singapore Nation. Acad. Sci., 22–24: 21 (1997): Eugenia nitidula.

oreophilum (Ridl.) I.M. Turner in J. Singapore Nation. Acad. Sci., 22–24: 22 (1997): Eugenia oreophila.

orites (Ridl.) I.M. Turner in J. Singapore Nation. Acad. Sci., 22–24: 22 (1997): Eugenia orites.

pahangense (Ridl.) I.M. Turner in J. Singapore Nation. Acad. Sci., 22–24: 22 (1997): Eugenia pahangensis.
  var. fraseri (M.R. Hend.) I.M. Turner in J. Singapore Nation. Acad. Sci., 22–24: 22 (1997): Eugenia pahangensis var. fraseri.

pallidulum (Ridl.) I.M. Turner in J. Singapore Nation. Acad. Sci., 22–24: 22 (1997): Eugenia pallidula.

paniense (Baker f.) J.W. Dawson in Fl. N. Caléd., 23: 37 (1999): Eugenia paniensis.

parvicarpum J.W. Dawson in Fl. N. Caléd., 23: 113 (1999)—New Caledonia.

pauper (Ridl.) I.M. Turner in J. Singapore Nation. Acad. Sci., 22–24: 22 (1997): Eugenia pauper.

pearsonianum (King) I.M. Turner in J. Singapore Nation. Acad. Sci., 22–24: 22 (1997): Eugenia pearsoniana.

pendens (Duthie) I.M. Turner in J. Singapore Nation. Acad. Sci., 22–24: 22 (1997): Eugenia pendens.

pendulinum J.W. Dawson in Fl. N. Caléd., 23: 117 (1999)—New Caledonia.

pennelii (Guillaumin) J.W. Dawson in Fl. N. Caléd., 23: 14 (1999): Acicalyptus pennelii.

perakense (King) I.M. Turner in J. Singapore Nation. Acad. Sci., 22–24: 22 (1997): Eugenia perakensis.

periyarensis Jomy Augustine & N. Sasidharan in Rheedea, 9(2): 155 (1999)—India (Kerala).

perryae I.M. Turner in J. Singapore Nation. Acad. Sci., 22–24: 22 (1997), nom. nov.: Eugenia benjamina King.

plumbeum (King) I.M. Turner in J. Singapore Nation. Acad. Sci., 22–24: 22 (1997): Eugenia plumbea.

politum (King) I.M. Turner in J. Singapore Nation. Acad. Sci., 22–24: 23 (1997): Eugenia polita.

polyanthum (Wight) Walp.
  var. sessile (M.R. Hend.) I.M. Turner in J. Singapore Nation. Acad. Sci., 22–24: 23 (1997): Eugenia polyantha var. sessilis.

porphyrantherum (Ridl.) I.M. Turner in J. Singapore Nation. Acad. Sci., 22–24: 23 (1997): Eugenia porphyranthera.

poyanum J.W. Dawson in Fl. N. Caléd., 23: 124 (1999)—New Caledonia.

praestigiosum (M.R. Hend.) I.M. Turner in J. Singapore Nation. Acad. Sci., 22–24: 23 (1997): Eugenia praestigiosa.

propinquum (Guillaumin) J.W. Dawson in Fl. N. Caléd., 23: 86 (1999): Caryophyllus propinquus.

pseudoclaviflorum (M.R. Hend.) I.M. Turner in J. Singapore Nation. Acad. Sci., 22–24: 23 (1997): Eugenia pseudoclaviflora.

pseudocrenulatum (M.R. Hend.) I.M. Turner in J. Singapore Nation. Acad. Sci., 22–24: 23 (1997): Eugenia pseudocrenulata.

pseudomolle (M.R. Hend.) I.M. Turner in J. Singapore Nation. Acad. Sci., 22–24: 23 (1997): Eugenia pseudomollis.

pulaiense I.M. Turner in J. Singapore Nation. Acad. Sci., 22–24: 23 (1997): Eugenia johorensis Ridl.

quadrangulare Guillaumin
  var. microsemmifolium (Guillaumin) J.W. Dawson in Fl. N. Caléd., 23: 98 (1999): S. microsemmifolium.

quadratum (King) I.M. Turner in J. Singapore Nation. Acad. Sci., 22–24: 24 (1997): Eugenia quadrata.

quadribracteatum (M.R. Hend.) I.M. Turner in J. Singapore Nation. Acad. Sci., 22–24: 24 (1997): Eugenia quadribracteata.

SYZYGIUM:-
rama-varmae *(Bourd.) V. Chithra in N.C. Nair & A.N. Henry, Fl. Tamil Nadu, India, Ser. 1,* 1: 157 (1983): Eugenia rama-varmae.
ramilepis *J.W. Dawson in Fl. N. Caléd.,* 23: 110 (1999)—New Caledonia.
rhomboideum *(Ridl.) I.M. Turner in J. Singapore Nation. Acad. Sci.,* 22–24: 24 (1997): Eugenia rhomboidea.
rostadonis *(Ridl.) I.M. Turner in J. Singapore Nation. Acad. Sci.,* 22–24: 24 (1997): Eugenia rostadonis.
rugosum Korth.
var. cordatum *(M.R. Hend.) I.M. Turner in J. Singapore Nation. Acad. Sci.,* 22–24: 24 (1997): Eugenia rugosa var. cordata.
salictoides *(Ridl.) I.M. Turner in J. Singapore Nation. Acad. Sci.,* 22–24: 24 (1997): Eugenia salictoides.
sarmentosum *J.W. Dawson in Fl. N. Caléd.,* 23: 116 (1999)—New Caledonia.
scalarinerve *(King) I.M. Turner in J. Singapore Nation. Acad. Sci.,* 22–24: 24 (1997): Eugenia scalarinervis.
schistaceum *J.W. Dawson in Fl. N. Caléd.,* 23: 48 (1999), nom. nov.: Jambosa neriifolia *Brongn. & Gris.*
scortechinii (King) Chantar. & J. Parn.
var. cuneatum *(M.R. Hend.) I.M. Turner in J. Singapore Nation. Acad. Sci.,* 22–24: 24 (1997): Eugenia scortechinii var. cuneata.
setosum *(King) I.M. Turner in J. Singapore Nation. Acad. Sci.,* 22–24: 24 (1997): Eugenia setosa.
spissifolium *(Ridl.) I.M. Turner in J. Singapore Nation. Acad. Sci.,* 22–24: 24 (1997): Eugenia spissifolia.
sriganesanii *K. Ravikumar & V. Lakshmanan in Rheedea,* 9(1): 59 (1999)—India (Tamil Nadu).
stapfianum *(King) I.M. Turner in J. Singapore Nation. Acad. Sci.,* 22–24: 25 (1997): Eugenia stapfiana.
subdecussatum *(Duthie) I.M. Turner in J. Singapore Nation. Acad. Sci.,* 22–24: 25 (1997): Eugenia subdecussata.
var. montanum *(King) I.M. Turner in J. Singapore Nation. Acad. Sci.,* 22–24: 25 (1997): Eugenia subdecussata var. montana.
swettenhamianum *(King) I.M. Turner in J. Singapore Nation. Acad. Sci.,* 22–24: 25 (1997): Eugenia swettenhamiana.
symingtonianum *(M.R. Hend.) I.M. Turner in J. Singapore Nation. Acad. Sci.,* 22–24: 25 (1997): Eugenia symingtoniana.
tahanense *(Ridl.) I.M. Turner in J. Singapore Nation. Acad. Sci.,* 22–24: 25 (1997): Eugenia tahanensis.
taipingense *(M.R. Hend.) I.M. Turner in J. Singapore Nation. Acad. Sci.,* 22–24: 25 (1997): Eugenia taipingensis.
tamilnadensis *Rathakr. & V. Chithra in N.C. Nair & A.N. Henry, Fl. Tamil Nadu, India, Ser. 1,* 1: 158 (1983), nom. nov.: Eugenia montana *Wight.*
tchambaense *J.W. Dawson in Fl. N. Caléd.,* 23: 38 (1999)—New Caledonia.
tectum *(King) I.M. Turner in J. Singapore Nation. Acad. Sci.,* 22–24: 25 (1997): Eugenia tecta.
tekuense *(M.R. Hend.) I.M. Turner in J. Singapore Nation. Acad. Sci.,* 22–24: 25 (1997): Eugenia tekuensis.
tetrapterum (Miq.) Chantar. & J. Parn.
var. pseudotetrapterum *(King) I.M. Turner in J. Singapore Nation. Acad. Sci.,* 22–24: 25 (1997): Eugenia pseudotetraptera.
thumra (Roxb.) Merr. & L.M. Perry
var. penangianum *(King) I.M. Turner in J. Singapore Nation. Acad. Sci.,* 22–24: 25 (1997): Eugenia thumra var. penangiana.
tiumanense *(Ridl.) I.M. Turner in J. Singapore Nation. Acad. Sci.,* 22–24: 26 (1997): Eugenia tiumanensis.
toninense *(Baker f.) J.W. Dawson in Fl. N. Caléd.,* 23: 34 (1999): Eugenia toninensis.
tontoutaense *J.W. Dawson in Fl. N. Caléd.,* 23: 100 (1999)—New Caledonia.
veillonii *J.W. Dawson in Fl. N. Caléd.,* 23: 108 (1999)—New Caledonia.
velutinum *A.P. Davis in Kew Bull.,* 52(3): 720 (1997)—Borneo (Brunei, Sarawak, Sabah).

SYZYGIUM:-
viridescens *(Ridl.) I.M. Turner in J. Singapore Nation. Acad. Sci.,* 22–24: 26 (1997): Eugenia viridescens.
virotii *J.W. Dawson in Fl. N. Caléd.,* 23: 40 (1999), nom. nov.: Cupheanthus microphyllus *Guillaumin.*
watsonianum *(M.R. Hend.) I.M. Turner in J. Singapore Nation. Acad. Sci.,* 22–24: 26 (1997): Eugenia watsoniana.
wrayi *(King) I.M. Turner in J. Singapore Nation. Acad. Sci.,* 22–24: 26 (1997): Eugenia wrayi.
xanthostemifolium *(Guillaumin) J.W. Dawson in Fl. N. Caléd.,* 23: 66 (1999): Caryophyllus xanthostemifolius.
zeylanicum (L.) DC.
var. magamalayanum *K. Ravikumar & V. Lakshmanan in Rheedea,* 9(1): 61 (1999)—India (Tamil Nadu).

T

TABEBUIA (Bignoniac.).
ochracea (Cham.) Standl.
subsp. heterotricha *(DC.) A.H. Gentry in Fl. Venezuela,* 8(4): 391 (1982): Tecoma heterotricha.

TABERNAEMONTANA (Apocynac.).
leeuwenbergiana *J.F. Morales in Novon,* 9(2): 238 (1999)—Colombia.

TACCA (Taccac.).
ankaranensis *M. Bardot-Vaucoulon in Adansonia,* 19(1): 154 (1997)—Madagascar.

TACHIA (Gentianac.).
longipes *L. Cobb & P.J.M. Maas in Brittonia,* 50(1): 11 (1998)—Surinam.

TACHIGALI (Legumin.).
albiflora *(Benoist) J.L. Zarucchi & P.S. Herendeen in Fl. Venezuelan Guayana (gen. eds. J.A. Steyermark et al.),* 4: 115 (1998): Sclerolobium albiflorum.
amplifolia *(Ducke) R.C. Barneby in Brittonia,* 48(2): 182 (1996): Sclerolobium amplifolium.
davidsei *J.L. Zarucchi & P.S. Herendeen in Fl. Venezuelan Guayana (gen. eds. J.A. Steyermark et al.),* 4: 116 (1998), nom. nov.: Sclerolobium aureum var. grandiflorum *Dwyer.*
dwyeri *(R.S. Cowan) J.L. Zarucchi & P.S. Herendeen in Fl. Venezuelan Guayana (gen. eds. J.A. Steyermark et al.),* 4: 116 (1998): Sclerolobium dwyeri.
micropetala *(Ducke) J.L. Zarucchi & J.J. Pipoly in Sida,* 16(4): 787 (1995): Sclerolobium micropetalum.
odoratissima *(Spruce ex Benth.) J.L. Zarucchi & P.S. Herendeen in Fl. Venezuelan Guayana (gen. eds. J.A. Steyermark et al.),* 4: 116 (1998): Sclerolobium odoratissimum.
paraensis *(Huber) R.C. Barneby in Brittonia,* 48(2): 182 (1996), as 'paranensis': Sclerolobium paraense.
physophora *(Huber) J.L. Zarucchi & P.S. Herendeen in Fl. Venezuelan Guayana (gen. eds. J.A. Steyermark et al.),* 4: 116 (1998): Sclerolobium physophorum.
pimichinensis *(R.S. Cowan) J.L. Zarucchi & P.S. Herendeen in Fl. Venezuelan Guayana (gen. eds. J.A. Steyermark et al.),* 4: 120 (1998): Sclerolobium pimichinense.
reticulosa *(Dwyer) J.L. Zarucchi & P.S. Herendeen in Fl. Venezuelan Guayana (gen. eds. J.A. Steyermark et al.),* 4: 120 (1998): Sclerolobium reticulosum.

TADEHAGI (Legumin.).
robustum *L. Pedley in Austrobaileya,* 5(2): 258 (1999)—Western Australia.

TADEHAGI:-
rodgeri *(Schindl.) B. Ghosh & G.G. Maiti in J. Pl. Anat. Morphol.,* 7(2): 112 (1997), isonym; *see H. Ohashi in J. Jap. Bot.,* 57(9): 266 (1982): Pteroloma rodgeri.

TAENIOPHYLLUM (Orchidac.).
arunachalense *A. Nageswara Rao & Jagdish Lal in Bioved,* 2(2): 225 (1991)—Assam (Arunachal Pradesh).
pusillum *(Willd.) G. Seidenfaden & P. Ormerod in G. Seidenfaden, Descr. Epidendrorum J.G. König 1791:* 23 (1995): Limodorum pusillum.
savaiiense *P.J. Cribb & W.A. Whistler, Orchids Samoa:* 131 (1996)—Western Samoa.
whistleri *P.J. Cribb in P.J. Cribb & W.A. Whistler, Orchids Samoa:* 130 (1996)—American Samoa, Western Samoa.

TAGETES (Compos.).
ser. Argentinae *J.A. Soule in D.J.N. Hind & H.J. Beentje (eds.), Compos.: Syst. (Proc. Internat. Compos.* Conf., Kew, 1994, 1): 442 (1996).
ser. Calcicolae *J.A. Soule in D.J.N. Hind & H.J. Beentje (eds.), Compos.: Syst. (Proc. Internat. Compos.* Conf., Kew, 1994, 1): 440 (1996).
ser. Campanulatae *J.A. Soule in D.J.N. Hind & H.J. Beentje (eds.), Compos.: Syst. (Proc. Internat. Compos.* Conf., Kew, 1994, 1): 441 (1996).
ser. Excelsae *J.A. Soule in D.J.N. Hind & H.J. Beentje (eds.), Compos.: Syst. (Proc. Internat. Compos.* Conf., Kew, 1994, 1): 441 (1996).
sect. Filifoliae *J.A. Soule in D.J.N. Hind & H.J. Beentje (eds.), Compos.: Syst. (Proc. Internat. Compos.* Conf., Kew, 1994, 1): 440 (1996).
ser. Foetidissimae *J.A. Soule in D.J.N. Hind & H.J. Beentje (eds.), Compos.: Syst. (Proc. Internat. Compos.* Conf., Kew, 1994, 1): 441 (1996).
subgen. Hydrotagetes *J.A. Soule in D.J.N. Hind & H.J. Beentje (eds.), Compos.: Syst. (Proc. Internat. Compos.* Conf., Kew, 1994, 1): 439 (1996).
ser. Inclusae *J.A. Soule in D.J.N. Hind & H.J. Beentje (eds.), Compos.: Syst. (Proc. Internat. Compos.* Conf., Kew, 1994, 1): 442 (1996).
subgen. Iya *J.A. Soule in D.J.N. Hind & H.J. Beentje (eds.), Compos.: Syst. (Proc. Internat. Compos.* Conf., Kew, 1994, 1): 439 (1996).
ser. Lacerae *J.A. Soule in D.J.N. Hind & H.J. Beentje (eds.), Compos.: Syst. (Proc. Internat. Compos.* Conf., Kew, 1994, 1): 440 (1996).
ser. Lemmoniae *J.A. Soule in D.J.N. Hind & H.J. Beentje (eds.), Compos.: Syst. (Proc. Internat. Compos.* Conf., Kew, 1994, 1): 440 (1996).
ser. Linifoliae *J.A. Soule in D.J.N. Hind & H.J. Beentje (eds.), Compos.: Syst. (Proc. Internat. Compos.* Conf., Kew, 1994, 1): 441 (1996).
ser. Minutae *J.A. Soule in D.J.N. Hind & H.J. Beentje (eds.), Compos.: Syst. (Proc. Internat. Compos.* Conf., Kew, 1994, 1): 442 (1996).
ser. Peruvianae *J.A. Soule in D.J.N. Hind & H.J. Beentje (eds.), Compos.: Syst. (Proc. Internat. Compos.* Conf., Kew, 1994, 1): 441 (1996).
ser. Sororiae *J.A. Soule in D.J.N. Hind & H.J. Beentje (eds.), Compos.: Syst. (Proc. Internat. Compos.* Conf., Kew, 1994, 1): 440 (1996).
arenicola *J.L. Panero & J.L. Villaseñor in Brittonia,* 48(1): 87 (1996)—Mexico Southwest (Guerrero).
persicifolia *(Benth.) B.L. Turner in Phytologia Mem.,* 10: 57 (1996), as 'persicaefolius': Adenopappus persicifolius.

TAHITIA (Tiliac.).
tahitensis *(Nadeaud) S.L. Welsh, Fl. Societensis:* 282 (1998): Entelia tahitensis.

TAINIA (Orchidac.).
emeiensis *(K.Y. Lang) Z.H. Tsi in Fl. Reipubl. Popul. Sin.,* 18: 236 (1999): Mischobulbum emeiense.
minor Hook. f.
var. laxiflora *(Makino) T. Hashimoto in Proc. 12th World Orchid Conf., 1987 (eds. K. Saito & R. Tanaka):* 124 (1987): T. laxiflora.
ruybarrettoi *(S.Y. Hu & Barretto) Z.H. Tsi in Fl. Reipubl. Popul. Sin.,* 18: 243 (1999), isonym; *see Aver., Opred. Orkhid. V'etnama:* 403 (1994): Ania ruybarrettoi.

TALAMANCALIA (Compos.).
fosbergii *(Cuatrec.) B. Nordenstam in Compositae Newsl.*, 29: 49 (1996): Senecio fosbergii.
putcalensis *(Hieron.) B. Nordenstam & J.F. Pruski in Compositae Newsl.*, 27: 35 (1995): Senecio putcalensis.

TALIERA (Palmae).
elata *(Roxb.) N. Wallich, Report on Calcutta Bot. Gard. to G.A. Bushby:* 29 (1840): Corypha elata.

TALINELLA (Portulacac.).
microphylla *U. Eggli in Adansonia*, 19(1): 56 (1997)—Madagascar.
pachypoda *U. Eggli in Adansonia*, 19(1) 57 (1997)—Madagascar.

TALISIA (Sapindac.).
chartacea *P. Acevedo-Rodríguez in Biollania*, ed. esp. 6 (J.J. Wurdack Festschr.): 144 (1997)—Venezuela.
laevigata *P. Acevedo-Rodríguez in Biollania*, ed. esp. 6 (J.J. Wurdack Festschr.): 145 (1997)—Venezuela.

TALTALIA *Ehr. Bayer in Sendtnera*, 5: 7 (1998). *ALSTROEMERIACEAE*
graminea *(Phil.) Ehr. Bayer in Sendtnera*, 5: 7 (1998): Alstroemeria graminea.

TAMAMSCHJANELLA *M.G. Pimenov & E.V. Kljuykov in Bot. Zhurn.*, 81(8): 75 (1996). *UMBELLIFERAE*
rhizomatica *(Hartvig) M.G. Pimenov & E.V. Kljuykov in Bot. Zhurn.*, 81(8): 78 (1996): Ligusticum rhizomaticum.
rubella *(E.A. Busch) M.G. Pimenov & E.V. Kljuykov in Bot. Zhurn.*, 81(8): 76 (1996): Eleutherospermum rubellum.

TAMARIX (Tamaricac.).
ramosissima Ledeb.
var. macedonica *K. Micevski in Prilozi Oddel. Biol. Med. Nauki Makedonska Akad. Nauk. Umet.*, 15(1–2): 38 (1994), contrary to Art. 37.5 ICBN (1994)—Yugoslavia.

TAMAYORKIS *D.L. Szlachetko in Fragm. Flor. Geobot.*, Supp. 3: 121 (1995). *ORCHIDACEAE*
ehrenbergii *(Rchb. f.) R. González Tamayo & D.L. Szlachetko in Ann. Bot. Fenn.*, 35(1): 22 (1998): Microstylis ehrenbergii.
hintonii *(Todzia) R. González Tamayo & D.L. Szlachetko in Ann. Bot. Fenn.*, 35(1): 22 (1998): Malaxis hintonii.
platyglossa *(B.L. Rob. & Greenm.) D.L. Szlachetko in Fragm. Flor. Geobot.*, Supp. 3: 122 (1995): Microstylis platyglossa.
wendtii *(Salazar) R. González Tamayo & D.L. Szlachetko in Ann. Bot. Fenn.*, 35(1): 22 (1998): Malaxis wendtii.

TAMIJIA *S. Sakai & H. Nagamasu in Edinburgh J. Bot.*, 57(2): 245 (2000). *ZINGIBERACEAE*
flagellaris *S. Sakai & H. Nagamasu in Edinburgh J. Bot.*, 57(2): 245 (2000)—Borneo (Sarawak, Brunei).

TAMRIDAEA *M. Thulin & B. Bremer in Pl. Syst. Evol.*, 211(1–2): 85 (1998). *RUBIACEAE*
capsulifera *(Balf. f.) M. Thulin & B. Bremer in Pl. Syst. Evol.*, 211(1–2): 87 (1998): Mussaenda capsulifera.

TANA *B.-E. van Wyk in Taxon*, 48(4): 743 (1999). *UMBELLIFERAE*
bojeriana *(Baker) B.-E. van Wyk in Taxon*, 48(4): 743 (1999): Peucedanum bojerianum

TANACETUM (Compos.).
musili *(Velen.) A. Soldano in Compositae Newsl.*, 32: 20 (1998): Pyrethrum musili.
tatsienense (Bureau & Franch.) K. Bremer & Humphries
var. tanacetopsis *(W.W. Sm.) A.J.C. Grierson in Edinburgh J. Bot.*, 57(3): 410 (2000): Chrysanthemum jugorum var. tanacetopsis

TANNODIA (Euphorbiac.).
sect. Amyreophylla *A. Radcliffe-Smith in Kew Bull.*, 53(1): 184 (1998).

TANNODIA:-
subgen. Domohinea *(Leandri) A. Radcliffe-Smith in Kew Bull.*, 53(1): 176 (1998): gen. Domohinea.
sect. Domohinea *(Leandri) A. Radcliffe-Smith in Kew Bull.*, 53(1): 177 (1998): gen. Domohinea.
sect. Neoholstia *(Rauschert) A. Radcliffe-Smith in Kew Bull.*, 53(1): 176 (1998): gen. Neoholstia.
grandiflora *A. Radcliffe-Smith in Kew Bull.*, 53(1): 180 (1998)—Madagascar.
var. myrtifolia *A. Radcliffe-Smith in Kew Bull.*, 53(1): 184 (1998)—Madagascar.
nitida *A. Radcliffe-Smith in Kew Bull.*, 53(1): 178 (1998)—Madagascar.
obovata *A. Radcliffe-Smith in Kew Bull.*, 53(1): 180 (1998)—Madagascar.
pennivenia *A. Radcliffe-Smith in Kew Bull.*, 53(1): 184 (1998)—Madagascar.
perrieri *(Leandri) A. Radcliffe-Smith in Kew Bull.*, 53(1): 177 (1998): Domohinea perrieri.
var. latifolia *A. Radcliffe-Smith in Kew Bull.*, 53(1): 178 (1998)—Madagascar.
var. ludiifolia *A. Radcliffe-Smith in Kew Bull.*, 53(1): 178 (1998)—Madagascar.

TAPEINOCHILOS (Costac.).
koordersianus *H.N. Ridley in J. Straits Branch Roy. Asiat. Soc.*, 34: 99 (1900)—Sulawesi.

TAPEINOSPERMA (Myrsinac.).
alatum *D.E. Holland & P.F. Stevens in Blumea*, 43(2): 348 (1998)—Moluccas.

TAPEINOSTEMON (Gentianac.).
longiflorum *B. Maguire & J.A. Steyermark in Mem. New York Bot. Gard.*, 32: 359 (1981)—Venezuela.
var. australe *B. Maguire & J.A. Steyermark in Mem. New York Bot. Gard.*, 32: 362 (1981)—Venezuela/Brazil border.
rugosum *B. Maguire & J.A. Steyermark in Mem. New York Bot. Gard.*, 32: 366 (1981)—Venezuela.
sessiliflorum *(Humb. & Bonpl. ex Schult.) J.F. Pruski & S.F. Smith in Brittonia*, 49(3): 347 (1997): Psychotria sessiliflora.

TAPHEOCARPA (Commelinac.).
calandrinioides *(F. Muell.) J.G. Conran in Austral. Syst. Bot.*, 9(4): 659 (1996): Aneilema calandrinioides.

TAPHROSPERMUM (Crucif.).
fontanum *(Maxim.) I.A. Al-Shehbaz & G. Yang in Harvard Pap. Bot.*, 5(1): 104 (2000): Dilophia fontana.
subsp. microspermum *I.A. Al-Shehbaz & G. Yang in Harvard Pap. Bot.*, 5(1): 105 (2000)—Qinghai, Xinjiang, Xizang.
himalaicum *(Hook. f. & Thomson) I.A. Al-Shehbaz, Kats. Arai & H. Ohba in Harvard Pap. Bot.*, 5(1): 102 (2000): Cochlearia himalaica.
lowndesii *(H. Hara) I.A. Al-Shehbaz in Harvard Pap. Bot.*, 5(1): 107 (2000): Glaribraya lowndesii.
tibeticum *(O.E. Schulz) I.A. Al-Shehbaz in Harvard Pap. Bot.*, 5(1): 107 (2000): Dipoma tibeticum.
verticillatum *(Jeffrey & W.W. Sm.) I.A. Al-Shehbaz in Harvard Pap. Bot.*, 5(1): 106 (2000): Cardamine verticillata.

TAPINANTHUS (Loranthac.).
cordifolius *R.M. Polhill & D. Wiens, Mistletoes of Africa:* 204 (1998)—Nigeria.
longiflorus *R.M. Polhill & D. Wiens, Mistletoes of Africa:* 191 (1998)—Angola.
praetexta *R.M. Polhill & D. Wiens, Mistletoes of Africa:* 205 (1998)—Ivory Coast.

TAPIPHYLLUM (Rubiac.).
cistifolium (Welw. ex Hiern) Robyns
var. latifolium *B. Verdcourt in Fl. Zambes.*, 5(2): 259 (1998)—Zambia.
rhodesiacum *(Tennant) D.M. Bridson in Kew Bull.*, 51(2): 351 (1996): Ancylanthos rhodesiacus.

TAPIRIRA (Anacardiac.).
rubrinervis *A.S. Barfod in Novon*, 9(4): 472 (1999)—Ecuador.

TAPURA (Dichapetalac.).
follii *G.T. Prance in Biollania*, ed. esp. 6 (J.J. Wurdack Festschr.): 492 (1997)—Brazil Southeast (Espírito Santo).
wurdackiana *G.T. Prance in Biollania*, ed. esp. 6 (J.J. Wurdack Festschr.): 492 (1997)—Brazil Southeast (Espírito Santo).

TARALEA (Legumin.).
cordata Ducke
var. rigida *(Schery) H.C. Lima in Fl. Venezuelan Guayana (gen. eds. J.A. Steyermark et al.)*, 5: 416 (1999): T. rigida.

TARAXACUM (Compos.).
subsect. Alpinae *(G.E. Haglund) N.N. Tzvelev in Novosti Sist. Vyssh. Rast.*, 31: 267 (1998): T. sect. Alpinae.
*subsect.*Arctica *B.A. Jurtzev in Bot. Zhurn.*, 82(1): 113 (1997).
subsect. Holmeniana *B.A. Jurtzev in Bot. Zhurn.*, 82(1): 112 (1997).
subsect. Phymatocarpa *B.A. Jurtzev in Bot. Zhurn.*, 82(1): 113 (1997).
subsect. Soczaviana *B.A. Jurtzev in Bot. Zhurn.*, 82(1): 113 (1997).
adamii *C. Claire in Bull. Soc. Bot. Rochelaise*, 12: 49 (1890 publ. 1891)—France.
amarellum *J. Kirschner & J. Štěpánek, Monogr. Taraxacum sect. Palustria:* 23 (1998)—Ireland.
ambitiosum *J. Kirschner & J. Štěpánek, Monogr. Taraxacum sect. Palustria:* 176 (1998)—Bulgaria.
ambrosium *J. Kirschner & J. Štěpánek, Monogr. taraxacum sect. Palustria:* 231 (1998)—Czechoslovakia.
amplexum *C.E. Sonck in Ann. Bot. Fenn.*, 35(2): 119 (1998)—Italy.
artificis *C.E. Sonck in Ann. Bot. Fenn.*, 35(3): 175 (1998)—France.
atrox *J. Kirschner & J. Štěpánek in Preslia*, 69(1): 55 (1997)—Czechoslovakia, Germany.
bibulum *J. Kirschner & J. Štěpánek, Monogr. Taraxacum sect. Palustria:* 85 (1998)—Czechoslovakia, Austria, Hungary.
bidentilobum *C.E. Sonck in Ann. Bot. Fenn.*, 35(2): 121 (1998)—Italy.
bufonium *J. Kirschner & J. Štěpánek, Monogr. Taraxacum sect. Palustria:* 257 (1998)—Czechoslovakia.
canaliculatum H. Lindb.
f. potens *(H. Lindb.) C.-F. Lundevall & H. Øllgaard in Preslia*, 71(1–2): 68 (1999): T. canaliculatum subsp. potens.
carbocapillum *C.E. Sonck in Ann. Bot. Fenn.*, 36(4): 292 (1999)—Pennsylvania.
cariciphilum *C.E. Sonck & H. Øllgaard in Ann. Bot. Fenn.*, 36(2): 146 (1999)—Finland, Sweden, Russia North.
catenatum *J. Kirschner & J. Štěpánek, Monogr. Taraxacum sect. Palustria:* 242 (1998)—Czechoslovakia.
clavatifrons *C.E. Sonck in Ann. Bot. Fenn.*, 35(3): 177 (1998)—France.
conturbatum *C.E. Sonck in Ann. Bot. Fenn.*, 35(3): 177 (1998)—France.
decrepitum *J. Kirschner & J. Štěpánek, Monogr. Taraxacum sect. Palustria:* 225 (1998)—Greece.
desideratum *C.E. Sonck in Ann. Bot. Fenn.*, 35(1): 31 (1998)—Finland.
fartoris *J. Kirschner & J. Štěpánek, Monogr. Taraxacum sect. Palustria:* 229 (1998)—Czechoslovakia.
fascinans *J. Kirschner, V. Mikoláš & J. Štěpánek in Preslia*, 69(1): 45 (1997)—Czechoslovakia, Hungary.
ferale *J. Kirschner & J. Štěpánek, Monogr. Taraxacum sect. Palustria:* 35 (1998)—France.
fluviatile *J. Kirschner & J. Štěpánek, Monogr. Taraxacum sect. Palustria:* 227 (1998)—Czechoslovakia, Hungary, Ukraina.
glossodon *C.E. Sonck & H. Øllgaard in Ann. Bot. Fenn.*, 36(4): 293 (1999)—Finland.
gotlandicum *(Dahlst.) H. Dahlstedt in Bot. Notiser*, 1909: 171 (1909), ref. in I.K. Supp.4 is to basionym: T. erythrospermum subsp. gotlandicum.†
haareanum *N.N. Tzvelev in Novosti Sist. Vyssh. Rast.*, 31: 267 (1998)—Russia Northwest.
haematicum *G.E. Haglund ex H. Øllgaard & H. Wittzell in Ann. Bot. Fenn.*, 32(4): 230 (1995)—Sweden, Denmark, Finland, Germany, Iceland, Norway, Baltic States (Latviya, Estoniya).

TARAXACUM:-
hatsuroi *H. Koidzumi in J. Pl. Iwateken,* 2(2): 103 (1937)—Japan.
heleocharis *J. Kirschner & J. Štěpánek, Monogr. Taraxacum sect. Palustria:* 108 (1998)—Germany.
hercynicum *J. Kirschner & J. Štěpánek in Preslia,* 69(1): 51 (1997)—Czechoslovakia, Germany.
hondoense Nakai ex H. Koidz.
f. yokouchii *Okuhara ex T. Shimizu, Fl. Nagano Pref.:* 1510 (1997)—Japan.
humidicola *J. Kirschner & J. Štěpánek, Monogr. Taraxacum sect. Palustria:* 236 (1998)—Czechoslovakia, Hungary.
illyricum *Dahlst. ex J. Kirschner & J. Štěpánek, Monogr. Taraxacum sect. Palustria:* 218 (1998)—Yugoslavia.
indigenum *J. Kirschner & J. Štěpánek, Monogr. Taraxacum sect. Palustria:* 143 (1998)—Czechoslovakia.
inundatum *J. Kirschner & J. Štěpánek, Monogr. Taraxacum sect. Palustria:* 240 (1998)—Czechoslovakia.
irroratum *J. Kirschner & J. Štěpánek, Monogr. Taraxacum sect. Palustria:* 253 (1998)—Czechoslovakia, Austria.
kalambakae *C.E. Sonck in Ann. Bot. Fenn.,* 36(3): 211 (1999)—Greece.
krasnoborovii *A.A. Krasnikov in Fl. Sibir. (Asterac.),* 13: 279 (1997 publ. 1998)—Altay.
krylovii *A.A. Krasnikov & V.M. Khanminchun in Fl. Sibir. (Asterac.),* 13: 280 (1997 publ. 1998)—Tuva.
lentum *J. Kirschner & J. Štěpánek, Monogr. Taraxacum sect. Palustria:* 216 (1998)—Bulgaria.
leucosquameum *C.E. Sonck in Ann. Bot. Fenn.,* 36(4): 291 (1999)—Belgium (Luxembourg).
limosicola *J. Kirschner & J. Štěpánek, Monogr. Taraxacum sect. Palustria:* 251 (1998)—Czechoslovakia.
litigiosum *J. Kirschner & J. Štěpánek, Monogr. Taraxacum sect. Palustria:* 236 (1998)—Czechoslovakia.
melancholicum *J. Kirschner & J. Štěpánek, Monogr. Taraxacum sect. Palustria:* 172 (1998)—Bulgaria.
memorabile *J. Kirschner & J. Štěpánek, Monogr. Taraxacum sect. Palustria:* 267 (1998)—Germany.
× mesohalobium *J. Kirschner & J. Štěpánek in Folia Geobot.,* 33(4): 411 (1998 publ. 1999)—Krym. (T. bessarabicum × salsum).
montellii *H. Lindb. ex C.E. Sonck in Ann. Bot. Fenn.,* 35(1): 29 (1998)—Finland.
nylandicum *C.E. Sonck & H. Øllgaard in Ann. Bot. Fenn.,* 36(4): 295 (1999)—Finland.
obuncum *J. Kirschner & J. Štěpánek, Monogr. Taraxacum sect. Palustria:* 165 (1998)—Bulgaria.
oddense *C.-F. Lundevall in Preslia,* 71(1–2): 118 (1999), nom. nov.: T. humile *G.E. Haglund.*
odiosum *J. Kirschner & J. Štěpánek, Monogr. Taraxacum sect. Palustria:* 139 (1998)—France.
paludem-ornans *J. Kirschner & J. Štěpánek, Monogr. Taraxacum sect. Palustria:* 183 (1998)—Czechoslovakia, Austria.
perenne *J. Kirschner & J. Štěpánek in Folia Geobot.,* 33(4): 405 (1998 publ. 1999)—Krym.
pindicum *J. Kirschner & J. Štěpánek in Folia Geobot.,* 33(4): 409 (1998 publ. 1999)—Greece.
porcellisinus *C.E. Sonck & H. Øllgaard in Ann. Bot. Fenn.,* 36(4): 296 (1999)—Finland.
portentosum *J. Kirschner & J. Štěpánek, Monogr. Taraxacum sect. Palustria:* 233 (1998)—Czechoslovakia.
potor *J. Kirschner & J. Štěpánek, Monogr. Taraxacum sect. Palustria:* 255 (1998)—Czechoslovakia, Hungary.
pratense *A.A. Krasnikov in Fl. Sibir. (Asterac.),* 13: 291 (1997 publ. 1998), homonym; *see Hegetschw., Fl. Schweiz:* 762 (1840)—Altay.
pseudoalaskanum *B.A. Jurtzev in Bot. Zhurn.,* 82(1): 114 (1997)—Magadan.
var. ulvelkotii *(Jurtzev) B.A. Jurtzev in Bot. Zhurn.,* 82(1): 114 (1997): T. alaskanum var. ulvelkotii.
pseudokamtschaticum *B.A. Jurtzev in Bot. Zhurn.,* 82(1): 114 (1997)—Alaska.
pseudopaucilobum *J. Kirschner & J. Štěpánek, Monogr. Taraxacum sect. Palustria:* 118 (1998)—Czechoslovakia, Hungary.
pseudoporrigens *C.E. Sonck in Ann. Bot. Fenn.,* 35(3): 179 (1998)—France.

TARAXACUM:-
pseudopulchrum *J. Kirschner & J. Štěpánek, Monogr. Taraxacum sect. Palustria:* 158 (1998)—Turkey.
pseudosuecicum *J. Kirschner & J. Štěpánek, Monogr. Taraxacum sect. Palustria:* 72 (1998)—Sweden.
pulchellum *J. Kirschner & J. Štěpánek, Monogr. Taraxacum sect. Palustria:* 115 (1998)—Hungary, Czechoslovakia.
radens *C.E. Sonck in Ann. Bot. Fenn.,* 35(3): 179 (1998)—France.
ranarium *J. Kirschner & J. Štěpánek, Monogr. Taraxacum sect. Palustria:* 247 (1998)—Hungary, Czechoslovakia.
rangiferinum *C.E. Sonck & H. Øllgaard in Ann. Bot. Fenn.,* 36(2): 144 (1999)—Finland, Norway, Sweden.
ranunculus *J. Kirschner & J. Štěpánek, Monogr. Taraxacum sect. Palustria:* 207 (1998)—Czechoslovakia.
rigidifolium *C.E. Sonck in Ann. Bot. Fenn.,* 36(3): 212 (1999)—Greece.
rigidipes *C.E. Sonck in Ann. Bot. Fenn.,* 35(2): 120 (1998)—Italy.
rupestre *A.A. Krasnikov in Fl. Sibir. (Asterac.),* 13: 291 (1997 publ. 1998), homonym; *see Greene in Pittonia,* 4: 229 (1901)—Altay.
rutilum *J. Kirschner & J. Štěpánek in Preslia,* 69(1): 47 (1997)—Germany.
sacrificatum *C.E. Sonck in Ann. Bot. Fenn.,* 35(1): 31 (1998)—Finland.
sajanense *A.A. Krasnikov in Fl. Sibir. (Asterac.),* 13: 292 (1997 publ. 1998)—Buryatiya/Irkutsk.
salsitatis *J. Kirschner, J. Štěpánek & Ş. Yıldırımlı in Folia Geobot.,* 33(4): 408 (1998 publ. 1999)—Turkey.
salsum *J. Kirschner & J. Štěpánek in Folia Geobot.,* 33(4): 406 (1998 publ. 1999)—Krym.
sertatum *J. Kirschner, H. Øllgaard & J. Štěpánek in Preslia,* 69(1): 49 (1997)—Czechoslovakia, Germany, Sweden, Denmark.
sinicum Kitag.
f. album *(Sato ex Liou) C.Y. Li in Clavis Pl. Chinae Bor.-Or. ed. 2 (ed. P.Y. Fu):* 729 (1995), basionym without latin descr. or type: T. sinicum var. album.
sordidum *J. Kirschner & J. Štěpánek, Monogr. Taraxacum sect. Palustria:* 222 (1998)—Turkey.
stenocephalum Boiss. & Kotschy
subsp. daralagesicum *(Schischk.) J. Kirschner & J. Štěpánek in Folia Geobot.,* 33(4): 402 (1998 publ. 1999): T. daralagesicum.
subsp. gumusanicum *(Soest) J. Kirschner & J. Štěpánek in Folia Geobot.,* 33(4): 403 (1998 publ. 1999): T. bessarabicum subsp. gumusanicum.
subsp. magnum *J. Kirschner & J. Štěpánek in Folia Geobot.,* 33(4): 402 (1998 publ. 1999)—North Caucasus (Krasnodar).
strictum *J. Kirschner & J. Štěpánek, Monogr. Taraxacum sect. Palustria:* 154 (1998)—Bulgaria.
sublimiforme *C.E. Sonck in Ann. Bot. Fenn.,* 36(3): 213 (1999)—Greece.
submicrocranum *C.E. Sonck in Ann. Bot. Fenn.,* 36(3): 216 (1999)—Greece.
subpolonicum *J. Kirschner & J. Štěpánek, Monogr. Taraxacum sect. Palustria:* 120 (1998)—Czechoslovakia, Poland.
subudum *J. Kirschner & J. Štěpánek, Monogr. Taraxacum sect. Palustria:* 225 (1998)—Bulgaria, Romania.
suspectum *J. Kirschner & J. Štěpánek, Monogr. Taraxacum sect. Palustria:* 162 (1998)—Bulgaria.
tenuifolium *H. Koidzumi in J. Pl. Iwateken,* 2(2): 102 (1937), homonym; *see W.D.J. Koch in Flora,* 23: 374 (1840)—Japan.
tragopogon *J. Kirschner & J. Štěpánek in Preslia,* 69(1): 43 (1997)—Czechoslovakia, Finland, Germany.
trigonense *C.E. Sonck in Ann. Bot. Fenn.,* 36(3): 214 (1999)—Greece.
tschuktschorum *(Tzvelev) B.A. Jurtzev & N.N. Tzvelev in Bot. Zhurn.,* 82(1): 113 (1997): T. soczavae var. tschuktschorum.
turbidum *C.E. Sonck in Ann. Bot. Fenn.,* 35(3): 181 (1998)—France.
uvidum *J. Kirschner & J. Štěpánek, Monogr. Taraxacum sect. Palustria:* 249 (1998)—Czechoslovakia, Austria, Hungary.

TARAXACUM:-
xinyuanicum *D.T. Zhai & Z.X. An in Acta Phytotax. Sin.,* 34(3): 318 (1996)—Xinjiang.
zagorae *C.E. Sonck in Ann. Bot. Fenn.,* 36(3): 215 (1999)—Greece.

TARAXIS *B.G. Briggs & L.A.S. Johnson in Telopea,* 7(4): 363 (1998). *RESTIONACEAE*
grossa *B.G. Briggs & L.A.S. Johnson in Telopea,* 7(4): 364 (1998)—Western Australia.

TARENNA (Rubiac.).
gilletii *(De Wild. & T. Durand) N. Hallé ex R.E. Gereau in Novon,* 8(2): 140 (1998): Chomelia gilletii.

TAUSCHIA (Umbellif.).
beruloides *L. Constance & J.M. Affolter in Brittonia,* 49(4): 460 (1997)—Mexico Southwest (Michoacan), Mexico Central (México State).
spellenbergii *L. Constance & J.M. Affolter in Brittonia,* 49(4): 458 (1997)—Mexico Northeast (Chihuahua).

TAVARESIA (Asclepiadac.).
thompsoniorum *E. van Jaarsveld & R. Nagel in Asklepios,* 76: 9 (1999)—Angola.

TAVERNIERA (Legumin.).
diffusa *(Cambess.) M. Thulin in Fl. Iran.,* 157: 465 (1984): Onobrychis diffusa.

TAXILLUS (Loranthac.).
delavayi (Tiegh.) Danser
var. barbatus *W.L. Cheng in Acta Bot. Yunnanica,* 20(4): 394 (1998)—Xizang.
var. yanjingensis *W.L. Cheng in Acta Bot. Yunnanica,* 20(4): 394 (1998)—Xizang.
liquidambaricola (Hayata) Hosok.
var. neriifolius *H.S. Kiu in Guihaia,* 17(4): 308 (1997), as 'liquidambaricolus'—China Southeast (Fujian, Guangdong).
lonicerifolius *(Hayata) S.T. Chiu in Taiwania,* 41(2): 157 (1996): Loranthus lonicerifolius.
var. longifolius *S.T. Chiu in Taiwania,* 41(2): 158 (1996)—Taiwan.
parasiticus *(L.) S.T. Chiu in Taiwania,* 41(2): 159 (1996): Scurrula parasitica.
renii *H.S. Kiu in Guihaia,* 17(4): 306 (1997)—China South-central (Yunnan).
rhododendricola *(Hayata) S.T. Chiu in Taiwania,* 41(2): 162 (1996), as 'rhododendricolus': Loranthus rhododendricola.
ritozanensis *(Hayata) S.T. Chiu in Taiwania,* 41(2): 163 (1996): Loranthus ritozanensis.
tsaii *S.T. Chiu in Taiwania,* 41(2): 164 (1996)—Taiwan.
wiensii *R.M. Polhill in R.M. Polhill & D. Wiens, Mistletoes of Africa:* 219 (1998)—Kenya.

TAXODIUM (Taxodiac.).
nuciferum *(L.) Brongn. ex J.O. Voigt, Hort. Suburb. Calcutt.:* 559 (1845): Taxus nucifera.

TAXUS (Taxac.).
cuspidata Siebold & Zucc.
var. caespitosa *(Nakai) Q.L. Wang in Clavis Pl. Chinae Bor.-Or. ed. 2 (ed. P.Y. Fu):* 73 (1995): T. caespitosa.
var. caespitosa *(Nakai) Y.N. Lee, Fl. Korea:* 1156 (1996), isonym; *see Q.L. Wang in Clavis Pl. Chinae Bor.-Or. ed. 2:* 73 (1995): T. caespitosa.
fuana *Nan Li & R.R. Mill in Novon,* 7(3): 263 (1997)—Xizang, Jammu-Kashmir, Nepal, Pakistan, India (Punjab).
wallichiana Zucc.
var. mairei *(Lemée & H. Lév.) L.K. Fu & Nan Li in Novon,* 7(3): 263 (1997): Tsuga mairei.

TEAGUEIA (Orchidac.).
alyssana *C.A. Luer & L. Jost in C.A. Luer, Ic. Pleurothallid. XX (Monogr. Syst. Bot. Missouri Bot. Gard.,* 79): 131 (2000)—Ecuador.
cymbisepala *C.A. Luer & L. Jost in C.A. Luer, Ic. Pleurothallid. XX (Monogr. Syst. Bot. Missouri Bot. Gard.,* 79): 132 (2000)—Ecuador.
jostii *C.A. Luer, Ic. Pleurothallid. XX (Monogr. Syst. Bot. Missouri Bot. Gard., 79):* 132 (2000)—Ecuador.

TEAGUEIA:-
sancheziae *C.A. Luer & L. Jost in C.A. Luer, Ic. Pleurothallid. XX (Monogr. Syst. Bot. Missouri Bot. Gard.*, 79): 133 (2000)—Ecuador.

TEDINGEA (Amaryllidac.).
pygmaea *(Snijman) D. Müller-Doblies & U. Müller-Doblies in Feddes Repert.*, 107(5–6): S.c.5 (1996): Strumaria pygmaea.

TEESDALIA (Crucif.).
conferta *(Lag.) O. Appel in Novon*, 8(3): 218 (1998): Iberis conferta.

TEHUANA *J.L. Panero & J.L. Villaseñor in Syst. Bot.*, 21(4): 555 (1996 publ. 1997). *COMPOSITAE*
calzadae *J.L. Panero & J.L. Villaseñor in Syst. Bot.*, 21(4): 555 (1996 publ. 1997)—Mexico Southwest (Oaxaca).

TELANTHOPHORA (Compos.).
cobanensis (J.M. Coult.) H. Rob. & Brettell
var. molinae *(H. Rob. & Brettell) B.L. Clark in Sida*, 19(2): 236 (2000): Telanthophora molinae.
grandifolia (Less.) H. Rob. & Brettell
var. serraquitchensis *(Greenm.) B.L. Clark in Sida*, 19(2): 236 (2000): Senecio serraquitchensis.
sublaciniata *(Greenm.) B.L. Clark in Sida*, 19(2): 236 (2000), as 'sublaciniatus': Senecio cobanensis var. sublaciniatus.

TELINE (Legumin.).
sect. Chronanthus *(DC.) S. Talavera & P.E. Gibbs in An. Jard. Bot. Madrid*, 57(1): 210 (1999): Cytisus sect. Chronanthus.
orientalis *(Loisel.) S. Talavera & P.E. Gibbs in An. Jard. Bot. Madrid*, 57(1): 210 (1999): Cytisus orientalis.
pallida (Poir.) G. Kunkel
subsp. silensis *M. del Arco Aguilar in Taxon*, 49(1): 21 (2000)—Canary Is.
patens *(DC.) S. Talavera & P.E. Gibbs in Lagascalia*, 18(2): 267 (1996): Genista patens.
tribracteolata *(Webb) S. Talavera & P.E. Gibbs in An. Jard. Bot. Madrid*, 57(1): 210 (1999): Cytisus tribracteolatus.

TELIPOGON (Orchidac.).
jucusbambae *C.H. Dodson & R. Escobar in Orquideologia*, 21(1): 65 (1998)—Peru.
maldonadoensis *C.H. Dodson & R. Escobar in Orquideologia*, 21(1): 53 (1998)—Ecuador.
puruantensis *C.H. Dodson & R. Escobar in Orquideologia*, 21(1): 56 (1998)—Ecuador.
sayakoae *D.E. Bennett & E.A. Christenson in Arnaldoa*, 6(1): 61 (1999 publ. 2000), as 'sayakoi'—Peru.
tungurahuae *C.H. Dodson & R. Escobar in Orquideologia*, 21(1): 58 (1998)—Ecuador.

TELOPEA (Proteac.).
aspera *M.D. Crisp & P.H. Weston in Fl. Australia*, 16: 498 (1995)—New South Wales.

TELOSIPHONIA *(Woodson) J. Henrickson in Aliso*, 14(3): 184 (1995 publ. 1996): Macrosiphonia subgen. Telosiphonia. (Apocynac.).
brachysiphon *(Torr.) J. Henrickson in Aliso*, 14(3): 186 (1995 publ. 1996): Echites brachysiphon.
hesperia *(I.M. Johnst.) J. Henrickson in Aliso*, 14(3): 191 (1995 publ. 1996): Macrosiphonia hesperia.
hypoleuca *(Benth.) J. Henrickson in Aliso*, 14(3): 185 (1995 publ. 1996): Echites hypoleuca.
lanuginosa *(M. Martens & Galeotti) J Henrickson in Aliso*, 14(3): 189 (1995 publ. 1996): Echites lanuginosa.
var. oaxacensis *J. Henrickson in Aliso*, 14(3): 191 (1995 publ. 1996)—Mexico Central (Puebla), Mexico Southwest (Oaxaca).
macrosiphon *(Torr.) J. Henrickson in Aliso*, 14(3): 187 (1995 publ. 1996): Echites macrosiphon.
nacapulensis *R.S. Felger & J. Henrickson in Aliso*, 14(3): 194 (1995 publ. 1996)—Mexico Northwest (Sonora).

TEMBURONGIA *S. Dransfield & K.M. Wong in Sandakania*, 7: 53 (1996). *GRAMINEAE*
simplex *S. Dransfield & K.M. Wong in Sandakania*, 7: 55 (1996)—Borneo (Brunei).

TEMNADENIA (Apocynac.).
odorifera *(Vell.) J.F. Morales in Novon*, 9(2): 240 (1999): Echites odorifera.

TENARIS (Asclepiadac.).
bikitaensis *(Peckover) J.E. Victor & A. Nicholas in S. Afr. J. Bot.*, 64(3): 207 (1998): Brachystelma bikitaense.
christianeae *(Peckover) J.E. Victor & A. Nicholas in S. Afr. J. Bot.*, 64(3): 207 (1998): Brachystelma christianeae.

TENGIA (Gesneriac.).
potiflora *S.Z. He in J. China Pharm. Univ.*, 23(5): 269 (1992)—China South-central (Guizhou).
scopulorum Chun
var. potiflora *(S.Z. He) W.T. Wang, A.L. Weitzman & L.E. Skog in Novon*, 7(4): 434 (1997 publ. 1998): T. potiflora.

TEPHROCACTUS (Cactac.).
aoracanthus (Lem.) Lem.
var. paediophilus *(A. Cast.) J.G. Lambert, Cactus d'Argentine, ed. 2*: 297 (1998): Opuntia paediophila.
pulcherrimus *J.J. Halda & L. Horáček in Acta Mus. Richnov. Sect. Nat.*, 7(2): 75 (2000)—Bolivia.

TEPHROSERIS (Compos.).
campestris (Retz.) Rchb.
f. spathulatus *(Miq.) R. Yin & C.Y. Li in Clavis Pl. Chinae Bor.-Or. ed. 2 (ed. P.Y. Fu)*: 694 (1995): Senecio aurantiacus var. spathulatus.
integrifolia (L.) Holub
var. spathulata *(Miq.) H. Ohba in J. Jap. Bot.*, 72(2): 125 (1997): Senecio aurantiacus var. spathulatus.
tundricola (Tolm.) Holub
subsp. lindstroemii *(Ostenf.) E.I. Wiebe in Fl. Sibir. (Asterac.)*, 13: 161 (1997 publ. 1998): Senecio integrifolius var. lindstroemii.
turczaninovii (DC.) Holub
subsp. reverdattoi *(Sobolevsk.) E.I. Wiebe in Fl. Sibir. (Asterac.)*, 13: 161 (1997 publ. 1998): Senecio reverdattoi.

TEPHROSIA (Legumin.).
carrollii *O. Téllez Valdés in Iselya*, 2(3): 106 (1985), nom. nov.: T. woodii *O. Téllez*.
macrophylla *N. Wallich, Report on Calcutta Bot. Gard. to G.A. Bushby*: 23 (1840), nom. nov.: Robinia macrophylla *Roxb.*
purpurea (L.) Pers.
var. piscatoria *(Aiton) F.R. Fosberg in Micronesica*, 25(2): 187 (1992): Galega piscatoria.
socotrana *M. Thulin in Nordic J. Bot.*, 17(4): 349 (1997)—Socotra.
tuitoensis *O. Téllez Valdés in Iselya*, 2(3): 104 (1985)—Mexico Southwest (Jalisco).
viridiflora *O. Téllez Valdés in Iselya*, 2(3): 101 (1985)—Mexico Southwest (Jalisco).

TERMINALIA (Combretac.).
glabrata
var. intonsa *F.R. Fosberg in Micronesica*, 23(1): 2 (1990)—Cook Is.
procera *W. Roxburgh, Pl. Coromandel*, 3: 18 (1811)—Andaman Is.†

TERNSTROEMIA (Theac.).
coniocarpa *L.K. Ling in Fl. Reipubl. Popul. Sin.*, 50(1): 185 (1998)—China Southeast (Guangdong, Guangxi, Hunan).
longipedicellata *L.K. Ling in Fl. Reipubl. Popul. Sin.*, 50(1): 184 (1998)—China South-central (Yunnan).
luteoflora *L.K. Ling in Fl. Reipubl. Popul. Sin.*, 50(1): 183 (1998)—China Southeast (Guangdong, Fujian, Guangxi, Hunan, Jiangxi), China South-central (Guizhou, Hubei, Yunnan), Hainan.
sichuanensis *L.K. Ling in Fl. Reipubl. Popul. Sin.*, 50(1): 186 (1998)—China South-central (Sichuan, Guizhou).
simaoensis *L.K. Ling in Fl. Reipubl. Popul. Sin.*, 50(1): 187 (1998)—China South-central (Yunnan).
yunnanensis *L.K. Ling in Fl. Reipubl. Popul. Sin.*, 50(1): 182 (1998)—China South-central (Yunnan).

TESSARIA (Compos.).
dodoneifolia (Hook. & Arn.) Cabrera
subsp. plucheoides *(Hassl.) S.E. Freire in Candollea*, 52(1): 214 (1997): T. plucheoides.
integrifolia Ruiz & Pav.
var. ambigua *(DC.) S.E. Freire in Candollea*, 52(1): 215 (1997): T. ambigua.

TETA (Convallariac.).
humilis *(Andrews) J.O. Voigt, Hort. Suburb. Calcutt.*: 651 (1845): Peliosanthes humilis.

TETRABERLINIA (Legumin.).
baregarum *J.J. Wieringa in Wageningen Agric. Univ. Pap.*, 99(4): 257 (1999)—Zaire.
korupensis *J.J. Wieringa in Wageningen Agric. Univ. Pap.*, 99(4): 267 (1999)—Cameroon.
longiracemosa *(A. Chev.) J.J. Wieringa in Wageningen Agric. Univ. Pap.*, 99(4): 272 (1999): Monopetalanthus longiracemosus.

TETRACERA (Dilleniac.).
macphersonii *G. Aymard C. in Novon*, 6(4): 325 (1996)—Panama.
xui *H. Zhu & H. Wang in Guihaia*, 19(4): 337 (1999)—China South-central (Yunnan).

TETRADENIA (Labiat.).
clementiana *P.B. Phillipson in Fl. Madagascar*, fam. 175: 87 (1998)—Madagascar.
cordata *P.B. Phillipson in Fl. Madagascar*, fam. 175: 77 (1998)—Madagascar.
falafa *P.B. Phillipson in Fl. Madagascar*, fam. 175: 81 (1998)—Madagascar.
herbacea *P.B. Phillipson in Fl. Madagascar*, fam. 175: 88 (1998)—Madagascar.
hildeana *P.B. Phillipson in Fl. Madagascar*, fam. 175: 78 (1998)—Madagascar.
isaloensis *P.B. Phillipson in Fl. Madagascar*, fam. 175: 90 (1998)—Madagascar.

TETRADIUM (Rutac.).
glabrifolium (Benth.) T.G. Hartley
var. glaucum *(Miq.) H. Ohba in Fl. Japan*, 2c: 35 (1999), isonym; *see T. Yamaz. in J. Jap. Bot.*, 72(4): 249 (1997): Euodia glauca.
var. glaucum *(Miq.) T. Yamazaki in J. Jap. Bot.*, 72(4): 249 (1997): Euodia glauca.
taiwanense *(T. Yamaz.) T. Yamazaki in J. Jap. Bot.*, 72(4): 249 (1997): Euodia taiwanensis.

TETRAGONIA (Aizoac.).
coronata *B.L. Rye & M.E. Trudgen in Nuytsia*, 10(3): 459 (1996)—Western Australia.
moorei *M. Gray in Telopea*, 7(2): 119 (1997)—South Australia, Queensland, New South Wales, Victoria, Western Australia.

TETRAMICRA (Orchidac.).
malpighiarum *J.A. Hernández & M.A. Díaz in Harvard Pap. Bot.*, 5(1): 189 (2000)—Cuba.
zanonii *M.A. Nir, Orchidaceae Antillanae*: 369 (2000)—Dominican Republic.

TETRANEMA (Scrophulariac.).
floribundum *B.E. Hammel & M.H. Grayum in Phytologia*, 79(4): 276 (1995 publ. 1996)—Costa Rica.
gamboanum *M.H. Grayum & B.E. Hammel in Phytologia*, 79(4): 270 (1995 publ. 1996)—Costa Rica.

TETRAPETALUM (Annonac.).
lambirense *K. Momose in Blumea*, 43(1): 117 (1998)—Borneo (Sarawak).

TETRAPLASANDRA (Araliac.).
flynnii *P.P. Lowry & K.R. Wood in Novon*, 10(1): 40 (2000)—Hawaiian Is.

TETRASIDA (Malvac.).
weberbaueri *(Ulbr.) P.A. Fryxell & F.J. Fuertes Aguilar in Contrib. Univ. Mich. Herb.*, 21: 193 (1997): Pseudabutilon weberbaueri.

TETRASTIGMA (Vitac.).
sect. Orbicularia *C.L. Li in Chinese J. Appl. Environ. Biol.*, 1(4): 329 (1995), without type.

TETRASTIGMA:-
ser. Palmata *C.L. Li in Chinese J. Appl. Environ. Biol.*, 1(4): 322 (1995), without type.
subgen. Palmicirrata *C.L. Li in Chinese J. Appl. Environ. Biol.*, 1(4): 331 (1995), without type.
ser. Trifoliolata *C.L. Li in Chinese J. Appl. Environ. Biol.*, 1(4): 314 (1995), without type.
apiculatum Gagnep.
var. pubescens *C.L. Li in Chinese J. Appl. Environ. Biol.*, 1(4): 314 (1995)—China South-central (Yunnan), Hainan.
baenzigeri *C.L. Li in Chinese J. Appl. Environ. Biol.*, 3(2): 139 (1997), as 'baenzigerii'—Thailand.
ceratopetalum *C.Y. Wu in Chinese J. Appl. Environ. Biol.*, 1(4): 327 (1995)—China South-central (Yunnan), China Southeast (Guangxi).
funingense *C.L. Li in Chinese J. Appl. Environ. Biol.*, 1(4): 316 (1995)—China South-central (Yunnan).
jingdongensis *C.L. Li in Acta Bot. Yunnanica*, 19(3): 220 (1997)—China South-central (Yunnan).
jinghongense *C.L. Li in Chinese J. Appl. Environ. Biol.*, 1(4): 324 (1995)—China South-central (Yunnan).
jinxiuense *C.L. Li in Chinese J. Appl. Environ. Biol.*, 1(4): 315 (1995)—China Southeast (Guangxi).
kwangsiense *C.L. Li in Chinese J. Appl. Environ. Biol.*, 1(4): 317 (1995)—China Southeast (Guangxi).
lincangense *C.L. Li in Chinese J. Appl. Environ. Biol.*, 1(4): 324 (1995)—China South-central (Yunnan).
lineare *W.T. Wang ex C.Y. Wu & C.L. Li in Chinese J. Appl. Environ. Biol.*, 1(4): 313 (1995)—China South-central (Yunnan).
longipedunculatum *C.L. Li in Chinese J. Appl. Environ. Biol.*, 1(4): 314 (1995)—China Southeast (Guangxi).
napaulense *(DC.) C.L. Li in Chinese J. Appl. Environ. Biol.*, 1(4): 331 (1995):Cissus napaulensis.
var. puberulum *(W.T. Wang) C.L. Li in Chinese J. Appl. Environ. Biol.*, 1(4): 331 (1995): T. serrulatum var. puberulum.
papillatum *(Hance) C.Y. Wu in Chinese J. Appl. Environ. Biol.*, 1(4): 312 (1995):Vitis papillata.
pilosum *C.L. Li in Chinese J. Appl. Environ. Biol.*, 3(2): 138 (1997)—Thailand.
pseudocruciatum *C.L. Li in Chinese J. Appl. Environ. Biol.*, 1(4): 318 (1995)—Hainan.
retinervium Planch.
var. pubescens *C.L. Li in Chinese J. Appl. Environ. Biol.*, 1(4): 316 (1995)—China South-central (Yunnan), China Southeast (Guangxi).
rumicispermum (Laws.) Planch.
var. lasiogynum *(W.T. Wang) C.L. Li in Chinese J. Appl. Environ. Biol.*, 1(4): 327 (1995): T. serrulatum var. lasiogynum.
serrulatum (Roxb.) Planch.
var. puberulum *(W.T. Wang & Z.Y. Cao) C.L. Li in Chinese J. Appl. Environ. Biol.*, 1(4): 331 (1995), synonym; *see W.T. Wang in Acta Phytotax. Sin.*, 17(3): 83 (1979): T. hypoglaucum var. puberulum.
var. pubinervium *C.L. Li in Fl. Reipubl. Popul. Sin.*, 48(2): 126 (1998), nom. nov.: T. hypoglaucum var. puberulum *W.T. Wang & Z.Y. Cao.*
sichouense *C.L. Li in Chinese J. Appl. Environ. Biol.*, 1(4): 322 (1995)—China South-central (Yunnan, Guizhou).
var. megalocarpum *C.L. Li in Chinese J. Appl. Environ. Biol.*, 1(4): 323 (1995)—China South-central (Yunnan, Guizhou), Xizang.
subtetragonum *C.L. Li in Chinese J. Appl. Environ. Biol.*, 1(4): 321 (1995)—China South-central (Yunnan).
taeniatum *C.L. Li in Chinese J. Appl. Environ. Biol.*, 3(2): 138 (1997)—Thailand.
tsaianum *C.Y. Wu in Chinese J. Appl. Environ. Biol.*, 1(4): 329 (1995)—China South-central (Yunnan).
venulosum *C.Y. Wu in Chinese J. Appl. Environ. Biol.*, 1(4): 320 (1995)—China South-central (Yunnan).
xishuangbannaense *C.L. Li in Chinese J. Appl. Environ. Biol.*, 1(4): 326 (1995)—China South-central (Yunnan).
xizangense *C.L. Li in Chinese J. Appl. Environ. Biol.*, 1(4): 324 (1995)—Xizang.
yiwuense *C.L. Li in Chinese J. Appl. Environ. Biol.*, 1(4): 315 (1995)—China South-central (Yunnan).

TETRATHECA (Tremandrac.).
chapmanii *J.J. Alford in Nuytsia*, 10(2): 144 (1995)—Western Australia.

TETRATHECA:-
paynterae *J.J. Alford in Nuytsia*, 10(2): 146 (1995)—Western Australia.

TETRODON *(Kraenzl.) M.A. Clements & D.L. Jones in Orchadian*, 12(7): 310 (1998): Eria sect. Tetrodon. (Orchidac.).
oppositifolius *(Kraenzl.) M.A. Clements & D.L. Jones in Orchadian*, 12(7): 310 (1998): Eria oppositifolia.
petrophilus *(Kraenzl.) M.A. Clements & D.L. Jones in Orchadian*, 12(7): 311 (1998): Eria petrophila.

TETRORCHIDIUM (Euphorbiac.).
duckei *A. Radcliffe-Smith & R. Govaerts in Kew Bull.*, 52(1): 189 (1997), nom. nov.: Adenophaedra minor *Ducke.*
gabonense *F.J. Breteler in Adansonia*, 21(1): 101 (1999)—Gabon.

TEUCRIUM (Labiat.).
capitatum L.
subsp. gypsicola *G. Mateo Sanz & V.J. Arán Redó in Flora Montiberica*, 9: 34 (1998), contrary to Art. 37.4 ICBN (1994)—Spain.
× carvalhoae *A.F. Carrillo, A. Hernández, E. Coy, J. Güemes & P. Sánchez-Gómez in Acta Bot. Malacitana*, 22: 221 (1997), as 'carvalhoi'—Spain. (T. balthazaris × capitatum subsp. gracillimum).
chlorostachyum Pau & Font Quer
subsp. melillense *(Maire) J. El Oualidi, J. Mathez & T. Navarro in Flora Medit.*, 7: 23 (1997): T. melillense.
clementiae *O. Ryding in Edinburgh J. Bot.*, 55(2): 210 (1998)—Ethiopia, Somalia.
conquense M.B. Crespo & Mateo
nothosubsp. siyasense *A.F. Carrillo & P. Sánchez-Gómez in Acta Bot. Malacitana*, 21: 285 (1996)—Spain. (T. capitatum subsp. gracillimum × gnaphalodes).
ekimii *H. Duman in Karaca Arbor. Mag.*, 4(3): 125 (1998)—Turkey.
× eloualidii *P. Sánchez Gómez & T. Navarro in An. Jard. Bot. Madrid*, 57(1): 167 (1999)—Spain. (T. freynii × lanigerum).
embergeri *(Sauvage & Vindt) J. El Oualidi, T. Navarro & A. Martin in Acta Bot. Malacitana*, 22: 198 (1997): T. mairei var. embergeri.
× guemesii *J.F. Jiménez, A.F. Carrillo, M.A. Carrión Vilches, P. Sánchez Gómez &* T. Navarro *in Acta Bot. Malacitana*, 24: 205 (1999)—Spain. (T. carolipaui subsp. fontqueri × lanigerum).
heterophyllum L'Hér.
subsp. brevipilosum *M. von Gaisberg in Willdenowia*, 30(2): 267 (2000)—Canary Is.
subsp. hierrense *M. von Gaisberg in Willdenowia*, 30(2): 267 (2000)—Canary Is.
joannis *(Sauvage & Vindt) J. El Oualidi, T. Navarro & A. Martin in Acta Bot. Malacitana*, 22: 198 (1997): T. mairei subsp. joannis.
leucocladum Boiss.
var. glandulosum *A. Danin in Willdenowia*, 27(1–2): 169 (1997)—Sinai.
subsp. jordanicum *A. Danin in Willdenowia*, 27(1–2): 172 (1997)—Palestine (Jordan).
subsp. sinaicum *A. Danin in Willdenowia*, 27(1–2): 171 (1997)—Sinai.
lusitanicum Schreb.
subsp. clementiae *T. Navarro, J. El Oualidi, Arnaud Martin & S. Puech in Acta Bot. Gallica*, 145(1): 64 (1998)—Spain.
martinii *S. Cirujano, R. Roselló, J.B. Peris & G. Stübing in An. Jard. Bot. Madrid*, 57(2): 407 (1999 publ. 2000)—Spain.
maximowiczii *N.S. Probatova in Sosud. Rast. Sovet. Dal'nego Vostoka*, 7: 301 (1995)—Japan.
× navarroi *P. Sánchez-Gómez, J. Güemes, A.F. Carrillo, E. Coy & A. Hernández in Acta Bot. Malacitana*, 21: 285 (1996)—Spain. (T. rivas-martinezii × thymifolium).
ozturkii *A.P. Khokhrjakov in Byull. Glavn. Bot. Sada (Moscow)*, 175: 54 (1997)—Turkey.
pilbaranum *B.J. Conn in Telopea*, 8(3): 299 (1999)—Western Australia.
polioides *O. Ryding in Edinburgh J. Bot.*, 55(2): 214 (1998)—Somalia.

TEUCRIUM:-
polium L.
var. aguilasense *(S. Puech) O. de Bolòs & J. Vigo, Fl. Països Catalans*, 3: 240 (1995), basionym not effectively publ.: T. polium subsp. aguilasense.
var. clapae *(S. Puech) O. de Bolòs & J. Vigo, Fl. Països Catalans*, 3: 240 (1995): T. polium subsp. clapae.
var. expassum *(Pau) O. de Bolòs & J. Vigo, Fl. Països Catalans*, 3: 238 (1995): T. expassum.
var. gracillimum *(Rouy) O. de Bolòs & J. Vigo, Fl. Països Catalans*, 3: 239 (1995): T. capitatum var. gracillimum.
subvar. homotrichum *(Font Quer) O. de Bolòs & J. Vigo, Fl. Països Catalans*, 3: 236 (1995): T. carthaginense var. homotrichum.
subsp. latifolium *(Willk.) O. de Bolòs & J. Vigo, Fl. Països Catalans*, 3: 236 (1995): T. aureum var. latifolium.
subsp. lutescens *(Coincy) O. de Bolòs & J. Vigo, Fl. Països Catalans*, 3: 235 (1995): T. eriocephalum var. lutescens.
var. pii-fontii *(P. Palau) O. de Bolòs & J. Vigo, Fl. Països Catalans*, 3: 239 (1995): T. polium subsp. pii-fontii.
× portusmagni *P. Sánchez Gómez, A.F. Carrillo, A. Hernández & T. Navarro in Acta Bot. Malacitana*, 24: 205 (1999), as 'portusmagnii'—Spain. (T. capitatum subsp. gracillimum × freynii).
× pseudothymifolium *P. Sánchez-Gómez, J. Güemes & A.F. Carrillo in Acta Bot. Malacitana*, 21: 283 (1996)—Spain. (T. rotundifolium × thymifolium).
rifanum *(Maire & Sennen) T. Navarro & J. El Oualidi in Acta Bot. Malacitana*, 22: 199 (1997): T. polium var. rifanum.
somalense *O. Ryding in Edinburgh J. Bot.*, 55(2): 217 (1998)—Somalia.
subspinosum Pourr. ex Willd.
subsp. spinescens *(Porta) A.M. Romo, Fl. Silvestres Baleares:* 253 (1994): T. marum var. spinescens.
taiwanianum *T.H. Hsieh & T.C. Huang in Taiwania*, 41(2): 86 (1996)—Taiwan.

THALICTRUM (Ranunculac.).
borderi *M. Gandoger, Dec. Pl. Nov.*, 1: 46 (1875)—France.
integratum *M. Gandoger, Dec. Pl. Nov.*, 1: 45 (1875)—France.
juranum *M. Gandoger, Dec. Pl. Nov.*, 1: 47 (1875)—France.
lancangense *Y.Y. Qian in Acta Phytotax. Sin.*, 35(3): 262 (1997)—China South-central (Yunnan).
minus L.
subsp. appendiculatum *(C.A. Mey.) I.A. Gubanov in Fl. Vostoch. Khangaya (MNR):* 131 (1983): T. appendiculatum.
subsp. thunbergii *(DC.) V.N. Voroschilov in A.K. Skvortsov (ed.), Florist. issl. v razn. raĭonakh SSSR:* 170 (1985): T. thunbergii.
obtusilobum *M. Gandoger, Dec. Pl. Nov.*, 1: 46 (1875)—France.
oxyphyllum *M. Gandoger, Dec. Pl. Nov.*, 1: 48 (1875)—France.
platyphyllum *M. Gandoger, Dec. Pl. Nov.*, 1: 46 (1875)—France.
pyrrha *M. Gandoger, Dec. Pl. Nov.*, 1: 47 (1875)—France.
rhynchocarpum Quart.-Dill. & A. Rich.
subsp. ruwenzoriense *K.A. Lye in Lidia*, 4(3): 89 (1998), nom. nov., as 'ruwenzoriensis': T. boivinianum *Staner & J. Léonard.*
simaoense *W.T. Wang & G.H. Zhu in Phytologia*, 79(5): 385 (1995 publ. 1996)—China South-central (Yunnan).
tenerifolium *M. Gandoger, Dec. Pl. Nov.*, 1: 47 (1875)—France.
yunnanense W.T. Wang
var. austroyunnanense *Y.Y. Qian in Acta Phytotax. Sin.*, 35(3): 262 (1997), as 'austro-yunnanense'—China South-central (Yunnan).

THAMNOCALAMUS (Gramin.).
crassinodus *(T.P. Yi) J.-P. Demoly in Bambou*, 21: 15 (1995): Fargesia crassinoda.
denudatus *(T.P. Yi) J.-P. Demoly in Bambou*, 21: 15 (1995): Fargesia denudata.
ibityensis *(A. Camus) D. Ohrnberger, Bamboos of the World Intro. 2:* 10 (1996): Arundinaria ibityensis.

THAMNOCALAMUS:-
nitidus *(Stapf) J.-P. Demoly, Bambous en France:* 76 (1996): Arundinaria nitida.

THAMNOSMA (Rutac.).
crenata *(Engl.) M. Thulin in Nordic J. Bot.*, 19(1): 7 (1999): T. africana var. crenata.
somalensis *M. Thulin in Nordic J. Bot.*, 19(1): 8 (1999)—Somalia.

THAPSIA (Umbellif.).
garganica L.
subsp. gymnesica *(Rosselló & Pujadas) A.M. Romo, Fl. Silvestres Baleares:* 197 (1994): T. gymnesica.
nitida Lacaita
var. meridionalis *A.J. Pujadas Salvà in An. Jard. Bot. Madrid*, 57(2): 465 (1999 publ. 2000)—Spain, Portugal.

THECACORIS (Euphorbiac.).
bussei *(Pax) A. Radcliffe-Smith in Kew Bull.*, 51(2): 330 (1996): Cyathogyne bussei.
grandifolia *(Pax & K. Hoffm.) R. Govaerts in R. Govaerts, D.G. Frodin & A. Radcliffe-Smith, World Checklist & Bibliogr.* Euphorbiaceae, 4: 1517 (2000): Cyathogyne grandifolia.
viridis (Müll. Arg.) G.L. Webster
subsp. dewevrei *(Pax) R. Govaerts in R. Govaerts, D.G. Frodin & A. Radcliffe-Smith, World Checklist & Bibliogr.* Euphorbiaceae, 4: 1520 (2000): Cyathogyne dewevrei.
subsp. glabra *(J. Léonard) R. Govaerts in R. Govaerts, D.G. Frodin & A. Radcliffe-Smith, World Checklist & Bibliogr.* Euphorbiaceae, 4: 1520 (2000): Cyathogyne viridis subsp. glabra.

THELLUNGIELLA (Crucif.).
parvula *(Schrenk) I.A. Al-Shehbaz & S.L. O'Kane in Novon*, 5(4): 309 (1995): Diplotaxis ? parvula.

THELOCACTUS (Cactac.).
ser. Ancistrocactoides *A.B. Doweld in Sukkulentÿ (Moscow)*, 1999(1): 31 (1999).
ser. Bicolores *(Kladiwa) A.B. Doweld in Sukkulentÿ (Moscow)*, 1999(1): 30 (1999): Thelocactus sect. Bicolores.
bicolor (Galeotti ex Pfeiff.) Britton & Rose
subsp. bolaensis *(Runge) A.B. Doweld in Sukkulentÿ (Moscow)*, 1999(1): 30 (1999): Echinocactus bolaensis.
subsp. commodus *(R. Haas) A.B. Doweld in Sukkulentÿ (Moscow)*, 1999(1): 30 (1999): T. bicolor var. commodus.
subsp. flavidispinus *(Backeb.) N.P. Taylor in Cactaceae Consensus Initiatives*, 5: 14 (1998): T. bicolor var. flavidispinus.
subsp. heterochromus *(F.A.C. Weber) A. Mosco & C. Zanovello in Bradleya*, 18: 62 (2000): Echinocactus heterochromus.
subsp. schwarzii *(Backeb.) N.P. Taylor in Cactaceae Consensus Initiatives*, 5: 14 (1998): T. schwarzii.
subsp. zwaki *J. Chvastek & J.J. Halda in Acta Mus. Richnov. Sect. Nat.*, 7(1): 38 (2000)—Mexico Northeast (Coahuila).
buekii (Klein) Britton & Rose
subsp. jarmilae *J.J. Halda & L. Horáček in Acta Mus. Richnov. Sect. Nat.*, 7(1): 38 (2000)—Mexico Northeast (Nuevo León).
subsp. matudae *(Sánchez-Mej. & A.B. Lau) A. Mosco & C. Zanovello in Bradleya*, 18: 62 (2000): Thelocactus matudae.
conothelos
subsp. argenteus *(Glass & R.A. Foster) C.E. Glass, Guía Identific. Cact. Amenazadas México*, 1: Th/con ssp. argenteus (1998 publ. 1997), as 'conothele': T. conothelos var. argenteus.
subsp. aurantiacus *(Glass & R.A. Foster) C.E. Glass, Guía Identific. Cact. Amenazadas México*, 1: Th/con ssp. aurantiacus (1998 publ. 1997), as 'conothele': T. conothelos var. aurantiacus.
subsp. flavus *(Mosco & Zanov.) A. Mosco & C. Zanovello in Bradleya*, 18: 63 (2000): Thelocactus flavus.
subsp. garciae *(Glass) A. Mosco & C. Zanovello in Bradleya*, 18: 63 (2000): Thelocactus garciae.
var. sniceri *J.J. Halda & J. Sladkovský in Acta Mus. Richnov. Sect. Nat.*, 7(1): 38 (2000)—Mexico Northeast (Nuevo León).

THELOCACTUS:-
flavus *A. Mosco & C. Zanovello in Cactus & Co.*, 3(1): 20, 22 (1999)—Mexico Northeast (San Luis Potosí).
garciae *C.E. Glass & M. Mendoza in C.E. Glass, Guía Identific. Cact. Amenazadas México*, 1: Th/ga (1998 publ. 1997)—Mexico Northeast (Tamaulipas).
hexaedrophorus (Lem.) Britton & Rose
var. decipiens *(A. Berger) J. Pilbeam, Thelocactus (Cactus File Handbook, 1):* 29 (1996): Basionym not stated.
var. droegeanus *(Hildm. ex K. Schum.) J. Pilbeam, Thelocactus (Cactus File Handbook, 1):* 29 (1996): Basionym not stated.
subsp. franci *J.J. Halda & J. Sladkovský in Acta Mus. Richnov. Sect. Nat.*, 7(1): 38 (2000)—Mexico Northeast (San Luis Potosí).
subsp. jarmilae *J.J. Halda & J. Chvastek in Acta Mus. Richnov. Sect. Nat.*, 7(1): 38 (2000)—Mexico Northeast (San Luis Potosí).
subsp. kvetae *J. Chvastek & J.J. Halda in Acta Mus. Richnov. Sect. Nat.*, 7(1): 39 (2000)—Mexico Northeast (San Luis Potosí).
var. labouretianus *(K. Schum.) J. Pilbeam, Thelocactus (Cactus File Handbook, 1):* 30 (1996): Basionym not stated.
subsp. lloydii *(Britton & Rose) N.P. Taylor in Cactaceae Consensus Initiatives*, 5: 14 (1998): T. lloydii.
var. paradensis *J. Pilbeam, Thelocactus (Cactus File Handbook, 1):* 31 (1996), without latin descr. or type—Cultivation.
leucacanthus
subsp. schmollii *(Werderm.) A. Mosco & C. Zanovello in Cactaceae Consensus Initiatives*, 7: 18 (1999): T. leucacanthus var. schmollii.
multicephalus *J.J. Halda & Panarotto in Acta Mus. Richnov. Sect. Nat.*, 5(1): 40 (1998)—Mexico Northeast (Nuevo León).
panarottoanus *J.J. Halda in Acta Mus. Richnov. Sect. Nat.*, 5(4): 161 (1998)—Mexico Northeast (San Luis Potosí).
rinconensis
subsp. freudenbergeri *(R. Haas) A. Mosco & C. Zanovello in Cactaceae Consensus Initiatives*, 7: 18 (1999): T. rinconensis var. freudenbergeri.
subsp. hintonii *J.M. Lüthy in Kakt. And. Sukk.*, 48(2): 39 (1997)—Mexico Northeast (Nuevo León).
subsp. icamolensis *J.J. Halda & P. Kupčák in Acta Mus. Richnov. Sect. Nat.*, 7(2): 75 (2000)—Mexico Northeast (Nuevo León).
subsp. multicephalus *(Halda & Panar.) J.M. Lüthy in Kakt. And. Sukk.*, 50(4): 80 (1999): T. multicephalus.
subsp. nidulans *(Quehl) C.E. Glass, Guía Identific. Cact. Amenazadas México*, 1: Th/rin ssp. hintonii (1998 publ. 1997): Echinocactus nidulans.
subsp. nidulans *(Quehl) A.B. Doweld in Sukkulentÿ (Moscow)*, 1999(1): 30 (1999), isonym; *see Glass, Guía Identific. Cact. Amazadas México*, 1: Th/rin subsp. hintonii (1998 publ. 1997): Echinocactus nidulans.
subsp. phymatothelos *(Poselg. ex Rümpler) C.E. Glass, Guía Identific. Cact. Amenazadas México*, 1: Th/rin ssp. hintonii (1998 publ. 1997), as 'phymatothele': Echinocactus phymatothelos.
subsp. phymatothelos *(Poselg. ex Rümpler) A.B. Doweld in Sukkulentÿ (Moscow)*, 1999(1): 30 (1999), isonym; *see Glass, Guía Identific. Cact. Amenazadas México*, 1: Th/rin ssp. hintonii (1998 publ. 1997): Echinocactus phymatothelos.
santaclarensis *J.J. Halda, P. Kupčák & J. Sladkovský in Acta Mus. Richnov. Sect. Nat.*, 7(1): 39 (2000)—Mexico Northeast (Nuevo León).
setispinus (Engelm.) E.F. Anderson
var. cachetianus *(Labour.) J. Pilbeam, Thelocactus (Cactus File Handbook, 1):* 42 (1996): Echinocactus setispinus var. cachetianus.
var. hamatus *(Muehlenpf.) J. Pilbeam, Thelocactus (Cactus File Handbook, 1):* 42 (1996): Echinocactus hamatus.
var. mierensis *(K. Schum.) J. Pilbeam, Thelocactus (Cactus File Handbook, 1):* 42 (1996): Echinocactus setispinus var. mierensis.

THELOCACTUS:-
setispinus
var. orcuttii *(K. Schum.) J. Pilbeam, Thelocactus (Cactus File Handbook, 1):* 42 (1996): Echinocactus setispinus var. orcuttii.
tulensis
subsp. buekii *(Klein) N.P. Taylor in Cactaceae Consensus Initiatives*, 5: 14 (1998): Echinocactus buekii.
subsp. huizachensis *J.J. Halda & J. Sladkovský in Acta Mus. Richnov. Sect. Nat.*, 7(1): 39 (2000)—Mexico Northeast (San Luis Potosí).
subsp. kupcaki *J.J. Halda & J. Sladkovský in Acta Mus. Richnov. Sect. Nat.*, 7(1): 40 (2000)—Mexico Northeast (San Luis Potosí).
subsp. matudae *(Sánchez-Mej. & A.B. Lau) N.P. Taylor in Cactaceae Consensus Initiatives*, 5: 14 (1998): T. matudae.
vaskoanus *J.J. Halda, L. Hovorka & V. Zatloukal in Acta Mus. Richnov. Sect. Nat.*, 7(2): 75 (2000)—Mexico Northeast (San Luis Potosí).

THELYMITRA (Orchidac.).
aggericola *D.L. Jones in Orchadian*, 12(11): 517 (1999)—Tasmania.
atronitida *J.A. Jeanes in Muelleria*, 14: 91 (2000)—Victoria.
erosa *D.L. Jones & M.A. Clements in Austral. Orchid Res.*, 3: 184 (1998)—Tasmania.
granitora *D.L. Jones & M.A. Clements in Orchadian*, 12(7): 326 (1998)—Western Australia.
gregaria *D.L. Jones & M.A. Clements in Orchadian*, 12(7): 327 (1998)—Victoria.
hiemalis *D.L. Jones & M.A. Clements in Orchadian*, 12(7): 330 (1998), nom. nov.: T. ixioides var. subdifformis *Nicholls*.
imbricata *D.L. Jones & M.A. Clements in Austral. Orchid Res.*, 3: 186 (1998), nom. nov.: T. nuda var. grandiflora *Lindl.*
improcera *D.L. Jones & M.A. Clements in Austral. Orchid Res.*, 3: 187 (1998)—Tasmania.
longiloba *D.L. Jones & M.A. Clements in Austral. Orchid Res.*, 3: 191 (1998)—Tasmania, Queensland, New South Wales, Victoria.
planicola *J.A. Jeanes in Muelleria*, 14: 94 (2000)—Victoria, New South Wales.
polychroma *D.L. Jones & M.A. Clements in Austral. Orchid Res.*, 3: 193 (1998)—Tasmania.
silena *D.L. Jones in Orchadian*, 12(11): 518 (1999)—Tasmania.
simulata *D.L. Jones & M.A. Clements in Austral. Orchid Res.*, 3: 195 (1998)—Tasmania, New South Wales (Australian Capital Territory, New South Wales).
spadicea *D.L. Jones & M.A. Clements in Austral. Orchid Res.*, 3: 196 (1998)—Tasmania.
sparsa *D.L. Jones & M.A. Clements in Austral. Orchid Res.*, 3: 197 (1998)—Tasmania.

THEMEDA (Gramin.).
triandra Forssk.
var. laxa *(Andersson) H.J. Noltie in Edinburgh J. Bot.*, 56(3): 402 (1999): Anthistiria laxa.

THEMISTOCLESIA (Ericac.).
alata *J.L. Luteyn in Fl. Ecuador*, 54(147): 371 (1996)—Colombia, Ecuador.
orientalis *J.L. Luteyn in Fl. Ecuador*, 54(147): 376 (1996)—Ecuador.

THENARDIA (Apocynac.).
chiapensis *J.K. Williams in Brittonia*, 47(4): 403 (1995)—Mexico Southeast (Chiapas).

THERMOPSIS (Legumin.).
alpina (Pall.) Ledeb.
var. licentiana *(E. Peter) Z.X. Peng & Y.M. Yuan ex L.Z. Shue in Fl. Loess-Plat. Sin.*, 2: 337 (1992): T. licentiana.

THESIUM (Santalac.).
caespitans *(Ledeb.) N.N. Tzvelev in Fl. Vostochnoĭ Evropÿ*, 9: 406 (1996): T. ramosum var. caespitans.
tuvense *I.M. Krasnoborov in Fl. Sibir. (Salicac.-Amaranthac.)*, 5: 86 (1992)—Tuva.

THEVETIA (Apocynac.).
alliodora *(Roem. & Schult.) L. Allorge in Succulentes*, 1998(1): 27 (1998): Cerbera alliodora.

THEVETIA:-
pinifolia *(Standl. & Steyerm.) J.K. Williams in Sida,* 17(1): 187 (1996): T. peruviana var. pinifolia.

THIBAUDIA (Ericac.).
gunnarii *J.L. Luteyn in Fl. Ecuador,* 54(147): 311 (1996)—Ecuador.
inflata *J.L. Luteyn in Fl. Ecuador,* 54(147): 299 (1996)—Ecuador.
litensis *J.L. Luteyn in Fl. Ecuador,* 54(147): 292 (1996)—Ecuador.
lugoi *J.L. Luteyn in Fl. Ecuador,* 54(147): 303 (1996)—Ecuador, Peru.
retusifolia *J.L. Luteyn in Fl. Ecuador,* 54(147): 313 (1996)—Ecuador.
sessiliflora *(A.C. Sm.) J.L. Luteyn in Fl. Ecuador,* 54(147): 295 (1996): Calopteryx sessiliflora.

THINOPYRUM (Gramin.).
intermedium *(Host) M.E. Barkworth & D.R. Dewey in Amer. J. Bot.,* 72(5): 772 (1985): Triticum intermedium.
subsp. barbulatum *(Schur) M.E. Barkworth & D.R. Dewey in Amer. J. Bot.,* 72(5): 772 (1985): Agropyron barbulatum.
ponticum *(Podp.) M.E. Barkworth & D.R. Dewey in Amer. J. Bot.,* 72(5): 772 (1985): Triticum ponticum.
pungens *(Pers.) M.E. Barkworth in Phytologia,* 83(4): 304 (1997 publ. 1998): Triticum pungens.
pycnanthum *(Godr.) M.E. Barkworth in Phytologia,* 83(4): 304 (1997 publ. 1998): Triticum pycnanthum.

THISMIA (Burmanniac.).
goodii *R. Kiew in Gard. Bull. Singapore,* 51(2): 179 (1999)—Borneo (Sabah).
lauriana *J.K. Jarvie in Blumea,* 41(1): 259 (1996)—Borneo (Kalimantan).

THLADIANTHA (Cucurbitac.).
palmatipartita *A.M. Lu & C. Jeffrey in Novon,* 10(4): 398 (2000)—China South-Central (Yunnan).

THLASPI (Crucif.).
apterocarpum *(Rech.f. & Aellen) V. Mozaffarian in Iranian J. Bot.,* 7(1): 139 (1996): Aethionema apterocarpum.
maassoumii *V. Mozaffarian in Iranian J. Bot.,* 7(1): 135 (1996)—Iran.
montanum L.
var. californicum *(S. Watson) P.K. Holmgren in Mem. New York Bot. Gard.,* 21(2): 80 (1971): T. californicum.
var. fendleri *(A. Gray) P.K. Holmgren in Mem. New York Bot. Gard.,* 21(2): 87 (1971): T. fendleri.
var. idahoense *(Payson) P.K. Holmgren in Mem. New York Bot. Gard.,* 21(2): 81 (1971): T. idahoense.
var. siskiyouense *P.K. Holmgren in Mem. New York Bot. Gard.,* 21(2): 84 (1971)—Oregon.
pulvinata *V. Mozaffarian in Iranian J. Bot.,* 7(1): 138 (1996)—Iran.
trinervium *(DC.) V. Mozaffarian in Iranian J. Bot.,* 7(1): 139 (1996): Hutchinsia trinervia.

× **THOMPSONARA** *hort. in Orchid Rev.,* 106(1222): 254 (1998). *ORCHIDACEAE* (Catasetum × Cymbidium × Grammatophyllum).

THOMPSONELLA (Crassulac.).
mixtecana *J. Reyes & L. López in Cact. Suc. Mex.,* 43(2): 28 (1998)—Mexico Southwest (Oaxaca).
xochipalensis *M. Gual Diaz, S. Peralta Gomez & E. Pérez-Calix in Acta Bot. Mex.,* 40: 38 (1997)—Mexico Southwest (Guerrero).

THONANDIA *H.P. Linder in Telopea,* 6(4): 612 (1996), nom. illegit., see Art. 52.1 ICBN (1994). *GRAMINEAE*
gracilis *(Hook. f.) H.P. Linder in Telopea,* 6(4): 612 (1996): Danthonia gracilis.
longifolia *(R. Br.) H.P. Linder in Telopea,* 6(4): 613 (1996): Danthonia longifolia.
nigricans *(Petrie) H.P. Linder in Telopea,* 6(4): 612 (1996): Danthonia semiannularis var. nigricans.

THONANDIA:-
semiannularis *(Labill.) H.P. Linder in Telopea,* 6(4): 613 (1996): Arundo semiannularis.
unarede *(Raoul) H.P. Linder in Telopea,* 6(4): 613 (1996): Danthonia unarede.

THOTTEA (Aristolochiac.).
abrahamii *M. Dan, P.J. Mathew, C.M. Unnithan & P. Pushpangadan in Kew Bull.,* 51(1): 179 (1996)—India (Kerala).
piperiformis *(Griff.) D.J. Mabberley in Blumea,* 44(2): 350 (1999): Asiphonia piperiformis.

THRIXSPERMUM (Orchidac.).
torajaense *P. O'Byrne in Malayan Orchid Rev.,* 32: 48 (1998)—Sulawesi.
walkeri *G. Seidenfaden & P. Ormerod in G. Seidenfaden, Descr. Epidendrorum J.G. König 1791:* 26 (1995)—Sri Lanka.
williamsianum *(Kores) P. Ormerod in Austral. Orchid Rev.,* 61(6): 37 (1996): Sarcochilus williamsianus.

THRYPTOMENE (Myrtac.).
remota *A.R. Bean in Austrobaileya,* 4(4): 647 (1997)—Northern Territory.

THUNBERGIA (Acanthac.).
colpifera *B. Hansen in Nordic J. Bot.,* 15(6): 583 (1995 publ. 1996)—Thailand.
laevis Wall. ex Nees
var. parviflora *(Trimen) L.H. Cramer in M.D. Dassanayake & W.D. Clayton, eds., Revised Handb. Fl. Ceylon,* 12: 13 (1998): T. fragrans var. parviflora.
var. vestita *(Nees) L.H. Cramer in M.D. Dassanayake & W.D. Clayton, eds., Revised Handb. Fl. Ceylon,* 12: 12 (1998): T. fragrans var. vestita.

THURANTHOS (Hyacinthac.).
aurantiacus *(H. Lindb.) F. Speta in Phyton (Austria),* 38(1): 81 (1998): Urginea aurantiaca.
congestus *(Wight) F. Speta in Phyton (Austria),* 38(1): 81 (1998): Urginea congesta.
coromandelianus *(Roxb.) F. Speta in Phyton (Austria),* 38(1): 84 (1998): Scilla coromandeliana.
govindappae *(G. Boraiah & Fathima) F. Speta in Phyton (Austria),* 38(1): 84 (1998): Urginea govindappae.
indicus *(Roxb.) F. Speta in Phyton (Austria),* 38(1): 84 (1998): Scilla indica.
macrocarpus *(Stedje) F. Speta in Phyton (Austria),* 38(1): 84 (1998): Drimia macrocarpa.
nagarjunae *(Hemadri & Swahari) F. Speta in Phyton (Austria),* 38(1): 84 (1998): Urginea nagarjunae.
noctiflorus *(Batt. & Trab.) F. Speta in Phyton (Austria),* 38(1): 84 (1998): Urginea noctiflora.
polyanthus *(Blatt. & McCann) F. Speta in Phyton (Austria),* 38(1): 84 (1998): Urginea polyantha.
polyphyllus *(Hook. f.) F. Speta in Phyton (Austria),* 38(1): 84 (1998): Urginea polyphylla.
razii *(Ansari) F. Speta in Phyton (Austria),* 38(1): 84 (1998): Drimia razii.
revolutus *(A.V. Duthie) F. Speta in Phyton (Austria),* 38(1): 84 (1998): Urginea revoluta.
senegalensis *(Kunth) F. Speta in Phyton (Austria),* 38(1): 85 (1998): Urginea senegalensis.
wightianus *(Hook. f.) F. Speta in Phyton (Austria),* 38(1): 85 (1998): Urginea wightiana.

THYMUS (Labiat.).
× aitanae Mateo, M.B. Crespo & Laguna
nothosubsp. dominguezii *(Socorro & Arrébola) R. Morales in An. Jard. Bot. Madrid,* 53(2): 200 (1995 publ. 1996): T. × dominguezii.
× almijarensis *J. Ruiz de la Torre & J. Ruiz del Castillo in Ecología (Madrid),* 6: 103 (1992)—Spain. (T. emmanuelis × longiflorus).
× aragonensis *G. Mateo Sanz, M.B. Crespo & N.E. Mercadal Ferreruela in Flora Montiberica,* 16: 23 (2000)—Spain. (T. loscosii × zygis subsp. zygis).
× beltraniae *O. Socorro, M.C. Espinar & M.L. Arrebola in Lagascalia,* 17(1): 186 (1993), as 'beltrani'—Spain. (T. membranaceus × orospedanus).
× benitorum *G. Mateo Sanz, N.E. Mercadal Ferreruela & J.M. Pisco García in Bot. Complutensis,* 20: 70 (1995), as 'benitoi'—Spain. (T. godayanus × pulegioides).

THYMUS:-
× brachychaetus (Willk.) Cout.
nothosubsp. sierranus *J. Peñas, J. Cabello, J.F. Mota & P. Sánchez-Gómez in Lagascalia,* 20(2): 291 (1998)—Spain. (T. mastichina subsp. mastichina × zygis subsp. gracilis).
× braunii Borbás
nothosubsp. celticus *(Heinr. Braun) P.A. Schmidt in Feddes Repert.,* 108(1–2): 109 (1997): T. × celticus.
nothosubsp. monteilii *(Sennen ex Machule) P.A. Schmidt in Feddes Repert.,* 108(1–2): 109 (1997): T. monteilii.
× celtibericus Pau
nothosubsp. bonichensis *(Mateo & M.B. Crespo) R. Morales in An. Jard. Bot. Madrid,* 53(2): 202 (1995 publ. 1996): T. × bonichensis.
× cremnophilus *J. Čáp in Čas. Nár. Muz. Řada Přír.,* 168(1–4): 116 (1999)—Greece. (T. jankae × sibthorpii).
× diazii *F. Alcaraz, S. Rivas-Martínez & P. Sánchez-Gómez in Itinera Geobot.,* 2: 118 (1989)—Spain. (T. hyemalis × moroderi).
× dominguezii *O. Socorro & M.L. Arrebola in Lagascalia,* 17(2): 355 (1994)—Spain. (T. serpylloides subsp. gadorensis × vulgaris subsp. aestivus).
dzalindensis *N.S. Probatova in Sosud. Rast. Sovet. Dal'nego Vostoka,* 7: 360 (1995)—Amur.
ebusitanus *(Font Quer) A.M. Romo, Fl. Silvestres Baleares:* 266 (1994): T. richardii var. ebusitanus.
emmanuelis *J. Ruiz de la Torre & J. Ruiz del Castillo in Ecología (Madrid),* 6: 101 (1992)—Spain.
funkii Coss.
subsp. burilloi *P. Sánchez-Gómez in P. Sánchez-Gómez et al., Fl. Murcia: Claves ident. pl. vasc.:* 347 (1996)—Spain.
subsp. sabulicola *(Coss.) P. Sánchez-Gómez in P. Sánchez-Gómez et al., Fl. Murcia: Claves ident. pl. vasc.:* 347 (1996): T. sabulicola.
granatensis Boiss.
subvar. hirsutus *O. de Bolòs & J. Vigo, Fl. Països Catalans,* 3: 321 (1995)—Spain.
hartvigii *R. Morales in Lagascalia,* 19(1–2): 251 (1997), nom. nov.: T. rechingeri *Hartvig.*
subsp. macrocalyx *(Hartvig) R. Morales in Lagascalia,* 19(1–2): 251 (1997): T. rechingeri subsp. macrocalyx.
herba-barona Loisel.
subsp. bivalens *M. Mayol, L. Sáez & J.A. Rosselló in Flora Montiberica,* 8: 61 (1998)—Baleares.
× herberoi *A. De la Torre, M. Vicedo, M.A. Alonso & J. Payá in Acta Bot. Malacitana,* 22: 224 (1997)—Spain. (T. moroderi × zygis subsp. gracilis).
jankae Čelak.
var. ilinicae *V. Matevski in Prilozi Oddel. Biol. Med. Nauki Makedonska Akad. Nauk. Umet.,* 9(1–2): 59 (1988 publ. 1991)—Yugoslavia.
karavaevii *V.M. Doronkin in Fl. Sibir. (Pyrolac.-Lamiac.),* 11: 213 (1997)—Yakutskiya.
kuminovianus *(Serg.) G.A. Peschkova, Stepnaya Fl. Baĭkal'skoĭ Sibiri:* 90 (1972): T. serpyllum var. kuminovianus.
× lainzii *P. Sánchez Gómez, S. Fernández Jiménez & F. Sáez in An. Jard. Bot. Madrid,* 54(1): 302 (1996)—Spain.(T. funkii × vulgaris subsp. vulgaris).
levitskyi *N.S. Probatova in Sosud. Rast. Sovet. Dal'nego Vostoka,* 7: 365 (1995)—Amur.
longicaulis C. Presl
var. rupestris *V. Matevski in Prilozi Oddel. Biol. Med. Nauki Makedonska Akad. Nauk. Umet.,* 9(1–2): 53 (1988 publ. 1991)—Yugoslavia.
× mercadalii *G. Mateo & J.M. Pisco in Flora Medit.,* 6: 86 (1996)—Spain. (T. izcoi × pulegioides).
moesiacus Velen.
f. albiflora *V. Matevski in Prilozi Oddel. Biol. Med. Nauki Makedonska Akad. Nauk. Umet.,* 9(1–2): 55 (1988 publ. 1991)—Yugoslavia.
f. orbelicus *(Ronniger) V. Matevski in Prilozi Oddel. Biol. Med. Nauki Makedonska Akad. Nauk. Umet.,* 9(1–2): 56 (1988 publ. 1991): T. stojanovii var. orbelicus.
mohamedii *B. Tahiri & M. Rejdali in Lagascalia,* 18(2): 289 (1996)—Morocco.
var. ebracteata *B. Tahiri & M. Rejdali in Lagascalia,* 18(2): 292 (1996)—Morocco.

THYMUS:-
mohamedii
var. rosatii B. Tahiri & M. Rejdali in Lagascalia, 18(2): 292 (1996)—Morocco.
subsp. rosatii B. Tahiri & M. Rejdali in Lagascalia, 18(2): 292 (1996)—Morocco.
mongolicus Ronniger
var. leucanthus H.L. Liu & D.Z. Ma in Fl. Ningxiaensis, 2: 521 (1988 publ. 1990)—Inner Mongolia (Ningxia).
× monrealensis Pau ex R. Morales
nothosubsp. conquensis G. Mateo Sanz & V.J. Arán Redó in Flora Montiberica, 4: 35 (1996)—Spain. (T. vulgaris subsp. vulgaris × zygis subsp. sylvestris).
× moralesii Mateo & M.B. Crespo
nothosubsp. navarroi (Mateo & M.B. Crespo) R. Morales in An. Jard. Bot. Madrid, 53(2): 208 (1995 publ. 1996): T. × navarroi.
nakhodkensis P.G. Gorovoĭ & R.V. Dudkin in Bot. Zhurn., 83(6): 109 (1998)—Primorye.
novograblenovii N.S. Probatova in Sosud. Rast. Sovet. Dal'nego Vostoka, 7: 363 (1995)—Kamchatka.
numidicus Poir.
var. kabylicus (Batt.) Maire ex A. Dubuis in Soc. Ech. Pl. Vasc. Eur. Bassin Médit. Bull., 25: 65 (1995): T. lanceolatus subsp. kabylicus.
× pastoris O. Socorro & M.L. Arrebola in Lagascalia, 17(2): 353 (1994)—Spain. (T. serpylloides subsp. gadorensis × zygis subsp. gracilis).
× porcii Borbás
nothosubsp. pilisiensis (Borbás) P.A. Schmidt in Feddes Repert., 108(1–2): 109 (1997): T. × pilisiensis.
praecox Opiz
var. arcticus (Durand) T. Karlsson in Svensk Bot. Tidskr., 91(5): 249 (1997 publ. 1998): T. serpyllum var. arcticus.
pulcherrimus Schur
subsp. carpathicus (Čelak.) P. Mártonfi in Thaiszia, 7(2): 156 (1997): T. carpathicus.
quinquecostatus Čelak.
f. albus Y.N. Lee, Fl. Korea: 1160 (1996), without type—Locality not stated.
sachalinensis N.S. Probatova in Sosud. Rast. Sovet. Dal'nego Vostoka, 7: 363 (1995)—Sakhalin.
schlothauerae N.S. Probatova in Sosud. Rast. Sovet. Dal'nego Vostoka, 7: 365 (1995), as 'schlothaueriae'—Khabarovsk.
× segurae G. Mateo & M.B. Crespo in An. Jard. Bot. Madrid, 55(1): 167 (1997)—Spain. (T. bracteatus × pulegioides).
serpyllum L.
var. paui (R. Morales) O. de Bolòs & J. Vigo, Fl. Països Catalans, 3: 323 (1995): T. leptophyllus subsp. paui.
talijevii Klokov & Des.-Shost.
subsp. rotundatus (Klokov) M.S. Knyazev in P.L. Gorchakovskiĭ et al., Opred. Sosud. Rast. Sredn. Urala: 377 (1994), without basionym ref.: T. rotundatus.
targuistii B. Tahiri & M. Rejdali in Lagascalia, 18(2): 287 (1996)—Morocco.
ternejicus N.S. Probatova in Sosud. Rast. Sovet. Dal'nego Vostoka, 7: 359 (1995)—Primorye.
tidighinii B. Tahiri & M. Rejdali in Lagascalia, 18(2): 293 (1996)—Morocco.
urussovii N.S. Probatova in Sosud. Rast. Sovet. Dal'nego Vostoka, 7: 365 (1995)—Primorye.
verchojanicus V.M. Doronkin in Fl. Sibir. (Pyrolac.-Lamiac.), 11: 220 (1997)—Yakutskiya.
× viciosoi Pau ex R. Morales in An. Jard. Bot. Madrid, 53(2): 210 (1995 publ. 1996)—Spain. (T. pulegioides × zygis subsp. zygis).

TIBETIA (Legumin.).
coelestis (Diels) H.P. Tsui in Fl. Reipubl. Popul. Sin., 42(2): 159 (1998): Astragalus coelestis.

TIBOUCHINA (Melastomatac.).
almedae C.A. Todzia in Brittonia, 51(3): 255 (1999)—Mexico Northwest (Sinaloa)
congestiflora C.A. Todzia in Brittonia, 51(3): 259 (1999)—Mexico Southwest (Guerrero).
connata Gleason ex C.A. Todzia in Brittonia, 51(3): 262 (1999)—Mexico Southwest (Guerrero).
hintonii Gleason ex C.A. Todzia in Brittonia, 51(3): 264 (1999)—Mexico Southwest (Guerrero, Jalisco, Oaxaca).

TIBOUCHINA:-
johnwurdackiana C.A. Todzia in Biollania, ed. esp. 6 (J.J. Wurdack Festschr.): 537 (1997)—Brazil West-central (Goiás, Brasília Distrito Federal).
macvaughii C.A. Todzia in Brittonia, 51(3): 267 (1999)—Mexico Southwest (Jalisco, Nayarit).
nanifolia C.A. Todzia in Brittonia, 51(3): 269 (1999)—Mexico Central (México State).
patens C.A. Todzia in Brittonia, 51(3): 270 (1999)—Mexico Southwest (Guerrero).
roseotincta C.A. Todzia in Brittonia, 51(3): 273 (1999)—Mexico Southwest (Guerrero).
serrana P.J.F. Guimarães & A.B. Martins in Rev. Brasil. Bot., 20(1): 15 (1997)—Brazil Southeast (São Paulo, Minas Gerais).
sinaloensis C.A. Todzia in Brittonia, 51(3): 275 (1999)—Mexico Northwest (Sinaloa).
thulia C.A. Todzia in Brittonia, 51(3): 276 (1999)—Mexico Northwest (Sinaloa).

TICOREA (Rutac.).
diandra J.A. Kallunki in Brittonia, 50(4): 504 (1998)—Ecuador, Peru.
froesii J.A. Kallunki in Brittonia, 50(4): 508 (1998)—Brazil Northeast (Maranhão), Brazil North (Pará).

TIGRIDIA (Iridac.).
vanhouttei (Baker) A. Espejo Serna & A.R. López-Ferrari in Acta Bot. Mex., 41: 2 (1997): Hydrotaenia vanhouttei.

TILESIA (Compos.).
baccata (L.) J.F. Pruski in Novon, 6(4): 414 (1996): Coreopsis baccata.
var. discoidea (S.F. Blake) J.F. Pruski in Novon, 6(4): 414 (1996): Wulffia baccata var. discoidea.
macrocephala (H. Rob.) J.F. Pruski in Novon, 6(4): 414 (1996): Wulffia macrocephala.
rubens (Alexander) J.F. Pruski in Novon, 6(4): 414 (1996): Wulffia rubens.

TILIA (Tiliac.).
corinthiaca Bosc ex K. Koch, Dendrologie, 1: 474 (1869)—Greece.†
dasystyla Steven
subsp. caucasica (V. Engl.) C.D. Pigott in Edinburgh J. Bot., 56(2): 171 (1999): T. rubra subsp. caucasica.
oblongifolia Rehder
var. sangzhiensis B.R. Liao & W.X. Wang in Acta Phytotax. Sin., 35(2): 192 (1997)—China Southeast (Hunan).
viridis (Bayer) L. Simonkai, Revis. Tiliarum Hungar. (Math. Term. Közlem., 22(7)): 320 (1888): Tilia alba var. viridis.
nothosubsp. moltkei (Dippel) C.C. Xifreda in Darwiniana, 35(1–4): 148 (1998): T. americana moltkei.

TILLAEA (Crassulac.).
likiangensis H. Chuang in Acta Bot. Yunnanica, 19(3): 225 (1997)—China South-central (Yunnan).
schimperi (C.A. Mey.) M.G. Gilbert, H. Ohba & K.T. Fu in Novon, 10(4): 366 (2000): Crassula schimperi.
subsp. phyturus (Mildbr.) M.G. Gilbert, H. Ohba & K.T. Fu in Novon, 10(4): 367 (2000): Crassula phyturus.
subsp. transvalensis (Kuntze) M.G. Gilbert, H. Ohba & K.T. Fu in Novon, 10(4): 367 (2000): Sedum transvalense.

TILLANDSIA (Bromeliac.).
atroviolacea R. Ehlers & P. Koide in J. Bromeliad Soc., 47(1): 21 (1997)—Mexico Southwest (Oaxaca).
bakiorum H.E. Luther in Selbyana, 20(1): 13 (1999)—Colombia.
× bergiana H. Takizawa & P. Koide in J. Bromeliad Soc., 50(6): 244 (2000)—Mexico Southeast (Chiapas). (T. beutelspacheri × flabellata).
borinquensis J.A. Cedeño-Maldonado & G.R. Proctor in Harvard Pap. Bot., 4(1): 113 (1999)—Puerto Rico.
calochlamys R. Ehlers & L. Hromadnik in Bromelie, 1998(2): 38 (1998)—Bolivia.
cathcartii (H. Luther) J.R. Grant in Phytologia, 79(3): 256 (1995 publ. 1996): Vriesea cathcartii.

TILLANDSIA:-
colganii R. Ehlers in J. Bromeliad Soc., 47(3): 103 (1997)—Bolivia.
crocata (E. Morren) N.E. Brown, Suppl. Johnson's Gard. Dict.: 1015 (1882): Phytarrhiza crocata.
delicata R. Ehlers in Bromelie, 1999(3): 64 (1999)—Mexico Southwest (Oaxaca).
erici R. Ehlers in Bot. Jahrb., 1998(1): 20 (1998)—Bolivia.
hasei R. Ehlers & L. Hromadnik in Bromelie, 1996(3): 76 (1996)—Bolivia.
heubergeri R. Ehlers in Bromelie, Sonderheft 3: 35 (1996)—Brazil Northeast (Bahia).
imporaensis R. Ehlers in Bromelie, 1998(3): 76 (1998)—Bolivia.
ionantha Planch.
f. fastigiata P. Koide in J. Bromeliad Soc., 43(4): 162 (1993)—Mexico Southwest (Oaxaca).
var. maxima R. Ehlers in Bromelie, 2000(1): 8 (2000)—Mexico Southwest (Oaxaca).
var. stricta P. Koide in J. Bromeliad Soc., 43(4): 162 (1993)—Mexico Southwest (Oaxaca).
jaguactalensis I.M. Ramírez M., G. Carnevali F.C. & F. Chi May in J. Bromeliad Soc., 50(2): 63 (2000)—Mexico Southeast (Quintana Roo).
kessleri H.E. Luther in Selbyana, 20(1): 10 (1999)—Bolivia.
kuzmae R. Ehlers in J. Bromeliad Soc., 50(3): 113 (2000)—Dominican Republic.
laxissima Mez
var. moorei H.E. Luther in Selbyana, 20(1): 13 (1999)—Peru.
lydiae R. Ehlers in Bromelie, 2000(2): 57 (2000)—Mexico Southwest (Guerrero).
× marceloi H. Takizawa & P. Koide in J. Bromeliad Soc., 49(5): 213 (1999)—Mexico Southwest (Oaxaca). (T. achyrostachys × bourgaei).
may-patii I.M. Ramírez M. & G. Carnevali F.C. in Harvard Pap. Bot., 4(1): 187 (1999)—Mexico Southeast (Quintana Roo).
mixtecorum R. Ehlers & P. Koide in J. Bromeliad Soc., 48(2): 52 (1998)—Mexico Southwest (Oaxaca).
moronesensis R. Ehlers in J. Bromeliad Soc., 50(3): 117 (2000)—Mexico Northeast (Zacatecas).
myosura Griseb. ex J.G. Baker in J. Bot., 16: 240 (1878)—Argentina Northwest.†
occulta H.E. Luther in Selbyana, 18(1): 99 (1997)—Mexico Northwest (Sinaloa).
organensis R. Ehlers in Bromelie, Sonderheft 3: 48 (1996)—Brazil Southeast (Rio de Janeiro).
paraisoensis R. Ehlers in J. Bromeliad Soc., 42(2): 55 (1992)—Mexico Southwest (Guerrero).
polzii R. Ehlers in Bromelie, 1997(1): 11 (1997)—Brazil South (Rio Grande do Sul).
portillae E. Gross & R. Wülfinghoff in J. Bromeliad Soc., 47(2): 52 (1997)—Ecuador.
pseudobaileyi C.S. Gardner
subsp. yucatanensis I.M. Ramírez, G. Carnevali & I. Olmstead in J. Bromeliad Soc., 50(2): 69 (2000); see also J. Bromeliad Soc., 50(4): 164 (2000) 'Errata'—Mexico Southeast (Campeche, Quintana Roo).
recurvifolia Hook.
var. subsecundifolia (W. Weber & Ehlers) W. Till in J. Bromeliad Soc., 45(6): 268 (1995): T. meridionalis var. subsecundifolia.
roseiflora R. Ehlers & W. Weber in Bromelie, Sonderheft 3: 52 (1996)—Brazil Southeast (Rio de Janeiro).
rubrispica R. Ehlers & P. Koide in J. Bromeliad Soc., 46(1): 7 (1996)—Mexico Southwest (Oaxaca).
salmonea R. Ehlers in Bromelie, 1995(3): 65 (1995)—Mexico Southeast (Chiapas).
samaipatensis W. Till in Bromelie, 1996(2): 36 (1996)—Bolivia.
scaposa (L.B. Sm.) R. Ehlers in Bromelie, 1999(3): 69 (1999), without basionym ref.: T. ionantha var. scaposa.
sphaerocephala Baker
var. tarijensis R. Ehlers & L. Hromadnik in Bromelie, 1998(2): 41 (1998)—Bolivia.
takizawae R. Ehlers & H. Luther in J. Bromeliad Soc., 50(5): 216 (2000)—Mexico Central (Puebla).
taxcoensis R. Ehlers in Bromelie, 1996(1): 26 (1996)—Mexico Central (Morelos).
tenuifolia L.
var. strobiliformis R. Ehlers in Bromelie, 1996(2): 41 (1996)—Brazil Southeast (Espírito Santo).

TILLANDSIA:-
thiekenii *R. Ehlers in J. Bromeliad Soc.*, 43(4): 151 (1993)—Brazil Southeast (Rio de Janeiro).
tillandsioides *(L.B. Sm.) J.R. Grant in Phytologia*, 75(2): 174 (1993): Vriesea tillandsioides.†
trigalensis *R. Ehlers in Bromelie*, 1997(2–3): 46 (1997)—Guatemala.
ulrici *R. Ehlers in Bromelie*, 2000(3): 66 (2000)—Mexico Southwest (Oaxaca).
xiphioides Ker Gawl.
subsp. prolata *H.E. Luther in J. Bromeliad Soc.*, 46(5): 211 (1996)—Bolivia.

TIMONIUS (Rubiac.).
elegans *S.P. Darwin in Syst. Bot.*, 22(1): 93 (1997)—New Guinea (Papua New Guinea).
hentyi *S.P. Darwin in Syst. Bot.*, 22(1): 92 (1997)—New Guinea (Papua New Guinea, Irian Jaya).
kostermansii *S.P. Darwin in Syst. Bot.*, 22(1): 95 (1997)—Moluccas.
morobensis *S.P. Darwin in Syst. Bot.*, 22(1): 94 (1997)—New Guinea (Papua New Guinea).
subcoriaceus Valeton
var. sessilis *S.P. Darwin in Syst. Bot.*, 22(1): 97 (1997)—New Guinea (Papua New Guinea).

TINTINABULUM (Polemoniac.).
campanulatum *(A. Gray) V. Grant in Phytologia*, 84(2): 78 (1998 publ. 1999): Gilia campanulata.
inyoense *(I.M. Johnst.) V. Grant in Phytologia*, 84(2): 78 (1998 publ. 1999), as 'inyoensis': Gilia inyoensis.
tenerrimum *(A. Gray) A.G. Day & V. Grant in Phytologia*, 84(5): 377 (1998 publ. 1999): Gilia tenerrima.

TINTINNABULARIA (Apocynac.).
gratissima *J.F. Morales in Novon*, 6(4): 392 (1996)—Mexico Gulf (Veracruz).
murallensis *J.K. Williams in Lundellia*, 2: 138 (1999)—Honduras.

TIRUCALLIA (Euphorbiac.).
aequoris *(N.E. Br.) P.V. Heath in Calyx*, 5(3): 87 (1996): Euphorbia aequoris.
alata *(Hook.) P.V. Heath in Calyx*, 5(3): 87 (1996): Euphorbia alata.
alluaudii *(Drake) P.V. Heath in Calyx*, 5(3): 87 (1996): Euphorbia alluaudii.
subsp. oncoclada *(Drake) P.V. Heath in Calyx*, 5(3): 87 (1996): Euphorbia oncoclada.
amarifontana *(N.E. Br.) P.V. Heath in Calyx*, 5(3): 87 (1996): Euphorbia amarifontana.
angrae *(N.E. Br.) P.V. Heath in Calyx*, 5(3): 87 (1996): Euphorbia angrae.
antisyphilitica *(Zucc.) P.V. Heath in Calyx*, 5(3): 87 (1996): Euphorbia antisyphilitica.
aphylla *(Willd.) P.V. Heath in Calyx*, 5(3): 87 (1996): Euphorbia aphylla.
arbuscula *(Balf. f.) P.V. Heath in Calyx*, 5(3): 87 (1996): Euphorbia arbuscula.
arceuthobioides *(Boiss.) P.V. Heath in Calyx*, 5(3): 87 (1996): Euphorbia arceuthobioides.
aspericaulis *(Pax) P.V. Heath in Calyx*, 5(3): 88 (1996): Euphorbia aspericaulis.
berotica *(N.E. Br.) P.V. Heath in Calyx*, 5(3): 88 (1996): Euphorbia berotica.
bottae *(Boiss.) P.V. Heath in Calyx*, 5(3): 88 (1996): Euphorbia bottae.
brachiata *(Klotzsch) P.V. Heath in Calyx*, 5(3): 88 (1996): Arthrothamnus brachiatus.
burmannii *(Klotzsch) P.V. Heath in Calyx*, 5(3): 88 (1996): Arthrothamnus burmannii.
calamiformis *(S. Carter) P.V. Heath in Calyx*, 5(3): 88 (1996): Euphorbia calamiformis.
cameronii *(N.E. Br.) P.V. Heath in Calyx*, 5(3): 88 (1996): Euphorbia cameronii.
caruncalifera *(L.C. Leach) P.V. Heath in Calyx*, 5(3): 88 (1996): Euphorbia carunculifera.
var. fasciata *P.V. Heath in Calyx*, 5(3): 88 (1996), nom. nov.: Euphorbia carunculifera subsp. subfastigiata *L.C. Leach*.
cassythoides *(Boiss.) P.V. Heath in Calyx*, 5(3): 88 (1996): Euphorbia cassythoides.
caterviflora *(N.E. Br.) P.V. Heath in Calyx*, 5(3): 88 (1996): Euphorbia caterviflora.
chersina *(N.E. Br.) P.V. Heath in Calyx*, 5(3): 89 (1996): Euphorbia chersina.

TIRUCALLIA:-
cibdela *(N.E. Br.) P.V. Heath in Calyx*, 5(3): 89 (1996): Euphorbia cibdela.
congestifolia *(L.C. Leach) P.V. Heath in Calyx*, 5(3): 89 (1996): Euphorbia congestifolia.
corymbosa *(N.E. Br.) P.V. Heath in Calyx*, 5(3): 89 (1996): Euphorbia corymbosa.
cuneata *(Vahl) P.V. Heath in Calyx*, 5(3): 89 (1996): Euphorbia cuneata.
curocana *(L.C. Leach) P.V. Heath in Calyx*, 5(3): 89 (1996): Euphorbia curocana.
damarana *(L.C. Leach) P.V. Heath in Calyx*, 5(3): 89 (1996): Euphorbia damarana.
densiflora *(Klotzsch) P.V. Heath in Calyx*, 5(3): 89 (1996): Arthrothamnus densiflorus.
dregeana *(Boiss.) P.V. Heath in Calyx*, 5(3): 89 (1996): Euphorbia dregeana.
enterophora *(Drake) P.V. Heath in Calyx*, 5(3): 89 (1996): Euphorbia enterophora.
ephedroides *(Boiss.) P.V. Heath in Calyx*, 5(3): 89 (1996): Euphorbia ephedroides.
var. debilis *(L.C. Leach) P.V. Heath in Calyx*, 5(3): 90 (1996): Euphorbia ephedroides var. debilis.
var. imminuta *(L.C. Leach & G. Will.) P.V. Heath in Calyx*, 5(3): 90 (1996): Euphorbia ephedroides var. imminuta.
fiherenensis *(Poiss.) P.V. Heath in Calyx*, 5(3): 90 (1996): Euphorbia fiherenensis.
gentilis *(N.E. Br.) P.V. Heath in Calyx*, 5(3): 90 (1996): Euphorbia gentilis.
giessii *(L.C. Leach) P.V. Heath in Calyx*, 5(3): 90 (1996): Euphorbia giessii.
goetzei *(Pax) P.V. Heath in Calyx*, 5(3): 90 (1996): Euphorbia goetzei.
gossypina *(Pax) P.V. Heath in Calyx*, 5(3): 90 (1996): Euphorbia gossypina.
var. coccinea *(Pax) P.V. Heath in Calyx*, 5(3): 90 (1996): Euphorbia gossypina var. coccinea.
gregaria *(Marloth) P.V. Heath in Calyx*, 5(3): 90 (1996): Euphorbia gregaria.
gummifera *(Boiss.) P.V. Heath in Calyx*, 5(3): 90 (1996): Euphorbia gummifera.
herrei *(A.C. White, R.A. Dyer & B. Sloane) P.V. Heath in Calyx*, 5(3): 90 (1996): Euphorbia herrei.
indecora *(N.E. Br.) P.V. Heath in Calyx*, 5(3): 90 (1996): Euphorbia indecora.
indurescens *(L.C. Leach) P.V. Heath in Calyx*, 5(3): 90 (1996): Euphorbia indurescens.
intisy *(Drake) P.V. Heath in Calyx*, 5(3): 90 (1996): Euphorbia intisy.
juttae *(Dinter) P.V. Heath in Calyx*, 5(3): 90 (1996), as 'judithae': Euphorbia juttae.
karooensis *(Boiss.) P.V. Heath in Calyx*, 5(3): 91 (1996): Euphorbia burmannii β karooensis.
lactiflua *(Phil.) P.V. Heath in Calyx*, 5(3): 91 (1996): Euphorbia lactiflua.
larica *(Boiss.) P.V. Heath in Calyx*, 5(3): 91 (1996): Euphorbia larica.
lateriflora *(Schumach.) P.V. Heath in Calyx*, 5(3): 91 (1996): Euphorbia lateriflora.
lignosa *(Marloth) P.V. Heath in Calyx*, 5(3): 91 (1996): Euphorbia lignosa.
macella *(N.E. Br.) P.V. Heath in Calyx*, 5(3): 91 (1996): Euphorbia macella.
mainty *(Leandri) P.V. Heath in Calyx*, 5(3): 91 (1996): Euphorbia mainty.
mauritanica *(L.) P.V. Heath in Calyx*, 5(3): 91 (1996): Euphorbia mauritanica.
mixta *(N.E. Br.) P.V. Heath in Calyx*, 5(3): 91 (1996): E. mixta.
muricata *(Thunb.) P.V. Heath in Calyx*, 5(3): 91 (1996): Euphorbia muricata.
negromontana *(N.E. Br.) P.V. Heath in Calyx*, 5(3): 91 (1996): Euphorbia negromontana.
nubica *(N.E. Br.) P.V. Heath in Calyx*, 5(3): 92 (1996): Euphorbia nubica.
paxiana *(Dinter) P.V. Heath in Calyx*, 5(3): 92 (1996): Euphorbia paxiana.
perpera *(N.E. Br.) P.V. Heath in Calyx*, 5(3): 92 (1996): Euphorbia perpera.
plagiantha *(Drake) P.V. Heath in Calyx*, 5(3): 92 (1996): Euphorbia plagiantha.
platyclada *(Rauh) P.V. Heath in Calyx*, 5(3): 92 (1996): Euphorbia platyclada.
var. hardyi *(Rauh) P.V. Heath in Calyx*, 5(3): 92 (1996): Euphorbia platyclada var. hardyi.

TIRUCALLIA:-
rectirama *(N.E. Br.) P.V. Heath in Calyx*, 5(3): 92 (1996): Euphorbia rectirama.
rhombifolia *(Boiss.) P.V. Heath in Calyx*, 5(3): 92 (1996): Euphorbia rhombifolia.
var. cymosa *(Klotzsch) P.V. Heath in Calyx*, 5(3): 92 (1996): Arthrothamnus cymosus.
rudolphi *(N.E. Br.) P.V. Heath in Calyx*, 5(3): 92 (1996): Euphorbia rudolphi.
sarcostemmatoides *(Dinter) P.V. Heath in Calyx*, 5(3): 92 (1996): Euphorbia sarcostemmatoides.
schimperi *(Presl) P.V. Heath in Calyx*, 5(3): 92 (1996): Euphorbia schimperi.
spartaria *(N.E. Br.) P.V. Heath in Calyx*, 5(3): 93 (1996): Euphorbia spartaria.
spicata *(Boiss.) P.V. Heath in Calyx*, 5(3): 93 (1996): Euphorbia spicata.
spinea *(N.E. Br.) P.V. Heath in Calyx*, 5(3): 93 (1996): Euphorbia spinea.
stapelioides *(Boiss.) P.V. Heath in Calyx*, 5(3): 93 (1996): Euphorbia stapelioides.
stenoclada *(Baill.) P.V. Heath in Calyx*, 5(3): 93 (1996): Euphorbia stenoclada.
stolonifera *(A.C. White, R.A. Dyer & B. Sloane) P.V. Heath in Calyx*, 5(3): 93 (1996): Euphorbia stolonifera.
tenax *(Burch.) P.V. Heath in Calyx*, 5(3): 93 (1996): Euphorbia tenax.
tirucalli *(L.) P.V. Heath in Calyx*, 5(3): 93 (1996): Euphorbia tirucalli.
transvaalensis *(Schltr.) P.V. Heath in Calyx*, 5(3): 93 (1996): Euphorbia transvaalensis.
verruculosa *(N.E. Br.) P.V. Heath in Calyx*, 5(3): 93 (1996): Euphorbia verruculosa.
virgata *(Haw.) P.V. Heath in Calyx*, 5(3): 93 (1996): Tithymalus virgatus.
var. berthelotii *(Boiss.) P.V. Heath in Calyx*, 5(3): 94 (1996): Euphorbia berthelotii.
var. desfoliata *(Menezes) P.V. Heath in Calyx*, 5(3): 94 (1996): Euphorbia obtusifolia var. desfoliata.
var. latibracteata *(G. Kunkel) P.V. Heath in Calyx*, 5(3): 94 (1996): Euphorbia obtusifolia f. latibracteata.
var. pseudodendroides *(H. Lindb.) P.V. Heath in Calyx*, 5(3): 94 (1996): Euphorbia pseudodendroides.
var. regis-jubae *(Webb & Berthel.) P.V. Heath in Calyx*, 5(3): 94 (1996): Euphorbia regis-jubae.
xylophylloides *(Lem.) P.V. Heath in Calyx*, 5(3): 94 (1996): Euphorbia xylophylloides.

TITHYMALUS (Euphorbiac.).
angustifrons *(Borbás) J. Chrtek & B. Křísa in Biológia, A (Czechoslovakia)*, 26(10): 768 (1971): Euphorbia angustifrons.
budensis *(T. Simon) J. Chrtek & B. Křísa in Biológia, A (Czechoslovakia)*, 26(10): 769 (1971): Euphorbia budensis.

TOCANTINIA *P. Ravenna in Onira, Bot. Leafl.*, 5(3): 9 (2000). *AMARYLLIDACEAE*
mira *P. Ravenna in Onira, Bot. Leafl.*, 5(3): 10 (2000)—Brazil North (Tocantins).

TOCOCA (Melastomatac.).
caryophyllaea *(DC.) S.S. Renner in Biollania*, ed. esp. 6 (J.J. Wurdack Festschr.): 499 (1997): Truncaria caryophyllaea.

TOFIELDIA (Melanthiac.).
coccinea Richardson
f. fusca *(Miyabe & Kudô) Q.S. Sun in Clavis Pl. Chinae Bor.-Or. ed. 2 (ed. P.Y. Fu)*: 764 (1995): T. fusca.

TOLUMNIA (Orchidac.).
sect. Gudrunia *(Braem) M.A. Nir, Orchidaceae Antillanae*: 370 (2000): gen Gudrunia.
sect. Olgasis *(Raf.) M.A. Nir, Orchidaceae Antillanae*: 370 (2000), with incorrect basionym ref.: gen. Olgasis.
subsect. Variegata *(Braem) M.A. Nir, Orchidaceae Antillanae*: 370 (2000): Tolumnia subgen. Variegata.
acunae *(M.A. Díaz) M.A. Nir, Orchidaceae Antillanae*: 372 (2000): Tolumnia tuerckheimii var. acunae.

TOLUMNIA:-
arizajuliana *(Withner & J. Jiménez Alm.) J.D. Ackerman in Lindleyana*, 12(3): 151 (1997): Oncidium arizajulianum.
borinquinensis *R.P. Sauleda & M.E. Ragan in Orchid Dig.*, 60(4): 185 (1996)—Puerto Rico, Leeward Is. (Virgin Is.).
gauntlettii *(Withner & H.P. Jesup) K. Senghas in Schlechter Orchideen, ed. 3*, I/C(33–36): 2114 (1997), isonym; *see M.A. Nir in Lindleyana*, 9(3): 149 (1994): Oncidium gauntlettii.
triquetra *(Sw.) K. Senghas in Schlechter Orchideen, ed. 3*, I/C(33–36): 2114 (1997), isonym; *see M.A. Nir in Lindleyana*, 9(3): 149 (1994): Epidendrum triquetrum.

TOMZANONIA *M.A. Nir in Lindleyana*, 12(4): 186 (1997). *ORCHIDACEAE*
filicina *(Dod) M.A. Nir in Lindleyana*, 12(4): 186 (1997): Dilomilis filicina.

TONGOLOA (Umbellif.).
fortunatii *(H. Boissieu) M.G. Pimenov & E.V. Kljuykov in Feddes Repert.*, 110(7–8): 489 (1999), as 'furtunatii': Pimpinella fortunatii.

TOONA (Meliac.).
ciliata M. Roem.
var. pilipetala *(C.DC) N.C. Nair & G.R. Kumari in N.C. Nair & A.N. Henry, Fl. Tamil Nadu, India, Ser. 1*, 1: 69 (1983): Cedrela toona var. pilipetala.
var. pilistila *(C.DC.) N.C. Nair & G.R. Kumari in N.C. Nair & A.N. Henry, Fl. Tamil Nadu, India, Ser. 1*, 1: 69 (1983): Cedrela toona var. pilistila.
hexandra (Wall. ex Roxb.) M. Roem.
var. candollei *(Bahadur) S.M. Almeida & M.R. Almeida in J. Bombay Nat. Hist. Soc.*, 91(3): 474 (1994 publ. 1995), without basionym ref.: Basionym not stated.
var. gamblei *(C.DC.) S.M. Almeida & M.R. Almeida in J. Bombay Nat. Hist. Soc.*, 91(3): 473 (1994 publ. 1995): Cedrela toona var. gamblei.
var. grandifolia *(C.DC.) S.M. Almeida & M.R. Almeida in J. Bombay Nat. Hist. Soc.*, 91(3): 473 (1994 publ. 1995): Cedrela microcarpa var. grandifolia.
var. griffithiana *(Pierre) S.M. Almeida & M.R. Almeida in J. Bombay Nat. Hist. Soc.*, 91(3): 473 (1994 publ. 1995): T. febrifuga var. griffithiana.
var. haslettii *(Haines) S.M. Almeida & M.R. Almeida in J. Bombay Nat. Hist. Soc.*, 91(3): 473 (1994 publ. 1995): Cedrela toona var. haslettii.
var. kingii *(C.DC.) S.M. Almeida & M.R. Almeida in J. Bombay Nat. Hist. Soc.*, 91(3): 474 (1994 publ. 1995): Cedrela kingii.
var. listeri *(C.DC.) S.M. Almeida & M.R. Almeida in J. Bombay Nat. Hist. Soc.*, 91(3): 473 (1994 publ. 1995): Cedrela toona var. listeri.
var. mollis *(Hand.-Mazz.) S.M. Almeida & M.R. Almeida in J. Bombay Nat. Hist. Soc.*, 91(3): 474 (1994 publ. 1995): Cedrela mollis.
var. parviflora *(Benth.) S.M. Almeida & M.R. Almeida in J. Bombay Nat. Hist. Soc.*, 91(3): 473 (1994 publ. 1995): Cedrela toona var. parviflora.
var. pilistila *(C.DC.) S.M. Almeida & M.R. Almeida in J. Bombay Nat. Hist Soc.*, 91(3): 473 (1994 publ. 1995): Cedrela toona var. pilistila.
var. pubinervis *(C.DC.) S.M. Almeida & M.R. Almeida in J. Bombay Nat. Hist. Soc.*, 91(3): 474 (1994 publ. 1995): Cedrela toona var. pubinervis.
subsp. velutina *(C.DC.) S.M. Almeida & M.R. Almeida in J. Bombay Nat. Hist. Soc.*, 91(3): 474 (1994 publ. 1995): Cedrela velutina.
rubriflora *C.J. Tseng in Acta Sci. Nat. Univ. Amoiensis*, 9(4): 303 (1962), with two types cited—China Southeast (Fujian).
sinensis (A. Juss.) M. Roem.
var. hupehana *(C.DC.) P.Y. Chen in Fl. Reipubl. Popul. Sin.*, 43(3): 39 (1997): Cedrela sinensis var. hupehana.

TOONA:-
sinensis
var. longifolia *(Wall. ex C.DC.) S.M. Almeida & M.R. Almeida in J. Bombay Nat. Hist. Soc.*, 91(3): 474 (1994 publ. 1995), as 'sinengii': Cedrela longifolia.
sureni Blume
var. glabrior *(C.DC.) S.M. Almeida & M.R. Almeida in J. Bombay Nat. Hist. Soc.*, 91(3): 474 (1994 publ. 1995): Cedrela febrifuga var. glabrior.
var. velutina *(Koord. & Valeton) S.M. Almeida & M.R. Almeida in J. Bombay Nat. Hist. Soc.*, 91(3): 474 (1994 publ. 1995), without basionym ref.: Cedrela febrifuga var. velutina.

TOPOBEA (Melastomatac.).
arboricola *F. Almeda in Proc. Calif. Acad. Sci.*, 52(9): 98 (2000)—Panama.
muricata *G. Lozano C. in Rev. Acad. Colomb. Cienc. Exact. Fis. Nat.*, 88: 342 (1999)—Colombia.

TORDYLIUM (Umbellif.).
ketenoglui *H. Duman & A. Duran in Edinburgh J. Bot.*, 56(1): 51 (1999)—Turkey.

TORENIA (Scrophulariac.).
concolor Lindl.
f. chingshuiana *S.S. Ying in Mem. Coll. Agric. Nation. Taiwan Univ.*, 28(2): 41 (1988)—Taiwan.

TORILIS (Umbellif.).
subsect. Sessiles *S.L. Jury in Soc. Éch. Pl. Vasc. Eur. Bassin Médit. Bull.*, 27: 73 (1998).
arvensis (Huds.) Link
subsp. recta *S.L. Jury in Lagascalia*, 18(2): 282 (1996)—Morocco.
ucranica Spreng.
var. aculeata *(Boiss.) V. Ciocârlan in Acta Horti Bot. Bucurest.*, 27: 158 (1998 publ. 1999): T. microcarpa var. aculeata.

× **TORMIMALUS** *J. Holub in Preslia*, 70(2): 115 (1998). *ROSACEAE* (Malus × Torminalis).
florentina *(Zuccagni) J. Holub in Preslia*, 70(2): 115 (1998): Crataegus florentina.

TORNABENEA (Umbellif.).
humilis *W. Lobin & K.H. Schmidt in Sommerfeltia*, 24: 83 (1997)—Cape Verde.
ribeirensis *K.H.A. Schmidt & W. Lobin in Feddes Repert.*, 110(1–2): 8 (1999)—Cape Verde.

TORREYCACTUS *A.B. Doweld in Sukkulentÿ (Moscow)*, 1998(1): 19 (1998). *CACTACEAE*
conothelos *(Regel & Klein) A.B. Doweld in Sukkulentÿ (Moscow)*, 1998(1): 19 (1998), as 'conothele': Echinocactus conothelos.
var. argenteus *(Glass & R.A. Foster) A.B. Doweld in Sukkulentÿ (Moscow)*, 1998(1): 19 (1998), as 'conothele': Thelocactus conothelos var. argenteus.
subsp. argenteus *(Glass & R.A. Foster) A.B. Doweld in Sukkulentÿ (Moscow)*, 3(1–2): 37 (2000): Thelocactus conothelos var. argenteus.
subsp. aurantiacus *(Glass & R.A. Foster) A.B. Doweld in Sukkulentÿ (Moscow)*, 3(1–2): 36 (2000): Thelocactus conothelos var. aurantiacus.
var. aurantiacus *(Glass & R.A. Foster) A.B. Doweld in Sukkulentÿ (Moscow)*, 1998(1): 19 (1998), as 'conothele': Thelocactus conothelos var. aurantiacus.
subsp. garciae *(Glass & Mend.-Garc.) A.B. Doweld in Sukkulentÿ (Moscow)*, 3(1–2): 36 (2000): Thelocactus garciae.

TORULARIA (Crucif.).
conferta *R.F. Huang in S.G. Wu (S.K. Wu) & Z.J. Feng (eds.), Biol. & Human Physiol. Hoh Xil Region*: 51 (1996), generic name illegitimate—Qinghai.

TOURNEFORTIA (Boraginac.).
delicatula *J.S. Miller in Novon*, 7(3): 265 (1997)—Peru.
gigantifolia *Killip ex J.S. Miller in Novon*, 10(1): 49 (2000)—Ecuador, Colombia.

TOURNEFORTIA:-
isabellina *J.S. Miller in Novon*, 9(2): 232 (1999)—Costa Rica.
luzonica I.M. Johnst.
subsp. angustissima *H. Riedl in Blumea*, 41(2): 443 (1996)—Philippines.
minutiflora *H. Riedl in Blumea*, 41(2): 443 (1996)—New Guinea (Papua New Guinea).
oppositifolia *H. Riedl in Blumea*, 41(2): 444 (1996)—Moluccas.
vasquezii *J.S. Miller in Novon*, 10(1): 45 (2000)—Peru.

TOVOMITA (Guttif.).
auriculata *N.L. Cuello A. in Novon*, 9(2): 150 (1999)—Venezuela, Peru.
foldatsii *N.L. Cuello A. in Novon*, 9(2): 151 (1999)—Venezuela.

TRACHYANDRA (Asphodelac.).
triquetra *M. Thulin in Nordic J. Bot.*, 15(6): 571 (1995 publ. 1996)—Somalia.

TRACHYCARPUS (Palmae).
latisectus *T.W. Spanner, H.J. Noltie & M. Gibbons in Edinburgh J. Bot.*, 54(2): 257 (1997)—India (West Bengal).
oreophilus *M. Gibbons & T.W. Spanner in Principes*, 41(4): 205 (1997)—Thailand.

TRACHYMENE (Umbellif.).
sect. Didiscus *(Hook.) B.L. Rye in Nuytsia*, 13(1): 199 (1999): gen. Didiscus.
sect. Dimetopia *(DC.) G.J. Keighery & B.L. Rye in Nuytsia*, 13(1): 41 (1999): gen. Dimetopia.
sect. Hemicarpus *(F. Muell.) B.L. Rye in Nuytsia*, 13(1): 224 (1999): gen. Hemicarpus.
anisocarpa (Turcz.) B.L. Burtt
var. trichocarpa *B.L. Rye in Nuytsia*, 13(1): 206 (1999)—Western Australia.
ceratocarpa *(W. Fitzg.) G.J. Keighery & B.L. Rye in Nuytsia*, 13(1): 42 (1999): Hydrocotyle ceratocarpa.
coerulea Graham
subsp. leucopetala *(F. Muell. ex Benth.) B.L. Rye in Nuytsia*, 13(1): 211 (1999): T. coerulea var. leucopetala.
grandis *(Turcz.) B.L. Rye in Nuytsia*, 13(1): 216 (1999): Dimetopia grandis.
oleracea (Domin) B.L. Burtt
subsp. sedimenta *B.L. Rye in Nuytsia*, 13(1): 219 (1999)—Western Australia.
pilbarensis *B.L. Rye in Nuytsia*, 13(1): 221 (1999)—Western Australia.
pyrophila *B.L. Rye in Nuytsia*, 13(1): 223 (1999)—Western Australia.

TRACHYNIA (Gramin.).
platystachya *(Balansa ex Coss.) H. Scholz in Willdenowia*, 28(1–2): 173 (1998): Brachypodium distachyon var. platystachyon.

TRACHYPOGON (Gramin.).
mayaensis *J.K. Wipff & S.D. Jones in Sida*, 18(1): 242 (1998)—Belize, Mexico Southwest (Oaxaca).
renvoizei *L. Catasús Guerra in Fontqueria*, 44: 144 (1996)—Cuba.

TRACHYSPERMUM (Umbellif.).
villosum *(Haines) P.K. Bhattacharyya & K. Sarkar, Fl. W. Champaran Distr. Bihar*: 253 (1998): Carum villosum.

TRACTEMA (Hyacinthac.).
liliohyacinthus *(L.) F. Speta in Phyton (Austria)*, 38(1): 124 (1998): Scilla liliohyacinthus.
merinoi *(S. Ortiz, Rodr. Oubiña & Izco) F. Speta in Phyton (Austria)*, 38(1): 124 (1998): Scilla merinoi.
monophyllos *(Link) F. Speta in Phyton (Austria)*, 38(1): 124 (1998): Scilla monophyllos.
odorata *(Link) F. Speta in Phyton (Austria)*, 38(1): 124 (1998): Scilla odorata.
ramburei *(Boiss.) F. Speta in Phyton (Austria)*, 38(1): 124 (1998): Scilla ramburei.
tingitana *(Schousb.) F. Speta in Phyton (Austria)*, 38(1): 124 (1998): Scilla tingitana.
umbellata *(Ramond) F. Speta in Phyton (Austria)*, 38(1): 124 (1998): Scilla umbellata.

TRACTEMA:-
verna *(Huds.) F. Speta in Phyton (Austria),* 38(1): 124 (1998): Scilla verna.

TRADESCANTIA (Commelinac.).
petricola *J.R. Grant in Novon,* 10(2): 121 (2000)—Costa Rica, Venezuela.

TRAGACANTHA (Legumin.).
balearica *(Chater) A.M. Romo, Fl. Silvestres Baleares:* 140 (1994): Astragalus balearicus.

TRAGIA (Euphorbiac.).
finalis *P.I. Forster in Austral. Syst. Bot.,* 10(6): 863 (1997)—Queensland.
jonesii *A. Radcliffe-Smith & R. Govaerts in Kew Bull.,* 52(2): 480 (1997), nom. nov.: T. scandens *M.E. Jones.*
lassia *A. Radcliffe-Smith & R. Govaerts in Kew Bull.,* 52(2): 480 (1997), nom. nov.: Lassia scandens *Baill.*
leptorrhachis *A. Radcliffe-Smith & R. Govaerts in Kew Bull.,* 52(2): 480 (1997), nom. nov.: Leptorrhachis hastata *Klotzsch.*

TRAGOPOGON (Compos.).
floccosus Waldst. & Kit.
subsp. borysthenicus *(Artemczuk) G. Dihoru in Acta Horti Bot. Bucurest.,* 1996–97: 93 (1998): T. borysthenicus.
heteropappus *C.H. An in Fl. Xinjiangensis,* 5: 479 (1999)—Xinjiang.
orientalis L.
var. latifolius *C.H. An in Fl. Xinjiangensis,* 5: 479 (1999)—Xinjiang.
pratensis L.
subsp. lamottei *(Rouy) O. de Bolòs & J. Vigo, Fl. Països Catalans,* 3: 986 (1995): T. lamottei.
pseudocastellanus *G. Blanca & C. Díaz de la Guardia in An. Jard. Bot. Madrid,* 54(1): 360 (1996)—Spain.
ruber S.G. Gmel.
var. leucocarpus *C.H. An in Fl. Xinjiangensis,* 5: 479 (1999)—Xinjiang.
verrucosobracteatus *C.H. An in Fl. Xinjiangensis,* 5: 479 (1999)—Xinjiang.

TRAPA (Trapac.).
bispinosa *W. Roxburgh, Pl. Coromandel,* 3: 29 (1815)—Bangladesh.†
grozdovii *V.N. Vassil. ex V.N. Tikhomirov in Byull. Mosk. Obshch. Ispȳt. Prir., Biol.,* 100(1): 89 (1995)—Russia Central.
natans L.
var. hamata *N.N. Tzvelev, Opred. Sosud. Rast. Severo-Zap. Rossii:* 680 (2000)—Northwest European Russia.

TRATTINNICKIA (Burserac.).
boliviana *(Swart) D.C. Daly in Kew Bull.,* 54(1): 130 (1999): T. lawrancei var. boliviana.
lancifolia *(Cuatrec.) D.C. Daly in Kew Bull.,* 54(1): 130 (1999): T. rhoifolia var. lancifolia.
mensalis *D.C. Daly in Kew Bull.,* 54(1): 132 (1999)—Brazil Southeast (Espírito Santo), Brazil Northeast (Bahia).

TREMBLEYA (Melastomatac.).
hatschbachii *Wurdack & E. Martins in Bol. Bot. (São Paulo),* 14: 40 (1995)—Brazil Southeast (Minas Gerais).

TREMULINA *B.G. Briggs & L.A.S. Johnson in Telopea,* 7(4): 361 (1998). *RESTIONACEAE*
tremula *(R. Br.) B.G. Briggs & L.A.S. Johnson in Telopea,* 7(4): 361 (1998): Restio tremulus.

TREPADONIA (Compos.).
oppositifolia *H. Robinson & H. Beltrán in Sida,* 19(1): 111 (2000)—Peru.

TRESANTHERA (Rubiac.).
condamineoides H. Karst.
var. thyrsiflora *(Steyerm.) P.G. Delprete in Fl. Neotrop.,* 77: 101 (1999): T. thyrsiflora.

TREVESIA (Araliac.).
lateospina *M.H.P. Jebb in Glasra (n.s.),* 3(2): 97 (1998)—Thailand.

TREWIA (Euphorbiac.).
nudiflora L.
var. dentata *S.R.M. Susila Rani & N.P. Balakrishnan in J. Econ. Taxon. Bot.,* 22(2): 352 (1998)—India (Maharashtra).
var. polycarpa *(Benth.) S.R.M. Susila Rani & N.P. Balakrishnan in J. Econ. Taxon. Bot.,* 22(2): 351 (1998): T. polycarpa.
var. tomentosa *S.R.M. Susila Rani & N.P. Balakrishnan in J. Econ. Taxon. Bot.,* 22(2): 351 (1998)—India (Andhra Pradesh, Gujarat, Maharashtra, West Bengal), Sikkim, Burma.

TRIARRHENA *(Maxim.) T. Nakai in J. Jap. Bot.,* 25(1–2): 7 (1950), as 'Tiarrhena': Imperata subgen. Triarrhena. (Gramin.).†
hackelii *(Nakai) T. Nakai in J. Jap. Bot.,* 25(1–2): 7 (1950), as 'Tiarrhena': Miscanthus hackelii.†
lutarioriparia L. Liou
f. altissima *L. Liou in Fl. Reipubl. Popul. Sin.,* 10(2): 23 (1997), nom. nov.: T. lutarioriparia f. yizhangqing *L. Liou.*
f. coccinea *L. Liou in Fl. Reipubl. Popul. Sin.,* 10(2): 23 (1997), nom. nov.: T. lutarioriparia f. yanzhihong *L. Liou.*
f. pendulifolia *L. Liou in Fl. Reipubl. Popul. Sin.,* 10(2): 23 (1997), nom. nov.: T. lutarioriparia f. chuiyeqing *L. Liou.*
f. purpureorosa *L. Liou in Fl. Reipubl. Popul. Sin.,* 10(2): 24 (1997), nom. nov.: T. lutarioriparia f. tiegangang *L. Liou.*
sacchariflora *(Maxim.) T. Nakai in J. Jap. Bot.,* 25(1–2): 7 (1950), as 'Tiarrhena': Imperata sacchariflora.†

× **TRIASPETALUM** *hort. in Orchid Rev.,* 104(1210): 220 (1996). *ORCHIDACEAE* (Cirrhopetalum × Trias).

× **TRIASPHYLLUM** *hort. in Orchid Rev.,* 104(1210): 220 (1996). *ORCHIDACEAE* (Bulbophyllum × Trias).

TRIBOLIUM (Gramin.).
sect. Acutiflorae *N.C. Visser & Spies ex H.P. Linder & G. Davidse in Bot. Jahrb.,* 119(4): 477 (1997).
sect. Uniolae *N.C. Visser & Spies ex H.P. Linder & G. Davidse in Bot. Jahrb.,* 119(4): 471 (1997).
pusillum *(Nees) H.P. Linder & G. Davidse in Bot. Jahrb.,* 119(4): 495 (1997): Urochlaena pusilla.

TRIBULOPIS (Zygophyllac.).
homalococca *(Domin) R.M. Barker in J. Adelaide Bot. Gard.,* 18(2): 91 (1998): Tribulus homalococcus.
var. alifer *(Domin) R.M. Barker in J. Adelaide Bot. Gard.,* 18(2): 92 (1998): Tribulus homalococcus var. alifer.

TRIBULUS (Zygophyllac.).
adelacanthus *R.M. Barker in Nuytsia,* 12(1): 17 (1998)—Western Australia.
macropterus Boiss.
var. arabicus *(Hosni) F. Al-Hemaid & Jacob Thomas in Arab Gulf J. Sci. Res.,* 14(2): 436 (1996), without basionym date: T. arabicus.
var. mollis *(Ehrenb. ex Schweinf.) F. Al-Hemaid & Jacob Thomas in Arab Gulf J. Sci. Res.,* 14(2): 435 (1996): T. mollis.
pentandrus
var. acanthopterus *(Kralik) H.A. Hosni in Candollea,* 51(1): 196 (1996): T. alatus var. acanthopterus.
var. bimucronatus *(Viv.) F. Al-Hemaid & Jacob Thomas in Arab Gulf J. Sci. Res.,* 14(2): 434 (1996): T. bimucronatus.
suberosus *H. Eichler ex R.M. Barker in J. Adelaide Bot. Gard.,* 17: 171 (1996), without type; *et in Nuytsia,* 12(1): 15 (1998), with type—Western Australia.
terrestris L.
var. parvispinus *(C. Presl) F. Al-Hemaid & Jacob Thomas in Arab Gulf J. Sci. Res.,* 14(2): 437 (1996): T. parvispinus.
var. rajasthanensis *(Bhandari & V.S. Sharma) F. Al-Hemaid & Jacob Thomas in Arab Gulf J. Sci. Res.,* 14(2): 437 (1996): T. rajasthanensis.

TRICARPELEMA (Commelinac.).
pumilum *(Hallier f.) R.B. Faden in Checklist Flow. Pl. & Gymn. Brunei Darussalam (ed. M.J.E. Coode et al.):* 435 (1996): Pollia pumila.

TRICERATOSTRIS *(Szlach.) D.L. Szlachetko & R. González Tamayo in Fragm. Flor. Geobot.,* 41(2): 1021 (1996): Oestlundorchis subgen. Triceratostris. (Orchidac.).
hio *(Szlach.) D.L. Szlachetko & R. González Tamayo in Fragm. Flor. Geobot.,* 41(2): 1023 (1996): Oestlundorchis hio.
rhombilabia *(Garay) D.L. Szlachetko & R. González Tamayo in Fragm. Flor. Geobot.,* 41(2): 1022 (1996): Deiregyne rhombilabia.
subsp. tamayoi *(Szlach.) D.L. Szlachetko & R. González Tamayo in Fragm. Flor. Geobot.,* 41(2): 1022 (1996): Oestlundorchis rhombilabia subsp. tamayoi.

TRICHANTHA (Gesneriac.).
fimbricalyx *(L.P. Kvist & L.E. Skog) H. Wiehler in Gesneriana,* 1(1): 122 (1995): Columnea fimbricalyx.
laevis *(L.P. Kvist & L.E. Skog) H. Wiehler in Gesneriana,* 1(1): 122 (1995): Columnea laevis.
minutiflora *(L.P. Kvist & L.E. Skog) H. Wiehler in Gesneriana,* 1(1): 122 (1995): Columnea minutiflora.
rubricalyx *(L.P. Kvist & L.E. Skog) H. Wiehler in Gesneriana,* 1(1): 122 (1995): Columnea rubricalyx.
tenella *H. Wiehler in Gesneriana,* 1(1): 95 (1995)—Ecuador, Colombia.

TRICHERA (Dipsacac.).
caroli-rechingeri *(Micevski) J. Holub in Preslia,* 70(2): 115 (1998): Knautia caroli-rechingeri.
catalaunica *(Szabó) J. Holub in Preslia,* 70(2): 115 (1998): Knautia catalaunica.
× hungarica *(Borbás) J. Holub in Preslia,* 70(2): 115 (1998): Knautia × hungarica.
lebrunii *(J. Prudhomme) J. Holub in Preslia,* 70(2): 115 (1998): Knautia lebrunii.
× leucantha *(Schur) J. Holub in Preslia,* 70(2): 115 (1998): Knautia × leucantha.
maxima *(Opiz) J. Holub in Preslia,* 70(2): 115 (1998): Scabiosa maxima.
× posoniensis *(Degen) J. Holub in Preslia,* 70(2): 116 (1998): Knautia × posoniensis.
× sambucifolia *(Godet) J. Holub in Preslia,* 70(2): 116 (1998): Knautia sylvatica var. sambucifolia.
slovaca *(Štěpánek) J. Holub in Preslia,* 70(2): 116 (1998): Knautia slovaca.
× speciosa *(Schur) J. Holub in Preslia,* 70(2): 116 (1998): Knautia × speciosa.
transsilvanica *(Schur) J. Holub in Preslia,* 70(2): 116 (1998): Knautia transsilvanica.

TRICHILIA (Meliac.).
carinata *M.E. Morales-Puentes in Rev. Acad. Colomb. Cienc. Exact. Fis. Nat.,* 84: 339 (1998)—Colombia.
cauliflora *J.-F. Leroy & M. Lescot in Bull. Mus. Nation. Hist. Nat., B, Adansonia,* Ser. 4, 18(1–2): 6 (1996)—Madagascar.
oligofoliolata *M.E. Morales-Puentes in Rev. Acad. Colomb. Cienc. Exact. Fis. Nat.,* 84: 343 (1998)—Colombia.
tsaratananensis *M. Lescot in Bull. Mus. Nation. Hist. Nat., B, Adansonia,* Ser. 4, 18(1–2): 4 (1996)—Madagascar.

TRICHLORA (Alliac.).
huascarana *P. Ravenna in Onira, Bot. Leafl.,* 4(9): 32 (2000)—Peru.
lactea *P. Ravenna in Onira, Bot. Leafl.,* 4(9): 31 (2000)—Peru.

TRICHLORIS (Gramin.).
mendocina *(Phil.) F. Kurtz in Mem. Fac. Ci. Exact. Fis. Nat. Univ. Cordoba,* 1896: 37 (1897), as 'mendozina': Chloris mendocina.†

TRICHOCENTRUM (Orchidac.).
sect. Saccatae *K. Senghas in Schlechter Orchideen,* I/B(31): 1939 (1995).
neudeckeri *W. Königer in Arcula,* 6: 175 (1996)—Bolivia.

TRICHOCENTRUM:-
obcordilabium *F. Pupulin in Novon,* 8(3): 283 (1998)—Ecuador.
popowianum *W. Königer in Arcula,* 6: 178 (1996)—Ecuador.
teaguei *E.A. Christenson in Orchid Dig.,* 61(1): 32 (1997)—Bolivia.
viridulum *F. Pupulin in Novon,* 8(3): 285 (1998)—Colombia.

TRICHOCEPHALUS *A.T. Brongniart, Mém. Fam. Rhamnées:* 67 (1826). *RHAMNACEAE*†
spicatus *(L.f.) A.T. Brongniart, Mém. Fam. Rhamnées:* 68 (1826): Phylica spicata.†
stipularis *(L.) A.T. Brongniart, Mém. Fam. Rhamnées:* 68 (1826): Phylica stipularis.†

TRICHOCEREUS (Cactac.).
walteri *(R. Kiesling) J.G. Lambert, Cactus d'Argentine, ed. 2:* 321 (1998): Lobivia walteri.

TRICHOGLOTTIS (Orchidac.).
jiewhoei *J.J. Wood, A. Lamb & C.L. Chan in Sandakania,* 11: 44 (1998)—Borneo (Sabah).
lasioglossa *(Schltr.) P. Ormerod in Austral. Orchid Rev.,* 62(3): 15 (1997): Pomatocalpa lasioglossa.
tinekeae *A. Schuiteman in Blumea,* 43(2): 492 (1998)—Borneo (Sabah, Sarawak).

TRICHOPHORUM (Cyperac.).
cespitosum (L.) Hartm.
nothosubsp. foersteri *G.A. Swan in Watsonia,* 22(3): 230 (1999)—Great Britain. (T. cespitosum subsp. cespitosum × subsp. germanicum).
rigidum *(Boeck.) P. Goetghebeur, A.M. Muasya & D.A. Simpson in Novon,* 10(2): 132 (2000): Scirpus rigidus.
subcapitatum *(Thwaites & Hook.) D.A. Simpson in Kew Bull.,* 53(1): 227 (1998): Scirpus subcapitatus.

TRICHOPILIA (Orchidac.).
aenigma *L.A. Garay in Lindleyana,* 11(4): 233 (1996)—Venezuela, Colombia.
undulatissima *D.E. Bennett & E.A. Christenson in Lindleyana,* 13(2): 95 (1998)—Peru.

TRICHOSALPINX (Orchidac.).
subgen. Pseudolepanthes *C.A. Luer, Ic. Pleurothallid. XV (Monogr. Syst. Bot. Missouri Bot. Gard., 64):* 5 (1997): T. sect Pseudolepanthes.
subgen. Xenia *C.A. Luer, Ic. Pleurothallid. XV (Monogr. Syst. Bot. Missouri Bot. Gard., 64):* 6 (1997).
amygdalodora *(Kraenzl.) C.A. Luer, Ic. Pleurothallid. XV (Monogr. Syst. Bot. Missouri Bot. Gard., 64):* 43 (1997): Pleurothallis amygdalodora.
atropurpurea *C.A. Luer & Hirtz in C.A. Luer, Ic. Pleurothallid. XV (Monogr. Syst. Bot. Missouri Bot. Gard.,* 64): 46 (1997)—Ecuador.
barbelifera *C.A. Luer & R. Vásquez in C.A. Luer, Ic. Pleurothallid. XV (Monogr. Syst. Bot. Missouri Bot. Gard.,* 64): 46 (1997)—Bolivia.
bricenoensis *C.A. Luer & R. Escobar in C.A. Luer, Ic. Pleurothallid. XV (Monogr. Syst. Bot. Missouri Bot. Gard.,* 64): 105 (1997)—Colombia.
calceolaris *C.A. Luer & R. Escobar in C.A. Luer, Ic. Pleurothallid. XV (Monogr. Syst. Bot. Missouri Bot. Gard.,* 64): 105 (1997)—Colombia.
caudata *C.A. Luer & R. Escobar in C.A. Luer, Ic. Pleurothallid. XV (Monogr. Syst. Bot. Missouri Bot. Gard.,* 64): 20 (1997)—Colombia, Panama.
dalstroemii *C.A. Luer, Ic. Pleurothallid. XV (Monogr. Syst. Bot. Missouri Bot. Gard., 64):* 52 (1997)—Ecuador.
deceptrix *G. Carnevali & I.M. Ramírez M. in Harvard Pap. Bot.,* 3(2): 250 (1998)—Venezuela.
decorata *C.A. Luer & R. Escobar in C.A. Luer, Ic. Pleurothallid. XV (Monogr. Syst. Bot. Missouri Bot. Gard.,* 64): 106 (1997)—Colombia.
dressleri *C.A. Luer, Ic. Pleurothallid. XV (Monogr. Syst. Bot. Missouri Bot. Gard., 64):* 54 (1997)—Panama.
dunstervillei *C.A. Luer, Ic. Pleurothallid. XV (Monogr. Syst. Bot. Missouri Bot. Gard., 64):* 55 (1997)—Venezuela.
fasciculata *C.A. Luer & Hirtz in C.A. Luer, Ic. Pleurothallid. XV (Monogr. Syst. Bot. Missouri Bot. Gard.,* 64): 58 (1997)—Ecuador.

TRICHOSALPINX:-
fissa *C.A. Luer, Ic. Pleurothallid. XV (Monogr. Syst. Bot. Missouri Bot. Gard., 64):* 59 (1997)—Ecuador, Venezuela.
fruticosa *C.A. Luer in Lindleyana,* 11(2): 105 (1996); *et in Ic. Pleurothallid. XV (Monogr. Syst. Bot. Missouri Bot. Gard., 64):* 59 (1997)—Costa Rica, Guatemala, El Salvador.
gentryi *C.A. Luer, Ic. Pleurothallid. XV (Monogr. Syst. Bot. Missouri Bot. Gard., 64):* 60 (1997)—Peru.
hamiltonii *C.A. Luer, Ic. Pleurothallid. XV (Monogr. Syst. Bot. Missouri Bot. Gard., 64):* 61 (1997)—Peru.
hirtzii *C.A. Luer, Ic. Pleurothallid. XV (Monogr. Syst. Bot. Missouri Bot. Gard., 64):* 61 (1997)—Ecuador.
jimburae *C.A. Luer & Hirtz in C.A. Luer, Ic. Pleurothallid. XV (Monogr. Syst. Bot. Missouri Bot. Gard.,* 64): 64 (1997)—Ecuador.
lamellata *C.A. Luer, Ic. Pleurothallid. XV (Monogr. Syst. Bot. Missouri Bot. Gard., 64):* 64 (1997)—Ecuador.
lankesteriana *C.A. Luer in Lindleyana,* 11(2): 108 (1996)—Costa Rica, Panama.
lenticularis *(Luer) C.A. Luer, Ic. Pleurothallid. XV (Monogr. Syst. Bot. Missouri Bot. Gard., 64):* 65 (1997): Pleurothallis lenticularis.
ligulata *C.A. Luer & Hirtz in C.A. Luer, Ic. Pleurothallid. XV (Monogr. Syst. Bot. Missouri Bot. Gard.,* 64): 65 (1997)—Ecuador.
lilliputalis *(Luer & Hirtz) C.A. Luer, Ic. Pleurothallid. XV (Monogr. Syst. Bot. Missouri Bot. Gard., 64):* 115 (1997): Lepanthopsis lilliputalis.
macphersonii *C.A. Luer, Ic. Pleurothallid. XV (Monogr. Syst. Bot. Missouri Bot. Gard., 64):* 66 (1997)—Panama.
metamorpha *C.A. Luer & Hirtz in C.A. Luer, Ic. Pleurothallid. XV (Monogr. Syst. Bot. Missouri Bot. Gard.,* 64): 67 (1997)—Ecuador.
parsonsii *C.A. Luer & Dod in C.A. Luer, Ic. Pleurothallid. XV (Monogr. Syst. Bot. Missouri Bot. Gard.,* 64): 72 (1997)—Costa Rica.
patula *C.A. Luer, Ic. Pleurothallid. XV (Monogr. Syst. Bot. Missouri Bot. Gard., 64):* 31 (1997), without type; *et Ic. Pleurothallid. XVI (Monogr. Syst. Bot. Missouri Bot. Gard., 65):* 82 (1998), with type—Ecuador, Bolivia, Venezuela.
psilantha *C.A. Luer, Ic. Pleurothallid. XV (Monogr. Syst. Bot. Missouri Bot. Gard., 64):* 32 (1997)—Bolivia.
purpurea *H. Seehawer in Orchidee,* 49(3): 128 (1998)—Brazil Southeast (Rio de Janeiro).
ramosii *C.A. Luer, Ic. Pleurothallid. XV (Monogr. Syst. Bot. Missouri Bot. Gard., 64):* 108 (1997)—Colombia.
ringens *C.A. Luer in Lindleyana,* 11(2): 108 (1996)—Panama, Costa Rica.
rotundata *(C. Schweinf.) R.L. Dressler in Novon,* 7(2): 124 (1997): Pleurothallis rotundata.
silverstonei *C.A. Luer, Ic. Pleurothallid. XV (Monogr. Syst. Bot. Missouri Bot. Gard., 64):* 108 (1997)—Colombia.
solomonii *C.A. Luer, Ic. Pleurothallid. XV (Monogr. Syst. Bot. Missouri Bot. Gard., 64):* 79 (1997)—Bolivia.
spathulata *C.A. Luer, Ic. Pleurothallid. XV (Monogr. Syst. Bot. Missouri Bot. Gard., 64):* 109 (1997)—Colombia.
steyermarkii *C.A. Luer, Ic. Pleurothallid. XV (Monogr. Syst. Bot. Missouri Bot. Gard., 64):* 79 (1997)—Venezuela.
strumifera *C.A. Luer, Ic. Pleurothallid. XV (Monogr. Syst. Bot. Missouri Bot. Gard., 64):* 80 (1997)—Ecuador.
tantilla *(Luer) C.A. Luer, Ic. Pleurothallid. XV (Monogr. Syst. Bot. Missouri Bot. Gard., 64):* 81 (1997): Pleurothallis tantilla.
teaguei *C.A. Luer, Ic. Pleurothallid. XV (Monogr. Syst. Bot. Missouri Bot. Gard., 64):* 82 (1997)—Bolivia.
teres *C.A. Luer, Ic. Pleurothallid. XV (Monogr. Syst. Bot. Missouri Bot. Gard., 64):* 83 (1997)—Ecuador.
todziae *C.A. Luer in Lindleyana,* 11(2): 111 (1996)—Costa Rica.
uvaria *C.A. Luer, Ic. Pleurothallid. XV (Monogr. Syst. Bot. Missouri Bot. Gard., 64):* 109 (1997)—Colombia.

TRICHOSALPINX:-
vasquezii *C.A. Luer, Ic. Pleurothallid. XV (Monogr. Syst. Bot. Missouri Bot. Gard., 64):* 35 (1997)—Bolivia.
wilhelmii *C.A. Luer, Ic. Pleurothallid. XV (Monogr. Syst. Bot. Missouri Bot. Gard., 64):* 86 (1997)—Ecuador.
zunagensis *C.A. Luer & Hirtz in C.A. Luer, Ic. Pleurothallid. XV (Monogr. Syst. Bot. Missouri Bot. Gard.,* 64): 110 (1997)—Ecuador.

TRICHOSANTHES (Cucurbitac.).
sect. Edulis *Rugayah in Reinwardtia,* 11(4): 232 (1999).
ser. Ovifolia *C.H. Yueh & Lu Q. Huang in Proc. Cross-strait Symp. Floristic Divers. & Conserv. (eds. S.T. Chiu &* C.I. Peng): 76 (1998).
subsect. Tetragonosperma *(C.Y. Cheng & C.H. Yueh) Rugayah in Reinwardtia,* 11(4): 229 (1999): T. sect. Tetragonosperma.
subsect. Villosae *C.H. Yueh & L.Q. Huang in Acta Phytotax. Sin.,* 35(2): 127 (1997); *et in Proc. Cross-strait Symp. Floristic Divers. & Conserv. (eds. S.T. Chiu &* C.I. Peng): 76 (1998):
auriculata *Rugayah in Reinwardtia,* 11(3): 216 (1998)—Borneo (Sarawak, Kalimantan, Sabah).
beccariana Cogn.
subsp. pusilla *Rugayah in Reinwardtia,* 11(3): 217 (1998)—Borneo (Kalimantan, Sabah, Sarawak).
densiflora *Rugayah in Reinwardtia,* 11(4): 252 (1999)—New Guinea (Irian Jaya).
dentifera *Rugayah in Reinwardtia,* 11(4): 253 (1999)—New Guinea (Papua New Guinea).
edulis *Rugayah in Reinwardtia,* 11(4): 254 (1999)—New Guinea (Papua New Guinea).
var. sativa *Rugayah in Reinwardtia,* 11(4): 256 (1999)—New Guinea (Papua New Guinea).
var. septemloba *Rugayah in Reinwardtia,* 11(4): 256 (1999)—New Guinea (Papua New Guinea).
emarginata *Rugayah in Reinwardtia,* 11(4): 258 (1999)—Sumatera.
floresana *Rugayah in Reinwardtia,* 11(4): 260 (1999)—Lesser Sunda Is.
homophylla Hayata
var. ishigakiensis *(E. Walker) T. Yamazaki in J. Jap. Bot.,* 73(1): 51 (1998): T. ishigakiensis.
kinabaluensis *Rugayah in Reinwardtia,* 11(4): 262 (1999), with two types cited; *et in Reinwardtia,* 11(5): 419 (2000)—Borneo (Sabah).
laeoica Cheng & Huang
f. sicyocarpa *Rugayah in Reinwardtia,* 11(4): 264 (1999)—New Guinea (Papua New Guinea).
f. sorongensis *Rugayah in Reinwardtia,* 11(4): 265 (1999)—New Guinea (Irian Jaya).
f. yapenensis *Rugayah in Reinwardtia,* 11(4): 265 (1999)—New Guinea (Irian Jaya).
leuserensis *Rugayah in Reinwardtia,* 11(4): 265 (1999)—Sumatera.
longispicata *Rugayah in Reinwardtia,* 11(4): 266 (1999)—Borneo (Sarawak).
montana *Rugayah in Reinwardtia,* 11(3): 218 (1998)—Jawa, Sumatera, Sulawesi, Moluccas, Borneo (Sabah).
mucronata *Rugayah in Reinwardtia,* 11(4): 268 (1999)—Borneo (Sabah).
obscura *Rugayah in Reinwardtia,* 11(4): 269 (1999)—Borneo (Sarawak).
ovigera Blume
var. boninensis *(Tuyama) H. Ohba in Fl. Japan,* 2c: 197 (1999): T. boninensis.
pendula *Rugayah in Reinwardtia,* 11(3): 219 (1998)—Borneo (Sabah).
philippinensis *Rugayah in Reinwardtia,* 11(4): 271 (1999)—Philippines.
planiglans *Rugayah in Reinwardtia,* 11(4): 273 (1999)—Philippines.
rotundifolia *Rugayah in Reinwardtia,* 11(3): 223 (1998)—Sumatera.
sepilokensis *Rugayah in Reinwardtia,* 11(4): 275 (1999)—Borneo (Sabah).
tricuspidata Lour.
f. asperifolia *Rugayah in Reinwardtia,* 11(4): 276 (1999)—Lesser Sunda Is.
f. seramensis *Rugayah in Reinwardtia,* 11(4): 277 (1999)—Moluccas.
f. siberutensis *Rugayah in Reinwardtia,* 11(4): 277 (1999)—Sumatera.

TRICHOSANTHES:-
tricuspidata
var. strigosa *S. Mitra & S. Bandyopadhyay in J. Bombay Nat. Hist. Soc.*, 95(2): 374 (1998)—India (West Bengal, Andhra Pradesh, Bihar, Karnataka, Kerala, Madhya Pradesh, Maharashtra, Orissa, Tamil Nadu), Assam (Arunachal Pradesh, Manipur, Meghalaya), Bangladesh, Burma.
var. tomentosa *(B. Heyne ex C.B. Clarke) G.R. Kumari in N.C. Nair & A.N. Henry, Fl. Tamil Nadu, India, Ser. 1*, 1: 174 (1983): T. palmata var. tomentosa.
valida *Rugayah in Reinwardtia*, 11(4): 277 (1999)—Sulawesi.
villosa Blume
subsp. mindorensis *Rugayah in Reinwardtia*, 11(4): 278 (1999)—Philippines.
vitifolia *(Naudin) N.E. Brown, Suppl. Johnson's Gard. Dict.*: 1017 (1882): Eopepon vitifolius.
wawrae Cogn.
f. hirsuta *Rugayah in Reinwardtia*, 11(4): 279 (1999)—Borneo (Sarawak).

TRICHOSPERMUM (Tiliac.).
stevensii *W. Takeuchi in Sida*, 18(4): 1077 (1999)—New Guinea (Papua New Guinea).

TRICYRTIS (Convallariac.).
viridula *Hiroshi Takahashi in Acta Phytotax. Geobot.*, 48(2): 123 (1997 publ. 1998)—China Southeast (Zhejiang, Jiangxi, Guangxi), China South-central (Guizhou, Yunnan).

TRIDACTYLE (Orchidac.).
trimikeorum *M. Dare in Excelsa*, 19: 84 (1999)—Zimbabwe.

TRIDAX (Compos.).
cajamarcensis *H. Robinson in Phytologia*, 82(1): 61 (1997)—Peru.
yecorana *B.L. Turner in Phytologia*, 79(4): 286 (1995 publ. 1996)—Mexico Northwest (Sonora).

TRIFOLIUM (Legumin.).
sect. Hiantia *Bobrov ex N.N. Tzvelev, Opred. Sosud. Rast. Severo-Zap. Rossii*: 680 (2000).
andricum *P. Lassen in Ann. Naturhist. Mus. Wien*, 98B(Suppl.): 301 (1996)—Greece.
gemellum Pourr. ex Willd.
var. tastetii *(Pau) J.A. Devesa in Lagascalia*, 21(2): 365 (2000): Trifolium striatum var. tastetii.
jokerstii *M.A. Vincent & R. Morgan in Novon*, 8(1): 91 (1998)—California.
leucotrichum *S. Petrović, Fl. Agr. Nyss.*: 228 (1882)—Yugoslavia.
lupinaster L.
subsp. pacificum *(Bobrov) R.V. Kamelin, Mater. Istorii Fl. Azii (Altaisk. Gorn. Strana)*: 122 (1998), without exact basionym page: T. pacificum.
subsp. popovii *(Roskov) R.V. Kamelin, Mater. Istorii Fl. Azii (Altaisk. Gorn. Strana)*: 122 (1998): Lupinaster popovii.
nigrescens Viv.
var. grandifolium *A.S. Ertekin & H. Akbayın in Israel J. Pl. Sci.*, 48(1): 71 (2000)—Turkey.
phitosianum *N. Böhling, W. Greuter & T. Raus in Bot. Chronika*, 13: 39 (2000)—Kriti.
plummerae *S. Watson, Bot. California*, 2: 440 (1880)—California.
wormskioldii Lehm.
var. siskiyouense *(J.M. Gillett) D. Isely, Native & Natural. Legumin. (Fabac.) U.S. (excl. Alaska & Hawaii)*: 936 (1998): T. siskiyouense.

TRIGLOCHIN (Juncaginac.).
protuberans *H.I. Aston in Muelleria*, 12(2): 215 (1999)—Western Australia.

TRIGONELLA (Legumin.).
aphanoneura *K.H. Rechinger in Fl. Iran.*, 157: 225 (1984)—Iran.
heratensis *K.H. Rechinger in Fl. Iran.*, 157: 215 (1984)—Afghanistan.
monantha C.A. Mey.
subsp. geminiflora *(Bunge) K.H. Rechinger in Fl. Iran.*, 157: 244 (1984): T. geminiflora.

TRIGONIA (Trigoniac.).
littoralis *J.R. Miguel & E.F. Guimarães in Bradea*, 8(32): 198 (2000)—Brazil Southeast (Rio de Janeiro).
nivea Cambess.
var. candida *(Warm.) J.R. Miguel & E.F. Guimarães in Daphne*, 3(4): 9 (1993): T. candida.

TRIGONOCHILUM (Orchidac.).
articulatum *W. Königer in Arcula*, 6: 181 (1996)—Ecuador.
dalessandroi *(Dodson & Garay) K. Senghas in Schlechter Orchideen, ed. 3*, I/C(33–36): 2211 (1997): Odontoglossum dalessandroi.
densiflorum *(Lindl.) K. Senghas in Schlechter Orchideen, ed. 3*, I/C(33–36): 2211 (1997): Odontoglossum densiflorum.
distans *(Rchb. f.) K. Senghas in Schlechter Orchideen, ed. 3*, I/C(33–36): 2211 (1997): Odontoglossum distans.
fidicularium *(Dalström) K. Senghas in J. Orchideenfreund*, 7(2): 135 (2000): Odontoglossum fidicularium.
flavescens *(Rolfe) K. Senghas in Schlechter Orchideen, ed. 3*, I/C(33–36): 2211 (1997): Odontoglossum flavescens.
fractiflexum *(Kraenzl.) W. Königer & H. Schildhauer in Arcula*, 5: 153 (1996): Odontoglossum fractiflexum.
funis *(F. Lehm. & Kraenzl.) K. Senghas in Schlechter Orchideen, ed. 3*, I/C(33–36): 2211 (1997): Oncidium funis.
graminifolium *(C. Schweinf.) K. Senghas in J. Orchideenfreund*, 7(1): 22 (20 Dec. 1999), preprint—Odontoglossum graminifolium.
hoeijeri *(Dalström) K. Senghas in J. Orchideenfreund*, 7(2): 135 (2000): Odontoglossum hoeijeri.
kopfianum *W. Königer in Arcula*, 8: 246 (1999)—Locality not stated.
myanthum *(Lindl.) K. Senghas in Schlechter Orchideen, ed. 3*, I/C(33–36): 2211 (1997): Odontoglossum myanthum.
myrianthum *(Rchb. f.) K. Senghas in Schlechter Orchideen, ed. 3*, I/C(33–36): 2211 (1997): Odontoglossum myrianthum.
robustum *(Rchb. f. & Warsz.) K. Senghas in Schlechter Orchideen, ed. 3*, I/C(33–36): 2211 (1997): Odontoglossum robustum.
rodrigoi *W. Königer in Arcula*, 10: 277 (2000)—Colombia.
scabiosum *(Kraenzl.) K. Senghas in Schlechter Orchideen, ed. 3*, I/C(33–36): 2211 (1997): Cyrtochilum scabiosum.

TRIGONOTIS (Boraginac.).
clarkei *R.R. Mill in Edinburgh J. Bot.*, 53(1): 122 (1996)—India (West Bengal).

TRILLIUM (Trilliac.).
camschatcense Ker Gawl.
var. tschonoskii *(Maxim.) V.N. Voroschilov in A.K. Skvortsov (ed.), Florist. issl. v razn. Raĭonakh SSSR*: 159 (1985): T. tschonoskii.
channellii *I. Fukuda, J.D. Freeman & M. Itou in Novon*, 6(2): 164 (1996)—Japan.
× crockerianum *J.J. Halda in Acta Mus. Richnov. Sect. Nat.*, 6(3): 234 (1999), as 'crockeranum'—California. (T. ovatum × rivale).
× klineanum *J.J. Halda in Acta Mus. Richnov. Sect. Nat.*, 6(3): 234 (1999)—California. (T. ovatum × rivale).
ovatum Pursh
f. maculosum *F.W. Case & R.B. Case, Trilliums*: 118 (1997)—California.

TRIMENIA (Trimeniac.).
nukuhivensis *W.L. Wagner & D.H. Lorence in Adansonia*, 21(2): 227 (1999)—Marquesas.

TRIMEZIA (Iridac.).
pardina *P. Ravenna in Pabstia*, 11(1): 1 (2000)—Brazil Southeast (Minas Gerais, Rio de Janeiro).

TRINIOCHLOA (Gramin.).
gracilis *M. Gómez-Sánchez & M. González-Ledesma in Novon*, 8(2): 144 (1998)—Mexico Southwest (Guerrero, Oaxaca).

TRIODIA (Gramin.).
acutispicula *M. Lazarides in Austral. Syst. Bot.*, 10(3): 400 (1997), nom. nov.: Plectrachne caroliniana *S.W.L. Jacobs*.
aeria *M. Lazarides in Austral. Syst. Bot.*, 10(3): 401 (1997)—Western Australia.
aristiglumis *(Lazarides) M. Lazarides in Austral. Syst. Bot.*, 10(3): 405 (1997): Plectrachne aristiglumis.
aurita *M. Lazarides in Austral. Syst. Bot.*, 10(3): 405 (1997)—Northern Territory.
biflora *M. Lazarides in Austral. Syst. Bot.*, 10(3): 412 (1997)—Western Australia.
bitextura *M. Lazarides in Austral. Syst. Bot.*, 10(3): 412 (1997), nom. nov.: Triraphis pungens *R. Br.*
bromoides *(F. Muell.) M. Lazarides in Austral. Syst. Bot.*, 10(3): 417 (1997): Triraphis bromoides.
bunglensis *(S.W.L. Jacobs) M. Lazarides in Austral. Syst. Bot.*, 10(3): 418 (1997): Plectrachne bunglensis.
bunicola *(S.W.L. Jacobs) M. Lazarides in Austral. Syst. Bot.*, 10(3): 419 (1997): T. scariosa subsp. bunicola.
bynoei *(C.E. Hubb.) M. Lazarides in Austral. Syst. Bot.*, 10(3): 422 (1997): Plectrachne bynoei.
contorta *(Lazarides) M. Lazarides in Austral. Syst. Bot.*, 10(3): 426 (1997): Plectrachne contorta.
danthonioides *(F. Muell.) M. Lazarides in Austral. Syst. Bot.*, 10(3): 427 (1997): Triraphis danthonioides.
desertorum *(C.E. Hubb.) M. Lazarides in Austral. Syst. Bot.*, 10(3): 429 (1997): Plectrachne desertorum.
dielsii *(C.E. Hubb.) M. Lazarides in Austral. Syst. Bot.*, 10(3): 431 (1997): Plectrachne dielsii.
helmsii *(C.E. Hubb.) M. Lazarides in Austral. Syst. Bot.*, 10(3): 434 (1997): Plectrachne helmsii.
integra *M. Lazarides in Austral. Syst. Bot.*, 10(3): 436 (1997)—Northern Territory.
latzii *M. Lazarides in Austral. Syst. Bot.*, 10(3): 444 (1997)—Northern Territory.
longiloba *M. Lazarides in Austral. Syst. Bot.*, 10(3): 446 (1997)—Northern Territory, Western Australia.
longipalea *M. Lazarides in Austral. Syst. Bot.*, 10(3): 448 (1997), nom. nov.: Plectrachne drummondii *C.E. Hubb.*
melvillei *(C.E. Hubb.) M. Lazarides in Austral. Syst. Bot.*, 10(3): 451 (1997): Plectrachne melvillei.
mollis *(Lazarides) M. Lazarides in Austral. Syst. Bot.*, 10(3): 456 (1997): Plectrachne mollis.
prona *M. Lazarides in Austral. Syst. Bot.*, 10(3): 463 (1997)—Western Australia.
rigidissima *(Pilg.) M. Lazarides in Austral. Syst. Bot.*, 10(3): 468 (1997): Triraphis rigidissima.
salina *M. Lazarides in Austral. Syst. Bot.*, 10(3): 471 (1997)—Northern Territory, Western Australia.
schinzii *(Henrard) M. Lazarides in Austral. Syst. Bot.*, 10(3): 474 (1997): Plectrachne schinzii.
triaristata *M. Lazarides in Austral. Syst. Bot.*, 10(3): 480 (1997)—Northern Territory, Queensland.
uniaristata *(Lazarides) M. Lazarides in Austral. Syst. Bot.*, 10(3): 482 (1997): Plectrachne uniaristata.
vella *M. Lazarides in Austral. Syst. Bot.*, 10(3): 483 (1997)—Queensland.

TRIPHYSARIA (Scrophulariac.).
chinensis *(D.Y. Hong) D.Y. Hong in Novon*, 6(4): 374 (1996): Orthocarpus chinensis.

TRIPLEUROSPERMUM (Compos.).
sect. Chionogeton *(Pobed.) N.S. Chandjian in Fl. Armenii*, 9: 574 (1995): T. subgen. Chionogeton.
sect. Gastrosulum *(Sch. Bip.) N.S. Chandjian in Fl. Armenii*, 9: 578 (1995), isonym; *see Tzvelev in Fl. Evropeĭskoĭ Chasti SSSR*, 7: 132 (1994): gen. Gastrosulum.

TRIPOGON (Gramin.).
sivarajanii *C.N. Sunil in Sida*, 18(3): 809 (1999)—India (Kerala).
vellarianus *A.K. Pradeep in Sida*, 18(3): 811 (1999)—India (Kerala).

TRIPORA *P.D. Cantino in Syst. Bot.*, 23(3): 380 (1998 publ. 1999). *VERBENACEAE*

TRIPORA:-
divaricata *(Maxim.) P.D. Cantino in Syst. Bot.*, 23(3): 382 (1998 publ. 1999): Caryopteris divaricata.

TRIPTEROSPERMUM (Gentianac.).
japonicum (Siebold & Zucc.) Maxim.
f. album *T. Shimizu, Fl. Nagano Pref.*: 1509 (1997), as 'albiflorum', see amendment sheet—Japan.
trinervium *(Thunb. ex Murray) Hiroyoshi Ohashi & H. Nakai in J. Jap. Bot.*, 71(2): 112 (1996): Convolvulus trinervius.
var. involubile *(N. Yonez.) Hiroyoshi Ohashi & H. Nakai in J. Jap. Bot.*, 71(2): 112 (1996): T. involubile.
volubile (D. Don) H. Hara
subsp. longipes *E. Aitken & D.G. Long in Edinburgh J. Bot.*, 53(1): 92 (1996)—India (West Bengal), Nepal.

TRIRAPHIS (Gramin.).
devia *T.S. Filgueiras & F.O. Zuloaga in Novon*, 9(1): 36 (1999)—Brazil West-central (Goiás).

TRISETARIA (Gramin.).
lapalmae *H. Scholz in Willdenowia*, 25(2): 571 (1996)—Canary Is.
michelii *(Savi) E. Banfi & F. Conti in Bocconea*, 10: 14, 220 (1998), isonym; *see D. Heller in Conspect. Fl. Orient.*, 6: 92 (1991): Bromus michelii.

TRISETELLA (Orchidac.).
lasiochila *F. Pupulin in Lindleyana*, 15(1): 30 (2000)—Costa Rica.

TRISETUM (Gramin.).
arduanum *E. Edgar & A.P. Druce in New Zealand J. Bot.*, 36(4): 545 (1998)—New Zealand North I., New Zealand South I.
drucei *E. Edgar in New Zealand J. Bot.*, 36(4): 548 (1998)—New Zealand North I., New Zealand South I.
lasiorhachis *(Hack.) E. Edgar in New Zealand J. Bot.*, 36(4): 549 (1998): T. antarcticum var. lasiorhachis.
lepidum *E. Edgar & A.P. Druce in New Zealand J. Bot.*, 36(4): 553 (1998)—New Zealand South I., New Zealand North I., Chatham Is.
serpentinum *E. Edgar & A.P. Druce in New Zealand J. Bot.*, 36(4): 554 (1998)—New Zealand South I., New Zealand North I.

× **TRISULOARA** *hort. in Orchid Rev.*, 106(1220): 126 (1998). *ORCHIDACEAE* (Barkeria × Brassavola × Cattleya × Epidendrum × Laelia × Sophronitis).

TRITICUM (Gramin.).
aestivum L.
subsp. yunnanense *King ex S.L. Chen in Novon*, 7(3): 230 (1997)—Qinghai.
baeoticum Boiss.
var. subaznaburticum *K. Hammer & A. Filatenko in Genet. Resourc. Crop Evol.*, 44(4): 286 (1997)—Transcaucasus (Armeniya).
var. subbuluchevskajae *K. Hammer & A. Filatenko in Genet. Resourc. Crop Evol.*, 44(4): 288 (1997)—Transcaucasus (Nakhichevan).
var. sublarionovii *K. Hammer & A. Filatenko in Genet. Resourc. Crop Evol.*, 44(4): 288 (1997)—Transcaucasus (Armeniya).
var. viridialbinigrescens *K. Hammer & A. Filatenko in Genet. Resourc. Crop Evol.*, 44(4): 288 (1997)—Iran.
var. viridihaussknechtii *K. Hammer & A. Filatenko in Genet. Resourc. Crop Evol.*, 44(4): 288 (1997)—Iraq.
var. viridinigrireuteri *K. Hammer & A. Filatenko in Genet. Resourc. Crop Evol.*, 44(4): 288 (1997)—Turkey.
dicoccon Schrank
var. italicum *K. Hammer in S. Padulosi, K. Hammer & J. Heller (eds.), Hulled Wheats:* 18 (1996)—Cultivation.
× duelongatum *L.V. Poleva in Byull. Glavn. Bot. Sada (Moscow)*, 171: 59 (1995), not validly publ., see Arts. 32.1(b), H.6 ICBN (1994)—Cultivation. (Elytrigia elongata × Triticum durum).

TRITICUM:-
monococcum L.
var. albanicum *K. Hammer & A. Filatenko in Genet. Resourc. Crop Evol.*, 44(4): 285 (1997)—Albania.
var. gaterslebense *K. Hammer & A. Filatenko in Genet. Resourc. Crop Evol.*, 44(4): 285 (1997)—Italy.
var. mansfeldii *K. Hammer & A. Filatenko in Genet. Resourc. Crop Evol.*, 44(4): 285 (1997)—Turkey.
var. reuteri *(Flaksb.) K. Hammer & A.T. Szabó in S. Padulosi, K. Hammer & J. Heller (eds.), Hulled Wheats:* 13 (1996): T. thaoudar var. reuteri.
var. viridivulgare *A. Filatenko & K. Hammer in Genet. Resourc. Crop Evol.*, 44(4): 285 (1997)—Locality not stated.
urartu Thumanjan ex Gandilyan
var. libanicum *K. Hammer & A. Filatenko in Genet. Resourc. Crop Evol.*, 44(4): 286 (1997)—Lebanon.
var. pubescentinigrum *K. Hammer & A. Filatenko in Genet. Resourc. Crop Evol.*, 44(4): 286 (1997)—Turkey.
var. pubescentiurartu *K. Hammer & A. Filatenko in Genet. Resourc. Crop Evol.*, 44(4): 286 (1997)—Turkey.
var. turcicum *K. Hammer & A. Filatenko in Genet. Resourc. Crop Evol.*, 44(4): 286 (1997)—Turkey.

TRITONIA (Iridac.).
marlothii M.P. de Vos
subsp. delpierrei *(M.P. de Vos) M.P. de Vos in Fl. S. Afr.*, 7(2:1): 121 (1999): T. delpierrei.

TRIUMFETTA (Tiliac.).
albida *(Domin) D.A. Halford in Austrobaileya*, 4(4): 539 (1997): T. micracantha var. albida.
antrorsa *D.A. Halford in Austrobaileya*, 4(4): 547 (1997)—Western Australia, Northern Territory.
aquila *D.A. Halford in Austrobaileya*, 4(4): 561 (1997)—Western Australia, Northern Territory.
arnhemica *D.A. Halford in Austrobaileya*, 4(4): 548 (1997)—Northern Territory.
aspera *D.A. Halford in Austrobaileya*, 4(4): 516 (1997)—Western Australia.
breviaculeata *D.A. Halford in Austrobaileya*, 4(4): 554 (1997)—Western Australia.
carteri *D.A. Halford in Austrobaileya*, 4(4): 541 (1997)—Western Australia.
centralis *D.A. Halford in Austrobaileya*, 4(4): 543 (1997)—Northern Territory, Western Australia.
cinerea *D.A. Halford in Austrobaileya*, 4(4): 535 (1997)—Northern Territory.
cladara *D.A. Halford in Austrobaileya*, 4(4): 563 (1997)—Western Australia.
clivorum *D.A. Halford in Austrobaileya*, 4(4): 545 (1997)—Western Australia.
subsp. brevipetala *D.A. Halford in Austrobaileya*, 4(4): 546 (1997)—Western Australia, Northern Territory.
coronata *D.A. Halford in Austrobaileya*, 4(4): 511 (1997)—Western Australia.
deserticola *D.A. Halford in Austrobaileya*, 4(4): 526 (1997)—Western Australia, Northern Territory.
echinata *D.A. Halford in Austrobaileya*, 4(4): 527 (1997)—Western Australia.
fissurata *D.A. Halford in Austrobaileya*, 4(4): 536 (1997)—Western Australia, Northern Territory.
hapala *D.A. Halford in Austrobaileya*, 4(4): 520 (1997)—Western Australia.
incana *D.A. Halford in Austrobaileya*, 4(4): 538 (1997)—Western Australia.
inermis *D.A. Halford in Austrobaileya*, 4(4): 559 (1997)—Northern Territory.
kenneallyi *D.A. Halford in Austrobaileya*, 4(4): 531 (1997)—Western Australia.
litticola *D.A. Halford in Austrobaileya*, 4(4): 518 (1997)—Northern Territory.
longipedunculata *D.A. Halford in Austrobaileya*, 4(4): 562 (1997)—Western Australia, Northern Territory.
maconochieana *D.A. Halford in Austrobaileya*, 4(4): 533 (1997)—Northern Territory, Western Australia.
marsupiata *D.A. Halford in Austrobaileya*, 4(4): 519 (1997)—Queensland.

TRIUMFETTA:-
mellina *D.A. Halford in Austrobaileya*, 4(4): 555 (1997)—Queensland, Western Australia, Northern Territory.
mitchellii *D.A. Halford in Austrobaileya*, 4(4): 540 (1997)—Western Australia.
monstrosa *D.A. Halford in Austrobaileya*, 4(4): 515 (1997)—Western Australia.
nutans *D.A. Halford in Austrobaileya*, 4(4): 549 (1997)—Western Australia.
oenpelliensis *D.A. Halford in Austrobaileya*, 4(4): 559 (1997)—Northern Territory.
pannosa *R.Br. ex D.A. Halford in Austrobaileya*, 4(4): 556 (1997)—Northern Territory, Queensland.
propinqua *D.A. Halford in Austrobaileya*, 4(4): 524 (1997)—Western Australia.
prostrata *D.A. Halford in Austrobaileya*, 4(4): 517 (1997)—Northern Territory.
pustulata *D.A. Halford in Austrobaileya*, 4(4): 565 (1997)—Western Australia.
rubiginosa *D.A. Halford in Austrobaileya*, 4(4): 530 (1997)—Western Australia.
rupestris *D.A. Halford in Austrobaileya*, 4(4): 557 (1997)—Queensland, Western Australia.
ryeae *D.A. Halford in Austrobaileya*, 4(4): 550 (1997)—Western Australia, Northern Territory.
subsp. brevipetala *D.A. Halford in Austrobaileya*, 4(4): 552 (1997)—Western Australia.
subsp. hirsuta *D.A. Halford in Austrobaileya*, 4(4): 552 (1997)—Western Australia.
saccata *D.A. Halford in Austrobaileya*, 4(4): 520 (1997)—Western Australia.
simplicifolia *(Sessé & Moç.) P.A. Fryxell in P. Mathew & M. Sivadasan (eds.), Divers. & Taxon. Trop. Flow. Pl.*: 187 (1998): Connarus simplicifolius.
simulans *D.A. Halford in Austrobaileya*, 4(4): 553 (1997)—Western Australia.
sylvicola *D.A. Halford in Austrobaileya*, 4(4): 529 (1997)—Northern Territory.
tenuiseta *D.A. Halford in Austrobaileya*, 4(4): 522 (1997)—Western Australia.
trisecta *D.A. Halford in Austrobaileya*, 4(4): 541 (1997)—Western Australia.
viridis *D.A. Halford in Austrobaileya*, 4(4): 544 (1997)—Northern Territory.

TRIVALVARIA (Annonac.).
macrophylla (Blume) Miq.
var. sumatrana *E.C.H. van Heusden in Nordic J. Bot.*, 17(2): 177 (1997)—Sumatera.

TRIXIS (Compos.).
antimenorrhoea (Schrank) Kuntze
subsp. discolor *(D. Don) L. Katinas in Darwiniana*, 34(1–4): 53 (1996): T. discolor.
divaricata (Kunth) Spreng.
subsp. discolor *(D. Don) L. Katinas in Candollea*, 53(1): 120 (1998): T. discolor.
nobilis *(Vell.) L. Katinas in Darwiniana*, 34(1–4): 74 (1996): Castra nobilis.
ophiorhiza Gardner
subsp. glandulifera *(Hassl.) L. Katinas in Darwiniana*, 34(1–4): 78 (1996): T. ophiorhiza var. glandulifera.
peruviana *L. Katinas in Darwiniana*, 34(1–4): 83 (1996)—Peru.
pruskii *D.J.N. Hind in Kew Bull.*, 55(2): 382 (2000)—Brazil Northeast (Bahia).
verbascifolia (Gardner) S.F. Blake
subsp. gigas *(Wawra) L. Katinas in Darwiniana*, 34(1–4): 96 (1996): T. gigas.

TROLLIUS (Ranunculac.).
vitalii *N.V. Stepanov, Florogenet. Analiz*, 1: 101 (1994)—Krasnoyarsk.
f. asiaticifolius *N.V. Stepanov, Florogenet. Analiz*, 1: 102 (1994)—Krasnoyarsk.
var. forficuloides *N.V. Stepanov, Florogenet. Analiz*, 1: 102 (1994)—Krasnoyarsk.
var. nadeshdae *N.V. Stepanov, Florogenet. Analiz*, 1: 102 (1994)—Krasnoyarsk.
yunnanensis (Franch.) Ulbr.
var. latilobus *Y. Fei in Fl. Yunnanica*, 11: 38 (2000)—China South-Central (Yunnan).

TROMMSDORFFIA (Amaranthac.).
cardenasii *(Standl.) T.M. Pedersen in Adansonia*, 19(2): 249 (1997): Iresine cardenasii.

TROMMSDORFFIA:-
costaricensis *(Standl.) T.M. Pedersen in Adansonia,* 19(2): 249 (1997): Iresine costaricensis.
macrophylla *(R.E. Fr.) T.M. Pedersen in Adansonia,* 19(2): 249 (1997): Iresine macrophylla.
weberbaueri *(Suess.) T.M. Pedersen in Adansonia,* 19(2): 249 (1997): Iresine weberbaueri.

TROMOTRICHE (Asclepiadac.).
sect. Caruncularia *(Haw.) P.V. Bruyns in S. Afr. J. Bot.*, 65(4): 306 (1999): gen. Caruncularia.
aperta *(Masson) P.V. Bruyns in S. Afr. J. Bot.*, 61(4): 208 (1995), as '(Masson) Sweet': Stapelia aperta.
pedunculata *(Masson) P.V. Bruyns in S. Afr. J. Bot.*, 61(4): 208 (1995), as '(Masson) Haw.': Stapelia pedunculata.

TROPAEOLUM (Tropaeolac.).
hookerianum Barnéoud
subsp. austropurpureum *J.M. Watson & A.R. Flores in Gayana, Bot.*, 57(1): 62 (2000)—Chile Central (Coquimbo).
subsp. pilosum *J.M. Watson & A.R. Flores in Gayana, Bot.*, 57(1): 63 (2000)—Chile Central (Coquimbo).
porifolium *(Cav.) L. Andersson & S. Andersson in Taxon,* 49(4): 733 (2000): Magallana porifolia.
trialatum *(Suess.) L. Andersson & S. Andersson in Taxon,* 49(4): 733 (2000): Magallana trialata.

TROPHIS (Morac.).
noraminervae *R. Cuevas Guzman & S. Carvajal Hernandez in Acta Bot. Mex.*, 47: 2 (1999)—Mexico Southwest (Jalisco).

TRYMALIUM (Rhamnac.).
monospermum *B.L. Rye in Nuytsia,* 13(2): 341 (2000)—Western Australia.
myrtillus
subsp. pungens *B.L. Rye in Nuytsia,* 11(1): 110 (1996)—Western Australia.

TSUGA (Pinac.).
chinensis (Franch.) Pritz.
var. patens *(Downie) L.K. Fu & Nan Li in Novon,* 7(3): 263 (1997): T. patens.
oblongisquamata *(W.C. Cheng & L.K. Fu) L.K. Fu & Nan Li in Novon,* 7(3): 263 (1997): T. chinensis var. oblongisquamata.

TUBERCULOCARPUS *J.F. Pruski in Novon,* 6(4): 415 (1996). *COMPOSITAE*
ruber *(Aristeg.) J.F. Pruski in Novon,* 6(4): 416 (1996): Aspilia rubra.

TUGARINOVIA (Compos.).
ovatifolia *(Ling & Ma) Y.Z. Zhao in Acta Bot. Bor.-Occid. Sin.*, 20(5): 875 (2000): Tugarinovia mongolica var. ovatifolia.

TULBAGHIA (Alliac.).
cameronii *J.G. Baker in J. Bot.*, 16: 321 (1878), in original IK as 'camerooni'—Tanzania.†

TULIPA (Liliac.).
cinnabarina *K. Persson in New Plantsman,* 7(4): 200 (2000)—Turkey.
heweri *L.W.D. van Raamsdonk in Nordic J. Bot.*, 18(1): 52 (1998)—Afghanistan.
sylvestris
var. mauritii *B. Valdés in Lagascalia,* 18(2): 326 (1996)—Morocco.
tarbagataica *D.Y. Tan & X. Wei in Acta Phytotax. Sin.*, 38(3): 302 (2000)—Xinjiang.
tianschanica Regel
var. sailimuensis *X. Wei & D.Y. Tan in Acta Phytotax. Sin.*, 38(3): 304 (2000)—Xinjiang.
wendelboi *F. Matin & M. Iranshahr in Iranian J. Bot.*, 7(2): 228 (1998)—Iran.

TUNILLA *D.R. Hunt & J. Iliff in Cactaceae Syst. Initiatives,* 9: 10 (2000). *CACTACEAE*
albisaetacens *(Backeb.) D.R. Hunt & J. Iliff in Cactaceae Syst. Initiatives,* 9: 10 (2000): Opuntia albisaetacens.
chilensis *(F. Ritter) D.R. Hunt & J. Iliff in Cactaceae Syst. Initiatives,* 9: 10 (2000): Platyopuntia chilensis.

TUNILLA:-
corrugata *(Salm-Dyck) D.R. Hunt & J. Iliff in Cactaceae Syst. Initiatives,* 9: 10 (2000): Opuntia corrugata.
erectoclada *(Backeb.) D.R. Hunt & J. Iliff in Cactaceae Syst. Initiatives,* 9: 10 (2000): Opuntia erectoclada.
ianthinantha *(F. Ritter) D.R. Hunt & J. Iliff in Cactaceae Syst. Initiatives,* 9: 10 (2000): Platyopuntia ianthinantha.
microdisca *(A.A. Weber) D.R. Hunt & J. Iliff in Cactaceae Syst. Initiatives,* 9: 11 (2000): Opuntia microdisca.
minuscula *(Backeb.) D.R. Hunt & J. Iliff in Cactaceae Syst. Initiatives,* 9: 11 (2000): Tephrocactus minusculus.
orurensis *(Cárdenas) D.R. Hunt & J. Iliff in Cactaceae Syst. Initiatives,* 9: 11 (2000): Opuntia orurensis.
picardoi *D.R. Hunt & J. Iliff in Cactaceae Syst. Initiatives,* 9: 11 (2000)—Argentina Northwest (Salta).
silvestris *(Backeb.) D.R. Hunt & J. Iliff in Cactaceae Syst. Initiatives,* 9: 11 (2000): Opuntia silvestris.
soehrensii *(Britton & Rose) D.R. Hunt & J. Iliff in Cactaceae Syst. Initiatives,* 9: 11 (2000): Opuntia soehrensii.
tilcarensis *(Backeb.) D.R. Hunt & J. Iliff in Cactaceae Syst. Initiatives,* 9: 11 (2000): Opuntia tilcarensis.

TUPISTRA (Convallariac.).
kwangtungiensis *S.S. Ying in Mem. Coll. Agric. Nation. Taiwan Univ.*, 31(1): 30 (1991), contrary to Arts. 37.4, 37.5 ICBN (1994)—China Southeast (Guangdong).
pingbianensis *J.L. Huang & X.Z. Liu in Acta Phytotax. Sin.*, 34(6): 592 (1996)—China South-central (Yunnan).

TURBINICARPUS (Cactac.).
ser. Lauani *A.B. Doweld in Sukkulentÿ (Moscow),* 3(1–2): 69 (2000).
sect. Normanbokea *(Kladiwa & Buxb.) A.B. Doweld in Sukkulentÿ (Moscow),* 1999(1): 27 (1999): gen. Normanbokea.
ser. Valdeziani *A.B. Doweld in Sukkulentÿ (Moscow),* 3(1–2): 69 (2000).
alonsoi *C. Glass & S. Arias in Kakt. And. Sukk.*, 47(2): 26 (1996)—Mexico Northeast (Guanajuato).
beguinii *(N.P. Taylor) A. Mosco & C. Zanovello in Bradleya,* 15: 81 (1997): Thelocactus beguinii.
subsp. hintoniorum *A. Hofer in Succulentes,* 2000(3): 21 (2000)—Mexico Northeast (Nuevo León).
booleanus *G.S. Hinton in Phytologia,* 80(1): 62 (1996)—Mexico Northeast (Nuevo León).
dickisoniae *(Glass & R.A. Foster) C.E. Glass & Hofer in C.E. Glass, Guía Identific. Cact. Amenazadas México,* 1: Tu/di (1998 publ. 1997): T. schmiedickeanus var. dickisoniae.
gracilis Glass & R.A. Foster
subsp. dickisoniae *(Glass & R.A. Foster) P. Panarotto in Piante Grasse,* 18(3): 120 (1998 publ. 1999): T. schmiedickeanus var. dickisoniae.
horripilus (Lem.) V. John & Ríha
subsp. wrobelianus *S. Nitzschke & Anaya Montes in Succulenta,* 79(2): 92 (2000)—Mexico Northeast (Hidalgo).
krainzianus
subsp. minimus *(Gerhart Frank) J. Lüthy & Hofer in Kakt. And. Sukk.*, 51(3): 70 (2000): T. krainzianus f. minimus.
× kubesi *J.J. Halda & P. Kupčák in Acta Mus. Richnov. Sect. Nat.*, 7(2): 76 (2000)—Cultivation. (T. krainzianus × pseudomacrochele).
macrochele (Werderm.) Buxb. & Backeb.
subsp. andersonii *(Mosco) A.B. Doweld in Sukkulentÿ (Moscow),* 3(1–2): 70 (2000): Turbinicarpus schmiedickeanus subsp. andersonii.
subsp. flaviflorus *(G. Frank & A.B. Lau) A.B. Doweld in Sukkulentÿ (Moscow),* 3(1–2): 69 (2000): Turbinicarpus flaviflorus.
subsp. klinkerianus *(Backeb. & H. Jacobsen) A.B. Doweld in Sukkulentÿ (Moscow),* 3(1–2): 70 (2000): Turbinicarpus klinkerianus.

TURBINICARPUS:-
macrochele
subsp. kupcakii *J.J. Halda & L. Horáček in Acta Mus. Richnov. Sect. Nat.*, 7(2): 76 (2000), as 'kupcaki'—Mexico Northeast (San Luis Potosí).
subsp. schwarzii *(Shurly) A.B. Doweld in Sukkulentÿ (Moscow),* 3(1–2): 70 (2000): Strombocactus schwarzii.
mandragora
subsp. beguinii *(N.P. Taylor) J.M. Lüthy in Kakt. And. Sukk.*, 50(11): 279 (1999): Thelocactus beguinii.
subsp. booleanus *(G.S. Hinton) J.M. Lüthy in Kakt. And. Sukk.*, 50(11): 279 (1999): T. booleanus.
subsp. pailanus *(Halda & Panar.) J.M. Lüthy in Kakt. And. Sukk.*, 50(11): 279 (1999): T. pailanus.
subsp. subterraneus *(Backeb.) J.M. Lüthy in Kakt. And. Sukk.*, 50(11): 279 (1999): Echinocactus subterraneus.
subsp. zaragozae *(Glass & R.A. Foster) J.M. Lüthy in Kakt. And. Sukk.*, 50(11): 279 (1999): Gymnocactus subterraneus var. zaragozae.
mombergeri *J. Říha in Kaktusy (Brno),* 32(2): 54 (1996)—Mexico Northeast (San Luis Potosí).
pailanus *J.J. Halda & P. Panarotto in Acta Mus. Richnov. Sect. Nat.*, 5(4): 162 (1998)—Mexico Northeast (Coahuila).
pseudomacrochele
subsp. krainzianus *(Gerhart Frank) C.E. Glass, Guía Identific. Cact. Amenazadas México,* 1: Tu/pseudom ssp. krainzianus (1998 publ. 1997): Toumeya krainziana.
subsp. lausseri *(Diers & G. Frank) C.E. Glass, Guía Identific. Cact. Amenazadas México,* 1: Tu/pseudom ssp. krainzianus (1998 publ. 1997): T. pseudomacrochele var. lausseri.
f. lausseri *(Diers & Gerhart Frank) P. Panarotto in Piante Grasse,* 18(3): 120 (1998 publ. 1999): T. pseudomacrochele var. lausseri.
f. minimus *(G. Frank) M. Zachar, R. Staník, A. Lux & I. Dráb, Rod Turbinicarpus:* 69 (1996), with incorrect basionym ref.: T. krainzianus f. minimus.
f. minimus *(G. Frank) C.E. Glass, Guía Identific. Cact. Amenazadas México,* 1: Tu/pseudom ssp. krainzianus (1998 publ. 1997), with incorrect basionym ref.: T. krainzianus f. minimus.
subsp. minimus *(Gerhart Frank) J.M. Lüthy & Hofer in Kakt. And. Sukk.*, 50(11): 280 (1999), basionym not validly publ. in cited ref.: T. krainzianus f. minimus.
pseudopectinatus (Backeb.) Glass & R.A. Foster
subsp. jarmilae *J.J. Halda & J. Chvastek in Acta Mus. Richnov. Sect. Nat.*, 7(1): 39 (2000)—Mexico Northeast (Nuevo León).
saueri
subsp. nelissae *J.J. Halda & P. Panarotto in Acta Mus. Richnov. Sect. Nat.*, 5(4): 161 (1998)—Mexico Northeast (Tamaulipas).
subsp. ysabelae *(Schlange) J.M. Lüthy in Kakt. And. Sukk.*, 50(11): 280 (1999): Thelocactus ysabelae.
schmiedickeanus
subsp. andersonii *A. Mosco in Cactaceae Consensus Initiatives,* 7: 18 (1999)—Mexico Northeast (San Luis Potosí).
subsp. bonatzii *(Gerhart Frank) P. Panarotto in Piante Grasse,* 18(3): 120 (1998 publ. 1999): T. bonatzii.
subsp. dickisoniae *(Glass & R.A. Foster) N.P. Taylor in Cactaceae Consensus Initiatives,* 5: 14 (1998): T. schmiedickeanus var. dickisoniae.
subsp. flaviflorus *(G. Frank & A.B. Lau) C.E. Glass, Guía Identific. Cact. Amenazadas México,* 1: Tu/sch ssp. flaviflorus (1998 publ. 1997): T. flaviflorus.
subsp. gracilis *(Glass & R.A. Foster) C.E. Glass, Guía Identific. Cact. Amenazadas México,* 1: Tu/sch ssp. gracilis (1998 publ. 1997): T. gracilis.
subsp. klinkerianus *(Backeb. & H. Jacobsen) N.P. Taylor in Cactaceae Consensus Initiatives,* 5: 14 (1998): T. klinkerianus.
subsp. klinkerianus *(Backeb. & H. Jacobsen) P. Panarotto in Piante Grasse,* 18(3): 120 (1998 publ. 1999), isonym; *see N.P. Taylor in Cactaceae Consensus Initiatives,* 5: 15 (1998): T. klinkerianus.

TURBINICARPUS:-
schmiedickeanus
subsp. macrochele *(Werderm.) N.P. Taylor in Cactaceae Consensus Initiatives,* 5: 14 (1998): Echinocactus macrochele.
subsp. macrochele *(Werderm.) P. Panarotto in Piante Grasse,* 18(3): 120 (1998 publ. 1999), without exact basionym page: Echinocactus macrochele.
var. panarottoi *J. Říha in Kaktusy (Brno),* 32(1): 25 (1996)—Mexico Northeast (San Luis Potosí).
subsp. panarottoi *(Říha) J.M. Lüthy in Kakt. And. Sukk.,* 50(11): 280 (1999): T. schmiedickeanus var. panarottoi.
var. polaskii *(Backeb.) M. Zachar, R. Staník, A. Lux & I. Dráb, Rod Turbinicarpus:* 47 (1996), basionym not validly publ.: T. polaskii.
subsp. rioverdensis *(Gerhart Frank) J.M. Lüthy in Kakt. And. Sukk.,* 50(11): 280 (1999), with incorrect basionym ref.: T. rioverdensis.
subsp. rubriflorus *(Gerhart Frank) P. Panarotto in Piante Grasse,* 18(3): 120 (1998 publ. 1999): T. schwarzii var. rubriflorus.
subsp. schwarzii *(Shurly) N.P. Taylor in Cactaceae Consensus Initiatives,* 5: 14 (1998): Strombocactus schwarzii.
f. schwarzii *(Shurly) P. Panarotto in Piante Grasse,* 18(3): 120 (1998 publ. 1999): Strombocactus schwarzii.
subterraneus (Backeb.) A.D. Zimmerman
var. zaragozae *(Glass & R.A. Foster) A.D. Zimmerman & E.F. Anderson in S. Arias Montes & N.P. Taylor, Threatened Cacti Mexico (Succ. Pl. Res.,* 2): 109 (1994), as 'zaragosae'; *see Glass, Guía Identific. Cact. Amenazadas México,* 1: Tu/za (1998 publ. 1997): Gymnocactus subterraneus var. zaragozae.†
valdezianus
var. albiflorus *(Pazout) M. Zachar, R. Staník, A. Lux & I. Dráb, Rod Turbinicarpus:* 42 (1996), basionym not validly publ.; Pelecyphora valdeziana var. albiflora.
viereckii
subsp. major *(Glass & R.A. Foster) C.E. Glass, Guía Identific. Cact. Amenazadas México,* 1: Tu/vi ssp. major (1998 publ. 1997): Gymnocactus viereckii var. major.
zaragozae *(Glass & R.A. Foster) C.E. Glass & Hofer in C.E. Glass, Guía Identific. Cact. Amenazadas México,* 1: Tu/za (1998 publ. 1997): Gymnocactus subterraneus var. zaragozae.

TURNERA (Turnerac.).
asymmetrica *M.M. Arbo in Bonplandia,* 10(1–4): 43 (2000)—Brazil Northeast (Bahia).
caatingana *M.M. Arbo in Bonplandia,* 10(1–4): 66 (2000)—Brazil Northeast (Bahia).
caerulea DC.
var. surinamensis *(Urb.) M.M. Arbo & Av. Fernández in Bonplandia,* 9(1–2): 95 (1996): T. ulmifolia var. surinamensis.
hatschbachii *M.M. Arbo in Bonplandia,* 10(1–4): 27 (2000)—Brazil Southeast (Espírito Santo).
var. miniata *M.M. Arbo in Bonplandia,* 10(1–4): 29 (2000)—Brazil Southeast (Espírito Santo).
hindsiana Benth.
subsp. brachyantha *M.M. Arbo in Bonplandia,* 9(3–4): 167 (1997)—Brazil Northeast (Maranhão), Brazil Southeast (Minas Gerais).
ignota *M.M. Arbo in Bonplandia* 9(3–4): 168 (1997)—Brazil Southeast (Minas Gerais).
maracasana *M.M. Arbo in Bonplandia,* 10(1–4): 29 (2000)—Brazil Northeast (Bahia).
princeps *M.M. Arbo in Bonplandia,* 10(1–4): 35 (2000)—Brazil Southeast (Minas Gerais).

TURPINIA (Staphyleac.).
dulongensis *H. Li in Fl. Dulongjian Reg.:* 196 (1993), without latin descr. Or type—China South-central (Yunnan).

TURRAEA (Meliac.).
africana *(Welw.) M. Cheek in Kew Bull.,* 51(4): 716 (1996): Naregamia alata var. africana.
alata *(Wight & Arn.) M. Cheek in Kew Bull.,* 51(4): 716 (1996): Naregamia alata.
rutilans *(Sm.) J. Bosser in Fl. Mascar.,* 69: 4 (1997): Gilibertia rutilans.

TWEEDIA (Asclepiadac.).
australis *(Malme) C. Ezcurra in Darwiniana,* 33(1–4): 367 (1995): Metastelma australe.

TWISSELMANNIA *I.A. Al-Shehbaz in Novon,* 9(2): 132 (1999). *CRUCIFERAE*
californica *I.A. Al-Shehbaz in Novon,* 9(2): 133 (1999)—California.

TYLECODON (Crassulac.).
bleckiae *G. Williamson in Cact. Succ. J. (U.S.A.),* 70(3): 127 (1998)—Namibia, Cape Province.
cordiformis *G. Williamson in Cact. Succ. J. (U.S.A.),* 70(5): 255 (1998)—Cape Province.
jarmilae *J.J. Halda in Acta Mus. Richnov. Sect. Nat.,* 5(1): 41 (1998)—Cape Province.
nigricaulis *G. Williamson & E. van Jaarsveld in Aloe,* 36(2–3): 43 (1999)—Cape Province.
peculiaris *E. van Jaarsveld in Aloe,* 35(1): 10 (1998)—Cape Province.
tribblei *E. van Jaarsveld in Bradleya,* 15: 68 (1997)—Cape Province.

TYLERIA (Ochnac.).
terrae-humilis *C. Sastre in Novon,* 8(3): 303 (1998)—Venezuela.

TYLOPHORA (Asclepiadac.).
cordigera *P.T. Li in J. S. China Agric. Univ.,* 14(1): 54 (1993), nom. nov.: Apocynum cordatum *Thunb.*
corollae *Bullock ex U. Meve & S. Liede in Edinburgh J. Bot.,* 53(3): 323 (1996)—Ethiopia.
erubescens *(Liede & Meve) S. Liede in Taxon,* 45(2): 206 (1996): Tylophoropsis erubescens.
ovata (Lindl.) Hook. ex Steud.
var. lanyuensis *(Liu & Lu) S.S. Ying in Mem. Coll. Agric. Nation. Taiwan Univ.,* 29(2): 54 (1989): T. lanyuensis.
urceolata *U. Meve in Kew Bull.,* 51(3): 585 (1996)—Cameroon, Gulf of Guinea Is. (Bioko).

TYPHA (Typhac.).
nothosect. Typhaolatae *E.V. Mavrodijev & Yu.E. Alekseev in Byull. Mosk. Obshch. Ispŷt. Prir., Biol.,* 103(6): 52 (1998). (T. sect. Bracteolatae × sect. Typha).
alekseevii *E.V. Mavrodiev in Feddes Repert.,* 110(1–2): 127 (1999)—North Caucasus (Dagestan).
valentinii *E.V. Mavrodiev in Feddes Repert.,* 111(7–8): 571 (2000)—Transcaucasus (Azerbaijan).

TYPHONIUM (Arac.).
brevipilosum *W.L.A. Hetterscheid & M. Sizemore in Aroideana,* 23: 52 (2000)—Sumatera.
circinnatum *W.L.A. Hetterscheid & J. Mood in Aroideana,* 23: 53 (2000)—Vietnam.
gaoligongense *(Z.L. Wang & H. Li) W.L.A. Hetterscheid & P.C. Boyce in Aroideana,* 23: 51 (2000): Sauromatum gaoligongense.
huense *Nguyen Van Dzu & T.B. Croat in Aroideana,* 20: 48 (1997 publ. 1998)—Vietnam.
johnsonianum *A. Hay & S.M. Taylor in Telopea,* 6(4): 563 (1996)—Northern Territory.
kunmingense H. Li
var. alatum *H. Li ex H. Peng & S.Z. He in Acta Bot. Yunnanica,* 19(1): 40 (1997)—China South-central (Guizhou).
var. cerebriforme *H. Li ex H. Peng & S.Z. He in Acta Bot. Yunnanica,* 19(1): 40 (1997)—China South-central (Guizhou).
mirabile *(A. Hay) A. Hay in Edinburgh J. Bot.,* 54(3): 334 (1997): Lazarum mirabile.
peltandroides *A. Hay, M.D. Barrett & R.L. Barrett in Nuytsia,* 13(1): 243 (1999)—Western Australia.
praetermissum *A. Hay in Edinburgh J. Bot.,* 54(3): 331 (1997)—Northern Territory.
smitinandii *D. Sookchaloem & J. Murata in Thai Forest Bull., Bot.,* 25: 57 (1997)—Thailand.
taylorii *A. Hay in Edinburgh J. Bot.,* 54(3): 334 (1997), as 'taylori'—Northern Territory.
venosum *(Dryand. ex Aiton) W.L.A. Hetterscheid & P.C. Boyce in Aroideana,* 23: 51 (2000): Arum venosum.

TYRBASTES *B.G. Briggs & L.A.S. Johnson in Telopea,* 7(4): 365 (1998). *RESTIONACEAE*
glaucescens *B.G. Briggs & L.A.S. Johnson in Telopea,* 7(4): 366 (1998)—Western Australia.

TZELTALIA *E. Estrada & M. Martínez in Brittonia,* 50(3): 289 (1998). *SOLANACEAE*
amphitricha *(Bitter) E. Estrada & M. Martínez in Brittonia,* 50(3): 291 (1998): Saracha amphitricha.
calidaria *(Standl. & Steyerm.) E. Estrada & M. Martínez in Brittonia,* 50(3): 292 (1998): Physalis calidaria.

U

ULEARUM (Arac.).
sagittatum Engl.
var. viridispadix *J. Bogner in Sendtnera,* 4: 10 (1997)—Brazil North (Acre).

ULEX (Legumin.).
airensis *M.D. Espírito Santo, P. Cubas, M.F. Lousã, C. Pardo & J.C. Costa in An. Jard. Bot. Madrid,* 55(1): 57 (1997)—Portugal.
australis Clemente
subsp. welwitschianus *(Planch.) M.D. Espírito Santo, P. Cubas, M.F. Lousã, C. Pardo & J.C. Costa in An. Jard. Bot. Madrid,* 55(1): 59 (1997): U. welwitschianus.

ULMUS (Ulmac.).
ismaelis *C.A. Todzia & J.L. Panero in Brittonia,* 50(3): 343 (1998)—Mexico Southwest (Oaxaca).

UMBILICUS (Crassulac.).
rupestris (Salisb.) Dandy
subsp. gaditanus *(Boiss.) O. Bolòs, X. Font & X. Pons, Atlas Corol. Fl. Vasc. Països Catalans,* 4: no. 583 (1994): U. gaditanus.

UNCARINA (Pedaliac.).
leandrii Humbert
var. rechbergeri *J.J. Lavranos in Haseltonia,* 3: 86 (1995)—Madagascar.
platycarpa *J.J. Lavranos in Haseltonia,* 3: 87 (1995), without type; *et in Cact. Succ. J. (U.S.A.),* 68(2): 82 (1996), with type—Madagascar.
roeoesliana *W. Rauh in Kakt. And. Sukk.,* 47(1): 15 (1996)—Madagascar.
turicana *J.J. Lavranos in Kakt. And. Sukk.,* 50(12): 317 (1999)—Madagascar.

UNCINIA (Cyperac.).
andina *G.A. Wheeler in Hickenia,* 2(45): 218 (1997)—Argentina South (Neuquén, Rio Negro), Chile South (Bíobío, La Araucania).
araucana *G.A. Wheeler in Aliso,* 15(1): 3 (1996 publ. 1997)—Chile South (La Araucania).
chilensis *G.A. Wheeler in Aliso,* 15(1): 1 (1996 publ. 1997)—Chile South (La Araucania, Bíobío).
ecuadorensis *G.A. Wheeler & P. Goetghebeur in Aliso,* 15(1): 10 (1996 publ. 1997)—Ecuador.
lacustris *G.A. Wheeler in Aliso,* 14(2): 141 (1995)—Ecuador.
obtusifolia *P.B. Heenan in New Zealand J. Bot.,* 34(1): 11 (1996)—New Zealand North I., New Zealand South I.
paludosa *G.A. Wheeler & P. Goetghebeur in Aliso,* 14(2): 142 (1995)—Ecuador.
scabriuscula *G.A. Wheeler in Hickenia,* 2(45): 215 (1997)—Argentina South (Neuquén, Rio Negro), Chile South (Aisén, Los Lagos).
subsacculata *G.A. Wheeler & P. Goetghebeur in Aliso,* 14(2): 145 (1995)—Ecuador.
tenuifolia *G.A. Wheeler & P. Goetghebeur in Aliso,* 14(2): 144 (1995)—Ecuador.

UNONA (Annonac.).
longiflora *W. Roxburgh, Pl. Coromandel,* 3: 87 (1820)—Bangladesh.†

UNONOPSIS (Annonac.).
stevensii *G.E. Schatz in Novon,* 8(4): 439 (1998)—Costa Rica.

URARIA (Legumin.).
floribunda *Wall. ex J.O. Voigt, Hort. Suburb. Calcutt.*: 221 (1845)—Locality not stated.
lagopodioides (L.) Desv.
var. alopecuroides *(Roxb.) A. Pramanik & Thoth. in J. Ind. Bot. Soc.*, 69(3–4): 468 (1990): Doodia alopecuroides.
rufescens (DC.) Schindl.
var. neglecta *(Prain) A. Pramanik & Thoth. in J. Ind. Bot. Soc.*, 69(3–4): 467 (1990): U. neglecta.

URCEOLA (Apocynac.).
malayana *D.J. Middleton in Blumea*, 41(1): 95 (1996)—Malaya (Peninsular Malaysia).

URGINAVIA *F. Speta in Phyton (Austria)*, 38(1): 86 (1998). *HYACINTHACEAE*
altissima *(L.f.) F. Speta in Phyton (Austria)*, 38(1): 87 (1998): Ornithogalum altissimum.
brachystachys *(Baker) F. Speta in Phyton (Austria)*, 38(1): 87 (1998): Urginea brachystachys.
micrantha *(A. Rich.) F. Speta in Phyton (Austria)*, 38(1): 87 (1998): Scilla micrantha.
porphyrantha *(Bullock) F. Speta in Phyton (Austria)*, 38(1): 87 (1998): Urginea porphyrantha.
viridula *(Baker) F. Speta in Phyton (Austria)*, 38(1): 87 (1998): Urginea viridula.

URGINEA (Hyacinthac.).
undulata (Desf.) Steinh.
subsp. caeculi *(Pau) M.B. Crespo & G. Mateo in Flora Montiberica*, 5: 58 (1997): U. undulata var. caeculi.

UROCHLOA (Gramin.).
brizantha (Hochst. ex A. Rich.) R.D. Webster
var. ciliata *(Basappa & Muniy.) V.N. Ashalatha & V.J. Nair in Bull. Bot. Surv. India*, 35(1–4): 28 (1993 publ. 1997): Brachiaria brizantha var. ciliata.
chennaveeraiana *(Basappa & Muniy.) V.N. Ashalatha & V.J. Nair in Bull. Bot. Surv. India*, 35(1–4): 28 (1993 publ. 1997): Brachiaria chennaveeraiana.
comata *(A. Rich.) M.S.M. Sosef in Fl. Gabon*, 5 bis: 64 (1999): Panicum comatum.
decidua *O. Morrone & F.O. Zuloaga in Novon*, 6(3): 315 (1996)—Brazil Northeast (Bahia).
deflexa *(Schumach.) V.N. Ashalatha & V.J. Nair in Bull. Bot. Surv. India*, 35(1–4): 28 (1993 publ. 1997), isonym; *see H. Scholz in Bull. Mus. Nation. Hist. Nat., B, Adansonia*, Ser. 4, 11(4): 443 (1989 publ. 1990): Panicum deflexum.
dictyoneura *(Fig. & De Not.) J.F. Veldkamp in Blumea*, 41(2): 418 (1996): Panicum dictyoneurum.
eruciformis *(Sm.) C.H. Nelson Sutherland & F.J. Fernández Casas in Fontqueria*, 51: 4 (1998): Panicum eruciforme.
extensa *(Chase) C.H. Nelson Sutherland & F.J. Fernández Casas in Fontqueria*, 51: 4 (1998): Brachiaria extensa.
fusiformis *(Reeder) J.F. Veldkamp in Blumea*, 41(2): 420 (1996): Brachiaria fusiformis.
glumaris *(Trin.) J.F. Veldkamp in Blumea*, 41(2): 420 (1996): Panicum glumare.
hybrida *(Basappa & Muniy.) V.N. Ashalatha & V.J. Nair in Bull. Bot. Surv. India*, 35(1–4): 28 (1993 publ. 1997): Brachiaria hybrida.
jubata *(Fig. & De Not.) M.S.M. Sosef in Fl. Gabon*, 5 bis: 64 (1999): Panicum jubatum.
munae *(Basappa) V.N. Ashalatha & V.J. Nair in Bull. Bot. Surv. India*, 35(1–4): 29 (1993 publ. 1997): Brachiaria munae.
nilagirica *(Bor) V.N. Ashalatha & V.J. Nair in Bull. Bot. Surv. India*, 35(1–4): 29 (1993 publ. 1997): Brachiaria nilagirica.
remota *(Retz.) V.N. Ashalatha & V.J. Nair in Bull. Bot. Surv. India*, 35(1–4): 29 (1993 publ. 1997): Panicum remotum.
semiundulata *(Hochst. ex A. Rich.) V.N. Ashalatha & V.J. Nair in Bull. Bot. Surv. India*, 35(1–4): 30 (1993 publ. 1997): Panicum semiundulatum.
var. lanata *(Basappa & Muniy.) V.N. Ashalatha & V.J. Nair in Bull. Bot. Surv. India*, 35(1–4): 30 (1993 publ. 1997): Brachiaria semiundulata var. lanata.
semiverticillata *(Rottler) V.N. Ashalatha & V.J. Nair in Bull. Bot. Surv. India*, 35(1–4): 30 (1993 publ. 1997): Panicum semiverticillatum.

UROCHLOA:-
stapfiana *(Basappa & Muniy.) V.N. Ashalatha & V.J. Nair in Bull. Bot. Surv. India*, 35(1–4): 30 (1993 publ. 1997): Brachiaria stapfiana.
supervacua *(C.B. Clarke) H.J. Noltie in Edinburgh J. Bot.*, 56(3): 394 (1999): Panicum supervacuum.
tanimbarensis *(Ohwi) J.F. Veldkamp in Blumea*, 41(2): 431 (1996): Brachiaria tanimbarensis.
villosa (Lam.) T.Q. Nguyen
var. barbata *(Bor) H.J. Noltie in Edinburgh J. Bot.*, 56(3): 395 (1999): Brachiaria villosa var. barbata.
var. glaberrima *(Basappa & Muniy.) V.N. Ashalatha & V.J. Nair in Bull. Bot. Surv. India*, 35(1–4): 30 (1993 publ. 1997): Brachiaria villosa var. glaberrima.

UROMYRTUS (Myrtac.).
rostrata *(Lauterb.) N. Snow & G.P. Guymer in Austrobaileya*, 5(2): 181 (1999): Myrtella rostrata.

UROPHYLLUM (Rubiac.).
parviflorum *F.C. How ex H.S. Lo in Bull. Bot. Res. North-East. Forest. Univ.*, 18(3): 278 (1998)—China South-central (Yunnan).
tsaianum *F.C. How ex H.S. Lo in Bull. Bot. Res. North-East. Forest. Univ.*, 18(3): 279 (1998)—China South-central (Yunnan).

URSIA (Legumin.).
eximia *(Stephan ex Ser.) A.P. Khokhrjakov in Novosti Sist. Vyssh. Rast.*, 31: 138 (1998): Trifolium eximium.

URSOPUNTIA *P.V. Heath in Calyx*, 6(2): 41 (1999), nom. nov.: Opuntia ser. Weberianae *Britton & Rose*. (Cactac.).
textoris *P.V. Heath in Calyx*, 6(2): 41 (1999), nom. nov., as 'comb. nov.', contrary to Art. 52.1 ICBN (1994): Opuntia weberi *Speg.*
var. deminuta *(Rausch) P.V. Heath in Calyx*, 6(2): 42 (1999): Tephrocactus weberi var. deminutus.
var. dispar *(A. Cast. & H.V. Lelong) P.V. Heath in Calyx*, 6(2): 42 (1999): Opuntia weberi var. dispar.
var. setiger *(Backeb.) P.V. Heath in Calyx*, 6(2): 42 (1999): Tephrocactus setiger.

URTICA (Urticac.).
lilloi *(Hauman) D.V. Geltman in Novon*, 8(1): 17 (1998): U. magellanica subsp. lilloi.
magellanica Poir.
subsp. bracteata *(Steud.) D.V. Geltman in Novon*, 8(1): 17 (1998): U. bracteata.
peruviana *D.V. Geltman in Novon*, 8(1): 15 (1998)—Peru.
pseudomagellanica *D.V. Geltman in Novon*, 8(1): 16 (1998)—Bolivia.

URVILLEA (Sapindac.).
paucidentata *M.S. Ferrucci in Bonplandia*, 9(1–2): 19 (1996)—Brazil West-central (Mato Grosso).
peruviana *M.S. Ferrucci in Bonplandia*, 9(3–4): 237 (1997)—Peru.

UTRICULARIA (Lentibulariac.).
australis R. Br.
var. tenuicaulis *(Miki) S. Hatusima in S. Hatusima & T. Amano, Fl. Ryukyus, S. of Amami Is., ed. 2*: 204 (1994), without basionym ref.: Basionym not stated.
paulineae *A. Lowrie in Nuytsia*, 12(1): 38 (1998)—Western Australia.

UVARIA (Annonac.).
borneensis *(Merr.) T.M.A. Utteridge in Blumea*, 45(2): 393 (2000), as 'borneense': Tetrapetalum borneense.
lambirensis *(K. Momose) T.M.A. Utteridge in Blumea*, 45(2): 393 (2000), as 'lambirense': Tetrapetalum lambirense.
puguensis *D.M. Johnson in Novon*, 9(1): 58 (1999)—Tanzania.
korupensis *R.E. Gereau & D. Kenfack in Adansonia*, 22(1): 41 (2000)—Cameroon.

V

VACCARIA (Caryophyllac.).
hispanica (Mill.) Rauschert
subsp. pyramidata *(Medik.) J. Holub in Preslia*, 70(2): 116 (1998): V. pyramidata.

VACCINIUM (Ericac.).
sect. Barandanum *S.P. Vander Kloet in Floristic Charact. & Diversity E. Asian Pl. (eds. A.L. Zhang & S.G. Wu)*: 117 (1998).
distichum *J.L. Luteyn in Fl. Ecuador*, 54(147): 321 (1996)—Ecuador.
dulongense *H. Li in Fl. Dulongjian Reg.*: 232 (1993), without latin descr. or type—China South-central (Yunnan).
hiepii *S.P. Vander Kloet in Bot. J. Linn. Soc.*, 134(4): 582 (2000)—Vietnam.
podocarpoideum *W.P. Fang & Z.H. Pan ex Q.Z. Lin in Guihaia*, 19(1): 28 (1999), homonym; *see W.P. Fang & Z.H. Pan in Acta Phytotax. Sin.*, 19(1): 107 (1981)—China Southeast (Hunan).
rubescens *R.C. Fang in Novon*, 9(2): 174 (1999)—China South-central (Yunnan).
tenuiflorum *R.C. Fang in Novon*, 9(2): 178 (1999)—Xizang.

VAILIA (Asclepiadac.).
salicina *(Decne.) G. Morillo in Acta Bot. Venezuel.*, 20(2): 21 (1997): Blepharodon salicinum.

VALERIANA (Valerianac.).
alternifolia Bunge
f. verticillata *(Kom.) S.X. Li in Clavis Pl. Chinae Bor.-Or. ed. 2 (ed. P.Y. Fu)*: 629 (1995): V. stubendorfii f. verticillata.
altoandina *A.L. Cabrera, Fl. Prov. Jujuy (Inst. Nac. Tecn. Agropec., 13)*, 9: 452 (1993)—Argentina Northwest (Jujuy, Salta).
corymbulosa *(Wedd.) A.L. Cabrera, Fl. Prov. Jujuy (Inst. Nac. Tecn. Agropec., 13)*, 9: 453 (1993): Phyllactis corymbulosa.
eupatoria *M. Sobral in Novon*, 10(2): 149 (2000)—Brazil South (Rio Grande do Sul).
excelsa
subsp. sambucifolia *(J.C. Mikan) J. Holub in Preslia*, 68(3): 286 (1996 publ. 1997): V. sambucifolia.
*subsp.*transiens *(E. Walther) J. Holub in Preslia*, 68(3): 286 (1996 publ. 1997): V. sambucifolia var. transiens.
petersenii *F. Weberling & H. Reese-Krug in Bot. Jahrb.*, 118(2): 160 (1996)—Bolivia, Peru, Chile North (Tarapacá).
tajuvensis *M. Sobral in Novon*, 9(1): 114 (1999)—Brazil South (Rio Grande do Sul, Santa Catarina).

VALERIANELLA (Valerianac.).
dentata (L.) Pollich
var. macrocarpa *(Velen.) M. Ančev, D. Delipavlov & M. Popova in Fl. Republ. Bŭlgariya*, 12: 384 (1995): V. morisonii var. macrocarpa.
eriocarpa Desv.
var. intermedia *M.A. Thiébaud, Valerianaceae: Compl. Prodr. Fl. Corse (ser. eds. D. Jeanmonod & H.M.* Burdet): 101 (1996)—Corse.
microcarpa Loisel.
var. glabra *M.-A. Thiébaud, Valerianaceae: Compl. Prodr. Fl. Corse (ser. eds. D. Jeanmonod & H.M.* Burdet): 86 (1996)—Corse.
pumila (L.) DC.
var. brachystephana *(Ten.) M. Ančev, D. Delipavlov & M. Popova in Fl. Republ. Bŭlgariya*, 10: 380 (1995), without basionym page: Fedia brachystephana.

VALIHA *S. Dransfield in Kew Bull.*, 53(2): 380 (1998). *GRAMINEAE*
diffusa *S. Dransfield in Kew Bull.*, 53(2): 381 (1998)—Madagascar.
perrieri *(A. Camus) S. Dransfield in Kew Bull.*, 53(2): 388 (1998): Ochlandra perrieri.

VALLESIA (Apocynac.).
aurantiaca *(M. Martens & Galeotti) J.F. Morales in Novon*, 8(3): 263 (1998): Neriandra aurantiaca.
sinaloensis *El. Meyer ex J.F. Morales in Novon*, 8(3): 263 (1998)—Mexico Northwest (Sinaloa).
spectabilis *E. Meyer ex J.F. Morales in Novon*, 8(3): 263 (1998)—Mexico Southwest (Jalisco).
vaupesana *J.F. Morales in Novon*, 8(3): 264 (1998)—Colombia.

VALLISNERIA (Hydrocharitac.).
annua *S.W.L. Jacobs & K.A. Frank in Telopea*, 7(2): 116 (1997)—Western Australia, Northern Territory, Queensland.
longipedunculata *X.S. Shen in Acta Bot. Bor.-Occid. Sin.*, 20(5): 889 (2000)—China Southeast (Anhui).
triptera *S.W.L. Jacobs & K.A. Frank in Telopea*, 7(2): 114 (1997)—Northern Territory, Western Australia.

VAN-TIEGHEMIA *A.V.F.Ch. Bobrov & A.P. Melikian in Bot. Zhurn.*, 85(7): 58 (2000). *PODOCARPACEAE*
densifolia *(Kunth) A.V.F.Ch. Bobrov & A.P. Melikian in Bot. Zhurn.*, 85(7): 59 (2000): Podocarpus taxifolius var. densifolius.
meridensis *(J. Buchholz & N.E. Gray) A.V.F.Ch. Bobrov & A.P. Melikian in Bot. Zhurn.*, 85(7): 59 (2000), as 'meridense': Podocarpus montanus var. meridensis.
montana *(Willd.) A.V.F.Ch. Bobrov & A.P. Melikian in Bot. Zhurn.*, 85(7): 59 (2000): Taxus montana.

VANDA (Orchidac.).
bidupensis *L.V. Averyanov & E.A. Christenson in Orchid Dig.*, 62(1): 27 (1998)—Vietnam.
taiwaniana *S.S. Ying in Mem. Coll. Agric. Nation. Taiwan Univ.*, 29(2): 65 (1989)—Taiwan.

VANGUERIA (Rubiac.).
soutpansbergensis *N. Hahn in Bothalia*, 27(1): 45 (1997)—Transvaal.

VANILLA (Orchidac.).
africana Lindl.
subsp. cucullata *(Kraenzl. ex J. Braun & K. Schum.) D.L. Szlachetko & T.S. Olszewski in Fl. Cameroun*, 34: 286 (1998): V. cucullata.
subsp. ramosa *(Rolfe) D.L. Szlachetko & T.S. Olszewski in Fl. Cameroun*, 34: 286 (1998): V. ramosa.
diabolica *P. O'Byrne in Malayan Orchid Rev.*, 30: 67 (1996)—Sulawesi.
hallei *D.L. Szlachetko & T.S. Olszewski in Fl. Cameroun*, 34: 282 (1998)—Gabon.
hirsuta *M.A. Clements & D.L. Jones in Lasianthera*, 1(1): 47 (1996)—New Guinea (Papua New Guinea).
ochyrae *D.L. Szlachetko & T.S. Olszewski in Fl. Cameroun*, 34: 291 (1998)—Cameroon.

VANWYKIA (Loranthac.).
rubella *R.M. Polhill & D. Wiens, Mistletoes of Africa:* 222 (1998)—Tanzania.

VASCONCELLEA (Caricac.).
crassipetala *(V.M. Badillo) V.M. Badillo in Ernstia*, 10(2): 76 (2000): Carica crassipetala.
cundinamarcensis *V.M. Badillo in Ernstia*, 10(2): 78 (2000), nom. nov.: Carica pubescens *Lenné & K. Koch*.
× heilbornii *(V.M. Badillo) V.M. Badillo in Ernstia*, 10(2): 76 (2000): Carica × heilbornii.
var. chrysopetala *(Heilborn) V.M. Badillo in Ernstia*, 10(2): 76 (2000), without basionym ref.: Carica chrysopetala.
var. fructifragrans *(García-Barr. & Hern. Cam.) V.M. Badillo in Ernstia*, 10(2): 76 (2000), as 'fructigragrans', without basionym ref.: Carica fructifragrans.
horovitziana *(V.M. Badillo) V.M. Badillo in Ernstia*, 10(2): 77 (2000): Carica horovitziana.
longiflora *(V.M. Badillo) V.M. Badillo in Ernstia*, 10(2): 77 (2000): Carica longiflora.
microcarpa (Jacq.) A.DC.
subsp. australis *(V.M. Badillo) V.M. Badillo in Ernstia*, 10(2): 77 (2000): Carica microcarpa subsp. australis.

VASCONCELLEA:-
microcarpa
subsp. baccata *(Heilborn) V.M. Badillo in Ernstia*, 10(2): 77 (2000): Carica baccata.
subsp. heterophylla *(Poepp. & Endl.) V.M. Badillo in Ernstia*, 10(2): 77 (2000): Carica heterophylla.
subsp. pilifera *(V.M. Badillo) V.M. Badillo in Ernstia*, 10(2): 77 (2000): Carica microcarpa subsp. pilifera.
omnilingua *(V.M. Badillo) V.M. Badillo in Ernstia*, 10(2): 77 (2000): Carica omnilingua.
palandensis *(V.M. Badillo, Van den Eynden & Van Damme) V.M. Badillo in Ernstia*, 10(2): 77 (2000): Carica palandensis.
pulchra *(V.M. Badillo) V.M. Badillo in Ernstia*, 10(2): 77 (2000): Carica pulchra.
sphaerocarpa *(García-Barr. & Hern. Cam.) V.M. Badillo in Ernstia*, 10(2): 78 (2000): Carica sphaerocarpa.
sprucei *(V.M. Badillo) V.M. Badillo in Ernstia*, 10(2): 78 (2000): Carica sprucei.
stipulata *(V.M. Badillo) V.M. Badillo in Ernstia*, 10(2): 78 (2000): Carica stipulata.
weberbaueri *(Harms) V.M. Badillo in Ernstia*, 10(2): 78 (2000): Carica weberbaueri.

VELARUM (Crucif.).
leiocarpum *(DC.) N.N. Tzvelev, Opred. Sosud. Rast. Severo-Zap. Rossii:* 679 (2000): Sisymbrium officinale var. leiocarpum.
officinale (L.) Rchb.
subsp. leiocarpum *(DC.) N.N. Tzvelev in Novosti Sist. Vyssh. Rast.*, 32: 184 (2000): Sisymbrium officinale var. leiocarpum.

VELASCOA *G. Calderón de Rzedowski & J. Rzedowski in Acta Bot. Mex.*, 39: 54 (1997). *CROSSOSOMATACEAE*
recondita *G. Calderón de Rzedowski & J. Rzedowski in Acta Bot. Mex.*, 39: 54 (1997)—Mexico Northeast (Querétaro).

VELLA (Crucif.).
bourgaeana *(Coss.) S.I. Warwick & I.A. Al-Shehbaz in Novon*, 8(3): 324 (1998): Euzomodendron bourgaeanum.

VELLOZIA (Velloziac.).
armata *R. de Mello-Silva in Bot. J. Linn. Soc.*, 120(3): 258 (1996)—Brazil Southeast (Minas Gerais).
auriculata *R. de Mello-Silva & N.L. de Menezes in Novon*, 9(4): 537 (1999)—Brazil Southeast (Minas Gerais).
gigantea *N.L. de Menezes & R. de Mello-Silva in Novon*, 9(4): 537 (1999)—Brazil Southeast (Minas Gerais).
sessilis *L.B. Sm. ex R. de Mello-Silva in Bol. Bot. (São Paulo)*, 16: 65 (1997)—Brazil West-central (Goiás).
tillandsioides *R. de Mello-Silva in Bot. J. Linn. Soc.*, 120(3): 261 (1996)—Brazil Southeast (Minas Gerais).

VENTRICULARIA (Orchidac.).
borneensis *J.J. Wood in Orchid Rev.*, 106(1221): 175 (1998)—Borneo (Sabah).

VEPRIS (Rutac.).
elegantissima *F. White & C.M. Pannell in Bull. Jard. Bot. Nation. Belg.*, 65(3–4): 419 (1996)—Malawi.
glaberrima *(Engl.) J.B. Hall ex D.J. Harris in Kew Bull.*, 55(2): 458 (2000): Oriciopsis glaberrima.

VERATRUM (Melanthiac.).
latifolium *(Desr.) W.B. Zomlefer in Harvard Pap. Bot.*, 2(2): 161 (1997): Melanthium latifolium.
maackii Regel
f. atropurpureum *(Honda) Hir. Takah. ex T. Shimizu, Fl. Nagano Pref.:* 1510 (1997): V. longebracteatum var. atropurpureum.
nigrum L.
subsp. ussuriense *(Loes.) V.N. Voroschilov in A.K. Skvortsov (ed.), Florist. issl. v razn. raĭonakh SSSR:* 158 (1985): V. nigrum var. ussuriense.
shuehshanarum *S.S. Ying in Mem. Coll. Agric. Nation. Taiwan Univ.*, 28(2): 48 (1988)—Taiwan.

VERBASCUM (Scrophulariac.).
faurei (Murb.) Hub.-Mor.
subsp. acanthifolium *(Pau) C. Benedí & J.M. Montserrat in Lagascalia*, 20(1): 169 (1997): Celsia acanthifolia.
lagurus Fisch. & C.A. Mey.
subsp. stefanovii *K. Sutorý in Phytologia Balcanica*, 6(1): 13 (2000), nom. nov.: Verbascum ponticum *Stef.*
prunellii *V. Rodríguez Gracia & E. Valdés Bermejo in An. Jard. Bot. Madrid*, 56(1): 185 (1998)—Spain.
pyrenaicum *M. Gandoger, Dec. Pl. Nov.*, 2: 9 (1876)—France.
rosellum *M. Gandoger, Dec. Pl. Nov.*, 2: 8 (1876)—France.
subalpestre *M. Gandoger, Dec. Pl. Nov.*, 2: 8 (1876)—France.
subbiserratum *M. Gandoger, Dec. Pl. Nov.*, 2: 9 (1876)—France.
thapsus L.
subsp. litigiosum *(Samp.) A. Galán de Mera & J.A. Vicente Orellana in An. Jard. Bot. Madrid*, 53(2): 261 (1995 publ. 1996): V. litigiosum.
subsp. martinezii *(Valdés) A. Galán de Mera & J.A. Vicente Orellana in An. Jard. Bot. Madrid*, 53(2): 261 (1995 publ. 1996): V. giganteum subsp. martinezii.
tzar-borisii *(Davidov ex Stoj.) B. Stefanova-Gateva in Fl. Republ. Bŭlgariya*, 10: 50 (1995): V. humile subsp. tzar-borisii.

VERBENA (Verbenac.).
africana *(R. Fern. & Verdc.) P.W. Michael in Telopea*, 7(3): 296 (1997): V. officinalis subsp. africana.
gaudichaudii *(Briq.) P.W. Michael in Telopea*, 7(3): 295 (1997): V. officinalis var. gaudichaudii.
incompta *P.W. Michael in Telopea*, 6(2–3): 181 (1995)—Victoria, Queensland, New South Wales, South Australia, Northern Territory, Brazil South, Transvaal, New Guinea (Papua New Guinea), Fiji, Norfolk I., New Zealand.

VERBESINA (Compos.).
albissima *A. Sagástegui Alva in Arnaldoa*, 4(1): 7 (1996)—Peru.
ampliatifolia *A. Sagástegui Alva & V. Quipuscoa Silvestre in Arnaldoa*, 5(2): 141 (1998)—Peru.
ancashensis *A. Sagástegui Alva & V. Quipuscoa Silvestre in Arnaldoa*, 6(1): 41 (1999 publ. 2000)—Peru.
aypatensis *A. Sagástegui Alva & V. Quipuscoa Silvestre in Arnaldoa*, 5(2): 143, 144 (1998)—Peru.
biserrata *H. Robinson & J. Panero in Rhodora*, 102(910): 129 (2000)—Ecuador.
brevilingua *A. Sagástegui Alva in Arnaldoa*, 5(1): 35 (1998)—Peru.
brunnea *A. Sagástegui Alva & V. Quipuscoa Silvestre in Arnaldoa*, 5(1): 38 (1998)—Peru.
calzadae *J.L. Panero & J.L. Villaseñor in Brittonia*, 48(4): 569 (1996 publ. 1997)—Mexico Southwest (Oaxaca).
capituliparva *A. Sagástegui Alva in Arnaldoa*, 4(1): 10 (1996)—Peru.
chachapoyensis *A. Sagástegui Alva & V. Quipuscoa Silvestre in Arnaldoa*, 5(1): 40 (1998)—Peru.
clarkiae *H. Robinson & J. Panero in Rhodora*, 102(910): 131 (2000), as 'clarkae'—Ecuador.
crassicephala *A. Sagástegui Alva & V. Quipuscoa Silvestre in Arnaldoa*, 6(1): 44 (1999 publ. 2000)—Peru.
huancabambae *A. Sagástegui Alva & V. Quipuscoa Silvestre in Arnaldoa*, 5(2): 146 (1998)—Peru.
juxtlahuacensis *J.L. Panero & J.L. Villaseñor in Brittonia*, 48(1): 89 (1996)—Mexico Southwest (Oaxaca).
leivae *A. Sagástegui Alva & V. Quipuscoa Silvestre in Arnaldoa*, 5(1): 42 (1998)—Peru.
ligulata *(Maguire & Wurdack) J.F. Pruski in Novon*, 6(4): 417 (1996): V. schomburgkii subsp. ligulata.
maldonadoensis *H. Robinson & J. Panero in Rhodora*, 102(910): 134 (2000)—Ecuador.
miahuatlana *B.L. Turner in Phytologia*, 79(4): 296 (1995 publ. 1996)—Mexico Southwest (Oaxaca).
monactioides *A. Sagástegui Alva, S. Leiva González & P. Lezama Asencio in Arnaldoa*, 3(2): 7 (1995 publ. 1996)—Peru.

VERBESINA:-

otuzcensis *A. Sagástegui Alva & V. Quipuscoa Silvestre in Arnaldoa,* 5(1): 44 (1998)—Peru.

pauciramea *A. Sagástegui Alva, S. Leiva González & P. Lezama Asencio in Arnaldoa,* 3(2): 9 (1995 publ. 1996)—Peru.

perijaensis *H. Robinson in Rhodora,* 102(910): 138 (2000)—Colombia.

perlanata *A. Sagástegui Alva & V. Quipuscoa Silvestre in Arnaldoa,* 5(1): 47 (1998)—Peru.

pichinchensis *H. Robinson in Rhodora,* 102(910): 136 (2000)—Ecuador.

spooneri *J.L. Panero in Brittonia,* 51(1): 92 (1999)—Bolivia.

VERNONANTHURA (Compos.).

hintoniorum *(B.L. Turner) H. Robinson in Smithsonian Contrib. Bot.,* 89: 85 (1999): Vernonia hintoniorum.

VERNONIA (Compos.).

subsect. Argenteae *Á.L. Cabrera in Candollea,* 54(1): 109 (1999).

bahiana *(H. Rob.) D.J.N. Hind in Kew Bull.,* 55(1): 185 (2000): Lepidaploa bahiana.

bellula *M. Dematteis in Candollea,* 55(2): 308 (2000)—Paraguay.

boquerona *B.L. Turner in Phytologia,* 81(2): 115 (1996 publ. 1997)—Mexico Southeast (Chiapas).

brunneri *(H. Rob.) Á.L. Cabrera in Candollea,* 54(1): 109 (1999): Mesanthophora brunneri.

bulo-burtiensis *Mesfin Tadesse in Kew Bull.,* 52(3): 579 (1997)—Somalia.

calulu Hiern

var. carinata *(Hutch. & Burtt) K. Kalanda in Fragm. Flor. Geobot.,* 40(2): 679 (1995): V. carinata.

dalettiensis *Mesfin Tadesse in Kew Bull.,* 52(3): 579 (1997)—Ethiopia.

didessana *Mesfin Tadesse in Kew Bull.,* 52(3): 584 (1997)—Ethiopia.

ganevii *D.J.N. Hind in Kew Bull.,* 54(4): 905 (1999 publ. 2000)—Brazil Northeast (Bahia).

gilbertii *Mesfin Tadesse in Kew Bull.,* 52(3): 575 (1997)—Ethiopia.

hystricosa *Á.L. Cabrera & M. Dematteis in Candollea,* 54(1): 103 (1999)—Paraguay.

lainzii *S. Ortiz in Canad. J. Bot.,* 77(6): 877 (1999)—Angola.

lanata *(Chodat) A.L. Cabrera in Hickenia,* 2(35): 155 (1995): V. glabrata var. lanata.

lanata *Mesfin Tadesse in Kew Bull.,* 52(3): 580 (1997), homonym; *see Cabrera in Hickenia,* 2(35): 155 (1995)—Somalia.

leonensis *Á.L. Cabrera in Candollea,* 54(1): 105 (1999)—Paraguay.

longibracteata *S. Ortiz & Rodr. Oubiña in Canad. J. Bot.,* 77(6): 879 (1999)—Angola.

novarae *A.L. Cabrera in Hickenia,* 2(54): 255 (1997)—Argentina Northwest (Salta).

ochyrae *S. Lisowski in Fragm. Flor. Geobot.,* 40(2): 572 (1995)—Zaire.

parvifolia *(Chodat) A.L. Cabrera in Hickenia,* 2(35): 156 (1995): V. glabrata var. parvifolia.

platyseta *S. Ortiz in Canad. J. Bot.,* 77(6): 881 (1999)—Angola.

plumosa *(O. Hoffm.) J.-P. Lebrun & A.L. Stork, Enum. Pl. Fleurs Afr. Trop.,* 4: 343 (1997): Gongrothamnus plumosus.

potrerillona *E.L. Ekman & I. Urban in Feddes Repert.,* 26: 98 (1929), in IK Supp. 8 as 'potretillona' in error—Cuba.†

profusa *M. Dematteis & Á.L. Cabrera in Candollea,* 54(1): 105 (1999)—Paraguay.

propinqua Hieron.

var. canescens *(Chodat) M. Dematteis in Candollea,* 54(1): 109 (1999): V. lepidifera var. canescens.

pseudoincana *(Hieron.) Á.L. Cabrera in Candollea,* 54(1): 110 (1999): V. rubricaulis var. pseudoincana.

pumilla *(Vell.) A.L. Cabrera in Hickenia,* 2(35): 158 (1995): Chrysocoma pumilla.

ramellae *A.L. Cabrera in Hickenia,* 2(35): 159 (1995), nom. nov.: V. sessilifolia var. auriculata *Chodat.*

raui *B.P. Uniyal in Fl. India,* 13: 382 (1995), nom. nov., as 'rauii': V. membranacea *Bedd. ex S. Moore.*

VERNONIA:-

schweinfurthii Oliv. & Hiern.

var. bukamaensis *(De Wild.) K. Kalanda & S. Lisowski in Fragm. Flor. Geobot.,* 40(2): 672 (1995): V. bukamaensis.

tayloriana *(Isawumi) J.-P. Lebrun & A.L. Stork, Enum. Pl. Fleurs Afr. Trop.,* 4: 345 (1997): Baccharoides tayloriana.

tewoldei *Mesfin Tadesse in Kew Bull.,* 52(3): 581 (1997)—Ethiopia.

thulinii *Mesfin Tadesse in Kew Bull.,* 52(3): 582 (1997)—Ethiopia.

vivekanathanii *B.P. Uniyal in Fl. India,* 13: 391 (1995), nom. nov.: V. monosis *Benth. ex C.B. Clarke.*

volkameriifolia DC.

subsp. siamica *(Hosseus) H. Koyama in Bull. Nation. Sci. Mus., B (Tokyo),* 23(4): 163 (1997): V. volkameriifolia var. siamica.

wendtiana *B.L. Turner in Phytologia,* 82(4): 275 (1997 publ. 1998)—Mexico Southwest (Oaxaca).

xiquexiquensis *D.J.N. Hind in Kew Bull.,* 55(1): 182 (2000)—Brazil Northeast (Bahia).

yabelloana *Mesfin Tadesse in Kew Bull.,* 52(3): 584 (1997)—Ethiopia.

VERNONIASTRUM *H. Robinson in Proc. Biol. Soc. Wash.,* 112(1): 233 (1999). *COMPOSITAE*

aemulans *(Vatke) H. Robinson in Proc. Biol. Soc. Wash.,* 112(1): 233 (1999): Vernonia aemulans.

ambiguum *(Kotschy & Peyr.) H. Robinson in Proc. Biol. Soc. Wash.,* 112(1): 233 (1999): Vernonia ambigua.

latifolium *(Steetz) H. Robinson in Proc. Biol. Soc. Wash.,* 112(1): 234 (1999): Crystallopollen latifolium.

musofense *(S. Moore) H. Robinson in Proc. Biol. Soc. Wash.,* 112(1): 234 (1999): Vernonia musofensis.

nestor *(S. Moore) H. Robinson in Proc. Biol. Soc. Wash.,* 112(1): 234 (1999): Vernonia nestor.

ugandense *(S. Moore) H. Robinson in Proc. Biol. Soc. Wash.,* 112(1): 234 (1999): Vernonia ugandensis.

uncinatum *(Oliv. & Hiern) H. Robinson in Proc. Biol. Soc. Wash.,* 112(1): 234 (1999): Vernonia uncinata.

viatorum *(S. Moore) H. Robinson in Proc. Biol. Soc. Wash.,* 112(1): 234 (1999): Vernonia viatorum.

VERONICA (Scrophulariac.).

sect. Scutellatae *(Benth.) N.N. Tzvelev, Opred. Sosud. Rast. Severo-Zap. Rossii:* 680 (2000), isonym; *see Ivanina in Sosud. Rast. Sovet. Dal'nego Vostoka,* 5: 320 (1991): Veronica subsect. Scutellatae.

austriaca L.

subsp. jacquinii *(Baumg.) Eb. Fischer in R. Wisskirchen & H. Haeupler, Standardliste Farn- und Blütenpfl.* Deutschl.: 538 (1998): V. jacquinii.

himalensis D. Don

subsp. yunnanensis *(P.C. Tsoong) D.Y. Hong in Novon,* 6(1): 23 (1996): V. himalensis var. yunnanensis.

krumovii *(Peev) D. Peev in Fl. Republ. Bŭlgariya,* 10: 159 (1995): V. chamaedrys subsp. krumovii.

langei

var. nevadensis *(Pau) M.M. Martínez Ortega & E. Rico in An. Jard. Bot. Madrid,* 57(2): 478 (1999 publ. 2000): V. repens var. nevadensis.

laxissima *D.Y. Hong in Novon,* 6(1): 23 (1996)—China South-central (Sichuan).

orsiniana

var. verdagueri *(Sennen) M.M. Martínez Ortega & E. Rico in An. Jard. Bot. Madrid,* 57(2): 478 (1999 publ. 2000): V. verdagueri.

pinnata L.

subsp. nana *A.V. Polozhij in Fl. Sibir. (Solanac.-Lobeliac.),* 12: 33 (1996 publ. 1997)—Altay.

rosea Desf.

subsp. atlantica *(Ball) I. Soriano in Lagascalia,* 18(2): 299 (1996): V. cuneifolia var. atlantica.

var. macrantha *Pau ex I. Soriano in Lagascalia,* 18(2): 300 (1996)—Morocco.

sennenii *(Pau) M.M. Martínez Ortega & E. Rico in An. Jard. Bot. Madrid,* 57(2): 477 (1999 publ. 2000): V. prostrata var. sennenii.

VERONICA:-

sergievskiana *A.V. Polozhij in Fl. Sibir. (Solanac.-Lobeliac.),* 12: 37 (1996 publ. 1997)—Chita.

stelleri Pall.

f. rufescens *Y.N. Lee, Fl. Korea:* 1161 (1996), without type—Locality not stated.

taigischensis *N.V. Stepanov, Florogenet. Analiz,* 1: 81 (1994); *et in Bot. Zhurn.,* 82(9): 93 (1997)—Krasnoyarsk.

tenuifolia

subsp. fontqueri *(Pau) M.M. Martínez Ortega & E. Rico in An. Jard. Bot. Madrid,* 57(2): 477 (1999 publ. 2000): V. fontqueri.

uralensis *M.S. Knjasev in Bot. Zhurn.,* 85(9): 118 (2000)—East European Russia.

yushanchienshanica *S.S. Ying in Mem. Coll. Agric. Nation. Taiwan Univ.,* 29(2): 59 (1989)—Taiwan.

VERONICASTRUM (Scrophulariac.).

axillare (Siebold & Zucc.) T. Yamaz.

var. simadai *(Masam.) H.Y. Liu in Fl. Taiwan, ed. 2,* 4: 642 (1998): Calorhabdos simadai.

brunonianum (Benth.) D.Y. Hong

subsp. sutchuenense *(Franch.) D.Y. Hong in Novon,* 6(1): 23 (1996): Calorhabdos sutchuenensis.

VEYRETELLA *D.L. Szlachetko & T.S. Olszewski in Fl. Cameroun,* 34: 100 (1998). *ORCHIDACEAE*

hetaerioides *(Summerh.) D.L. Szlachetko & T.S. Olszewski in Fl. Cameroun,* 34: 102 (1998): Habenaria hetaerioides.

VEYRETIA *D.L. Szlachetko in Fragm. Flor. Geobot.,* Supp. 3: 115 (1995). *ORCHIDACEAE*

aphylla *(Ridl. ex Oliv.) D.L. Szlachetko in Fragm. Flor. Geobot.,* Supp. 3: 116 (1995): Pelexia aphylla.

cogniauxiana *(Barb. Rodr. ex Cogn.) D.L. Szlachetko in Fragm. Flor. Geobot.,* Supp. 3: 116 (1995): Spiranthes cogniauxiana.

hassleri *(Cogn.) D.L. Szlachetko in Fragm. Flor. Geobot.,* Supp. 3: 116 (1995): Spiranthes hassleri.

lindmaniana *(Kraenzl.) D.L. Szlachetko in Fragm. Flor. Geobot.,* Supp. 3: 116 (1995): Spiranthes lindmaniana.

neuroptera *(Rchb. f. & Warm.) D.L. Szlachetko in Fragm. Flor. Geobot.,* Supp. 3: 116 (1995): Spiranthes neuroptera.

var. camposnovensis *(Hoehne) D.L. Szlachetko in Fragm. Flor. Geobot.,* Supp. 3: 116 (1995): Spiranthes camposnovensis.

rupestris *D.L. Szlachetko in Fragm. Flor. Geobot.,* Supp. 3: 116 (1995), nom. nov. as 'comb. nov.': Spiranthes rupestris *Barb. Rodr.*

sagittata *(Rchb. f. & Warm.) D.L. Szlachetko in Fragm. Flor. Geobot.,* Supp. 3: 116 (1995): Spiranthes sagittata.

var. depauperata *(Cogn.) D.L. Szlachetko in Fragm. Flor. Geobot.,* Supp. 3: 116 (1995): Spiranthes sagittata var. depauperata.

simplex *(Griseb.) D.L. Szlachetko in Fragm. Flor. Geobot.,* Supp. 3: 116 (1995): Spiranthes simplex.

sincorensis *(Schltr.) D.L. Szlachetko in Fragm. Flor. Geobot.,* 41(2): 862 (1996): Spiranthes sincorensis.

undulata *D.L. Szlachetko in Fragm. Flor. Geobot.,* 41(2): 862 (1996)—Brazil South (Santa Catarina).

VIBURNUM (Caprifoliac.).

farreri Stearn

var. stellipilum *D.Z. Ma & H.L. Liu in Fl. Ningxiaensis,* 2: 521 (1988 publ. 1990)—Inner Mongolia (Ningxia).

grandiflorum Wall. ex DC.

f. foetens *(Decne.) N.P. Taylor & D. Zappi in Curtis's Bot. Mag.,* 14(3): 122 (1997): V. foetens.

versatile *R.H. Miao, W.B. Liao & Q.Y. Sun in Acta Sci. Nat. Univ. Sunyatseni,* 38(6): 129 (1999)—China North-central (Shaanxi).

VICIA (Legumin.).

sect. Amurense *Y. Endo & H. Ohashi in J. Jap. Bot.,* 71(5): 259 (1996).

subsect. Grandiflorae *(Radzhi) M.M. Fedoronchuk in Ukr. Bot. Zhurn.,* 53(5): 593 (1996): V. ser. Grandiflorae.

VICIA:-
sect. Laticarpa (Stank.) M.M. Fedoronchuk in Ukr. Bot. Zhurn., 53(5): 595 (1996): V. subsect. Laticarpae.
subsect. Sativae (Buchenau) M.M. Fedoronchuk in Ukr. Bot. Zhurn., 53(5): 593 (1996): V. IV. Sativa.
angustifolia
subsp. pseudoangustifolia (Rouy) F.J. Tardío Pato in An. Jard. Bot. Madrid, 56(2): 266 (1998): V. communis var. pseudoangustifolia.
chianschanensis (P.Y. Fu & Y.A. Chen) Z.D. Xia in Acta Phytotax. Sin., 34(4): 431 (1996): V. ramuliflora f. chianschanensis.
erzurumica N. Demirkuş & S. Erik in Willdenowia, 28(1–2): 154 (1998)—Turkey.
graminea Sm.
var. nigricarpa N.R. Bastos & S.T.S. Miotto in Pesquisas, Bot., Bras., 46: 146 (1996)—Brazil South (Santa Catarina, Rio Grande do Sul).
japonica A. Gray
subsp. pallida (Turcz.) V.N. Voroschilov in A.K. Skvortsov (ed.), Florist. issl. v razn. raĭonakh SSSR: 179 (1985): V. pallida.
longidentata Z.D. Xia in Acta Phytotax. Sin., 34(3): 431 (1996)—China South-central (Sichuan).
monantha Retz.
subsp. calcarata (Desf.) C. Romero Zarco in An. Jard. Bot. Madrid, 57(1): 223 (1999): V. calcarata.
multijuga Z.D. Xia in Acta Phytotax. Sin., 34(4): 424 (1996), homonym; see Rech. f. in Bot. Jahrb. Syst., 75: 340 (1951)—China North-central (Shaanxi, Gansu).
narbonensis L.
f. joannis (Tamamasch.) M.M. Fedoronchuk & Dubovik in Ukr. Bot. Zhurn., 53(5): 595 (1996): V. joannis.
nipponica J. Matsumura in Bot. Mag. Tokyo, 16: 81 (1902)—Japan.
var. ramosa H. Nakam. ex T. Shimizu, Fl. Nagano Pref.: 1509 (1997)—Japan.
sepium L.
subsp. eriocalyx (Čelak.) G.H. Loos in Flor. Rundbr., 30(2): 153 (1996), isonym; see Holub in Folia Geobot. Phytotax., 8(2): 177 (1973): V. sepium β eriocalyx.
var. hartii J.R. Akeroyd in Watsonia, 21(1): 72 (1996)—Ireland, Great Britain.
ternata Z.D. Xia in Acta Phytotax. Sin., 34(4): 425 (1996)—China South-central (Sichuan)
unijuga A. Braun
f. albiflora Y.N. Lee, Fl. Korea: 1159 (1996), without type—Locality not stated.
subsp. minor (Nakai) Y.N. Lee, Fl. Korea: 1159 (1996): V. unijuga var. minor.
var. trifoliolata Z.D. Xia in Acta Phytotax. Sin., 34(4): 433 (1996)—China South-central (Sichuan), China North-central (Shaanxi, Shanxi).
wushanica Z.D. Xia in Acta Phytotax. Sin., 34(4): 430 (1996)—China North-central (Gansu).

VIDALASIA D.D. Tirvengadum in Biogeographica, 74(4): 164 (1998). RUBIACEAE
fusca (Craib) D.D. Tirvengadum in Biogeographica, 74(4): 165 (1998): Randia fusca.
morindifolia (Elmer) D.D. Tirvengadum in Biogeographica, 74(4): 166 (1998), as 'morindaefolia': Gardenia morindifolia.
murina (Craib) D.D. Tirvengadum in Biogeographica, 74(4): 165 (1998): Randia murina.
pubescens (Tirveng. & Sastre) D.D. Tirvengadum in Biogeographica, 74(4): 166 (1998): Fosbergia pubescens.
tonkinensis (Pit.) D.D. Tirvengadum in Biogeographica, 74(4): 165 (1998): Gardenia tonkinensis.

VIETNAMIA P.T. Li in J. S. China Agric Univ., 15(4): 72 (1994). ASCLEPIADACEAE
inflexa P.T. Li in J. S. China Agric. Univ., 15(4): 73 (1994)—Vietnam.

VIGNA (Legumin.).
spectabilis (Standl.) A. Delgado-Salinas in Novon, 8(4): 852 (1998): Phaseolus spectabilis.
unguiculata (L.) Walp.
subsp. aduensis R.S. Pasquet in Kew Bull., 52(4): 840 (1997)—Ethiopia.

VIGUIERA (Compos.).
ayutlana B.L. Turner in Phytologia, 79(5): 369 (1995 publ. 1996)—Mexico Southwest (Jalisco).
benthamii V.M. Badillo, Enum. Compuestas (Asterac.) Venezuela (Rev. Fac. Agron. (Maracay), 45): 128 (1994), nom. nov. superfl.: Hymenostephium angustifolium Benth.
grammatoglossa DC.
var. huajuapana J.L. Panero & J.L. Villaseñor in Brittonia, 48(4): 571 (1996 publ. 1997)—Mexico Southwest (Oaxaca).
inegii M.S. González Elizondo, M. González Elizondo & J. Rzedowski in Acta Bot. Mex., 53: 39 (2000)—Mexico Northeast (Durango).
meridensis (S.F. Blake) V.M. Badillo, Enum. Compuestas (Asterac.) Venezuela (Rev. Fac. Agron. (Maracay), 45): 129 (1994): Hymenostephium meridense.
oligantha M.S. González Elizondo, M. González Elizondo & J. Rzedowski in Acta Bot. Mex., 53: 37 (2000)—Mexico Northeast (Durango).
paranensis (Malme) J.U. Moreira dos Santos in Rodriguésia, 42–44(68–70): 47 (1990–92 publ. 1996)—Aspilia paranensis.

VILLADIA (Crassulac.).
cucullata Rose
subsp. apiculata R. Moran & C.H. Uhl in Cact. Succ. J. (U.S.A.), 70(4): 202 (1998)—Mexico Northeast (San Luis Potosí, Hidalgo, Nuevo León, Tamaulipas).
recurva R.V. Moran, M. Kimnach & C.H. Uhl in Cact. Suc. Mex., 43(1): 3 (1998)—Mexico Southwest (Oaxaca, Guerrero).

VILLARSIA (Menyanthac.).
goldblattiana R. Ornduff in Novon, 9(3): 407 (1999)—Cape Province.

VILLASENORIA B.L. Clark in Sida, 18(3): 632 (1999). COMPOSITAE
orcuttii (Greenm.) B.L. Clark in Sida, 18(3): 633 (1999): Senecio orcuttii.

VINCETOXICUM (Asclepiadac.).
assadii M. Zaeifi in Iranian J. Bot., 8(1): 106 (1999)—Iran.
brachystelmoides (P.I. Forst.) S. Liede in Taxon, 45(2): 205 (1996): Cynanchum brachystelmoides.
calcareum (H. Ohashi) S. Liede in Taxon, 45(2): 205 (1996): Cynanchum calcareum.
changqinianum P.T. Li in J. S. China Agric. Univ., 12(3): 40 (1991), nom. nov.: Asclepias paniculata Bunge.
christinae (P.I. Forst.) S. Liede in Taxon, 45(2): 205 (1996): Cynanchum christinae.
grandifolium (Hemsl.) P.T. Li in J. S. China Agric. Univ., 12(3): 41 (1991): Cynanchum grandifolium.
liebianum (F. Muell.) S. Liede in Taxon, 45(2): 205 (1996): Tylophora liebiana.
linearisepalum (P.T. Li) P.T. Li in J. S. China Agric. Univ., 12(3): 41 (1991): Cynanchum linearisepalum.
linearisepalum (P.T. Li) S. Liede in Taxon, 45(2): 205 (1996), isonym; see P.T. Li in J. S. China Agric. Univ., 12(3): 41 (1991): Cynanchum linearisepalum.
mozaffarianii M. Zaeifi in Iranian J. Bot., 8(1): 108 (1999)—Iran.
nanum (Buch.-Ham. ex Wight) S. Liede in Taxon, 45(2): 205 (1996): Cynanchum nanum.
otophyllum (C.K. Schneid.) P.T. Li in J. S. China Agric. Univ., 12(3): 41 (1991): Cynanchum otophyllum.
pingshanicum (M.G. Gilbert & P.T. Li) S. Liede in Taxon, 45(2): 206 (1996): Cynanchum pingshanicum.
rockii (M.G. Gilbert & P.T. Li) S. Liede in Taxon, 45(2): 206 (1996): Cynanchum rockii.
thailandense (P.T. Li) S. Liede in Taxon, 45(2): 206 (1996): Cynanchum thailandense.
tsiangii (P.T. Li) P.T. Li in J. S. China Agric. Univ., 12(3): 41 (1991): Cynanchum tsiangii.
utriculosum (Costantin) S. Liede in Taxon, 45(2): 206 (1996): Cynanchum utriculosum.
wangii (P.T. Li & W. Kittr.) S. Liede in Taxon, 45(2): 206 (1996): Cynanchum wangii.

VIOLA (Violac.).
sect. Adnatae (W. Becker) V.V. Nikitin in Bot. Zhurn., 83(3): 132 (1998); et in Novosti Sist. Vyssh. Rast., 31: 222 (1998), isonym; see C.J. Wang in Fl. Reipubl. Popul. Sin., 51: 41 (1991): V. grex Adnatae.
sect. Arction Juz. ex V.V. Zuev in Fl. Sibir. (Geraniac.-Cornac.), 10: 96 (1996), without type.
sect. Arosulatae (Borbás) V.V. Zuev in Fl. Sibir. (Geraniac.-Cornac.), 10: 88 (1996): V. subsect. Arosulatae.
subser. Arvenses (Drabble) M. Espeut in Monde Pl., 464: 37 (1999): V. sect. Arvenses.
subsect. Boreali-americanae (W. Becker) N.L. Gil-ad in Boissiera, 53: 42 (1997): V. "N." Boreali-Americanae.
ser. Calcaratae (W. Becker) M. Espeut in Monde Pl., 464: 37 (1999): V. [unranked] Calcaratae.
ser. Cuneatifoliae V.V. Nikitin in Bot. Zhurn., 83(3): 130 (1998), as 'Cuneatofoliae'.
ser. Dacicae V.V. Nikitin in Bot. Zhurn., 83(3): 135 (1998).
ser. Declinatae Klokov ex V.V. Nikitin in Bot. Zhurn., 83(3): 135 (1998).
ser. Denticulatae V.V. Nikitin in Bot. Zhurn., 83(3): 129 (1998).
supersect. Dischidium (Ging.) V.V. Nikitin in Bot. Zhurn., 83(3): 133 (1998); et in Novosti Sist. Vyssh. Rast., 31: 223 (1998): V. sect. Dischidium.
ser. Ebracteatae (Kupffer) V.V. Nikitin in Bot. Zhurn., 83(3): 135 (1998): V. subsect. Ebracteatae.
subsect. Eflagellatae (Kitt. ex Juz.) M. Espeut in Monde Pl., 464: 37 (1999), basionym not validly publ.: V. [unranked] Eflagellatae.
ser. Elatiores V.V. Nikitin in Bot. Zhurn., 83(3): 130 (1998).
sect. Estolonosae (Kupffer) V.V. Nikitin in Bot. Zhurn., 83(3): 132 (1998); et in Novosti Sist. Vyssh. Rast., 31: 222 (1998): V. grex Estolonosae.
ser. Fimbriatae V.V. Nikitin in Bot. Zhurn., 83(3): 129 (1998).
ser. Hirtae V.V. Nikitin in Bot. Zhurn., 83(3): 131 (1998).
ser. Hymettianae Klokov ex V.V. Nikitin in Bot. Zhurn., 83(3): 135 (1998).
sect. Icmasion Juz. ex N.N. Tzvelev, Opred. Sosud. Rast. Severo-Zap. Rossii: 679 (2000), without latin descr.
subsect. Infudibulares V.V. Nikitin in Bot. Zhurn., 83(3): 136 (1998).
ser. Lacteae V.V. Nikitin in Bot. Zhurn., 83(3): 130 (1998).
ser. Melanium (Ging.) V.V. Nikitin in Bot. Zhurn., 83(3): 135 (1998): V. sect. Melanium.
subsect. Melanium (Ging.) V.V. Nikitin in Bot. Zhurn., 83(3): 135 (1998): V. sect. Melanium.
sect. Mirabiles (Nyman) V.V. Zuev in Fl. Sibir. (Geraniac.-Cornac.), 10: 86 (1996), with incorrect basionym ref.: V. subsect. Mirabiles.
sect. Mirabiles (Nyman ex Borbás) V.V. Nikitin in Bot. Zhurn., 83(3): 130 (1998); et in Novosti Sist. Vyssh. Rast., 31: 217 (1998): V. grex Mirabiles.
ser. Occultae V.V. Nikitin in Bot. Zhurn., 83(3): 136 (1998).
subser. Parvulae (Klokov) M. Espeut in Monde Pl., 464: 37 (1999), basionym not validly publ.: V. ser. Parvulae.
supersect. Plagiostigma (Godr.) V.V. Nikitin in Bot. Zhurn., 83(3): 132 (1998); et in Novosti Sist. Vyssh. Rast., 31: 221 (1998): V. sect. Plagiostigma.
ser. Pyrenaicae V.V. Nikitin in Bot. Zhurn., 83(3): 131 (1998).
sect. Rosulantes (Borbás) V.V. Zuev in Fl. Sibir. (Geraniac.-Cornac.), 10: 87 (1996): V. subsect. Rosulantes.
ser. Rupestres Juz. ex V.V. Nikitin in Bot. Zhurn., 83(3): 129 (1998).
ser. Scotophyllae Juz. ex V.V. Nikitin in Bot. Zhurn., 83(3): 132 (1998).
ser. Subpubescens V.V. Nikitin in Bot. Zhurn., 83(3): 136 (1998).
ser. Tomentosae V.V. Nikitin in Bot. Zhurn., 83(3): 136 (1998).
ser. Tricolores (W. Becker) M. Espeut in Monde Pl., 464: 37 (1999): V. [unranked] Tricolores.
ser. Vericaninae Juz. ex V.V. Nikitin in Bot. Zhurn., 83(3): 130 (1998).

VIOLA:-
alba Besser
var. lactea *M. Espeut in Monde Pl.*, 464: 17 (1999), nom. nov.: V. dehnhardtii var. albiflora *W. Becker.*
var. picta *(Moggr.) M. Espeut in Monde Pl.*, 464: 16 (1999): V. hirta var. picta.
× albovii *V.V. Nikitin in Novosti Sist. Vÿssh. Rast.*, 31: 225 (1998)—Transcaucasus (Adzhariya). (V. oreades × orthoceras).
argenteria *B. Moraldo & G. Forneris in Allionia*, 28: 51, 52 (1987–88 publ. 1988), nom. nov.: V. nummularifolia *All.*
awagatakensis *T. Yamazaki, I. Ito & Ageishi in J. Jap. Bot.*, 72(1): 60 (1997)—Japan.
brevistipulata (Franch. & Sav.) W. Becker
var. crassifolia *(Koidz.) F. Maek. ex S. Akiyama, H. Ohba & S. Tabuchi in Fl. Japan*, 2c: 187 (1999): V. glabella var. crassifolia.
var. yezoana *Toyok. ex H. Nakai, H. Igarashi & Hiroyoshi Ohashi in J. Jap. Bot.*, 71(6): 357 (1996), nom. nov.: V. hidakana var. ciliata *Nakai ex Tatew.*
brittoniana Pollard
f. pectinata *(E.P. Bicknell) N.L. Gil-ad in Boissiera*, 53: 54 (1997): V. pectinata.
canina L.
var. contracta *V.V. Nikitin in Fl. Vostochnoĭ Evropÿ*, 9: 193 (1996)—Russia North.
castillonii *(W. Becker) C.C. Xifreda & A.M. Sanso in Zuloaga & Morrone (eds.), Cat. Pl. Vasc. Rep. Arg. 2 (Monogr. Syst. Bot.* Missouri Bot. Gard. 74): 1246 (1999): Viola exigua var. castillonii.
× chinoi F. Maek.
f. sordida *E. Hama ex T. Shimizu, Fl. Nagano Pref.*: 1505 (1997)—Japan.
donetzkiensis *M.V. Klokov in Novosti Sist. Vÿssh. Nizsh. Rast. (Kiev)*, 1974: 102 (1974)—Ukraina.
× eizasieboldii *Sugim. ex T. Shimizu, Fl. Nagano Pref.*: 1506 (1997)—Japan. (V. eizanensis × sieboldii).
fuscoviolacea *(L.G. Adams) T.A. James in Muelleria*, 9: 35 (1996): V. hederacea subsp. fuscoviolacea.
× glabrescens *(Neuman) N.N. Tzvelev, Opred. Sosud. Rast. Severo-Zap. Rossii*: 679 (2000): Viola arenaria var. glabrescens.
glandularis *H.E. Ballard & P.M. Jørgensen in Cat. Vasc. Pl. Ecuador (Monogr. Syst. Bot. Missouri Bot. Gard., 75)*: 956 (1999), nom. nov.: V. glandulifera *Hook.*
grypoceras A. Gray
var. ripensis *N. Yamada & M. Okamoto in Bull. Osaka Mus. Nat. Hist.*, 50: 2 (1996)—Japan.
× hyperborea *(Rupr.) V.V. Nikitin in Fl. Vostochnoĭ Evropÿ*, 9: 201 (1996): V. epipsila var. hyyperborea.
× hyrcanica *V.V. Nikitin in Novosti Sist. Vÿssh. Rast.*, 31: 213 (1998)—Transcaucasus (Azerbaydzhan). (V. sieheana × sylvestris).
× interposita *M. Kitagawa in Rep. Inst. Sci. Res. Manchoukuo*, 6(4): 123 (1942)—Manchuria. (V. prionantha × variegata).
ircutiana Turcz.
subsp. tenuicornis *(W. Becker) V.N. Voroschilov in A.K. Skvortsov (ed.), Florist. issl. v razn. raĭonakh SSSR*: 182 (1985): V. tenuicornis.
× jettmarii *Hand.-Mazz. in Oesterr. Bot. Z.*, 88(4): 302 (1939)—China North-Central (Shaanxi). (V. pinnata × prionantha).
kitaibeliana Schult.
subsp. trimestris *(DC.) M. Espeut in Monde Pl.*, 464: 20 (1999): V. tricolor var. trimestris.
kusanoana Makino
f. subrosea *H. Igarashi in J. Jap. Bot.*, 75(3): 187 (2000)—Japan.
lainzii *P. Montserrat Recoder in An. Jard. Bot. Madrid*, 54(1): 237 (1996)—Spain.
lehmannii *W. Becker ex H.E. Ballard & P.M. Jørgensen in Cat. Vasc. Pl. Ecuador (Monogr. Syst. Bot. Missouri Bot. Gard., 75)*: 956 (1999)—Ecuador.
× litoralis Spreng.
var. litvinovii *V.V. Nikitin in Fl. Vostochĭ Evropÿ*, 9: 194 (1996)—Russia North.
mandshurica W. Becker
var. patrinii *(DC.) Y.N. Lee, Fl. Korea*: 1159 (1996): V. patrinii.

VIOLA:-
× menitzkii *V.V. Nikitin in Novosti Sist. Vÿssh. Rast.*, 31: 230 (1998)—Transcausasus (Gruziya). (V. arvensis × occulta).
minuta M. Bieb.
subsp. dagestanica *(Rupr.) V.V. Nikitin in Novosti Sist. Vÿssh. Rast.*, 31: 224 (1998): V. minuta var. dagestanica.
subsp. meyeriana *(Rupr.) V.V. Nikitin in Novosti Sist. Vÿssh. Rast.*, 31: 225 (1998): V. minuta var. meyeriana.
mirabilis L.
subsp. brachysepala *(Maxim.) V.N. Voroschilov in A.K. Skvortsov (ed.), Florist. issl. v razn. raĭonakh SSSR*: 182 (1985), as 'brachisepala': V. brachysepala.
subsp. subglabra *(Ledeb.) V.V. Zuev in Fl. Sibir. (Geraniac.-Cornac.)*, 10: 87 (1996): V. mirabilis var. subglabra.
× miyajiana Koidz.
f. sordida *E. Hama ex T. Shimizu, Fl. Nagano Pref.*: 1506 (1997)—Japan.
nantouensis *S.S. Ying in Mem. Coll. Agric. Nation. Taiwan Univ.*, 28(2): 45 (1988)—Taiwan.
nemoralis Kütz.
var. abbreviata *V.V. Nikitin in Fl. Vostochnoĭ Evropÿ*, 9: 194 (1996)—Russia North.
novae-angliae House
subsp. grisea *(Fernald) N.L. Gil-ad in Boissiera*, 53: 68 (1997): V. septentrionalis var. grisea.
× obtusoacuminata *T. Hashim. ex T. Shimizu, Fl. Nagano Pref.*: 1506 (1997)—Japan. (V. acuminata × obtusa).
odorata L.
var. incompta *(Jord.) M. Espeut in Monde Pl.*, 464: 15 (1999): V. incompta.
× ogawae Nakai
f. variegata *E. Hama ex T. Shimizu, Fl. Nagano Pref.*: 1506 (1997)—Japan.
× okuharae *F. Maek. ex T. Shimizu, Fl. Nagano Pref.*: 1506 (1997)—Japan. (V. hirtipes × violacea var. makinoi).
parnonia *Kit Tan, G. Sfikas & G. Vold in Ann. Bot. Fenn.*, 34(3): 149 (1997)—Greece.
patrinii DC.
f. toyokoroensis *Koji Ito in J. Jap. Bot.*, 71(5): 301 (1996)—Japan.
phalacrocarpa Maxim.
f. plena *Okuhara ex T. Shimizu, Fl. Nagano Pref.*: 1506 (1997)—Japan.
polycephala *H.E. Ballard & P.M. Jørgensen in Novon*, 7(1): 13 (1997), nom. nov.: V. parvifolia *Benth.*
× popovae *V.V. Nikitin in Novosti Sist. Vÿssh. Rast.*, 31: 216 (1998)—North Caucasus (Stavropol). (V. nemoralis subsp. nemoralis × sieheana).
primorskajensis *(W. Becker) V.N. Voroschilov in A.K. Skvortsov (ed.), Florist. issl. v razn. raĭonakh SSSR*: 182 (1985): V. tenuicornis subsp. primorskajensis.
× pseudomakinoi *M. Mizush. ex T. Shimizu, Fl. Nagano Pref.*: 1507 (1997)—Japan. (V. sieboldii f. variegata × violacea var. makinoi).
rupestris F.W. Schmidt
subsp. glaberrima *(Murb.) V.V. Nikitin in Fl. Vostochnoĭ Evropÿ*, 9: 190 (1996): V. rupestris var. glaberrima.
nothosubsp. glabrescens *(Neuman) V.V. Nikitin in Fl. Vostochnoĭ Evropÿ*, 9: 190 (1996): V. arenaria var. glabrescens.
subsp. orioli-bolosii *J. Molero, L. Sáez & L. Villar in Acta Bot. Barcin.*, 45: 376 (1998)—Spain.
subsp. sacchalinensis *(H. Boissieu) V.N. Voroschilov in A.K. Skvortsov (ed.), Florist. issl. v razn. raĭonakh SSSR*: 182 (1985): V. sacchalinensis.
× savatieri Makino
f. variegata *E. Hama ex T. Shimizu, Fl. Nagano Pref.*: 1507 (1997)—Japan.
sergievskiae *N.N. Tzvelev, Opred. Sosud. Rast. Severo-Zap. Rossii*: 679 (2000), nom. nov.: Viola rupestris var. glaberrima *Murb.*
serresiana *M. Erben in Bot. Chronika*, 13: 51 (2000)—Greece.
sieboldii Maxim.
f. variegata *(S. Nagas. ex F. Maek.) F. Maek. & T. Hashim. ex T. Shimizu, Fl. Nagano Pref.*: 1507 (1997), homonym; *see E. Hama in J. Jap. Bot.*, 51(11): 342 (1976): V. pumilio f. variegata.

VIOLA:-
sororia Willd.
subsp. affinis *(Leconte) R.J. Little in Phytologia*, 80(4): 295 (1996): V. affinis.
suavis M. Bieb.
var. catalonica *(W. Becker) M. Espeut in Monde Pl.*, 464: 16 (1999): V. catalonica.
subandina *J.M. Watson in A. Hoffmann, M.K. Arroyo, F. Liberona, M. Muñoz & J. Watson, Pl.* Altoandinas Fl. Silv. Chile: 66 (1998), nom. nov.: V. pusilla *Hook. & Arn.*
subatlantica *(Maire) M. Ibn Tattou in Bocconea*, 8: 189 (1998): V. tricolor var. subatlantica.
× taradakensis Nakai
var. eizalacea *F. Maek. & T. Hashim. ex T. Shimizu, Fl. Nagano Pref.*: 1507 (1997)—Japan.
f. variegata *E. Hama ex T. Shimizu, Fl. Nagano Pref.*: 1507 (1997)—Japan.
tineorum *M. Erben & F.M. Raimondo in Giorn. Bot. Ital.*, 129(1): 80 (1995)—Sicilia.
ucriana *M. Erben & F.M. Raimondo in Giorn. Bot. Ital.*, 129(1): 85 (1995)—Sicilia.
uniflora L.
subsp. lasczinskyi *V.V. Zuev in Fl. Sibir. (Geraniac.-Cornac.)*, 10: 98 (1996)—West Siberia.
vaginata Maxim.
f. purpureocalcarata *H. Igarashi in J. Jap. Bot.*, 75(3): 187 (2000)—Japan.
× wilhelmii *V.V. Nikitin in Novosti Sist. Vÿssh. Rast.*, 31: 226 (1998)—North Caucasus (Krasnodar). (V. oreades × saxatilis).
woosanensis *Y.N. Lee & J. Kim in Korean J. Pl. Taxon.*, 28(1): 30 (1998)—South Korea.
yuzufelensis *A.P. Khokhrjakov in Byull. Glavn. Bot. Sada (Moscow)*, 175: 53 (1997), as 'yzufelensis'—Turkey.

VIORNA *É. Spach, Hist. Nat. Vég.*, 7: 268 (1838), homonym; *see Rchb., Handb. Nat. Pfl.-Syst.*: 277 (1837). *RANUNCULACEAE*

× **VIRAPHANDHUARA** *hort. in Orchid Rev.*, 108(1231): new orchid hybrids 15 (2000). *ORCHIDACEAE* (Aerides × Ascocentrum × Neofinetia × Rhynchostylis × Vanda).

VIROLA (Myristicac.).
aequatorialis *P. Muriel & H. Balslev in Nordic J. Bot.*, 20(4): 443 (2000)—Ecuador.
kwatae *D. Sabatier in Adansonia*, 19(2): 273 (1997)—French Guiana.

VISCUM (Viscac.).
calvinii *R.M. Polhill & D. Wiens, Mistletoes of Africa*: 295 (1998)—Tanzania.
chyuluense *R.M. Polhill & D. Wiens, Mistletoes of Africa*: 299 (1998)—Kenya.
congdonii *R.M. Polhill & D. Wiens in Novon*, 7(3): 278 (1997)—Tanzania, Malawi, Zimbabwe.
cylindricum *R.M. Polhill & D. Wiens in Novon*, 7(3): 279 (1997)—Malawi, Tanzania.
exile *B.A. Barlow in Blumea*, 41(2): 343 (1996)—Sulawesi.
gracile *R.M. Polhill & D. Wiens, Mistletoes of Africa*: 295 (1998)—Tanzania.
grandicaule *R.M. Polhill & D. Wiens, Mistletoes of Africa*: 304 (1998)—Gulf of Guinea Is. (Annobón).
griseum *R.M. Polhill & D. Wiens, Mistletoes of Africa*: 300 (1998)—Kenya.
hoolei *(Wiens) R.M. Polhill & D. Wiens, Mistletoes of Africa*: 298 (1998): V. capense subsp. hoolei.
iringense *R.M. Polhill & D. Wiens, Mistletoes of Africa*: 301 (1998)—Tanzania.
littorum *R.M. Polhill & D. Wiens, Mistletoes of Africa*: 299 (1998)—Mozambique.
loranthicola *R.M. Polhill & D. Wiens in Novon*, 7(3): 279 (1997)—Tanzania, Malawi, Zambia.
luisengense *R.M. Polhill & D. Wiens, Mistletoes of Africa*: 290 (1998)—Tanzania.
petiolatum *R.M. Polhill & D. Wiens, Mistletoes of Africa*: 283 (1998)—Zaire.
scurruloideum *B.A. Barlow in Blumea*, 41(2): 344 (1996)—Jawa.
subverrucosum *R.M. Polhill & D. Wiens, Mistletoes of Africa*: 300 (1998)—Malawi.

VISENIA (Sterculiac.).
velutina *(DC.) J.O. Voigt, Hort. Suburb. Calcutt.*: 109 (1845): Riedlea velutina.

VITEX (Verbenac.).
negundo L.
var. sichuanensis *J.L. Liu in Acta Phytotax. Sin.*, 33(5): 501 (1995)—China South-central (Sichuan).

VITIS (Vitac.).
ser. Adstrictae *F.S. Wang in J. Trop. Subtrop. Bot.*, 8(1): 7 (2000).
ser. Davidianae *F.S. Wang in J. Trop. Subtrop. Bot.*, 8(1): 7 (2000).
sect. Labruscoideae *(Planch.) F.S. Wang in J. Trop. Subtrop. Bot.*, 8(1): 3 (2000): V. ser. Labruscoideae.
ser. Piasezkianae *F.S. Wang in J. Trop. Subtrop. Bot.*, 8(1): 6 (2000).
sect. Romanetianae *F.S. Wang & C.S. Zhu in J. Trop. Subtrop. Bot.*, 8(1): 8 (2000).
sect. Sinocineriae *F.S. Wang & C.S. Zhu in J. Trop. Subtrop. Bot.*, 8(1): 4 (2000).
sect. Wuhanenses *F.S. Wang & C.S. Zhu in J. Trop. Subtrop. Bot.*, 8(1): 9 (2000).
amurensis Rupr.
var. funiushanensis *F.S. Wang in J. Trop. Subtrop. Bot.*, 8(1): 5 (2000)—China Southeast (Henan).
baihensis *L.X. Niu in Acta Univ. Agric. Bor.-Occid.*, 23(5): 122 (1995)—China North-central (Shaanxi).
baileyana *T.V. Munson, Vitis baileyana, Munson (leaflet)*: [1] (1893), nom. nov.; *see Bull. Torrey Bot. Club*, 20: 500 (1893): V. virginiana *Munson*.
balanseana Planch.
var. ficifolioides *(W.T. Wang) C.L. Li in Chinese J. Appl. Environ. Biol.*, 2(2): 240 (1996): V. ficifolioides.
var. tomentosa *C.L. Li in Chinese J. Appl. Environ. Biol.*, 2(3): 240 (1996)—China Southeast (Guangxi).
bashanica *P.C. He in Acta Univ. Agric. Bor.-Occid.*, 23(5): 121 (1995)—China North-central (Shaanxi).
bellula (Rehder) W.T. Wang
var. pubigera *C.L. Li in Chinese J. Appl. Environ. Biol.*, 2(3): 251 (1996)—China Southeast (Guangdong, Guangxi, Hunan).
bryoniifolia Bunge
var. ternata *(W.T. Wang) C.L. Li in Chinese J. Appl. Environ. Biol.*, 2(3): 253 (1996): V. adstricta var. ternata.
cinerea (Engelm.) Engelm. ex Millardet
var. tomentosa *(Planch.) B.L. Comeaux in Sida*, 19(1): 129 (2000): Vitis berladieri var. tomentosa.
davidiana *(Carrière) N.E. Brown, Suppl. Johnson's Gard. Dict.*: 1021 (1882): Cissus davidiana.
davidii
var. hispida *X.D. Wang & S.C. Chen in Acta Phytotax. Sin.*, 37(6): 603 (1999)—China Southeast (Fujian).
var. ningqiangensis *L.X. Niu in Acta Univ. Agric. Bor.-Occid.*, 23(5): 123 (1995)—China North-central (Shaanxi).
fengqinensis *C.L. Li in Chinese J. Appl. Environ. Biol.*, 2(3): 245 (1996)—China South-central (Yunnan).
flexuosa Thunb.
var. wallichii *(DC.) F.S. Wang in J. Trop. Subtrop. Bot.*, 8(1): 5 (2000): V. wallichii.
hekouensis *C.L. Li in Chinese J. Appl. Environ. Biol.*, 2(3): 246 (1996)—China South-central (Yunnan).
heyneana Roem. & Schult.
subsp. ficifolia *(Bunge) C.L. Li in Chinese J. Appl. Environ. Biol.*, 2(3): 250 (1996): V. ficifolia.
lanceolatifoliosa *C.L. Li in Chinese J. Appl. Environ. Biol.*, 2(3): 253 (1996), as 'laceolatifoliosa'—China Southeast (Hunan, Jiangxi, Guangdong, Guangxi).
luochengensis W.T. Wang
var. tomentosinerva *C.L. Li in Chinese J. Appl. Environ. Biol.*, 2(3): 241 (1996), as 'tomentoso-nerva'—China Southeast (Guangdong).
menghaiensis *C.L. Li in Chinese J. Appl. Environ. Biol.*, 2(3): 250 (1996)—China South-central (Yunnan).

VITIS:-
mengziensis *C.L. Li in Chinese J. Appl. Environ. Biol.*, 2(3): 245 (1996)—China South-central (Yunnan).
qinlingensis *P.C. He in Acta Univ. Agric. Bor.-Occid.*, 23(5): 122 (1995)—China North-central (Shaanxi).
ruyuanensis *C.L. Li in Chinese J. Appl. Environ. Biol.*, 2(3): 244 (1996)—China Southeast (Guangdong).
shenxiensis *C.L. Li in Chinese J. Appl. Environ. Biol.*, 2(3): 239 (1996)—China North-central (Shaanxi).
tiubaensis *L.X. Niu in Acta Univ. Agric. Bor.-Occid.*, 23(5): 123 (1995)—China North-central (Shaanxi).
wuhanensis *C.L. Li in Chinese J. Appl. Environ. Biol.*, 2(3): 243 (1996)—China South-central (Hubei), China Southeast (Henan, Jiangxi).
var. arachnoidea *X.D. Wang & C.L. Li in Acta Phytotax. Sin.*, 37(6): 604 (1999)—China Southeast (Jiangxi).
xunyangensis *P.C. He in Acta Univ. Agric. Bor.-Occid.*, 23(5): 121 (1995)—China North-central (Shaanxi).
yunnanensis *C.L. Li in Acta Bot. Yunnanica*, 19(3): 217 (1997)—China South-central (Yunnan).

VOCHYSIA (Vochysiac.).
gentryi *L. Marcano-Berti in Pittieria*, 20: 113 (1993)—Colombia.
lopezpalacioi *L. Marcano-Berti in Pittieria*, 20: 112 (1993)—Colombia.
ortegae *L. Marcano-Berti & J. Bautista Bautista in Pittieria*, 24: 63 (1996)—Venezuela.
penae *L. Marcano-Berti in Pittieria*, 20: 111 (1993)—Venezuela.

VOLKAMERIA (Verbenac.).
bicolor *T. Hardwicke in Asiat. Res.*, 6: 366 (1799)—India.

VOLUTARIA (Compos.).
djiboutensis *G. Wagenitz in Ann. Naturhist. Mus. Wien*, 98B(Suppl.): 178 (1996)—Djibouti.

VOTOMITA (Melastomatac.).
ventuarensis *T. Morley in Novon*, 9(2): 241 (1999)—Venezuela.

VOUARANA (Sapindac.).
anomala *(Steyerm.) P. Acevedo-Rodríguez in Biollania*, ed. esp. 6 (J.J. Wurdack Festschr.): 146 (1997): Toulicia anomala.

VOYRIA (Gentianac.).
subgen. Leiphaimos *(Schltdl. & Cham.) V.A. Albert & L. Struwe in Brittonia*, 49(4): 473 (1997). gen. Leiphaimos.
aurantiaca Splitg.
subsp. monantha *J.S. Pringle in Fl. Ecuador*, 53(159A): 122 (1995)—Peru.

× **VRIECANTAREA** *J.R. Grant in Phytologia*, 79(3): 256 (1995 publ. 1996). *BROMELIACEAE* (Alcantarea × Vriesea).

VRIESEA (Bromeliac.).
bahiana *E.M.C. Leme in J. Bromeliad Soc.*, 45(4): 150 (1995)—Brazil Northeast (Bahia).
barii *J.F. Morales in Novon*, 9(3): 401 (1999)—Costa Rica.
calimaniana *E.M.C. Leme & W. Till in Bromélia*, 4(2): 9 (1997)—Brazil Southeast (Espírito Santo).
capixabae *E.M.C. Leme in Harvard Pap. Bot.*, 4(1): 150 (1999)—Brazil Southeast (Espírito Santo).
chapadensis *E.M.C. Leme in Harvard Pap. Bot.*, 4(1): 157 (1999)—Brazil Northeast (Bahia).
diamantinensis *E.M.C. Leme in J. Bromeliad Soc.*, 47(4): 168 (1997)—Brazil Southeast (Minas Gerais).
exaltata *E.M.C. Leme in Harvard Pap. Bot.*, 4(1): 159 (1999)—Brazil Northeast (Bahia).
fabioi *E.M.C. Leme in Harvard Pap. Bot.*, 4(1): 153 (1999)—Brazil Northeast (Bahia).
fidelensis *E.M.C. Leme in Bradea*, 8(21): 118 (1999)—Brazil Southeast (Rio de Janeiro).
fradensis *And. Costa in Bromélia*, 2(4): 7 (1995)—Brazil Southeast (Rio de Janeiro).

VRIESEA:-
garlippiana *E.M.C. Leme in J. Bromeliad Soc.*, 47(3): 99 (1997)—Brazil Southeast (Rio de Janeiro).
grandiflora *E.M.C. Leme in J. Bromeliad Soc.*, 50(2): 52 (2000)—Brazil Southeast (Rio de Janeiro).
gruberi *R. Ehlers in Bromelie*, 1999(1): 10 (1999)—Colombia.
haberi *J.F. Morales in Novon*, 9(3): 402 (1999), as 'haberii'—Costa Rica.
hasei *R. Ehlers in Bromelie*, 1998(2): 55 (1998)—Venezuela.
jonesiana *E.M.C. Leme in Harvard Pap. Bot.*, 4(1): 154 (1999)—Brazil Southeast (São Paulo).
lutheri *J.M. Manzanares & W. Till in J. Bromeliad Soc.*, 50(4): 169 (2000)—Ecuador.
minor *(L.B. Sm.) E.M.C. Leme in J. Bromeliad Soc.*, 46(6): 245 (1996): V. bituminosa var. minor.
minuta *E.M.C. Leme in Bromélia*, 2(4): 24 (1995)—Brazil Northeast (Bahia).
mollis *E.M.C. Leme in Harvard Pap. Bot.*, 4(1): 156 (1999)—Brazil Southeast (Rio de Janeiro).
nanuzae *E.M.C. Leme in J. Bromeliad Soc.*, 47(4): 172 (1997)—Brazil Southeast (Minas Gerais).
oleosa *E.M.C. Leme in Harvard Pap. Bot.*, 4(1): 160 (1999)—Brazil Northeast (Bahia).
osaensis *J.F. Morales in Novon*, 9(3): 403 (1999)—Costa Rica.
pseudoatra *E.M.C. Leme in Harvard Pap. Bot.*, 4(1): 162 (1999)—Brazil Southeast (Rio de Janeiro).
rafaelii *E.M.C. Leme in Harvard Pap. Bot.*, 4(1): 163 (1999)—Brazil Southeast (Minas Gerais).
repandostachys *E.M.C. Leme in Harvard Pap. Bot.*, 4(1): 151 (1999)—Brazil Southeast (Espírito Santo).
sandrae *E.M.C. Leme in J. Bromeliad Soc.*, 45(4): 147 (1995)—Brazil Northeast (Bahia).
sazimae *E.M.C. Leme in Bromélia*, 2(4): 26 (1995)—Brazil Southeast (São Paulo).
simulans *E.M.C. Leme in J. Bromeliad Soc.*, 47(4): 169 (1997)—Brazil Southeast (Minas Gerais).
simulans *J.F. Morales in Novon*, 9(3): 404 (1999), homonym; *see Leme in J. Bromeliad Soc.*, 47(4): 169 (1997)—Costa Rica.
tillii *J.M. Manzanares in J. Bromeliad Soc.*, 48(4): 169 (1998)—Ecuador.
tiquirensis *J.F. Morales in Novon*, 9(3): 405 (1999)—Costa Rica.
vulcanicola *J.F. Morales in Novon*, 9(3): 405 (1999)—Costa Rica.
ynghii *(Carrière) N.E. Brown, Suppl. Johnson's Gard. Dict.*: 1021 (1882): Encholirion ynghii.

VRYDAGZYNEA (Orchidac.).
albida (Blume) Blume
var. formosana *(Hayata) T. Hashimoto in Proc. 12th World Orchid Conf., 1987 (eds. K. Saito & R. Tanaka)*: 122 (1987): V. formosana.
brassii *P. Ormerod in Austral. Orchid Rev.*, 63(1): 13 (1998)—New Guinea (Papua New Guinea).
sessilifolia *P. Ormerod in Austral. Orchid Rev.*, 63(1): 13 (1998)—New Guinea (Papua New Guinea).

VULPIA (Gramin.).
muralis (Kunth) Nees
var. concinna *(Paunero) J. do Amaral Franco & M.L. Rocha Afonso in Silva Lusitana*, 5(1): 141 (1997): V. broteri var. concinna.
var. spicata *J. do Amaral Franco & M.L. Rocha Afonso in Silva Lusitana*, 5(1): 141 (1997)—Portugal.

W

WACHENDORFIA (Haemodorac.).
multiflora *(Klatt) J.C. Manning & P. Goldblatt in Strelitzia*, 9: 709 (2000): Babiana multiflora.

WAHLENBERGIA (Campanulac.).
akaroa *J.A. Petterson in New Zealand J. Bot.*, 35(1): 24 (1997)—New Zealand South I.

WAHLENBERGIA:-
albomarginata Hook.
subsp. decora *J.A. Petterson in New Zealand J. Bot.*, 35(1): 39 (1997)—New Zealand South I.
subsp. flexilis *(Petrie) J.A. Petterson in New Zealand J. Bot.*, 35(1): 37 (1997): W. flexilis
subsp. laxa *(G. Simpson) J.A. Petterson in New Zealand J. Bot.*, 35(1): 38 (1997): W. laxa.
subsp. olivina *J.A. Petterson in New Zealand J. Bot.*, 35(1): 36 (1997)—New Zealand South I.
brehmeri *T.G. Lammers in Novon*, 8(1): 35 (1998), nom. nov.: W. rotundifolia *Brehmer.*
celata *P.I. Forster in Austrobaileya*, 5(4): 661 (2000)—Queensland.
congesta (Cheeseman) N.E. Br.
subsp. haastii *J.A. Petterson in New Zealand J. Bot.*, 35(1): 32 (1997)—New Zealand South I.
denticulata (Burch.) A.DC.
var. transvaalensis *(Adamson) W.G. Welman in Bothalia*, 26(2): 157 (1996): Lightfootia denticulata var. transvaalensis.
littoricola
subsp. vernicosa *(J.A. Petterson) P.J. de Lange & E.K. Cameron in New Zealand J. Bot.*, 37(3): 435 (1999): W. vernicosa.
longifolia (A.DC.) Lammers
var. corymbosa *(Adamson) W.G. Welman in Bothalia*, 26(2): 157 (1996): Lightfootia longifolia var. corymbosa.
pygmaea Colenso
subsp. drucei *J.A. Petterson in New Zealand J. Bot.*, 35(1): 44 (1997); *see also New Zealand J. Bot.*, 35 (2): 267 (1997)—New Zealand North I.
subsp. tararua *J.A. Petterson in New Zealand J. Bot.*, 35(1): 43 (1997)—New Zealand North I.
rubens (H. Buek) Lammers
var. brachyphylla *(Adamson) W.G. Welman in Bothalia*, 26(2): 157 (1996): Lightfootia rubens var. brachyphylla.
rubioides (Banks ex A.DC.) Lammers
var. stokoei *(Adamson) W.G. Welman in Bothalia*, 26(2): 157 (1996): Lightfootia rubioides var. stokoei.
subulata (L'Hér.) Lammers
var. congesta *(Adamson) W.G. Welman in Bothalia*, 26(2): 157 (1996): Lightfootia subulata var. congesta.
var. tenuifolia *(Adamson) W.G. Welman in Bothalia*, 26(2): 157 (1996): Lightfootia subulata var. tenuifolia.
tenella (L.f.) Lammers
var. palustris *(Adamson) W.G. Welman in Bothalia*, 26(2): 157 (1996): Lightfootia diffusa var. palustris.
var. stokoei *(Adamson) W.G. Welman in Bothalia*, 26(2): 157 (1996): Lightfootia diffusa var. stokoei.
tenerrima (H. Buek) Lammers
var. montana *(Adamson) W.G. Welman in Bothalia*, 26(2): 157 (1996): Lightfootia tenella var. montana.
uitenhagensis (H. Buek) Lammers
var. debilis *(Sond.) W.G. Welman in Bothalia*, 26(2): 157 (1996), without basionym ref.; *et in Bothalia*, 27(2): 140 (1997), with basionym ref.: Lightfootia ciliata var. debilis.
var. filifolia *(Adamson) W.G. Welman in Bothalia*, 26(2): 157 (1996): Lightfootia divaricata var. filifolia.
vernicosa *J.A. Petterson in New Zealand J. Bot.*, 35(1): 26 (1997)—New Zealand North I., Chatham Is., Kermadec Is.
violacea *J.A. Petterson in New Zealand J. Bot.*, 35(1): 17 (1997)—New Zealand North I., New Zealand South I.

× **WAILAIARA** *hort. in Orchid Rev.*, 107(1226): 126 (1999). *ORCHIDACEAE* (Aerides × Arachnis × Ascocentrum × Rhynchostylis × Vanda).

× **WAIRONARA** *hort. in Orchid Rev.*, 105(1215): 190 (1997). *ORCHIDACEAE* (Aerides × Renanthera × Rhynchostylis × Vanda).

WALLENIA (Myrsinac.).
maestrensis *C. Panfet Valdés & I. Ventosa Rodríguez in Willdenowia*, 30(1): 141 (2000)—Cuba.

WALSURA (Meliac.).
trifolia (A. Juss.) Harms
var. ternata *(Roxb.) H.O. Saxena in H.O. Saxena & M. Brahmam, Fl. Orissa*, 1: 284 (1994): W. ternata.

WALTHERIA (Sterculiac.).
laxa *M. Thulin in Nordic J. Bot.*, 19(1): 13 (1999)—Somalia.

WARMINGIA (Orchidac.).
buchtienii *(Schltr.) Schltr. ex Garay & E.A. Christenson in Lindleyana*, 11(1): 25 (1996): Macradenia buchtienii.

WARNOCKIA *M.W. Turner in Pl. Syst. Evol.*, 203(1–2): 78 (1996), nom. nov.: Brazoria § Stachyastrum *Engelm. & A. Gray.* (Labiat.).
scuttelarioides *(Engelm. & A. Gray) M.W. Turner in Pl. Syst. Evol.*, 203(1–2): 79 (1996): Brazoria scuttelarioides.

WASABIA (Crucif.).
japonica (Miq.) Matsum.
var. koreana *(Nakai) Y.N. Lee, Fl. Korea:* 1158 (1996): W. koreana.

WATSONIA (Iridac.).
meriana (L.) Mill.
var. bulbillifera *(J.W. Mathews & L. Bolus) D.A. Cooke in J. Adelaide Bot. Gard.*, 18(1): 6 (1998): W. bulbillifera.

WEDELIA (Compos.).
africana Pers.
var. ambigua *(C.D. Adams) M.A. Isawumi in Feddes Repert.*, 108(7–8): 558 (1997): Aspilia africana var. ambigua.
var. guineensis *(O. Hoffm. & Muschl.) M.A. Isawumi in Feddes Repert.*, 108(7–8): 558 (1997): Aspilia guineensis.
var. minor *(C.D. Adams) M.A. Isawumi in Feddes Repert.*, 108(7–8): 558 (1997): Aspilia africana var. minor.
angustifolia *(Oliv. & Hiern) M.A. Isawumi in Feddes Repert.*, 108(7–8): 559 (1997), homonym; *see B.L. Turner in Phytologia*, 72(5): 391 (1992): Aspilia angustifolia.
bracteosa *(C.D. Adams) M.A. Isawumi in Feddes Repert.*, 108(7–8): 558 (1997): Aspilia bracteosa.
bussei *(O. Hoffm. & Muschl.) M.A. Isawumi in Feddes Repert.*, 108(7–8): 560 (1997): Aspilia bussei.
chevalieri *(O. Hoffm. & Muschl.) M.A. Isawumi in Feddes Repert.*, 108(7–8): 561 (1997): Aspilia chevalieri.
ciliata *(Schumach.) A. Soldano in Compositae Newsl.*, 32: 20 (1998): Verbesina ciliata.
helianthoides *(Schumach.) M.A. Isawumi in Feddes Repert.*, 108(7–8): 560 (1997), without basionym page: Coronocarpus helianthoides.
subsp. ciliata *(Schumach.) M.A. Isawumi in Feddes Repert.*, 108(7–8): 560 (1997), without basionym page: Verbesina ciliata.
subsp. papposa *(O. Hoffm. & Muschl.) M.A. Isawumi in Feddes Repert.*, 108(7–8): 560 (1997), species name not validly publ.: Aspilia helianthoides var. papposa.
subsp. prieuriana *(DC.) M.A. Isawumi in Feddes Repert.*, 108(7–8): 560 (1997), without basionym page: Blainvillea prieuriana.
kotschyi *(Sch. Bip.) M.A. Isawumi in Feddes Repert.*, 108(7–8): 559 (1997), without basionym page: Dipterotheca kotschyi.
var. alba *(Berhaut) M.A. Isawumi in Feddes Repert.*, 108(7–8): 559 (1997), species name not validly publ.: Aspilia kotschyi var. alba.
kotschyi *(Sch. Bip.) A. Soldano in Compositae Newsl.*, 32: 20 (1998): Dipterotheca kotschyi.
linearifolia *(Oliv. & Hiern) M.A. Isawumi in Feddes Repert.*, 108(7–8): 559 (1997), homonym; *see Baker in Mart., Fl. Bras.*, 6(3): 182 (1884): Aspilia linearifolia.
loxensis *H. Robinson in Phytologia*, 84(5): 345 (1998 publ. 1999)—Ecuador.
meridensis *(V.M. Badillo) V.M. Badillo, Enum. Compuestas (Asterac.) Venezuela (Rev. Fac. Agron. (Maracay), 45):* 131 (1994): Aspilia meridensis.

WEDELIA:-
mortonii *(C.D. Adams) M.A. Isawumi in Feddes Repert.*, 108(7–8): 559 (1997): Aspilia mortonii.
paludosa *(Berhaut) M.A. Isawumi in Feddes Repert.*, 108(7–8): 559 (1997), homonym; *see DC., Prodr.*, 5: 538 (1836): Aspilia paludosa.
rudis *(Baker) H. Robinson in Phytologia*, 82(1): 61 (30 Sept. 1997): Zexmenia rudis.
rudis *(Oliv. & Hiern) M.A. Isawumi in Feddes Repert.*, 108(7–8): 561 (Nov. 1997), homonym; *see H. Rob. in Phytologia*, 82(1): 61 (30 Sept. 1997): Aspilia rudis.
subsp. fontinaloides *(C.D. Adams) M.A. Isawumi in Feddes Repert.*, 108(7–8): 561 (1997): Aspilia rudis subsp. fontinaloides.
spenceriana *(Muschl.) M.A. Isawumi in Feddes Repert.*, 108(7–8): 561 (1997): Aspilia spenceriana.
zuliana *S. Jiménez H.in Acta Bot. Venezuel.*, 21(2): 107 (1998 publ. 1999)—Venezuela.

WEHRAUHIA (Bromeliac.).
millennia *J.R. Grant in J. Bromeliad Soc.*, 50(1): 3 (2000)—Panama.

WEINGARTIA (Cactac.).
jarmilae *J.J. Halda & L. Horáček in Acta Mus. Richnov. Sect. Nat.*, 7(1): 40 (2000)—Bolivia.
neumanniana
subsp. kargliana *(Rausch) R. Slaba in Kaktusy (Brno)*, 35(3): 71 (1999): W. kargliana.

WEINMANNIA (Cunoniac.).
sect. Fasciculata *Bernardi ex R.D. Hoogland & H.C. Hopkins in Adansonia*, 20(1): 21 (1998).
coodei *H.C. Hopkins in Adansonia*, 20(1): 52 (1998)—Sulawesi.
croftii *H.C. Hopkins in Adansonia*, 20(1): 76 (1998)—Bismarck Archipelago.
devogelii *H.C. Hopkins in Adansonia*, 20(1): 48 (1998)—Sulawesi.
eymana *H.C. Hopkins in Adansonia*, 20(1): 50 (1998), as 'eymaeana'—Sulawesi.
furfuracea *H.C. Hopkins in Adansonia*, 20(1): 49 (1998)—Sulawesi, Moluccas.
hooglandii *H.C. Hopkins & J.C. Bradford in Adansonia*, 20(1): 37 (1998)—Malaya (Peninsular Malaysia).
loxensis *G. Harling in Fl. Ecuador*, 61(78): 44 (1999)—Ecuador.
marquesana F. Br.
var. myrsinites *(Fosberg & Sachet) H.C. Hopkins & J. Florence in Adansonia*, 20(1): 122 (1998): W. parviflora var. myrsinites.
ouaiemensis *(Guillaumin & Virot) R.D. Hoogland in Adansonia*, 20(1): 102 (1998): Cunonia ouaiemensis.
tremuloides *H.C. Hopkins & J. Florence in Adansonia*, 20(1): 123 (1998)—Marquesas.

WELLSTEDIA (Boraginac.).
laciniata *M. Thulin & A.N.B. Johansson in Biodiversity African Pl. (Proc. XIV AETFAT Congr.), eds. L.J.G. van der* Maesen et al.: 81 (1996)—Somalia.
robusta *M. Thulin in Nordic J. Bot.*, 18(6): 663 (1998 publ. 1999)—Somalia.
somalensis *M. Thulin & A.N.B. Johansson in Biodiversity African Pl. (Proc. XIV AETFAT Congr.), eds. L.J.G. van der* Maesen et al.: 83 (1996)—Somalia.

WERAUHIA *J.R. Grant in Trop. Subtrop. Pflanzenwelt*, 91: 28 (1995). *BROMELIACEAE*
sect. Jutleya *J.R. Grant in Trop. Subtrop. Pflanzenwelt*, 91: 39 (1995).
acuminata *(Mez & Wercklé) J.R. Grant in Trop. Subtrop. Pflanzenwelt*, 91: 30 (1995): Vriesea acuminata.
ampla *(L.B. Sm.) J.R. Grant in Trop. Subtrop. Pflanzenwelt*, 91: 30 (1995): Vriesea ampla.
apiculata *(L.B. Sm.) J.R. Grant in Trop. Subtrop. Pflanzenwelt*, 91: 30 (1995): Vriesea apiculata.
attenuata *(L.B. Sm. & Pittendr.) J.R. Grant in Trop. Subtrop. Pflanzenwelt*, 91: 39 (1995): Vriesea attenuata.
balanophora *(Mez) J.R. Grant in Trop. Subtrop. Pflanzenwelt*, 91: 39 (1995): Guzmania balanophora.

WERAUHIA:-
bicolor *(L.B. Sm.) J.R. Grant in Trop. Subtrop. Pflanzenwelt,* 91: 30 (1995): Vriesea bicolor.
boliviana *H.E. Luther in Selbyana,* 19(2): 218 (1998 publ. 1999)—Bolivia.
bracteosa *(Mez & Wercklé) J.R. Grant in Trop. Subtrop. Pflanzenwelt,* 91: 40 (1995): Thecophyllum bracteosum.
breedloveana *(L.B. Sm.) J.R. Grant in Trop. Subtrop. Pflanzenwelt,* 91: 31 (1995): Vriesea breedloveana.
broadwayi *(L.B. Sm.) J.R. Grant in Phytologia,* 79(3): 255 (1995 publ. 1996): Vriesea broadwayi.
brunei *(Mez & Wercklé) J.R. Grant in Trop. Subtrop. Pflanzenwelt,* 91: 31 (1995): Vriesea brunei.
capitata *(Mez & Wercklé) J.R. Grant in Trop. Subtrop. Pflanzenwelt,* 91: 40 (1995): Thecophyllum capitatum.
chiapensis *(Matuda) J.R. Grant in Trop. Subtrop. Pflanzenwelt,* 91: 31 (1995): Vriesea chiapensis.
comata *(Mez & Wercklé) J.R. Grant in Trop. Subtrop. Pflanzenwelt,* 91: 40 (1995): Thecophyllum comatum.
cornus-cervi *(Rohweder) J.R. Grant in Trop. Subtrop. Pflanzenwelt,* 91: 31 (1995): Vriesea cornus-cervi.
cowellii *(Mez & Britton) J.R. Grant in Trop. Subtrop. Pflanzenwelt,* 91: 40 (1995): Tillandsia cowellii.
diantha *(H. Luther) J.R. Grant in Trop. Subtrop. Pflanzenwelt,* 91: 40 (1995): Vriesea diantha.
dodsonii *(L.B. Sm.) J.R. Grant in Phytologia,* 79(3): 255 (1995 publ. 1996): Vriesea dodsonii.
dressleri *(Rauh) J.R. Grant in Trop. Subtrop. Pflanzenwelt,* 91: 42 (1995): Vriesea dressleri.
gibba *(L.B. Sm.) J.R. Grant in Phytologia,* 79(3): 255 (1995 publ. 1996): Vriesea gibba.
gigantea *(Mart. ex Schult. f.) J.R. Grant in Trop. Subtrop. Pflanzenwelt,* 91: 31 (1995): Vriesea gigantea.
gladioliflora *(H. Wendl.) J.R. Grant in Trop. Subtrop. Pflanzenwelt,* 91: 31 (1995): Tillandsia gladioliflora.
graminifolia *(Mez & Wercklé) J.R. Grant in Trop. Subtrop. Pflanzenwelt,* 91: 32 (1995): Vriesea graminifolia.
greenbergii *(Utley) J.R. Grant in Trop. Subtrop. Pflanzenwelt,* 91: 42 (1995): Vriesea greenbergii.
guadelupensis *(Baker) J.R. Grant in Phytologia,* 79(3): 255 (1995 publ. 1996) Tillandsia guadelupensis.
hainesiorum *(L.B. Sm.) J.R. Grant in Trop. Subtrop. Pflanzenwelt,* 91: 42 (1995): Vriesea hainesiorum.
haltonii *(H. Luther) J.R. Grant in Trop. Subtrop. Pflanzenwelt,* 91: 42 (1995): Vriesea haltonii.
haplostachya *(Sauvalle) J.R. Grant in Phytologia,* 79(3): 255 (1995 publ. 1996) Tillandsia haplostachya.
hygrometrica *(André) J.R. Grant in Trop. Subtrop. Pflanzenwelt,* 91: 42 (1995): Caraguata hygrometrica.
kathyae *(Utley) J.R. Grant in Trop. Subtrop. Pflanzenwelt,* 91: 42 (1995): Vriesea kathyae.
kupperiana *(Suess.) J.R. Grant in Trop. Subtrop. Pflanzenwelt,* 91: 32 (1995): Vriesea kupperiana.
latissima *(Mez & Wercklé) J.R. Grant in Trop. Subtrop. Pflanzenwelt,* 91: 43 (1995): Thecophyllum latissimum.
laxa *(Mez & Wercklé) J.R. Grant in Trop. Subtrop. Pflanzenwelt,* 91: 43 (1995): Thecophyllum laxum.
leucophylla *(L.B. Sm.) J.R. Grant in Trop. Subtrop. Pflanzenwelt,* 91: 43 (1995): Vriesea leucophylla.
luis-gomezii *(Utley) J.R. Grant in Trop. Subtrop. Pflanzenwelt,* 91: 43 (1995): Vriesea luis-gomezii.
lutheri *S. Pierce & J.E. Aranda in J. Bromeliad Soc.,* 50(5): 206 (2000)—Panama.
lyman-smithii *(Utley) J.R. Grant in Trop. Subtrop. Pflanzenwelt,* 91: 43 (1995): Vriesea lyman-smithii.
marnier-lapostollei *(L.B. Sm.) J.R. Grant in Trop. Subtrop. Pflanzenwelt,* 91: 33 (1995): Vriesea marnier-lapostollei.
nephrolepis *(L.B. Sm. & Pittendr.) J.R. Grant in Trop. Subtrop. Pflanzenwelt,* 91: 44 (1995): Vriesea nephrolepis.

WERAUHIA:-
nocturna *(Matuda) J.R. Grant in Trop. Subtrop. Pflanzenwelt,* 91: 33 (1995): Vriesea nocturna.
notata *(L.B. Sm. & Pittendr.) J.R. Grant in Trop. Subtrop. Pflanzenwelt,* 91: 44 (1995): Vriesea notata.
nutans *(L.B. Sm.) J.R. Grant in Phytologia,* 79(3): 255 (1995 publ. 1996): Vriesea nutans.
ochracea *(Rauh & E. Gross) J.R. Grant in Trop. Subtrop. Pflanzenwelt,* 91: 33 (1995): Vriesea ochracea.
orjuelae *(L.B. Sm.) J.R. Grant in Trop. Subtrop. Pflanzenwelt,* 91: 44 (1995): Vriesea orjuelae.
ororiensis *(Mez) J.R. Grant in Trop. Subtrop. Pflanzenwelt,* 91: 44 (1995): Guzmania ororiensis.
ovandensis *(Matuda) J.R. Grant in Trop. Subtrop. Pflanzenwelt,* 91: 33 (1995): Vriesea ovandensis.
panamaensis *(E. Gross & Rauh) J.R. Grant in Trop. Subtrop. Pflanzenwelt,* 91: 33 (1995): Vriesea panamaensis.
paniculata *(Mez & Wercklé) J.R. Grant in Trop. Subtrop. Pflanzenwelt,* 91: 45 (1995), as 'panniculata': Thecophyllum paniculatum.
patzeltii *(Rauh) J.R. Grant in Trop. Subtrop. Pflanzenwelt,* 91: 45 (1995): Vriesea patzeltii.
subsp. panamaensis *(Rauh) J.R. Grant in Trop. Subtrop. Pflanzenwelt,* 91: 47 (1995): Vriesea patzeltii var. panamaensis.
paupera *(Mez & Sodiro) J.R. Grant in Trop. Subtrop. Pflanzenwelt,* 91: 47 (1995): Thecophyllum pauperum.
pectinata *(L.B. Sm.) J.R. Grant in Trop. Subtrop. Pflanzenwelt,* 91: 33 (1995): Vriesea pectinata.
pedicellata *(Mez & Wercklé) J.R. Grant in Trop. Subtrop. Pflanzenwelt,* 91: 47 (1995): Thecophyllum pedicellatum.
picta *(Mez & Wercklé) J.R. Grant in Trop. Subtrop. Pflanzenwelt,* 91: 47 (1995): Thecophyllum pictum.
pittieri *(Mez) J.R. Grant in Trop. Subtrop. Pflanzenwelt,* 91: 33 (1995): Vriesea pittieri.
proctorii *J.A. Cedeño-Maldonado in Harvard Pap. Bot.,* 4(1): 115 (1999)—Puerto Rico.
ringens *(Griseb.) J.R. Grant in Trop. Subtrop. Pflanzenwelt,* 91: 35 (1995): Tillandsia ringens.
rubra *(Mez & Wercklé) J.R. Grant in Trop. Subtrop. Pflanzenwelt,* 91: 47 (1995): Thecophyllum rubrum.
rugosa *(Mez & Wercklé) J.R. Grant in Phytologia,* 79(3): 255 (1995 publ. 1996): Vriesea rugosa.
sanguinolenta *(Linden ex Cogn. & Marchal) J.R. Grant in Trop. Subtrop. Pflanzenwelt,* 91: 35 (1995): Vriesea sanguinolenta.
singuliflora *(Mez & Wercklé) J.R. Grant in Trop. Subtrop. Pflanzenwelt,* 91: 48 (1995): Thecophyllum singuliflorum.
sintenisii *(Baker) J.R. Grant in Trop. Subtrop. Pflanzenwelt,* 91: 48 (1995): Caraguata sintenisii.
stenophylla *(Mez & Wercklé) J.R. Grant in Trop. Subtrop. Pflanzenwelt,* 91: 48 (1995): Thecophyllum stenophyllum.
subsecunda *(Wittm.) J.R. Grant in Trop. Subtrop. Pflanzenwelt,* 91: 35 (1995): Vriesea subsecunda.
tarmaensis *(Rauh) J.R. Grant in Trop. Subtrop. Pflanzenwelt,* 91: 35 (1995): Vriesea tarmaensis.
tonduziana *(L.B. Sm.) J.R. Grant in Trop. Subtrop. Pflanzenwelt,* 91: 35 (1995): Vriesea tonduziana.
umbrosa *(L.B. Sm.) J.R. Grant in Trop. Subtrop. Pflanzenwelt,* 91: 48 (1995): Vriesea umbrosa.
urbaniana *(Mez) J.R. Grant in Phytologia,* 79(3): 255 (1995 publ. 1996): Guzmania urbaniana.
uxoris *(Utley) J.R. Grant in Trop. Subtrop. Pflanzenwelt,* 91: 48 (1995): Vriesea uxoris.
vanhyningii *(L.B. Sm.) J.R. Grant in Trop. Subtrop. Pflanzenwelt,* 91: 35 (1995): Vriesea vanhyningii.
verrucosa *(L.B. Sm.) J.R. Grant in Trop. Subtrop. Pflanzenwelt,* 91: 38 (1995): Vriesea verrucosa.
vietoris *(Utley) J.R. Grant in Trop. Subtrop. Pflanzenwelt,* 91: 48 (1995): Vriesea vietoris.
viridiflora *(Regel) J.R. Grant in Trop. Subtrop. Pflanzenwelt,* 91: 38 (1995): Pitcairnia viridiflora.
viridis *(Mez & Wercklé) J.R. Grant in Trop. Subtrop. Pflanzenwelt,* 91: 49 (1995): Thecophyllum viride.
vittata *(Mez & Wercklé) J.R. Grant in Trop. Subtrop. Pflanzenwelt,* 91: 38 (1995): Thecophyllum vittatum.
werckleana *(Mez) J.R. Grant in Trop. Subtrop. Pflanzenwelt,* 91: 38 (1995): Vriesea werckleana.

WERAUHIA:-
williamsii *(L.B. Sm.) J.R. Grant in Trop. Subtrop. Pflanzenwelt,* 91: 49 (1995): Vriesea williamsii.
woodsoniana *(L.B. Sm.) J.R. Grant in Trop. Subtrop. Pflanzenwelt,* 91: 39 (1995): Vriesea woodsoniana.

WHALLEYA *K.E. Wills & J.J. Bruhl in Austral. Syst. Bot.,* 13(3): 462 (2000). *GRAMINEAE*
proluta *(F. Muell.) K.E. Wills & J.J. Bruhl in Austral. Syst. Bot.,* 13(3): 465 (2000): Panicum prolutum.
pungens *K.E. Wills & J.J. Bruhl in Austral. Syst. Bot.,* 13(3): 463 (2000)—New South Wales.
subxerophila *(Domin) K.E. Wills & J.J. Bruhl in Austral. Syst. Bot.,* 13(3): 465 (2000): Panicum subxerophilum.

WHITEHEADIA (Hyacinthac.).
etesionamibensis *U. Müller-Doblies & D. Müller-Doblies in Feddes Repert.,* 108(1–2): 82 (1997)—Namibia.

WHYTOCKIA (Gesneriac.).
bijieensis *Yin Z. Wang & Z.Yu Li in Acta Phytotax. Sin.,* 35(1): 67 (1997)—China South-central (Guizhou).
gongshanensis *Yin Z. Wang & H. Li in Acta Phytotax. Sin.,* 38(2): 189 (2000)—China South-central (Yunnan).

WIGANDIA (Hydrophyllac.).
wurdackiana *R. Ferreyra in Biollania,* ed. esp. 6 (J.J. Wurdack Festschr.): 328 (1997)—Peru.

WIGGINSIA (Cactac.).
sect. Eowigginsia *(Doweld) A.B. Doweld in Sukkulentÿ (Moscow),* 3(1–2): 61 (2000): Peronocactus sect. Eowigginsia.
langsdorfii (Lehm.) D.M. Porter
subsp. multiceps *(Hofacker & K. Herm) A.B. Doweld in Sukkulentÿ (Moscow),* 1999(2): 25 (1999): Parodia langsdorfii subsp. multiceps.
subsp. pulvinatus *(Vliet) A.B. Doweld in Sukkulentÿ (Moscow),* 1999(2): 25 (1999): Notocactus pulvinatus.
nothohorstii *A.B. Doweld in Sukkulentÿ (Moscow),* 3(1–2): 62 (2000), nom. nov.: Notocactus horstii *F. Ritter.*
pauciareolata *(Arechav.) R. Kiesling in Zuloaga & Morrone (eds.), Cat. Pl. Vasc. Rep. Arg. 2 (Monogr. Syst. Bot.* Missouri Bot. Gard. 74): 1246 (1999): Echinocactus pauciareolatus.
werdermanniana *(Herter) A.B. Doweld in Sukkulentÿ (Moscow),* 3(1–2): 62 (2000): Notocactus werdermannianus.

WIKSTROEMIA (Thymelaeac.).
dolichantha Diels
var. effusa *(Rehder) B. Peterson in Fl. Thailand,* 6(3): 242 (1997), isonym; *see C.Y. Chang in Fl. Sichuanica,* 9: 233 (1989): W. effusa.
johnplewsii *W.L. Wagner & D.H. Lorence in Novon,* 8(3): 318 (1998)—Marquesas.
subcyclolepidota *L.P. Liu & Y.S. Lian in Acta Phytotax. Sin.,* 34(4): 440 (1996)—China North-central (Gansu).

WILDPRETIA *U. Reifenberger & A. Reifenberger in Vieraea,* 25: 203 (1996), nom. nov.: Sonchus subgen. Lactucosonchus *Sch. Bip.* (Compos.).
beltraniae *U. Reifenberger & A. Reifenberger in Vieraea,* 25: 204 (1996)—Canary Is.
webbii *(Sch. Bip.) U. Reifenberger & A. Reifenberger in Vieraea,* 25: 204 (1996): Sonchus webbii.

WILLDAMPIA *A.S. George in West. Austral. Nat.,* 22(3): 192 (1999). *LEGUMINOSAE*
formosa *(G. Don) A.S. George in West. Austral. Nat.,* 22(3): 192 (1999): Donia formosa.

WILLEMETIA *A.T. Brongniart, Mém. Fam. Rhamnées:* 63 (1826). *RHAMNACEAE*†
africana *(L.) A.T. Brongniart, Mém. Fam. Rhamnées:* 64 (1826): Ceanothus africana.†
stipitata (Jacq.) Dalla Torre
subsp. albanica *(Kümmerle & Jáv.) L. Kirschnerová in Taxon,* 45(4): 629 (1996): W. stipitata f. albanica.

× **WILLIAMARA** *hort. in Orchid Rev.*, 107(1230): 391 (1999). *ORCHIDACEAE* (Colax × Neogardneria × Promenaea × Zygopetalum).

WILLISIA (Podostemac.).
arekaliana *G.R. Shivamurthy & K.B. Sadanand in Kew Bull.*, 52(1): 243 (1997)—India (Kerala).

WILLUGHBEIA (Apocynac.).
edulis *W. Roxburgh, Pl. Coromandel*, 3: 77 (1820)—Bangladesh.†
lunduensis *D.J. Middleton in Blumea*, 41(1): 123 (1996)—Borneo (Sarawak).

WIMMERELLA *L. Serra, M.B. Crespo & T.G. Lammers in Novon*, 9(3): 415 (1999). *CAMPANULACEAE*
arabidea *(C. Presl) L. Serra, M.B. Crespo & T.G. Lammers in Novon*, 9(3): 416 (1999): Rapuntium arabideum.
bifida *(Thunb.) L. Serra, M.B. Crespo & T.G. Lammers in Novon*, 9(3): 416 (1999): Lobelia bifida.
frontidentata *(E. Wimm.) L. Serra, M.B. Crespo & T.G. Lammers in Novon*, 9(3): 417 (1999): Laurentia frontidentata.
giftbergensis *(E. Phillips) L. Serra, M.B. Crespo & T.G. Lammers in Novon*, 9(3): 417 (1999): Lobelia giftbergensis.
hederacea *(Sond.) L. Serra, M.B. Crespo & T.G. Lammers in Novon*, 9(3): 417 (1999): Laurentia hederacea.
hedyotidea *(Schltr.) L. Serra, M.B. Crespo & T.G. Lammers in Novon*, 9(3): 417 (1999): Laurentia hedyotidea.
longitubus *(E. Wimm.) L. Serra, M.B. Crespo & T.G. Lammers in Novon*, 9(3): 417 (1999): Laurentia longitubus.
mariae *(E. Wimm.) L. Serra, M.B. Crespo & T.G. Lammers in Novon*, 9(3): 417 (1999): Laurentia mariae.
pygmaea *(Thunb.) L. Serra, M.B. Crespo & T.G. Lammers in Novon*, 9(3): 417 (1999): Lobelia pygmaea.
secunda *(L.f.) L. Serra, M.B. Crespo & T.G. Lammers in Novon*, 9(3): 417 (1999): Lobelia secunda.

WIMMERIA (Celastrac.).
microphylla *L. Radlkofer in Beih. Bot. Centralbl.*, 15: 359 (1903)—Mexico Central (Puebla).

WINIKA *M.A. Clements, D.L. Jones & B.P.J. Molloy in Orchadian*, 12(5): 214 (1997). *ORCHIDACEAE*
cunninghamii *(Lindl.) M.A. Clements, D.L. Jones & B.P.J. Molloy in Orchadian*, 12(5): 215 (1997): Dendrobium cunninghamii.

WITHANIA (Solanac.).
chevalieri *A.E. Gonçalves in Garcia de Orta, Ser. Bot.*, 14(1): 149 (1999)—Cape Verde.

WITHERINGIA (Solanac.).
wurdackiana *C. Benítez de Rojas in Biollania*, ed. esp. 6 (J.J. Wurdack Festschr.): 247 (1997)—Venezuela.

WITTROCKIA (Bromeliac.).
corallina *E.M.C. Leme in J. Bromeliad Soc.*, 42(2): 51 (1992)—Brazil Southeast (São Paulo).
cyathiformis *(Vell.) E.M.C. Leme, Canistrum - Bromeliads Atlantic Forest:* 67 (1997): Tillandsia cyathiformis.
gigantea *(Baker) E.M.C. Leme, Canistrum - Bromeliads Atlantic Forest:* 70 (1997): Nidularium giganteum.
paulistana *E.M.C. Leme, Nidularium - Bromeliads of Atlantic Forest:* 224 (2000)—Brazil Southeast (São Paulo).
tenuisepala *(Leme) E.M.C. Leme, Canistrum - Bromeliads Atlantic Forest:* 65 (1997): Nidularium tenuisepalum.

WOLFFIA (Lemnac.).
borealis *(Engelm. ex Hegelm.) E. Landolt & O. Wildi in Ber. Geobot. Inst. E.T.H. Stiftung Rübel*, 44: 137 (1977): W. brasiliensis var. borealis.

WOLFFIELLA (Lemnac.).
caudata *E. Landolt in Ber. Geobot. Inst. Eidgenöss. Techn. Hochsch. Stift. Rübel*, 58: 121 (1992)—Bolivia.

WOLLASTONIA (Compos.).
lifuana *(Hochr.) F.R. Fosberg in Allertonia*, 7(2): 80 (1993): Lipochaeta lifuana.

WOLLEMIA *W.G. Jones, K.D. Hill & J.M. Allen in Telopea*, 6(2–3): 173 (1995). *ARAUCARIACEAE*
nobilis *W.G. Jones, K.D. Hill & J.M. Allen in Telopea*, 6(2–3): 174 (1995)—New South Wales.

WOODIELLANTHA (Annonac.).
grandifolia *(Airy Shaw) K. Mat-Salleh & A. Latiff in Sandakania*, 11: 52 (1998): Woodiella sympetala var. grandifolia.

× **WOODWARDARA** *hort. in Orchid Rev.*, 105(1218): 382 (1997). *ORCHIDACEAE* (Colax × Neogardneria × Zygopetalum).

WOONYOUNGIA *Y.W. Law in Bull. Bot. Res. North-East. Forest. Univ.*, 17(4): 354 (1997). *MAGNOLIACEAE*
septentrionalis *(Dandy) Y.W. Law in Bull. Bot. Res. North-East. Forest. Univ.*, 17(4): 356 (1997): Kmeria septentrionalis.

WURDASTOM *B. Wallnöfer in Ann. Naturhist. Mus. Wien*, 98B(Suppl.): 461 (1996). *MELASTOMATACEAE*
bullata *(Wurdack) B. Wallnöfer in Ann. Naturhist. Mus. Wien*, 98B(Suppl.): 461 (1996): Alloneuron bullatum.
cuatrecasasii *(Wurdack) B. Wallnöfer in Ann. Naturhist. Mus. Wien*, 98B(Suppl.): 461 (1996): Alloneuron cuatrecasasii.
dorrii *(Wurdack) B. Wallnöfer in Ann. Naturhist. Mus. Wien*, 98B(Suppl.): 461 (1996): Alloneuron dorrii.
dudleyi *(Wurdack) B. Wallnöfer in Ann. Naturhist. Mus. Wien*, 98B(Suppl.): 461 (1996): Alloneuron dudleyi.
ecuadorensis *(Wurdack) B. Wallnöfer in Ann. Naturhist. Mus. Wien*, 98B(Suppl.): 461 (1996), as 'ecuadorense': Alloneuron ecuadorense.
hexamera *(Wurdack) B. Wallnöfer in Ann. Naturhist. Mus. Wien*, 98B(Suppl.): 461 (1996): Alloneuron hexamerum.
sneidernii *(Wurdack) B. Wallnöfer in Ann. Naturhist. Mus. Wien*, 98B(Suppl.): 461 (1996): Alloneuron sneidernii.
subglabra *(Wurdack) B. Wallnöfer in Ann. Naturhist. Mus. Wien*, 98B(Suppl.): 461 (1996): Alloneuron subglabrum.

WURMBEA (Colchicac.).
saccata *T.D. Macfarlane & S.J. van Leeuwen in Nuytsia*, 10(3): 429 (1996)—Western Australia.

X

XANTHISMA (Compos.).
texanum DC.
f. orientale *J.C. Semple in Rhodora*, 76(805): 15 (1974)—Texas.

XANTHOCERCIS (Legumin.).
rabiensis *L.J.G. van der Maesen in Bull. Jard. Bot. Nation. Belg.*, 66(1–2): 20 (1997)—Gabon.

XANTHOGALUM (Umbellif.).
sachokianum *I.I. Karyagin in Dokl. Akad. Nauk Az. SSR*, 2(10): 425 (1946)—North Caucasus (Dagestan), Transcaucasus (Azerbaydzhan).

XANTHOPHYLLUM (Polygalac.).
flavescens *W. Roxburgh, Pl. Coromandel*, 3: 82 (1820)—Bangladesh.†

XANTHOPHYLLUM:-
virens *W. Roxburgh, Pl. Coromandel*, 3: 82 (1820)—Bangladesh.†

XANTHOPHYTUM (Rubiac.).
bullatum *C. Tange in Nordic J. Bot.*, 15(6): 578 (1995 publ. 1996)—Borneo (Sarawak).
kinabaluense *C. Tange in Nordic J. Bot.*, 15(6): 575 (1995 publ. 1996), nom. nov.: Myrioneuron borneense *Stapf.*

XANTHOSIA (Umbellif.).
eichleri *J.M. Hart & M.J. Henwood in Nuytsia*, 12(2): 186 (1998)—Western Australia.
scopulicola *J.M. Hart & M.J. Henwood in Austral. Syst. Bot.*, 13(2): 250 (2000)—New South Wales.
stellata *J.M. Hart & M.J. Henwood in Austral. Syst. Bot.*, 13(2): 252 (2000)—New South Wales, Victoria, Queensland.
ternifolia *J.M. Hart & M.J. Henwood in Austral. Syst. Bot.*, 13(2): 256 (2000), nom. nov.: X. tasmanica *Gand.*

XANTHOSOMA (Arac.).
dealbatum *M.H. Grayum in Phytologia*, 82(1): 55 (1997)—Costa Rica.
pottii *E.G. Gonçalves in Aroideana*, 22: 3 (1999)—Brazil West-central (Mato Grosso do Sul).
riparium *E.G. Gonçalves in Novon*, 10(1): 26 (2000)—Brazil West-central (Goiás).
weeksii *M. Madison in Aroideana*, 1(1): 24 (1978), in Supp. 17 as 'wieksii'—Ecuador.†

XANTHOSTEMON (Myrtac.).
fruticosus *Peter G. Wilson & L.L. Co in Sida*, 18(1): 283 (1998)—Philippines.

XANTHOXALIS (Oxalidac.).
corniculata (L.) Small
subsp. repens *(Thunb.) N.N. Tzvelev in Fl. Vostochnoĭ Evropy̆*, 9: 368 (1996): Oxalis repens.
stricta (L.) Small
subsp. villicaulis *(Wiegand) N.N. Tzvelev in Fl. Vostochnoĭ Evropy̆*, 9: 367 (1996): Oxalis europaea f. villicaulis.

XENOPHYLLUM *V.A. Funk in Novon*, 7(3): 235 (1997). *COMPOSITAE*
acerosum *(Cuatrec.) V.A. Funk in Novon*, 7(3): 238 (1997): Werneria acerosa.
amblydactylum *(S.F. Blake) V.A. Funk in Novon*, 7(3): 238 (1997): Werneria amblydactyla.
ciliolatum *(A. Gray) V.A. Funk in Novon*, 7(3): 239 (1997): Werneria ciliolata.
crassum *(S.F. Blake) V.A. Funk in Novon*, 7(3): 239 (1997): Werneria crassa.
dactylophyllum *(Sch. Bip.) V.A. Funk in Novon*, 7(3): 239 (1997): Werneria dactylophylla.
decorum *(S.F. Blake) V.A. Funk in Novon*, 7(3): 239 (1997): Werneria decora.
digitatum *(Wedd.) V.A. Funk in Novon*, 7(3): 239 (1997): Werneria digitata.
esquilachense *(Cuatrec.) V.A. Funk in Novon*, 7(3): 239 (1997): Werneria esquilachensis.
fontii *(Cuatrec.) V.A. Funk in Novon*, 7(3): 239 (1997): Werneria fontii.
humile *(Kunth) V.A. Funk in Novon*, 7(3): 239 (1997): Werneria humilis.
incisum *(Phil.) V.A. Funk in Novon*, 7(3): 239 (1997): Werneria incisa.
var. pubescens *(Rockh.) Á.L. Cabrera & S.E. Freire in Zuloaga & Morrone (eds.), Cat. Pl. Vasc. Rep. Arg. 2 (Monogr. Syst. Bot.* Missouri Bot. Gard. 74): 1245 (1999): Werneria incisa var. pubescens.
lycopodioides *(S.F. Blake) V.A. Funk in Novon*, 7(3): 240 (1997): Werneria lycopodioides.
marcidum *(S.F. Blake) V.A. Funk in Novon*, 7(3): 240 (1997): Werneria marcida.
poposum *(Phil.) V.A. Funk in Novon*, 7(3): 240 (1997): Werneria poposa.
pseudodigitatum *(Rockh.) V.A. Funk in Novon*, 7(3): 240 (1997): Werneria pseudodigitata.
rigidum *(Kunth) V.A. Funk in Novon*, 7(3): 240 (1997): Werneria rigida.
rosenii *(R.E. Fr.) V.A. Funk in Novon*, 7(3): 240 (1997): Werneria rosenii.

XENOPHYLLUM:-
roseum *(Hieron.) V.A. Funk in Novon,* 7(3): 240 (1997): Werneria rosea.
sotarense *(Hieron.) V.A. Funk in Novon,* 7(3): 240 (1997): Werneria sotarensis.
staffordiae *(Sandwith) V.A. Funk in Novon,* 7(3): 240 (1997): Werneria staffordiae.
weddellii *(Phil.) V.A. Funk in Novon,* 7(3): 240 (1997): Werneria weddellii.

XENOSTEGIA (Convolvulac.).
tridentata (L.) D.F. Austin & Staples
subsp. angustifolia *(Jacq.) A.D.J. Meeuse in Bothalia,* 26(1): 49 (1996), isonym; *see Lejoly & Lisowski in Fragm. Flor. Geobot.,* 38(2): 379 (1993): Ipomoea angustifolia.

XERONEMA (Phormiac.).
callistemon
f. bracteosa *(L.B. Moore) P.J. de Lange & E.K. Cameron in New Zealand J. Bot ,* 37(3): 435 (1999): X. callistemon var. bracteosa.

XEROPHYTA (Velloziac.).
connata *G. McPherson & H. van der Werff in Novon,* 7(4): 387 (1997 publ. 1998)—Madagascar.
schnizleinia (Hochst.) Baker
var. somalensis *(A. Terracc.) K.A. Lye in Lidia,* 3(5): 143 (1995): Vellozia schnizleinia var. somalensis.

× **XERRIARA** *hort. in Orchid Rev.,* 106(1223): 318 (1998). *ORCHIDACEAE* (Arachnis × Ascocentrum × Rhynchostylis).

XIPHOTHECA (Legumin.).
sect. Congestae *A.L. Schutte in Ann. Missouri Bot. Gard.,* 84(1): 94 (1997).

XIZANGIA (Scrophulariac.).
bartsioides *(Hand.-Mazz.) C.Y. Wu & D.D. Tao in Novon,* 9(2): 288 (1999): Pterygiella bartsioides.

XYLOBIUM (Orchidac.).
subpulchrum *R.L. Dressler in Orquideologia,* 21(3): 310 (2000)—Peru.

XYLOPIA (Annonac.).
involucrata *M.C. Dias & L.S. Kinoshita in Kew Bull.,* 53(2): 471 (1998)—Brazil Northeast (Bahia).
mwasumbii *D.M. Johnson in Novon,* 9(1): 55 (1999)—Tanzania.
orinocensis *K.J. Bagstad & D.M. Johnson in Contrib. Univ. Mich. Herb.,* 22: 26 (1999)—Venezuela.
rigidiflora *K.J. Bagstad & D.M. Johnson in Contrib. Univ. Mich. Herb.,* 22: 24 (1999)—Brazil North (Amazonas, Roraima).

XYLOPODIA *M. Weigend, Nasa & Conquest S. America - Syst. Rearrang. in Loasac.:* 215 (1997). *LOASACEAE*
klaprothioides *M. Weigend, Nasa & Conquest S. America - Syst. Rearrang. in Loasac.:* 215 (1997): Peru.

XYLOSMA (Flacourtiac.).
longipedicellata *A. Pool in Novon,* 7(1): 75 (1997)—Nicaragua, Honduras.
racemosa (Siebold & Zucc.) Miq.
var. caudata *(S.S. Lai) S.S. Lai in Fl. Reipubl. Popul. Sin.,* 52(1): 40 (1999): X. congesta var. caudata.
smithiana *F.R. Fosberg in Allertonia,* 7(2): 66 (1993)—Tonga.

XYLOTHAMIA (Compos.).
palmeri A. Gray
var. pachylepis *(H.M. Hall) G.L. Nesom ex M.A. Lane & R.L. Hartman in Amer. J. Bot.,* 83(3): 364 (1996): Haplopappus palmeri subsp. pachylepis.

XYRIS (Xyridac.).
amorimii *R. Kral in Novon,* 8(4): 388 (1998)—Brazil Northeast (Bahia).
anceps Lam.
var. minima *(Steud.) J.M. Lock in Kew Bull.,* 53(4): 890 (1998): X. minima.

XYRIS:-
asterotricha *J.M. Lock in Kew Bull.,* 54(2): 312 (1999)—Zimbabwe.
atrovirida *A.N.L. Doust & B.J. Conn in Austral. Syst. Bot.,* 10(2): 222 (1997)—Western Australia.
bampsii *S. Lisowski in Syst. Geogr. Pl.,* 69(2): 205 (1999)—Zaire.
bissei *A.J. Urquiola Cruz & R. Kral in Fontqueria,* 48: 176 (1997 publ. 1998)—Cuba.
boliviana *R. Kral in Novon,* 8(4): 388 (1998)—Bolivia.
dilatatiscapa *R. Kral & M. Jansen-Jacobs in Novon,* 8(4): 399 (1998)—Guyana.
ednae *J.M. Lock in Kew Bull.,* 53(4): 883 (1998)—Uganda.
erosa *J.M. Lock in Kew Bull.,* 53(4): 885 (1998)—Tanzania.
exilis *A.N.L. Doust & B.J. Conn in Austral. Syst. Bot.,* 10(2): 228 (1997)—Western Australia.
ferreirae *R. Kral in Novon,* 8(4): 391 (1998)—Brazil North (Amazonas).
glandacea *L.A. Nilsson in Kongl. Svenska Vetenskapsakad. Handl.,* 24(14): 50, pl. 3 (1892), in I.K. Supp. 1 as 'glandulacea' in error—Brazil Southeast.†
gongylospica *R. Kral in Novon,* 8(4): 393 (1998)—Venezuela.
guillenii *R. Kral in Novon,* 8(4): 395 (1998)—Bolivia.
itatiayensis *(Malme) M. das Graças Lapa Wanderley & M. das Graças Sajo in Acta Bot. Brasil.,* 10(2): 380 (1996): X. obtusiuscula var. itatiayensis.
kibaraensis *S. Lisowski in Syst. Geogr. Pl.,* 69(2): 207 (1999)—Zaire.
laniceps *J.M. Lock in Kew Bull.,* 54(2): 323 (1999)—Zambia.
laxifolia Mart.
var. iridifolia *(Chapm.) R. Kral in Novon,* 9(2): 214 (1999): X. iridifolia.
lejolyanus *S. Lisowski in Syst. Geogr. Pl.,* 69(2): 209 (1999)—Zaire.
mantuensis *A.J. Urquiola Cruz & R. Kral in Fontqueria,* 48: 175 (1997 publ. 1998)—Cuba.
maxima *A.N.L. Doust & B.J. Conn in Austral. Syst. Bot.,* 10(2): 242 (1997)—Western Australia.
popeanus *S. Lisowski in Syst. Geogr. Pl.,* 69(2): 211 (1999)—Zaire.
porcata *J.M. Lock in Kew Bull.,* 54(2): 317 (1999)—Zambia.
rhodolepis *(Malme) J.M. Lock in Kew Bull.,* 53(4): 886 (1998): X. humpatensis var. rhodolepis.
stricta Chapm.
var. obscura *R. Kral in Novon,* 9(2): 218 (1999)—Alabama.
subasperula *R. Kral in Novon,* 8(4): 395 (1998)—Bolivia.
subtilis *J.M. Lock in Kew Bull.,* 53(4): 892 (1998)—Uganda, Tanzania.
symoensii *B. Brylska & S. Lisowski in Syst. Geogr. Pl.,* 69(2): 212 (1999)—Zaire.
tasmanica *(D.I. Morris) A.N.L. Doust & B.J. Conn in Austral. Syst. Bot.,* 10(2): 212 (1997): X. gracilis subsp. tasmanica.

Y

× **YEEPENGARA** *hort. in Orchid Rev.,* 106(1220): 126 (1998). *ORCHIDACEAE* (Aerides × Phalaenopsis × Rhynchostylis × Vanda).

YINSHANIA (Crucif.).
acutangula (O.E. Schulz) Y.H. Zhang
subsp. microcarpa *(K.C. Kuan) I.A. Al-Shehbaz, G. Yang, L.L. Lu & T.Y. Cheo in Harvard Pap. Bot.,* 3(1): 83 (1998): Cochlearia microcarpa.
subsp. wilsonii *(O.E. Schulz) I.A. Al-Shehbaz, G. Yang, L.L. Lu & T.Y. Cheo in Harvard Pap. Bot.,* 3(1): 83 (1998): Cochlearia henryi var. wilsonii.
albiflora Ma & Y.Z. Zhao
var. gobica *Z.X. An in Fl. Xinjiangensis,* 2(2): 381 (1995)—Xinjiang.

YINSHANIA:-
hunanensis *(Y.H. Zhang) I.A. Al-Shehbaz, G. Yang, L.L. Lu & T.Y. Cheo in Harvard Pap. Bot.,* 3(1): 92 (1998): Hilliella hunanensis.
lichuanensis *(Y.H. Zhang) I.A. Al-Shehbaz, G. Yang, L.L. Lu & T.Y. Cheo in Harvard Pap. Bot.,* 3(1): 90 (1998): Hilliella lichuanensis.
rivulorum *(Dunn) I.A. Al-Shehbaz, G. Yang, L.L. Lu & T.Y. Cheo in Harvard Pap. Bot.,* 3(1): 87 (1998): Nasturtium rivulorum.
rupicola *(D.C. Zhang & J.Z. Shao) I.A. Al-Shehbaz, G. Yang, L.L. Lu & T.Y. Cheo in Harvard Pap. Bot.,* 3(1): 91 (1998): Cochlearia rupicola.
subsp. shuangpaiensis *(Z.Yu Li) I.A. Al-Shehbaz, G. Yang, L.L. Lu & T.Y. Cheo in Harvard Pap. Bot.,* 3(1): 92 (1998): Hilliella shuangpaiensis.
sinuata *(K.C. Kuan) I.A. Al-Shehbaz, G. Yang, L.L. Lu & T.Y. Cheo in Harvard Pap. Bot.,* 3(1): 88 (1998): Cochlearia sinuata.
subsp. qianwuensis *(Y.H. Zhang) I.A. Al-Shehbaz, G. Yang, L.L. Lu & T.Y. Cheo in Harvard Pap. Bot.,* 3(1): 89 (1998): Hilliella sinuata var. qianwuensis.
yixianensis *(Y.H. Zhang) I.A. Al-Shehbaz, G. Yang, L.L. Lu & T.Y. Cheo in Harvard Pap. Bot.,* 3(1): 84 (1998): Hilliella yixianensis.

YOUNGIA (Compos.).
longiflora *(Babc. & Stebbins) C. Shih in Fl. Reipubl. Popul. Sin.,* 80(1): 150 (1997): Y. japonica subsp. longiflora.
pseudosenecio *(Vaniot) C. Shih in Fl. Reipubl. Popul. Sin.,* 80(1): 157 (1997): Lactuca pseudosenecio.
silhetensis (DC.) Babc. & Stebbins
subsp. bhutanica *A.J.C. Grierson & L.S. Springate in Edinburgh J. Bot.,* 57(3): 410 (2000)—Bhutan.

YUA (Vitac.).
chinensis *C.L. Li in Chinese J. Appl. Environ. Biol.,* 2(1): 47 (1996)—China South-central (Sichuan, Guizhou, Yunnan), China Southeast (Guangxi).

YUCCA (Agavac.).
angustissima
subsp. avia *(Reveal) F. Hochstätter in Cactaceae Rev.,* 1(2): 21 (1999); *cf. Repert. Pl. Succ. (I.O.S.),* 50: 6 (1999 publ. 2000): Y. angustissima var. avia.
subsp. kanabensis *(McKelvey) F. Hochstätter in Cactaceae Rev.,* 1(2): 21 (1999); *cf. Repert. Pl. Succ. (I.O.S.),* 50: 6 (1999 publ. 2000): Y. kanabensis.
subsp. toftiae *(S.L. Welsh) F. Hochstätter in Cactaceae Rev.,* 1(2): 21 (1999); *cf. Repert. Pl. Succ. (I.O.S.),* 50: 6 (1999 publ. 2000): Y. toftiae.
arkansana *W. Trelease in Rep. (Annual) Missouri Bot. Gard.,* 13: 63 (1902)—Arkansas, Texas.
subsp. freemanii *(Shinners) F. Hochstätter, Yucca I:* 61 (2000): Yucca freemanii.
subsp. louisianensis *(Trel.) F. Hochstätter in Succulenta,* 78(3): 122 (1999): Y. louisianensis.
baileyi
subsp. intermedia *(McKelvey) F. Hochstätter in Cactaceae Rev.,* 1(2): 21 (1999); *cf. Repert. Pl. Succ. (I.O.S.),* 50: 6 (1999 publ. 2000): Y. intermedia.
capensis *L.W. Lenz in Cact. Succ. J. (U.S.A.),* 70(6): 289 (1998)—Mexico Northwest (Baja California Sur).
declinata *J.E. Laferrière in Cact. Succ. J. (U.S.A.),* 67(6): 347 (1995)—Mexico Northwest (Sonora).
elata
subsp. utahensis *(McKelvey) F. Hochstätter in Cactaceae Rev.,* 1(2): 21 (1999); *cf. Repert. Pl. Succ. (I.O.S.),* 1(2): 7 (1999 publ. 2000): Y. utahensis.
glauca
subsp. albertana *F. Hochstätter in Piante Grasse,* 20(1): 2 (2000)—Alberta.
subsp. stricta *(Sims) F. Hochstätter in Cactaceae Rev.,* 1(2): 21 (1999); *cf. Repert. Pl. Succ. (I.O.S.),* 50: 7 (1999 publ. 2000): Y. stricta.
harrimaniae Trel.
subsp. gilbertiana *(Trel.) F. Hochstätter, Yucca I:* 36 (2000): Yucca harrimaniae var. gilbertiana.

YUCCA:-
harrimaniae
subsp. neomexicana *(Wooton & Standl.) F. Hochstätter in Cactaceae Rev.*, 1(2): 21 (1999); *cf. Repert. Pl. Succ. (I.O.S.)*, 50: 7 (1999 publ. 2000): Y. neomexicana.
subsp. sterilis *(Neese & S.L. Welsh) F. Hochstätter in Cactaceae Rev.*, 1(2): 21 (1999); *cf. Repert. Pl. Succ. (I.O.S.)*, 50: 7 (1999 publ. 2000): Y. harrimaniae var. sterilis.
linearifolia *K.H. Clary in Brittonia*, 47(4): 394 (1995)—Mexico Northeast (Nuevo León, Coahuila).
linearis *(Trel.) D.J. Ferguson in Cact. Succ. J. (U.S.A.)*, 68(3): 130 (1996)—Y. rostrata var. linearis.
mixtecana *A. García-Mendoza in Acta Bot. Mex.*, 42: 1 (1998)—Mexico Southwest (Oaxaca), Mexico Central (Puebla).
nana *F. Hochstätter in Succulenta*, 77(2): 72 (1998)—Utah.
whipplei Torr.
subsp. newberryi *(McKelvey) F. Hochstätter in Succulenta*, 79(1): 39 (2000): Y. newberryi.

YUNNANOPILIA *C.Y. Wu & D.Z. Li in Acta Bot. Yunnanica*, 22(3): 249 (2000). *OPILIACEAE*
longistaminea *(W.Z. Li) C.Y. Wu & D.Z. Li in Acta Bot. Yunnanica*, 22(3): 250 (2000): Melientha longistaminea.

YUSHANIA (Gramin.).
ailuropodina *T.P. Yi in J. Bamboo Res.*, 15(3): 6 (1996)—China South-central (Sichuan).
ambositrensis *(A. Camus) D. Ohrnberger, Bamboos of the World Intro. 3:* 14 (1996): Arundinaria ambositrensis.
baviensis *(Balansa) J.J.N. Campb. ex D. Ohrnberger, Bamboos of the World Intro. 3:* 14 (1996): Arundinaria baviensis.
boliana *J.-P. Demoly in Bambou*, 35: 3 (2000)—China South-Central (Sichuan).
dafengdingensis *T.P. Yi in J. Bamboo Res.*, 15(3): 9 (1996)—China South-central (Sichuan).
elegans *(Kurz) J.J.N. Campb. ex D. Ohrnberger, Bamboos of the World Intro. 2:* 10 (1996), isonym; *see R.B. Majumdar in Karth. et al., Fl. Ind. Enumerat. - Monocot.*: 282 (1989): Arundinaria elegans.
ferax *(Keng) J.-P. Demoly, Bambous en France:* 76 (1996): Arundinaria ferax.
fungosa *(T.P. Yi) J.-P. Demoly in Bambou*, 21: 15 (1995): Fargesia fungosa.
hirsuta *(Munro) J.J.N. Campb. ex D. Ohrnberger, Bamboos of the World Intro. 2:* 10 (1996), isonym; *see R.B. Majumdar in Karth. et al, Fl. Ind. Enumerat. - Monocot.*: 282 (1989): Arundinaria hirsuta.
humbertii *(A. Camus) D. Ohrnberger, Bamboos of the World Intro. 3:* 14 (1996): Arundinaria humbertii.
longissima *K.F. Huang in Wuyi Sci. J.*, 2: 20 (1982)—China Southeast (Fujian).
madagascariensis *(A. Camus) D. Ohrnberger, Bamboos of the World Intro. 3:* 15 (1996): Arundinaria madagascariensis.
marojejyensis *(A. Camus) D. Ohrnberger, Bamboos of the World Intro. 3:* 15 (1996): Arundinaria marojejyensis.
microphylla *(Munro) D. Ohrnberger, Bamboos of the World Intro. 2:* 10 (1996), isonym; *see R.B. Majumdar in Karth. et al., Fl. Ind. Enumerat. - Monocot.*: 283 (1989): Arundinaria microphylla.
papillosa *(W.T. Lin) D. Ohrnberger, Bamboos of the World Intro. 4:* 20 (1997): Sinarundinaria papillosa.
perrieri *(A. Camus) D. Ohrnberger, Bamboos of the World Intro. 3:* 15 (1996): Arundinaria perrieri.
pingshanensis *T.P. Yi in J. Bamboo Res.*, 19(1): 21 (2000)—China South-central (Sichuan).
pubescens *T.P. Yi in J. Bamboo Res.*, 19(1): 24 (2000)—China South-central (Sichuan).
rigidula *(E.G. Camus) D. Ohrnberger, Bamboos of the World Intro. 4:* 20 (1997): Arundinaria rigidula.
schmidiana *(A. Camus) D. Ohrnberger, Bamboos of the World Intro. 2:* 10 (1996): Arundinaria schmidiana.

YUSHANIA:-
wardii *(Bor) D. Ohrnberger, Bamboos of the World Intro. 3:* 15 (1996): Arundinaria wardii.

Z

ZAGROSIA *F. Speta in Phyton (Austria)*, 38(1): 124 (1998). *HYACINTHACEAE*
persica *(Hausskn.) F. Speta in Phyton (Austria)*, 38(1): 124 (1998): Scilla persica.

ZAMIA (Zamiac.).
gentryi *C.H. Dodson in Novon*, 8(1): 12 (1998)—Ecuador.
lacandona *B. Schutzman & A.P. Vovides in Novon*, 8(4): 441 (1998)—Mexico Southeast (Chiapas).
polymorpha *D.W. Stevenson, A. Moretti & M. Vázquez Torres in Delpinoa*, 37–38: 4 (1995–96 publ. 1998)—Belize, Mexico Southeast (Campeche, Quintana Roo, Yucatán).
urep *B. Wallnöfer in Linz. Biol. Beitr.*, 28(2): 1056 (1996)—Peru.
vazquezii *D.W. Stevenson, S. Sabato & P. De Luca in Delpinoa*, 37–38: 14 (1995–96 publ. 1998)—Mexico Gulf (Veracruz).

ZANTEDESCHIA (Arac.).
sect. Aestivae *Y. Singh in S. Afr. J. Bot.*, 62(6): 324 (1996).
valida *(Letty) Y. Singh in S. Afr. J. Bot.*, 62(6): 323 (1996): Z. albomaculata subsp. valida.

ZANTHOXYLUM (Rutac.).
ailanthoides Siebold & Zucc.
var. boninshimae *(Koidz.) T. Yamaz. ex H. Ohba in Fl. Japan*, 2c: 39 (1999): Fagara boninshimae.
f. inermis *(Hatus.) H. Ohba in Fl. Japan*, 2c: 39 (1999): Fagara ailanthoides var. inermis.
f. pubescens *(Hatus.) H. Ohba in Fl. Japan*, 2c: 39 (1999): Fagara ailanthoides var. pubescens.
gentryi *C. Reynel in Novon*, 5(4): 362 (1995)—Colombia.
lenticulare *C. Reynel in Novon*, 5(4): 364 (1995)—Colombia, Venezuela.
lepidopteriphilum *C. Reynel in Novon*, 5(4): 365 (1995)—Peru, Ecuador.
piperitum (L.) DC.
f. rotundatum *Yokouchi ex T. Shimizu, Fl. Nagano Pref.*: 1509 (1997)—Japan.
schinifolium Siebold & Zucc.
var. okinawense *(Nakai) Hatus. ex K. Simabuku in Island Stud. Okinawa*, 7: 56 (1989): Fagara okinawensis.

ZAPOTECA (Legumin.).
gracilis *(Griseb.) M. Bässler in Fl. Repúbl. Cuba*, Ser. A, 2: 156 (1998): Calliandra gracilis.
quichoi *H.M. Hernández & A.M. Hanan-Alipi in Brittonia*, 50(2): 211 (1998)—Mexico Southeast (Tabasco).

ZEA (Gramin.).
mays L.
subsp. huehuetenangensis *(H.H. Iltis & Doebley) J. Doebley in Maydica*, 35(2): 148 (1990): Z. mays var. huehuetenangensis.
nicaraguensis *H.H. Iltis & B.F. Benz in Novon*, 10(4): 382 (2000)—Nicaragua.

ZEHNERIA (Cucurbitac.).
formosana *(Hayata) S.S. Ying in Mem. Coll. Agric. Nation. Taiwan Univ.*, 30(3): 101 (1990), as 'Zehenaria': Melothria formosana.
maysorensis (Wight & Arn.) Arn.
var. umbellata *(Chakrav.) G.R. Kumari in N.C. Nair & A.N. Henry, Fl. Tamil Nadu, India, Ser. 1*, 1: 175 (1983): Melothria mucronata var. umbellata.
perpusilla (Blume) Bole & M.R. Almeida
var. deltifrons*(Ohwi) H. Ohba in Fl. Japan*, 2c: 200 (1999): Melothria perpusilla var. deltifrons.

ZEHNERIA:-
repanda *(Blume) C.M. Simmons in Blumea*, 45(1): 240 (2000): Bryonia repanda.
ridens *B. Verdcourt in Kew Bull.*, 52(3): 729 (1997)—Tanzania.
somalensis *M. Thulin in Nordic J. Bot.*, 16(3): 297 (1996 publ. 1997)—Somalia.

ZEPHYRANTHES (Amaryllidac.).
amoena *P. Ravenna in Onira, Bot. Leafl.*, 3(16): 54 (1999)—Brazil South (Paraná).
breviscapa *P. Ravenna in Onira, Bot. Leafl.*, 3(16): 54 (1999)—Bolivia.
dichromantha *T.M. Howard in Herbertia*, 51: 38 (1996)—Mexico Northeast (San Luis Potosí).
diluta *P. Ravenna in Onira, Bot. Leafl.*, 3(16): 55 (1999)—Argentina Northwest (San Juan).
elegans *P. Ravenna in Onira, Bot. Leafl.*, 3(16): 53 (1999)—Bolivia.
fragrans *P. Ravenna in Onira, Bot. Leafl.*, 3(16): 52 (1999)—Bolivia.
hondurensis *P. Ravenna in Onira, Bot. Leafl.*, 3(16): 54 (1999)—Honduras, Guatemala.
moctezumae *T.M. Howard in Herbertia*, 51: 40 (1996)—Mexico Northeast (San Luis Potosí).
modesta *P. Ravenna in Onira, Bot. Leafl.*, 3(17): 62 (1999)—Argentina Northwest (Jujuy).

ZERAVSCHANIA (Umbellif.).
scabrifolia *M.G. Pimenov in Opred. Rast. Sred. Azii*, 7: 375 (1983)—Tadzhikistan.

ZEUKTOPHYLLUM (Aizoac.).
calycinum *(L. Bolus) H.E.K. Hartmann in Bradleya*, 16: 88 (1998): Ruschia calycina.

ZEUXINE (Orchidac.).
assamica *I. Barua & K. Barua in J. Econ. Taxon. Bot.*, 21(2): 491 (1997)—Assam.
debrajiana *Sud. Chowdhury in Indian Forester*, 122(1): 87 (1996)—Assam.
lancifolia *(Ames) P. Ormerod in Austral. Orchid Rev.*, 63(4): 11 (1998): Hetaeria lancifolia.
odzalaensis *D. Geerinck & J. Lejoly in Bull. Jard. Bot. Nation. Belg.*, 66(1–2): 154 (1997)—Congo.
shuishiehensis *S.S. Ying in Mem. Coll. Agric. Nation. Taiwan Univ.*, 29(2): 57 (1989), as 'shuishiehense'—Taiwan.

ZEYLANIDIUM (Podostemac.).
maheshwarii *C.J. Mathew & V.K. Satheesh in Aquatic Bot.*, 54(1): 74 (1996)—India (Kerala).

ZHUKOWSKIA *D.L. Szlachetko, R. González Tamayo & P. Rutkowski in Adansonia*, 22(2): 236 (2000). *ORCHIDACEAE*
cerina *(Lindl.) D.L. Szlachetko, R. González Tamayo & P. Rutkowski in Adansonia*, 22(2): 236 (2000): Spiranthes cerina.
lobata *(Lindl.) D.L. Szlachetko, R. González Tamayo & P. Rutkowski in Adansonia*, 22(2): 236 (2000): Spiranthes lobata.
smithii *(Rchb. f.) D.L. Szlachetko, R. González Tamayo & P. Rutkowski in Adansonia*, 22(2): 236 (2000): Spiranthes smithii.

ZIGADENUS (Melanthiac.).
neglectus *A. Espejo, A.R. López-Ferrari & Ceja in Acta Bot. Mex.*, 48: 47 (1999)—Mexico Northeast (Guanajuato, Querétaro).
virescens (Kunth) J.F. Macbr.
var. porrifolius *(Greene) O.S. Walsh ex A. Espejo & A.R. López-Ferrari in Acta Bot. Mex.*, 48: 49 (1999): Z. porrifolius.

ZINGIBER (Zingiberac.).
argenteum *J. Mood & I. Theilade in Nordic J. Bot.*, 17(4): 341 (1997)—Borneo (Sarawak).
aurantiacum *(Holttum) I. Theilade in Gard. Bull. Singapore*, 48(1–2): 232 (1996 publ. 1998): Z. gracile var. aurantiacum.
bisectum *D. Fang in Guihaia*, 16(1): 5 (1996)—China Southeast (Guangxi).
chlorobracteatum *J. Mood & I. Theilade in Nordic J. Bot.*, 19(5): 515 (1999)—Borneo (Sabah).
collinsii *J. Mood & I. Theilade in Nordic J. Bot.*, 19(5): 525 (1999), without latin descr.; *et in Nordic J. Bot.*, 20(1): 32 (2000), with latin descr.—Vietnam.

ZINGIBER:-
eborinum *J. Mood & I. Theilade in Sandakania,* 9: 21 (1997)—Borneo (Sabah).
ellipticum *(S.Q. Tong & Y.M. Xia) Q.G. Wu & T.L. Wu in Acta Phytotax. Sin.,* 34(4): 416 (1996): Plagiostachys elliptica.
ellipticum *(S.Q. Tong & Y.M. Xia) S.Q. Tong & Y.M. Xia in Fl. Yunnanica,* 8: 548 (1997), isonym; *see Q.G. Wu & T.L. Wu in Acta Phytotax. Sin.,* 34(4): 416 (1996): Plagiostachys elliptica.
flagelliforme *J. Mood & I. Theilade in Nordic J. Bot.,* 19(5): 517 (1999)—Borneo (Sabah).
flammeum *I. Theilade & J. Mood in Sandakania,* 9: 24 (1997)—Borneo (Sabah).
flavovirens *I. Theilade in Nordic J. Bot.,* 19(4): 397 (1999)—Thailand.
fraseri *I. Theilade in Gard. Bull. Singapore,* 48(1–2): 214 (1996 publ. 1998)—Malaya (Peninsular Malaysia).
georgeae *J. Mood & I. Theilade in Nordic J. Bot.,* 19(5): 521 (1999), as 'georgei'—Borneo (Sarawak).
gulinense *Y.M. Xia in Acta Bot. Yunnanica,* 18(1): 51 (1996)—China South-central (Yunnan).
kelabitianum *I. Theilade & H. Christensen in Edinburgh J. Bot.,* 55(2): 239 (1998)—Borneo (Sarawak).
koshunense *C.T. Moo in Novon,* 10(1): 88 (2000)—Taiwan.
lambii *J. Mood & I. Theilade in Nordic J. Bot.,* 17(4): 339 (1997)—Borneo (Sabah).
larsenii *I. Theilade in Nordic J. Bot ,* 19(4): 404 (1999)—Thailand.
latifolium *I. Theilade & J. Mood in Nordic J. Bot.,* 17(4): 343 (1997)—Borneo (Sabah).
longibracteatum *I. Theilade in Nordic J. Bot.,* 19(4): 408 (1999)—Thailand.
longiglande *D. Fang & D.H. Qin in Guihaia,* 16(1): 7 (1996)—China Southeast (Guangxi).
montanum *(J. König) I. Theilade in Gard. Bull. Singapore,* 48(1–2): 225 (1996 publ. 1998): Amomum montanum.
neotruncatum *T.L. Wu, K. Larsen & N.J. Turland in Novon,* 10(1): 91 (2000), nom. nov.: Z. truncatum *S.Q. Tong.*
newmanii *I. Theilade & J. Mood in Nordic J. Bot.,* 19(4): 407 (1999)—Thailand.
officinale Roscoe
var. rubrum *I. Theilade in Gard. Bull. Singapore,* 48(1–2): 218 (1996 publ. 1998)—Cultivation.
paucipunctatum *D. Fang in Guihaia,* 16(1): 6 (1996)—China Southeast (Guangxi).
pendulum *J. Mood & I. Theilade in Nordic J. Bot.,* 17(4): 342 (1997)—Borneo (Sabah).
peninsulare *I. Theilade in Nordic J. Bot.,* 19(4): 401 (1999)—Thailand.
petiolatum *(Holttum) I. Theilade in Gard. Bull. Singapore,* 48(1–2): 232 (1996 publ. 1998): Z. gracile var. petiolatum.
phillippsiae *J. Mood & I. Theilade in Nordic J. Bot.,* 19(5): 520 (1999), as 'phillippsii'—Borneo (Sabah).
simaoense *Y.Y. Qian in Bull. Bot. Res. North-East. Forest. Univ.,* 18(3): 284 (1998)—China South-central (Yunnan).
velutinum *J. Mood & I. Theilade in Nordic J. Bot.,* 19(5): 519 (1999)—Borneo (Sabah).
villosum *I. Theilade in Nordic J. Bot.,* 19(4): 396 (1999), homonym; *see Stokes, Bot. Mat. Med.,* 1: 63 (1812)—Thailand.
vinosum *J. Mood & I. Theilade in Nordic J. Bot.,* 17(4): 337 (1997)—Borneo (Sabah).
viridiflavum *J. Mood & I. Theilade in Nordic J. Bot.,* 19(5): 514 (1999)—Borneo (Sabah).

ZINNIA (Compos.).
acerosa (DC.) A. Gray
var. guanajuatensis *G. Calderón de Rzedowski & J. Rzedowski in Acta Bot. Mex.,* 36: 78 (1996)—Mexico Northeast (Guanajuato).
zamudiana *G. Calderón de Rzedowski & J. Rzedowski in Acta Bot. Mex.,* 36: 79 (1996)—Mexico Northeast (Querétaro).

ZIZIPHORA (Labiat.).
clinopodioides Lam.
subsp. tomentosa *(Juz.) R.V. Kamelin & I.A. Gubanov in I.A. Gubanov, Konsp. Fl. Vneshnei Mongolii (Sosud. Rast.):* 90 (1996): Z. tomentosa.

ZIZIPHUS (Rhamnac.).
acidojujuba *C.Y. Cheng & M.J. Liu in J. Hebei Agric. Univ.,* 17(4): 3 (1994), nom. nov.: Z. vulgaris var. spinosa *Bunge.*
f. granulata *C.Y. Cheng & M.J. Liu in J. Hebei Agric. Univ.,* 17(4): 5 (1994)—China North-central (Hebei).
f. infecunda *C.Y. Cheng & M.J. Liu in J. Hebei Agric. Univ.,* 17(4): 5 (1994)—China North-central (Hebei).
f. trachysperma *C.Y. Cheng & M.J. Liu in J. Hebei Agric. Univ.,* 17(4): 5 (1994)—China North-central (Hebei).
hajarensis *D.W. Duling, S.A. Ghazanfar & H.D.V. Prendergast in Kew Bull.,* 53(3): 733 (1998)—Oman.
jujuba Mill.
f. allochroa *C.Y. Cheng & M.J. Liu in J. Hebei Agric. Univ.,* 17(4): 7 (1994)—China North-central (Hebei, Shanxi), China Southeast (Henan).
f. apyrena *C.Y. Cheng & M.J. Liu in J. Hebei Agric. Univ.,* 17(4): 8 (1994)—China North-central (Shanxi, Hebei), China Southeast (Henan).
f. carnosicalycis *C.Y. Cheng & M.J. Liu in J. Hebei Agric. Univ.,* 17(4): 8 (1994)—China North-central (Shanxi).
f. heteroformis *C.Y. Cheng & M.J. Liu in J. Hebei Agric. Univ.,* 17(4): 8 (1994)—China North-central (Shanxi, Shandong).
f. tortuosa *C.Y. Cheng & M.J. Liu in J. Hebei Agric. Univ.,* 17(4): 7 (1994)—China North-central (Shanxi, Hebei, Shandong), China Southeast (Henan).
robertsoniana *H.J. Beentje in Kew Bull.,* 51(1): 197 (1996)—Kenya.

ZOLLERNIA (Legumin.).
cowanii *V. de F. Mansano in Brittonia,* 51(2): 155 (1999)—Brazil Southeast (Minas Gerais).

ZOMICARPELLA (Arac.).
amazonica *J. Bogner in Sendtnera,* 4: 8 (1997)—Brazil North (Amazonas).

ZORNIA (Legumin.).
amaralaensis *M. Brandão in Daphne,* 4(3): 5 (1994)—Brazil Southeast (Minas Gerais).
brasiliensis Vogel
var. adenocarpa *L.P. de Queiroz in Bol. Bot. (São Paulo),* 16: 108 (1997)—Brazil Southeast (Minas Gerais).
confusa *R. Vanni in Brittonia,* 48(4): 495 (1996 publ. 1997)—Brazil Southeast (Minas Gerais), Mexico Central, Mexico Northeast, Mexico Gulf, Mexico Southwest, Mexico Southeast, Guatemala, Honduras, Costa Rica, Venezuela, Panama.
crinita *(Mohlenbr.) R. Vanni in Darwiniana,* 33(1–4): 8 (1995): Z. pardina var. crinita.
gavilanesii *M. Brandão & N.M. Sousa Costa in Daphne,* 1(1): 5 (1990)—Brazil Southeast (Minas Gerais).

ZOTOVIA *E. Edgar & H.E. Connor in New Zealand J. Bot.,* 36(4): 569 (1998), nom. nov.: gen. Petriella *Zotov.* (Gramin.).
acicularis *E. Edgar & H.E. Connor in New Zealand J. Bot.,* 36(4): 569 (1998)—New Zealand South I.
colensoi *(Hook. f.) E. Edgar & H.E. Connor in New Zealand J. Bot.,* 36(4): 571 (1998): Ehrharta colensoi.
thomsonii *(Petrie) E. Edgar & H.E. Connor in New Zealand J. Bot.,* 36(4): 573 (1998): Ehrharta thomsonii.

ZOYSIA (Gramin.).
japonica Steud.
f. macrostachya *H.D. Dong & L.J. Gong in Bull. Bot. Res. North-East. Forest. Univ.,* 17(4): 364 (1997)—Manchuria (Liaoning).

ZYGIA (Legumin.).
sect. Barticaea *R.C. Barneby & J.W. Grimes, Silk Tree, Guanacaste, Monkey's Earring (Mem. New York Bot. Gard., 74(2)):* 82 (1997).
sect. Codonocalyx *R.C. Barneby & J.W. Grimes, Silk Tree, Guanacaste, Monkey's Earring (Mem. New York Bot. Gard., 74(2)):* 128 (1997).

ZYGIA:-
sect. Ingopsis *R.C. Barneby & J.W. Grimes, Silk Tree, Guanacaste, Monkey's Earring (Mem. New York Bot. Gard., 74(2)):* 130 (1997).
sect. Marmaroxylon *(Killip) R.C. Barneby & J.W. Grimes, Silk Tree, Guanacaste, Monkey's Earring (Mem. New York Bot. Gard., 74(2)):* 71 (1997): gen. Marmaroxylon.
sect. Nothellobium *R.C. Barneby & J.W. Grimes, Silk Tree, Guanacaste, Monkey's Earring (Mem. New York Bot. Gard., 74(2)):* 73 (1997).
sect. Parazygia *R.C. Barneby & J.W. Grimes, Silk Tree, Guanacaste, Monkey's Earring (Mem. New York Bot. Gard., 74(2)):* 75 (1997).
sect. Pseudocojoba *R.C. Barneby & J.W. Grimes, Silk Tree, Guanacaste, Monkey's Earring (Mem. New York Bot. Gard., 74(2)):* 83 (1997).
sect. Zygiopsis *R.C. Barneby & J.W. Grimes, Silk Tree, Guanacaste, Monkey's Earring (Mem. New York Bot. Gard., 74(2)):* 64 (1997).
bangii *R.C. Barneby & J.W. Grimes, Silk Tree, Guanacaste, Monkey's Earring (Mem. New York Bot. Gard., 74(2)):* 95 (1997), nom. nov.: Pithecellobium laxiflorum *Rusby.*
basijuga *(Ducke) R.C. Barneby & J.W. Grimes, Silk Tree, Guanacaste, Monkey's Earring (Mem. New York Bot. Gard., 74(2)):* 68 (1997): Pithecellobium basijugum.
claviflora *(Benth.) R.C. Barneby & J.W. Grimes, Silk Tree, Guanacaste, Monkey's Earring (Mem. New York Bot. Gard., 74(2)):* 65 (1997): Pithecellobium claviflorum.
coccinea (G. Don) L. Rico
var. macrophylla *(Benth.) R.C. Barneby & J.W. Grimes, Silk Tree, Guanacaste, Monkey's Earring (Mem. New York Bot. Gard., 74(2)):* 91 (1997): Pithecellobium macrophyllum.
var. oriunda *(J.F. Macbr.) R.C. Barneby & J.W. Grimes, Silk Tree, Guanacaste, Monkey's Earring (Mem. New York Bot. Gard., 74(2)):* 90 (1997): Pithecellobium oriundum.
codonocalyx *R.C. Barneby & J.W. Grimes, Silk Tree, Guanacaste, Monkey's Earring (Mem. New York Bot. Gard., 74(2)):* 128 (1997)—Colombia.
collina *(Sandwith) R.C. Barneby & J.W. Grimes, Silk Tree, Guanacaste, Monkey's Earring (Mem. New York Bot. Gard., 74(2)):* 80 (1997): Pithecellobium collinum.
dinizii *(Ducke) D.A. Neill, G.P. Lewis & B.B. Klitgaard in Cat. Vasc. Pl. Ecuador (Monogr. Syst. Bot. Missouri Bot. Gard., 75):* 956 (1999): Pithecellobium dinizii.
dissitiflora *R.C. Barneby & J.W. Grimes, Silk Tree, Guanacaste, Monkey's Earring (Mem. New York Bot. Gard., 74(2)):* 92 (1997)—Colombia.
eperuetorum *(Sandwith) R.C. Barneby & J.W. Grimes, Silk Tree, Guanacaste, Monkey's Earring (Mem. New York Bot. Gard., 74(2)):* 83 (1997): Pithecellobium eperuetorum.
garcia-barrigae *(Barbosa) R.C. Barneby & J.W. Grimes, Silk Tree, Guanacaste, Monkey's Earring (Mem. New York Bot. Gard., 74(2)):* 123 (1997): Pithecellobium latifolium subsp. garcia-barrigae.
heteroneura *R.C. Barneby & J.W. Grimes, Silk Tree, Guanacaste, Monkey's Earring (Mem. New York Bot. Gard., 74(2)):* 91 (1997)—Ecuador, Peru.
inundata *(Ducke) H.C. Lima ex R.C. Barneby & J.W. Grimes, Silk Tree, Guanacaste, Monkey's Earring (Mem. New York Bot. Gard., 74(2)):* 130 (1997): Inga inundata.
lathetica *R.C. Barneby & J.W. Grimes, Silk Tree, Guanacaste, Monkey's Earring (Mem. New York Bot. Gard., 74(2)):* 69 (1997)—Ecuador, Colombia.
latifolia (L.) Fawc. & Rendle
var. communis *R.C. Barneby & J.W. Grimes, Silk Tree, Guanacaste, Monkey's Earring (Mem. New York Bot. Gard., 74(2)):* 119 (1997)—Brazil North (Roraima).
var. controversa *R.C. Barneby & J.W. Grimes, Silk Tree, Guanacaste, Monkey's Earring (Mem. New York Bot. Gard., 74(2)):* 121 (1997)—Brazil North (Rondônia, Acre, Amazonas), Bolivia.
var. glabrata *(Mart.) R.C. Barneby & J.W. Grimes, Silk Tree, Guanacaste, Monkey's Earring (Mem. New York Bot. Gard., 74(2)):* 122 (1997): Pithecellobium glabratum.

ZYGIA:-
latifolia
var. lasiopus *(Benth.) R.C. Barneby & J.W. Grimes, Silk Tree, Guanacaste, Monkey's Earring (Mem. New York Bot. Gard., 74(2)):* 120 (1997): Pithecellobium lasiopus.
morongii *R.C. Barneby & J.W. Grimes, Silk Tree, Guanacaste, Monkey's Earring (Mem. New York Bot. Gard., 74(2)):* 108 (1997)—Paraguay, Argentina Northeast (Chaco, Corrientes, Formosa, Misiones).
multipunctata *R.C. Barneby & J.W. Grimes, Silk Tree, Guanacaste, Monkey's Earring (Mem. New York Bot. Gard., 74(2)):* 94 (1997)—Colombia.
ocumarensis *(Pittier) R.C. Barneby & J.W. Grimes, Silk Tree, Guanacaste, Monkey's Earring (Mem. New York Bot. Gard., 74(2)):* 74 (1997): Klugiodendron ocumarense.
palustris *R.C. Barneby & J.W. Grimes, Silk Tree, Guanacaste, Monkey's Earring (Mem. New York Bot. Gard., 74(2)):* 66 (1997)—Venezuela.
pithecolobioides *(Kuntze) R.C. Barneby & J.W. Grimes, Silk Tree, Guanacaste, Monkey's Earring (Mem. New York Bot. Gard., 74(2)):* 127 (1997): Feuilleea pithecolobioides.
potaroensis *R.C. Barneby & J.W. Grimes, Silk Tree, Guanacaste, Monkey's Earring (Mem. New York Bot. Gard., 74(2)):* 68 (1997)—Guyana, Venezuela.
racemosa *(Ducke) R.C. Barneby & J.W. Grimes, Silk Tree, Guanacaste, Monkey's Earring (Mem. New York Bot. Gard., 74(2)):* 71 (1997): Pithecellobium racemosum.
ramiflora *(Benth.) R.C. Barneby & J.W. Grimes, Silk Tree, Guanacaste, Monkey's Earring (Mem. New York Bot. Gard., 74(2)):* 76 (1997), homonym; *see Kosterm. in Bull. Org. Sci. Res. Indonesia,* 20: 26 (1954): Pithecellobium ramiflorum.
sabatieri *R.C. Barneby & J.W. Grimes, Silk Tree, Guanacaste, Monkey's Earring (Mem. New York Bot. Gard., 74(2)):* 83 (1997)—French Guiana.
steyermarkii *(Schery) R.C. Barneby & J.W. Grimes, Silk Tree, Guanacaste, Monkey's Earring (Mem. New York Bot. Gard., 74(2)):* 79 (1997): Pithecellobium steyermarkii.
tetragona *R.C. Barneby & J.W. Grimes, Silk Tree, Guanacaste, Monkey's Earring (Mem. New York Bot. Gard., 74(2)):* 80 (1997)—French Guiana.
transamazonica *R.C. Barneby & J.W. Grimes, Silk Tree, Guanacaste, Monkey's Earring (Mem. New York Bot. Gard., 74(2)):* 78 (1997)—Brazil North (Amazonas).
turneri *(McVaugh) R.C. Barneby & J.W. Grimes, Silk Tree, Guanacaste, Monkey's Earring (Mem. New York Bot. Gard., 74(2)):* 117 (1997): Pithecellobium turneri.

ZYGOPETALUM (Orchidac.).
euglossum *(Rchb. f.) N.E. Brown, Suppl. Johnson's Gard. Dict.:* 1026 (1882): Pescatorea euglossa.
fimbriatum *(Regel) N.E. Brown, Suppl. Johnson's Gard. Dict.:* 1026 (1882): Pescatorea fimbriata.
russelianum *(Rchb. f.) N.E. Brown, Suppl. Johnson's Gard. Dict.:* 1026 (1882): Pescatorea russeliana.
sincoranum *V.P. Castro Neto & M.A. Campacci in Orquidário,* 14(2): 36 (2000)—Brazil Northeast (Bahia).

ZYGOPHYLLUM (Zygophyllac.).
applanatum *L. Van Zyl in Bothalia,* 29(2): 233 (1999)—Namibia.
aurantiacum (Lindl.) F. Muell.
subsp. cuneatum *H. Eichler ex R.M. Barker in J. Adelaide Bot. Gard.,* 17: 163 (1996)—South Australia.
subsp. simplicifolium *H. Eichler ex R.M. Barker in J. Adelaide Bot. Gard.,* 18(1): 61 (1998)—South Australia, Northern Territory.
subsp. verticillatum *H. Eichler ex R.M. Barker in J. Adelaide Bot. Gard.,* 17: 164 (1996)—South Australia.
botulifolium *L. Van Zyl in Bothalia,* 27(2): 131 (1997)—Cape Province.
eichleri *R.M. Barker in J. Adelaide Bot. Gard.,* 18(1): 50 (1998)—Western Australia, South Australia, Northern Territory.
flavum *H. Eichler ex R.M. Barker in J. Adelaide Bot. Gard.,* 17: 166 (1996)—South Australia.

ZYGOPHYLLUM:-
fuscatum *L. Van Zyl in Bothalia,* 27(2): 129 (1997)—Cape Province.
halophilum *R.M. Barker in J. Adelaide Bot. Gard.,* 18(1): 55 (1998)—Western Australia, South Australia.
hirticaule *L. Van Zyl in Bothalia,* 29(2): 235 (1999)—Namibia.
lobulatum *(Benth.) H. Eichler in J. Adelaide Bot. Gard.,* 17: 168 (1996): Z. iodocarpum var. lobulatum.
marliesiae *R.M. Barker in J. Adelaide Bot. Gard.,* 18(1): 63 (1998)—South Australia.
pamiricum *V.I. Grubov in Novosti Sist. Vÿssh. Rast.,* 31: 175 (1998)—Xinjiang.
pinnatum Cham.
subsp. chakassicum *G.A. Peschkova in Fl. Sibir. (Geraniac.-Cornac.),* 10: 32 (1996)—Krasnoyarsk.
pterocarpum Bunge
subsp. tuvinicum *G.A. Peschkova in Fl. Sibir. (Geraniac.-Cornac.),* 10: 33 (1996)—Tuva.
pterocaule *L. Van Zyl in Bothalia,* 29(2): 231 (1999)—Cape Province, Namibia.
reticulatum *H. Eichler ex R.M. Barker in J. Adelaide Bot. Gard.,* 18(1): 57 (1998)—Western Australia, South Australia.
rowelliae *R.M. Barker in J. Adelaide Bot. Gard.,* 18(1): 53 (1998)—South Australia, Northern Territory, Queensland.
tetrapterum *H. Eichler ex R.M. Barker in J. Adelaide Bot. Gard.,* 17: 164 (1996)—Western Australia.

ZYGOSEPALUM (Orchidac.).
revolutum *L.A. Garay & G.A. Romero-Gonzalez in Harvard Pap. Bot.,* 4(2): 487 (1999)—Colombia.

ZYGOSTATES (Orchidac.).
leptosepala *A.L.V. Toscano de Brito & L. Kollman in Kew Bull.,* 52(1): 239 (1997)—Brazil Southeast (Espírito Santo).
ligulata *A.L.V. Toscano de Brito & M.C.M. Werkhoven in Lindleyana,* 13(1): 3 (1998)—Guyana, Surinam.

NEW AND OVERLOOKED GENERIC NAMES LISTED UNDER FAMILIES

A

ACANTHACEAE
CYNAROSPERMUM *K. Vollesen*
ICHTHYOSTOMA *M. Hedrén & K. Vollesen*

AIZOACEAE
CAULIPSOLON *C. Klak*
HAMMERIA *P.M. Burgoyne*
× HARBOTRYA *S.A. Hammer*
HARTMANTHUS *S.A. Hammer*
× LAPIDERMA *S.A. Hammer*

ALISMATACEAE
ASTONIA *S.W.L. Jacobs*

ALOACEAE
× ALGASTOLOBA *D.M. Cumming*
× BAYERARA *D.M. Cumming*
× CUMMINGARA *G.D. Rowley*
× MAYSARA *D.M. Cumming*
× ROWLEYARA *D.M. Cumming*
× SMITHARA *D.M. Cumming*

ALSTROEMERIACEAE
TALTALIA *Ehr. Bayer*

AMARANTHACEAE
HEBANTHODES *T.M. Pedersen*
LECOSIA *T.M. Pedersen*
PEDERSENIA *J. Holub*

AMARYLLIDACEAE
CEARANTHES *P. Ravenna*
× CRIMOCHARIS *D.J. Lehmiller*
TOCANTINIA *P. Ravenna*

ANNONACEAE
CLEISTOPETALUM *H. Okada*
KLAROBELIA *L.W. Chatrou*
MOSANNONA *L.W. Chatrou*
PSEUDOMALMEA *L.W. Chatrou*
SANRAFAELIA *B. Verdcourt*

APOCYNACEAE
ECUA *D.J. Middleton*
HYLAEA *J.F. Morales*
TELOSIPHONIA *(Woodson) J. Henrickson*

ARALIACEAE
RAUKAUA *B.C. Seemann*

ARAUCARIACEAE
WOLLEMIA *W.G. Jones, K.D. Hill & J.M. Allen*

ARISTOLOCHIACEAE
ASIPHONIA *W. Griffith*

ASCLEPIADACEAE
AUSTRALLUMA *D.C.H. Plowes*
BALLYANTHUS *P.V. Bruyns*
BAYNESIA *P.V. Bruyns*
BOREALLUMA *D.C.H. Plowes*
CALYPTRANTHERA *J. Klackenberg*

ASCLEPIADACEAE:-
CAUDANTHERA *D.C.H. Plowes*
CRENULLUMA *D.C.H. Plowes*
CRYPTOLLUMA *D.C.H. Plowes*
CYLINDRILLUMA *D.C.H. Plowes*
LARRYLEACHIA *D.C.H. Plowes*
MADANGIA *P.I. Forster, D.J. Liddle & I.M. Liddle*
MONOLLUMA *D.C.H. Plowes*
SANGUILLUMA *D.C.H. Plowes*
SAUROLLUMA *D.C.H. Plowes*
SINOMARSDENIA *P.T. Li & J.J. Chen*
SOMALLUMA *D.C.H. Plowes*
SPIRALLUMA *D.C.H. Plowes*
SULCOLLUMA *D.C.H. Plowes*
VIETNAMIA *P.T. Li*

B

BORAGINACEAE
ANCHUSELLA *M. Bigazzi, E. Nardi & F. Selvi*
CEBALLOSIA *G. Kunkel ex H. Förther*
EXARRHENA *(A.DC.) O.D. Nikiforova*
GYMNOMYOSOTIS *(A.DC.) O.D. Nikiforova*
HILGERIA *H. Förther*
PONTECHIUM *U.-R. Böhle & H.H. Hilger*
SAURIA *M.S. Bajtenov*
SETULOCARYA *R.R. Mill & D.G. Long*
STOMOTECHIUM *J.G.C. Lehmann*

BROMELIACEAE
× AECHOPSIS *D. Butcher*
× ANAGELIA *E.L. Smith*
× ANANANANAS *D.A. Beadle*
× ANDROLAECHMEA *C. Chevalier*
× BILLMEA *K. Williams*
CANISTROPSIS *(Mez) E.M.C. Leme*
× CANMEA *Racine C. Foster & M.B. Foster*
× CANULARIUM *D.A. Beadle*
× CRYPTANANAS *D.A. Beadle*
× CRYPTMEA *E.L. Smith*
× DYCKCOHNIA *G.H. Anderson ex J.R. Grant*
× DYCKTIA *D.A. Beadle*
EDMUNDOA *E.M.C. Leme*
× HECHCOHNIA *G.H. Anderson ex J.R. Grant*
× HOHENTEA *D.A. Beadle*
× NEOBERGIA *E.L. Smith*
× NEODSTROPSIS *D. Butcher*
× NEOISTRUM *D.A. Beadle*
× NEOLARIUM *Racine C. Foster & M.B. Foster*
× NEOROCKIA *D. Butcher*
× ORTHOLARIUM *Racine C. Foster & M.B. Foster*
× ORTHOMEA *E.L. Smith*
× PITINIA *J. Irvin ex E. Baskerville*
× PORTEMEA *L. Ariza-Julia*
× PUCOHNIA *D.A. Beadle*
× QUESMEA *E. Knobloch*
× QUESREGELIA *J.F. Carrone*
× STREPTOLARIUM *D.A. Beadle*

BROMELIACEAE
× STREPTOMEA *E.L. Smith*
× VRIECANTAREA *J.R. Grant*
WERAUHIA *J.R. Grant*

BURSERACEAE
PSEUDODACRYODES *R. Pierlot*

C

CACTACEAE
ACHARAGMA *(N.P. Taylor) A. D. Zimmerman ex C.E. Glass*
× APOROHELIOCEREUS *hort. ex H.K. Airy Shaw*
BOLIVICACTUS *A.B. Doweld*
BRAVOCACTUS *A.B. Doweld*
CINTIA *K. Kníže & J. Říha*
EMORYCACTUS *A.B. Doweld*
ESCOBARIOPSIS *A.B. Doweld*
ESCOBRITTONIA *A.B. Doweld*
ESCOCORYPHANTHA *A.B. Doweld*
GRISEOCEREUS *(P.V. Heath) P.V. Heath*
× GRISULATOCEREUS *P.V. Heath*
GYMNOCALYCIUM *Pfeiff. ex L. Mittler*
KADENICARPUS *A.B. Doweld*
LODIA *A. Mosco & C. Zanovello*
MEYEROCACTUS *A.B. Doweld*
NEONAVAJOA *A.B. Doweld*
NIGELLICEREUS *(P.V. Heath) P.V. Heath*
× OREOBIVIA *M. Lowry*
PARRYCACTUS *A.B. Doweld*
PERONOCACTUS *A.B. Doweld*
PIERREBRAUNIA *E. Esteves Pereira*
PLUTONOPUNTIA *P.V. Heath*
PUEBLOA *A.B. Doweld*
× RITTELLICEREUS *P.V. Heath*
RITTEROCACTUS *A.B. Doweld*
SALMONOPUNTIA *P.V. Heath*
× STENILLOCACTUS *P.V. Heath*
TORREYCACTUS *A.B. Doweld*
TUNILLA *D.R. Hunt & J. Iliff*
URSOPUNTIA *P.V. Heath*

CAMPANULACEAE
CALCARATOLOBELIA *R.L. Wilbur*
WIMMERELLA *L. Serra, M.B. Crespo & T.G. Lammers*

CARICACEAE
HOROVITZIA *V.M. Badillo*

CARYOPHYLLACEAE
PSAMMOSILENE *W.C. Wu & C.Y. Wu*
PSEUDOCERASTIUM *C.Y. Wu, X.H. Guo & X.P. Zhang*

CELASTRACEAE
GLOVERIA *M. Jordaan*
ROBSONODENDRON *R.H. Archer*

COMPOSITAE
ANDERBERGIA *B. Nordenstam*
× ASTERIGERON *N.N. Tzvelev*
BATOPILASIA *G.L. Nesom & R.D. Noyes*
BRENANDENDRON *H. Robinson*
CAATINGANTHUS *H. Robinson*
CABOBANTHUS *H. Robinson*
CARLQUISTIA *B.G. Baldwin*
CASTRILANTHEMUM *R. Vogt & C. Oberprieler*
CATOLESIA *D.J.N. Hind*
CAUCASALIA *B. Nordenstam*
CAXAMARCA *M.O. Dillon & A. Sagástegui Alva*
CHARADRANAETES *J.P. Janovec & H. Robinson*
CONSTANCEA *B.G. Baldwin*
CRINITINA *J. Soják*
CURIO *P.V. Heath*
ELEPHANTOPSIS *(Sch. Bip.) C.F. Baker*
EUTHAMIA *(Nutt.) Cassini*
FABERIOPSIS *C. Shih & Y.L. Chen*
× FILFIA *J. Holub*
HARMONIA *B.G. Baldwin*
HILLIARDIELLA *H. Robinson*
× JACONECIO *P.V. Heath*
× JACORARIA *P.V. Heath*
JENSIA *B.G. Baldwin*
KOYAMASIA *H. Robinson*
KYHOSIA *B.G. Baldwin*
LAMPROPAPPUS *(O. Hoffm.) H. Robinson*
LIGULARIOPSIS *Y.L. Chen*
MANYONIA *H. Robinson*
MICROPAPPUS *(Sch. Bip.) C.F. Baker*
MISBROOKEA *V.A. Funk*
MONTICALIA *C. Jeffrey*
MTONIA *H.J. Beentje*
MYANMARIA *H. Robinson*
OOCEPHALA *(S.B. Jones) H. Robinson*
ORBIVESTUS *H. Robinson*
OSMITES *L.*
PARAMIFLOS *J. Cuatrecasas*
PHITOSIA *G. Kamari & W. Greuter*
PSECTRA *(Endl.) P. Tomšovic*
PSILOTHONNA *(E. Mey. ex DC.) E.P. Phillips*
RACHELIA *J.M. Ward & I. Breitwieser*
RAYJACKSONIA *R.L. Hartman & M.A. Lane*
ROBINSONECIO *T.M. Barkley & J.P. Janovec*
SEMIRIA *D.J.N. Hind*
SIAPAEA *J.F. Pruski*
SOLANECIO *(Sch.Bip.) W.G. Walpers*
TEHUANA *J.L. Panero & J.L. Villaseñor*
TUBERCULOCARPUS *J.F. Pruski*
VERNONIASTRUM *H. Robinson*
VILLASENORIA *B.L. Clark*
WILDPRETIA *U. Reifenberger & A. Reifenberger*
XENOPHYLLUM *V.A. Funk*

CONVALLARIACEAE
HETEROPOLYGONATUM *M.N. Tamura & M. Ogisu*

CONVOLVULACEAE
CLEIOSTOMA *Raf.*

CORSIACEAE
CORSIOPSIS *D.X. Zhang, R.M.K. Saunders & C.M. Hu*

CRASSULACEAE
BALFOURIA *(H. Ohba) H. Ohba*
KUNGIA *K.T. Fu*
OHBAEA *V.V. Byalt & I.V. Sokolova*

CROSSOSOMATACEAE
VELASCOA *G. Calderón de Rzedowski & J. Rzedowski*

CRUCIFERAE
BAIMASHANIA *I.A. Al-Shehbaz*
BARBAMINE *A.P. Khokhrjakov*
CRUCIHIMALAYA *I.A. Al-Shehbaz, S.L. O'Kane & R.A. Price*
GOERKEMIA *Ş. Yıldırımlı*
IANHEDGEA *I.A. Al-Shehbaz & S.L. O'Kane*
OLIMARABIDOPSIS *I.A. Al-Shehbaz, S.L. O'Kane & R.A. Price*
PSEUDOARABIDOPSIS *I.A. Al-Shehbaz, S.L. O'Kane & R.A. Price*
SIBAROPSIS *S. Boyd & T.S. Ross*
SINOSOPHIOPSIS *I.A. Al-Shehbaz*
TWISSELMANNIA *I.A. Al-Shehbaz*

CUCURBITACEAE
BAYABUSUA *W.J.J.O. de Wilde*
BORNEOSICYOS *W.J.J.O. de Wilde*

CYCADACEAE
EPICYCAS *D.J. de Laubenfels*

CYPERACEAE
CAPEOBOLUS *J. Browning*
KRECZETOVICZIA *N.N. Tzvelev*

D

DILLENIACEAE
NEODILLENIA *G.A. Aymard C.*

DIPSACACEAE
BASSECOIA *B.L. Burtt*
PTEROCEPHALODES *V. Mayer & F. Ehrendorfer*

DROSERACEAE
FREATULINA *J. Chrtek & Z. Slavíková*

E

ECDEIOCOLEACEAE
GEORGEANTHA *B.G. Briggs & L.A.S. Johnson*

EPACRIDACEAE
LEPTECOPHYLLA *C.M. Weiller*
PLANOCARPA *C.M. Weiller*

ERIOCAULACEAE
PAEPALANTHUS *Mart.*
PHILODICE *Mart.*

EUPHORBIACEAE
ARCHILEPTOPUS *P.T. Li*
AUBLETIANA *J. Murillo-A.*
BALAKATA *H.-J. Esser*
DISTICHIRHOPS *R.M.A.P. Haegens*
NEOSHIRAKIA *H.-J. Esser*
NOTHOBACCAUREA *R.M.A.P. Haegens*
RHODOTHYRSUS *H.-J. Esser*
SHIRAKIOPSIS *H.-J. Esser*

G

GENTIANACEAE
ARIPUANA *L. Struwe, P.J.M. Maas & V.A. Albert*
CELIANTHA *B. Maguire*

GESNERIACEAE
EMARHENDIA *R. Kiew, A. Weber & B.L. Burtt*
HOVANELLA *A. Weber & B.L. Burtt*
PARAISOMETRUM *W.T. Wang*
RIDLEYANDRA *A. Weber & B.L. Burtt*
SENYUMIA *R. Kiew, A. Weber & B.L. Burtt*
SPELAEANTHUS *R. Kiew, A. Weber & B.L. Burtt*

GRAMINEAE
ACIDOSASA *C.D. Chu & C.S. Chao ex P.C. Keng*
ANATHEROSTIPA *(Hack. ex Kuntze) P. Peñailillo B.*
ANEMANTHELE *J.F. Veldkamp*
AUSTRODANTHONIA *H.P. Linder*
AUSTROSTIPA *S.W.L. Jacobs & J. Everett*
CATHARIOSTACHYS *S. Dransfield*

GRAMINEAE:-
CYRTOCHLOA *S. Dransfield*
DIACHROA *T. Nuttall*
DIANDRANTHUS *L. Liou*
× DUPOA *J. Cayouette & S.J. Darbyshire*
× ELYTESION *M.E. Barkworth & D.R. Dewey*
EREMIUM *O. Seberg & I. Linde-Laursen*
FIMBRIBAMBUSA *E.A. Widjaja*
GAOLIGONGSHANIA *D.Z. Li, C.J. Hsueh & N.H. Xia*
JOYCEA *H.P. Linder*
× LEYDEUM *M.E. Barkworth*
NEOLOLEBA *E.A. Widjaja*
NICORAELLA *M.A. Torres*
PARABAMBUSA *E.A. Widjaja*
PATELLOCALAMUS *W.T. Lin*
× PHYLLOSASA *J.-P. Demoly*
PINGA *E.A. Widjaja*
PSEUDORAPHIS *Griff. ex R. Pilger*
QIONGZHUEA *(T.H. Wen & Ohrnb.) J.R. Xue & T.P. Yi*
REHIA *F. Fijten*
RHEOCHLOA *T.S. Filgueiras, P.M. Peterson & Y. Herrera-Arrieta*
RUBIMONS *B.S. Sun*
× SASINARIA *J.-P. Demoly*
× SCHEDOLIUM *J. Holub*
× SCHEDOLOLIUM *R.J. Soreng & E.E. Terrell*
SELLULOCALAMUS *W.T. Lin*
SHIBATAEA *Makino ex T. Nakai*
TEMBURONGIA *S. Dransfield & K.M. Wong*
THONANDIA *H.P. Linder*
TRIARRHENA *(Maxim.) T. Nakai*
VALIHA *S. Dransfield*
WHALLEYA *K.E. Wills & J.J. Bruhl*
ZOTOVIA *E. Edgar & H.E. Connor*

GROSSULARIACEAE
× RIBULARIA *J. Holub*

H

HYACINTHACEAE
AUTONOE *(Webb & Berthel.) F. Speta*
AVONSERA *F. Speta*
CHARYBDIS *F. Speta*
CHOUARDIA *F. Speta*
EBERTIA *F. Speta*
FESSIA *F. Speta*
IGIDIA *F. Speta*
MERWILLA *F. Speta*
NAMOPHILA *U. Müller-Doblies & D. Müller-Doblies*
PFOSSERIA *F. Speta*
PSEUDOPROSPERO *F. Speta*
RHADAMANTHOPSIS *(Oberm.) F. Speta*
SCHNARFIA *F. Speta*
URGINAVIA *F. Speta*
ZAGROSIA *F. Speta*

I

IRIDACEAE
ALATAVIA *G.I. Rodionenko*

L

LABIATAE
BRACHYSOLA *B.L. Rye*
CALCHAS *P.V. Heath*

LABIATAE:-
× CLINOMICROMERIA *R. Govaerts*
ISODON *(Schrad. ex Benth.) Spach*
MADLABIUM *I.C. Hedge*
OBTEGOMERIA *A. Doroszenko & P.D. Cantino*
WARNOCKIA *M.W. Turner*

LAURACEAE
SEXTONIA *H. van der Werff*

LEGUMINOSAE
ALYSISTYLES *N.E. Br. ex R.A. Dyer*
BALIZIA *R.C. Barneby & J.W. Grimes*
BARNEBYDENDRON *J.H. Kirkbride*
BIKINIA *J.J. Wieringa*
BLANCHETIODENDRON *R.C. Barneby & J.W. Grimes*
BOBGUNNIA *J.H. Kirkbride & J.H. Wiersema*
BOBROVIA *A.P. Khokhrjakov*
BRENANIODENDRON *J. Léonard*
ECUADENDRON *D.A. Neill*
FALCATARIA *(I.C. Nielsen) R.C. Barneby & J.W. Grimes*
GUINETIA *L. Rico & M. Sousa S.*
HESPERALBIZIA *R.C. Barneby & J.W. Grimes*
HYDROCHOREA *R.C. Barneby & J.W. Grimes*
HYLODESMUM *H. Ohashi & R.R. Mill*
ICURIA *J.J. Wieringa*
LACKEYA *R.H. Fortunato, L.P. de Queiroz & G.P. Lewis*
LEUCOCHLORON *R.C. Barneby & J.W. Grimes*
MONTIGENA *P.B. Heenan*
OHWIA *Hiroyoshi Ohashi*
ORYXIS *A. Delgado Salinas & G.P. Lewis*
OTTLEYA *D.D. Sokoloff*
PELTIERA *D.J. Du Puy & J.-N. Labat*
SPHINGA *R.C. Barneby & J.W. Grimes*
STONESIELLA *M.D. Crisp & P.H. Weston*
SYLVICHADSIA *D.J. Du Puy & J.-N. Labat*
WILLDAMPIA *A.S. George*

LEMNACEAE
LANDOLTIA *D.H. Les & D.J. Crawford*

LENTIBULARIACEAE
LECTICULA *J.H. Barnhart*
SETISCAPELLA *J.H. Barnhart*

LOASACEAE
AOSA *M. Weigend*
CHICHICASTE *M. Weigend*
NASA *M. Weigend*
PRESLIOPHYTUM *(Urb. & Gilg) M. Weigend*
XYLOPODIA *M. Weigend*

LOGANIACEAE
PHYLLANGIUM *C.R. Dunlop*
SCHIZACME *C.R. Dunlop*

M

MAGNOLIACEAE
WOONYOUNGIA *Y.W. Law*

MALPIGHIACEAE
EXCENTRADENIA *W.R. Anderson*

MALVACEAE
FUERTESIMALVA *P.A. Fryxell*
RHOEADIA *C. Lemaire*

MARCGRAVIACEAE
MARCGRAVIASTRUM *(Wittm. ex Szyszył.) A.C. de Roon & S. Dressler*
SARCOPERA *H.G. Bedell*

MELASTOMATACEAE
WURDASTOM *B. Wallnöfer*

MELIACEAE
LEPIDOTRICHILIA *(Harms) T.D. Pennington & B.T. Styles*

MENISPERMACEAE
CIONOMENE *B.A. Krukoff*

MYRISTICACEAE
MAULOUTCHIA *(Baill.) O. Warburg*

MYRTACEAE
ALUTA *B.L. Rye & M.E. Trudgen*
ANETHOLEA *Peter G. Wilson*
CORYMBIA *K.D. Hill & L.A.S. Johnson*
LITHOMYRTUS *F. Muell.*
PETRAEOMYRTUS *L.A. Craven*
STENOSTEGIA *A.R. Bean*

O

OLACACEAE
ENGOMEGOMA *F.J. Breteler*

OPILIACEAE
YUNNANOPILIA *C.Y. Wu & D.Z. Li*

ORCHIDACEAE
× ADAPASIA *hort.*
× AGANANTHES *hort.*
× AGANAX *hort.*
ALA *D.L. Szlachetko*
ANDINIA *(Luer) C.A. Luer*
APPENDICULOPSIS *(Schltr.) D.L. Szlachetko*
ARCHIVEA *E.A. Christenson & R. Jenny*
ASCOCENTROPSIS *K. Senghas & H. Schildhauer*
× BALAGUERARA *hort.*
× BAPTICIDIUM *hort.*
× BIFRANISIA *hort.*
× BLETIAGLOTTIS *hort.*
× BOGARDARA *hort.*
BRASILOCYCNIS *G. Gerlach & M.W. Whitten*
× CALOMITRA *hort.*
CANNAEORCHIS *M.A. Clements & D.L. Jones*
× CARMICHAELARA *hort.*
CHAMAELEORCHIS *K. Senghas & Lückel*
× CHAUBARDIANTHES *hort.*
CHELYORCHIS *R.L. Dressler & N.H. Williams*
× CHONDRANTHES *hort.*
× CHUATIANARA *hort.*
× CLARKEARA *hort.*
COLLARE-STUARTENSE *K. Senghas & L. Bockemühl*
COOKTOWNIA *D.L. Jones*
CYANICULA *S.D. Hopper & A.P. Brown*
× CYCNANDRA *hort.*
× CYCSELLIA *hort.*
DANHATCHIA *L.A. Garay & E.A. Christenson*
× DEROSAARA *hort.*
DIDYMOPLEXIOPSIS *G. Seidenfaden*
DISKYPHOGYNE *D.L. Szlachetko & R. González Tamayo*
× DOREENHUNTARA *hort.*
DRACONANTHES *(Luer) C.A. Luer*
× ELECALTHUSA *hort.*
× ELIARA *hort.*
× ELYTHODIA *hort.*
× ESTELARA *hort.*
EUCHILE *(Dressler & G.E. Pollard) C.L. Withner*
× EUCLADES *hort.*
EXALARIA *L.A. Garay & G.A. Romero-Gonzalez*
FINGARDIA *D.L. Szlachetko*
GARAYANTHUS *D.L. Szlachetko*
× GERBERARA *hort.*
× GIGARA *hort.*
GLOSSOCHILOPSIS *D.L. Szlachetko*
GRACIELANTHUS *R. González Tamayo & D.L. Szlachetko*
GUANCHEZIA *G.A. Romero & G. Carnevali*
HALLEORCHIS *D.L. Szlachetko & T.S. Olszewski*
× HOULLORA *hort.*
HYLAEORCHIS *G. Carnevali & G.A. Romero*
HYSTERIA *Reinw.*

ORCHIDACEAE:-
× IACOVIELLOARA *hort.*
INDIA *A. Nageswara Rao*
JEJEWOODIA *D.L. Szlachetko*
JENNYELLA *E. Lückel & H. Fessel*
JOSTIA *C.A. Luer*
× KEFOREA *hort.*
× KIPPENARA *hort.*
KORNASIA *D.L. Szlachetko*
× LAIPENCHIHARA *hort.*
LISOWSKIA *D.L. Szlachetko*
LUECKELIA *R. Jenny*
× MACCOLMARA *hort.*
× MALCOLMCAMPBELLARA *hort.*
× MEECHAIARA *hort.*
× MONKHOUSEARA *hort.*
× MOOREARA *hort.*
× NEOAERISTYLIS *hort.*
× NEOSTYLOPSIS *hort.*
NEZAHUALCOYOTLIA *R. González Tamayo*
NOTHERIA *P. O'Byrne & J.J. Vermeulen*
NOTYLIOPSIS *P. Ortiz V.*
OBERONIOIDES *D.L. Szlachetko*
OCHYRELLA *D.L. Szlachetko & R. Gonzalez Tamayo*
× ODDYARA *hort.*
OPHIOGLOSSELLA *A. Schuiteman & P. Ormerod*
× ORCHIOPHRYS *hort.*
× PALERMOARA *hort.*
× PANCZAKARA *hort.*
× PERISTERCHILUS *hort.*
PFITZERIA *K. Senghas*
PODANDRIELLA *D.L. Szlachetko*
× POMATOCHILUS *hort.*
× PORTERARA *hort.*
PORTILLIA *W. Königer*
PRAECOXANTHUS *S.D. Hopper & A.P. Brown*
× PROMELLIA *hort.*
× PROMOSEPALUM *hort.*
PSEUDOPERISTYLUS *(P.F. Hunt) D.L. Szlachetko & T.S. Olszewski*
RACIBORSKANTHOS *D.L. Szlachetko*
RENZORCHIS *D.L. Szlachetko & T.S. Olszewski*
× RITTERSHAUSENARA *hort.*
RODRIGOA *L.A. Braas*
× ROHRLARA *hort.*
× RUPPARA *hort.*
× SAKABAARA *hort.*
SARMENTICOLA *K. Senghas & L.A. Garay*
× SARPARIZA *hort.*
SCELOCHILOPSIS *C.H. Dodson & M.W. Chase*
× SCHILLIGERARA *hort.*
× SCHOMBOLAELIOPSIS *hort.*
SCHUNKEA *K. Senghas*
SEEGERIELLA *K. Senghas*
SEIDENFADENIELLA *C. Sathish Kumar*
SEIDENFIA *D.L. Szlachetko*
× SENGHASARA *hort.*
× SHEEHANARA *hort.*
× SIEGERISTARA *hort.*
× SIGMACIDIUM *hort.*
× STENIELLA *hort.*
STILIFOLIUM *W. Königer & D. Pongratz*
× SUSANPERREIRAARA *hort.*
× SYMMONSARA *hort.*
TAMAYORKIS *D.L. Szlachetko*
TETRODON *(Kraenzl.) M.A. Clements & D.L. Jones*
× THOMPSONARA *hort.*
TOMZANONIA *M.A. Nir*
× TRIASPETALUM *hort.*
× TRIASPHYLLUM *hort.*
TRICERATOSTRIS *(Szlach.) D.L. Szlachetko & R. González Tamayo*
× TRISULOARA *hort.*
VEYRETELLA *D.L. Szlachetko & T.S. Olszewski*
VEYRETIA *D.L. Szlachetko*
× VIRAPHANDHUARA *hort.*
× WAILAIARA *hort.*
× WAIRONARA *hort.*
× WILLIAMARA *hort.*
WINIKA *M.A. Clements, D.L. Jones & B.P.J. Molloy*
× WOODWARDARA *hort.*
× XERRIARA *hort.*
× YEEPENGARA *hort.*
ZHUKOWSKIA *D.L. Szlachetko, R. González Tamayo & P. Rutkowski*

P

PALMAE
× MAXIMBIGNYA *S.F. Glassman*

PAPAVERACEAE
EHRENDORFERIA *T. Fukuhara & M. Lidén*
ICHTYOSELMIS *M. Lidén & T. Fukuhara*

PITTOSPORACEAE
AURANTICARPA *L.W. Cayzer, M.D. Crisp & I.R.H. Telford*

PLUMBAGINACEAE
CERATOLIMON *M.B. Crespo & M.D. Lledó*
SAHARANTHUS *M.B. Crespo & M.D. Lledó*

PODOCARPACEAE
BRACTEOCARPUS *A.V. Bobrov & A.P. Melikyan*
CORNERIA *A.V.F.Ch. Bobrov & A.P. Melikian*
GAUSSENIA *A.V.F.Ch. Bobrov & A.P. Melikian*
LAUBENFELSIA *A.V.F.Ch. Bobrov & A.P. Melikian*
MANOAO *B.P.J. Molloy*
MARGBENSONIA *A.V. Bobrov & A.P. Melikyan*
METADACRYDIUM *Baum.-Bod. ex A.P. Melikian & A.V.F.Ch. Bobrov*
STACHYPITYS *A.V.F.Ch. Bobrov & A.P. Melikian*
VAN-TIEGHEMIA *A.V.F.Ch. Bobrov & A.P. Melikian*

POLEMONIACEAE
BRYANTIELLA *J.M. Porter*
DAYIA *J.M. Porter*
LATHROCASIS *L.A. Johnson*
MACULIGILIA *V. Grant*
MICROGILIA *J.M. Porter & L.A. Johnson*
SALTUGILIA *(V.E. Grant) L.A. Johnson*

POLYGONACEAE
× REYLLOPIA *J. Holub*

PORTULACACEAE
DENDROPORTULACA *U. Eggli*
LEWISIOPSIS *R. Govaerts*
PARAKEELYA *M.A. Hershkovitz*

POTAMOGETONACEAE
COLEOGETON *(Rchb.) D.H. Les & R.R. Haynes*

PROTEACEAE
CATALEPIDIA *P.H. Weston*
EIDOTHEA *A.W. Douglas & B.P.M. Hyland*
MEGAHERTZIA *A.S. George & B.P.M. Hyland*

R

RANUNCULACEAE
CORIFLORA *W.A. Weber*
VIORNA *É. Spach*

RESTIONACEAE
ACION *B.G. Briggs & L.A.S. Johnson*
APODASMIA *B.G. Briggs & L.A.S. Johnson*
CATACOLEA *B.G. Briggs & L.A.S. Johnson*
CHORDIFEX *B.G. Briggs & L.A.S. Johnson*
CYTOGONIDIUM *B.G. Briggs & L.A.S. Johnson*
DAPSILANTHUS *B.G. Briggs & L.A.S. Johnson*
EURYCHORDA *B.G. Briggs & L.A.S. Johnson*
GURINGALIA *B.G. Briggs & L.A.S. Johnson*
KULINIA *B.G. Briggs & L.A.S. Johnson*
MELANOSTACHYA *B.G. Briggs & L.A.S. Johnson*
PLATYCHORDA *B.G. Briggs & L.A.S. Johnson*
SAROPSIS *B.G. Briggs & L.A.S. Johnson*

RESTIONACEAE:-
STENOTALIS *B.G. Briggs & L.A.S. Johnson*
TARAXIS *B.G. Briggs & L.A.S. Johnson*
TREMULINA *B.G. Briggs & L.A.S. Johnson*
TYRBASTES *B.G. Briggs & L.A.S. Johnson*

RHAMNACEAE
COLUBRINA *Rich. ex A.T. Brongniart*
GRANITITES *B.L. Rye*
RETANILLA *(DC.) A.T. Brongniart*
SAGERETIA *A.T. Brongniart*
SCUTIA *(Comm. ex DC.) A.T. Brongniart*
SOULANGIA *A.T. Brongniart*
TRICHOCEPHALUS *A.T. Brongniart*
WILLEMETIA *A.T. Brongniart*

ROSACEAE
PLEIOSORBUS *L.H. Zhou & C.Y. Wu*
× TORMIMALUS *J. Holub*

RUBIACEAE
ALEISANTHIOPSIS *C. Tange*
ANTHOSPERMOPSIS *(K. Schum.) J.H. Kirkbride*
CYANONEURON *C. Tange*
EVERISTIA *S.T. Reynolds & R.J.F. Henderson*
FOSBERGIA *D.D. Tirvengadum & C. Sastre*
GANGUELIA *E. Robbrecht*
GUIHAIOTHAMNUS *H.S. Lo*
LANDIOPSIS *Capuron ex J. Bosser*
RIODOCEA *P.G. Delprete*
RUBOVIETNAMIA *D.D. Tirvengadum*
STENOSEPALA *C. Persson*
TAMRIDAEA *M. Thulin & B. Bremer*
VIDALASIA *D.D. Tirvengadum*

RUTACEAE
ANDREADOXA *J.A. Kallunki*
CROSSOSPERMA *T.G. Hartley*
DINOSPERMA *T.G. Hartley*
DUTAILLIOPSIS *T.G. Hartley*
LEIONEMA *(F. Muell.) Paul G. Wilson*
PERRYODENDRON *T.G. Hartley*
PITAVIASTER *T.G. Hartley*

S

SAPINDACEAE
HAPLOCOELOPSIS *F.G. Davies*
NAMATAEA *D.W. Thomas & D.J. Harris*

SAXIFRAGACEAE
DARMERA *A. Voss*

SCROPHULARIACEAE
AMEROGLOSSUM *Eb. Fischer, S. Vogel & A.V. Lopes*
BARTHLOTTIA *Eb. Fischer*
BARTSIELLA *M. Bolliger*
HELIOHEBE *P.J. Garnock-Jones*
PHILCOXIA *P. Taylor & V.C. Souza*
PSEUDOMELASMA *Eb. Fischer*
PSEUDOMISOPATES *J. Güemes*
SIEVERSANDREAS *Eb. Fischer*

SOLANACEAE
ARCHIHYOSCYAMUS *A.M. Lu*
DARCYA *A.T. Hunziker*
DARCYANTHUS *A.T. Hunziker*
ERIOLARYNX *(Hunz.) A.T. Hunziker*
TZELTALIA *E. Estrada & M. Martínez*

STILBACEAE
KOGELBERGIA *J.P. Rourke*

STYRACACEAE
CHANGIOSTYRAX *Tao Chen*
NEOSTYRAX *G.S. Fan*

T

TECOPHILAEACEAE
KABUYEA *R.K. Brummitt*

TRIGONIACEAE
ISIDODENDRON *J.L. Fernández-Alonso, J.A. Pérez-Zabala & A. Idarraga-Piedrahita*

U

UMBELLIFERAE
ANDRIANA *B.-E. van Wyk*
BALINOTELLA *J. Soják*
BRACHYSCIAS *J.M. Hart & M.J. Henwood*
CANNABOIDES *B.-E. van Wyk*
EKIMIA *H. Duman & M.F. Watson*
GONGYLOTAXIS *M.G. Pimenov & E.V. Kljuykov*
HUEGELIA *Rchb.*
OREOCOMOPSIS *M.G. Pimenov & E.V. Kljuykov*
PSEUDOCANNABOIDES *B.-E. van Wyk*
PSEUDOTRACHYDIUM *(Kljuykov, Pimenov & V.N. Tikhom.) M.G. Pimenov & E.V. Kljuykov*
SINOCARUM *H. Wolff ex R.H. Shan & F.T. Pu*
TAMAMSCHJANELLA *M.G. Pimenov & E.V. Kljuykov*
TANA *B.-E. van Wyk*

URTICACEAE
HAROLDIELLA *J. Florence*

V

VERBENACEAE
ALOYSIA *Palau*
DISCRETITHECA *P.D. Cantino*
PSEUDOCARYOPTERIS *(Briq.) P.D. Cantino*
TRIPORA *P.D. Cantino*

W

WINTERACEAE
AUSTRODRIMYS *A.B. Doweld*
PSEUDODRIMYS *A.B. Doweld*

Z

ZINGIBERACEAE
CORNUKAEMPFERIA *J. Mood & K. Larsen*
SIAMANTHUS *K. Larsen & J. Mood*
TAMIJIA *S. Sakai & H. Nagamasu*

ZYGOPHYLLACEAE
IZOZOGIA *G. Navarro*